Enhanced Oil Recovery

Second Edition

Enhanced Oil Recovery

Second Edition

Don W. Green
*Emeritus Distinguished Professor of Chemical and
Petroleum Engineering University of Kansas*

G. Paul Willhite
*Ross H. Forney Distinguished Professor of Chemical
and Petroleum Engineering University of Kansas*

Henry L. Doherty Memorial Fund of AIME

Society of Petroleum Engineers
Richardson, Texas, USA

Disclaimer

This book was prepared by members of the Society of Petroleum Engineers and their well-qualified colleagues from material published in the recognized technical literature and from their own individual experience and expertise. While the material presented is believed to be based on sound technical knowledge, neither the Society of Petroleum Engineers nor any of the authors or editors herein provide a warranty either expressed or implied in its application. Correspondingly, the discussion of materials, methods, or techniques that may be covered by letters patents implies no freedom to use such materials, methods, or techniques without permission through appropriate licensing. Nothing described within this book should be construed to lessen the need to apply sound engineering judgment nor to carefully apply accepted engineering practices in the design, implementation, or application of the techniques described herein.

ISBN 978-1-61399-494-8

18 19 20 21 22 23 / 10 9 8 7 6 5 4 3 2 1

Society of Petroleum Engineers
222 Palisades Creek Drive
Richardson, TX 75080-2040 USA

http://store.spe.org
service@spe.org
1.972.952.9393

Dedication

This book is dedicated to the memory of Fred H. Poettmann in recognition of
his contributions and encouragement.

About the Authors

Don W. Green is Emeritus Distinguished Professor of Chemical and Petroleum Engineering at the University of Kansas. His career began in 1962 in the Production Research Division of Continental Oil Company before he joined the University of Kansas in 1964. At the University of Kansas, Green was chair of his department from 1970 to 1974 and from 1996 to 2000. He was codirector of the Tertiary Oil Recovery Project with G. Paul Willhite from 1974 to 2007. Green has authored or coauthored 70 refereed publications, more than 100 technical meeting presentations, is editor of the 6th, 7th and 8th editions of Perry's *Chemical Engineers' Handbook,* and is currently editing the 9th edition. He holds a BS degree in petroleum engineering from the University of Tulsa, and MS and PhD degrees in chemical engineering from the University of Oklahoma. Green has won numerous teaching awards at the University of Kansas, including the Honors for Outstanding Progressive Educator (HOPE) Award and the Chancellor's Club Career Teaching Award. He has also been featured as an outstanding educator in the American Society for Engineering Education journal *Chemical Engineering Education.* Green was an SPE Distinguished Lecturer, recipient of the SPE Distinguished Achievement Award for Petroleum Engineering Faculty, the Improved Oil Recovery (IOR) Pioneer Award and was named an Honorary Member of SPE in 2007. He is also a Fellow of the American Institute of Chemical Engineers. Green received the University of Kansas School of Engineering Distinguished Engineering Service Award (DESA) in 2015, and was named to the University of Tulsa College of Engineering and Natural Sciences Hall of Fame in 2017.

G. Paul Willhite is the Ross H. Forney Distinguished Professor of Chemical and Petroleum Engineering at the University of Kansas, located in Lawrence, Kansas. He has been a member of the faculty since 1969, and has served as chair of the department from 1988 to 1996 and as interim chair from 2003 to 2004. In 1974, Willhite cofounded the Tertiary Oil Recovery Project and served as codirector from 1974 to 2009. From 1962 to 1969, he worked in the Production Research Division of Continental Oil Company, Ponca City, Oklahoma. Willhite is the author of the SPE textbook *Waterflooding*, published in 1986, and the coauthor of the SPE textbook *Enhanced Oil Recovery*, originally published in 1998. He holds a BS degree from Iowa State University (1959) and a PhD degree from Northwestern University (1962), both in chemical engineering. Willhite is a Distinguished Member of SPE. He is a recipient of the 1981 SPE Distinguished Achievement Award for Petroleum Engineering Faculty, the 1986 Lester C. Uren Award, and the 2001 SPE John Franklin Carll Award. In 1995, Willhite received the Professional Achievement Citation in Engineering from the College of Engineering, Iowa State University. In 2004, he received the IOR Pioneer Award at the SPE/DOE Improved Oil Recovery Symposium. Willhite was elected to the National Academy of Engineering in 2006, and received the Anson Marston Medal from the College of Engineering at Iowa State University in 2009. He was elected as an Honorary Member of SPE-AIME in 2012, and was inducted into the Alumni Hall of Fame in the Department of Chemical and Biological Engineering at Iowa State University in 2013.

Preface

The development of enhanced-oil-recovery (EOR) processes has been ongoing since the end of World War II, when operators who owned reservoirs with declining reserves recognized that significant quantities of oil remained in their reservoirs after primary and secondary recovery (primarily waterflooding). Research and field activity increased as production from major reservoirs declined, worldwide consumption of oil increased, and discoveries of major new reservoirs became infrequent. Intense interest in EOR processes was stimulated in response to the oil embargo of 1973 and the following energy "crisis." The period of high activity lasted until the collapse of worldwide oil prices in 1986.

Over the years, interest in EOR has been tempered by the increase in oil reserves and production. The discovery of major oil fields in the North Slope of Alaska, the North Sea, regions such as Indonesia and South America, and the Athabasca oil sands of Canada have added large volumes of oil to the worldwide market. The development of horizontal drilling combined with hydraulic fracturing in shale oil reservoirs has added large volumes of light crude oil to the worldwide market. In addition, estimates of reserves from reservoirs in the Middle East increased significantly, leading to the expectation that the oil supply will be plentiful and that the oil price would remain in the vicinity of USD 20 to 25/bbl (constant dollars) for many years.

Although large volumes of oil remain in mature reservoirs, the oil will not be produced in large quantities by EOR processes unless these processes can compete economically with the cost of oil production from conventional sources. Thus, as reservoirs age, a dichotomy exists between the desire to preserve wells for potential EOR processes and the lack of economic incentive because of the existence of large reserves of oil in the world.

Enhanced Oil Recovery describes technologies that can be applied to recover oil that cannot be produced by primary recovery or waterflooding or to recover oil that remains after application of these processes. While many of the technologies were economical at the oil prices that existed in 2013 and most of 2014, they may not be at the oil prices of 2015 through 2017. Development of these processes represents significant technological advances in our understanding of oil recovery from petroleum reservoirs and may be the stimulus for future technological developments.

Approach

This text is written as an introduction to EOR processes, which are processes normally applied after waterflooding. These include polymer, micellar-polymer, and CO_2 flooding and thermal-recovery processes that are typically implemented following primary production. Written for seniors and first-year graduate students in petroleum engineering, we assume that those using this text have a basic understanding of petrophysics (porosity and permeability, saturation), fluid properties (viscosity, density, formation volume factor, and phase behavior), and material balances (volumetrics and elementary depletion calculations). We also assume that students have some grasp of the complexity of reservoirs through exposure to geology courses. These topics can be found in other texts.

We have included three background, or review, chapters that cover microscopic (pore-level)-displacement efficiency, linear-displacement theory, and macroscopic (volumetric) -displacement efficiency, respectively. These chapters can be used by petroleum engineering students for review or by those in other engineering or science disciplines as background information for the study of the different EOR processes treated in the book. The text has been used in a one-semester graduate course in our master's degree program taken by students majoring in both petroleum and chemical engineering. The text contains more material than can be covered in a one-semester course, allowing the instructor to place more emphasis on some processes than others.

Chapter 1 introduces EOR processes and methods of screening reservoirs that are candidates for potential application. Chapter 2 reviews fundamental concepts for oil recovery from porous rocks at the microscopic or pore scale. Chapter 3 develops linear-displacement theory on the basis of fractional-flow concepts. In Chapter 4, we introduce volumetric-displacement efficiency of processes. Chapter 5 covers polymer flooding, and Chapter 6 introduces miscible-displacement processes, including CO_2 miscible flooding. Chapter 7 presents chemical flooding, and Chapter 8 covers thermal recovery. A number of EOR commercial field applications have been in operation since publication of the first edition and several of these are described in the chapters covering the different processes.

In describing the different EOR processes, we focus on the fundamental concepts of each process. However, we also present methods of predicting oil recovery when the processes are applied to oil reservoirs. Many methods are available to calculate displacement performance, ranging from simple models based on volumetric sweep to sophisticated reservoir simulators. The use of reservoir simulators is beyond the scope of this text. We chose a middle course that reinforces fundamental mechanisms but requires the mathematical skills expected of students taking this as a first course on EOR. In some cases, the computations are tedious, but they can be done easily with short computer programs. Selected programs are included in the Appendices.

While this text was being written, important developments in EOR technology took place in laboratories and oil fields throughout the world. We have included those developments where appropriate. This was possible because we had access to numerous high-quality technical publications prepared by our colleagues in universities and the petroleum industry.

This book began as a comprehensive text on oil-recovery processes authorized by the SPE Textbook Committee. The chapter on waterflooding in the original outline was expanded into the text *Waterflooding* (published in 1986); writing of this text resumed following completion of *Waterflooding*. In the years that followed, development of micellar-polymer-flooding technology was phased out as a direct result of the collapse of oil prices in 1986 and the development of new oil supplies throughout the world, which led to projections of oil prices in the vicinity of USD 20 to 25/bbl (in constant dollars) for many years. We attempted to preserve the important parts of this technology in the text, even though at initial writing it appeared unlikely that the technology would be applied for many years.

By 2012, there was a steady rise in oil price from USD 20/bbl when the first edition was prepared to more than USD 100/bbl, which stimulated application of EOR processes throughout the world. The SPE Textbook Committee requested the preparation of a second edition with emphasis on field applications of EOR processes. As the preparation of the second edition was nearing completion, oil prices declined to the vicinity of USD 35 to 50/bbl, as the effect of large volumes of oil produced from horizontal wells impacted the worldwide oil market.

Thermal-recovery processes continue to be the major contributor to production from EOR processes. The chapter on thermal-recovery processes is extensive and could be used for a single course. In Canada, the extensive deposits of tar sands has stimulated the development and application of steam-assisted gravity drainage (SAGD) and cyclic steam stimulation (CSS). These topics have been added to the text.

The development of CO_2 miscible flooding in west Texas created increased application of this process. We have added several field case histories from major reservoirs to the chapter on miscible-displacement processes. There will be continued development and application of this technology in addition to the material covered in this text. We anticipate that a wealth of field case histories will be developed from ongoing projects; therefore, students and instructors should look for additional material as they use the text.

Extensive field application of polymer flooding is occurring in the Daqing field in China, where oil production from polymer flooding is estimated to be in excess of 1.5 Bbbl. Polymer flooding by use of horizontal wells is under development in some heavy-oil reservoirs in Canada.

Although there have been substantial developments in surfactant formulations since the first edition and some pilot field tests, no information was available on field tests. Consequently, the revision of the chapter on chemical flooding focuses primarily on the development and testing of surfactant formulations that are effective over a wide range of reservoir conditions in laboratory tests.

During the past 10 years, laboratory research has demonstrated that waterflood recovery from oil-wet and intermediately wetted cores can be increased by injecting water containing low salinity (Low Sal). Mechanisms contributing to this increase in oil recovery are not well understood and continue to be a major area of research. Although field tests are in progress, the topic of Low Sal is in an early stage of development and is not covered in this revision.

Acknowledgements

We would like to thank several colleagues for their contributions to the development of this text. Tom Hewitt (now at Stanford University) introduced us to the usefulness of fractional-flow concepts in understanding immiscible-displacement processes while Paul Willhite was on sabbatical leave at Chevron Oil Field Research Corporation (now Chevron Production Technology Company). Fred Poettmann, one of our editors from the SPE Textbook Committee, provided problems from his EOR course at the Colorado School of Mines and also critical comments on each chapter. He passed away before he had an opportunity to use the text in his EOR class, a goal that he reminded us of through continual encouragement to finish the book. Fred Stalkup, the second editor for the text, provided valuable information for the chapter on miscible displacement. Several unknown external reviewers reviewed each chapter. We appreciate their comments, which contributed much to the text.

We are also indebted to many on the staffs at the University of Kansas and at the Society of Petroleum Engineers (SPE) in Dallas, who helped us in the completion of the text. At the University of Kansas, Ruth Sleeper and Shari Gladman prepared some of the text and Vera Sehon, Megan Gannon, and Jim Busse of the Center for Research Incorporated prepared graphics for several chapters and provided emergency repairs to other figures. The writing of this book spanned several years and several editors at SPE. We appreciate the editorial work of Flora Cohen, Carla Atwal, and Holly Hargadine. We also appreciate the review by our colleague, Shapour Vossoughi, and a number of students who took the course at the University of Kansas.

For the second edition, revisions made to Chapters 5 through 8 were reviewed by several colleagues at other institutions. These include Randy Seright (New Mexico Institute of Mining and Technology), Gary Pope (University of Texas at Austin), George Hirasaki (Rice University), and Sayed Farouq Ali (University of Calgary). Field case histories of application of miscible-flooding processes (Chapter 6) were provided by several individuals. These include Lanny Schoeling, Mark Linroth, Steve Pennell, and Reza Barati (Kinder Morgan Company) for assistance in data gathering and reviewing the material for the Scurry Area Canyon Reef Operations Committee (SACROC) Unit field case; Gary F. Teletzke (Exxon Mobil Upstream Research Company) for his review and guidance of the Means San Andres Unit field case; Charles A. Peterson and Bob Schwager (Devon Energy Company), who provided more-recent data and information resulting from field monitoring in discussion of the

Madison reservoir CO_2 project, which is based almost exclusively on Peterson et al. (2012); and Nick Valenti and Travis Melster (Quantum Resources) for review of the write-up and for providing current field data for the Jay/Little Escambia Creek (LEC) Field N_2 Miscible-Flood Project.

Finally, we especially acknowledge the SPE staff responsible for managing and editing the book throughout the process leading to publication. Jane Eden, Editorial Service Manager, oversaw and guided the project. Her extensive interaction with the authors was very helpful. Judith Mathis, Senior Staff Editor, who was responsible for editing the manuscript, did a superb job in requiring consistency and accuracy in the text, and clear imaging of figures and tables. She and Jane were clearly dedicated to producing a book of high quality. Recognition is also given to David Grant, Digital Publishing Manager, for the excellent cover design.

G. Paul Willhite
Don W. Green
November 2017

Table of Contents

Chapter 1

Introduction to EOR Processes

1.1 Definition of EOR

Oil recovery operations traditionally have been subdivided into three stages: primary, secondary, and tertiary. Historically, these stages described the production from a reservoir in a chronological sense. Primary production, the initial production stage, resulted from the displacement energy naturally existing in a reservoir. Secondary recovery, the second stage of operations, usually was implemented after primary production declined. Traditional secondary recovery processes are waterflooding, pressure maintenance, and gas injection, although the term secondary recovery is now almost synonymous with waterflooding. Tertiary recovery, the third stage of production, was that obtained after waterflooding (or whatever secondary process was used). Tertiary processes used miscible gases, chemicals, and/or thermal energy to displace additional oil after the secondary recovery process became uneconomical.

The drawback to consideration of the three stages as a chronological sequence is that many reservoir production operations are not conducted in the specified order. A well-known example is production of the heavy oils that occur throughout much of the world. If the crude is sufficiently viscous, it may not flow at economic rates under natural energy drives, so primary production would be negligible. For such reservoirs, waterflooding would not be feasible; therefore, the use of thermal energy might be the only way to recover a significant amount of oil. In this case, a method considered to be a tertiary process in a normal, chronological depletion sequence would be used as the first, and perhaps final, method of recovery.

In other situations, the so-called tertiary process might be applied as a secondary operation instead of waterflooding. This action might be dictated by such factors as the nature of the tertiary process, availability of injectants, and economics. For example, if a waterflood before application of the tertiary process would diminish the overall effectiveness, then the waterflooding stage might reasonably be bypassed.

Because of such situations, the term "tertiary recovery" fell into disfavor in petroleum engineering literature and the designation of "enhanced oil recovery" (EOR) became more accepted. This latter term is used throughout this book. Another descriptive designation commonly used is "improved oil recovery" (IOR), which includes EOR but also encompasses a broader range of activities (e.g., reservoir characterization, improved reservoir management, and infill drilling). The term IOR is not used in this book.

Because of the difficulty of chronological oil-production classification, classification based on process description is more useful and is now the generally accepted approach, although the naming of the processes still incorporates the earlier scheme based on chronology. Oil recovery processes now are classified as primary, secondary, and EOR processes. A classification scheme is clearly useful in that it establishes a basis for communication among technical persons. However, it also has a pragmatic utility in the implementation of tax laws and accounting rules.

Primary recovery results from the use of natural energy present in a reservoir as the main source of energy for the displacement of oil to producing wells. These natural energy sources are solution gas drive, gas-cap drive, natural waterdrive, fluid and rock expansion, and gravity drainage. The particular mechanism of lifting oil to the surface, once it is in the wellbore, is not a factor in the classification scheme.

Secondary recovery results from the augmentation of natural energy through injection of water or gas to displace oil toward producing wells. Gas injection, in this case, is either into a gas cap for pressure maintenance and gas-cap expansion or into oil-column wells to displace oil immiscibly according to relative permeability and volumetric sweepout considerations. Gas processes that are based on other mechanisms, such as oil swelling, oil viscosity reduction, or favorable phase behavior, are considered EOR processes. An immiscible gas displacement is not as efficient as a waterflood and is used infrequently as a secondary recovery process today. (Its use in earlier times was much more prevalent.) Today, waterflooding is almost synonymous with the secondary recovery classification.

EOR results principally from the injection of gases or liquid chemicals and/or the use of thermal energy. Hydrocarbon gases, carbon dioxide (CO_2), nitrogen, and flue gases are among the gases used in EOR processes. In this book, the use of a gas is considered an EOR process if the recovery efficiency significantly depends on a mechanism other than immiscible frontal displacement characterized by high-interfacial-tension (IFT) permeabilities. A number of liquid chemicals are commonly used, including polymers, surfactants, and hydrocarbon solvents. Thermal processes typically consist of the use of steam or hot water, or rely on the in-situ generation of thermal energy through oil combustion in the reservoir rock.

EOR processes involve the injection of a fluid or fluids of some type into a reservoir. The injected fluids and injection processes supplement the natural energy present in the reservoir to displace oil to a producing well. In addition, the injected fluids interact with the reservoir rock/oil system to create conditions favorable for oil recovery. These interactions might, for example, result in lower IFTs, oil swelling, oil viscosity reduction, wettability modification, or favorable phase behavior. The interactions are attributable to physical and chemical mechanisms and to the injection or production of thermal energy. Simple waterflooding and the injection of dry gas for pressure maintenance or oil displacement are excluded from the definition.

EOR processes often involve the injection of more than one fluid. In a typical case, a relatively small volume of an expensive chemical (primary slug) is injected to mobilize the oil. This primary slug is displaced with a larger volume of a relatively inexpensive chemical (secondary slug). The purpose of the secondary slug is to displace the primary slug efficiently with as little deterioration as possible of the primary slug. In some cases, additional fluids of even lower unit cost are injected after a secondary slug to reduce expenses. In such a case of multiple fluid injection, all injected fluids are considered to be part of the EOR process, even though the final chemical slug might be water or dry gas that is injected solely to displace volumetrically the fluids injected earlier in the process.

1.2 Target Oil Resource for EOR Processes

Several studies (Energy Research and Development Administration 1976; National Petroleum Council 1976, 1984; US Office of Technology Assessment 1978; US DOE 1989, 1990) in the US have estimated the potential oil recovery through the application of EOR processes. Part of the objectives of these studies was estimating the target oil resource for EOR (i.e., the amount of oil that would remain after exhaustion of recovery through primary and secondary processes). **Fig. 1.1,** which shows the total US oil resources, is a recent example. In the US as of the end of 1993, approximately 536×10^9 bbl of oil had been discovered. The cumulative production through 1993 was approximately 162×10^9 bbl, and the proven reserves amounted to 23×10^9 bbl (US DOE 1989, 1990). Proven reserves is the oil remaining in known reservoirs that can be expected to be recovered through application of current proven technology at economic conditions on the specified date. Thus, the proven reserves at the end of 1993 include primary and waterflood recovery. A small amount of EOR oil is also included in the proven reserves and is principally oil expected to be recovered through the application of steam processes in California.

As Fig. 1.1 shows, cumulative production plus oil reserves accounts for approximately one-third of the original oil in place (OOIP). Thus, the total target for EOR processes is large, amounting to approximately 351×10^9 bbl in the US alone. If this one-third recovery fraction for primary plus secondary production holds worldwide, then the EOR target approaches 2×10^{12} bbl for the world, not including countries that formerly had centrally planned economies.

The physical/chemical characteristics of the target oil are varied and range from high-API-gravity, volatile crudes of low viscosity to low-API-gravity, heavy crudes of very high viscosity. Significant amounts of oil exist across this physical/chemical spectrum, and, therefore, EOR technology cannot focus on a particular oil type without eliminating a large fraction of the target resource. Clearly, no single EOR process will be applicable to all crudes, and a number of different processes will have to be developed.

A parallel difficulty is that the oil resource exists in reservoirs of widely varying characteristics. Oil reservoir types range from very thick carbonate reef formations at significant depths to relatively shallow, thin sandstone bodies. Subsequent chapters will show that reservoir rock type and structure have an effect on most EOR processes and are important variables. Willhite (1986) describes the role of geology and its significance for displacement processes.

Finally, the saturation, distribution, and physical state of the oil in a reservoir as a result of past production operations are important factors in the implementation of an EOR process. Typically, a reservoir will undergo primary production followed by waterflooding. Recovery by those processes in individual reservoirs might have approached 35 to 50% OOIP when the waterflood reached an economic limit. The residual oil in the part of the reservoir swept by the waterflood remains largely

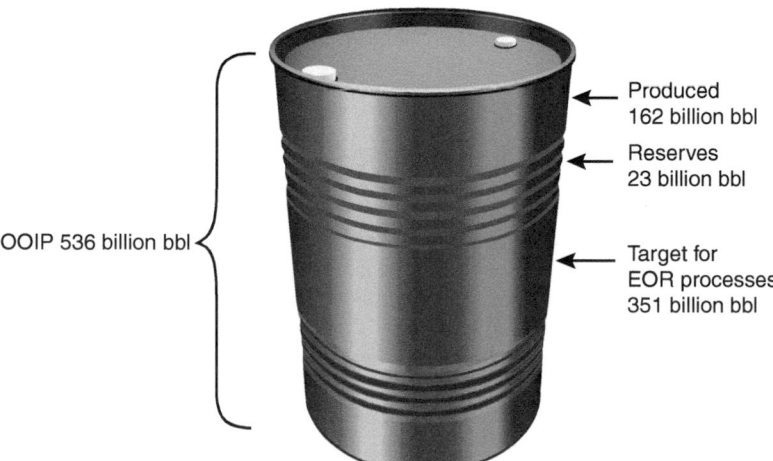

OOIP 536 billion bbl

Produced
162 billion bbl

Reserves
23 billion bbl

Target for
EOR processes
351 billion bbl

Fig. 1.1—US oil barrel showing oil produced, proven reserves, and target for EOR at end of 1993 (after US DOE 1996[1])

[1] Personal communication, BDM/US DOE, Bartlesville, Oklahoma, June 1996.

as isolated, trapped droplets (or ganglia) in the pores or films around the rock particles, depending on the rock wettability. Residual saturation typically is approximately 20 to 35% in swept regions. An EOR process must be able to mobilize the droplets or oil blobs and to create an oil bank that can be efficiently propagated to producing wells.

In other circumstances, an EOR process might be applied after primary production (i.e., as a secondary recovery operation in a chronological sense). In this case, the oil may exist at a relatively high saturation, perhaps approximately 50 to 60%, and may still exist in a connected state with a relative permeability to oil significantly greater than zero. A gas saturation might also be present, depending on the primary recovery mechanism. In this case, the EOR process would be expected to build an oil bank, much in the manner of a waterflood. The displacement efficiency, however, must be better than for a waterflood in that only small amounts of trapped oil should be left behind. That is, the residual oil saturation following the displacement must be low relative to waterflooding because the economic attractiveness of an EOR process applied as a secondary operation normally is compared with the alternative of waterflooding [i.e., recovery (and costs) greater than that expected from a standard waterflood].

In some instances, EOR processes are implemented as the initial or primary production stage. The usual situation is a viscous oil that would not be produced in economic quantities by primary mechanisms or waterflooding. The use of thermal energy, which decreases oil viscosity, is generally the preferred process in such cases. Again, the EOR process must mobilize the oil and displace it efficiently toward production wells.

In summary, the target oil resource is very large and occurs under diverse conditions. Oil type, reservoir rock, and formation type, as well as the oil's distribution, saturation, and physical state resulting from past operations, must all be considered in the design of an EOR process for a particular reservoir. This diversity has led to the development of several different EOR processes that can be considered for implementation.

1.3 Idealized Characteristics of an EOR Process

1.3.1 Efficient Microscopic and Macroscopic Displacement. The overall displacement efficiency of any oil recovery displacement process can be considered conveniently as the product of microscopic and macroscopic displacement efficiencies. In equation form,

$$E = E_D E_V, \dots\dots\dots\dots\dots\dots\dots\dots\dots\dots\dots\dots\dots\dots\dots\dots\dots\dots\dots (1.1)$$

where E = overall displacement efficiency (oil recovered by process/oil in place at start of process), E_D = microscopic displacement efficiency expressed as a fraction, and E_V = macroscopic (volumetric) displacement efficiency expressed as a fraction. Microscopic displacement relates to the displacement or mobilization of oil at the pore scale. That is, E_D is a measure of the effectiveness of the displacing fluid in moving (mobilizing) the oil at those places in the rock where the displacing fluid contacts the oil. E_D is reflected in the magnitude of the residual oil saturation, S_{or}, in the regions contacted by the displacing fluid.

Macroscopic displacement efficiency relates to the effectiveness of the displacing fluid(s) in contacting the reservoir in a volumetric sense. Alternative terms conveying the same general concept are sweep efficiency and conformance factor. E_V is a measure of how effectively the displacing fluid sweeps out the volume of a reservoir, both areally and vertically, as well as how effectively the displacing fluid moves the displaced oil toward production wells. Both areal and vertical sweeps must be considered, and it is often useful to further subdivide E_V into the product of areal and vertical displacement efficiencies. E_V is reflected in the magnitude of average or overall residual oil saturation, S_{or}, because the average is based on residual oil in both swept and unswept parts of the reservoir.

Consider the magnitude of these efficiencies in a typical waterflood. For an example in which initial oil saturation, S_{oi}, is 0.60 and S_{or} in the swept region is 0.30,

$$E_D = \frac{S_{oi} - S_{or}}{S_{oi}} = \frac{0.60 - 0.30}{0.60} = 0.50. \dots\dots\dots\dots\dots\dots\dots\dots\dots\dots\dots\dots\dots\dots (1.2)$$

A typical waterflood sweep efficiency, E_V, at the economic limit is 0.7. Therefore,

$$E = E_D E_V = 0.50 \times 0.70 = 0.35. \dots\dots\dots\dots\dots\dots\dots\dots\dots\dots\dots\dots\dots\dots (1.3)$$

Thus, for a typical waterflood, the overall displacement efficiency is on the order of one-third, which also represents the oil recovery efficiency (neglecting volume changes associated with pressure changes). This one-third figure is by no means a universal result applicable to all reservoirs. Individual reservoirs yield higher or lower recovery efficiencies, depending on the oil and reservoir characteristics. The result, however, does indicate that significant amounts of oil remain following the completion of a waterflood. And it is seen that this oil remains as a result of two factors. First, a residual oil saturation remains in those places swept by the water. Second, a large portion of the reservoir is not contacted by the injected water and thus oil has not been displaced from these regions and has not been displaced to production wells. In addition, some oil from the swept region may be displaced into unswept regions and increase the oil saturation in those regions over what it was before the flood began.

It is desirable in an EOR process that the values of E_D and E_V, and consequently E, approach 1.0. An idealized EOR process would be one in which the primary displacing fluid (primary slug) removed all oil from the pores contacted by the fluid ($S_{or} \rightarrow 0$), and in which the displacing fluid contacted the total reservoir volume and displaced oil to production wells. A secondary fluid slug used to displace the primary slug would behave in a similar manner in that it would displace the primary slug efficiently both microscopically and macroscopically. As will be seen, the development of a "magic" displacing fluid or fluids having properties that will yield this result and still be economical is a monumental and unfulfilled task.

Several physical/chemical interactions occur between the displacing fluid and oil that can lead to efficient microscopic displacement (low S_{or}). These include miscibility between the fluids, decreasing the IFT between the fluids, oil volume expansion, and reducing oil viscosity. The maintenance of a favorable mobility ratio between displaced and displacing fluids also contributes to better microscopic displacement efficiency. EOR processes are thus developed with consideration of these factors. Fluids used as primary displacing slugs have one or more of the favorable physical/chemical interactions with the oil. Fluids used to displace the primary slug ideally should also have similar favorable interactions with the primary slug. The goal with an acceptable EOR fluid is to maintain the favorable interaction(s) as long as possible during the flooding process.

Macroscopic displacement efficiency is improved by maintenance of favorable mobility ratios between all displacing and displaced fluids throughout a process. Favorable ratios contribute to improvement of both areal and vertical sweep efficiencies. An ideal EOR fluid then is one that maintains a favorable mobility ratio with the fluid being displaced. Another factor important to good macroscopic efficiency is the density difference between displacing and displaced fluids. Large density differences can result in gravity segregation (i.e., the underriding or overriding of the fluid being displaced). The effect is to bypass fluids at the top or bottom of a reservoir, reducing E_V. If density differences do exist between fluids, this might be used to advantage by flooding in an updip or downdip direction. Reservoir geology, and in particular geologic heterogeneity, is an important factor in the consideration of macroscopic displacement efficiency. The effects of mobility and density differences can be amplified or diminished by the nature of the geology. An ideal EOR fluid thus is one that has a favorable mobility ratio with the fluid(s) being displaced and, further, maintains this favorable condition throughout the process. In addition, the density of an ideal EOR fluid should be comparable with that of the displaced fluid unless flooding can be performed in an updip or downdip direction.

1.3.2 Practical Considerations. Fluids that possess the properties required for good microscopic and macroscopic displacement efficiencies are certainly known or can be developed. A practical concern, however, is that the fluids are expensive, or for the case of thermal processes, the cost of developing the thermal energy that the fluids carry is high.

As described later, the nature of flow in porous media and rock/fluid interactions lead to the diminished effectiveness of injected fluid slugs. For example, fluid/fluid mixing causes injected fluid concentrations to change and physical adsorption causes the loss of certain chemical components. For thermal processes, heat conduction to overburden and underburden rocks results in a loss of thermal energy from the process. Such chemical losses, changes in composition, or losses of thermal energy mean that the injected fluid slug size must be large enough to sustain the losses or changes and still operate effectively. Thus, the size of the fluid slugs that are injected and their unit costs become major considerations in the design of an EOR process. In fact, injected fluid cost and crude oil price (and instability in price) are the two most important factors controlling the economic implementation of EOR processes.

Another consideration is the ease of handling an EOR fluid and its general compatibility with the physical injection/production system. Highly toxic or corrosive fluids, or fluids that are not readily injected, are not very amenable to use in EOR processes. While such fluids might be used with installation of special equipment, the cost is usually prohibitive.

The availability of an EOR fluid is also a consideration. If an EOR process is applied in a major reservoir, the fluid requirements for that single reservoir can be quite large. And if that process is widely accepted for application across the country or the world, the volume requirements can become an important limiting factor for its application. CO_2 is an example of this. A study by the US Office of Technology Assessment (1978) indicated that the total CO_2 required could ultimately reach 50 $\times$ 10^{12} scf. While CO_2 occurs naturally in underground reservoirs and is a byproduct of some commercial operations, such as fertilizer production, this projected demand could be difficult to satisfy, especially when geographic factors are considered.

The implementation and success or failure of an EOR process are always affected by reservoir geology and reservoir geologic heterogeneities. Processes that are well-understood in a laboratory environment and properly designed for the reservoir fluids may fail when implemented in the reservoir because of geologic factors. Reports of such failures are numerous in petroleum engineering literature. Geologic factors may lead to unexpected losses of chemicals or bypassing of fluids because of channeling in high-permeability zones or fractures. Similarly, fluid movement may be very nonuniform because of variations in rock properties. Unexpected chemical adsorption can sometimes occur, causing a deterioration of fluid slugs. Factors of this type, unless properly identified and understood before the start of a process, will likely cause a project failure. A number of procedures exist that can be used before implementation of an EOR process in an attempt to describe the reservoir geology. These procedures include geologic evaluations of well cores and logs, single-well and well-to-well tracer tests, pressure-transient analysis, and seismic surveys.

1.4 General Classifications and Description of EOR Processes

EOR processes can be classified into five categories: mobility-control, chemical, miscible, thermal, and other processes, such as microbial EOR.

Mobility-control processes, as the name implies, are those that are based primarily on maintaining favorable mobility ratios to improve the magnitude of E_V. Examples are thickening of water with polymers and reducing gas mobility with foams. Chemical processes are those in which certain chemicals, such as surfactants or alkaline agents, are injected to use a combination of phase behavior and IFT reduction to displace oil, thereby improving E_D. In some cases, mobility control is also a part of the chemical process, providing the potential of improving both E_V and E_D. In miscible processes, the objective is to inject fluids that are directly miscible with the oil or that generate miscibility in the reservoir through composition alteration. Examples are injection of hydrocarbon solvents or CO_2. Phase behavior is a major factor in the application of such processes.

Thermal processes rely on the injection of thermal energy or the in-situ generation of heat to improve oil recovery. Steam injection and in-situ combustion from air or oxygen injection are examples. Alteration of oil viscosity, favorable phase behavior, and in some cases, chemical reaction are the primary mechanisms leading to improved oil recovery. "Other processes" is a catch-all category. Examples of processes in this category are microbial-based techniques, immiscible CO_2 injection, and mining of resources at shallow depths. Such methods are not considered in this book.

The classification scheme is not altogether satisfactory in that there is a certain lack of precision in the terms used. For example, chemical processes is one category but chemicals clearly are used in all the processes. Also, there is some overlap in mechanisms between the categories. For example, the chemical processes rely on phase behavior and at least a limited solubility between the different fluids, which is similar to the miscible processes. Despite these shortcomings, the indicated names are used throughout this book, principally because they are deeply embedded in the petroleum engineering literature and are quite useful, with proper clarification, to divide the discussions in the book.

The following discussion of the processes acquaints the reader with the methods in general. This overview will be particularly useful for the material discussed in Chapters 2 through 4 on microscopic and macroscopic displacement efficiencies and linear displacement processes.

1.4.1 Mobility-Control Processes. A widely applied mobility-control process is the polymer-augmented waterflood shown schematically in **Fig. 1.2.** In a typical application, a solution of partially hydrolyzed polyacrylamide polymer in brine, at a concentration of a few hundred to several hundred ppm of polymer, is injected to displace oil (and associated water) toward production wells. The size of the polymer slug might be as much as 50 to 100% pore volume (PV) and might be varied in composition. That is, the highest polymer concentration used is injected for a period of time followed by slugs at successively lower concentrations. The final fluid injected is water or brine.

Polymer solutions are designed to develop a favorable mobility ratio between the injected polymer solution and the oil/water bank being displaced ahead of the polymer. The purpose is to develop a more uniform volumetric sweep of the reservoir, both vertically and areally, as illustrated in Fig. 1.2 for one-quarter of a five-spot pattern. In a conventional waterflood, if the mobility ratio is unfavorable, the water tends to finger by the oil and to move by the shortest path to the production well. This effect is amplified by reservoir geologic heterogeneities.

A polymer solution moves in a more uniform manner, as Fig. 1.2 shows. While flow still tends to be greatest in high-permeability zones and along the shortest path between the injection and production wells, the effect is damped because polymer solution mobility is less than water mobility. Thus, at the economic limit, E_V is larger for a polymer flood than for a waterflood. It is generally accepted that polymer solutions do not significantly affect final, or endpoint, residual oil saturation. But, depending on the nature of the fractional flow curve and the volume of water injected, the "effective" residual oil saturation at the economic level of a flood may be lower for polymer displacing a viscous oil than for a waterflood. The primary mechanism in a polymer flood, however, is an increase in the macroscopic sweep efficiency.

Partially hydrolyzed polyacrylamide polymers affect mobility in two ways. First, solutions of polymers have apparent viscosities that are larger than that of water. The polymer solutions are non-Newtonian, however, and can exhibit significant sensitivity to shear (i.e., apparent viscosity can be a function of the shear rate to which a solution is subjected). The solutions are also sensitive to brine type and concentration, which can affect the apparent viscosity. Second, polyacrylamide polymers adsorb on porous media and/or are mechanically entrapped as a result of their large physical size. This polymer retention reduces the amount of polymer in solution but also causes a decrease in the effective permeability of the porous medium. The mobility of a polyacrylamide polymer solution is thus reduced to less than that of the displaced oil/water bank by a combination of viscosity and effective permeability reduction.

Polymer types other than partially hydrolyzed polyacrylamides may also be used. The most common alternatives are called bio-polymers and are produced by fermentation manufacturing processes. These polymers affect the apparent solution viscosity but have little effect on apparent rock permeability because retention is much smaller.

The most serious limitation to polymer-augmented waterflooding is that projected ultimate recoveries are small compared with those of other EOR processes. Polymer flooding works primarily to improve macroscopic efficiency rather than microscopic efficiency. The process also is affected by the production operations that preceded the polymer flood. A previous successful waterflood, for example, can result in a polymer flood having only a minimal effect.

Other processes exist that are based on the application of foams, relative permeability alteration, or permeability blockage in high-permeability zones in an attempt to increase oil recovery. As discussed later, mobility-control processes also are used extensively with other EOR methods to improve overall process efficiency.

1.4.2 Chemical Processes. Chemical processes involve the injection of specific liquid chemicals that effectively displace oil because of their phase-behavior properties, which result in decreasing the IFT between the displacing liquid and oil. The surfactant/polymer

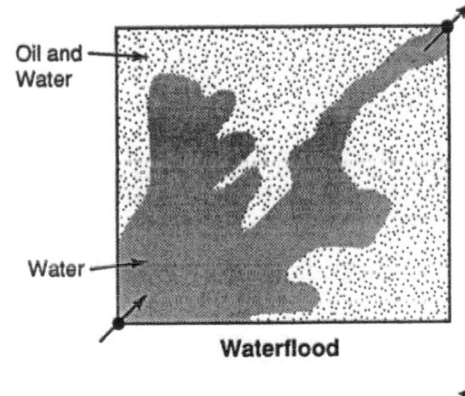

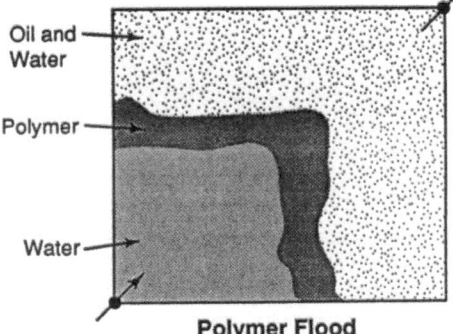

Fig. 1.2—Schematic of macroscopic displacement efficiency improvement with polymer-augmented waterflooding (one-quarter of a five-spot pattern).

process illustrated in **Fig. 1.3** has been demonstrated to have the most potential in terms of ultimate oil recovery in this category of EOR methods. In this process, the primary displacing liquid slug is a complex chemical system called a micellar solution. This solution contains a surfactant (usually a petroleum sulfonate), cosurfactant (an alcohol), oil, electrolytes, and water. The surfactant slug is relatively small, typically 10% PV. The surfactant slug is followed by a mobility buffer, a solution that contains polymer at a concentration of a few hundred ppm. This polymer solution is often graded in concentration, becoming more dilute in polymer as more of the solution is injected. The total volume of the polymer solution is typically approximately 1 PV.

The micellar solution has limited solubility with the oil and is designed to have an ultralow IFT with the oil phase. When this solution contacts residual oil drops, the drops, under a pressure gradient, are deformed as a result of the low IFT and are displaced through the pore throats. Coalescence of oil drops results in an oil bank that, along with water, moves ahead of the displacing chemical slug. The micellar slug also is designed to have a favorable mobility ratio with the oil bank and the water flowing ahead of the slug to prevent viscous fingering of the slug into the oil bank and to increase the macroscopic displacement efficiency.

The polymer-solution mobility buffer is injected to displace the micellar solution efficiently. The IFT between the polymer and micellar solutions is quite low, and only a small residual saturation of the micellar slug is trapped. The existence of a favorable mobility ratio between the polymer and micellar solutions also contributes to an efficient displacement.

In this process, the displacements are immiscible; that is, complete solubility does not exist between the micellar solution and oil or between the micellar and polymer solutions. A low IFT between displacing fluids is desirable at both ends of the micellar slug. A low IFT between the micellar solution and oil is required to mobilize discontinuous oil drops or films. At the back of the micellar slug, a low IFT results in minimal trapping and bypassing of the micellar solution. Clearly, if the micellar solution were not efficiently displaced by the polymer solution, then the micellar slug would deteriorate rapidly.

The surfactant processes have significant potential because of the possibility of designing a process where both E_V and E_D increase. There are important problems, however. The process is complex technologically and can be justified only when oil prices are relatively high and when residual oil after waterflooding is substantial. The chemical solutions, which contain surfactant, cosurfactant, and sometimes oil, are expensive. Chemical losses can be severe. Such losses can occur as a result of adsorption, phase partitioning and trapping, and bypassing owing to fingering if mobility control is not maintained. These losses must be compensated for by increasing the volume of micellar solution injected. The stability of surfactant systems in general is known to be sensitive to high temperatures and high salinity. Systems that can withstand these conditions must be developed if the process is to have wide applicability. For example, early applications have essentially excluded carbonate reservoirs, in part because of the high salinity usually associated with such formations and high concentrations of divalent ions.

There are several variations to the surfactant process, and some of these will be described later in this book. Other chemical methods have also been developed. Alkaline flooding is a process in which injected alkaline chemicals react with certain components in the oil to generate a surfactant in situ. The process has potential but apparently is limited in scope of application. Various alcohol processes also have been tested under laboratory conditions, but these have not been attempted in the field. These processes will be discussed but only in a limited manner.

1.4.3 Miscible Processes. The primary objective in a miscible process is to displace oil with a fluid that is miscible with the oil (i.e., forms a single phase when mixed at all proportions with the oil) at the conditions existing at the interface between

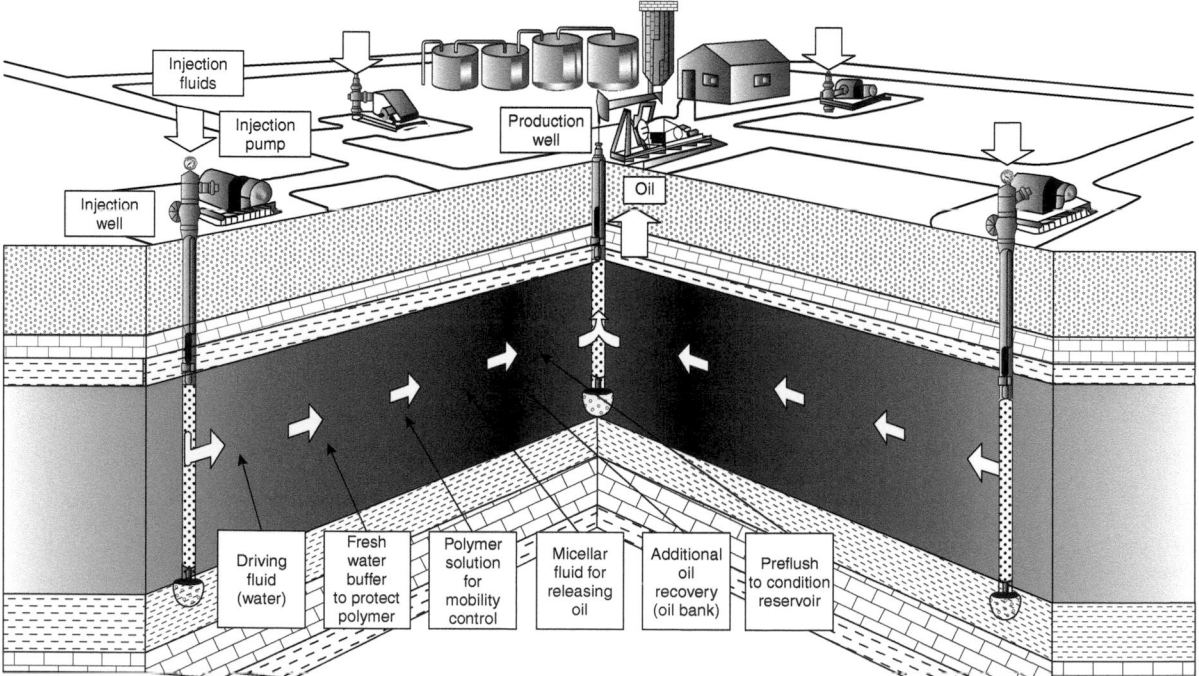

Fig. 1.3—Surfactant/polymer process (after US DOE).

the injected fluid and the oil bank being displaced. There are two major variations in this process. In one, called a first-contact-miscible (FCM) process, the injected fluid is directly miscible with the reservoir oil at the conditions of pressure and temperature existing in the reservoir. **Fig. 1.4** illustrates the FCM process. A relatively small slug of a hydrocarbon fluid, such as liquefied petroleum gas (LPG), is injected to displace the oil. The primary slug size would be approximately 10 to 15% PV. The LPG slug, in turn, is displaced by a larger volume of a less-expensive gas that is high in methane concentration (dry gas). In some cases, water may be used as the secondary displacing fluid.

The process is effective primarily because of miscibility between the primary slug and the oil phase. Primary-slug/oil interfaces are eliminated, and oil drops are mobilized and moved ahead of the primary slug. Miscibility between the primary slug and the secondary displacing fluid (dry gas in Fig. 1.4) is also desirable. Otherwise, the primary slug would be trapped as a residual phase as the process progresses.

The other variation of the miscible processes is the multiple-contact-miscible (MCM) process. In this, the injected fluid is not miscible with the reservoir oil on first contact. Rather, the process depends on the

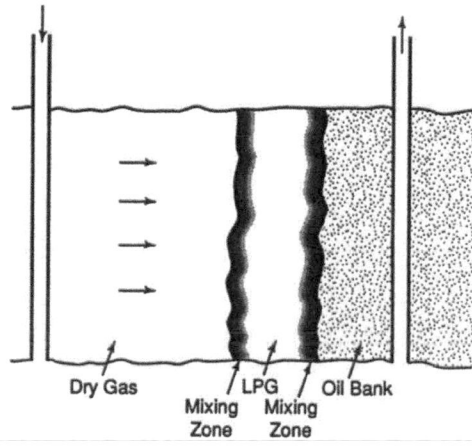

Fig. 1.4—FCM process with LPG and dry gas.

modification of composition of the injected phase, or oil phase, through multiple contacts between the phases in the reservoir and mass transfer of components between them. Under proper conditions of pressure, temperature, and composition this composition modification will generate miscibility between the displacing and displaced phases in situ.

The CO_2 miscible process illustrated in **Fig. 1.5** is one such process. A volume of relatively pure CO_2 is injected to mobilize and displace residual oil. Through multiple contacts between the CO_2 and oil phase, intermediate- and higher-molecular-weight hydrocarbons are extracted into the CO_2-rich phase. Under proper conditions, this CO_2-rich phase will reach a composition that is miscible with the original reservoir oil. From that point, miscible or near-miscible conditions exist at the displacing front interface. Under ideal conditions, this miscibility condition will be reached very quickly in the reservoir and the distance required to establish multiple-contact miscibility initially is negligible compared with the distance between wells. CO_2 volumes injected during a process are typically approximately 25% PV.

The critical temperature of CO_2 is 87.8°F, and thus, in most cases it is injected as a fluid above its critical temperature. The viscosity of CO_2 at injection conditions is small, approximately 0.06 to 0.10 cp, depending on reservoir temperature and pressure. Oil and water are therefore displaced by CO_2 under unfavorable-mobility-ratio conditions in most cases. As described earlier, this leads to fingering of the CO_2 through the oil phase and also to poor macroscopic displacement efficiency.

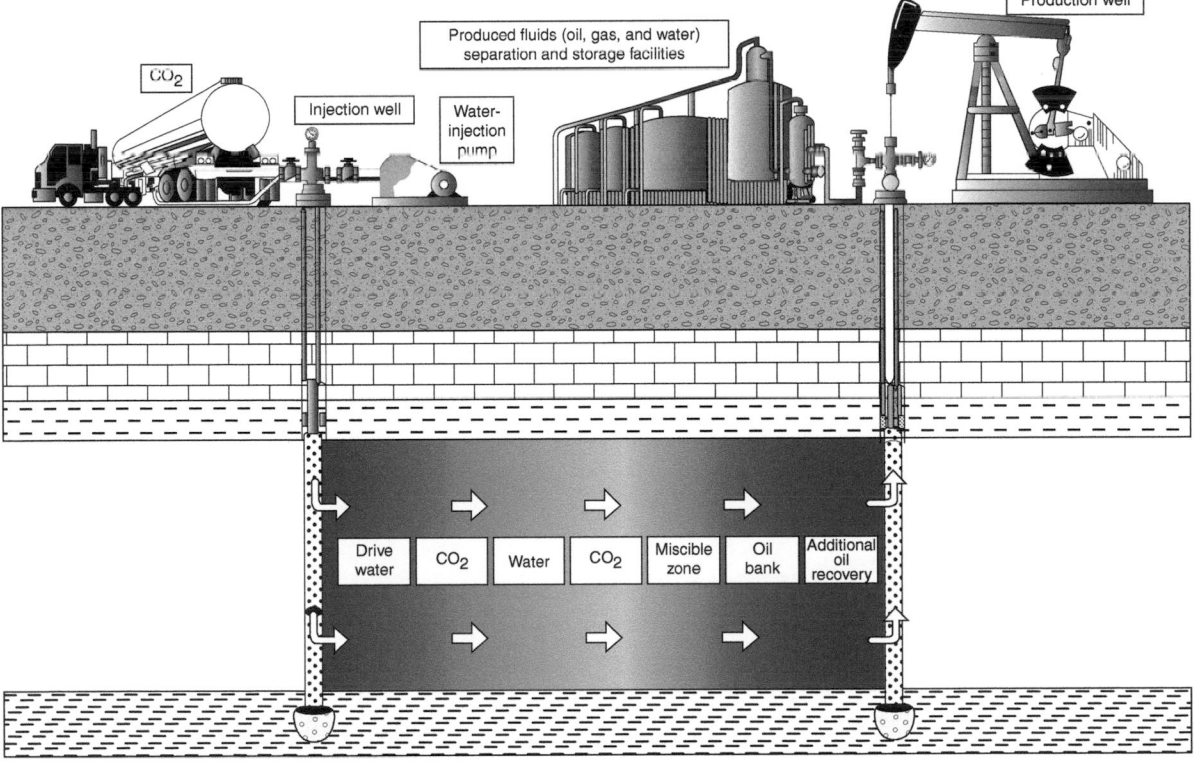

Fig. 1.5—CO₂ miscible process (after US DOE).

One approach to overcoming this difficulty has been to inject slugs of CO_2 and water alternately. This method is called the water-alternating-gas (WAG) process. The purpose of the water injection is to reduce the relative permeability to CO_2 and thereby to reduce its mobility. Another advantage of the WAG process is that it spreads the demand for CO_2 over time. Other methods of mobility control are being tested. These include the use of foams and polymers in conjunction with CO_2 injection. Another problem with the CO_2 process results from the density difference between CO_2 and water and sometimes between CO_2 and the oil. At injection conditions, CO_2 has a specific gravity of approximately 0.4 (again depending on the specific reservoir conditions). Depending on oil density, CO_2 can tend to move to the top of the formation and to override the displaced fluids. In some cases, this gravity effect is exploited by flooding from the top of the reservoir and displacing fluids downdip, but this can be done only where the reservoir structure is suitable.

For the reasons cited, CO_2 often channels in a reservoir and breaks through at production wells relatively early in the process. Because the fuel value of CO_2 is zero, it is usually separated from other produced gases, recompressed, and reinjected. Recycling of CO_2 adds to the cost of a project, but is typically less expensive than purchasing all new CO_2. The separated natural gas has its normal fuel value and is thus salable.

Other gases are suitable for application as MCM displacement fluids in a manner similar to that described for CO_2. These include relatively dry hydrocarbon gases (high CH_4 content), nitrogen, or flue gases. The difference is that these gases usually require much higher pressures to achieve miscibility than CO_2. These other gases are more suitable for deep reservoirs where high pressures can be achieved without fracturing the reservoir rock. A rough rule of thumb for fracturing pressures is 0.6 psi/ft of depth. If fracture pressure is exceeded in the process, the reservoir rock will fail and injected fluids will channel through the fractures, bypassing most of the oil. Thus, the process design and choice of displacing fluid depend on operating pressure, which in turn depends on reservoir depth.

Another modification of the MCM process uses a hydrocarbon fluid that is rich in components such as ethane and propane. In this process, these injected components condense into the oil phase, enriching the oil with the lighter components. Again, under proper conditions, the oil-phase composition can be modified so that it becomes miscible with the injected fluid and in-situ generation of miscibility occurs.

Problems with the miscible processes are primarily those described for the CO_2 MCM process. The miscible fluids generally have small viscosities and therefore fingering and poor volumetric sweeps result. Reservoir heterogeneities magnify this problem. The development of methods to control mobility has proved to be a difficult task. Density differences also contribute to poor volumetric contact because of gravity override unless these density differences can be used to advantage in dipping reservoirs. Finally, the fluids applicable at moderate reservoir pressures are expensive and, in some cases, in limited supply.

1.4.4 Thermal Processes. Thermal processes may be subdivided into hot waterfloods, steam processes, and in-situ combustion. The hot waterflood has been used only sparingly and with limited success and will not be considered here. Steam is used in two different ways: cyclic steam stimulation and steamdrive (steamflood). **Fig. 1.6** shows steam stimulation, sometimes called steam soak or the huff 'n' puff process. This is a single-well method in which steam is injected into a production well for a specified period. The well is then closed in for a while, the so-called "soak" part of the process. The well

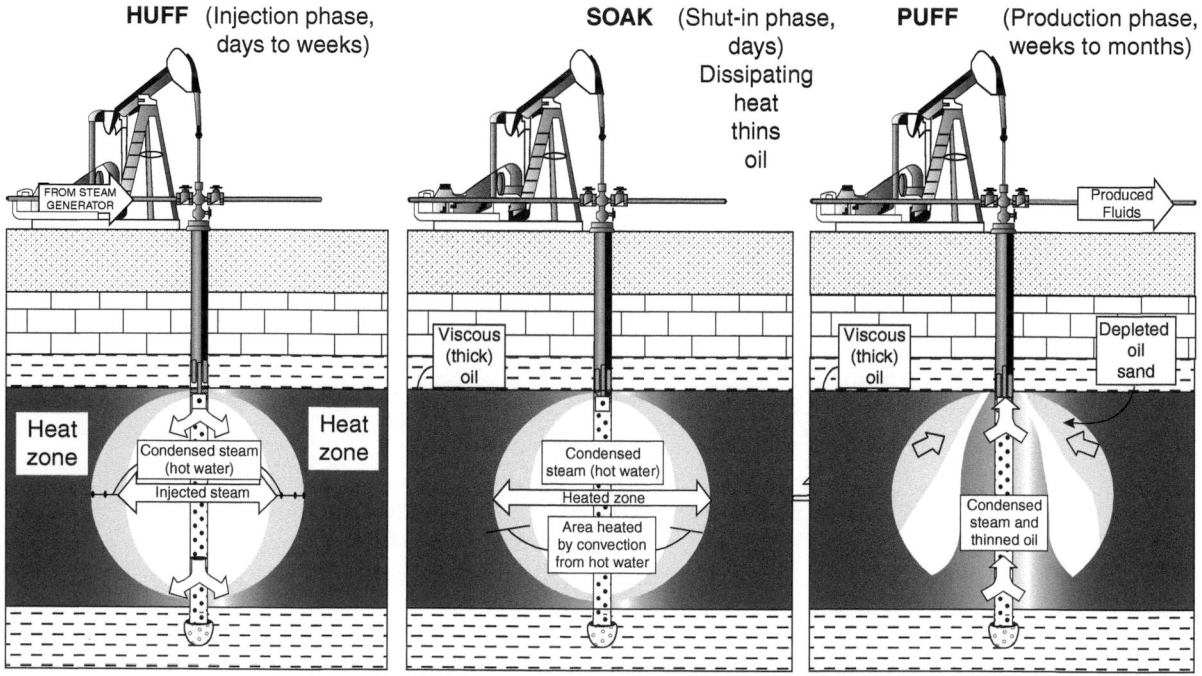

Fig. 1.6—Cyclic-steam-stimulation process (after US DOE).

is next opened for production, which continues until flow rates diminish to a point when the entire procedure is repeated. A typical well may go through several cycles, with the effect of the steam gradually diminishing with continued applications.

Production is increased through a combination of mechanisms, including viscosity reduction, steam flashing, oil swelling, and steam stripping. The cumulative effect of these mechanisms is greatest on heavier (low-API-gravity) oils with high viscosities. Steam injection therefore tends to be used on viscous, low-API-gravity oils.

The second general method of steam application is the steamdrive, or steamflood, process shown in **Fig. 1.7.** In this method, steam is injected through injection wells and the fluids are displaced toward production wells that are drilled in specified patterns.

Recovery mechanisms in this method also are based on viscosity reduction, oil swelling, steam stripping, and steam-vapor drive. As the steam loses energy in its movement through the reservoir, condensation to liquid water occurs. Therefore, the process consists of a hot waterflood in the region of condensation followed by steam displacement. The process has been applied primarily to low-API-gravity, high-viscosity oils but is also applicable to lighter crudes.

A major problem with steam processes is that the steam density is much lower than that of oil and water and therefore the steam tends to move to the top of a reservoir, overriding a large part of the oil body. This is compensated for partially by heat conduction away from the zone of actual contact by the steam, however, and the heated portion of a reservoir can be a high percentage of total reservoir volume. The heated volume depends significantly on the reservoir structure. Mobility control is also a problem with the steamdrive process because steam viscosity is small compared with the viscosities of liquid water and oil. Other points of concern include heat losses, equipment problems from operating at high temperatures, and pollutant emissions resulting from surface steam generation.

In Canada, the extensive heavy oil and tar sands deposits represent a hydrocarbon resource estimated to contain 1.845 Bbbl of hydrocarbon (Alberta Energy Regulator 2014, Table 1). The resource is immobile at reservoir temperature. The capability to drill and complete long horizontal wells in these deposits led to the development of steam-assisted gravity drainage (SAGD). SAGD is based on reservoir heating through a set of parallel horizontal wells, as shown in **Fig. 1.8.** In this process, well pairs are completed so that they are approximately 5 m apart in the vertical plane. The upper well is used for steam injection and the lower well is the production well. Gravity segregation of steam and condensate occur because of the differences in densities, and the steam eventually rises to the top of the steam chamber, as shown in Fig. 1.8 (Butler 1994). As the reservoir heats, bitumen is mobilized and flows by gravity to the production well.

In-situ combustion, shown schematically in **Fig. 1.9,** is another thermal process. In this process, thermal energy is generated in the reservoir by combustion, which may be initiated with either an electric heater or gas burner or may be spontaneous. Oxygen, as air or in a partially purified state, is compressed at the surface and continuously injected (dry process), often together with water (wet process). In the heating and combustion that occur, the lighter components of the oil are vaporized and moved ahead. Depending on the peak temperature attained, thermal cracking may occur, and vapor products from this reaction also move downstream. Part of the oil is deposited as a coke-like material on the reservoir rock, and this solid material serves as the fuel in the process. Thus, as oxygen injection is continued, a combustion front slowly propagates through the reservoir, with the reaction components displacing vapor and liquids ahead toward production wells.

Recovery mechanisms include viscosity reduction from heating, vaporization of fluids, and thermal cracking. Injected gases and water pick up energy as they pass through the burned zone and move toward the combustion front. Ahead of the combustion front, a steam plateau exists (i.e., a region of condensing steam in which the temperature is almost constant at

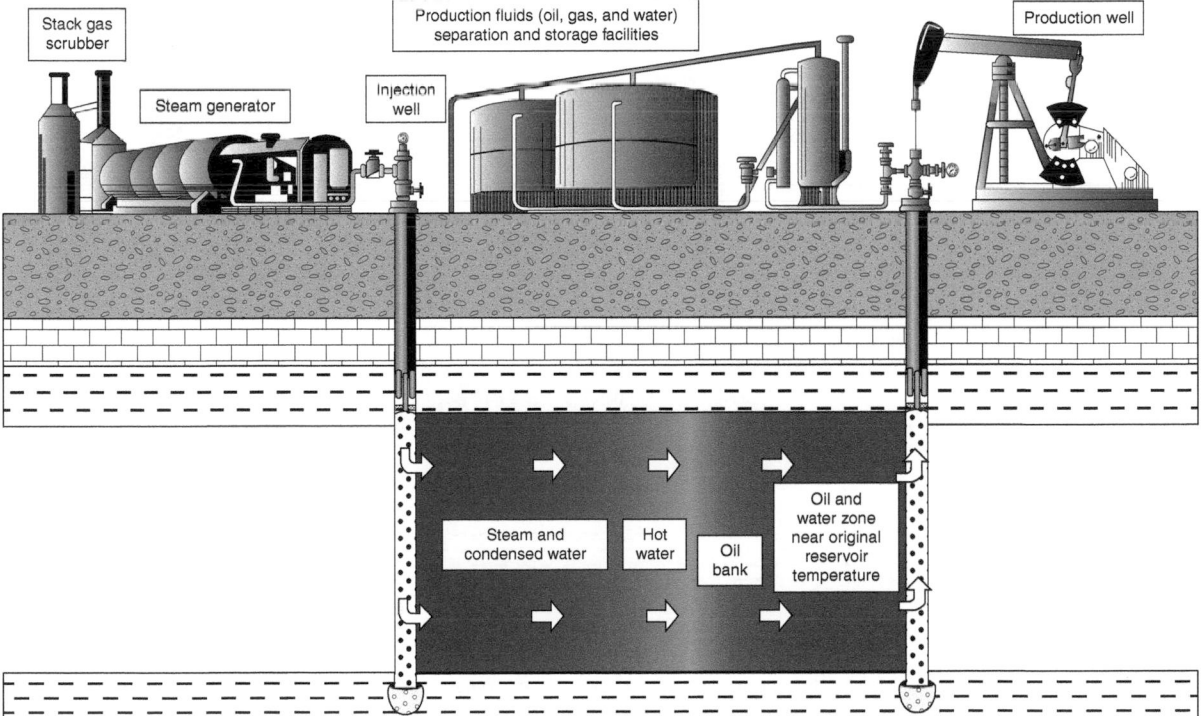

Fig. 1.7—Steamflooding process (after US DOE).

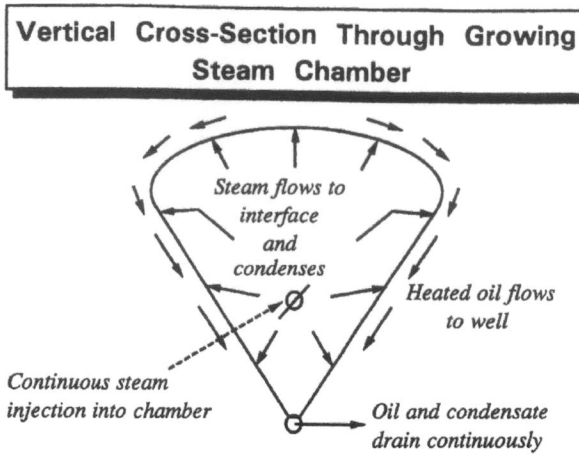

Vertical Cross-Section Through Growing Steam Chamber

Steam flows to interface and condenses

Heated oil flows to well

Continuous steam injection into chamber

Oil and condensate drain continuously

Fig. 1.8—Schematic of the SAGD process (Butler 1994). *Printed in* Horizontal Wells for the Recovery of Oil, Gas and Bitumen, *Petroleum Society Monograph No. 2, CIM, Calgary. Reproduced with permission from the Canadian Institute of Mining, Metallurgy and Petroleum.*

the steam saturation temperature corresponding to the reservoir pressure). A hot waterflood essentially exists in this region, much in the same manner as in a steamdrive process. Ahead of the steam plateau, the temperature decreases to the original reservoir temperature.

There are variations to the in-situ combustion process. In wet combustion, water is injected along with air. The water effectively picks up energy in the burned zone behind the front. It also has beneficial effects on the combustion process and reduces the combustion-zone temperature. In another variation, not often applied, the combustion is carried out in a reverse manner. Combustion is started at the production wells. Oxygen is still injected at injection wells and so the combustion zone moves in the direction opposite to the fluid flow.

The in-situ combustion process effectively displaces oil in the regions contacted. Approximately 30% of the oil in place is required as fuel in the burning. This percentage varies, of course, depending on the oil composition and saturation, combustion conditions, and rock properties.

A major problem with this method is control of the movement of the combustion front. Depending on reservoir characteristics and fluid distributions, the combustion front may move in a nonuniform manner through the reservoir, with resulting poor volumetric contact. Also, if proper conditions are not maintained at the combustion front, the combustion reaction can weaken and cease completely. The process effectiveness is lost if this occurs. Finally, because of the high temperatures generated, significant equipment problems can occur at the wells. Pollutant emission control also can be of concern in some cases.

1.5 Potential of the Different Processes

As mentioned previously, the potential recoveries in the US by the different EOR processes were estimated in a number of relatively early studies (Energy Research and Development Administration 1976; National Petroleum Council 1976, 1984; US Office of Technology Assessment 1978; US DOE 1989, 1990). While specific recovery estimates for the different processes vary, the general conclusions are similar. Results from the study by the National Petroleum Council (NPC) (1984) are summarized in **Figs. 1.10 through 1.12.** Data shown in these figures indicate the general magnitude of oil recovery predicted from the use of the different processes, giving the reader a feel for their relative predicted importance. In the NPC study, mobility-control processes were included within the chemical-process category. Most of the recovery within this category, however,

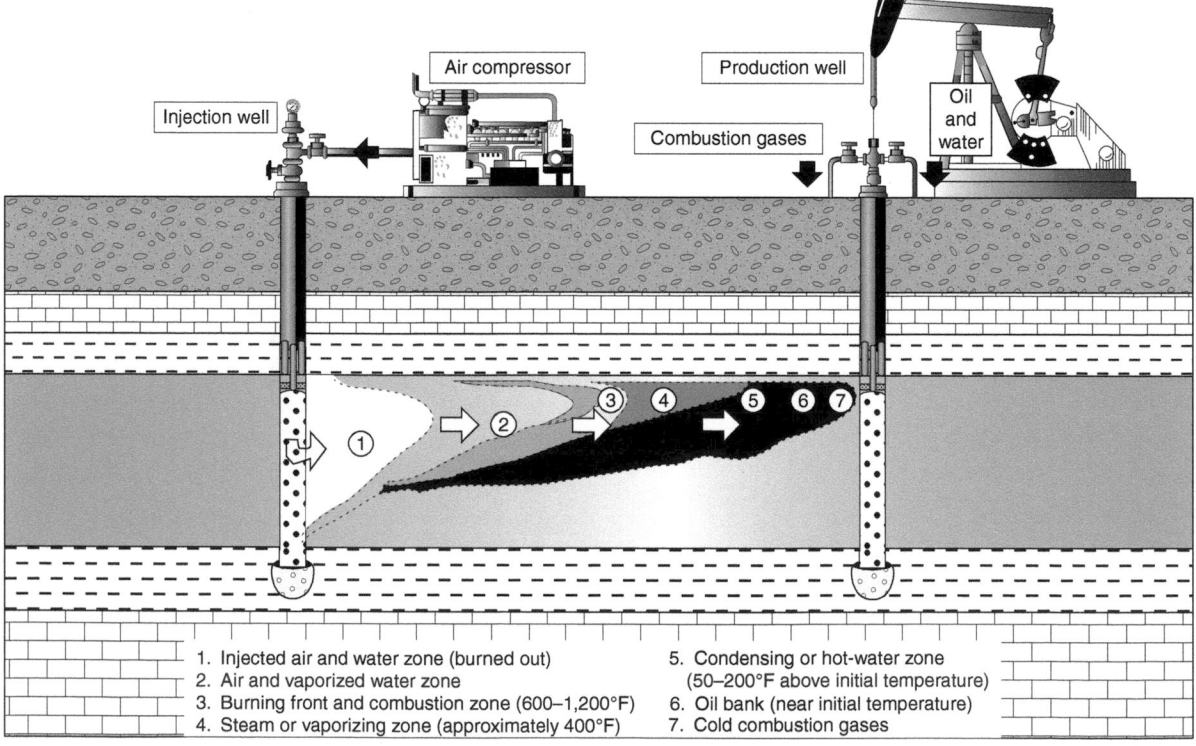

1. Injected air and water zone (burned out)
2. Air and vaporized water zone
3. Burning front and combustion zone (600–1,200°F)
4. Steam or vaporizing zone (approximately 400°F)
5. Condensing or hot-water zone (50–200°F above initial temperature)
6. Oil bank (near initial temperature)
7. Cold combustion gases

Fig. 1.9—In-situ combustion process (after US DOE).

results from the use of chemicals that decrease the IFT between the displacing fluid and the oil.

The NPC study was based on examination of reservoirs having more than 50×10^6 bbl OOIP. This data base encompassed approximately 67% of the total OOIP in the US as of 1980. Fig. 1.10 shows projected ultimate recoveries for reservoirs in the data base for each of the major processes and cumulative ultimate recoveries for two different technology scenarios, both at an assumed oil price of USD 30/bbl (constant 1983 dollars) and a minimum rate of return of 10%. The implemented-technology case was based on application of technology in existence at the time of the study. In the advanced-technology case, use of new technology that might be developed over the next 30 years was assumed (which is the reason for the distinction in recovery through and after 2013 in Fig. 1.11).

The effect of price on recovery is shown in Fig. 1.11, where crude oil prices from USD 20 to 50/bbl were assumed for the two technology scenarios. Oil price has a significant effect in the USD 20 to 30/bbl range, but little effect above approximately USD 40/bbl. Predicted potential production rates for the different processes and cumulative rates are presented in Fig. 1.12 for the advanced-technology case at an oil price of USD 30/bbl.

Other projections (US Office of Technology Assessment 1978; US DOE 1990) were more optimistic in terms of ultimate recovery. All the studies projected that a significant amount of incremental oil could be recovered by EOR processes under favorable economic conditions. Recovery rates could increase over the next several years beyond the dates of the studies. However, ultimate recoveries will not be achieved until well into the 21st century.

Actual cumulative production rates for all EOR projects in the US for the years 1984–95 are shown in Fig. 1.12 (Moritis 1990, 1996) along with the NPC projections. As of the beginning of 1996, thermal processes contributed approximately 424,000 B/D, miscible and immiscible gas processes contributed approximately 299,000 B/D, and chemical processes (including polymer processes) added less than 1,000 B/D. Moritis (1996) reports data on the status of EOR projects around the world as of the beginning of 1996.

The *Oil & Gas Journal* has presented biennial summaries of EOR projects in the US and around the world since 2000. **Fig. 1.13** summarizes the data from the 2014 report for the US (Koottungal 2014). Comparison of the data presented in Fig. 1.13 with the projections in Fig. 1.12 shows that the predictions made in the 1970s and 1980s for the US, both for the individual process and the total production rates, were overly optimistic. Production from miscible/immiscible gas projects has come the closest to the projections for the US, while production

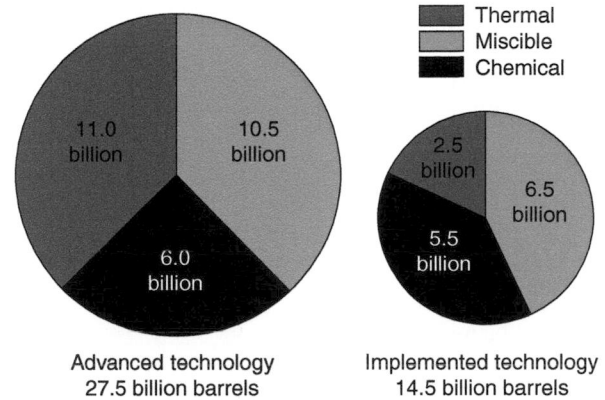

Fig. 1.10—Potential EOR ultimate recovery (National Petroleum Council 1984).

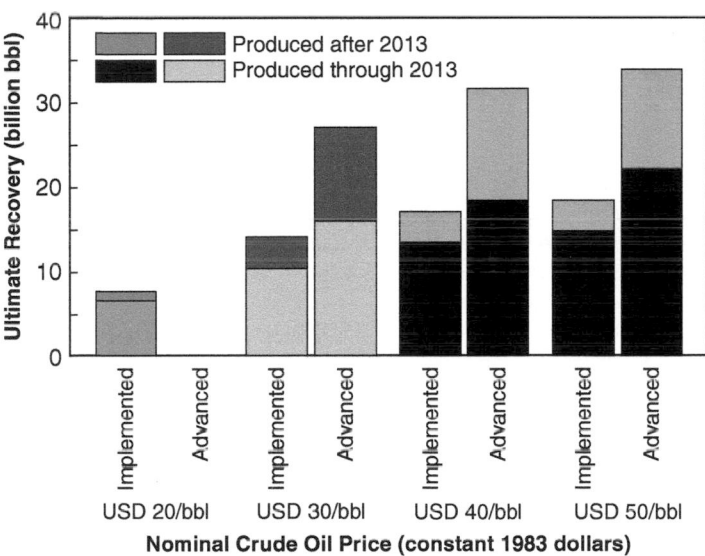

Fig. 1.11—Potential EOR ultimate recovery as a function of oil price (National Petroleum Council 1984).

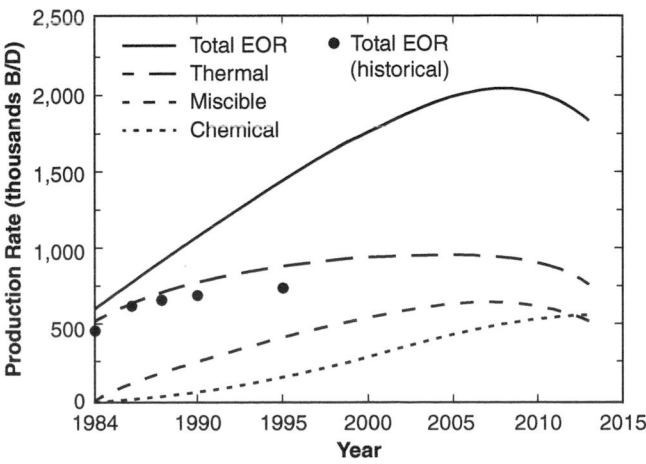

Fig. 1.12—Potential EOR production rates (National Petroleum Council 1984).

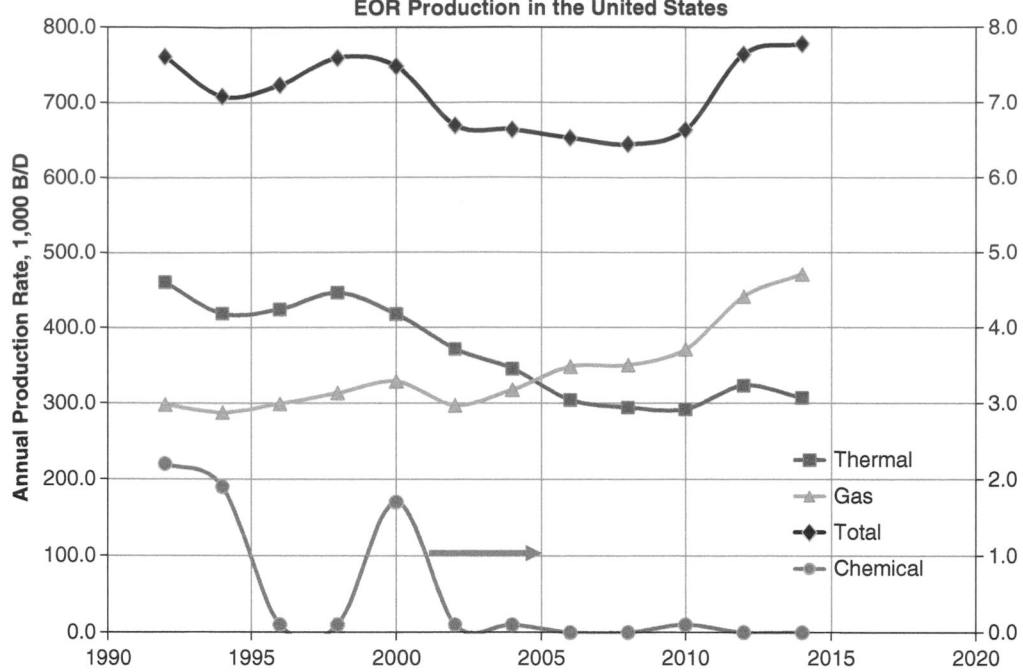

Fig. 1.13—EOR recoveries in the US, 1992–2014 (Koottungal 2014).

from thermal processes has fallen short and is declining, and major chemical projects have not materialized. This shortfall is attributed, in part, to large fluctuations in oil price over this period and the development of technologies (other than EOR) that compete for development of expanded production. However, for thermal and chemical processes, there are major projects outside the US, including, for example, SAGD in Canada and polymer flooding in China. In later chapters of this book, a number of industrial-scale field projects for the different processes are described in some detail. While the US EOR production has not met the earlier predictions, the magnitude of production has been very significant, accounting for approximately 10% of total US crude production as of 2014.

1.6 Screening Criteria for Process Applicability

The US national studies (Energy Research and Development Administration 1976; National Petroleum Council 1976, 1984; US Office of Technology Assessment 1978; US DOE 1989, 1990) used general, or rule-of-thumb, technical screening criteria in the process evaluations. These criteria reflect current estimates of the range of oil and reservoir properties over which the different processes are applicable. **Table 1.1** gives a more recent set of such criteria (Taber et al. 1996). Restrictions on the application of the processes exist. For example, the CO_2 miscible process is limited to reservoirs with sufficient depth to obtain the miscibility pressure and to oils that have relatively high API gravity because of miscibility pressure and/or mobility problems. Steamdrive has reservoir depth limitations because of heat losses and the steam temperatures obtainable. Surfactant/polymer processes are generally limited because of salinity and temperature and the associated difficulty of designing stable surfactant/polymer systems. The screening criteria shown in Table 1.1 are only approximate. In specific cases, successful designs may be developed to exceed the published criteria, and thus, each potential reservoir situation should be considered individually. Also, as the technology develops, the limitations will be relaxed to reflect new knowledge about known processes, variations of known processes, or even new processes.

1.7 Organization of the Textbook

The first four chapters present a general overview of the major EOR processes and a discussion of reservoir engineering principles that relate to oil recovery. The topics have been subdivided into principles that relate to displacement on a microscopic scale (Chapter 2), to linear displacement (Chapter 3), and to macroscopic (or volumetric) displacement (Chapter 4). For those familiar with trapping and mobilization processes in porous media, Chapters 2 through 4 will be a review.

The different classifications of processes are described in the remainder of the book. For each general type of process, a chapter deals with the fundamental displacement mechanisms for the process type. EOR technology is advancing relatively rapidly. A difficulty in writing a book is determining which technologies are reasonably well-established and which are speculative and largely unproved. The purpose of each chapter about the different process types is to describe the principles established to relate to the recovery mechanisms. In the different chapters, certain specific processes are described in more detail, including current design procedures. Practical problems associated with the processes are discussed. Because of space limitations, all the variations of the different process types cannot be described in detail. For example, Chapter 6 emphasizes a description of the CO_2 miscible process. Other miscible processes such as condensing-gas drive and nitrogen displacement, however, are presented in less detail. The discussions given are intended to provide the reader with a basis for proceeding to other, related processes.

EOR Method	Oil Properties			Reservoir Characteristics					
	Gravity °API	Viscosity (cp)	Composition	Oil Saturation (% PV)	Formation Type	Net Thickness (ft)	Average Permeability (md)	Depth (ft)	Temperature (°F)
Gas Injection Methods (Miscible)									
Nitrogen (& Flue Gas)	>35 / 48 /[1]	<0.4 / 0.2	High % of C_1–C_7	>40 / 75 /	Sandstone or carbonate	Thin unless dipping	N.C.[2]	>6,000	N.C.
Hydrocarbon	>23 / 41 /	<3, 0.5	High % of C_2–C_7	>30 / 80 /	Sandstone or carbonate	Thin unless dipping	N.C.	>4,000	N.C.
Carbon Dioxide	>22 / 36 /	<10 / 1.5	High % of C_5–C_{12}	>20 / 55 /	Sandstone or carbonate	(Wide range)	N.C.	>2,500	N.C.
Chemical									
Micellar-/Polymer, Alkaline/Polymer (ASP), and Alkaline Flooding	>20 / 35 /	<35 / 13	Light, intermediate; some organic acids for alkaline floods	>35 / 53 /	Sandstone preferred	N.C.	>10 / 450 /	<9,000 / 3,250	<200 / 80
Polymer Flooding	>15 ≤ 40	<150, >10	N.C.	>70 / 80 /	Sandstone preferred	N.C.	>10[3] / 800 /	<9,000	<200 / 140
Thermal									
Combustion	>10 / 16→?	<5,000→1,200	Some asphaltic components	>50 / 72 /	High porosity sand/ sandstone	>10	>50[4]	<11,500 / 3,500	>100 / 135
Steam	>8–13.5→?	<200,000 / 4,700	N.C.	>40 / 66 /	High porosity sand/ sandstone	>20	>200[5]	<4,500 / 1,500	N.C.

1. Underlined values represent the approximate mean or average for current field projects. / indicates that higher value of parameter is better.
2. N.C. = not critical.
3. >5 md from some carbonate reservoirs.
4. Transmissibility > 20 md-ft/cp.
5. Transmissibility > 50 md-ft/cp.

Table 1.1—Summary of screening criteria for enhanced recovery methods (Taber et al. 1996).

Nomenclature

Parameter definitions followed by dimensions and typical units used in text.

D = depth, L, ft

E = overall displacement efficiency, L³/L³, volume fraction

E_D = microscopic displacement efficiency, L³/L³, volume fraction

E_V = macroscopic (volumetric) displacement efficiency, L³/L³, volume fraction

h = pay-zone thickness, L, ft

k = permeability, L², md

p_R = reservoir pressure, mL/t², psi

S_o = oil saturation, L³/L³, volume fraction

S_{oi} = initial oil saturation, L³/L³, volume fraction

S_{or} = residual oil saturation, L³/L³, volume fraction

$\overline{S}_{or}$ = average residual oil saturation (averaged over entire reservoir volume), L³/L³, volume fraction

T_R = reservoir temperature, T, °F

ϕ = porosity, L³/L³, volume fraction

μ = viscosity, mL/t, cp

References

Alberta Energy Regulator. 2014. ST98-2014: Alberta's Energy Reserves 2013 and Supply/Demand Outlook 2014–2023. Report, Alberta Energy Regulator, Calgary (May 2014).

Butler, R. M. 1994. *Horizontal Wells for the Recovery of Oil, Gas and Bitumen*, Petroleum Society Monograph No. 2, CIM, Calgary, 171.

Enhanced Oil Recovery Potential in the United States. 1978. NTIS Order # PB-276594, Office of Technology Assessment, US Congress, Washington, DC (January 1978), Library of Congress Catalog Card No. 77-600063.

Koottungal, L. 2014. 2014 Worldwide EOR Survey. *Oil & Gas Journal*, **112** (4).

Moritis, G. 1990. CO_2 and HC Injection Lead EOR Production Increase. *Oil & Gas J.* **88:** 49–82.

Moritis, G. 1996. New Technology, Improved Economics Boost EOR Hopes. *Oil & Gas J.* **94**B: 39–61.

National Petroleum Council. 1976. *Enhanced Oil Recovery*. National Petroleum Council, US DOE, Washington, DC (December 1976), Library of Congress Catalog Card No. 76-62538.

National Petroleum Council. 1984. *Enhanced Oil Recovery*. National Petroleum Council, US DOE, Washington, DC (June 1984), Library of Congress Catalog Card No. 84-061296.

Research and Development in Enhanced Oil Recovery. 1976. Energy Research and Development Administration, Washington, DC (1976).

Taber, J. J., Martin, F. D., and Seright, R. S. 1996. EOR Screening Criteria Revisited. Presented at the SPE Improved Oil Recovery Symposium, Tulsa, 21–24 April. SPE-35385-MS. https://doi.org/10.2118/35385-MS.

US DOE. 1990. Oil Research Program Implementation Plan. DOE/FE-0188P, US DOE, Washington, DC (April 1990).

US DOE. 1989. Major Program Elements for an Advanced Geoscience Oil and Gas Recovery Research Initiative. DOE/BC—89/9/SP, Program Study Summary Report, Geoscience Institute for Oil and Gas Recovery Research, University of Texas at Austin, Austin, Texas (October 1989).

Willhite, G. P. 1986. *Waterflooding*, Vol. 3. Richardson, Texas: Textbook Series, SPE.

SI Metric Conversion Factors

bbl	× 1.589 873	E−01	=	m³
cp	× 1.0*	E−03	=	Pa·s
ft	× 3.048*	E−01	=	m
ft³	× 2.381 685	E−02	=	m³
°F	(°F − 32)/1.8		=	°C
psi	× 6.894 757	E+00	=	kPa

*Conversion factor is exact.

Chapter 2

Microscopic Displacement of Fluids in a Reservoir

2.1 Introduction

An important aspect of any enhanced-oil-recovery (EOR) process is the effectiveness of process fluids in removing oil from the rock pores at the microscopic scale. Microscopic displacement efficiency, E_D, largely determines the success or failure of a process. For crude oil, E_D is reflected in the magnitude of S_{or} [i.e., the residual oil saturation (ROS) remaining in the reservoir rock at the end of the process] in places contacted by the displacing fluids. Because EOR processes typically involve the injection of multiple fluid slugs, the efficiency of displacement of these slugs through the reservoir is also of interest. Poor efficiency leads to early deterioration and breakdown of the slugs, which in turn leads to poor project performance.

Capillary and viscous forces govern phase trapping and mobilization of fluids in porous media and thus microscopic displacement efficiency. An understanding and appreciation of the magnitude of these forces is required to understand the recovery mechanisms involved in EOR processes.

This chapter discusses forces related to phase trapping and mobilization in multiphase fluid systems, including descriptions of the roles of interfacial tension (IFT), rock wettability, and capillary pressure. Simple models of trapping and mobilization, experimental data, and empirical expressions that correlate trapping and mobilization data with system parameters are presented. Also discussed are displacement under immiscible conditions when two or more phases flow simultaneously, the role of phase behavior in displacement when solubility of the components in the different phases is significant, and the application of pseudoternary diagrams for phase-behavior representation.

This chapter is a review of the important factors relating to microscopic displacement behavior. Willhite (1986) presents additional discussion of the subject.

2.2 Capillary Forces

2.2.1 Surface Tension and IFT. Whenever immiscible phases coexist in a porous medium as in essentially all processes of interest, surface energy related to the fluid interfaces influences the saturations, distributions, and displacement of the phases. As **Fig. 2.1** shows, water coexists with oil in a reservoir even when the reservoir has not been waterflooded or flooded by a natural waterdrive. Even though the water may be immobile in this case, interfacial forces can still influence performance of subsequent flow processes. If a reservoir has been waterflooded or there is a natural waterdrive, then water saturations will be high and the water phase will be mobile. Most EOR processes use fluids that are not completely miscible with the oil phase and/or the water phase. Interfacial forces must then be examined to determine their significance for oil recovery.

A free liquid surface is illustrated in **Fig. 2.2,** where A, B, and C represent molecules of the liquid. Molecules that are well below the surface, such as A, are, on average, attracted equally in all directions owing to cohesive forces, and their movement therefore tends to be unaffected by cohesive forces. Molecules B and C, however, which are at or near the liquid/air interface, are acted on unequally. A net downward force tends to pull these molecules back into the bulk of the liquid. The surface thus acts like a stretched membrane, tending to shorten as much as possible.

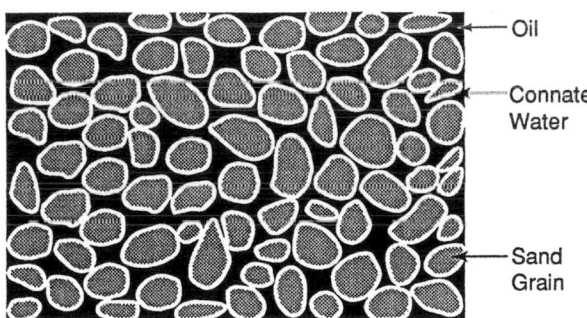

Fig. 2.1—Closeup of oil and water between grains of rock.

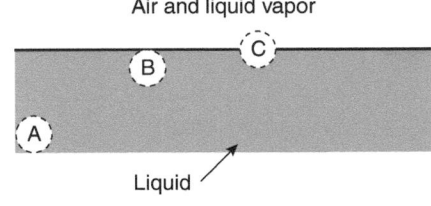

Fig. 2.2—Free liquid surface indicating molecular positions.

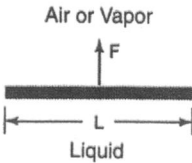

Fig. 2.3—Free liquid surface, force and length used in definition of surface tension.

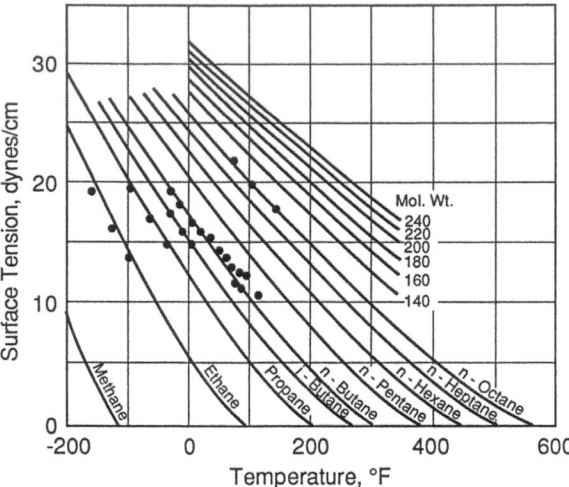

Fig. 2.4—Surface tension of paraffin hydrocarbons (Katz and Saltman 1939; Amyx et al. 1960. Reproduced with permission of McGraw-Hill Inc.).

This surface force, which is a tensile force, is quantified in terms of surface tension, σ, the force acting in the plane of the surface per unit length of the surface. Surface tension can be visualized as shown in **Fig. 2.3,** where a force F is acting normal to a liquid surface of length L. The force per unit length, F/L, required to create additional surface area (that is, to extend the surface) is the surface tension, usually expressed in dynes/cm. Surface tension is related to the work required to create new surface area. Assume that the force F in Fig. 2.3 moves a distance dx, creating new surface in the amount of Ldx. The work done can be expressed as

$$W = Fdx \dots\dots\dots\dots\dots\dots\dots\dots\dots\dots\dots\dots (2.1)$$

or $W = \sigma dA, \dots\dots\dots\dots\dots\dots\dots\dots\dots\dots\dots\dots (2.2)$

where F = force applied to surface, dynes; L = length over which force is applied, cm; σ = IFT, F/L, dynes/cm; and dA = new surface area, Ldx, cm². Thus, the work needed to create additional surface is directly proportional to σ. From this argument, σdA also is seen to represent a surface-energy term.

The term "surface tension" usually is reserved for the specific case in which the surface is between a liquid and its vapor or air. If the surface is between two different liquids, or between a liquid and a solid, the term "interfacial tension" is used. The surface tension of water in contact with its vapor at room temperature is approximately 73 dynes/cm. IFTs between water and pure hydrocarbons are approximately 30 to 50 dynes/cm at room temperature. Mixtures of hydrocarbons such as crude oils will have lower IFTs that depend on the nature and complexity of the liquids. IFTs and surface tensions are relatively strong functions of temperature. Surface tensions of several paraffin hydrocarbons are shown in **Fig. 2.4** (Katz and Saltman 1939; Amyx et al. 1960), and IFTs between water and a number of crude oils are given in **Table 2.1** (Donaldson et al. 1969).

One of the simplest ways to measure the surface tension of a liquid is to use a capillary tube, as shown in **Fig. 2.5.** When a capillary tube of radius r is placed in a beaker of water, the water will rise in the capillary tube to a certain height, h, as a result of the force difference created across the curvature of the meniscus. (The water level will rise if it wets the capillary-tube material. Solid wettability will be discussed later.) At static conditions, the force owing to surface tension (vertical component of surface tension multiplied by the wetted perimeter) will be balanced by the force of gravity acting on the column of fluid: upward vertical force of surface tension × wetted perimeter = downward gravitational force acting on fluid column, or

$$\sigma \cos\theta 2\pi r = \pi r^2 h\left(\rho_w - \rho_a\right)g, \dots\dots\dots\dots\dots\dots\dots\dots\dots\dots\dots\dots\dots\dots\dots\dots\dots\dots\dots\dots\dots\dots (2.3)$$

where r = capillary-tube radius, cm; h = height of water rise in the capillary, cm; ρ_w = water density, g/cm³; ρ_a = air density, g/cm³; g = gravity acceleration constant, 980 cm/s²; and θ = contact angle between water and capillary tube.

Solving for σ gives the familiar equation for capillary rise,

$$\sigma = \frac{rh\left(\rho_w - \rho_a\right)g}{2\cos\theta}. \dots\dots\dots\dots\dots\dots\dots\dots\dots\dots\dots\dots\dots\dots\dots\dots\dots\dots\dots\dots\dots\dots\dots\dots\dots\dots (2.4)$$

Thus, if one carefully measures θ (measured through the liquid) and the height of the column for a given capillary radius, the surface tension can be determined.

Example 2.1—Calculation of Surface Tension From Rise in a Capillary Tube. Calculate the surface tension of water at 77°F if $\theta = 38°$, the capillary radius is 100 μm, and the height of the water column is 12 cm. (Neglect the density of air compared with the density of water.)

Solution.

$$\sigma = \frac{\left(100 \times 10^{-6} \text{ m}\right)\left(100 \text{ cm/m}\right)\left(12 \text{ cm}\right)\left(1 \text{ g/cm}^3\right)\left(980 \text{ cm/s}^2\right)}{2 \times 0.788}$$

$$= 74.6 \text{ g/s}^2 = 74.6 \text{ g} \cdot \text{cm/}\left(\text{s}^2 \cdot \text{cm}\right)$$

$$= 74.6 \text{ dynes/cm}.$$

Field	Oil Formation	Location	Oil Viscosity (cp)	IFT (dynes/cm)
West Delta	Offshore	Louisiana	30.4	17.9
Cayuga	Woodbine	Texas	82.9	17.9
Fairport	Lansing	Kansas	5.3	20.8
Bayou	Choctaw	Louisiana	16.1	15.6
Chase-Silica	Kansas City	Kansas	6.7	19.6
Hofra	Paleocene	Libya	6.1	27.1
Black Bay	Miocene	Louisiana	90.8	17.7
Bar-Dew	Bartlesville	Oklahoma	9.0	21.4
Bar-Dew	Bartlesville	Oklahoma	6.8	21.4
Eugene Island	Offshore	Louisiana	7.4	16.2
Cambridge	Second Berea	Ohio	15.3	14.7
Grand Isle	Offshore	Louisiana	10.3	16.1
Bastían Bay	Uvigerina	Louisiana	112.2	24.8
Oklahoma City	Wilcox	Oklahoma	6.7	20.1
Glenpool	Glen	Oklahoma	5.1	24.7
Cumberland	McLish	Oklahoma	5.8	18.5
Allen District	Allen	Oklahoma	22.0	25.9
Squirrel	Squirrel	Oklahoma	33.0	22.3
Berclair	Vicksberg	Texas	44.5	10.3
Greenwood-Waskom	Wacatoch	Louisiana	5.9	11.9
Ship Shoal	Miocene	Louisiana	22.2	17.3
Gilliland	–	Oklahoma	12.8	17.8
Clear Creek	Upper Bearhead	Louisiana	2.4	17.3
Ray	Arbuckle	Kansas	21.9	25.3
Wheeler	Ellenburger	Texas	4.5	18.2
Rio Bravo	Rio Bravo	California	3.8	17.8
Tatums	Tatums	Oklahoma	133.7	28.8
Saturday Island	Miocene	Louisiana	22.4	31.5
North Shongaloo-Red	Takio	Louisiana	5.2	17.7
Elk Hills	Shallow Zone	California	99.2	12.6
Eugene Island	Miocene	Louisiana	27.7	15.3
Fairport	Reagan	Kansas	31.8	23.4
Long Beach	Alamitos	California	114.0	30.5
Colgrade	Wilcox	Louisiana	360.0	19.9
Spivey Grabs	Mississippi	Kansas	26.4	24.5
Elk Hills	Shallow Zone	California	213.0	14.2
Trix-Liz	Woodbine A	Texas	693.8	10.6
St. Teresa	Cypress	Illinois	121.7	21.6
Bradford	Devonian	Pennsylvania	2.8	9.9
Huntington Beach	South Main Area	California	86.2	16.4
Bartlesville	Bartlesville	Oklahoma	180.0	13.0
Rhodes Pool	Mississippi Chat	Kansas	43.4	30.5
Toborg	–	Texas	153.6	18.0

Table 2.1—IFT between water and various crude oils (Donaldson et al. 1969).

In practice, the contact angle is rather difficult to measure. A simpler technique is the ring tensiometer method shown in **Fig. 2.6** being used to measure IFT between oil and water. In this method, a carefully cleaned platinum/iridium ring is pulled through the interface. As this occurs, the interfacial area increases (Fig. 2.6a), and the surface is stretched further until it finally breaks (Fig. 2.6b). The force on the ring is measured throughout the process. At the breaking point of the surface, the force divided by the circumference of the ring (corrected for a geometric factor) is the σ_{ow} value at the experimental temperature. This method is reliable and convenient over the range of surface and IFT values commonly encountered for water, hydrocarbons,

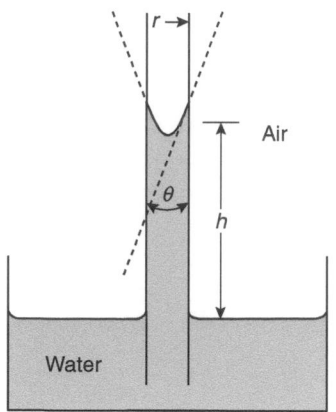

Fig. 2.5—Use of a capillary tube to illustrate a method of measuring surface tension.

$$\sigma = \frac{\mu g}{2L} \cdot F$$

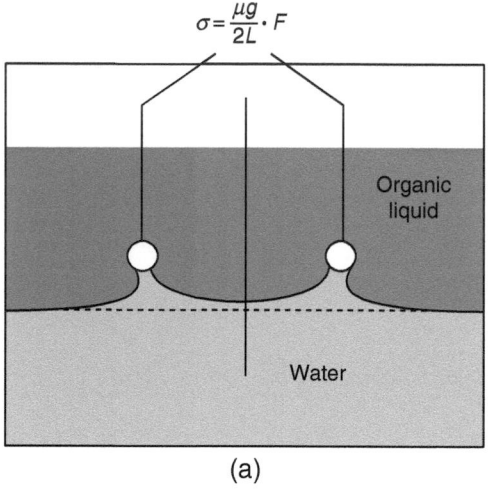

(a)

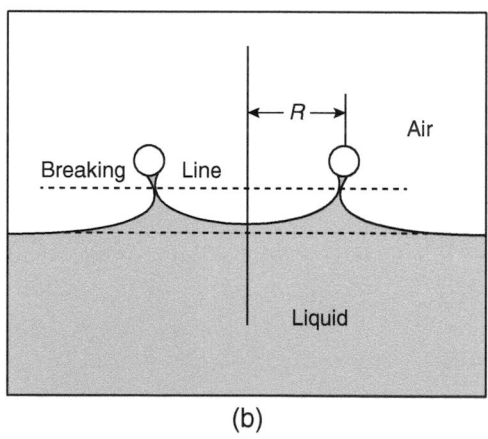

(b)

Fig. 2.6—Use of ring tensiometer to measure IFT: (a) distension of interfacial film during IFT measurements and (b) condition of surface at breaking point.

water/hydrocarbon systems, and some water/hydrocarbon/surfactant systems ($\sigma > 1.0$ dyne/cm).

For ultralow IFTs such as those encountered in surfactant processes (Chapter 7), other means of measurement must be used. These include the spinning-drop and the pendant-drop methods.

2.2.2 Solid Wettability. Fluid distributions in porous media are affected not only by the forces at fluid/fluid interfaces, but also by forces at fluid/solid interfaces. Wettability is the tendency of one fluid to spread on or adhere to a solid surface in the presence of a second fluid. When two immiscible phases are placed in contact with a solid surface, one phase usually is attracted to the solid more strongly than the other phase. The more strongly attracted phase is called the wetting phase.

Rock wettability affects the nature of fluid saturations and the general relative permeability characteristics of a fluid/rock system. A simple example of the effect on saturations is shown in **Fig. 2.7,** which shows ROSs in a strongly water-wet and a strongly oil-wet rock. The location of a phase within the pore structure depends on the wettability of that phase. Considering the effect of wettability on fluid distributions, it is easy to rationalize that relative permeability curves are strong functions of wettability, as Willhite (1986) discusses in some detail. Rocks are also known to have intermediate and/or mixed wettability, depending on the physical/chemical makeup of the rock and the composition of the oil phase. Intermediate wettability occurs when both fluid phases tend to wet the solid, but one phase is only slightly more attracted than the other. Mixed wettability results from a variation or heterogeneity in chemical composition of exposed rock surfaces or cementing-material surfaces in the pores. Because of this mixed chemical exposure, the wettability condition may vary from point to point. In fact, water sometimes wets the solid over part of the surface and oil over the remaining part.

Wettability can be quantitatively treated by examining the interfacial forces that exist when two immiscible fluid phases are in contact with a solid. **Fig. 2.8** shows a water drop placed in contact with a homogeneous rock surface in the presence of an oil phase. The water drop spreads on the solid until forces are balanced, as shown in Fig. 2.8. A force balance at the line of intersection of solid, water, and oil yields

$$\sigma_{os} - \sigma_{ws} = \sigma_{ow} \cos\theta, \dots\dots\dots\dots\dots\dots\dots\dots\dots\dots (2.5)$$

where σ_{os}, σ_{ws}, and σ_{ow} = IFTs between solid and oil, water and solid, and water and oil, respectively, dynes/cm, and θ = contact angle, measured through the water.

While σ_{ow} can be measured as described earlier, σ_{os} and σ_{ws} have never been measured directly. Experimental methods have not been developed to make these determinations. Therefore, the contact angle, θ, is used to measure wettability. For Fig. 2.8, the solid is water-wet if $\theta < 90°$ and oil-wet if $\theta > 90°$. A contact angle approaching 0° indicates a strongly water-wet system and an angle approaching 180° indicates a strongly oil-wet rock. By convention, contact angles are measured through the water phase; if contact angles were measured through the oil phase, the inverse of the preceding rule defining wettability would apply. Intermediate wettability, as described earlier, occurs when θ is close to 90°. Willhite (1986) presents additional discussion of intermediate and mixed wettability, and describes the effect of liquid composition on wettability.

In the example of IFT measurement with a capillary tube discussed earlier (see Fig. 2.5), water was the fluid used and was assumed to wet the glass capillary. If mercury were used as the liquid, the mercury would be depressed in the capillary as shown in **Fig. 2.9.** Here, mercury does not wet the glass. If as before, θ is measured through the liquid, then its value exceeds 90° and cos θ is a negative number. The value of h is also a negative quantity, and thus Eq. 2.4 yields a positive value for the surface tension of mercury. The mercury surface is in tension because a force acts upward in the mercury in the capillary because of the liquid head differential between the mercury in the container and capillary.

2.2.3 Capillary Pressure. Because interfaces are in tension in the systems described, a pressure difference exists across the interface. This pressure, called capillary pressure, can also be illustrated by fluid rise in a capillary tube. **Fig. 2.10** shows water

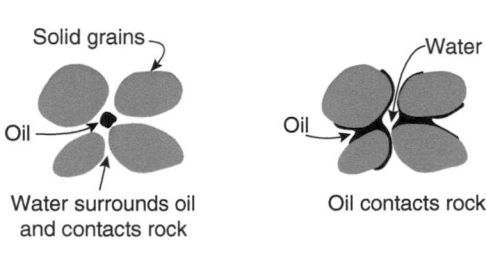

Water-wet system   **Oil-wet system**

Fig. 2.7—Effect of wettability on saturation.

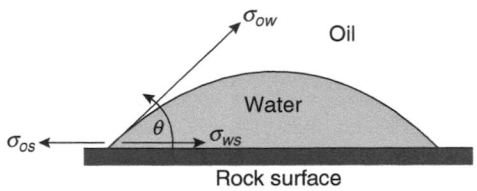

Fig. 2.8—Interfacial forces at an interface between two immiscible fluids and a solid.

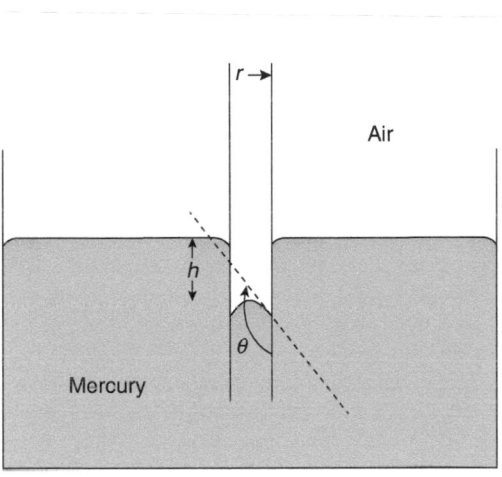

Fig. 2.9—Capillary depression of a nonwetting liquid in a capillary tube.

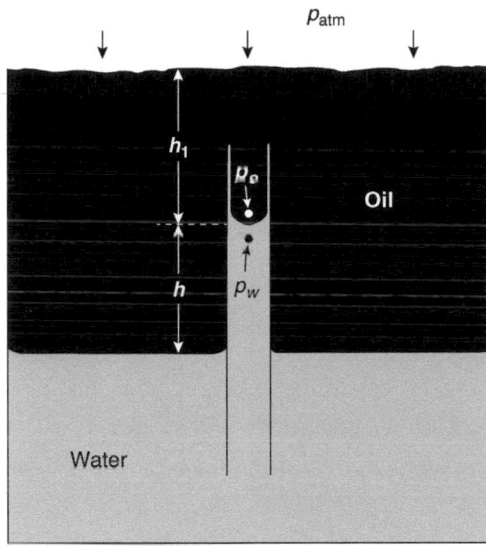

Fig. 2.10—Capillary pressure resulting from interfacial forces in a capillary tube.

rise in a glass capillary. The fluid above the water is an oil, and because the water preferentially wets the glass of the capillary, there is a capillary rise. Two pressures, p_o and p_w, are identified in the figure; p_o is oil-phase pressure at a point just above the oil/water interface, and p_w is water-phase pressure just below the interface.

Force balances yield the following:

$$p_o = p_{atm} + \rho_o g h_1 \dots \dots \dots \dots \dots \dots \dots \dots \dots \dots \dots \dots \dots \dots \dots \dots \dots (2.6)$$

and $$p_w = p_{atm} + \rho_o g (h_1 + h) - \rho_w g h, \dots \dots \dots \dots \dots \dots \dots \dots \dots \dots \dots \dots \dots (2.7)$$

where p_{atm} = atmospheric pressure, dynes/cm²; h and h_1 = fluid heights defined in Fig. 2.10, cm; ρ_o and ρ_w = oil and water densities, g/cm³; and g = gravity acceleration constant, 980 cm/s².

The water pressure is calculated by summing pressure heads through the oil and subtracting the water head in the capillary. The pressure at the oil/water interface in the container is assumed the same as the water pressure at the same elevation in the capillary. Subtracting Eq. 2.7 from Eq. 2.6 gives

$$p_o - p_w = h(\rho_w - \rho_o) g = P_c. \dots \dots \dots \dots \dots \dots \dots \dots \dots \dots \dots \dots \dots \dots \dots (2.8)$$

The result indicates that a pressure difference exists across the interface, which is designated the capillary pressure, P_c. Note that the larger pressure exists in the nonwetting phase.

Recalling that

$$\sigma = \frac{rh(\rho_w - \rho_o) g}{2 \cos \theta}, \dots \dots \dots \dots \dots \dots \dots \dots \dots \dots \dots \dots \dots \dots \dots \dots (2.4)$$

there results

$$\sigma_{ow} = \frac{rP_c}{2 \cos \theta} \dots \dots \dots \dots \dots \dots \dots \dots \dots \dots \dots \dots \dots \dots \dots \dots \dots \dots \dots \dots (2.9)$$

or $P_c = \dfrac{2\sigma_{ow}\cos\theta}{r}$. . . . . (2.10)

The capillary pressure is thus related to the fluid/fluid IFT, the relative wettability of the fluids (through θ), and the size of the capillary, r. The capillary pressure may be positive or negative; the sign merely expresses in which phase the pressure is lower. The phase with the lower pressure will always be the phase that preferentially wets the capillary. Notice that P_c varies inversely as a function of the capillary radius and increases as the affinity of the wetting phase for the rock surface increases. This concept is extremely important to the discussion that follows.

The straight capillary is an idealistic and simplistic approximation to capillary phenomena in oil-bearing rocks. The complexity of the pore structure prohibits rigorous analytical examination. One successful treatment (Leverett 1941), however, involves a configuration made up of uniform spherical particles of definite sizes on the order of magnitude found in oil-bearing rocks. For this system of unconsolidated uniform spheres, an expression for the capillary pressure was developed by Plateau (1863).

$$P_c = \sigma\left(1/R_1 + 1/R_2\right),$$ . . . . (2.11)

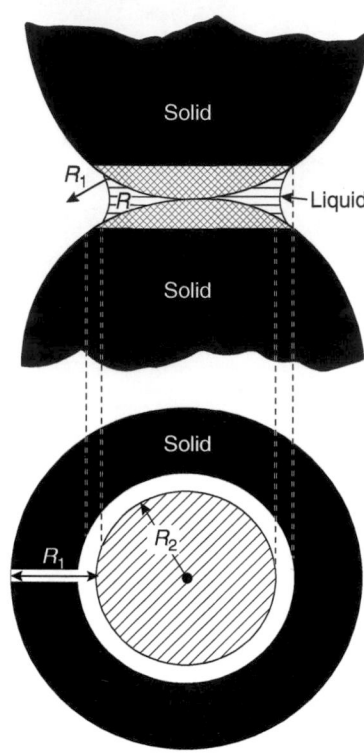

where R_1 and R_2 are the radii of curvature shown in **Fig. 2.11** (Leverett 1941). Eq. 2.11 has been called the Laplace equation and shown to be a general relationship if R_1 and R_2 are taken as the principal radii of curvature of the fluid/fluid interface at the localized position of capillary pressure determination. In a straight capillary, $1/R_1$ and $1/R_2$ are equal and are given by the radius of the capillary divided by the cosine of the contact angle, $r/\cos\theta$.

The values of R_1 and R_2 are related to the saturation of the wetting-phase fluid within a porous medium. Therefore, the capillary pressure depends on the saturation of the fluid phase that wets the system, although the exact nature of the dependence may not be simply stated because the variation of R_1 and R_2 with the saturation is quite complex.

There are a number of methods to measure capillary pressures in reservoir rocks, and numerous papers exist on the interpretation of capillary pressure curves. Willhite (1986) discusses capillary pressure data with implications for fluid distributions in reservoirs. **Fig. 2.12** (Craig Jr. 1971) shows typical drainage and imbibition curves for a reservoir rock. Hysteresis of the type shown is a common phenomenon discussed by several authors (Willhite 1986; Craig Jr. 1971). Also, the drainage curves generally exhibit a threshold pressure, as shown in Fig. 2.12, that is related to the size of the larger pores in the rock. The steeply rising portion of the curve occurs at a water saturation generally corresponding to the connate water saturation.

Fig. 2.11—Wetting of spheres showing radii of curvature (Leverett 1941).

2.3 Viscous Forces

Viscous forces in a porous medium are reflected in the magnitude of the pressure drop that occurs as a result of flow of a fluid through the medium. One of the simplest approximations used to calculate the viscous force is to consider a porous medium as a bundle of parallel capillary tubes. With this assumption, the pressure drop for laminar flow through a single tube is given by Poiseuille's law (Craft and Hawkins 1959).

$$\Delta p = -\frac{8\mu L\bar{v}}{r^2 g_c},$$ . . . . (2.12)

where Δp = pressure drop across the capillary tube, $p_2 - p_1$, lbf/ft²; L = capillary-tube length, ft; r = capillary-tube radius, ft; $\bar{v}$ = average velocity in the capillary tube, ft/sec; μ = viscosity of flowing fluid, lbm/(ft-sec); and g_c = conversion factor.

For an alternative set of units,

$$\Delta p = -\left(6.22\times10^{-8}\right)\left(\frac{\mu L\bar{v}}{r^2 g_c}\right),$$ . . . . (2.13)

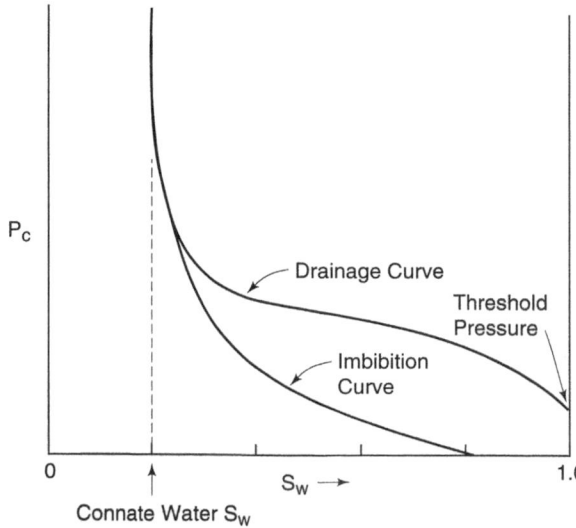

Fig. 2.12—Typical capillary pressure curves for a reservoir rock, water-wet system.

where Δp is in psi, r is in in., $\bar{v}$ is in ft/D, μ is in cp, and L is in ft.

Viscous forces in a porous medium can be expressed in terms of Darcy's law,

$$\Delta p = -(0.158)\left(\frac{\overline{v}\mu L\phi}{k}\right), \dots\dots\dots\dots\dots\dots\dots\dots\dots\dots\dots\dots\dots (2.14)$$

where Δp = pressure drop across the porous medium $p_2 - p_1$, psi; $\overline{v}$ = average velocity of fluid in the pores of the porous medium, ft/D; μ = fluid viscosity, cp; L = length of the porous medium, ft; ϕ = porous medium; and k = permeability of the porous medium, darcies. For a bundle of capillaries of equal size, the permeability is given by

$$k = 20 \times 10^6 d^2 \phi, \dots\dots\dots\dots\dots\dots\dots\dots\dots\dots\dots\dots\dots\dots (2.15)$$

where k = permeability of the bundle of capillary tubes, darcies; d = diameter of the capillary tubes, in.; and ϕ = effective porosity of the bundle of capillaries.

The magnitude of viscous forces can be illustrated with example calculations.

Example 2.2—Calculation of Pressure Gradient for Water Flow in a Capillary. Calculate the pressure gradient, $\Delta p/L$, for water flow at a typical reservoir rate of 1.0 ft/D through a straight capillary with a diameter of 0.004 in. and a water viscosity of 1.0 cp.

Solution. Eq. 2.12 is applicable.

$$\frac{\Delta p}{L} = -\frac{8\mu\overline{v}}{r^2 g_c}$$

$$\frac{\Delta p}{L} = -\frac{8\left(0.672 \times 10^{-3}\,\text{lbm/ft-sec}\right)\left(1.157 \times 10^{-5}\,\text{ft/sec}\right)}{\left(1.667 \times 10^{-5}\,\text{ft}\right)^2\left(32.2\,\text{lbm-ft/lbf-sec}^2\right)}$$

$$= -6.95\,\text{lbf/(ft}^2\text{-ft)} = -0.048\,\text{lbf/(in.}^2\text{-ft)}$$

$$= -0.048\,\text{psi/ft}.$$

The effective permeability of this single capillary is given by

$$k = 20 \times 10^6 \times 1.0 \times (0.0004)^2$$

$$= 3.2\,\text{darcies}.$$

Example 2.3—Calculation of Pressure Gradient for Viscous Oil Flow in a Rock. Calculate the pressure gradient for flow of an oil (viscosity of 10 cp) at an interstitial flow rate of 1.0 ft/D. The rock permeability is 250 md, and the porosity is 0.20.

Solution. Eq. 2.14 is applicable.

$$\frac{\Delta p}{L} = -\frac{0.158 \times 1.0\,\text{ft/D} \times 10\,\text{cp} \times 0.2}{0.250\,\text{darcies}}$$

$$= -1.264\,\text{psi/ft}.$$

Examples 2.2 and 2.3 illustrate that viscous forces yield pressure gradients in reservoir rocks on the order of 0.1 to > 1 psi/ft. These are typical values for the bulk of the reservoir volume. Values can be significantly higher for the regions around injection and production wells.

2.4 Phase Trapping

Trapping of oil or other fluids in a porous medium such as a reservoir rock is not understood completely and cannot be rigorously described mathematically. The trapping mechanism, however, is known to depend on (1) the pore structure of the porous medium, (2) fluid/rock interactions related to wettability, and (3) fluid/fluid interactions reflected in IFT and sometimes in flow instabilities. Trapping and mobilization are related to these factors in a complex way.

While a rigorous analysis is not possible, a number of models partly describe the forces involved in phase trapping and mobilization. In addition, considerable experimental data on trapping have been reported and correlated with various parameter groups. A few of the simple models and important data correlations are presented in this section. The closely related phenomenon of mobilization of trapped phases is discussed in the following section.

2.4.1 Trapping in a Single Capillary. *Trapping in a Single Capillary—Jamin Effect.* It has been recognized for some time that the pressure required to force a nonwetting phase through a capillary system, such as a porous rock, can be quite high. This phenomenon, called the Jamin effect, has been discussed by several authors (Gardescu 1930; Muskat 1949; Bethel and Calhoun 1953; Taber 1969). The phenomenon can be described most easily by analyzing a trapped oil droplet or gas bubble in a preferentially water-wet capillary, as shown in **Figs. 2.13 and 2.14.** The physical conditions for these cases are described, and then the cases are treated quantitatively. The analyses use the IFT, wettability, and capillary pressure concepts discussed earlier.

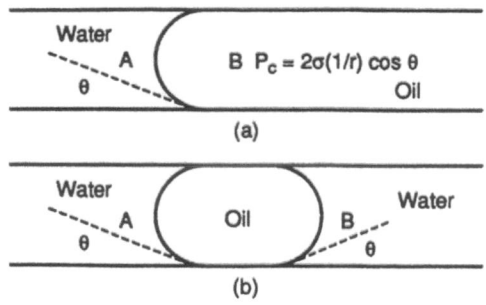

Fig. 2.13—Illustration of oil/water interfaces: continuous phases vs. trapped drop.

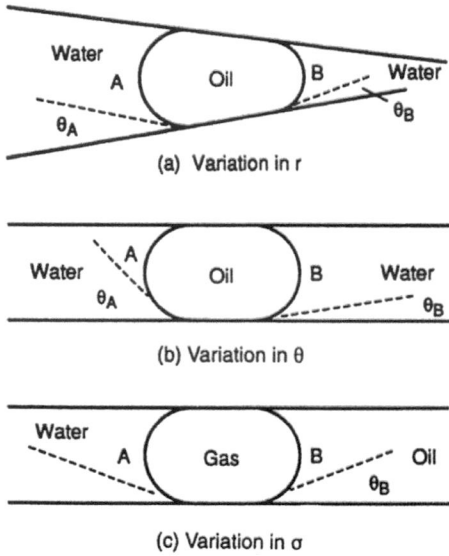

Fig. 2.14—Different conditions of trapping of a droplet in a capillary.

Fig. 2.13 shows two possible conditions of drop size. In Fig. 2.13a, a very long, continuous water filament is in contact with a continuous oil filament. The system is static, with different pressures existing at Points A and B because of capillary forces. The static pressure difference must be exceeded for flow to occur. In Fig. 2.13b, water contacts both sides of a finite oil drop. Again the static pressure difference, $p_B - p_A$, if one exists, must be overcome to initiate flow.

Fig. 2.14 shows three variations of the case shown in Fig. 2.13b, the finite drop in contact with water on both sides of the drop. In Fig. 2.14a, the capillary tube size varies and therefore the radius is smaller on one side of the drop than on the other. Fig. 2.14b shows a situation where the contact angle is different on the two sides of the drop, which could result, for example, if the drop were displaced in one direction, causing an advancing contact angle different from the receding angle. Finally, in Fig. 2.14c, a gas drop is trapped between water on one side and oil on the other. Again, determining the static pressure differences between Points A and B is important because this value must be exceeded to initiate flow.

For the condition in Fig. 2.13a, the pressure across the interface is just the capillary pressure, P_c:

$$p_B - p_A = P_c = \frac{2\sigma_{ow}\cos\theta}{r}. \qquad\qquad (2.16)$$

For the other cases in Figs. 2.13 and 2.14, a generalized expression can be written by simply adding pressure drops across each interface. An assumption is made that the pressure within the oil or gas drop is constant from one end of the drop to the other.

$$p_B - p_A = \left(\frac{2\sigma_{ow}\cos\theta}{r}\right)_A - \left(\frac{2\sigma_{ow}\cos\theta}{r}\right)_B \qquad (2.17)$$

Subscripts A and B indicate that the values are determined for the interfaces at Points A and B, respectively. Application of Eq. 2.17 to the cases in Figs. 2.13b and 2.14a through 2.14c yields the following forms, respectively.

Fig. 2.13b.

$$p_B - p_A = \left(\frac{2\sigma_{ow}\cos\theta}{r}\right)_A - \left(\frac{2\sigma_{ow}\cos\theta}{r}\right)_B = 0, \qquad (2.18)$$

because the conditions at Point A are the same as at Point B. Pressure in the oil phase would exceed the pressure in the water phase by the value of P_c, but there would be no net pressure change across the drop.

Fig. 2.14a.

$$p_B - p_A = 2\sigma_{ow}\cos\theta\left(\frac{1}{r_A} - \frac{1}{r_B}\right). \qquad (2.19)$$

Assuming that $\theta_A = \theta_B$, the pressure difference at static conditions is directly proportional to the difference, across the oil drop, of the inverse of the capillary radii. If $r_B < r_A$, then $p_A > p_B$ and a pressure drop exists in the direction from Point A to Point B. This gradient would have to be exceeded to induce flow into the narrower part of the capillary constriction. The drop is trapped at a finite pressure difference of $(p_B - p_A)$.

Fig. 2.14b.

$$p_B - p_A = \frac{2\sigma_{ow}}{r}\left(\cos\theta_A - \cos\theta_B\right). \qquad (2.20)$$

For an advancing contact angle at Point B and a receding angle at Point A, $\theta_A > \theta_B$ and $\cos\theta_A < \cos\theta_B$. This situation occurs when the drop is on the verge of moving to the right in the figure. Again, $p_A > p_B$ and a pressure gradient exists in the potential direction of flow at static, or trapped, conditions.

Fig. 2.14c.

$$p_B - p_A = \frac{2}{r}\left(\sigma_{ow}\cos\theta_A - \sigma_{go}\cos\theta_B\right). \qquad (2.21)$$

In this case, IFT and contact angles are different at the two interfaces because the fluid systems are different. Again, if σ_{go} cos θ_B > σ_{gw} cos θ_A, a pressure drop exists from Point A to Point B when this system is static.

For any of the situations, if the parameters are known or can be estimated, then the pressure drop required to initiate flow (i.e., to overcome the trapped conditions) can easily be calculated.

Example 2.4—Pressure Required To Force an Oil Drop Through a Pore Throat. Calculate the threshold pressure necessary to force an oil drop through a pore throat that has a forward radius of 6.2 μm and a rear radius of 15 μm. Assume that the wetting contact angle is zero and the IFT is 25 dynes/cm. Express the answer in dynes/cm² and psi. What would be the pressure gradient in psi/ft if the drop length were 0.01 cm?

Solution.

$$p_A - p_B = 2 \times 25 \text{ dynes/cm} \left(\frac{1}{0.00062} - \frac{1}{0.0015} \right) 1/\text{cm}$$

$$= -47,300 \text{ dynes/cm}^2$$

$$= -47,300 \text{ dynes/cm}^2 \times 1.438 \times 10^{-5} \text{ psi/(dynes/cm}^2)$$

$$= 0.68 \text{ psi}$$

$$\Delta p/L = -\frac{0.68 \text{ psi}}{0.01 \text{ cm}} \times \frac{30.48 \text{ cm}}{\text{ft}} = -2,073 \text{ psi/ft.}$$

As Example 2.4 illustrates, the pressure gradient required to move an oil drop through a constriction can indeed be quite large. It is misleading, however, to assume that the pressure gradient calculated for the single capillary exists throughout the length of the flow path in a reservoir. This assumption would be true if a reservoir consisted of a single capillary path with pore constrictions in series. In fact, numerous alternative paths for fluid flow in a permeable rock exist, and this mitigates the effect of single-capillary trapping. Nonetheless, the Jamin effect and calculations like that in Example 2.4 illustrate that trapping forces are large compared with typical viscous forces and that they will not be overcome by normal waterflood conditions. That is, oil drops trapped as residual oil in a waterflood are not likely to be displaced by additional waterflooding at conditions of injection rates and wellhead pressures normally attainable. The effect of altering parameters, such as the IFT, to reduce trapping forces is discussed later.

Trapping in a Single Capillary With Fluid Bypassing. Pore channels in reservoir rocks are not straight, smooth capillaries but irregularly shaped channels. Isolated oil drops in a channel do not ordinarily seal the channel; bypassing by a second phase is possible because of channel geometry. The calculations of the Jamin effect in the previous section assumed static conditions with no bypassing of oil drops by the water phase.

A simplification of the situation that exists in a rock is shown in **Fig. 2.15,** which is based on an experimental study by Arriola (1980). The capillary used in the study was approximately 100 μm square in cross section. The constriction was also approximately square and was approximately 10 μm at its narrowest point. An oil drop (nonane) isolated in the capillary was displaced by flowing water. Because the capillary was square in cross section, the oil drop could not fill the entire cross section and water was able to bypass the oil drop at the corners of the capillary. At the constriction of the capillary, the oil drop became trapped as a result of the Jamin effect discussed earlier. Water continued to bypass the trapped drop, as illustrated in Fig. 2.15.

A force balance indicates that the oil drop should move into the constriction until capillary forces are balanced by viscous forces in the flowing water. At that condition, the oil drop will be trapped. Viscous forces consist of form drag (resulting from the geometry) and interfacial shear.

Fig. 2.16 shows experimental results for the system (Arriola 1980). Measured pressure drop across the constriction is plotted as a function of the distance of the drop-front interface from the throat of the constriction (point of minimum diameter of the constriction). Results of several experiments for a number of different water flow rates are shown. Fig. 2.16 also contains a plot of the calculated pressure drop based on the Laplace equation and the radius of curvature of the front interface of the drop. This calculation is basically the Jamin-effect calculation described in the previous section. Drop-interface curvature was determined with a microscope that measured capillary dimensions and drop position. Water was assumed to wet the capillary completely (cos θ = 1.0).

Laplace's equation fits the experimental data over part of the range, but yields a pressure drop that is a few percent low when the drop interface is less than approximately 50 μm from the throat. Arriola et al. (1983) attributed the difference to optical distortion in microscope readings, unidentified wall effects, and viscous losses in the drop resulting from internal fluid circulation. In general, however, the agreement is quite good, illustrating the balance of viscous and capillary forces existing when one fluid phase flows by and around a trapped phase.

2.4.2 Pore-Doublet Model. Another relatively simple model for oil trapping is the pore doublet. In this model, the complexity of the porous medium is extended beyond that of a single capillary by considering flow in two connected parallel capillaries, as illustrated in **Fig. 2.17.** Although this model still lacks the complexity of a real reservoir rock, it allows the concept of differential flows in different flow channels in a rock to be introduced.

Analysis of flow in a pore doublet contributes to the understanding of one of the ways in which oil drops can be isolated and trapped. The following analysis is taken from Willhite (1986). Stegemeier (1977) also presents an analysis of the pore doublet.

In Fig. 2.17, water displaces oil from two pores with radii of r_1 and r_2, respectively. Pore 1 is smaller than Pore 2, and the two pores are connected at Points A and B to form a pore doublet. The pores are assumed to be water-wet, and for purposes of this

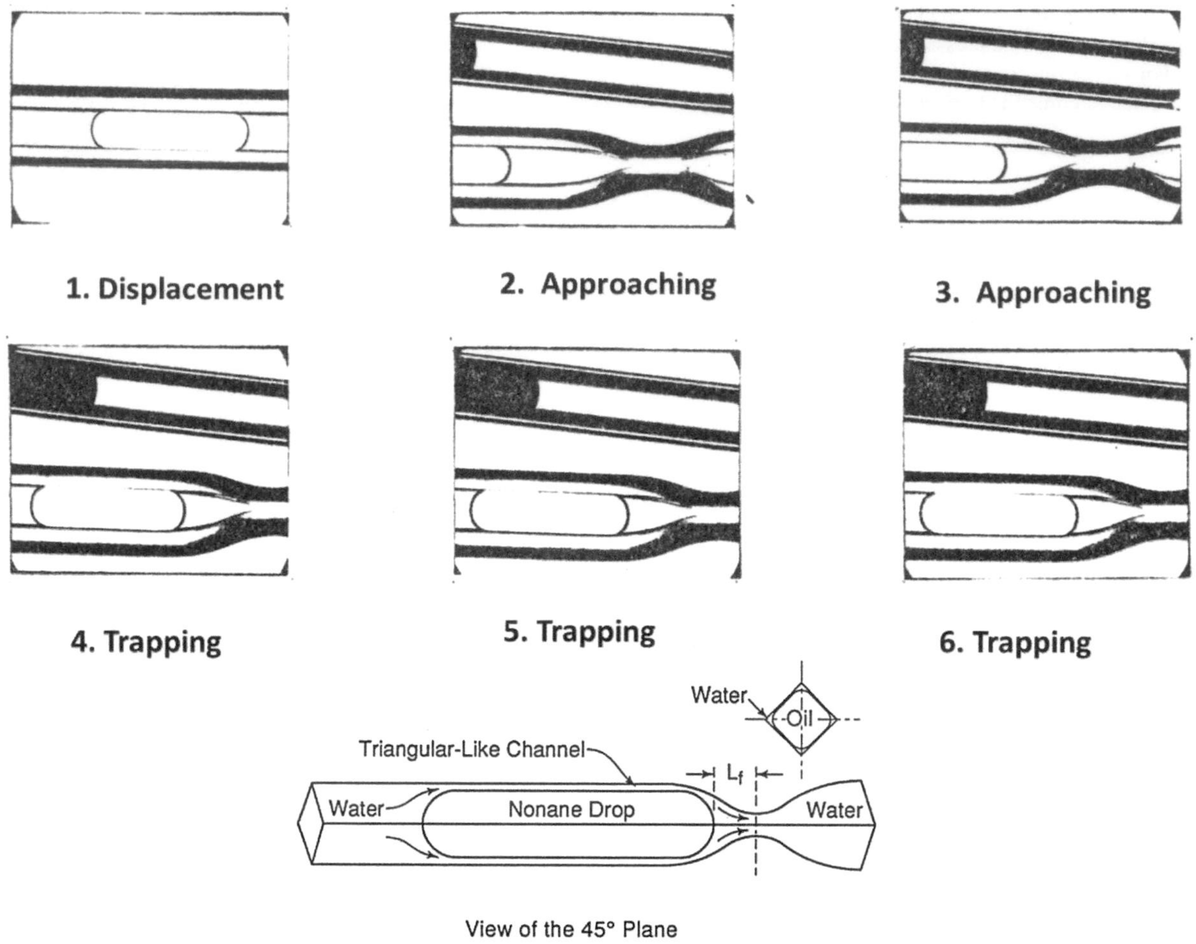

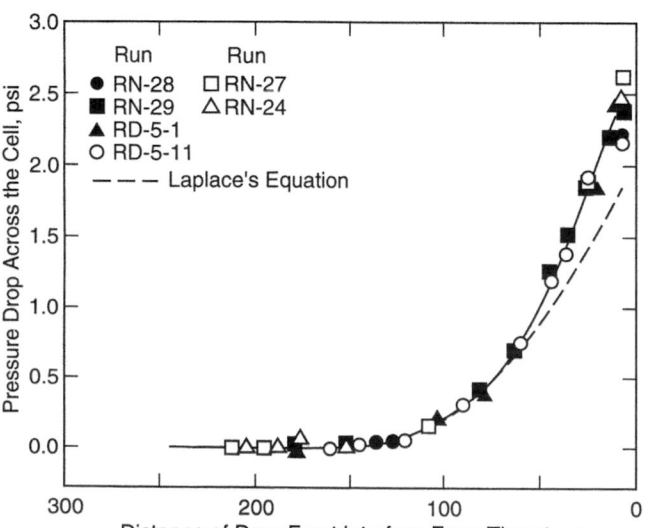

1. Displacement **2. Approaching** **3. Approaching**

4. Trapping **5. Trapping** **6. Trapping**

View of the 45° Plane

Fig. 2.15—Trapping of nonane drops in square capillary constrictions (Arriola 1980).

development, the viscosities and densities of the oil and water phases are assumed to be equal. Oil will be trapped if displacement proceeds faster in one pore than in the other and if there is insufficient pressure difference between Points A and B to displace the isolated oil drop from the pore with the slower displacement rate. Pressures at Points A and B are assumed to remain constant until trapping occurs.

The trapping process can be simulated by estimating the velocity of the water in each pore from elementary models of fluid flow and capillary forces. If the densities of both phases are constant, the flow of each phase will be steady and the flow rate can be computed from Poiseuille's equation for laminar flow in a circular tube (Craft and Hawkins 1959). When the velocity is $\bar{v}_1$, the pressure drop caused by viscous forces between the flowing fluid and the pore walls is given by Eq. 2.12 presented earlier and rewritten here for cm-g-s units.

$$\Delta p_1 = -\frac{8\mu L_1 \bar{v}_1}{r_1^2}, \quad\ldots\ldots\ldots\ldots\ldots (2.22)$$

where L_1 = length of the pore filled with the particular phase.

Because the pores are preferentially wetted by water, a difference in pressure exists between the water and the oil across an oil/water interface, as discussed earlier. The pressure in the oil phase is higher than the pressure in the water phase, as given by Eq. 2.23.

$$\Delta P_c = p_{oi} - p_{wi} = \frac{2\sigma\cos\theta}{r}, \quad\ldots\ldots\ldots (2.23)$$

Fig. 2.16—Pressure drop during the trapping of nonane drops in square capillary constrictions (Arriola 1980).

where ΔP_c is used for capillary pressure to emphasize that it is a pressure change and subscript i denotes the pore channel, either 1 or 2.

Considering the pressure difference between Points A and B after water enters Pore 1, for either pore there results

$$p_A - p_B = \underbrace{p_A - p_{wi}}_{\substack{\text{pressure drop caused by}\\\text{viscous forces in water}\\\text{phase}}} + \underbrace{p_{wi} - p_{oi}}_{\substack{\text{pressure change}\\\text{across interface}\\\text{resulting from}\\\text{capillary force}}} + \underbrace{p_{oi} - p_B}_{\substack{\text{pressure drop in}\\\text{oil phase caused}\\\text{by viscous forces}}}. \quad \ldots (2.24)$$

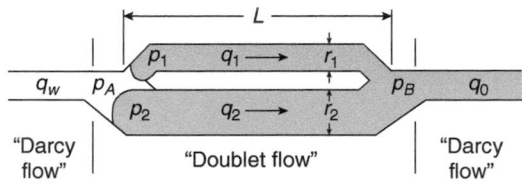

Fig. 2.17—Pore-doublet model for illustration of displacement and trapping of oil filament (Willhite 1986).

Substitution of Eqs. 2.22 and 2.23 into Eq. 2.24 for Pore 1 yields Eq. 2.25 for downstream minus upstream pressure.

$$p_B - p_A = -\frac{8\mu_w L_w \bar{v}_1}{r_1^2} + \frac{2\sigma\cos\theta}{r_1} - \frac{8\mu_o L_o \bar{v}_1}{r_1^2}. \quad \ldots\ldots\ldots(2.25)$$

Because $\mu_o = \mu_w = \mu$ and $L \approx L_w + L_o$,

$$\Delta p_{BA} = p_B - p_A = -\underbrace{\frac{8\mu L \bar{v}_1}{r_1^2}}_{\substack{\text{viscous}\\\text{pressure}\\\text{drop}}} + \underbrace{\Delta P_{c1}}_{\substack{\text{capillary}\\\text{pressure}}} \quad \ldots\ldots\ldots(2.26)$$

The magnitude of the pressure terms on the right side of Eq. 2.26 is illustrated by an example calculation.

Example 2.5—Pressure Change Along a Capillary, Two-Phase Flow. Consider the oil displacement by water in a single pore of radius r at a velocity of 1 ft/D. The length of the pore is 0.02 in., the viscosity is 1 cp, and the IFT is 30 dynes/cm. The contact angle, θ, is zero. Calculate the pressure difference, $p_B - p_A$ for different radius values.

Solution. Eq. 2.26 applies. Numerical values of $p_B - p_A$ for different pore radii are shown in **Table 2.2.**

As Table 2.2 shows, $p_B - p_A$ is always positive for the displacement of a nonwetting phase at rates expected in reservoir rocks (i.e., downstream pressure, p_B, is always larger than upstream pressure, p_A, because of the dominance of capillary forces over viscous forces).

In the pore-doublet model, the same pressure difference $(p_B - p_A)$ exists across both pores until water breaks through from one pore. Thus, for Pore 2,

$$p_B - p_A = -\frac{8\mu L \bar{v}_2}{r_2^2} + \frac{2\sigma\cos\theta}{r_2}. \quad \ldots\ldots\ldots(2.27)$$

Eqs. 2.26 and 2.27 can be used to determine the relative flow velocities in Pores 1 and 2 in the doublet model. **Table 2.3** summarizes the possibilities for $\bar{v}_1$ and $\bar{v}_2$ for different values of Δp_{BA}. For displacement of oil to occur from both pores in the doublet, both $\bar{v}_1$ and $\bar{v}_2$ must be positive.

Table 2.3 shows that for this to occur, two conditions must exist: $\Delta p_{BA} < \Delta P_{c_1}$ and $\Delta p_{BA} < \Delta P_{c_2}$. It also follows that $\Delta P_{c_2} < \Delta P_{c_1}$ because $r_2 > r_1$. Thus, to have $\Delta p_{BA} < \Delta P_{c_2}$,

Pore Radii (µm)	$-8\mu L \bar{v}/r^2$ dynes/cm²	psi × 10⁴	ΔP_c dynes/cm²	psi	$p_B - p_A$ dynes/cm²	psi
2.5	−22.6	−3.25	240,000	3.45	+ 239,977	3.45
5	−5.6	−0.81	120,000	1.73	+ 119,994	1.73
10	−1.41	−0.20	60,000	0.86	+ 59,999	0.86
25	−0.23	−0.033	24,000	0.35	+ 24,000	0.35
50	−0.056	−0.008	12,000	0.17	+ 12,000	0.17
100	−0.014	−0.002	6,000	0.086	+ 6,000	0.086

Table 2.2—Capillary and viscous forces for different sizes of pore radii, Example 2.5.

$\Delta p_{BA} = +\Delta P_{c1}$	$\bar{v}_1 = 0$	$\Delta p_{BA} = +\Delta P_{c2}$	$\bar{v}_2 = 0$
$\Delta p_{BA} < +\Delta P_{c1}$	$\bar{v}_1 > 0$	$\Delta p_{BA} < +\Delta P_{c2}$	$\bar{v}_2 > 0$
$\Delta p_{BA} > +\Delta P_{c1}$	$\bar{v}_1 < 0$	$\Delta p_{BA} > +\Delta P_{c2}$	$\bar{v}_2 < 0$

Table 2.3—Conditions for flow in pore-doublet model.

$$-\frac{8\mu L\bar{v}_1}{r_1^2} + \Delta P_{c_1} < \Delta P_{c_2} \quad \dots\dots\dots\dots\dots\dots\dots\dots\dots\dots\dots\dots \quad (2.28a)$$

$$\text{or} -\frac{8\mu L\bar{v}_1}{r_1^2} < \Delta P_{c_2} - \Delta P_{c_1}. \quad \dots\dots\dots\dots\dots\dots\dots\dots\dots\dots\dots \quad (2.28b)$$

Applying Eq. 2.23 for the capillary pressure terms, we see that a value of $\bar{v}_1$ for which

$$\bar{v} > \frac{\sigma\cos\theta r_1^2}{4\mu L}\left(\frac{1}{r_1} - \frac{1}{r_2}\right). \quad \dots\dots\dots\dots\dots\dots\dots\dots\dots\dots\dots\dots \quad (2.29)$$

is required for $\bar{v}_2$ to be positive (i.e., to move the oil in Pore 2). Eq. 2.29 was used with an equality sign replacing the less-than sign to calculate values of $\bar{v}_1$ corresponding to $\bar{v}_2 = 0$ for different r_2/r_1 ratios for the same example used to prepare Table 2.2. The value of r_1 is 2.5 µm. **Table 2.4** shows results of the calculation.

The $\bar{v}_1$ values in Table 2.4 are orders of magnitude larger than normal reservoir velocities (i.e., to displace any oil from Pore 2, very large velocities are required). According to the pore-doublet model, oil is displaced only from the small pore at this stage of the displacement process. In fact, for velocities that could be present in a reservoir during displacement in Pore 1, the computed velocity in Pore 2 would be negative. Negative or reverse flow is limited and probably does not occur. Probably, the radius of curvature of the oil/water interface becomes smaller, preventing flow from Pore 2 into Pore 1.

At the instant that all oil in Pore 1 has been displaced, the pressure at Point B drops and p_A becomes larger than p_B. For a constant velocity through Pore 1, the pressure difference, $p_B - p_A$, caused by frictional losses in that pore is now available to force the isolated oil globule from Pore 2, as depicted in **Fig. 2.18.** As the figure shows, movement of the oil drop creates a difference in the contact angles (receding vs. advancing contact angles). According to the Jamin effect described earlier, this would result in phase trapping. Also, any narrowing of the pore channel would result in trapping because of differences in capillary radii between the front and back sides of the drop. Example 2.6 illustrates the magnitude of the parameters of contact angle and pore-channel narrowing required to trap the oil drop by the Jamin effect.

Example 2.6—Jamin Effect. Water flowing in Pore 1 of the pore-doublet model has reached Point B, isolating the oil in Pore 2. The radius of Pore 1 is 2.5 µm and the velocity is 1.0 ft/D (0.3 m/d). Find the radius of curvature at the downstream end of the drop that would hold the drop motionless if the radius of Pore 2 is 10 µm. Also, if the capillary were straight, find the advancing contact angle that would hold the drop motionless. (Assume the water is wetting with $\theta = 0$ when the drop is at rest.)

Solution. The pressure drop owing to viscous flow in Pore 1 was computed from Eq. 2.22 in an earlier example and is given in Column 2 of Table 2.2. From the table, $\Delta p_{BA} = -22.6$ dynes/cm².

The capillary pressure at the upstream interface of Pore 2 was computed from Eq. 2.23 and appears in Column 3 of Table 2.2. By substitution into Eq. 2.19, $-22.6 = 60,000 - 2\sigma/r_B$.

$$\frac{1}{r_B} = \frac{60,022.6}{2\sigma} = \frac{60,022.6 \text{ dynes/cm}}{2 \times 30 \text{ dynes/cm}}$$

$$= 1000.38 \text{ cm}^{-1}.$$

$$r_B = 9.996 \text{ µm}.$$

r_2/r_1	$\bar{v}_1$ (ft/D)
2	5,315
4	7,940
10	9,580
20	10,090
40	10,600

Table 2.4—Pore-doublet model, required velocity in small pore to maintain zero velocity in large pore ($r_1 = 2.5$ µm).

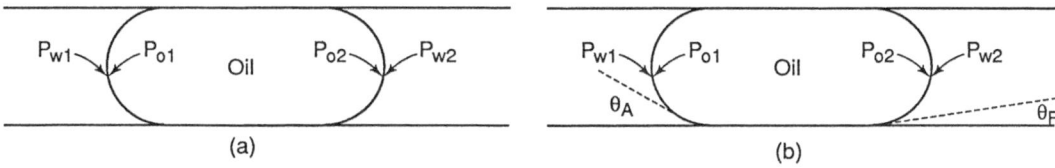

Fig. 2.18—Trapped oil drop in larger pore of pore doublet after displacement of oil from smaller pore (Willhite 1986): (a) isolated oil globule in Pore 2 after water breakthrough in Pore 1, equal contact angles, and (b) advancing and receding contact angles for displacement of isolated globule in Pore 2 after water breakthrough in Pore 1.

Therefore, a decrease of only 0.004 µm in the radius of curvature on the downstream side would be sufficient to trap the isolated oil in the 10-µm pore.

The isolated oil drop could also be trapped in a uniform pore by hysteresis in contact angles. If the receding contact angle were 0°, the advancing contact angle for the motionless globule can be computed from Eq. 2.20:

$$\frac{2\sigma}{r_2}\left(\cos\theta_B - \cos\theta_A\right) = 22.6 \text{ dynes/cm}^2,$$

$$\frac{(2)(30 \text{ dynes/cm})}{10 \text{ µm} \times 10^{-4} \text{ cm/µm}}\left(\cos\theta_B - \cos\theta_A\right)$$

$$= 22.6 \text{ dynes/cm}^2,$$

where $\cos\theta_B - \cos\theta_A = 3.767 \times 10^{-4}$. Because $\theta_B = 0°$, $\cos\theta_A = 0.99962$ and $\theta_A = 1.57°$.

From this example, either a small change in contact angle caused by hysteresis or a small change in pore radius, r_2, would be sufficient to trap the oil globule. In systems composed of water, crude oils, and reservoir rocks, oil/water contact angles exhibit much greater hysteresis than calculated in this example.

The pore-doublet model is by no means an exact representation of a porous medium. It does, however, incorporate the mechanism of competing flows in parallel flow channels that exists in reservoir rocks. The influence of capillary forces on the movement of fluids in the parallel channels is demonstrated. Nonuniform flow resulting from capillary forces is shown to lead to isolation of oil in the larger pore (where water is the wetting phase). Once the oil drop is isolated, it becomes trapped by capillary forces. The example calculations clearly indicate the dominance of capillary forces over viscous forces for typical reservoir-rock conditions.

2.4.3 Experimental Data in Reservoir Rocks—Capillary Number Correlation. Considerable experimental data exist on the trapping of residual oil in rocks and other porous media. Most of these data consist of measurements of residual saturations when a nonwetting phase (oil) is displaced by a wetting phase (water). Fewer data exist for trapping of a wetting phase displaced by a nonwetting phase. The list below gives typical experimental steps in the collection of such data. For this discussion, water will be used as the displacing fluid and oil as the displaced fluid, although the procedure is applicable for any two immiscible phases.

1. A core is first saturated with the water phase to be evaluated, part of which will remain as an initial water saturation after Step 2.
2. The core next is flooded with oil to establish an initial oil saturation. This flood leaves a residual saturation of water comparable to that found as connate water in reservoirs.
3. The core is then flooded with the water phase being evaluated at a specific constant rate. This flood establishes ROS for the particular flood conditions. The water phase in Step 3 is the same as in Step 1.
4. Material balances on all fluids are maintained for each step of the experiment to determine saturations. Pressure drops are also measured.

This procedure leaves a residual saturation that results from trapping of a phase that initially was at a relatively high saturation and was continuous (i.e., the phase flowed initially as connected stringers or ganglia). Trapping is a result of saturation reduction that leads, for a nonwetting phase, to isolation of drops or ganglia, as described earlier for the doublet model. For a displaced wetting phase, thin films that cover the surface lose hydraulic connectivity, leaving the wetting phase distributed in the smallest pores and crevices.

Experimental studies of the type described yield data on residual saturations of a displaced phase for the conditions and parameters of the particular experiments. A method of correlating the data is desirable. The use of dimensionless groupings of variables involving the ratio of viscous to capillary forces has proved to be reasonably successful. An analysis of the pore-doublet model yields the following dimensionless grouping of parameters (Moore and Slobod 1956), which is a ratio of the viscous to capillary forces in flow through a capillary.

$$\frac{F_v}{F_c} = \frac{v\mu_w}{\sigma_{ow}\cos\theta}, \dotfill (2.30)$$

where F_v and F_c = viscous and capillary forces, respectively, and v = interstitial velocity.

Subscript w denotes displacing phase, and σ_{ow} is the IFT between the displaced and displacing phases. The dimensionless group, or variations of the group, are called the capillary number. In this text, the capillary number is defined as in Eq. 2.30, but without the $\cos\theta$ term. Therefore,

$$N_{ca} = \frac{v\mu_w}{\sigma_{ow}}. \dotfill (2.31)$$

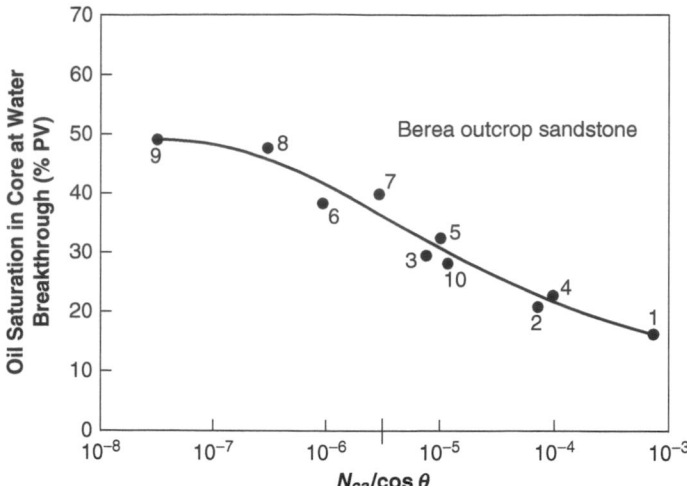

Fig. 2.19—Correlation of $N_{ca}/\cos\theta$ term with oil saturation in core at water breakthrough (Moore and Slobod 1956; Abrams 1975). Abrams converted μ to poise for dimensional consistency.

Any set of consistent units can be used because the group is dimensionless. Some investigators use interstitial velocity in the definition and others use Darcy velocity. Eqs. 2.30 and 2.31 use interstitial velocity. The relationship between the two forms is $N_{ca}^* = \phi N_{ca}$, where N_{ca}^* is based on Darcy velocity.

Figs. 2.19 and 2.20 (Moore and Slobod 1956; Abrams 1975) show two experimental data sets that have been correlated by use of the dimensionless group in Eq. 2.31. The plots show ROS as a function of $N_{ca}/\cos\theta$ for a large number of experiments. The data collectively represent experiments on cores of various lengths and in which velocity, viscosity, and IFT were varied over significant ranges. Water was the wetting phase for all experiments shown. The data of Moore and Slobod (1956) (Fig. 2.19) were measured with the procedure described earlier. However, the reported residual saturations are values at the time of water breakthrough rather than final saturations. Oil recovery after water breakthrough was small for the water-wet system investigated.

The data in Fig. 2.20 (Abrams 1975) came from a strongly water-wet porous medium with $\cos\theta \approx 1.0$. The velocity used by Abrams was $u/[\phi(S_{oi} - S_{or})]$.

The capillary number, N_{ca}, as defined in the preceding does a reasonable job of correlating the data, although there is significant scatter. Basically, however, the results show that at capillary numbers less than approximately 10^{-6}, the residual oil is relatively constant and is not a function of the magnitude of N_{ca}. Waterfloods typically operate at conditions where $N_{ca} < 10^{-6}$, and N_{ca} values on the order of 10^{-7} are probably most common. This implies, for example, that waterflood recoveries should be independent of injection rate over the range of values that can be accomplished in practice. Moore and Slobod (1956) showed that waterflood recoveries from laboratory cores in which waterflood water was imbibed into the cores were just as good as those when water was injected at typical rates used in the field. This result agrees with the experience of other investigators.

The correlation also indicates that if the value of N_{ca} could be increased to more than approximately 10^{-5} in a flood, then the magnitude of residual oil would decrease. As seen, the projected decrease in S_{or} is a smooth function of N_{ca}. At values on the order of 10^{-2}, virtually all oil is recovered.

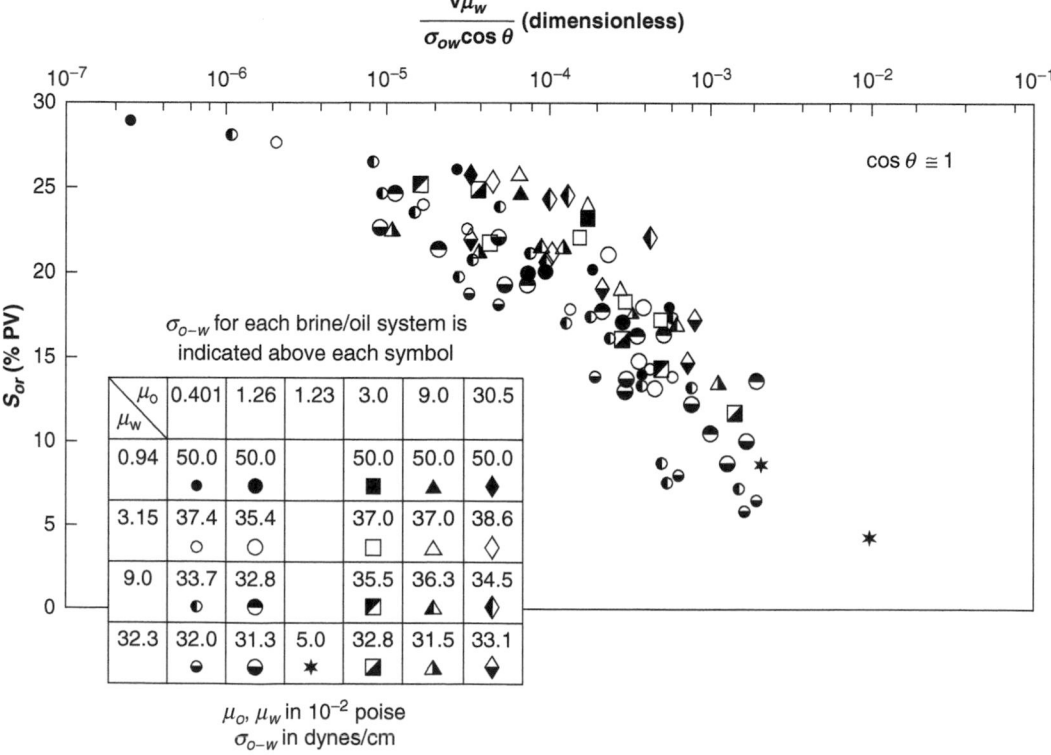

Fig. 2.20—Correlation of N_{ca} term with residual oil saturation (Abrams 1975).

The correlation shows that N_{ca} can be increased by either (1) increasing the flow rate of the displacing fluid, (2) increasing the viscosity of the displacing fluid, or (3) reducing IFT between the displaced and displacing fluids. This implication has been verified by experiments, as indicated by the data in Fig. 2.20. All three variables were changed in different experiments within the experimental set, and yet, with the use of N_{ca}, the data are grouped along one curve.

Abrams (1975) suggested that the correlation would be improved if the water/oil viscosity ratio were considered. His improved correlation is shown in **Fig. 2.21**, where N_{ca} is multiplied by $(\mu_w/\mu_o)^{0.4}$. These are the same data used for Fig. 2.20, but the viscosity ratio term has been added in the correlation. As can be seen, the improved correlation reduces the data scatter.

Abrams (1975) also investigated the effect of rock type. For a number of sandstones, the behavior was approximately the same as that shown in Fig. 2.21. **Fig. 2.22** shows the effect of rock type. The sandstones all showed a change of slope at values of the correlating parameter, $N_{ca}(\mu_w/\mu_o)^{0.4}$, of approximately 10^{-4} to 10^{-5}. A limestone sample investigated exhibited a linear decrease in S_{or} in the range of the parameter of 10^{-6} to 10^{-2}. The correlations of Figs. 2.19 through 2.21 contain $\cos\theta$ in the denominator of the capillary number term. The implication is that if the fluids were close to neutral wettability ($\theta \to 90°$), the term would increase significantly and residual saturation would be reduced markedly, but this has not been tested. The magnitude of water-wettability has been varied in the different experiments, but over a moderate range. Thus, the validity of the correlation has not been established at conditions close to neutral wettability. The general effect of wettability on phase trapping will be discussed later.

The validity of the capillary number as a correlating parameter has also been demonstrated in a single-capillary model (Arriola et al. 1983). The physical model was described earlier and is shown in Fig. 2.15, in which a capillary with an approximately square cross section was used. In the experiments, nonane drops were displaced through the capillary by a second immiscible liquid. The nonane drops were trapped at a constriction in the capillary and held there by viscous forces resulting from drag and interfacial shear caused by the other flowing liquid. In different experiments, either water, heptyl-alcohol/water, or isobutyl-alcohol/water systems were used as the displacing fluids. IFTs between these liquids and the nonane were 32.1, 8.16, and 2.31 dynes/cm, respectively. Flow velocities also were varied in different experiments.

A modified capillary number was defined as (Arriola 1980)

$$N_{ca,m} = \frac{\Delta p}{\sigma b_c}, \dotfill (2.32)$$

where Δp = pressure drop across the capillary, σ = IFT between the immiscible liquid pairs, and b_c = minimum width of the capillary in the constriction. The term is dimensionless, so any consistent set of units is applicable.

Results for the experiments are shown in **Fig. 2.23**, in which the distance of the front end of a trapped nonane drop from the throat (point of minimum constriction width) is plotted vs. $N_{ca,m}$. The data lie along one smooth curve, indicating that $N_{ca,m}$ is a viable correlating group. Under conditions where the nonane drop approached within approximately 8 μm of the throat, the drop passed through the constriction. Again, various values of $N_{ca,m}$ were achieved in the experiments by variation of Δp (through flow-rate variation) or σ. The initial nonane drop length was the same in all runs.

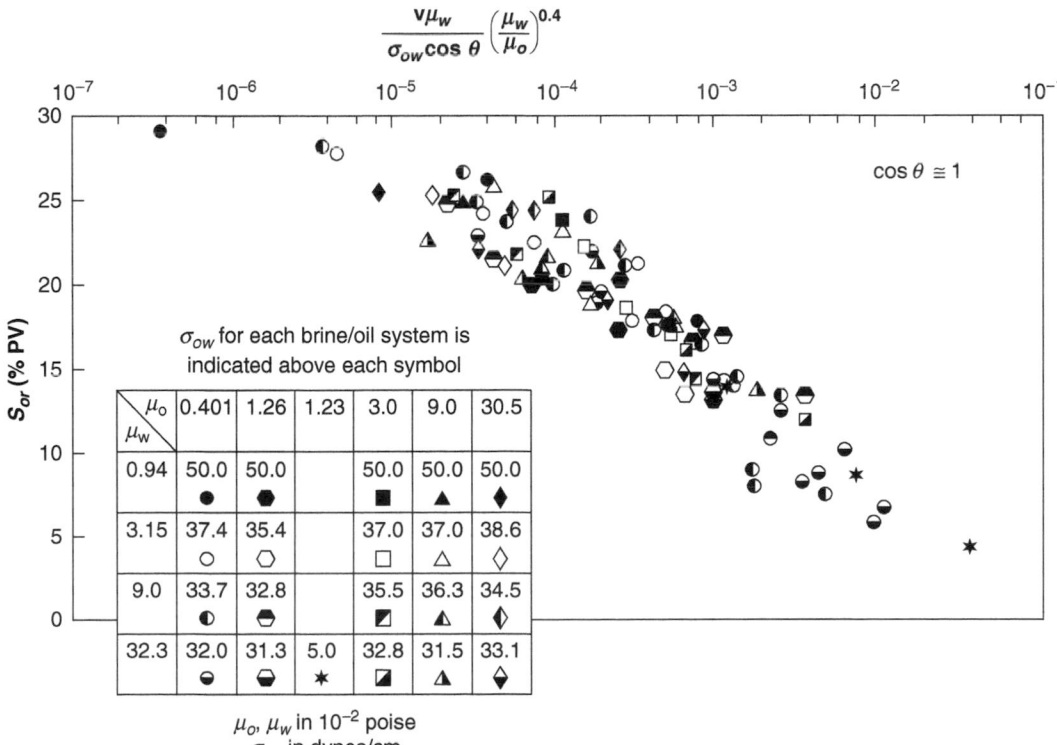

Fig. 2.21—Correlation of N_{ca} $(\mu_w/\mu_o)^{0.4}$ with residual oil saturation (Abrams 1975).

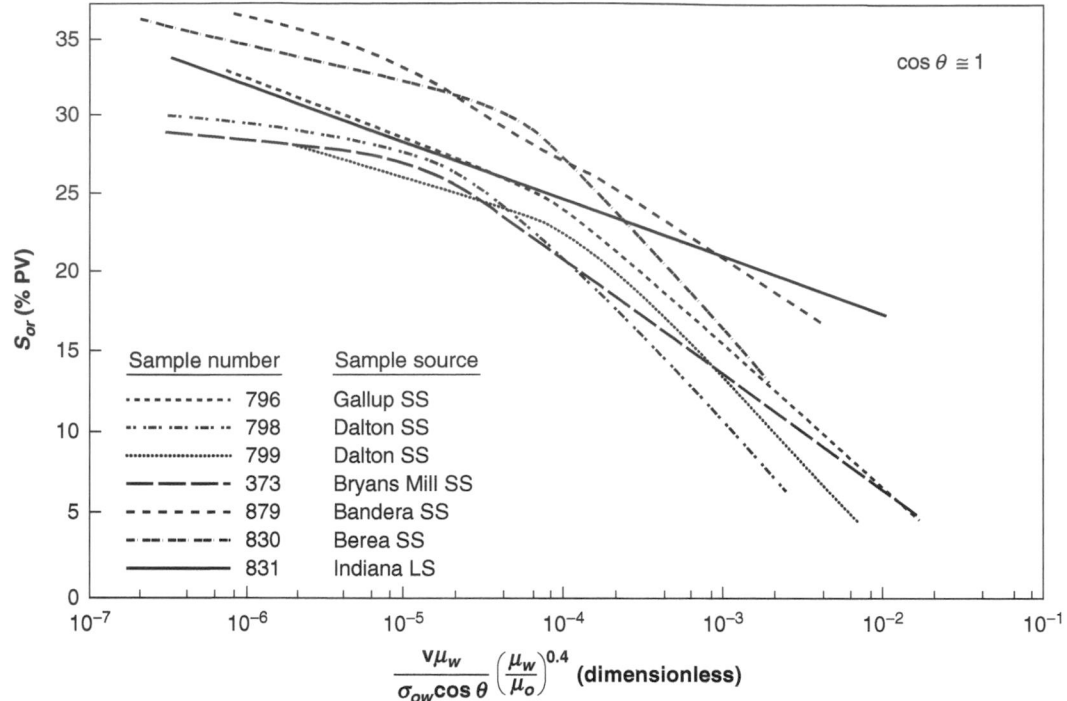

Fig. 2.22—Correlation of N_{ca} $(\mu_w/\mu_o)^{0.4}$ with residual oil saturation for different rocks (Abrams 1975).

Various forms of a capillary number have been defined by different investigators (Stegemeier 1977). If Darcy's law is introduced into Eq. 2.31 and Darcy velocity is used,

$$N_{ca}^* = \frac{k_{rw} k \nabla \Phi}{\sigma_{ow}}, \quad \dots \dots \dots \dots \dots \dots \dots \dots \dots \dots \dots \dots \dots \dots \dots \dots \dots \dots \dots \dots (2.33)$$

where k_{rw} = relative permeability of the displacing phase, k = absolute permeability of the porous medium, and $\nabla \Phi$ = gradient of the flow potential. Any consistent set of units can be used because the group is dimensionless. Eq. 2.33 is analogous to Eq. 2.31. Brownell and Katz (1947) suggested an alternative group in which Eq. 2.33 was divided by k_{rw}. This removes the saturation dependence of the dimensionless group.

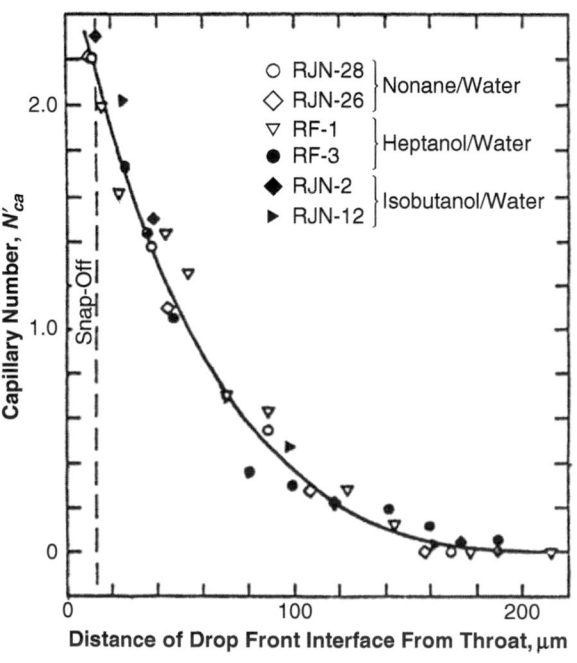

Fig. 2.23—Correlation of $N_{ca,m}$ in a single capillary with a constriction (Arriola et al. 1983).

A more generalized expression for a capillary number, which incorporates porous medium geometrical factors in a manner similar to Dullien (1969) and Melrose and Bradner (1974), is presented by Stegemeier (1977).

2.4.4 Effect of Rock Wettability on Trapping. The model and experimental data described earlier were based on trapping of a nonwetting phase. Solid wettability of a phase affects the nature and, to some extent, the magnitude of trapping.

A prime example of the effect of wettability is the asymmetry of relative permeability curves. **Fig. 2.24** shows typical curves for strongly water-wet and strongly oil-wet systems. At a given saturation of a fluid, the relative permeability to that fluid will be larger if it is the nonwetting rather than the wetting fluid. This is clearly seen, for example, by comparing the relative permeabilities of water at a water saturation of 60% in Figs. 2.24a and 2.24b. The water relative permeability value is much larger when water is the nonwetting phase because of the location of the wetting and nonwetting phases in the pore structure relative to the solid boundaries. Even though relative permeability curves are affected by wettability, for many rocks the residual saturations at which phases cease to flow altogether are not strong functions of wettability.

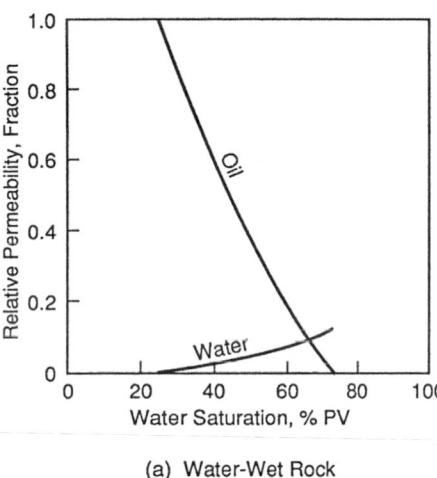
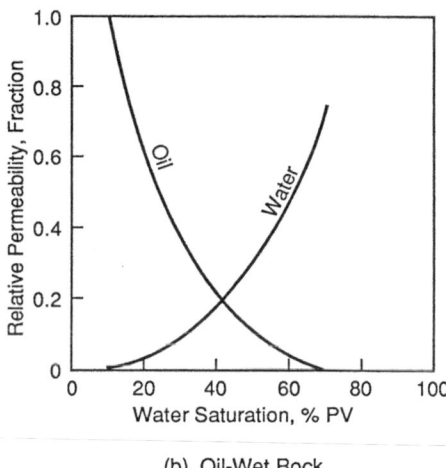

(a) Water-Wet Rock

(b) Oil-Wet Rock

Fig. 2.24—Effect of wettability on relative permeability curves: (a) typical water/oil relative permeability characteristics, strongly water-wet rock, and (b) typical water/oil relative permeability characteristics, strongly oil-wet rock. PV = pore volume.

As discussed earlier, a nonwetting phase tends to be trapped as isolated drops when the nonwetting phase is displaced by a wetting phase. The isolated drops are held by strong capillary forces that cannot be overcome by the relatively small viscous forces, and the trapped phase tends to reside in the larger pores. Flooding at successively larger capillary numbers displaces the trapped phase from smaller to larger pores successively.

When a wetting phase is trapped, it is held in small cracks and crevices interconnected by thin fluid layers around the solid (Reisberg and Doscher 1956). One result of the nature of the wettability and the physical location of a trapped phase is the length or distance over which trapping occurs in porous media. When a nonwetting phase is being displaced, the trapping occurs as isolated drops or ganglia, which reside in the larger pores. The competition of viscous and capillary forces that leads to trapping occurs over very short distances, probably on the order of a few pore lengths. When a wetting phase is trapped by a displacing nonwetting phase, however, the trapping occurs over relatively larger distances in the porous medium. In linear waterfloods in which oil wets the medium, this is reflected in early water breakthrough followed by continued oil production for long periods of time. That is, the oil and water flow together in the porous medium over long distances. The wetting-phase saturation is reduced rather slowly to a point where capillary forces dominate viscous forces and flow ceases. As previously indicated, however, the final residual saturation for a displaced wetting phase is often not too different from the value for a displaced nonwetting phase. The efficiency with which this final residual saturation is accomplished is quite different and is poorer when a wetting phase is the displaced phase.

Phases that display intermediate wettability are those for which $\theta \to 90°$. Ideally, the simple models described earlier apply, regardless of wettability and, conceptually at least, apply for fluids of intermediate wettability. As previously indicated, however, the applicability of the correlation has not been established for systems where $\cos \theta$ closely approaches zero.

The simple capillary models do not incorporate all the complexity of a reservoir rock. For example, assuming no effect of advancing or receding contact angles, a true neutrally wetting fluid ($\theta = 90°$) would not be trapped in a converging capillary. Even for this ideal case of neutral wetting, however, trapping has been demonstrated (Purcell 1950) to occur in rocks because of geometry effects not considered in the simple capillary or pore-doublet models already described. Purcell (1950) considered a "doughnut-shaped pore" where the pore is like the doughnut hole. He showed that, depending on the position of the fluid/fluid interface, the pressure in one phase could be greater than, equal to, or less than the pressure in the other phase, even though the wetting properties and contact angle were the same. Thus, the capillary pressure that must be overcome by viscous forces is a strong function of pore geometry and not a function of wettability and IFT only.

Such arguments have led to the belief that the magnitude of final residual saturation often is not a strong function of wettability. Certainly, however, the location of a trapped phase, the manner in which trapping occurs, and the efficiency of approach to residual saturation are functions of wettability. Mobilization, as will be discussed, also depends on wettability.

2.5 Mobilization of Trapped Phases—Alteration of Viscous/Capillary Force Ratio

2.5.1 Mobilization vs. Trapping. The models and experimental data described earlier provide a description of the trapping phenomenon and insight into the magnitude of forces involved in phase trapping in a porous rock. The experiments discussed, which were correlated with the capillary number, were for physical situations in which the displaced phase was initially at a relatively high saturation. The phase was essentially in continuous contact throughout the porous medium initially. Such a condition would exist, for example, at the initiation of a typical waterflood.

The question arises as to whether a capillary number correlation is applicable for the mobilization of oil or another phase once it has been trapped. The answer is that the same type of correlation is applicable; however, for strongly water-wet systems, curves representing data taken under conditions of trapping will not necessarily coincide with curves representing data taken under conditions of mobilization. Experience indicates that once the nonwetting phase is trapped, it is somewhat more difficult to mobilize. There is hysteresis in the trapping/mobilization process.

Moore and Slobod (1956) present data illustrating this hysteresis. They conducted numerous experiments in which an initially continuous nonwetting phase was trapped by a displacing wetting phase. These data, referenced earlier (Fig. 2.19), showed that the amount of trapping was dependent on the conditions that existed at the front end of the flood (i.e., at the interface between the displacing and displaced phases). That is, the trapping depended on conditions at the specific point of trapping and not on conditions that might occur later as additional fluids moved by the point of trapping. **Table 2.5** gives the results of Moore and Slobod (1956).

The effect of flood rate on residual oil for three different flooding conditions in three separate cores is shown in the nine results at the top of the table. A flooding rate of a constant 2.0 ft/D left ROSs of 41.6, 48.2, and 49.5% in three different cores. This trapped oil was presumably left as isolated oil drops. When corresponding floods were conducted at a much higher rate of 200 ft/D, residual saturation was reduced significantly, as Table 2.5 shows. This result is to be expected from the N_{ca} correlation if viscous forces have been increased by a factor of 100. In the third set of experiments, the initial flood rate was again 2.0 ft/D. After trapping occurred at the lower rate, the rate was increased to 200 ft/D. As seen, residual saturations were somewhat lower than for the run made at a rate of only 2.0 ft/D, but significantly larger than for the run made at the high rate (200 ft/D).

The experiments show that once the oil has been trapped, the process is not completely reversible. In the two-rate experiment, the higher rate mobilized some of the oil trapped during the preceding low-rate flood, but the final residual saturation was not as low as it would have been if the high rate had existed at the point of trapping at the time of trapping. Similar results were obtained when viscosity ratio was varied and when the IFT was reduced, as Table 2.5 shows. Thus, mobilization of a trapped phase apparently does not occur until an N_{ca} value higher than the value that existed when an initially continuous phase was trapped is achieved. This result has been substantiated by other investigators (Chatzis and Morrow 1984).

2.5.2 Mobilization by Alteration of the Viscous/Capillary Force Ratio. The difficulty of mobilizing trapped oil was demonstrated in a set of displacement experiments conducted by Taber (1969). In these experiments, oil at typical residual waterflood saturation in Berea sandstone cores was displaced by increasing the pressure drops in the displacement or by reducing IFT. The sequence of operations was as follows.

Step 1. A core was saturated with a 2.5% brine solution.

Step 2. The core was flooded with an oil phase (Soltrol™) until no more water could be displaced.

Step 3. The core was then very slowly waterflooded to create an ROS. Sometimes this waterflooding step was performed by simple capillary imbibition of water by placing the oil-saturated core in a beaker of water. ROSs after the waterflood were in the range of 37 to 44%.

Step 4. The final displacement of the residual oil thus created was accomplished by increasing the flooding rate (by increasing the pressure drop across the core) until oil was observed to emerge from the core. The pressure drop was then held constant until oil production ceased. At this point, the pressure drop was again increased slightly and the oil production allowed to proceed. This process was repeated with pressure increases occurring in small steps and oil production recorded for each step. For these final displacements, different alcohols and surfactant systems were used (as well as water) so that IFT could be varied in addition to pressure drop.

The result of this work was that no residual oil was produced from the cores unless a "specific" critical value of $\Delta p/L\sigma$ was exceeded. This value was on the order of 6.0 (psi/ft)/(dynes/cm) and was considered a minimum value for the indicated parameter group. A value of approximately 30.0 was required for reasonable production rates of residual oil from the cores.

The magnitude of the pressure gradient required to mobilize residual oil far exceeds values obtainable under normal field waterflood conditions of pressure drop and well spacing. The implication of Taber's work for field application clearly is that if residual oil

	Residual Oil (%PV)		
	Torpedo	Elgin	Berea
Effect of flood rate			
2 ft/D at front	41.6	48.2	49.5
200 ft/D at front	33.8	32.3	39.5
2 ft/D at front followed by 200 ft/D behind front	38.1	44.5	42.6
Effect of favorable viscosity ratio			
μ_o/μ_w = 1.0 at flood front	41.6	48.2	49.5
μ_o/μ_w = 0.055 at flood front	19.3	22.2	22.1
μ_o/μ_w = 1.0 at flood front followed by 0.055 behind front	41.0	47.5	48.8
Effect of reducing IFT			
σ = 30 dynes/cm at flood front	41.6	48.2	49.5
σ = 1.5 dynes/cm at flood front	28.5	27.5	31.5
σ = 30 dynes/cm at flood front followed by 1.5 dynes/cm behind front	41.0	46.0	48.0

Table 2.5—Effect on residual oil of changing capillary or viscous forces at the front and behind the front (Moore and Slobod 1956).

is to be recovered by processes involving fluid displacement, the oil/water IFT must be very significantly reduced to achieve production at reasonable pressure gradients. This conclusion can be shown from the result of Taber's experiments by a simple calculation.

Example 2.7—IFT Required for Oil Mobilization, Taber Experiments. Calculate the IFT required for a Taber number, $\Delta p/L\sigma$, of 30.0 (psi/ft)/(dynes/cm) in a sandstone reservoir if the maximum pressure gradient obtainable is 0.5 psi/ft.

Solution.

$$\frac{\Delta p}{L\sigma} = 30.0 \text{ (psi/ft)/(dynes/cm)}.$$

$$\sigma = \frac{\Delta p}{L(30.0)} = \frac{0.5 \text{ psi/ft}}{30.0 \text{ (psi/ft)/(dynes/cm)}}$$

$$= 0.016 \text{ dynes/cm}.$$

The value of σ calculated in the example is three orders of magnitude smaller than the normal IFT between oil and water. The significance is that surface-active agents, which reduce oil/water IFTs, or miscible processes, which eliminate interfaces entirely, must be used to recover residual oil. The only alternatives would be to alter geometrics and/or to use very close well spacing, which is generally not practical.

Considerable experimental data have been published on the mobilization and removal of residual oil. Stegemeier (1977) summarizes and reviews these results. The porous media used included unconsolidated sand, glass beads, sandstones, and sintered polytetrafluoroethylene particles. Oil and water were the fluids typically used, but in some cases, alcohols and surfactants were added to reduce IFT.

Residual saturation data are correlated as a function of capillary number N_{ca}^* in **Fig. 2.25** (Stegemeier 1977). The ordinate in the figure is the fraction of normal waterflood residual oil that remains trapped following a displacement at the indicated N_{ca}^*. The data represent a diverse set of conditions. The Abrams (1975), Moore and Slobod (1956), and Wagner and Leach (1966) data are for trapping of an initially continuous phase. Two of these data sets were discussed earlier. The other data are for physical situations in which a trapped phase was mobilized and then displaced at the indicated capillary number. More recent data (Chatzis and Morrow 1984; Garnes et al. 1990), both for trapping of a continuous oil phase and mobilization of a trapped phase, are in general agreement with the results in Fig. 2.25.

The following points are emphasized relative to the correlation.

1. Oil that is initially continuous exhibits a partial oil trapping at lower values of N_{ca}^* than those required to mobilize trapped, discontinuous oil. At higher values of N_{ca}^*, however, trapping and mobilization curves appear to be indistinguishable.
2. Variations in the correlating curves, resulting from the use of different types of porous media, are significant. Nonetheless, the general trends of the data are consistent, making the N_{ca}^* a useful way of correlating the data.
3. Wettability is quite important, as illustrated by the data of Dombrowski and Brownell (1954), which are for mobilization of oil trapped as pendular rings.

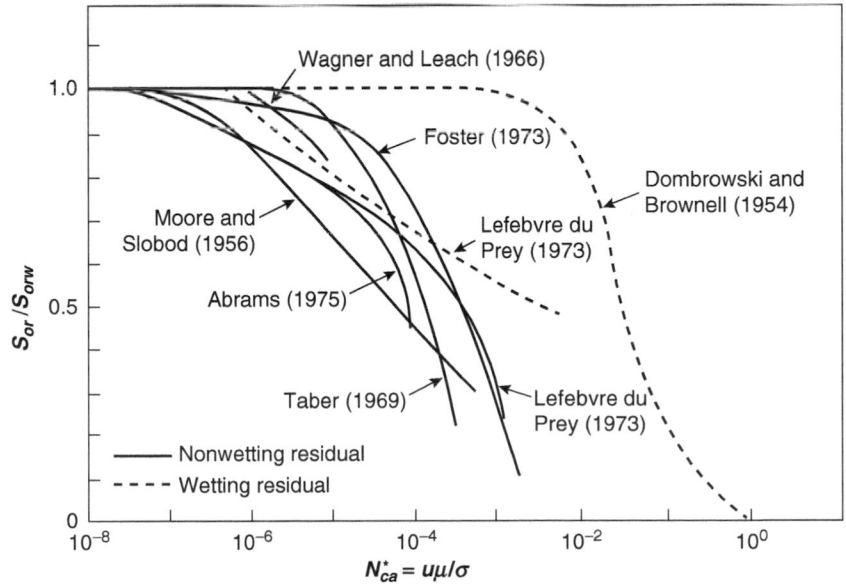

Fig. 2.25—Correlation of recoveries of residual phases as a function of N_{ca}^* (Stegemeier 1977).

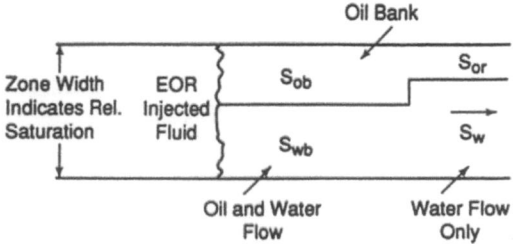

Fig. 2.26—Development of an oil bank in an EOR process initially at ROS.

The data and resulting correlation clearly show that reduction of oil saturation to a value significantly below normal waterflood residual saturation, if accomplished by alteration of the viscous/capillary force ratio, can be achieved only by altering the ratio by several orders of magnitude. In the application of waterflooding, it generally will not be possible to increase the flow rate enough to recover additional oil. Likewise, the use of polymers or other viscosity-enhancing chemicals cannot be used to increase viscous forces to the desired level in field applications. In both cases, allowable pressure drop through the reservoir between injection and production wells is the limiting factor. While both actions—flow-rate increase and/or viscosity increase—might improve recovery to a small extent, the magnitude of increase that can be accomplished in practical application severely limits the utility of this approach.

The only practical alternative to increase microscopic displacement efficiency significantly over that of a waterflood is to use miscible fluids or to create a very large reduction in IFT between the displacing phase and the trapped or mobilized oil. What is required is a "magic" fluid that is miscible (eliminates IFT completely) or has an ultralow IFT with the oil phase. This fluid could then mobilize trapped oil and, having mobilized the oil, could displace it toward producing wells with a microscopic displacement efficiency approaching unity. It is clear that the miscibility or low IFT must be maintained throughout the process. Otherwise, oil that has been mobilized in one part of the reservoir might be trapped at another part where displacement efficiency deteriorates. This latter concern is important because chemicals, such as surfactants, that reduce IFT can be removed from a displacement front through adsorption onto the rock, partitioning into phases that are trapped, and dilution through mixing.

2.5.3 Mobilization of Residual Oil—Formation of an Oil Bank. In preceding sections, the roles of viscous and capillary forces were described for (1) trapping of an oil phase that initially was connected and flowing and (2) mobilization of oil drops or ganglia that were initially trapped. Both processes, trapping and mobilization, may occur in an EOR process. If reservoir oil is at a relatively high saturation at the start of an EOR process, then the displacement must be efficient to minimize trapping of the flowing oil. Alternatively, in processes at or near ROS at their initiation, the displacement must first mobilize residual oil to form a flowing oil bank and then displace this bank efficiently.

Fig. 2.26 illustrates the development and displacement of an oil bank in a reservoir initially at ROS. In this example, the injected EOR fluid has a relatively low IFT, both with water and oil, so that both fluids are displaced efficiently. At the initiation of such a process, trapped oil drops or ganglia are mobilized and connected with other drops. A stabilized oil bank of relatively constant saturation forms that flows ahead of the injected fluid front, as illustrated. Fig. 2.26 shows phase saturations as a function of position at some arbitrary, fixed time.

Two-phase flow exists in the stabilized oil-bank zone ahead of the injected EOR fluid. This oil bank grows in volume (i.e., the length increases as the process proceeds). At the leading edge of the bank, oil is mobilized and solubilized into the bank. Thus, once the bank is established, it becomes the mobilizing fluid. Note that no oil production would occur from the reservoir until the oil bank arrived at the producing point.

At the trailing edge of the oil bank, fluids are displaced efficiently by the injected fluid as long as a relatively large N_{ca} is maintained. If the capillary numbers between the injected fluid and both water and oil are large, no significant trapping of either phase will occur.

Thus, at the initiation of a displacement process involving residual oil, the injected fluid must mobilize oil that was trapped and immobile (i.e., an oil bank must be developed). Once the oil bank is formed, additional oil will be mobilized at the leading edge of the bank. The injected fluid must now displace an oil phase that is continuous (i.e., the injected fluid must prevent trapping of a flowing continuous phase).

Capillary number correlations presented earlier indicate that there is a difference in mobilization and trapping behaviors at relatively large ROSs, not too different from saturations obtained in waterflooding. However, at small final residual saturations, the two phenomena are apparently governed by N_{ca}^* values that are comparable. The implication for process design is that it is not generally necessary to consider one N_{ca}^* value for mobilization and another for trapping.

The development and growth of an oil bank is governed by displacement efficiency, phase behavior, the material balance on the system, and relative mobility relationships. The development of such banks will be illustrated throughout the book and, in some cases, described by simple mathematical models.

2.6 Mobilization of Trapped Phases—Role of Phase Behavior

Mobilization of trapped oil and displacement of oil can be accomplished by use of favorable phase-behavior relationships between the oil and a displacing fluid. Phase-behavior relationships, for example, can result in solubilization of a displacing fluid into the oil with resulting swelling of oil volume. Relative permeability considerations then lead to improved oil recovery. Alternatively, extraction of components from the oil phase into a displacing phase can result in oil components being transported through a rock. And, as described in Chapter 6, alteration of composition through repeated contacts between injected fluids and original reservoir oil can create miscibility between displacing and displaced phases with attendant high displacement efficiency.

The application of ternary phase diagrams has proved useful for description of displacement processes in which phase behavior is important. The following section discusses general concepts that relate to the use of ternary diagrams and to the role of phase behavior in displacement processes.

2.6.1 Phase Behavior—Ternary Diagrams. The description of displacement processes involving liquid/liquid or liquid/vapor equilibrium is facilitated by the use of ternary diagrams to describe the phase behavior. Ternary, or triangular, phase diagrams can be used to plot the phase behavior of systems consisting of three components. In some cases, for systems containing more than three components, certain components can be grouped to form pseudocomponents. A common example is the decomposition of crude oil into CH_4, C_2–C_6 components, and C_{7+} compounds. The phase behavior on a ternary diagram is plotted at fixed pressure and temperature.

A ternary diagram for hypothetical Components A, B, and C is shown in **Fig. 2.27.** Compositions are in weight percent. If these three components were miscible, then no multiphase region would appear on the diagram. The vertexes represent the pure components, and the sides of the equilateral triangle are scaled to represent the binary compositions of the three possible pairs. Systems consisting of all three components are represented by points interior to the triangle. Either weight, mole, or volume percentage may be used, although the last should be used only if there are no significant volume changes on mixing. An equilateral triangle is used in Fig. 2.27. Alternatively, a right triangle may be used, as illustrated by Benham et al. (1960).

As an example of the representation of concentrations, Point M on Side AC in Fig. 2.27 represents a composition of 67.5% of Component A and 32.5% of Component C. This can be determined by measurement along Side AC; Point M is 67.5% of the distance from Point C to Point A. Alternatively, the distance can be measured along a line drawn perpendicular to the base that reaches from the base to Vertex A. The altitude of the triangle represents 100% of a single component, so Point M is 67.5% of the distance from the base to Vertex A.

Point P in the interior of the triangle consists of a mixture of 40% of Component A, 40% of Component B, and 20% of Component C. Point P is located so that the weight of Component A is to the weight of Component B is to the weight of Component C as the ratio of the perpendicular distances Pa is to Pb is to Pc from Point P to the respective sides. The weight fraction of Component C is given by the ratio of Pc to the altitude of the triangle. Lines $\overline{Pa}$, $\overline{Pb}$, and $\overline{Pc}$ sum to equal the altitude of the triangle. In triangular-coordinate graph paper, ruled lines are included so that percentages may be read directly from the paper.

An important property of ternary diagrams is that mixtures of the different components can be easily represented. For example, all mixtures of Components A and C lie along Line AC. Recall that Point M is a mixture of 67.5% of Component A and 32.5% of Component C. Likewise, all ternary mixtures of Binary Mixture M with Component B lie along Line $\overline{BM}$. Point P represents a mixture of 60% of Mixture M and 40% of Component B. Its location is determined by the so-called inverse-lever-arm rule. The Distance MP is to the Distance MB as the amount of Component B is to the total weight of the mixture. Conversely, the Length $\overline{PB}$ is to the Length $\overline{MB}$ as the amount of Mixture M is to the total amount of the mixture. In equation form,

$$\overline{MP}/\overline{MB} = \text{amount of B/total amount of mixture.} \dots\dots\dots\dots\dots\dots\dots\dots\dots\dots\dots\dots\dots\dots (2.34)$$

These same mixing rules apply to the combination of any two mixtures represented on the ternary. All mixtures of Component P with another system defined by Point U on the plot would be along Line $\overline{PU}$. The point representing the composition of the final mixture would again be determined by applying the inverse-lever-arm rule along the line.

Phase relationships may also be represented on a triangular phase diagram, as shown in **Fig. 2.28.** The diagram shows the phase conditions at equilibrium (constant temperature and pressure) for a system consisting of hypothetical Components A, B, and C. Compositions are in mole percent. The plot is typical of hydrocarbon systems in which vapor/liquid equilibrium exists over regions of the concentration domain. All concentrations represented by points lying inside the two-phase envelope would separate into two phases, while systems with concentrations lying outside the two-phase envelope would exist as a single phase. The curve defining this boundary of the two-phase region is called the binodal curve.

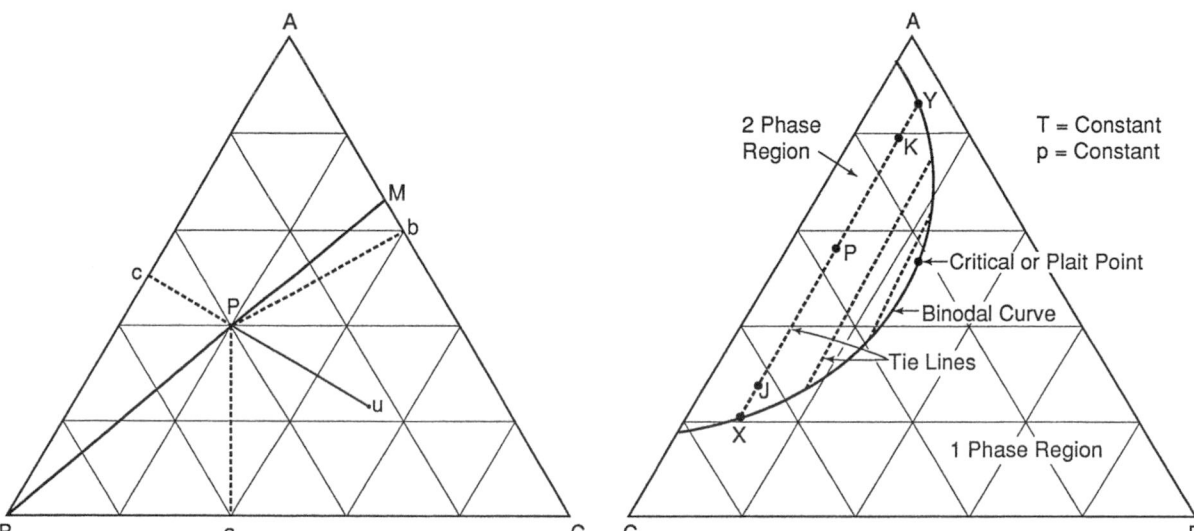

Fig. 2.27—Ternary phase diagram for a system consisting of Components A, B, and C which are miscible in all proportions.

Fig. 2.28—Ternary phase diagram for a system consisting of Components A, B, and C, which have limited solubility.

Consider a system with an overall composition P that is in the two-phase region. If the system were a typical hydrocarbon system, then it would separate into a vapor phase, denoted by Y, and a liquid phase, denoted by X. The compositions of these two phases are determined by measurement, as described previously, or by reading directly from the graph paper if it is ruled. The relative amounts of the two phases can also be calculated using the inverse-lever-arm rule. That is, the Distance $\overline{YP}$ is to the Distance $\overline{YX}$ as the amount of liquid is to the total amount of material. Similarly, $\overline{PX}$ is to $\overline{YX}$ as the amount of vapor is to the total.

Lines connecting equilibrium concentrations such as X and Y are called tie-lines. There are, in fact, an infinite number of these, but for practical purposes, several usually are drawn on a diagram and interpolation is used for concentrations lying between tie-lines. The ends of the tie-lines converge at a single point on the two-phase envelope. This point is the critical, or plait, point. This point of convergence is the point at which the phase properties become indistinguishable and therefore is the critical concentration at the specified temperature and pressure conditions. The IFT between the phases approaches zero in the vicinity of the plait point.

As indicated, mixtures that lie within the two-phase region are in thermodynamic equilibrium. If one applies the phase rule to a ternary system of two phases in equilibrium for a given pressure and temperature, the number of additional independent variables that must be set to define the system is one. For example, in the two-phase region, if the concentration of only one of the components in either phase is set, the system is defined. In the single-phase region, the composition of two of the components must be set. The tie-lines drawn between liquid and vapor curves connect vapor- and liquid-phase compositions corresponding to the equilibrium. Along any tie-line, the compositions of the gas and liquid phases in equilibrium are invariant. The ratio of the amount of gas phase to the amount of liquid phase does vary. For example, the liquid and vapor phases at Point K have the same composition as Point J, while the liquid mole fraction is 90% at Point J and 10% at Point K.

The tie-lines for a given system are by no means parallel and may swing significantly upward or downward in slope as the amount of any component in the mixture increases or decreases. The point of convergence (the critical, or plait, point) does not necessarily lie at the summit (midpoint) of the binodal curve, but frequently lies well to one side or the other. Its location and the consequent position of the tie-lines may be such that a very small change in the percentage of one phase results in a very large change in the other conjugate phase. The plait point for liquid/vapor hydrocarbon systems is that point at which the bubblepoint and dewpoint curves merge. It is a point of fixed composition corresponding to a definite ratio of any two components. At constant pressure, the position of the plait point varies with temperature.

Reservoir fluids are complex mixtures of hydrocarbons with components ranging from methane to C_{40+}. In miscible displacement processes, a fluid that is, or eventually will become, miscible with the reservoir fluid is injected into the reservoir. The injection of this fluid alters the chemical composition of the total system and consequently the thermodynamic properties. Rigorous thermodynamic analysis of such a process is only possible if all chemical constituents are identified, compositions are known, and thermodynamic properties are available. These conditions are never met in practice. As mentioned earlier, experience has shown that complex hydrocarbon systems can be represented with groups of hydrocarbons that preserve many of the important properties of the system. A typical representation of the process is a pseudoternary diagram with C_1, C_2–C_6, and C_{7+}, as the pseudocomponents, as shown in **Fig. 2.29.** Temperature and pressure are constant. Thus, concentrations of two components are sufficient to define any point on the diagram. Phase Boundary VO is the saturated vapor curve, and Boundary LO is the saturated liquid curve. An important assumption in using pseudocomponents and a pseudoternary diagram is that the composition of a pseudocomponent does not change in the different phases. For example, the relative compositions of components that make up the C_{7+} pseudocomponent must be approximately the same in the liquid phase as in the equilibrium vapor phase. If this does not hold, then use of the pseudoternary diagram can lead to serious errors in calculations of phase compositions and amounts. The oil represented by Point E in Fig. 2.29 is one that contains mostly heavy components. It would be a relatively low-API-gravity oil. Conversely, an oil having the composition of Point B would be a fairly volatile, high-API-gravity crude.

For systems that separate into two or more liquid phases, equilibrium compositions can also be shown on a ternary diagram.

An example is the alcohol/water/oil system shown in **Fig. 2.30.** Systems with overall compositions lying in the region under the binodal curve will separate into two phases, an oil-rich phase and a water-rich phase. For example, a system with overall composition of O will separate into Phases Y and X. The tie-lines that connect the equilibrium phases are generally obtained experimentally. Distances AA′ and BB′ represent the solubility of water in the oil and the solubility of the oil in water, respectively. Alcohol is completely soluble in both the water and the oil for the system illustrated. Note that the diagram is a pseudoternary diagram in that oil is a pseudocomponent. In such a diagram, electrolytes (e.g., NaCl, $CaCl_2$) are typically included with the water component, which is then designated as a brine and is also a pseudocomponent.

2.6.2 Mobilization and Displacement Through Favorable Phase Behavior.
Alcohol/oil/water systems will be used to illustrate the manner in which phase behavior between displacing and displaced fluids can result in mobilization and displacement of oil. Alcohol systems have been studied as displacement agents in the laboratory. However, they have not been applied successfully in the field for various practical and economic reasons. Nonetheless, the system is useful to

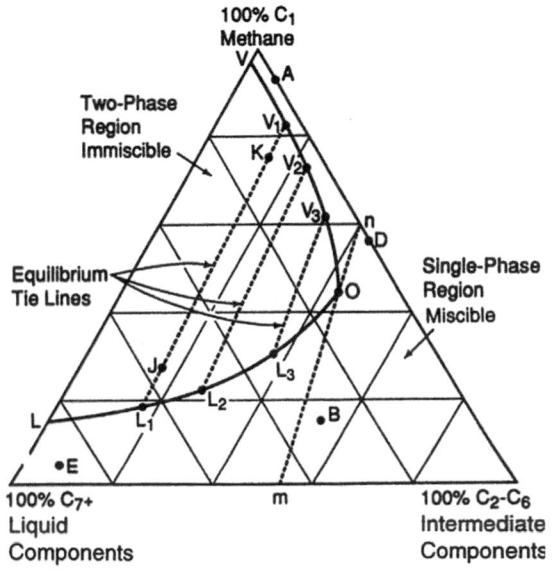

Fig. 2.29—Pseudoternary diagram.

demonstrate mechanisms associated with mobilization through favorable phase behavior. Mechanisms of miscible displacement, oil swelling, and oil extraction are illustrated in the following sections.

Miscible Displacement. **Fig. 2.31** shows a system consisting of a hypothetical alcohol, oil, and water. The alcohol is miscible with either oil or water. Over much of the concentration domain, however, two liquid phases exist in equilibrium when all three components are present.

Consider the injection of pure alcohol in a displacement process. Assume that at the point of contact with the alcohol, oil and water in the reservoir were at an overall volumetric composition represented by Point A in Fig. 2.31. Mixing of the alcohol with oil and water at the designated composition would occur along Line $\overline{AC}$. At high alcohol compositions, above the binodal curve, the mixed system would be a single phase. For fluid flow occurring in this regime, oil and water would be displaced completely. Any trapped oil contacted by the miscible system would be solubilized and thereby displaced, and trapping of a residual phase would not occur.

In effect, for such a miscible process, interfaces are eliminated (i.e., IFT is reduced to zero). Interfaces will be eliminated even though local equilibrium (complete mixing) may not necessarily be achieved as the process proceeds. Zero IFT corresponds to an infinite value of N_{ca}. Thus, a miscible process can be viewed as one in which the viscous/capillary force ratio is increased to infinity. The process would mobilize oil and generate an oil bank that would flow ahead of the alcohol front as described earlier. Two-phase flow would exist in the region of the oil bank. Single-phase flow would occur in the alcohol zone as long as concentrations were at or above the binodal curve.

Processes that operate within a miscible phase-behavior regime displace oil by total solubilization. The flow is single phase, and a residual fluid is not trapped. Such a process can solubilize both oil and water, as in the alcohol example. Alternatively, a displacing fluid that is miscible with oil but immiscible with water would solubilize only oil. In this latter case, two-phase flow could exist at the displacement front, but oil would still be mobilized and displaced through solubilization. Oil trapping would not occur as long as the oleic phase volume remained above residual saturation.

EOR fluid systems that exhibit phase behavior of the type shown for the hypothetical alcohol generally would require extremely large alcohol slugs if the displacement were to maintain miscibility throughout because, as mixing occurs at the alcohol front, alcohol concentration is reduced and overall concentration moves along Line $\overline{AC}$ toward the two-phase region (Taber et al. 1961). For slugs of practical volume, concentrations would in fact move into the two-phase region at some point and the process would degenerate to an immiscible process. General behavior in this immiscible mode will be discussed in the following sections.

Mobilization Through Increase in Oleic Phase Volume. Consider the tertiary butyl alcohol (TBA)/Soltrol/water system displayed in the pseudoternary diagram of **Fig. 2.32** (Taber et al. 1961). TBA is preferentially soluble in the hydrocarbon phase, which is a light oil, and the plait point lies on the left side of the diagram. A mixing line from pure TBA to a reservoir-oil/water composition at Point A passes well to the right of the plait point.

If a large volume of TBA were injected (as opposed to a relatively small slug), oil would be mobilized and a stabilized oil bank would form, as previously described. Upstream of the stabilized oil bank, a composition-transition zone would exist. At relatively high concentrations of alcohol, above the binodal curve, the system would be single phase and very efficient displacement would occur. The concentration would vary along the mixing line that runs from Point A on the brine/Soltrol base line to Point C at the

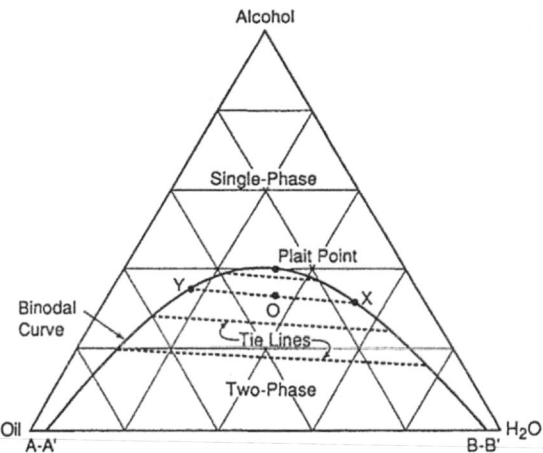

Fig. 2.30—Ternary phase diagram for liquid/liquid equilibrium.

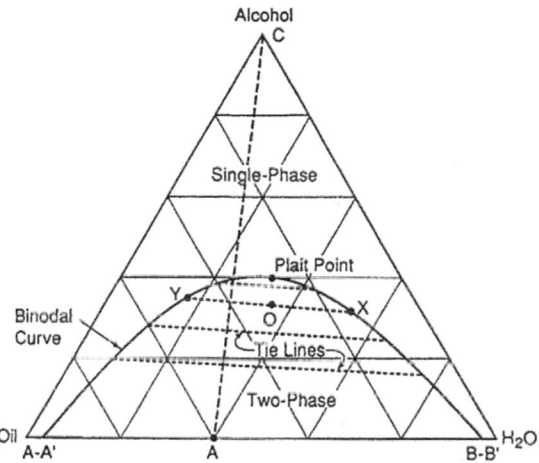

Fig. 2.31—Ternary phase diagram for an alcohol displacement process.

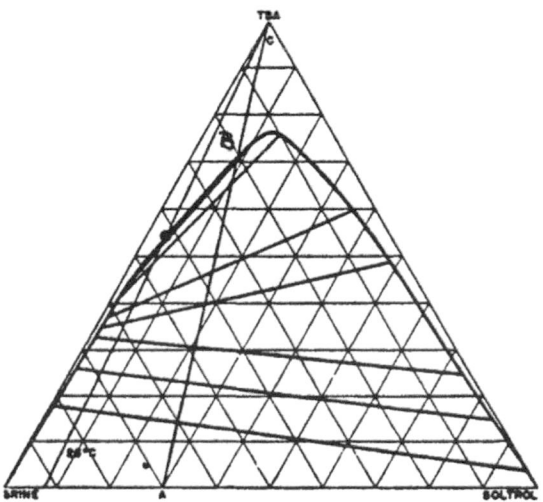

Fig. 2.32—Pseudoternary diagram for a TBA/Soltrol/brine system (Taber et al. 1961).

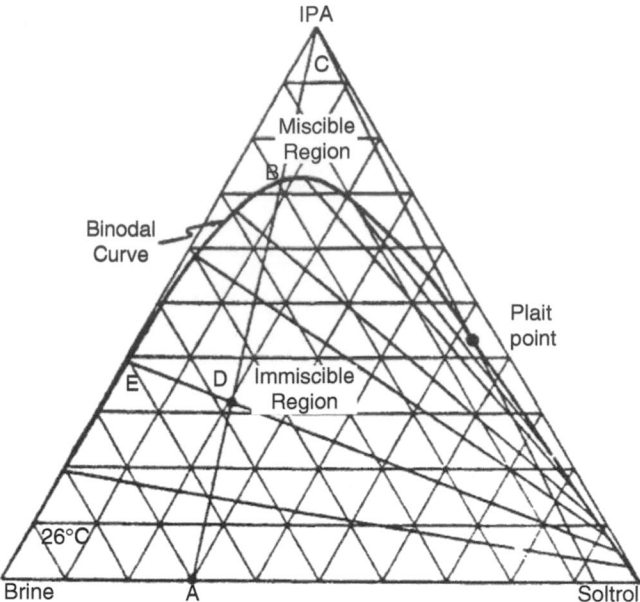

Fig. 2.33—Pseudoternary diagram for an IPA/Soltrol/brine system (Taber et al. 1961).

alcohol apex. When alcohol concentration falls below the binodal curve, there would be a phase separation into oleic and aqueous phases, resulting in two-phase flow. When the plait point is to the left of the mixing line, as is the case with TBA and the oil considered, the oleic phase would increase in volume along this mixing path. This result can be seen by applying the inverse-level-arm rule along the path. For the particular system shown, because of the tie-line slopes, the oil phase decreases in volume at low alcohol concentration but then increases in volume as the binodal curve is approached.

The increase in the oleic-phase volume in the transition zone promotes continuity of the oil phase and a favorable fractional flow relative to the aqueous phase. The oleic phase would tend to move faster than the aqueous phase. If the alcohol slug were large enough to maintain a miscible zone throughout the displacement, trapped residual oil would not be encountered by this miscible zone for the illustrated TBA process. If the alcohol slug were of such a size that a breakdown of miscibility occurred, then some residual oil would be left, but the swelling of the oleic-phase volume that occurred would diminish this residual. Thus, the relative increase in oleic-phase volume that would occur because of the phase behavior would promote recovery of the oil.

As the plait point is approached in a three-component system, the IFT between equilibrium phases approaches zero because the phases tend to become indistinguishable. If the mixing path carried the system concentration very near the plait point, then an increase in N_{ca} might be expected as a result of reduction in capillary forces. Were this to occur, displacement efficiency would be improved. This usually does not occur, however, because concentrations are sufficiently removed from the plait point that reduction in IFT is negligible. As previously discussed, the N_{ca} must be increased a few orders of magnitude above the value at normal waterflooding conditions before residual saturation is affected.

If a small slug of alcohol were injected (rather than a large volume), followed by brine injection, there would be concentration-transition zones at both the leading and trailing ends of the alcohol slug. The relative increase in oleic-phase volume and favorable fractional flow would still occur until the slug volume was diluted by the aqueous phase to the point where the aqueous phase became continuous and the oleic phase was left as a residual.

Mobilization Through Oleic-Phase Extraction. **Fig. 2.33** shows another alcohol system, isopropyl alcohol (IPA)/Soltrol/ brine. IPA is preferentially soluble in the aqueous phase and thus the plait point lies to the right of the binodal curve. The IPA/ Soltrol/water mixing Line $\overline{AC}$ passes to the left of the plait point.

As with the other alcohol system discussed, if a large volume of IPA were injected, the oil would be mobilized and an oil bank would form. At high alcohol concentration, a composition-transition zone would form in the miscible region along Line $\overline{AC}$ on the ternary diagram. In contrast to the TBA displacement, however, when overall composition fell within the two-phase region, the aqueous phase would increase in volume.

The implication of the reducing oleic-phase volume is that a residual oleic-phase saturation would be deposited at some point in the transition zone. Where the saturation is above a residual value, the flow rate of the oil would be less than that of the aqueous phase. If the IPA slug were large enough to maintain miscibility, then the trapped residual oil would be solubilized when the miscible zone arrived. If miscibility were lost at some point in the process, however, then a residual saturation would be left. The flowing aqueous phase would extract some of the oil from the trapped phase. This process is not as effective as the oil-swelling mechanism that occurred with the TBA system, and ultimate recovery generally would not be as good. As with the TBA system, the concentration path would not approach sufficiently close to the plait point to derive any benefit from IFT reduction.

If a slug of IPA were injected followed by brine, composition-transition zones would form at the leading and trailing edges of the slug. When the slug was sufficiently diluted, trapped oleic phase would not be displaced.

Summary Statement. Different mechanisms involving system phase behavior have been discussed. As described, the mechanisms of increase in oleic-phase volume and extraction do not depend on modification of viscous/capillary forces to increase oil recovery. If by virtue of phase behavior a significant reduction in IFT occurs, however, then the mechanisms of alteration of viscous/capillary forces and favorable use of phase behavior are complementary. Overall recovery is improved if both mechanisms are operative. Miscible displacement can be viewed as involving either a phase-behavior mechanism or an alteration of capillary forces. As discussed in Chapter 6, multicontact-miscible processes are dependent on phase behavior to develop a miscible condition in situ. Again, the different mechanisms are complementary. As discussed in Chapter 7, surfactant processes depend on both favorable phase behavior and IFT reduction. The thermal processes, especially steamflooding, are also shown to use both mechanisms (Chapter 8).

In a very general sense, understanding viscous/capillary force relationships and system phase behavior is necessary to comprehend recovery mechanisms for most of the EOR processes. Such an understanding in detail is difficult to obtain for complex EOR fluid systems operating in complex rock structures.

Problems

2.1 **Fig. 2.34** (Jordan et al. 1956) contains a set of data illustrating the effect of pressure gradient on the recovery of discontinuous residual oil. **Table 2.6** summarizes selected data.

Compute the capillary number for each data point as N_{ca}^*. For the purpose of this problem, the IFT is 30 dynes/cm, the viscosity of water is 1 cp, and the absolute permeability of the rock is 156 md. Porosity is 0.20.

Relative permeability data for water are given by the following equations, where $S_{wi} = 0.29$:

$$S_w^* = \frac{S_w - S_{wi}}{1 - S_{wi}}$$

and $k_{rw} = S_w^{*2.5}$.

The capillary number is dimensionless.

Compare the ratio of S_{or}/S_{orw} at your values of the capillary number with the correlations of **Fig. 2.35.**

2.2 The water/air capillary pressure data in **Table 2.7** were measured for a sandstone core from the Jayhawk formation. The oil-saturated part of the Jayhawk formation is underlain by a water table at a depth of 1,500 ft ($S_w = 100\%$ below this depth). Estimate and plot the water saturation above the 1,500-ft level. Assume that only water and oil exist in the formation (no gas saturation).

Additional data: p (at 1,500 ft) = 750 psia, $\rho_w = 1.02$ g/cm^3, $\rho_o = 0.80$ g/cm^3, $\sigma_{ow} = 25$ dynes/cm, and $\sigma_{aw} = 70$ dynes/cm.

Assume that the rock is strongly water-wet (i.e., the contact angle in the presence of air is approximately the same as the contact angle in the presence of oil).

2.3 **Fig. 2.36** shows a gas bubble confined in a 0.0002-in.-diameter capillary tube. The tube is wetted by water in the presence of gas and by oil in the presence of gas. The bubble is motionless. The contact angles, measured through the wetting phases, are 10 and 30° for the water and oil, respectively.

The pressure in the oil phase is 6.9×10^6 dynes/cm^2. The IFT between the water and gas, σ_{gw}, is 70 dynes/cm and that between the oil and gas, σ_{go}, is 33 dynes/cm.

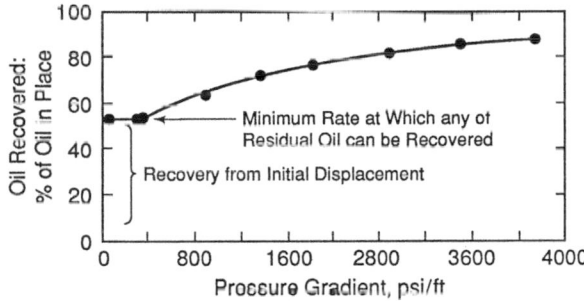

S_o (PV)	Pressure Gradient (psi/ft)
0.355	70
0.355	370
0.284	600
0.213	1,200
0.142	2,400
0.114	3,200

Table 2.6—Oil saturation as a function of applied pressure gradient, Problem 2.1.

Fig. 2.34—Pressure gradient required to recover disconnected residual oil, Problem 2.1 (Jordan et al. 1956).

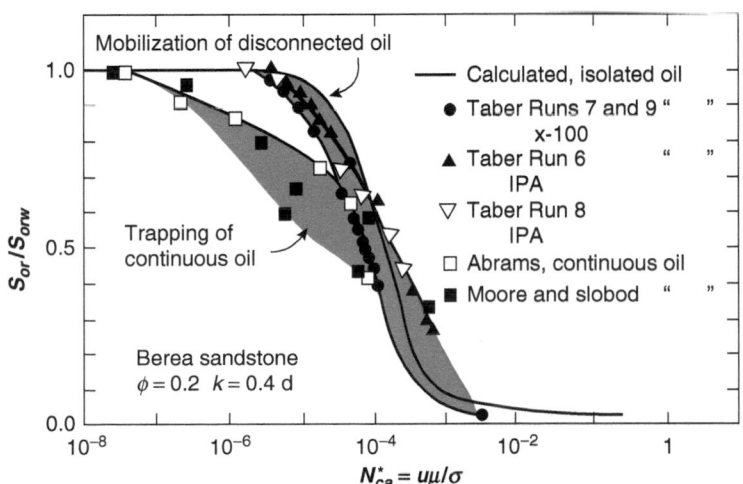

Fig. 2.35—Correlation of residual oil saturation with capillary number at trapping and mobilization, Problem 2.1 (Stegemeier 1977).

$$P_c = P_a - P_w$$

P_c (psi)	S_w (%)	P_c (psi)	S_w (%)	P_c (psi)	S_w (%)
2	100	10	54	24	34
3	93	12	48	28	32
4	86	14	43	32	31
5	78	16	40	36	31
6	73	18	38	40	31
7	67	20	36	44	31

Table 2.7—Air/water capillary pressure: Jayhawk sandstone, Problem 2.2.

1. Calculate the *minimum* pressure in the system and indicate the phase in which this pressure exists.
2. What is the pressure in the water phase that will just cause the gas bubble to be displaced to the left (←)? At this condition, what is the pressure difference between the oil and water phases (i.e., what is $p_w - p_o$)?

2.4 To determine water/oil IFT, water is being displaced through a small tube into an oil bath as shown in **Fig. 2.37.** The water forms a spherical drop, which breaks away from the water in the injection tube when it is large enough and falls to the bottom of the container.

For one experiment, the following data are given:

r_t = inside radius of discharge tube = 0.04 in.

r_d = drop radius (at time of breaking away) = 0.104 in.

ρ_w = water density = 1.0 g/cm³

ρ_o = oil density = 0.75 g/cm³

With these data, estimate the water/oil IFT.

2.5 A researcher assembles a bundle of capillary tubes to simulate a porous medium. The tubes are of uniform size and are 0.004 in. in diameter. It is known that water wets the capillary tube glass with a contact angle of approximately 30°.

The tubes are placed in a horizontal position and completely filled with water. The water is at atmospheric pressure. The tubes are then flooded with oil (i.e., oil is injected, displacing the water). The IFT between the oil and water is 30 dynes/cm.

1. What is the minimum pressure required in the oil phase to displace the water?
2. Assume the capillaries are constructed such that they varied in diameter from one end to the other, as shown in **Fig. 2.38.** For this condition, what pressure would be required in the oil phase to displace the water? That is, at what pressure would the displacement start, and what pressure would be required to displace the water completely?

Assume that the water is in contact with a water reservoir at the outlet end and that capillary forces do not exist at that point.

2.6 Magichem Chemical Co. has announced the development of a new chemical that can be used in displacement processes to recover more oil than waterflooding. The company reports the following displacement test results to substantiate their claim:

- Core length = 2.0 in.
- Initial oil saturation = 70%.

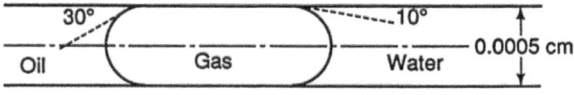

Fig. 2.36–Trapped gas bubble, Problem 2.3.

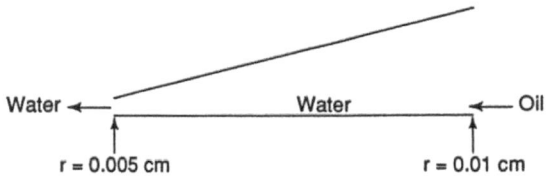

Fig. 2.38—Capillary of varying diameter, Problem 2.5.

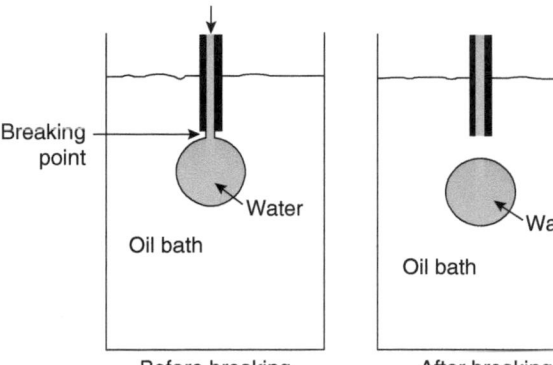

Fig. 2.37—Oil/water IFT measurement, Problem 2.4.

- Residual saturation after typical waterflood = 37%.
- Residual saturation after flooding with new chemical = 20%.
- Pressure drop during displacement test with new chemical (across total core length) = 50 psi.
- Viscosity of new chemical = 3.5 cp.
- IFT (new-chemical/oil) = 15 dynes/cm.
- Cost of the new chemical is relatively low.

On the basis of these results, would you agree that the new chemical shows potential as an oil recovery agent? Justify and document, as completely as you can, the reason for your answer.

2.7 One of the petroleum sulfonates being used in a commercial laboratory has the trade name TRS 10-80™. The spinning-drop method was used to measure the IFT between a solution of TRS 10-80 in brine and a light hydrocarbon. The IFT at 80°F was 0.03 dynes/cm.

Assume this surfactant solution is to be used to displace the light hydrocarbon from a 4.0-ft Berea sandstone linear core (k = 200 md). The core initially has a water saturation of 20% and a hydrocarbon saturation of 80%.

The core is waterflooded before injection of the TRS 10-80 solution. The ROS after waterflooding is 40%. The TRS 10-80 solution is then injected. The experiment is conducted at a temperature of 80°F.

1. If it is desired to recover 90% of the original oil in place through a combination of both floods, what pressure drop do you predict would be required across the core during the TRS 10-80 displacement?
2. If the temperature were increased to 100°F, would the required pressure drop be larger or smaller? Why?
3. TRS 10-80 is adsorbed on Berea sandstone. What effect would adsorption have on the experiment?

This problem can be solved using Fig. 2.35.

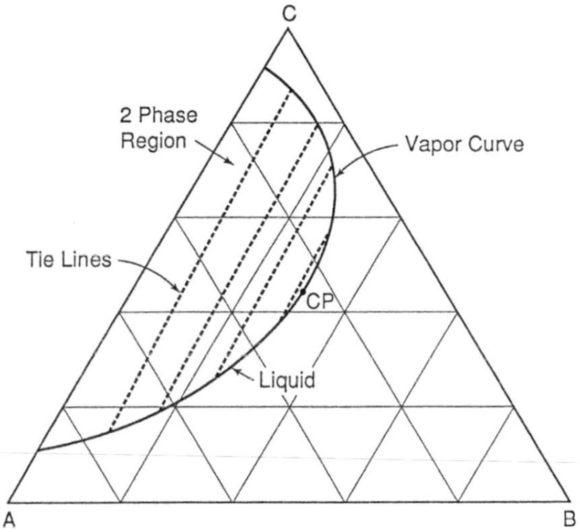

2.8 The pore-doublet described in Section 2.4.2 assumes that oil and water viscosities are equal. In many cases, oil viscosity is considerably larger than water viscosity.

For typical waterflooding conditions, and assuming that the pore-doublet model is applicable, describe the general effect on trapping of a relatively large oil viscosity. (It is not necessary to derive a precise relationship.)

2.9 **Fig. 2.39** shows the phase behavior of mixtures of Chemicals A, B, and C. The diagram is on a pound-mass basis and is for a fixed T and p. Mixtures of Components A, B, and C are contained in three vessels as shown in **Fig. 2.40.**

The contents of the three vessels are mixed together in a fourth vessel, which is held at the p and T corresponding to the ternary diagram.

1. Describe the resulting mixture as follows: (a) Show the overall composition on the diagram; (b) indicate whether the mixture is one phase or two phases at the specified p and T; and (c) if the mixture is two phase, specify the composition and amount of each phase.
2. Assume that a long porous-media bed is originally saturated with a liquid mixture which is 70% A, 25% B, and 5% C. This mixture is displaced by pure C, which is a vapor. When the injected vapor first breaks through at the end of the bed, what will be the composition of this effluent vapor phase? Use the ternary diagram to show the basis for your answer.

Fig. 2.39—Ternary diagram, Problem 2.9.

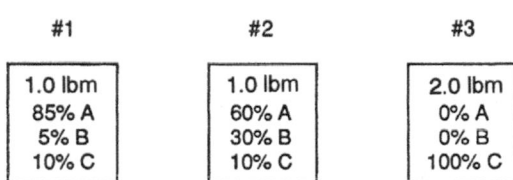

Fig. 2.40—Components to be mixed, Problem 2.9.

#1	#2	#3
1.0 lbm	1.0 lbm	2.0 lbm
85% A	60% A	0% A
5% B	30% B	0% B
10% C	10% C	100% C

Nomenclature

Parameter definitions are followed by dimensions and typical units used in text.

A = area, L^2, ft^2

b_c = width of capillary construction, L, cm

d = diameter, L, ft

E_D = microscopic displacement efficiency, L^3/L^3, volume fraction

r_d = drop radius, L, cm

r_t = tube radius, L, cm

r_1 = radius of Pore 1 in doublet model, L, cm

r_2 = radius of Pore 2 in doublet model, L, cm

F = force, mL/t^2, dynes

F_c = capillary forces, mL/t^2, dynes

F_v = viscous forces, mL/t^2, dynes

g = gravity acceleration constant, L/t^2, 32 ft/sec²

g_c = conversion factor [32.2 lbm·ft/(lbf·sec²)]

h = height, L, cm

k = permeability, L^2, darcies

k_{rw} = relative permeability to water

L = length, L, ft or cm

L_f = distance from oil drop front to capillary construction, L, cm

L_o = length of oil phase in doublet model, L, cm

L_w = length of water phase in doublet model, L, cm

L_1 = length of Pore 1 in doublet model, L, cm

L_2 = length of Pore 2 in doublet model, L, cm

N_{ca} = capillary number based on interstitial (pore) velocity, $N_{ca}^* = \phi N_{ca}$

N_{ca}^* = capillary number based on Darcy velocity

$N_{ca,m}$ = modified capillary number, Eq. 2.32

p = pressure, m/Lt^2, psi

p_{atm} = atmospheric pressure, m/Lt^2, dynes/cm²

Δp = pressure change, m/Lt^2, psi

P_c = capillary pressure, m/Lt^2, dynes/cm²

q = flow rate, L^3/t, ft/D

r = radius, L, cm

R = radius of curvature, L, cm

S = saturation, L^3/L^3, volume fraction

S_D = dimensionless saturation

S_{ob} = saturation of oil in oil bank, L^3/L^3, volume fraction

S_{oi} = initial oil saturation, L^3/L^3, volume fraction

S_{or} = ROS, L^3/L^3, volume fraction

S_{orw} = ROS at end of a waterflood, L^3/L^3, volume fraction

S_{wb} = saturation of water in oil bank, L^3/L^3, volume fraction

S_{wi} = initial water saturation, L^3/L^3, volume fraction

T = temperature, T, °F

u = Darcy velocity in a porous medium, L/t, ft/hr

v = interstitial pore velocity, L/t, ft/hr

$\bar{v}$ = average velocity in a capillary tube, L/t, cm/s

$\bar{v}_1$ = average velocity in Pore 1 of doublet model, L/t, cm/s

$\bar{v}_2$ = average velocity in Pore 2 of doublet model, L/t, cm/s

W = work, mL^2/t^2

x = distance in x-direction, L, cm

θ = contact angle, degrees

μ = viscosity, m/Lt, cp

ρ = fluid density, L^3, g/cm³

σ = IFT, m/t^2, dynes/cm

ϕ = porosity

Φ = flow potential, mL/t^2

Subscripts

a = air phase

atm = atmospheric conditions

aw = air/water interface

A = indicates position

B = indicates position

g = gas phase

go = gas/oil interface

gw = gas/water interface

i = pore channel index

o = oil phase

os = oil/solid interface

ow = oil/water interface

w = water phase

ws = water/solid interface

1 = indicates position

2 = indicates position

References

Abrams, A. 1975. The Influence of Fluid Viscosity, Interfacial Tension, and Flow Velocity on Residual Oil Saturation Left by Waterflood. *SPE J.* **15** (5): 437–447. SPE-5050-PA. https://doi.org/10.2118/5050-PA.

Amyx, J. W., Bass, D. M. Jr., and Whiting, R. L. 1960. *Petroleum Reservoir Engineering.* New York City: McGraw-Hill Book Co.

Arriola, A. 1980. *An Experimental Study of the Effects of Viscous and Capillary Forces on the Trapping and Mobilization of Oil Drops in Capillary Constrictions.* PhD dissertation, University of Kansas, Lawrence, Kansas.

Arriola, A., Willhite, G. P., and Green, D. W. 1983. Trapping of Oil Drops in a Noncircular Pore Throat and Mobilization Upon Contact With a Surfactant. *SPE J.* **23** (1): 99–114. SPE-9404-PA. https://doi.org/10.2118/9404-PA.

Benham, A. L., Dowden, W. E., and Kunzman, W. J. 1960. Miscible Fluid Displacement—Prediction of Miscibility. In *Petroleum Transactions*, AIME, Vol. 219, 229-37. SPE-1484-G.

Bethel, F. T. and Calhoun, J. C. 1953. Capillary Desaturation in Unconsolidated Beads. *J Pet Technol* **5** (8): 197–202. SPE-953197-G. https://doi.org/10.2118/953197-G.

Brownell, L. E. and Katz, D. L. 1947. Flow of Fluids Through Porous Media—Part II: Simultaneous Flow of Two Homogeneous Phases. *Chem. Eng. Prog.* **43**: 601–612.

Chatzis, I. and Morrow, N. R. 1984. Correlation of Capillary Number Relationships for Sandstone. *SPE J.* **24** (5): 555–562. SPE-10114-PA. https://doi.org/10.2118/10114-PA.

Craft, B. C. and Hawkins, M. F. 1959. *Applied Petroleum Engineering.* Englewood Cliffs, New Jersey: Prentice-Hall.

Craig, F. F. Jr. 1971. *The Reservoir Engineering Aspects of Waterflooding*, Vol. 3. Richardson, Texas: Monograph Series, SPE.

Dombrowski, H. S. and Brownell, L. E. 1954. Residual Equilibrium Saturation of Porous Media. *Ind. Eng. Chem.* **46** (6): 1207–1219. https://doi.org/10.1021/ie50534a037.

Donaldson, E. C., Thomas, R. D., and Lorenz, P. B. 1969. Wettability Determination and Its Effect on Oil Recovery. *SPE J.* **9** (1): 13–20. SPE-2338-PA. https://doi.org/10.2118/2338-PA.

Dullien, F. A. L. 1969. Determination of Pore Accessibilities—An Approach. *J Pet Technol* **21** (1): 14–15. SPE-2114-PA. https://doi.org/10.2118/2114-PA.

Foster, W. R. 1973. A Low-Tension Waterflooding Process. *J Pet Technol* **25** (2): 205–210. SPE-3803-PA. https://doi.org/10.2118/3803-PA.

Gardescu, I. I. 1930. Behavior of Gas Bubbles in Capillary Spaces. In *Transactions of the Society of Petroleum Engineers*, Vol. 86, Part 1, SPE-930351-G, 351–370. Richardson, Texas: SPE.

Garnes, J. M., Mathisen, A. M., Scheie, A. et al. 1990. Capillary Number Relations for Some North Sea Reservoir Sandstones. Presented at the SPE/DOE Enhanced Oil Recovery Symposium, Tulsa, 22–25 April. SPE-20264-MS. https://doi.org/10.2118/20264-MS.

Jordan, J. K., McCardell, W. M., and Hocott, C. R. 1956. Effect of Rate on Oil Recovery by Waterflooding. Report, Humble Oil & Refining Co., Houston.

Katz, D. L. and Saltman, W. 1939. Surface Tension of Hydrocarbons. *Ind. Eng. Chem.* **31** (1): 91–94. https://doi.org/10.1021/ie50349a019.

Lefebvre du Prey, E. G. 1973. Factors Affecting Liquid-Liquid Relative Permeabilities of a Consolidated Porous Medium. *SPE J.* **13** (1): 39–47. SPE-3039-PA. https://doi.org/10.2118/3039-PA.

Leverett, M. C. 1941. Capillary Behavior in Porous Media. In *Transactions of the Society of Petroleum Engineers,* Vol. 142, Part 1, SPE-941152-G, 152–169. Richardson, Texas: SPE.

Melrose, J. C. and Brandner, C. F. 1974. Role of Capillary Forces in Determining Microscopic Displacement Efficiency for Oil Recovery by Waterflooding. *J Can Pet Technol* **13** (4): 54–62. PETSOC-74-04-05. https://doi.org/10.2118/74-04-05.

Moore, T. F. and Slobod, R. C. 1956. The Effect of Viscosity and Capillarity on the Displacement of Oil by Water. *Producers Monthly* (August 1956): 20–30.

Muskat, M. 1949. *Physical Principles of Oil Production.* New York City: McGraw-Hill Book Co.

Plateau, T. A. F. 1863. Experimental and Theoretical Researches on the Figures of Equilibrium of a Liquid Mass Withdrawn From the Actions of Gravity. In *The Annual Report of the Board of Regents of the Smithsonian Institution*, 270–285. Washington, DC: Grover Printing Office.

Purcell, W. R. 1950. Interpretation of Capillary Pressure Data. *J Pet Technol* **2** (8): 11–12. SPE-950369-G. https://doi.org/10.2118/950369-G.

Reisberg, J. and Doscher, T. M. 1956. Interfacial Phenomena in Crude Oil-Water Porous Media. *Producers Monthly* **21** (1): 43–50.

Stegemeier, G. L. 1977. Mechanisms of Entrapment and Mobilization of Oil in Porous Media. In *Improved Oil Recovery by Surfactant and Polymer Flooding*, eds. D. O. Shah and R. S. Schechter, 55–91. New York City: Academic Press.

Taber, J. J. 1969. Dynamic and Static Forces Required To Remove a Discontinuous Oil Phase From Porous Media Containing Both Oil and Water. *SPE J.* **9** (1): 3–12. SPE-2098-PA. https://doi.org/10.2118/2098-PA.

Taber, J. J., Kamath, I. S. K., and Reed, R. L. 1961. Mechanism of Alcohol Displacement of Oil From Porous Media. *SPE J.* **1** (3): 195–212. SPE-1536-G. https://doi.org/10.2118/1536-G.

Wagner, O. R. and Leach, R. O. 1966. Effect of Interfacial Tension on Displacement Efficiency. *SPE J.* **6** (4): 335–344. SPE-1564-PA. https://doi.org/10.2118/1564-PA.

Willhite, G. P. 1986. *Waterflooding*, Vol. 3. Richardson, Texas: Textbook Series, SPE.

SI Metric Conversion Factors

cp	× 1.0*	E–03	=	Pa·s
dyne/cm	× 1.0*	E+00	=	mN/m
ft	× 3.048*	E–01	=	m
°F	(°F–32)/1.8		=	°C
psi	× 6.894 757	E+00	=	kPa

* Conversion factor is exact.

Chapter 3

Displacement in Linear Systems

3.1 Introduction

This chapter covers displacement mechanisms described by frontal-advance theory (Claridge and Bonder 1974; Patton et al. 1971; Pope 1980). The objective is to introduce fundamental concepts of enhanced-oil-recovery (EOR) processes using simple mathematical models that retain important features of more-complex models presented later. The application of frontal-advance theory to predict waterflooding performance in a linear system is reviewed (Willhite 1986). Then, frontal-advance theory is applied to viscous waterflooding and chemical flooding processes, such as polymer and surfactant flooding. Injection of chemicals as slugs is introduced. Finally, dispersion (or mixing)—when one fluid displaces a second, miscible fluid—is described, as are viscous fingering and its effect on displacement.

3.2 Waterflood Performance—Frontal-Advance Equations

3.2.1 Frontal Advance and Related Equations. This section describes the application of frontal-advance theory to oil displacement by water in a linear system (Willhite 1986). For the discussion, the rock is considered homogeneous with porosity ϕ, permeability k, length L, and cross-sectional area A. The water in the rock is initially at interstitial saturation, S_{iw}. Interstitial water saturation is defined as the saturation at which the water is immobile (i.e., the relative permeability to water, k_{rw}, is zero). Also, there is no gas saturation. When water is injected into this linear system at a sufficient rate for the frontal-advance assumptions to apply, each water saturation, S_w, travels at a constant velocity through the system given by Eq. 3.1. This equation, called the frontal-advance or Buckley-Leverett equation, is derived by Willhite (1986) as

$$\frac{dx_{S_w}}{dt} = \frac{q_t}{A\phi}\left(\frac{\partial f_w}{\partial S_w}\right)_{S=S_w}, \quad \dots\dots\dots\dots\dots\dots\dots\dots (3.1)$$

where x_{S_w} = location of water saturation, S_w, measured from $x = 0$; A = cross-sectional area; ϕ = porosity; q_t = injection rate; f_w = fractional flow of water; t = time from the beginning of injection; and any consistent set of units may be used.

When the system is horizontal and gravity and capillary forces are negligible, the fractional flow of water can be computed from Eq. 3.2.

$$f_w = \left(k_w/\mu_w\right)/\left(k_o/\mu_o + k_w/\mu_w\right), \quad \dots\dots\dots\dots\dots\dots (3.2)$$

where k_w = permeability of rock to water, k_o = permeability of rock to oil, μ_w = viscosity of water, μ_o = viscosity of oil, and f_w is dimensionless.

Fig. 3.1 is a typical plot of fractional flow vs. saturation. The derivative, $\partial f_w/\partial S_w$, can be evaluated graphically by constructing tangents to the $f_w - S_w$ curve at a given saturation or numerically if the relative permeabilities, $k_{ro}(S_w)$ and $k_{rw}(S_w)$, are available. For many combinations of fluid and rock properties, the frontal-advance solution is characterized by a saturation discontinuity at the flood front where the water saturation jumps from S_{iw} to S_{wf}, the flood-front saturation. This discontinuity occurs because the velocities of low water saturations ($< S_{wf}$) are less than the velocity of the flood-front saturation and are overtaken by this saturation.

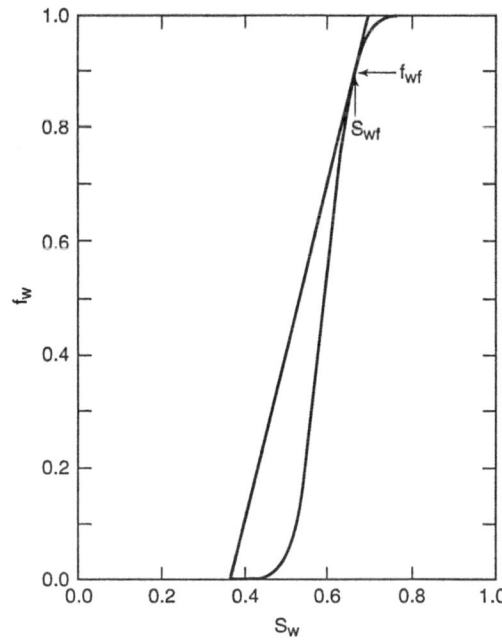

Fig. 3.1—Fractional flow vs. saturation.

The flood-front saturation is found by constructing a tangent to the fractional-flow curve from S_{iw}, when the water in the rock is initially at interstitial saturation (i.e., immobile, as depicted in Fig. 3.1). The slope of the tangent is given by

$$f'_{wf} = \left(f_{wf} - f_{iw}\right)\big/\left(S_{wf} - S_{iw}\right), \dots\dots\dots\dots\dots\dots\dots\dots\dots\dots\dots\dots\dots (3.3)$$

where f_{wf} = fractional flow of water at the flood-front saturation, f_{iw} = fractional flow of water at interstitial water saturation equal to zero, S_{wf} = flood-front saturation, and S_{iw} = interstitial water saturation.

The flood-front or saturation discontinuity moves at a velocity given by

$$v_{wf} = \left(q_t/A\phi\right)\left(\partial f_w/\partial S_w\right)_{S_{wf}}. \dots\dots\dots\dots\dots\dots\dots\dots\dots\dots\dots\dots (3.4)$$

When the interstitial water saturation is immobile, as depicted in Fig. 3.1, $f_{iw} = 0$ and v_{wf} is given by

$$v_{wf} = \left(q_t/A\phi\right)\left[f_{wf}\big/\left(S_{wf} - S_{iw}\right)\right]. \dots\dots\dots\dots\dots\dots\dots\dots\dots\dots (3.5)$$

All saturations less than S_{wf} travel at the flood-front velocity.

Eq. 3.3 is also an expression of the conservation of mass across a saturation discontinuity as the flood front, sometimes called a "shock wave," travels through the porous rock. This is discussed in Section 3.3.1.

The location of a particular saturation is found by integrating Eq. 3.1 with respect to time to obtain

$$x_{S_w} = \left(q_t t/A\phi\right)\left(\partial f_w/\partial S_w\right)_{S_w}. \dots\dots\dots\dots\dots\dots\dots\dots\dots\dots\dots (3.6)$$

Because the velocity of every saturation is constant, the graph of saturation location vs. time is a set of straight lines starting from the origin. This graph is often plotted in dimensionless form by introducing the following terms into Eq. 3.6. Let

$$x_D = x/L, \dots\dots\dots\dots\dots\dots\dots\dots\dots\dots\dots\dots\dots\dots\dots\dots\dots\dots\dots (3.7)$$

where x_D = dimensionless distance from origin.

$$t_D = q_t t/A\phi L, \dots\dots\dots\dots\dots\dots\dots\dots\dots\dots\dots\dots\dots\dots\dots\dots (3.8)$$

where t_D = dimensionless time [the same as Q_i, the number of pore volumes (PVs) of fluid injected]. Eq. 3.6 becomes

$$x_{DS_w} = t_D f'_w. \dots\dots\dots\dots\dots\dots\dots\dots\dots\dots\dots\dots\dots\dots\dots\dots\dots\dots (3.9)$$

Fig. 3.2 is a graph of dimensionless distance/time for the movement of water saturations predicted by the frontal-advance equation. Saturations $S_{iw} < S_w < S_{wf}$ travel at the same velocity and are located on the flood-front path. The region ahead of the flood front has a uniform saturation. Saturations greater than S_{wf} travel at progressively slower velocities, as indicated by the decreasing slopes in Fig. 3.2 and the fan-like distribution on the x_D/t_D graph. This region is sometimes called a spreading wave.

Saturation profiles or saturation histories can be constructed by making cross sections through the time/distance graph. A saturation profile is a graph of the locations of all saturations along a cross section of fixed time, as illustrated by the dashed line at $t_D = 0.15$ in Fig. 3.2. **Fig. 3.3** displays the saturation profile at $t_D = 0.15$ that was obtained from Fig. 3.2. The saturation history is the graph of saturation vs. time at a particular value of x_D. A plot of water saturation vs. t_D for $x_D = 1$, shown in **Fig. 3.4,** illustrates the arrival of water saturations at the end of the linear system.

Displacement performance is obtained by determining the average saturation of the region displaced by water. Before breakthrough (the arrival of the flood-front saturation at $x_D = 1$), only oil is produced and the volume of oil is equal to the water injected until breakthrough. At breakthrough, the volume of oil produced, expressed as a fraction of PV, is t_{Dbt}, given by

$$t_{Dbt} = 1/f'_{wf}. \dots\dots\dots\dots\dots\dots (3.10)$$

At and after breakthrough, the volume of oil displaced is given by

$$N_p = \left(\overline{S}_w - S_{iw}\right)A\phi L/B_o, \dots\dots\dots\dots (3.11)$$

where N_p = stock-tank barrels of oil displaced, $\overline{S}_w$ = average water saturation in the linear system, and B_o = oil formation volume factor (FVF).

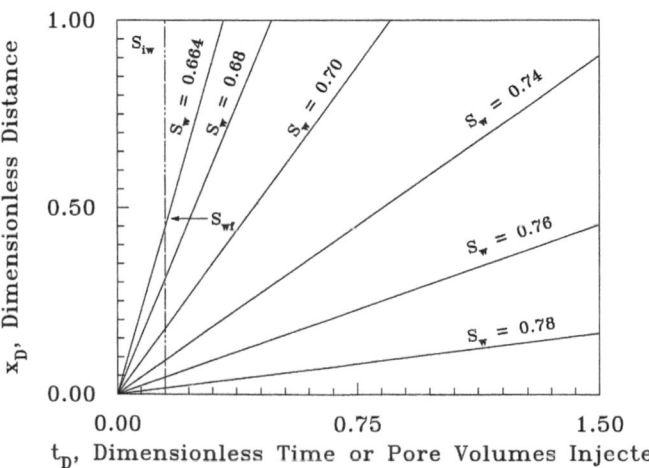

Fig. 3.2—x_D/t_D diagram for a linear waterflood.

The average water saturation at any time after breakthrough is computed from the Welge (1952) equation given by

$$\overline{S}_w = S_{w2} + t_{D2}\left(1 - f_{w2}\right), \dots\dots\dots\dots (3.12)$$

where S_{w2} = water saturation at $x_D = 1$, f_{w2} = fractional flow of water at $x_D = 1$, and t_{D2} = dimensionless time required to propagate saturation S_{w2} from the inlet of the system ($x_D = 0$) to the end of the system ($x_D = 1$). The value of t_{D2} is obtained from

$$t_{D2} = 1/f'_{w2}. \dots\dots\dots\dots\dots\dots (3.13)$$

$\overline{S}_w$ may also be determined graphically, as shown by Willhite (1986).

Example 3.1 illustrates the solution of the frontal-advance equation for a waterflood of a linear core initially at interstitial water saturation.

Example 3.1—Application of Frontal-Advance Equations: Linear Waterflood. A core is saturated with oil and water at interstitial water saturation. **Table 3.1** gives the properties of the core, fluids, and saturations. Prepare the following:

1. A dimensionless-distance, dimensionless-time x_D/t_D graph showing the displacement until water/oil ratio (WOR) = 50 is reached at the end of the system.
2. A saturation profile when the flood front is located at $x_D = 0.75$.
3. The volume of oil displaced from the beginning of the waterflood to WOR = 50.

Relative permeability relationships are given by

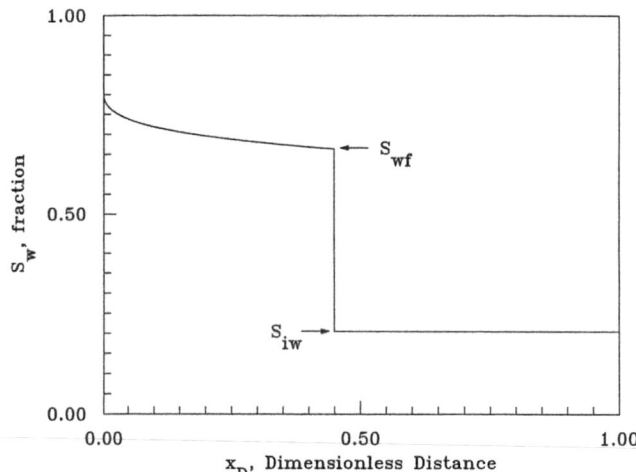

Fig. 3.3—Saturation profile at $t_D = 0.15$.

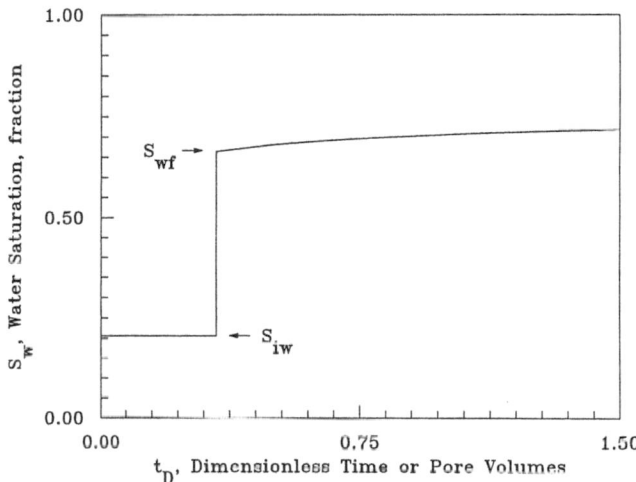

Fig. 3.4—Saturation history at $x_D = 1$ (outlet of the system).

$$k_{ro} = \alpha_1 \left(1 - S_{wD}\right)^m. \dots\dots\dots\dots\dots\dots\dots\dots\dots\dots\dots\dots\dots\dots\dots\dots\dots\dots\dots\dots (3.14)$$

and $k_{rw} = \alpha_2 S_{wD}^n, \dots\dots\dots\dots\dots\dots\dots\dots\dots\dots\dots\dots\dots\dots\dots\dots\dots\dots\dots\dots\dots\dots\dots\dots\dots\dots\dots (3.15)$

where $\alpha_1 = 0.8$, $\alpha_2 = 0.2$, $m = 2$, $n = 2$, and

$$S_{wD} = \left(S_w - S_{iw}\right)/\left(1 - S_{or} - S_{iw}\right), \dots\dots\dots\dots\dots\dots\dots\dots\dots\dots\dots\dots\dots\dots\dots\dots\dots\dots\dots\dots (3.16)$$

where S_{or} = waterflood residual oil saturation (ROS). Assume that $B_o = B_w = 1.0$.

Solution. Construction of the distance/time diagram, saturation profile, and displacement performance requires determination of the flood-front saturation and derivatives of the fractional-flow curve at various saturation values. These were determined numerically with programs presented in Appendix C (Willhite 1986). For this example, $S_{wf} = 0.4206$ and $f_{w2} = 0.65076$. The slope of the tangent to the fractional-flow curve from $f_w = 0$, $S_{iw} = 0.30$ is found with Eq. 3.3:

Property	Value
ϕ	0.20
S_{iw}	0.30
S_{or}	0.30
μ_o (cp)	40
μ_w (cp)	1
B_o (bbl/STB)	1

Table 3.1—Rock and fluid properties, Example 3.1.

$$f'_{wf} = \left(f_{wf} - f_{iw}\right) / \left(S_{wf} - S_{iw}\right)$$
$$= (0.65076 - 0.0)/(0.4206 - 0.3)$$
$$= 5.396.$$

When the flood reaches the end of the linear system ($x_D = 1.0$),

$$S_{w2} = S_{wf},$$

$$t_{D2} = 1/f'_w$$
$$= 1/5.396$$
$$= 0.185.$$

The average water saturation at $t_{D2} = 0.185$ is computed with Eq. 3.12:

$$\overline{S}_w = S_{w2} + t_{D2}\left(1 - f_{w2}\right)$$
$$= 0.4206 + 0.185(1 - 0.6508)$$
$$= 0.485.$$

Table 3.2 shows other saturations and computed parameters.

S_{w2}	f_{w2}	f'_{w2}	t_{D2}	$\overline{S}_w$
0.4206	0.65076	5.39578	0.18533	0.4853
0.4262	0.67991	5.03890	0.19846	0.4897
0.4318	0.70708	4.68777	0.21332	0.4943
0.4374	0.73232	4.34687	0.23005	0.4989
0.4430	0.75569	4.01949	0.24879	0.5037
0.4485	0.77727	3.70789	0.26970	0.5086
0.4541	0.79716	3.41350	0.29295	0.5136
0.4597	0.81545	3.13708	0.31877	0.5185
0.4653	0.83225	2.87883	0.34736	0.5236
0.4709	0.84766	2.63857	0.37899	0.5286
0.4765	0.86177	2.41585	0.41393	0.5337
0.4821	0.87469	2.20995	0.45250	0.5388
0.4877	0.88650	2.02007	0.49503	0.5438
0.4932	0.89729	1.84530	0.54192	0.5489
0.4988	0.90715	1.68468	0.59358	0.5540
0.5044	0.91614	1.53725	0.65051	0.5590
0.5100	0.92435	1.40505	0.71324	0.5640
0.5156	0.93183	1.27817	0.78237	0.5689
0.5212	0.93865	1.16472	0.85857	0.5739
0.5268	0.94487	1.06086	0.94263	0.5787
0.5324	0.95053	0.96581	1.03541	0.5836
0.5380	0.95568	0.87882	1.13789	0.5884
0.5435	0.96036	0.79922	1.25123	0.5931
0.5491	0.96462	0.72637	1.37671	0.5978
0.5547	0.96849	0.65970	1.51585	0.6025
0.5603	0.97200	0.59866	1.67039	0.6071
0.5659	0.97519	0.54278	1.84237	0.6116
0.5715	0.97808	0.49160	2.03418	0.6161
0.5771	0.98069	0.44471	2.24865	0.6205

Table 3.2—Summary of average water saturation calculations, Example 3.1, breakthrough to WOR > 50.

1. The distance/time graph is constructed by determining the time that each saturation arrives at the end of the system ($x_D = 1.0$) and constructing a straight line from that point to the origin. Because Eq. 3.13 is applicable,

$$t_{D2} = 1/f'_{w2}.$$

At $S_{w2} = 0.538$, $f'_{w2} = 0.87882$. Thus,

$$\begin{aligned} t_{D2} &= 1/f'_{w2} \\ &= 1/0.87882 \\ &= 1.13789. \end{aligned}$$

The distance/time graph shown in **Fig. 3.5** was prepared by drawing a line from the origin ($x_D = 0$, $t_D = 0$) to the end of the system ($x_D = 1$) for the arrival-time-selected water saturations ($t_D = t_{D2}$).

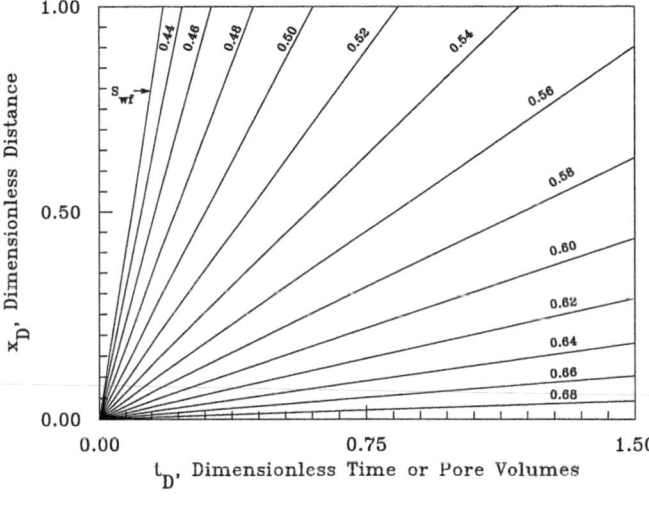

Fig. 3.5—x_D/t_D graph for parameters of Example 3.1.

2. The saturation profile when the flood front is at $x_D = 0.75$ is the locus of all saturations at the corresponding value of t_D. From the frontal-advance equation,

$$\begin{aligned} t_D &= x_{Df}/f'_{wf} \\ &= 0.75/f'_{wf} \\ &= 0.139. \end{aligned}$$

Locations of other saturations ($S_{wf} < S_w < 1 - S_{or}$) at $t_D = 0.139$ are found from Eq. 3.9, $x_{DS_w} = t_D f'_w$. At $S_w = 0.560$, $f'_w = 0.60183$. Thus,

$$\begin{aligned} x_{Dw} &= (0.139)(0.60183) \\ &= 0.0837. \end{aligned}$$

The computed saturation profile is given in **Table 3.3** and presented in **Fig. 3.6**. In Table 3.3, distances were determined at increments of 0.02 saturation units from $S_w = 1 - S_{or}$ to S_{wf}.

3. Displacement performance is obtained by computing the average water saturation with Eq. 3.12. **Table 3.4** summarizes the results with the displacement performance expressed as a fraction of the PV, $N_p/(A\phi L)$, determined from Eq. 3.11:

$$N_p = \left[\left(\overline{S}_w - S_{iw}\right)/B_o\right] A\phi L.$$
Thus, $N_p/A\phi L = \left(\overline{S}_w - S_{iw}\right)/B_o$.
From Table 3.2 at $S_{w2} = 0.5547$, $\overline{S}_w = 0.6025$, and

$$\begin{aligned} N_p/A\phi L &= 0.6025 - 0.3 \\ &= 0.3025 \text{ when } B_o = 1.0. \end{aligned}$$

F_{wo}, the producing WOR, is calculated by rearrangement of the fractional-flow equation (Eq. 3.2). Because

$$f_w = q_w/(q_w + q_o), \dotfill (3.17)$$

where q_w = water flow rate and q_o = oil flow rate, and

$$F_{wo} = (q_w/q_o)(B_o/B_w), \dotfill (3.18)$$

then $F_{wo} = \left[f_w/(1 - f_w)\right](B_o/B_w). \dotfill (3.19)$

3.2.2 Displacement of Interstitial Water. During a waterflood, the injected water displaces the interstitial water as well as oil. The movement of interstitial water during a waterflood can be followed by use of a simple model. This model, with some

S_w	f'_w	x_D
0.70	0.00000	0.000
0.68	0.02914	0.00405
0.66	0.06842	0.00951
0.64	0.12137	0.01687
0.62	0.19289	0.02681
0.60	0.28982	0.04028
0.58	0.42169	0.05861
0.56	0.60183	0.08365
0.54	0.84880	0.11798
0.52	1.18799	0.16513
0.50	1.65289	0.22975
0.48	2.28437	0.31752
0.46	3.12370	0.43419
0.44	4.19083	0.58251
0.4206*	5.3958	0.75
0.30	–	0.75
0.30	–	1.00

*S_{wf}

Table 3.3—Saturation profile at $x_D = 0.75$, Example 3.1, $t_D = 0.139$.

modification, also will be useful in following the movement of chemicals used to improve oil recovery by viscosity increase or alteration of interfacial tension (IFT). The interstitial water is chemically distinct from the injected water, but is displaced miscibly. No mixing is assumed to occur between the injected and interstitial fluids, so that a distinct boundary exists between the injected and interstitial water as if the displacement process were piston-like. Under these assumptions, the interstitial water will be distributed in the saturation profile, as depicted in **Fig. 3.7.**

The location of the boundary between the injected and interstitial water, x_{Db}, at any time can be found by applying a material balance on the interstitial water. Because the volume of interstitial water is conserved and no water is displaced from the linear system under the conditions being considered,

$$A\phi x_f S_{iw} = \int_{xb}^{xf} A\phi S_w dx$$
$$= A\phi \overline{S}_{wb}\left(x_f - x_b\right), \dots\dots\dots\dots\dots\dots\dots\dots\dots\dots\dots\dots\dots\dots\dots\dots\dots\dots (3.20)$$

where $\overline{S}_{wb}$ = average water saturation in the interval between x_b and x_f.

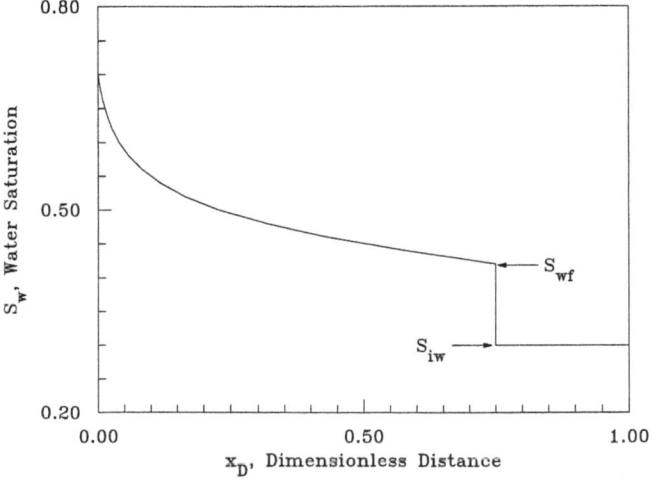

Fig. 3.6—Saturation profile when $x_D = 0.75$, Example 3.1.

S_{w2}	t_D	F_{wo} (bbl/bbl)	$N_p/A\phi L$(PV)
0.30	0.00000	0.00000	0.00000
0.30	0.01853	0.00000	0.01853
0.30	0.03707	0.00000	0.03707
0.30	0.05560	0.00000	0.05560
0.30	0.07413	0.00000	0.07413
0.30	0.09266	0.00000	0.09266
0.30	0.11120	0.00000	0.11120
0.30	0.12973	0.00000	0.12973
0.30	0.14826	0.00000	0.14826
0.4206	0.18533	1.86332	0.18533
0.4262	0.19846	2.12410	0.18972
0.4318	0.21332	2.41390	0.19427
0.4374	0.23005	2.73576	0.19895
0.4430	0.24879	3.09308	0.20374
0.4485	0.26970	3.48969	0.20861
0.4541	0.29295	3.92990	0.21356
0.4597	0.31877	4.41858	0.21855
0.4653	0.34736	4.96123	0.22358
0.4709	0.37899	5.56411	0.22863
0.4765	0.41393	6.23434	0.23370
0.4821	0.45250	6.98001	0.23878
0.4877	0.49503	7.81043	0.24385
0.4932	0.54192	8.73625	0.24891
0.4988	0.59358	9.76974	0.25395
0.5044	0.65051	10.92510	0.25897
0.5100	0.71324	12.21878	0.26397
0.5156	0.78237	13.66995	0.26893
0.5212	0.85857	15.30101	0.27386
0.5268	0.94263	17.13829	0.27874
0.5324	1.03541	19.21285	0.28359
0.5380	1.13789	21.56150	0.28839
0.5435	1.25123	24.22817	0.29313
0.5491	1.37671	27.26550	0.29783
0.5547	1.51585	30.73700	0.30248
0.5603	1.67039	34.71980	0.30707
0.5659	1.84237	39.30822	0.31160
0.5715	2.03418	44.61852	0.31607
0.5771	2.24865	50.79517	0.32048

Table 3.4—Estimated displacement performance, Example 3.1, linear water-flood at interstitial water saturation.

$\overline{S}_{wb}$ may be computed by use of the expanded version of the Welge equation developed by Craig (1971). From Willhite (1986), the average water saturation within the interval $x_1 \leq x \leq x_2$ obtained with the frontal-advance solution is given by

$$\overline{S}_w = \frac{x_2 S_{w2} - x_1 S_{w1}}{x_2 - x_1} - \left(\frac{q_t t}{A\phi}\right)\left(\frac{f_{w2} - f_{w1}}{x_2 - x_1}\right). \quad\dots\dots\dots\dots\dots\dots\dots\dots\dots\dots\dots\dots\dots\dots (3.21)$$

In this case,

$$\overline{S}_{wb} = \frac{x_f S_{wf} - x_b S_{wb}}{x_f - x_b} - \left(\frac{q_t t}{A\phi}\right)\left(\frac{f_{wf} - f_{wb}}{x_f - x_b}\right). \quad\dots\dots\dots\dots\dots\dots\dots\dots\dots\dots\dots\dots\dots (3.22)$$

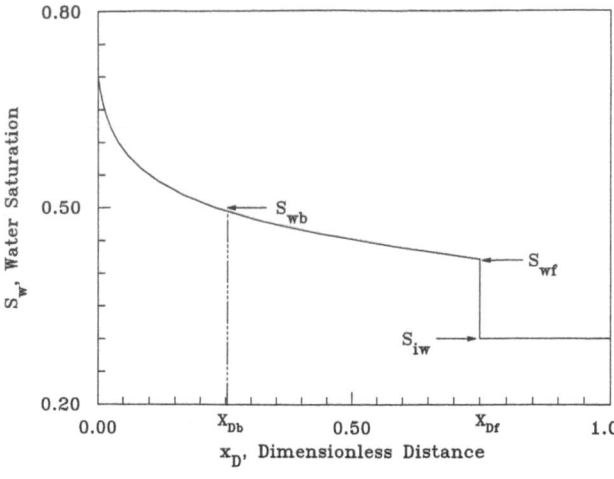

Fig. 3.7—Saturation profile showing interstitial water displaced by the injected fluid.

Substituting into Eq. 3.20 and rearranging gives

$$x_f S_{iw} = x_f S_w - x_b S_{wb} - (q_t t/A\phi)(f_{wf} - f_{wb})$$

$$\dots \dots \dots \dots \dots \dots \dots \dots \dots \dots \text{(3.23)}$$

or $x_f \left(S_{wf} - S_{iw}\right) = x_b S_{wb} + (q_t t / A\phi)(f_{wf} - f_{wb})$

$$\dots \dots \dots \dots \dots \dots \dots \dots \dots \dots \text{(3.24)}$$

From the frontal-advance equation,

$$x_f = (q_t t/A\phi)\left[f_{wf}/\left(S_{wf} - S_{iw}\right)\right]. \dots \dots \text{(3.25)}$$

Substituting Eq. 3.25 into Eq. 3.24 gives

$$x_b = (q_t t/A\phi)(f_{wb}/S_{wb}). \dots \dots \dots \dots \text{(3.26)}$$

Because the location of S_{wb} must also satisfy the frontal-advance solution,

$$xb = (q_t t/A\phi)(\partial f_w/\partial S_w)_{S_{wb}}. \dots \dots \dots \dots \dots \dots \dots \dots \dots \dots \dots \dots \dots \dots \dots \text{(3.27)}$$

Comparison of Eq. 3.26 and 3.27 gives

$$(\partial f_w/\partial S_w)_{S_{wb}} = f_{wb}/S_{wb}. \dots \dots \dots \dots \dots \dots \dots \dots \dots \dots \dots \dots \dots \dots \dots \text{(3.28)}$$

Thus, the derivative of the fractional-flow curve with a slope of f_{wb}/S_{wb} is the slope of a line from the origin ($f_w = 0$, $S_w = 0$) that is tangent to the fractional-flow curve, as shown in **Fig. 3.8**. The values of f_{wb} and S_{wb} are determined from the intersection of the tangent. For the problem in Example 3.1, $f_{wb} = 0.8989$ and $S_{wb} = 0.4941$.

The location of the boundary between the interstitial water and the injected water can now be plotted because this boundary also travels at a constant velocity. Fig. 3.9 shows the boundary for the parameters of Example 3.1. The shaded area is the expanding region occupied by the interstitial water. Expansion occurs because interstitial water is added to this region as the flood front advances at a constant velocity.

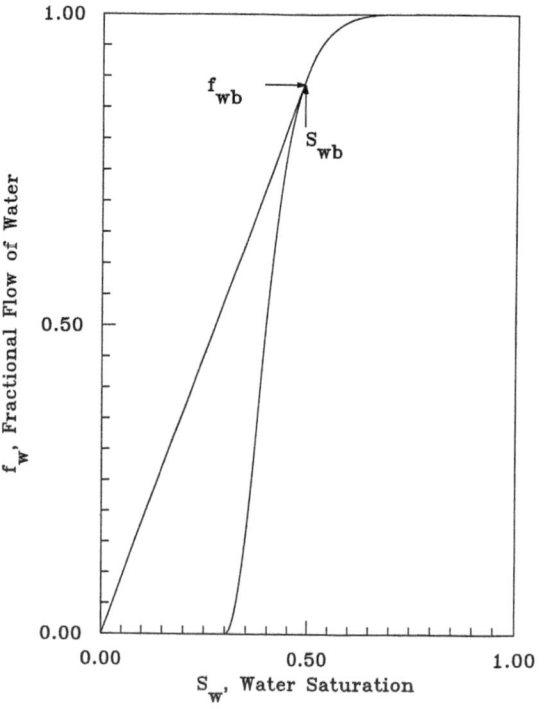

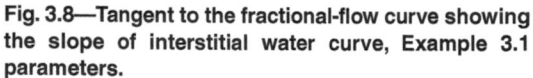

Fig. 3.8—Tangent to the fractional-flow curve showing the slope of interstitial water curve, Example 3.1 parameters.

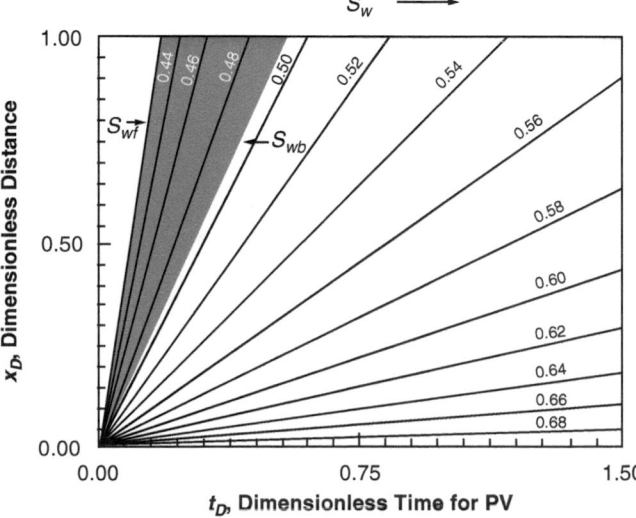

Fig. 3.9—x_D/t_D diagram showing region occupied by the interstitial water, Example 3.1 parameters.

3.3 Viscous Waterflood in a Linear System

Waterflood displacement efficiency is affected by the viscosity ratio of the displaced to the displacing fluid. This can be shown simply by altering the viscosity in a displacement calculation. For example, if the water viscosity in Example 3.1 were 40 cp (both interstitial and injected water) rather than 1 cp, the waterflood performance presented in **Fig. 3.10** would be predicted by the frontal-advance solution. Also reproduced in Fig. 3.10 is the predicted performance for the waterflood in Example 3.1 with $\mu_w = 1$ cp.

This example suggests that the injection of a viscous fluid is an attractive possibility to improve waterflood displacement efficiency, particularly in reservoirs containing viscous oil. However, it does not properly consider the role of interstitial or previously injected water that must be displaced.

In this section, frontal-advance techniques are used to estimate waterflood performance when the injected fluid is viscous

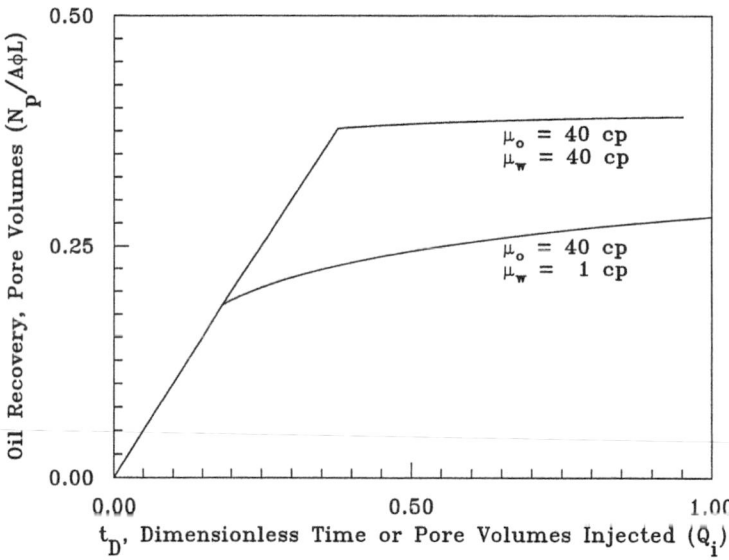

Fig. 3.10—Waterflood performance, frontal-advance solution, $\mu_w = 40$ cp vs. $\mu_w = 1$ cp and $\mu_o = 40$ cp vs. $\mu_w = 40$ cp, Example 3.1 parameters.

but still miscible with the interstitial or previously injected water, which has a low viscosity. The viscous fluid is assumed not to be adsorbed on the rock, an assumption that will be modified in Section 3.5. As in Section 3.2, no mixing occurs between the viscous fluid and the low-viscosity resident water. Thus, a boundary exists between the viscous and displaced water where there is a step change, or jump, in viscosity from μ_w to u_w^*.

The displacement process can be described as a waterflood in which a viscous fluid displaces both oil and low-viscosity resident water. The resident water is miscibly displaced by the injected fluid. It is the resident water, rather than the injected viscous fluid, that forms the leading flood front. Because there is a discontinuity in the viscosity between the viscous and resident fluids, a second discontinuity in saturation, or shock front, must form at the viscous-water/resident-water boundary. In Section 3.3.1, a relationship between the velocity of this discontinuity and fractional flows is derived on the basis of mass conservation.

3.3.1 Determination of Velocity and Saturations of the Viscous Shock. Fig. 3.11 illustrates the locations of a saturation discontinuity or shock wave moving at a constant velocity v_t when the flow rate q_t is constant. The shock front moves from x_{ft} to $x_{f(t+\Delta t)}$ in the time increment Δt, but is confined within the interval $x_2 - x_1$. Saturations and fractional flow are assumed to be uniform on either side of the saturation discontinuity because the distance can be made arbitrarily small by the choice of Δt. From Fig. 3.11, the volume of water in the volume element $x_2 - x_1$ at time t is

$$\text{Vol}/t = \left(x_{ft} - x_1\right)A\phi S_{w3}^* + \left(x_2 - x_{ft}\right)A\phi S_{w1}, \dots \dots \dots (3.29)$$

where S_{w3}^* = saturation on the upstream side of the discontinuity and S_{w1} = saturation on the downstream side of the discontinuity. At time $t + \Delta t$, the volume of water in the volume element is

$$\text{Vol}/_{t+\Delta t} = \left[x_{f(t+\Delta t)} - x_1\right]A\phi S_{w3}^* + \left[x_2 - x_{f(t+\Delta t)}\right]A\phi S_{w1}. \dots \dots \dots (3.30)$$

A volume balance on water across the volume element during the time Δt thus yields (by subtraction of Eq. 3.29 from Eq. 3.30)

$$\left\{\left[x_{f(t+\Delta t)} - x_1\right]A\phi S_{w3}^* + \left[x_2 - x_{f(t+\Delta t)}\right]A\phi S_{w1}\right\}$$

$$-\left[\left(x_{ft} - x_1\right)A\phi S_{w3}^* + \left(x_2 - x_{ft}\right)A\phi S_{w1}\right]$$

$$= \left(q_t\Delta t\right)\left(f_{w3}^* - f_{w1}\right). \dots \dots \dots (3.31)$$

After some rearrangement,

$$\left[x_{f(t+\Delta t)} - x_{ft}\right]\left(S_{w3}^* - S_{w1}\right) = q_t\Delta t/A\phi\left(f_{w3}^* - f_{w1}\right). \dots \dots (3.32)$$

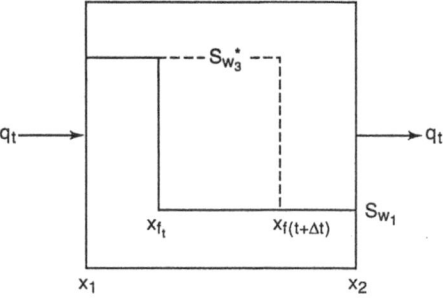

Fig. 3.11—Schematic of shock front at two times.

$$\text{or } \frac{\left[x_{f(t+\Delta t)} - x_{ft}\right]}{\Delta t} = \frac{q_t}{A\phi}\left(\frac{f^*_{w3} - f_{w1}}{S^*_{w3} - S_{w1}}\right). \quad \dots \dots \dots \dots \dots \dots \dots \dots \dots \dots \dots \dots (3.33)$$

The velocity of the saturation discontinuity is obtained by letting $\Delta t \to 0$ to obtain

$$\frac{dx_f}{dt} = \frac{q_t}{A\phi}\left(\frac{f^*_{w3} - f_{w1}}{S^*_{w3} - S_{w1}}\right). \quad \dots \dots \dots \dots \dots \dots \dots \dots \dots \dots \dots \dots (3.34)$$

Saturations in the viscous waterflood also satisfy the frontal-advance solution. Thus,

$$\frac{dx_{S^*_{w3}}}{dt} = \frac{q_t}{A\phi}\left(\frac{\partial f^*_w}{\partial S^*_w}\right)_{S^*_{w3}}. \quad \dots \dots \dots \dots \dots \dots \dots \dots \dots \dots \dots \dots (3.35)$$

Because the saturation S^*_{w3} moves at the same velocity as the saturation discontinuity,

$$\left(\frac{\partial f^*_w}{\partial S^*_w}\right)_{S^*_{w3}} = \frac{f^*_{w3} - f_{w1}}{S^*_{w3} - S_{w1}}. \quad \dots \dots \dots \dots \dots \dots \dots \dots \dots \dots \dots \dots (3.36)$$

To complete the specification of the saturation discontinuity, use is made of the miscibility between the viscous solution and the resident water. At the saturation discontinuity, the velocity of the viscous phase must be equal to the velocity of the displaced water because of miscibility. The velocity of the resident water phase is given by

$$v_1 = f_{w1}q_t / A\phi S_{w1}, \quad \dots \dots \dots \dots \dots \dots \dots \dots \dots \dots \dots \dots \dots \dots (3.37)$$

where $A\phi S_{w1}$ is the cross-sectional area that the water flows through.

By analogy,

$$v^*_3 = f^*_{w3}q_t / A\phi S^*_{w3}. \quad \dots \dots \dots \dots \dots \dots \dots \dots \dots \dots \dots \dots \dots \dots (3.38)$$

At the boundary between the viscous solution and the displaced water, the velocities of the viscous phase and the displaced water must be equal (i.e., $v_1 = v^*_3$). Thus,

$$f_{w1}/S_{w1} = f^*_{w3}/S^*_{w3}. \quad \dots \dots \dots \dots \dots \dots \dots \dots \dots \dots \dots \dots \dots \dots (3.39)$$

It is convenient to express these velocities in terms of dimensionless parameters by introducing the specific velocity defined as

$$v^*_{D3} = v^*_3 / (q_t / A\phi). \quad \dots \dots \dots \dots \dots \dots \dots \dots \dots \dots \dots \dots \dots \dots (3.40)$$

Converting Eq. 3.34 to dimensionless form by introducing x_D and t_D yields

$$\left(\frac{dx_D}{dt_D}\right)_{S^*_{w3}} = \frac{f^*_{w3} - f_{w1}}{S^*_{w3} - S_{w1}}. \quad \dots \dots \dots \dots \dots \dots \dots \dots \dots \dots \dots \dots (3.41)$$

Thus $v^*_{D3} = \left(f^*_{w3} - f_{w1}\right) / \left(S^*_{w3} - S_{w1}\right). \quad \dots \dots \dots \dots \dots \dots \dots \dots \dots \dots \dots \dots (3.42)$

Because the specific water velocity must be equal to the velocity of the discontinuity,

$$v^*_{D3} = v_{D1}. \quad \dots \dots \dots \dots \dots \dots \dots \dots \dots \dots \dots \dots \dots \dots \dots \dots (3.43)$$

and $\left(\frac{\partial f^*_w}{\partial S^*_w}\right)_{S^*_{w3}} = \frac{f^*_{w3} - f_{w1}}{S^*_{w3} - S_{w1}} = \frac{f^*_{w3}}{S^*_{w3}} = \frac{f_{w1}}{S_{w1}}. \quad \dots \dots \dots \dots \dots \dots \dots \dots (3.44)$

Inspection of Eq. 3.44 shows that f^*_{w3} and S^*_{w3} can be found by constructing a tangent from the origin to the $f^*_w - S^*_w$ curve for u^*_w, as shown in **Fig. 3.12.** The intersection of this tangent with the fractional-flow curve for μ_w gives the values of f_{w1} and S_{w1}. With these values, the displacement of the shock caused by the differences in viscosities is completely defined.

Example 3.2 illustrates the determination of flood-front saturations for a viscous waterflood.

Example 3.2—Viscous Waterflood. The linear reservoir in Example 3.1 is to be flooded with a viscous solution that does not adsorb. The viscosity of the solution is 4 cp, and no mixing occurs between the viscous solution and the interstitial water. All other parameters used in Example 3.1 remain unchanged (see Table 3.1 and Eqs. 3.14 through 3.16). Find the flood-front saturations.

Solution. Fractional-flow curves for the viscous and nonviscous waterfloods are shown in Fig. 3.12. **Table 3.5** contains values of fractional flow and saturations used to prepare Fig. 3.12. A tangent was constructed to the u_w^* fractional-flow curve to determine f_{w3}^* (0.926) and S_{w3}^* (0.576). The intersection of the tangent with the μ_w fractional-flow curve is at $f_{w1} = 0.687$ and $S_{w1} = 0.428$. These values were obtained numerically with a root-finding program and verified graphically. Note that S_{wf} from Example 3.1 is 0.4206 and that $S_{w1} > S_{wf}$.

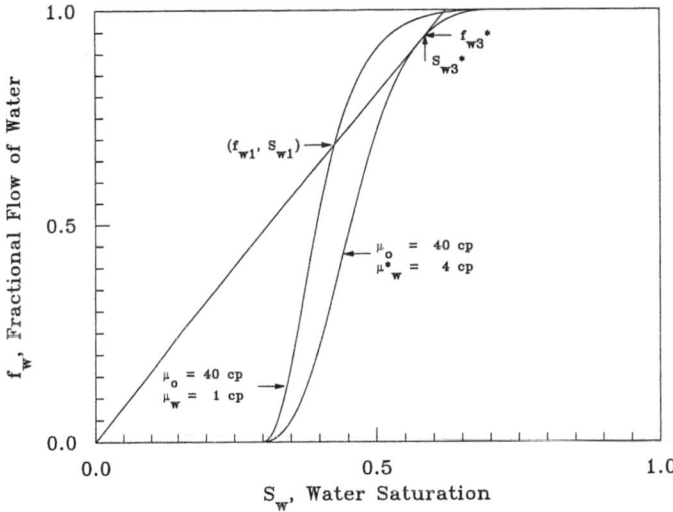

Fig. 3.12—Construction of tangent to find f_{w3}^*, S_{w3}^*, f_{w1}, and S_{w1}, Example 3.2.

The region behind the viscous shock behaves as a viscous waterflood and is described by the frontal-advance solution. Thus, the specific velocity of each saturation $S_w^*(S_w^* \geq S_{w3}^*)$ is obtained from the derivative at the viscous fractional-flow curve, as in Eq. 3.45:

$$v_{Dv}^* = \left(\partial f_w^* / \partial S_w^*\right)_{S_w^*}. \dots\dots\dots\dots\dots\dots\dots\dots\dots\dots\dots\dots\dots\dots\dots\dots\dots\dots\dots (3.45)$$

S_w, S_w^*	f_w^* $u_w^* = 4.0$ cp	f_w $\mu_w = 1.0$ cp
0.30	0.00000	0.00000
0.32	0.00688	0.02695
0.34	0.02994	0.10989
0.36	0.07223	0.23747
0.38	0.13514	0.38462
0.40	0.21739	0.52632
0.42	0.31469	0.64748
0.44	0.42024	0.74355
0.46	0.52632	0.81633
0.48	0.62597	0.87003
0.50	0.71429	0.90909
0.52	0.78879	0.93726
0.54	0.84906	0.95745
0.56	0.89608	0.97182
0.58	0.93156	0.98196
0.60	0.95745	0.98901
0.62	0.97561	0.99379
0.64	0.98770	0.99690
0.66	0.99509	0.99877
0.68	0.99889	0.99972
0.70	1.00000	1.00000

Table 3.5—Fractional-flow data from Fig. 3.12, Example 3.2.

The formation of the viscous shock when viscous water is injected after the beginning of a waterflood is complicated. Appendix A presents a conceptual model of this process.

3.3.2 Oil Recovery During a Viscous Waterflood. The volume of oil displaced during a viscous waterflood is determined by computing the average water saturation in the system at various points in time, as done for waterflooding calculations. When the initial oil saturation is $1 - S_{iw}$, the displaced oil, expressed in PVs, is given by

$$N_P/A\phi L = \left(\overline{S}_w - S_{iw}\right)/B_o. \dots\dots\dots\dots\dots\dots\dots (3.46)$$

Because the saturation profile for a viscous waterflood may have several discontinuities, the average water saturation must be determined by integrating the saturation distribution at discrete times. When $S_{w1} > S_{wf}$, the saturation profile is depicted by **Fig. 3.13** as long as the flood front is in the system. In Fig. 3.13, $S_{w1} = 0.428$ and $S_{wf} = 0.4206$, so there is a small change in the saturation profile between the waterflood front, x_f, and the viscous flood front, x_3. This difference is small, but is retained in the average water saturation calculations. The average water saturation is given by

$$S_w = \frac{\int_0^{x_3} S_w^* \, dx + \int_{x_3}^{x_1} S_{w1} \, dx = \int_{x_1}^{x_f} S_w \, dx + \int_{x_f}^{L} S_{iw} \, dx}{L}$$

$$= \overline{S}_{w3}^* \frac{x_3}{L} + \left(\frac{x_1 - x_3}{L}\right) S_{w1} + \left(\frac{x_f - x_1}{L}\right) \overline{S}_{w1} + S_{iw}\left(\frac{L - x_f}{L}\right). \dots\dots\dots (3.47)$$

In dimensionless form, Eq. 3.47 becomes

$$\overline{S}_w = x_{D3}\overline{S}_{w3}^* + \left(x_{D1} - x_{D3}\right)S_{w1} + \left(x_{Df} - x_{D1}\right)\overline{S}_{w1} + \left(1 - x_{Df}\right)S_{iw}, \dots\dots\dots (3.48)$$

where $\overline{S}_w^*$ and $\overline{S}_{w1}$ are the average water saturations in the respective regions. Both $\overline{S}_w^*$ and $\overline{S}_{w1}$ can be computed from the modified Welge equation when the appropriate fractional-flow curves are used. The viscous shock is assumed to form immediately when the viscous fluid is injected. According to the computations in Example A-1, this appears to be a reasonable assumption. When viscous water injection begins at $t_{Do} = 0$,

$$\overline{S}_{w3}^* = S_{w3}^* - \left(t_D/x_{D3}\right)\left(f_{w3}^* - 1\right) \dots\dots\dots\dots\dots\dots\dots (3.49)$$

$$\text{and } \overline{S}_{w1} = \frac{x_{Df}S_{wf} - x_{D1}S_{w1}}{x_{Df} - x_{D1}} - t_D \frac{f_{wf} - f_{w1}}{x_{Df} - x_{D1}}. \dots\dots\dots\dots\dots (3.50)$$

Because the viscous shock forms immediately, the oil bank also forms at the same time.

To compute the average saturations, it is necessary to know the locations of saturations S_{wf}, S_{w1}, and S_{w3}^*. These saturations travel at different velocities given by their fractional-flow relationships. By starting the injection of viscous water at $t_D = 0$, the following equations give the location of each region during the viscous waterflood:

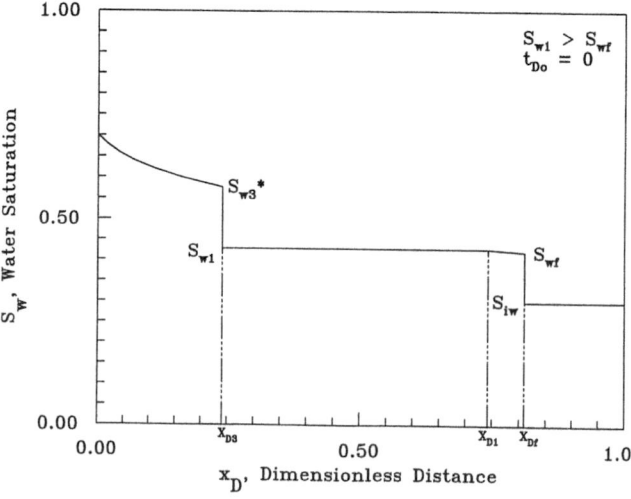

Fig. 3.13—Saturation profile during a viscous waterflood at interstitial water saturation when $S_{w1} > S_{wf}$.

$$x_{Df} = f_{wf}' t_D, \dots\dots\dots\dots\dots (3.51)$$

$$x_{D1} = f_{w1}' t_D, \dots\dots\dots\dots\dots (3.52)$$

$$\text{and } x_{D3} = f_{w3}'^* t_D. \dots\dots\dots\dots (3.53)$$

Before breakthrough, oil recovery is a linear function of t_D. At breakthrough of the flood front,

$$N_P/A\phi L = \overline{S}_w - S_{iw} = t_{Df}$$

$$= 1/f_{wf}'. \dots\dots\dots\dots\dots (3.54)$$

The oil bank $(S_w - S_{w1})$ arrives at the end of the system when $x_{D1} = 1$ or

$$t_{D1} = 1/f_{w1}'. \dots\dots\dots\dots\dots (3.55)$$

Between breakthrough and arrival of the oil bank, x_{D1},

$$\overline{S}_w = x_{D3}\overline{S}_{w3}^* + \left(x_{D1} - x_{D3}\right)S_{w1} + \left(1 - x_{D1}\right)\overline{S}_{w1}. \quad\text{.........................}(3.56)$$

After breakthrough, Eq. 3.56 can be expressed in terms of S_{w2}, the saturation at the end of the system, by substituting Eqs. 3.49 and 3.50 for $\overline{S}_{w3}^*$ and $\overline{S}_{w1}$ respectively, to obtain

$$\overline{S}_w = S_{w2} + t_{D2}\left(1 - f_{w2}\right), \quad\text{.........................}(3.57)$$

with $t_D = \dfrac{1}{f'_{w2}}$

where S_{w2} = water saturation at the end of the linear system. During the time period $t_{Df} \le t_D \le t_{D1}$, S_{w2} will increase from S_{wf} to S_{w1}. When the oil bank arrives at the end of the system $(x_{D1} = 1.0)$, the average water saturation is given by

$$\overline{S}_w = x_{D3}\overline{S}_{w3}^* + \left(1 - x_{D3}\right)S_{w1}. \quad\text{.........................}(3.58)$$

Substituting Eq. 3.49 for $\overline{S}_{w3}^*$ and Eq. 3.53 for x_{D3} gives

$$\overline{S}_w - S_{w1} + t_D\left(1 - f_{w1}\right). \quad\text{.........................}(3.59)$$

The viscous waterflood arrives at the end of the system when $x_{D3} - 1$ or $t_D = 1/f_{w3}^*$. Therefore, for $t_D \ge 1/f_{w3}^*$,

$$\overline{S}_w^* = S_w^* + t_D\left(1 - f_w^*\right). \quad\text{.........................}(3.60)$$

3.4 Viscous Waterflood in a Linear System Initially at Interstitial Water Saturation

In this section, a viscous waterflood in a linear reservoir at interstitial water saturation (initial oil saturation) is considered. As in Section 3.3, the injected viscous fluid is miscible with the interstitial water and is not retained on the porous rock by adsorption or other mechanisms. Dispersion is neglected so that a sharp boundary is maintained between the viscous solution and the interstitial water. A viscous shock forms instantaneously when the viscous solution is injected into the reservoir. This shock has the same properties as in Section 3.3. Thus,

$$\left(\frac{dx_D}{dt_D}\right)_{S_{w3}^*} = \frac{f_{w3}^* - f_{w1}}{S_{w3}^* - S_{w1}}. \quad\text{.........................}(3.61)$$

and $x_{D3} = t_D \dfrac{f_{w3}^* - f_{w1}}{S_{w3}^* - S_{w1}}.$ $\quad\text{.........................}(3.62)$

When $S_{w1} < S_{wf}$, the oil bank forms immediately, overtakes S_{wf}, and has a uniform water saturation, S_{w1}. **Fig. 3.14** illustrates the saturation profile for this flood. The velocity of the oil bank is given by

$$\left(\frac{dx}{dt}\right)_{S_{wi}} = \frac{q_t}{A\phi}\frac{f_{w1}}{\left(S_{w1} - S_{iw}\right)} \quad\text{.........................}(3.63)$$

or, in dimensionless form,

$$\left(\frac{dx_D}{dt_D}\right)_{S_{wi}} = \frac{f_{w1}}{S_{w1} - S_{iw}}. \quad\text{.........................}(3.64)$$

Water saturations greater than S_{w3}^* travel at velocities given by

$$\left(\frac{dx}{dt}\right)_{S_w^*} = \frac{q_t}{A\phi}\left(\frac{\partial f_w^*}{\partial S_w^*}\right)_{S_w^*}. \quad\text{.........................}(3.65)$$

The oil-bank region has a uniform saturation, S_{w1}, while the saturations greater than S_{w3}^* form a fan-shaped region like that described in Section 3.3.

Breakthrough of water is obtained from

$$t_{D1} = \left(S_{w1} - S_{iw}\right)/f_{w1}, \quad\text{.........................}(3.66)$$

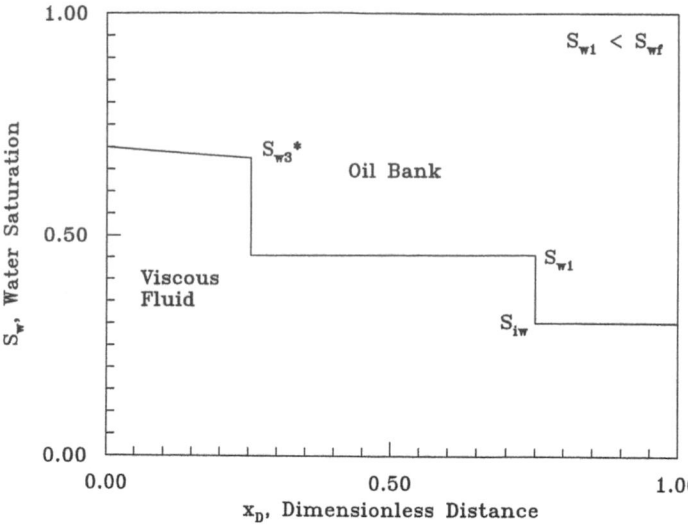

Fig. 3.14—Saturation profile for viscous waterflood at interstitial water saturation when $S_{w1} < S_{wf}$.

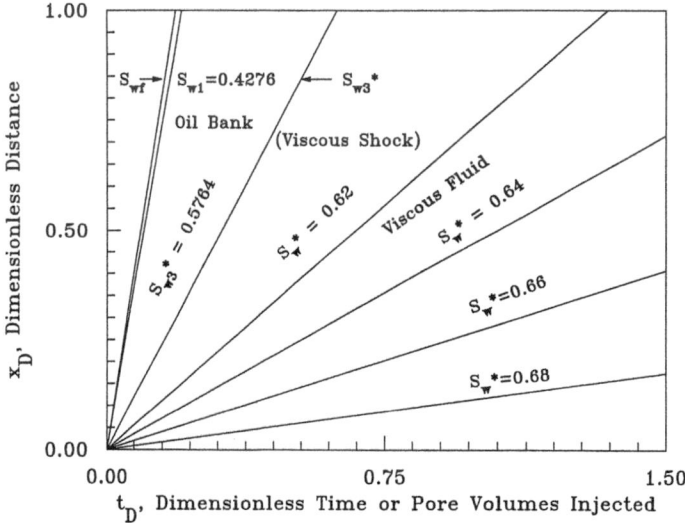

Fig. 3.15—Distance/time diagram for viscous waterflood at interstitial water saturation when $S_{w1} > S_{wf}$, Example 3.3.

while the viscous solution breaks through at

$$t_{D3} = \left(S_{w3}^* - S_{w1}\right)\big/\left(f_{w3}^* - f_{w1}\right). \quad \dots \text{(3.67)}$$

In some cases, $S_{w1} > S_{wf}$ and the oil bank cannot overtake the flood-front saturation. A saturation profile for this case is shown in Fig. 3.13. The displacement is characterized by arrival of the waterflood front, a zone of increasing water saturation, a region of constant water saturation, the viscous shock, and a region of increasing water saturation. Example 3.3 illustrates the computation of viscous waterflood performance assuming that the initial water saturation is uniform, as in Example 3.2. **Fig. 3.15** illustrates the distance/time diagram for this waterflood.

Example 3.3—Performance of a Viscous Waterflood—Initial Interstitial Water Saturation. Estimate the performance of a viscous waterflood with the parameters of Example 3.2 when the viscous waterflood starts with the reservoir at interstitial water saturation. Fluid and rock properties are the same as in Example 3.2.

Solution. When the reservoir is at interstitial water saturation, the saturation profile develops immediately and propagates through the system. **Table 3.6** summarizes saturation values for the various zones determined in Examples 3.1 and 3.2.

Because S_{w1} is slightly larger than S_{wf}, it never overtakes S_{wf} and the saturation profile shown in Fig. 3.13 is obtained. Thus, there are three distinct banks in the flood: water bank, oil bank, and viscous shock.

The x_D/t_D diagram can be generated from the fractional-flow data. From Eqs. 3.51 through 3.53, $x_{Df} = 5.3958 \, t_D$, $x_{D1} = 4.9499 \, t_D$, and $x_{D3}^* = 1.606 \, t_D$. The saturation paths are presented in Fig. 3.15. The region between S_{wf} and S_{w1} is narrow in this case (0.007 saturation units), which in practice would not be worth separating, but it is done here for understanding. In this region, each saturation travels at a different velocity, as indicated by several paths drawn in Fig. 3.15. Paths for saturations behind the viscous shock are obtained from Eq. 3.6: $x_D^* = t_D f_w^{'*}$. Values of $f_w^{'*}$ are given in **Table 3.7** for several values of S_w^*. Also included in Table 3.7 are the arrival time of S_w^* at $x_D = 1.0$ and the location of x_D^* when $t_D = 1.5$.

Fig. 3.13 is a saturation profile constructed from the x_D/t_D diagram at $t_D = 0.15$.

Oil recovery is obtained by applying Eqs. 3.57 through 3.60 to the appropriate zones. A convenient way to compute oil recovery is to choose the saturation at the end of the linear system and then compute the arrival time t_D when $x_D = 1.0$. **Table 3.8** summarizes the oil recovery calculations.

Displacement performance of the viscous waterflood is compared with a normal waterflood in **Fig. 3.16.**

Waterflood	Front of Oil Bank	Viscous Shock
$S_{wf} = 0.4206$	$S_{w1} = 0.4276$	$S_{w3}^* = 0.5764$
$f_{wf} = 0.6508$	$f_{w1} = 0.6869$	$f_{w3}^* = 0.9259$
$f_{wf}' = 5.3958$	$f_{w1}' = 4.9499$	$f_{w3}' = 1.6064$

Table 3.6—Saturation for various zones, Example 3.3.

S_w^*	$f_w'^*$	x_{Dw} at $t_D = 1.5$	t_D at $x_{Dw} = 1.0$
0.5764	1.6064	2.410	0.622
0.60	1.0865	1.630	0.920
0.64	0.4766	0.715	2.098
0.66	0.2717	0.408	3.680
0.68	0.1164	0.175	8.591
0.70	0	0	∞

Table 3.7—Location of saturations behind the viscous shock when $t_D = 1.5$, Example 3.3.

3.5 Chemical Flooding in a Linear System

The viscous waterflood discussed in Sections 3.3 and 3.4 enhances displacement performance because fractional-flow curves are altered. However, one or more chemical species must be added to the injected water to make it viscous. In most cases, the added chemical is polymer. Also, adding surfactants to the injected water can alter the shape of the fractional-flow curves by decreasing IFT and can change the end points of the curves by reducing the ROS. In this section, the frontal-advance equation that describes the transport of chemical species in oil displacement processes is developed. Then, the use of frontal-advance theory to estimate displacement performance of chemical floods, such as polymer and surfactant floods, is illustrated.

3.5.1 Transport of Chemical Species in Porous Rocks. The transport of chemical species in porous media can be described by applying material-balance concepts to each species. Attention is restricted to 1D, isothermal, two-phase flow to simplify the mathematical model. A single chemical species is added to the injected fluid. Dispersion, or fluid mixing, is neglected. Effects of dispersion are discussed in Section 3.8. Other assumptions include neglecting gravity and capillary forces and viscous fingering. Fluids are

S_w	f_w	Event at End of System	t_D	$\bar{S}_w$	$\bar{S}_w - S_{iw}$
0.4206	0.6508	←Water bank	0.1853	0.4853	0.1853
0.4276	0.6869	←Arrival of oil bank	0.202	0.4909	0.1909
0.5764	0.9259	←Arrival of viscous shock	0.623	0.623	0.323
0.60	0.9575		0.920	0.639	0.339
0.62	0.9756		1.345	0.653	0.353
0.64	0.9877	{Region behind viscous shock}	2.098	0.666	0.366
0.66	0.9951		3.681	0.678	0.378
0.68	0.9989		8.591	0.689	0.389

Table 3.8—Summary of recovery calculations, Example 3.3, viscous waterflood at interstitial water saturation.

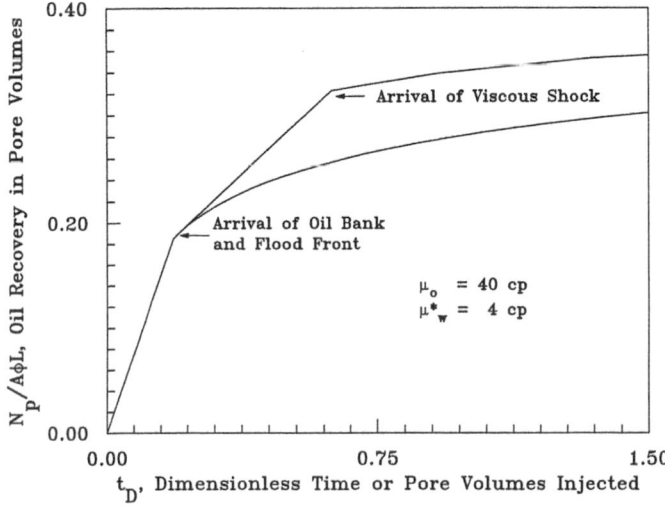

Fig. 3.16—Comparison of displacement performance for a viscous waterflood at interstitial water saturation with a normal waterflood, Example 3.3.

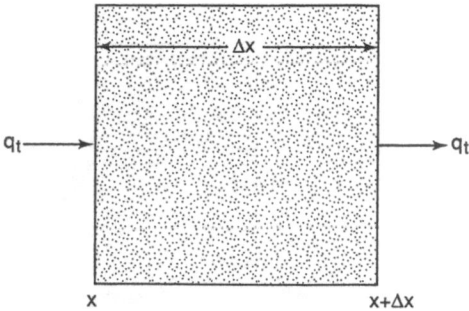

Fig. 3.17—Incremental element of porous rock.

considered incompressible. There is no mass transfer between oil and water phases.

Consider the flow of oil and water through the incremental element of porous rock shown in **Fig. 3.17**. The water contains one chemical species that adsorbs on the surface of the rock. A material balance on this chemical species entering and leaving the incremental element over a small time increment Δt yields

$$q_t f_w C_i \big|_x \Delta t - q_t f_w C_i \big|_{x+\Delta x} \Delta t = \left[C_i S_w \big|_{t+\Delta t} \Delta x A - C_i S_w \big|_t \Delta x A \right] \phi$$
$$+ A_i \rho_{gr} (1-\phi) \Delta x \big|_{t+\Delta t} A - A_i \rho_{gr} (1-\phi) \Delta x \big|_t A, \ \ldots \ldots \ldots \ (3.68)$$

where the vertical bars indicate that the term is evaluated at the indicated position or time; the first two terms represent the net amount of Species i that enters the differential element by fluid flow during Δt; the second two terms represent the net retention of Species i in the element resulting from changes in concentration, C_i, or water saturation, S_w, in the pore space ($A\phi\Delta x$) that can be occupied by Species i; and the third pair of terms is the net retention of Species i by the rock in Δt. In Eq. 3.68, C_i = concentration of Species i, ϕ = porosity fraction of PV, A_i = amount of Species i retained by the rock, and ρ_{gr} = grain density of the rock. It is assumed that the porosity occupied by Species i is ϕ.

Dividing both sides of Eq. 3.68 by $A\Delta x\Delta t$, gives

$$-\frac{\left[(q_t f_w C_i)\big|_{x+\Delta x} - (q_t f_w C_i \big|_x)\right]}{A} = \frac{\left[C_i S_w \big|_{t+\Delta t} - C_i S_w \big|_t\right]\phi}{\Delta t}$$
$$+\frac{\left\{\left[A_i \rho_{gr}(1-\phi)\right]_{t+\Delta t} - \left[A_i \rho_{gr}(1-\phi)\right]_t\right\}}{\Delta t}. \ \ldots\ldots\ldots\ldots\ldots\ldots\ldots \ (3.69)$$

The limit as Δx and Δt tend to zero in Eq. 3.69 is Eq. 3.70. Recalling that q_t and ϕ are constant,

$$-\frac{q_t}{A\phi} \frac{\partial(f_w C_i)}{\partial x} = \frac{\partial(C_i S_w)}{\partial t} + \frac{1}{\phi} \frac{\partial\left[A_i \rho_{gr}(1-\phi)\right]}{\partial t}. \ \ldots\ldots\ldots\ldots\ldots\ldots\ldots \ (3.70)$$

Eq. 3.70 can be simplified as follows. A material balance considering water as a chemical constitutent is given by

$$-\frac{q_t}{A\phi} \frac{\partial(f_w C_w)}{\partial x} = \frac{\partial(C_w S_w)}{\partial t}. \ \ldots\ldots\ldots\ldots\ldots\ldots\ldots\ldots\ldots\ldots\ldots\ldots \ (3.71)$$

Retention of water on the rock is neglected. Because C_i is present in small concentrations, C_w is essentially constant, and Eq. 3.71 becomes

$$-\frac{q_t}{A\phi} \frac{\partial(f_w)}{\partial x} = \frac{\partial(S_w)}{\partial t}. \ \ldots\ldots\ldots\ldots\ldots\ldots\ldots\ldots\ldots\ldots\ldots\ldots\ldots\ldots \ (3.72)$$

It is convenient to define a new variable $\hat{C}_i$ to represent the retention of Species i on the rock in terms of the PV of the rock:

$$\hat{C}_i = \frac{A_i \rho_{gr}(1-\phi)}{\phi}. \ \ldots\ldots\ldots\ldots\ldots\ldots\ldots\ldots\ldots\ldots\ldots\ldots\ldots\ldots\ldots \ (3.73)$$

Then, Eq. 3.70 becomes

$$-\frac{q_t}{A\phi} \frac{\partial(f_w C_i)}{\partial x} = \frac{\partial(S_w C_i)}{\partial t} + \frac{\partial\hat{C}_i}{\partial t}. \ \ldots\ldots\ldots\ldots\ldots\ldots\ldots\ldots\ldots\ldots \ (3.74)$$

Next, Eq. 3.74 is expanded by differentiating the products. Before carrying out the differentiation, Eqs. 3.72 and 3.74 are converted to dimensionless form. Recalling that $x_D = x/L$ and $t_D = (q_t t)/(A\phi L)$, then

$$\frac{\partial(S_w C_i)}{\partial t_D} + \frac{\partial\hat{C}_i}{\partial t_D} + \frac{\partial(f_w C_i)}{\partial x_D} = 0. \ \ldots\ldots\ldots\ldots\ldots\ldots\ldots\ldots\ldots\ldots\ldots \ (3.75)$$

and $\partial S_w/\partial t_D + \partial f_w/\partial x_D = 0.$ $\ \ldots\ldots\ldots\ldots\ldots\ldots\ldots\ldots\ldots\ldots\ldots \ (3.76)$

Expanding Eq. 3.75 gives

$$S_w \frac{\partial C_i}{\partial t_D} + C_i \frac{\partial S_w}{\partial t_D} + \frac{\partial \hat{C}_i}{\partial t_D} + fw \frac{\partial C_i}{\partial x_D} + C_i \frac{\partial f_w}{\partial x_D} = 0, \quad \dots\dots\dots\dots\dots\dots\dots\dots\dots\dots\dots\dots (3.77)$$

and gathering terms gives

$$S_w \frac{\partial C_i}{\partial t_D} + C_i \left(\frac{\partial S_w}{\partial t_D} + \frac{\partial f_w}{\partial x_D} \right) + \frac{\partial \hat{C}_i}{\partial t_D} + f_w \frac{\partial C_i}{\partial x_D} = 0. \quad \dots\dots\dots\dots\dots\dots\dots\dots\dots\dots\dots\dots (3.78)$$

The term in parentheses in Eq. 3.78 is identically equal to zero. Eq. 3.78 becomes

$$S_w \frac{\partial C_i}{\partial t_D} + \frac{\partial \hat{C}_i}{\partial t_D} + f_w \frac{\partial C_i}{\partial x_D} = 0. \quad \dots\dots\dots\dots\dots\dots\dots\dots\dots\dots\dots\dots\dots\dots (3.79)$$

To proceed further, a relationship must be developed between C_i and $\hat{C}_i$. If retention is assumed to be instantaneous and reversible, an equilibrium retention isotherm such as that shown in **Fig. 3.18** relates A_i to C_i and thus $\hat{C}_i$ to C_i. The isotherm in Fig. 3.18 has negative curvature, which is necessary for the flow solutions.

Because $\hat{C}_i = f(C_i)$,

$$\frac{\partial \hat{C}_i}{\partial t_D} = \left(\frac{\partial C_i}{\partial t_D} \right) \left(\frac{\partial \hat{C}_i}{\partial C_i} \right). \quad \dots\dots\dots\dots\dots\dots\dots\dots\dots\dots\dots\dots\dots\dots\dots\dots (3.80)$$

Let

$$D_i = \partial \hat{C}_i / \partial C_i, \quad \dots\dots\dots\dots\dots\dots\dots\dots\dots\dots\dots\dots\dots\dots\dots\dots\dots\dots\dots (3.81)$$

where D_i = slope of the equilibrium isotherm for C_i. Then,

$$\frac{\partial \hat{C}_i}{\partial t_D} = D_i \frac{\partial C_i}{\partial t_D}. \quad \dots\dots\dots\dots\dots\dots\dots\dots\dots\dots\dots\dots\dots\dots\dots\dots\dots\dots\dots (3.82)$$

and Eq. 3.79 becomes

$$\left(S_w + D_i \right) \frac{\partial C_i}{\partial t_D} + f_w \frac{\partial C_i}{\partial x_D} = 0. \quad \dots\dots\dots\dots\dots\dots\dots\dots\dots\dots\dots\dots\dots\dots\dots (3.83)$$

Eq. 3.83 describes the movement of a sharp concentration front through the porous rock. There is no dispersion or mixing of the fluids, and the concentration jumps from C_{i0} to C_i at the concentration front.

The concentration front can be followed by deriving an expression for the concentration velocity in a manner similar to that for the development of the frontal-advance equation for waterflooding.

Consider a path of constant composition where

$$C_i = C_i \left(x_D, t_D \right). \quad \dots\dots\dots\dots\dots\dots\dots\dots\dots (3.84)$$

Then, $$dC_i = \left(\frac{\partial C_i}{\partial x_D} \right)_{t_D} dx_D + \left(\frac{\partial C_i}{\partial t_D} \right)_{x_D} dt_D. \quad \dots\dots\dots (3.85)$$

for a path of constant composition $dC_i = 0$ and

$$\left(\frac{dx_D}{dt_D} \right)_{C_i} = - \left(\frac{\partial C_i}{\partial t_D} \right)_{x_D} \Big/ \left(\frac{\partial C_i}{\partial x_D} \right)_{t_D} = v_{ci}. \quad \dots\dots\dots (3.86)$$

Eq. 3.86 gives the specific velocity of the concentration shock. Combining Eqs. 3.83 and 3.86 gives

$$v_{ci} = \frac{dx_D}{dt_D} = \frac{f_w}{S_w + D_i}. \quad \dots\dots\dots\dots\dots\dots\dots\dots (3.87)$$

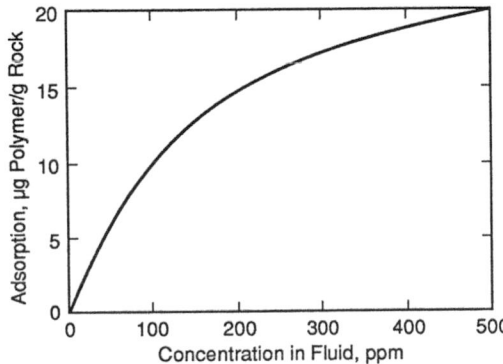

Fig. 3.18—Typical adsorption isotherm for chemical species on porous rock.

When the concentration jumps from C_{i0} (usually zero) to C_{ii} (the injected concentration), there is a change in viscosity or relative permeability curves associated with C_{ii}. This creates a chemical shock front analogous to the viscous waterflood example. In this case, the chemical shock is retarded by retention of chemical species on the rock. For the sharp-front approximation,

$$D_i = \hat{C}_{ii} / C_{ii}. \quad\dotfill\quad (3.88)$$

In Eq. 3.88, $\hat{C}_{ii}$ is the retention of Species i on the rock in units of mass per unit pore volume and is computed from

$$\hat{C}_{ii} = \left[A_i \rho_{gr} (1-\phi) \right] / \phi, \quad\dotfill\quad (3.89)$$

where A_i = chemical retention and ρ_{gr} = density of sand grain.

Because the concentration front causes the saturation shock and these shocks must travel at the same specific velocity, the specific velocity of the saturation shock is

$$\left(\frac{dx_D}{dt_D} \right)_{S_{w3}^*} = \left(\frac{\partial f_w}{\partial S_w} \right)_{S_{w3}^*} = \frac{f_{w3}^* - f_{w1}}{S_{w3}^* - S_{w1}}. \quad\dotfill\quad (3.90)$$

Thus,

$$\frac{dx_D}{dt_D} = \frac{f_{w1}}{S_{w1} + D_i} = \frac{f_{w3}^* - f_{w1}}{S_{w3}^* - S_{w1}} = \frac{f_{w3}^*}{S_{w3}^* + D_i}. \quad\dotfill\quad (3.91)$$

Inspection of Eq. 3.91 shows that the values of f_{w3}^* and S_{w3}^* can be found by drawing a tangent to the $f_w^* - S_w^*$ fractional-flow curve from the point $f_w^* = 0$, $S_w^* = -D_i$. The intersection of this tangent with the (f_w, S_w) curve for the original oil/water system gives (f_{w1}, S_{w1}). **Fig. 3.19** shows the construction procedure. **Fig. 3.20** shows the saturation profile generated from the injection of Species i. The flood-front saturation is S_{wf}, and the flood front moves at a specific velocity given by

$$\left(\frac{dx_D}{dt_D} \right) = \frac{f_{wf}}{S_{wf} - S_{iw}}. \quad\dotfill\quad (3.92)$$

Because $S_{w1} > S_{wf}$, the oil-bank front travels at a specific velocity given by

$$\frac{dx_D}{dt_D} = \left(\frac{\partial f_w}{\partial S_w} \right)_{S_{w1}} = f_w' \,|\, S_{w1}. \quad\dotfill\quad (3.93)$$

and arrives at the end of the system when

$$t_{D1} = \left(1 / f_w' \right) |\, S_{w1}. \quad\dotfill\quad (3.94)$$

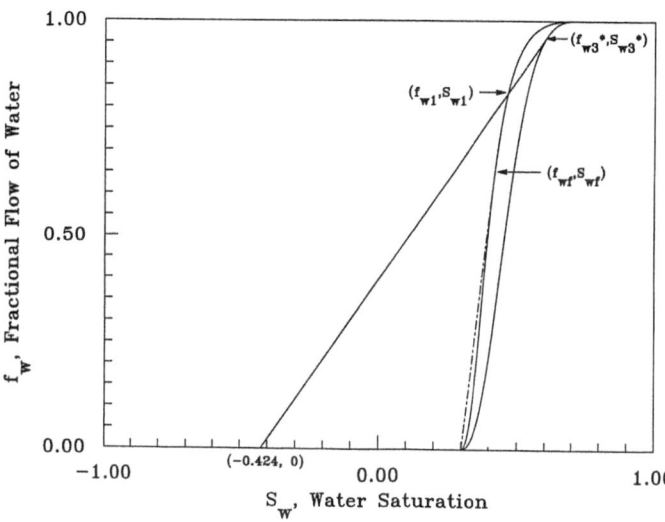

Fig. 3.19—Construction procedure to determine f_{w3}^* and S_{w3}^* for when adsorption occurs, Example 3.4.

The chemical shock arrives at the end of the system when

$$t_{D3} = \frac{S_{w3}^* + D_i}{f_{w3}^*} = \frac{S_{w1} + D_i}{f_{w1}}. \quad\dots\dots\quad (3.95)$$

Saturations greater than S_{w3}^* move at the velocity determined by the properties of the fractional-flow curve behind the chemical shocks, as discussed in Section 3.3. Thus, for $1 - S_{or} > S_w^* \geq S_{w3}^*$,

$$x_D^* = t_D \left(\frac{\partial f_w^*}{\partial S_w^*} \right)_{S_w^*}. \quad\dotfill\quad (3.96)$$

It can be demonstrated that Eq. 3.96 is correct by performing an overall material balance on Species i. If Species i is injected at $t_D = 0$, the location of C_{ii} is given by

$$x_{D3} = t_D f_{w3}'^*. \quad\dotfill\quad (3.97)$$

Fig. 3.21 shows the concentration profile after injection for t_D. At time t, the amount of Species i injected must be equal to the amount retained plus the amount in the pore space where $C_i = C_{ii}$. That is,

$$q_t t C_{ii} = \int_0^{x3} A\phi S_w C_{ii}\,dx + \int_0^{x3} A\phi \hat{C}_{ii}\,dx, \quad .. \; (3.98)$$

where the second term represents the amount of Species i in the water behind the chemical shock and the third term represents the quantity of Species i retained on the rock.

Rearranging Eq. 3.98 gives

$$q_t t C_{ii} = A C_{ii} \int_0^{x3} \phi S_w\,dx + A\phi \hat{C}_{ii} x_3, \quad .... \; (3.99)$$

and converting to a dimensionless form by introducing x_D and t_D gives

$$t_D C_{ii} = C_{ii} x_{D3} \overline{S}^*_{wD3} + \hat{C}_{ii} x_{D3}. \quad .... \; (3.100)$$

When the Welge equation (Eq. 3.12) is applied,

$$x_{D3} \overline{S}^*_{wD3} = x_{D3} S^*_{w3} - t_D\left(f^*_{w3} - 1\right). \quad ..... \; (3.101)$$

Substituting into Eq. 3.100 gives

$$t_D C_{ii} = C_{ii}\left[x_{D3} S^*_{w3} - t_D\left(f^*_{w3} - 1\right)\right] + \hat{C}_{ii} x_{D3}.$$
$$.................................. \; (3.102)$$

Recall that $x_{D3} = t_D f'^*_{w3}$. Substituting for x_{D3} and rearranging yield

$$f'^*_{w3}\left(C_{ii} S^*_{w3} + \hat{C}_{ii}\right) = C_i f^*_{w3}. \quad .... \; (3.103)$$

or $f'^*_{w3} = \dfrac{f^*_{w3}}{S^*_{w3} + \left(\hat{C}_{ii}/C_{ii}\right)} = \dfrac{f^*_{w3}}{S^*_{w3} + D_i}, \quad .... \; (3.104)$

which is equivalent to Eq. 3.90. This shows that the chemical species is conserved within the region occupied by the moving chemical shock.

3.5.2 Movement of Interstitial and Injected Water.

A consequence of the assumptions made to develop the fractional-flow model for Species i is that the chemical species travels slower than the water in which it was injected. For simplicity, consider the injection of water containing Species i into a porous rock completely saturated with water. If Species i is not retained, the concentration profile at t_D is given by the dashed line in **Fig. 3.22.** When Species i is retained as in the models assumed in this section, the concentration profile is given by the solid line. The area between the two curves represents the amount of species adsorbed by the rock.

Let x_{Db} = location of the interface between the injected and resident water and x_{Di} = location of the concentration shock for Species i. In this case,

$$\hat{C}_{ii} A\phi x_{Di} = \left(x_D - x_{Di}\right) A\phi C_{ii}. \quad .... \; (3.105)$$

Introducing the definition of D_i,

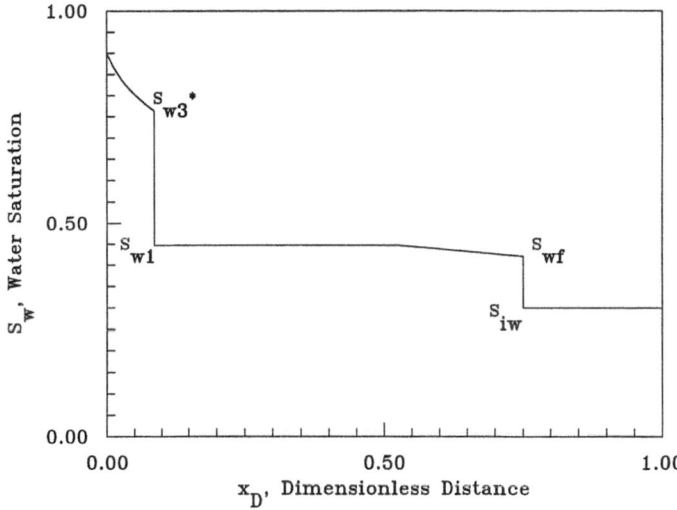

Fig. 3.20—Saturation profile for chemical flood started at interstitial water saturation.

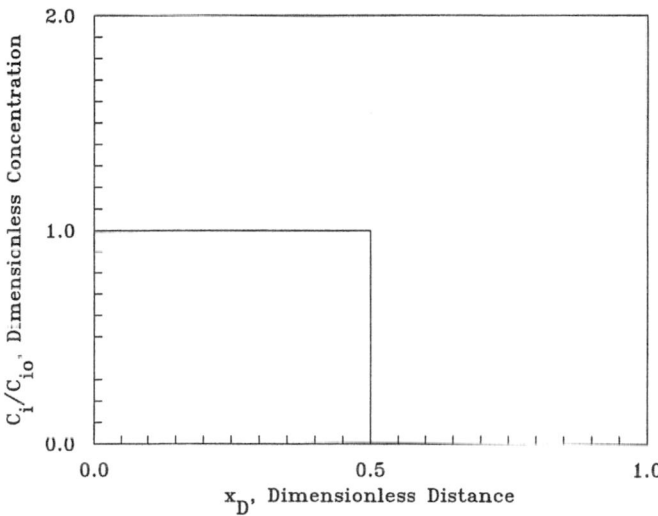

Fig. 3.21—Concentration profile after injection of t_D PV of solution.

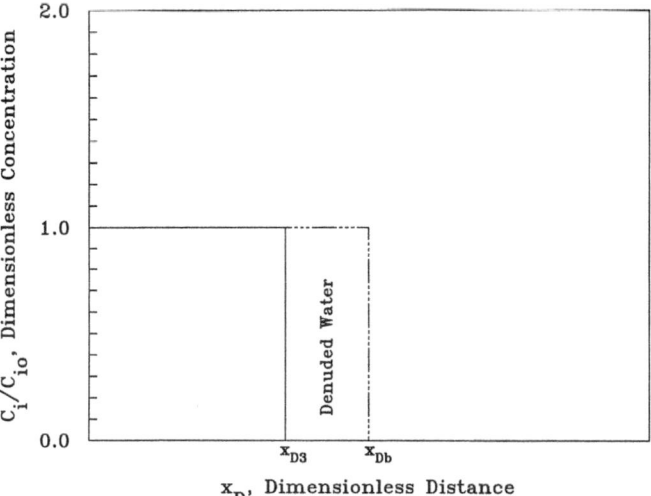

Fig. 3.22—Concentration profiles showing the effect of retention in porous rock for step-function concentration changes.

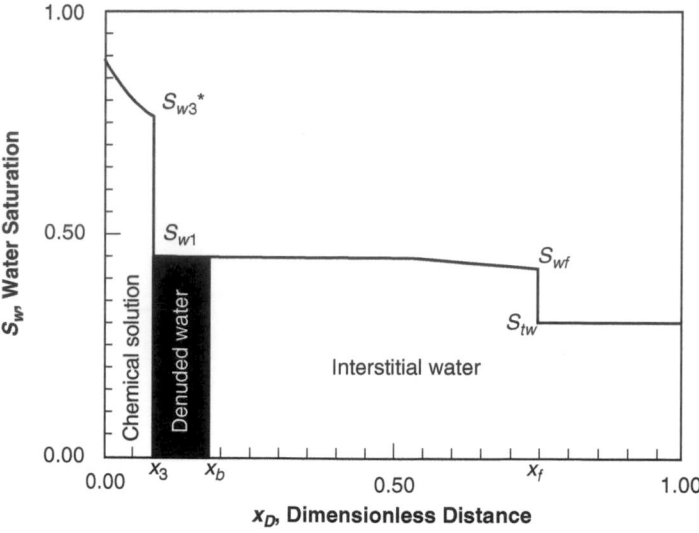

Fig. 3.23—Locations of denuded and interstitial water during a chemical flood initiated at interstitial water saturation.

$$D_i x_{Di} = x_D - x_{Di}, \quad \dots \dots \dots \dots (3.106)$$

gives $x_D = x_{Di}(1+D_i) \dots \dots \dots (3.107)$

or $x_{Di} = x_D/(1+D_i). \quad \dots \dots \dots (3.108)$

The model assumes that Species i is removed instantaneously as the chemical front advances, and thus "denuded" water flows ahead of the chemical front where it displaces resident water.

The location of the boundary between the denuded water and the resident water can be found easily in frontal-advance models by making a material balance on the denuded water. Refer to **Fig. 3.23,** where the location of the chemical shock x_3 and the boundary x_b are identified. In the region behind the chemical shock, the amount of chemical retained is $= A_i \rho_{gr} x_3 A(1-\phi)$.

Because the concentration of chemical in the injected solution is C_{ii}, the volume of denuded water released is given by

$$V_w = \left[A_i \rho_{gr} x_3 A(1-\phi)\right]/C_{ii}, \quad \dots \dots \dots \dots \dots \dots \dots \dots \dots \dots \dots \dots (3.109)$$

where V_w = volume of denuded water. The denuded water is confined to the interval $x_b - x_3$, where the water saturation is S_{w1}. Thus,

$$A\phi(x_b - x_3)S_{w1} = \frac{A_i \rho_{gr} x_3 A(1-\phi)}{C_{ii}}. \quad \dots \dots \dots \dots \dots \dots \dots \dots (3.110)$$

Solving for x_b gives

$$x_b = x_3 \left[1 + \frac{A_i \rho_{gr}(1-\phi)}{C_{ii}\phi S_{w1}}\right] \dots \dots \dots \dots \dots \dots \dots \dots \dots \dots (3.111)$$

or $x_{Db} = x_{D3}(1+D_i/S_{w1}). \quad \dots \dots \dots \dots \dots \dots \dots \dots (3.112)$

Recall from Eq. 3.53 that

$$x_{D3} = f''^*_{w3} t_D.$$

Substituting Eq. 3.91 for f''^*_{w3} yields

$$x_{Db} = t_D\left(\frac{f_{w1}}{S_{w1}+D_i}\right)\left(1+\frac{D_i}{S_{w1}}\right). \quad \dots (3.113)$$

Eq. 3.113 shows that the location of the boundary separating the denuded and interstitial water is a linear function of t_D. Thus, the specific velocity is constant and is given by

$$v_{Db} = \frac{dx_{Db}}{dt_D} = \frac{f_{w1}}{S_{w1}}. \quad \dots \dots \dots \dots (3.114)$$

Fig. 3.24 is a distance/time diagram showing the movement of the denuded and interstitial water.

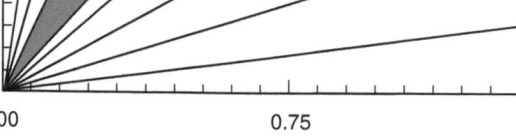

Fig. 3.24—Distance/time diagram showing movement of denuded and interstitial water in a chemical flood.

It is also possible to develop relationships for the location of the boundary between denuded and resident water when a waterflood precedes a chemical flood by making a material balance on the chemical species.

3.6 Applications of the Chemical Flooding Model

The chemical flooding model developed in Section 3.5 is useful for understanding basic displacement processes for both polymer and surfactant floods. It also may be extended to predict the performance of carbonated waterfloods (Claridge and Bonder 1974). In this section, the chemical flooding model is used to estimate displacement performance for a polymer flood at interstitial water saturation, a low-tension chemical flood at interstitial water saturation, and a low-tension chemical flood at ROS.

3.6.1 Polymer Flood of a Linear System Initially at Interstitial Water Saturation. Certain high-molecular-weight polymers increase the viscosity of water significantly when concentrations on the order of a few hundred parts per million are dissolved in the water. Characteristics of these polymers are discussed in more detail in Chapter 5. Because polymer solutions are not believed to alter relative permeability curves, the improvement in oil recovery is a result of an increase in viscosity, as discussed for viscous waterflooding. Because all polymers adsorb, or are retained on porous rocks to some degree, the relationship between polymer retention on the rock and the concentration of the polymer in solution must be known. As long as the adsorption isotherm has negative curvature, as in Fig. 3.18, the sharp-front assumption used in the chemical flooding model applies.

Estimation of polymer flood performance when the initial water saturation is uniform but immobile follows the general procedure outlined in Sections 3.3 and 3.4. The specific velocity of the polymer shock is computed from Eq. 3.44 or Eq. 3.90 after values of (f_{w3}^*, S_{w3}^*) and (f_{w1}, S_{w1}) are found from the intersection of a line from $(0, -D_p)$ to $f_w' - S_w''$, as depicted in Fig. 3.19.

$$v_{D3}^* = \frac{\partial f_w^{*}}{\partial S_w^*} = \frac{f_{w3}^* - f_{w1}}{S_{w3}^* - S_{w1}}$$

$$= \frac{f_{w3}^*}{S_{w3}^* + D_i} = \frac{f_{w1}}{S_{w1} + D_i}. \dotfill (3.115)$$

Eq. 3.115 is also an expression of the requirement that the specific velocities of the water and polymer solutions are equal at a miscible boundary.

If $S_{w1} < S_{wf}$, the second shock travels at a specific velocity given by

$$v_{D1} = f_{w1}/(S_{w1} - S_{iw}). \dotfill (3.116)$$

The saturation profile before water breakthrough is similar to that in Fig. 3.14. Displacement performance is computed following the same procedure described in Section 3.4. Example 3.4 illustrates the computation of polymer flood performance at interstitial water saturation.

Example 3.4—Polymer Flood in a Linear System. A polymer flood is to be conducted in a linear system. Properties of the rocks and fluids from Example 3.1 will be used. The oil viscosity is 40 cp. A concentration of 300 ppm polymer is used to raise the viscosity of the injected water to 4 cp, and the adsorption isotherm of Fig. 3.18 represents the retention of polymer on the rock. Estimate the oil recovery as a function of PVs injected. The density of the rock is 2.65 g/cm³, and the porosity is 0.267.

Solution. The value of D_i must be estimated from the adsorption isotherm before (f_{w3}^*, S_{w3}^*) and (f_{w1}, S_{w1}) can be determined. From Fig. 3.18, $A_i = 17.5$ µg polymer/g rock.

$$C_{ii} = 300 \text{ ppm}$$
$$= 300 \times 10^{-6} \text{ g/cm}^3.$$

From Eq. 3.89.

$$\hat{C}_{ii} = \left(\frac{17.5 \times 10^{-6} \text{ g}}{\text{g rock}}\right)\left(\frac{2.65 \text{ g rock}}{\text{cm}^3 \text{ rock volume}}\right)$$

$$\times \left[\frac{(1-0.267) \text{ cm}^3 \text{ rock volume}}{0.267 \text{ cm}^3 \text{ PV}}\right]$$

$$= 1.27 \times 10^{-4} \text{ g/cm}^3 \text{ PV}.$$

Thus, D_i is computed with Eq. 3.88:

$$D_i = \hat{C}_{ii} / C_{ii}$$
$$= (1.275 \times 10^{-4})/(300 \times 10^{-6})$$
$$= 0.424.$$

The graphs of $f_w^* - S_w^*$ and $f_w - S_w$ are shown in Fig. 3.19 with the tangent drawn from $(-0.424, 0)$ to the $f_w^* - S_w^*$ curve. From this construction, $f_{w3}^* = 0.9657$, $f_{w1} = 0.8319$, $f_{w3}^* = 0.6082$, $S_{w1} = 0.462$, and $f_{w3}' = 0.9356$. In this case, $S_{w1} > S_{wf}$, so that two shocks are formed separated by a region of constant water saturation at S_{w1} and a region where S_w decreases from S_{w1} to S_{wf}.

The polymer shock travels at a specific velocity equal to

$$v_{D3}^* = \frac{f_{w3}^* - f_{w1}}{S_{w3}^* - S_{w1}}$$

$$= 0.9356.$$

The waterflood shock travels at the same specific velocity (5.3958) as in Example 3.3, while the oil-bank front travels at the velocity of saturation S_{w1} on the $f_w - S_w$ fractional-flow curve. This specific velocity is f_{w1}'. The remainder of the solution follows Example 3.3. The following equations give the locations of the three regions as a function of t_D:

$$x_{Df} = 5.3958 \, t_D,$$

$$x_{D1} = 2.8846 \, t_D,$$

and $x_{D3} = 0.9356 \, t_D$.

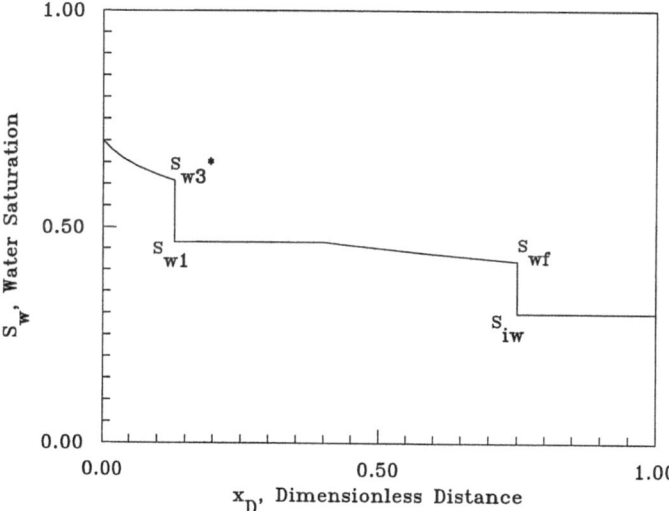

Fig. 3.25—Saturation profile when x_{Df} = 0.75, Example 3.4.

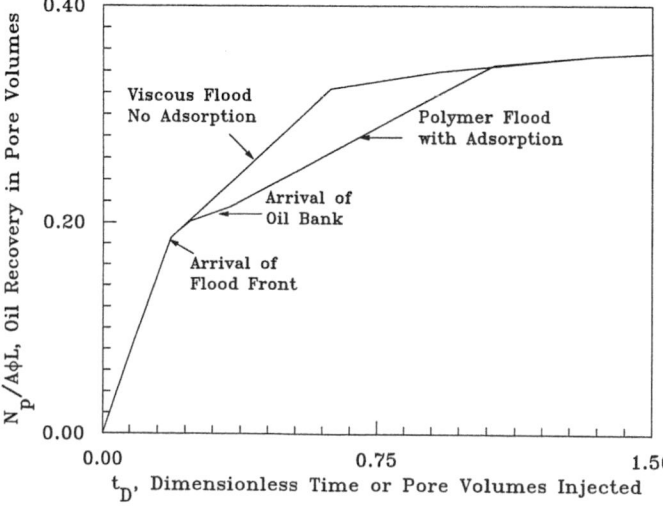

Fig. 3.26—Comparison of oil recovery from a viscous waterflood and a polymer flood initiated at interstitial water saturation, Example 3.4.

Fig. 3.25 is the saturation profile when x_{Df} = 0.75. Because $S_{w1} > S_{wf}$, there is a small variable saturation region where $S_{w1} > S_w > S_{wf}$.

Fig. 3.26 compares oil recovery from the polymer flood with oil recovery from the viscous waterflood in Example 3.3. Recall that the viscosities of the injected solutions are identical. The polymer flood recovery lags the viscous waterflood because the polymer shock travels at a slower specific velocity because of polymer retention. At large t_D, the recoveries are identical.

3.6.2 Low-Tension Flood of a Linear System Initially at Interstitial Water Saturation. Addition of certain chemicals to the injected water can reduce the IFT between the injected fluid and the oil. A low-tension flood is one in which the IFT is on the order of 10^{-3} dynes/cm or lower. The required chemicals, usually mixtures of surfactants and cosurfactants, are discussed in Chapter 7. The mixture of chemicals is treated as a single component.

The reduction of the IFT between two fluids in porous rock has two effects on the relative permeability curves (Talash 1976). First, the relative permeability curves have less curvature. In the relative permeability functions used in this text, the values of m and n would decrease, approaching a limiting value of 1.0 where the fluids are completely miscible. The second effect is a reduction of the ROS from S_{or}, the waterflood residual oil, to S_{orc} the residual saturation to the low-tension flood. Therefore, the fractional-flow curve representing the low-tension oil system will shift toward higher water saturations, as shown in **Fig. 3.27,** and will have a different curvature.

Chemicals that reduce IFT have one characteristic in common with polymers: Adsorption and retention occur as a result of rock/fluid interactions, as discussed in Chapter 7. Adsorption of the chemical constituents, primarily surfactants, is often much larger than for polymers. Adsorption isotherms are similar in shape to Fig. 3.18, with negative curvature, and the sharp-front assumptions apply.

Estimation of displacement performance of a low-tension flood at interstitial water saturation is analogous to the approach outlined in Section 3.6.1. First, the equilibrium adsorption isotherm is necessary to compute D_i. Relative permeability functions corresponding to the low-tension flood are used if available. Otherwise, the use of the relative permeability curves for a normal waterflood will give conservative results. The fractional-flow curves are drawn in Fig. 3.27. The tangent from $(0, -D_i)$ to the $f_w^* - S_w^*$ curve gives values of $\left(f_{w3}^*, S_{w3}^*\right)$ and (f_{w1}, S_{w1}), as in Section 3.5.1. The remainder of the computations are identical to those outlined in Section 3.3.2 and Example 3.3.

As in polymer flooding, two shocks form. The waterflood shock at the front moves at the largest velocity and is followed by the chemical shock. If $S_{w1} > S_{wf}$, the two shocks will be separated by a region of constant saturation, S_{w1}, with the flood-front velocity given by Eq. 3.116. Example 3.5 illustrates the application of frontal-advance theory to a low-tension flood.

Example 3.5—Low-Tension Flood in a Linear System at Interstitial Water Saturation. A chemical system has been found that reduces the IFT sufficiently to obtain an ROS of 0.10 in laboratory

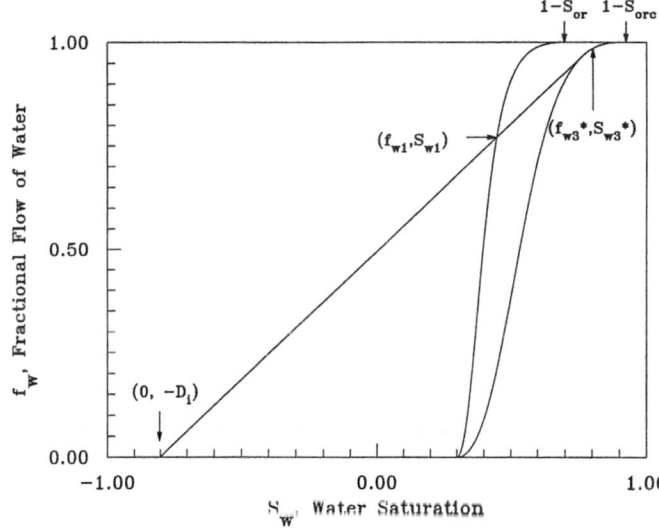

Fig. 3.27—Fractional-flow diagram for a low-tension system, Example 3.5.

displacement tests for the reservoir rock and oil system presented in Example 3.1. The relative permeability curves for the chemical flood are obtained by using the same values for α_1, α_2, m, and n, as in Example 3.1, but shifting the ROS from 0.30 to 0.10. The viscosity of the injected solution is 4 cp. Prepare a x_D/t_D diagram and determine the oil recovery ($N_p/A\phi L$) as a function of PV injected to $t_D = 1.5$.

Solution. In this example, the chemical is assumed to be strongly adsorbed with a value of $D_i = 0.8$. Values of (f_{w3}^*, S_{w3}^*) and (f_{w1}, S_{w1}) must be found by constructing the tangent to the chemical fractional-flow curve, as shown in Fig. 3.27. From this construction, $f_{w3}^* = 0.96619$, $f_{w1} = 0.77070$, $S_{w3}^* = 0.76304$, $S_{w1} = 0.44680$, $f_{w3}'^* = 0.61815$, and $f_{w1}' = 3.8043$.

Because $S_{w1} > S_{wf}$, $f_{wf} = 0.65076$, $S_{wf} = 0.4206$, and $f_{wf}' = 5.3958$.

The following equations give the dimensionless locations of S_{wf}, S_{w1}, and S_{w3}^* as a function of t_D:

$$x_{Df} = 5.3958 t_D,$$

$$x_{D1} = 3.8043 t_D,$$

and $x_{D3}^* = 0.31815 t_D.$

Construction of the x_D/t_D diagram is straightforward. Locations of saturations $S_{wf} \leq S_w \leq S_{w1}$ are found from the frontal-advance equation:

$$x_D = f_w' t_D.$$

Saturations $S_w^* \geq S_{w3}$ are found with Eq. 3.9 from the $f_w^* - S_w^*$ fractional-flow curve:

$$x_D^* = f_w'^* t_D.$$

The x_D/t_D graph is presented in **Fig. 3.28.**

Oil recovery calculations follow the procedure outlined in Section 3.3.2 and illustrated in Example 3.3. **Table 3.9** summarizes the results and **Fig. 3.29** plots the oil recovery vs. PVs injected, t_D.

3.6.3 Low-Tension Flood of a Linear System Initially at ROS. A principal application of low-tension floods is in reservoirs that are nearly at waterflood ROS. In these reservoirs, there is little oil flow in the regions swept by large volumes of water. Injection of a low-tension chemical system can mobilize the residual oil, creating a growing oil bank that is displaced through the system by a chemical shock.

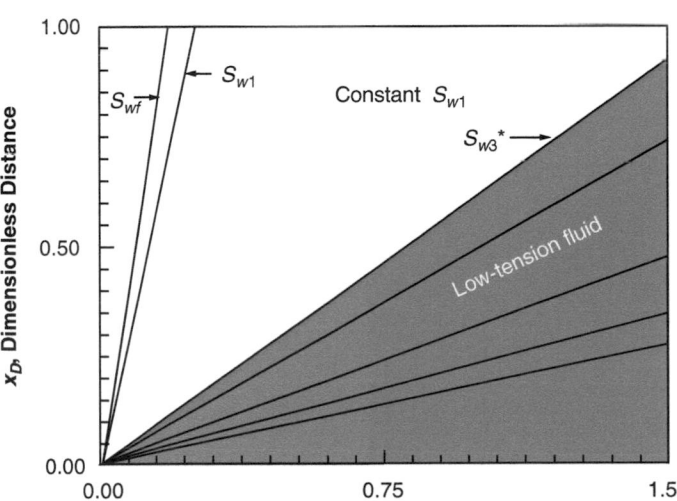

Fig. 3.28—Distance/time diagram for a low-tension flood at interstitial water saturation, Example 3.5.

S_{w2}	f_{w2}	Region	t_D	$\bar{S}_w$	$S_w - S_{iw}$
0.4206	0.6508	←Arrival of flood front	0.1853	0.4853	0.1853
0.4468	0.7707	← Arrival of oil bank	0.26286	0.5071	0.2071
0.7630	0.9662	← Arrival of low-tension shock	1.6177	0.8177	0.5177
0.78	0.9756		2.0172	0.8292	0.5292
0.81	0.9877		3.1476	0.8487	0.5487
0.828	0.9926	{Low-tension flood}	4.3228	0.8600	0.5600
0.84	0.9951		5.5215	0.8671	0.5671
0.87	0.9989		12.8899	0.8701	0.5701
0.90	1.00		∞	–	–

Table 3.9—Summary of oil recovery calculations, Example 3.5, low-tension flood at interstitial water saturation.

The analysis of displacement performance is analogous to that in Section 3.6.2 with the exception that there is no flow of oil when chemical injection starts. **Fig. 3.30** shows the fractional-flow diagram for the mobilization of residual oil. A chemical shock forms with the same values of $\left(f_{w3}^*, S_{w3}^*\right)$ and (f_{w1}, S_{w1}), forming an oil bank with water saturation S_{w1}. Because the oil in front of this bank is immobile, a second shock is created where S_{w1} increases abruptly to $1 - S_{or}$. The specific velocity of this oil-bank shock is the slope of the line connecting S_{w1} with $1 - S_{or}$. Note that the oil-bank shock moves faster than all other saturations in the saturation profile shown in **Fig. 3.31.**

Displacement performance is computed by determining the arrival times of the various banks. The oil-bank specific velocity is given by Eq. 3.117 while the location of the oil bank is expressed by Eq. 3.118

$$v_{Do} = \frac{1 - f_{w1}}{1 - S_{or} - S_{w1}}. \quad \ldots\ldots\ldots\ldots (3.117)$$

$$x_{Do} = v_{Do} t_D. \quad \ldots\ldots\ldots\ldots\ldots (3.118)$$

The oil bank arrives at the end of the system when

$$t_D = 1/v_{Do}. \quad \ldots\ldots\ldots\ldots\ldots (3.119)$$

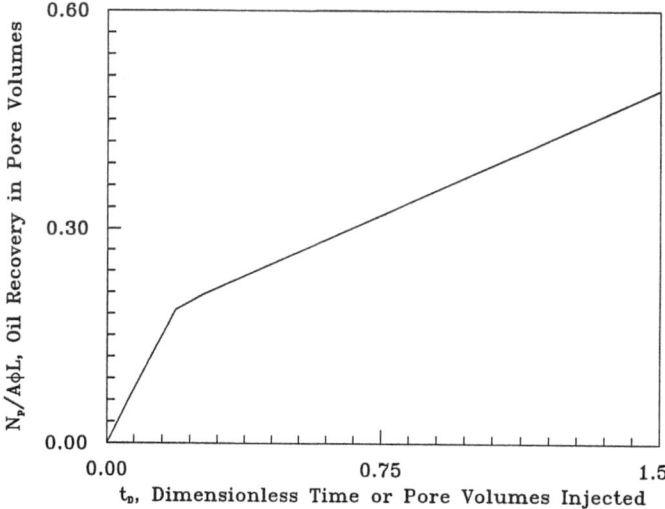

Fig. 3.29—Oil recovery from low-tension flood at interstitial water saturation, Example 3.5.

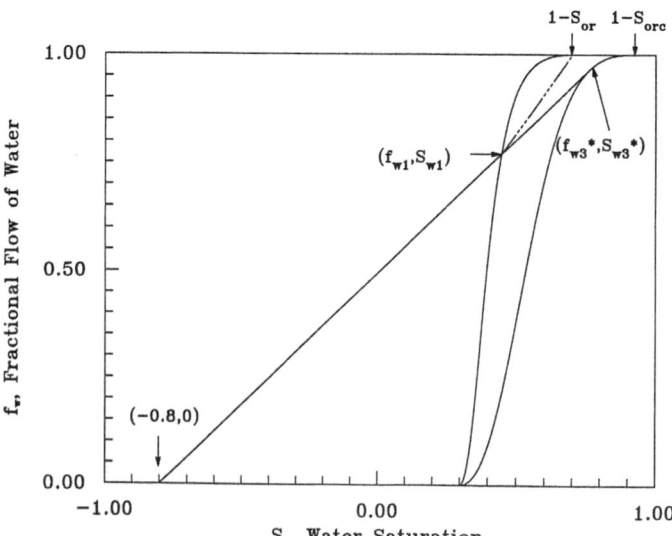

Fig. 3.30—Fractional-flow diagram for mobilization of residual oil by a low-tension flood, Example 3.6.

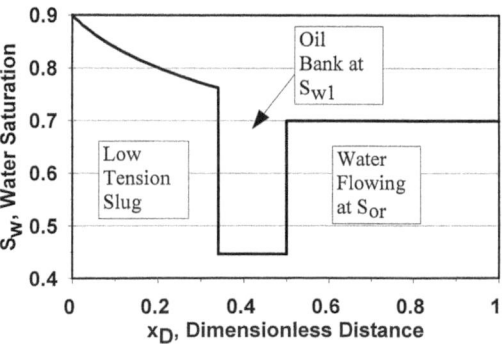

Fig. 3.31—Saturation profile for a low-tension flood at ROS.

The chemical shock travels at a specific velocity given by Eq. 3.91,

$$v_{D3}^* = f_{w3}^* / (S_{w3}^* + D_i),$$

and arrives at the end of the system when (Eq. 3.95)

$$t_D = (S_{w3}^* + D_i) / f_{w3}^*.$$

Oil displacement is computed in the same way as for the polymer flood except that there is no oil recovery until the oil bank breaks through. Thereafter, the average water saturation in the linear system is given by Eqs. 3.57 and 3.58. Example 3.6 illustrates prediction of low-tension flood performance at ROS.

Example 3.6—Low-Tension Flood in a Linear System at Waterflood ROS. A low-tension flood will be conducted in a linear system at ROS. The chemical system has the same properties as in Example 3.5. That is, the chemical solution leaves an ROS after chemical flood of 0.10, is retained by the rock to give $D_i = 0.8$, and has a viscosity of 4 cp. The waterflood ROS is 0.30. Relative permeability relationships are given by Eqs. 3.14 through 3.16. Estimate the oil recovery as a function of PVs injected.

Solution. All properties are the same as in Example 3.5 except the initial oil saturation. The flood-front saturation is now S_{w1}, and the oil bank corresponding to this saturation travels at a specific velocity given by Eq. 3.117.

$$\begin{aligned} v_{Do} &= (1 - f_{w1}) / (1 - S_{or} - S_{w1}) \\ &= (1 - 0.7707) / (1 - 0.30 - 0.4468) \\ &= 0.9056. \end{aligned}$$

The oil bank arrives at the end of the linear system at

$$\begin{aligned} t_D &= 1/v_{Do} \\ t_D &= 1/0.9056 \\ &= 1.1043. \end{aligned}$$

Before this time, water is produced from the system. The long time required for arrival of the oil bank results from the large value of D_i assumed in this example. In most systems of commercial interest, breakthrough occurs from 0.2 to 0.5 PV injected. **Fig. 3.32** presents the distance/time diagram for this example.

Oil is produced at a constant oil cut until the chemical shock arrives at the end of the system when $t_D = 1.6177$. When this happens, the average water saturation is given by Eq. 3.60.

$$\begin{aligned} \overline{S}_w^* &= S_{w3}^* + t_D (1 - f_{w3}^*) \\ &= 0.7630 + (1.6177)(1 - 0.9662) \\ &= 0.8177. \end{aligned}$$

Because the initial water saturation S_{wi} was 0.70, the volume of oil displaced when the chemical shock arrives is $0.8177 - 0.70$, or 0.1177.

Average saturations when $S_{w2} > S_{w3}^*$ are identical to those in Table 3.9. The oil displaced is $\overline{S}_w - S_{wi}$. **Table 3.10** summarizes the results of these computations. Oil recovery is plotted vs. PVs injected in **Fig. 3.33**.

Note that the form of the relative permeability functions yields a zero derivative of the fractional-flow curve at S_{or}. Thus, $S_{orc} = 0.10$ cannot be attained in a finite time.

3.7 Displacement of Slugs

Chemicals used in EOR processes are expensive, and thus continuous injection of polymer, surfactant, or a miscible solvent is not economically possible. Frontal-advance theory can be applied to investigate the use of a chemical slug rather than

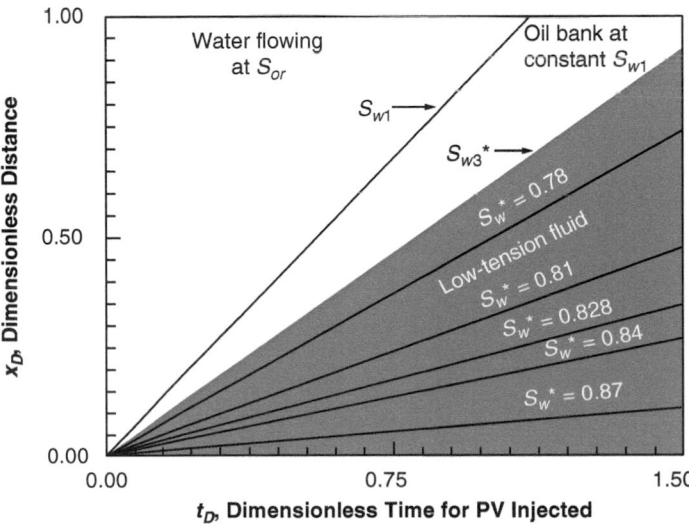

Fig. 3.32—Distance/time diagram for a low-tension flood at ROS, Example 3.6.

| | | | | | Oil Displaced |
S_{w2}	f_{w2}	Region	t_D at $x_D = 1.0$	$\bar{S}_w$	$\bar{S}_w - S_{wi} = N_p/A\phi L$
0.4468	0.77070	←Arrival of oil bank	1.1043	0.70	0
0.7630	0.9662	← Arrival of low-tension shock	1.6177	0.8177	0.1177
0.78	0.9756		2.0172	0.8292	0.1292
0.81	0.9877		3.1476	0.8487	0.1487
0.828	0.9926	{Low-tension flood}	4.3228	0.8600	0.1600
0.84	0.9951		5.5215	0.8671	0.1670
0.87	0.9989		12.8899	0.8701	0.1701
0.90	1.0000		∞	0.9000	0.2000

*$S_{wi} = 0.7$.

Table 3.10—Oil recovery calculations for a low-tension chemical flood at ROS, Example 3.6*.

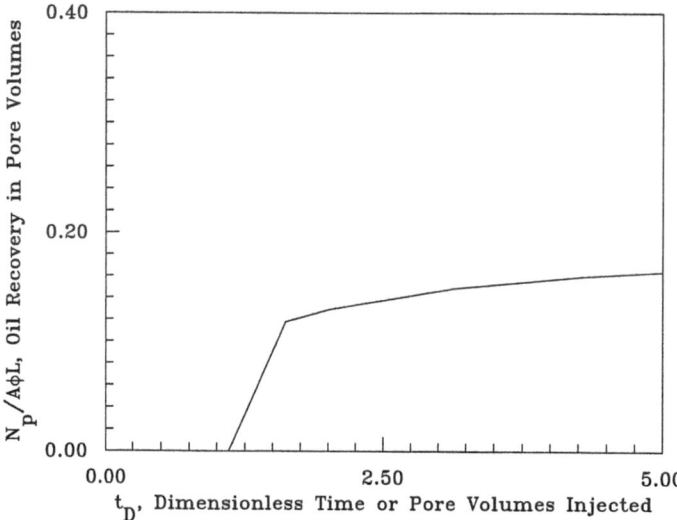

Fig. 3.33—Oil recovery from low-tension flood at ROS.

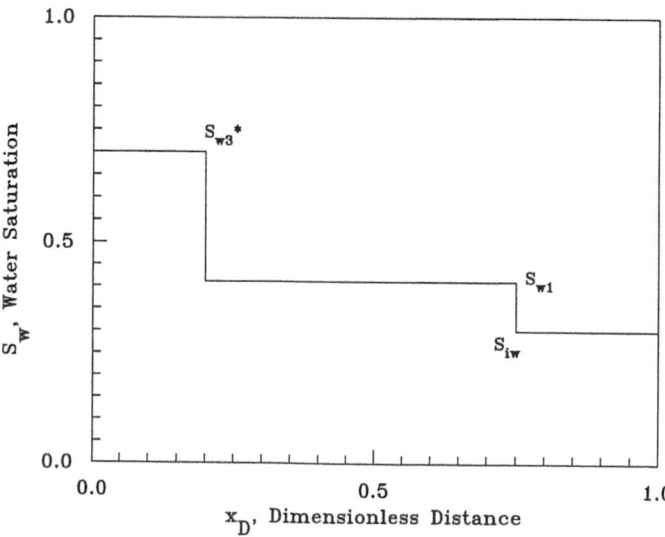

Fig. 3.34—Saturation profile for a viscous waterflood in piston-like displacement.

continuous injection. The chemical slug is displaced by a drive fluid that is assumed to be miscible with the slug. For example, viscous water or polymer can be displaced by water containing no injection chemical. As in previous use of frontal-advance theory, mixing between miscible fluids is neglected. Thus, viscous fingering or unstable displacement caused by unfavorable mobility ratios is not considered by the approach presented here. Mixing, or dispersion, is presented later in this chapter.

3.7.1 Piston-Like Displacement. A simple example is used to illustrate the displacement of a slug through a linear system as a viscous waterflood with a nonadsorbing chemical species. The viscosity of the viscous fluid is large enough so that $f_{w3}^* \cong 1.0$ and $S_{w3}^* \cong 1 - S_{or}$. Thus, the saturation profile during injection of the viscous fluid is depicted in **Fig. 3.34** when the flood begins at interstitial water saturation and $S_{w1} < S_{wf}$. The dimensionless velocities of the flood front and the saturation shocks are given by Eqs. 3.63 and 3.120, respectively.

$$v_{D3}^* = 1/(1 - S_{or}). \qquad \dots \dots \dots \dots \dots (3.120)$$

At t_{Do}, injection is switched to low-viscosity drive water. The drive water is miscible with the viscous water and at the boundary between these fluids $v_{Dw} = v_{Dv}$ or

$$f_w/S_w = f_w^*/S_w^*. \qquad \dots \dots \dots \dots \dots \dots (3.121)$$

With the properties of the viscous fluid described earlier,

$$f_{w3}^* = f_w^* = 1 \qquad \dots \dots \dots \dots \dots \dots (3.122)$$

and $S_{w3}^* = S_w^* = 1 - S_{or}. \qquad \dots \dots \dots \dots (3.123)$

Therefore, the velocity of the miscible boundary, or the slope of the $x_D - t_D$ trajectory, is given by

$$\mathrm{d}x_D/\mathrm{d}t_D = v_{Db} = 1/(1 - S_{or}). \quad \ldots (3.124)$$

The drive water arrives at the end of the linear system when

$$t_D = t_{Do} + 1 - S_{or}. \quad \ldots \ldots \ldots \ldots (3.125)$$

Fig. 3.35 shows the trajectory of the slug. **Fig. 3.36** shows the concentration profile corresponding to the nonadsorbing chemical species in the viscous slug. Because there is no mixing between the drive water and the viscous slug, a small slug is as effective as a large slug in this example. In practice, this will never be the case.

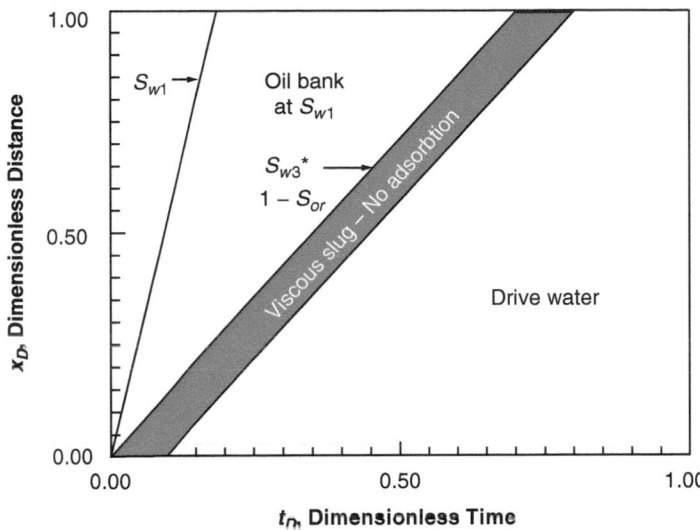

Fig. 3.35—Trajectory of slug when a nonadsorbing slug is displaced through a porous rock.

Displacement of an Adsorbing Slug. When a chemical species, such as polymer or surfactant, adsorbs or is retained on the rock by an equilibrium process, it will desorb when contacted with drive water containing no chemical. The desorption process is identical to the adsorption process. Thus, the concentration profile of the chemical species at various times is a square wave that travels through the linear system at a specific velocity given by Eq. 3.91:

$$v_{Db} = f_{w1}/(S_{w1} + D_i)$$
$$= 1/(1 - S_{or} + D_i). \quad \ldots \ldots \ldots \ldots \ldots \ldots \ldots \ldots \ldots \ldots \ldots \ldots (3.126)$$

The drive water initially travels at a specific velocity given by Eq. 3.124 that is faster than the velocity of the rear of the chemical slug but is slowed down because the chemical species desorbs into the drive water. **Fig. 3.37** is the distance/time diagram of the chemical flood showing the motion of the slug.

Displacement of a Chemical Slug That Does Not Desorb. Some chemical species retained by the rock desorb at such a slow rate that the retention process may be considered irreversible. When a slug containing this species is displaced through the porous rock, desorption of the chemical into the drive water is negligible. The porous rock behind the chemical shock contains the injected concentration, C_u. Because retention is assumed to be irreversible, the slug decreases continuously in size (but not concentration) as it is displaced through the rock. This is depicted in **Fig. 3.38**, where concentration profiles are shown as a function of t_D for a chemical slug equivalent to 0.424 PV at $x_D = 0$.

If the chemical slug is small enough, it will disappear before it reaches the end of the linear system, and the displacement returns to the original waterflood fractional-flow curves at that point. Design of an ideal slug would propagate the slug just to the end of the system before it vanished. It is possible to estimate this ideal slug size from fractional-flow theory by visualizing the displacement process on the distance/time diagram.

Section 3.5 showed that the chemical slug traveled at a specific velocity given by Eq. 3.91:

$$v_{D3} = f_{w3}^*/(S_{w3}^* + D_i). \quad \ldots \ldots (3.127)$$

When chemical injection begins at $t_D = 0$, the leading edge of the chemical slug arrives at the end of the linear system at a dimensionless time given by

$$t_D = (S_{w3}^* + D_i)/f_{w3}^*. \quad \ldots \ldots (3.128)$$

For the piston-like displacement conditions assumed in this section,

$$t_D = 1 - S_{or} + D_i. \quad \ldots \ldots \ldots (3.129)$$

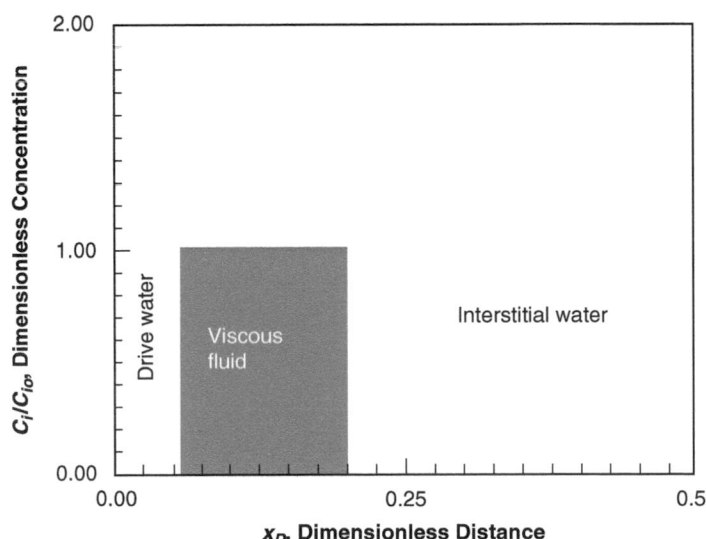

Fig. 3.36—Concentration profile during displacement of a slug of nonadsorbing chemical through porous rock.

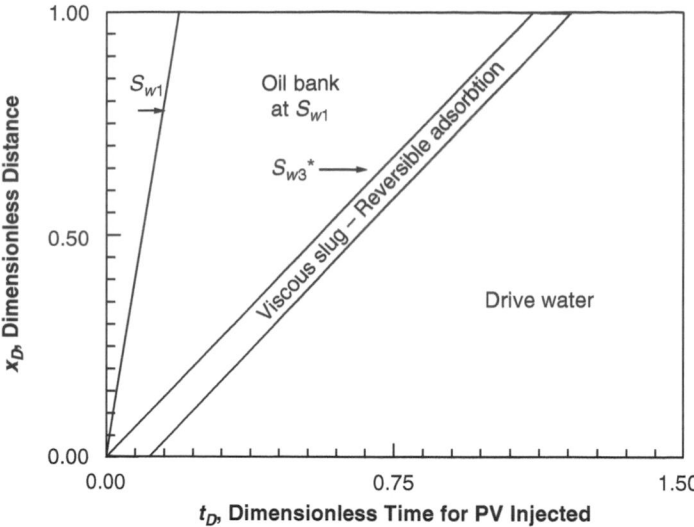

Fig. 3.37—Distance/time diagram for displacement of an adsorbing slug through porous rock.

The rear of the chemical slug is displaced by drive water. Because of piston-like displacement, miscibility, and the assumption of no mixing or viscous fingering, the drive-water boundary moves at a velocity given by

$$v_{Dw} = 1/(1 - S_{or}). \quad\ldots\ldots\ldots\ldots (3.130)$$

This velocity is higher than the velocity of the chemical shock and tracks the rear of the chemical slug when retention is irreversible. The drive water arrives at the end of the linear system at a dimensionless time given by

$$t_D = t_{Do} + 1 - S_{or}. \quad\ldots\ldots\ldots\ldots (3.131)$$

The chemical slug vanishes as it just reaches the end of the system when Eqs. 3.129 and 3.131 have the same value of t_D. Thus, when $t_{Do} = D_i$ the chemical slug will just reach the end of the system when it vanishes. **Fig. 3.39** is a distance/time diagram illustrating the movement of the chemical slug.

3.7.2 Nonpiston-Like Displacement. In many systems, $S_{w3}^* \neq 1 - S_{or}$. Consequently, a two-phase flow region exists behind the viscous flood front, as illustrated in Section 3.4. When the chemical is injected as a slug, the drive fluid displaces this region of variable saturation. The efficiency of slug displacement depends on the characteristics of the drive fluid. The methods discussed in Appendix A must be used to follow the various fronts.

If the mobility of the drive fluid is equal to the mobility of the chemical slug and adsorption or desorption do not affect viscosities and relative permeabilities, the slug will be displaced through the porous rock as if the slug were infinite. This is an ideal situation that does not exist. Consequently, displacement of a chemical slug when there are marked differences between the fractional-flow curves of the chemical slug and the drive fluid is investigated.

Consider a viscous waterflood with a nonadsorbing chemical like the one discussed in Section 3.4. Assume that a slug of viscous solution equal to t_{Do} is injected before switching to low-viscosity drive water. Viscous fingering is neglected, and the drive water and viscous fluid are assumed to be miscible. Viscous fingering and dispersion are assumed to be negligible to simplify the development of conceptual models presented in this section. When viscous fingering and/or dispersion are incorporated into the models, displacement calculations become complex. Results predicted by neglecting these terms are always optimistic and thus set upper bounds on displacement performance.

When the drive-water injection begins, the saturation profile is given by Fig. 3.25. Fractional-flow curves for the two fluids are represented by Fig. 3.12. As discussed in Appendix A, a saturation discontinuity develops between the drive water and the viscous fluid because of the differences in fluid viscosities. The specific velocity of the drive water at any saturation is larger than the specific velocity of the saturations behind the viscous flood front. Thus, the drive water cuts into the rear of the saturation profile, forming a saturation discontinuity between the drive water and the viscous fluid, as shown in **Fig. 3.40.** The specific velocity of this discontinuity is given by Eq. A-2, where (f_w, S_w) and (f_w^*, S_w^*) are the fractional-flow/saturation pairs that satisfy the material-balance requirements across a discontinuity, as shown in Fig. A-1:

$$v_{Dv} = (f_w - f_w^*)/(S_w - S_w^*).$$

Because the drive water and the viscous water are miscible,

$$v_{Dv} = f_w/S_w = f_w^*/S_w^*.$$

The path followed by the saturation discontinuity between the drive water and the viscous fluid is estimated with the front-tracking method presented in Appendix A—Following the Oil-Bank Shock by Front Tracking. In this case, the advancing

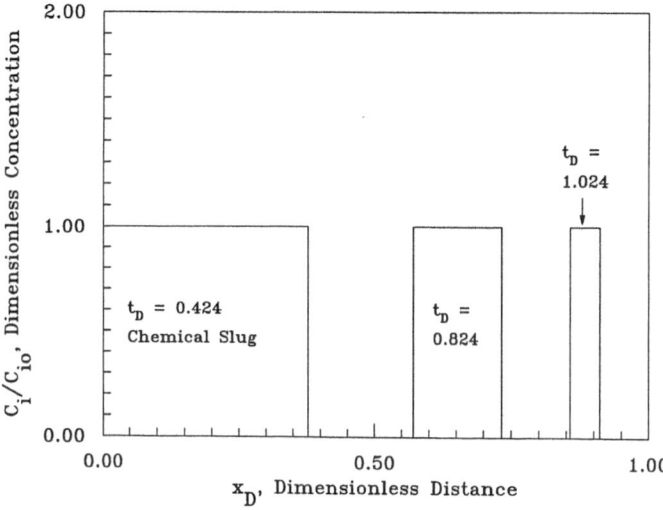

Fig. 3.38—Concentration profile for displacement of a chemical slug that does not desorb.

drive-water shock intersects paths of constant saturation, S_w^*, eliminating these saturations, while saturations S_w evolve from the saturation discontinuity.

The front-tracking method starts when drive-water injection begins and determines the time required for the viscous shock to intersect a smaller value of s_w^* in the saturation profile.

The oil-bank and flood-front saturations for the viscous waterflood are assumed to be established instantaneously, as described in Section 3.4, when a nonadsorbing viscous fluid is injected into a linear system at interstitial water saturation. The locations of saturations S_w^* are obtained from the frontal-advance equation and given by

$$x_D^* = t_D f_w'^*. \quad \dots \dots \dots \dots \dots (3.132)$$

The drive-water shock forms at $t_D = t_{D1}$ when injection is switched from viscous water to drive water. This shock travels at an average velocity of $\overline{v}_{Dv}^{n+1}$ between timesteps n and $n+1$, where

$$\overline{v}_{Dv}^{n+1} = \frac{1}{2}\left[\left(\frac{f_w^*}{S_w^*}\right)^{n+1} + \left(\frac{f_w^*}{S_w^*}\right)^n\right]. \quad \dots (3.133)$$

As noted earlier, the velocity of this shock is greater than v_{Dw}^* and overtakes S_w^*. Following the development in Appendix A—Following the Oil-Bank Shock by Front Tracking,

$$x_{Dv}^{n+1} = x_{Dv}^n + \overline{v}_{Dv}^{n+1}\left(t_D^{n+1} - t_D^n\right). \quad \dots \dots (3.134)$$

Because S_w^* is overtaken by the shock,

$$x_{Dv}^{n+1} = t_D^{n+1} f_w'^{*\,n+1} \quad \dots \dots \dots \dots \dots (3.135)$$

and $x_{Dv}^n = t_D^n f_w'^{*\,n}, \quad \dots \dots \dots \dots \dots (3.136)$

where $S_w^{*\,n}=$ saturation overtaken by the viscous shock at timestep n and $S_w^{*\,n+1}=$ saturation overtaken by the viscous shock at timestep $n+1$. The time when saturation $S_w^{*\,n+1}$ is overtaken by the drive water shock is given by

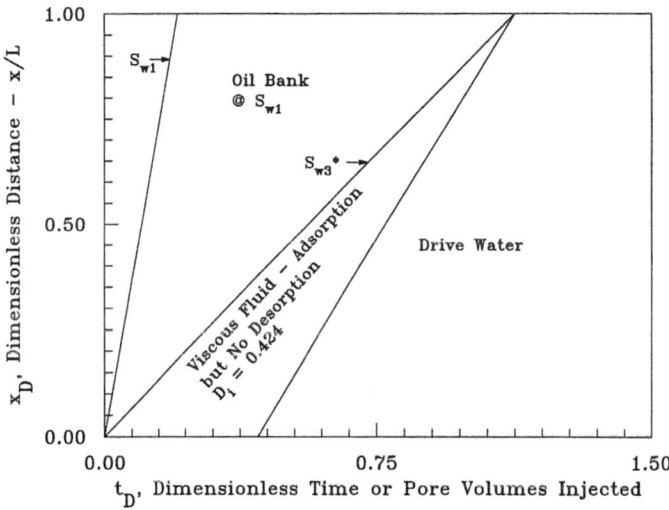

Fig. 3.39—Distance/time diagram for displacement of a chemical slug that does not desorb.

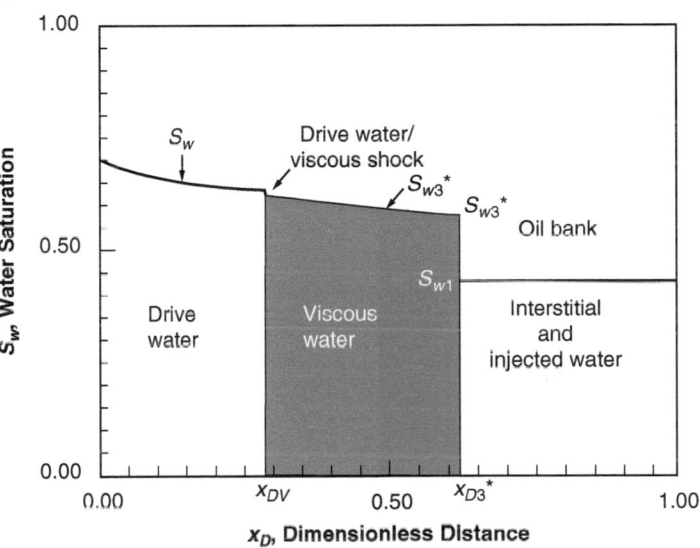

Fig. 3.40—Saturation profile for displacement of a nonadsorbing viscous slug by drive water.

$$t_D^{n+1} = t_D^n\left[\left(f_w'^{*\,n} - \overline{v}_{Dv}^{n+1}\right)/\left(f_w'^{*\,n+1} - \overline{v}_{Dv}^{n+1}\right)\right]. \quad \dots \dots \dots \dots \dots \dots \dots \dots \dots \dots \dots \dots \dots \dots \dots \dots \dots \dots \dots \dots (3.137)$$

Successive solution for pairs of t_D^{n+1} and x_{Dv}^{n+1} yields the trajectory of the drive-water shock. **Fig. 3.41** shows the distance/time diagram for the viscous waterflood of Example 3.3 when drive-water injection begins at $t_{Do} = 0.2$. Supporting calculations associated with the development of Fig. 3.41 are presented in Example 3.7.

Example 3.7—Injection of a Viscous Water Slug. The viscous waterflood described in Example 3.3 will be converted to drive-water injection after 0.2 PV of viscous water are injected. Prepare the distance/time diagram for this flood.

Solution. The saturation profile is computed when drive-water injection begins at $t_D = 0.2$. From Table 3.8, the water bank is seen to be already out of the system ($t_D = 0.1853$) and the oil bank is about to arrive at the end of the system ($t_D = 0.202$). Locations of other saturations, summarized in **Table 3.11,** are determined from the frontal-advance equation.

A saturation discontinuity forms immediately when the drive water is injected. Table A-1 gives computed (f_w, S_w) and $\left(f_w^*, S_w^*\right)$ saturation pairs that satisfy the material-balance requirement across the saturation discontinuity. Following Example A.1, the location of the saturation discontinuity is estimated starting at $S_w^* = 0.7$. In Example A.1, the average specific velocity as the

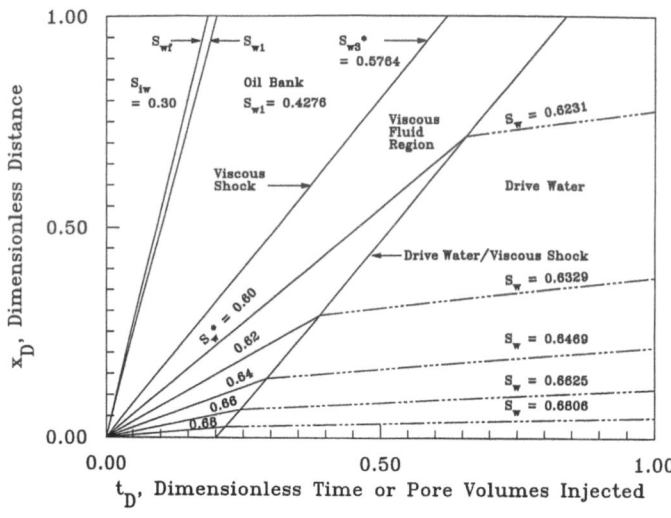

Fig. 3.41—Distance/time diagram for displacement of a 0.2-PV viscous slug by drive water when the viscous fluid does not adsorb, Example 3.7.

saturation discontinuity moves from $S_w^* = 0.70$ to $S_w^* = 0.68$ is 1.4488. As this happens, new saturation $S_w = 0.6806$ evolves from the saturation profile.

The time corresponding to the point where saturation $S_w = 0.6806$ evolves from the saturation discontinuity can be found by substitution into Eq. 3.137. Applying Eq. 3.137 gives

$$t_D^{(1)} = t_D^{(0)} \left[\frac{f_w'^{*(0)} - \overline{v}_{Dv}^{(1)}}{f_w'^{(1)} - \overline{v}_{Dv}^{(1)}} \right]$$

$$= (0.2)(0. - 1.4488) / (0.1164 - 1.4488)$$

$$= 0.2175,$$

and $x_{Dv}^{(1)} = (0.2175)(0.1164)$

$$= 0.0253.$$

Table 3.12 summarizes the remaining computations. The drive-water shock approaches v_{D3}^* asymptotically because $\overline{v}_{Dv} < v_{D3}^*$.

3.8 Dispersion During Miscible Displacement

3.8.1. General Description of the Dispersion Process. In all miscible displacement processes, mixing occurs between the displacing and displaced fluids. That is, there is dispersion between the different fluids. This dispersion dilutes the displacing fluid with the displaced fluid and thereby affects the phase behavior. Thus, understanding the dispersion phenomenon is important in the design of miscible displacement processes. This section emphasizes a description of the dispersion process, the physical mechanisms that cause dispersion, and a mathematical treatment of dispersion in relatively simple systems.

S_w	f_w	f_{Sw}	x_D	Comments
0.4276	0. 6868	4.9505	0.9901	←Front of oil bank
0.4276	0.6868	1.6064	0.3213	←Viscous saturation
0.5764	0.9259	1.6064	0.3213	
0.60	0.9575	1.08646	0.2173	
0.62	0.9756	0.74360	0.1487	
0.64	0.9877	0.47655	0.0953	{Shock displacement by viscous region}
0.66	0.9951	0.27166	0.0543	
0.68	0.9989	0.11638	0.0233	
0.70	1.0000	0	0	

Table 3.11—Saturation profile when drive-water injection begins at $t_{D1} = 0.2$, Example 3.7.

Timestep	S_w^*	$\overline{v}_{Dv}$	$f_{Sw}'^*$	t_D	x_{Dv}
0	0.70	1.4286	0	0.20	0
1	0.68	1.4488	0.1164	0.2175	0.0253
2	0.66	1.4884	0.2717	0.2452	0.0662
3	0.64	1.5255	0.4766	0.2931	0.1397
4	0.62	1.5585	0.7436	0.3892	0.2894
5	0.60	1.5847	1.0865	0.6570	0.7139
6	0.5764	1.6011	1.6064	∞	–

Table 3.12—Summary of calculations and parameters for determining the trajectory of the drive-water shock, Example 3.7.

Consider a linear porous medium such as shown in **Fig. 3.42.** The porous medium initially contains Fluid A, either a gas or a liquid. At a designated time, Fluid B is injected into the system at the left face at a constant velocity. Fluid B, which is miscible with Fluid A, displaces Fluid A from the system. The lower part of Fig. 3.42 shows a plot of the concentration of Fluid B vs. time at the inlet face of the porous medium (at $x = 0$). This is a so-called step change in concentration. A normalized concentration of Fluid B has been plotted so that the concentration varies between 0 and 1.0.

If injection of Fluid B were continued and a concentration-monitoring device were placed at some fixed position downstream, the concentration profile for Fluid B would be similar to that shown in Fig. 3.42. The curve of concentration of Fluid B vs. time would be S-shaped. Fluid B would first appear sometime before the time required for injection of 1.0 PV of Fluid B. The concentration would increase very slowly at first, followed by a much sharper increase and a concentration tail, until no more Fluid A was present in the effluent. The 0.50 concentration point would occur near the time required to inject 1.0 PV. The S-shaped curve in Fig. 3.42 is the classic concentration profile for miscible displacement of one fluid by a second in a linear system in which the mobility ratio is favorable and gravity effects are negligible. The transition from the step-change input to the S-shaped curve is the result of mixing, or dispersion, of Fluids A and B in the porous medium as the displacement occurs. If absolutely no mixing of the two fluids occurred, then the concentration response at the fixed downstream position would be a step change, as at the inlet, and would occur at a time corresponding to injection of 1.0 PV of Fluid B.

As another illustration, consider the injection of a small slug of Fluid B followed again by Fluid A, rather than changing totally to Fluid B at the inlet. The inlet concentration might appear as in **Fig. 3.43.** If this Fluid B slug were followed as it moved downstream and concentration profiles were measured at different fixed times, the profiles would appear as shown in Fig. 3.43. The concentration profiles would tend to spread as the fluid moved downstream, and the amplitude (maximum concentration) would decrease. Again, this would occur because of the dispersion of Fluids A and B.

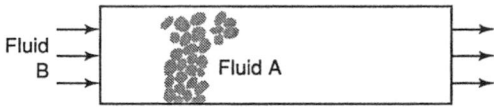

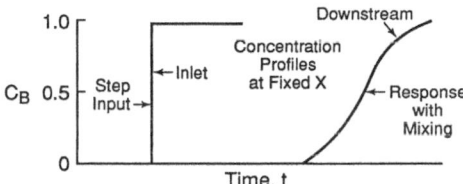

Fig. 3.42—Miscible displacement of Fluid A by Fluid B, step change in concentration at inlet.

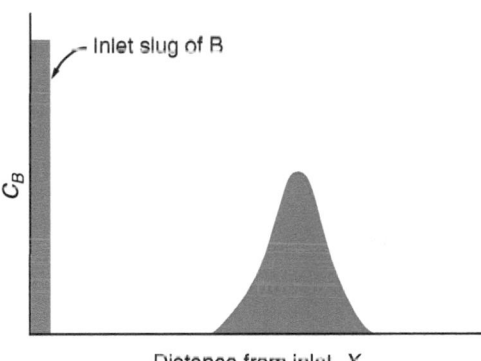

Fig. 3.43—Concentration profiles for injection of a slug of Fluid B to displace Fluid A.

The phenomenon described is longitudinal dispersion in that it occurs in the direction of primary flow (i.e., along the axis of flow). Dispersion also occurs in a transverse direction—that is, in a direction perpendicular to flow. This is illustrated by considering the injection of Fluid B to occur at a single point in the porous medium rather than across the entire inlet face. In this case, Fluid B would be dispersed in a direction normal to the flow direction as well as along the axis of flow. This mixing is called transverse dispersion.

3.8.2 Mechanisms and Models of Longitudinal Dispersion Phenomena. The dispersion of miscible fluids may be attributed to several different physical phenomena. This section discusses these phenomena. Simplified porous media models are used to illustrate the ways in which dispersion occurs.

Molecular Diffusion. Molecular diffusion is present in all systems in which miscible fluids are brought into physical contact. If two fluids are placed in contact in a straight capillary, the diffusion process may be described quantitatively using Fick's first law. For the diffusion of Fluid B into Fluid A, the diffusional flow across any plane may be expressed as

$$m_{Bx} = -D_{BA}A(\partial C_B / \partial x), \dots\dots\dots\dots\dots\dots\dots\dots\dots\dots\dots\dots\dots\dots (3.138)$$

where m_{Bx} = rate of Fluid B diffusing in the x-direction across a plane, A = cross-sectional area open for diffusion, C_B = concentration of Fluid B, D_{BA} = molecular diffusion coefficient for Fluid B diffusing into Fluid A, and x = position along the diffusion path. That is, the diffusion of Fluid B into Fluid A is proportional to the concentration gradient of Fluid B. The negative sign is imposed because the diffusion is in the direction of a decreasing concentration (i.e., from a higher to a lower concentration of Fluid A).

Diffusion coefficients, as defined in Eq. 3.138, typically are functions of concentration, temperature, and the particular chemical species present. For a rigorous analysis, the dependence on concentration and/or temperature should be included before integration of the equation. Assuming an average value for D_{BA} over the concentration and temperature ranges involved is often satisfactory, however, and this is what generally is done in reservoir analysis. In addition, the expression for diffusion in Eq. 3.138 requires an assumption of ideal fluid behavior, an assumption that is often valid for systems encountered in EOR processes.

If a reservoir rock were a bundle of straight capillary tubes, as is sometimes assumed in simple models, then Eq. 3.138 would be directly applicable to describe the molecular diffusion process. However, a straight capillary is not a very good representation of a porous medium. Fluids must move through a rock in a tortuous path. For example, if these fluids were assumed to move on average at approximately 45° to the basic flow direction, then the diffusion coefficient would need to be modified as follows (Perkins and Johnston 1963).

$$D_{aBA}/D_{BA} = 1/\sqrt{2} = 0.707, \dots\dots\dots\dots\dots\dots\dots\dots\dots\dots\dots\dots\dots\dots\dots\dots\dots\dots \quad (3.139)$$

where D_{aBA} = apparent diffusion coefficient in a porous medium.

Eq. 3.139 has been shown to be a fair representation of the molecular diffusion coefficient in a porous medium. However, several investigations (Perkins and Johnston 1963; Brigham et al. 1961) have suggested that a more sophisticated approach is to use the formation electrical resistivity factor. This approach recognizes the analogy between electrical conductivity and mass diffusion in porous media. That is, the path of electric current flow through a rock is essentially the path of mass movement by diffusion. The expression for the apparent diffusion coefficient is

$$D_{aBA}/D_{BA} = 1/F_R\phi, \dots\dots\dots\dots\dots\dots\dots\dots\dots\dots\dots\dots\dots\dots\dots\dots\dots\dots\dots \quad (3.140)$$

where F_R = formation electrical resistivity defined as R/R', (R = electrical resistivity of a porous medium saturated with a liquid that conducts electricity and R' = electricity resistivity of the liquid in the porous medium). The factor $1/F_R\phi$ typically ranges from 0.6 to 0.7, although it may be significantly lower, depending on the porous medium (Carman 1956).

The diffusion process is molecular in nature (i.e., it results from the random motion of molecules in solution). Diffusion is the dominant dispersion mechanism if flow rates are very low in a porous medium. At rates that commonly exist in reservoir displacement processes, however, dispersion also results from bulk flow or convective phenomena. The models that follow illustrate some ways in which dispersion can occur because of bulk movement of the fluids.

Velocity Profile (Taylor) Effect. Taylor (1953) showed that if one fluid (Fluid B) displaces a second (Fluid A) in a straight capillary, there will be a dispersion of Fluid B into Fluid A because of the velocity profile in the capillary as illustrated in **Fig. 3.44.** Assume that Fluid A is initially in the capillary and that Fluid A is displaced by Fluid B. Assume further that the flow rate is relatively low so that flow is laminar. The velocity profile radially across the capillary would have the parabolic form shown in Fig. 3.44. The profile would be such that

$$\bar{v} = (\tfrac{1}{2})v_{max}, \dots\dots\dots\dots\dots\dots\dots\dots\dots\dots\dots\dots\dots\dots\dots\dots\dots\dots\dots\dots\dots \quad (3.141)$$

and the maximum velocity would occur in the center of the capillary. If the effluent from the capillary were collected and the concentration measured, breakthrough of Fluid B would occur when a volume equal to one-half of the capillary volume had been injected. With continued collection and measurement of the effluent, the concentration of Fluid B would increase until there was no more Fluid A in the effluent. Thus, Fluids A and B would mix in the collected effluent because of the velocity profile that develops in laminar flow in a capillary.

Molecular diffusion also would occur across the Fluid A/Fluid B boundaries in the capillary, and this would affect the effluent concentration. However, this does not alter the basic conceptual idea that mixing occurs because of velocity gradients in the capillary.

Series of Mixing Cells. Another model for the mixing of miscible fluids is based on the assumption that a porous medium is a series of mixing cells or tanks in which fluids are perfectly mixed (Aris and Amundson 1957). **Fig. 3.45** illustrates this concept. Fluid B, which is displacing Fluid A, enters a pore (Tank 1) from the left. Fluids in the tank are perfectly mixed (i.e., there are no concentration gradients in the tank). This assumption means that, at any instant, the fluid being withdrawn from the tank has the same composition as the fluid in the tank. The mixed fluids from the first pore then enter the second pore (Tank 2), where the fluids are again perfectly mixed. The process is repeated as Fluid B moves through the system.

The assumption of perfect mixing in each pore is mathematically convenient and allows a quantitative expression to be developed for the mixing, or dispersion, process. Conceptually, it is clear that such a model, with or without perfect mixing in each pore, will lead to dispersion of Fluid B into Fluid A as the displacement process continues.

Stagnant Pockets. Another contribution to dispersion can be attributed to the flow behavior around stagnant pockets or dead-end pores (Aris 1959). **Fig. 3.46** illustrates this concept. Again, assume that Fluid B is displacing Fluid A, which was originally in the porous medium. The part of Fluid A that is in the main flow channel is displaced directly by Fluid B. However,

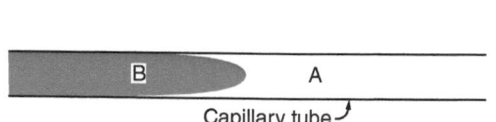

Fig. 3.44—Dispersion resulting from laminar flow in a straight capillary (Taylor effect).

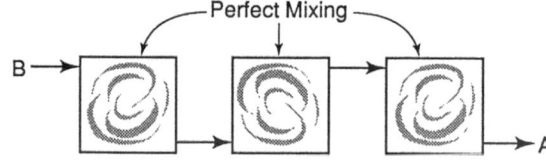

Fig. 3.45—Dispersion based on porous medium being viewed as a series of mixing tanks.

some of Fluid A is in stagnant pockets or dead-end pores (i.e., pore spaces that are connected to the main channel but through which there is no flow). This quantity of Fluid A is not displaced directly but is initially bypassed by Fluid B. Displacement of this bypassed fluid does occur slowly, however, as a result of molecular diffusion between the main flow channel and the stagnant pocket. Thus, Fluid A, which was in the pocket, is slowly removed. The overall result of this process is to cause mixing of Fluids A and B in the medium as the displacement progresses through the system.

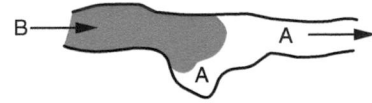

Fig. 3.46—Dispersion resulting from bypassing of fluid trapped in stagnant pockets.

Variation in Flow Paths. Dispersion of Fluid B into Fluid A can result from the variation of flow paths through the media encountered by different fluid particles (Raimondi et al. 1959; Keulemans 1957). **Fig. 3.47** shows this model. As before, Fluid A is being displaced by Fluid B. Visualize two separate particles (very small quantities) of Fluid B located at Point 1 in Fig. 3.47. As flow progresses, the particles move downstream to Point 2 but by slightly different paths through the medium. Because the flow paths are different, the particles, which were originally side by side, arrive downstream at different times. That is, dispersion of the fluid has resulted directly from the tortuosity inherent in the porous medium. Fluids become mixed because flow paths are varied as flow progresses through the system.

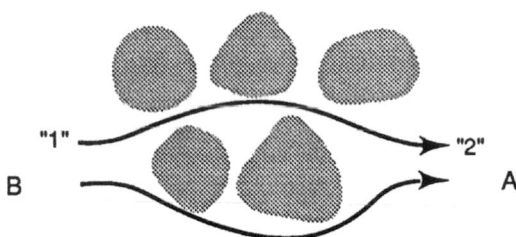

Fig. 3.47 Dispersion caused by variation of flow paths in a porous medium.

Summary Discussion of Dispersion Models. The simplified models described illustrate different ways in which dispersion can occur in a porous medium. The models are described in a qualitative way. Mathematical formulations have been developed to describe the diffusion process, the Taylor model, and the perfectly mixed tank model. None of these formulations is adequate to describe the dispersion process in the porous media of reservoir rocks, however, so the mathematical relationships are not presented. The dispersion process in reservoir rocks represents some combination of all the different mechanisms, and a precise mathematical description is not available. Simplified mathematical models have been developed and found to be useful, and Section 3.8.3 presents an introduction to this approach. Perkins and Johnston (1963) provide an excellent review of dispersion models.

3.8.3 Mathematical Description of Longitudinal Dispersion Process. Even though the dispersion process in reservoir rock is complicated, numerous experimental investigations (Perkins and Johnston 1963) have shown that the process can be approximated with a relatively simple mathematical model. Probably the most useful approach is based on the introduction of a dispersion coefficient, a parameter analogous to the molecular diffusion coefficient. The partial-differential equation that describes dispersion in porous media, which is based on use of the dispersion coefficient, is derived below.

The following primary assumptions are made in the derivation.

1. Fluid B is displacing Fluid A, and the two fluids are miscible.
2. Flow is single phase.
3. The fluids are incompressible.
4. The mobility ratio is unity (i.e., there is no viscous fingering).
5. The fluids are of equal density.
6. Flow is in only one direction (x-direction).
7. The fluid velocity is constant.
8. Flow is through a porous medium of constant porosity and of constant cross-sectional area, A.

The system is sketched in Fig. 3.42. Consider a mass balance on Fluid B over the volume element $A\Delta x$ and over a time period Δt.

$$m_{\text{end}} - m_{\text{begin}} = \sum m_{\text{in}} - \sum m_{\text{out}}, \quad\dots\dots\dots\dots\dots\dots\dots\dots\dots\dots\dots\dots\dots\dots (3.142)$$

where m_{end} and m_{begin} = amounts of Fluid B in the element at the end and beginning of Δt, respectively, and $\sum m_{\text{in}}$ and $\sum m_{\text{out}}$ = amounts of Fluid B that flow in and out of the element during Δt, respectively. Eq. 3.142 is a symbolic representation and states that the net accumulation of Fluid B in the volume element during Δt is equal to the net flow of Fluid B into the element during the same Δt.

Eq. 3.142 must now be rewritten in terms of fluid densities, concentrations, and velocities. In doing this it must be recognized that the net movement of Fluid B into the element occurs by two mechanisms. The first is by convection (i.e., by bulk flow of the fluid stream) and is directly dependent on the fluid velocity. The second mechanism is dispersion, which in the context used here includes all the different mixing phenomena discussed earlier. It is assumed that an expression for dispersion can be used that is analogous to Fick's law:

$$m_{Bx} = -K_\ell A (\partial C_B / \partial x), \quad\dots\dots\dots\dots\dots\dots\dots\dots\dots\dots\dots\dots\dots\dots\dots\dots (3.143)$$

where m_{Bx} = rate of dispersion of Fluid B across an arbitrary plane, C_B = concentration of Fluid B, K_l = longitudinal dispersion coefficient, and x = distance along the dispersion path.

The net dispersional flux of Fluid B across any plane in the system is proportional to the negative of the concentration gradient.

With these concepts, Eq. 3.142 can now be rewritten as

$$A\phi C_B \Delta x\big|_{t+\Delta t} - A\phi C_B \Delta x\big|_t = vA\phi C_B\big|_x \Delta t - vA\phi C_B\big|_{x+\Delta x}\Delta t$$
$$-K_\ell A\phi(\partial C_B/\partial x)\big|_x \Delta t + K_\ell A\phi(\partial C_B/\partial x)\big|_{x+\Delta x}\Delta t, \quad\ldots\ldots\ldots\ldots\ldots \quad (3.144)$$

where ϕ = porosity, t = time, and v = pore velocity (this is the Darcy velocity divided by ϕ).

The vertical bar in Eq. 3.144 denotes that the quantity to the left is evaluated at the time or position specified with the bar. Simplifying yields

$$C_B\big|_{t+\Delta t} - C_B\big|_x \Delta x = vC_B\big|_x - vC_B\big|_{x+\Delta x}\Delta t$$
$$+\left[-K_\ell(\partial C_B/\partial x)\big|_x + K_\ell(\partial C_B/\partial x)\big|_{x+\Delta x}\right]\Delta t. \quad\ldots\ldots\ldots\ldots\ldots \quad (3.145)$$

Dividing through by Δx and Δt gives

$$\frac{C_B\big|_{t+\Delta t} - C_B\big|_t}{\Delta t} = \frac{-(vC_B\big|_{x+\Delta x} - vC_B\big|_x)}{\Delta x}$$
$$+\frac{K_\ell(\partial C_B/\partial x)\big|_{x+\Delta x} - K_\ell(\partial C_B/\partial x)\big|_x}{\Delta x}. \quad\ldots\ldots\ldots\ldots\ldots \quad (3.146)$$

Now let $\Delta t \to 0$ and $\Delta x \to 0$, and let v and K_l be constant:

$$\frac{\partial C_B}{\partial t} = -v\frac{\partial C_B}{\partial x} + K_\ell\frac{\partial^2 C_B}{\partial x^2}. \quad\ldots\ldots\ldots\ldots\ldots \quad (3.147)$$

Eq. 3.144 reduces to the classic convection-diffusion equation in one space dimension. The diffusion coefficient has been replaced by the dispersion coefficient, which is an empirical parameter. That is, the coefficient must be determined from experimental data and is not known a priori. Empirical correlations for K_l and the method of calculating it from displacement data are presented in Section 3.8.4.

Solutions to Eq. 3.147 are available for various boundary conditions. For the case of a step change in input concentration, the solution (Brigham et al. 1961) is

$$C_B = \tfrac{1}{2}\left\{1 - \mathrm{erf}\left[(x - vt)/2\sqrt{K_\ell t}\right]\right\}, \quad\ldots\ldots\ldots\ldots\ldots \quad (3.148)$$

with boundary and initial conditions given by $C_B = 0$, $0 < x$, $t = 0$; $C_B = 1.0$, $x = 0$, $t > 0$; and $C_B = 0$, $x \to \infty$, $t > 0$.

Defining C_B as having values between 0 and 1.0 is analogous to defining a normalized concentration

$$C_B = \left(C_B^* - C_{B0}\right)/\left(C_{Bi} - C_{B0}\right), \quad\ldots\ldots\ldots\ldots\ldots \quad (3.149)$$

where C_B^* = actual concentration of Fluid B, C_{B0} = initial concentration of Fluid B in the system, and C_{Bi} = injected or maximum concentration of Fluid B. The error function (erf) is a tabulated function defined as

$$\mathrm{erf}(\zeta) = \frac{2}{\sqrt{\pi}}\int_0^\zeta e^{-\zeta^2}\,d\zeta \ldots\ldots\ldots\ldots\ldots\ldots \quad (3.150)$$

$$\text{or } \mathrm{derf}(\zeta)/d\zeta = 2/\sqrt{\pi}\,e^{-\zeta^2}. \quad\ldots\ldots\ldots\ldots\ldots \quad (3.151)$$

Tabulated values are given in Appendix B. Eq. 3.148 provides a means of calculating a concentration response in a linear system when the input concentration of a particular chemical species is changed in a step-wise manner. A sample calculation is shown in Example 3.8.

Example 3.8—Calculation of Concentration Profile in a Linear Miscible Displacement. This example illustrates the calculation of a concentration profile in a linear Berea sandstone core during a miscible displacement. The core is 0.165 ft in diameter and 4.01 ft long. Porosity is 0.206. The core is initially saturated with brine of concentration 30,000 ppm of

NaCl (30 g NaCl/L). Brine flow rate is 2.12×10^{-4} ft³/hr. Flow interstitial velocity is 1.155 ft/D. The concentration of the brine injected is suddenly changed to 20,000 ppm NaCl (step-function change). Calculate the concentration (normalized) at the effluent end of the core (i.e., at $x = L = 4.01$ ft), assuming the dispersion coefficient is given by $K_l = 3.46 \times 10^{-4}$ ft²/hr.

Solution. The calculation is made by first substituting all known values into Eq. 3.148. The only unspecified independent variable is t. Thus, different values of t are specified, and the error function solution is calculated to obtain C_B as a function of time. Consistent units must be used. Results are given in **Table 3.13** and **Fig. 3.48,** which show the normalized concentrations.

Notice that the time of 83.4 hours corresponds to the time required to inject 1 PV of brine and that $C_B = 0.5$ at this time. To obtain concentration in parts per million of NaCl, simply use the definition of the normalized concentration:

$$C_B = \left(C_B^* - C_{B0}\right)/\left(C_{Bi} - C_{B0}\right)$$

or $C_B = (-30{,}000)/(20{,}000 - 30{,}000),$

where C_B^* = concentrations in parts per million of NaCl.

3.8.4 Calculation of Longitudinal Dispersion Coefficient From Experimental Data. It is not possible to predict the dispersion coefficient for a given system from fundamental principles. However, K_l can be determined for a specified porous medium and specified flow conditions by conducting an experimental miscible displacement and empirically fitting concentration data to the appropriate differential equation solution. A number of investigators (Perkins and Johnston 1963) have done this and have shown that K_l is a function of several system parameters. In particular, K_l depends on the structure of the porous medium, the particle size of the medium, the velocity of the displacement, and the miscible fluids. On the basis of experimental results, empirical correlations have been developed that relate K_l to different system parameters. The following sections give the method of calculating K_l from experimental data. This is followed by a discussion of some of the available correlations for K_l.

Plot of Data on Probability Graph Paper. The nature of the solution to the describing partial-differential equation for the case of a step concentration input (Eq. 3.148) is such that a plot of C_B vs. $(x - vt)/\sqrt{t}$ yields a straight line on probability paper. This leads to a simple method for calculating K_l. To arrive at the method, the solution is first changed to a slightly different form (Brigham et al. 1961). Consider the argument of the error function,

$$\frac{x - vt}{2\sqrt{K_\ell}\sqrt{t}} = \frac{1}{2\sqrt{K_\ell}}\left(\frac{x - vt}{\sqrt{t}}\right). \quad \dots\dots\dots\dots\dots\dots\dots\dots\dots\dots\dots\dots\dots\dots\dots\dots\dots\dots\dots\dots (3.152)$$

Set $x = L$, the measuring-point location, which is usually the core effluent, to obtain

$$\frac{1}{2\sqrt{K_\ell}}\left(\frac{L - vt^*}{\sqrt{t^*}}\right) = \frac{1}{2\sqrt{K_\ell}}\left(\frac{LA\phi - vA\phi t^*}{\sqrt{A\phi}\sqrt{A\phi t^*}}\right), \quad \dots\dots\dots\dots\dots\dots\dots\dots\dots\dots\dots\dots\dots\dots\dots\dots\dots\dots (3.153)$$

where A = cross-sectional area for flow. Now, introducing V_p = volume in 1 PV, V_i = volume of fluid injected, and t^* = time to inject 1 PV gives

$$\text{arg} = \frac{1}{2\sqrt{K_\ell}}\left(\frac{V_p - V_i}{\sqrt{A\phi t^* v}}\right)\frac{\sqrt{v}}{\sqrt{A\phi}}$$

t (hours)	Error Function Argument	erf Argument	C_B^*
76.7	0.9851	0.8364	0.082
79.2	0.6062	0.6087	0.196
81.3	0.2971	0.3256	0.337
83.4	0.0000	0.0000	0.500
85.9	−0.3527	−0.3821	0.692
88.0	−0.6379	−0.6330	0.817
92.6	−1.2399	−0.9205	0.960
95.4	−1.5923	−0.9757	0.988

*C_B is a normalized concentration.

Table 3.13—Calculated concentration profile in a dispersion process, Example 3.8.

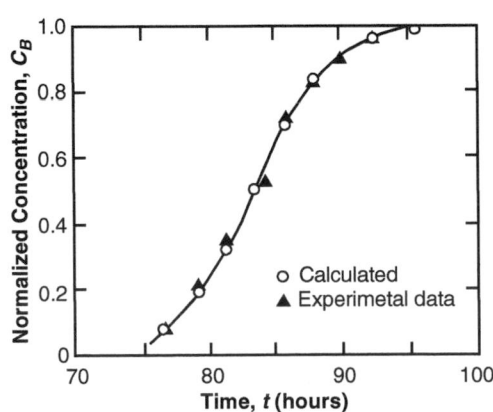

Fig. 3.48—Calculated and experimental concentration profiles in a linear system, Examples 3.9 and 3.10.

$$= \frac{1}{2\sqrt{K_\ell}}\left(\frac{v_p - v_i}{\sqrt{v_i}}\right)\sqrt{L/t^*}\left(\frac{1}{\sqrt{A\phi}}\right). \quad\dots\dots\dots\dots\dots\dots\dots\dots\dots (3.154)$$

Because $t^* = A\phi L/vA\phi$, the argument may be written

$$\arg = \frac{1}{2\sqrt{K_\ell}}\left(\frac{v_p - v_i}{\sqrt{v_i}}\right)\frac{L}{\sqrt{V_p t^*}} = \frac{1}{2\sqrt{K_\ell}}U\frac{L}{\sqrt{V_p t^*}}. \quad\dots\dots\dots\dots\dots (3.155)$$

The solution can now be written as

$$C_B = \frac{1}{2}\left[1 - \mathrm{erf}\left(\frac{UL}{2\sqrt{K_\ell}\sqrt{V_p t^*}}\right)\right], \quad\dots\dots\dots\dots\dots\dots\dots\dots (3.156)$$

where $U = \left(V_p - V_i\right)/\sqrt{V_i}$.

Probability graph paper may now be used. This graph paper is designed such that a plot of the dependent variable C_B vs. the error-function argument yields a straight line. Because U is the only parameter in the error-function argument that varies with C_B, U vs. C_B should result in a straight line, as stated earlier. Example 3.9 illustrates the plotting method. The calculation of K_ℓ from the data is described in Example 3.10.

Example 3.9—Plotting Experimental Dispersion Data. Experimental miscible-displacement data were taken for a system having the same properties as described in Example 3.8. The data are shown in Fig. 3.48 and tabulated in **Table 3.14.** Rock properties, fluids, and flow rates are as given in Example 3.8. Plot the data on probability paper.

Solution. **Fig. 3.49** shows the data, plotted on probability paper, with the best straight line through the data indicated.

The results shown in Fig. 3.49 are for an experiment run under nearly ideal conditions. The rock was relatively homogeneous and the properties of the displaced and displacing liquids were similar. For this system, the probability graph plot is essentially linear over the range $C_B = 0.1$ to 0.9, which is typical for such an ideal case. For some systems, however, linearity exists over a smaller C_B range. Reasons for deviations from the theory are discussed in Section 3.8.8.

Calculation of **K**$_\ell$. The equations previously presented in this section may be used with data, as shown in Fig. 3.49, to calculate the dispersion coefficient for the conditions of the experiment. Applying Eq. 3.156 at $C_B = 0.9$ gives

$$0.9 = \frac{1}{2}\left[1 - \mathrm{erf}\left(U_{90}\frac{1}{2\sqrt{K_\ell}}\frac{L}{\sqrt{V_p t^*}}\right)\right], \quad\dots\dots\dots\dots\dots\dots\dots\dots (3.157)$$

where U_{90} = value of U evaluated when $C_B = 0.9$.

L (ft)	4.01	
V_p (ft³)	0.01766	
v (ft/hr)	0.0481	

t	C_B	V_i/V_p	U
76.7	0.085	0.920	0.0111
78.4	0.160	0.940	0.0082
79.2	0.205	0.950	0.0068
80.5	0.275	0.965	0.0047
81.3	0.350	0.975	0.0034
82.6	0.445	0.990	0.0013
84.3	0.525	1.000	0.0
85.1	0.650	1.020	−0.00263
85.9	0.725	1.030	−0.00393
87.2	0.770	1.045	−0.00585
88.0	0.820	1.055	−0.00712
90.1	0.890	1.080	−0.0102
92.6	0.960	1.110	−0.0139

Table 3.14—Experimental data for a linear miscible displacement, Example 3.9.

$$0.90622 = -U_{90}\frac{1}{2\sqrt{K_\ell}}\frac{L}{\sqrt{V_p t^*}}. \quad\ldots\ldots\ldots\ldots \quad (3.158)$$

At $C_B = 0.1$, where $U = U_{10}$,

$$0.10 = \frac{1}{2}\left[1 - \text{erf}\left(U_{10}\frac{1}{2\sqrt{K_\ell}}\frac{L}{\sqrt{V_p t^*}}\right)\right] \quad\ldots\ldots \quad (3.159)$$

and $$0.90622 = U_{10}\frac{1}{2\sqrt{K_\ell}}\frac{1}{\sqrt{V_p t^*}}. \quad\ldots\ldots\ldots \quad (3.160)$$

Adding Eqs. 3.158 and 3.160 yields

$$1.8124 = \frac{1}{2\sqrt{K_\ell}}\frac{L}{\sqrt{V_p t^*}}(U_{10} - U_{90}). \quad\ldots\ldots\ldots \quad (3.161)$$

Now, squaring both sides and solving for K_ℓ gives

$$K_\ell = \left[\frac{L(U_{10} - U_{90})}{3.625}\right]^2 \frac{1}{V_p t^*}. \quad\ldots\ldots\ldots \quad (3.162)$$

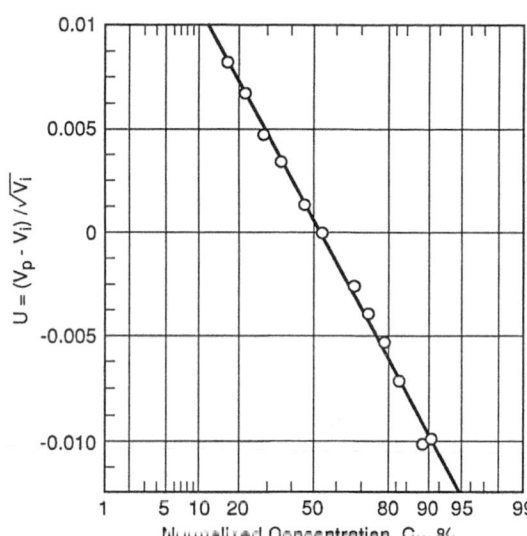

Fig. 3.49—Plot of experimental dispersion data on probability graph paper, Examples 3.9 and 3.10.

Thus, K_ℓ may be calculated from the probability-paper plot with U values at 10 and 90% concentrations. The following procedure can be used to calculate K_ℓ from dispersion data in a linear displacement experiment.

1. Measure C_B at the exit of the core. Measure the PVs of fluid displaced.
2. Plot C_B vs. U on probability paper.
3. Place the best straight line through the data.
4. Select U_{10} and U_{90} from the plot.
5. Calculate K_ℓ from Eq. 3.162.

An alternative procedure may be used in which U values are selected at other concentrations, such as 0.20 and 0.80. In this case, an appropriate equation to replace Eq. 3.162 must be developed with the above approach.

The method illustrated above is used in Example 3.10 to calculate the dispersion coefficient for the experimental data of Example 3.8.

Example 3.10—Calculation of K_ℓ From Experimental Data. Fig. 3.49 is a probability-paper plot of the experimental data. Calculate the value of K_ℓ.

Solution. The required U values are $U_{10} = 0.0105$ ft$^{3/2}$ and $U_{90} = -0.0099$ ft$^{3/2}$. From Eq. 3.162,

$$K_\ell = \left[\frac{L(U_{10} - U_{90})}{3.625}\right]^2 \frac{1}{V_p t^*}$$

$$= \left[\frac{(4.01 \text{ ft})(0.0105 + 0.0099)(\text{ft}^{3/2})}{3.625}\right]^2 \frac{1}{(0.01766\text{ft}^3)(83.4 \text{ hr})}$$

$$= 3.46 \times 10^{-4} \text{ ft}^2/\text{hr}.$$

This is equal to the dispersion coefficient used in Example 3.8 to calculate the theoretical concentration profile. Fig. 3.48 shows the experimental and theoretical profiles.

3.8.5 Empirical Correlations for the Longitudinal Dispersion Coefficient. The equation development and examples presented in Section 3.8.4 illustrate the manner in which longitudinal dispersion coefficients are determined experimentally. Numerous studies of dispersion have shown that K_ℓ is a function of the porous medium properties, fluid properties, and flow rate (Perkins and Johnston 1963). Empirical correlations have been presented to relate K_ℓ to the appropriate parameters.

One premise in the correlations is that K_ℓ can be represented as the sum of molecular-diffusion and convective-dispersion components. That is,

$$K_\ell = D_a + K_c, \quad\ldots\ldots\ldots\ldots\ldots\ldots\ldots\ldots\ldots\ldots\ldots\ldots\ldots\ldots\ldots\ldots\ldots\ldots\ldots\ldots\ldots \quad (3.163)$$

where D_a = apparent molecular-diffusion component and K_c = convective-dispersion component. At relatively low flow rates, the convective component is negligible and diffusion controls. At high flow rates, diffusion is negligible and convective dispersion is

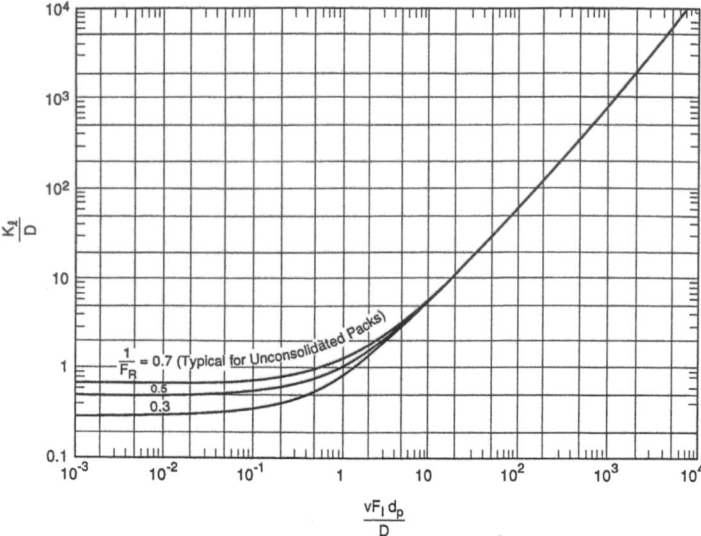

Fig. 3.50—Longitudinal dispersion coefficients for porous media (Perkins and Johnston 1963).

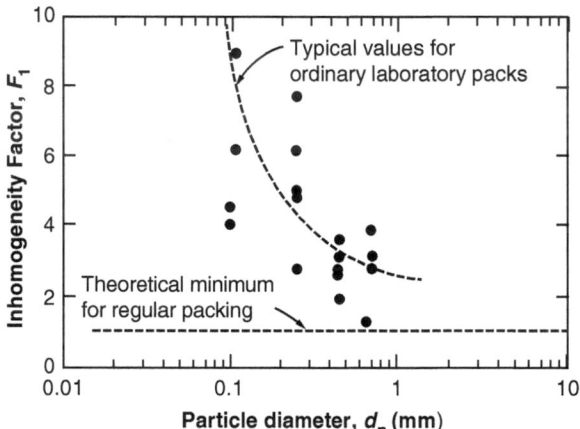

Fig. 3.51—Inhomogeneity factor for random packs of spheres (Perkins and Johnston 1963).

dominant. Between these extremes, both components contribute to the overall dispersion process, and this is the regime commonly encountered in reservoir flow processes.

Perkins and Johnston (1963) present the following correlation for the longitudinal dispersion coefficient:

$$K_\ell/D = 1/F_R\phi + 0.5\left(vF_I d_p/D\right) \quad .. \text{ (3.164)}$$

for $vF_I d_p/D < 50$,

where v = interstitial velocity (i.e., Darcy velocity divided by ϕ), d_p = average particle diameter, D = molecular diffusion coefficient, and F_I = inhomogeneity factor for the porous medium. Eq. 3.164 is dimensionless and any consistent set of units is applicable. Eq. 3.164 is plotted in **Fig. 3.50** along with a suggested correlating curve for $vF_I d_p/D < 50$.

In general, the inhomogeneity factor must be determined experimentally. **Fig. 3.51** shows results for random packs of spheres of various sizes (Perkins and Johnston 1963). For consolidated porous media, F_I cannot easily be separated from d_p, and the product $F_I d_p$ is often used. **Table 3.15** gives suggested values for a few outcrop cores.

Over the range of applicability of Eq. 3.164, the convective component of the dispersion coefficient is proportional to the first power of the velocity. At higher velocities, however, experimental data suggest that K_ℓ/D varies with $(vF_I d_p/D)^{1.2}$. This is reflected in the slope of the correlating curve in Fig. 3.50.

Example 3.11—Application of Empirical Correlation to Calculate K_ℓ. Consider the data of Example 3.8. Use the correlating equation of Perkins and Johnston (1963),

$$K_\ell/D = 1/F_R\phi + 0.5(vF_I d_p/D), \text{ to estimate } K_\ell.$$

Solution. In Berea core, $v = 0.0481$ ft/hr, $F_I d_p = 0.146$ in. $= 0.0121$ ft (average value of Berea cores, Table 3.15) $1/F_R\phi = 0.6$ (estimated for relatively homogeneous rock), $D = 3.87\times10^{-5}$ ft²/hr (assumed for brine system), $K_\ell/D = 0.60 + [(0.5)(0.0481$ ft/hr)(0.0121 ft)/3.87 × 10^{-5} ft²/hr] $= 0.60 + 7.52 = 8.12$, and thus $K_\ell = 3.14 \times 10^{-4}$ ft²/hr. This compares to $K_\ell = 3.46 \times 10^{-4}$ ft²/hr determined experimentally. The error is

$$100(3.46 - 3.14 / 3.46) = 9.2\%.$$

This is certainly an acceptable agreement. The correlations are no better than ±20% in general. Note that, for the conditions of the experiment, the molecular diffusion component is less than 10% of the total.

Source	Dispersion	Rock	$F_I d_p$ (in.)
Crane and Gardner (1961)	Transverse	Berea	0.098
Brigham et al. (1961)	Longitudinal	Berea	0.154
	Longitudinal	Torpedo	0.067
Raimondi et al. (1959)	Longitudinal	Berea	0.181
Handy (1959)	Longitudinal	Boise	0.217
		Average	0.143

Table 3.15—Values of $F_I d_p$ for outcrop sandstones.

3.8.6 Dispersion-Zone Width. The width of the dispersion zone when one fluid displaces a second, miscible fluid is an important parameter to examine because it is a reflection of the amount of mixing between the displaced and displacing fluids. In a miscible displacement process, the mixing-zone width directly relates to the miscible slug size that must be injected.

If the width is arbitrarily defined as the distance between positions at which the dimensionless concentration is 10 and 90%, it can be calculated from

$$x_{10} - x_{90} = 3.625\sqrt{K_\ell t}, \quad\dots\dots\dots\dots\dots\dots\dots\dots\dots\dots\dots\dots\dots\dots\dots \quad (3.165)$$

where x_{10} and x_{90} = distances to positions at which concentration is 10 and 90%, respectively. Eq. 3.165 can be derived from Eq. 3.148 by use of an approach analogous to the derivation of Eq. 3.162. If the width were defined as the distance between positions at which the concentrations have different values (such as 1 and 99%), the width expression would have the same form, but the constant 3.625 would change.

The result indicates that the width is proportional to $\sqrt{t}$. Because $x = vt$, the width of the dispersion zone is proportional to the square root of the mean distance traveled. Thus, the relative width of the dispersion zone, compared with mean travel distance, decreases as the front progresses through the porous medium, as illustrated in the following example.

Example 3.12—Calculation of Dispersion-Zone Width. The objective of this example is to calculate the dispersion-zone width after the front has traveled different distances through the porous medium. Width is defined as the distance between positions at which normalized concentrations are 10 and 90%. The data from Example 3.8 are used.

Solution. In Example 3.8, $v = 0.0481$ ft/hr, $K_\ell = 3.46 \times 10^{-4}$ ft²/hr, and $x_{10} - x_{90} = 3.625\sqrt{K_\ell t}$. Let $x = 4.0$ ft.

$$x_{10} - x_{90} = 3.625\left[\left(3.46 \times 10^{-4} \text{ ft}^2/\text{hr}\right)(4.0 \text{ ft})/(0.0481 \text{ ft/hr})\right]^{\frac{1}{2}}$$

$$= 0.61 \text{ ft.}$$

Therefore, the relative width is

$$(0.61/4.0) \times 100 = 15\%.$$

For $x = 100$ ft,

$$x_{10} - x_{90} = 3.625\left[\left(3.46 \times 10^{-4} \text{ ft}^2/\text{hr}\right)(100\text{ft})/(0.0481 \text{ ft/hr})\right]^{\frac{1}{2}}$$

$$= 3.07 \text{ ft.}$$

The relative width is

$$(3.07/100)100 = 3.1\%$$

For $x = 400$ ft,

$$x_{10} - x_{90} = 3.625\left[\left(3.46 \times 10^{-4} \text{ ft}^2/\text{hr}\right)(400 \text{ ft})/(0.0481 \text{ ft/hr})\right]^{\frac{1}{2}}$$

$$= 6.15 \text{ ft.}$$

The relative width is

$$(6.15/400)100 = 1.5\%$$

Example 3.12 indicates that relative width decreases as the interface moves through the porous medium, as described earlier. For normal reservoir well spacings, calculated dispersion-zone widths are relatively small fractions of the distances between wells. The implication for design is that miscible slug sizes can be relatively small compared with the reservoir volume. However, widths calculated with the dispersion theory must be considered minimum widths. As described later in this chapter, several factors increase the amount of mixing and cause deviations from ideal behavior. These factors generally lead to much larger apparent dispersion and hence larger dispersion-zone widths.

3.8.7 Empirical Correlations for the Transverse Dispersion Coefficient, K_t. Dispersion also occurs in a direction transverse to the principal flow direction. The molecular-diffusion component is not affected by flow and thus is the same for longitudinal and transverse dispersion. The convective component of transverse dispersion, however, is an order of magnitude smaller than the corresponding component of longitudinal dispersion. A correlation given by Perkins and Johnston (1963) is

$$K_t/D = 1/F_R\phi + 0.0157(vF_l d_p/D) \quad\dots\dots\dots\dots\dots\dots\dots\dots\dots\dots\dots\dots\dots \quad (3.166)$$

for $vF_l d_p/D < 10^4$.

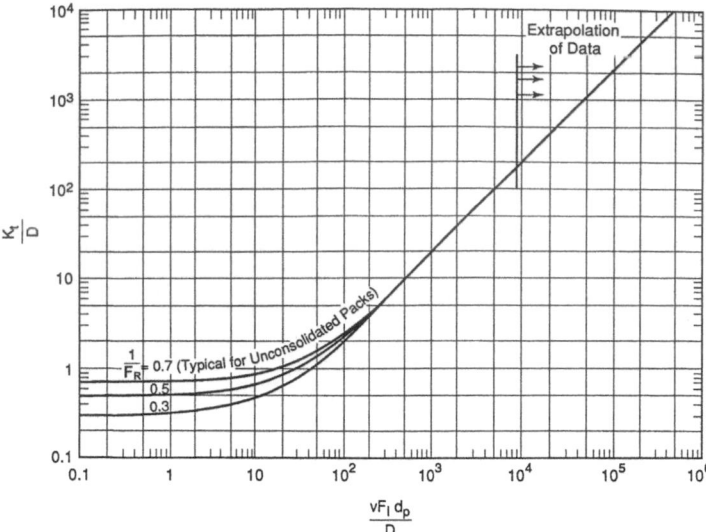

Fig. 3.52—Transverse dispersion coefficients for porous media (Perkins and Johnston 1963).

The correlation is plotted in **Fig. 3.52.** The inhomogeneity factor, F_I, is generally assumed to have the same value in correlations for both K_ℓ and K_r.

K_t generally is used to model dispersion when more than one space dimension is considered. A typical example is the release of a component at a single point in a porous medium through which a fluid is flowing in a specified direction. In this case, dispersion occurs in the direction of flow and also in directions transverse to flow. A 3D spread of the traced component occurs even though bulk flow is in only one direction. This spread (dispersion) is not equal in all directions, however, as indicated by the correlations for K_ℓ and K_r. Further, the traced component would move with the bulk flow; i.e., the dispersion is superimposed on the bulk transport.

3.8.8 Deviations From Ideal Behavior. The data used in Examples 3.9 through 3.12 were taken under nearly ideal conditions. The rock was relatively homogeneous, and displaced and displacing liquids had properties that were quite similar. A number of factors cause deviations from idealized behavior, resulting in measured concentration profiles that do not agree with the solution to Eq. 3.147.

Viscous Fingering. The idealized behavior is not followed if the viscosity of the displacing Fluid B is less than that of the displaced Fluid A (i.e., if $\mu_A/\mu_B > 1.0$). In this event, so-called viscous fingering occurs in the displacement process, as shown schematically in **Fig. 3.53.** Fluid B tends to channel, or finger, into Fluid A. When this happens in a linear displacement experiment, the apparent value of K_ℓ increases significantly because, in most cases, the average concentration of the effluent is measured. Flow in fingers is mixed with bypassed fluid, creating a much longer mixing zone.

The effect is illustrated in **Fig. 3.54,** which shows average effluent concentrations from linear experiments in which μ_A/μ_B was varied over a wide range (Brigham et al. 1961). The amount of dispersion and the general appearance of the concentration profiles change dramatically when μ_A/μ_B is increased to greater than 1.0. Viscosity ratios are called unfavorable or favorable depending on whether viscous fingering occurs.

An unfavorable viscosity ratio results in an increase in the apparent dispersion, the magnitude becoming larger as the viscosity ratio becomes more unfavorable. A second effect of an unfavorable ratio is to cause deviation from the idealized theory, as **Fig. 3.55** (Brigham et al. 1961) shows. In this figure, error-function plots are shown for unfavorable and favorable ratios. When μ_A/μ_B increases to only slightly more than unity (1.002), the error-function plot is no longer linear. A straight line cannot be drawn through the data points. It follows that the solution of Eq. 3.147 could no longer be fit to the concentration profile shown in Fig. 3.48. A dispersion coefficient cannot reasonably be determined for the experiment.

The reason for growth of viscous fingers may be described intuitively by considering **Fig. 3.56a.** Fluid B, which is displacing Fluid A, has a larger viscosity than Fluid A ($\mu_A/\mu_B < 1.0$). When a perturbation in the displacement front occurs as a result of flow in the tortuous path of the medium, the perturbation tends to dampen out because the flow resistance through the displacing liquid is greater than the resistance through the displaced fluid. However, when the viscosity ratio is made less than unity, as in Fig. 3.56b, perturbations

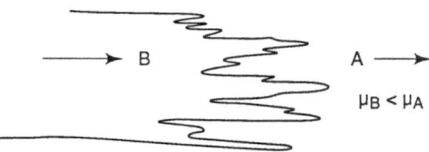

Fig. 3.53—Viscous fingering in a displacement process.

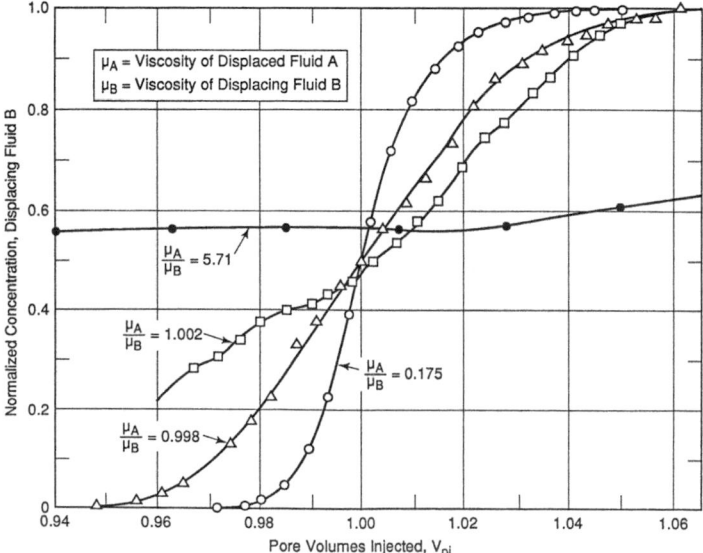

Fig. 3.54—Effect of viscosity ratio on effluent-concentration curve (Brigham et al. 1961).

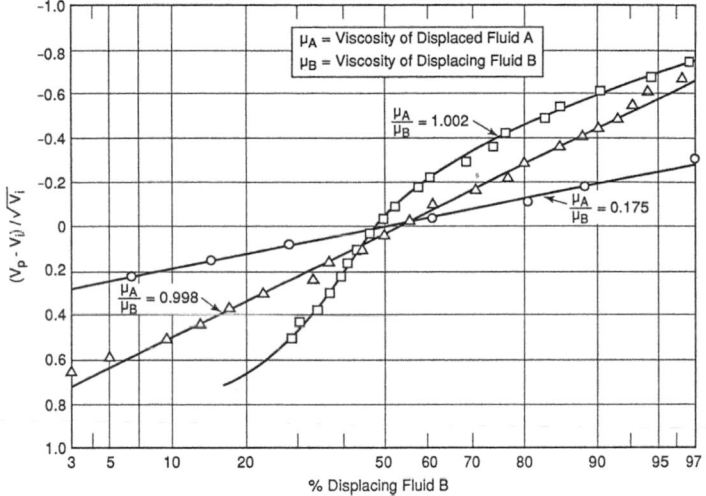

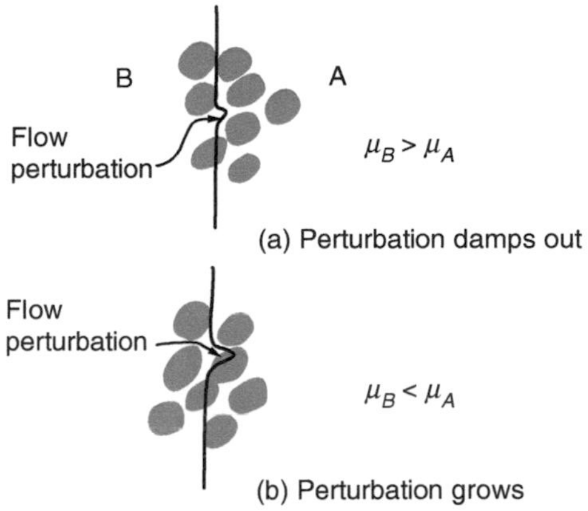

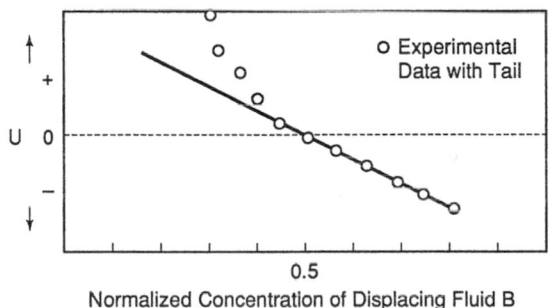

Fig. 3.56—Conditions promoting growth or dampening of flow perturbation.

Fig. 3.55—Effect of viscosity ratio on error function plot, Bead Pack 113-1 (Brigham et al. 1961).

Fig. 3.57—Asymmetry in the concentration profile.

tend to grow. In this case, the resistance through the displacing fluid is less than through the displaced fluid and more Fluid B flows through the path of least resistance, creating a finger that grows. It is interesting to consider that transverse dispersion will diminish fingering because it decreases concentration differences in a direction normal to finger growth. Viscous fingering is discussed in greater detail in Section 3.9, where a model for dispersion is described (Koval 1963).

Asymmetry of Tail. An experimental result commonly encountered is the asymmetry of the dispersion tail, illustrated in **Fig. 3.57.** In this case, the data fit a straight line on an error-function plot over part of the concentration range, but then deviate from a straight line at higher or lower concentrations. Several factors can cause such behavior.

One of these factors is deadend PV (i.e., a fraction of the pore volume has openings to the pores but no exits). There is no bulk flow through such pores, and fluid trapped in the pores can be removed only by molecular diffusion. If these pores form a significant fraction of the total PV, then a long concentration tail occurs because diffusion is a slow process. The concept is essentially the stagnant-pocket models discussed earlier, but the effect is more pronounced when an asymmetric concentration tail occurs. Deadend PV has been treated mathematically by Coats and Smith (1964).

Other factors that cause asymmetry in the concentration profile include the presence of a second, immiscible or partially soluble phase and permeability heterogeneities. Either of these, because of their effect on flow, can result in an apparent large dispersion coefficient and/or a deviation from the simple dispersion model. Perkins and Johnston (1963) discuss some of these factors.

Multiphase Systems. The treatment of dispersion presented earlier was based on the assumption that a single phase existed in the porous medium and that the species being considered were miscible. The presence of a second phase significantly affects the dispersion process (Thomas et al. 1963). As the saturation of a phase decreases, the dispersion of miscible species in that phase increases. That is, the presence of a second phase increases the apparent magnitude of the dispersion coefficient over values obtained in a single-phase system. The amount of dispersion in a phase also is affected by the wettability of the phase.

Thomas et al. (1963) also determined that the dispersion coefficient is not an adequate measure of the mixing process in multiphase flow. Their results indicate that the mixing is not well-described by the theory and differential-equation solutions based on an empirical dispersion coefficient.

3.8.9 Dispersion in Cylindrical or Field-Pattern Geometries.
The dispersion experiments and mathematical models described earlier were for linear systems. Most reservoir flow patterns, however, involve flow in cylindrical geometry (i.e., radial flow into or out of a well). In typical well patterns, such as the five-spot pattern, the flow is diverging/converging radial flow. Dispersion in linear flow has been studied extensively because of the simplicity of experimentation and ease of mathematical description. The study of the empirical dependence of K_ℓ on velocity would be difficult in a cylindrical geometry.

For cylindrical geometries, the describing differential equation is the cylindrical form of Eq. 3.147. Solution of the equation would yield C_B as a function of radial position and time. A difficulty in obtaining a solution is that K_ℓ is generally a function of fluid velocity, which, in a cylindrical system, is a function of radial position.

In a well-pattern geometry such as a five-spot pattern, the task of obtaining a solution is even more difficult. Baldwin (1966) and Brigham and Smith (1965) have obtained approximate solutions for this case. The solutions were used to describe chemical tracer tests to obtain information on reservoir flow patterns.

3.9 Viscous Fingering—Instability in Displacement Fronts

3.9.1 General Description. The nature of the front in a displacement process differs markedly depending on whether the mobility ratio is greater or less than unity. For example, if a solvent displaces an oil phase with which it is miscible at $M \leq 1.0$ and there are no gravity effects, the displacement process is efficient. A uniform front results, with little penetration of the displacing solvent into the displaced oil other than by molecular diffusion and dispersion.

Displacements carried out at conditions where the mobility ratio is greater than unity behave in a quite different manner. In this case, the displacement front becomes unstable and numerous fingers of displacing solvent penetrate into the displaced oil. **Fig. 3.58** (Habermann 1960) shows a schematic of this behavior for four different M values. Fig. 3.58 shows a displacement in a quarter of a five-spot pattern representative of data reported for miscible displacements (Habermann 1960). Similar results were reported for immiscible water/oil displacements, as shown in **Fig. 3.59** (van Meurs and van der Poel 1958), although concern has been raised about the effect of fluid wettability on fingering in immiscible systems (van Meurs 1957).

Viscous instability begins at conditions where $M > 1.0$. The magnitude of the fingering becomes more pronounced as the mobility ratio increases in magnitude, as indicated for the four M values in Fig. 3.58.

3.9.2 Criterion for Onset of Viscous Fingering. Collins (1961) described a simple model to quantify the criterion for the occurrence of viscous instabilities. Consider the linear, miscible displacement of oil by a solvent shown in **Fig. 3.60.** Flow is single phase, and there is no effect of gravity on the flow. At the time of consideration of the system, the solvent front is located at position x_f along the flow path.

In the flow region bounded by the dashed lines, a small perturbation or protrusion of the solvent front has occurred such that the front location is at position $x_f + \varepsilon$. The length parameter, ε, represents a length that is small relative to x_f. Perturbations, or small displacements, of the type shown clearly would occur in a displacement process in a porous medium in which flow paths are tortuous.

The focus of the analysis is the determination of conditions under which ε grows in time because, if ε does grow in time, then the front will be unstable (i.e., viscous fingers will form along the front). At conditions where ε does not grow or even diminishes in size, the front stability or uniformity will be maintained.

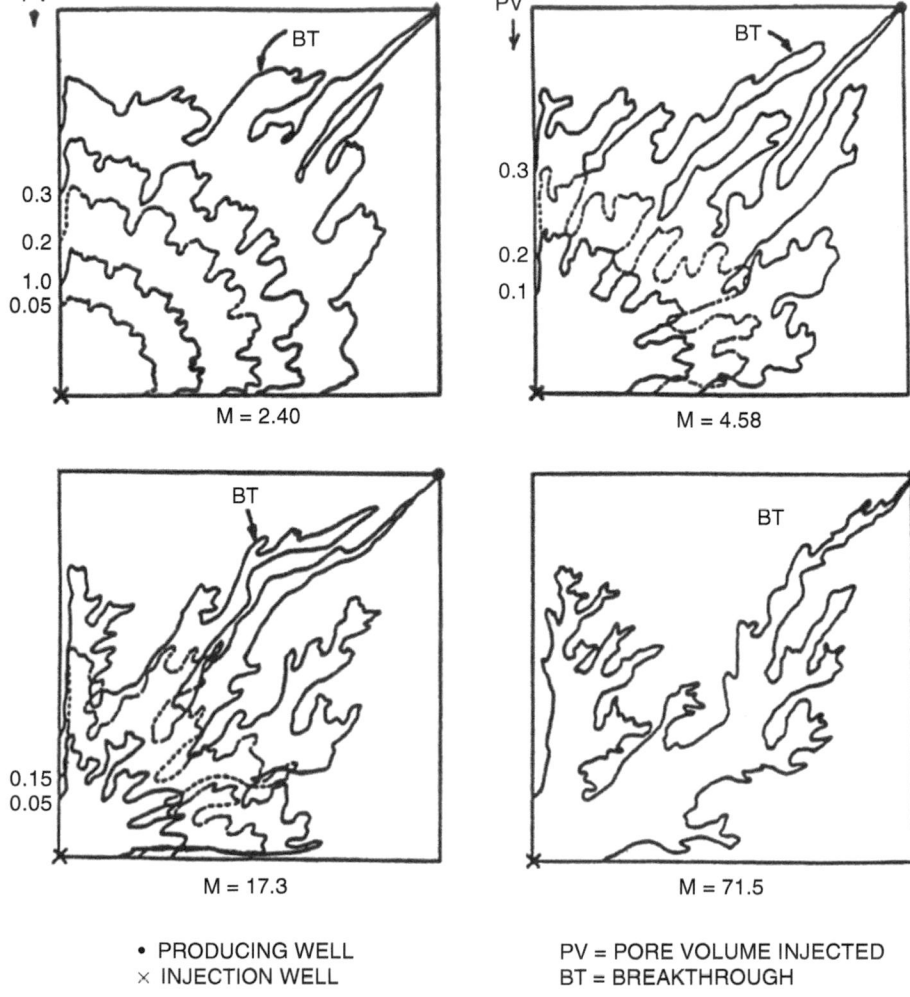

• PRODUCING WELL	PV = PORE VOLUME INJECTED
× INJECTION WELL	BT = BREAKTHROUGH

Fig. 3.58—Miscible displacement in one-quarter of a five-spot pattern at mobility ratios greater than unity; viscous fingering (Habermann 1960).

The analysis proceeds by examining the flow resistances along the different flow regions. If the solvent and oil resistances are assumed to be in series, application of Darcy's equation in the unperturbed region yields

$$(\Delta p)_{xf} + (\Delta p)_{L-xf} = \frac{u\mu_s x_f}{k} - \frac{u\mu_o (L - x_f)}{k}, \quad \ldots(3.167)$$

where $(\Delta p)_{xf}$ = pressure drop from the entrance to position x_f, $(\Delta p)_{L-xf}$ = pressure drop from position x_f to position L, u = superficial (Darcy) front velocity, k = permeability of the porous medium, μ_s = solvent viscosity, μ_o = oil viscosity, and appropriate units are used.

Recognizing that

$$u = \phi\left(dx_f/dt\right), \ldots\ldots\ldots\ldots\ldots\ldots (3.168)$$

and solving for the velocity of the front gives

$$\frac{dx_f}{dt} = \frac{-k\Delta p}{\phi\mu_s x_f + \phi\mu_o\left(L - x_f\right)}. \ldots\ldots\ldots\ldots (3.169)$$

Δp is the total pressure drop across the system, defined as $(p_L - p_o)$. With the definition $M = \mu_o/\mu_s$,

$$\frac{dx_f}{dt} = \frac{-k\Delta p}{\phi\mu_s\left[ML + (1 - M)x_f\right]}. \ldots\ldots\ldots\ldots(3.170)$$

A similar application of Darcy's equation in the region of perturbed flow yields

$$\frac{d\left(x_f + \varepsilon\right)}{dt} = \frac{-k\Delta p}{\phi\mu_s\left[ML + (1 - M)\left(x_f + \varepsilon\right)\right]}. \quad \ldots(3.171)$$

Subtracting Eq. 3.170 from Eq. 3.171 and carrying out some algebraic manipulation give

$$\frac{d\varepsilon}{dt} = \frac{-k\Delta p(1 - M)\varepsilon}{\phi\mu_s\left[ML + (1 - M)x_f\right]^2}, \ldots\ldots\ldots\ldots (3.172)$$

provided that $\varepsilon \ll x_f$, which was an initial assumption. Eq. 3.172 is an ordinary-differential equation with ε as the dependent variable (under the additional assumption that x_f is a constant). The solution is

$$\varepsilon = \varepsilon_0 e^{Ct}, \ldots\ldots\ldots\ldots\ldots\ldots\ldots\ldots (3.173)$$

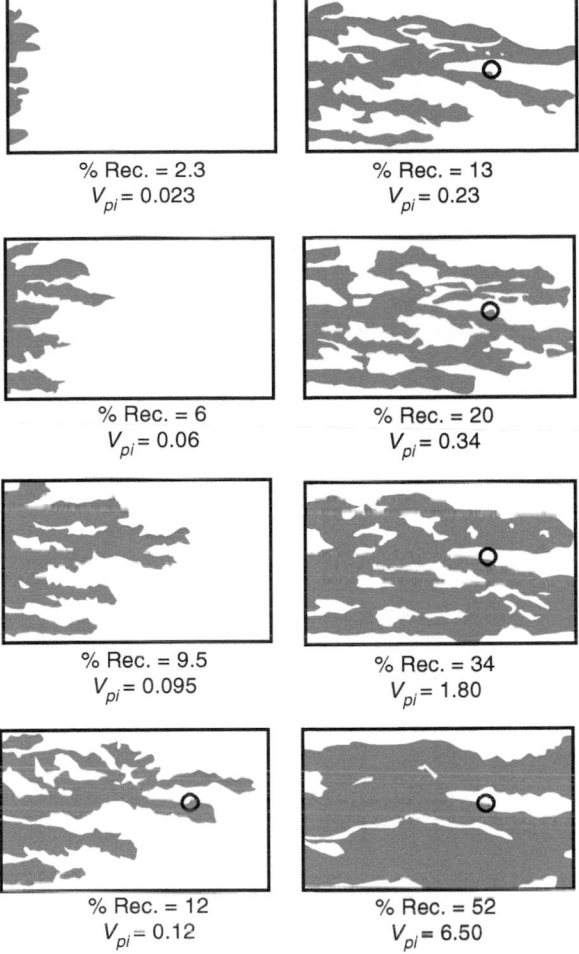

% Rec. = 2.3
V_{pi} = 0.023

% Rec. = 13
V_{pi} = 0.23

% Rec. = 6
V_{pi} = 0.06

% Rec. = 20
V_{pi} = 0.34

% Rec. = 9.5
V_{pi} = 0.095

% Rec. = 34
V_{pi} = 1.80

% Rec. = 12
V_{pi} = 0.12

% Rec. = 52
V_{pi} = 6.50

Fig. 3.59—Immiscible displacement in a linear system at an unfavorable mobility ratio; viscous fingering (van Meurs and van der Poel 1958).

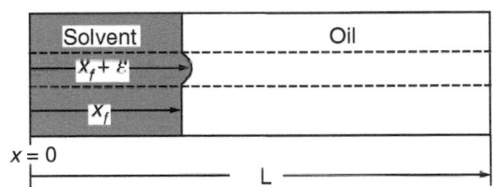

Fig. 3.60—Flow model for quantitative determination of the criterion for viscous instability.

$$\text{where } C = \frac{-k\Delta p(1 - M)}{\phi\mu_s\left[ML + (1 - M)x_f\right]^2} \ldots\ldots\ldots\ldots\ldots\ldots\ldots\ldots\ldots\ldots\ldots\ldots\ldots\ldots\ldots\ldots\ldots\ldots\ldots\ldots\ldots\ldots (3.174)$$

and ε_0 = initial length of the perturbation (i.e., the length at zero time). Inspection of Eqs. 3.173 and 3.174 shows that ε will grow exponentially if $M > 1.0$ because Δp is a negative quantity. If $M < 1.0$, the perturbation length ε will decay exponentially. The latter case leads to stable, nonfingering flow, while the former case leads to viscous finger growth.

The tortuosity of the path in a porous medium will lead to localized small flow perturbations. The analysis indicates that these perturbations, once formed, will grow or decay depending on the magnitude of the mobility ratio, with the value of M relative to unity being the critical factor. Fingers will grow if the displacing fluid is more mobile than the displaced fluid, and they will decay if the converse is true. While not considered in the analysis, capillary and gravity forces may affect the stability of the displacement front and may reduce fingering.

3.9.3 Dispersion in the Presence of Viscous Fingering. The model for dispersion described in Section 3.9 was shown to be invalid when the mobility ratio is unfavorable (i.e., when $M = \mu_o/\mu_s > 1.0$). In this case, viscous fingers form, and K_t, as defined in Eq. 3.147, is not a good description of the dispersion process. Miscible displacement in the presence of viscous fingers is called an unstable displacement.

Koval (1963) developed a model to describe an unstable miscible displacement in homogeneous or heterogeneous porous media for single-phase 1D flow. In this model, the frontal-advance equation is assumed to apply to the displacing phase (solvent).

$$\mathrm{d}x/\mathrm{d}t\,|_{S_s} = \left(q_t/A\phi\right)\left(\partial f_s/\partial S_s\right)|_{S=S_s}, \quad\dotfill (3.175)$$

where q_t = total volumetric flow rate, A = cross-sectional area normal to flow, and S_s = saturation or volume fraction occupied by solvent.

In an immiscible displacement process, such as a waterflood, fractional flow of the displacing phase, f_D, is given by

$$f_D = \frac{1}{1+\left(k_o/k_D\right)\left(\mu_D/\mu_o\right)}, \quad\dotfill (3.176)$$

where k_o = effective permeability of displaced oil phase, k_D = effective permeability of displacing water phase, μ_o = viscosity of oil phase, and μ_d = viscosity of water phase. For a miscible process with a solvent displacing oil, Koval expressed the fractional flow of solvent, f_s, in a general functional form as

$$f_s = f(S_s, H, E), \quad\dotfill (3.177)$$

where, in addition to solvent saturation, S_s, f_s is dependent on a heterogeneity factor, H, and a viscosity ratio factor, E. Koval then proceeded to determine the specific form of f_s.

The two miscible fluids are taken to be ideal in that they either do not mix or retain their individual identities and properties if they mix. From this assumption,

$$k_o = kS_o = k\left(1-S_s\right) \quad\dotfill (3.178)$$

and $k_s = kS_s$, $\quad\dotfill (3.179)$

where k = permeability of the porous media and k_s = effective permeability of the solvent displacing phase. It follows that

$$k_o/k_s = \left(1-S_s\right)/S_s. \quad\dotfill (3.180)$$

By analogy with Eq. 3.176,

$$f_s = \frac{1}{1+\left[\left(1-S_s\right)/S_s\right]\left(1/E\right)\left(1/H\right)}, \quad\dotfill (3.181)$$

or $f_s = KS_s/\left[1+S_s\left(K-1\right)\right]$, $\quad\dotfill (3.182)$

where $K = HE$. $\quad\dotfill (3.183)$

In the development that follows, the porous medium is assumed to be homogeneous with $H = 1.0$; thus, $K = E$. The parameter K (or E) is a function of the viscosity ratio, μ_o/μ_s, and is defined specifically in Eq. 3.195.

Expressions for the performance of the miscible displacement are obtained by integrating Eq. 3.175 and combining the result with Eq. 3.182. Integration of Eq. 3.175 yields

$$xS_s = \left(q_t t/A\phi\right)\left(\partial f_s/\partial S_s\right)|\,S_s. \quad\dotfill (3.184)$$

If $x = L$, the porous medium length,

$$1/\left(q_t t/A\phi L\right) = \left(\partial f_s/\partial S_s\right)|\,S_s. \quad\dotfill (3.185)$$

or $1/V_{pi}\left(\partial f_s/\partial S_s\right)|\,s_s = f_s'$, $\quad\dotfill (3.186)$

where V_{pi} = number of PVs of solvent injected for S_s to reach the end of the porous medium.

An expression for f_s' is obtained by differentiating Eq. 3.182:

$$\frac{1}{V_{pi}} = \frac{K}{\left[1+\left(K-1\right)S_s\right]^2}. \quad\dotfill (3.187)$$

Eq. 3.182 is used to eliminate S_s in terms of f_s:

$$V_{pi} = \frac{K}{\left[K - f_s(K-1)\right]^2} \quad \dots \dots \dots \dots \dots \dots \dots \dots \dots \dots \dots \dots \dots (3.188)$$

$$\text{or } f_s = \frac{K - \left(K/V_{pi}\right)^{\frac{1}{2}}}{K-1}. \quad \dots \dots \dots \dots \dots \dots \dots \dots \dots \dots \dots (3.189)$$

Eqs. 3.188 and 3.189 are valid at and after breakthrough. The equations specify the relationship between a specific fractional flow value, f_s, and the number of PVs of solvent injected to reach that fractional flow at the porous medium effluent.

At the point of initial breakthrough of solvent

$$f_s = 0. \quad \dots \dots \dots \dots \dots \dots \dots \dots \dots \dots \dots \dots \dots \dots \dots \dots (3.190)$$

Substituting into Eq. 3.189 yields

$$\left(V_{pi}\right)_{bt} = 1/K, \quad \dots \dots \dots \dots \dots \dots \dots \dots \dots \dots \dots \dots \dots (3.191)$$

where $(V_{pi})_{bt}$ = PVs injected at solvent breakthrough.

All the oil will have been produced when $f_s = 1.0$. Again, from Eq. 3.189,

$$\left(V_{pi}\right)_{comp} = K, \quad \dots \dots \dots \dots \dots \dots \dots \dots \dots \dots \dots \dots \dots (3.192)$$

where $(V_{pi})_{comp}$ = PVs injected when all the oil has been produced.

Between the point of solvent breakthrough and total oil recovery, oil produced is given by

$$N_p = \left(V_{pi}\right)_{bt} + \int_{(V_{pi})_{bt}}^{V_{pi}} \left\{ 1 - \left[\frac{K - \left(K/V_{pi}\right)^{\frac{1}{2}}}{K-1} \right] \right\} dV_{pi}, \quad \dots \dots \dots \dots (3.193)$$

where N_p = oil recovery in PVs (the oil FVF is assumed to equal unity for solvent and oil).

Integration of Eq. 3.193 yields

$$N_p = \frac{2\left(K/V_{pi}\right)^{\frac{1}{2}} - 1 - V_{pi}}{K-1}, \quad \dots \dots \dots \dots \dots \dots \dots \dots \dots (3.194)$$

which gives N_p as a function of V_{pi} and K.

The parameter K is yet to be specified. If the porous medium is assumed to be homogeneous ($H = 1.0$), then $K = E$. Koval postulated that E was not simply the ratio of viscosities because of the mixing of oil and solvent. To obtain an expression for E, Eq. 3.194 was matched against the experimental data of Blackwell et al. (1959). The expression for E proposed, based on this match, was

$$E = \left[0.78 + 0.22\left(\mu_o/\mu_s\right)^{\frac{1}{4}} \right]^4. \quad \dots \dots \dots \dots (3.195)$$

The expression for E incorporates, in part, the commonly used one-quarter power mixing rule for determination of viscosities of mixtures of solvents.

Fig. 3.61 compares Koval's model with the data of Blackwell et al. (1959). The agreement is quite good, partly because the data were used to determine the expression for E. Nonetheless, the Koval method, sometimes called the K-factor method, is an accepted approach to the description of performance of miscible displacement when the mobility ratio is unfavorable.

Example 3.13—Application of K-Factor Method for Calculating Performance of a Linear Miscible Displacement. An oil in a homogeneous sandpack is to be displaced miscibly with a solvent. Properties of the system and fluids are $\mu_o = 1.2$ cp, $\mu_s = 0.10$ cp, and $K = 500$ md. Calculate the fractional flow of solvent in the effluent stream as a function of the number of PVs of solvent injected.

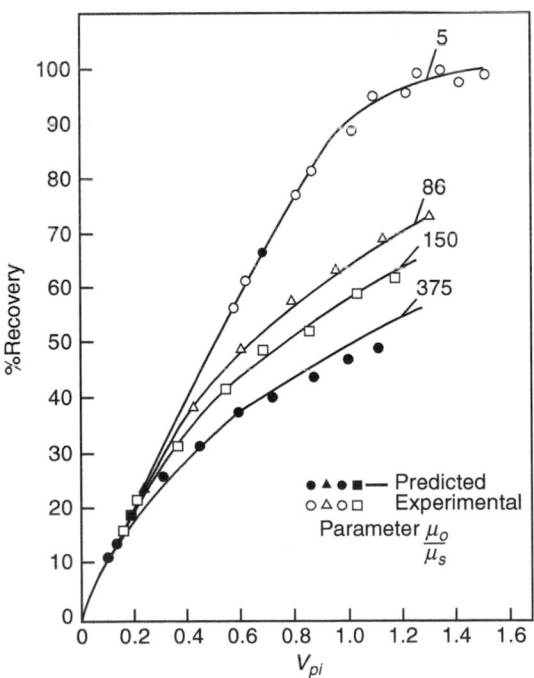

Fig. 3.61—Comparison of the experimental data of Blackwell et al. (1959) with predictions based on K-factor method (Koval 1963).

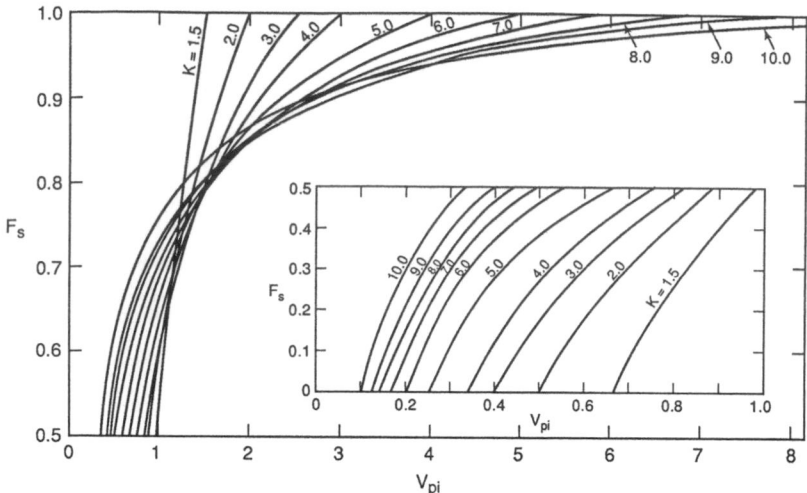

Fig. 3.62—Solvent cut as a function of PVs injected with K as a parameter, Example 3.13 (Koval 1963).

Solution. Eq. 3.189 is applicable after solvent breakthrough. K must be calculated first with Eq. 3.195.

$$K = E = \left[0.78 + 0.22\left(\mu_o/\mu_s\right)^{\frac{1}{4}}\right]^4.$$

$$= \left[0.78 + 0.22\left(1.2/0.1\right)^{\frac{1}{4}}\right]^4$$

$$= 2.00.$$

First, determine $(V_{pi})_{bt}$ and then substitute different values of $V_{pi}\left[V_{pi} > \left(V_{pi}\right)_{bt}\right]$ into Eq. 3.189 to calculate f_s.

$$\left(V_{pi}\right)_{bt} = 1/K = 1/2.0 = 0.5.$$

V_{pi}	f_s
0.6	0.17
0.8	0.42
1.0	0.59
1.5	0.85
2.0	1.00

Fig. 3.62 (Koval 1963) plots the results along with curves for other values of K.

Problems

3.1 Consider the linear waterflood calculation of Example 3.1. Assume that the calculation can be applied to a linedrive flood where the rows of injectors and producers are 660 ft apart. The width of a single pattern may be considered to be 330 ft, and reservoir thickness is 20 ft. Other data are given in Example 3.1.
   1. Calculate the volume of water injected (barrels) up to the time water breakthrough occurs at the producing end of the pattern.
   2. Calculate the volume of oil produced (barrels) up to the time the WOR reaches 50.
   3. Calculate and plot the volume of oil recovered (barrels) vs. water injected (barrels) out to a WOR of 50.

3.2 Examples 3.2 through 3.5 consider linear floods for the same conditions as in Example 3.1 with a different injected fluid. In Examples 3.2 and 3.3, viscous water with $\mu = 4.0$ cp is injected. In Example 3.4, adsorbing polymer solution with $\mu = 4.0$ cp is injected. In Example 3.5, low-tension liquid with $\mu = 4.0$ cp is injected. Assume that the example results can be applied to the linedrive pattern in the reservoir of Problem 3.1. Calculate and plot the volume of oil recovered (barrels) vs. water injected (barrels) out to a WOR of 50. Compare the results from the different cases.

3.3 A waterflood is planned in a reservoir that has the fluid and rock properties in **Table 3.16.** Relative permeability relationships are given by Eqs. 3.14 through 3.16. Determine the following using frontal-advance relationships.
   1. Flood-front saturation.
   2. Oil recovery (PV) at breakthrough.

Property	Value
ϕ	0.097
S_{iw}	0.20
S_{or}	0.38
μ_w (cp)	0.65
μ_o (cp)	1.25
α_1	1.0
α_2	0.5
m	3
n	1.5

Table 3.16—Rock and fluid properties, Problem 3.3.

3. Oil recovery (PV) when the F_{wo} = 10, 25, and 50.
4. Saturation profile, $S_w - r_D$, when $t_D = 0.25$.
5. Location of the boundary between the injected water and the resident water when t_D = 0.25.
6. Prepare a plot of x_D vs. t_D for these saturations: S_{wf}, S_w = 0.45, 0.5, 0.54, and S_{wh}.

3.4 A reservoir contains an oil (°API = 25) that has a viscosity of 20 cp at reservoir temperature. The operator would like to know whether the reservoir is a candidate for waterflooding. **Table 3.17** presents the rock and fluid properties for a reservoir to be waterflooded, and Eqs. 3.14 through 3.16 represent relative permeabilities for oil and water. Using frontal-advance relationships, determine the following.
1. Oil recovery (in PV) as a function of PVs, t_D, of water injected for $0 < t_D < 1.5$. Plot oil recovery vs. t_D.
2. Oil recovery (PV) as a function of PVs of fluid injected for $0 < t_D < 1.5$ assuming that a nonadsorbing, viscous fluid was injected. The viscosity of the viscous fluid is 10 cp.

3.5 A chemical flood is being considered for the reservoir described in Problem 3.4. This chemical viscosifies the water but does not reduce the ROS. Sufficient chemical will be added to the injected water to increase the viscosity to 10 cp. Relative permeability relationships are not affected by the chemical; however, the chemical adsorbs on the reservoir rock and is retained irreversibly. Laboratory measurements indicate that adsorption follows an equilibrium isotherm similar to that in Fig. 3.18. The slope of the equilibrium isotherm, D_i, has a value of 1.00 at the injected concentration.
1. Determine the saturation profile when 0.25 PV of chemical has been injected in the reservoir. Plot the saturation vs. x_D. Identify the regions on the saturation profile occupied by the resident brine and the viscous chemical.
2. Estimate the oil recovery (PV) as a function of PVs of chemical injected for the interval $0 < t_D < 1.5$.
3. Determine the incremental oil resulting from the chemical flood.

3.6 A low-tension waterflood is to be evaluated for the reservoir described in Problem 3.3. Laboratory data indicate that the ROS is reduced to 0.15 when Chemical A is added to the injected water and is injected continuously. Chemical A adsorbs on the rock and is retained irreversibly. The amount of adsorption is not known but will be studied by varying the value of D_i. Laboratory data indicate that the relative permeability relationships are altered by the presence of Chemical A but can be represented by Eqs. 3.14 through 3.16 with the values of α_1, α_2, m, and n given in **Table 3.18.** The viscosity of the low-tension fluid is 3.0 cp. Determine the following for the case when $D_i = 0.50$.
1. Breakthrough saturation and the WOR at breakthrough.
2. Water saturation in the oil bank, S_{w1}.
3. Water saturation at the chemical flood front, S^*_{w3}.

Property	Value
ϕ	0.20
S_{iw}	0.30
S_{or}	0.20
μ_w (cp)	1.0
μ_o (cp)	20.0
α_1	1.0
α_2	0.72
m	3.72
n	2.56

Table 3.17—Rock and fluid properties, Problem 3.4.

Property	Value
α_1	1.0
α_2	1.0
m	1.5
n	1.2

Table 3.18—Relative permeability parameters, Problem 3.6—Chemical A.

4. Saturation profile when $t_D = 0.2$.
5. PVs of chemical injected when the chemical flood front reaches the end of the linear system.
6. Oil recovery as a function of PVs of chemical injected.
7. Incremental oil from low-tension flooding for values of $t_D \le 1/f'_{sw3}$.

3.7 Assume that the chemical flood in Problem 3.6 is conducted late in the life of the waterflood where the oil saturation in the swept region is approximately at S_{or}. Determine the following.
1. Water saturation of the oil bank, S_{wl}, and the WOR at breakthrough of the oil bank.
2. Water saturation at the chemical flood front, S^*_{w3}.
3. Saturation profile when the chemical flood front is at $x_D = 0.5$.
4. PVs of chemical injected when the chemical flood front reaches the end of the linear system.
5. Oil recovery as a function of PVs of chemical injected.
6. Fractional flow of water as a function of PVs of chemical injected.
7. Incremental oil from low-tension flooding for the interval between zero and the time that the chemical flood front reaches the end of the linear system as determined in 4.

3.8 A reservoir containing a viscous oil ($\mu_o = 47.8$ cp at reservoir temperature of 125°F) is to be waterflooded. Relative permeability data were obtained on preserved reservoir core from linear displacement tests run at constant injection rate. The k_{rw}/k_{ro} data obtained from the laboratory corefloods can be correlated with Eq. 3.196 for the saturation range of $0.356 \le S_w \le 0.78$. The viscosity of water at reservoir temperature is 0.59 cp; therefore, an unfavorable mobility ratio would be expected.

$$\ln\left(\frac{k_{rw}}{k_{ro}}\right) = a + b\frac{\ln(S_w)}{S_w}, \dots\dots\dots\dots\dots\dots\dots\dots\dots\dots\dots\dots\dots\dots (3.196)$$

where $a = 3.544877$ and $b = 4.742485$. Note that k_{rw}/k_{ro} in Eq. 3.196 has a finite value when $S_w = 1 - S_{or}$.

Consider a chemical flood to which a low concentration of chemical is added to the injected water to increase the viscosity and reduce the mobility ratio. The ROS to waterflood is not affected by the chemical. Estimate the displacement performance of an ideal chemical flood by assuming adsorption losses are negligible and the viscosity of the injected solution is 10 cp at reservoir temperature. **Table 3.19** gives other reservoir properties.
1. Determine the flood-front saturation, if any.
2. Determine the saturation of the oil bank.
3. Determine the saturation of the chemical flood front and sketch the saturation profile after 0.1 PV of chemical has been injected into this reservoir.
4. Estimate the number of PVs injected when the chemical flood front just reaches the end of the linear system (i.e., t_D at $X_{D3} = 1.0$).
5. Estimate oil recovery (PV) as a function of t_D to a WOR of 25. What fraction of the oil is recovered after the chemical flood front reaches the end of the linear system?
6. Determine the incremental oil recovery as a function of t_D.

3.9 A thin reservoir zone (8 ft thick) is to be opened in several wells to production. The oil from this zone is expected to have a viscosity of 2.36 cp at reservoir temperature. Because the oil is essentially dead, it will be necessary to supplement the reservoir energy to produce oil at economical rates. A waterflood is under consideration. Relative permeability data from an adjacent interval have been correlated with water saturation by use of Eqs. 3.197 and 3.198. Log calculations indicate that the initial water saturation should be 0.136. An ROS of 0.325 was obtained on cores from a geologically similar interval.
1. Calculate the waterflood recovery (PV) as a function of PVs, t_D, of fluid injected for $0 \le t_D$ to a WOR of $F_{wo} = 50$. The viscosity of water at reservoir temperature is 0.63 cp. Relative permeability relationships are given by Eqs. 3.197 and 3.198.

Property	Value
ϕ	0.21
μ_o (cp)	47.8
μ_w (cp)	0.59
S_{iw}	0.356
S_{or}	0.22

Table 3.19—Reservoir and fluid properties, Problem 3.8.

$$k_{ro} = \left(a_o + b_o S_{wD}^{0.5}\right)^2 \quad \text{(3.197)}$$

$$\text{and } \ln k_{rw} = a_w + b_w \frac{\ln S_w}{S_w}, \quad \text{(3.198)}$$

where Eq. 3.16 gives S_{wD} and **Table 3.20** gives the coefficients a_o, b_o, a_w, and b_w for the waterflood.

2. Estimate the potential for a low-tension chemical flood if the ROS to chemical flooding is $S_{orc} = 0.1$ for the interval from $t_D = 0$ until the chemical flood front breaks through. Determine the incremental oil when the viscosity of the low-tension solution is 10 cp and the retention factor, D_i, is 0.25. Assume that Eqs. 3.198 and 3.199 represent the relative permeability relationships for the low-tension flood,

$$k_{ro} = \left(a_o + b_o S_{wD}^{0.5}\right)^2, \quad \text{(3.199)}$$

where S_{wD} is $\left(S_w - S_{iw}\right)/\left(1 - S_{orc} - S_{iw}\right)$ for the low-tension flood.

3.10 Estimate the incremental oil in Problem 3.9 when the low-tension flood begins at the end of the waterflood when the ROS is assumed to be 0.325. Compare your estimation of oil recovery vs. PVs of chemical injected with the results of Problem 3.9.

3.11 Show that the erf(ζ) solution (Eq. 3.148) given for the convection-diffusion partial-differential equation does indeed satisfy the equation.

3.12 Consider Example 3.9. Assume that a time of 50 hours has elapsed since the inlet brine concentration was changed. At this specific point in time, calculate the concentration of NaCl (in parts per million) as a function of position x in the core.

3.13. Hall and Geffen (1957) reported an 8.5-ft mixing-zone length for the following conditions in their dispersion experiments: core length = 62 ft, v = 2.0 ft/D, unfavorable viscosity ratio, and Torpedo sandstone core. The mixing-zone length is measured between the 95 and 5% concentration levels. How does this mixing-zone length compare with that which would be predicted by the solution to the diffusivity equation for a favorable mobility ratio?

3.14. Consider a linear core that originally contains Fluid A. Starting at time zero, Fluid B is injected at a constant rate of 0.0413 ft/hr. After a fixed period of constant injection, the injected fluid is changed to Fluid C and injection is continued at the same rate. The system is shown in **Fig. 3.63.**

Fluid A is miscible with Fluid B, and Fluid C is miscible with Fluid B. Fluids A and C are not miscible. Also, if Fluids A and C are mixed with Fluid B the system can become immiscible. In particular, if the concentrations of Fluids A and C each exceed 1% when mixed with Fluid B, then the system becomes immiscible.

It is desired to inject a slug of Fluid B of sufficient size such that the system will stay miscible throughout the displacement process from the entrance to the end at $x = 50$ ft [i.e., until the concentration of Fluid B is 50% at $x = 50$ ft (mean front position)].

Fluids A, B, and C have similar physical properties. The dispersion coefficient is assumed to be a constant value of $K = 3.1 \times 10^{-4}$ ft²/hr for all three fluids.

With this information, estimate the minimum slug size of Fluid B so that miscibility will not be lost. Define the size in terms of the total time of injection of Fluid B or in terms of the distance between the mean positions of the front and back of the slug.

3.15. Consider flow in a linear sandstone core. An experiment is run in which Chemical A is displaced by Chemical B, with which it is miscible. A step change in concentration is made at the inlet ($x = 0$) at $t = 0$. The velocity of the displacement is 6.6×10^{-5} in./sec.

When the mean position of the front has traveled 96 in. ($t = 1.461 \times 10^6$ seconds), the width of the mixing zone is 19.5 in. The width is defined as the distance between the points where the concentrations of Chemical B are 0.02 and 0.98 (normalized concentrations).

a_o	1.000677
b_o	−0.99625
a_w	−0.55874189
b_w	1.2140403
a_{o1}	1.281874
b_{o1}	−9.42322

Table 3.20—Coefficients for relative permeability correlations, Eqs. 3.197 through 3.199.

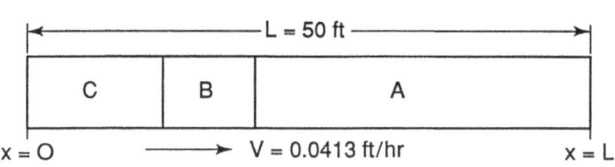

Mean Positions of Fluids Shown

Fig. 3.63—Injection of a slug, Problem 3.14.

1. Calculate the dispersion coefficient. Assume that the dispersion process fits the theory perfectly.
2. Assume that a second experiment is run in the same system but at a velocity three times the original. Calculate the width of the mixing zone when the mean position of the front has traveled 96 in. (same position as in the original experiment).
3. If a third experiment were run, but at a velocity of 10 times the original velocity, would the width of the mixing zone be larger than determined in Part 2? Again assume that the width is to be measured when the mean position of the front has traveled 96 in. Explain the reasons for your answer.

Additional information (applies to all questions):

1. Assume that the Perkins and Johnston (1963) correlation, Eq. 3.164, applies and that the contribution of the molecular diffusion term is completely negligible.
2. The following parameter values may be used if required: $F_I d_p = 0.24$ in. and $D = 1.0 \times 10^{-5}$ cm^2/s.

3.16. Assume that a propane miscible displacement is to be carried out in a field test. The minimum propane slug size required in the test to ensure miscibility throughout the displacement process needs to be estimated. Pressure and temperature conditions are such that propane is miscible with the crude oil and with methane, which will be used to displace the propane slug.

For simplicity, assume that the reservoir is linear and is 500 ft long (distance between injection and production well). $\phi = 20\%$ and $S_{or} = 37\%$.

Determine the minimum slug size, assuming that the porous medium is homogeneous. For this purpose, define the minimum slug size to be that which will just maintain a zone of propane of at least 95% of original composition between the oil/propane zone and the trailing propane/methane zone. Give your answer as a percent of the total reservoir hydrocarbon pore volume.

Additional data: $\mu_o = 3.0$ cp, $\mu_{c3} = 0.10$ cp, and $\mu_{c1} = 0.015$ cp.

3.17. A miscible displacement is to be carried out in a given reservoir with Chemical B. Chemical B is miscible with the crude oil at all compositions.

The flood is to be conducted in a line drive. The injection well and an observation well are 400 ft apart. The flood-front velocity is to be 1.5 ft/D. Assume that the reservoir rock is similar to Berea sandstone. The viscosity of Chemical B is approximately the same as the crude oil viscosity (slightly favorable mobility ratio).

1. If a slug of Chemical B is injected as a step change in concentration, how long will it take for breakthrough at the observation well? Assume that a 1% concentration constitutes breakthrough.
2. What is the width (in feet) of the mixed zone at this time? Assume that the width is the distance between the points where the Chemical B concentrations are 99 and 1%.

Additional data: D (Chemical B in oil) $= 5.43 \times 10^{-5}$ ft^2/hr, $1/(F_R\phi) \approx 0.5$, and $F_I d_p = 0.15$ in. Use the Perkins and Johnston (1963) correlation, Eq. 3.164, to estimate K_I for this calculation.

3.18. A laboratory experiment on dispersion is conducted on an outcrop core. In the experiment, water is displaced by water containing alcohol as a tracer. The frontal velocity for the experiment is 5.0 ft/D. The measured dispersion coefficient, K_I, is 1.63×10^{-3} ft^2/hr.

1. Estimate the dispersion coefficient at a velocity of 1.0 ft/D.
2. Estimate the dispersion coefficient at a velocity of 250 ft/D.
3. For the dispersion coefficient of Part 1, calculate the width of the dispersion zone at 48 hours after the start of injection. Assume that the width of the dispersion zone is defined to be the distance between points where the concentrations are 0.01 and 0.99.

Additional data: D (alcohol in water) $= 4.26 \times 10^{-5}$ ft^2/hr and $F_R\phi = 1.2$.

3.19. The dispersion data shown in **Fig. 3.64** were taken in the laboratory. **Table 3.21** shows the run conditions.

1. Determine the dispersion coefficient for this run.
2. Plot the experimental data against the analytical solution to show the "goodness" of fit between theory and data. Plot concentration of the core effluent vs. time.
3. How well do these results compare with those obtained with the Perkins and Johnston (1963) correlation? That is, how does your answer to Part 1 compare to the dispersion coefficient value calculated with an appropriate correlation presented by Perkins and Johnston?

3.20. A dispersion experiment is conducted in the laboratory in a linear core, as shown in

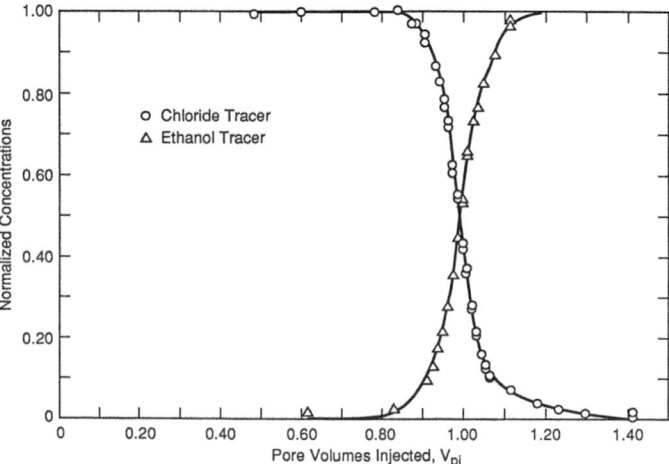

Fig. 3.64—Glycerol displacement run in Berea sandstone core, Problem 3.19.

Experimental Conditions	
Flow rate (cm³/h)	6
Flow velocity (in./D)	13.8
Collection rate (test tube/hr)	1
Permeability before run (md)	303
Average temperature (°F)	77
Temperature range (°F)	70 to 85
Average pressure in core (psig)	40*
Core Dimensions	
Length (in.)	48.07
Diameter (in.)	1.981
Area (in.²)	3.081
PV (cm³)	500.53
Porosity	0.2062
Displacement Fluids	
Fluid initially in core: brine, 30 g NaCl/L, 650 g glycerol/L	
Viscosity at 22.5 s⁻¹ and 77°F (cp)	7.4
Density at 77°F (g/cm³)	1.1551
Fluid injected: brine, 20g NaCl/L, 700 g glycerol/L, 2.5 vol% ethanol	
Viscosity at 22.5 s⁻¹ and 77°F (cp)	11.3
Density at 77°F (g/cm³)	1.1609

*Pressure caused by effluent valve restricting flow.

Table 3.21—Glycerol displacement run, Problem 3.19, Berea sandstone core.

Fig. 3.65. The core properties are $\phi = 0.19$ and $V_p = 455.7$ cm³. The liquid is injected at a constant rate. The linear (interstitial) velocity is 1.39×10^{-4} in./sec. In the dispersion experiment, a dye is suddenly injected with the liquid and the concentration of the dye, C_B, in the exit stream is measured continuously. **Table 3.22** presents the resulting data, also plotted in **Figs. 3.66 and 3.67.**

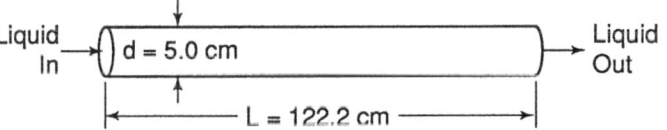

Fig. 3.65—Dispersion experiment, Problem 3.20.

Assume that the experiment is to be repeated at a higher velocity. The new liquid velocity will be 3.33×10^{-3} in./sec. This is 24 times greater than the first velocity. The dispersion coefficient is assumed to be related to velocity by

$$K_\ell = \alpha v^{1.2},$$

where K_ℓ = dispersion coefficient, v = velocity, and α = constant to be determined.

For this experiment, calculate and plot the predicted normalized exit concentration vs. V_i and U.

Volume Injected, V_i (cm³)	Brigham Parameter,* U (cm³/²)	Normalized Concentration of Exit Stream, C_b
410	2.257	0.05
420	1.742	0.10
440	0.748	0.29
455.7	0	0.50
470	−0.660	0.68
490	−1.550	0.87
500	−1.981	0.92

*Equal to $\left(V_p - V_i\right)/\sqrt{V_i}$.

Table 3.22—Dispersion data, Problem 3.20.

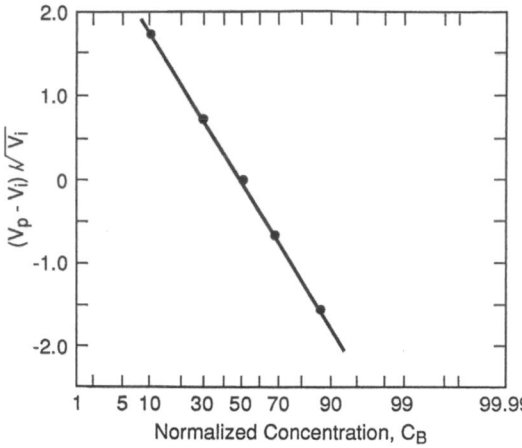

Fig. 3.66—Probability-paper plot of data, Problem 3.20.

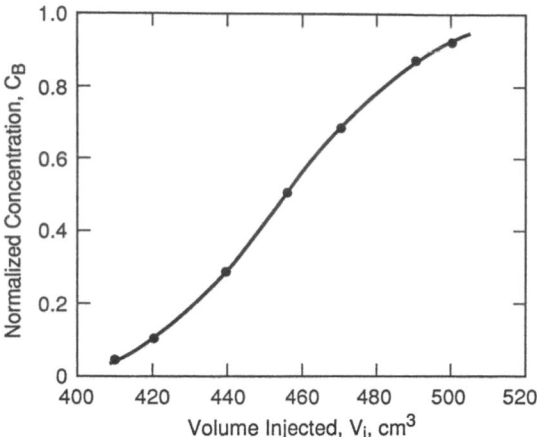

Fig. 3.67—Experimental dispersion data, Problem 3.20.

3.21 Consider the miscible displacement of an oil by a solvent in a homogeneous porous medium. For the conditions of each case specified below, calculate the normalized concentration of the solvent in the effluent as a function of number of PVs of solvent injected. Use the Koval K-factor method.
1. $\mu_o = 5.0$ cp and $\mu_s = 1.0$ cp.
2. $\mu_o = 5.0$ cp and $\mu_s = 0.2$ cp.
Plot the results as C vs. V_{pi}.

3.22 Consider the miscible displacement of an oil by a solvent in a homogeneous porous medium. $\mu_o = \mu_s = 1.0$ cp; the dispersion coefficient, K_ℓ, is 5×10^{-5} cm²/s; $V_p = 550$ cm³; core length, L, is 48 in.; and the time to inject 1 PV, t, is 3.5×10^5 seconds.

Calculate the normalized concentration of solvent in the effluent as a function of the number of PVs injected. Plot the results as C vs. V_{pi} on the same graph as Problem 3.21.

3.23 Brigham et al. (1961) present miscible displacement dispersion data for cases in which the viscosity ratio is unfavorable. Some of the data are given in Figs. 3.55 and 3.56. Compare the data for $\mu_A/\mu_B > 1.0$ with the calculated concentrations obtained from the Koval model for viscous fingering.

3.24 Consider the model by Collins (1961) model that describes the growth of viscous fingers for an unfavorable mobility ratio (Eq. 3.173). Calculate the increase in ε as a function of time when $\varepsilon_0 = 0.001$ m, $L = 5.0$ m, $x_f = 1.0$ m, $M = 1.2$, $\phi = 0.20$, $\mu_s = 0.5$ cp $= 0.5 \times 10^{-3}$ Pa·s, $\Delta P = 11\,310$ kPa, and $K = 200$ md $= 0.197 \times 10^{-12}$ m². Refer to Fig. 3.60.

Nomenclature

a	=	constant in relative permeability correlation in Eq. 3.196
a_o	=	constant in relative permeability correlation in Eq. 3.197
a_{o1}	=	constant in relative permeability correlation in Eq. 3.199
a_w	=	constant in relative permeability correlation in Eq. 3.198
A	=	cross-sectional area, L², ft²
A_i	=	amount of Species i retained at C_{ii}, m/m, μg/g rock
b	=	constant in relative permeability correlation in Eq. 3.196
b_o	=	constant in relative permeability correlation in Eq. 3.197
b_{o1}	=	constant in relative permeability correlation in Eq. 3.199

b_w	=	constant in relative permeability correlation in Eq. 3.198
B	=	FVF, L³/L³, bbl/STB
C	=	constant defined in Eq. 3.174
C_B	=	actual or normalized concentration of Component B, m/L³ or fraction, lbm/ft³ or fraction
C_B^*	=	actual concentration of Component B, m/L³, lbm/ft³
C_{Bi}	=	injected or maximum concentration of Component B, m/L³, lbm/ft³
C_{B0}	=	initial concentration of Component B in the system, m/L³, lbm/ft³
C_i	=	concentration of Species i, m/L³, lbm/ft³
C_{ii}	=	injected concentration of Species i, m/L³, lbm/ft³
$\hat{C}_{ii}$	=	retained Species i or polymer at concentration C_{i0} in fluid, m/L³, lbm/PV

C_{i0}	=	initial concentration of Species i, m/L^3, lbm/ft^3
d_p	=	average particle diameter, L, ft
D	=	molecular diffusion coefficient, L^2/t, ft^2/hr
D_a	=	apparent diffusion coefficient in porous medium, L^2/t, ft^2/hr
D_{BA}	=	molecular diffusion coefficient, L^2/t, ft^2/hr
D_i	=	ratio of injected to retained chemical concentrations, fraction
E	=	Koval viscosity ratio factor
f	=	fractional flow, L^3/L^3, volume fraction
f_D	=	fractional flow of displacing phase
f_{iw}	=	fractional flow of water at interstitial water saturation, $f_{iw}=0$ in this text, L^3/L^3, volume fraction
f_s'	=	derivative of fractional flow of solvent with respect to saturation or volume fraction of solvent
f_w'	=	derivative of water fractional-flow curve with respect to S_w
f_w^*	=	fractional flow of viscous or chemical solution having a different fractional-flow curve than water, L^3/L^3, volume fraction
f_{wb}	–	fractional flow of water at interface between interstitial and injected water at saturation S_{wb}, L^3/L^3, volume fraction
f_{wf}	=	fractional flow of water at flood front at saturation S_{wf}, L^3/L^3, volume fraction
f_{wf}'	=	derivative of water fractional-flow curve at saturation S_{wf}
f_{w1}	=	fractional flow of water in oil bank preceding viscous or chemical flood saturation discontinuity, L^3/L^3, volume fraction
f_{w2}	=	fractional flow of water corresponding to saturation S_{w2}, L^3/L^3, volume fraction
f_{w3}	=	fractional flow of water corresponding to saturation S_{w3}, L^3/L^3, volume fraction
f_{w1}'	=	derivative of water fractional-flow curve at saturation S_{w1}
f_{w2}'	=	derivative of water fractional-flow curve at saturation S_{w2}
f_{w3}^*	=	fractional flow of water at saturation S_{w3}^*, saturation of viscous or chemical shock front, L^3/L^3, volume fraction
$f_{w3}'^*$	=	derivative of fractional-flow curve with respect to S_w^* at saturation, saturation of chemical or viscous-shock front
F_I	=	inhomogeneity factor
F_R	=	formation electrical resistivity, R/R
F_{wo}	=	WOR, L^3/L^3, bbl/bbl
H	=	Koval heterogeneity factor

k	=	permeability, L^2, md
k_D	=	effective permeability of displacing phase, L^2, md
k_r	=	relative permeability, dimensionless
K	=	Koval factor = HE
K_c	=	convective contribution to longitudinal dispersion coefficient, L^2/t, ft^2/hr
K_ℓ	=	longitudinal dispersion coefficient, L^2/t, ft^2/hr
K_t	=	transverse dispersion coefficient, L^2/t, ft^2/hr
L	=	length, L, ft
m	=	exponent in Eq. 3.14
m_{begin}	=	amount of Fluid B in differential volume element at beginning of time period Δt, m, lbm
m_{Bx}	=	rate of Fluid B diffusing across plane in x-direction, m/t, lbm/hr
m_{end}	=	amount of Fluid B in differential volume element at end of time period Δt, m, lbm
m_{in}	=	amount of Fluid B flowing into element, Eq. 3.142
m_{out}	=	amount of Fluid B flowing out of element, Eq. 3.142
M	=	mobility ratio, dimensionless
n	▬	exponent in Eq. 3.15
N_p	=	oil displaced by a flood, L^3, STB
p	=	pressure, m/Lt^2, psi
Δp	=	pressure drop, m/Lt^2, psi
p_l	=	pressure at downstream position, m/Lt^2, psi
p_o	=	pressure at upstream position, m/Lt^2, psi
q	=	flow rate, L^3/t, B/D
q_t	=	total injection rate, L^3/t, B/D
Q_i	=	pore volumes injected, L^3/L^3
R	=	electrical resistivity of porous medium saturated with liquid that conducts electricity
R'	=	electrical resistivity of liquid in porous medium
S_{iw}	=	interstitial water saturation, L^3/L^3, volume fraction
S_{or}	=	waterflood ROS, L^3/L^3, volume fraction
S_{orc}	=	low-tension flood ROS, L^3/L^3, volume fraction
S_s	=	saturation of solvent, L^3/L^3, volume fraction
S_w	=	water saturation, L^3/L^3, volume fraction
S_w^*	=	water saturation in region where flowing fluid system is viscous fluid or chemical solution, L^3/L^3, volume fraction
$\overline{S}_w$	=	average water saturation of the linear system, L^3/L^3, volume fraction

S_{wb} = water saturation at miscible boundary between injected and interstitial water, L^3/L^3, volume fraction

$\overline{S}_{wb}$ = average water saturation in region behind miscible boundary between injected and interstitial water, L^3/L^3, volume fraction

S_{wD} = dimensionless water saturation

S_{wf} = flood-front saturation, L^3/L^3, volume fraction

S_{wr} = water saturation in water bank at oil-bank shock, L^3/L^3, volume fraction

S_{w1} = water saturation in oil bank preceding chemical flood saturation discontinuity, L^3/L^3, volume fraction

S_{w2} = water saturation at end of linear system, L^3/L^3, volume fraction

S_{w3} = water saturation on μ_w fractional-flow curve in developing viscous or chemical shock when chemical flood begins after waterflood, L^3/L^3, volume fraction

S_{w3}^* = water saturation of shock front associated with/f_w^*, fractional flow of viscous or chemical solution, L^3/L^3, volume fraction

t = time, t, days

t^* = time to inject 1 PV, t, days

t_D = dimensionless time or PVs injected

t_{Dbt} = dimensionless time when flood front breaks through at end of linear system

t_{Df} = dimensionless time of flood-front breakthrough

t_{Do}^* = dimensionless time when S_w^* evolves from saturation profile

t_{D1} = dimensionless time when S_{w1} breaks through at end of linear system

t_{D2} = dimensionless time when water saturation S_{w2} arrives at end of linear system

Δt = time increment, t, seconds

u = Darcy velocity, L/t, ft/D

U = parameter in solution of diffusivity equation, $\left(V_p - V\right)/\sqrt{V}$, $L^{3/2}$, $ft^{3/2}$

U_{10} = value of U evaluated when $C_B = 0.1$, $L^{3/2}$, $ft^{3/2}$

U_{90} = value of U evaluated when $C_B = 0.9$, $L^{3/2}$, $ft^{3/2}$

v = pore velocity, u/ϕ, L/t, ft/D

$\overline{v}$ = average velocity of fluid flowing in circular conduit, L/t, ft/D

v_{ci} = specific velocity of concentration shock

v_D = specific velocity, dimensionless

v_{Do} = specific velocity of oil bank

v_{Dv} = specific velocity of viscous shock

$\overline{v}_{Dv}$ = average specific velocity of viscous shock

v_{Dv}^* = specific velocity associated with saturation S_w in flood

v_{D1} = specific velocity of water saturation S_{w1}

v_{D3}^* = specific velocity of S_{w3}^*, chemical solution saturation discontinuity

v_{max} = maximum velocity of fluid flowing in circular conduit, L/t, ft/D

v_{wf} = velocity of water saturation S_{wf} L/t, ft/D

v_1 = velocity of resident water phase, L/t, ft/D

v_3 = velocity of saturation S_{w3}^*, Eq. 3.38, L/t, ft/D

v_3^* = velocity of shock front associated with viscous or chemical flood, L/t, ft/D

V_i = volume of fluid injected, L^3, ft^3

V_p = volume in 1 PV, $A\phi L$, L^3, ft^3

V_{pi} = number of PVs of solvent injected for S_s to reach the end of porous medium

$(V_{pi})_{bt}$ = PVs of solvent injected at time of solvent breakthrough, L^3/L^3, volume fraction

$(V_{pi})_{comp}$ = PVs of solvent injected at time that all oil has been recovered in miscible displacement, L^3/L^3, volume fraction

x = linear distance, L, ft

x_b = location of miscible boundary between injected and interstitial water, L, ft

x_D = dimensionless distance in x-direction

x_{Db} = dimensionless location of miscible boundary between injected and interstitial water

x_{Df} = dimensionless distance to flood front, L, ft

x_{Di} = dimensionless location of chemical tracer front when adsorption occurs

x_{Do} = dimensionless location of oil-bank saturation discontinuity for low-tension flood at ROS

x_{Dw} = dimensionless distance to point of saturation S_w

x_{D1} = dimensionless location of leading edge of viscous or chemical shock

x_{D3} = shock dimensionless distance to location of saturation S_{w3}^*

x_f = location of flood-front saturation in x-direction, L, ft

x_{ft} = location of flood-front saturation at time t, L, ft

x_{Ss} = location of solvent saturation S_s in linear system

x_{Sw} = location of water saturation S_w in x-direction, L, ft

x_{wf}	=	location of the flood front, L, ft
x_1, x_2, x_3	=	positions in x-direction
x_{10}	=	distance to position where concentration is 10% of injected concentration, L, ft
x_{90}	=	distance to position where concentration is 90% of injected concentration, L, ft
α	=	constant in Problem 3.10
α_1, α_2	=	constants in Eq. 3.15
ε	=	length parameter
ε_0	=	initial length of perturbation
ζ	=	dummy integration variable
μ	=	viscosity, m/Lt, cp
μ^*	=	viscosity of injected viscous water or chemical solution, m/Lt, cp
μ_A	=	viscosity of Fluid A, m/Lt, cp

μ_B	=	viscosity of Fluid B, m/Lt, cp
μ_D	=	viscosity of displacing phase, m/Lt, cp
μ_1	=	viscosity of Fluid 1, m/Lt, cp
μ_2	=	viscosity of Fluid 2, m/Lt, cp
ρ	=	fluid density, m/L^3, lbm/ft^3
ρ_{gr}	=	grain density, m/L^3, lbm/ft^3
ϕ	=	porosity, fraction
ϕ_i	=	porosity occupied by Species i, fraction

Subscripts

g	=	gas phase
o	=	oil phase
s	=	solvent
w	=	water phase

References

Aris, R. 1959. The Longitudinal Diffusion Coefficient in Flow Through a Tube With Stagnant Pockets. *Chem. Eng. Sci.* **11** (3): 194–98. https://doi.org/10.1016/0009-2509(59)80086-5.

Aris, R. and Amundson, N. R. 1957. Some Remarks on Longitudinal Mixing and Diffusion in Fixed Beds. *AIChE J.* **3** (2): 280–82. https://doi.org/10.1002/aic.690030226.

Baldwin, D. E. Jr. 1966. Prediction of Tracer Performance in a Five-Spot Pattern. *J Pet Technol* **18** (4): 513–517. SPE-1230-PA. https://doi.org/10.2118/1230-PA.

Blackwell, R. J., Rayne, J. R., and Terry, W. M. 1959. Factors Influencing the Efficiency of Miscible Displacement. In *Petroleum Transactions*, AIME, Vol. 217, 1–8. SPE-1131-G.

Brigham, W. E. and Smith, D. H. 1965. Prediction of Tracer Behavior in Five-Spot Flow. Presented at the SPE Conference on Production Research and Engineering, Tulsa, 3–4 May. SPE-1130-MS. https://doi.org/10.2118/1130-MS.

Brigham, W. E., Reed, P. W., and Dew, J. N. 1961. Experiments on Mixing During Miscible Displacement in Porous Media. *SPE J.* **1** (1): 1–8. SPE-1430-G. https://doi.org/10.2118/1430-G.

Carman, P. C. 1956. *Flow of Gases Through Porous Media.* New York City: Reinhold Press.

Claridge, E. L. and Bonder, P. L. 1974. A Graphical Method for Calculating Linear Displacement With Mass Transfer and Continuously Changing Mobilities. *SPE J.* **14** (6): 609–18. SPE 4673-PA. https://doi.org/10.2118/4673-PA.

Coats, K. H. and Smith, B. D. 1964. Dead-End Pore Volume and Dispersion in Porous Media. *SPE J.* **4** (1): 73–84. SPE-647-PA. https://doi.org/10.2118/647-PA.

Collins, R. E. 1961. *Flow of Fluids Through Porous Materials*, Vol. 196. Tulsa: PennWell Publishing Co.

Craig, F. F. Jr. 1971. *The Reservoir Engineering Aspects of Waterflooding*, Vol. 3. Richardson, Texas: Monograph Series, SPE.

Grane, F. E. and Gardner, G. H. F. 1961. Measurements of Transverse Dispersion in Granular Media. *J. Chem. Eng. Data* **6** (2): 283–287. https://doi.org/10.1021/je60010a031.

Habermann, R. 1960. The Efficiency of Miscible Displacement as a Function of Mobility Ratio. In *Petroleum Transactions*, AIME, Vol. 219, 264–272. SPE-1540-G.

Hall, H. N. and Geffen, T. M. 1957. Laboratory Study of Solvent Flooding. In *Petroleum Transactions*, AIME, Vol. 210, 48–57. SPE-711-G.

Handy, L. L. 1959. An Evaluation of Diffusion Effects in Miscible Displacement. *J Pet Technol* **11** (3): 61–63. SPE-1130-G. https://doi.org/10.2118/1130-G.

Keulemans, A. I. M. 1957. *Gas Chromatography.* New York City: Reinhold Press.

Koval, E. J. 1963. A Method for Predicting the Performance of Unstable Miscible Displacement in Heterogeneous Media. *SPE J.* **3** (2): 145–154. SPE-450-PA. https://doi.org/10.2118/450-PA.

Patton, J. T., Coats, K. H., and Colegrove, G. T. 1971. Prediction of Polymer Flood Performance. *SPE J.* **11** (1): 72–84. SPE-2546-PA. https://doi.org/10.2118/2546-PA.

Perkins, T. K. and Johnston, O. C. 1963. A Review of Diffusion and Dispersion in Porous Media. *SPE J.* **3** (1): 70–81. SPE-480-PA. https://doi.org/10.2118/480-PA.

Pope, G. A. 1980. The Application of Fractional Flow Theory to Enhanced Oil Recovery. *SPE J.* **20** (3): 191–205. SPE-7660-PA. https://doi.org/10.2118/7660-PA.

Raimondi, P., Gardner, G. H. F., and Petrick, C. B. 1959. Effects of Pore Structure and Molecular Diffusion on the Mixing of Miscible Liquids in Porous Media. Preprint 43, presented at the AIChE/SPE Joint Symposium, San Francisco, 6–9 December.

Talash, A. W. 1976. Experimental and Calculated Relative Permeability Data for Systems Containing Tension Additives. Presented at the SPE Improved Oil Recovery Symposium, Tulsa, 22–24 March. SPE-5810-MS. https://doi.org/10.2118/5810-MS.

Taylor, G. I. 1953. Dispersion of Soluble Water in Solvent Flowing Slowly Through a Tube. *Proc,* Royal Society, Vol. 219, 186.

Thomas, G. H., Countryman, G. R., and Fatt, I. 1963. Miscible Displacement in a Multiphase System. *SPE J.* **3** (3): 189–196. SPE-538-PA. https://doi.org/10.2118/538-PA.

van Meurs, P. and van der Poel, C. 1958. A Theoretical Description of Water-Drive Processes Involving Viscous Fingering. In *Petroleum Transactions,* AIME, Vol. 213, 103–112. SPE-931-G.

van Meurs, P. V. 1957. The Use of Transparent Three-Dimensional Models for Studying the Mechanism of Flow Processes in Oil Reservoirs. In *Petroleum Transactions,* AIME, Vol. 210, 295–301. SPE-678-G.

Welge, H. J. 1952. A Simplified Method for Computing Oil Recovery by Gas or Water Drive. *J Pet Technol* **4** (4): 91–98. SPE-124-G. https://doi.org/10.2118/124-G.

Willhite, G. P. 1986. *Waterflooding,* Vol. 3. Richardson, Texas: Textbook Series, SPE.

SI Metric Conversion Factors

bbl	×	1.589 873	E–01	=	m^3
cp	×	1.0*	E–03	=	Pa·s
dynes/cm	×	1.0*	E+00	=	mN/m
ft	×	3.048*	E–01	=	m
ft^2	×	9.290 304*	E–02	=	m^2
ft^3	×	2.831 685	E–02	=	m^3
°F		(°F–32)/1.8		=	°C
in.	×	2.54*	E+00	=	cm
$in.^2$	×	6.451 6*	E+00	=	cm^2
md	×	9.869 233	E–04	=	μm^2
psi	×	6.894 757	E+00	=	kPa

*Conversion factor is exact.

Chapter 4

Macroscopic Displacement of Fluids in a Reservoir

4.1 Introduction

Oil recovery in any displacement process depends on the volume of reservoir contacted by the injected fluid. A quantitative measure of this contact is the volumetric displacement (sweep) efficiency, E_V. Volumetric sweep is a macroscopic efficiency defined as the fraction of reservoir (or project) pore volume (PV) invaded by the injected fluid, or stated another way, the fraction of PV that has been contacted or affected by the injected fluid. Clearly, E_V is a function of time in a displacement process.

Overall displacement efficiency in a process can be viewed conceptually as a product of the volumetric sweep, E_V, and the microscopic efficiency, E_D, defined in Chapter 1.

$$E = E_V E_D, \dots\dots\dots\dots\dots\dots\dots\dots\dots\dots\dots\dots\dots\dots\dots\dots\dots\dots (4.1)$$

where E = overall hydrocarbon displacement efficiency, the volume of hydrocarbon displaced divided by the volume of hydrocarbon in place at the start of the process measured at the same conditions of pressure and temperature; E_V = macroscopic (volumetric) displacement efficiency; and E_n = microscopic (volumetric) hydrocarbon displacement efficiency.

This chapter concerns the volumetric sweep efficiency of displacement processes. Four factors generally control how much of a reservoir will be contacted by a displacement process: (1) the properties of the injected fluids, (2) the properties of the displaced fluids, (3) the properties and geological characteristics of the reservoir rock, and (4) the geometry of the injection and production well pattern. Over a number of years, a significant understanding of how these factors affect volumetric sweep has developed, and this is the principal focus of this chapter.

Parameters that control volumetric sweep and how those parameters affect displacement performance are described. General principles relating to volumetric sweep that must be considered in planning enhanced-oil-recovery (EOR) processes are discussed. This chapter is not intended to prescribe specific design procedures, but to provide an overview of the principles that underlie most design approaches.

4.2 Volumetric Displacement Efficiency and Material Balance

Volumetric displacement, or sweep efficiency, is often used to estimate oil recovery by use of material-balance concepts. For example, consider a displacement process that reduces the initial oil saturation to a residual saturation in the region contacted by the displacing fluid. If the process is assumed to be piston-like, the oil displaced is given by

$$N_p = \left(\frac{S_{o1}}{B_{o1}} - \frac{S_{o2}}{B_{o2}} \right) V_p E_V, \dots\dots\dots\dots\dots\dots\dots\dots\dots\dots\dots\dots\dots\dots (4.2)$$

where N_p = oil displaced, S_{o1} = oil saturation at the beginning of the displacement process, S_{o2} = residual oil saturation (ROS) at the end of the process in the volume of reservoir contacted by the displacing fluid, B_{o1} = formation volume factor (FVF) at initial conditions, B_{o2} = FVF at the end of the process, and V_p = reservoir PV.

Dividing both sides of Eq. 4.2 by the oil in place (OIP) at the start of the process gives the fractional recovery as a product of the microscopic and macroscopic displacement efficiencies.

$$N_p / N_1 = E_D E_V, \dots\dots\dots\dots\dots\dots\dots\dots\dots\dots\dots\dots\dots\dots\dots\dots\dots (4.3)$$

where N_1 = OIP at the beginning of the displacement process. If displacement performance data are available, Eq. 4.2 also can be used to estimate volumetric sweep. For example, if waterflood recovery data are available, the equation can be rearranged to solve for E_V:

$$E_V = \frac{N_p}{V_p \left(\dfrac{S_{o1}}{B_{o1}} - \dfrac{S_{o2}}{B_{o2}} \right)}, \dotfill (4.4)$$

where N_p = oil produced in the waterflood.

In a waterflood of a reservoir that has been producing under solution-gas drive, resaturation of volume occupied by trapped gas may occur (Willhite 1986). In this case, the material-balance equation, solved in terms of E_V and assuming piston-like displacement, is given by

$$E_V = \frac{1 - \left(1 - \dfrac{N_p}{N} \right) \dfrac{B_o}{B_{oi}}}{1 - \dfrac{S_{orw}}{S_{oi}}}, \dotfill (4.5)$$

where N_p = total oil produced by primary recovery plus waterflooding, N = original OIP, S_{oi} = initial oil saturation, S_{orw} = ROS in the swept system at the end of the waterflood, B_{oi} = initial FVF, and B_o = FVF at conditions of the waterflood.

Volumetric sweep efficiencies of waterfloods range from only a few percent to almost 100%. Because of the assumptions made in the derivation of Eqs. 4.4 and 4.5, it cannot be determined whether the entire vertical cross section of the reservoir has been swept, leaving some parts of the reservoir or the edges unswept, or whether one or more vertical zones were thoroughly swept while other zones received little injected fluid.

4.3 Volumetric Displacement Efficiency Expressed as the Product of Areal and Vertical Displacement Efficiencies

Volumetric sweep efficiency can be considered conceptually as the product of the areal and vertical sweep efficiencies. Consider a reservoir that has uniform porosity, thickness, and hydrocarbon saturation, but that consists of several layers. For a displacement process conducted in the reservoir, E_V can be expressed as

$$E_V = E_A E_I, \dotfill (4.6)$$

where E_A = areal sweep (displacement) efficiency in an idealized or model reservoir, area swept divided by total reservoir area; and E_I = vertical sweep (displacement) efficiency, pore space invaded by the injected fluid divided by the pore space enclosed in all layers behind the location of the leading edge (leading areal location) of the front. All efficiencies are expressed as fractions. E_I is the volumetric sweep efficiency of the region confined by the largest areal sweep efficiency in the system.

For a real reservoir, in which porosity, thickness, and hydrocarbon saturation vary areally, E_A is replaced by a pattern sweep efficiency, E_P:

$$E_V = E_p E_I, \dotfill (4.7)$$

where E_P = pattern sweep (displacement) efficiency, hydrocarbon pore space enclosed behind the injected-fluid front divided by total hydrocarbon pore space in the pattern or reservoir.

In essence, E_P is an areal sweep efficiency that has been corrected for variations in thickness, porosity, and saturation. In either case, overall hydrocarbon recovery efficiency in a displacement process may be expressed as

$$E = E_p E_I E_D. \dotfill (4.8)$$

While Eqs. 4.6 and 4.7 are conceptually correct, application to practical problems generally is difficult. To use Eqs. 4.6 and 4.7 to determine E_V, independent estimates of E_P (or E_A) and E_I are required. Such estimates are difficult to obtain because, for displacements in 3D systems, E_A and E_I typically are not independent. In the absence of vertical effects, areal sweep can be approximated from correlations developed from scaled physical models or mathematical models. Additionally, there are methods to estimate vertical sweep efficiency, but the availability of model studies is more limited. In practice, E_V usually is determined by application of appropriate correlations or mathematical models that are based on 3D systems and not by independent calculation of E_A and E_I. Nonetheless, it is useful to consider E_V as the product of E_A and E_I to understand the parameters that affect volumetric sweep. Thus, E_A and E_I and the parameters that affect these efficiencies are considered separately in this chapter.

Studies have shown that both E_A and E_I are strongly influenced by the mobility ratio in either a miscible or immiscible displacement process. Therefore, the concept of mobility ratio will be reviewed before the discussion of the factors that affect E_A and E_I.

4.4 Definition and Discussion of Mobility Ratio

Mobility of a fluid phase flowing in a porous medium is defined on the basis of the Darcy equation:

$$u_i = -\left(k_i / \mu_i \right)\left(dp/dx \right), \dotfill (4.9)$$

where u_i = superficial (Darcy) velocity of Phase i, k_i = effective permeability of Phase i, μ_i = viscosity of Phase i, p = pressure, and x = length, and, where appropriate, units should be used with each term. For single-phase flow, k_i is the absolute permeability of the porous medium. For multiphase flow, it is the effective permeability of the flowing phase and is, therefore, a function of the saturation of the phase. Mobility of the fluid phase, λ_i, is given by

$$\lambda_i = (k_i/\mu_i). \quad\dotfill \quad (4.10)$$

In calculations involving a displacement process, a useful concept is the mobility ratio, M, of the displacing and displaced fluid phases.

$$M = \lambda_D/\lambda_d, \quad\dotfill \quad (4.11)$$

where λ_D = mobility of the displacing fluid phase and λ_d = mobility of the displaced fluid phase. Note that M is a dimensionless quantity. The mobility ratio is an extremely important parameter in any displacement process. It affects both areal and vertical sweep, with sweep decreasing as M increases for a given volume of fluid injected. Further, M affects the stability of a displacement process, with flow becoming unstable (nonuniform displacement front) when $M > 1.0$. This unstable flow is called viscous fingering and was discussed in Chapter 3. Because a value of M relative to unity is so significant, a value > 1.0 is referred to as an unfavorable mobility ratio. Conversely, a value < 1.0 is a favorable mobility ratio.

M can be defined in a variety of ways, depending on the flow conditions in a specific process. For example, when one solvent is displacing a second solvent with which the first solvent is completely miscible and only one phase is flowing,

$$M = \mu_d/\mu_D. \quad\dotfill \quad (4.12)$$

Eq. 4.12 holds because the permeability to each solvent is the absolute porous-medium permeability.

For a waterflood in which piston-like flow is assumed, with only water flowing behind the front and only oil flowing ahead of the front,

$$M = \left(\frac{k_{rw}}{\mu_w}\right)_{S_{or}} \left(\frac{\mu_o}{k_{ro}}\right)_{S_{iw}} . \quad\dotfill \quad (4.13)$$

Relative permeabilities, k_{rw} and k_{ro}, are measured at ROS and interstitial (immobile) water saturation, respectively, because of the assumption of piston-like displacement.

Generally, two (or more) phases are flowing both ahead of and behind a displacement front. In this case, phase saturations may change with position and time. Craig (1971) defined a mobility ratio for waterflooding (or any immiscible displacement process) as

$$M_{\bar{S}} = \left(\frac{k_{rD}}{\mu_D}\right)_{\bar{S}_D} \left(\frac{\mu_d}{k_{rd}}\right)_{\bar{S}_d} . \quad\dotfill \quad (4.14)$$

$$\text{or } M_{\bar{S}} = \left(\lambda_D\right)_{\bar{S}_D} \big/ \left(\lambda_d\right)_{\bar{S}_d} , \quad\dotfill \quad (4.15)$$

where $\left(\lambda_D\right)_{\bar{S}_D}$ = mobility of the displacing phase measured at the average displacing-phase saturation at breakthrough of that phase and $\left(\lambda_D\right)_{\bar{S}_D}$ = mobility of the displaced-phase measured at the average saturation ahead of the displacement front.

Yet another definition of mobility ratio is based on total mobility, λ_t, ahead of and behind the displacement front:

$$M_t = (\lambda_{tD})_{\bar{S}_D} / (\lambda_{td})_{\bar{S}_d} . \quad\dotfill \quad (4.16)$$

Here, M_t is the mobility ratio based on total mobilities, which are given by

$$\left(\lambda_{tD}\right)_{\bar{S}_D} = \sum_i \left(\frac{k_i}{\mu_i}\right)_D . \quad\dotfill \quad (4.17)$$

That is, $\left(\lambda_{tD}\right)_{\bar{S}_D}$ is the sum of mobilities of all phases flowing behind the displacement front measured at the average saturation behind the front. An analogous definition holds for $\left(\lambda_{td}\right)_{\bar{S}_d}$. Stalkup (1983) recommended this definition "... for characterizing mobility ratio between an oil bank and the solvent displacing the oil bank when mobile water is present in either region."

Stalkup also points out that in many oil recovery processes there will be more than one displacement front. For example, many EOR processes involve the injection of multiple slugs of different fluids. The flow behavior of any specific displacement front is affected not only by the mobilities of the fluids immediately ahead of and behind that front, but also by the mobilities of fluids in regions around the other fronts. For this more general case, there is no uniquely defined mobility ratio that allows prediction of such important parameters as sweep efficiency.

As pointed out, there are a number of different definitions of mobility ratio. Clearly, caution must be expressed in the application of this concept. In general, when empirical correlations based on mobility ratio are applied, the user must ensure that the mobility ratio is defined in a manner consistent with the definition used by the author of the correlation.

4.5 Areal Displacement Efficiency

As discussed in Section 4.3, expressing volumetric sweep as a product of areal and vertical sweep efficiencies is useful because it allows correlations that are based on independent studies of the different efficiencies (i.e., vertical or areal sweeps) to be applied. One must recognize, however, that this approach is an approximation because the areal sweep is not truly independent of vertical sweep and vice-versa. Nonetheless, a number of studies of areal sweep based on the use of physical models have been reported. Such studies have yielded areal displacement efficiencies, E_A, as functions of various parameters for idealized or homogeneous reservoirs. For a real reservoir with variable properties, E_A must be replaced with pattern displacement efficiency, E_P.

4.5.1 Parameters Affecting Areal Displacement Efficiency. Areal displacement efficiency is controlled by four main factors: injection/production well pattern, reservoir permeability heterogeneity, mobility ratio, and relative importance of gravity and viscous forces. Studies conducted with physical models have focused principally on the first two factors. These factors involve parameters that can be controlled in the laboratory and that have a significant effect on areal sweep. Typically, the models used have been thin, which minimizes the role of gravity segregation and allows areal sweep to be determined independent of vertical sweep (i.e., vertical sweep is essentially 100%). Results and correlations for E_A on the basis of physical models are described in Section 4.5.2.

A number of different injection/production well patterns have been used in reservoir displacement processes. **Fig. 4.1** (Craig 1971) shows several of these patterns. The five-spot pattern has been the most commonly used for both waterflooding and EOR processes, and most of the results in this book are for this pattern. The principles and effects illustrated in the studies based on five-spot patterns generally are applicable to other patterns, even though the specific correlations resulting from the studies are pattern dependent. Craig (1971) presents correlations for several pattern types and gives a summary of references to modeling studies for different patterns.

Permeability heterogeneity often has a marked effect on areal sweep. This effect may be quite different from reservoir to reservoir, however, and thus it is difficult to develop generalized correlations. Anisotropy in permeability has been considered and some of these results are shown in Section 4.5.2. Willhite (1986) presents a lengthy discussion of the role of geology on waterflood performance.

4.5.2 Correlations of Areal Displacement Efficiency Based on Modeling Studies. Physical models have been widely used for studies of areal sweep efficiency. Specific well patterns are represented by model configurations that are appropriate fractions of the patterns being investigated (e.g., one-quarter or one-eighth of a five-spot pattern with an injection well on one corner and a production well on the opposite corner of the model). Gravity effects are eliminated by adjusting the densities of the different fluids or by using thin models so that gravity override or underride is minimized.

Because laboratory models are small in size compared with a reservoir, physical scaling laws should be used in constructing and operating the models. These scaling laws are described elsewhere (Craig et al. 1955). Basically, scaling requires that injection rates, interfacial tension (IFT) values, and physical size be appropriately adjusted in the models to represent property conditions in a displacement process in a reservoir. These scaling laws have been followed, to the extent possible, in some modeling work, but have not been considered carefully in other work. Claridge (1972) gives a good summary and extensive bibliography for these modeling studies. Results from unscaled models, while not as quantitatively reliable as those from scaled models, are nonetheless useful and show trends in behavior and performance of displacement processes.

Most of the reported modeling studies (Slobod and Caudle 1952; Dyes et al. 1954; Habermann 1960; Kimbler et al. 1964) are based on the use of fluid systems that are completely miscible. In some cases, however, immiscible fluid systems have been used (Craig et al. 1955; Douglas et al. 1959). The fronts, or interfaces, between displaced and displacing fluids were monitored by use of dyed fluids that can be photographed or by X-ray shadowgraph techniques. In the latter approach, one of the fluids is tagged with a substance that absorbs X-rays, while the other fluid is relatively invisible to X-rays. In either case, the experimental approach is to monitor the displacement front as a function of the volume of displacing fluid injected. Usually, material-balance measurements also are conducted during a displacement to corroborate the front-tracking results.

The two parameters of primary concern in studies of the type described are the injection/production well pattern geometry and the mobility ratio. Typical results follow.

Correlations Based on Miscible Fluids, Five-Spot Pattern. **Figs. 4.2 and 4.3** (Habermann 1960) show fluid fronts at different points in a flood for different mobility ratios. These results are based on photographs taken during displacements of one colored liquid by a second, miscible colored liquid in a scaled model. The viscosity ratio was varied in different floods and, because only one phase was present, M is given by Eq. 4.12.

Fig. 4.4 gives areal sweep efficiency at breakthrough of the displacing fluid as a function of M. The upper curve is based on a measurement of area invaded shown in the photographs obtained during a flood, while the lower curve is based on a material-balance calculation made from the known volume of injected liquid and assuming piston-like displacement. Because of mixing along the liquid/liquid interface, the area measurement yields a slightly larger areal sweep value than the material-balance result. As indicated in Fig. 4.4, the difference in the two curves is a measure of the mixing that occurred at the interface.

As seen, E_A at breakthrough is a strong function of M. At $M = 1.0$, areal sweep is approximately 70%. E_A increases slightly at smaller, favorable mobility ratios and decreases very sharply as M is increased. The poorer performance at larger M values

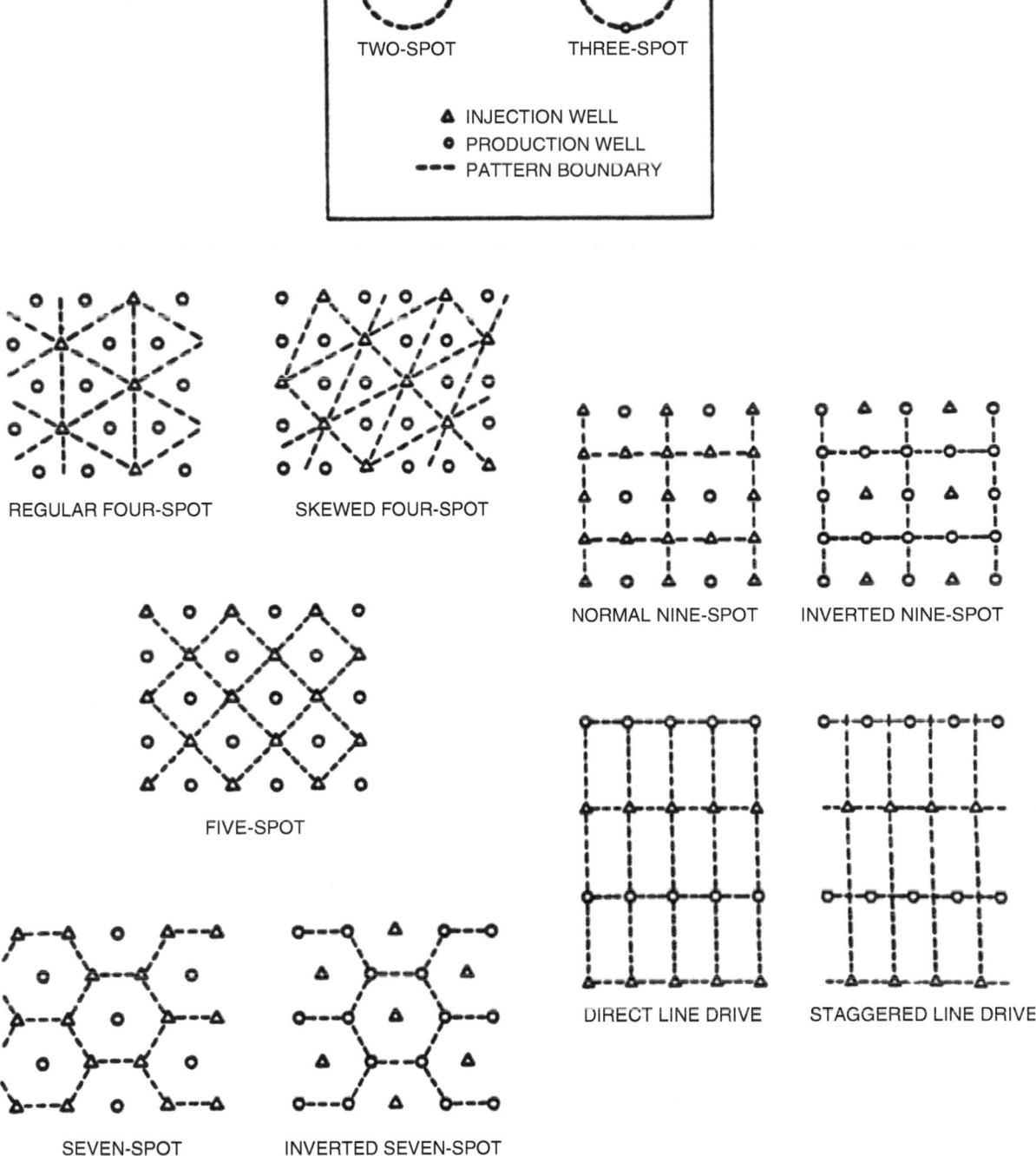

Fig. 4.1—Flooding patterns (Craig 1971).

occur for two reasons. First, viscous fingering occurs at $M > 1.0$, and becomes more pronounced as M increases. Second, for $M > 1.0$, the geometry contributes to the early breakthrough. Because of the geometry, the smallest flow resistance, and therefore the largest flow velocity, is along the center line connecting the injection and production wells. Fluid flowing along this line breaks through first in a homogeneous reservoir. When $M > 1.0$, the resident fluid of a certain mobility is replaced by the injected fluid, which has a higher mobility. Because the largest flow is along the centerline path, the resistance along that path is reduced more than along any other flow path. Proportionately, a larger volume of injected fluid flows along the centerline, reducing the resistance even more. The result is that the larger the value of M, the earlier breakthrough occurs. The argument is very similar to the one used in the analysis of viscous fingering presented in Section 3.9.

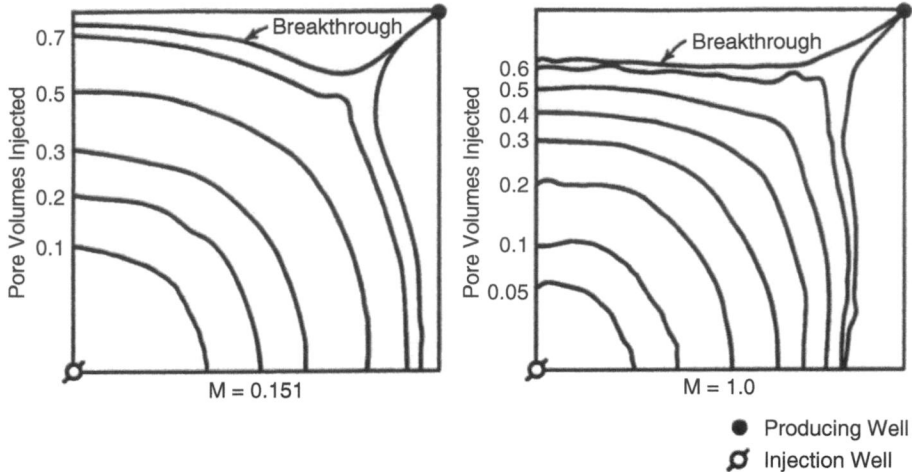

Fig. 4.2—Miscible displacement in one-quarter of a five-spot pattern at mobility ratios ≤ 1.0 (Habermann 1960).

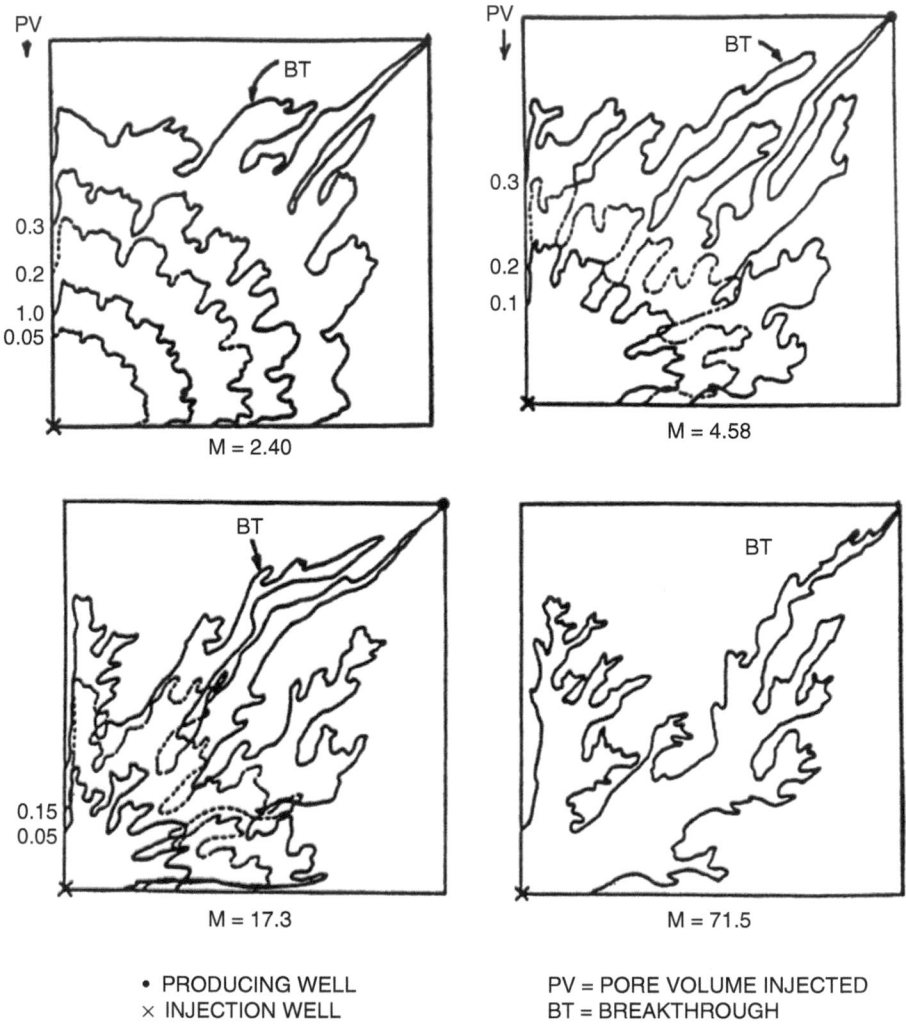

Fig. 4.3—Miscible displacement in one-quarter of a five-spot pattern at mobility ratios > 1.0, viscous fingering (Habermann 1960).

Habermann (1960) also presented values of E_A as a function of dimensionless PVs injected, V_i/V_p, after breakthrough, as shown in **Fig. 4.5.** Results are given for $M = 0.216$ (favorable) to 71.5 (unfavorable). As seen, flood performance becomes poorer as M increases. For unfavorable mobility ratios, not only is areal sweep at breakthrough poorer, but performance continues to be poor as the flood progresses. As Fig. 4.5 shows, at larger M values, areal sweep will not approach 100% for a reasonable injection volume even in idealized laboratory models.

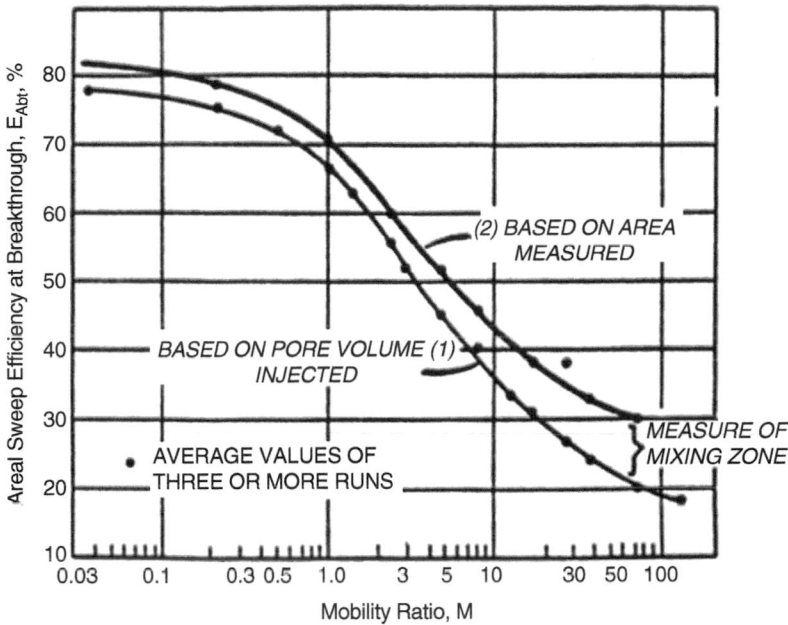

Fig. 4.4—Areal sweep efficiency at breakthrough as a function of mobility ratio; miscible fluid displacement, five-spot pattern (Habermann 1960).

Fig. 4.6 compares the work of Habermann (1960) with that of two other reported studies (Dyes et al. 1954; Caudle and Witte 1959). These investigations were also performed with miscible liquids; however, the fronts were tracked with the X-ray shadowgraph technique. Agreement between investigators is good in the region around unit mobility ratio, but is relatively poor at small and large M values. The differences are not understood and must be attributed to differences in experimental systems and scaling. The variation in reported results indicates that caution should be used in application of the correlations to determine specific values of areal sweep. Trends in sweep as a function of M are consistent among different investigators.

Correlations Based on Miscible Fluids, Other Patterns. Numerous modeling studies for patterns other than a five-spot have been reported. Craig (1971) gives a summary listing of references. As an example of such studies, **Fig. 4.7** (Kimbler et al. 1964) shows one reported result of areal sweep as a function of mobility ratio for one-eighth of a nine-spot pattern. This study was conducted with miscible liquids and the X-ray shadowgraph method. Shown is E_A at breakthrough and for various volumes of fluid injected after breakthrough. V_i/V_p is the number of dimensionless PVs injected.

In a nine-spot pattern, behavior is affected by the relative producing rates of corner wells (those wells on the corner of the pattern) and side wells (those wells in the middle of each side of the pattern). Fig. 4.1 shows a nine-spot pattern. The effect of the production rates described by Kimbler et al. (1964) shows that sweep efficiency can be improved for a given volume injected by properly balancing the production rates of side and corner wells. Comparison of Fig. 4.7 with Figs. 4.4 and 4.5 illustrates the effect of geometry for the one particular case of producing-rate ratios shown in Fig. 4.7.

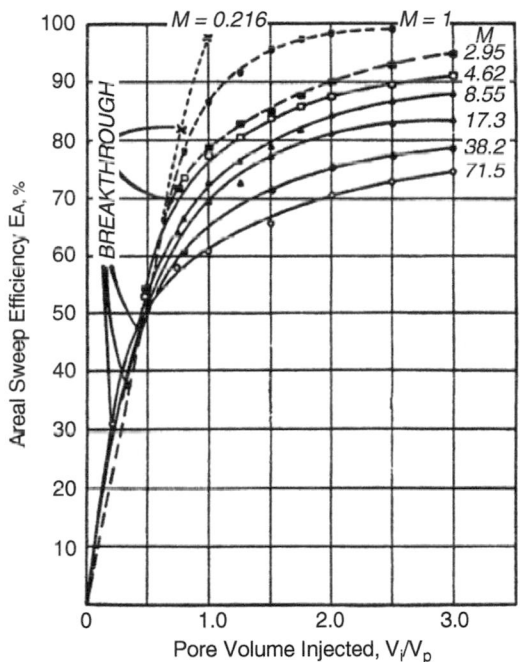

Fig. 4.5—Areal sweep efficiency after breakthrough as a function of mobility ratio and PVs injected; miscible fluid displacement, five-spot pattern (Habermann 1960).

Correlations Based on Immiscible Fluids, Five-Spot Pattern. Craig et al. (1955) conducted an experimental study of areal displacement efficiency for immiscible fluids consisting of oil, gas, and water. A laboratory model was scaled to field conditions by adjusting flow rates and IFTs between the fluids. The study was conducted in consolidated sandstone cores, and fronts were monitored with the X-ray shadowgraph technique.

Fig. 4.8 compares areal sweep efficiency at breakthrough as a function of mobility ratio to the data of Dyes et al. (1954), which were obtained with miscible fluids. In a displacement process involving immiscible fluids, the definition of mobility ratio that should be used and the saturation at which the mobility ratio should be computed are questionable. The problem occurs because saturation generally will vary with position ahead of and behind the displacement front.

In a waterflood in which oil ahead of the front flows at connate water saturation, the oil saturation ahead of the front may be assumed to be constant. Behind the front, however, water saturation varies with position and oil and water flow simultaneously. Craig

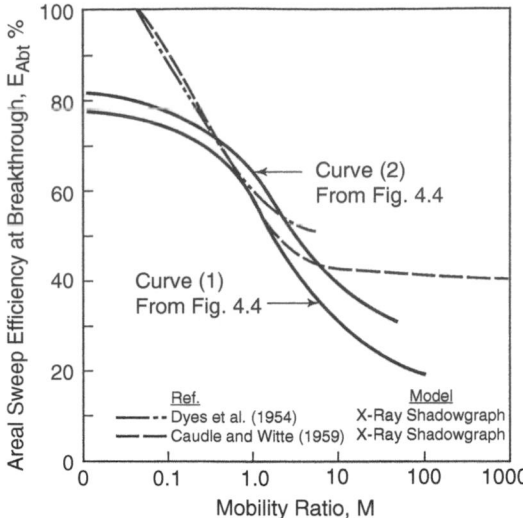

Fig. 4.6—Comparison of results from laboratory five-spot models; miscible systems (Habermann 1960).

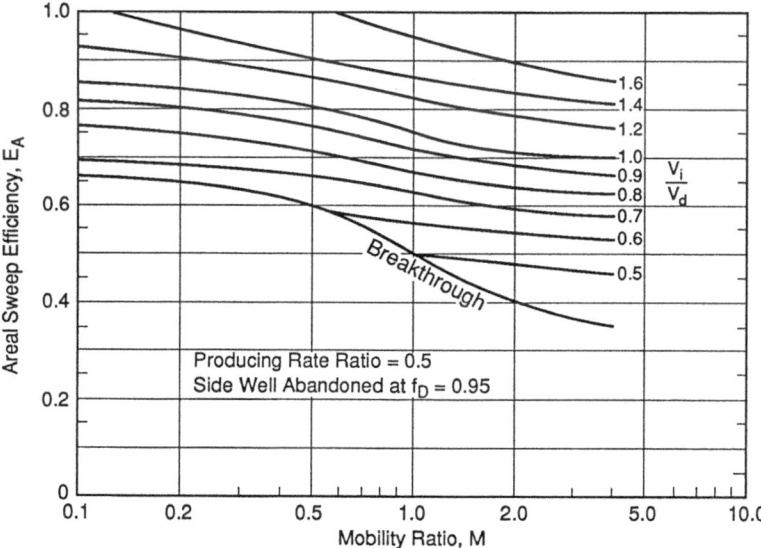

Fig. 4.7—Areal sweep efficiency as a function of mobility ratio; miscible fluid displacement, nine-spot pattern. Producing rate ratio is the production rate of a corner well ratioed to the production rate of a side well (Kimbler et al. 1964).

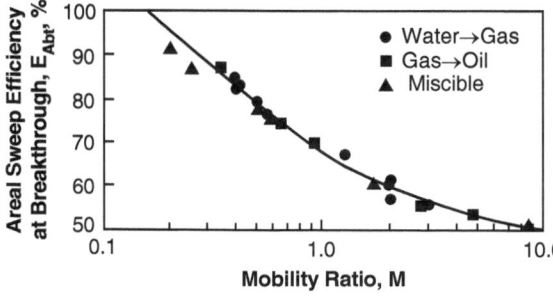

Fig. 4.8—Areal sweep efficiency at breakthrough as a function of mobility ratio; immiscible fluid displacement, five-spot pattern (Craig et al. 1955; Craig 1971).

et al. (1955) determined that the best data correlation was obtained when the mobility ratio was based on water relative permeability at the average saturation behind the front at water breakthrough. The mobility ratio in Fig. 4.8 was defined in this manner.

4.5.3 Prediction of Areal Displacement Performance on the Basis of Modeling Studies. For some investigations in which physical models were used, sufficient results have been reported to allow performance calculations to be conducted. This approach is illustrated in this section. Performance calculations based on physical model results should be viewed as approximations at best. In all cases, models used to obtain the data have been small in size and the scaling of these models to field dimensions is questionable. As noted previously, significantly different results have been obtained by different investigators, especially at relatively large or small mobility ratios. Additionally, Dougherty and Sheldon (1964) have presented a mathematical argument indicating that results based on physical model studies with miscible fluids are misleading at $M > 2$ to 3. They attribute this to the effect of mixing on mobility ratio at the fluid/fluid interface. Despite the inherent problems with modeling studies, however, performance calculations based on the studies are useful in that they illustrate the manner in which important parameters affect flood behavior.

Prediction Based on Piston-Like Displacement. Caudle and Witte (1959) published results from laboratory models of a five-spot pattern in which displacements were conducted with miscible liquids. Scaling rules were not closely followed. Their results are presented in a form that can be used to make performance calculations. Because miscible liquids were used in the experiments, however, the performance calculations are restricted to those floods in which piston-like displacement is a reasonable assumption (i.e., the displacing phase flows only in the swept region and the displaced phase flows in the unswept region). No production of displaced phase occurs from the region behind the front.

Figs. 4.9 through 4.11 show data from the experiments. In Fig. 4.9, E_A is given as a function of M (defined as in Eq. 4.12) for various values of injected PVs. The ratio V_i/V_{pd} is a dimensionless injection volume defined as injected volume divided by displaceable PV, V_{pd}. For a waterflood, V_{pd} is given by

$$V_{pd} = Ah\phi\left(S_{oi} - S_{or}\right). \quad \ldots \ldots (4.18)$$

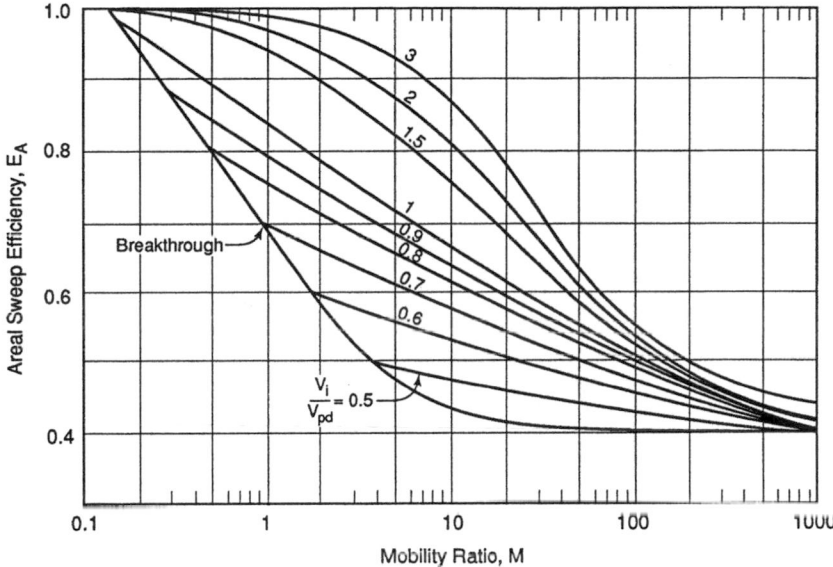

Fig. 4.9—Areal sweep efficiency as a function of mobility ratio and injected volume, five-spot pattern (Caudle and Witte 1959).

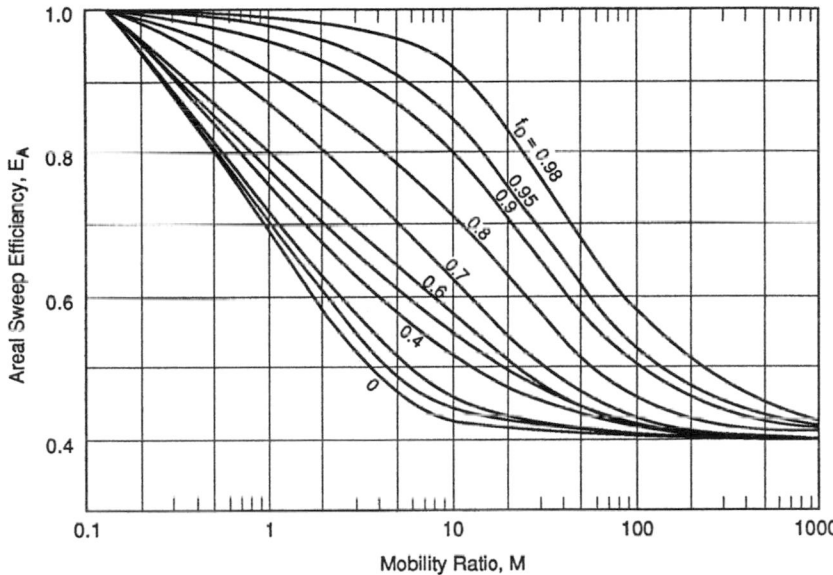

Fig. 4.10—Areal sweep efficiency as a function of mobility ratio and fractional flow at displacing phase, five-spot pattern (Caudle and Witte 1959).

Fig. 4.10 gives E_A as a function of M for different values of the fractional flow of the displacing phase, f_D, at the producing well. As previously stated, production of the displaced phase is assumed to come entirely from the unswept region of the pattern.

Finally, Fig. 4.11 presents the conductance ratio, γ, as a function of M for various values of E_A, but only for values of M between 0.1 and 10. Conductance is defined as injection rate divided by the pressure drop across the pattern, $q/\Delta p$. At any mobility ratio other than $M = 1.0$, conductance will change as the displacement process proceeds. For a favorable mobility ratio, conductance will decrease as the area swept, E_A, increases. The opposite will occur for unfavorable M values. The conductance ratio, shown in Fig. 4.11, is the conductance at any point of progress in the flood divided by the conductance at that same point for a displacement in which the mobility ratio is unity (referenced to the displaced phase).

By combining Figs. 4.9 through 4.11, performance calculations can be performed. Areal sweep, as a function of volume injected, is available from Fig. 4.9. Fractional production of either phase can be determined with Fig. 4.10. Rate of injection may be determined as a function of E_A from Fig. 4.11. To apply Fig. 4.11, however, it is also necessary to use the appropriate expression for initial injection rate. This is given by Craig (1971) for a five-spot pattern using parameters for the displaced phase:

$$i = \frac{0.001538 k k_{rd} h \Delta p}{\mu_d \left(\log \dfrac{d}{r_w} - 0.2688 \right)}, \quad\quad\quad\quad\quad\quad\quad\quad\quad\quad\quad\quad\quad\quad\quad\quad (4.19)$$

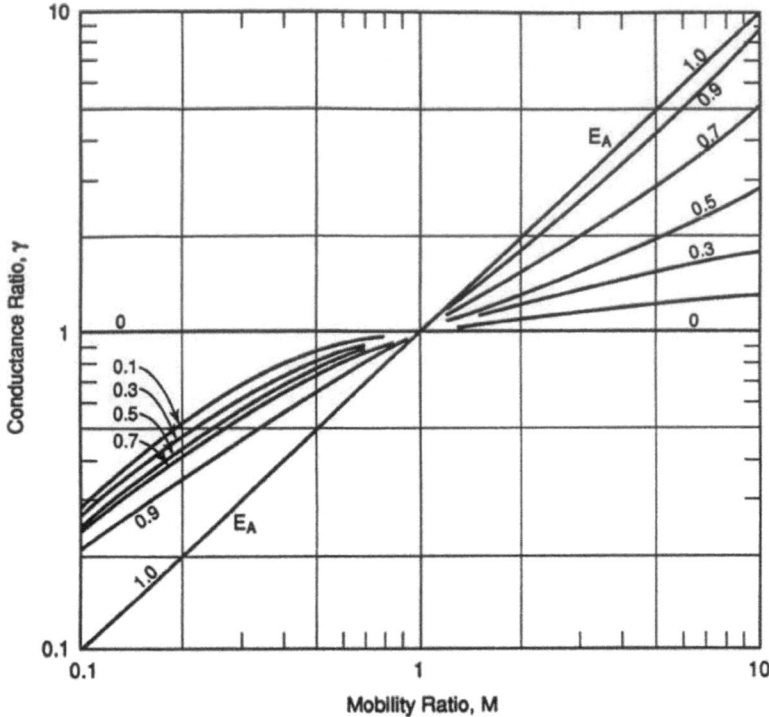

Fig. 4.11—Conductance ratio as a function of mobility ratio and areal sweep, five-spot pattern (Caudle and Witte 1959).

where i = injection rate at start of a displacement process, B/D; k = absolute rock permeability, md; k_{rd} = relative permeability of displaced phase; h = reservoir thickness, ft; Δp = pressure drop, psi; μ_d = viscosity of displaced phase, cp; d = distance measured between injection and production wells, ft; and r_w = wellbore radius, ft.

Eq. 4.19 is applicable where the displaced phase is the only phase flowing or where the mobility ratio is unity. At complete floodout (100% areal sweep), Eq. 4.19 is applicable, but with displaced-phase properties replaced by the properties of the displacing phase $(k_{rD}$ and $\gamma_D)$. The displacing-phase rate at complete floodout may also be obtained by multiplying the result of Eq. 4.19 by M. Craig (1971) presents corresponding equations for other flooding patterns. To determine flood progress as a function of time, a step-wise procedure is required because injection rate will change with time (other than for floods at unit mobility ratio). At any point in the flood, the flow rate is given by

$$q = i\gamma, \dots\dots\dots\dots\dots\dots\dots\dots\dots\dots\dots\dots\dots\dots\dots\dots\dots\dots\dots\dots\dots\dots\dots\dots\dots\dots\dots\dots\dots\dots\dots\dots (4.20)$$

where q = total flow rate at any specific time in the displacement process, and i = injection rate at the start of the displacement process as given by Eq. 4.19. Because fluids are assumed to be incompressible, the total flow rate, q, at any time is equal to the injection rate, which is equal to the production rate at that same point in time.

Example 4.1—Performance Calculations Based on Physical Modeling Results. A waterflood is conducted in a five-spot pattern for which the pattern area is 20 acres. Reservoir properties are

h	=	20 ft
ϕ	=	0.20
S_{oi}	=	0.80
S_{or}	=	0.25
μ_o	=	10 cp
μ_w	=	1.0 cp
B_o	=	1.0 RB/STB
k	=	50 md
k_{rw}	=	0.27 (at ROS)
k_{ro}	=	0.94 (at interstitial water saturation)
Δp	=	1,250 psi
r_w	=	0.5 ft

Use the method of Caudle and Witte (1959) to calculate (1) the barrels of oil recovered at the point in time at which the producing water/oil ratio (WOR) = 20, (2) the volume of water injected at the same point, (3) the rate of water injection at the same point in time, and (4) the initial rate of water injection at the start of the waterflood.

Solution. Apply the correlations presented in Figs. 4.9 through 4.11.

1. Calculate oil recovered.

$$M = (k_{rw}/\mu_w)/(k_{ro}/\mu_o).$$

If we assume piston-like displacement, $M = (0.27/1.0 \text{ cp})(10.0 \text{ cp}/0.94) = 2.9$ and $f_D = 20/21 = 0.95$. From Fig. 4.10, $E_A = 0.94$. Oil recovered is then

$$N_p = Ah\phi\left(S_{oi} - S_{or}\right)E_A \frac{1}{5.615 \text{ ft}^3/\text{bbl}} \frac{1}{B_o}$$

$$= \left[\left(43,560 \text{ ft}^2/\text{acre} \times 20 \text{ acre}\right) \times 20 \text{ ft} \times 0.20\right.$$

$$\left. \times (0.80 - 0.25) \times 0.94\right]/\left(5.615 \text{ ft}^3/\text{bbl} \times 1.0 \text{ RB/STB}\right)$$

$$= 321,000 \text{ STB}.$$

2. Calculate total water injected. From Fig. 4.9, $V_i/V_{pd} = 2.5$ (at $E_A = 0.94$).

$$V_{pd} = V_p\left(S_{oi} - S_{or}\right)$$

$$= Ah\phi\left(S_{oi} - S_{or}\right)(1/5.615)$$

$$= (43.560 \times 20) \times 20 \times 0.20 \times \frac{(0.80 - 0.25)}{5.615}$$

$$= 341,300 \text{ bbl}.$$

$$V_i = V_{pd} \times 2.5$$

$$= 341,300 \times 2.5$$

$$= 853,300 \text{ bbl}.$$

3. Calculate water injection rate at the same point in time. From Eq. 4.19,

$$i = \frac{0.001538 k k_{ro} h \Delta p}{\mu_o\left(\log\dfrac{d}{r_w} - 0.2688\right)}$$

$$= \frac{0.001538 \times 50 \times 0.94 \times 20 \times 1250}{10\left(\log\dfrac{660}{0.5} - 0.2688\right)}$$

$$= 63.4 \text{ B/D}.$$

From Fig. 4.11, $\gamma = 2.7$. From Eq. 4.20,

$$q = i\gamma$$

$$= 63.4 \times 2.7$$

$$= 171 \text{ B/D},$$

which is the injection rate because flow is constant through the system.

4. Calculate initial water injection rate. This is the injection rate, i, calculated from Eq. 4.19 with the properties of the oil: $i = 63.4$ B/D.

Fig. 4.12 (Mahaffey et al. 1966) shows another set of miscible displacement data for a five-spot pattern. This study was conducted in a parallel-plate glass model scaled so that dispersion effects were at or near the molecular diffusion level. Differences with the Caudle and Witte (1959) data exist, particularly for the breakthrough curve at large mobility ratio values. Because of better scaling in the Mahaffey et al. (1966) work, their data are probably more representative of behavior in a reservoir. Mahaffey et al. (1966) compare their results to those of other investigators and present data for an inverted five-spot pattern.

Claridge (1972) developed an improvement of the areal sweep correlation for a five-spot pattern. Basically, he observed that the sweep efficiencies reported by Caudle and Witte (1959), which were based on unscaled models, were too high compared with scaled-model results. Their reported sweep efficiencies, however, agreed with the fractional area lying inside a curve drawn through the tips of the viscous fingers in photographs of displacements taken in more properly scaled models. The X-ray

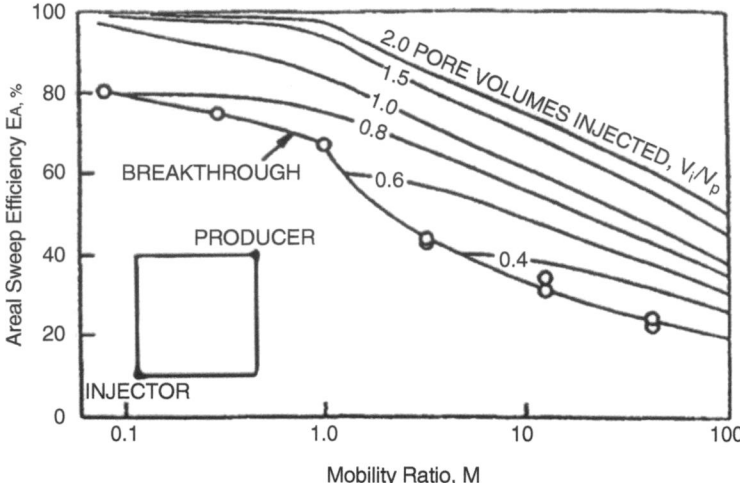

Fig. 4.12—Areal sweep efficiency for a miscible displacement in a five-spot pattern with dispersion scaled (Mahaffey et al. 1966).

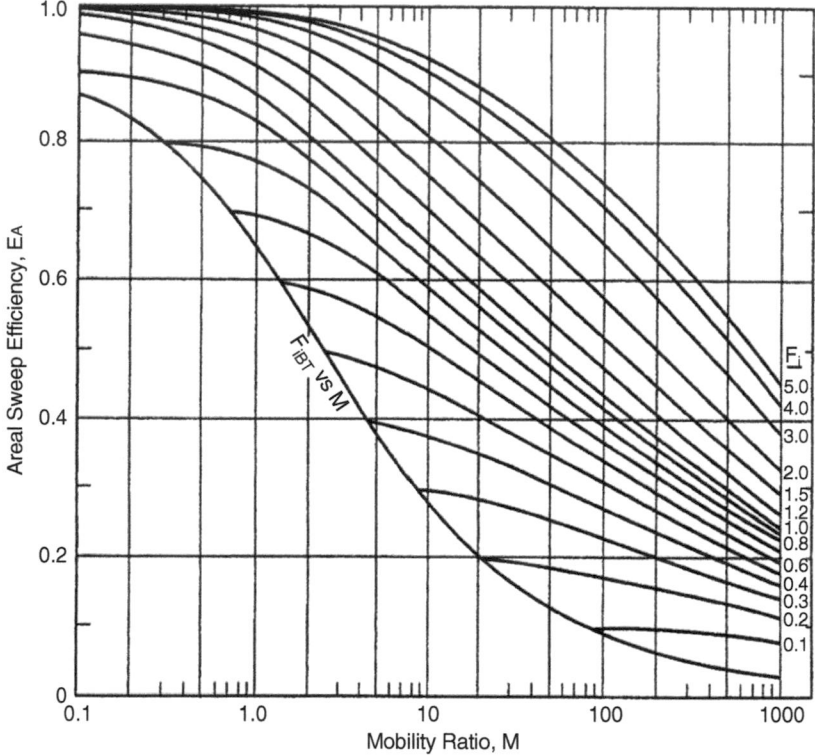

Fig. 4.13—Claridge correlation for areal sweep efficiency (Claridge 1972).

shadowgraph work of Caudle and Witte thus was assumed by Claridge to predict "the fraction of total pattern area invaded by fingers at breakthrough and subsequent levels of throughput."

Claridge (1972) developed a correlation by combining the Caudle and Witte (1959) data with the model of viscous fingering derived by Koval (1963) (see Section 3.8). Even though the Koval model was derived for a linear flow system, its application was justified on the basis of observations that it approximately fit the fingering in a five-spot geometry.

The correlation is shown in **Fig. 4.13,** where recovery efficiency is given in displaceable PVs of oil produced as a function of displaceable PVs of fluid injected, F_i, and mobility ratio, M. For a miscible flood of a porous medium that originally is at interstitial water saturation (water immobile), and for which $S_{or} = 0$, the displaceable PV is the original hydrocarbon PV. Alternatively, for a waterflood, the displaceable PV is given by Eq. 4.18.

The curves of Fig. 4.13 are represented by the following equations:

$$F_{ibt} = \left[0.9/(M+1.1) \right]^{\frac{1}{2}} \quad \dots \dots \dots \dots \dots \dots \dots \dots \dots \dots \dots \dots \dots \dots \dots \dots \dots \dots \dots \dots \dots \dots \dots \dots (4.21)$$

and

$$\frac{N_p/V_{pd} - F_{ibt}}{1.0 - N_p/V_{pd}} = \left(\frac{1.6}{F_\mu^{0.61}}\right)\left(\frac{F_i - F_{ibt}}{1.0 - F_{ibt}}\right)^{1.28/F_\mu^{0.26}}, \quad \dots \dots \dots \dots \dots \dots \dots \dots \dots \dots \dots \dots \dots \dots (4.22)$$

where F_μ is the Koval (1963) effective mobility ratio given by

$$F_\mu = \left[0.78 + 0.22\left(\frac{\mu_o}{\mu_s}\right)^{\frac{1}{4}}\right]^4 \quad \dots \dots \dots \dots \dots \dots \dots \dots \dots \dots \dots \dots \dots (4.23)$$

and where F_i = dimensionless displaceable PVs injected, V_i/V_{pd}, and F_{ibt} = dimensionless displaceable PVs of solvent injected at the time of solvent breakthrough at the producing well. The FVFs of the fluids were assumed to equal unity.

Example 4.2—Application of Claridge Correlation To Calculate Areal Displacement Efficiency. A miscible displacement is to be conducted in 20-acre five-spot pattern in a reservoir with the following properties.

h	=	20 ft
ϕ	=	0.20
S_{oi}	=	0.75
μ_o	=	2.0 cp
μ_s	=	0.04 cp
$B_o = B_S$	=	1.0 RB/STB (assumed for simplicity)

A very large solvent slug is to be injected. Calculate oil recovery out to a solvent injection of 1.0 PV. Compare this to expected recovery if the mobility ratio were unity.

Solution. Calculate M for the solvent injection:

$$M = \mu_o/\mu_s = 2.0/0.04 = 50.$$

Apply Eqs. 4.21 through 4.23 to calculate N_p:

$$F_{ibt} = \left(\frac{0.9}{M + 1.1}\right)^{\frac{1}{2}} = \left(\frac{0.9}{50 + 1.1}\right)^{\frac{1}{2}}$$
$$= 0.133 \text{ PV.}$$

$$F_\mu = \left[0.78 + 0.22\left(\frac{\mu_o}{\mu_s}\right)^{\frac{1}{4}}\right]^4$$
$$= \left[0.78 + 0.22(50)^{\frac{1}{4}}\right]^4$$
$$= 3.47.$$

$$V_{pd} = Ah\phi S_{oi}$$
$$= \left(20 \text{ acre} \times 43,560 \text{ ft}^2/\text{acre}\right)(20 \text{ ft})0.20 \times 0.75$$
$$= 2.61 \times 10^6 \text{ ft}^3$$
$$= 4.65 \times 10^5 \text{ bbl.}$$

Calculate N_p for various values of F_i with Eq. 4.22. Make the corresponding calculation for $M = \mu_o/\mu_s = 1.0$.
$\quad F_{ibt} = 0.655.$
$\quad F_\mu = 1.0 = M$ (true only for $M = 1.0$).
For both cases, before breakthrough, $N_p = F_i V_{pd}$. **Fig. 4.14** plots the results which are given in **Tables 4.1 and 4.2.**

This approach can also be used for other patterns for which areal sweep data are available. Data are available for four-spot (Caudle et al. 1968), seven-spot (Guckert 1961), nine-spot (Kimbler et al. 1964; Guckert 1961), and linedrive patterns (Dyes et al. 1954). An alternative set of data for a five-spot pattern is also available in Dyes et al. (1954). In general, calculations based on the Dyes et al. (1954) data are more favorable (i.e., yield higher efficiencies) than calculations that are based on the Caudle and Witte (1959) results.

Prediction Considering a Mobile Displaced Phase Behind the Displacement Front. In an immiscible displacement such as a waterflood or a gas/liquid displacement, there is typically two-phase flow and a saturation gradient behind the front. In this

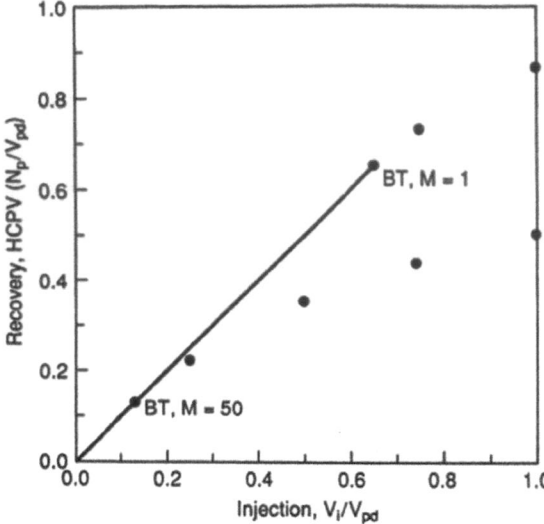

Fig. 4.14—Calculations of areal sweep, Example 4.2.

case, after breakthrough, the major part of the displaced phase production comes from the unswept region but part comes from the swept region. The methods based on modeling results for miscible systems will not yield correct results for this situation.

Craig (1971), Claridge (1972), and Willhite (1986) developed a calculation procedure to handle a displacement process in which there is a significant amount of displaced phase production from the region behind the front. The procedure is based on correlations for areal sweep and on frontal-advance theory.

Willhite (1986) describes the method as used in the prediction of waterflood performance in detail. Chapter 4 of Willhite (1986) describes the method and gives a detailed example calculation. Areal sweep correlations based on laboratory models are available for this method for five-spot patterns but generally not for other geometries.

4.5.4 Calculations of Areal Displacement Efficiency When a Directional Permeability Exists. As discussed by Craig (1971), limited data are available for determination of areal sweep when a directional permeability exists in the flood pattern. One such study is based on a mathematical analysis of results obtained from a potentiometric model (Landrum and Crawford 1960). Results are limited to steady-state conditions with $M = 1.0$. As in other studies of areal sweep, gravity effects are negligible.

The calculations are summarized in **Table 4.3,** which shows areal sweep efficiency at breakthrough for five-spot and line-drive patterns as a function of permeability ratio, k_x/k_y. The x- and y-direction permeabilities are normal to each other. For the results given in the top half of the table, the permeability direction k_x is parallel to a line running from the injection to the production wells. In the results shown in the bottom part of the table, the permeability direction k_x is rotated 45° from a line running from the injection well.

As indicated, directional permeability can have a marked effect on areal sweep. Proper orientation of the well pattern can compensate for the directional permeability and actually work to improve sweep over a case where no directional permeability exists.

4.5.5 Calculation of Areal Displacement Efficiency With Mathematical Modeling. Displacement processes can be modeled mathematically, and numerous papers exist in the literature that describe the models and their applications. Most of these models are based on the use of numerical analysis methods and digital computers. In general, it is beyond the scope of this book to describe the mathematical modeling procedures and results; however, two examples of mathematical modeling that relate to areal sweep efficiency are referenced.

Douglas et al. (1959) described a method of mathematically simulating a 2D immiscible displacement. The method is based on the numerical solution of the partial-differential equations that describe the flow of two immiscible phases in two space dimensions. The general approach is that used in most of the numerical simulation studies performed over the past several years.

F_i	N_p/V_{pd}	N_p (bbl)
0.133	0.133	6.19×10^4
0.15	0.150	6.96×10^4
0.25	0.224	1.04×10^5
0.50	0.353	1.64×10^5
0.75	0.439	2.04×10^5
1.00	0.503	2.34×10^5

Table 4.1—Oil production as a function of volume injected, Example 4.2.

F_i	N_p/V_{pd}	N_p (bbl)
0.655	0.655	3.05×10^5
0.75	0.735	3.42×10^5
1.0	0.867	4.03×10^5

Table 4.2—Oil production as a function of volume injected, unit mobility ratio, Example 4.2.

k_x Parallel to Straight Line Connecting Injector and Producer		
	Areal Sweep Efficiency at Breakthrough (%)	
k_x/k_y	Five-Spot	Line Drive
0.1	12	10
1	72	56
3	43	80
10	15	95
100	1	100

k_x Rotated 45° From Straight Line Connecting Injector and Producer		
	Areal Sweep Efficiency at Breakthrough (%)	
k_x/k_y	Five Spot	Line Drive
0.1	90	60
1	72	56
3	77	59
10	90	59
100	100	—

Table 4.3—Sweep efficiency at breakthrough with directional permeability (Landrum and Crawford 1960).

A comparison was made between the calculated results and experimental displacements conducted in a laboratory model study of waterflooding in a five-spot pattern. The results of that comparison (**Fig. 4.15**) show that agreement is excellent. Mobility ratio is not reported; rather, the oil/water viscosity ratios are given for the different runs. The viscosity ratios correspond to a wide range of mobility ratios from favorable to unfavorable.

Note that the mathematical model does not account for viscous fingering, yet the agreement with experimental data is good. The effects of mobility ratio and geometry on resistance along flow paths are treated, and these effects apparently dominate in the physical system modeled.

Higgins and Leighton (1962) developed a mathematical model based on frontal-advance theory and the concept of streamtubes. This model can be used to simulate immiscible displacements in uniform or irregular patterns and is very useful for modeling areal displacement processes. The model does not account for viscous fingering. Willhite (1986) describes application of this model in waterflooding. Higgins and Leighton (1962) compared their model results to those of Douglas et al. (1959) and found good agreement for waterflooding in a five-spot pattern.

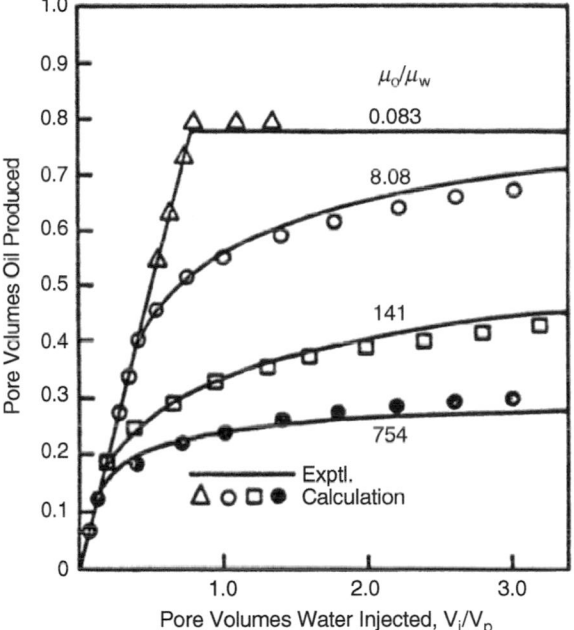

Fig. 4.15—Comparison of calculated and experimental results, waterflood in a five-spot pattern (Douglas et al. 1959).

4.6 Vertical Displacement Efficiency

Areal sweep efficiency, must be combined in an appropriate manner with vertical sweep to determine overall volumetric displacement efficiency. It is useful, however, to examine the factors that affect vertical sweep in the absence of areal displacement factors. To accomplish this, sample models and modeling studies based on linear reservoirs and linear displacement processes are useful and are considered in this section.

4.6.1 Factors Affecting Vertical Displacement Efficiency. Vertical sweep efficiency is controlled primarily by four factors: gravity segregation caused by differences in density, mobility ratio, vertical-to-horizontal permeability variation, and capillary forces. The effect of these factors on vertical sweep has been studied by a number of investigators. Both physical and mathematical models based on numerical simulation have been used. The effects of these parameters on vertical and volumetric sweep are illustrated in the following section. Relatively simple ways to estimate vertical displacement efficiency with simple mathematical models and correlations based on physical modeling results are also discussed.

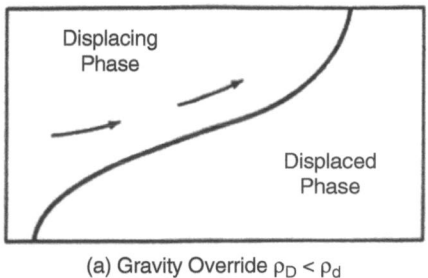 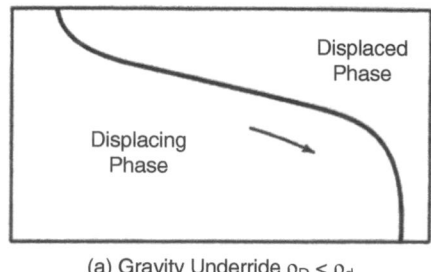

(a) Gravity Override $\rho_D < \rho_d$ (a) Gravity Underride $\rho_D < \rho_d$

Fig. 4.16—Gravity segregation in displacement processes.

4.6.2 Effect of Gravity Segregation and Mobility Ratio on Vertical Displacement Efficiency. Gravity segregation occurs when density differences between injected and displaced fluids are large enough to induce a significant component of fluid flow in the vertical direction even when the principal direction of fluid flow is in the horizontal plane. When the injected fluid is less dense than the displaced fluid, gravity segregation occurs and the displacing fluid overrides the displaced fluid, as shown in **Fig. 4.16a.** Gravity override is observed in steam displacement, in-situ combustion, CO_2 flooding, and solvent flooding processes. Gravity segregation also occurs when the injected fluid is more dense than the displaced fluid, as Fig. 4.16b shows for a waterflood. Gravity segregation leads to early breakthrough of the injected fluid and reduced vertical sweep efficiency.

Gravity Segregation in Horizontal Reservoirs. Scaled models or numerical simulators must be used to define when gravity forces become important and to describe their effects on displacement efficiency. Much of the readily available information on gravity segregation was obtained from displacement studies in scaled horizontal laboratory models that were homogeneous and isotropic. Other information is based on calculations made with numerical simulators. Results from these two sources are the basis of the vertical-sweep-efficiency correlations discussed in this section. As predictive tools, the correlations presented should be considered as approximations. The correlations correctly indicate the dependence of sweep on the different parameters used, but they do not allow precise calculation of E_I for actual reservoirs.

Craig et al. (1957) studied vertical sweep efficiency by conducting a set of scaled experiments in linear systems and five-spot models. Scaling laws used were of the type based on dimensional analysis described in the literature (Geertsma et al. 1956). Both consolidated and unconsolidated sands were used. Data taken in the linear systems are discussed in this section, while data taken in the five-spot model are described in Section 4.7.

The linear models used were from 10 to 66 in. long with length/height ratios ranging from 4.1 to 66. Experiments were conducted with miscible and immiscible liquids having mobility ratios from 0.057 to 200. Immiscible waterfloods were conducted at $M < 1$. In this case, oil flow behind the displacement fronts was essentially zero. Vertical sweep, E_I, was determined at breakthrough by material balance and visual observation of produced effluent.

Results of the linear displacements are shown in **Fig. 4.17,** where E_I at breakthrough is given as a function of a dimensionless group called a viscous/gravity ratio (Stalkup 1983).

$$R_{v/g} = \left(\frac{u\mu_d}{kg\Delta\rho}\right)\left(\frac{L}{h}\right), \dotfill (4.24)$$

where u = linear Darcy velocity, μ_d = viscosity of displaced phase, k = porous media permeability, g = gravity acceleration constant, $\Delta\rho$ = density difference between displacing and displaced phases, L = length of system, and h = height of system and consistent units should be used. Expressed in customary units, $R_{v/g}$ is given by

$$R_{v/g} = \left(\frac{2050u\mu_d}{k\Delta\rho}\right)\left(\frac{L}{h}\right), \dotfill (4.25)$$

where u is in B/(D-ft²), μ_d is in centipoise, k is in millidarcies, $\Delta\rho$ is in grams per cubic centimeter, and L and h are in feet.

The magnitude of viscous forces relative to gravity forces increases with increasing $R_{v/g}$ values. At small $R_{v/g}$ values, the displaced phase tends to override or underride, depending on the magnitude of the liquid densities, which leads to early breakthrough of the displacing phase, even for $M = 1.0$. As $R_{v/g}$ becomes relatively large in magnitude, with $M = 1.0$, E_I approaches 100%. Data correlated in Fig. 4.17 were obtained for the case in which the horizontal permeability, k_H, is equal to vertical permeability, k_V. If these permeabilities are not equal, Stalkup (1983) suggests the permeability be approximated as $k = \sqrt{k_V k_H}$.

The application of the correlation to estimate the importance of gravity effects is illustrated in Example 4.3.

Example 4.3—Relative Importance of Gravity Segregation in a Displacement Process. A miscible displacement process will be used to displace oil from a linear reservoir having the following properties:

L = 300 ft

h = 10 ft

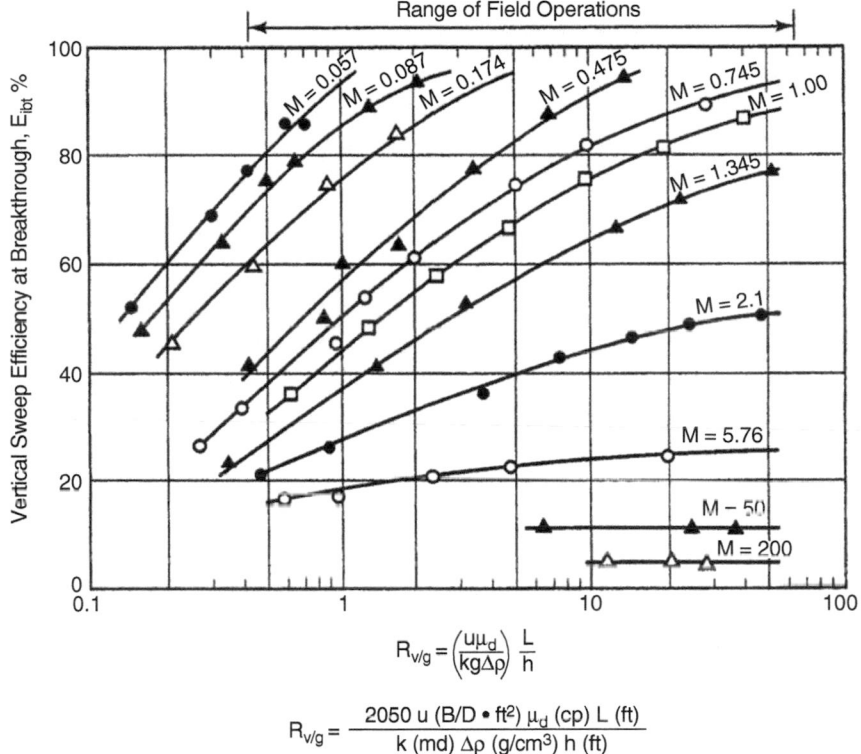

$$R_{v/g} = \left(\frac{u\mu_d}{kg\Delta\rho}\right)\frac{L}{h}$$

$$R_{v/g} = \frac{2050\ u\ (B/D \bullet ft^2)\ \mu_d\ (cp)\ L\ (ft)}{k\ (md)\ \Delta\rho\ (g/cm^3)\ h\ (ft)}$$

Fig. 4.17—Volumetric (vertical) sweep efficiency at breakthrough as a function of the ratios of viscous/gravity forces, linear system (Craig et al. 1957).

$$\phi \quad = \quad 0.2$$

$$S_{oi} \quad = \quad 0.75$$

$$S_{iw}^{'} \quad = \quad 0.25$$

$$k_o \quad = \quad 200\ md\ \text{(effective permeability to oil at interstitial water saturation)}$$

Determine the effect of gravity segregation on the vertical sweep efficiency if the oil is displaced miscibly by a solvent with a density of 0.7 g/cm³ and a viscosity of 2.3 cp at reservoir temperature. The density of the oil is 0.85 g/cm³ and the viscosity is 2.3 cp.

For this example, consider displacement at a frontal advance rate of 0.375 ft/D. This is equivalent to a Darcy velocity of 0.075 ft/D.

Solution. Calculate the viscous/gravity ratio with Eq. 4.25.

$$u = 0.075\ ft/D\left(\frac{1}{5.615\ ft^3/bbl}\right)$$

$$= 0.0134\ B/(D-ft^2)$$

$$\text{and } R_{v/g} = \frac{(2,050)(0.0134)(2.3)(300)}{(200)(0.85-0.70)(10)} = 63.$$

Calculate M.

$$M = \frac{k_D}{\mu_D}\frac{\mu_d}{k_d} = \frac{\mu_d}{\mu_D}$$

$$= 2.3/2.3 = 1.0.$$

From Fig. 4.17, E_I (at breakthrough) = 0.86. As discussed in Chapter 3, mixing (dispersion) in miscible flow leads to breakthrough in a linear system before 1.0 PV has been injected. Thus, the importance of gravity override in the proposed miscible displacement is relatively small and probably not significant.

If the frontal advance rate were 0.0075 ft/D, the value of $R_{v/g}$ would be 6.3. In this case, E_I would be approximately 0.70 and gravity segregation becomes an important mechanism in the displacement. Any change in a parameter in $R_{v/g}$ that reduces its

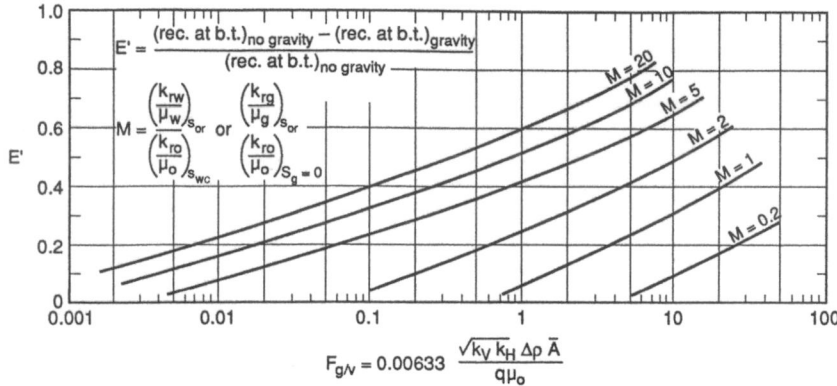

Fig. 4.18—Gravity segregation in two-phase water/oil or gas/oil displacements, linear systems (Spivak 1974).

numerical value contributes to gravity effects. For example, increasing the permeability from 200 to 4,000 md has the same effect as reducing the frontal advance rate from 0.075 to 0.00375 ft/D.

Similar studies of gravity segregation have been conducted with mathematical models. This approach has the advantage of producing the complete displacement performance rather than the more limited results obtainable from physical models. Spivak (1974) used a 2D and 3D numerical model to study gravity effects during waterflooding and gasflooding. **Fig. 4.18** presents a correlation of the results. The correlation is in terms of a gravity/viscous ratio, $F_{g/v}$, similar to that used by Craig et al. (1957). The effect of having a vertical permeability that is different from the horizontal permeability is reflected in the definition of $F_{g/v}$. This difference in permeabilities was not considered in the correlation of Craig et al.

$$F_{g/v} = \frac{0.00633\sqrt{k_V k_H}\,\Delta\rho A}{q\mu_d}, \dots\dots\dots\dots\dots\dots\dots\dots\dots\dots\dots\dots\dots\dots\dots\dots\dots\dots\dots\dots\dots\dots\dots\dots\dots\dots (4.26)$$

where k_V and k_H = vertical and horizontal permeabilities, respectively, md; A = average cross-sectional area for flow, ft^2; q = flow rate, ft^3/D; and $\Delta\rho$ = density difference, psi/ft.

The area term, A, in the definition of $F_{g/v}$ is average cross-sectional area normal to flow. Using an average area makes the correlation applicable to linear displacements as well as displacements in five-spot patterns.

The recovery factor, E', in the correlation of Spivak (1974) is defined in terms of recovery at breakthrough with gravity effects compared with recovery at breakthrough in the absence of gravity effects. E' is not equivalent to vertical displacement efficiency, E_I, which is used in the correlation of Craig et al. (1957). Additionally, the factor $R_{v/g}$ in the Craig correlation contains a length/height ratio, L/h. Thus, the two correlations are not directly comparable. It should be noted that if the same permeability is used in the two correlations and if adjustments are made for units, then $R_{v/g} = L/(hF_{g/v})$.

The correlations of Craig et al. (1957) and Spivak (1974) indicate the following effects of various parameters on gravity segregation, as summarized by Spivak:

1. Gravity segregation increases with increasing horizontal and vertical permeability.
2. Gravity segregation increases with increasing density difference between the displacing and displaced fluids.
3. Gravity segregation increases with increasing mobility ratio.
4. Gravity segregation increases with decreasing rate. (This effect can be reduced by viscous fingering, as discussed in a following section.)
5. Gravity segregation decreases with increasing level of viscosity for a fixed viscosity ratio.

The effect of mobility ratio in a displacement is to accelerate or to retard the gravity effect, depending on whether M is greater than or less than 1.0. If the displacing phase is less mobile than the displaced phase, then the override or underride is hindered owing to the greater viscous flow resistance through the injected phase. If $M > 1.0$, then viscous resistance in the displacing phase is smaller and gravity override or underride is enhanced.

When M > 1.0, viscous fingering can occur along with gravity segregation. Thus, at conditions where gravity effects are important and the mobility ratio is unfavorable, vertical sweep can be affected by both the tendency of the displacing fluid to flow into the gravity tongue and the superposition of viscous fingers onto the gravity tongue.

Flow Regions in Miscible Displacement at Unfavorable Mobility Ratios. Flow experiments in a vertical cross section in horizontal porous media have shown that four flow regions, depicted in **Fig. 4.19,** are possible when the mobility ratio is unfavorable (Stalkup 1983; Crane et al. 1963). The onset of the different regimes is governed by the magnitude of the viscous/gravity ratio defined in Eqs. 4.24 and 4.25.

Region I occurs at very low $R_{v/g}$ values and is characterized by a single gravity tongue, with the displacing liquid either underriding or overriding the displaced liquid. Vertical sweep is a strong function of $R_{v/g}$. At larger $R_{v/g}$ values, in Region II, a single gravity tongue still exists, but vertical sweep is relatively insensitive to the value of the viscous/gravity ratio. The transition to Region III occurs at a particular critical $R_{v/g}$ value. In Region III, viscous fingers are formed along the gravity tongue and appear as secondary fingers along the primary gravity tongue. Vertical sweep is improved by the formation of the viscous fingers in this region. In Region IV, flow is dominated by the viscous forces and by viscous fingering. A gravity tongue does not form because of the strong viscous fingering. The vertical sweep in this region is relatively insensitive to $R_{v/g}$.

The different flow regions are shown in **Fig. 4.20** (Stalkup 1983) as a function of $R_{v/g}$ and for different values of M. Stalkup points out that the flow regions illustrated in Fig. 4.20 were estimated from several different literature sources and should be considered only as approximations. A generalized correlation of flow regions for immiscible displacements has not been developed.

Viscous fingering likely occurred in the physical modeling studies of gravity effects in which miscible liquids were used, as it did in the areal displacement models. Thus, results shown in Fig. 4.17 reflect the influence of gravity forces, mobility ratio effects, and in some cases, viscous fingering. At unfavorable mobility ratios, the importance of viscous fingering depends on the existing flow region. The numerical modeling studies of Spivak (1974) did not consider viscous fingering.

Gravity Segregation in Dipping Reservoirs. Density differences between displacing and displaced fluids can have a pronounced effect on displacement processes in dipping reservoirs. The different flow regimes previously described can occur, depending on the relative importance of gravity and viscous forces. When the reservoir has a dip and the general flow direction is not horizontal, however, the specific correlations shown in Figs. 4.17 through 4.20 are not directly applicable. Corresponding correlations for dipping reservoirs do not generally exist.

When the reservoir has a dip, gravity can be used to advantage to improve displacement performance. For example, if an oil were displaced from a reservoir by injecting a less dense, more mobile solvent updip, gravity would tend to stabilize the displacement front. That is, if the displacement velocity were sufficiently low, gravity forces would act to prevent the formation of fingers at the solvent/oil interface. Similarly, in a waterflood, downdip injection of water can work to stabilize the interface between the water and oil bank.

The criterion for stable displacement in a dipping reservoir, for the cited conditions of density and mobility, can be determined by examining a relatively simple model of flow at the interface, as shown in **Fig. 4.21.**

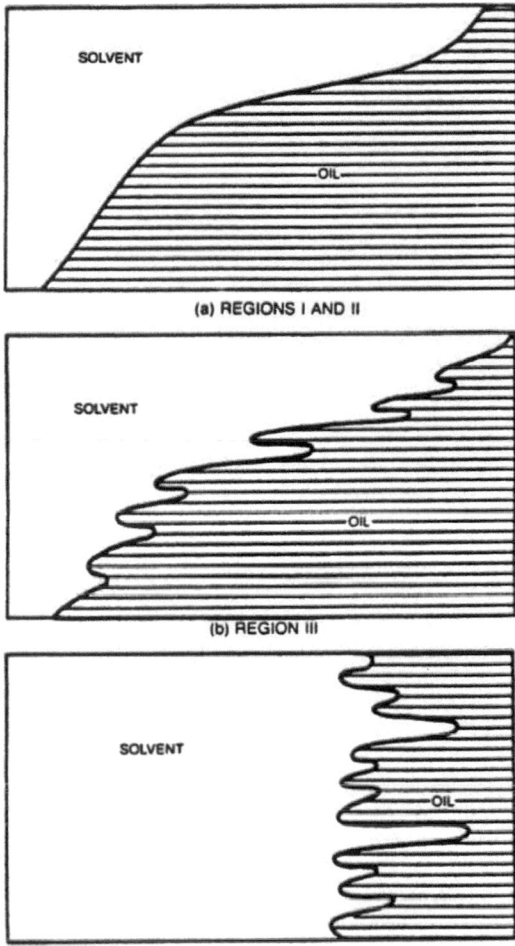

Fig. 4.19—Flow regimes in miscible displacement of unfavorable mobility ratios (Stalkup 1983).

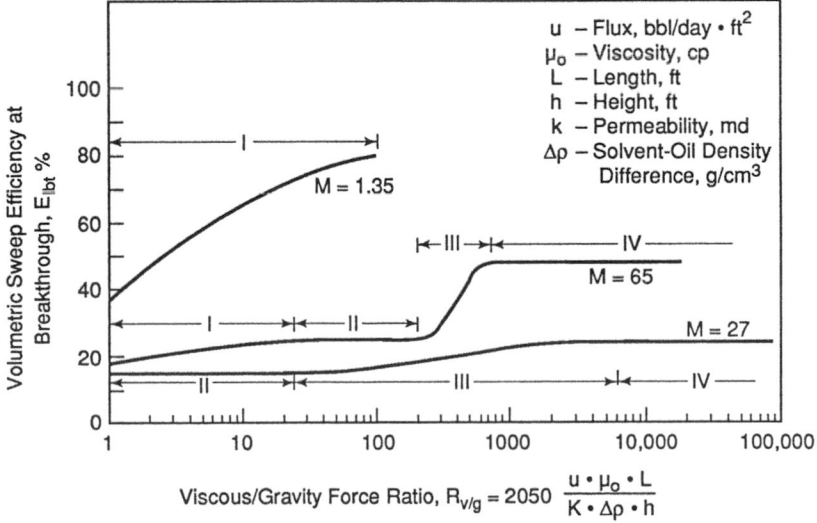

Fig. 4.20—Flow regimes in miscible displacement in linear systems as a function of viscous/gravity and mobility ratios, $R_{v/g}$ defined in Eq. 4.25 (Stalkup 1983).

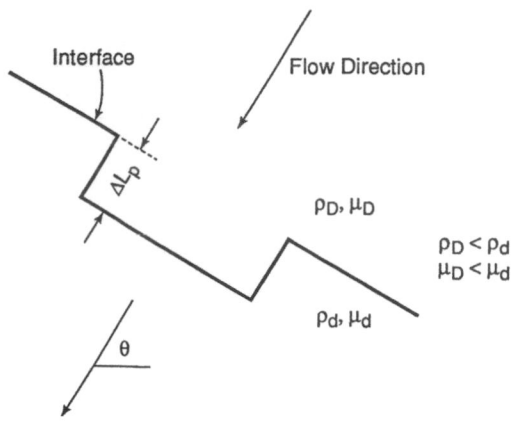

Fig. 4.21—Model for determination of stability criterion in a dipping reservoir.

Consider a displacement in a downdip direction in which the reservoir dip is at an angle θ to the horizontal. Assume that the interface is sharp with only the displacing fluid flowing ahead of the front. This would occur in an immiscible displacement under the assumption of piston-like flow or in a miscible displacement under the assumption of negligible mixing at the interface. Consider further that a small perturbation, finger, or protrusion has formed at the interface. The objective of the analysis is the determination of conditions under which the protrusion will remain stable or, alternatively, will grow in magnitude, leading to an unstable interface.

According to Hill (1952), the finger will remain stable if just across the interface of the finger the pressure in the displaced phase is greater than the pressure in the displacing phase (i.e., $p_d > P_D$). Darcy's equation can be applied to obtain expressions for p_d and p_D across the interface. From Fig. 4.21,

$$p_d = p_o + \rho_d g \Delta L_p \sin\theta - \frac{\mu_d u \Delta L_p}{k_d} \qquad \ldots\ldots\ldots\ldots (4.27)$$

and

$$p_D = p_o + \rho_D g \Delta L_p \sin\theta - \frac{\mu_D u \Delta L_p}{k_D}, \qquad \ldots\ldots\ldots\ldots\ldots\ldots (4.28)$$

where p_o = reference pressure at the point of the unperturbed interface, ρ_d = density of the displaced phase, ρ_D = density of the displacing phase, μ_d = viscosity of the displaced phase, μ_D = viscosity of the displacing phase, k_d = effective permeability of the displaced phase, k_D = effective permeability of the displacing phase, u = Darcy velocity, g = gravity acceleration constant, and ΔL_p = length of the perturbation (Fig. 4.21). For the interface to remain stable,

$$p_d - p_D > 0 \qquad \ldots\ldots\ldots\ldots\ldots\ldots\ldots\ldots\ldots\ldots\ldots\ldots\ldots\ldots\ldots\ldots (4.29)$$

or, subtracting Eq. 4.28 from Eq. 4.27,

$$g\left(\rho_d - \rho_D\right)\Delta L_p \sin\theta - \left(\frac{\mu_d}{k_d} - \frac{\mu_D}{k_D}\right)u\Delta L_p > 0. \qquad \ldots\ldots\ldots\ldots\ldots\ldots\ldots (4.30)$$

Eq. 4.30 establishes the maximum velocity, called the critical velocity, u_c, at which the interface will remain stable. If, as stated, $\lambda_D > \lambda_d$ and $\rho_D < \rho_d$, then,

$$u_c = \frac{g\left(\rho_d - \rho_D\right)\sin\theta}{\left(\dfrac{\mu_d}{k_d} - \dfrac{\mu_D}{k_D}\right)}. \qquad \ldots\ldots\ldots\ldots\ldots\ldots\ldots\ldots\ldots\ldots\ldots\ldots\ldots (4.31)$$

In customary units, Eq. 4.31 is

$$u_c = \frac{0.0439\left(\rho_d - \rho_D\right)\sin\theta}{\left(\dfrac{\mu_d}{k_d} - \dfrac{\mu_D}{k_D}\right)}, \qquad \ldots\ldots\ldots\ldots\ldots\ldots\ldots\ldots\ldots\ldots\ldots (4.32)$$

where ρ is in lbm/ft³, μ is in centipoise, k is in darcies, and u_c is in feet per day. If $k_d = k_D = k$ and if the mobility ratio, $M = \mu_d/\mu_D$, is introduced,

$$u_c = \frac{0.0439 k \left(\rho_d - \rho_D\right)\sin\theta}{\mu_D \left(M - 1\right)}. \qquad \ldots\ldots\ldots\ldots\ldots\ldots\ldots\ldots\ldots\ldots\ldots (4.33)$$

Eqs. 4.31 through 4.33 establish the maximum velocity at which the interface will remain stable for the stated relative properties of the displacing and displaced phases.

The same analysis is applicable for other relative property values of the fluids and for displacements in the updip direction. For example, if properties were such that $\rho_d > \rho_D$, $k_d = k_D = k$, and $\mu_D < \mu_d$, then examination of Eq. 4.30 shows that the interface would be unconditionally stable if flow were in the downdip direction, as shown in Fig. 4.21.

The derivation of the expression for u_c was based on the assumption of a sharp concentration, or phase, transition at the interface between the displacing and displaced fluids. Dumore (1964) considered a miscible displacement in which mixing occurred at the interface, resulting in a concentration-transition zone. For this case, both viscosity and density are a function of position because of the concentration gradient. In one form of the derivation it was assumed that both density and $\ln \mu$ were linear functions of concentration:

$$\rho = C_D \rho_D + (1 - C_D)\rho_d, \ \rho_D < \rho_d \ \dots\dots\dots\dots\dots\dots\dots\dots\dots\dots (4.34)$$

and $\ln \mu = C_D \ln \mu_D + (1 - C_D)\ln \mu_d, \ \mu_D < \mu_d, \ \dots\dots\dots\dots\dots\dots\dots\dots (4.35)$

where C_D = concentration of the displacing phase, ρ = density of the mixture, and μ = viscosity of the mixture. The assumption about the dependence of $\ln \mu$ on concentration is an approximation based on commonly used mixing rules for viscosity.

Dumore defined a minimum stable Darcy velocity, u_{st}, which he showed to be related to u_c by

$$\frac{u_{st}}{u_c} = \frac{\mu_d - \mu_D}{\mu_d (\ln \mu_d - \ln \mu_D)} \ \dots\dots\dots\dots\dots\dots\dots\dots\dots\dots\dots\dots (4.36)$$

$$\text{or } \frac{u_{st}}{u_c} = \frac{\left(1 - \dfrac{1}{M}\right)}{\ln M}. \ \dots\dots\dots\dots\dots\dots\dots\dots\dots\dots\dots\dots\dots (4.37)$$

Because $\mu_D < \mu_d (M > 1.0)$ and $\rho_D < \rho_d$ for the conditions of the derivation and flow is downdip, u_{st}/u_c is always less than unity. Thus, Eqs. 4.36 and 4.37 for u_{st} provide a more stringent and more reliable condition for stable flow than the results for u_c based on a sharp transition zone.

Stalkup (1983) presented a relationship between u_c and u_{st} assuming a quarter-power blending rule for viscosity:

$$\mu = \left[\frac{C_D}{\mu_D^{1/4}} + \frac{(1 - C_D)}{\mu_d^{1/4}}\right]^{-4} \ \dots\dots\dots\dots\dots\dots\dots\dots\dots\dots (4.38)$$

The resulting equation relating u_{st} and u_c is

$$\frac{u_{st}}{u_c} = \frac{\left(1 - \dfrac{1}{M}\right)}{4(M^{1/4} - 1)}. \ \dots\dots\dots\dots\dots\dots\dots\dots\dots\dots\dots (4.39)$$

The choice of application of Eqs. 4.35 and 4.37 or Eqs. 4.38 and 4.39 should be based on the dependence of viscosity on concentration. In the absence of data, Eqs. 4.38 and 4.39 are recommended.

The application of the criterion for the critical flow rate, u_c, is illustrated in Example 4.4.

Example 4.4—Calculation of Criterion for Gravity-Stabilized Flow. A miscible displacement is to be conducted in a laboratory experiment in which one glycerol/brine solution is displaced vertically downward by a second solution having a different concentration of glycerol. Liquid properties are as follows.

Liquid 1, Displacing Liquid:

       30.00 g NaCl/L
       650 g glycerol/L
       $\rho_D = 1.1551$ g/cm³ at 77°F
       $\mu_D = 7.4$ cp at 77°F

Liquid 2:

       20.00 g NaCl/L
       700 g glycerol/L
       $\rho_d = 1.1609$ g/cm³ at 77°F
       $\mu_d = 11.3$ cp at 77°F

The liquid velocity is to be 0.237 ft/D (Darcy velocity). Porosity of the porous medium is 0.206 and permeability is 303 md. Determine whether the flow will be stable (i.e., whether viscous fingering will occur).

Solution. Eq. 4.31 is applicable for the calculation of u_c.

$$u_c = \frac{0.0439(\rho_d - \rho_D)\sin\theta}{\left(\dfrac{\mu_d}{k_d} - \dfrac{\mu_D}{k_D}\right)}.$$

$k_d = k_D = 303$ md or 0.303 darcies, $\rho_d = 72.440$ lbm/ft³, and $\rho_D = 72.078$ lbm/ft³. Therefore,

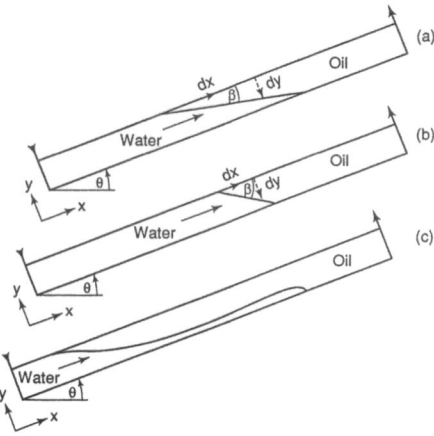

Fig. 4.22—Stable and unstable displacements in an immiscible flood conducted in the updip direction (Dake 1978).

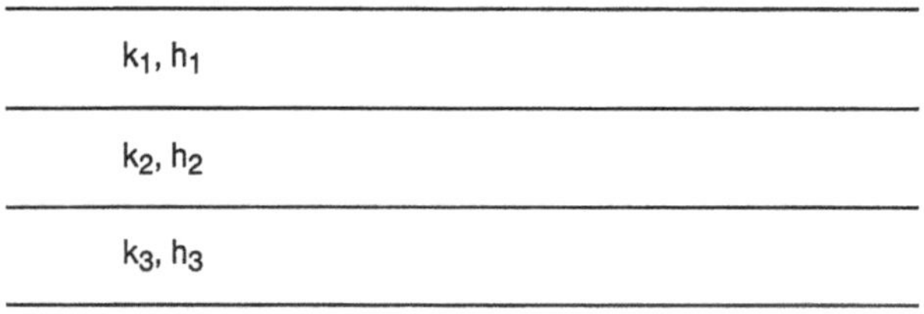

---
k_1, h_1
---
k_2, h_2
---
k_3, h_3

Fig. 4.23—Vertical cross section with layers of different permeabilities and thicknesses.

$$u_c = \frac{0.0439\left(72.440 - 72.078\right)1.0 \times 0.303}{\left(11.3 - 7.4\right)}$$

$$= 0.00123 \text{ ft/D}.$$

Because the planned velocity of 0.237 ft/D $>> u_c$, the flow will be unstable. Viscous fingering will occur.

Calculate the ratio u_{st}/u_c. Eq. 4.34 is applicable.

$$\frac{u_{st}}{u_c} = \frac{\left[1-\left(1/M\right)\right]}{\ln M}.$$

$$M = \frac{\mu_d}{\mu_D} = \frac{11.3}{7.4} = 1.527.$$

$$\frac{u_{st}}{u_c} = \frac{\left[1-\left(1/1.527\right)\right]}{\ln 1.527}$$

$$= 0.815$$

Fluid mixing results in an even smaller maximum velocity for stable flow.

When a stable displacement occurs in either the updip or downdip direction, the interface stabilizes at a specific angle relative to the dip of the formation, as illustrated in **Fig. 4.22** for a waterflood (Dake 1978).

Expressions relating the angle of the interface to flow rate, angle of dip, mobility ratio, and fluid densities are given by Dake (1978), Hawthorne (1960), and Stalkup (1983). Conditions depicted in Figs. 4.22a and 4.22b are for stable flow, while Fig. 4.22c depicts an unstable flow in which u_c is exceeded.

4.6.3 Effect of Vertical Heterogeneity and Mobility Ratio on Vertical Displacement Efficiency. Vertical variation in permeability in reservoirs is relatively common. **Fig. 4.23** shows a vertical cross section of such a reservoir in which the vertical section has been divided into layers of different thicknesses and permeabilities. Such a geologic model is an idealization because permeability typically is not constant over a significant thickness with abrupt changes to a different value (i.e., actual reservoir layering is not as idealized as shown in the figure).

The vertical variation in permeability will lead to a reduction of vertical sweep efficiency at breakthrough in a displacement process owing to uneven flow in the different layers. This would occur at idealized conditions of unit mobility ratio and in the absence of gravity segregation. Different models of flow in layered reservoirs are described in the following sections. Crossflow between layers and gravity segregation are assumed to be negligible, although crossflow is discussed briefly.

Unit-Mobility-Ratio Displacement in a Layered Linear Reservoir With No Crossflow. Consider a displacement involving incompressible fluids that is carried out in a system like the one shown in Fig. 4.23. Permeabilities and thicknesses of the layers may be different, but porosities are assumed to be equal. Overall pressure drop is constant and the same for all layers. Consider further that there is no crossflow between layers and that the flow is piston-like in each layer. If the displacement is conducted at $M = 1.0$, then by Darcy's equation

$$\frac{q_j}{q_t} = \frac{\lambda_j h_j}{\sum\limits_{k=1}^{n} \lambda_k h_k} = \frac{k_j h_j}{\sum\limits_{k=1}^{n} k_k h_k}, \dots\dots\dots\dots\dots\dots\dots\dots\dots\dots\dots\dots\dots\dots (4.40)$$

where subscripts j and k refer to the layers, q_j = flow rate in Layer j, q_t = total flow rate in the vertical section, λ_j = mobility of injected fluid in Layer j, k_j = permeability of Layer j, h_j = thickness of the Layer j, and the sum in the denominator applies to all n layers in the system. Eq. 4.40 shows that for a system in which the overall pressure drop is the same for all layers, the highest fluid velocity will occur in the layer with the largest permeability. After some algebraic manipulation, application of the equation allows calculation of sweep efficiency at breakthrough.

$$E_{Ibt} = \frac{\sum\limits_{k=1}^{n} \lambda_k h_k}{\lambda_j \sum\limits_{k=1}^{n} h_k} = \frac{\sum\limits_{k=1}^{n} k_k h_k}{k_j \sum\limits_{k=1}^{n} h_k}, \dots\dots\dots\dots\dots\dots\dots\dots\dots\dots\dots\dots\dots\dots (4.41)$$

where breakthrough occurs in Layer *j*, which has the largest mobility. If all layers were of equal thickness, Eq. 4.41 would simplify to

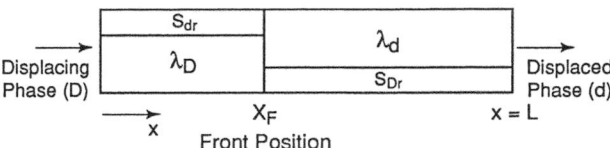

$$E_{Ibt} = \frac{\sum_{k=1}^{n} k_k}{n k_j}, \quad\quad\quad (4.42)$$

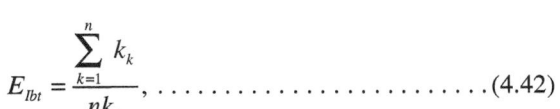

Fig. 4.24—Schematic of piston-like displacement in a linear system.

where, again, breakthrough occurs in Layer *j*, which has the largest mobility. Note that the vertical sweep efficiency is equivalent to the volumetric sweep efficiency of the vertical cross section.

If piston-like displacement is assumed, Darcy's equation can be applied, as was done in the development of Eqs. 4.40 through 4.42, along with a material balance to calculate recovery of the displaced phase. Recovery can be calculated as a function of fluid injected both before and after breakthrough of the displacing phase.

Displacement at a Nonunit Mobility Ratio in a Layered, Linear Reservoir With No Crossflow. As in the previous section, consider a linear, layered reservoir like that shown in Fig. 4.23. Assume that an immiscible displacement is conducted at a constant pressure drop across the system and that in any given layer the displacement is piston-like, as shown in **Fig. 4.24.** Crossflow between layers does not exist. Single-phase flow exists in each zone (i.e., both ahead of and behind the displacing phase front). The mobility ratio is defined by Eq. 4.13 and may be equal to, less than, or greater than unity. At nonunit mobility ratios, total resistance across the system varies as a function of volume injected and thus the rate of injection varies at constant pressure drop.

Expressions describing the displacement in a single layer can be derived by application of Darcy's equation. In a single layer,

$$\frac{k_{rD}k}{\mu_D}\frac{dp}{dx}\bigg|_D = \frac{k_{rd}k}{\mu_d}\frac{dp}{dx}\bigg|_d \quad\quad\quad (4.43)$$

$$\text{or } M\frac{dp}{dx}\bigg|_D = \frac{dp}{dx}\bigg|_d. \quad\quad\quad (4.44)$$

Because

$$\Delta p = \Delta p_D + \Delta p_d, \quad\quad\quad (4.45)$$

$$\Delta p = X_f \frac{dp}{dx}\bigg|_D + (L - X_f)\frac{dp}{dx}\bigg|_d, \quad\quad\quad (4.46)$$

where Δp = overall pressure drop between the entrance and exit, X_f = location of the displacing phase fluid front (Fig. 4.24), and L = total length of the system.

From Eqs. 4.44 and 4.46,

$$\Delta p = \left[X_f + (L - X_f)M \right]\frac{dp}{dx}\bigg|_D \quad\quad\quad (4.47)$$

$$\text{or } \frac{dp}{dx}\bigg|_D = \frac{\Delta p}{\left[X_f + (L - X_f)M \right]}. \quad\quad\quad (4.48)$$

Again, from Darcy's equation,

$$v_D = \frac{dX_f}{dt} = \frac{-k_{rD}k}{\mu_D}\frac{dp}{dx}\bigg|_D \left[\frac{1}{\phi(1 - S_{dr} - S_{Dr})}\right], \quad\quad\quad (4.49)$$

where v_D = linear velocity of displacing phase front, S_{dr} = residual saturation of the displaced phase (Fig. 4.24), and S_{Dr} = residual saturation of the displacing phase in the region ahead of the displacing phase front (e.g., interstitial water saturation). Substitution of Eq. 4.48 into Eq. 4.49 gives

$$\frac{dX_f}{dt} = \frac{\lambda_{rD}k\Delta p}{\left[ML + (1 - M)X_f \right]\phi(1 - S_{dr} - S_{Dr})}. \quad\quad\quad (4.50)$$

In field units,

$$\frac{dX_f}{dt} = \frac{6.327\lambda_{rd}k\Delta p}{\left[ML+(1-M)X\right]f\left(1-S_{dr}-S_{Dr}\right)}, \quad \dots \dots \dots (4.51)$$

where X_f is in feet, t is in days, L is in feet, k is in darcies, Δp is in pounds per square inch, λ_{rD} is in cp^{-1}, and M, S_{dr}, and S_{Dr} are dimensionless parameters.

Eq. 4.51 is a differential equation that may be integrated to determine the position of the displacing phase front, X_f, as a function of the length of time of injection, t. Separation of variables and integration results in

$$MLX_f +(1-M)\frac{X_f^2}{2} = \frac{-6.327\lambda_{rD}k\Delta p}{\phi\left(1-S_{dr}-S_{Dr}\right)}. \quad \dots \dots \dots (4.52)$$

From Eq. 4.52,

$$t = \frac{\phi\left(1-S_{dr}-S_{Dr}\right)}{6.327\lambda_{rD}k\Delta p}\left[MLX_f +(1-M)\frac{X_f^2}{2}\right] \quad \dots \dots \dots (4.53)$$

and

$$X_f = \frac{-ML\pm\left[(ML)^2 +\dfrac{2\times6.327(1-M)\lambda_{rD}k\Delta pt}{\phi\left(1-S_{dr}-S_{Dr}\right)}\right]^{\!1/2}}{(1-M)}. \quad \dots \dots \dots (4.54)$$

There are two roots to Eq. 4.54, and the proper one must be selected.

At any time up to breakthrough in the layer, the volume of displacing phase fluid injected is given by

$$V_i = A\left(1-S_{dr}-S_{Dr}\right)X_f \times\frac{1}{5.615}\phi \text{ ft}^3\text{/bbl,} \quad \dots \dots \dots (4.55)$$

where V_i = volume injected, bbl; and A = cross-sectional area of the layer, ft^2. The injection rate after breakthrough is given by (in units of barrels per day)

$$i_{abt} = 1.127\frac{k_{rD}kA}{\mu_D}\frac{\Delta p}{L}, \quad \dots \dots \dots (4.56)$$

and the volume injected at any time after breakthrough is

$$V_{iabt} = V_{ibt} + i_{abt}\left(t-t_{bt}\right), \quad \dots \dots \dots (4.57)$$

where t = time from the beginning of the injection process, days; and t_{bt} = time of breakthrough, days.

Eqs. 4.53 through 4.57 may be applied to a multilayer linear system to calculate volumes injected and recovered as a function of time by application of the equations to each layer in the system. The application is described in Example 4.5.

Example 4.5—Displacement Performance Calculation in a Linear, Layered Reservoir, M = 1.0. Consider the reservoir shown in **Fig. 4.25.** It is a horizontal, linear system having a width of 100 ft and a length of 350 ft. The reservoir consists of two layers, each 5.0 ft thick. Layer 1 has a permeability of 20 md, while Layer 2 has a permeability of 100 md. Porosity is 0.18. There is no cross-flow between layers.

The reservoir has an initial oil saturation, S_{oi}, of 80% and a water saturation, S_{iw} of 20%. Water is injected at the end at $X = 0$, displacing oil to the production end at $X = L$ (350 ft). A constant pressure drop of 500 psi is maintained across the system.

In the calculations to be made, it is assumed that the displacement is by plug or piston-like flow. That is, only oil flows ahead of the front ($S_w = S_{iw}$) and only water flows behind the front ($S_o = S_{or}$).

Additional data are as follows.

μ_w = 1.0 cp at T_R

μ_o = 20 cp at T_R

B_o = 1.0 RB/STB

B_w = 1.0 RB/STB

k_{rw} = 0.27 at $S_{or} = 0.25$

k_{ro} = 0.94 at $S_{iw} = 0.20$

The mobility ratio is given by

$$M = (k_{rw}/\mu_w)(\mu_w/k_{ro})$$
$$= (0.27/1.0)(20/0.94)$$
$$= 5.74 \text{ (unfavorable).}$$

Calculate the following:

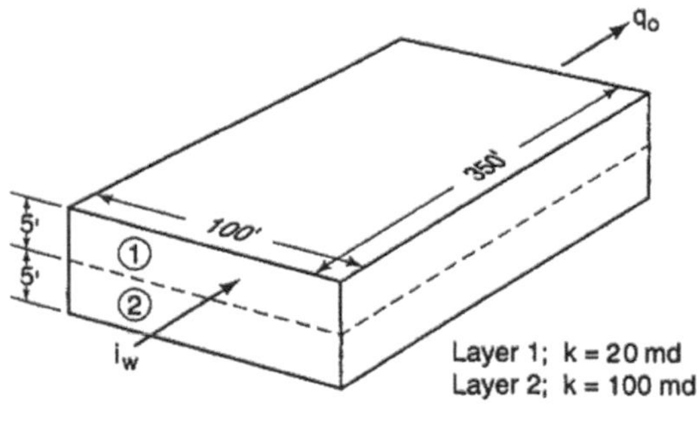

Fig. 4.25—Illustration of reservoir for Example 4.5.

Layer 1; k = 20 md
Layer 2; k = 100 md

1. Oil recovery (barrels) vs. water injected (barrels), assuming the water mobility is equal to the oil mobility [i.e., $M = 1.0$ (based on oil phase)].
2. Oil recovery (barrels) vs. water injected (barrels) at the specified mobility ratio, $M = 5.74$.
3. Oil recovery (barrels) vs. water injected (barrels) assuming the reservoir consists of a single 10-ft-thick layer and an average permeability of 60 md.
4. Vertical displacement efficiencies at water breakthrough for conditions of Parts 2 and 3.
5. For $M = 5.74$, show the relative locations of the water fronts in the two layers for at least three different points in time.

Solution.

1 and 2. The calculation procedure is as follows:
   a. The position of the front is set at certain discrete values in each layer.
   b. The volume of water injected corresponding to each front position is calculated.
   c. Oil recovery corresponding to each front position is calculated.
   d. The time corresponding to each front position in each layer is calculated with Eq. 4.52.
   e. Calculations for each individual layer are combined to yield the overall performance.

Eq. 4.53 is applicable.

$$t = \frac{\phi(1-S_{or}-S_{iw})}{6.327\lambda_{rw}k\Delta p}\left[MLX_f + (1-M)\frac{X_f^2}{2}\right]$$
$$= \frac{0.18(1-0.20-0.25)}{6.327\times0.27k\times500}\left[350MX_f + (1-M)\frac{X_f^2}{2}\right].$$

Solution of the equation yields the location of the water front as a function of time up to the time of water breakthrough in the layer. After breakthrough, which will occur first in Layer 2, the water injection rate in Layer 2 is given by

$$i_{abt} = 1.127\lambda_{rw}kA\frac{\Delta p}{L}$$
$$= 1.127\times0.27\times0.100\times5\times100\times\frac{(500)}{350}$$
$$= 21.74 \text{ B/D.}$$

   Tables 4.4 and 4.5 show calculations for Part 1 for $M = 1.0$ and Part 2 for $M = 5.74$. Results are plotted in **Fig. 4.26.** Total performance for the two layers was determined by combining volumes injected and oil produced for the two layers at equal points in time obtained from the calculations in Tables 4.4 and 4.5.
   The results show the general effect of mobility ratio on recovery as a function of water injected in a layered reservoir.
3. Calculation of oil recovery vs. water injected, averaged properties. If the two-layer system is considered as a single layer with an average permeability of 60 md, then water injected is equal to oil produced. The result of the calculation is also shown in Fig. 4.26.
4. Calculation of vertical displacement efficiency, E_I, at water breakthrough. From Fig. 4.26 for $M = 1.0$, $N_p = 3,700$ bbl at breakthrough. Total mobile oil is 6,170 bbl. Therefore, $E_I = 3,700/6,170 = 0.60$. Alternatively, from Eq. 4.42,

$$E_I = 20+100/2\times100 = 0.60.$$

   From Fig. 4.26 for $M = 5.74$, $N_p = 3,475$ bbl. Vertical sweep at breakthrough is reduced as M increases.
5. Relative location of the water front at different points in time, $M = 5.74$. **Fig. 4.27** shows the locations of the water front at times of 368, 478 (breakthrough), and 1,839 days. Data were obtained from calculations in Table 4.5.

Dykstra-Parsons Model for Vertical Heterogeneity. As described in previous sections, the effects of reservoir heterogeneity on vertical sweep efficiency can be estimated with simple models by assuming that the reservoir is represented by

X (ft)	V, for Each Layer (bbl)	N_p for Each Layer (bbl)	Layer 1* t_1 (days)	Layer 2** t_2 (days)
70	617	617	815.6	163.1
140	1,234	1,234	1,631	326.3
210	1,851	1,851	2,447	489.4
280	2,468	2,468	3,263	652.2
350	3,085	3,085	4,078	815.6
–	3,782	–	–	1,000
–	5,672	–	–	1,500
–	7,062	–	–	2,000
–	8,452	–	–	2,500
–	9,842	–	–	3,000
–	11,232	–	–	3,500
–	12,839	–	–	4,078

*k = 20 md.

**k = 100 md.

Table 4.4—Results for Example 4.5, Part 1; M = 1.0 and λ_{rw} = 0.047 cp^{-1}.

X (ft)	V, for Each Layer (bbl)	N_p for Each Layer (bbl)	Layer 1* t_1 (days)	Layer 2** t_2 (days)
70	617	617	747.6	149.5
140	1,234	1,234	1,360.7	272.2
210	1,851	1,851	1,839.2	367.9
280	2,468	2,468	2,183.0	436.7
350	3,085	3,085	2,392.3	478.5
–	3,782	–	–	750
–	8,987	–	–	1,000
–	14,422	–	–	1,250
–	19,857	–	–	1,500
–	25,292	–	–	1,750
–	30,727	–	–	2,000
–	36,162	–	–	2,392
–	44,684	–	–	–

*k = 20 md.

**k = 100 md.

Table 4.5—Results for Example 4.5, Part 2; M = 5.74 and λ_{rw} = 0.27 cp^{-1}.

noncommunicating layers and by neglecting gravity segregation. Such a model was developed by Dykstra and Parsons for piston-like displacement in a linear reservoir flooded at constant pressure drop (Willhite 1986; Stalkup 1983; Dykstra and Parsons 1950). The Dykstra-Parsons model is based on subdividing the reservoir into n layers of equal thickness that have different permeabilities. Layers are arranged in order of descending permeability, as shown in **Fig. 4.28.**

When the displacement is piston-like, the vertical sweep efficiency is given by Eqs. 4.58 and 4.59. For $M = 1.0$,

$$E_I = \frac{n_j + \sum_{k-j+1}^{n} \frac{k_k}{k_j}}{n}, \dots\dots\dots\dots\dots\dots\dots\dots\dots\dots\dots\dots\dots\dots\dots\dots\dots \quad (4.58)$$

where n = total number of layers, n_j = number of layers flooded out, and k = permeability. Eq. 4.58 is a special case of Eqs. 4.41 and 4.42 and is applicable at the specific points in time at which the successive layers are flooded out. For M other than unity, the displacement efficiency is given by

$$E_I = \frac{n_j + \frac{(n-n_j)M}{M-1} - \frac{1}{M-1}\sum_{k=j+1}^{n}\left[M^2 + \frac{k_k}{k_j}\left(1-M^2\right)\right]^{\frac{1}{2}}}{n}. \dots\dots\dots\dots\dots\dots\dots\dots\dots \quad (4.59)$$

In piston-like displacement, only displaced fluid is produced before breakthrough and no displaced fluid is produced after breakthrough in a particular layer. Thus, the ratio of displacing fluid to displaced fluid at the producing well can be determined from this model and is given by Eqs. 4.60 and 4.61. This is the WOR for a waterflood, F_{wo}. Again, the equations apply when n_j layers have been flooded out. For $M = 1.0$,

$$F_{wo} = \frac{\sum\limits_{k=1}^{j} k_k}{\sum\limits_{k=j+1}^{n} k_k}. \quad \ldots\ldots\ldots\ldots\ldots (4.60)$$

For $M > 1.0$ or < 1.0,

$$F_{wo} = \frac{\sum\limits_{k=1}^{j} k_k}{\sum\limits_{k=j+1}^{n} \frac{k_k}{\left[M^2 + \frac{k_k}{k_j}\left(1 - M^2\right)\right]^{1/2}}}. \quad \ldots\ldots (4.61)$$

Dykstra and Parsons also developed a correlation for vertical sweep efficiency as a function of a parameter that describes permeability variation. Permeability tends to be log-normally distributed (Law 1944). On the basis of this tendency, calculations were made for a hypothetical series of reservoirs that were each assumed to consist of 50 layers. Permeabilities were assigned to each layer in each reservoir according to a log-normal distribution. The distribution was characterized by a parameter V, the permeability variation defined as follows:

$$V = \frac{k_{84.1} - k_{50}}{k_{50}}, \quad \ldots\ldots\ldots\ldots\ldots (4.62)$$

where k_{50} = permeability value at the 50th percentile and $k_{84.1}$ = permeability value at the 84.1 percentile.

The Dykstra-Parsons model can be used to obtain E_I as a function of permeability variation, V, and mobility ratios, M, at various values of WOR, F_{wo}. Results of the calculations for F_{wo} values of 1.0 and 25 for a linear reservoir are shown in **Figs. 4.29 and 4.30.** The correlations show that both reservoir heterogeneity and mobility ratio strongly affect vertical sweep efficiency. Figs. 4.29 and 4.30 give optimistic estimates of vertical sweep efficiency for layered systems because viscous fingering is not modeled and most displacement processes are not piston-like.

Dykstra and Parsons (1950), Craig (1971), and Willhite (1986) discuss methods of calculating permeability variation for particular reservoirs. A perfectly homogeneous reservoir would have a permeability variation of zero, while a totally heterogeneous reservoir would have a variation of 1.0. $V = 0.7$ is typical for many reservoirs and is often assumed when there are no data available. Willhite (1986) gives examples of the application of the Dykstra-Parsons model.

Effect of Crossflow in a Layered Reservoir. The models previously discussed were based on the assumption that crossflow does not occur between the layers of the reservoir. This would be strictly true only where permeability barriers existed between the layers. In the absence of such barriers, factors exist that lead to crossflow between layers, resulting in the introduction of error into the models discussed in the previous sections.

The effects of crossflow are difficult to treat mathematically, but may be modeled with numerical simulators. In an early paper Douglas et al. (1959) modeled an immiscible oil displacement in a stratified reservoir in which crossflow was considered. Goddin et al. (1966) later conducted a similar study. In these works, the effects of viscous and capillary forces on crossflow

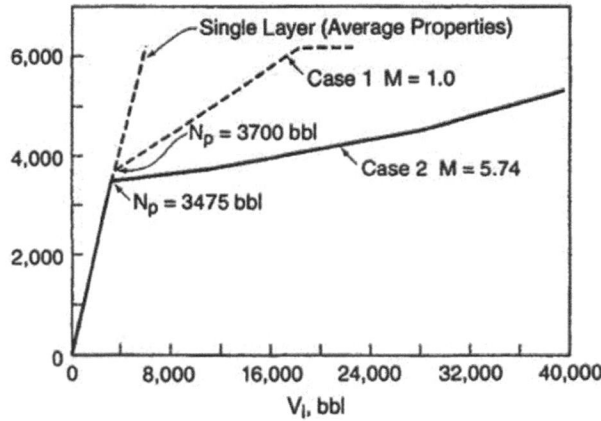

Fig. 4.26—Results of Example 4.5, oil recovery as a function of water injected.

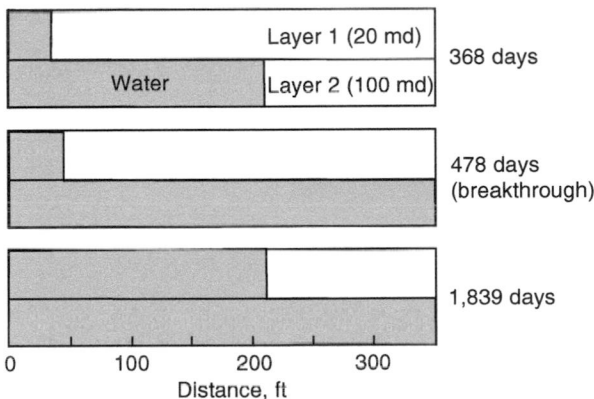

Fig. 4.27—Results of Example 4.5, relative location of the water front at different times.

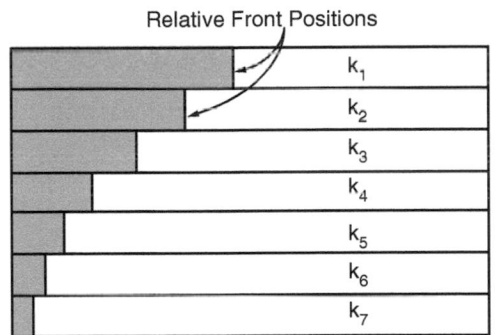

Fig. 4.28—Dykstra-Parsons model, reservoir layering.

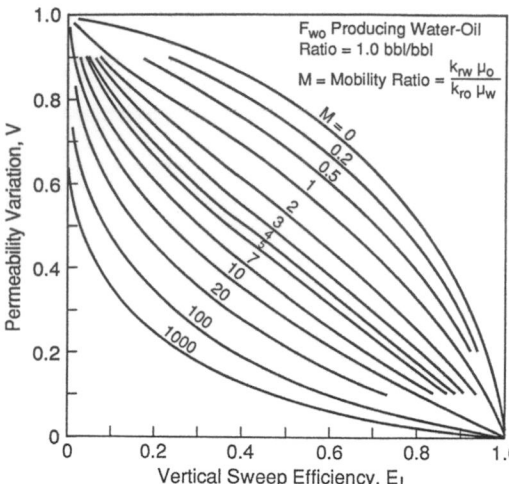

Fig. 4.29—Vertical sweep efficiency as a function of permeability variation and mobility ratio for a WOR of 1.0, linear system (Dykstra and Parsons 1950).

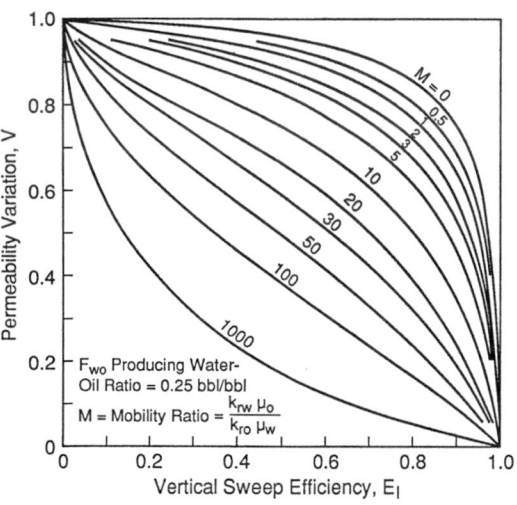

Fig. 4.30—Vertical sweep efficiency as a function of permeability variation and mobility ratio for a WOR of 25.0, linear system (Dykstra and Parsons 1950).

were analyzed. Gravity forces were considered to be negligible. Goddin et al. (1966) characterized their results in terms of a crossflow index. They showed that under certain conditions, crossflow caused by viscous and capillary forces significantly affected waterflood performance. Capillary pressure gradients clearly affected the amount and location of crossflow. They found that under conditions of favorable mobility ratio, "computed oil recovery with crossflow was always intermediate between that predicted for a uniform reservoir and that for a layered reservoir with no crossflow." The results of Goddin et al. (1966) and Douglas et al. (1959) are described in detail by Willhite (1986). Craig (1971) also reviewed published studies of crossflow. He concluded that crossflow acts to improve sweep efficiency at favorable mobility ratios, but that the reverse is true at adverse mobility ratios.

4.7 Volumetric Displacement Efficiency

As discussed in previous sections, areal and vertical sweep efficiencies generally have been determined from 2D physical or mathematical models. Areal sweep efficiencies have been determined by use of areal models that simulated various well patterns (e.g., five-spot or linedrive patterns). In these studies, vertical sweep effects were minimized by the use of homogeneous, relatively thin, porous media and/or fluids of matched density. Vertical sweep efficiencies generally have been determined by use of linear models with multiple layers (vertical heterogeneities) or a large vertical dimension. Areal effects have been minimized by the use of linear flow (i.e., by the negation of the influence of a third space dimension).

Using 3D models as a basis of correlation of either E_A or E_I is difficult because these efficiencies are not truly independent. For example, the influence of gravity override or underride on vertical sweep in a five-spot areal pattern would be affected by the 3D geometry of the reservoir model and by any vertical heterogeneities, such as layers with different properties, that exist. Thus, where 3D models have been used in studies of displacement efficiency, volumetric sweep, E_V, generally has been the efficiency of concern. One exception is the study of Withjack and Akervoll (1988) in which volumetric, areal, and vertical sweep efficiencies were reported for miscible displacements.

Likewise, use of correlations of E_A or E_I based on 2D models to calculate E_V by application of Eq. 4.6 should be undertaken with care. Geometry effects and the interdependence of factors that control E_A and E_I make this approach to calculation of E_V subject to significant error.

Volumetric displacement efficiency should be determined on the basis of the application of an approach that considers the 3D nature of a reservoir. Methods of estimating volumetric displacement efficiency in a 3D reservoir fall into two classifications (Willhite 1986). One classification is based on direct application of 3D models. While a limited number of such studies has been conducted with physical models, the application of numerical simulators has a greater potential. In this latter approach, the entire 3D reservoir is modeled by solving the fluid-flow equations in three spatial coordinates. The solution of the equation set yields volumetric displacement efficiency. This approach is limited in that the number of gridblocks that can reasonably be used in a model is limited and this affects the accuracy of the calculations. This limitation is especially true for displacements with miscible or near-miscible fluids.

A second classification is based on a layered-reservoir model. The reservoir is divided into a number of noncommunicating layers. Displacement performance is calculated in each layer with correlations or calculations based on a 2D areal model. Performances in individual layers are summed to obtain overall, or volumetric, displacement efficiency.

These two approaches to determining volumetric sweep efficiency are discussed briefly in this section.

4.7.1 Volumetric Sweep Calculations Based on Physical Modeling Studies, Five-Spot Pattern. Craig et al. (1957) conducted a study of volumetric displacement efficiency as a function of viscous/gravity ratio. Their results for a linear system were presented in Section 4.6.2. They also used physical models in a study of a five-spot pattern.

The models consisted of unconsolidated sandpacks representing one-eighth of a five-spot pattern. The larger of the models was 46 in. between wellbores and 4 in. thick. The scaling factors were such that this represented a five-spot with 10-acre spacing. Immiscible gas and liquid flooding were simulated in the five-spot models.

In another study, Withjack and Akervoll (1988) used computed tomography scanning to investigate volumetric displacement efficiency in a 3D model of a five-spot pattern. Their model was 8 in. wide, 8 in. long, and 6 in. thick and represented one-quarter of a five-spot pattern. The model was physically scaled with dimensionless scaling groups, and unconsolidated beads composed the porous medium. Miscible displacements were conducted at different mobility ratios and with fluids of different densities.

Results of both studies are consolidated and given in **Fig. 4.31,** where E_V at breakthrough of the displacing phase is correlated as a function of viscous/gravity ratio, $R_{v/g,5}$. In customary units, the ratio is defined as

$$R_{v/g,5} = \frac{512i\mu d}{k_H \Delta \rho y^2}, \quad \dotsi \quad (4.63)$$

where i is in barrels per day, μ_d is in centipoise, k is in

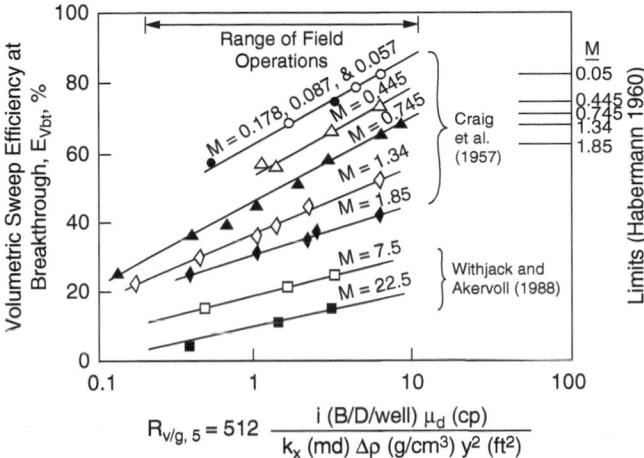

$$R_{v/g,5} = 512 \frac{i \text{ (B/D/well) } \mu_d \text{ (cp)}}{k_x \text{ (md) } \Delta\rho \text{ (g/cm}^3) \ y^2 \text{ (ft}^2)}$$

Fig. 4.31—Volumetric displacement efficiency at breakthrough as a function of the ratio of viscous/gravity forces, five-spot pattern (after Craig et al. 1957; Withjack and Akervoll 1988).

millidarcies, $\Delta\rho$ is in grams per cubic centimeter, and y is in feet. Volumetric sweep at breakthrough decreases with increasing mobility ratio and with decreasing ratio of viscous to gravity forces.

Volumetric sweep in the absence of gravity forces is not 100% at breakthrough, as discussed in Section 4.6. That is, the straight lines in Fig. 4.31 do not extrapolate to 100% sweep efficiency. The upper limits of the extrapolations may be approximated from correlations of E_A based on areal-sweep data taken in the absence of gravity effects. Data taken by Habermann (1960) were used for this purpose, and the resulting limits of various values of M are shown along the right side of Fig. 4.31.

Example 4.6—Calculation of Volumetric Sweep Efficiency in a Homogeneous Five-Spot Pattern in Which Gravity Segregation Is Possible. Consider the problem of a waterflood in a five-spot pattern shown in **Fig. 4.32** with the following data:

$h \quad - \quad 20 \text{ ft}$

$\phi \quad = \quad 0.20$

$S_{oi} \quad = \quad 0.80$

$\mu_o \quad = \quad 5 \text{ cp}$

$\mu_w \quad = \quad 1.0 \text{ cp}$

$k \quad = \quad 50 \text{ md}$

$B_o \quad = \quad 1.0 \text{ RB/STB}$

$\rho^o \quad = \quad 0.875 \text{ g/cm}^3$

$\rho_w \quad - \quad 1.00 \text{ g/cm}^3$

$k_{rw} \quad = \quad 0.27 \text{ (at } S_{or} \text{ of } 0.25)$

$k_{ro} \quad = \quad 0.94 \text{ (at } S_{iw} \text{ of } 0.20)$

$i \quad = \quad 120 \text{ B/D}$

1. Estimate the barrels of oil to be recovered at water breakthrough assuming that gravity effects are negligible.
2. Estimate the barrels of oil recovered at water breakthrough taking gravity underride into consideration.

Solution.

1. Calculate oil recovered at breakthrough with no gravity effects. Apply Fig. 4.4, the correlation of Habermann (1960), to calculate M:

$$M = \left(k_{rw}/\mu_w\right)\left(\mu_o/k_{ro}\right)$$
$$= 0.27 \, / \, 1.0 \times 5.0 \, / \, 0.94$$
$$= 1.4.$$

From the upper curve of Fig. 4.4, $E_{Abt} = 0.65$ (sweep efficiency at breakthrough).

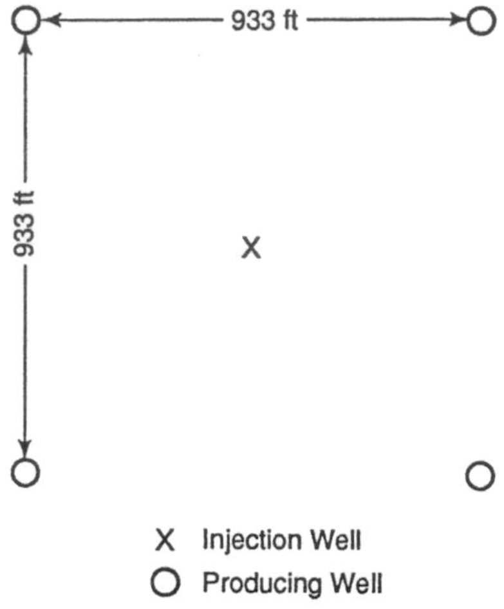

Fig. 4.32—Well pattern, Example 4.6.

$$N_p = V_p \left(S_{oi} - S_{or} \right) E_{abt}$$

$$= \left(933 \text{ ft}\right)^2 \times 20 \text{ ft} \times 0.20 \left(\frac{1}{5.615 \text{ ft}^3/\text{bbl}} \right) (0.80 - 0.25)0.65$$

$$= 221,700 \text{ bbl.}$$

2. Calculate oil recovered at breakthrough with gravity underride. Apply Fig. 4.31, the correlation of Craig et al. (1957), to calculate $R_{v/g,5}$:

$$R_{v/g,5} = 512 i \mu_o / k \Delta \rho y^2$$

$$= \frac{512(120)(5.0)}{50(1.00 - 0.875)(20)^2}$$

$$= 123.$$

In Fig. 4.31, $R_{v/g,5} = 123$ is off the chart to the right. Extrapolation of an interpolated straight line for $M = 1.4$, however, indicates that the upper limit of E_A would be reached at a value of $R_{v/g,5} < 123$. Thus, gravity underride is insignificant; $E_{Abt} = 0.65$. If the injection rate were reduced by a factor of two, so that $R_{v/g,5} = 61.5$, there would be a slight decrease of E_{Abt}.

As indicated in Example 4.6, the correlations of Craig et al. (1957) and Withjack and Akervoll (1988) can be used to estimate volumetric displacement efficiency in a homogeneous reservoir. The correlations are based on data taken in 3D physical models. The correlations are not applicable where either areal or vertical heterogeneities exist in a reservoir. Section 6.9 gives additional results from the Withjack and Akervoll study.

It is possible to use data taken in 2D models to estimate behavior in 3D systems. As an example, the existence of certain flow regimes, governed by gravity fingers and/or viscous fingering, was discussed in Section 4.6. The data for the correlations were taken in 2D models. Stalkup (1983) suggested that the correlation can be used for 3D systems by adjusting the flow velocity. For a five-spot pattern, the suggested velocity to use is

$$u = 1.25i/hd, \dotfill (4.64)$$

where i = injection rate, h = formation thickness, d = distance between production wells, and u = Darcy average velocity. The formula is based on the assumption that the proper width for calculating the average linear velocity is one-fifth the distance between wells. Stalkup (1983) cautions, however, that Eq. 4.64 is an approximation that is based on limited data and that the underlying assumptions could be substantially in error.

4.7.2 Calculations of Volumetric Sweep With Numerical Simulators. Three-dimensional volumetric displacement efficiency may be determined by numerical simulation (i.e., mathematical modeling that is based on computer solutions). This approach requires the formulation of differential equations that mathematically represent the physical displacement process. The equations then must be solved by use of mathematical numerical-analysis techniques, which require a computer. Because of the complexity of the equations and the calculation procedures, significant effort often is required to obtain a solution algorithm. Once an algorithm has been developed, however, a number of cases can be computed and the effect of different parameters on displacement performance can be evaluated.

Harpole (1980) described the application of a 3D, three-phase numerical model (black-oil model) to simulate waterflood performance in the West Seminole field of west Texas. In this reservoir, after an initial primary stage of production that lasted several years, a peripheral waterflood was established as a pressure-maintenance project. This was followed by the development of a waterflood on 40-acre five-spot patterns. **Fig. 4.33** (Harpole 1980) illustrates the field outline and flood patterns.

The numerical model first was used to match the performance history of the field. On the basis of this history-matching process, geologic and flow models of the reservoir, which served as a basis for subsequent decisions, were developed. For example, an important operating consideration throughout the field's life, and especially during the pattern waterflooding period, was the extent of vertical communication within the reservoir. There was particular concern about communication between the oil zone and the overlying gas cap. The model study indicated the degree of sensitivity of the waterflood performance to vertical communication. On the basis of the history match with the model, it was concluded that a small, but important, vertical communication existed between different layers. Subsequent management and operating decisions were based on this conclusion.

Volumetric waterflood displacement efficiency, in effect, was calculated for various proposed operating and development schemes. On the basis of the results of those calculations and corresponding economic considerations, reservoir management plans were implemented.

Some 30 months after the original simulation, model predictions for the management plan implemented were compared with field performance for that same 30-month period (**Fig. 4.34**). Agreement is quite good, indicating the utility of the modeling approach.

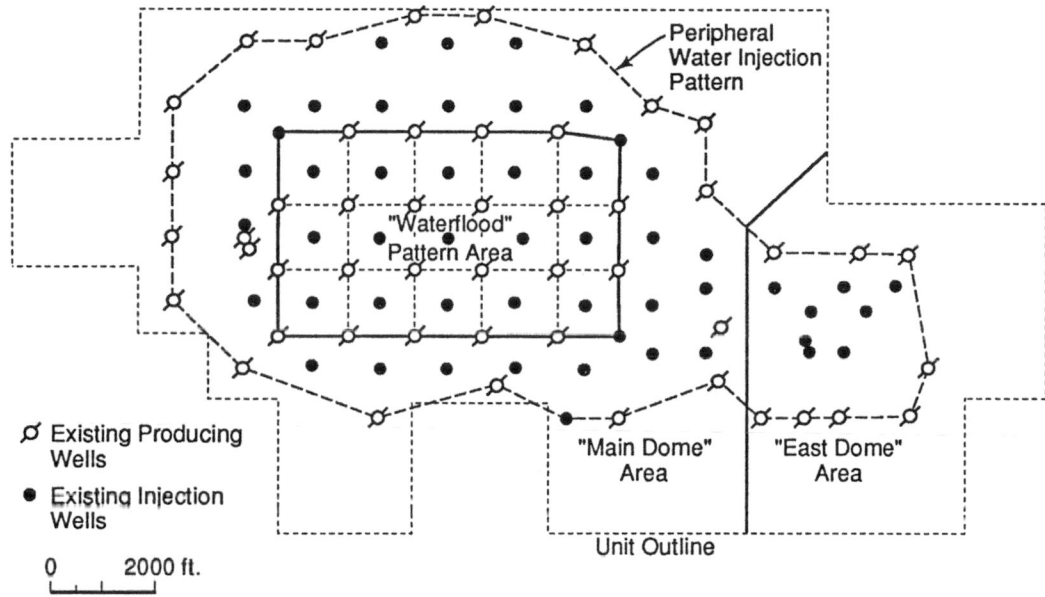

Fig. 4.33—Unit outline, West Seminole field (Harpole 1980).

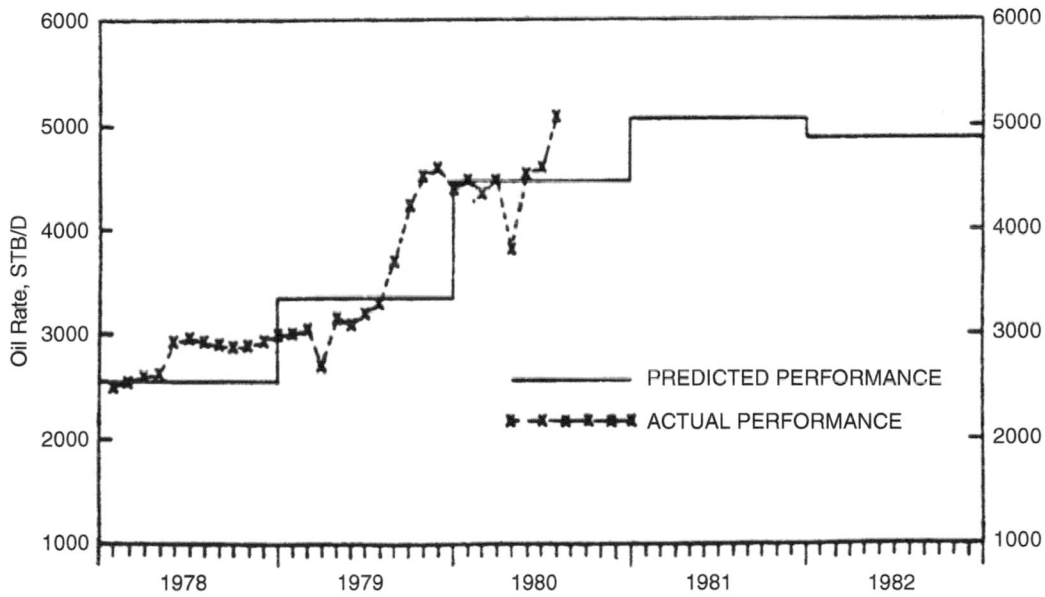

Fig. 4.34—Comparison of predicted vs. actual performance, West Seminole field (Harpole 1980).

4.7.3 Calculations of Volumetric Sweep Based on Layered Models. An approach to the simulation of performance in a 3D reservoir is to subdivide the reservoir into vertical layers with the assumption that crossflow between layers does not occur. The performance in each of the 2D layers is evaluated, and results from all layers are combined to obtain the 3D result. The key assumption in this approach is that no communication exists between layers, or if it does, any errors introduced by the crossflow are acceptable.

As discussed in Section 4.7, the layers may be described in different ways. For example, the layers and layer properties may be set on the basis of what is known about the reservoir geology. This method is valid when discrete layers are identifiable from well logs or core data. Layers also may be set from the results of a permeability-variation model such as that described by Dykstra and Parsons (1950). In this approach, the layers are hypothetical and the number of layers to be used in a model must be determined. This determination was addressed by Craig (1970).

Example 4.7—Application of a Layered Model To Calculate Volumetric Sweep Efficiency at Breakthrough, Five-Spot Pattern. Consider a 20-acre, five-spot pattern such as that illustrated in Example 4.5. The reservoir is thought to consist of two noncommunicating layers. In Layer 1, $h_1 = 8$ ft, $\phi = 0.18$, and $k_1 = 150$ md, and in Layer 2, $h_2 = 12$ ft, $\phi = 0.18$, and $k_2 = 60$ md. Other properties are the following:

S_{oi} = 0.75

k_{rw} = 0.40 (at S_{or} of 0.30)

k_{ro} = 0.98 (at S_{iw} of 0.25)

μ_w = 1.0 cp

μ_o = 2.5 cp

For piston-like displacement,

$$M = \left(\frac{k_{rw}}{\mu_w}\right)_{S_{or}} \Big/ \left(\frac{k_{ro}}{\mu_o}\right)_{S_{iw}}$$

$$= \left(\frac{0.40}{1.0}\right) \Big/ \left(\frac{0.98}{2.5}\right)$$

$$= 1.02.$$

Gravity effects are negligible. A constant pressure drop of 500 psi is maintained across the system (i.e., from injection-well to production-well bottomhole pressure).

From the Caudle and Witte (1959) results for areal displacement efficiency in a five-spot pattern (Figs. 4.9 through 4.11), calculate the volumetric sweep efficiency at water breakthrough.

Calculate relative injection rates in the two layers. The flow rate in each layer is constant because $M \approx 1.0$. Assume piston-like displacement in each layer.

Solution. Eq. 4.19 is applicable:

$$i = \frac{0.001538 k k_{ro} h \Delta p}{\mu_o \left(\log \dfrac{d}{r_w} - 0.2688 \right)}.$$

The ratio of the injection rates into Layers 1 and 2 is

$$i_1 / i_2 = (k_1 / k_2)(h_1 / h_2)$$

$$= (150)(8)/(60)(12)$$

$$= 1.67.$$

Breakthrough occurs in Layer 1. From Fig. 4.9, $E_{A1} = 0.68$ (at breakthrough). The volume injected in Layer 1 at breakthrough can be calculated from Fig. 4.9:

$$V_{i1} = A_1 h_1 \phi \left(S_{oi} - S_{or}\right) \frac{1}{5.615} \times E_{A1}$$

$$= (933 \text{ ft})^2 (8 \text{ ft})(0.18)(0.75 - 0.30)\left(\frac{1}{5.615 \text{ ft}^3/\text{bbl}}\right)0.68$$

$$= 68,300 \text{ bbl.}$$

The volume injected in Layer 2 at this same point in time is calculated as

$$V_{i2} = V_{i1}(i_2 / i_1)$$

$$= 68,000(1/1.67)$$

$$= 40,900 \text{ bbl.}$$

The volumetric sweep efficiency is determined by the ratio of the total volume injected to the total displaceable volume.

$$E_V = \frac{V_{i1} + V_{i2}}{A\left(h_1 + h_2\right)\phi\left(S_{oi} - S_{or}\right)\left(\dfrac{1}{5.615}\right)}$$

$$= \frac{(68,300 + 40,900 \text{ bbl})}{(933 \text{ ft})^2 (20 \text{ ft})0.18(0.75 - 0.30)\left(\dfrac{1}{5.615 \text{ ft}^3/\text{bbl}}\right)}$$

$$= 109,200/251,100$$

$$= 0.435.$$

Note that a vertical displacement efficiency could be calculated for this example:

$$E_V = E_{A1}E_I$$

$$0.435 = 0.68E_I$$

$$E_I = 0.64.$$

Alternatively, E_I could be calculated directly on the basis of its definition and then E_V could be determined as the product of E_{A1} and E_I.

E_V as the product of E_{A1} and E_I is calculated as follows for a two-layer five-spot pattern. At breakthrough $E_{A1} = 0.68$. At this same point in time,

$$E_{A2} = 40,900/150,688$$
$$= 0.27.$$

$$E_I = \frac{E_{A2}}{E_{A1}}\frac{(h_1 + h_2)}{(h_1 + h_2)} + \left[\frac{(E_{A2} - E_{A1})}{E_{A1}}\right]\frac{h_1}{(h_1 + h_2)}$$

$$= \frac{\dfrac{E_{A2}}{E_{A1}}(h_1 + h_2) + \dfrac{(E_{A2} - E_{A1})}{E_{A1}}h_1}{h_1 + h_2}$$

$$= \frac{\dfrac{0.27}{0.68} \times 20 + \dfrac{(0.68 - 0.27)}{0.68} \times 8 \text{ ft}}{20}$$

$$= 0.64.$$

$$E_V = E_{A1}E_1$$
$$= 0.68 \times 0.64$$
$$= 0.435 \text{ (agrees with previous calculation)}.$$

While this approach to calculation of E_V is valid, it is a more complex approach and is of limited utility. Note that this example does not establish the validity of calculating E_V as the product of E_{A1} and E_I. The calculation is valid only within the constraints of the layered, noncommunicating model applied.

The example calculation was made much simpler by use of a value of $M = 1.0$. For any other mobility ratio at constant pressure drop, the rate in each layer would be a function of time and would require a numerical integration technique to determine volume injected.

Willhite (1986) gives an example of a waterflood calculation in a linear layered reservoir in Chapter 5. That example is based on the application of the Dykstra-Parsons model (Dykstra and Parsons 1950).

Craig (1970) showed that results obtained with the layered-model approach are dependent on the number of layers used in the calculation. Where separate layers are identifiable from geologic considerations, this dependence is appropriate. When a model based on permeability variation is applied, however, a dependence on the number of layers is questionable because the layers are hypothetical (i.e., they are constructs for the model). Craig (1970) addressed this problem and developed a series of tables that show the minimum number of equal-thickness layers required to match performance of a 100-layer reservoir at specified values of M and producing WORs. This guide should be used in conjunction with the Dykstra-Parsons model.

Problems

4.1 Consider a dispersion experiment with data shown in **Table 4.6.** Assume that the experiments were conducted with the core oriented in the vertical direction. Two liquids were used in the experiment: Liquid 1 is a brine with 30.00 g NaCl/L and 650 g glycerol/L. $\rho_1 = 1.1551$ g/cm^3 and $\mu_1 = 7.4$ cp at 77°F.

Liquid 2 is a brine with 20.00 g NaCl/L and 700 g glycerol/L. $\rho_2 = 1.1609$ g/cm^3 and $\mu_2 = 11.3$ cp at 77°F.

Analyze the flow in terms of the Dumore model (Dunmore 1964) for gravity stabilized fingering for four possible displacement conditions: (1) Liquid 1 displacing Liquid 2, flow downward; (2) Liquid 1 displacing Liquid 2, flow upward; (3) Liquid 2 displacing Liquid 1, flow downward; and (4) Liquid 2 displacing Liquid 1, flow upward.

4.2 Consider the reservoir shown in **Fig. 4.35,** a horizontal, linear system having a width of 100 ft and a length of 350 ft. The reservoir consists of two layers, each 5.0 ft in thickness. Layer 1 has a permeability of 20 md, while Layer 2 has a permeability of 100 md. Porosity is 0.18.

The reservoir has an initial oil saturation, S_{oi}, of 80% and an initial water saturation, S_{iw}, of 20%. Water is injected at the end at $x = 0$, displacing oil to the production end at $x = L$ (350 ft). A constant pressure drop of 500 psi is maintained across the system.

Experimental Conditions

Flow rate (cm³/h)	6
Flow velocity (in./D)	13.8
Collection rate per hour	One test tube
Permeability before run (md)	303
System compressibility (cm³)	
Before run	3.5
After run	3.0
Average temperature (°F)	77
Temperature range (°F)	70 to 85
Average pressure in core* (psig)	40 to 115

Core Dimensions

Length (in.)	48.1
Diameter (in.)	2 to 0.2
Area (in.²)	3.1
PV (cm³)	500.53
Porosity	0.2062

Displacement Fluids

	Brine	Viscosity** (cp)	Density† (g/cm³)
Fluid initially in core	30 g/L NaCl	7.4	1.1551
	650 g/L glycerol		
Fluid injected	20 g/L NaCl	11.3	1.1609
	700 g/L glycerol		
	2.5 vol% ethanol		

Time sequence: Injection began at 1:30 a.m. on 15 June 1976, with Test Tube 141. The run ended at 1:30 p.m. on 22 June 1976.

*Pressure caused by effluent valve restricting flow.

**At 22.5 sec⁻¹ and 77°F.

†At 77°F.

Table 4.6 —Displacement Run 4, glycerol displacement in Berea Sandstone Core A, Problem 4.1.

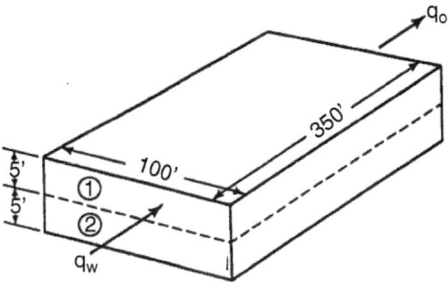

Fig. 4.35—Layered reservoir, Problem 4.2.

In the calculations to be made, it is assumed that the displacement is by plug flow. That is, only oil flows ahead of the front $(S_w = S_{iw})$ and only water flows behind the front $(S_o = S_{or})$. Calculate and plot the following:

1. Oil recovery (barrels) vs. water injected (barrels) assuming that the reservoir consists of a single layer with an average permeability of 60 md.
2. Oil recovery (barrels) vs. time (days) for the condition of Part 1.
3. Oil recovery (barrels) vs. water injected (barrels), assuming that the reservoir consists of two noncommunicating layers as described.
4. Oil recovery (barrels) vs. time (days) for the condition of Part 3.
5. For Part 4, show the relative location of the water front in the two layers for at least three different points in time.

Additional data are

μ_w = 1.0 cp at T_R,

μ_o = 20 cp at T_R,

B_o = 1.0 RB/STB,

B_w = 1.0 RB/STB,

S_{or} = 0.25

k_{rw} = 0.27 at $S_w = 1 - S_{or}$, and

k_{ro} = 0.94 at $S_o = 1 - S_{iw}$.

4.3 Consider the analysis of the linear waterflood in the two-layer reservoir of Problem 4.2. For that problem, the mobility ratio was unfavorable ($M = 5.74$). Recalculate Parts 3 and 4 of Problem 4.2 for $M = 1.0$ and 0.5. Assume that, in each case, M is modified by changing the viscosity of the injected water. Prepare two graphs to present your results: Graph 1, oil recovery vs. water injected, and Graph 2, oil recovery vs. time. Put the results for all three M values (the two above plus the original value, $M = 5.74$) on each graph.

4.4 Consider the problem of a waterflood in a five-spot pattern (Fig. 4.32) with the following data:

h = 20 ft

ϕ = 0.2

S_{oi} = 0.8

μ_o = 10 cp

μ_w = 1.0 cp

k = 50 md

B_o = 1.0 RB/STB

ρ_o = 0.875 g/cm^3

ρ_w = 1.00 g/cm^3

i = 120 B/D

1. Estimate the barrels of oil recovered at water breakthrough assuming that gravity effects are negligible.
2. Estimate the barrels of oil recovered at water breakthrough taking gravity underride into consideration. For Part 1, use an appropriate correlation presented in the text. For Part 2, use the correlation presented by Craig (1971) (Fig. 4.31).
3. Recalculate Parts 1 and 2 for $M = 1.0$.
4. Calculate oil produced as a function of water injected out to a producing WOR of 20.
5. Calculate the WOR as a function of water injected out to a ratio of 20.
6. Plot your results. Use the correlations of Caudle and Witte (1959) (Figs. 4.9 through 4.11) for your calculations.

4.5 Consider Example 4.5, which concerns linear displacement in a layered reservoir (two layers). Data are the same except for the oil viscosity. Consider Case 1, $\mu_o = 1.74$ cp, and Case 2, $\mu_o = 3.48$ cp. For both cases calculate oil recovery (barrels) vs. water injected, oil recovery (barrels) vs. time, and vertical displacement efficiency at water breakthrough. Compare the results of the first two calculations with Example 4.5 for $M = 5.74$.

Nomenclature

Parameter definitions are followed by dimensions and typical units used in text.

A =	area, L^2, ft^2	
B =	FVF, L^3/L^3, RB/STB	
B_{oi} =	initial oil FVF, L^3/L^3, RB/STB	
B_{o1} =	oil FVF at the beginning of process, L^3/L^3, RB/STB	
B_{o2} =	oil FVF at the end of process, L^3/L^3, RB/STB	
C_i =	concentration, L^3/L^3, vol/vol	
d =	distance between wells, L, ft	
E =	overall hydrocarbon recovery efficiency	
E' =	recovery efficiency in Spivak (1974) correlation, Fig. 4.18	
E_A =	areal sweep efficiency	
E_{Abt} =	areal sweep efficiency at breakthrough	

E_D = microscopic displacement efficiency

E_I = vertical displacement efficiency

E_{Ibt} = vertical displacement efficiency at breakthrough

E_P = pattern sweep efficiency

E_V = macroscopic (volumetric) recovery efficiency

E_{Vbt} = volumetric sweep efficiency at breakthrough

f = fractional flow of a phase, L^3/L^3, volume fraction

$F_{g/v}$ = gravity/viscous ratio defined in Eq. 4.26

F_i = displaceable PVs of fluid injected, Claridge (1972) correlation

F_{ibt} = displaceable PVs of fluid injected at breakthrough of injected fluid at producing well, Claridge (1972) correlation

F_{wo} = producing WOR, L^3/L^3, vol/vol

F_μ = effective viscosity ratio, Koval (1963) model, Eq. 4.23

g = gravity acceleration constant L/t^2, 32.174 ft/sec^2

h = formation thickness, L, ft

h_j = thickness of Layer j, L, ft

h_k = thickness of Layer k, L, ft

i = volumetric injection rate, L^3/t, B/D

i_{abt} = injection rate after breakthrough, L^3/t, B/D

k = absolute permeability, L^2, md

k_i = effective permeability of Phase i, L^2, md

k_j = absolute permeability of Layer j, L^2, md

k_k = absolute permeability of Layer k, L^2, md

k_r = relative permeability

k_x = permeability in x-direction, L^2, md

k_y = permeability in y-direction, L^2, md

k_{50} = permeability at 50th percentile, log-normal distribution, Eq. 4.62, L^2, md

$k_{84.1}$ = permeability at 84.1 percentile, log-normal distribution, Eq. 4.62, L^2, md

L = length, L, ft

L_p = length of flow perturbation, L, ft

M = mobility ratio

$M_{\bar{s}}$ = mobility ratio based on average saturation, Eq. 4.14

M_t = mobility ratio based on total mobilities, Eq. 4.16

n = number of layers

n_j = number of layers flooded out

N = original OIP, L^3, STB

N_p = oil displaced or produced, L^3, STB

N_1 = OIP at beginning of process, L^3, STB

p = pressure, m/Lt^2, psi

p_r = reference pressure, m/Lt^2, psi

Δp = pressure drop, m/Lt^2, psi

q = volumetric flow rate, L^3/t, B/D

q_j = total volumetric injection or flow rate in Layer j, L^3/t, B/D

q_t = total volumetric injection or flow rate, L^3/t, B/D

r_w = wellbore radius, L, ft

$R_{v/g}$ = viscous/gravity ratio defined in Eq. 4.24

$\bar{S}$ = average saturation, L^3/L^3, volume fraction

S_{dr} = residual saturation of displaced phase, L^3/L^3, volume fraction

S_{Dr} = residual saturation of displacing phase, L^3/L^3, volume fraction

S_{iw} = interstitial water saturation, L^3/L^3, volume fraction

S_{oi} = initial oil saturation, L^3/L^3, volume fraction

S_{or} = residual oil saturation, L^3/L^3, volume fraction

S_{o1} = oil saturation at beginning of displacement process, L^3/L^3, volume fraction

S_{o2} = oil saturation at end of displacement process, L^3/L^3, volume fraction

t = time, t, days

T_R = reservoir temperature, T, °F

u = Darcy velocity, L/t, B/(D-ft^2)

u_c = critical velocity defined in Eq. 4.31, L/t, ft/D

u_i = Darcy velocity of Phase i, L/t, STB/(D-ft^2)

u_{st} = critical velocity defined in Eq. 4.36, L/t, ft/D

V = Dykstra-Parsons coefficient, Eq. 4.62

V_i = total volume injected, L^3, bbl

V_p = PV, L^3, bbl

V_{pd} = displaceable PV, L^3, bbl

x = distance, L, ft

X_f = location of a displacement front, L, ft

y = vertical dimension, L, ft

γ = conductance ratio

θ = angle of reservoir dip

λ = mobility, L^3t/m, md/cp

λ_i = mobility of Phase i, L^3t/m, md/cp

λ_j = mobility of fluid in Layer j, L^3t/m, md/cp

λ_k = mobility of fluid in Layer k, L^3t/m, md/cp

λ_r = relative mobility, k_r/μ, L^3t/m, cp^{-1}

λ_t = total mobility; sum of mobilities of all flowing phases, L^3t/m, md/cp

μ = viscosity, m/Lt, cp

μ_i = viscosity of Phase i, m/Lt, cp

ρ = density, m/L^3, g/cm^3

$\Delta\rho$ = density difference, m/L^3, g/cm^3

ϕ = porosity

Subscripts

abt = after breakthrough

bt = breakthrough

d = displaced phase

D = displacing phase

H = horizontal

o = oil

s = solvent

V = vertical

w = water

References

Caudle, B. H. and Witte, M. D. 1959. Production Potential Changes During Sweep-Out in a Five-Spot System. *J Pet Technol* **12** (12): 63–65. SPE-1334-G. https://doi.org/10.2118/1334-G.

Caudle, B. H., Hickman, B. M., and Silberberg, I. H. 1968. Performance of the Skewed Four-Spot Injection Pattern. *J Pet Technol* **20** (11): 1315–1319. SPE-2128-PA. https://doi.org/10.2118/2128-PA.

Claridge, E. L. 1972. Prediction of Recovery in Unstable Miscible Flooding. *SPE J.* **12** (2): 143–155. SPE-2930-PA. https://doi.org/10.2118/2930-PA.

Craig, F. F. Jr. 1970. Effect of Reservoir Description on Performance Predictions. *J Pet Technol* **22** (10): 1239–1245. SPE-2652-PA. https://doi.org/10.2118/2652-PA.

Craig, F. F. Jr. 1971. *The Reservoir Engineering Aspects of Waterflooding*, Vol. 3. Richardson, Texas: Monograph Series, SPE.

Craig, F. F. Jr., Geffen, T. M., and Morse, R. A. 1955. Oil Recovery Performance of Pattern Gas or Water Injection Operations From Model Tests. In *Petroleum Transactions*, AIME Vol. 204, 7–15. SPE-413-G.

Craig, F. F. Jr., Sanderlin, J. L., Moore, D. W. et al. 1957. A Laboratory Study of Gravity Segregation in Frontal Drives. In *Petroleum Transactions*, AIME, Vol. 210, 275–282. SPE-676-G.

Crane, F. E., Kendall, H. A., and Gardner, G. H. F. 1963. Some Experiments of the Flow of Miscible Fluids of Unequal Density Through Porous Media. *SPE J.* **3** (4): 277–280. SPE-535-PA. https://doi.org/10.2118/535-PA.

Dake, L. P. 1978. *Fundamentals of Reservoir Engineering*. New York City: Elsevier Science Publishing Co. Inc..

Dougherty, E. L. and Sheldon, J. W. 1964. The Use of Fluid-Fluid Interfaces To Predict the Behavior of Oil Recovery Processes. *SPE J.* **4** (2): 171–182. SPE-781-PA. https://doi.org/10.2118/781-PA.

Douglas, J. Jr., Peaceman, D. W., and Rachford, H. H. 1959. A Method for Calculating Multi-Dimensional Immiscible Displacement. In *Petroleum Transactions*, AIME, Vol. 216, 297–308. SPE-1327-G.

Dumore, J. M. 1964. Stability Considerations in Downward Miscible Displacements. *SPE J.* **4** (4): 356–362. SPE-961-PA. https://doi.org/10.2118/961-PA.

Dyes, A. B., Caudle, B. H., and Erickson, R. A. 1954. Oil Production After Breakthrough as Influenced by Mobility Ratio. *J Pet Technol* 6 (4): 27–32. SPE-309-G. https://doi.org/10.2118/309-G.

Dykstra, H. and Parsons, R. L. 1950. The Prediction of Oil Recovery by Waterflood. In *Secondary Recovery of Oil in the United States*, second edition, 160. New York City: American Petroleum Institute.

Geertsma, J., Croes, G. A., and Schwarz, N. 1956. Theory of Dimensionally Scaled Models of Petroleum Reservoirs. In *Petroleum Transactions*, AIME, Vol. 207, 118–127. SPE-539-G.

Goddin, C. S. Jr., Craig, F. F. Jr., Wilkes, J. O. et al. 1966. A Numerical Study of Waterflood Performance in a Stratified System With Crossflow. *J Pet Technol* **18** (6): 765–771. SPE-1223-PA. https://doi.org/10.2118/1223-PA.

Guckert, L. G. 1961. *Areal Sweepout Performance of Seven- and Nine-Spot Patterns*. MS thesis, Pennsylvania State University, University Park, Pennsylvania.

Habermann, R. 1960. The Efficiency of Miscible Displacement as a Function of Mobility Ratio. In *Petroleum Transactions*, AIME, Vol. 219, 264–72. SPE-1540-G.

Harpole, K. J. 1980. Improved Reservoir Characterization—A Key to Future Reservoir Management for the West Seminole San Andres Unit. *J Pet Technol* **32** (11): 2009–2019. SPE-8274-PA. https://doi.org/10.2118/8274-PA.

Hawthorne, R. G. 1960. Two-Phase Flow in Two-Dimensional Systems—Effects of Rate, Viscosity and Density or Fluid Displacement in Porous Media. In *Petroleum Transactions*, AIME, Vol. 219, 81–87. SPE-1323-G.

Higgins, R. V. and Leighton, A. J. 1962. A Computer Method To Calculate Two-Phase Flow in Any Irregularly Bounded Porous Media. *J Pet Technol* **14** (6): 679–683. SPE-243-PA. https://doi.org/10.2118/243-PA.

Hill, S. 1952. Channeling in Packed Column. *Chemical Engineering Science* **1** (6): 247–253. https://doi.org/10.1016/0009-2509(52)87017-4.

Kimbler, O. K., Caudle, B. H., and Cooper, H. E. Jr. 1964. Areal Sweepout Behavior in a Nine-Spot Injection Pattern. *J Pet Technol* **16** (2): 199–202. SPE-784-PA. https://doi.org/10.2118/784-PA.

Koval, E. J. 1963. A Method for Predicting the Performance of Unstable Miscible Displacement in Heterogeneous Media. *SPE J.* **3** (2): 145–154. SPE-450-PA. https://doi.org/10.2118/450-PA.

Landrum, B. L. and Crawford, P. B. 1960. Effect of Directional Permeability on Sweep Efficiency and Production Capacity. *J Pet Technol* **12** (11): 67–71. SPE-1597-G. https://doi.org/10.2118/1597-G.

Law, J. 1944. A Statistical Approach to the Interstitial Heterogeneity of Sand Reservoirs. In *Transactions*, AIME, Vol. 155, No. 1, 202–222. SPE-944202-G. https://doi.org/10.2118/944202-G.

Mahaffey, J. L., Rutherford, W. M., and Matthews, C. S. 1966. Sweep Efficiency by Miscible Displacement in a Five-Spot. *SPE J.* **6** (1): 73–80. SPE-1233-PA. https://doi.org/10.2118/1233-PA.

Slobod, R. L. and Caudle, B. H. 1952. X-Ray Shadowgraph Studies of Areal Sweepout Efficiencies. In *Petroleum Transactions*, AIME, Vol. 195, 265–270. SPE-211-G.

Spivak, A. 1974. Gravity Segregation in Two-Phase Displacement Processes. *SPE J.* **14** (6): 619–632. SPE-4630-PA. https://doi.org/10.2118/4630-PA.

Stalkup, F. I. 1983. *Miscible Displacement*, Vol. 8. Richardson, Texas: Monograph Series, SPE.

Willhite, G. P. 1986. *Waterflooding*, Vol. 3. Richardson, Texas: Textbook Series, SPE.

Withjack, E. M. and Akervoll, I. 1988. Computed Tomography Studies of 3-D Miscible Displacement Behavior in a Laboratory Five-Spot Model. Presented at the SPE Annual Technical Conference and Exhibition, Houston, 2–5 October. SPE-18096-MS. https://doi.org/10.2118/18096-MS.

SI Metric Conversion Factors

bbl	$\times$	1.589 873	E–01	=	m³
cp	$\times$	1.0*	E–03	=	Pa·s
ft	$\times$	3.048*	E–01	=	m
ft²	$\times$	9.290 304*	E–02	=	m²
°F		(°F–32)/1.8		=	°C
in.	$\times$	2.54*	E+00	=	cm
in.²	$\times$	6.451 6*	E+00	=	cm²
md	$\times$	9.869 233	E–04	=	μm²
psi	$\times$	6.894 757	E+00	=	kPa

* Conversion factor is exact.

Chapter 5

Mobility-Control Processes

5.1 Introduction

Mobility control is a generic term describing any process by which an attempt is made to alter the relative rates at which injected and displaced fluids move through a reservoir. The objective of mobility control is to improve the volumetric sweep efficiency of a displacement process. In some processes, there is also an improvement in microscopic displacement efficiency at a specified volume of fluid injected. Mobility control is usually discussed in terms of the mobility ratio, M, and a displacement process is considered to have mobility control if $M \le 1.0$. Volumetric sweep efficiency generally increases as M is reduced, and it is sometimes advantageous to operate at a mobility ratio considerably less than unity, especially in reservoirs with substantial variation in the vertical or areal permeability.

Examination of Eq. 4.14 shows how the mobility ratio can be changed in a displacement process. Inspection of Eq. 4.14, reproduced here as Eq. 5.1, indicates that the mobility ratio can be modified by any combination of changes in the permeabilities of the rock to the displacing and displaced fluids and/or changes in fluid viscosities.

$$M = \left(k_{rD} / \mu_D\right)_{S_D} \left(\mu_d / k_{rd}\right)_{S_d}, \dots\dots\dots\dots\dots\dots\dots\dots\dots\dots\dots\dots\dots\dots\dots\dots(5.1)$$

where k_{rD} = relative permeability of the displacing phase, k_{rd} = relative permeability of the displaced phase, μ_D = viscosity of the displacing phase, μ_d = viscosity of the displaced phase, S_D – average saturation of the displacing phase in the region behind the displacing-phase front, S_d = average saturation of the displaced phase in the region ahead of the displacing-phase front, and consistent units are used.

Because it is often not feasible to change the properties of the displaced fluid when it is oil or the permeabilities of the rock to the displaced fluids, most mobility control processes of current interest involve addition of chemicals to the injected fluid. These chemicals increase the apparent viscosity of the injected fluid and/or reduce the effective permeability of rock to the injected fluid. The chemicals used are primarily polymers when the injected fluid is water and surfactants that form foams when the injected fluid is a gas. In some cases, mobility control is attained in gas-injection processes by the injection of alternate slugs of gas and water.

This chapter introduces the principles of mobility control applicable to the development of enhanced-oil-recovery (EOR) processes. Mobility control is essential to the effectiveness of such processes as micellar/polymer flooding and offers much potential to improve the effectiveness of other processes, such as miscible gasflooding and steam displacement.

5.2 Process Description

5.2.1 Polymer-Augmented Waterflooding. High-molecular-weight water-soluble polymers in dilute concentrations [on the order of a few hundred parts per million (ppm)] increase the viscosity of water significantly. Two types of polymers are commonly used for mobility control in waterfloods: partially hydrolyzed polyacrylamide and xanthan biopolymers. Properties of these polymers are discussed later, but here note that the mobility of polymer solutions containing polyacrylamide is reduced by a combination of increased solution viscosity and reduced rock permeability caused by polymer retention. In contrast, xanthan polymer solutions reduce the mobility of the injected solution by increasing the solution viscosity. When the initial mobility ratio is unfavorable for a waterflood, mobility control can increase the microscopic displacement efficiency at a specified water/oil ratio (WOR). In most cases, polymer flooding is used to increase volumetric sweep efficiency.

In a polymer-augmented waterflood, polymer is injected continuously at the initial polymer concentration for a limited period. Reducing the polymer concentration systematically as more pore volumes (PVs) are injected (as depicted in **Fig. 5.1**) is the most cost-effective method to conduct a flood. After sufficient polymer has been injected, the polymer slug is displaced through the reservoir by injecting water. Polymer selection, injected concentration, polymer volume injected, and the method of reducing the polymer concentration with PVs of fluid injected are determined from data obtained in laboratory experiments and by simulating the polymer flood with computer models (Sorbie 1991a).

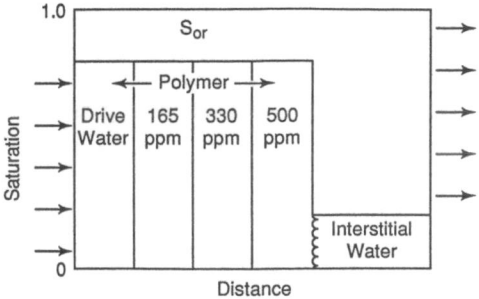

Fig. 5.1—Injection schedule for a continuous polymer flood.

5.2.2 Mobility Control To Complement Other EOR Processes. *Chemical Flooding.* Chemical flooding processes are based on the injection of a surfactant solution or the in-situ generation of surfactants by injected-solution/crude-oil reaction to mobilize oil. Because the injected chemicals are expensive, a small chemical slug (approximately 5 to 40% PV) is injected in the process. This slug is displaced through the reservoir by a polymer bank, which in turn is displaced by drive water. **Fig. 5.2** (Reppert et al. 1990) illustrates a typical chemical flooding process. Mobility control is essential for three steps in the chemical flooding process. In the chemical slug, mobility control is needed to prevent the chemical slug from fingering into the oil/water bank where it would

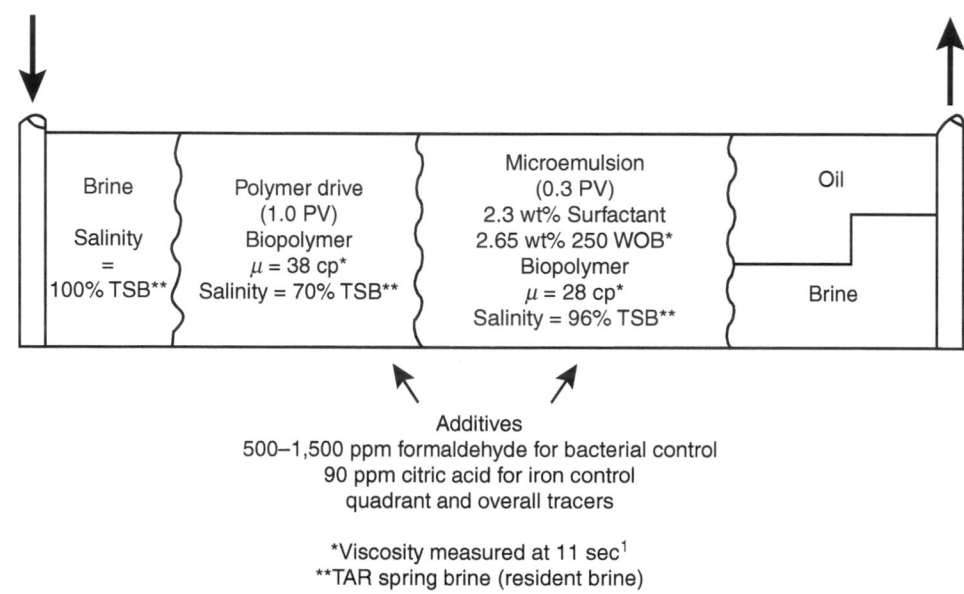

Additives
500–1,500 ppm formaldehyde for bacterial control
90 ppm citric acid for iron control
quadrant and overall tracers

*Viscosity measured at 11 sec^1
**TAR spring brine (resident brine)

Fig. 5.2—Mobility control of a chemical flood process (Reppert et al. 1990).

dissipate by dispersive mixing. Mobility control between the mobility buffer (polymer bank) and chemical slug minimizes dissipation of the slug by mixing with the mobility buffer. Finally, mobility control is needed to prevent the drive water from fingering through the mobility buffer into the chemical slug.

Miscible Gas-Injection Processes. In miscible flooding by gaseous solvents, a slug of solvent is injected that is either miscible with the crude oil on first contact or develops miscibility through repeated contacts with the oil. Because the viscosities of most solvents are much lower than oil or water viscosities, the mobility ratio is typically very unfavorable. This leads to such negative effects as viscous fingering, channeling through high-permeability zones, and generally reduced sweep efficiency. Mobility control improves the performance of such miscible processes.

The water-alternating-gas (WAG) process often is used for mobility control in miscible gas-injection processes. Mobility control is obtained by choosing a water/gas injection ratio that minimizes gas bypassing. Although the process is based in principle on simultaneous flow of water and gas in porous rock, as depicted in **Fig. 5.3**[2], gas and water are injected as slugs in practice. Water and gas injection slugs are sized so that two-phase flow is obtained in the mixing zone as gas fingers through the water slug. WAG ratios can be computed from relative permeability data for linear displacement but also may be determined empirically in field applications from interpretation of field data.

Steamflooding. Mobility control is also a concern in steamflooding. The injected steam rises to the top of the reservoir as a result of gravity segregation. Areal sweep is often quite high in this region. Oil displacement is confined primarily to the upper portion of a reservoir, leaving a transition zone of hot water above the cold oil, as shown in **Fig. 5.4** (Blevins et al. 1969). The region contacted by steam may have residual oil saturations (ROSs) as low as 5 to 10%. The steam mobility in this region of steam displacement is high, so injection rates must be reduced to avoid excessive steam production at production wells. Vertical expansion of the steam zone is slow relative to lateral movement. Mobility control is needed to reduce steam movement laterally across the pattern, thereby promoting faster vertical expansion of the steam zone.

In-situ generation of foam has potential as a mobility-control process for both miscible and steamflooding processes. Foam forms when a gas phase displaces a liquid solution containing a surfactant in porous rock. The foam is a viscous, compressible

[2] Personal communication with Larry W. Lake, University of Texas at Austin, 1986.

dispersion of gas in water separated by a surfactant film at the interface. Mobility control is obtained in the foamed region because the permeability of gas in the foamed region is reduced markedly. The properties of the foam are governed by the porous medium, the characteristics of the surfactant, and the foam quality. These properties must be determined from laboratory testing. In practice, the surfactant solution may be injected as a slug between gas-injection cycles or concurrently. An inert gas, such as nitrogen or methane, must be injected with the steam to develop foams in steamflooding applications.

5.3 Physical and Chemical Characteristics of Polymers

5.3.1 Polymer Types. Two types of polymers, often called macromolecules, are used widely in EOR processes: polyacrylamides and polysaccharides. Polyacrylamides can be manufactured by polymerization of the acrylamide monomer, shown in the brackets in **Fig. 5.5** (Willhite and Dominguez 1977), to produce a polymer

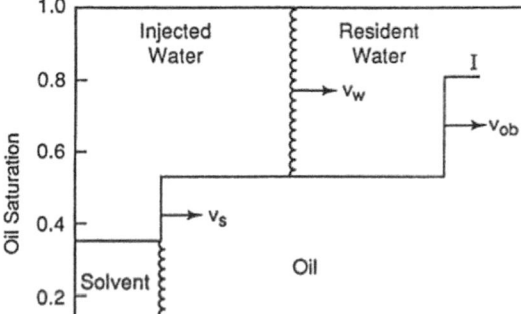

Fig. 5.3—Mobility control with alternate injection of water and gas[2].

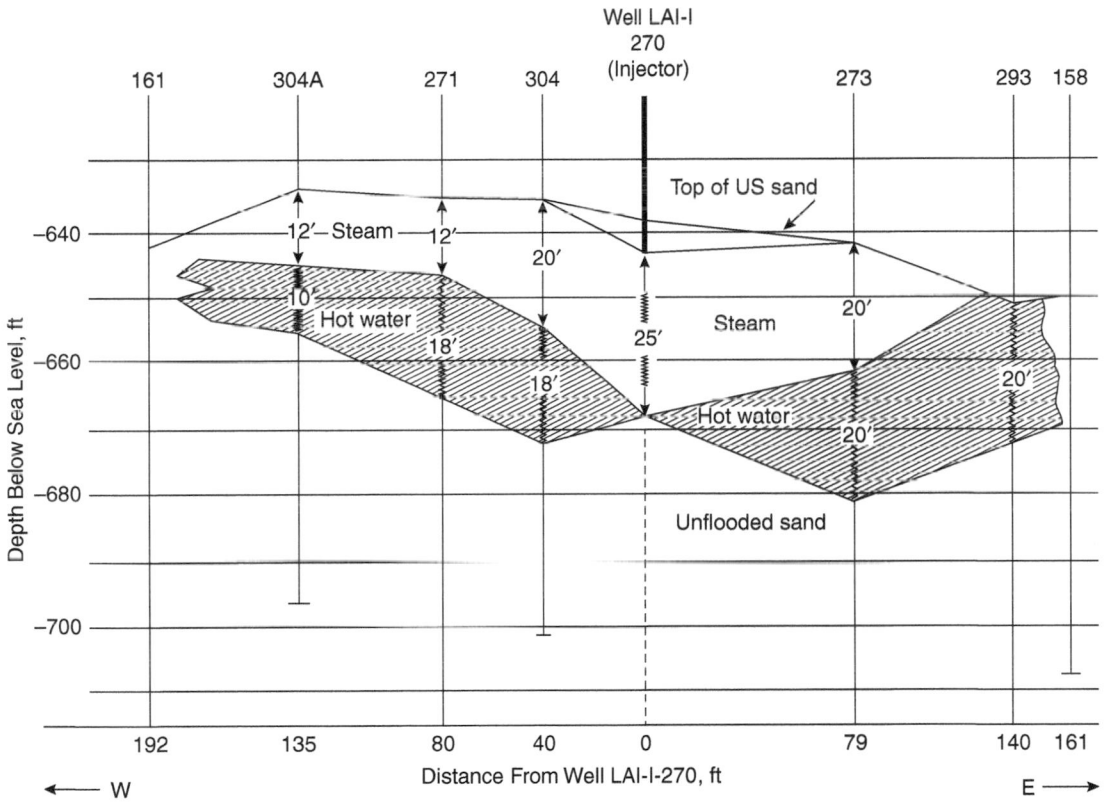

Fig. 5.4—Gravity segregation in a steam displacement process (Blevins et al. 1969).

that resembles a flexible coil. Polymerization produces macromolecules with average molecular weights ranging from 0.5 million to 30 million, depending on the extent of polymerization. Molecular weights commonly used range from 1 million to 10 million. Polyacrylamide adsorbs strongly on mineral surfaces. Thus, the polymer is partially hydrolyzed to reduce adsorption by reacting polyacrylamide with a base, such as sodium hydroxide or potassium hydroxide or sodium carbonate. Hydrolysis converts some of the amide groups (NH_2) to carboxyl groups (COO^-), as shown in Fig. 5.5. The degree of hydrolysis is the fraction of amide groups that are converted by hydrolysis and ranges from 15 to 35% in commercial products. Polyacrylamide also is used in the "unhydrolyzed" form in some applications. Even "unhydrolyzed" polyacrylamide will have small amounts (2 to 4%) of hydrolyzed groups unless exceptional precautions are taken in the manufacturing process. Partially hydrolyzed polyacrylamides are also produced by copolymerization. Polyacrylamides are supplied as a dry polymer or as liquid emulsions with oil- or water-external systems.

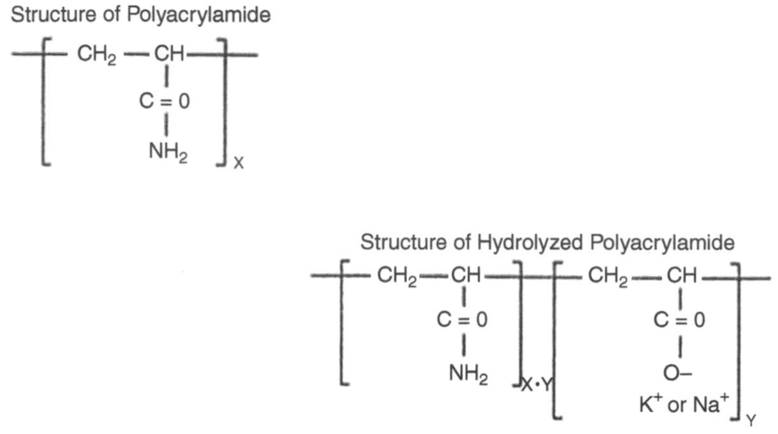

Structure of Polyacrylamide

Structure of Hydrolyzed Polyacrylamide

Fig. 5.5—Molecular structure of high-molecular-weight polyacrylamides (Willhite and Dominguez 1977).

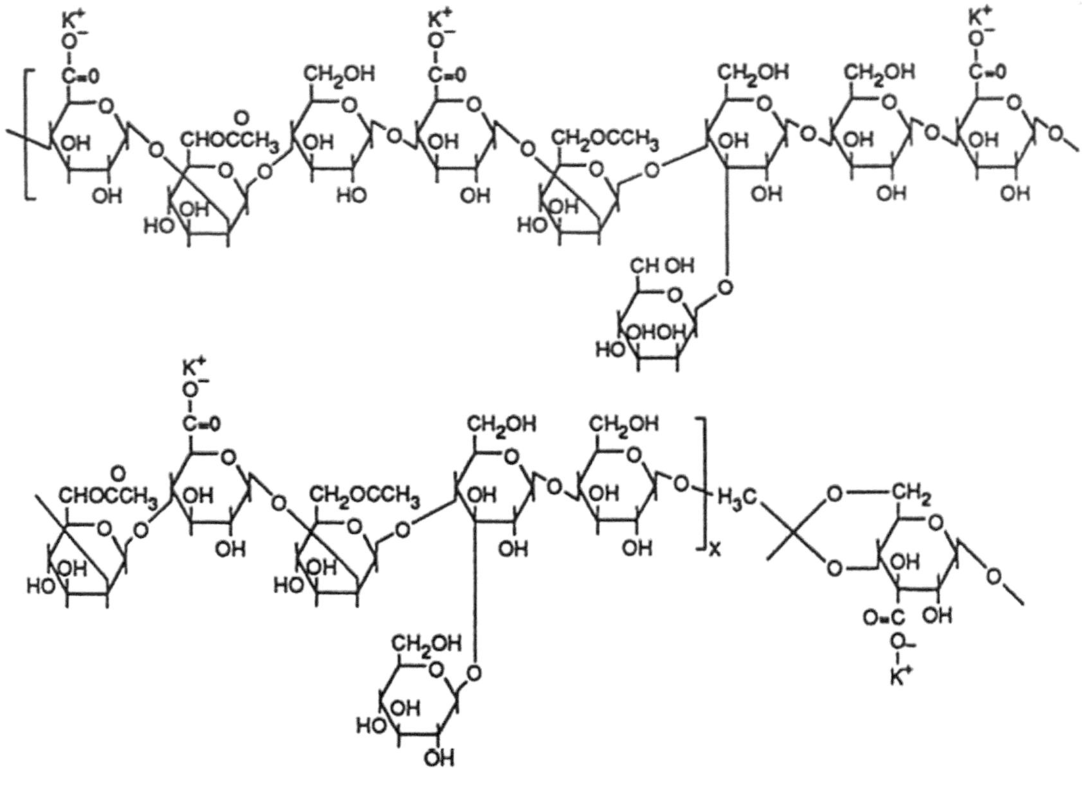

Fig. 5.6—Structure of xanthan biopolymer (Xanthan Gum/Keltrol/Kelzan 1974).

The most widely used polysaccharide is xanthan gum, which is a biopolymer produced commercially by microbial action of the organism *Xanthomonas campestris* (Lipton 1974) on a carbohydrate feed stock. Typical structure of the xanthan biopolymer is shown in **Fig. 5.6** (Xanthan Gum/Keltrol/Kelzan Tech. Bull. 1974). The polymer acts like a semirigid rod and is quite resistant to mechanical degradation. Average reported molecular weights of xanthan biopolymers used in EOR processes range from 1 million to 15 million, depending on the method used to determine the molecular weight. The properties of a particular biopolymer depend to a large measure on the organism used to manufacture the polymer. There are many strains of *Xanthomonas campestris,* and different biopolymers consequently have been developed from these strains. Xanthan biopolymers are supplied as a dry powder or as a concentrated broth. Unless special precautions are taken in the manufacturing process, the biopolymer product contains cellular debris that must be removed by filtration before it can be injected into porous rocks. Dry xanthan biopolymers are also susceptible to formation of microgels, which have plugging tendencies (Chauveteau and Kohler 1984, Kohler and Chauveteau 1981). Kolodziej (1987) proposed that the formation of microgels in solutions made from xanthan was dependent on the presence of denatured proteins and salt.

Unlike ordinary chemical reactions, where the products are discrete, identifiable molecular species, all polymerization reactions yield polymers with a broad distribution of molecular weights. This characteristic is called polydispersivity. **Fig. 5.7** (Herr and

Routson 1974) shows an example of the size distribution for a high-molecular-weight polyacrylamide. Molecular weight distributions are difficult to obtain, so product specifications report an average molecular weight based on weight average or number average. Viscosity-averaged molecular weights often are determined from correlations of intrinsic viscosity with molecular weight (*API RP 63* 1990).

Other polymers have been developed for particular applications. For example, polyvinylpyrrolidone (Doe et al. 1987) was developed for high-temperature applications in harsh environments where polyacrylamides and biopolymers were found not to be applicable.

5.3.2 Polymer Stability. The property that makes polymers useful for EOR applications is that small concentrations of polymer, on the order of a few hundred to a few thousand ppm (by weight), increase the viscosity of an aqueous solution significantly. The rheology of these solutions is examined later. This section addresses the critical question of polymer selection for a particular reservoir environment.

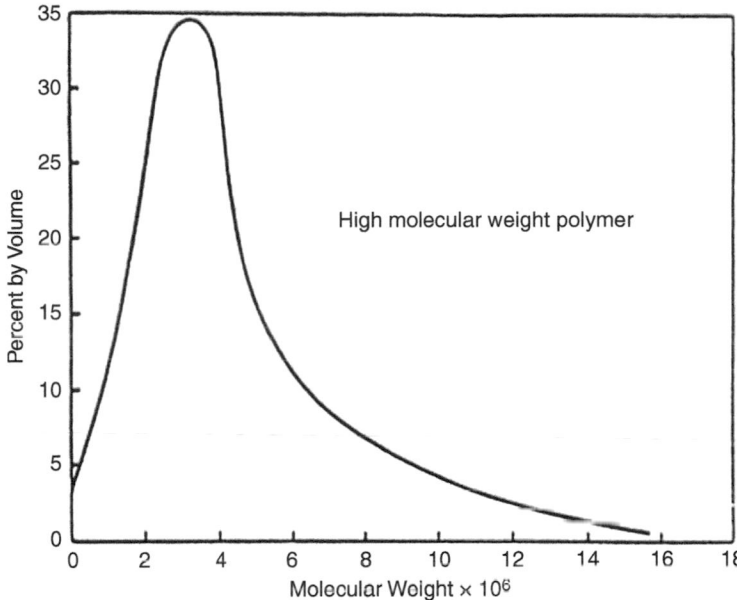

Fig. 5.7—Frequency distribution of high-molecular-weight polyacrylamide at 1-ppm concentration (Herr and Routson 1974).

For a polymer to be useful in EOR applications, it must be stable at reservoir conditions for the expected residence time in the reservoir rock. Because polymers can degrade under certain conditions, short-time laboratory tests can be misleading. Polymer stability at reservoir temperature and in the presence of reservoir brine is essential to EOR applications.

Several papers (Knight 1973; Wellington 1983; Yang and Treiber 1985) describe methods of evaluating polymer stability and present experimental data defining principal effects. The important variables are summarized here.

It is well-established that both polyacrylamides and biopolymers are susceptible to oxidative attack by dissolved oxygen in the injected water. Degradation is detected by the loss of solution viscosity with time (Knight 1973; Wellington 1983; Yang and Treiber 1985). At low temperatures, the reaction rate is slow and can go undetected in short tests. The degradation rate increases as temperature increases, which is consistent with chemical reaction kinetics. The oxidative degradation reaction is catalyzed by dissolved metal ions, such as Fe^{+++} (Shupe 1981). Degradation by oxidative attack can be prevented or minimized by reducing the oxygen content of the water or brine to less than a few parts per billion.

This usually is done by use of oxygen scavengers or deaeration. Sodium dithonite is used to stabilize polyacrylamides (Pye 1967). Yang and Treiber (1985) provide guidelines for the use of oxygen scavengers to prevent degradation of polyacrylamides in field brines under simulated reservoir conditions. A mixture of thiourea, isopropyl alcohol, and sodium bisulfate was found to retard oxidative attack on xanthan biopolymers at temperatures up to 207°F (Wellington 1983).

Oxygen scavengers are not typically used in field applications because most reservoirs have a reducing environment and because dissolved oxygen is consumed rapidly after the injected fluid containing oxygen enters the reservoir.

The thermal stability of polymers (i.e., stability at higher temperatures) is a second important consideration. Laboratory tests indicate that the carbon/carbon backbone of polyacrylamides is stable in the absence of oxygen and divalent ions to temperatures up to 194°F (Ryles 1983). During incubation at high temperatures, however, polyacrylamides undergo hydrolysis by reaction of the amide groups with water. This is reflected in an increase in solution viscosity, as shown in **Fig. 5.8.** The behavior of xanthan polymers at elevated temperatures is complex (Seright and Henrici 1990). Acetyl groups in this polymer are susceptible to base-catalyzed hydrolysis. For example, unbuffered xanthan solutions exhibit a decrease in pH when exposed to elevated temperatures. The increase in hydrogen-ion concentration is attributed to the generation of H+ when O-acetyl groups are hydrolyzed. The xanthan molecule has a helical structure that appears to take on different configurations, depending upon salinity, divalent-ion concentrations, and temperature. Changes in structural configuration can be correlated in terms of a transition temperature. The transition temperature increases with salinity and divalent-ion content (Seright and Henrici 1990). In the absence of dissolved oxygen, Seright and Henrici (1990) estimate that a xanthan solution could maintain at least half its original viscosity for 5 years if the temperature does not exceed 167 to 176°F. This estimate is consistent with current practice, in which most xanthan biopolymers are limited to applications where the temperature is ≤140°F. Scleroglucan polymers have been reported to be more stable than xanthan at elevated temperatures (Davison and Mentzer 1982; Kalpakci et al. 1990).

It is usually necessary to prepare the injected polymer solution in reservoir brine. Reservoir brines often contain high concentrations of divalent cations, in particular Ca^{++} and Mg^{++}. The solution viscosity of each polymer is affected by the presence of divalent cations, as discussed in Section 5.3.3. Both polyacrylamides and biopolymers are stable in high concentrations of divalent cations at low reservoir temperatures (Davison and Mentzer 1982; Zaitoun and Potie 1983; Moradi-Araghi and Doe 1987). Ferric ion (Fe^{+++}) will cause gelation of polyacrylamide and must be excluded or chelated within reservoir brines.

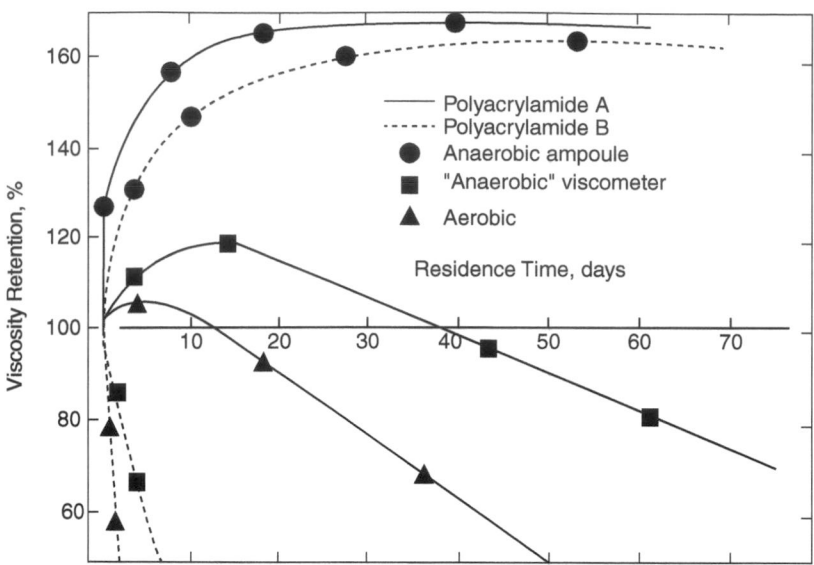

Fig. 5.8—Viscosity/time history indicating thermal degradation of polyacrylamide at 194°F (Ryles 1983).

The presence of divalent cations causes stability problems for polyacrylamides at elevated temperatures. As the degree of hydrolysis increases as a result of polyacrylamide reaction with water, the solubility of the polymer decreases in the presence of calcium and magnesium. **Fig. 5.9** presents approximate guidelines for the stability of polyacrylamide exposed to divalent cations as a function of temperature (Moradi-Araghi and Doe 1987).

Polymer degradation also results from bacterial attack. Biopolymers are susceptible to biological attack resulting in the loss of solution viscosity from the destruction of the carbohydrate backbone, which can be rapid. For this reason, the polymer broth usually contains a bactericide, such as formaldehyde, to control bacterial growth. At one time, it was thought that bacterial attack would not be a problem in petroleum reservoirs because the organisms could not be transported through the porous rock. Unfortunately, bacterial attack has been observed in at least two field tests (Van Horn 1981; Bragg et al. 1982). Specimens of bacteria were recovered from core material from the Loudon reservoir, demonstrating that bacteria could be transported appreciable distances through porous rock. In the case of the Loudon field test (Reppert et al. 1990), high concentrations of formaldehyde (1,500 ppm) successfully controlled the bacterial attack. When H_2S is present, no known bactericide is effective.

Polyacrylamides are perceived to be less susceptible to biological attack than biopolymers, but no data to support this perception exist in the published literature.

5.3.3 Rheological Properties. *Effect of Shear Rate.* Polymers are of interest in EOR applications because of their rheological properties in dilute solutions. Aqueous solutions of polyacrylamides and xanthan biopolymers often exhibit non-Newtonian rheological behavior. A Newtonian fluid has a linear relationship between shear stress and shear rate given in Eq. 5.2. The proportionality constant in this relationship is the viscosity of the fluid.

$$\tau = \mu\dot{\gamma}, \dotfill (5.2)$$

where τ = shear stress, μ = solution viscosity, and $\dot{\gamma}$ = shear rate.

A more general expression relating shear stress to shear rate is

$$\tau = K\dot{\gamma}^n, \dotfill (5.3)$$

where K and n are constants that characterize the fluid. If $n \neq 1.0$, then shear stress does not vary linearly with shear rate and the fluid is non-Newtonian. For this case, an apparent viscosity, μ_a, may be defined by

$$\tau = \mu_a\dot{\gamma}s. \dotfill (5.4)$$

The apparent viscosity varies with shear rate. The subscript on μ_a is usually dropped when discussing non-Newtonian fluids.

Normally, the apparent viscosity of polymer solutions used in EOR processes decreases as shear rate increases. Fluids with this rheological characteristic are said to be shear thinning. The apparent viscosity decreases because the polymer molecules are able to align themselves with the shear field to reduce internal friction. Often it is possible to represent the rheological

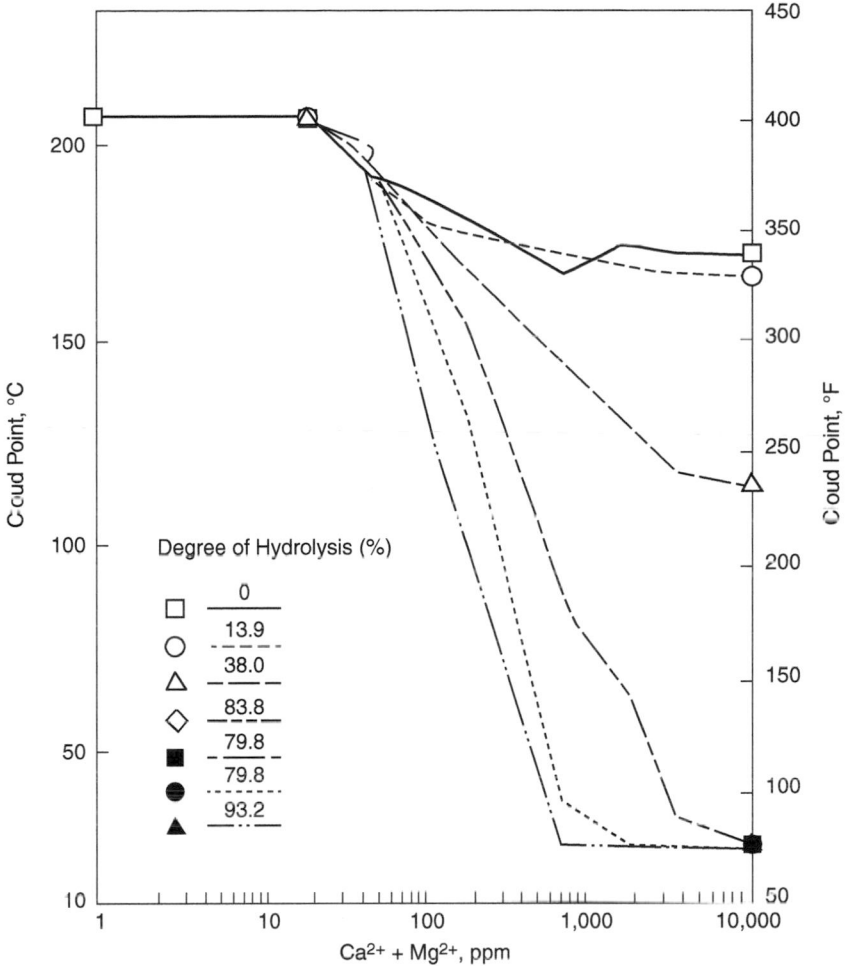

Fig. 5.9—Approximate stability guidelines for polyacrylamides exposed to divalent cations (Moradi-Araghi and Doe 1987).

properties of a shear-thinning fluid by the power-law model given by Eq. 5.5, which results from combining Eqs. 5.3 and 5.4. The constants K and n depend on the concentration of the polymer.

$$\mu = K\dot{\gamma}^{(n-1)}, \quad \dots\dots\dots\dots\dots\dots\dots\dots\dots (5.5)$$

where μ = apparent viscosity, K = power-law constant, n = power-law exponent, $\dot{\gamma}$ = shear rate, and consistent units should be used.

Fig. 5.10 shows the correlation of K and n with polymer concentration for a xanthan biopolymer (Willhite and Uhl 1988). The data can be correlated as a function of polymer concentration by use of curve-fitting techniques to obtain Eqs. 5.6 and 5.7 (Hejri 1989). Similar correlations can be developed for other polymers if the rheological data are available.

$$n = 1.0/(1.0 + 0.002C^{0.943}) \quad \dots\dots\dots\dots\dots (5.6)$$

and $K = 5.435 + 2.362 \times 10^{-5} C^{2.286}, \quad \dots\dots\dots(5.7)$

where K has units of $mPa \cdot s^n$ and C = polymer concentration (ppm). The shear-thinning behavior may encompass

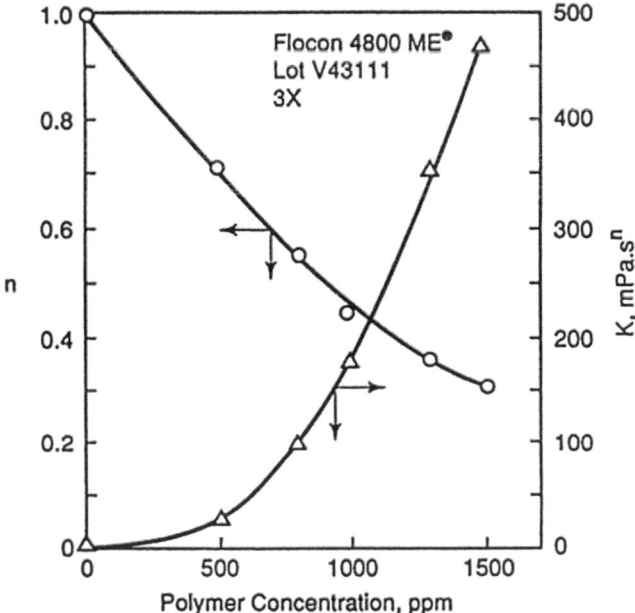

Fig. 5.10—Variation of power-law parameters with polymer concentration, Flocon 4800 ME (Willhite and Uhl 1988).

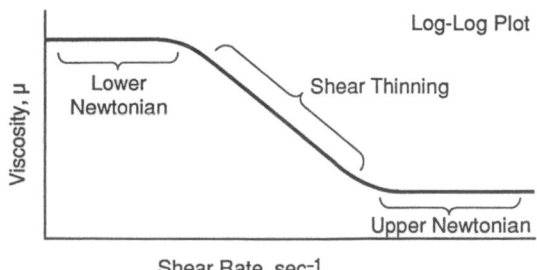

Fig. 5.11—Rheology of a shear-thinning fluid (Willhite and Uhl 1988).

a wide range of shear rates and, in many cases, may be the only behavior that is measurable with available viscometers. However, shear thinning is often just one part of the rheological behavior. **Fig. 5.11** (Willhite and Uhl 1988), a plot of apparent viscosity vs. shear rate, represents a typical rheogram for a shear-thinning fluid. At low shear rates, the fluid behaves as a Newtonian fluid in that the apparent viscosity is constant. This region is called the lower Newtonian region. As shear rate increases, there is a transition to the shear-thinning behavior represented by the power-law model. At high shear rates, there is another transition from shear-thinning behavior to Newtonian behavior. This region is called the upper Newtonian flow region. **Fig. 5.12** (Hughes et al. 1990) shows rheograms for a biopolymer demonstrating portions of the shear-thinning behavior.

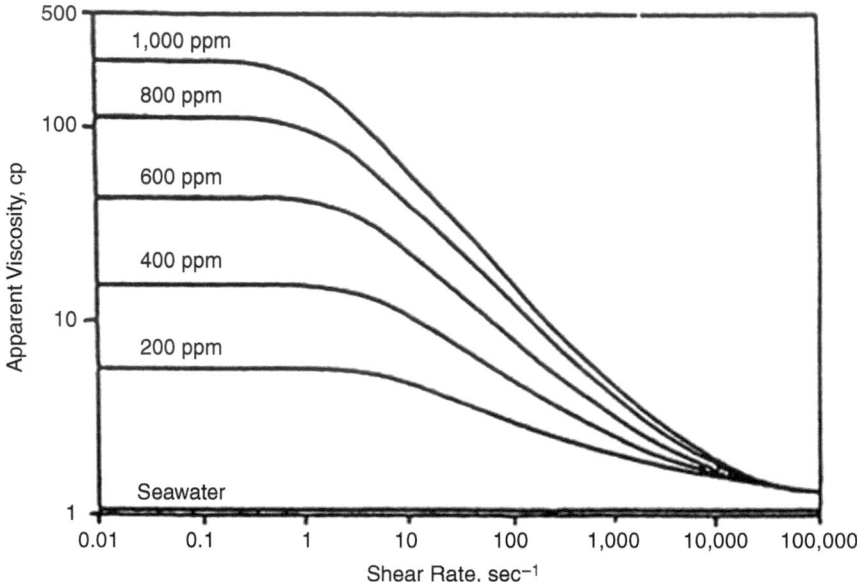

Fig. 5.12—Rheograms of a biopolymer showing a Newtonian region at low shear rates (Hughes et al. 1990).

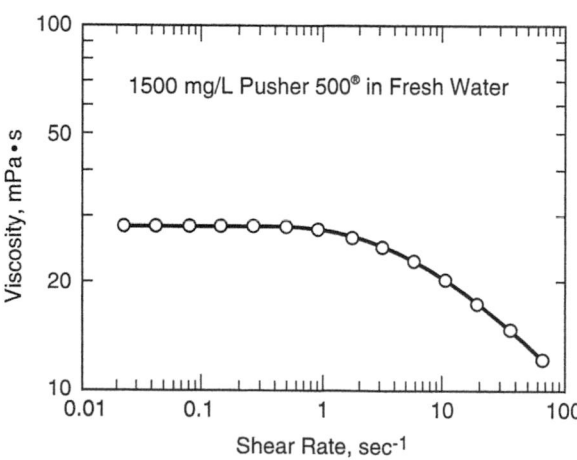

Fig. 5.13—Rheogram of a 1500-mg/L solution of Pusher 500 in fresh water (Martin and Kantamukkala 1984).

Eq. 5.8 gives a model developed by Carreau (1968, 1972) describing the complete rheological behavior of a shear-thinning fluid. The parameters n and τ_r are determined by fitting the model to experimental data. The parameter μ_{pN} is the viscosity in the lower Newtonian region, and μ_∞ is the viscosity of the polymer at very high shear rates. At high shear rates, μ_∞ is approximated by the viscosity of the solvent.

$$\mu - \mu_\infty = (\mu_{pN} - \mu_\infty)[1 + (\dot{\gamma}/\tau_r)^2]^{[(n-1)/2]} \ldots \ldots \ldots (5.8)$$

Fig. 5.13 (Martin and Kantamukkala 1984) shows the rheological behavior of a 1,500-ppm solution of Pusher 500®, a partially hydrolyzed polyacrylamide, in fresh water. The line through the data points represents the correlation obtained from Eq. 5.8. **Table 5.1** contains representative parameters obtained for several solutions of Pusher 500 (Martin 1986).

Example 5.1 illustrates the use of the Carreau model to fit experimental data.

Example 5.1—Application of Carreau Model. Viscometric data are obtained to evaluate polymer solutions over a range of shear rates. A data set for a 1,500-ppm solution of Pusher 500 in 53 meq/L NaCl brine at 72.3°F is presented in **Table 5.2**[3]. Fit these data to the Carreau model (Eq. 5.8) and determine τ_r and n. Assume that the viscosity of the polymer solution at high shear rates, μ_∞, is the viscosity of the solvent. In this case, $\mu_\infty = 1.0$ cp.

[3] Personal communication with H. W. Gao, NIPER, Bartlesville, Oklahoma, USA (23 December 1988).

Concentration (mg/L)	μ_o (cp)	K (mPa·s)	n	τ_r (sec^{-1})	F_s
750 in fresh water	7.5	10.9	0.818	6.0	11.2
1500 in fresh water	28.2	40.6	0.712	3.1	17.3
2000 in fresh water	64.5	80.4	0.660	1.7	22.8
2500 in fresh water	138.0	142.5	0.619	1.0	27.3
1500 in 3% NaCl	6.9	10.1	0.854	9.6	14.1

*μ_∞ is assumed to be 1 cp.

Table 5.1—Properties of Pusher 500 solutions (Martin 1986)*.

Shear Rate (seconds^{-1})	Viscosity (cp)
0.945	23.74
1.285	23.36
1.747	22.56
2.37	21.90
3.23	21.22
4.39	20.52
5.96	19.58
8.11	18.67
11.02	17.64
14.98	16.50
20.4	15.41
27.7	14.26
37.6	13.18
51.2	12.12
69.5	11.12
94.5	10.21
128.5	9.34

*Personal communication with H. W. Gao, NIPER, Bartlesville, Oklahoma, USA (23 December 1988).

Table 5.2—Solution viscosity for 1,500-ppm Pusher 500 in 53 meq/L NaCl at 72.3°F, Example 5.1*.

Solution. **Fig. 5.14** presents a graph of $(\mu - \mu_\infty)$ vs. shear rate. In the absence of data at lower shear rates, it is assumed that the shear rate of 0.945 seconds^{-1} is the end of the lower Newtonian region. The shear-thinning region is present from approximately 15 to 128.4 seconds^{-1}, as indicated by the linear relationship between μ and $\dot{\gamma}$. A long transition region is present between $\dot{\gamma} = 0.945$ and 15 seconds^{-1} for this solution.

The Carreau model requires determination of τ_r and n. The value of τ_r is found from the intersection of the lines extrapolated from the Newtonian and the power-law regions, as shown in Fig. 5.14. From this intersection, the value of τ_r is 4.25 seconds^{-1}. The power-law constant n is found from the slope of the graph in the power-law region. That is,

$$\log(\mu - \mu_\infty) = \left(\frac{n-1}{2}\right)\log\left[1 + \left(\frac{\dot{\gamma}}{\tau_r}\right)^2\right] + \log(\mu_{pN} - \mu_\infty).$$

In this region, $(\dot{\gamma}/\tau_r)^2 \gg 1$ at high shear rates so that

$$\log(\mu - \mu_\infty) = (n-1)\log(\dot{\gamma}) - (n-1)\log\tau_r + \log(\mu_{pN} - \mu_\infty). \dotfill (5.9)$$

Eq. 5.9 applied to any two points on the shear-thinning region, yields

$$n - 1 = \log\left(\frac{\mu_1 - \mu_\infty}{\mu_2 - \mu_\infty}\right)\Big/\log\left(\frac{\dot{\gamma}_1}{\dot{\gamma}_2}\right). \dotfill (5.10)$$

With shear rates of 37.6 and 128.5 seconds^{-1},

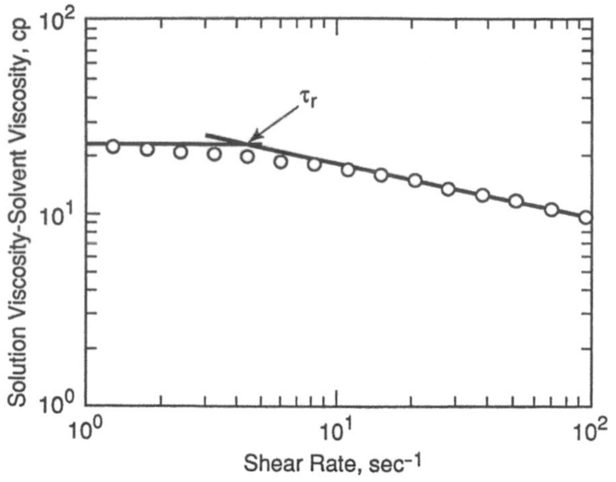

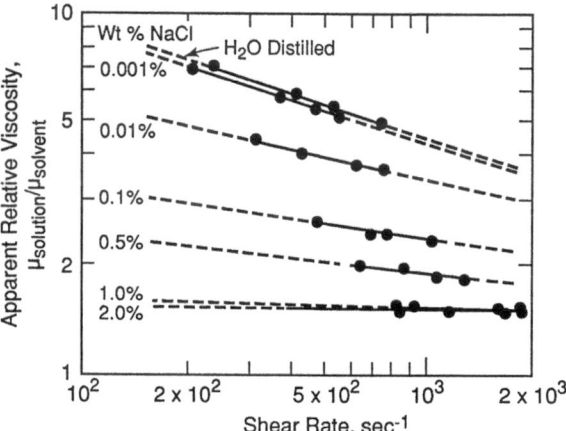

Fig. 5.14—Graph of viscosity/shear-rate data in the form to determine Carreau model parameters, Example 5.1.

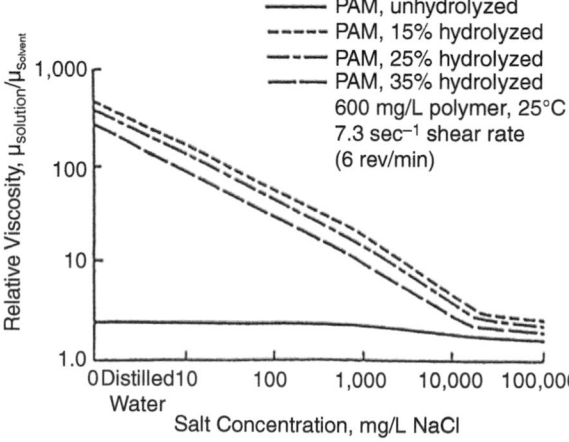

Fig. 5.15—Variation in relative viscosity with salinity and shear rate for 250-ppm Pusher 700 solutions at 25°C (Mungan 1969a). (Reprinted with permission from the Canadian Inst. of Mining, Metallurgy, and Petroleum.)

Fig. 5.16—Relative viscosity of hydrolyzed polyacrylamides in NaCl waters (Martin and Sherwood 1975).

$$n - 1 = \log\left(\frac{12.18}{8.34}\right) \Big/ \log\left(\frac{37.6}{128.5}\right)$$
$$= -0.308$$
$$n = 0.692.$$

The Carreau model parameters are $\mu_{pn} = 23.74$ cp, $\mu_\infty = 1.00$ cp, $\tau_r = 4.2$ seconds^{-1}, and $n = 0.692$. A least-squares fit also could be used to determine the power-law constant n. The line through the data points in Fig. 5.14 represents the fitted data.

Effect of Salinity. The rheological behavior of polymer solutions also may be affected by salinity and divalent-ion content. The effects are specific to polymer type, and the largest effects are observed with polyacrylamides.

Because it is not possible to produce unhydrolyzed polyacrylamide commercially, the discussion concerning polyacrylamides refers to partially hydrolyzed polyacrylamides. Hydrolysis of polyacrylamide introduces negative charges on the backbone of the polymer chain that have a large effect on the rheological properties of the polymer solution. At low salinities, the negative charges on the polymer backbone repel each other and cause the polymer chain to stretch. Each polymer molecule occupies more space in solution, and the apparent viscosity of a dilute solution increases accordingly. For example, the apparent viscosity of a dilute solution (250 ppm) of Pusher 500 at a shear rate of 200 seconds^{-1} in distilled water is about seven times the viscosity of water. Larger differences are observed at lower shear rates.

When an electrolyte, such as NaCl, is added to a polymer solution, the repulsive forces are screened by a double layer of electrolytes and extension is reduced. As the electrolyte concentration increases, the extension of the polymer chain decreases and the solution viscosity declines. **Fig. 5.15** (Mungan 1969a) illustrates the effect of salinity on the relative viscosity (apparent solution viscosity/solvent viscosity) of 250-ppm solutions of Pusher 700®, a partially hydrolyzed polyacrylamide (Martin and Sherwood 1975). Chain extension also is controlled by the degree of hydrolysis. This is shown in **Fig. 5.16,** where rheograms are compared for four polymers with the same average molecular weight but different degrees of hydrolysis. Salinity has little effect on the relative viscosity of the "unhydrolyzed" polymer. At each salinity, the relative viscosity decreases as the degree of hydrolysis increases. The relative viscosity of all solutions decreases rapidly with salinity, however, reaching values of 3 to 5 for salinities greater than approximately 20,000 ppm. Thus, much of the increase in solution viscosity anticipated from rheological data taken on polymer solutions prepared in distilled water is not attainable at salinities expected in reservoir brines.

Divalent ions (Ca^{++}, Mg^{++}) bond readily to the negatively charged macro-ion in preference to a monovalent ion, such as sodium. The effect of divalent-ion concentration on relative viscosity is more pronounced than sodium-ion concentration because the divalent ions locate themselves in such a way as to screen the negative charges on the backbone more effectively (Oosawa 1971).

Compared with solutions of partially hydrolyzed polyacrylamides, viscosities of xanthan solutions are much less affected by changes in salinity or divalention content. **Figs. 5.17 and 5.18** illustrate this by plotting the solution viscosity at various shear rates vs. salinity and divalent-ion content[4].

---

[4] Technical Information FLOCON® Biopolymer 4800. 1983. Pfizer Chemical Company.

5.4 Flow of Polymers Through Porous Media

5.4.1 Polymer Retention. When a polymer flows through a porous sandpack or rock, there is usually a measurable amount of polymer retention. Retention is caused primarily by adsorption on the surface of the porous material and mechanical entrapment in pores that are small relative to the size of the polymer molecule in solution (Willhite and Dominguez 1977; Gogarty 1967a, b). In most cases, retention of polymers used in EOR applications is considered instantaneous and irreversible. This is not exactly true because small amounts of polymer can be removed from porous rock by prolonged exposure to water or brine injection. Usually, however, the rate of release is so small that it is not possible to measure the concentrations accurately. It is thus more accurate to state that the rate of polymer retention is much greater than the rate of polymer removal.

Retention also may occur when flow rates are suddenly increased after polymer has been injected at a constant rate until a steady-state condition has been attained (i.e., until the effluent concentration has reached the injected concentration). This type of retention, called hydrodynamic retention, is characterized by expulsion of the polymer when the flow rate is reduced suddenly. Thus, it is possible to obtain polymer concentrations in the effluent of linear displacement experiments that are larger than the injected concentration (Chauveteau and Kohler 1974; Maerker 1973; Dominguez and Willhite 1977). Hydrodynamic retention appears to be reversible because the amount of polymer retained after an increase in flow rate is approximately the same as the amount of polymer recovered when the rate is reduced.

The amount of polymer retained when a polymer solution is displaced through a porous medium must be determined by experimental measurement. If the porous material is unconsolidated and the permeability is on the order of 1 darcy or larger, polymer adsorption can be estimated with batch adsorption experiments. In these experiments, a polymer solution of known concentration is contacted with a known mass of sand grains until no further change in polymer concentration is detected. The concentration of the equilibrated polymer solution is determined, and adsorption is computed by material balance. **Fig. 5.19** (MacWilliams et al. 1973) presents adsorption data for partially hydrolyzed polyacrylamides on unconsolidated Miocene sand. Adsorption ranges from approximately 20 μg/g rock at 38% hydrolysis to approximately 700 μg/g rock for the sample with minimum hydrolysis. The minimum in the adsorption is related to a charge interaction between the negatively charged silica surface on the sand and the negatively charged carboxyl group on the polymer, which acts to reduce adsorption. Adsorption data in Fig. 5.19 indicate that polyacrylamides used in mobility-control processes must be partially hydrolyzed to reduce adsorption to acceptable levels.

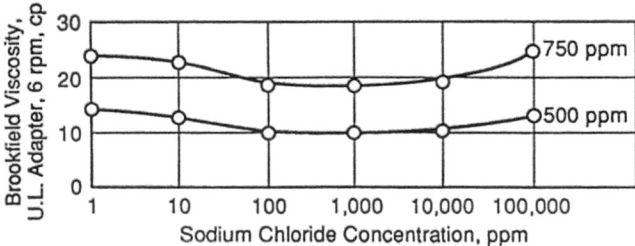

Fig. 5.17—Effect of salinity on viscometric properties of xanthan biopolymer (Technical Information FLOCON® Biopolymer 4800 1983).

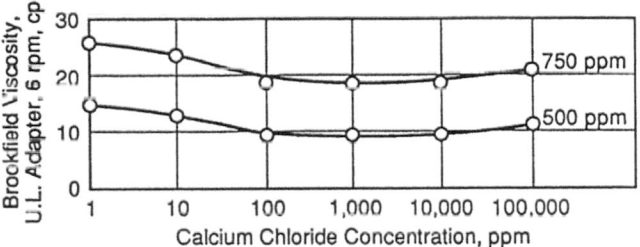

Fig. 5.18—Effect of divalent-ion content on viscometric behavior of xanthan biopolymer (Technical Information FLOCON® Biopolymer 4800 1983).

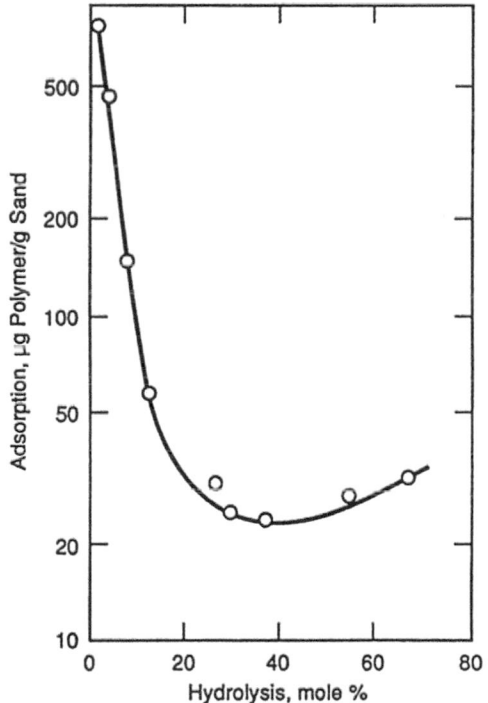

Fig. 5.19—Adsorption of hydrolyzed polyacryla-mide in 2.2% NaCl solutions on Miocene sand (MacWilliams et al. 1973).

Polymer retention in consolidated porous media cannot be determined with batch adsorption techniques because the process of disaggregation to obtain representative granular material generates significant amounts of new surface area and polymer adsorption is usually excessive.

Polymer retention in porous media is determined primarily by flow experiments. Two methods are commonly used. In the first method, a polymer solution is injected at a constant frontal-advance rate into a linear core plug or series of linear core plugs until the polymer concentration in the effluent is equal to the injected polymer concentration (radial core segments are preferred by some investigators). **Fig. 5.20** shows a representative concentration profile for the displacement of Pusher 700 through

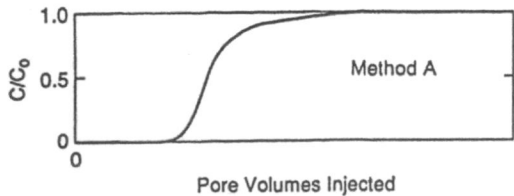

Fig. 5.20—Concentration profile for the constant displacement of a polymer through a porous rock for determination of polymer retention.

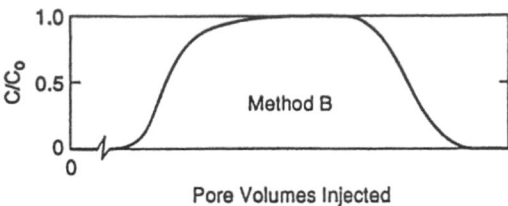

Fig. 5.21—Concentration profile for the displacement of a slug of polymer through a porous rock for determination of polymer retention.

Berea sandstone core material (Pusher 700 is a partially hydrolyzed polyacrylamide with a higher molecular weight than Pusher 500). It is often necessary to inject several PVs of polymer into the porous material because the concentration profile typically has a long "tail." In the second method, injection is switched to brine or water after the effluent concentration reaches the injected concentration and the mobile polymer is displaced from the pore space. **Fig. 5.21** is a concentration profile used to determine polymer retention by the second method. Polymer retention in each method is determined by material balance.

Table 5.3 summarizes some polymer retention data from displacement experiments for partially hydrolyzed polyacrylamides. Retention varies from 35 to approximately 1,000 lbm/acre-ft over a wide range of fluid and rock properties. Information on converting retention values from pounds per acre-foot to micrograms per gram of rock is given as a footnote in Table 5.3. Several trends are present in the limited amount of retention data in the literature. **Fig. 5.22** shows the variation of polymer retention with brine permeability at ROS (Vela et al. 1976). The retention at low permeabilities is large and is probably a result of excessive mechanical entrapment of polymer molecules in small pores. Another possible explanation is high clay content. Polymer concentration appears to have little effect on retention for the data shown in Fig. 5.22. The weak concentration dependence in Fig. 5.22 is reinforced by data from Shah (1978) for the retention of partially hydrolyzed polyacrylamide on Berea core material shown in **Fig. 5.23.** Retention at 50 ppm polymer concentration is 77% of the retention at 1,070 ppm.

Limited data, summarized in **Table 5.4,** have been published for the retention of biopolymers on porous media. Values of approximately 38 to 78 lbm/acre-ft have been reported.

Retention of xanthan biopolymer depends on the effective permeability of the porous rock. Retention data from Huh et al. (1990)[5] plotted in **Fig. 5.24** show the same trend as Fig. 5.22. Polymer retention increases as the effective permeability decreases. The uncertainty range indicated in Fig. 5.24 shows the effect of different values of inaccessible PV (IPV) on polymer retention determined by material-balance calculations (IPV is discussed in Section 5.4.2). In experiments conducted in 10- and 33-md porous media, the injected concentration was not reached when the experiment ended. Some uncertainty exists in these values, possibly because of the effects of sandface plugging. Retention of biopolymers is generally less than that of polyacrylamides at comparable concentrations. Also, retention of biopolymer may be lower when oil is present in the porous media. Kolodziej (1988) reported retention of xanthan biopolymer in Berea sandstone to be 75 lbm/acre-ft-PV for 100% brine corefloods and 38 lbm/acre-ft-PV in corefloods at ROS.

Polymer retention in porous media may be correlated by use of the Langmuir isotherm model, which is given by

$$\hat{C} = a_1 b_1 C / (1 + b_1 C), \dots\dots\dots\dots\dots\dots\dots\dots\dots\dots\dots\dots\dots\dots\dots\dots\dots\dots\dots\dots\dots\dots\dots (5.11)$$

where $\hat{C}$ = polymer adsorption, C = polymer concentration in solution, and a_1, b_1 = constants.

The Langmuir model is an equilibrium relationship, and its application assumes retention is instantaneous. The constants a_1 and b_1 are determined by fitting the data. If the Langmuir retention model applies, the graph of $1/\hat{C}$ vs. $1/C$ on a linear scale is a straight line with slope $1/a_1 b_1$ and intercept $1/a_1$. **Fig. 5.25** is a typical retention isotherm for high-molecular-weight polymers (Dawson and Lantz 1972). In the Langmuir model, retention is reversible. Thus, when polymer retention is considered to be irreversible, the Langmuir model cannot be used when the polymer concentration is decreasing.

Section 5.1 pointed out that high-molecular-weight polymers used in EOR applications have a wide range of average sizes, expressed by the average molecular weight, and an often unknown size distribution. Porous rocks also have characteristic pore sizes. Because mechanical retention is a significant contributor to polymer loss in porous media, a relationship between polymer retention in a specific porous medium and average molecular size should be expected. However, general relationships of this kind have not been developed because of the difficulty in characterizing both polymers and porous rocks. When polymers from various suppliers are screened for a particular reservoir rock, it is necessary to conduct flow tests to determine which polymers can be transported through a porous rock with acceptable retention.

5.4.2 Inaccessible PV. Polymer molecules are larger than water molecules and are large relative to some pores in a porous rock. Because of this, polymers do not flow through all the pore space contacted by the brine. The fraction of the pore space not contacted by the polymer solution is called the IPV. The concept of IPV is illustrated from the results of an experiment shown in **Fig. 5.26** (Dawson and Lantz 1972). In this experiment, a polymer solution containing 2% NaCl was displaced through a Berea sandstone core until no further polymer was retained. Then the polymer and NaCl concentration of the injected fluid were reduced for a period to create a "pulse" change in NaCl and polymer concentrations. The concentration profiles shown in Fig. 5.26 are effluent profiles of polymer and NaCl. The midpoint of the change in salt concentration arrived at approximately 1 PV injected, as expected from displacement theory discussed in Chapter 3, assuming complete contact with the PV. The polymer pulse, however, arrived approximately 0.24 PV earlier than expected and thus did not contact all the PV in the core. Approximately 24% of the pore space was not accessible to the polymer.

[5] Personal communication with C. Huh, Exxon USA, Houston, 9 October 199

Polymer	Concentration (ppm)	Total Dissolved Solids (ppm)	Porous Medium	Permeability (md)	Retention* (lbm/acre-ft)	Reference
HPAM						
$M_w = 3\times10^6$	500	Brine	Miocene sand	53 at S_{or}	684	Jennings et al. (1971)
HPAM						
$M_w = 3\times10^6$	500	0	Muffled Berea		201**	Mungan (1969a)
	500	0	Berea		316**	Mungan (1969a)
	500	0	Ottawa sand		747†	Mungan (1969a)
HPAM						
Pusher 700						
$M_w = 5\times10^6$	500	Brine	Reservoir core	359	34.9	Hirasaki and Pope (1974)
	500	Surfactant	Reservoir core	117	44.0	Hirasaki and Pope (1974)
	400	Surfactant	Reservoir core	97	46.9	Hirasaki and Pope (1974)
	500	Surfactant	Reservoir core	80	75.4	Hirasaki and Pope (1974)
HPAM						
$M_w = 5\times10^6$	300	133,000	Reservoir core	45 at S_{or}	224	Vela et al. (1976)
	300	133,000	Reservoir core	30 at S_{or}	561	Vela et al. (1976)
	300	13,340	Reservoir core	17 at S_{or}	580	Vela et al. (1976)
	300	20,000	Reservoir core	13 at S_{or}	64	Vela et al. (1976)
HPAM						
Pusher 500	750	1,000 NaCl	Berea sandstone	350	99	Martin et al. (1983)
	750	20,000 NaCl	Berea sandstone	550	147	Martin et al. (1983)
Pusher 700	750	1,000 NaCl	Berea sandstone	550	70	Martin et al. (1983)
	750	20,000 NaCl	Berea sandstone	550	135	Martin et al. (1983)
Pusher 1000®	750	1,000 NaCl	Berea sandstone	550	107	Martin et al. (1983)
	750	20,000 NaCl	Berea sandstone	550	160	Martin et al. (1983)
Betz Hi Vis®	750	1,000 NaCl	Berea sandstone	550	68	Martin et al. (1983)
	750	20,000 NaCl	Berea sandstone	550	130	Martin et al. (1983)
Cyanatrol 960 S®	750	1,000 NaCl	Berea sandstone	550	103	Martin et al. (1983)
	750	20,000 NaCl	Berea sandstone	550	155	Martin et al. (1983)
Nal-flo®	750	1,000 NaCl	Berea sandstone	550	95	Martin et al. (1983)
	750	20,000 NaCl	Berea sandstone	550	149	Martin et al. (1983)
HPAM		1,000	Wyoming reservoir core	200	40	Martin et al. (1983)
		70,000	Oklahoma reservoir core	300	415	Martin et al. (1983)
		1,270	South American reservoir core	1,500	151	Martin et al. (1983)

* Assuming 20% porosity.

** Conversion from micrograms per gram to pounds per acre-foot is achieved by multiplying micrograms per gram by $2.717(1 - \phi)\rho_g$, where ρ_g is the grain density of the rock in grams per cubic centimeter. For quartz, $\rho_g = 2.65$ g/cm³.

† Assuming 35% porosity.

Table 5.3—Retention of polyacrylamides during flow through porous media.

IPV has been observed in all types of porous media for both polyacrylamides and biopolymers and is considered to be a general characteristic of polymer flow in porous media. The magnitude of the IPV can range from 1 to 2% to as much as 25 to 30%, depending on the polymer and porous medium. **Table 5.5** presents representative data on IPV for a xanthan and polyacrylamide in porous materials. A slight decrease in IPV with concentration is indicated by the polyacrylamide data. Several models have been offered to explain why IPV occurs (Chauveteau and Kohler 1984; Kolodziej 1987; DiMarzio and Guttman 1970; Chauveteau 1982), but none has gained universal acceptance.

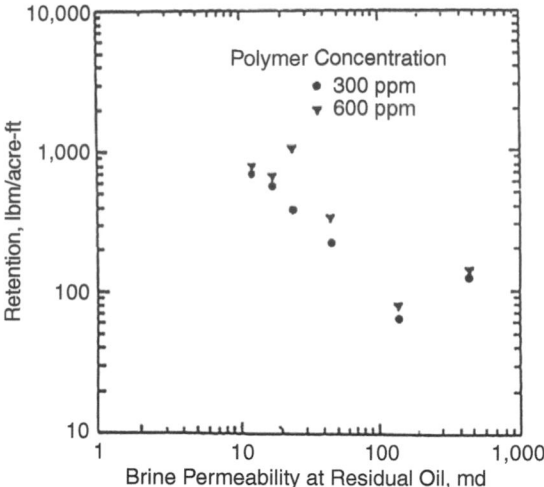

Fig. 5.22—Variation of polymer retention with permeability in sandstone containing ROSs (Vela et al. 1976).

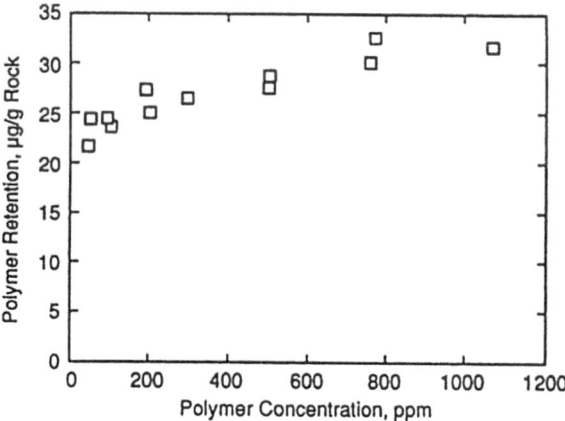

Fig. 5.23—Retention of Pusher 700 by Berea sandstone (Shah 1978).

The impact of IPV on polymer transport in porous rocks often is concealed by polymer retention. In displacement experiments where a constant polymer concentration is injected into a porous medium that has not been contacted previously by polymer, retention causes the effluent concentration to lag, as seen in **Fig. 5.27** (Hughes et al. 1990). However, IPV offsets part or all of this lag so that polymer breakthrough can be essentially the same as tracer breakthrough even though significant polymer retention occurs.

5.4.3 Flow Characteristics. *Permeability Reduction.* Polymer retention reduces the apparent permeability of the rock. Permeability reduction depends on the type of polymer, the amount of polymer retained, the pore-size distribution, and the average size of the polymer relative to the pores in the rock. Permeability reduction is determined experimentally by first displacing polymer solution through a porous medium and then displacing the polymer with brine and measuring the permeability to brine after all mobile polymer has been displaced. **Fig. 5.28** (Smith 1970) illustrates the effect of initial rock permeability on the permeability reduction of Berea sandstone cores by partially hydrolyzed polyacrylamide in 3% NaCl. The trend in permeability reduction in Fig. 5.28 is consistent with the trend of increased retention as permeability decreases shown in Fig. 5.22.

It is convenient to describe the permeability reduction in terms of the initial brine permeability. In practice, this is done by defining the residual resistance factor (Eq. 5.12) as the ratio of the brine mobility before contact with polymer, λ_w, to the brine mobility after all mobile polymer has been displaced from the pore space, λ_{wp}. Residual resistance factors are shown in **Tables 5.6 and 5.7** for several xanthan biopolymer and polyacrylamide solutions.

$$F_{rr} = \lambda_w / \lambda_{wp} \quad \dots\dots\dots\dots\dots\dots\dots\dots\dots\dots\dots (5.12)$$

and $F_{rr} = k_w / k_{wp}, \quad \dots\dots\dots\dots\dots\dots\dots\dots\dots\dots\dots (5.13)$

where F_{rr} = residual resistance factor for the porous matrix after contact with a particular polymer solution, k_{wp} = permeability of the porous matrix to brine after contact with polymer solution, and k_w = initial brine permeability. The permeability to brine after the mobile polymer has been displaced, k_{wp}, is assumed to be the same as the permeability of the porous medium to the flow of polymer, k_p.

Retention of xanthan biopolymer is relatively small when bacterial debris are removed by filtration. The permeability to brine after contact with polymer may be reduced from 10 to 30%. Table 5.6 compares the brine permeability before and after contact of Berea sandstone and unconsolidated sandpacks with xanthan biopolymer. Compared with xanthans, polyacrylamides usually cause larger reductions in brine permeability. At high salinities or divalent-ion content, permeability reduction is decreased. Table 5.7 contains typical permeability data before and after contact of porous rocks with polyacrylamides.

The permeability reduction usually persists for a large number of PVs of fluid throughput. In laboratory tests with relatively low fluid throughput, little change in brine permeability occurs, as shown in **Table 5.8.** However, prolonged fluid injection eventually erodes the permeability reduction, as indicated in **Fig. 5.29,** where the residual resistance factor declines markedly with PVs of throughput (Needham et al. 1974).

Prediction of the permeability reduction from properties of the porous rock and the polymer is not possible at this time. Experimental measurement with the rock and polymer of interest is necessary. It is often possible, however, to correlate permeability reduction for the same polymer in the same type of porous medium and use the resulting correlation for interpolation and extrapolation. Gogarty (1967b) correlated the permeability of consolidated porous rocks after contact with polyacrylamide by use of an empirical relationship.

Willhite and Uhl (1988) determined the permeability of Berea sandstone cores to brine following contact with xanthan biopolymer. The empirical correlation of the data is

$$k_{wp} = a k_w^b, \quad \dots\dots\dots\dots\dots\dots\dots\dots\dots\dots\dots\dots\dots\dots\dots\dots\dots\dots\dots\dots\dots\dots\dots (5.14)$$

Polymer	Concentration (ppm)	Total Dissolved Solids	Porous Medium	Permeability (md)	Retention*	Reference
Kelzan M		Brine	Nevada sand	6,000	38 lbm/acre-ft**	Patton et al. (1971)
Xanflood	750	1,000 ppm NaCl	Berea sandstone	350	48 lbm/acre-ft	Martin et al. (1983)
	750	20,000 ppm NaCl	Berea sandstone	550	75 lbm/acre-ft	
Xanthan broth	750	1,000 ppm NaCl	Berea sandstone	550	36 lbm/acre-ft	Martin et al. (1983)
(Abbott)	750	20,000 ppm NaCl	Berea sandstone	550	77 lbm/acre-ft	
Biopolymer 1035	750	1,000 ppm NaCl	Berea sandstone	550	41 lbm/acre-ft	Martin et al. (1983)
(Pfizer)	750	20,000 ppm NaCl	Berea sandstone	550	46 lbm/acre-ft	
Biopolymer		Brine	Berea sandstone		78 lbm/acre-ft†	Trushenski et al. (1974)
Xanthan	580		Bentheim sandstone		70 µg/g	Lötsch (1988)
Xanthan	630				83 µg/g	
Xanthan	1,350				151 µg/g	
Xanthan	2,450				114 µg/g	
Scleroglucan	100		Sand		17 µg/g	Lötsch (1988)
Scleroglucan	310		Sand		35 µg/g	
Scleroglucan	560		Sand		58 µg/g	
Scleroglucan	850		Sand		117 µg/g	
Scleroglucan	920		Sand		126 µg/g	
Scleroglucan	1,450		Sand		149 µg/g	
Xanthan	400		Reservoir core at S_{oi}		31 µg/g	Hughes et al. (1990)
Xanthan	500		Reservoir core at S_{or}		35 µg/g	
Xanthan	500		Reservoir core at S_{or}		55 µg/g	
Xanthan	500		Reservoir core at $S_o = 0$		76 µ/g	
Xanthan	325		Reservoir core at $S_o = 0$		70 µ/g	

* Assuming 20% porosity.

** Assuming 35% porosity.

† Conversion from micrograms per gram to pounds per acre-foot is achieved by multiplying micrograms per gram by $2.717(1-\phi)\rho_g$, where ρ_g is the grain density of the rock in grams per cubic centimeter. For quartz, $\rho_g = 2.65$ g/cm³.

Table 5.4—Retention of biopolymer during flow through porous media.

where a and b are parameters fitted from experimental data in Table 5.6 and k_w is in md. Hejri et al. (1991) measured the permeability of brine in sandpacks after displacement with xanthan biopolymer and correlated the data with

$$k_{wp} = 0.377 k_w^{1.088} . \dots\dots\dots\dots\dots\dots\dots\dots\dots\dots\dots\dots \quad (5.15)$$

Polymer Mobility in Porous Media. Polymers are non-Newtonian fluids. Consequently, the flow characteristics in porous media for some polymers are related to the rheological properties described in Section 5.3.3. For shear-thinning fluids with rigid structures, such as xanthan biopolymers, the flowing polymer exhibits Newtonian characteristics at low frontal-advance rates, a shear-thinning region at intermediate frontal-advance rates, and Newtonian characteristics at high frontal-advance rates. Because the shear-thinning region often includes the range of most reservoir frontal-advance rates, most experimental data have been taken in this region.

Data describing the flow of polymers in porous media can be obtained by conducting steady-state flow tests in core plugs or sandpacks over the range of frontal-advance rates anticipated in the bulk of the reservoir and in the vicinity of the wellbore. In these

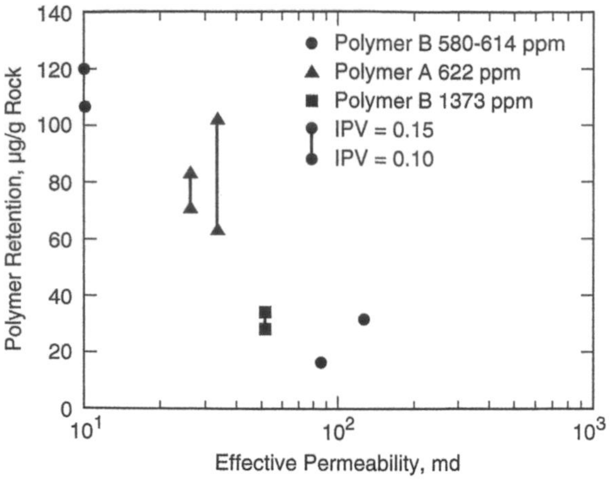

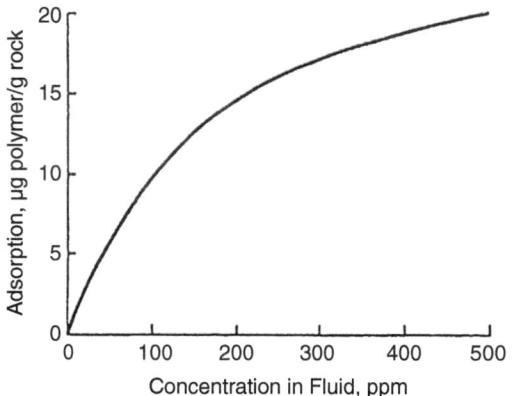

Fig. 5.24—Retention of xanthan biopolymer in Berea sandstone (Huh et al. 1990; personal communication with C. Huh, Exxon USA, Houston, 19 October 1990).

Fig. 5.25—Typical adsorption isotherm for high-molecular-weight polymers (Dawson and Lantz 1972).

tests, polymer of a specific concentration is injected at a constant rate. Pressure drops are measured across the entire length of the porous medium and between measuring ports spaced along the porous medium, as depicted in **Fig. 5.30.** A constant rate is maintained until the pressure drop reaches a steady state. A series of measurements of flow rate vs. pressure drop is taken to determine the flow properties of the polymer in the porous material.

The data can be analyzed by assuming that Darcy's law applies to the flow of polymer in porous media:

$$u = \lambda_p \left(\Delta p / L \right), \quad \dots\dots\dots\dots\dots\dots\dots\dots\dots\dots\dots (5.16)$$

where u = Darcy velocity, λ_p = mobility of the polymer in the porous rock, Δp = pressure drop, L = length over which the pressure drop is measured, and consistent units are used. The mobility of the polymer is defined by

$$\lambda_p = k_p / \mu_p, \quad \dots\dots\dots\dots\dots\dots\dots\dots\dots\dots\dots\dots (5.17)$$

where k_p = permeability of the porous medium to polymer and μ_p = apparent viscosity of the polymer at the average shear rate existing in the porous medium.

The polymer mobility, λ_p, can be calculated from data of the type described. In general, however, neither k_p nor μ_p is independently known and the apparent viscosity of the polymer cannot be calculated unless a value for k_p is assumed. When k_p is not known, an effective viscosity may be defined by assuming that $k_p = k_w$:

$$\mu_e = k_w / \lambda_p, \quad \dots\dots\dots\dots\dots\dots\dots\dots\dots\dots\dots\dots (5.18)$$

where μ_e = effective viscosity.

Fig. 5.31 presents effective viscosity data as a function of frontal-advance rate for the flow of xanthan biopolymer through Frannie reservoir core (Castagno et al. 1987). The effective viscosity decreases as the frontal-advance rate increases because the biopolymer is shear-thinning in this range of frontal-advance rates. Another common practice is to assume that k_p is the permeability of the porous rock to brine, k_{wp}, after the mobile polymer has been displaced and compute the apparent viscosity of the polymer.

Polymer	Concentration (ppm)	Inaccessible PV (%)	Comments	Reference
Xanthan	500	31	Reservoir core at S_{or}	Hughes et al. (1990)
Xanthan	500	29	Reservoir core at S_{or}	Hughes et al. (1990)
Xanthan	500SQ*	25	Reservoir core at $S_w = 1.0$	Hughes et al. (1990)
Xanthan	325	20	Reservoir core at $S_w = 1.0$	Hughes et al. (1990)
Pusher 700	51.5	0.24	Berea sandstone core	Shah (1978)
2% NaCl	106	0.28	$\phi = 0.206$, $k = 277$ md	
2% NaCl	201	0.26		
2% NaCl	297	0.25		
2% NaCl	502	0.21		
2% NaCl	760	0.195		
2% NaCl	1,070	0.187		

*SQ = super quality.

Table 5.5—Inaccessible PV.

In studies of polymer flow in porous media, experimental data are often reported in terms of a flow resistance called the resistance factor. The resistance factor, F_r, is the ratio of the brine mobility in the porous medium before polymer contact to the polymer mobility in the same porous medium and is defined by

$$F_r = \lambda_w / \lambda_p. \dots\dots\dots\dots\dots (5.19)$$

The resistance factor also is equivalent to the ratio of the pressure drop in the porous medium when polymer is flowing to the pressure drop when water flows at the same rate before exposure of the porous medium to polymer.

Example 5.2 illustrates the calculation of the polymer mobility from a set of experimental data.

Example 5.2—Calculation of Polymer Mobility. Data for the flow of a partially hydrolyzed polyacrylamide through a 10-in.-long Berea sandstone core were obtained by injecting polymer solution at a constant frontal-advance rate and measuring the pressure drop between pressure taps located 8 in. apart. The following data are available (Gao and French 1988): k_w = 576 md; Δp = 1.18 psi; distance between pressure taps is 0.667 ft; frontal-advance rate, v, is 1.85 ft/D; and porosity is 0.21. Determine the polymer mobility in md/cp from these data.

Solution. From Eq. 5.16,

$$\lambda_p = uL / \Delta p$$
$$= \phi vL / \Delta p.$$

Because units are mixed, a unit conversion factor is needed:

$$\lambda_p = \frac{\phi vL}{\Delta p} \frac{ft^2}{D\text{-}psi} \left[\frac{1}{(3.281 \text{ ft/m})^2}\right]\left[\frac{1}{(86,400 \text{ sec/D})}\right]$$
$$\times \left[\frac{1}{(6,894.8 \text{ Pa/psi})}\right]\left[\frac{(10^{-3} \text{ Pa·s})}{cp}\right]$$
$$\times \left[\frac{1}{(0.98692 \times 10^{-15} \text{ m}^2 / \text{md})}\right]$$
$$= 158.005(\text{md/cp})(\text{psi-D/ft}^2)(\phi vL / \Delta p)(\text{ft}^2 / \text{D-psi})$$
$$= 158.005[(0.21)(1.85)(0.667) / 1.18] \text{ md/cp}$$
$$= 34.7 \text{ md/cp.}$$

Assuming $\mu_w = 1.0$ cp,

$$\lambda_w = 576 \text{ md/cp}$$

and $F_r = (576 \text{ md/cp})/(34.7 \text{ md/cp})$
$$= 16.6.$$

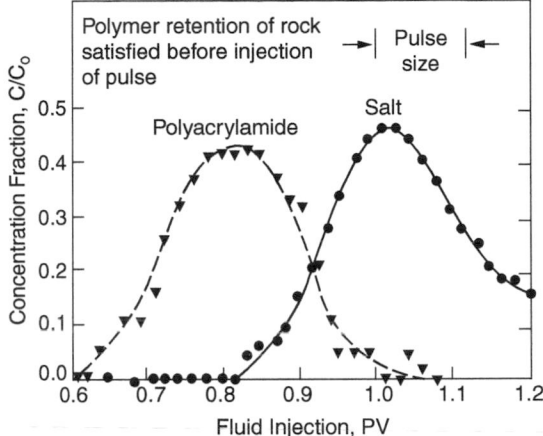

Fig. 5.26—Early arrival of polymer front caused by IPV (Dawson and Lantz 1972).

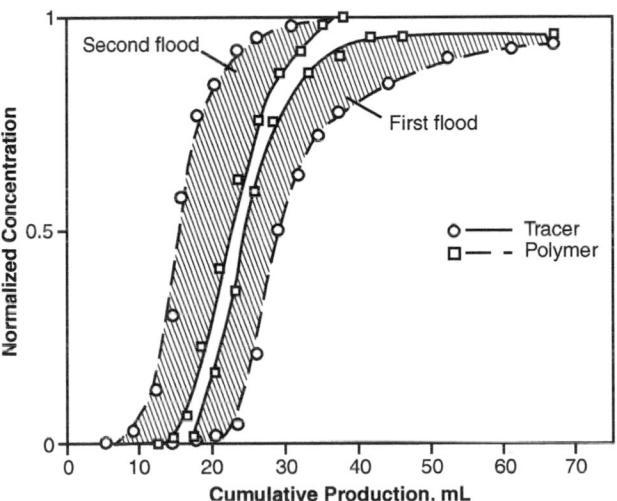

Fig. 5.27—Comparison of tracer and polymer concentration profiles in the effluent when IPV is significant (Hughes et al. 1990).

Flow Regimes. The lower Newtonian region is recognized when the polymer mobility, or resistance factor, is independent of flow rate or frontal-advance rate at low rates. Chauveteau and Zaitoun (1981) presented data for xanthan biopolymer demonstrating that the lower Newtonian region can be reached at low shear rates (low Darcy velocities). **Fig. 5.32** illustrates the variation of the apparent viscosity with the estimated in-situ shear rate for polymer flow through Fontainbleau sandstone. Also plotted in Fig. 5.32 is the apparent viscosity of the biopolymer. In the porous media used in these experiments, Chauveteau and Zaitoun found that permeability reduction caused by polymer retention was negligible, so that $k_p = k_w$. The apparent viscosity of the polymer was constant over an extensive range of shear rates, indicating that Newtonian flow was present. However, the apparent viscosity in the lower Newtonian regime was less than the corresponding viscosity in the lower Newtonian regime from rheological data. A model based on the concept of excluded PV was proposed to explain the observed phenomena (Chauveteau and Kohler 1984; Chauveteau 1982).

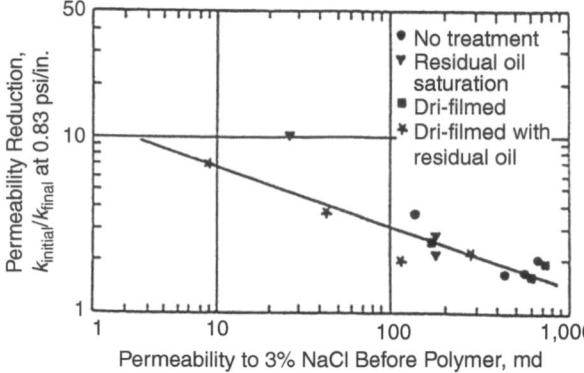

Fig. 5.28—Comparison of tracer and polymer concentration profiles in the effluent when IPV is significant (Smith 1970).

In the shear-thinning region, most polymers are represented by Darcy's law modified for a power-law fluid:

$$u^{n_c} = \lambda_p^* (\Delta p / L), \quad \dots \dots \dots \dots \dots \dots \dots \dots \dots \dots \dots (5.20)$$

where λ_p^* = polymer mobility constant, n_c = polymer flow constant, and consistent units are used. Analysis of Eq. 5.20 shows that a plot of pressure drop vs. flow rate (or Darcy velocity) should be a straight line on log-log graph paper when the power-law model is valid. Such a data set is shown in **Fig. 5.33** for the flow of a xanthan biopolymer in Berea sandstone core material. In Fig. 5.33, a linear relationship is observed over a range of frontal-advance rates from 0.23 to 117 ft/D. Values of λ_p^* and n_c are obtained from the graph or by analysis of the data. There is little evidence of either upper or lower Newtonian flow regimes in these data. Values of λ_p^* and n_c obtained from the analysis of the data in Fig. 5.33 are presented in **Table 5.9.**

Fig. 5.34 is a data set for the flow of xanthan biopolymer (1,000 to 2,000 ppm) in a sandpack with a permeability of 15.26 darcies (Hejri et al. 1991). The graph of pressure drop vs. Darcy velocity begins to depart from linear behavior at Darcy velocities less than 0.5 ft/D. The curves asymptotically approach the Newtonian region where the graph of pressure drop vs. flow rate is a straight line with slope = 1.0 because n_c = 1.0 for a Newtonian fluid.

Polyacrylamides exhibit similar flow characteristics at low and moderate frontal-advance rates. A lower Newtonian regime exists at low frontal-advance rates. As frontal-advance rate increases, there is a transition to a shear-thinning region. **Fig. 5.35** illustrates pressure-drop vs. flow-rate data for the flow of polyacrylamide in Berea core material (Gao and French 1988)[3]. This polymer is shear-thinning because $n_c<1.0$. However, polyacrylamides are not as shear-thinning as biopolymers. The mobility of polyacrylamide in porous rocks is strongly influenced by the reduction of permeability caused by the retained polymer. The computation of λ_p^* and n_c from these data is illustrated in Example 5.3.

Example 5.3—Calculation of λ_p^* and n_c From Flow Data. The pressure-gradient vs. frontal-advance rate data in Fig. 5.35 were obtained from Gao and French (1988)[3]. These data are part of the experimental data from a study of the flow through Berea core material of a 1,500-ppm solution of Pusher 500, a partially hydrolyzed polyacrylamide, in 53 meq/L NaCl.

Porous Material	k_w (md)	k_{wp} (md)	F_{rr}	Comments and References
Ottawa sandpack	893	525	1.7	After contact with 2,000 ppm
Ottawa sandpack	1,768	1,573	1.12	Pfizer 4800ME™
Ottawa sandpack	8,255	6,522	1.27	Hejri (1989)
Ottawa sandpack	17,394	15,260	1.14	
Fired Berea	909	818 to 847	1.08 to 1.11	Willhite and Uhl (1988)
Fired Berea	876	668	1.32	
Fired Berea	412	352	1.18	
Fired Berea	425	352	1.21	
Fired Berea	177	153	1.16	
Fired Berea	192	155	1.24	
Berea				Martin et al. (1983)
Berea			1.2	750 mg/L Abbott broth, 1,000 ppm NaCl
Berea			1.2	20,000 ppm NaCl
Berea			5.3	750 mg/L Xanflood™, 1,000 ppm NaCl
Berea			6.7	20,000 ppm NaCl
Berea			3.2	750 mg/L Pfizer 1035™, 1,000 ppm NaCl
Berea			2.3	20,000 ppm NaCl

Table 5.6—Reduction in brine permeability after contact with xanthan biopolymer.

[3] Personal communication with H. W. Gao, NIPER, Bartlesville, Oklahoma, USA, 23 December 1988.

Polymer Concentration (ppm)	Porous Material	k_w (md)	k_{wp} (md)	F_{rr}	Reference
1,200		156	87	1.79	Gogarty (1967b)
1,200 to 200 NaOH		203	91	2.23	
1,800		140	56	2.50	
1,400		466	114	4.09	
1,400 to 400 NaCl		438	104	4.21	
1,400 to 400 NaCl		140	12	11.67	
750 mg/L	Berea	350 to 550 md			
750 mg/L	Polymer				
1,000 ppm NaCl	Sweepaid 102®			2.26	Martin et al. (1983)
1,000 ppm NaCl	Betz Hi Vis®			2.1	
1,000 ppm NaCl	Cyanatrol 960S®			14.8	
1,000 ppm NaCl	Pusher 500			2.7	
1,000 ppm NaCl	Pusher 700			3.3	
1,000 ppm NaCl	Pusher 1000			4.2	
1,000 ppm NaCl	Nal-flo			2.2	
20,000 ppm NaCl	Sweepaid 102			1.1	
20,000 ppm NaCl	Betz Hi Vis			1.3	
20,000 ppm NaCl	Cyanatrol 960			3.9	
20,000 ppm NaCl	Pusher 500			1.4	
20,000 ppm NaCl	Pusher 700			1.1	
20,000 ppm NaCl	Pusher 1000			1.5	
20,000 ppm NaCl	Nal-flo			3.2	

Table 5.7—Reduction in brine permeability after contact with partially hydrolyzed polyacrylamide.

Table 5.10 summarizes pressure gradients and frontal-advance rates. These data are to be analyzed to obtain values of λ_p^* and n_c. Porosity of the sandstone is 0.21.

Solution. Because the plot of $\Delta p/L$ vs. v is linear on the log-log graph in Fig. 5.35 for frontal-advance rates within the range listed in Table 5.10, the data can be fitted to Eq. 5.20 by use of least squares. Eq. 5.20 can be rearranged in the form

$$\Delta p / L = (\phi^{n_c} / \lambda_p^*)v^{n_c}$$
$$= a_2 v^{n_c} . \quad\dotfill (5.21)$$

From a least-squares analysis of the data, $a_2 = 1.005$ and $n_c = 0.918$, with a correlation coefficient of 0.995 and the units used in Table 5.10. Thus, this polymer is slightly shear-thinning in the Berea core over the range of frontal-advance rates studied.

For the data in Table 5.10, a_2 has the units of $(psi/ft)/(ft/D)^{n_c}$. Oilfield units of λ_p^* are $(md/cp)/(ft/D)^{1-n_c}$. The conversion factor developed in Example 5.2 for oilfield units was 158.005 md/cp = 1(ft²/psi-D). λ_p^* is computed from the following relationship:

Run	Polymer Solution Injected (PV)	Solvent Solution Injected (PV)	Permeability (md)
1	15.7	67.3	89.1
		92.9	88.8
2	14.1	57.5	92.4
		83.3	92.1
3	11.8	54.5	88.3
		86.5	88.3

*Original permeability = 203 md.

Table 5.8—Persistence of permeability reduction in Berea sandstone cores* treated with polyacrylamide (Gogarty 1967b).

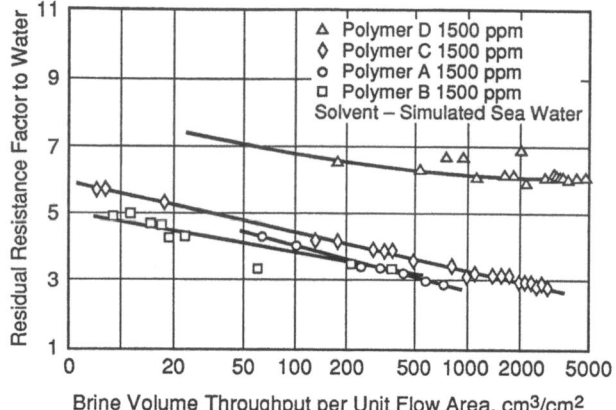

Fig. 5.29—Degradation of permeability reduction of poly-acrylamide after prolonged brine injection (Needham et al. 1974).

$$\lambda_p^* = 158.005 \left(\frac{\text{md}}{\text{cp}} \right) \left(\frac{\text{psi-D}}{\text{ft}^2} \right) \left(\frac{\phi^{n_c}}{a_2} \right) \frac{(\text{ft/D})^{n_c}}{(\text{psi/ft})}$$

$$= 158.005 \frac{\phi^{n_c}}{a_2} \left(\frac{\text{md}}{\text{cp}} \right) / \left(\frac{\text{ft}}{\text{D}} \right)^{1-n_c}$$

$$= 158.005(0.21)^{0.918}/1.005$$

$$= 37.52(\text{md/cp})/(\text{ft/D})^{1-n_c}.$$

For frontal-advance rates, v, between 0.03 and 5.56 ft/D, the Darcy velocity is related to the pressure gradient by

$$u^{0.918} = \frac{37.52(\text{md/cp})/(\text{ft/D})^{1-n_c}}{158.005(\text{md/cp})(\text{psi-D/ft}^2)} (\Delta p / L)(\text{psi/ft}) \quad \dots\dots\dots\dots\dots\dots\dots\dots\dots\dots\dots\dots\dots\dots\dots\dots \quad (5.22)$$

$$= 0.2375(\Delta p/L)(\text{ft/D})^{n_c}.$$

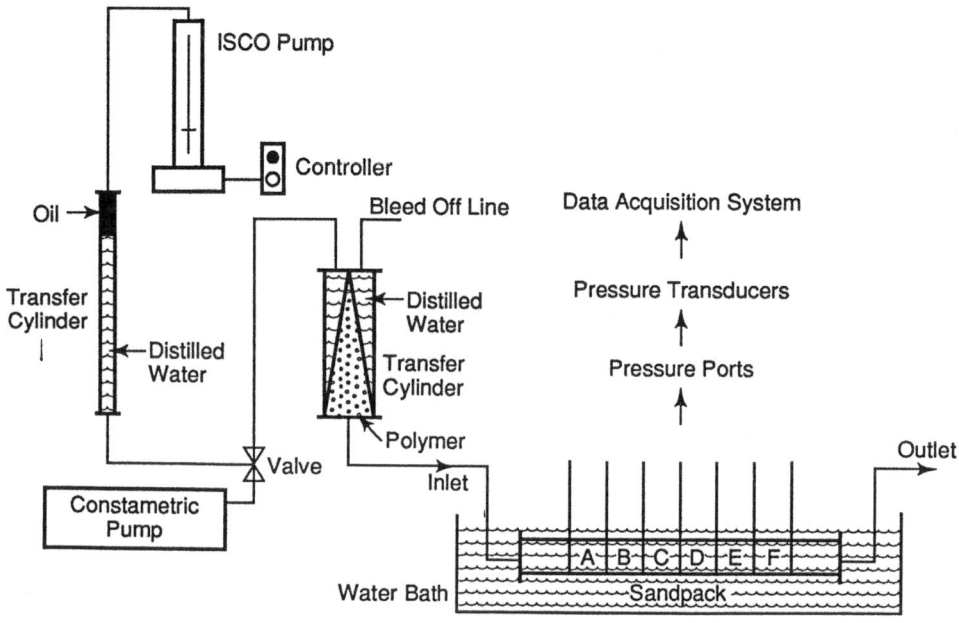

Fig. 5.30—Core layout with pressure taps for determination of polymer mobility.

Prediction of Polymer Mobility in the Shear-Thinning Region. Prediction of polymer mobility in porous media for conditions other than those measured in laboratory corefloods is sometimes desirable. Experimental data can be correlated to allow prediction of polymer mobility from rheological measurements for particular polymer/rock systems. This section presents examples of correlations developed for xanthan biopolymers.

Correlations based primarily on theoretical models of polymer flow in porous media assume that the power-law index for polymer flow in porous media, n_c, is identical to the power-law index determined from rheological measurements. This is not a good assumption, and n_c for flow in porous media must be determined from analysis of experimental data. Several polymer/rock systems have been studied in which $n_c \neq n$ (Willhite and Uhl 1988; Hejri 1989; Gogarty 1967b). When sufficient experimental data are taken, correlations may be developed relating polymer mobility to polymer and rock properties. Willhite and Uhl (1988) correlated λ_p^* and n_c with k_{wp} for three xanthan concentrations with Eqs. 5.23a through 5.23g. Units for λ_p^* are $(\text{md/cp})/(\text{ft/D})^{1-n_c}$. For a 500-ppm xanthan concentration,

$$\lambda_p^* = 0.783 k_{wp}^{0.708} \quad \dots \quad (5.23a)$$

and $n_c = 0.618 k_{wp}^{0.009}. \quad \dots \quad (5.23b)$

For a 1,000-ppm concentration,

$$\lambda_p^* = 0.746 k_{wp}^{0.578} \quad \dots \quad (5.23c)$$

and $n_c = 0.659 k_{wp}^{-0.035}. \quad \dots \quad (5.23d)$

For a 1,500-ppm concentration,

$$\lambda_p^* = 0.679 k_{wp}^{0.488}, \quad \dots \quad (5.23e)$$

and $n_c = 0.710 k_{wp}^{-0.073}, \quad \dots \quad (5.23f)$

$$k_{wp} = 0.83 k_w. \quad \dots \quad (5.23g)$$

In Eq. 5.23g, k_w is the permeability to water (md) before the core is contacted with polymer. If the core is at ROS, k_w is the permeability to water at ROS. The polymer mobility is computed from

$$\lambda_p = \lambda_p^* u^{1-n_c}. \quad \dots \quad (5.24)$$

At high shear rates, the polymer mobility approaches the mobility of water (or brine) as the upper Newtonian flow region is encountered. Under these conditions, the polymer mobility is limited by

$$\lambda_p = k_{wp}/\mu_\infty, \quad \dots \quad (5.25)$$

where μ_∞ = viscosity of the brine.

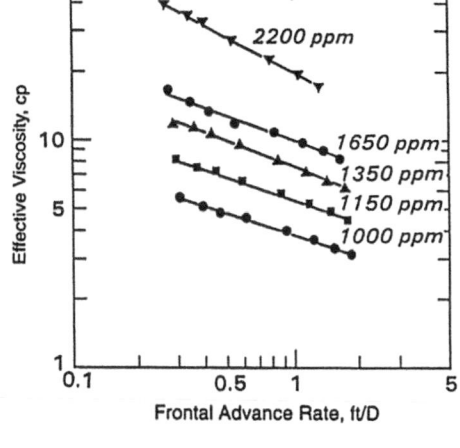

Fig. 5.31 Effective viscosity as a function of frontal-advance rate (Castagno et al. 1987).

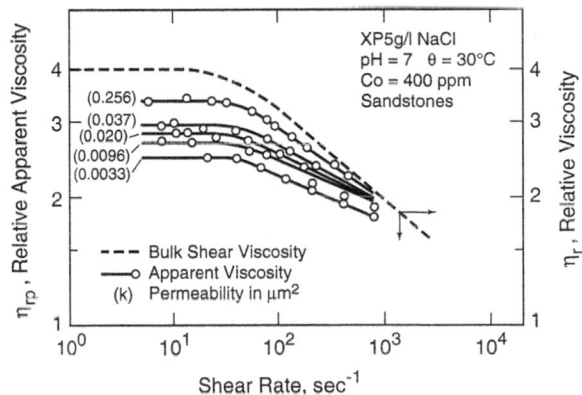

Fig. 5.32—Comparison of apparent viscosity in Fontainbleau sandstone and the bulk viscosity in the upper Newtonian and shear-thinning regions (Chauveteau and Zaitoun 1981).

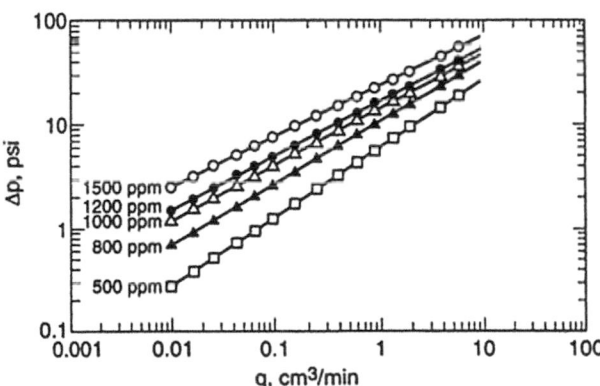

Fig. 5.33—Pressure-drop/flow-rate data for concentrations during the flow of Flocon 4800ME through Berea sandstone core (Willhite and Uhl 1988).

Correlations developed from experimental data generally are specific to the polymer/rock system used in their development. Hejri et al. (1991) developed correlations for the flow of xanthan biopolymer through unconsolidated sandpacks over a range of permeability from 0.53 to 15.3 darcies for polymer concentrations of 1,000, 1,500, and 2,000 ppm. Flow behavior was shear-thinning over a range of frontal-advance rates from 1 to 117 ft/D.

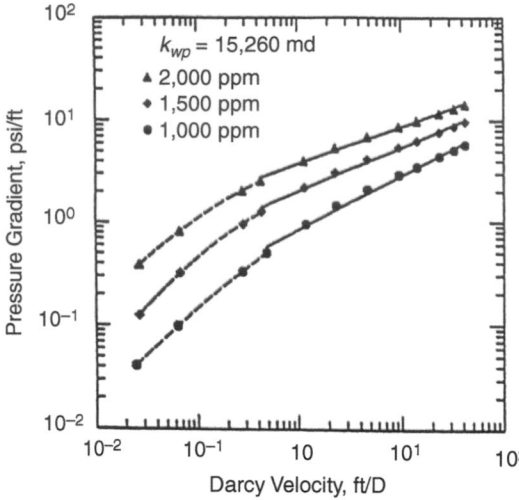

Fig. 5.34—Pressure-drop/flow-rate data for xanthan in an unconsolidated sandpack showing onset of Newtonian region at low frontal-advance rates (Hejri et al. 1991).

Table 5.11 presents rheological properties of the polymer. Correlations are presented in Eqs. 5.26a through 5.26e.

$$k_{wp} = 0.377 k_w^{1.088}. \quad\dots\dots\dots\dots\dots\dots\dots\dots \text{(5.26a)}$$

$$n_c = 0.37n + 0.132. \quad\dots\dots\dots\dots\dots\dots \text{(5.26b)}$$

For 1,000 ppm,

$$\lambda_p^* = 0.212 k_{wp}^{0.700}. \quad\dots\dots\dots\dots\dots\dots\dots \text{(5.26c)}$$

For 1,500 ppm,

$$\lambda_p^* = 0.187 k_{wp}^{0.621}. \quad\dots\dots\dots\dots\dots\dots\dots \text{(5.26d)}$$

For 2,000 ppm,

$$\lambda_p^* = 0.190 k_{wp}^{0.552}. \quad\dots\dots\dots\dots\dots\dots\dots \text{(5.26e)}$$

The correlations represented by Eqs. 5.26a through 5.26e can be extended to interpolate for polymer concentrations between 1,000 and 2,000 ppm by use of a correlation based on the modified Blake-Kozeny model for the flow of non-Newtonian fluids (Christopher and Middleman 1965). Eq. 5.27 is an expression for λ_{pBK}^* derived from the Blake-Kozeny model. Note that all parameters are either properties of the porous medium or rheological measurements. Eq. 5.27 underestimates λ_p^* by approximately 50%. However, Hejri et al. (1991) were able to correlate λ_{pBK}^* and λ_p^* for the unconsolidated sandpack data with Eq. 5.28. Eqs. 5.27 and 5.28, along with Eq. 5.24, predict polymer mobility for polymer concentrations ranging from 1,000 to 2,000 ppm within approximately 7%.

$$\lambda_{pBK}^* = k_{wp}^{(1+n)/2} \Big/ \left[\left(\frac{K}{12}\right)\left(\frac{9n+3}{n}\right)^n (150\phi)^{(1-n)/2} \right] \quad\dots\dots\dots\dots\dots\dots\dots\dots\dots\dots \text{(5.27)}$$

and $\lambda_p^* = 2.322 \lambda_{pBK}^{*0.947}$, $\quad\dots\dots\dots\dots\dots\dots\dots\dots\dots\dots\dots\dots\dots\dots\dots\dots\dots\dots\dots\dots\dots\dots \text{(5.28)}$

where λ_{pBK}^* is in $(md/cp)/(ft/D)^{1-n}$, K is in cp^n, and k_{wp} is in md.

In some polymer/rock systems, $n_c = n$. For example, Cannella et al. (1988) and Teeuw and Hesselink (1980) report good agreement between n_c and n for xanthan biopolymers in porous media used in their studies. Cannella et al. report permeability reduction from biopolymer retention was less than a factor of 0.9. They were able to correlate the apparent viscosity with a modified power-law expression of the form given by

$$\mu_p = \mu_\infty + K\dot{\gamma}_a^{n-1} \quad\dots\dots\dots\dots\dots\dots\dots\dots\dots\dots\dots\dots\dots\dots\dots\dots\dots\dots\dots\dots\dots\dots\dots \text{(5.29)}$$

and $\mu_p = \mu_\infty + K\left(\frac{3n+1}{4n}\right)^n \left(\frac{6u}{\sqrt{k\phi}}\right)^{n-1}$, $\quad\dots\dots\dots\dots\dots\dots\dots\dots\dots\dots\dots\dots\dots\dots\dots\dots\dots\dots \text{(5.30)}$

where consistent units are used.

In Eq. 5.29, the apparent shear rate is given by

$$\dot{\gamma}_a = 6[(3n+1)/4n](u/\sqrt{k\phi}). \quad\dots\dots\dots\dots\dots\dots\dots\dots\dots\dots\dots\dots\dots\dots\dots\dots\dots\dots\dots \text{(5.31)}$$

Polymer Concentration (ppm)	λ_p^* [(md/cp)/(ft/D)]	n_c
500	45.7	0.65
800	24.3	0.579
1,000	17.7	0.524
1,200	14.2	0.507
1,500	10.05	0.471

Table 5.9—Parameters for flow of xanthan biopolymer in Berea sandstone core represented by Fig. 5.33 (Willhite and Uhl 1988) data.

In Eq. 5.31, the units of k must be ft² or m² and be consistent with the Darcy velocity. Units of the Darcy velocity may be ft/sec or m/s.

The constant 6 was determined empirically by matching the data for particular rock and fluid systems. A representative correlation is illustrated in **Fig. 5.36. Table 5.12** presents rheological parameters determined by Cannella et al. (1988) for the xanthan biopolymer used to develop the correlation.

Polymer mobility in the shear-thinning region can be predicted from correlations of experimental data. However, predictions from theoretical models based solely on rheological data are not dependable. Some experimental data are required for the particular polymer/rock system of interest to develop a suitable correlation or to verify the applicability of a particular correlation.

Prediction of Polymer Mobility in the Shear-Thickening Region. At high frontal-advance rates, polyacrylamides exhibit an unusual flow behavior in porous rocks. The flowing fluid appears to become more viscous as the flow rate increases. This behavior is called shear thickening. **Fig. 5.37** shows the development of shear-thickening behavior (i.e., the sharp increase in the resistance factor) as flow rate increases for a 500-ppm solution of partially hydrolyzed polyacrylamide flowing in a sandstone (MacWilliams et al. 1973).

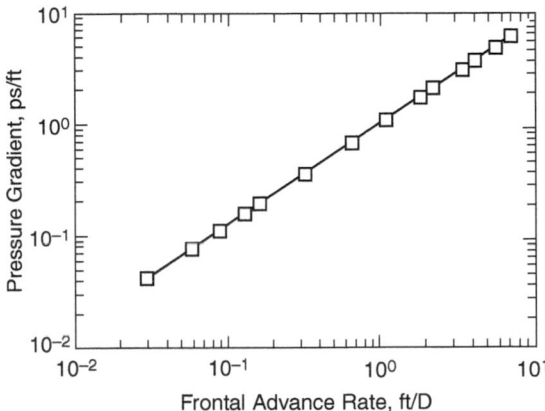

Fig. 5.35—Pressure-drop/flow-rate data for polyacrylamide in Berea core material (Gao and French 1988).

Shear thickening is caused by the viscoelastic nature of polyacrylamide. Polyacrylamide has a flexible coil conformation in solution. When the flexible polyacrylamide molecule flows from pore to pore, it deforms (i.e., stretches) to adjust to the flow field. If the average flow time from one constriction to the next is large relative to the time required for the polymer molecule to relax and assume the random coil configuration, the polymer remains shear thinning. The characteristic time required for the polymer molecule to relax is called the relaxation time and can be measured with specially designed rheogoniometers. At high flow rates, however, the transit time between pore throats (i.e., successive deformations) is of the same order of magnitude as the relaxation time of the polymer and the polymer chains remain elongated during flow, increasing the apparent viscosity of the flowing fluid. Shear thickening of polyacrylamide is a characteristic of flow in porous materials and is not observed in rheological measurements of polyacrylamide at comparable shear rates.

The transition from shear thinning to shear thickening often occurs over a narrow range of frontal-advance rates. **Fig. 5.38** shows experimental data (Heemskerk 1984) for the flow of polyacrylamides in porous rocks. Shear-thinning behavior

Frontal-Advance Rate (ft/D)	Pressure Gradient (psi/ft)
0.0295	5.10
0.0577	3.84
0.0883	3.18
0.129	2.12
0.159	1.77
0.316	0.689
0.652	0.242
1.090	0.078
1.85	0.161
2.22	1.095
3.43	0.194
4.14	0.042
5.56	0.110

Table 5.10—Flow behavior of 1,500 ppm Pusher 500® in Berea core (576 md) at 72.3°F, Example 5.3 (Gao and French 1988).

C (ppm)	n	K (mPa·s^n)
1,000	0.418	181
1,500	0.325	438
2,000	0.273	835

Table 5.11—Power-law parameters of Flocon 4800 MX® (Hejri et al. 1991).

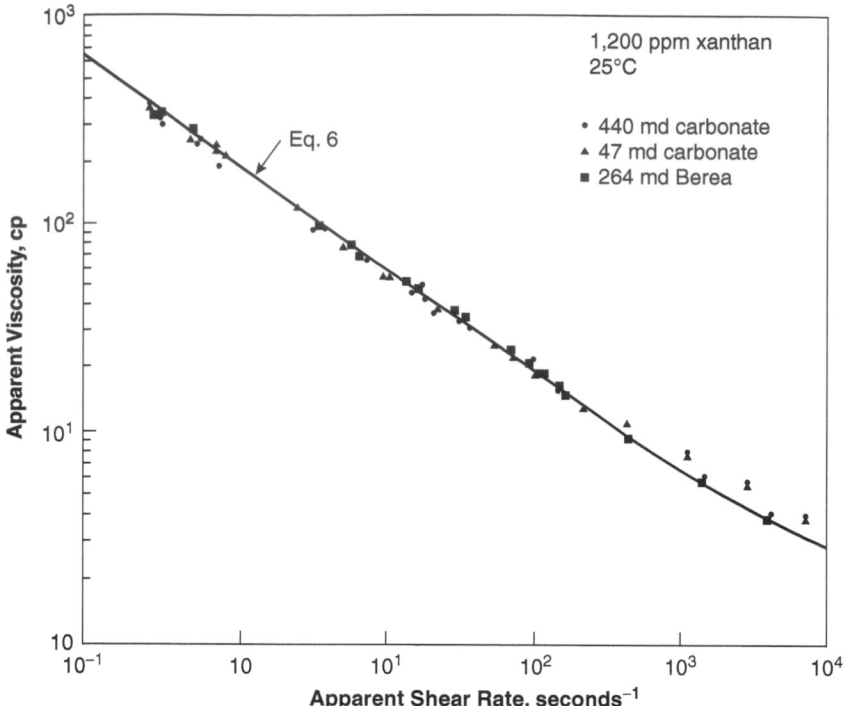

Fig. 5.36—Correlation of apparent viscosity with apparent shear rate for 1,200-ppm xanthan biopolymer in reservoir rocks (Cannella et al. 1988).

was observed in each porous rock until a critical frontal-advance rate, v_c, was attained. Thereafter, shear-thickening behavior was observed. Shear thickening causes an additional pressure drop in the flowing fluid. This pressure drop, Δp_E, is shown in **Fig. 5.39,** which also shows the pressure drop, Δp_v, that would have been observed if the flow remained shear thinning. When shear thickening occurs, the slope of pressure drop vs. frontal-advance rate (or volumetric flow rate) on log-log graph paper, n_c, is larger than unity. In the shear-thinning region where $v < v_c$, the flow index, n_c, is less than 1.0.

Example 5.4 illustrates the analysis of polymer flow data in the shear-thickening region.

Example 5.4—Analysis of Polymer Flow in the Shear-Thickening Region. This example is a continuation of Example 5.3. Additional data were taken at frontal-advance rates up to 35.34 ft/D on the same Berea core (K_w=576 md and $\phi = 0.21$) (Gao and French 1988). **Table 5.13** presents the data. A plot of pressure gradient vs. frontal-advance rate is shown in **Fig. 5.40,** including the data used in Example 5.3. There are two linear regions in Fig. 5.40, the shear-thinning region previously discussed and a shear-thickening region from approximately 14 to 35.34 ft/D. The transition from shear thinning to shear thickening begins at a frontal-advance rate of approximately 6 ft/D. Analyze the data to determine the critical velocity.

Solution. The shear-thickening region of the data can be fitted to the power-law model of Eq. 5.20 by determining the values of λ_p^* and n_c with the same procedure described in Example 5.3. Regression analysis was used to obtain the following values: $a_2 = 0.297$, $n_{c2} = 1.481$, $R_2 = 0.999$ (correlation coefficient), $\lambda_{p2}^* = 158.005\phi^{n_c}/a_2 = 158.005(0.21)^{1.481}/0.297$, and $\lambda_{p2}^* = 52.74$ (md/cp)/(ft/D)$^{1-n_c}$.

An approximate model of the flow of 1,500 ppm Pusher 500 in 53 meq/L NaCl in Berea sandstone can be developed for the entire range of frontal-advance rates by use of the two linear regions in Fig. 5.40, which appear to intersect at $v_{c2} \approx 10$ ft/D. The value of v_{c2} can be determined by solving Eq. 5.20 at the intersection of the two regions. In the shear-thinning region,

$$u^{n_{c1}} = \lambda_{p1}^* (\Delta p / L).$$

C (ppm)	n	K (cp-sec^{n-1})	μ_{pN} (cp)
300	0.75	17	8.6
600	0.60	43	26.0
1,200	0.48	195	102.0
1,600	0.35	620	1,000.0

Table 5.12—Xanthan rheological parameters (Cannella et al. 1988).

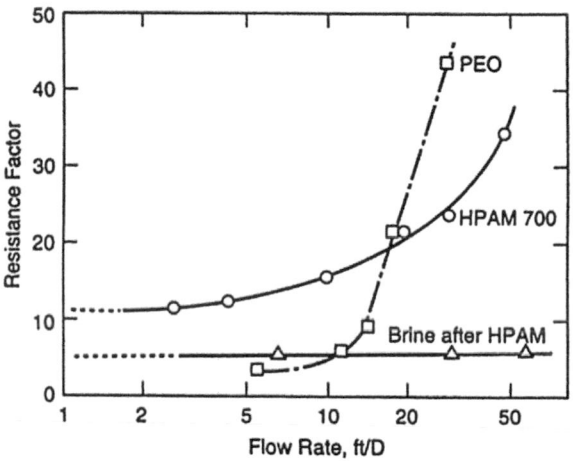

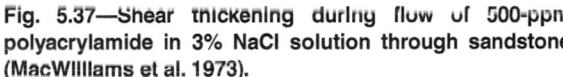

Fig. 5.37—Shear thickening during flow of 500-ppm polyacrylamide in 3% NaCl solution through sandstone (MacWilliams et al. 1973).

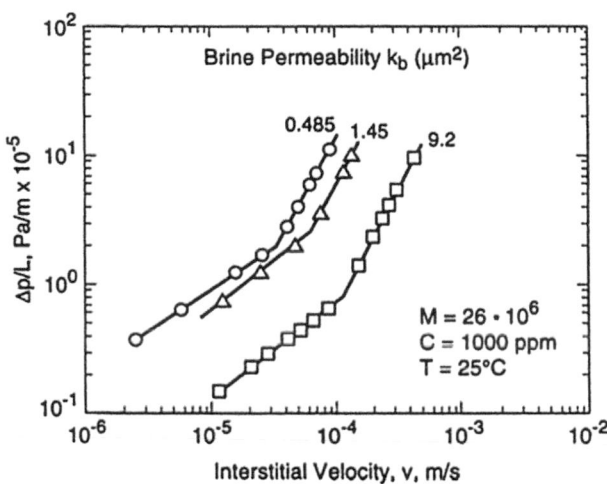

Fig. 5.38—Shear-thickening behavior of polyacrylamide in sandstone (Heemskerk et al. 1984).

Frontal-Advance Rate (ft/D)	Pressure Gradient (psi/ft)
35.34	59.19
28.32	42.03
24.79	34.26
21.33	26.79
17.71	20.82
14.01	15.12
10.66	11.01
10.47	10.59
6.95	6.60
5.56	5.10

Table 5.13—Flow behavior of 1,500 ppm Pusher 500® in Berea core at 72.3°F, Example 5.4 (Gao and French 1988).

In the shear-thickening region,

$$u^{n_{c2}} = \lambda_{p2}^* (\Delta p / L).$$

At the critical velocity, both velocity and pressure gradients are equal. Thus,

$$u^{n_{c2}} = (\lambda_{p2}^* / \lambda_{p1}^*)^{1/(n_{c2} - n_{c1})}. \quad \dots \dots \dots \dots \dots \dots \dots (5.32)$$

Solving for u_c with Eq. 5.32 yields

$$n_c = 0.659 k_{wp}^{-0.035}$$

$$= (0.659)(33.2)^{-0.035}$$

$$= 0.583.$$

Thus, $v_{c2} = 1.831 / 0.21$

$$= 8.72 \text{ ft/D}.$$

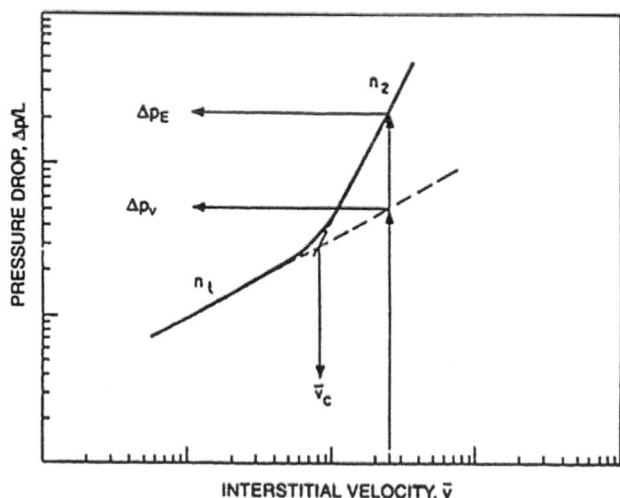

Fig. 5.39—Determination of critical velocity for onset of shear-thickening flow for polyacrylamide in cores (Heemskerk et al. 1984).

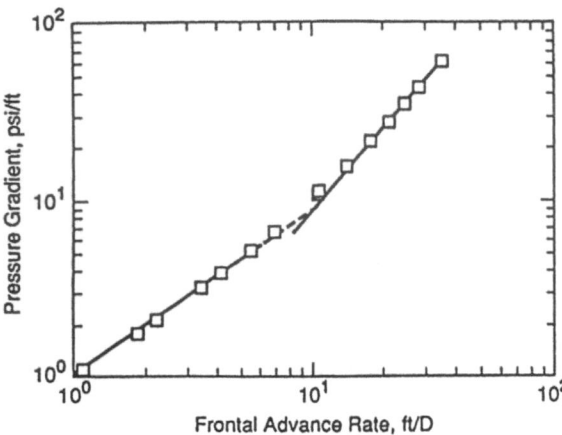

Fig. 5.40—Determination of critical velocity for onset of shear-thickening flow in Example 5.4 (Gao and French 1988).

The onset of viscoelastic or shear-thickening flow is a function of permeability, temperature, salinity, molecular weight, and polymer concentration. Flow parameters must be determined experimentally. **Fig. 5.41** (Heemskerk et al. 1984) presents experimental data showing the effect of average molecular weight on polymer flow in sandstones with permeabilities ranging from 912 to 115 md. **Table 5.14** summarizes power-law parameters obtained from these data.

In Table 5.14, n_{c1} is the flow index for the shear-thinning region and n_{c2} is the flow index for the shear-thickening region. The value of v_{c2} can be computed from the value of the critical Deborah number, N_{Dec}, as described in the next section.

In oilfield applications, the viscoelastic character of polyacrylamide solutions is evaluated with a simple apparatus called the screen viscometer. The screen viscometer (MacWilliams et al. 1973), shown in **Fig. 5.42,** consists of a bulb with known volume connected to a glass tube. A set of five 100-mesh stainless-steel screens is inserted into a fitting at the end of the glass tube. The screen viscometer is used by determining the times required for known volumes of polymer and brine to flow through the viscometer. The screen factor, F_s, is defined as the ratio of the flow time for the polymer solution, Δt_p, to the flow time for the brine, Δt_b:

$$F_s = \Delta t_p / \Delta t_b. \dots\dots\dots\dots\dots\dots\dots\dots\dots\dots\dots\dots\dots\dots\dots\dots\dots\dots\dots\dots\dots\dots\dots\dots (5.33)$$

Table 5.15 presents values of screen factors for some polyacrylamide solutions. The screen factor was initially used to correlate flow resistance in porous rocks. However, as discussed in the next section, the screen factor commonly is used to evaluate

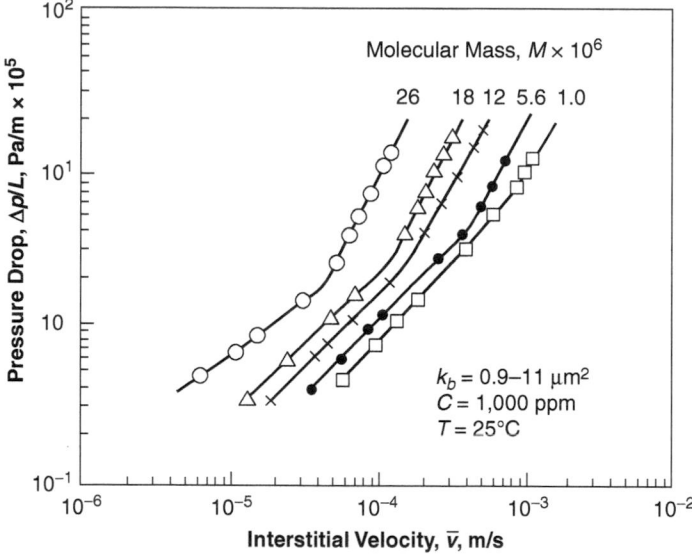

Fig. 5.41—Pressure drop as a function of interstitial velocity for solutions with polyacrylamides of different molecular weights (Heemskerk et al. 1984).

M (× 10^{-6})	k_b (md)	ϕ	n_{c1}	n_{c2}	τ_f	N_{Dec}
26	1,095	0.19	0.70	1.96	1.1	1.09
18	963	0.19	0.88	1.88	0.32	0.94
12	1,034	0.19	0.93	1.76	0.24	1.04
5.6	916	0.19	0.95	1.60	0.10	0.98
1.0	917	0.19	0.96	1.43	0.06	1.11

Table 5.14—Parameters for core flow tests in Bentheim sandstones (Heemskerk et al. 1984).

shear degradation of the polymer solution. The screen factor has been shown to be a direct measure of the viscoelastic characteristics of a polymer solution (Lim et al. 1986).

Shear Degradation. Flexible polymer molecules such as polyacrylamide are quite susceptible to shear degradation (Maerker 1975, 1976). They can be easily degraded if subjected to high shear rates, as would occur if a large pressure drop is taken across a valve or an orifice. Shear degradation causes a rupture of the polymer chain with corresponding changes in the average molecular weight, screen factor, and solution viscosity. The screen factor is more sensitive to shear degradation than the solution viscosity. Thus, the screen factor is used widely to assess the rheological quality of a polyacrylamide solution. Biopolymers do not degrade under a similar shear history.

Shear degradation of partially hydrolyzed polyacrylamides may occur when polymer solutions are injected into the reservoir. The injected solution is sheared when it passes through perforations in the well casing. Degradation also may occur when a polymer solution flows through a porous matrix at high frontal velocities, as might exist in the vicinity of the injection wellbore. **Fig. 5.43** (Maerker 1975) shows degradation of a 600-ppm solution of polyacrylamide in 3.3% brine (3% NaCl and 0.3% $CaCl_2$) forced through consolidated sandstone plugs over a wide range of flow rates. Flow rates are plotted as fluxes [i.e., ft^3/(ft^2-D)], which are equivalent to the Darcy velocity of the fluid.

The data in Fig. 5.43 show that shear degradation occurs over a wide range of flow rates. At the highest fluxes, 80 to 90% of the screen factor was lost. Loss of solution viscosity was less than 20% in the range of fluxes studied. The onset of shear degradation occurred after viscoelastic flow began. These results show that loss of viscoelastic properties can be significant at high fluxes in the vicinity of the injection wellbore. Thus, shear degradation must be considered in the flow of partially hydrolyzed polyacrylamides in porous media (Maerker 1975, 1976).

Solvent properties affect shear degradation, as shown in **Figs. 5.44 and 5.45** (Maerker 1975). Degradation decreases as ionic strength decreases. For example, the onset of degradation increases from 18 to 100 ft^3/(ft^2-D) when the salinity is decreased from 130,000 to 3,000 ppm NaCl. Causes of mechanical degradation are not fully understood, and explanations based on solution properties conflict. There should be more entanglement of polymer molecules at 3,000 ppm because the polymer should be able to expand and thus more degradation would be expected at low salinities. However, the polymer has a tighter coiled configuration and thus less flexibility at 130,000 ppm than at 3,000 ppm. The degradation experiments in porous media suggest that polymers are more susceptible to degradation when tightly coiled at high salinities. Calcium causes more degradation than expected from changes in ionic strength. Effects of polymer concentration on degradation are shown in **Fig. 5.46.** For concentrations of 300 to 600 ppm, screen factor was not affected by concentration. Solution viscosity losses were identical up to a flux of approximately 100 ft^3/(ft^2-D).

Data from studies by Seright (1983) support the concept that degradation occurs primarily within a short distance after the fluid enters the porous matrix. He presented flow data showing that an extra pressure drop is observed at the entrance of the porous medium when degradation begins. **Table 5.16** contains a set of runs in a 6-in. Berea core over a wide range of fluxes. Polymer was injected at constant rate until the pressures stabilized. Pressure is plotted vs. length in **Fig. 5.47** for two different polymer flow rates and for brine flow. The pressure from Pressure Tap 2 to the end of the core is a linear function of distance, indicating undetectable degradation in that region. Extrapolation of the linear pressure profile to the core entrance gives the value of the pressure (p_{int}) that would have been measured if all degradation occurred at the sandface as the polymer enters the core. The difference between the actual measured entrance pressure, p_i, and p_{int} is the estimated pressure drop ($\Delta p_{md} = p_i - p_{int}$) associated with polymer degradation. These data suggest that, at the higher flow rates, degradation occurs in the entrance

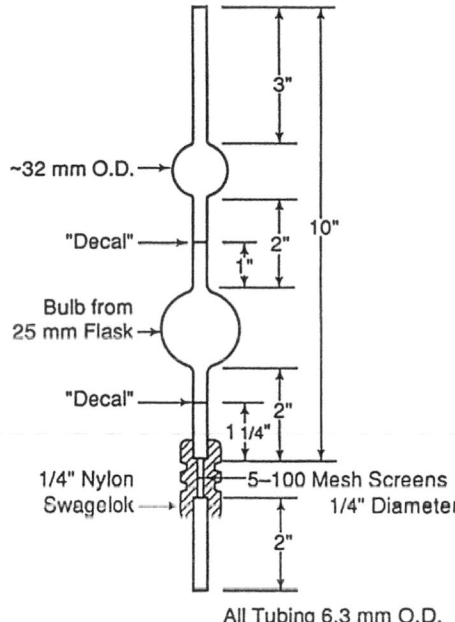

Fig. 5.42—Screen viscometer.

Polymer	Concentration (ppm)	F_s	Reference
A	600	17.8	Seright (1983)
A	600	18.9	Seright (1983)
A	600	18.1	Seright (1983)
Pusher 500®	750 in fresh water	11.2	Martin and Kantamukkala (1984)
Pusher 500®	1,500 in fresh water	17.3	Martin and Kantamukkala (1984)
Pusher 500®	2,000 in fresh water	22.8	Martin and Kantamukkala (1984)
Pusher 500®	2,500 in fresh water	27.3	Martin and Kantamukkala (1984)
Pusher 500®	1,500 in 3% NaCl	14.1	Martin and Kantamukkala (1984)

Table 5.15—Screen factors for selected polyacrylamide solutions.

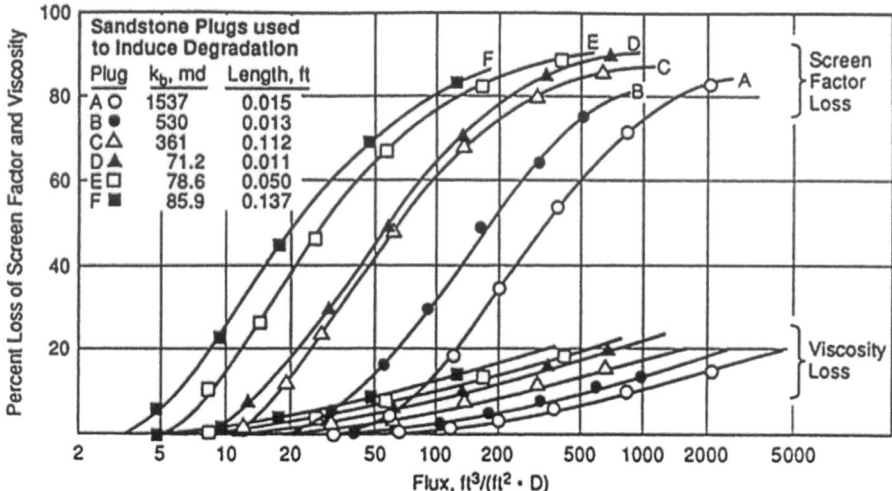

Fig. 5.43—Laboratory degradation data: percent losses of screen factor and viscosity as functions of flux through consolidated sandstone plugs for 600-ppm polyacrylamide solutions in 3.3% brine (Maerker 1975).

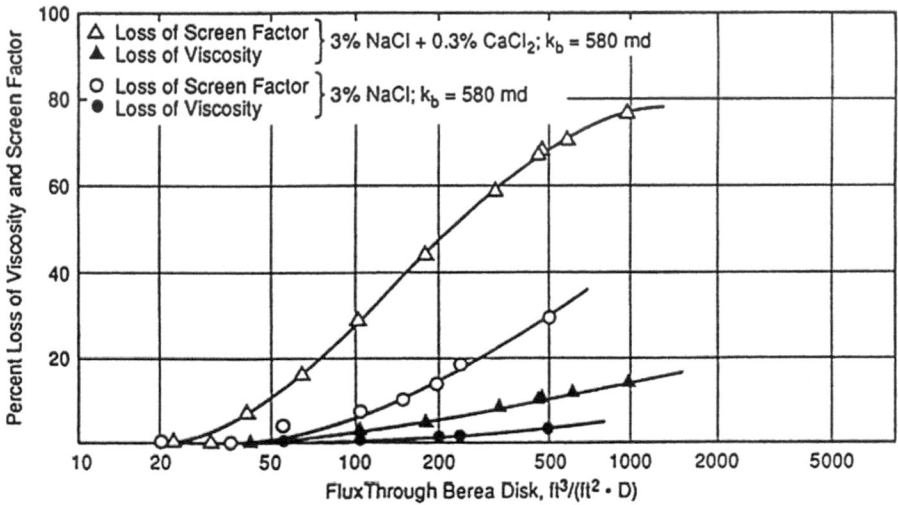

Fig. 5.44—Effect of CaCl$_2$ addition on mechanical degradation of 600-ppm polyacrylamide solutions in Berea sandstone (Maerker 1975).

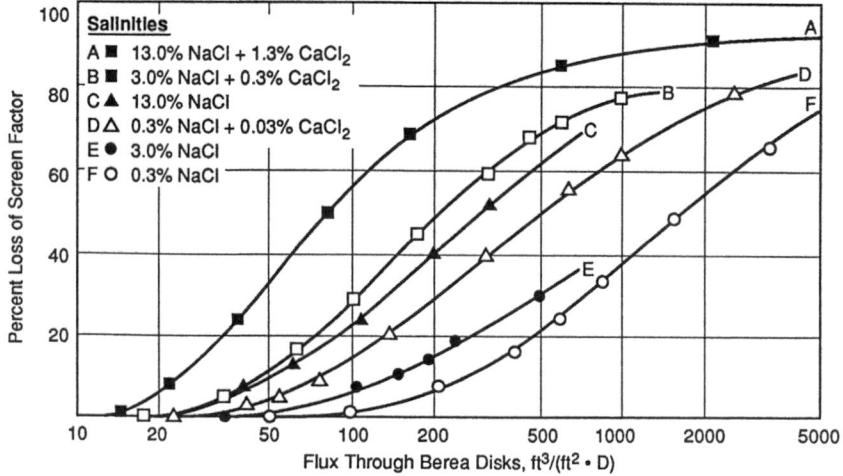

Fig. 5.45—Effect of NaCl and CaCl$_2$ concentrations on mechanical degradation of 600-ppm polyacrylamide solutions in Berea sandstone (Maerker 1975).

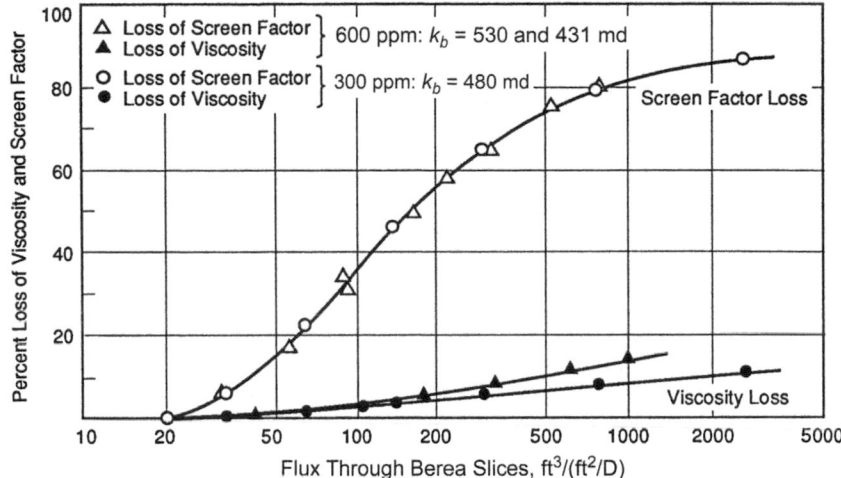

Fig. 5.46—Effect of polyacrylamide concentration on mechanical degradation in Berea sandstone (Maerker 1975).

region and that no further degradation occurs after this region. The onset of degradation was detected when the screen factor began to drop from its initial value of 17.8. Note that changes in the solution viscosity were much smaller than changes in the screen factor.

The data in Table 5.16 may be examined to determine flow characteristics of the polymer solution in the porous matrix. **Fig. 5.48** plots pressure drop vs. u for the data in Table 5.16. This pressure drop is the total pressure drop across the 6-in. core after the pressure drop caused by shear degradation, Δp_{md}, is deducted. The graph is linear from $u = 3.7$ to 10.7 ft/D and the slope is 1.49, indicating that the fluid is shear-thickening in this interval. The onset of shear degradation was indicated by the value of Δp_{md} determined at $u = 10.7$ ft/D. Thus, there was a substantial region of shear-thickening flow before significant shear degradation was observed.

Seright also presented data for the flow of a polymer solution that was degraded before flow studies. **Table 5.17** shows results obtained when a polymer solution had been degraded by injecting the solution into a 150-md Berea core at a flux of 42.1 ft³/(ft²-D). The screen factor of the polymer solution decreased from 18.2 to 6.5, while the solution viscosity decreased from 2.56 to 2.10 cp. There were no shear-thinning effects on this polymer because the salinity was 3.3%. Table 5.17 summarizes the displacement data when this solution was used for fluxes ranging from 7.1 to 56.4 ft³/(ft²-D). Measurements of the screen factor indicate that no further degradation occurred until the flux exceeded 42.1 ft³/(ft²-D).

These data suggest that shear degradation is inevitable in some situations. This has led some operators to flow polyacrylamide solutions through shear plates to degrade the polymer slightly before injection to reduce the pressure drop caused by

Pressure* (psi)					Correlation Coefficient	Flux (ft/D)	F_r	Δp_{md} (psi)	Screen Factor	Viscosity (cp)
$p_{0\%}$	$p_{19\%}$	$p_{57\%}$	$p_{84\%}$	$p_{intercept}$						
						0	—	—	17.8	2.44
0.53	—	—	—	0.53	—	0.38	4.9	0	17.8	2.44
8	6.6	3.7	1.8	8	0.998	3.7	6.4	0	17.8	2.44
20	16.2	8.6	3.3	20	1.000	6.8	8.5	0	17	2.43
40	31.7	16.5	5.9	39	1.000	10.7	10.6	1	16.1	2.41
75	57.9	28.8	9.5	71	0.000	17.7	11.6	4	13.1	2.39
177	123	59	15.5	150	0.996	43.6	10.0	27	7.9	2.13
313	207	98	23	252	0.995	105	6.9	61	4.5	1.88
432	283	133	—	343	0.998	171	5.8	89	3.3	1.70
586	382	183	—	464	0.998	289	4.6	122	2.5	1.67

*$p_{x\%}$ = pressure at x% core length and $p_{100\%} = 0$ psig. $p_{intercept}$ and correlation coefficients were determined by linear regression with all points except $x = 0$.

Table 5.16—Linear coreflood results for 600 ppm Polymer A in 3.3% brine (229-md Berea core, 6-In. length, $\phi = 0.20$) (Seright 1983).

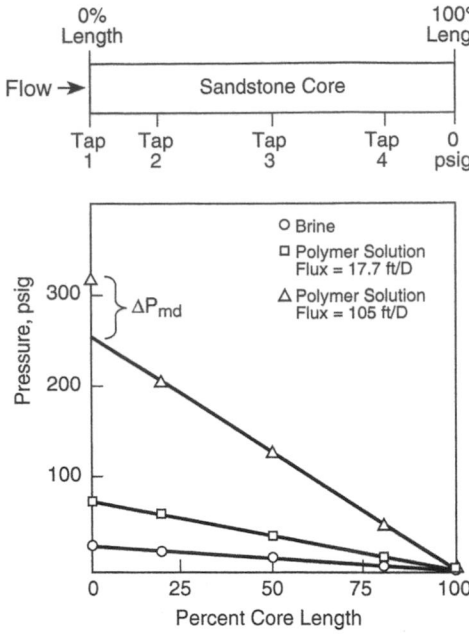

Fig. 5.47—Pressure vs. percent core length for polymer-degradation studies (Seright 1983).

mechanical degradation that occurs in the vicinity of the wellbore. By reducing the pressure drop, injecting at higher rates becomes possible.

The onset of viscoelastic effects can be correlated with properties of the fluid and rock. For unconsolidated porous media, viscoelastic effects begin at a Deborah number of approximately 0.5 (Metzner et al. 1966; Marshall and Metzner 1967). The Deborah number is dimensionless and is defined by

$$N_{De} = \dot{\varepsilon}\tau_f, \dotfill (5.34)$$

where N_{De} = Deborah number, $\dot{\varepsilon}$ = stretch rate of the fluid, and τ_f = relaxation time of the fluid.

The stretch rate of the fluid (time^{-1}) is approximated by

$$\dot{\varepsilon} = \overline{v}/\left(\overline{d}_p/2\right), \dotfill (5.35)$$

where $\overline{d}_p$ = average grain diameter, $\overline{v}$ = average interstitial velocity, and consistent units are used.

For consolidated rocks, the average grain diameter cannot be obtained easily. An equivalent average grain diameter is estimated from

$$\overline{d}_p = \left[(1-\phi)/\phi\right]\sqrt{150k_w/(1-\phi)}, \dotfill (5.36)$$

where k_w = permeability to water at 100% saturation in units of ft^2 or m^2 to be consistent with Eq. 5.35.

Relaxation times are determined from rheological experiments. Table 5.14 includes representative values for polyacrylamides (Heemskerk etl. 1984) and critical values of the Deborah number for the onset of viscoelasticity. The Deborah numbers were in the vicinity of unity for these data.

The effects of mechanical degradation for uniform materials can be correlated from the stretch rate of the fluid. **Fig. 5.49** is the correlation (Maerker 1975) between screen-factor loss and a group $\dot{\varepsilon}L_D^{1/3}$ determined as follows. The average grain diameter is computed from Eq. 5.36 with appropriate modification for units. When k_w is in md and v is in ft/D, the average grain diameter and stretch are computed from

$$\dot{\varepsilon} = (2v/86,400\,\overline{d}_p)\text{ seconds}^{-1}. \dotfill (5.37)$$

$$\text{and } \overline{d}_p = \frac{1-\phi}{\phi}\sqrt{\frac{150k_w\,1.0623\times10-14}{1-\phi}}\text{ft.} \dotfill (5.38)$$

$$L_D = L/\overline{d}_p, \dotfill (5.39)$$

where L_D = dimensionless flow distance in which degradation occurs in number of grain diameters and L = distance where degradation takes place.

In radial geometry, mechanical degradation occurs under a diminishing shear rate as the injected polymer flows away from the wellbore. Maerker (1975) showed that the total degradation during radial flow can be estimated by computing the stretch rate at the sandface radius and the distance where degradation takes place at one-half the wellbore radius. Eq. 5.40 gives the stretch rate as a function of radial distance.

$$n_c = 0.659k_{wp}^{-0.035}$$

$$= (0.659)(33.2)^{-0.035}$$

$$= 0.583, \dotfill (5.40)$$

where q = injection rate, B/D; h = depth of perforation, ft; and $\dot{\varepsilon}_i$ = stretch rate, seconds^{-1}, at radius r_i, ft.

When degradation occurs in a porous rock, the region is confined to a small interval (0.5 to 1 in.) after the polymer

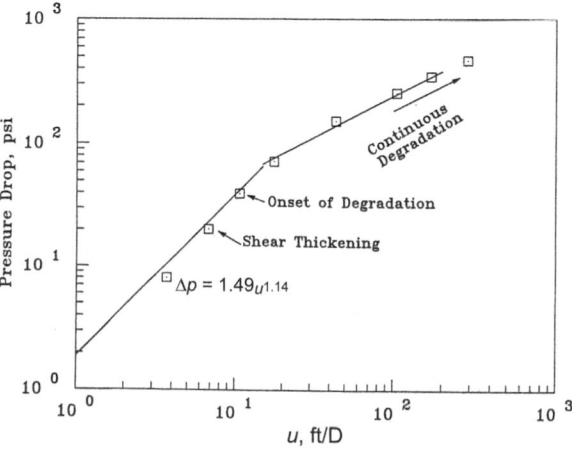

Fig. 5.48—Onset of shear thickening and degradation in Berea sandstone core (Seright 1983).

Pressure (psi)					Correlation Coefficient	Flux [ft³/(D–ft²)]	F_r	Δp_{md} (psi)	Screen Factor	Viscosity (cp)
$p_{0\%}$	$p_{19\%}$	$p_{49\%}$	$p_{81\%}$	$p_{intercept}$						
						0	—	—	18.1	2.56
*253	178	107	29	218	0.997	42.1	9.6	35	6.5	2.10
7	5.9	3.8	1.5	7	1.000	7.1	1.8	0	6.2	2.14
220	16.2	10.3	3.8	20	1.000	15.3	2.4	0	6.2	2.14
55	44	30	10.0	55	0.998	25.5	4.0	0	6.1	2.11
110	88	59	18.7	110	0.998	34.1	6.0	0	6.0	2.08
157	126	83	26	157	0.999	41.6	7.0	0	5.6	2.05
204	164	105	31	204	0.999	49.2	7.7	0	5.2	2.08
248	193	121	34	239	0.999	56.4	7.9	9	4.8	2.02

*Data in this row concern a freshly prepared polymer solution forced once through a core at high flux. This solution was then reinjected into the same core to obtain the data in the next row.

Table 5.17—Linear coreflood results for 600 ppm Polymer A in 3.3% brine (150-md Berea core, 6-in. length, $\phi = 0.215$) (Seright 1983).

enters the porous rock. Example 5.5 illustrates the use of this correlation to estimate degradation of polyacrylamide in an injection well.

Example 5.5—Application of Maerker Correlation To Estimate Degradation. A polyacrylamide solution is to be injected into a well at a rate of 5 B/(D-ft). The well is completed open hole with a diameter of 6 in. Average permeability to the flow of polymer solution is 50 md, and the average porosity is 15%. Determine whether mechanical degradation is likely to occur.

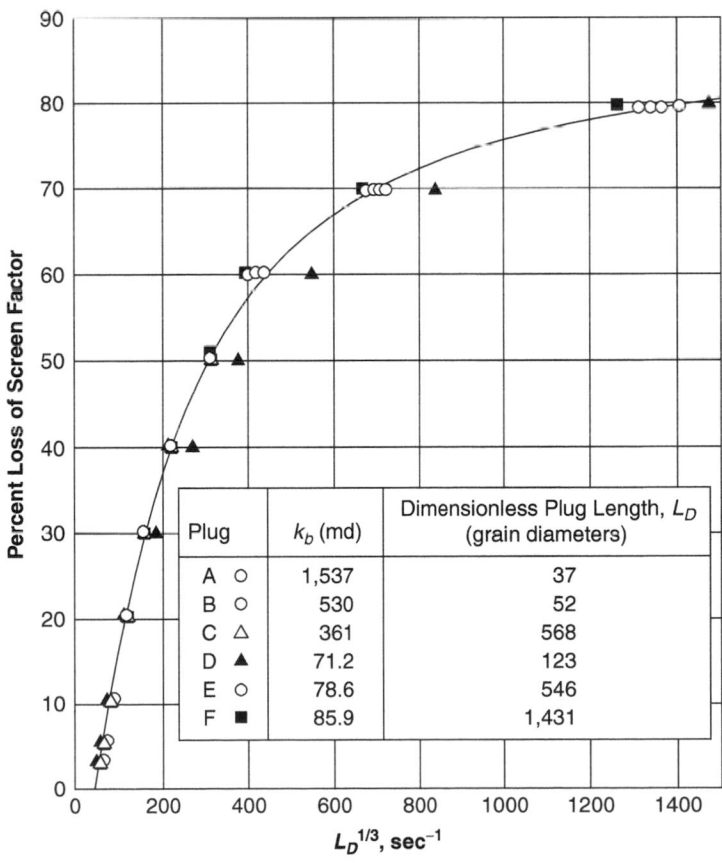

Fig. 5.49—Correlation of screen-factor loss caused by mechanical degradation in consolidated porous media (Maerker 1975).

Solution. Begin by computing the parameters necessary to evaluate the stretch rate in radial geometry with Eq. 5.40. The average grain diameter is estimated with Eq. 5.38.

$$\bar{d}_p = \frac{1-\phi}{\phi} \sqrt{\frac{150 k_w 1.0623 \times 10^{-14}}{1-\phi}}$$

$$= \frac{1-0.15}{0.15} \sqrt{\frac{(150)(50)(1.0623 \times 10^{-14})}{1-0.15}}$$

$$= 5.49 \times 10^{-5} \text{ ft.}$$

The stretch rate is evaluated at the wellbore radius with Eq. 5.40.

$$\dot{\varepsilon}_i = 5.615 q / (86,400 \pi \phi h \bar{d}_p r_i)$$

$$= \frac{5.615(5)}{86,400\pi(0.15)(5.49 \times 10^{-5})(0.25)}$$

$$= 50 \text{ seconds}^{-1}.$$

The dimensionless distance, L_D, is computed by assuming $L = \frac{1}{2} r_w$ at the sandface radius.

$$LD = r_w / (2\bar{d}_p)$$

$$= 0.25/(2)(5.49 \times 10^{-5})$$

$$= 2,277.$$

Thus, $\dot{\varepsilon}_i L_D^{0.33} = (50)(2,277)^{0.333}$

$$= 641 \text{ seconds}^{-1}.$$

From Fig. 5.49, the screen-factor loss caused by mechanical degradation is approximately 67%. Thus, significant mechanical degradation of the polymer is expected at this injection rate. Because solution viscosity is not as sensitive to mechanical degradation as the screen factor, additional data are needed to evaluate the effects on solution viscosity.

Southwick and Manke (1988) studied the injection of polyacrylamide solutions into perforated completions and found that degradation is not described completely by the Deborah number. A permeability dependence was observed. This reinforces the position that experimental data are required to evaluate the particular porous-rock/ polymer solution being considered.

Mechanical degradation is more extensive and cannot be correlated as well in heterogeneous porous matrices such as carbonate rocks (Martin and Kantamukkala 1984). This is demonstrated in **Fig. 5.50,** where rheograms are displayed for the injected solution and effluent solutions of a series of flow tests conducted at interstitial velocities ranging from 24 to 278 ft/D. The core plug used in these experiments was from the Phosphoria formation and had a permeability of 85 md (Martin 1986).

There is substantial loss in solution viscosity caused by mechanical degradation. **Table 5.18** summarizes Carreau-model parameters and screen factors for these solutions. The reduction in solution viscosity is a clear indication that the average molecular weight of the polymer was reduced by mechanical degradation. This reduction was confirmed by size-exclusion chromatography when a decrease in the molecular weight of approximately 68% was determined for the effluent from the flow experiment at a frontal-advance rate of 278 ft/D.

Holzwarth et al. (1988) determined the molecular weight distribution of partially hydrolyzed polyacrylamides before and after injection through at 2-in.-thick Berea core (570 md) at a frontal-advance rate of 600 ft/D. The molecular-weight distributions in **Fig. 5.51** show that shear degradation caused nearly a complete loss of the high-molecular-weight tail of the polyacrylamide molecular weight distribution.

5.4.4 Estimation of Injection Rates/Pressure Drop for Polymer Flooding. Injection rate is a critical variable in EOR processes. In this section, an approximate model is developed to predict the injection rate or the pressure drop during injection of polymer solution into a well. The well is assumed to be completed open hole or to have sufficient perforations to neglect the pressure drop across the perforations. The polymer solution flows radially away from the wellbore. Fluids are considered to be incompressible. Single-phase flow is used to simplify model development. These assumptions can be removed for specific applications.

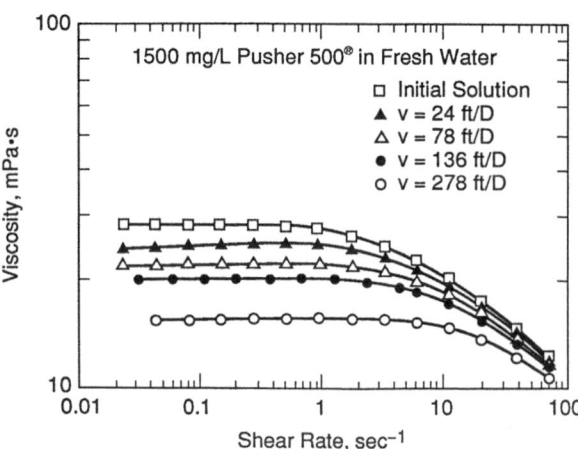

Fig. 5.50—Effect of mechanical degradation in carbonate core on solution viscosity (Martin and Kantamukkala 1984).

V (ft/D)	μ_o (cp)	K (mPa·s^n)	n	τ_r (seconds^{-1})	F_s
*	28.2	40.6	0.712	3.1	17.3
24.4	24.7	38.8	0.724	4.4	17.9
58.5	22.9	38.1	0.720	5.2	11.1
78.0	21.9	37.3	0.723	6.0	10.2
136.0	20.0	34.4	0.735	6.8	9.2
278.0	15.5	29.1	0.762	12.8	6.8

^aSolution properties before injection.

Table 5.18—Carreau-model parameters for 1500 mg/L Pusher 500® in fresh water degraded in Core P-3 (Martin and Kantamukkala 1984).

When injection begins at a constant rate, the leading edge of the polymer zone moves radially away from the wellbore and the pressure drop changes continuously with time. The location of the polymer front can be determined by material balance if dispersion and IPV are neglected. Eq. 5.41 gives the radius of the polymer front, r_p, after W_p barrels of polymer have been injected into a formation of thickness h and porosity ϕ.

$$r_p = \sqrt{\frac{5.615 \rho_p W_p C}{\pi h \left[\phi S_w C \rho_p + \hat{C}(1-\phi)\rho_r \right]}}, \quad \ldots\ldots\ldots\ldots\ldots\ldots\ldots\ldots\ldots\ldots(5.41)$$

where r_p = radius of polymer front, ft; ρ_p = density of polymer solution, fbm/ft³; ρ_r = grain density of rock, lbm/ft³; C = polymer concentration, $\dfrac{\text{lbm polymer}}{\text{lbm solution}}$;

$\hat{C}$ = polymer retention, $\dfrac{\text{lbm polymer retained}}{\text{lbm rock}}$ or $\dfrac{\text{g pol}}{\text{g rock}}$;

h = formation thickness, ft; and W_p – cumulative volume of polymer solution injected, bbl. The polymer displaces a Newtonian fluid with mobility λ_N. Flow is assumed to be radial from the injection wellbore to an effective radius r_e, where the reservoir pressure is maintained constant at p_e.

Shear-Thinning Fluids. Polymer flow is represented by three flow regimes, depicted in **Fig. 5.52**, following the Carreau-type model for shear-thinning fluids. In the lower Newtonian region, the polymer mobility is constant. The transition from Newtonian flow to shear-thinning flow is assumed to be abrupt when the Darcy velocity is u_{c1}, the critical Darcy velocity for the onset of shear-thinning flow. In most reservoir applications, flow rates are well above u_{c1} and laboratory data are confined to the shear-thinning region. Thus, the value of u_{c1} must be estimated. In the shear-thinning region, the polymer mobility is given by

$$\lambda_p = \lambda_p^* / u^{n_c - 1}. \quad \ldots\ldots\ldots\ldots\ldots\ldots(5.42)$$

The shear-thinning region ends abruptly upon the onset of the upper Newtonian region, where the polymer mobility approaches the mobility of the solvent, as shown in Fig. 5.52. The limiting value of λ_p is given by

$$\lambda_{puN} = k_{wp} / \mu_w. \quad \ldots\ldots\ldots\ldots\ldots\ldots (5.43)$$

The critical velocity for the onset of the upper Newtonian region is found by noting that, on the graph of $\Delta p/L$ vs. u or v, the shear-thinning and upper Newtonian regions intersect at the critical value of $\Delta p/L$ and u_{c2}. Thus, from Eqs. 5.16 and 5.20, $\Delta p/L = u_{c2}$,

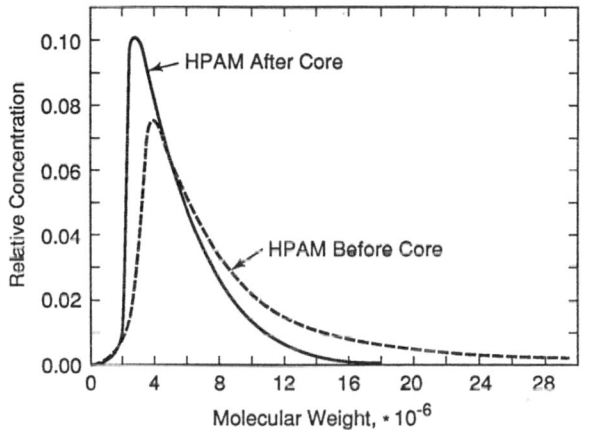

Fig. 5.51—Molecular weight distribution of partially hydrolyzed polyacrylamide after passage through a core (Holzwarth et al. 1988).

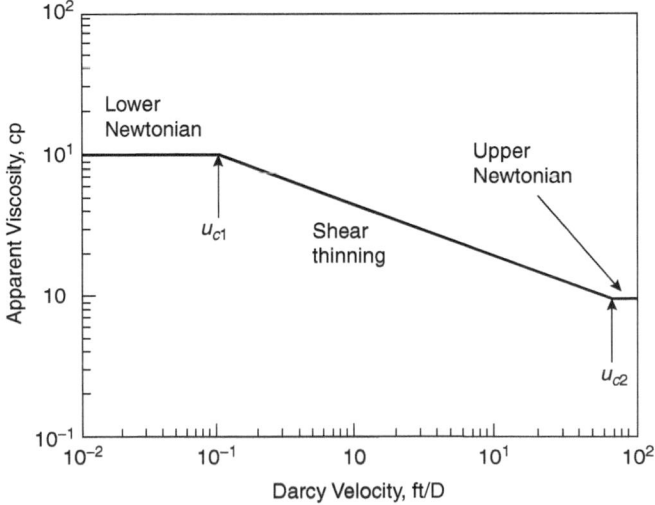

Fig. 5.52—General model for the variation of polymer mobility with Darcy velocity for the flow of shear-thinning fluids in a porous material.

$$u_{c2}^{1-n_{c1}} = \lambda_{puN}/\lambda_p^* \text{ or } u_{c2} = \left(\lambda_{puN}/\lambda_p^*\right)^{1/(1-n_{c1})}, \dots \tag{5.44}$$

where u_{c2} is the critical velocity where the upper Newtonian region begins. The three flow regions are now defined. Parameters are, for the lower Newtonian region,

$$0 < u < u_{c1}. \dots \tag{5.45a}$$

and $\lambda_p = \lambda_{pN};$ $\dots \tag{5.45b}$

for the shear-thinning region,

$$u_{c1} < u < u_{c2} \dots \tag{5.46a}$$

and $\lambda_p = \lambda_p^* u^{1-n_c};$ $\dots \tag{5.46b}$

and for the upper Newtonian region,

$$u_{c2} < u. \dots \tag{5.47a}$$

and $\lambda_p = \lambda_{puN}.$ $\dots \tag{5.47b}$

The pressure drop in the radial system is the sum of the pressure drops across each flow region present in the system. In an injection well, the progression of flow regimes (that could possibly be present) with increasing distance from the wellbore is upper Newtonian if $u_{c2} < q_t/(2\pi r_w h)$, shear-thinning for $u_{c1} < u < u_{c2}$, and lower Newtonian if $u < u_{c1}$, where $q_t =$ injection rate, $h =$ thickness of interval taking fluid, and $r_w =$ wellbore radius.

When fluid flow is considered to be incompressible, the expressions that relate injection rate to pressure drop may be derived easily because transients are propagated through the system instantaneously. The Darcy velocity at any radial position is given by

$$u = q_t / 2\pi rh. \dots \tag{5.48}$$

For Newtonian fluids, Darcy's law is given by

$$u = -\lambda_N \left(dp/dr\right), \dots \tag{5.49}$$

where λ_N is the mobility of the Newtonian fluid. The pressure drop between Locations r_1 and r_2 is obtained by setting Eq. 5.48 equal to Eq. 5.49 and integrating to obtain

$$\Delta p_{12} = \frac{q_t}{2\pi\lambda_N h}\ln\left(\frac{r_2}{r_1}\right). \dots \tag{5.50}$$

In the power-law region, Eq. 5.20 is the appropriate form for Darcy's law, where the polymer mobility is given by Eq. 5.42. It can be shown that the pressure drop between r_1 and r_2 is given by

$$\Delta p_{12} = \left(\frac{q_t}{2\pi h}\right)^{n_c} \frac{\left(r_2^{1-n_c} - r_1^{1-n_c}\right)}{\lambda_p^*\left(1-n_c\right)}. \dots \tag{5.51}$$

Depending on the flow rate, one or more of the flow regimes may be encountered as the polymer moves away from the injection well. It is necessary to check the Darcy velocity as a function of radial position to determine which flow regimes are present. If all flow regimes are present in the system, the expression for the pressure drop is given by Eq. 5.52. Eq. 5.52 can be adjusted for a particular application if a region cannot exist on the basis of the criteria established in the preceding section.

$$\Delta p = \frac{q_t}{2\pi\lambda_{puN} h}\ln\left(\frac{r_{c2}}{r_w}\right) + \left(\frac{q_t}{2\pi h}\right)^{n_{c1}}\left[\frac{r_{c1}^{1-n_{c1}} - r_{c2}^{1-n_{c1}}}{\lambda_{p1}^*\left(1-n_{c1}\right)}\right]$$

$$+ \frac{q_t}{2\pi h\lambda_{pN}}\ln\left(\frac{r_p}{r_{c1}}\right) + \frac{q_t}{2\pi h\lambda_N}\ln\left(\frac{r_e}{r_p}\right) \dots \tag{5.52}$$

where r_w = wellbore radius, r_p = radius to position of polymer front location, r_{c2} = radius where upper Newtonian region ends and shear-thinning region begins, and r_{c1} = radius where shear-thinning region ends and lower Newtonian region begins and consistent units are used.

Example 5.6 illustrates the use of this model to estimate the pressure drop for the injection of a shear-thinning fluid into a reservoir.

Example 5.6—Estimation of Pressure Drop Through a Reservoir. A polymer flood is being designed that uses 1,000 ppm xanthan biopolymer. The polymer solution will be injected into a sandstone reservoir at a rate of 10 B/D-ft. The permeability of the reservoir is 200 md, porosity is 0.19, and thickness is 50 ft. Initial oil saturation is 0.70, and ROS is 0.30. A five-spot pattern is planned on 10-acre spacing. Radius of the injection well is 3.25 in., and there is no wellbore damage. At the beginning of the polymer flood, the reservoir is at interstitial water saturation and contains a crude oil with a viscosity of 1.0 cp at reservoir temperature and pressure. For this example, the displacement process is considered piston-like so that a sharp displacement front will form between the displaced oil and the injected polymer solution. Polymer retention and IPV are neglected. Determine the bottomhole pressure (BHP) in the injection well when 1,000 bbl of polymer solution has been injected. The average reservoir pressure at the effective radius of the five-spot pattern is 400 psi.

Solution. The effective radius, r_e, for the region of radial flow is the radius of a circle that has one-half the area of the five-spot pattern. That is,

$$2\pi r_e^2 = A_p$$

$$r_e = \sqrt{A_p / 2\pi}$$

or

$$= \sqrt{\frac{(10 \text{ acres})(43{,}560 \text{ ft}^2/\text{acre})}{2\pi}}$$

$$= 263.3 \text{ ft.}$$

$$r_w = 3.25 \text{ in.}/12 \text{ in./ft}$$

$$= 0.271 \text{ ft.}$$

When 1,000 bbl of polymer has been injected, the polymer front is located at a distance computed with Eq. 5.41. Because polymer retention is neglected in this example,

$$r_p = \sqrt{\frac{5.615 W_p}{\pi h \phi (1 - S_{or})}}$$

$$= \sqrt{\frac{5.615 \text{ ft}^3/\text{bbl}(1{,}000 \text{ bbl})}{\pi (50 \text{ ft})(0.19)(0.7)}} \quad \dotfill \quad (5.53)$$

$$= 16.4 \text{ ft.}$$

Flow properties of the polymer in the porous rock are obtained from Willhite and Uhl (1988).

The value of $k_{wp} = 0.83 \, k_w$. In this case, k_w is the permeability of water at ROS. Assume that $k_{rw} = 0.2$ at S_{or}, so that $k_w = 40$ md. Then, $k_{wp} = 33.2$ md. Ahead of the polymer front, k_{ro} at $S_{iw} = 1.0$, so that $k_o = 200$ md.

From Willhite and Uhl, at 1,000 ppm,

$$\lambda_p^* = 0.746 \, k_{wp}^{0.578}$$

$$= 0.746(33.2)^{0.578} \quad \dotfill \quad (5.54)$$

$$= 5.65 \text{(md/cp)/(ft/D)}^{(1-n_c)}$$

and $n_c = 0.659 k_{wp}^{-0.035}$

$$= (0.659)(33.2)^{-0.035}$$

$$= 0.583. \quad \dotfill \quad (5.55)$$

In this sandstone, Newtonian flow is assumed to begin when the frontal-advance rate is less than 0.1 ft/D. Thus,

$$u_{c1} = v_{c1}\phi$$

$$= (0.1 \text{ ft/D})(0.19) \quad \dotfill \quad (5.56)$$

$$= 0.019 \text{ ft/D.}$$

The lower Newtonian region will be encountered when the Darcy velocity is 0.019 ft/D. If flow is radial,

$$u = q_t / 2\pi r h. \quad \dots \dots \dots \dots (5.57)$$

The radius where Newtonian flow begins is found by solving Eq. 5.57 with $u = u_{c1}$. Thus,

$$r_{c1} = q_t / 2\pi h u_{c1}$$
$$= \frac{[10 \text{ B/(D-ft)}](5.615 \text{ ft}^3 / \text{bbl})}{2\pi(0.019 \text{ ft/D})}$$
$$= 470.3 \text{ ft.}$$

Because $r_{c1} > r_e$, the lower Newtonian flow region is not present.

The existence of an upper Newtonian region can be determined by first finding the critical Darcy velocity where flow changes from shear-thinning to Newtonian flow. The viscosity of the water (solvent) in the polymer solution is 1.0 cp. When the upper Newtonian region is encountered, the polymer mobility is given by

$$\lambda_{puN} = k_{wp} / \mu_w$$
$$= 33.2 \text{ md/1.0 cp} \quad \dots \dots \dots \dots (5.58)$$
$$= 33.2 \text{ md/cp.}$$

It is necessary to determine whether the upper Newtonian flow region is present in the vicinity of the injection well. From Eq. 5.44, the critical velocity for the transition to Newtonian flow is

$$u_{c2} = \left(\lambda_{puN} / \lambda_p^*\right)^{1/(1-n_{c2})}$$
$$= [(33.2 \text{ md/cp})/5.65]^{1/(1-0.582)}$$
$$= (5.876)^{2.392}$$
$$= 69.87 \text{ ft/D.}$$

At an injection rate of 10 B/(D-ft) into a well with a diameter of 6.5 in., the maximum Darcy velocity occurs at the sandface. Thus,

$$u_{max} = q_t / 2\pi h r_w$$
$$= \frac{[10 \text{ B/(D-ft)}](5.615 \text{ ft}^3 / \text{bbl})}{2\pi(0.271 \text{ ft})}$$
$$= 33 \text{ ft/D.}$$

For this example, flow is entirely in the shear-thinning region. Thus, Eq. 5.52 is modified for the flow of a shear-thinning fluid to obtain

$$\Delta p = \left(\frac{q_t}{2\pi h}\right)^{n_c} \left[\frac{r_p^{1-n_{c1}} - r_w^{1-n_{c1}}}{\lambda_{p1}^*(1-n_{c1})}\right] + \frac{q_t}{2\pi h \lambda_N} \ln\left(\frac{r_e}{r_p}\right). \quad \dots \dots \dots (5.59)$$

The value of λ_N is the mobility of oil at interstitial water saturation. Assuming piston-like displacement ahead of the polymer bank,

$$\lambda_N = k_o / \mu_o$$
$$= 200 \text{ md/cp.}$$

To compute the pressure drop with Eq. 5.59, it is necessary to introduce conversion factors to keep units consistent. When the units are barrels, days, feet, millidarcies, and centipoise, Eq. 5.59 becomes

$$\Delta p = 158.005 \left(\frac{5.615 q_t}{2\pi h}\right)^{n_{c1}} \left[\frac{r_p^{1-n_{c1}} - r_w^{1-n_{c1}}}{\lambda_p^*(1-n_{c1})}\right]$$
$$+ \frac{q_t}{0.007081 h \lambda_N} \ln\left(\frac{r_e}{r_p}\right) \quad \dots \dots \dots (5.60)$$

In Eq. 5.60, the conversion factor 0.007081 has units of B/D/[(psi)(md/cp)(ft)].

$$\Delta p = 158.005 \left[\frac{(5.615)(10)}{2\pi} \right]^{0.583}$$

$$\times \left[\frac{16.4^{1-0.583} - (0.271)^{1-0.583}}{(5.65)(1-0.583)} \right]$$

$$+ \frac{10}{(0.007081)(200)} \ln \left(\frac{263.3}{16.4} \right)$$

$$= 158.005(3.585) \left[\frac{3.21 - 0.580}{(5.65)(0.417)} \right] + 7.061 \ln(16.05)$$

$$= 632.3 + 19.6$$

$$= 651.9 \text{ psi.}$$

Thus, the BHP in the injection well is (652 + 400), or 1,052 psi. Because the injection rate is 10 B/D-ft, the injection time can be computed directly:

$$t = W_p / q_i$$

$$= 1,000 \text{ bbl}/(10 \text{ B/D-ft})(50 \text{ ft}) \dots \dots \dots \dots \dots \dots \dots \dots \dots \dots \dots \dots \dots \dots \dots \dots (5.61)$$

$$= 2 \text{ days.}$$

The flow rate could be computed when the pressure drop is constant. Eq. 5.60 is nonlinear in flow rate, and thus a root-finding scheme is necessary to find q_i when Δp is given.

Shear-Thickening Fluids. The model developed for shear-thinning fluids can be adapted for shear-thickening fluids. In this case, the flow regimes shown in **Fig. 5.53** may be present as the fluid flows radially from the wellbore into the formation. The upper Newtonian region is not present because polyacrylamide undergoes mechanical degradation when the Darcy velocity exceeds some maximum value. Properties of the flow regimes are, with mechanical degradation (first 0.5 in. from sandface), $u_{max} < u$; for shear thickening, $u_{c2} < u < u_{max}$ and $\lambda_p = \lambda_{p2}^* u^{1-n_{c2}}$; for shear thinning, $u_{c1} < u < u_{c2}$ and $\lambda_p = \lambda_{p1}^* u^{1-n_{c1}}$; and for the lower Newtonian region, $0 < u < u_{c1}$ and $\lambda_p = \lambda_N$. u_{max}, the Darcy velocity where mechanical degradation begins, usually must be determined experimentally.

When all regions are present between the sandface and the polymer front, the pressure drop is given by

$$\Delta p = \left(\frac{q}{2\pi h} \right)^{n_{c2}} \left[\frac{r_2^{1-n_{c2}} - r_w^{1-n_{c2}}}{\lambda_{p2}^* \left(1 - n_{c2} \right)} \right] + \left(\frac{q}{2\pi h} \right)^{n_{c1}}$$

$$\times \left[\frac{r_1^{1-n_{c1}} - r_2^{1-n_{c1}}}{\lambda_{p1}^* \left(1 - n_{c1} \right)} \right] + \frac{q}{2\pi h \lambda_{pN}} \ln \left(\frac{r_p}{r_1} \right) + \frac{q}{2\pi h \lambda_N} \ln \left(\frac{r_e}{r_p} \right) , \dots \dots \dots \dots \dots \dots \dots \dots \dots (5.62)$$

where consistent units are used.

Eq. 5.62 does not account for the pressure drop caused by mechanical degradation. In general, this pressure drop must be determined experimentally. Seright (1983) was able to correlate this pressure loss with u_{max} for a specific polymer/rock system.

5.4.5 Two-Phase Flow for Polymer Systems.
Two-phase flow of polymer and oil occurs during a polymer flood in the region behind the zone where interstitial water (original water in place) displaces the oil. Two-phase flow also occurs when the polymer slug used for polymer flooding or mobility control is displaced by brine. Questioning whether relative permeability curves developed for the flow of oil and brine are affected by polymer flow and polymer retention is reasonable. Limited experimental data are available for when oil is displaced by polymer in a polymer flood that begins when there is a mobile oil saturation. In displacement calculations, relative permeabilities to oil for polymer displacing oil are assumed to be identical to those for water displacing oil. If permeability is not reduced by polymer

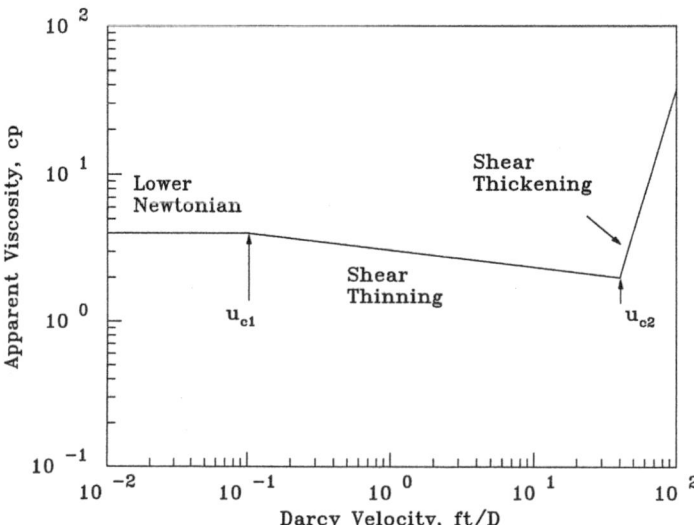

Fig. 5.53—General model for the variation of polymer mobility with Darcy velocity for the flow of a shear-thickening fluid in a porous material.

retention, the relative permeabilities for polymer are assumed to be the same as those for water. The relative permeability for polymer must be adjusted when the permeability is reduced by polymer retention. The endpoint of the relative permeability curve must be reduced by the ratio of k_{rwp}/k_{rw} at S_{or}. This adjustment also will change the shape of the relative permeability curve for polymer.

Schneider and Owens (1982) conducted an extensive study to determine the effect of polymer on relative permeability in a displacement sequence in which polymer solution was injected into a reservoir that was at waterflood ROS. Steady-state relative permeability data were obtained for Berea sandstone and reservoir cores having a range of permeabilities and wettabilities. All the tests were conducted with polyacrylamides.

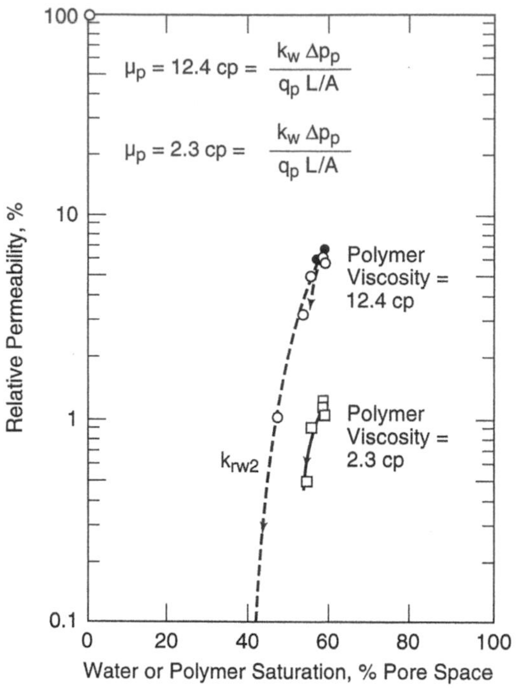

Two-phase flow of oil and polymer was studied in water-wet cores on the secondary drainage path. That is, the relative permeability curves were determined as the polymer saturation was decreased from $1 - S_{or}$ to S_{wip}. Permeability to oil was essentially unaffected by the flow of polymer; however, significant changes were observed in relative permeability curves for polymer. **Fig. 5.54** shows relative permeability curves for the displacement of polymer by oil and the displacement of water by oil before the rock was contacted with polymer (Schneider and Owens 1982).

Interpretation of Fig. 5.54 requires consideration of polymer flow characteristics. When steady-state displacement tests are conducted with a polymer solution, as discussed in Section 5.4, the polymer mobility is extracted from the experimental data. Permeability to polymer can be calculated if the apparent viscosity of the polymer solution is known at the Darcy velocity of the polymer phase. For the core in Fig. 5.54, the apparent viscosity was determined with Eq. 5.18 to be 2.3 cp at S_{or} from the polymer mobility with $k_p = k_{wp}$. Because the effective polymer viscosity at S_{or} did not vary significantly with flow rate, the apparent viscosity for relative permeability computations was assumed to be constant throughout the steady-state tests. The relative permeability curve for polymer solution is significantly less than the corresponding relative permeability curve for the displacement of water before contact of the core with polymer solution.

Relative permeability curves also were determined for the displacement of oil by water following the polymer/oil tests. **Fig. 5.55** compares the relative permeability data for the oil and water phases before and after the rock was contacted with polymer. In water-wet rocks, there was little difference between the ROS obtained before and after polymer contacted the rock, as would be expected. Oil relative permeabilities are relatively unaffected. The relative permeability to water after polymer contact, k_{rwp}, is reduced significantly, as seen in Fig. 5.55. The displacement process is analogous to what happens at the rear of a polymer slug that is being displaced through the reservoir by drive water.

Fig. 5.56 presents typical data for the same displacement sequence in an oil-wet rock. The same relative permeability trends are observed in oil-wet rock, although careful inspection is necessary because k_{rw1} and k_{rw2} are quite different owing to hysteresis. The ROS decreased after exposure to polymer because the oil saturation in an oil-wet rock exists as thin films and in small pores. Injection of a viscous liquid will decrease the oil saturation in oil-wet rocks, as discussed in Chapter 2. The relative permeability of water following polymer contact, k_{rwp}, is significantly lower than the relative permeability to water before polymer contact, k_{rw1}, because the water saturation increased.

These differences in relative permeability characteristics have significant effects on displacement calculations.

5.5 Polymer-Augmented Waterflood

Aqueous solutions of high-molecular-weight polymers are used widely to augment waterflooding. Polymer-augmented waterflooding can be divided into two broad classifications. When the mobility ratio of a waterflood is unfavorable, continuous injection of polymer solution increases the microscopic displacement efficiency at a particular

$$\mu_p = 12.4 \text{ cp} = \frac{k_w \, \Delta p_p}{q_p \, L/A}$$

$$\mu_p = 2.3 \text{ cp} = \frac{k_w \, \Delta p_p}{q_p \, L/A}$$

Fig. 5.54—Water and estimated Pusher 700 relative permeabilities when effective polymer viscosities are used during displacement of polymer by oil (Sample Berea-4) (Schneider and Owens 1982).

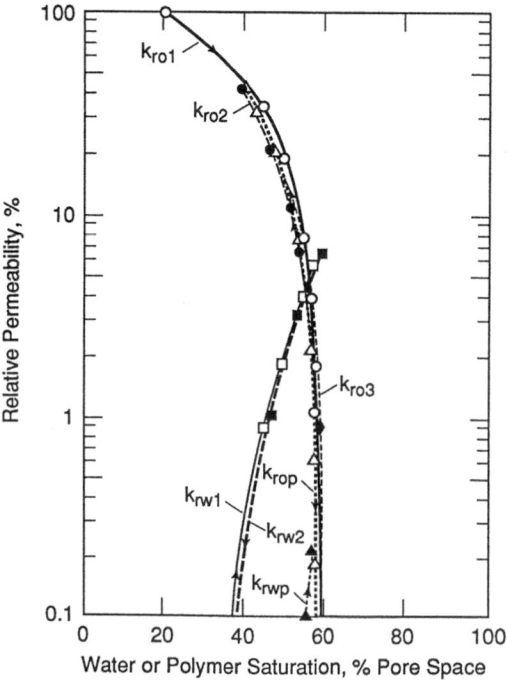

Fig. 5.55—Water/oil/Pusher 700 relative permeabilities (Sample Berea-4) (Schneider and Owens 1982).

WOR and increases volumetric sweep efficiency in the reservoir. Even when the mobility ratio is favorable, reservoir heterogeneity primarily in the vertical direction may cause poor volumetric sweep. In this case, polymer-augmented waterflooding is used to reduce the water mobility in the high-permeability layers, so that oil can be displaced from lower-permeability layers.

Polymers also are used to improve volumetric sweep efficiency in reservoirs that are fractured or have considerable permeability variation. In this application, commonly called in-situ permeability modification, the polymer is crosslinked in situ to form a gel network. In-situ gelation reduces the permeability in the region contacted by the polymer. By selection of chemical systems, treatments can be designed to penetrate only the near-wellbore region or to extend considerable distances into the reservoir. Water injection resumes after the polymer treatment.

This section illustrates the use of polymer to improve the microscopic and volumetric sweep of a waterflood by continuous polymer injection.

5.5.1 Design Considerations.
The first step in the design process is to determine whether a reservoir is a candidate for polymer-augmented waterflooding. Reservoirs that are good candidates can be recognized by poor volumetric sweep efficiency of the waterflood, low waterflood recovery compared with similar reservoirs, rapid breakthrough of injected water into production wells, and high WOR throughout the flood life.

A mobility-control problem can be recognized at the microscopic level by determining the mobility ratio of the waterflood. If the mobility ratio is much larger than unity, both microscopic and volumetric sweep efficiency decrease. At high mobility ratios, viscous fingering occurs, leading to poor sweep efficiency. Effects of reservoir heterogeneity on volumetric sweep efficiency can be assessed by determining the Dykstra-Parsons coefficients from core analysis (Willhite 1986) and from displacement calculations to predict whether the volumetric displacement efficiency is likely to be low.

The critical question that must be addressed in design is whether polymer flooding in a particular reservoir is economically attractive. In most cases, the economics of current operations is compared with economics based on projections of the displacement performance (Argabright et al. 1986; Sengul et al. 1984) with a polymer flood. Usually, some type of mathematical model must be used to predict displacement performance when a polymer flood is installed. Available models have varying degrees of complexity. Relatively simple models were chosen for this text to emphasize concepts and design approaches. Section 5.5.2 examines polymer flooding in a linear system.

5.5.2 Polymer-Augmented Waterflooding in a Linear Reservoir. *Continuous Polymer Injection.*
A linear displacement model represents a porous rock of length, L, width, w, and thickness, h. Porosity and permeability are uniform throughout the linear system. For the purposes of this discussion, the reservoir has a uniform initial oil saturation. The viscosities of the oil and water, as well as the relative permeability relationships, are known. It is assumed that various polymers have been evaluated through core testing, so that flow characteristics, retention, and IPV are known as functions of concentration.

A polymer-augmented waterflood in a linear reservoir is one case presented for the displacement model for chemical flooding in Chapter 3, where an adsorbing viscous chemical was displaced through a linear system. The displacement model introduced in Chapter 3 did not account for IPV. Section 5.4.2 demonstrated that a part of the PV, called the IPV, is not accessible to polymer molecules. This portion is accessible to water and other low-molecular-weight solvents. IPV is incorporated into the displacement equations by introducing the following concepts.

Let ϕ_{IPV} represent the porosity that is not accessible to polymer. Then, at every point in the porous rock, the PV that is accessible to polymer is $(\phi S_w - \phi_{IPV})$. A material balance over the polymer following the derivation of Section 3.5 yields

$$C\frac{\partial f_w}{\partial x_D} + \left(S_w - \frac{\phi_{IPV}}{\phi} + D_p \right)\frac{\partial C}{\partial t_D} = 0, \dots\dots\dots\dots\dots\dots\dots\dots\dots\dots\dots\dots(5.63)$$

where C = concentration, f_w = fractional flow of water, S_w = water saturation, x_D = dimensionless distance, and t_D = dimensionless time. Eq. 5.63 is solved as discussed in Section 3.4. From this solution, the specific velocity of the polymer front is given by

$$f'^*_{w3} = v^*_D = \frac{f^*_w}{S^*_w - (\phi_{IPV}/\phi) + D_p}, \dots\dots\dots\dots\dots\dots\dots\dots\dots\dots\dots\dots\dots\dots(5.64)$$

where the asterisk denotes conditions at and behind the polymer front.

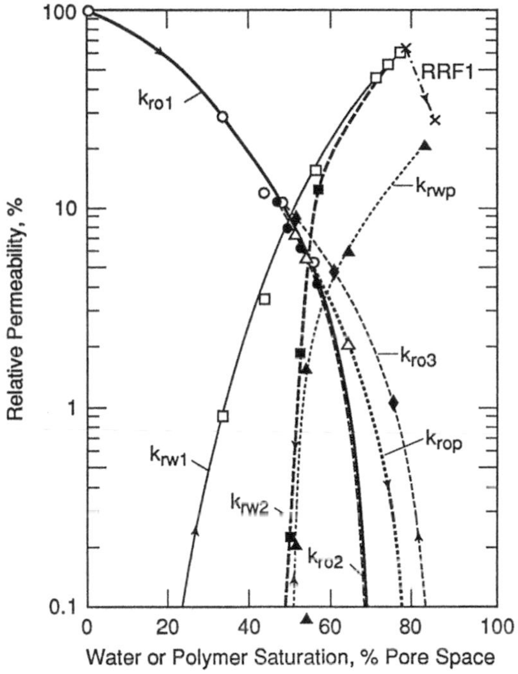

Fig. 5.56—Water/oil/Pusher 500 relative permeabilities in an oil-wet rock (Sample Tensleep-1) (Schneider and Owens 1982).

The effective IPV is defined as the fraction of the total PV that is not accessible to polymer. It is convenient to define $\phi_e = \phi_{IPV}/\phi$ so that Eq. 5.64 becomes

$$v_D^* = f_w^* \Big/ \left(S_w^* - \phi_e + D_p \right), \dots\dots\dots\dots\dots\dots\dots\dots\dots\dots\dots\dots\dots\dots\dots\dots\dots (5.65)$$

where ϕ_e = fraction of the total PV that is inaccessible. Eq. 5.65 shows that the specific velocity of the polymer front is retarded by polymer retention (through the term D_p) but is accelerated by IPV. In some polymer systems, $\phi_e = D_p$ and the polymer front appears to be displaced through the linear system as if there were no retention. Retention still occurs, but the polymer moves at a higher velocity because of IPV.

In Example 3.5, the displacement performance of a polymer flood was calculated in terms of the PVs of fluid injected for a particular reservoir-rock/fluid system, neglecting IPV. Now the displacement model is used to compare the benefits of waterflooding a reservoir with a polymer-augmented waterflood.

Example 5.7 illustrates the comparison of waterflood and polymer flood performance in a linear system when the injection rate is constant.

Example 5.7—Polymer Flood in a Linear Reservoir Originally at Interstitial (Immobile) Water Saturation. The potential of using polymer-augmented waterflooding to increase oil recovery from a uniform reservoir must be evaluated. For the purposes of this example, consider a linear reservoir segment that is 500 ft wide and 20 ft thick. Production and injection wells are 1,000 ft apart. Properties of the reservoir rock and fluids, summarized in **Table 5.19,** are identical to those used in Example 3.1. Injection rate is constant at 200 B/D. Relative permeability relationships are

$$k_{ro} = 0.8 \left(1 - S_d \right)^2, \dots\dots\dots\dots\dots\dots\dots\dots\dots\dots\dots\dots\dots\dots\dots\dots (5.66)$$

$$k_{rw} = 0.2 S_{wD}^2, \dots\dots\dots\dots\dots\dots\dots\dots\dots\dots\dots\dots\dots\dots\dots\dots\dots (5.67)$$

$$\text{and } S_{wD} = \left(S_w - S_{iw} \right) \Big/ \left(1 - S_{or} - S_{iw} \right). \dots\dots\dots\dots\dots\dots\dots\dots\dots\dots\dots\dots\dots (5.68)$$

Solution. Evaluation of a polymer flood begins with an assessment of the waterflood displacement efficiency. The mobility ratio is a universal guide to waterflood performance. The mobility ratio, $M_{\bar{s}}$, is

$$M_{\bar{s}} = \left(k_{rw} \right)_{\bar{S}_{wf}} \mu_o \Big/ \left(k_{ro} \right)_{S_{wi}} \mu_w. \dots\dots\dots\dots\dots\dots\dots\dots\dots\dots\dots\dots\dots (5.69)$$

In Example 3.1, the waterflood displacement performance was estimated for this reservoir. The flood-front saturation was 0.4206, and the average water saturation behind the flood front, $\bar{S}_{wf}$, was determined to be 0.485. The relative permeability of water is given by Eq. 5.67.

Substituting $\bar{S}_{wf} = 0.485$ into Eq. 5.68 with $S_{iw} = 0.30$ and $S_{or} = 0.30$ gives

$$S_{wD} = (0.485 - 0.30) / (1 - 0.30 - 0.30)$$

$$= 0.463.$$

Then, $k_{rw} = 0.2(0.463)^2$

$$= 0.043$$

and $M_{\bar{s}} = (0.043)(40)/(0.8)(1.0)$

$$= 2.15.$$

The mobility ratio is greater than unity, which suggests that increased displacement efficiency could be obtained by using a polymer-augmented waterflood.

The endpoint mobility ratio also is used to describe waterflood displacement efficiency because endpoint relative permeability data are often more readily available than relative permeability curves. In this case, the endpoint mobility ratio is

Porosity	0.267
S_{oi}	0.70
S_{or}	0.30
μ_o (cp)	40.0
μ_w (cp)	1.0
B_o (bbl/STB)	1.1
B_w (bbl/STB)	1.0

Table 5.19—Reservoir rock and fluid properties, Example 5.7.

$$M = (k_{rw})_{S_{or}} \mu_o / (k_{ro})_{S_{iw}} \mu_w$$

$$= (0.20 / 1.0)(0.8 / 40.0) \cdots \cdots (5.70)$$

$$= 10.0$$

The endpoint mobility ratio is much larger than $M_{\overline{S}}$ and exaggerates the effect of the high oil viscosity on the displacement efficiency.

Consider the effect of adding sufficient polymer to the injected water so that the apparent viscosity of the polymer solution is 4 cp. Assume that no reduction in water relative permeability is caused by the polymer. The mobility ratio, $M_{\overline{S}}$, would be expected to decrease by a factor of four to approximately 0.54, which is clearly favorable. In Example 3.5, the displacement performance of a polymer flood was estimated for a polymer solution containing 300 ppm polymer with an apparent viscosity of 4.0 cp. Retention is 17.5 μg/g at 300 ppm, so that $D_p = 0.424$. Effective IPV for this system is estimated to be 0.25. Thus, $-\phi_e + D_p = 0.174$.

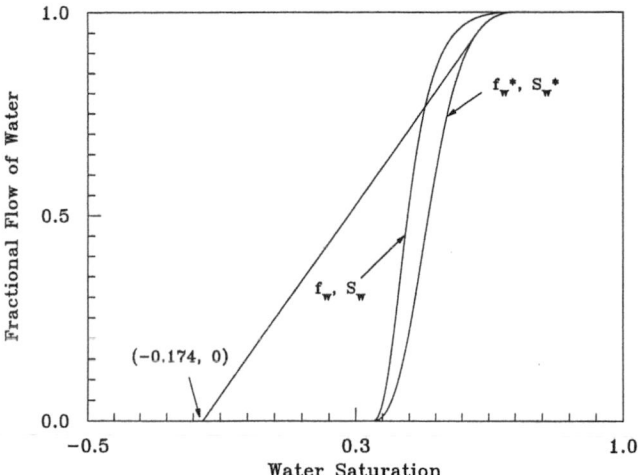

Fig. 5.57—Fractional-flow curve for the polymer flood in Example 5.7.

The polymer-flood-front properties S_{w3}^* and f_{w3}^* are found by drawing a tangent from the point $(0, -0.174)$ to the fractional-flow curve with polymer as the displacing phase. The tangent, shown in **Fig. 5.57,** intersects the fractional-flow curve for the waterflood at f_{w1}, S_{w1}. Locations of S_{w3}^*, f_{w3}^* and S_{w1}, f_{w1} are shown in Fig. 5.57. The value of S_{wf} is the same as in Example 3.5.

Two flood fronts form in this case because $S_{w1} > S_{wf}$. The saturation profile is similar to Fig. 3.34 and is characterized by a flood front with saturation S_{wf}, an oil bank where the water saturation is constant, S_{w1}, and a polymer flood front with saturation S_{w3}^*. **Table 5.20** presents saturations and fractional flows corresponding to each saturation. Velocities of the three distinct banks calculated from Eqs. 3.142 and 3.143 are also included. Oil recovery during a continuous polymer flood is computed by tracking the three regions as they are displaced through the linear system and then making a material balance as described in Section 3.2.7.

Eqs. 5.71 through 5.74 give the locations of these fronts in terms of dimensionless time and the dimensionless time when each region reaches the end of the linear system.

$$x_D = f_w' t_D, \dots \dots (5.71)$$

$$x_{Df} = 5.3958 t_D, \dots \dots (5.72)$$

$$x_{D1} = 3.8530 t_D, \dots \dots (5.73)$$

$$\text{and } x_{D3}^* = 1.2377 t_D. \dots \dots (5.74)$$

Arrival time at the end of the linear system $(x_D = 1.0)$ is easily computed from these equations:

$$t_D = 1 / f_w', \dots \dots (5.75)$$

with $t_{Df} = 0.1853$, $t_{D1} = 0.2595$, and $t_{D3}^* = 0.8079$.

There is a small difference (0.0754 PV) between the arrival of the waterflood front and the oil bank. Locations of saturations behind the polymer front are found with

$$x_{Dw}^* = f_w' * t_D. \dots \dots (5.76)$$

	Water Saturation	Fractional Flow	$f_w' = V_D$
Waterflood front	0.4206	0.6508	5.3958
Front of oil bank	0.4459	0.7673	3.8530
Front of polymer bank	0.5925	0.9687	1.2377

Table 5.20—Computed parameters for polymer flood, Example 5.7.

The displacement performance is calculated next. The PV of the linear segment is computed as

$$V_p = hwL\phi$$

$$= \frac{(20 \text{ ft})(500 \text{ ft})(1{,}000 \text{ ft})(0.267)}{5.615 \text{ ft}^3/\text{bbl}} \quad \dotfill \quad (5.77)$$

$$= 475{,}512 \text{ bbl}.$$

Oil and water rates at reservoir conditions are obtained directly from the fractional flow at the end of the system:

$$q_o = (1.0 - f_w)q_t \dotfill (5.78)$$

and $q_w = f_w q_t$. $\dotfill$ (5.79)

The displacement performance is estimated in terms of t_D, the number of PVs injected. When the injection rate is constant, the time since the beginning of the displacement is determined from

$$t = (V_p/q_t)t_D$$

$$= [475{,}512 \text{ bbl}/(200 \text{ B/D})]t_D \dotfill (5.80)$$

$$= 2{,}377.6 t_D \text{ days}.$$

At any time during the displacement,

$$N_p = [(S_{oi}/B_{oi}) - (S_o/B_o)]V_p$$

$$= \left(\frac{\bar{S}_w}{B_o} - \frac{S_{wi}}{B_{oi}} + \frac{1}{B_{oi}} - \frac{1}{B_o} \right) V_p. \quad \dotfill \quad (5.81)$$

When $B_o = B_{oi}$,

$$N_p = \left[\left(\bar{S}_w - S_{wi} \right) / B_o \right] V_p. \quad \dotfill \quad (5.82)$$

Arrival of Waterflood Front, t_{Df}. Because there is no initial gas saturation and the system is assumed to be incompressible, at reservoir conditions,

$$B_o N_p / V_p = t_{Df}$$
$$\dotfill (5.83)$$
$$= 0.1853$$

and $N_p = \dfrac{(0.1853)(475{,}512 \text{ bbl})}{1.1 \text{ bbl/STB}}$

$$= 80{,}102 \text{ STB}. \quad \dotfill \quad (5.84)$$

Before breakthrough,

$$q_o = q_t/B_o$$

$$= (200 \text{ B/D})/(1.1 \text{ bbl/STB}) \quad \dotfill \quad (5.85a)$$

$$= 181.8 \text{ STB/D}$$

and $q_w = 0$. $\dotfill$ (5.85b)

At breakthrough,

$$q_o = [q_t(1 - f_w)]/B_o$$

$$= [200(1 - 0.6508)]/1.1$$

$$= 63.5 \text{ STB/D},$$

$$q_w = q_t f_{wf}/B_w$$

$$= 200(0.6508)$$

$$= 130.2 \text{ B/D},$$

$$t = 2{,}377.6 t_D$$

$$= 2{,}377.6(0.1853)$$

$$= 440.6 \text{ days},$$

and $F_{wo} = (q_w / q_o)$
$= 2.05.$

Arrival of Oil Bank, t_{DI}. The oil bank arrives after 0.2595 PV has been injected, t_{D1}, and the water saturation, S_{w1}, at the end of the system remains constant at 0.4459 until the polymer front arrives. During this time, the average water saturation in the linear system is given by Eq. 3.87, which is rewritten as

$$\overline{S}_w = S_{w1} + t_D (1 - f_{w1}). \dots\dots\dots\dots\dots\dots\dots\dots\dots\dots\dots\dots\dots\dots (5.86)$$

At arrival of the oil bank ($t_{D1} = 0.2595$),

$$\overline{S}_w = 0.4459 + 0.2595(1 - 0.7673)$$
$$= 0.5063,$$
$$B_o N_p / V_p = 0.5063 - 0.30$$
$$= 0.2063 \text{ (at reservoir conditions)},$$
$$N_p - [(0.2063)(475,512 \text{ bbl})]/(1.1 \text{ bbl/STB})$$
$$= 89,181 \text{ STB},$$
$$q_o = [(200 \text{ B/D})(1 - 0.7673)]/(1.1 \text{ bbl/STB})$$
$$= 42.31 \text{ STB/D},$$
$$q_w = (200 \text{ B/D})(0.7673)$$
$$= 153.5 \text{ B/D},$$
$$t = (2,377.6)(0.2595)$$
$$= 617.0 \text{ days},$$

and $F_{wo} = 3.63$ bbl water/STB oil.

Note that because the water saturation in the oil bank is constant, a linear relationship exists between cumulative production and t_D until the polymer front arrives. Oil and water rates remain constant during this period.

Arrival of Polymer Front. Breakthrough of the polymer flood front, S_{w3}^*, occurs after 0.8079 PV has been injected. The average water saturation at and after polymer breakthrough is given by Eq. 3.88, rewritten as

$$\overline{S}_{w3}^* = S_w^* + t_D (1 - f_w^*). \dots\dots\dots\dots\dots\dots\dots\dots\dots\dots\dots\dots\dots\dots (5.87)$$

At breakthrough $f_{w3}^* = 0.9487$, $t_D = 0.8079$, and $S_{w3}^* = 0.5925$.

$$\overline{S}_{w3}^* = 0.5925 + 0.8079(1 - 0.9487)$$
$$= 0.6339,$$
$$B_o N_p / V_p = 0.6339,$$
$$N_p = [(0.339)(475,512 \text{ bbl})]/(1.1 \text{ bbl/STB})$$
$$= 144,383 \text{ STB},$$
$$q_o = 9.33 \text{ STB/D},$$
$$q_w = 189.7 \text{ B/D},$$
$$t = 1,921 \text{ days},$$

and $F_{wo} = 20.3$ bbl water/STB oil.

The remainder of the displacement performance was computed by assuming the arrival of different polymer saturations at the end of the system. Fractional-flow data and derivatives for the polymer solution with an apparent viscosity of 4 cp, presented in **Table 5.21,** are identical to those developed for a viscous waterflood in Table 3.11. **Table 5.22** presents the displacement performance for the polymer-augmented waterflood.

The performance of the polymer-augmented waterflood can be compared with that of a waterflood. Waterflood performance (at reservoir conditions) was estimated for this fluid/rock system in Example 3.1 and was presented in Table 3.4 in terms of t_D, $(N_p/A\phi L)$, and F_{wo}. With these values, the waterflood performance for a constant injection rate of 200 B/D is presented in **Table 5.23** with $B_o = 1.1$ bbl/STB.

A composite table **(Table 5.24)** was constructed using linear interpolation to compare the polymer-augmented waterflood with the waterflood.

S_w^*	f_w^*	$f_w'^*$	t_D	$\overline{S}_w$
0.5925	0.9487	1.238	0.8079	0.634
0.62	0.9756	0.743	1.345	0.653
0.64	0.9877	0.477	2.098	0.666
0.66	0.9951	0.272	3.681	0.678
0.68	0.9989	0.116	8.591	0.689

Table 5.21—Fractional-flow calculations for the polymer region, Example 5.7.

Fluid at End of System	t_D	t (days)	$\overline{S}_{w2}$ Reservoir Condition	N_p (bbl)	q_o (STB/D)	q_w (B/D)	F_{wo} (bbl/STB)
Start injection	0^+	0^+	0.30	0	181.8	0	0
Water-bank arrival	0.185	440	0.485	79,972	63.5	130.2	2.05
Oil-bank arrival	0.2595	617	0.506	89,181	42.3	153.5	3.63
Polymer-front arrival	0.8079	1,921	0.634	144,303	9.33	189.7	20.3
	1.345	3,198	0.653	152,596	4.44	195.1	43.9
Polymer injection	2.098	4,988	0.666	158,216	2.24	197.5	88.3
	3.691	8,752	0.678	163,403	0.89	199.0	22.3

Table 5.22—Estimated oil displacement from a polymer-augmented waterflood, Example 5.7.

t_D (PV)	t (days)	Reservoir $\overline{S}_{w2}$	N_p (STB)	q_o (STB/D)	q_w (B/D)	F_{wo} (bbl/STB)
0^+	0^+	0.30	0	181.8	0	0
0.185	440	0.485	79,972	63.5	130.2	2.05
0.249	592	0.504	88,186	44.4	151.1	3.40
0.347	825	0.524	96,832	30.5	166.4	5.46
0.594	1,412	0.554	109,800	16.9	181.4	10.74
0.650	1,545	0.559	111,961	15.2	183.2	12.0
0.943	2,242	0.579	120,607	10.0	189.0	19.0
1.25	2,972	0.593	126,659	7.2	192.1	26.7
1.377	3,274	0.598	128,820	6.43	192.9	30.0
2.034	4,836	0.616	136,602	3.97	195.6	49.3
2.249	5,347	0.620	138,331	3.51	196.1	55.9

Table 5.23—Waterflood performance of linear reservoir, Example 5.7.

The original oil in place (OOIP) is 302,599 STB. This comparison shows that, at polymer breakthrough, the incremental oil recovery is approximately 9.2% OOIP. To obtain this recovery, it is necessary to inject 200 B/D of polymer solution continuously for 1,921 days (5.26 years). The incremental oil recovery peaks in the vicinity of polymer breakthrough.

Examining the economics of continuous polymer injection is useful. Let the polymer cost be USD 1.50/lbm of active polymer. Daily polymer cost is computed below for injection of 200 B/D at a polymer concentration of 300 ppm.

$$m_p = (200 \text{ B/D})\left(\frac{350 \text{ lbm water}}{\text{bbl}}\right)\left(\frac{300 \text{ lbm polymer}}{10^6 \text{ lbm water}}\right)$$

$$= 21 \text{ lbm/D}$$

and $C_p =$ USD 31.50/D, where $m_p =$ amount of polymer required and $C_p =$ cost of polymer.

When the polymer front arrives at the end of the linear system, the total amount of polymer injected would be 40,341 lbm. Thus, if polymer were injected continuously, the estimated incremental oil would be 0.686 STB/lbm polymer injected. Polymer would cost USD 2.18/STB incremental oil.

Time (days)	Waterflood (bbl)	Polymer-Augmented Waterflood (STB)	Incremental Oil Recovery (STB)
0	0	0	0
440	79,972	79,927	−45
592	88,186	88,106	−80
617	89,149	89,187	38
825	96,832	97,991	1,159
1,412	109,800	122,838	13,038
1,545	111,961	128,468	16,507
1,921*	116,976	144,383	27,407
2,242	120,607	146,448	25,841
2,972	126,659	149,106	22,447
3,274	128,820	152,835	24,105
4,836	136,602	157,739	21,137
5,347	138,331	158,711	20,380

*Polymer breakthrough.

Table 5.24—Comparison of oil recovery for a polymer-augmented waterflood and a waterflood, Example 5.7.

Under many conditions, continuous injection of polymer solutions is not economical. Consequently, various schemes based on the injection of polymer slugs are used in polymer-augmented waterfloods. Section 3.6 introduced the theory of injecting a slug of viscous fluid. Section 5.5.3 covers application of these concepts to a polymer-augmented waterflood.

5.5.3 Slug Injection. Section 5.5.2 stated that continuous injection of polymer during a polymer-augmented waterflood would not be economically attractive under many conditions. In practice, smaller volumes of polymer solution, called slugs, are injected. In most cases, the polymer is not injected at a constant concentration but rather in a staged sequence of concentration reduction. The schedule for polymer injection is empirically derived, often from laboratory experiments or from computer simulations of the process (Mungan 1969b, 1971; Uzoigwe et al. 1974). The polymer slug is displaced by drive water. **Table 5.25** (DeHekker et al. 1986) shows a typical injection plan.

The objectives of this concentration sequence are to reduce the total amount of polymer used and to prevent, or at least to reduce, viscous fingering of low-concentration fluid into regions of higher concentrations. Viscous fingering occurs because each reduction in polymer concentration is accompanied by a reduction in the apparent solution viscosity. It is assumed that mixing in the reservoir will dampen out the viscous fingers.

Methods to predict the effect of viscous fingering on the displacement process are available. Most are complex and involve approximations of viscous fingering by a dispersion process. A model developed by Claridge (1978) and Stoneberger and Claridge (1988) can be used for simple linear displacement calculations.

Section 3.6.2 introduced the theory of slug displacement when viscous fingering is not considered. This section considers the design of a slug of constant polymer concentration to replace continuous polymer injection. The slug is displaced through the porous rock by drive water. Viscous fingering and dispersive mixing are neglected.

The size of the polymer slug is based on displacing a region of constant concentration through the reservoir. The minimum amount of polymer is determined from retention data by material balance. Let A_i = polymer retention in µg/g rock. Then,

$$P_r = \left(V_p/\phi\right)\left(1-\phi\right)\rho_r A_i = \hat{C}_o V_p \times 1,000. \dotfill (5.88a)$$

and $P_i = t_{Dp} C_o V_p,$ $\dotfill (5.88b)$

where P_r = polymer retained, P_i = polymer injected, and C_o and $\hat{C}_o$ have units of g/cm³ of PV.

Polymer Concentration (ppm)	PV Injected
500	0.3
330	0.1
165	0.1
0	Remainder of flood

Table 5.25—Schedule for polymer concentrations for slug injection scheme (DeHekker et al. 1986).

When retention is satisfied at C_o,

$$t_{Dp} C_o V_p = \hat{C}_o V_p \dotfill (5.89)$$

$$\text{or } t_{Dp} = \hat{C}_o / C_o \dotfill (5.90)$$

$$\text{and } t_{Dp} = D_p. \dotfill (5.91)$$

Therefore, the minimum polymer slug size to satisfy retention is D_p PV. Ideally, when D_p PV are injected, the polymer flood front will just reach the end of the linear system but no polymer will be produced.

The path followed by the rear of the polymer bank can be determined with the front-tracking methods introduced in Section 3.2.6. This section develops a method for locating the rear of the polymer bank that is based on a material balance on the polymer, assuming that retention is irreversible.

Consider a polymer slug that is being displaced through a linear system by injecting drive water when $t_D = D_p$ (i.e., sufficient polymer has been injected to satisfy retention). As long as the polymer flood front, x_3^*, has not arrived at the end of the system, the material balance on the polymer is

$$q_t t_o C_o = A x_3^* (1 - \phi) \rho_r A_i + \left(x_3^* - x_r^* \right) A \left(\phi \overline{S}_{wr}^* - \phi_{IPV} \right) C_o. \dotfill (5.92)$$

The first term represents the mass of polymer injected, while the second term is the amount of polymer retained on the rock that has been contacted by polymer. The third term is the amount of polymer in the accessible PV at concentration C_o. In Eq. 5.92, $x_r^* =$ location of the rear of the polymer bank and $\overline{S}_{wr}^* =$ average water saturation in the region between x_r^* and x_3^*. $\overline{S}_{wr}^*$ is obtained from the Welge equation (Willhite 1986):

$$\overline{S}_{wr}^* = \frac{x_{D3}^* S_{w3}^* - x_{Dr}^* S_{wr}^*}{x_{D3}^* - x_{Dr}^*} - t_{Dr} \frac{\left(f_{w3}^* - f_{wr}^* \right)}{\left(x_{D3}^* - x_{Dr}^* \right)}, \dotfill (5.93)$$

where subscript D represents dimensionless distances (i.e., $x_D = x/L$.)

A solution is developed to find x_{Dr}^* for each value of S_{wr}^* as the drive water displaces the polymer slug through the linear system. Dividing Eq. 5.92 by $A \phi L C_{io}$ and expressing it in dimensionless terms yield

$$t_{Do} = x_{D3}^* D_p + \left(x_{D3}^* - x_{Dr}^* \right) \left(\overline{S}_{wr}^* - \phi_e \right), \dotfill (5.94)$$

where $D_p = \hat{C}_o / C_o$,

$$\phi_e = \phi_{IPV} / \phi,$$

$$x_{Dr}^* = x_r^* / L,$$

and $x_{D3}^* = x_3^* / L$.

All saturations within the polymer slug travel at specific velocities determined by the frontal-advance equation. Thus,

$$x_{Dr}^* = f'^*_{wr} t_{Dr} \dotfill (5.95)$$

$$\text{and } x_{D3}^* = f'^*_{w3} t_D. \dotfill (5.96)$$

Substituting these equations for x_{Dr}^* and x_{D3}^* into Eq. 5.92 and solving for t_D yield

$$t_{Dr} = \frac{t_{Do}}{f'^*_{w3} \left(\overline{S}_{wr}^* + D_p - \phi_e \right) - f'^*_{wr} \left(\overline{S}_{wr}^* - \phi_e \right)}. \dotfill (5.97)$$

The solution proceeds by computing the value of t_D corresponding to each value of S_{wr}^* from $1 - S_{or}$ to S_{w3}^*. Then, x_{Dr}^* is computed from Eq. 5.95.

Note that substitution of Eqs. 5.95 and 5.96 into Eq. 5.94 gives

$$\overline{S}_{wr}^* = \frac{f'^*_{w3} S_{w3}^* - f'^*_{wr} S_{wr}^*}{f'^*_{w3} - f'^*_{wr}} - \frac{f_{w3}^* - f_{wr}^*}{f'^*_{w3} - f'^*_{wr}}. \dotfill (5.98)$$

Example 5.8 illustrates the application of this solution technique to predict the displacement performance of a polymer-augmented waterflood where D_p PV of polymer at concentration C_o is followed by drive water.

Example 5.8—Displacement of a Polymer Slug in a Linear System. Estimate the performance of the polymer-augmented waterflood when a slug of D_p PV is injected into the linear reservoir described in Example 5.7.

Solution. The polymer-augmented waterflood performance to $t_D = 0.424(D_p)$ is identical to that estimated in Example 5.7. Values of S_{wf}, S_{w1}, and S_{w3}^* remain unaffected by the shift from continuous to slug injection. When drive-water injection begins, the polymer flood front is located at x_{D3}^*.

$$x_{D3}^* = f_{w3}' t_D$$
$$= (1.2377)(0.424) \ldots\ldots\ldots\ldots\ldots \quad (5.99)$$
$$= 0.525.$$

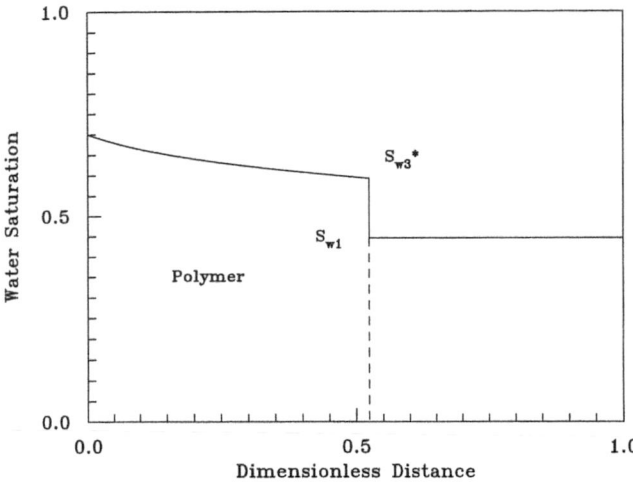

Fig. 5.58—Saturation profile when drive-water injection begins in Example 5.8.

Fig. 5.58 presents the saturation profile corresponding to this time.

As the drive water is injected, it displaces the polymer slug. Next, the location of the rear of the polymer slug as it is displaced through the porous rock is computed with the material-balance method. **Table 5.26** contains values of S_w^*, f_w^*, and $f_w'^*$ from the polymer fractional-flow curve. In this example, x_{Dr}^* for $S_{wr}^* = 0.6581$ is computed to illustrate the procedure.

From Table 5.26, $f_{wr}^* = 0.9945$ and $f_{wr}'^* = 0.2891$. Also from Table 5.26, $S_{w3}^* = 0.5925$, $f_{w3}^* = 0.9487$, and $f_{wr}'^* = 1.2377$. $\overline{S_{wr}^*}$ is determined from Eq. 5.98:

$$\overline{S_{wr}^*} = \frac{f_{w3}'^* S_{w3}^* - f_{wr}'^* S_{wr}^*}{f_{w3}'^* - f_{wr}'^*} - \frac{f_{w3}^* - f_{wr}^*}{f_{w3}'^* - f_{wr}'^*}$$
$$= \frac{(1.2377)(0.5925) - (0.2891)(0.6581) - 0.9487 + 0.9945}{1.2377 - 0.2891}$$
$$= 0.5887/0.9486$$
$$= 0.6206.$$

S_w^*	f_w^*	$f_w'^*$
0 70000	1.00000	0.00000
0.69462	0.99993	0.02799
0.68925	0.99970	0.05831
0.68387	0.99929	0.09111
0.67849	0.99871	0.12659
0.67312	0.99793	0.16491
0 66774	0.99693	0.20629
0.66237	0.99570	0.25092
0.65699	0.99423	0.29903
0.65162	0.99248	0.35084
0.64624	0.99045	0.40659
0.64086	0.98810	0.46653
0.63549	0.98542	0.53091
0.63011	0.98239	0.60000
0.62473	0.97896	0.67405
0.61936	0.97513	0.75335
0.61398	0.97085	0.83816
0.60861	0.96611	0.92876
0.60323	0.96086	1.02539
0.59785	0.95507	1.12831
0.59248	0.94871	1.23776

Table 5.26—Fractional-flow data, polymer flood, Example 5.8.

Recall from Example 5.7 that $\phi_e = 0.25$. The time when the rear of the polymer bank is at saturation S_{wr}^* is found by solving Eq. 5.97 for t_{Dr}. In Eq. 5.97, $t_{Do} = 0.424$.

$$\begin{aligned}
t_{Dr} &= \frac{t_{Do}}{f_{w3}'^* \left(\overline{S_{wr}^*} + D_p - \phi_e\right) - f_{wr}'^* \left(\overline{S_{wr}^*} - \phi_e\right)} \\
&= \frac{0.424}{1.2377(0.6206 + 0.424 - 0.25) - 0.2891(0.6206 - 0.25)} \\
&= 0.424/(0.9834 - 0.1071) \\
&= 0.4839.
\end{aligned}$$

The location of the rear of the polymer bank is found from the frontal-advance solution:

$$\begin{aligned}
x_{Dr}^* &= f_{wr}'^* t_{Dr} \\
&= (0.2891)(0.4839) \dots\dots\dots\dots\dots\dots\dots\dots\dots\dots\dots\dots\dots\dots\dots\dots\dots\dots\dots \quad (5.100) \\
&= (0.1399).
\end{aligned}$$

The remaining calculations of t_D for this example were performed with a short computer program. Columns 2 and 3 of **Table 5.27** contain pairs of coordinates on the distance/time diagram (x_{Dr}^*/t_D) that correspond to each saturation S_{wr}^* at the rear of the polymer slug. The drive water is moving faster than the saturations in the polymer bank and gradually overtakes the polymer bank. The drive water arrives at the end of the linear system at the same time as the polymer flood front, x_{D3}^*. **Fig. 5.59** shows the path traced by the rear of the polymer bank. The location of the rear of the polymer slug is almost a linear function of time for this example. Thus, for this case, the rear of the polymer slug appears to travel at a constant velocity. Fig. 5.59 also shows the waterflood front, oil bank, polymer flood front, and paths of selected saturations in the polymer slug. Saturations in the drive-water region are discussed later.

When the polymer flood front arrives at the end of the linear system, the displacement process becomes a waterflood. The WOR jumps from 3.53 in the oil/water bank to 27.2 at the polymer flood front and then continues to increase. The remainder of the oil will be produced at high WOR. Oil recovery when the polymer front reaches the end of the linear system is identical to that in Table 5.22 at $t_D = 0.8079$, corresponding to polymer-flood-front breakthrough. Remember that in the case of slug injection where the PV of polymer injected equals D_p, the polymer flood front disappears just as the polymer reaches the end

t_D	S_w^*	x_{Dr}^*
0.42400	0.70000	0.00000
0.42938	0.69462	0.01202
0.43528	0.68925	0.02538
0.44178	0.68387	0.04025
0.44893	0.67849	0.05683
0.45682	0.67312	0.07534
0.46555	0.66774	0.09604
0.47521	0.66237	0.11924
0.48595	0.65699	0.14531
0.49790	0.65162	0.17468
0.51125	0.64624	0.20787
0.52622	0.64086	0.24549
0.54306	0.63549	0.28831
0.56210	0.63011	0.33725
0.58372	0.62473	0.39346
0.60843	0.61936	0.45836
0.63685	0.61398	0.53378
0.66977	0.60861	0.62206
0.70825	0.60323	0.72623
0.75367	0.59785	0.85037
0.80790	0.59248	1.00000

Table 5.27—Location of rear of polymer bank, Example 5.8.

of the linear system. Incremental oil displaced at this time is 27,658 STB from the injection of 0.424 PV of polymer solution. Polymer required in the slug is

$$m_p = (0.424 \ \text{PV})(474,472 \ \text{bbl/PV})(350 \ \text{lbm/bbl})$$

$$\times \left(300 \times 10^{-6} \ \text{lbm polymer/lbm}\right)$$

$$= 21,170 \ \text{lbm}.$$

Incremental oil at this point is 1.31 STB/lbm polymer. This estimate is optimistic because it does not consider dispersion of the polymer or viscous fingering.

Drive-Water Region. The drive-water region contains all saturations $S_{w3}^* \le S_w \le 1 - S_{or}$. These saturations evolve at the rear of the polymer bank as the displacement proceeds through the system in a manner similar to that discussed in Section 3.6.2. At every saturation S_{wr}^*, the velocity of the rear of the polymer bank is given by Eq. 5.101. The retention term, D_p, is omitted at the rear of the polymer bank because retention is assumed to be irreversible.

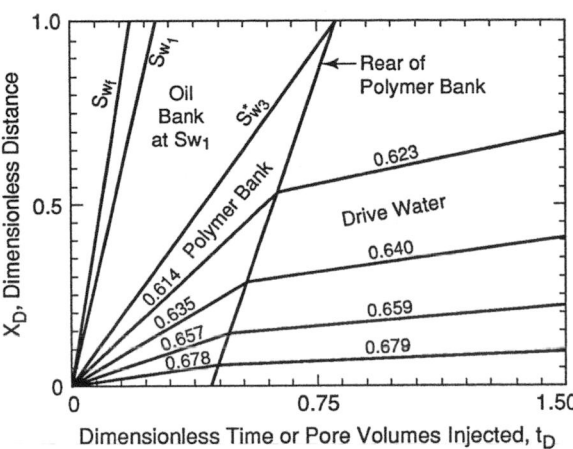

Fig. 5.59—Distance/time diagram showing displacement of the polymer bank by drive water in Example 5.8.

$$v_{Dr}^* = f_{wr}^* / \left(S_{wr}^* - \phi_e\right). \quad \dots \dots \dots \dots \dots \dots \dots \dots \dots \dots \dots \dots \dots \dots (5.101)$$

Because the drive water and the polymer solution are miscible, the specific velocity of the polymer-free water must be equal to the specific velocity of the rear of the polymer bank. That is,

$$\frac{f_{wr}}{S_{wr} - \phi_e} = \frac{f_{wr}^*}{S_{wr}^* - \phi_e}. \quad \dots \dots \dots \dots \dots \dots \dots \dots \dots \dots \dots \dots \dots \dots \dots \dots \dots \dots (5.102)$$

A step-function change in polymer concentration occurs at the rear of the polymer slug. A discontinuity in saturation also occurs because, at the polymer rear, the specific velocity in Eq. 5.102 must also satisfy

$$v_{Dr}^* = \left(f_{wr}^* - f_{wr}\right) / \left(S_{wr}^* - S_{wr}\right), \quad \dots \dots \dots \dots \dots \dots \dots \dots \dots \dots \dots \dots \dots \dots \dots \dots \dots (5.103)$$

where S_{wr} = water saturation in the polymer-free water and f_{wr} = fractional flow of water in the polymer-free water.

The water just behind the rear of the polymer bank is called polymer-free water. Because polymer travels faster through the porous rock than the water with which it was injected, a denuded region of "polymer water" is present between the drive water and the rear of the polymer slug.

At every value of S_{wr}^*, there is a corresponding value of S_{wr} at the saturation discontinuity. Thus, as the polymer-free water displaces the polymer slug, saturations S_{wr}^* are overtaken by the drive water and new saturations, S_{wr} evolve at the rear of the polymer bank. The saturations S_{wr} can be found graphically for each value of S_{wr}^* by observing that f_{wr}, S_{wr} is the intersection of a line from $f_w = 0$, $S_w = \phi_e$ to f_{wr}^*, S_{wr}^* with the f_w, S_w fractional-flow curve. **Fig. 5.60** shows one line for the fractional-flow curves used in Example 5.8 when $\phi_e = 0.25$. At high fractional flows in the polymer bank, the difference between saturations S_{wr}^* and S_{wr} is difficult to determine graphically. A root-function program was used to find pairs of f_{wr}^*, S_{wr}^* and f_{wr}, S_{wr} for the problem in Example 5.8, and values of these parameters are summarized in **Table 5.28.** The saturation differences are quite small.

Paths of saturations for the drive water displacing the polymer slug can be determined by noting that each saturation S_{wr} evolves from the rear of the polymer bank at x_{Dr}, t_{Dr}. Thus, at any time $t_D > t_{Dr}$, the location of each saturation is given by applying Eq. 5.71 to obtain

$$x_D = x_{Dr} + \left(t_D - t_{Dr}\right) f_{wr}'. \quad \dots \dots \dots (5.104)$$

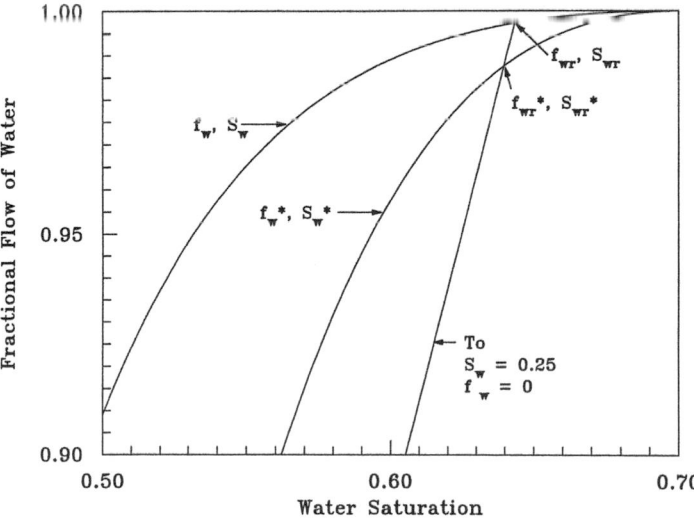

Fig. 5.60—Determination of f_{wr}, S_{wr} from the fractional-flow curve in Example 5.8.

S_w^*	f_{wr}^*	S_{wr}	f_{wr}	$f_{wr}'^*$
0.70000	1.00000	0.70000	1.00000	2.22223
0.69462	0.99993	0.69464	0.99998	2.24893
0.68925	0.99970	0.68934	0.99993	2.27593
0.68387	0.99929	0.68409	0.99983	2.30321
0.67849	0.99871	0.67890	0.99969	2.33074
0.67312	0.99793	0.67377	0.99951	2.35850
0.66774	0.99693	0.66872	0.99928	2.38647
0.66637	0.99570	0.66372	0.99901	2.41461
0.65699	0.99423	0.65881	0.99868	2.44287
0.65162	0.99248	0.65396	0.99831	2.47123
0.64624	0.99045	0.64921	0.99789	2.49962
0.64086	0.98810	0.64454	0.99742	2.52800
0.63549	0.98542	0.63996	0.99689	2.55631
0.63011	0.98239	0.63548	0.99632	2.58447
0.62473	0.97896	0.63112	0.99569	2.61242
0.61936	0.97513	0.62688	0.99502	2.64007
0.61398	0.97085	0.62277	0.99431	2.66731
0.60861	0.96611	0.61878	0.99355	2.69406
0.60323	0.96086	0.61495	0.99276	2.72020
0.59785	0.95507	0.61127	0.99194	2.74561
0.59248	0.94871			

Table 5.28—Fractional-flow parameters for rear of polymer bank, Example 5.8.

Example 5.9 gives a sample calculation of a saturation path.

Example 5.9—Saturation Path in Drive-Water Region Following Injection of a Polymer Slug. Generate saturation paths in the drive-water region for the polymer-augmented waterflood using a 0.424-PV slug of polymer described in Example 5.8.

Solution. One path in the drive-water region is calculated in this section from the results of Example 5.8. From Example 5.8 at $S_{wr}^* = 0.6581$, $t_{Dr} = 0.4839$ and $x_{Dr} = 0.1399$.

From a root-finding program, a line from $S_w = 0.25$, $f_w = 0$ to $f_{wr}^* = 0.9945$, $S_{wr}^* = 0.6581$ intersects the f_w, S_w fractional-flow curve at $f_{wr} = 0.9988$, $S_{wr} = 0.6598$, and $f_{wr}' = 0.0689$. Substitution into Eq. 5.104 yields, for $S_{wr} = 0.6598$,

$$x_D = x_{Dr} + (t_D - t_{Dr})f_{wr}'$$
$$= 0.1399 + (t_D - 0.4839)(0.0689).$$

Selected values of S_{wr}^* were chosen for construction of saturation paths. Fig. 5.59 shows paths for saturations in the drive-water region.

Fig. 5.59 displays a complete picture of the saturation paths in the example linear displacement of a polymer slug where dispersion and viscous fingering are neglected. Displacement performance after water breakthrough is estimated by determining the average water saturation along lines of constant t_D. Because water saturations travel at slow specific velocities, the remaining oil will be produced at high WORs. Saturation profiles may be constructed from this diagram because a saturation profile is the locus of saturations along a line of constant t_D.

Displacement performance after the polymer flood front breaks through and disappears is found by determining the average saturation of the linear system at specified values of t_D. After the polymer front just reaches the end of the system, a saturation profile ranges from $1-S_{or}$ at the beginning of the system to S_{w2} at the end of the system. The location of every saturation is found with Eq. 5.104. In Eq. 5.104, x_{Dr} is the location where the saturation S_{wr} evolved from the rear of the polymer bank at t_{Dr}.

The average water saturation in the linear system is found by integrating the saturation distribution over the length of the system with

$$\overline{S}_w = \int_0^{1.0} S_w \, dx_D. \quad \dots \dots \dots \dots \dots \dots \dots \dots \dots \dots \dots \dots \dots \dots \dots \dots (5.105)$$

If there are m saturations within the system (i.e., $x_{DSwr} \leq 1.0$), the integral in Eq. 5.105 can be evaluated numerically with a short computer program. Using the trapezoidal rule for integrating Eq. 5.105 results in

$$\overline{S}_w = 0.5 \sum_{j=1}^{m-1} \left(S_{wrj} + S_{wrj+1} \right) \left(x_{Dj+1} - x_{Dj} \right), \dots\dots\dots\dots\dots\dots\dots\dots\dots\dots\dots\dots\dots\dots(5.106)$$

where x_{Dj} = location of saturation S_{wrj} when t_D PV of fluid has been injected.

The oil displaced at t_D is computed by material balance:

$$N_P = [(\overline{S}_w - S_{iw})V_p] / B_o. \dots\dots\dots\dots\dots\dots\dots\dots\dots\dots\dots\dots\dots\dots\dots\dots(5.107)$$

Insufficient Polymer To Satisfy Retention. If the amount of polymer injected is not sufficient to satisfy retention in the porous rock, the drive water will overtake the polymer front at some point in the linear system. Recall that in Examples 5.8 and 5.9, sufficient polymer was injected so that the polymer front just reached the end of the system.

The time and location at which the polymer front is overtaken by the drive water are determined by material balance, as in Eqs. 5.92 and 5.94. If t_{Dp} = PVs of polymer injected, then the polymer front will be overtaken by the drive water when $x_{D3b}^* = t_{Dp} / D_p$. The time when this happens can be estimated from the frontal-advance solution. Let t_{Db} = dimensionless time when the polymer front is overtaken by the drive water. Then,

$$t_{Db} = x_{D3b}^* / f_{w3}^{\prime *}. \dots\dots\dots\dots\dots\dots\dots\dots\dots\dots\dots\dots\dots\dots\dots\dots\dots\dots\dots\dots\dots(5.108)$$

The path followed by the rear of the polymer slug is predicted in the same manner as described earlier and illustrated in Example 5.8.

When the polymer front is overtaken by the drive water, the process becomes a waterflood. For example, when $t_{Dp} = D_p/2$ (in Example 5.8), the polymer front is overtaken at $t_D = 0.5/1.2377 = 0.404$. **Fig. 5.61** shows the saturation distribution at this instant. Note that the oil bank created by the polymer front is present, with a saturation discontinuity from S_{wr} to S_{w1} at the rear of the oil bank. This saturation discontinuity is not stable. Because of miscibility, the velocity of this discontinuity is given by

$$v_{Dr} = f_{w1} / S_{w1} = f_{wr} / S_{wr} \dots\dots\dots\dots\dots\dots\dots\dots\dots\dots\dots\dots\dots\dots\dots\dots(5.109)$$

or $v_{Dr} = \left(f_{wr} - f_{w1} \right) / \left(S_{wr} - S_{w1} \right). \dots\dots\dots\dots\dots\dots\dots\dots\dots\dots\dots\dots\dots(5.110)$

The velocity at the rear of the oil bank would be constant and stable if $v_{Dr} = f_{wr}'$. Inspection of the fractional-flow curve or tables of f_w' at the saturation between S_{wr} and S_{w1} shows that $v_{Dr} > f_{wr}'$. Therefore, the drive water cuts into the rear of the oil bank until $S_{wr} = S_{w1}$.

The velocity at the rear of the oil bank increases until $S_{wr} = S_{w1}$. The location of the rear of the oil bank at t_D is determined by making a material balance on the water phase. Fig. 5.61 is a generalized saturation profile for the displacement of the oil bank by the drive water after the polymer flood front has been overtaken. The total volume of water injected (polymer water and drive water) is given by

$$t_D = \underbrace{\int_0^{x_{Dr}} S_w dx_D}_{\substack{\text{water behind} \\ \text{oil bank}}} + \underbrace{S_{w1}\left(x_{D1} - x_{Dr} \right)}_{\substack{\text{water in} \\ \text{oil bank}}} + \underbrace{\int_{x_{D1}}^{x_{Df}} S_w dx_D}_{\substack{\text{water in region} \\ \text{between} \\ S_{w1} \text{ and } S_{wf}}} - \underbrace{S_{iw} x_{Df}}_{\substack{\text{interstitial} \\ \text{water}}}. \dots\dots\dots\dots\dots\dots\dots\dots\dots\dots\dots(5.111)$$

The first integral on the right side of Eq. 5.111 represents the water in the region behind the oil bank. This region contains all saturations S_{wrj} that evolved from the rear of the polymer or oil bank. Let x_{Dr} represent the location at which saturation S_{wrj} evolved at t_{Dj} as the drive water displaced the polymer or oil bank. Thus, the location of saturation S_{wrj} after t_D PV of fluid has been injected is found from

$$x_{Dj} = x_{Drj} + \left(t_D - t_{Dj} \right) f_{wrj}'. \dots\dots\dots\dots(5.112)$$

The location of the rear of the oil bank, x_{Dr}, is found by integrating the velocity of the saturation discontinuity. Recall that

$$dx_{Dr} / dt = v_{Dr}. \dots\dots\dots\dots\dots\dots\dots\dots(5.113)$$

Then, for the timestep from t_D^{n-1} to t_D^n,

$$x_{Dr}^n = x_{Dr}^{n-1} + \frac{\left(v_{Dr}^n + v_{Dr}^{n-1} \right)}{2}\left(t_D^n - t_D^{n-1} \right). \dots\dots\dots(5.114)$$

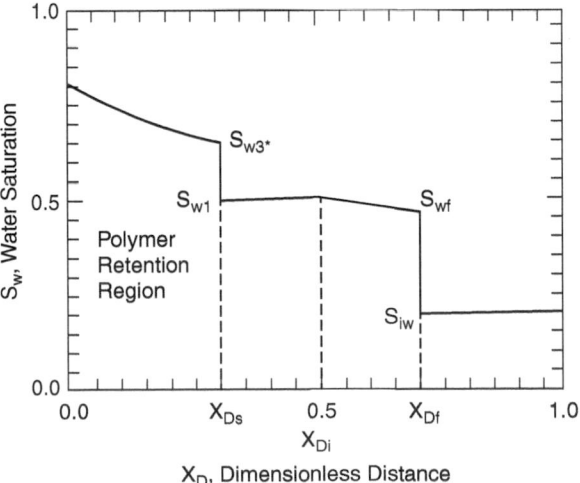

Fig. 5.61—Saturation profile when polymer front is overtaken by drive water.

When there are m saturations ($m - 1$ saturation increments) from $x_D = 0$ to x_{Dr}^n, the first integral is approximated by

$$\int_0^{x_{Dr}} S_{wd} x_D = 0.5 \sum_{j=1}^{m-1} \left(S_{wrj} + S_{wrj+1} \right)\left(x_{Dj+1} - x_{Dj} \right), \dotsfill (5.115)$$

where $x_{Drm} = x_{Dr}^n$.

Note that

$$x_{D1} = t_D f_{w1}' \dotsfill (5.116)$$

and $x_{Df} = t_D f_{wf}'. \dotsfill (5.117)$

The second integral in Eq. 5.111 is evaluated by applying the Welge equation. Thus,

$$\int_{x_{D1}}^{x_{Df}} S_w \, dx = \overline{S}_{wf1} \left(x_{Df} - x_{D1} \right), \dotsfill (5.118)$$

where $\overline{S}_{wf1} = \dfrac{f_{wf}' S_{wf} - f_{w1}' S_{w1} - f_{wf} + f_{w1}}{f_{wf}' - f_{w1}'}. \dotsfill (5.119)$

$\overline{S}_{wf1}$ is independent of t_D.

Inspection of Eqs. 5.111 through 5.119 shows that x_{Dr} and t_D are dependent variables once a value of S_{wr} is chosen. In principle, x_D can be eliminated from Eq. 5.111 by use of Eqs. 5.108 and 5.110 through 5.119. Then a value of t_D can be found for each value of S_{wr} between S_{wrb} and S_{w1}. However, this approach is sensitive to small numerical errors. Example 5.10 illustrates the solution of Eq. 5.111 when an approximation technique is used.

Example 5.10—Polymer Slug With Insufficient Polymer To Satisfy Retention. The amount of polymer to be injected as a slug in a linear system is chosen so that the polymer front disappears before reaching the end of the system. The properties of the system in Example 5.7 are used in this example, where a polymer solution with an apparent viscosity of 4 cp displaces oil with a viscosity of 40 cp. IPV is 0.25 and $D_p = 0424$. In this example, the polymer slug will be $D_p/2 = 0.212$ PV injected. Determine the displacement performance of the polymer flood.

Solution. During polymer injection ($t_D \le 0.212$), the polymer flood performs exactly as described in Examples 5.7 and 5.8. A waterflood front forms at saturation S_{wf}, followed by an oil bank that has constant water saturation S_{w1}. The oil bank is displaced by a polymer flood front, S_{w3}^*. Table 5.20 presents properties of these fronts.

The first part of the solution is to determine the displacement performance to the time when the polymer flood front is overtaken by the drive water. When the volume of polymer injected is $D_p/2$, the polymer front will disappear when $x_{D3}^* = 0.5$. That is, $D_p/2$ is the amount of polymer required to satisfy retention for one-half of the linear system. Then,

$$t_{Db} = x_{D3}^* / f_{w3}'^*$$
$$= 0.5 / f_{w3}'^*$$
$$= 0.5 / 1.2377$$
$$= 0.404.$$

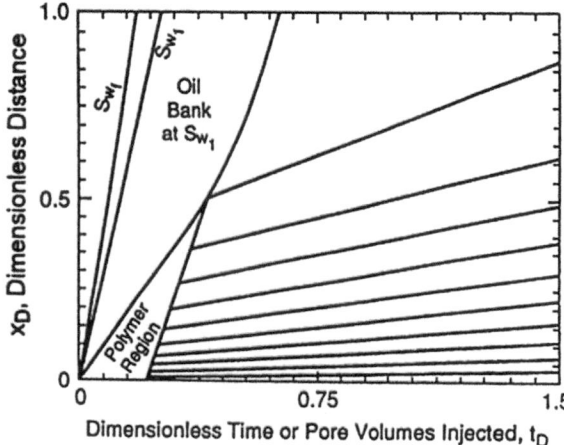

Fig. 5.62—Distance/time diagram showing polymer being overtaken by drive water, Example 5.10.

The path of the rear of the polymer front and the saturation paths in the drive water can be determined with the method presented in Section 5.7.2. Program PREAR in Appendix C was used to determine the values of S_{wr}, x_{Dr}, and t_D when the saturation interval between $1 - S_{or}$ and S_{w3}^* was divided into 100 increments. **Fig. 5.62** presents the distance/time diagram for $0 < t_D \le 0.404$.

After the polymer front is overtaken, the path of the rear of the oil bank (x_{Dr}, t_D) is determined by solving for t_D and x_{Dr} for each saturation between S_{wrb} and S_{w1}. For the purposes of this example, the saturation interval $S_{wrb} - S_{w1}$ was divided into 50 increments. Eq. 5.111 is solved in the following manner.

1. t_D^n is estimated by assuming that $t_D^n - t_D^{n-1} \approx t_D^{n-1} - t_D^{n-2}$ beginning with the last saturation increment.
2. Locations of all saturations are computed at t_D with Eqs. 5.113 through 5.116.
3. Eq. 5.115 is used to compute the first integral in Eq. 5.111.

1. The value of t_D is computed from Eq. 5.111 and called t_{Dm}.
2. If $|t_{Dm} - t_D^n|$ is less than some value ε, calculations cease and the value of t_D^n is accepted. Otherwise, an iterative solution is used.

If the number of saturation increments is large, the two values are in good agreement after the first computation.

The computations are too complicated to conduct by hand but are easily performed with a short computer program. Program PSLUG in Appendix C computes the location of the rear of the oil bank (x_{Dr1}, t_d) using the procedure described above. Computed results are presented in **Table 5.29.** Values of t_D and t_{Dm} are within 1.5×10^{-4}, which shows that the calculation technique is valid.

Fig. 5.63 is the distance/time diagram for the displacement process. After the polymer slug is overtaken by the drive water, the oil bank is eroded continually by the drive water until the drive-water saturation at the rear of the oil bank decreases to S_{w1}. The curved path at the rear of the oil bank depicts this process. An oil bank of water saturation S_{w1} is produced until breakthrough of the drive water at $t_D \approx 0.626$. Thereafter, oil production declines as higher water saturations reach the end of the system. Paths of water saturations in the drive-water region $(1 - S_{or} > S_w \geq S_{w1})$ were computed with Eq. 5.112 with the values of x_{Drj} and t_{Dj} obtained from Table 5.29.

Displacement performance for the linear system $(0 \leq x_D \leq 1.0)$ is obtained by computing the average water saturation in this region for various t_D. The average water saturation in the linear system is given by

$$\bar{S}_w = \int_0^1 S_{wD} \, dx_D. \dots\dots\dots\dots\dots\dots\dots\dots\dots\dots\dots\dots\dots\dots \text{(5.120)}$$

The volume of oil displaced at reservoir conditions is found by subtracting the volume of water initially present at the beginning of the flood. Thus,

$$N_p = [(\bar{S}_w - S_{wi})V_p] / B_o. \dots\dots\dots\dots\dots\dots\dots\dots\dots\dots\dots\dots\dots \text{(5.121)}$$

The integral in Eq. 5.120 is evaluated in the same manner as Eq. 5.106.

t_D	t_{Dm}	x_{Dr}	S_{wr}
0.40381*	0.40381	0.50000	0.59248
0.41525	0.41514	0.51740	0.58662
0.42669	0.42658	0.53535	0.58075
0.43814	0.43802	0.55389	0.57489
0.44958	0.44946	0.57305	0.56903
0.46102	0.46090	0.59285	0.56316
0.47246	0.47234	0.61334	0.55730
0.48390	0.48379	0.63454	0.55144
0.49535	0.49523	0.65650	0.54557
0.50679	0.50667	0.67925	0.53971
0.51823	0.51811	0.70285	0.53385
0.52967	0.52955	0.72732	0.52799
0.54112	0.54099	0.75273	0.52212
0.55256	0.55244	0.77912	0.51626
0.56400	0.56388	0.80654	0.51040
0.57544	0.57532	0.83506	0.50453
0.58688	0.58676	0.86473	0.49867
0.59833	0.59820	0.89562	0.49281
0.60977	0.60964	0.92779	0.48694
0.62121	0.62108	0.96131	0.48108
0.63265	0.63253	0.99625	0.47522
0.63384**	0.63372	1.00000	0.47461

*Polymer bank is overtaken by drive water.

**Rear of oil bank leaves system

Table 5.29—Summary of calculation to locate rear of oil bank, Example 5.10.

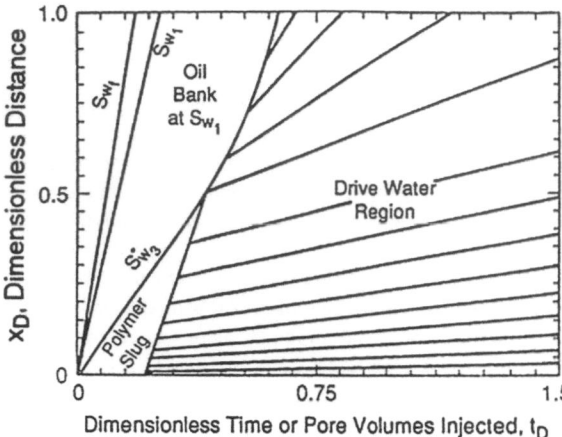

Fig. 5.63—Distance/time diagram—0.212-PV polymer slug displaced by drive water, Example 5.10.

Table 5.30 presents the displacement performance for the polymer flood when enough polymer is injected to satisfy polymer retention for one-half the length of the system. **Fig. 5.64** compares the displacement efficiency of the polymer flood (t_{DP} = 0.212) with a waterflood of the same linear reservoir. **Fig. 5.65** presents graphs of WOR vs. cumulative production for the same floods.

In Example 5.8, the incremental oil recovered as a result of the injection of a polymer slug (0.424 PV) was found to be 27,658 STB at t_D = 0.8079, corresponding to arrival of the polymer slug at the end of the system. When the polymer slug is reduced to 0.212 PV, as in Example 5.10, cumulative oil production is estimated to be 132,625 STB at t_D = 0.8079. Incremental oil above the waterflood is 15,739 STB for an oil/polymer ratio of 1.49 STB/lbm of polymer. Reduction of the size of the polymer slug reduces the cost of polymer and the incremental oil recovery. If economic parameters were known, it would be possible to determine an optimum slug size by conducting a series of these calculations.

t_D (PV)	Time (days)	$\overline{S}_{w2}$ (fraction)	N_P (STB)	q_o (STB/D)	q_w (B/D)	F_{wo} (B/D)
0.0	0.0	0.300	0	181.8	0.0	0.0
0.185*	440.0	0.485	79,972	63.5	130.2	2.05
0.21200	504.04	0.494	83,813	53.68	140.96	2.63
0.21764	517.45	0.496	84,518	51.99	142.81	2.75
0.22447	533.68	0.497	85,347	50.11	144.88	2.89
0.23277	553.43	0.500	86,315	48.00	147.20	3.07
0.24297	577.68	0.502	87,452	45.67	149.77	3.28
0.25562	607.76	0.505	88,787	43.07	152.62	3.54
0.26001**	618.19	0.506	89,227	42.31	153.46	3.63
0.27153	645.58	0.509	90,386	42.31	153.46	3.63
0.29186	693.92	0.514	92,431	42.31	153.46	3.63
0.31842	757.07	0.520	95,103	42.31	153.46	3.63
0.35413	841.95	0.528	98,693	42.31	153.46	3.63
0.40381	960.08	0.540	103,692	42.31	153.46	3.63
0.46102	1,096.10	0.553	109,412	42.31	153.46	3.63
0.51823	1,232.13	0.566	115,167	42.31	153.46	3.63
0.57544	1,368.15	0.580	120,923	42.31	153.46	3.63
0.63384†	1,507.00	0.593	126,798	25.97	171.43	6.60
0.68293	1,623.71	0.598	128,993	15.22	183.26	12.04
0.83371	1,982.19	0.608	133,021	8.74	190.39	21.79
0.17602	2,796.06	0.620	138,191	4.79	194.73	40.69
1.17664	2,797.53	0.620	138,196	4.79	194.73	40.69
1.89771	4,511.93	0.633	143,956	2.42	197.33	81.39
3.32138	7,896.79	0.646	149,563	1.32	198.55	150.86
4.35441	10,352.87	0.653	152,427	1.03	198.86	192.15

* Water breakthrough.

** Oil bank arrives.

† Oil bank ends.

Table 5.30—Displacement performance of polymer flood, Example 5.10 (0.212-PV polymer slug followed by drive water, injection rate 200 B/D, B_o = 1.1).

Note that the comparisons of incremental oil recovery in Examples 5.8 through 5.10 assume that the water and polymer injection rates are constant throughout the flood. Sufficient pressure drop is assumed to exist across the system to maintain the rates at the values in the examples. This is often not the case. Section 5.5.4 addresses the problem of estimating pressure drop during a polymer flood in a linear reservoir.

5.5.4 Estimation of Pressure Drop During a Polymer Flood at Constant Injection Rate.

When the injection rate is maintained constant during a polymer flood, as assumed in Example 5.7, the pressure drop must change in response to the changes in fluid saturations and mobilities along the length of the system. The pressure drop can be computed at any point in the displacement process when the saturation profile is known. **Fig. 5.66** depicts the saturation profile and distribution of fluid components during a polymer flood. This figure, adapted from Fig. 3.34, is referenced in Example 5.11. Because the fluids were assumed to be incompressible in the development of the polymer flood model, the total flow rate, q_t, is invariant with respect to x at a specific time. The relationship between flow rate and the pressure gradient at every location is given by Darcy's law. That is,

$$q_t = -\left(\lambda_{ro} + \lambda_{rw}\right)k_b A\left(dp/dx\right), \dots\dots\dots\dots (5.122)$$

where λ_{ro} = relative mobility of the oil phase, λ_{rw} = relative mobility of the aqueous phase, k_b = base permeability, A = cross-sectional area perpendicular to the direction of flow, p = pressure, and consistent units are used.

The relative mobility of the aqueous phase in the region contacted by polymer is the relative mobility of the polymer flowing in the rock. Because λ_{ro} and λ_{rw} are functions of the saturation profile and fluid properties, when the saturation profile is known, as in Fig. 5.66, the mobilities of both phases can be computed at every value of x. Eq. 5.122 can be integrated to find the pressure drop corresponding to the value of t_D. Thus,

$$\Delta p = \frac{q_t}{k_b A}\int_0^L \left(\frac{1}{\lambda_{ro} + \lambda_{rw}}\right)dx. \dots\dots\dots\dots (5.123)$$

When expressed in terms of dimensionless distance, x_D, Eq. 5.123 becomes

$$\Delta p = \frac{q_t L}{k_b A}\int_0^1 \left(\lambda_r^{-1}\right)dx_D, \dots\dots\dots\dots (5.124)$$

where λ_r^{-1} = reciprocal relative mobility,

$$\lambda_r^{-1} = 1/\left(\lambda_{ro} + \lambda_{rw}\right). \dots\dots\dots\dots (5.125)$$

The value of the integral in Eq. 5.124 is analogous to a viscosity in the single-phase flow of a homogeneous fluid through a linear system. For this reason, this integral is defined as $\overline{\lambda}^{-1}$, the average apparent viscosity. That is,

$$\overline{\lambda}^{-1} \equiv \int_0^1 \left(\lambda_r^{-1}\right)dx_D, \dots\dots\dots\dots (5.126)$$

where $\overline{\lambda}^{-1}$ = average apparent viscosity of a linear system corresponding to the saturation profile that is present at t_D. With this definition,

$$q_t = k_b A\Delta p/\overline{\lambda}^{-1}L \dots\dots\dots\dots (5.127a)$$

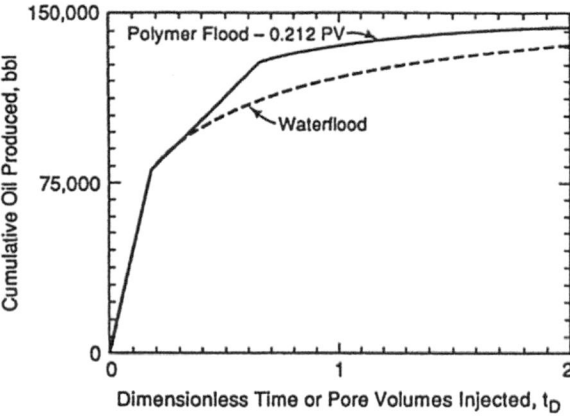

Fig. 5.64—Comparison of displacement performance for 0.212-PV polymer slug with a waterflood, Example 5.10.

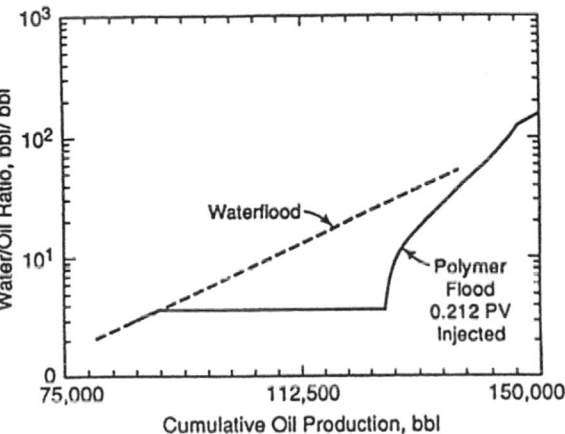

Fig. 5.65—WORs for 0.212-PV polymer flood and a waterflood in Example 5.10.

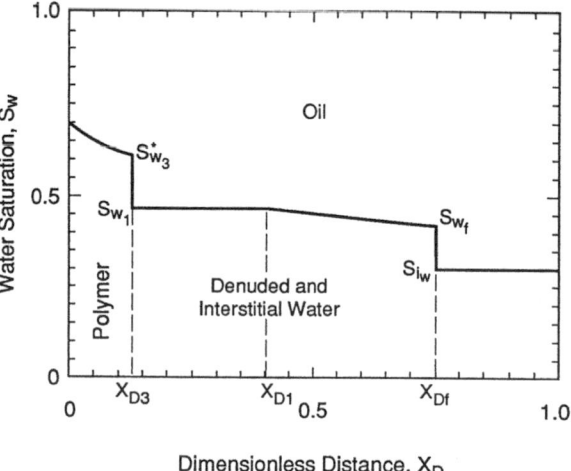

Fig. 5.66—Saturation profile in Example 5.11 for computation of the average apparent viscosity.

or $\Delta p = q_i \bar{\lambda}^{-1} L / k_b A.$ .... (5.127b)

Saturation profiles and fluid distributions change during the displacement but remain functions of t_D in all cases. Thus, the average apparent viscosity is also a unique function of t_D for a particular polymer flood. Once the relationship between $\bar{\lambda}^{-1}$ and t_D is determined by integrating Eq. 5.126, it is possible to compute the injection rate or pressure drop during a linear polymer flood when one of these parameters is known.

Computation of the average apparent viscosity is a straightforward application of numerical integration techniques. The procedure is illustrated by determining the average apparent viscosity when the saturation profile is given by Fig. 5.66. In this case, there are four discrete regions where fluid saturations and/or properties change with dimensionless distance, x_D: (1) the region where $0 \leq x_D \leq x_{D3}^*$ and polymer and oil are flowing, (2) the region where $x_{D3}^* \leq x \leq x_{D1}$ and oil and water are flowing and the water saturation is constant, (3) the region where $x_{D1} \leq x \leq x_{Df}$ and the water saturation decreases from S_{w1} to S_{wf}, and (4) the region where $x_{Df} \leq x \leq 1$ and oil is flowing at interstitial water saturation.

Applying Eq. 5.126 to the saturation profile in Fig. 5.66 yields

$$\bar{\lambda}^{-1} = \int_0^{x_{D3}^*} \lambda_r^{-1*} \, dx_D + \left(x_{D1} - x_{D3}^*\right)\lambda_{rSw1}^{-1} + \int_{x_{D1}}^{x_{Df}} \lambda_r^{-1} \, dx_D$$
$$+ \left(1 - x_{Df}\right)\lambda_{rSiw}^{-1}.$$
.... (5.128)

Eq. 5.128 may be viewed as the sum of the contribution of each region to the apparent viscosity of the total system.

After the polymer shock breaks through the end of the system, the average apparent viscosity for $S_{w2} \geq S_{w3}^*$ is computed from

$$\bar{\lambda}^{-1} = \int_0^1 \lambda_r^{-1*} \, dx_D.$$ .... (5.129)

The dimensionless distances x_{D3}^*, x_{D1}, and x_{Df} are determined from the frontal-advance solution:

$$x_{D3}^* = \left(\frac{f_{w3}^*}{S_{w3}^* - \phi_e + D_i}\right)t_D,$$ .... (5.130)

$$x_{D1} = f_{w1}' t_D,$$ .... (5.131)

and $x_{Df} = f_{wf}' t_D.$ .... (5.132)

Saturation profiles are found by use of the following relationships. For saturations in the region $0 \leq x_D \leq x_{D3}^*$ (i.e., $1 - S_{or} \geq S_w \geq S_{w3}^*$),

$$x_D^* = f_w^* t_D,$$ .... (5.133)

and for saturations in the region $x_{D1} \leq x_D \leq x_{Df}$ (i.e., $S_{w1} \geq S_w \geq S_{wf}$),

$$x_D = f_w t_D.$$ .... (5.134)

The evaluation of the average apparent viscosity for the saturation profile of Fig. 5.66 is illustrated in Example 5.11.

Example 5.11—Estimation of Pressure Drop During a Continuous Polymer Flood in a Linear Reservoir. Determine the pressure drop for the polymer flood in Example 5.7 when the waterflood front, x_{Df}, is located at a distance of 0.75 from the entrance of the system. Base permeability, the permeability to oil at interstitial water saturation, is 250 md. The injection rate is constant at 200 B/D. Recall that the linear segment of the reservoir being simulated is 500 ft wide and 20 ft thick. Injection and production wells are 1,000 ft apart.

Solution. The first step is to determine the saturation profile that corresponds to $x_{Df} = 0.75$. It is useful to summarize flood-front parameters from Example 5.7.

$$S_{wf} = 0.4206, \ S_{w1} = 0.4459, \ S_{w3}^* = 0.5925,$$

$$f_{wf} = 0.6508, \ f_{w1} = 0.7673, \ f_{w3}^* = 0.9687,$$

$$f_{wf}' = 5.3958, \ f_{w1}' = 3.8530, \ f_{w3}'^* = 1.2377$$

When the waterflood front is located at $x_{Df} = 0.75$, the dimensionless injection time is given by

$$t_D = x_{Df} / f_{wf}'$$
$$= 0.75 / 5.3958$$ .... (5.135)
$$= 0.139.$$

Also, x_{D1}, x_{Df}, x_{D3}^* can be located:

$$x_{D1} = f_{w1}' t_D$$
$$= (0.139)(3.8530)$$
$$= 0.536,$$

$$x_{Df} = f_{wf}' t_D,$$

and $x_{D3}^* = (0.139)(1.2377)$
$$= 0.172.$$

A numerical method is required to find $\bar{\lambda}_r^{-1}$ because $\bar{\lambda}_r^{-1}$ is a nonlinear function of x_D. The following relationships are used to compute λ_r^{-1}.

$$\lambda_{ro} = k_{ro}/\mu_o$$
$$= \left[0.8\left(1 - S_{wD}\right)^2\right]/\mu_o \quad\dots\dots\dots\dots\dots\dots\dots\dots\dots\dots\dots\dots(5.136)$$

$$\lambda_{rw} = k_{rw}/\mu_w$$
$$= 0.2 S_{wD}^2/\mu_w. \quad\dots\dots\dots\dots\dots\dots\dots\dots\dots\dots\dots\dots\dots(5.137)$$

The viscosity of the water is replaced by the apparent viscosity of the polymer solution in the region contacted by polymer.

There are two integrals to be evaluated in Eq. 5.128 when flood fronts from both waterflood and polymer floods are in the system. The regions are treated separately with the same approach.

The integral for the region behind the polymer shock can be rewritten as an equivalent integration with respect to the derivative of the fractional-flow curve (Willhite 1986). Recall that $x_D^* = t_D f_w'^*$, $0 \le x_D^* \le x_{D3}^*$. Then,

$$\bar{\lambda}_r^{-1*} = \frac{\displaystyle\int_0^{x_D^*} \lambda_r^{-1} dx_D}{x_D^*}$$

$$= \frac{\displaystyle\int_0^{f_w'^*} \lambda_r^{-1*} df_w^*}{f_w'^*} \quad\dots\dots\dots\dots\dots\dots\dots\dots\dots\dots\dots\dots(5.138)$$

$$\int_0^{x_D^*} \lambda_r^{-1} dx_D = \bar{\lambda}_r^{-1*} x_{D^*}. \quad\dots\dots\dots\dots\dots\dots\dots\dots\dots\dots\dots\dots(5.139)$$

The saturation profile in Fig. 5.66 was constructed when $x_{df} = 0.75$. Because $f_{wf}' = 5.3958$, the value of $t_D = 0.139$. Also, for the set of relative permeability curves used in this example, $f_w' = 0$ at $S_w^* = 1 - S_{or}$.

Values of λ_r^{*-1} are plotted vs. $f_w'^*$ in **Fig. 5.67.** The integral is the area under the curve between $f_{wo}'^*$ and $f_{w3}'^*$. Actual calculations were performed numerically with the trapezoidal rule by dividing the saturation interval between $S_w^* = 1 - S_{or}$ and S_{w3}^* into 100 increments. **Table 5.31** gives selected values of $\bar{\lambda}_r^{*-1}$ and $f_w'^*$ for each saturation S_w^*.

The average apparent viscosity, $\bar{\lambda}_r^{-1*}$, for the region behind the polymer flood front is 28.66 cp. From Eq. 5.139,

$$\int_0^{x_{D3}} \lambda_r^{-1} dx_D = (28.66)(0.172)$$
$$= 4.93 \; cp.$$

The second integral includes the waterflood region where the saturation changes from S_{w1} to S_{wf}. Numerical integration could be used to compute the value of the integral in the same manner as for the polymer region. In this case, however, there is not a large difference between S_{w1} and S_{wf} at $S_{w1} = 0.4459$, $k_{ro} = 0.3228$, $k_{rw} = 0.0266$, and the value of $\lambda_{rSw1}^{-1} = 28.84$ cp. At $S_{wf} = 0.4206$, $k_{ro} = 0.3903$, $k_{rw} = 0.0182$, and $\lambda_{rSwf}^{-1} = 35.77$ cp. The trapezoidal rate is applied to the interval $x_{D1} \le x_D \le x_{Df}$ to give

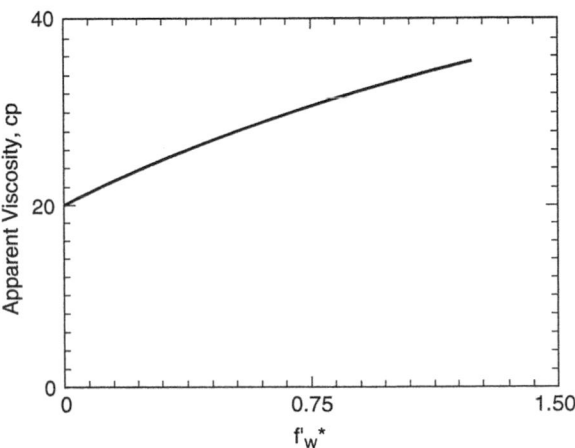

Fig. 5.67—Apparent viscosity vs. fractional flow, polymer region, Example 5.11.

S_w^*	$\overline{\lambda}_r^{*-1}$(cp)	$f_w^{'*}$
0.592478	28.662727	1.237757
0.597854	28.050756	1.128313
0.603230	27.458923	1.025389
0.608606	26.887281	0.928756
0.613982	26.335779	0.838164
0.619358	25.804293	0.753352
0.624735	25.292606	0.674053
0.630111	24.800457	0.599995
0.635487	24.327520	0.530910
0.640863	23.873444	0.466529
0.646239	23.437824	0.406590
0.651615	23.020252	0.350839
0.656991	22.620281	0.299028
0.662367	22.237457	0.250921
0.667743	21.871311	0.206288
0.673119	21.521368	0.164913
0.678495	21.187138	0.126587
0.683871	20.868097	0.091115
0.689247	20.563623	0.058309
0.694623	20.272432	0.027994
0.699999	20.000088	0.000004

Table 5.31—Average apparent viscosity computed for Example 5.11, polymer region.

$$\int_{x_{D1}}^{x_{Df}} \lambda_r^{-1}\, dx_D = \frac{\left(\lambda_{rSw1}^{-1} + \lambda_{rSwf}^{-1}\right)}{2}\left(x_{Df} - x_{D1}\right)$$

$$= \frac{(28.84 + 35.77)}{2}(0.75 - 0.536). \quad\dots\dots\dots\dots\dots\dots\dots\dots\dots\dots\dots\dots\dots\dots\dots\dots\dots (5.140)$$

$$= 6.91 \text{ cp.}$$

Collecting terms yields

$$\overline{\lambda}^{-1} = 4.93 + 28.84(0.536 - 0.172) + 6.91 + 50(1.0 - 0.75)$$
$$= 34.84 \text{ cp.}$$

The pressure drop at $t_D = 0.139$ is computed with Eq. 5.141 when parameters are in oilfield units (psi, cp, darcies, ft, and B/D). Constant 1.127 is a unit conversion factor.

$$q_1 = \frac{1.127 k_b A(p_i - p_l)}{\overline{\lambda}^{-1} L} \quad\dots\dots\dots\dots\dots\dots\dots\dots\dots\dots\dots\dots\dots\dots\dots\dots\dots\dots\dots\dots\dots\dots\dots (5.141)$$

and $$\Delta p = \frac{(200 \text{ B/D})(34.84 \text{ cp})(1{,}000 \text{ ft})}{(1.127)(0.250 \text{ darcies})(10{,}000 \text{ ft}^2)} \quad\dots\dots\dots\dots\dots\dots\dots\dots\dots\dots\dots\dots\dots\dots\dots (5.142)$$

$$= 2{,}473 \text{ psi.}$$

As the polymer flood progresses, the average apparent viscosity changes with the saturation profile. When the polymer flood front breaks through, $\overline{\lambda}^{-1} = 28.66$ cp. If the injection rate remains constant, the pressure drop at this time will be 2,034 psi. Because there is a unique saturation profile at every value of t_D, there is a unique average apparent viscosity. Values of $\overline{\lambda}^{-1}$ computed for the polymer flood are presented in **Table 5.32** and plotted in **Fig. 5.68** as functions of t_D. Pressure drops corresponding to an injection rate of 200 B/D are also included in Table 5.32.

5.5.5 Estimation of Injection Rate With a Constant Pressure Drop. When the pressure is constant during a polymer-augmented waterflood, the injection rate changes as the saturation profiles develop and are displaced through the linear system. Section 5.5.4 showed that the average apparent viscosity is a function of t_D. Thus, at any value of t_D, the injection rate can be predicted at any point in the displacement process when the pressure drop is specified. This computation is illustrated in Example 5.12.

Example 5.12—Estimation of Injection Rate in a Polymer Flood When the Pressure Drop Is Constant. The continuous polymer flood described in Example 5.7 is to be operated at a constant pressure drop of 1,000 psi between injection and production wells. Determine the injection rate (and thus the production rate) at $t_D = 0.479$.

Solution. The flow rate for a linear displacement is computed from Eq. 5.142. Table 5.32 contains values of the average apparent viscosity, $\bar{\lambda}^{-1}$, as a function of t_D. At $t_D = 0.479$, the average apparent viscosity is 28.708 cp. Substitution into Eq. 5.141 yields

$$q_i = \frac{1.127 k_b A\left(p_i - p_p\right)}{\bar{\lambda}^{-1} L},$$

at $t_D = 0.479$ gives

$$q_i = \frac{(1.127)(0.250 \text{ darcies})(10,000 \text{ ft}^2)(1,000 \text{ psi})}{(28.708 \text{ cp})(1,000 \text{ ft})}$$

$$= 98.1 \text{ B/D}.$$

t_D	Average Apparent Viscosity (cp)	Pressure Drop at 200 B/D (psi)
0.000	50.000	3,549
0.021	47.753	3,390
0.062	43.260	3,071
0.103	38.767	2,752
0.144	34.274	2,433
0.185	29.781	2,114*
0.200	29.578	2,100
0.215	29.375	2,085
0.230	29.172	2,071
0.245	28.969	2,056
0.260	28.766	2,042**
0.369	28.737	2,040
0.479	28.708	2,038
0.589	28.679	2,036
0.698	28.651	2,034
0.808	28.622	2,032†
0.886	28.006	1,988
0.975	27.409	1,946
1.077	26.833	1,905
1.193	26.275	1,865
1.327	25.737	1,827
1.484	25.217	1,790
1.667	24.716	1,754
1.884	24.232	1,720
2.143	23.764	1,687
2.459	23.313	1,655
2.850	22.875	1,624

* Waterflood breakthrough.

**Oil-bank breakthrough.

† Polymer breakthrough.

Table 5.32—Average apparent viscosity and pressure drop calculations, Example 5.11, continuous polymer flood at 200 B/D.

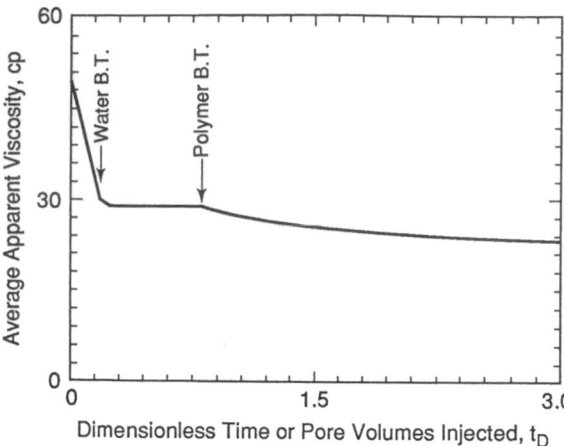

Fig. 5.68—Variation of the average apparent viscosity with PV injected, Example 5.11.

It is possible to calculate the injection rate for every value of t_D in the displacement process with this procedure.

To complete the solution for a linear polymer flood at constant pressure drop, it is necessary to predict rates and cumulative production as functions of time. This task is easy when the injection rate is constant. When the pressure drop is constant, however, injection rates can vary over a wide range. For example, the injection rate would be 72.7 B/D at $t_D = 0.103$ when the pressure drop is constant at 1,000 psi, as in Example 5.12. It is necessary to develop a procedure for determining the time when injection rates vary.

When the pressure drop is constant, the flow rate varies with time. The solution developed in this section gives the flow rate at all values of dimensionless time, t_D, or PVs injected, Q_i. The time can be extracted from pairs of q_t, t_D data by recalling that

$$t_D^{n+1} = \int_0^{t^{n+1}} \frac{q_t dt}{V_p} \qquad (5.143)$$

$$\text{and } t_D^{n} = \int_0^{t^{n}} \frac{q_t dt}{V_p}. \qquad (5.144)$$

Subtracting Eq. 5.143 from Eq. 5.144 gives

$$t_D^{n+1} - t_D^{n} = \int_{t^{n}}^{t^{n+1}} \frac{q_t dt}{V_p}, \qquad (5.145)$$

which becomes

$$t_D^{n+1} - t_D^{n} \approx \left(\frac{q_t^{n+1} + q_t^{n}}{2V_p} \right)(t^{n+1} - t^{n}) \qquad (5.146)$$

when q_t is approximated by $(q_t^{n+1} + q_t^{n})/2$. Thus,

$$t^{n+1} = t^{n} + \frac{2(t_D^{n+1} - t_D^{n})V_p}{q_t^{n+1} + q_t^{n}} \qquad (5.147)$$

when $n = 0$, $t^n = 0$, and $t_D{}^n = 0$. For this case, t^1 is estimated with

$$t^1 = t^0 + \frac{2(t_D^1)V_p}{(q_t^1 + q_t^0)}. \qquad (5.148)$$

The value q_t^0 is the flow rate of the oil and water at the initial water saturation. If the interstitial water is immobile,

$$q_t^0 = \frac{1.127 k_b A (p_i - p_p)}{\lambda_{roS_{wi}} L}. \qquad (5.149)$$

Example 5.13 illustrates computation of the flow-rate/time relationships for the polymer flood in Example 5.7.

Example 5.13—Estimation of Injection-Rate/Time History During a Polymer Flood in a Linear Reservoir at Constant Pressure Drop. Predict the injection rate as a function of time for the polymer flood in Example 5.7 when the pressure drop is maintained constant at 1,000 psi.

Solution. Injection rates are computed at every value of t_D in Table 5.32 for a constant pressure drop of 1,000 psi with the procedure in Example 5.12. **Table 5.33** gives rates for each value of t_D.

Times corresponding to each value of t_D are computed sequentially. Eq. 5.149 is used for the first time increment, and then Eq. 5.148 is used. Computed times also are presented in Table 5.33.

5.5.6 Polymer-Augmented Waterflood in a Layered Reservoir. Reservoirs are usually heterogeneous, particularly with respect to permeability. The largest changes in permeability typically occur in the vertical direction. When there is a wide range of permeabilities in the vertical cross section, the injected water flows through the most-permeable layers where the displacement process occurs rapidly. Water breakthrough from these layers may cause high WORs and abandonment of the waterflood while substantial portions of the low-permeability zones are being waterflooded at a slower rate.

Injecting polymer into a linear system slows the flow of injected fluids in *all* zones contacted by the polymer. The amount of polymer flowing into each zone of a uniform, layered reservoir will be proportional to the permeability of the layer. It is tempting to assume that polymer injection will change the vertical distribution of fluids flowing in the system and thus change the displacement efficiency. To be an effective mobility-control agent, however, the polymer solution must not be slowed as much in low-permeability zones as in high-permeability zones.

t_D	Average Apparent Viscosity (cp)	Injection Rate (B/D)
0.000	50.000	56.35
0.021	47.753	59.00
0.062	43.260	65.13
0.103	38.767	72.68
0.144	34.274	82.21
0.185	29.781	94.61
0.200	29.578	95.26
0.215	29.375	95.91
0.230	29.172	96.58
0.245	28.969	97.26
0.260	28.766	97.95
0.369	28.737	98.04
0.479	28.708	98.14
0.589	28.679	98.24
0.698	28.651	98.34
0.808	28.622	98.44
0.886	28.006	100.60
0.975	27.409	102.8
1.077	26.833	105
1.193	26.275	107.2
1.327	25.737	109.5
1.484	25.217	111.7
1.667	24.716	114
1.884	24.232	116.3
2.143	23.764	118.6
2.459	23.313	120.9
2.850	22.875	123.2

Table 5.33—Injection rate during continuous polymer flood, pressure drop maintained constant at 1,000 psi, Example 5.12.

The impact of polymer flooding on the displacement performance of a linear reservoir with noncommunicating layers can be predicted by several techniques. If a polymer-flood model, such as that used to generate the polymer-flood performance in Examples 5.7 through 5.10, is available, the displacement performance can be simulated easily when the pressure drop is constant. A similar calculation technique can be developed for constant injection rate, but the computations are complex and are not included in this text.

When no crossflow occurs between layers, the performance of a multilayered, linear reservoir flooded under a constant pressure drop can be simulated by computing the displacement performance of each layer separately as a function of time, as described in Section 5.5.5. Then, the predicted performance of the composite reservoir is determined by combining the results from individual layers at the same time. Composite results are computed for predetermined time increments by linear interpolation. When crossflow between layers occurs, a numerical model must be used to simulate polymer flood performance (Gao et al. 1993; Sorbie 1991b). Consideration of crossflow in polymer flooding is beyond the scope of this text.

5.5.7 Polymer-Augmented Waterflood in Patterns or Other Arrangements of Production and Injection Wells.
Estimation of polymer flood performance in pattern floods or other flooding programs where the flow is not linear requires the use of more-complex models or reservoir simulators. An adaptation of a black-oil reservoir simulator is available in the public domain (Ray and Diaz Munoz 1986). Most reservoir software companies have polymer flood simulators of various degrees of complexity. Application of reservoir simulators is beyond the scope of this text.

Polymer-augmented waterflood performance in pattern floods can be simulated with streamtube modeling with the approach originally introduced for waterflooding by Higgins and Leighton (Willhite 1986; Higgins and Leighton 1962; Leighton and Higgins 1975). In these models, the flooding pattern is represented by a collection of streamtubes connecting the injection and production wells, as depicted in **Fig. 5.69.** Fluid flow is 1D in each stream-tube along the principal axis

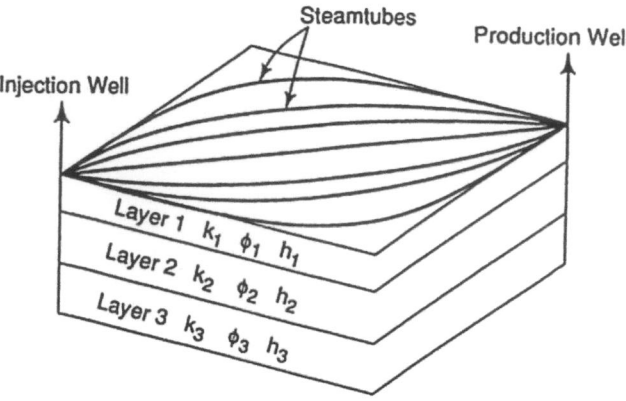

Fig. 5.69—Streamtube model for simulation of displacement in one-quarter of a layered five-spot pattern (Ray and Diaz Munoz 1986).

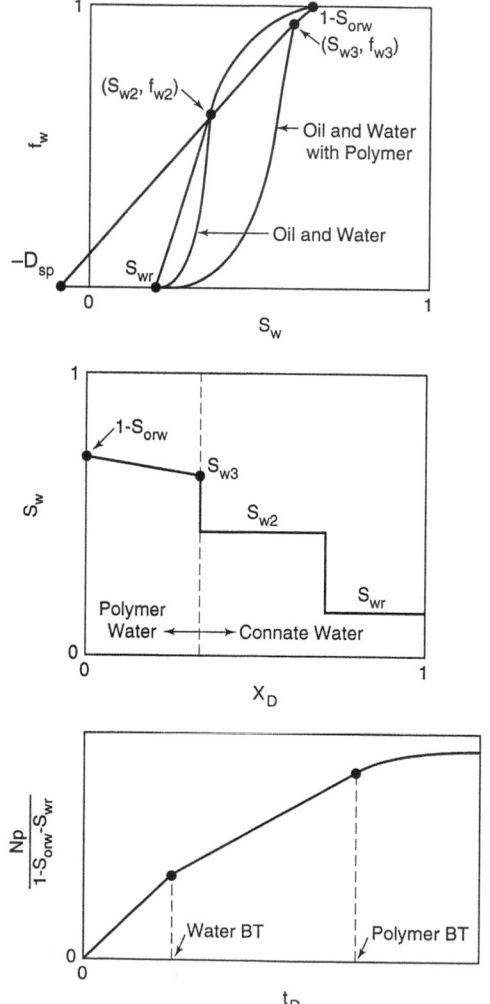

Fig. 5.70—Fractional-flow, saturation profile, and cumulative production relationships for the DOE streamtube model (Ray and Diaz Munoz 1986).

of the streamtube. There is no crossflow across streamtube boundaries. The polymer flooding models developed in Chapter 3 and in this chapter also are valid for each streamtube.

In principle, these models simulate the polymer-augmented waterflood performance in each streamtube at constant pressure drop. **Fig. 5.70** illustrates the fractional-flow, saturation, and production profiles used in each streamtube. Performance of the pattern is determined by combining the displacement performance of each stream-tube at the same point in time. A model based on streamtube concepts, developed by the US Department of Energy (DOE), is available to the public and documented in a report (Ray and Diaz Munoz 1986).

Example 5.14 illustrates the use of the DOE model to predict the performance of a polymer-augmented waterflood in a five-spot pattern.

Example 5.14—Simulation of a Polymer-Augmented Waterflood in a Five-Spot Pattern With a Streamtube Model. A polymer-augmented waterflood is to be conducted in a reservoir on 20-acre spacing. The reservoir is located at a depth of 2,500 ft. OOIP in the 20-acre pattern was 1,475,800 STB. Oil recovery from primary depletion and the current waterflood is 421,700 STB, leaving an estimated 1,054,100 STB of oil in place at the beginning of the polymer flood. Average oil saturation at this point is 0.5. The ROS is 0.25, so the mobile oil left in the reservoir is approximately 527,250 STB.

Tables 5.34 and 5.35 summarize the formation, rock, and fluid properties.

The reservoir has a Dykstra-Parsons coefficient of 0.6 and is assumed to be represented by five noncommunicating layers with properties given in Table 5.35. Relative permeability curves for oil/water and oil/polymer systems are represented by

$$k_{ro} = 0.8\left(1 - S_{wD}\right)^{2.0} \dots \dots \dots \dots \dots \dots (5.150)$$

$$\text{and } k_{rw} = 0.2 S_{wD}^{2.0}, \dots \dots \dots \dots \dots \dots \dots (5.151)$$

where S_{wD} is defined by

$$S_{wD} = \frac{S_w - S_{iw}}{1 - S_{or} - S_{iw}}. \dots \dots \dots \dots \dots \dots (5.152)$$

The plan is to inject 0.3 PV of polymer at a concentration of 900 ppm followed by a maximum of 3 PV of drive water. The effective viscosity of the polymer solution is 10.9 cp.

Solution. **Table 5.36** summarizes the predicted polymer-augmented waterflood performance. A summary of injection history is presented in **Table 5.37.** In the simulation, 218,420 lbm of polymer is injected in the first 4.7 years. Oil displaced at a WOR of 115.9 was 430,900 STB for an oil/polymer ratio of 1.97 STB/lbm. This example illustrates the capabilities of this polymer flood predictive model. Results from such a prediction form the basis for economic analysis of the polymer flood.

For example, **Fig. 5.71** illustrates the effect of slug size on cumulative oil recovery for a polymer-augmented waterflood in a five-spot pattern with a single layer. An economic analysis could be made to determine the optimum slug size on the basis of simulated displacement results. Further information on the model and assumptions are found in Ray and Diaz Munoz (1986).

Formation Properties and Pattern Volumes

Formation depth (ft)	2,500.0
Formation temperature (°F)	125.0
Formation average permeability (md)	200.0
Formation porosity	0.3
Total net thickness (ft)	50.0
Total PV (10^3 bbl)	2,327.5
OOIP (10^3 STB)	1,475.8
OIP at start of flood (10^3 STB)	1,054.1
Dykstra-Parsons coefficient	0.60
Pattern area (acres)	20.00
Wellbore radius (ft)	0.50
Injectivity coefficient (psi/ft)	0.51
Water/oil mobility ratio	2.08
Polymer/oil mobility ratio	0.17
S_{oi}	0.70
S_{or}	0.25

Fluid Properties

Oil FVF (RB/STB)	1.104
Water FVF (RB/STB)	1.008
API oil gravity (°API)	25.0
Oil viscosity (cp)	5.000
Water viscosity (cp)	0.600
Formation pressure (psi)	1,000.0
Solution gas/oil ratio (scf/STB)	175.0
Specific gravity of gas	0.700
Polymer concentration (ppm)	900.0
Polymer adsorption (lbm/acre-ft)	100.000
Polymer adsorption parameter (vol slug/PV)	0.041
Polymer viscosity (cp)	10.900
Resistance factor	12.00
Residual resistance factor	1.101
Power-law coefficient (cp-sec^{n-1})	4.774
Power-law exponent	0.600
Polymer slug size (PV)	0.300
Maximum injected PV, polymer + chase water (PV)	3.000

Table 5.34—Input data for simulation of polymer-augmented waterflood (Ray and Diaz Munoz 1986), Example 5.14.

Layer	φ	k (md)	h (ft)	S_{wi}
1	0.30	529.76	10.00	0.50
2	0.30	216.56	10.00	0.50
3	0.30	132.61	10.00	0.50
4	0.30	81.69	10.00	0.50
5	0.30	39.38	10.00	0.50

Table 5.35—Layer properties for simulation of polymer-augmented waterflood (Ray and Diaz Munoz 1986), Example 5.14.

Pattern Production Summary

Pattern life ratio	3.20
Pattern area (acres)	20.00
Pattern PV (10^3 bbl)	2,327.5
OOIP (10^3 STB)	1,475.8
OIP at start of flood (10^3 STB)	1,054.1
Total oil production (10^3 STB)	430.9
Oil recovery (fraction OOIP)	0.2920
Oil recovery (fraction OIP at start of flood)	0.4088
Total polymer injected (10^3 bbl)	218.4
Oil/polymer ratio (STB/lbm)	1.97

Time (years)	Oil Rate (STB/D)	Gas Rate (Mscf/D)	Water Rate (STB/D)	Cumulative Oil (10^3 STB)	Cumulative Gas (MMscf)	Cumulative Water (10^3 STB)	Oil Recovery (fraction oil at start)	WOR
0.00	0.0	0.0	0.0	0.0	0.00	0.00	0.0000	0.00
0.40	309.4	54.1	451.8	45.18	7.91	65.97	0.0429	1.46
1.17	147.4	25.8	247.8	86.76	15.18	135.88	0.0823	1.68
1.98	128.1	22.4	250.2	124.63	21.81	209.86	0.1182	1.95
2.83	112.5	19.7	252.0	159.25	27.87	287.39	0.1511	2.24
3.70	90.6	15.9	261.7	188.22	32.94	371.11	0.1786	2.89
4.70	70.6	12.4	239.7	213.94	37.44	458.39	0.2030	3.39
5.77	64.0	11.2	226.3	238.86	41.80	546.55	0.2266	3.54
6.44	109.5	19.2	352.0	265.64	46.49	632.67	0.2520	3.22
7.05	108.0	18.9	396.3	289.88	50.73	721.58	0.2750	3.67
7.73	92.2	16.1	362.9	312.83	54.75	811.89	0.2968	3.93
8.46	77.3	13.5	348.3	333.46	58.35	904.75	0.3163	4.50
9.22	72.1	12.6	337.9	353.42	61.85	998.34	0.3353	4.69
9.98	71.4	12.5	342.0	373.03	65.28	1,092.31	0.3539	4.79
10.76	56.1	9.8	343.9	389.02	68.08	1,190.25	0.3690	6.13
11.30	39.3	6.9	539.6	396.80	69.44	1,297.18	0.3764	13.74
11.72	38.2	6.7	701.7	402.74	70.48	1,406.13	0.3821	18.37
12.17	31.0	5.4	684.3	407.72	71.35	1,516.13	0.3868	22.09
12.60	28.5	5.0	690.4	412.28	72.15	1,626.58	0.3911	24.20
12.95	21.1	3.7	879.8	414.98	72.62	1,739.08	0.3937	41.76
13.30	17.3	3.0	882.4	417.19	73.01	1,852.11	0.3958	50.98
13.66	17.2	3.0	878.5	419.41	73.40	1,965.13	0.3979	51.02
14.01	17.1	3.0	873.0	421.62	73.78	2,078.16	0.4000	51.01
14.37	17.0	3.0	866.5	423.84	74.17	2,191.18	0.4021	50.87
14.73	17.0	3.0	869.5	426.06	74.56	2,304.20	0.4042	51.03
15.08	17.0	3.0	879.0	428.25	74.94	2,417.25	0.4063	51.58
15.35	16.6	2.9	1,127.1	429.92	75.24	2,530.87	0.4078	67.97
15.63	9.7	1.7	1,130.2	430.91	75.41	2,645.24	0.4088	115.94

Table 5.36—Pattern production report, Example 5.14.

Time (years)	Fraction (PV)	Water Rate (STB/D)	Polymer Rate (lbm/D)	Injection (B/D-psi)	Cumulative Water (10³ STB)	Cumulative Polymer (10³ lbm)
0.00	0.000	0.0	0.0	0.000	0.00	0.00
0.40	0.050	409.3	129.0	0.319	115.45	36.40
1.17	0.100	390.4	123.1	0.304	230.90	72.81
1.98	0.150	375.2	118.3	0.293	346.35	109.21
2.83	0.200	361.0	113.8	0.281	461.81	145.61
3.70	0.250	317.1	100.0	0.247	577.26	182.02
4.70	0.300	296.4	93.4	0.231	692.71	218.42
5.77	0.350	471.9	0.0	0.368	808.16	218.42
6.44	0.400	514.6	0.0	0.401	923.61	218.42
7.05	0.450	463.9	0.0	0.362	1,039.06	218.42
7.73	0.500	433.0	0.0	0.338	1,154.52	218.42
8.46	0.550	416.9	0.0	0.325	1,269.97	218.42
9.22	0.600	420.1	0.0	0.328	1,385.42	218.42
9.98	0.650	405.4	0.0	0.316	1,500.87	218.42
10.76	0.700	582.6	0.0	0.454	1,616.32	218.42
11.30	0.750	743.5	0.0	0.580	1,731.77	218.42
11.72	0.800	718.2	0.0	0.560	1,847.23	218.42
12.17	0.850	721.6	0.0	0.563	1,962.68	218.42
12.60	0.900	902.8	0.0	0.704	2,078.13	218.42
12.95	0.950	901.3	0.0	0.703	2,193.58	218.42
13.30	1.000	897.3	0.0	0.700	2,309.03	218.42
13.66	1.050	891.8	0.0	0.695	2,424.48	218.42
14.01	1.100	885.2	0.0	0.690	2,539.93	218.42
14.37	1.150	888.0	0.0	0.692	2,655.39	218.42
14.73	1.200	897.6	0.0	0.700	2,770.84	218.42
15.08	1.250	1,145.20	0.0	0.893	2,886.29	218.42
15.35	1.300	1,140.90	0.0	0.890	3,001.74	218.42
15.63	1.350	1,207.30	0.0	0.941	3,117.19	218.42

Table 5.37—Pattern injection report, Example 5.14.

5.6 In-Situ Permeability Modification

The amount of oil that is recoverable from a reservoir by a displacement process depends on (1) the effectiveness with which the injected fluid displaces oil from the pores in the rock (microscopic displacement efficiency) and (2) the volumetric fraction of the reservoir contacted by the injected fluid (macroscopic sweep efficiency). This latter efficiency is governed by the mobility ratio but also in large measure by the geologic heterogeneity of the reservoir rock. Permeabilities vary both areally and vertically, and large changes typically occur in the vertical direction in a single well. As an example, **Fig. 5.72** shows permeability variation with depth for a shallow sandstone reservoir in eastern Kansas (Willhite and Jordan 1981).

When a displacement process is implemented in a reservoir with large variations in vertical permeability, injected fluid tends to flow through the zones with the highest permeabilities; thus, low-permeability zones may receive only a small fraction of the injected fluid. Example 4.4 gave an illustration of the effect of a variation in vertical permeability for a linear reservoir consisting of two layers and for displacement processes conducted under different mobility ratios.

Bypassing part of the reservoir by the injected fluid can lead to production of relatively large volumes of injected fluid per barrel of recovered oil. The result can be that the displacement process reaches the economic limit at a time when a large volume of oil remains in the bypassed or unswept regions of the reservoir.

The objective of an in-situ permeability-modification process is to treat the reservoir in such a way that the effective permeability of the high-permeability zones is significantly reduced. Conceptually, it would be desirable to reduce permeability across the entire reservoir (i.e., from injection wells to production wells). Practically, treatment tends to be limited to the

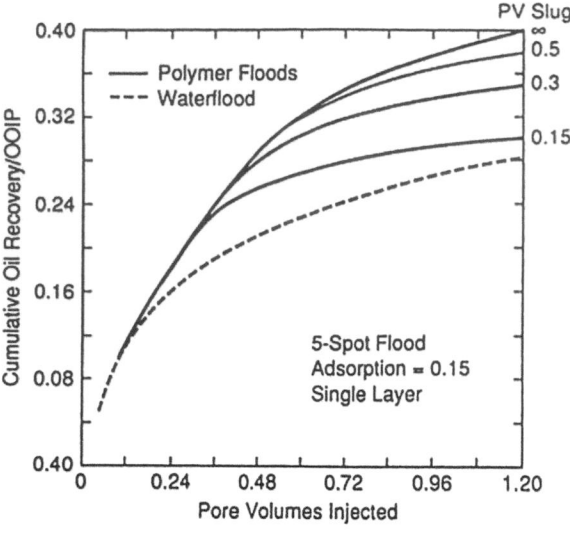

Fig. 5.71—Effect of polymer slug size on oil recovery in a five-spot pattern with a single layer (Ray and Diaz Munoz 1986).

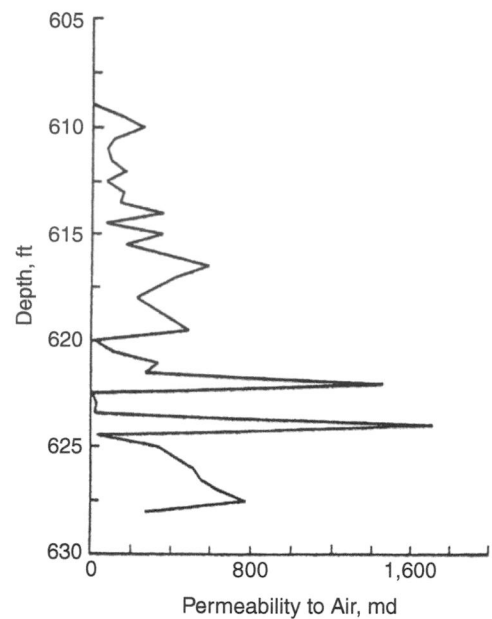

Fig. 5.72—Example of vertical variation in permeability, eastern Kansas reservoir (Willhite and Jordan 1981).

region relatively near the wellbore. Also, provisions must be made to prevent the gelling agents from entering and damaging the lower-permeability zones.

Various methods of permeability modification have been applied. These include crosslinked polymer (gelled polymer), microbial-based processes, and precipitation processes. This chapter describes only the application of crosslinked polymers. In-situ permeability modification is not strictly a mobility-control process in that volumetric sweep efficiency is improved through modification of rock permeability rather than through mobility adjustment of an injected fluid. The processes are related, however, because of the use of polymer systems that are similar and because the result (i.e., improved volumetric sweep) is similar. Thus, permeability modification is described in this chapter.

5.6.1 General Description of the Process. In the crosslinked, or gelled, polymer process, an aqueous solution or solutions of moderate viscosity containing polymer and a crosslinking agent are injected into high-permeability or fractured zones. Polymer and the crosslinker react to form a viscous gel (i.e., a fluid with a very high apparent viscosity). The gel is essentially immobile and thus acts to reduce the apparent permeability of the rock matrix or fracture.

Depending on the chemical system, the procedure used to mix the reacting chemicals is different. In some applications,

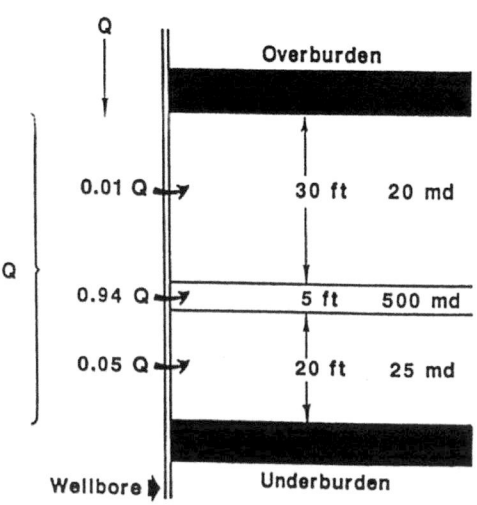

Fig. 5.73—Flow distribution during waterflood of a three-layer five-spot pattern with no crossflow between layers (Abdo et al. 1984).

the reacting chemicals are mixed at the surface by passing the different solutions through an in-line mixer before they enter the wellbore. Reaction to form a gel occurs in the reservoir. Another approach, often used with biopolymers, is to mix the solutions in tanks at the surface before injection. Reaction starts at the surface and continues after injection into the reservoir. Yet another procedure is to inject the different solutions as slugs. In this case, mixing occurs in the reservoir as a result of dispersion and chromatographic transport of the different chemical species.

The general problem addressed by the permeability-modification process and the potential effect on production are illustrated in **Figs. 5.73 through 5.75.** Waterflooding a reservoir consisting of three noncommunicating layers is considered (Abdo et al. 1984). A computer model was used to simulate a waterflood in a 10-acre five-spot pattern. After 500 days of waterflooding at an injection rate of 2,000 B/D, the relative injection rates were as shown in Fig. 5.73. The high-permeability streak took approximately 94% of the injected water, while it comprised only 9% of the reservoir net pay. A very small relative oil recovery was obtained from the two lower-permeability zones after 500 days, and the WOR was large.

Permeability modification was then implemented to reduce the permeability in the 500-md layer by a factor of 10 out to a radius of 75 ft around the injection well. Five hundred days of additional waterflooding at the same rate following the permeability-modification treatment yielded

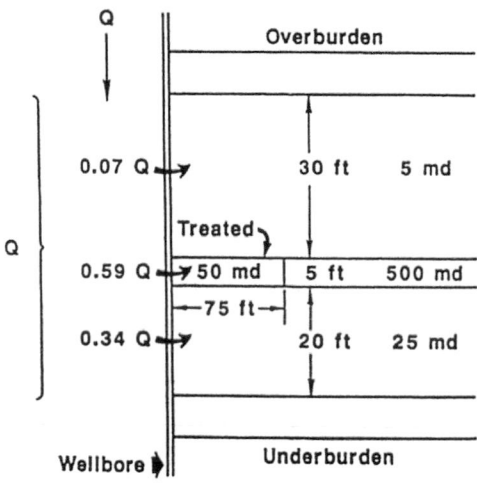

Fig. 5.74—Effect of permeability-modification treatment of 500-md zone on the flow distribution during a waterflood of a three-layer five-spot pattern (Abdo et al. 1984).

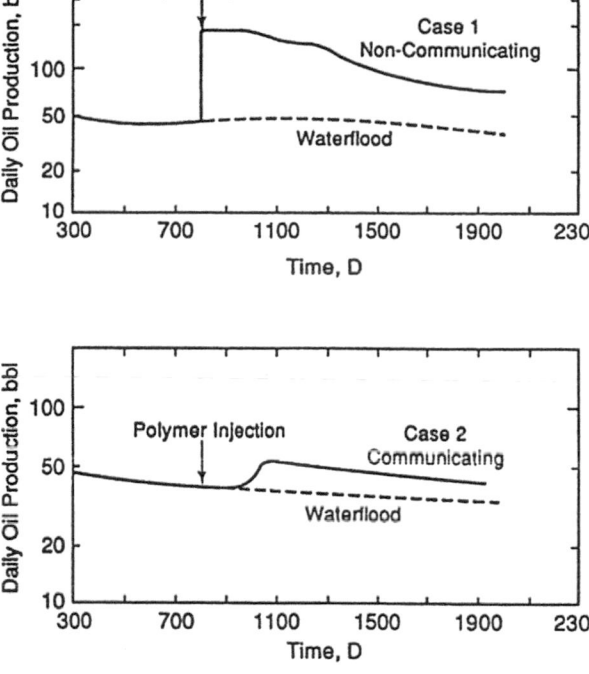

Fig. 5.75—Comparison of oil production response from permeability modification of 500-md zone during a waterflood of a three-layer five-spot pattern (Abdo et al. 1984).

the calculated results shown in Figs. 5.74 and 5.75. The injectivity profile was modified significantly, with much improved oil recoveries occurring in the low-permeability zones.

Abdo et al. (1984) also investigated the effect of interlayer communication on the permeability-modification treatment. Results for a calculation carried out for the same reservoir as previously described, but assuming that the ratio of vertical to horizontal permeability is 0.10 for all layers, are shown in Fig. 5.75. Crossflow reduced the effectiveness of the post-treatment waterflood compared with the case with no interlayer communication. Both the time response of the oil production and the amount of the production were adversely affected by interlayer flow. Still, the treatment did result in increased oil recovery and reduced WOR compared with the case with no treatment.

Permeability modification also is applicable to fractured systems and to reservoirs with areal heterogeneities, although the latter case is more difficult to address. Basically, the process requires application of suitable chemical systems and techniques to place these systems selectively in the reservoir rock. Calculations show that some gel solution will flow into the low-permeability zone unless provision is made to isolate the different zones mechanically during treatment (Todd et al. 1991). This must be considered in the design. Calculations of the type described also indicate the desirability of conducting permeability modifications that extend a considerable distance into the reservoir from the treated wells.

Permeability modification has been applied most commonly to injection wells but also is applicable to production wells. Production well applications typically involve use of a gelled polymer to shut off or to reduce water production where, for example, water coning is occurring. Production well treatments are not considered in this book, but descriptions are available in the literature (Hessert and Fleming 1979; Sloat 1979).

5.6.2 Crosslinked Polymer Chemical Systems. A number of in-situ permeability-modification processes are based on the gelation of such high-molecular-weight polymers as partially hydrolyzed polyacrylamides and polysaccharides. In these processes, the polymer molecules are bound together into a gel structure through the use of a crosslinking chemical agent, such as a trivalent cation [Cr(III) or Al(III)] or an organic crosslinker. The gel structure may be formed at the surface and injected subsequent to gelation, or the system may be designed so that the gel structure does not form until the chemicals are placed in the reservoir.

Polyacrylamides and partially hydrolyzed polyacrylamides form gels in the presence of Cr(III) and other trivalent cations (Hessert and Fleming 1979; Clampitt and Hessert 1974; Schoeling et al. 1989). In one process, polyacrylamide at a concentration of a few thousand ppm is mixed with sodium dichromate at a concentration of a few hundred ppm and a reducing agent, such as sodium bisulfite, sodium thiosulfate, or thiourea. Cr(VI) in the dichromate is reduced to Cr(III), which subsequently reacts with the polymer to form crosslinks and a 3D gel structure. The chemical process is summarized by

$$\text{reducing agent} + M^{+6} \rightarrow M^{+3}, \dots\dots\dots\dots\dots\dots\dots\dots\dots\dots\dots (5.153)$$

$$\text{polymer} + M^{+3} \rightarrow M^{+3} - \text{polymer}, \dots\dots\dots\dots\dots\dots\dots\dots\dots\dots\dots\dots (5.154)$$

$$\text{and polymer+polymer} - M^{+3} \rightarrow \text{polymer} - M^{+3} - \text{polymer}, \dots\dots\dots\dots\dots\dots\dots\dots\dots\dots\dots\dots\dots\dots\dots \text{(5.155)}$$

where M represents the metal ion. Eqs. 5.153 through 5.155 represent a simplified model of the gelation process. In fact, the kinetics of the process are complex and only partly understood. Gel formation may take minutes to a few months, depending on reducing-agent type, component compositions, and other system variables, such as temperature. **Table 5.38** shows typical ranges of component concentrations and conditions used with polyacrylamide/Cr(III) gels. In the implementation of a treatment of this gel type, system components are mixed at the surface, usually with an in-line mixer. The gel solution is then pumped into the reservoir, where the reaction proceeds until a gel is formed. Gel slug size must be designed to be compatible with the gelation time at reservoir conditions. For this process, if the chemical system were mixed and allowed to gel at the surface before it was pumped, the shear imposed during injection would destroy the gel. Premature gelation may cause plugging of the wellbore.

In another process, which uses Cr(III) as the crosslinker, the polymer is crosslinked with a Cr(III) carboxylate-complex crosslinking agent. Cr(III) can be introduced directly as chromium acetate (Sydansk 1990). The complexation of the Cr(III) prevents chromium precipitation at high pH, and the gelation rate can be controlled to give gel times from minutes to days to weeks. The other advantages of this chemical system are that the process is less sensitive to pH (gelation will occur at high pH); the gelation is not adversely affected by H_2S; and the use of dichromate, which is classified as a hazardous chemical, is not necessary. This process also is reported to produce gels at temperatures exceeding 200°F (Sydansk 1990).

Biopolymers (polysaccharides) also react with Cr(III) to form gels (Abdo et al. 1984). Until the gel structure is well-developed, these gels can be sheared and then pumped into porous rock. When subjected to moderate shear, the gel solution will reheal when the shear is removed to form a gel. Thus, these systems are often mixed and held for a few hours to gel in surface tanks before they are pumped into the porous rock. It is assumed that the gel solution will reheal after flow is stopped. The solution will not reheal if the gel solution is subjected to prolonged shear beyond the nominal beaker gel time (Dolan 1989). Alternatively, the chemical system can be mixed at the surface and designed to gel in situ. Because biopolymer/Cr(III) gels are not necessarily designed for in-situ gelation, the Cr(III) may be added directly to the system in the form of chromic chloride. The use of an oxidation-reduction reaction to convert Cr(VI) to Cr(III) and thereby to control gelation time is not necessary. Low Cr(III) concentrations, on the order of 25 ppm, are suitable for gelation of biopolymers at the polymer concentrations in the range of 2,000 to 6,000 ppm shown in Table 5.38.

Al(III) also can be used as a crosslinking agent for polymers and is typically applied in a slug-type process referred to as the aluminum-citrate process (Bruning et al. 1983). This process has been described as one in which a gel layer is built along the walls of the porous medium; however, there is significant uncertainty as to how the process actually works (Parmeswar and Willhite 1988).

The process is implemented as a slug process, often involving two different types of polyacrylamide polymer. In a typical application, a cationic polyacrylamide is first injected at a concentration of a few hundred ppm. The polymer is designed to adsorb onto the porous medium. Next, a larger slug of an anionic polyacrylamide is injected, and this polymer undergoes additional adsorption/entrapment in the porous medium. A slug of aluminum citrate is then injected; the citrate is used to keep aluminum in solution. Al(III) reacts with the retained polymer to form the first step of the crosslinking process. Additional anionic polyacrylamide is injected next and crosslinks through the Al(III) with the polymer retained earlier. In this manner, a gel is formed in situ. The aluminum citrate process is purported to achieve greater in-depth treatment around a well than the Cr(III)/polymer processes. This is accomplished through the systematic development of a gel in situ as opposed to the injection of bulk gel solution (Thomas 1976). The development of in-depth treatment has been questioned, however (Parmeswar and Willhite 1988).

Other polymer-crosslinking systems also have been used. One system that has shown promise is described by Mumallah (1988). Other polymers, lignosulfonate and phenoformaldehyde, for example, have been used. In a slightly different application, a polymer is generated in situ by causing a polymerization reaction to occur within the reservoir rock (Woods et al. 1986; Hess et al. 1971).

5.6.3 Gelation Time and Gel Strength. The time required for gelation to occur is an important design variable for the Cr(III)/polyacrylamide system. With this system, chemicals are mixed at the surface and injected as a viscous solution. The solution is designed to react and gel at a designated time after being placed in the formation. Gel time is an ill-defined variable because gelation is a continuous kinetic process and there is no clearly defined point at which one can state that a gel has been formed. The literature is replete with techniques for the measurement of gel time, and many provide a consistent measurement within the technique, but results do not necessarily correlate with other methods.

Polymer (ppm)	2,000 to 6,000
Sodium dichromate (source of Cr) (ppm)	500 to 1,000
Reducing agent (ppm)	500 to 1,000
Initial pH	4 to 6
Brine salinity (ppm)	0 to 100,000
Temperature (°F)	<160

Table 5.38—Typical ranges of component concentrations and conditions for the Cr^{+3}/polyacrylamide gel systems.

One method is based on the measurement of viscosity (Terry et al. 1981; Purkaple and Summers 1988). **Fig. 5.76** (Jordan et al. 1982) shows an example measurement with a Brookfield viscometer. In this procedure, the gel solution is mixed in a thermally jacketed beaker and then the viscometer is turned on with a specified spindle at a set shear rate. Apparent viscosity is measured as a function of time and typically yields a curve like that shown in Fig. 5.76. There is no apparent increase in viscosity for a period of time, and then a period of rapid increase occurs. Gel time may be arbitrarily defined as the time at which viscosity starts to increase, the time at which the extensions of the two approximately straight lines intersect, or the time at which the apparent viscosity reaches a specified value. This method has been found to give reproducible results for gel times of up to 10 days or more (Terry et al. 1981).

Another approach is to use a sophisticated viscometer, a rheogoniometer, to monitor a parameter referred to as G', the storage modulus (Sydansk 1990). In this method, properties are measured under an imposed oscillatory shear. Alternatively, for screening purposes, simple bottle tests can be used in which the gel solution behavior is observed when a bottle holding a gel is tilted in a specified way or inverted (Sydansk 1990).

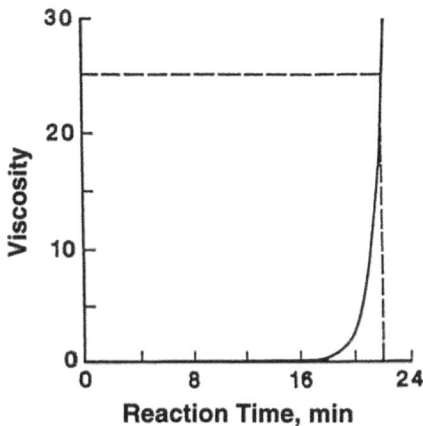

Fig. 5.76—Viscosity of a polyacrylamide gel during gelation (Jordan et al. 1982).

Gel strength also is a variable that is related to process performance, but it too is not defined precisely. Methods of measurement based on the use of shear viscosity, dynamic viscosity, a penetrometer device, gel breakdown pressure, or visually observed flow characteristics have been proposed. These give a relative empirical indication of strength. **Table 5.39** (Sydansk 1990) gives a typical qualitative measurement based on visual observation.

Phenomena that relate to gel strength and that are relatively easy to measure are syneresis and swelling (Gales et al. 1988). Swelling occurs when gel volume increases when the gel is brought into contact with excess solvent. Syneresis is a reduction in volume (gel shrinkage) resulting from a separation of solvent from the gel. Syneresis appears as a breaking up of the gel. Gels of various types can be formulated that are stable in beakers at controlled conditions for long periods. The long-term stability in reservoir environments, however, has not been widely investigated.

A number of factors affect gel time and strength. **Tables 5.40 and 5.41** list several of these effects. These parameters affect all the different gel systems described but not necessarily in the same direction or to the same degree. Gel time generally decreases (gelation rate increases) with increasing polymer concentration, degree of hydrolysis, and molecular weight. An increase in crosslinker concentration also accelerates gelation, and for the redox process, an increase in reducing agent concentration and strength accelerates the rate.

Gel time is shorter at high temperatures. For example, for the redox-reaction-based process gel time is dependent on temperature in the manner described by the Arrhenius equation used in chemical kinetics (Jordan et al. 1982). Gel time decreases with increasing temperature, and a plot of the logarithm of gel time vs. $1/T$ yields a straight line. The gel time, t_g, decreases by approximately a factor of two for each 18°F change in temperature, T. This relationship has been found to hold when gel time is controlled by the rate of generation of Cr(III) through an oxidation-reduction reaction.

System pH also affects gelation time, as shown in **Fig. 5.77** for a Cr(III)/polyacrylamide gel system in which thiourea was used as the reducing agent (Huang et al. 1986). Gel time increased significantly as initial pH was increased above a value of approximately 4. In this case, gels did not readily form outside the pH range of 3 to 5.5. The effect in Huang's data results from the effect of H^+ on the rate of the oxidation-reduction reaction. Just the opposite effect has been found in systems in which Cr(III) was added directly to the system (i.e., gelation rate increases with increasing pH) (Hunt et al. 1989).

A = No detectable gel formed. Gel appears to have the same viscosity (fluidity) as the original polymer solution, and no gel is visually detectable.

B = Highly flowing gel. Gel appears to be only slightly more viscous than the initial polymer solution.

C = Flowing gel. Most of the obviously detectable gel flows to the bottle cap upon inversion.

D = Moderately flowing gel. A small portion (5 to 15%) of the gel does not readily flow to the bottle cap upon inversion—usually characterized as a "tonguing" gel (i.e., after hanging out of the bottle, gel can be made to flow back into the bottle by slowly turning the bottle upright).

E = Barely flowing gel. Gel slowly flows to the bottle cap and/or a significant portion (>15%) of the gel does not flow upon inversion.

F = Highly deformable nonflowing gel. The gel does not flow to the bottle cap upon inversion (gel flows to just short of reaching the bottle cap).

G = Moderately deformable nonflowing gel. The gel flows about halfway down the bottle upon inversion.

H = Slightly deformable nonflowing gel. Only the gel surface deforms slightly upon inversion.

I = Rigid gel. There is no gel-surface deformation upon inversion.

J = Ringing rigid gel. A tuning-fork-like mechanical vibration can be felt after the bottle is tapped.

Table 5.39—Bottle-test gel strength codes (Sydansk 1990).

Polymer concentration and type

Crosslinker concentration

Reducing agent strength and concentration

Temperature

Salinity

pH

Contaminants

  Fe^{+2}

  Fe^{+3}

  H_2S

  Microorganisms

  Divalent ions, Ca^{++}/Mg^{++}

Table 5.40—Factors affecting gel time and strength.

Partially hydrolyzed polyacrylamide/Cr(III) carboxylate gels are less sensitive to pH than gels made by reducing Cr(VI) to Cr(III) (Sydansk 1990). Stable gels formed over a pH range of 4.0 to 12.5. Gelation rate increases slightly with pH for the Cr(III) carboxylate gels up to a pH of 10.5.

Salinity also affects gelation rate. Sydansk (1990) reported that the rate increased slightly for salinities between 0 and 1,000 ppm and was relatively insensitive to salinities between 1,000 and 30,000 ppm. Polyacrylamide gels produced with the Cr(VI) reduction have been reported to gel faster with increasing salinity at low salt concentrations but to gel slower with increasing salinity at high salt concentrations (Jordan et al. 1982).

The contaminants listed in Table 5.40 affect the gelation rate. H_2S, which is often present in oilfield brines, is a strong reducing agent and will accelerate the reduction of Cr(VI) to Cr(III). H_2S is not a severe problem for the Cr(III) carboxylate gels or for gels made by crosslinking biopolymers with Cr(III). Iron and divalent cations (Ca^{++} and Mg^{++}) can interfere in the process because of ionic effects and promotion of crosslinking. Biopolymers, and to a lesser degree polyacrylamide, are susceptible to destruction by microorganisms.

Gel strength, swelling, and syneresis are dependent on the same parameters that affect gel time. Xanthan/Cr(III) gels have been shown to synerese at low pH (< approximately 4.0) and swell at higher pH (Gales et al. 1988). Both xanthan and polyacrylamide gels have been shown to synerese rather rapidly at 194°F (Nagra et al. 1986). The tendency of xanthan/Cr(III) gels to synerese with increasing temperature has been shown in other studies (Gales et al. 1988). However, polyacrylamide/Cr(III) carboxylate gels have been reported to be stable for long periods at temperatures as high as 255°F. Lignosulfonate and phenoformaldehyde gels have also been reported to be stable at high temperatures (Nagra et al. 1986). Gel strength increases with an increase in concentrations of polymer and crosslinker. However, at high concentration ratios of crosslinking agent to polymer, syneresis is promoted. In this case, the gel apparently is "overcrosslinked" as the reaction continues and solvent is forced out of the gel. This is illustrated in **Fig. 5.78** for a xanthan/Cr(III) gel where degree of swelling is plotted as a function of Cr(III)/polymer ratio. A negative degree of swelling indicates syneresis (Gales et al. 1988).

Gel Type	pH	T	Polymer Concentration	Hydrolysis (%)	Polymer M_w	Metal Ion Concentration	Salinity
Partially hydrolyzed** polyacrylamide/ Cr(III), Cr(VI) → Cr(III) reduction	Decrease	Increase*	Increase	Increase	Increase	Increase	Increase at low concentration, decrease at high concentration
Partially hydrolyzed polyacrylamide/ Cr(III), carboxylate	Increase	Increase	Increase	Increase	Increase	Increase	Increase at low concentration, little effect at high concentration
Polysaccharide/Cr(III)†	Increase	Increase	Increase	—	Increase	Increase	

* Increase indicates the gelation rate increases with an increasing value of the system parameters.

** Susceptible to H_2S.

† Susceptible to biodegradation.

Table 5.41—Effect of system parameters on gelation rate.

Salinity also affects gel stability, and gels can synerese at very low or very high salt concentrations. Gels prepared in fresh water have been observed to synerese markedly when exposed to a brine (Nagra et al. 1986). However, the stability of gels prepared in brine has been noted to be relatively insensitive to change in salinity or to divalent-ion concentration (Sydansk 1990; Gales et al. 1988).

5.6.4 Behavior of Gel Systems in Porous Media.
The crosslinked polymer systems mixed at the surface are injected into a formation, where reaction occurs to form a gel. Ideally, once formed, the gel has sufficient strength or viscosity to be immobile. The effective permeability of the rock matrix or fractures in which the gel resides is reduced or, in some cases, essentially eliminated. For processes based on injection of slugs in which the different reactants reside, mixing followed by reaction occurs within the rock. The objective here also is to form a gel that is not mobile.

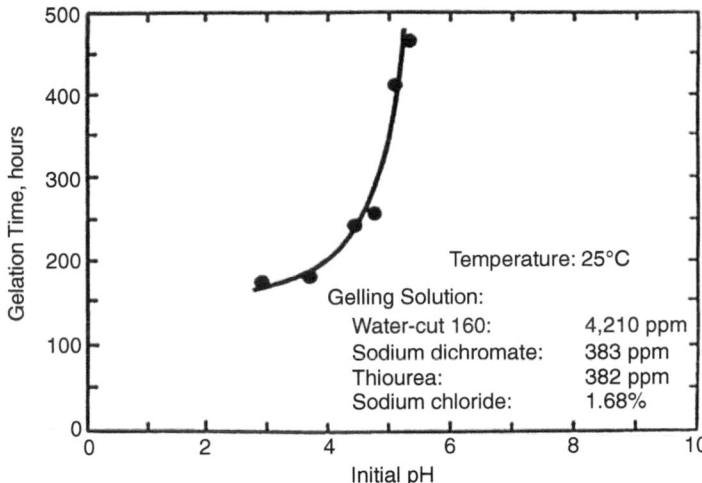

Fig. 5.77—Effect of pH on gelation time for a redox system (Huang et al. 1986).

Stability Under Flow Conditions. Gel stability under flow conditions in a porous medium depends on gel strength and the reservoir environment. Stiff or very viscous gels can withstand large pressure gradients without being moved through a porous medium. Sydansk (1990) reported applying a gradient of 1,000 psi/ft at temperatures up to 255°F without gel breakdown. On the other hand, Martin et al. (1988) reported breakdown of polyacrylamide/Cr(III) and xanthan/Cr(III) gels used in conjunction with CO_2 miscible displacements.

Properly designed gel systems will remain stable and essentially immobile under pressure gradients imposed at reservoir conditions. One series of tests of gel stability, for a xanthan/Cr(III) gel, is represented by the data in **Fig. 5.79** (Eggert et al. 1992). The tests were conducted in porous media consisting of unconsolidated sandpacks having original permeabilities between 3,700 and 4,900 md. A sandpack was flooded with a gel system and then flow was stopped for a period to allow the gel to form in situ. Flow of brine was then imposed at a pressure gradient of 13.5 psi/ft and continued for up to 4 months. Permeability was calculated as a function of time, as illustrated in the figure.

The sandpack permeability was reduced to less than 1% of its original value, and the recovery of permeability was small over the test period. Some Cr(III) and a very small amount of polymer were flushed out of the sandpacks during brine flow. An interesting result is that gels exhibiting up to 71% volume loss (syneresis) in beaker tests were quite effective and stable in the corefloods. Recovery of permeability during flushing actually was greater in gels that exhibited swelling in beaker tests. Changing the pH of the injected brine had some effect on sandpack permeability but not to the degree inferred from pH effects observed in beaker experiments.

Gel Solution Behavior During Injection. Gel solution behavior during the period a solution is being injected is of interest because it relates to gel placement in the porous medium. It has been shown that, at least for some systems, gelation time during flow in a porous medium is not the same as gelation time measured in a beaker or viscometer (Huang et al. 1986; McCool et al. 1991; Hejri et al. 1989; Jousset et al. 1990; Marty et al. 1991).

Fig. 5.80 shows results illustrating behavior in porous medium for a polyacrylamide/Cr(III)/redox system. The data were taken by flowing the gel system at a steady rate through an unconsolidated sandpack. The gel system was mixed at the inlet of the sandpack with an in-line mixer. Pressure drop was measured along the sandpack and converted to apparent viscosity by use of

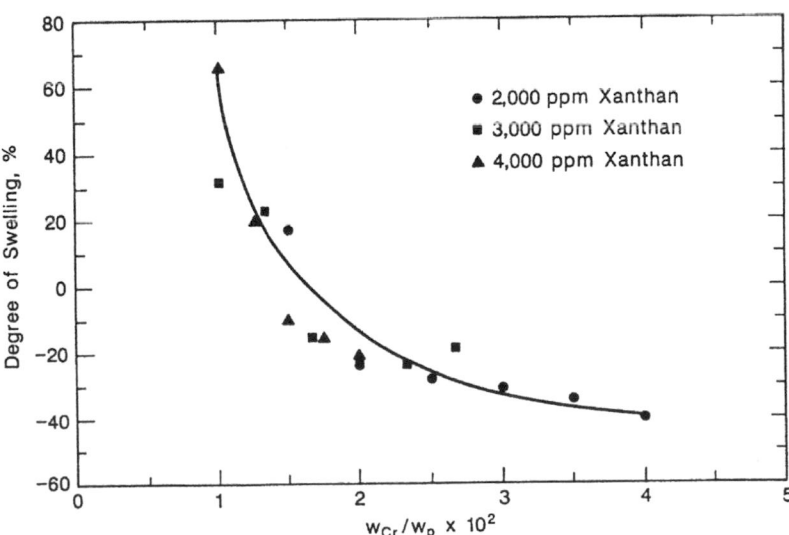

Fig. 5.78—Variation of the degree of swelling of Cr/xanthan gels with chromium/polymer ratio (Gales et al. 1988).

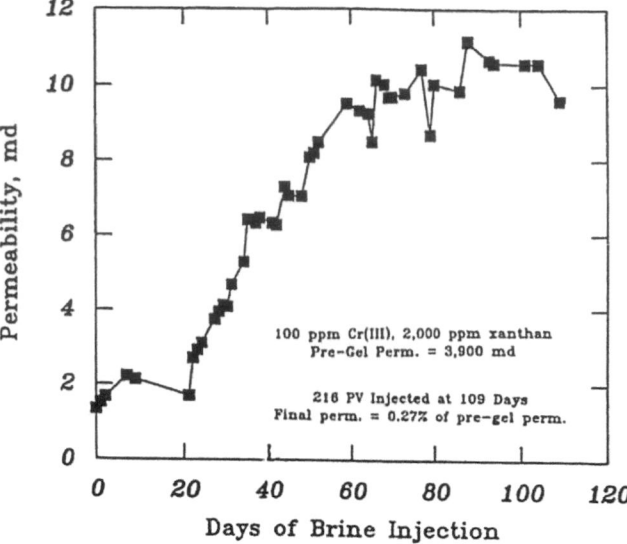

Fig. 5.79—Effect of brine injection on permeability of a sandpack/ gel system treated with 100-ppm-Cr(III)/2,000-ppm-xanthan gel (Eggert et al. 1992).

Darcy's law. Apparent viscosity is plotted vs. distance at different times for the gel system up to approximately 240 hours.

As seen in Fig. 5.80, a region of high resistance built up in the interior of the sandpack. The time required for this resistance to develop was less then the beaker gel time. In fact, at an elapsed time of 240 hours, the leading edge of the injected gel system had exited the sandpack. The age of the gel system at the point of buildup of resistance was about 125 hours.

The behavior is attributed to the filtration of gel particles that formed during the gelation reaction (McCool et al. 1991). Adsorption of polymer and reaction between adsorbed polymer and Cr(III)/polymer in the flowing stream also contributed to the development of the resistance. The location of the point of buildup of resistance has been shown to be a function of flow rate and shear in the porous medium (Marty et al. 1991). Similar behavior was observed for xanthan/Cr(III) gels (Hejri et al. 1989; Jousset et al. 1990).

Analysis of these and similar data indicates that subjecting the gel solution to a shear causes earlier gelation than observed in simple beaker tests. That is, subjecting the gel solution to a shear, such as flowing through porous media, increases the gelation rate. This result was confirmed in experiments conducted in a viscometer (Aslam et al. 1984). It also was determined that the magnitude of the imposed shear rate affects the ultimate gel strength; the larger the imposed shear rate, the weaker the final gel.

5.6.5 Design Considerations for Gel Treatments. Design of a gel treatment consists of two parts: analysis of the reservoir problem and design of the gel system to be applied. The reservoir problem is typically analyzed with standard reservoir engineering calculations combined with field testing. Injectivity profiles conducted to determine the zones in which fluids are leaving the wellbore in an injection well are useful. Radioactive tracers, spinner surveys, and shut-in temperature profiles are commonly used. These do not define flow in the region away from a wellbore, however. Chemical or radioactive tracer tests can be used to obtain information on flow patterns over a wider region of a reservoir. Depending on flow conditions, tracer tests may require several months for completion of field monitoring and analysis.

Design of the gel typically consists of conducting beaker tests to determine gelation time and gel strength, followed by corefloods to evaluate the magnitude of permeability reduction. Sydansk (1990), Purkaple and Summers (1988), and Nagra et al. (1986) give examples of the approach. As part of the tests, it is useful to evaluate long-term stability of a gel system.

The design slug volume is calculated by estimating the volume of reservoir to be treated radially around the wellbore. When the objective is to reduce fracture permeability, an estimate of the fracture volume is required. This approach is an approximate calculation at best and is illustrated in Example 5.15.

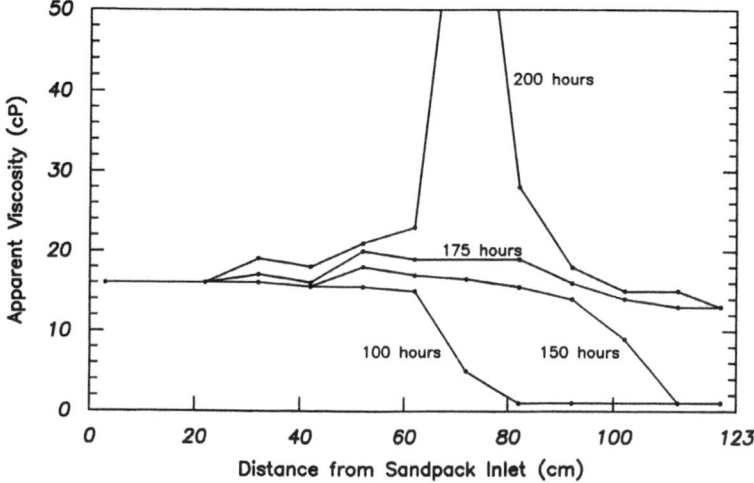

Fig. 5.80—Apparent viscosity profiles in a long sandpack/gel solution displacement (McCool et al. 1991).

Example 5.15—Calculation of Gel Volume. A five-spot pattern on 10-acre spacing has been under waterflood for several years. The reservoir consists of two sandstone layers separated by a 2-to 3-ft shale streak that is considered impermeable. The upper layer is 20 ft thick and has an absolute permeability of 50 md. Porosity of the reservoir rock in this layer is 20%, and the initial oil saturation is estimated to be 60%. The lower layer is 10 ft thick and has an average absolute permeability of 250 md. The average porosity of this interval is 25%, and the initial oil saturation was 70%. Waterflood tests indicate that residual saturations are 0.35 and 0.30, respectively, for the low- and high-permeability layers. Oil viscosity is 2 cp, and the water viscosity is 1 cp at reservoir temperature.

The WOR is high, and the possibility of reducing this ratio by carrying out a permeability-modification treatment on the lower layer needs to be evaluated. It is thought that the treated interval should extend to a radius of 50 ft into the reservoir. Determine the approximate volume of gel solution that must be injected into the high-permeability zone.

Solution. Because the high-permeability zone has been under waterflood for many years, it is reasonable to assume that the oil saturation in the immediate vicinity of the wellbore is at S_{or}. If all the gel solution enters the high-permeability zone, the volume of gel solution required is determined by material balance:

$$W_p = \pi r_p^2 h\phi(1 - S_{or}).$$

For this case,

$$W_p = \frac{\pi(50 \text{ ft})^2(10 \text{ ft})(0.25)(1 - 0.30)}{5.615 \text{ ft}^3 / \text{bbl}}$$
$$= 2,448 \text{ bbl}.$$

This assumes that the gel solution displaces the resident water in piston-like flow and that the amount of gel solution entering the low-permeability zone is negligible. This assumption is examined in Example 5.16.

One concern during the placement of a gel system is penetration of the gel into zones other than the targeted high-permeability zones. Unless the targeted zones are isolated mechanically by packers, some of the gel system will flow into all permeable rock that is open to the injection fluid. The effect of this is detrimental because subsequent injectivity will be reduced, not only in the high-permeability, targeted zones, but over the entire injection interval.

The problem of zonal penetration has been addressed (Seright 1988). The calculations are based on application of Darcy's law in the vicinity of the treated well, as illustrated in Example 5.16.

Example 5.16—Penetration of Gel Solution in Reservoir Layers. The five-spot pattern described in Example 5.15 is to be treated with a gel solution to reduce the water flow in the high-permeability layer at the bottom of the reservoir. Gel solution will be injected into the high-permeability layer at a constant BHP of 1,850 psi. The operator prefers not to isolate the two layers during the treatment so that gel solution may enter both layers. The mean reservoir pressure is 750 psi at an effective radius of 263.3 ft. The effective wellbore radius is 1.0 ft.

The gel solution contains 1,500 ppm xanthan biopolymer. Flow characteristics of this gel solution will be assumed to be the same as those for xanthan flowing in Berea sandstone (Willhite and Uhl 1988). Find the injection time and volume of gel solution injected when the polymer reaches a distance of 50 ft from the well in the high-permeability zone. It will be assumed that the high-permeability zone is at S_{or} in the vicinity of the injection wellbore, while the low-permeability zone is at initial oil saturation until polymer injection begins. Relative permeability relationships for this reservoir rock are given by Eqs. 5.66 through 5.68. Base permeability for these relationships is the absolute permeability to liquid.

Solution. It is necessary to determine what flow regions are present in the region contacted by polymer. This was done in Example 5.6, where the injection rate was specified and the pressure drop was determined. In this example, an iterative process is used because the flow rate is not known a priori and must be calculated indirectly. Xanthan biopolymer is shear-thinning, so possible additional flow regions include upper Newtonian and lower Newtonian.

The transition from the shear-thinning to the upper Newtonian region occurs when the Darcy velocity exceeds u_{c2} given by Eq. 5.44. Because there is a single non-Newtonian region, $n_c = n_{c1}$ and

$$u_{c2} = \left(\lambda_{puN}/\lambda_p^*\right)^{1/(1-n_{c1})}.$$

In the upper Newtonian region, from Eq. 5.43,

$$\lambda_{puN} = k_{wp}/\mu_w.$$

Evaluation of u_{c2} requires values of λ_p^* and λ_{puN} for both layers.

Because the high-permeability zone is at ROS in the vicinity of the wellbore, $k_w = k_w$ at S_{or}. From Eq. 5.67, $k_{rw} = 0.2\, S_{wD}^2$. Because $S_{wD} = 1.0$ at S_{or}, $k_{rw} = 0.2$ and $k_w = (0.2)(250) = 50$ md. In the low-permeability zone, $k_b = 50$ md, so $k_w = (0.2)(50) = 10$ md.

Flow properties of the xanthan biopolymer in Berea core material are given by Eqs. 5.23e through 5.23g:

$$\lambda_p^* = 0.679\, k_{wp}^{0.488},$$
$$n_c = 0.710\, k_{wp}^{-0.073},$$

and $k_{wp} = 0.83\, k_w$.

Rheological properties of the xanthan polymer at this concentration are $K = 0.426$ Pa·s^n and $n = 0.307$. Xanthan biopolymer is shear-thinning because $n < 1.0$.

For the high-permeability zone, $k_{wp} = (0.83)(50) = 41.5$ md. The viscosity of brine is taken as 1 cp. Therefore,

$$\lambda_p^* = 0.679 \, k_{wp}^{0.488}$$
$$= (0.679)(41.5)^{0.488}$$
$$= 4.18 \text{ (md/cp)/(ft/D)}^{1-n_c}$$

and $n_c = 0.710(41.5)^{-0.073}$
$$= 0.54.$$

For the low-permeability zone,

$$\mu_c = 0.63 / (0.0423 \text{ cp}^{-1})$$
$$= 14.9 \text{ cp.}$$

and $n_c = 0.71(8.3)^{-0.073}$
$$= 0.61.$$

Table 5.42 summarizes the polymer flow parameters.

High-Permeability-Zone Calculations. Because $k_{wp} = 41.5$ md at S_{or},

$$\lambda_{puN} = 41.5 \text{ md}/1.0 \text{ cp}$$
$$= 41.5 \text{ md/cp.}$$

Thus, $u_{c2} = (41.5 / 4.18)^{1/1-0.54}$
$$= 146.9 \text{ ft/D.}$$

The maximum Darcy velocity occurs at the injection wellbore. The injection rate corresponding to u_{c2} is

$$q = 2\pi r_w u_{c2} h / 5.615$$
$$= \frac{2(3.1416)(1.0 \text{ ft})(146.9 \text{ ft/D})(10 \text{ ft})}{5.615 \text{ ft}^3 / \text{bbl}}$$
$$= 1,644 \text{ B/D.}$$

Thus, as long as the injection rate is less than 1,644 B/D in the high-permeability zone, the upper Newtonian region will not be encountered.

The lower Newtonian flow region is not encountered until Darcy velocities of approximately 0.1 ft/D are attained using the correlations of Willhite and Uhl (1988). Thus,

$$u_{cl} = 0.1 \text{ ft/D.}$$

This corresponds to an injection rate of

$$q = \frac{2\pi(50 \text{ ft})(0.1 \text{ ft/D})(10 \text{ ft})}{5.615 \text{ ft}^3 / \text{bbl}}$$
$$= 55.95 \text{ B/D}$$

at a radius of 50 ft. Thus, as long as the polymer injection rate into the high-permeability zone is between 56 and 1,644 B/D, flow will be shear thinning and the following equation gives the relationship between flow rate and pressure drop where

	Zone	
Thickness (ft)	20	10
ϕ	0.20	0.25
k_b (md)	50	250
λ_p^* (md-cp)/(ft-D)$^{1-n_c}$	1.91	4.18
n_c	0.61	0.54
k_{wp} (md)	8.3	41.5
λ_{pN} (md/cp)	0.78	–
λ_N (md/cp)	20.0	41.5

Table 5.42—Flow parameters, Example 5.16.

$$\Delta p = 158.005 \left(\frac{5.615 q_t}{2\pi h} \right)^{n_{c1}} \left[\frac{r_p^{1-n_{c1}} - r_w^{1-n_{c1}}}{\lambda_p^* (1 - n_{c1})} \right]$$

$$+ \frac{q_t}{(0.007081)\lambda_N h} \ln\left(\frac{r_e}{r_p} \right),$$

where r_p = radius of polymer-contacted region; r_w = wellbore radius, ft; r_e = effective radius, ft; h = formation thickness, ft; q = injection rate, B/D; Δp = pressure drop between r_w and r_e, psi; and λ_N = mobility of fluid in region between r_p and r_e, md/cp.

This equation is solved by assuming values of r_p from r_w to r_p = 50 ft and computing the value of q_t for each r_p. An example calculation is outlined for r_p = 50 ft. The value of $\lambda_N = \lambda_w$ = 41.5 md/cp:

$$1{,}100 = 158.005 \frac{(5.615 q_t)^{0.54}}{[2\pi(10)]^{0.54}} \left[\frac{(50)^{1-0.54} - 1}{(4.18)(1 - 0.54)} \right]$$

$$+ \frac{q_t \ln(263.3/50)}{(0.00708)(41.5)(10)}$$

$$= 112.56 q_t^{0.54} + 0.565\, q_t.$$

This equation is solved by trial and error or by using a root-finding program to determine the value of q_t. For q_t = 64.04 B/D, Δp = 1,100 psi.

Check

$$\Delta p = 112.56(64.04)^{0.54} + 0.565(64.04)$$

$$= 1{,}063.82 + 36.18$$

$$= 1{,}100 \text{ psi.}$$

The injection rate varies with the location of the polymer front and thus with time. The mobility of the polymer is less than the mobility of the fluids being displaced, so the injection rate will always decrease with time. The relationship between time and injection rate is found from the cumulative fluid injection. Assuming piston-like displacement,

$$W_p = \pi(r_p^2 - r_w^2)\phi h (1 - S_{or}).$$

The cumulative volume of polymer can be expressed as the integral of the injection rate over time. Recalling that

$$W_p^{n+1} \int_0^{t^{n+1}} q_t dt$$

and $W_p^n = \int_0^{t^n} q_t dt.$

Thus, successive times are related by

$$t_c^{n+1} = t_c^n + \frac{2\left(W_p^{n+1} - W_p^n\right)}{q_t^{n+1} + q_t^n}.$$

Times corresponding to successively larger values of time are found by starting at $t = 0$ where $t^0 = 0$ and

$$q_t^{(0)} = \frac{\lambda_w h \Delta p}{\ln(r_e / r_w)}.$$

Thus, $t^{(1)} = \dfrac{2 W_p^{(1)}}{q_t^{(1)} + q_t^0}.$

Table 5.43 contains values of q_t, W_p, and t corresponding to the movement of the polymer front to the radius indicated. Values of q_t for radius r_p were found using a short computer program. Then values of W_p and t were computed as described above.

Low-Permeability-Zone Calculations. The flow regions must be defined for the low-permeability zones. From Eq. 5.44,

$$u_{c2} = \left(\frac{\lambda puN}{\lambda_p^*} \right)^{1/(1-n_{c1})},$$

λ_{puN} = 8.3 md/cp,

$u_{c2} = (8.3 / 1.91)^{1/(1-0.61)}$

$= 43.3$ ft/D.

Radius of Polymer Front (ft)	Pressure Drop (psi)	Injection Rate (B/D)	Volume of Polymer Injected (bbl)	Average Injection Rate (B/D)	Time (days)	Injectivity (B/D-ft)
1	1,100	579.99	0.00	–	0.00	58.00
1.5	1,100	547.22	1.22	563.60	< 0.00	54.72
2	1,100	515.38	2.94	531.30	0.01	51.54
5	1,100	372.20	23.50	443.79	0.05	37.22
10	1,100	247.65	96.93	309.92	0.29	24.76
15	1,100	183.46	219.33	215.55	0.86	18.35
20	1,100	145.10	390.67	164.28	1.90	14.51
25	1,100	119.82	610.98	132.46	3.56	11.98
30	1,100	102.00	880.24	110.91	5.99	10.20
35	1,100	88.79	1,198.46	95.40	9.33	8.88
40	1,100	78.63	1,565.63	83.71	13.71	7.86
45	1,100	70.58	1,981.76	74.60	19.29	7.06
50	1,100	64.04	2,446.85	67.31	26.20	6.40

Table 5.43—Injection rate for high-permeability layer, Example 5.16.

Thus, the upper Newtonian flow region will be reached at the wellbore when

$$q_t = \frac{2\pi(1.0 \text{ ft})(43.3 \text{ ft/D})(20 \text{ ft})}{5.615 \text{ ft}^3 / \text{bbl}}$$

$$= 968 \text{ B/D}.$$

This rate is not likely to be obtained in the low-permeability layer; thus, the upper Newtonian region will not be present.

The lower Newtonian region will be reached in the low-permeability zone if the Darcy velocity decreases to u_{cl} in the region contacted by polymer. The injection rate at any time is related to the radius and the Darcy velocity by Eq. 5.57 modified by the factor 5.615 ft³/bbl with q_t in B/D.

$$q_t = 2\pi r h u / 5.615.$$

If the injection time is long enough, there will be some radius r_{cl} where q_t in Eq. 5.57 is satisfied with $u = u_{cl}$. The pressure drop at this instant is given by

$$\Delta p = 158.005 \left(r_{cl} u_{cl} \right)^{n_{cl}} \left[\frac{r_{cl}^{1-n_{cl}} - r_w^{1-n_{cl}}}{\lambda_p^* (1 - n_{cl})} \right]$$

$$+ \frac{158.028 r_{cl} u_{cl}}{\lambda_N} \ln\left(\frac{r_e}{r_{cl}} \right),$$

which was obtained by replacing q_t with $(2\pi r_{cl} u_{cl} h)/5.615$. In the low-permeability zone, the region ahead of the polymer front is assumed to be at initial oil saturation. Thus, $\lambda_N = k_o/\mu_o = (40 \text{ md})/(2 \text{ cp})$ or 20 md/cp.

When the value of u_{cl} is specified, this equation is solved to find the value of r_{cl} for a given pressure drop. Substituting the parameters for the low-permeability zone into this equation results in the following equation, which is solved for r_{cl}.

$$1,100 = 158.005 \left(r_1 u_{cl} \right)^{0.61} \left[\frac{r_p^{1-0.61} - r_w^{1-0.61}}{(1.91)(1-0.61)} \right]$$

$$+ \frac{158.028 r_1 u_{cl} \ln(263.3/r_1)}{20}.$$

The value of r_{cl} when $u_{cl} = 0.1$ ft/D, assumed for this example, is found by a root-finding program to be 27.77 ft. Thus, for $r_p < r_{cl}$ the polymer will be in the shear-thinning region and the pressure drop is given by

$$1,100 = 158.005 \frac{(5.615 q_t)^{0.61}}{\left[2\pi(20)\right]^{0.61}} \left[\frac{r_p^{1-0.61} - r_w^{1-0.61}}{(1.91)(1-0.61)} \right]$$

$$+ \frac{q_t \ln\left(263.3/r_p\right)}{(0.007081)(20)(20)}.$$

Injection rates into the low-permeability zone are computed from this equation in the same manner as for the high-permeability zone. Values of q_t corresponding to each $r_p < r_{cl}$ are computed using a root-finding program. **Table 5.44** summarizes the values. Injection times are determined as discussed in the high-permeability calculations.

Note that the injection time required for the radius of the polymer zone to reach 50 ft in the high-permeability zone is 26.2 days. In the low-permeability zone, the lower Newtonian region of polymer flow is encountered ($r_p = r_{cl}$) after 12.58 days of injection. Thus, it is necessary to consider Newtonian flow of the polymer in the low-permeability zone for times greater than 12.58 days (i.e., $r_p > r_{cl}$).

Eq. 5.52 (with the upper Newtonian region deleted) gives the pressure drop when there is Newtonian flow of polymer in the region between r_{cl} and r_p.

$$\Delta p = 158.005 \left(\frac{q_t}{2\pi h} \right)^{n_{cl}} \left[\frac{r_{cl}^{1-n_{cl}} - r_w^{1-n_{cl}}}{\lambda_p^* (1 - n_{cl})} \right]$$

$$+ \frac{q_t}{2\pi h \lambda_{pN}} \ln \left(\frac{r_p}{r_{cl}} \right) + \frac{q_t}{2\pi h \lambda_N} \ln \left(\frac{r_e}{r_p} \right).$$

Because the polymer mobility is less than the mobility of the fluids being displaced, the injection rate will continue to decline as the polymer moves farther into the formation. This also means that the location of r_{cl} changes for each r_p greater than the initial value of r_{cl}. Therefore, the injection rate must be expressed in terms of r_{cl} in the following equation.

$$\Delta p = 158.005 \left(r_{cl} u_{cl} \right)^{n_{cl}} \left[\frac{r_{cl}^{1-n_{cl}} - r_w^{1-n_{cl}}}{\lambda_p^* (1 - n_{cl})} \right]$$

$$+ \frac{158.028 r_{cl} u_{cl}}{\lambda_{pN}} \ln \left(\frac{r_p}{r_{cl}} \right) + \frac{158.028 r_{cl} u_{cl}}{\lambda_N} \ln \left(\frac{r_e}{r_p} \right).$$

In the low-permeability zone, the polymer mobility, λ_{pN}, is computed from

$$\lambda_{pN} = \lambda_p^* u_{cl}^{(1-n_{cl})}$$

$$= 1.91[(md/cp)/(ft/D)^{(1-0.61)}](0.1 \ ft/D)^{(1-0.61)}$$

$$= 0.78 \ md/cp.$$

Substitution of the appropriate values into the equation for pressure drop gives

$$1,100 = 158.005 \left(0.1 r_{cl} \right)^{0.61} \left[\frac{r_p^{1-0.61} - r_w^{1-0.61}}{(1.91)(1 - 0.61)} \right]$$

$$+ \frac{158.028 r_{cl} (0.1) \ln \left(r_p / r_{cl} \right)}{0.78} + \frac{158.028 r_{cl} (0.1) \ln \left(263.3 / r_p \right)}{(20)}.$$

Radius of Polymer Front (ft)	Pressure Drop (psi)	Injection Rate (B/D)	r_{cl} (ft)	Volume of Polymer Injected (bbl)	Average Injection Rate (B/D)	Time (days)	Injectivity (B/D-ft)
1	1,100	558.96	–	0.00	–	0.00	27.95
1.5	1,100	474.51	–	1.82	516.73	<0.00	23.73
2	1,100	412.37	–	4.36	443.44	0.01	20.62
5	1,100	233.15	–	34.91	322.76	0.10	11.66
10	1,100	138.88	–	144.02	186.01	0.69	6.94
15	1,100	101.06	–	325.85	119.97	2.21	5.05
20	1,100	80.52	–	580.43	90.79	5.01	4.03
25	1,100	67.52	–	907.73	74.02	9.43	3.38
27.77	1,100	62.16	27.77	1,120.73	71.34	12.58	3.11
30	1,100	58.68	26.22	1,307.78	60.42	15.68	2.93
35	1,100	53.05	23.70	1,780.55	55.86	24.14	2.65
40	1,100	49.14	21.96	2,326.07	51.09	34.82	2.46
45	1,100	46.23	20.66	2,944.31	47.68	47.79	2.31
50	1,100	43.95	19.64	3,635.30	45.09	63.11	2.20

Table 5.44—Injection rate for low-permeability layer, Example 5.16.

This equation is solved to obtain a value of r_{cl} for each value of r_p assumed. The injection rate corresponding to each value of r_{cl} is $q_t = (2\pi r_{cl} u_{cl} h)/5.615$. Table 5.44 summarizes computed rates and times.

Comparison of injection rates at similar times in Tables 5.43 and 5.44 shows that substantial amounts of polymer enter *both* layers. Approximately 26.2 days is required to displace the polymer front to a distance of 50 ft in the high- (250-md) permeability layer. However, when the low- (50-md) permeability layer is open in the injection wellbore, injection rates into this zone are substantial, as Table 5.44 indicates. By 26.2 days, the polymer front in the low-permeability zone is located 36.0 ft from the injection well. Polymer volume in this zone is approximately 1,886 bbl, so the total volume of polymer required to treat the high-permeability layer to a radius of 50 ft is 2,446 + 1,886 = 4,332 bbl.

Example 5.16 assumes that the flow properties of the gel solution are identical to the flow properties of the polymer solution. This is probably valid for a short time after the crosslinking constituents are added to the polymer solution. In the absence of other correlations, this example shows that selective treatment of the high-permeability layer with the polymer/rock properties selected for this example is not possible without isolating this layer from lower-permeability layers.

5.7 Field Experience

Numerous polymer-augmented waterfloods and gelled polymer conformance projects have been implemented on an economic (fieldwide) scale since the early 1960s. Additionally, a large number of pilot floods and tests have been conducted.

Manning et al. (1983) identified 273 polymer projects that had been instituted or planned as of 1983. According to a statistical analysis of the fieldwide projects, median recovery of oil was 2.91% OOIP, or 4.02% of the remaining oil in place (OIP) at the start of the projects. This amounted to a median recovery of 24.9 STB/acre-ft and 2.34 STB oil recovered per 1 lbm of polymer injected. Incremental oil recovery by polymer-augmented waterflooding is not large on a percentage basis. Nonetheless, the incremental recovery can be sufficient to yield a satisfactory economic return.

Manning et al. also reported that approximately 90% of the fieldwide and pilot polymer projects were in sandstone reservoirs and approximately 10% were in carbonate lithologies. **Table 5.45** (Manning et al. 1983) gives a summary of data for selected projects.

Schurz et al. (1989) summarized results from 99 projects initiated during 1980–89. **Table 5.46** compares median values from these projects with those reported by Manning et al. In most cases, the oil/water viscosity ratio at reservoir temperature is between 5.4 and 6.8, which would yield a slightly to moderately unfavorable mobility ratio when relative permeability effects are considered. Average permeability ranges from 54 to 101 md. Partially hydrolyzed polyacrylamide was used in most polymer floods, probably because it has the lowest cost. There is a trend toward higher polymer concentrations in floods initiated during 1980–89. Projected median incremental oil recovery ranges between 3.7 and 4.8% OOIP at a ratio of 0.64 to 1.3 bbl oil/lbm polymer. Most of the polymer-augmented waterfloods were initiated while the WOR was between 2.4 and 5.0. Most floods have been in sandstone reservoirs.

Table 5.46 also shows parameters from selected projects in which reduction of the mobility ratio to improve displacement efficiency was a significant design parameter. High concentrations of partially hydrolyzed polyacrylamide were used in each project to increase the viscosity of the injected polymer solution. Permeability in each reservoir was significantly higher than the average of the projects summarized in Cols. 1 and 2 in Table 5.46, favoring high injection and production rates. Projected incremental oil was 13 to 25% OOIP, considerably higher than expected from the other reservoirs surveyed by Manning et al. (1983) and Schurz (1989).

Christopher et al. (1988) described a polymer-augmented waterflood in the Reagan sandstone in Nebraska. The formation contains an upper light-oil zone underlain by a heavy-oil zone. The light oil, which is the target of the flood, has a 31 °API gravity and a viscosity of 24 cp. Porosity and permeability of the sandstone are relatively high at 24% and 2,580 md, respectively.

The unit had been under waterflood since the late 1960s, and the flood had been successful in spite of an unfavorable mobility ratio of approximately 8.0 (based on k_w at ROS). A polymer flood was initiated in 1985 to increase oil recovery. Ten inverted nine-spot patterns covering approximately 1,685 acres were used.

Design procedures included laboratory tests on polymer rheology, relative permeability, shear degradation, screen factor, stability, and salinity effects. Computer simulations were performed to predict recoveries and to examine optimum polymer concentration. Field injectivity tests were conducted to examine injectivity behavior with time and to gain experience in surface handling of the polymer. Pressure-falloff tests were conducted in conjunction with the injectivity tests. Finally, the preparation involved design of the polymer-injection plant and analysis of costs.

The polymer selected was a partially hydrolyzed polyacrylamide. Over the course of polymer injection, concentration varied from 1,000 to 800 ppm. The polymer was mixed in available fresh water.

Fig. 5.81 shows performance of the polymer flood through October 1987. The oil production responded to the polymer after a few months of injection. WOR started to decline almost immediately after initiation of injection. Christopher et al. (1988) report that, on the basis of an assumed exponential decline curve for continued waterflooding, the incremental oil recovery from polymer flooding was 1.64% OOIP as of October 1987. The project was judged to be a technical and economic success.

Krebs (1976) presents an example of a polymer-augmented waterflood that was not successful. The flood was conducted in the Ranger zone of the Wilmington Field, California. This zone is an unconsolidated sand at depths from 2,500 to 4,000 ft and is made up of beds that are approximately 10 to 50 ft thick. Porosities are high, approximately 30 to 40%, because the

Project	Field	State/Country	Oil Recovery (10^3 STB)	Formation	Lithology	Depth (ft)	Temperature (°F)	Area (acres)
Algyo	Algyo	Hungary	—	Algyo 2, Szoreg 1	—	—	—	—
Anderson-Kerr	—	Texas	60	—	Sandstone	—	—	—
Big Sinking	Big Sinking	Kentucky	—	Corniferous	Limestone	1,000	70	40
Bond	Bond	Oklahoma	238	Allen	Sandstone	850	78	261
Bone Pile	Bone Pile	Wyoming	—	Mini	—	10,800	207	1,760
Brelum	Brelum	Texas	250	Government Wells	Sandstone	1,950	115	264
Byerly	Byerly	Kansas	—	Squirrel	Sandstone	600	77	—
Caprock	Caprock	New Mexico	—	Queen	Sandstone	2,775	100	—
Caprock	Caprock	New Mexico	—	Queen	—	—	—	—
Cement	Cement	Oklahoma	28	—	Sandstone	3,554	98	316
Cushing	Cushing	Oklahoma	—	Layton	Sandstone	1,500	95	2,140
Cushing	Cushing	Oklahoma	287	Prue	Sandstone	2,500	105	2,140
Hamm	Hamm	Wyoming	1,160	Minnelusa	Sandstone	7,925	147	380
Hankensbuettal	Hankensbuettal	West Germany	—	Dogger-Beta	Sandstone	4,900	136	—
Kuehne Ranch Southeast Range	Kuehne Ranch Southeast Range	Wyoming	784	Minnelusa	Sandstone	7,945	130	662
Kummerfeld	Kummerfeld	Wyoming	820	Minnelusa	Sandstone	7,600	130	712

Table 5.45—Database for technical survey of polymer flooding projects (Manning et al. 1983).

Project	Number of Injectors	Number of Producers	Pattern Size (acres)	Net Pay (ft)	Permeability (md)	Permeability Range (md)	Dykstra-Parsons Permeability Variation	ϕ
Algyo	—	—	—	—	—	—	—	—
Anderson-Kerr	—	—	—	—	600	—	—	0.19
Big Sinking	8	3	5	24	40	30 to 50	0.16	0.22
Bond	—	—		—	48	—	0.55	0.2
Bone Pile	—	—	—	—	—	—	—	—
Brelum	12	8	264	10	399	36 to 1,490	0.69	0.29
Byerly	—	—	—	30	17.1	—	—	0.206
Caprock	—	—	—	10	175	5 to 1,500	0.86	0.19
Caprock	—	—	—	—	—	—	—	—
Cement	10	33	40	133	34	1 to 1,168	0.91	0.17
Cushing	27	8	20	15	—	—	0.93	0.14
Cushing	27	8	20	12	—	—	0.82	0.12
Hamm	2	4	—	28	239	—	0.76	0.225
Hankensbuettal	2	7	—	59	2,828	2,000 to 4,000	0.21	0.28
Kuehne Ranch Southeast Range	3	5	—	18	110	—	0.714	0.167
Kummerfeld	4	6	80	22	135	15 to 950	0.71	0.165

Table 5.45—Database for technical survey of polymer flooding projects (Manning et al. 1983) (continued).

Project	Oil (°API)	Oil Viscosity (cp)	Previous Production	Polymer Solution Injected (PV)	Polymer Solution Injected (10³ bbl)	Polymer Type	Polymer Concentration (ppm)	m
Algyo	—	—	—	—	—	—	—	—
Anderson-Kerr	—	—	—	0.022	400	Synthetic	—	—
Big Sinking	40	4	Waterflood	—	—	Synthetic	600	—
Bond	—	20	—	—	—	Synthetic	—	8
Bone Pile	32	—	—	—	—	—	—	—
Brelum	22	10	Primary	0.25	2,700	Synthetic	389 to 75	5.8
Byerly	28.9	24.2	—	—	—	Synthetic	250	3.8
Caprock	34.1	4.85	—	—	—	Synthetic	250	2.7
Caprock	—	—	—	—	—	Synthetic	—	—
Cement	35	6	—	—	—	Biopolymer	—	—
Cushing	39	5	Waterflood	—	—	Synthetic	—	—
Cushing	43	4	Waterflood	—	—	Synthetic	—	—
Hamm	—	24.8	—	0.284	5,285	Synthetic	137 to 625	17.3
Hankensbuettal	—	11	—	0.41	—	Synthetic	1,500 to 2,000, 2,000 to 0	—
Kuehne Ranch Southeast Range	—	18	—	0.147	2,013	Synthetic	252 to 345	17.3
Kummerfeld	23	38	Waterflood	0.171	2,400	Synthetic	200	18

Table 5.45—Database for technical survey of polymer flooding projects (Manning et al. 1983) (continued).

Project	λ_p/λ_o	Initial WOR	Projected Life (months)	Polymer Recovery (% OOIP)	Incremental Oil (% OIP)	Incremental Oil (STB/acre-ft)	Incremental Oil (STB/lbm polymer)	Projected Ultimate Incremental Oil (10³ STB)
Algyo	—	—	—	—	—	—	—	—
Anderson-Kerr	—	6	—	—	—	4.8	1.4	—
Big Sinking	—	—	39	0	0	0	0	—
Bond	—	—	—	—	—	—	—	238
Bone Pile	—	—	—	—	—	—	—	—
Brelum	0.9	10.2	180	—	9.83	—	—	250
Byerly	0.63	—	219	—	—	—	—	—
Caprock	0.34	—	36	—	—	—	—	—
Caprock	—	—	—	—	—	—	—	—
Cement	—	—	—	—	—	—	—	28
Cushing	—	—	—	—	—	—	—	287
Cushing	—	—	—	—	—	—	—	125
Hamm	—	0.5	—	8.56	—	108.92	5.2	—
Hankensbuettal	—	5.67	—	7.22	11.20	117.58	—	—
Kuehne Ranch Southeast Range	—	0.015	—	2.33	2.47	30.94	2.25	920
Kummerfeld	—	0.32	—	6.14	6.52	52	6.44	1,427

Table 5.45—Database for technical survey of polymer flooding projects (Manning et al. 1983) (continued).

	Median Values From Surveys		Selected Individual Projects		
	1960–82[1] (226 projects)	1980–89[2] (99 projects)	Marmul[3]	Oerrel[4]	Courtenay[5]
Reservoir and fluid properties					
Oil/water viscosity ratio at reservoir temperature	6.8	5.4	114	39	55
Reservoir temperature (°F)	115	115	115	136	86
Permeability (md)	101	54	15,000	2,000	2,500
OOIP present at startup (%)	74.6	74	≈92	81.5	≈69
Producing WOR at startup	2.4	5.0	1	4	4
Lithology (sand : carbonate)	10:1	2.4:1	sand	sand	sand
Polymer injection					
Polymer (HPAM : xanthan)	19:1	11:1	HPAM	HPAM	HPAM
Polymer concentration (ppm)	250	500	1,000	1,500	1,000
Polymer retention (lbm polymer/acre-ft)	10.2	32	373	162	536
Projected incremental recovery over waterflooding					
% OOIP	4.8	3.7	25	≈13	≈14
bbl oil/lbm polymer	1.3	0.64	1.2	≈1.4	≈0.5
bbl oil/acre-ft	25	20.5	461	≈230	≈290

HPAM = hydrolyzed polyacrylamide.

[1]Manning et al. (1983)

[2]Schurz et al. (1989)

[3]Koning et al. (1988)

[4]Maitin et al. (1988); Maitin and Volz (1981)

[5]Putz et al. (1988)

Table 5.46—Properties of polymer flooding field projects (Schurz et al. 1989).

sand is unconsolidated. Permeabilities are on the order of darcies. The oil gravity averages 18 °API, and the oil viscosity is approximately 31 cp at the reservoir temperature of 135°F.

Waterflooding was initiated in 1964 in the Ranger zone area that was later to be polymer flooded. Response to the waterflood was generally favorable, but WORs showed a steady increase. The waterflood mobility ratio was approximately 14, indicating that improved sweep efficiency might be expected from polymer flooding. Tests with a polyacrylamide indicated that the mobility ratio could be reduced to just slightly above 1.0 with a polymer concentration of 250 ppm. Resistance factor measured in the laboratory was 10.7, and polymer adsorption was measured and predicted to be 85 lbm polymer/acre-ft.

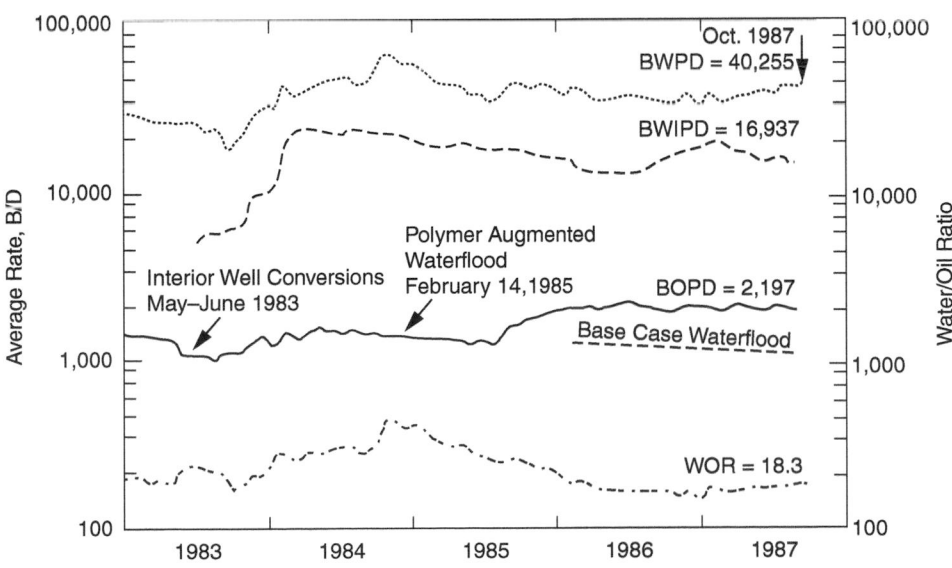

Fig. 5.81—Fluid production rates, Reagan sandstone (Christopher et al. 1988).

Polymer injection began in 1969. The injected concentration was reduced to 150 ppm because of injectivity and polymer-handling problems but later was increased to the design value of 250 ppm.

The performance of the flood was predicted at various stages. **Fig. 5.82** shows a prediction made in 1970. The figure shows actual performance data and predicted results from a reservoir simulator. Also shown is the predicted performance for continued waterflooding.

As Fig. 5.82 shows, the reservoir did not respond to the polymer flood. WORs were not significantly changed, and oil production did not increase. In fact, by 1973 oil production had dropped below the rate predicted for continued waterflooding. The polymer flood was discontinued in 1972, 2.5 years after initiation, because of the poor performance.

Several reasons were cited for the poor behavior (Krebs et al. 1976). One was that the polymer concentration used was not high enough. The apparent polymer slug viscosity was very sensitive to polymer concentration, and reductions in viscosity and resistance factor in the reservoir might have contributed to the poor behavior. The loss of injection rate early in the flood also was detrimental. This loss resulted from plugging by scale, undissolved polymer, and possibly polymer adsorption. Additionally, heterogeneities in the reservoir led to channeling of the polymer slug (i.e., the concentration of polymer was not sufficient to reduce flow in high-permeability zones).

Hochanadel et al. (1990) analyzed a number of waterfloods and polymer floods conducted in the Minnelusa sandstone formation of the Powder River basin, Wyoming. The formation exhibits high permeability variation, with the Dykstra-Parsons coefficient typically ranging between 0.6 and 0.8. Primary recoveries have varied significantly because of the reservoir heterogeneity (5 to 15% OOIP). Low primary recovery makes the reservoirs good candidates for waterflooding. However, unfavorable waterflood mobility ratios, ranging between 2 and 25, and reservoir heterogeneity reduce waterflood efficiency.

A large number of polymer floods have been conducted in the formation during 1975–90. The floods typically involved a combination of polymer-augmented waterflooding and the use of an aluminum citrate crosslinker. An accepted approach has been to inject a cationic polymer followed by an anionic polymer and aluminum citrate crosslinker. The approach is based on the assumption that the cationic polymer will adsorb onto the reservoir rock and that the polymer will be crosslinked by aluminum to anionic polymer subsequently injected. The crosslinking process leads to reduced permeability in those zones penetrated by the polymer (i.e., to in-situ permeability modification).

Polymer slug sizes have ranged from 10 to 40% PV (Hochanadel et al. 1990). The volume of anionic polymer injected has been much larger than the volume of cationic polymer. For example, in one project, the amounts of cationic and anionic polymer injected were 10,000 lbm and 629,115 lbm, respectively (Doll and Hanson 1987). In this project, the average anionic polymer concentration was 233 ppm, which is typical of the Minnelusa floods. Injection of polymer is thought to have had two positive effects: (1) polymer crosslinking reduced permeability variation and (2) the mobility ratio improved. Any increased recovery of polymer flooding over waterflooding results from a combination of both mechanisms.

Table 5.47 (Hochanadel et al. 1990) summarizes conditions and performance of 31 polymer floods. Comparison of the polymer flood results to waterflood results in the same formation led Hochanadel et al. to several conclusions. Polymer floods, on average, recovered an additional 7.5% OOIP at a cost of USD 1.69/bbl of incremental oil. At equal injection volumes, polymer floods recovered more oil and produced less water than waterfloods.

Sydansk and Smith (1988) describe implementation of a successful gel treatment technology to improve conformance. In this technology, acrylamide polymers are crosslinked with Cr(III). The crosslinking agent is "a mixture of complex Cr(III) ions containing low-molecular-weight carboxylate anions." (Sydansk and Smith 1988) The carboxylate anion used was acetate.

Nine field tests were conducted in the Big Horn basin, Wyoming, during 1985–86 (Sydansk 1988). Tests were conducted in both sandstones and carbonate formations, but all the formations contained significant natural fracturing, which adversely affected waterflood performance.

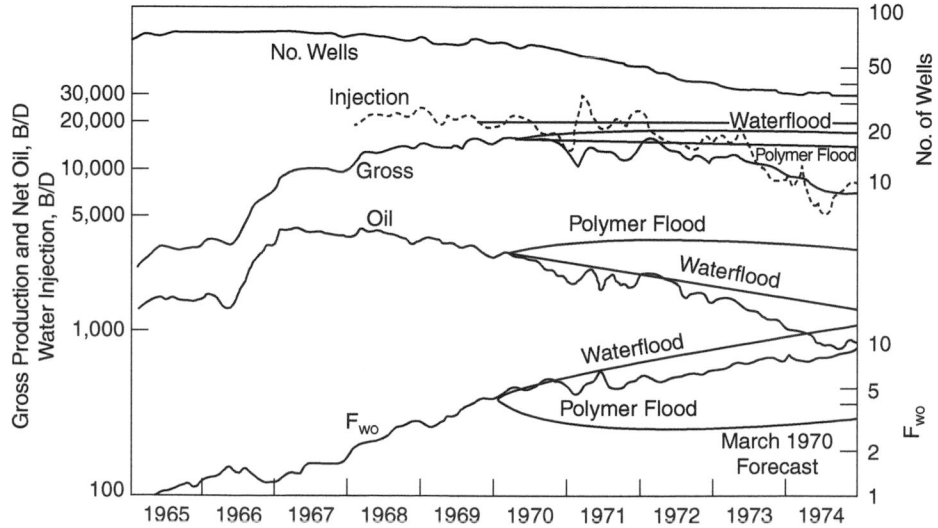

Fig. 5.82—Predicted and actual performance of Ranger zone polymer flood, Wilmington field (Krebs 1976).

Field	Operator	Gravity (°API)	Porosity (%)	Ah (acre-ft)	OOIP (10³ STB)	PV (10³ bbl)	Cumulative Oil (10³ bbl)	Cumulative Water (10³ bbl)	Cumulative Injected (10³ bbl)
Hamm	Presidio	20	22.5	10,650	13,547	18,590	6,531	6,705	17,558
Sharp	Anderson	26	13.4	3,645	2,180	3,789	1,049	1,511	4,245
Shepherd Ranch	Lario	19	14.3	5,710	4,812	7,600	2,248	778	2,981
Simpson Ranch	Pacific Enterprises	20	19.0	1,705	1,825	2,513	826	997	1,678
Wagonspoke	Fancher	28	17.0	6,280	6,095	8,282	2,722	3,428	8,020
OK	URC	30	18.6	5,935	6,300	8,565	2,747	3,206	6,784
Kummerfeld	Pacific Enterprises	20	16.5	14,288	12,019	18,290	5,170	3,845	19,734
Edsel	Apache	21	19.6	8,152	8,673	12,886	3,340	2,296	5,895
Semlek, West	Pacific Enterprises	24	19.4	9,593	10,322	14,438	3,877	2,328	9,827
Tholson	Osborn HRS	23	17.3	9,904	9,566	13,392	3,554	5,056	6,528
Rozet Southeast	Presidio	23	11.2	8,067	4,665	7,009	1,692	1,754	3,035
Kuehne Ranch	Samedan	23	14.4	9,280	9,250	14,725	3,211	494	9,561
Stewart R, East	Gallagher	21	15.7	2,720	2,009	3,313	679	589	1,457
Spring	Presidio	21	15.1	6,721	4,582	7,807	1,536	945	2,920
Rozet Northeast	Bridge Oil		17.1	4,800	3,723	5,475	1,089	367	1,274
Kuehne Ranch Southeast	Presidio	23	16.7	10,591	9,468	13,721	2,629	3,634	12,007
Deadman Creek	Meridian	23	17.2	6,560	5,871	8,754	1,356	971	3,438
North Rainbow R	Apache	26	18.4	9,260	8,130	12,428	1,706	1,044	3,023
Victor	Quintana	21	14.2	4,050	3,167	4,453	617	18	640
Big Mac	Powder River	22	19.3	2,651	2,646	3,970	514	17	504
Kiehl, West	Pacific Enterprises	24	23.0	1,165	1,519	2,079	292	2	283
Right a Way	WEM	20	14.4	3,731	2,094	4,176	378	1	513
Pownall Ranch	Pacific Enterprises	25	15.1	5,293	5,270	6,201	816	27	871
Lone Cedar	Apache	26	17.0	7,128	7,724	9,034	1,189	25	1,313
Glo, North	Anderson	22	16.9	4,932	4,528	6,084	682	69	440
Kiehl	Pet Inc.	21	22.4	8,840	9,042	15,364	1,300	71	1,225
Swartz Draw	PG&E	23	15.4	5,914	4,423	7,066	624	47	873
Candy Draw	Santa Fe	25	15.1	9,181	6,991	10,755	901	19	710
Bracken	Anadarko	24	16.5	1,909	1,979	2,443	234	80	184
Lily	Vintage	21	19.8	4,546	5,205	6,974	507	118	1,118
American	Ladd	22	14.3	2,194	1,624	2,429	89	24	111

Table 5.47—Minnelusa polymer floods less than 15,000 acre-ft as of 1 March 1989 (Hochanadel et al. 1990).

Polymer concentration in the injected gel systems ranged from 3,000 to 8,500 ppm. The volumes of gel injected ranged from 4,000 to 37,000 bbl/well treated. Seven of the wells treated were waterflood injection wells and two were production wells.

Fig. 5.83 shows the response of offset production wells to the gel treatment of a waterflood injection well. Oil production increased rapidly and markedly in response to the injection well treatment.

The favorable response presumably resulted from the sealing of fractures with the gel and subsequent diversion of injected water into previously unswept zones. Sydansk and Smith concluded that the gel system could be used in fractured reservoirs to reduce WORs and to increase oil recovery and that the system was cost-effective.

5.7.1 Large-Scale Polymer Floods. There have been several large-scale commercial applications of polymer flooding (Putz et al. 1988, 1994; Abib et al. 1991; Delamaide et al. 1994; Chauveteau et al. 1988; Wang et al. 1993, 1995, 2000, 2002a, 2002b 2008,

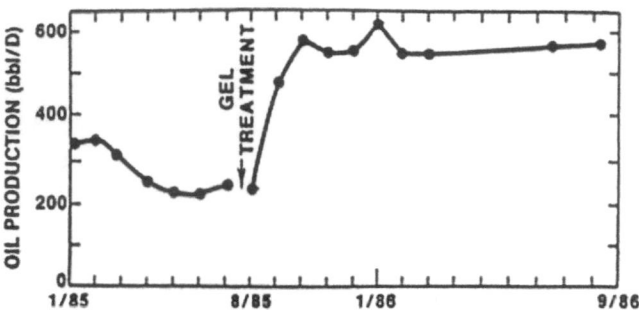

Fig. 5.83—Combined offset production, well response to the Injection Well 0-17 field test (Sydansk and Smith 1988).

2009, 2011, 2013; Guo et al. 2000; Liang et al. 2011). The information on these projects was prepared early in the project life; consequently, overall performance data were never published for most of these projects. Three field applications are described in this section.

Courtenay Field. The Courtenay Field (Putz et al. 1988), shown in **Fig. 5.84,** is an appendix of the Chateaurenard Field located in the Paris Basin approximately 100 km southeast of Paris. Polymer flooding in this field was initiated by conducting a pilot flood from 1985 to 1987 (Putz et al. 1988), which was expanded to a commercial flood from 1989 to 1993 (Putz et al. 1994).

Pilot Flood (1985–87) (Putz et al. 1988). The isopach map of the field is also shown in Fig. 5.84. The Courtenay Field, found at a depth of 600 m, is described (Putz et al. 1988) as an elongated layer composed of unconsolidated Griselles sands, which formed on a monocline dipping 2° to the northwest. Average thickness is 3.2 m. Permeability ranges from 500 md to 5 darcies, with an average permeability of 2 darcies. Highly permeable zones are confined to channels containing clean sand, which are thought to be distributed randomly in the field. Lateral continuity of these sands on the well spacing of the pilot was apparently believed to be present. Shale content in the reservoir varies from 2 to 15%. Average porosity is 30%. The reservoir contains a paraffinic oil (SG = 0.89) with a viscosity of 40 cp at the reservoir temperature of 30°C. The oil is undersaturated with a bubblepoint of 5.2 bar (≈75 psi). Initial water saturation was 30% and was close to fresh. The water viscosity at reservoir temperature was 0.73 cp. Original oil in place was estimated to be 1,335 kt (9.4 million bbl). The primary recovery mechanism was an edgewater drive from the North West aquifer coupled with some water influx from the south. The field had not been waterflooded. The endpoint mobility ratio for a waterflood was approximately 3.7.

The pilot area was located in the middle of the field where both the pressure gradients and the reservoir pressure were minimal. An inverted 10-acre five-spot pilot polymer flood was conducted in the field from October 1985 to October 1987. The pilot pattern consisted of one injection well and four production wells, as shown on the isopach map in **Fig. 5.85** (Putz et al. 1988). Well spacing for the pilot was 200 m to enable completion of the pilot within an acceptable time period. Pore volume of the pattern was approximately 60 000 m³, so the OOIP was approximately 42 000 m³. The sand is unconsolidated. Lateral continuity of these channels was not well-defined, as became apparent during the flood when Well CY40 had no response to the polymer flood. Injection and production wells were completed openhole, underreamed followed by installation of a gravel pack. In addition to sand control, the well completion minimized mechanical degradation of the injected and produced polymer. Other details regarding well completion, surface equipment, and operations are found in Putz et al. (1988). The oil production rate from the four pattern wells had been stable before polymer injection at approximately 8 m³/d, with water production at 33 m³/d.

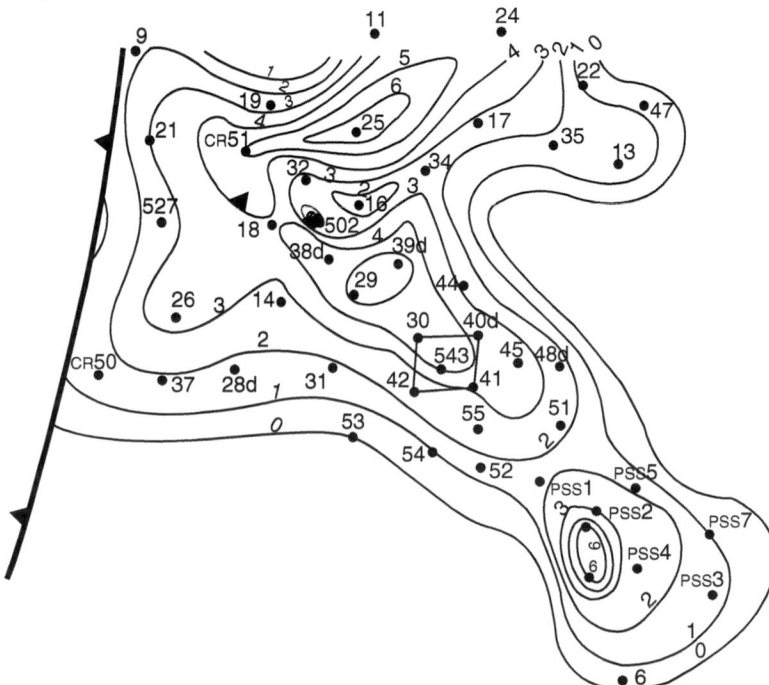

The injection program was based on oversizing the polymer slug to (a) verify that a larger slug would be more profitable, (b) produce a large amount of polymer that could be matched by reservoir simulation, and (c) analyze produced polymer to determine if polymer molecules were degraded as they flowed through the reservoir. The polymer slug size was based on a computer simulation that accounted for polymer retention (20 µg/g) and a residual resistance factor of 1.3 measured in laboratory experiments. Polyacrylamide polymer with a molecular weight of 6.5 million Daltons and 30% hydrolyzed was prepared in fresh water and injected into Well 543 at a concentration of 1,000 ppm, starting at 50 m³/d and increasing to a rate of 100 m³/d (600 B/D) from 1 October 1985 to 7 February 1987. The volume of polymer injected at 1,000 ppm was 24 600 m³.

Polymer concentration was reduced in steps until 13 October 1987, when chase water injection began at a rate of 100 m³/d. Approximately 22 600 m³ of

Fig. 5.84—Isopach of Griselles interval, Courtenay Field (Putz et al. 1988).

polymer solution was injected at decreasing concentrations. Total polymer solution injected was 0.81 PV.

Fig. 5.86 (Putz et al. 1988) shows the wellhead pressures and injection rates of Well 543 and production rates from the pattern. Putz et al. (1988) noted that after September 1986, the wellhead pressure was constant even when rates or polymer concentration changed. It was suggested that this behavior was consistent with injection above the parting pressure. Oil cuts increased significantly in Wells CY30, CY 48, and CY 41. A small increase in oil cut was observed in Well CY40, indicating a minimal response of that well to the polymer flood.

For the period from 1 October 1986 to 31 May 1988, cumulative oil produced was estimated from Fig. 5.86 at 19 400 m³. Incremental oil attributed to polymer injection was approximately 11 700 m³ (73,600 bbl), which is equivalent to 28% of the OOIP. The amount of polymer injected was 38.2 t (84,216 lbm). Incremental oil was approximately 0.87 bbl/lbm polymer. Analysis of the produced polymer indicated that there was no significant mechanical or chemical degradation of the polymer after flow through the formation. The analysis also indicated an increase in molecular weight with time, suggesting that fractionation of the polymer occurred during transport through the reservoir.

Commercial Polymer Flood (1989–93) (Putz et al. 1994). In 1989, the pilot flood area was included in a commercial flood. This flood was conducted in three patterns, as shown in **Fig. 5.87** (Putz et al. 1994). The pore volume of these regions was estimated to be 640 000 m³ (4.03 million bbl). Original oil in place in the pilot region was 448 000 m³ (2.82 million bbl). The project area consisted of four injection wells and 18 producing wells. Well 543 was shut in and Wells 30 and 41, pilot flood producers, were converted to injection wells. There were three contiguous zones within the project area. Injection of polyacrylamide polymer began on 1 July 1989 at an average rate of 380 m³/d (2,390 B/D) and continued through November 1993, when injection was converted to water at an average rate of 420 m³/d. Polymer concentration was maintained at 900 ppm until October 1991, when the concentration was decreased in steps (as shown in **Fig. 5.88**) (Putz et al. 1994) to zero when water injection began. A total of 409 t of polymer was injected during this period of time.

Responses of individual wells are described in Arshad et al. (1994). The field production history is shown in **Fig. 5.89** (Putz et al. 1994). Cumulative oil production as of May 1994 was 122.1 kt (860,000 bbl). Incremental oil attributed to polymer injection was estimated to be 47.8 kt (336,572 bbl), which was equivalent to 116.9 kt oil/kt polymer (0.374 bbl/lbm

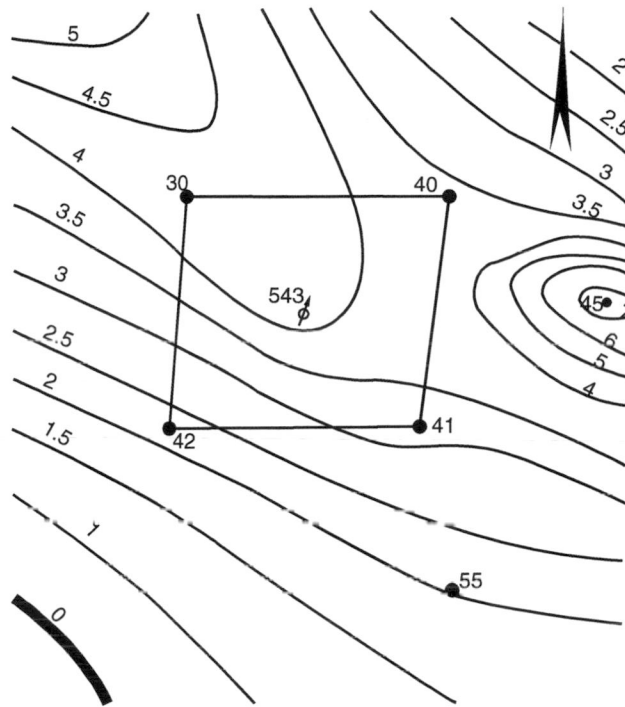

Fig. 5.85—Isopach of five-spot pattern—Courtenay polymer pilot (Putz et al. 1988).

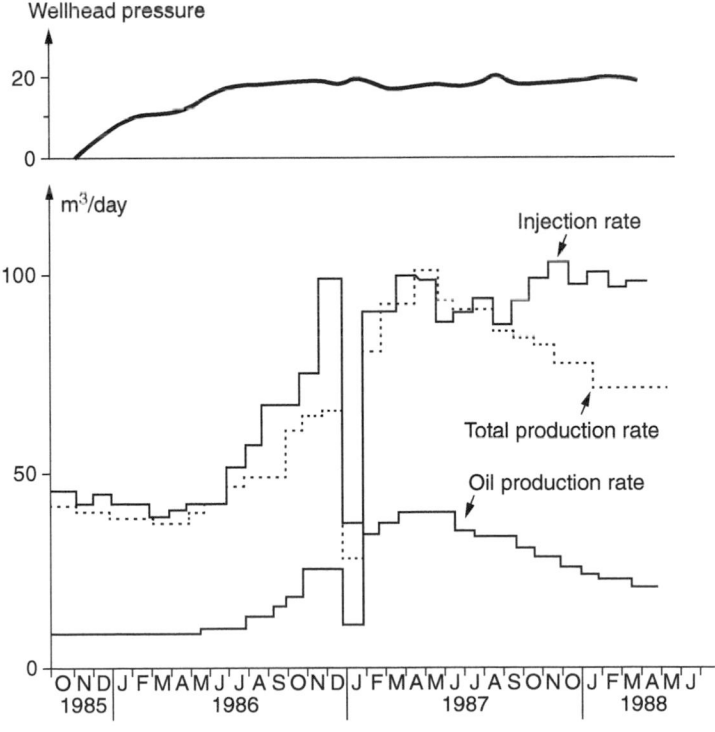

Fig. 5.86—Wellhead pressure and injection rate of Well 543 and total production and oil production rates from the Courtenay polymer pilot (Putz et al. 1988).

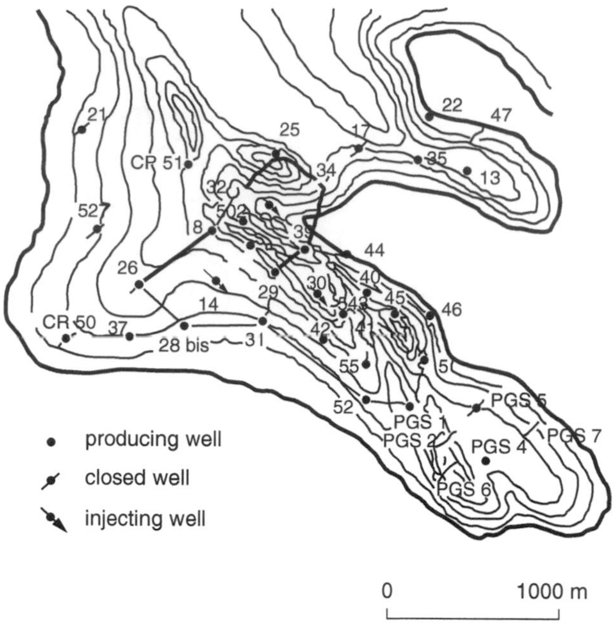

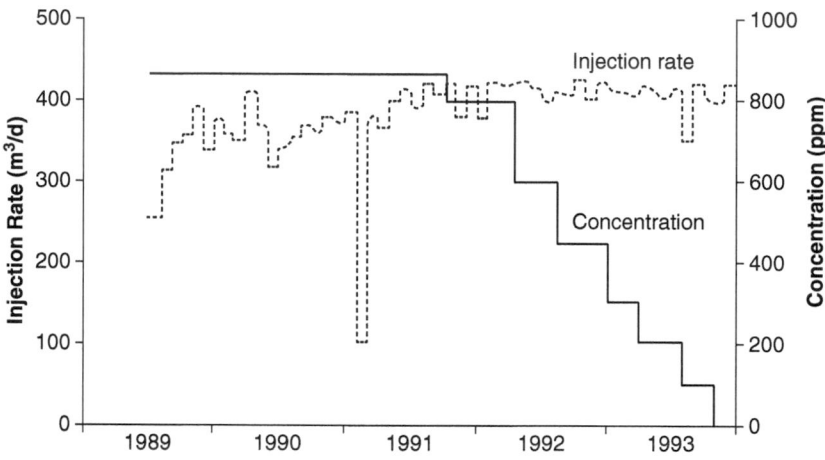

Fig. 5.87—1322 Isopach of Courtenay Field showing commercial polymer-injection areas (Putz et al. 1994).

polymer). Incremental oil resulting from the polymer flood was approximately 12% of OOIP.

During the flood, 22.2 t of polymer was produced from 11 wells, with amounts varying based on the location of the well. Three wells (28-bis, 34, and PGS 1) did not respond to polymer injection. Analysis of the produced polymer confirmed that polymer propagated through the reservoir retained good rheological properties with minimal degradation. Chromatographic separation of polymer molecules was observed. These results were consistent with observations during the pilot flood.

Polymer Floods—Daqing Oil Field. The Daqing Oil Field (Abib et al. 1991), shown in **Fig. 5.90,** is the largest oil field in the People's Republic of China and contained an OOIP in excess of 2 billion tons (≈14 billion bbl). The oil field is located 1000 km north of Beijing in a sedimentary basin and was discovered in 1959 (Delamaide et al. 1994). It consists of three reservoirs [Goataizi, Putaohua(P) and Saertu(S)], which were deposited along an anticline that extends from 700 to 1200 m deep along a N-NE/S-SW trend.

Properties of the Daqing reservoir fluids and rocks are summarized in **Table 5.48** (Chauveteau et al. 1988). The reservoirs are heterogeneous, containing sandstone bodies of various sizes that are (1) sheet sands interbedded with shales, but with good continuity, and (2) channel deposits with poor lateral

Fig. 5.88—Injection-rate and polymer-concentration schedule for a commercial flood (Putz et al. 1994).

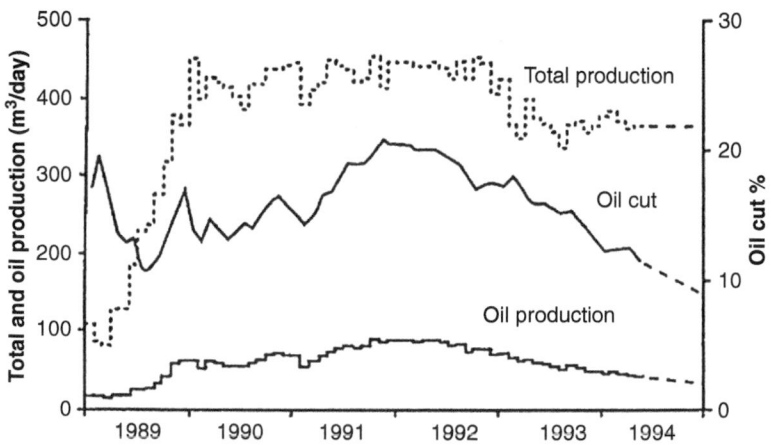

Fig. 5.89—Field production from the Courtenay commercial polymer flood (Putz et al. 1994).

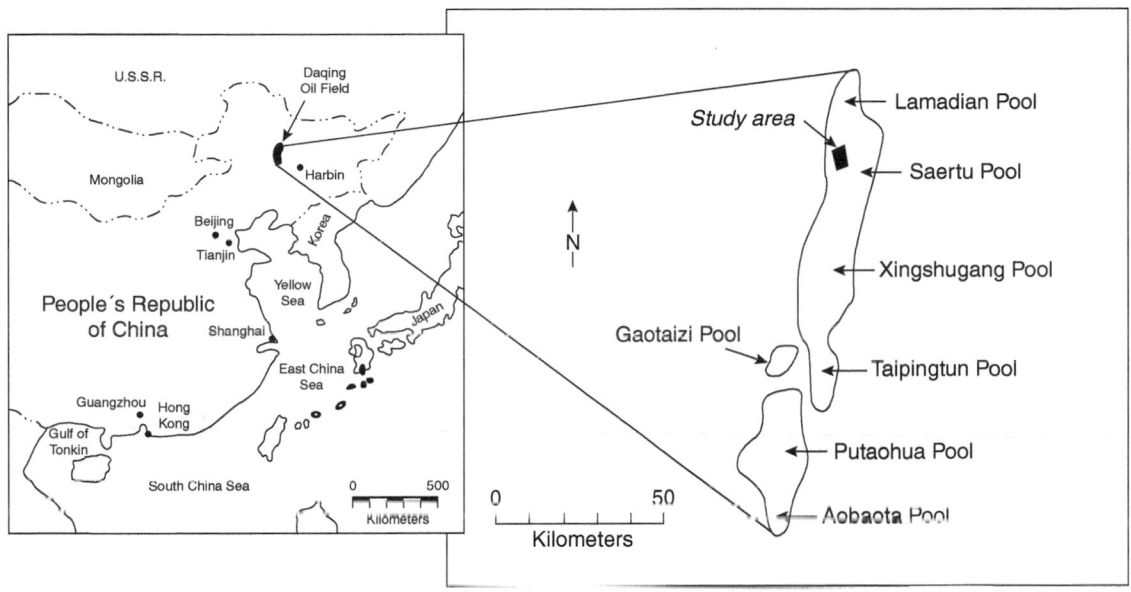

Fig. 5.90—Outline of Daqing Oil Field (Abib et al. 1991).

continuity. **Fig. 5.91** (Chauveteau et al. 1988) is a cross section through the Putaohua interval depicting the reservoir heterogeneity. Poor lateral continuity of channel deposits suggests that sweeping these deposits effectively by water or polymer may require infill drilling to develop enough interwell communication to displace the oil to production wells, or accepting low volumetric sweep efficiency.

Water injection began in 1963 to maintain reservoir pressure. Since 1979, more than 10,000 infill wells have been drilled to help maintain production at 1 million B/D. Water cuts in producing wells were on the order of 90% or higher. Average oil recovery at water cuts of 90 to 95% was approximately 34.6% (Wang et al. 2008).

At Daqing, the endpoint relative permeabilities for the oil/water/reservoir rock were $k_{ro} = 0.8$ and $k_{rw} = 0.5$, which is characteristic of a weakly oil-wet environment (Wang et al. 2008). Water viscosity at reservoir temperature is 0.6 cp. The endpoint mobility ratio based on the oil viscosity of 9.5 cp is 9.9, which is unfavorable. Viscous fingering and poor volumetric sweep efficiency occurred during waterflooding (Wang et al. 2009). Injecting a polymer solution with an apparent viscosity of 35 to 40 cp would reduce the endpoint mobility ratio to 0.17 to 0.15, which would reduce viscous fingering, increase volumetric sweep efficiency, and reduce the water cut. This potential was the basis for the development of polymer flooding at Daqing.

Polymer flooding in Daqing began in the 1980s with a pilot project in the central part of the field. The results of this pilot project led to an expansion to a multiwell project on larger well spacing (Wang et al. 2002). Expansion to a large commercial project began in 1996. By 2001, annual production attributed to polymer flooding was more than 70 million bbl (Wang et al. 2013). Polymer flooding involved more than 3,000 wells, 1,300 injectors, and 1,700 production wells in an area of 44,000 acres. By the end of 2011, annual production from polymer flooding in Daqing was more than 10 million tons per year

Net thickness of individual layers [m (ft)]	1 to 20 (3 to 65)
Permeability (md)	50 to 5,000
Porosity (%)	25 to 30
Original saturation (%)	60 to 80
Wax content (%)	25
Pour point [°C (°F)]	25 (77)
Original gas/oil ratio [m³/m³ (ft³/bbl)]	40 (225)
Formation volume factor	1.12
Saturation pressure [bar (psig)]	90 (1,300)
Specific gravity	0,86 (33 °API)
Oil viscosity (at initial reservoir conditions) (cp)	9.5
Original reservoir temperature [°C (°F)]	40 to 45 (105 to 115)
Original reservoir pressure (average) [bar (psig)]	100 (1,450)
Average water salinity (currently in place) (g/1)	<7

Table 5.48—Properties of Daqing reservoirs and fluids (Chauveteau et al. 1988).

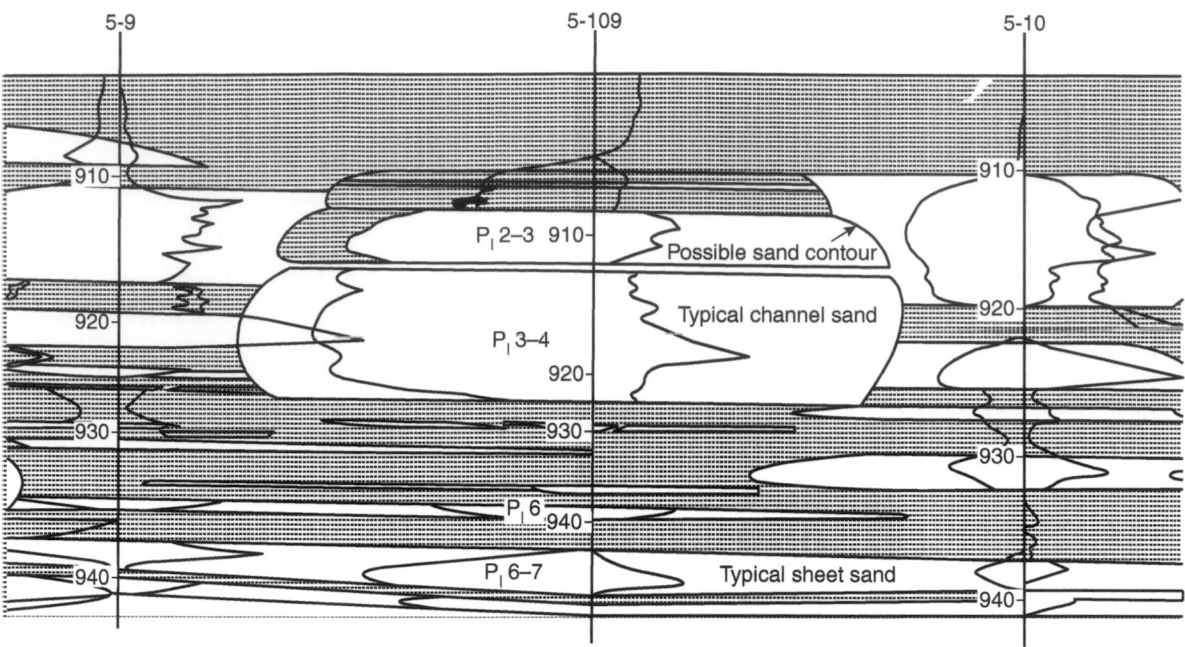

Fig. 5.91—Cross section through the P interval in the West Central Saertu area of Daqing (Chauveteau et al. 1988).

(73 million bbl) and had been sustained for 10 years, with cumulative production by the end of 2011 (Wang et al. 2013) at 1.58×10^8 ton (1.15 billion bbl) (Wang et al. 1993). Average incremental recovery resulting from polymer flooding is 13.3%. This project is the largest commercial-scale application of polymer flooding in the world. The following section summarizes field results from some of the pilot projects that led to commercial-scale operation.

Pilot Polymer Floods—West Central Saertu (1989–92). Two pilot floods were planned and completed for the West Central Saertu area of the field (Delamaide 1994; Wang et al. 1993). The results of the polymer flooding are presented in Wang et al. (1993). The pilot patterns shown in **Fig. 5.92** (Delamaide et al. 1994) consisted of four inverted five-spots with spacing between injection and production of 106 m. The distance between producers was 150 m, and the area of each pilot was 90 000 m^2 (22.2 acres). Pilot One (PO) was completed only in the P formation in P I_{1-4}. Pilot Two (PT) was conducted in P(P I_{1-4}) and S(S I_{1-3}) intervals. All wells were new and were cored and logged. Correlation of sand bodies on the basis of these data was difficult because of the heterogeneity of the deposits. The PV of PO was 319 368 m^3 (2,008,505 bbl), and the PV of PT was 581 988 m^3 (3,660,122 bbl).

Properties of the formation and fluids are summarized in **Table 5.49** (Delamaide 1994). Details of the well completion and operation of the pilots are found in Delamaide et al. (1994).

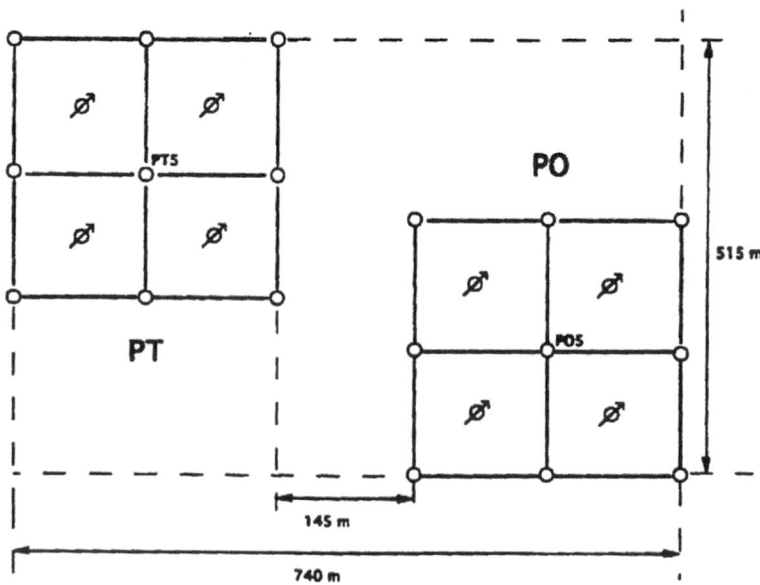

Fig. 5.92—Inverted five-spot patterns in the West Central Saertu polymer pilot (Delamaide et al. 1994).

The P and S formations were waterflooded beginning in June 1960 by injection into multiple layers. Early breakthrough of water led to development of multilayer injection in 1965. Interpretation of well logs from 30 wells drilled in 1988 indicated that the vertical sweep of the waterflood for S I_{1-3} was 30 and 50% for the P I_{1-4} interval and low volumetric sweep of the reservoirs, respectively.

Preflush Waterflood. Water injection was conducted in both pilots to condition the reservoir before polymer injection. Goals included (1) stabilizing the saturation distribution after infill drilling in a region that had been waterflooded for many years, (2) establishing communication between injection and production wells, and (3) flushing saline formation water (7,000 ppm) from the formation to take advantage of the high viscosity of HPAAM solutions at low salinity.

The PO pattern waterflush began on 9 December 1989 and ended on 4 August

Properties	P I$_{1-4}$	S I$_{1-3}$
Net thickness (m)	6.6 to 19.2	2.3 to 11.6
Average thickness (m)	15	6.1
Average air permeability (md)	1,400—PO area	1,100
	3,800—PT area	–
Dykstra-Parsons coefficient	0.6 to 0.8	0.5 to 0.7
Salinity formation water (ppm)	7,000	7,000
Salinity injected water (ppm)	800 to 1,300	
Salinity produced water (ppm)	2,000 to 4,000	
Calcium and magnesium (ppm)	<20	

Table 5.49—Properties of P I$_{1-4}$ and S I$_{1-3}$ intervals in pilot areas (Delamaide et al. 1994).

1990. Cumulative water injected was 28.5% PV at an average rate of 100 m³/d (629 B/D). Cumulative oil produced was 16,124 t (≈20,999 m³, assuming B_o = 1.12 m³/m³), which is equivalent to 0.066 saturation units. Average water cut at the end of the waterflood was 95%.

The PT pattern waterflush began on 5 December 1989 and ended on 6 November 1990. Cumulative water injected was 28.3% PV at an average rate of 200 m³/d (1,258 B/D). Cumulative oil produced was 28 818 t (≈37 530 m³, assuming B_o = 1.12 m³/m³), which is 0.064 saturation units. Preflushes were not used on subsequent polymer floods.

Polymer Floods. The polymer flood plan was based on an injection of 0.5 PV of polymer solution containing 915 ppm polymer. A partially hydrolyzed polyacrylamide (HPAAM) with a molecular weight of 10 million Daltons was used. The degree of hydrolysis was chosen to minimize polymer retention and give a high intrinsic viscosity (Chauveteau et al. 1988). Polymer concentration was selected to give a mobility ratio of approximately unity for the polymer flood. Polymer concentration was stepped down after the main slug was injected to minimize channeling of the chase water.

Polymer injection began on 5 August 1990 in the PO pilot at a rate of 100 m³/d for each injection well. Wellhead pressure rose 2 to 3 MPa (290–435 psi) compared to water injection, with average injectivity decreasing to approximately 36% of the water injectivity. **Table 5.50** (Wang et al. 1993) summarizes the injection data for the PO pilot.

In the PT pilot, polymer injection began on 7 July 1990 and was completed on 24 February 1992. **Table 5.51** summarizes the polymer injection schedule (Wang et al. 1993).

Polymer injection in both pilots was completed when the water cut reached 95% in the central wells. **Fig. 5.93** (Delamaide et al. 1994) illustrates the oil rate and water cut of the PO pilot. The oil rate attributed to water injection was approximately 50 m³/d. Oil rate increased from ≈50 m³/d to 150 m³/d approximately 150 days after polymer injection began and was accompanied by a decrease in the water cut from 92.6 to 76.6%. Water cuts returned to 90% as oil rates decreased to prepolymer-injection rates at the end of the flood. Cumulative oil recovered was 73 120 m³ from the injection of 161 tons of polymer. This is a ratio of 1.3 bbl oil/lbm polymer. Cumulative oil corresponding to the initial waterflood was approximately 40 000 m³ (800 days × 50 m³/d), so that the incremental oil resulting from polymer flooding was approximately 33 100 m³, which is equivalent to 0.59 bbl/lbm polymer.

Oil rates and water cuts from the PT pilot are shown in **Fig. 5.94** (Delamaide et al. 1994). In this pattern, the water cut increased from 80 to 92% near the end of the preflush period, which was the response to water injection. The oil rate increased sharply after polymer injection began, rising from approximately 50 m³/d at the end of the preflush period to approximately 225 m³/d before declining toward the prepolymer-flood production rate. Individual well response was mixed, with some wells having increased oil rates and decreased water cuts while other wells did not respond. Cumulative oil produced from the PT pattern was 118 950 m³ (748,171 bbl) from the injection of 285 tons (628,311 lbm) of polymer. The oil/polymer ratio for the PT pattern was 1.19 bbl/lbm polymer.

In both patterns, the production of substantial volumes of polymer solution at concentrations as high as 400 ppm was observed and attributed to vertical heterogeneity in each reservoir interval. Overall performance of the pilot floods is presented in **Table 5.52** (Delamaide et al. 1994). The results of these polymer pilot floods led to the decision to implement polymer flooding on a large scale in 1992.

Injectivity. **Fig. 5.95** (Delamaide et al. 1994) is a plot of bottomhole flowing pressure in an injection well during the pilot flood when water injection was converted to polymer injection, as indicated earlier. Injection of polymer solutions with viscosities on the order of 25 to 35 cp should increase the bottomhole pressure by more than the 2 to 3 MPa reported by Wang[6].

According to Wang[6], the overburden pressure is approximately 2.3 MPa/100 m. Because depths of injection zones range from 800 to 1200 m, the bottomhole fracturing pressure ranges from 18.4 to 27.6 MPa. Thus, the wellhead fracturing pressures (less hydrostatic head) are in the range of 10.56 to 15.84 MPa, depending on the depth of the injection zones, which vary across the field. The mean subsurface depth of Well P01 was 920 m, which projects to a fracturing pressure of 21.2 MPa for this well. Injection pressure is controlled to be at least 1 MPa below the fracturing pressure. The cause of the small pressure increase of 2 to 3 MPa observed following the switch to polymer injection is not understood.

Commercial-Scale Polymer Pilot Test (1993–95) (Wang et al. 1995). The formation selected for the first commercial-scale polymer pilot test in Daqing was the Pu 1$^{1-4}$ intervals located in the northern part of the Saertu oil field (Wang et al. 1995). This

---

[2] Personal communication with D. Wang, Daqing Oil Company, 5 December 2013.

Injection Period	Solution Injected		Polymer Injected (t)	Average Polymer Concentration (ppm)	PV Injected (ppm)
	m³	PV			
5 August 1990 to 31 October 1991	168 502	0.527	140.69	835	440
1–30 November 1991	12 482	0.039	8.51	682	27
1 December 1991 to 14 February 1992	29 462	0.092	11.17	379	35
15–20 February 1992	2 573	0.008	0.63	245	2
Cumulative	213 019	0.667	161	756	504

Table 5.50—Polymer injection data–PO pilot (Wang et al. 1993).

Injection Period	Solution Injected		Polymer Injected (t)	Average Polymer Concentration (ppm)	PV Injected (ppm)
	m³	PV			
7 August 1990 to 31 December 1991	493 621	10.504	269.1	916	462
1 January to 21 February 1992	41 604	0.072	16.6	400	29
Cumulative	335 225	0.576	285. 7	852	491

Table 5.51—Polymer injection data for Pilot Two (Wang et al. 1993).

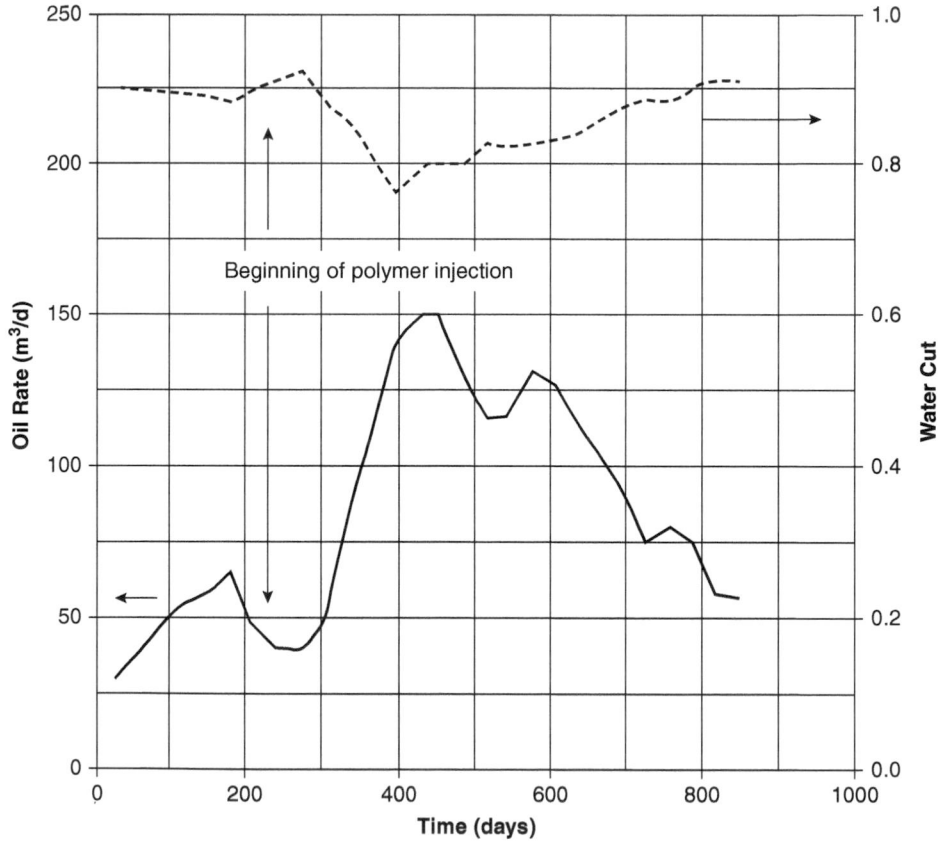

Fig. 5.93—Oil rate and water cut from the PO pilot polymer flood (Delamaide et al. 1994).

area was waterflooded using line-drive patterns beginning in 1963. Fifty new wells were drilled for the pilot test from the end of 1990 to the beginning of 1991 for the polymer pilot test. **Fig. 5.96** [Fig. 1 of Wang et al. (1995)] shows the well layout for this polymer flood. The layout consisted of 25 five-spot patterns with approximately 250 m spacing between injection and production wells. The test area contained 36 production and 25 injection wells; one well (6-27) was completed with reinforced

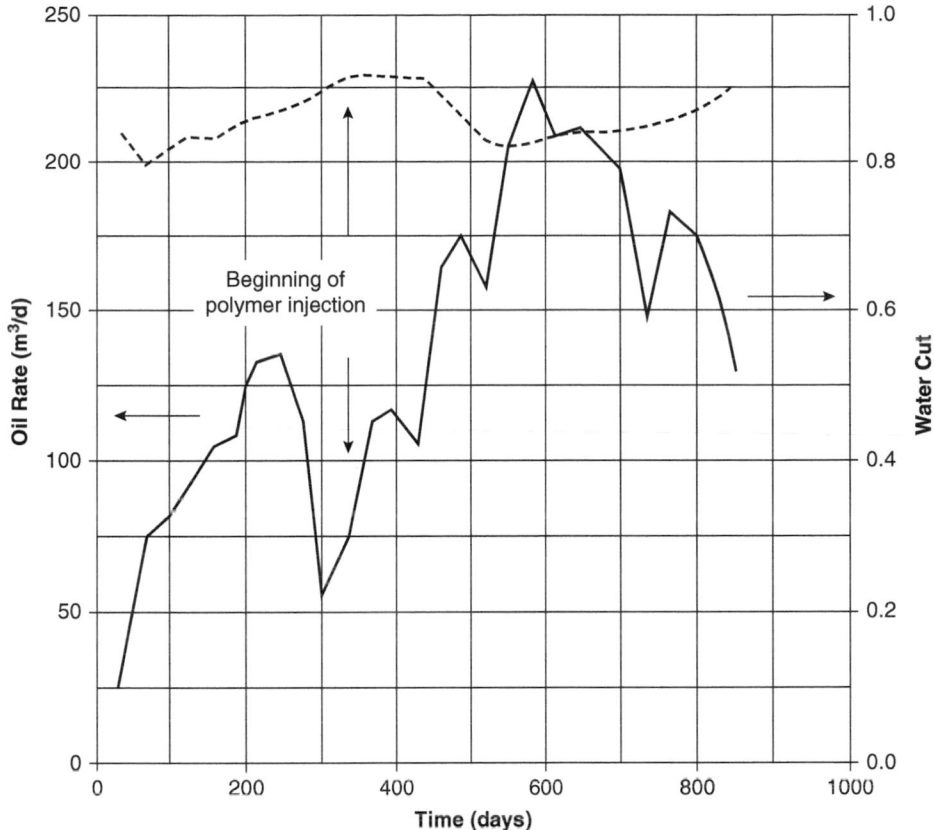

Fig. 5.94—Oil rate and water cut from the PT polymer pilot (Delamaide et al. 1994).

	PO	PT
Pore volume (m³)	313 369	581 988
Oil produced during water flush (t)	16 124	28 818
Oil produced during water flush (SG = 0.86) (std m³)	18 749	33 509
Reservoir volume of oil produced during water flush (B_o – 1.12) (m³)	20 999	37 750
Change in oil saturation because of water flush	0.067	0.064
Polymer injected (m³)	213 019	335 225
Polymer injected (t)	161	285.7
Oil attributed to polymer injection (t)	38 804	59 767
Oil attributed to polymer injection, (SG = 0.86) (std m³)	45 120	76 949
Oil attributed to polymer injection (B_o = 1.12) (m³)	50 535	77 836
Change in oil saturation attributed to polymer injection	0.161	0.134
Oil/polymer (t/t)	241	209
Oil/polymer (bbl/lbm)	0.82	0.77

Table 5.52—Overall performance of pilot polymer floods (Delamaide et al. 1994).

glass fiber to permit logging during the polymer flood for observation of fluid movement. The central well area, outlined in Fig. 5.96, contained nine injection wells and 16 production wells.

The reservoir is quite heterogeneous in terms of both thickness and permeability. A cross section through the Pu 1$^{1-4}$ test interval is shown in **Fig. 5.97** (Wang et al. 1995). This interval was subdivided into five sedimentary units; representative properties of the interval are summarized in **Table 5.53** (Wang et al. 1995).

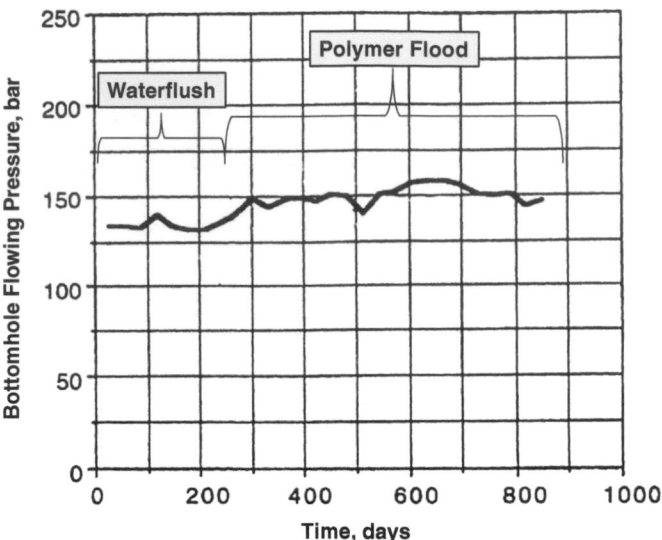

Fig. 5.95—Injection pressure during water flush and polymer flood (Delamaide et al. 1994).

This interval had been waterflooded from 1963 to 1990 in various layers. Several regions in the Pu 1^{1-4} interval were considered to be watered out. For example, **Fig. 5.98** (Wang et al. 1995) is an E/W cross section through the line of wells indicated by the dashed line on Fig. 5.96.

It is clear from Fig. 5.98 that the waterflood did not sweep the vertical cross section uniformly in spite of the high reservoir permeability. Good volumetric sweep appears to be present areally only at the bottom of each interval. This figure reinforces the interpretation that this unit has a high degree of heterogeneity both areally and vertically. Average residual oil saturation in the water-swept regions was 61.2%, and the watered-out thickness averaged 62.8%.

The pilot area was waterflooded for 15 months, from October 1991 to December 1992. The average water-injection rate in each well was 239 m^3/d; average liquid production was 220 t at a water cut of 88.7%. Cumulative oil production from the central well area was 16.72×10^4 t, which was 4.3% of the OOIP (390×10^4 t). Additional details can be found in Wang et al. (1995).

The polymer injection plan was based on injection of a 0.357 PV slug of 1,000 ppm polymer, followed by 0.03 PV slugs containing 500 ppm and 250 ppm polymer, respectively; however, 0.748 PV of polymer was injected at an average polymer concentration of 791 mg/L Viscosity of the solution ranged from 30 to 35 mPa.s; average daily injection rate was 239 m^3. Chase water was injected after polymer injection was completed. Polymer injection began on 8 January 1993 and was completed in April 1997 when water injection began (Guo et al. 2000). Water injection ended in October 1998. **Fig. 5.99** summarizes the injection data for the project (Guo et al. 2000).

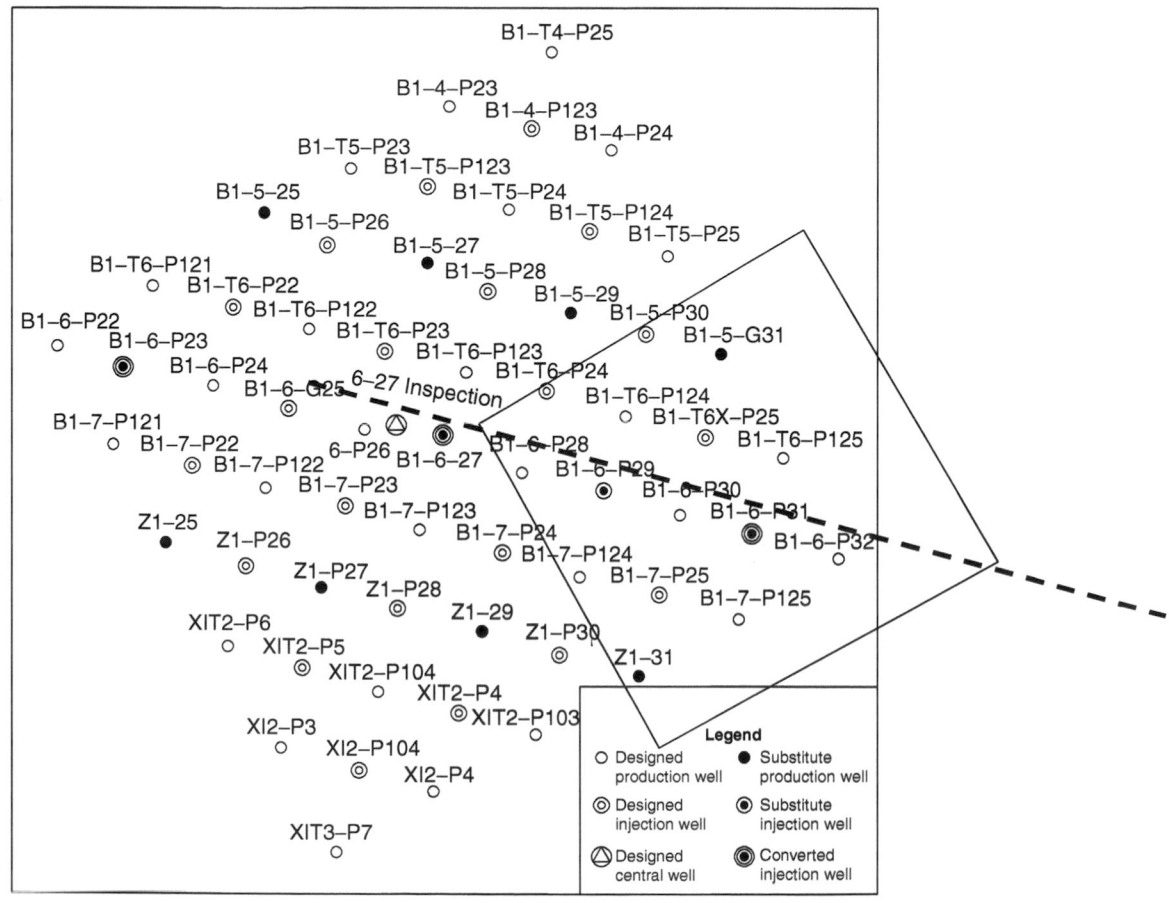

Fig. 5.96—Commercial-scale pilot polymer flood with 25 five-spot patterns (Wang et al. 1995).

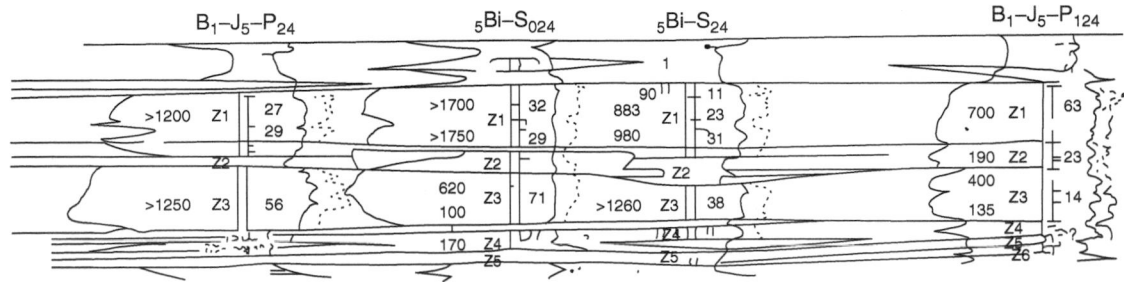

Fig. 5.97—Cross section showing five subzones of the Pu 1$^{1\text{-}4}$ interval (Wang et al. 1995).

Area of pilot test (m²)	3 130 000
Pore volume (m²)	10.86×10^6
Reservoir depth (m)	1036–1117
Reservoir thickness (m)	5.1–20.9
Average thickness (m)	13.2
Average porosity (fraction)	0.263
Permeability (μm²)	342×10^{-3} to 1166×10^{-3}
Average permeability (μm²)	870×10^{-3}
Dykstra-Parsons permeability variation coefficient	0.753–0.812
Reservoir temperature (°C)	45
Oil viscosity (cp)	9–10
Salinity of formation water (mg/L)	7000
Salinity of produced water (mg/L)	3000–4000

Table 5.53—Reservoir and fluid properties (Wang et al. 1995).

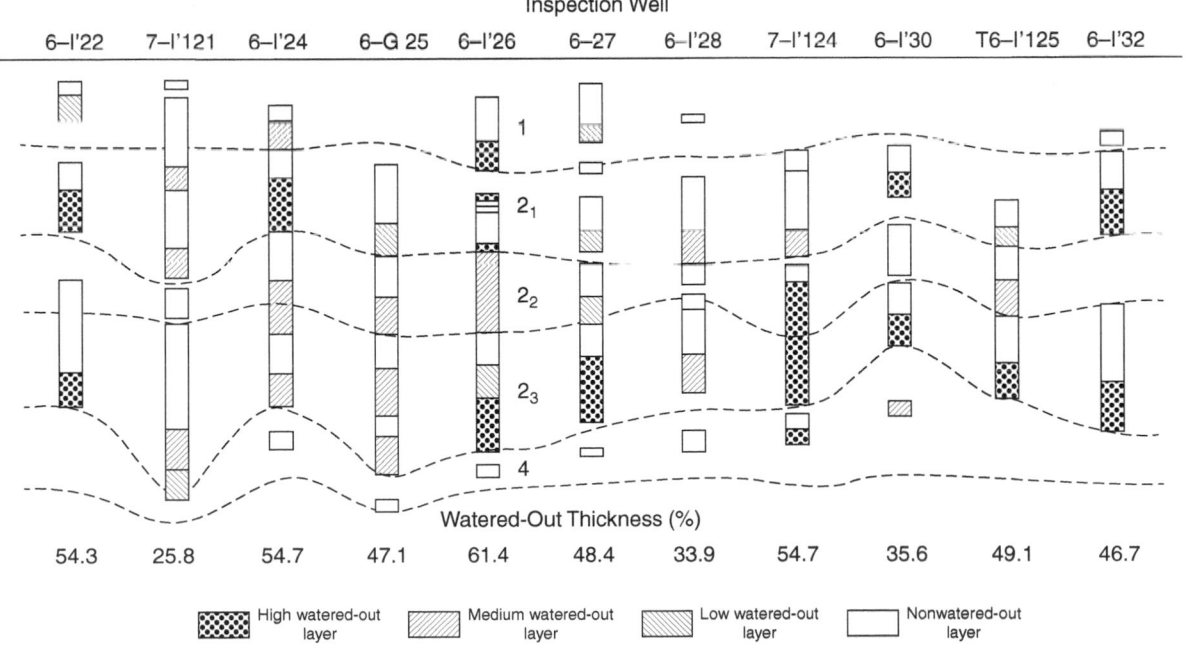

Fig. 5.98—Estimated watered-out regions in cross section in Fig. 5.96 (Wang et al. 1995).

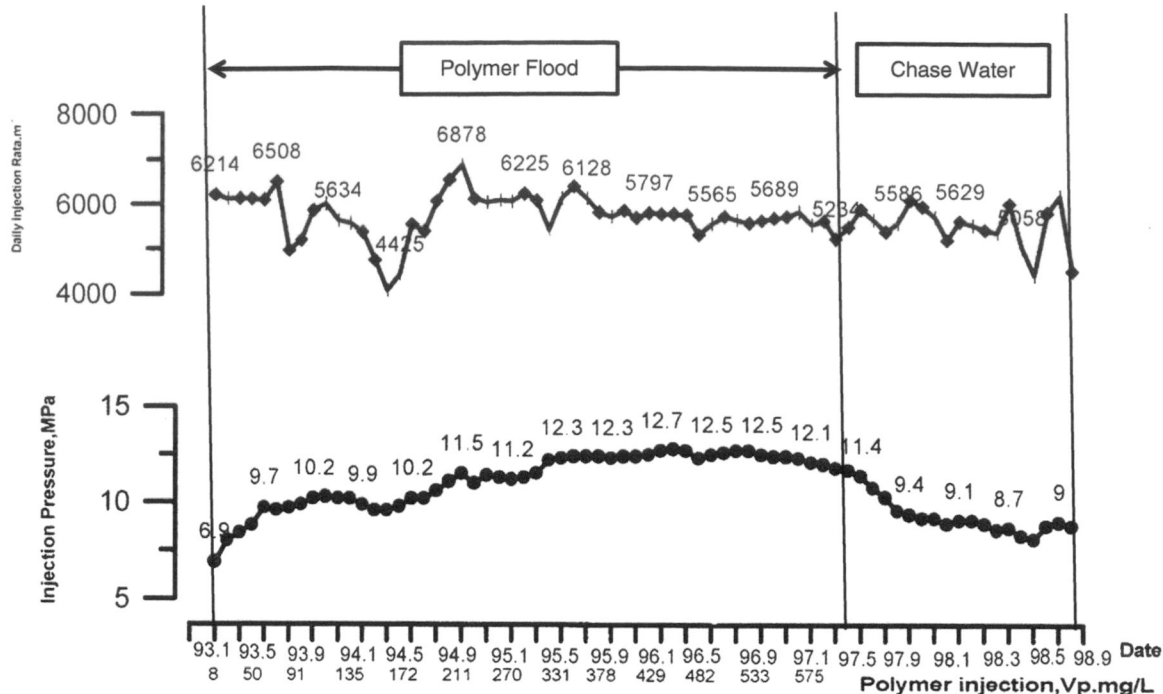

Fig. 5.99—Injection data (Guo et al. 2000).

Production response of the total and central well areas is shown in **Fig. 5.100** during the first 3 years of the pilot (Guo et al. 2000). The water cut decreased approximately 6 months after polymer injection, from approximately 90.75 to 76.1% in December 1994. Oil rate from the pilot region increased from 900 t/d during water injection to 1200 t/d at the end of 1994. Cumulative oil production from the beginning of water injection was 96.59 × 10⁴ t, with 59.9 × 10⁴ t attributed to polymer

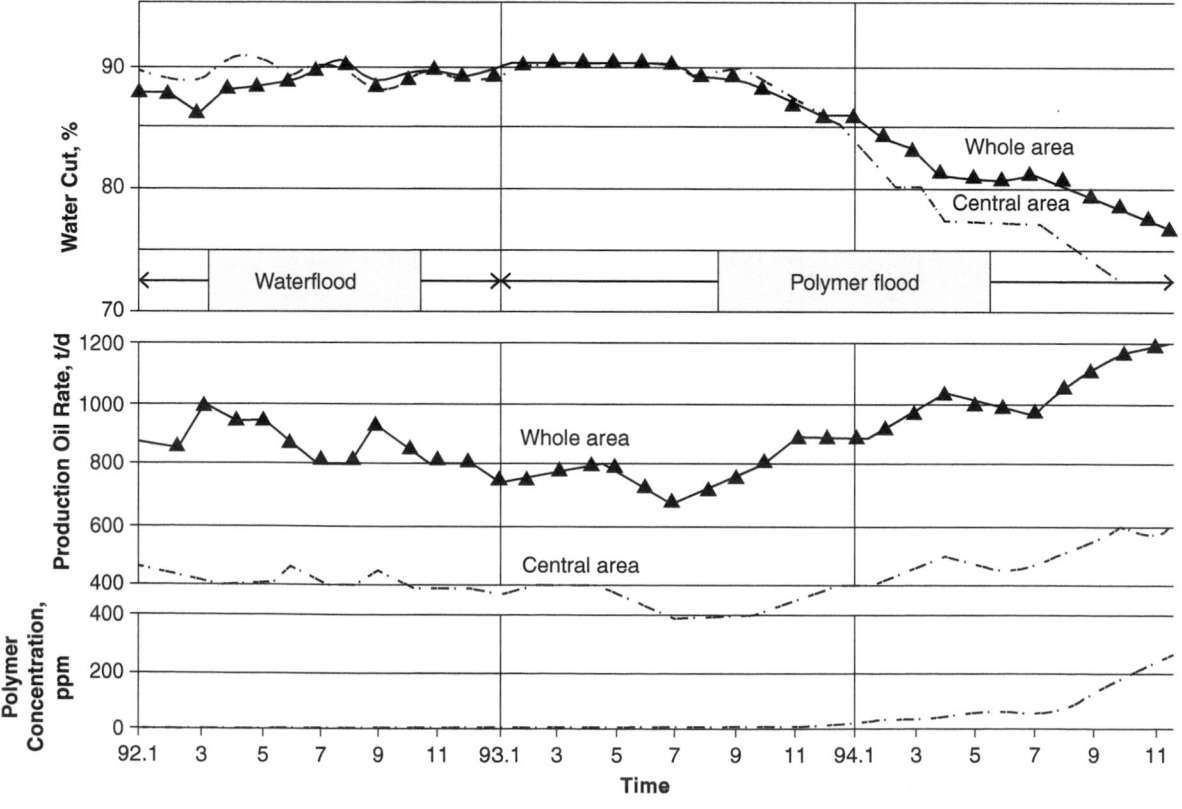

Fig. 5.100—Production response of a commercial-scale polymer flooding test (Wang et al. 1995).

flooding. Although the flood was still in progress when these data were presented, overall recovery was 12.6% OOIP (Wang et al. 2002). The 16 central production wells produced 70.62 × 10⁴ t of oil, which was 18.1% OOIP.

Fig. 5.101 presents the water cut and polymer concentration data in the produced fluids from the project (Guo et al. 2000). Assessment of the effectiveness of the polymer flood is difficult in a heterogeneous reservoir, where the volumetric sweep cannot be determined other than by estimating vertical sweep efficiency from limited data. In this project, analysis of cored wells before and after the polymer flood demonstrated that polymer flooding increased the effective flooded thickness after waterflooding (Guo et al. 2000). **Table 5.54** shows the distribution of flushed regions in each reservoir layer in Well 6-c27, which was cored in April 1991 before the polymer flood, and Well 6-c26, which was cored in June 1998, well after the polymer flood was completed.

High-Concentration Polymer Flooding. Laboratory studies of high-concentration polymer flooding have been interpreted to mean that increased oil recovery occurred because of the elasticity of the polymer (Wang et al. 2000). Wang et al. (2011) report that two field pilot tests of high-concentration polymer flooding, begun in 2002 in the Daqing oil field, began by converting 100 wells from conventional polymer flooding to high-concentration polymer flooding. In conventional polymer flooding, the injected polymer concentration was approximately 1,000 ppm, and the molecular weight of the polymer was approximately 15 to 18 million Daltons. Incremental recovery above waterflooding was 10 to 12% OOIP. In the high-concentration polymer flooding, the injected polymer concentration was approximately 2000 mg/L, and the molecular weight was 35 million Daltons. Incremental oil recovery resulting from polymer flooding was more than 21% OOIP.

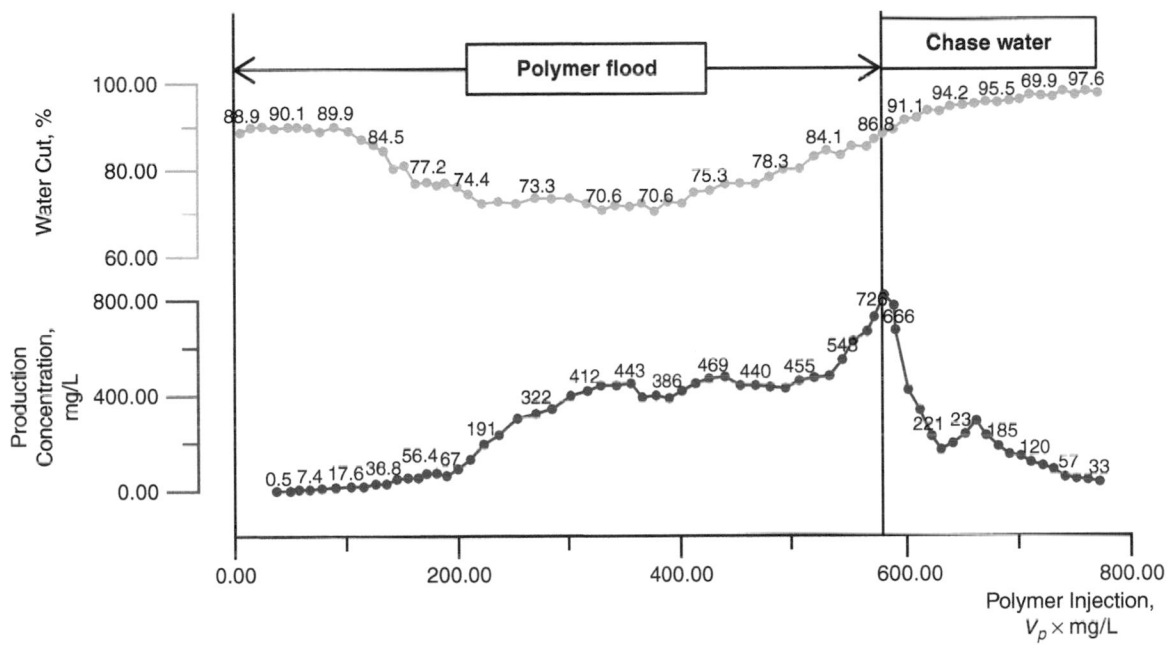

Fig. 5.101—Production and water-cut data as a function of polymer injection (mg/L×Vp) (Guo et al. 2000).

Well	Date Well Drilled	Reservoir Layer	Effective Thickness (m)	Thickness of Flush (m)					Oil Displacement Efficiency (%)	Recovery (%)
				Strong Flush	Medium Flush	Weak Flush	Total	%		
6-c27	Post WF (91.4)	P11²	2	0.54	1.37	–	1.91	95.5	50	47.8
		P12	7.3		2.66	0.49	3.15	43.2	38.9	16.8
		P13	5.1	3.82	1.1	–	4.92	96.5	66.3	64
		P14	0.5		0.15	–	0.15	30,0	38.3	11.5
		Total	14.9	4.36	5.28	0.49	10.13	68	54,2	36.8
6-c26	Post PF (98.6)	P11²	2.1	1.41	0.64	–	2.05	97.6	63.4	61.9
		P12	7.9	6.61	1.29	–	7.9	100	61.4	61.4
		P13	6.2	5.75	0.28	–	6.03	97.3	67.5	65.7
		P14	0.4	–	0.11	–	0.11	27.5	39.8	10.9
		Total	16.6	13.77	2.32	–	16.09	96.9	63.7	61.7

Table 5.54—Vertical sweep efficiency before and after pilot flood (Guo et al. 2000).

Application of high-concentration polymer injection was expanded to more than 5,000 wells, replacing conventional polymer flooding. Based on 1,000 wells with sufficient production history to predict total recovery, estimated incremental recovery is more than 20% OOIP. Because of these results, total recovery is estimated to be 42% OOIP for waterflooding and more than 60% OOIP for high-concentration polymer flooding. In-situ viscosity of the polymer solution was estimated to be at least 100 cp. Under these conditions, the endpoint mobility ratio would be 0.06, ensuring high volumetric sweep. The volumetric sweep efficiency for recovery of 60% OOIP would be approximately 80%.

Sweep Improvement. The successful application of polymer flooding at Daqing led to other developments to improve volumetric sweep efficiency (Wang et al. 2008; Wang et al. 2013). Polymer solutions lose viscosity because of mechanical/shear degradation in pumps and at the sandface, where polymer enters the porous rock at high velocities (Wang et al. 2009). Viscosity loss can be compensated by increasing the polymer concentration/molecular weight. For example, if the viscosities of injected solutions were 35 to 40 cp (Wang et al. 2008), loss of 50% of the viscosity because of degradation would result in a polymer solution with an in-situ viscosity ranging from 17.5 to 20 cp. Even with polymer degradation, the endpoint mobility ratio for polymer flooding would be 0.34 and 0.3, which remain favorable for effective volumetric sweep. The favorable mobility ratio also enhances crossflow in zones with different permeabilities (Wang et al. 2008).

Individual zones at Daqing have different permeabilities. Consequently, the molecular weight of the polymer that is effective for one zone may not allow entry into a zone of lower permeability. **Fig. 5.102** is a correlation of polymer molecular weight with molecular weight developed from laboratory data and reservoir cores (Wang et al. 2013).

This correlation provides guidance on the selection of the molecular weight for a particular interval. Even when controlling injection with molecular weight and concentration of the polymer solution, high-permeability channels dominate the vertical sweep in many wells. This has led to the development of a separate layer injection of polymer at Daqing. **Fig. 5.103** shows an estimate of the relationship between the pressure increases in the injection well vs. polymer concentration for different permeabilities (Wang et al. 2013). Distance between injection and production wells is 250 m. The molecular weight was 12 to 16 million Daltons for the calculations in Fig. 5.103.

Some layers at Daqing are isolated by barriers to vertical flow. The layers differ in permeability in the vertical direction as well as in the plane of flow. This led to the development of eccentric multilayer injection technology, which is used

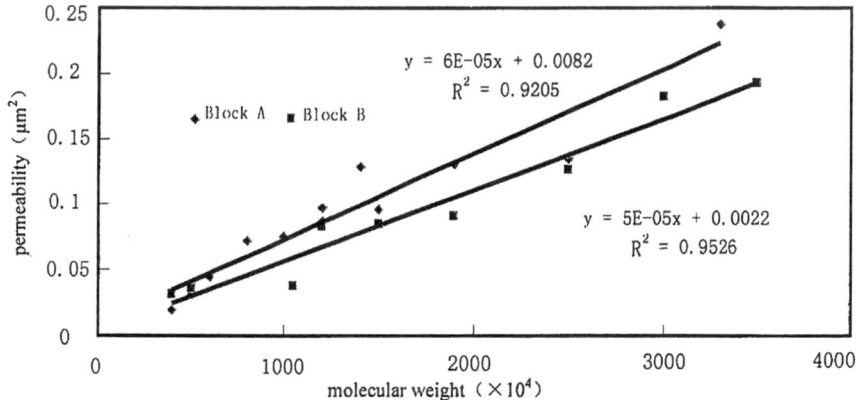

Fig. 5.102—Correlation of polymer molecular weight with permeability of reservoir cores from Block A and Block B (Wang et al. 2013).

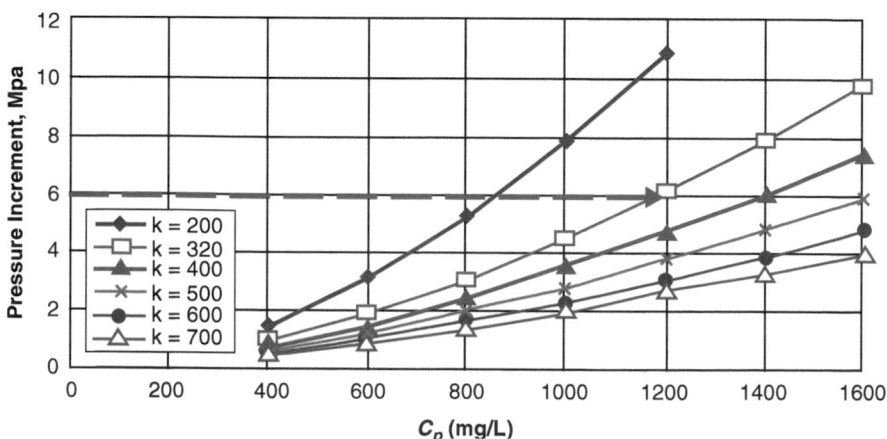

Fig. 5.103—Effect of polymer solution concentration on pressure under different conditions (Wang et al. 2013).

at Daqing. Injection into multiple zones is done through an eccentric injection string, which consists of a series of packers and eccentric injection mandrels as shown in **Fig. 5.104** (Liang et al. 2011). Each zone is isolated by packers when the injection string is run.

The injection string can be fitted with two types of blanking plugs. A pressure blanking plug is used to control the injection rate into high-permeability layers. A molecular-weight blanking plug is used for low-permeability layers. This blanking plug is designed to degrade the injected polymer to the molecular weight that will increase injectivity in the low-permeability layer. The amount of degradation is determined by the flow rate and the nozzle size (Liang et al. 2011). Both blanking plug and nozzle can be changed by wireline. **Table 5.55** is a comparison of injection rates in Well N1 based on the interpretation of oxygen-activation log data. The data show that this tool redistributed the injection rates, increasing the vertical sweep in this well. This technology has been applied in more than 300 wells in Daqing (Liang et al. 2011). The string can be used in wells where the distance between layers is greater than 8 m (He et al. 2013).

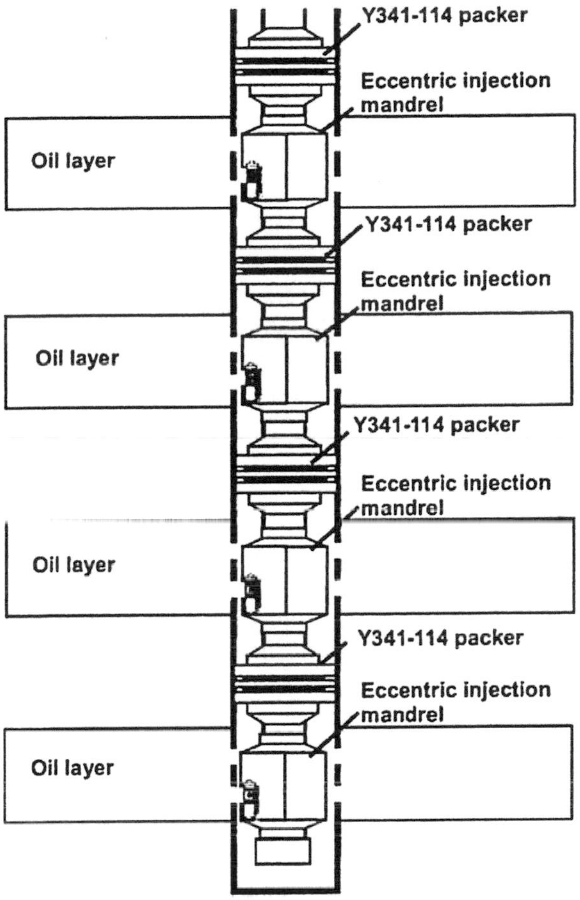

Polymer Flooding Using Horizontal Wells—Pelican Lake Field, Alberta, Canada (Delamaide et al. 2013). The Pelican Lake heavy oil field (also called the Brintnell) in Alberta, Canada, shown in **Fig. 5.105,** is a thin reservoir (average thickness is approximately 5 m) at a shallow depth (300–450 m), which contains more than 6 billion bbl of oil (Delamaide et al. 2013). Crude-oil viscosities range from 600 to 80,000 cp. The reservoir is composed of unconsolidated sands, which extend over an area of approximately 177 000 ha. Reservoir energy is low, with little dissolved gas (4–6 m³/m³) and limited waterdrive. Properties for the reservoir and fluids are presented in **Table 5.56** (Delamaide et al. 2013).

Early attempts to develop production from the reservoir using vertical wells were uneconomic. In 1988, CS Resources

Fig. 5.104—Eccentric separate injection string for rate and molecular weight control in each layer (Liang et al. 2011).

began a program to develop production using horizontal wells. Horizontal wells enabled more access to the reservoir volume and performed significantly better than vertical wells. As the technology for drilling horizontal wells developed over a period of several years, multilateral wells were developed (**Fig. 5.106**), leading to development of the entire field for primary production using horizontal wellbores (Delamaide et al. 2013).

Even with horizontal wells, primary production was estimated to be approximately 5 to 7% OOIP. Other methods of increasing recovery were investigated because the reservoir was too shallow and too thin to apply thermal methods.

Polymer flooding is generally not considered as a recovery process for viscous oil reservoirs developed using vertical wells because injectivity is limited because of the high pressure drop around the wellbore and the limited vertical exposure of the reservoir. Development of horizontal drilling in thin formations with wellbores on the order of 1250 m offered the potential of increasing recovery by injecting a viscous polymer solution.

The first polymer flood was a pilot project that was developed and implemented in July 1997. The pilot flood consisted of three 1250-m-long horizontal wells, shown in **Fig. 5.107** (Delamaide et al. 2013). The middle well was used as the

| | | | | Commingled Injection | | | Different-Molecular-Weights Polymer Injection | | | |
Layer	Type	Effective Thickness (m)	Effective Permeability (μm²)	Absolute Intake (m³/d)	Relative Intake (%)	Molecular Weight (10⁴)	Absolute Intake (m³/d)	Relative Intake (%)	Designed Molecular Weight (10⁴)	Actual Molecular Weight (10⁴)
SII10-12 SII13	Enhanced	8.50	0.72	15.9	14.84		20.92	18.66	800	785
SII14-16 SIII 1-3	Restricted	13.80	>1.2	91.30	85.17	1200	74.66	66.66	1200	1200
SII5+6 SIII8-9	Enhanced	1.80	0.21	0	0		15.42	600	600	532

Table 5.55—Comparison of oxygen-activation log data for Well N1 (Liang et al. 2011).

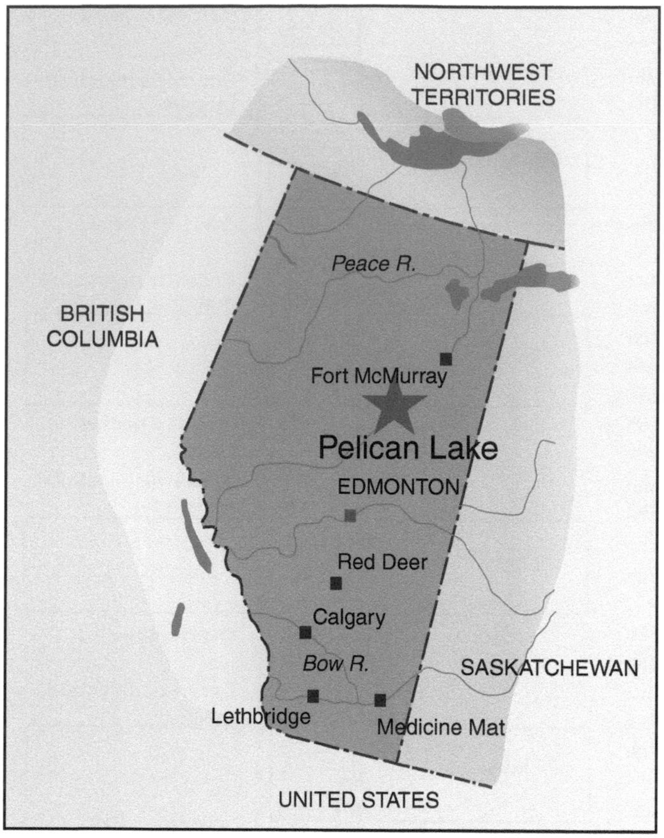

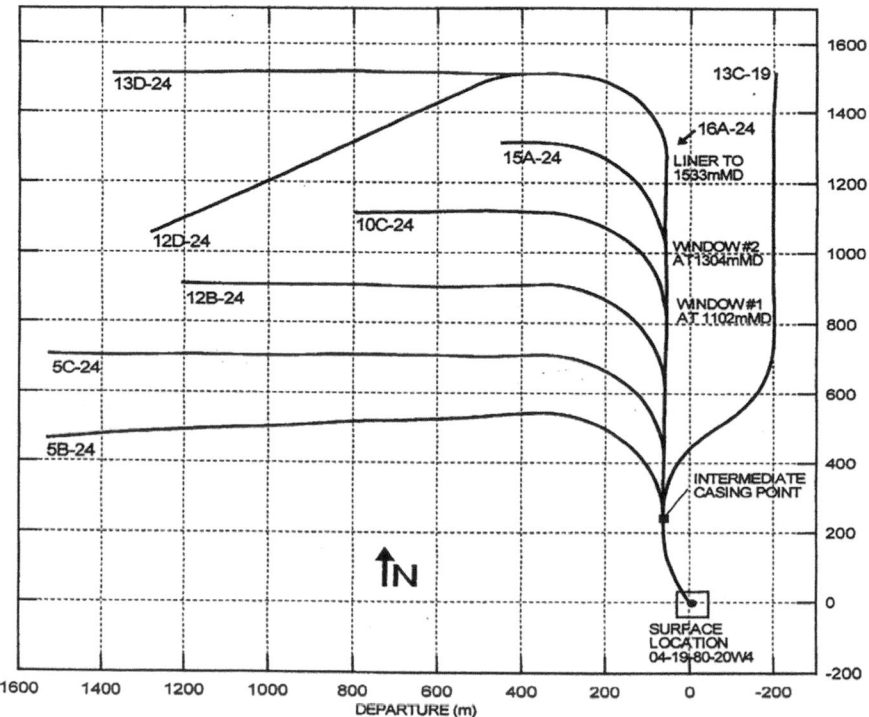

Fig. 5.105—Location of Pelican Lake Heavy Oil Field (Delamaide et al. 2013).

Depth (m)	300–450
Thickness (m)	1–9
Average thickness (m)	5
Porosity (fraction)	0.28–0.32
Permeability (md)	300–3,000
Oil saturation (volume fraction)	0.6–0.7
Temperature (°C)	12–17
Initial pressure (kPa)	1,800–2,600
Oil gravity (°API)	11.5–16.5
Dead oil viscosity (cp)	600–80,000

Table 5.56—Reservoir and fluid properties– Pelican Lake heavy-oil field (Delamaide et al. 2013).

Fig. 5.106—Multilateral horizontal well plan for Well 13D-24-80-21W4M (Delamaide et al. 2013).

injection well. Numerous operational difficulties were encountered, and the desired injectivity could not be achieved. The pilot was suspended in 1998 (Zaitoun et al. 1998). The property changed hands through a series of mergers and acquisitions; CS Resources was overtaken by PanCanadian in July 1997, and PanCanadian's interest in the field was sold to Canadian Natural Resources (CRNL). PanCanadian then merged with the Alberta Energy Company (AEC), the other major operator in Pelican Lake, to form EnCana. Later on, Cenovus became the corporate successor to EnCana. These activities delayed further evaluation of polymer flooding for several years.

Waterflooding pilots using horizontal wells were begun near the end of 2000 by AEC. EnCana (EnCana 2005) initiated waterfloods in other parts of the field in 2002. Further development of polymer flooding was delayed by implementation of waterflood pilots. CRNL initiated a waterflood pilot in 2003 consisting of 12 horizontal injection wells, as shown in **Fig. 5.108** (Delamaide et al. 2013). Response to water injection occurred within 3 months, as shown in **Fig. 5.109** (Delamaide et al. 2013). Incremental recovery attributed to waterflooding was estimated to be 12%.

The successful implementation of waterflooding in a viscous oil reservoir led to fieldwide expansion by both CRNL and other operators in a series of projects approved by the regulatory agency (Energy Resources Conservation Board, succeed by Alberta Energy Regulator in 2013). Waterflood recovery was estimated to be on the order of 7.5 to 10% OOIP.

Polymer flooding was revisited on the basis of laboratory work and re-evaluation of prior experience in 2004 (Delamaide et al. 2013). CRNL developed a polymer flood pilot in pad HTLP 6 consisting of five horizontal wells (1500 m in length), shown in **Fig. 5.110** (Delamaide et al. 2013). Three production wells (14-34, 15-34, and 16-34) were separated by injection wells 2/15 34 and 2/16 34, as shown in **Fig. 5.111.**

Polymer injection began in May 2005 at an initial rate of 930 B/D per well. Polymer concentration was 600 ppm, giving a viscosity of 20 cp in the injected water. Wellhead pressure increases led to a reduction of the injection rate and polymer concentration (13 cp) in August 2005, as shown in **Fig. 5.112** (Delamaide et al. 2013). Maximum wellhead injection pressure (MAWHIP) was limited by regulatory approval in 2005 to 7650 kPa to avoid fracturing the formation.

The response of the pilot to polymer injection is shown in **Figs. 5.113 through 5.115** (Delamaide et al. 2013) for each of the three horizontal production wells.

Average oil-production rate for the HTLP 6 pilot from the beginning of polymer injection was approximately 500 B/D, with the average water cut approximately 50%. Incremental oil recovery in the HTLP 6 pilot was estimated to be 15 to 21% OOIP (Delamaide et al. 2013).

The success of the HTLP 6 polymer pilot led to expansion of polymer flooding by CRNL in a multilateral pilot. Well locations and timing of injection are shown in **Fig. 5.116.** Pilot production data shown in **Fig. 5.117** average approximately 750 B/D, with the water cut increasing from less than 10% to approximately 30%.

Successful pilot projects in Pelican Lake led to commercial expansion of polymer flooding in 2008 to 51% of CRNL properties (Zaitoun et al. 1998) by 2013. Polymer flooding has expanded slowly by permit areas. Cenovus has also expanded polymer flooding to a significant portion of their field. Estimated ultimate recovery factors (primary, waterflooding, and polymer flooding) range from 15 to 32% in areas under flood by CRNL (Delamaide et al. 2013). Summaries of waterflood and polymer flood performance are found in EnCana (2002, 2003, 2005); Moriarty (2008); and Canadian Natural (2010, 2013).

5.8 Mobility Control To Maintain Chemical Slug Integrity

5.8.1 Design Mobility. Mobility-control design for chemical flooding processes is based on prevention of viscous fingering between the chemical slug and the oil/water bank and between the mobility buffer and the chemical slug. This can be achieved by selecting the properties of injected fluids so that the mobility ratio is unity. The method developed by Gogarty et al. (1970) is used to determine the design mobility for the chemical slug. The design procedure begins by defining the total relative mobility of the oil/water bank as

$$\lambda_n = k_{ro} / \mu_o + k_{rw} / \mu_w \dots \dots \dots \dots \dots (5.156)$$

and $\lambda_{rt} = \lambda_{ro} + \lambda_{rw}, \dots \dots \dots \dots \dots (5.157)$

where k_{ro} = relative permeability of the oil phase, k_{rw} = relative permeability of the water phase, μ_o = oil viscosity, μ_w = water viscosity, λ_{rt} = total relative mobility, λ_{ro} = relative mobility of the oil phase, and λ_{rw} = relative mobility of the water phase.

The total relative mobility is equal to the total mobility of the oil/water bank divided by the base permeability, which is usually the absolute liquid permeability or the permeability to oil at connate water saturation. Total relative mobility varies with saturation because relative permeabilities vary with saturations, as depicted in **Fig. 5.118.**

Because oil- and water-bank saturations are not known a priori, there is uncertainty as to what saturation should be used to evaluate the total relative mobility. Gogarty et al. (1970) proposed that the design mobility should be the minimum total relative mobility encountered over the saturation range. Their approach is conservative because the actual relative total mobility of the oil/water bank may be greater than the design mobility.

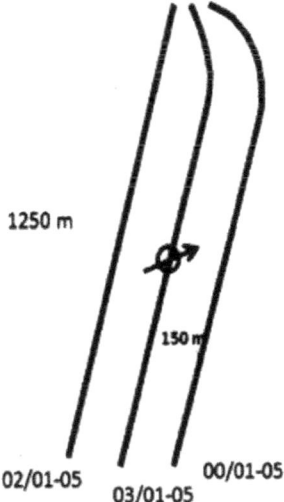

1250 m

150 m

02/01-05   00/01-05
       03/01-05

Fig. 5.107—Layout of horizontal wells for the first polymer pilot in the Pelican Lake Field in Township 81, Range 22 W4M (Delamaide et al. 2013).

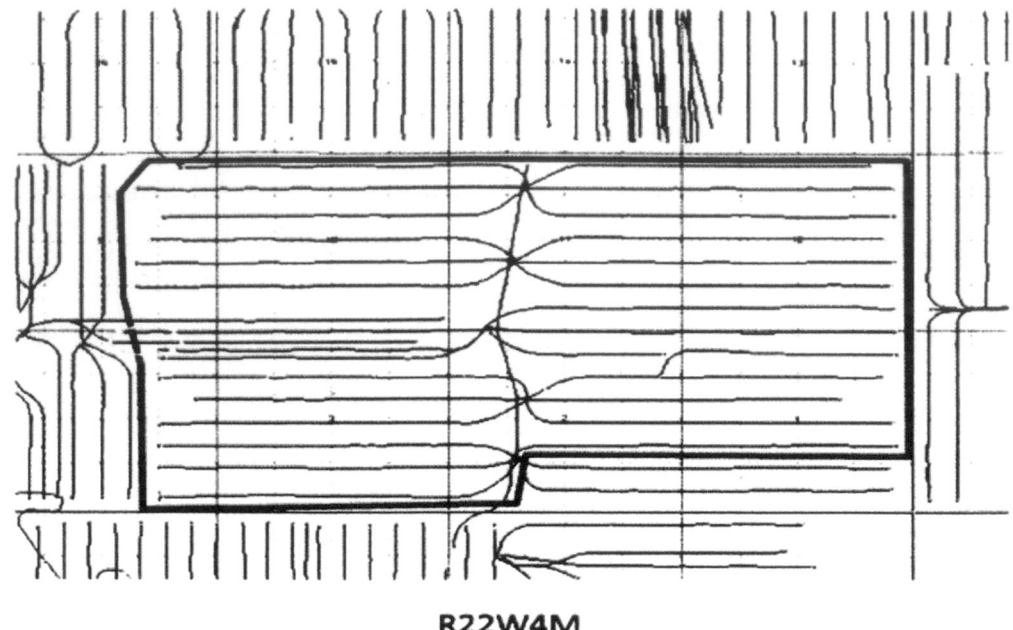

Fig. 5.108—CRNL North Horsetail waterflood pilot map (Delamaide et al. 2013).

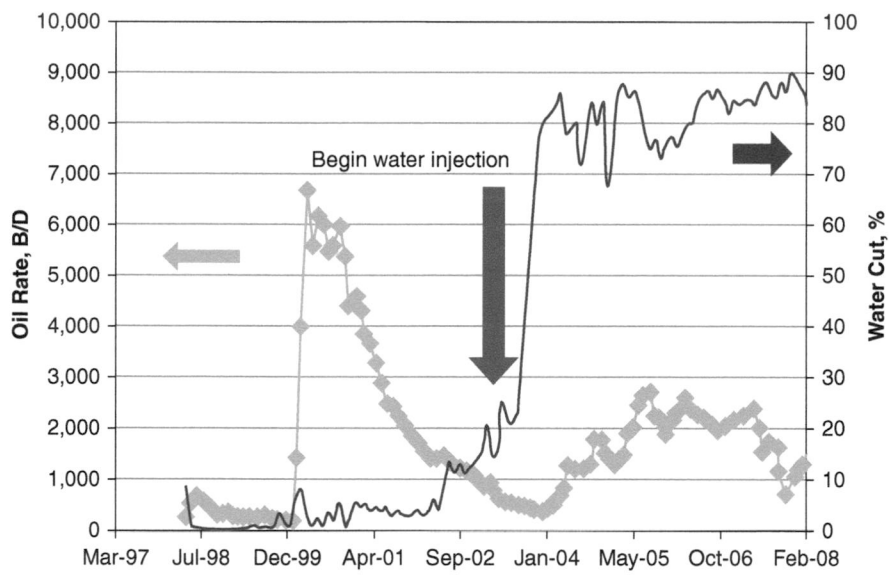

Fig. 5.109—CRNL North Horsetail horizontal waterflood response (Delamaide et al. 2013).

Two methods are used to determine the design mobility. In the first method, the design mobility is estimated from relative permeability data for the reservoir rock. To find this minimum, total relative mobilities are calculated for all saturations and plotted vs. saturation, as shown in **Fig. 5.119.** The minimum total relative mobility in Fig. 5.119 is the design mobility for the oil/water bank, λ_{td}. The reciprocal of the design relative mobility is the maximum apparent viscosity of the oil/water bank.

Estimation of λ_{td} from relative permeability data must consider the history or path followed in building the oil/water bank. For example, consider the creation of an oil bank in a strongly water-wet porous rock. When the chemical slug is injected, the oil saturation increases and fluid flow at the front of the oil bank is on the drainage path (Willhite 1986). Total relative mobilities should be computed from drainage relative permeability data. At the rear of the oil bank, the water saturation increases and the displacement process follows the imbibition path. Imbibition relative permeability data describe fluid flow in this region. Somewhere in the oil bank, the flow must switch from drainage to imbibition paths. Determination of the design mobility with the Gogarty et al. procedure is straightforward for a single rock sample when imbibition and drainage relative permeability curves coincide. When there is hysteresis in the relative permeability curves, however, as shown in **Fig. 5.120** for reservoir

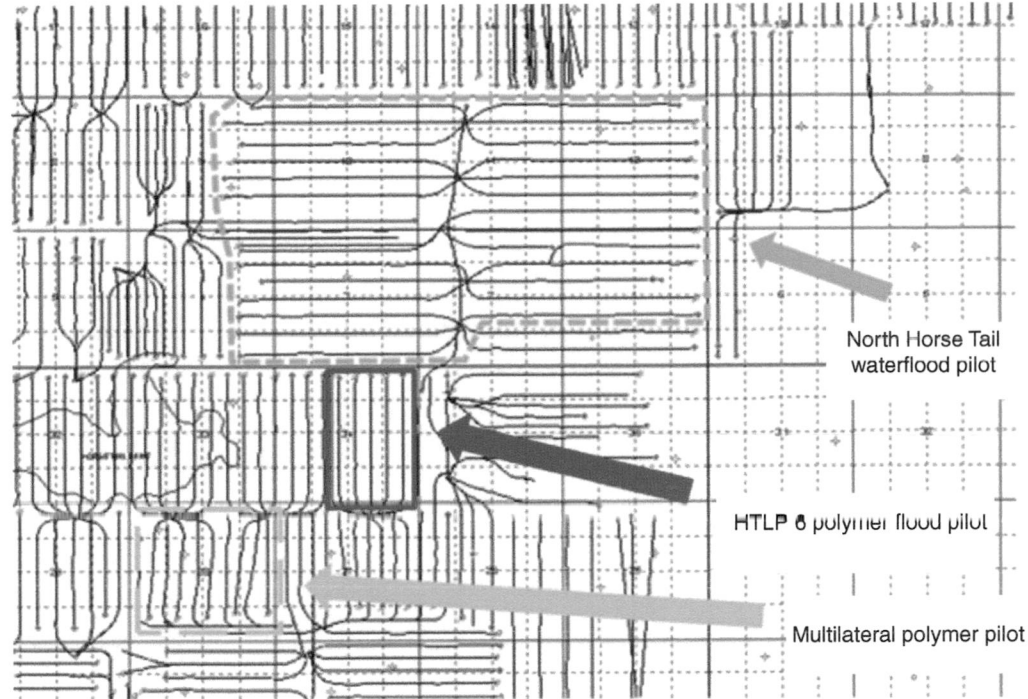

Fig. 5.110—Map showing locations of the North Horse Tail waterflood pilot, HTLP 6 polymer flood pilot and multilateral pilot (Delamaide et al. 2013).

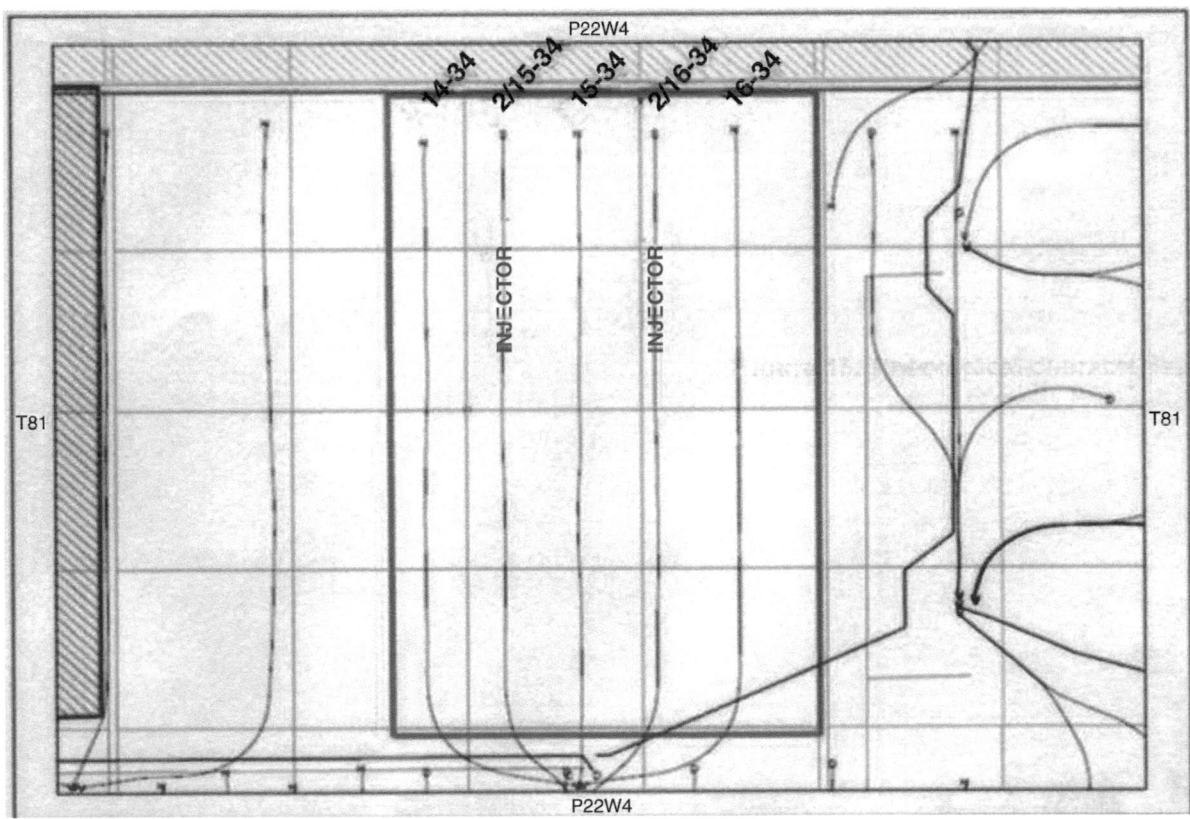

Fig. 5.111—CRNL HTLP 6 polymer flood (modified from CRNL 2006) (Delamaide et al. 2013).

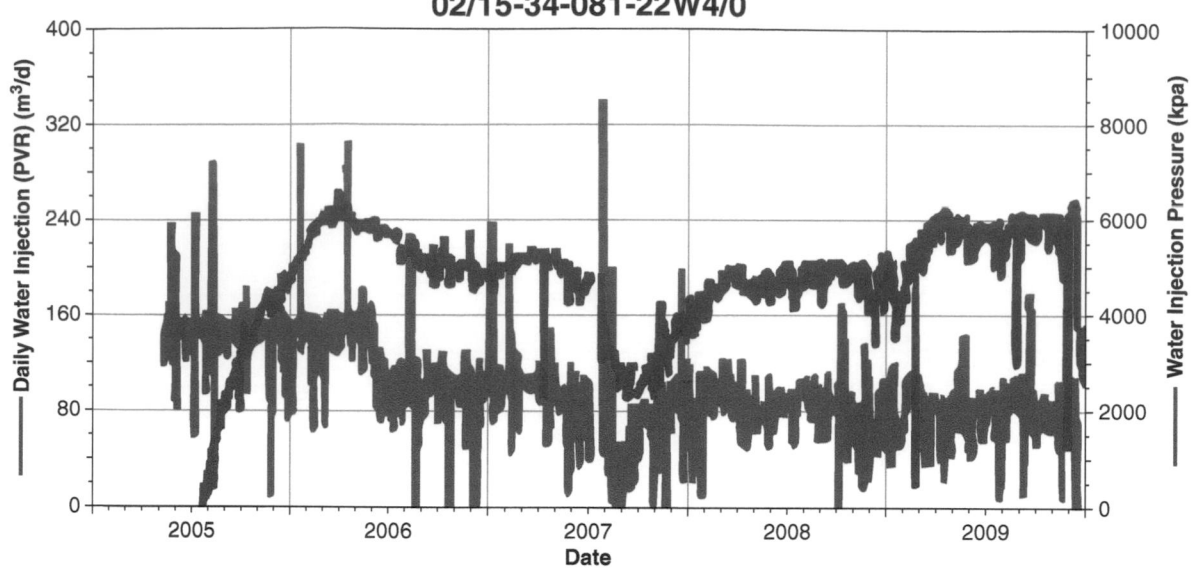

Fig. 5.112—Wellhead pressure and rate for one of the injection wells in the HTLP 6 polymer pilot (Delamaide et al. 2013).

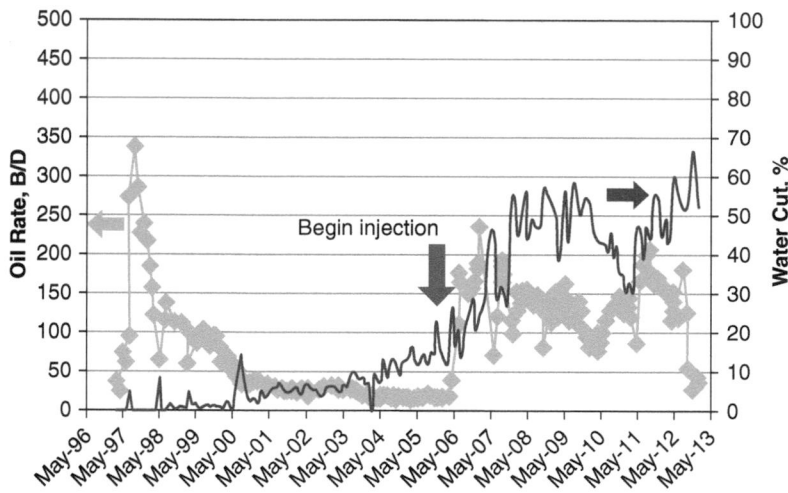

Fig. 5.113—HTLP 6 polymer pilot—Production response and water cut from Well 00/14-34-081-22W4 (Delamaide et al. 2013).

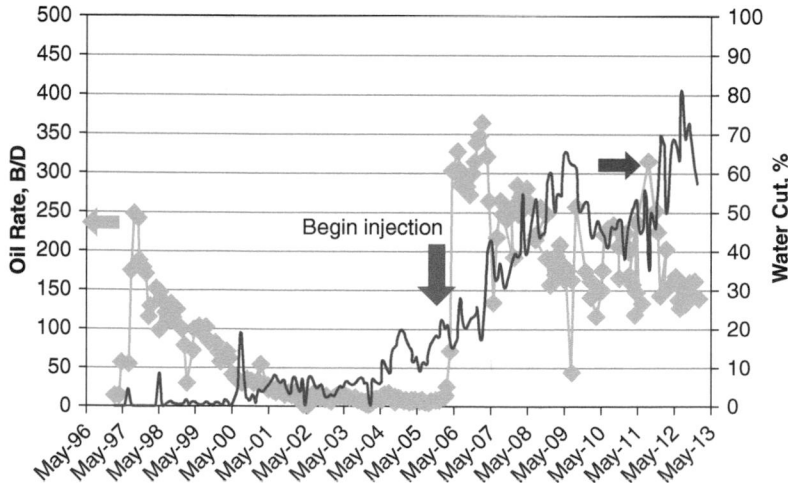

Fig. 5.114—6 polymer pilot—production response and water cut from Well 00/15-34-081-22W4 (Delamaide et al. 2013).

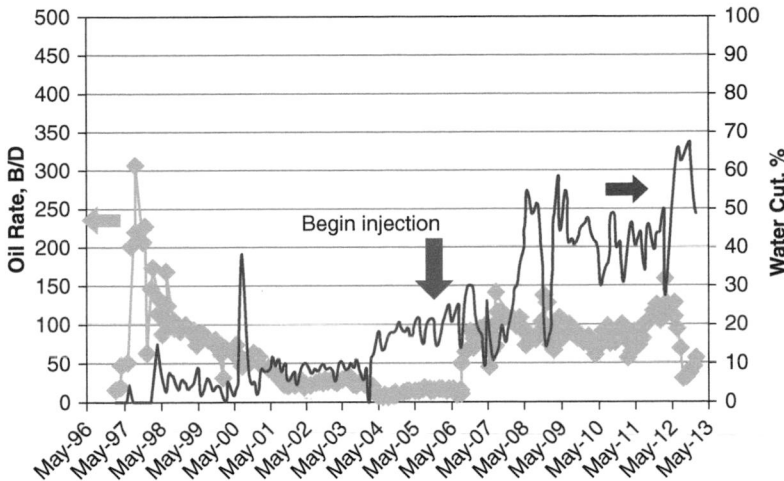

Fig. 5.115—HTLP 6 polymer pilot—production response and water cut from Well 00/16-34-081-22W4 (Delamaide et al. 2013).

rock from the El Dorado Admire reservoir (Chang et al. 1978), a design mobility can be computed from each set of data, as shown in **Fig. 5.121.** The design mobility should be the lowest total relative mobility, which in this case is taken from the drainage curve (Chang et al. 1978).

Selection of the design mobility is complicated by variation of reservoir properties. Relative permeabilities of rock samples from different parts of the same reservoir are often different, as **Fig. 5.122** shows (Gogarty et al. 1970). **Fig. 5.123** presents total relative mobilities for these rocks. The design mobility from Sample 4 is the conservative value for this reservoir but may not be representative of the reservoir as a whole. In this case, other information was used to confirm the selection of Sample 4. The value of λ_{rt} was estimated from transient well testing to be 0.05 cp^{-1}. Water saturation in the region surrounding the test well

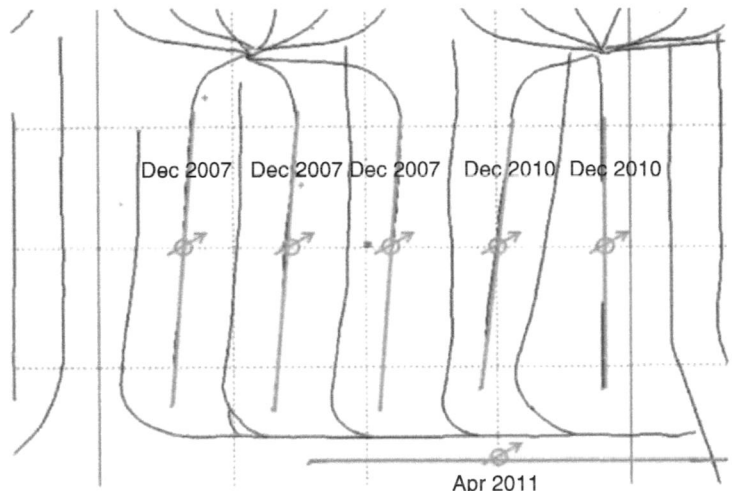

Fig. 5.116—Horizontal well arrangement for a multiwall lateral polymer pilot (Delamaide et al. 2013).

was estimated to be 0.60 by material balance. Because this value was in closest agreement with the data for Sample 4, the design mobility was based on Sample 4. Chang et al. (1978) also present an example illustrating the use of transient pressure testing to determine the set of relative permeability data used to represent the field average.

The second method used to determine the design mobility is to measure the mobility of the oil/water bank produced by the chemical flood in native-state cores (Gogarty et al. 1970; Chang et al. 1978). The displacement process must produce a stable oil/water bank in a region where the interfacial tension (IFT) is not altered. **Fig. 5.124** (Chang et al. 1978) presents the results of an oil displacement test in a 4-ft core in which an aqueous surfactant system was injected. A stabilized oil bank formed in the interval where the IFT was not affected by the surfactant. This region is shaded in Fig. 5.124. Total relative mobility was calculated from pressure data for the stabilized oil/water bank with

$$\lambda_t = q_t \Delta L / A \Delta p \dots\dots\dots\dots\dots\dots\dots\dots\dots\dots\dots\dots\dots\dots\dots\dots\dots\dots (5.158)$$

$$\text{and } \lambda_{rt} = \lambda_t / k_b, \dots\dots\dots\dots\dots\dots\dots\dots\dots\dots\dots\dots\dots\dots\dots\dots\dots\dots (5.159)$$

where λ_t = total mobility of the oil bank, λ_{rt} = total relative mobility of the oil bank, q_t = flow rate, Δp = pressure drop between pressure ports, ΔL = distance between pressure ports, A = cross-sectional area of core, k_b = base permeability for relative permeability data, and appropriate units are used.

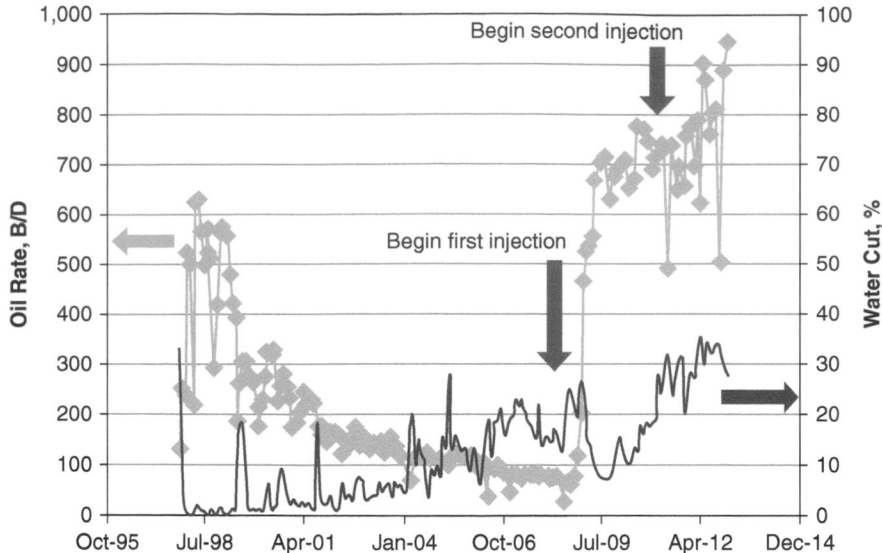

Fig. 5.117—Production response to a multilateral pilot (Delamaide et al. 2013).

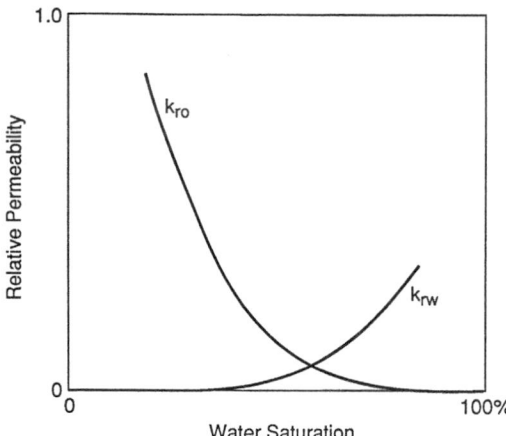

Fig. 5.118—Typical water/oil relative permeability curves.

Pressure taps must be carefully installed and monitored to ascertain that each port is measuring the pressure of the same phase to eliminate capillary pressure effects. McCool et al. (1983) describe procedures for obtaining and interpreting pressure data from laboratory core tests.

5.8.2 Chemical Slug Mobility. Determination of the design mobility for the oil/water bank establishes the mobility requirements for the chemical slug and the mobility buffer. The relative mobility of the chemical slug must be less than or equal to the design mobility to have mobility control at the chemical/oil-bank interface. The mobility of the slug is controlled primarily by adjusting the slug viscosity. Viscosities in some formulations can be controlled by altering the concentrations of the components that make up the chemical slug. In other systems, it is necessary to add a viscosifier, such as a polymer, to the chemical slug to obtain mobility control. When the chemical slug behaves as a Newtonian fluid, its viscosity does not depend on frontal-advance rate. Chemical slug viscosity can be determined from viscometric measurements. The relative mobility of the chemical slug, λ_{rc}, is given by

$$\lambda_{rc} = k_{rc}/\mu_c, \dots\dots\dots\dots\dots\dots\dots\dots\dots\dots\dots\dots\dots\dots\dots\dots\dots\dots\dots\dots\dots\dots \text{(5.160)}$$

where k_{rc} = relative permeability of the chemical slug at the anticipated ROS to chemical flood and μ_c = viscosity of the chemical slug at the anticipated frontal-advance rate.

Because the chemical slug does not displace all the oil in systems of practical interest, the permeability of the chemical slug will approximate the brine permeability at S_{orc}, the ROS to chemical flood. Relative mobility of the chemical slug is computed directly from Eq. 5.160 from relative permeability correlations of k_{rw} with S_w assuming k_{rw} (at S_{orc}) = k_{rc} (at S_{orc}).

Water permeability is a strong function of ROS. **Fig. 5.125** illustrates experimental data for Berea sandstone obtained by Morrow et al. (1985) for the endpoint relative permeability of water as a function of ROS. An empirical correlation between k_{rw} and S_w developed by Negahban[7] for $S_w \le 0.80$ is

$$k_{rw} = \frac{-0.173 + 0.312 S_w}{0.848 - 0.901 S_w}. \dots\dots\dots\dots\dots\dots\dots\dots\dots\dots\dots\dots\dots\dots\dots\dots\dots\dots\dots \text{(5.161)}$$

When the chemical slug behaves as a non-Newtonian fluid, its viscosity varies with shear rate and thus with frontal-advance rate. Non-Newtonian behavior is observed in chemical slugs with and without polymer content. In principle, the viscosity of the chemical slug can be determined by measuring the viscosity at the shear rate anticipated in the core displacement tests. As discussed in Section 5.4.3, however, methods of estimating in-situ shear rate from rock and fluid properties are not adequate. It

[7] Personal communication with S. Negahban, Amoco Production Co., Tulsa, 1984.

is necessary to determine chemical slug mobility by direct measurement in laboratory core experiments.

Fig. 5.126 (Gogarty et al. 1970) presents results of such a laboratory program for three similar chemical slugs with slightly different compositions. Note that the three slugs are non-Newtonian fluids because the relative mobility increases with frontal-advance rate. Slug 1 does not meet the requirements of the design mobility and would not maintain mobility control between the chemical slug and the oil/water bank. Slugs 2 and 3 are adequate for frontal-advance rates below approximately 10 to 20 ft/D. Because reservoir rates are about 1 ft/D, these chemical slugs would yield mobility control except in the immediate region of the wellbore, where frontal-advance rates could exceed 10 ft/D. In this case, a frontal-advance velocity cutoff may be established for mobility control. Viscous fingering is accepted in the region within a few feet of the injection wellbore to improve injectivity and project economics.

Displacement experiments also provide a means to correlate viscometric data with mobility at selected frontal-advance rates in the core. Chapter 7 discusses this further.

Example 5.17 illustrates the application of these concepts to choose the viscosities of the chemical slug and the mobility at the leading edge of the polymer buffer for a chemical flooding process.

Fig. 5.119—Total relative mobility vs. water saturation (Gogarty et al. 1970).

Example 5.17—Mobility Control in a Chemical Flood. A chemical flood is being designed for a reservoir. The current chemical formulation reduces the ROS to 0.20 at a frontal-advance rate of 1.0 ft/D when the chemical is continuously injected into a sample of the reservoir rock. **Table 5.57** gives properties of the reservoir rock, oil, and water. Relative permeabilities are represented by Eqs. 5.162 and 5.163. For this example, assume that k_{rw} (at S_{orc}) is represented by Fig. 5.118 and that Eqs. 5.162 through 5.164 are applicable.

$$k_{ro} = \left(1 - S_{wD}\right)^{2.5}, \quad\dots\dots\dots\dots\dots\dots\dots\dots\dots\dots\dots\dots\dots\dots\dots\dots\dots\dots (5.162)$$

where $k_{rw} = 0.209 \, S_{wD}^2. \quad\dots\dots\dots\dots\dots\dots\dots\dots\dots\dots\dots\dots\dots\dots\dots\dots\dots\dots\dots\dots (5.163)$

and $S_{wD} = \dfrac{S_w - S_{iw}}{1 - S_{or} - S_{iw}}. \quad\dots\dots\dots\dots\dots\dots\dots\dots\dots\dots\dots\dots\dots\dots\dots\dots\dots (5.164)$

Measurements indicate that the viscosity of the chemical slug is 15 cp at flow rates expected in the reservoir. Determine whether the laboratory experiments have been run under conditions where there is mobility control between the chemical slug and the oil/water bank. If mobility control was not obtained, indicate what the viscosity should be.

Solution. The relative-total-mobility/water-saturation graph must be prepared to determine the design mobility. **Table 5.58** contains relative mobilities computed at saturation increments of 0.016.

The minimum total relative mobility is 0.0423 cp^{-1}, which is equivalent to an apparent viscosity of 23.6 cp. Thus, the chemical slug must exhibit an apparent viscosity of 23.6 cp to obtain mobility control between the slug and the oil/water bank.

Next, determine the viscosity of the chemical slug. For this example, assume the chemical slug is a Newtonian fluid so that variations of viscosity with frontal-advance rate do not have to be considered. That is, the viscosity is a function of composition but does not change with shear rate. The viscosity of the chemical slug is obtained from the definition of mobility. It is first necessary to determine the relative permeability of the chemical slug in the presence of residual oil. The ROS to chemical flooding is 0.2. From Fig. 5.125, k_{rw} at S_{orc} is approximately 0.63. Substituting into the expression for mobility and solving for μ_c yield

$$\lambda_{rc} = k_{rc} / \mu_c$$

and $\mu_c = 0.63 / (0.0423 \text{ cp}^{-1})$

$\qquad = 14.9 \text{ cp}.$

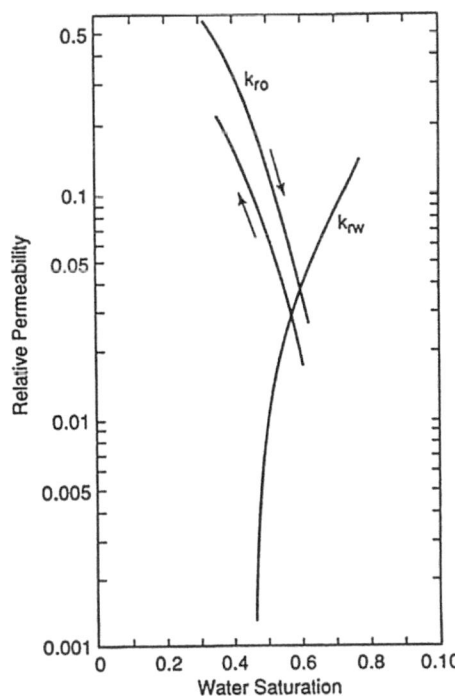

Fig. 5.120—Steady-state relative permeability data, Run 2 (Chang et al. 1978).

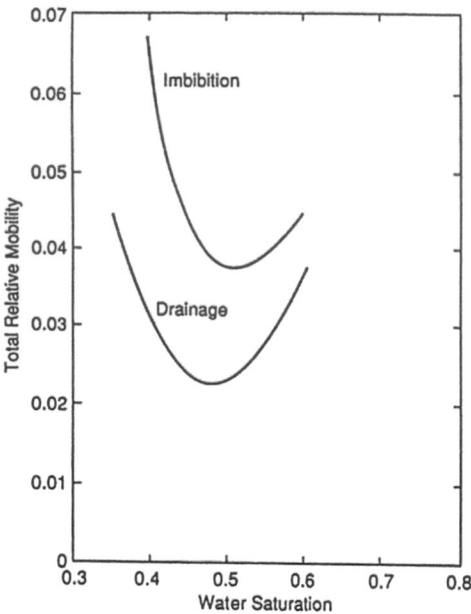

Fig. 5.121—Total relative mobility data, Run 2 (Chang et al. 1978).

Therefore, the viscosity of the chemical slug should be adequate for mobility control if $S_{orc} = 0.2$.

Selection of the correct mobility of the chemical slug is also affected by the oil viscosity. **Fig. 5.127** illustrates the change in the apparent mobility of the chemical slug required to maintain unit mobility ratio as a function of oil viscosity.

The displacement efficiency of the chemical slug is 33.3% of S_{orw}. If the displacement efficiency were higher, the viscosity of the slug would need to be increased to provide mobility control. The required chemical slug viscosity is a function of microscopic displacement efficiency because permeability of the chemical slug varies with chemical flood oil residual, as noted in Fig. 5.125. **Fig. 5.128** is a plot of chemical slug viscosity vs. oil viscosity at different values of S_{orc}. The data in Fig. 5.125 were used to determine k_{rc}. Higher chemical slug viscosity is required as the displacement efficiency increases because the chemical slug relative permeability is higher at a lower ROS.

Example 5.17 evaluated mobility control when the relative permeability of the chemical slug resulted only from ROS. If the chemical slug contains polymer, or if the chemical slug mobility varies with frontal-advance rate, the mobility of the chemical systems must be evaluated in the porous rock at residual saturations anticipated from the chemical slug over the frontal-advance rates expected.

5.8.3 Mobility Buffer. The mobility buffer must be miscible with the chemical slug to avoid trapping of the slug as a residual saturation. There must be mobility control between the mobility buffer and the chemical slug. The mobility buffer is water thickened by adding small amounts of polymer to the injected water. Mobility control is obtained by increasing the polymer concentration until unit mobility exists between the leading edge of the mobility buffer and the chemical slug. Therefore, for mobility control,

$$\lambda_p \leq \lambda_{rt} k_b, \dots\dots\dots\dots\dots\dots\dots\dots\dots\dots\dots\dots\dots\dots\dots\dots\dots\dots\dots\dots\dots (5.165)$$

where λ_p = polymer mobility.

Polymers used for mobility control are non-Newtonian fluids, and thus, the polymer mobility in porous rock varies with frontal-advance rate, as illustrated in **Fig. 5.129** (Gogarty et al. 1970). By selecting polymer concentration, it is usually

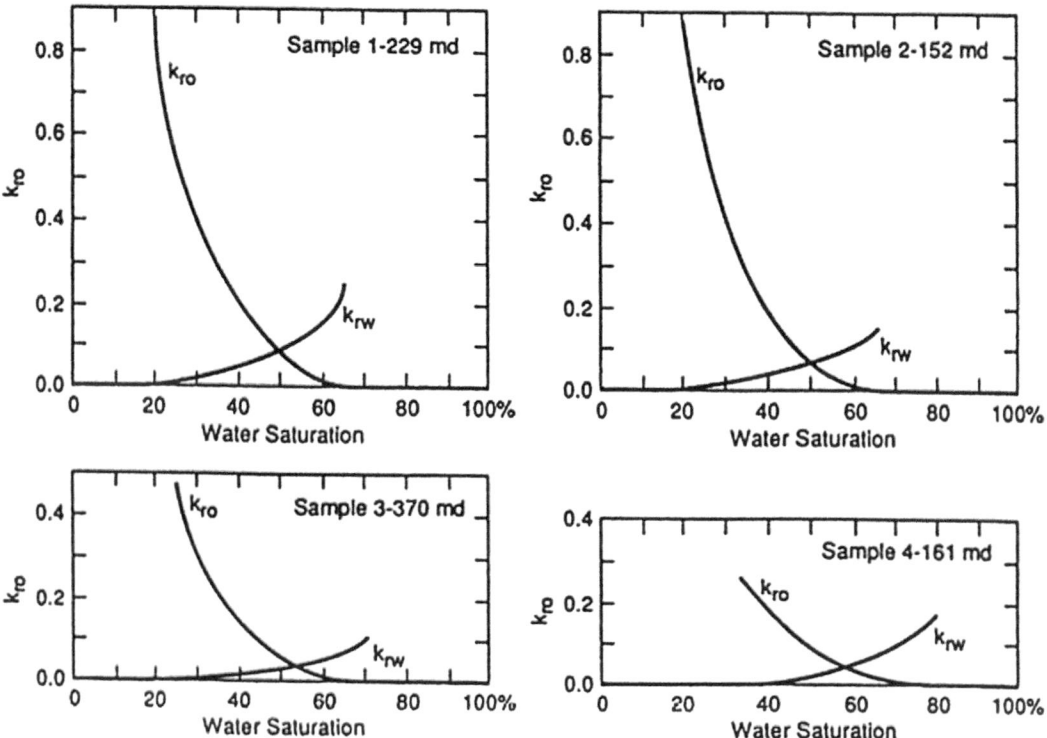

Fig. 5.122—Relative permeability curves for samples of the same reservoir (Gogarty et al. 1970).

possible to meet the design mobility. Data presented in Fig. 5.129 were obtained from laboratory tests on small reservoir core plugs. Polymer mobility tests should be conducted at the ROS expected for chemical flooding to include the effects of permeability reduction resulting from the chemical flood ROS.

When laboratory data are not available, it is necessary to estimate polymer mobility. The mobility of polymer in porous rock at a particular flow rate is given by

$$\lambda_p = k_p / \mu_p, \quad\dots\dots\dots\dots\dots\dots\dots\dots (5.166)$$

where λ_p = polymer mobility, k_p = permeability of the polymer at ROS to chemical flooding, and μ_p = viscosity of the polymer at the average frontal-advance rate. The relative polymer mobility is expressed by

$$\lambda_{rp} = \lambda_p / k_b. \quad\dots\dots\dots\dots\dots\dots\dots\dots (5.167)$$

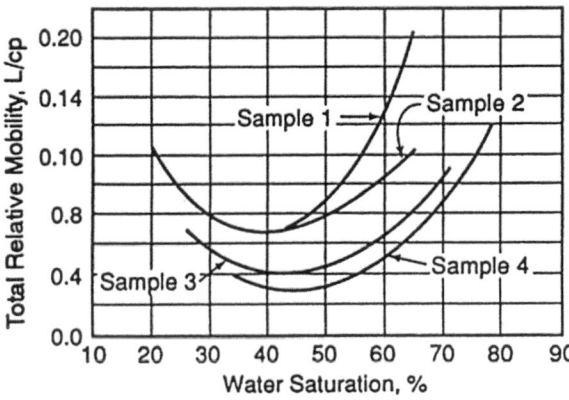

Fig. 5.123—Total relative mobilities for samples of the same reservoir (Gogarty et al. 1970).

There are two difficulties in estimating λ_p. Because polymers are non-Newtonian fluids, the viscosity varies with shear rate, as discussed in Section 5.4.3. Viscosities can be easily measured at various shear rates with a standard viscometer. A relationship is needed between shear rate and frontal-advance rate to convert viscometric data to equivalent core data. Eq. 5.168 is an empirical expression for shear rate during flow in porous media (Chauveteau and Zaitoun 1981). Although this model has been widely used for computations, it does not estimate shear rates correctly for most polymer/rock systems of practical interest. At best, Eq. 5.168 may be correlated against experimental data to find the shear rate that yields the apparent viscosity observed in the rock when the frontal-advance rate is specified.

$$\dot{\gamma} = 4a_3 u / \sqrt{k\phi}, \quad\dots\dots\dots\dots\dots\dots\dots\dots\dots\dots\dots\dots\dots\dots\dots\dots (5.168)$$

where a_3 = an adjustable constant characterizing the rock, u = Darcy velocity, k = absolute liquid permeability, ϕ = porosity, $\dot{\gamma}$ = in-situ shear rate, and consistent units are used.

The second difficulty is estimation of the effective permeability of the polymer at ROS to chemical flooding. This permeability is approximated by the permeability to brine at S_{orc} for some polymers. However, polymers are retained in porous rock, and brine permeability is reduced owing to polymer retention. As discussed in Section 5.4, it is usually necessary to determine the amount of permeability reduction by conducting laboratory tests on small cores. This can be done for screening purposes. Candidates identified in screening tests should be evaluated thoroughly by determining polymer mobility over the range of frontal-advance velocities of interest. Section 5.4 discusses determination and correlation of polymer mobility data further.

Costs prohibit continuous injection of the mobility buffer. At some point in the displacement process, sufficient polymer has been injected to prevent the drive water from fingering through the mobility buffer into the chemical slug. A region of variable concentration forms owing to mixing between the drive water and the polymer. If the polymer concentration in the mobility buffer remains constant, the size of the mobility buffer needed to protect the chemical slug can be estimated from experimental displacement runs or from mathematical models.

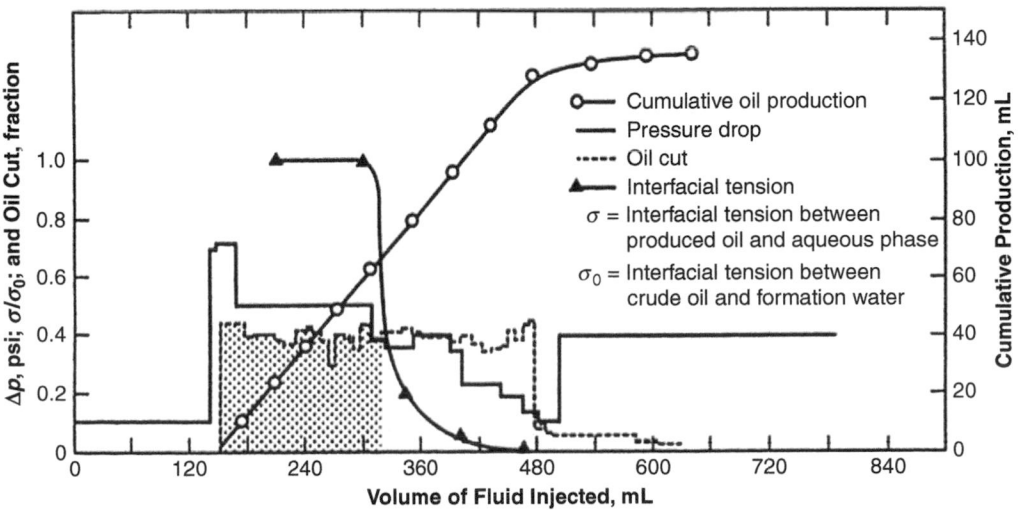

Fig. 5.124—Laboratory flow test results of Run 23.

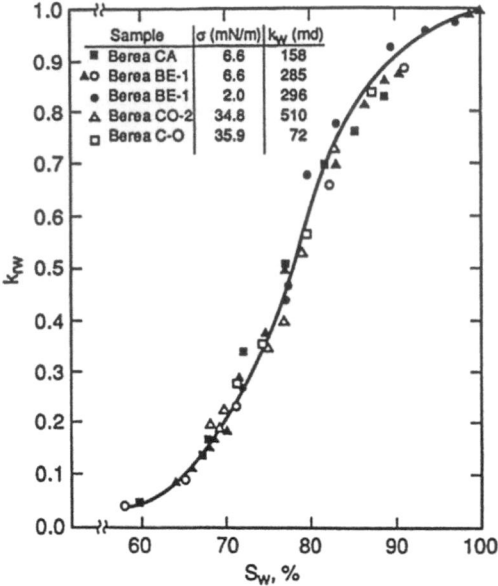

Sample	σ (mN/m)	k_w (md)
■ Berea CA	6.6	158
▲○ Berea BE-1	6.6	285
● Berea BE-1	2.0	296
△ Berea CO-2	34.8	510
□ Berea C-O	35.9	72

Fig. 5.125—Relative permeability at reduced ROS in Berea sandstone samples at various IFTs (Morrow et al. 1985).

Bank size may be determined experimentally. Results of a mixing-zone study conducted in a 16-ft Berea sandstone core (Gupta and Trushenski 1978) illustrate one approach. In this study, glycerine, biopolymer, and polyacrylamide solutions of various mobilities were displaced through the core by drive water. The length of the mixing zone was determined from effluent concentrations to be the volume between the 5 and 95% concentrations. **Fig. 5.130** summarizes the results. The mixing-zone volume is a function of mobility ratio, as would be expected. For example, the mobility buffer must be 0.5 PV to prevent reduction of the polymer concentration at the leading edge of the mobility buffer below 95% of the injected concentration in the effluent when the mobility ratio between the mobility buffer and the chase water is five.

Bank size also is estimated with numerical models that simulate mixing and viscous fingering between the drive water and the mobility buffer. For chemical flooding, the size of the mobility buffer must be increased to compensate for the fact that polymer molecules move faster through the porous rock than water injected at the same time. This phenomenon, called IPV, was discussed in detail in Section 5.4. The effect of IPV can be seen in the concentration profiles shown in **Fig. 5.131** (Gupta and Trushenski 1978) for three displacements conducted at similar mobility ratios. Both biopolymer and polyacrylamide mixing zones arrive at the end of the core approximately 0.17 PV before the glycerine mixing zones.

The mobility buffer usually is not injected at constant concentration. A tapered slug is used to reduce the total amount of polymer required by varying the polymer concentration from design mobility at the polymer/chemical-slug interface to zero at the rear of the mobility buffer where drive water pushes the mobility buffer. Although the drive water fingers into the mobility buffer, the mixed fluid encounters progressively more-viscous regions and thus mobility is reduced.

Fig. 5.132 (Trantham 1977) shows the injection schedule for polymer concentration in the mobility buffer for the North Burbank surfactant field test. The total amount of polymer injected was 0.7 PV at an average concentration of 710 ppm. The quantity of polymer is equivalent to a 0.2-PV slug of polymer at a concentration of 2,500 ppm, which is at the leading edge of the mobility buffer.

Design of tapered, or graded, viscosity slugs is based on empirical models. The simplest model assumes that the polymer concentration declines exponentially from the front of the slug to the lowest concentration at the rear of the slug. Other models include simulation of viscous fingering in some manner. Claridge (1978) and Stoneberger and Claridge (1988) developed a method based on Koval's method to design graded viscosity banks. The preferred approach is to evaluate the proposed slug design in laboratory experiments. This requires use of scaled displacements in linear core or sandpack systems.

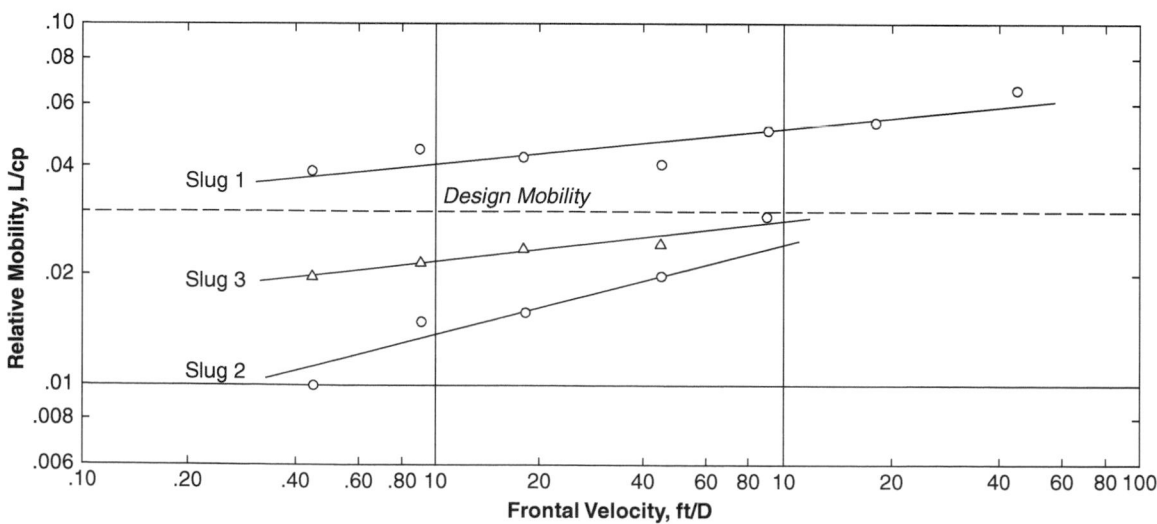

Fig. 5.126—Effect of slug composition on mobility-control design (Gogarty et al. 1970).

Porosity	0.23
S_{oi}	0.70
S_{or}	0.30
S_{gi}	0.0
μ_o(cp)	20.0
μ_w(cp)	0.5

Table 5.57—Properties of reservoir rock and fluids, Example 5.17.

S_w	λ_{ro} (cp^{-1})	λ_{rw} (cp^{-1})	λ_{rt} (cp^{-1})
0.300	0.0500	0.0000	0.0500
0.316	0.0451	0.0007	0.0458
0.332	0.0406	0.0027	0.0433
0.348	0.0363	0.0060	0.0423
0.364	0.0323	0.0107	0.0430
0.380	0.0286	0.0167	0.0453
0.396	0.0252	0.0241	0.0493
0.412	0.0220	0.0328	0.0548
0.428	0.0191	0.0428	0.0619
0.444	0.0164	0.0542	0.0706
0.460	0.0139	0.0669	0.0808
0.476	0.0117	0.0809	0.0927
0.492	0.0097	0.0963	0.1061
0.508	0.0087	0.1130	0.1230
0.524	0.0064	0.1311	0.1375
0.540	0.0051	0.1505	0.1555
0.556	0.0039	0.1712	0.1751
0.572	0.0029	0.1933	0.1962
0.588	0.0021	0.2167	0.2188
0.604	0.0014	0.2414	0.2428
0.620	0.0009	0.2675	0.2684
0.636	0.0005	0.2949	0.2955
0.652	0.0002	0.3237	0.3239
0.668	0.0001	0.3538	0.3539
0.684	0.0000	0.3852	0.3852
0.700	0.0000	0.4180	0.4180

Table 5.58—Computation of total relative mobility, Example 5.17.

5.9 Foam as an EOR Agent

A foam is a dispersion of a relatively large volume of gas in a small volume of liquid (Raza 1970), as illustrated in the photograph of a foam shown in **Fig. 5.133** (Patton et al. 1983). A foam is produced when a liquid that contains a small concentration of foaming agent (surfactant) comes into contact with a gas and sufficient mechanical energy is provided to cause the liquid to foam.

Foams are useful in EOR processes because they have a relatively high resistance to flow when displaced through a porous medium. The resistance typically is significantly higher than that of the individual phases that make up the foam. Thus, they are potentially suitable for improving displacement efficiency in a process or for blocking the flow of condensed fluids.

The following are possible applications (Raza 1970).

1. Blocking or restricting flow of undesired fluids, such as the coning of gas or water in a production well.
2. Blocking or restricting flow of injected fluids in high-permeability streaks or fractures (profile modification).
3. Improving the mobility ratio in displacement processes by reducing the mobility of the injected phase.

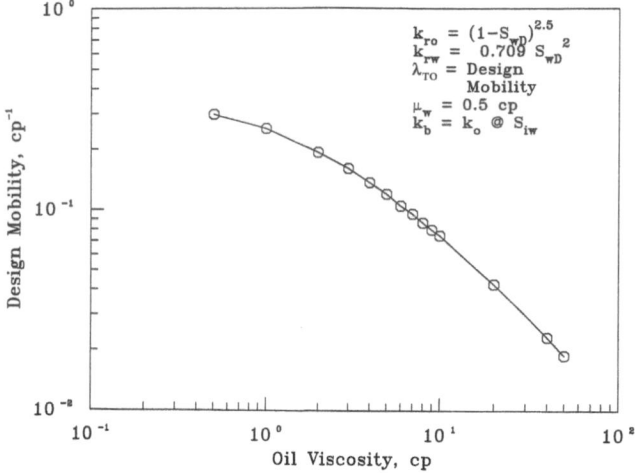

Fig. 5.127—Variation of design mobility with oil viscosity, Example 5.17.

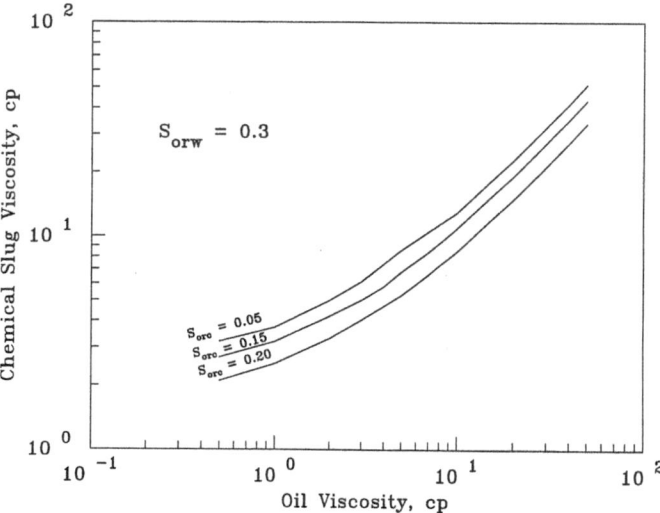

Fig. 5.128—Effect of S_{orc} on chemical-slug viscosity, Example 5.17.

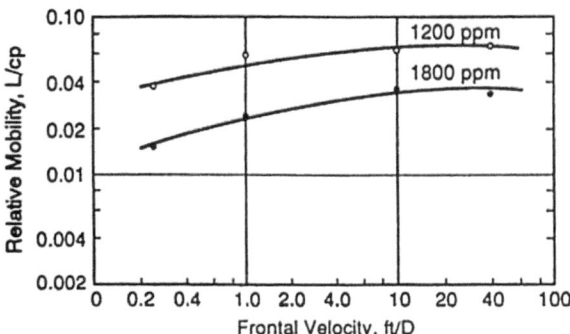

Fig. 5.129—Effect of thickened water composition on mobility (Gogarty et al. 1970).

The schematic of a foam displacement process shown in **Fig. 5.134** (Raza 1970) illustrates the last mechanism. An unconsolidated sandpack ($k = 2,760$ md) was initially saturated with a surfactant solution (water containing 2% foaming agent). Air was injected at a constant pressure difference of 10 psia across the sandpack. Pressure drops across individual sections of the sandpack were monitored through pressure taps located along the sandpack as shown in the top sketch. A foam was generated as a result of the contact between the air and the surfactant solution, and the volume of the foam continued to grow as air injection proceeded. As Fig. 5.134 indicates, the foam's resistance to flow was much larger than that for the surfactant solution. Essentially all the pressure drop occurred over the volume occupied by foam. The foam mobility was much lower than that of the surfactant solution, which was essentially equivalent to water. This is characteristic of the flow behavior of foams at reservoir conditions.

Foams have been tested as diverting agents in waterfloods (Holm 1970), as steam-diverting agents to reduce gravity override or steam channeling in steamfloods (Duerksen 1986), and as mobility-control fluids in CO_2 displacements (Heller et al. 1985).

5.9.1 Properties That Characterize Foam. Foams typically are characterized by quantifying foam quality and bubble size (Lake 1989). Foam quality is expressed as the fraction or percentage of total volume that is gas. Bubble size is specified in terms of average size or diameter (foam texture) and the range of bubble sizes present.

Foam quality typically ranges from 75 to 90% but may be higher, approaching 97% (Lake 1989), or significantly lower (Raza 1970). Quality is a function of pressure and temperature, as well as foam constituents and physical conditions associated with the production and handling of the foam. **Table 5.59** (Raza 1970) shows an example of data on quality for foams generated in different porous media. In these experiments, the porous medium was saturated with a 0.25 N (normality) brine containing 2% of a surfactant called O.K. Liquid. The liquid was then displaced with gas, generating a foam. Raza (1970) also investigated the effect of a second liquid phase on foam formation. **Table 5.60** shows the effect of oil on the quality of a water foam. Raza concluded that a second liquid phase tends to suppress the foaming action of the initial phase. That is, water foamability is reduced by the presence of an oil phase and vice versa.

Average bubble size and distribution of sizes may vary significantly. Bubble size is affected by the foam quality and the variables that control quality. **Fig. 5.135** (David and Marsden 1969) shows typical data on size distributions. Foams with a relatively large distribution of bubble sizes are more likely to be unstable (Lake 1989). Pore sizes in reservoir rocks are usually smaller than the foam-bubble sizes shown in Fig. 5.135. The foam flow in porous media might therefore be expected to be affected by foam quality and bubble size, and this is the case.

5.9.2 Foaming Agents. A foaming agent (i.e., a surfactant) is required to produce a foam. Numerous surfactants are suitable for this purpose. Surfactants generally are selected for specific applications on the basis of laboratory test procedures and empirical relationships. There are no theoretical quantitative relationships available to make reliable, detailed predictions of foam properties. Several investigators (Holm 1970; Duerksen 1986; Heller et al. 1985; Dellinger et al. 1984) have studied foaming agents, and illustrations of this work are discussed below.

Heller et al. (1985) and Dellinger et al. (1984) studied the behavior of a number of surfactants for possible use in mobility control with the CO_2 miscible process. Because CO_2 is generally injected above its critical temperature and pressure, it is a relatively dense fluid at displacement conditions, typically having a density greater than 0.6 g/cm². Because of this, Heller et al. (1985) studied surfactants known to be good liquid emulsifying agents.

A parameter used to rank surfactants is an empirical number called the hydrophile/lipophile balance (HLB) number. Heller et al. used surfactants with HLB numbers of 8 to 13. They also required that the surfactants be soluble in an NaCl/CaCl$_2$ brine and that they be anionic or nonanionic. This latter requirement was imposed because cationic surfactants probably would adsorb strongly onto reservoir rock.

The screening procedure consisted of putting a few milliliters of foam solution into a test tube filled with isooctane. Isooctane was used to simulate a CO_2 phase because the density of CO_2 at typical reservoir displacement conditions is similar to that of isooctane at room conditions (room temperature and atmospheric pressure). The test tube was stoppered and shaken vigorously. The degrees of foaming and foam stability were measured and recorded. **Table 5.61** (Heller et al. 1985) shows results of these screening tests. The surfactants shown are roughly ranked from good to poor (from top to bottom in the table) in terms of their potential for use in a CO_2 displacement. Heller et al. evaluated the foams further in flow tests in porous media and described flow properties in terms of foam mobility (discussed in Section 5.9.3).

Foam also has been tested as a way to modify the permeability profile in injection wells in a waterflood (Holm 1970). More than 100 surfactants were tested for this purpose, and it was found that certain ethoxylated alkyl sulfates containing amide stabilizers were efficient foaming agents. For the field test, a modified ammonium lauryl sulfate called O.K. Liquid was used.

Foaming agents are applicable in steamdrives as steam-diverting agents (Duerksen 1986). Gravity override and steam channeling are important problems in steam displacement. By creating a high flow resistance, foams can be used as diverting agents to make steam move more uniformly through a reservoir. Surfactants used in conjunction with steam displacement must be capable of generating flow resistance and remaining stable at steamflood temperatures, which are commonly approximately +400°F.

Duerksen (1986) screened approximately 50 surfactants by measuring flow resistance in stainless-steel-wool packs and various porous media. He found that a number of sulfonate surfactants had good foamability and stability at the required temperature.

5.9.3 Rheology of Foams—Flow in a Tube. Rheological behavior of foams has been measured in a viscometry apparatus **(Fig. 5.136)** (Patton et al. 1983; Raza and Marsden 1967). Typically, the foam constituents are mixed and passed through a foam generator, such as a packed bed. The foam is then displaced through a small-diameter tube, and flow rates, pressure drop, and temperature are measured.

Experiments of this type have determined that a foam behaves as a non-Newtonian fluid. A foam typically has the characteristics of a pseudoplastic fluid (i.e., the apparent viscosity of the foam decreases as the shear rate increases) (Patton et al. 1983).

For a Newtonian fluid, the relationship between shear stress, τ and shear rate, $\dot{\gamma}$ is given by

$$\tau = \mu \dot{\gamma}, \dotfill (5.169)$$

where μ = fluid viscosity and is not a function of the shear rate. Viscosity is dependent on temperature and pressure, however. Shear rate is expressed in terms of a velocity gradient, such as dv/dr, in flow in a tube.

As previously discussed, Eq. 5.169 does not hold for non-Newtonian fluids because the relationship between τ and $\dot{\gamma}$ is nonlinear. A useful model to describe non-Newtonian fluids is the power-law model attributed to Ostwald and de Waele (Ostwald 1925, 1929). The model is given by Eq. 5.3, which is rewritten here as

$$\tau = K \dot{\gamma}^n, \dotfill (5.170)$$

where K = power-law constant or consistency index and n = power-law exponent or flow-behavior index. The shear stress, τ, is related to pressure drop, and shear rate, $\dot{\gamma}$, is related to volumetric flow rate for flow in a tube as (Mooney 1932)

$$q = \Delta p d / 4L, \dotfill (5.171)$$

where Δp = pressure drop along a length L of the tube, d = tube diameter, L = tube length, q = volumetric flow rate, and consistent units are used.

Figs. 5.137 and 5.138 (Patton et al. 1983) show foam data. Fig. 5.137 plots shear stress vs. shear rate for a foam formed of nitrogen and brine. The different curves are for capillary tubes of different lengths. The curves are all slightly concave downward, indicating that the shear stress is not increasing linearly with the imposed shear rate. On this plot, the apparent viscosity of the foam at any point is the ratio $\tau / \dot{\gamma}$. As seen, the apparent viscosity of the foam decreases as shear rate increases, a behavior called pseudoplastic. Comparison of Eqs. 5.169 and 5.170 shows that the apparent viscosity, μ_a, also can be expressed as

$$\mu_a = K \dot{\gamma}^{n-1} \dotfill (5.172)$$

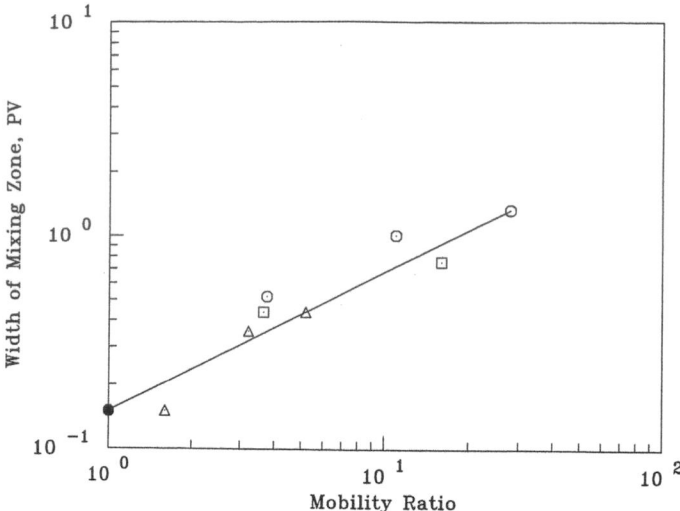

Fig. 5.130—Effect of mobility ratio on mixing-zone width (Gupta and Trushenski 1978).

Power-law fluids should yield a straight line when log τ is plotted vs. log $\dot{\gamma}$ (Eq. 5.170). Fig. 5.138 shows several data sets for tubes of different lengths and diameters. The slope of each line is the flow-behavior index n. When the slope is unity ($n = 1$), the fluid is Newtonian and K is the Newtonian viscosity. Fluids exhibit greater non-Newtonian behavior as n deviates more from unity. For foams, $n < 1.0$, indicating pseudoplastic behavior.

The foam indices in Fig. 5.138 are functions of capillary-tube diameter and length, a characteristic of foams observed by other investigators (Raza and Marsden 1967). Foam-behavior indices are also known to be functions of the foam composition and quality, as well as temperature and pressure.

Foam rheological behavior is somewhat like that of the polymers described earlier in that they also are pseudoplastic. However, the foam flow is more complex and is dependent on a larger number of variables, as discussed.

Experimental work of the type discussed indicates that foams can be described in a useful way by a power-law model. Foams are complex systems, however, and the model parameters are not unique but depend on the system geometry, foam properties, pressure, and temperature.

5.9.4 Flow of Foams in Porous Media. Foam rheology depends on the geometry of the flow system. This is particularly true when a foam flows through a porous medium. Heller et al. (1985) discussed two important requirements for the rheological characterization of a foam to be geometrically independent. First, the foam-bubble (cell) size should be at least 20 times smaller than the characteristic flow dimension (pore size). Otherwise, the effect of the pore walls on the foam macroscopic or bulk behavior will be significant (Mooney 1932). Bulk properties, such as apparent viscosity or even quality, will be affected by the geometry at low ratios of bubble size to flow dimension. Dietz et al. (1985) pointed out that foam flow at the pore level may be such that rheology data taken in a standard viscometer may be meaningless. Second, pressure variations along this flow path should be such that fluid compressibility effects are negligible. Standard viscometry measurements are based on incompressible fluids, and large compressibility effects during flow can introduce significant error (Heller et al. 1985).

The rheological description of foam flow in porous media has been treated in different ways (Lake 1989). One approach has been to use the single-phase fluid viscosities to calculate relative permeabilities to each fluid on the basis of experimental measurements of flow rates and pressure drop in foam flow through a porous medium.

Fig. 5.139 (Bernard and Holm 1964) shows data of this type. The data represent a series of displacements through a sandpack with a permeability of 3,190 md. In the displacements, water or foaming-agent solution was displaced through the sandpack at different rates, holding the pressure gradient constant. A constant gas pressure was maintained at the injection face of the sandpack. The relative permeability to the gas and the gas saturation were measured.

Gas saturation was determined to be the same at a given flow rate, regardless of whether a foam was present. The relative permeability to the gas phase was very dependent on the presence of foam, however, and was more than two orders of magnitude smaller when foam was flowing in the sandpack than when water was flowing with no foaming agent. This approach to describing rheology in effect treats the flow of foam as two-phase flow.

Another approach to describing foam rheology has been to calculate an apparent foam viscosity from flow-rate and pressure-drop measurements made during flow through a porous medium (Minssieux 1974). Darcy's law is used with the rock permeability, or water relative permeability if an oil phase is present, to calculate apparent foam viscosity. Results show that the apparent viscosity calculated in this manner is a strong function of foam quality, decreasing approximately linearly as foam quality increases. This approach essentially treats the foam as a single phase.

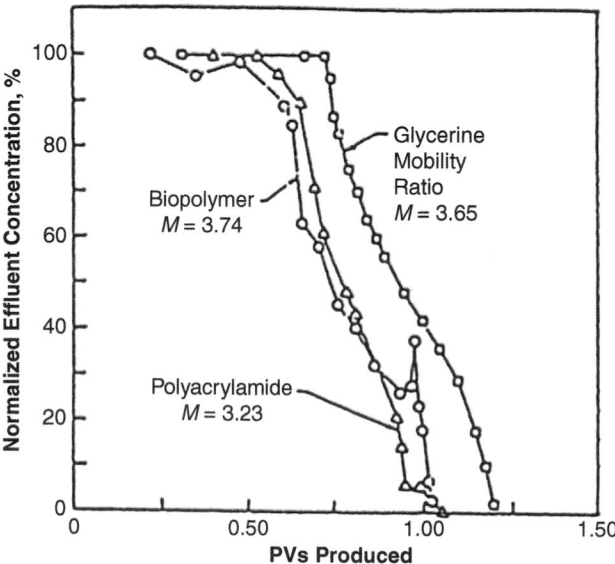

Fig. 5.131—Mixing with chase water at similar mobility ratio (16-ft Berea core, 408 md, 110°F, 5 ft/D) (Gupta and Trushenski 1978).

A third approach is to describe the foam flow in terms of foam mobility, λ (Heller et al. 1985). The quantities typically measured in a displacement or flow experiment in porous media are pressure drop and flow rate. These results can be used with Darcy's law to calculate a mobility.

$$\lambda = k/\mu = (q_t/A)/(\Delta p/L), \quad \ldots \ldots (5.173)$$

where q_t = total volumetric flow rate, A = cross-sectional area normal to flow direction, Δp = pressure drop across the porous medium, L = length of the porous medium, λ = foam mobility, and consistent units are used. A relative mobility, λ_r, also can be defined as

$$\lambda_r = \lambda/k, \quad \ldots \ldots \ldots \ldots \ldots (5.174)$$

where k = permeability of the porous medium.

Heller et al. (1985) analyzed a number of experimental results reported in the literature. They restricted their analyses to experiments that met the following conditions.

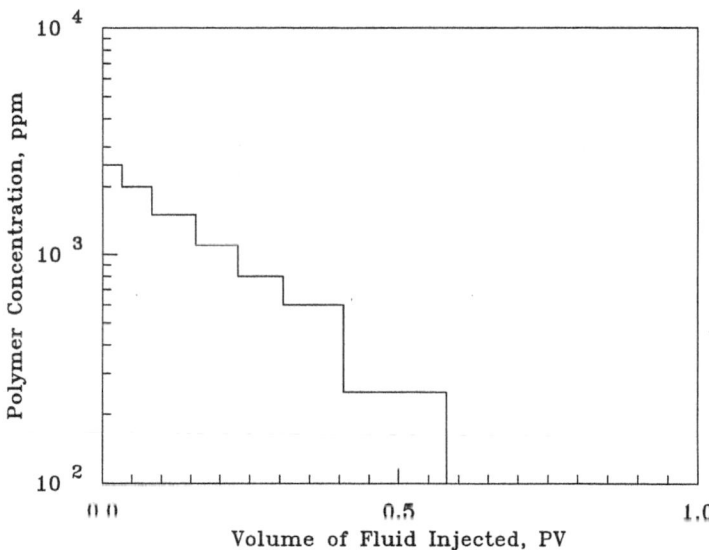

Fig. 5.132—Injection schedule for graded mobility buffer, North Burbank surfactant test (Trantham 1977).

1. The experiment was conducted under steady-state conditions. The results would thus represent conditions well behind a foam-displacement front where properties are changing.
2. The foam quality was thought to be fairly uniform along the length of the system. Compressibility effects should therefore be small.

Most of the experiments were performed in sandpacks with air/liquid foam. **Table 5.62** shows the results. Foam relative mobilities, λ_r, ranged from 0.001 to 0.636 in the data they analyzed.

Heller et al. (1985) also conducted foam-flow experiments and determined relative mobilities for the foams. **Figs. 5.140 and 5.141** show typical results. Relative mobility was found to be a function of the average foam velocity, as shown (i.e., mobility increased with increasing velocity). This finding is consistent with the pseudoplastic nature of the foams determined in viscometers. No clear dependence of λ_r on foam quality was observed, however. Their analyses of literature results also indicated that λ_r is not a strong function of quality over the range of 75 to 90% quality.

Example 5.18—Calculation of Foam Relative Permeabilities and Mobilities. Raza (1970) measured flow resistance with foam flowing in a sandpacked tube with a diameter of 1.5 in. and a length of 12.0 in. In the experiments, a liquid foaming solution (brine containing surfactant) and nitrogen were simultaneously injected into the sandpack at fixed rates. Pressure drop across the sandpack was monitored, and the injection was continued until steady-state conditions were reached. Gas saturation at the end of the experiment was measured by weighing the tube. Temperature was 25°C.

Table 5.63 (Heller et al. 1985) summarizes data for one series of runs. With these data, calculate the relative permeability to gas and liquid and the foam mobility and relative mobility as defined by Heller et al. (1985) for the last data entry in Table 5.63.

Solution. Relative permeabilities must be calculated by application of Darcy's law.

$$q_t = \frac{kk_r A}{\mu}\frac{\Delta p}{L},$$

where k = permeability of the sandpack, darcies; k_r = relative permeability to a specific fluid; A = cross-sectional area of the sandpack, cm^2; μ = fluid viscosity; Δp = pressure drop, atm; L = length of sandpack, cm; and q_t = volumetric flow rate of the fluid cm^3/s.

Substituting the experimental values, for gas,

$$\frac{76.8 \text{ cm}^3/\text{h}}{3,600 \text{ sec/hr}} = \left(\frac{3.0 \text{ darcies } k_{rg}}{0.0178 \text{ cp}}\right)\left[\frac{\pi(1.5 \text{ in.})^2(2.54 \text{ cm/in.})^2}{4}\right]$$
$$\times \left(\frac{1.63 \text{ atm}}{12 \text{ in.} \times 2.54 \text{ cm/in.}}\right).$$

Solving for k_{rg},

$$k_{rg} = 2.08 \times 10^{-4}.$$

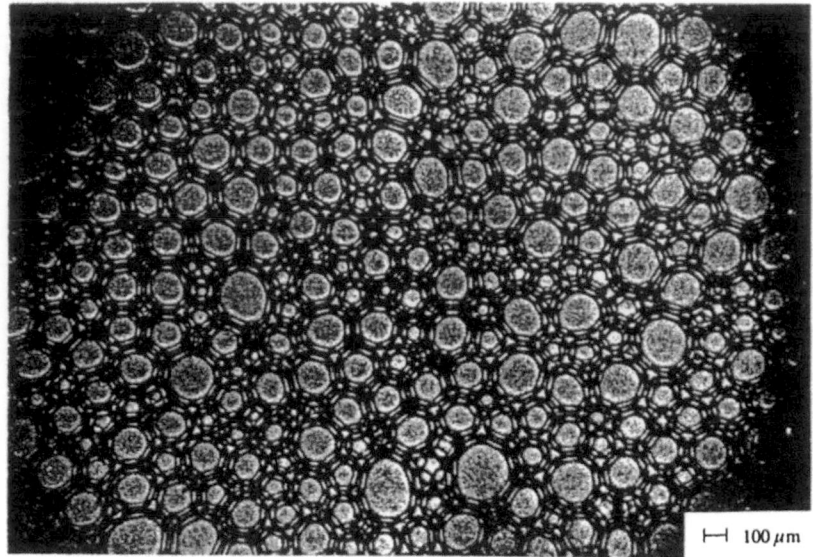

Fig. 5.133—Foam (bubbles flattened, 0.107-mm thick) (Patton et al. 1983).

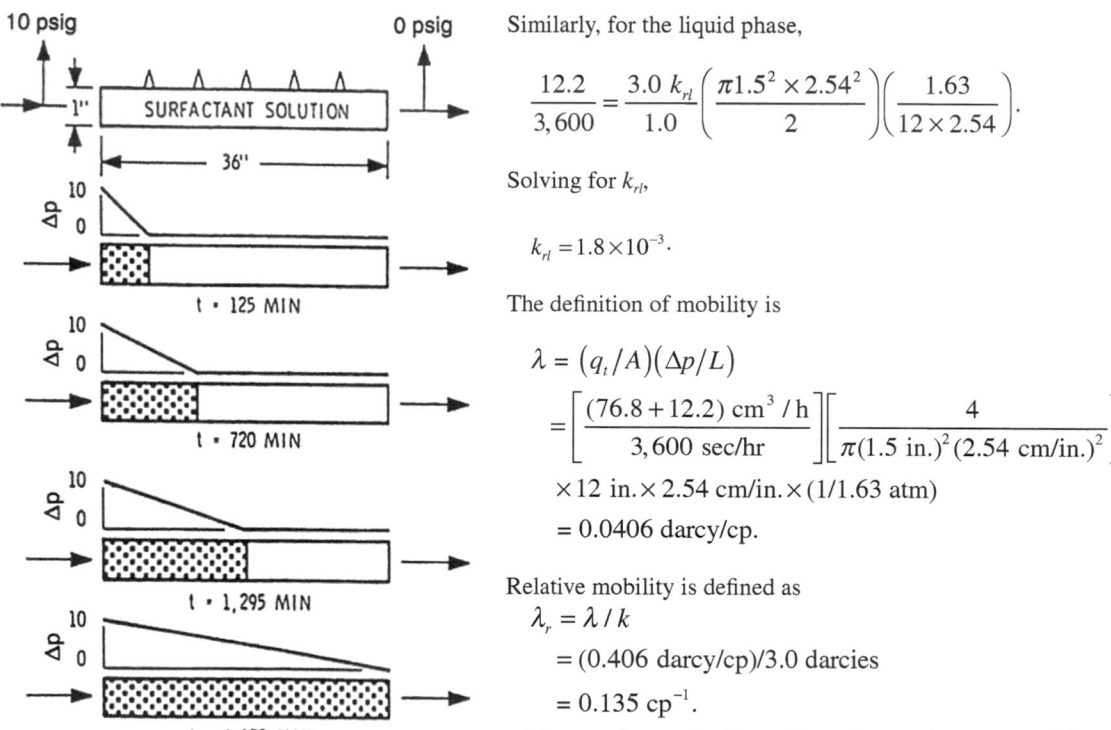

Fig. 5.134—In-situ generation and propagation of foam in a 2,760-md unconsolidated sandpack (Raza 1970).

Similarly, for the liquid phase,

$$\frac{12.2}{3,600} = \frac{3.0 \, k_{rl}}{1.0} \left(\frac{\pi 1.5^2 \times 2.54^2}{2} \right) \left(\frac{1.63}{12 \times 2.54} \right).$$

Solving for k_{rl},

$$k_{rl} = 1.8 \times 10^{-3}.$$

The definition of mobility is

$$\lambda = (q_t/A)(\Delta p/L)$$

$$= \left[\frac{(76.8 + 12.2) \, \text{cm}^3/\text{h}}{3,600 \, \text{sec/hr}} \right] \left[\frac{4}{\pi(1.5 \, \text{in.})^2 (2.54 \, \text{cm/in.})^2} \right]$$

$$\times 12 \, \text{in.} \times 2.54 \, \text{cm/in.} \times (1/1.63 \, \text{atm})$$

$$= 0.0406 \, \text{darcy/cp}.$$

Relative mobility is defined as
$$\lambda_r = \lambda / k$$

$$= (0.406 \, \text{darcy/cp})/3.0 \, \text{darcies}$$

$$= 0.135 \, \text{cp}^{-1}.$$

Flow resistance in foam flow also can be expressed by presenting results in terms of a resistance factor, F_r (Duerksen 1986). This factor is defined as the pressure drop measured across a foam generator through which a mixture of water, surfactant, nitrogen, and water vapor was flowing divided by the pressure drop for the same system without the presence of surfactant. **Table 5.64** shows typical data taken at three different temperatures.

A number of variables affect foam performance in developing flow resistance. For steamflood applications, Duerksen (1986) observed that (1) increasing temperature decreases foamability and performance, (2) increasing noncondensable gas-phase composition (N_2) increases performance, (3) increasing salt concentration in the brine adversely affects performance, and (4) performance is relatively insensitive to foam quality.

Foams in steam systems can be generated at the low fluid velocities that occur at relatively large distances away from a wellbore. That is, very little mechanical energy is required to develop a foam if a system is properly designed. In a foam displacement, however, the foam requires continued regeneration to maintain high resistance.

Porous Medium					Experimental Conditions		Foam Quality (%)	Gas Saturation at Breakthrough, Gas/Brine Test (%)
Type	Permeability (md)	Description	Length (in.)	Diameter (in.)	Mean Pressure (psig)	Δp (psi/ ft)		
Unconsolidated sandstone	2,200	Uniform	6.5	6.0	495	2.5	72.1	26.0
Unconsolidated sandstone	3,000	Uniform	124.0	3.0	795	1.0	75.1	22.6
Unconsolidated sandstone	3,000	Uniform	124.0	3.0	95	1.0	81.5	20.5
Unconsolidated sandstone	3,000	Uniform	124.0	3.0	5	1.0	85.5	19.0
Unconsolidated sandstone	3,000	Uniform	48.0	1.5	95	2.5	89.5	22.6
Unconsolidated sandstone	3,000	Uniform	18.0	1.5	45	10.0	86.1	22.2
Unconsolidated sandstone	3,000	Uniform	18.0	1.5	2	2.5	79.0	18.7
Unconsolidated sandstone	15,700	Uniform	18.0	1.5	2	2.5	89.0	21.2
Unconsolidated sandstone	35,500	Uniform	18.0	1.5	2	2.5	92.5	23.2
Unconsolidated sandstone	13,700	Nonuniform	18.0	1.5	45	10.0	73.5	26.0
Unconsolidated sandstone	39,300	Nonuniform	18.0	1.5	45	10.0	72.0	26.0
Unconsolidated sandstone	14,500	Multilayer system	18.0	1.5	87.5	3.4	49.5	7.4
Unconsolidated sandstone	14,500	Multilayer system	18.0	1.5	89.5	0.7	55.3	15.8
Torpedo sandstone	500	Natural	12.0	1.88	495	10.0	40.5	19.5
Torpedo sandstone	500	Natural	12.0	1.88	95	10.0	55.0	20.2
Torpedo sandstone	500	Natural	12.0	1.88	45	10.0	58.2	20.4
Torpedo sandstone	500	Natural	12.0	1.88	5	10.0	59.1	20.6
Torpedo sandstone	500	Natural	12.0	1.88	25	50.5	38.2	18.4
Torpedo sand	500	Natural	34.5	1.88	93	2.0	61.0	25.4
Sandstone	1.5	Natural	5.7	1.97	250	21.0	23.6	16.4
Berea sandstone	1,000	Fired	28.5	1.94	5	5.0	60.8	15.2
Limestone	63	Intergranular	8.0	3.32	98	6.0	39.6	23.6
Limestone	344	Vugular	1.0	3.3	3	2.0	8.1	5.7
Limestone	11	Vugular	26.9	4.33	100	3.1	11.0	7.0
Conglomerate	525	Nonuniform	13.1	1.97	250	10.0	31.6	14.8

Table 5.59—Effect of porous medium, pressure level, and pressure differential on foam quality (Raza 1970).

5.10 WAG Process

The mobility ratio between injected gas and the displaced oil bank in CO_2 and other miscible gas-displacement processes is typically very unfavorable because of the relatively low viscosity of the injected phase. For example, the viscosity of CO_2 at 110°F is about 0.03 cp at 1,500 psia and 0.06 cp at 2,500 psia (Stalkup 1983). Viscosity ratios, μ_o/μ_{CO_2}, for floods conducted in west Texas have been reported to be between 8 and 50 (Stalkup 1983). As discussed in Chapters 3 and 4, an unfavorable mobility ratio results in viscous fingering and reduced volumetric sweep efficiency.

A technique developed to overcome this problem is to inject specified volumes, or slugs, of water and gas alternately. Simultaneous flow of the two fluids results in reduction of the mobility of each phase. The combined mobility of the two phases is less than that of the injected gas alone, and thus the mobility ratio in the process is improved. The process is called the water-alternating-gas (WAG) process. Caudle and Dyes (1958) proposed this general method to improve oil recovery.

In WAG injection, water/gas injection ratios have ranged from 0.5 to 4 volumes of water per volume of gas at reservoir conditions. The sizes of the alternate slugs range from 0.1% to 2% PV (Huang and Holm 1988). Total or cumulative slug sizes of CO_2 in reported field projects typically have been 15% to 30% HCPV (Stalkup 1983; Huang and Holm 1988), although smaller and larger slugs have been used.

A problem in the WAG process is that injected water blocks contact between the injected gas phase and resident oil. This reduces displacement efficiency at the pore scale (i.e., it results in a larger ROS). This effect has been found to be a strong function of rock wettability and more detrimental in water-wet rocks (Stalkup 1970).

Porous Medium				Oil		Foam Quality (%)		Time Until Gas Breakthrough (minutes)	
Type	Length (in.)	Diameter (in.)	Permeability (md)	Description	Type	Saturation	Oil	No Oil	Oil	No Oil
Unconsolidated sand	18	1.5	3,000	Uniform	Core test fluid	16.0	31.0	78.0	28	2,225
Unconsolidated sand	18	1.5	3,000	Uniform	Crude oil	15.4	22.0	78.0	16	2,225
Unconsolidated sand	124	3.0	4,000	Uniform	Crude oil	25.0	25.9	87.5	190	15,890
Unconsolidated sand	18	1.5	14,500	Multilayer	Core test fluid	13.7	31.6	55.3	22	300
Torpedo sandstone	12	1.875	500	Native state	Crude oil	22.6	11.9	40.5	6	18
Torpedo sandstone	12	1.875	500	Native state	Crude oil/ pentane mixture	47.3	18.5	40.5	8	18

* Foaming solution = 2% OK liquid in ¼ N—brine.

Table 5.60—Effect of oil on the quality and nature of water foam (Raza 1970)*.

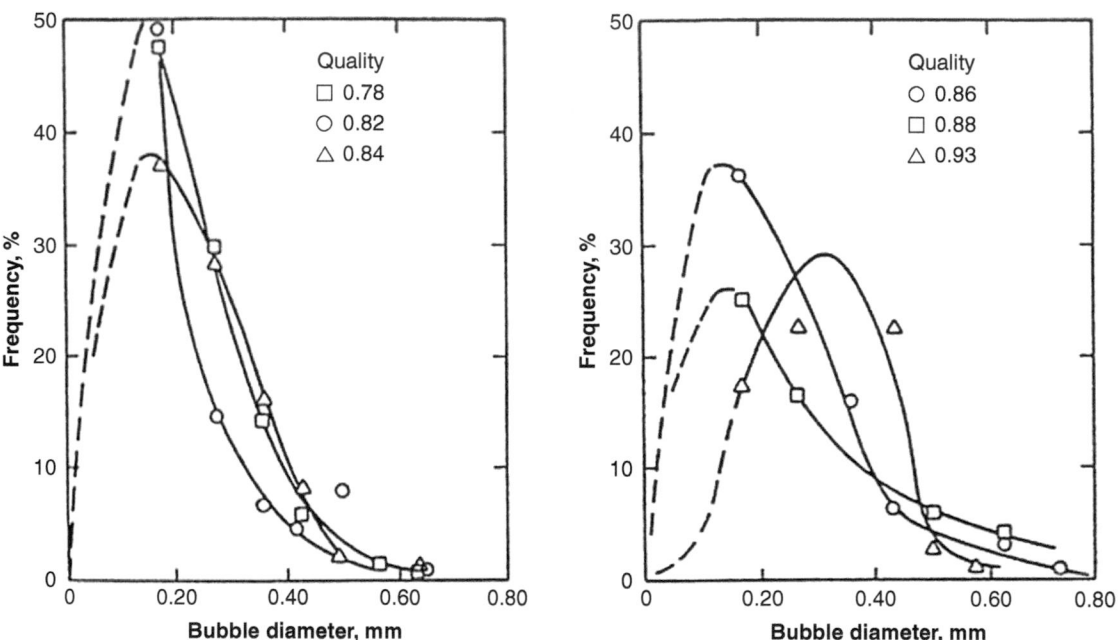

Fig. 5.135—Bubble-size frequency distribution (David and Marsden 1969).

The WAG process is discussed in Sections 5.10.1 through 5.10.4. The water-blocking phenomenon is described first, followed by discussions of flow in one space dimension and volumetric sweep efficiency in two and three space dimensions. Finally, field experiences are briefly summarized.

5.10.1 Effect of Water Blocking on Displacement Efficiency. A number of investigators have shown that the presence of a mobile water phase can adversely affect the displacement of oil by a solvent. As an example, Tiffin and Yellig (1983) conducted displacement experiments in linear Berea cores. The cores in their natural state were strongly water-wet. In some cases, the Berea cores were treated chemically to make them oil-wet.

The experimental procedure basically consisted of first creating immobile, or connate, water saturation by injecting an oil into a core that was 100% saturated with water to reduce the water saturation to approximately 31%. Saturation in the core was next reduced to ROS by waterflooding. Then, CO_2 and water were injected simultaneously at a specified water/CO_2 ratio. Conditions were such that the CO_2 phase would develop miscibility with the oil phase in the manner described in Chapters 1 and 6.

	Surfactant	HLB	Ionic Type**	Manufacturer[†]	
Foamed in <1 minute to full height of tube {	Emulphogene BC-720		N	e	
	Aerosol 30 Alipal CD-128		Z	a	
			A	e	
	Arfoam 2213		A	b	
	Igepal CO-630	13.0	N	e	Refoamed easily after aging several days
Foamed in <2 minutes to full height of tube {	Igepal CO-710	13.6	N	e	}
	Makon 14	15.0	N	h	
	Monateric ADFA		Z	g	
	Sulfotex PAI		A	f	
	Sulfotex RIF		A	f	
	Trycol TDA-8	12.5	N	d	
	Alipal CO-436		A	e	
	Deriphat BAW		Z	f	
Foamed in <2 minutes to full height of tube {	Emsorb 6900	15.0	N	d	
	Plurafac C-17	16.0	N	c	
	Witcolate 1247-H		A	i	
	Witcolate 1259		A	i	Did not refoam easily after aging several days
	Arnox 930-70		N	b	}
Foamed in >2 minutes to full height of tube {	Emersal 6462		A	d	
	Pluronic F-68	30.5	N	c	
	BioSoft EA-4	≈12.0	N	h	
Shaken >2 minutes; very little foam produced {	Deriphat 160-C		Z	f	
	Emsorb 6903	10.2	N	d	

* Mixed brine = 0.5% $CaCl_2$ and 0.5 wt% NaCl in distilled water.

** Ionic type: A = anionic, N = nonionic, Z = zwitterionic (or amphoteric).

[†]Manufacturer: a = American Cyanamid, b = Arjay Inc., c = BASF, d = Emery Industries, e = GAF, f = Henkel Chemicals, g = Mona Industries, h = Stepan Chemical Co., and i = Witco Chemical Co.

Table 5.61—Foaming results of 0.05% (active) surfactant in mixed brine* with isooctane (Heller et al. 1985).

Oil recoveries were determined by material balance. The process was a tertiary recovery process in that the displacements were conducted after a waterflood.

Figs. 5.142 and **5.143** (Tiffin and Yellig 1983) show results for the water-wet and oil-wet cores. In the water-wet cores, the simultaneous injection of water with CO_2 resulted in a significant decrease in recovery (i.e., the water caused oil trapping to occur). The trapped oil was not miscibly displaced. As shown in Fig. 5.143, the effect was much less pronounced in oil-wet Berea cores. Tiffin and Yellig (1983) showed that, for tertiary CO_2 flooding, mobile water from a previous waterflood did not change overall recovery.

Similar results have been presented by several researchers (Huang and Holm 1988; Stalkup 1970; Raimondi and Torcaso 1964; Lin and Huang 1990), showing that oil recovery is adversely affected by the presence of mobile water in water-wet cores, but the effect of mobile water is small to negligible in mixed-wet and preferentially oil-wet cores. In these last cases, the roles of prolonged contact and diffusion of the solvent to the trapped oil were shown to be important (Huang and Holm 1988; Lin and Huang 1990).

A correlation of the trapping phenomenon is given by

$$S_{or,wb} = \frac{S_{or}}{1 + \alpha k_{ro} / k_{rw}}, \dotfill (5.175)$$

where S_{or} = ROS following waterflooding, $S_{or,wb}$ = ROS following miscible displacement in the presence of mobile water, α = an empirical constant, and k_{ro} and k_{rw} are oil and water relative permeabilities, respectively. A value of $\alpha = 1.0$ represents strong

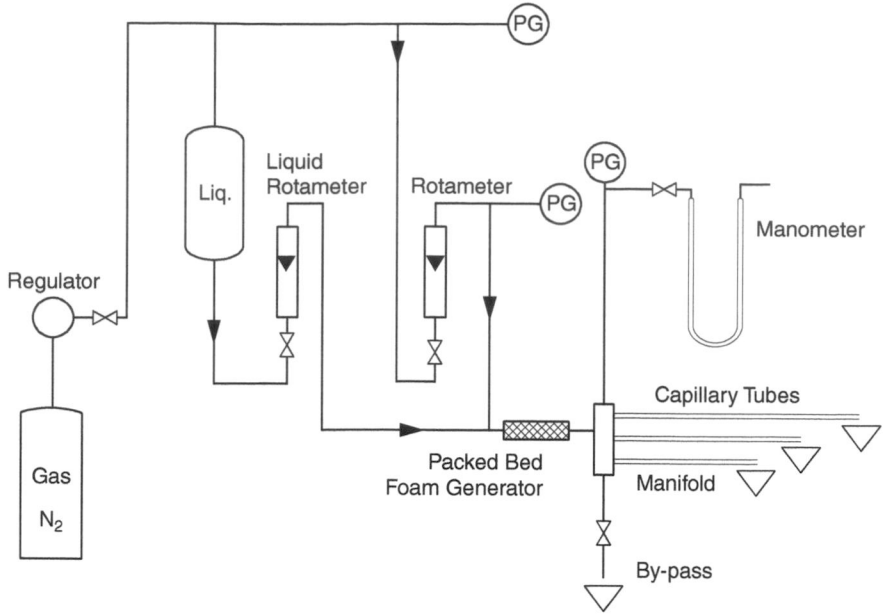

Fig. 5.136—Capillary viscometry apparatus (Patton et al. 1983).

oil trapping, while values on the order of 100 represent relatively weak trapping (Lin and Huang 1990). The correlation was first presented by Raimondi and Torcaso (1964) and later modified by Chase and Todd (1984). **Fig. 5.144** shows an example of $S_{or,wb}$ as a function of S_w for different values of α (Lin and Poole 1991). The curves also illustrate the effect of water saturation, manifested by k_{ro}/k_{rw} in the correlation.

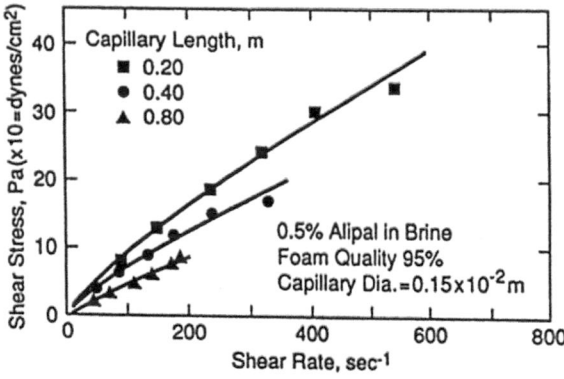

Fig. 5.137—Pseudoplastic characteristics of foam (Patton et al. 1983).

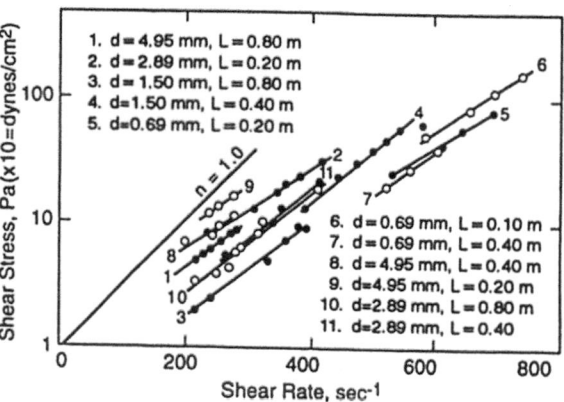

Fig. 5.138—Foam data represented by Ostwald-de Waele model (Patton et al. 1983).

The correlation given by Eq. 5.175 does not account for the recovery by the diffusion process that occurs in mixed-wet and oil-wet rocks. Lin and Huang (1990) give a correlation to account for this time-dependent effect.

Clearly, oil trapping in the presence of mobile water, when it occurs, works to the disadvantage of the WAG process. Positive effects from improved mobility ratio may be offset by trapping.

Example 5.19—Oil Trapped by Mobile Water in the WAG Process. Consider flow through a linear Berea core that is strongly water-wet. A solvent is used to displace oil that is at waterflood ROS ($S_{or} = 35\%$). The solvent is injected simultaneously with water. The water/solvent ratio is 1:1. Fluid viscosities are $\mu_w = 1.0$ cp and $\mu_o = 2.0$ cp. Calculate the residual saturation caused by water blocking.

Solution. From Eq. 5.175,

$$S_{or,wb} = \frac{S_{or}}{1 + \alpha k_{ro}/k_{rw}}$$

or $S_{or,wb} = \dfrac{S_{or}}{1 + \alpha (f_o/f_w)(\mu_o\mu_w)}.$

Assuming that $\alpha = 1.0$,

$$S_{or,wb} = \frac{0.35}{1 + 1.0 \times 1.0/1.0 \times 2.0/1.0}$$
$$= 0.35/(1+2) = 0.12.$$

Water blocking is estimated to result in a residual saturation of 12% after miscible displacement.

5.10.2 Displacement in One Space Dimension. In this discussion, flow is considered as though water and solvent were injected simultaneously. This assumption simplifies calculations because injection in field practice occurs as alternate injections of discrete slugs (Stalkup 1983). When discrete slugs are injected, the solvent fingers into the water slug because of the unfavorable mobility ratio. Thus, small slugs dissipate relatively rapidly, approaching a condition of simultaneous injection.

Stalkup (1983)[8] developed equations to describe the condition of equal-velocity flow of solvent and water phases. He chose this condition as an optimum rate to take advantage of the increased mobility resulting from WAG injection. If too little water is injected, so that the solvent velocity is larger, then a solvent bank will form ahead of the water. This will result in an unfavorable mobility ratio at the solvent/oil interface, with resulting fingering of solvent into the oil bank, which would reduce the efficiency of the process.

If too much water is injected, the water will move faster than the solvent, resulting in a high water saturation at the solvent/oil interface. Oil trapping is likely to be increased by the higher water saturation. The magnitude of oil trapping depends on rock wettability, as discussed in Section 5.10.1.

Equations are derived below that allow calculation of the water/solvent injection ratio required for equal water/solvent velocities. The model is for ID flow and is based on fractional-flow theory. Equations and graphical solutions are presented to describe WAG injection for both secondary and tertiary recovery conditions. Walsh (1991) gives a more comprehensive treatment using fractional-flow theory.

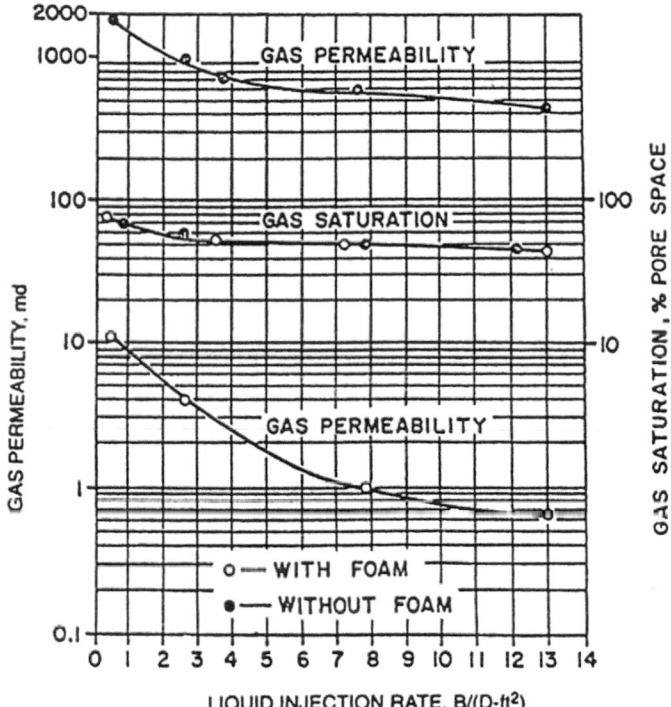

Fig. 5.139—Effect of liquid rate on gas saturation and permeability (Bernard and Holm 1964).

Secondary Recovery—Water at Immobile Saturation. Consider the flow condition shown in **Fig. 5.145**. Oil is displaced by a solvent/water bank in which water and solvent are flowing at equal velocities. The problem is to determine the injection ratio at which the equal-velocity condition holds (i.e., $v_s = v_w$). Volume balances on solvent and water are made on a differential element located at the interface between the solvent/water bank and the oil bank, as shown in Fig. 5.145. It is assumed here that no oil is bypassed by the solvent/water bank and that there is no solubility of one phase into another. Modified equations that incorporate these effects could be derived easily. Water is assumed to be immobile at the initial saturation, S_{wi}.

A balance on solvent over Δt, the time required for the water/solvent bank to advance a length Δx, yields

$$V_s \big|_{t+\Delta t} - V_s \big|_t = (V_s)_{\text{in}} - (V_s)_{\text{out}} \quad\dots\dots\dots\dots\dots\dots\dots\dots (5.176)$$

$$\text{or } A\phi S_s \Delta x \big|_{t+\Delta t} - 0 \big|_t = q_t f_s \Delta t \big|_{\text{in}} - 0 \big|_{\text{out}}, \quad\dots\dots\dots\dots\dots\dots\dots\dots (5.177)$$

where q_t = total flow rate, solvent plus water; f_s = fractional flow of solvent; A = cross-sectional area; ϕ = porosity; S_s = solvent saturation; x = distance; and t = time.

Dividing by Δt and Δx, letting $\Delta t \to 0$ and $\Delta x \to 0$, and rearranging gives

$$\frac{dx_s}{dt} = v_s = \frac{q_t f_s}{A\phi S_s}. \quad\dots\dots\dots\dots\dots\dots\dots\dots (5.178)$$

A similar balance on water yields

$$\frac{dx_w}{dt} = v_w = \frac{q_t f_w}{A\phi\left(S_w - S_{wi}\right)}, \quad\dots\dots\dots\dots\dots\dots\dots\dots (5.179)$$

where S_{wi} = initial water saturation. Equating $v_s = v_w$,

$$\frac{f_s}{S_s} = \frac{f_w}{S_w - S_{wi}}. \quad\dots\dots\dots\dots\dots\dots\dots\dots (5.180)$$

$$\text{or } \frac{1 - f_w}{1 - S_w} = \frac{f_w}{S_w - S_{wi}}. \quad\dots\dots\dots\dots\dots\dots\dots\dots (5.181)$$

[8] Personal communication with F. I. Stalkup, Arco Oil & Gas Co., Plano, Texas, USA, 1992.

	Bernard and Holm (1964)	Marsden and Khan (1966)	Albrecht and Marsden (1970)	Raza (1970)	Chiang et al. (1980)	Bernard et al. (1980)
Porous medium	Sandpack	Sandpack (25 to 120 mesh) Berea core Alundum™	Sandpack (100 to 200 mesh	Sandpack	Sandpack (60 mesh)	Carbonate core
Length (ft)	1 to 30	0.2 0.125, 0.13 0.125	2	1.5, 1	3.8	3, 2.7
Diameter (in.)	0.5	1 0.75 1.06	1.5	1.5	11.375×0.25	2×2
k (darcies)	3 to 150	5.3 to 58 0.26 to 0.48 4.41	5	3 to 15.7	32.7	0.086 0.114 0.052
Δp (psi)	5 150	— —	10 14	24 to 84	3.1	50 100
P/P_o*	1.34 11.20	— –	1.68 1.95	— —	1.21	1.02 1.03
Foamer	1% Solution	0.1 or 1% Aerosol MA™	0.1% O.K.™ (1.5% NaCl)	2.0% O.K.™ (0.25 N NaCl)	1% Suntech IV™	1% Alipal CD-128™
λ (darcies/cp)	—	0.006 to 3.5 0.013 to 0.1 0.35 to 0.75	0.075 to 0.35	0.013 to 0.041	3.0 to 3.7	0.003 to 0.063
λ_r (cp⁻¹)	—	0.01 to 0.06 0.05 to 0.20 0.08 to 0.17	0.015 to 0.069	0.001 to 0.007	0.09 to 0.11	0.056 to 0.636

*P/P_o is the ratio of (absolute) inlet to outlet pressures.

Table 5.62—Selected foam studies (Heller et al. 1985).

Solution of Eq. 5.181 gives a value of f_w that is the injected fraction of water required to yield equal water and solvent velocities. Because S_w is unknown, Eq. 5.181 must be solved simultaneously with the solvent/water fractional-flow curve. This is done graphically as shown in **Fig. 5.146.**

Eq. 5.181 indicates that a straight line drawn from the point $(1, 1)$ to $(0, S_{wi})$ will intersect the point (f_w, S_w) on the fractional-flow curve, which satisfies both Eq. 5.181 and the fractional-flow curve. The value of f_w at this intersection is the desired fractional water flow.

Tertiary Recovery—Oil Initially at Residual Saturation to Waterflooding, S_{orw}. **Fig. 5.147** shows the displacement for this case. Oil is originally at S_{orw}, the residual saturation following a waterflood. Simultaneous injection of solvent and water creates an oil bank that flows ahead of the solvent/water bank. Again, the problem is to determine the fractional flow of water that will yield equal solvent and water velocities.

Volume balances are made on an element located at the interface, as previously shown. As before, it is assumed that no oil is bypassed and that there is no solubility of one phase into another. Stalkup (1983) gives corresponding equations with these assumptions removed.

A volume balance on solvent over Δt required for the front to advance a distance Δx is given by

$$A\phi S_s \Delta x = q_t f_s \Delta t. \dotfill (5.182)$$

Rearranging and letting $\Delta x \to 0$ and $\Delta t \to 0$ give

$$\frac{dx_s}{dt} = v_s = \frac{q_t f_s}{A\phi S_s}.$$

In the tertiary recovery case, there is water flow ahead of the solvent/water bank and thus the water balance is

$$A\phi S_w \Delta x |_{t+\Delta t} - A\phi S_{wob} \Delta x |_t$$
$$= q_t f_w \Delta t |_{in} - q_t f_{wob} \Delta t |_{out}, \dotfill (5.183)$$

where S_{wob} = water saturation in the oil bank at the solvent/water and oil-bank interface and f_{wob} = fractional flow of water in the oil bank at the interface. Dividing by Δt and Δx, letting $\Delta t \to 0$ and $\Delta x \to 0$, and rearranging give

$$\frac{dx_w}{dt} = v_w = \frac{q_t \left(f_w - f_{wob} \right)}{A\phi \left(S_w - S_{wob} \right)} \dots (5.184)$$

Setting $v_s = v_w$,

$$\frac{f_s}{S_s} = \frac{1-f_w}{1-S_w} = \frac{f_w - f_{wob}}{S_w - S_{wob}} \dots (5.185)$$

It also must be true that

$$v_w = v_s = v_{ob}, \dots (5.186)$$

where v_{ob} is the velocity of the trailing edge of the oil bank. The saturation of the water in the oil bank, S_{wob}, and the water fractional flow, f_{wob}, must satisfy the fractional-flow equation. That is,

$$v_{ob} = \left(\frac{dx}{dt} \right)_{S_{wob}} = \frac{q_t}{A\phi} \left(\frac{\partial f_w}{\partial S_w} \right)_{S_{wob}} \dots (5.187)$$

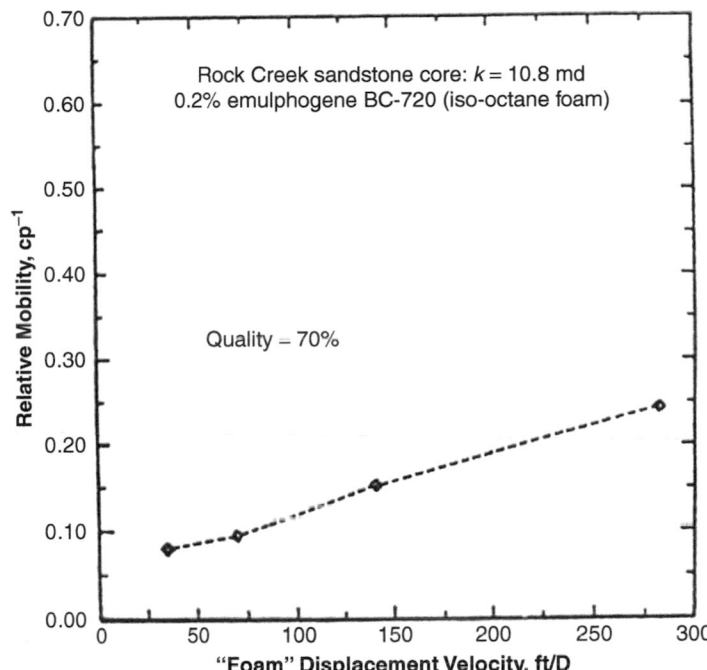

Fig. 5.140—Relative mobility for isooctane foam (Heller et al. 1985).

Eqs. 5.185 and 5.187, combined with the solvent/water and oil/water fractional-flow curves, make possible a graphical solution for f_w and S_w, as shown in **Fig. 5.148**. From Eq. 5.185, a line drawn from the point $(1,1)$ through (f_w, S_w) on the solvent/water fractional-flow curve must pass through (f_{wob}, S_{wob}) on the oil/water fractional-flow curve. Thus,

$$\frac{1-f_w}{1-S_w} = \frac{f_w - f_{wob}}{S_w - S_{wob}} = \frac{1-f_{wob}}{1-S_{wob}}. \dots (5.188)$$

Eq. 5.187 requires that this line also be tangent to the oil/water fractional-flow curve at the point (f_{wob}, S_{wob}), as shown in Fig. 5.148. The required water fractional injection rate to attain equal water and solvent velocities is given by the point (f_w, S_w) on the solvent/water fractional-flow curve.

Application of this solution is illustrated in Example 5.20.

Example 5.20—Calculation of Relative Water Injection Rate in a WAG Process, Tertiary Recovery. A WAG tertiary recovery process is to be implemented. The reservoir has previously been waterflooded to S_{or}. Relative oil and water permeability data are given by (Willhite 1986)

$$k_{ro} = \left(1 - S_{wD} \right)^{2.56} \dots (5.189)$$

and $k_{rw} = 0.785\, S_{wD}^{3.72}, \dots (5.190)$

where $S_{wD} = \dfrac{S_w - S_{iw}}{1 - S_{or} - S_{iw}}. \dots (5.191)$

Table 5.65 gives the k_{ro} and k_{rw} values for S_{iw} = 0.363 and S_{or} = 0.205.

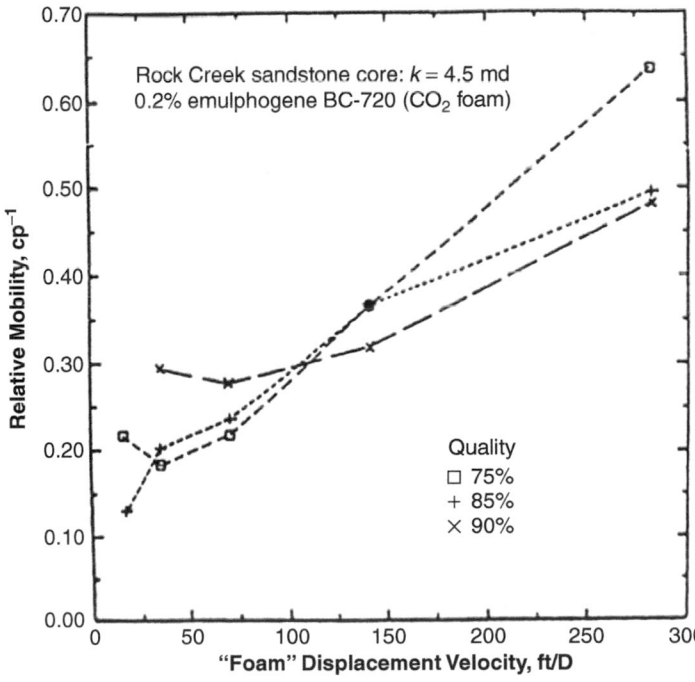

Fig. 5.141—Relative mobility for dense CO_2 foam, Core A (Heller et al. 1985).

k Sandpack (md)	q_g (cm³/h)	q_l (cm³/h)	Δp (atm)	S_g
3,000	35.7	48.8	5.36	0.789
	34.3	48.8	5.73	0.790
	62.8	48.8	5.72	0.798
	30.2	12.2	1.68	0.886
	76.8	12.2	1.63	0.886

Table 5.63—Data for a series of low-resistance runs (Heller et al. 1985), Example 5.18.

Water fractional flow is given by

$$f_w = \frac{1}{1 + (k_{ro}/k_{rw})(\mu_w/\mu_o)}. \quad\dots\dots\dots\dots\dots\dots\dots\dots\dots\dots\dots\dots\dots\dots\dots \quad (5.192)$$

and also is tabulated in Table 5.65 for $\mu_o/\mu_w = 2.0$.

The residual oil is to be displaced by a WAG injection in which the solvent viscosity, μ_s, is 0.06 cp at reservoir conditions. The relative injection rates are to be set such that the solvent and water flow at equal velocities in the solvent/water bank that displaces the oil bank. Calculate the fractional water injection rate.

Solution. Assume that the relative solvent and water permeabilities are the same as for oil and water. The fractional flow data for solvent and water are in Table 5.65, and $\mu_s/\mu_w = 0.06$.

Fig. 5.148 plots fractional-flow curves for oil/water and solvent/water. To determine the required f_w, draw a line from the point (1,1) tangent to the oil/water fractional-flow curve, as shown in Fig. 5.148. This line crosses the solvent/water fractional-flow curve at (f_w, S_w). The values read from the figure are $f_w = 0.42$ and $S_w = 0.694$. The required injection ratio is thus 42% water and 58% solvent or 0.72 volumes of water per volume of solvent at reservoir conditions.

It is also interesting to determine the improvement in mobility ratio that results from the application of the WAG process. This is illustrated in Example 5.21.

Example 5.21—Mobility Ratio Improvement From Application of the WAG Process. With the data and conditions of Example 5.19, determine the improvement in mobility ratio that results from use of the WAG process compared with pure solvent injection.

Solution. For the WAG process,

$$M_{WAG} = \frac{(\lambda_s + \lambda_w)_{WAG}}{(\lambda_o + \lambda_w)_{ob}}, \quad\dots\dots\dots\dots\dots\dots\dots\dots\dots\dots\dots\dots\dots\dots \quad (5.193)$$

	Resistance Factor**		
Surfactant	At 375°F	At 400°F	At 425°F
Stepanflo 30	27.4	25.1	21.2
Suntech IV	21.7	17.5	14.3
CRC Sulfonate 6	22.3	18.6	14.6
CRC Sulfonate 13	27.8	18.9	19.3
CRC Sulfonate 8	28.6	24.3	23.4
CRC Sulfonate 1	19.1	16.3	—
CRC Sulfonate 3a	20.4	17.0	16.3
CRC Sulfonate 2	20.0	17.9	—
CRC Sulfonate 3b	21.4	19.9	16.6
CRC Sulfonate 4	17.0	11.0	1.1

*Brine contains 1% NaCl and 500 ppm $CaCl_2$.

** Resistance factor = (Δp surfactant + brine + nitrogen)/(Δp brine + nitrogen).

Table 5.64—Effect of temperature on resistance factor of Chevron Research Co. (CRC) surfactants in brine (Duerksen 1986)*.

where the numerator is the total relative mobility in the water/solvent WAG bank and the denominator is the total relative mobility in the oil bank. From Example 5.20, the water saturation in the WAG solvent/water bank is $S_w = 0.694$. Thus, from Eqs. 5.189 through 5.191,

$$\begin{aligned} S_{wD} &= \frac{S_w - S_{iw}}{1 - S_{or} - S_{iw}} \\ &= \frac{0.694 - 0.363}{1 - 0.205 - 0.363} \\ &= 0.766 \end{aligned}$$

and $k_{rw} = 0.785\, S_{wD}^{3.72}$

$$\begin{aligned} &= 0.785(0.766)^{3.72} \\ &= 0.291. \end{aligned}$$

Assuming that the solvent relative permeability/saturation relationship is the same as the oil relative permeability/saturation relationship,

$$\begin{aligned} k_{rs} &= k_{ro} = \left(1 - S_{wD}\right)^{2.56} \\ &= (1 - 0.766)^{2.56} \\ &= 0.024. \end{aligned}$$

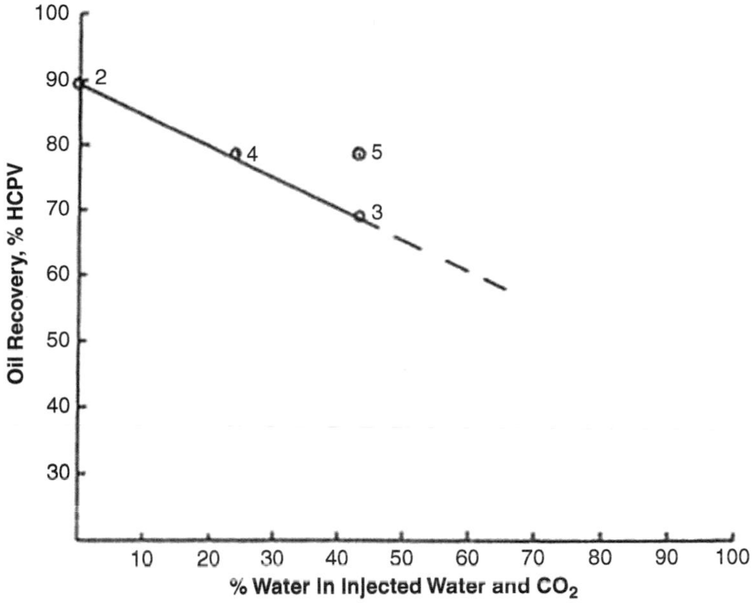

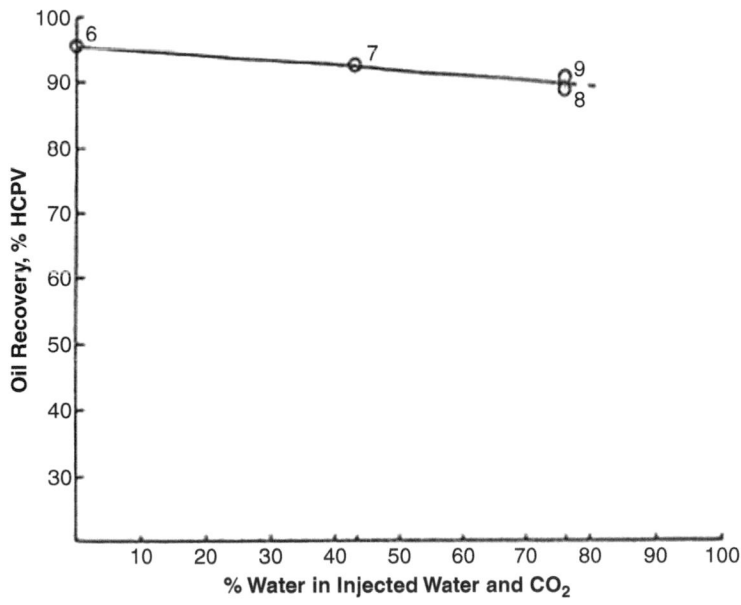

Fig. 5.142—Effect of mobile water on oil recovery for CO_2 displacements of reservoir crude oil (multiple-contact miscible process), tertiary displacements, water-wet core (Tiffin and Yellig 1983).

Thus, $(\lambda_s + \lambda_w)_{WAG} = k_{rs} / \mu_s + k_{rw} / \mu_w$

$$= (0.024 / 0.06) + (0.291 / 1.0) \quad (5.194)$$

$$= 0.691.$$

Fig. 5.143—Effect of mobile water on oil recovery for CO_2 displacements of reservoir crude oil (multiple-contact miscible process), tertiary displacements, oil-wet core (Tiffin and Yellig 1983).

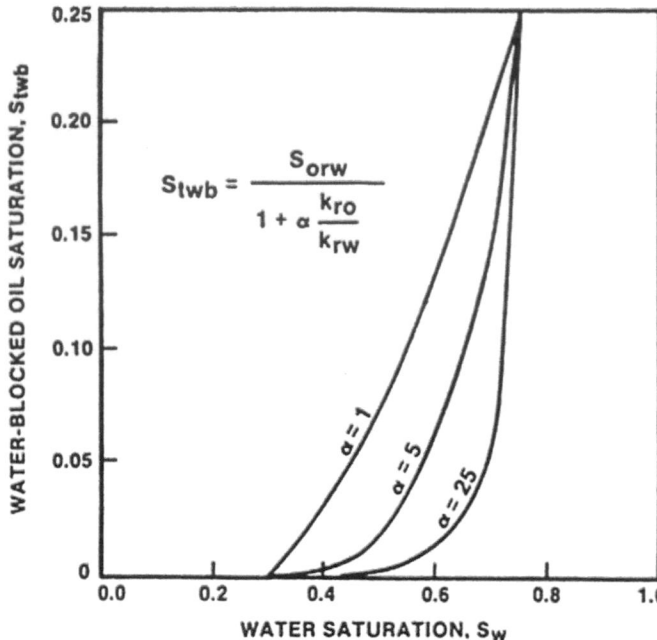

Fig. 5.144—Water-blocking function used in simulation study (Lin and Poole 1991).

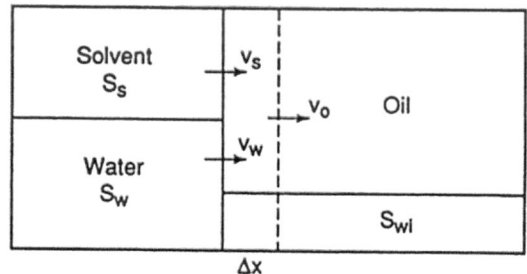

Fig. 5.145—Secondary WAG displacement.

In the oil bank, from Fig. 5.148,

$$S_w = 0.50,$$

$$S_{wD} = \frac{0.50 - 0.363}{1 - 0.205 - 0.363}$$
$$= 0.317,$$

$$k_{rw} = 0.785(0.317)^{3.72}$$
$$= 0.011,$$

$$k_{ro} = (1 - 0.317)^{2.56}$$
$$= 0.377,$$

and $(\lambda_o + \lambda_w)_{ob} = k_{ro}/\mu_o + k_{rw}/\mu_w$

$$= (0.377/2.0) + (0.011/1.0)$$

$$= 0.200. \ \dots\dots\dots\dots\dots \ (5.195)$$

The mobility ratio in the WAG process is

$$M_{WAG} = 0.691/0.200 = 3.46.$$

If the displacing fluid were pure solvent, then the total relative mobility of the displacing fluid would be

$$\lambda_s = k_{rs}/\mu_s$$
$$= 1.0/0.06$$
$$= 16.67.$$

The oil-bank total relative mobility would be the same. Thus,

$$M_s = \lambda_s / \left[(\lambda_o + \lambda_w)_{ob} \right]$$
$$= 16.67/0.20$$
$$= 83.3, \qquad \dots\dots\dots\dots\dots \ (5.196)$$

where M_s is the mobility ratio with pure solvent displacing the oil/water bank. Use of the WAG displacement thus results in an improvement of the mobility ratio by a factor of 83.3/3.46 = 24.

5.10.3 Volumetric Sweep Efficiency. As discussed in Chapter 4, reduction of the mobility ratio in a displacement process results in an improvement of volumetric sweep efficiency. However, calculation of sweep from the empirical correlations presented in Chapter 4 is probably not justified in a WAG process because of the complex nature of the flow in the region behind the oil bank. In application, the process usually is modeled with computer-based mathematical models (Lin and Poole 1991; Hsie and Moore 1988; Claridge 1982). Limited results from properly scaled physical models have also been reported (Jackson et al. 1985).

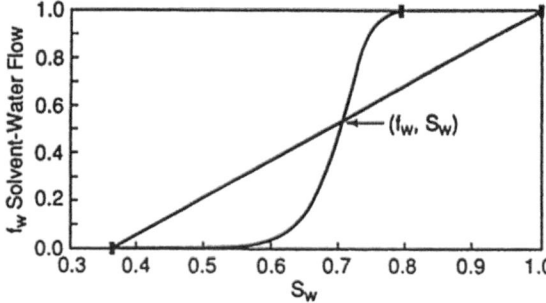

Fig. 5.146—Secondary WAG process, calculation of water fractional flow.

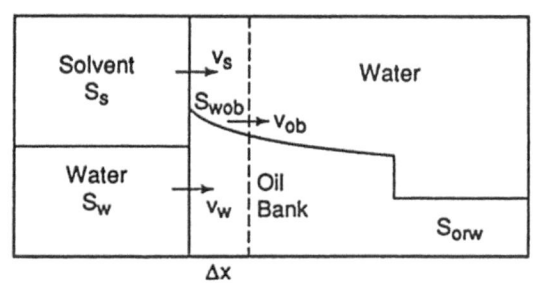

Fig. 5.147—Tertiary WAG displacement.

As an illustration, a study of the Dollarhide Field in west Texas is briefly described (Lin and Poole 1991). The field has two productive zones separated by a limestone barrier. The lower zone contained approximately 75% of the OOIP. The field is divided into three fault blocks. OOIP was estimated to be approximately 138×10^6 bbl. The field was developed on 40-acre spacing. Recovery was initially by primary production and then by waterflood, which was quite successful. Ultimate recovery by a combination of primary and secondary production was estimated to be 43.1% OOIP.

The reservoir oil has a gravity of 40 °API and a viscosity of 12.2 cp at reservoir conditions. Initial reservoir pressure was 3,300 psia, and temperature was 120°F. The bubblepoint pressure was approximately 2,830 psia. CO_2 miscibility pressure (discussed in Chapter 6) was 1,650 psia, and thus, a CO_2 miscible flood was considered to be an appropriate EOR process to be applied.

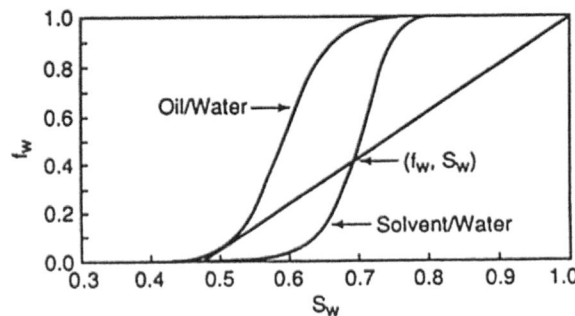

Fig. 5.148—Tertiary WAG process, calculation of water fractional flow.

CO_2 miscible flooding began in 1985 in the south fault block of the field. A computer simulation was conducted to test various CO_2 injection schemes. These included single-slug CO_2 injection, WAG injection, and a third method called a hybrid process.

A computer simulation of the type described by Todd and Longstaff (1972) was conducted. In the simulated WAG injection, water blocking was modeled with Eq. 5.175 with $\alpha = 5.0$ on the basis of core experiments. Other parameters used in the model were based on laboratory measurements and history matching of primary and secondary recovery.

In the single-slug injection runs, CO_2 slug sizes ranging from 8.8 to 50% HCPV were investigated. An optimum value of 30% was determined because, for larger volumes, the incremental oil recovered dropped below 0.1 bbl/Mcf of CO_2 injected. This CO_2 slug size was used in the WAG simulations.

S_w	S_{wD}	k_{ro}	k_{rw}	S_w	f_w Oil/Water	f_w Solvent/Water
0.36	0	1	0	0.363	0	0
0.38	0.039351852	0.902328627	4.65736×10^{-6}	0.38	1.03229×10^{-5}	3.0969×10^{-7}
0.40	0.085648148	0.795152032	8.40583×10^{-5}	0.4	0.000211382	6.34277×10^{-6}
0.42	0.131944444	0.69611606	0.000419493	0.42	0.001203787	3.61558×10^{-5}
0.44	0.178240741	0.604987208	0.00128415	0.44	0.004227267	0.00012734
0.46	0.224537037	0.52152642	0.003031524	0.46	0.01149198	0.000348646
0.48	0.270833333	0.445488632	0.006088657	0.48	0.02660743	0.00081937
0.50	0.31712963	0.376622246	0.010951425	0.5	0.054959775	0.001741642
0.52	0.363425926	0.314668525	0.018180961	0.52	0.103580232	0.003454712
0.54	0.409722222	0.259360879	0.028400806	0.54	0.179659555	0.006527297
0.56	0.456018519	0.210424031	0.042294557	0.56	0.286729983	0.011916102
0.58	0.502314815	0.167573008	0.060603866	0.58	0.419722354	0.021238526
0.60	0.548611111	0.13051193	0.084126722	0.6	0.563162462	0.037235325
0.62	0.594907407	0.098932507	0.113715938	0.62	0.696864879	0.064516348
0.64	0.641203704	0.072512146	0.150277811	0.64	0.805632778	0.11059487
0.66	0.6875	0.050911519	0.194770917	0.66	0.884411099	0.186688029
0.68	0.733796296	0.033771323	0.248205027	0.68	0.936302321	0.306025325
0.70	0.780092593	0.020707804	0.311640111	0.7	0.967844428	0.474504057
0.72	0.826388889	0.011306246	0.386185428	0.72	0.985572825	0.67206763
0.74	0.872685185	0.005110814	0.472998694	0.74	0.994626464	0.84739618
0.76	0.918981481	0.001606852	0.573285297	0.76	0.99860052	0.955370151
0.80	1	0	0.785	0.795	1	1
1.00	1.474537037	0	3.328655169	1	1	1

Table 5.65—Fractional-flow data, Example 5.20.

Results of the single-slug runs are shown in **Table 5.66,** where recoveries are expressed as percent OOIP. The table also shows predicted recoveries for continuation of the waterflood, which was in place, and for continued waterflooding with infill drilling. The infill drilling converted the flood spacing from 80 to 40 acres. The 40-acre spacing was used in all CO_2 flooding calculations. All computer runs were terminated at an economic limit corresponding to a producing WOR of 97% water cut.

In the WAG runs, predicted behavior was first determined to be relatively insensitive to WAG ratios between 0.5 to 2.0 (volume water/volume CO_2 at reservoir conditions) as long as the total CO_2 slug size was kept constant. A ratio of 1:1 was selected for the simulations. Table 5.66 and **Fig. 5.149** show the results of the WAG run at these conditions.

The WAG process yielded an incremental ultimate recovery of 1.9% OOIP above the single-slug process. As shown in Fig. 5.149, however, oil production response to the injected CO_2 was delayed in the WAG process compared with the single-slug process. Incremental oil in the WAG injection resulted because production was extended approximately 4 years before the economic limit was reached. The behavior in the WAG injection, compared with the single-slug injection, results from a combination of the increased sweep in the WAG process, the extension in time required to inject the volume of CO_2, and oil trapping in the WAG process. Although oil recovery is predicted to be higher in the WAG process, the economics may not be as favorable because of the delayed production.

Guided by the results of the single-slug and WAG simulations, a third approach, a hybrid CO_2 injection, was examined. This approach used a pre-WAG CO_2 injection of 8.8% HCPV. This was followed by WAG injection at a 1:1 ratio until a total of 30% HCPV of CO_2 was injected (i.e., a total WAG injection of 42.4% HCPV). Table 5.66 and Fig. 5.149 show the results.

	Waterflood (End of History Match)	Continuous Waterflood	Waterflood With Infill	8.8% HCPV CO_2 Slug	20% HCPV CO_2 Slug	30% HCPV CO_2 Slug	40% HCPV CO_2 Slug	50% HCPV CO_2 Slug	30% HCPV CO_2 1:1 WAG	30% HCPV CO_2 Hybrid
Cumulative oil recovery (% OOIP)	42.5	48.0	50.8	57.5	61.5	64.1	66.2	67.9	66.0	64.8
Incremental oil recovery over infill waterflood (% OOIP)	NA	NA	0.0	6.7	10.7	13.3	15.4	17.1	15.2	14.0
Solvent efficiency (bbl/Mcf)	NA	NA	NA	0.187	0.132	0.109	0.095	0.084	0.215	0.110

*OOIP = 16.0×10^6 bbl. Oil recoveries calculated at an economic limit of 90% water cut or 97% water cut in the field.

NA = not applicable.

Table 5.66—Summary of simulation prediction results for the Dollarhide field study (Lin and Poole 1991).

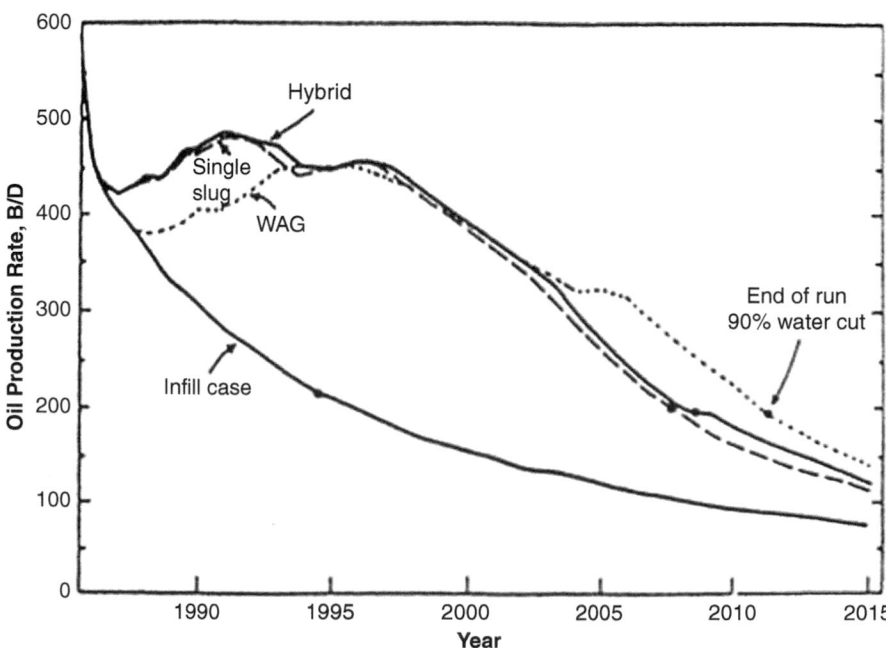

Fig. 5.149—Comparison of predicted oil rates for various CO_2 injection processes at Dollarhide field (Lin and Poole 1991).

Single-Slug CO₂ Projects	% OOIP	WAG CO₂ Projects	% OOIP
Little Creek—sandstone	18	SACROC (Phase III)—limestone	7.6
Twofreds—sandstone	10	SACROC (total)—limestone	7.0
Garber—sandstone	14	SACROC (pilot)—limestone	6.0
Mead-Strawn—sandstone	15	Slaughter Estate—dolomite	21.0
Shannon, West Sussex—sandstone	13	South Welch—dolomite	5.2
Maljamar—dolomitic sandstone	18	Shannon—sandstone North Meadow Creek	2.0
		Levelland—dolomite	8.9
Average	15	Average	8.2

Table 5.67—Incremental oil recovery from field CO_2 projects (Huang and Holm 1988).

The hybrid injection scheme yielded a recovery of 1.2% OOIP less than the WAG injection, but 0.7% OOIP more than the single-slug process. The oil recovery response was essentially identical to that of the single-slug injection. A study to determine the optimum size of the pre-WAG slug yielded a value of 5% HCPV. An economic analysis conducted in conjunction with the simulation results showed the hybrid process to be the most attractive of the three schemes.

Another approach to the evaluation of the WAG process, but one that has been much less widely used, is to conduct displacement experiments in a dimensionally scaled physical model. Jackson et al. (1985) conducted displacements in a single-layer bead pack that simulated one-quarter of a five-spot pattern. The pack was $15 \times 15 \times 1.0$ in. They examined effects of pack wettability, WAG injection, and total CO_2 slug size on recovery in secondary and tertiary miscible displacements.

Results indicated that tertiary displacements in oil-wet packs tended to be controlled by viscous fingering. The WAG process performed slightly better than single-slug injection, and the optimum WAG ratio was 1:1. In water-wet packs, the displacements tended to be controlled by gravity tonguing and maximum recovery was obtained with the single-slug process.

In summary, both physical modeling and computer simulation have shown that WAG performance is dependent on rock wettability. Oil trapping in water-wet rocks is a significant negative factor. Volumetric sweep can be improved with WAG injection, but a corresponding delay in production response can adversely affect the economics of the process.

5.10.4 Field Experience. Table 5.67 (Huang and Holm 1988) summarizes oil recoveries in several field projects. The field, rock lithology, and recovery, expressed as percent OOIP, are shown for several single-slug and WAG CO_2 miscible displacement projects. Rock wettability was not known. CO_2 slug sizes varied in the different projects but were greater than 15% HCPV in all cases. Average recovery in the single-slug projects was 15% OOIP, while it was only 8.2% OOIP in the WAG projects. The Slaughter Estate project was conducted somewhat differently in that a large slug of CO_2/H_2S solvent was injected and this was followed by an equal size N_2 slug. Both solvent and gas were injected in WAG cycles.

The results shown in Table 5.67 raise questions about the efficiency of the WAG process. Data are reported for only a small number of projects, however, and such parameters as slug size, lithology, and waterflood ROS varied significantly among the different projects. Thus, the results shown in Table 5.67 should not be generalized to all WAG applications but should serve as a warning in consideration of the process.

There are reported successes, such as the Quarantine Bay pilot tertiary project in a US Gulf Coast low-dip reservoir (Hsie and Moore 1988). Most US Gulf Coast miscible projects have been in relatively steeply dipping reservoirs, and single-slug processes have been used to develop gravity-stable floods. WAG injection was used in Quarantine Bay because the dip is small. Computer simulation predicted that the WAG process would perform better than continuous CO_2 injection. An optimum design, based on simulation, was a WAG process with a total CO_2 slug volume of 19.5% HCPV and a 1:1 WAG ratio. In the actual pilot, a slug size of 18.9% HCPV of CO_2 was used and the WAG ratio was raised to 2:1 during the process on the basis of later simulation results. Through October 1987, the reported pilot recovery was 16.9% OOIP, compared with a predicted value of 18.3% OOIP. The WAG process was judged to be successful (Hsie and Moore 1988).

Single-slug CO_2 displacement was compared with the WAG process in the Denver Unit of the Wasson (San Andres) Field of west Texas (Tanner et al. 1992). The Denver Unit had been successfully waterflooded. ROS after waterflooding was relatively high at 40%, leaving a large target for CO_2 flooding.

Side-by-side areas were established for CO_2 miscible flooding. Continuous CO_2 injection was used in one area and a 1:1 WAG ratio process in the other. In both areas, an ultimate solvent slug size of 40% HCPV was used.

The early oil production response in the area of continuous solvent injection was much more favorable than in the WAG area. As the flood progressed, however, some producing wells in the continuous area tended to high gas/oil ratios and had to be shut in. This led the operators to consider a combination, or hybrid, process that consisted of continuous CO_2 slug injection for 4 to 6 years followed by a conventional 1:1 WAG process. They called this the Denver Unit WAG (DUWAG).

Fig. 5.150 compares results of a computer simulation of continuous injection, conventional 1:1 WAG injection, and 1:1 DUWAG injection. The advantage of the combined approach (DUWAG) is shown. Tanner et al. (1992) reported that this

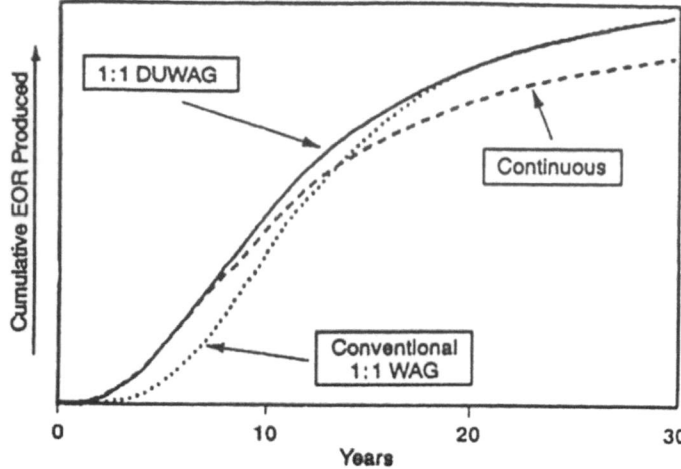

Fig. 5.150—Calculated cumulative incremental EOR recovery vs. time.

process is being applied in the Denver Unit, although specific parameters, such as slug size, WAG ratio, and time at which conversion is made from continuous injection to WAG injection, may vary from location to location within the unit.

Problems

5.1 The viscosity of a 5,000-ppm solution of partially hydrolyzed polyacrylamide was measured at several shear rates with a cone-and-plate rheometer. **Table 5.68** gives experimental data as a function of shear rate. Plot the viscosity vs. shear rate on log-log paper and determine the power-law exponent, n, and the power-law constant, K, from the experimental data using least-squares analysis and correlate the viscosity as a function of shear rate. Compare values of the viscosity from the correlation with experimental data.

5.2 **Table 5.69** presents viscosity data for a 1,500-ppm biopolymer solution obtained with a cone-and-plate rheometer. The data are ordered in the sequence of shear rates measured. Plot the viscosity vs. shear rate on log-log paper and identify the region of shear-thinning behavior. Notice that the last two data points were obtained at low shear rates after the sample was subjected to a shear rate of 297 seconds^{-1}. Is there any evidence of degradation of the polymer

Shear Rate (seconds^{-1})	Viscosity (cp)
1.75	6.921
2.083	7.364
2.733	7.111
3.546	6.938
4.661	6.676
6.353	6.205
8.624	5.785
11.95	5.298
16.56	4.838
23.31	4.351
33.04	3.883
47.22	3.441
68.04	3.026
97.68	2.669
139.5	2.363
199.5	2.094

Table 5.68—Rheological data for 5,000-ppm Calgon 834 in 3% NACl.

Viscosity (cp)	Shear rate (seconds^{-1})
385.7	0.9417
272.4	1.8709
205.8	2.9696
129.1	5.9204
93.37	9.417
57.48	18.71
41.5	29.696
25.79	59.2
19.28	94.17
13.11	187.1
10.44	297
624.3	0.2976
847.1	0.0942

Table 5.69—Rheology data for 1,500-ppm biopolymer solution.

Darcy Velocity (ft/D)	Apparent Viscosity (cp)
0.66	3.23
0.92	3.2
1.9	2.9
3.8	2.9
5.9	2.9
8	3.23
10	2.9
22	2.9
50	4.21
95	6.83
140	9.44
200	14.97
300	22.24
370	26.74
440	34.88

Table 5.70—Apparent viscosity data.

Flow Rate (cm³/s)	Pressure Drop (psi/in.)
0.000138	0.034
0.00028	0.066
0.00116	0.175
0.00306	0.31
0.00604	0.443
0.0102	0.69
0.0224	1.0
0.0483	1.63
0.10166	2.56
0.1533	3.44
0.2005	4.24
0.2566	5.06
0.3666	6.56
0.4666	7.84

Berea core: $\phi = 0.197$, $k = 312$ md, and dimensions are 2×2×12 in.

Table 5.71—Flow data for 1,000-ppm xanthan in 1% NaCl in Berea sandstone.

caused by shear rate? Fit the data to the Carreau (1968, 1972) model (Eq. 5.8) assuming that the viscosity of the polymer at high shear rates is equal to the viscosity of the solvent (2% NaCl at 77°F). Compare the measured data with the correlation developed using the Carreau model.

5.3 **Table 5.70** gives apparent viscosity data obtained for a 500-ppm solution of partially hydrolyzed polyacrylamide flowing through a 804-md sandpack for Darcy velocity (q/A) ranging from 0.66 to 440 ft/D. The porosity of the sandpack is 42%. The polymer is shear thinning with a power-law constant of $K = 4.76$ mPa·s $^{-0.19}$ and a power-law exponent of $n = 0.815$. Examine the data and determine the flow regimes that are present. Correlate the data using Eq 5.20 and determine the polymer-mobility constant, λ_p^*, and n_c for each flow regime identified. Compare values of the apparent viscosity from the correlation with experimental data.

5.4 Flow experiments were conducted in a Berea sandstone core to determine the flow characteristics of a 1,000-ppm solution of xanthan in 1% NaCl. The core was $2 \times 2 \times 12$ in., porosity was 0.197, and permeability to brine was 312 md. Polymer was injected at a constant injection rate until the pressure drop stabilized. Flow rates were varied between 0.000138 and 0.4666 cm³/s. **Table 5.71** summarizes the experimental data. Determine the apparent viscosity as a function of Darcy velocity and plot the data. Correlate the data with the power-law model given by Eq. 5.20. Compare the correlation with the experimental pressure drops assuming that the flow rate is the independent variable.

5.5 A displacement experiment was conducted where a 50-ppm xanthan solution in 15,500 ppm NaCl was injected into a Clashach sandstone core to determine the amount of polymer retained. The Clashach sandstone is more than 99.5% quartzitic (Sorbie et al. 1987) and has a low concentration of clays. Consequently, retention by adsorption is expected to be low. **Table 5.72** summarizes polymer concentration data from this run when 0.995 PV of polymer solution was injected followed by 0.938 PV of brine. The PV of the core was 303 cm³, and the porosity was 0.175. Assuming that polymer retention was irreversible, use material-balance calculations to estimate the amount of polymer retained. The length of the core was 100 cm, and the cross-sectional area was 17.35 cm². Express the polymer retention as mg/g of rock. The density of the solid matrix may be taken as 2.65 g/cm³.

5.6 **Table 5.73** contains effluent data from an experiment where a 250-ppm solution of partially hydrolyzed polyacrylamide was injected into a sandstone core to determine polymer retention. In this experiment, 570 cm³ of polymer was injected followed by 138 cm³ of brine. At the end of the experiment, concentration of the polymer in the effluent was below limits of measurement. The core was 15.2 cm long, and the diameter was 5.08 cm. PV of the core was 62.5 cm³. Initial oil saturation was 0.29. Analyze the effluent data to determine the polymer retention in mg/g rock. Assume that the density of the matrix is 2.65 g/cm³. Express the polymer retention in lbm/acre-ft. Use the retention data to estimate the value of D_p for this sandstone.

5.7 **Table 5.74** summarizes dispersion data for the Clashach sandstone core described in Problem 5.5. Compare these data with the polymer concentration in the effluent from Problem 5.5. Estimate the IPV for this polymer in Clashach sandstone.

5.8 A 500-ppm solution of polyacrylamide is to be injected into a reservoir through a slimhole completion. This completion is 2⅞-in. tubing cemented to the surface in a 5⅜-in. hole. The productive formation was completed open hole. The polymer injection rate is expected to be 1 B/D-ft. The reservoir is a consolidated sandstone with a porosity of 0.20

PV Injected	C/C_o	PV Injected	C/C_o
0.516	0.000	1.451	0.980
0.659	0.007	1.495	1.000
0.681	0.054	1.528	1.000
0.714	0.108	1.550	0.980
0.736	0.196	1.572	0.959
0.758	0.297	1.605	0.926
0.791	0.385	1.627	0.865
0.813	0.486	1.649	0.824
0.835	0.568	1.682	0.743
0.868	0.622	1.704	0.662
0.890	0.689	1.726	0.601
0.912	0.743	1.759	0.520
0.956	0.784	1.792	0.459
0.978	0.804	1.803	0.385
1.000	0.838	1.836	0.324
1.022	0.838	1.858	0.264
1.044	0.858	1.880	0.203
1.077	0.899	1.913	0.196
1.099	0.899	1.935	0.149
1.132	0.919	1.968	0.108
1.154	0.959	1.990	0.088
1.187	0.939	2.023	0.074
1.209	0.959	2.045	0.054
1.242	0.953	2.067	0.061
1.264	0.959	2.100	0.041
1.286	0.980	2.122	0.041
1.319	0.980	2.144	0.027
1.363	0.959	2.166	0.020
1.396	1.000	2.199	0.020
1.418	1.000	2.232	0.020
1.440	0.980	2.254	0.000

Table 5.72—Polymer concentration data.

and a permeability to brine at S_{or} of 10 md. Using the Maerker (1975) correlation, estimate the loss in screen factor and solution viscosity resulting from shear degradation of this polymer.

5.9 A xanthan biopolymer (C = 1,500 ppm) is to be used for mobility control in a waterflood in a sandstone reservoir. Flow characteristics of the biopolymer are given by the correlations developed by Willhite and Uhl (1988) in Eqs. 5.23e through 5.23g; **Table 5.75** gives properties of the reservoir. An injectivity test is planned in a watered-out five-spot pattern with a pattern area of 5 acres. Predict the pressure drop between the injection well and the effective radius of the pattern as a function of volume of polymer injected when 5,000 bbl of polymer is injected at a constant rate of 10 B/D-ft. Plot pressure drop as a function of volume of polymer injected (barrels). Assume that the region around the wellbore is at ROS and the permeability to water at ROS is 50 md. The polymer is shear thinning until the Darcy velocity is less than 0.1 ft/D. Polymer retention is 40 lbm/ acre-ft. The density of the water is 62.4 lbm/ft³ at reservoir temperature. Find the fraction of polymer retained. Determine the time required for the injectivity test.

5.10 A test is planned to evaluate the injectivity of a polyacrylamide solution into a 20-ft reservoir zone. **Table 5.76** gives properties of the reservoir and rock. The region around the injection well is at waterflood ROS. Transient well testing indicated that there was no skin damage to the well from prior testing. The reservoir pressure is 480 psi, and the injection well is on 10-acre spacing. Laboratory data are available on the injectivity of a 1,900-ppm polyacrylamide solution (at reservoir temperature) into core taken from the reservoir zone; **Table 5.77** presents these data. Estimate the BHP in the injection well after 30 days of continuous polymer injection at a rate of 250 B/D.

Sample	Polymer Concentration (ppm)	Cumulative Effluent Volume (cm³)	Sample	Polymer Concentration (ppm)	Cumulative Effluent Volume (cm³)
1	0.0	6.0	61	215.0	366.0
6	0.0	36.0	62	215.0	372.0
7	3.2	42.0	63	212.0	378.0
8	9.6	48.0	65	244.0	390.0
9	30.6	54.0	66	215.0	396.0
10	72.5	60.0	67	215.0	402.0
11	110.8	66.0	68	215.0	408.0
12	133.0	72.0	69	230.0	414.0
13	150.0	78.0	70	208.0	420.0
14	162.0	84.0	71	215.0	426.0
15	170.8	90.0	72	210.0	432.0
16	179.0	96.0	73	210.0	438.0
17	183.0	102.0	74	199.0	444.0
18	187.5	108.0	77	199.0	462.0
19	191.6	114.0	78	197.0	468.0
20	193.2	120.0	80	197.0	480.0
21	196.3	126.0	81	190.0	486.0
22	196.3	132.0	83	190.0	498.0
23	203.0	138.0	84	181.0	504.0
24	205.2	144.0	88	181.0	528.0
25	205.2	150.0	89	179.0	534.0
26	207.0	156.0	90	179.0	540.0
27	207.4	162.0	91	172.0	546.0
36	207.4	216.0	93	167.0	558.0
37	209.0	222.0	94	167.0	564.0
38	212.0	228.0	95	167.0	570.0
39	212.0	234.0	96	172.5	576.0
40	217.0	240.0	97	176.2	582.0
41	212.0	246.0	98	161.1	588.0
42	217.0	252.0	99	123.8	594.0
43	226.0	258.0	100	107.0	600.0
44	217.0	264.0	101	73.3	606.0
45	221.0	270.0	102	56.4	612.0
46	221.0	276.0	103	47.2	618.0
47	221.0	282.0	104	36.7	624.0
48	217.0	288.0	105	28.8	630.0
49	217.0	294.0	106	21.0	636.0
50	217.0	300.0	107	15.7	642.0
51	221.0	306.0	108	10.5	648.0
52	221.0	312.0	109	7.8	654.0
53	226.0	318.0	110	5.2	660.0
54	226.0	324.0	113	5.2	678.0
55	250.0	330.0	114	2.6	684.0
56	56.0	336.0	115	2.6	690.0
57	212.0	342.0	116	2.0	696.0
58	212.0	348.0	117	1.0	702.0
59	212.0	354.0	118	0.0	708.0
60	212.0	360.0	–	–	–

Table 5.73—Effluent data for 250-ppm partially hydrolyzed polyacrylamide injected into sandstone core.

PV Injected	C/C_o	PV Injected	C/C_o
0.593	0.000	1.550	1.000
0.670	0.007	1.583	0.993
0.692	0.007	1.605	1.000
0.714	0.007	1.627	1.000
0.736	0.007	1.649	1.000
0.769	0.014	1.682	0.986
0.791	0.020	1.704	0.980
0.813	0.041	1.737	0.980
0.846	0.074	1.759	0.959
0.868	0.128	1.781	0.932
0.901	0.196	1.803	0.905
0.923	0.277	1.836	0.865
0.945	0.365	1.858	0.811
0.978	0.459	1.913	0.676
1.000	0.554	1.935	0.595
1.022	0.635	1.968	0.520
1.044	0.709	1.990	0.446
1.077	0.770	2.023	0.372
1.099	0.824	2.045	0.311
1.132	0.872	2.067	0.250
1.154	0.899	2.089	0.203
1.187	0.926	2.122	0.162
1.220	0.946	2.155	0.122
1.242	0.959	2.177	0.095
1.264	0.966	2.210	0.068
1.286	0.980	2.232	0.047
1.319	0.986	2.254	0.034
1.341	0.993	2.287	0.027
1.363	0.986	2.309	0.020
1.418	1.000	2.342	0.020
1.462	1.000	2.364	0.007
1.495	1.000	2.386	0.000
1.528	1.014	–	–

Table 5.74—Dispersion data.

Wellbore radius (ft)	0.5
Porosity (fraction)	0.20
Water viscosity at reservoir temperature (cp)	0.9
Thickness (ft)	20
ROS (fraction)	0.25

Table 5.75—Reservoir rock and fluid properties.

Porosity (fraction)	0.31
Permeability to air (md)	570
ROS (fraction)	0.30
Oil gravity (°API)	27
Oil viscosity (cp)	8
Brine viscosity at 150°F (cp)	0.45
Brine mobility at S_{or} (md/cp)	73
Effective wellbore radius (ft)	0.37

Table 5.76—Reservoir and Rock Properties

Darcy Velocity (ft/D)	Mobility (md/cp)	Resistance Factor*
200	9.1	8.0
100	11.0	6.6
50	13	5.6
10	10	7.3

*Based on brine mobility of 73 md/cp at ROS

Table 5.77—Laboratory injectivity data for 1,900-ppm polymer at reservoir temperature.

Porosity (fraction)	0.32
Thickness (ft)	10
Viscosity of water at reservoir temperature (cp)	0.5
Permeability to brine at residual oil saturation (md)	804
Residual oil saturation	0.3
Distance to observation well (ft)	216
Wellbore radius (ft)	0.333

Table 5.78—Reservoir rock and fluid properties.

	Layer 1	Layer 2
Oil viscosity (cp)	10	10
Water viscosity (cp)	1	1
k_o at S_{iw} (md)	200	20
k_w at S_{or} (md)	20	1
S_{or}	0.30	0.35
S_{iw}	0.3	0.40
Thickness (ft)	10	20
Porosity (fraction)	0.25	0.15

Table 5.79—Reservoir, rock, and fluid properties.

5.11 The expressions for pressure drop in an injection well developed in Section 5.4.4 do not account for the presence of wellbore damage (positive skin) or wellbore enhancement (negative skin). Conventionally, the skin is introduced into fluid-flow calculations by assuming that there is a hypothetical damaged region of radius r_s with permeability k_s in the immediate vicinity of the wellbore. The skin is defined by Eq. 5.197 when the viscosity of the flowing fluid is the same in the reservoir and the damaged region.

$$ s = \left(\frac{k}{k_s} - 1 \right) \ln \frac{r_s}{r_w}. \dots\dots\dots\dots\dots\dots\dots\dots\dots\dots\dots\dots\dots\dots\dots\dots\dots (5.197) $$

Develop the appropriate form of Eq. 5.60 when there is a skin. What assumptions do you need to make regarding the properties of polymer flowing through the region represented by the "skin"?

5.12 The polymer tested in laboratory experiments in Problem 5.3 is to be injected into an unconsolidated sand formation down casing. **Table 5.78** gives the reservoir data. Injection of polymer into this well at a rate of 500 B/D is desired. Compare pressure drop between the injection well and an observation well located 216 ft away. The reservoir is watered out and at ROS. Estimate the pressure drop as a function of volume of polymer injected for the first 15,000 bbl of polymer. Plot pressure drop vs. volume of polymer injected and time of injection. Polymer retention is neglected, and there is no skin in the well.

5.13 Use the properties in Problem 5.12 and estimate the injection rate as a function of time for 30 days when the pressure drop between the injection well and the observation well is maintained at 500 psi.

5.14 Consider the injection of a 500-ppm xanthan solution into a 2.5-acre five-spot pattern in a reservoir that has two layers with properties given in **Table 5.79.** To simplify the calculations, assume that polymer displaces oil in a piston-like manner. There is no communication between layers. Polymer injection is to be performed by maintaining a constant pressure drop of 500 psi between the injection well and the effective radius of the pattern. Using the flow properties of xanthan in Berea sandstone core developed by Willhite and Uhl (1988) (Eqs. 5.23c, 5.23d, and 5.23g), estimate the injection rate for each layer for the first 30 days of injection. Estimate the injection rate for the well by combining the injection rates for each layer at the same point in time. Assume that the polymer flow is Newtonian when the Darcy velocity is less than 0.1 ft/D. Retention of polymer is neglected.

5.15 A reservoir has two productive zones that are separated by an impermeable shale stringer. **Table 5.80** gives zone properties. An estimate of the injection rate into a watered-out well in the central part of the field is wanted in connection with preliminary studies of a polymer flood. The region around the well is at ROS. A xanthan polymer will be used at a concentration of 600 ppm. Table 5.12 gives the rheological properties of xanthan biopolymer. The residual resistance factor is 1.10. Retention of polymer is neglected.

1. Derive a relationship to predict the injection rate of a shear-thinning fluid in radial flow using the correlation of Cannela et al. (1988) (Eqs. 5.29 through 5.31) to estimate the apparent viscosity of the polymer solution as a function of the Darcy velocity. The pressure drop, Δp, is known between the injection well and the effective radius of the pattern, r_e. Assume piston-like displacement of the resident brine by the polymer. Limit your relationship to the shear-thinning flow regime.
2. Estimate the injection rate (in barrels per day) into each zone when the radius of the polymer front is 20 ft from the wellbore. $r_e = 300$ ft, and Δp between the injection well and r_e is 750 psi. The effective wellbore radius is 0.75 ft.
3. Estimate the time (in days) for the polymer front to reach a radius of 20 ft in each zone.

5.16 A polymer flood is being proposed for a thin reservoir that contains a 20-cp oil. **Table 5.81** summarizes properties of the reservoir, rock, and fluids. The reservoir energy was limited, and primary recovery was negligible. Because of the high oil viscosity, a polymer flood is under consideration. Laboratory tests indicate that a polymer concentration of 300 ppm will increase the viscosity of field brine to 5 cp at the frontal-advance rates expected in the reservoir. Polymer retention at the 300-ppm injected polymer concentration is 20 mg/g, and the density of the rock matrix is 2.65 g/cm³. IPV is 0.20. Estimate the oil recovery for a linear polymer flood in this reservoir as a function of PVs of polymer injected to a WOR of 20. Plot oil recovery vs. PVs of polymer injected. Eqs. 3.14 through 3.16 give relative permeabilities for the rock and fluid pairs. Assume that oil and water FVFs are 1.0.

5.17 Consider the possibility of injecting a 0.25-PV slug of polymer in Problem 5.16 followed by continuous injection of brine. The amount of polymer is not sufficient to satisfy retention. Thus, the brine will overtake the polymer bank somewhere in the system and the polymer flood will revert to a waterflood. **Table 5.82** contains the location of each water saturation when the saturation evolved as the drive water displaces the polymer region. This table was prepared by running PREAR in Appendix C and choosing selected values from the output. Table 5.82 also includes values of f_{wr} and f'_{wr} corresponding to each saturation S_{wr}.

1. Draw the saturation profile when polymer injection ends and just before drive-water injection begins. Label all regions, including polymer, oil bank, and the region at initial water saturation.
2. Draw the saturation profile when the polymer bank disappears. Label all regions that are still within the linear system (i.e., $0 \le x_D \le 1$).
3. Determine the oil recovery (in PVs) at this point in the polymer flood.
4. How much incremental oil is from polymer flooding?
5. How many PVs of polymer and brine must be injected when $S_{w2} = 0.54624$?
6. Draw the saturation profile when $S_{w2} = 0.54624$ and determine the oil recovery (in PVs). Compare with the waterflood recovery at the same number of PVs of water injected.

5.18 The polymer flood in Problem 5.16 will not begin until late in the life of the waterflood, when the WOR is 10.

1. Find the number of PVs of water injected when the WOR is 10.
2. Determine the saturation profile at a WOR of 10 and plot the profile as a function of dimensionless distance for the linear system ($0 \le x_D \le 1$).
3. Determine the oil recovery by waterflooding and the remaining mobile oil that is potentially recoverable by polymer flooding.

	Layer 1	Layer 2
Porosity (fraction)	0.2	0.18
Thickness (ft)	5	50
k_w (md)	500	50
k_w at S_{or} (md)	50	5.0
μ_w (cp)	1.0	1.0
μ_∞ (cp)	1.0	1.0

Table 5.80—Rock and fluid properties.

Oil viscosity (cp)	20
Water viscosity (cp)	0.9
Initial water saturation (fraction)	0.30
ROS (fraction)	0.35
Porosity (fraction)	0.20
α_1	1.0
α_2	0.5
m	3
n	2

Table 5.81—Fluid and rock properties.

Timestep	x_{Dpr}	t_D	S_{wr}	f_{wr}	f'_w
1	0	0.25	0.6499997	1.0000000	0.0000000
10	0.0008209	0.2503635	0.6406955	0.9999982	0.0005859
20	0.0039957	0.2517437	0.6302759	0.9999819	0.0028607
30	0.0102422	0.2543966	0.6200511	0.9999325	0.0071753
40	0.0205573	0.2586734	0.6098636	0.9998268	0.0140557
50	0.0363768	0.2650743	0.599849	0.9996394	0.0239688
60	0.0598302	0.2743342	0.5900613	0.9993423	0.0374292
70	0.0941957	0.2875752	0.5805492	0.9989067	0.0549568
80	0.1447629	0.3065976	0.5714161	0.998309	0.0768899
90	0.2205935	0.3344691	0.5628135	0.9975387	0.1031648
100	0.33855	0.3768792	0.5548559	0.9966018	0.133363
101**	0.3537569	0.3822314	0.54624	0.9952849	0.1738021
106	0.384759	0.4089426	0.5348689	0.9929364	0.2424807
111	0.4186434	0.4357038	0.5234978	0.9896862	0.3334123
116	0.4558527	0.4624649	0.5121267	0.985244	0.4534099
121	0.4969143	0.4892261	0.5007557	0.9792309	0.6114262
126	0.542457	0.5159873	0.4893846	0.9711508	0.8191773
131	0.5932317	0.5427485	0.4780135	0.9603554	1.0918664
136	0.6501344	0.5695095	0.4666424	0.9460003	1.4488981
141	0.7142298	0.5962704	0.4552714	0.9269941	1.914335
146	0.7867724	0.6230313	0.4439003	0.9019462	2.5164993
151†	0.8692164	0.6497921	0.432529	0.8691297	3.2855868

* When 0.25 PV of polymer is displaced by drive water.

** Polymer flood front overtaken by drive water.

† Saturation discontinuity eliminated at rear of oil bank.

Table 5.82—Evolution of drive-water saturations and fractional-flow properties.*

4. Continuous injection of polymer is to be considered with the same polymer and rock characteristics used in Problem 5.16. Will an oil bank form when polymer injection is started? Why or why not?

5.19 A reservoir consists of two layers separated by a permeability barrier so that there is no flow between layers except at the well-bores. The permeability of the upper layer and lower layers is 1,400 and 33 md, respectively. **Table 5.83** gives other properties of the reservoir. A determination of the potential of a polymer flood for this reservoir is desired assuming that displacement is linear. The wells are on 10-acre spacing so that the distance between rows of injection and production wells is 660 ft. Polymer is retained by the reservoir rock; Table 5.83 gives the values of polymer retention at the injected concentration of 250 ppm. Retention is assumed to be irreversible, and the density of the rock matrix is taken as 2.65 g/cm³. The effective IPV is 0.25. Relative permeability relationships given by Eq. 3.197 and 3.198 are assumed to be applicable for both layers. Use constants in Table 3.20 for Layer 1. $a_w = -3$ for Layer 2.

1. Estimate the oil recovery (in STB) as a function of volume of polymer injected into each layer until the WOR exceeds 100 in each layer. Compare the displacement performance with that expected from a waterflood under the same conditions.

2. Estimate injection and production rates as a function of time if the pressure difference between the injection and production wells is maintained constant at 1,000 psi.

5.20 Estimate the oil recovery from polymer flooding in Problem 5.19 using the Dykstra-Parsons model to describe fluid displacement. What assumptions must be made to use the Dykstra-Parsons model? What is the anticipated effect of these assumptions on the projected oil recovery?

5.21 The DOE has placed in the public domain a computer program to calculate performance of a polymer flood in a five-spot pattern. Program documentation and diskettes are available from the DOE, Bartlesville, Oklahoma, USA (Ray and Diaz Munoz 1986). The program name is polymer flood predictive model (PFPM). The program can be used to simulate a standard waterflood as well as a polymer-augmented waterflood. Example problems are presented as a part of the documentation. The program is designed to be run on a personal computer. Use the PFPM program to

	Layer 1	Layer 2
Porosity	0.248	0.146
Permeability (md)	1,400	33
Thickness (ft)	8	20
Initial water saturation	0.136	0.410
ROS (fraction)	0.325	0.325
Viscosity of water (cp)	0.63	0.63
Viscosity of oil (cp)	2.36	2.36
Oil FVF (bbl/STB)	1.05	1.05
Polymer retention (mg/g)	30	60
Effective IPV (fraction)	0.25	0.25
Apparent viscosity of polymer solutions (cp)	6.3	6.3

Table 5.83—Reservoir rock and fluid properties.

make the calculation described in Example 5.14, a polymer-augmented waterflood. Run the comparable waterflood and compare the results.

5.22 Investigate the effect of polymer retention on the polymer flood recovery in Example 5.14 by varying the amount of polymer retained between 50 and 200 lbm/acre-ft. Compare the predicted oil recovery by plotting oil recovery in barrels vs. volume of fluid injected.

5.23 A permeability modification treatment with a gelled-polymer system is planned for a well that contains three zones with properties given in **Table 5.84**. The plan is to treat Zone A with a gelling solution by isolating this zone. Zone A is watered out and is at waterflood ROS. The gelling solution contains 5,000 ppm polyacrylamide and a crosslinking agent. A treatment radius of 75 ft is desired. Estimate the injection rate as a function of time when a $\Delta p = 1,000$ psi is maintained between the injection wellbore (downhole) and the mean reservoir pressure of 500 psi (at a radius of 500 ft from the injection well). The data in Problem 5.1 give the apparent viscosity of the polymer solution. Because the gel time of this gelling solution is several days, assume that the gelling solution has the same apparent viscosity as the polymer solution. The following data were obtained from a single flow test in formation core from Zone A at reservoir temperature. ROS was 0.325, brine mobility at S_{or} was 399 md/cp, and polymer mobility was 28.5 md/cp at a Darcy velocity of 1 ft/D. Use Eq. 5.24 to estimate polymer mobility by assuming that $n = n_c$. Because core data are limited, shear thickening cannot be considered. The maximum polymer mobility is 399 md/cp in the upper Newtonian region. Neglect retention in the porous rock. The effective radius of the well is 1.0 ft.

5.24 **Fig. 5.151** shows an injection well in a reservoir with two intervals of significantly different permeability; **Table 5.85** gives reservoir rock and fluid properties. The well was hydraulically fractured with a fracture extending to a distance of 150 ft on both sides from the injection **(Fig. 5.152)**. Both layers are exposed to the fracture. Water now flows preferentially in the high-permeability zone, leading to high WORs in surrounding production wells and incomplete sweep of the low-permeability zone. It is desired to treat the well with a gelled polymer solution to reduce the permeability in the high-permeability zone. For this problem, assume that the fracture width is large so that the flow resistance in the fracture is neglected. Both layers will be exposed to the gelling solution. To simplify the problem, assume that both layers are at ROS. The pressure drop between the injection well and the production wells is maintained at 250 psi.

1. Develop the expression for the injection rate of gelling solution into each layer as a function of the depth of penetration assuming linear flow between the injection and the production well.
2. Find the distance that the gelling solution penetrated in both layers if the gelling solution is injected for a period of 7 days.
3. Estimate the volume of polymer injected into each layer in the 7 days of injection.
4. Determine the water-injection rate into each layer if the permeability to water is reduced by a factor of 500 in the region contacted by the gelling solution.
5. Compare the water-injection rates before and after treatment and determine the percentage of the injected fluid that enters each layer.

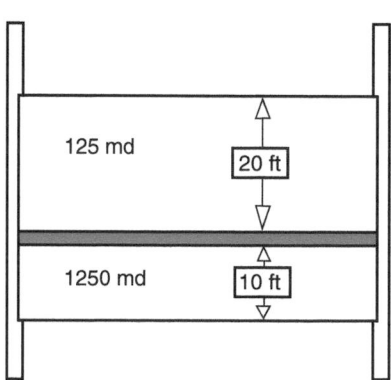

Fig. 5.151—Cross section of reservoir between injection and production wells.

	Zone A	Zone B	Zone C
Porosity (fraction)	0.248	0.187	0.146
Thickness (ft)	8	18.8	20.2
Permeability (md)	1,400	319	33

Table 5.84—Productive-zone properties.

	Layer 1	Layer 2
Porosity (fraction)	0.3	0.35
ROS (fraction)	0.2	0.2
Permeability to water at ROS (darcies)	0.125	1.25
Water viscosity (cp)	1	1
Gelling-solution viscosity (cp)	50	50
Thickness (ft)	20	10

Table 5.85—Reservoir rock and fluid properties.

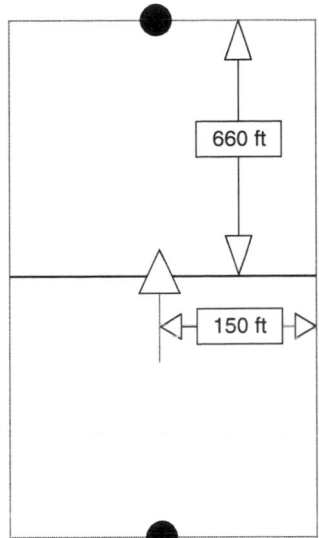

Fig. 5.152—Plan view of well layout and orientation of vertical fracture in the injection well.

5.25 A micellar-polymer flood is to be designed for a reservoir which has been waterflooded to ROS. Viscosities of the oil and formation water at reservoir temperature are 3 and 0.9 cp, respectively. Eqs. 3.14 through 3.16 give relative permeability relationships, and Table 5.86 gives relative permeability parameters. Determine the mobility of the micellar-polymer solution required to obtain mobility control in the displacement.

5.26 The micellar-polymer solution in Problem 5.25 reduces the ROS from 0.35 to 0.10. A xanthan biopolymer is to be used for mobility control. If the Darcy velocity for the displacement is 1 ft/D, what should the concentration of the polymer solution be to maintain mobility control between the polymer buffer and the micellar-polymer solution? Permeability of the reservoir rock to oil at interstitial water saturation and the permeability at 100% oil saturation are the same: 100 md. Eq. 5.161 gives the permeability of water as a function of maximum water saturation. Assume that the Willhite and Uhl (1988) correlation (Eqs. 5.23a through 5.23g) gives the flow properties of the polymer in reservoir rock.

5.27 A WAG process is to be implemented on a reservoir that was previously waterflooded to an ROS, S_{or}. Eqs. 5.189 and 5.190 give relative oil and water permeabilities, respectively, of $S_{iw} = 0.363$ and $S_{or} = 0.205$. Residual oil is to be displaced by a WAG process where the solvent viscosity is 0.04 cp. Water and oil viscosities are 1.0 and 3.0 cp, respectively. Water and solvent injection rates are to be set so that the water and solvent flow at equal velocities in the solvent/water bank that displaces the oil bank. Calculate the relative water and solvent injection rates. Assume linear flow, as in Example 5.19.

5.28 Refer to Problem 5.27.

1. Calculate the mobility ratio for the WAG process.
2. Compare the mobility ratio for the WAG process with the ratio that would hold if residual oil were displaced by pure solvent.

α_1	1.0
α_2	0.3
m	3
n	2

Table 5.86—Relative permeability parameters.

Nomenclature

a	=	correlation constant, permeability to water after polymer contact
a_w	=	constant in relative correlation in Eq. 3.198
a_1	=	correlation constant, Langmuir isotherm model
a_2	=	correlation constant, non-Newtonian flow in porous media
a_3	=	correlation constant, Eq. 5.168
A	=	cross-sectional area, L^2, ft^2
A_i	=	retention of polymer, m/m, μg/g rock
A_p	=	area of five-spot pattern, L^2, acres
b	=	correlation constant, permeability to polymer after polymer contact

b_1	=	correlation constant, Langmuir isotherm model
b_2	=	correlation constant, non-Newtonian flow in porous media
B	=	FVF, L^3/L^3, bbl/STB
B_{oi}	=	initial oil FVF, L^3/L^3, bbl/STB
C	=	polymer concentration, m/m, ppm
$\hat{C}$	=	retention of polymer, m/m, g polymer/g rock
C_o	=	injected polymer concentration, m/m, g/cm^3
$\hat{C}_o$	=	retention of polymer at C_o, m/m, g polymer/g rock

C_p = polymer cost, M/L^3 (USD/bbl)

d = diameter, L, ft

$\bar{d}_p$ = mean sand grain diameter, L, ft

D_p = retention factor, dimensionless

f = fractional flow, L^3/L^3, volume fraction

f'_w = derivative of f_w with respect to S_w

f^*_w = fractional flow of water in polymer region, L^3/L^3 (volume fraction)

f'^*_w = derivative of f^*_w with respect to S^*_w

f'^*_w = derivative of f_w in polymer region with respect to S^*_w

f_{wf} = fractional flow of water at S_{wf}

f'_{wf} = derivative of f_w with respect to S_w at S_{wf}

f'^*_{wo} = derivative of f^*_w with respect to S^*_w at $S^*_w = 1-S_{or}$, L^3/L^3 (volume fraction)

f_{wob} = fractional flow of water in oil bank ahead of solvent/water bank, L^3/L^3, volume fraction

f_{wr} = fractional flow of drive water at rear of polymer bank, L^3/L^3, volume fraction

f'_{wr} = derivative of f_{wr} with respect to S_w at S_{wr}

f^*_{wr} = fractional flow of water at rear of polymer bank, L^3/L^3, volume fraction

f'^*_{wr} = derivative of f_{wr} with respect to S_w at S_{wr}

f_{w1} = fractional flow of water at S_{w1}, L^3/L^3, volume fraction

f'_{w1} = derivative of f_w with respect to S_w at S_{w1}

f'_{w3} = derivative of f_w with respect to S_w at S_{w3}

f^*_{w3} = fractional flow of water at polymer-flood front, L^3/L^3, volume fraction

f'^*_{w3} = derivative of fractional flow of water with respect to S^*_w at polymer-flood front

F_r = resistance factor for porous matrix during flow of polymer solution, dimensionless

F_{rr} = residual resistance factor for porous matrix after contact with polymer solution, dimensionless

F_s = screen factor, dimensionless

F_{wo} = WOR, L^3/L^3, bbl/bbl

G' = storage modulus, m/Lt^2, Pa

h = formation thickness, L, ft

k = absolute permeability, L^2, md or μm^2

k_b = base permeability for relative permeability correlation, L^2, md

k_p = permeability of porous matrix to polymer solution, L^2, md

k_r = relative permeability

k_{r1} = relative permeability to liquid

k_{re} = relative permeability of chemical slug at anticipated ROS to chemical flood

k_{rop} = relative permeability to oil after rock was contacted with polymer, S_w increasing

k_{ro1} = relative permeability to oil before contact with polymer, S_w increasing

k_{ro2} = relative permeability to oil before contact with polymer, S_w decreasing

k_{ro3} = relative permeability to oil during displacement by polymer

k_{rwp} = relative permeability to water in polymer-contacted rock, S_w increasing

k_{rw1} = relative permeability to water before contact with polymer, S_w increasing

k_{rw2} = relative permeability to water before contact with polymer, S_w decreasing

k_{wp} = permeability of porous medium to water after contact with polymer, L^2, md

K = power-law constant or consistency index, mt^{n-2}/L, $mPa \cdot s^n$

L = length, L, ft

L_D = dimensionless flow distance in which polymer degradation occurs

m_p = polymer requirement, m/t, lbm/D

M = mobility ratio

$M_{\bar{s}}$ = mobility ratio based on average water saturation behind flood front

M_t = total mobility ratio

M_w = molecular weight of polymer

n = power-law index for non-Newtonian fluid

n_c = power-law constant for porous media

n_{c1} = power-law constant for polymer flow in porous media in shear-thinning region

n_{c2} = power-law constant for polymer flow in porous media in shear-thickening region

N_{De} = Deborah number, dimensionless

N_{Dec} = critical Deborah number, dimensionless

N_p = cumulative oil production, L^3, STB

p = pressure, m/Lt^2, psi

p_i = inlet pressure, m/Lt^2, psi

p_{int} = extrapolated inlet pressure, m/Lt^2, psi

p_o = pressure at core exit, m/Lt^2, psi

p_l = pressure at end of linear system, m/Lt^2, psi

Δp = pressure drop, m/L^2, t, psi

Δp_E = pressure drop in viscoelastic region, m/Lt^2, psi

Δp_L = pressure at $x = L$, m/Lt^2, psi

Δp_{Pmd} = pressure drop from mechanical degradation, m/Lt^2, psi

Δp_v = pressure drop in viscous region, m/Lt^2, psi

Δp_{12} = pressure drop between r_1 and r_2, m/Lt^2, psi

P_i = polymer injected, m

P_r = polymer retained, m

q = flow or injection rate, L³/t, B/D

q_{it} = initial total flow rate, L³/t, B/D

q_t = total volumetric flow rate, L³/t, ft/D

$q_t^{(0)}$ = injection rate at beginning of polymer injection, L³/t, B/D

Qi = number of PV injected

r = radius, L, ft

r_e = radius of external boundary for complete injection rate, L, ft

r_i = radius at Location i, L, ft

r_p = radius of polymer front, L, ft

r_w = wellbore radius, L, ft

r_1 = radius of Region 1 boundary, L, ft

r_2 = radius of Region 2 boundary, L, ft

R^2 = correlation coefficient

S = saturation, L³/L³, volume fraction

S_{iw} – interstitial immobile water saturation, L³/L³, volume fraction

S_{oi} = initial oil saturation, L³/L³, volume fraction

S_{or} = ROS to waterflood, L³/L³, volume fraction

S_{orc} = ROS to chemical flood, L³/L³, volume fraction

S_{orw} = ROS to waterflood, L³/L³, volume fraction

$S_{or,wb}$ = ROS following miscible displacement in the presence of mobile water, L³/L³, volume fraction

S_{twb} = water-blocked oil saturation, L³/L³

S_w^* = water saturation in polymer region, L³/L³, volume fraction

S_{wD} = dimensionless water saturation

S_{wf} = water saturation at flood front, L³/L³, volume fraction

$\overline{S}_{wf}$ = average water saturation behind flood front, L³/L³, volume fraction

$\overline{S}_{wf1}$ = average water saturation between S_{wf} and S_{w1} in a linear displacement, L³/L³, volume fraction

S_{wi} = initial water saturation, L³/L³, volume fraction

S_{wip} = initial water saturation in presence of polymer, L³/L³, volume fraction

S_{wob} = water saturation in oil bank ahead of solvent/water bank, L³/L³, volume fraction

S_{wr} = Drive-water saturation at rear of polymer bank, L³/L³, volume fraction

S_{wr}^* = water saturation at rear of polymer bank, L³/L³ (volume fraction)

$\overline{S}_{wr}^*$ = average water saturation in polymer bank, L³/L³ (volume fraction) time, t, days

S_{wrb} = water saturation at rear of polymer bank at instant polymer disappears, L³/L³, volume fraction

S_{wrj} = drive-water saturation at dimensionless location x_D, L³/L³, volume fraction

S_{w2} = water saturation at end of linear system, L³/L³, volume fraction

$\overline{S}_{w2}$ = average water saturation when water saturation at end of system is S_{w2}, L³/L³, volume fraction

S_{w3}^* = water saturation on polymer region, L³/L³, volume fraction

$\overline{S}_{w3}^*$ = average water saturation behind polymer-flood front, L³/L³, volume fraction

t = time, t, days

$t^{(0)}$ = time corresponding to the beginning of injection, t, days

$t^{(1)}$ = time corresponding to the end of time increment 1, t, days

t_D = dimensionless time, PV injected

t_{Db} = dimensionless time when polymer front is overtaken by drive water

t_{Df} = dimensionless time for waterflood front to reach end of linear system

t_{Dm} = dimensionless time estimated from material balance on total volume of water injected

t_{Do} = dimensionless polymer injection time, PVs of polymer injected for slug process

t_{Dp} = PVs of polymer injected, L³/L³, volume fraction

t_{Dr} = dimensionless time that S_{wr} has existed

t_{D1} = dimensionless time for S_{w1} to reach end of linear system

t_{D3}^* = dimensionless time when saturation S_{w3}^* arrives at the end of a linear system

t_g = gelation time, t, hours

t_0 = time for polymer injection, t, days

Δt = time increment, t, days

Δt_b = time for brine to flow through screen viscometer, t, seconds

Δt_p = time for polymer solution to flow through screen viscometer, t, seconds

T = temperature, T, °F

u = Darcy velocity, L/t, ft/D

u_{c1} = Darcy velocity for onset of shear-thinning flow, L/t, ft/D

u_{c2} = Darcy velocity for onset of upper Newtonian or shear-thickening flow, L/t, ft/D

u_{max} = Darcy velocity where mechanical degradation begins, L/t, ft/D

v = interstitial velocity, L/t, ft/D

$\overline{v}$ = average interstitial velocity, L/t, ft/D

v_c	=	critical velocity where flow regimes change, L/t, ft/D
v_{c2}	=	critical velocity for onset of shear-thickening flow, L/t, ft/D
v_D^*	=	specific velocity of S_w^*, dimensionless
v_{Dr}	=	specific velocity of drive water at rear of polymer bank, dimensionless
v_{Dr}^*	=	specific velocity of rear of polymer bank, dimension
v_{ob}	=	oil-bank velocity, L/t, ft/D
v_s	=	solvent velocity in WAG water/solvent bank, L/t, ft/D
v_w	=	water velocity in WAG water/solvent bank, L/t, ft/D
V	=	volume, L³, bbl
V_p	=	PV, L³, bbl
V_s	=	volume of solvent, L³, bbl
w	=	formation width, L (ft)
w_{cr}	=	mass of chromium in gel, m, g
w_p	=	mass of polymer in gel, m, g
W_p	=	cumulative polymer injected, L³, bbl
$W_p^{(1)}$	=	volume of polymer injected at time increment 1, L³, bbl
x	=	position or distance, L, ft
x_D^*	=	dimensionless location of saturation
x_{Df}	=	dimensionless location of waterflood front
x_{Do}	=	dimensionless location where S_{wr} evolved from saturation profile at rear of polymer bank
x_{Dr}	=	dimensionless location of S_{wr}
x_{Dr}^*	=	dimensionless location of rear of polymer bank
x_{Drm}	=	dimensionless distance x_{Dr} at end of nth increment in Eq. 5.115
x_D	=	dimensionless distance
x_{D1}	=	dimensionless location of S_{w1}
x_{D3}^*	=	dimensionless location of polymer flood front at saturation S_{w3}^*
x_{D3b}^*	=	dimensionless location of polymer flood front when overtaken by drive water
x_r^*	=	location of rear of polymer bank, L, ft
x_3^*	=	location of polymer flood front, L, ft
Δx	=	distance increment, L, ft
α	=	parameter in correlation for oil trapping in miscible displacement with a WAG process
α_1	=	parameter in generalized oil relative permeability correlation
α_2	=	parameter in generalized water relative permeability correlation

$\dot{\gamma}$	=	shear rate, 1/t, seconds⁻¹
$\dot{\gamma}_a$	=	apparent shear rate, 1/t, seconds⁻¹
$\dot{\gamma}_r$	=	relaxation shear rate in Carreau model, 1/t, seconds⁻¹
$\dot{\varepsilon}$	=	stretch rate of fluid, 1/t, seconds⁻¹
$\dot{\varepsilon}_i$	=	rate of fluid stretch at Radius i, 1/t, seconds⁻¹
η	=	relative apparent viscosity (apparent viscosity of polymer solution/viscosity of brine)
η_r	=	relative viscosity (viscosity of polymer solution/viscosity of brine)
$\dot{\gamma}_2$	=	shear rate at Point 1, seconds⁻¹
γ_r^{-1}	=	shear rate at Point 2, seconds⁻¹
λ	=	foam mobility, mL/t, md/cp
λ_N	=	mobility of Newtonian fluid in porous rock, mL/t, md/cp
λ_p	=	mobility of polymer solution in porous rock, mL/t, md/cp
λ_p^*	=	polymer mobility constant defined in Eq. 5.20
λ_{pBK}^*	=	polymer mobility constant derived from Blake-Kozeny model
λ_{pN}	=	polymer mobility in lower Newtonian region, mL/t, md/cp
λ_{puN}	=	polymer mobility in upper Newtonian region, mL/t, md/cp
λ_{p1}^*	=	polymer mobility constant for Region 1 (shear thinning)
λ_{p2}^*	=	polymer mobility constant for Region 2 (shear thickening)
λ_r	=	relative mobility, Lt/m, cp⁻¹
λ_r^{-1}	=	reciprocal relative mobility, m/Lt, cp
λ_r^{*-1}	=	reciprocal relative mobility in polymer region, m/Lt, cp
λ_{rc}	=	relative total mobility of chemical slug, Lt/m, cp⁻¹
λ_{ro}/S_{iw}	=	oil relative mobility at interstitial water saturation, Lt/m, cp⁻¹
$\lambda_r^1 \bar{S}_{w1}^1$	=	reciprocal relative mobility at S_{w1}, m/Lt, cp
$\lambda_r \bar{S}_{wf}^1$	=	reciprocal relative mobility at S_{wf}
λ_{rt}	=	relative total mobility of oil bank, Lt/m, cp⁻¹
λ_t	=	total mobility, mL/t, md/cp
λ_{td}	=	design mobility of chemical slug to maintain unit mobility ratio, defined as minimum total relative mobility in Fig. 5.119, Lt/m, cp⁻¹
λ_w	=	water-phase mobility, mL/t, md/cp
λ_{wp}	=	water-phase mobility after rock has been exposed to polymer solution, mL/t, md/cp
$\bar{\lambda}^{-1}$	=	average apparent viscosity for linear system, m/Lt, cp

$\bar{\lambda}_r^{*-1}$	=	average apparent viscosity in polymer region, m/Lt, cp
μ	=	viscosity, m/Lt, cp
μ_a	=	apparent viscosity, m/Lt, cp
μ_c	=	viscosity of chemical slug at anticipated frontal-advance rate, m/Lt, cp
μ_{CO2}	=	CO_2 viscosity, m/Lt, cp
μ_e	=	effective viscosity of polymer solution when permeability of polymer solution is assumed to be k_w, m/Lt, cp
μ_p	=	polymer-solution viscosity at average shear rate in medium, m/Lt, cp
μ_{pN}	=	polymer viscosity in lower Newtonian region, m/Lt, cp
μ_∞	=	polymer viscosity in upper Newtonian region, usually taken as brine viscosity, m/Lt, cp
μ_1	=	viscosity at shear rate $\dot{\gamma}_1$, m/Lt, cp
μ_2	=	viscosity at shear rate $\dot{\gamma}_2$, m/Lt, cp
ρ_g	=	grain density, m/L³, g/cm³
ρ_p	=	polymer-solution density, m/L³, lbm/ft³
ρ_r	=	bulk density of rock, m/L³, lbm/ft³
Σ	=	interfacial tension, m/t², dynes/cm
τ	=	shear stress, m/Lt², dynes/cm²
τ_f	=	relaxation time of fluid, t, seconds
τ_r	=	shear rate in Carreau model for intersection of lower Newtonian and shear-thinning regions as in Fig. 5.14, 1/t, seconds^{-1}
ϕ	=	porosity, fraction
ϕ	=	inaccessible fraction of porosity
ϕ_{Ipv}	=	inaccessible PV

Subscripts

d	=	displaced phase
D	=	displacing phase
g	=	solvent phase
j	=	index of the number of saturation increments in a linear system
L	=	liquid phase
o	=	oil phase
ob	=	oil bank
S	=	solvent phase
w	=	water phase
WAG	=	WAG process

Superscripts

$-$	=	average

References

Abdo, M. K., Chung, H. S., Phelps, C. II. ct al. 1984. Field Experience With Floodwater Diversion by Complexed Biopolymers. Presented at the SPE/DOE Enhanced Oil Recovery Symposium, Tulsa, 15–18 April. SPE-12642-MS. https://doi.org/10.2118/12642-MS.

Abib, O., Moretti, F. J., Mei, C. et al. 1991. Application of Geological Modeling and Reservoir Simulation to the West Saertu Area of the Daqing Oil Field. *SPE Res Eng* **6** (1): 99–106. SPE-17560-PA. https://doi.org/10.2118/17560-PA.

Albrecht, R. A. and Marsden, S. S. 1970. Foams as Blocking Agents in Porous Media. *SPE J.* **10** (1): 51–55. SPE-2357-PA. https://doi.org/10.2118/2357-PA.

API RP 63, Recommended Practices for Evaluation of Polymers Used in Enhanced Oil Recovery Operations. 1990. Washington, DC: American Petroleum Institute.

Argabright, P. A., Rhudy, J. S., and Trujillo, E. M. 1986. Anatomy of a Full-Field Polymer-Augmented Waterflood Project in Water Soluble Polymers. In *Water-Soluble Polymers*, ed. J. E. Glass, Chapter 15, 269–311, Advances in Chemistry Series, Vol. 213. Washington, DC: American Chemical Society. https://doi.org/10.1021/ba-1986-0213.ch015.

Arshad, A. M., Manan, M. A., and Ismail, A. R. 1994. The Application of Pressure Drop Through a Horizontal Well Correlation to Oil Well Production Performance. Presented at the SPE Asia Pacific Oil and Gas Conference, Melbourne, Australia, 7–10 November. SPE-28801-MS. https://doi.org/10.2118/28801-MS.

Aslam, S., Vossoughi, S., and Willhite, G. P. 1984. Viscometric Measurement of Chromium(III)-Polyacrylamide Gels by Weissenberg Rheogoniometer. Presented at the SPE/DOE Enhanced Oil Recovery Symposium, Tulsa, 15–16 April. SPE-12639-MS. https://doi.org/10.2118/12639-MS.

Bernard, G. G. and Holm, L. W. 1964. Effect of Foam on Permeability of Porous Media to Gas. *SPE J.* **4** (3): 267–274. SPE-983-PA. https://doi.org/10.2118/983-PA.

Bernard, G. G., Holm, L. W., and Harvey, C. P. 1980. Use of Surfactant to Reduce CO_2 Mobility in Oil Displacement. *SPE J.* **20** (4): 281–292. SPE-8370-PA. https://doi.org/10.2118/8370-PA.

Blevins, T. R., Aseltine, R. J., and Kirk, R. S. 1969. Analysis of a Steam Drive Project, Inglewood Field, California. *J Pet Technol* **21** (9): 1141–1150. SPE-2291-PA. https://doi.org/10.2118/2291-PA.

Bragg, J. R., Gale, W. W., McElhannon, W. A. et al. 1982. Loudon Surfactant Flood Pilot Test. Presented at the SPE/DOE Enhanced Oil Recovery Symposium, Tulsa, 4–7 April. SPE-10862-MS. https://doi.org/10.2118/10862-MS.

Bruning, D. D., Hedges, J. H., and Zornes, D. R. 1983. Use of the Aluminum Citrate Process in the Commercial Scale North Burbank Unit Polymerflood. *Proc*, Fifth Tertiary Oil Recovery Conference, University of Kansas, Lawrence, Kansas, 111–130.

Canadian Natural. 2010. Annual ERCB Performance Presentation, In Situ Oil Sands Schemes, Approvals: 9673/9467/9572/ 10147/10423/10787 (24 March 2010), 1–45. http://aer.ca/data-and-publications/activity-and-data/in-situ-performance-presentations.

Canadian Natural. 2013. Annual Performance Presentation, In Situ Oil Sands Schemes 9673/10147/10423/10787 (March 2013). http://aer.ca/data-and-publications/activity-and-data/in-situ-performance-presentations.

Cannella, W. J., Huh, C., and Seright, R. S. 1988. Prediction of Xanthan Rheology in Porous Media. Presented at the SPE Annual Technical Conference and Exhibition, Houston, 2–5 October. SPE-18089-MS. https://doi.org/10.2118/18089-MS.

Carreau, J. P. 1968. *Rheological Equations from Molecular Network Theories*. PhD dissertation, University of Wisconsin, Madison, Wisconsin.

Carreau, J. P. 1972. Rheological Equations from Molecular Network Theories. *Trans. Soc. Rheol.* **16** (1): 99–127. https://doi. org/10.1122/1.549276.

Castagno, R. E., Shupe, R. D., and Gregory, M. D. 1987. Method for Laboratory and Field Evaluation of a Proposed Polymer Flood. *SPE Res Eng* **2** (4): 452–460. SPE-13124-PA. https://doi.org/10.2118/13124-PA.

Caudle, B. H. and Dyes, A. B. 1958. Improving Miscible Displacement by Gas-Water Injection. In *Petroleum Transactions, AIME*, Vol. 213, SPE-911-G, 281–283. Richardson, Texas: Society of Petroleum Engineers.

Chang, H. L., Al-Rikabi, H. M., and Pusch, W. H. 1978. Determination of Oil/Water Bank Mobility in Micellar-Polymer Flooding. *J Pet Technol* **30** (7): 1055–1060. SPE-6048-PA. https://doi.org/10.2118/6048-PA.

Chase, C. A. Jr. and Todd, M. R. 1984. Numerical Simulation of CO_2 Flood Performance. *SPE J.* **24** (6): 597–605. SPE-10514-PA. https://doi.org/10.2118/10514-PA.

Chauveteau, G. 1982. Rodlike Polymer Solution Flow Through Fine Pores: Influence of Pore Size on Rheological Behavior. *J. Rheol.* **26** (2): 111–142. https://doi.org/10.1122/1.549660.

Chauveteau, G. and Kohler, N. 1974. Polymer Flooding: The Essential Elements for Laboratory Evaluation. Presented at the SPE Improved Oil Recovery Symposium, Tulsa, 22–24 April. SPE-4745-MS. https://doi.org/10.2118/4745-MS.

Chauveteau, G. and Kohler, N. 1984. Influence of Microgels in Polysaccharide Solutions on Their Flow Behavior Through Porous Media. *SPE J.* **24** (3): 361–368. SPE-9295-PA. https://doi.org/10.2118/9295-PA.

Chauveteau, G. and Zaitoun, A. 1981. Basic Rheological Behavior of Xanthan Polysaccharide Solutions in Porous Media: Effects of Pore Size and Polymer Concentration. *Proc*, European Symposium on Enhanced Oil Recovery, Bournemouth, 197–212.

Chauveteau, G., Combe, J., and Dong, H. 1988. Preparation of Two Polymer Pilot Tests in Daqing Oil Field. Presented at the SPE International Meeting on Petroleum Engineering, Tianjin, China, 1–4 November. SPE-17632-MS. https://doi.org/10.2118/17632-MS.

Chiang, J. C., Sawyal, S. K., Castanier, L. M. et al. 1980. Foam as a Mobility Control Agent in Steam Injection Processes. Presented at the SPE California Regional Meeting, Los Angeles, California, USA, 9–11 April. SPE-8912-MS. https://doi.org/10.2118/8912-MS.

Christopher, C. A., Clark, T. J., and Gibson, D. H. 1988. Performance and Operation of a Successful Polymer Flood in the Sleepy Hollow Reagan Unit. Presented at the SPE/DOE Enhanced Oil Recovery Symposium, Tulsa, 22–27 April. SPE-17395-MS. https://doi.org/10.2118/17395-MS.

Christopher, R. H. and Middleman, S. 1965. Power-Law Flow Through a Packed Tube. *Ind. Eng. Chem. Fundamen.* **4** (4): 422–426. https://doi.org/10.1021/i160016a011.

Clampitt, R. L. and Hessert, J. E. 1974. Method for Controlling Formation Permeability. U.S. Patent No. 3,785,437 (15 January).

Claridge, E. L. 1978. A Method for Designing Graded Viscosity Banks. *SPE J.* **18** (5): 315–324. SPE-6848-PA. https://doi.org/10.2118/6848-PA.

Claridge, E. L. 1982. CO_2 Flooding Strategy in a Communicating Layered Reservoir. *J Pet Technol* **34** (12): 2746–2756. SPE-10289-PA. https://doi.org/10.2118/10289-PA.

David, A. and Marsden, S. S. Jr. 1969. The Rheology of Foam. Presented at the Fall Meeting of the Society of Petroleum Engineers of AIME, Denver, 28 September–1 October. SPE-2544-MS. https://doi.org/10.2118/2544-MS.

Davison, P. and Mentzer, E. 1982. Polymer Flooding in North Sea Reservoirs. *SPE J.* **22** (3): 353–362. SPE-9300-PA. https://doi.org/10.2118/9300-PA.

Dawson, R. and Lantz, R. B. 1972. Inaccessible Pore Volume in Polymer Flooding. *SPE J.* **12** (5): 448–452. SPE-3522-PA. https://doi.org/10.2118/3522-PA.

DeHekker, T. G., Bowzer, J. L., Coleman, R. V. et al. 1986. A Progress Report on Polymer-Augmented Waterflooding in Wyoming's North Oregon Basin and Byron Fields. Presented at the SPE/DOE Enhanced Oil Recovery Symposium, Tulsa, 20–23 April. SPE-14953-MS. https://doi.org/10.2118/14953-MS.

Delamaide, E., Corlay, P., and Wang, D. 1994. Daqing Oil Field: The Success of Two Pilots Initiates First Extension of Polymer Injection in a Giant Oil Field. Presented at the SPE/DOE Improved Oil Recovery Symposium, Tulsa, 17–20 April. SPE-27819-MS. https://doi.org/10.2118/27819-MS.

Delamaide, E., Zaitoun, A., Renard, G. et al. 2013. Pelican Lake Field: First Successful Application of Polymer Flooding in a Heavy Oil Reservoir. Presented at the SPE Enhanced Oil Recovery Conference, Kuala Lumpur, Malaysia, 2-4, July 2013. SPE-165234-MS. https://doi.org/10.2118/165234-MS.

Dellinger, S. E., Patton, J. T., and Holbrook, S. T. 1984. CO_2 Mobility Control. *SPE J.* **24** (2): 191–196. SPE-9808-PA. https://doi.org/10.2118/9808-PA.

Dietz, D. N., Bruining, J., and Heijna, H. B. 1985. Foamdrive Seldom Meaningful. *J Pet Technol* **37** (5): 921–922. SPE-11274-PA. https://doi.org/10.2118/11274-PA.

DiMarzio, E. A. and Guttman, C. M. 1970. Separation by Flow. *Macromolecules* **3** (2): 131–146. https://doi.org/10.1021/ma60014a005.

Doe, P. H., Moradi-Araghi, A., Shaw, J. E. et al. 1987. Development and Evaluation of EOR Polymers Suitable for Hostile Environments—Part 1: Copolymers of Vinylpyrrolidone and Acrylamide. *SPE Res Eng* **2** (4): 461–467. SPE-14233-PA. https://doi.org/10.2118/14233-PA.

Dolan, D. M. 1989. *An Experimental Study of the Effects of pH and Shear on the Gelation of a Xanthan-Chromium(III) Solution.* MS thesis, University of Kansas, Lawrence, Kansas.

Doll, T. E. and Hanson, M. T. 1987. Performance and Operation of the Hamm Minnelusa Sand Unit, Campbell County, Wyoming. *J Pet Technol* **39** (12): 1565–1570. SPE-15162-PA. https://doi.org/10.2118/15162-PA.

Dominguez, J. G. and Willhite, G. P. 1977. Retention and Flow Characteristics of Polymer Solutions in Porous Media. *SPE J.* **17** (2): 111–121. SPE-5835-PA. https://doi.org/10.2118/5835-PA.

Duerksen, J. H. 1986. Laboratory Study of Foaming Surfactants as Steam-Diverting Agents. *SPE Res Eng* **1** (1): 44–52. SPE-12785-PA. https://doi.org/10.2118/12785-PA

Eggert, R. W. Jr., Willhite, G. P., and Green, D. W. 1992. Experimental Measurement of the Persistence of Permeability Reduction in Porous Media Treated With Xanthan/Cr(III) Gel Systems. *SPE Res Eng* **7** (1): 29–35. SPE-19630-PA. https://doi.org/10.2118/19630-PA.

EnCana Corporation 2005. In Situ Oil Sands Scheme Approval 9404 Brintnell Sector Enhanced Recovery of Crude Bitumen by Water Injection: Semi-Annual Presentation. EnCana Corporation, Calgary (8 June 2005).

EnCana Corporation. 2002. 2002 Annual Report EnCana[2]: Building a Best-In-Class Enterprise. Encanca Corporation, Calgary.

EnCana Corporation. 2003. 2003 Annual Report to Shareholders: What Matters. EnCana Corporation, Calgary.

Gales, J. R., Young, T. S., Willhite, G. P. et al. 1988. Equilibrium Swelling and Syneresis Properties of Xanthan Gum-Cr(III) Gels. Presented at the SPE/DOE Enhanced Oil Recovery Symposium, Tulsa, 17 20 April. SPE-17328-MS.

Gao, H. W. and French, T. R. 1988. Optimal Rheological Character in Dynamic Stability of Polymer Flow Through Porous Media. NIPER-303 (DE88001225), US Department of Energy, Washington, DC (April 1988).

Gao, H. W., Chang, M. M., Burchfield, T. E. et al. 1993. Studies of the Effects of Crossflow and Initiation Time of a Polymer Gel Treatment on Oil Recovery in a Waterflood Using a Permeability Modification Simulator. Presented at the SPE/DOE Enhanced Oil Recovery Symposium, Tulsa, 22–25 April. SPE-20216-MS.

Gogarty, W. B. 1967a. Rheological Properties of Pseudoplastic Fluids in Porous Media. *SPE J.* **7** (2): 149–160. SPE-1566-PA. https://doi.org/10.2118/1566-PA.

Gogarty, W. B. 1967b. Mobility Control With Polymer Solutions. *SPE J.* **7** (2): 161–173. SPE-1566-B. https://doi.org/10.2118/1566-B.

Gogarty, W. B., Meabon, H. P., and Milton, H. W. Jr. 1970. Mobility Control Design for Miscible-Type Waterfloods Using Micellar Solutions. *J Pet Technol* **22** (2): 141–147. SPE-1847-E-PA. https://doi.org/10.2118/1847-E-PA.

Guo, W., Cheng, J., Yuming, Y. et al. 2000. Commercial Pilot Test of Polymer Flooding in Daqing Oil Field. Presented at the SPE/DOE Improved Oil Recovery Symposium, Tulsa, 3–5 April. SPE-59275-MS. https://doi.org/10.2118/59275-MS.

Gupta, S. P. and Trushenski, S. P. 1978. Micellar Flooding—The Propagation of the Polymer Mobility Buffer Bank. *SPE J.* **18** (1): 5–12. SPE-6204-PA. https://doi.org/10.2118/6204-PA.

He, L., Yang, G., Sun, F. et al. 2013. Overview of Key Zonal Water Injection Technologies in China. Presented at the International Petroleum Technology Conference, Beijing, 26–28 March. IPTC-16868-MS. https://dx.doi.org/10.2523/IPTC-16868-MS.

Heemskerk, J., Janssen-van Rosmalen, R., Holtslag, R. J. et al. 1984. Quantification of Viscoelastic Effects of Polyacrylamide Solutions. Presented at the SPE/DOE Enhanced Oil Recovery Symposium, Tulsa, 15–18 April. SPE-12652-MS. https://doi.org/10.2118/12652-MS.

Hejri, S. 1989. *An Experimental Investigation into the Flow and Rheological Behavior of Biopolymer and Biopolymer/Cr(III) Systems in Porous Media.* PhD dissertation, University of Kansas, Lawrence, Kansas.

Hejri, S., Green, D. W., and Willhite, G. P. 1989. In-Situ Gelation of a Xanthan/Cr(III) Gel System in Porous Media. Presented at the SPE Annual Technical Conference and Exhibition, San Antonio, Texas, USA, 8–11 October. SPE-19634-MS.

Hejri, S., Willhite, G. P., and Green, D. W. 1991. Development of Correlations To Predict Biopolymer Mobility in Porous Media. *SPE Res Eng* **6** (1): 91–98. SPE-17396-PA. https://doi.org/10.2118/17396-PA.

Heller, J. P., Cheng, L. L., and Kuntamukkula, M. S. 1985. Foamlike Dispersions for Mobility Control in CO_2 Floods. *SPE J.* **25** (4): 603–613. SPE-11233-PA. https://doi.org/10.2118/11233-PA.

Herr, J. W. and Routson, W. G. 1974. Polymer Structure and Its Relationship to the Dilute Solution Properties of High Molecular Weight Polyacrylamide. Presented at the Fall Meeting of the Society of Petroleum Engineers of AIME, Houston, 6–9 October. SPE-5098-MS. https://doi.org/10.2118/5098-MS.

Hess, P. H., Clark, C. O., Haskin, C. A. et al. 1971. Chemical Method for Formation Plugging. *J Pet Technol* **23** (5): 559–564. SPE-3045-PA. https://doi.org/10.2118/3045-PA.

Hessert, J. E. and Fleming, P. D. III. 1979. Gelled Polymer Technology for Control of Water in Injection and Production Wells. *Proc*, Third Tertiary Oil Recovery Conference, University of Kansas, Lawrence, Kansas, 58–70.

Higgins, R. V. and Leighton, A. J. 1962. Computer Prediction of Water Drive of Oil and Gas Mixtures Through Irregularly Bounded Porous Media—Three-Phase Flow. *J Pet Technol* **14** (9): 1048–1054. SPE-283-PA. https://doi.org/10.2118/283-PA.

Hirasaki, G. J. and Pope, G. A. 1974. Analysis of Factors Influencing Mobility and Adsorption in the Flow of Polymer Solution Through Porous Media. *SPE J.* **14** (4): 337–346. SPE-4026-PA. https://doi.org/10.2118/4026-PA.

Hochanadel, S. M., Lunceford, M. L., and Farmer, C. W. 1990. A Comparison of 31 Minnelusa Polymer Floods With 24 Minnelusa Waterfloods. Presented at the SPE/DOE Enhanced Oil Recovery Symposium, Tulsa, 22–25 April. SPE-20234-MS. https://doi.org/10.2118/20234-MS.

Holm, L. W. 1970. Foam Injection Test in the Siggins Field, Illinois. *J Pet Technol* **22** (12): 1499–1506. SPE-2750-PA. https://doi.org/10.2118/2750-PA.

Holzwarth, G., Soni, L., Schulz, D. N. et al. 1988. Absolute MWDs of Polyacrylamides by Sedimentation and Light Scattering. In *Water-Soluble Polymers for Petroleum Recovery*, eds. G. A. Stahl and D. N. Schultz, 215–229. New York, New York: Plenum Press.

Hsie, J. C. and Moore, J. S. 1988. The Quarantine Bay 4RC CO_2 WAG Pilot Project: A Postflood Evaluation. *SPE Res Eng* **3** (3): 807–814. SPE-15498-PA. https://doi.org/10.2118/15498-PA.

Huang, C. G., Green, D. W., and Willhite, G. P. 1986. An Experimental Study of the In-Situ Gelation of Chromium(+3)/Polyacrylamide Polymer in Porous Media. *SPE Res Eng* **1** (6): 583–592. SPE-12638-PA. https://doi.org/10.2118/12638-PA.

Huang, E. T. S. and Holm, L. W. 1988. Effect of WAG Injection and Rock Wettability on Oil Recovery During CO_2 Flooding. *SPE Res Eng* **3** (1): 119–129. SPE-15491-PA. https://doi.org/10.2118/15491-PA.

Hughes, D. S., Teeuw, D., Cottrell, C. W. et al. 1990. Appraisal of the Use of Polymer Injection To Suppress Aquifer Influx and To Improve Volumetric Sweep in a Viscous Oil Reservoir. *SPE Res Eng* **5** (1): 33–40. SPE-17400-PA. https://doi.org/10.2118/17400-PA.

Huh, C., Lange, E. A., and Cannella, W. J. 1990. Polymer Retention in Porous Media. Presented at the SPE/DOE Enhanced Oil Recovery Symposium, Tulsa, 22–25 April. SPE-20235-MS. https://doi.org/10.2118/20235-MS.

Hunt, J. A., Young, T. S., Green, D. W. et al. 1989. A Study of Cr(III)-Polyacrylamide Reaction Kinetics by Equilibrium Dialysis. *AIChE J.* **35** (2): 250–258. https://doi.org/10.1002/aic.690350209.

Jackson, D. D., Andrews, G. L., and Claridge, E. L. 1985. Optimum WAG Ratio vs. Rock Wettability in CO_2 Flooding. Presented at the SPE Annual Technical Conference and Exhibition, Las Vegas, Nevada, USA, 22–25 September. SPE-14303-MS. https://doi.org/10.2118/14303-MS.

Jennings, R. R., Rogers, J. H., and West, T. J. 1971. Factors Influencing Mobility Control by Polymer Solutions. *J Pet Technol* **23** (3): 391–401. SPE-2867-PA. https://doi.org/10.2118/2867-PA.

Jordan, D. S., Green, D. W., Terry, R. E. et al. 1982. The Effect of Temperature on Gelation Time for Polyacrylamide/Chromium(III) Systems. *SPE J.* **22** (4): 463–471. SPE-10059-PA. https://doi.org/10.2118/10059-PA.

Jousset, F., Green, D. W., Willhite, G. P. et al. 1990. Effect of High Shear Rate on In-Situ Gelation of a Xanthan/Cr(III) System. Presented at the SPE/DOE Enhanced Oil Recovery Symposium, Tulsa, 22–25 April. SPE-20213-MS. https://doi.org/10.2118/20213-MS.

Kalpakci, B., Jeans, Y. T., Magri, N. F. et al. 1990. Thermal Stability of Scleroglucan at Realistic Reservoir Conditions. Presented at the SPE/DOE Enhanced Oil Recovery Symposium, Tulsa, 22–25 April. SPE-20237-MS. https://doi.org/10.2118/20237-MS.

Knight, B. L. 1973. Reservoir Stability of Polymer Solutions. *J Pet Technol* **25** (5): 618–626. SPE-4167-PA. https://doi.org/10.2118/4167-PA.

Kohler, N. and Chauveteau, G. 1981. Xanthan Polysaccharide Plugging Behavior in Porous Media—Preferential Use of Fermentation Broth. *J Pet Technol* **33** (2): 349–358. SPE-7425-PA. https://doi.org/10.2118/7425-PA.

Kolodziej, E. J. 1987. Mechanism of Microgel Formation in Xanthan Biopolymer Solutions. Presented at the SPE Annual Technical Conference and Exhibition, Dallas, Texas, USA, 27–30 September. SPE-16730-MS. https://doi.org/10.2118/16730-MS.

Kolodziej, E. J. 1988. Transport Mechanisms of Xanthan Biopolymer Solutions in Porous Media. Presented at the SPE Annual Technical Conference and Exhibition, Houston, 2–5 October. SPE-18090-MS. https://doi.org/10.2118/18090-MS.

Koning, E. J. L., Mentzer, E., and Heemskerk, J. 1988. Evaluation of a Pilot Polymer Flood in the Marmul Field, Oman. Presented at the SPE Annual Technical Conference and Exhibition, Houston, 2–5 October. SPE-18092-MS. https://doi.org/10.2118/18092-MS.

Krebs, H. J. 1976. Wilmington Field, California, Polymer Flood—A Case History. *J Pet Technol* **28** (12): 1473–1480. SPE-5828-PA. https://doi.org/10.2118/5828-PA.

Lake, L. W. 1989. *Enhanced Oil Recovery*. Englewood Cliffs, New Jersey: Prentice-Hall.

Leighton, A. J. and Higgins, R. V. 1975. Improved Method To Predict Multiphase Waterflood Performance for Constant Rates or Pressures. Report, RI 8055, US Bureau of Mines, Washington, DC.

Liang, Y., Zhang, S., Pei, X. et al. 2011. Practice and Understanding of Separate-Layer Polymer Injection in Daqing Oil Field. *SPE Prod & Oper* **26** (3): 224–228. SPE-128103-PA. https://doi.org/10.2118/128103-PA.

Lim, T., Uhl, J. T., and Prud'homme, R. K. 1986. The Interpretation of Screen-Factor Measurements. *SPE Res Eng* **1** (3): 272–276. SPE-12285-PA. https://doi.org/10.2118/12285-PA.

Lin, E. C. and Huang, E. T. S. 1990. The Effect of Rock Wettability on Water Blocking During Miscible Displacement. *SPE Res Eng* **5** (2): 205–212. SPE-17375-PA. https://doi.org/10.2118/17375-PA.

Lin, E. C. and Poole, E. S. 1991. Numerical Evaluation of Single-Slug, WAG, and Hybrid CO_2 Injection Processes, Dollarhide Devonian Unit, Andrews County, Texas. *SPE Res Eng* **6** (4): 415–420. SPE-20098-PA. https://doi.org/10.2118/20098-PA.

Lipton, D. 1974. Improved Injectability of Biopolymer Solutions. Presented at the Fall Meeting of the Society of Petroleum Engineers of AIME, Houston, 6–9 October. SPE-5099-MS. https://doi.org/10.2118/5099-MS.

Lötsch, T. 1988. Untersachungen zum Retentionsverhalten von Polymeren beim viskosen Fluten. In *Developments in Petroleum Science*, ed. W. Littman, 24. New York, New York: Elsevier Science Publishers.

MacWilliams, D. C., Rogers, J. H., and West, T. J. 1973. Water Soluble Polymers in Petroleum Recovery. In *Water Soluble Polymers*, ed. N. M. Bikales, 105–126. New York, New York: Plenum Press.

Maerker, J. M. 1973. Dependence of Polymer Retention on Flow Rate. *J Pet Technol* **25** (11): 1307–1308. SPE-4423-PA. https://doi.org/10.2118/4423-PA.

Maerker, J. M. 1975. Shear Degradation of Partially Hydrolyzed Polyacrylamide Solutions. *SPE J.* **15** (4): 311–322. SPE-5101-PA. https://doi.org/10.2118/5101-PA.

Maerker, J. M. 1976. Mechanical Degradation of Partially Hydrolyzed Polyacrylamide Solutions in Unconsolidated Porous Media. *SPE J.* **16** (4): 172–174. SPE-5672-PA. https://doi.org/10.2118/5672-PA.

Maitin, B. K. and Volz, H. 1981. Performance of Deutsche Texaco AG's Oerrel and Hankensbuettel Polymer Floods. Presented at the Offshore Technology Conference, Houston, 4–7 May. SPE 9794 MS.

Maitin, B. K., Daboul., B., and Sohn, W. O. 1988. Numerical Simulation for Planning and Evaluation of Polymer Flood Process: A Field Performance Analysis. Presented at the SPE International Meeting on Petroleum Engineering, Tianjin, China, 1–4 November. SPE-17631-MS. https://doi.org/10.2118/17631-MS.

Manning, R. K., Pope, G. A., Lake, L. W. et al. 1983. A Technical Survey of Polymer Flooding Projects. Report DOE/ET/10327-19, US Department of Energy, Bartlesville, Oklahoma (September 1983).

Marsden, S. S. and Khan, S. A. 1966. The Flow of Foam Through Short Porous Media and Apparent Viscosity Measurements. *SPE J.* **6** (1): 17–25. SPE-1319-PA. https://doi.org/10.2118/1319-PA.

Marshall, R. J. and Metzner, A. B. 1967. Flow of Viscoelastic Fluids Through Porous Media. *Ind. Eng. Chem. Fundamen.* **6** (3): 393–400. https://doi.org/10.1021/i160023a012.

Martin, F. D. 1986. Mechanical Degradation of Polyacrylamide Solutions in Core Plugs From Several Carbonate Reservoirs. *SPE Form Eval* **1** (2): 139–150. SPE-12651-PA. https://doi.org/10.2118/12651-PA.

Martin, F. D. and Kantamukkala, M. S. 1984. The Influence of Mechanical Degradation on the Viscous, Elastic and Elongational Flow Properties of Polymer Solutions Used in Enhanced Oil Recovery. In *Proc.*, Ninth International Congress on Rheology, Mexico City, 411–419.

Martin, F. D. and Sherwood, N. S. 1975. The Effect of Hydrolysis of Polyacrylamide on Solution Viscosity, Polymer Retention and Flow Resistance Properties. Presented at the SPE Rocky Mountain Regional Meeting, Denver, 7–9 April. SPE-5339-MS. https://doi.org/10.2118/5339-MS.

Martin, F. D., Hatch, M. J., Shepitka, J. S. et al. 1983. Improved Water-Soluble Polymers for Enhanced Recovery of Oil. Presented at the SPE International Symposium on Oilfield and Geothermal Chemistry, Denver, 1–3 June. SPE-11786-MS. https://doi.org/10.2118/11786-MS.

Martin, F. D., Kovarik, F. S., Chang, P. W. et al. 1988. Gels for CO_2 Profile Modification. Presented at the SPE/DOE Enhanced Oil Recovery Symposium, Tulsa, 17–20 April. SPE-17330-MS. https://doi.org/10.2118/17330-MS.

Marty, L., Green, D. W., and Willhite, G. P. 1991. The Effect of Flow Rate on In-Situ Gelation of a Chrome/Redox/Polyacrylamide System. *SPE Res Eng* **6** (2): 219–224. SPE-18504-PA. https://doi.org/10.2118/18504-PA.

McCool, C. S., Green, D. W., and Willhite, G. P. 1991. Permeability Reduction Mechanisms Involved in In-Situ Gelation of a Polyacrylamide/Chromium(VI)/Thiourea System. *SPE Res Eng* **6** (1): 77–83. SPE-17333-PA. https://doi.org/10.2118/17333-PA.

McCool, C. S., Parmeswar, R., and Willhite, G. P. 1983. Interpretation of Differential Pressure in Laboratory Surfactant/Polymer Displacements. *SPE J.* **23** (5): 791–803. SPE-10713-PA. https://doi.org/10.2118/10713-PA.

Metzner, A. B., White, J. L., and Denn, M. M. 1966. Behavior of Viscoelastic Materials in Short Time Processes. *Chem. Eng. Prog.* **62:** 81–92.

Minssieux, L. 1974. Oil Displacement by Foams in Relation to Their Physical Properties in Porous Media. *J Pet Technol* **26** (1): 100–108. SPE-3991-PA. https://doi.org/10.2118/3991-PA.

Mooney, M. 1932. Explicit Formulas for Slip and Fluidity. *Rheology* **2**: 210–222.

Moradi-Araghi, A. and Doe, P. H. 1987. Hydrolysis and Precipitation of Polyacrylamides in Hard Brines at Elevated Temperatures. *SPE Res Eng* **2** (2): 189–198. SPE-13033-PA. https://doi.org/10.2118/13033-PA.

Moriarty, M. 2008. Enhanced Recovery of Crude Bitumen by Injection. Presentation, Britnell Field In Situ Oil Sands Schemes, Approvals: 9673/9467/9572/10147/10423/10787 (15 February, 2008). http://www.aer.ca/documents/oilsands/insitu-presentations/2008AthabascaCNRLBrintnell9673_9467_9572_10147.pdf.

Morrow, N. R., Chatzis, I., and Lim, H. 1985. Relative Permeabilities at Reduced Residual Saturations. *Can. Pet. Tech.* **24:** 62–69.

Mumallah, N. A. 1988. Chromium (III) Propionate: A Crosslinking Agent for Water-Soluble Polymers in Hard Oilfield Brines. *SPE Res Eng* **3** (1): 243–250. SPE-15906-PA. https://doi.org/10.2118/15906-PA.

Mungan, N. 1969a. Rheology and Adsorption of Aqueous Polymer Solutions. *J Can Pet Technol* **8** (2): 45–50. PETSOC-69-02-01. https://doi.org/10.2118/69-02-01.

Mungan, N. 1969b. Le Contrôle de la Mobilité dans les Injections de Polymères. *Rev. IFP* **24:** 232–250.

Mungan, N. 1971. Improved Waterflooding Through Mobility Control. *Cdn. J. Chem. Eng.* **49** (1): 32–37. https://doi. org/10.1002/cjce.5450490107.

Nagra, S. S., Batycky, J. P., Nieman, R. E. et al. 1986. Stability of Waterflood Diverting Agents at Elevated Temperatures in Reservoir Brines. Presented at the SPE Annual Technical Conference and Exhibition, New Orleans, 5–8 October. SPE-15548-MS. https://doi.org/10.2118/15548-MS.

Needham, R. B., Threlkeld, C. B., and Gall, J. W. 1974. Control of Water Mobility Using Polymers and Multivalent Cations. Presented at the SPE Improved Oil Recovery Symposium, Tulsa, 22–24 April. SPE-4747-MS. https://doi.org/10.2118/4747-MS.

Oosawa, F. 1971. *Polyelectrolytes*. New York, New York: Marcel Dekker Inc.

Ostwald, W. 1925. Uber die Geschwindigkeitsfunktion der Viskosität disperser Système (in German). *Kolloid-Zeitschrift* **36:** 99–117.

Ostwald, W. 1929. Ueber die rechnerische Darstellung des Strukturgebietes der Viskosität (in German). *Kolloid-Zeitschrift* **47** (2): 176–187. https://doi.org/10.1007/BF01496959.

Parmeswar, R. and Willhite, G. P. 1988. A Study of the Reduction of Brine Permeability in Berea Sandstone With the Aluminum Citrate Process. *SPE Res Eng* **3** (3): 959–966. SPE-13582-PA. https://doi.org/10.2118/13582-PA.

Patton, J. T., Coats, K. H., and Colegrove, G. T. 1971. Prediction of Polymer Flood Performance. *SPE J.* **11** (1): 72–84. SPE-2546-PA. https://doi.org/10.2118/2546-PA.

Patton, J. T., Holbrook, S. T., and Hsu, W. 1983. Rheology of Mobility-Control Foams. *SPE J.* **23** (3): 456–460. SPE-9809-PA. https://doi.org/10.2118/9809-PA.

Purkaple, J. D. and Summers, L. E. 1988. Evaluation of Commercial Crosslinked Polyacrylamide Gel Systems for Injection Profile Modification. Presented at the SPE/DOE Enhanced Oil Recovery Symposium, Tulsa, 17–20 April. SPE-17331-MS. https://doi.org/10.2118/17331-MS.

Putz, A. G., Bazin, B., and Pedron, B. M. 1994. Commercial Polymer Injection in the Courtenay Field, 1994 Update. Presented at the SPE Annual Technical Conference and Exhibition, New Orleans, 25–28 September. SPE-28601-MS. https://doi. org/10.2118/28601-MS.

Putz, A. G., Lecourtier, J. M., and Bruckert, L. 1988. Interpretation of High Recovery Obtained in a New Polymer Flood in the Chateaurenard Field. Presented at the SPE Annual Technical Conference and Exhibition, Houston, 2–5 October. SPE-18093-MS. https://doi.org/10.2118/18093-MS.

Pye, D. J. 1967. Water Flooding Process. U.S. Patent No. 3,343,601.

Raimondi, P. and Torcaso, M. A. 1964. Distribution of the Oil Phase Obtained Upon Imbibition of Water. *SPE J.* **4** (1): 49–55. SPE-570-PA. https://doi.org/10.2118/570-PA.

Ray, R. M. and Diaz Munoz, J. 1986. Polymer Predictive Model. Report DOE/BC-86/10/SP (DE87001207), US Department of Energy, Washington, DC (December 1986).

Raza, S. H. 1970. Foam in Porous Media: Characteristics and Potential Applications. *SPE J.* **10** (4): 328–335. SPE-2421-PA. https://doi.org/10.2118/2421-PA.

Raza, S. H. and Marsden, S. S. 1967. The Streaming Potential and the Rheology of Foam. *SPE J.* **7** (4): 359–368. SPE-1748-PA. https://doi.org/10.2118/1748-PA.

Reppert, T. R., Bragg, J. R., Wilkinson, J. R. et al. 1990. Second Ripley Surfactant Flood Pilot Test. Presented at the SPE/DOE Enhanced Oil Recovery Symposium, Tulsa, 22–25 April. SPE-20219-MS. https://doi.org/10.2118/20219-MS.

Ryles, R. G. 1983. Elevated Temperature Testing of Mobility Control Reagents. Presented at the SPE Annual Technical Conference and Exhibition, San Francisco, California, USA, 5–8 October. SPE-12008-MS. https://doi.org/10.2118/12008-MS.

Schneider, F. N. and Owens, W. W. 1982. Steady-State Measurements of Relative Permeability for Polymer/Oil Systems. *SPE J.* **22** (1): 79–86. SPE-9408-PA. https://doi.org/10.2118/9408-PA.

Schoeling, L. G., Green, D. W., and Willhite, G. P. 1989. Introducing EOR Technology to Independent Operators. *J Pet Technol* **41** (12): 1344–1350. SPE-17401-PA. https://doi.org/10.2118/17401-PA.

Schurz, G. F., Martin, F. D., Seright, R. S. et al. 1989. Polymer-Augmented Waterflooding and Control of Reservoir Heterogeneity. Presented at the New Mexico Tech Centennial Symposium, Socorro, New Mexico, 16–19 October. NMT 890029.

Sengul, M. M., Skinner, T. K., Kirchner, W. D. et al. 1984. Simulation of Polymer Flood With Propagating Hydraulic Fracture. Presented at the SPE Annual Technical Conference and Exhibition, Houston, 16–19 September. SPE-13125-MS.

Seright, R. S. 1983. The Effects of Mechanical Degradation and Viscoelastic Behavior on Injectivity of Polyacrylamide Solutions. *SPE J.* **23** (3): 475–485. SPE-9297-PA. https://doi.org/10.2118/9297-PA.

Seright, R. S. 1988. Placement of Gels To Modify Injection Profiles. Presented at the SPE/DOE Enhanced Oil Recovery Symposium, Tulsa, 17–20 April. SPE-17332-MS. https://doi.org/10.2118/17332-MS.

Seright, R. S. and Henrici, B. J. 1990. Xanthan Stability at Elevated Temperatures. *SPE Res Eng* **5** (1): 52–60. SPE-14946-PA. https://doi.org/10.2118/14946-PA.

Shah, B. 1978. *An Experimental Study of Inaccessible Pore Volume as a Function of Polymer Concentration During Flow Through Porous Media*. MS thesis, University of Kansas, Lawrence, Kansas.

Shupe, R. D. 1981. Chemical Stability of Polyacrylamide Polymers. *J Pet Technol* **33** (8): 1513–1529. SPE-9299-PA. https://doi.org/10.2118/9299-PA.

Sloat, B. 1979. Producing Well Water-Oil Ratio Control—Process Design and Evaluation. *Proc*, Third Tertiary Oil Recovery Conference, University of Kansas, Lawrence, Kansas, 38–57.

Smith, F. W. 1970. The Behavior of Partially Hydrolyzed Polyacrylamide Solutions in Porous Media. *J Pet Technol* **22** (2): 148–156. SPE-2422-PA. https://doi.org/10.2118/2422-PA.

Sorbie, K. S. 1991a. *Polymer-Improved Oil Recovery*, 312–40. London: Blackie & Son Ltd.

Sorbie, K. S. 1991b. *Polymer-Improved Oil Recovery*, 274–297. London: Blackie & Son Ltd.

Sorbie, K. S., Parker, A., and Clifford, P. J. 1987. Experimental and Theoretical Study of Polymer Flow in Porous Media. *SPE Res Eng* **2** (3): 281–304. SPE-14231-PA. https://doi.org/10.2118/14231-PA.

Southwick, J. G. and Manke, C. W. 1988. Molecular Degradation, Injectivity, and Elastic Properties of Polymer Solutions. *SPE Res Eng* **3** (4): 1193–1201. SPE-15652-PA. https://doi.org/10.2118/15652-PA.

Stalkup, F. I. 1970. Displacement of Oil by Solvent at High Water Saturation. *SPE J.* **10** (4): 337–348. SPE-2419-PA. https://doi.org/10.2118/2419-PA.

Stalkup, F. I. 1983. *Miscible Displacement*, 8, Monograph Series. Richardson, Texas: Society of Petroleum Engineers.

Stoneberger, M. W. and Claridge, E. L. 1988. Graded-Viscosity-Bank Design With Pseudoplastic Fluids. *SPE Res Eng* **3** (4): 1221–1232. SPE-14230-PA. https://doi.org/10.2118/14230-PA.

Sydansk, R. D. 1988. A New Conformance-Improvement-Treatment Chromium(III) Gel Technology. Presented at the SPE/DOE Enhanced Oil Recovery Symposium, Tulsa, 17–20 April. SPE-17329-MS. https://doi.org/10.2118/17329-MS.

Sydansk, R. D. 1990. A Newly Developed Chromium(III) Gel Technology. *SPE Res Eng* **5** (3): 346–352. SPE-19308-PA. https://doi.org/10.2118/19308-PA.

Sydansk, R. D. and Smith, T. B. 1988. Field Testing of a New Conformance-Improvement-Treatment Chromium(III) Gel Technology. Presented at the SPE/DOE Enhanced Oil Recovery Symposium, Tulsa, 17–20 April. SPE-17383-MS. https://doi.org/10.2118/17383-MS.

Tanner, C. S., Baxley, P. T. Crump, J. G. III et al. 1992. Production Performance of the Wasson Denver Unit CO_2 Flood. Presented at the SPE/DOE Enhanced Oil Recovery Symposium, Tulsa, 22–24 April. SPE-24156-MS. https://doi.org/10.2118/24156-MS.

Teeuw, D. and Hesselink, F. T. 1980. Power-Law Flow and Hydrodynamic Behaviour of Biopolymer Solutions in Porous Media. Presented at the SPE International Symposium on Oilfield and Geothermal Chemistry, Stanford, California, USA, 28–30 May. SPE-8982-MS. https://doi.org/10.2118/8982-MS.

Terry, R. E., Huang, C., Green, D. W. et al. 1981. Correlation of Gelation Times for Polymer Solutions Used as Sweep Improvement Agents. *SPE J.* **21** (2): 229–235. SPE-8419-PA. https://doi.org/10.2118/8419-PA.

Thomas, C. P. 1976. The Mechanism of Reduction of Water Mobility by Polymers in Glass Capillary Arrays. *SPE J.* **16** (3): 130–136. SPE-5556-PA. https://doi.org/10.2118/5556-PA.

Tiffin, D. L. and Yellig, W. F. 1983. Effects of Mobile Water on Multiple-Contact Miscible Gas Displacements. *SPE J.* **23** (3): 447–455. SPE-10687-PA. https://doi.org/10.2118/10687-PA.

Todd, B. J., Willhite, G. P., and Green, D. W. 1991. Radial Modeling of In-Situ Gelation in Porous Media. Presented at the SPE Production Operations Symposium, Oklahoma City, Oklahoma, USA, 7–9 April. SPE-21650-MS. https://doi.org/10.2118/21650-MS.

Todd, M. R. and Longstaff, W. J. 1972. The Development, Testing, and Application of a Numerical Simulator for Predicting Miscible Flood Performance. *J Pet Technol* **24** (7): 874–882. SPE-3484-PA. https://doi.org/10.2118/3484-PA.

Trantham, J. C. 1977. North Burbank Unit Tertiary Recovery Pilot Test, Second Annual Report, May 1976-1977. Report BERC/TPR-77/5, Energy Resources Development Administration, Bartlesville, Oklahoma (August 1977).

Trushenski, S. P., Dauben, D. L., and Parrish, D. R. 1974. Micellar Flooding—Fluid Propagation, Interaction, and Mobility. *SPE J.* **14** (6): 633–645. SPE-4582-PA. https://doi.org/10.2118/4582-PA.

Uzoigwe, A. C., Scanlon, F. C., and Jewett, R. L. 1974. Improvements in Polymer Flooding: The Programmed Slug and the Polymer-Conserving Agent. *J Pet Technol* **26** (1): 33–41. SPE-4024-PA. https://doi.org/10.2118/4024-PA.

Van Horn, L. E. 1981. El Dorado Micellar-Polymer Demonstration Project, Sixth Annual Report for the Period September 1979–August 1980. Report DOE/ET/130070-63, US Department of Energy, Washington, DC.

Vela, S., Peaceman, D. W., and Sandvik, E. I. 1976. Evaluation of Polymer Flooding in a Layered Reservoir With Crossflow, Retention, and Degradation. *SPE J.* **16** (2): 82–96. SPE-5102-PA. https://doi.org/10.2118/5102-PA.

Walsh, M. P. 1991. An Analysis of Miscible Flooding Chase-Fluid Strategies. 1991. *SPE Res Eng* **6** (4): 437–444. SPE-19866-PA. https://doi.org/10.2118/19866-PA.

Wang, D., Cheng, J., Wu, J. et al. 2002a. Experiences Learned after Production more than 300 Million Barrels of Oil by Polymer Flooding in Daqing Oil Field. Presented at the SPE Annual Technical Conference and Exhibition, San Antonio, Texas, USA, 29 Septermber–2 October. SPE-77693-MS. https://doi.org/10.2118/77693-MS.

Wang, D., Cheng, J., Wu, J. et al. 2002b. Producing by Polymer Flooding more than 300 Million Barrels of Oil, What Experiences Have Been Learnt? Presented at the SPE Asia Pacific Oil and Gas Conference and Exhibition, Melbourne, Australia, 8–10 October. SPE-77872-MS. https://doi.org/10.2118/77872-MS.

Wang, D., Cheng, J., Yang, Q. et al. 2000. Viscous-Elastic Polymer Can Increase Microscale Displacement Efficiency in Cores. Presented at the SPE Annual Technical Conference and Exhibition, Dallas, Texas, USA, 1–4 October. SPE-63227-MS. https://doi.org/10.2118/63227-MS.

Wang, D., Dong, H., Lv, C. et al. 2009. Review of Practical Experience by Polymer Flooding at Daqing. *SPE Res Eng* **12** (3): 470–476. SPE-114342-PA. https://doi.org/10.2118/114342-PA.

Wang, D., Han, P., Shao, Z. et al. 2008. Sweep-Improvement Options for the Daqing Oil Field. *SPE Res Eng* **11** (1): 18–26. SPE-99441-PA. https://doi.org/10.2118/99441-PA.

Wang, D., Hao, Y., Delamaide, E. et al. 1993. Result of Two Polymer Flooding Pilots in the Central Area of Daqing Oil Field. Presented at the SPE Annual Technical Conference and Exhibition, Houston, 3–6 October. SPE-26401-MS. https://doi.org/10.2118/26401-MS.

Wang, D., Wang, G., Xia, H. et al. 2011. Incremental Recoveries in the Field of Large Scale High Viscous-Elastic Fluid Flooding are Double that of Conventional Polymer Flooding. Presented at the SPE Annual Technical Conference and Exhibition, Denver, 30 October–2 November. SPE-146473-MS. https://doi.org/10.2118/146473-MS.

Wang, D., Zhang, J., Fanru, M. et al. 1995. Commercial Test of Polymer Flooding in Daqing Oil Field Daqing Petroleum Administrative Bureau. Presented at the SPE International Meeting on Petroleum Engineering, Beijing, 14–17 November. SPE-29902-MS. https://doi.org/10.2118/29902-MS.

Wang, Y., Pang, Y., Shao, Z. et al. 2013. The Polymer Flooding Technique Applied at High Water Cut Stage in Daqing Oilfield. Presented at the North Africa Technical Conference and Exhibition, Cairo, Egypt, 15–17 April. SPE-164595-MS. https://doi.org/10.2118/164595-MS.

Wellington, S. L. 1983. Biopolymer Solution Viscosity Stabilization—Polymer Degradation and Antioxidant Use. *SPE J.* **23** (6): 901–912. SPE-9296-PA. https://doi.org/10.2118/9296-PA.

Willhite, G. P. 1986. *Waterflooding,* Vol. 3. Richardson, Texas: Textbook Series, SPE.

Willhite, G. P. and Dominguez, J. G. 1977. Mechanisms of Polymer Retention in Porous Media. In *Improved Oil Recovery by Surfactant and Polymer Flooding*, ed. D. O. Shah, 511–514. New York, New York: Academic Press.

Willhite, G. P. and Jordan, D. S. 1981. Alteration of Permeability in Porous Rocks With Gelled Polymers. In *Polymer Reprints*, Vol. 22, No. 2, 53–58, American Chemical Society, Washington, DC (August 1981).

Willhite, G. P. and Uhl, J. T. 1988. Correlation of the Flow of Flocon® 4800 Biopolymer with Polymer Concentration and Rock Properties in Berea Sandstone. In *Water-Soluble Polymers for Petroleum Recovery*, eds. G. A. Stahl and D. N. Schultz, 101–119. New York, New York: Plenum Press. https://doi.org/10.1007/978-1-4757-1985-7_5.

Woods, P., Schramko, K., Turner, D. et al. 1986. In-Situ Polymerization Controls CO_2/Water Channeling at Lick Creek. Presented at the SPE/DOE Enhanced Oil Recovery Symposium, Tulsa, 20–23 April. SPE-14958-MS. https://doi.org/10.2118/14958-MS.

Xanthan Gum/Keltrol/Kelzan. 1974. Technical Bulletin X474, Kelco Company, San Diego, California, USA.

Yang, S. H. and Treiber, L. E. 1985. Chemical Stability of Polyacrylamide Under Simulated Field Conditions. Presented at the SPE Annual Technical Conference and Exhibition, Las Vegas, Nevada, USA, 22–25 September. SPE-14232-MS. https://doi.org/10.2118/14232-MS.

Zaitoun, A. and Potie, B. 1983. Limiting Conditions for the Use of Hydrolyzed Polyacrylamides in Brines Containing Divalent Ions. Presented at the SPE International Symposium on Oilfield and Geothermal Chemistry, Denver, 1–3 June. SPE-11785-MS. https://doi.org/10.2118/11785-MS.

Zaitoun, A., Tabary, R., Fossey, J. P. et al. 1998. Implementing a Heavy-Oil Horizontal-Well Polymer Flood in Western Canada. Presented at the Seventh UNITAR International Conference on Heavy Crude and Tar Sands, Beijing, 27–30 October. Manuscript No. 191.

SI Metric Conversion Factors

acre	×	4.046 873	E-01	=	ha
°API		141.5/(131.5 +°API)		=	g/cm³
bbl	×	1.589 873	E-01	=	m³
cp	×	1.0*	E-03	=	Pa·s
dynes/cm	×	1.0*	E+00	=	mN/m
ft	×	3.048*	E-01	=	m
ft²	×	9.290 304*	E-02	=	m²
ft³	×	2.831 685	E-02	=	m³
°F		(°F-32)/1.8		=	°C
in.	×	2.54*	E+00	=	cm
lbm	×	4.535 924	E-01	=	kg
md	×	9.869 233	E-04	=	µm²
psi	×	6.894 757	E+00	=	kPa
scf/bbl	×	1.801 175	E-01	=	std m³/m³

* Conversion factor is exact.

Chapter 6

Miscible Displacement Processes

6.1 Introduction

Miscible displacement processes are defined here as processes where the effectiveness of the displacement results primarily from miscibility between the oil in place and the injected fluid. Displacement fluids, such as hydrocarbon solvents, carbon dioxide (CO_2), flue gas, and nitrogen, are considered. Miscibility plays a role in surfactant processes, but is not the primary recovery mechanism for these processes, so the use of surfactants is not discussed in this chapter. Miscibility also plays a role in other processes that are basically immiscible, such as polymer-augmented waterflooding. These displacements also are not treated in this chapter. This organization is consistent with Stalkup (1983a).

The displacement processes treated here may be conveniently classified as first-contact miscible (FCM) or multiple-contact miscible (MCM) on the basis of the manner in which miscibility is developed. As implied, the former process involves injection of a displacement fluid that is miscible with the crude oil (i.e., it forms only a single phase upon first contact when mixed in all proportions with the crude oil). In the latter process, miscible conditions are developed in situ through composition alteration of the injected fluid or crude oil as the fluids move through the reservoir.

This chapter first discusses phase-behavior considerations related to miscible displacement processes. Requirements for an FCM process are presented, followed by a discussion of MCM processes. Ternary diagrams are used to describe conceptually the manner in which miscibility is achieved in the MCM processes. It will be shown that, for a specified fluid/reservoir system, the so-called minimum miscibility pressure (MMP) is an important parameter for these processes. Methods of determining this pressure, which rely on experimental measurements, empirical correlations, and equation-of-state (EOS) calculations, are presented. Finally, oil recovery mechanisms, methods of modeling the processes, general design principles, and several examples of commercial field applications are discussed.

6.2 General Description of Miscible Displacement

In an immiscible displacement process, such as a waterflood, the microscopic displacement efficiency, E_D, is generally much less than unity. Part of the crude oil in the places contacted by the displacing fluid is trapped as isolated drops, stringers, or pendular rings, depending on the wettability. When this condition is reached, relative permeability to oil is reduced essentially to zero and continued injection of the displacing fluid is ineffective because the fluid simply flows around the trapped oil. The oil does not move in the flowing stream because of capillary forces, which prevent oil deformation and passage through constrictions in the pore passages.

This limitation to oil recovery may be overcome by the application of miscible displacement processes in which the displacing fluid is miscible with the displaced fluid at the conditions existing at the displacing-fluid/displaced-fluid interface. Interfacial tension (IFT) is eliminated. If the two fluids do not mix in all proportions to form a single phase, the process is called immiscible.

Fig. 6.1 shows schematically an idealized FCM displacement process that involves the injection of a specified volume, or slug, of a solvent that is miscible with the crude oil. The solvent in Fig. 6.1 is a mixture of low-molecular-weight hydrocarbons [liquefied petroleum gases (LPGs)]. If the process is operated as a secondary recovery process, the oil is displaced efficiently ahead of the LPG slug, leaving little or no residual oil. Mixing and dispersion will occur at the solvent/oil interface, as described in Chapter 3, and a mixing zone with a concentration gradient develops.

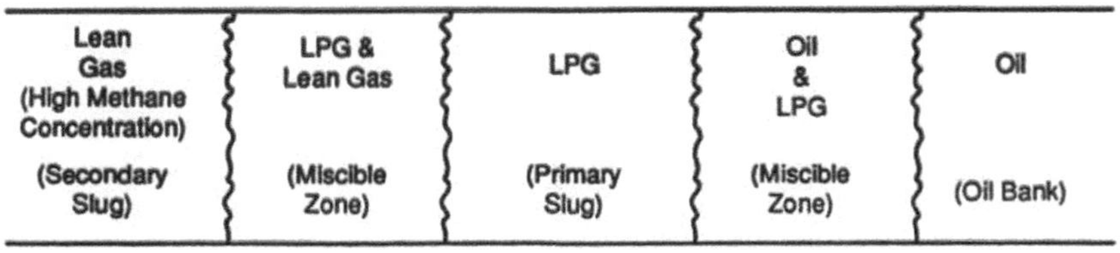

Lean Gas (High Methane Concentration)	LPG & Lean Gas	LPG	Oil & LPG	Oil
(Secondary Slug)	(Miscible Zone)	(Primary Slug)	(Miscible Zone)	(Oil Bank)

Fig. 6.1—Miscible displacement.

If the process is used as a tertiary process with the oil at waterflood residual saturation, the injected solvent must displace enough of the water phase to contact the residual oil and then displace the oil as a single-phase mixture with the solvent. Mixing of solvent and oil results in a mixture with a higher viscosity than the pure solvent, which makes water displacement more efficient. The mixing leads to the development of an oil bank, followed by a solvent/oil mixture bank that is rich in oil at the front end and rich in solvent at the back. As the displacement proceeds, the oil bank continues to grow, and oil is displaced through the reservoir as long as the integrity of the injected solvent slug is maintained; that is, as long as the fluid bank displacing the oil is miscible with the oil. The resulting microscopic displacement efficiency is much greater than for immiscible processes and can approach 100%.

In practice, solvents that are miscible with crude oil are more expensive than water or dry gas, and thus an injected solvent slug must be relatively small for economic reasons. For this situation, the primary (solvent) slug may be followed by a larger volume of a less-expensive fluid, such as water or a lean gas, as shown in Fig. 6.1. Ideally, this secondary slug should be miscible with the primary slug, thus yielding an efficient displacement of the primary slug. Under proper conditions of pressure, temperature, and composition, a lean gas that is high in CH_4 concentration will be miscible with LPG, as Fig. 6.1 illustrates. Again, a mixing zone will develop at the interface between the primary slug (LPG) and the secondary slug (lean gas).

When water, which is not miscible with the solvent, is used as the chase fluid a residual solvent saturation will be retained in the rock and the primary slug will deteriorate as it is displaced toward a producing well by the water. The displacing water typically has a relatively low mobility, which improves sweepout efficiency over that obtained with the solvent. For this reason, alternate injection of water and solvent, called a water-alternating-gas (WAG) process, often is used. This process was discussed in Chapter 5.

Fig. 6.1 indicates that only hydrocarbon is flowing. If the process is applied as a tertiary process after waterflooding, however, then water saturation will be above the waterflood residual saturation in at least part of the reservoir and water also will be flowing in that part of the reservoir.

In the MCM process, the oil and injected solvent are not miscible upon first contact at reservoir conditions. Rather, the process depends on modification of the oil or injected solvent compositions to such a degree that the fluids become miscible as the solvent moves through the reservoir. For example, under the proper conditions, injected methane, which is not miscible in all proportions with oil, might extract certain hydrocarbon components from the oil as flow occurs through the reservoir. The enriched methane slug could thus become miscible with oil. In this process, miscibility does not exist initially but is dynamically developed as the process continues. Such processes are sometimes called dynamic miscible processes because of the manner in which miscibility develops.

Various gases and liquids are suitable for use as miscible displacement agents in either FCM or MCM processes. These include low-molecular-weight hydrocarbons, mixtures of hydrocarbons (LPGs), CO_2, nitrogen, or mixtures of these. The particular application depends on the reservoir pressure, temperature, and compositions of the crude oil and the injected fluid. Miscibility development depends on the phase behavior of the system, which in turn depends on the factors cited.

6.3 Principles of Phase Behavior Related to Miscibility

Various methods exist to represent the vapor/liquid phase behavior of multicomponent systems. These methods include the use of pressure/temperature, pressure/composition, and ternary diagrams, which are discussed in this chapter. These diagrams provide a convenient way to present the boundaries of the single- and multiphase regions typically determined from experimental data or from calculations with an EOS.

A problem in dealing with crude oils is that they are complex fluids made up of numerous chemical components. Further, the exact chemical composition usually is not known because of the technical difficulty and expense involved in obtaining such information.

The Gibbs phase rule defines the number of degrees of freedom for a mixture of chemical components as (Glasstone 1946)

$$F = n - P + 2, \dots\dots\dots\dots\dots\dots\dots\dots\dots\dots\dots\dots\dots\dots\dots\dots\dots\dots\dots\dots\dots\dots (6.1)$$

where F = degrees of freedom, n = number of different chemical components, and P = number of phases.

According to the rule, F degrees of freedom (e.g., temperature, pressure, composition) must be specified before the phase behavior can be described completely. For a typical crude oil, F is a large number and thus rigorous specification of phase behavior is difficult or impossible. Therefore, the common practice is to use approximate methods, such as the pseudoternary diagram described in Chapter 2.

In what follows, simplified mixtures of two or three components are used in a number of instances to illustrate principles of phase behavior that relate to understanding miscible displacement. Stalkup (1983a), McCain (1973), and Sage and Lacey (1939) contain additional discussions of phase behavior.

6.3.1 Pressure/Temperature Diagrams. A pressure/temperature (p-T) phase diagram for a multicomponent hydrocarbon mixture, such as a reservoir oil, provides useful information. **Fig. 6.2** shows an idealized p-T diagram. Note that the diagram is for a fixed overall composition.

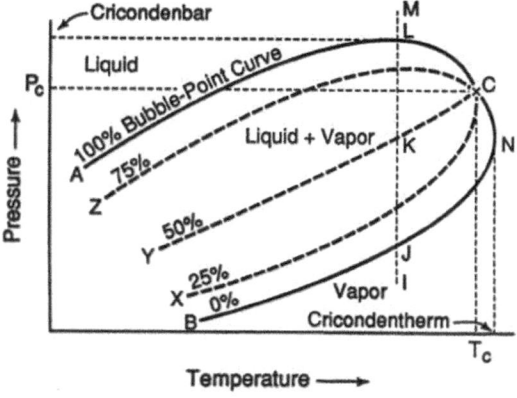

Fig. 6.2—Typical pressure/temperature diagram for a multicomponent fixed-composition system.

The bubblepoint and dewpoint curves meet at the critical point (Point C). The critical point is also the point at which properties of the liquid and gas become identical (McCain 1973). All points within Curve ACB consist of two phases. Two phases can exist at a pressure greater than the critical pressure and at a temperature greater than the critical temperature, unlike in a single-component system. For a multicomponent system, the maximum pressure at which two phases can exist in equilibrium is called the cricondenbar and the maximum temperature at which two phases can exist is called the cricondentherm.

If the system originally exists at Point M in Fig. 6.2 and undergoes an isothermal pressure reduction, the following phase changes occur.

1. The system originally exists as a single fluid phase with essentially liquid properties.
2. At Point L, the system changes to a saturated liquid phase with a single "bubble" of vapor present at negligible volume compared with that of the liquid (bubblepoint pressure).
3. As the pressure is reduced further, more vapor is formed until, at Point K, the system is a 50/50 vol% mixture of liquid and vapor.
4. Further pressure reduction to the Point J causes all the liquid to vaporize except for a single drop of negligible volume compared with that of the vapor (dewpoint pressure).
5. Still further pressure reduction causes the system to exist as a vapor without any liquid present.

Fig. 6.2 shows the general behavior for a fixed composition of a system consisting of two or more components. If the composition were changed, then the position of the two-phase envelope on the pressure/temperature plot would shift. This shift is shown in **Fig. 6.3,** which is a composite pressure/temperature diagram showing several different compositions of an ethane/n-heptane binary system. The behavior is typical of other binary systems. The dashed line is the locus of the critical point conditions for the different compositions. **Fig. 6.4** shows the critical point loci for various pairs of hydrocarbons in the n-paraffin series.

The locus of critical points, as Fig. 6.4 shows, lies at a higher pressure than the cricondenbar values taken from the different phase-boundary curves at constant composition. Thus, the locus of critical points on the p-T diagram specifies, at each temperature, the maximum pressure at which two phases can exist.

Data such as those in Figs. 6.3 and 6.4 indicate conditions at which mixtures of the binary pairs are miscible. For example, at a temperature of 300°F, a pressure greater than approximately 1,250 psia is required before all possible mixtures of ethane and n-heptane are miscible. At a lower pressure, there are possible concentrations at which the system will separate into two phases. Similar minimum pressures at other temperatures can be determined from Fig. 6.4.

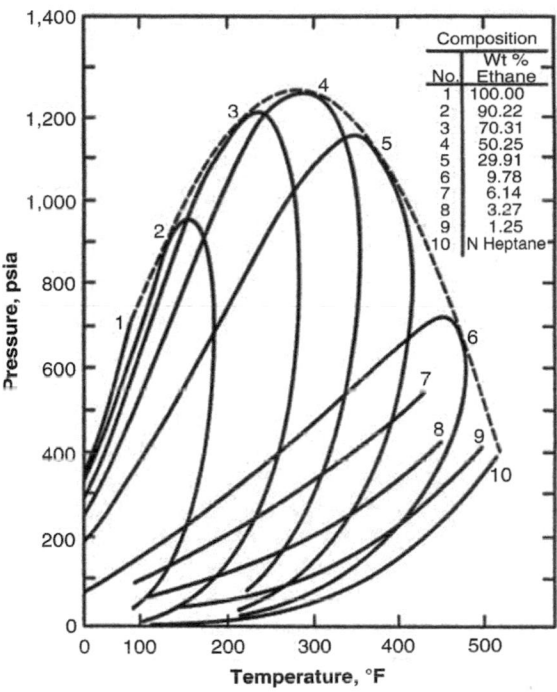

Fig. 6.3—Pressure/temperature phase diagrams for mixtures of ethane and n-heptane (Kay 1938).

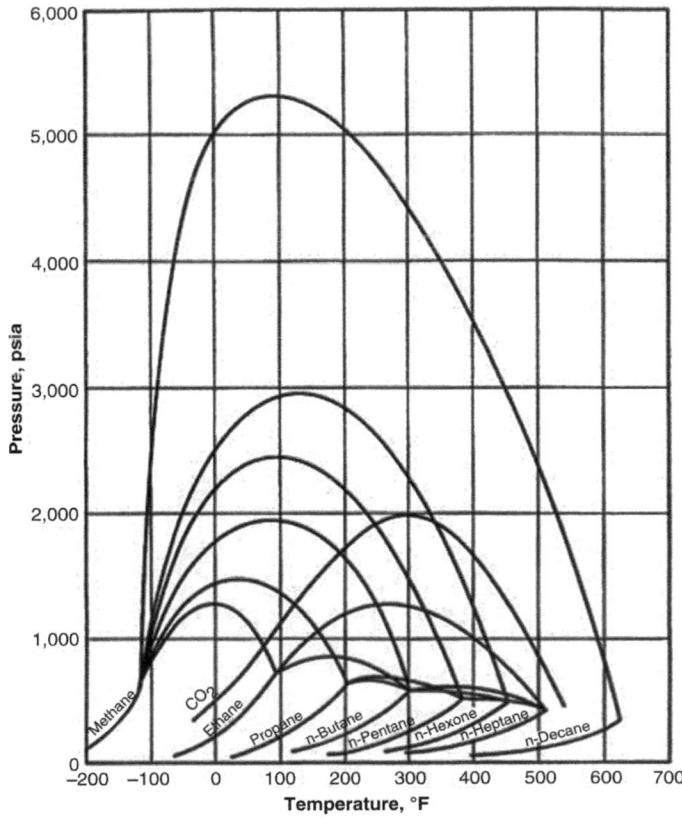

Fig. 6.4—Critical loci of binary n-paraffin systems (Brown et al. 1948).

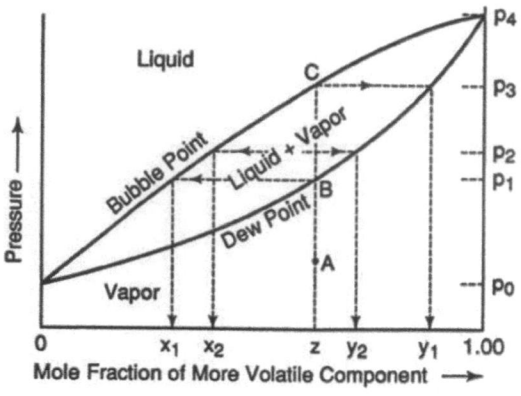

Fig. 6.5—Pressure/composition diagram illustrating isothermal compression.

For systems containing a larger number of components, the description is more complex because of the larger number of degrees of freedom and no maps of critical points typically are available.

6.3.2 Pressure/Composition Diagrams. Phase behavior information also can be presented on a pressure/composition (*p-x*) diagram such as that shown in **Fig. 6.5** for an idealized binary system. On the diagram, the composition is expressed as a mole fraction of the more volatile component. To illustrate the use of the diagram, the behavior of a system initially in the vapor phase subjected to an isothermal compression through the two-phase region is described.

If the pressure is increased on a system represented by Point A, no phase change occurs until the dewpoint (Point B) is reached at pressure p_1. At the dewpoint, an infinitesimal amount of liquid forms with a composition given by x_1. The composition of the vapor is still equal to the original composition, z. As the pressure is increased, more liquid forms and the compositions of the coexisting liquid and vapor phases are given by projecting the ends of the straight, horizontal line through the two-phase region of the composition axis. For example, at p_2, both liquid and vapor are present and the compositions are given by x_2 and y_2. At p_3, bubblepoint (Point C) is reached. The composition of the liquid is equal to the original composition, z. The infinitesimal amount of vapor still present at the bubblepoint has a composition given by y_3. Upon further compression, the system becomes totally liquid.

As previously indicated, the extremities of a horizontal line through the two-phase region represent the compositions of coexisting phases. However, the composition of a phase and the amount of a phase present in a two-phase system should not be confused. At the dewpoint, for example, only an infinitesimal amount of liquid is present, but it consists of finite mole fractions of the two components. An equation for the relative amounts of liquid and vapor in a two-phase system may be derived by applying the lever-arm rule in a manner similar to that discussed in Chapter 2.

For the system shown in Fig. 6.5, the binary pair would be miscible over all possible concentrations at pressures greater than p_4 (liquid) or less than p_0 (vapor). Between these pressures, the system would be single phase only over limited concentration ranges.

For a binary system at a temperature above the critical temperature of the more volatile component, two phases do not form over all compositions (0 to 100%) of the more volatile component. **Fig. 6.6** (Hutchinson and Braun 1961) shows data for one such system (methane/butane) at several temperatures. For this system at 100°F (lowest temperature in Fig. 6.6), two phases are not formed above a methane composition of approximately 87 mol%. The composition of methane above which only one phase exists decreases with increasing temperature. This accounts for the difference in appearance of the pressure/composition plots in Fig. 6.6. Comments about Fig. 6.5 are applicable to Fig. 6.6. The cricondenbars have been indicated on each phase envelope in Fig. 6.6. For a binary system, the cricondenbar pressure on a *p-x* diagram is also the critical pressure. At each specified temperature, the cricondenbar pressure is the maximum pressure at which two phases can exist. Above this pressure, all methane/*n*-butane mixtures will be a single phase (i.e., complete miscibility will exist).

Fig. 6.7 shows a pressure/composition diagram for a binary system consisting of *i*-butane and CO_2 at temperatures between 100 and 250°F (Besserer and Robinson 1975). At 160°F, for example, the system would be single phase at all compositions above approximately 1,090 psi.

A pressure/composition diagram for a three-component system consisting of methane, *n*-butane, and decane at a fixed temperature of 160°F is shown in **Fig. 6.8** (Stalkup 1983a; Hutchinson and Braun 1961). The abscissa is the mol% of methane, C_1, added to a reservoir-fluid mixture originally containing 30.8% C_1, 55.4% *n*-C_4, and 13.8% C_{10}. The loci of the bubble-point and dewpoint pressures are shown along with the locations of the cricondenbar pressure and plait point. As previously

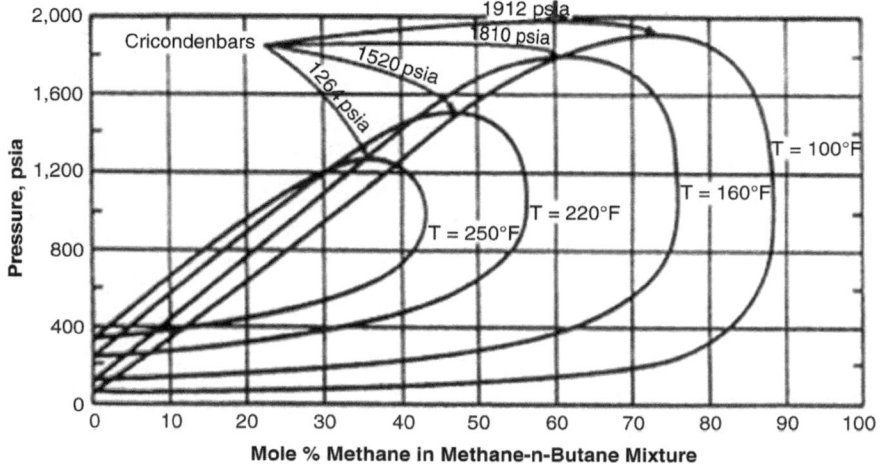

Fig. 6.6—Pressure/composition isotherms for the methane/*n*-butane system (Hutchinson and Braun 1961).

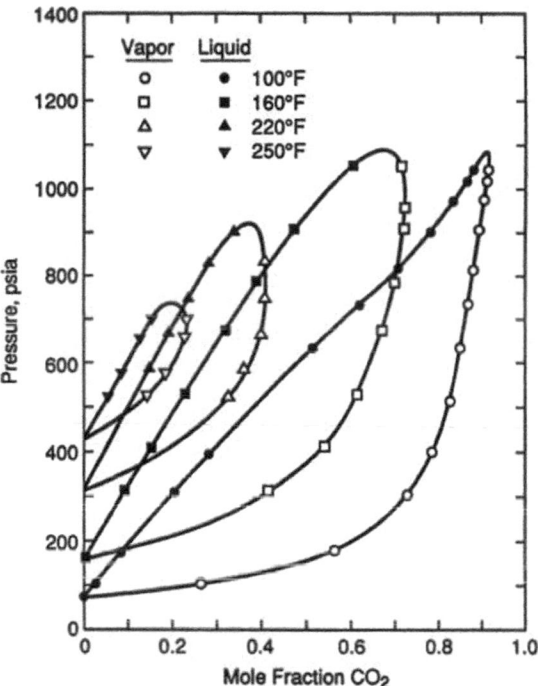

Fig. 6.7—Pressure-equilibrium phase composition diagram for *i*-butane/CO₂ system (Besserer and Robinson 1975).

stated, the plait point or critical point is that point where the bubblepoint and dewpoint curves meet and where the properties of the liquid and vapor phases become indistinguishable.

A single phase exists above the phase-boundary lines, while two phases exist below the phase boundaries. Conditions for miscibility to exist between C_1 and the reservoir fluid can be determined from the diagram. For example, if incremental amounts of C_1 were added to the reservoir fluid at a constant pressure of 2,000 psia (and a temperature of 160°F), the mixture of fluids would be single phase as long as the concentration of added C_1 was less than approximately 25 mol%. Further C_1 addition would cause a gas phase to separate from the liquid phase. However, if addition of C_1 continued and enough was added for the overall composition of gas to exceed approximately 97 mol%, then a single phase would form again as a gas. At the low-gas-concentration end, gas goes into solution in the liquid, while at the high-gas-concentration end, the liquid is vaporized into the gas.

If the pressure on the system were raised to a value equal to or higher than the cricondenbar pressure (3,250 psig), then addition of C_1 to the reservoir fluid would not cause two phases to form. The gas (C_1) and reservoir fluid would be single phase over all possible concentrations of added C_1.

Fig. 6.9 shows additional data on this system. Bubblepoint and dewpoint loci are shown for the addition of two different gases containing different amounts of C_4 to the original reservoir fluid. As can be seen, if a heavier component, such as C_4, is a component in the added gas phase, the bubblepoint and dewpoint curves shift to lower pressures. Miscibility between the added gas and reservoir fluid could be achieved at a lower pressure when C_4 is in the gas than for the gas that is 100% C_1.

Pressure/composition diagrams also can be developed for more-complex systems containing a larger number of components. The use of these diagrams is illustrated in Section 6.5.3.

6.3.3 Ternary Diagrams. Chapter 2 described the use of ternary diagrams to describe the phase behavior of three-component systems at constant temperature and pressure. The extension to multi-component systems and the use of pseudoternary diagrams also were discussed. **Fig. 6.10** gives an example of a true ternary system, and **Fig. 6.11** (Hutchinson and Braun 1961) gives an example of a pseudoternary system.

Concentration ranges over which a system is miscible at a specified temperature and pressure can be determined directly from the diagram. The overall concentration of mixtures of any two solutions lies along the straight line connecting the concentrations of the two solutions. Therefore, the inverse lever-arm rule can be used to determine the composition of the resulting mixture.

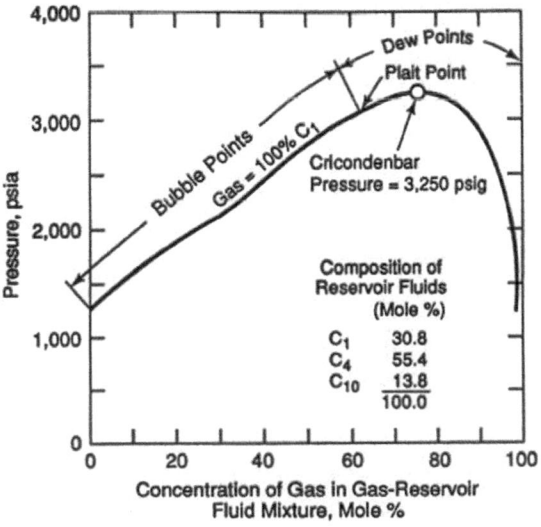

Fig. 6.8—Pressure/composition diagram for mixtures of C_1 with a C_1/n-C_4/C_{10} liquid (Hutchinson and Braun 1961).

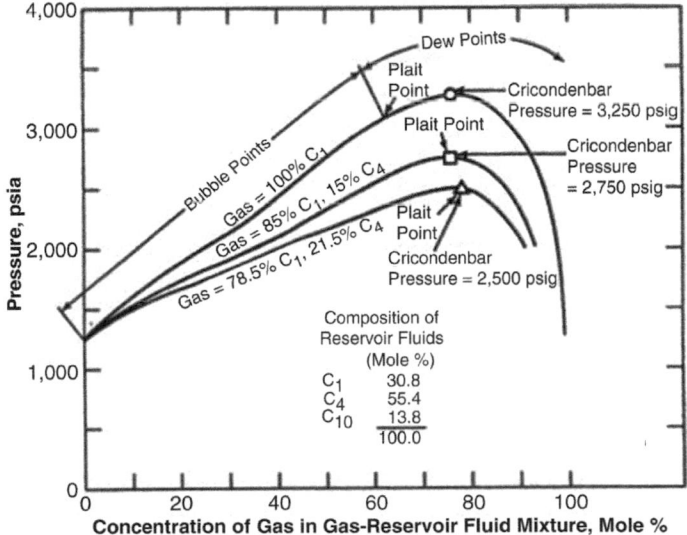

Fig. 6.9—Effect of gas composition on the cricondenbar of the gas (C_1 to n-C_4)/decane system at 160°F (Hutchinson and Braun 1961).

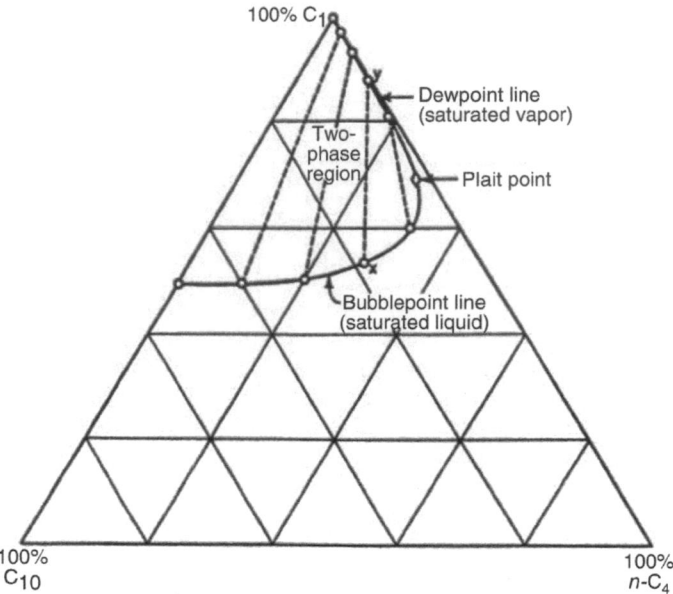

Fig. 6.10—Phase relations (mol%) for methane/n-butane/decane system at 160°F and 2,500 psia (Hutchinson and Braun 1961).

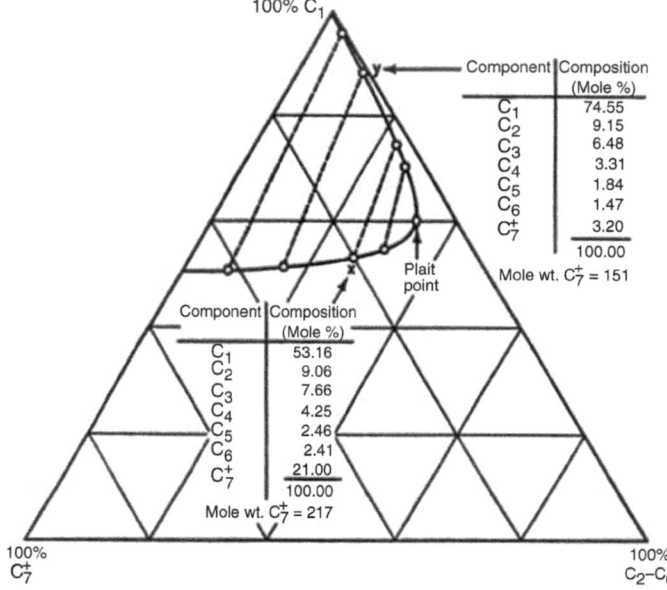

Fig. 6.11—Phase relations (mol%) of the University Block 31 field reservoir fluid at 140°F and 4,000 psia (Hutchinson and Braun 1961).

Fig. 6.10 shows the phase behavior for the ternary system of C_1, n-C_4, and C_{10} at 2,500 psia and 160°F. At these conditions, all C_1/n-C_4 mixtures are miscible, as are all n-C_4/C_{10} mixtures. Note that this is consistent with the data given in Fig. 6.4.

Mixtures of C_1 and C_{10} are not miscible over the total concentration range and neither are mixtures of all three components. Any mixture of components that yields an overall composition within the two-phase region is an immiscible mixture.

Similar conclusions can be made for a multicomponent system such as that shown in Fig. 6.11 as long as the pseudoternary representation is valid. Representing a system of more than three components on a ternary diagram is an approximation, however, as can be seen by noting the concentrations for the equilibrium vapor and liquid phases given by Points x and y on the binodal curve of Fig. 6.11. Ideally, a pseudocomponent, such as C_2 through C_6 should have the same relative composition in both phases, but this did not occur. The lighter components in the C_2 through C_6 pseudocomponent tended to go to the vapor phase, while the heavier components moved to the liquid phase. This movement can be seen, for example, by noting the ratio of C_2/C_6 in the vapor and liquid phases. The ratio is significantly larger in the vapor phase than in the liquid. Nonetheless, a pseudoternary diagram is often a useful approximation, especially for describing how miscibility is achieved in displacement processes.

Changing either temperature or pressure of a system modifies the binodal curve (i.e., changes the bubblepoint and dewpoint curves). **Fig. 6.12** (Hutchinson and Braun 1961) illustrates this alteration for the same ternary system previously discussed (Fig. 6.10). An increase in pressure from 2,500 to 3,250 psia at a constant temperature of 160°F causes a reduction in the overall concentration area encompassed by the two-phase envelope. At the higher pressure, the system is single phase over a broader range of possible concentrations (i.e., a higher pressure is favorable for the development of miscibility between different components).

6.4 FCM Process

An FCM process normally consists of injecting a relatively small primary slug that is miscible with the crude oil, followed by injection of a larger, less expensive secondary slug. Economic considerations are important in determining the slug sizes. Ideally, the secondary slug should be miscible with the primary slug; thus, phase behavior must be considered at both leading and trailing edges of the primary slug. If the different slug materials are not miscible, then a residual saturation of the primary slug material will be trapped in the displacement process.

Clark et al. (1958) illustrated the general conditions of first-contact miscibility and immiscibility, as shown in **Figs. 6.13a through 6.13c.** Methane and crude oil are partially soluble in one another. At conditions typically encountered in a reservoir, however, they do not mix in all proportions and two phases exist. The displacement of crude oil by methane at the reservoir conditions of Fig. 6.13a would thus be an immiscible flood.

On the other hand, higher-molecular-weight hydrocarbons, such as propane or LPG, are completely soluble with oil at most reservoir conditions. As shown in Fig. 6.13b, oil exists as a liquid and propane exists as a gas at atmospheric conditions. If the temperature and pressure are increased to the indicated reservoir conditions, however, the propane exists as a liquid (see Fig. 6.4), as does the crude oil. These two liquids will mix completely and are thus miscible. The displacement of oil with propane at the specified conditions would be a miscible process.

Fig. 6.13c gives a third example considering propane and methane. Both exist as a gas at atmospheric conditions. At the reservoir conditions of 150°F and 2,000 psi, methane still appears to be a gas (actually, it is a fluid above its critical point), while propane is a liquid. If the two hydrocarbons are brought into contact, they will mix in all proportions and appear to be a single-phase gas. The mixture is more properly called a fluid because it is above its critical point. Thus, the displacement of propane by methane at the specified conditions would be a miscible displacement. This behavior is verified by noting the locus of the critical points for mixtures of methane and propane in Fig. 6.4.

The determination of reservoir pressures required for FCM displacements is further illustrated in Examples 6.1 and 6.2.

Example 6.1—Determination of Pressure Required for First-Contact Miscibility With a Butane Primary Slug. An FCM displacement is to be designed in which a slug of butane is the primary displacing solvent. The butane slug is to be displaced by dry gas consisting essentially of methane. Assume that the crude oil is represented, in a phase-behavior sense, by n-decane and reservoir conditions are 160°F and 2,500 psi. The data presented in Figs. 6.4 and 6.10 are assumed to apply.

Determine whether miscible conditions would exist at the front and back of the solvent (butane) slug.

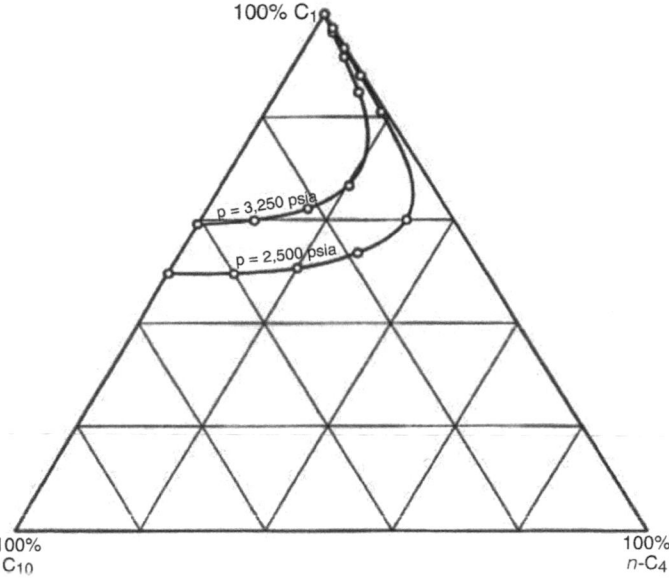

Fig. 6.12—Effect of pressure on the phase behavior for the methane/ n-butane/decane system (Hutchinson and Braun 1961).

Solution. The primary slug is butane (C_4) displacing oil (C_{10}). Refer to Fig. 6.4. The locus of critical points for C_4/C_{10} is not given. At 160°F and 2,500 psi, however, C_4 is in a liquid state. C_4 and C_{10} will be miscible. Miscibility is also indicated on the ternary phase-behavior diagram (Fig. 6.10). The secondary slug is methane (C_1) displacing butane (C_4). In Fig. 6.4, at 160°F, a pressure of 2,500 psia is well above the locus of critical points for C_1/C_4 mixtures. Therefore C_1 and C_4 will be miscible. Again, miscibility also is indicated in Fig. 6.10.

At 2,500 psia and 160°F, the proposed displacement will be miscible at both the leading and trailing edges of the butane slug. If the reservoir pressure were 1,500 psi at 160°F, then the displacement of butane by methane would be at conditions below the critical locus. At some point in the process at these conditions, two phases would form and the dry gas would

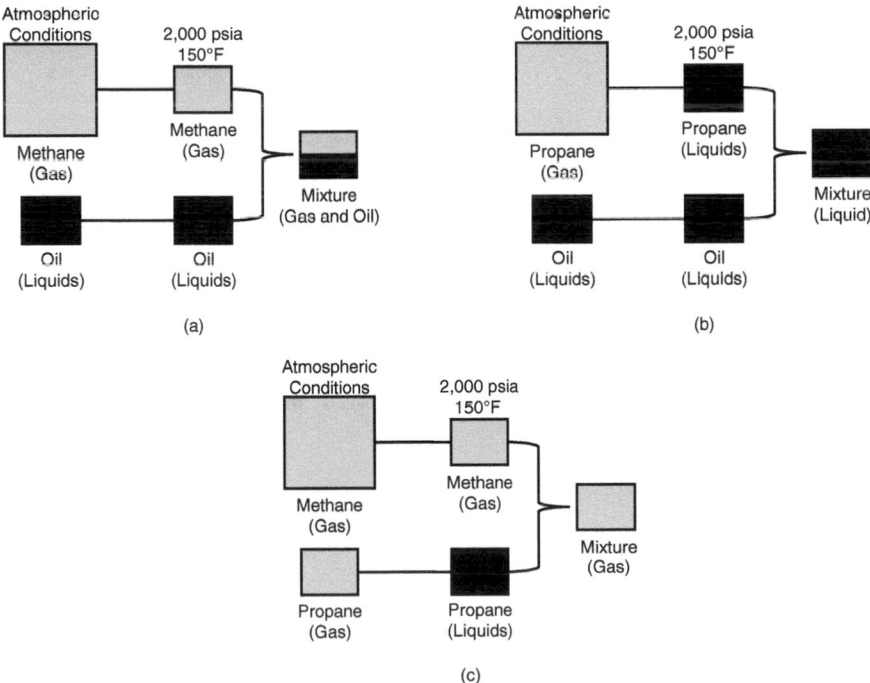

Fig. 6.13—(a) Immiscible two-phase mixture of methane gas and oil liquid at typical reservoir conditions (Clark et al. 1958). (b) Miscibility of propane (or LPG) liquid and oil liquid at reservoir temperature and pressure conditions. Here, propane (or LPG) liquid is a liquid in the presence of a liquid (Clark et al. 1958). (c) Miscibility of methane gas and propane (or LPG) liquid at reservoir temperature and pressure conditions. Here, propane (or LPG) is a gas in presence of gas above the critical point (Clark et al. 1958).

displace the butane immiscibly at a reduced microscopic displacement efficiency (i.e., butane would be bypassed by the methane and trapped in the reservoir). This would result in a relatively rapid degradation of the integrity of the primary butane slug and dramatically influence oil production.

Example 6.2—Determination of First-Contact Miscibility Pressure With Methane as Primary Slug. Suppose the oil of Example 6.1 (decane) is to be displaced by methane at a reservoir temperature of 160°F. What pressure would be required to achieve first-contact miscibility? The data presented in Fig. 6.4 are assumed to apply.

Solution. The primary slug is methane (C_1) displacing decane (C_{10}). Refer to Fig. 6.4. At 160°F, the critical pressure is approximately 5,200 psia. A higher pressure would be required for the C_1/C_{10} system to be single phase over the total concentration range from 100% C_1 to 100% C_{10} and thus to be miscible.

The results of Example 6.2 indicate why methane is not usually applied as the primary slug in an FCM displacement. The pressure required to achieve miscibility typically exceeds a pressure that can be reached in the reservoir (because of the lower pressure at which formation rock would fracture).

Phase behavior and conditions for first-contact miscibility can also be shown on a ternary or pseudoternary diagram (Stalkup 1983a). Three different example conditions are shown in **Fig. 6.14.** Assume that the primary slug is a mixture of intermediate-molecular-weight hydrocarbons (C_2 through C_6). The secondary slug is a light hydrocarbon (C_1) or nitrogen (N_2) and the crude oil is a mixture, as indicated on the pseudoternary diagram.

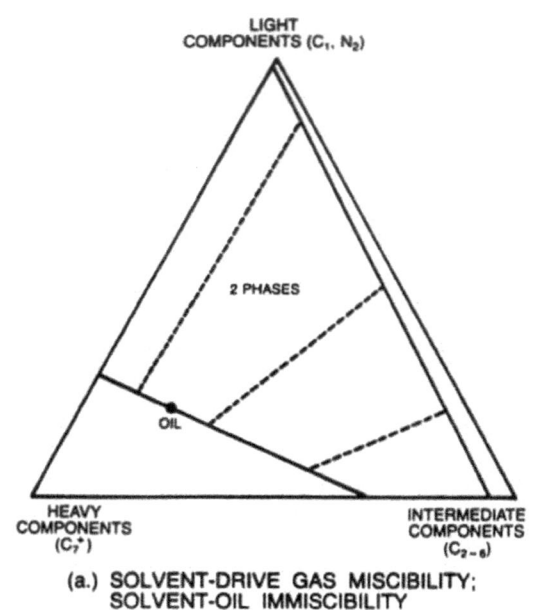

(a.) SOLVENT-DRIVE GAS MISCIBILITY;
SOLVENT-OIL IMMISCIBILITY

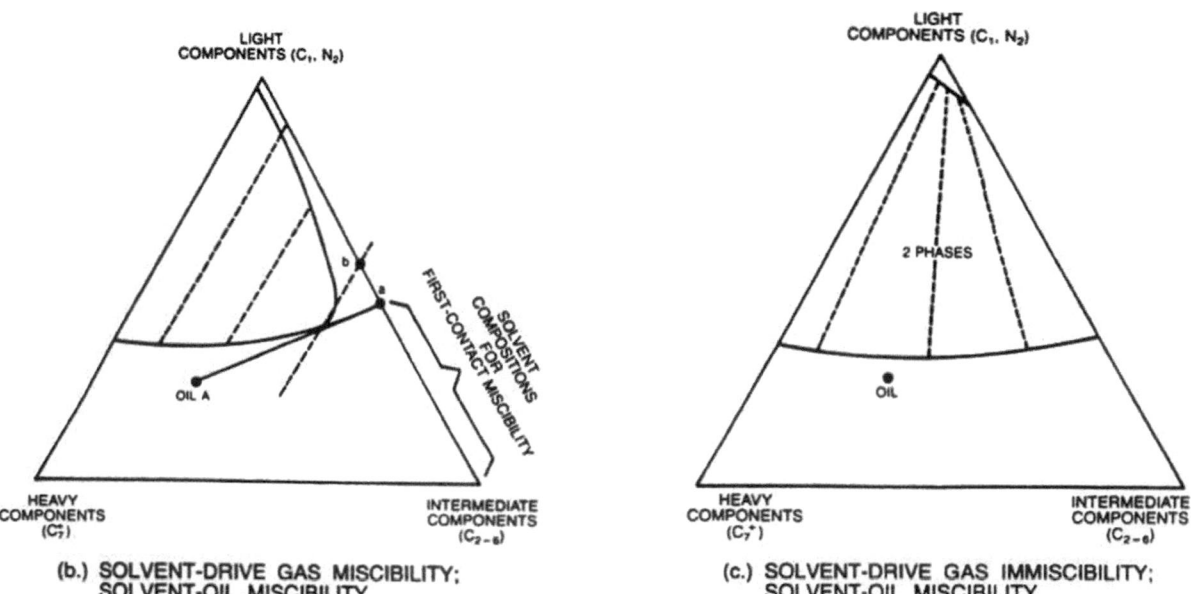

(b.) SOLVENT-DRIVE GAS MISCIBILITY;
SOLVENT-OIL MISCIBILITY

(c.) SOLVENT-DRIVE GAS IMMISCIBILITY;
SOLVENT-OIL MISCIBILITY

Fig. 6.14—Phase-behavior considerations for first-contact miscibility; pseudoternary diagrams (Stalkup 1983a).

In Fig. 6.14a, miscibility exists between the primary and secondary slug materials but not between the primary slug and crude oil. The two-phase boundaries intersect the base of the triangle, indicating that some mixtures of (C_2 through C_6) and C_{7+}, and C_2 through C_6 and the reservoir oil will form two phases.

Fig. 6.14c shows the phase boundaries for a system in which miscibility exists between the primary slug and the reservoir oil. However, mixtures of the primary and secondary slugs form two phases over a wide range of compositions.

Finally, Fig. 6.14b shows a phase diagram for which miscibility exists between the primary slug and crude oil and between the primary and secondary slugs. This condition is required for an ideal first-contact displacement process. Also, for the system described by Fig. 6.14b, the primary slug consisting of a mixture of C_2 through C_6 could be diluted significantly with the light component and still be miscible with the crude oil. Point a on the diagram gives the maximum composition of light component in a primary slug that would be miscible with Oil A. Fig. 6.14b shows the range of primary slug compositions that would be miscible with Oil A.

As described in Example 6.1, a basic concern in the design of a process is the phase behavior between the primary slug and the crude oil and between the primary slug and the secondary slug fluid that displaces the primary slug. In propane, butane, or LPG displacements, the phase behavior of the latter is generally the controlling factor. **Fig. 6.15** shows loci of cricondenbars as functions of temperature for methane and four potential primary-slug materials. As previously stated, at a specified reservoir temperature and for a given fluid pair, the cricondenbar pressure is the minimum reservoir pres-

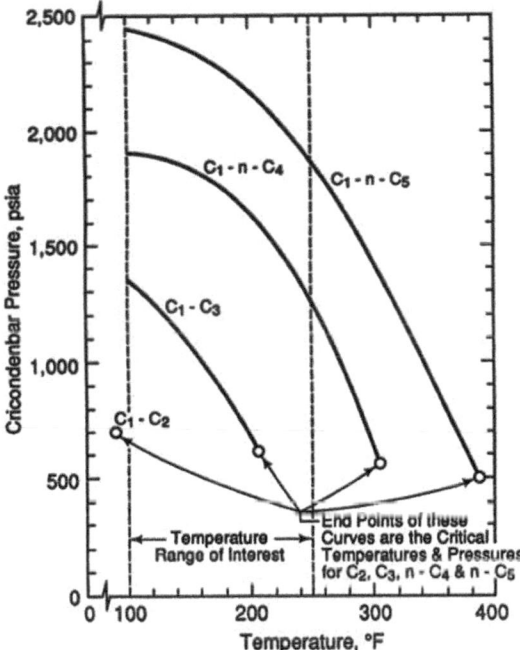

Fig. 6.15—Loci of the cricondenbars for methane and potential slug materials C_2, C_3, n-C_4, and n-C_5 (Hutchinson and Braun 1961).

sure that would ensure miscibility between the primary- and secondary-slug materials. If LPG consisting of a mixture of hydrocarbons were used as the primary slug, data for the specific LPG fluid and methane would be required to determine the cricondenbar pressure at the reservoir temperature.

Although the pressure required for miscibility between the primary slug and crude oil usually is not the controlling pressure, the phase behavior should still be considered. When the reservoir temperature is below the critical temperature of the primary-slug material, the pressure must be just high enough to liquefy the primary slug. If the hydrocarbon slug is liquid at reservoir conditions, it will be miscible with the oil. (Asphaltene deposition, which is sometimes a problem, is discussed below.) Typically, the hydrocarbon slug is liquid, as Fig. 6.4 shows. The critical temperature of propane is 206°F and that of butane is 304°F.

When the reservoir temperature is above the critical temperature of the primary slug solvent, the pressure required for complete miscibility between the slug solvent and the reservoir oil becomes more difficult to estimate. Under these conditions, the solvent cannot be liquefied and the pressure must be above the cricondenbar for the primary-slug/reservoir-oil system at the reservoir temperature. The cricondenbar depends on the reservoir temperature and the composition of the crude oil. Cricondenbar data for such systems as propane, butane, or LPG and crude oils are typically not available over all composition ranges of interest. Neither are data to prepare a pseudoternary diagram, such as that shown in Fig. 6.11, usually available. Clearly, if such data are available, they can be used to determine whether the system will be miscible at reservoir conditions. For the system shown in Fig. 6.11, if the primary slug were represented by the C_2 through C_6 component (LPG) at that apex of the diagram, then the primary slug would be miscible with both the crude oil and a secondary slug of 100% methane.

According to Stalkup (1983a), when phase-behavior data are required for the primary-slug/crude-oil system, it is usual to obtain pressure/composition (p-x) data at the reservoir temperature, and p-x diagrams are determined by use of high-pressure visual cells. In the experiments, measured amounts of the slug material are added to the reservoir oil and dewpoints and bubblepoints are measured. Vol% liquid also can be determined as a function of pressure at each composition. The application of such pressure/composition diagrams is illustrated later in this chapter.

A primary slug that is a liquid is not always completely miscible with a reservoir oil. Such solvents as propane and butane are known to cause the precipitation of asphalts from certain crude oils. Propane has been used as a deasphalting agent in refineries (Hutchinson and Braun 1961). If a significant amount of the oil displaced miscibly in a reservoir is deasphalted by the slug material, the asphalt could plug some of the pores and reduce the effective rock permeability. When the oil is displaced by the primary-slug solvent in a porous medium, however, a transition-zone fluid is formed between the main body of the slug and the undeasphalted reservoir oil. This fluid is miscible with the slug solvent at the rear of the transition zone and with the reservoir oil at the front of this zone. Under this condition, the miscible displacement process should proceed without appreciable deasphalting. The possibility of permeability loss from plugging by asphaltene should be recognized, however. Asphaltene deposition in the vicinity of production or injection wells, in particular, can be a serious problem.

Another consideration that relates to phase behavior is the possibility that mixing will occur between all three fluids involved in a displacement—i.e., between reservoir oil, the primary slug, and the secondary slug. If no mixing or dispersion of reservoir fluids occurs, then ideally an infinitesimally small volume of primary solvent is required. Significant

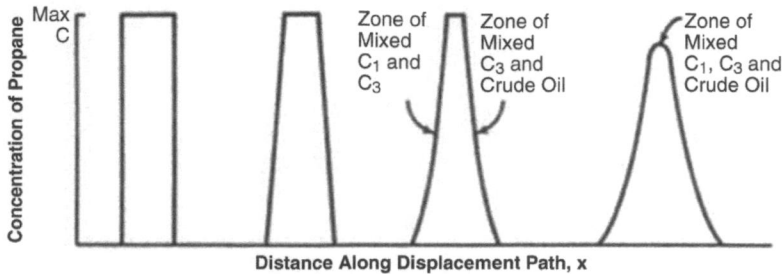

Fig. 6.16—Idealized concentration profiles for a miscible displacement process as the miscible slug advances through the reservoir.

dispersion does occur in reservoirs, however, and the process design must consider this. The effect of dispersion is to dilute the primary slug by both the reservoir oil and the secondary slug. **Fig. 6.16** shows an idealized concentration profile in which propane is displacing oil and is being displaced by methane. The mixed zone grows as flow proceeds through the reservoir. At some point, methane could "break through" the propane slug and a zone containing three components would be formed (assuming oil can be treated as a single pseudocomponent). The result could be the formation of two phases and deterioration of the process to an immiscible displacement. Viscous instabilities also contribute significantly to mixing and slug deterioration. Methods of quantitatively treating the mixing and dispersion phenomena are described in Chapter 3.

6.5 MCM Process

An MCM displacement process is one in which the condition of miscibility is generated in the reservoir through in-situ composition changes resulting from multiple contacts and mass transfer between reservoir oil and injected fluid. The multiple-contact processes are classified as vaporizing-gas (lean-gas) displacements, condensing and condensing/vaporizing-gas (enriched-gas) displacements, and CO_2 displacements.

In the vaporizing-gas process, the injected fluid is generally a relatively lean gas; i.e., it contains mostly methane and other low-molecular-weight hydrocarbons (or sometimes inert gases, such as nitrogen). In this approach, the composition of the injected gas is modified as it moves through the reservoir so that it becomes miscible with the original reservoir oil. That is, the injected fluid is enriched in composition through multiple contacts with the oil, during which intermediate components in the oil are vaporized into the injected gas. Under proper conditions, this enrichment can be such that the injected fluid of modified composition will become miscible with the oil at some point in the reservoir. From that point on, under idealized conditions, a miscible displacement will occur.

In the condensing, or enriched-gas, process the injected fluid generally contains larger amounts of intermediate-molecular-weight hydrocarbons and thus is more expensive. In this approach, reservoir oil near the injection well is enriched in composition by contact with the injected fluid first put into the reservoir. Hydrocarbon components are condensed from the injected fluid into the oil and thus the process is called a condensing process. Under proper conditions, the oil will be sufficiently modified in composition to become miscible with additional injected fluid and a miscible displacement will ensue. The enriched-gas process typically can be operated at a lower pressure than the vaporizing process.

The enriched-gas process was long thought to operate mechanistically, as described in the preceding paragraph. More recently it has been recognized that the process is often one of a combination of condensing and vaporizing mechanisms (Zick 1986). The light intermediate components in the injected gas (C_2 through C_4) condense into the reservoir oil as previously described. However, middle intermediate components (C_{4+}) are vaporized from the oil into the gas phase, and this prevents the development of miscibility between fresh injected gas and enriched oil at the entry point of the injection process (the oil becomes heavier). As the injection continues, the condensation of the light intermediates contained in the injected gas into the first oil contacted diminishes because this oil becomes saturated. But, the vaporization of the middle intermediates continues, with the result that the injected gas is slightly enriched (as in a vaporizing process). As this condensation/vaporization process moves downstream, the gas becomes enriched to a greater extent because it contacts more oil. The enrichment occurs to the point that the gas "nearly" becomes miscible with the original oil and results in an efficient displacement even though miscibility is never completely developed (i.e., the two phases are never miscible in all proportions).

CO_2 is not FCM with most crude oils at normal reservoir temperatures and pressures. However,

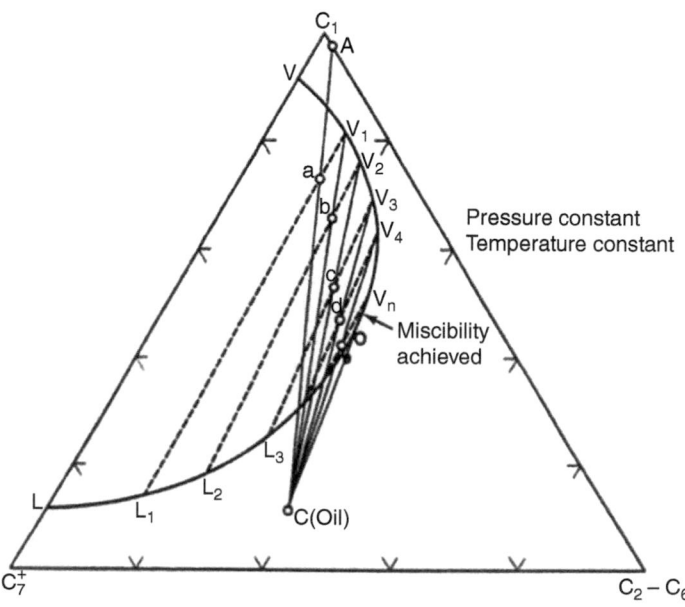

Fig. 6.17—Representation of the vaporizing-gas displacement process on a pseudoternary diagram, development of miscibility.

in-situ development of miscibility will occur under proper conditions of pressure, temperature, and composition. In general, the process behavior is analogous to the vaporizing process. Under some conditions, however, the phase behavior is more complex and two liquid phases or two liquid phases and a vapor phase form.

6.5.1 Vaporizing-Gas (Lean-Gas) Displacement Process. The vaporizing process may be described on a pseudoternary diagram, as shown in **Fig. 6.17.** The apexes of the diagram represent 100% concentration of methane (C_1), intermediate hydrocarbons (C_2 through C_6), and heavy hydrocarbons (C_{7+}), respectively. Temperature and pressure are constant and equal to the displacement conditions in the reservoir. Phase Boundary VO is the saturated vapor curve, and Phase Boundary LO is the saturated liquid curve. Point O is the critical or plait point.

Fig. 6.17 is a pseudoternary diagram, as described in Chapter 2, because certain components in the system are combined to form pseudocomponents. These are C_2 through C_6 and C_{7+}. As previously discussed, for this approximation to be valid, the individual components in each pseudocomponent should partition into the different phases in the same proportions as they exist in the pseudocomponent. This does not happen precisely, but the approach to this behavior is often sufficiently close that the use of the pseudoternary diagram is justified. In particular, a pseudoternary diagram is useful in describing the development of miscibility in a multiple-contact process.

The process descriptions that follow are based on the assumption that local thermodynamic equilibrium exists between the different phases. This assumption is generally thought to be valid at reservoir displacement conditions where advance rates are very low (except in the immediate vicinity of wellbores). As discussed later, at high water saturations, the water can block contact between the solvent and crude oil in strongly water-wet rocks and mass-transfer effects can be important (Stalkup 1970).

Assume that the reservoir crude oil is represented by Point C in Fig. 6.17. The injected fluid is shown as Point A and is a mixture consisting mostly of methane and much smaller amounts of the intermediate components. The injected fluid would be classified as a dry gas. From the locus of critical points in Fig. 6.4, we see that very high pressures would be required to generate first-contact miscibility at typical reservoir temperatures. However, miscibility may be created in a dynamic fashion in the reservoir. The process operates conceptually as follows (see Fig. 6.17).

1. At the leading edge of the displacement front, Gas A mixes with Oil C. The resulting composition of the mixture is along Line AC, say at Point a.
2. Mixture a is in the two-phase region and separates into a vapor, V_1, and a liquid, L_1.
3. Vapor V_1 moves ahead of Liquid L_1 and contacts fresh oil of Composition C. The resulting mixture is along Line V_1C, say at Point b.
4. Mixture b separates into Vapor V_2 and Liquid L_2.
5. The process continues with the vapor-phase composition changing along the saturated vapor curve, V_3, V_4...etc.
6. Finally, at Point e, the vapor becomes miscible with Oil C because the mixing line lies entirely in the single-phase region.
7. From this point, as the displacement continues the process will be miscible for an ideal process. In fact, as a result of mixing in the reservoir, miscibility may be successively lost and redeveloped.

Miscibility will not be generated for all combinations of injected gases and crude oils. An example in which miscibility does not result is shown in **Fig. 6.18.** For this example, the reservoir oil is assumed to have the composition at Point B. As before, assume Gas A mixes with Oil B along Line AB, yielding Mixture a, which gives Vapor V_1 and Liquid L_1. Vapor V_1 moves ahead and mixes with the new Oil B, which gives V_2 and L_2. In this case, however, as the process continues the vapor will change by successive mixing only to Point V_3 and then further enrichment will not occur. The composition change ceases because Line V_3L_3, when extended, passes through Point B, the oil composition. That is, mixtures of V_3 and B will be along the tie-line and further enrichment of the vapor by the intermediate and heavy components will not occur. The process will continue as an immiscible process.

The distinction between processes that will and will not be miscible is determined by the so-called limiting tie-line, illustrated by Line mn in Fig. 6.18. This line is tangent to the binodal curve at Critical Point O.

The process described is called a vaporizing-type process. The vaporizing term derives from the fact that the injected dry gas is enriched by vaporizing intermediate and heavy components from the oil. The general effect of pressure on the process is illustrated in Fig. 6.12. As pressure is increased, the two-phase envelope becomes progressively smaller and the critical point shifts toward lower concentrations of intermediates. The limiting tie-line thus shifts toward lower concentrations of intermediates in the

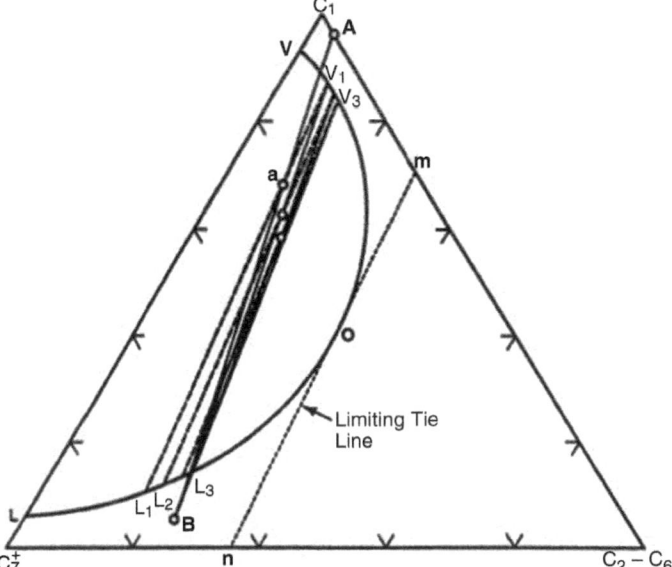

Fig. 6.18—Representation of the high-pressure process on a pseudoternary diagram; miscibility does not develop.

oil, and more oils develop miscibility with a dry gas of specified composition, such as Gas A. The minimum pressure at which the limiting tie-line just passes through the reservoir oil composition is called the MMP. That is, the MMP is the minimum pressure at which in-situ miscibility can be achieved in the MCM process for a specified fluid system. In the vaporizing MCM process, the injected gas composition must lie to the left of and the reservoir oil composition must lie on or to the right of the limiting tie-line, as shown in Fig. 6.17. Relatively high pressures are required for most oils and thus the process is sometimes designated a high-pressure process. Typical applications of this process are at pressures well above 3,000 psi.

This described development of miscibility is an idealized picture of the events at the leading edge of the displacement front. Behind the front, the situation is essentially the same. In the region of developing miscibility, the oil-phase composition at a fixed point moves along the lower part of the binodal curve toward higher concentrations of the C_{7+} components, as shown in Fig. 6.17. Thus, the oil at a fixed position is being depleted of its lighter components.

There is a transition zone in MCM displacement processes because a finite distance is required for the gas-phase composition to be modified to create a miscible condition. Data from laboratory experiments suggest that this distance is short (no more than a few feet in laboratory cores or sandpacks) (Negahban et al. 1990; Yellig 1982) relative to reservoir dimensions.

There can be a liquid-phase dropout in the vaporizing process. Gas-phase composition behind the front (i.e., between the miscible front and the injection point) changes along the binodal curve $V_1V_2V_3$. Mixing of any two points yields compositions that are just slightly inside the two-phase envelope, as illustrated in **Fig. 6.19.** Application of the inverse lever-arm rule illustrates that the liquid dropout is relatively small.

Thus, even in the most idealized processes, phase-behavior considerations indicate that liquid will remain in a reservoir for at least two reasons. First, some liquid will not be displaced in the region in which compositions are changing and miscibility is being generated. In practice, this is a negligibly small volume. Second, a small amount of liquid will be lost as a result of liquid dropout behind the front. In practice, this loss also should be small and is even reduced somewhat by subsequent vaporization in the process.

6.5.2 Condensing and Condensing/Vaporizing-Gas (Enriched-Gas) Displacement Process. In the condensing process, the injected fluid contains significant amounts of intermediate components (C_2 through C_6), rather than being a dry gas. The process depends on the condensation of these components into the reservoir oil, thereby modifying the oil composition. The modified oil then becomes miscible with the injected fluid.

The process is illustrated in **Fig. 6.20.** The fluid to be injected is represented by Point A and the reservoir oil by Point C. Note that injection of a relatively dry gas would not generate miscibility because the oil lies well to the left of the limiting tie-line.

1. The idealized sequence of events for the condensing process is as follows, where again local thermodynamic equilibrium is assumed to exist.
2. The injected Fluid A mixes with Oil C. The resulting composition is along Line AC, say at Point a.
3. Mixture a is in the two-phase region and separates into Vapor V_1 and Liquid L_1.

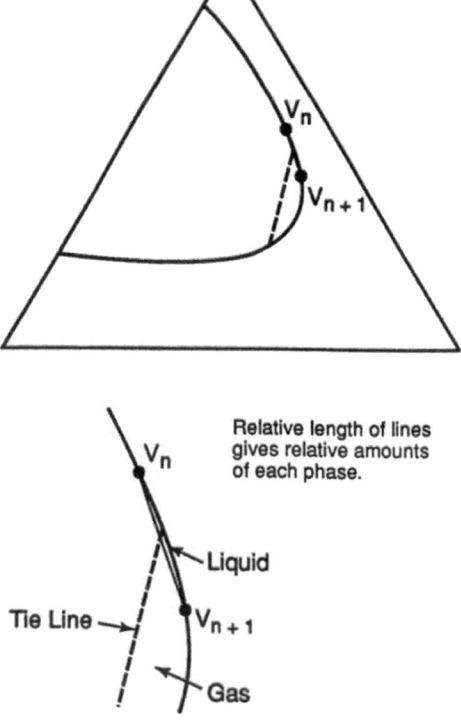

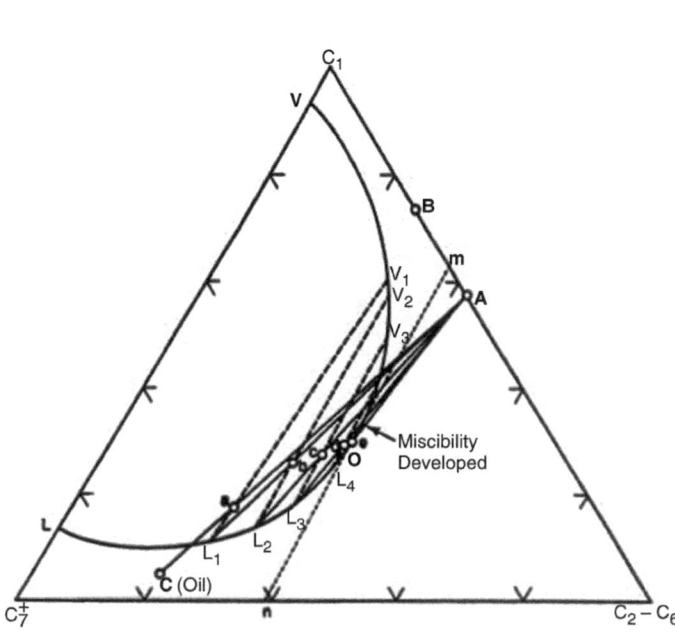

Fig. 6.19—Phase behavior behind the point of miscibility indicating liquid-phase dropout.

Fig. 6.20—Representation of the enriched-gas (condensate) process on a ternary diagram.

4. Vapor V_1 moves ahead of Liquid L_1. For ease of conceptualization, it might be viewed that L_1 is immobile at this point in the process. Liquid L_1 mixes with additional injected Gas A and the resulting fluid is along Mixing Line AL_1, say at Point b.
5. Mixture b separates into Vapor V_2 and Liquid L_2.
6. The process continues with the oil-phase composition changing from L_2, L_3, $L_4...L_n$.
7. For the tie-line through Mixture e the immobile oil, L_n, becomes miscible with the injected fluid because Mixing Line AL_n, will lie entirely in the single-phase region.
8. From this time, the process will become miscible and will continue as a miscible process because the mixture resulting from mixing along Line AL_n will be miscible with the fluids ahead of this point.

For this ideal conceptualization of the condensing-type process, a composition transition zone exists ahead (downstream) of the point where miscibility is achieved. This results because the injected gas continues to condense partially into the oil as the injected gas moves downstream. Once miscibility is generated, the transition zone stabilizes. Analogous to behavior behind the front in the vaporizing-gas process, where a small amount of liquid condensation occurs, in the condensing-gas process, there is a small amount of vaporization in the transition zone downstream of the point of miscibility.

The limiting Tie-Line mn is shown in Fig. 6.20. In the example, Fluids A and C are on opposite sides of Tie-Line mn. If the injection fluid were at Point B, to the left of this line, miscibility would not be achieved. This is because, at some point in the process, extension of one of the L_n V_n lines would pass through Point B and further modification of the oil-phase composition would not occur. The process would then continue as an immiscible displacement.

If pressure were changed, the two-phase region would be modified and the limiting tie-line would shift. For a given injection-fluid/crude-oil pair, the MMP in a condensing-gas process is the minimum pressure at which the limiting tie-line just passes through the injection fluid composition; i.e., the injection fluid composition is on the limiting tie-line. Thus, in the condensing MCM process, the injected fluid composition must lie on or to the right of and the oil composition must lie to the left of the limiting tie-line.

In the condensing-gas process, there is an alternative to increasing the pressure in that the injection gas composition can be enriched to achieve miscibility. At a fixed pressure, the minimum enrichment at which the limiting tie-line passes through the injection gas composition is called the minimum miscibility enrichment (MME).

More-recent investigations of the enriched-gas process have shown that an alternative mechanism, one that involves both condensation and vaporization, is often responsible for the efficient displacement in this process (Zick 1986; Stalkup 1987; Mansoori and Gupta 1988). Zick (1986) was the first to present a detailed description of the mechanism, although Stalkup (1983a) indicated earlier that there were problems in trying to describe the enriched-gas process on a pseudoternary diagram.

The problem of description arises from the assumption that a pseudoternary diagram is a valid representation of a multicomponent system having many more than three components. As discussed in Chapter 2, if specific components making up a pseudocomponent do not partition into the different phases in the same proportions as specified in the pseudocomponent, then the pseudoternary diagram will not describe the system correctly. To the extent that the pseudocomponent concept fails, the description of the condensing process on a pseudoternary diagram for a multicomponent system may be incorrect.

Zick (1986) proposed the following mechanism, which has been generally substantiated by later studies (Stalkup 1987; Mansoori and Gupta 1988). The oil/gas system is assumed to be composed of four groups of components: (1) lean components, such as C_1, N_2, and CO_2; (2) light intermediate components, such as C_2 through C_4, which are the enriching components; (3) middle intermediates, which may range from C_4 to C_{10} on the low-molecular-weight side up to C_{30} on the high side (these components are generally not in the injected gas but are in the reservoir oil and may be vaporized from the oil); and (4) high-molecular weight components that cannot be vaporized from the oil in significant amounts.

When the enriched gas, containing components from Groups 1 and 2, contacts reservoir oil, light intermediates condense into the oil, making it lighter. The gas moves faster than the oil, so the depleted gas moves ahead. Additional fresh injected gas contacts the oil, continuing to decrease its density. If this continued until the oil became miscible with the injected gas, it would be the condensing-gas process as previously described. However, there is a countereffect.

The middle intermediates in the oil are stripped from the oil into the gas because these components are not originally in the gas. Thus, the oil at the upstream location tends to become saturated with light intermediates but depleted of middle intermediates. This prevents the development of miscibility between injected gas and reservoir oil, as previously described for the condensing-gas process.

If this were all that occurred, the process would not be very efficient. However, downstream of the injection point is a positive mechanism. After some period of injection, oil downstream "sees" a gas that is rich both in light and middle intermediates. This gas occurs because oil upstream becomes saturated with light intermediates (and therefore these components do not condense out of the gas into the oil) but the gas strips oil middle intermediates. Thus, gas downstream actually becomes a little richer. As this process continues and moves downstream, the condensing/vaporizing mechanism results in a gas that is almost miscible with reservoir oil. While miscibility (single phase) may never be reached, the displacement process can be very efficient.

Zick (1986) presented experimental evidence to support the proposed mechanism. One type of experiment consisted of making multiple contacts between an injection gas and an oil in a pressure/volume/temperature (PVT) cell. Reservoir oil was placed in the cell, a specified amount of gas was injected, and the system was allowed to equilibrate. The gas phase and a small amount of liquid phase were removed and analyzed. Fresh injectant gas was then charged to the cell and the process was repeated. Approximately seven contacts of this type were made.

If the process was a condensing-gas process only, oil density would decrease monotonically and injected gas-phase density would increase monotonically. If the mechanism were a condensing/vaporizing one, however, then the densities would go through a minimum and maximum, respectively. This latter behavior is what occurred, as shown in **Fig. 6.21.** Zick (1986), Stalkup (1987), and Mansoori and Gupta (1988) have also shown that calculations made with an EOS support the condensing/vaporizing process for many systems.

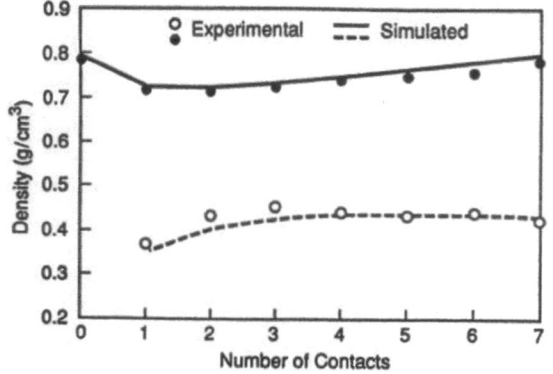

Fig. 6.21—Experimental and simulated densities from multicontact experiment at 3,600 psig (Zick 1986).

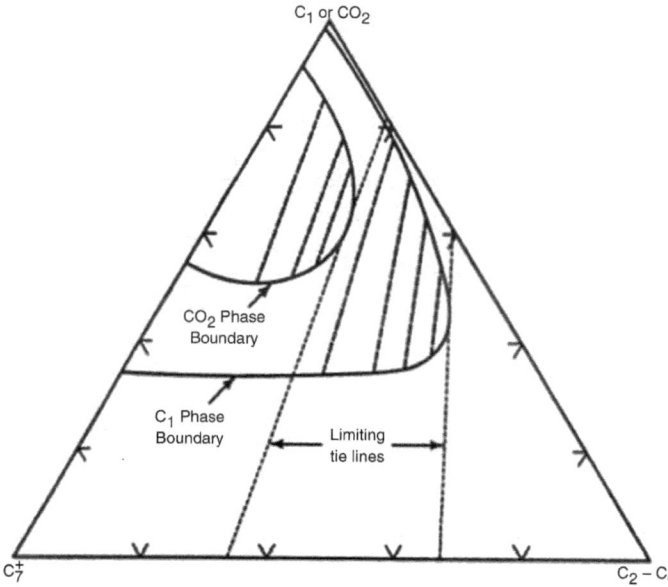

Fig. 6.22—Comparison of two-phase envelopes for methane/hydrocarbon and CO₂/hydrocarbon systems.

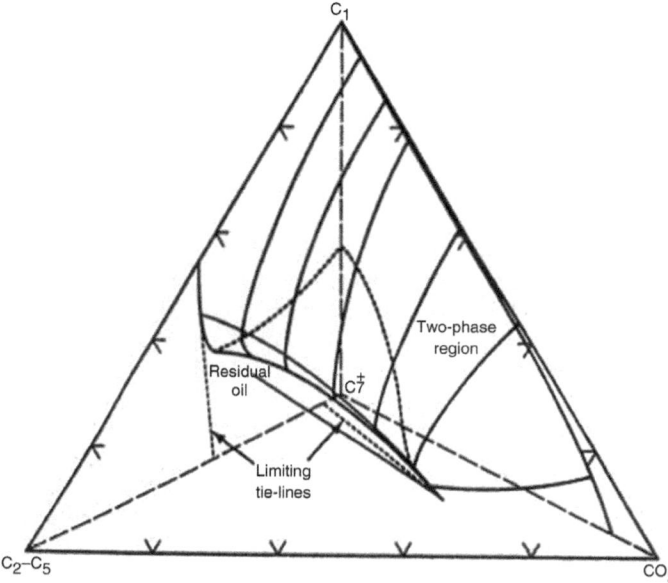

Fig. 6.23—Representation of the phase behavior of a methane/CO₂/hydrocarbon system on a pseudoquaternary diagram.

The condensing/vaporizing process is not presented as easily on a pseudoternary diagram as the straight condensing or vaporizing processes. Zick (1986) has shown that, for such a process, multiple contacts do not lead to a convergence at a critical point on a pseudoternary diagram as occurs for a true ternary system. This behavior of converging as concentrations approach the critical point but then diverging before the critical point is reached is described by Stalkup (1987). The effect of dispersion on the process is also discussed by Stalkup (1987).

6.5.3 CO₂ Miscible Displacement Process. An idealized description of the CO₂ miscible displacement process on a pseudoternary diagram is essentially the same as for the high-pressure vaporizing process. At temperatures above approximately 120°F, the phase behavior diagram for $CO_2/(C_2$ through $C_6)/C_{7+}$ is similar in appearance to the $C_1/(C_2$ through $C_6)/C_{7+}$ system. The primary difference is that, at the same temperature and pressure, the two-phase envelope is much smaller for the CO₂ system than for the CH_4 system, as illustrated in **Fig. 6.22.** Also, the limiting tie-line for the CO₂ system tends to have a slope that is more parallel to the CO_2/C_{7+} side of the ternary diagram than the C_1 system. Thus, miscibility can be generated between CO₂ and reservoir oils at lower pressures than between methane and reservoir oils. This is partly illustrated by reference to Fig. 6.4, which shows the locus of binary critical points for CO₂ and octane. This characteristic of CO₂, the ability to generate miscibility at much lower pressures, provides its primary advantage over methane.

If methane were added to the CO₂/hydrocarbon system, the phase behavior would be changed. **Fig. 6.23** shows a pseudoquaternary diagram at reservoir pressure and temperature. This representation of a 3D phase diagram shows the behavior of a system consisting of CO_2, C_1, C_2 through C_6, and C_{7+}. Arguments relative to the generation of miscibility, such as those used for the ternary diagrams, could be developed. The result would be to show that to develop dynamic miscibility, the advancing gas phase would move along the 3D surface of the two-phase region until the mixing tie-line between the gas phase and reservoir oil was entirely in the single-phase region.

The effect of adding methane is to increase the miscibility pressure. That is, for a given oil, a higher pressure is required for miscibility if C_1 is present in the displacing fluid. This is also true for the presence of other noncondensable gases, such as N_2 or O_2. These impurities in the CO₂ stream will generally be detrimental to miscibility development. For this reason, stack or flue gases are not directly applicable as miscible displacement agents at reservoir pressures commonly encountered. LPG components (such as C_2 and C_3) typically present in CO₂ produced from a CO₂ flood, however, are helpful and offset the effect of C_1 and the other noncondensable gases. The presence of H_2S also is helpful for the development of miscibility.

At lower temperatures (approximately 120°F or less), CO₂ phase behavior is often such that two liquid phases or two liquid phases and a vapor phase are formed. The description of the process on a pseudoternary diagram is more complex for such conditions. It has been pointed out by Metcalfe and Yarborough (1979) and Shelton and Yarborough (1977) that there are two general types of phase behavior in CO_2/hydrocarbon systems. These are indicated in **Figs. 6.24a and 6.24b** (Stalkup 1983a), which show the different types on p-x diagrams.

The first and simplest type of behavior is shown in Fig. 6.24a. The abscissa is concentration of CO_2 in the CO_2/hydrocarbon mixture. In this type of behavior, the two-phase region consists of a liquid and vapor in equilibrium. Bubblepoint (100% liquid) and dewpoint (100% vapor) curves meet at the critical point. Loci of constant percents of liquid or vapor exist between the bubble- and dewpoint curves, as indicated in Fig. 6.24a. At a constant temperature and pressure, the phase boundaries on a pseudoternary diagram generally would be of the form in Fig. 6.22. This first type of behavior usually exists at temperatures of 120°F or above.

At temperatures below approximately 120°F, the behavior can be of the type seen in Fig. 6.24b. At lower pressures, a two-phase region consisting of a liquid and vapor exists, as previously discussed and shown in **Fig. 6.25a** on a pseudoternary diagram. At higher pressures, however, a liquid/liquid two-phase region exists. Further, at intermediate pressures, there is a relatively small composition region in which three phases, two liquid phases and a vapor phase, exist in equilibrium. On a ternary diagram, the three-phase region appears as in Figs. 6.25b and 6.25c (Stalkup 1983a). The pseudoternary diagrams in Fig. 6.25 are at the corresponding pressures, as indicated in Fig. 6.24b. When the overall composition falls within the three-phase region on a ternary diagram, the compositions of the three equilibrium phases are at the apexes of the three-phase triangle. Stalkup (1983a) further divided the second type of phase behavior into three subcategories, Types IIa, IIb, and IIc. He discusses these different types and the effect on development of miscibility in detail.

The existence of regions of two liquid phases and three phases undoubtedly affects the recovery mechanisms in a displacement. With multiple liquid phases, a mechanism of liquid/liquid extraction occurs that is analogous to the vaporization mechanism previously described. Whatever the specific mechanisms, it has been shown that multiple-contact miscibility can be achieved (Gardner et al. 1981) at temperatures below 120°F.

6.6 Experimental Verification of the Role of Phase Behavior in Miscible Displacement

The relationship between the phase behavior and oil recovery in an experimental linear displacement test was demonstrated by Metcalfe and Yarborough (1979). They conducted displacements in which the hydrocarbon to be recovered was a binary mixture consisting of 40 mol% butane and 60 mol% decane. The displacing fluid was pure CO_2. Thus, the system could be represented accurately on a ternary diagram with CO_2, butane, and decane at the apexes. The porous medium consisted of an 8-ft-long, 2-in.-diameter Berea sandstone core. There was no water phase present in the displacement, and local thermodynamic equilibrium was assumed to exist. Experimental results are shown in **Figs. 6.26 through 6.28**. The first

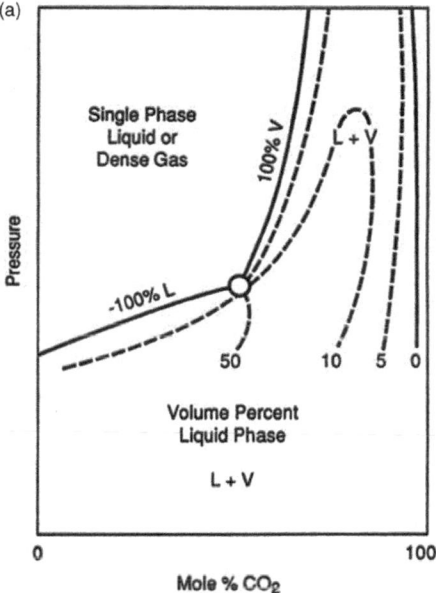

(a)

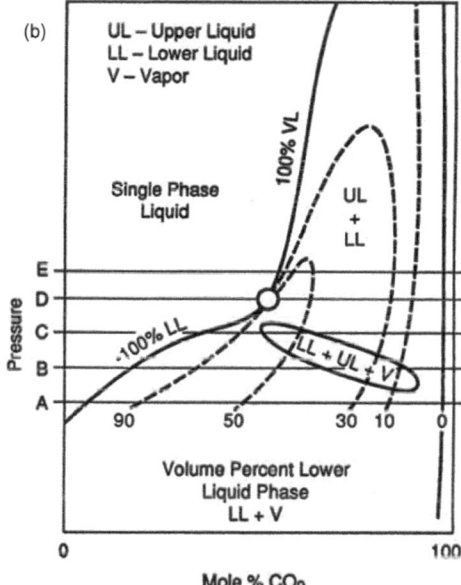

(b)

Fig. 6.24—(a) CO_2/hydrocarbon phase behavior at temperatures above approximately 120°F: Type I phase behavior (Stalkup 1983a). **(b)** CO_2/hydrocarbon phase behavior at temperatures above approximately 120°F: Type II phase behavior (Stalkup 1983a).

experiment, shown in Fig. 6.26, was at a pressure of 1,900 psia. At this pressure and 160°F, the three components are miscible over all compositions. Thus, the displacement was an FCM displacement. As shown in Fig. 6.26, the reduced liquid concentration essentially followed the ideal mixing line from the oil composition (40% butane and 60% decane) to the pure CO_2 apex. Oil recovery was 99% of original oil in place (OOIP).

A second displacement, shown in Fig. 6.27, was conducted at 1,700 psia with all other variables held constant. At this pressure, a small two-phase envelope exists, but the oil composition lies to the right of the limiting tie-line. The produced-fluid composition changed in essentially a linear manner (after CO_2 was first detected in the effluent) from the initial oil composition to the vicinity of the critical point. Compositions then "moved" around the two-phase envelope and finally to pure CO_2. Thus, the concentration change was consistent with the description of an MCM process. Oil recovery was 90% OOIP.

The third experiment, shown in Fig. 6.28, was conducted at a pressure of 1,500 psia. At this pressure, both oil and injected fluid are to the left of the limiting tie-line, and thus, achievement of miscibility would not be expected. The produced-fluid concentration changed almost linearly from the initial composition to a point on the two-phase envelope. The effluent then became two phase, as seen in Fig. 6.28, indicating an immiscible displacement. The behavior was thus consistent with the theoretical behavior described on a ternary diagram. Recovery was 81% OOIP, significantly less than for the MCM and FCM displacements.

The Metcalfe and Yarborough (1979) data show the manner in which the development of miscibility affects oil recovery for a relatively simple system. An FCM process, in which the displacing fluid is miscible in all proportions with the displaced phase, has essentially 100% efficiency at the microscopic scale. All the oil is displaced at those places contacted by the displacing miscible fluid. Recovery efficiency is also high in an MCM process. There are factors, however, that contribute to a reduced efficiency: (1) miscibility must be developed in situ, (2) complete miscibility may not exist across the entire miscible zone, and (3) dispersion may lead to a temporary or, at some locations, permanent loss of miscibility. Thus, recovery in an MCM process is usually less than in an FCM process. Finally, displacements conducted at conditions where miscibility is not achieved yield recovery efficiencies smaller than for MCM processes. The reduction in efficiency depends on process conditions and their nearness, in a thermodynamic sense, to the state required for miscibility.

6.7 Measurement and Prediction of the MMP or MME in a Multiple-Contact Process

An important design consideration is the determination of the conditions at which dynamic miscibility will be achieved for specified reservoir fluids and reservoir characteristics. The temperature of the process is set by reservoir conditions, but the pressure may be controlled within certain limits. Likewise, the injected-gas composition can be set. Thus, the problem becomes one of determining the minimum pressure (MMP) or minimum gas enrichment (MME) at which miscibility will be achieved in a multiple-contact process for a specified reservoir fluid composition and the reservoir temperature.

6.7.1 Experimental Measurement of MMP or MME. Miscibility pressure may be measured by conducting displacement tests under idealized conditions. The slimtube test has become a generally accepted procedure, although precise test conditions and interpretations have not been standardized by the petroleum industry. Another method, which was introduced later, uses a rising-bubble apparatus (RBA) and has been shown to yield comparable results and to take less time than a slimtube test.

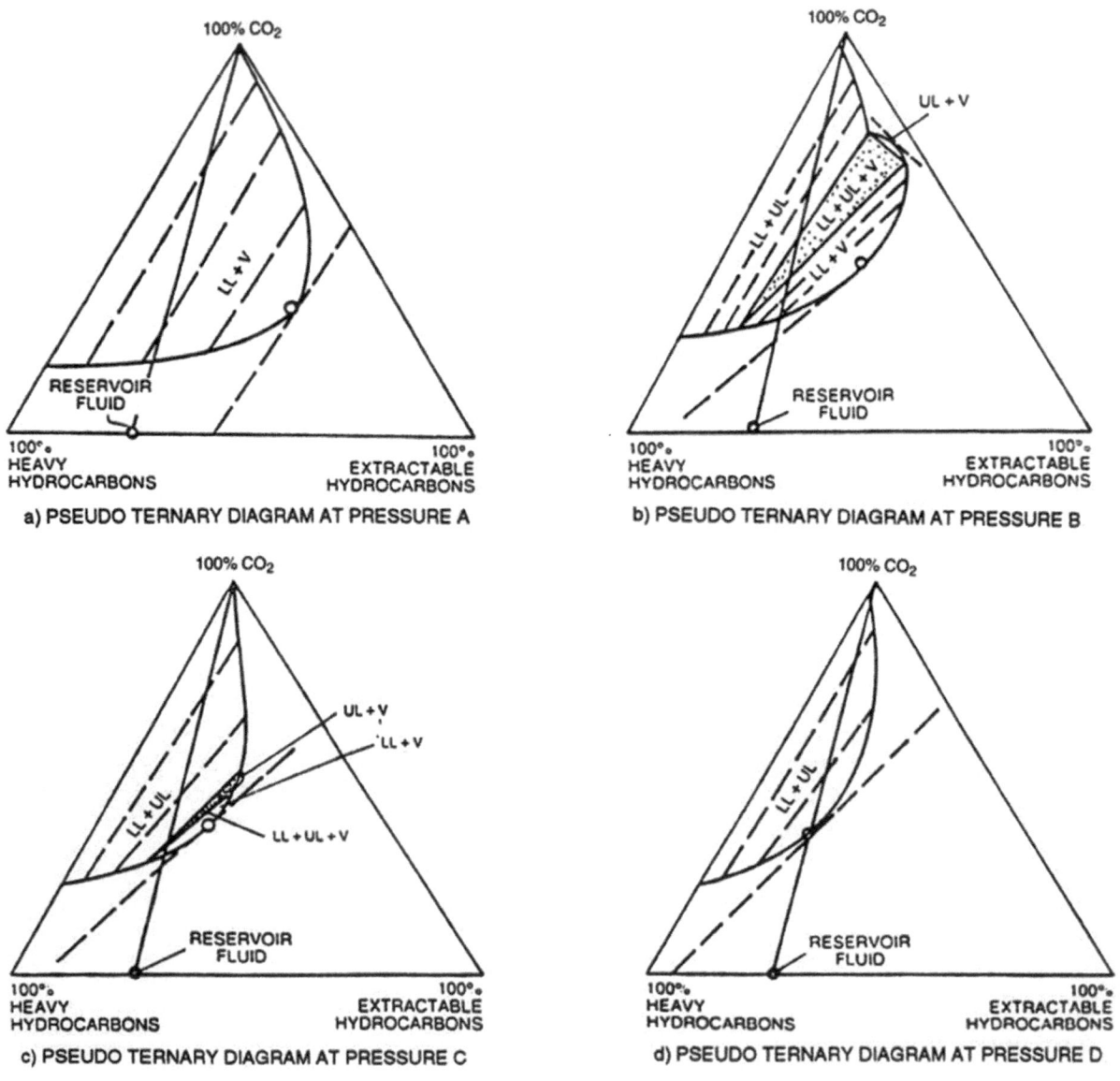

Fig. 6.25—Pseudoternary diagrams corresponding to the CO_2/hydrocarbon system of Fig. 6.24b (Stalkup 1983a).

Fig. 6.29 is a schematic of typical slimtube test equipment. The slimtube typically consists of a stainless-steel tube that has an approximate $5/16$-in. inside diameter and that is approximately 40 ft long. The tube is packed uniformly with fine-grade sand or glass beads of a size on the order of 100 mesh. The ratio of particle size to tubing diameter is sufficiently small that wall effects are negligible. The tube is coiled in a manner so that flow is basically horizontal and gravity effects are insignificant. A pump system is provided to force fluids through the porous medium pack, and pressure is controlled by a backpressure regulator. The coiled tube and certain auxiliary equipment are placed inside a constant-temperature bath, usually an air bath. At the effluent end of the tube, fluid collection and measurement systems are provided. This equipment may range from a simple graduated cylinder and wet test motor to a more complicated system involving a gas chromatograph. A small visual cell also is included at the effluent end so that the fluid product can be observed.

To conduct a test, the porous medium in the tubing is filled with the hydrocarbon to be displaced. This is typically a sample of the crude oil from the reservoir being considered for miscible displacement. The system is brought to test temperature, and the backpressure regulator is set at the desired displacement pressure. The displacing fluid is then injected at a constant rate. A linear rate of advancement considerably higher than what might be expected in an actual reservoir usually is used to complete an experiment in a reasonable amount of time. The pressure drop across the system is generally a small fraction of the average absolute pressure level in the tube. The hydrocarbon recovery at displacing-fluid breakthrough, recovery at the time of injection of a specified number of PVs, and/or ultimate hydrocarbon recovery are recorded.

The entire experiment is repeated several times at different pressures, but with all other variables held constant. Recoveries are plotted as a function of displacement pressure, as illustrated in **Fig. 6.30.** The MMP is typically assumed to be the pressure at the "break" in the curve (i.e., the pressure above which very little additional recovery occurs). A miscibility pressure of approximately 1,800 psi is estimated from the data in the example in Fig. 6.30. There is no universally accepted method of defining MMP from slimtube experiments. Klins (1984) summarizes the methods that have been used. The different methods generally yield results that are sufficiently close to each other for engineering purposes.

An alternative approach that has been used, especially for condensing-gas drive processes, is to measure the minimum gas enrichment necessary to achieve miscibility (MME) at a fixed pressure. Repeated displacements are run with increasingly higher concentrations of an intermediate compound such as C_{2+} in the displacing solvent. **Fig. 6.31** (Mansoori and Gupta 1988) shows typical results from such slimtube displacements for two different pressures.

The slimtube test is appropriate for the condensing/vaporizing displacement process, even though it is thought that miscibility is never completely developed (Zick 1986). Efficiency is high in the process, and recovery performance behaves generally in the manner of the vaporizing or condensing processes.

The basis of the slimtube test is that the small-diameter tube filled with an unconsolidated porous medium serves as an idealized multiple contactor for the displaced and displacing fluids. The displacing and displaced fluids are assumed to be in local thermodynamic equilibrium at all points in the porous medium. The length/diameter ratio of the slimtube is set so that nonidealities, such as fingering caused by

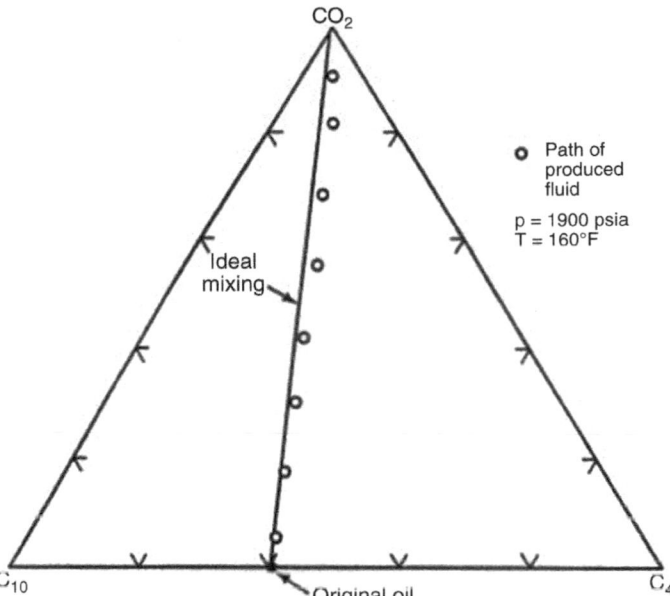

Fig. 6.26—FCM displacement of a C_4/C_{10} mixture by pure CO_2, representation on a ternary diagram (Metcalfe and Yarborough 1979).

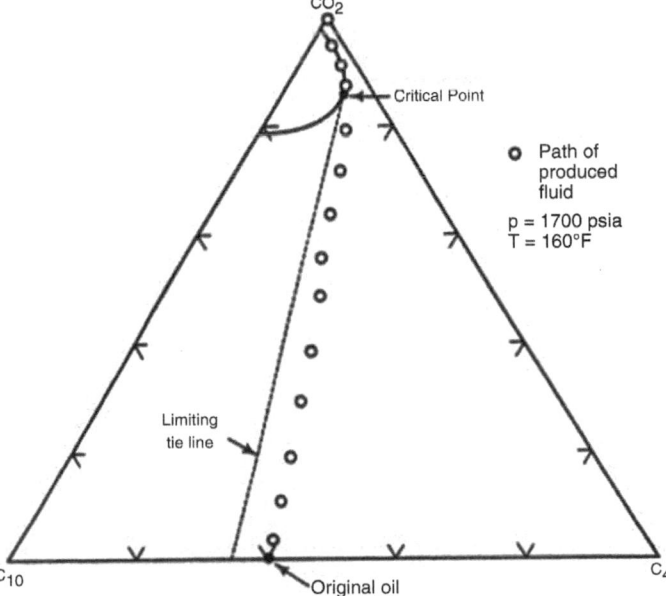

Fig. 6.27—Dynamic miscible displacement of a C_4/C_{10} mixture by pure CO_2, representation on a ternary diagram (Metcalfe and Yarborough 1979).

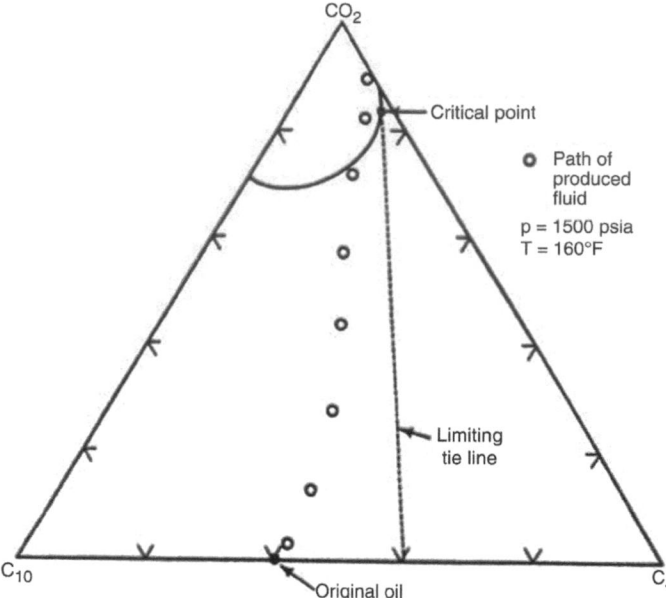

Fig. 6.28—Dynamic immiscible displacement of a C_4/C_{10} mixture by pure CO_2, representation on a ternary diagram (Metcalfe and Yarborough 1979).

porous-medium heterogeneities, unfavorable viscosity ratios, or gravity effects, are negligible. Thus, the efficiency of the displacement is taken to be the result of the thermodynamic phase behavior of the system and not a function of particular reservoir rock characteristics. As indicated in Fig. 6.30, oil recovery in an MCM process conducted in a slimtube typically is somewhat lower than in an FCM process.

In some cases, visual observations of the fluids flowing within the tube are used to confirm the existence of single- or multiphase flow. Chromatographic analyses of the effluents can provide additional information about the nature of the displacement.

Slimtube experiments are relatively simple to conduct (although they can require significant time) and typically have been used in the design of a miscible process or for the screening of reservoirs to determine candidates for miscible processes. The assumption that nonideal effects, such as viscous fingering and gravity segregation, are negligible may not always be valid; thus, it is recommended that slimtube conditions be checked by use of an FCM displacement as a control.

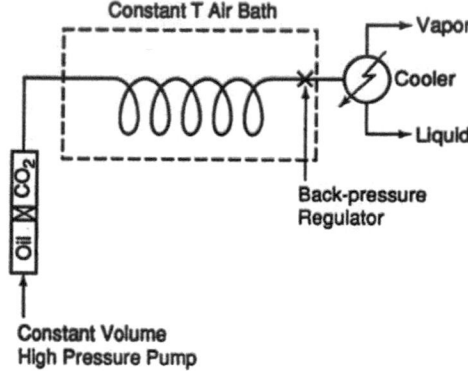

Fig. 6.29—Schematic of slimtube apparatus for experimental measurement of miscibility pressure.

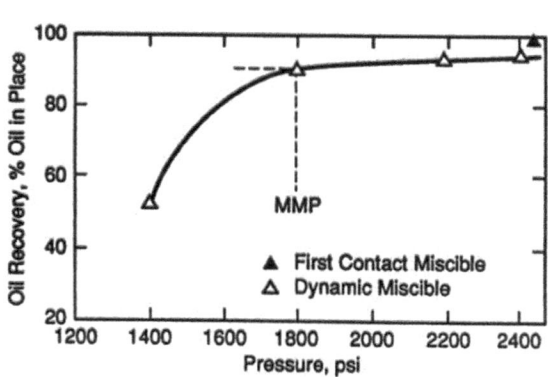

Fig. 6.30—Slimtube data.

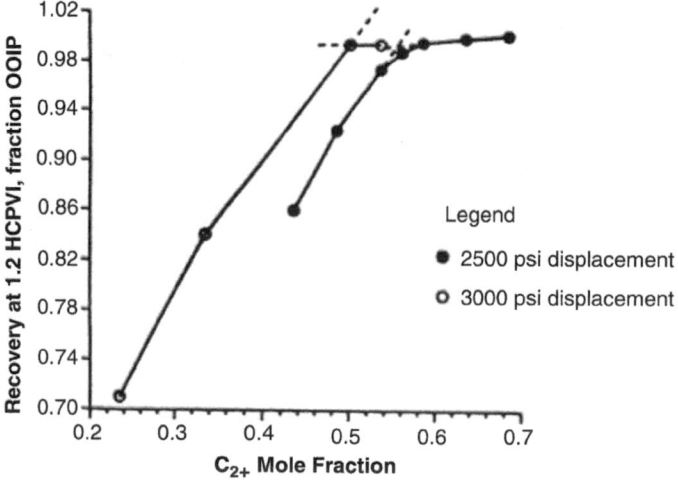

Fig. 6.31—Slimtube recovery as a function of solvent enrichment level (Mansoori and Gupta 1988).

The RBA for measuring MMP consists of "a flat glass tube mounted vertically in a high-pressure sight gauge in a temperature-controlled bath" (Christiansen and Haines 1987). A back light is used on the sight gauge so that behavior can be observed. The tube and gauge are first filled with distilled water. Then, the water is displaced with the oil to be tested except for a short column at the bottom. The temperature is held at a desired value, usually the temperature of the reservoir of interest, and pressure is held at a specified test value.

Next, a small bubble of gas is injected into the bottom of the tube. The gas composition is that of the injection gas to be tested in the MCM process. The gas bubble rises first through the water column and then through the oil. Its behavior is observed as it rises through the region of the sight glass.

The behavior of the bubble is quite distinctive at or slightly above the MMP. As the bubble rises, it changes shape and disperses into the oil. Far below the MMP, the bubble retains its near-spherical shape as it rises but decreases in size as a result of mass transfer between the gas and oil. Far above the MMP, the bubble disperses into the oil quite rapidly.

The inventors of the process presented data showing that values of MMP determined with the RBA agreed within acceptable accuracy with values obtained with the slimtube apparatus (Christiansen and Haines 1987). This agreement of the two methods also was demonstrated in other work (Sibbald et al. 1991). Investigators who have used the RBA concluded that the method is significantly faster than the slimtube test, and thus more data on a system can be obtained in a given time.

Zhou and Orr (1998) analyzed the RBA method and showed that changes in bubble shape as the bubble moved upward through the oil phase were controlled by changes in the interfacial tension driven by mass transfer between the phases. They also concluded on the basis of experimental work that the method was reasonably reliable when used to determine MMP for ternary vaporizing gas drives. However, they concluded that use of the method was not a reliable way to determine MME for ternary condensing systems. Instead, for such systems, they proposed use of a falling drop method in which a drop of oil is allowed to fall through the gas phase. "Whether either a falling-drop or a rising-bubble experiment could be used to determine the MMP accurately in condensing/vaporizing gas drives has not been determined" (Jessen and Orr 2008).

A method of estimating MMP for CO_2 systems containing contaminates of N_2, CH_4, C_2, C_3, and H_2S up to concentrations of 20% mole fraction was developed by Johns et al. (2010). The method requires values of MMP for pure CO_2 and for binary mixtures of CO_2 and each contaminate. These values can be obtained from experiments, such as slimtube experiments, or from accurate analytical calculations. The number of MMP values required is equal to the number of components in the injection gas. The MMP values of the different binary mixtures were shown to be linear functions of the contaminate concentrations at small concentrations. The MMP for the actual contaminated CO_2 gas mixture was then shown (to a good approximation) to be a linear function of "pseudoMMP" values obtained for the MMP values of the different binary mixtures. When the procedure was tested using a west Texas crude oil and various CO_2 contaminated gases, it was found to give MMP values within ±15% of values obtained from slimtube experiments or reliable simulations.

Methods to estimate MMP as a function of fluid composition and temperature are discussed in Section 6.7.2. The experimental basis for the predictive methods is usually the MMP as measured in a slimtube. That is, the MMP determined in a slimtube experiment usually is accepted as the correct value against which a predicted value is compared.

6.7.2 MMP Prediction. MMP is commonly predicted from (1) empirical correlations based on experimental results and (2) phase-behavior calculations based on an EOS and computer modeling. The first approach, the use of empirical correlations, is relatively simple to apply. A predicted value can be significantly in error, however, especially if a correlation is applied at conditions far removed from the experimental conditions on which the correlation was based. Empirical correlations typically are used to obtain first-pass estimates or as a screening tool.

The second approach, which involves the application of an EOS, can be used to obtain more-reliable results. However, this approach requires availability of a significant amount of composition data for the reservoir fluids. Such data often are not available, although they can be obtained from laboratory analyses that are somewhat tedious. Further, the calculations are complex and involve the application of a solution algorithm and a computer. Such algorithms are not universally available throughout the industry. Also, because equilibrium constants required in the calculations are not known precisely or because the EOS is not sufficiently accurate, the calculations must be calibrated against experimental PVT data. This can be a laborious task.

As stated in Section 6.7.1, the experimental basis for MMP calculations typically is the miscibility pressure measured in slimtube displacements.

Empirical Correlations. A number of empirical correlations exist that can be used to estimate the MMP for hydrocarbon systems (vaporizing or enriched gas) and CO_2/hydrocarbon systems. Examples of these correlations are presented next. Stalkup (1983a) and Klins (1984) discuss additional correlations.

Condensing-Gas Drive. An early correlation developed by Benham et al. (1960) is applicable to condensing-gas displacements. Their correlation is based on a study of the calculated phase behavior of various mixtures of five different reservoir fluids with intermediates (C_2 through C_4 mixtures) and methane (C_1). Phase behavior was represented on pseudoternary diagrams with C_1, C_2 through C_4, and C_{5+} as the pseudocomponents. Right triangles rather than equilateral triangles were used for the diagrams.

Benham et al. (1960) observed that, in the region of the critical point, calculated tie-lines connecting equilibrium compositions in the two-phase region tended to have negative slopes (on a right-triangular diagram) that were close to being parallel to the C_1/C_{5+} edge of the diagram. This led them to assume that the limiting tie-line (tie-line tangent to the critical point) is, in fact, parallel to the C_1/C_{5+} edge.

For dynamic miscibility to be achieved in a condensing-gas process, sufficient intermediates must be added to the methane in the displacing fluid to make the displacing fluid composition lie to the rich-gas side of the composition given by extension of the limiting tie-line (see Fig. 6.20). Thus, for the assumption made by Benham et al. (1960), the composition of intermediates in the critical point fluid is the minimum composition that can exist in a C_1/C_2 through C_4 mixture to ensure the dynamic development of miscibility in a condensing-gas process.

Benham et al. (1960) calculated the critical pressures and temperatures for a number of mixtures of reservoir fluids, methane, and intermediates. They correlated their calculated results by plotting the maximum composition of C_1 in the displacing fluid as a function of pressure, temperature, C_{5+} molecular weight of the reservoir fluid, and the average molecular weight of the intermediate components in the displacing fluid. The following ranges of parameters were considered: pressure, 1,500 to 3,000 psia; temperature, 70 to 260°F; average molecular weight of C_{5+} in the reservoir fluid, 180 to 240; and average molecular weight of C_2 through C_4 components in the displacing fluid, 34 to 58.

Stalkup (1983a) crossplotted the Benham et al. (1960) results to present the MMP as a function of the same variables. **Figs. 6.32 through 6.34** show his plots for temperatures of 100, 150, and 200°F, respectively. Stalkup (1983a) also reported a comparison of experimentally measured miscibility pressures with those calculated with the correlation. For 10 different reservoir conditions, the average error in the predicted miscibility pressure was 340 psia and the maximum error was 800 psia. Yarborough and Smith (1970) also reported comparisons of predicted and measured miscibility pressures for a number of reservoir oils.

An example calculation with the Benham et al. (1960) correlation is given in Example 6.3.

Example 6.3—Calculation of the MMP for a Condensing-Gas Displacement Process. Assume that a reservoir that is to be flooded with a condensing-gas process is at 150°F. The average molecular weight of the C_{5+} fraction of the reservoir oil is 200. A displacement gas is available with the composition 60 mol% CH_4, 30 mol% C_3H_8, and 10 mol% C_4H_{10}. The maximum pressure at which the displacement can be conducted is 2,000 psia. Determine whether multiple-contact miscibility can be achieved at the specified conditions.

Solution. The molecular weight of the intermediate components (C_3 and C_4) must be calculated first. The molar composition of intermediates is 75% C_3 and 25% C_4. The average molecular weight of the intermediates is

$$\bar{M}_w = 0.75 \times 44 + 0.25 \times 58 = 47.5$$

The intermediates make up 40 mol% of the composition. From Fig. 6.33 (by interpolation), we can see that the molecular weight of C_2 through C_4 is 49. The MMP is 2,070 psia. This pressure is well above the maximum operating pressure, and miscibility could not be achieved. If the composition of intermediate components in the displacement gas were increased to 50 mol%, the predicted MMP would be < 2,000 psia.

Vaporizing-Gas Drive. A very limited amount of information is available for the prediction of MMP in vaporizing-gas displacement processes other than for CO_2 miscible displacement. Stalkup (1983a) presents one correlation for estimating miscibility pressure with a lean hydrocarbon gas based on data taken over the following range of conditions: temperature, 140 to 265°F; saturation pressures of reservoir oil, 596 to 4,035 psia; average molecular weight of C_{7+} in the reservoir oil, 149 to 216; and MMP, 3,250 to 4,750 psia.

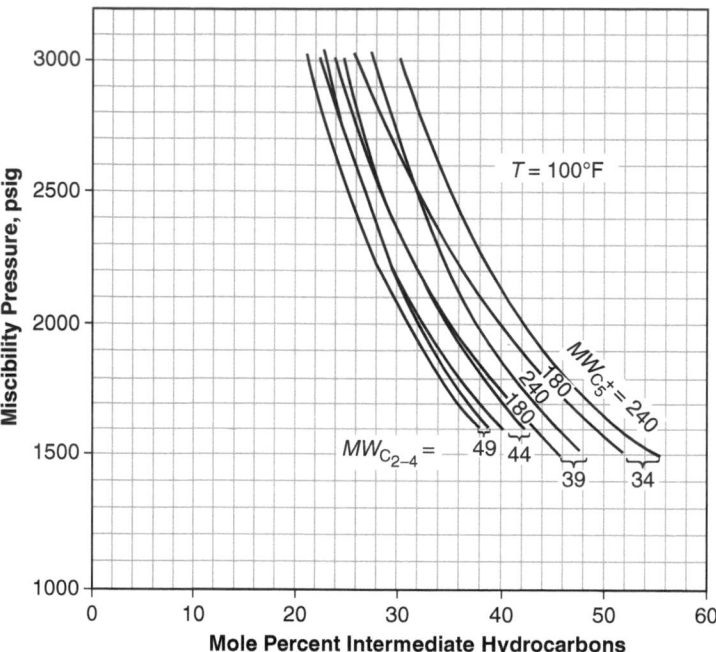

Fig. 6.32—Condensing-gas-drive miscibility-pressure correlation, $T = 100°F$ (Benham et al. 1960; Stalkup 1983a).

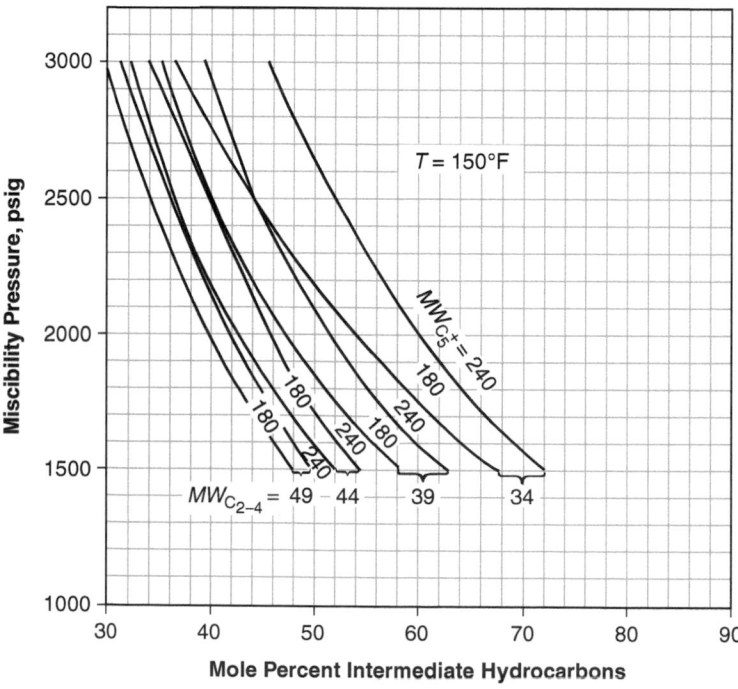

Fig. 6.33—Condensing-gas-drive miscibility-pressure correlation, $T = 150°F$ (Benham et al. 1960; Stalkup 1983a).

The correlation is shown in **Fig. 6.35**, where the ratio of miscibility pressure, p_m, is plotted as a function of a correlating parameter that characterizes the reservoir oil. The correlation is expressed algebraically as

$$\frac{P_{so}}{P_m} = \frac{2.252X^*}{X^* + 4.091}, \quad \ldots\ldots\ldots \text{(6.2)}$$

where $X^* = \dfrac{W_i W_1 T_R^{\frac{1}{3}}}{W_{C_{7+}}}, \quad \ldots\ldots \text{(6.3)}$

where W_i = amount of C_2 through C_6 hydrocarbons + CO_2 + H_2S in the oil, W_1 = amount of methane (C_1) + N_2 in the oil, $W_{C_{j+}}$ = amount of C_{7+} hydrocarbons in the oil, and T_r = reduced temperature of the oil (see Eqs. 6.9 and 6.11). W_1, W_i, and $W_{C_{j+}}$ are expressed in units of lbm/lbm mol.

The correlation was developed from data on nine oils. The average deviation between the value predicted by the correlation and the experimental MMP was 260 psi, and the maximum deviation was 640 psi.

CO₂ Miscible Displacement. A number of empirical correlations exist for predicting MMP in CO_2 miscible displacement. Three of these are presented along with results that indicate the accuracy of the predictions, but without a detailed discussion of the underlying assumptions. All the correlations presented are based on experimental data from slimtube experiments, although the experimental conditions and definition of MMP are not precisely the same in all cases.

Holm and Josendal (1974, 1980) developed a correlation that is an extension of that of Benham et al. (1960). The correlation, shown in **Fig. 6.36**, gives MMP as a function of temperature and molecular weight of the C_{5+} fraction in the oil. Mungan (1981) extended the Holm and Josendal (1974, 1980) correlation to cover additional molecular weights of the C_{5+} fraction of the reservoir oil; this extension

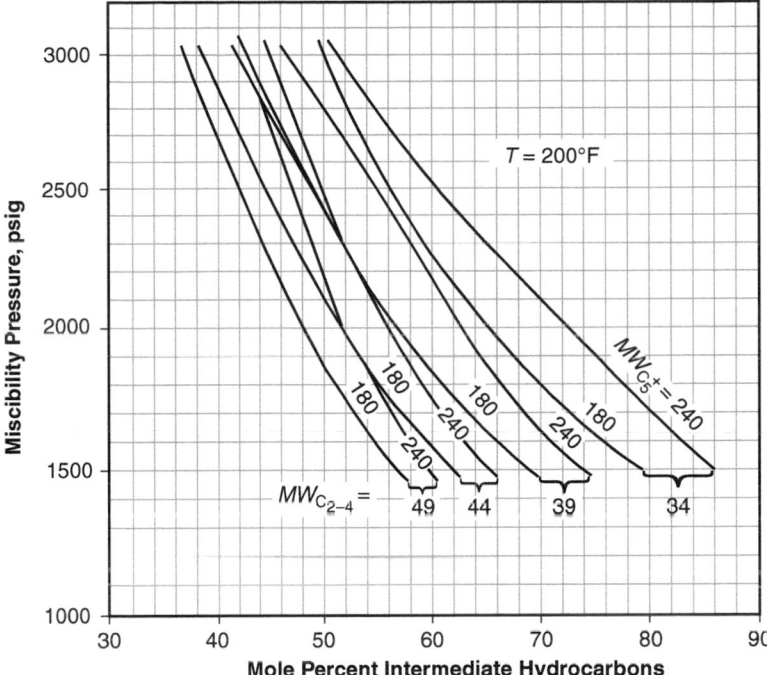

Fig. 6.34—Condensing-gas-drive miscibility-pressure correlation, $T = 200°F$ (Benham et al. 1960; Stalkup 1983a).

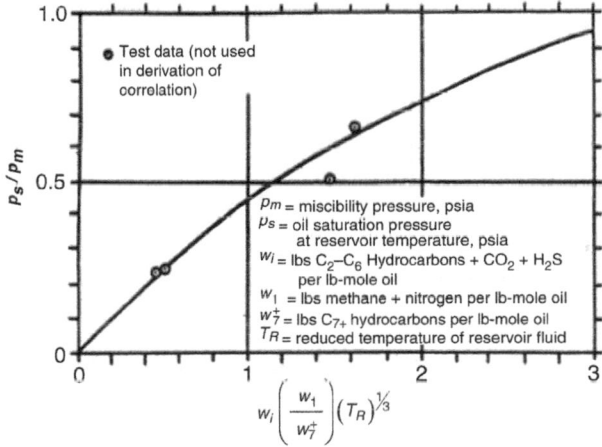

Fig. 6.35—Vaporizing-gas-drive miscibility-pressure correlation (Stalkup 1983a).

is also shown in Fig. 6.36. Holm and Josendal (1982) later showed that the development of miscibility with CO_2 was related to the CO_2 density. They correlated CO_2 density at the MMP value with the C_5 through C_{30} content of the reservoir oil. The Holm and Josendal correlation shown in Fig. 6.36 is applicable for a displacing fluid that is pure CO_2.

Yellig and Metcalfe (1980) developed an even simpler correlation, shown in **Fig. 6.37**. MMP is correlated as a single curve as a function of temperature. However, if the bubblepoint pressure of the oil is larger than the MMP from the curve, then the bubblepoint pressure is taken as the MMP. This accounts for the possible formation of two phases when the pressure is below the bubblepoint of the reservoir oil. The Yellig and Metcalfe (1980) correlation is applicable for a pure CO_2 displacing phase.

Yellig and Metcalfe (1980) compared MMP values predicted with their correlation and the Holm and Josendal (1974) correlation with experimental values measured in slimtube experiments. **Fig. 6.38** shows typical results of this comparison for the Yellig and Metcalfe (1980) correlation. While the general agreement is good, predicted values for specific oils are in error by magnitudes of approximately 500 psia. The error analysis indicates that empirical correlations provide estimates of MMP, but should be used in actual design calculations with a great deal of caution.

Johnson and Pollin (1981) and Alston et al. (1985) developed correlations that account for impurities in the CO_2. The Alston et al. (1985) correlation applies to pure CO_2 streams as well as streams with impurities. For pure CO_2, the MMP, P_{CO_2}, is given by

$$P_{CO_2} = 8.78 \times 10^{-4} \left(T\right)^{1.06} \left(\bar{M}_{wC_{5+}}\right)^{1.78} \left(\frac{X_{vol}}{X_{int}}\right)^{0.136}, \quad \ldots\ldots\ldots\ldots\ldots\ldots\ldots\ldots\ldots\ldots \text{(6.4)}$$

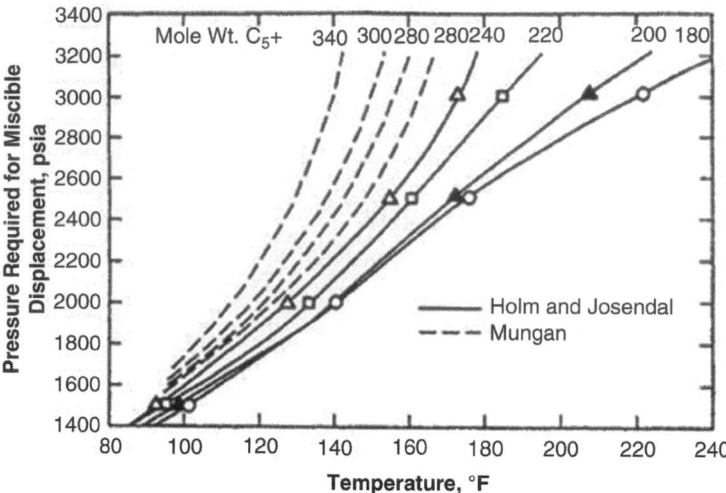

Fig. 6.36—Pressure required for miscible displacement in CO_2 flooding (Mungan 1981; Holm and Josendal 1974).

where P_{CO_2} = MMP for pure CO_2 injection; T = reservoir temperature; $\bar{M}_{wC_{5+}}$ = molecular weight of the C_{5+} fraction in the oil; X_{vol} = mole fraction of the volatile component in the oil, assumed to consist of C_1+N_2; and X_{int} = mole fraction of the intermediate component in the oil, assumed to consist of C_2 through C_4, CO_2, and H_2S. For oils with nearly equal fractions of the volatile and intermediate components, the correlation factor $(X_{vol}/X_{int})^{0.136}$ is very nearly unity.

When the CO_2 stream is contaminated with other components, the MMP is affected. The addition of C_1 or N_2 to the CO_2 increases MMP, while the addition of C_2, C_3, C_4, or H_2S reduces the MMP (Alston et al. 1985). Alston et al. (1985) developed a correlation based on a pseudocritical temperature of the impure CO_2 stream defined with a weight-fraction mixing rule:

$$T'_{cm} = \sum_{i=1}^{n} X_i T_{ci} - 459.7, \dotfill (6.5)$$

where T'_{cm} = weight-average critical temperature, X_i = mass fraction of Component i, and T_{ci} = critical temperature of Component i. For components C_1, C_3, C_4, CO_2, and N_2, the critical temperature, T_{ci}, was taken to be the true critical temperature. For both C_2 and H_2S, however, the true critical temperature values were replaced by an apparent value of T_{ci} = 585°R. This substitution of the apparent value for C_2 and H_2S gave results that improved the agreement between predicted and experimentally measured values of MMP.

T'_{cm} is used to calculate a correction factor applied to Eq. 6.4:

$$F_{imp} = (87.8 / T'_{cm})1.935 \times 87.8 / T'_{cm}. \dotfill (6.6)$$

The MMP is then given by

$$P_{CO_2} - imp = P_{CO_2} F_{imp}. \dotfill (6.7)$$

Alston et al. (1985) point out that the correction factor can be applied to other literature MMP values based on pure CO_2. They also suggest use of the bubblepoint correction suggested by Yellig and Metcalfe (1980). When the calculated MMP is less than the bubblepoint of the reservoir oil, the bubblepoint pressure should be taken as the MMP.

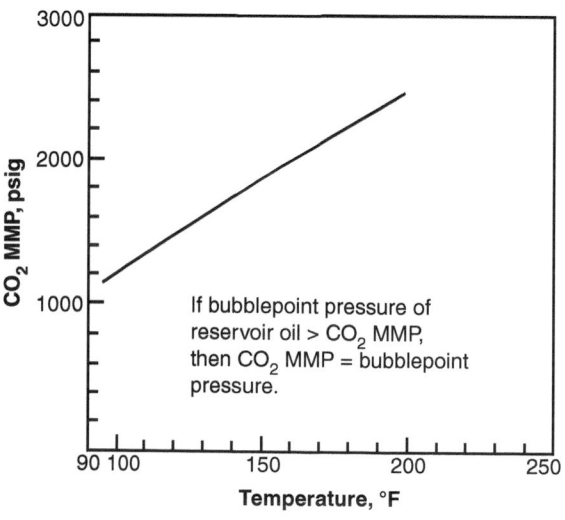

Fig. 6.37—Temperature/bubblepoint pressure of CO_2 MMP correlation (Yellig and Metcalfe 1980).

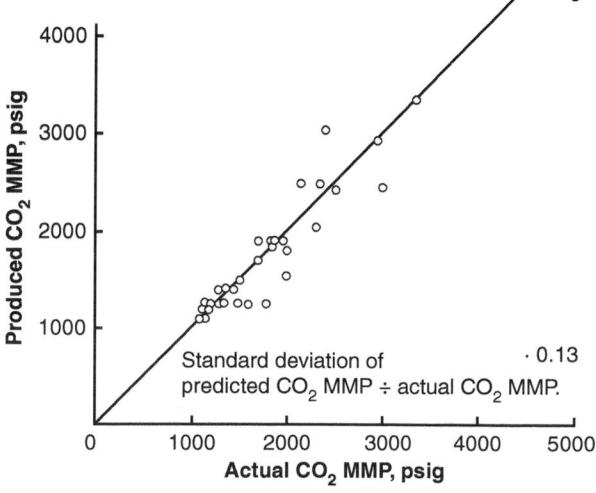

Fig. 6.38—Accuracy of temperature/bubblepoint pressure for predicting CO_2 MMP (Yellig and Metcalfe 1980).

Alston et al. (1985) compared the predicted value obtained with their correlations to 68 experimental values. The systems used consisted of their data plus available literature data. Pure CO_2 and contaminated CO_2 streams were included in the data set. The comparison is shown in **Fig. 6.39.** The average error in MMP is 6.9%, and the standard deviation is 8.7%. The maximum error is approximately 700 psi.

Sebastian et al. (1985) developed another correlation for impure CO_2 streams. The correlation relates MMP to the MMP for pure CO_2 and the pseudocritical temperature of the drive gas. The correlation is expressed algebraically as

$$\frac{P_{CO_2}-\text{imp}}{P_{CO_2}}=1.0-2.13\times10^{-2}\left(T_{cp}-304.2\right)+2.51\times10^{-4}\times\left(T_{cp}-304.2\right)^2-2.35\times10^{-7}\left(T_{cp}-304.2\right)^3,\ \ldots\ldots(6.8)$$

where $P_{CO_2}-\text{imp}$ and P_{CO_2} are MMPs for CO_2 with impurities and pure CO_2, respectively. T_{cp} is the pseudocritical temperature of the injected drive gas given by

$$T_{cp}=\sum_i X_i T_{ci},\ \ldots\ldots\ldots\ldots\ldots\ldots\ldots\ldots\ldots\ldots\ldots\ldots\ldots\ldots\ldots\ldots\ldots\ldots\ldots\ldots(6.9)$$

where T_{ci} = critical temperature of components in the mixture and X_i = mole fraction of components in the mixture. **Table 6.1** gives critical temperatures of selected gases. The critical temperature of H_2S was adjusted from its true value of 373 K to 325 K to improve the fit of the correlation to experimental data. Note that for a mixture with a pseudocritical temperature of 304.2 K, the value for pure CO_2, the ratio of MMPs in Eq. 6.8 reduces to unity. Sebastian et al. (1985) used the correlation of Yellig and Metcalfe (1980) to calculate the MMP of pure CO_2, P_{CO_2}.

The correlation of Sebastian et al. (1985) was based mainly on west Texas oils. The average absolute deviation from experimental data was 12%, with a standard deviation of 17% when P_{CO_2} was calculated with the Yellig and Metcalfe (1980) correlation. When experimental data were used for P_{CO_2}, the average deviation decreased to 8%. Sebastian et al. (1985) noted that the correlation was in general agreement with that of Alston et al. (1985).

Stalkup (1983a) compared the predictions from four correlations [not including the Alston et al. (1985) correlation] against 19 experimental MMP values. The average error in MMP was 580 psi, and the maximum error was 2,370 psi. The empirical correlations are useful for calculating first estimates of MMP and as a screening device. The correlations cannot be relied on to give accurate values for specific oils. Where an accurate value is required, experimental measurements should be made with a slimtube apparatus.

The empirical correlations also are useful to indicate the manner in which various parameters affect the MMP. As summarized by Stalkup (1983a), Holm and Josendal (1974) concluded the following relative to development of miscibility in CO_2 displacements.

1. Dynamic miscibility occurs when the CO_2 density is sufficiently great that the dense gaseous CO_2 or liquid CO_2 solubilizes the C_5 through C_{30} hydrocarbons contained in the reservoir oil. For the oils examined in their study, Holm and Josendal (1974) observed that miscible displacement occurred at CO_2 densities ranging from 0.4 to 0.65 g/cm^3, depending on both the total amount of C_5 through C_{30} hydrocarbons in the C_{5+} fraction of the reservoir oil and the distribution of hydrocarbons in this carbon-number range.

2. Reservoir temperature is an important variable affecting MMP because of its effect on the pressure required to achieve the CO_2 density necessary for miscible displacement. A higher temperature results in a higher miscibility pressure requirement when other factors are equal.

3. MMP is inversely related to the total amount of C_5 through C_{30} hydrocarbons present in the crude oil. The more of these hydrocarbons contained in the oil, the lower the MMP.

4. MMP is affected by the molecular-weight distribution of the individual C_5 through C_{30} hydrocarbons in the reservoir oil. Low-molecular-weight, gasoline-range hydrocarbons are particularly effective in promoting miscibility and result in a lower MMP requirement when other factors are equal.

5. MMP also is affected, but to a much smaller degree, by the types of hydrocarbons present in the C_5 through C_{30} fraction. For example, aromatics result in lower MMP.

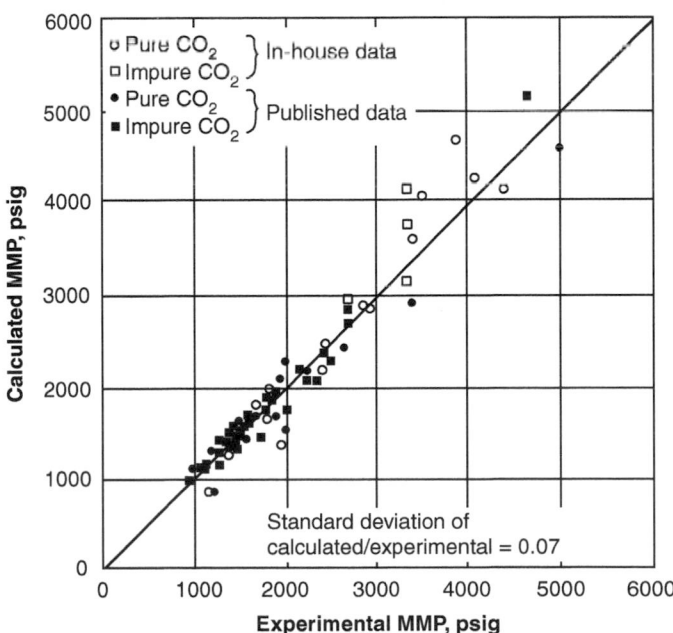

Fig. 6.39—Accuracy of correlation for predicting pure and impure CO_2 MMP (Alston et al. 1985).

Constituent	Molecular Weight	Critical Temperature (°R)	Critical Pressure (psia)
Methane	16.04	343.3	673.1
Ethane	30.07	549.8	708.3
Propane	44.09	666.0	617.4
Isobutane	58.12	765.3	550.7
n-Butane	58.12	765.3	550.7
Isopentane	72.15	829.8	483.0
n-Pentane	72.15	845.6	489.5
Isohexane, 121*	86.17	880.9	450.5
n-Hexane, 155*	86.17	914.1	439.7
Isoheptane, 174*	100.20	937.6	417.0
Isoheptane, 194*	100.20	955.9	400.0
n-Heptane, 209*	100.20	972.3	396.9
Isooctane, 229*	114.22	999.0	379.0
Isooctane, 240*	114.22	1,019.0	391.0
n-Octane, 258*	114.22	1,024.9	362.1
n-Nonane	128.25	1,071.0	331.0
n-Decane	142.88	1,114.0	306.0
Helium	4.0	9.4	33.2
Air	29.0	238.4	547.0
Nitrogen	28.0	226.9	492.0
Oxygen	32.0	277.9	730.0
CO_2	44.01	547.7	1,073.0
H_2S	34.08	672.4	1,306.0

* Boiling point in degrees Farenheit.

Table 6.1—Critical properties of natural-gas constituents (Katz et al. 1959; Rossini 1953).

6. Properties of the heavy fraction (i.e., > C_{30} hydrocarbons) also influence MMP, although they are not as important as the total quantity of C_{30+} material.
7. Development of dynamic miscibility does not require the presence of C_2 through C_4 hydrocarbons.
8. The presence of methane in the reservoir oil does not change the MMP appreciably.

Stalkup (1983a) also points out that there is not universal agreement on these conclusions. For example, Yellig and Metcalfe (1980) are not in agreement with Holm and Josendal (1974, 1982) on the effect of C_{5+} components in the oil phase. Most of the conclusions, however, are reflected in the different correlations that have been discussed.

Conclusion 8 is not reflected in the correlations. In fact, the Alston et al. (1985) correlation predicts that the presence of C_1 will give a slightly higher MMP. The conclusion of Holm and Josendal (1974) was based on the observation that C_1 in the oil tends to be stripped out by the initial contacts of displacing gas. In subsequent contacts, the C_1 content of the oil has been reduced to the point that the MMP value is not markedly affected.

Example 6.4—Calculation of MMP With Different Correlations. It is desired to estimate the MMP for a specified crude oil. Two different displacing fluids are to be considered. The first is 100% CO_2, and the second is a mixture of 92.5 mol% CO_2 and 7.5 mol% C_1.

The MMP values are to be estimated with the correlations of Holm and Josendal (1974), Yellig and Metcalfe (1980), Alston et al. (1985), and Sebastian et al. (1985). The reservoir temperature is 130°F, the C_{5+} molecular weight for oil components is 185.8, the volatiles make up 5.0 mol% of the oil and the intermediates 7.5 mol%, and the ratio of mole fractions of volatiles to intermediates in the oil is 0.667.

Solution. Note that volatiles are C_1 and N_2 and intermediates are C_2, C_3, C_4, CO_2, and H_2S. The CO_2 critical temperature is 87.8°F, and the C_1 critical temperature is –116.4°F.

1. 100% CO_2 displacing fluid. The Holm and Josendal (1974) correlation (Fig. 6.36) gives a C_{5+} molecular weight of 185.8 and an MMP of 1,880 psia. The Yellig and Metcalfe (1980) correlation (Fig. 6.37) gives an MMP of 1,630 psia (assuming that the bubblepoint pressure is less than this value). The Alston et al. (1985) correlation (Eq. 6.4) gives

$$P_{CO_2} = 8.78 \times 10^{-4} \times (130°F)^{1.06} \times (185.8)^{1.78} \times (0.667)^{0.136}$$

$$= 8.78 \times 10^{-4} \times 174.1 \times 10,930 \times 0.946$$

and $p_{mm} = P_{CO_2} = 1,581\,\text{psia}.$

The Sebastian et al. (1985) correlation is not applicable to pure CO_2 streams.
The measured value for this oil made with a slimtube apparatus was 1,500 psia. The Holm and Josendal (1974) error was

$$E_r = (1,880 \text{ psia} - 1,500 \text{ psia})/(1,500 \text{ psia}) \times 100 = 25\%.$$

The Yellig and Metcalfe (1980) error was

$$E_r = (1,630 \text{ psia} - 1,500 \text{ psia})/(1,500 \text{ psia}) \times 100 = 8.7\%.$$

The Alston et al. (1985) error was 5.4%.

2. 92.5 mol% CO_2 and 7.5 mol% C_1 displacing fluid. The correlations of Holm and Josendal (1974) and Yellig and Metcalfe (1980) are for 100% CO_2 and do not predict the effect of adding impurities to the CO_2 stream. For the Alston et al. (1985) correlation, Eqs. 6.2 through 6.5 are applicable. First, calculate the weight fraction of CO_2 and C_1: $CO_2 = $ 92.5 mol% and 97.1 wt% and $C_1 = 7.5$ mol% and 2.9 wt%. The weight percent is given by

$$CO_2 = \left(\frac{0.925 \times 44}{0.925 \times 44 + 0.075 \times 16} \right) \times 100 = 97.1 \text{ wt\%}.$$

Apply Eq. 6.5 for T'_{cm}:

$$T'_{cm} = 0.029 \times 343.3 + 0.971 \times 547.7 - 459.7 = 82.1°F.$$

Apply Eq. 6.6:

$$F_{imp} = \left(\frac{87.8}{82.1} \right)^{1.93 \times 87.8/82.1} = 1.15.$$

Apply Eq. 6.7:

$$p_{mm} = P_{CO_2} - \text{imp} - 1,581 \text{ psia} \times 1.15 = 1,818 \text{ psia}.$$

For the Sebastian et al. (1985) correlation apply Eq. 6.9 to obtain T_{cp}:

$$T_{cp} = 0.925 \times 304.2 + 0.075 \times 190.6 = 295.7 \text{ K}.$$

Apply Eq.6.8:

$$\frac{P_{CO_2} - \text{imp}}{P_{CO_2}} = 1.0 - 2.13 \times 10^{-2} (295.7 - 304.2)$$
$$+ 2.51 \times 10^{-4} (295.7 - 304.2)^2$$
$$- 2.35 \times 10^{-7} (295.7 - 304.2)^3$$
$$= 1.0 + 0.181 + 0.018 + 0.00014$$
$$= 1.199.$$

From the Yellig and Metcalfe (1980) correlation results in Part 1, $P_{CO_2} = 1,630$ psia. Therefore,

$$p_{mm} = P_{CO_2} - \text{imp} = 1,630 \times 1.199 = 1.954 \text{ psia}.$$

This value is 136 psia higher than the result obtained with the Alston et al. (1985) correlation. The measured value is not known; however, the MMP would be expected to increase with the addition of C_1 to the CO_2 stream.

Note that both the Alston et al. (1985) and the Sebastian et al. (1985) correlations indicate that the presence of intermediate hydrocarbons (C_2, C_3, etc.) and H_2S decreases the MMP. For example, in Eq. 6.8, components that have a critical temperature higher than CO_2 will tend to increase the pseudocritical temperature above that of CO_2 (304.2 K), leading, in some cases, to a $P_{CO_2} - \text{imp}/P_{CO_2}$ ratio less than unity.

EOS Phase-Behavior Calculations. EOSs are used to predict behavior by predicting fugacity coefficients, which in turn are used to predict phase-equilibrium *K* values. Some excellent work (Redlich and Kwong 1949; Fussell and Yanosik 1978) along these lines has been performed, with much of the later work (Turek et al. 1984; Henry and Metcalfe 1983) focusing on CO_2/hydrocarbon systems. A typical approach is to apply the generalized Redlich-Kwong EOS (Redlich and Kwong 1949) with appropriate mixing rules developed for CO_2/hydrocarbon systems (Fussell and Yanosik 1978; Turek et al. 1984). The success of the method depends on the correct predictions of the number and compositions of phases present at a given temperature, pressure, and overall reservoir fluid composition. Work such as that of Henry and Metcalfe (1983), Metcalfe and Yarborough (1979), and Turek et al. (1984) has shown that CO_2/hydrocarbon systems are complex and that, under certain conditions, three or more phases may exist in equilibrium. Thus, sophisticated EOS modeling with computer technology is required to describe adequately the compositional effects that occur in enhanced-oil-recovery (EOR) processes involving CO_2 injection.

Solution techniques for phase behavior vary depending on the EOS used, but three thermodynamic restrictions imposed on the system must be satisfied: (1) the material balance must be preserved, (2) the chemical potential in each phase must be equal for each component, and (3) the system of phases predicted must have the lowest possible Gibbs free energy at the given temperature and pressure.

Reliable phase behavior for CO_2/hydrocarbon systems can be predicted with EOS-type calculations, but this method is not for the novice and is not considered in detail in this text. Also, there are limitations in using an EOS. Experimental data are needed for calibration of an equation (i.e., an EOS does not do a good job of predicting phase behavior a priori). Also, for systems of the type considered here, data close to the plait point are required for calibration; otherwise the EOS may not predict the MMP well.

Examples of different approaches to calculation of MMP, generally on the basis of EOS models, include numerical simulation (Stalkup 1987), application of analytical theory (Wang and Orr 1998), application of the method of characteristics (Yuan and Johns 2005), and the multiple-mixing-cell method (Ahmadi and Johns 2008).

6.8 Fluid Properties in Miscible Displacement

The performance of a miscible displacement process depends on fluid physical properties that affect flow behavior in a reservoir. Two important properties are density and viscosity, although other properties, such as compressibility, solubility in an aqueous phase, and IFT, are sometimes required in calculations.

Example properties of fluids used in miscible displacement are presented in this section to provide a background for discussions of flow factors that affect miscible displacement in subsequent sections and for calculations of performance.

Properties based on rock/fluid interactions, such as relative permeability and capillary pressure, are not discussed here. Additional data may be found in Stalkup (1983a) and Katz et al. (1959) and engineering and scientific handbooks.

6.8.1 Fluid Density. The result in a displacement process can be gravity override, underride, or fingering, depending on the difference in density between the displaced and displacing fluids, as discussed in Chapter 4. Knowledge of the relative densities of the fluids and fluid mixtures is important for the process design.

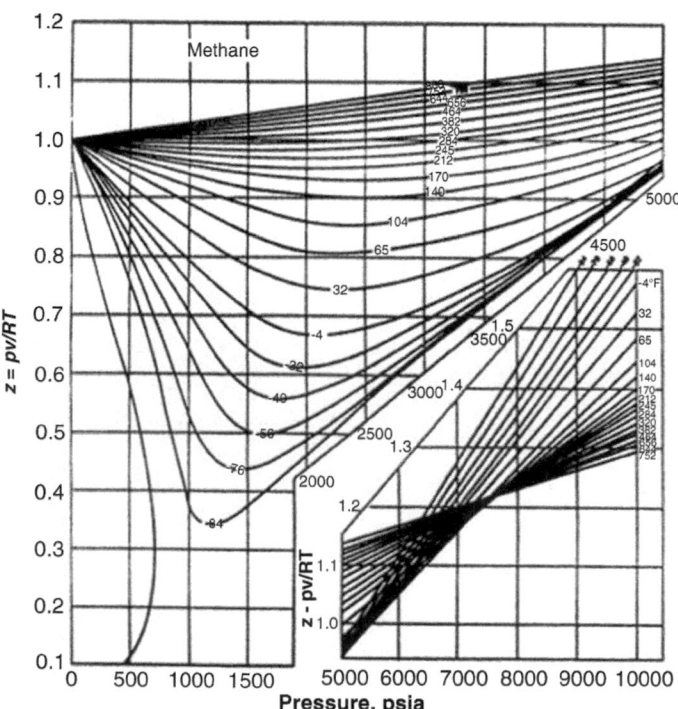

Fig. 6.40—Compressibility factors for methane (Katz et al. 1959; Brown et al. 1948).

Crude oils tend to have specific gravities (60/60°F) in the 0.80 to 0.95 range (ρ = 50 to 60 lbm/ft³). Densities at reservoir temperature and pressure differ from those at surface conditions, and the change in density as a function of pressure and temperature can be estimated. McCain (1973) describes the application of ideal solution behavior to make such estimates. Mixing rules based on ideal solution behavior also can be used to estimate the effect of concentration on density (McCain 1973).

The densities of fluids used in miscible processes generally are significantly smaller than the densities of the crude oils displaced, except in the case of CO_2 at moderately low reservoir temperatures (90 to 110°F).

Application of Compressibility Factors for Gas Density. Gas density can be calculated with the EOS.

$$pv = ZRT, \quad \dots\dots\dots\dots\dots\dots (6.10)$$

where p = pressure; v = specific volume; T = temperature; R = gas constant, 10.73 (psi-ft³)/(°R-lbm-mol); and Z = compressibility factor, dimensionless. The compressibility factor is a measure of the deviation from the ideal-gas law.

Fig. 6.40 (Brown et al. 1948; Katz et al. 1959) gives the compressibility factor for methane as a function of temperature and pressure. **Fig. 6.41** (Katz et al. 1959) gives compressibility factors for a number of hydrocarbons as functions of reduced temperature and pressure. The reduced properties are defined as

$$T_r = T / T_c \quad \ldots \ldots \ldots \ldots \ldots \ldots (6.11)$$

and $p_r = p/p_c,$ $\ldots \ldots \ldots \ldots \ldots (6.12)$

where T_r = reduced temperature, p_r = reduced pressure, T_c = critical temperature, and p_c = critical pressure. Table 6.1 lists critical temperatures and pressures for several hydrocarbons.

Example 6.5—Calculation of C_1 and C_3 Density.
Consider propane, C_3, at a temperature of 270°F and a pressure of 1,500 psia. Calculate the density.

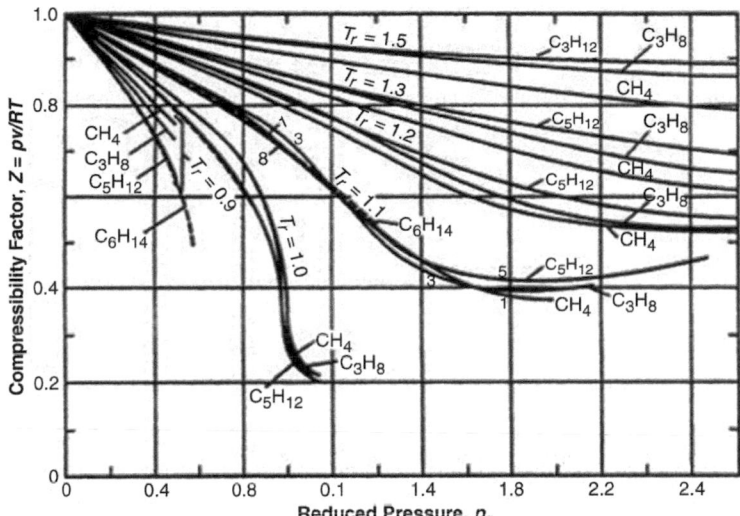

Fig. 6.41—Compressibility factors for various hydrocarbons (Katz et al. 1959).

Solution.

$$T_r = 730°R \: / \: 666°R = 1.10.$$

$$p_r = 1,500 \text{ psia} \: / \: 617.4 \text{ psia} = 2.43.$$

At these conditions, C_3 is above the critical temperature and pressure. From Fig. 6.40, $Z \approx 0.4$. From Eq. 6.10,

$$v = \frac{0.4 \times \left[10.73 (\text{psi-ft}^3)/(°R\text{-lbm-mol}) \right] \times 730°R}{1,500 \text{ psia}}$$

$$= 2.09 \text{ ft}^3 \text{/lbm mol.}$$

and $p = \dfrac{1}{2.09 \text{ ft}^3 \text{/lbm mol}} = 0.48 \text{ lbm mol/ft}^3$

$$= 0.48 \text{ lbm mol/ft}^3 \times 44 \text{ lbm/lbm mol}$$

$$= 21.1 \text{ lbm/ft}^3.$$

The density is approximately 40 to 50% of the density of a typical crude oil. At this same temperature and pressure, the density of C_1 is approximately 3.2 lbm/ft³ (from Z determined from Fig. 6.40).

The correlation of the Z-factor data in terms of reduced properties (i.e., the theorem of corresponding states) is expressed in a generalized Z-factor chart like that shown in **Fig. 6.42** (Katz et al. 1959; Standing and Katz 1942). In this chart, temperature and pressure are expressed as pseudoreduced temperature and pseudoreduced pressure. These are calculated with Eqs. 6.11 and 6.12, but for mixtures, T_c and p_c are replaced by pseudocritical properties, T_{cp} and p_{cp}, calculated on a molar-average basis.

$$T_{cp} = \sum_i X_i T_{ci} \quad \ldots \ldots \ldots \ldots \ldots \ldots \ldots \ldots \ldots \ldots \ldots \ldots \ldots \ldots \ldots \ldots \ldots \ldots \ldots \ldots \ldots (6.9)$$

and $P_{cp} = \sum_i X_i p_{ci},$ $\ldots \ldots \ldots \ldots \ldots \ldots \ldots \ldots \ldots \ldots \ldots \ldots \ldots \ldots \ldots \ldots \ldots \ldots \ldots \ldots \ldots (6.13)$

where T_{cp} = pseudocritical temperature of a mixture, P_{cp} = pseudocritical temperature of a mixture, T_{ci} = critical temperature of Component i, p_{ci} = critical pressure of Component i, and X_i = mole fraction of Component i.

Liquid Density—Intermediate Hydrocarbons. Density data for propane are shown in **Fig. 6.43** (Brown et al. 1948; Katz et al. 1959). Note that at the conditions given in Example 6.5 (270°F and 1,500 psia), the density is approximately 0.34 g/cm³ or 21.2 lbm/ft³, the same value as obtained from the compressibility-factor correlation.

Liquid densities are read at temperatures below the critical temperature of 206.4°F and at pressures sufficiently high to liquefy the C_3. As seen, the liquid-phase compressibility is relatively small. At 150°F and 1,500 psia, the density is approximately 0.47 g/cm³ (29.3 lbm/ft³). As indicated earlier, this is significantly less than the density of most crude oils at reservoir conditions.

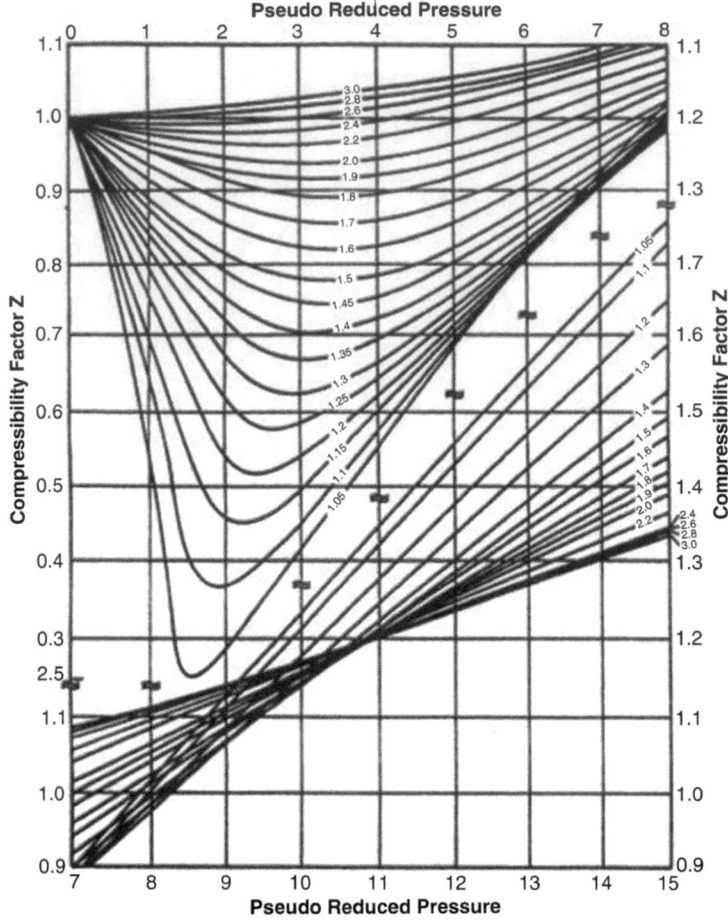

Fig. 6.42—Compressibility factor for natural gases (Katz et al. 1959; Standing and Katz 1942).

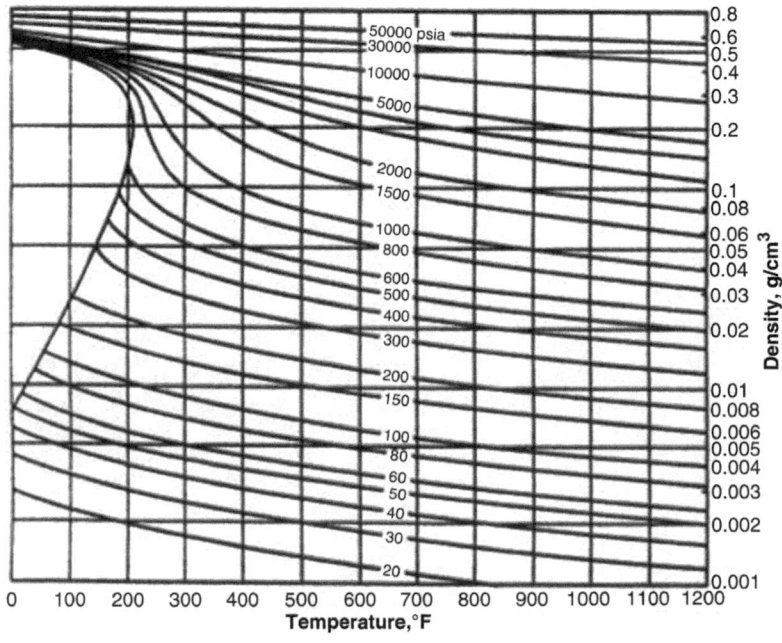

Fig. 6.43—Density of propane (Brown et al. 1948; Katz et al. 1959).

Alani and Kennedy (1960) give data on liquid molal volumes of several hydrocarbons over a wide range of temperatures and pressures. They also present information on temperatures of hydrocarbons for temperatures up to 250°F and pressures up to 5,000 psia.

Density of CO₂. Data on compressibility factors and densities for CO_2 given by Sage and Lacey (1955) are reproduced by Stalkup (1983a). **Fig. 6.44** shows compressibility factors at four temperatures. **Fig. 6.45** gives density as a function of temperature and pressure.

Example 6.6—Calculation of CO₂ Density. Calculate the CO_2 density at 150°F and 1,500 psia. As seen from Table 6.1, these conditions are above the critical temperature and pressure.

Solution. Apply the Z-factor from Fig. 6.44, $Z \approx 0.54$.

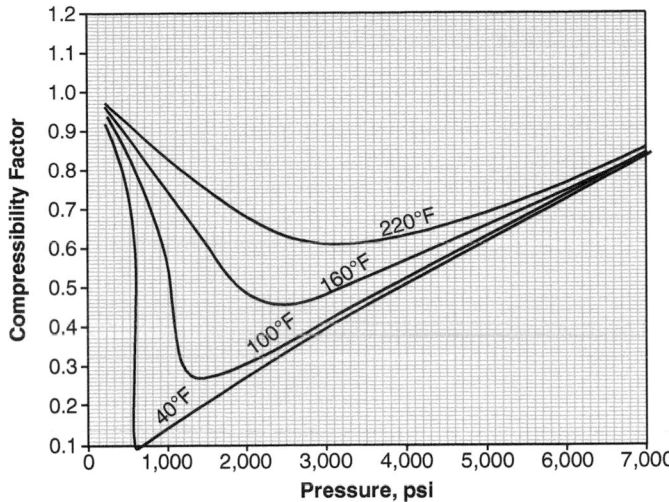

Fig. 6.44—Compressibility factors for CO₂ (Sage and Lacey 1955; Stalkup 1983a).

$$v = ZRT/p$$

$$= \frac{0.54 \times \left[10.73(\text{psi-ft}^3)/(°\text{R-lbm-mol})\right] \times 610°\text{R}}{1,500 \text{ psia}}$$

$$= 2.36 \text{ ft}^3/\text{lbm mol} \times \frac{1}{44 \text{ lbm/lbm mol}}$$

$$= 0.0536 \text{ ft}^3/\text{lbm}.$$

$$\rho = 18.6 \text{ lbm/ft}^3.$$

The density of C_3 at these same conditions is 29.3 lbm/ft³.

From Fig. 6.45, $\rho = 0.30$ g/cm³ or 18.7 lbm/ft³. At 270°F and 1,500 psia, the density is approximately 10.0 lbm/ft³. The density of C_3 at these same conditions is 21.1 lbm/ft³ (Example 6.5).

Note that at 95°F and 1,500 psia, CO_2 density (Fig. 6.45) is 0.73 g/cm³ (45.5 lbm/ft³), which is close to the density of light crude oils at the same conditions. The density of C_3 at 95°F and 1,500 psia is only 0.52 g/cm³ (32.4 lbm/ft³), indicating an important difference in the compressibility behavior of CO_2 and C_3.

Density of CO₂-Saturated Liquid Hydrocarbon Phases. Densities of liquid mixtures of hydrocarbons and of CO_2 and hydrocarbons can be calculated with an EOS (Simon et al. 1978). An estimate of liquid density can be calculated on the basis of a mole-averaged density of pure components as follows:

$$\frac{X_1}{\rho_1} + \frac{X_2}{\rho_2} + \frac{X_3}{\rho_3} + ... = \frac{1}{\rho_{mix}}, \quad .... (6.14)$$

where X_i = mole fraction of Component i in the mixture, ρ_i = density of Component i, and ρ_{mix} = density of the mixture. ρ_{mix} can be converted to units of lbm/ft³ by multiplication by the average molecular weight of the mixture.

A problem, however, is that the condensed-phase density of a component such as CO_2, normally a vapor or fluid above the critical point at the temperature and pressure of the mixture, is not known or predicted easily. In this case, an "apparent" liquid or condensed-phase density can be used if experimental data are available. For example, Orr and Silva (1983) and Orr et al. (1983) conducted experiments with both a multiple-contact

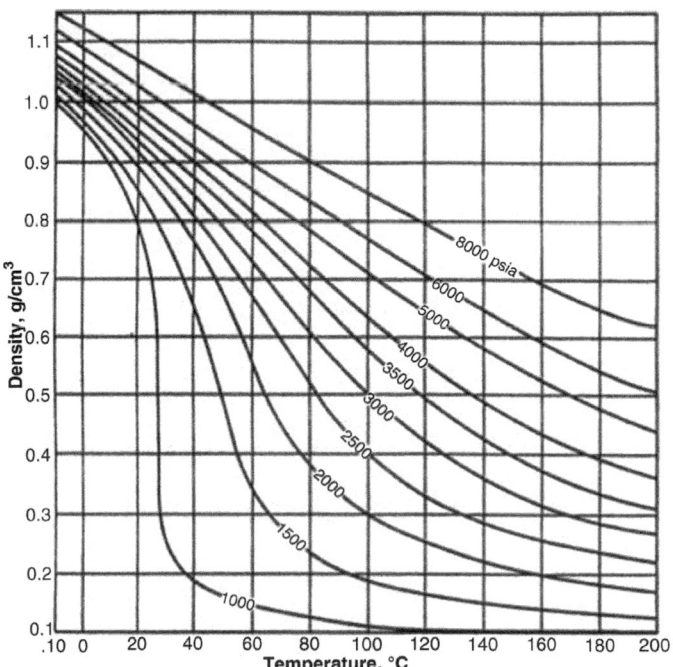

Fig. 6.45—Density of CO₂ (Ruska Instrument Corp.).

device and a slimtube in which CO_2/crude-oil phase behavior and displacement behavior were measured. Measurements were made at 32°C and various pressures.

The observed experimental data from the slimtube were simulated with a relatively simple mathematical model in which the different components were allowed to have different densities in the vapor and liquid phases. The system was assumed to consist of three pseudocomponents: CO_2, C_5 through C_{12}, and C_{13+}. **Table 6.2** shows apparent densities used in the calculations for the different components in the different phases at four pressures. The performance data were well-matched, assuming that the hydrocarbons had the same apparent density in each phase. However, CO_2 had a significantly larger apparent density in the liquid phase than in the vapor phase. And at pressures where two liquid phases existed in equilibrium, the CO_2 apparent density was different in each phase. Green and Swift (1985) give similar density data.

6.8.2 Fluid Viscosity. Mobility ratio in a displacement process is a direct function of the viscosities of displaced and displacing fluids. For miscible displacements, assuming the relative permeability relationships of the different nonaqueous fluids are the same,

$$M = \mu_d/\mu_D, \dotfill (6.15)$$

where μ_d = viscosity of the displaced phase and μ_D = viscosity of the displacing phase.

The viscosities of crude oils vary over a wide range, from less than the value for water (≈ 1.0 cp) to very large values for heavy oils. Viscosities of miscible solvents tend to be significantly smaller in value; thus, the mobility ratio is usually unfavorable in a miscible process.

Viscosity of Hydrocarbon Gases. Katz et al. (1959) present viscosity data for natural gases and intermediate-molecular-weight hydrocarbons. **Fig. 6.46** gives viscosity as a function of temperature and pressure for C_3. A typical value, which corresponds to temperature and pressure conditions used in examples in the previous section ($T = 270$°F and $p = 1,500$ psia) is $\mu = 0.036$ cp. C_3 is a fluid above the critical point at these conditions.

Viscosities of gases and fluids above critical conditions have been correlated as functions of reduced properties (Stalkup 1983a; Turek et al. 1984). **Fig. 6.47** presents a correlation by Gonzalez and Lee (1968). The correlation is applicable to single components and mixtures of hydrocarbons.

Component	Molecular Weight	800 psi		1,000 psi			1,200 psi		1,400 psi	
		Vapor	Liquid	Vapor	CO_2-Rich Liquid	Oil-Rich Liquid	CO_2-Rich Liquid	Oil-Rich Liquid	CO_2-Rich Liquid	Oil-Rich Liquid
CO_2	44	0.143	0.918	0.2313	0.780	0.898	0.74,0.82	0.895	0.79,0.83	0.893
C_5 through C_{12}	119	0.689	0.689	0.696	0.696	0.696	0.702	0.702	0.708	0.708
C_{13+}	323	0.978	0.978	0.978	0.978	0.978	0.978	0.978	0.978	0.978

Density (g/cm³)

Table 6.2—Component densities for simulations of slimtube displacements (Orr et al. 1983).

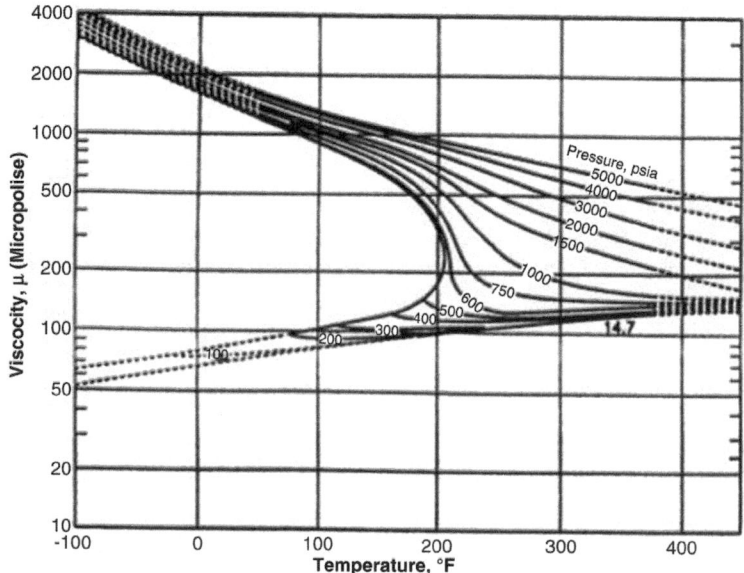

Fig. 6.46—Viscosity of propane (Katz et al. 1959).

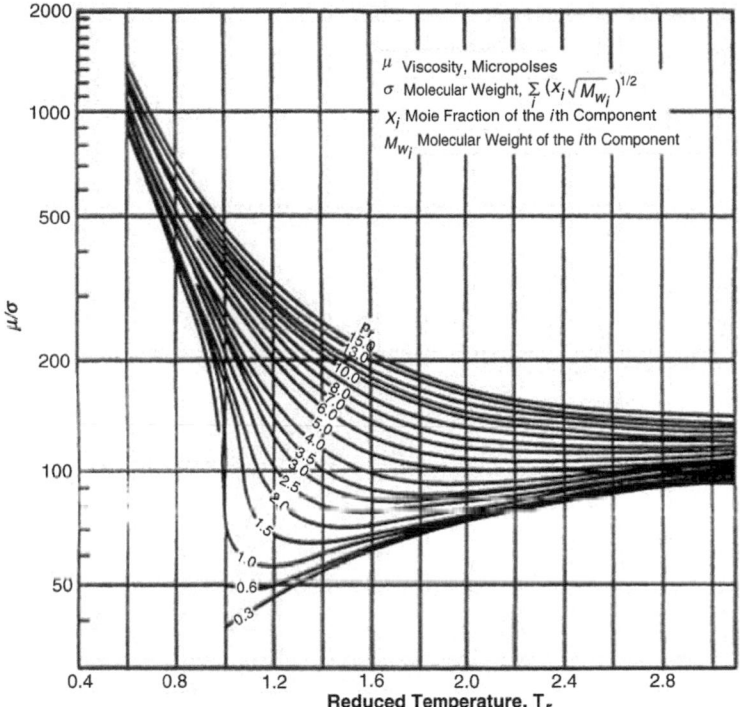

Fig. 6.47—Viscosities of hydrocarbon systems as functions of reduced temperature (Gonzalez and Lee 1968).

Example 6.7—Viscosity Calculation for a Gas Mixture. Calculate the viscosity of a mixture that is 70 mol% C_3 and 30 mol% C_2 at 150°F and 1,500 psia using the correlation of Gonzalez and Lee (1968).

Solution. Calculate pseudoreduced properties.

$$T_{cp} = \sum X_i T_{ci}$$
$$= 0.70 \times 666°R + 0.30 \times 549.8°R = 631.1°R.$$

$$P_{cp} = \sum X_i P_{ci}$$
$$= 0.70 \times 617.4 \text{ psia} + 0.30 \times 708.3 \text{ psia}$$
$$= 644.7 \text{ psia.}$$

The critical properties of the pure components were taken from Table 6.1

$$T_r = T/T_{cp} = 610°R/631.1°R = 0.97.$$
$$p_r = p/P_{cp}$$
$$= 1,500 \text{ psia}/644.7 \text{ psia} = 2.33.$$

Apply Fig. 6.47.

$$\mu/\sigma = 230$$

$$\sigma = \left(\sum_i X_i \sqrt{M_{wi}} \right)^{\frac{1}{2}}$$
$$= \left(0.7 \times \sqrt{44} + 0.3\sqrt{30} \right)^{\frac{1}{2}}$$
$$= 2.51.$$
$$\mu = 577 \mu p$$
$$= 0.058 \text{ cp.}$$

Reference to Fig. 6.4 indicates that the mixture is well above the locus of critical points at the specified conditions.

At 270°F and 1,500 psia, the viscosity (Fig. 6.47) is $\mu = 0.027$ cp. This value is less than the viscosity of pure C_3 at the same temperature and pressure (0.035 cp). The value of viscosity for pure C_3 at 270°F and 1,500 psia, obtained from Fig. 6.47, is in agreement with that read from Fig. 6.46.

Viscosity of Hydrocarbon Liquids. Methods and charts for estimating liquid viscosities are presented by Katz et al. (1959). For a few intermediate hydrocarbons, charts such as Fig. 6.46 are available.

A typical value for C_3 at $T = 150$°F and $p = 1,500$ psia, where C_3 is a liquid, is $\mu = 0.083$ cp (Fig. 6.46).

Fig. 6.48 (Katz et al. 1959; Rossini 1953) gives viscosities for a number of pure hydrocarbons at a pressure of 1.0 atm and a range of temperatures. For crude oils and mixtures of hydrocarbons, the method of Lohrenz et al. (1964) is commonly used.

CO_2 Viscosity. The CO_2 viscosity as a function of temperature and pressure is given in **Fig. 6.49** (Stalkup 1983a; Kennedy and Thodos 1961). A typical value at $T = 150$°F and $p = 1,500$ psia, where CO_2 is above the critical point, is $\mu = 0.025$ cp.

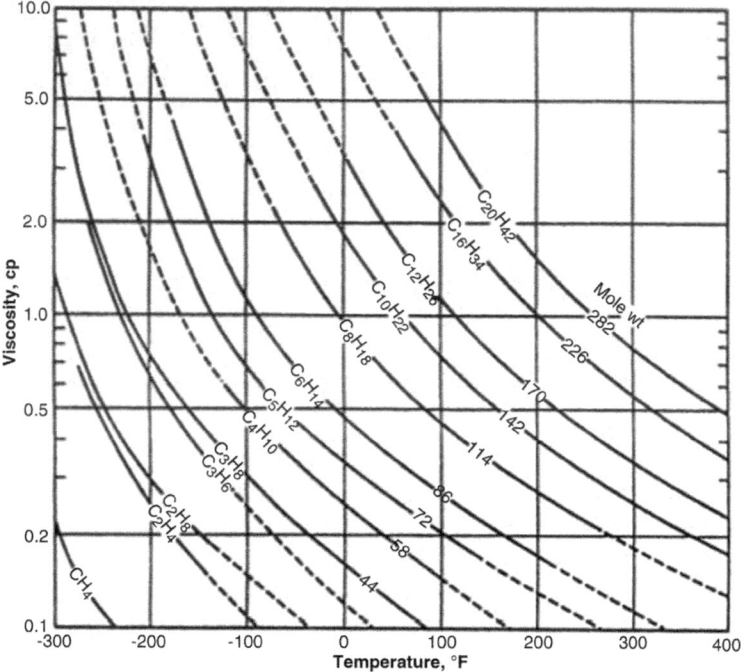

Fig. 6.48—Viscosities of hydrocarbon liquids at 1.0 atm (Katz et al. 1959; Rossini 1953).

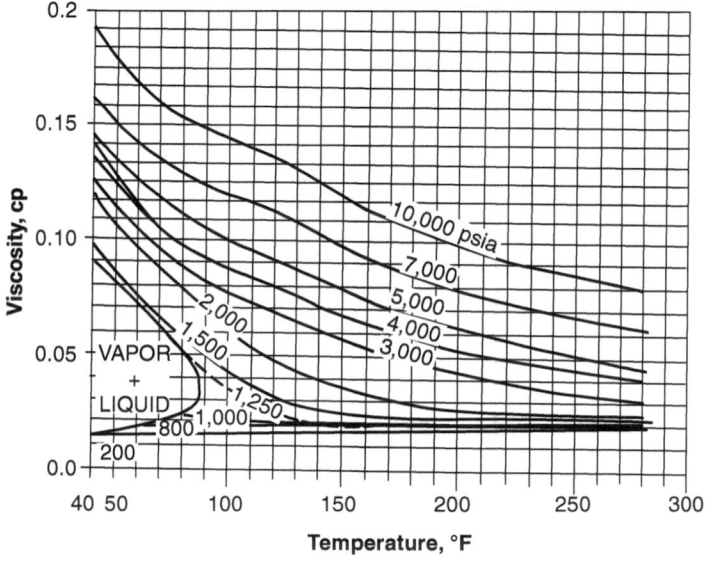

Fig. 6.49—Viscosity of CO_2 (Kennedy and Thodos 1961; Stalkup 1983a).

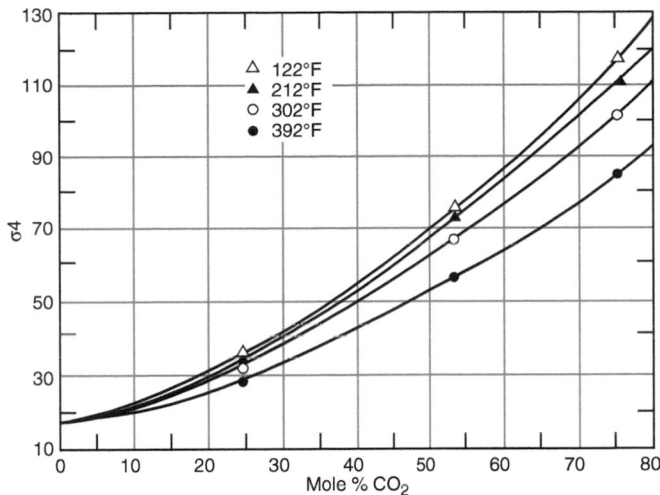

Fig. 6.50—Correction chart for mixtures containing CO_2 (Gonzalez and Lee 1968).

Gonzalez and Lee (1968) present a correction-factor chart for mixtures of hydrocarbon gases that contain up to 80 mol% CO_2. This chart, given in **Fig. 6.50,** should be used in conjunction with Fig. 6.47. The σ value calculated from the σ^4 value read from Fig. 6.50 should be added to the σ value obtained when the procedure in Fig. 6.47 is used.

6.8.3 Additional Properties. Properties other than density and viscosity are required in design. For example, such data include diffusion coefficients, surface tension, additional transport properties, and thermodynamic properties. Data can be found in engineering and scientific handbooks, such as that by Katz et al. (1959).

In the design of a CO_2 displacement process, it is useful to estimate the amount of CO_2 that is dissolved in a water phase. The gas loss owing to solubility in water is significant. **Fig. 6.51** (Dodds et al. 1956) shows CO_2 solubility in fresh water as a function of temperature and pressure. The effect of salinity on solubility is important and is shown in **Fig. 6.52** as a function of pressure at a temperature of 77°F (Stewart and Munjal 1970).

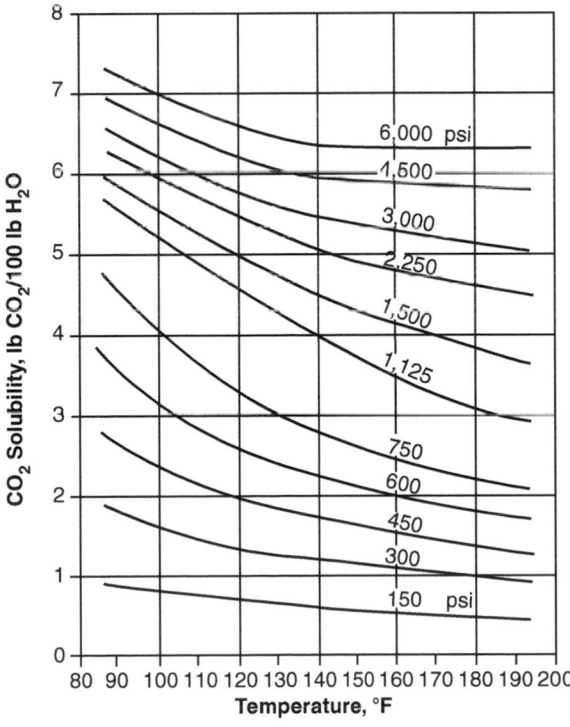

Fig. 6.51—CO_2 solubility in fresh water (Dodds et al. 1956; Stalkup 1983a).

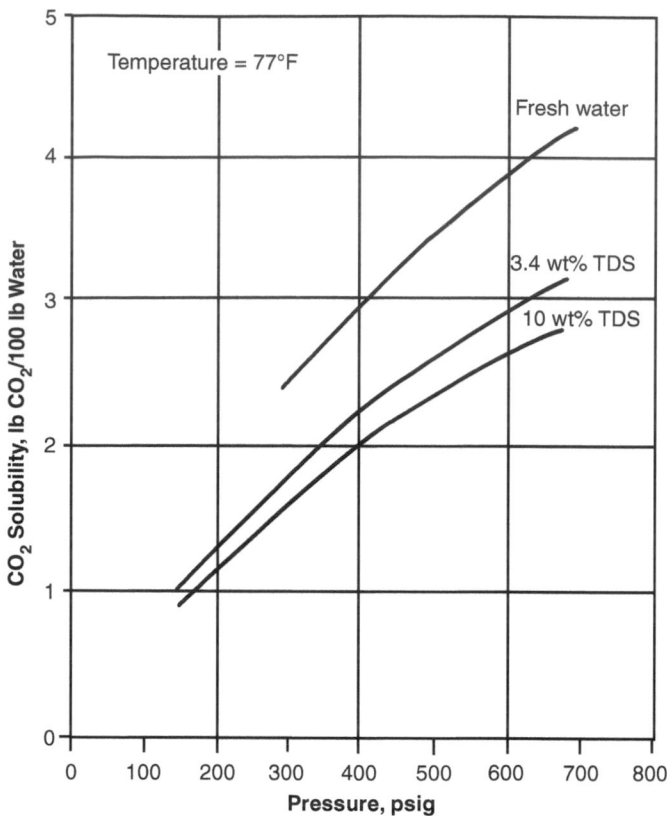

Fig. 6.52—Effect of salinity on CO_2 solubility in water at a temperature of 77°F (Stewart and Munjal 1970; Stalkup 1983a).

Example 6.8—CO_2 Solubility in Water at Reservoir Conditions. Assume that CO_2 is injected into a reservoir at 1,500 psia and 150°F. Estimate the percentage loss of CO_2 by solubility into water in the reservoir, assuming that 1.0×10^6 scf of CO_2 contacts a residual water phase that is at a 25% saturation. The rock porosity is 0.20, and oil saturation is negligible. Calculate the loss assuming that the water is fresh and that the water in contact with the CO_2 becomes saturated with CO_2.

Solution. Calculate the volume of reservoir invaded by the CO_2. From Fig. 6.45,

$$\rho_{CO_2} = 0.31 \text{ g/cm}^3 \left[\left(62.4 \text{ lbm/ft}^3 \right) / \left(\text{g/cm}^3 \right) \right]$$

$$= 19.3 \text{ lbm/ft}^3,$$

$$m_{CO_2} = 1.10^6 \text{ scf} \left[(1/379 \text{ scf}) / (\text{lbm mol}) \right]$$

$$= 1.16 \times 10^5 \text{ lbm},$$

$$V_{iCO_2} \text{ (at reservoir conditions)} = 1.16 \times 10^5 \text{ lbm} \left(\frac{1}{19.3 \text{ lbm/ft}^3} \right)$$

$$= 6,010 \text{ ft}^3,$$

$$V_R = 6,010 \text{ ft}^3 / \phi S_{CO_2}$$

$$= 6,010 \text{ft}^3 / \left[0.20 (1 - 0.25) \right]$$

$$= 4.01 \times 10^4 \text{ ft}^3,$$

and $M_w = 4.01 \times 10^4 \text{ ft}^3 \phi S_w \rho_{H_2O}$

$$= 4.01 \times 10^4 \text{ ft}^3 \times 0.20 \times 0.25 \times 62.4 \text{ lbm/ft}^3$$

$$= 1.25 \times 10^4 \text{ lbm H}_2\text{O}.$$

From Fig. 6.51,

$$CO_2 \text{ solubility} \approx 4.2 \text{ lbm CO}_2 / 100 \text{ lbm H}_2\text{O}$$

and CO_2 loss $= \dfrac{4.2 \text{ lbm } CO_2}{100 \text{ lbm } H_2O} \times 1.25 \times 10^5 \text{ lbm } H_2O.$

$$= 5,250 \text{ lbm}$$

$$= 4.5\% \text{ of } CO_2 \text{ injected.}$$

Reference to Fig. 6.52 indicates that if the water had a salinity of approximately 3 to 10% total dissolved solids, the loss of CO_2 would be approximately 25 to 30% less. Note that at a lower pressure and temperature (e.g., 1,000 psia and 100°F), the relative loss of CO_2 would be significantly larger.

6.9 Factors Affecting Microscopic and Macroscopic Displacement Efficiency of Miscible Processes

Displacement efficiencies at the microscopic and macroscopic levels in a miscible process are less than 100%. The magnitudes of the efficiencies depend on a number of factors, including whether a displacement is conducted as a secondary or a tertiary process. In a secondary recovery process, this text assumes that there is no mobile water unless water is injected as a part of the process. In a tertiary process, both oil and water will be displaced and will be mobile. The existence of a flowing water phase may affect process performance.

6.9.1 Microscopic Displacement Efficiency—No Mobile Water. In a miscible displacement conducted as a secondary recovery process, the IFT between displacing (solvent) and displaced (oil) phases is zero. According to N_c correlations discussed in Chapter 2, because $N_c \rightarrow \infty$, residual saturation in portions of the rock contacted by the displacing phase should be essentially zero.

Experimental studies of the FCM process show that the residual saturation of the displaced phase is very small. In laboratory experiments in relatively homogeneous porous media in which solvent is continuously injected, typical oil recoveries are 97 to 100% of the OOIP (Metcalfe and Yarborough 1979). When the solvent is injected as a small primary slug followed by a secondary slug (e.g., C_3 followed by C_1), however, recoveries can be poorer as a result of dispersion and mixing of the different slug materials (Koch and Slobod 1957). Significant mixing can result in loss of miscibility at either the leading or the trailing edge of the primary slug displacement.

Laboratory studies of MCM processes have shown that recoveries are somewhat poorer than for first-contact processes (Metcalfe and Yarborough 1979; Holm and Josendal 1974; Kremesec and Sebastian 1988). Displacements conducted in slimtubes at pressures in excess of the MMP often yield recoveries on the order of 90 to 97% OOIP. Displacements in well-designed slimtube experiments should not be affected by such factors as viscous fingering or gravity override (Orr et al. 1982). Efficiencies should thus be representative of microscopic efficiencies.

There are different reasons for the reduced recoveries at the microscopic level in the MCM processes. One is that a finite distance of travel is required in the process before miscibility is achieved (i.e., a sufficient number of contacts between the different fluids must be made). It is thought that the distance required to achieve miscibility is relatively small. For example, in one reported laboratory study (Campbell and Orr 1985) in which MCM displacements were carried out in a small "micromodel," miscibility was developed over a distance equivalent to a few pore lengths. In laboratory studies (Negahban et al. 1990; Yellig 1982) in linear cores in which displacements were conducted at unfavorable mobility ratios, however, distances of up to 8 ft were required to obtain a displacement efficiency corresponding to a miscible displacement. The longer distance required was attributed to instabilities in the flow caused by viscous fingering or gravity tonguing. These effects are discussed further in Section 6.9.2. The fact that a finite distance is required to develop miscibility or to dampen viscous instabilities is the rationale for using slimtubes that are at least 20 ft long. Tubes are designed to be sufficiently long to minimize length effects. In a field application, this entrance-length effect on overall recovery is insignificant, although mixing and flow instabilities do affect recovery.

As discussed in Section 6.5.1 and illustrated in Fig. 6.19, in a vaporizing-type MCM process, a small amount of liquid-phase dropout can occur as a result of mixing effects when compositions are in the vicinity of the binodal curve. Bypassing in the flow process at the microscopic level owing to small-scale heterogeneities or dead-end pores, for example (Gardner et al. 1981), can cause mixing that is analogous to what would occur in a number of finite mixing cells placed in series (as opposed to mixing cells of infinitesimal size). The mixing process can result in overall compositions that are within the two-phase region, which would lead to the trapping of a residual liquid phase, though at a small saturation.

Mass transfer by dispersion in a displacement process also can lead to reduced displacement efficiencies, as described by Gardner et al. (1981). They used a numerical simulator to show the effect of dispersion on CO_2 displacement efficiency. **Fig. 6.53** shows one of their results on a pseudoternary diagram.

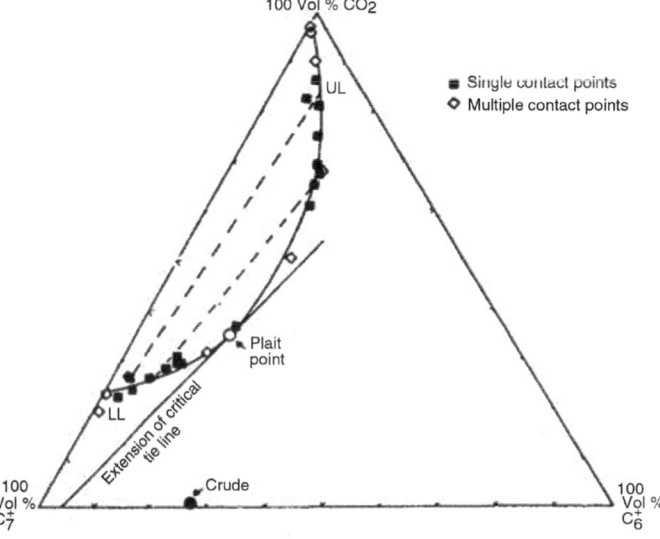

Fig. 6.53—Composition routes calculated by 1D simulator for displacements of Wasson crude by CO_2 at 2,000 psia and 105°F (Gardner et al. 1981).

Dispersion causes the composition path that occurs during a displacement to move into the multiphase region of the phase-behavior diagram. This results in the formation of a liquid phase, which remains as a trapped phase because it is at a relatively low saturation. Gardner et al. (1981) also showed that the higher the dispersion level, the poorer the calculated recovery efficiency. They concluded that displacement efficiencies achieved in their slimtube experiments at pressures well above the MMP were limited by the levels of dispersion in the experiments.

There can be a partial recovery of a liquid phase that is dropped out by phase behavior and trapped (Gardner et al. 1981). Part of the liquid will be revaporized or stripped by subsequent contact with a flowing gas phase, such as the CO_2 phase. The recovery by a revaporization process occurs at a relatively slow rate, however, and is probably small in most displacement processes.

Dispersion and mixing at the microscopic level, combined with the associated phase behavior, are probably the major reasons that microscopic displacement efficiencies in MCM processes conducted in the absence of mobile water are not 100%. Efficiencies typically range from 90 to 97% even in idealized experiments, such as those conducted in a slimtube apparatus.

6.9.2 Macroscopic Displacement Efficiency—No Mobile Water. Four major factors affect recovery efficiency at the macroscopic level in a miscible process in which there is no mobile water: mobility ratio, viscous fingering, gravity segregation, and reservoir heterogeneity. Chapters 3 and 4 discuss the general effects of these factors on any displacement process. This section, discusses mobility ratio, viscous fingering, and gravity segregation specifically in miscible displacements.

Effect of Mobility Ratio. The mobility ratio for a miscible process is given by Eq. 6.15:

$$M = \mu_d / \mu_D = \mu_o / \mu_s. \dots\dots\dots\dots\dots\dots\dots\dots\dots\dots\dots\dots\dots\dots\dots\dots (6.15)$$

As described in Section 6.8, the viscosities of miscible solvents are typically small (< 0.1 cp) and thus the mobility ratios are unfavorable. Stalkup (1983a) lists data for a number of field projects involving the condensing-gas-drive process, the vaporizing-gas process, and the CO_2 miscible process. Viscosity ratios, or mobility ratios as defined by Eq. 6.15, range from 4 to 86 in the different MCM processes. Stalkup (1983a) also reports data on a number of FCM projects. For these, the viscosity ratio ranges from 4 to 40.

An adverse ratio results in relatively poor areal sweep, as discussed in Chapter 4. For example, Fig. 4.4 (lower curve) shows sweep efficiency at breakthrough in a homogeneous five-spot pattern decreases from 67% for $M = 1.0$ to 35% for $M = 10.0$. Reservoir heterogeneities compound the problem of poor volumetric sweep, as illustrated in Example 4.2.

Effect of Viscous Fingering. An adverse viscosity ratio in a miscible process results in viscous fingering, which leads to reduced volumetric sweep. The Koval (1963) model defines an effective viscosity ratio, E, that characterizes the effect of viscous fingering:

$$E = \left[0.78 + 022 \left(\mu_d / \mu_D \right)^{\frac{1}{4}} \right]^4. \dots\dots\dots\dots\dots\dots\dots\dots\dots\dots\dots\dots\dots\dots (6.16)$$

Table 6.3 gives values of E over the range of viscosity ratios in the field tests reported by Stalkup (1983a).

Koval (1963) determined breakthrough volumetric displacement efficiency in homogeneous linear systems as a function of E **(Fig. 6.54)**. Breakthrough efficiencies taken from the curve in Fig. 6.54 are also given in Table 6.3.

The effect of viscous fingering on displacement efficiency in an MCM process is more complicated than described by the Koval model. Gardner and Ypma (1984) showed that recovery efficiency is affected by mixing caused by transverse dispersion between viscous fingers and by the size of the physical system.

Fig. 6.55 shows results from two sets of experiments on a linear system. In the first set, CO_2 was used to displace an oil (Soltrol™) at conditions at which the process was FCM. The viscosity ratio, μ_o/μ_s, was 16, which is unfavorable. In the second set, CO_2 displaced a crude oil (Wasson crude) in an MCM process conducted at a pressure above the MMP. For this displacement, the viscosity ratio was 21. Also shown in Fig. 6.55 is a numerical simulation of the CO_2/Wasson crude displacement. There was no mobile water present in the experiments.

In both sets of experiments, the oil recovery was delayed because of the adverse mobility ratio (i.e., there was early breakthrough of CO_2). In the FCM process, ultimate recovery approached 100% with continued injection of CO_2. In the MCM process, however, the recovery leveled off, indicating that final recovery would not approach 100% with continued CO_2 injection.

μ_o/μ_s	E	Breakthrough Recovery Linear System (%)
4	1.42	70
10	1.88	53
20	2.40	40
50	3.47	30
80	4.28	23

Table 6.3—Breakthrough volumetric displacement in a homogeneous linear system, Koval method (Koval 1963).

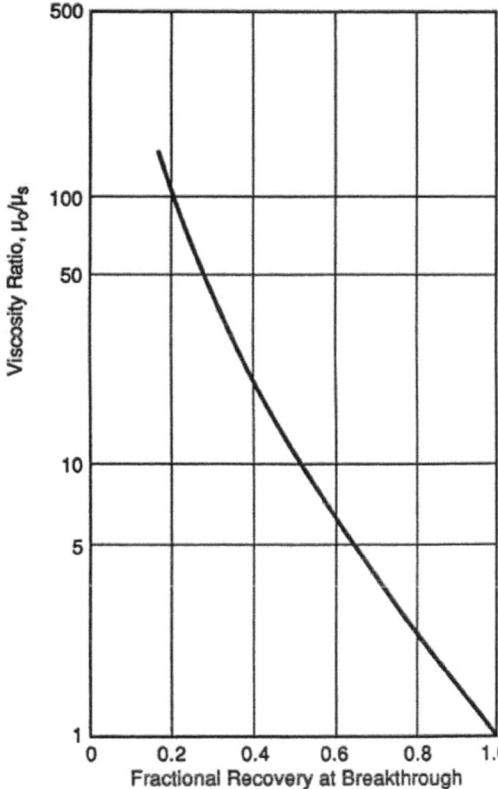

Fig. 6.54—Estimated breakthrough recovery as a function of viscosity ratio by the *K*-factor method (Koval 1963).

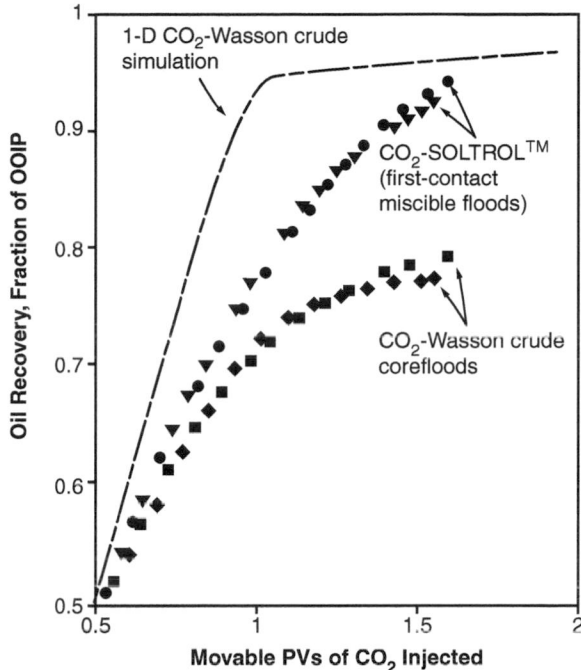

Fig. 6.55—Impact of viscous instability on secondary CO_2 flood oil recovery efficiency (Gardner and Ypma 1984).

For the MCM process, the viscous fingering not only delayed the oil recovery but also worked to reduce total recovery. The implication is that an interaction or synergism occurs between viscous fingering and phase behavior to cause liquid-phase dropout and trapping.

Gardner and Ypma (1984) also correlated available literature data on CO_2 MCM displacements conducted above the MMP by plotting measured residual oil saturation (ROS) vs. residence time in the core, as shown in **Fig. 6.56.** Residence time is defined as the core length, L, divided by average pore velocity of the displacing phase. The reported experimental data were taken in 2.0-in.-diameter cores. Flood velocities varied from 1 to more than 15 ft/D. Gardner and Ypma (1984) found that residual saturation was independent of residence time up to approximately $L/v = 0.35$ days. Above this value, the ROS decreased, indicating improved displacement efficiency.

To provide additional insight, they conducted another set of experiments in which CO_2 displaced oil in FCM displacements. ROS after 1.1 PV injected was determined as a function of residence time, as shown in **Fig. 6.57.** In this correlation, recovery at

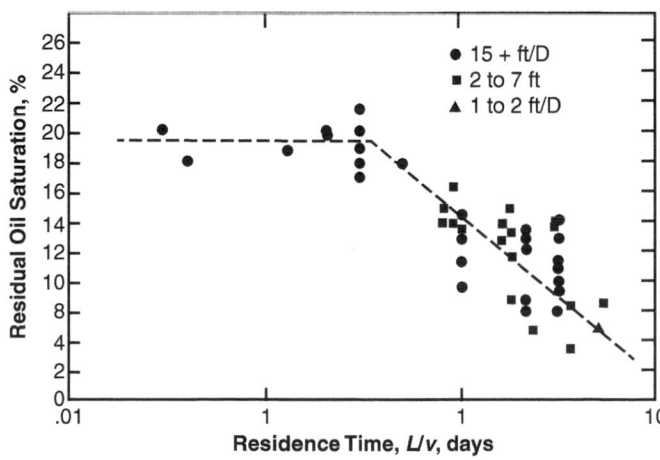

Fig. 6.56—Apparent impact of residence time on ROS for 2-in.-diameter CO_2/crude corefloods (Gardner and Ypma 1984).

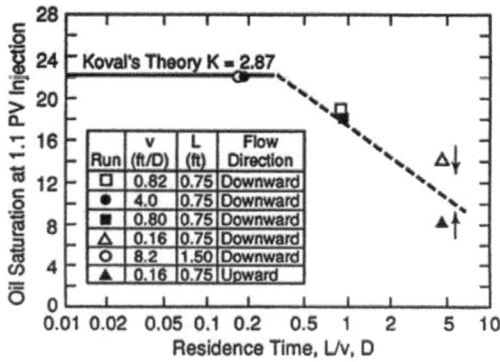

Fig. 6.57—Apparent impact of residence time on remaining oil saturation for 2-in.-diameter secondary FCM CO_2/ Soltrol™ corefloods (Gardner and Ypma 1984).

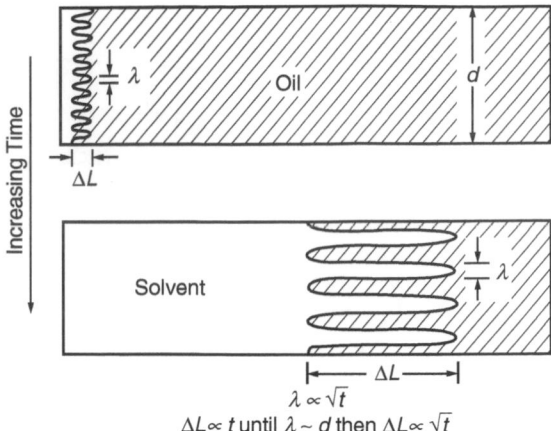

$$\lambda \propto \sqrt{t}$$

$\Delta L \propto t$ until $\lambda \sim d$ then $\Delta L \propto \sqrt{t}$

Fig. 6.58—Schematic of finger growth in unstable linear miscible displacements (Gardner and Ypma 1984).

1.1 PV injected was used as the measure of oil recovery because all the oil eventually will be recovered in an FCM process if injection is continued. Again, recovery was independent of residence time up to a value of approximately 0.35 days and increased at larger values. As Fig. 6.57 indicates, the Koval theory predicted a constant recovery that is independent of rate.

The effect of residence time on recovery when there is an adverse viscosity ratio is explained by examining the effect of transverse dispersion on the viscous fingers that form. **Fig. 6.58** shows schematically the growth of fingers in a linear displacement. The diameter of the core is assumed to be sufficiently large to allow viscous fingers to form initially. Once formed, the lengths of the fingers grow in direct proportion to time because the lengths are related to convection (i.e., the bulk flow rate). As discussed in Chapter 3, the widths of the fingers grow in proportion to the square root of time because this growth is governed by dispersion (Perkins et al. 1965).

If the core system is sufficiently long, the finger widths will begin to overlap (i.e., to mix into one another) before the fluids in the region of the fingers are produced. The number and sizes of the fingers will be affected, and if transverse dispersion has sufficient time to work over the entire width of the core, viscous fingers can be eliminated totally. The elimination of the fingers will be enhanced by longer residence times in the core. The improved recovery indicated in Figs. 6.56 and 6.57 is thus postulated to be a direct result of transverse dispersion, which works to damp out the fingers (i.e., to stabilize the flow).

Transverse dispersion effects are described by the transverse Péclet number, N_{Pe_t}, given by

$$N_{Pe_t} = vd^2/K_t L = t_{\text{disp}}/t_{\text{con}}, \dots\dots\dots\dots\dots\dots\dots\dots\dots\dots\dots\dots\dots\dots\dots\dots\dots\dots\dots\dots\dots\dots (6.17)$$

where v = average pore (interstitial) velocity, d = core diameter, L = core length, K_t = transverse dispersion coefficient, t_{disp} = dispersion time, t_{con} = convection time, and appropriate units should be used to make the group dimensionless.

It has been reasoned (Campbell and Orr 1985) that in a general way ultimate recovery should be related to $1/N_{Pe}$ in the manner shown in **Fig. 6.59.** At small values of $K_t L/vd^2$, the convection time is small relative to dispersion time. That is, L/v, the time for a fluid particle to travel the length of the core, is small compared with d^2/K_t, the time for a particle to disperse over the width of the core. Under these conditions, flow is governed by the viscous fingers (unstable displacement). At large values of $K_t L/vd^2$, dispersion eliminates the fingers and flow is stable. Between these limits, a transition zone exists. The expression in Fig. 6.59 that

$$\lambda_c < d, \dots\dots\dots\dots\dots\dots\dots\dots\dots\dots\dots\dots\dots\dots\dots\dots\dots\dots\dots\dots\dots\dots\dots\dots\dots\dots\dots\dots\dots\dots (6.18)$$

where λ_c = critical wave length for viscous fingers, is a statement that the core diameter must be sufficiently large to allow fingers to form initially.

The introduction of the core diameter or width, d, into the correlation expressed what Gardner and Ypma (1984) called the lateral boundary effect. The physical size of the system affects the behavior. They tested this by conducting FCM displacements with CO_2 and oil (Soltrol) in cores of different diameters and lengths. Results are shown in **Fig. 6.60,** where experimental data

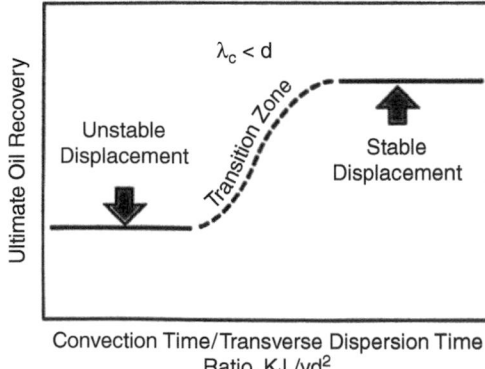

Fig. 6.59—Idealized impact of transverse dispersion on "ultimate" crude oil recovery by CO_2 flooding (Gardner and Ypma 1984).

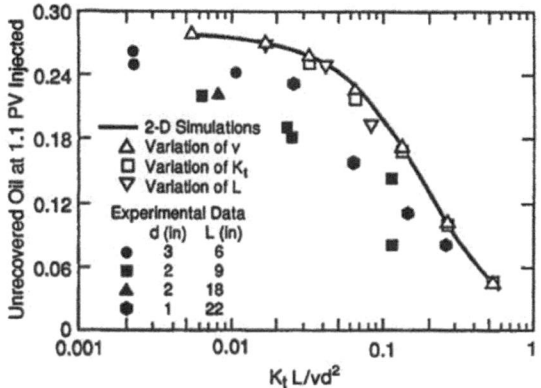

Fig. 6.60—Comparison of lateral boundary effects: 2D simulation vs. experiment (Gardner and Ypma 1984).

as well as results of numerical simulations are presented. While the agreement between the simulations and data is not good in terms of quantitative values, the agreement in the trend is excellent. The results indicate that the recovery behavior in an unstable miscible process is related to the transverse Péclet number. Heller et al. (1987) used the same dimensionless group in an analysis that can be applied to calculate the maximum length to which transverse dispersion will allow the fingers to grow.

The reduced recovery efficiency that occurs in an MCM process with unstable flow (i.e., when viscous fingers are present) results from interaction between the fingering and the phase behavior. According to Gardner and Ypma (1984), "in the case of a stable displacement (no fingers), the only place pure CO_2 and original crude oil meet is near the injection. But in the unstable case, this unfavorable encounter takes place over the entire length of the system as a result of bypassing (by viscous fingers) and transverse mixing. And since CO_2 by itself is not an effective crude-oil removal agent, a lower ultimate recovery in the unstable core is to be expected. Such a mechanism should come into play in any unstable vaporizing-gas-drive-type process."

While not specifically discussed, it seems clear that viscous fingering in a condensing-gas process also would lead to adverse mixing and phase behavior, which could reduce recovery efficiency.

The results of viscous fingering are important in that they indicate the mechanism whereby oil is trapped in a miscible process. The results, however, cannot be applied directly to a displacement in a reservoir. The dimensionless group $K_t L/vd^2$ would tend to be much smaller in a reservoir application because of the larger transverse dimension compared with a laboratory core. Viscous fingering would be more important in a reservoir displacement and would have a more significant adverse effect on recovery because of poorer volumetric sweepout. Gardner and Ypma (1984) point out that laboratory floods conducted at relatively low values of $K_t L/vd^2$ would yield displacement results more representative of reservoir conditions than experiments conducted in long, narrow cores where viscous fingers were damped out. Stalkup (1983a) also points out the danger of applying laboratory results involving longitudinal and transverse dispersion directly to field applications.

Effect of Gravity. The effect of density differences in a displacement process is discussed in Section 4.6.2. A dimensionless group used to characterize gravity effects is the viscous/gravity ratio, $R_{v/g}$, defined in field units as

$$R_{v/g} = (2,050u\mu_d/k\Delta\rho)(L/h), \dotfill (6.19)$$

where u = linear Darcy velocity, B/D-ft^2; k = permeability, md; μ_d = viscosity of displaced phase, cp; $\Delta\rho$ = density difference between displacing and displaced phases, g/cm^3; h = height of system, ft; and L = length of system, ft. This equation was previously presented as Eq. 4.25. Fig. 4.17 gives vertical sweep efficiency for a linear displacement at breakthrough as a function of $R_{v/g}$ and M.

The magnitude of $R_{v/g}$ also determines the flow regime, as discussed in Section 4.6.2 and shown in Fig. 4.19 (Stalkup 1983a). When the mobility ratio is unfavorable ($M > 1.0$) at relatively small values of $R_{v/g}$, flow is dominated by a single gravity tongue (Regions I and II). At larger values (Region III), viscous fingers form along the gravity tongue, with a resulting increase in displacement efficiency. At still larger values (Region IV), flow is dominated by viscous fingers and gravity effects are relatively unimportant. The particular values of $R_{v/g}$ over which these flow regions exist depend on the value of M, as shown in Fig. 4.19.

The use of $R_{v/g}$ and M to determine flow conditions and performance is illustrated in Example 6.9.

Example 6.9—Vertical Sweep Efficiency at Breakthrough—CO_2 Displacement. Consider the displacement of an oil by CO_2 in a linear system. The effective length of the horizontal reservoir is 500 ft and the thickness is 40 ft. **Table 6.4** lists other properties and conditions.

Calculate the sweep efficiency at breakthrough at the downstream end of the reservoir. Assume that the correlation of Fig. 4.17 is applicable. Also calculate the flow regime that exists in the displacement.

Solution. Calculate $R_{v/g}$.

$$R_{v/g} = \frac{2,050 \times 0.06 \text{ B/D-ft}^2 \times 1.0 \text{ cp} \times 500 \text{ ft}}{120 \text{ md } (0.78 - 0.68) \text{ g/cm}^3 \, 40\text{ft}} = 128.$$

T (°F)	120
p (psia)	2,000
ρ_o (g/cm^3)	0.78
ρ_{CO_2} (g/cm^3)	0.68*
μ_o at 120°F (cp)	1.0
μ_{CO_2} (cp)	0.049**
k (md)	120†
u (B/D-ft^2)	0.06

*From Fig. 6.45

**From Fig. 6.49

†Vertical and horizontal

Table 6.4—Example 6.9 properties and conditions.

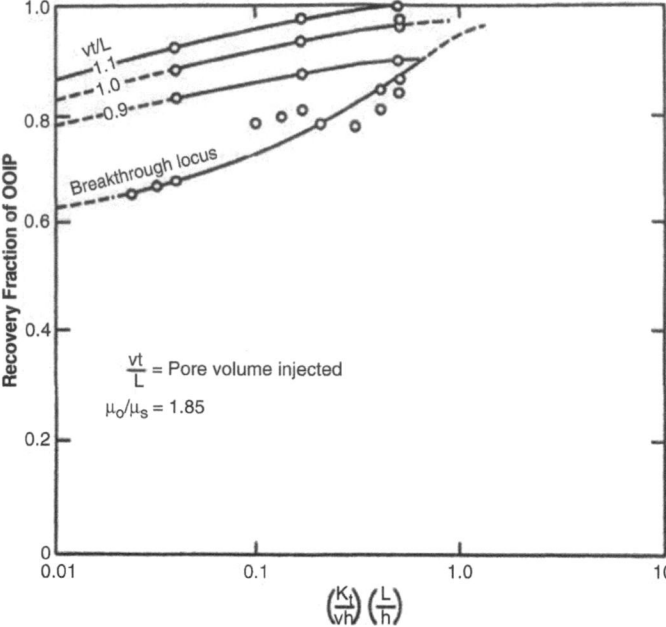

Fig. 6.61—Vertical sweep efficiency correlation for a viscosity ratio of 1.85 when transverse mixing is by molecular diffusion (Pozzi and Blackwell 1963).

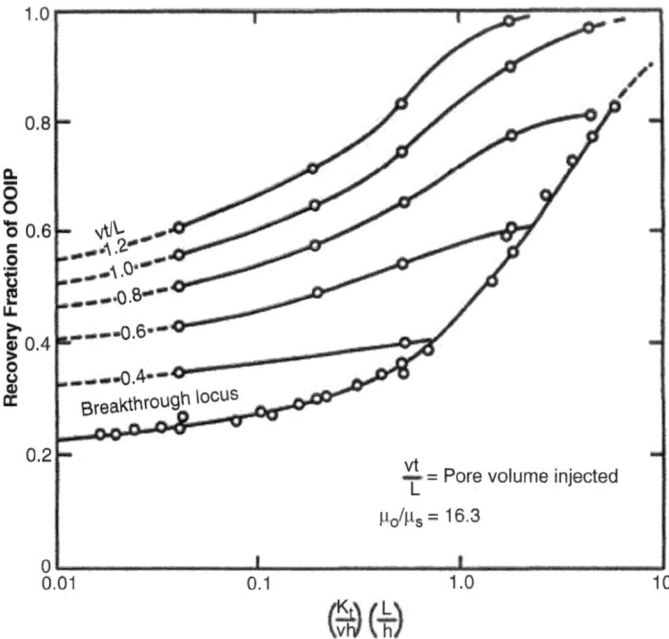

Fig. 6.62—Vertical sweep efficiency correlation for a viscosity ratio of 16.3 when transverse mixing is by molecular diffusion (Pozzi and Blackwell 1963).

Calculate M (viscosity ratio).

$$M = \mu_o / \mu_s = 1.0 \text{ cp}/0.049 \text{ cp} = 20.$$

From Fig. 4.17 (the curves must be extrapolated), $E_I \approx 20\%$ at breakthrough. From Fig. 4.19, flow is in Regions II and III and is dominated by the gravity tongue, with viscous fingers superimposed. In this range of $R_{v/g}$ and M, the recovery efficiency at breakthrough is poor.

Vertical sweep efficiency depends on transverse dispersion as well as on mobility ratio and gravity effects, $R_{v/g}$. Pozzi and Blackwell (1963) conducted a series of FCM displacements in linear systems in which they considered the effect of transverse dispersion (by molecular diffusion) on recovery efficiency. By proper scaling, they designed the experiments such that flow was in Region II (Fig. 4.19), where there is no dependence on $R_{v/g}$. Flow is dominated by a single gravity tongue. The effect of M was considered by varying the viscosity ratio in different sets of experiments.

Pozzi and Blackwell (1963) presented their results as a series of correlative plots of fractional recovery as a function of the dimensionless group

$$G = \left(K_t / vh\right)\left(L/h\right), \quad \ldots \ldots \ldots \ldots \ldots (6.20)$$

where K_t = transverse dispersion, v = pore (interstitial) or frontal velocity, h = model height, and L = model length and where consistent units should be used. This is the same correlating group Gardner and Ypma (1984) used in their studies of the effects of viscous fingering and transverse dispersion on recovery.

The correlation of Pozzi and Blackwell (1963) is given in **Figs. 6.61 through 6.63** for $M = 1.85$, 16.3, and 69, respectively. In the experiments, scaling was such that transverse dispersion was controlled by molecular diffusion, the parameter vt/L, or the number of PVs injected. In each core, breakthrough recovery is along the bottom curves in Figs. 6.61 through 6.63. Examination of the curves shows that transverse mixing acts to increase recovery efficiency when all other parameters are held constant.

The application of the Pozzi and Blackwell (1963) correlation is illustrated in Example 6.10.

Example 6.10—Effect of Transverse Dispersion With Gravity Tonguing. Consider Example 6.9 and assume that the Pozzi and Blackwell (1963) correlation applies. Using the same flow conditions, calculate the recovery efficiency at breakthrough. Use a porosity of 0.15 to calculate interstitial velocity.

Solution. The calculation in Example 6.9 showed that flow was in or very near Region II. The Pozzi and Blackwell (1963) correlation is applicable. $M = 20$ (use Fig. 6.62 for $\mu_o/\mu_s = 16.3$). Calculate the value of the dimensionless group given in Eq. 6.20. Let $K_t = 1.8 \times 10^{-4}$ ft²/hr.

$$u = 0.06 \text{ B/D-ft}^2 \times 5.615 \text{ ft}^3/\text{bbl}\left(\frac{1}{24 \text{ hr/day}}\right)\left(\frac{1}{0.15}\right)$$

$$= 0.094 \text{ ft/hr.}$$

$$\left(\frac{K_t}{uh}\right)\left(\frac{L}{h}\right) = \left(\frac{1.8 \times 10^{-4} \text{ ft}^2/\text{hr}}{0.094 \text{ ft/hr} \times 40 \text{ ft}}\right)\left(\frac{500 \text{ ft}}{40 \text{ ft}}\right) = 0.00060.$$

From an extrapolation of Fig. 6.62, fractional recovery at breakthrough ≈ 20%. This value agrees with the result of Example 6.9. Transverse dispersion does not have a significant effect at the conditions specified when flow is in Region II.

The results of Pozzi and Blackwell (1963) for a linear system were extended to displacements in a five-spot pattern by Stalkup (1983a). The result is shown in **Fig. 6.64,** where sweepout efficiency is plotted as a function of displaceable PVs injected, V_{pD}.

$$V_{pD} = V_{si} / \left[V_p \left(1 - S_{orm} - S_{wi} \right) \right], \quad \ldots (6.21)$$

where V_{si} = volume of solvent injected, V_p = PV in the pattern area, S_{orm} = ROS at the end of the miscible displacement, and S_{wi} = irreducible water saturation.

Fig. 6.64 is applicable only at relatively small values of $K_t L / v h^2$ where transverse dispersion has a negligible effect on the displacement. The correlation applies in Region II of the different flow regimes, as does the Pozzi and Blackwell (1963) correlation. Stalkup (1983a) points out that the correlation in Fig. 6.64 should be used only as an approximation. In a five-spot pattern, both the velocity and the transverse dimension vary over a large range during a flood. This makes the extrapolation from a linear system to a five-spot pattern subject to significant error. Stalkup also discusses the fact that if the solvent is soluble in the reservoir brine or miscible flood residual oil to any significant extent, then V_{pD} should be corrected for this "loss" of solvent.

Stalkup (1983a) made several key assumptions in the calculation of Fig. 6.64. It was assumed that the Dietz theory (Dietz 1953; Crane et al. 1963) was applicable for describing vertical sweep efficiencies, E_I, in a linear system (2D, length times height).

$$E_I = \left[1/(M-1) \right] \left[2 \left(M V_{pD} \right)^{1/2} - 1 - V_{pD} \right],$$
$$\ldots \ldots \ldots \ldots \ldots \ldots \ldots \ldots \ldots \ldots \ldots \ldots (6.22)$$

where all terms are dimensionless. Eq. 6.22 applies in Region II and was derived under the assumption that transverse dispersion is negligible. Stalkup (1983a) also assumed that

$$E_V = E_A E_I, \quad \ldots \ldots \ldots \ldots \ldots \ldots (6.23)$$

where E_V = volumetric sweep efficiency, E_A = areal sweep efficiency, and E_I = vertical sweep efficiency for a linear system (Eq. 6.19). E_A was determined from areal sweepout data taken in a five-spot model at $M = 1.0$.

Withjack and Akervoll (1988) used tomography to measure volumetric sweep efficiency in a 3D model of a five-spot pattern (see Section 4.7) at various M and $R_{v/g}$ values. Their breakthrough data are given in Fig. 4.31, along with the data of Craig et al. (1957), where breakthrough is plotted as a function of M and $R_{v/g,5}$.

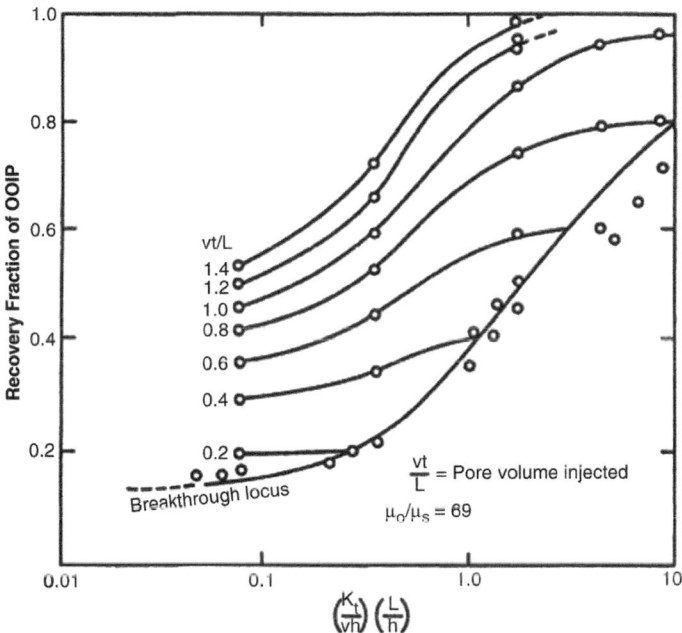

Fig. 6.63—Vertical sweep efficiency correlation for a viscosity ratio of 69 when transverse mixing is by molecular diffusion (Pozzi and Blackwell 1963).

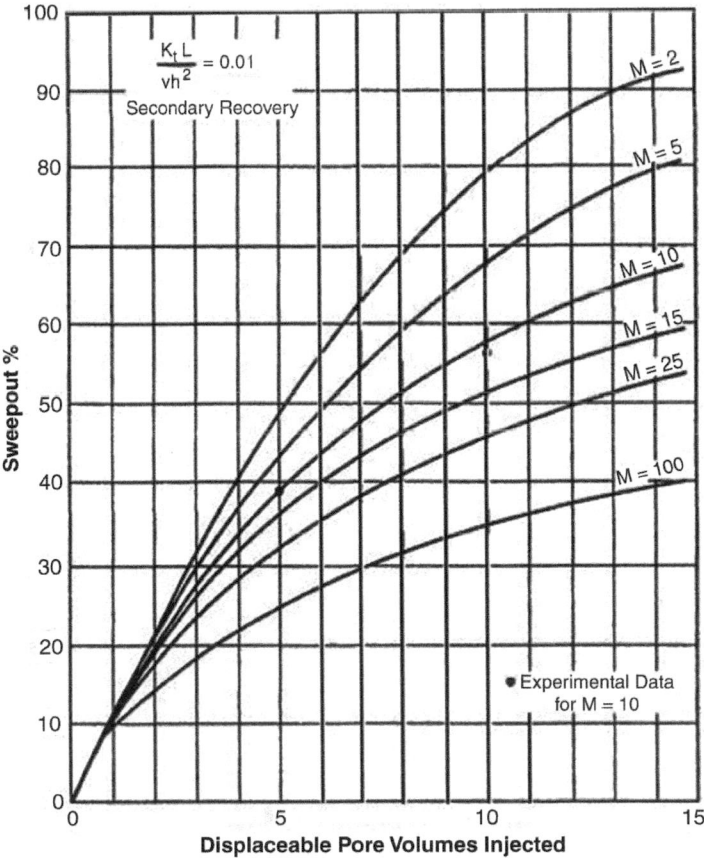

Fig. 6.64—Volumetric sweepout in a normal five-spot pattern for Region II flow (Stalkup 1983a).

Figs. 6.65 and 6.66 give volumetric sweepouts as functions of $R_{v/g}$ for $M = 7.5$ and 22.4, respectively. $R_{v/g}$ is defined by use of the "average" velocity in a five-spot pattern (in customary units) and is given by

$$R_{v/g} = (2,050 \times 1.25 i \mu_o / hLk\Delta\rho)(L/h), \dots\dots\dots\dots\dots\dots\dots\dots\dots\dots\dots\dots\dots\dots\dots\dots (6.24)$$

where i = injection rate per well, B/D. Other units are as given for Eq. 6.19. Note that $R_{v/g}$ for a five-spot pattern defined in Eq. 6.24 differs from $R_{v/g,5}$ used in the correlation of Fig. 4.31 in that the velocity terms (flow terms) are different.

Fig. 6.67 compares the sweepout data of Withjack and Akervoll (1988) with the correlation of Stalkup (1983a). Stalkup's correlation is based on vertical displacement efficiency in a 2D linear system (height times length). Flow is assumed to be in Region II, where sweepout is not a strong function of $R_{v/g}$. Withjack and Akervoll (1988) determined that, for a 3D system, sweepout increased with increasing $R_{v/g}$ (i.e., sweepout was not independent of $R_{v/g}$). Furthermore, the onset of Region III occurred at lower values of $R_{v/g}$ in a 3D system than in a 2D system. This is reflected in the data points in Fig. 6.67. At the lower values of $R_{v/g}$, recovery measured in the 3D model is considerably poorer than predicted by the Stalkup correlation. As the flow pattern changes from Region II to Region III flow, where viscous fingers are superimposed on a gravity tongue, sweepout improves owing to the formation of the viscous fingers.

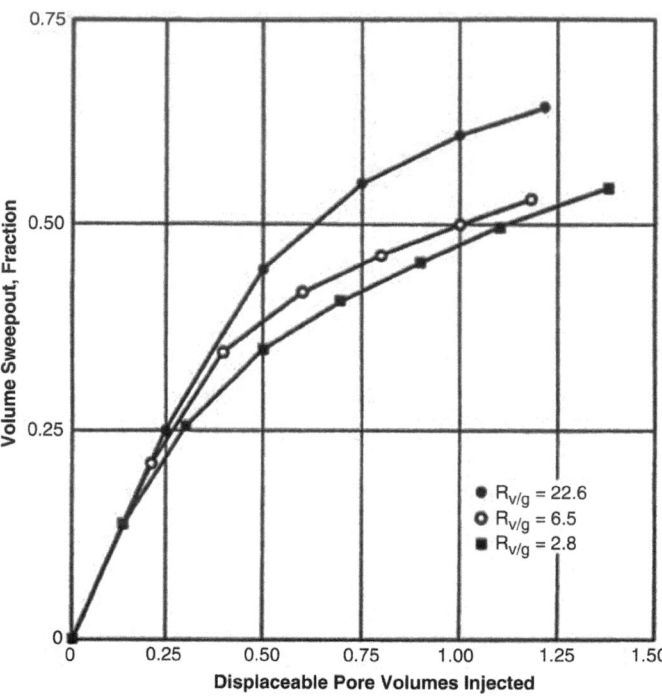

Fig. 6.65—Volumetric sweepout in a five-spot pattern with $M = 7.5$ (Withjack and Akervoll 1988).

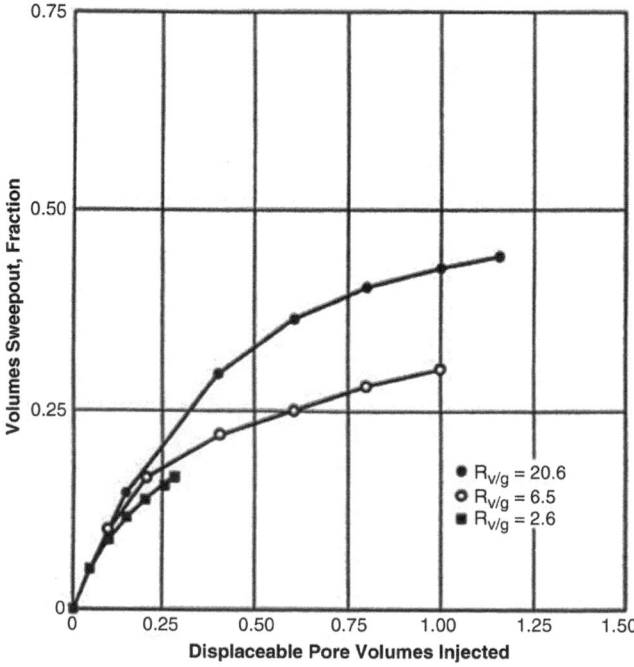

Fig. 6.66—Volumetric sweepout in a five-spot pattern with $M = 22.4$ (Withjack and Akervoll 1988).

6.9.3 Displacement Efficiency When Mobile Water Is Present.

Mobile water is present in a miscible displacement process in two general situations. If the displacement is conducted after a waterflood, as a tertiary process, then water will be displaced along with oil. Two-phase flow will occur ahead of the displacement front. If the injected solvent (C_3, CO_2, etc.) does not effectively displace water, simultaneous solvent/water flow also will occur over a region of the displacement.

In some applications, water is injected alternately with a solvent to improve the mobility ratio in the displacement. The presence of a flowing water phase in the region of the system in which solvent is also flowing reduces effective permeability to the solvent, thereby reducing its mobility. The method is called a WAG process (see Chapter 5).

The presence of a water phase, flowing or stagnant, has no significant effect on phase behavior in either an FCM process or an MCM process. Miscibility is developed in basically the same manner whether water is present or not (Tiffin and Yellig 1983).

The presence of a flowing water phase, however, has a negative effect on displacement efficiency, although in many situations the effect is relatively small. The different behaviors discussed in Sections 6.9.1 and 6.9.2 will still be present (Doscher and Gharib 1983), but made more complex by the presence of mobile water. The major additional problem created by flowing water is that the water blocks part of the oil away from the solvent (i.e., it reduces the ability of the solvent to contact and mobilize the oil). This occurs in both FCM and MCM processes.

Tertiary Recovery—No Water Injected. In a tertiary process in which only solvent is injected, the extent to which the waterflood water reduces displacement efficiency is unclear. Tiffin and Yellig (1983) reported that the presence of water from a prior waterflood had negligible effect on ultimate recovery in both water-wet and oil-wet cores when pure CO_2 was the displacing fluid. ROSs in their experiments were comparable with values obtained in the absence of water.

In another study, however, in which a flow cell was used to visualize the displacement (Campbell and Orr 1985), miscible solvents were reported to be relatively inefficient at displacing oil in a tertiary process. Water from a prior waterflood effectively blocked the waterflood residual oil from contact with injected solvent. CO_2 was somewhat more efficient than other solvents because CO_2 can readily diffuse through a water phase and contact blocked oil. The relatively high solubility of CO_2 in water compared with other solvents enhances this mass transfer to the trapped oil. It has also been reported by Stalkup (1970), however, that diffusion of solvents, such as C_3, can be effective in displacing oil trapped by a water phase given sufficient contact time. Diffusion is a relatively slow process.

Campbell and Orr (1985) demonstrated the role of CO_2 diffusion in the recovery of oil trapped in dead-end pores with a physical model (**Fig. 6.68**). Fig. 6.68a shows oil trapped in a pore. CO_2 is flowing in the main channel, but water in the rest of the pore prevents direct contact between oil and CO_2. The CO_2 solubilizes in the water, diffuses through the water, and is solubilized in the oil. The dissolved CO_2 causes the oil to swell in volume, forcing the water from the pore throat (Fig. 6.68b). This eventually leads to direct contact of the oil by the CO_2 (Fig. 6.68c), and given sufficient time, the oil will be partially recovered from the pore.

The relative wettability of the rock to fluids also affects the displacement efficiency. In the tertiary recovery experiments Campbell and Orr (1985) conducted in a visual cell, two different oils were used. One way that the oils differed was the degree to which they wet the porous-medium system (glass). The oil that wet the glass the most strongly was displaced more efficiently by the CO_2. The role of wettability is discussed later in this section.

Stalkup (1983a) suggested that a model can be used to make rough estimates of recovery efficiency when mobile water is present. He suggests that a modified mobility be used in the calculation of $R_{v/g}$:

$$R_{v/g,\text{tert}} = \left(\frac{2,030u}{\Delta\rho}\right)\left[\frac{1}{(k_H k_V)^{1/2}\left(\dfrac{k_{roi}}{\mu_o} + \dfrac{k_{rwi}}{\mu_w}\right)}\right]\left(\frac{L}{h}\right), \quad \dots\dots\dots\dots (6.25)$$

where $R_{v/g,\text{tert}}$ = tertiary (mobile water) viscous-to-gravity ratio; u = Darcy velocity, B/D-ft²; $\Delta\rho$ = density difference between solvent and water, g/cm³; k_H = horizontal permeability, md; k_V = vertical permeability, md; k_{roi} = relative oil permeability at start of the solvent injection; k_{rwi} = relative water permeability at start of the solvent injection; L = length of system, ft; h = thickness of system, ft; and μ_o and μ_w = viscosities of oil and water, respectively, cp.

Also, the mobility ratio should be defined as

$$M_{\text{tert}} = \frac{k_{rs}}{\mu_s}\bigg/\left(\frac{k_{roi}}{\mu_o} + \frac{k_{rwi}}{\mu_w}\right), \quad \dots\dots\dots\dots\dots\dots\dots\dots\dots\dots (6.26)$$

where M_{tert} = tertiary (mobile water) mobility ratio, k_{rs} = relative permeability to solvent, and μ_s = solvent viscosity. Stalkup (1983a) emphasizes that this approach should be used only for rough estimates of recovery efficiency.

Simultaneous or Alternate Solvent/Water Injection. Literature reports of laboratory studies in water-wet porous media (Stalkup 1970) clearly indicate that the presence of water at high saturations blocks oil from contact with an injected

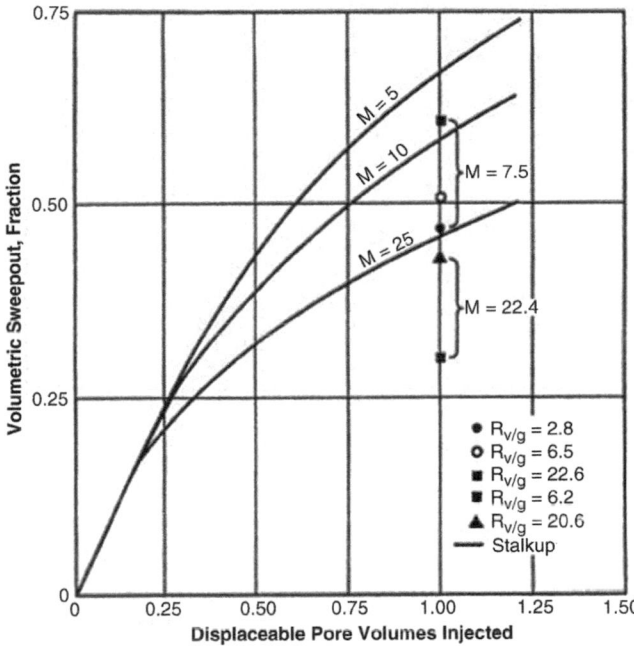

Fig. 6.67—Volumetric sweepout in a five-spot pattern, Region II (Stalkup 1983a; Withjack and Akervoll 1988).

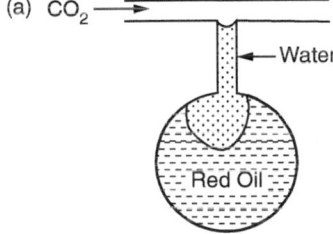

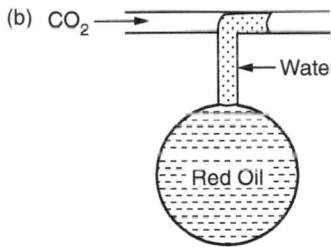

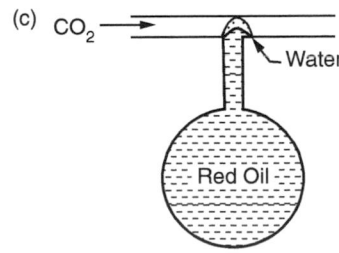

Fig. 6.68—Recovery of red Soltrol™ isolated by water in a dead-end pore by CO_2 at 1,200 psi and 77°F: (a) at start of CO_2 injection; (b) position of water barrier after 18 hours; and (c) position of water barrier after 26.5 hours (Campbell and Orr 1985).

solvent. This trapped oil is bypassed by the injected solvent front and is trapped by the water. Continued injection, however, can recover much of this oil. The diffusion of solvent through the water phase probably plays a major role in the recovery process, which is controlled by the mass-transfer rate (Stalkup 1970; Tiffin and Yellig 1983). As stated earlier, the rate of CO_2 diffusion through water is relatively high compared with that of hydrocarbon solvents or N_2. Therefore, CO_2 should be more efficient than other solvents when simultaneous or alternate water slugs are injected with the solvent.

Stalkup (1983a, 1970) conducted FCM displacements in long, strongly water-wet cores in which the water saturation was varied in different experiments. The solvent was C_3. Water saturations were first established in the cores by simultaneous injection of oil and water at a fixed ratio. After steady conditions were achieved, C_3 and water were injected simultaneously at the same ratio used to establish the initial saturation. Water and solvent were injected for a number of PVs.

Results in **Fig. 6.69** show a normalized recovery factor, S_{orm}/S_{or}, as a function of the water saturations in the core. The ratio represents the fraction of waterflood residual oil that is trapped and not recovered by the solvent/waterflood. The fraction increases markedly with increasing water saturation.

Raimondi and Torcaso (1964) conducted similar experiments in water-wet cores that resulted in a correlation expressing trapped oil as a function of saturation and relative permeability. This correlation, modified by Todd et al. (1982), is given by

$$S_{orm}/S_{or} = 1/\left[1 + \alpha\left(k_{ro}/k_{rw}\right)\right], \quad\quad\quad\quad\quad\quad\quad\quad\quad\quad\quad\quad\quad\quad\quad (6.27)$$

where α is an empirical parameter between 1.0 and ∞. A value of 1.0 corresponds to the original correlation. At larger values of α, the amount of trapped oil is reduced and becomes negligible as $\alpha \to \infty$. The parameter α is a function of wettability, increasing as the rock becomes more oil-wet.

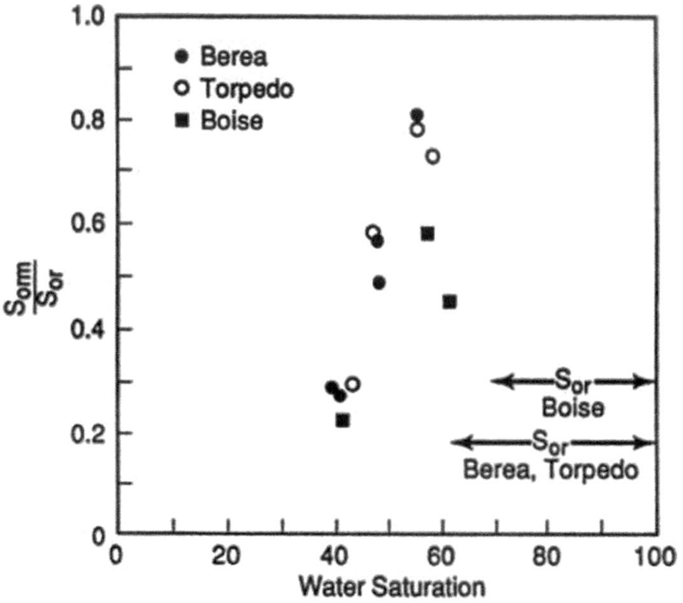

Other studies have reported the effect of mobile water on recovery efficiency. Tiffin and Yellig (1983) investigated the WAG process in a CO_2 MCM process. They reported that recovery efficiency decreased as the ratio of injected water to injected CO_2 increased. The decrease, however, was a strong function of the wettability, as shown in **Figs. 6.70 and 6.71**. Oil recovery in water-wet cores decreased much more than in oil-wet cores. Apparently, the water blocks the oil from the CO_2 much more effectively when the water is the wetting phase. This same conclusion was reached by others (Campbell and Orr 1985; Shelton and Schneider 1975).

Tiffin et al. (1991) examined the effect of water blocking in the enriched-gas displacement process in water-wet systems. They observed the same negative effect as in the CO_2 WAG and FCM processes. They also observed that diffusion through the water contributed little to increased recovery.

The interaction between viscous fingering, transverse dispersion, and phase behavior

Fig. 6.69—Fraction of waterflood ROS trapped by water (Stalkup 1970, 1983a).

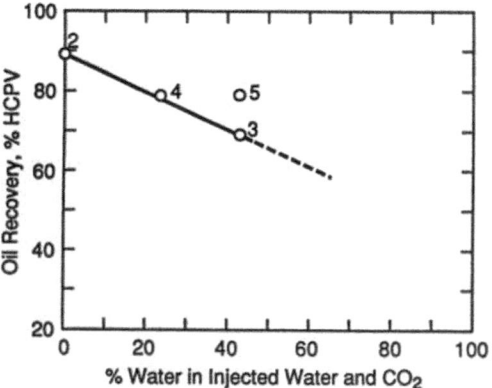

Fig. 6.70—Effect of mobile water on oil recovery for CO_2 displacements of reservoir crude oil (MCM process), tertiary displacements, water-wet core (Tiffin and Yellig 1983).

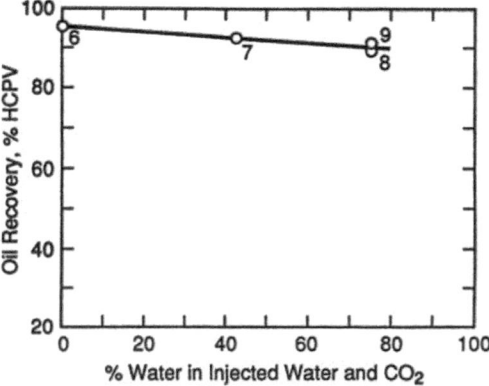

Fig. 6.71—Effect of mobile water on oil recovery for CO_2 displacements of reservoir crude oil (MCM process), tertiary displacements, oil-wet core (Tiffin and Yellig 1983).

discussed in Section 6.9.2 (Gardner and Ypma 1984) also occurs in the presence of high water saturations (Campbell and Orr 1985). However, the magnitudes of dispersion coefficients apparently are larger when a water phase is present than when pores are filled only with miscible fluid (Stalkup 1970).

Other effects also come into play that are less well-understood when multiple phases are present. Campbell and Orr (1985) reported that decreasing IFT between CO_2-rich and oil-rich phases when the two phases are not miscible probably contributes to oil recovery. They also discussed the role of capillary effects, which can be important even when a water phase is not present. They describe an interesting experiment in which oil was completely displaced from a dead-end pore by CO_2 as a result of capillary forces.

6.10 Miscible Displacement Performance Modeling

A number of procedures are available for calculation of performance of a miscible displacement process. These range from simple calculations that may be considered as rough estimates to complex numerical simulations that require large computers (Kremesec and Sebastian 1988). This section presents methods of making relatively simple calculations that can be performed by hand. Most of these are based on results from physical models, such as those discussed in Chapter 4. The calculations should be considered as rough estimates of performance. Although they do serve to illustrate the effects of different variables on process performance, they generally are not sufficient for final design calculations.

The calculations considered in Section 6.10.1 focus on ultimate recovery and recovery as a function of solvent (and other drive fluids) injected. The rates of injection and recovery are considered as well. The methods are illustrated for single-layer reservoirs, but can be extended to multilayer systems. Effects of mobility ratio and gravity segregation also are considered.

The calculation techniques are illustrated with example problems. Generic "solvents" are used in most of the calculations, but the methods generally are applicable to the miscible fluids and processes discussed in this chapter.

6.10.1 Recovery by Material Balance. A very simple estimate of recovery from a process can be made by application of a material balance. For example, for a displacement conducted as a tertiary process following a waterflood, recovery is given by

$$N_p = \left(\frac{V_p S_{orw}}{B_o} - \frac{V_p S_{orm}}{B_o} \right) E_{RW} \left(\frac{E_{RM}}{E_{RW}} \right), \quad \dots\dots\dots\dots\dots\dots\dots\dots\dots\dots\dots\dots\dots\dots\dots\dots \quad (6.28)$$

where N_p = oil recovery, V_p = reservoir PV, B_o = formation volume factor (FVF) at conditions of the miscible flood, S_{orw} = average ROS at end of waterflood in regions contacted by water, S_{orm} = ROS at end of miscible flood in regions contacted by solvent, E_{RW} = volumetric hydrocarbon sweep efficiency in waterflood, and E_{RM} = volumetric hydrocarbon sweep efficiency in miscible displacement. The ratio E_{RM}/E_{RW} is used to relate waterflood sweep to miscible fluid sweep.

Eq. 6.28 can be rewritten as

$$N_p = \frac{NB_{oi}}{S_{oi}} \left(\frac{S_{orw}}{B_o} - \frac{S_{orm}}{B_o} \right) E_{RW} \left(\frac{E_{RM}}{E_{RW}} \right), \quad \dots\dots\dots\dots\dots\dots\dots\dots\dots\dots\dots\dots\dots\dots \quad (6.29)$$

where N = OOIP at discovery of reservoir, S_{oi} = initial oil saturation, and B_{oi} = FVF at initial conditions of reservoir.

Eq. 6.29 as a fractional recovery becomes

$$\frac{N_p}{N} = \frac{B_{oi}}{B_o S_{oi}} \left(S_{orw} - S_{orm} \right) E_{RW} \left(\frac{E_{RM}}{E_{RW}} \right). \quad \dots\dots\dots\dots\dots\dots\dots\dots\dots\dots\dots\dots\dots \quad (6.30)$$

For typical values of the different parameters, Eq. 6.30 yields recoveries of 12 to 18% OOIP.

Example 6.11—Estimate of Oil Recovery by Overall Material Balance. Apply Eq. 6.30 to calculate the fraction of OOIP that would be recovered by a CO_2 MCM process for the conditions listed in **Table 6.5**.

S_{oi}	0.75
S_{orw}	0.30
S_{orm}	0.08
E_{rw}	0.70
E_{RM}/E_{RW}	0.7
B_{oi} (RB/STB)	1.1
B_o (RB/STB)	1.05

Table 6.5—Example 6.11 conditions.

The ratio of efficiencies was reported as a reasonable estimate for displacements conducted in field tests in west Texas (US Government 1978). This ratio is dependent on CO_2-slug size. In the field experience referenced, slug size was approximately 40% HCPV.

Solution. Calculate N_p/N

$$\frac{N_p}{N} = \frac{1.10 \text{ RB/STB}}{1.05 \text{ RB/STB} \times 0.75}(0.30 - 0.08)0.70 \times 0.70 = 0.15.$$

Recovery is estimated to be 15% OOIP. This is within the range of expectation for a CO_2 miscible process (Stalkup 1983a).

6.10.2 Calculations Based on Physical Models—Single-Layer Reservoir. Physical models discussed in Chapter 4 and in this chapter can be used as the basis for performance estimates. The calculations in this section are based on a single layer.

Secondary Recovery From a Five-Spot Pattern—Continuous Solvent Injection. Results from an appropriately scaled physical model can be used in a straightforward manner to determine oil production as a function of displaceable PVs of miscible solvent injected. Displaceable PVs injected was defined in Eq. 6.21. In this approach, M is first calculated and values of production as functions of injection are read directly from the selected correlation. Most of the available correlations are *for* $R_{v/g} \to \infty$ (i.e., gravity effects are negligible).

Chapter 4 discussed results from several model studies. The different studies are in good agreement for mobility ratios in the vicinity of unity. At large M values, however, significant differences exist, reflecting differences in scaling. Claridge (1972) developed a set of curves giving displacement efficiency as a function of M and PVs injected for $R_{v/g} \to \infty$. His correlation was based on the physical model results of Caudle and Witte (1959) and the viscous fingering model of Koval (1963). Claridge's correlation is used here to illustrate this approach. A FORTRAN computer program that uses the Claridge correlation is given in Appendix D.

Example 6.12—Secondary Recovery From a Five-Spot Pattern—Application of Claridge Correlation. Assume that a large slug of a solvent is to be used to displace oil miscibly in a secondary recovery process. A five-spot pattern is to be used. Calculate oil recovery as a function of solvent injected with the Claridge (1972) correlation. **Table 6.6** lists data for this example. Gravity effects are assumed to be negligible (i.e., $R_{v/g} \to \infty$).

Because the slug is large, assume continuous solvent injection. The reservoir is completely liquid filled; i.e., there is no initial gas saturation requiring liquid fill-up.

Solution. The OOIP is

$$N = Ah\phi S_{oi} \frac{1}{B_o}\left(\frac{1}{5.615 \text{ ft}^3/\text{bbl}}\right)$$

$$= 20 \text{ acres} \times 43,650 \text{ ft}^2/\text{acre} \times 20 \text{ ft} \times 0.18$$

$$\times 0.75\left(\frac{1}{1.15 \text{ RB/STB}}\right)(1/5.615)$$

$$= 364,000 \text{ STB}.$$

Well spacing, A (acres)	20
Thickness, h (ft)	20
Porosity, ϕ	0.18
Permeability, k (md)	120
Initial oil saturation, S_{oi}	0.75
Connate water saturation, S_{ow}	0.25
ROS at end of miscible displacement, S_{orm}	0.05
Oil viscosity at reservoir temperature, μ_o (cp)	1.5
Solvent viscosity at reservoir temperature, μ_s (cp)	0.06
Oil FVF, B_o (RB/STB)	1.15
Reservoir temperature, T (°F)	120
Average reservoir pressure, p (psia)	2,500

Table 6.6—Example 6.12 data.

The displaceable reservoir volume of solvent injected is

$$V_{pDs} = V_{si} / \left[Ah\phi (1 - S_{cw} - S_{orm}) \left(\frac{1}{5.615 \text{ ft}^2/\text{bbl}} \right) \right]$$

$$= V_{si} / \left[20 \text{ acres} \times 43{,}560 \text{ ft}^2/\text{acre} \times 20 \text{ ft} \right.$$

$$\left. \times 0.18 (1 - 0.25 - 0.05) \left(\frac{1}{5.615 \text{ ft}^3/\text{bbl}} \right) \right]$$

$$- V_{si} / 391{,}000 \text{ bbl}.$$

Calculate M.

$$M = \mu_o / \mu_s$$

$$= 1.5 \text{ cp} / 0.06 \text{ cp}$$

$$= 25.$$

Calculate the areal sweep efficiency, E_{As}, vs. displaceable PVs injected, V_{pD}, from **Fig. 6.72** (Claridge 1972). V_{pD} is defined in Eq. 6.21. Note that for a single-layer reservoir, F_{As} is equivalent to volumetric sweep efficiency. The dimensionless PVs injected also could be expressed in terms of HCPVs injected as

$$V_{ph} = V_{si} / \left[V_p (1 - S_{cw}) \right]$$

$$= V_{pDs} \left[(1 - S_{orm} - S_{cw}) / (1 - S_{cw}) \right].$$

Calculate the oil produced.

$$N_p = E_{As} Ah\phi (S_{oi} - S_{orm})(1/B_o)$$

$$= E_{As} 20 \text{ acres} \times 43{,}560 \text{ ft}^2/\text{acre} \times 20 \text{ ft} \times 0.18 (0.75 - 0.05)$$

$$\times \left(\frac{1}{1.15 \text{ RB/STB}} \right) \left(\frac{1}{5.615 \text{ ft}^3/\text{bbl}} \right)$$

$$= 340{,}000 \, E_{As}.$$

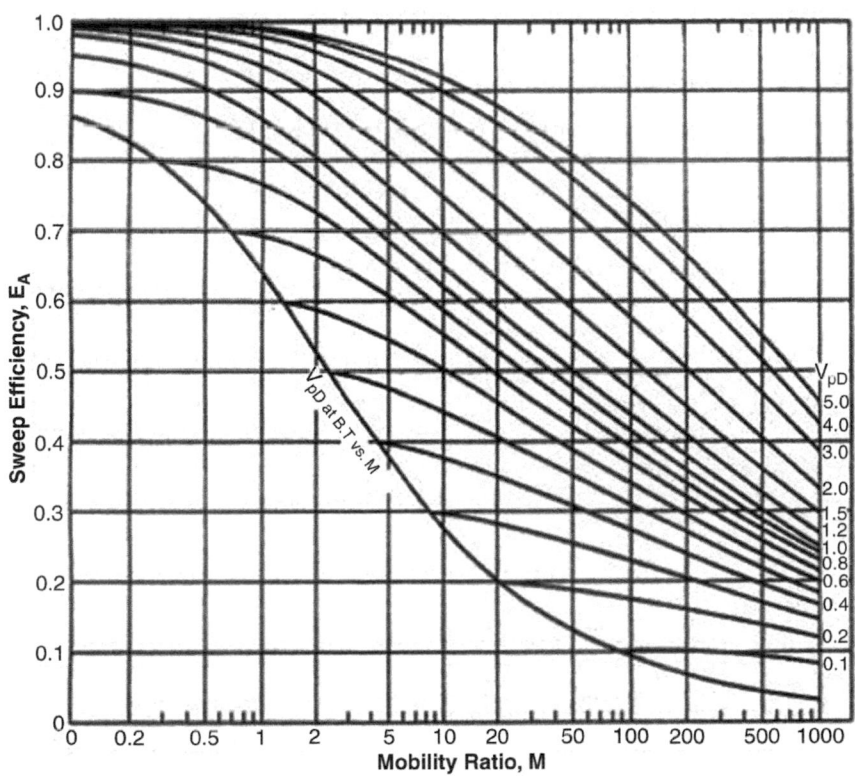

Fig. 6.72—Oil recovery in miscible flooding for five-spot well patterns (Claridge 1972).

Results are shown in **Table 6.7** and **Fig. 6.73.** At the very unfavorable mobility ratio, breakthrough of solvent occurs relatively early, at 0.18 displaceable PV injected. The process becomes less efficient as the injection continues, as indicated by the change in slope of the recovery curve (Fig. 6.73).

The same general approach can be used with results from physical models other than the Claridge correlation. For example, if calculation of the viscous/gravity ratio group, $R_{v/g}$, indicated that flow was in Region II, then use of Stalkup's correlation (Stalkup 1983a) would be appropriate. It is necessary to calculate or to estimate the average velocity of flow for the calculation of $R_{v/g}$.

Tertiary Recovery From a Five-Spot Pattern—Continuous Solvent Injection. In a miscible displacement conducted as a tertiary recovery process, the oil resource target is assumed to exist in two different saturation conditions. In the part of the reservoir previously swept by water or other secondary recovery fluid, oil is at a residual saturation. In the remaining part of the reservoir, oil saturation is relatively high and may exist at a saturation of $(1.0—S_{cw})$.

The mobility ratio in a miscible displacement is typically more unfavorable than the value in a waterflood. Thus, sweep efficiency will be poorer in a tertiary miscible process than in the preceding secondary waterflood. The swept zone in the miscible process thus may reasonably be assumed to exist initially at ROS.

Stalkup (1983a) proposed a procedure for making a rough performance calculation for a tertiary miscible displacement in which solvent injection is continuous. The process is shown schematically in **Fig. 6.74,** where piston-like displacement

V_{pDs}	Volume Injected, V_{Si} (RB)	E_{As}	N_p (STB)	N_p/N
0.185	72,300	0.185	62,900	0.173
0.20	78,200	0.197	67,000	0.184
0.30	117,300	0.270	91,800	0.252
0.40	156,400	0.345	117,300	0.322
0.50	195,500	0.395	134,300	0.369
0.60	234,600	0.440	149,600	0.411
0.70	273,700	0.480	163,200	0.448
0.80	312,800	0.515	175,100	0.481
0.90	351,900	0.540	183,600	0.504
1.00	391,000	0.570	193,800	0.532
1.20	469,200	0.615	209,100	0.574
1.50	586,500	0.670	227,800	0.626
2.0	782,000	0.725	246,500	0.677
3.0	1,173,000	0.800	272,000	0.747
4.0	1,564,000	0.845	287,300	0.789
5.0	1,955,000	0.870	295,800	0.813

*Breakthrough of solvent occurs at $V_{pDs} \approx 0.185$.

Table 6.7—Calculated results, Example 6.12*.

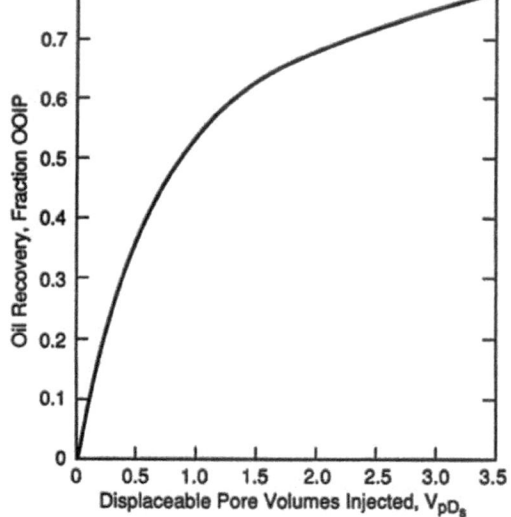

Fig. 6.73—Calculated recovery, Example 6.12.

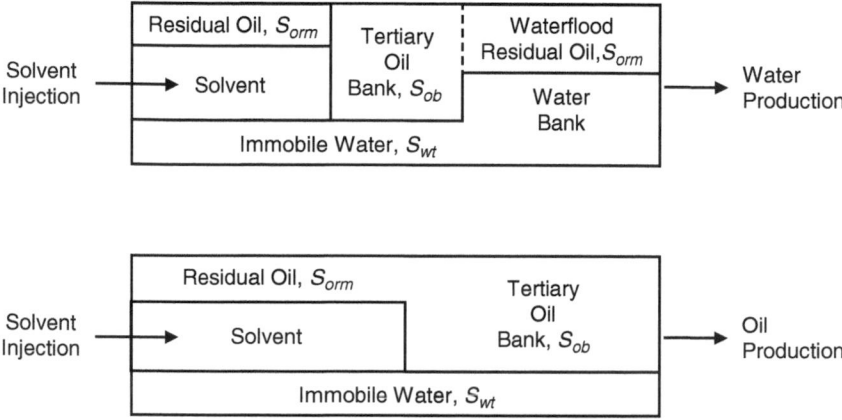

Fig. 6.74—Tertiary recovery, continuous solvent injection.

is assumed. Fractional-flow behavior and mixing of miscible fluids would cause fluid interfaces to be "smeared;" thus, the piston-like displacement is an idealization. In this procedure the following assumptions and equations are applied.

1. Residual oil from the prior waterflood is banked up into a tertiary oil bank by injected miscible solvent. At the leading edge of the oil bank, waterflood residual oil is mobilized and water is displaced with a sharp front.

2. The tertiary oil bank is displaced by solvent with a sharp front. There may be an ROS, S_{orm}, after the solvent displacement. The water saturation through the oil and solvent banks is assumed to be a constant value, S_{wt}. The saturation would, in fact, vary across these banks, and the average value could be estimated from frontal-advance theory.

3. The mobility ratio governing sweepout is the ratio of mobility in the solvent-invaded region to that in the region of water flow ahead of the tertiary oil bank.

4. Sweepout correlations used for the oil bank and solvent are the same.

5. A pseudodisplaceable PV injected is calculated that includes the oil bank:

$$V_{pDob} = V_{pDs} + \Delta V_{pDob}, \quad\dotfill (6.31)$$

where V_{pDob} = pseudodisplaceable PVs injected and accounts for solvent plus the oil bank, V_{pDs} = displaceable PVs of solvent injected defined in Eq. 6.21 with S_{wi} replaced by S_{wt}, and ΔV_{pDob} = incremental contribution to the pseudodisplaceable PVs from the oil bank. The quantity ΔV_{pDob} is determined by calculating the number of PVs occupied by the oil bank:

$$\Delta V_{pDob} = \frac{\left(\text{oil displaced by the solvent bank}\right)}{\text{displaceable PV in the oil bank}}.$$

$$\Delta V_{pDob} = E_{Vs}V_p\left(S_{orw} - S_{orm}\right) / \left[V_p\left(1 - S_{wt} - S_{orw}\right)\right]$$

$$= E_{Vs}\left[\left(S_{orw} - S_{orm}\right) / \left(S_{ob} - S_{orw}\right)\right], \quad\dotfill (6.32)$$

where E_{Vs} = volumetric sweep efficiency of the solvent, S_{orw} = waterflood ROS, S_{orm} = ROS in zone invaded by solvent, S_{ob} = oil saturation in the oil bank, S_{wt} = water saturation in the oil and solvent banks, and V_p = reservoir PV. The saturations are indicated in Fig. 6.74.

The saturation, S_{orw}, can be determined from the relative permeability curves. Oil-bank saturation, S_{ob}, can be estimated from frontal-advance theory (Chapter 3) and the Welge tangent method (Stalkup 1983a). Alternatively, as a simplification, S_{ob} can be assumed equal to $(1 - S_{wr})$, a value that is somewhat too large.

6. Oil recovery is given by

$$N_p = \left[V_p E_{Vs}\left(S_{orw} - S_{orm}\right) - V_p\left(E_{Vob} - E_{Vs}\right)\left(S_{ob} - S_{orw}\right)\right]\left(1/B_o\right), \quad\dotfill (6.33)$$

where E_{Vob} = volumetric sweep efficiency of the oil bank. Saturations and the oil FVF, B_o, are at the conditions of the solvent displacement.

Example 6.13 illustrates the procedure using the Claridge (1972) correlation, as in the previous example. A FORTRAN computer program to make the calculation is in Appendix D.

Example 6.13—Tertiary Recovery From a Five-Spot Pattern—Application of Claridge Correlation. Assume that a large slug of a solvent is to be used to displace oil miscibly in a tertiary recovery process. The unit has previously been waterflooded to S_{orw}. **Table 6.8** lists data for this example. Other data are as given in Example 6.12. There are no gravity effects or vertical heterogeneities. Calculate oil recovery as a function of PVs of solvent injected.

Solution. Calculate OOIP:

$$N = Ah\phi S_{orw} \frac{1}{B_o}\left(\frac{1}{5.615 \text{ ft}^3/\text{bbl}}\right)$$

$$= 20 \text{ acres} \times 43,560 \text{ ft}^2/\text{acre} \times 20 \text{ ft} \times 0.18 \times 0.30$$

$$\times \left(\frac{1}{1.15 \text{ RB/STB}}\right)\left(\frac{1}{5.615 \text{ ft}^3/\text{bbl}}\right)$$

$$= 145,700 \text{ STB}.$$

The total displaceable volume, V_{pD}, is 391,000 RB (Example 6.12). Calculate the effective mobility ratio (solvent-slug mobility/water-zone mobility):

$$M_e = \frac{k_{rs}/\mu_s}{k_{rw}/\mu_w}$$

$$= \frac{0.80/0.06 \text{ cp}}{0.20/0.70 \text{ cp}}$$

$$= 46.7.$$

Waterflood ROS, S_{orw}		0.30
Oil-bank saturation, S_{ob}		0.75
Water saturation in oil bank, S_{wt}		0.25
Water viscosity, μ_w (cp)		0.70
Relative water permeability at waterflood ROS, k_{rw}		0.20
Relative solvent permeability at water saturation in oil bank, k_{rs}		0.80

Table 6.8—Example 6.13 data.

The following procedure is used until V_{pDs} reaches the maximum desired value:

1. Set V_{pDs}.
2. Calculate ΔV_{pDob} and V_{pDob} from Eqs. 6.32 and 6.31.
3. Determine E_{As} and E_{Aob} from Fig. 6.72, the Claridge correlation, or equations in Section 4.5. Areal sweep in the Claridge correlation is equal to volumetric sweep.
4. Calculate N_p with Eq. 6.33.
5. Return to Step 1.

Results are shown in **Table 6.9** and **Fig. 6.75.** Displaceable PVs injected, V_{pDs}, in Fig. 6.75 could be converted to HCPVs injected as was done in Example 6.12:

$$V_{ph} = V_{pDs}\left[\left(1 - S_{orm} - S_{wt}\right)/\left(1 - S_{wc}\right)\right].$$

The results are presented in terms of fractional recovery, N_p/N, where N in this case is the oil in place at the start of the tertiary recovery project. The fractional recovery also could be expressed relative to OOIP. The recovery at 2.0 displaceable PV of solvent injected is calculated to be 14.6% OOIP.

In a tertiary project, oil-bank breakthrough occurs after a certain amount of solvent is injected ($\approx$33,200 RB in the example). Also, if solvent injection is continuous, as in this example, the solvent/oil produced ratio is large, which adversely affects the economics of the project.

Stalkup (1983a) discusses the fact that the procedure is a rough approximation. The overall average mobility and controlling mobility ratio change as the displacement proceeds. The waterflooded area is relatively large in the early part of the process, and consequently, the water mobility has a large effect. The size of the oil bank and the relative importance of the oil-bank

V_{pDs}	ΔV_{pDob}	V_{pDob}	E_{As}	E_{Aob}	Volume Solvent Injected, V_{si} (RB)	N_p	N_p/N
0.07	0.039	0.109	0.07	0.109	27,400	0	0
0.087	0.048	0.135	0.087	0.135	34,000	0	0*
0.10	0.056	0.156	0.100	0.156	39,100	120	—
0.20	0.111	0.311	0.190	0.270	78,200	5,600	0.038
0.30	0.167	0.467	0.260	0.340	117,300	14,100	0.097
0.40	0.222	0.622	0.315	0.405	156,400	18,600	0.128
0.50	0.278	0.778	0.360	0.450	195,900	24,000	0.165
0.60	0.333	0.933	0.397	0.495	234,600	26,800	0.184
0.70	0.389	1.089	0.425	0.520	273,700	30,800	0.211
0.80	0.444	1.244	0.460	0.560	312,800	34,000	0.233
0.90	0.500	1.400	0.485	0.590	351,900	35,900	0.246
1.00	0.556	1.556	0.510	0.605	391,000	41,200	0.283
1.20	0.667	1.867	0.550	0.650	469,200	44,900	0.308
1.50	0.833	2.333	0.600	0.690	586,500	53,200	0.365
2.0	1.111	3.111	0.665	0.750	782,000	62,200	0.427
3.0	1.667	4.667	0.740	0.810	1,173,000	74,600	0.512
4.0	2.222	6.222	0.785	0.825	1,564,000	86,600	0.594

*Oil bank breakthrough.

Table 6.9—Calculated results, Example 6.13.

mobility increase with continued injection. For typical flooding conditions, these factors lead to a calculated oil-bank breakthrough that is too early and oil production that is too optimistic in the early period. Stalkup (1983a) has shown, however, that the calculated behavior is generally consistent with results obtained from more-sophisticated mathematical models.

Tertiary Recovery—Displacement With Solvent Slugs Driven by Water. Stalkup (1983a) extended the calculation procedure to describe displacement of solvent slugs by water or gas. The case of solvent slugs driven by water is considered here. When water displaces a solvent slug, breakdown of the slug occurs from immiscible displacement of the slug by water; i.e., there will be a trapped residual solvent phase. The procedure is as follows.

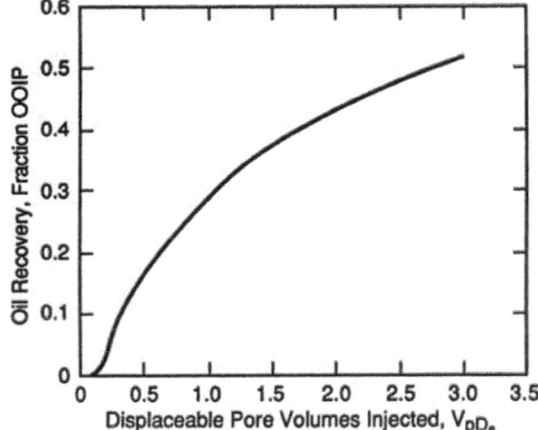

Fig. 6.75—Calculated results, Example 6.13.

1. The procedure for continuous solvent injection is used until all the solvent has been injected.
2. During water injection, until all the solvent is left as a residual phase or has been produced (i.e., until the last of the solvent slug has been left as a residual phase), use the same value of M as in Step 1. This means that the same sweepout curve will be used for this period. Breakthrough of injected water is assumed not to occur until the last of the solvent slug has been left as a residual.

The sweepout of the drive-water front is given by

$$V_{pDw} = W_i / \left[V_p \left(1 - S_{orm} - S_{srw} - S_{wt} \right) \right], \dots \dots \dots (6.34)$$

where V_{pDw} = displaceable PVs of water injected, S_{srw} = residual solvent saturation remaining after immiscible displacement of solvent by water, and W_i = volume of water injected. Before water breakthrough,

$$V_{pDw} = E_{Vw}, \dots \dots \dots (6.35)$$

where E_{Vw} = volumetric sweep efficiency of the water.

Calculate a pseudodisplaceable PV injected for the solvent phase that accounts for the displacement of solvent by water.

$$V_{pDs} = V'_{pDs} + \Delta V_{pDs}, \dots \dots \dots (6.36)$$

where $V'_{pDs} = V_{si} / \left[V_p \left(1 - S_{orm} - S_{wt} \right) \right]. \dots \dots \dots (6.37)$

V'_{pDs} = total (maximum) solvent-slug size expressed as displaceable PVs, and

$$\Delta V_{pDs} = \Delta V_{pDw} \left(\frac{S_s - S_{srw}}{1 - S_{orm} - S_{wt}} \right)$$

$$= \Delta V_{pDw} \left(\frac{1 - S_{orm} - S_{wt} - S_{srw}}{1 - S_{orm} - S_{wt}} \right), \dots \dots \dots (6.38)$$

where S_s = solvent saturation in the solvent bank.

The pseudodisplaceable PV for the leading edge of the oil-bank sweepout is given by

$$V_{pDob} = V_{pDs} + \Delta V_{pDob}, \dots \dots \dots (6.39)$$

where ΔV_{pDob} is calculated from Eq. 6.32. Eq. 6.33 gives oil recovery for a tertiary flood during this period.

3. Once the slug has been dissipated (i.e., produced and/or left as a residual phase) continued oil recovery results from further injection of water. With water displacing only oil, the mobility ratio is more favorable than with solvent displacement. Therefore, it is appropriate to use a different mobility ratio, one based on water displacing the oil bank. The transition to this different sweepout curve on the correlation for volumetric sweep efficiency must be estimated, and Stalkup recommends that a smooth sweepout curve be drawn from the first curve to the second. This is clearly only a rough approximation. During this phase of the process the sweepout of the leading edge of the injected water bank is given by

$$V_{pDw} = V'_{pDw} + \Delta V_{pDw}, \dots \dots \dots (6.40)$$

where V'_{pDw} = displaceable PVs of water injected at the point in time that the last of the solvent slug is left as a residual phase and ΔV_{pDw} = displaceable PVs of water injected beyond the point in time that the last of the solvent slug is left as a residual phase.

The pseudodisplaceable PV for the leading edge of the tertiary oil bank is given by

$$V_{pDob} = V'_{pDob} + \Delta V_{pDw}, \dotfill (6.41)$$

where V'_{pDob} = pseudodisplaceable PVs of oil bank at the point in time that the last of the solvent slug is left as a residual phase. After the solvent slug has dissipated, oil production is given by

$$N_p = \left[V_p E_{Vs,max} \left(S_{orw} - S_{orm} \right) - V_p \left(E_{Vob} - E_{Vw} \right) \left(S_{ob} - S_{orw} \right) \right] 1/B_o, \dotfill (6.42)$$

where $E_{Vs,max}$ = volumetric sweep efficiency of leading edge of the solvent bank at the point in time that the last of the solvent was left as a residual phase.

The procedure is illustrated in Example 6.14. A FORTRAN computer program to make the calculation is given in Appendix D.

Example 6.14—Tertiary Recovery From a Five-Spot Pattern—Solvent Slug Driven by Water. It is desired to examine the effect of solvent-slug size on miscible displacement performance in a five-spot pattern that has previously been waterflooded. The Stalkup model combined with the Claridge correlation is used for the calculation. Gravity effects and reservoir heterogeneities are negligible. **Table 6.10** gives data for this example.

Calculate oil recovery as a function of HCPV of fluid injected for solvent-slug sizes ranging from 10% to 100% HCPV. Water is to be used to displace the solvent. HCPV is based on connate water saturation, S_{cw}.

Solution. M_e is calculated as the ratio of mobility of the solvent to mobility of the water in the waterflooded area:

$$M_e = \frac{k_{rs}/\mu_s}{k_{rw}/\mu_w}$$
$$= \frac{0.8/0.06}{0.2/0.7}$$
$$= 46.7.$$

Oil in place at start of process is

$$N = AL\phi S_{orw}/B_o$$
$$= 20 \text{ acres} \times 43,560 \text{ ft}^2/\text{acre} = 20 \text{ ft} \times 0.18 \times 0.30$$
$$\times \left(\frac{1}{5.615 \text{ft}^3/\text{bbl}} \right) \left(\frac{1}{1.15 \text{ RB/STB}} \right)$$
$$= 145,700 \text{ STB}.$$

Well spacing, A (acres)	20
Thickness, h (ft)	20
Porosity, ϕ	0.18
Permeability, k (md)	120
Oil viscosity, μ_o (cp)	1.5
Water viscosity, μ_w (cp)	0.7
Solvent viscosity, μ_s (cp)	0.06
Oil FVF, B_o (RB/STB)	1.15
Reservoir temperature, T (°F)	120
Reservoir pressure, p (psia)	2,500
Connate water saturation, S_{cw}	0.25
Water saturation in oil and solvent banks, S_{wt}	0.35
ROS in solvent bank, S_{orm}	0.10
Oil saturation in oil bank, S_{ob}	0.65
Waterflood ROS, S_{orw}	0.30
Relative permeability to water at S_{orw}, k_{rw}	0.20
Relative permeability to solvent and oil at S_w, k_{rs}	0.80

Table 6.10—Example 6.14 data.

The solution procedure is as follows.

1. Solvent injection phase—Set total HCPV of solvent to be injected. Calculate total V_{pDs} to be injected. $V_{ph} = 0.10$.

$$V'_{pDs} = \frac{V_{ph}(1 - S_{cw})}{1 - S_{orm} - S_{wt}}$$

$$= \frac{0.10(1 - 0.25)}{1 - 0.10 - 0.35}$$

$$= 0.1364 \text{ (maximum value)},$$

or in general,

$$V'_{pDs} = 1.364 \times V_{ph}.$$

Follow the procedure in Example 6.13 until the specified total V_{pDs} has been reached

a. Set V_{pDs}.
b. Calculate ΔV_{pDob} and V_{pDob}.
c. Calculate E_{As} and E_{Aob} with the Claridge correlation. Use the equations given in Section 4.5. Areal sweep is equal to volumetric sweep.
d. Calculate N_p.
e. If $V_{pDs} < V'_{pDs}$, return to Step 1a and repeat the calculation for the next V_{pDs}. If $V_{pDs} = V'_{pDs}$, proceed to Step 2.

2. Water injection phase to point in time that last of solvent is left as a residual.

a. Set V_{pDw}.
b. Calculate ΔV_{pDs} and V_{pDs}.
c. Calculate ΔV_{pDob} and V_{pDob}.
d. Calculate E_{Aw}, E_{As}, and E_{Aob} with the Claridge correlation.
e. Check to determine if $E_{Aw} > E_{As}$ (i.e., whether the water bank has overtaken the solvent bank). If it has, reduce the V_{pDw} value and return to Step 2a. If $E_{Aw} \leq E_{As}$, proceed to Step 2f.
f. Calculate N_p.
g. If $E_{Aw} < E_{As}$, return to Step 2a and repeat the calculation for the next V_{pDw}. If $F_{Aw} = F_{As}$, proceed to Step 3.

3. Water injection phase—waterflood the remaining oil.

a. Calculate a new M_e that is the ratio of mobility of the water in the waterflood zone to the mobility of the oil bank. This will require shifting to a new M_e value in the Claridge correlation, as indicated in Step 3d.
b. Set ΔV_{pDw} and calculate V_{pDw}.
c. Calculate V_{pDob}.
d. Calculate E_{Aw} and E_{Aob}. M_e is assumed to change linearly over 1 displaceable PV of water injected from the value used in Steps 1 and 2 to the value calculated for Step 3.
e. Calculate N_p.
f. If the maximum amount of water has not been injected, return to Step 3b and repeat the calculation for the next ΔV_{pDw}.

A FORTRAN computer program written to make the calculations is given in Appendix D. The calculation was made for slug sizes of 10, 25, 50, and 100% HCPV.

Results for the 10% slug are given in **Table 6.11.** Oil recoveries as functions of HCPV injected for all four slug sizes are shown in **Fig. 6.76.**

Example 6.14 illustrates the effect of solvent-slug size on recovery. The model used to make the calculations is based on a number of assumptions and should be viewed only as a rough approximation of actual reservoir performance. Nonetheless, trends and general behavior exhibited are indicative of miscible displacement performance. The Claridge (1972) correlation was used in the calculations to obtain sweep efficiency. Other correlations, such as Stalkup (1983a) or Withjack and Akervoll (1988), could also be applied.

6.11 Design Procedures and Criteria

A number of factors must be considered in the design of a miscible displacement process. This section discusses those that relate primarily to the laboratory and modeling studies that precede field testing. In addition to the design work discussed here, significant reservoir analysis is required. The reservoir analysis should focus on geology, fluid distributions, performance of primary production, analysis of any prior displacement process (such as waterflooding), injection rates, and pattern-geometry effects. Gaining an understanding of reservoir heterogeneity is especially critical for the final design. An economic study, including a sensitivity analysis, also should be conducted.

V_{pDs}	ΔV_{pDob}	V_{pDob}	E_{As}	E_{Aob}	Voume Injected (bbl)	V_{ph}	N_p (STB)	N_p/N^*
0.01	0.0057	0.0157	0.0100	0.057	3,072	0.0073	–	–
0.03	0.0171	0.0471	0.0300	0.0471	9,216	0.0220	–	–
0.05	0.0286	0.0786	0.0500	0.0786	15,360	0.0367	–	–
0.07	0.0400	0.1100	0.0700	0.1100	21,505	0.0513	–	–
0.09	0.0514	0.1414	0.0700	0.1414	27,649	0.0660	–	–
0.10	0.0571	0.1571	0.1000	0.1563	30,721	0.0733	139	0.0010
0.12	0.686	0.1886	0.1200	0.1820	36,865	0.0880	1,110	0.0076
0.1364	0.0779	0.2143	0.1364	0.2012	41,892	0.1000	2,232	0.0153

End Slug Injection

V_{pDw}	E_w	V_{pDob}	E_{As}	E_{Aob}	Volume Injected (bbl)	V_{ph}	N_p (STB)	N_p/N
0.01	0.01	0.2245	0.1431	0.2084	45,522	0.1087	2,797	0.0192
0.05	0.05	0.2629	0.1657	0.2340	60,045	0.1433	4,498	0.0309
0.10	0.10	0.3090	0.1907	0.2620	78,198	0.1867	6,408	0.0440
0.15	0.15	0.3538	0.2134	0.2869	96,352	0.2300	8,246	0.0566
0.20	0.20	0.3976	0.2345	0.3094	114,505	0.2733	10,037	0.0689
0.25	0.25	0.4406	0.2540	0.3299	132,658	0.3167	11,785	0.0809
0.26	0.26	0.4491	0.2578	0.3338	136,289	0.3253	12,129	0.0832

Slug Left as Residual

V_{pDw}	E_w	V_{pDob}		E_{Aob}	Volume Injected (bbl)	V_{ph}	N_p (STB)	N_p/N
0.36	0.3292	0.5491		0.3970	172,595	0.4120	13,515	0.0928
0.46	0.3984	0.6491		0.4602	208,902	0.4987	14,529	0.0997
0.56	0.4675	0.7491		0.5234	245,208	0.5854	15,543	0.1067
0.66	0.5367	0.8491		0.5866	281,515	0.6720	16,557	0.1136
1.06	0.8134	1.249		0.8394	426,741	1.019	20,614	0.1415

*N = Oil in place at start of process.

Table 6.11—Calculated results for Example 6.14, slug size of 10% HCPV.

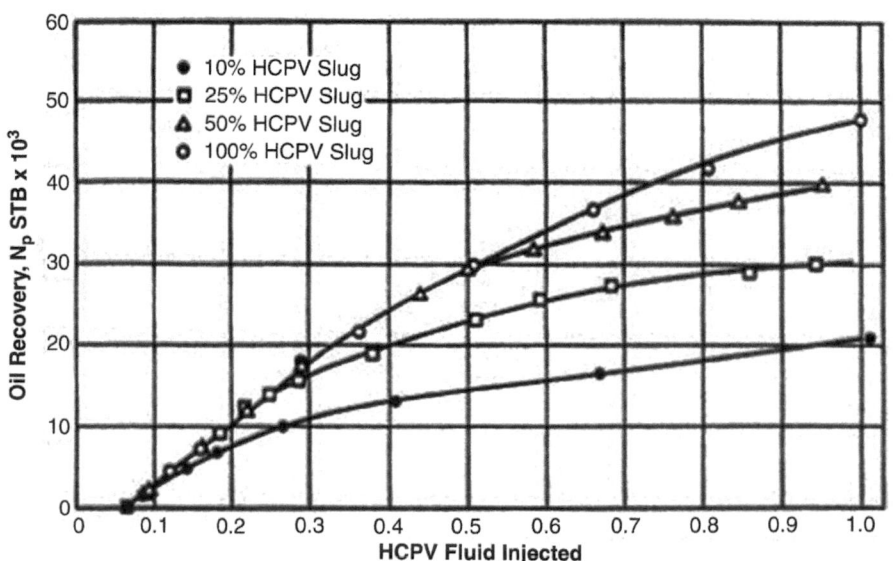

Fig. 6.76—Calculated results for Example 6.14; effect of slug size on recovery.

In a miscible displacement process, a solvent slug that efficiently displaces the reservoir oil at the microscopic level must be used. Typically, ROS to the solvent should be 10% or less. Depending on reservoir fluids and conditions (p and T) and the solvents available, either a first-contact or multiple-contact process can be applied. Volumetric sweepout (which is dependent on mobility ratio, gravity effects, reservoir heterogeneities, and well pattern) must be such that the overall recovery efficiency is sufficient to yield a satisfactory economic return. The solvent-slug size should be designed so that the miscible and immiscible portions (after solvent-slug breakdown) of the process yield this acceptable recovery.

Meeting these criteria requires that certain laboratory and modeling studies be conducted in the design work. Phase-behavior studies relating to miscibility must be conducted. Fluid mobilities and methods of exercising mobility control should be considered, as should the role of gravity forces. This probably will require conducting laboratory displacements. Mathematical modeling should be used to examine such factors as optimum slug size and volumetric sweepout.

6.11.1 Phase Behavior—Selection of a Solvent. The selection of a solvent is governed by reservoir pressure and temperature, reservoir oil composition, and solvent availability and cost. Reservoir temperature and especially pressure are, in turn, related to reservoir depth. Pressures in excess of the fracture pressure cannot be used because fracturing would lead to poor sweep efficiency. A rule of thumb for fracture pressure is 0.6 psi/ft of depth.

Miscibility of a solvent/oil system and efficiency of displacement in a phase-behavior sense are typically determined from slimtube experiments, as discussed in Section 6.7. These experiments should be designed so that viscous-fingering, gravity-segregation, and heterogeneity effects are minimized. Under these conditions, the slimtube apparatus is basically a multiple-contact device in an idealized porous medium. Recovery from an experiment represents the maximum displacement efficiency at the microscopic level that can be expected on the basis of the phase behavior of the solvent/oil system.

Slimtube experiments or RBA experiments can be used to measure the MMP for a specified solvent or the MME of a solvent at a specified pressure. The experiments are relatively simple and inexpensive to conduct, making the testing of several systems feasible.

An alternative or complementary experiment to the slimtube test is the continuous multiple-contact test described by Orr and Silva (1983) and Orr et al. (1983). In the device used for this test, multiple contacts between oil and solvent are made in a continuous process (as opposed to a batch process) but a porous medium is not used. Phase-behavior data that relate to the development of miscibility are obtained.

Classic phase-behavior data taken in a PVT cell also can be useful. These experiments consist of allowing fixed amounts of solvent and oil to equilibrate in a cell at fixed pressure and temperature. Samples of each phase then can be used for measurements of compositions and other properties such as viscosity and density. PVT data are relatively time-consuming and expensive to obtain. Such data are useful to calibrate an EOS. The data also can be used to identify potential problems, such as asphaltene deposition.

6.11.2 Mobility-Control Considerations. As discussed in Section 6.8, viscosities of solvents are relatively small compared with those of reservoir oils. For example, viscosities of hydrocarbon gases or CO_2 are on the order of 0.05 to 0.10 cp at usually encountered flooding conditions. Thus, the mobility ratio between the solvent slug and the displaced oil bank is typically very unfavorable.

Research is under way on the use of additives to increase hydrocarbon or CO_2 solvent viscosity. The application of mobility-control agents, such as foams or gels, also has been examined. To date, however, these approaches are experimental and are not widely accepted technology in solvent flooding.

The most widely used method of mobility control is to alternate water and gas injection—the WAG injection process described in Chapter 5. In this process, slugs of solvent and water are alternately injected through the injection wells. The presence of a high water saturation reduces the effective permeability of the solvent and thus its mobility.

An evaluation of the improved sweep resulting from improving the effective mobility ratio by use of the WAG process is probably best accomplished with a reservoir simulator. Relative permeability data for the reservoir rock are required for these calculations. Research has shown that high water saturations can lead to water blocking [i.e., to poor contact between oil and solvent (Shelton and Schneider 1975; Tiffin et al. 1991)] in strongly water-wet rocks. An assessment of the residual oil resulting from this blocking in strongly water-wet rocks can be estimated with the approach of Raimondi and Torcaso (1964) and Todd et al. (1982). However, for most reservoir rocks (which are not strongly water-wet), experimental data are required to determine the parameters in the correlation.

6.11.3 Gravity Force Considerations. Differences in density between solvent and oil can lead to overriding of the oil by a gravity tongue. Correlations that relate gravity effects to recovery were discussed in Sections 4.6 and 6.9. The correlations use a dimensionless group, R_{vg}, a ratio of viscous/gravity forces. The role of the gravity force also can be examined with a reservoir simulator.

Under suitable conditions, a gravity-stable miscible displacement can be considered. Section 4.6 discusses methods of estimating this rate. In this process, the solvent is injected above the oil zone to be displaced, driving the oil downward. Drive gas is then injected after and above the primary solvent slug, forcing it downward through the reservoir. Use of a WAG process in this circumstance would be questionable because the higher-density water would tend to finger downward through the solvent and oil banks.

6.11.4 Corefloods. Reservoir corefloods are useful to estimate displacement efficiency at the microscopic level. Effects on displacement of such phenomena as water blocking at high water saturations can be investigated. Data on relative permeabilities in cores can be used to estimate injectivities under field conditions if hysteresis in the relative permeabilities is carefully modeled.

Corefloods need to be designed carefully. Samples of reservoir rock suitable for making long cores for laboratory displacement investigations are difficult to obtain. Miscible displacements in short, small-diameter cores cannot be readily extrapolated to field dimensions. Quarried rocks, such as Berea sandstone, can be used, but the microscopic characteristics, especially wettability, would be quite different from actual reservoir rocks.

6.11.5 Mathematical Modeling. Mathematical modeling, or simulation, of a miscible displacement process can be achieved with a variety of models ranging from those that are relatively simple to complex, computer-based, numerical finite-difference simulators. The calculation described in Section 6.10, which is based on data from laboratory physical models, is an example of a relatively simple approach. Results from such a model can be used to describe general performance behavior and as a screening tool. The results, however, should be considered only approximate at best.

The most comprehensive models are numerical finite-difference simulators based on numerical solutions of partial-differential equations that describe the fluid system. The literature (Todd el at. 1982; Moore and Clark 1988; Nagel et al. 1990; Fussell and Fussell 1979) contains many examples of the use of such simulators to describe miscible displacement processes.

In a numerical finite-difference simulator, the reservoir is subdivided into gridblocks. The describing partial-differential equations are approximated by algebraic [finite-difference equations, and the dependent variables (e.g., pressure, composition, saturation) are defined at each gridblock. The reservoir can be approximated in two dimensions or, more rigorously, in three dimensions. The latter typically requires a significantly larger number of gridblocks, and therefore computation time often is increased markedly. In addition to the finite-difference equations, certain auxiliary equations are required. These include an EOS that allows computation of equilibrium phases and compositions and relative permeability functions. The model solution involves simultaneous solution of a very large set of nonlinear algebraic equations. The results of the computation are pressure, saturations, and in some cases, phase compositions as functions of time and position throughout the reservoir.

Two types of models that have been widely used are modified black-oil simulators and compositional simulators. These differ in the treatment of phase behavior and composition of each phase. Black-oil simulators basically treat the different phases as individual components (i.e., oil, gas, or water). Compositional simulators account for the different components in each phase and use a more rigorous EOS. Compositional simulators are more complex and thus generally more expensive and time-consuming to apply. Stalkup (1983a) gives a summary of different model types and their application.

One problem commonly encountered in numerical finite-difference simulators used to describe miscible displacement processes is numerical dispersion. This is an artificial and often large mixing or dispersion that results from the solution of the equations in the model. However, dispersion caused by reservoir heterogeneity is significant, and simulators often incorporate numerical dispersion to approximate dispersion in a reservoir. Stalkup (1983a) gives a good discussion of the effect of numerical dispersion.

An example of a result from application of a numerical finite-difference simulator is shown in **Fig. 6.77** (Todd et al. 1982). The simulator was used to evaluate alternative designs of a proposed CO_2 tertiary food in a west Texas reservoir. The two designs examined were a CO_2 slug followed by water injection and a WAG process followed finally by water. In both cases, 0.4 HCPV of CO_2 was injected. The effects of such factors as gravity-driven crossflow between reservoir layers and reservoir stratification were examined.

Fig. 6.77 shows oil recovery from a 30-acre five-spot pattern predicted for the two different designs. Also shown is oil recovery assuming that the waterflood was continued. In the calculation, ultimate oil recovery was slightly greater and recovery rate

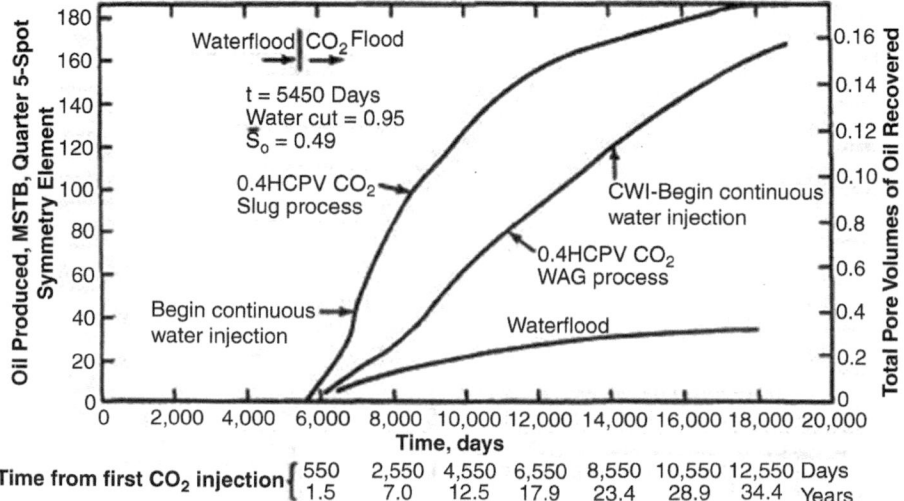

Fig. 6.77—Predicted recovery performance, 30-acre five-spot, alternative process results (Todd et al. 1982).

was faster when the simple CO_2 slug was used. There is an advantage, however, to the WAG process in that the demand for CO_2 is spread over a much longer period, which can improve the economics. Also, other calculations not shown indicated that in highly stratified reservoirs with significant gravity-driven crossflow, the WAG process performed better than the single-slug process.

A very useful model and one that is relatively simple to run is the CO_2 miscible food predictive model (CO_2 PM) published by the US Department of Energy (US DOE 1986). This model is not based on a solution of finite-difference equations, but rather uses a method-of-characteristics solution to the fractional-flow equation. The CO_2 PM describes miscible displacement in a layered five-spot pattern. CO_2 injection or a WAG process, in either a secondary recovery or tertiary recovery mode, can be modeled. A manual is available from the US DOE and several example calculations are described in the manual.

Examples of reservoir simulation to field-scale applications are discussed briefly in the following section, which includes brief descriptions of four major miscible flooding projects. Modeling details for the specific applications can be found in the references associated with each project.

6.12 Field Experience
A large number of field trials and commercial applications of miscible flooding have been implemented. Stalkup (1983b) indicated that by 1983 more than 50 field tests of the FCM process, 19 condensing-gas projects, 11 vaporizing-gas-drive projects, and at least 36 miscible CO_2 tests had been reported. Stalkup (1983a) summarizes field application of the different processes. Later, Brock and Bryan (1989) summarized 1972–87 field-test results with CO_2.

Table 6.12 gives a summary of early CO_2 miscible floods presented by Brock and Bryan (1989). The projects are divided into three categories: field scale, producing pilots, and nonproducing pilots. Field-scale projects involved multiple patterns

Field	State	Reservoir	Lithology	Depth (ft)	Reservoir Temperature (°F)	Porosity (%)	Permeability (md)	Net Pay (ft)
Field scale								
Dollarhide	TX	Devonian	Trip. chert	7,800	120	17.0	9.0	48
East Vacuum	NM	San Andres	Oolitic dolomite	4,400	101	11.7	11.0	71
Ford Geraldine	TX	Delaware	Sandstone	2,680	83	23.0	64.0	23
Means	TX	San Andres	Dolomite	4,400	100	9.0	20.0	54
North Cross	TX	Devonian	Trip. chert	5,400	106	22.0	5.0	60
Northast Purdy	OK	Springer	Sandstone	8,200	148	13.0	44.0	40
Rangely	CO	Weber	Sandstone	6,500	160	15.0	5 to 50.0	110
SACROC (17 pattern)	TX	Canyon Reef	Carbonate	6,400	130	9.4	3.0	139
SACROC (4 pattern)	TX	Canyon Reef	Carbonate	6,400	130	9.4	3.0	139
South Welch	TX	San Andres	Dolomite	4,850	92	12.8	13.9	132
Twofreds	TX	Delaware	Sandstone	4,820	104	20.3	33.4	18
Wertz	WY	Tensleep	Sandstone	6,200	165	10.7	16.0	185
Producing pilots								
Garber	OK	Crews	Sandstone	1,950	95	17.0	57.0	21
Little Creek	MS	Tuscaloosa	Sandstone	10,400	248	23.4	75.0	30
Maljamar	NM	San Andres	Anhydrous dolomite	4,050	90	10.0	11.2	49
Maljamar	NM	Grayburg	Dolomitic sandstone	3,700	90	11.0	13.9	23
North Coles Levee	CA	Stevens	Sandstone	9,200	235	15.0	9.0	136
Quarantine Bay	LA	Sand 4 (RC)	Sandstone	8,180	183	26.4	230.0	15
Slaughter Estate	TX	San Andres	Dolomite	4,985	105	12.0	8.0	75
Weeks Island	LA	South S and R (B)	Sandstone	13,000	225	26.0	1,200.0	186
West Sussex	WY	Shannon	Sandstone	3,000	104	19.5	28.5	22
Nonproducing pilots								
Little Knife	ND	Mission Canyon	Sucr. dolomite	9,800	245	21.0	30.0	16
South Pine	MT	Red River	Cryst. dolomite	9,000	205	17.0	10.0	11

(1) S_{or} was reduced from 38% after waterflood to 20% after CO_2.

Table 6.12—Summary of selected CO_2 miscible flood projects (from Brock and Bryan 1989).

and were typically commercial projects. Producing pilots were pilot floods that used a producing well, while nonproducing pilots were pilot floods with observation wells only. General characteristics of the floods are given. The incremental recoveries reported are recoveries over and above the expected recovery had prior operations been continued. Also, some of the incremental recoveries are compared with primary recovery, while most are compared with waterflooding recovery. The incremental recovery, however, is an extrapolated value often based on data taken relatively early in the project. The last two columns of Table 6.12 are gross and net amounts of CO_2 required per barrel of oil produced. Gross CO_2 is the total amount injected, while net CO_2 is the purchased CO_2 (i.e., the gross CO_2 less recycled CO_2). Both continuous CO_2-slug and WAG injection are represented in the projects, although not identified in the table.

The projects reported were generally successful in terms of incremental oil recovery. However, Brock and Bryan (1989) did not make economic analyses. They do report that it was not possible to locate reports of projects that were outright failures.

Stalkup's summaries of hydrocarbon miscible floods for tertiary FCM, condensing-gas-drive, and vaporizing-gas-drive projects (Stalkup 1983b) are reported in **Tables 6.13 through 6.15,** respectively. Note that most of the FCM floods were conducted in the 1950s and 1960s, while more of the MCM processes were conducted later. The CO_2 projects reported in Table 6.11 tended to be conducted later yet. This timing reflects both the cost of solvents and the development of the technologies.

As reported in the 2014 *Oil and Gas Journal* biennial survey of EOR projects (Kuuskraa and Wallace 2014), the number of CO_2-EOR commercial projects increased steadily from the early 1980s. This growth, which occurred almost entirely within the US, is shown in **Fig. 6.78.** Kruuskraa and Wallace (2014) describe the historical growth in terms of two phases: (1) the development of large natural CO_2 sources in Colorado and New Mexico and (2) construction of a CO_2 pipeline into the Permian Basin in west Texas. Additional development of CO_2 supply sources, including both natural and industrial sources, was responsible for the second-phase growth that began in the early years of the 21st Century. **Fig. 6.79** presents a map of CO_2 sources and data on projects as of May 2014. There is potential for a third phase of growth on the basis of the development of additional CO_2 sources and pipelines; however, oil price will be a major factor in such development (Kruuskraa and Wallace 2014).

Denny (2009) conducted a study to identify US reservoirs in which CO_2-EOR projects could potentially be developed (Mohan et al. 2008). They located 1,673 target reservoirs on the basis of the following criteria:

- Gravity > 22 °API
- Pressure > MMP
- Depth > 2,500 ft
- Viscosity > 10 cp
- Saturated oil > 20% PV
- Sandstone or carbonate

The following are descriptions of four commercial projects—three CO_2 miscible and the fourth an N_2 miscible project that requires a much higher operating pressure than a CO_2 flood to obtain miscibility.

6.12.1 Scurry Area Canyon Reef Operations Committee (SACROC) Unit. *Early History and Geology.* The SACROC unit, located in the Kelly-Snyder field in the Permian Basin near the town of Snyder, Texas, was discovered in 1948. It is the largest of a chain of canyon-reef fields of Pennsylvanian age and covers an area of approximately 50,000 acres (Langston et al. 1988; Pipes and Schoeling 2014). The discovery well was drilled to a depth of 6,700 ft and yielded an initial production of 530 B/D (Langston et al. 1988). Reservoir thickness is quite variable, ranging from as little as 10 ft on the flanks to as much as 900 ft along the crest (Langston et al. 1988; Pipes and Schoeling 2014). The oil was initially undersaturated with a solution gas/oil ratio (GOR) of approximately 1,000 scf/bbl and a bubblepoint pressure of 1,820 psia. Reservoir and fluid properties are summarized in **Tables 6.16 and 6.17** (Langston et al. 1988; Linroth and Rickard 2014; Dicharry et al. 1973). The most current estimate of OOIP is 2.8 billion STB (Pipes and Schoeling 2014).

A computer-generated structure of SACROC is shown in **Fig. 6.80** (Barati et al. 2016). The unit has been described as a chain of limestone mounds making up what is called the Horseshoe Atoll. **Fig. 6.81** (Barati et al. 2016) shows a type log of the canyon-reef complex, which has been divided into four major zones with significantly different reservoir characteristics, termed the Cisco, green zone (GZ), upper middle canyon, and lower middle canyon. A transition zone has been identified in parts of the unit. Limestone is the dominant mineral in the reef complex, with thin shale members of 1 to 10 ft in thickness making up less than 3% of the total unit. The shale members are important stratigraphic markers.

The SACROC unit has a long and complex development history, as illustrated in **Fig. 6.82** (Barati et al. 2016), where the production/injection timeline is summarized through 2001. Between the date of the discovery well (1948) and the end of 1950, on the order of 1,600 production wells were drilled. During this period, reservoir pressure in regions of the reservoir dropped by as much as 50%, from 3,122 psi at discovery to 1,650 psi (Raines 2005). Ultimate recovery at the time was predicted to be approximately 19%, typical of a solution-gas-drive primary-production mechanism (Langston et al. 1988). Most of the Kelly-Snyder field was drilled on irregular 40-acre spacing by the end of 1950 (Raines 2005).

The unit was formed in 1953, following which, a centerline water-injection program was implemented for pressure maintenance (Dicharry et al. 1973). Large volumes of water were injected into the thickest portion of the reservoir, repressurizing the reservoir and increasing oil production, as seen in **Figs. 6.82 and 6.83** (Barati et al. 2016; Pennell et al. 2015). By the end of 1971, SACROC was producing 134,000 BOPD, had a cumulative oil production of 528 million STB, and an estimated

Field	Operator	Year Started	Type Project	Oil Gravity (°API)	Oil Viscosity (cp)	Viscosity Ratio μ_o/μ_s	Depth (ft)	Thickness (ft)	Area (acres)	Slug Size* (% HCPV)	Incremental Recovery* (% OOIP)	Oil/Gross Slug (STB/RB)	Oil/Net Slug (STB/RB)
Burkett Unit (KS)	Phillips	1958	LPG←G←W**	42	–	25	2,100	30	10	10	7.0	0.67	–
Johnson (NE)	Ohio	1958	LPG←G	37	1.1	18	4,600	10	164	5.5	5 to 34**	>0.9	–
South Ward (TX)	Atlantic	1959	LPG←G←W	35	4	40	2,400	32	10	7.5	11.5	1.5	2.2
Adena Clar A (CO)	Union	1962	LPG←G	44	0.42	6.5	5,500	28		7	–	0.46	2.2
Adena Hough A (CO)	Union	1963	LPG←G/W	44	0.42	6.5	5,500	28	80	12.5	–	0.6	1.2
Hibberd Pool South Cuyama (CA)	Atlantic	1963	LPG←G←W	35	1.7	23	4,300	60	80	7.4	13.5	1.6	>3
Phegly Unit	Mobil	1964	LPG←G/W	37	3	30	4,900	8	785	4	3.7	0.85	–

*Treated area.

**LPG = liquefied petroleum gas, G = dry gas, and W = water.

Table 6.13—Summary of selected tertiary FCM flood projects (Stalkup 1983b).

Field	Operator	Year Started	Project Type	Oil Gravity (°API)	Oil Viscosity (cp)	Viscosity Ratio, μ_o/μ_s	Depth (ft)	Thickness (ft)	Area (acres)	Slug Size (%HCPV)	Recovery (%OOIP)	Status
Haynesville (LA)	Haynesville Operators Committee	1953	LPG←RG←G←W	–	–	–	–	–	3,200	–	–	–
Bronte (TX)	Exxon U.S.A.	1956	RG	–	–	–	–	–	1,300	–	–	–
Bronte (TX)	Exxon U.S.A.	1956	RG	–	–	–	–	–	460	–	–	–
Seeligson Zone 20B-07 (TX)	Exxon U.S.A.	1957	RG→G/W	40	–	12	6,000		877	52	5 to 10 (incremental) 50 to 54 (ultimate)	NC, disc.
Stratton (TX)	Southern Minerals Corp.	1957	RG	–	–	–	–	–	–	–	–	–
South Coles Levee (CA)	Ohio	1957	RG	–	–	–	–	–	–	–	–	–
Elk Basin (WY)	Amoco Production Co.	–	–	–	–	–	–	–	–	–	–	–
Midland Farms (TX)	Amoco Production Co.	1960	RG→G→G/W	41	0.3	–	8,350	–	600	2	–	term., disc, prof.
Neale Lilliedoll (LA)	Arco Oil & Gas Co.	1962	tertiary RG←G/W	–	–	11	10,100	13.5	640	15	–	term.
Golden Spike (Alberta)	Imperial Oil Co.	1964	vertical RG←G	37	0.8	–	5,672	550	1,305	7	3.1 (incremental)	comp., disc, prof.
Pembina Bear Lake Cardium (Alberta)	Cities Service Co.	1968	RG←G←G/W	37.6	1.5	52	4,857	50	4,160	3	–	–
Ante Creek (Alberta)	Amoco Production Co.	1968	RG/W	42	0.13	4.7	11,000	29.5	6,000	cont. rich gas	–	HF,NE
Intisar D Reef (Libya)	Occidental Petroleum Corp.	1969	vertical RG	40	0.46	20	8,950	888	3,300	–	70* (ultimate)	HF, succ, prof.
Levelland (TX)	Amoco Production Co.	1972	RG←RG/W←G/W	30	1.9	86	4,900	250	1,190	14	>27*	HF, succ, prof.
South Swan Hills (Alberta)	Amoco Production Co.	1973	RG/W←G/W	42	0.39	18	8,500	73	830	15	20* (incremental) 65* (ultimate)	HF, prom., prof.
Central Mallett Slaughter (TX)	Amoco Production Co.	1972	–	30	2	–	4,800	–	12	20 and increasing	6.0 (incremental) and increasing	JS, TETT, unprof.

Table 6.14—Summary of selected condensing-gas-drive miscible flood projects (Stalkup 1983a).

Field	Operator	Year Started	Project Type	Oil Gravity (°API)$_i$	Oil Viscosity (cp)	Viscosity Ratio, μ_o/μ_s	Depth (ft)	Thickness (ft)	Area (acres)	Slug Size (%HCPV)	Recovery (%OOIP)	Status
Camp Sand Haynesville (TX)	Marathon Oil Co.	1966	RG←G←W	46	1.05	–	8,000	5	≈2,200	≈10	–	–
Rainbow area (Alberta)	various	–	vertical	–	–	–	–	–	–	–	–	–

*Estimated with simulators

Symbol Key:
  LPG = liquefied natural gas.
  RG = rich gas
  G = lean gas
  RG←G/W = rich gas displaced by alternate gas/water
  NC = nearing completion
  disc. = discouraging

  term. = terminated
  NE = not elevated
  †ETT = too early to tell
  JS = just started
  HF = half finished
  prof. = profitable

Table 6.14—Summary of selected condensing-gas-drive miscible flood projects (Stalkup 1983a). (Continued)

Field	Operator	Year Started	Project Type	Oil Gravity (°API)	Viscosity Ratio μ_o/μ_s	Miscibility Pressure (psi)	Depth (ft)	Thickness (ft)	Acreage	Status	Recovery (%OOIP)
Block 31 Devonian (TX)	Arco Oil & Gas Co.	1949 1952	G, N	46	10	3,500	8,400	125 to 150	7,840	NC, succ, prof.	>60*
Neale 10,400-ft (LA)	Arco Oil & Gas Co.	1956	G	47	4	4,000	10,400	10	–	term.	36**
University Block 9 (TX)	Arco Oil & Gas Co.	1960	LPG slug by G by G/W	38	29	4,500	5,300	4 to 45, 10	4,000	–	40
Raleigh (MS)	Chevron U.S.A. Inc.	1960	G	45.5	3.5	<5,400	12,600	18	812	–	>52
Hassi-Massaoud (Algeria)	SONATRACH	1964	G and G/W	44	7	3,700	10,800	300	18,000	succ, prof.	–
Fairway (TX)	Hunt Oil Co.	1966	G/W	48	5.5	4,800	10,000	55	22,600	HF,succ, prof.	50*
Bridger Lake (WY/UT)	Phillips Petroleum Co.	1970	G	40	19	5,300	16,500	20	3,800	NC,succ, prof.	32*
East Binger (OK)	Phillips Petroleum Co.	1977	N	38	–	–	10,000	–	13,000	JS, TETT	–
Fordoche (LA)	Sun Gas Co.	1971	G + N	45	5	–	13,500	25 to 35	6,100	HF.succ.	50*
Jay-Little (FL)	Exxon U.S.A.	1982	N/W (20% HCPV)	51	6	3,600	15,000	350	14,000	–	54*
Lake Barre† (LA)	Texaco Inc.	1978	N	40–43	–	5,100(G) 5,800(N)	17,500	57	1,194	–	53* (6.5* incremental)

*Projected.

**Recovery in the east and west sections of the sand where miscible displacement definitely was achieved was estimated by the operator to be approximately 50%.

† Full miscibility apparently has not been achieved; reservoir pressure is several hundred pounds per square inch below dynamic miscibility pressure.

Symbol Key:

G = natural gas

N = flue gas or nitrogen

G/W = alternate gas/water injection

succ. = successful

prof. = profitable

TETT = too early to tell

NC = nearing completion

term. = terminated

HF = half finished

JS = just started

Table 6.15—Summary of selected vaporizing-gas-drive miscible flood projects (Stalkup 1983a).

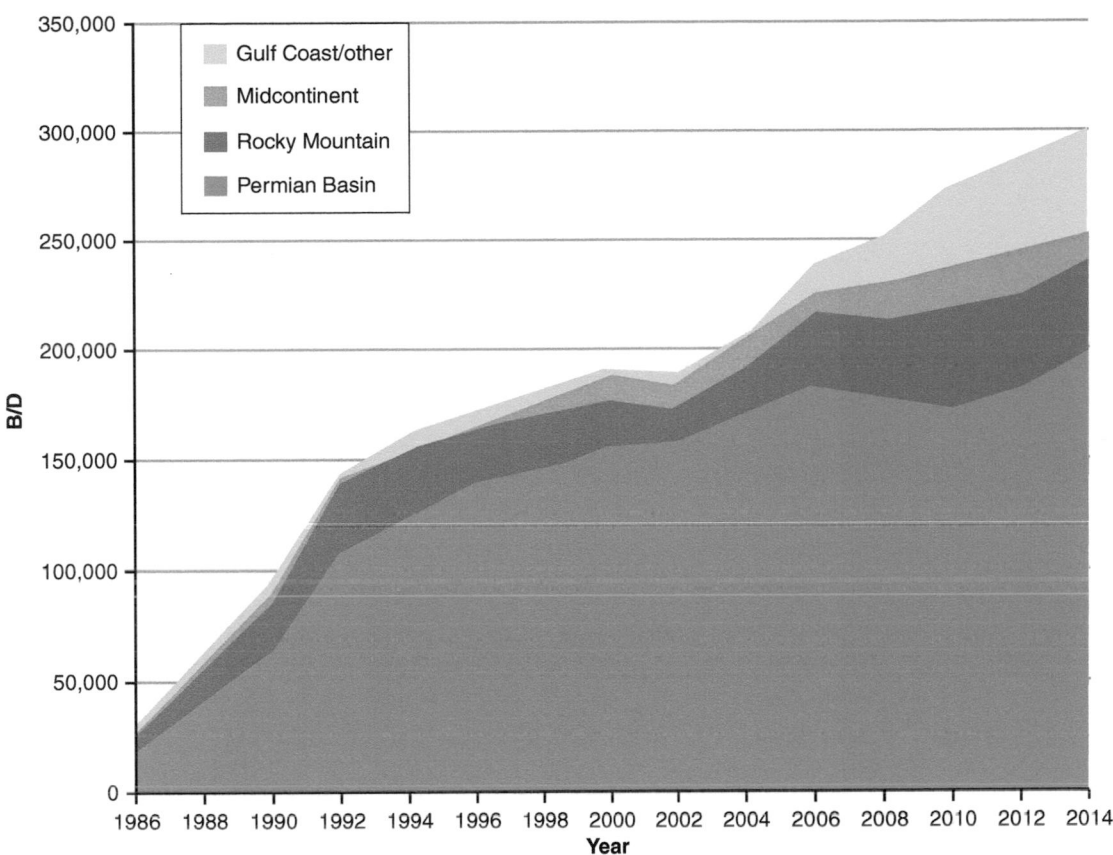

Fig. 6.78—Historical CO₂-EOR production in the US (Kuuskraa and Wallace 2014).

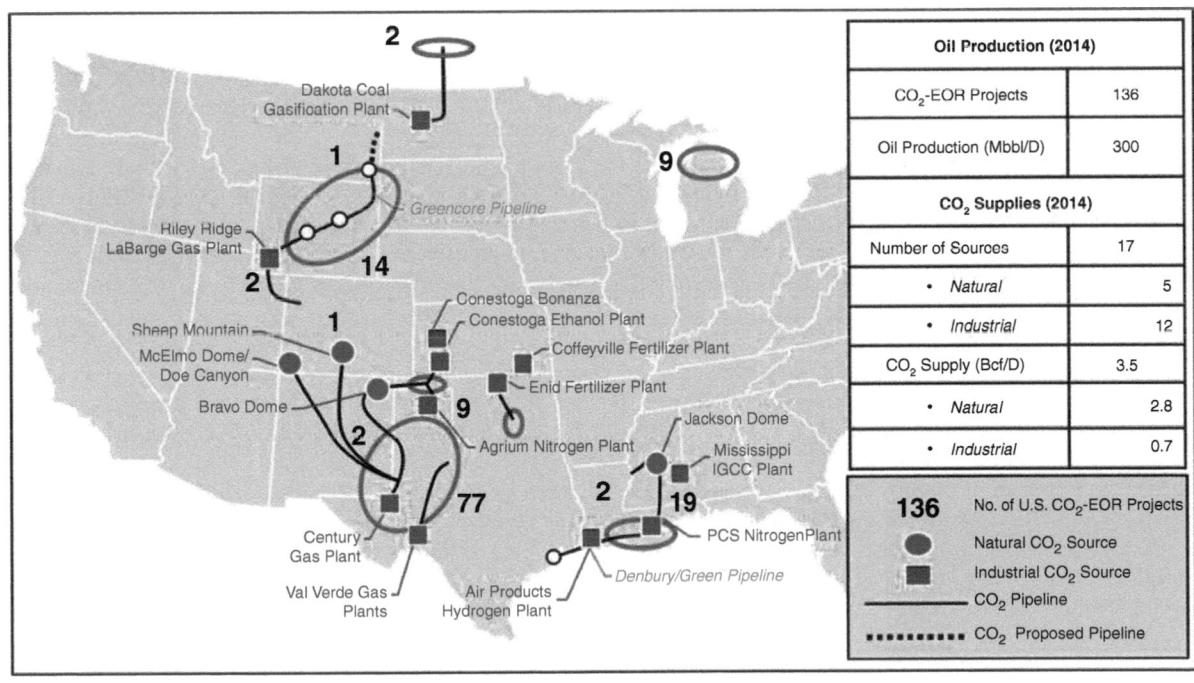

Fig. 6.79—CO₂-EOR operations and sources, 2014 (Kuuskraa and Wallace 2014).

Discovery date	November 1948
Formation	Canyon Reef limestone
Depth (ft)	6,200 to 7,000
OOIP (BSTB)	2.8
Average porosity (%)	7.6
Average permeability (md)	19
Reservoir temperature (°F)	130
Pressure (psig)	3,300
R_{si} (scf/bbl)	1,000
B_o (RB/STB)	1.65
Viscosity (cp)	0.33
S_{oi} (%)	80
S_{orw} (%)	36
P_b (psig)	1,800
MMP (psig)	1,850
Drive mechanism	Solution gas
API gravity	42
Gas gravity	1.13
Salinity (ppm)	59,000
Sulfur content	0.32
Water/oil mobility ratio	0.3
CO_2/oil mobility ratio	8
CO_2 flood GOR (scf/STBO)	2,600
NGL (bbl/million scf HC gas)	230

Table 6.16—SACROC fluid and rock properties (Langston et al. 1988; Linroth and Rickard 2014).

recovery factor of 19% (Linroth 2012). During the first 7.5 years of injection, the percentage of reservoir area with a bottom-hole pressure greater than bubblepoint pressure increased from a value of 1% to a value of 80% (Dicharry et al. 1973).

Initial CO_2 Flooding, 1970s–1990s. Planning for initial CO_2 flooding began in 1968. Decisions were made to use a WAG process and, because of a limited supply of CO_2, to divide the injection program into three phases, as shown in **Fig. 6.84** (Langston et al. 1988). CO_2 was first injected in Phase 1 in early 1972, following injection of water to increase reservoir pressure to an MMP of 2,200 to 2,300 psia, as had been estimated as of that date (although much of the Phase 1 area did not reach these higher pressures) (Langston et al. 1988). The injection was successful, increasing production from 40,000 to 108,500 STB/D over a period of approximately 18 months (Langston et al. 1988; Kane 1979). Breakthrough of CO_2 occurred in approximately 6 months.

For Phase 2, CO_2 was first injected in 1974 following pattern waterflooding from 1972. Pressures over much of the Phase 2 region were maintained at values of 2,000 to 2,200 psia. Rather than the dramatic increase of oil production observed in Phase 1, oil production increased gradually from a value of 70,000 STB/D to a peak value of 85,600 STB/D over a period of 18 months. Breakthrough of CO_2 at a producing well occurred in 6 months, similar to that for Phase 1. Less than 15% of the injected CO_2 was produced (Kane 1979).

In the Phase 3 region, a pattern waterflood preceded CO_2 injection and was successful, increasing oil production by approximately 18,000 STB/D. CO_2 injection was initiated in late 1976, and the result was basically a stabilization of the oil-production rate. CO_2 breakthrough and production performance were similar to those of Phases 1 and 2 (Kane 1979). Issues affecting production throughout the three phases of production were restricted CO_2 availability, early CO_2 breakthrough, limited fluid-handling facilities, various technical problems, and existence of regions of the reservoir that were not pressured to the MMP.

CO_2 injection with high WAG ratios was continued through the 1990s and was considered successful, although the overall unit production continued to decrease during this period, as shown in Fig. 6.83. The flooding included three projects targeted to relatively small areas in the northern, central, and southern parts of the unit. Fig. 6.83 is a plot of fluid-production rates over time for the entire SACROC unit, and reflects the oil-production rates discussed, but does not show rates for specific parts of the unit.

CO_2 Flooding Development After 2000. Kinder Morgan purchased the SACROC unit in 2000 and expanded on Penn-zoil's successful centerline CO_2 floods of the 1990s. A new centerline pipeline was installed, which delivered an additional 300 MMscf/D of CO_2 to the SACROC area by 2003. CO_2 provided by the pipeline was essentially pure, while recycled CO_2

Reservoir Fluid Composition*	
Component	Mol (%)
CO_2	0.32
N_2	0.83
C_1	28.65
C_2	11.29
C_3	12.39
$i\text{-}C_4$	1.36
$n\text{-}C_4$	6.46
$i\text{-}C_5$	1.98
$n\text{-}C_5$	2.51
C_6	4.06
C_7^+	<u>30.15</u>
Total	100.00
Molecular weight of C_7^+	197.4
Specific gravity of C_7^+	0.841
Bubblepoint pressure at 130°F (psia)	1,820
Reservoir fluid viscosity at 1,820 psia and 130°F (cp)	0.38
Reservoir fluid density at 1,820 psia and 130°F (lbm/ft³)	41.8

Flash Separation Data		
First-Stage Separator Conditions	25 psia and 95°F	31 psia and 75°F
Solution GOR (scf/STB)	990	910
Stock-tank oil gravity (°API)	41.0	42.7
Casinghead gas gravity	1.087	1.030
FVF** at 3,137 psia (bbl/STB)	1.528	1.472
FVF** at 1,820 psia (bbl/STB)	1.557	1.500

*Average of 11 samples

**Formation volume factor

Table 6.17—SACROC crude-oil composition and properties (Dicharry et al. 1973).

contained a small percentage of light hydrocarbons. The company developed several new areas and expanded existing areas. The development consisted of expanding the flooding, phase by phase, from the central area outward (Pennell et al. 2015). Planning for the CO_2 flooding was based on an MMP value of 1,865 psia (Linroth and Rickard 2014). The phases, referred to as "expansion areas," are listed in order of development between 1996 and 2015 in **Table 6.18** and locations of each area are shown in **Fig. 6.85** (Bohn 2014). In general, the first wave of expansion projects were traditional pattern CO_2 floods. However, in an effort to improve sweep efficiency and target bypassed pay zones, waves of horizontal-well patterns were implemented. Additionally, harvesting projects to recover CO_2 that had been "buried" were conducted successfully. As shown in Fig. 6.83, oil production resulting from the projects increased from approximately 8,000 STB/D in July 2001 to 37,000 STB/D in the spring of 2015. CO_2-injection rate reached a maximum of 1,100 MMscf/D, while the recycled-CO_2 rate increased over time to 900 MMscf/D. The CO_2 purchase rate for the unit reached 370 MMscf/D in early 2004, but by early 2013, purchases decreased to approximately 150 MMscf/D. This decrease in purchased CO_2 was caused by a combination of increased WAG ratios in the more-mature flood areas and efforts to recover produced CO_2 from the earlier project areas (Pennell et al. 2015). Estimates indicated that purchases will be eliminated by 2018.

Relative performance of the different projects is shown in **Fig. 6.86,** which is a plot of oil recovery (expressed as % OOIP) vs. total reservoir barrels of fluid injected (CO_2 plus water), expressed as number of PVs. The recoveries range

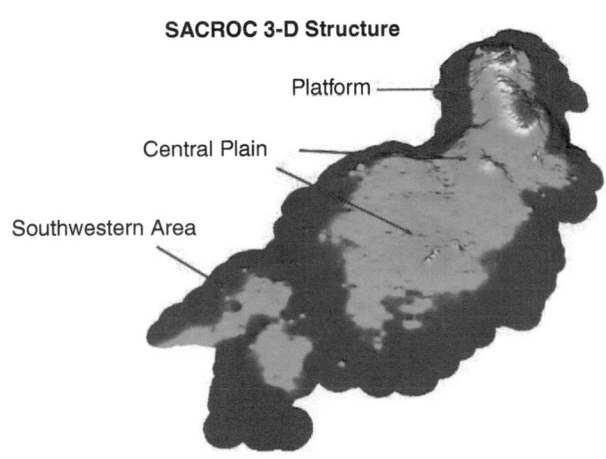

SACROC 3-D Structure

Platform

Central Plain

Southwestern Area

Fig. 6.80—SACROC 3D structure (Barati et al. 2016).

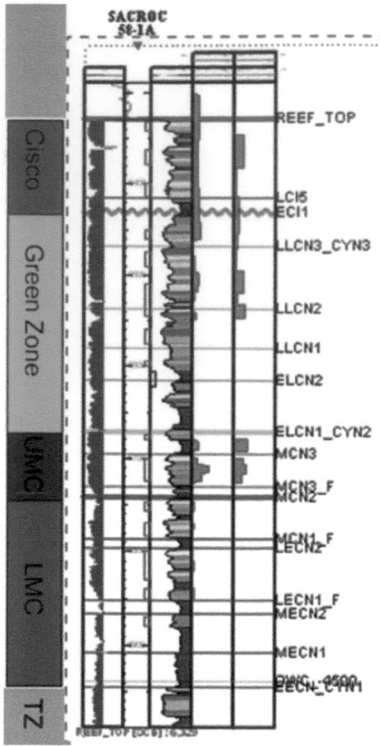

Fig. 6.81—Type log of the canyon-reef complex divided into four zones (Barati et al. 2016).

from a high of 18.5% OOIP for the Center Line 3 (CL3) project down to approximately 6.5% OOIP for the Bull's Eye project (Pennell et al. 2015). The high recovery in the CL3 project has been attributed to the relatively large volume of CO_2 injected, 134% HCPV. For the Gilligan's Island (GI), Center Ring 1 (CR1), and Center Line 4 (CL4) projects, the recovery-curve slopes increase toward the end of the production periods. This increase in recovery rate is attributed to different factors, including conformance improvement, infill drilling, and use of horizontal wells to recover oil from pay zones otherwise bypassed. As discussed earlier, there were CO_2 projects in SACROC before Kinder Morgan acquired the property and implemented CO_2 flooding. The earlier projects would be expected to have some effect on the projects conducted since 2000. Analysis by Kinder Morgan has concluded that there is likely a small effect, but the magnitude is difficult to determine (Pennell et al. 2015).

Platform-Area Projects. The platform area, composing the northern part of the unit, generally has the richest pay of the SACROC. A number of separate CO_2 projects have been conducted in this region, starting in 2008, as shown in Table 6.18 and Fig. 6.85. Several 40-acre five-spot patterns contain more than 10 million STB OOIP, and in much of the area, there is significant permeability heterogeneity, which was considered in the project design. Generally, in the projects, CO_2 was injected continuously until gas breakthrough occurred, and then a WAG-injection process was initiated. Oil-production response occurred early after start of gas injection. Recoveries as of early 2015 were on the order of 7 to 8% OOIP, with project lives of 7 to 8 years. Relative performance of the platform-area projects is shown in **Fig. 6.87** (Pennell et al. 2015). Two of the specific platform-area projects are described in the following subsections.

South Platform Project. The south platform project was initiated in March 2008. The reservoir thickness was relatively large in the project area, with an OOIP of approximately 105 million STB (Barati et al. 2016). Historical injection-profile surveys had shown that injected fluids had preferentially gone into the higher-permeability Cisco zone and GZ. Thus, it was thought necessary to design an injection program that would get acceptable fluid injection into the lower-permeability middle canyon reservoir, as well as the higher-permeability layers. While polymer-gel systems had been somewhat successful in redirecting injected fluids in other project areas, in this project, a decision was made to use mechanical isolation of layers by implementing dual-injection completions. This allowed CO_2/WAG-injection design to balance injection rates into the different layers: the Cisco/GZ and the middle canyon zone (Pennell et al. 2015).

The project area initially contained 24 40-acre regular five-spot patterns **(Fig. 6.88).** Oil production peaked at the end of 2009 at approximately 6,000 STB/D, and then went on a relatively steep decline, which was mitigated by the addition of two partial patterns on the east side of the project area. Performance of the project is summarized in Fig. 6.88, which shows the project-area patterns, injection and production rates, and cumulative oil and CO_2 production. As of March 2015, the project recovery was 8.8% OOIP (Pennell et al. 2015).

P1B Project. The P1B project, initiated in June 2009, contained a relatively rich reservoir with an estimated OOIP of 173 million STB. The project was originally designed to have 24 40-acre five-spot patterns; however, an experimental infill project, designated as Infill Development Assessment (IDEA), was included to assess the effect of using 10-acre patterns. IDEA consisted of four 10-acre five-spot patterns, as shown in **Fig. 6.89** (Pennell et al. 2015). As shown in Fig. 6.87, the smaller spacing did result in a 2 to 3% higher OOIP recovery. However, a downside was that the shorter well spacing increased interwell connectivity in high-permeability layers, resulting in very early breakthrough of CO_2, which caused problems with the

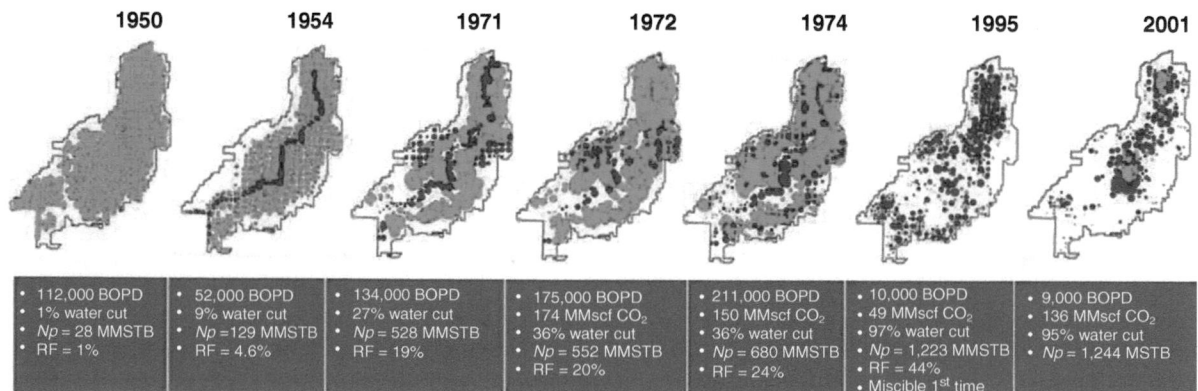

Fig. 6.82—SACROC history (Barati et al. 2016).

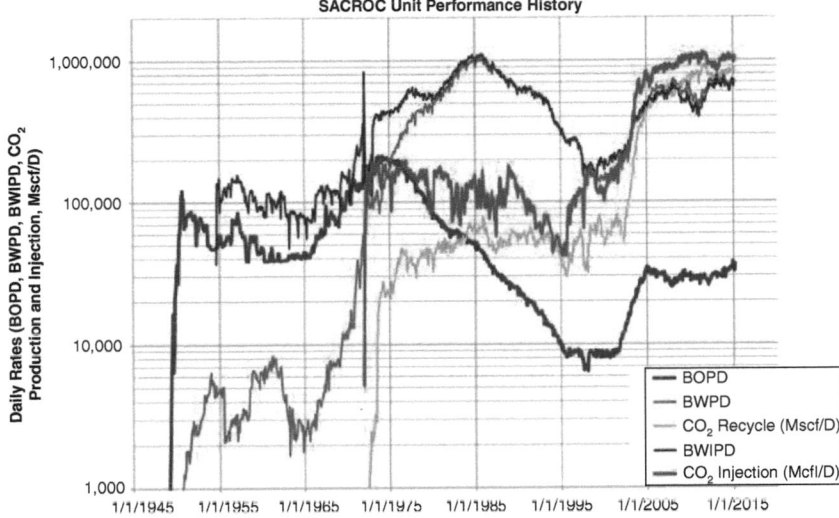

Fig. 6.83—Oil, water, and CO₂ injection and production rates (Pennell et al. 2015).

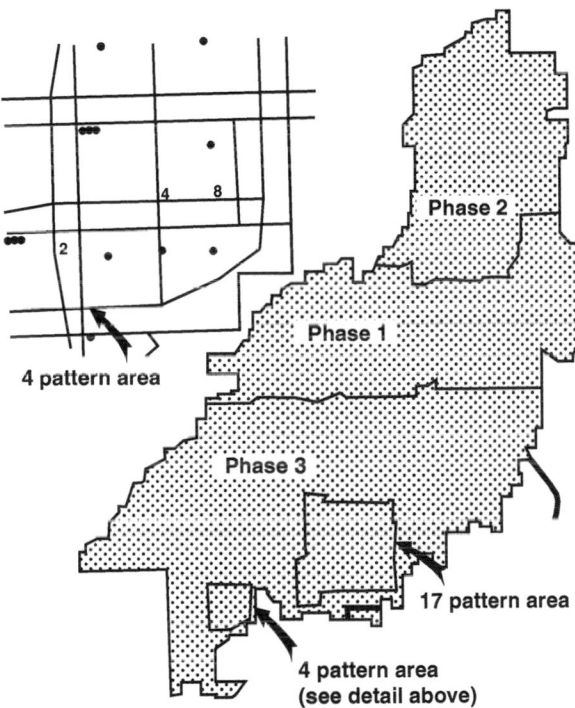

Fig. 6.84—Areas of three phases of initial CO₂ flooding (Langston et al. 1988).

submersible pumps. Producing GOR increased from 20,000 scf/bbl to more than 100,000 scf/bbl in December 2009. The problem was addressed by use of large-volume crosslinked-polymer systems (see subsection Operating Issues), which eventually resulted in GOR reduction to 20,000 scf/bbl by April 2010 (Pennell et al. 2015).

A map of the P1B project area and project performance is shown in Fig. 6.89. The oil rate peaked at approximately 7,500 STB/D in the middle of 2012, followed by a decline to 5,000 STB/D by March 2015. Total recovery as of that date was 6.8% OOIP (Pennell et al. 2015).

Operating Issues. *Conformance Improvement Approaches.* Early breakthrough of injected fluids and poor sweep efficiency have occurred in much of the SACROC unit. While the average permeability is reported to be 19 md (Table 6.16), there are known to be very-high-permeability "streaks," features judged to be laterally extensive. Where these features are present, they have led to early CO₂ breakthrough, poor conformance, and high CO₂ gross-usage ratios (CO₂ injected/bbl oil recovery), especially in the GZ (Fig. 6.81). The conformance problem has not been uniform, varying from pattern to pattern (Pipes and Schoeling 2014), and has been addressed in three ways: (1) by fluid rate and pressure balancing (Linroth and Rickard 2014), (2) by application of relatively large gelled-polymer treatments (Pipes and Schoeling 2014), and (3) by use of dual completions in injection wells (Pennell et al. 2015). In all approaches, the purpose was to modify CO₂ flow in the reservoir in such a way as to improve volumetric sweep efficiency.

The first approach involved extensive and continuous monitoring of intake pressures and producing GOR at production wells, combined with calculation of flow streamlines. On the basis of these data, producing-well pump-intake pressures or pump speeds were altered to effect changes in CO₂ flow behavior that would improve sweep. One case study described in Linroth and Rickard (2014) was conducted in the southwest centerline project area (Fig. 6.85). The western region of this area had been part of a water curtain and was overpressured, resulting in essentially no oil recovery because CO₂ flow into the region was impeded by the high pressure. Initial flooding recovery was extremely poor, only approximately 2.5% of OOIP. With alteration of pump-intake pressures and rates, recovery in the western region was significantly improved. The processing rate in the region was very high (on the order of 50% PV/yr), which is a favorable condition because the effects on flow from pump changes could be detected quickly and additional changes made as required. Overall, this approach to improvement of conformance has been judged to be favorable (Linroth and Rickard 2014).

Starting in 2007, conformance control was addressed by the application of large gelled-polymer treatments. The method of choice involved use of Cr(III) carboxylate/acrylamide-polymer gels tailed with either high concentrations of polymer or squeezed cement (Pipes and Schoeling 2014). Candidate wells were selected on the basis of published guidelines (Sydansk

Year	Project
1996	Center Line 1 & 2 Expansions (CL1 & CL2)
2001	Center Line 3 (CL3)
2002	Center Line 4 (CL4)
2002	Center Line 5 (CL5)
2003	Bull's Eye (BE)
2004	Center Ring 1 (CR1)
2005	Center Ring 2 (CR2)
2007	Southwest Center Line 1 & 2 (SWCL1 & 2)
2008	Southwest Center Line 3 (SWCL3)
2008	Gilligan's Island (GI)
2008	South Platform (SP)
2009	Platform Phase 1 (P1)
2010	Chiquita
2011–12	South Shore
2012–13	Platform Phase 2 (P2)
2013–14	Platform Phase 3 South (P3S)
2014	Chiquita Expansion

Table 6.18—SACROC Unit CO$_2$ flood development projects (Barati et al. 2016).

SACROC Unit Project Area Map

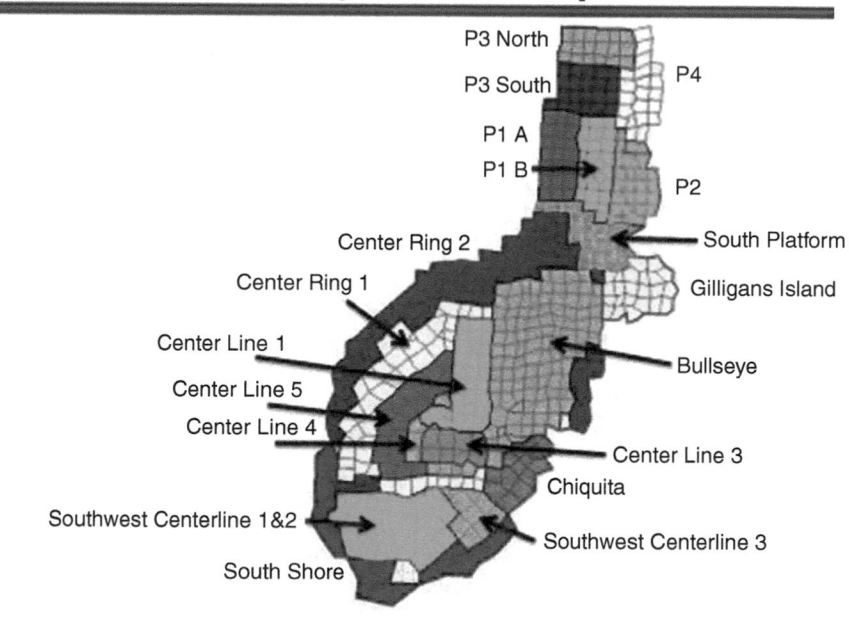

Fig. 6.85—Project areas (Bohn 2014).

and Southwell 2000), measured injection profiles, and step-rate tests. Any well having an injective index of 20 (B/D)/psi or greater was considered for treatment with gelled polymer (Pipes and Schoeling 2014). The best outcomes resulted, in general, when 20,000 bbl or more of polymer gel was injected. The gel consisted of Cr-crosslinked medium and high-molecular-weight (M_w) polyacrylate polymer, with a concentration of 5,000 ppm at the start of injection, increasing to 12,000 ppm at the latter stages. A final tail-in stage of 30,000 ppm was used. The polymer system was designed to not crosslink until it was deep in the formation (Pennell et al. 2015). The treatments were judged to be most effective when the gel treatment was implemented before initial CO$_2$ injection. Results of a treatment for one pattern in the P1 platform area are illustrated in **Fig. 6.90** (Pipes and Schoeling 2014). Gel treatments in the GZ were successful in that the GOR was reduced from a value of more than 100,000 scf/bbl to a value of 20,000 scf/bbl by April 2010. There were problems with the polymer treatments. Polymer breakthrough occurred at a significant number of injection wells, causing those wells to be shut in. Where one interval of a dual-completion injector was to be treated, both completion intervals had to be shut in during the treatment period.

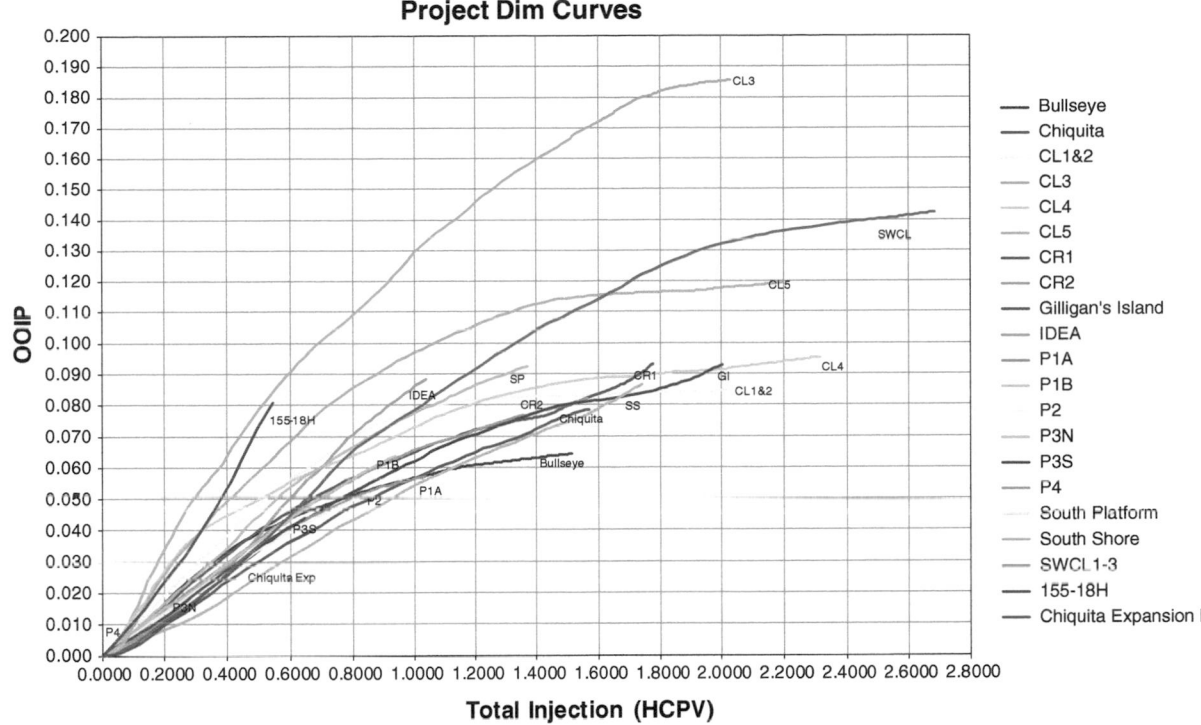

Fig. 6.86—Dimensionless EOR vs. PV of fluid injected for all projects (Schoeling 2015).[9]

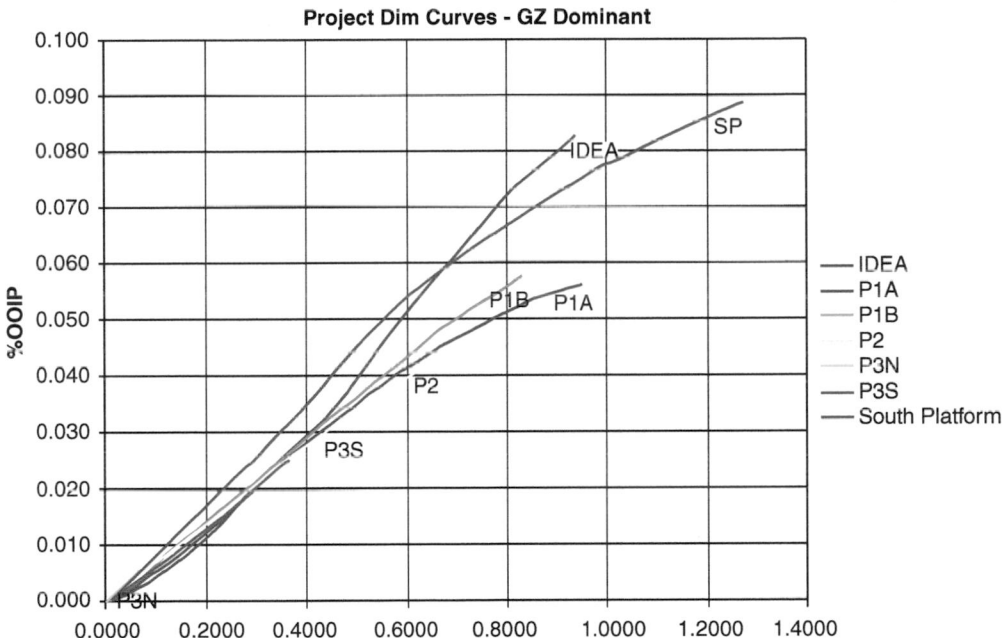

Fig. 6.87—Dimensionless EOR vs. PV of fluid injected for the platform areas (Pennell et al. 2015).

The third approach involved mechanical isolation of high-permeability zones from low-permeability zones by use of dual-injection completions (e.g., high-permeability Cisco/GZ formations isolated from the low-permeability middle Canyon). This allowed higher injection rates into the tighter middle-canyon formation. In some cases, two individual "twin" injection wells were drilled (Pennell et al. 2015).

In the fourth approach, short and long lateral wells were drilled in the tight middle canyon and fractured with gelled-acid stages, which were spaced to optimize sweep efficiency between injectors and offset producers. To enhance injection, fractures were designed to have a half-length of approximately 100 ft, with a 20-ft growth zone below the wellbore and an 80-ft zone above.

---

[9] Personal communication with L. Schoeling, Kinder-Morgan Company, Houston (October 2015).

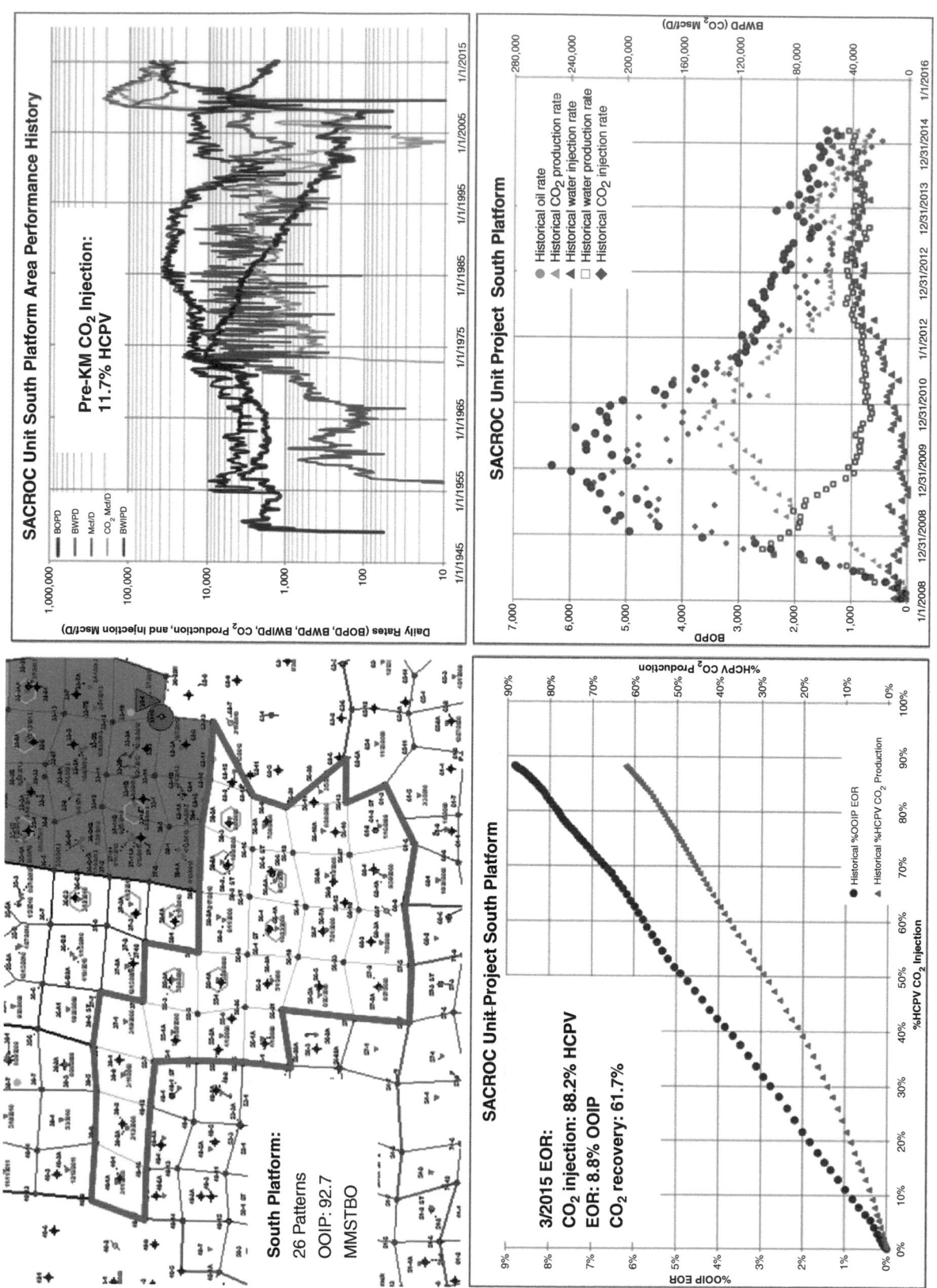

Fig. 6.88—South platform project map and performance (Pennell et al. 2015).

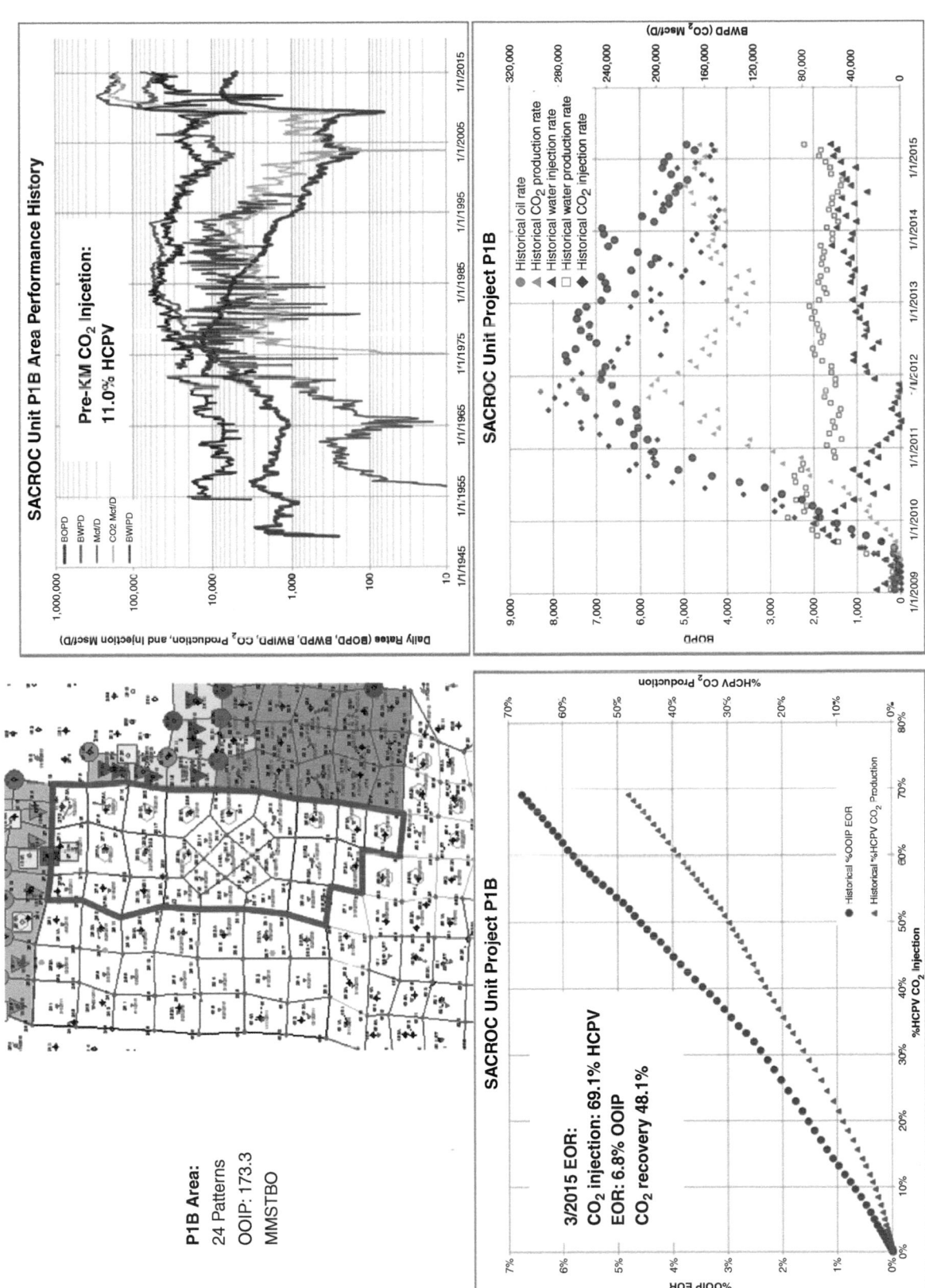

Fig. 6.89—P1B project map and performance (Pennell et al. 2015).

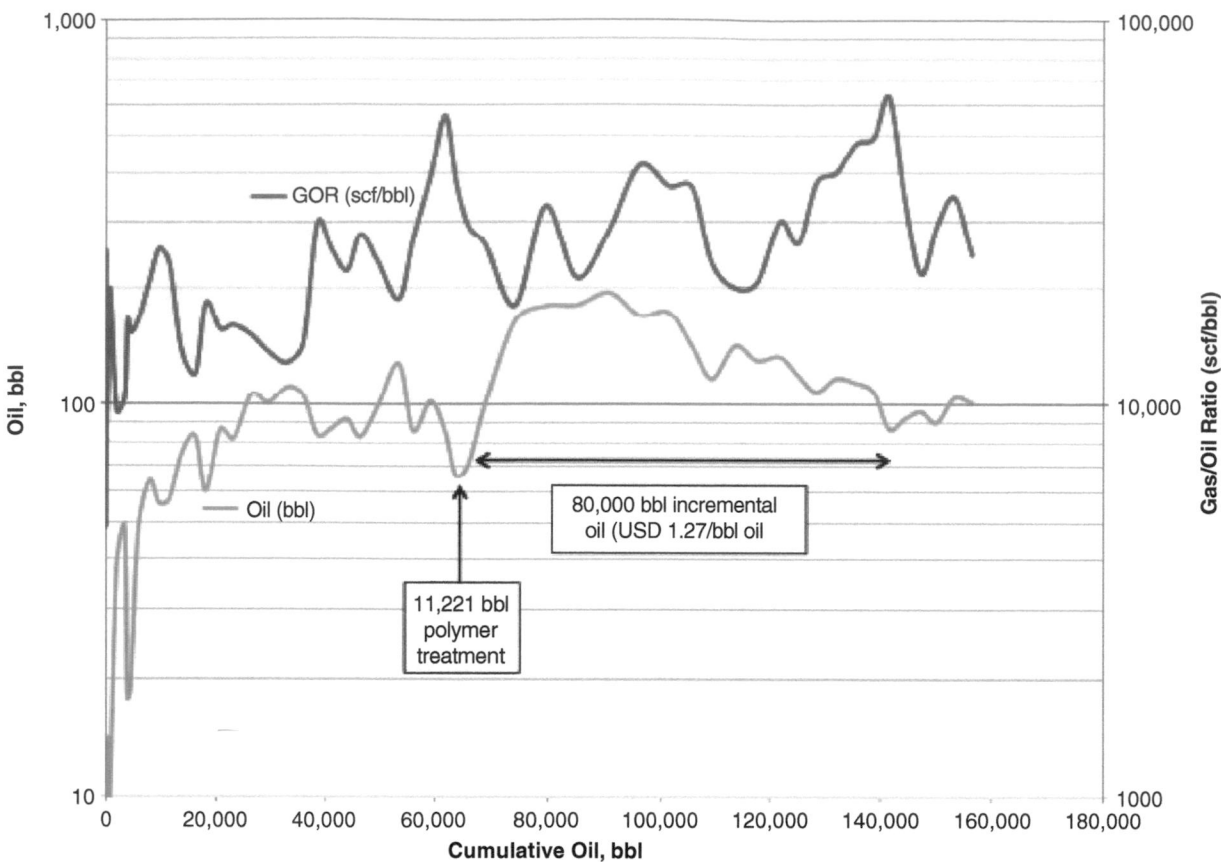

Fig. 6.90—Pattern 38-1A, Project P1, production plot showing effect of polymer treatment.

CO₂-Processing Rates and CO₂ Use. Processing rates (% HCPV processed /year) varied a great deal between the different project areas, with values as low as 8% to as high as 50% (Linroth and Rickard 2014; Barati et al. 2016). As discussed in the Conformance Improvement Approaches subsection, the magnitude of the processing rate had an important impact on determining the effects of changes in operating procedures. CO_2-usage values in the different project areas also differed by large amounts, resulting mostly from differences in % HCPV of CO_2 injected. As examples, the net-usage values, expressed as Mcf CO_2/bbl oil, varied from 5.6 at 152% HCPV CO_2 injected in the southwest centerline region, to 9.0 at 57% HCPV CO_2 injected in the bull's-eye area, to 14.4 at 68% HCPV CO_2 injected in the Platform 2 area (see Fig. 6.85). Gross-usage factors in these three areas, at the same % HCPV CO_2 injected, were 35.0, 29.2, and 43.7 Mcf CO_2/bbl oil, respectively (Schoeling 2015)[9].

Asphaltene Deposition. Asphaltene deposition has not been a concern in SACROC; however, paraffin-wax deposition has occurred in the unit, resulting in pump failures. Analysis of SACROC crude oil has shown high paraffin wax and low asphaltene concentrations (Barati et al. 2016).

Loss of Fluid Injectivity. Significant injectivity loss was observed in more than 150 injection wells, a sizable fraction of the total wells, as a result of the use of the WAG process (Barati et al. 2016). Vertical heterogeneity is large in the SACROC unit. One study showed that wells with a Dykstra-Parsons coefficient on the order of 0.83 exhibited major injectivity loss, while wells with a coefficient on the order of 0.76 showed no significant loss. Vertical-injection profiles were measured on a large number of wells before, during, and after WAG implementation. These showed that both water- and CO_2-injection profiles were altered in those wells with injectivity loss (Barati et al. 2016).

6.12.2 Means San Andres Unit. *Early History and Geology.* The Means field (discovered and originally produced in 1934) is located approximately 50 miles northwest of Midland, Texas. It was discovered by Humble Oil Company and has been operated by Exxon and ExxonMobil over the years. It produces primarily from the San Andres formation, located along the northeastern edge of the Central Basin platform. Reservoir rock is a dolomite with minor amounts of shale and anhydrite, and a producing horizon that is at a depth of 4,200 to 4,800 ft subsurface. The field also includes the associated Grayburg reservoir, which makes only a minor contribution to total production. The San Andres is a north/south trending anticline made up of north and south domes separated by a dense structural saddle that runs east to west, as shown in **Fig. 6.91** (Goodwin 1989; Magruder et al. 1990; Stiles and Magruder 1992). It has a thickness of several hundred feet, of which the upper 200 to 300 ft are productive (Stiles and Magruder 1992). On the basis of the analysis of some 5,000 ft of core material, permeability was determined to be extremely variable, with significant lateral discontinuity in the producing zones (Goodwin 1989). Reservoir and fluid properties are shown in **Table 6.19** (Magruder et al. 1990).

MEANS SAN ANDRES UNIT

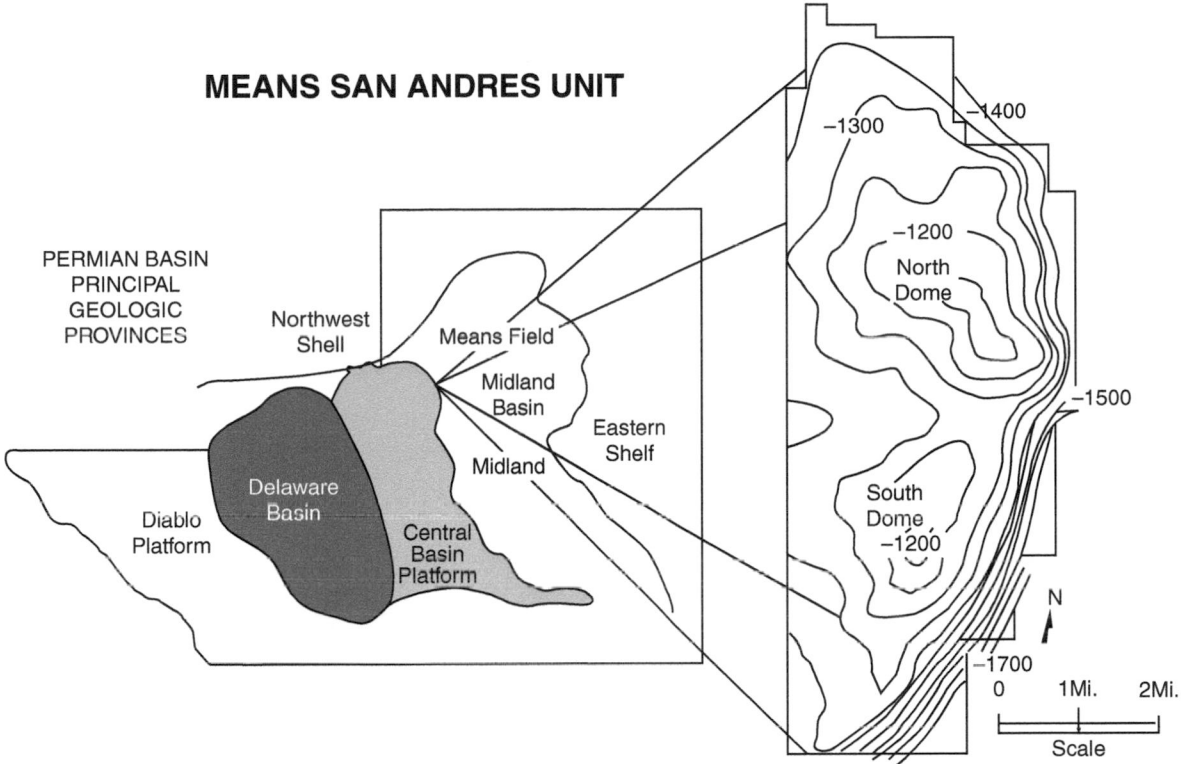

Fig. 6.91—Means San Andres Unit (Goodwin 1989).

Development of the field occurred slowly over the 1930s and 1940s, with a significant expansion in the 1950s caused by the implementation of hydraulic fracturing. Primary production was from fluid expansion aided by a weak waterdrive. The field reached a peak production of more than 10,000 STB/D in the mid-1950s, followed by a steep decline, as seen in **Fig. 6.92,** which shows the field production from 1934 to 2011 (Pathak et al. 2011). By 1963, reservoir pressure had declined to approximately 800 psi, less than one-half of original pressure. In that year, the field was unitized for the purpose of implementing a waterflood.

A waterflood was begun in 1963 by use of a peripheral pattern on the basis of 40-acre spacing. The mobility ratio for water, because of the relatively high oil viscosity of 6 cp, was unfavorable, with a value on the order of 2. The flood was successful, increasing reservoir pressure to near initial pressure and increasing oil production to the maximum allowable, as specified by

Formation	San Andres
Lithology	Dolomite
Area (acres)	14,328
Depth (ft)	4,400
Gross thickness (ft)	300
Average net pay (ft)	54
Average porosity (%)	9.0 to 25
Average permeability (md)	20 to 1,000
Average interstitial water saturation (%)	29
Primary drive	Weak waterdrive
Average original pressure (psig)	1,850
Stock-tank oil gravity (°API)	29
Oil viscosity (cp)	6
FVF (RB/STB)	1.04
Saturation pressure (psi)	310

Table 6.19—Means San Andres Unit reservoir and fluid properties (Magruder et al. 1990).

the Texas Railroad Commission (Goodwin 1989). However, by 1970, the allowable production rate could not be sustained, and a different pattern (a 3:1 line pattern) was implemented, which increased production for the flooded area from approximately 14,000 to 17,000 STB/D, as shown in **Figs. 6.92 and 6.93.** By late 1973, production was again in decline. The geological model was revised, and with higher oil prices, 20-acre infill wells were drilled, increasing the production rate and adding an estimated 22 million STB to the ultimate recovery. Production decline resumed again into the early 1980s, as shown Figs. 6.92 and 6.93.

Tertiary-Recovery Project Planning. Early screening work to determine the feasibility of implementing a CO_2-recovery project began in early 1980. It was known that a number of successful CO_2-flood projects had been conducted in Permian San Andres reservoirs, which provided an incentive. However, there were concerns about flooding the Means unit because of the relatively high oil viscosity, vertical layering, and lateral discontinuity of the reservoir. In addition, the MMP had been determined to be in the range of 1,850 to 2,300 psi, which was close to the formation parting pressure of approximately 2,700 psi. Thus, a decision was made to conduct a pilot test (Stiles et al. 1983).

The pilot test consisted of five wells covering an area of 1.5 acres. The test did not include a producing well, but rather depended on logging and fluid sampling. There was one injector, three logging/observation wells, and one fluid-sampling well. Gas injection, consisting of 97% CO_2 and 3% HC, began in 1981. The test was successful in determining that the 6-cp oil could be displaced, CO_2-WAG injectivity was satisfactory at a pressure between the MMP and the parting pressure, and CO_2 override or channeling through high-permeability streaks did not present major problems, nor did the lateral discontinuities over the distances in the test. Additionally, the presence of a small concentration of HC in the injected-gas stream did not appear to adversely affect the displacement (Stiles et al. 1983; Magruder et al. 1990).

Laboratory testing on core material indicated that the S_{orw} to waterflooding was approximately 34%, and the residual saturation after a CO_2-WAG flood was approximately 9%, data which allowed good estimates of expected recovery in areas swept by the flood. Planned injection volume was 40% HCPV purchased CO_2 plus approximately 15% HCPV recycled gas (Magruder et al. 1990).

To improve conformance and flood-response time, as part of the project, a major infill-drilling program was planned, which involved drilling more than 200 producers and 158 injectors. The initial plan consisted of 167 patterns on 10-acre spacing covering 6,700 acres and including 82% of the OOIP (Magruder et al. 1990). However, the area was later increased to 8,500 acres and 172 patterns. An inverted-nine-spot pattern with infill drilling to 10-acre spacing was used. Operating pressure was designed to be on the order of 2,000 psi. All produced gases were to be compressed, mixed with makeup CO_2, and reinjected because economics did not favor building and operating a plant to separate CO_2 from the produced-gas stream (Magruder et al. 1990; Stiles and Magruder 1992; Pathak et al. 2011).

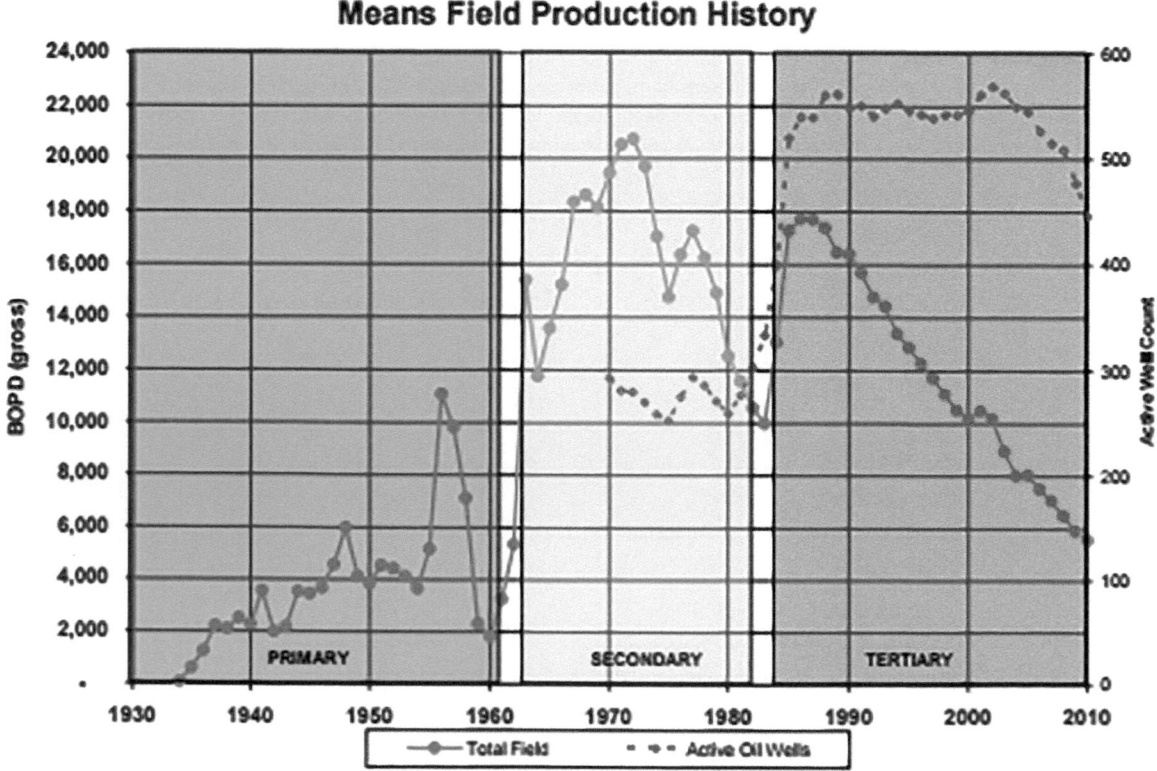

Fig. 6.92—Production of the Means San Andres Unit, 1934–2011 (Pathak et al. 2011).

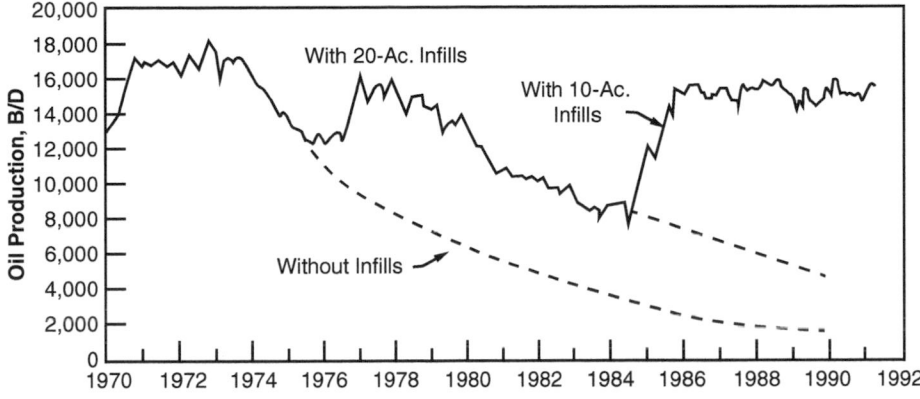

Fig. 6.93—Means San Andres Unit production, 1970–90 (Stiles and Magruder 1992).

Final design parameters are shown in **Table 6.20.**

Project Implementation and Performance. The final integrated-reservoir-development plan consisted of CO_2-WAG injection and a pattern-modification/infill-drilling program, as described in the preceding subsection (Magruder et al. 1990; Stiles and Magruder 1992). A map of the total Means unit area and the smaller area encompassed by the CO_2-WAG process is shown in **Fig. 6.94.** CO_2-WAG injection was begun in November 1983. The integrated nature of the project, involving both CO_2-WAG injection and extensive infill drilling, meant that oil recovery would result from both miscible-gas displacement and additional waterflooding. One concern was the possible loss of injectivity from WAG injection; however, a field study of 15 injection wells established that this was not a significant problem (Magruder et al. 1990). Reservoir modeling with continuous updating of the geologic model was conducted to improve performance (Stiles and Magruder 1992).

Fig. 6.93, previously referenced to show the effect of moving to 20-acre spacing, also shows the effect of the 10-acre infill drilling. The plot includes production from the entire Means unit, thereby including effects of CO_2-WAG injection and waterflooding outside of the CO_2-project area. The dashed lines on the figure show the estimated recoveries had the changes in operation not been made. Thus, the combination of miscible displacement and infill drilling account for an incremental production rate of approximately 10,000 STB/D as of 1990 (Stiles and Magruder 1992). Oil production in the CO_2-project area only, as well as gas injection and production, are shown in **Fig. 6.95** (Stiles and Magruder 1992) for the same time period. After 0.2 HCPV had been injected, CO_2 retention in the reservoir was 72%, which was deemed reasonable (Stiles and Magruder 1992).

Initial predicted recoveries were 16.6 million STB from CO_2-WAG and 21.4 million STB from additional waterflooding, although it was difficult to determine production to be assigned to the different recovery mechanisms. These values would result in an ultimate recovery of 50% of OOIP, including primary, secondary, and tertiary (Magruder et al. 1990). However, as of 1990, results exceeded initial project predictions, indicating that ultimate recovery would exceed these projections (Magruder et al. 1990; Stiles and Magruder 1992).

Additional evidence to support the effectiveness of CO_2-WAG displacement can be seen in **Fig. 6.96,** which shows oil-rate increase following initiation of CO_2 injection for a single pattern. Many wells experienced increases in rate on the order of 50 STB/D because of the CO_2 mobilization of waterflood residual oil.

The field-production rate to 2011 is shown in Fig. 6.92. In 2011, the average oil rate was 5,821 STB/D with a water cut of approximately 96%. Most of the field production (approximately 87%) came from the CO_2-WAG project area, with

Process	CO_2 miscible
CO_2 supply	Sheep Mountain (southeast Colorado)
Operating pressure (psi)	2,000
WAG ratio	2:1
CO_2 bank (slug size) (HCPV)	0.4
Ultimate CO_2 injection (with CO_2 recycle) (HCPV)	0.55
Flood pattern (with all wells active)	40-acre inverted nine-spot (10-acre well spacing)
Maximum injection rate (MMscf/D)	70
Maximum CO_2 production rate (MMscf/D)	30
Incremental tertiary recovery (10^6 STB)	16.6 (7.1% OOIP)
CO_2 usage (Mscf/tertiary STB)	15.2

Table 6.20—Means San Andres Unit Original CO_2 project design (Magruder et al. 1990).

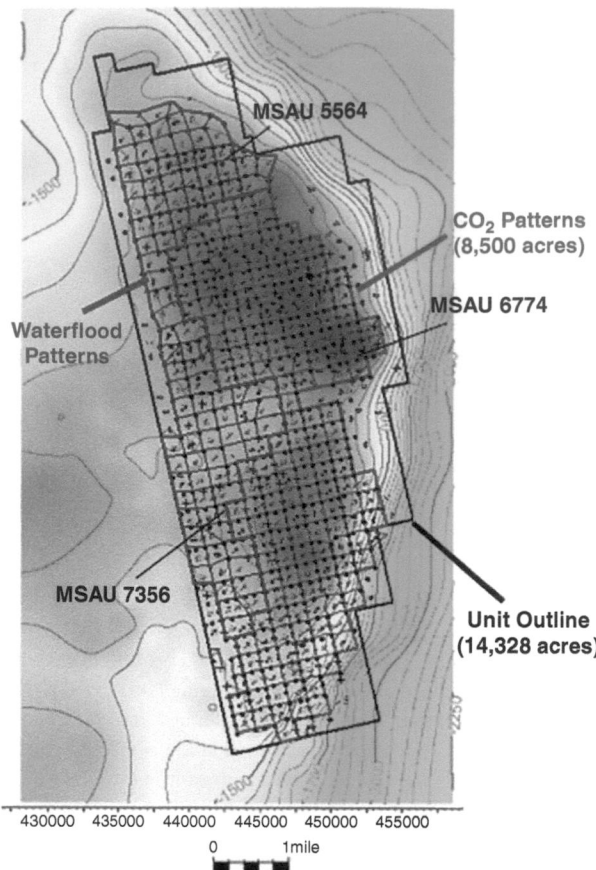

Fig. 6.94—The Means field structure map (Pathak et al. 2011).

the remainder from the waterflood area. Water production was approximately 140,000 B/D, which was reinjected back into the San Andres, and average GOR was 9,000 scf/STB. Approximately 50 MMscf/D of produced gas was compressed, combined with 20 MMscf of CO_2 makeup, and reinjected. The WAG ratio was 4 bbl water/Mscf gas. As of 2011, there were 465 active producing wells, 175 CO_2-WAG injectors, and 105 water injectors. At the end of that year, cumulative oil production was 286.8 million STB, 63.9% of OOIP (C and C Reservoirs and DAKS 2015). **Fig. 6.97** shows historical recoveries from the CO_2 and waterflood project areas through 2004, as well as projections to 2030 (Wilkinson et al. 2004).

6.12.3 Beaver Creek Madison Reservoir. *Early History and Geology.* The Madison reservoir, discovered in 1954, is one of 12 separate formations in the Beaver Creek field, located in the Wind River basin of central Wyoming. Devon Energy Corporation acquired the Beaver Creek field in December 2000, and has been the owner/operator since that date. The Madison is a Mississippian reservoir located at an average depth of 11,235 ft and is the deepest producing formation in the Beaver Creek field. The reservoir has been described as a fault-bounded asymmetrical anticline; however, it is bounded on the northwest flank by a water/oil contact. The formation consists of carbonate rock, ranging from limestone to dolomite, with wettability that runs from neutral to oil wet. Geologic and oil characteristics are summarized in **Table 6.21,** and a computer-generated-structure map is shown in **Fig. 6.98** (Peterson et al. 2012).

The producing interval varies from approximately 450 to 475 ft in gross thickness across the reservoir, which covers

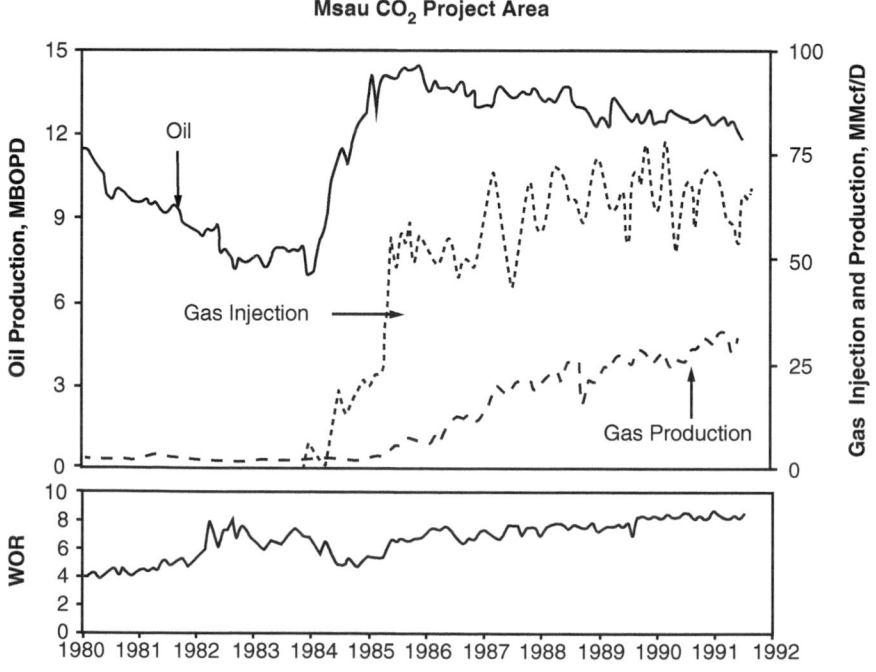

Fig. 6.95—Performance of tertiary project area (Stiles and Magruder 1992).

an area just greater than 900 acres. Six distinct porosity intervals have been identified (A through F), within which the net pay averages 212 ft. The reservoir had a relatively low initial water saturation, estimated to be 8 to 10% on average. On the basis of an assumed S_{wi} of 10%, OOIP was calculated to be 109 million bbl (formation water is extremely fresh, making log-derived water-saturation estimates questionable).

The initial reservoir pressure was 5,301 psig, significantly greater than the bubblepoint pressure of 673 psig. The reservoir initially produced under liquid expansion and a partial waterdrive. A waterflood was initiated in 1959, before primary production had started to decline, which maintained reservoir pressure at greater than the bubblepoint pressure (Pollock 1960). A maximum production rate of 8,800 B/D was reached in late 1962, and with the addition of injection wells through the 1960s, the maximum water-injection rate reached 18,000 B/D. Total recovery from the relatively short primary period and the waterflood up to the time of initiation of a CO_2 flood was 44.6 million bbl or 41% of OOIP. The production history before the CO_2 project, including water injection and production, is shown in **Fig. 6.99** (Peterson et al. 2012).

CO₂ Project Planning. Evaluation of a proposed CO_2-flooding project was initiated in January 2005 and led to implementation of CO_2 injection in July 2008. The planning was conducted in three phases. In the first phase, comparisons were made

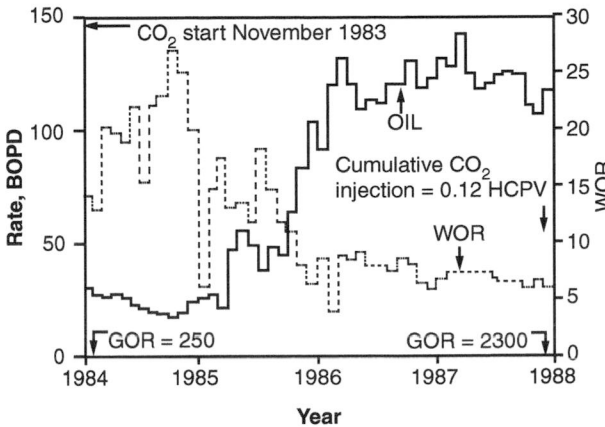

Fig. 6.96—Pattern 184 monthly performance data (Magruder et al. 1990).

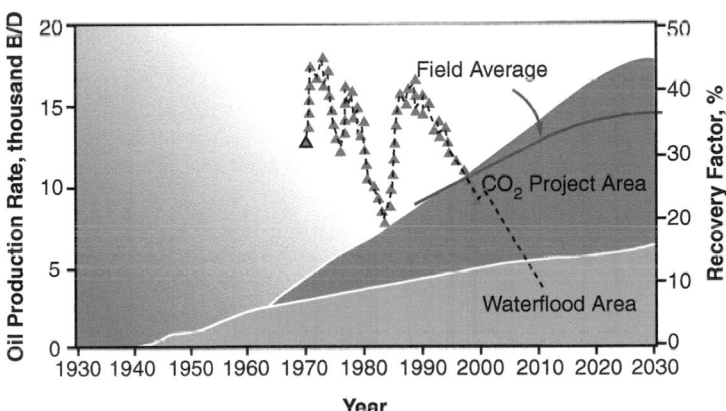

Fig. 6.97—Means field recovery factor by process (Wilkinson et al. 2004).

Matrix	Limestone/Dolomite
Productive area (acres)	914
Oil column height (ft)	828
Gross interval (ft)	450 to 475
Net pay average (ft)	212
Average depth (ft TVD)	11,235
Average porosity (%)	10
Average S_w (%)	8 to 10
Permeability (md)	9
Reservoir temperature (°F)	234
Original pressure (psig)	5,301
Bubblepoint pressure (psig)	673
Oil gravity (°API)	39.5
Oil viscosity (cp)	0.64
GOR (initial) (scf/bbl)	288
Mean miscibility pressure (psig)	2,650

Table 6.21—Summary of geologic and reservoir characteristics (Peterson et al. 2012).

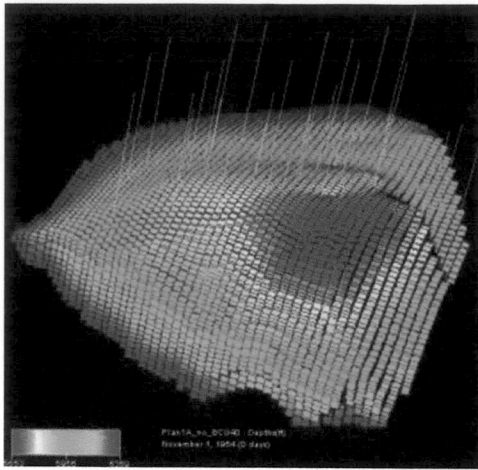

Fig. 6.98—3D view of the structure and grid of the Beaver Creek Madison numerical model (Peterson et al. 2012).

between the reservoir and fluid characteristics of the Madison and those of other active CO_2 projects. The general range of recoveries was estimated through analytical modeling, and preliminary economics determined from this study were positive. Phase 2 involved laboratory core analyses to determine rock wettability, capillary pressure, and relative permeability. CO_2 corefloods were conducted. An MMP value of 2,650 psig was determined by use of a recombined-crude-oil and MCM test. Phase 2 work also included the use of a fully compositional sector model that was based on laboratory PVT analysis to study both gravity-stable displacement and viscous lateral displacement in a representative cross section of the reservoir.

In Phase 3, a fully compositional 3D simulation was carried out on the entire reservoir. A history match was performed on a well-by-well basis by use of liquid rates and comparison of calculated oil rate, water rate, gas rate, and well pressures with actual data. Several possible injection/production scenarios were evaluated. Modeling results led to a conclusion that both gravity drainage and lateral viscous displacements would be successful. The final design selected included "continuous upstructure and peripheral CO_2 injection of an initial slug volume prior to initiating downstructure WAG injection" (Peterson et al. 2012). Well locations are shown in **Fig. 6.100.** The work in Phase 3 also included design of facilities and planning for construction and project implementation. The project was predicted to recover an additional 11.1 million bbl over its life.

CO_2-Flood Performance. The initial response occurred from the peripheral flood in November 2008, with an increase in oil and gas rates in an outer-ring producer. As the flood proceeded, increase in production moved from the outer-ring wells toward the center wells. No significant early breakthrough of injected gas was observed, indicating relatively good conformance. Conversion of two wells to injection wells to maintain reservoir operating pressure in the center of the reservoir, along with an overall increase in reservoir operating pressure, improved sweep efficiency and producing-well productivity, resulting in the improvement of the peak rate above modeled predictions. Production peaked in November 2012 at a rate of 5,541 B/D. Production and injection histories for all fluids are shown in **Fig. 6.101.**

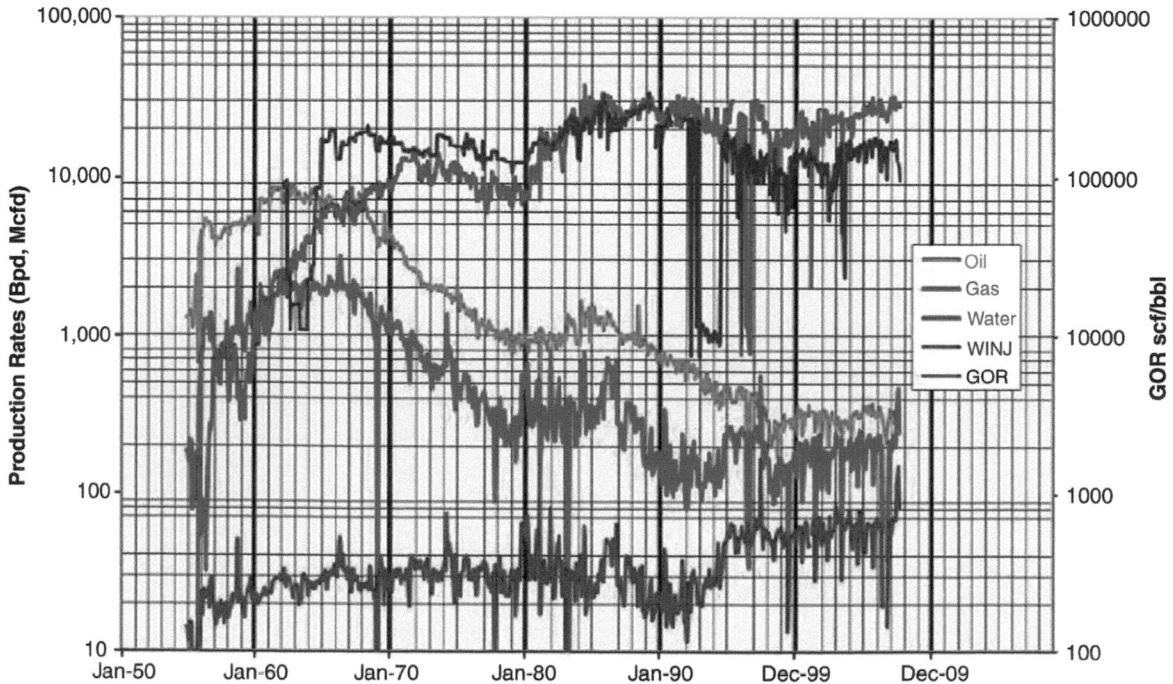

Fig. 6.99—Beaver Creek Madison production history before CO_2 flood project (Peterson et al. 2012).

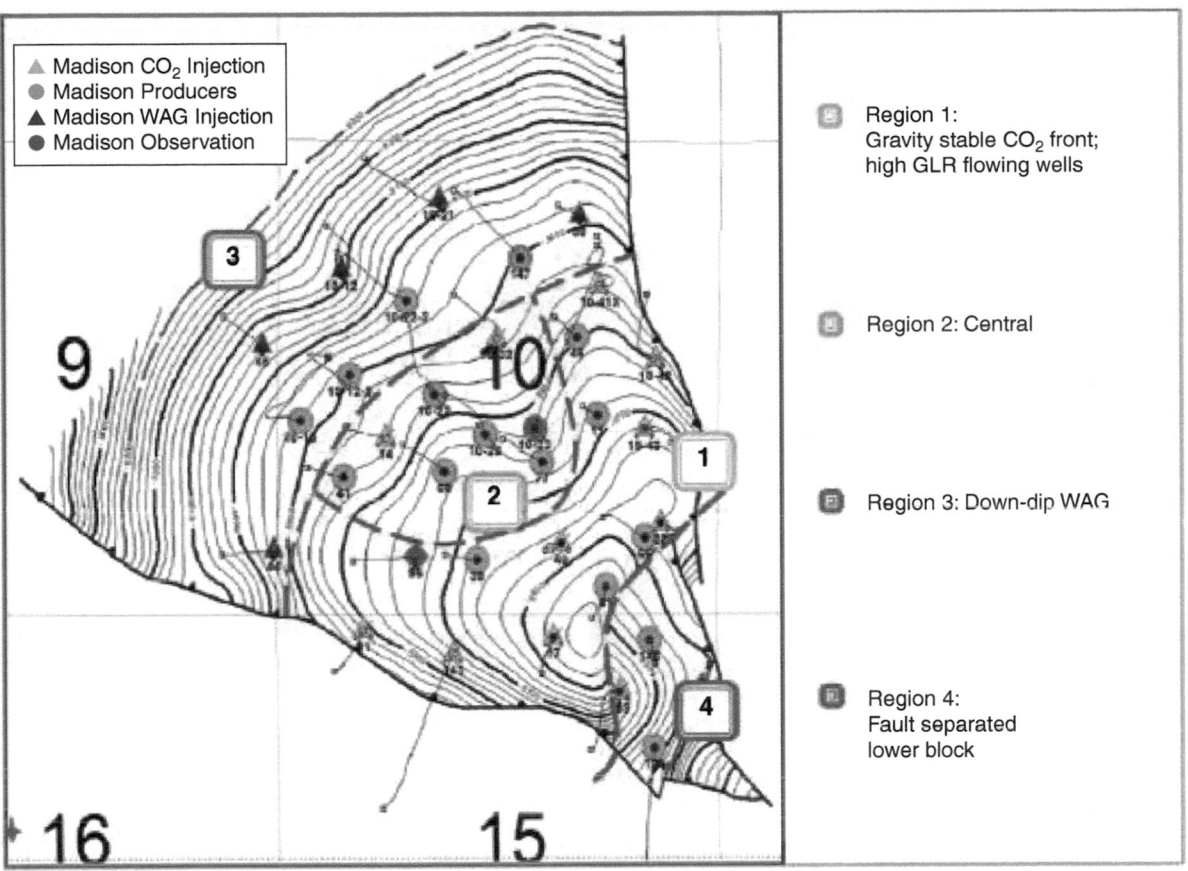

Fig. 6.100—Madison structure map indicating region analysis patterns (Peterson et al. 2012).

CO₂ Usage. Purchased CO_2 is provided through a 47.1-mile-long pipeline, which extends from a distribution pipeline near Jeffrey City, Wyoming. The pipeline was completed in the spring of 2008. Gas-compression facilities for CO_2 recycling were installed and made operational in the spring of 2009, before significant CO_2 breakthrough. **Fig. 6.102** shows predicted oil recovery vs. CO_2 injection for different probability scenarios. The curve for actual production through the end of 2014 is also shown. Note that performance to date exceeds even the optimistic forecast (P10). At the time of writing, this is still

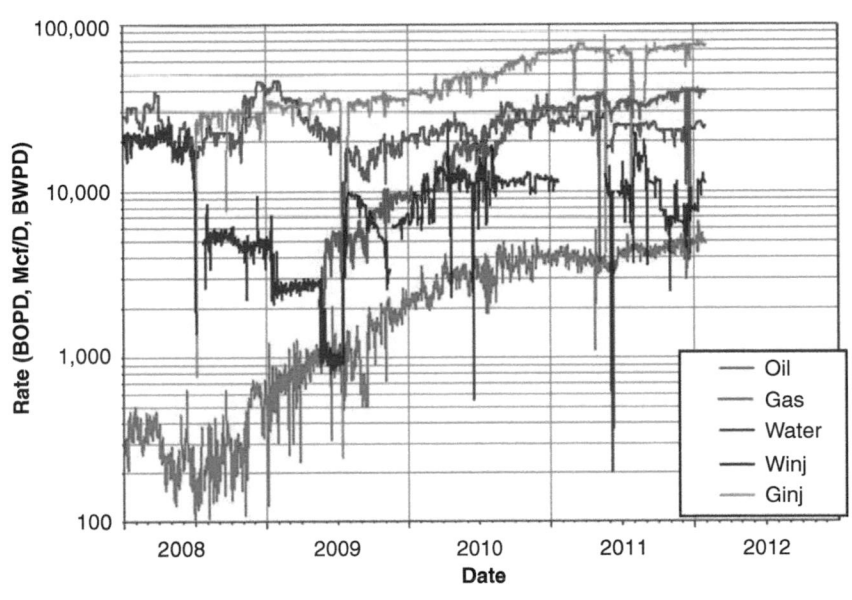

Fig. 6.101—Madison production and injection history since CO_2 flood initiation (Peterson et al. 2012).

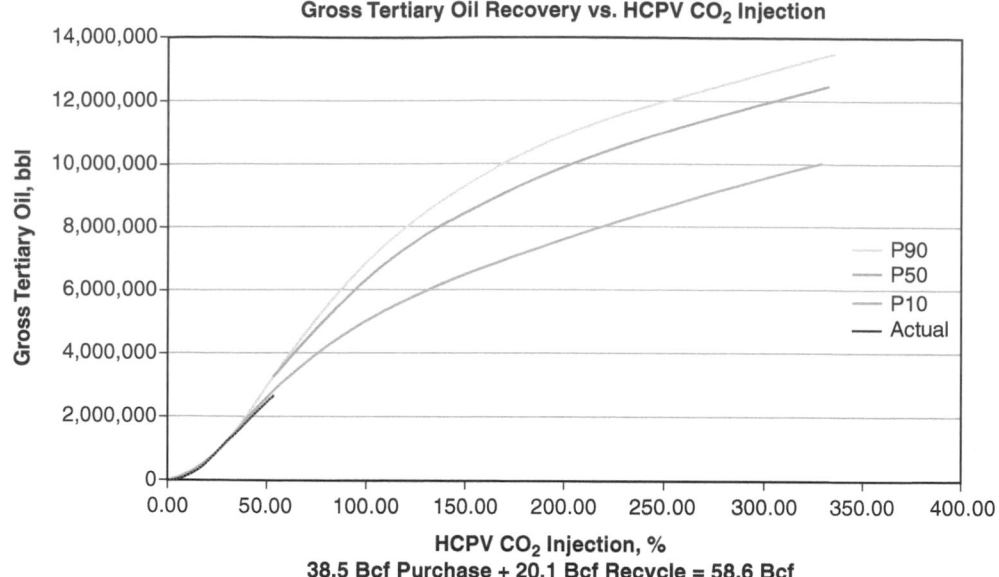

Fig. 6.102—Beaver Creek Unit Madison incremental tertiary oil recovery vs. HPV of CO₂ injected (Peterson et al. 2012).

being evaluated to determine the flood estimated ultimate recovery (EUR); however, the current estimate is that the EUR will exceed 15 million bbl. Several factors are likely contributing to better-than-anticipated recovery, including improved sweep because of the addition of midreservoir injectors, higher operating reservoir pressures, and the drilling/completion of several infill-production wells. On the basis of an oil recovery of 7.5 million bbl as of year-end 2014, and the CO_2 injected volume listed on the abscissa of the plot, the gross CO_2 usage was 16.6 Mcf/bbl (**Fig. 6.103),** and the net was 7.7 Mcf/bbl (**Fig. 6.104),** indicating very efficient CO_2 usage.

Operational Issues. Regular testing and data-gathering procedures were followed; for example, monitoring downhole pressures in essentially all wells to ensure that reservoir pressure did not drop below the MMP. Four different regions of the reservoir (shown in Fig. 6.100) were designated for focused monitoring because of geologic and elevation differences on the structure. Region 1 was the gravity-stable flood area, Region 2 the center of the peripheral flood, Region 3 the downdip WAG area, and Region 4 was a lower faulted area. Within these regions, localized injection and withdrawal rates were balanced to ensure that the pressures were maintained above the MMP. Limitation of CO_2 supply was partly responsible for pressure decline, resulting in a need to modify injection-well locations and to inject water downdip to keep pressure above the MMP. As of year-end 2014, pressures throughout the reservoir were still well above the MMP; however, there are some significant pressure differentials in the reservoir and more balancing work was needed. These pressure differentials are caused (at least in part) by ongoing injectivity issues such that the center of the flood is at lower pressures than the surrounding areas (Peterson 2015).

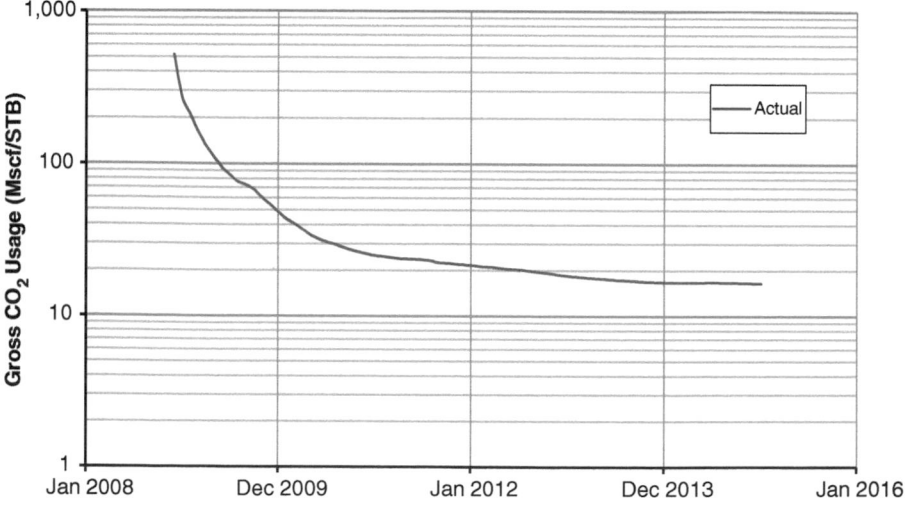

Fig. 6.103—Historical gross CO₂ usage (cumulative CO₂ purchases + cumulative CO₂ recycled)/cumulative tertiary oil.

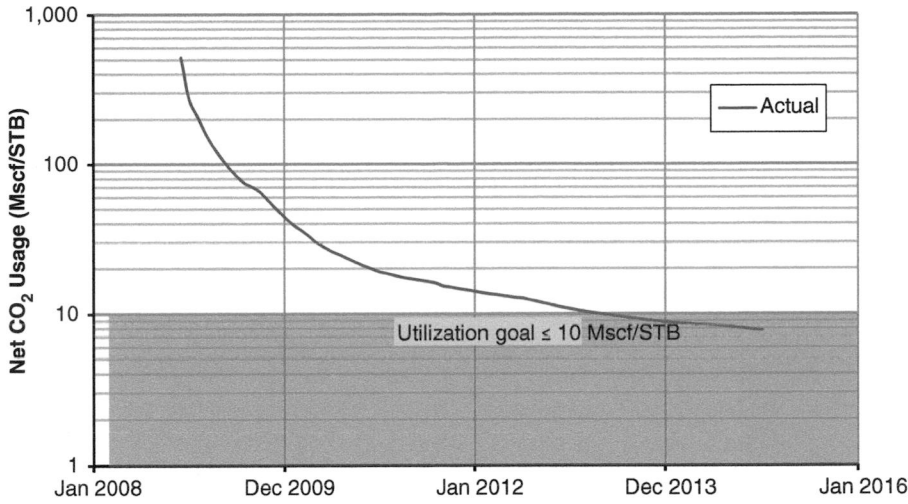

Fig. 6.104—Historical net CO₂ usage: cumulative CO₂ purchases/cumulative tertiary oil.

The conformance of CO₂ injection is being monitored through production logging of injection profiles over time. Plots of these profiles at four different times are given as **Fig. 6.105,** indicating that the "D" zone took most of the injected fluid. This was as expected on the basis of the properties of the different zones. While it was originally thought that there could be loss of injectivity because of relative permeability changes induced by the alternating of different injected fluids, a significant loss did not occur. There does, however, seem to be some loss of injectivity observed in mid-2014, which early indications suggest may be scale deposition. This was to be studied with the belief that acid treatments may correct the issue (Peterson 2015).

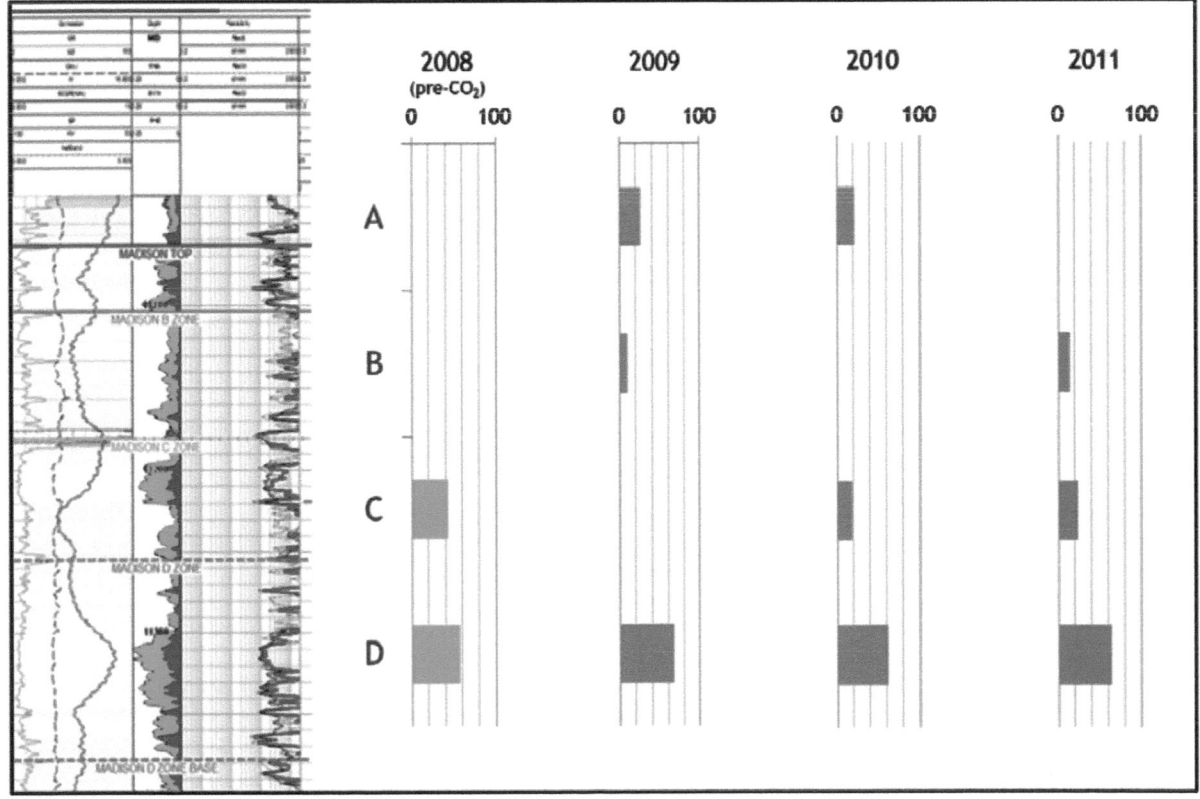

Fig. 6.105—Typical injection-well injection profile log illustrating variations in zonal injection over time (Peterson et al. 2012).

Rock Properties	
Average porosity (%)	14
Average permeability (md)	20
Average thickness (ft)	350
Average oil saturation (%)	87
S_{or} to water (%)	25
S_{or} to N_2 (miscible) (%)	7
Fluid Properties	
Oil gravity (°API)	50
Solution GOR (scf/bbl)	1,830
Bubblepoint pressure (psia)	2,830
MMP at reservoir temperature (psia)	3,600
Initial reservoir pressure (psia)	7,850
Oil viscosity at reservoir temperature (cp)	0.18
FVF (RB/STB)	1.76
Water/oil mobility ratio	0.3
Reservoir Properties	
Area (acres)	14,400
Depth (datum ss) (ft)	15,400
Reservoir temperature (°F)	285
Peak number of wells	138
Original well spacing (acres)	160

Table 6.22—Jay/LEC rock, fluid, and reservoir properties (Langston et al. 1981; Christian et al. 1981; Valenti 2014).

Other issues reported were handling of supercritical CO_2, asphaltene precipitation/deposition, and selection of proper materials for the piping associated with injection wells (Peterson et al. 2012).

6.12.4 Jay/Little Escambia Creek (LEC) Field N$_2$ Miscible-Flood Project. *Early History and Geology.* The Jay/LEC field was discovered in June 1970 and straddles the Florida/Alabama border approximately 40 miles north of Pensacola, Florida. The field is approximately 7 miles long by 3 miles wide (14,400 productive acres) and produces primarily from the Smackover dolomite formation located at an average depth of approximately 15,500 ft. Average thickness is 350 ft. Original reservoir pressure was 7,850 psia with a reservoir temperature of 285°F. The oil has a 50 °API gravity with a bubblepoint pressure of 2,830 psia, significantly undersaturated with associated gas that was sour, containing 8.8 mol% H$_2$S (Langston et al. 1981; Christian et al. 1981; Lawrence et al. 2002). Early in the field life, the OOIP was estimated to be 728 million STB (Langston et al. 1981). Later studies resulted in a revised OOIP estimate of 830 million STB (Lawrence et al. 2002). In 2014, the OOIP was further revised to be 920 million STB. The primary reason for the increase in the OOIP estimate is the inclusion of additional pay intervals in the lower Smackover formation. A summary of fluid, rock, and reservoir properties is given in **Table 6.22.**

The reservoir structure has been described as a southeast plunging "nose" with little faulting. Reservoir rock consists of many low- and high-permeability zones of significant thickness that are continuous over large areas. There also exist numerous thin zones characterized by the presence of stylolites with cemented material, above and/or below the feature, ranging in thickness from a few millimeters to several centimeters. These thin zones were studied in detail during the 1990s by examination of some 28,750 ft of core taken from 138 wells. The zones have very low permeability, with a horizontal extent varying typically from 100 to 300 ft, and thus restrict vertical flow where present (Lawrence et al. 2002).

During the initial development period (1970 to 1974), 102 wells were drilled on 160-acre spacing, resulting in 89 producers and 13 dry holes. Production increased during primary production, reaching a value of approximately 100,000 B/D by early 1974, as shown in **Fig. 6.106** (Melster et al. 2014). The field was unitized in 1974, and a 3:1 staggered line-drive waterflood was implemented. The flood was successful, maintaining the maximum allowable production rate of just more than 100,000 B/D through the 1970s. Additional wells were drilled during the period from 1977 to 1979, reducing well spacing to approximately 136 acres. As of January 1981, cumulative oil production from the field was 296 million STB, approximately 32% of

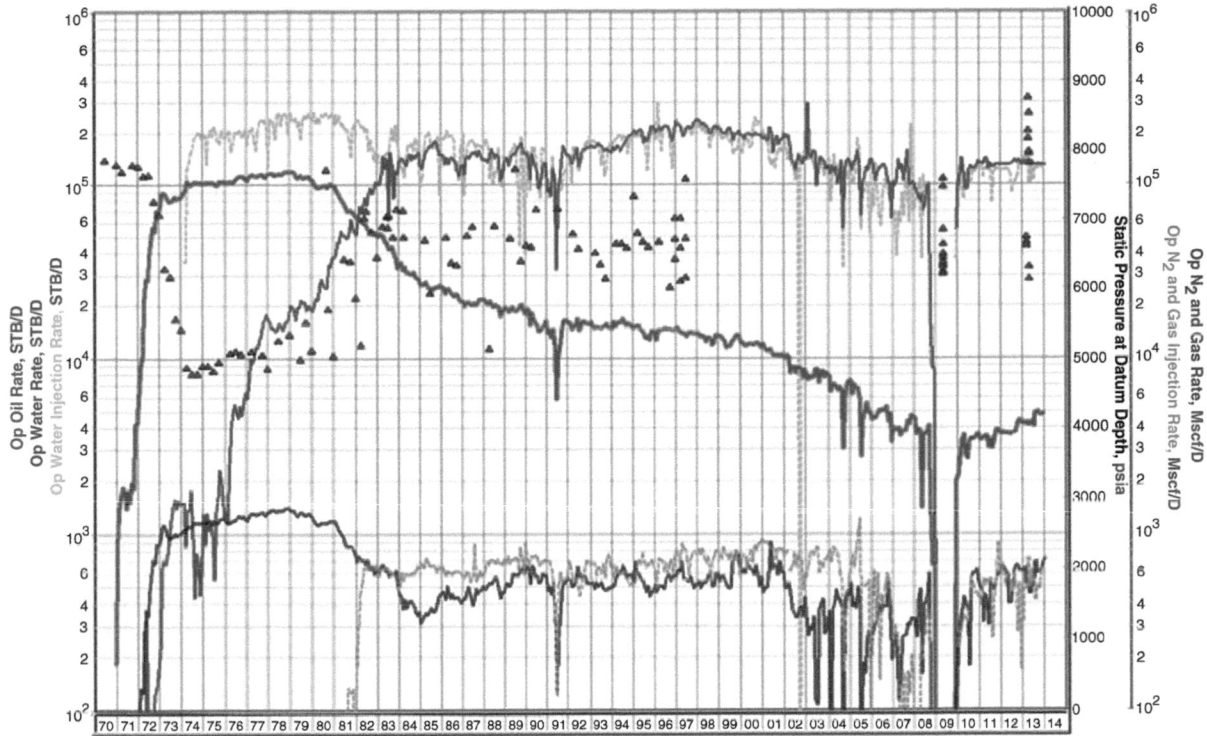

Fig. 6.106—Jay Field production and injection history (Melster et al. 2014).

the OOIP estimate as of that date. Cumulative water injection at that point was 524 million bbl (Christian et al. 1981). Production and injection rates are shown in Fig. 6.106 (Melster et al. 2014).

Nitrogen-WAG EOR Project: Initial Project Planning. EOR studies were begun by Exxon in late 1977, and the N_2-WAG project was approved in June 1980. Project evaluation included core analyses, measurement of MMP, corefloods to determine microscopic displacement efficiency by miscible displacement, field testing to examine possible loss of fluid injectivity, and simulation by use of 2D and 3D models.

Traditional slimtube experiments were conducted to measure MMP values of CO_2, N_2, and CH_4 at reservoir temperature. The MMP determined for all three fluids was 3,600 psia, well below the reservoir pressure. N_2 was selected as the injection fluid because a third-party supplier was available, while there was essentially no economically viable access to a CO_2 supply. CH_4 was rejected because the cost exceeded that of N_2. Even with reduced reservoir temperatures as a result of waterflooding, the MMP of N_2 was found to be significantly below the expected reservoir pressure, thus confirming its suitability as the displacement gas. Early coreflood experiments by use of a hydrocarbon gas that was miscible with the oil demonstrated that miscible displacement of waterflood residual oil was very efficient, driving the S_{or} value to almost 0% (Christian et al. 1981). Later tests that used N_2 as the displacing fluid yielded S_{or} values on the order of 7% (Lawrence et al. 2002). Field testing showed that water-injectivity loss on the order of 40% could occur following gas injection (Christian et al. 1981), although this did not prove to be a significant issue during the project.

On the basis of laboratory studies and extensive modeling, the initial plan was to inject N_2 at a rate of 67 MMscf/D with a WAG ratio of unity or more. Total injected gas was to be 20% of the HCPV, with a lifetime of 15 years. Predicted incremental recovery was 47 million STB or 6.5% of OOIP.

Project Implementation and Performance. The N_2-WAG process was implemented in late 1981. As shown in Fig. 6.106, oil production had started to decline in 1980, and this decline continued following initiation of N_2 injection, but at a rate slower than what was expected to occur with continuation of the waterflood only. The initial N_2 injection rate was approximately 160,000 MMcf/D and increased with time to values on the order of 170,000 to 180,000 MMcf/D. Early water injection was approximately 160,000 B/D, but was also increased later, as shown in Fig. 6.106.

In the 1990s, ExxonMobil undertook a major Jay/LEC unit study focused on a refined geological and reservoir-simulation model (Lawrence et al. 2002). As a result of the study, initiatives were implemented that included optimization of N_2 injection among field patterns, conversion of some producers to injectors, increased water-injection rate, and installation of equipment for increased capture and reinjection of N_2. A major result was the recognition of the importance of the stylolites and associated thin cement zones that restricted vertical flow. These flow restrictions limited gravity segregation. As a result, high-permeability intervals between the flow restrictions had very high sweep efficiency, while other less-permeable layers were poorly swept. Early modeling, performed before implementation of the changes resulting from the study, had predicted a field-production rate of 2,000 STB/D in the year 2000. With the improvements, the rate was

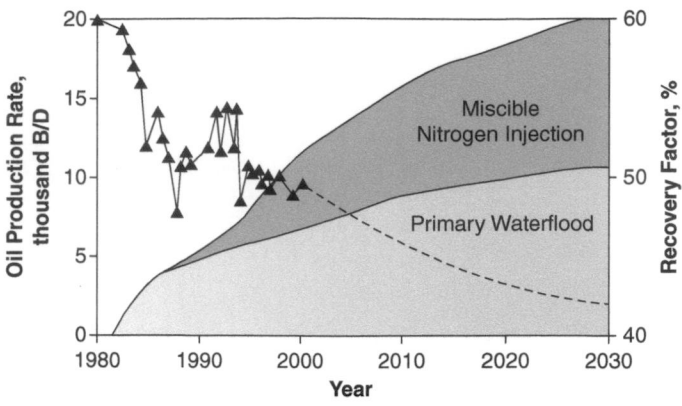

Fig. 6.107—Incremental WAG benefits over waterflooding.

approximately 10,000 STB/D as of that date. The change in slope of the decline curve in the 1990s is clearly seen in Fig. 6.106.

Cumulative production as of 2002 was approximately 440 million STB, which was 48% of the OOIP estimate as of that date. Ultimate cumulative incremental production was estimated to be 7 to 10% of OOIP. **Fig. 6.107** shows the actual production rate through 2002, as well as a comparison between predicted ultimate incremental production under N_2 injection and under waterflood (Lawrence et al. 2002). As of 2015, cumulative production was 464 million STB, with a recovery factor of 50%. The latest EUR is approximately 3 to 5% of OOIP (Valenti 2014).

Surface Facilities and 2009 Retrofitting of Facilities. During the field life, the Jay plant processed produced fluids for sale or reinjection. Up until the mid-2000s, surface facilities included N_2-rejection units (NRUs) operated at cryogenic conditions (–298°F) to separate light hydrocarbon molecules from N_2 for sale and reinjection, respectively. Equipment for gas sweetening, sulfur recovery, natural-gas-liquids (NGL) separation, and compression was also required. In the late 1990s, as decline of production accelerated, technical problems related to aging of facilities increased, with the issues exacerbated by the low oil price and associated poor economic conditions. By the early 2000s, only approximately one-half of the original wells were still active. In 2009, Quantum Resources Management, then the operating company for the field, implemented a major program of retrofitting and "right sizing" the field equipment (Melster et al. 2014). The plan entailed short-term, midterm and long-term projects to scale costs up or down relative to production volumes. For example, because the cost of operating the NRUs was excessive, the units were shut down, and a less-expensive gas-recovery system was installed that captured NGLs and left some light hydrocarbons in the N_2 gas stream, which were reinjected. Computer simulations involving phase-behavior calculations indicated that the presence of the hydrocarbons would not significantly affect the miscibility of the gas with the crude oil (Lawrence et al. 2002). The project was successful in reducing operating cost per barrel by 40% and increasing profitability, which was critical during a low-oil-price environment.

The operating-cost improvements in the Jay field afforded more attention to restoring production in the field, which subsequently climbed from 2,000 STB/D in 2010 to 5,000 STB/D in 2014 (Fig. 6.106). The contributing factors to increased oil

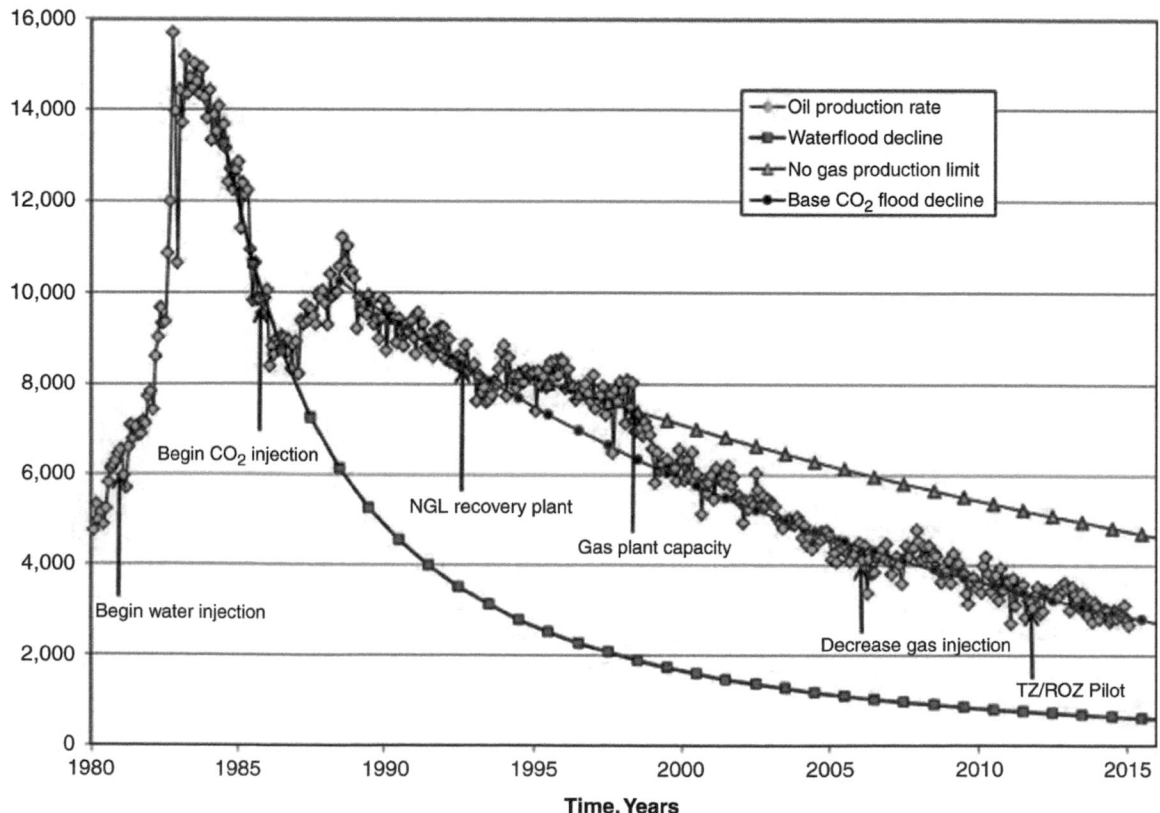

Fig. 6.108—East Vacuum oil-production rate, 1980–2015 (Moffitt et al. 2015).

production include reactivating shut-in producers, cleaning fill from injector-well completions, and converting wells to injection. The increased oil production during this time frame was also matched with declining water/oil and gas/oil production ratios. This, in turn, further improved the profitability of the Jay field and extended the expected life of the field.

6.12.5 References to Selected Additional Field Projects. This subsection contains brief descriptions and references to selected miscible-flood field projects. Two of the areas referenced— Prudhoe Bay and the North Sea—are very expensive regions in which to operate, creating a challenge for EOR processes.

East Vacuum Grayburg San Andres Unit. This unit is located in southeast New Mexico, not far from the west Texas location of the Means San Andres unit discussed in Section 6.12.2. The two units produce from the same formation (mainly the San Andres); however, the properties of the crude oil are significantly different. The East Vacuum was unitized in 1978, waterflooding was initiated in 1980, and CO_2 flooding was started in 1985. As of 2015, the CO_2-EOR project had been operating for 30 years. A plot of oil production vs. time over this period is shown in **Fig. 6.108** (Moffitt et al. 2015). Total recovery to that date was estimated at 12.5% OOIP. Major events occurring during the time period covered are indicated on the plot and discussed in Moffitt et al. (2015). Brownlee and Sugg (1987) describe the project implementation and results for the first 2 years of the project, while Harpole and Hallenbeck (1996) present a summary of the flood performance for the first 10 years of the project.

Miscible Gasflood at Prudhoe Bay. The Prudhoe Bay reservoir has a large gas cap with the major production occurring as gravity drainage. The waterflooding and EOR operations are mainly confined to the down structure and peripheral areas of the reservoir. EOR, consisting of a hydrocarbon first-contact miscible drive with inverted nine-spot patterns on 80-acre spacing, was implemented in 1982 (McGuire et al. 2000). McGuire et al. (2000) primarily describes detailed analysis of the flood performance on the basis of physical and chemical analyses of core data taken at a well located 300 ft from a producing well, in addition to simulator history matching of the core results.

Survey of North Sea EOR Projects. Awan et al. (2008) summarize a survey of EOR projects initiated in the North Sea between 1975 and 2005. Projects discussed include FCM hydrocarbon gasflooding, WAG, simultaneous WAG injection processes, and foam-assisted WAG. CO_2 flooding was not implemented during the time frame covered because of lack of availability of a CO_2 supply, although studies indicated significant possibilities for future application. Almost all of the implemented EOR processes were deemed to be successful, with WAG the most-successful operating approach.

Problems

6.1 Consider the pressure/composition diagram shown in **Fig. 6.109** for a binary mixture of CO_2 and a hydrocarbon. The diagram is for a system at 90°F.

1. Assume that a mixture of 85% CO_2 and 15% hydrocarbon is placed in a PVT cell at 90°F. Show the phases and the relative amounts of the phases (qualitatively) at pressures of 300, 600, 1,250, 1,800, 2,500, and 3,500 psia. Use cell "blanks," as shown in **Fig. 6.110** for your answer.

2. If the same mixture were placed in a cell at 130°F, what would you expect the phase behavior to be? Again, use cell "blanks" shown in Fig. 6.110 for your answer.

6.2 The pseudoternary diagram for the system consisting of CO_2 and Jayhawk crude is shown in **Fig. 6.111**. The diagram is at reservoir temperature and pressure and is given on a mass-composition basis (pounds mass component/pounds mass total). The composition of the crude oil is shown.

Assume that this crude oil is to be displaced in a slimtube test with pure CO_2 as

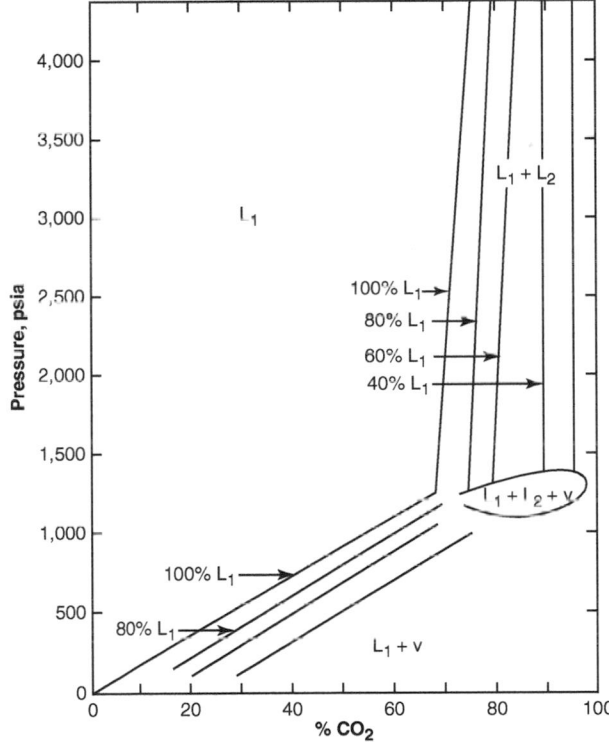

Fig. 6.109—Pressure/composition diagram, Problem 6.1.

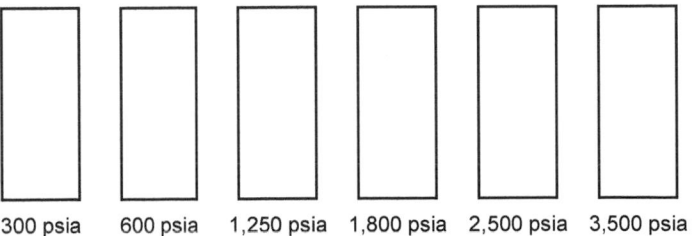

300 psia   600 psia   1,250 psia   1,800 psia   2,500 psia   3,500 psia

Fig. 6.110—Cell "blanks," Problem 6.1.

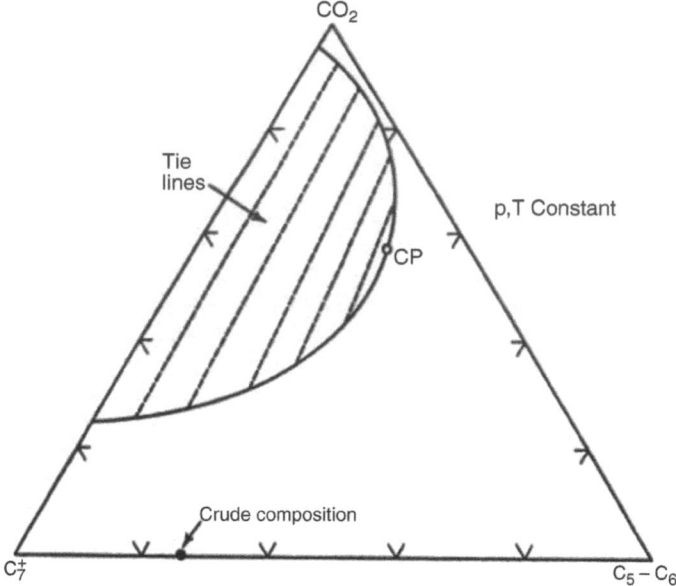

Fig. 6.111—Ternary diagram, CO₂/Jayhawk crude, Problem 6.2 [compositions are on a mass fraction basis (1 lbm component/ 1 lbm total)].

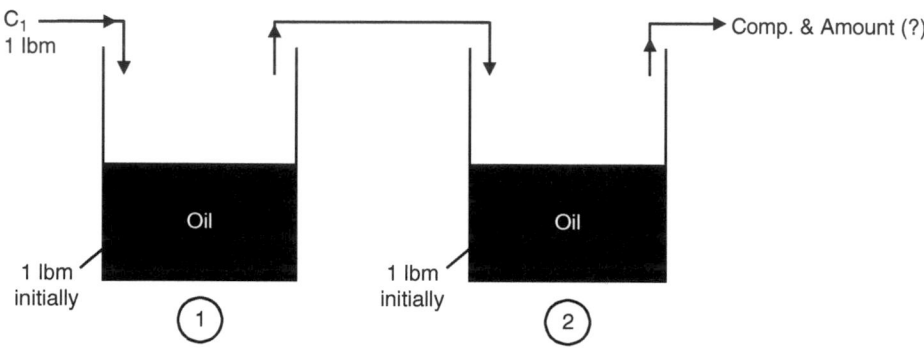

Fig. 6.112—Mixing cells in series, Problem 6.4.

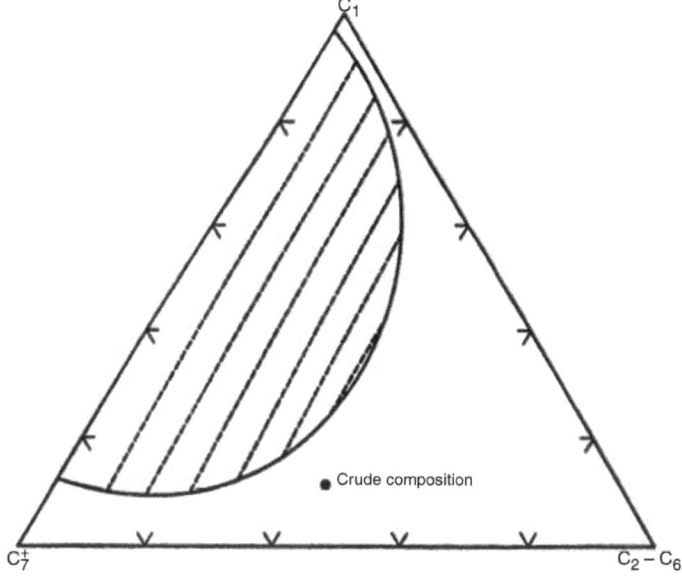

Fig. 6.113—Pseudoternary diagram, Problem 6.4.

the displacing fluid. The test is to be conducted at the reservoir T and p. The slimtube is to be saturated with the crude oil (no water) before CO_2 is injected.

Using the pseudoternary diagram and the additional data provided below, estimate the fractional recovery of crude expected from the slimtube after a reasonable amount of CO_2 has been injected (approximately 1.2 PV of CO_2).

Relative permeability data indicate that, in an immiscible drive, the residual saturation of the hydrocarbon-rich phase will be 32%. $\rho_o = 0.75$ g/cm³ and $\rho_{CO_2} = 0.6$ g/cm³ (apparent density of the CO_2 in the hydrocarbon-rich phase).

6.3 An oil reservoir initially contained 1.5×10^6 STB oil in place. Other properties were $S_{oi} = 0.75$ and $B_{oi} = 1.2$ RB/STB. The reservoir was never produced under primary production, but was waterflooded initially. Recovery under waterflood was 715,000 STB. It was estimated that the waterflood sweep efficiency was 65%.

The reservoir is being considered for a CO_2 miscible flood. Estimate the recovery in stock-tank barrels that might be expected in the CO_2 flood. The waterflood was conducted at essentially initial reservoir pressure, which is also the assumed CO_2 flood pressure. State any assumptions that you make.

6.4 A dynamic miscible displacement process is to be modeled with a series of perfectly mixed cells as shown in **Fig. 6.112.** The displacing fluid is to be pure methane. The composition of the reservoir oil is shown on the pseudoternary diagram in **Fig. 6.113.**

The mixing cells each initially contain 1.0 lbm of reservoir oil. Then, 1 lbm of methane is fed to Cell 1. After equilibrium is reached, all the vapor is drawn off and injected into Cell 2. Again, after equilibrium is reached, the vapor is drawn off and injected into the next cell, etc.

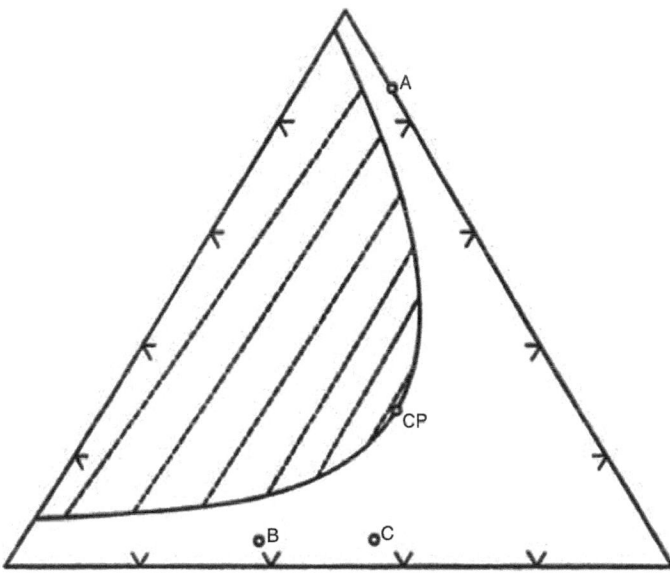

Fig. 6.114—Ternary diagram, Problem 6.5.

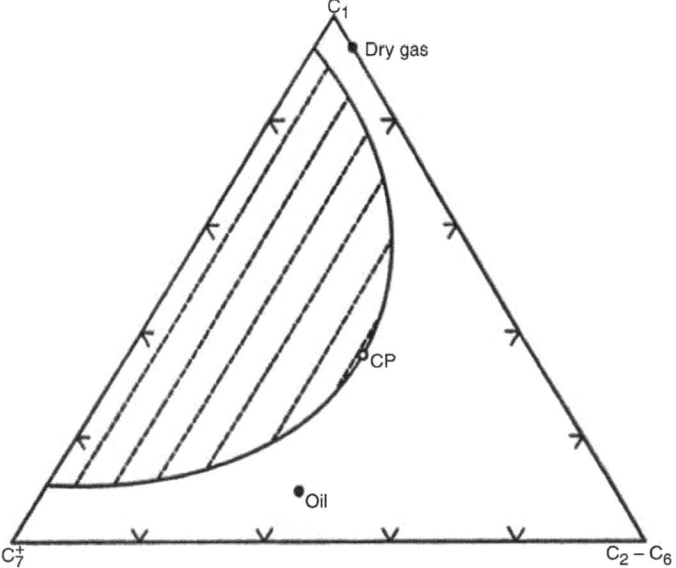

Fig. 6.115—Pseudoternary diagram, Problem 6.7.

Determine the composition and amount of vapor withdrawn for Cell 2.

6.5 Consider the pseudoternary diagram in **Fig. 6.114.** The reservoir oil composition is shown at Point C. Injection of a dry gas of Composition A is planned to create a miscible displacement process in the reservoir.

 1. Use the ternary diagram to illustrate the manner in which a miscible displacement will be formed in the reservoir.
 2. If the reservoir oil composition were at Point B, could a miscible condition be created in the reservoir by injection of gas of Composition A? Explain and show the reason for your answer.

6.6 A laboratory experiment on CO_2 miscible displacement is required. In one run, the plan is to conduct an FCM displacement. The "crude oil" in this experiment is to be *n*-octane. The displacement temperature will be 200°F.

At what pressure would you run this experiment to ensure first-contact miscibility in the displacement process? Indicate the basis for your answer.

6.7 A pseudoternary diagram for a dynamic miscible displacement process is shown in **Fig. 6.115.**

A dry gas of the composition indicated would develop dynamic miscibility with an oil of the composition indicated. Assume that miscibility is developed at a certain point in the reservoir.

 1. Describe the fluid conditions that exist both ahead of and behind the "point" of miscibility.
 2. Once miscibility with the reservoir oil is achieved at a certain position, will there be miscibility behind this position (i.e., between the point at which miscibility exists with the oil and the injection well)? Explain your answer. Use the ternary diagram as necessary.

6.8 A CO_2 displacement test is conducted in a slimtube apparatus. A pseudoternary phase diagram for the fluid system is shown in **Fig. 6.116.** The oil composition is indicated on the diagram.

A displacement test is conducted with a fluid that is 96% CO_2 and 4% C_2 through C_6.

 1. Will miscibility be achieved in this displacement?
 2. When the injected gas first breaks through at the exit end of the core, what will its composition be? Show this point on the ternary diagram. (Neglect any effects of dispersion.)
 3. Consider a location near the entrance of the slimtube. Several PVs of gas will have flowed through this position by the end of the run. What will be the final oil composition at this position near the entrance? Show the composition point on the ternary diagram.

6.9 A miscible displacement is to be considered for the system described on the pseudoternary diagram in **Fig. 6.117.** The oil composition is shown on the diagram. The injection fluids being considered are C_1 (methane) and $C_1/(C_2$ through $C_6)$ (methane enriched with C_2 through C_6 components).

The injected-fluid cost increases with increasing amounts of C_2 through C_6. Pressure is fixed.

 1. Specify the composition of the minimum-cost injection fluid that could be used in an MCM (dynamic) process.
 2. Describe how your proposed process would work (i.e., describe the mechanism for the development of miscibility). Use the ternary diagram as a part of your description.
 3. Suppose that a slimtube test were conducted in which the oil was displaced by a gas of your recommended composition. Assume further that at some relatively early time in the displacement, a gas breaks through at the outlet end of the tube. Using the phase-behavior data, estimate the composition of this initial effluent gas.

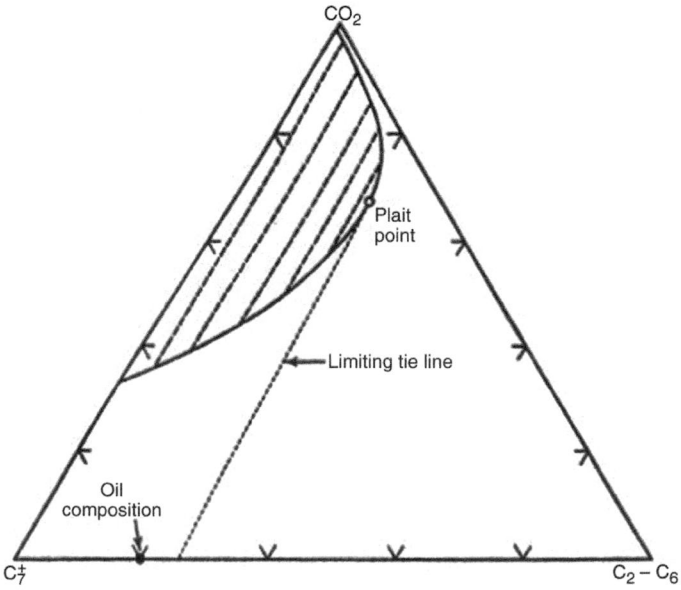

Fig. 6.116—Pseudoternary diagram, Problem 6.8.

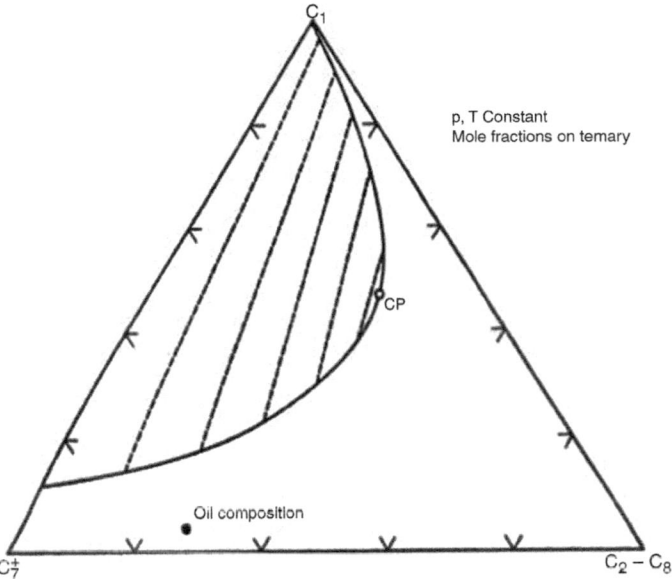

Fig. 6.117—Pseudoternary diagram, Problem 6.9.

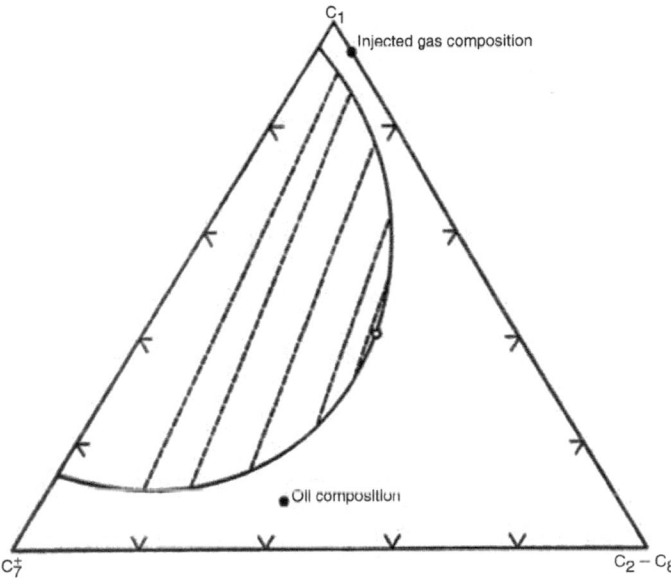

Fig. 6.118—Pseudoternary diagram, Problem 6.12.

6.10 The Jayhawk reservoir is at a depth of 4,000 ft. It contains a 35 °API crude oil of known composition. The composition is such that the reservoir might be suitable for CO_2 flooding. The reservoir temperature is 120°F.

A laboratory study is needed to determine whether CO_2 flooding would be suitable. You are to design the laboratory investigation.

1. What kinds of experiments would you conduct and what would your experimental system(s) consist of? Describe in a general way.
2. What variables would you investigate (i.e., what parameters are of particular importance in the experiments)?
3. What experimental data would you take and what would be your indications that CO_2 either is or is not suitable?

6.11 A laboratory experiment on CO_2 miscible displacement is required. In one run, the plan is to conduct an FCM displacement. The "crude oil" in this experiment is to be n-octane. The displacement temperature will be 200°F.

At what pressure would you run this experiment to ensure first-contact miscibility in the displacement process? Indicate the basis for your answer.

6.12 A dynamic miscible displacement test is conducted in the laboratory for the system shown on the ternary diagram in **Fig. 6.118.** A dry gas with the indicated composition is injected into a long core to displace the resident oil with the composition shown. The oil phase is originally at residual saturation and is not mobile.

1. Will miscibility be achieved? Explain.

2. What will be the composition of the gas exiting from the core when the injected gas first breaks through (arrives) at the exit point of the core? Give the composition and show its location on the ternary diagram.

3. Assume that a large number of PVs of dry gas are injected over a long period of time. After this is done, what is your estimate of the composition of the oil that remains in the core? Give the composition and show its location on the ternary diagram.

6.13 In Problem 6.4, high-pressure (vaporizing) dynamic miscible displacement is simulated by multiple batch contacts as shown in Fig. 6.112.

Crude oil in Cell 1 is contacted with a specified amount of vapor. After equilibrium, the vapor was removed and put in contact with "fresh" oil in Cell 2, where equilibrium again is established. The process can be repeated through any number of cells.

You are to design a similar experiment for the rich-gas (condensing) miscible displacement process. **Fig. 6.119** shows the pseudoternary phase diagram. Also shown are the rich-gas composition and the reservoir crude oil composition.

The rich gas introduced into Cell 1 is to be one-half the amount of the oil originally in the cell. You are to decide on the method of subsequent contacts (i.e., what kinds of vapor/liquid contacts should be made in cells following the first). The process is to simulate the rich-gas, condensing-type process.

Whatever your contact sequence, the amount of vapor introduced into a cell should be one-half the amount of liquid originally in the cell. This injection ratio (one part gas and two parts oil) should be maintained for all cells.

1. Describe the sequence of vapor/liquid contacts (i.e., the procedure of your experiment). Use a sketch to illustrate.
2. Determine the number of total contacts required before a single phase exists at equilibrium in the final cell.
3. Determine the number of total contacts required before dynamic miscibility is achieved.

Assume that the phase diagram is in terms of moles and that "amount" in the problem statement also refers to moles.

6.14 Consider Example 6.13 in which the Claridge correlation was used to predict oil recovery in a five-spot pattern. The process was a miscible displacement used for tertiary recovery.

Use the same data but assume that flow is in Region II and that the Stalkup correlation is applicable (Fig. 6.64).

Calculate N_p/N and N as a function of the volume of solvent injected. You will have to interpolate in Fig. 6.64. Use **Fig. 6.120** with the interpolation shown.

6.15 A reservoir oil has the following composition: 10% C_1, 20% C_2 through C_6, and 70% C_7. A miscible displacement is being considered for this reservoir. Three potential systems—rich gas (30% C_2 through C_6 and 70% C_1) followed by C_1, dry gas (95% C1 and 5% C_2 through C_6), and C_3 followed by C_1—are to be considered because of the availability of fluids.

The maximum pressure that can be applied to the reservoir is 1,200 psi. The reservoir temperature is 100°F. **Figs. 6.121 and 6.122** give the phase-behavior data.

Which, if any, of these fluid systems would give a miscible displacement? Clearly state the reason for your decision relative to each fluid system.

6.16 The CO_2 dynamic miscible process was stated previously to resemble the high-pressure, vaporizing-gas, dynamic miscible displacement process.

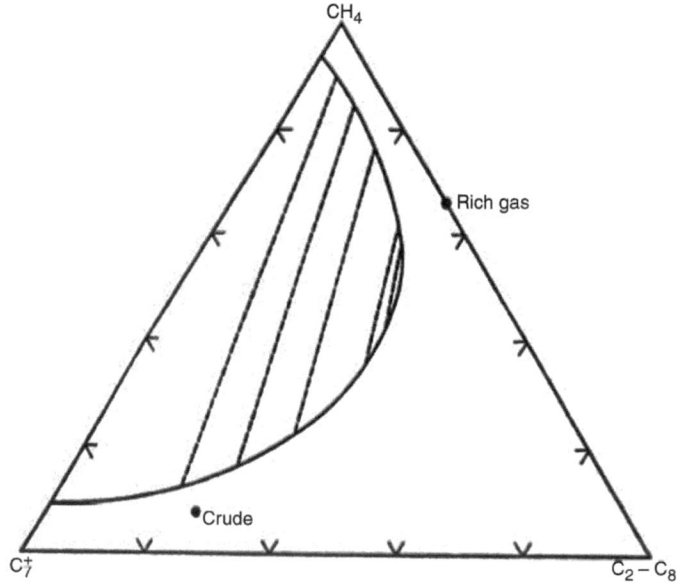

Fig. 6.119—Pseudoternary diagram, Problem 6.13.

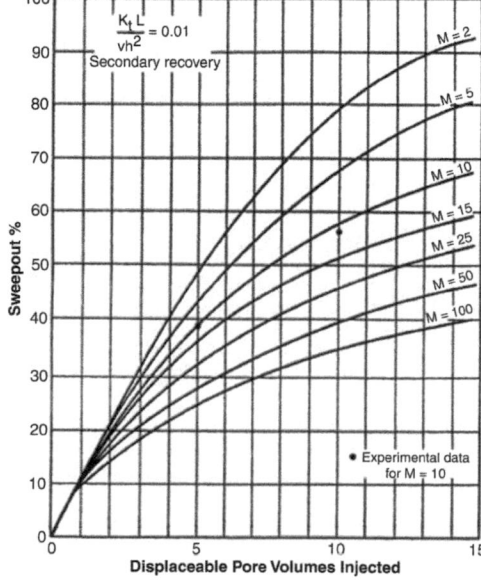

Fig. 6.120—Stalkup correlation, Problem 6.14.

1. Describe the similarities that exist in the two processes. Use diagrams to answer, as necessary.
2. In what way(s) are the processes different?
3. Because the two processes are similar, would it be advantageous to mix CH_4 and CO_2 to form a displacement gas? Explain.

6.17 Consider the miscibility pressure of a synthetic hydrocarbon of the composition shown in **Table 6.23** when displaced by pure CO_2. The MMP for a CO_2 displacement in a slimtube apparatus was determined to be 1,220 psia at a temperature of 150°F. You want to estimate the MMP at this same temperature if the amount of $C_{20}H_{42}$ is doubled, while leaving the amounts of the other components constant.
   1. Which correlation (of those discussed in the text) is the most appropriate for making this estimate? Why?
   2. Use your recommended correlation to estimate the MMP for the modified composition.

6.18 Consider the use of CO_2 as a dynamic miscible displacement fluid. On the basis of what you can determine about the properties of CO_2, would you expect the following factors that affect displacement efficiency to be of concern on typical applications?

   1. Mobility control
   2. Gravity override or underride.
     Explain the basis for your answer.

6.19 Consider the pressure/composition diagrams shown in **Fig. 6.123** for the binary system consisting of n-hexane and CO_2. Assume that laboratory experiments are to be conducted in which n-hexane is displaced by CO_2.
     Using the information in **Table 6.24**, prepare a plot of MMP vs. temperature for this system.

6.20 Data on four different crude oils are shown in **Tables 6.25 through 6.28**. The data include oil composition, reservoir temperature, and specific gravity or API gravity. With these data, estimate MMP with the following methods, if they are applicable.

   1. National Petroleum Council (1976)
   2. NHolm and Josendal (1974)
   3. NHolm and Josendal (1982)
   4. NYellig and Metcalfe (1980)
   5. NAlston et al. (1985)

6.21 Consider a slimtube displacement of Johanning B crude oil at a temperature of 102°F. The displacing fluid is pure CO_2, and the displacing pressure is 1,150 psia. **Fig. 6.124** is a pseudoternary diagram for the crude oil and CO_2 at 102°F and 1,150 psia. The Johanning B oil has a gravity of 36.8 °API and a molecular weight of 223.

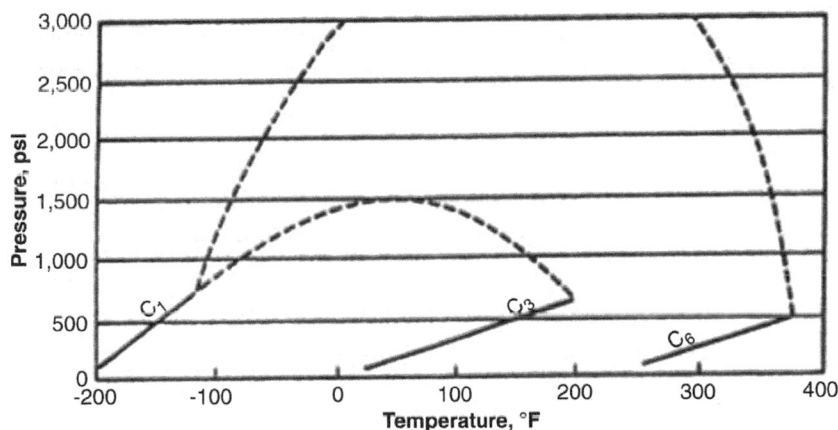

Fig. 6.121—Critical loci of binary mixtures, Problem 6.15.

C_5H_{12}	0.2
$C_{10}H_{22}$	0.3
$C_{20}H_{42}$	0.3
$C_{30}H_{62}$	0.1
$C_{35}H_{72}$	0.1
Total	1.0

Table 6.23—Composition (mole fraction) of synthetic hydrocarbon, Problem 6.17.

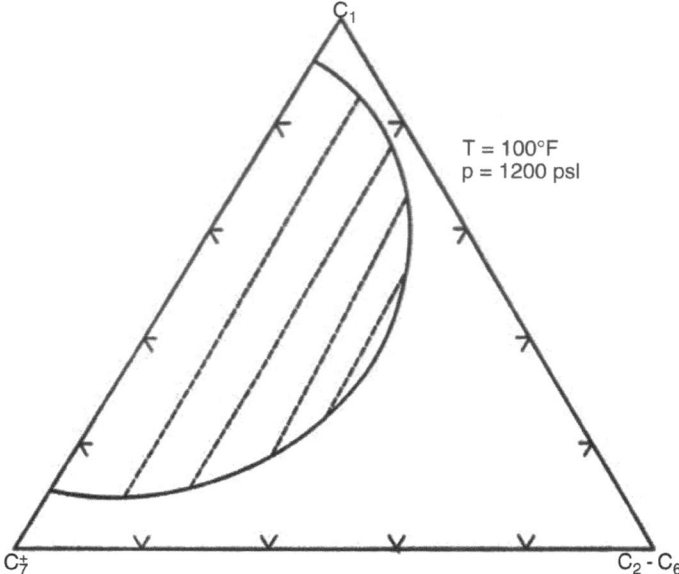

Fig. 6.122—Pseudoternary diagram, Problem 6.15.

Relative permeability data on a similar crude-oil/CO_2 system indicate that the residual hydrocarbon-phase saturation is 30% for two-phase flow. The apparent CO_2 density, dissolved in the oil, is estimated to be approximately the same as the oil density. The slimtube displacement yielded an oil recovery of approximately 78% (ultimate recovery after many PVs of throughput).

Determine whether this recovery is consistent with the ternary diagram.

6.22   Consider a linear miscible displacement in a reservoir for which the following data are given: $L = 400$ ft (length of linear system), $h = 30$ ft (thickness of reservoir), $T = 120°F$, $p = 1,500$ psia, and $k = 150$ md ($k_V = k_H$). Oil properties are $\rho_o = 0.75$ g/cm^3 at 120°F and $\mu_o = 1.2$ cp at 120°F. The solvent to be injected is 20 mol% C_3 and 80 mol% C_4.

Assuming that a large slug of solvent is injected, calculate the volumetric sweep efficiency at breakthrough. Assume that the correlation in Fig. 6.72 is applicable.

The injection rate is such that the Darcy flow velocity is $u = 0.10$ B/D-ft^2.

6.23   Assume that a five-spot pattern is to be flooded by miscible displacement. The following properties exist: well spacing = 20 acres, $h = 20$ ft, $\phi = 18\%$, and $k = 120$ md. The pattern has been waterflooded and is at ROS to water.

Using the FORTRAN computer programs for miscible displacement at tertiary conditions for continuous solvent injection and finite slug injection, calculate the effect of slug size on oil recovery. **Table 6.29** gives the rock and fluid properties.

Calculate oil recovery for solvent slug sizes ranging from 25% HCPV to continuous solvent injection.

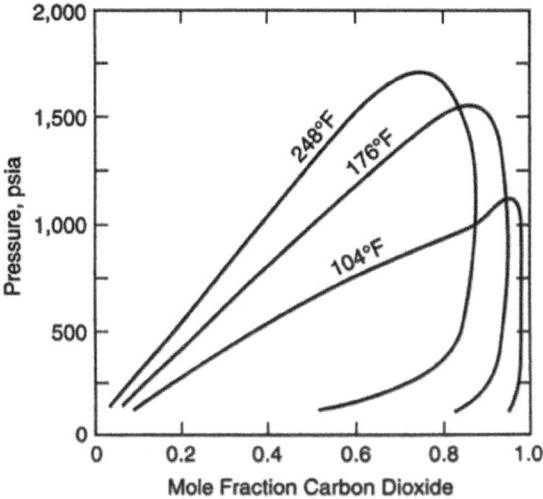

Fig. 6.123—Isothermal pressure/composition data for CO2/n-hexane, Problem 6.19.

Pressure (psia)	CO$_2$ (mole fraction)		K-Values	
	Liquid	Vapor	CO$_2$	n-C$_6$H$_{14}$
At 40°C				
113	0.080	0.949	11.85	0.055
240	0.170	0.972	5.73	0.034
354	0.252	0.977	3.87	0.031
480	0.356	0.982	2.76	0.027
600	0.450	0.982	2.18	0.032
733	0.574	0.984	1.71	0.037
844	0.687	0.985	1.44	0.048
973	0.829	0.984	1.19	0.091
1,039	0.882	0.982	1.11	0.149
1,085	0.915	0.981	1.07	0.221
At 80°C				
125	0.052	0.815	15.83	0.195
237	0.110	0.894	8.15	0.119
355	0.167	0.923	5.52	0.092
444	0.207	0.934	4.50	0.084
593	0.287	0.943	3.29	0.080
719	0.353	0.948	2.69	0.080
854	0.422	0.947	2.24	0.092
973	0.486	0.947	1.95	0.102
1,084	0.541	0.949	1.75	0.112
1,203	0.599	0.945	1.58	0.137
1,327	0.683	0.930	1.36	0.220
1,458	0.752	0.918	1.22	0.330
1,518	0.805	0.906	1.13	0.480
1,546	0.821	0.886	1.08	0.639
At 120°C				
130	0.028	0.507	17.83	0.508
251	0.080	0.715	8.93	0.310
383	0.135	0.808	5.98	0.223
511	0.180	0.843	4.67	0.191
644	0.243	0.853	3.51	0.194
769	0.289	0.862	2.99	0.194
885	0.339	0.873	2.58	0.193
1,008	0.380	0.875	2.30	0.202
1,125	0.433	0.877	2.03	0.216
1,259	0.486	0.874	1.80	0.245
1,379	0.538	0.866	1.61	0.290
1,464	0.570	0.853	1.50	0.341
1,593	0.632	0.828	1.31	0.467
1,661	0.676	0.801	1.19	0.615
1,682	0.693	0.793	1.15	0.675

Table 6.24—CO$_2$/n-hexane vapor/liquid equilibrium data, Problem 6.19.

Single Carbon Number	Mole Fraction	Cumulative Mole Fraction	Weight Fraction	Cumulative Weight Fraction	Volume Fraction	Cumulative Volume Fraction	Weight	Volume
5	0.0390	0.0390	0.016	0.016	0.021	0.021	2.82	4.46
6	0.0937	0.1327	0.044	0.060	0.053	0.074	7.87	11.41
7	0.1027	0.2354	0.055	0.115	0.063	0.174	9.86	13.56
8	0.1354	0.3708	0.081	0.196	0.089	0.226	14.49	19.34
9	0.1191	0.4899	0.081	0.277	0.087	0.313	14.41	18.76
10	0.0750	0.5649	0.056	0.333	0.059	0.372	10.05	12.85
11	0.0559	0.6208	0.046	0.380	0.048	0.420	8.22	10.36
12	0.0432	0.6640	0.039	0.419	0.040	0.460	6.96	8.65
13	0.0393	0.7033	0.039	0.457	0.039	0.499	6.88	8.44
14	0.0329	0.7362	0.035	0.492	0.035	0.534	6.25	7.57
15	0.0273	0.7635	0.032	0.524	0.031	0.565	5.62	6.73
16	0.0226	0.7861	0.028	0.552	0.028	0.593	5.02	5.95
17	0.0223	0.8084	0.030	0.582	0.029	0.622	5.29	6.21
18	0.0134	0.8218	0.019	0.601	0.018	0.640	3.36	3.93
19	0.0107	0.8325	0.016	0.617	0.015	0.655	2.81	3.27
20	0.0136	0.8461	0.021	0.638	0.020	0.675	3.74	4.32
21	0.0110	0.8571	0.018	0.656	0.017	0.692	3.20	3.68
22	0.0077	0.8648	0.013	0.669	0.012	0.704	2.31	2.64
23	0.0096	0.8744	0.017	0.686	0.016	0.720	3.00	3.40
24	0.00988	0.8832	0.016	0.702	0.015	0.735	2.85	3.22
25	0.0068	0.8900	0.013	0.715	0.012	0.747	2.29	2.58
26	0.0038	0.8938	0.007	0.722	0.007	0.754	1.33	1.49
27	0.0059	0.8997	0.0121	0.734	0.011	0.765	2.12	2.37
28	0.0038	0.9035	0.008	0.742	0.007	0.772	1.41	1.57
29	0.0040	0.9075	0.009	0.751	0.008	0.780	1.53	1.69
30	0.0040	0.9115	0.009	0.76	0.008	0.788	1.58	1.75
31	0.0040	0.9155	0.009	0.769	0.008	0.796	1.62	1.78
32	0.0040	0.9195	0.009	0.778	0.008	0.804	1.66	1.82
33	0.0013	0.9208	0.003	0.781	0.003	0.807	0.55	0.61
34	0.0014	0.9222	0.003	0.784	0.003	0.810	0.61	0.67
35	0.0014	0.9236	0.004	0.788	0.003	0.813	0.62	0.68
36	0.0014	0.9250	0.004	0.792	0.003	0.816	0.64	0.69
+37	0.0750	1.0000	0.208	1.000	0.184	1.000	37.13	39.88
							178.1	216.3

*M_w = 178.1, γ = 0.823, K_w = 11.6, and reservoir T = 90°F.

Table 6.25—Reported composition of Maljamar crude oil*, Problem 6.20.

Single Carbon Number	Mole Fraction	Cumulative Mole Fraction	Weight Fraction	Cumulative Weight Fraction	Volume Fraction	Cumulative Volume Fraction	Weight	Volume
6	0.014	0.0214	0.006	0.006	0.007	0.007	1.18	1.70
7	0.256	0.270	0.122	0.128	0.144	0.151	24.58	33.8
8	0.125	0.395	0.067	0.195	0.076	0.227	13.38	17.86
9	0.084	0.479	0.051	0.246	0.056	0.283	10.16	13.23
10	0.069	0.548	0.046	0.292	0.05	0.333	9.25	11.82
11	0.06	0.608	0.044	0.336	0.047	0.380	8.82	11.12
12	0.046	0.654	0.037	0.373	0.039	0.419	7.41	9.21
13	0.037	0.691	0.032	0.405	0.034	0.453	6.48	7.94
14	0.031	0.722	0.029	0.434	0.030	0.483	5.89	7.13
15	0.033	0.755	0.034	0.468	0.035	0.518	6.80	8.13
16	0.026	0.781	0.029	0.497	0.029	0.547	5.77	6.85
17	0.018	0.799	0.021	0.518	0.021	0.568	4.27	5.01
18	0.014	0.813	0.017	0.535	0.018	0.586	3.51	4.11
19	0.011	0.824	0.014	0.549	0.014	0.600	2.89	3.36
20	0.009	0.833	0.012	0.561	0.012	0.612	2.48	2.86
21	0.006	0.839	0.009	0.570	0.009	0.621	1.75	2.00
22	0.006	0.845	0.009	0.579	0.009	0.630	1.80	2.05
23	0.005	0.850	0.008	0.587	0.008	0.638	1.56	1.77
24	0.005	0.855	0.008	0.595	0.008	0.646	1.62	1.83
+ 25**	0.145	1.000	0.405	1.000	0.354	1.000	81.20	82.94
25 to 30	0.070							
							200.8	234.7

*M_w = 200.8, γ = 0.856, K_w = 11.6, and reservoir T = 118°F.

**CCN ı25 properties based on M_w = 560 and γ = 0.070.

Table 6.26—Adjusted composition of west Texas stock-tank oil* of Yellig and Metcalfe (1980), Problem 6.20.

Single Car-bon Number	Mole Fraction	Cumulative Mole Fraction	Volume Fraction	Cumulative Volume Fraction	Weight Fraction	Cumulative Weight Fraction	Weight
5	–	–	–	–	–	–	–
6	–	–	–	–	–	–	–
7	0.056	0.056	0.030	0.030	0.024	0.024	5.6
8	0.122	0.178	0.070	0.100	0.059	0.083	13.91
9	0.128	0.306	0.080	0.180	0.070	0.153	16.38
10	–	–	–	–	–	–	–
11	0.096	0.402	0.065	0.245	0.059	0.211	14.98
12	0.054	0.456	0.040	0.285	0.037	0.248	9.18
13	0.067	0.523	0.055	0.340	0.051	0.299	12.33
14	–	–	–	–	–	–	–
15	0.068	0.591	0.060	0.400	0.058	0.357	14.42
16	–	–	–	–	–	–	–
17	0.057	0.648	0.055	0.455	0.054	0.411	13.68
18	–	–	–	–	–	–	–
19	0.062	0.710	0.065	0.520	0.065	0.475	16.62
20	–	–	–	–	–	–	–
21	–	–	–	–	–	–	–
22	0.060	0.770	0.070	0.590	0.070	0.546	18.60
23	–	–	–	–	–	–	–
24	–	–	–	–	–	–	–
25	0.094	0.864	0.120	0.710	0.121	0.667	33.09
+26	0.136	1.000	0.29	1.000	0.333	1.000	57.39**
							226.18

*γ_{API} = 34.6 °API and reservoir T = 126°F.

**Assumed C_{26}^+ = C_{30}.

Table 6.27—Abernaty-Collins crude* compositional summary based on *ASTM D-86* distillation, Problem 6.20.

Single Carbon Number	Mole Fraction	Cumulative Mole Fraction	Volume Fraction	Cumulative Volume Fraction	Weight Fraction	Cumulative Weight Fraction	Weight
5	–	–	–	–	–	–	–
6	–	–	–	–	–	–	–
7	–	–	–	–	–	–	–
8	0.052	0.052	0.030	0.030	0.026	0.026	5.93
9	0.143	0.195	0.090	0.120	0.079	0.105	18.30
10	0.137	0.332	0.095	0.215	0.086	0.191	19.45
11	0.087	0.419	0.065	0.280	0.061	0.251	13.57
12	–	–	–	–	–	–	–
13	0.074	0.493	0.060	0.340	0.057	0.308	13.62
14	0.079	0.572	0.070	0.410	0.068	0.377	15.64
15	–	–	–	–	–	–	–
16	0.072	0.644	0.070	0.480	0.069	0.446	16.27
17	–	–	–	–	–	–	–
18	0.070	0.714	0.075	0.555	0.075	0.521	17.28
19	–	–	–	–	–	–	–
20	–	–	–	–	–	–	–
21	0.081	0.795	0.095	0.650	0.097	0.617	23.98
22	–	–	–	–	–	–	–
23	–	–	–	–	–	–	–
24	0.095	0.890	0.125	0.775	0.127	0.745	32.11
25	–	–	–	–	–	–	–
+26	0.110	1.000	0.225	1.000	0.255	1.000	46.42**
							223.1

*γ_{API} = 38.8 °API and reservoir T = 102°F.

**Assumed C_{26}' = C_{30}.

Table 6.28—Johanning B crude* compositional summary based on *ASTM D-86* distillation, Problem 6.20.

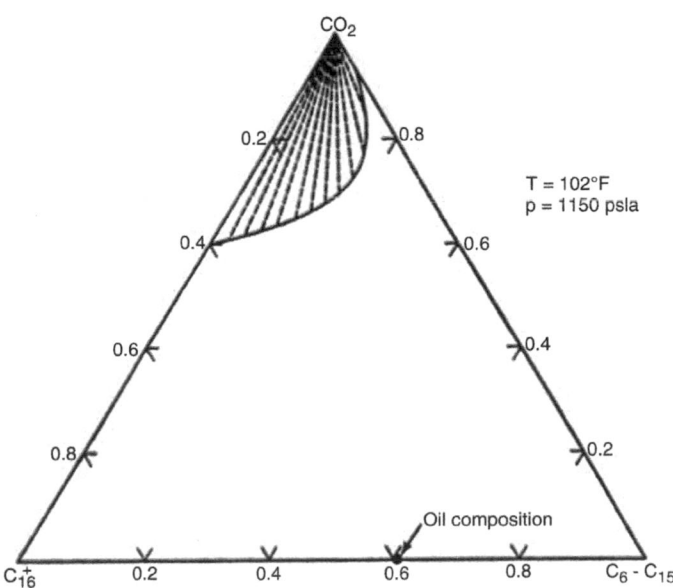

Fig. 6.124—Calculated ternary phase diagram for Johanning B crude oil (p = 1,150 psia and T = 102°F), Problem 6.21 (compositions in mole fractions).

μ_o (cp)	2.0
μ_w (cp)	0.7
μ_s (cp)	0.25
Saturations	
Connate water saturation, S_{cw}	0.2
ROS to miscible flood, S_{orm}	0.1
ROS to waterflood, S_{orw}	0.27
Oil saturation in oil bank, S_{ob}	0.65
Water saturation in oil bank, S_{wt}	0.35
Residual water saturation to solvent, S_{srw}	0.17
Relative Permeabilities	
K_{rw} at S_{orw}	0.30
k_{rs} at S_{orm}	0.70
k_{ro} at S_{cw}	0.50

Table 6.29—Rock and fluid properties, Problem 6.23.

Nomenclature

A = area spacing of pattern (i.e., area covered by pattern), L^2, acres

B_o = oil FVF, L^3/L^3, RB/STB

B_{oi} = oil FVF at original conditions, L^3/L^3, RB/STB

C = concentration, m/L^3, lbm mol/ft^3

d = core diameter, L, ft

E_r = error, %

E = effective viscosity ratio, Koval model, Eq. 6.16

E_A = areal displacement efficiency

E_{Aob} = areal sweep efficiency of mobile oil bank ahead of injected solvent or water bank

E_{As} = areal sweep efficiency of injected solvent

E_{Aw} = areal sweep efficiency of injected water

E_D = microscopic displacement efficiency

E_I = vertical displacement efficiency

E_{RM} = volumetric hydrocarbon displacement efficiency in miscible flood

E_{RW} = volumetric hydrocarbon displacement efficiency in waterflood

E_V = volumetric displacement efficiency

E_{Vob} = volumetric sweep efficiency of mobile oil bank ahead of injected solvent bank

E_{Vs} = volumetric sweep efficiency of injected solvent

E_{Vsmax} = volumetric sweep efficiency of leading edge of solvent bank at time that last of solvent is left as residual phase

E_{Vw} = volumetric sweep efficiency of injected water bank

F = degrees of freedom, phase behavior

F_{imp} = correction factor for MMP, Eqs. 6.6 and 6.7

G = dimensionless group defined by Eq. 6.20

h = thickness of core, model, or reservoir, L, ft

i = injection rate, L^3/t, B/D

k = permeability, L^2, md

k_H = horizontal permeability, L^2, md

k_r = relative permeability

k_{roi} = relative permeability of oil phase at start of solvent injection, Eq. 6.25

k_{rwi} = relative permeability of water phase at start of solvent injection, Eq. 6.25

k_V = vertical permeability, L^2, md

K_t = transverse dispersion coefficient, L^2/t, ft^2/hr

K_w = Watson characterization factor

L = liquid composition on binodal curve or a liquid phase

L = length of core or reservoir, L, ft

ΔL = length of viscous finger, L, ft

L_n = specific liquid composition on binodal curve

m = mass injected, m, lbm

M = mobility ratio; reduces to viscosity ratio in miscible displacement

M_e = effective mobility ratio

M_{tert} = mobility ratio for tertiary miscible displacement, Eq. 6.26

M_w	=	molecular weight
$\bar{M}_w$	=	average molecular weight
$\bar{M}_{wC_{5+}}$	=	average molecular weight of C_{5+} component in oil
M_{wi}	=	molecular weight of Component i in mixture, m/m, lbm/lbm mol
n	=	number of different chemical components in mixture
N	=	OOIP, or oil at initiation of process, L^3, STB
N_{ca}	=	capillary number, dimensionless group
N_p	=	oil produced, L^3, STB
N_{Pet}	=	transverse Péclet number, dimensionless group, Eq. 6.17
p	=	pressure, m/Lt^2, psi
p_c	=	critical pressure, m/Lt^2, psi
p_{ci}	=	critical pressure of Component i in mixture, m/Lt^2, psia
P_{CO_2}	=	MMP for pure CO_2, Eq. 6.4, m/Lt^2, psia
$P_{CO_2}-imp$	=	MMP for CO_2 solvent that contains impurities, Eq. 6.8, m/Lt^2, psia
P_{cp}	=	pseudocritical pressure of mixture, m/Lt^2, psi
p_m	=	miscibility pressure, vaporizing-gas displacement, Eq. 6.2, m/Lt^2, psi
p_{mm}	=	MMP, m/Lt_2, psi
p_r	=	reduced pressure, dimensionless
p_s	=	saturation pressure, m/Lt_2, psi
p_{so}	=	saturation pressure of oil, Eq. 6.2, m/Lt_2, psi
P	=	number of phases
R	=	gas constant, 10.73 $(psi-ft^3)/(°R-lbm-mol)$
$R_{v/g}$	=	viscous-to-gravity ratio, dimensionless group, Eq. 6.19
S	=	saturation, L^3/L^3, volume fraction
S_{cw}	=	connate water saturation, immobile water, L^3/L^3, volume fraction
S_{ob}	=	oil saturation in mobile oil bank, L^3/L^3, volume fraction
S_{oi}	=	original oil saturation, L^3/L^3, volume fraction
S_{or}	=	ROS, L^3/L^3, volume fraction
S_{orm}	=	ROS at end of miscible displacement, L^3/L^3, volume fraction
S_{orw}	=	ROS at end of waterflood, L^3/L^3, volume fraction
S_s	=	saturation of solvent in solvent bank, L^3/L^3, volume fraction
S_{srw}	=	residual solvent saturation remaining after immiscible displacement of solvent by water, L^3/L^3, volume fraction

S_{wi}	=	irreducible water saturation, L^3/L^3, volume fraction
S_{wt}	=	water saturation in mobile oil or solvent bank, L^3/L^3, volume fraction
t	=	time, t, hours or days
T	=	temperature, T, °F or °R
T_c	=	critical temperature, T, °R or K
T_{ci}	=	critical temperature of Component i in mixture, T, °R or K
T'_{cm}	=	weight-averaged critical temperature of mixture, Eq. 6.5, T, °F
T_{cp}	=	pseudocritical temperature of mixture, T, °R or K
T_r	=	reduced temperature, dimensionless
u	=	Darcy velocity, L/t, ft/hr, ft/D, or B/D-ft^2
v	=	average pore (interstitial) velocity, L/t, ft/hr or ft/D
V	=	vapor composition on binodal curve or vapor phase
V	=	volume, L^3, ft^3 or bbl
V_i	=	volume injected, L^3, ft^3
V_n	=	specific vapor composition on binodal curve
V_p	=	PV, L^3, ft^3 or bbl
V_{pD}	=	displaceable reservoir PVs or displaceable PV injected, L^3/L^3, volume fraction
V_{pDob}	=	pseudodisplaceable PVs occupied by mobile oil bank, Stalkup model, L^3/L^3, volume fraction
V'_{pDob}	=	pseudodisplaceable PVs of oil bank at time that last of solvent slug is left as residual phase, Stalkup model, L^3/L^3, volume fraction
V_{pDs}	=	displaceable or pseudodisplaceable PVs of solvent injected, Stalkup model, L^3/L^3, volume fraction
V'_{pDs}	=	maximum displaceable PVs of solvent injected, Stalkup model, L^3/L^3, volume fraction
V_{pDw}	=	displaceable PVs of water injected, Stalkup model, L^3/L^3, volume fraction
V'_{pDw}	=	displaceable PVs of water injected at time that last of solvent slug is left as residual phase, Stalkup model, L^3/L^3, volume fraction
ΔV_{pDvob}	=	incremental contribution to pseudodisplaceable PVs from oil bank, Stalkup model, L^3/L^3, volume fraction
V_{ph}	=	HCPV
V_R	=	reservoir volume contacted, L^3, ft^3
V_{si}	=	total volume of miscible solvent injected, L^3, bbl

w_i	=	volume of water injected, L^3, bbl
$W_{C_{7+}}$	=	amount of C_{7+} hydrocarbons in an oil per lbm mol of oil, Eq. 6.3, m/m, lbm/lbm mol
W_i	=	amount of C_2 through C_6 hydrocarbon+CO_2 + H_2S in an oil per lbm mol of oil, Eq. 6.3, m/m, lbm/lbm mol
W_I	=	amount of C_1 hydrocarbon+N_2 in an oil per lbm mol of oil, Eq. 6.3, m/m, lbm/lbm mol
x	=	distance along displacement path, L, ft
X	=	mass or mole fraction
X_i	=	mole or mass fraction of Component i in mixture
X_{int}	=	mole fraction of intermediate components in an oil; C_2, C_3, C_4, CO_2, and H_2S, Eq. 6.4
X_{vol}	=	mole fraction of volatile components in oil; C_1+N_2, Eq. 6.4
y	=	vapor composition, m/m, mole fraction

z	=	composition, m/m, mole fraction
Z	=	gas compressibility factor
α	=	empirical parameter, Eq. 6.27
γ	=	specific gravity
γ_{API}	=	API gravity, °API
λ	=	width of viscous finger, L, ft
λ_c	=	critical wavelength for viscous fingers, L, ft
μ	=	viscosity, m/Lt, cp
v	=	specific volume, L^3/m, ft^3/lbm mol
ρ	=	density, m/L^3, lbm/ft^3 or g/cm^3
$\Delta\rho$	=	density difference between displacing and displaced phases, m/L^3, g/cm^3
ρ_i	=	density of Component i in mixture, m/L^3, lbm/ft^3
σ	=	parameter in viscosity correlation, Fig. 6.47
ϕ	=	porosity

Subscripts

con	=	convection	i	=	Component i
CO_2	=	carbon dioxide	mix	=	mixture of components
d	=	displaced phase	o	=	oil phase
disp	=	dispersion	s	=	solvent phase
D	=	displacing phase	w	=	water phase

References

Ahmadi, K. and Johns, R. T. 2008. Multiple Mixing-Cell Method for MMP Calculations. Presented at the SPE Annual Technical Conference and Exhibition, Denver, 21–24 September. SPE-116823-MS. https://doi.org/10.2118/116823-MS.

Alani, G. H. and Kennedy, H. T. 1960. Volumes of Liquid Hydrocarbons at High Temperatures and Pressures. In *Petroleum Transactions*, AIME, Vol. 219, 288–292. SPE-1399-G.

Alston, R. B., Kokolis, G. P., and James, C. F. 1985. CO_2 Minimum Miscibility Pressure: A Correlation for Impure CO_2 Streams and Live Oil Systems. *SPE J.* **25** (2): 268–274. SPE-11959-PA. https://doi.org/10.2118/11959-PA.

Awan, A. R., Teigland, R., and Kleppe, J. 2008. A Survey of North Sea Enhanced-Oil-Recovery Projects Initiated During the Years 1975 to 2005. *SPE Res Eval & Eng* **11** (3): 497–512. SPE-99546-PA. https://doi.org/10.2118/99546-PA.

Barati, R. B., Pennel, S., Matson, M. et al. 2016. Overview of CO_2 Injection and WAG Sensitivity in SACROC. Presented at the SPE Improved Oil Recovery Conference, Tulsa, 11–13 April. SPE-179569-MS. https://doi.org/10.2118/179569-MS.

Benham, A. L., Dowden, W. E., and Kunzman, W. J. 1960. Miscible Fluid Displacement—Prediction of Miscibility. In *Petroleum Transactions*, AIME, Vol. 219, 229–237. SPE-1484-G.

Besserer, G. J. and Robinson, D. B. 1975. Equilibrium-Phase Properties of Isopentane-Carbon Dioxide System. *J. Chem. Eng. Data* **20** (1): 93–96. https://doi.org/10.1021/je60064a011.

Bohn, S. 2014. SACROC Project Area Production and Economic Analysis. Internal Report, Kinder Morgan Company, Houston.

Brock, W. R. and Bryan, L. A. 1989. Summary Results of CO_2 EOR Field Tests, 1972–1987. Presented at the SPE Joint Rocky Mountain Regional/Low Permeability Reservoirs Symposium and Exhibition, Denver, 6–8 March. SPE-18977-MS. https://doi.org/10.2118/18977-MS.

Brown, G. G., Katz, D. L., Oberfell, G. G. et al. 1948. *Natural Gasoline and the Volatile Hydrocarbons*. Tulsa: Natural Gas Assn. of America.

Brownlee, M. H. and Sugg, L. A. 1987. East Vacuum Grayburg-San Andres Unit CO_2 Injection Project: Development and Results to Date. Presented at the SPE Annual Technical Conference and Exhibition, Dallas, Texas, USA, 27–30 September. SPE-16721-MS. https://doi.org/10.2118/16721-MS.

Campbell, B. T. and Orr, F. M. Jr. 1985. Flow Visualization for CO_2/Crude Oil Displacements. *SPE J.* **25** (5): 665–678. SPE-11958-PA. https://doi.org/10.2118/11958-PA.

Caudle, B. H. and Witte, M. D. 1959. Production Potential Changes During Sweep-Out in a Five-Spot System. *J Pet Technol* **12** (12): 63–65. SPE-1334-G. https://doi.org/10.2118/1334-G.

Christian, L. D., Shirer, J. A., Kimbel, E. L. et al. 1981. Planning a Tertiary Oil-Recovery Project for Jay/LEC Fields Unit. *J Pet Technol* **33** (8): 1535–1544. SPE-9805-PA. https://doi.org/10.2118/9805-PA.

Christiansen, R. L. and Haines, H. K. 1987. Rapid Measurement of Minimum Miscibility Pressure With the Rising-Bubble Apparatus. *SPE Res Eng* **2** (4): 523–527. SPE-13114-PA. https://doi.org/10.2118/13114-PA.

Claridge, E. L. 1972. Prediction of Recovery in Unstable Miscible Flooding. *SPE J* **12** (2): 143–155. SPE-2930-PA. https://doi.org/10.2118/2930-PA.

Clark, N. J., Shearin, H. M., Schultz, W. P. et al. 1958. Miscible Drive—Its Theory and Application. *J Pet Technol* **10** (6): 11–20. SPE-1036-G. https://doi.org/10.2118/1036-G.

Craig, F. F. Jr., Sanderlin, J. L., More, D. W. et al. 1957. A Laboratory Study of Gravity Segregation in Frontal Drives. In *Petroleum Transactions*, AIME, Vol. 2010, 275–282. SPE-676-G.

Crane, F. E., Kendall, H. A., and Gardner, G. H. F. 1963. Some Experiments on the Flow of Miscible Fluids of Unequal Density Through Porous Media. *SPE J.* **3** (4): 277–280. SPE-535-PA. https://doi.org/10.2118/535-PA.

Denny, D. 2009. Potential for Additional CO_2-Flood Projects in the US. *J Pet Technol* **61** (1): 55–57. SPE-0109-0055-JPT. https://doi.org/10.2118/0109-0055-JPT.

Dicharry, R. M., Perryman, T. L., and Ronquille, J. D. 1973. Evaluation and Design of a CO_2 Miscible Flood Project, SACROC Unit, Kelly-Snyder Field. *J Pet Technol* **25** (11): 1309–1318. SPE-4083-PA. https://doi.org/10.2118/4083-PA.

Dietz, D. N. 1953. A Theoretical Approach to the Problem of Encroaching by Bypassing Edge Water. *Proc.*, Academy Science Amsterdam, Vol. B56, 83.

Dodds, W. S., Stutzman, L. F., and Sollami, B. J. 1956. Carbon Dioxide Solubility in Water. *Ind. Eng. Chem. Chem. Eng. Data Series* **1** (1): 92–95. https://doi.org/10.1021/i460001a018.

Doscher, T. M. and Gharib, S. 1983. Physically Scaled Model Studies Simulating the Displacement of Residual Oil by Miscible Fluids. *SPE J.* **23** (3): 440–446. SPE-8896-PA. https://doi.org/10.2118/8896-PA.

Fussell, D. D. and Yanosik, J. L. 1978. An Iterative Sequence for Phase-Equilibria Calculations Incorporating the Redlich-Kwong Equation of State. *SPE J.* **18** (3): 173–182. SPE-6050-PA. https://doi.org/10.2118/6050-PA.

Fussell, L. T. and Fussell, D. D. 1979. An Iterative Technique for Compositional Reservoir Models. *SPE J.* **19** (4): 211–220. SPE-6891-PA. https://doi.org/10.2118/6891-PA.

Gardner, J. W. and Ypma, J. G. J. 1984. An Investigation of Phase-Behavior/Macroscopic-Bypassing Interaction in CO_2 Flooding. *SPE J.* **24** (5): 508–520. SPE-10686-PA. https://doi.org/10.2118/10686-PA.

Gardner, J. W., Orr, F. M., and Patel, P. D. 1981. The Effect of Phase Behavior on CO_2-Flood Displacement Efficiency. *J Pet Technol* **33** (11): 2067–2081. SPE-8367-PA. https://doi.org/10.2118/8367-PA.

Glasstone, S. 1946. *Textbook of Physical Chemistry*, second edition, 791. New York City, New York: Van Nostrand Reinhold Co. Inc.

Gonzalez, M. H. and Lee, A. L. 1968. Graphical Viscosity Correlation for Hydrocarbons. *AIChE J.* **14** (2): 242–44. https://doi.org/10.1002/aic.690140208.

Goodwin, J. M. 1989. Infill Drilling and Pattern Modification in the Means San Andres Unit. *Proc.*, 8th IOR Conference, Tertiary Oil Recovery Project, University of Kansas, Wichita, Kansas, 8–9 March, 84–94.

Green, D. W. and Swift, G. W. 1985. Development of a Method for Evaluating Carbon Dioxide Miscible Flooding Prospects. Final report, DOE/BC/10122-50, US Department of Energy, Bartlesville Project Office, Bartlesville, Oklahoma (March 1985).

Harpole, K. J. and Hallenbeck, L. D. 1996. East Vacuum Grayburg San Andres Unit CO2 Flood Ten Year Performance Review: Evolution of a Reservoir Management Strategy and Results of WAG Optimization. Presented at the SPE Annual Technical Conference and Exhibition, Denver, 6–9 October. SPE-36710-MS. https://doi.org/10.2118/36710-MS.

Heller, J. P., Kovarik, F. S., and Taber, J. J. 1987. Improvement of CO_2 Flood Performance. Annual report, DOE/MC/21136-10, US Department of Energy Morgantown Energy Technology Center, Morgantown, West Virginia (May 1987).

Henry, R. L. and Metcalfe, R. S. 1983. Multiple-Phase Generation During Carbon Dioxide Flooding. *SPE J.* **23** (4): 595–601. SPE-8812-PA. https://doi.org/10.2118/8812-PA.

Holm, L. W. and Josendal, V. A. 1974. Mechanisms of Oil Displacement by Carbon Dioxide. *J Pet Technol* **26** (12): 1427–1438. SPE-4736-PA. https://doi.org/10.2118/4736-PA.

Holm, L. W. and Josendal, V. A. 1980. Discussion of Determination and Prediction of CO_2 Minimum Miscibility Pressures. *J Pet Technol* **32** (1): 870–871 (discussion follows paper). SPE-7477-PA. https://doi.org/10.2118/7477-PA.

Holm, L. W. and Josendal, V. A. 1982. Effect of Oil Composition on Miscible-Type Displacement by Carbon Dioxide. *SPE J.* **22** (1): 87–98. SPE-8814-PA. https://doi.org/10.2118/8814-PA.

Hutchinson, C. A. and Braun, P. H. 1961. Phase Relations of Miscible Displacement in Oil Recovery. *AIChE J.* **7** (1): 64–72. https://doi.org/10.1002/aic.690070117.

Jessen, K, and Orr, F. M. Jr. 2008. On Interfacial-Tension Measurements To Estimate Minimum Miscibility Pressures. *SPE Res Eval & Eng* **11** (5): 933–939. SPE-110725-PA. https://doi.org/10.2118/110725-PA.

Johns, R. T., Ahmadi, K., and Zhou, D. et al. 2010. A Practical Method for Minimum-Miscibility-Pressure Estimation of Contaminated CO_2 Mixtures. *SPE Res Eval & Eng* **13** (5): 764–772. SPE-124906-PA. https://doi.org/10.2118/124906-PA.

Johnson, J. P. and Pollin, J. S. 1981. Measurement and Correlation of CO_2 Miscibility Pressures. Presented at the SPE/DOE Enhanced Oil Recovery Symposium, Tulsa, 5–8 April. SPE-9790-MS. https://doi.org/10.2118/9790-MS.

Kane, A. V. 1979. Performance Review of a Large-Scale CO_2-WAG Enhanced Recovery Project, SACROC Unit—Kelly-Snyder Field. *J Pet Technol* **31** (2): 217–231. SPE-7091-PA. https://doi.org/10.2118/7091-PA.

Katz, D. L., Cornell, D., Vary, J. A. et al. 1959. *Handbook of Natural Gas Engineering*. New York City, New York: McGraw-Hill Book Co.

Kay, W. B. 1938. Liquid-Vapor Phase Equilibrium Relations in the Ethane—*n*-Heptane System. *Ind. Eng. Chem.* **30** (4): 459–465. https://doi.org/10.1021/ie50340a023.

Kennedy, J. T. and Thodos, G. 1961. Transport Properties of Carbon Dioxide. *AIChE J.* **7** (4): 625–631. https://doi.org/10.1002/aic.690070419.

Klins, M. A. 1984. *Carbon Dioxide Flooding*. Boston, Massachusetts: Intl. Human Resources Development Corp.

Koch, H. A. Jr. and Slobod, R. L. 1957. Miscible Slug Process. In *Petroleum Transactions*, AIME Vol. 210, 40–47. SPE-714-G.

Koval, E. J. 1963. A Method for Predicting the Performance of Unstable Miscible Displacement in Heterogeneous Media. *SPE J.* **3** (2): 145–154. SPE-450-PA. https://doi.org/10.2118/450-PA.

Kremesec, V. J. Jr. and Sebastian, H. M. 1988. CO_2 Displacements of Reservoir Oils From Long Berea Cores: Laboratory and Simulation Results. *SPE Res Eng* **3** (2): 496–504. SPE-14306-PA. https://doi.org/10.2118/14306-PA.

Kuuskraa, V. and Wallace, M. 2014. CO_2-EOR Set for Growth as New CO_2 Supplies Emerge. *Oil and Gas Journal* **112** (5): 92–105.

Langston, E. P., Shirer, J. A., and Nelson, D. E. 1981. Innovative Reservoir Management - Key to Highly Successful Jay/LEC Waterflood. *J Pet Technol* **33** (5): 783–791. SPE-9476-PA. https://doi.org/10.2118/9476-PA.

Langston, M. V., Hoadley, S. F., and Young, D. N. 1988. Definitive CO_2 Flooding Response in the SACROC Unit. Presented at the SPE/DOE Enhanced Oil Recovery Symposium, Tulsa, 17–20 April. SPE-17321-MS. https://doi.org/10.2118/17321-MS.

Lawrence, J. J., Maer, N. K., Stern, D. et al. 2002. Jay Nitrogen Tertiary Recovery Study: Managing a Mature Field. Presented at the Abu Dhabi International Petroleum Exhibition and Conference, Abu Dhabi, 13–18 October. SPE-78527-MS. https://doi.org/10.2118/78527-MS.

Linroth, M. 2012. Rejuvenating a Mature EOR Asset: Miscible CO_2 Flooding at SACROC. Presented at the 18th Annual CO_2 Flooding Conference, Midland, Texas, USA, 3–7 December.

Linroth, M. A. and Rickard, A. E. 2014. Pressure and Rate Rebalancing to Improve Recovery in a Miscible CO_2 EOR Project. Presented at the SPE Improved Oil Recovery Symposium, Tulsa, 12–16 April. SPE-169175-MS. https://doi.org/10.2118/169175-MS.

Lohrenz, J., Bray, B. G., and Clark, C. R. 1964. Calculating Viscosities of Reservoir Fluids From Their Compositions. *J Pet Technol* **16** (10): 1171–1176. SPE-915-PA. https://doi.org/10.2118/915-PA.

Magruder, J. B., Stiles, L. H., and Yelverton, T. D. 1990. Review of the Means San Andres Unit CO_2 Tertiary Project. *J Pet Technol* **42** (5): 638–644. SPE-17349-PA. https://doi.org/10.2118/17349-PA.

Mansoori, J. and Gupta, S. P. 1988. An Interpretation of the Displacement Behavior of Rich Gas Drives Using an Equation-of-State Compositional Model. Presented at the SPE Annual Technical Conference and Exhibition, Houston, 2–5 Houston. SPE-18061-MS. https://doi.org/10.2118/18061-MS.

McCain, W. D. 1973. *The Properties of Petroleum Fluids*. Tulsa: PennWell Publishing Co.

McGuire, P. L., Spence, A. P., and Redman, R. S. 2000. Performance Evaluation of a Mature Miscible Gas Flood at Prudhoe Bay. Presented at the SPE/DOE Improved Oil Recovery Symposium, Tulsa, 3–5 April. SPE-59326-MS. https://doi.org/10.2118/59326-MS.

Melster, T., Valenti, N., Geiger, P. et al. 2014. Right-Sizing the Jay/LEC Field - Commercial 30 Year EOR Project. Presented at the SPE Improved Oil Recovery Symposium, Tulsa, 12–16 April. SPE-169080-MS. https://doi.org/10.2118/169080-MS.

Metcalfe, R. S. and Yarborough, L. 1979. The Effect of Phase Equilibria on the CO_2 Displacement Mechanism. *SPE J.* **19** (4): 242–252. SPE-7061-PA. http://dx.doiorg/10.2118/7061-PA.

Moffitt, P., Pecore, D., Trees, M. et al. 2015. East Vacuum Grayburg San Andres Unit, 30 Years of CO2 Flooding: Accomplishments, Challenges and Opportunities. Presented at the SPE Annual Technical Conference and Exhibition, Houston, 28–30 September. SPE-175000-MS. https://doi.org/10.2118/175000-MS.

Mohan, H., Carolus, M. J., and Biglarbigi, R. 2008. The Potential for Additional Carbon Dioxide Flooding Projects in the United States. Presented at the SPE Symposium on Improved Oil Recovery, Tulsa, 19–23 April. SPE-113975-MS. https://doi.org/10.2118/113975-MS.

Moore, J. S. and Clark, G. C. 1988. History Match of the Maljamar CO_2 Pilot Performance. Presented at the SPE Enhanced Oil Recovery Symposium, Tulsa, 16–21 April. SPE-17323-MS. https://doi.org/10.2118/17323-MS.

Mungan, N. 1981. Carbon Dioxide Flooding—Fundamentals. *J Can Pet Technol* **20** (1): 87–92. PETSOC-81-01-03. https://doi.org/10.2118/81-01-03.

Nagel, R. G., Hunter, B. E., Peggs, J. K. et al. 1990. Tertiary Application of a Hydrocarbon Miscible Flood: Rainbow Keg River B Pool. *SPE Res Eng* **5** (3): 301–08. SPE-17355-PA. https://doi.org/10.2118/17355-PA.

National Petroleum Council. 1976. Enhanced Oil Recovery. Washington, DC (December 1976).

Negahban, S., Shiralkar, G. S., and Gupta, S. P. 1990. Simulation of the Effects of Mixing in Gasdrive Core Tests of Reservoir Fluids. *SPE Res Eng* **5** (3): 402–408. SPE-17377-PA. https://doi.org/10.2118/17377-PA.

Orr, F. M. Jr. and Silva, M. K. 1983. Equilibrium Phase Compositions of CO_2/Hydrocarbon Mixtures—Part 1: Measurement by a Continuous Multiple-Contact Experiment. *SPE J.* **23** (2): 272–280. SPE-10726-PA. https://doi.org/10.2118/10726-PA.

Orr, F. M. Jr., Silva, M. K., and Cheng-Li, L. 1983. Equilibrium Phase Compositions of CO_2/Crude Oil Mixtures—Part 2: Comparison of Continuous Multiple-Contact and Slim-Tube Displacement Tests. *SPE J.* **23** (2): 281–291. SPE-10725-PA. https://doi.org/10.2118/10725-PA.

Orr, F. M. Jr., Silva, M. K., Lien, C. L. et al. 1982. Laboratory Experiments To Evaluate Field Prospects for CO_2 Flooding. *J Pet Technol* **34** (4): 888–898. SPE-9534-PA. https://doi.org/10.2118/9534-PA.

Pathak, P., Fitz, D. E., and Babcock, K. P. 2011. Residual Oil Saturation Determination for EOR Projects in a Mature West Texas Carbonate Field. Presented at the SPE Enhanced Oil Recovery Conference, Kuala Lampur, Malaysia, 19–21 July. SPE-145229-MS. https://doi.org/10.2118/145229-MS.

Pennell, S., Linroth, M., Schoeling. L. et al. 2015. SACROC Unit. Internal Report, Kinder Morgan Company, Houston.

Perkins, T. K., Johnston, O. C, and Hoffman, R. N. 1965. Mechanics of Viscous Fingering in Miscible Systems. *SPE J.* **5** (4): 301–317. SPE-1229-PA. https://doi.org/10.2118/1229-PA.

Peterson, C. A. 2015. Personal communication (April 2015).

Peterson, C. A., Pearson, E. J., Chodur, V. T. et al. 2012. Beaver Creek Madison CO_2 Enhanced Oil Recovery Project Case History; Riverton, Wyoming. Presented at the SPE Improved Oil Recovery Symposium, Tulsa, 14–18 April. SPE-152862-MS. https://doi.org/10.2118/152862-MS.

Pipes, J. W. and Schoeling, L. G. 2014. Performance Review of Gel-Polymer Treatments in a Miscible CO_2 Enhanced Recovery Project, SACROC Unit, Kelly-Snyder Field. Presented at the SPE Improved Oil Recovery Symposium, Tulsa, 12–16 April. SPE-169175-MS. https://doi.org/10.2118/169176-MS.

Pollock, C. B. 1960. Beaver Creek Madison, Wyoming's Deepest Water Injection Project. *J Pet Technol* **12** (1): 39–41. SPE-1226-G. https://doi.org/10.2118/1226-G.

Pozzi, A. L. and Blackwell, R. J. 1963. Design of Laboratory Models for Study of Miscible Displacement. *SPE J.* **3** (1): 28–40. SPE-445-PA. https://doi.org/10.2118/445-PA.

Raimondi, P. and Torcaso, M. A. 1964. Distribution of the Oil Phase Obtained Upon Imbition of Water. *SPE J.* **4** (1): 49–55. SPE-570-PA. https://doi.org/10.2118/570-PA.

Raines, M. 2005. Tertiary Recovery at the SACROC Unit. In *Oil and Gas Fields in West Texas*, Vol VIII. Midland, Texas: West Texas Geological Society.

Redlich, O. and Kwong, J. N. S. 1949. On the Thermodynamics of Solutions. V. An Equation of State. Fugacities of Gaseous Solutions. *Chem. Rev.* **44** (1): 233–244. https://doi.org/10.1021/cr60137a013.

Rossini, F. D. (ed.) 1953. *Selected Values of Physical and Thermodynamic Properties of Hydrocarbons*. Pittsburgh, Pennsylvania: Carnegie Press.

Sage, B. H. and Lacey, W. N. 1939. *Volumetric and Phase Behavior of Hydrocarbons*. Stanford, California: Stanford University Press.

Sage, B. H. and Lacey, W. N. 1955. *Some Properties of the Lighter Hydrocarbons, Hydrogen Sulfide, and Carbon Dioxide*, Monograph on API Research Project 37. Dallas, Texas: American Petroleum Institute.

Sebastian, H. M., Wenger, R. S., and Renner, T. A. 1985. Correlation of Minimum Miscibility Pressure for Impure CO_2 Streams. *J Pet Technol* **37** (11): 2076–2082. SPE-12648-PA. https://doi.org/10.2118/12648-PA.

Shelton, J. L. and Schneider, F. N. 1975. The Effects of Water Injection on Miscible Flooding Methods Using Hydrocarbons and Carbon Dioxide. *SPE J.* **15** (3): 217–226. SPE-4580-PA. https://doi.org/10.2118/4580-PA.

Shelton, J. L. and Yarborough, L. 1977. Multiple Phase Behavior in Porous Media During CO_2 or Rich-Gas Flooding. *SPE J.* **29** (9): 1171–1178. SPE-5827-PA. https://doi.org/5827-PA.

Sibbald, L. R., Novosad, Z., and Costain, T. G. 1991. Methodology for the Specification of Solvent Blends for Miscible Enriched Gas Drives (includes associated papers 23836, 24319, 24471, and 24548). *SPE Res Eng* **6** (3): 373–378. SPE-20205-PA. https://doi.org/10.2118/20205-PA.

Simon, R., Rosman, A., and Zana, E. 1978. Phase-Behavior Properties of CO_2-Reservoir Oil Systems. *SPE J.* **18** (1): 20–26. SPE-6387-PA. https://doi.org/10.2118/6387-PA.

Stalkup, F. I. 1970. Displacement of Oil by Solvent at High Water Saturation. *SPE J.* **10** (4): 337–348. SPE-2419-PA. https://doi.org/10.2118/2419-PA.

Stalkup, F. I. 1983a. *Miscible Displacement*, Vol. 8. Monograph Series. Richardson, Texas: SPE.

Stalkup, F. I. 1983b. Status of Miscible Displacement. *J Pet Technol* **35** (4): 815–826. SPE-9992-PA. https://doi.org/10.2118/9992-PA.

Stalkup, F. I. 1987. Displacement Behavior of the Condensing/Vaporizing Gas Drive Process. Presented at the SPE Annual Technical Conference and Exhibition, Dallas, Texas, 27–30 September. SPE-16715-MS. https://doi.org/10.2118/16715-MS.

Standing, M. B. and Katz, D. L. 1942. Density of Natural Gases, In *Transactions of the AIME*, Vol. 146, 140–149. SPE-942140-G. https://doi.org/10.2118/942140-G.

Stewart, P. B. and Munjal, P. K. 1970. Solubility of Carbon Dioxide in Pure Water, Synthetic Sea Water, and Synthetic Sea Water Concentrates at -5° to 25°C and 10- to 45-Atm. Pressure. *J. Chem. Eng. Data* **15** (1): 67–71. https://doi.org/10.1021/je60044a001.

Stiles, L. H. and Magruder, J. B. 1992. Reservoir Management in the Means San Andres Unit. *J Pet Technol* **44** (4): 469–475. SPE-20751-PA. https://doi.org/10.2118/20751-PA.

Stiles, L. H., Chiquito, R. M., George, C. J. et al. 1983. Design and Operation of a CO_2 Tertiary Pilot: Means San Andres Unit. Presented at the SPE Annual Technical Conference and Exhibition, San Francisco, California, USA, 5–8 October. SPE-11987-MS. https://doi.org/10.2118/11987-MS.

Sydansk, R. D. and Southwell, G. P. 2000. More Than 12 Years' Experience With a Successful Conformance-Control, Polymer-Gel Technology. *SPE Prod & Fac* **15** (4): 270–278. SPE-66558-PA. https://doi.org/10.2118/66558-PA.

Tiffin, D. L. and Yellig, W. F. 1983. Effects of Mobile Water on Multiple-Contact Miscible Gas Displacement. *SPE J.* **23** (3): 447–455. SPE-10687-PA. https://doi.org/10.2118/10687-PA.

Tiffin, D. L., Sebastian, H. M., and Bergman, D. F. 1991. Displacement Mechanism and Water Shielding Phenomena for a Rich-Gas/Crude-Oil System. *SPE Res Eng* **6** (2): 193–199. SPE-17374-PA. https://doi.org/10.2118/17374-PA.

Todd, M. R., Cobb, W. M., and McCarter, E. D. 1982. CO_2 Flood Performance Evaluation for the Cornell Unit, Wasson San Andres Field. *J Pet Technol* **34** (10): 2271–2282. SPE-10292-PA. https://doi.org/10.2118/10292-PA.

Turek, E. A., Metcalfe, R. S., Yarborough, L. et al. 1984. Phase Equilibria in CO_2— Multicomponent Hydrocarbon Systems: Experimental Data and an Improved Prediction Technique. *SPE J.* **24** (3): 308–324. SPE-9231-PA. https://doi.org/10.2118/9231-PA.

US Department of Energy (US DOE). 1986. Fossil Energy: Supporting Technology for Enhanced Oil Recovery—CO_2 Miscible Flood Predictive Model. DOE/BC-86/12/SP, US DOE, Bartlesville, Oklahoma, and Ministry of Energy of Mines, Venezuela (December 1986).

US Government. 1978. Enhanced Oil Recovery Potential in the United States. NTIS Order #PB-276594, Congress of the United States, Office of Technology Assessment, Washington, DC (January 1978).

Valenti, N. 2014. Personal communication. Quantum Resources Management LLC (16 October 2014).

Wang, Y. and Orr, F. M. Jr. 1998. Calculation of Minimum Miscibility Pressure. Presented at the SPE/DOE Improved Oil Recovery Symposium, Tulsa, 19–22 April. SPE-39683-MS. https://doi.org/10.2118/39683-MS.

Wilkinson, J. R., Genetti, D. B., Henning, G. T. et al. 2004. Lessons Learned from Mature Carbonates for Application to Middle East Fields. Presented at the Abu Dhabi International Petroleum Exhibition and Conference, Abu Dhabi, 10–13 October. SPE-88770-MS. https://doi.org/10.2118/88770-MS.

Withjack, E. M. and Akervoll, I. 1988. Computed Tomography Studies of 3D Miscible Displacement Behavior in a Laboratory Five-Spot Model. Presented at the SPE Annual Technical Conference and Exhibition, Houston, 2–5 October. SPE-18096-MS. https://doi.org/10.2118/18096-MS.

Yarborough, L. and Smith, L. R. 1970. Solvent and Driving Gas Compositions for Miscible Slug Displacement. *SPE J.* **10** (3): 298–310. SPE-2543-PA. https://doi.org/10.2118/2543-PA.

Yellig, W. F. 1982. Carbon Dioxide Displacement of a West Texas Reservoir Oil. *SPE J.* **22** (6): 805–815. SPE-9785-PA. https://doi.org/10.2118/9785-PA.

Yellig, W. F. and Metcalfe, R. S. 1980. Determination and Prediction of CO_2 Minimum Miscibility Pressures (includes associated paper 8876). *J Pet Technol* **32** (1): 160–168. SPE-7477-PA. https://doi.org/10.2118/7477-PA.

Yuan, H. and Johns, R. T. 2005. Simplified Method for Calculation of Minimum Miscibility Pressure or Enrichment. *SPE J.* **10** (4): 416–425. SPE-77381-PA. https://doi.org/10.2118/77381-PA.

Zhou, D. and Orr, F. M Jr. 1998. Analysis of Rising-Bubble Experiments to Determine Minimum Miscibility Pressures. *SPE J.* **3** (1): 19–25. SPE-30786-PA. https://doi.org/10.2118/30786-PA.

Zick, A. A. 1986. A Combined Condensing/Vaporizing Mechanism in the Displacement of Oil by Enriched Gases. Presented at the SPE Annual Technical Conference and Exhibition, New Orleans, 5–8 October. SPE-15493-MS. https://doi.org/10.2118/15493-MS.

SI Metric Conversion Factors

acre	×	4.046 873	E–01	=	ha
°API		141.5/(131.5 + °API)		=	g/cm³
atm	×	1.013 250*	E+05	=	Pa
bbl	×	1.589 873	E–01	=	m³
cp	×	1.0*	E–03	=	Pa·s
ft	×	3.048*	E–01	=	m
ft³	×	2.831 685	E–02	=	m³
°F		(°F–32)/1.8		=	°C
in.	×	2.54*	E+00	=	cm
lbm	×	4.535 924	E–01	=	kg
psi	×	6.894 757	E+00	=	kPa
°R		°R/1.8		=	K

*Conversion factor is exact.

Chapter 7

Chemical Flooding

7.1 Introduction

This chapter discusses processes to improve recovery efficiency primarily through the use of a displacing fluid that has a low interfacial tension (IFT) with the displaced crude oil. The effect of IFT on recovery in a displacement process is shown in **Fig. 7.1** (Stegemeier 1977), where residual oil is correlated as a function of capillary number, N_{ca}^*. (Section 2.4.3 describes this relationship.) As covered in Chapter 2, IFT must be reduced from 10 to 30 dynes/cm in a typical waterflood to approximately 10^{-3} dynes/cm before a large reduction in the waterflood residual oil saturation (ROS) is achieved. Significant reduction in S_{or} is possible with an IFT of approximately 10^{-2} dynes/cm.

The process treated in depth, called the micellar/polymer process, is based on injection of a chemical system that contains surface-active agents (i.e., surfactants). The chemical system, usually called a micellar or microemulsion solution, typically contains the following components: surfactant, cosurfactant (which may be alcohol or another surfactant), hydrocarbon, water, and electrolytes. A water soluble polymer is added to increase viscosity and to attain mobility control during the displacement process. **Table 7.1** (Poettmann 1974) shows example ranges of composition. In the process, injection of the surfactant or micellar solution usually is followed by injection of an aqueous solution to which polymer has been added to maintain mobility control—thus the name micellar/polymer process.

Variations of the surfactant/polymer process have gone by different names. When the surfactant solution contains a hydrocarbon, it was called a micellar solution, microemulsion, surfactant, low-tension, soluble-oil, and chemical flooding. The differences are in the chemical composition and the volume of the primary slug injected. For example, the low-tension flood uses a relatively small concentration of surfactant (approximately 2 to 3 wt%) and usually no hydrocarbon. An alkaline chemical often is added to promote IFT reduction and to reduce surfactant adsorption. The slug volume usually is much larger than the volume for a process that uses a relatively high surfactant concentration.

A different but related process is alkaline polymer flooding. Some crude oils contain acidic constituents (high acid number), which react with alkaline solutions to create surfactants, sometimes referred to as soaps. In this process, the high-pH injected solution reacts with the crude oil to form surfactants in situ. Here, several mechanisms may be effective in promoting oil recovery. Decreasing the IFT between the displacing solution and the oil is one of those mechanisms. In standard alkaline flooding, no surfactant is injected; however, a variation on the process includes a surfactant injected with the alkaline chemical to improve the process, and polymer is added for mobility control. This process is called an alkaline/surfactant/polymer (ASP) flood. When the crude oil does not contain acidic constituents (low acid number), the alkaline solution may reduce surfactant retention. In this case, the differences between alkaline flooding and micellar/polymer flooding become blurred.

This chapter emphasizes the micellar/polymer process. This process obtains ultralow IFTs ($< 10^{-3}$ dynes/cm) for a specific crude-oil/reservoir-brine by formulating a microemulsion system. A surfactant is the principal component that causes the system to exhibit low IFT. The limited range of compositions where ultralow IFTs can be obtained is controlled by the surfactant type, cosurfactant (if needed), brine salinity, crude oil composition, and reservoir temperature. A unique formulation must be developed for each reservoir environment.

This chapter includes a brief introduction to surfactants, with emphasis on those that have potential use in chemical flooding. As noted earlier, we are specifically interested in formulating surfactant systems that exhibit ultralow IFTs. Some interesting phase-behavior characteristics of these systems simplify the search for regions of ultralow IFT. Ultralow IFTs occur in regions where a large amount of oil and/or water is solubilized by the phase containing most of the surfactant. Phase-behavior studies can locate these regions; thus, we will discuss phase behavior of microemulsion systems, parameters affecting phase behavior, and the relationship between phase behavior and IFT in some detail.

Maintaining regions of low IFT at appropriate locations in the displacement process governs process performance. Phase behavior and displacement mechanisms are important parameters here. Interactions with the rock and its mineral constituents affect maintenance of these regions. Simple models that can be used to simulate the process are discussed. Finally, design procedures and field experience in application of the process are described.

As mentioned in Chapter 1, the term "chemical flooding" is somewhat ambiguous because chemicals are used in all enhanced-oil-recovery (EOR) processes. However, the term describes surfactant-based processes that reduce IFT and is used here to encompass such processes.

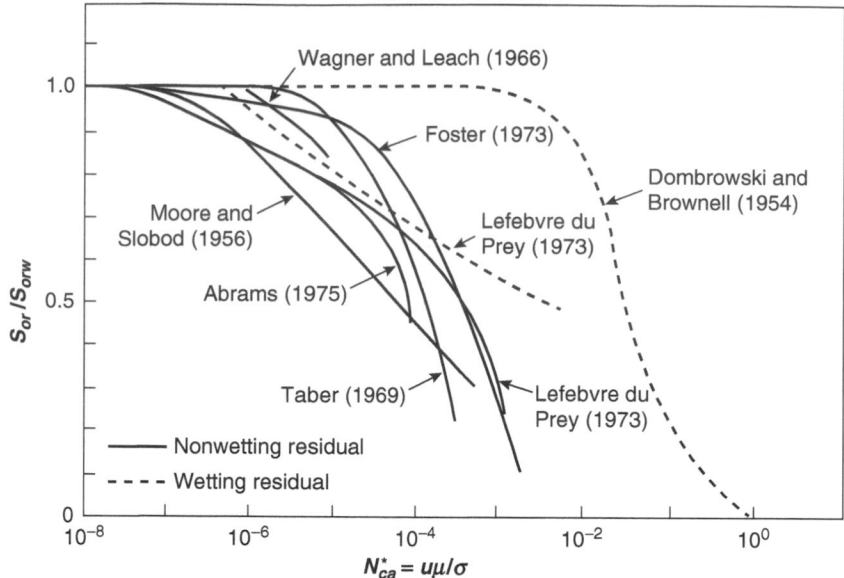

Fig. 7.1—ROS as a function of N_{ca}^* (Stegemeier 1977).

Component	Vol%
Hydrocarbon	0 to 80
Water	10 to 95
Surfactant	<1 to 15
Cosurfactant	0 to 10
Electrolyte	<1 to 10

Table 7.1—Ranges of micellar solution compositions (from Poettmann 1974).

7.2 Description of the Micellar/Polymer Process

Fig. 7.2 shows the micellar/polymer process. In most situations, the micellar/polymer process is implemented as a tertiary displacement near the end of a waterflood. Fig. 7.2a shows a tertiary process where the initial oil saturation is S_{orw}, ROS after waterflooding. A specified volume, or primary slug, of micellar solution is injected. The volume of the slug is one of the process variables; however, typical volumes are approximately 3 to 30% of the flood pattern pore volume (PV) (Gogarty 1976; Reppert et al. 1990). The micellar solution has a very low IFT with the residual crude oil and mobilizes the trapped oil, forming an oil bank ahead of the slug. As discussed later, the micellar slug also has a relatively low IFT with the brine and thus displaces brine as well as oil. Both oil and water flow in the oil bank.

Because oil is initially at residual saturation in a tertiary flood, no oil production occurs until the oil bank breaks through at the end of the flow system. **Fig. 7.3** shows results from a tertiary flood conducted in a laboratory linear core. Fractional flow of oil in the core effluent is plotted as a function of PVs of micellar solution injected (Davis and Jones 1968). In this experiment, oil saturation was reasonably constant over most of the oil bank, as indicated by the fractional flow remaining approximately constant until approximately 0.8 PV had been injected.

Fig. 7.2b shows an idealized displacement conducted as a secondary flood where water is immobile at the initial water saturation. Oil is produced immediately upon injection of the chemical slug because it is initially at a high saturation. Both oil and water flow in the oil bank because some of the interstitial water is displaced by the micellar slug.

The micellar solution must be designed so that a favorable mobility ratio exists between the micellar slug and the oil bank. The viscosity of the micellar solution is adjusted to accomplish this. Polymer often is added to the micellar solution to increase its apparent viscosity. Thus, the process has the potential to increase both volumetric sweep efficiency and microscopic displacement efficiency. This improvement is an advantage of the micellar/polymer process over mobility-control (Chapter 5) and miscible (Chapter 6) processes.

In some cases, a preflush is injected ahead of the micellar solution to adjust brine salinity or pH. The preflush solution may contain a sacrificial adsorbent (i.e., a relatively inexpensive chemical that will be adsorbed on the rock and occupy adsorption sites). The purpose is to reduce adsorption and loss of the surfactant contained in the micellar solution that will follow. In pilot tests, the volume of the preflush may be as large as 100% of the flood-pattern PV. This amount is not practical in a full-scale field application where much smaller volumes must be used. Field experience has shown that preflushes are often

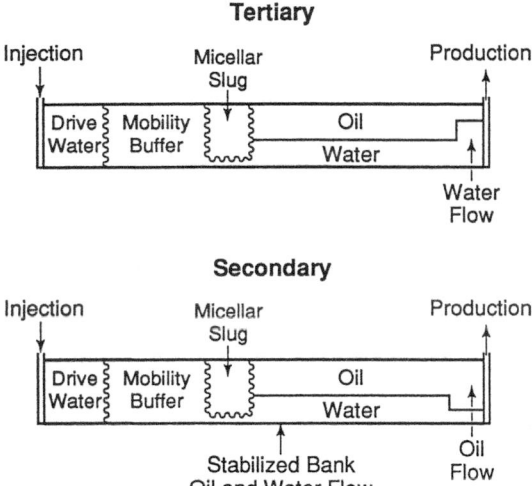

Fig. 7.2—Micellar/polymer displacement process (Poettmann 1974).

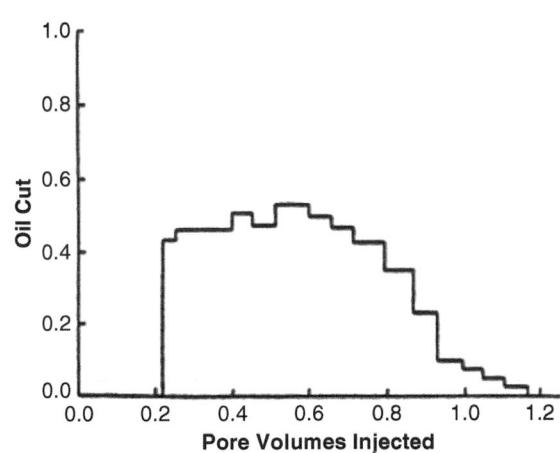

Fig. 7.3—Oil cut, laboratory micellar/polymer displacement test (Davis and Jones 1968).

not effective in controlling brine salinity and divalent-ion content (Pursley et al. 1973). Thus, a micellar solution generally must be designed to tolerate resident brine.

A micellar solution is relatively expensive, so a limited volume, or slug size, is used. The micellar slug must be displaced with a less expensive fluid. Water is not a suitable displacing fluid because an unfavorable mobility ratio usually exists between water (brine) and a micellar solution. Therefore, a mobility buffer slug that consists of a solution of polymer in water is used. The mobility buffer slug, which may range up to 100% of the flood-pattern PV, is then followed by water.

The polymer mobility buffer typically contains polymer at a concentration from 250 to 2,500 ppm. Both polyacrylamide and biopolymer of the type described in Chapter 5 are used. The salinity of the polymer solution is a design variable. The solution often contains a biocide, especially when a biopolymer is used, because microorganisms can degrade the polymer and reduce the viscosity of the solution.

As Chapter 5 discusses, the polymer concentration in the polymer slug often is tapered (i.e., the concentration of polymer is graded from the original injected concentration to 100% water). Tapering reduces cost and minimizes the effects of an adverse mobility ratio.

The polymer mobility buffer is such an integral concern that the overall process usually is called a micellar/polymer flood, as indicated in Fig. 7.2. When the polymer and micellar solutions are not miscible, there should be a low IFT between the liquids. Some of the micellar solution will be trapped as a residual phase and lost from the primary micellar slug. Physical processes, such as adsorption of surfactant and other constituents, degrade or diminish slug volume as the process proceeds.

In laboratory corefloods, recovery efficiency can be quite high, approaching 90% or more of the oil present at the start of displacement. Fig. 7.4 (Davis and Jones 1968) shows a recovery curve from a laboratory tertiary displacement. These data are for the same displacement shown in Fig. 7.3.

7.3 Surfactants

Surface-active agents, or surfactants, are chemical substances that adsorb on or concentrate at a surface or fluid/fluid interface when present at low concentrations in a system (Rosen 1978). They alter the interfacial properties significantly; in particular, they decrease the surface tension, or IFT. This section presents an introduction to surfactants and their effect on IFT. A more thorough treatment of surfactants and microemulsions is found in Bourrel and Schechter (1988).

In their most common form, surfactants consist of a hydrocarbon portion (nonpolar) and a polar, or ionic, portion. Fig. 7.5 (Ottewill 1984) is a simplified sketch of the molecule. The hydrocarbon portion is often called the "tail" and the ionic portion the "head" of the molecule. The hydrocarbon portion can be either a straight chain or a branched chain. The nonpolar and polar portions are called lipophilic and hydrophilic moieties, respectively. (A moiety is simply a part or portion.) The entire molecule is sometimes called an amphiphile because it contains the nonpolar and polar moieties. Fig. 7.6 (Lake 1989) shows examples of surfactants.

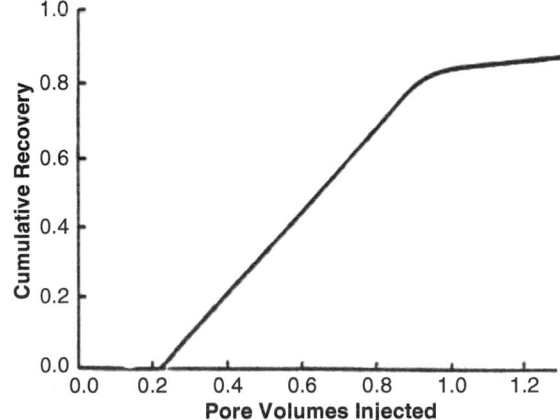

Fig. 7.4—Cumulative recovery curve, laboratory micellar/polymer displacement test (Davis and Jones 1968).

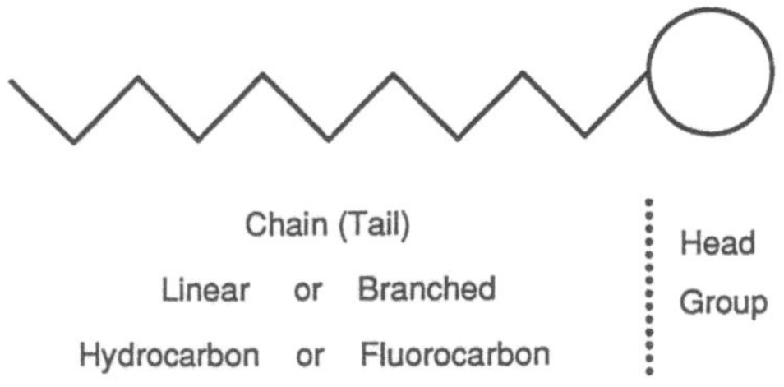

Chain (Tail)

Linear or Branched

Hydrocarbon or Fluorocarbon

Head Group

Fig. 7.5—Schematic of surface-active molecule (Ottewill 1984).

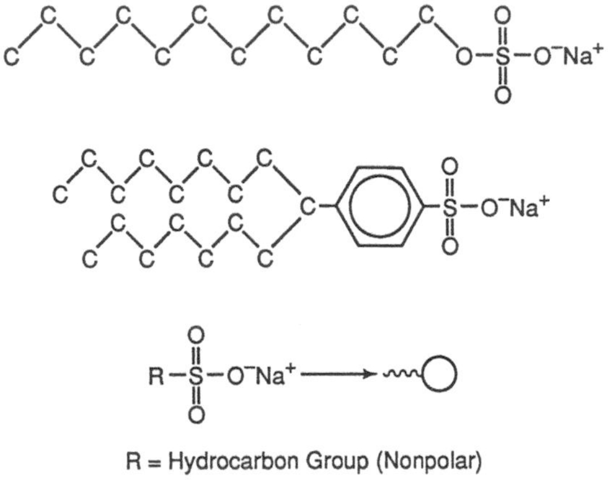

R = Hydrocarbon Group (Nonpolar)

Fig. 7.6—Representative surfactant molecular structures: (a) sodium dodecyl sulfate, (b) Texas No. 1 sulfonate, and (c) commercial petroleum sulfonates. (Lake, *Enhanced Oil Recovery*, © 1988, p.356. Reprinted by permission of Prentice Hall, Upper Saddle River, New Jersey.)

Texas Number 1

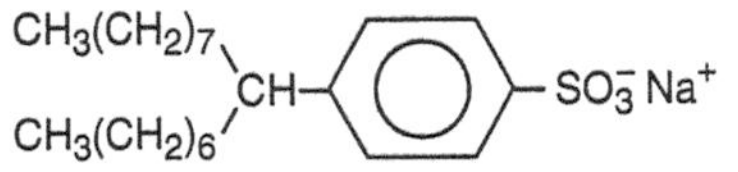

Fig. 7.7—Example of a hydrocarbon sulfonate.

In surfactants of the type shown in Fig. 7.6, the hydrocarbon portion interacts very weakly with water molecules in aqueous solution. In fact, the water molecules act to try to "squeeze" the hydrocarbon out of the water. Thus, the tail is called hydrophobic. On the other hand, the head or polar portion interacts strongly with water molecules, undergoing solvation. This part of the surfactant is hydrophilic.

Basically, it is the balance between the hydrophilic and hydrophobic parts of a surfactant that gives it the characteristics associated with a surface-active agent. An empirical number that has sometimes been used to characterize surfactants is the hydrophilic/lipophilic balance (HLB) (Garrett 1972). This number indicates relatively the tendency to solubilize in oil or water and thus the tendency to form water-in-oil or oil-in-water emulsions. Low HLB numbers are assigned to surfactants that tend to be more soluble in oil and to form water-in-oil emulsions.

7.3.1 Classification and Structure of Surfactants. Surfactants may be classified according to the ionic nature of the head group (Ottewill 1984) as anionic, cationic, nonionic, and zwitterionic. Examples of these types follow (Ottewill 1984).

Anionic: sodium dodecyl sulfate ($C_{12}H_{25}SO_4^-Na+$). In aqueous solution, the molecule ionizes, and thus the surfactant has a negative charge. This surfactant is classified as anionic because of the negative charge on its head group.

Cationic: dodecyltrimethylammonium bromide ($C_{12}H_{25}N^+Me_3$ Br^-). In aqueous solution, ionization occurs and the surfactant head group has a positive charge and is cationic.

Nonionic: dodecylhexaoxyethylene glycol monoether ($C_{12}H_{25}$ $[OCH_2CH_2]_6OH$). In this particular molecule, which does not ionize, the head group is larger than the tail group.

Zwitterionic: 3-dimethyldodecylamine propane sulfonate,

$$C_{12}H_{25} - \overset{+}{\underset{Me_2}{N}} - CH_2 - CH_2 - CH_2 - SO_3^-.$$

This surfactant has two groups of opposite charge.

Anionics and nonionics have been used as surfactants in EOR processes. Anionic surfactants have been the most widely used because they have good surfactant properties, are relatively stable, exhibit relatively low adsorption on reservoir rock, and can be manufactured economically. Nonionics have been used primarily as cosurfactants to improve the behavior of surfactant systems. Nonionics are much more tolerant of high-salinity brine, but their surface-active properties (reduction of IFT) are not generally as good as those of anionics. Cationics are usually not used because they adsorb strongly on reservoir rocks.

The most common surfactants used in micellar/polymer flooding are sulfonated hydrocarbons. **Fig. 7.7** shows an example of a hydrocarbon sulfonate produced by sulfonating a relatively pure organic structure to form an organic acid followed by neutralization with sodium hydroxide.

The term "crude oil sulfonates" refers to the product when a crude oil is sulfonated after it has been topped. "Petroleum sulfonates" are sulfonates produced when an intermediate-molecular-weight refinery stream is sulfonated, while "synthetic sulfonates" are the product when a relatively pure organic compound is sulfonated. Crude oil and petroleum sulfonates have been used for low-salinity applications (<2 to 3 wt% NaCl). These surfactants have been widely used because they are effective at attaining low IFT, relatively inexpensive, and reported to be chemically stable (Salter 1986). Surfactants considered practical for EOR applications have some water solubility. Petroleum sulfonates are soluble because of the ionic sulfonate group SO_3^-. Thus, they tend to precipitate or become primarily oil-soluble in brines that have high salinity or hardness (high calcium or magnesium ion content). Calcium and magnesium sulfonates are quite oil-soluble. The hydrocarbon tail also affects solubility. Ottewill (1984) stated that, as a rough approximation, compounds with a hydrocarbon chain length of approximately C_{16} or less are soluble, while those with a length greater than C_{16} are not.

The characteristics and structure of petroleum sulfonates suitable for EOR applications depend on the chemical composition of the feedstock, degree of sulfonation, and the average number of sulfonate groups attached to each molecule. Refinery streams that have been sulfonated include white oil, fractionated gas oil, vacuum gas oil, and lube-oil extract (Salter 1986). **Table 7.2** presents properties of typical petroleum sulfonates (Salter 1986). The sulfonated stream contains unreacted oil, the sodium salt of the sulfonated hydrocarbon (if neutralized with sodium hydroxide), and sodium sulfate from neutralization of the excess sulfuric acid used in the sulfonation process.

Stream Sulfonated	White Oil	Fractionated Gas Oil	Vacuum Gas Oil	Lube-Oil Extract
Equivalent weight	422	382	400	450
Di- and tri-sulfonate (%)	1.6	17.1	34.8	11.5
		Unreacted Oil		
Aromatics	14.5	7.5	13.3	64.7
Monoaromatics	54.5	46.6	39.8	21.7
Diaromatics	29.5	22.7	32.3	28.5
Triaromatics	7.6	5.3	9.8	18.2
Tetra-aromatics	1.4	5.3	7.5	11.7
Penta-aromatics	0.7	2.7	1.5	2.6
Thiophenoaromatics	5.5	6.7	4.5	8.3

Table 7.2—Properties of various petroleum sulfonates (Salter 1986).

Molecular weights of the feedstock typically range from 350 to 450 but may vary from these values significantly on both the high and low sides. When the stream to be sulfonated contains polynuclear aromatics (more than one aromatic ring on each molecule), these materials are preferentially sulfonated, leading to polysulfonates. Petroleum sulfonates are characterized by two parameters: the equivalent weight, which is the molecular weight divided by the number of sulfonate groups, and the percentage of the sulfonated material that is polysulfonates (i.e., has more than one sulfonate group per molecule). In general, petroleum sulfonates with high polysulfonate content are not good EOR candidates. Also, as a rule of thumb relative to solubility, Gale and Sandvik (1973) state that petroleum sulfonates with equivalent weights above 450 are normally oil-soluble and not water-soluble. Lower-equivalent-weight sulfonates tend to be water-soluble.

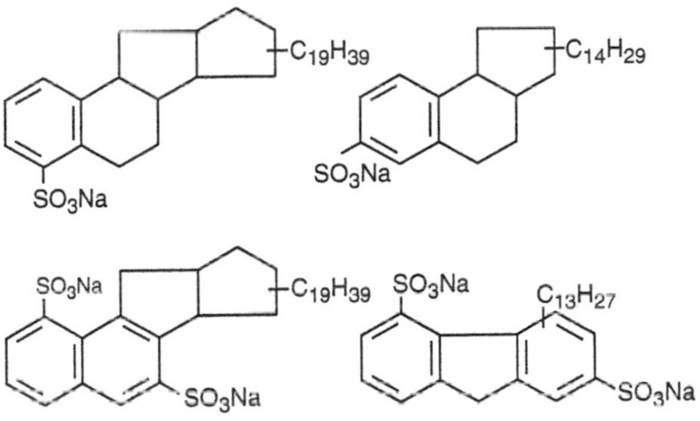

Fig. 7.8—Examples of compounds resulting from sulfonation.

Sulfonation of crude oil has also been used to reduce the cost of surfactants for field applications. Because crude oils contain many compounds, sulfonation yields a complex mixture of compounds of different molecular weights and structures. **Fig. 7.8** shows some examples of compounds that have been identified. Note that two of the compounds have undergone disulfonation and have two sulfonate groups.

Table 7.3 (Malmberg et al. 1982) shows examples of synthetic petroleum sulfonates. The sulfonates are described in terms of the olefin and aromatic hydrocarbons that were combined to make the alkylate that was then sulfonated. These surfactants are called alkyl-aryl sulfonates. Broad and narrow cut refer to the boiling-point range of the portion of the alkylate used for sulfonation. The equivalent weight and percent active sulfonate for each of the samples are also reported. The concentration of a petroleum sulfonate often is reported in terms of the percent active sulfonate, which is simply the percent of the total material supplied that is sulfonated (i.e., other components in the mixture are considered inert). The lower section of Table 7.3 shows representative molecular-weight distributions of the three olefins used.

Straight-chain sulfonates also may be manufactured by sulfonating an olefin. An example is an internal alpha olefin sulfonate (Baviére et al. 1988):

$$R - CH = CH - (CH_2)_n - SO_3Na,$$

where R is typically a straight chain of 13 to 20 carbons. The sulfonation process proceeds by reacting an alkene with sulfur trioxide, SO_3.

High salinity adversely affects the water solubility of surfactants and can hamper the performance. Baviére et al. (1988) state that the alpha olefin sulfonates are more tolerant of salinity than typical petroleum sulfonates.

Another surfactant type that has been developed for high salinity tolerance is given by the general expression (Maerker and Gale 1992)

$$RO(C_3H_6O)_m(C_2H_4O)_nSO_3Na,$$

where R is an iso-tridecyl alcohol radical and m and n have values ranging from 1 to 6. These compounds were reported to be manufactured by first reacting propylene oxide and then ethylene oxide with iso-tridecyl alcohol in a two-step process.

Project Sample	Components of Alkylate*	Equivalent Weight	Active Sulfonate (%)
1	Propylene tetramer + mixed xylenes	369	65
2	Propylene pentamer + mixed xylenes	425	51
4	Normal C_{15} through C_{18} olefin**+ toluene	427	69
5	Propylene pentamer + toluene, narrow cut	400	59
6	Propylene pentamer + toluene, broad cut	400	55
7	Propylene pentamer + benzene tower feed, broad cut	403	52
8	Propylene pentamer + benzene tower feed, narrow cut	401	58
9	Propylene tetramer + xylene tower bottoms	380	65
13	Alkylate bottoms	475	52

Molecular Weight Distribution in Alkylbenzene Fraction (Normalized)

Project Sample	C_{13}	C_{14}	C_{15}	C_{16}	C_{17}	C_{18}	C_{19}	C_{20}	C_{21}	C_{22}	C_{23}	C_{24}	C_{25}	C_{26}	C_{27}	C_{28}
1	0.7	0.8	1.0	1.6	1.6	7.3	25.9	46.2	12.8	1.8	0.4	—	—	—	—	—
4	—	—	—	—	—	—	—	0.3	0.5	11.9	23.4	38.6	19.6	4.2	1.5	0.1
7	—	0.2	0.6	0.9	1.7	3.0	5.2	7.6	13.3	22.0	20.7	13.8	6.8	2.7	0.5	0.2

*Aromatic components are derived from the petrochemical BTX operation. Benzene, toluene, xylenes, and a residue of higher homologues (primarily three alkyl carbons) are coproduced and then separated by successive distillation.
**Straight-chain alkylate was made from commercially available wax-cracked alpha olefin with approximately equal amounts of C_{15}, C_{16}, and C_{17} chains. The position of the double bond was randomized before the alkylation. Another compound with lower molecular weight was originally planned, but the olefin was not available when the alkylate was produced, hence no Sample 3.

Table 7.3—Description of example synthetic sulfonates (Malmberg et al. 1982).

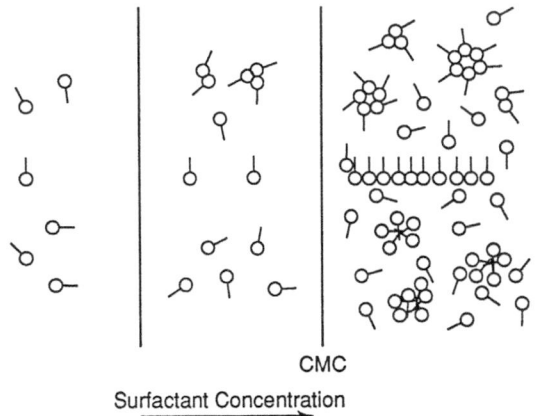

CMC

Surfactant Concentration

Fig. 7.9—Formation of micelles.

The products were then sulfated and finally neutralized with sodium hydroxide. Section 7.6.5 discusses the application of this type of surfactant. Recent advances in surfactant structure are described in Section 7.11.

7.3.2 Micelles and Microemulsions. From the early days of the study of surface-active agents, it was recognized that their properties indicated that colloidal aggregates were in solution (Rosen 1978). **Fig. 7.9** depicts the formation of these aggregates, which are called micelles.

When a surfactant is added to a solvent at very low concentrations, the dissolved surfactant molecules are dispersed as monomers, as Fig. 7.9 shows. As the concentration of surfactant increases, the molecules tend to aggregate. Above a specific concentration, called the critical micelle concentration (CMC), further addition of surfactant results in the formation of micelles. The concentration of surfactant as monomers essentially remains constant above the CMC. That is, surfactant added at concentrations above the CMC results in formation of additional micelles but relatively little change in monomer concentration.

If the solvent is water, the micelles form with the tail portion directed inward and the head (polar) portion outward. Water would be the continuous phase, as shown in the lower right side of Fig. 7.9. For a hydrocarbon solvent, the orientation of the surfactant molecules is reversed, as indicated in the upper right side of Fig. 7.9, and the hydrocarbon would be the continuous phase.

The structure and properties of micelles were first described in a classic paper by G. S. Hartley in 1936, in which he proposed the following properties as summarized by Ottewill (1984).

1. Micelles have a spherical shape, as shown in **Fig. 7.10.** The micelle's radius is approximately as long as the hydrocarbon chain in the surfactant.
2. Micelles contain approximately 50 to 100 monomer units.
3. The process of micellization occurs over a very narrow concentration range.
4. For surfactants in aqueous solution, the interior of the micelle is formed by the association of hydrocarbon chains. They have many of the properties of a liquid hydrocarbon, including the ability to solubilize organic compounds.

Many of Hartley's proposals were correct. However, it is now known that the range of micelle properties is broader than originally proposed (Ottewill 1984). Other micelle shapes, depicted in Fig. 7.10, have been postulated on the basis of

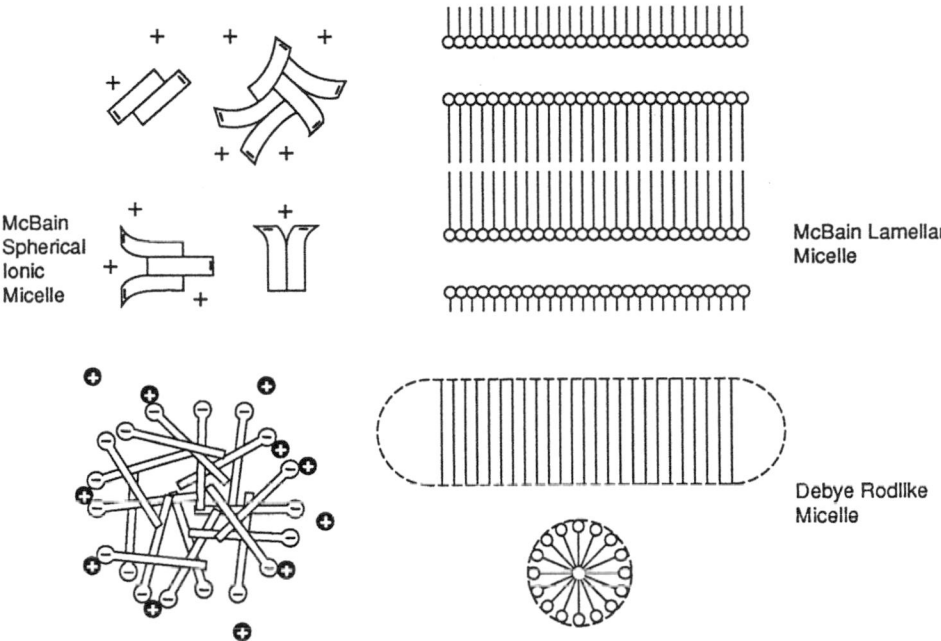

Fig. 7.10—Various models of micelle structures (McBain spherical ionic micelle, McBain lamellar micelle, Hartley spherical micelle, Debye cylindrical micelle). (Ottewill, R.H.: "Introduction," *Surfactants*, T.F. Tadros (ed.), Academic Press, 1984. Reprinted by permission of John Wiley & Sons, Inc.)

evidence from light-scattering and neutron-scattering experiments. These shapes include the rodlike and lamellar shapes shown in Fig. 7.10.

The number of monomer units in a micelle also can be significantly outside the range Hartley proposed. Rosen (1978) presents an extended list of aggregation numbers (number of monomer units per micelle) for several surfactants. For a given system, micellization occurs over a narrow concentration range. This concentration is small, approximately 1.0×10^{-5} to 1.0×10^{-4} gmol/L for surfactants used in EOR. For example, for a surfactant with a molecular weight of 450, this lower limit of CMC corresponds to a concentration of 0.0045 g/L. Rosen (1978) also presents an extended list of CMC values for common surface-active agents.

A number of solution properties change in trend rather abruptly as a function of surfactant concentration at the CMC. These include such properties as electrical conductivity, surface tension, and osmotic pressure.

The cores of spherically shaped micelles formed in aqueous solution are capable of solubilizing organics. Under the right conditions, significant amounts of oil can be solubilized into the micelles. Conversely, in hydrocarbon solvents containing micelles, water can be solubilized into the interior of the micelle. Thus, while oil and water each have very limited solubility for the other phase, the addition of a surfactant at a concentration above the CMC significantly increases the apparent solubility.

Micelles that have solubilized a phase that is immiscible with the solvent often are called swollen micelles or the solutions are called microemulsions. One definition of a microemulsion used in EOR processes is "...a stable, translucent micellar solution of oil and water that may contain electrolytes and one or more amphiphilic compounds" (Healy and Reed 1974). The swollen micelles typically range from 10 to 200 μm, much larger than micelles that have not solubilized a phase. In this book, the term "microemulsion" or "micellar solution" is used for liquid phases that contain surfactant, oil, and water (brine).

Depending on aggregate size, an aqueous-phase microemulsion may appear transparent or semitransparent to the eye, even though a significant amount of hydrocarbon is solubilized in the micelle.

The term microemulsion is used to contrast these systems with macroemulsions, where particle sizes are much larger and the systems are opaque to visible light. Microemulsions are thermodynamically stable, while macroemulsions are not, although macroemulsions may persist for long periods.

Microemulsions can be of the water-external or hydrocarbon-external type, as **Fig. 7.11** illustrates. The external phase is the continuous phase. The nature of the physical configuration is determined by system parameters, such as hydrocarbon/water ratio, surfactant, and temperature. In some cases, intermediate or lamellar structures, such as shown in Fig. 7.11,

Definition

A microemulsion is a stable, translucent micellar solution of oil, water that may contain electrolytes, and one or more amphiphilic compounds.

(a) Water-External

(c) Oil-External

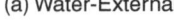

(b) Lamellar

Fig. 7.11—Definition and structure of microemulsion (Healy and Reed 1974).

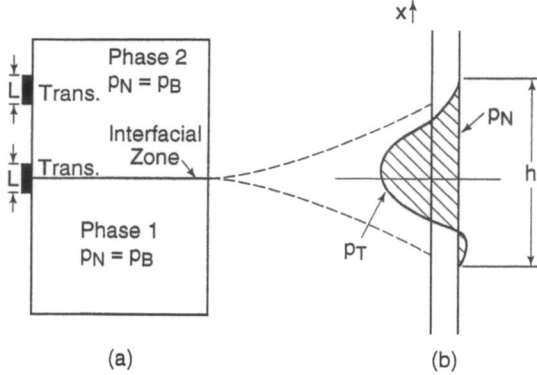

Fig. 7.12—(a) Two-phase system separated by a planar interface. (b) Normal, p_N, and transverse, p_T, pressure profiles across the interfacial zone (Davis and Scriven 1982).

are also thought to exist. The notion of an external phase is less clear in this situation. Scriven (1976) developed a model of a bicontinuous structure.

7.3.3 Mechanism for IFT Reduction by Surfactants. IFT is the force per unit length required to create new surface area at the interface between two immiscible fluids. IFT also is a condition of mechanical equilibrium at an interface. The description in this section was developed by Davis and Scriven (1982) by analyzing stresses in the interfacial region between two fluids.

Consider the two-phase system separated by a planar interface shown in **Fig. 7.12a.** Phase 1 is on the bottom because it has a higher density than Phase 2. The interfacial zone has a thickness h_I of 10 to several hundred angstroms. The system is at hydrostatic equilibrium.

The region shown in Fig. 7.12a is assumed to be sufficiently thin that the bulk pressures in Phases 1 and 2 are equal to p_b. This neglects any effect of fluid density on pressures above and below the interfacial zone.

The pressure distribution in the region shown in Fig. 7.12b is affected by the presence of the interface. The interfacial zone is considered to be inhomogeneous because the densities and compositions of the fluids within the interfacial zone vary with Position x. Because of this variation in fluid density, the stresses acting on molecules are not uniform within the interfacial zone and vary with both direction and position. These differences in stresses within the interfacial zone cause the pressure to vary with position and direction. For this geometry, the pressure has two components, a component p_N that is normal to the interface (x direction) and a second component, p_T, that is transverse and lies in the plane of the interface. The magnitude of p_T depends on position in the interfacial zone, as depicted in Fig. 7.12b. At $x \gg h_I$, $P_t = P_n = P_B$.

Assume that it is possible to place a transducer of height L and width w (into the figure) centered on the interface and that $L > h_I$. (This is a conceptual transducer because no real transducer exists with these capabilities.) The force exerted on the transducer face by the fluid tension is given by

$$F_I = -\int_{-L/2}^{L/2} p_T w \, dx. \dots\dots\dots\dots\dots\dots\dots\dots\dots\dots\dots\dots\dots\dots\dots (7.1)$$

The force exerted on a similar transducer face in the bulk phase is

$$F_B = -p_N wL$$
$$= \int_{-L/2}^{L/2} p_N w \, dx. \dots\dots\dots\dots\dots\dots\dots\dots\dots\dots\dots\dots\dots\dots\dots (7.2)$$

The increase in force caused by the presence of the interfacial zone is

$$F_I - F_B = w \int_{-L/2}^{L/2} \left(p_N - p_T \right) dx. \dots\dots\dots\dots\dots\dots\dots\dots\dots\dots\dots (7.3)$$

The IFT is defined as the force/unit length. Thus,

$$\sigma = \left(F_I - F_B \right) / w \dots\dots\dots\dots\dots\dots\dots\dots\dots\dots\dots\dots\dots\dots\dots (7.4)$$

$$\text{or } \sigma = \int_{-L/2}^{L/2} \left(p_N - p_T \right) dx. \dots\dots\dots\dots\dots\dots\dots\dots\dots\dots\dots\dots\dots (7.5)$$

Because $P_N = P_T$ for $L > h_I$, Eq. 7.5 can be written as

$$\sigma = \int_{-\infty}^{\infty} \left(p_N - p_T \right) dx. \dots\dots\dots\dots\dots\dots\dots\dots\dots\dots\dots\dots\dots\dots (7.6)$$

It is possible to compute the IFT, or surface tension, for pure fluids with the mechanical equilibrium model. However, the computations are quite complex.

Consider that two immiscible liquids, heptane and water, are brought into contact. When a surfactant is added to this system, surfactant molecules adsorb at the interface, displacing some of the heptane and water molecules there. The surfactant molecules orient themselves such that the hydrophilic part is directed into the water phase and the hydrophobic part into the heptane phase. Accumulation of the surfactant in the interfacial zone disrupts the fluid structure in this region and increases p_T. This is reflected in the rapid decrease in the IFT as the surfactant concentration increases up to the CMC.

The IFT between an aqueous surfactant solution and a hydrocarbon phase is a function of the salinity, temperature, the surfactant concentration, surfactant type and purity, and the nature of the hydrocarbon phase. **Fig. 7.13** illustrates the general behavior of IFT between a relatively pure surfactant solution (containing a single surfactant species) and a hydrocarbon phase. The IFT decreases rather sharply as surfactant concentration increases until the CMC is reached. Beyond the CMC, little change in IFT occurs. Surfactant added in excess of the CMC contributes to the formation of micelles and does not increase the concentration at the water/ hydrocarbon interface. Thus, there is only a small incremental effect on IFT.

The IFT properties of petroleum sulfonates, which are mixtures, generally are similar to those of single-component surfactant systems. There is, however, a difference in that a sharp CMC is not usually observed (Gale and Sandvik 1973). The IFT between an aqueous surfactant system and a hydrocarbon phase may decrease significantly at concentrations well above the CMC, the point of onset of formation of micelles.

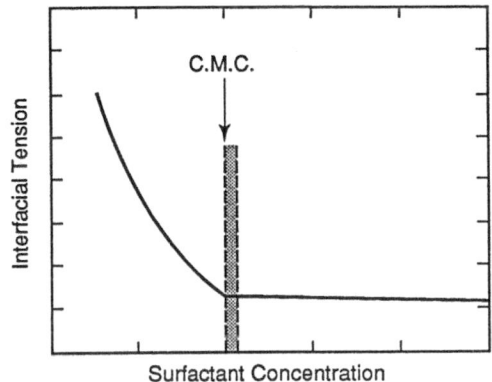

Fig. 7.13—IFT as a function of concentration— pure (single-component) surfactant.

7.3.4 Surfactants and IFTs in EOR Systems. Surfactant systems that contain no alcohol or limited amounts of alcohol or cosurfactant exhibit ultralow IFTs (Gale and Sandvik 1973; Foster 1973; Hill et al. 1973; Wilson et al. 1976). Puig et al. (1979) and Hall (1980) show that ultralow IFTs observed under these conditions are caused by the presence of a finely dispersed phase at the interface between the oil and brine. This mesophase is liquid crystalline, surfactant-rich, and often exhibits birefringence under polarized light. Puig et al. found that ultralow tensions were observed only when the particles were large enough to cause turbidity in the solution. Manning et al. (1983) present an excellent report on the measurement of ultralow IFT. These systems have not been used successfully in field applications and are not considered further.

Micellar solutions, or microemulsions, the type of solutions used in EOR processes, are complex. Characteristics of these solutions are discussed in the next section.

7.4 Phase Behavior of Microemulsions

Microemulsion systems can be designed that have ultralow IFT values with either aqueous or hydrocarbon phase (approximately 10^{-3} dynes/cm). This property makes micellar solutions, or microemulsions, attractive for use as oil recovery agents. Ultralow IFTs correlate with high solubilization of oil and water by the microemulsion system (Healy et al. 1976). Thus, regions of low IFT are usually found by studying the phase behavior of microemulsion systems to locate regions of high solubilization. This section discusses phase behavior, and Section 7.5 discusses the relationship between phase behavior and IFT.

The phase behavior of microemulsions is complex and dependent on a number of parameters, including the types and concentrations of the surfactants, cosurfactants, hydrocarbons, and brine; temperature; and to a much lesser degree, pressure. There are no universal equations of state for even simple microemulsions. Thus, phase behavior for a particular system has to be measured experimentally, and results typically are presented in graphical form. Where desired, results can be put in equation form for application in a mathematical model.

7.4.1 Phase-Behavior Representations on a Ternary Diagram. A microemulsion usually is composed of at least five components: a surfactant, cosurfactant, hydrocarbon, water, and NaCl. Cosurfactant and NaCl are not essential. To study phase behavior rigorously, the effects of each component would need to be determined over a wide range of compositions. Time and economic constraints prohibit extensive phase-behavior studies of each system. Consequently, the number of components must be reduced by combining one or more components into pseudocomponents. For example, water and NaCl are commonly represented by the brine pseudocomponent and the hydrocarbon phase may be a mixture of hydrocarbons. When brine is a pseudocomponent, the microemulsion system has four components and phase behavior can be represented on a quaternary diagram, as depicted in **Fig. 7.14.** In this diagram, the apexes are 100% surfactant, cosurfactant, brine, and hydrocarbon. All points within the diagram represent overall compositions of the four components.

Determining the entire phase behavior of a microemulsion system on a quaternary diagram is not necessary because the region of interest for surfactant flooding is usually at relatively low surfactant concentrations. A few systems have been studied on the quaternary diagram to understand their fundamental concepts (Salter 1978, 1983; Blevins et al. 1981), but another pseudocomponent is needed to reduce the region of investigation to a pseudoternary diagram. In most cases, the surfactant and cosurfactant are treated as a pseudocomponent and subsequently called the "surfactant." Thus, one apex of the pseudoternary diagram represents 100% "surfactant." The surfactant is composed of a fixed ratio of surfactant to cosurfactant. Fig. 7.14 shows a pseudoternary triangular diagram representing a plane of a quaternary diagram. The surfactant pseudocomponent has a specific surfactant/alcohol ratio. Therefore, every point on the pseudoternary diagram, except the brine and hydrocarbon apices, has a surfactant composition with the same surfactant/alcohol ratio.

The use of pseudoternary diagrams to represent phase behavior has other consequences. While the overall composition of every mixture of the three components must lie on or within the boundaries of the triangle, the compositions of phases that are in equilibrium will not lie on the pseudoternary diagram unless the components behave as true pseudocomponents. A component is a true pseudocomponent if that component has the same composition in every phase. In the material that follows,

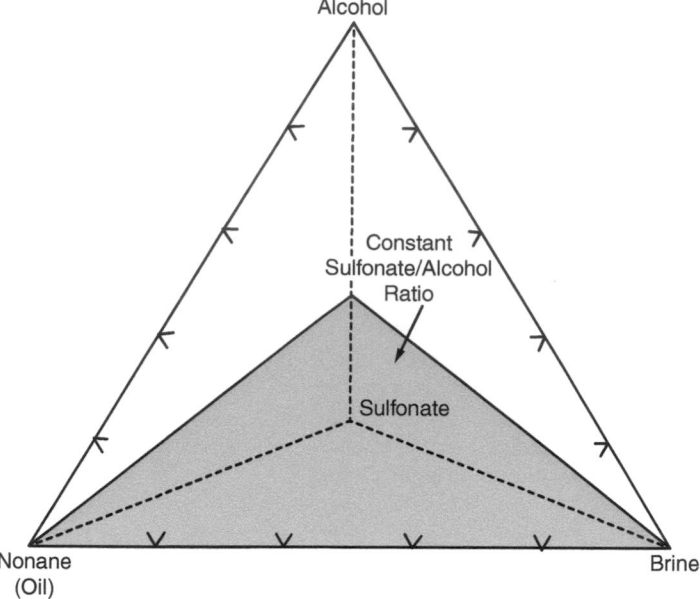

Fig. 7.14—Quaternary diagram with pseudoternary diagram at constant sulfonate/alcohol ratio.

pseudoternary diagrams are presented and discussed as if components behave as true pseudocomponents. Phase boundaries are drawn on ternary diagrams, implying that the actual compositions of equilibrium phases lie on the ternary diagram. For some systems, this is a good assumption. For most systems, however, some partitioning of the components that make up a pseudocomponent occurs. NaCl, cosurfactant, and surfactant do not always partition equally between phases. In spite of this, pseudoternary diagrams are one tool for the study of phase behavior of surfactant systems.

Fig. 7.15 shows a general representation of "ideal" phase behavior of a microemulsion on a ternary diagram. The apex locations on the equilateral triangle represent surfactant, brine, and oil, the three components of the solution. Concentrations may be expressed on a mass or volume basis.

In the single-phase region of the diagram, the solution is a microemulsion or micellar solution over most of the concentration range (other than at concentrations of surfactant below the CMC). At high water concentrations, as Point S_1 on Fig. 7.15,

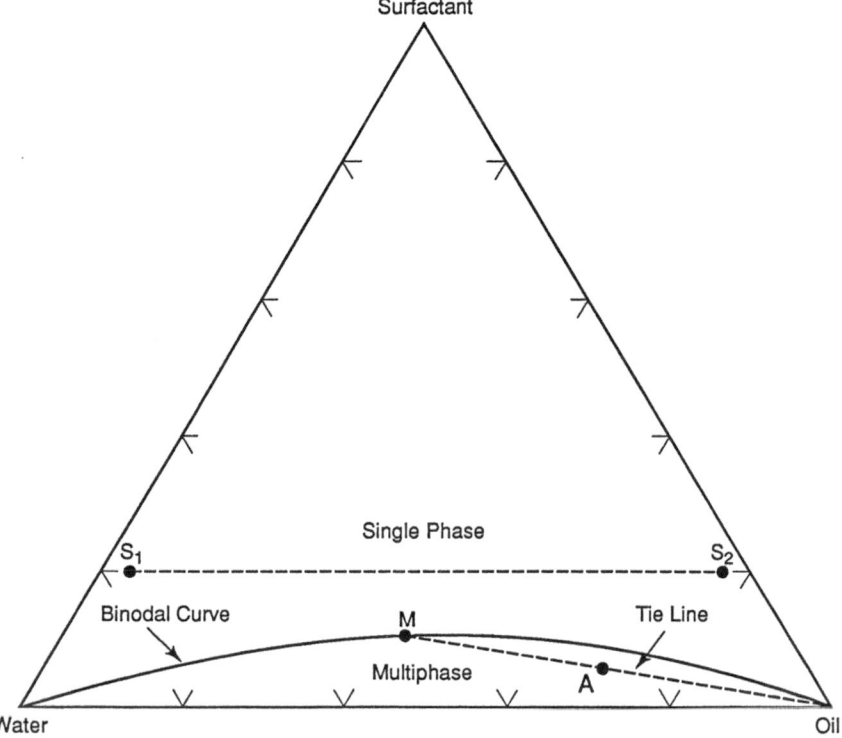

Fig. 7.15—Pseudoternary phase diagram for a micellar (microemulsion) solution.

the microemulsion would be a water-external system with oil solubilized in the cores of the micelles. A microemulsion with high oil concentration, such as Point S_2, would be an oil-external system with water solubilized in the cores of the micelles. The dashed line connecting Points S_1 and S_2 represents all mixtures of these two compositions. Although the mixtures remain single phase and thermodynamically stable, the microemulsion structure changes through a series of intermediate states (Bourrel and Schechter 1988). The structures of these intermediate states are not well-understood. However, the solutions are thermodynamically stable and isotropic.

At any concentration in the multiphase region, such as Point A in Fig. 7.15, the solution will separate into two or sometimes three phases. For a concentration that yields two phases, one phase is typically a microemulsion lying on the binodal curve and the second phase is relatively pure oil or water. System A separates into a microemulsion at Point M on the binodal curve and an oil at the pure-oil apex. The rules that apply to tie-lines and the lever-arm rule (described in Chapter 2) are applicable.

7.4.2 Effect of Brine Salinity on Phase Behavior. In general, increasing salinity of an aqueous phase (brine) decreases the solubility of an ionic surfactant. Surfactant is driven out of a brine as the electrolyte concentration increases. Thus, brine salinity has a significant effect on phase behavior.

Healy et al. (1976) described the phase behavior of a "simple" or ideal microemulsion system and the effect of the brine salinity on the phase behavior. They found that, for their ideal system, the multiphase behavior divides into three basic classes illustrated in **Fig. 7.16.**

At relatively low brine salinity, solutions at concentrations within the multiphase region divide into a water-external microemulsion and an excess-oil phase. The microemulsion is saturated with oil at that composition and temperature. The system would appear as shown on the left of Fig. 7.16 below the ternary diagrams. Because the microemulsion is the aqueous phase and is denser than the oil phase, it resides below the oil phase and is call a lower-phase microemulsion. At high salinity, the system separates into an oil-external microemulsion (hydrocarbon or oleic phase) and an excess, more dense, water (brine) phase. In this case, the microemulsion is an upper-phase microemulsion (Healy et al. 1976).

At intermediate salinity, the system is more complex. At lower surfactant concentrations, a three-phase region exists. Solutions with overall concentrations within this region separate into microemulsion, water, and oil phases, as Fig. 7.16 indicates. For a simple (ideal) system at fixed salinity, the microemulsion composition is invariant for any concentration within the three-phase triangle. Because the density of the microemulsion is intermediate to the oil and brine densities, it is called a middle-phase (or midphase) microemulsion and is designated by M^* in Fig. 7.16. A middle-phase microemulsion is saturated with both oil and water at the temperature and overall composition of the system. The three-phase region is of particular interest because ultralow IFTs against both water and oil usually are found in this region.

The surfactant on a pseudoternary diagram is composed of surfactant plus cosurfactant (alcohol) at a fixed ratio. Changing the surfactant/cosurfactant (S/A) ratio at constant salinity will change the phase behavior and the composition region encompassed by the three-phase triangle. Consequently, an infinite number of pseudoternary diagrams with surfactant as the pseudocomponent exist that will have three-phase regions.

Two-phase lobes exist to the upper right and upper left of the three-phase triangle. Systems with overall concentrations within these lobes separate into two equilibrium phases, as indicated by tie-lines in Fig. 7.16. There is a third two-phase region located at very low surfactant concentrations below the three-phase region. This region typically is quite small and therefore is not included on the diagram.

The microemulsion concentration point, Point M^*, on the three-phase triangle will shift toward the water or oil apex as salinity is increased or decreased around the optimal salinity value. **Fig. 7.17** (Nelson and Pope 1978) illustrates this behavior.

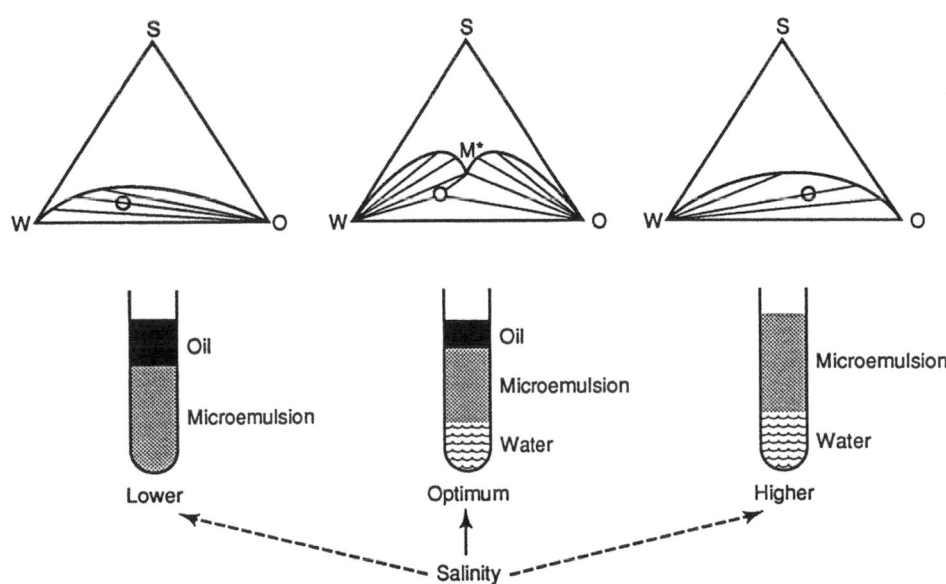

Fig. 7.16—Effect of salinity on microemulsion phase behavior (Healy et al. 1976).

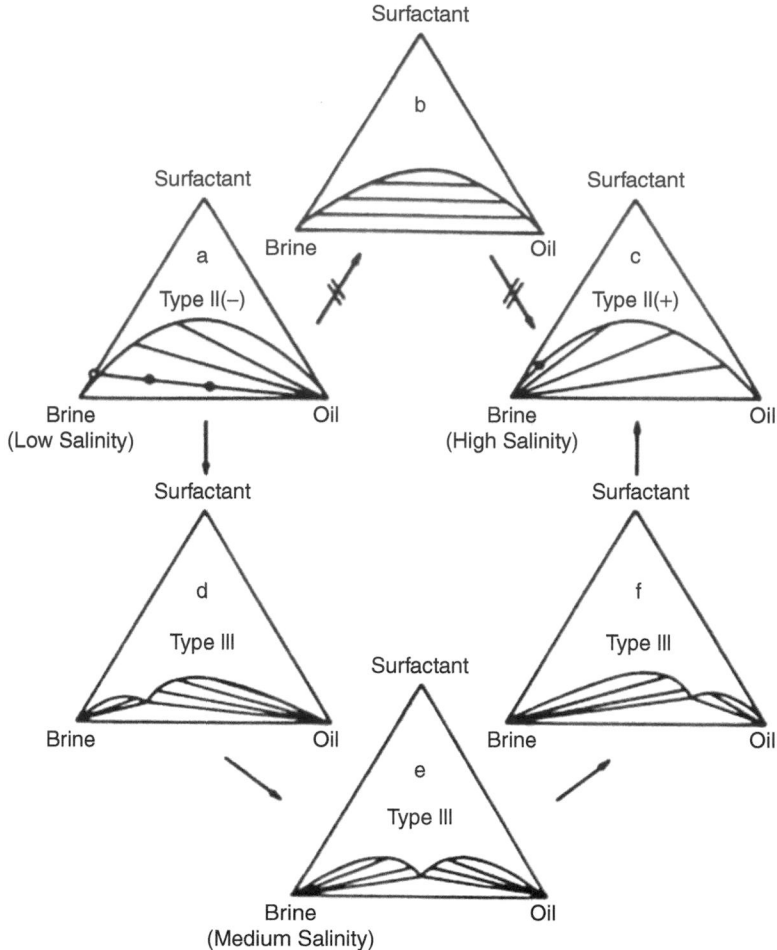

Fig. 7.17—Generalized phase diagrams illustrating the effect of changing salinity (Nelson and Pope 1978).

Figs. 7.17 and 7.18 use an alternative terminology to that of Healy et al. (1976). Lower-, middle-, and upper-phase systems are called Type II(–), III, and II(+) phase environments, respectively (Nelson and Pope 1978). This follows the terminology originally used by Windsor (1954).

In the Type II(–) environment, a maximum of two phases exists. Tie-line slopes are negative, which is the basis for the Type II(–) designation. Windsor referred to the microemulsions formed in the Type II(–) phase environment as Type I microemulsions. In the Type II(+) environment, again only two phases exist. Tie-lines have a positive slope, hence the designation of Type II(+). Windsor called the microemulsions formed in this environment Type II microemulsions.

In the Type III phase environment, three phases exist in equilibrium. Windsor called these microemulsions, which are in equilibrium with water and oil, Type III. As previously indicated, however, Type II(–) and Type II(+) lobes also may exist in this environment. Nelson and Pope (1978) point out that the behavior illustrated at the top of Fig. 7.17 has not been observed for microemulsions used in EOR processes. That is, the transition from Type II(–) to Type II(+), or vice versa, always occurs through the Type III environment (by Path a-d-e-f-c and not by Path a-b-c).

The application of the lever-arm rule to determine amounts of equilibrium phases is illustrated in Example 7.1. Systems that separate into three phases are discussed in the following example.

Example 7.1—Phase Behavior, Ternary Diagram. Consider the ternary phase diagram shown in **Fig. 7.19.** This is for a Type III simple, or ideal, system discussed by Healy et al. (1976).

1. Consider 100 cm³ of a mixture at the overall composition at Point A on the diagram. Assuming that equilibrium is reached, what are the amounts and compositions of the equilibrium phases? Compositions are in volume percent.
2. Repeat Part 1 for 100 cm³ of a mixture at the overall composition given by Point B on the diagram.

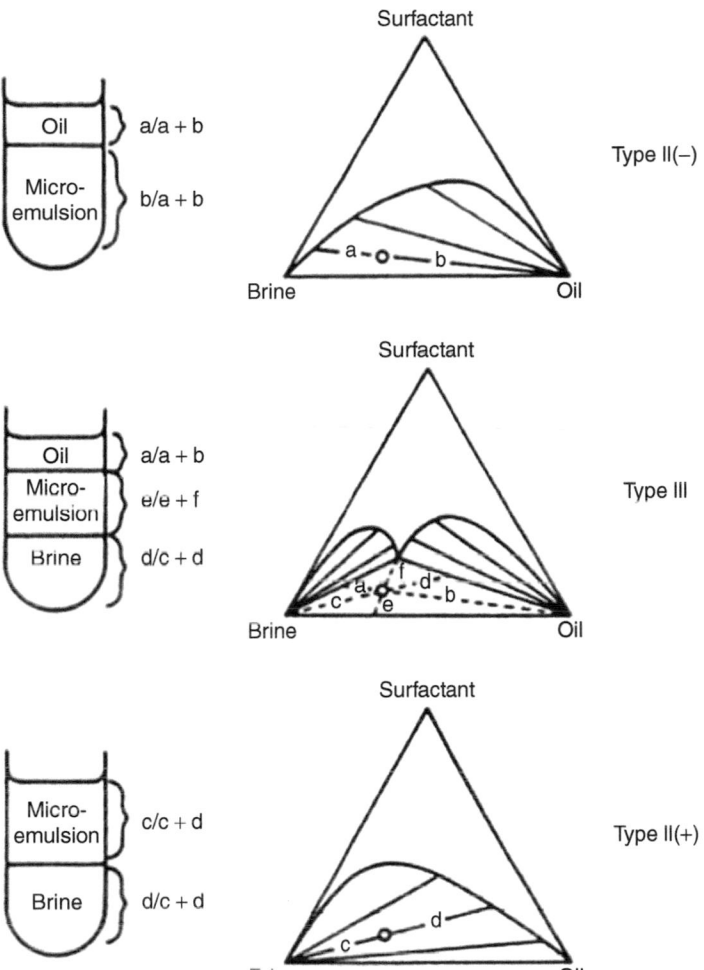

Fig. 7.18—Ternary representation of phase diagrams (Nelson and Pope 1978).

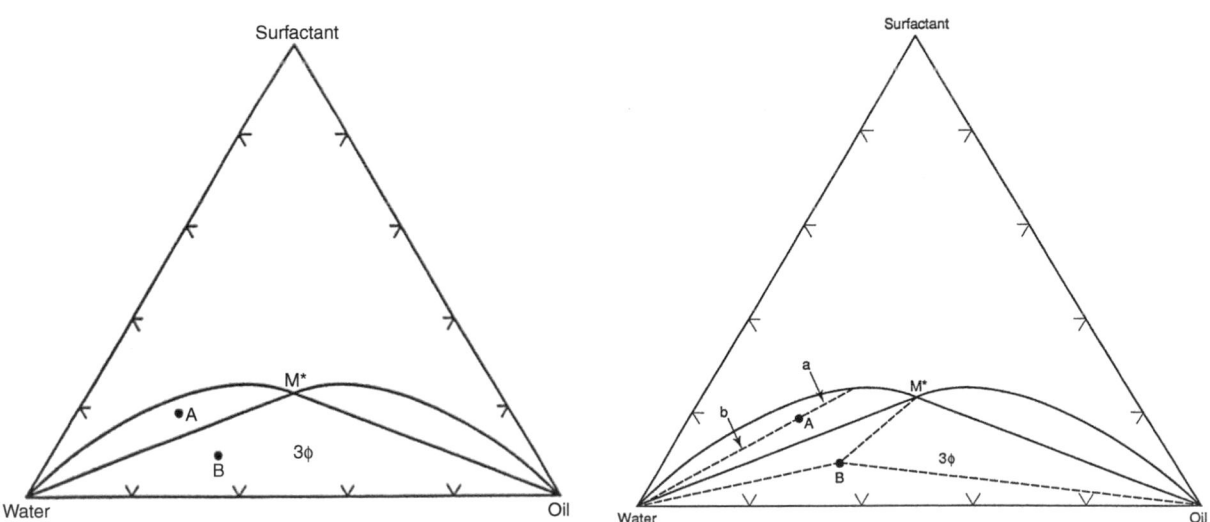

Fig. 7.19—Ternary diagram, Example 7.1.

Fig. 7.20—Ternary diagram, solution of Example 7.1.

Solution. Refer to **Fig. 7.20.**

1. The overall composition at Point A is 19% (19 cm³) surfactant, 20% (20 cm³) oil, and 61% (61 cm³) water. The equilibrium tie-line is assumed to intersect the pure-water apex. Water is in equilibrium with a microemulsion. The amount of the water phase is determined by measurement along the tie-line:

$$a / (a+b) = 6/23 \text{ (by measurement)},$$

$$V_{H_2O} = 6/23 \times 100 = 26 \text{ cm}^3.$$

The composition of the microemulsion, determined by material balance, is 19 cm³ (25.7%) surfactant, 20 cm³ (27.0%) oil, and 35 cm³ (47.3%) water. All the oil and surfactant were assumed to be in the microemulsion phase. Also,

$$V_{me} = 100 - 26$$
$$= 74 \text{ cm}^3.$$

The values of concentrations also could have been read directly from the diagram.

2. The overall composition at Point B is 9% (9 cm³) surfactant, 31% (31 cm³) oil, and 60% (60 cm³) water. The system equilibrates with three phases; a microemulsion at Point M* and pure-oil and -water phases. The composition and amounts of the phases can be determined by material balance, assuming that all the surfactant is in the microemulsion:

$$V_s = 9 \text{ cm}^3.$$

The composition of surfactant in the microemulsion is 23% (from the diagram):

$$\text{Thus, } 0.23 V_{me} = 9 \text{ cm}^3$$
$$V_{me} = 39 \text{ cm}^3.$$

Also from the diagram,

$$V_o = V_w.$$

This is true because the compositions of water and oil are equal in the microemulsion, according to the compositions on the diagram:

$$V_o + V_w + V_s = 39 \text{ cm}^3,$$
$$2V_o = 39 - 9 = 30 \text{ cm}^3,$$

and $V_o = V_w = 15 \text{ cm}^3.$

The composition of the microemulsion is 9 cm³ (23%) surfactant, 15 cm³ (38.5%) oil, and 15 cm³ (38.5%) water. For the lower phase (water),

$$V = 60 - 15 = 45 \text{ cm}^3.$$

For the upper phase (oil),

$$V = 31 - 15 = 16 \text{ cm}^3.$$

Fig. 7.21 illustrates an alternative way to present the phase change that Lake (1989) called a tent diagram. Fig. 7.21 also illustrates the existence of plait points on the binodal curve where the two phases are indistinguishable and the IFT approaches zero. In some systems, the excess phases are not necessarily pure water or oil as shown in the figure. Both of the equilibrium phases in a two-phase system may be micellar solutions. The plait point shifts with salinity, as Fig. 7.21 indicates.

The three-phase region is a characteristic of the thermodynamic properties of liquid systems and is not unique to surfactant/oil/brine systems. Because of this, an understanding of the evolution of the three-phase region with changing variables, such as salinity, can be developed by studying a pure-component system. Knickerbocker et al. (1979) obtained phase-behavior data for brine/hydrocarbon/alcohol systems and demonstrated that three liquid phases would form for many hydrocarbons. Because these systems do not form micelles, the presence of the three-phase region is not caused by micelle formation.

Knickerbocker et al. (1982) developed a model for the evolution of the three-phase region as one of the principal variables, such as salinity, increases. The elements of this model are depicted in **Fig. 7.22,** where a series of pseudoternary diagrams is shown with hydrocarbon, alcohol, and brine as the apices. Each pseudoternary diagram represents constant brine salinity. It is assumed that NaCl does not partition between phases. At low salinities, a two-phase system is present in which an oil phase is in equilibrium with an alcohol-rich brine. A plait point is shown where the two liquid phases are indistinguishable. As salinity increases, the lower alcohol-rich phase becomes saturated with respect to NaCl and a third phase erupts from this critical tie-line. Note that there is little solubility of NaCl in the hydrocarbon phase except near the plait point. The critical endpoint (CEP)

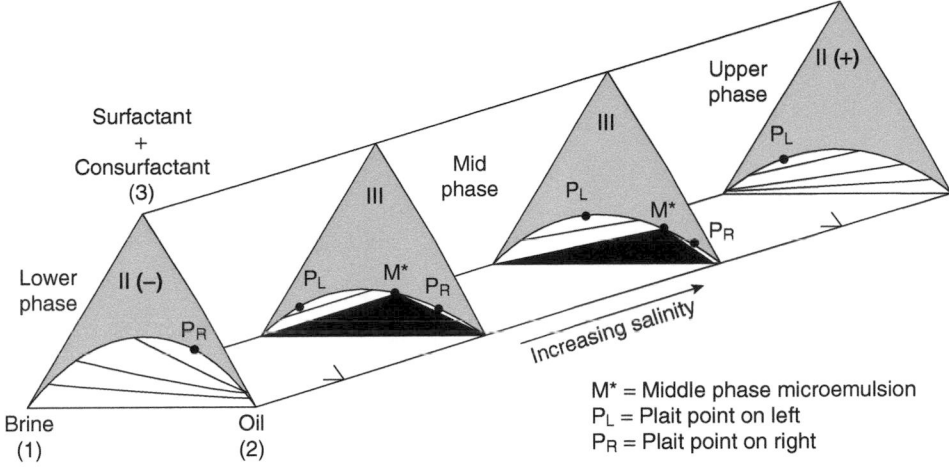

Fig. 7.21—Phase behavior, effect of salinity (Lake 1989).

of this tie-line is indicated on Fig. 7.22. The three-phase region is shaded and is surrounded by three two-phase regions. One two-phase region has alcohol-rich phases, while the other two-phase region has hydrocarbon-rich phases. There are two plait points on the pseudoternary diagram.

One corner of the new three-phase region is the middle-phase composition, Point M*, a phase that is saturated with respect to alcohol, brine, and hydrocarbon at a specific total salinity. As salinity increases, the middle-phase composition moves toward the brine baseline as the capability of the middle phase to solubilize brine is reduced. At a specific salinity, the three-phase region merges into a two-phase region where an alcohol-rich hydrocarbon phase is in equilibrium with an alcohol-poor brine phase. The three-phase region disappears at the CEP indicated in the upper part of Fig. 7.22, where the alcohol-rich hydrocarbon phase becomes indistinguishable from the middle phase. Distances have been exaggerated on Fig. 7.22 to illustrate concepts. CEPs are difficult to find, and the transition may occur over a narrow concentration range. However, the concepts introduced by Fig. 7.22 are sound and provide the basis for understanding phase behavior in microemulsion systems.

7.4.3 Phase-Behavior Representation on a Volume-Fraction Diagram.
Phase-behavior data also may be represented on a volume-fraction diagram, as **Fig. 7.23** (Malmberg et al. 1982) illustrates. In this diagram, the relative volume of each phase is plotted as a function of salinity. This data set is often called a "salinity scan." The overall composition of surfactant and cosurfactant is fixed. Phases are plotted as they would appear in a test tube (i.e., the upper phase is at the top of the diagram). As salinity increases, the system proceeds from lower- to middle- to upper-phase microemulsion.

7.4.4 Solubilization Parameters.
The volume of oil and brine that can be solubilized by a microemulsion is of interest in characteriz-

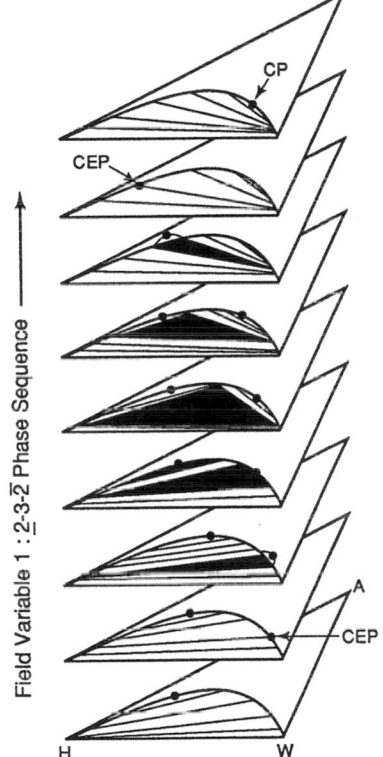

Fig. 7.22—Effect of salinity on phase behavior for an alcohol/brine/hydrocarbon system. (Reprinted with permission from Knickerbocker, B.M. et al., "Patterns of here-Liquid Phase Behavior Illustrated by alcohol-Hydrocarbon-Water-Salt Mixtures," *J. Phys. Chem.* Copyright 1982 American Chemical Society.)

ing a surfactant system. Healy et al. (1976) expressed the amounts of oil and water solubilized by a unit of surfactant in terms of solubilization parameters. Solubilization parameters are defined as follows (Bourrel and Schechter 1988).

$$P_o = \frac{V_o}{V_s} = \frac{\text{volume of oil in microemulsion phase}}{\text{volume of surfactant in microemulsion phase}}. \quad \dots \dots \dots \dots \dots \dots \dots \dots \dots (7.7)$$

$$P_w = \frac{V_w}{V_s} = \frac{\text{volume of water in microemulsion phase}}{\text{volume of surfactant in microemulsion phase}}. \quad \dots \dots \dots \dots \dots \dots \dots \dots \dots (7.8)$$

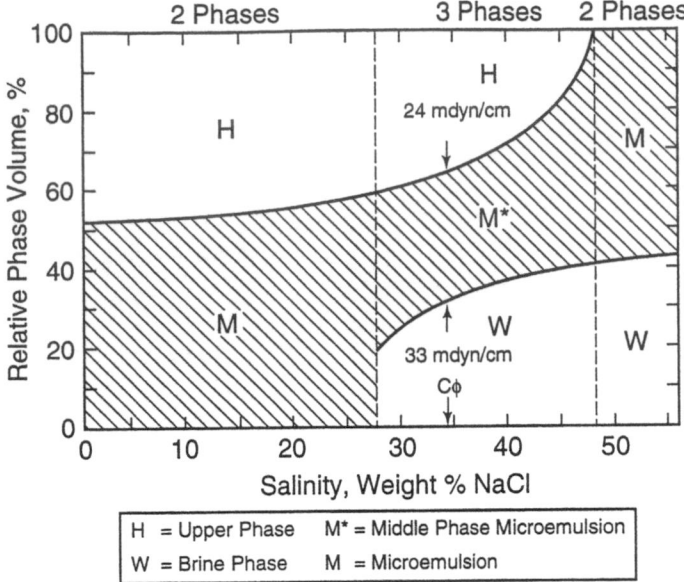

Fig. 7.23—Volume-fraction diagram (2% TAA and 1% iso-octane) (Malmberg et al. 1982).

V_s is the volume of surfactant in the system and includes no cosurfactant. The surfactant is assumed to be in the microemulsion phase and not in the excess-oil or -water phase.

Fig. 7.24 is a typical plot showing the solubilization parameters as functions of salinity. These data and much of the data to follow were presented by Healy et al. (1976). The surfactants used were anionic and were monoethanol amine salts of alkylorthoxylene sulfonic acid (MEACNOXS). The N in MEACNOXS represents the carbon number or the alkyl side chain and is 9, 12, or 15 for their data. The neutralized form of the molecule is a sulfonate. The surfactants were supplied by Exxon Chemical Co. USA. The number of carbon atoms in the side chain of the surfactant used for Fig.7.24 is 12.

The microemulsion systems were made with an alcohol cosolvent, either tertiary amyl alcohol (TAA) or tertiary butyl alcohol (TBA). The surfactant/alcohol ratio was typically 67 vol% surfactant to 33 vol% alcohol. The oil used was a mixture of paraffinic oil (denoted by I) and aromatic oil (denoted by N). The ratio was 90 vol% paraffinic to 10 vol% aromatic. Finally, equal volumes of brine and oil were used in the mixtures, usually 48.5% of each. The difference was the surfactant and the alcohol cosolvent.

In Fig. 7.24, data are shown for V_o/V_s in the lower- and middle-phase microemulsions, while data for V_w/V_s were taken in the middle- and upper-phase systems. In the upper-phase system, all the oil present is solubilized in the microemulsion phase. Thus, the solubilization parameter, V_o/V_s, is constant, assuming that all surfactant is in the upper phase. The corresponding reasoning holds for V_w/V_s and lower-phase microemulsions.

The solubilization parameters of the component in the excess phase increase or decrease monotonically with salinity but are equal at one point in the middle-phase region. The salinity at which the parameters are equal is called the optimal salinity for phase behavior. The manner in which these parameters are related to IFT will be discussed in Section 7.5.

Example 7.2—Optimal Salinity for Phase Behavior. Phase-behavior data are taken on a particular system. For these data, overall composition is held constant except for brine salinity. The overall composition is 47% brine, 47% oil, 4% surfactant, and 2% alcohol. Eight 100-mL samples with different salinities are mixed and allowed to equilibrate. The phase volumes presented in **Table 7.4** are then measured. **Fig. 7.25** shows the appearance of Samples 1 and 5 after equilibration.

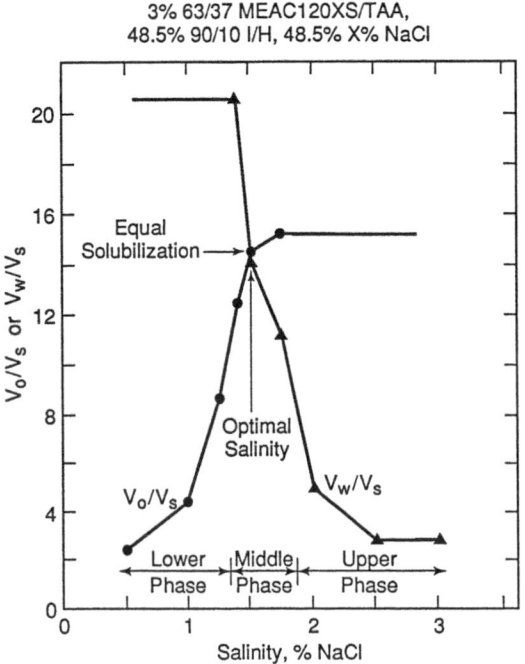

Fig. 7.24—Solubilization parameters as functions of salinity (Healy et al. 1976).

Sample	Salinity (wt%)	Microemulsion Type	Microemulsion Volume (%)	Excess-Oil Volume (%)	Excess-Water Volume (%)
1	0.50	Lower	65.0	35.0	0
2	0.75	Lower	67.4	32.6	0
3	1.00	Lower	72.2	27.8	0
4	1.25	Lower	78.6	21.4	0
5	1.50	Middle	79.6	13.4	7.0
6	1.75	Middle	72.0	8.6	19.4
7	2.00	Upper	69.0	0	31.4
8	2.25	Upper	65.8	0	34.2

Table 7.4—Phase volume data (Example 7.2).

1. Determine the optimal salinity on the basis of phase behavior (as defined by Healy and Reed 1974).
2. Calculate and plot on a ternary diagram the composition of the middle-phase microemulsion at optimal salinity. Also, show the three-phase region at optimal salinity on the ternary diagram.

Solution.

1. A plot of V_o/V_s and V_w/V_s is required. **Table 7.5** shows the volumes. V_s includes only the surfactant volume. The data are plotted in **Fig. 7.26**. The optimal salinity is 1.6% at $V_o/V_s = 8.8$.
2. At optimal salinity,

$$V_o/V_s = V_w/V_s = 8.8,$$

$$V_s = 4.0 \text{ cm}^3,$$

$$V_o = 8.8 \times 4 = 35.2 \text{ cm}^3,$$

and $V_w = 35.2 \text{ cm}^3$.

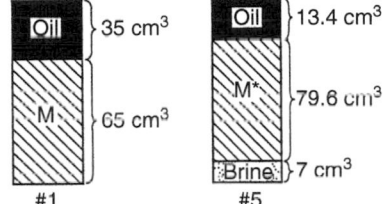

Fig. 7.25—Phase behavior, Samples 1 and 5, Example 7.2.

The composition of the microemulsion phase is 6.0 cm³ (7.9%) surfactant/alcohol, 35.2 cm³ (46.1%) oil, and 35.2 cm³ (46.1%) brine. Note that percent compositions sum to 100.1% as a result of rounding. **Fig. 7.27** plots the microemulsion composition. All the surfactant is assumed to be in the microemulsion and the upper and lower phases are assumed to be pure oil and brine, respectively.

7.4.5 Real Phase Behavior. Phase-behavior characteristics introduced in Sections 7.4.1 and 7.4.2 represent ideal behavior. Real micellar systems used in practice deviate from the ideal behavior presented by Healy et al. (1976). This section briefly summarizes several kinds of real phase behavior.

For a system that exhibits ideal behavior, the phase diagrams of lower-, middle-, and upper-phase systems are given by Fig. 7.16, which implies that multiple-phase regions are uniquely defined (i.e., there is a single-phase region composed of isotropic microemulsions above the multiphase regions and two-phase regions are developed). **Fig 7.28** is an example of phase behavior in a real system where several middle-phase compositions were found rather than the single Point M* indicated in Fig. 7.16.

Salinity	Type	Microemulsion Volume (%)	V_w(%)	V_o(%)	V_w/V_s	V_o/V_s
0.50	Lower	65.0	47.0	12.0	—	3.0
0.75	Lower	67.4	47.0	14.4	—	3.6
1.00	Lower	72.2	47.0	19.2	—	4.8
1.25	Lower	78.6	47.0	25.6	—	6.4
1.50	Middle	79.6	40.0	33.6	10.0	8.4
1.75	Middle	72.0	27.6	38.4	6.9	9.6
2.00	Upper	69.0	16.0	47.0	4.0	—
2.25	Upper	65.8	12.8	47.0	3.2	—

Table 7.5—Solubilization parameters for Example 7.2.

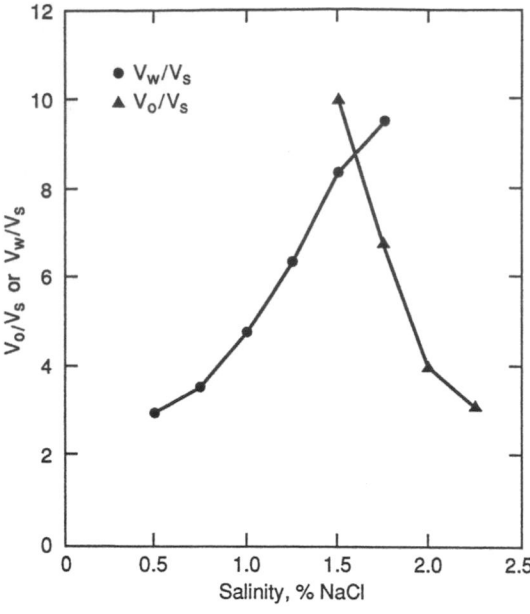

Fig. 7.26—Plot of V_o/V_s and V_w/V_s for Example 7.2.

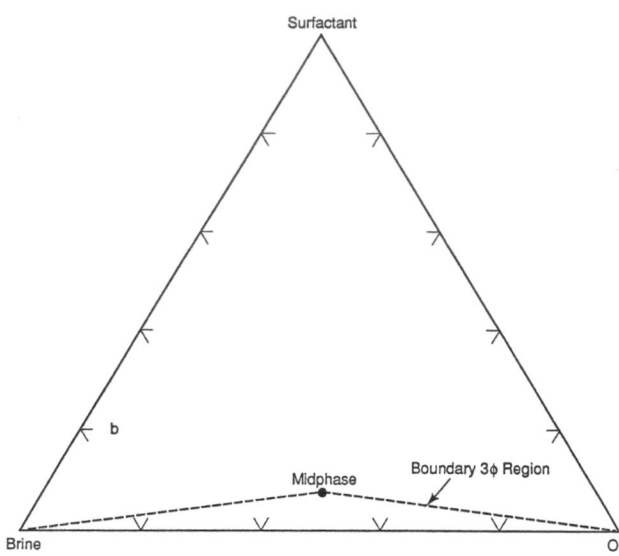

Fig. 7.27—Three-phase triangle, Example 7.2.

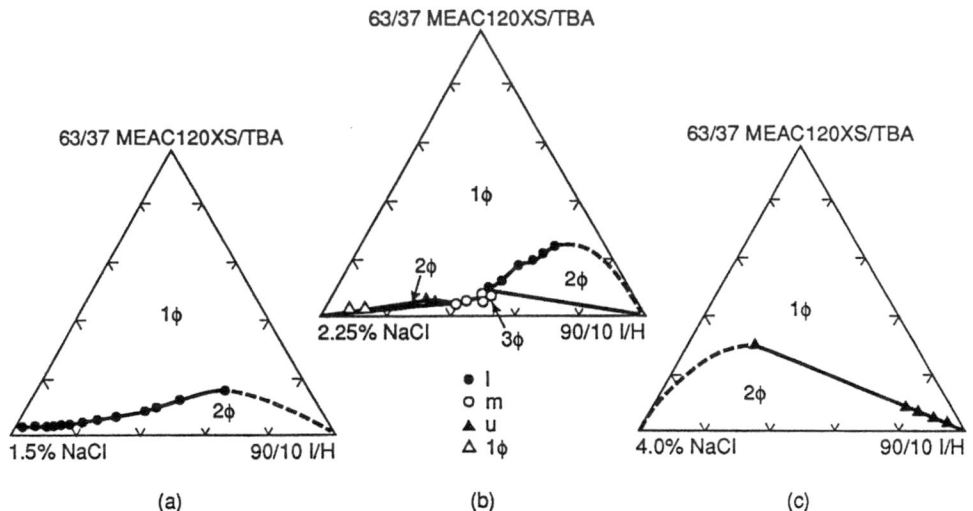

Fig. 7.28—Real phase behavior (Healy et al. 1976).

The two-phase region on the left of the three-phase region is flat and narrow. This behavior probably indicates fractionation of one or more of the components into a pseudocomponent.

Phase behavior often is more complex in the remainder of the phase diagram than is revealed in simple salinity scans where compositions are limited to relatively small regions of the phase diagram (usually at low surfactant concentrations). Salter (1983) developed partial pseudoternary diagrams for a surfactant system consisting of C_{11} paraethyl benzene sulfonate, 2.2% NaCl, and isooctane. There was no cosurfactant in this system. **Figs. 7.29a through 7.29c** show the structures, middle-phase compositions, and phase diagram for the lower part of this system. The phase behavior is quite complex.

Three regions are shown in Fig. 7.29a. At high surfactant concentrations (10 to 20 wt%), high oil concentrations (40 to 85 wt%), and relatively low brine concentrations (<7 wt%), a two-phase region forms containing a precipitate in equilibrium with an oil-rich microemulsion. This is not ideal phase behavior. A large multiphase formed just above the oil/brine boundary of the pseudoternary diagram. An isotropic microemulsion separated the precipitate region from the multiphase region.

The multiphase region contains several middle-phase compositions (Fig. 7.29b) and thus does not behave as an ideal system. This is particularly interesting because the surfactant is essentially monosulfonate, although there is a distribution of the sulfonate group between ortho and meta sites on the benzene ring. High-performance liquid chromatography analysis showed that the two surfactant species did not fractionate. Salter (1983) attributes part of the non-pseudocomponent behavior to fractionation of brine between equilibrium phases.

Several phases are birefringent and thus have some type of structure. Surfactant systems that have liquid crystalline structures are birefringent, which means that polarized light passes through the solution when viewed through crossed polars. Liquid crystals and precipitates may be removed in some systems by addition of a cosurfactant or by changing the temperature. In any case, real phase behavior is considerably different from ideal phase behavior and must be considered in designing a surfactant system for a particular reservoir. These results show that determination of the phase behavior of systems from a single scanned parameter, such as salinity, may underestimate the complexity of the phase diagram. Salter also observed that small amounts of fractionation of a surfactant blend had substantial effects on the phase behavior.

Pseudoternary Diagram Representation. Representation of phase behavior of a micellar solution on a ternary diagram requires the use of pseudocomponents. Typically, the cosurfactant and surfactant are lumped together as a single component and brine and oil are considered to be single components. This is only an approximation because the constituents of a pseudocomponent do not usually partition into the different phases in the exact ratio that they exist in the pseudocomponent. In some cases, salt partitions between surfactant and brine phases and is not a good pseudocomponent. In these cases, the total composition is represented by a single point on the pseudoternary diagram, but the compositions of the equilibrium phases are not on the same pseudoternary diagram because the salt concentration in the brine has changed. Phase behavior can be treated in a more complex way. Quaternary and pseudoquaternary diagrams have been used, for example (Salter 1978; Blevins et al. 1981).

Presence of Other Phases. Under some conditions, phases other than isotropic solutions have been observed. These phases typically have been highly viscous, sometimes consisting of liquid crystals (Healy and Reed 1974; Scriven 1976; Willhite et al. 1980). Some tend to occur at high surfactant concentrations and low temperatures.

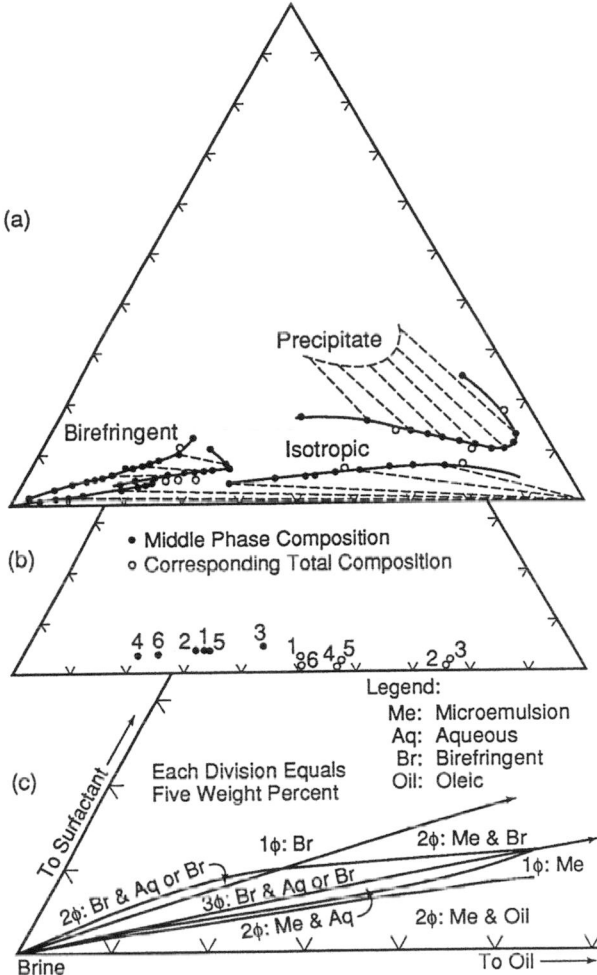

Fig. 7.29—Example of complex phase behavior: 2.2% brine/iso-octane system (Salter 1983).

Others are found in all parts of the phase diagram. Gel-like phases also have been observed when polymer is added to a micellar solution (Pope et al. 1982).

7.5 Phase Behavior and IFT

A strong correlation exists between phase behavior of a microemulsion system and the IFT between the equilibrium phases. Because the phase behavior progresses in an orderly way as brine salinity is increased, a reasonable conclusion is that IFT also would be a function of salinity, as has been demonstrated to be the case (Healy et al. 1976).

This section describes the relationships between IFT values of equilibrium phases and phase behavior, which depends on salinity. Section 7.6 describes the manner in which system parameters, such as temperature, pressure, and chemical composition, affect phase behavior and IFT. Later sections show that the performance of a microemulsion displacement is optimal when the flood is operated in the vicinity of optimal salinity. The consideration of optimal salinity is critical for design of a microemulsion flood.

7.5.1 IFT as a Function of Salinity.
Fig. 7.30 is a typical plot of IFT between equilibrium phases as a function of salinity. The quantity σ_{mo} is the IFT between the microemulsion phase and the excess-oil phase, while σ_{mw} is the IFT between the microemulsion and excess-brine phases. σ_{mo} decreases significantly as salinity increases as phase behavior progresses from a lower-phase system [Type II(−)], into a middle-phase system (Type III), and toward an upper-phase system [Type II(+)]. Change in this direction corresponds to increasing solubilization of oil from the excess phase into the microemulsion (Fig. 7.24), yielding a lower-density microemulsion.

Correspondingly, σ_{mw} increases as the progression from middle-phase to upper-phase occurs and water is driven out of the microemulsion phase. Also, σ_{mw} increases as the volume of water solubilized by the microemulsion decreases. Thus, a relationship exists between solubilization of the excess phases and the IFT between the microemulsion and the excess phases.

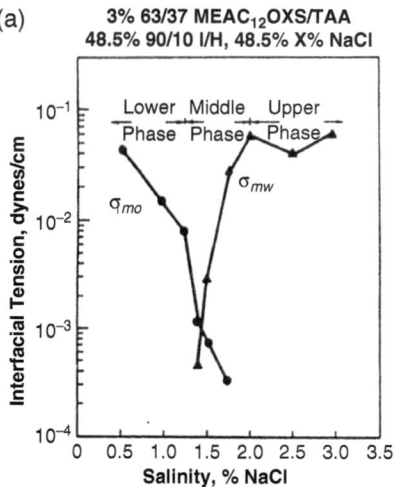

Fig. 7.30—IFT as a function of salinity (Healy et al. 1976).

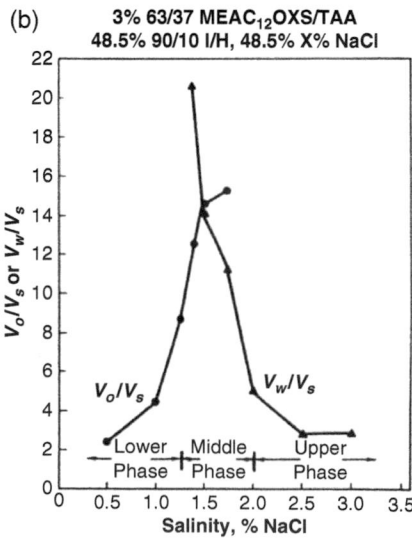

Fig. 7.31—Optimal salinities for IFT and phase behavior (Healy et al. 1976).

The IFT values σ_{mo} and σ_{mw} are not defined in the absence of an excess phase. For example, σ_{mo} is not defined in an upper-phase system because all the oil has been solubilized. There is no excess-oil phase in equilibrium with the microemulsion. Both σ_{mo} and σ_{mw} are defined in the middle-phase environment.

The IFT values shown in Fig. 7.30 are typical values measured in EOR systems. Recall that the IFT for an oil/water system (without surfactant) is typically approximately 30 dynes/cm. Thus, the values shown are three and four orders of magnitude below values encountered in a standard waterflood. In a flood with systems of the type shown in Fig. 7.30, the capillary number could be dramatically increased above that in a waterflood and trapped oil would be mobilized.

The value of salinity at which $\sigma_{mo} = \sigma_{mw}$ is called the optimal salinity for IFT (Healy et al. 1976). This salinity is usually very close to the optimal salinity for phase behavior previously defined as the salinity for which $V_o/V_s = V_w/V_s$, as is shown in **Fig. 7.31.**

The fact that optimal salinity for phase behavior is essentially equal to optimal salinity for IFT has an important practical result. IFT is relatively difficult to measure when tensions are ultralow. Such instruments as the spinning-drop apparatus or the pendant-drop apparatus (Cayias et al. 1975) must be used. However, measurement of solubilization parameters is relatively easy. Thus, for a specific system under consideration, one can first determine optimal salinity by the relatively easy phase measurements. It is generally believed that solubilization parameters must be on the order of 10 or greater to obtain ultralow IFTs suitable for EOR applications (Salter 1986). When the solubilization parameters are in the right range, IFT values can be measured in the vicinity of optimal salinity to determine whether the tensions are sufficiently small. This is a useful design procedure that is discussed later in the chapter.

7.5.2 Correlation of IFT and Solubilization Parameters. Section 7.5.1 demonstrated that the IFTs, σ_{mo} and σ_{mw}, are related to the solubilization parameters. Fig. 7.31a plots IFT vs. solubilization parameter for the same data in Fig. 7.31b. The data can be correlated empirically. One correlation that describes the data of Healy and Reed (1977) discussed in this section is of the form

$$\log\left(\sigma_{mo}/\sigma'_{mo}\right) = \frac{a}{m_o\left(V_o/V_s\right)+1} \quad\ldots\ldots\ldots\ldots\ldots\ldots (7.9)$$

and $$\log\left(\sigma_{mw}/\sigma'_{mw}\right) = \frac{b}{m_w\left(V_w/V_s\right)+1}, \quad\ldots\ldots\ldots\ldots\ldots (7.10)$$

where a, b, m_o, m_w are constants and $\log \sigma'_{mw}$ and $\log \sigma'_{mo}$ are intercept values obtained from experimental data. Healy and Reed (1977) reported the following values for the constants and intercepts: $a = 6.285$, $m_o = 0.04477$, $\log \sigma'_{mo} = -7.058$, $b = 12.167$, $m_w = 0.01280$, and $\log \sigma'_{mw} = -12,856$. The constants in Eqs. 7.9 and 7.10 are specific to the surfactant and oil used to generate the phase behavior and interfacial data and to the temperature.

A similar equation has been proposed by Nelson (1982) to be more general than Eqs. 7.9 and 7.10:

$$\log \sigma_{mo,mw} = \frac{4.80}{1+0.210\left(V_{o,w}/V'_s\right)^{-5.40}}, \quad\ldots\ldots\ldots\ldots (7.11)$$

where $\sigma_{mo,mw}$ = IFT at the microemulsion/oil or the microemulsion/water interface, dynes/cm; $V_{o,w}$ = volume of oil or water in the microemulsion phase, cm^3; and V_s' =volume of surfactant plus alcohol in the microemulsion phase, cm^3. Note that V_s' is defined differently from V_s, which has been used previously. By definition, $V_o = V_w$ at optimal salinity and $\sigma_{mo} \approx \sigma_{mw}$. Eq. 7.11 may be used with phase-behavior solubilization data to estimate σ_{mo} and σ_{mw} at optimal salinity. It is emphasized that Eq. 7.11 is an empirical relationship based on correlating a limited amount of data. As noted earlier, empirical correlations such as Eq. 7.11 are specific to the data used to obtain the parameters and should not be extrapolated beyond the data used to establish the correlation.

Huh (1979) developed a theoretical relationship between the solubilization parameter and IFT for a middle-phase microemulsion. Huh's model envisions the middle phase to consist of alternating layers of oil and water with surfactant at the interfaces. For the oil/microemulsion interface,

$$\frac{P_o^2 \sigma_{mo}}{\cos\left[(\pi/2)\phi_1\right]} = a_H. \dots\dots\dots\dots\dots\dots\dots\dots\dots\dots\dots\dots\dots\dots (7.12)$$

For the water/microemulsion interface,

$$\frac{P_w^2 \sigma_{mw}}{\cos\left[(\pi/2)\phi_2\right]} = a_H, \dots\dots\dots\dots\dots\dots\dots\dots\dots\dots\dots\dots\dots (7.13)$$

where σ_{mo} = IFT tension between the microemulsion and the oil phase, dynes/cm; σ_{mw} = IFT tension between the microemulsion and the excess-water phase, dynes/cm; ϕ_1 = volume fraction of oil in the middle phase on a surfactant-free basis $[P_o/(P_o+P_w)]$; ϕ_2 = volume fraction of water in the microemulsion on a surfactant-free basis $[P_w/(P_o+P_w)]$; and a_H = empirical constant usually determined experimentally, dynes/cm.

At optimal salinity, $P_o = P_w = P^*$ and $\sigma_{mo} = \sigma_{mw} = \sigma^*$. Because $\phi_1 = \phi_2 = \frac{1}{2}$, Eqs. 7.12 and 7.13 become

$$\frac{\left(P^{*2}\right)\sigma^*}{\cos(\pi/4)} = a_H, \dots\dots\dots\dots\dots\dots\dots\dots\dots\dots\dots\dots\dots (7.14)$$

where σ^* = IFT at optimal salinity, dynes/cm. For regions near optimal, Eqs. 7.12 and 7.13 may be used to estimate IFT in the two-phase region.

Values of a_H have been determined experimentally for alkylbenzene sulfonates (0.48 ± 0.05 dynes/cm) and ethoxylated alkylphenols (0.34 ± 0.06 dynes/cm) (Graciaa et al. 1982). Barakat et al. (1983b) reported a value of 0.40 ± 0.15 dynes/cm for alkane and alpha-olefin sulfonates. Verkruyse and Salter (1985) obtained similar values of a_H for nonionic surfactants (alkylbenzene sulfonates) after accounting for the surfactant that was in the oil phase in the computation of the solubilization parameters.

Example 7.3—Phase Behavior and IFT. Fig. 7.32 shows the phase diagram for a surfactant system under investigation. All compositions are given in volume fractions. The surfactant consists of a mixture of petroleum sulfonate and alcohol in a 2:1 ratio.

1. Determine the number, type, and composition of phases formed when a mixture consisting of 70 mL brine and 30 mL oil is added to 600 mL of solution containing 0.15 surfactant and 0.85 oil.
2. IFT data for this phase diagram are correlated with Eqs. 7.9 and 7.10 with specific constants.

$$\log(\sigma_{mo}) = -7.058 + \frac{6.285}{0.04477(V_o/V_s)+1}$$

$$\text{and } \log(\sigma_{mw}) = -12.856 + \frac{12.167}{0.01280(V_w/V_s)+1},$$

where σ = IFT, dynes/cm; V_o = oil solubilized in the microemulsion volume, cm^3; V_w = water solubilized in the microemulsion volume, cm^3; and V_s = volume of sulfonate solubilized in the microemulsion, cm^3. Determine the IFT(s) between equilibrium phases (if any) found in Part 1.

3. Use Eq. 7.11 to calculate the IFT for the equilibrium phases. Compare the results of Parts 2 and 3.

Solution.

1. The overall composition is as follows.

Oil, 30 + 0.85 × 600	=	540 cm³	77.1 vol%
Brine	=	70 cm³	10.0 vol%
Surfactant, 0.15 × 600	=	90 cm³	12.9 vol%

The surfactant includes petroleum sulfonate and alcohol. Locate overall composition on the ternary diagram. It is in the upper two-phase lobe. Construct a tie-line through the overall composition and the 100% oil apex. The equilibrium phases are (refer to **Fig. 7.33**) a microemulsion phase with 50% oil, 22% brine, and 28% surfactant and a 100% oil phase.

2. To find the IFT between phases, determine σ_{mo} (no-excess-brine phase). V_o and V_s (petroleum sulfonate only) are needed.

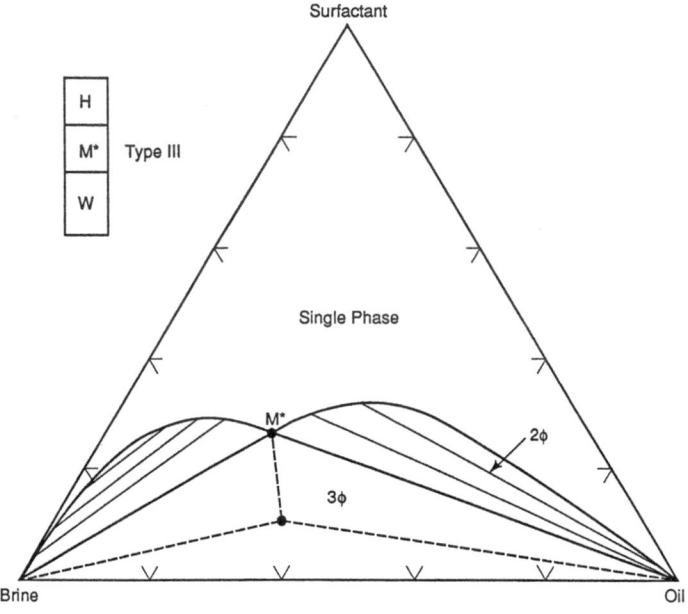

Fig. 7.32—Ternary diagram, Example 7.3 (representative tie-lines shown).

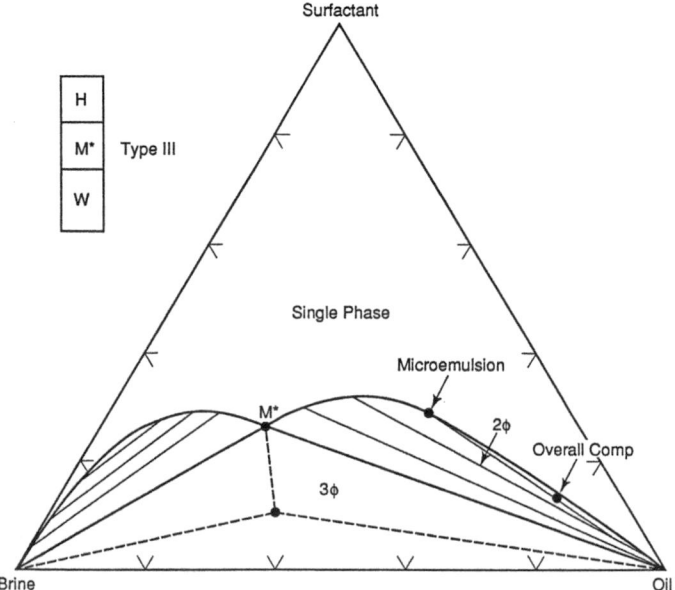

Fig. 7.33—Phase behavior, Example 7.3.

3. $0.28 \, V_{me} = 9 \text{ cm}^3$

    and $V_{me} = 321.4 \text{ cm}^3$,

    where V_{me} = volume of the microemulsion phase, cm^3.

    $$V_s = \tfrac{2}{3} \times 90 = 60 \text{ cm}^3.$$

    The surfactant/brine ratio is specified to be 2:1.

    $$V_o = 0.50 \times 321.4 = 160.7 \text{ cm}^3.$$
    $$V_o / V_s = 160.7 / 60 = 2.68.$$

    $$\log \sigma_{mo} = -7.058 + \frac{6.285}{0.04477 \left(V_o / V_s' \right) + 1}$$
    $$= -7.058 + \frac{6.285}{0.04477 \times 2.68 + 1}$$
    $$= -1.446.$$

    $\sigma_{mo} = 0.036 \text{ dynes/cm.}$

4. From Eq. 7.11

    $$\log \sigma_{mo} = \frac{4.80}{1 + 0.21 \left(V_o / V_s' \right) - 5.40}$$
    $$= \frac{4.80}{1 + 0.21 \left(160.7 / 90 \right)^{-5.40}}$$
    $$= -1.909.$$

    $\sigma_{mo} = 0.012 \text{ dynes/cm.}$

The agreement is poor, indicating the approximate nature of Eq. 7.11.

7.5.3 Reasons for Ultralow IFT in Microemulsion Systems. The causes of ultralow IFTs have been extensively investigated. Low IFTs are known to be closely associated with phase behavior near critical points. For example, at the plait point of a liquid/liquid system, two phases become indistinguishable and the IFT between two equilibrium phases goes to zero. The plait point is one of the critical points for a solution of a given composition. The fact that microemulsion systems exhibit ultralow IFTs over wide ranges of salinities, surfactant concentrations, and temperatures suggests that a critical phenomenon is involved. Several papers provide support for this interpretation (Knickerbocker et al. 1982; Davis and Scriven 1980). The general explanation offered from current research is that surfactant/oil/brine systems at optimal salinity are near the tricritical point where the three phases become chemically indistinguishable and thus exhibit ultralow IFTs between all phases.

7.6 Variables Affecting Phase Behavior and IFT

A number of variables affect phase behavior and solubilization parameters, and thus IFT, including temperature; types of ions in the brine phase, alcohol, and oil; water/oil ratio (WOR); surfactant structure; addition of polymer to the solution; and pressure. Examples of the effects of changing several of these parameters will be given. These examples provide information to guide the development of design criteria for the selection of a surfactant system for a particular reservoir oil. A number of the examples are from Healy et al. (1976) discussed in Section 7.4.3. Their examples are for systems at a WOR of 1:1; however, the systems were not very sensitive to this parameter.

7.6.1 Effect of Oil Type. The system shown in Fig. 7.31 contained 10 vol% aromatics in the oil constituent, as discussed in Section 7.4.3. When this aromatic fraction is replaced by additional paraffin oil, optimal salinity is increased, as shown in **Fig. 7.34** (Healy et al. 1976). For this system, increasing aromaticity causes optimal salinity and the IFT value at optimal salinity to decrease. If the aromaticity is decreased while holding salinity constant, the phase behavior shifts from lower- to middle- to upper-phase microemulsions. Values of σ_{mo} and V_w / V_s decrease while σ_{mw} and V_o / V_s increase.

Hydrocarbon/brine/surfactant systems have been studied extensively to develop a systematic approach to correlate phase behavior with hydrocarbon composition. These studies are also useful in selecting a surfactant for a particular hydrocarbon/brine system. Cayias et al. (1976) observed that IFT minima for a series of pure-alkane/0.2%-surfactant/1%-NaCl-brine systems occurred at an alkane carbon number (ACN) of 8. Mixtures of pure alkane hydrocarbons also had IFT minima at an equivalent ACN (EACN) of 8. The EACN is the sum of the mole-fraction-weighted ACN of each pure species. This concept was extended to crude oils by determining the EACN of several crude oil systems. It was proposed that a single EACN would

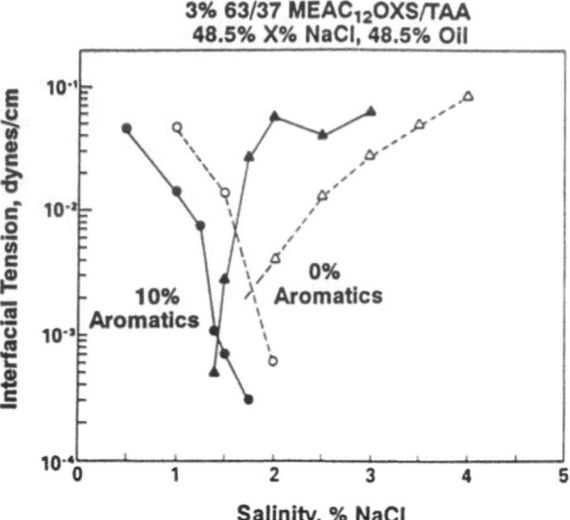

Fig. 7.34—IFT, effect of oil (Healy et al. 1976).

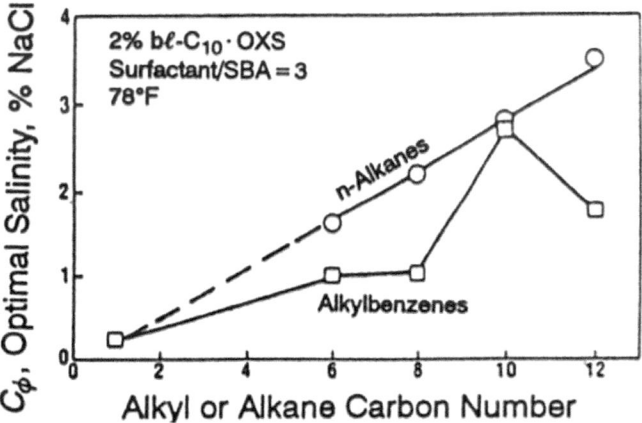

Fig. 7.35—Divergence of alkanes from alkylbenzenes (Puerto and Reed 1983).

be found that characterized a particular crude oil. In subsequent research (Salager et al. 1979), a correlation was developed between optimal salinity and EACN. This concept has been used in the design and evaluation of oils for field tests (Glinsmann 1979) and for determining the effect of using live crude oil on surfactant flooding efficiency (Nelson 1983).

The concept of EACN is not universally applicable. Variations in EACN with alcohol cosolvent type, total WOR of the sample, and crude oil composition have been reported (Tham and Lorenz 1981). Reproducibility of IFT minima was questioned by Shah et al. (1981). Significant differences were observed between alkanes and alkylbenzenes when optimal salinity was correlated with ACN (Puerto and Reed 1983), as shown in **Fig. 7.35.** Puerto and Reed (1983) concluded that there was no simple relationship between optimal salinity and the carbon numbers of n-alkanes and n-alkylbenzenes or n-alkylcyclohexanes.

Puerto and Reed (1983) developed a three-parameter correlation of microemulsion phase behavior. They found that three parameters—optimal salinity, C_ϕ; solubilization parameter, V_o/V_s; and oil molar volume, V_{mo}—improved correlation of microemulsion phase behavior with the type of oil for a wide range of conditions. Such a correlation is shown in **Fig. 7.36,** which represents the same data presented in Fig. 7.35. In Fig. 7.36, nC_i indicates normal alkane and $nC_{i\phi}$ is a normal alkylbenzene. Microemulsions at optimal salinities form within the region defined by the three-parameter system. Dashed lines of constant solubilization parameter are also shown. It is clear from Fig. 7.36 that it is necessary to supplement the ACN with the molar volume to account for the complex phase behavior of mixtures of hydrocarbons in microemulsions. A unique three-parameter representation was found for a given surfactant system when oils of high or low molar volume were excluded.

Comparison of three-parameter correlations for different surfactant systems and temperatures revealed important trends that can have significant impact on the design of microemulsion systems. Effects of temperature and alcohol cosolvent are shown in **Fig. 7.37** for n-alkanes for three different surfactant systems. One surfactant system, bl-C12BTXS, formed middle-phase microemulsions with high solubilization parameters at 140 and 200°F without use of alcohol. The optimal salinity, C_ϕ, for this surfactant was independent of temperature. The other two surfactants required use of secondary butyl alcohol (SBA) to obtain middle-phase microemulsions over the same range of oil molar volumes. For these surfactants, increasing temperature increased the optimal salinity at the expense of reduced solubility parameters.

Three-parameter correlations lead to the possibility of identifying equivalent oils for modeling live crude oil. Equivalent oils have the same optimal salinity, solubilization parameter, and molar volume. Puerto and Reed (1983) present mixing rules and examples of the use of a three-parameter correlation.

The effect of oil type on phase behavior can be very dramatic. **Fig. 7.38** (Maerker and Gale 1992) shows solubilization parameters vs. salinity for two different oils and for a particular surfactant mixture studied for use in a field test (see Section 7.13). Salinity is represented as percent of a Tar Springs brine (TSB), a brine of approximately 10% salinity with significant concentrations of Ca++ and Mg++ ions. One hundred percent TSB on the abscissa corresponds to approximately 10% salinity.

As Fig. 7.38a shows, the behavior was classic when a diesel oil was used as the oleic phase. When Loudon crude oil was used, however, a nonideal behavior resulted. V_o/V_s increased with increasing salinity at lower salinities, but then suddenly decreased. An optimal salinity, as previously defined, did not exist for this system.

The existence of this and other nonidealities discussed in this chapter should be kept in mind.

7.6.2 Effect of Cosurfactant Type. Alcohol cosurfactants were originally added to surfactant systems to increase solubility of certain surfactants and to alter the viscosity of the system. **Table 7.6** contains compositions of several micellar systems

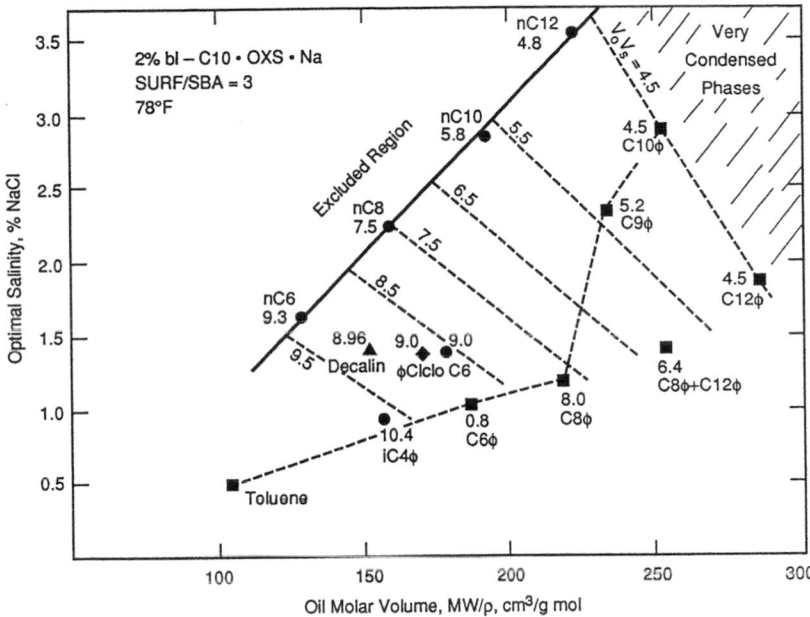

Fig. 7.36—Three-parameter representation of optimal salinity region (Puerto and Reed 1983).

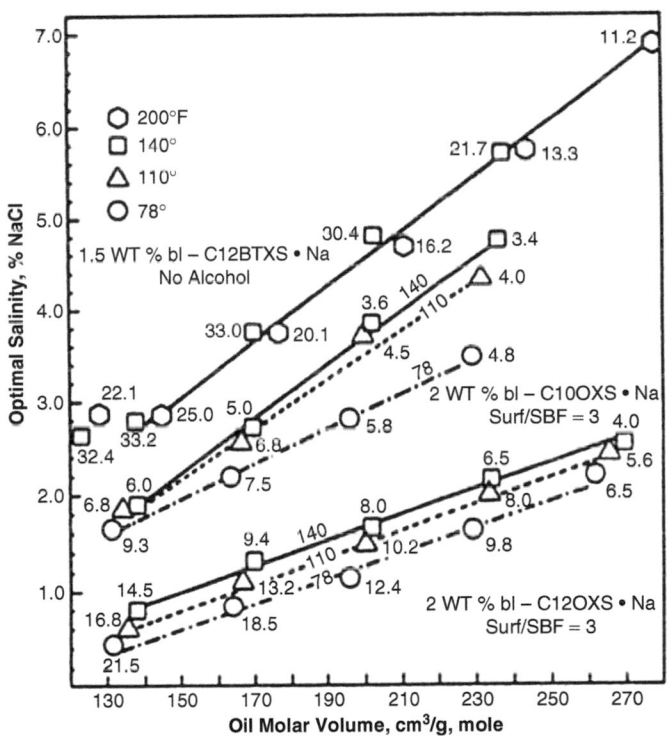

Fig. 7.37—Effect of temperature on alkane line: three surfactants (Puerto and Reed 1983). Solubility parameters are indicated adjacent to each data point.

before the addition of alcohol (Jones and Dreher 1976). Included in Table 7.6 are alcohols Jones and Dreher (1976) studied to determine the effect of alcohol on these micellar systems. Water-soluble alcohols (lower molecular weight) make a microemulsion more hydrophilic (i.e., to increase in capacity to solubilize water but decrease in capacity to solubilize oil). Alcohols with low water solubility (higher molecular weight), such as pentanol and hexanol, have the opposite effect: oil solubilization increases and water solubilization decreases.

It was soon realized that the cosurfactant type and concentration affect the phase behavior, IFT, and viscosity of a system (Wade et al. 1978). **Fig. 7.39** shows salinity scans for two different alcohol cosolvents, TAA and TBA. The surfactant is

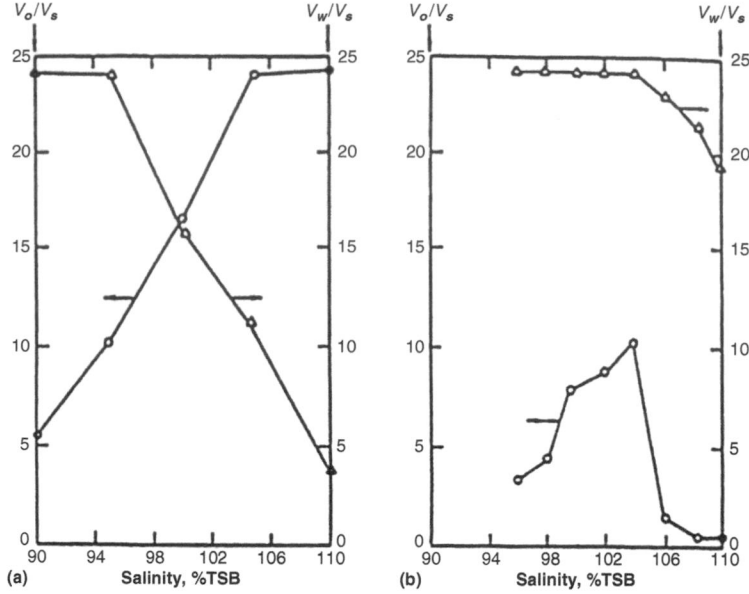

Fig. 7.38—Solubilization parameters vs. salinity at 78°F (2% of a 60/40 blend of i-C₁₃H₂₇O(PO)₄(EO)₂SO3Na/i-C₁₃H₂₇O(PO)₃(EO)₄SO₃Na in a 50/50 mixture of oil and brine where the oil component is (a) diesel oil or (b) Loudon crude oil (Maerker and Gale 1992).

	A	B	C	D	E	F
Composition (vol%)						
Petroleum sulfonate[1]	12.0	10.0	11.7	10.0	11.5	11.9
Added hydrocarbon[2]	63.1	40.0	22.8	40.0	33.7	53.1
Added water[3]	24.9	50.0	65.5	50.0	54.8	35
Average equivalent weight of sulfonate	470	440	420	440	424	397
Sulfonate activity (wt%)	62.0	60.8	61.6	60.8	62.0	49.8
Added hydrocarbon	LSRG[4]	60% IC[5]/40% HN[6]	IC	IC	IC	PC[7]
Total water content of micellar slug (wt%)	28.1	54.5	70.0	54.5	60.0	41.7
Electrolyte concentration in slug water (ppm)[8]	2,400	3,900	10,000	3,900	13,100	10,600
Electrolyte	Na₂SO₄	(NH₄)₂SO₄	(NH₄)₂SO₄	(NH₄)₂SO₄	(NH₄)₂SO₄	Na₂SO₄

Alcohols Used as Cosurfactants		
Cosurfactant	Source	Composition
2-propanol	Fisher Scientific Co.	A.C.S. grade
1-pentanol	Mallinckrodt	Analytical reagent
p-pentanol	Union Carbide Corp.	1-pentanol—62.3%
		2- and 3-methyl—1-butanol—37.1%
		2,2-dimethyl—1–propanol—0.6%
1-hexanol	J.T. Baker Chemical Co.	A.C.S. grade
p-hexanol	Union Carbide Corp.	1-hexanol—93.6%
		2-ethyl—1-butanol—5.9%
		1-butanol—0.1%
		Unknowns—0.4%
2-hexanol	Matheson, Coleman & Bell	Reagent grade
p-nonylphenol	K&K Laboratories	95% to 99% pure

[1] Except for Composition A, which contains sodium sulfonates (manufactured by Shell Chemical Co.), and Composition F, which contains a 44:56 (wt/wt) blend of sodium sulfonates (TRS-16/TRS-40), all surfactants are ammonium petroleum sulfonates made by Marathon Oil Co.
[2] Does not include the unsulfonated hydrocarbon contribution from the sulfonate.
[3] Does not include water from the sulfonate.
[4] Light, straight-run gasoline.
[5] Illinois crude: 37 °API; 7- to 9-cp viscosity; 1,474 refractive index.
[6] Heavy naphtha.
[7] Platformer charge.
[8] Does not include the approximately 400-ppm TDS contained in the added water.

Table 7.6—Micellar slug compositions before addition of alcohol (Jones and Dreher 1976).

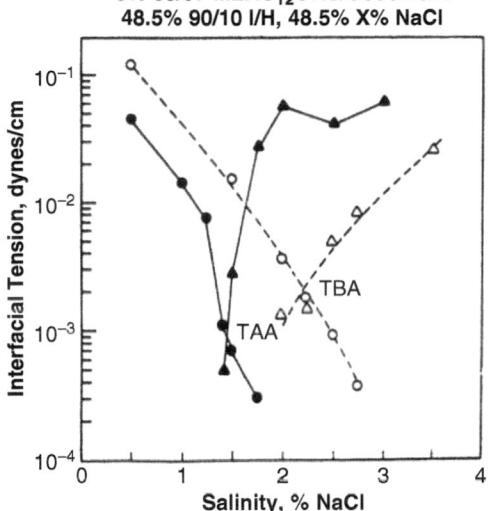

3% 63/37 MEAC$_{12}$OXS/Cosolvent
48.5% 90/10 I/H, 48.5% X% NaCl

Fig. 7.39—IFT, effect of alcohol (Healy et al. 1976).

the same in both cases and is the same as for the system in Fig. 7.31. Decreasing the molecular weight of the alcohol (making the alcohol more water-soluble) shifted optimal salinity to a higher value and increased the IFT value at optimal salinity. From this it can be inferred that the optimal solubilization parameter decreased. With TBA, the optimal salinity is 2.2% NaCl and the IFT at this point is 2.2×10^{-3} dynes/cm. For TAA, the corresponding values are 1.4% and 9×10^{-4} dynes/cm.

At a fixed salinity, an increase in molecular weight of the alcohol (making the alcohol more oil-soluble) causes the phase behavior to tend in the direction of lower to middle to upper phase. Corresponding to this, σ_{mo} and V_w/V_s decrease while σ_{mw} and V_o/V_s increase (Healy et al. 1976).

Salter (1977) conducted extensive studies on systems containing alcohols and mixtures of alcohols (C_3 to C_{14}) by using salinity scans.

In general, optimal salinity was affected by the type and amount of alcohol present in the system. **Fig. 7.40** illustrates the effect of alcohol concentration on optimal salinity and optimal IFT for a system in which the surfactant was Amoco Mahogany AA® sulfonate and the alcohol was TBA. Because TBA is quite water-soluble, the optimal salinity increased with increasing alcohol concentration. **Fig. 7.41** summarizes optimal salinities and optimal IFTs for a series of alcohols ranging from isopropanol (IPA) to normal hexanol (NHA) with the same system. Highly oil-soluble alcohols like isoamyl alcohol, normal amyl alcohol, and NHA cause the optimal salinity to decrease with increasing alcohol concentration.

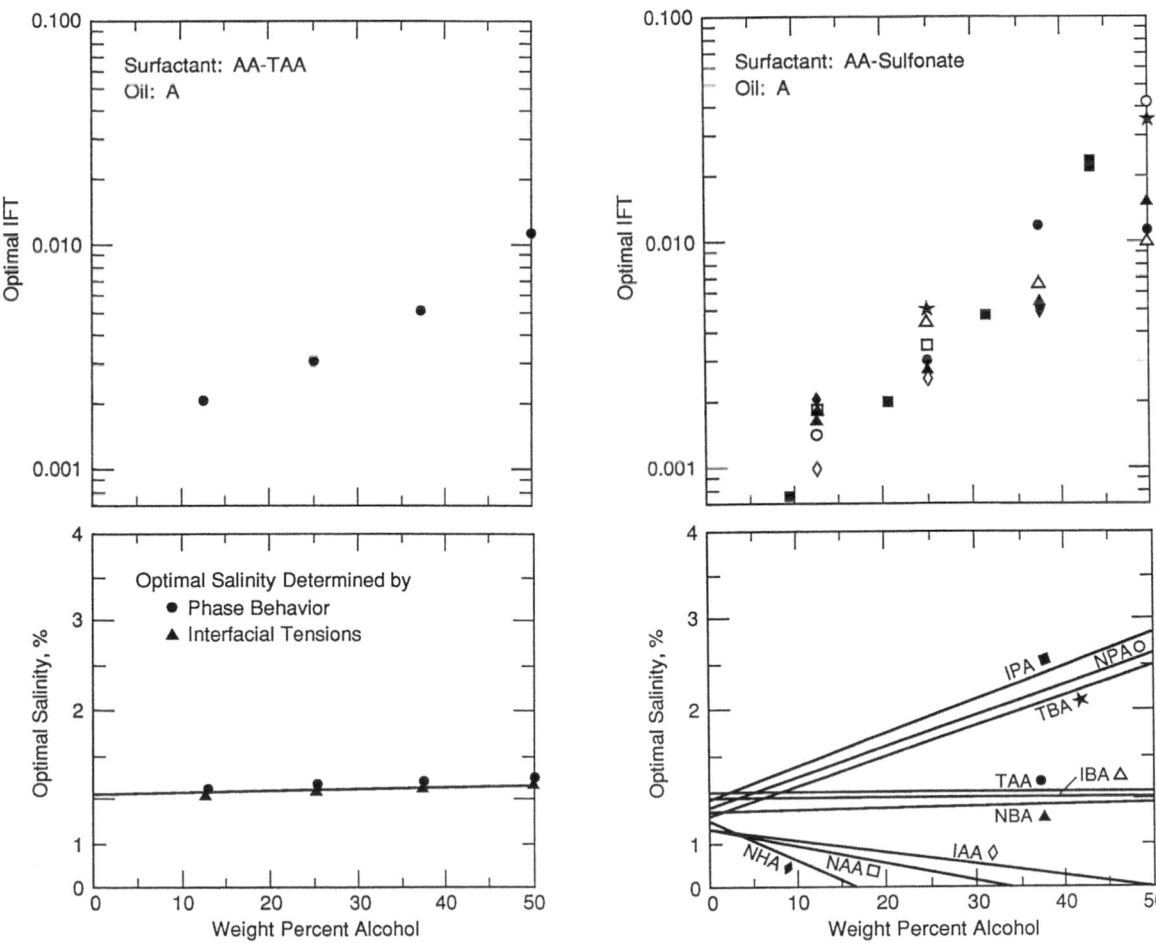

Fig. 7.40—Variation of optimal salinity and optimal IFT for AA/TBA system (Salter 1977).

Fig. 7.41—Variation of optimal salinity and optimal IFT for AA-9 alcohols (Salter 1977).

The discussion of microemulsion systems up to this point has assumed that an alcohol is the cosurfactant. The surfactant component shown on the pseudoternary diagrams is composed of a surfactant and alcohol. These systems are effective for oil displacement as long as the surfactant and cosurfactant do not separate as they are displaced through the porous rock and the designed phase environment can be maintained. Some separation is inevitable because retention of surfactant usually is larger than retention of the cosurfactant when the cosurfactant is alcohol. This was confirmed in laboratory (Willhite et al. 1980; Gupta 1984) and field tests (Miller et al. 1980).

7.6.3 Effect of Temperature. An increase in temperature causes solubilization parameters V_o/V_s and V_w/V_s to decrease at optimal salinity, increasing the IFT and shifting the optimal salinity for a given system to a higher value. This is shown in **Fig. 7.42,** where the optimal solubilization parameter decreased from 14.5 to 7.5 as the temperature increased from 74 to 150°F and the optimal salinity increased from 1.6 to 2.1%. IFTs at optimal salinity (**Fig. 7.43**) increased with increasing temperature, as expected from the change in solubilization parameters.

Fig. 7.44 (Novosad 1982) shows the effect of temperature for two additional systems. The figure shows the relative amounts of the different phases that would exist at different temperatures. Texas No. 1® and PDM 337® are commercial surfactants. Concentrations of propanol cosurfactant and NaCl brine are given in the caption.

As temperature is increased for a system at a specified concentration, the phase shifts from upper- to middle- to lower-phase environments. This is consistent with the data shown in Figs. 7.42 and 7.43.

Additional insight into the effect of temperature on phase behavior and optimal salinity can be obtained from salinity scans for pure alkanes at different temperatures. **Figs. 7.45 through 7.47** (Puerto and Reed 1983) show phase maps for bℓ-C$_{10}$·OXS for pure alkane hydrocarbons at temperatures of 78, 110, and 140°F. Lower-, middle-, and upper-phase microemulsions are denoted by the indicated symbols. Tagged symbols ($\bullet$, $\square$) indicate the presence of very condensed phases (VCPs), which are gels, viscous phases, or precipitates. Note that an increase in temperature from 78 to 110°F removed the VCP. Puerto and Reed report the effect of temperature on VCP to resemble "melting" of VCPs. The dashed lines in Figs. 7.46 and 7.47 represent C_ϕ, the optimal salinity for phase behavior. The three-phase region widens and C_ϕ increases as temperature increases. This trend with temperature is not general. For example, the opposite trend occurs with oxylkylated sulfate surfactants used in the Loudon system. VCPs also can be removed by addition of an alcohol, but the solubilization parameter decreases.

7.6.4 Effect of Divalent Ions. Oilfield brines typically contain divalent ions Ca^{++} and Mg^{++}. These ions contribute to the hardness of a brine (i.e., the tendency to precipitate and an increased incompatibility with a surfactant). Divalent ions are present in the porous matrices of many reservoir rocks. Injected fluids can pick up divalent ions by dissolution and/or ion exchange, as discussed in Section 7.9.3. The divalent-ion content of a surfactant system may change markedly as it flows through the reservoir rock.

The presence of divalent ions shifts the optimal salinity to a lower value, as shown in **Fig. 7.48.** The original system contained NaCl as the salt in the brine and is the same system shown in Fig. 7.42. The brine was then modified by replacing the NaCl by a 10:1 mixture of NaCl/CaCl$_2$ · 2H$_2$O. Phase behavior and the values of σ_{mo} and σ_{mw} were affected.

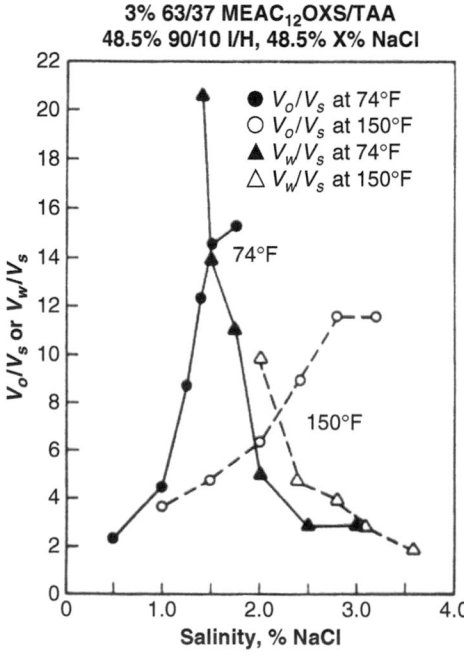

Fig. 7.42—Effect of temperature on solubilization parameters (Healy et al. 1976).

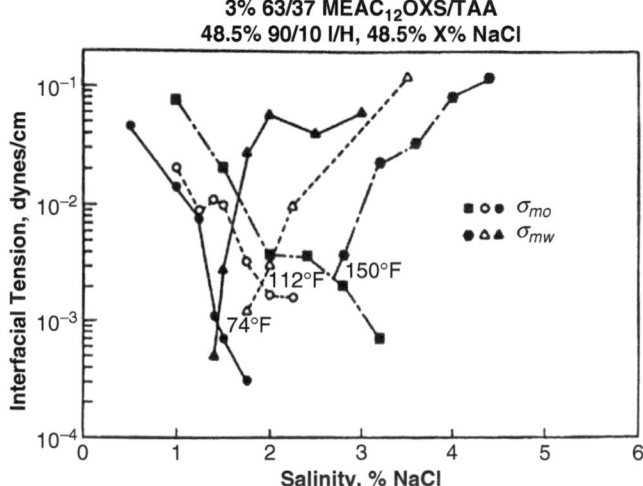

Fig. 7.43—Effect of temperature on IFT (Healy et al. 1976).

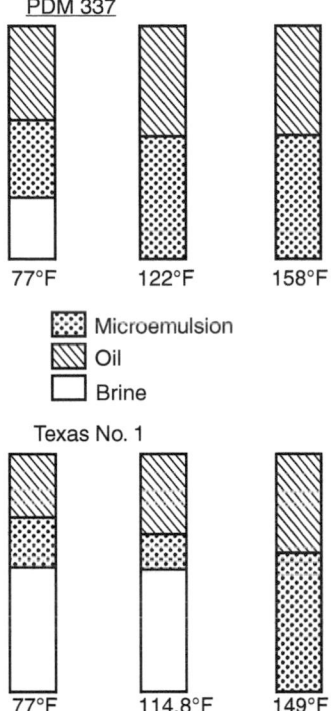

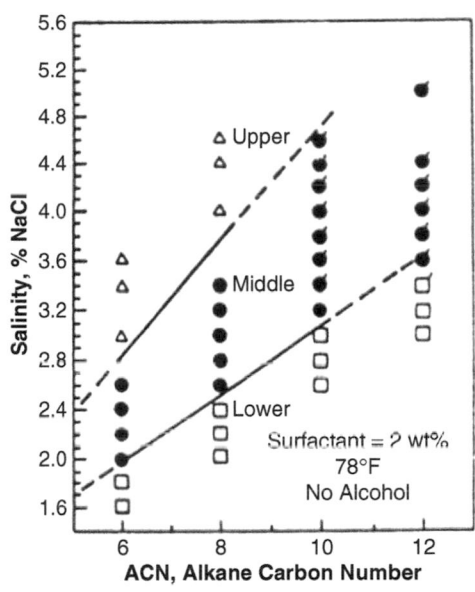

Fig. 7.44—Effect of temperature on phase behavior (2% 1:6 Texas No. 1/n-propanol in 1.5% NaCl; 3% 1: 1 PDM 337/SBA in 1.5% NaCl) (Novosad 1982).

Fig. 7.45—Phase map for bℓ-C10·OXS against pure alkanes; T = 78°F (Puerto and Reed 1983).

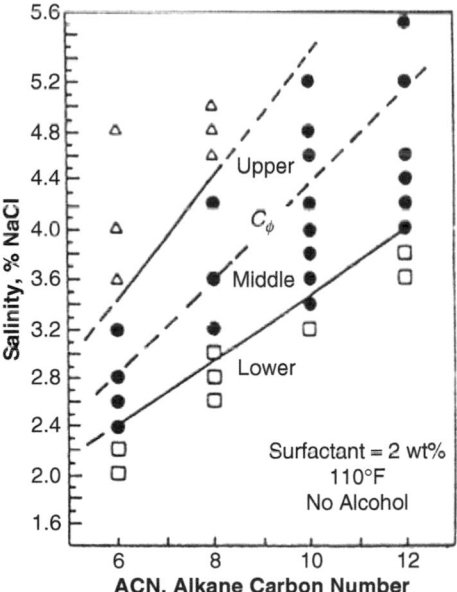

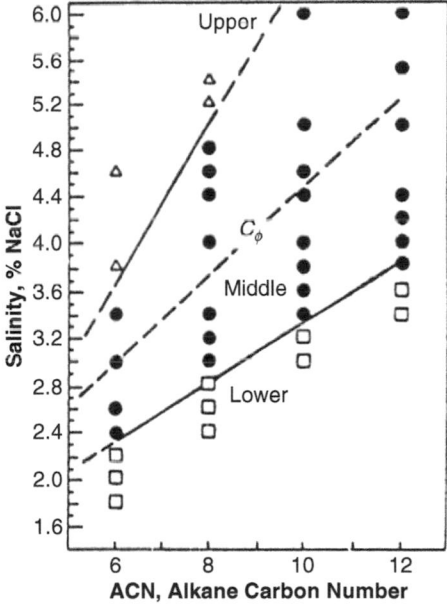

Fig. 7.46—Phase map for bℓ-C10·OXS against pure alkanes; T = 110°F (Puerto and Reed 1983).

Fig. 7.47—Phase map for bℓ-C10·OXS against pure alkanes; T = 110°F (Puerto and Reed 1983).

The IFT curves denoted by NaCl, Ca++ in Fig. 7.48 have been shifted to a lower salinity value. Optimal salinity has been reduced from approximately 1.5 to 1.0% total dissolved solids (TDS). The IFT at optimal salinity has increased from approximately 9×10^{-4} to 1×10^{-3} dynes/cm. This shift reflects the fact that the surfactants are less compatible with divalent ions in the aqueous phase. The divalent ions tend to drive the system toward an upper-phase system.

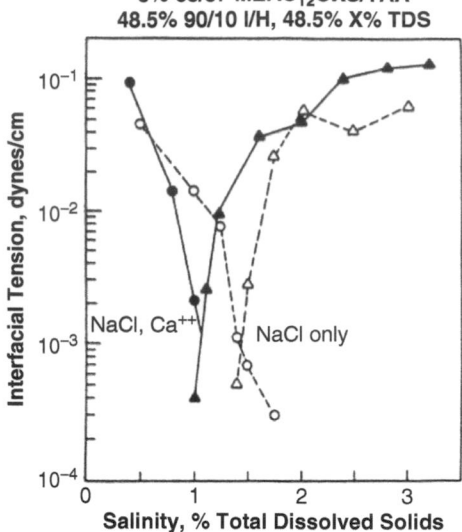

Fig. 7.48—IFT, effect of Ca²⁺ (Healy et al. 1976).

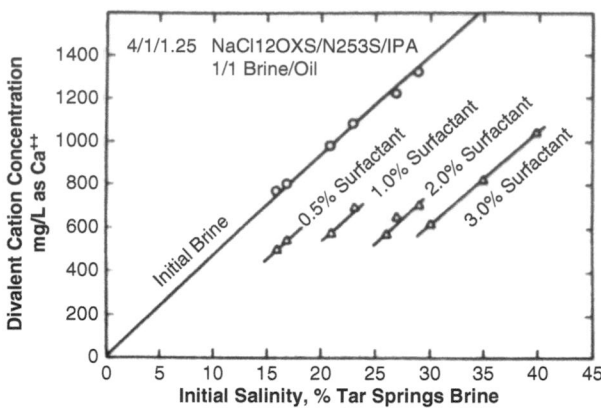

Fig. 7.49—Divalent-cation content of equilibrium brine phases (Glover et al. 1979).

The depletion of divalent cations from the equilibrium brine phase is a consequence of the association of the divalent cations with the surfactant (Glover et al. 1979). Support for this interpretation is shown in **Fig. 7.49,** where calcium concentrations of the initial brine and brine phases in equilibrium with middle- and upper-phase microemulsions are plotted vs. initial brine salinity. Divalent-cation/surfactant complexes tend to be oil-soluble and to promote formation of upper-phase systems. Glover et al. (1979) describe the complexation of calcium with sodium orthoxylene sulfonate (NaC12OXS, where C12 indicates the predominant carbon in the alkyl chain) as an equilibrium relationship for divalent-cation monosulfonate:

$$Na(C12OXS) + Ca^{++} \rightleftharpoons Ca(C12OXS)^+ + Na^+. \dots \dots \dots \dots \dots \dots (7.15)$$

The equilibrium relationship for the formation is given by

$$K_1 = \left(\frac{X_{2S}}{X_{1S}}\right)\left(\frac{C_1}{C_2}\right)$$

$$= \frac{\text{moles divalent-cation sulfonate}}{\text{moles monovalent-cation sulfonate}}, \dots \dots \dots \dots \dots \dots \dots \dots (7.16)$$

where K_1 = equilibrium constant; X_{1S} = moles of monovalent-cation sulfonate per mole of total sulfonate; X_{2S} = moles of divalent-cation sulfonate complex per mole of total sulfonate; C_1 = monovalent-cation concentration, g mol/L; and C_2 = divalent-cation concentration, g mol/L. Calcium may complex two sulfonate molecules, forming divalent-cation disulfonate complexes according to the equilibrium relationship

$$2NaC12OXS + Ca^{++} \rightleftharpoons Ca(C12OXS)_2 + 2Na^+, \dots \dots \dots \dots \dots \dots \dots (7.17)$$

where the concentration at equilibrium satisfies the equilibrium relationship

$$K_2 = \left(\frac{X_{2SS}}{X_{1S}^2}\right)\left(\frac{C_1^2}{C_2}\right) \dots \dots \dots \dots \dots \dots \dots \dots \dots \dots \dots \dots (7.18)$$

and X_{2SS} = moles of divalent-cation disulfonate complex per mole of total sulfonate.

The equilibrium constants K_1 and K_2 are determined by fitting the experimental data. A decrease in divalent-cation content of an excess-water phase compared with original brine composition (Fig. 7.49) is assumed to be a direct measure of the quantity of divalent-cation complex formed. **Fig. 7.50** compares predicted excess divalent cation per mole sulfonate as a function of sulfonate concentration for K_1 = 6.18 and K_2 = 12.15 for the data set in Fig. 7.49. Fig. 7.50 also shows the fit of an empirical relationship given by

$$K_1^* = \left(\frac{X_{2S}}{X_{1S}}\right)^{0.426}\left(\frac{C_1}{C_2}\right). \dots \dots \dots \dots \dots \dots \dots \dots \dots \dots \dots (7.19)$$

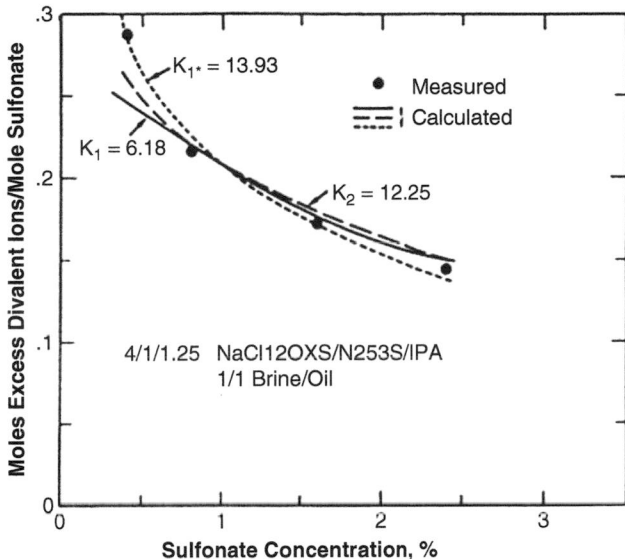

Fig. 7.50—Excess-divalent-cation/sulfonate ratio vs. sulfonate concentration (Glover et al. 1979).

Glover et al. (1979) correlated the optimal salinity for this system with the mole fraction of divalent cation, X_{2S}, for a fixed ratio of sulfonate to N25-3S (ethoxylated) cosurfactant. According to this relationship, the optimal salinity for phase behavior, C_ϕ, decreases with increasing divalent-cation association with the surfactant:

$$C_\phi = 42.6 - 89 X_{2S}. \dotfill (7.20)$$

There is error in the calculation of associated surfactant because there is some association of calcium with the N25-3S cosurfactant (Hirasaki 1982a).

Hirasaki (1982a) developed a model that can be used to estimate the fraction of divalent ions associated with the surfactant in the microemulsion. This model shows that the optimal salinity is a function of the alcohol associated with the surfactant and the divalent-ion fraction of the total cations associated with the surfactant.

7.6.5 Effect of Surfactant Structure. Petroleum sulfonates were initially investigated for oil recovery because they were relatively inexpensive and exhibited several desirable properties. Much of the initial research focused on correlating IFT and oil recovery from laboratory corefloods with the equivalent weight of the surfactant (Gale and Sandvik 1973). Lowest IFTs were found when petroleum sulfonates with equivalent weights in the vicinity of 400 to 450 were used. As discussed earlier, the production process for petroleum sulfonates typically yields a mixture of compounds of different equivalent weights. **Fig. 7.51** shows an example distribution. In this example, compounds with equivalent weights ranging from <300 to >600 are present in the mixture.

Gale and Sandvik (1973) showed that the IFT properties of a particular surfactant system being considered for a field application were governed by the higher-equivalent-weight molecules. Their system consisted of a mixture of surfactant molecules with a wide range of equivalent weights. They examined the effect on IFT with an oil by measuring IFT as a function of the surfactant equivalent weight. **Fig. 7.52** shows a plot of IFT vs. equivalent weight for one system they examined. They also conducted IFT experiments in which they varied the percentages of high- and low-equivalent-weight surfactants in the mixture. For example, they doubled the amount of the 25% of the compounds with the highest equivalent weights and found that IFT was reduced by a factor of three. Conversely, they removed the 25% of the compounds with the highest equivalent weights and found that the IFT increased by an order of magnitude. In both cases, adjustments were made so that the average equivalent weight was not significantly altered. Their conclusion was that the high-equivalent-weight molecules dominated the surfactant properties.

The molecular structure of the surfactant affects phase behavior, solubilization parameters, and consequently, IFTs. For example, **Fig. 7.53** shows the effect of changing the length of the hydrocarbon tail on IFT for the same basic system described by Healy et al. (1976). The N designation is the carbon number of the side chain on the surfactant described in Section 7.4.3. Increasing N from 9 to 12 to 15

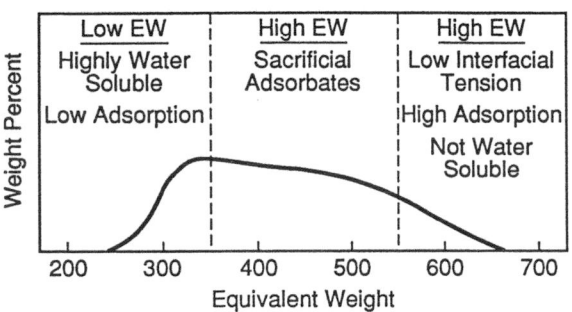

Fig. 7.51—Example distribution of equivalent weights of a petroleum sulfonate.

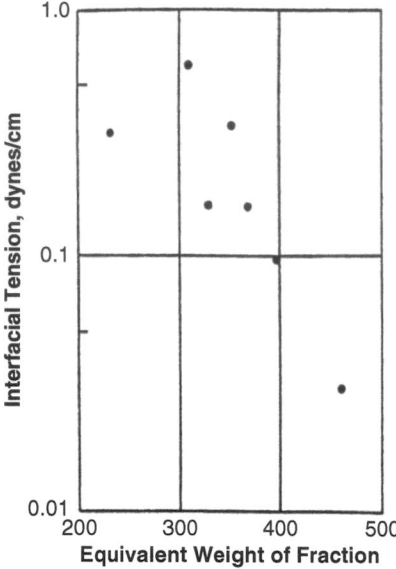

Fig. 7.52—IFT vs. equivalent weight (0.5% surfactant in 2% Na₂SO₄ brine equilibrated with Borregos crude oil) (Gale and Sandvik 1973).

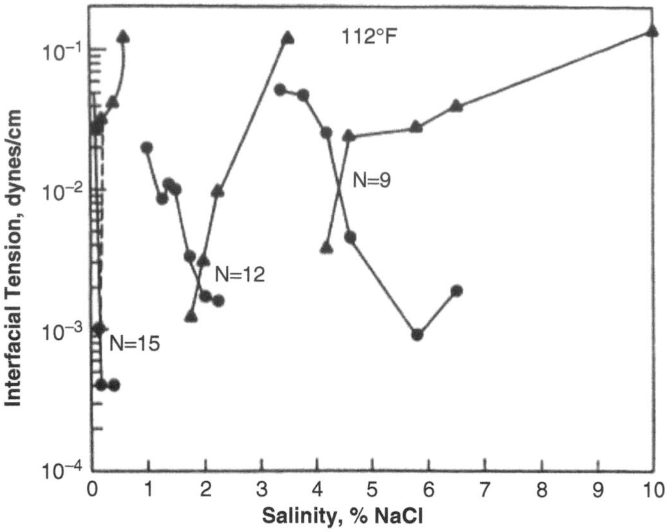

Fig. 7.53—IFT, effect of surfactant structure (Healy et al. 1976).

reduced optimal salinity from 4.4 to 0.2% NaCl, with a corresponding reduction in IFT and increase in the optimal solubilization parameters. The data shown in Fig. 7.53 are at 112°F, but Healy et al. (1976) point out that the same general behavior was observed at lower and higher temperatures.

There have been extensive investigations of surfactant structure to discover which structures give highest oil and water solubilization (Salter 1986). This effort was driven by economic and technical considerations. First, an analysis of the white-oil supply showed that the quantity of the white-oil sulfonates (largely monosulfonated) tested extensively in laboratories could not meet the demand if micellar/polymer flooding was adopted on a large scale.

Petroleum sulfonates and crude oil sulfonates have poor salinity tolerance and generally have been used in reservoirs where the brine salinity is 2 wt% or less. However, many reservoirs have temperatures and salinities beyond the range where petroleum sulfonates may be used. A second characteristic of petroleum sulfonates is that the fraction of the material that is disulfonated can range from <1% for white-oil sulfonates to >30% in crude oil sulfonates. Disulfonates have low equivalent weights and thus are quite water-soluble. Petroleum sulfonates have been found to fractionate into mono- and disulfonated components owing to preferential adsorption and to separate chromatographically as the surfactant is displaced through porous rock (Willhite et al. 1980; Gupta 1984).

These problems provided an incentive to develop new molecules that gave high solubilization parameters under harsh conditions. Finally, it became evident that a single surfactant could not be used in all reservoir-oil/brine types.

Synthetic sulfonates can be produced economically in large quantities. Because many surfactant molecules can be made, it is of interest to determine which molecules give high solubilization parameters. Barakat et al. (1983a) studied the phase behavior of a series of alkyl benzene sodium sulfonates to determine factors that enhanced oil and water solubilization in microemulsions. Most of the surfactants were pure monoisomeric species. These species are described by the notation $m\phi C_n SO_3 Na$, where n is the length of the alkyl chain and m is the carbon where the benzene ring is attached. For example, the structure of $5\phi C_{12} SO_3 Na$, with the benzene ring attached to the fifth carbon of the dodecyl chain, is shown in **Fig. 7.54.**

Fig. 7.55 shows the correlation of solubilization parameter with ACN for families of pure alkyl benzene sodium sulfonates and some mixtures at a salinity of 1% NaCl and a concentration of 0.0227 M. Cosurfactants were 3% 2-butanol and 2% isopentanol. Each line represents a surfactant of the same molecular weight. The ACN corresponds to the carbon number of the pure alkane that exhibited optimal solubilization with that particular surfactant. For example, decane has an ACN of 10. The solubilization parameters in Fig. 7.55 have the same slope and are correlated by

$$P^* = -1.5 N_{AC}^* + 2.5 C_t - 1.4, \dots\dots\dots\dots\dots\dots\dots\dots\dots\dots\dots\dots\dots\dots\dots\dots\dots\dots\dots\dots\dots\dots \quad (7.21)$$

where C_t = total number of carbons in the surfactant's hydrocarbon tail and N_{AC}^* = ACN at optimal solubilization. For a particular surfactant, solubilization is enhanced by increasing the length of the hydrocarbon tail.

The data also show that for a given molecular weight, solubilization parameters increase as the length of the short chain, C_{sc}, at the point where the benzene ring is attached decreases. For $5\phi C_{12} SO_3 Na$, $C_{sc} = 4$. Highest solubilization parameters were correlated for alkyl benzene sulfonates with

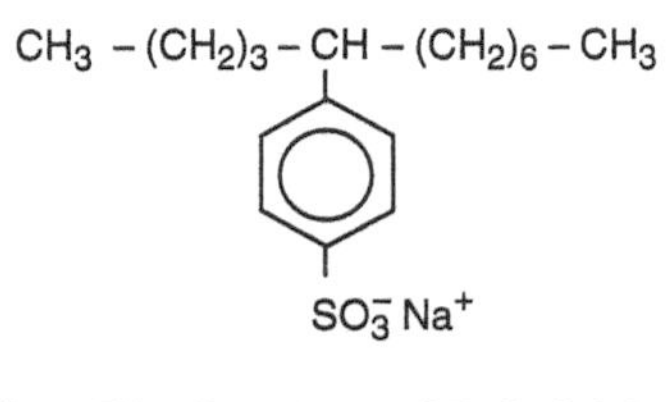

$$CH_3 - (CH_2)_3 - CH - (CH_2)_6 - CH_3$$

$$SO_3^- Na^+$$

Fig. 7.54—Structure of $5\phi C_{12} SO_3 Na$.

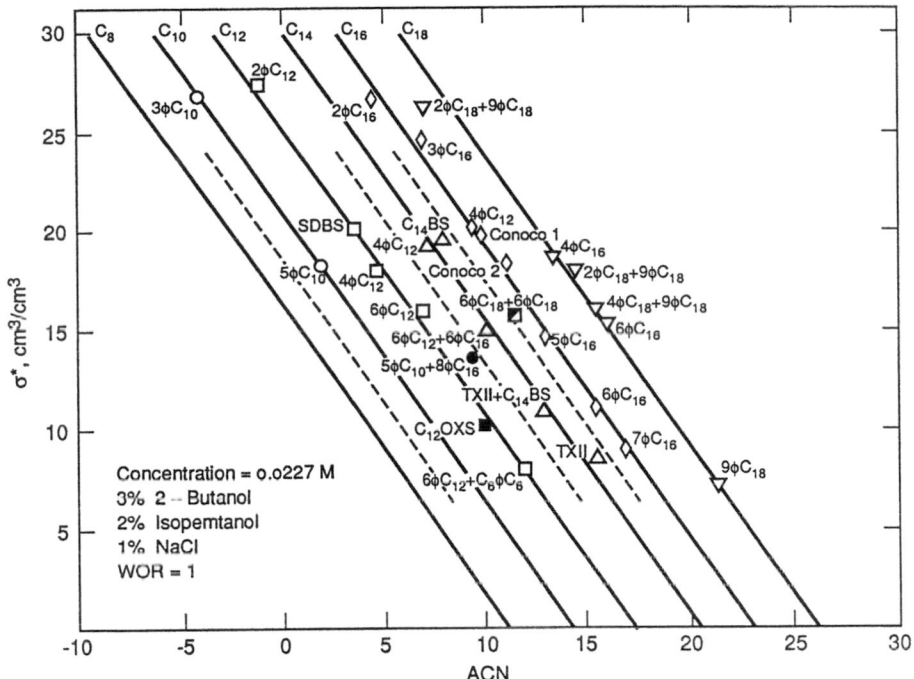

Fig. 7.55—Correlation of optimum solubilization parameter, σ*, vs. ACN for several families of pure alkyl benzene sulfonates and some mixtures (Barakat et al. 1983b).

$$P^* = -3.0C_{SC} + 30. \dots \dots (7.22)$$

Salter (1986) presents examples showing that the location of the sulfonate on the benzene ring also influences solubilization parameters. Monoisometrically pure paraethylbenzene sulfonate (C_{12}PEBS) forms the two species when sulfonated. **Fig. 7.56** shows the structures of C_{12}PEBS. Sulfonation at the meta position adjacent to the ethyl side chain is the dominant species (Salter 1983). This sulfonate was separated into two species, and the optimal salinity was determined for mixtures of these sulfonates. **Fig. 7.57** shows that the highest solubilization parameters were observed when the sulfonate was in the ortho position for C_{12}PEBS. At this value, the optimal salinity would be approximately 2 wt% NaCl.

Considerable research has been directed at developing surfactant structures that are tolerant of high salinity and hardness (Graciaa et al. 1982; Carmona et al. 1985; Gale et al. 1981). One of these surfactants was developed by Exxon and was used in an extensive series of field tests in the Loudon field (Maerker and Gale 1992). The Loudon surfactant system was a blend of two similar surfactants that provided flexibility for adjustments to compensate for uncontrollable variations in manufacturing or variations in makeup-water salinity. There was no cosurfactant. Also, a surfactant system was desired that would have optimal salinities near the resident brine salinity because of the failure of preflushes in field tests to condition a reservoir to a lower salinity adequately.

A family of surfactants (Gale et al. 1981) was developed with the general formula

$$R_1O(C_3H_6O)_m (C_2H_4O)_n YX,$$

where R_1 is an isotridecyl alcohol radical, C_3H_6O is a propylene oxide group, C_2H_4O is an ethylene oxide group, m and n have values between 1 and 6, X is a monovalent sodium cation, and Y is a sulfate group. This family of surfactants has relatively high optimal salinity. It was possible to formulate microemulsions by mixing two surfactants without use of alcohol as a cosurfactant. Because the two surfactants were similar on a molecular basis, chromatographic separation of the surfactant was eliminated by process design. **Fig. 7.58** shows the optimal salinity map constructed from phase-behavior data with a large number of single surfactants. Values of m and n are averages of distributions in a single product. Lines of constant optimal salinity are expressed as percentages of the resident brine, TSB. Fig. 7.58 suggests that

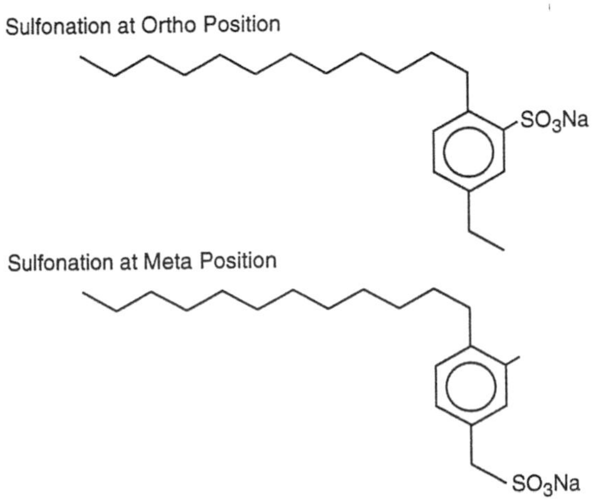

Sulfonation at Ortho Position

Sulfonation at Meta Position

Fig. 7.56—Structures of C_{12} PEBS.

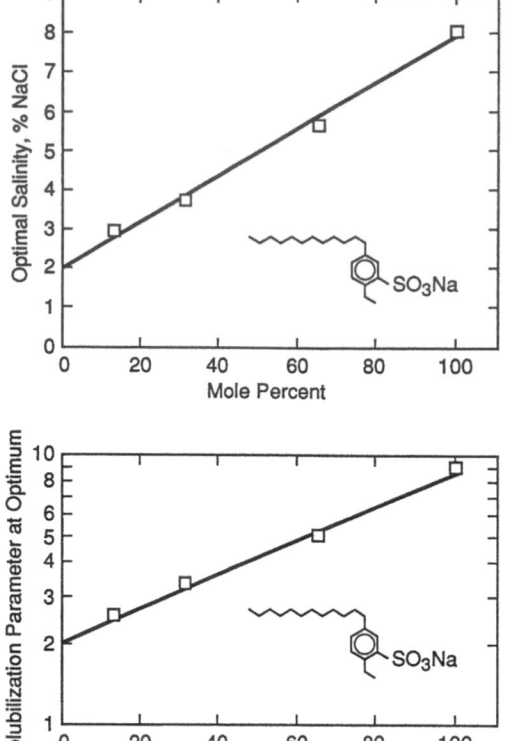

Fig. 7.57—Effect of point of sulfonation on optimum parameters for $C_{12}PEBS$ (Salter 1986).

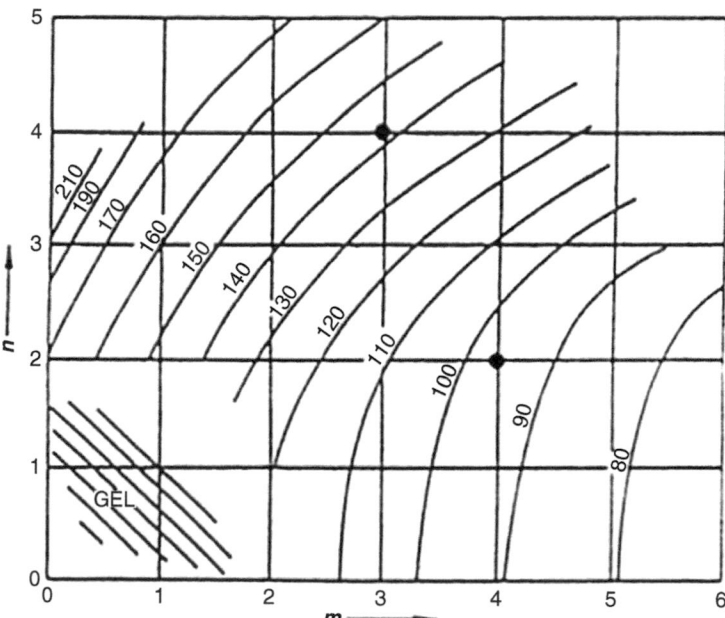

Fig. 7.58—Optimal salinity contours for i-$C_{13}H_{27}O(PO)_m (EO)n SO_3 Na$ as functions of m and n for 1% surfactant in 50:50 diesel/X% TSB at 78°F (Maerker and Gale 1992).

an infinite number of combinations of m and n, or blends of surfactants with different average values of m and n, will yield formulations that have low IFTs. The Loudon formulation was composed of a mixture of two surfactants, i-$C_{13}H_{27}O(PO)_4(EO)_2 SO_3 Na$ and i-$C_{13}H_{27}O(PO)_3(EO)_4 SO_3 Na$. The compositions of these surfactants are indicated in Fig. 7.58.

Surfactant systems other than those discussed here have been investigated and are described in the literature (Salter 1986; Chou and Bae 1989). Design of surfactants that have high solubilization parameters in harsh environments and in the presence of divalent ions appears to be possible, although further development may be needed. It also is necessary to determine whether these surfactant systems displace oil effectively and have tolerable adsorption losses.

7.6.6 Effect of Pressure. As with liquid systems, the general effect of pressure on phase behavior is small. In one set of reported experiments (Nelson 1983), the effect of pressure alone was negligible. When different oils were pressurized with methane so that some methane dissolved in the oil, however, there was some effect. Nelson (1983) concluded that for live crude oils—i.e., crude oils containing significant amounts of gas (C1, C2, etc.)—the possible influence of pressure on behavior should be considered in the design process.

Another study (Skauge and Fotland 1990) reported that increasing the pressure caused a shift in phase behavior toward a lower-phase microemulsion. For a given system, optimal salinity increased as pressure on the system increased.

7.6.7 Effect of Polymer Addition. In earlier formulations of micellar slugs, polymer typically was not added directly to a micellar solution. Because a polymer solution displaces the micellar slug, however, dispersion causes mixing at the micellar-slug/polymer-slug interface. This mixing could affect phase behavior and IFT and thus influence process performance. More recently, polymer has been added to micellar solutions to increase the solution viscosity for the purpose of mobility control.

The effect of polymer on the phase behavior of a micellar solution has been investigated (Pope et al. 1982; Trushenski 1977; Szabo 1979). Behavior is complex, but it has been shown that polymer can be added to a micellar slug under controlled conditions to increase slug viscosity without adversely affecting phase behavior or IFT. Pope et al. (1982) concluded that "all combinations of anionic and nonionic surfactants with all anionic and nonionic polymers studied showed regions of composition that were compatible and regions of composition that were not compatible." They also concluded that anionic surfactants were somewhat more compatible with nonionic polymers.

In aqueous surfactant/polymer systems (i.e., in the absence of oil) over certain concentration ranges, two aqueous phases form (Pope et al. 1982; Szabo 1979). One is a surfactant-rich phase and the other is a polymer-rich phase. IFT between the phases often is quite low. This phase separation is a strong function of salinity in a manner analogous to the phase behavior of micellar solutions. Pope et al. (1982) defined a critical electrolyte concentration (CEC) above which the phase separation occurred. The CEC was found to be a function of the surfactant/cosurfactant combination but not a strong function of polymer type or concentration.

When oil is present, the effect on phase behavior and IFT of adding polymer to a micellar solution is small in many cases. Examples are shown in **Figs. 7.59 and 7.60** (Pope et al. 1982). Fig. 7.59 is a volume-fraction diagram for a specific micellar solution with and without the addition of polymer (Xanflood is a biopolymer). The effect of polymer addition is to shift the phase boundaries along the salinity axis. Fig. 7.60 shows IFT behavior for another example system. Again, polymer addition causes a shift along the salinity axis. The magnitude of IFT at optimal salinity is not changed significantly by the polymer addition. Pope et al. (1982) point out that the most dramatic effect of adding polymers is a shift in the three-phase boundaries. However, some anionic surfactants are reported to precipitate in the presence of partially hydrolyzed polyacrylamide.

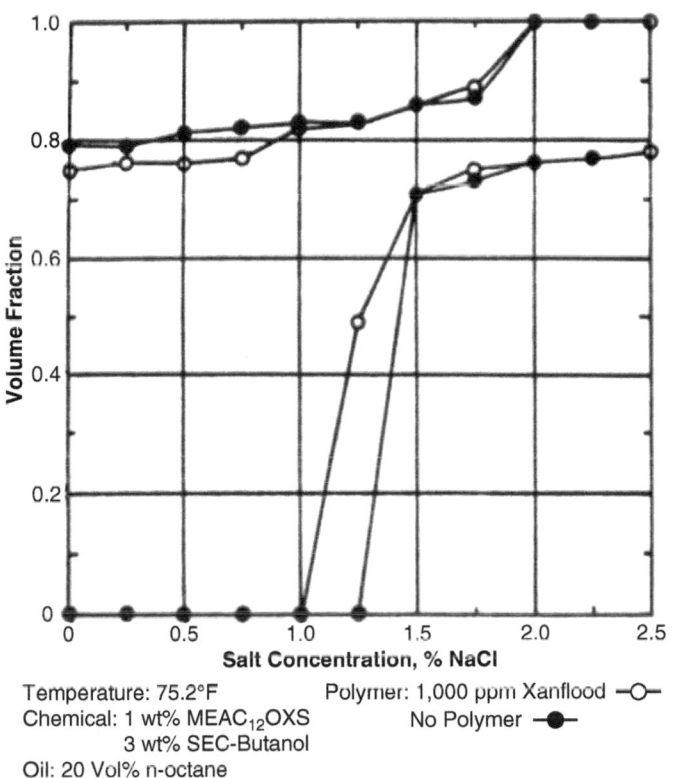

Temperature: 75.2°F   Polymer: 1,000 ppm Xanflood —O—
Chemical: 1 wt% MEAC$_{12}$OXS   No Polymer —●—
　　　　　3 wt% SEC-Butanol
Oil: 20 Vol% n-octane

Fig. 7.59—Effect of polymer on phase behavior (MEAC$_{12}$OXS behavior at 24°C) (Pope et al. 1982).

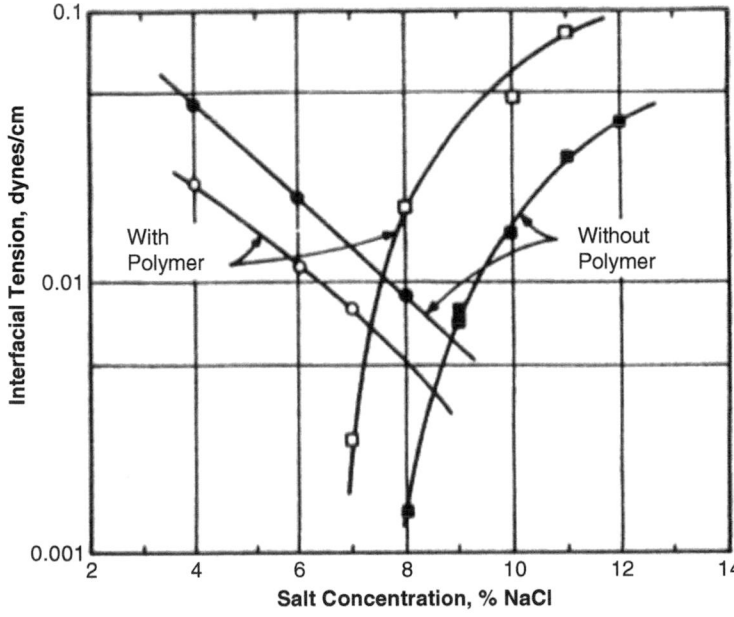

Fig. 7.60—Effect of polymer on IFT for CO-610 formulation (Pope et al. 1982).

When polymer is added to a micellar solution, viscosity is increased. In some cases, the effect is predictable in that increased viscosity corresponds to that expected from the polymer concentration. In other cases, gels or very viscous solutions form (Pope et al. 1982).

Thus, when polymer is used with a micellar solution the behavior should be checked experimentally for the specific system.

7.7 Viscosity and Density of Microemulsions

As discussed in Chapter 4, the magnitudes of the density and viscosity of the displacing fluid relative to the displaced fluid are important design variables that affect volumetric displacement efficiency. The tendency for gravity override or underride to occur is determined by the relative densities of the displaced and displacing fluids. Areal and vertical sweep efficiencies are in large measure determined by the mobility ratio in the displacement process, which is inversely proportional to the displacing-fluid viscosity.

Both density and viscosity are functions of microemulsion composition. Viscosity, in particular, can be varied over a wide range by proper adjustment of composition and/or by polymer addition.

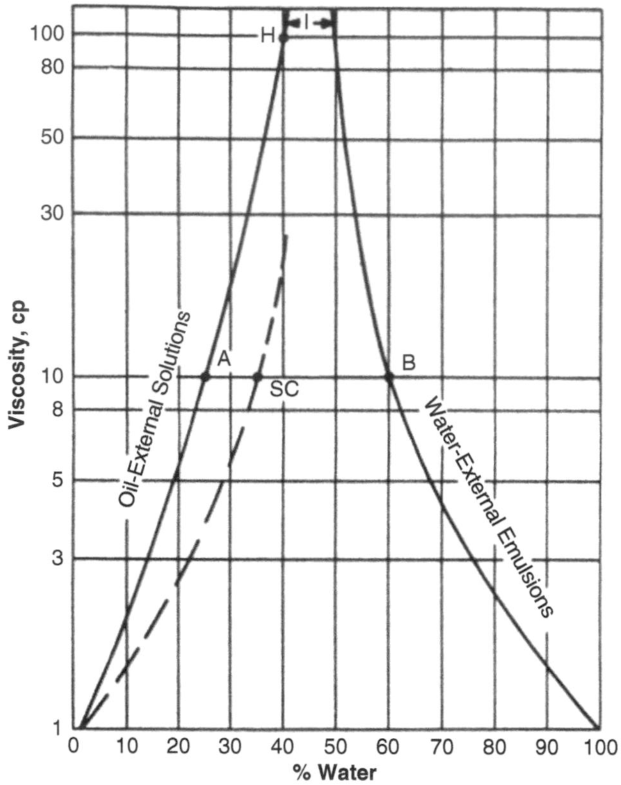

7.7.1 Viscosity of Microemulsions.
Viscosity of microemulsions varies from values on the order of magnitude of water to significantly larger values. The structure of a microemulsion (i.e., whether it is water- or oil-external) has a large effect, as shown in **Fig. 7.61** (Gogarty and Tosch 1968). At low water content, the system is oil-external. Viscosity increases as water is added, creating swollen micelles. For this system, viscosity increased two orders of magnitude between its initial state with very little water and the state near 50% water content. At higher water content, after inversion to a water-external system, the viscosity decreases with further addition of water.

A = Oil-external slug
H = Oil-external solution
B = Water-external emulsion
SC = Salt-containing micellar solution
I = Inversion point between oil- and water-external slugs

Fig. 7.61—Viscosity vs. percent water (Gogarty and Tosch 1968).

In practice, viscosity usually is adjusted by adding alcohol cosurfactant and/or polymer to the microemulsion. **Figs. 7.62 and 7.63** (Jones and Dreher 1976) show the typical effects of alcohol concentration and type on viscosity. These data are for Systems A and B in Table 7.6. As shown, viscosity can change by a large magnitude over a relatively narrow range of alcohol concentration. The existence of a maximum or minimum in the viscosity/alcohol concentration relationship is common. The enhanced viscosities are shear-sensitive and are disrupted easily by high shear rates. Solutions recover initial viscosities slowly after high shear. While the change in viscosity is believed to be related to structural changes of the microemulsion, these structures are not well understood.

The large sensitivity of viscosity to concentration is undesirable because of the mixing that occurs in a displacement process in porous media. Also, alcohol type and concentration affect other properties of the system, such as IFT. A system that is optimal in an overall sense usually is designed by trial and error.

Microemulsion solution viscosity also can be increased by addition of a polymer, such as polyacrylamides or biopolymers of the types discussed in Chapter 5. Polymer can affect phase behavior and IFT, although, at least in some cases, this has been shown to be a relatively small effect (Healy et al. 1976) (see Section 7.6.7). The effect of polymer addition on viscosity can be quite significant, as shown in **Fig. 7.64** (Pope et al. 1982), especially at lower salinities.

7.7.2 Density.
Densities of surfactants and alcohols are approximately 1.2 and 0.8 g/cm^3, respectively. As discussed, a microemulsion typically consists of a mixture of surfactant, alcohol, oil, and brine. The volumes of oil and water are much larger than the surfactant/alcohol volume, however, so the microemulsion density usually lies between the oil and brine density values. Thus, in most displacements with microemulsions, the density of the displacing phase lies between the densities of the oil and water being displaced. Difference in densities can contribute to gravity override at the front of the surfactant bank and gravity underride at the rear of the surfactant bank (Tham et al. 1983).

7.8 Displacement Mechanisms

The micellar/polymer, or microemulsion, flooding process was described in general in Section 7.2 and illustrated in Figs. 7.2a and 7.2b. Figs. 7.3 and 7.4 show results of displacements conducted as tertiary floods in laboratory cores. This section discusses microscopic-level displacement mechanisms in more detail. Specific topics include miscible vs. immiscible

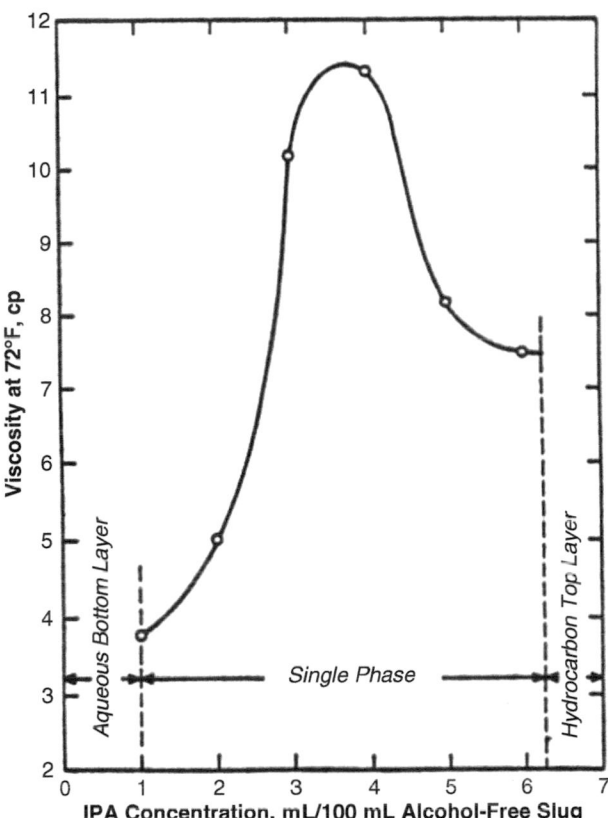

Fig. 7.62—Viscosity of System A (Table 7.4) in the single-phase region (Jones and Dreher 1976).

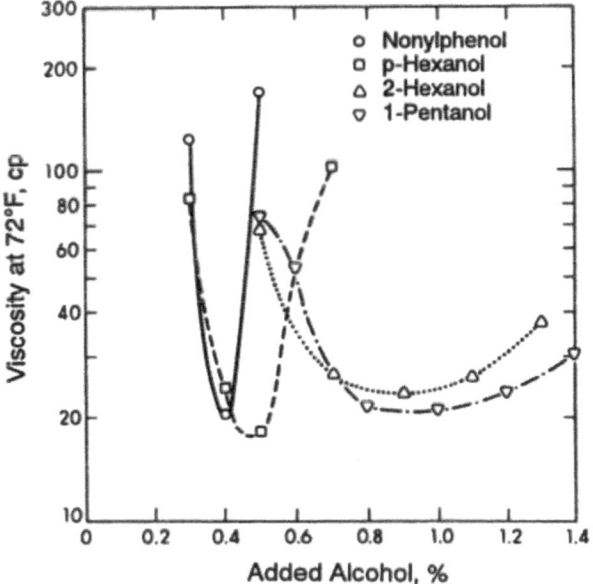

Fig. 7.63—Effect of alcohol type on viscosity of System B (Table 7.4) (Jones and Dreher 1976).

displacement, the relationship between oil recovery and IFT, the role of phase behavior, and gravity segregation. Generally accepted criteria for an efficient displacement are described. Additional design criteria are discussed in Sections 7.9 through 7.11.

7.8.1 Miscible vs. Immiscible Displacement. Microemulsion systems used in practice have had markedly different compositions. As shown in **Fig. 7.65,** compositions of some systems have been in the single-phase region, while compositions of others have been on or near the binodal curve boundary of the multiphase region. When a microemulsion slug, at whatever initial composition, is displaced through reservoir rock, the composition will change owing to mixing and surfactant retention. The effects of mixing can be estimated by considering the process shown in Fig. 7.2a. At the leading edge of the chemical slug, mixing with oil and water phases that are being displaced will occur. At the back edge of the slug, mixing with the polymer solution, the mobility buffer, will occur. The dilution path followed by the microemulsion slug as it propagates through the rock is of interest.

Fig. 7.66 shows a simplistic view of the dilution process. The microemulsion slug composition for this example is in the single-phase region. For a tertiary process, the original reservoir fluid composition is taken to be at a water saturation of 0.65, with a corresponding waterflood ROS, S_{orw}, of 0.35. The mobility buffer is an aqueous phase represented by the pure-water apex of the ternary diagram. Dilution paths for the leading and trailing edges of the microemulsion slug are shown. The dilution paths shown are simplistic in that no bypassing or differential bank velocities are considered (i.e., only mixing between fluids at the three indicated concentrations is considered). Note that because of multiphase fluid flow, the actual oil saturation at the leading edge of the mixing zone would correspond to that in the flowing water/oil bank and is larger than S_{or}. Dilution paths should be constructed by use of "flowing saturations." These saturations can be estimated with the calculation methods described in Section 7.10.2 and illustrated in Examples 7.9 and 7.10.

At the leading edge of the bank, a miscible displacement occurs initially as oil and water are solubilized into the microemulsion. If the slug size is finite and of a volume that is practical in field applications, dilution will occur as the process proceeds, taking the overall composition into the multiphase region. Once the overall composition is in the multiphase part of the diagram, equilibrium phases will separate and the displacement will be immiscible. Efficiency of oil and water displacement at the microscopic level is then dependent on the IFT between the distinct phases.

For this example, a similar dilution behavior will occur at the trailing edge of the bank, where there is mixing between the mobility buffer and microemulsion. The dilution path will take the overall composition into the multiphase region of the diagram and from a miscible to an immiscible displacement.

Fig. 7.67 is a schematic of the process. Fig. 7.67a shows a core that has been waterflooded and has reached an ROS to waterflooding. In Figs. 7.67a and 7.67b, a microemulsion bank has been injected followed by a mobility buffer, viscous water in this example. As long as the microemulsion slug is able to solubilize oil and water and to remain a single phase, the process

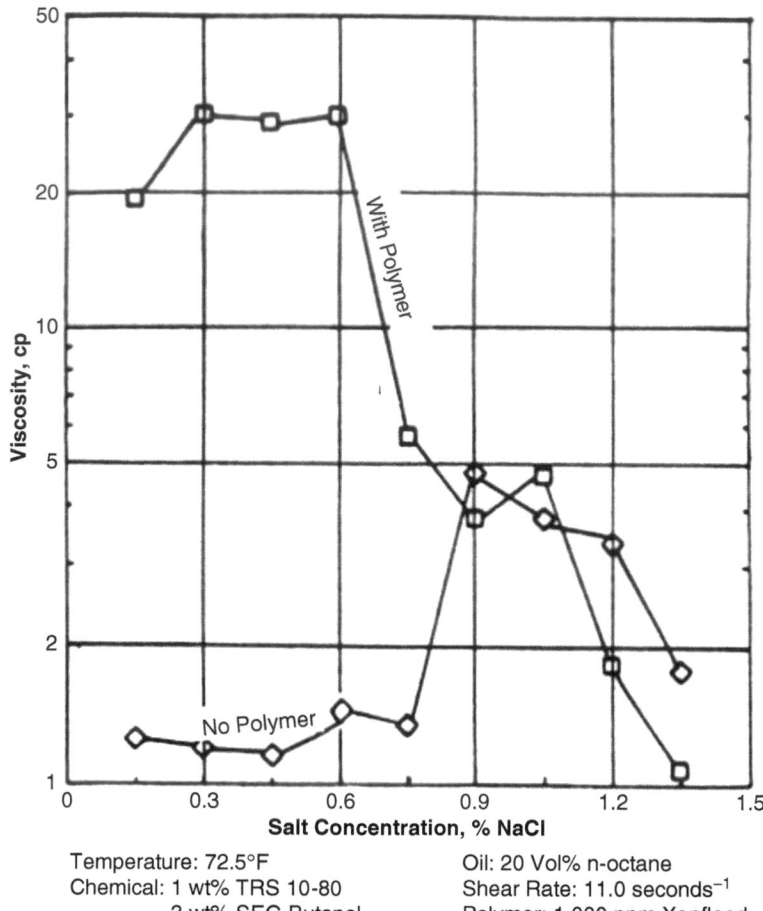

Temperature: 72.5°F
Chemical: 1 wt% TRS 10-80
3 wt% SEC-Butanol

Oil: 20 Vol% n-octane
Shear Rate: 11.0 seconds^{-1}
Polymer: 1,000 ppm Xanflood

Fig. 7.64—Viscosity of TRS 10-80 with oil at 75°F (Pope et al. 1982).

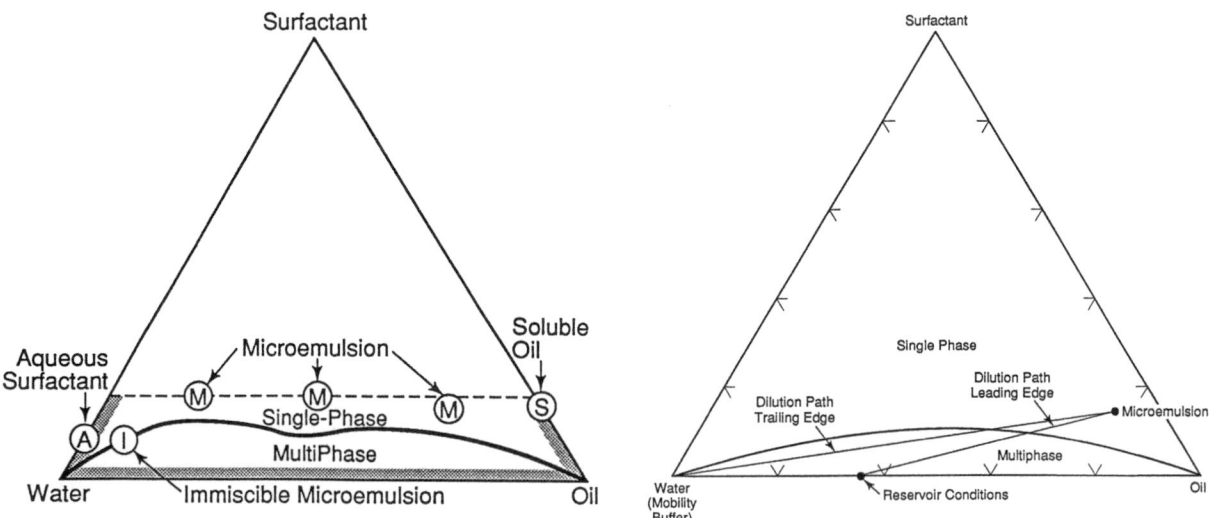

Fig. 7.65—Examples of microemulsion compositions used in practice.

Fig. 7.66—Dilution paths on a ternary diagram, ideal mixing.

behaves as a miscible displacement. The efficiency is high for a micellar displacement, often approaching 100%, as indicated in Figs. 7.67b and 7.67c.

When the microemulsion bank has been sufficiently diluted, phase separation will occur, as indicated on the ternary diagram of Fig. 7.66. The closer the initial composition is to the boundary of the multiphase region, the sooner this separation will occur. The process converts to an immiscible displacement, as shown in Fig. 7.67d. Oil displacement efficiency is

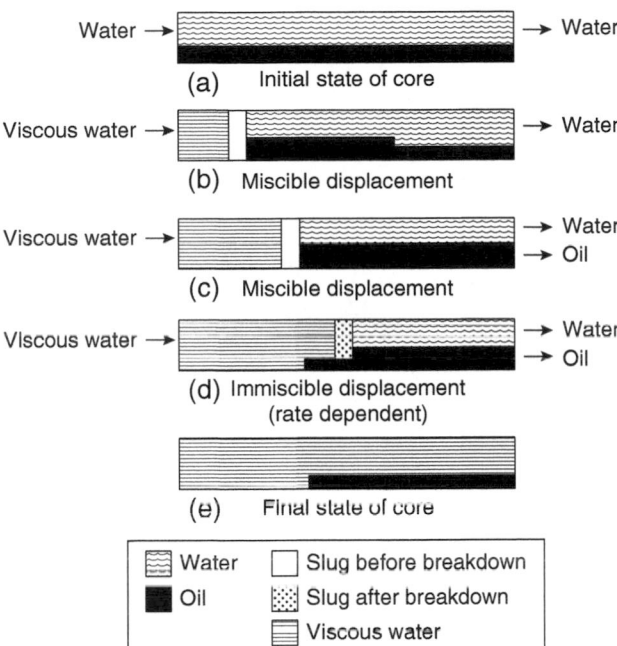

Fig. 7.67—Conceptualization of a microemulsion flood as a miscible/immiscible displacement.

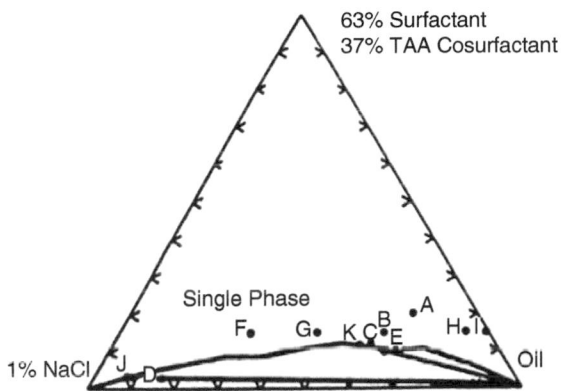

Fig. 7.68—Phase behavior and injection composition experiments to demonstrate relative importance of miscible/immiscible displacement (Healy et al. 1975).

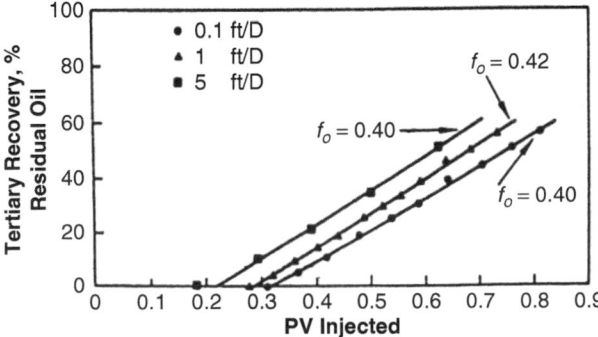

Fig. 7.69—Oil recovery from locally miscible displacement (Healy et al. 1975).

governed by the IFT between the microemulsion and oil phases and may be affected by the flow rate as well. The dependence on rate is indicated by capillary number correlations, as shown in Fig. 7.1.

At ultralow IFT values, the efficiency may approach that of a miscible process. An ROS will be left by the microemulsion slug, even at low IFT. Although not shown in Fig. 7.67, a residual microemulsion phase can be left by the displacing viscous water if dilution causes multiple phases to flow and if the IFT value between phases is not sufficiently small.

Healy et al. (1975) addressed the question of the relative importance of miscible vs. immiscible displacement. A series of microemulsion displacements was conducted in laboratory cores that had been previously waterflooded (i.e., the microemulsion floods were tertiary recovery floods). Injected microemulsion composition and the flow rate were varied.

Three corefloods were made with each core at waterflood ROS. The injected slug composition was the same in each run and is shown as Point A in **Fig. 7.68.** This is a lower-phase system. The microemulsion slug (chemical slug) contained approximately 20% surfactant/cosurfactant. In each run, slug injection was continuous (i.e., an "infinite" slug was used). Flow rate was varied between runs so that interstitial velocities were 0.1, 1.0, and 5.0 ft/D. The surfactant and oil used were the same as those discussed in Section 7.6 (Healy et al. 1976).

The displacements were anticipated to behave as a miscible process because the composition of the injected slug was well above the binodal curve and an infinite slug size was used. The microemulsion slug mobility was designed to be less than the minimum oil mobility expected in the oil bank. Thus, an efficient displacement was expected.

Fig. 7.69 shows the cumulative recovery (expressed as percent oil in place) as a function of PV injected up to the time of surfactant breakthrough. The slope of the recovery curve is proportional to oil fractional flow, f_o, in the effluent. As seen, f_o was essentially independent of rate. An effect of flow rate on oil-bank breakthrough was attributed to slug bypassing at the higher rates. Cumulative recovery at a specific PV injected was strongly dependent on flow rate, with highest recoveries at the largest interstitial velocity of 5 ft/D. The appearance of surfactant in the effluent also was weakly dependent on rate.

The primary conclusion drawn from the runs was that the process behaved as a miscible displacement and that $f_o \approx 0.4$ for this system represented a maximally efficient produced oil cut. In addition, displacement efficiency increased with flow rate.

Two additional series of runs were made corresponding to Compositions J and K on Fig. 7.68. Both solutions were lower-phase microemulsions in equilibrium with an upper phase that was essentially oil. These microemulsions are saturated with oil at the injection conditions and cannot solubilize additional oil. Compositions D and E are the overall compositions corresponding to the equilibrium systems. Because Compositions J and K are on the binodal curve, any dilution caused by mixing in the core would result in a multiphase system. Thus, displacements with Compositions J and K, even with continuous microemulsion injection, were immiscible throughout a coreflood.

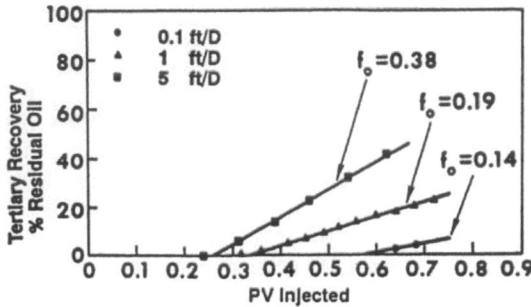

Fig. 7.70—Oil recovery from immiscible displacement (σ_{mo} = 0.03 dynes/cm) (Healy et al. 1975).

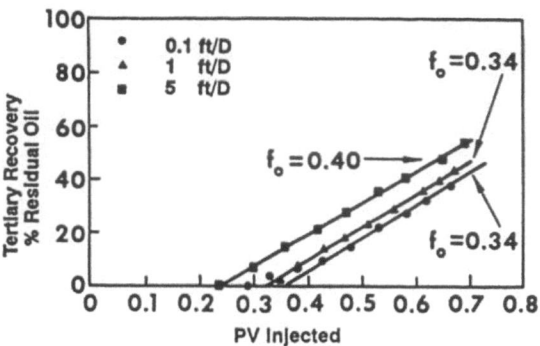

Fig. 7.71—Oil recovery from immiscible displacement (σ_{mo} = 0.002 dynes/cm) (Healy et al. 1975).

IFT values were measured between equilibrium phases. For Composition D, σ_{mo} = 0.03 dynes/cm between Composition J and the excess-oil phase, and for System E, the value was 0.002 dynes/cm between Composition K and the excess-oil phase. Rates were set at 0.1, 1.0, and 5.0 ft/D in the different displacements. Results for injection of Compositions J and K are given in **Figs. 7.70 and 7.71**, respectively. For Composition J, there was a large effect of rate. Fractional oil flow and recovery increased markedly with rate up to the time of surfactant breakthrough. Recovery efficiency was very poor at the lowest rate. The effect of rate was much less pronounced when Composition K was injected, where σ_{mo} was an order of magnitude smaller than for Composition J. In both systems, the micro-emulsions displaced brine miscibly because the microemulsions could solubilize additional brine on the dilution path.

Fig. 7.72 summarizes the results by plotting fractional oil flow as a function of rate for the series of three experiments. At relatively high IFT values, the process is rate-sensitive, with recovery efficiency, expressed by f_o, approaching that of a miscible process at high rates. At a low IFT value, the process is weakly dependent on rate and efficiency is only slightly less than for a miscible process. This is consistent with the capillary number correlations discussed in Chapter 2 and shown in **Fig. 7.73** for these experiments.

Healy et al. (1975) concluded from these runs that it was not possible, on the basis of displacement efficiency, to differentiate between miscible and immiscible floods conducted at sufficiently high capillary numbers.

Additional experiments were then conducted with microemulsion slugs at Compositions A, B, and C in Fig. 7.68. In these experiments, conducted in 4-ft cores, 0.05-PV slugs were used. The slugs were followed by a biopolymer solution to maintain mobility control. The rates were constant at specified values between 0.1 and 5.0 ft/D. Again, f_o was monitored as a function of PV injected. The maximum PV injected at which f_o remained constant (at a value >0.35) was recorded. This PV injected corresponded to an inferred maximum length of travel of a slug in the core for which the displacement was maximally efficient. The length depended on the injected slug composition and, at equal flow rate, was largest for Composition A and smallest for Composition C. This was consistent with the idea that high surfactant concentration prolonged the miscible displacement.

Finally, experiments were conducted with the same systems (Compositions A, B, and C) to determine the length of travel before slug breakdown occurred (i.e., before a slug was diluted to a condition where phase separation occurred). Runs were conducted with Compositions A, B, and C at a rate of 1.0 ft/D with a slug size corresponding to the volume equivalent of 0.05 PV injected into a 4.0-ft core. Lengths of the cores varied between 4.0 and 12.0 in. Core effluents were observed visually for evidence of slug breakdown. Slug breakdown in the short cores occurred well before a length of travel that corresponded to the maximum length of travel inferred from the measured f_o in the 4.0-ft cores. The slugs were rapidly diluted, modifying

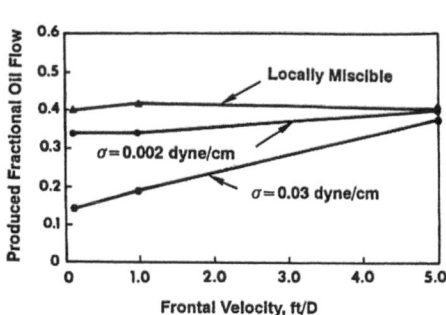

Fig. 7.72—Effect of rate on fractional oil flow (Healy et al. 1975).

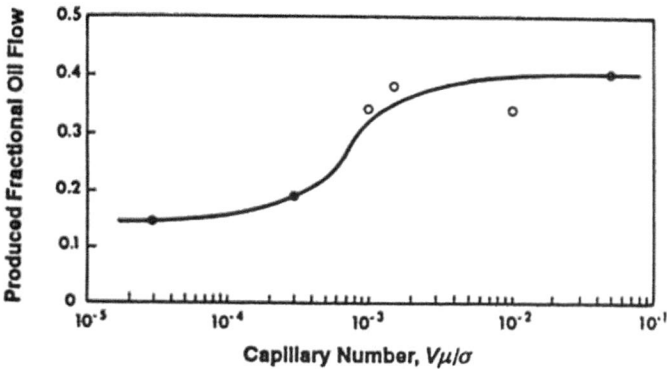

Fig. 7.73—Fractional oil flow as a function of capillary number (Healy et al. 1975).

concentrations and moving them into the multiphase region, but displacement efficiency was still high for a significant period after slug breakdown because of low IFT between the slug and the oil.

The results of these experiments indicate that microemulsion flooding is principally an immiscible process in field applications, where slug size and surfactant concentration are limited by economics. Miscible displacement may occur in the very early stages of a flood, but the chemical slug quickly breaks down (multiple phases form) and the process becomes immiscible. Consequently, laboratory and process design must be based on simulating conditions where the process is immiscible.

7.8.2 Phase Behavior, Optimal Salinity, and Oil Recovery. Optimal salinity for phase behavior and IFT were previously defined and shown to be approximately equal. This section discusses the relationship between those optimal salinities and oil recoveries in a microemulsion flood. Data taken by Healy and Reed (1977) demonstrate the relationship. The study described here used the surfactant/cosurfactant/oil system described in Section 7.5.1.

Several experiments were conducted in which microemulsions of different compositions were used in floods of 4.0-ft sandstone cores. Before microemulsion flooding, the cores were waterflooded to ROS (i.e., the cores contained discontinuous oil and a continuous water phase at the start of a chemical flood). A slug of microemulsion was injected with a slug size specified by

$$B \times C_s = 32, \dots\dots\dots\dots\dots\dots\dots\dots\dots\dots\dots\dots\dots\dots\dots\dots\dots\dots\dots\dots (7.23)$$

where B = slug size, % PV; and C_s = surfactant in the slug, vol%. The application of Eq. 7.23 was somewhat arbitrary but ensured that the same volume of surfactant was used in the different displacements.

The chemical slug was followed by a mobility buffer consisting of a biopolymer solution. Care was taken that mobility control existed at the trailing edge of the microemulsion. Polymer was not added to the surfactant solution. Consequently, mobility control between the surfactant and the oil bank was not ensured.

Table 7.7 shows experimental conditions. As indicated, brine salinity in the microemulsion was a variable in the different runs (salinity was the same in the waterflood brine as in the chemical slug). This caused the phase behavior and IFT values to vary, as described in Section 7.4.2 and shown in Table 7.7. Rates also varied, but over a narrow range between 0.5 and 2.2 ft/D. The number N in Table 7.7 is the carbon number of the alkyl side chain on the surfactant. The cosurfactant was TAA. The term

N	V (ft/D)	Salinity (% NaCl)	Microemulsion Type	Microemulsion Viscosity at 23 seconds⁻¹ (cp)	σ_{mo} (dynes/cm)	σ_{mw} (dynes/cm)	σ_c (dynes/cm)	$v\mu/\sigma_c$
9	0.8	2.0	Lower	2	2.7×10^1	0.0	2.7×10^{-1}	2.1×10^{-5}
9	0.5	7.0	Upper	3	0.0	6.0×10^{-2}	6.0×10^{-2}	8.8×10^{-5}
9	1.0	1.0	Lower	2	6.5×10^{-1}	0.0	6.5×10^{-1}	1.1×10^{-5}
9	1.1	2.0	Lower	2	2.7×10^{-1}	0.0	2.7×10^{-1}	2.9×10^{-5}
9	1.0	3.0	Lower	4	9.0×10^{-2}	0.0	9.0×10^{-2}	1.6×10^{-4}
9	1.2	3.8	Middle	12	2.2×10^{-2}	1.0×10^{-2}	2.2×10^{-2}	2.3×10^{-3}
9	1.1	5.0	Middle	26	1.2×10^{-2}	2.2×10^{-2}	2.2×10^{-2}	4.6×10^{-3}
9	1.0	7.0	Upper	3	0.0	6.0×10^{-2}	6.0×10^{-2}	1.8×10^{-4}
9	1.0	8.0	Upper	5	0.0	1.2×10^{-1}	1.2×10^{-1}	1.5×10^{-4}
9	2.3	2.0	Lower	2	2.7×10^{-1}	0.0	2.7×10^{-1}	6.0×10^{-5}
9	2.0	7.0	Upper	3	0.0	6.0×10^{-2}	6.0×10^{-2}	3.5×10^{-4}
12	0.5	0.5	Lower	3	4.5×10^{-2}	0.0	4.5×10^{-2}	1.2×10^{-4}
12	0.5	1.25	Lower	12	8.0×10^{-3}	0.0	8.0×10^{-3}	2.6×10^{-3}
12	0.5	2.5	Upper	6	0.0	4.0×10^{-2}	4.0×10^{-2}	2.6×10^{-4}
12	0.9	0.5	Lower	3	4.5×10^{-2}	0.0	4.5×10^{-2}	2.1×10^{-4}
12	1.0	1.0	Lower	3	1.5×10^{-2}	0.0	1.5×10^{-2}	7.0×10^{-4}
12	1.0	1.25	Lower	12	8.0×10^{-3}	0.0	8.0×10^{-3}	5.3×10^{-3}
12	1.0	1.4	Middle	11	1.0×10^{-3}	5.0×10^{-4}	1.0×10^{-3}	3.9×10^{-2}
12	1.0	1.5	Middle	10	7.0×10^{-4}	3.0×10^{-3}	3.0×10^{-3}	1.2×10^{-2}
12	1.0	1.75	Middle	13	3.0×10^{-4}	2.7×10^{-2}	2.7×10^{-2}	1.7×10^{-3}
12	1.0	2.5	Upper	6	0.0	4.0×10^{-2}	4.0×10^{-2}	5.3×10^{-4}
12	1.0	3.0	Upper	4	0.0	6.1×10^{-2}	6.1×10^{-2}	2.3×10^{-4}
12	2.2	0.5	Lower	3	4.5×10^{-2}	0.0	4.5×10^{-2}	5.1×10^{-4}

Table 7.7—Continuous microemulsion injection floods (Healy and Reed 1977).

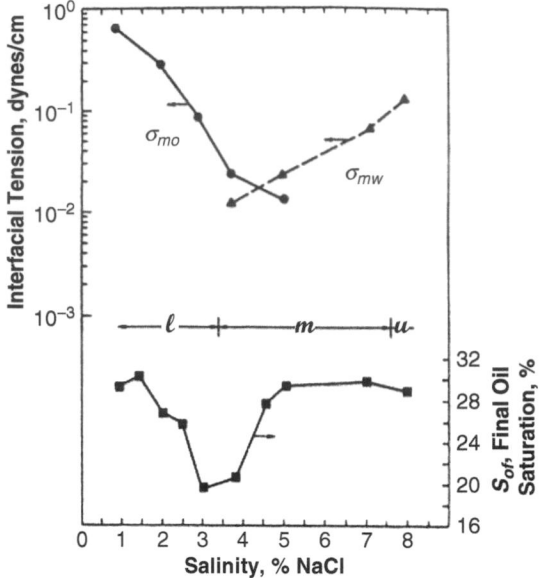

Fig. 7.74—IFT and oil recovery ($N = 9$) (Healy and Reed 1977).

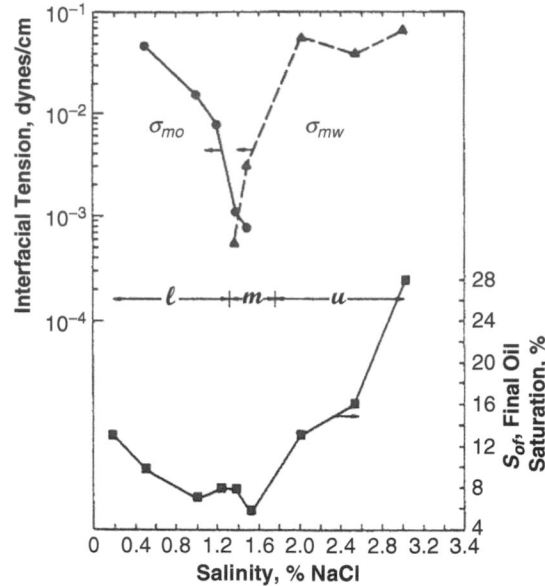

Fig. 7.75—IFT and oil recovery ($N = 12$) (Healy and Reed 1977).

σ_c is the larger of the two IFT values, σ_{mo} and σ_{mw}, and was called the critical IFT by Healy and Reed. The last column is the capillary number, N_{ca}, based on σ_c.

Parameters measured for each flood were fractional oil and water flow, f_o and f_w, in the oil bank and the final ROS, S_{of}, at the end of a displacement.

The most striking result from the experiments was the correlation between optimal salinity for IFT and oil recovery expressed by S_{of}. **Figs. 7.74 and 7.75** show two examples. In essentially all cases, the maximum oil recovery (minimum S_{of}) occurred at a salinity at or very near optimal salinity. That is, the best recovery resulted when the microemulsion system was in a three-phase environment, at or near the point where $\sigma_{mo} = \sigma_{mw}$ and $V_o/V_s = V_w/V_s$.

The effects on the recovery behavior of a number of variables—temperature, surfactant structure, alcohol cosurfactant type, and WOR in the microemulsion—were investigated. The results were consistent with the minimum S_{of} occurring at or near optimal salinity for IFT.

The data show that a displacement with a chemical slug is most efficient when the IFT between phases is low at both the leading and trailing edges of a microemulsion slug. The salinity at which both σ_{mo} and σ_{mw} are small is in the vicinity of optimal salinity. If σ_{mo} is too large, oil will not be mobilized and displaced efficiently by the slug. However, if σ_{mw} is too large, then a relatively large residual saturation of microemulsion will be trapped at the trailing edge of the slug and the slug will deteriorate as it is transported through the rock.

Healy and Reed (1977) also postulated that a flood was controlled by the value of σ_c, the larger value of σ_{mo} and σ_{mw}. There was a correlation between the capillary number based on σ_c and S_{of}. However, this correlation was not as strong or consistent as the dependence of S_{of} on optimal salinity.

Example 7.4—Displacement Efficiency as a Function of Salinity. **Table 7.8** gives phase behavior and IFT data for a given sulfonate/oil/water system (3% surfactant, 48.5% oil, 48.5% brine of X% NaCl).

Assume a series of linear displacements is to be conducted displacing an oil of 5.0-cp viscosity. The displacement velocity in all cases is to be 1.0 ft/D. Microemulsions at the salinities indicated in Table 7.8 are to be used in the different displacements.

Salinity (% NaCl)	V_o/V_s	V_w/V_s	σ_{mo} (dynes/cm)	σ_{mw} (dynes/cm)	Microemulsion
0.5	2.5	—	5×10^{-2}	—	Lower
1.0	4.0	—	2×10^{-2}	—	Lower
1.25	8.0	—	4×10^{-3}	—	Lower
1.5	14.0	16.5	6×10^{-4}	3×10^{-4}	Middle
1.75	16.0	13.0	4×10^{-4}	4.5×10^{-4}	Middle
2.0	—	5.5	—	3×10^{-2}	Upper
2.5	—	2.5	—	8×10^{-2}	Upper

Table 7.8—Phase behavior and IFT data, Example 7.4.

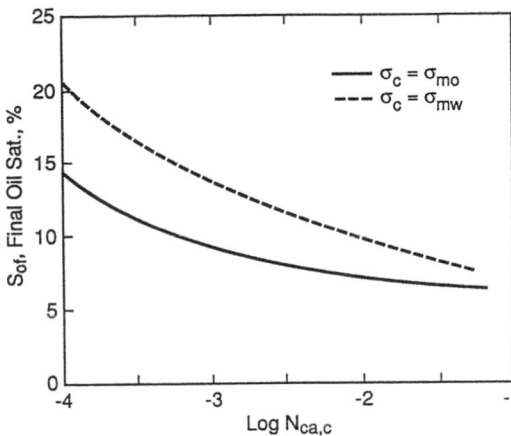

Fig. 7.76—Capillary number correlation for Example 7.4 (Healy and Reed 1977).

Using the Healy and Reed (1977) correlation (**Fig. 7.76**), prepare a plot of final oil saturation, S_{of}, vs. salinity (percent NaCl) that you would expect to result from the experiments. Note that 1.0 cp = 10^{-2} g/(cm·s).

Solution. Prepare a table of the controlling capillary number, $N_{ca,c}$ from data provided. Here, $N_{ca,c}$ is defined using interstitial velocity rather than Darcy velocity (Healy and Reed 1977).

$$N_{ca,c} = v\mu/\sigma_c,$$

where μ = viscosity of the displacing phase, cp.

$$N_{ca,c} = \frac{0.000353 \text{ cm/s} \times 0.01 \text{ g/(cm·s)}}{\sigma_c (g \cdot cm)/(s^2 \cdot cm)}$$

$$= \frac{0.00000353}{\sigma_c}.$$

Table 7.9 shows the controlling capillary number for each salinity. $N_{ca,c}$ is then used with Fig. 7.76 to obtain S_{of}. The minimum S_{of} corresponding to the maximum oil recovery occurs very near optimal salinity (**Fig. 7.77**).

In some experimental work reported in the literature, oil recovery does not correlate directly with IFT measurements made on equilibrated phases (Gerbacia and McMillen 1982). Other factors, such as loss of mobility control and phase partitioning, can dominate in a flow process in porous media.

7.8.3 Effect of a Salinity Gradient on Displacement Efficiency.
Displacement efficiency is affected by phase behavior and by brine salinity, as described previously. In a field application, at least three distinct fluid zones exist, each of which could have a different salinity. Ahead of the microemulsion slug is a zone containing brine that remains as a result of a waterflood or as a result of a preflush of a specific brine/chemical system to condition the reservoir. The microemulsion slug contains a brine of a certain salinity. And finally, behind the microemulsion slug is a mobility-buffer zone, which is an aqueous phase at a particular salinity. The brine type (type of ions) and concentration are not necessarily the same in the three zones.

Consideration of the effect of the salinity of the different fluid zones on displacement efficiency has led to a design concept in which a salinity gradient is used (Nelson 1982; Glover et al. 1979). Because displacement efficiency is highest at optimal salinity, Nelson (1982) postulated that a flood should be conducted, to the degree possible, in the three-phase environment near optimal salinity. (Nelson used "midpoint salinity" instead of optimal salinity because this is the condition at which $V_o/V_s = V_w/V_s$.) Nelson's work led to the conclusion that this condition is satisfied and that a flood is most efficient when conducted in a salinity gradient (i.e., when the salinities of the different fluid zones are fixed in a particular order).

An alternative way to look at the displacement process under a salinity gradient is to express phase behavior over the ranges of the different phase environments as a function of the surfactant/cosurfactant concentration and salinity (Nelson 1982, 1981). **Fig. 7.78** shows an example of this approach using a salinity requirement diagram.

Nelson (1982) obtained data of this type by equilibrating samples of systems that were 80% brine; 5.0, 2.0, and 0.8% surfactant/cosurfactant; and the remainder oil. He varied the percent NaCl in the brine phase over a range of approximately 0 to 2.5 wt%. Nelson recorded the concentration ranges over which Types II(−), III, and II(+) phase environments were observed and the midpoint salinities (optimal salinity for phase behavior). These regions are indicated in Fig. 7.78.

Fig. 7.78 indicates that the region of the three-phase environment and the midpoint salinity are functions of the surfactant/cosurfactant concentration. Midpoint, or optimal, salinity

Salinity (% NaCl)	Controlling σ_c (dynes/cm)	$N_{ca,c}$	log $N_{ca,c}$	S_{of} (%)
0.50	5×10^{-2} oil	3.5×10^{-4}	−3.46	11.0
1.00	2×10^{-2} oil	8.8×10^{-4}	−3.06	9.5
1.25	4×10^{-3} oil	4.4×10^{-3}	−2.36	7.8
1.50	6×10^{-4} oil	2.9×10^{-2}	−1.54	6.7
1.75	4.5×10^{-4} water	3.9×10^{-2}	−1.41	8.0
2.00	3×10^{-2} water	5.9×10^{-4}	−3.23	15.0
2.50	8×10^{-2} water	2.2×10^{-4}	−3.66	17.5

Table 7.9—Capillary numbers, Example 7.4.

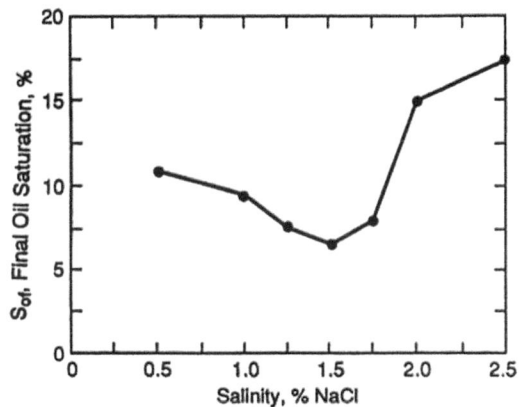

Fig. 7.77—ROS as a function of salinity, Example 7.4.

typically decreases as the surfactant concentration is reduced, as shown in Fig. 7.78. This reflects the fact that the phase behavior expressed on a pseudoternary diagram is only an approximation (i.e., a multicomponent system is approximated by three pseudocomponents). This "real" behavior is reflected in the salinity requirement diagram (Fig. 7.78) by the "bending over" of the upper Type III environment boundary and the narrowing of the Type III environment at the lower surfactant concentrations. For an ideal system, of the type discussed in Section 7.4.2, the upper and lower boundaries between Types III, II(+), and II(–) environments would be horizontal lines. Frequently, the behavior of systems that do not contain divalent ions will approximate ideal systems. (The general effect of divalent ions is discussed in Section 7.6.4.)

Nelson (1982) correlated a series of microemulsion displacement experiments reported in the literature (Gupta and Trushenski 1979) in which the salinity of the different fluid zones was varied. The chemical system consisted of a commercial surfactant, IPA, brine containing NaCl in distilled water, and decane as the oil. The runs were conducted in 1-ft-long Berea cores at an interstitial velocity of 2.0 ft/D. Each core was waterflooded to ROS before the microemulsion flood. Chemical slug size was 10% PV. The chemical slug was followed by a polymer solution consisting of a 1,000-ppm biopolymer in an NaCl brine. The mobility ratio was favorable at both the leading and trailing edges of the microemulsion slug. The only variables changed in the different runs were the salinities of the waterflood brine, chemical slug, and mobility buffer.

The salinity requirement diagram and the paths followed by eight different floods are shown in **Fig. 7.79.** The shaded area is the three-phase region. The band running through this region is the midpoint salinity. Numbers in circles in the midpoint salinity band are the volume percents of active surfactant/cosurfactant in the equilibrium microemulsion phase at midpoint salinity.

Dashed lines and run numbers indicate concentration paths followed by different experiments, and arrows indicate the direction of change of concentration. For example, in Run 6, the waterflood salinity was 2.2% (no surfactant), the microemulsion salinity was 0.2%, and the polymer salinity was 0.2%. The path through the displacement thus proceeded from 2.2% brine and 0% surfactant to 0.2% brine and 8.8% surfactant to 0.2% brine and 0% surfactant. The concentration of the full-strength chemical slug was 8.8% surfactant/cosurfactant.

As shown in Fig. 7.79, several concentration paths were examined. **Table 7.10** gives results of eight experiments and the phase environments for the different fluid zones. ROS at the end of the displacements and the percent of the surfactant retained in the core also are listed.

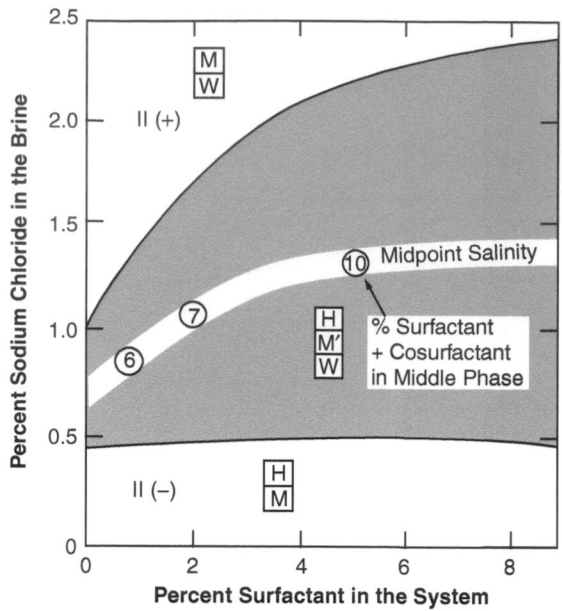

Fig. 7.78—Salinity requirement diagram (Nelson 1982).

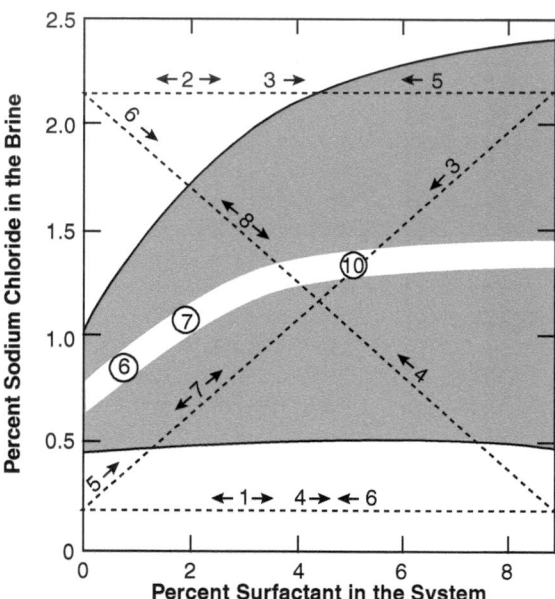

Fig. 7.79—Salinity requirement diagram and initial composition paths (Nelson 1982).

The most obvious result is that recovery efficiency was high in Runs 3, 6, and 7, where the salinity of the polymer mobility-control solution was low, and in the Type II(–) phase environment. Surfactant retention was also relatively small in these runs. The leading waterflood salinity and/or the microemulsion salinity was in either the Type II(+) or III phase environment. The salinity in all three cores passed through the three-phase environment, where IFT values are very small. Run 1 had a poor recovery efficiency, but also had a polymer solution salinity in the Type II(–) environment and a small surfactant retention. In this core, however, all three zones were at salinities in the Type II(–) environment. Evidently, the low IFT required for efficient displacement was not achieved.

Other runs, such as Runs 4, 5, and 8, passed through the three-phase environment but recovery efficiencies were poor. Also, all the injected surfactant was retained in the core at the end of these displacements. The data indicate that a salinity gradient in the different fluid zones, ending with a low-salinity mobility-control solution in the Type II(–) environment, achieves the best results.

The explanation of the results relates to the phase behavior as affected by salinity and the corresponding IFT values between equilibrium phases. Runs 3, 6, and 7 passed through the Type III environment, where IFT is low. At the trailing edge of the

| | Phase Type Promoted | | | | |
Chemical Flood	Waterflood Brine	Chemical Slug	Polymer Drive	ROS After Chemical Flood (% PV)	Injected Surfactant Retained by Core (%)
1	II(−)	II(−)	II(−)	29.1*	52
2	II(+)/III	II(+)/III	II(+)/III	25.2*	100*
3	II(+)/III	II(+)/III	II(−)	2.0**	61*
4	II(−)	II(−)	II(+)/III	17.6*	100*
5	II(−)	II(+)/III	II(+)/III	25.0	100
6	II(+)/III	II(−)	II(−)	5.6**	59**
7	II(−)	II(+)/III	II(−)	7.9*	73**
8	II(+)/III	II(−)	II(+)/III	13.7**	100*

*Average of duplicates.
**Average of triplicates.

Table 7.10—Phase environment type and chemical flood performance data associated with Fig. 7.79 (Nelson 1982).

| | Phase Type Promoted | | | ROS After Chemical Flood (% PV) | Injected Surfactant Retained by Core (%) |
Chemical Flood Description	Waterflood Brine	Chemical Slug	Polymer Drive		
Constant Salinity	III	III	III	12.3*	100
Salinity gradient	II(+)	III	II(−)	3.5	55

*Average of duplicates.

Table 7.11—Phase environment type and chemical flood performance data associated with Fig. 7.80 (Nelson 1982).

floods, the systems were in the Type II(−) environment (i.e., a lower-phase microemulsion environment in which surfactant was in the aqueous phase). The aqueous phase was at high saturation and therefore propagated through the core, carrying the surfactant. This tended to maintain conditions for oil displacement at or near the Type III, or three-phase, environment.

Conversely, in Runs 4, 5, and 8, the salinity was high at the trailing edge of the floods and a Type II(+), or upper-phase, system existed. Even though IFT was probably very low during part of the flood, the surfactant partitioned in the oleic phase toward the rear. This phase is at relatively low saturation and is either slow moving or trapped. Thus, the surfactant was left in the core and not carried to the location where it was required to displace the oil bank.

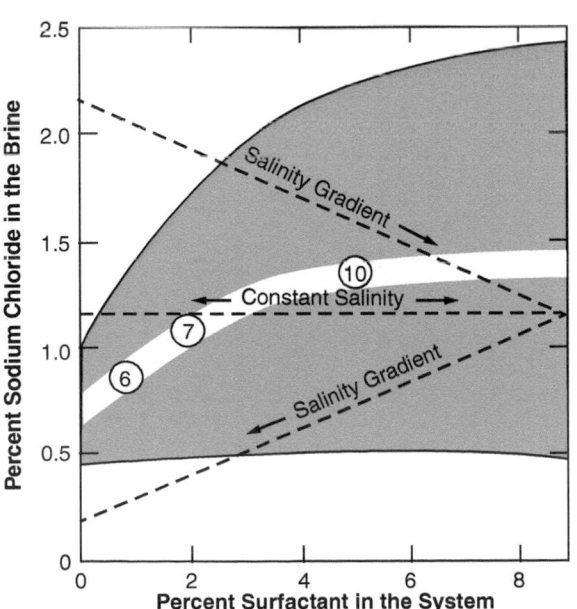

Fig. 7.80—Comparison of constant-salinity and salinity-gradient designs (Nelson 1982).

Runs 1 and 2 were conducted at constant salinity, in the Types II(−) and II(+) environments, respectively. A low IFT was probably not achieved in either case. There was a difference in surfactant retention in that all the surfactant was retained in the core in Run 2, where surfactant partitioned into the oleic phase at the trailing edge.

As stated by Nelson (1982), the floods are excellent examples of how flood effectiveness depends on the compositions and phase environments developed in the mixing zones. In all the floods discussed, the oil, rock, temperature, flow rate, slug size, and surfactant composition were the same. In Runs 1, 4, 6, and 7, the chemical slugs were identical, but the recovery efficiency ranged from 4 to 82% of waterflood ROS. Runs 2, 3, 5, and 7 had identical chemical slugs, and efficiency ranged from 19 to 94%. The utility of specifying the salinity of the drive mobility buffer is clearly indicated by the results.

Table 7.11 gives results from two additional runs that demonstrate the validity of the above conclusion. The concentration paths for the two runs are given in **Fig. 7.80.** The constant-salinity run was conducted by Gupta and Trushenski (1979) and the salinity-gradient run by Nelson (1982). The constant-salinity flood was at a salinity of 1.17%, which is optimal salinity for IFT for a chemical slug at full strength (8.8% surfactant/cosurfactant). Even though the constant-salinity displacement was almost entirely within the Type III environment, the performance was significantly poorer than for the flood conducted with a salinity gradient. The fact that the polymer solution was in the Type II(+) environment had a negative effect on the constant-salinity flood and probably trapped the microemulsion.

Example 7.5—Salinity-Gradient Concept. The salinity-gradient concept is to be tested in laboratory displacement runs. To determine the salinity requirement diagram, solubility data are measured over a range of salinities for four different surfactant concentrations. Phase behavior also is observed and recorded. All data are for a single surfactant system and at a constant temperature. **Figs. 7.81a through 7.81d** show the data.

1. With these data, construct a salinity requirement diagram for this system. Include the midpoint salinity locus in the diagram.
2. On the diagram, plot the displacement paths for the following two systems: (a) waterflood, 2% NaCl; chemical slug, 0.5% NaCl; and polymer slug, 2% NaCl; and (b) waterflood, 1.5% NaCl; chemical slug, 0.3% NaCl; and polymer slug, 0.3% NaCl. Both systems will use a chemical slug that contains 6% surfactant.
3. Describe the expected displacement recovery efficiency for the two systems. Briefly explain the reason(s) for your answer.

Solution.

1. The salinity requirement diagram is determined by plotting the boundaries of the different phase environments: Types II(–), III, and II(+). **Fig. 7.82** shows the data points with smooth curves drawn through them.
2. The composition paths followed by Runs a and b are given on the salinity requirement diagram. Arrows indicate the direction of changing composition.

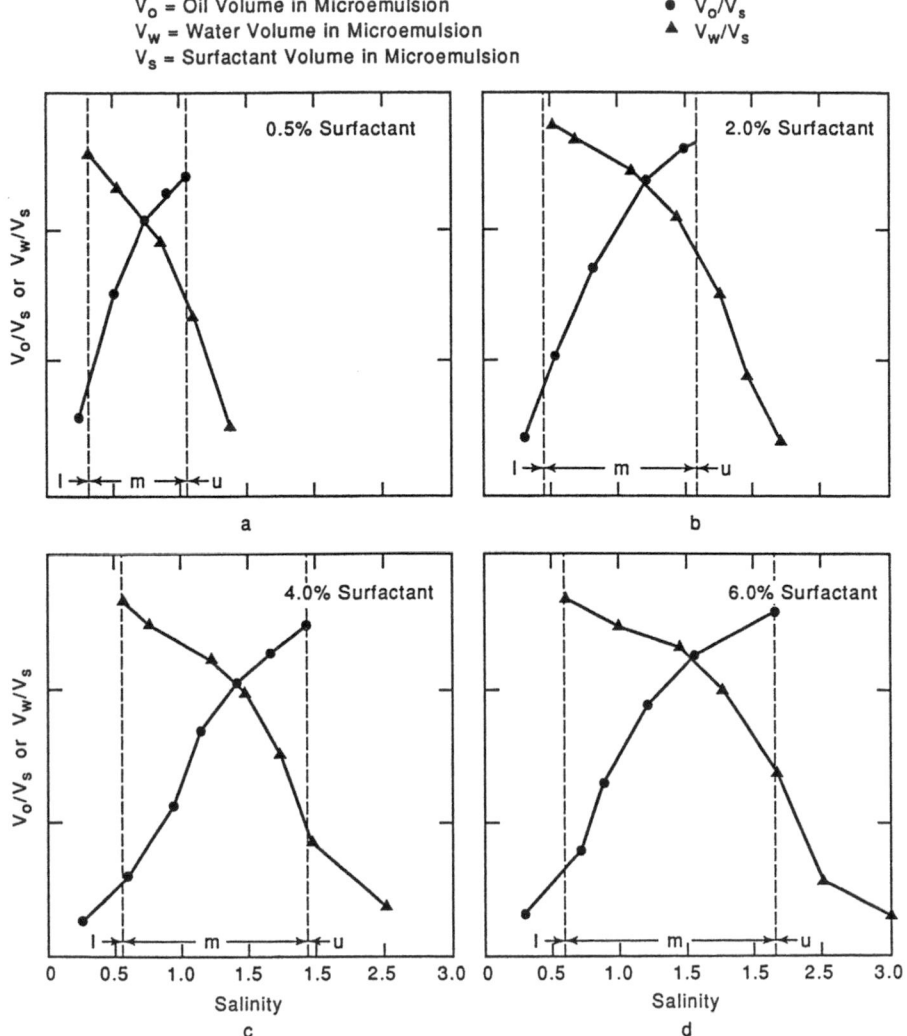

Fig. 7.81—Solubility data for Example 7.5.

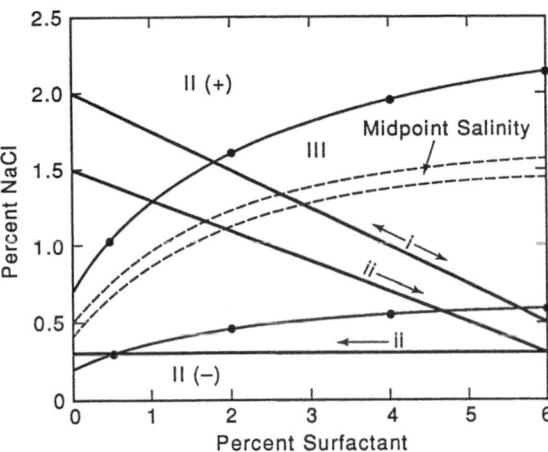

Fig. 7.82—Salinity requirement diagram and displacement paths, Example 7.5.

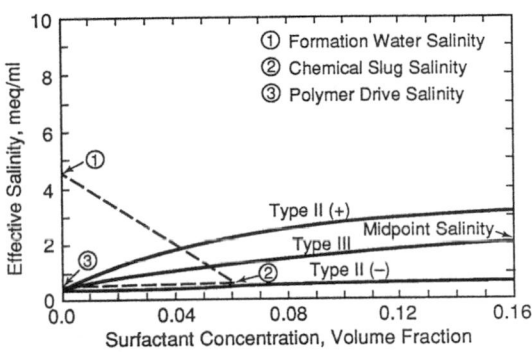

Fig. 7.83—Salinity requirement diagram for wedge-forming experiment (Tham et al. 1983).

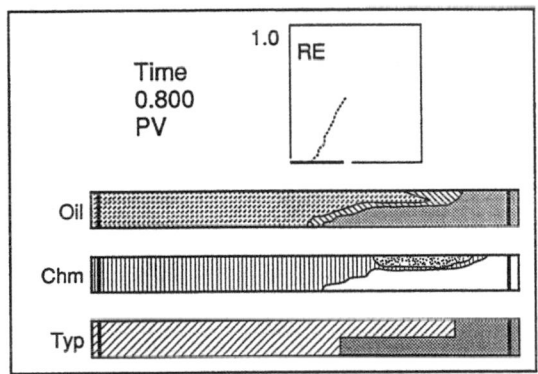

Square Frame – Recovery Curves
········· Fraction of Waterflood Residual Oil Recovered
—— Fraction of Surfactant Recovered

Upper Rectangular Frame – Oil Saturation
▨ Chemical Flood Residual Oil
▧ Oil Saturations Intermediate Between Chemical Flood Residual and Waterflood Residual
▩ Oil Saturations Equal to, or Greater than, Waterflood Residual

Middle Rectangular Frame – Surfactant Concentration
‖‖‖ Surfactant Concentration Equal to, or Less than, Adsorbed or Trapped Concentration
▨ Active Surfactant Concentration
☐ No Surfactant

Lower Rectangular Frame – Phase Environment Type
▨ Type III
▧ Type II (+)

Fig. 7.84—Computed profiles during wedge-forming case at 0.8 PV injected (Tham et al. 1983).

3. Displacement a would be inefficient because a Type II(+) phase is the final displacing fluid. Surfactant partitions into a slow-moving oleic phase and is "phase trapped." Displacement b would be an efficient displacement with relatively low surfactant loss and high recovery (assuming that σ_{mo} and σ_{mw} are small). Surfactant partitions into an aqueous phase at the rear of the flood.

7.8.4 Gravity Segregation in Microemulsion Flooding. Under certain conditions, a wedge of residual oil is left on the bottom of a core in laboratory chemical flooding experiments in which the chemical flooding system has a narrow Type III salinity range (Nelson 1982). Analysis of this region indicated that little surfactant was present in the liquids or adsorbed on the rock. Thus, this region was bypassed by the chemical flooding system.

This region of bypassed oil, called an "oil wedge," is formed by gravity segregation (Tham et al. 1983). By using a chemical flood simulator, Tham et al. (1983) were able to show that an oil wedge forms when there is gravity override at the front of the chemical slug and gravity underride at the back of the chemical slug.

Fig. 7.83 shows the salinity requirement diagram for an experiment in which an oil wedge formed. There is a large mixing zone at the front of the chemical slug, which is in the Type II(+) environment (Path 1-2). Because the density of the microemulsion phase is smaller than that of the formation water, gravity override occurs, as depicted in **Fig. 7.84,** produced by the simulator. At the back of the chemical slug, the mixing path between the mobility buffer and the microemulsion is along Path 2-3. These microemulsions are in the Type III phase environment and contain large oil concentrations. Because the polymer drive water is denser than the microemulsions, the polymer drive water underruns the rear of the chemical slug (Tham et al. 1983). Surfactant is lost as the Type III microemulsion is trapped with the residual oil behind the chemical slug.

Chemical flood systems that have Type III regions over a broad range of salinities are less likely to form an oil wedge. Results (Tham et al. 1983) from a chemical flood simulator show that these situations can be overcome by design of the slug. Gravity override at the surfactant front can be reduced by selecting a chemical flooding system that has a high upper Type III boundary [minimizing the Type II(+) mixing region] on the salinity requirement diagram. Gravity underride at the back of the chemical slug can be reduced by increasing the injection rate (possible only in laboratory corefloods), increasing the viscosity of the drive fluid, and decreasing the salinity of the drive fluid to move the rear mixing zone into a Type II(−) phase environment. Some of these alternatives may conflict with choosing a chemical system that gives maximum oil recovery.

7.8.5 Additional Comments on Displacement Mechanisms. A number of investigations of displacement mechanisms in addition to those previously discussed have been reported in the literature (Stegemeier 1977; Davis and Jones 1968; Willhite et al. 1980; Trushenski et al. 1974; Cash et al. 1975; Gladfelter and Gupta 1980; Fleming et al. 1981; Pope and Nelson 1978).

Macroemulsions frequently are produced during a micellar/polymer displacement process. Macroemulsions can have a high mobility and thus adversely affect the process by causing an unfavorable mobility ratio at the oil-bank interface. It also has been suggested that macroemulsions can be efficient displacement agents and can enhance recovery under some circumstances (Willhite et al. 1980; Cash et al. 1975).

Chromatographic-like separation of the chemicals that make up a micellar slug is known to occur as the slug propagates through a porous medium. Examples of this separation are given in Stegemeier (1977), Davis and Jones (1968), Willhite et al. (1980), Trushenski et al. (1974), Cash et al. (1975), Gladfelter and Gupta (1980), Fleming et al. (1981), and Pope and Nelson (1978). Mechanisms that cause this separation are discussed in Section 7.9.

7.9 Surfactant Loss From Rock/Fluid Interactions and Phase Partitioning

Surfactant loss from an injected micellar slug can occur by at least three processes: precipitation, adsorption onto the porous medium, and phase partitioning into a static or slow-moving phase. These mechanisms all result in retention of surfactant in a porous medium and deterioration of the composition of the chemical slug, leading to poor displacement efficiency (Novosad 1981; Meyers and Salter 1981; Bae and Petrick 1977; Pursley and Graham 1975; Bae et al. 1974; Smith and Malmberg 1975; Smith et al. 1975; Gilliland and Conley 1976; Hurd 1976; Boneau and Clampitt 1977; Widmyer et al. 1977; Strange and Talash 1977; Lawson and Dilgren 1978; Malmberg and Smith 1976; Lawson 1978; Somasundaran and Hanna 1979, 1985; Holm 1978; Rathmell et al. 1978; Wanosik et al. 1978; Kellerhals 1982; Putz et al. 1981; Somasundaran et al. 1984; Ziegler and Handy 1981; Hedges and Glinsmann 1979; Satter et al. 1980; Chan and Gupta 1984; Wang 1993). The negative effects of retention and phase splitting were illustrated in Section 7.8.3, which dealt with displacement under different salinity gradients.

Additionally, the brine composition of the different slugs in the process can be altered during flow through a porous medium by cation exchange between the fluids and rock. This can alter phase behavior and IFT and contribute to a reduced displacement efficiency.

7.9.1 Precipitation. The presence of multivalent ions (Somasundaran et al. 1984) can lead to surfactant precipitation as a result of phase separation of surfactant and cosurfactant (Meyers and Salter 1981) and chromatographic separation of the different surfactant species present (Gale and Sandvik 1973). Because precipitation results in retention in a porous medium, differentiating between precipitation and other mechanisms that retain surfactant is often difficult in experiments (Trushenski et al. 1974).

Figs. 7.85 and 7.86 (Somasundaran et al. 1984) show example effects of surfactant concentration, salt concentration, and ion type on precipitation of surfactant. The surfactants used in these studies were a sodium dodecylbenzenesulfonate (NaD-DBS) and a sodium benzenesulfonate (Texas No. 2®). Brines containing K+, Ca++, or Al+++ ions were used. The solutions did not contain an alcohol cosurfactant or an oil. Figs. 7.85 and 7.86 plot percent light transmission through a solution as a function of salt or surfactant concentration. Reduction in light transmission indicates turbidity and precipitation of surfactant.

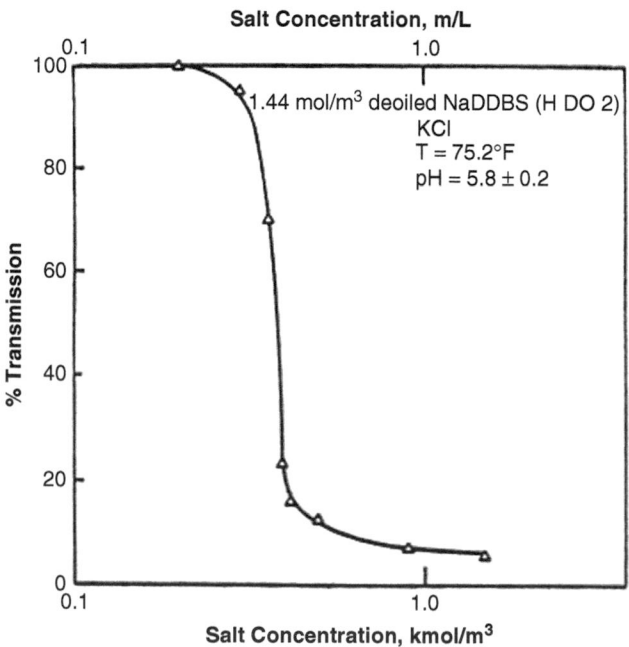

Fig. 7.85—Light transmission of dodecylbenzenesulfonate solutions as a function of added salt concentration.

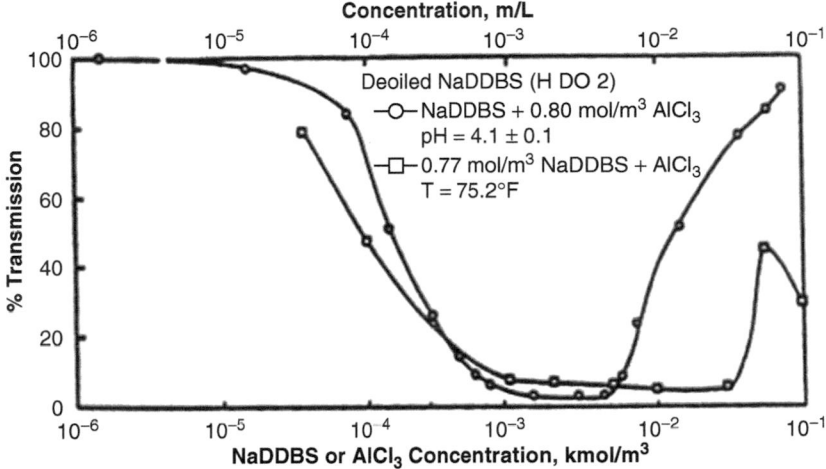

Fig. 7.86—Light transmission of sodium dodecylbenzenesulfonate solutions as a function of dodecylbenzenesulfonate and AlCl₃ concentrations (Somasundaran et al. 1984).

The system in Fig. 7.85 contained a fixed amount of NaD-DBS to which a KCl solution was added. The light transmission through the system decreased sharply at a KCl concentration of approximately 0.5 kmol/m³, indicating precipitation. Solubility of the surfactant decreased as salinity increased. Fig. 7.86 shows a similar effect in the lower curve, where the added salt is AlCl₃. Precipitation occurred at much lower salt concentrations when the ion was Al⁺⁺⁺ than when it was K⁺. The behavior with Ca⁺⁺ was consistent in that divalent ions decreased solubility but not to the extent of Al⁺⁺⁺.

Fig. 7.86 also shows the effect of increasing surfactant concentration at fixed salinity. As NaDDBS concentration increased above approximately 10^{-5} kmol/m³, precipitation occurred, increasing with surfactant concentration. However, at higher surfactant concentrations above approximately 5×10^{-3} kmol/m³, precipitated surfactant redissolved. This behavior also was observed in the presence of divalent ions. Some apparent redissolution of surfactant also is shown at higher concentrations of AlCl₃. Somasundaran et al. (1984) point out that redissolution did not occur at higher surfactant concentrations.

An alcohol cosurfactant may increase the solubility of a surfactant (Novosad 1981). This characteristic depends on alcohol type. The alcohol must be matched with the surfactant. **Figs. 7.87 and 7.88** show example data illustrating this effect. Surfactants in these solutions were a commercial surfactant (TRS10-80®) and sodium 8-phenyl-n-hexadecyl-p-sulfonate (Texas No. 1®). Brine was 1% NaCl. No oil was added to the system. The alcohol used was SBA. Precipitated sediment was measured as a function of the alcohol/surfactant ratio. As Figs. 7.87 and 7.88 show, sediment either remained constant over a range of concentration ratios and then declined significantly or decreased steadily with increasing concentration ratios. In either case, precipitation was not present when the alcohol/surfactant ratio was sufficiently high.

Gale and Sandvik (1973) pointed out that chromatographic separation of surfactant species can occur in a porous medium because of differential adsorption of the molecular species. When a surfactant consists of a broad range of equivalent weights, such as might be present in a petroleum sulfonate or crude oil sulfonate, separation leads to modified and different concentrations of the chemical species in different zones of a chemical slug. If a

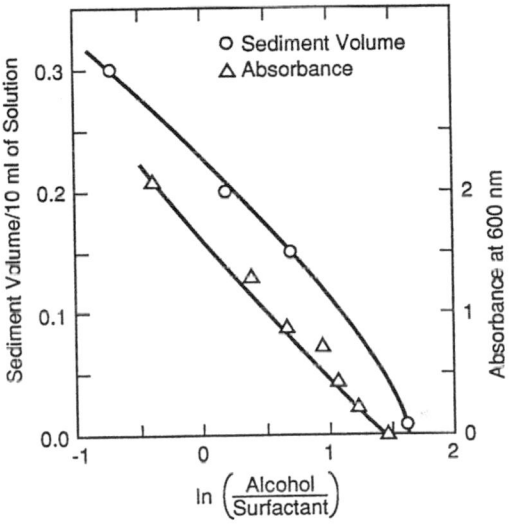

Fig. 7.87—Solubility of 1% Texas 1/sec-butylalcohol in 1% NaCl at 22°C (Novosad 1981).

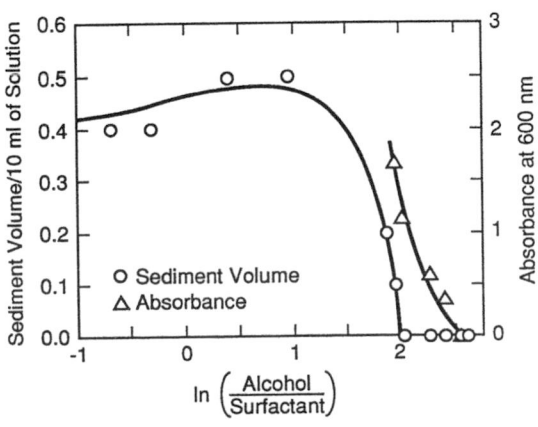

Fig. 7.88—Solubility of 1% TRS 10-80/sec-butyl-alcohol in 1% NaCl at 22°C (Novosad 1981).

solution that initially is at a concentration at which the surfactant is soluble undergoes a change in concentration, then precipitation is possible. This might happen if, for example, the concentrations of the higher-molecular-weight surfactants are increased relative to the lower-molecular-weight surfactants.

Surfactant can precipitate under a number of different conditions. The process is complex and extremely difficult to predict. In addition to the variables discussed, precipitation is dependent on temperature. It also is a kinetic process [i.e., there is a certain reaction time involved that may range from minutes to hours (Somasundaran et al. 1984)]. A surfactant system being considered for field application should be tested for precipitation under conditions as close as possible to those in the field. Even so, in experimental work, it often is difficult to isolate the specific cause of precipitation or even to be certain that this is the reason for surfactant loss rather than an entirely different mechanism, such as adsorption or phase partitioning. Surfactant retention in the porous medium by any of the mechanisms of loss is generally very detrimental to the micellar/polymer flooding process.

7.9.2 Adsorption. Adsorption of surfactant on reservoir rock is determined by batch equilibrium tests on crushed core material and from core displacement tests. The amount of material adsorbed may be expressed in three different units: mass of surfactant adsorbed/mass of rock (mg/g), mass of surfactant adsorbed per unit of PV (mg/mL PV), and molecules of surfactant adsorbed per unit surface area (μeq/m^2). Conversion between these units is possible if surface area/mass of rock, bulk density of the rock, porosity of the rock, and the equivalent weight of the surfactant are known. In core displacements, surface areas are usually not known. As a result, adsorption is reported in mg/g or mg/mL PV.

In batch equilibrium tests on crushed core, a known volume of surfactant solution at a known concentration is mixed with a specified mass of crushed rock in a sealed container. Fluid samples are withdrawn at intervals and analyzed. When the concentration remains constant with time, the system is at equilibrium and the test is completed. Adsorption is computed from a material balance on the surfactant. When concentrations are expressed in μeq/m^2, the adsorption is given by (Meyers and Salter 1981)

$$A_S = \left(\frac{C_{AS}^i - C_{AS}^f}{m_a S_a} \right) \rho V_a, \dots\dots\dots\dots\dots\dots\dots\dots\dots\dots\dots\dots\dots\dots\dots\dots\dots\dots (7.24)$$

where A_S = adsorption, μeq/m^2; C_{AS}^i = active surfactant concentration in the initial solution, meq/g; C_{AS}^f = active surfactant concentration in the final solution, meq/g; m_a = mass of adsorbent, g; S_a = surface area of adsorbent, m^2/g; ρ = density of initial fluid, g/cm^3; and V_a = volume of surfactant solution added, cm^3.

When the surface area is not known, adsorption is reported per unit PV or per mass of adsorbent. Eqs. 7.25 and 7.26 give conversions between the units of adsorption.

To convert adsorption from μeq/m^2 to mg/mL PV,

$$A_S \left(\frac{mg}{mL\ PV} \right) = A_S \left(\frac{\mu eg}{m^2} \right) \left(\frac{10^{-6} eq}{\mu eq} \right) M_s \left(\frac{g}{eq} \right) S_a \left(\frac{m^2}{g\ rock} \right)$$

$$\times \rho_b \left(\frac{g\ rock}{cm^3\ bulk\ volume} \right) \frac{1}{\phi}$$

$$\times \left(\frac{cm^3\ bulk\ volume}{cm^3\ PV} \right) \left(\frac{cm^3}{mL} \right) \left(\frac{10^3\ mg}{g} \right)$$

$$= 10^{-3} \frac{M_s S_a \rho_b}{\phi} A_S \left(\frac{\mu eq}{m^2} \right), \dots\dots\dots\dots\dots\dots\dots\dots\dots\dots (7.25)$$

where M_s = equivalent weight of the surfactant, g/eq.

By analogy,

$$A_S \left(\frac{mg}{g\ rock} \right) = 10^{-3} M_s S_a A_S \left(\frac{\mu eq}{m^2} \right). \dots\dots\dots\dots\dots\dots\dots\dots\dots\dots\dots\dots (7.26)$$

Fig. 7.89 illustrates the approach to equilibrium during adsorption of a petroleum sulfonate on crushed Bell Creek and Berea core material (Meyers and Salter 1981). Adsorption is a rate-dependent process in these materials. Equilibrium was reached in approximately 3 weeks for the samples in Fig. 7.89. This is a relatively long time for laboratory tests but a short time in a field application.

Adsorption in core material is often achieved by injecting a surfactant solution of known composition into a core until the effluent concentration is equal to the injected concentration. If the concentrations of all effluent samples are analyzed, the retained surfactant can be determined by material balance. This requires many analyses. The work can be reduced by extracting the adsorbed surfactant from the core with an appropriate solvent, such as methanol/chloroform or IPA, and analyzing the extract for surfactant. Adsorption is calculated as the difference between the total surfactant recovered and the surfactant in the pore space of the rock, if any.

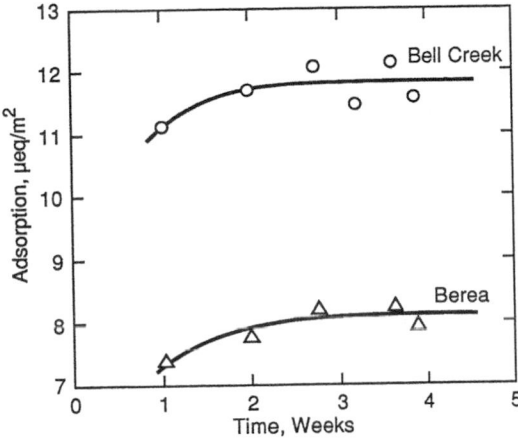

Fig. 7.89—Adsorption vs. time—Fluid 1 (Meyers and Salter 1981).

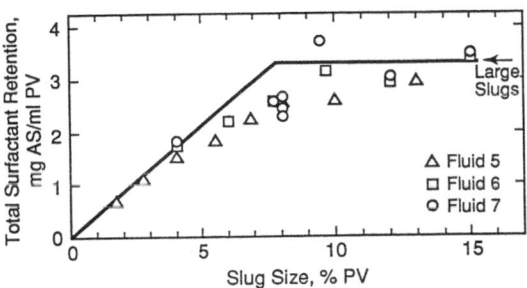

Fig. 7.90—Surfactant retention during coreflood experiments (Meyers and Salter 1981).

Adsorption can also be determined by injecting surfactant slugs of increasing size into cores until retention reaches a maximum. **Fig. 7.90** shows adsorption data for one system as a function of surfactant slug size (Meyers and Salter 1981).

The adsorption of surfactants onto reservoir rock materials is a function of the surfactant type, equivalent weight, and concentration; rock minerals; temperature; clay content; redox condition; and the flow rate of the solution. **Table 7.12** gives typical values of surfactant adsorption on Berea sandstone core materials along with references to the data sources where specific conditions for the experiments are described. **Table 7.13,** although not comprehensive, summarizes some of the adsorption data available for several systems (Meyers and Salter 1981).

Early adsorption studies with petroleum sulfonates discovered that equivalent weight had an effect on retention, as illustrated in **Fig. 7.91,** where data are shown for adsorption of different equivalent weight fractions of a petroleum sulfonate on a calcium montmorillonite clay (Gale and Sandvik 1973). A similar dependence on equivalent weight was found for adsorption onto a crushed reservoir rock, although the adsorption (expressed in mg/g) was considerably smaller for rock materials than for pure clays. Adsorption is higher on clays because the surface area is considerably larger than on crushed reservoir rock.

When a surfactant solution that contains a broad mixture of different equivalent weight compounds flows through a porous medium, differential adsorption will lead to separation of the different surfactant species. Higher-equivalent-weight compounds will tend to be preferentially adsorbed and thus will move more slowly than lower-equivalent-weight compounds. For example, Willhite et al. (1980), Gupta (1984), and Chan and Gupta (1984) have documented chromatographic separation of mono- and disulfonated species.

Gale and Sandvik (1973) point out that middle-equivalent-weight compounds will adsorb to some extent and thereby occupy adsorption sites on the rock material. Assuming adsorption is not reversible, this will reduce the adsorption of the higher-equivalent-weight compounds, which are the most effective at reducing IFT in petroleum sulfonate solutions. The middle-equivalent-weight compounds thus act as sacrificial adsorbents. The problem of chromatographic separation within surfactant systems is reduced when the surfactants are chemically similar, as in the Exxon surfactant used in the Loudon field test (Maerker and Gale 1992).

Adsorption is a function of concentration, as indicated for the system shown in **Fig. 7.92** (Gale and Sandvik 1973). The amount adsorbed typically increases with surfactant concentration in the solution. For pure surfactants, the amount adsorbed reaches a maximum at the CMC. For surfactant mixtures, adsorption continues to increase well beyond the point of onset of micelle formation. Typically, the amount adsorbed reaches a maximum (i.e., plateau) at a sufficiently high concentration of surfactant. For some surfactant systems, the adsorption isotherm is reasonably approximated by the Langmuir model discussed in Chapter 3.

A number of investigators (Novosad 1982; Pope and Nelson 1978; Bae and Petrick 1977; Somasundaran and Hanna 1985) have observed that the amount adsorbed passes through a maximum and decreases in magnitude beyond this maximum. Several reasons for this behavior have been proposed.

Reference	Observed Adsorption (mg/g rock)
Healy et al. (1975)	0.28 to 0.40
Novosad (1982)	0.1 to 1.2
Gale and Sandvik (1973)	0.4 to 0.7
Pursley and Graham (1975)	0.37 to 0.72

Table 7.12—Surfactant adsorption on Berea core material.

Reference	Study Type	Method*	Adsorbent	Surfactant	Plateau Value	Factors Considered**	Conclusions
Gogarty and Tosch (1968)	Laboratory	FC	Berea	Potassium	~0.1 mg/mL PV	0,RA	Presence of oil and water decreases adsorption; physical entrapment rather than adsorption controls loss.
Hill et al. (1973)	Laboratory	S,FC	Berea	70/30 M470 and M380	S(1.3 µeq/m²)D (0.8 mg/mL PV)	FWD,SA	Sacrificial agents decrease adsorption.
Foster (1973)	–	–	–	Petroleum sulfonate	–	FWD,SA	Sacrificial agents decrease adsorption.
Gale and Sandvik (1973)	Laboratory	S,FC	Berea	Several sulfonates	S(0.8 to 1.9 µeq/m²) D(2.6 to 7.7 mg AS/mL PV)	F	Fractionation important; desorption minimal; high pH reduces loss.
Pursley et al. (1973)	Field, laboratory		Loudon	Feed D petroleum sulfonate	(4 mg/mL PV)	F,SA,FWD, DI	No fractionation observed; salinity gradient used but not effective; dual base protected against divalent ions.
Pursley and Graham (1975)	Field, laboratory	PF,FC	Borregos	380 AEW sulfonate	(9.1 to 11.7 mg AS/mL PV)	F	Fractionation very important.
Trushenski et al. (1974)	Laboratory	FC	Berea	AA-sulfonate	(3.1 to 9.3 mg AS/mL PV)		Retrograde isotherms; no desorption.
Bae et al. (1974)	Laboratory	S,FC	Silica gel, Berea	450 AEW sulfonate	S(0.2 to 0.4 µeq/m²) laboratory D(3.8 mg AS/mL PV)	0	Adsorption isotherm maximum; oil-external microemulsions adsorb more; presence of residual oil negligible, has negligible effect on loss.
Healy et al. (1975)	Laboratory	FC	Berea	C12OXS	Ads (4 mg/mL PV) Ret. (5.4 mg/mL PV)	RA	Retention increases with slug breakdown.
Smith and Malmberg (1975)	Laboratory	FC	Berea	TRS 10 and Neodol 25-3S	(6.6 to 14.9 mg AS/mL PV)	F,0	See Gilliland and Conley (1976).
Smith et al. (1975)	Laboratory	FC	Berea	TRS 10 and Neodol 25-3S	(6.2 mg AS total/mL PV)	F,0	High-equivalent-weight sulfonate adsorbs preferentially; sulfate and sulfonate do not separate; residual oil has no effect.
Gilliland and Conley (1976)	Laboratory	FC	Big Muddy	427 AEW sulfonate	(3.3 mg AS/mL PV)	0	Retention increases with salinity, decreases with presence of residual oil and increased alcohol.
Hurd (1976)	Laboratory	S,FC	Loma Novia	Several sulfonates	S(0.5 to 4.3 µeq/m²) D(0.3 to 3.1 mg AS/mL PV)	SA	Adsorption increases with salinity and equivalent weight; adsorption decreases as pH increases; sacrificial agents can reduce adsorption.
Bae and Petrick (1977)	Laboratory	FC	Berea	TRS 18 and 40	D(4 to 8 mg AS/mL PV)	SA,K	Sacrificial agents reduce adsorption; time to reach equilibrium increases with surfactant concentration.
Boneau and Clampitt (1977)	Laboratory	FC	Berea, Burbank	TRS 10-80	Water-wet (Berea 0.7 mg/mL PV) Oil-wet (Berea 3.1/Burbank 4.0 mg/mL PV)	DI,SA,FWD	Oil-wet cores adsorb more.
Widmyer et al. (1977)	Field	PF	Salem	TRS 10-80	(14.9 mg AS/mL PV)	–	–
Strange and Talash (1977)	Field, laboratory	PF,?	Salem	TRS 10-80	Field (3.2 mg AS/mL PV) Laboratory (2.9 mg AS/mL PV)	DI,SA,FWD	Field gave higher loss than laboratory measurements.

Reference	Type	Method*	Formation	Surfactant	Retention	Variables**	Comments
Lawson and Diigren (1978)	Laboratory	S	Berea, Benton	Sipponate 10	(3 to 4.6 μeq/m²)	F,0	No preferential adsorption; replacing oil in deoiled surfactant decreases adsorption.
Malmberg and Smith (1976)	Laboratory	FC	Berea, Seeligson, Cottage Grove	Synthetic sulfonate	(16.6, 21.3, 22.1 mg AS/mL PV)	F,0	Smaller slugs and residual oil decrease adsorption.
Lawson (1978)	Laboratory	S	Berea	Bryton 430	(4.2 μeq/m²)	DI,F	No selective adsorption vs. sulfate; divalent ions increase adsorption; large anions decrease adsorption.
Glover et al. (1979)	Laboratory	FC	Berea	MEAC$_{12}$OXS/NaC$_{12}$OXS + N25-3S	(1.6 to 6.5 mg AS/mL PV)	RA,DI	System containing divalent ions had severe phase trapping, much larger retention.
Somasundaran and Hanna (1979)	Laboratory	S	Various	AA-sulfonate	(2 to 10 μeq/m²)	pH,DI	Structure-making and structure-breaking ions.
Gupta and Trushenski (1979)	Laboratory	FC	Berea	AA-sulfonate	(1.8 to 3.0 mg AS/mL PV)	FWD	Retention decreases with decreased drive-water salinity.
Holm (1978)	Laboratory	FC	Berea	Stepan		DI	Silicate preflush increases surfactant production.
Rathmell et al. (1978)	Laboratory	FC	Berea	Petrostep 465	(1.6 mg AS/mL PV)	—	Live crude decreases adsorption over stock-tank oil.
Wanosik et al. (1978)	Laboratory	FC	Sloss	AA-sulfonate	(2.0 to 4.8 mg/mL PV)	—	Clean core adsorbs more than fresh core.
Kellerhals (1982)	Laboratory	FC	El Dorado	Aqueous system	(4.5 mg AS/mL PV)	FWD	—
Putz et al. (1981)	Laboratory	FC	Chateaurenard	Petroleum sulfonate	(6.0 mg AS/mL PV)	—	—
Somasundaran et al. (1984)	Laboratory	S	Na kaolinite	C$_{12}$ LABS	(1.4 μeq/m²)	0	Precipitation is a major cause of loss; increase in surfactant concentration or added oil redissolves precipitated surfactant.
Ziegler and Handy (1981)	Laboratory	S	Fired Berea	C$_{12}$ LABS	(3 μeq/m²)	—	—
Malmberg et al. (1982)	Laboratory	FC	Berea	DOE synthetics	(3.1 to 3.5 mg AS/mL PV)	—	—
Hedges and Glinsmann (1979)	Laboratory	FC	Berea	TRS 10-410	(1 to 3 mg AS/mL PV)	—	—
Glinsmann (1979)	Laboratory	FC	Berea	TRS 10-410	(0 to 3.2 mg AS/mL PV)	SA	Retention is a function of brine salinity (≈2 mg AS/mL PV in three-phase region).
Gupta (1984)	Laboratory	FC	Berea, Sloss	AA-sulfonate	Berea (3 to 5 mg AS/mL PV) Sloss (5.0)	DI,0,FWD,K	Calcium increases retention; oil has no effect on small micellar slugs; larger slugs, higher salts, caused phase trapping; FWD decreases retention; time effects evident.
Satter et al. (1980)	Field, laboratory	T	Berea, Manvel	Sulfonate and ethoxylate	Berea (13.1 to 17.8 mg/mL PV), Manvel (54.3 mg/mL)	—	—

*S = static, D = dynamic corefloods, PC = field core, and T = tracer test.
**RA = separate retention from adsorption, F = surfactant fractionation, K = adsorption kinetics, DI = effects of divalent ions, FWD = freshwater-drive employed SA = sacrificial agents, and 0 = effect of oil.

Table 7.13—Summary of surfactant retention studies (Meyers and Salter 1981).

Adsorption is a rate-dependent process and thus can vary with flow rate in coreflooding tests. Bae and Petrick (1977) measured higher amounts adsorbed when the rate was 2.0 ft/D than when the rate was 36 ft/D. At reservoir rates typical of those away from a wellbore, the rate dependency should not be a major concern. Adsorption also has been shown to be a function of temperature, decreasing slightly with increasing temperature (Novosad 1982).

Surfactant adsorption is strongly affected by the redox condition of the system (Wang 1993). Laboratory cores typically have been exposed to oxygen and are in an aerobic state. Wang (1993) showed that when laboratory adsorption experiments were conducted in cores that were restored to anaerobic conditions, surfactant adsorption was reduced significantly compared with results from experiments conducted with cores at aerobic conditions.

Effects of other variables on adsorption have been studied. For example, Meyers and Salter (1981) examined the effect of brine/oil mass ratio on adsorption from a petroleum-sulfonate/TAA solution on crushed Berea and Bell Creek core material. They found no effect of brine/oil mass ratio in the surfactant on the amount of surfactant adsorbed by the rock. When a middle-phase microemulsion was injected into waterflooded Berea cores, total surfactant retention was found to be independent of the brine/oil mass ratio in the microemulsion. Adsorption for this surfactant system was 3.3 mg active surfactant/mL PV. Thus, adsorption was a function of the total surfactant concentration in the injected fluid.

The significance of adsorption on a chemical flood is illustrated in Example 7.6.

Example 7.6—Estimation of Significance of Adsorption in a Chemical Flood. Consider the micellar displacement process in a system with 5-acre spacing. A 5% PV microemulsion slug is to be injected, and the slug will contain 5 vol% petroleum sulfonate. Assuming that the average adsorption is 0.4 mg/g of rock, calculate the fraction of the injected surfactant that will be adsorbed. (This adsorption amount is consistent with data presented in Table 7.12.)

Additional data: p_r (solid rock) = 2.7 g/cm³, p (sulfonate) = 1.1 g/cm³, and $\phi = 30\%$.

Solution. Consider a 5-acre pattern, 1.0 ft in thickness.

$$V_b = 5 \text{ acres} \times 43{,}560 \text{ ft}^2/\text{acre} \times 1.0 \text{ ft}$$

$$= 217{,}800 \text{ ft}^3 \text{ rock (bulk)}.$$

$$\rho_b = 2.7 \times 62.4 \text{ lbm/ft}^3 \times (1-\phi)$$

$$= 117.9 \text{ lbm rock/ft}^3 \text{ bulk}.$$

$$m_r = 217{,}800 \text{ ft}^3 \text{ bulk} \times 117.9 \text{ lbm rock/ft}^3 \text{ bulk}$$

$$= 25.7 \times 10^6 \text{ lbm rock}.$$

$$A_S = 0.4 \text{ mg/g rock} \times 0.001 \text{ g/mg} \times 454 \text{ g/lbm}$$

$$\times 25.7 \times 10^6 \text{ lbm rock}$$

$$= 4.67 \times 10^6 \text{ g sulfonate}.$$

$$= 10{,}280 \text{ lbm sulfonate}.$$

Calculate mass of sulfonate injected.

$$V_{slug} = 5\% \text{ PV}$$

$$= 5 \text{ vol\% sulfonate in slug}.$$

$$V_{slug} = 0.05 \times 217{,}800 \text{ ft}^3 \times 0.30$$

$$= 3{,}267 \text{ ft}^3 \text{ of slug}.$$

$$V_s = 0.05 \times 3{,}267 \text{ ft}^3$$

$$= 163 \text{ ft}^3 \text{ sulfonate}.$$

$$m = (1.1 \times 62.4) \text{ lbm/ft}^3 \times 163 \text{ ft}^3$$

$$= 11{,}200 \text{ lbm sulfonate}.$$

$$f_a = 10{,}280/11{,}200$$

$$= 0.92.$$

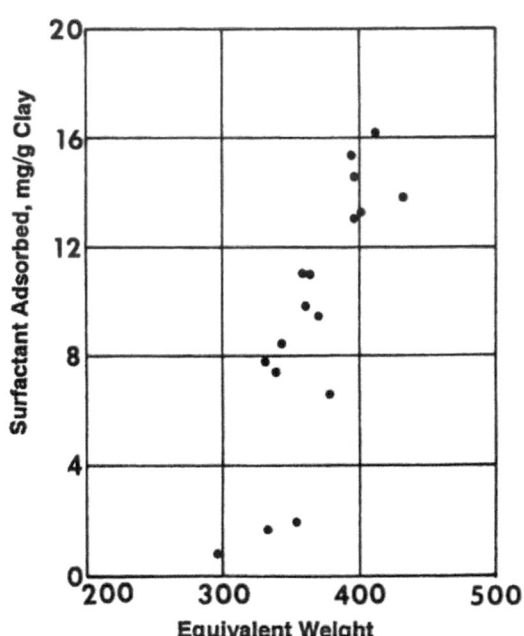

Fig. 7.91—Surfactant adsorption on Ca montmorillonite (25 mL of 0.5% surfactant in 2% Na₂SO₄ equilibrated with 7 g of API No. 23 standard Ca montmorillonite) (Gale and Sandvik 1973).

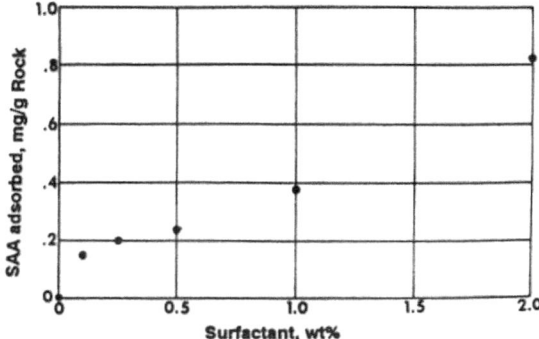

Fig. 7.92—Surfactant adsorption isotherm (surfactant dissolved in 0.75% Na₂CO₃, 0.5% NH₃ brine, ≈500-md Berea rock, crushed-rock adsorption technique) (Gale and Sandvik 1973).

The adsorption amount would vary with position in the flood because of changing compositions as the flood progressed. Nonetheless, the calculation indicates that adsorption can be quite significant.

Adsorption is important for design and always should be considered. The magnitude of expected adsorption should be estimated so that the slug size and composition can be set properly. Steps can be taken to reduce the amount of adsorption. Gale and Sandvik (1973) showed that the middle-equivalent-weight surfactant compounds act as sacrificial adsorbents. Bae and Petrick (1977) reduced adsorption on Berea rock by using preflushes of Na_2CO_3. The use of alkaline additives or alkaline preflushes to increase pH to reduce adsorption in dilute surfactant systems also has been discussed (Krumrine et al. 1982). Novosad (1982) showed that the addition of low-molecular-weight alcohols to the surfactant solution reduces adsorption. Such steps as these should be considered where necessary to reach acceptable adsorption levels.

Surfactant retention can vary between laboratory and field tests. The Loudon field tests (described in Section 7.6.2) were designed to use a mixture of sulfonates. Surfactant-retention data from a series of laboratory core tests averaged 0.55 mg/g rock for chemical slug sizes of approximately 0.4 PV. Surfactant retention in cores taken from a post-pilot evaluation well in the second reported test averaged 0.07 mg/g rock, which was in good agreement with a value of 0.08 mg/g rock determined by overall material balance (Reppert et al. 1990). This value was lower than those observed in laboratory tests with the same drive-water salinity and in the first Loudon field test (Bragg et al. 1982) with this surfactant system. This discrepancy has been attributed to the difference in redox condition between the laboratory cores and the reservoir (Wang 1993). Conventional laboratory tests are conducted under aerobic (oxidizing) conditions, whereas reservoir conditions tend to be anaerobic (reducing). The work of Wang (1993) showed that adsorption at anaerobic conditions is smaller than adsorption at aerobic conditions. Laboratory data taken before the work of Wang, such as reported in Tables 7.12 and 7.13, may indicate a larger adsorption level than will occur in most reservoirs.

7.9.3 Cation Exchange. Ion exchange is a reaction between an electrolyte in solution and an electrolyte attached to a solid surface. A general reaction involving cations may be expressed as

$$M^-A^+ + B^+ \rightleftharpoons M^-B^+ + A^+, \dots\dots\dots\dots\dots\dots\dots\dots\dots\dots\dots\dots\dots\dots (7.27)$$

where M represents a solid material, such as a clay, and A^+ and B^+ are cations. Ion exchange resembles adsorption and, for engineering purposes, often is considered to be a special case of adsorption.

Clays found in reservoir rocks have the interesting property such that, through an ion-replacement process involving Al^{+++}, the clay acquires a negative charge during diagenesis. To preserve electroneutrality, cations from the associated fluid are incorporated within the clay structure. When this clay later comes into contact with a different liquid, as occurs during a displacement process, the cations within the clay structure are available for exchange with cations in the contacting liquid (Amphlett 1964). This cation-exchange process can result in a significant change in the composition of a displacing fluid.

The capacity for cation exchange for a given rock is expressed in terms of the cation-exchange capacity, Q_m, usually given in units of milliequivalents per kilogram of rock.

$$Q_m \text{ (meq/kg rock)} = \frac{\text{gmol of ion available} \times \text{charge on ion}}{1000 \times \text{mass of rock (kg)}}$$

$$= \frac{\text{mass of ion available (g)}}{1000 \times \text{equivalent weight of ion} \times \text{mass of rock (kg)}}. \dots\dots\dots\dots\dots\dots\dots (7.28)$$

The capacity also can be expressed in terms of unit PV as

$$Q_v = Q_m \rho_s \frac{(1-\phi)}{\phi}, \dots\dots\dots\dots\dots\dots\dots\dots\dots\dots\dots\dots\dots\dots\dots\dots\dots\dots (7.29)$$

where Q_v = cation-exchange capacity expressed on a unit PV basis, meq/m³ PV; ρ_s = density of rock, kg/m³; and ϕ = porosity. **Table 7.14** (Crocker et al. 1983) gives cation-exchange capacities for several rocks, including those commonly used in petroleum laboratories. **Table 7.15** (Amphlett 1964) shows typical ranges of cation-exchange capacities for common clays.

Different ions have different affinities for clays (i.e., there is a selectivity in the reactions between the ions and clay sites). The order of affinity for several ions is

$$Li < Na^+ < K^+ < Mg^{2+} < Ca^{2+}. \dots\dots\dots\dots\dots\dots\dots\dots\dots\dots\dots\dots\dots\dots (7.30)$$

Li is held less tightly than Na, which is held less tightly than K, etc.

Relative affinities of ions for solids can be expressed in terms of equilibrium constants. Consider, for example, the reaction

$$Ca^{2+} + 2(Na^+ - Clay^-) \rightleftharpoons (Ca^{2+} - Clay^{2-}) + 2Na^+. \dots\dots\dots\dots\dots\dots\dots\dots\dots (7.31)$$

Sandstone	Porosity (fraction)	Permeability (md)	Density (g/mL)	Surface Area (m²/g)	Clay Dispersion Classification	Cation-Exchange Capacity (meq/kg)
Bandera	0.174	12	2.18	5.50	Pore lining	11.99
Berea	0.192	302	2.09	0.93	Grain cementing	5.28
Coffeyville	0.228	62	2.09	2.85	Pore bridging	23.92
Cottage Grove	0.261	284	1.93	2.30	Pore bridging	17.96
Noxie	0.270	421	1.85	1.43	Pore lining	10.01
Torpedo	0.245	94	1.98	2.97	Pore bridging	29.27

Table 7.14—Physical properties and cation-exchange capacity of selected rocks (Crocker et al. 1983).

Type	Mineral	Capacity* (meq/kg)
Kaolinite group	Kaolinite	20 to 100
Illite group	Muscovite	105
	Illite	130 to 420
Fibrous clays	Attapulgite	180 to 220
Montmorillonite group	Nontronite	570 to 640
Micaceous derivatives	Saponite	690 to 810
	Montmorillonite	800 to 1,500
	Biotite	30
	Vermiculite (pure)	1,000 to 1,500

* These values are representative of a number of samples, but wide variations are possible for a given mineral depending on its source and precise composition. In some cases where edge and corner exchange is appreciable, the capacity depends markedly on particle size (e.g., kaolinite and illite; where exchange principally involves interlayer cations, the effect of particle size is small).

Table 7.15—Cation-exchange capacities of some clay minerals (Amphlett 1964).

Assuming that the law of mass action holds and that the system displays ideal behavior at equilibrium,

$$K_{Ca^{2+}-Na^+} = \frac{\left[Ca^{2+}-Clay^{2-}\right]\left[Na^+\right]^2}{\left[Ca^{2+}\right]\left[Na^+-Clay^{2-}\right]^2}, \quad\dots\dots\dots\dots\dots\dots\dots\dots\dots\dots\dots\dots\dots\dots\dots\dots\dots \quad (7.32)$$

where the brackets denote appropriate equilibrium concentrations. Similar equations could be written for other ion pairs, such as Mg^{2+} and Na^+, for example. The magnitude of $K_{Ca^{2+}-Na^+}$ indicates the relative tendency of the two ions to react with the sites on the clay. The larger the value of $K_{Ca^{2+}-Na^+}$ in Eq. 7.32, the greater the tendency of Ca^{2+} to attach to the clay compared with the tendency of Na^+. The $K_{A^+-B^+}$ value for any ion pair (A^+, B^+) is a function of the type of ions and the nature of the solid. This constant is sometimes called the exchange constant (Tan 1982).

Eq. 7.32 holds at equilibrium. When the fluid concentration is changed for a system originally at equilibrium, there will be a transient period before the system comes to a new equilibrium. The length of this period depends on the rate of ion exchange, which in turn depends on such system parameters as ion diffusion rates in the bulk fluid and within the solid and the rate of the exchange reaction of the solid site. In petroleum processes, it is often acceptable to assume local equilibrium because rates of fluid flow are relatively small in the bulk of the reservoir system.

The calculations shown in **Figs. 7.93 and 7.94** (Pope et al. 1978) were compared with corefloods that were conducted in Berea sandstone cores in the absence of an oil phase. Fig. 7.93 shows a case in which the residual brine in the core contained 0.010 meq/mL Ca^{2+} and 0.050 meq/mL Na^+. This brine was displaced with a brine containing 0.010 meq/mL Ca^{2+} and 0.030 meq/mL Na^+; i.e., the Na^+ concentration was reduced in the injected liquid.

Cation exchange is indicated by the reduction of Ca^{2+} concentration in the effluent from 0.010 to approximately 0.005 meq/mL. The Ca^{2+} was exchanged for Na^+ on the rock. Once the exchange capacity of the rock was reached, the Ca^{2+} concentration increased to the injected value.

Fig. 7.94 shows initial and injected compositions for the second example. In this example, the injected Na^+ composition was increased. A corresponding increase in Ca^{2+} concentration in the effluent occurred as Na^+ replaced Ca^{2+} on the rock.

The solid and dashed lines in Figs. 7.93 and 7.94 are concentration profiles calculated with a model developed by Pope et al. (1978).

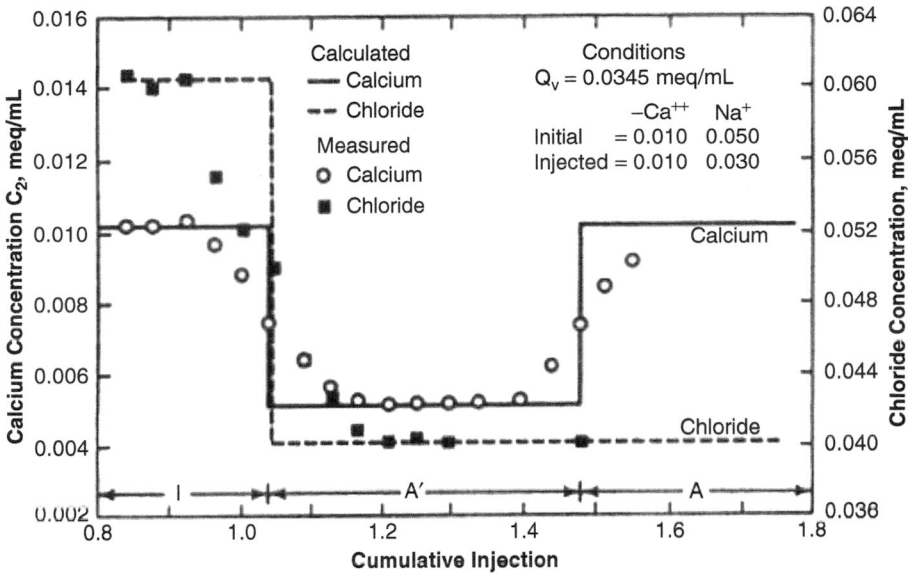

Fig. 7.93—First case, composition route (schematic) distance/time diagram and production history calculated for the two-cation case and stated conditions (Pope et al. 1978).

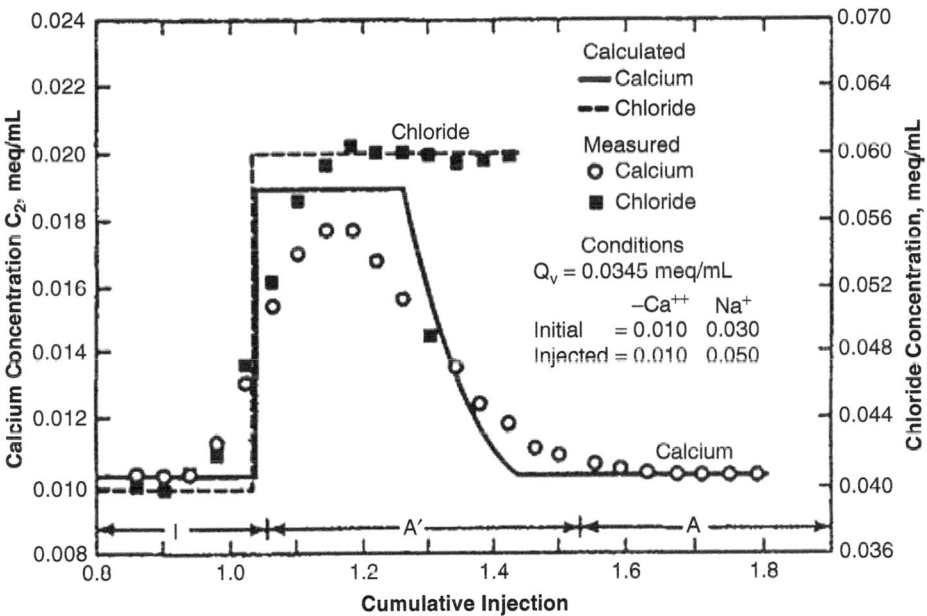

Fig. 7.94—Second case, composition route (schematic) distance/time diagram and production history calculated for the two-cation case and stated conditions (Pope et al. 1978).

Example 7.7—Cation-Exchange Capacity. Consider the coreflood shown in Fig. 7.93. The initial concentrations of Na^+ and Ca^{2+} are 0.050 and 0.010 meq/mL, respectively. Final concentrations are 0.030 and 0.010 meq/mL, respectively. Assume that the PV of the rock was 100 mL and that the amount of Ca^{2+} exchanged with Na^+ was 0.215 meq. The value of $K_{Ca^{2+}-Na^+}$ is 3.0. Calculate the amount of Ca^{2+} originally on the rock.

Solution. Let Ca_{Clay} = amount of Ca^{2+} adsorbed on the rock initially, meq, and Na_{clay} = amount of Na^+ adsorbed on the rock initially, meq. Applying Eq. 7.32 for the initial condition gives

$$3.0 = \frac{Ca_{clay} \times (0.05)^2}{(0.01) \times (Na_{clay})^2}$$

and $12.0 = \dfrac{Ca_{clay}}{\left(Na_{clay}\right)^2}.$

Next, apply Eq. 7.32 for the final equilibrium condition. The amount of Ca^{2+} and Na^+ exchanged was 0.215 meq, which corresponds to 0.00215 meq/mL PV. Therefore,

$$3.0 = \frac{\left[\left(Ca_{clay}\right)+0.00215\right](0.03)^2}{(0.01)\left[\left(Na_{clay}\right)-0.00215\right]^2}$$

and $33.33 = \dfrac{\left[\left(Ca_{clay}\right)+0.00215\right]}{\left[\left(Na_{clay}\right)-0.00215\right]^2}.$

Solving the two equations simultaneously yields Na_{clay} = 0.01360 meq/mL and Ca_{clay} = 0.00222 meq/mL. These are the concentrations of Na^+ and Ca^{2+} originally adsorbed on the rock.

Example 7.7 illustrates the manner in which the composition of salts can be changed in an injected fluid. This change can affect the performance of a chemical slug and should be considered in the design process. For example, Ca picked up by the surfactant slug can shift a lower-phase microemulsion, which has effective oil displacement properties, to an upper-phase microemulsion that can be trapped by the mobility buffer. Chan and Kremesec (1985) and Hirasaki (1982b) give additional examples of cation exchange in corefloods.

Ca associates with micelles to form a complex in which an equilibrium exists between the Ca in the clay and the Ca associated with the micelles (Hirasaki 1982b). Thus, in systems where the preflood, surfactant, mobility buffer, and drive water have the same salinity and Ca ion content, the Ca concentration of the surfactant will increase. This can shift the surfactant system from a middle- or lower-phase environment to an upper-phase environment, where surfactant loss can be large if the IFT between the mobility buffer and the surfactant is not ultralow. Use of low-salinity polymer solution and drive water can offset these effects if a lower-phase environment occurs.

An example of Ca association with micellar fluid is shown in **Fig. 7.95** (Gupta 1984), where Ca concentration in the effluent is plotted vs. PVs injected. In this experiment, 2.0 PV of micellar solution were displaced by 2.0 PV of low-salinity polymer water in oil-free Berea core. Shown in Fig. 7.95 is a dashed curve representing the Ca concentration in the effluent based on ion-exchange theory. The Ca content of the micellar slug increased from 50 to more than 300 ppm in the surfactant bank, indicating uptake of Ca by the micelles. The same effect was observed when a 0.1-PV micellar slug was displaced through oil-free Berea core by a low-salinity polymer water, as shown in **Fig. 7.96**. Hirasaki (1982b) showed that the observed Ca concentrations in Figs. 7.95 and 7.96 are consistent with predictions based on a theory of Ca/micellar solution association.

7.9.4 Phase Partitioning/Trapping. Surfactant may partition into all liquid phases present and contiguous in a system. The amount of partitioning must be determined experimentally from analyses of equilibrium phases. The importance of this was shown in Section 7.8.3, which described the effect of a salinity gradient. When the mobility buffer salinity is such that the system is in the Type II(+) phase environment, surfactant partitions to the oleic phase. This phase is relatively slow-moving in

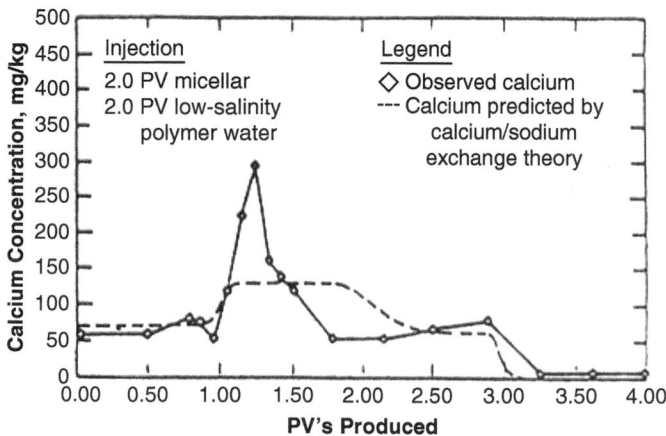

Fig. 7.95—Produced Ca behavior for a large-slug ion-exchange test (2.0-PV slug, 2.0-PV polymer water, oil-free Berea core) (Gupta 1984).

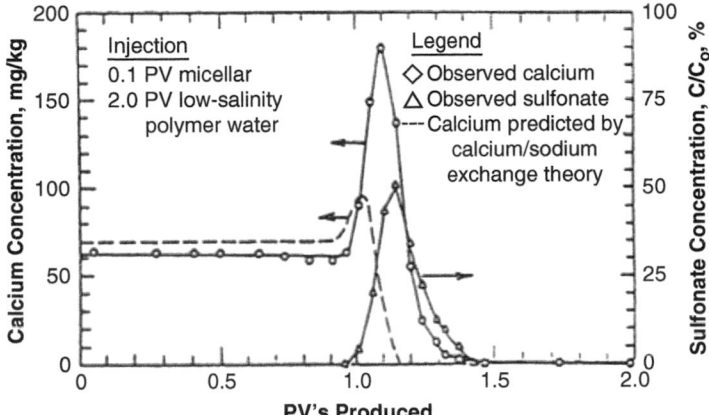

Fig. 7.96—Produced Ca and sulfonate behavior for a small-slug ion-exchange test (0.10-PV slug, 2.0-PV polymer water, oil-free Berea core) (Gupta 1984).

a displacement, and thus, when surfactant is transferred to this phase, it is removed from the displacement front because this phase can be trapped as an ROS by the mobility buffer. The result can be a very poor efficiency. Phase trapping also can occur as a result of shifts in phase boundaries caused by ion exchange when Ca is added to the slug.

Surfactant retention by phase partitioning often is of the same order of magnitude as adsorption, according to displacement experiments by several investigators (Gale and Sandvik 1973; Nelson 1982; Novosad 1982; Healy et al. 1975). As is the case for other mechanisms of surfactant retention, loss by phase partitioning is complex and difficult to calculate in the absence of experimental data on the particular system of interest.

7.10 Modeling Chemical Flood Displacement

In this section, oil recovery by a chemical flood is first estimated with a simple material-balance calculation. The purpose of the calculation is to provide an order-of-magnitude approximation of recovery that might be expected from a flood.

Beyond a material balance, there are basically two approaches to modeling micellar/polymer floods. The first approach presented in this section uses the frontal-advance theory developed in Chapter 3 to predict displacement performance. The second approach involves solution of a system of equations describing the transport of each chemical species through the porous rock. This system of equations is solved with finite-difference techniques. Such a program is called a chemical flood simulator (Hirasaki et al. 2006). Use of chemical flood simulators is beyond the scope of this text.

7.10.1 Estimating Oil Recovery by Material Balance. Oil recovery by the surfactant/polymer process can be approximated by application of a simple material balance. Because a favorable mobility ratio is maintained in the process, volumetric sweep efficiency is assumed to be the same as for a waterflood preceding the surfactant process. Recovery is then given by

$$N_p = \frac{Ah\phi\left(S_{orw} - S_{orc}\right)}{B_o 5.615} E_{vw}, \dots\dots\dots\dots\dots\dots\dots\dots\dots\dots\dots\dots\dots\dots\dots\dots (7.33)$$

where N_p = oil recovered in process, STB; A = pattern area, ft^2; h = reservoir thickness, ft; ϕ = porosity; S_{orw} = ROS at termination of waterflood (corresponds to one endpoint saturation of the waterflood relative permeability curve); S_{orc} = ROS left by surfactant slug; B_o = oil FVF, RB/STB; and E_{vw} = volumetric sweep efficiency of waterflood preceding chemical flood.

ROSs left by chemical floods, S_{orc}, typically range from 0.05 to 0.15 PV in laboratory corefloods. These residuals may be considerably higher (0.15 to 0.25 PV) when a chemical system is optimized for economics of field-scale operation. The material balance is applied in Example 7.8.

Example 7.8—Oil Recovery by Material Balance, Surfactant/Polymer Process. Consider application of a surfactant/polymer process to a shallow sandstone reservoir having the following properties.

A = 20-acre spacing, five-spot pattern
h = 20 ft
ϕ = 0.18
S_{orw} = 0.30
S_{orc} = 0.08
B_o = 1.05 RB/STB
E_{vw} = 0.70

Calculate ultimate oil recovery as a fraction of original oil in place (OOIP) assuming an initial oil saturation, S_{oi}, of 0.75.

$$N = Ah\phi S_{oi} / B_o$$

$$= \frac{20 \text{ acres} \times 43{,}560 \text{ ft}^2 / \text{acre} \times 20 \text{ ft} \times 0.18 \times 0.75}{1.05 \text{ bbl/STB} \times 5.615 \text{ ft}^3 / \text{bbl}}$$

$$= 399{,}000 \text{ STB}.$$

Solution. Apply Eq. 7.33:

$$N_p = \frac{20 \text{ acres} \times 43{,}560 \text{ ft}^2 / \text{acre} \times 20 \text{ ft} \times 0.18(0.30 - 0.08)0.70}{1.05 \text{ bbl/STB} \times 5.615 \text{ ft}^3 / \text{bbl}}$$

$$= 81{,}900 \text{ STB},$$

and $N_p / N = 0.205$,

where N_p = oil displaced by the chemical flood, STB, and N = OOIP, STB.

The combined oil recovery by waterflooding and the surfactant/polymer process is (assuming resaturation of oil in the unswept region)

$$N_p = \frac{20 \text{ acres} \times 43{,}560 \text{ ft}^2 / \text{acre} \times 20 \text{ ft} \times 0.18(0.75 - 0.08)0.70}{1.05 \text{ bbl/STB} \times 5.615 \text{ ft}^3 / \text{bbl}}$$

$$= 249{,}491 \text{ STB}.$$

$N_p/N = 0.625$ (total recovery after waterflooding and micellar/ polymer flooding).

7.10.2 Estimating Oil Recovery by Frontal-Advance Theory: Low-Tension Chemical Flood at Interstitial (Immobile) Water Saturation, Linear System. Chapter 3 presented the development of a model for a chemical flood based on a single chemical species. To use this model to predict displacement performance for a micellar/polymer flood, the surfactant/cosurfactant is assumed to be represented by a single species. Only two-phase flow can be approximated by the frontal-advance model developed in Chapter 3. The surfactant remains in the aqueous phase (i.e., a lower-phase system) so that the surfactant/ oil phase behavior is approximated by a set of relative permeability curves that reflect a reduction in both IFT and ROS. Displacement of the chemical species is pistonlike, and thus dispersion and viscous fingering are neglected. Mass transfer between the aqueous and oil phases is neglected. The chemical species is adsorbed according to a Langmuir isotherm, and adsorption is irreversible. The exchange of divalent ions between the rock and the surfactant solution, as well as the change of phase behavior resulting from interaction of divalent cations with the surfactant, cannot be accounted for in this model.

When a chemical flood is initiated in a linear system that is at interstitial water saturation, two flood fronts form, as depicted by the saturation profile in **Fig. 7.97.** The first flood front represents the displacement of oil by the chemical slug. The velocity of this flood front is given by Eq. 3.94, which is rewritten here as

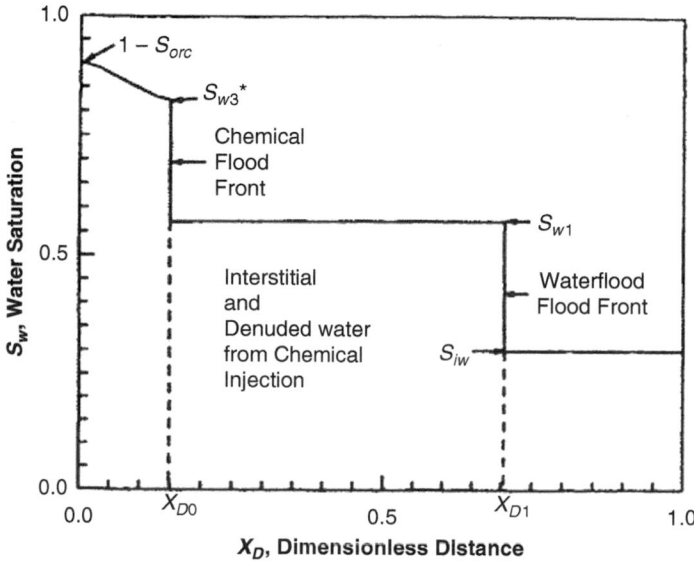

Fig. 7.97—Saturation profile during a chemical flood in a reservoir at interstitial (immobile) water saturation, S_{iw}.

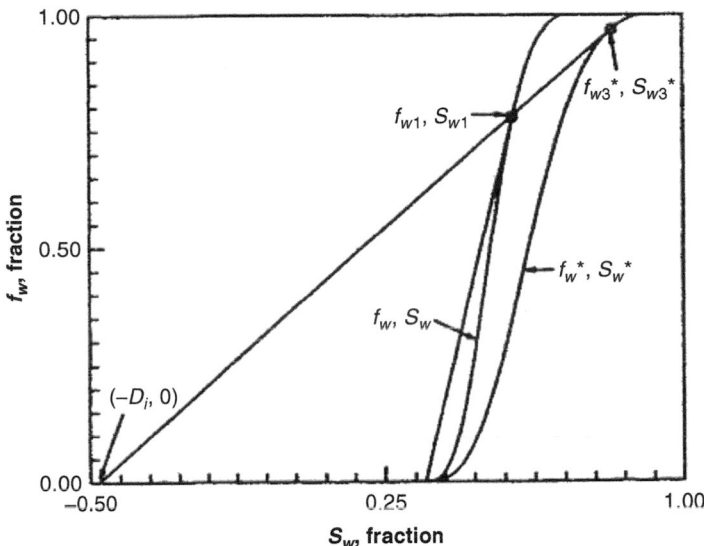

Fig. 7.98—Fractional-flow curves for a chemical flood in a reservoir at interstitial (immobile) water saturation, Example 7.9.

$$\left(\frac{dx_D}{dt_D}\right)_{S_{w3}^*} = \frac{f_{w1}}{S_{w1}+D_i} = \frac{f_{w3}^*-f_{w1}}{S_{w3}^*-S_{w1}} = \frac{f_{w3}^*}{S_{w3}^*+D_i}. \qquad (7.34)$$

D_i is defined by Eq. 3.88 and is determined from the adsorption isotherm for the particular surfactant/rock system. Because displacement of the chemical slug is piston-like, the displacement process is represented by fractional-flow curves, f_w^*/S_w^* for the chemical flood with endpoint saturations at S_{iw} and $1-S_{orc}$ and f_w/S_w for a normal waterflood. **Fig. 7.98** presents these fractional-flow curves. The saturation at the chemical flood front, S_{w3}^*, is found by constructing a tangent to the f_w^*/S_w^* fractional-flow curve for the chemical flood from the point $S_{w3}^* = -D_i$, $f_w^* = 0$. The tangent to the f_w^* curve intersects at S_{w3}^*. The injection of the chemical caused an oil bank to form with a constant saturation S_{w1}. The water saturation, S_{w1}, is found from the intersection of the tangent to the f_w^*/S_w^* fractional-flow curve with the f_w/S_w fractional-flow curve. The velocity of this flood front must also satisfy Eq. 7.34.

The second flood front represents the displacement of oil by interstitial water as in a normal waterflood. The waterflood front saturation normally is found by constructing a tangent to the fractional-flow curve from $S_w = S_{iw}$, $f_w = 0$ to the f_w/S_w fractional-flow curve, as shown in **Fig. 7.98**. The point of tangency is f_{wf}, S_{wf}. If $S_{wf} < S_{w1}$, the flood-front saturation will be S_{wf} and there will be a saturation gradient from S_{wf} to S_{w1}, as shown in Fig. 3.20. If $S_{wf} > S_{w1}$, the flood-front saturation will be S_{w1}.

Displacement performance is estimated by advancing the flood-front saturations through the linear system. The waterflood front will advance at a specific velocity given by Eqs. 3.92 or 3.93, depending on S_{w1} and S_{wf} while the chemical flood front will advance at a specific velocity given by Eq. 7.34. The arrival time for these saturations at the end of the system is determined directly from the frontal-advance solution. Recall that

$$x_D = t_D f_w'. \qquad (7.35)$$

Thus, from Eqs. 3.84 through 3.86,

$$t_{Df} = x_{Df}/f_{wf}', \qquad (7.36)$$

$$t_{D1} = x_{D1}/f_{w1}', \qquad (7.37)$$

and $t_{D3} = x_{D3}/f_{w3}'^*. \qquad (7.38)$

Displacement performance is computed by determining the average water saturation during the displacement process with Eqs .3.87 and 3.93.

$$N_p = \frac{(\bar{S}_w - S_{iw})V_p}{B_o} \qquad (7.39)$$

if $S_{wf} > S_{wl}$.
   For $t_D < t_{D1}$,

$$\bar{S}_w - S_{iw} = t_D; \dots\dots\dots\dots\dots\dots\dots\dots\dots\dots\dots\dots\dots\dots\dots\dots\dots\dots\dots(7.40)$$

for $t_{D1} < t_D < t_{D3}$,

$$\bar{S}_w = S_{w1} + t_D\left(1 - f_{w1}\right); \dots\dots\dots\dots\dots\dots\dots\dots\dots\dots\dots\dots\dots\dots\dots\dots(7.41)$$

and for $t_D > t_m$,

$$\bar{S}_w = S_{w2}^* + t_D\left(1 - f_{w2}^*\right). \dots\dots\dots\dots\dots\dots\dots\dots\dots\dots\dots\dots\dots\dots\dots(7.42)$$

Example 7.9 illustrates the calculation of displacement performance of a surfactant flood at interstitial water saturation.

Example 7.9—Linear Surfactant Flood in a Reservoir With Initial Immobile Water Saturation. A reservoir is 1,320 ft long, 660 ft wide, and 20 ft thick. The initial oil saturation is 65%. A low-tension flood is to be considered as a secondary recovery process. Using frontal-advance theory, estimate the oil production as a function of PVs injected and time. The injection rate is constant at 2,000 B/D. The parameters of the flood are $\phi = 0.20$, $k_{ro} = \alpha_1(1 - S_{wD})^m$, and $k_{rw} = \alpha_2(S_{wD})^n$, with $\alpha_1 = 0.8$, $\alpha_2 = 0.2$, $m = 1.5$, $n = 2.5$, and

$$S_{wD} = \frac{S_w - S_{iw}}{1 - S_{or} - S_{iw}},$$

where $S_{iw} = 0.35$. $S_{orw} = 0.30$ (waterflood ROS), $S_{orc} = 0.10$ (low-tension flood ROS), $\mu_o = 10$ cp, $\mu_w = 1.0$ cp, $\mu_c = 1.0$ cp (surfactant solution), $B_o = 1.0$ bbl/STB, and $C_{io} = 1.5$ wt% (injected surfactant concentration).

Solution. Construct fractional-flow curves with Eq. 3.2:

$$f_w = \frac{k_{rw} / \mu_w}{k_{ro} / \mu_o + k_{rw} / \mu_w}.$$

There are two fractional-flow curves required, one with $S_{orw} = 0.30$ and another with $S_{orc} = 0.10$ (Fig. 7.98). The f_w/S_w curve is for a waterflood and the f_w^* / S_w^* curve is for a low-tension flood. The surfactant fractional-flow curve was constructed with the same values of m and n. Relative permeabilities are affected by IFT, and values of m and n for the surfactant/oil system should be obtained experimentally.

Next, consider adsorption of surfactant on the rock. Assume that the low-tension flood has the adsorption isotherm shown in Fig. 3.18. At $C_{io} = 1.5$ wt%, $A_{io} = 0.68$ mg adsorbed/g rock. From Eq. 3.73,

$$\hat{C}_{io} = \frac{A_{io}\rho_r\left(1 - \phi\right)}{\phi}$$

$$= 0.68 \times 10^{-3} \text{ g/g rock} \times 2.65 \text{ g rock/cm}^3 \text{ rock}$$

$$\times \left[\left(1 - 0.2\right) / 0.2\right]$$

$$= 7.21 \times 10^{-3} \text{ g adsorbed/cm}^3 \text{ rock.}$$

Also, $C_{io} = 1.5$ wt%=15,000 ppm

$$\approx 15,000 \times 10^{-6} \text{ g/cm}^3.$$

The density of the chemical solution was assumed to be 1.0 g/cm³ for this example. The solution density should be determined experimentally. From Eq. 3.122,

$$D_i = \hat{C}_{io} / C_{io}$$

$$= \left(7.21 \times 10^{-3}\right) / \left(15 \times 10^{-3}\right) = 0.48.$$

The problem is now solved with the fractional-flow curves. Construct a line from $(-D_i, 0)$ tangent to the low-tension flood fractional-flow curve f_w^* / S_w^* (Fig. 7.98). The parameters f_{w3}^* and S_{w3}^* are obtained from the intersection of this tangent with the f_w^* / S_w^* fractional-flow curve. The slope of the tangent is $f_{w3}^{\prime\prime*}$. Parameters f_{w1} and S_{w1} are obtained from the intersection of the tangent line and the waterflood fractional-flow curve. The value of f_{w1}' is the slope of the f_w/S_w fractional-flow curve at S_{w1}.

$$
\begin{aligned}
f_{w3}^* &= 0.972 \\
S_{w3}^* &= 0.825 \\
f_{w3}^{\prime*} &= 0.745 \\
f_{w1} &= 0.783 \\
S_{w1} &= 0.572 \\
f_{w1}' &= 4.23
\end{aligned}
$$

The waterflood front is obtained by drawing a line from $(S_{iw}, 0)$ tangent to the f_w/S_w curve.

$$
\begin{aligned}
f_{wf} &= 0.80 \\
S_{wf} &= 0.575 \\
f_{wf}' &= 3.53
\end{aligned}
$$

The relative positions of the fronts are obtained by using the dimensionless velocity relationships. From Eq. 7.35,

$$ x_D = f_w' t_D $$

and $v_D = dx_D / dt_D = f_w'$.

From the fractional-flow parameters, $v_{Df} = 3.53$, $v_{DI} = 4.23$, and $v_{D3}^* = 0.745$. In all floods, interstitial water is displaced at S_{wf} with a velocity of v_{Df}. However, because $v_{DI} > v_{Df}$, the front at S_{wf} will be overtaken quickly by the front at S_{w1}. The original waterflood front is therefore ignored. Thus, in this example, two fronts will form. The first is a jump from S_{iw} to S_{w1}. The second is from S_{w1} to S_{w3}^*.

The saturation profile at a dimensionless time of $t_D = 0.2$ is shown in **Fig. 7.99** for this example. Breakthrough occurs when $x_D = 1.0$ in Eq. 7.35.

$$ t_{D3} = \frac{1}{0.745} = 1.342 \text{ PV injected.} $$

$$ t_{D1} = \frac{1}{4.23} = 0.236 \text{ PV injected.} $$

From $t_D = 0$ to $t_D = 0.236$, only oil is produced at a rate equal to the injection rate. From $t_D = 0.236$ to $t_D = 1.342$, oil is produced at a cut based on the low-tension flood fractional-flow curve. Cumulative oil recovery is based on average water saturation from Eq. 7.41,

$$ \overline{S}_w = S_{w1} + t_D \left(1 - f_{w1}\right). $$

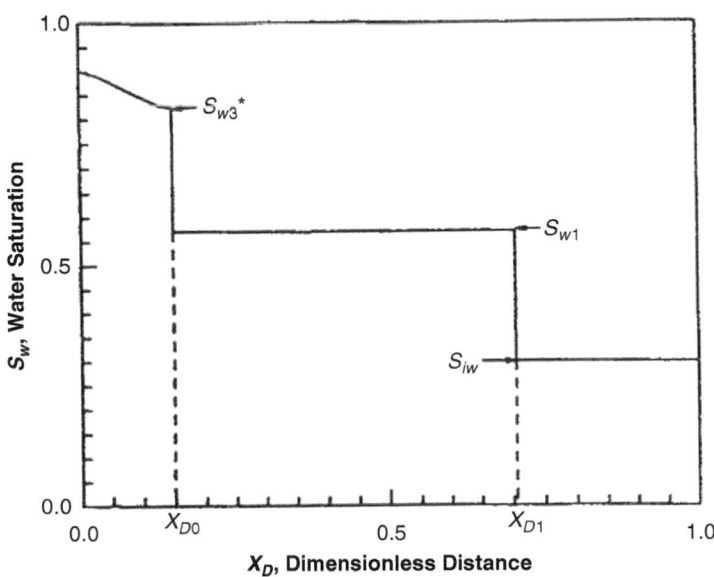

Fig. 7.99—Saturation profile for Example 7.9 at $t_D = 0.2$.

$\bar{S}_w$	f_w	f_w'	t_D (PV)	t (D)	N_p^* (bbl)	Mobile Oil Recovered, $S_{orc} = 0.1$ (%)
0.35	0	–	0	0	0	–
0.572	0.783	4.23	0.236**	73.2	146,470	43
0.825	0.972	0.745	1.342†	416.4	318,120	93
0.85	0.986	0.440	2.273	705	330,061	97
0.879	0.996	0.183	5.474	1,700	340,436	99.7

*N_p is a linear function of t_D before breakthrough.
**Oil bank (S_{w_1}) arrives at end of system.
†Chemical flood front arrives at end of system.

Table 7.16—Calculated displacement performance, Example 7.9.

For $t_D > 1.342$,

$$\bar{S}_w^* = S_{w2}^* + t_D\left(1 - f_{w2}^*\right).$$

Oil produced is

$$N_p = \left(\bar{S}_w^* - S_{iw}\right)\left(A\phi L / 5.615\right).$$

Table 7.16 shows cumulative oil produced. The fraction of mobile oil recovered was calculated as the ratio of oil recovered to original mobile oil in place, where

$$N_m = \left(S_{oi} - S_{orc}\right)\left(A\phi L / 5.615\right)$$
$$= \frac{(0.65 - 0.10)660 \text{ ft} \times 20 \text{ ft} \times 1,320 \text{ ft} \times 0.20}{5.615 \text{ ft}^3 / \text{bbl}}.$$

Fig. 7.100 plots cumulative oil produced vs. real time. Real time is determined from dimensionless time by recalling that

$$t_D = qt / A\phi L.$$

7.10.3 Mobility Control in a Chemical Flood, Linear System. Effective displacement of oil in a chemical flood requires a favorable mobility ratio between the chemical slug and the oil bank displaced by the slug. When the mobility ratio is unfavorable, the chemical slug may finger into the oil bank, where dispersion and mixing could render it ineffective.

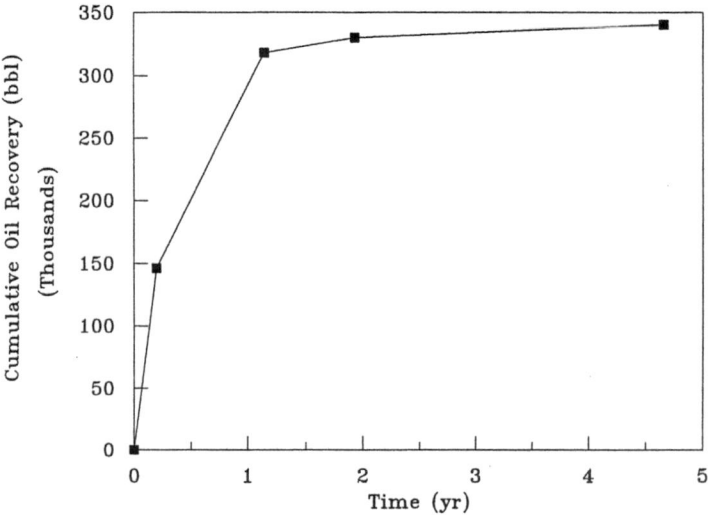

Fig. 7.100—Cumulative oil production vs. time for Example 7.9.

Section 5.7 introduced the concept of mobility control, including a method of estimating the mobility of a chemical solution displacing an oil bank. The design mobility for the chemical slug was found as the minimum of the total relative mobility for the oil/water system from Eq. 5.154, rewritten here as

$$\lambda_{rd} = \min\left(\lambda_{rw} + \lambda_{ro}\right), \dots\dots\dots\dots\dots\dots\dots\dots\dots\dots\dots\dots\dots\dots\dots\dots\dots\dots(7.43)$$

where λ_{rd} = design mobility, cp^{-1}; $\lambda_{rw} = k_{rw}/\mu_w$; and $\lambda_{ro} = k_{ro}/\mu_o$ for $S_{iw} \leq S_w \leq 1 - S_{or}$.

In most cases, it is necessary to adjust the viscosity of the chemical slug so that its effective mobility is equal to or less than the design mobility. If the average water saturation in the chemical slug is close to $1 - S_{orc}$, then the mobility of the chemical slug is approximated by

$$\lambda_c = \frac{k_{rw} \text{ at } 1 - S_{orc}}{\mu_c}, \dots\dots\dots\dots\dots\dots\dots\dots\dots\dots\dots\dots\dots\dots\dots\dots\dots\dots\dots\dots(7.44)$$

where μ_c = apparent viscosity of the chemical slug, cp.

The viscosity of the chemical slug is obtained from Eq. 7.44 when the design mobility is known. If the average water saturation in the chemical slug is much lower than $1 - S_{orc}$, the average mobility of the region behind the chemical flood front must be computed to estimate the required viscosity of the chemical slug. The selection of the viscosity of a chemical slug is illustrated in Example 7.10 for a chemical flood in a linear reservoir at ROS following a waterflood.

Example 7.10—Mobility-Control Design for the Chemical Slug in a Reservoir at ROS. Consider the reservoir in Example 7.9 with the oil saturation at a waterflood ROS of 0.30. A chemical flood is to be designed that will reduce the oil saturation to 0.10 in core tests. For this example, it is assumed that the viscosity of the injected chemical can be altered by slight changes in composition or by addition of a nonadsorbing viscosifier. Determine the design viscosity of the chemical solution and the displacement performance as a function of time. Reservoir and fluid properties used in Example 7.9 will be used, with the exception that the relative permeability curves for the chemical flood change in response to changing oil saturation and low IFT. **Table 7.17** gives the values used in the expression for relative permeabilities. Values of m and n decrease for the chemical flood because of the ultralow IFT values between the oil and chemical solution. The value of α_2 increases because the reduction in oil saturation increases k_{rw} at S_{orc}. The injection rate is constant at 2,000 B/D.

Solution. The first step is to determine the design mobility to maintain mobility control between the chemical slug and the oil/water bank ahead of the slug. Total relative mobilities were computed for a range of saturations ($S_{iw} < S_w < 1 - S_{orc}$) with the waterflood properties and selected values presented in **Table 7.18** and plotted in **Fig. 7.101.** Inspection of Table 7.18 and Fig. 7.101 shows that the minimum mobility of the oil/water bank is approximately 0.0564 cp^{-1}.

To find the flood-front saturation of the chemical shock, an average water saturation must be assumed behind the chemical flood front to compute the mobility of the chemical slug. An initial estimate of the mobility of the chemical flood obtained from Eq. 7.43 where k_{rw} at $1 - S_{orc} = \alpha_2 = 0.4$ is $0.0564 = 0.4/\mu_c$. Thus, $\mu_c = 7.09$ cp. The average apparent viscosity of the chemical slug is 7.09 cp but will be increased to 8.0 cp to cover uncertainties in the calculations.

Fig. 7.102 shows the fractional-flow curves for the chemical flood ($\mu_c = 8$ cp) and the waterflood ($\mu_w = 1$ cp). The retention of chemical in this system is the same as in Example 7.9, so $D_i = 0.48$. A tangent from $(-0.48,0)$ intersects the f_w^* / S_w^* fractional-flow curve at $f_w^* = 1.0$ and $S_w^* = 0.9$ (after rounding). Thus, there will be no oil production after breakthrough of the chemical slug and the average mobility of the chemical slug for this example was estimated correctly. The tangent to f_w^* / S_w^* intersects the f_w/S_w curve at $f_{w1} = 0.758$ and $S_{w1} = 0.566$. Recall that S_{w1} is the water saturation in the oil bank ahead of the chemical shock. Fig. 7.102 also shows the waterflood front saturation S_{wf}. In this case, $S_{wf} > S_{w1}$ and a single oil bank forms with water saturation of S_{w1}, as depicted in Fig. 7.97. Values of saturations and fractional flows obtained from Fig. 7.102 are $S_{w3}^* = 0.90$, $f_{w3}^* = 1.00$, $f_w^* = 0.725$, $S_{w1} = 0.566$, $f_{w1} = 0.758$, $f_{wf}' = 3.537$, $S_{wf} = 0.580$, $f_{wf} = 0.814$.

Because the initial oil saturation is S_{orw}, an oil bank is formed that has a water saturation S_{w1}, as noted above. The front of the oil bank travels at a velocity given by Eq. 3.17. No oil is produced until the oil bank reaches the end of the linear system.

	Waterflood	Chemical Flood
α_1	0.8	0.8
α_2	0.2	0.4
m	1.5	1.2
n	2.5	1.1
S_{iw}	0.35	0.35
S_{orw} or S_{orc}	0.30	0.10

Table 7.17—Values used in expression for relative permeabilities, Example 7.10.

s_w	k_{ro}	λ_{ro} (cp⁻¹)	k_{rw}	λ_{rw} (cp⁻¹)	λ_t (cp⁻¹)
0.3500	0.8000	0.0800	0.0000	0.0000	0.0800
0.3640	0.7525	0.0752	0.0001	0.0001	0.0753
0.3780	0.7059	0.0706	0.0004	0.0004	0.0710
0.3920	0.6604	0.0660	0.0010	0.0010	0.0670
0.4060	0.6159	0.0616	0.0020	0.0020	0.0636
0.4200	0.5724	0.0572	0.0036	0.0036	0.0608
0.4340	0.5300	0.0530	0.0056	0.0056	0.0586
0.4480	0.4888	0.0489	0.0083	0.0083	0.0572
0.4620	0.4486	0.0449	0.0116	0.0116	0.0564
0.4760	0.4096	0.0410	0.0156	0.0156	0.0565
0.4900	0.3718	0.0372	0.0202	0.0202	0.0574
0.5040	0.3353	0.0335	0.0257	0.0257	0.0592
0.5180	0.3000	0.0300	0.0319	0.0319	0.0619
0.5320	0.2660	0.0266	0.0390	0.0390	0.0656
0.5460	0.2335	0.0233	0.0469	0.0469	0.0703
0.5600	0.2024	0.0202	0.0558	0.0558	0.0760
0.5740	0.1728	0.0173	0.0655	0.0655	0.0828
0.5880	0.1448	0.0145	0.0763	0.0763	0.0907
0.6020	0.1185	0.0119	0.0880	0.0880	0.0998
0.6160	0.0941	0.0094	0.1007	0.1007	0.1101
0.6300	0.0716	0.0072	0.1145	0.1145	0.1216
0.6440	0.0512	0.0051	0.1293	0.1293	0.1345
0.6580	0.0333	0.0033	0.1453	0.1453	0.1486
0.6720	0.0181	0.0018	0.1624	0.1624	0.1642
0.6860	0.0064	0.0006	0.1806	0.1806	0.1812
0.7000	0.0000	0.0000	0.2000	0.2000	0.2000

Table 7.18—Summary of computations to determine mobility of chemical slug for mobility-control design, Example 7.10.

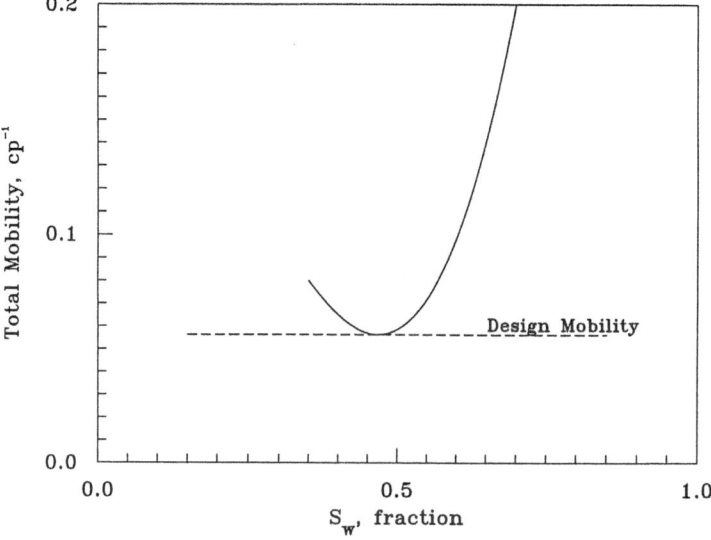

Fig. 7.101—Variation of total relative mobility with saturation for the oil/water bank in Example 7.10.

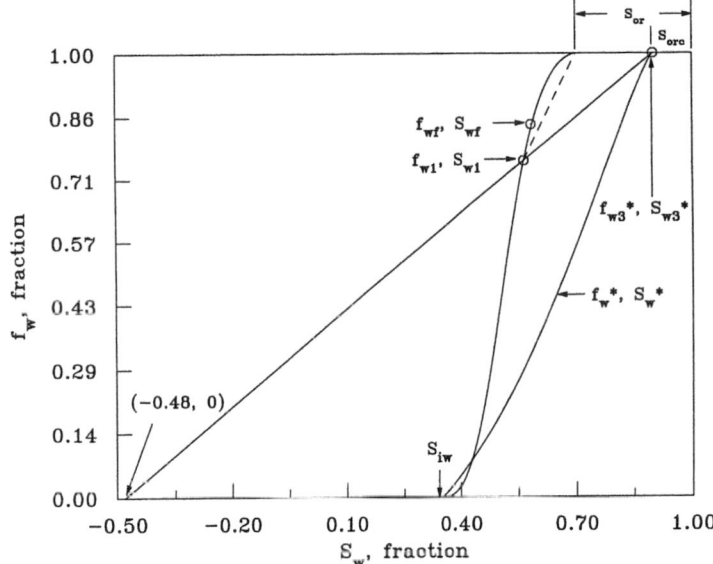

Fig. 7.102—Fractional-flow curves for a chemical flood in a reservoir at waterflood ROS, Example 7.10.

$$v_{Do} = \frac{1 - f_{w1}}{1 - S_{or} - S_{w1}}$$

$$= \frac{1 - 0.758}{1 - 0.3 - 0.566}$$

$$- 1.806.$$

From Eq. 3.152, the oil bank arrives at the end of the linear system when

$$t_{D1} = 1 / v_{Do}$$

$$= 0.554.$$

Water is produced for $0 < t_D < 0.554$. The oil bank is produced at a constant oil cut. The WOR is

$$F_{wo} = f_{w1} / (1 - f_{w1})$$

$$= 0.758 / (1 - 0.758)$$

$$= 3.13.$$

The chemical slug travels at a specific velocity equal to $f_{w3}'^*$. Thus, the chemical slug reaches the end of the system when

$$t_{D3} = 1 / f_{w3}'^*$$

$$= 1 / 0.725$$

$$= 1.379.$$

Fig. 7.103 shows a saturation profile at $t_D = 0.4$. Because $S_{w3}^* = 1 - S_{orc}$ in this example, there is no oil production after arrival of the chemical slug. Cumulative oil recovery at this point is

$$N_p = \frac{(S_{or} - S_{orc}) V_p}{B_o}.$$

From Example 7.9, $B_o = 1.0$,

$$V_p = \frac{(660 \text{ ft})(20 \text{ ft})(1,320 \text{ ft})(0.2)}{5.615 \text{ ft}^3 / \text{bbl}}$$

$$= 620,623 \text{ bbl,}$$

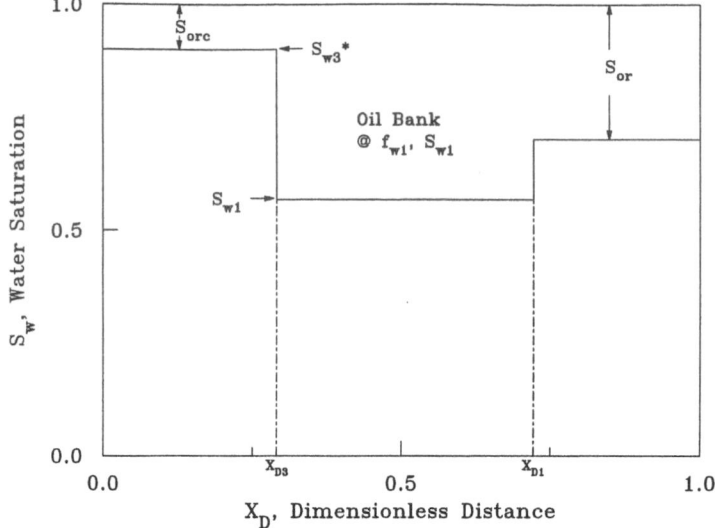

Fig. 7.103—Saturation profile for Example 7.10 at $t_D = 0.4$.

and $N_p = (0.3 - 0.1)(620,623)$
  $= 124,125$ bbl.

In this example, the cumulative oil production is a linear function of t_D for $t_D > t_{D1}$. Thus, for $t_{D1} < t_D < t_{D3}$,

$$N_p = \frac{(S_{or} - S_{orc})V_p}{B_o}\left(\frac{t_D - t_{D1}}{t_{D3} - t_{D1}}\right).$$

Table 7.19 summarizes the production performance for this chemical flood.

7.11 Advances in Surfactant Technology

7.11.1 Development of Low-Cost Effective Surfactant Systems. Loudon surfactants were expensive to produce. When interest in chemical flooding resumed in the early 2000s, other routes were investigated to develop low-cost effective surfactant systems. These efforts, pioneered by Hirasaki et al. (2006), began with the development of alcohol propoxy sulfates with the general formula of (branched alcohol)-(propoxyl groups) n-sulfate. Several alcohol propoxy sulfates were developed by propoxylating ether alcohols and sulfating the product. Two of these surfactants were $C_{16-17}(PO)_7 SO_4^-$ (Petrostep S1, manufactured by Stepan) and $C_{13}(PO)_{13} SO_4^-$ (Petrostep S13D, also manufactured by Stepan). Adding the PO group to the surfactant

t_D	t (days)	N_p (bbl)	q_o (B/D)	q_w (B/D)	F_{wo}
0 to 0.554−	0 to 1,719	0	0	200	∞
0.554+	1,719	0	48.4	156.6	3.13
0.6	1,862	6,921	48.4	156.6	3.13
0.7	2,172	21,966	48.4	156.6	3.13
0.8	2,482	37,011	48.4	156.6	3.13
0.9	2,793	52,057	48.4	156.6	3.13
1.0	3,103	67,103	48.4	156.6	3.13
1.1	3,413	82,148	48.4	156.6	3.13
1.2	3,724	97,194	48.4	156.6	3.13
1.3	4,034	112,239	48.4	156.6	3.13
1.379	4,279.2	124,125	0	200	∞

Table 7.19—Displacement performance, chemical flood in a reservoir at waterflood residual saturation, Example 7.10.

increased the oil solubilization and divalent-ion tolerance (Levitt et al. 2006; Liu et al. 2008). It was necessary to use a cosurfactant such as an internal olefin sulfonate (IOS) C_{15-18} (Petrostep S2) to obtain the desired optimal salinity and solubility in brine at optimum salinity. The branching of the internal olefin was thought to reduce the tendency for the formation of viscous phases such as liquid crystals and similar structures, reducing the amount of alcohol required as a solvent (Levitt et al. 2006; Liu et al. 2008). However, in most formulations, a cosolvent such as isobutyl alcohol (IBA) or diethylene glycol butyl ether (DGBE) was used to control viscous phases. A water-soluble polymer is part of the formulation to provide mobility control (Levitt and Pope 2008).

Many formulations use alkali (sodium carbonate or sodium hydroxide) to reduce adsorption of anionic surfactants on sandstones and carbonates (Hirasaki and Zhang 2004; Zhang et al. 2006; Hirasaki et al. 2008). Some crude oils contain acids that react with the alkali to create soaps. Alkali constituents also permit some flexibility to adjust the electrolyte composition of the surfactant slug so that the aqueous-phase stability is greater than or equal to the optimal salinity.

It is well-known based on earlier studies (Nelson 1982; Pope et al. 1979) that effective displacement requires use of a salinity gradient to ensure that the displacement process remains in a Type III environment as long as possible, followed by a Type I phase environment where displacement of the mobilized oil by the polymer bank is effective. Consequently, surfactant formulations are usually designed so that the salinity of the polymer drive is substantially less than that of the surfactant slug. For maximum robustness of the process design, the initial salinity should be higher than optimum salinity (Type II). Various uncertain reservoir conditions have less impact on the process with a salinity gradient going from Type II to Type III to Type I.

There are many examples of surfactant systems developed with this approach to design. **Fig. 7.104** (Flaaten 2007, Fig 4.151) shows solubility data as a function of salinity for a surfactant system that contained 0.625 wt% C_{16-17}(7PO)SO_4^-, 0.375 wt% IOS_{15-18}, and 0.25 wt% SBA. The surfactant slug was clear at 52°C at a salinity of 6 wt% NaCl after 2,000 ppm Flopaam 3330 S polymer was added to provide mobility control and did not contain alkali.

Optimal salinity was 4.9 wt% NaCl, and the solubilization parameter was 16. IFT based on the Huh relationship (Huh 1979) was estimated to be 1.17×10^{-3} dynes/cm at optimal salinity. The IFTs are usually not measured because they can be reliably calculated from the Huh equation. Fig. 7.103 shows the Type I, II, and III regions identified from the phase-behavior data.

Fig. 7.105 (Flaaten 2007, Fig. 4.152) shows the results of a linear core test in which a Berea core at ROS was flooded with a surfactant slug containing polymer for a mobility control flood (Flaaten 2007). The core was saturated with synthetic brine for the waterflood containing 156,525 mg/L TDS with 8667 mg/L Ca^{++} + Mg^{++}. The waterflood ROS was 0.294.

The flood consisted of injecting 0.3 PV of surfactant slug at a salinity of 4.9 wt% NaCl, followed by 2.3 PV of polymer solution. The salinity of the polymer slug was 3 wt%. Injection of the surfactant solution mobilized the residual oil, forming an oil bank with a fractional flow of approximately 0.4. The oil bank arrived at the end of the system after 0.25 PV was injected. Surfactant arrival at the end of the core corresponds to the decline of the fractional flow of oil, as dispersion of the surfactant slug through mixing with the oil bank occurred at 0.83 PV. Overall recovery was approximately 97% of the residual oil to waterflood, leaving an ROS of 0.009. Approximately 17% of the oil was recovered as an oil dispersion following surfactant breakthrough. These results demonstrated that the mixing between the formation brine and the surfactant slug did not prevent the efficient displacement of the residual oil, which was interpreted to mean that the surfactant could tolerate mixing with hard brine.

Fig. 7.106 (Flaaten 2007, Fig. 4.153) shows the salinity and surfactant concentrations of effluent samples collected during the surfactant flood. The shape of the salinity concentration data indicates that the hard resident brine was displaced miscibly by the leading edge of the surfactant front and did not damage the surfactant. Fig. 7.104 shows a favorable salinity gradient starting in Type II and decreasing to Type III, and most importantly, ending up in Type I. Surfactant did not arrive until the salinity decreased to Type III.

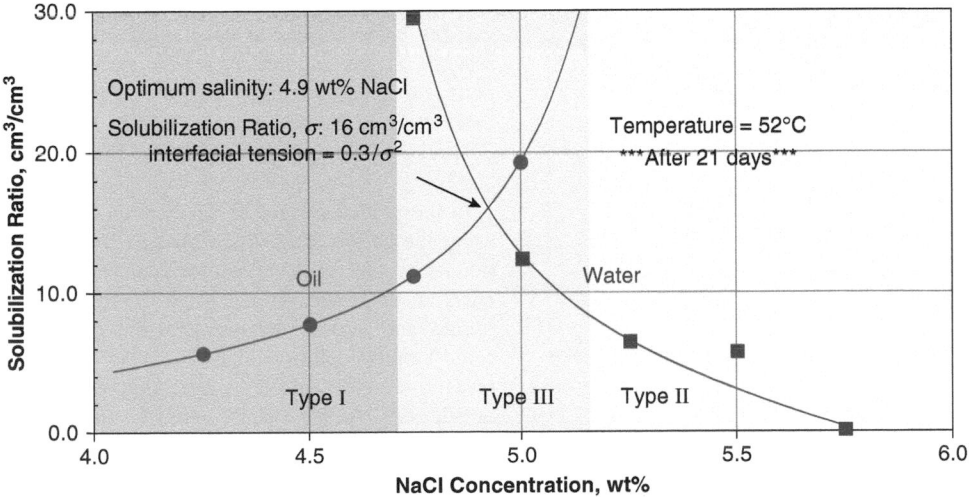

Fig. 7.104—Solubilization ratio as a function of salinity for Crude D at 52°C. The surfactant system contained 0.625 wt% C_{16-17}(7PO)SO_4^-, 0.375 wt% IOS_{15-18}, and 0.25 wt% SBA (Flaaten 2007, Fig. 4.151).

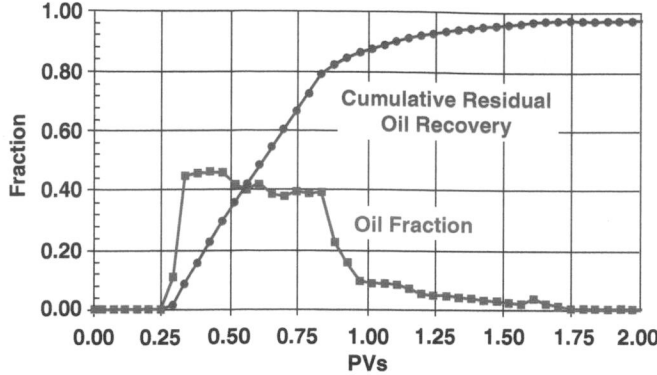

Fig. 7.105—Oil recovery and fractional flow of oil from surfactant flood of core containing Crude D at 52°C (Flaaten 2007, Fig. 4.152).

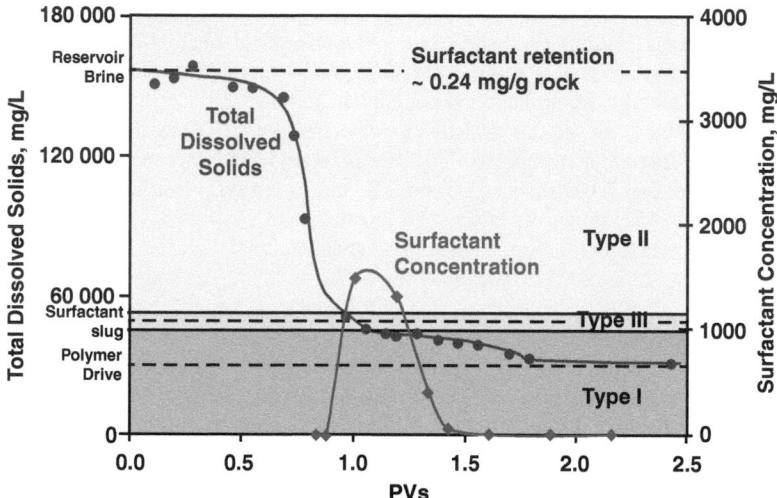

Fig. 7.106—TDS (from conductivity measurements) and surfactant concentrations from analysis of effluent samples for the surfactant flood in Figure 7.105. (Flaaten 2007, Fig. 4.153).

The results of this experiment demonstrate that a properly formulated surfactant system displaced most of the residual oil from a homogeneous linear core containing hard brine. However, the surfactant system was prepared in soft water with an aqueous stability limit of approximately 6 wt%.

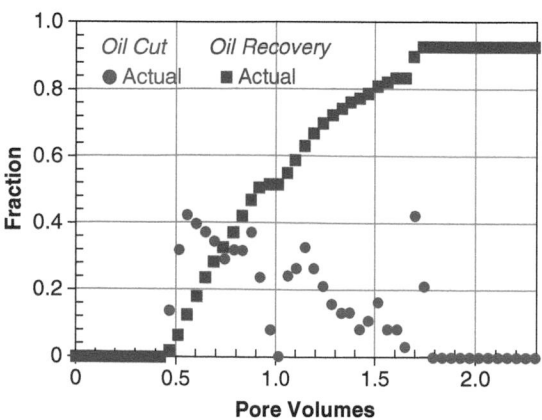

Fig. 7.107—Oil recovery and oil cut for Core Flood 1a by use of the surfactant system formulated in diluted hard brine (98 000 mg/L TDS, 5500 mg/L divalent ions) (Flaaten et al. 2008b, Fig. 6).

Many surfactants systems have been developed supported by displacement performance in linear cores. This section summarizes key advances in developing cost-effective surfactant systems that are suitable for a wide variety of reservoir conditions. There are a number of papers describing different systems that have been developed (Levitt and Pope 2008; Flaaten 2007; Solairaj et al. 2012; Yang et al. 2010; Zhao et al. 2008; Ahmad 2012).

In the previous example, application of surfactant flooding using the system studied by Flaaten was limited by optimal salinity as well as low tolerance for divalent ions. In a subsequent study, Flaaten et al. (2008b) demonstrated that optimal salinity and hardness tolerance for this base surfactant system could be increased by adding an ethoxy sulfate (C_8-3EO-SO_3) as a cosurfactant and increasing the cosolvent concentration of SBA or IBA. From phase-behavior tests at a temperature of 52°C, optimal salinity for a surfactant system formulated with hard brine containing 5500 mg/L Ca^{++}, Mg^{++} was 98 000 mg/L with a solubilization ratio of 10.

Performance of a surfactant flood in a Berea core at 52°C is shown in **Fig. 7.107** (Flaaten et al. 2008b, Fig. 6). In this flood, a

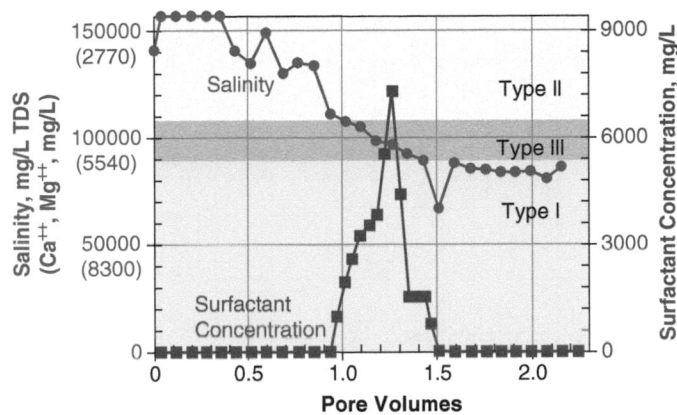

Fig. 7.108—TDS and surfactant concentrations in effluent samples of Core Flood 1a (Flaaten et al. 2008b, Fig. 10).

Berea core containing brine TDS of 157 000 mg/L, with 8700 mg/L Ca++, Mg++ at ROS (0.294) was flooded with a surfactant system formulated in diluted hard brine. The TDS of the polymer drive was 70 650 mg/L (3900 mg/L Ca++, Mg++). In the displacement, a 0.3 PV surfactant slug [5000 mg/L C16-17-7PO-SO4, 5000 mg/L C15-18 IOS, 5000 mg/L C3-3EO-SO3, 10,000 mg/L IBA, and 2,000 ppm Flopaam 3330S formulated in diluted hard brine (98 000 mg/L TDS, 5500 mg/L divalent ions)] was displaced by 2 PV of polymer solution containing 2,000 ppm Flopaam 3330S. Oil recovery was 95% of the waterflood ROS at approximately 1.8 PV of injection.

Fig. 7.108 (Flaaten et al. 2008b, Fig. 10) shows the salinity and surfactant concentrations of effluent samples with Type I, II, and III phase boundaries indicated (Flaaten et al. 2008b). The delay in the arrival of surfactant in effluent samples was interpreted as retention of surfactant in a Type II environment in the oil phase that was remobilized as a Type III microemulsion.

7.11.2 Reducing Surfactant Retention. Retention of surfactant is a major challenge in designing a chemical flood. The primary minerals in sandstone reservoirs are silica (quartz) and clays. Silica has a negative surface charge, so adsorption of anionic surfactants should be limited; however, clays are positively charged at reservoir conditions and adsorb the negatively charged anionic surfactant. Liu et al. (2008) point out that the use of alkali in a surfactant/polymer flood changes the charge of clays from positive to negative, which has been shown to reduce adsorption. Thus, the addition of alkali to a surfactant/polymer flood is effective even when the crude oil does not contain acidic components. As noted earlier, addition of alkali to a surfactant formulation requires the use of soft water to prevent precipitation of calcium and magnesium. The following example illustrates the development of an ASP system applied in a field test in southern Illinois, USA (Sharma et al. 2013).

Sharma et al. (2013) describe the design and implementation of an ASP pilot test in the Bridgeport sandstone reservoir in Illinois (Dean 2011). The reservoir contains a nonreactive oil. The development of the surfactant formulation is described by Dean (2011). The surfactant formulation consisting of 0.75% TDA-13PO-SO4, 0.25% C20-24 IOS, 0.75% IBA, 1.6% NaCl, 1% Na_2CO_3, and 2,200 ppm FP 3330S was prepared in softened water. Optimal salinity was 1.6% NaCl for this formulation. **Fig. 7.109** (Dean 2011, Fig. 2.22) shows the oil recovery from Bridgeport Core B-5 as function of pore volumes injected for the ASP flood, where 0.3 PV of surfactant was displaced by a 1.1 PV polymer drive prepared with 2,100 ppm FP3330S and 1.5% NaCl in soft water. Recovery was 93% of the waterflood ROS.

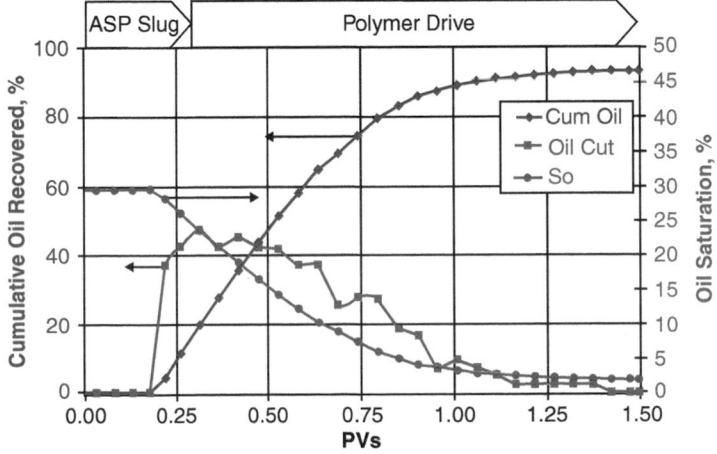

Fig. 7.109—Oil recovery from ASP flood of Bridgeport Core B-5 (Dean 2011, Fig. 2.22).

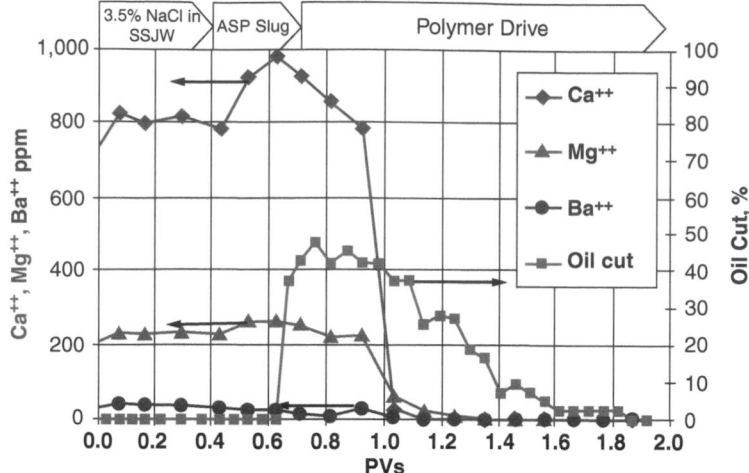

Fig. 7.110—Divalent-ion concentrations displaced by 0.4-PV slug of 3.5% NaCl solution superimposed on oil-recovery data (Dean 2011, Fig. 2.21).

The waterflood was conducted using synthetic hard brine, containing 6,267 ppm NaCl, 880 ppm Ca^{++}, 220 ppm Mg^{++}, and 140 ppm Ba^{++}. TDS was 19,442 ppm. It is well-known that sodium displaces divalent ions from clays. Injection of the surfactant slug was preceded by the injection of a soft water slug containing 3.5% NaCl. The primary purpose of the preflood was to displace the Ba^{++} ions to prevent precipitation of $BaSO_4$. **Fig. 7.110** (Dean 2011, Fig. 2.21) shows the calcium and magnesium concentration superimposed on the recovery curve (Dean 2011). The increase in calcium and magnesium concentration in the effluent at 0.4 PV injected was caused by the displacement of Ca^{++} and Mg^{++} from the clays by ion exchange. The surfactant front broke through at 0.74 PV of chemical slug injection (1.14 PV in Dean 2011, Fig. 2.21). In this flood, most of the calcium and magnesium was displaced from the core before the divalent ions could mix with the injected alkaline solution and surfactant. Surfactant retention was 0.24 mg/g.

7.11.3 Surfactant Floods in Carbonate Reservoirs. At one time, it was thought that carbonate reservoirs were not good candidates for surfactant flooding because the charge on the carbonate surface is positive, leading to high retention of anionic surfactants, which are negatively charged. However, Zhang et al. (2006) pointed out that the surface charge of carbonate minerals changed from positive to negative if the pH was increased above 9. This opened the possibility of applying ASP flooding techniques developed for sandstone reservoirs to carbonate reservoirs, which has been demonstrated in several laboratory studies. ASP flooding in carbonate reservoirs will be discussed in Section 7.14.5 because it is usually studied in reservoirs in which the oil has some level of reactivity, creating soaps when contacted with alkaline solutions. ASP formulations have been developed for carbonate reservoirs containing nonreactive oils (Liu et al. 2008). ASP floods are not good candidates for a carbonate reservoirs that contain anhydrite ($CaSO_4$) because calcium dissolves in the alkaline solution and precipitates as $CaCO_3$ (Lopez-Salinas et al. 2011), consuming the alkali.

Tolerance of high-divalent-ion content remains an issue with sulfates and sulfonates. Calcium and magnesium, the principal divalent cations, associate with anionic surfactants and increase oil solubility, decreasing the optimal salinity. In some cases, divalent cations may precipitate an anionic surfactant or prevent the surfactant solution from meeting the aqueous-phase salinity limit (APSL) for the reservoir conditions. Although the use of additives to increase tolerance to divalent ions such as sodium metaborate or ethylenediaminetetraacetic acid (EDTA) has been studied (Yang et al. 2010), they are generally not cost-effective, and other surfactant systems must be considered if the injection water cannot be softened economically.

7.11.4 Temperature Stability of Surfactant Systems. Use of ether sulfates for surfactant flooding is limited to reservoirs with temperatures less than approximately 65°C. Above this temperature, hydrolysis of the ether sulfate to base alcohols occurs, and surfactant activity declines with time of exposure to elevated temperatures (Talley 1988). However, when sodium carbonate is added to the surfactant and the pH is maintained in the range of 10 to 11, then the hydrolysis of ether sulfates is much slower, which extends their use to higher temperatures (Adkins et al. 2010).

A new family of surfactants was developed by ethyoxylating and propoxylating Guerbet alcohols to increase salinity and temperature tolerance. Guerbet alcohols are large branched alcohols produced by dimerization of linear alcohols (Adkins et al. 2012) to produce a molecule with twice the molecular weight. **Fig. 7.111** shows the structure of a Guerbet alcohol (O'Lenick and Bilbo 1987). Adkins et al. (2012) demonstrated that salinity and temperature tolerance could be increased by attaching large numbers of EO and PO groups to the Guerbet alcohol before adding a surface active group such as a sulfate or sulfonate. This family of surfactants is termed Guerbet alkoxy sulfates.

Guerbet alkoxy sulfates have high salinity tolerance. The structure of the surfactant is adjusted by controlling the number of EO and PO groups that are attached to the Guerbet alcohol before sulfating. **Fig. 7.112** (Adkins et al. 2010, Fig. 4) shows solubilization data for Crude Oil #2 as a function of sodium carbonate concentration for an oil concentration of 50% at 100°C.

Optimal salinity is 27,000 pm Na_2CO_3. APSL for this system is 35,000 ppm Na_2CO_3. The solubilization ratio at optimal salinity was 17. The surfactant formulation was 0.25% C_{32}-7PO-14EO sulfate, 0.25% C_{20-24} IOS, and 0.5% TEGBE (triethylene glycol monobutyl ether).

This surfactant system was tested in a Bentheimer core that had been waterflooded to ROS. The surfactant formulation included 2,000 ppm Flopaam 3630 for mobility control and 680 ppm $Na_2S_2O_3$ (500 ppm $S_2O_3^-$) for oxygen scavenging. **Fig. 7.113** (Adkins et al. 2010, Fig. 5) is a graph of a surfactant flood of a Bentheimer core using this surfactant formulation. The flood was a 0.5 PV slug followed by a polymer solution containing 900 ppm Flopaam 3630S and 1,000 ppm Na_2CO_3 to induce a salinity gradient. Average oil cut in the oil bank was 65%. Total oil recovery was 94.7% of the ROS after waterflood.

In general, Guerbet alkoxy sulfate systems require the use of soft water for reservoirs that have high-divalent-ion content or chelating agents such as $EDTANa_4$ where cost effective. Alcohol propoxylate ether sulfates and large hydrophobe surfactants prepared from Guerbet alcohols are not stable at temperatures above 60 to 65°C except for special pH conditions. At elevated temperatures, these sulfates hydrolyze returning to their alcohol precursors.

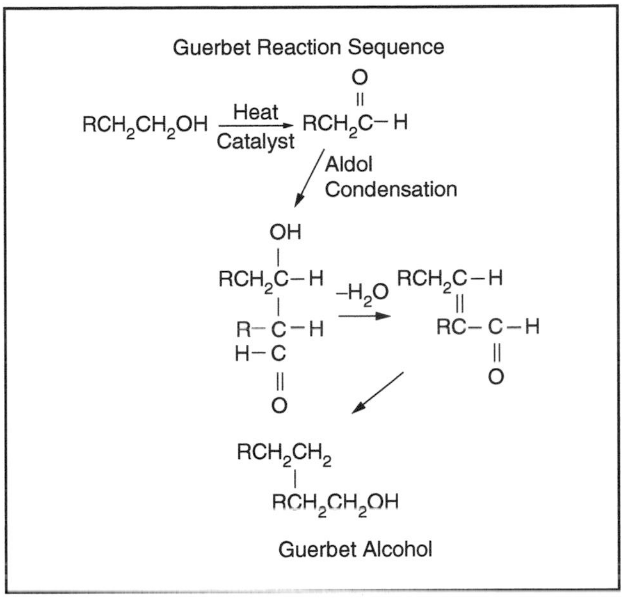

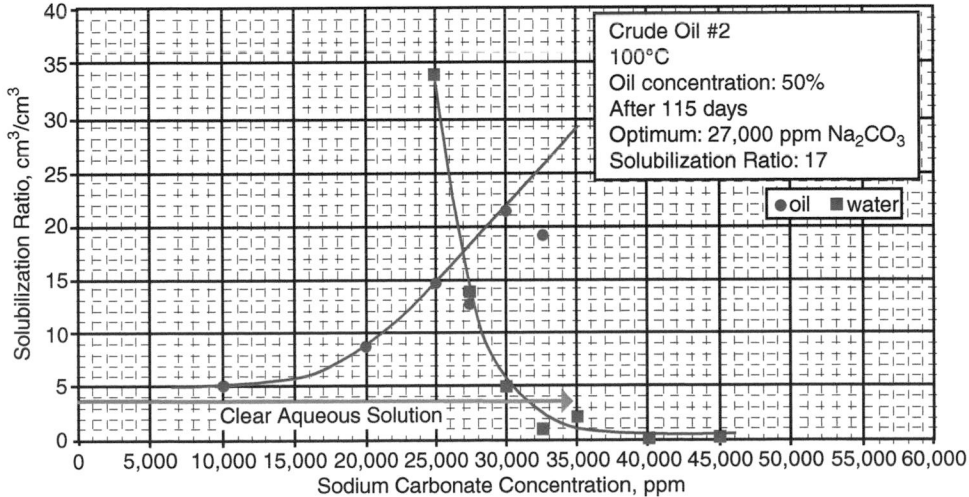

Fig. 7.111—Structure of a Guerbet alcohol (O'Lenick and Bilbo 1987).

Adkins et al. (2012) discovered that thermally and chemically stable surfactants could be prepared by carboxylating alkoxy Guerbet alcohols. This family of surfactant was developed by adding large numbers of EO and PO groups to Guerbet alcohols, followed by carboxylating the Guerbet alkoxy alcohol. Some examples of Guerbet alkoxy carboxylate surfactants are the molecules C_{24}-25PO-18EO carboxylate, C_{28}-35PO-10EO carboxylate, and C_{28}-45PO-60EO carboxylate (Lu et al. 2012). Surfactant formulations often require the use of a cosurfactant and/or a cosolvent.

Data were presented that demonstrated the surfactants were chemically stable at elevated temperatures and had a high tolerance for divalent-ion concentrations with and without alkali. **Fig. 7.114** (Lu et al. 2012, Fig. 18) shows the phase behavior of a surfactant formulation containing 0.5 wt% C_{28}-45PO-60EO carboxylate and 0.5 wt% C15-18 IOS with Oil #2 at 120°C. Oil recovery from a Silurian dolomite core following injection of 0.25 PV of this surfactant solution followed by 4,500 ppm FP 3330S polymer is shown in **Fig. 7.115** (Lu et al. 2012, Fig. 21).

7.12 Design Procedures and Criteria

This section focuses on the general criteria required for a micellar/polymer flood and on the design of the chemical slug. A design procedure to determine the viscosity of the chemical and polymer slugs to maintain mobility control will be described. Chapter 5 discussed other factors involved in polymer design.

Fig. 7.112—Solubilization ratios as a function of sodium carbonate concentration from phase-behavior data by use of 50% oil ratio (Adkins et al. 2010, Fig. 4).

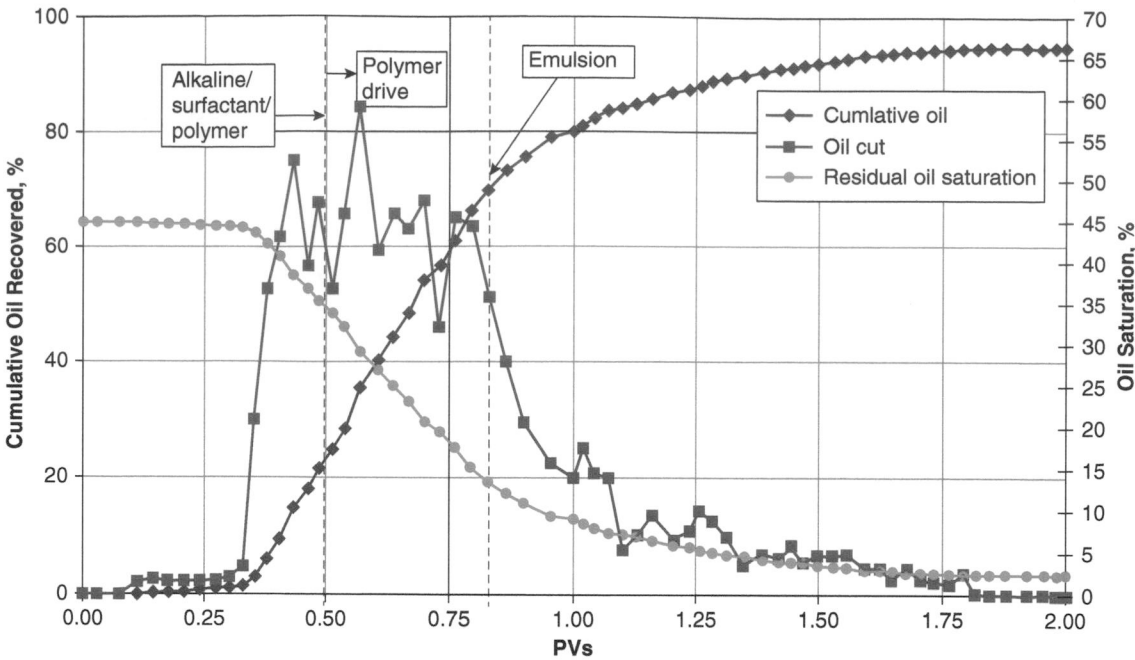

Fig. 7.113—Oil recovery for a Guerbet alkoxy sulfate surfactant flood in a Bentheimer core at 100°C by use of the surfactant system shown in Adkins et al. (2010), Fig. 5.

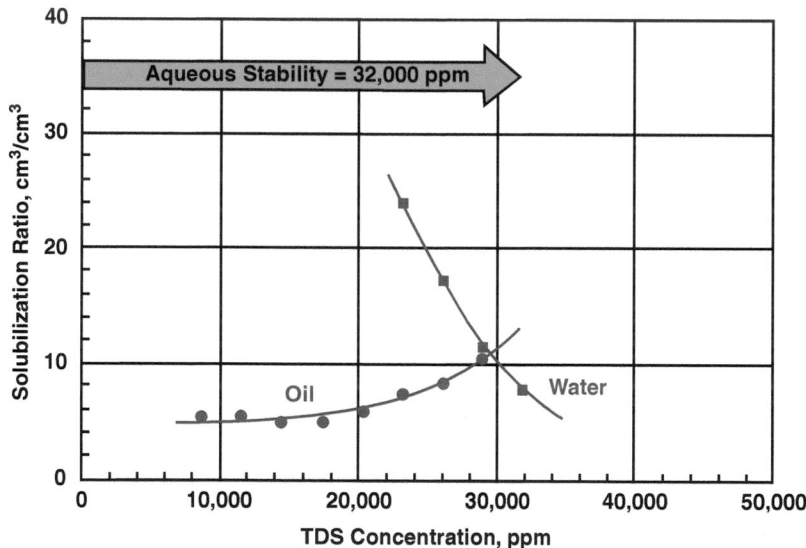

Fig. 7.114—Oil recovery from a Silurian dolomite core containing waterflood residual oil at 100°C (Lu et al. 2012, Fig. 18). The resident brine contained 66,800-ppm TDS, including 38,000-ppm salinity, 4,400-ppm Ca^{++}, and 900 ppm Mg^{++}.

It is clearly necessary to have a good understanding of the reservoir geology and the primary and secondary recovery processes that might have preceded implementation of a micellar/polymer flood. Gaining that understanding is part of the overall design process, but is not treated in this section. Knight (1977) present an example of such an analysis.

7.12.1 General Criteria for an Efficient Micellar/Polymer Flood. A number of general criteria must be met for a chemical flood to perform at a high and acceptable efficiency.

1. Low IFT between the primary chemical slug and the oil bank.
2. Low IFT between the mobility buffer and the primary chemical slug.
3. Favorable mobility ratio between the chemical slug and the oil bank.
4. Favorable mobility ratio between the mobility buffer and the primary chemical slug.
5. Maintenance of the integrity of the chemical slug (i.e., prevention of surfactant losses that seriously degrade the slug).

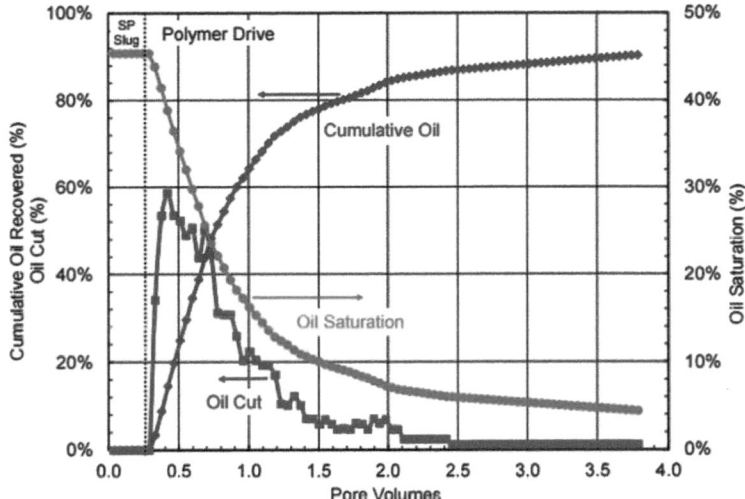

Fig. 7.115—Oil recovery from Silurian dolomite core by injecting 0.25 PV surfactant solution containing 2% surfactant (1% C₃₂-7PO-32EO carboxylate, 1% C₁₉₋₂₃ IOS, and 4,500 ppm FP 3330S polymer) (Lu et al. 2012, Fig. 21).

Design of a process considering these factors typically involves significant laboratory work with the specific rock/chemical system of interest. Mathematical modeling of the process and consideration of the economics usually are required.

7.12.2 General Approach in the Design. Design of a micellar/polymer flood is a complex process. The process begins with the reservoir crude oil [or an equivalent oil (Puerto and Reed 1983)], the composition of the reservoir brine, and the composition of possible injection waters at reservoir temperature. Selection of a surfactant and cosurfactant (if needed) that will give high solubilization parameters at reservoir salinity is the critical step. This selection is required to obtain ultralow IFTs necessary for effective displacement. The composition of the surfactant/cosurfactant solution depends on a number of parameters, including, at least, surfactant/cosurfactant types, reservoir oil, brine type and composition, polymer compatibility, pH, and temperature (Salter 1986). No equations are available to calculate a priori the effects of these parameters. Optimal conditions must be determined through a series of laboratory experiments involving phase behavior and IFT measurements.

It is also important to understand the behavior of a chemical system as it flows through the reservoir rock. The integrity of the chemical slugs must be maintained for some minimum acceptable flow period. Again, in the absence of data, no equations are readily available, and thus it is desirable to conduct laboratory corefloods.

Salter (1986) discusses surfactant selection for chemical floods for low- and high-salinity regions. Selection of a surfactant involves identifying those surfactants that have the potential of giving high solubilization parameters. It is also necessary to determine which structures are commercially available or might be manufactured. Once candidates have been identified, extensive phase-behavior experiments are conducted over the range of parameters of interest as an initial screen. IFT measurements are made on the most likely systems, and finally, corefloods are conducted with the most promising systems that satisfy both phase-behavior and IFT criteria. Once data on phase behavior, IFT, and coreflood performance are available, mathematical models can be used to optimize the design and to provide information for economic calculations.

In developing the experimental program for the design of the chemical slug, optimal salinity and the manner in which it is to be attained in the reservoir of interest should be considered. Three approaches can be taken.

1. Optimal salinity of the micellar slug can be adjusted so that it equals, or approximately equals, the salinity of the resident brine in the reservoir, considering the divalent-ion content of the brine and the divalent ions that may be added to the solution by cation exchange or dissolution. This may be difficult if salinity is high or there are significant quantities of divalent ions. In this case, it would be necessary to use a salinity-tolerant synthetic surfactant. An example of this approach was discussed in Section 7.6 (Maerker and Gale 1992).

2. A preflush ahead of the micellar slug can be used to adjust reservoir salinity or divalent-ion concentrations to a desired level. High-pH preflushes will precipitate divalent ions. Preflushes have been used and typically require a large volume. While the concept of a preflush is appealing, a preflush does not always prevent mixing of a micellar slug with original reservoir brine (Pursley et al. 1973) and generally is avoided as a method to control salinity in process design. If it is known when a waterflood is conducted that a micellar/polymer flood will follow, then the waterflood brine can be adjusted to the desired salinity.

3. The salinity-gradient concept can be applied by considering the salinity of the resident reservoir brine and setting the salinities of the micellar slug and mobility-control slug to optimize recovery (Nelson and Pope 1978; Nelson 1982).

If one of these approaches can be established early in the design, experimental work can be focused in terms of salinity requirements, which will reduce time and effort.

Another difference in design philosophy is the use of a relatively large slug of low surfactant concentration vs. a smaller slug of higher surfactant concentration (Gogarty 1976). There are proponents of both approaches and examples of field applications of each will be discussed in Section 7.13. Again, an early decision on the approach in the design process will reduce the work required to design the chemical slug.

7.12.3 Phase-Behavior and IFT Measurements. Once general salinity requirements are established, phase-behavior experiments can be conducted to screen candidate systems. Because optimal salinity for phase behavior is very close to the optimal salinities for IFT and oil recovery, measurement of solubilization parameters can be used as the initial screen (Healy et al. 1976).

One procedure is to set the micellar slug composition and measure solubilization parameters as salinity is varied. If optimal salinity for the system is near the desired salinity, the system becomes a candidate for further study. This can be repeated for any desired number of combinations of slug compositions. Salinity can be varied by use of a simple salt, such as NaCl, but at some point actual reservoir brine or a synthetic brine that simulates the reservoir brine should be used.

7.12.4 Formulation of Surfactant Systems. Formulation of a surfactant system for a specific oil, reservoir conditions, and rock type requires extensive phase-behavior experiments followed by linear displacement tests to verify oil recovery at reservoir conditions. A few guidelines have been developed that are based on the research program at the University of Texas (Levitt et al. 2006; Flaaten et al. 2008a; Solairaj et al. 2012; Yang et al. 2010) and in other laboratories (Ahmad 2012). The following guidelines are helpful in developing a surfactant system in phase behavior studies:

- The aqueous surfactant solution (surfactant, cosurfactant, cosolvent, polymer, and/or alkali) must be a clear, one-phase solution at reservoir temperature at or above the optimal salinity determined for the oil/water/surfactant system. The salinity at which a second phase occurs (also termed the cloud point) for a given formulation is called the APSL.
- The solubilization ratio must be higher than 10 mL/mL at optimal salinity to have an acceptable IFT for mobilizing oil.
- The microemulsion formed must be low-viscosity and free from macroemulsions and liquid crystalline phases, particularly near the optimum salinity.
- Coalescence of the microemulsion phase observed by equilibration of the pipettes near the optimum salinity should be fast.
- The final parameter, which evolved through numerous experiments, is the expectation that an effective surfactant system will recover at least 90% of the waterflood ROS in linear corefloods at reservoir conditions, with a final oil saturation less than 3% (a metric independent of the starting oil saturation).

IFT can be measured on micellar systems that pass the phase-behavior screen. However, IFT can be estimated with sufficient accuracy using the Huh equation in the vicinity of optimal salinity.

Because micellar solutions generally include polymer for mobility control, the effect of polymer compatibility, concentration, and molecular weight are evaluated during the phase-behavior studies. Viscosity measurements should be made on systems that are to be used in corefloods. Because the fluids are non-Newtonian, the effect of shear rate on viscosity should be considered.

7.12.5 Corefloods. For those systems that pass the phase-behavior and IFT screens, corefloods should be conducted to evaluate displacement behavior in porous media. It is recommended that cores from the reservoir of interest be used in at least some of these tests. Gogarty (1983a) reported that recoveries that differed by as much as a factor of three were noted when the same chemical system was used in different sandstone cores.

Corefloods can be used to measure such factors as adsorption, cation exchange, effect of micellar slug size, and effectiveness of mobility control. Also, if the salinity-gradient concept is used in the design, this can be tested in corefloods. The floods should be conducted at reservoir temperature.

7.12.6 Mobility Control. The system of chemical slugs should be designed so that there is a favorable mobility ratio between the micellar slug and the oil bank and between the leading edge of the polymer slug and the micellar slug. The method of setting the mobilities in the different slugs suggested by Gogarty et al. (1970) was used in Example 7.10.

Typically, relative permeability data are available, but the saturations in the oil bank are not known and the total mobility cannot be calculated. Gogarty et al. (1970) suggest using the relative permeability data to plot total mobility as a function of saturation, as shown in **Fig. 7.116.** As illustrated, the curve will pass through a minimum at some saturation.

The minimum total relative mobility is taken as the design relative mobility for the micellar slug. If the micellar slug has a relative mobility equal to or less than this minimum, the mobility ratio will be favorable.

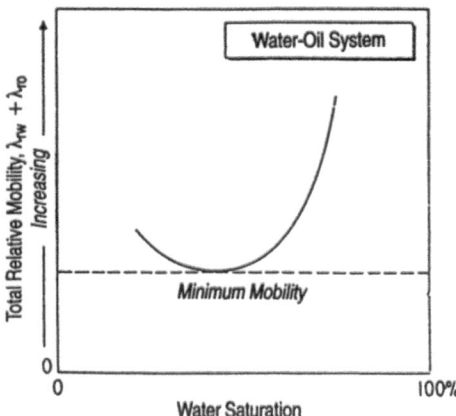

Fig. 7.116—Total relative mobility vs. water saturation (Gogarty et al. 1970).

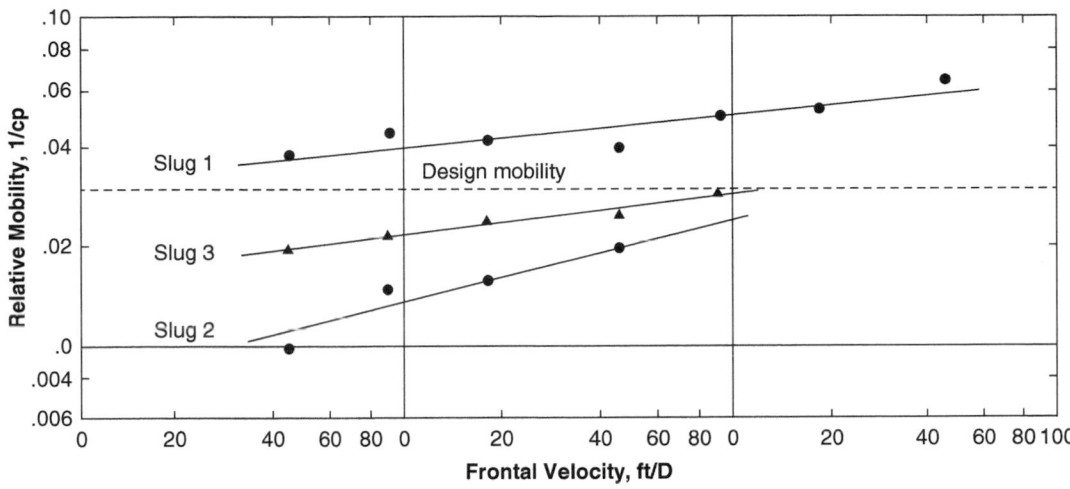

Fig. 7.117—Effect of slug composition on mobility-control design (Gogarty et al. 1970).

Once the micellar slug mobility is fixed, the polymer slug mobility is set at a smaller value to maintain mobility control at the polymer-slug/micellar-slug interface. Because polymer solutions are non-Newtonian, shear rate should be considered. The effect of shear rate and other factors that affect polymer mobility are discussed in Chapter 5.

Gogarty et al. (1970) tested three different chemical slugs by measuring relative mobility of the slugs as a function of frontal velocity. **Fig. 7.117** shows the results. Relative mobility is a function of flow rate, indicating the non-Newtonian nature of the slugs. Slugs 2 and 3 satisfy the design criterion at velocities expected in the reservoir except at locations very near the wellbore.

Gogarty et al. (1970) also studied relative mobilities of polymer under several conditions. Mobility was found to be controlled by a combination of solution viscosity and mechanical entrapment, as discussed in Chapter 5. The use of a mobility buffer that was graded (i.e., decreased in polymer concentration as injection was continued) was recommended. The viscosity of the polymer solution was graded so that the viscosity ratio between adjacent polymer slugs was constant (Poettmann 1974).

7.12.7 Modeling of the Process. The models described in Section 7.10 can be used to describe mechanisms and general behavior but are of limited use in design. Typically, relatively sophisticated computer-based simulations are used in a design. Gogarty (1983) references a number of computer simulators. The description of such simulators is beyond the scope of this book.

The US Department of Energy (US DOE) has published a series of simulators that can be used as screening tools in chemical flooding design (Ray and Munoz 1986). These are available upon request. UTCHEM is a 3D, three-phase multicomponent chemical flooding simulator developed by the University of Texas at Austin and is available for download (UTCHEM-9.0 Technical Documentation 2000). Applications of UTCHEM are presented in the references (Pope and Nelson 1978; Pope et al. 1979, 1987; Saad et al. 1989; Bhuyan et al. 1990; Delshad et al. 1998, 2013; Anderson et al. 2006; Mohammadi et al. 2009).

7.13 Field Experience

The micellar/polymer process has been widely field tested, but has had only limited commercial application (Knight 1977; Ferrell et al. 1984, 1988; Cole 1988a, b; Huh et al. 1990a, b; Troiani et al. 1979). **Table 7.20** summarizes data from three large-scale field implementations. The complexity and high cost of the process slowed application, especially after the oil price drop in 1986. The development of new surfactant systems that tolerate wider ranges of reservoir conditions and the higher oil prices between 2006 and 2014 led to several pilot tests of selected surfactant systems. As of this writing, results from these tests have not been published. A critical fact in chemical flooding is the absence of performance from large-scale field tests.

7.13.1 Marathon M-1 Project. Marathon Oil Company's M-1 project was a field-scale commercial application of micellar/polymer flooding in the depleted Robinson Sand reservoir. The project involved an area of 407 acres (Cole 1988b). Approximately 60% of the project area was developed on 2.5-acre five-spot patterns. The remaining area was developed with 5-acre five-spot displacement patterns. The M-1 project was designed to determine the effect of pattern spacing on displacement performance. The Marathon surfactant system was formulated with crude-oil sulfonates and other cosurfactants. Injection began in 1977 with a 0.1-PV slug of surfactant into each pattern followed by 1.05 PV of mobility-control solution, which was displaced by drive water. **Fig. 7.118** (Stover 1988) shows the production response from the Marathon M-1 project.

Ultimate oil recovery from the project area was estimated at 1,397,000 bbl of oil, or 21% of the OOIP at the beginning of the project. Oil recovery was 807,000 bbl (22.3%) from the 2.5-acre development and 590,000 bbl (19.4%) from the 5.0-acre development. Recoveries were significantly less than predicted by laboratory studies. The lower recoveries were attributed to poor volumetric sweep efficiency and salinity/hardness effects (Cole 1988b).

	Marathon	Exxon	Conoco
Reservoir name	Robinson (M-1 Project)	Weller Sand (Loudon field)[1]	Second Wall Creek (Big Muddy field)
Lithology	SS	SS	SS
Area in flood (acres)	407	0.71	90
Pattern type	Five spot	Five spot (center well producer)	Five spot
Spacing (acres)	2.5; 5.0	–	10
Tracer study	Yes	Yes	Yes
Permeability (md)	103	67 to 189	56[2]
Porosity (%)	18.9	19.5	$\approx$ 20.0
Thickness (ft)	0 to 60; average 27.8	8 to 28; average 15.6	65
Depth (ft)	< 1,000	1,400 to 1,600	3,100
Temperature (°F)	72	78	115
Crude Oil			
Viscosity (cp)	5 to 6	5.0	5.0
API gravity (°API)	36	–	–
Geology	Stacked and isolated sand lenses; meandering river, migrating point bars	Deltaic deposit, fine to very fine grain sand	Highly jointed with low closure pressure
Heterogeneity	Lorenz coefficient = 0.44	Significant thickness variation	Fracture joint system; Dykstra-Parsons VDP = 0.01
Tertiary or secondary flood	Tertiary	Tertiary	Tertiary
Oil saturation at start of flood, swept zone (% PV)	40	24.1	32
Chemical slug			
Surfactant type	Crude oil sulfonate	$RO(C_3H_6O)m(C_2H_4O)n$ SO_3Na[3]	Blend of synthetic sulfonates
Surfactant concentration, active (wt%)	10[4]	2.3	3
Cosurfactant type	Hexanol	–	Isobutyl alcohol
Cosurfactant concentration	0.8 vol%[5]	–	5 wt%
Oil (wt%)	7.5[6]	2.65; 250 white oil base[7]	–
Water (wt%)	80[8]	96 of resident salinity	–
Salts (wt%)	2.5[9]	96 of resident salinity	0.6
Polymer in slug	No	Yes, biopolymer	Yes, polyacrylamide, 2,200 ppm
pH	6.5 to 7.5	5.2	–
Other additives	Citric acid, 500 ppm[10]	Formaldehyde, citric acid, 90 mg/L[11]	–
Viscosity (cp)	< 40	28[12]	12[13]
Slug size (% PV)	10	30	10.2
Formation water (mg/L)	16 575	104 000	Brine preflush
Ca	166	2,840	–
Mg	118	1,210	–
Mobility-control buffer			
Polymer type	Polyacrylamide	Biopolymer	Polyacrylamide
Biocide	No	Yes; formaldehyde; 90 mg/L	–
Polymer concentration (ppm)	1,156	40 cp[14]	1,400; 12cp[13]
Slug size (% PV)	11	100	18[15]
Polymer concentration (ppm)	800	–	–
Slug size (% PV)	19	–	–

Table 7.20—Data on selected micellar/polymer field applications.

	Marathon	Exxon	Conoco
Polymer concentration (ppm)	625	–	–
Slug size (% PV)	32	–	–
Polymer concentration (ppm)	411	–	–
Slug size (% PV)	12	–	–
Polymer concentration (ppm)	200	–	–
Slug size (% PV)	11	–	–
Polymer concentration (ppm)	50	–	–
Slug size (% PV)	10	–	–
Polymer concentration (ppm)	0	–	–
Slug size (% PV)	35	–	–
Salinity	–	70% of formation salinity	0.4 wt%
Date injection started	February 1977	August 1982	1980 (preflush) January 1981 (chemical)
Tertiary oil recovery			
Date	September 1983	November 1983	14% of oil in place at start of project
Oil recovery	See Fig. 7.118	68% of waterflood residual oil; see Figs. 7.119 and 7.120.	–
Problems	–	Production of an emulsion.	Low matrix k; fractures at p less than hydrostatic p; lack of containment of injected fluids.
References	Knight (1977); Garrett (1972)	Reppert et al. (1990); Maerker and Gale (1992); Huh et al. (1990b)	Ferrell et al. (1984, 1988); Troiani et al. (1979); Cole (1988a)

[1] This is the second of two pilot tests in the Loudon field in which a high-salinity-tolerant surfactant was used.
[2] k_w = 1 to 2 md at ROS. Fracturing occurs at less than hydrostatic pressure, leading to larger effective permeability.
[3] R represents i-C$_{13}$H$_{27}$; m = 3 or 4; n = 2 or 4; mixture of two surfactants used.
[4] Target value reported; actual value 9.2 to 10.8.
[5] Cosurfactant added at field after chemical slug made up at refinery. Refinery chemical slug consisted of surfactant, oil, water, and salt; target value reported; actual value 0.4 to 1.0.
[6] Target value reported; actual value 5 to 15.
[7] Surfactant, 250 white oil base and biopolymer broth mixed with formation brine.
[8] Water contained 501 mg/L salts.
[9] Specified as inorganic salt.
[10] Citric acid added at field as chelating agent.
[11] Formaldehyde added as a biocide.
[12] Viscosity measured at a shear rate of 11 seconds^{-1}.
[13] Measured at 12 rev/min on Brookfield viscometer.
[14] Concentration not specified. Sufficient polymer added to make viscosity equal to 38 cp measured at 11 seconds^{-1} shear rate.
[15] Polymer slug injection terminated earlier than planned because of low injectivities and production. Planned volume was 0.4 PV.

Table 7.20—Data on selected micellar/polymer field applications. (Continued)

7.13.2 Conoco Big Muddy Field Low-Tension Flood. Conoco Incorporated's Big Muddy field project used a so-called low-tension flood process and was a commercial-scale demonstration project (Borah and Gregory 1988). The Wall Creek reservoir was depleted from extensive waterflooding before initiation of the low-tension flood. The 90-acre project area was flooded in nine 10-acre five-spot patterns. Formation water was fresh, with a relatively low divalent-ion concentration. Consequently, the surfactant system was formulated with a petroleum sulfonate and isobutyl alcohol. A small salinity gradient was used between the surfactant and the mobility buffer. A pilot project preceded the field demonstration project.

Oil production was approximately 290,000 bbl, which was approximately one-fourth of the projected recovery. Lack of fluid containment was the primary cause of low oil recovery. Displacing fluids migrated from the project area as a result of poor completion techniques in old wells, natural fractures in the reservoir, and unintentional pressure parting.

7.13.3 Exxon Loudon Field Tests. The Exxon Loudon field test is the third in a series of four scaled field tests to evaluate a surfactant system for the Loudon field (Maerker and Gale 1992). The first field test, described by Pursley et al. (1973), showed that a preflush would not protect the surfactant slug and that a salinity-tolerant surfactant would be required.

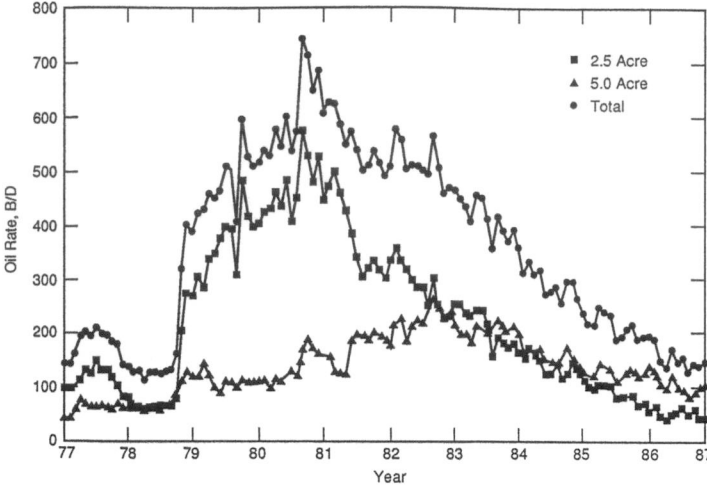

Fig. 7.118—M-1 project performance, 2.5- and 5.0-acre spacing (Stover 1988).

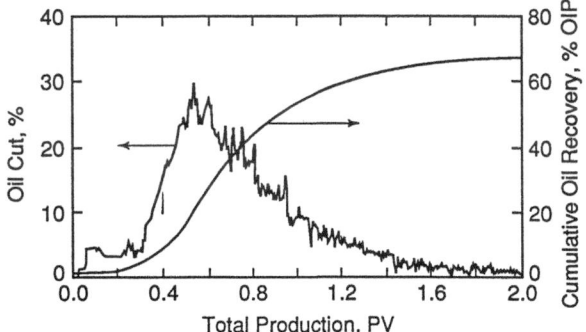

Fig. 7.119—Second Ripley pilot oil cut and cumulative recovery (Reppert et al. 1990).

The salinity-tolerant system developed by Exxon, described in Section 7.6 (Maerker and Gale 1992), was used in three tests. All tests involved injection of 0.3 PV of surfactant slug followed by a polymer drive and drive water. The first field test was conducted on a 0.68-acre five-spot pattern to determine the oil recovery efficiency and surfactant loss (Krumrine et al. 1982). The second field test in a 0.71-acre confined five-spot pattern was designed to determine the effect of slug size on oil recovery, to demonstrate that formaldehyde was an effective biocide for preventing biodegradation of the polysaccharide biopolymer, and to develop cost-effective methods for breaking produced-oil/brine/surfactant emulsions (Reppert et al. 1990).

Fig. 7.119 shows the oil production response of this project. Oil recovery was 68% of the waterflood residual oil initially present. **Fig. 7.120** shows production of the different slugs in the center well of the Exxon project. The data are presented as concentrations normalized to injected concentrations. Formaldehyde was present in the surfactant and polymer slugs. Methanol was used as a tracer for the surfactant slug. The lag in breakthrough of surfactant reflects retention in the porous media. However, adsorption was estimated to be approximately 0.08 mg/g rock or less, which is quite low and very favorable. The reason for this low adsorption is thought to be the anaerobic (reducing) condition of the reservoir, as discussed in Section 7.9.2 (Somasundaran 1984). There is a significant lag in breakthrough of polymer, reflecting a relatively high retention in the reservoir.

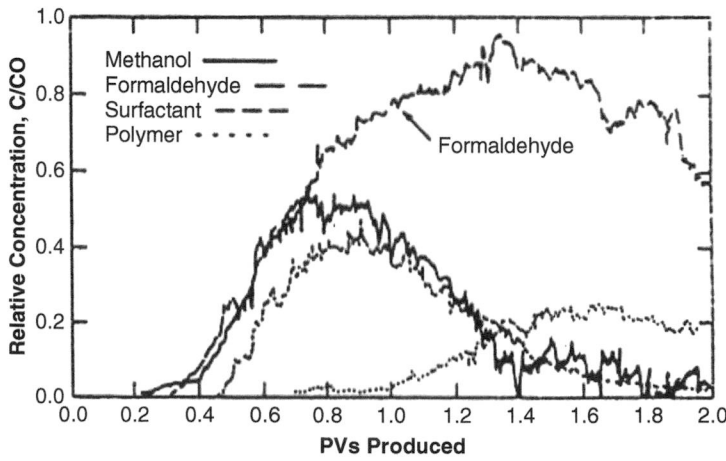

Fig. 7.120—Well 20 production and chemical transport, second Ripley pilot (Reppert et al. 1990).

The third field test was designed to evaluate this surfactant system on 2.5- and 5.0-acre spacings (Huh et al. 1990a). A 40-acre test consisted of nine contiguous 2.5-acre five-spot patterns. Oil recovery averaged 27% of the OOIP at the beginning of the project. The 80-acre test consisted of nine contiguous 5-acre five-spot patterns. Average oil recovery from the 5-acre project region was 33% of the OOIP at the beginning of the project.

Oil recovery was significantly less than anticipated. This was attributed to the failure of the polymer drive bank to propagate through the reservoir as designed (Huh et al. 1990a, b).

7.13.4 Project Results. The Marathon M-1 demonstration was not economical. Economic data were not provided for the Conoco and Exxon tests, but would not be positive at oil prices that existed during 1986–1990 (less than USD 20/bbl). It was generally accepted that an oil price of USD 25 to 30/bbl (1990 dollar value) would be required for the process to be attractive economically.

All three of the applications summarized were in sandstone reservoirs. Because of the problems associated with high salinity and especially high-divalent-ion concentrations, the micellar/polymer process has not been used in carbonate reservoirs. Also, carbonate reservoirs tend to be more heterogeneous and to have lower permeabilities than sandstone reservoirs, which adversely affects the process.

7.14 Alkaline Flooding

In alkaline flooding, a high-pH chemical system is injected. If the reservoir crude oil has sufficient "saponifiable components," a reaction will occur in which surfactants are formed in situ. In most of the literature, these saponifiable components are described as petroleum acids, even though their structure is not known. We adopt this practice in this section, recognizing that not all saponifiable components that may form surfactants are petroleum acids. Several mechanisms, including reduction of IFT, contribute to increased oil displacement efficiency as a result of the formation of the surfactant. The process generally has been applied with crude oils of relatively low API gravity.

The high costs of micellar/polymer systems and the relatively low cost of alkaline agents stimulated other variations of the process. Cosurfactant-enhanced alkaline flooding is a modification of the basic process in which a surfactant (called a cosurfactant to the surfactant formed in situ) is injected with the alkaline chemical (Nelson et al. 1984). The complementary effect of the cosurfactant improves process performance.

Mobility control can improve displacement efficiency in alkaline floods. In most cases, polymer is used as a mobility buffer to displace the primary slug. However, polymer has been injected with the alkaline fluid in two field tests (Manji and Stasiuk 1987; Meyers et al. 1992). In addition, a preflush is commonly performed to condition the reservoir before injection of the primary slug.

This section presents the formation of surfactants in situ, IFT reduction, mechanisms that affect recovery, and rock/fluid interactions. The cosurfactant-enhanced alkaline process is described, and references to field applications are given.

7.14.1 Chemicals Used and In-Situ Formation of Surfactants. Several different alkaline agents have been used, including sodium hydroxide, sodium orthosilicate, sodium carbonate, ammonium hydroxide, and ammonium carbonate (Gogarty 1983b). The first three have been the most widely considered. **Fig. 7.121** compares several commonly used alkaline materials (Mayer et al. 1983). **Table 7.21** compares selected properties of alkalis. The process is dependent on alkali reacting with petroleum acids in a crude oil to form surfactants in situ. These petroleum acids are many in number and varied in composition (Mayer et al. 1983).

Addition of the alkali chemicals results in a high pH because of the dissociation in the aqueous phase. NaOH, for example, dissociates to yield [OH⁻] as follows:

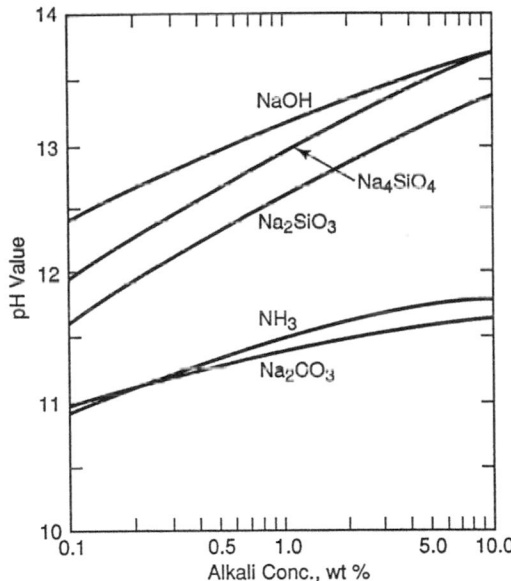

Fig. 7.121—pH comparison of several commonly used alkaline chemicals (Mayer et al. 1983).

$$NaOH \rightarrow Na^+ + OH^-. \dots\dots\dots\dots\dots\dots\dots(7.45)$$

Equilibrium dissociation of water is given by

$$K = \frac{\left[OH^-\right]\left[H^+\right]}{\left[H_2O\right]}, \dots\dots\dots\dots\dots\dots\dots\dots\dots\dots\dots(7.46)$$

where brackets indicate molar concentrations and an increase in [OH⁻] causes a decrease in [H⁺]. Water concentration is essentially constant. Because pH is defined as

$$pH = -\log_{10}\left[H^+\right], \dots\dots\dots\dots\dots\dots\dots\dots\dots\dots\dots\dots\dots(7.47)$$

| | Formula | Molecular Weight | pH of 1-wt% Alkaline Solution | | | Solubility | |
			0% NaCl	1% NaCl	Na$_2$O (%)	Cold Water (g/100 cm³)	Hot Water (g/100 cm³)
Sodium hydroxide	NaOH	40	13.15	12.5	0.775	42	347
Sodium orthosilicate	Na$_4$SiO$_4$	184	12.92	12.4	0.674	15	56
Sodium metasilicate	Na$_2$SiO$_3$	122	12.60	12.4	0.508	19	91
Ammonia	NH$_3$	17	11.45	11.37	–	89.9	7.4
Sodium carbonate	Na$_2$CO$_3$	106	11.37	11.25	0.585	7.1	45.5

Table 7.21—Selected properties of alkaline agents.

the pH of the solution increases ($\log[H^+]$ is a negative number). Note that for water at pH=7.0, the concentration of $[H^+]$ is 10^{-7}. Sodium carbonate dissociates as

$$Na_2CO_3 \rightarrow 2Na^+ + CO_3^{2-}, \dots\dots\dots\dots\dots\dots\dots\dots\dots (7.48)$$

followed by the hydrolysis reaction

$$CO_3^{2-} + H_2O \rightarrow HCO_3^- + OH^-. \dots\dots\dots\dots\dots\dots\dots\dots (7.49)$$

The dissociation of sodium silicate compounds is complex, involving formation of oligomeric species. Consequently, dissociation cannot be represented by a single chemical equation.

The hydroxide ion must react with a petroleum acid from the crude oil to form a surfactant. A general mechanism is shown in **Fig. 7.122** and described by the following (deZabala et al. 1982).

Some of the petroleum acid in the crude oil partitions into the aqueous phase according to the solubility expression

$$K_D = [HA_o] / [HA_w], \dots\dots\dots\dots\dots\dots\dots\dots\dots\dots (7.50)$$

where K_D = distribution or partition coefficient and HA_o and HA_w denote petroleum acid in the oil and water phases, respectively. This also can be expressed as

$$HA_o \rightleftharpoons HA_w, \dots\dots\dots\dots\dots\dots\dots\dots\dots\dots\dots (7.51)$$

as shown in Fig. 7.122.

The petroleum acid dissociates in the aqueous phase according to the expression

$$HA_w \rightleftharpoons H^+ + A^-, \dots\dots\dots\dots\dots\dots\dots\dots\dots\dots (7.52)$$

as governed by the equilibrium relationship

$$K_A = \frac{[H^+][A^-]}{[HA_w]}. \dots\dots\dots\dots\dots\dots\dots\dots\dots\dots\dots (7.53)$$

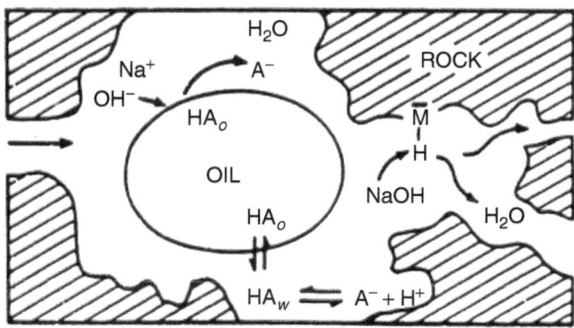

Fig. 7.122—Schematic of alkali recovery process (deZabala et al. 1982).

The species A⁻ is an anionic surface-active agent. In effect, caustic (OH⁻) consumes the hydrogen ion by the reaction

$$HA_w + OH^- \rightleftharpoons A^- + H_2O. \dotfill (7.54)$$

According to Eq. 7.54, this results in an increase in [A⁻]. The net effect of all the reactions is shown in the upper left corner of Fig. 7.122. The alkali ions shown in Fig. 7.122 are discussed later.

A measure of the potential of a crude oil to form surfactants is given by the acid number. This is the amount of KOH, usually given in milligrams, required to neutralize 1 g of the petroleum acid in the crude oil. Unfortunately, acid number does not always correlate with oil recovery.

7.14.2 Chemical Formulation and IFT. Several investigators (Campbell 1982; Jennings 1975; Burk 1987; Jennings et al. 1974) have measured the effect of alkaline chemical type and concentration on IFT between aqueous and oil phases. A typical result is shown in **Fig. 7.123** (Campbell 1982), where IFT is plotted as a function of alkali concentration. The minimum IFT value occurs in the concentration range of 0.05 to 0.10 wt%, and the minimum value is approximately 0.01 dyne/cm. There is little difference between the two alkaline chemicals tested. Other work has also shown that sodium carbonate, sodium hydroxide, and sodium orthosilicate are equally effective at reducing IFT (Burk 1987). The result that a fairly sharp minimum in IFT occurs over a narrow concentration range is typical.

Hardness in the water (i.e., the presence of Ca⁺⁺ or Mg⁺⁺) has an adverse effect on IFT reduction because Ca associates with the petroleum surfactant in the same manner as discussed in micellar/polymer flooding and becomes preferentially oil-soluble. **Fig. 7.124** (Campbell 1982) shows plots of IFT vs. alkali concentration for the same systems shown in Fig. 7.123, except that a simulated reservoir brine has been used instead of a brine of 1% NaCl (Campbell 1982). The simulated reservoir brine had a combined Ca⁺⁺ and Mg⁺⁺ concentration of 0.41%. The minimum IFT values occur at a higher concentration of alkaline chemical for the simulated brine, and the IFT reduction is diminished.

Jennings (1975) reported IFT measurements for 164 crude oils. The data were expressed in terms of a caustic coefficient and were correlated against acid number, crude density, and crude viscosity. The caustic coefficient is an empirical number

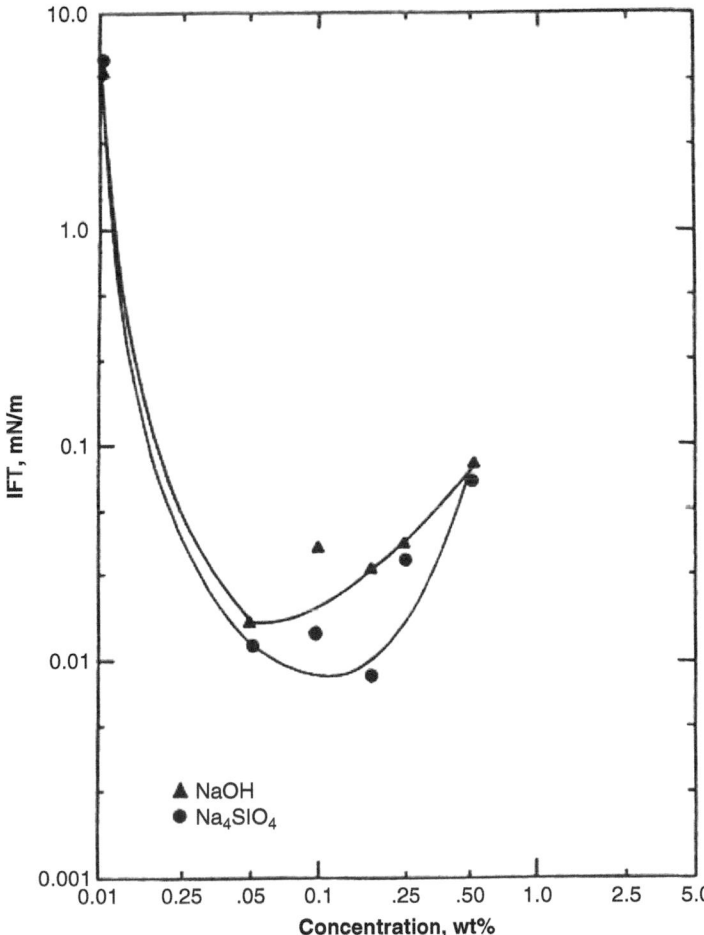

Fig. 7.123—IFT comparison in soft water. Spinning-drop IFT values for C-331 crude oil (32 cp) vs. sodium hydroxide and sodium orthosilicate in softened water with 1% NaCl; values are measured after 5 minutes of spinning (Campbell 1982).

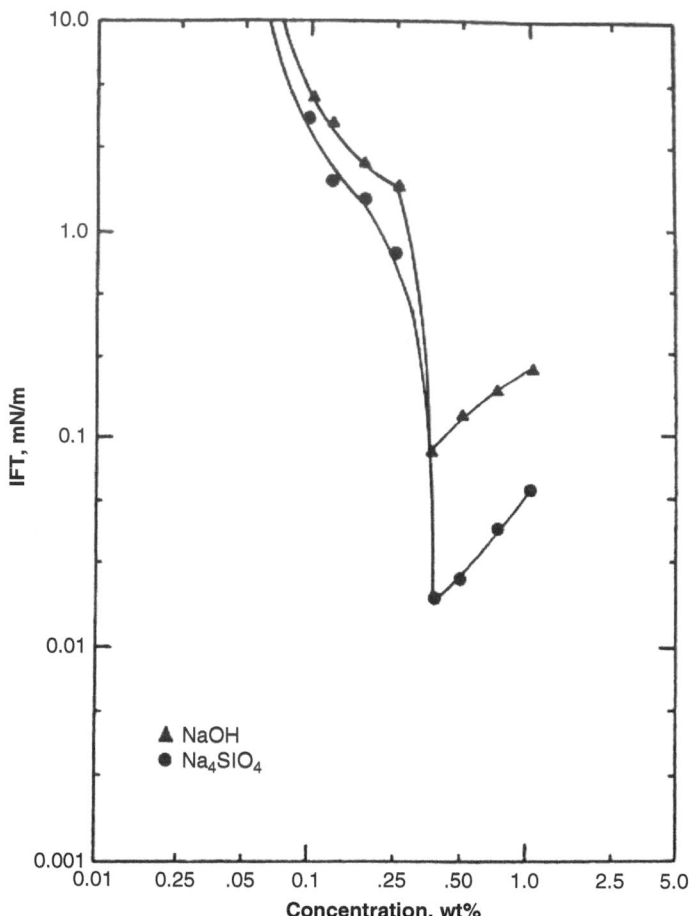

Fig. 7.124—IFT comparison in hard water. Spinning-drop IFT values for C-331 crude oil (32 cp) vs. sodium hydroxide and sodium orthosilicate in produced brine; values are measured after 5 minutes of spinning (Campbell 1982).

related to a log-log plot of IFT vs. concentration. IFT did not correlate particularly well against any of the parameters. However, almost all oils with gravities of 20 °API or less produced IFT values of less than 0.01 dynes/cm. Acid numbers for the oils ranged from essentially 0 to 5.25 mg KOH/g oil. Sodium hydroxide was the alkaline chemical. In 90% of the cases, the minimum IFT occurred in the vicinity of 0.01 wt% caustic.

IFTs for aqueous solutions of alkaline chemical and oil are functions of oil type, salinity ions, alkaline chemical concentration, and alkaline chemical type. IFT also depends on the system temperature, usually decreasing in magnitude as temperature increases.

7.14.3 Recovery Mechanisms. The effectiveness of alkaline flooding is attributed to a number of mechanisms. The four most commonly referenced are (Johnson 1976) emulsification and entrainment, wettability reversal (oil-wet to water-wet), wettability reversal (water-wet to oil-wet), and emulsification and entrapment. Other mechanisms mentioned include emulsification with coalescence, wettability gradients, oil-phase swelling, disruption of rigid films, low IFT (deZabala et al. 1982), and improved sweep resulting from precipitates altering flow (Campbell 1982). Johnson (1976) describes the first four mechanisms listed in detail with references to the original sources.

The first mechanism, emulsification and entrainment, results from reduction of IFT and the formation of an emulsion in which oil is entrained. If the emulsion is mobile, the oil saturation will decrease and oil will move through the reservoir.

Introduction of alkaline chemicals can cause a reversal of wettability from either oil-wet to water-wet or vice versa. The change in wettability and subsequent readjustment of fluids within the pores favorably affects the relative permeability to the oil phase. Discontinuous residual oil can be reconnected and caused to flow. When this wettability reversal is coupled with IFT reduction, the ROS can decrease significantly.

Jennings et al. (1974) described recovery by emulsification and entrapment. In this mechanism, an emulsion formed by decreasing the IFT is subsequently trapped by the pore throats. This, in effect, causes a reduction of flow in high-permeability zones and results in an improvement of the effective mobility ratio between displacing and displaced fluids. Viscous fingering is diminished. A series of laboratory experiments in glass-bead packs, where the displacement process could be photographed, was used to demonstrate this mechanism. In these experiments, no significant reduction in oil saturation below that resulting from a waterflood occurred. Improved oil recovery resulted from the improvement in mobility ratio.

In any specific alkaline chemical displacement, one or more of the different mechanisms may dominate the recovery efficiency. This depends on the chemical/rock system, but the process is not sufficiently understood to predict this behavior. Results from laboratory core experiments generally have shown that improved oil recovery is not directly correlated to the acid number of the crude oil. This suggests that other saponifiable constituents must exist that react with alkali to produce surface-active agents. The magnitude of IFT reduction is a better criterion (Mayer et al. 1983). IFT values between the alkaline chemical solution and the oil of approximately 0.01 dynes/cm or less are required for significant recovery. Some disagreement exists in the literature about whether the pH of the alkaline system must be high (> 11 to 12) (Campbell 1982) or can be lower (Burk 1987; French and Burchfield 1990).

Addition of polymer to an alkaline chemical solution to increase the viscosity and to provide mobility control has been shown to be effective (Burk 1987). Burk presented a series of coreflooods using sodium carbonate, sodium hydroxide, and sodium orthosilicate, which were shown to be equally effective in decreasing IFT. In alkaline floods without added polymer, tertiary recoveries of oil were approximately 8 to 17% of residual oil after waterflooding. With the addition of polymer to the alkaline solutions, recoveries increased to approximately 88 to 97% of residual oil. The increase is attributed to improved mobility ratios with polymer.

7.14.4 Rock/Fluid Interactions and Loss of Alkaline Chemical.

Alkaline chemical is consumed during a displacement process. A small amount of the chemical is used in the reaction with petroleum acids in the crude oil to form surfactants. Much of the chemical, however, can be lost through ion exchange with clays in the rock, reaction with ions (particularly in hard water), and reaction with rock mineral, resulting in mineral dissolution (Cheng 1986).

Fig. 7.122 illustrates ion exchange for a sodium hydroxide alkaline solution. The reversible ion exchange between H^+ and Na^+ is represented by

$$\overline{M}H + Na^+ + OH^- \rightleftharpoons \overline{M}Na + H_2O, \quad\quad\quad\quad\quad\quad\quad\quad\quad\quad\quad\quad\quad\quad\quad\quad\quad (7.55)$$

where $\overline{M}$ denotes an ion-exchange site on the rock (deZabala et al. 1982). For example, the hydrogen exchange capacity for a Wilmington reservoir sand was reported to be approximately 1.0 meq/100 g sand. Ion exchange is reversible and occurs relatively rapidly (Cheng 1986). Generally, it is satisfactory to assume that equilibrium exists in the exchange process. **Fig. 7.125** (Bunge and Radke 1982) shows example exchange isotherms. Adsorption by ion exchange has been shown to be larger in magnitude on montmorillonite than on illite or kaolinite (Huang et al. 1986).

Ion exchange has the same effect on transit time through a core as adsorption. The exchange process causes a delay in breakthrough that can be significant, as illustrated in Example 7.11.

Example 7.11—Delay in Breakthrough Caused by Ion Exchange. A 4-ft core is to be flooded with a caustic solution (NaOH) that is 0.1 wt%. The hydrogen-exchange capacity is 1.0 meq/100 g sand. Assuming that ion exchange occurs and that the system is always at local equilibrium, calculate the delay in breakthrough of NaOH in PV. The sand porosity is 22%, and the grain density is 2.6 g/cm³.

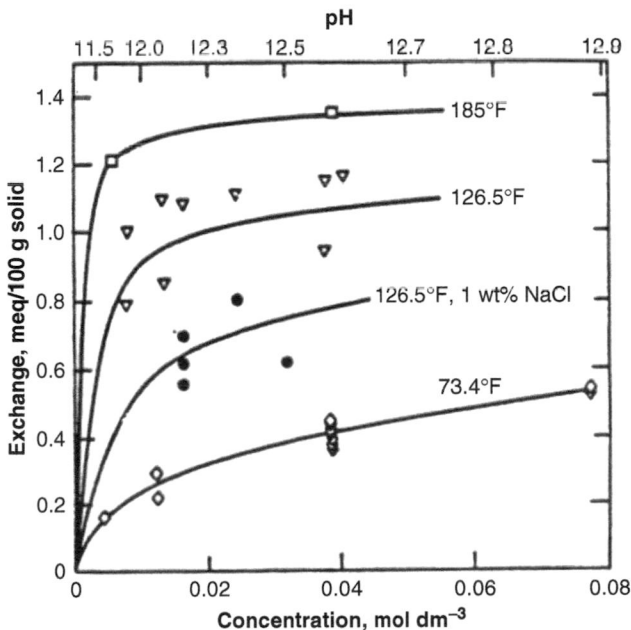

Fig. 7.125—Hydroxide-exchange isotherms for Wilmington sand with NaOH at three temperatures and two salinities (Bunge and Radke 1982).

Solution. Assume a 5.0-cm core diameter and a 100-cm core length. (The answer in PV will be independent of the assumed diameter and length.)

$$\rho_s = \frac{\pi d^2}{4} \times L \times (1-\phi)$$

$$= \frac{\pi \times (5.0)^2 \times 100 \times (1.0-0.22)(2.6)}{4}$$

$$= 3982 \text{ g sand.}$$

Total ion-exchange capacity is

$$Q_t = 3982 \text{ g sand} \times \frac{1.0 \text{ meq}}{100 \text{ g sand}}$$

$$= 39.82 \text{ meq.}$$

Total Na^+ absorbed is

$$m_{Na} = 39.82 \times 10^{-3} \text{ eq} \times M_{Na}$$

$$= 39.80 \times 10^{-3} \times 23$$

$$= 0.916 \text{ g Na.}$$

$$V_p = 431.8 \text{ cm}^3.$$

$$C_{NaOH} = 0.1 \text{ wt\%.}$$

$$m_{NaOH} = 432 \text{ cm}^3 \times 1.0 \frac{\text{g}}{\text{cm}^3} \times 0.001$$

$$= 0.432 \text{ g NaOH.}$$

$$m_{Na} = 0.432 \text{ g NaOH} \times \frac{23 \text{ g Na}}{40 \text{ g NaOH}}$$

$$= 0.248 \text{ g Na.}$$

Delay in breakthrough is

$$\Delta t_D = \frac{0.916 \text{ g adsorbed}}{0.248 \text{ g/PV}}$$

$$= 3.7 \text{ PV.}$$

Alkali reacts with Ca^{++} and Mg^{++} and can precipitate. This reaction can be advantageous and is sometimes used in chemical flooding to soften the brine. Preflushing with alkaline chemicals is common, but the reaction results in loss of alkaline chemicals (Pursley et al. 1973; Rivenq et al. 1985). Further, the precipitates can cause pore plugging. While this can be used as an in-situ permeability modification approach, it often has negative effects. For example, injectivity can be adversely affected.

Alkali can irreversibly dissolve minerals from a rock, particularly silica and anhydrite (Mayer et al. 1983; French and Burchfield 1990). Calcium and magnesium silicates are formed that remove alkali from the aqueous phase. These silicates also precipitate and can cause severe plugging. Cheng (1986) gives example effects.

7.14.5 Cosurfactant-Enhanced Alkaline Flooding. As discussed in Section 7.13.1, naphthenic acids in the crude oil can react with alkali to form soaps. The soaps usually have a low optimum salinity[10]. Consequently, it is possible to formulate surfactant systems that are under-optimum when injected, but can mix with alkaline soaps to create solutions that pass through near-optimum conditions at the alkaline surfactant front. Alkaline flooding is improved when a cosurfactant is added to the injected alkaline solution (Nelson et al. 1984; Johnson 1976); the added cosurfactant is an active, preformed surfactant.

[10]Personal communication with G. Hirasaki, 3 May 2015.

As previously discussed, IFT values for an alkaline system are lowest at low concentrations of alkaline chemical (≈ 0.10 wt%). However, losses of chemical by reactions in the aqueous phase and by rock/fluid interactions prevent propagation of a dilute alkaline fluid through the reservoir at desired rates. Alkaline loss can be offset by using higher concentrations of alkali to propagate the slug through the reservoir, but the IFT decreases as alkaline concentration increases. Thus, a compromise must be made between selecting an optimal concentration for IFT reduction and one that will propagate through the reservoir.

Nelson et al. (1984) resolved this dilemma by adding a cosurfactant to the alkaline flood and called the process "cosurfactant-enhanced surfactant flooding." They said that "at concentrations of alkali above that required for minimum IFT, the systems become overoptimal" (by analogy with micellar systems). "Excess alkali plays the same role as excess salt." They proposed that a cosurfactant can be used to increase the salinity requirement of an alkaline chemical system in a manner analogous to a micellar system. Thus, the system with a cosurfactant can be made to have an optimal alkali concentration that is larger than that without a cosurfactant present.

Nelson et al. (1984) used phase-behavior data for cosurfactant-enhanced systems to identify regions in which optimal interfacial activity and oil displacement were observed as functions of the concentration of alkaline material and cosurfactant. **Fig. 7.126** is an example of an activity map.

The term "activity" was introduced because the phase behavior with increasing salinity is not as distinct as phase boundaries in ASP phase tests in nonreactive oils. Phase behavior is screened by mixing crude oil with alkaline solutions containing fixed amounts of cosurfactant. Emulsions form as soaps are created, which have varying degrees of stability. **Table 7.22** describes the appearance of sample tubes from screening tests reported by Nelson et al. (1984). In terminology used in surfactant phase-behavior studies, over-optimum corresponds to Type II, optimum corresponds to Type III, and under-optimum corresponds to Type I.

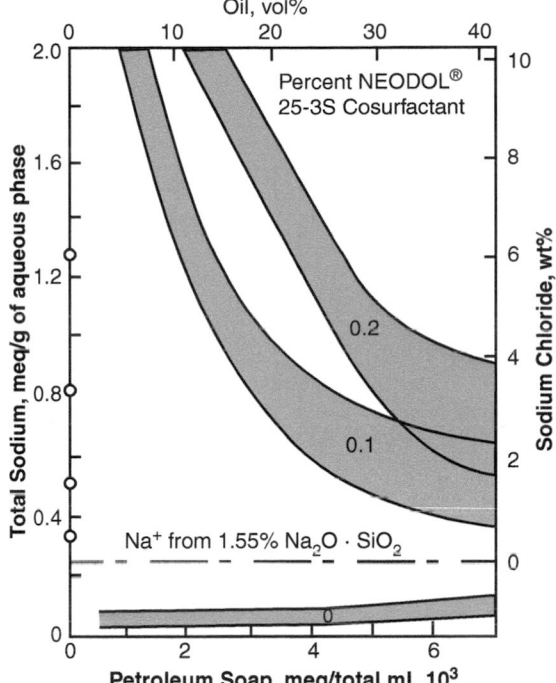

Fig. 7.126—Activity maps for 1.55% Na₂O·SiO₂ and 0, 0.1, and 0.2% NEODOL® 25-3S with a US Gulf Coast crude at 168°F (Nelson et al. 1984).

The map shows the region of greatest oil displacement activity as a function of two variables selected from among those desired to be studied. In Fig. 7.126, the two variables are total sodium present and concentration of cosurfactant. Three active regions are shown for 0, 0.10, and 0.20 wt% cosurfactant (NEODOL). The shaded regions of high activity are the concentration regions in which the process should operate for efficient oil recovery. Phase-behavior data also can be displayed on salinity requirement diagrams (Johnson 1976).

The description of phases as emulsions is commonly used. Photos of salinity scans by Liu et al. (2008) and Stoll et al. (2011) illustrate the difficulty in identifying some phase boundaries. Liu et al. (2008) were able to identify the optimal salinity by viewing phase behavior tubes through cross-polarized light. Stoll et al. (2011) identified the Type II phase boundaries readily but found other phase boundaries more difficult to determine accurately.

Table 7.22 was taken from Table 1 of Nelson et al. (1984). In subsequent research, Liu et al. (2008) demonstrated that the variation of optimal salinity with surfactant and WOR could be represented by collapsing salinity scan data into a single correlation when optimal salinity was plotted against the molar ratio of petroleum soap to cosurfactant. **Fig. 7.127**

Over-Optimum (Type II)	Viscous, water-in-oil emulsions form dark color during shaking. Oleic phase wets wall of sample tube.
	Color lightens during shaking, but darkens quickly after shaking is stopped. Oleic and aqueous phases stream separately down wall of sample tube when tube is inverted.
Optimum (Type III)	Light-colored, stable emulsion forms. May show a silvery sheen during shaking; only a trace, if any, of unsolubilized oil or water.
	Lightest-colored, most-stable emulsion *Sorm*. No unsolubilized oil or water.
	Light-colored, stable emulsions form. Only a trace of unsolubilized oil.
Under-Optimum (Type I)	Color lightens during shaking, then fades slowly after shaking is stopped. Oil drops stream as they run down the wall of sample tube when tube is inverted. Most stable foams usually form here.
	Color lightens during shaking, but fades quickly after shaking is stopped. Some solubilized oil. Remainder of oil is small drops that elongate as they run down wall of sample tube when tube is inverted.
	Dark color during shaking. Oil is dispersed in large, spherical drops.

Table 7.22—Appearance of sample tubes from screening tests reported by Nelson et al. (1984).

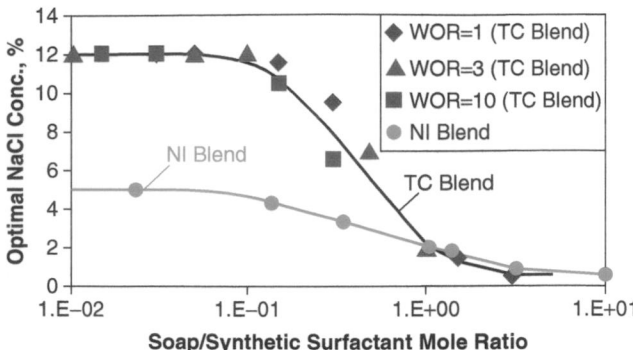

Fig. 7.127—Optimal salinity as a function of the molar ratio of petroleum soap to cosurfactant blends with MY4 crude oil at ambient temperature (Liu et al. 2008, Fig. 2).

(Liu et al. 2008, Fig. 2) illustrates data obtained by Liu et al. from salinity scans on MY4 crude oil using two surfactant blends. Blend NI contained 1 wt% Na_2CO_3, 4:1 weight ratio of N67(S1), and IOS. Blend TC was a 1:1 weight mixture of C_{12} (3EO)SO_4 and i-C_{13}(4PO) SO_4. The MY4 crude oil had an acid number of 0.2 mg KOH/g crude oil.

In Fig. 7.127, the optimal salinity of the TC cosurfactant blend is 12% NaCl and was 5% for the NI blend. The optimal salinity of the petroleum soap was 0.5%. Liu et al. (2008) point out that if cosurfactant was not added to the alkaline solution, the optimum salinity for the petroleum soap of 0.5% would be less than the salinity of most reservoirs so that over-optimum (Type II) behavior would occur. Adding cosurfactant to the alkaline solution permits adjustment of the salinity to a region in which optimum conditions could be attained during an ASP flood.

Stoll et al. (2011) show that the approach of Liu can be simulated by a thermodynamic mixing rule by Eq. 7.56, where p_{opt} is the optimal salinity of the mixture, p_A is the optimal salinity of the petroleum soaps, p_S is the optimal salinity of the cosurfactant blend, and R is the ratio of moles of petroleum soap to moles of cosurfactant:

$$P_{opt}(R) = p_A^{\frac{R}{1+R}} p_S^{\frac{1}{1+R}}. \dots\dots\dots\dots\dots\dots\dots\dots\dots\dots\dots\dots (7.56)$$

The correlation of the TC blend from Fig. 7.127 using the Salager et al. (1979) mixing rules is shown in **Fig. 7.128** (Stoll et al. 2011, Fig. 7).

An important contribution of the work of Liu et al. (2008) was the demonstration that an ASP flood could be completed in a carbonate sandpack. For this demonstration, an ASP displacement was carried out in a dolomite sandpack that was 1 in. in diameter and 1 ft long. The sandpack was saturated with 2 wt% NaCl brine and flooded with MY4 oil (19 cp) to establish the initial oil saturation. The sandpack containing MY4 oil and interstitial water saturation was aged at 60°C for 60 hours to change the wettability of the sand from water-wet to some level of oil-wetness. Then, the sandpack was waterflooded at ambient temperature with 2 wt% NaCl brine to a ROS of 0.18. An ASP formulation was prepared containing 0.2% NI blend, 1 wt% Na_2CO_3, 2 wt% NaCl, and 0.5 wt% Flopaam 3330S. The flood consisted of injecting 0.5 PV slug followed by 1 PV of polymer drive solution with the same NaCl and polymer concentrations. Viscosity of the polymer drive was approximately the same as the surfactant formulation (45 mPa·s or 45 cp).

The ASP flood **Fig. 129** (Liu et al. 2008, Fig. 20) recovered 98% of the residual oil, with 80% recovered as clean oil. Surfactant breakthrough was estimated to occur at 0.99 PV. This is the first documented ASP flood in a dolomite porous medium. Additional details are found in Liu et al. (2008).

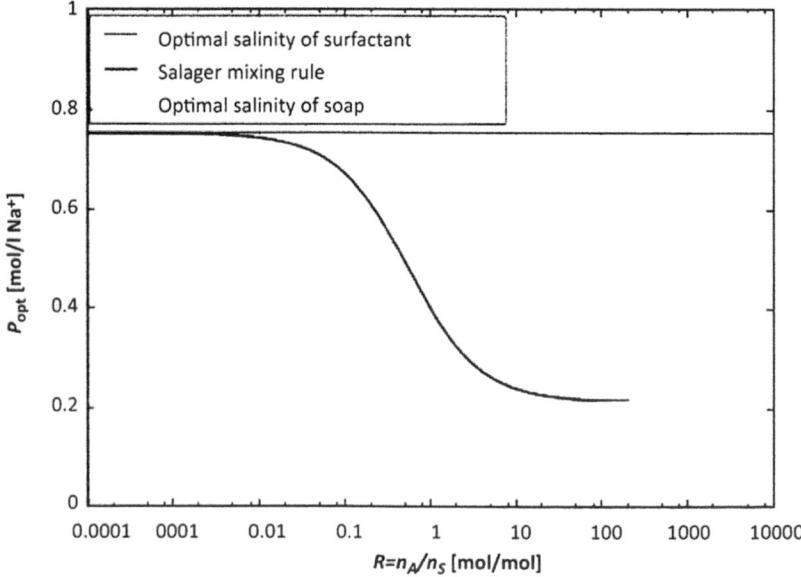

Fig. 7.128—Correlation of optimal salinity from data of Liu et al. (2006) with the ratio of petroleum soap to cosurfactant (Stoll et al. 2011, Fig. 7).

Stoll et al. (2011) describe the development of an ASP flood from laboratory to field testing in a sandstone reservoir containing a viscous crude oil (90 cp) and a high acid number (0.8 mg KOH/g oil). Reservoir temperature is 46°C, and divalent-ion content is low (Mg 16 ppm, Ca 26 ppm.)

An ASP surfactant formulation was developed from phase-behavior tests that contained 1 wt% Na_2CO_3, 0.3 wt% surfactant, 0.3 wt% Flopaam 20 million molecular weight, and 1 wt% cosolvent. The surfactant consisted of 0.225 wt% Shell ENORDET alcohol propoxy sulfate and 0.075 wt% Shell ENORDET internal olefin sulfonate. **Fig. 7.130** (Stoll et al. 2011, Fig. 4) shows data obtained during an ASP flood on a 5-cm-diameter core (30 cm in length) obtained from an outcrop of Bentheim sandstone. The surfactant flood was preceded by a 2.2-PV waterflood. In the flood 0.3 PV of ASP surfactant formulation was followed by 2.6 PV of polymer drive. The viscosity of the ASP and polymer drives was 27 mPa·s.

The data show that the oil bank arrived at approximately 2.6 PV and ended at approximately 3.4 PV, with recovery of 98% of the waterflood residual oil.

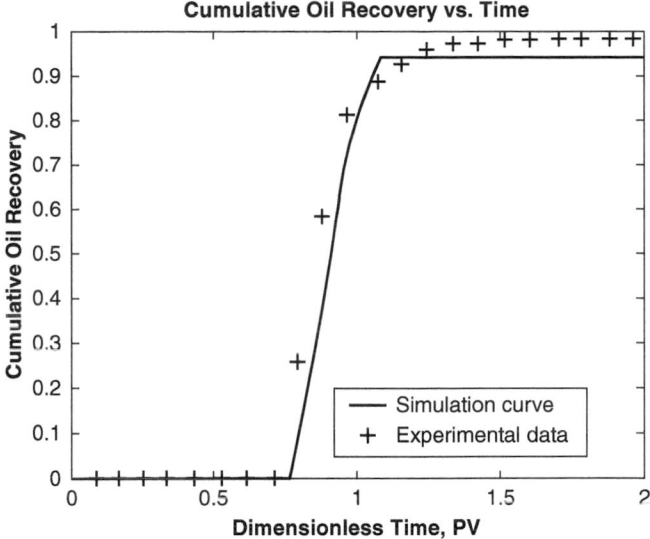

Fig. 7.129—Observed and predicted recovery from ASP flood in dolomite sandpack containing MY4 crude oil at 25°C (Liu et al. 2008, Fig. 20).

Analysis of other measurements indicated that approximately half of the oil bank was produced as clean oil, with the remaining oil produced as an emulsion. Retention of surfactant was estimated to be 45 mg/g. The solid lines in Fig. 7.130 represent results from a simulation of the displacement.

Formulation of surfactant systems for some crude oils with high wax content and viscosity is difficult to obtain high oil recovery. These crude oils typically have high acid numbers, which must be considered when designing ASP floods. Zhao et al. (2008) present results from surfactant solutions formulated for five crude oils with high wax/asphaltene/paraffin content and high viscosity at reservoir temperatures varying from 46 to 119°C. Surfactant systems containing high-carbon-number internal olefin sulfonates formulated with cosurfactants, cosolvents, and alkali have been developed, which result in high oil recovery—generally accepted to be 90% of the waterflood ROS.

The base surfactants were high-molecular-weight IOSs, which were combined with cosurfactants, cosolvents, and alkali to obtain recoveries of nearly 100% of the waterflood ROS. The IOS surfactants have the general formulas of

R—CH()H)–CH2–CH(SO3–)R', R'–CH=CH–CJ(SO3–)–R',
where R+R' = C_{12-C15} for C_{15-18} IOS,
R+R' = C_{11-23} for C_{14-26} IOS,
R+R' = C_{17-21} for C_{21-25} IOS,
R+R' = C_{21-25} for C_{24-28} IOS.

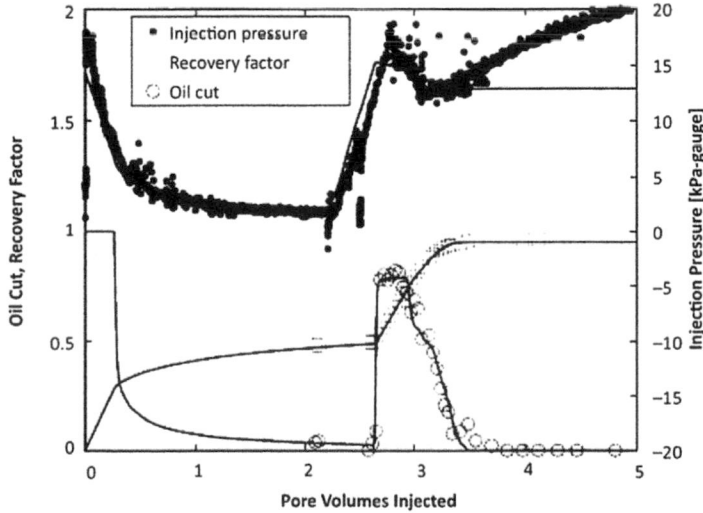

Fig. 7.130—Oil recovery of other parameters from ASP flood in Bentheim outcrop core (Stoll et al. 2011, Fig. 4).

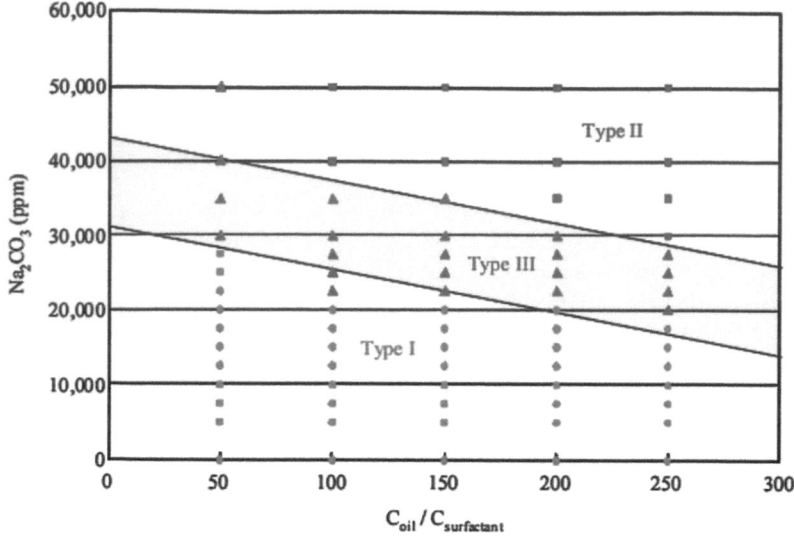

Fig. 7.131—Activity map for Crude Oil 5 at 62°C at five oil/surfactant ratios (Zhao et al. 2008, Fig. 6).

Surfactant formulations prepared using cosolvents and alkali were used to prepare activity maps before linear core floods. **Fig. 7.131** (Zhao et al. 2008, Fig. 6) is an activity map for mixtures of 2% C20-24 IOS with and without cosolvent (2% t-pentanol), with 1% $NaBO_2$ for Crude 2 at a reservoir temperature of 104°C at five ratios of oil to surfactant. Shown in Fig. 7.131 are microemulsion Regions I, II, and III, which were encountered at different Na_2CO_3 concentrations and oil/surfactant ratios. Although the map is plotted in terms of the ratio of oil/surfactant, this is equivalent to the ratio of moles of petroleum soap/moles of surfactant because the acid number of the crude oil is constant and crude-oil concentration can be expressed in terms of moles of petroleum soap, assuming that there is sufficient alkali to convert all petroleum acids to petroleum soaps. This activity map indicates that the optimal salinity of the surfactant (0 petroleum soap) is between 31,000 and 43,000 ppm Na_2CO_3.

 Fig. 7.132 (Zhao et al. 2008, Fig 9) shows the oil recovery from an ASP flood in a Berea core containing Crude 5 at water-flood ROS. The ASP surfactant composition was 0.2 wt% surfactant containing 0.1 wt% C20-24 IOS, 0.1 wt% C16 branched akylbenzene sulfonate, 1 wt% DGBE, 27,500 ppm Na_2CO_3, and 3,000 ppm Flopaam 3630S (SNF Floerger). A 0.5-PV slug was injected, followed by a polymer slug containing 2,000 ppm Flopaam 3630S. Oil recovery was 95%, leaving a ROS of 2.3% following the ASP flood.

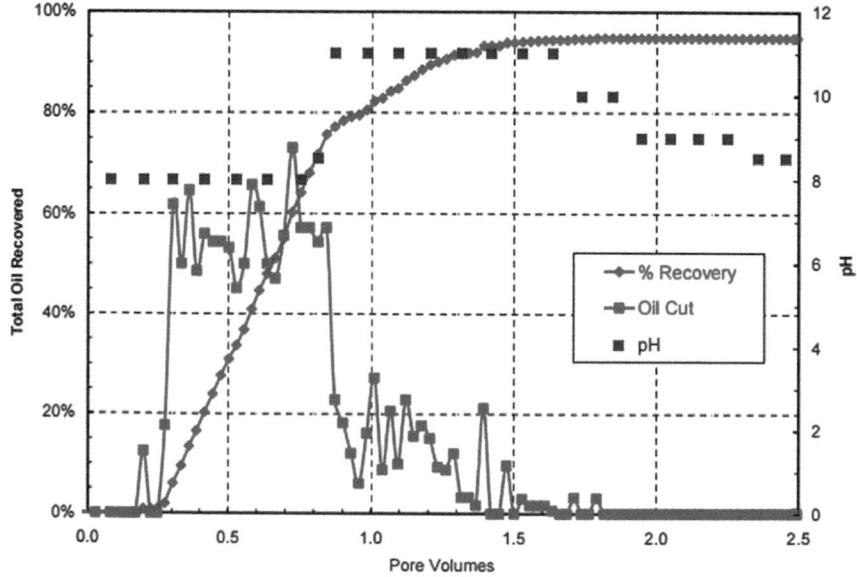

Fig. 7.132—Oil recovery and pH measurements from a surfactant flood in a Berea sandstone core containing Crude Oil 5 at ROS. The 0.5-PV ASP slug (0.1 wt% C_{20-24} IOS, 0.1 wt% C_{16} branched alkylbenzene sulfonate, 1 wt% DGBE, 27,500 ppm Na_2CO_3, and 3,000 ppm Flopaam 3630S) was followed by a polymer drive containing 2,000 ppm Flopaam 3630S (Zhao et al. 2008, Fig. 9).

Design of ASP Formulations-IFT Maps. ASP flooding is being applied in the Daqing reservoir in China, with substantial oil recovery. A different approach was used to develop ASP formulations for this field. Surfactant and polymer systems are screened to identify those compositions that appear to exhibit low IFT judging from the results of many individual samples prepared, sealed, shaken, and observed. From this process, systems that appear to be candidates are tested in detail to develop activity maps showing the variation of IFT with surfactant, alkali, and salt concentration. Multiple samples are prepared over a grid that spans the ranges of surfactant concentration, alkaline concentrations, or salinity determined by previous screening tests.

Fig. 7.133 (Gao et al. 1995, Fig. 5) shows a salinity requirement map, developed for the PS-D$_2$ sulfonate dissolved in injection water containing 1 wt% NaOH and 1000 mg/L Flopaam 3330S for an ASP system developed for use in Daqing (Gao et al. 1995). IFTs on the order of 10^{-3} were found over a large range of surfactant concentrations and salinities. PS-D$_2$ is a petroleum sulfonate.

The effectiveness of this formulation was tested in a linear ASP flood in Daqing core. The coreflooding procedure is summarized as follows from Gao et al. (1995).

1. Saturate clean Daqing core with 7000-mg/L produced water.
2. Displace produced water with dead Daqing crude oil.
3. Waterflood with 1200-mg/L injection water. Continue injection until 98% water cut is reached.
4. Inject ASP formulation for some fraction of a pore volume.
5. Inject polymer drive solution.
6. Inject 1200-mg/L injection water.

Fig. 7.134 (Gao et al. 1995, Fig. 7) shows production response and supporting data from a linear coreflood in which a 0.3-PV ASP slug was injected that contained 1 wt% NaOH, 1 wt% PS-D$_2$, and 1000 mg/L Flopaam 3330S followed by a polymer-drive solution assumed to be 1000 mg/L. Cumulative oil recovery at 98% water cut was approximately 85%. Surfactant retention was estimated to be 0.17 mg/g, and NaOH consumption was 0.30 mg/g.

A more extensive activity map is shown in **Fig. 7.135** (Li et al. 2008, Fig. 1), showing contours bounding regions of low IFT for two different surfactants. The surfactants were Sa, a proprietary surfactant made in China, and ORS-41, a linear alkylbenzene sulfonated obtained from Oil Chemical Technologies Incorporated. The alkali used was NaOH.

Polymer is always required for mobility control. Activity maps are prepared from surfactant solutions containing sufficient polymer for mobility control. Laboratory displacement tests were not reported in the paper. On the basis of this activity map, the design presented in **Table 7.23** (Li et al. 2008) was developed for implementation in the Xing2 Area.

Alkaline Cosolvent Flooding. Crude oils that have high viscosity often have acidic components that will react with alkali to form petroleum soaps. These oils have been overlooked as candidates for alkaline flooding because viscous macroemulsions form following saponification that inhibit effective displacement in porous media. Fortenberry et al. (2013) demonstrate that the use of selected cosolvents leads to the development of emulsions that have low microemulsion viscosities. The addition of a cosolvent allows phase behavior to be tailored for a particular reservoir situation. This type of alkaline flooding is designated

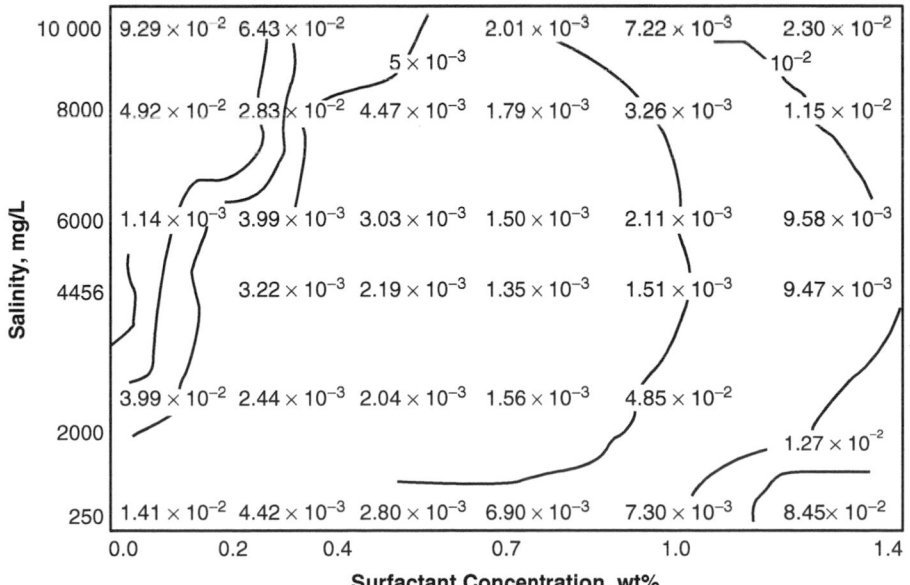

Fig. 7.133—Salinity requirement map with IFT contours for surfactant formulations containing PS-D$_2$, 1 wt% NaOH, and 1000 mg/L Flopam 3330S (Gao et al. 1995, Fig. 5).

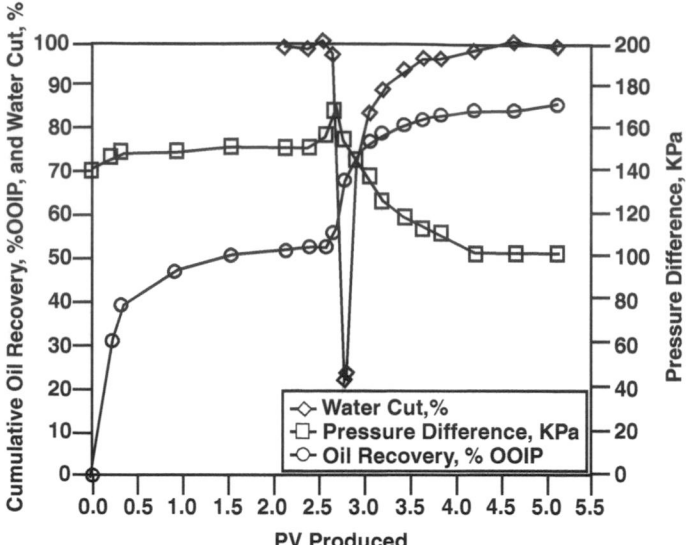

Fig. 7.134—Oil recovery, water cut, and pressure data from ASP flood in a linear Daqing core (Gao et al. 1995, Fig. 7).

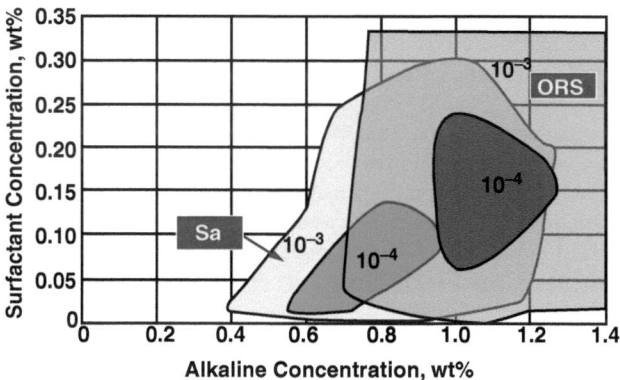

Fig. 7.135—Activity map of low IFTs for mixtures of crude oil and surfactant solution containing surfactant and alkaline (Li et al. 2008, Fig. 1).

Slug	Volume (PV)	Composition	Viscosity (mPa·s)
Preflush polymer	0.0375	1400 mg/L Polymer	40
ASP main slug	0.35	1.0% NaOH+0.2%S+650 mg/L Polymer	40
ASP sub slug	0.1	1.0% NaOH+0.1%S+1500 mg/L Polymer	35
Post-polymer	0.1	1000 mg/L Polymer	30
Post-polymer	0.1	630 mg/L Polymer	15
Waterflood		Until water cut reaches 98%	

Table 7.23—Injection plan for ASP in Xing2 Area.

as alkaline cosolvent polymer (ACP) flooding because effective formulations can be developed by adding a cosolvent to the injected alkaline fluid without a surfactant. Polymer is included in the formulation to provide effective mobility control. Fortenberry et al. (2013) present four examples of ACP systems that were formulated and demonstrated to displace more than 90% of the waterflood residual oil in Berea and Bentheim corefloods. One example based on Crude Oil #1 is included in this section.

Crude Oil #1 has a gravity of 19 °API and a viscosity of 180 cp at the reservoir temperature of 38°C. The activity diagram for Crude Oil #1 at 38°C is shown in **Fig. 7.136** (Fortenberry et al. 2013, Fig. 1). The activity diagram for Crude Oil #1 when 3% IBA-10EO cosolvent was added to the alkaline solution is shown in **Fig. 7.137** (Fortenberry et al. 2013, Fig. 2). Including cosolvent expands the Type III region, enhancing the capability of a broad region in which low IFT is present. A critical factor is the effect of the cosolvent on the emulsion viscosity at optimum salinity, shown in **Fig. 7.138** (Fortenberry et al. 2013, Fig. 3) at 30% oil concentration. Emulsion viscosity was reduced by a factor of 10. **Fig. 7.139** (Fortenberry et al. 2013, Fig. 9) shows the results of an ACP flood in reservoir core when 0.3 PV of an alkaline solution containing 30,000 ppm Na$_2$CO$_3$, 3 wt% IBA-10EO cosolvent, and 4,000 ppm FP 3630S is displaced through the core with a 1.6-PV polymer drive containing 4,000 ppm FP 3630S in 10,000-ppm Na$_2$CO$_3$.

7.14.6 Design and Field Experience. The design approach is similar to that for a micellar/polymer process, particularly if the cosurfactant-enhanced alkaline process is to be applied (Nelson et al. 1984). The acid number of the crude oil is the only direct measure of the surfactant-forming tendency of a crude oil, although it is recognized that the acid number does not correlate directly with oil recovery (French and Burchfield 1990). The reasons for this are not understood. In general, the process is not recommended if CO$_2$ content is high, if gypsum content is greater than approximately 0.1 wt%, or if the montmorillonite clay

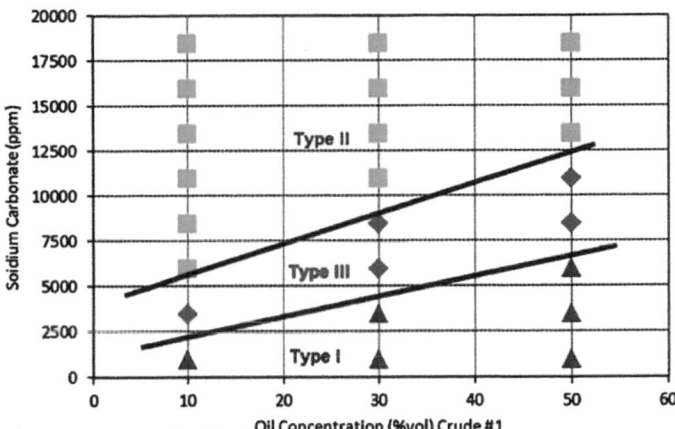

Fig. 7.136—Activity diagram for Crude Oil 1 without cosolvent at 38°C (Fortenberry et al. 2013, Fig. 1).

content is high (French and Burchfield 1990). Low-pH alkalis should be considered for carbonate formations (French and Burchfield 1990). Phase behavior and IFT as functions of concentrations should be measured. Plots of IFT vs. alkali concentration, salinity requirement diagrams, or activity maps can be used as appropriate.

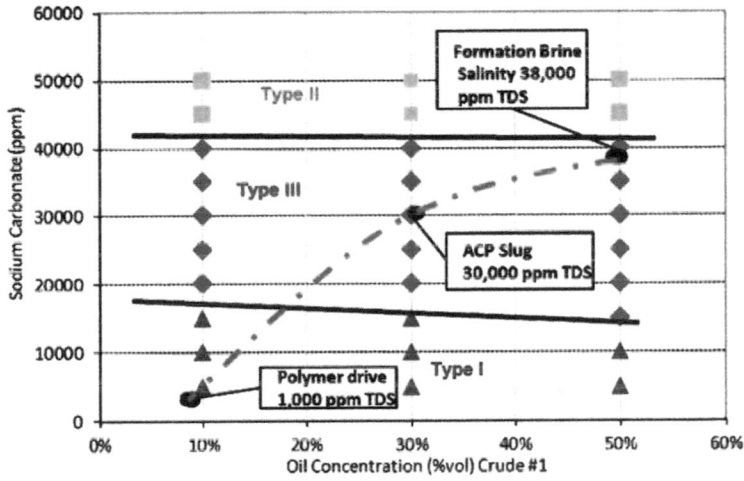

Fig. 7.137—Activity diagram for Crude Oil 1 with 3% IVBA-10EO cosolvent at 38°C (Fortenberry et al. 2013, Fig. 2).

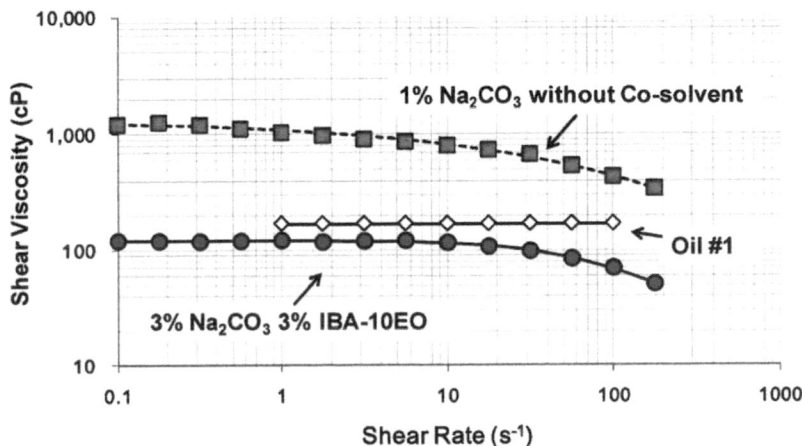

Fig. 7.138—Emulsion (30% oil) viscosity at optimum salinity and 38°C with and without 3% IBA-10EO cosolvent, Crude Oil 1 (Fortenberry et al. 2013, Fig. 3).

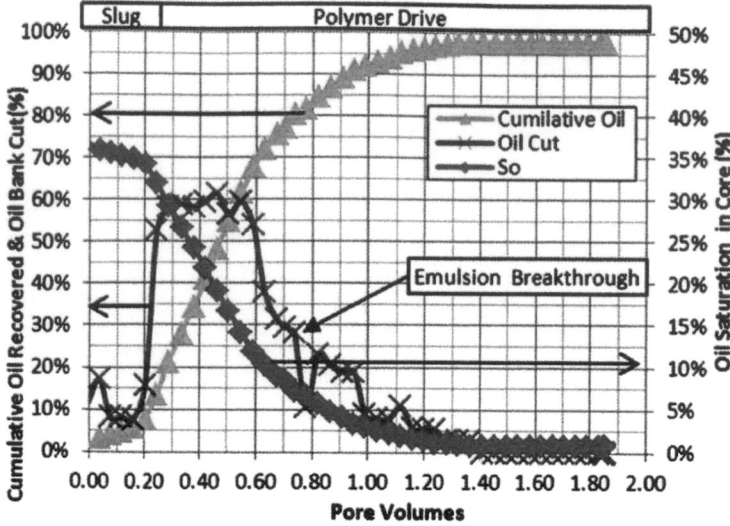

Fig. 7.139—ACP-1.2 oil recovery-reservoir core (Fortenberry et al. 2013, Fig. 9).

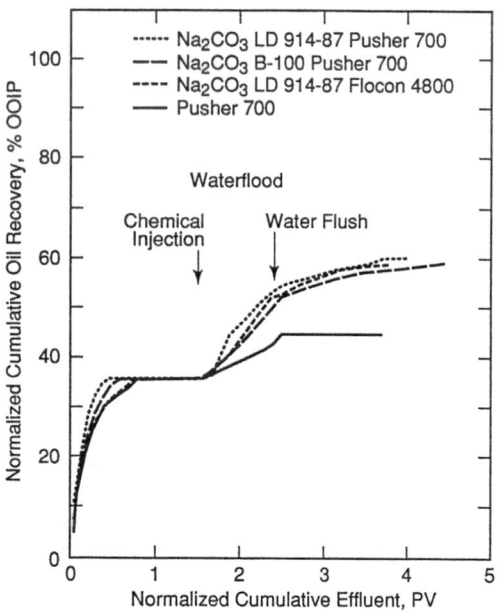

Fig. 7.140—West Kiehl radial sandpack coreflood comparison of normalized recoveries (Clark et al. 1993).

Design steps should include measurements of alkali consumption by the reservoir rock and brine. Actual preserved core samples from the reservoir to be flooded should be used in laboratory corefloods. Corefloods should be run at reservoir temperature, and pressure drop per unit length should be of a magnitude close to that which will occur in the field (Johnson 1976). The next subsection summarizes early alkaline flooding tests in the United States. Extensive application of alkaline flooding has occurred in the Daqing oil field in China. These applications are described in the following subsection.

Field Tests in the United States. **Fig. 7.140** shows displacement performance for a waterflood followed by a cosurfactant-enhanced polymer flood in Berea sandstone for the West Kiehl field (Clark et al. 1993). Waterflood recovery was approximately 36% OOIP. Three cosurfactant formulations tested averaged 23% OOIP beyond the waterflood recovery, giving a total recovery of approximately 59% OOIP.

A number of field tests have been described in the literature (Gogarty 1983b; Mayer et al. 1983; Johnson 1976; Doll 1988a, b; Graue and Johnson 1974; Dauben et al. 1987). **Table 7.24** presents properties and selected operating data for three alkaline projects.

Field test results have been reported for two cosurfactant-enhanced alkaline floods. Both tests used a single well for injection. A cosurfactant-enhanced alkaline flood began in the West Kiehl field in December 1987 (Meyers et al. 1992; Clark et al. 1993). This flood was conducted in the Minnelusa Lower B sand, which has an average thickness of 11 ft, average porosity of 23%, and average permeability of 350 md. Locations of the injection and production wells are shown on the net porosity-foot isopach map in **Fig. 7.141** (Meyers et al. 1992). Injection of alkaline water began at the end of primary recovery, so the cosurfactant-enhanced alkaline flood was in fact a secondary recovery process in this field. From December 1987 to July 1990, approximately 501,000 bbl of alkaline/surfactant solution was injected. This was followed by 122,926 bbl of polymer solution, which was tapered at the end of the injection period to drive water injection. Incremental recovery from the cosurfactant-enhanced alkaline polymer flood was estimated by subtracting the estimated waterflood response from the total oil production. The ultimate incremental oil from the ASP flood was estimated to be 256,508 bbl, or 12.9% OOIP. Recovery is higher in the swept regions because only part of the reservoir volume was contacted by the ASP solutions.

Shell Oil Company conducted the second test of cosurfactant-enhanced alkaline flooding in the White Castle field, Louisiana, USA (Falls et al. 1994). The test was conducted in the 1.5-acre Q sand of this field, which was at residual saturation (S_{or} = 0.20) following waterflooding. This test was planned to evaluate injection facility design, to determine the injectivity of process slugs, to establish the degree of gravity segregation during the process, to evaluate the effectiveness of this process to recover residual waterflood oil, to measure alkali consumption and surfactant retention, and to obtain information about

	Reservoir Name (Field)		
	Second and Third Zones (Whitter)	Ranger (Wilmington)	Almy (Isenhour)
Lithology	Sandstone	Sandstone	Sandstone
Area in flood (acres)	63	–	173
Pattern type	Line drive	Staggered line	–
Spacing (acres)	1 to 2	≈ 12	29
Tracer study	Yes	–	–
Permeability (md)	320 to 495	240	21
Porosity (%)	30	25	15.5
Thickness (ft)	137	320	15
Depth (ft)	1,500 to 2,100	2,225 to 2,800	3,570
Temperature (°F)	120	125	97
Brine	Fresh, hardness <1	Fresh, hardness <1	–
Crude Oil	–	–	2.8
Viscosity (cp)	40	23	43
API gravity (°API)	20	17	–
Acid number (mg KOH/g)	–	2.05	–
Geology	Steep-dip fault isolates flooded area	–	–
Heterogeneity (Dykstra-Parsons)	0.66 to 0.74	0.70	–
Tertiary or secondary flood	Secondary	Tertiary	Secondary
Oil saturation at start of flood (%)	51	51	35
Chemical(s) injected			
Type	NaOH	Orthosilicate	Na_2CO_3 polymer
Concentration (wt%)	0.2	0.4	–
Slug size (% PV)	20	67	37
Recovery (incremental) (%)	5 to 7	–	–
Incremental oil/chemical injected (bbl/lbm chemical)	0.32 to 0.43	–	0.13 (estimated)
Date started	October 1966	January 1979	September 1980
References	Mayer et al. (1983); Graue and Johnson (1974)	Mayer et al. (1983); Dauben et al. (1987)	Doll (1988a, b)

Table 7.24—Data on selected alkaline flooding field applications.

treating produced fluids by use of conventional production facilities. Polymer was not used for mobility control in this test. The cosurfactant-enhanced alkaline slug was injected into Well 264, shown in **Fig. 7.142,** which is downdip from two producing wells, Wells 267 and 269. Falls et al. (1994) give details of the slug compositions. A 0.269-PV slug of alkaline solution was injected followed by drive water. Nearly 80% of the oil was produced from Well 267. Oil recovery from the reservoir volume swept by preflood tracers was 57% of the residual waterflood oil. Incremental oil recovery above the 0.5% oil cut that existed in producing wells at the beginning of the flood was 38% of the waterflood residual oil in the swept region. Shahin and Thigpen (1996) describe a companion polymer injection test.

ASP Field Tests—Daqing Oil Field. Daqing oil field, a sandstone deposit (Chang et al. 2006) with multiple sand intervals, was found in a structure that is 90 miles long, 6 miles wide, and 2,300 to 3,900 ft deep. The reservoir's temperature is 45°C (113°F), and it contains oils with live viscosities of approximately 9 to 10 cp at reservoir temperature. Dead-oil viscosity is 25 cp at reservoir temperature (Gao et al. 1995). Oil density is 0.91 g/cm^3 (Gao et al. 1995) but varies as low as 0.834 to 0.851 in different zones in the Daqing oil field (Wang et al. 1999). TDS in the reservoir brine range from 5,000 to 7,000 ppm. The acid content is 0.1 mg KOH/g of oil (Gao et al. 1995; Wang et al. 1999). Estimated OOIP is 36 billion bbl.

ASP flooding has been carried out in the Lamadian (L), Saertu (S), and Xinshungang (X) reservoirs. Eight ASP pilot tests have been completed in the Daqing oil field (Wang et al. 1999; Chang et al. 2006). **Fig. 7.143** (Chang et al. 2006, Fig. 2) is an outline of the field showing the location of the ASP pilots.

Table 7.25 contains a summary of the pilot tests. In Table 7.25, the well spacing is the distance between the injection and production wells. In Table 7.25, Slug 1 is a preflush with a polymer solution, and Slugs 2 and 3 are ASP solutions with different

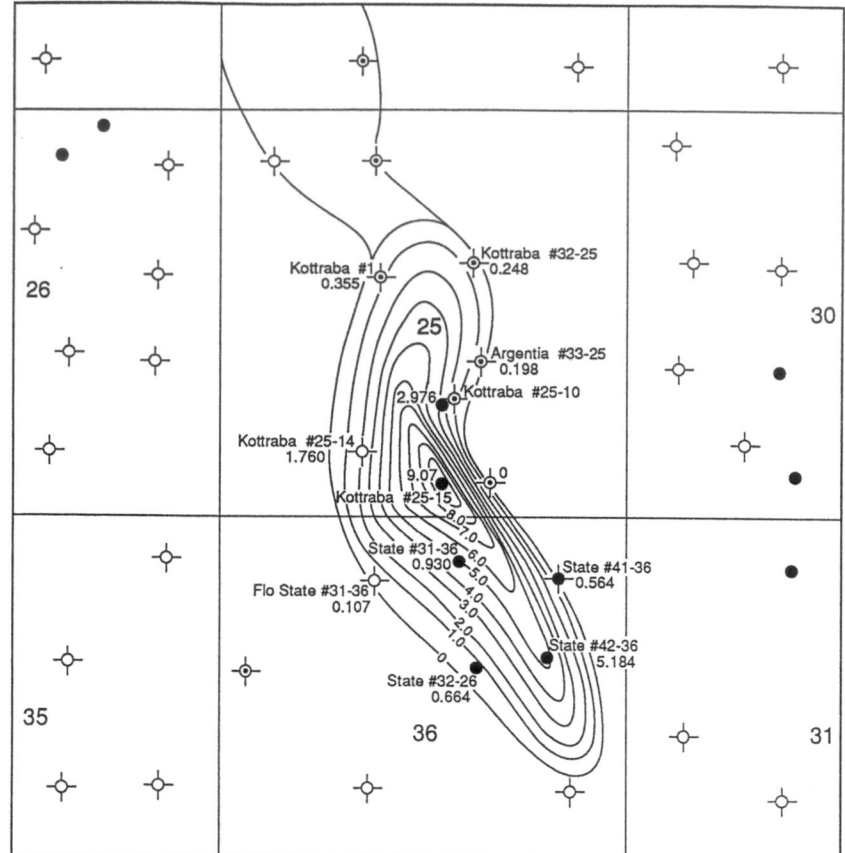

Fig. 7.141—West Kiehl net porosity-foot isopach (Meyers et al. 1992).

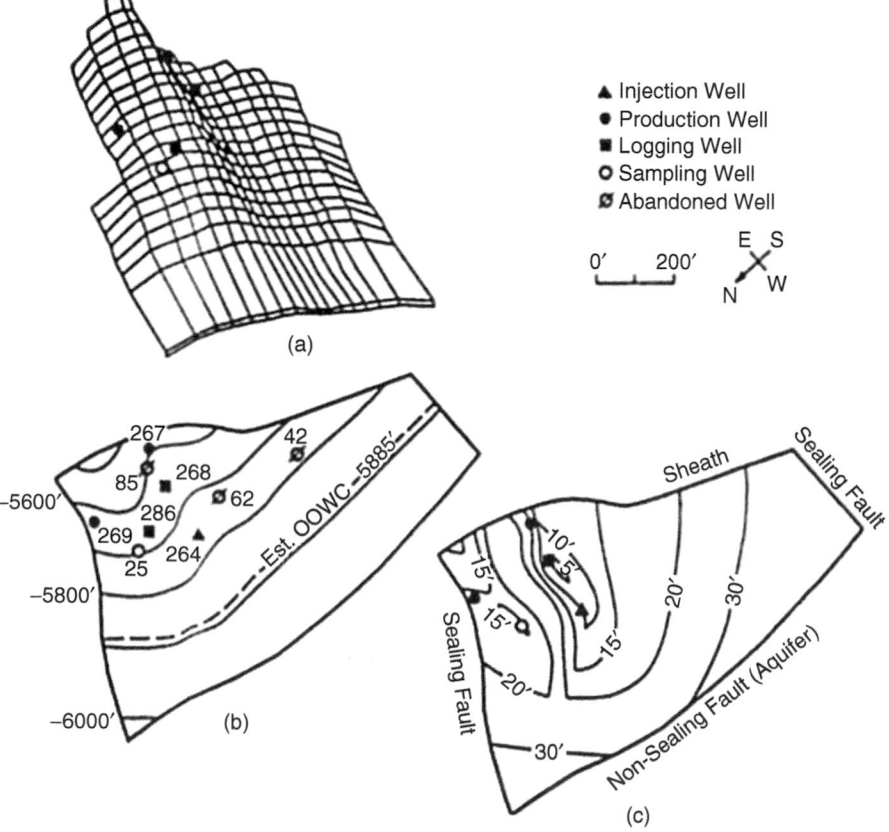

Fig. 7.142—White Castle Q sand reservoir: (a) 3D representation of structure, (b) elevation contours and well identifications, and (c) stratigraphic net pay (Falls et al. 1994).

chemical compositions. Slug 4 is the polymer-drive solution, which is not used in some pilot tests.

ASP Pilot in West Central Saertu, Daqing Oil Field. As described in the previous subsection, eight ASP floods have been completed in the Daqing oil field. Some are continuing as commercial applications. The ASP pilot in the West Central Saertu is well-documented and was the precursor of five additional ASP pilots (Wyatt et al. 2004), followed by three industrial tests (Zhu et al. 2012). The pilot is located on the west section of the Saertu anticline (Gao et al. 1996) and contains several productive intervals. Average depth of the reservoir is 814 m; the reservoir has a gentle dip from east to west. The pilot area (Po) is shown in **Fig. 7.144** (Gao et al. 1996, Fig. 1). Reservoir and fluid properties in the Po Pilot Area are summarized in **Table 7.26** (Gao et al. 1996).

The target interval for the ASP pilot is the SII_{1-3}, which has an average thickness of 10.5 m and a net sand thickness of 8.6 m. The SII_{1-3} interval contains three different strata (SII_1, SII_2, and SII_3), which are noncommunicating except at wells where all intervals are open. The net thickness of each layer is summarized in **Table 7.27**. The Dykstra-Parsons permeability variation is between 0.5 and 0.8 based on cores from Wells Po4 and Po5.

Before development of the ASP pilot area, the reservoir was waterflooded by a line drive, with wells Zhong 3-6, 3-8, 3-10, and 3-12 forming the upper line of injectors and wells Zhong 7-6, 7-8, 7-10, and 7-12 forming the lower line of injectors. The effectiveness of the line-drive waterflood was estimated from the evaluation of cores and logs from the 13 wells in the pilot area. Table 7.27 summarizes the analysis. The flooded intervals were located at the bottom of the interval. Volumetric sweep of the line-drive waterflood was approximately 38% on the basis of the water-contacted thickness in **Table 7.28**.

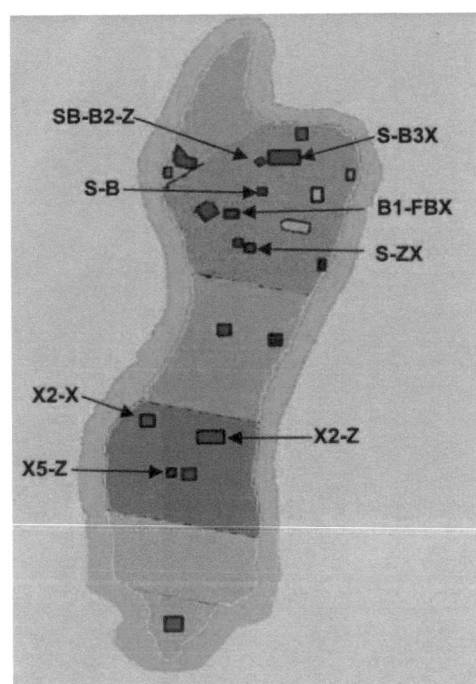

Fig. 7.143—ASP pilot test locations in the Daqing oil field (Chang et al. 2006, Fig. 2).

Fig. 7.145 (Gao et al. 1996, Fig. 2) shows the net thickness contours of the Po area, with the results of tracer tests superimposed on the contour map. Tracers were injected into wells Po1–Po4 and produced at wells Po5–Po13. Breakthrough of tracers occurred from 28 to 49 days after injection in wells in which tracer was detected, indicating rapid fluid movement and good communication between injection and responding wells. Estimated tracer velocities between wells ranged from 2.5 and 3.5 m/d, which is consistent with the average air permeability of 1.43 μm^2. The tracer injected into Well Po4 was not detected in Wells Po5 and Po12, indicating poor transmissibility between those wells. Tracer injected into Well Po3 was not detected in Well Po10. Other properties of the reservoir and reservoir fluids are summarized in Table 7.25.

The pilot area was waterflooded from 25 June 1993 to 23 September 1994. Incremental oil attributed to the pilot waterflood was 8400 m³ based on reservoir simulation. Total oil produced before the ASP pilot began was 42 000 m³. This left an average oil saturation of 0.52 in the pilot area as the target for ASP flooding.

The pilot area was waterflooded from 25 June 1993 to 25 September 1994 at four inverted five-spot patterns with Wells Po1, Po2, Po3, and Po4 as injectors, shown in **Fig. 7.146** (Gao et al. 1996, Fig. 2a). Because the pilot area had been subjected to the line-drive waterflood, the water cut in the test area was 68% in July 1993 and increased to 82.7% in September 1994. Additional oil recovered from the waterflood was estimated to be 8000 m³ on the basis of reservoir simulation (Gao et al. 1996). Total oil produced from the pilot area before the ASP pilot began was 42 000 m³, or 30.6% OOIP. The average oil saturation remaining in the pilot area was 0.52.

Number	Location	Spacing (ft)	Wells (Injector/ Producer)	Start Date	Slug 1 (Vp)	Slug 2 (Vp)	Slug 3 (Vp)	Slug 4 (Vp)	Incremental Recovery (%OOIP)
ASP 1	S-ZX	348	4/9	9/1994	0.30	0.29	–	–	21.40
ASP 2	X5-Z	462	1 /4	1/1995	0.30	0.30	0:18	–	25.00
ASP 3	X2-X	656	4/12	9/1996	0.04	0.35	0.10	0.25	19.40
ASP 4	S-B	246	3/4	12/1997	0.33	0.15	0.25.	–	23.24
ASP 5	B1-FBX	820	6/12	3/1997	0.30.	0.15	0.20	–	20.63
ASP 6	X2-Z	820	17/27	4/2000	0.04	0.35	0.10	0.20	Ongoing
ASP 7	SB-B2-Z	246	3/4	10/2004	0.04	0.35	0.15	0.20	Ongoing
ASP 8	S-B3X	820	–/13	8/2002	0.04	0.35	0.10	0.20	Ongoing

Table 7.25—Summary of ASP pilot tests conducted in Daqing oil field (Chang et al. 2006).

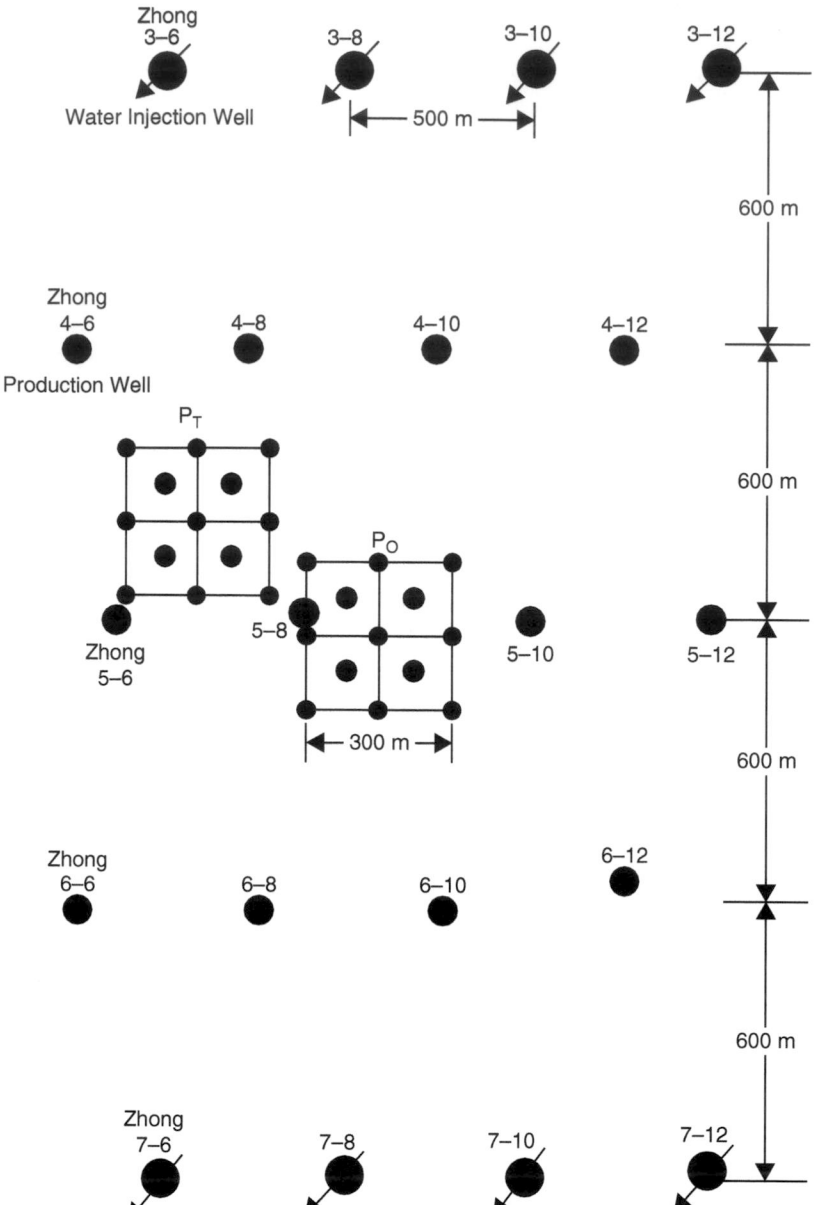

Fig. 7.144—Well layout in Saertu ASP pilot area (Gao et al. 1996, Fig. 1).

The ASP flood began in September 1994 with the injection of 0.327 PV of ASP solution containing 1.25% Na₂CO₃, 0.3% (active) Petrostep B-100, and 1200 mg/L Alcoflood 1275A in fresh water. The ASP slug was followed by 0.273 PV of polymer drive containing 800 mg/L in fresh water. The polymer drive was followed by water injection. Details and quantities are given in Gao et al. (1996).

Production wells responded to ASP injection within 0.05 PV of fluid injected. **Fig. 7.147** (Gao et al. 1996, Fig. 3) shows the oil rate and oil cut from Well Po5. Oil rate increased from approximately 4 m³/d and peaked at approximately 27 m³/d. **Table 7.29** shows the injection profile from Well Po3 measured after ASP flooding (Wang et al. 1999), showing good distribution of injected fluids between the three sublayers. Wells Po7 and Po12 had little response to the ASP flood, indicating that the areas around those wells were not contacted by the injected ASP fluid.

All other wells produced additional oil. Production data are presented in Gao et al. (1996). **Fig. 7.148** (Gao et al. 1996, Fig. 11) shows the oil rate and oil cut from the Po pilot area during the project.

Total pilot production as of March 1996 was 25 300 m³ at an oil cut of 29.7%. This is equivalent to 18.3% OOIP with an estimated oil cut of 29.7%. **Fig. 7.149** (Wyatt et al. 2004, Fig. 15) is the projection of total recovery from the Po pilot estimating the incremental recovery at 31% OOIP. Actual recovery was reported to be 21.4% (Zhu et al. 2012) or 29 532 m³. **Table 7.30** summarizes the oil production from the Po area.

Sandstone thickness (m)	10.5
Net thickness (m)	8.6
Depth (m)	814
Temperature (°C)	45
Oil density (g/cm³)	0.85
Oil viscosity (mPa·s)	11.5
Oil acid number (mg KOH/g)	0.01
Connate water salinity (mg/L)	6000
Freshwater salinity (mg/L)	600
Average air permeability (µm²)	**1.43**
Average effective porosity (%)	26
Total pilot PV (m³)	203 300
OOIP (stock tank) (m³)	138 000
Formation volume factor (res m³/std m³)	1.102
Oil volume (res m³)	152 076
Original oil saturation (%)	74.8

Table 7.26—Reservoir and fluid properties, West Central Saertu.

Layer	Thickness (m)	% Total Net Thickness
SII₁	2.8	23.6
SII₂	5.2	60.4
SII₃	0.6	7.0
Total	8.6	100

Table 7.27—Net thickness in pilot area.

Interval	Thickness (m)
Unflooded thickness	5.3
Weakly flooded	1.8
Moderately flooded	1.0
Strongly flooded	0.5

Table 7.28—Assessment of waterflood effectiveness on the basis of core and well-log analysis for 13 pilot pattern wells.

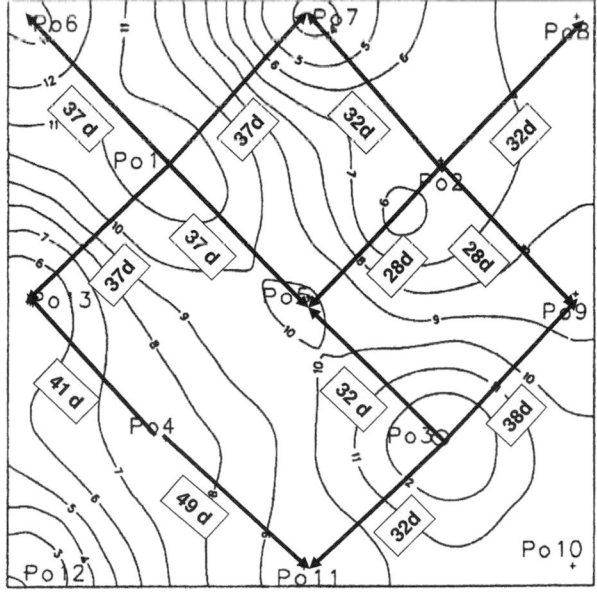

Fig. 7.145—Isopach of pilot area with tracer times between wells superimposed (after Gao et al. 1996, Fig. 2).

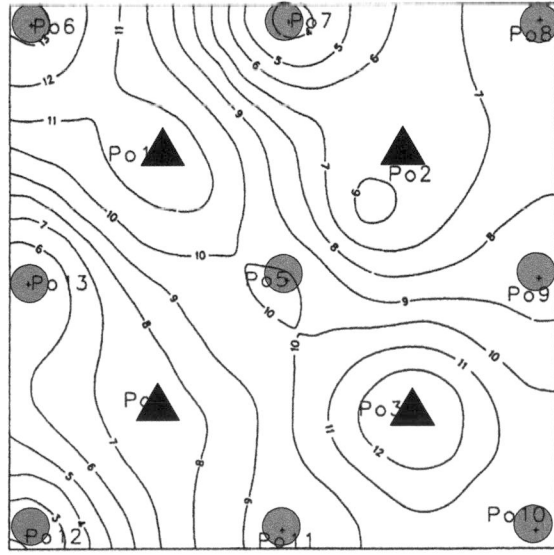

Fig. 7.146—Four inverted five-spot patterns for pilot waterflood and ASP pilot (after Gao et al. 1996, Fig. 2).

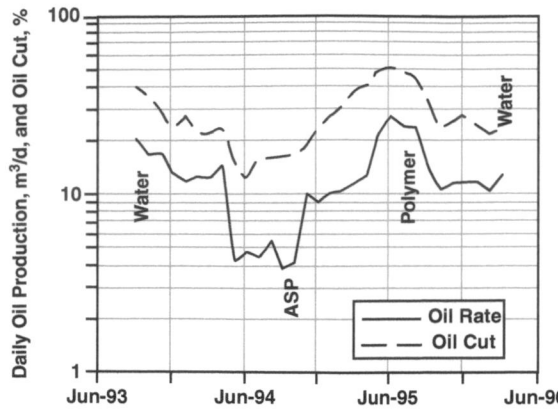

Fig. 7.147—Production response of Well Po5 to waterflood and ASP flood (Gao et al. 1996, Fig. 10).

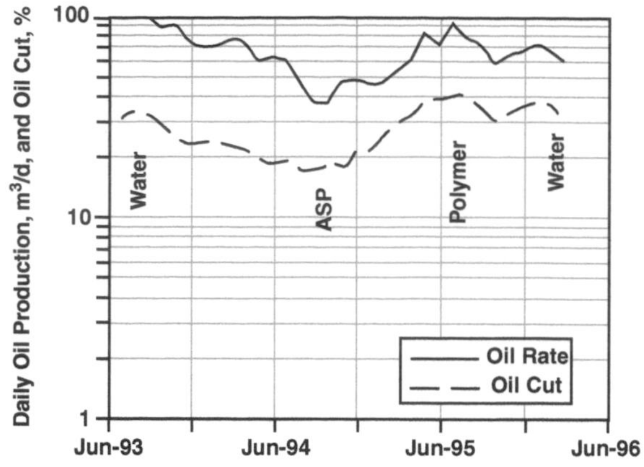

Fig. 7.148—Oil rate and oil cut from Po pilot (Gao et al. 1996, Fig. 11).

Sublayer	Percentage of Fluid Intake (%)
SII1	38.5
SII2	34.2
SII3	28.2

Table 7.29—Fluid distribution in Injection Well Po3 after ASP flooding.

The Po pilot demonstrated that the amount of incremental oil attributed to ASP was 21.4% above the recovery by waterflooding. Substantial incremental oil was recovered using an ASP flood, which led to five additional pilots and, ultimately, three industrial-scale field trials (Wyatt et al. 2004).

ASP Field Tests—PetroChina. Successful ASP pilot tests described in the previous subsections resulted in the testing of ASP flooding in large-scale field tests, followed by commercial application of ASP flooding in four blocks in the Daqing oil field. **Table 7.31** (Zhu et al. 2012) summarizes the results of four large-scale field tests. The first large-scale test began in May 2000; subsequent large-scale tests followed in 2005, 2006, and 2008.

Commercial application of ASP flooding was implemented in the Xin 1-2 Region, the South 6 Region, the East 1 Block of the Xin 6 Region, and the East II Block of the Xin 6 Region during the period from 2007–2009. **Table 7.32** summarizes the available data for these projects.

These projects represent the largest application of ASP in the world and provide operating experience to guide the development of chemical flooding.

Operating Problems. *Scale and Corrosion.* In small pilot projects, operating problems such as scaling and the presence of emulsions in produced fluids are resolved on a small scale to avoid interference with the primary objective of the test, which is usually to determine the amount of incremental oil that can be produced by the process. These problems are magnified as the size of the projects increases, transitioning from 10 to 15 producing wells in small pilots to 44 to 102 producing wells in large-scale pilots to 214 to 203 producing wells in PetroChina's commercial-scale operations.

Scaling and corrosion caused by alkaline flooding are major operating problems (Zhu et al. 2012). Both affect the performance of the pumps, which results in frequent maintenance and increased costs. **Table 7.33** summarizes the history of

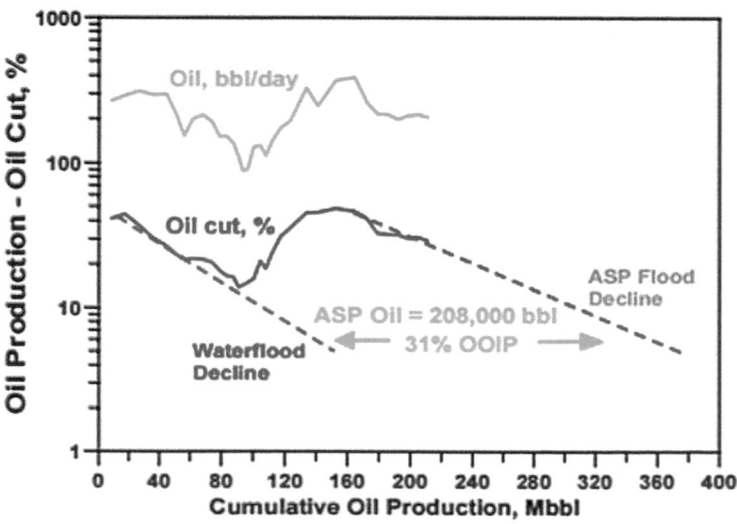

Fig. 7.149—Projection of Po pilot area production (Wyatt et al. 2004, Fig. 15).

Production Period	Oil Produced (m³)	Oil Produced (rm³)	% OOIP	Average Remaining Oil Saturation
Initial reservoir	–	–	–	0.748
Line-drive waterflood (estimated oil displaced from pilot area before drilling pilot wells)	33,800	37,248	24.5	0.565
Five-spot waterflood (estimated by simulation)	8,400	9.257	6.1	0.52
ASP pilot (Zhu et al. 2012)	29,532	32,544	21.4	0.36
Total	71,732	79,049	52.0	–

Table 7.30—Summary of oil production from Po Pilot Area, West Central Saertu.

Projects	Start Time	Well Number and Pattern	Well Spacing (m)	Effective Thickness (m)	Effective Permeability (μm²)	ASP Formula	Predicted Improved Recovery (%OOIP)
Strong alkali ASP flooding of middle part of Xing-2 Zone	2000.5	17 injectors 27 producers Five spots	250	7.8	0.85	HABS + NaOH +HPAM	18.0
Strong alkali ASP flooding of South-5 Zone	2005.7	29 injectors 39 producers Five spots	175	10.0	0.867	HABS + NaOH +HPAM	18.6
Strong alkali ASP flooding of North-I East Zone in Type II layer	2006.6	49 injectors 63 producers Five spots	125	7.7	0.670	HABS +NaOH + HPAM	22.5
Weak alkali ASP flooding of North-2 West Zone in Type II layer	2008.11	35 injectors 44 producers Five spots	125	6.6	0.533	PS +Na₂CO₃ +HPAM	19.3

Table 7.31—ASP flooding extensive field tests in PetroChina (Zhu et al. 2012).

Blocks	Start Time	Well Number	Well Spacing (m)	Thickness of Sandstone (m)	Effective Thickness (m)	Effective Permeability (μm²)	Geological Reserves (10⁴ m³)
Xin 1-2 Region	2007.8	112 injectors/ 143 producers	150	10.4	8.6	0.517	666.5
South 6 Region	2009.1	144 injectors/ 160 producers	175	13.2	10.7	0.539	1273.6
East I Block of Xin 6 Region	2009.6	102 injectors/ 129 producers	141	6.8	5.7	0.550	469.0
East II Block of Xin 6 Region	2009.10	105 injectors/ 109 producers	141	7.3	5.7	0.515	452.3

Table 7.32—ASP flooding commercial-application blocks in PetroChina (Zhu et al. 2012).

	2007			2008			Jan-Sept 2009		
Classification	Times	Pump Checking Rate (%)	Checking Period (day)	Times	Pump Checking Rate (%)	Checking Period (day)	Times	Checking Rate (%)	Checking Period (day)
Scaling	13	20.6	155	42	66.7	126	23	36.5	162
Not scaling	49	77.8	349	35	55.6	227	17	27.0	264
Total	62	98.4	308	77	122.2	158	40	63.5	205

Table 7.33—Statistics of pump-checking situations in Narrth-1 East Test Zone (Zhu et al. 2012).

pump-checking maintenance from January 2007–September 2009 in the North-1 East test zone, where there are 63 production wells.

Emulsions. Emulsions accompany alkaline chemical flooding and contribute to the displacement mechanism, as discussed in Section 7.14.3. In reactive oils, emulsions form easily in ASP floods because of the tendency of alkali with reactive constituents of oils to form soaps. Emulsions range from oil in water (o/w) to water in oil (w/o) and multiple types (Kang and Wang 2001). Emulsions create problems because they are viscous and are difficult to produce in some cases. Some emulsions are quite stable, which causes difficulties in reducing the sediment and water content to levels required for pipeline or refinery specifications. Table 7.33 (Zhu et al. 2012) summarizes the operating problems associated with oil produced in the ASP flood in the North-1 East zone of the Daqing oil field, which were not completely resolved.

Demulsifying oil/water emulsions is a field in itself and is beyond the scope of this text. Kang and Wang (2001) studied Daqing oil/water emulsions and concluded that the de-emulsification mechanism involves partial replacement of emulsifying components at the oil/water interface with de-emulsifier molecules. The challenge is to find the appropriate de-emulsifier from the many chemicals available.

The usual approach is to combine chemical additives with electrostatic treatment and, in some cases, heat to destabilize the emulsion. This is a trial-and-error, iterative process for a particular emulsion. Identifying the chemical additive that is effective is challenging and varies with the reservoir oil and fluids. **Table 7.34** summarizes the problems experienced in the ASP flood in the North-1 East zone of the Daqing oil field.

Problems

7.1 Consider the ternary phase diagram in **Fig. 7.150** for a Type III simple, or ideal, system.

1. Consider 100 mL of a mixture at the overall Composition A on the diagram. Assuming that equilibrium is reached, what are the amounts and compositions of the equilibrium phases?
2. Repeat Part 1 for 100 mL of a mixture at the overall Composition B on the diagram. Assume volume percents for all components (i.e., the diagram compositions are by volume).
3. What is the effect on the phase behavior of significantly increasing the salinity of the brine? To answer, show the diagram at the higher salinity.

7.2 1. In their paper on "Immiscible Microemulsion Flooding," Healy and Reed (1977) introduce the concept of a controlling IFT, σ_c. They argue that the magnitude of σ_c determines the outcome of a displacement and that one desires to minimize σ_c. What experimental evidence supports this argument?
2. Consider a microemulsion displacement with a very large slug of middle-phase microemulsion, as performed by Healy and Reed (1977). The following parameters are given: salinity = 1.5% NaCl, μ_{me} = 6 cp, 1 cp = 0.01 g/cm·s, σ_{mo} = 2.0 × 10^{-2} dynes/cm, and σ_{mw} = 0.8 × 10^{-2} dynes/cm. The surfactant system resembles that used by Healy and Reed with N = 12.

   a. What flooding velocity should be used to obtain a fractional oil flow of 0.4?
   b. At that velocity, what final oil saturation, S_{of}, should result from the flood?
      Use results in Healy and Reed (1977) for analysis of this problem.

7.3 A phase-behavior diagram has been determined for a microemulsion system at a relatively low, constant salinity (**Fig. 7.151**). At the salinity used, the system forms lower-phase microemulsions. The binodal curve is shown in the figure.

Stage	Date	Dehydration System	Sewage Treatment System
First	2008.12.18–2009.3.21	Water content before and after dehydration in exported oil reached the standard. As content of polymer and surfactant and pH value increased, dehydration electrical field turned unstable with a high frequency of disabled electrical field.	Suspended solid content exceeded standard
Second	2009.3.21–2009.5.12	Water content in oil exceeded standard before dehydration and could not recover after crossing electrical field. It was dehydrated by use of the thermal-chemical deposition method. Water content in exported oil was overproof in most cases.	Suspended solid content exceeded standard
Third	2009.5.12–2009.5.28	Oil, gas, and water could not be separated in produced liquid that was full of foam. Ten wells were shut down. The system was decompressed manually, but the liquid could not be exported.	Exceeded standard–seriously
Fourth	2009.5.28-2009.12.31	Demulsifying agents and antifoam agents produced by HLX Corporation were added. Oil and water could be separated obviously. The electrical field was stable. Water content in exported oil reached the standard.	Exceeded standard–little

Table 7.34—Treating problems in ASP Flood-North-1 East zone of Daqing oil field (Zhu et al. 2012).

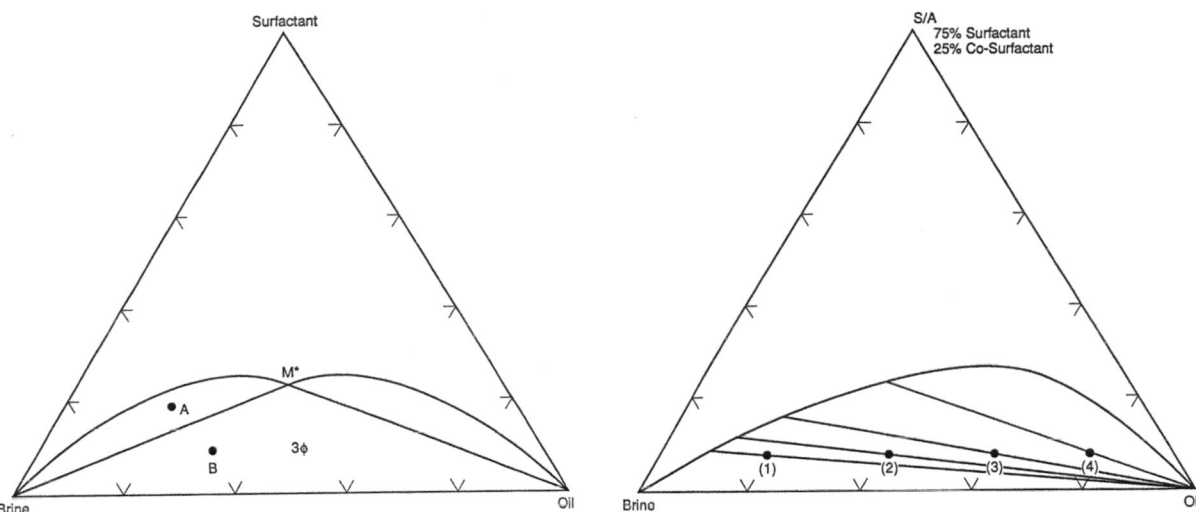

Fig. 7.150—Ternary diagram, Problem 7.1.   Fig. 7.151—Ternary diagram, Problem 7.3.

The surfactant, S/A, is 75% sulfonate and 25% alcohol (cosurfactant). The system behaves as a "simple system," as described in Fig. 1 of Healy et al. (1976). Assume that four samples are made up at overall Compositions 1 through 4 in Fig. 7.151. Assume further that the IFT, σ_{mo}, is given by

$$\log\left(\sigma_{mo}\right) = -7.058 + \frac{6.285}{0.04477\left(V_o / V_s\right) + 1}.$$

1. Which sample (1, 2, 3, or 4) will yield the lowest value of σ_{mo}?
2. Calculate the value of σ_{mo} for this sample.

7.4 Nelson (1982) reports that the relationship between σ_{mo} or σ_{mw} and V_o/V_s or V_w/V_s may be correlated by

$$\log_{10}\sigma_{mo,mw} = \frac{4.80}{1 + 0.10\left(V_{o,w} / V_s\right)} - 5.40,$$

where $\sigma_{mo,mw}$ = IFT between microemulsion and oil or water, dynes/cm; $V_{o,w}$ = volume of oil or water in microemulsion phase, cm³; and V_s = volume of surfactant in microemulsion phase, cm³. That is, the same equation describes IFTs between microemulsion/excess oil and microemulsion/excess water.

Using the equation and the solubility vs. salinity data presented in **Table 7.35**, calculate σ_{mo} and σ_{mw} at optimum salinity.

7.5 Consider the phase-behavior and IFT data shown in **Table 7.36.** With these data, do the following.

1. Develop an equation that relates σ_{mo} to V_o/V_s.
2. Estimate σ_{mo} at $V_o/V_s = 10.0$.
3. Determine the optimum salinity.
4. Estimate σ_{mw} at $V_w/V_s = 10.0$.

Salinity	V_o/V_s	V_w/V_s	Phase
0.5	2.2	—	Lower
1.0	4.2	—	Lower
1.25	8.7	21.0	Middle
1.50	14.8	16.0	Middle
1.75	16.0	12.0	Middle
2.0	—	5.2	Upper
2.5	—	3.0	Upper

Table 7.35—Solubility data, Problem 7.4.

σ_{mo} (dynes/cm)	σ_{mw} (dynes/cm)	V_o/V_s	V_w/V_s	Salinity (% NaCl)	Phase
0.0398	–	2.0	–	0.5	Lower
0.00759	–	7.2	–	1.0	Lower
0.000676	–	12.0	–	1.5	Middle
0.000316	–	15.0	–	1.8	Middle
–	0.0502	–	6.0	2.0	Middle
–	0.00803	–	–	1.8	Middle
–	0.000805	–	–	1.6	Middle

Table 7.36—Phase behavior and IFT data, Problem 7.5.

V_o/V_s	V_w/V_s	Salinity	Microemulsion Type
2.0	–	0.5	Lower
7.2	–	1.0	Lower
11.0	14.5	1.5	Middle
14.0	10.0	1.8	Middle
–	5.0	2.2	Upper

Table 7.37—Solubility data, Problem 7.6.

5. Assume that a displacement is to be conducted with a microemulsion at the optimum salinity. The displacement rate is to be 1.0 ft/D. Fluid viscosities are 12 cp. On the basis of results of Healy and Reed (1977), would you expect this to be an efficient displacement?

7.6 Consider the solubility behavior for the microemulsion system shown in **Table 7.37.** The overall composition is 48.5% water, 48.5% oil, and 3.0% surfactant.

1. What is the optimal salinity, C_ϕ?
2. Assuming the salinity is 1.5%, estimate the boundary of the three-phase region.

7.7 Healy and Reed (1977) show that for their system, and for an alkyl chain length $N = 12$, S_{of} roughly correlates with $N_{ca,c}$, as shown in Fig. 2 of Healy and Reed (1977). Use the data of Table 3 of Healy and Reed (1977) to determine whether a similar correlation exists for the surfactant with $N = 9$.

Healy and Reed show that the data in their Tables 2 and 3 yield values of optimal salinity for oil recovery, C^*, that are approximately equal to optimal salinity for IFT, C_σ (Figs. 3 and 4 of Healy and Reed 1977). Can this same conclusion be drawn for the data taken at 150°F [Table 5 of Healy and Reed (1977)]?

7.8 In a micellar flood, it is desired to set the micellar slug viscosity so that its mobility is less than the total mobility in the oil bank ahead of the slug. That is, for a favorable mobility ratio

$$\frac{k}{\mu_{me}} < \left(\frac{k_{ro}}{\mu_o} + \frac{k_{rw}}{\mu_w} \right) k$$

where k = formation permeability, md; k_{ro}, k_{rw} = relative permeability to oil and water, respectively; μ_o, μ_w = viscosity of oil and water, respectively, cp; and μ_{me} = viscosity of micellar slug, cp. Consider that a Berea core, originally containing oil of 5.0-cp viscosity, has been waterflooded to an ROS of 25%. **Fig. 7.152** shows the oil/water relative permeability curves. Water viscosity is 1.0 cp. What micellar slug viscosity would guarantee that the mobility ratio will be favorable for all flow conditions? Show your calculations.

7.9 Many eastern Kansas sandstone reservoirs are approaching the economic limit of waterflooding and are candidates for EOR by a surfactant/polymer process. Preliminary screening tests in the laboratory identified several surfactant systems that will tolerate the salinity and divalent-ion concentrations found in the original connate fields.

As the project engineer for a small company, you have been asked to evaluate the feasibility of conducting a surfactant/polymer flood in an old reservoir. Because your company has no laboratory facilities, all testing and evaluation work must be contracted out to a commercial laboratory. You have been requested to write the specifications for the contract.

Prepare a comprehensive laboratory program that will lead to the design and testing of an acceptable surfactant/polymer system for an eastern Kansas reservoir. For the purposes of this program you may assume the following.

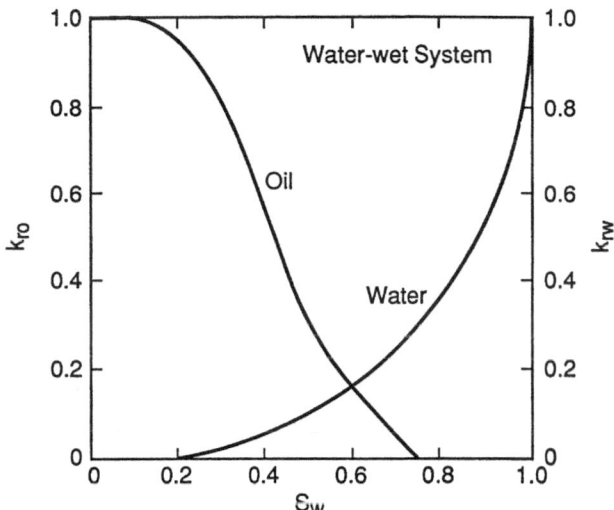

Fig. 7.152—Water/oil relative permeability (water-wet system), Problem 7.8.

1. Rock and fluid properties are known, but relative permeability data will have to be obtained on new cores.
2. The connate brine contains 6 wt% NaCl, 1 wt% CaCl$_2$, and 0.5 wt% MgCl$_2$.
3. The supply water for the waterflood obtained from the Arbuckle formation contains 3 wt% NaCl, 0.5 wt% CaCl$_2$, and 0.3 wt% MgCl$_2$.
4. The surfactants are effective in the range of 1 to 5 wt%.
5. The commercial laboratory is fully equipped to carry out all required testing and analyses.

In developing your program, identify technical papers in the literature that support the design approach you select. Use specific examples of each laboratory test you specify. Include a clear statement of the objective and expected result of each laboratory program.

7.10 A reservoir is at waterflood ROS and is a candidate for a low-tension surfactant flood. Waterflood ROS, S_{orw}, is 35%. Relative permeability functions are as follows: waterflood, $k_{ro}=0.9\,(1-S_{wD})^{3.0}$ and $K_{rw}=0.3(S_{wD})^{2.5}$; low-tension flood, $k_{ro}=0.9(1-S_{wD})^{2.0}$, $k_{rw}=0.3(1-S_{wD})^{1.5}$, $S_{iw}=0.30$ (initial water saturation before waterflood), $S_{orw}=0.35$ (ROS following waterflood), $S_{orc}=0-15$ (ROS following low-tension flood), and

$$S_{wD} = \frac{S_w - S_{iw}}{1-S_{or}-S_{iw}}.$$

Oil and aqueous phase viscosities are $\mu_o = 15$ cp and $\mu_w = 2$ cp. The reservoir PV is 900,000 ft^3.

Assuming a linear flood and no adsorption on the rock, calculate oil production as a function of time. The injection rate is constant at 1,500 B/D. Use frontal-advance theory.

7.11 While designing a microemulsion for a field project, three microemulsions were formulated with the rheological properties indicated on **Fig. 7.153.** The figure also shows approximate frontal velocities, v. Corefloods indicated that the reciprocal relative mobility of the oil bank was on the order of 52 cp. Laboratory corefloods indicated that all three microemulsions gave reasonable oil recoveries. The injection wells are open hole and shot, resulting in an average wellbore diameter of 1 ft. The design injectivity for these wells is 5 B/D-ft. Formation porosity is 20%.

1. Which microemulsion would you choose for this project: A, B, or C? Why?
2. For the microemulsion you chose, what is the microemulsion viscosity in the reservoir away from the wellbore?
3. What is the approximate viscosity ratio between the stabilized oil bank and the microemulsion you chose?
4. Assuming that everything else is equal, what can you tell about the injectivity of Slugs A, B, and C? Support your statement with calculations.

7.12 The phase data in **Table 7.38** are for a Type II(–) system. The components are oil, brine, and surfactant, and the compositions are in weight percent.

1. Plot the data on a pseudoternary phase diagram. What is the composition of the plait point? Plot the plait point.
2. What is the composition and weight fraction of the equilibrium separated phases for 200 g of mixture with total composition of 4% surfactant and 77% oil? Calculate and plot on the pseudoternary diagram.
3. What weight of surfactant must be added to 100 g of a 20% oil-in-brine mixture to make it single phase? What is the composition of this final mixture?
4. What is the composition of the mixture when 150 g of a solution containing 10% oil and 40% surfactant is added to 150 g of a solution of 50% oil and 40% surfactant?

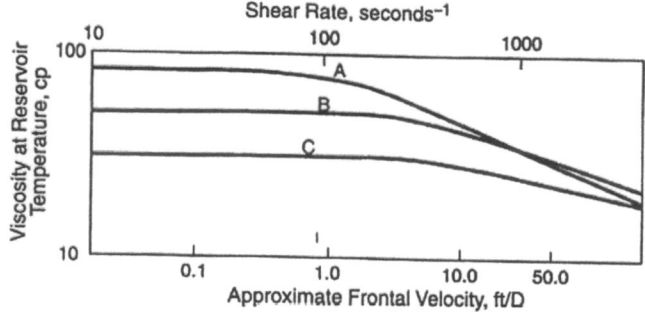

Fig. 7.153—Rheology data for Problem 7.11.

Phase A		Phase B	
Surfactant	Oil	Surfactant	Oil
0	2	0	98
9	9	2	97
18	18	4	95
25	32	6	92
27	40	11	84
26	55	17	74

Table 7.38—Phase-behavior data.

    5. What is the composition of the resulting phases when 100 g of a solution composed of 12% surfactant and 5% oil is added to a 100-g solution composed of 29% surfactant and 77% oil?

7.13 Magichem Chemical Company has announced the development of a new chemical that can be used in displacement processes to recover more oil than with waterflooding. **Table 7.39** shows the results reported by the company to substantiate its claim. The cost of the new chemical is relatively low. On the basis of these coreflood results, would you continue testing this new chemical as an oil-recovery agent? Justify and document the rationale for your answer.

7.14 **Table 7.40** contains oil/water relative permeability data for a reservoir rock. This curve is representative of the reservoir and is to be used to determine the minimum mobility for the design of a micellar slug. The viscosities of the oil and water at reservoir conditions are 5 and 1 cp, respectively.

    1. What is the minimum relative mobility of the water/oil bank?
    2. What is the design viscosity for the micellar solution if mobility control is attained?

7.15 Example 7.9 assumes that the retention of surfactant during a low-tension flood is 0.68 mg/g rock. Determine the effect of discovering a surfactant that has a retention of 0.25 mg/g rock on oil recovery up to a WOR of 50. Compare with Example 7.9 by plotting oil displaced (in barrels) vs. time (in days).

7.16 Solve Example 7.10 with the same assumptions as in Problem 7.15.

7.17 Suppose that the viscosity of the chemical solution in Problem 7.16 is 5 cp. What effect will this have on oil recovery if adsorption of surfactant is 0.25 mg/g?

7.18 What is the minimum size of a chemical slug that will just satisfy retention in Example 7.10?

7.19 A chemical-flood predictive model for a five-spot layered pattern was developed by the US DOE to screen reservoirs as potential candidates for surfactant flooding (Ray and Munoz 1986). The model is in the public domain. Table 4.12 of Ray et al. (1986) gives the data for the base-case micellar/polymer project; these data are also included in the files available from the US DOE. Determine the oil recovery as a function of time using the chemical-flood predictive model for the base-case data. Investigate the effect on oil recovery of varying the clay content from 0.05 to 0.25 wt%.

7.20 Determine the effect of the Dykstra-Parson coefficient on oil recovery as a function of time using the base-case chemical-flood predictive model (Problem 7.19). Vary the Dykstra-Parsons coefficient between 0.4 and 0.8.

7.21 Use the US DOE chemical-flood predictive model to explore the effect of pattern spacing on the base case. Vary the pattern spacing between 2.5 and 40 acres (20 acres is the base case). Compare oil recovery as a function of time for a total flood area of 40 acres (i.e., sixteen 2.5-acre five-spots, eight 5-acre five-spots, four 10-acre five-spots, two 20-acre five-spots and one 40-acre five-spot). Identify the factors to use to determine which flooding pattern gives the largest rate of return.

Core length (in.)	2.0
Sor	0.37
Sorc	0.20
Viscosity of chemical (cp)	3.5
IFT (chemical/oil) (dynes/cm)	15
Pressure drop during displacement (psi)	50

Table 7.39—Displacement test results, Problem 7.13.

S_w	k_{rw}	k_{ro}
0.2	0	0.88
0.3	0.011	0.425
0.4	0.03	0.22
0.5	0.075	0.09
0.6	0.133	0.03
0.65	0.20	0.018

Table 7.40—Relative permeability data, Problem 7.14.

7.22 An oil reservoir at waterflood ROS is to be considered as a candidate for a low-tension flood. The reservoir is a narrow, linear sand body that can be flooded with a linear pattern. Estimated PV between each injection and production well is approximately 160,000 bbl. The viscosity of the oil is 15 cp, and the viscosity of the reservoir brine is 0.9 cp. Laboratory tests indicate that the ROS following the chemical flood is 15%.

It is desired to conduct the flood at a mobility ratio of unity by adding polymer to the injected surfactant solution. The polymer viscosifies the solution but does not alter oil-recovery capability. Because there are no data on retention of surfactant, the base case will be run under the assumption that surfactant retention is negligible. Assume that the chemical solution is injected continuously at a rate of 100 B/D. **Table 7.41** gives the relative permeability parameters for the waterflood and the low-tension flood.

1. Determine the viscosity (to the nearest centipoise) of the chemical solution required to obtain mobility control between the oil bank and the chemical solution.
   For Questions 2 through 4, assume that the viscosity of the chemical solution is 9 cp.
2. Investigate the effect of surfactant retention on the fractional flow of oil in the oil bank and the dimensionless arrival time of the oil bank, t_{D1L}, by determining f_{o1} and t_{D1L} at values of $D_i = 0, 0.2, 0.6$, and 1.0. Plot the fractional flow of oil in the oil bank as a function of D_i. Plot t_{D1L} as a function of D_i.
3. Compare the fractional flow of oil in the effluent (oil cut) of the linear system as a function of t_D (PVs of fluid injected) for $D_i = 0$ and 0.6.
4. Determine the oil recovery (in barrels) at breakthrough of the chemical solution, and plot the recovery as a function of time for $D_i = 0.6$.

Parameter	Waterflood	Low-Tension Flood
S_{iw}	0.3	0.3
α_1	0.9	0.9
m	3.0	2.0
α_2	0.3	0.5
n	2.5	1.5
S_{or}	0.35	0.15

Table 7.41—Relative permeability data.

Nomenclature

a	=	length indicated on Fig. 7.20	A_{io}	=	amount of chemical adsorbed on porous medium from adsorption isotherm, m/m, mg/g rock
a_H	=	empirical constant in Eqs. 7.12 and 7.13			
A	=	cross-sectional area, L^2, ft^2	A_s	=	surfactant adsorption, m/L^2, µeq/m^2

b	=	length indicated on Fig. 7.20
B	=	slug size, % PV
B_o	=	oil FVF, L^3/L^3, bbl/STB
C	=	concentration, m/L^3 or L^3/L^3, g/cm^3 or fraction
C_{AS}^f	=	active sulfonate concentration in the final solution, m/L^3, meq/g
C_{AS}^i	=	active sulfonate concentration in the initial solution, m/L^3, meq/g
Ci	=	chemical (surfactant) concentration, m/L^3, g/cm^3
C_{io}	=	injected chemical (surfactant) concentration, m/L^3, g/cm^3
$\hat{C}_{io}$	=	adsorbed chemical (surfactant) concentration (Eq. 3.107), m/L^3, g adsorbed/cm^3
$nC_{i\phi}$	=	normal alkylbenzene
nC_i	=	normal alkane
C_n	=	number of carbons in an alkyl chain of length n
C_s	=	concentration of surfactant in slug, vol%
C_{NaOH}	=	concentration of sodium hydroxide in caustic solution, wt%
C_{sc}	=	carbons in short chain in hydrocarbon tail
C_t	=	carbons in the surfactant's hydrocarbon tail
$C*$	=	optimal salinity for oil recovery, m/m, ppm
C_σ	=	optimal salinity for IFT, m/m, ppm
C_ϕ	=	optimal salinity for phase behavior, m/m, ppm
C_1	=	monovalent-cation concentration, m/L^3, g mol/L
C_2	=	divalent-cation concentration, m/L^3, g mol/L
d	=	diameter, L, ft
D_i	=	ratio of injected chemical concentration to that retained on rock, m/m, fraction
E_{vw}	=	volumetric sweep efficiency of waterflood preceding a chemical flood, L^3/L^3, fraction
f_a	=	fraction of chemical adsorbed, L^3/L^3, fraction
f_o	=	fractional flow of oil phase, L^3/L^3, fraction
f_{o1}	=	fractional flow of oil when the oil bank arrives at the end of a linear system = $1 - f_{w1}$
f_w	=	fractional flow of water on fractional-flow curve for water or brine, L^3/L^3, fraction
f_w'	=	slope of fractional-flow curve for water at S_w
f_w^*	=	fractional flow of water on fractional-flow curve of chemical solution, L^3/L^3, fraction
f_{wf}	=	fractional flow of water at S_{wf}, water saturation at water discontinuity in normal waterflood, L^3/L^3, fraction
f_{wf}'	=	slope of fractional-flow curve for water at S_{wf}
f_{w1}	=	fractional flow of water in oil bank preceding chemical flood saturation discontinuity, L^3/L^3, fraction
f_{w1}'	=	slope of fractional-flow curve for water at S_{w1}
f_{w2}^*	=	fractional flow of chemical solution at effluent end of linear system, L^3/L^3, fraction
f_{w3}^*	=	fractional flow of chemical solution at effluent end of linear system, L^3/L^3, fraction
$f_{w3}'^*$	=	slope of fractional-flow curve for chemical solution at S_{w3}^*
F_B	=	force exerted by fluid in bulk phase, mL/t^2, dynes
F_I	=	force exerted by fluid in interfacial zone, mL/t^2, dynes
F_{wo}	=	WOR, L^3/L^3, bbl/bbl
h	=	thickness, L, ft
h_I	=	thickness of interfacial zone, L, cm
k	=	permeability, L^2, md
HA_o	=	concentration of petroleum acid in the oil phase, moles/L
HA_W	=	concentration of petroleum acid in the water phase, moles/L
k_{ro}	=	relative permeability to oil
k_{rw}	=	relative permeability to water
K, K_1 K_1', K_2	=	equilibrium constants
K_1^*	=	equilibrium constant representing the ratio of divalent-cation sulfonate to monovalent-cation sulfonate in the middle-phase microemulsion, $M*$, moles/moles
K_A	=	equilibrium constant
K_D	=	distribution or partition coefficient
L	=	length, L, ft
m	=	mass, m, g
m_a	=	mass of absorbent, m, g
m_{Na}	=	amount of sodium ion (Na^+) adsorbed, g
m_{NaOH}	=	amount of sodium hydroxide in one pore volume of solution, g
m_o	=	constant in Eq. 7.9
m_r	=	weight of rock matrix, lbm
m_s	=	mass of rock matrix, g
m_w	=	constant in Eq. 7.0
$M*$	=	middle-phase composition
M_{Na}	=	molecular weight of Na
M_s	=	equivalent weight of surfactant
n	=	constant

N	=	carbon number of alkyl side chain on surfactant
N	=	oil in place originally or at start of process, L^3, STB
N_{AC}^*	=	ACN at optimal salinity
N_{ca}	=	capillary number based on interstitial velocity, $v\mu/\sigma$, dimensionless
N_{ca}^*	=	capillary number based on Darcy velocity, $u\mu/\sigma$, dimensionless
$N_{ca,c}$	=	critical capillary number, $v\mu/\sigma_c$, dimensionless
$N_{ca,c}^*$	=	critical capillary number based on interstitial velocity, dimensionless
N_m	=	mobile oil originally in place, L^3, STB
N_p	=	oil produced, L^3, STB
p	=	pressure, m/Lt^2, psi
p_B	=	bulk pressure in Fig. 7.10, m/Lt^2, dynes/cm²
p_N	=	normal component of pressure, m/Lt^2, dynes/cm²
p_T	=	transverse pressure in interfacial zone, m/Lt^2, dynes/cm²
P^*	=	optimum solubilization parameter, $P_o = P_w$
P_o	=	oil solubilization parameter, volume of oil/volume of surfactant, L^3/L^3, fraction
P_w	=	water solubilization parameter, volume of water/volume of surfactant, L^3/L^3, fraction
q	=	volumetric flow rate, L^3/t, B/D
Q_m	=	cation-exchange capacity, m/m, g/g
Q_t	=	total ion exchange capacity, meq
Q_v	=	cation-exchange capacity, m/L^3, g/cm³
S_u	=	surface area of adsorbent per unit mass, L^2/m, m²/g
S_{iw}	=	interstitial water saturation, L^3/L^3, fraction
S_o	=	oil saturation, L^3/L^3, fraction
S_{of}	=	ROS to chemical flood, L^3/L^3, fraction
S_{oi}	=	initial oil saturation, L^3/L^3, fraction
S_{or}	=	ROS, L^3/L^3, fraction
S_{orc}	=	ROS at end of chemical flood, L^3/L^3, fraction
S_{orw}	=	ROS following waterflood, L^3/L^3, fraction
S_w	=	water saturation, L^3/L^3, fraction
S_w^*	=	saturation of water in region of chemical solution flow, L^3/L^3, fraction
$\overline{S}_w$	=	average water saturation, L^3/L^3, fraction
$\overline{S}_w^*$	=	average chemical solution saturation in region behind chemical flood saturation discontinuity, L^3/L^3, fraction
S_{wD}	=	dimensionless water saturation

S_{wf}	=	water saturation at water discontinuity (shock) in normal waterflood, L^3/L^3, fraction
S_{wi}	=	initial water saturation, L^3/L^3, fraction
S_{w1}	=	water saturation in oil bank preceding chemical flood saturation discontinuity, L^3/L^3, fraction
S_{w2}^*	=	chemical solution saturation at effluent end of linear system, L^3/L^3, fraction
S_{w3}^*	=	saturation of water at chemical front shock, L^3/L^3, fraction
t	=	time, t, days
t_D	=	dimensionless time, $qt/A\phi L$
t_{Df}	=	dimensionless time for S_{wf} to reach x_{Df}
t_{D1}	–	dimensionless time at which S_{wf} reaches x_{D1}
t_{D1L}	=	dimensionless time when the oil bank arrives at the end of a linear system
t_{D3}	=	dimensionless time at which S_{w3}^* reaches x_{D3}
Δt_D	=	dimensionless time delay in breakthrough, PV
T	=	temperature, T, °F
u	=	Darcy velocity, L/t, ft/D
v	=	interstitial (frontal) velocity, L/t, ft/D
v_D	=	specific velocity, dimensionless
v_{Df}	=	specific velocity of S_{wf}, dimensionless
v_{Do}	=	dimensionless velocity of oil bank preceding chemical solution front
v_{D1}	=	specific velocity of S_{w1}, dimensionless
v_{D3}^*	=	specific velocity of S_{w3}^*, dimensionless
V	=	volume, L^3, bbl
V_a	=	volume of surfactant solution added, m³, cm³
V_b	=	bulk volume of reservoir rock, L^3, ft³
V_{DP}	=	Dykstra-Parsons coefficient
V_{H_2O}	=	volume of water phase, cm³
V_{me}	=	volume of microemulsion phase, cm³
V_{mo}	=	molar volume of oil, m/L^3, g mol/L
V_o	=	volume of oil or oleic phase, L^3, bbl
$V_{o,w}$	=	volume of water or oil in the microemulsion phase, cm³
V_p	=	PV, L^3, bbl
V_s	=	volume of surfactant, L^3, bbl
V_s'	=	volume of surfactant plus alcohol in microemulsion phase, L^3, cm³
V_{slug}	=	volume of sulfonate injected in slug, % PV
V_w	=	volume of aqueous phase, L^3, bbl
w	=	width, L, cm
x	=	distance or length, L, ft
x_D	=	dimensionless distance

x_{Df}	=	dimensionless distance to location of S_{wf}	ρ_b	=	bulk density of porous rock, g/cm^3 or lbm/ft^3
x_{D1}	=	dimensionless distance to location of S_{wl}	ρ_r	=	rock density, m/L^3, g/cm^3
x_{D3}	=	dimensionless distance to location of S_{w3}^*	ρ_s	=	density of solid rock, kg/m^3
X_{1S}	=	moles of monovalent-cation sulfonate per mole of total sulfonate, m/m, mol/mol	σ	=	IFT, m/t^2, dynes/cm
			σ^*	=	IFT at optimal salinity, m/t^2, dynes/cm
X_{2S}	=	moles of divalent-cation sulfonate complex per mole of total sulfonate, m/m, mol/mol	σ_c	=	critical IFT, m/t^2, dynes/cm
			σ_{mo}	=	IFT between microemulsion and oleic phase, m/t^2, dynes/cm
X_{2SS}	=	moles of divalent-cation/sulfonate complex per mole of total sulfonate, m/m, mol/mol	σ'_{mo}	=	intercept value of experimental data, IFT between microemulsion and oleic phase, m/t^2, dynes/cm
α_1, α_2	=	constants in relative permeability correlation			
			$\sigma_{mo,mw}$	=	IFT at microemulsion/oil or microemulsion/water interface, m/t^2, dynes/cm
λ_c	=	mobility of the chemical slug, Lt/m, 1/cp			
λ_{rd}	=	design relative mobility, Lt/m, 1/cp	σ_{mw}	=	IFT between microemulsion and aqueous phases, m/t^2, dynes/cm
λ_{ro}	=	relative mobility of oil phase, Lt/m, 1/cp			
λ_{rw}	=	relative mobility of water phase, Lt/m, 1/cp	σ'_{mw}	=	intercept value of experimental data, IFT between microemulsion and aqueous phase, m/t^2, dynes/cm
μ	=	viscosity, m/Lt, cp			
μ_c	=	apparent viscosity of chemical slug, cp			
μ_{me}	=	viscosity of micellar slug, m/Lt, cp	φ	=	porosity
μ_o	=	oil viscosity, m/Lt, cp	φ_1	=	volume fraction of oil in middle phase on surfactant-free basis, L^3/L^3, fraction
μ_w	=	water viscosity, m/Lt, cp			
ρ	=	density of initial fluid, m/L^3, g/cm^3	φ_2	=	volume fraction of water in middle phase on surfactant-free basis, L^3/L^3, fraction

Subscripts

b	=	bulk
me	=	microemulsion
o	=	oil
s	=	surfactant
t	=	total
w	=	water

References

Adkins, S., Liyange, P. J., Pinnawala Arachchilage, G. W. P. et al. 2010. A New Process for Manufacturing and Stabilizing High-Performance EOR Surfactants at Low Cost for High-Temperature, High-Salinity Oil Reservoirs. Presented at the SPE Improved Oil Recovery Symposium, Tulsa, 24–28 April. SPE-129923-MS. https://doi.org/10.2118/129923-MS.

Adkins, S., Pinnawala Arachchilage, G. W. P., Solairaj, S. et al. 2012. Development of Thermally and Chemically Stable Large-Hydrophobe Alkoxy Carboxylate Surfactants. Presented at the SPE Improved Oil Recovery Symposium, Tulsa, 14–18 April. SPE-154256-MS. https://doi.org/10.2118/154256-MS.

Ahmad, M. S. 2012. *Methodology for Designing and Evaluating Chemical Systems for Improved Oil Recovery*. MS thesis, University of Kansas, Lawrence, Kansas.

Amphlett, C. B. 1964. *Inorganic Ion Exchanges*. New York City: Elsevier Publishing Co.

Anderson, G. A., Delshad, M., King, C. B. et al. 2006. Optimization of Chemical Flooding in a Mixed-Wet Dolomite Reservoir. Presented at the SPE/DOE Symposium on Improved Oil Recovery, Tulsa, 22–26 April. SPE-100082-MS. https://doi.org/10.2118/100082-MS.

Bae, J. H. and Petrick, C. B. 1977. Adsorption/Retention of Petroleum Sulfonates in Berea Cores. *SPE J.* **17** (5): 353–357. SPE-5819-PA. https://doi.org/10.2118/5819-PA.

Bae, J. H., Petrick, C. B., and Ehrlich, R. 1974. A Comparative Evaluation of Microemulsions and Aqueous Surfactant Systems. Presented at the SPE Improved Oil Recovery Symposium, Tulsa, 22–24 April. SPE-4749-MS. https://doi.org/10.2118/4749-MS.

Barakat, Y., Fortney, L. N., LaLanne-Cassou, C. et al. 1983a. The Phase Behavior of Simple Salt-Tolerant Sulfonates. *SPE J.* **23** (6): 913–918. SPE-10679-PA. https://doi.org/10.2118/10679-PA.

Barakat, Y., Fortney, L. N., Schechter, R. S. et al. 1983b. Criteria for Structuring Surfactants to Maximize Solubilization of Oil and Water: II. Alkyl Benzene Sodium Sulfonates. *Journal of Colloid Interface Science* **92** (2): 561–574. https://doi.org/10.1016/0021-9797(83)90177-7.

Baviére, M., Bazin, B., and Noik, C. 1988. Surfactants for EOR: Olefin Sulfonate Behavior at High Temperature and Hardness. *SPE Res Eng* **3** (2): 597–603. SPE-14933-PA. https://doi.org/10.2118/14933-PA.

Bhuyan, D., Lake, L. W., and Pope, G. A. 1990. Mathematical Modeling of High-pH Chemical Flooding. *SPE Res Eng* **5** (2): 213–220. SPE-17398-PA. https://doi.org/10.2118/17398-PA.

Blevins, C. E., Willhite, G. P., and Michnick, M. J. 1981. Investigation of Three-Phase Regions Formed by Petroleum Sulfonate Systems. *SPE J.* **21** (5): 581–592. SPE-8260-PA. https://doi.org/10.2118/8260-PA.

Boneau, D. F. and Clampitt, R. L. 1977. A Surfactant System for the Oil-Wet Sandstone of the North Burbank Unit. *J Pet Technol* **29** (5): 501–506. SPE-5820-PA. https://doi.org/10.2118/5820-PA.

Borah, M. T. and Gregory, M. D. 1988. A Summary of the Big Muddy Field Low-Tension Flood Demonstration Project. Presented at the SPE Rocky Mountain Regional Meeting, Casper, Wyoming, USA, 11–13 May. SPE-17536-MS. https://doi.org/10.2118/17536-MS.

Bourrel, M. and Schechter, R. S. 1988. *Microemulsions and Related Systems.* New York City: Marcel Dekker Inc.

Bragg, J. R., Gale, W. W., McElhannon, W. A. Jr. et al. 1982. Loudon Surfactant Flood Pilot Test. Presented at the SPE Enhanced Oil Recovery Symposium, Tulsa, 4–7 April. SPE-10862-MS. https://doi.org/10.2118/10862-MS.

Bunge, A. L. and Radke, C. J. 1982. Migration of Alkaline Pulses in Reservoir Sands. *SPE J.* **22** (6): 998–1012. SPE-10288-PA. https://doi.org/10.2118/10288-PA.

Burk, J. H. 1987. Comparison of Sodium Carbonate, Sodium Hydroxide, and Sodium Orthosilicate for EOR. *SPE Res Eng* **2** (1): 9–16. SPE-12039-PA. https://doi.org/10.2118/12039-PA.

Campbell, T. C. 1982. The Role of Alkaline Chemicals in the Recovery of Low-Gravity Crude Oils. *J Pet Technol* **34** (11): 2510–2516. SPE-8894-PA. https://doi.org/10.2118/8894-PA.

Carmona, I., Schecter, R. S., Wade, W. H. et al. 1985. Ethoxylated Oleyl Sulfonates as Model Compounds for Enhanced Oil Recovery. *SPE J.* **25** (3): 351–357. SPE-11771-PA. https://doi.org/10.2118/11771-PA.

Cash, R. L. Jr., Cayias, J. L., Hayes, M. et al. 1975. Spontaneous Emulsification—A Possible Mechanism for Enhanced Oil Recovery. Presented at the Fall Meeting of the Society of Petroleum Engineers of AIME, Dallas, 28 September–1 October. SPE-5562-MS. https://doi.org/10.2118/5562-MS.

Cayias, J. L., Schechter, R. S., and Wade, W. H. 1975. The Measurement of Low Interfacial Tension via the Spinning Drop Technique. In *ACS Symposium Series*, ed. L. K. Mittal, Vol. 8, Chapter 17, 234–247. Washington, DC: American Chemical Society.

Cayias, J. L., Schechter, R. S., and Wade, W. H. 1976. Modeling Crude Oils for Interfacial Tension. *SPE J.* **16** (6): 351–357. SPE-5813-PA. https://doi.org/10.2118/5813-PA.

Chan, A. F. and Gupta, S. P. 1984. The Propagation of Oil-Moving and Solubilizing Components of Broad Equivalent-Weight Sulfonate Systems in Micellar Floods. *SPE J.* **24** (4): 435–446. SPE-10200-PA. https://doi.org/10.2118/10200-PA.

Chan, A. F. and Kremesec, V. J. Jr. 1985. Cation Exchange in Porous Media With Broad-Equivalent-Weight Sulfonate Micellar Fluids. *SPE J.* **25** (4): 580–586. SPE-12126-PA. https://doi.org/10.2118/12126-PA.

Chang, H. L., Zhang, Z. Q., Wang, Q. M. et al. 2006. Advances in Polymer Flooding and Alkaline/Surfactant/Polymer Processes as Developed and Applied in the People's Republic of China. *J Pet Technol* **58** (2): 84–89. SPE-89175-JPT. https://doi.org/10.2118/89175-JPT.

Cheng, K. H. 1986. Chemical Consumption During Alkaline Flooding: A Comparative Evaluation. Presented at the SPE/DOE Enhanced Oil Recovery Symposium, Tulsa, 20–23 April. SPE-14944-MS. https://doi.org/10.2118/14944-MS.

Chou, S. I. and Bae, J. H. 1989. Using Oligomeric Surfactants To Improve Oil Recovery. *SPE Res Eng* **4** (3): 373–380. SPE-16725-PA. https://doi.org/10.2118/16725-PA.

Clark, S. R., Pitts, M. J., and Smith, S. M. 1993. Design and Application of an Alkaline-Surfactant-Polymer Recovery System to the West Kiehl Field. *SPE Advanced Technology Series* **1** (1): 172–179. SPE-17538-PA. https://doi.org/10.2118/17538-PA.

Cole, E. L. 1988a. An Evaluation of the Big Muddy Field Low-Tension Flood Demonstration Project. Report DOE/BC/10830-9, US Department of Energy, Bartlesville, Oklahoma (December 1988).

Cole, E. L. 1988b. An Evaluation of the Robinson M-1 Commercial Scale Demonstration Project of Enhanced Oil Recovery by Micellar-Polymer Flood. Report DOE/BC/10830-10, US Department of Energy, Bartlesville, Oklahoma (December 1988).

Crocker, M. E., Donaldson, E. C., and Marchin, L. M. 1983. Comparison and Analysis of Reservoir Rocks and Related Clays. Presented at the SPE Annual Technical Conference and Exhibition, San Francisco, 5–8 October. SPE-11973-MS. https://doi.org/10.2118/11973-MS.

Dauben, D. L., Easterly, R. A., and Western, M. M. 1987. An Evaluation of the Alkaline Waterflooding Demonstration Project, Ranger Zone, Wilmington Field, California. Report DOE/BC/10830-5, US Department of Energy, Bartlesville, Oklahoma (May 1987).

Davis, H. T. and Scriven, L. E. 1980. The Origins of Low Interfacial Tensions for Enhanced Oil Recovery. Presented at the SPE Annual Technical Conference and Exhibition, Dallas, 21–24 September. SPE-9278-MS. https://doi.org/10.2118/9278-MS.

Davis, H. T. and Scriven, L. E. 1982. Stress and Structure in Fluid Interfaces. In *Advances in Chemical Physics*, eds. I. Prigogine and S. A. Rice, Vol. 49, Chapter 6, 357–454. Hoboken, New Jersey: John Wiley & Sons, Inc. https://doi.org/10.1002/9780470142691.ch6.

Davis, J. A. and Jones, S. C. 1968. Displacement Mechanisms of Micellar Solutions. *J Pet Technol* **20** (12): 1415–1428. SPE-1847-2-PA. https://doi.org/10.2118/1847-2-PA.

Dean, R. M. 2011. *Selection and Evaluation of Surfactants for Field Pilots.* MS thesis, University of Texas at Austin, Austin, Texas (May 2011).

Delshad, M., Han, C., Veedu, F. K. et al. 2013. A Simplified Model for Simulations of Alkaline-Surfactant-Polymer Floods. *Journal of Petroleum Science and Engineering* **108**: 1–9. https://doi.org/10.1016/j.petrol.2013.04.006.

Delshad, M., Han, W., Pope, G. A. 1998. Alkaline/Surfactant/Polymer Flood Predictions for the Karamay Oil Field. Presented at the SPE/DOE Improved Oil Recovery Symposium, Tulsa, 19–22 April. SPE-39610-MS. https://doi.org/10.2118/39610-MS.

DeZabala, E. F., Vislocky, J. M., Rubin, E. et al. 1982. A Chemical Theory for Linear Alkaline Flooding. *SPE J.* **22** (2): 245–258. SPE-8997-PA. https://doi.org/10.2118/8997-PA.

Doll, T. E. 1988a. An Update of the Polymer-Augmented Alkaline Flood at the Isenhour Unit, Sublette County, Wyoming. *SPE Res Eng* **3** (2): 604–608. SPE-14954-PA. https://doi.org/10.2118/14954-PA.

Doll, T. E. 1988b. Performance Data Through 1987 of the Isenhour Unit, Sublette County, Wyoming, Polymer-Augmented Alkaline Flood. Presented at the SPE Rocky Mountain Regional Meeting, Casper, Wyoming, USA, 11–13 May. SPE-17801-MS. https://doi.org/10.2118/17801-MS.

Falls, A. H., Thigpen, D. R., Nelson, R. C. et al. 1994. Field Test of Cosurfactant-Enhanced Alkaline Flooding. *SPE Res Eng* **9** (3): 217–223. SPE-24117-PA. https://doi.org/10.2118/24117-PA.

Ferrell, H. H., Gregory, M. D., and Borah, M. T. 1984. Progress Report: Big Muddy Field Low-Tension Flood Demonstration Project With Emphasis on Injectivity and Mobility. Presented at the SPE/DOE Fourth Symposium on Enhanced Oil Recovery, Tulsa, 15–18 April. SPE-12682-MS. https://doi.org/10.2118/12682-MS.

Ferrell, H. H., King, D. W., and Sheely, C. Q. Jr. 1988. Analysis of the Low-Tension Pilot at Big Muddy Field, Wyoming. *SPE Form Eval* **3** (2): 315–321. SPE-12683-PA. https://doi.org/10.2118/12683-PA.

Flaaten, A. K. 2007. *Experimental Study of Microemulsion Characterization and Optimization in Enhanced Oil Recovery: A Design Approach for Reservoirs with High Salinity and Hardness.* MS thesis, University of Texas at Austin, Austin, Texas (December 2007).

Flaaten, A., Nguyen, Q. P., Pope, G. A. et al. 2008a. A Systematic Laboratory Approach to Low-Cost High-Performance Chemical Flooding. Presented at the SPE/DOE Improved Oil Recovery Symposium, Tulsa, 19–23 April. SPE-113469-MS. https://doi.org/10.2118/113469-MS.

Flaaten, A., Nguyen, Q. P., Zhang, J. et al. 2008b. ASP Chemical Flooding Without the Need for Soft Water. Presented at the SPE Annual Technical Conference and Exhibition, Denver, 21–24 September. SPE-116754-MS. https://doi.org/10.2118/116754-MS.

Fleming, P. D. III, Thomas, C. P., and Winter, W. K. 1981. Formulation of a General Multiphase, Multicomponent Chemical Flood Model. *SPE J.* **21** (1): 63–76. SPE-6727-PA. https://doi.org/10.2118/6727-PA.

Fortenberry, R. P., Kim, D. H., Nizamidin, N. et al. 2013. Use of Co-Solvents to Improve Alkaline-Polymer Flooding. Presented at the SPE Annual Technical Conference and Exhibition, New Orleans, 30 September–2 October. SPE-166478-MS. https://doi.org/10.2118/166478-MS.

Foster, W. R. 1973. A Low-Tension Waterflooding Process. *J Pet Technol* **25** (2): 205–210. SPE-3803-PA. https://doi.org/10.2118/3803-PA.

French, T. R. and Burchfield, T. E. 1990. Design and Optimization of Alkaline Flooding Formulations. Presented at the SPE/DOE Enhanced Oil Recovery Symposium, Tulsa, 22–25 April. SPE-20238-MS. https://doi.org/10.2118/20238-MS.

Gale, W. W. and Sandvik, E. I. 1973. Tertiary Surfactant Flooding: Petroleum Sulfonate Composition Efficacy Studies. *SPE J.* **13** (4): 191–199. SPE-3804-PA. https://doi.org/10.2118/3804-PA.

Gale, W. W., Puerto, M. C., Ashcraft, T. et al. 1981. Propoxylated Ethoxylated Surfactants and Method of Recovering Oil Therewith. US Patent No. 4,293,428 (6 October 1981).

Gao, S., Li, H., and Li, H. 1995. Laboratory Investigation of Combination of Alkali/Surfactant/Polymer Technology for Daqing EOR. *SPE Res Eng* **10** (3): 194–197. SPE-27631-PA. https://doi.org/10.2118/27631-PA.

Gao, S., Li, H., Yang, Z. et al. 1996. Alkaline/Surfactant/Polymer Pilot Performance of the West Central Saertu, Daqing Oil Field. *SPE Res Eng* **11** (3): 181–188. SPE-35383-PA. https://doi.org/10.2118/35383-PA.

Garrett, H. E. 1972. *Surface Active Chemicals.* New York City: Pergamon Press.

Gerbacia, W. and McMillen, T. J. 1982. Oil-Recovery Surfactant Formulation Development Using Surface Design Experiments. *SPE J.* **22** (2): 237–244. SPE-8878-PA. https://doi.org/10.2118/8878-PA.

Gilliland, H. E. and Conley, F. R. 1976. Pilot Flood Mobilizes Residual Oil. *Oil & Gas J.*: 43–48.

Gladfelter, R. E. and Gupta, S. P. 1980. Effect of Fractional Flow Hysteresis on Recovery of Tertiary Oil. *SPE J.* **20** (6): 508–520. SPE-7577-PA. https://doi.org/10.2118/7577-PA.

Glinsmann, G. R. 1979. Surfactant Flooding With Microemulsions Formed In Situ—Effect of Oil Characteristics. Presented at the SPE Annual Technical Conference and Exhibition, Las Vegas, Nevada, USA, 23–26 September. SPE-8326-MS. https://doi.org/10.2118/8326-MS.

Glover, C. J., Puerto, M. C., Maerker, J. M. et al. 1979. Surfactant Phase Behavior and Retention in Porous Media. *SPE J.* **19** (3): 183–193. SPE-7053-PA. https://doi.org/10.2118/7053-PA.

Gogarty, W. B. 1976. Status of Surfactant or Micellar Methods. *J Pet Technol* **28** (1): 93–102. SPE-5559-PA. https://doi.org/10.2118/5559-PA.

Gogarty, W. B. 1983a. Enhanced Oil Recovery Through the Use of Chemicals–Part 1. *J Pet Technol* **35** (9): 1581–1590. SPE-12367-PA. https://doi.org/10.2118/12367-PA.

Gogarty, W. B. 1983b. Enhanced Oil Recovery Through the Use of Chemicals–Part 2. *J Pet Technol* **35** (10): 1767–1775. SPE-12368-PA. https://doi.org/10.2118/12368-PA.

Gogarty, W. B. and Tosch, W. C. 1968. Miscible-Type Waterflooding: Oil Recovery With Micellar Solutions. *J Pet Technol* **20** (12): 1407–1414. SPE-1847-1-PA. https://doi.org/10.2118/1847-1-PA.

Gogarty, W. B., Meabon, H. P., and Milton, H. W. Jr. 1970. Mobility Control Design for Miscible-Type Waterfloods Using Micellar Solutions. *J Pet Technol* **22** (2): 141–147. SPE-1847-E-PA. https://doi.org/10.2118/1847-E-PA.

Graciaa, A., Fortney, L. N., Schechter, R. S. et al. 1982. Criteria for Structuring Surfactants To Maximize Solubilization of Oil and Water: Part 1—Commercial Nonionics. *SPE J.* **22** (5): 743–749. SPE-9815-PA. https://doi.org/10.2118/9815-PA.

Graue, D. J. and Johnson, C. E. Jr. 1974. Field Trial of Caustic Flooding Process. *J Pet Technol* **26** (12): 1353–1358. SPE-4740-PA. https://doi.org/10.2118/4740-PA.

Gupta, S. P. 1984. Compositional Effects on Displacement Mechanisms of the Micellar Fluid Injected on the Sloss Field Test. *SPE J.* **24** (1): 38–48. SPE-8827-PA. https://doi.org/10.2118/8827-PA.

Gupta, S. P. and Trushenski, S. P. 1979. Micellar Flooding—Compositional Effects on Oil Displacement. *SPE J.* **19** (2): 116–128. SPE-7063-PA. https://doi.org/10.2118/7063-PA.

Hall, A. C. 1980. Interfacial Tension and Phase Behavior in Systems of Petroleum Sulfonate/Brine/*n*-Alkane. *Colloids and Surfaces* **1** (2): 209–228. https://doi.org/10.1016/0166-6622(80)80007-2.

Healy, R. N. and Reed, R. L. 1974. Physiochemical Aspects of Microemulsion Flooding. *SPE J.* **14** (5): 491–501. SPE-4583-PA. https://doi.org/10.2118/4583-PA.

Healy, R. N. and Reed, R. L. 1977. Immiscible Microemulsion Flooding. *SPE J.* **17** (2): 129–139. SPE-5817-PA. https://doi.org/10.2118/5817-PA.

Healy, R. N., Reed, R. L., and Carpenter, C. W. Jr. 1975. A Laboratory Study of Microemulsion Flooding (includes associated papers 6395 and 6396). *SPE J.* **15** (1): 87–103. SPE-4752-PA. https://doi.org/10.2118/4752-PA.

Healy, R. N., Reed, R. L., and Stenmark, D. G. 1976. Multiphase Microemulsion Systems. *SPE J.* **16** (3): 147–160. SPE-5565-PA. https://doi.org/10.2118/5565-PA.

Hedges, J. H. and Glinsmann, G. R. 1979. Compositional Effects on Surfactant Flood Optimization. Presented at the SPE Annual Technical Conference and Exhibition, Las Vegas, Nevada, 23–26 September. SPE-8324-MS. https://doi.org/10.2118/8324-MS.

Hill, H. J., Reisberg, J., and Stegemeier, G. L. 1973. Aqueous Surfactant Systems for Oil Recovery. *J Pet Technol* **25** (2): 186–194. SPE-3798-PA. https://doi.org/10.2118/3798-PA.

Hirasaki, G. J. 1982a. Interpretation of the Change in Optimal Salinity With Overall Surfactant Concentration. *SPE J.* **22** (6): 971–982. SPE-10063-PA. https://doi.org/10.2118/10063-PA.

Hirasaki, G. J. 1982b. Ion Exchange With Clays in the Presence of Surfactant. *SPE J.* **22** (2): 181–192. SPE-9279-PA. https://doi.org/10.2118/9279-PA.

Hirasaki, G. J. and Zhang, D. L. 2004. Surface Chemistry of Oil Recovery From Fractured, Oil-Wet, Carbonate Formations. *SPE J.* **9** (2): 151–162. SPE-88365-PA. https://doi.org/10.2118/88365-PA.

Hirasaki, G. J., Miller, C. A. and Pope, G.A. 2006. Surfactant Based Enhanced Oil Recovery and Foam Mobility Control. Third Semi-Annual Technical Report, Department of Energy Contract DE-FC26-03NT15406 (February 2006).

Hirasaki, G. J., Miller, C. A. and Puerto, M. 2008. Recent Advances in Surfactant EOR. Presented at the SPE Annual Technical Conference and Exhibition, Denver, 21–24 September. SPE-115386-MS. https://doi.org/10.2118/115386-MS.

Holm, L. W. 1978. Correlation of Oleic and Aqueous Micellar Processes for Tertiary Oil Recovery. Presented at the SPE Symposium on Improved Methods for Oil Recovery, Tulsa, 16–19 April. SPE-7066-MS. https://doi.org/10.2118/7066-MS.

Huang, L., Yang, P., and Qin, T. 1986. A Study of Caustic Consumption by Clays. Presented at the SPE/DOE Enhanced Oil Recovery Symposium, Tulsa, 20–23 April. SPE-14945-MS. https://doi.org/10.2118/14945-MS.

Huh, C. 1979. Interfacial Tensions and Solubilizing Ability of a Microemulsion Phase That Coexists With Oil and Brine. *Journal of Colloid and Interface Science* **71** (2): 408–426. https://doi.org/10.1016/0021-9797(79)90249-2.

Huh, C., Landis, L. H., Maer, N. K. Jr. et al. 1990a. Simulation To Support Interpretation of the Loudon Surfactant Pilot Tests. Presented at the SPE Annual Technical Conference and Exhibition, New Orleans, 23–26 September. SPE-20465-MS. https://doi.org/10.2118/20465-MS.

Huh, C., Lange, E. A., and Cannella, W. J. 1990b. Polymer Retention in Porous Media. Presented at the SPE/DOE Enhanced Oil Recovery Symposium, Tulsa, 22–25 April. SPE-20235-MS. https://doi.org/10.2118/20235-MS.

Hurd, B. G. 1976. Adsorption and Transport of Chemical Species in Laboratory Surfactant Waterflooding Experiments. Presented at the SPE Improved Oil Recovery Symposium, Tulsa, 22–24 March. SPE-5818-MS. https://doi.org/10.2118/5818-MS.

Jennings, H. Y. Jr. 1975. A Study of Caustic Solution-Crude Oil Interfacial Tensions. *SPE J.* **15** (3): 197–202. SPE-5049-PA. https://doi.org/10.2118/5049-PA.

Jennings, H. Y. Jr., Johnson, C. E. Jr., and McAuliffe, C. E. 1974. A Caustic Waterflooding Process for Heavy Oils. *J Pet Technol* **26** (12): 1344–1352. SPE-4741-PA. https://doi.org/10.2118/4741-PA.

Johnson, C. E. Jr. 1976. Status of Caustic and Emulsion Methods. *J Pet Technol* **28** (1): 85–92. SPE-5561-PA. https://doi.org/10.2118/5561-PA.

Jones, S. C. and Dreher, K. D. 1976. Cosurfactants in Micellar Systems Used for Tertiary Oil Recovery. *SPE J.* **16** (3): 161–167. SPE-5566-PA. https://doi.org/10.2118/5566-PA.

Kang, W. and Wang, D. 2001. Emulsification Characteristic and De-emulsifiers Action for Alkaline/Surfactant/Polymer Flooding. Presented at the SPE Asia Pacific Improved Oil Recovery Conference, Kuala Lumpur, Malaysia, 8–9 October. SPE-72138-MS. https://doi.org/10.2118/72138-MS.

Kellerhals, G. E. 1982. Laboratory Core Floods To Support the El Dorado Micellar/Polymer Project. *J Pet Technol* **34** (6): 1378–1388. SPE-8197-PA. https://doi.org/10.2118/8197-PA.

Knickerbocker, B. M., Pesheck, C. V., Davis, H. T. et al. 1982. Patterns of Three-Liquid-Phase Behavior Illustrated by Alcohol-Hydrocarbon-Water-Salt Mixtures. *J. Phys. Chem.* **86** (3): 393–400. https://doi.org/10.1021/j100392a022.

Knickerbocker, B. M., Pesheck, C. V., Scriven, L. E. et al. 1979. Phase Behavior of Alcohol-Hydrocarbon-Brine Mixtures. *J. Phys. Chem.* **83** (15): 1984–1990. https://doi.org/10.1021/j100478a012.

Knight, B. L. (ed.) 1977. Commercial Scale Demonstration, Enhanced Oil Recovery by Micellar-Polymer Flooding. M-1 Project, Design Report. US Department of Energy, Bartlesville, Oklahoma (April 1977). http://www.osti.gov/scitech/biblio/7101620/.

Krumrine, P. H., Falcone, J. S. Jr., and Campbell, T. C. 1982. Surfactant Flooding 1: The Effect of Alkaline Additives on IFT, Surfactant Adsorption, and Recovery Efficiency. *SPE J.* **22** (4): 503–513. SPE-8998-PA. https://doi.org/10.2118/8998-PA.

Lake, L. W. 1989. *Enhanced Oil Recovery*. Englewood Cliffs, New Jersey: Prentice-Hall Inc.

Lawson, J. B. 1978. The Adsorption of Non-Ionic and Anionic Surfactants on Sandstone and Carbonate. Presented at the SPE Fifth Symposium on Improved Methods for Oil Recovery, Tulsa, 16–19 April. SPE-7052-MS. https://doi.org/10.2118/7052-MS.

Lawson, J. B. and Dilgren, R. E. 1978. Adsorption of Sodium Alkyl Aryl Sulfonates on Sandstone. *SPE J.* **18** (1): 75–82. SPE-6121-PA. https://doi.org/10.2118/6121-PA.

Levitt, D. and Pope, G. A. 2008. Selection and Screening of Polymers for Enhanced-Oil Recovery. Presented at the SPE/DOE Improved Oil Recovery Symposium, Tulsa, 19–23, April. SPE-113845-MS. https://doi.org/10.2118/113845-MS.

Levitt, D., Jackson, A., Heinson, C. et al. 2006. Identification and Evaluation of High-Performance EOR Surfactants. Presented at the SPE/DOE Symposium on Improved Oil Recovery, Tulsa, 22–26 April. SPE-100089-MS. https://doi.org/10.2118/100089-MS.

Li, H., Xu, D., Jiang, J. et al. 2008. Performance and Effect Analysis of ASP Commercial Flooding in Central Xing 2 Area of Daqing Oil Field. Presented at the SPE/DOE Improved Oil Recovery Symposium, Tulsa, 19–23 April. SPE-114348-MS. https://doi.org/10.2118/114348-MS.

Liu, S., Zhang, D., Yan, W. et al. 2008. Favorable Attributes of Alkaline-Surfactant-Polymer Flooding. *SPE J.* **13** (1): 5–16. SPE-99744-PA. https://doi.org/10.2118/99744-PA.

Lopez-Salinas, J. L., Hirasaki, G. J., and Miller, C. A. 2011. Determination of Anhydrite in Reservoirs for EOR. Presented at the SPE International Symposium on Oilfield Chemistry, The Woodlands, Texas, USA, 11–13 April. SPE-141420-MS. https://doi.org/10.2118/141420-MS.

Lu, J., Britton, C., Solairaj, S. et al. 2012. Novel Large-Hydrophobe Alkoxy Carboxylate Surfactants for Enhance Oil Recovery. Presented at the SPE Improved Oil Recovery Symposium, Tulsa, 14–18 April. SPE-154261-MS. https://doi.org/10.2118/154261-MS.

Maerker, J. M. and Gale, W. W. 1992. Surfactant Flood Process Design for Loudon. *SPE Res Eng* **7** (1): 36–44. SPE-20218-PA. https://doi.org/10.2118/20218-PA.

Malmberg, E. W. and Smith, L. 1976. The Adsorption Losses of Surfactants in Tertiary Recovery Systems. In *Improved Oil Recovery by Surfactant and Polymer Flooding*, eds. D. O. Shah and R. S. Schechter, 275–291. New York City: Academic Press.

Malmberg, E. W., Gajderowicz, C. C., Martin, F. D., et al. 1982. Characterization and Oil Recovery Observations on a Series of Synthetic Petroleum Sulfonates. *SPE J.* **22** (2): 226–236. SPE-8323-PA. https://doi.org/10.2118/8323-PA.

Manji, K. H. and Stasiuk, B. W. 1987. Design Considerations for David Alkali/Polymer Waterflood During Uncertain World Oil Prices. Presented at the Petroleum Society of CIM Annual Technical Meeting, Calgary, 7–10 June. CIM 87-38-29.

Manning, C. D., Pesheck, C. V., Puig, J. E. et al. 1983. Measurement of Interfacial Tension. Technical Report, DOE/BC/10116-12, US Department of Energy, Washington, DC (1 November 1983).

Mayer, E. H., Berg, R. L., Carmichael, J. D. et al. 1983. Alkaline Injection for Enhanced Oil Recovery—A Status Report. *J Pet Technol* **35** (1): 209–221. SPE-8848-PA. https://doi.org/10.2118/8848-PA.

Meyers, J. J., Pitts, M. J., and Wyatt, K. 1992. Alkaline-Surfactant-Polymer Flood of the West Kiehl, Minnelusa Unit. Presented at the SPE/DOE Enhanced Oil Recovery Symposium, Tulsa, 22–24 April. SPE-24144-MS. https://doi.org/10.2118/24144-MS.

Meyers, K. O. and Salter, S. J. 1981. The Effect of Oil/Brine Ratio on Surfactant Adsorption From Microemulsion. *SPE J.* **21** (4): 500–512. SPE-8989-PA. https://doi.org/10.2118/8989-PA.

Miller, R. J., Rosenwald, G. W., and Howell, W. D. 1980. El Dorado Micellar-Polymer Demonstration Project: Fifth Annual Report for the Period September 1978–August 1979. Report DOE/ET/13070-53, US Department of Energy, Bartlesville Energy Technology Center, Bartlesville, Oklahoma (February 1980).

Mohammadi, H., Delshad, M., and Pope, G. A. 2009. Mechanistic Modeling of Alkaline/Surfactant/Polymer Floods. *SPE Res Eval & Eng* **12** (4): 512–527. SPE-110212-PA. https://doi.org/10.2118/110212-PA.

Nelson, R. C. 1981. Further Studies on Phase Relationships in Chemical Flooding. In *Surface Phenomena in Enhanced Oil Recovery*, ed. D. O. Shah, 73–104. New York City: Plenum Press. https://doi.org/10.1007/978-1-4757-0337-5_4.

Nelson, R. C. 1982. The Salinity-Requirement Diagram—A Useful Tool in Chemical Flooding Research and Development. *SPE J.* **22** (2): 259–270. SPE-8824-PA. https://doi.org/10.2118/8824-PA.

Nelson, R. C. 1983. The Effect of Live Crude on Phase Behavior and Oil-Recovery Efficiency of Surfactant Flooding Systems. *SPE J.* **23** (3): 501–510. SPE-10677-PA. https://doi.org/10.2118/10677-PA.

Nelson, R. C. and Pope, G. A. 1978. Phase Relationships in Chemical Flooding. *SPE J.* **18** (5): 325–338. SPE-6773-PA. https://doi.org/10.2118/6773-PA.

Nelson, R. C., Lawson, J. B., Thigpen, D. R. et al. 1984. Cosurfactant-Enhanced Alkaline Flooding. Presented at the SPE Enhanced Oil Recovery Symposium, Tulsa, 15–18 April. SPE-12672-MS. https://doi.org/10.2118/12672-MS.

Novosad, J. 1981. Adsorption of Pure Surfactant and Petroleum Sulfonate at the Solid-Liquid Interface. In *Surface Phenomena in Enhanced Oil Recovery*, ed. D. O. Shah, 675–694. New York City: Plenum Publishing.

Novosad, J. 1982. Surfactant Retention in Berea Sandstone—Effects of Phase Behavior and Temperature. *SPE J.* **22** (6): 962–970. SPE-10064-PA. https://doi.org/10.2118/10064-PA.

O'Lenick, A. J. and Bilbo, R. E. 1987. Guerbet Alcohols: A Versatile Hydrophobe. *Soap/Cosmetics/Chemical Specialties* (April): 52–55,115.

Ottewill, R. H. 1984. Introduction. In *Surfactants*, ed. T. F. Tadros, 1–18. San Francisco: Academic Press.

Poettmann, F. H. 1974. Microemulsion Flooding. In *Secondary and Tertiary Oil Recovery Processes*, 67–93. Oklahoma City, Oklahoma: Interstate Oil Compact Commission.

Pope, G. A. and Nelson, R. C. 1978. A Chemical Flooding Compositional Simulator. *SPE J.* **18** (5): 339–354. SPE-6725-PA. https://doi.org/10.2118/6725-PA.

Pope, G. A., Lake, L. W., and Helfferich, F. G. 1978. Cation Exchange in Chemical Flooding: Part I—Basic Theory Without Dispersion. *SPE J.* **18** (6): 418–434. SPE-6771-PA. https://doi.org/10.2118/6771-PA.

Pope, G. A., Lake, L. W., and Sepehrnoori, K. 1987. Modeling and Scale-Up of Chemical Flooding—Final Report. Report DOE/BC/10846-6, US Department of Energy, Washington, DC (March 1987).

Pope, G. A., Tsaur, K., Schechter, R. S. et al. 1982. The Effect of Several Polymers on the Phase Behavior of Micellar Fluids. *SPE J.* **22** (6): 816–830. SPE-8826-PA. https://doi.org/10.2118/8826-PA.

Pope, G. A., Wang, B., and Tsaur, K. 1979. A Sensitivity Study of Micellar/Polymer Flooding. *SPE J.* **19** (6): 357–368. SPE-7079-PA. https://doi.org/10.2118/7079-PA. https://doi.org/10.2118/7079-PA.

Puerto, M. C. and Reed, R. L. 1983. A Three-Parameter Representation of Surfactant/Oil/Brine Interaction. *SPE J.* **23** (4): 669–682. SPE-10678-PA. https://doi.org/10.2118/10678-PA.

Puig, J. E., Franses, E. I., Davis, H. T. et al. 1979. On Interfacial Tensions Measured With Alkyl Aryl Sulfonate Surfactants. *SPE J.* **19** (2): 71–82. SPE-7055-PA. https://doi.org/7055-PA.

Pursley, S. A. and Graham, H. L. 1975. Borregos Surfactant Pilot Test. *J Pet Technol* **27** (6): 695–700. SPE-4084-PA. https://doi.org/10.2118/4084-PA.

Pursley, S. A., Healy, R. N., and Sandvik, E. I. 1973. A Field Test of Surfactant Flooding, Loudon, Illinois. *J Pet Technol* **25** (7): 793–802. SPE-3805-PA. https://doi.org/10.2118/3805-PA.

Putz, A., Chevalier, J. P., Stock, G. et al. 1981. A Field Test of Microemulsion Flooding, Chateaurenard Field, France. *J Pet Technol* **33** (4): 710–718. SPE-8198-PA. https://doi.org/10.2118/8198-PA.

Rathmell, J. J., Smith, F. W., Salter, S. J. et al. 1978. Evaluation of the Optimal Salinity Concept for Design of a High Water Content Micellar Fluid. Presented at the SPE Symposium on Improved Methods for Oil Recovery, Tulsa, 16–19 April. SPE-7067-MS. https://doi.org/10.2118/7067-MS.

Ray, R. M. and Munoz, J. D. 1986. Chemical Flood Predictive Model. Report DOE/BC-86/11/SP(DE7001208), US Department of Energy, Washington, DC (December 1986).

Reppert, T. R., Bragg, J. R., Wilkinson, J. R. et al. 1990. Second Ripley Surfactant Flood Pilot Test. Presented at the SPE/DOE Enhanced Oil Recovery Symposium, Tulsa, 22–25 April. SPE-20219-MS. https://doi.org/10.2118/20219-MS.

Rivenq, R., Sardin, M., Scheich, D. et al. 1985. Sodium Carbonate Preflush: Theoretical Analysis and Application to Chateaurenard Field Test. Presented at the SPE Annual Technical Conference and Exhibition, Las Vegas, 22–25 September. SPE-14294-MS. https://doi.org/10.2118/14294-MS.

Rosen, M. J. 1978. *Surfactants and Interfacial Phenomena*. New York City: John Wiley & Sons Inc.

Saad, N., Pope, G. A., and Sepehrnoori, K. 1989. Simulation of Big Muddy Surfactant Pilot. *SPE Res Eng* **4** (1): 24–34. SPE-17549-PA. https://doi.org/10.2118/17549-PA.

Salager, J. L., Morgan, J. C., Schechter, R. S. et al. 1979. Optimum Formulation of Surfactant/Water/Oil Systems for Minimum Interfacial Tension or Phase Behavior. *SPE J.* **19** (2): 107–115. SPE-7054-PA. https://doi.org.10.2118/7054-PA.

Salter, S. J. 1977. The Influence of Type and Amount of Alcohol on Surfactant-Oil-Brine Phase Behavior and Properties. Presented at the SPE Annual Fall Technical Conference and Exhibition, Denver, 9–12 October. SPE-6843-MS. https://doi.org/10.2118/6843-MS.

Salter, S. J. 1978. Selection of Pseudo-Components in Surfactant-Oil-Brine-Alcohol Systems. Presented at the SPE Symposium on Improved Methods of Oil Recovery, Tulsa, 16–19 April. SPE-7056-MS. https://doi.org/10.2118/7056-MS.

Salter, S. J. 1983. Optimizing Surfactant Molecular Weight Distribution: I. Sulfonate Phase Behavior and Physical Properties. Presented at the SPE Annual Technical Conference and Exhibition, San Francisco, 5–8 October. SPE-12036-MS. https://doi.org/10.2118/12036-MS.

Salter, S. J. 1986. Criteria for Surfactant Selection in Micellar Flooding. Presented at the SPE International Meeting on Petroleum Engineering, Beijing, 17–20 March. SPE-14106-MS. https://doi.org/10.2118/14106-MS.

Satter, A., Shum, Y-M., and Widmyer, R. H. 1980. Single Well Cyclic Method for Determining In-Situ Chemical Retention. Presented at the SPE/DOE Enhanced Oil Recovery Symposium, Tulsa, 20–23 April. SPE-8837-MS. https://doi.org/10.2118/8837-MS.

Scriven, L. E. 1976. Equilibrium Bi-Continuous Structures. In *Micellization, Solubilization and Microemulsions*, ed. K. L. Mittal. New York City: Plenum Press.

Shah, K. D., Green, D. W., Michnick, M. J. et al. 1981. Phase Boundaries of Microemulsions as a Way of Characterizing Hydrocarbon Mixtures. *SPE J.* **21** (6): 763–770. SPE-8986-PA. https://doi.org/10.2118/8986-PA.

Shahin, G. T. and Thigpen, D. R. 1996. Injecting Polyacrylamide Into Gulf Coast Sands: The White Castle Q Sand Polymer-Injectivity Test. *SPE Res Eng* **11** (3): 174–180. SPE-24119-PA. https://doi.org/10.2118/24119-PA.

Sharma, A., Azizi-Yarand, A., Clayton, B. et al. 2013. The Design and Execution of an Alkaline/Surfactant/Polymer Pilot Test. *SPE Res Eval & Eng* **16** (4): 423–431. SPE-154318-PA. https://doi.org/10.2118/154318-PA.

Skauge, A. and Fotland, P. 1990. Effect of Pressure and Temperature on the Phase Behavior of Microemulsions. *SPE Res Eng* **5** (4): 601–608. SPE-14932-PA. https://doi.org/10.2118/14932-PA.

Smith, L. and Malmberg, E. W. 1975. Measurement of Surfactant Loss in Porous Media. Presented at the SPE International Symposium on Oilfield Chemistry, Dallas, 16–17 January. SPE-5306-MS. https://doi.org/10.2118/14932-PA.

Smith, L., Malmberg, E. W., Kelley, H. W. et al. 1975. The Quantitative Analysis of Adsorbed Petroleum Sulfonates. Presented at the SPE California Regional Meeting, Ventura, California, USA, 2–4 April. SPE-5369-MS. https://doi.org/10.2118/5369-MS.

Solairaj, S., Britton, C., Lu, J. et al. 2012. New Correlation To Predict the Optimum Surfactant Structure for EOR. Presented at the SPE Improved Oil Recovery Symposium, Tulsa, 14–18 April. SPE-154262-MS. https://doi.org/10.2118/154262-MS.

Somasundaran, P. and Hanna, H. S. 1979. Adsorption of Sulfonates on Reservoir Rocks. *SPE J.* **19** (4): 221–232. SPE-7059-PA. https://doi.org/10.2118/7059-PA.

Somasundaran, P. and Hanna, H. S. 1985. Adsorption/Desorption of Sulfonates by Reservoir Rock Minerals in Solutions of Varying Sulfonate Concentrations. *SPE J.* **25** (3): 343–350. SPE-10603-PA. https://doi.org/10.2118/10603-PA.

Somasundaran, P., Celik, M., Goyal, A. et al. 1984. The Role of Surfactant Precipitation and Redissolution in the Adsorption of Sulfonate on Minerals. *SPE J.* **24** (2): 233–239. SPE-8263-PA. https://doi.org/10.2118/8263-PA.

Stegemeier, G. L. 1977. Mechanism of Entrapment and Mobilization of Oil in Porous Media. In *Improved Oil Recovery by Surfactants and Polymer Flooding*, eds. D. O. Shah and R. S. Schechter, 55–91. New York City: Academic Press Inc. https://doi.org/10.1016/B978-0-12-641750-0.50007-4.

Stoll, W. M., al Shureqi, H., Finol, J. et al. 2011. Alkaline/Surfactant/Polymer Flood: From the Laboratory to the Field. *SPE Res Eval & Eng* **14** (6): 702–712. SPE-129164-PA. https://doi.org/10.2118/129164-PA.

Stover, D. F. 1988. Commercial Scale Demonstration Enhanced Oil Recovery by Micellar-Polymer Flood. Report DOE/ET/13077-130, US Department of Energy, Bartlesville, Oklahoma (November 1988).

Strange, L. K. and Talash, A. W. 1977. Analysis of Salem Low-Tension Waterflood Test. *J Pet Technol* **29** (11): 1380–1384. SPE-5885-PA. https://doi.org/10.2118/5885-PA.

Szabo, M. T. 1979. The Effect of Sulfonate/Polymer Interaction on Mobility Buffer Design. *SPE J.* **19** (1): 4–14. SPE-6201-PA. https://doi.org/10.2118/6201-PA.

Talley, L. D. 1988. Hydrolytic Stability of Alkylethoxy Sulfates. *SPE Res Eng* **3** (1): 235–242. SPE-14912-PA. https://doi.org/10.2118/14912-PA.

Tan, K. H. 1982. *Principles of Soil Chemistry.* New York City: Mercel Dekker Inc.

Tham, M. J., Nelson, R. C., and Hirasaki, G. J. 1983. Study of the Oil Wedge Phenomenon Through the Use of a Chemical Flood Simulator. *SPE J.* **23** (5): 746–758. SPE-10729-PA. https://doi.org/10.2118/10729-PA.

Tham, M. K. and Lorenz, P. B. 1981. The EACN of a Crude Oil: Variations With Cosurfactant and Water Oil Ratio. In *Enhanced Oil Recovery: Proceedings of the third European Symposium on Enhanced Oil Recovery, Bournemouth, UK, 21–23 September, 1981*, ed. F. J. Fayers, Chapter 6, 123–134. New York City: Elsevier.

Troiani, L. R., Davis, J. G., Ferrell, H. H. et al. 1979. Big Muddy Field Low Tension Flood Demonstration Project. First Annual Report, April 1978–March 1979, DOE/SF /01424-13, US Department of Energy, Bartlesville, Oklahoma (August 1979).

Trushenski, S. P. 1977. Micellar Flooding: Sulfonate-Polymer Interaction. In *Improved Oil Recovery by Surfactant and Polymer Flooding*, ed. D. O. Shah and R. S. Schechter, 555–575. New York City: Academic Press.

Trushenski, S. P., Dauben, D. L., and Parrish, D. R. 1974. Micellar Flooding—Fluid Propagation, Interaction, and Mobility. *SPE J.* **14** (6): 633–645. SPE-4582-PA. https://doi.org/10.2118/4582-PA.

UTCHEM-9.0 Technical Documentation, Vol. II. 2000. The University of Texas at Austin, Austin, Texas (July 2000).

Verkruyse, L. A. and Salter, S. J. 1985. Potential Use of Nonionic Surfactants in Micellar Flooding. Presented at the SPE Oilfield and Geothermal Chemistry Symposium, Phoenix, Arizona, USA, 9–11 April. SPE-13574-MS. https://doi.org/10.2118/13574-MS.

Wade, W. H., Morgan, J. C., Schechter, R. S., et al. 1978. Interfacial Tension and Phase Behavior of Surfactant Systems. *SPE J.* **18** (4): 242–252. SPE-6844-PA. https://doi.org/10.2118/6844-PA.

Wang, D., Cheng, J., Wu, J. et al. 1999. Summary of ASP Pilots in Daqing Oil Field. Presented at the SPE Asia Pacific Improved Oil Recovery Conference, Kuala Lumpur, Malaysia, 25–26 October. SPE-57288-MS. https://doi.org/10.2118/57288-MS.

Wang, F. H. L. 1993. Effects of Reservoir Anaerobic, Reducing Conditions on Surfactants Retention in Chemical Flooding. *SPE Res Eng* **8** (2): 108–116. SPE-22646-PA. https://doi.org/10.2118/22648-PA.

Wanosik, J. L., Treiber, L. E., Myal, F. R. et al. 1978. Sloss Micellar Pilot: Project Design and Performance. Presented at the SPE Symposium on Improved Methods for Oil Recovery, Tulsa, 16–19 April. SPE-7092-MS. https://doi.org/10.2118/7092-MS.

Widmyer, R. H., Satter, A., Frazier, G. D. et al. 1977. Low-Tension Waterflood Pilot at the Salem Unit, Marion County, Illinois—Part 2: Performance Evaluation. *J Pet Technol* **29** (8): 933–938. SPE-5833-PA. https://doi.org/10.2118/5833-PA.

Willhite, G. P., Green, D. W., Okoye, D. M. et al. 1980. A Study of Oil Displacement by Microemulsion Systems—Mechanisms and Phase Behavior. *SPE J.* 20 (6): 459–472. SPE-7580-PA. https://doi.org/10.2118/7580-PA.

Wilson, P. M., Murphy, C. L., and Foster, W. R. 1976. The Effects of Sulfonate Molecular Weight and Salt Concentration on the Interfacial Tension of Oil-Brine-Surfactant Systems. Presented at the SPE Improved Oil Recovery Symposium, Tulsa, 22–24 March. SPE-5812-MS. https://doi.org/10.2118/5812-MS.

Windsor, P. A. 1954. *Solvent Properties of Amphiphillic Compounds.* London: Butterworth Scientific Publication.

Wyatt, K., Pitts, M. J., and Surkalo, H. 2004. Field Chemical Flood Performance Comparison with Laboratory Displacement in Reservoir Core. Presented at the SPE/DOE Fourteenth Symposium on Improved Oil Recovery, Tulsa, 17–21 April. SPE-89385-MS. https://doi.org/10.2118/89385-MS.

Yang, H. T., Britton, C., Liyange, P. J. et al. 2010. Low-Cost, High-Performance Chemicals for Enhanced Oil Recovery. Presented at the SPE Improved Oil Recovery Symposium, Tulsa, 24–28 April. SPE-129978-MS. https://doi.org/10.2118/129978-MS.

Zhang, D. L., Liu, S., Puerto, M. et al. 2006. Wettability Alteration and Spontaneous Imbibition in Oil-Wet Carbonate Formations. *Journal of Petroleum Science and Engineering* **52** (1–4): 213–226. https://doi.org/10.1016/j.petrol.2006.03.009.

Zhao, P., Jackson, A. C., Britton, C. et al. 2008. Development of High-Performance Surfactants to Difficult Oils. Presented at the SPE/DOE Improved Oil Recovery Symposium, Tulsa, 19–23 April. SPE-113432-MS. https://doi.org/10.2118/113432-MS.

Zhu, Y., Hou, Q., Liu, W. et al. 2012. Recent Progress and Effects Analysis of ASP Flooding Field Tests. Presented at the SPE Improved Oil Recovery Symposium, Tulsa, 14–18 April. SPE-151285-MS. https://doi.org/10.2118/151285-MS.

Ziegler, V. M. and Handy, L. L. 1981. Effect of Temperature on Surfactant Adsorption in Porous Media. *SPE J.* **21** (2): 218–228. SPE-8264-PA. https://doi.org/10.2118/8264-PA.

Chapter 8

Thermal Recovery Processes

8.1 Introduction

Thermal recovery processes rely on the use of thermal energy in some form both to increase the reservoir temperature, thereby reducing oil viscosity, and to displace oil to a producing well. Four processes have evolved over the past 50 years to the point of commercial application. These are cyclic steam stimulation (CSS), steam-assisted gravity drainage (SAGD), steamdrive, and forward in-situ combustion. The history of the development of thermal recovery processes is well-documented by Prats (1982) and Butler (1991).

The motivation for developing thermal recovery processes was the existence of major reservoirs all over the world that were known to contain billions of barrels of heavy oil and tar sands that could not be produced with conventional techniques. In many reservoirs, the oil viscosity was so high that primary recovery on the order of a few percent of original oil in place (OOIP) was common. In some reservoirs, primary recovery was negligible.

This chapter introduces mechanisms contributing to oil displacement by thermal recovery processes. Simple models are used to describe the displacement processes. Including all such models is not possible. Extensive treatment of thermal models with comparisons are found in Prats (1982), Boberg (1988), Burger et al. (1985), and Butler (1991). Field results are introduced where appropriate.

Thermal recovery processes are the most advanced enhanced-oil recovery (EOR) processes and contribute significant amounts of oil to daily production. Most thermal oil production is the result of cyclic steam injection and steamdrive. In 2014, worldwide production from cyclic steam, SAGD, and steamdrive was more than 1,500,000 B/D. Consequently, more emphasis is placed on these processes. Reservoirs under treatment by these processes are shallow (generally < 3,500 ft in depth).

In-situ combustion has been tested extensively in a variety of reservoirs, with mixed results. Daily production from in-situ combustion in 2014 was approximately 22,000 B/D. Although steam injection and in-situ combustion have been used successfully in the same reservoir (Ramey et al. 1992), steam injection has been the process of choice for reasons other than process efficiency. In-situ combustion is the only thermal recovery process that can be used in deep, high-pressure reservoirs.

Thermal recovery processes are applicable to a wide range of reservoirs. Table 1.1 presents screening criteria for EOR processes that were introduced in Chapter 1. **Table 8.1** summarizes the criteria for thermal recovery processes. These criteria are to be used as a guide in selecting candidates for thermal recovery processes. Exceptions to the criteria may be found in specific reservoirs.

Many of the criteria are identical for steam and in-situ combustion. Consequently, it is not uncommon to find that a reservoir will satisfy the criteria for both processes. Three criteria where there are significant differences will be discussed in this section: depth, reservoir pressure, and average reservoir permeability.

Steam processes are limited to depths on the order of 3,000 ft because wellbore heat losses can become excessive. As discussed later in this chapter, insulated injection tubing can be used to reduce heat losses and increase this depth. Although the depth limitation for in-situ combustion is suggested to be 11,500 ft, this is not a process limitation if air can be injected at the reservoir pressure and the crude oil deposits sufficient fuel to sustain the combustion front.

Reservoir pressure is the second criterion where the two processes differ. In steamdrive projects, the fraction of energy transported as latent heat decreases as pressure increases. The temperature of steam increases with pressure, as does the heat loss to the surroundings. Consequently, under the same conditions, the volume of the reservoir that can be contacted by steam decreases with injection pressure. Reservoir pressure becomes a limiting factor in the application of steamdrive processes. The maximum injection pressure is easy to establish. It is not practical to inject steam in the field near the critical pressure of steam, which is 3,206.2 psi (critical temperature is 705.4°F). Although steam has been injected at pressures of 2,500 psi in field projects, most successful steam-injection projects operate at pressures on the order of 1,500 psi or lower. Reservoir pressure is not limited by a similar mechanism in the in-situ-combustion process. Reservoir pressure for in-situ-combustion projects is affected by compression costs and injection rates. Projects are under consideration that have reservoir pressures significantly higher than the 2,000 psi in the screening criteria.

Screening Parameters	Thermal Recovery	
	Steam	In-Situ Combustion
Oil gravity (°API)	10 to 34	10 to 35
In-situ oil viscosity, μ (cp)	$\leq 15{,}000$	$\leq 5{,}000$
Depth, D (ft)	$\leq 3{,}000$	$\leq 11{,}500$
Pay-zone thickness, h (ft)	≥ 20	≥ 20
Reservoir temperature, T_r (°F)	—	—
Porosity, ϕ (fraction)	≥ 0.20	≥ 0.20
Average permeability, k (md)	250	35
Transmissibility, kh/μ (md-ft/cp)	≥ 5	≥ 5
Reservoir pressure, p_r (psi)	$\leq 1{,}500$	$\leq 2{,}000$
Minimum oil content at start of process, $S_o \times \phi$ (fraction)	≥ 0.10	≥ 0.08
Salinity of formation brine, total dissolved solids (ppm)	—	—
Rock type	Sandstone or carbonate	Sandstone or carbonate

Table 8.1—Screening parameters for thermal recovery processes.

Permeability of the reservoir is the third criterion where there are substantial differences between steamdrive and in-situ combustion. In-situ combustion can be applied in reservoirs that have lower permeability than the permeability limit for steam-drive because the air-injection rates are sufficient to sustain the combustion front. In contrast, the steam zone can advance only as long as heat losses from the steam zone to the surrounding formations can be maintained by the steam-injection rate. In low-permeability reservoirs, it is not possible to inject steam at sufficient rates to propagate a steam zone appreciable distances into the reservoir. Models to predict the minimum rate of steam injection are presented later in this chapter.

Thermal recovery processes rely on the fact that crude oil viscosity decreases markedly with temperature, as illustrated for selected crude oils in **Fig. 8.1** (Braden 1966). Thus, reservoir heating is an essential part of all thermal recovery processes. We begin our examination of thermal recovery methods by discussing wellbore heat losses when hot fluid is injected from the surface to heat a reservoir. Fundamental concepts of reservoir heating by hot-fluid injection are introduced. Then, we focus on oil recovery processes, beginning with CSS, SAGD, and steam displacement. We conclude with in-situ combustion, in which reservoir heating is accomplished by injection of air or oxygen to react with the reservoir oil. We note that in-situ combustion is an effective displacement process and is not limited to reservoirs with viscous oils.

8.2 Heat Losses During Steam Injection

Steam for cyclic steam, SAGD, and steamdrive processes is usually supplied by once-through steam generators that produce 80%-quality steam at the steam generator outlet. In some cases, steam is generated at a cogeneration plant, where natural gas is used to fire steam generators and part of the steam is used to produce electricity. Economy of scale dictates use of large generators located at central sites. Steam is distributed to individual wells through a series of insulated lines.

Fig. 8.2 depicts the route taken by steam leaving the generator. Some heat is lost in the distribution lines. Usually, steam is injected into tubing set on a packer with the annulus between the tubing and the packer boiled dry. Heat is lost in the wellbore by convection and radiation between the tubing string and the casing. The rate of heat lost can be reduced by use of insulated tubing if economically justified.

The temperature and quality of the steam that reaches the injection point in a well are determined by a complex interaction between heat losses and frictional losses as the steam flows. This section presents methods of estimating heat losses during steam injection.

8.2.1 Heat-Loss Rate From Distribution Lines. Most steam-distribution lines are insulated with such material as calcium silicate wrapped with an aluminum cover, as shown in **Fig. 8.3.** Heat transfer occurs by conduction through the insulation to the aluminum cover, where heat is lost by radiation and a combination of forced or natural convection to the surroundings. Heat-transfer resistances caused by scale on the surface of the steam line and conduction through the steel are usually negligible. In addition, heat-transfer coefficients for condensation are so large that the temperature of the distribution line is essentially the same as the steam temperature.

Heat transfer through the insulation occurs by conduction and is given by

$$\dot{Q}_{\ell p} = \frac{2\pi k_{hins}\left(T_s - T_{is}\right)}{\ln\left(\dfrac{r_{is}}{r_{do}}\right)}, \dotfill (8.1)$$

where $\dot{Q}_{\ell p}$ = heat-loss rate per unit length of pipe, Btu/(hr-ft); r_{is} = radius of outside surface of insulation, ft; r_{do} = radius of outside surface of pipe, ft; k_{hins} = thermal conductivity of insulation, Btu/(hr-ft-°F); T_s = temperature of the steam, °F; and T_{is} = temperature of the surface of the insulation, °F.

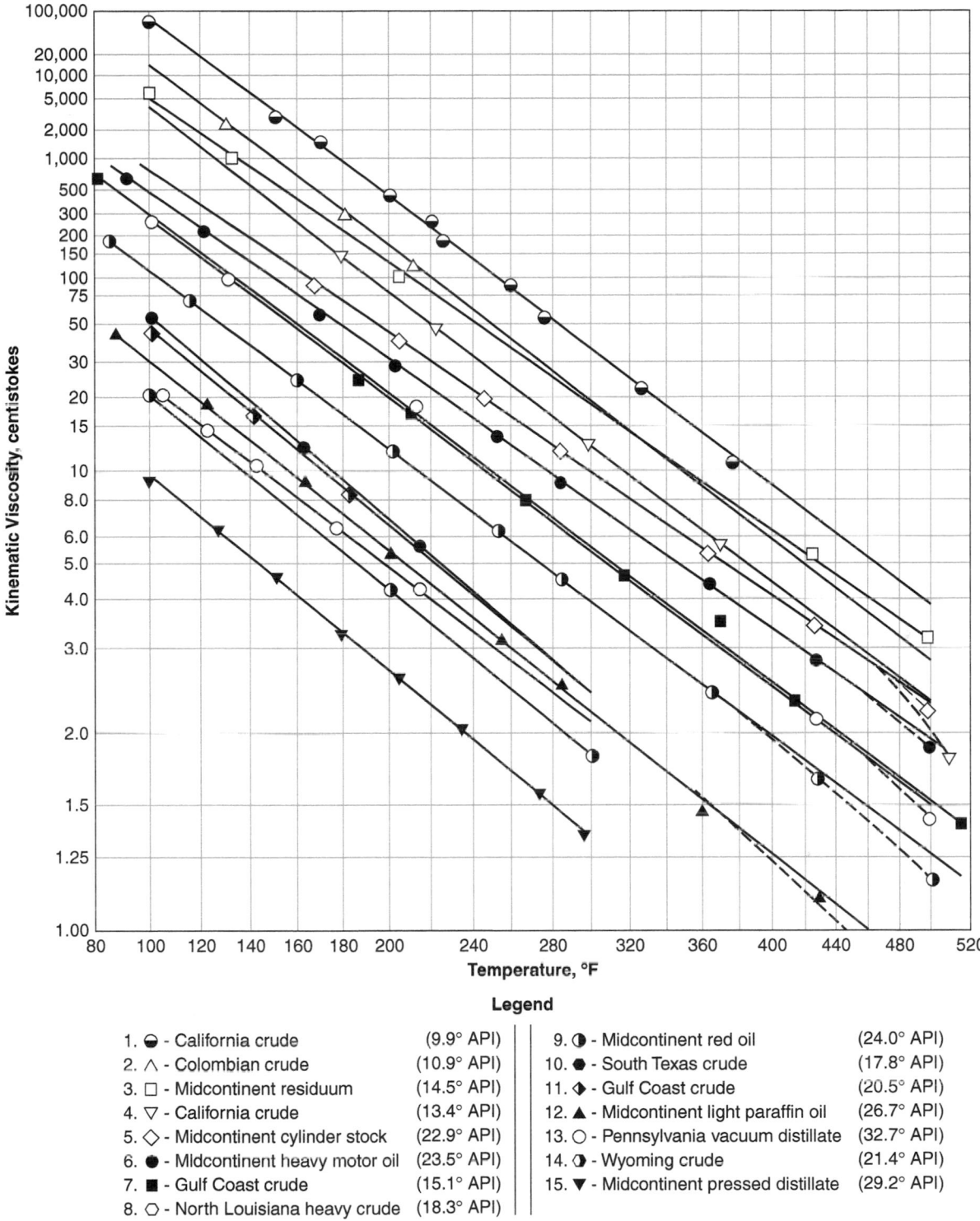

Fig. 8.1—Typical viscosity/temperature relationship for a heavy oil (Braden 1966).

Legend

1. ◓ - California crude   (9.9° API)
2. △ - Colombian crude   (10.9° API)
3. □ - Midcontinent residuum   (14.5° API)
4. ▽ - California crude   (13.4° API)
5. ◇ - Midcontinent cylinder stock   (22.9° API)
6. ● - Midcontinent heavy motor oil   (23.5° API)
7. ■ - Gulf Coast crude   (15.1° API)
8. ○ - North Louisiana heavy crude   (18.3° API)

9. ◑ - Midcontinent red oil   (24.0° API)
10. ● - South Texas crude   (17.8° API)
11. ◆ - Gulf Coast crude   (20.5° API)
12. ▲ - Midcontinent light paraffin oil   (26.7° API)
13. ○ - Pennsylvania vacuum distillate   (32.7° API)
14. ◑ - Wyoming crude   (21.4° API)
15. ▼ - Midcontinent pressed distillate   (29.2° API)

The heat that flows through the insulation is exchanged with the surroundings through the mechanisms of convection and radiation. These heat-transfer mechanisms act in parallel. Thus,

$$\dot{Q}_{\ell p} = \dot{Q}_{\ell r} + \dot{Q}_{\ell c}, \dotfill (8.2)$$

where $\dot{Q}_{\ell r}$ = heat-loss rate per unit length of pipe owing to radiation, Btu/(hr-ft), and $\dot{Q}_{\ell c}$ = heat-loss rate per unit length of pipe as a result of convection, Btu/(hr-ft).

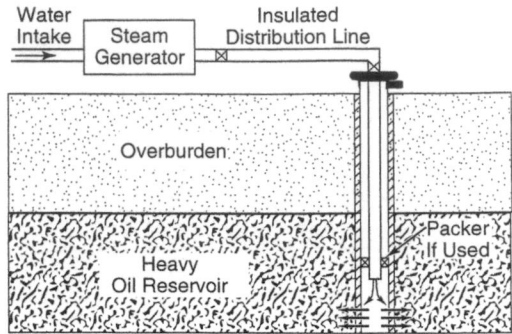

Fig. 8.2—Steam flow in cyclic-steam-injection and steamdrive projects.

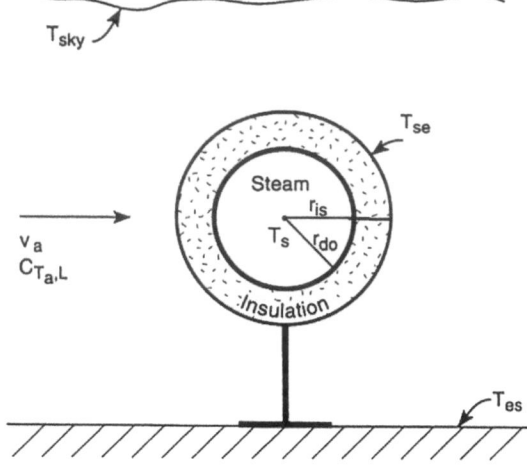

Fig. 8.3—Insulated steam-distribution line.

Convective heat losses from a heated pipe are quite sensitive to the velocity and direction of the wind. When the wind velocity is zero, heat losses occur as a result of free convection currents generated by the temperature difference between the surface of the heated pipe and the ambient air. At high wind velocities, forced convection dominates the heat exchange.

Heat transfer from convection is described in terms of a heat-transfer coefficient defined by

$$\dot{Q}_{\ell c} = 2\pi r_{is} h_c \left(T_{is} - \overline{T}_a\right), \dotfill (8.3)$$

where h_c = heat-transfer coefficient for convection based on the outer surface area of the insulation $(2\pi r_{is}L)$ and the temperature difference between the surface of the insulation and the ambient air, Btu/(hr-ft²-°F), and $\overline{T}_a$ = average ambient air temperature, °F.

The heat-transfer coefficient for convection can be estimated from correlations developed for certain geometrical configurations. It is convenient to use a correlation developed for heat transfer from a circular cylinder to wind flowing at velocity v_w normal to the axis of the cylinder. Correlations are available that cover a wide range of wind velocities. When forced convection dominates the heat-transfer process, the following correlation from McAdams (1954) may be used.

$$h_{fc} = \frac{0.12 k_{ha}}{r_{ec}} N_{Re}^{0.6}, \dotfill (8.4)$$

where h_{fc} = heat-transfer coefficient for forced convection, Btu/(hr-ft²-°F); k_{ha} = thermal conductivity of air, Btu/(hr-ft²-°F/ft); r_{ec} = external radius of conduit exposed to air, ft; and N_{Re} = modified Reynolds number for air flow normal to the pipe.

$$N_{Re} = 4,365 \frac{r_{ec} v_w \rho_a}{\mu_a}, \dotfill (8.5)$$

where v_w = wind velocity normal to the pipe, miles/hr; ρ_a = density of air evaluated at ambient air temperature, lbm/ft³; and μ_a = viscosity of air, cp. The properties of the air $(K_{ha}$ and $\mu_a)$ are evaluated at the average film temperature $\left(T_{sa} + \overline{T}_a\right)/2$. This leads to an iterative solution because T_{se}, the outside surface temperature, is not known a priori. Eq. 8.4 is valid for $1,000 \leq N_{Re} \leq 50,000$.

Heat loss from the cylindrical surface depicted in Fig. 8.3 owing to radiation is estimated with

$$\dot{Q}_{\ell r} = \pi r_r \varepsilon \sigma \left[\left(T_{sea}^4 - T_{skya}^4\right) + \left(T_{sea}^4 - \overline{T}_{esa}^4\right)\right], \dotfill (8.6)$$

where ε = emissivity of the surface; σ = Stefan-Boltzman constant, 1.713×10^{-9} Btu/hr-ft²-°R⁴; r_r = radius of the surface that is radiating, ft; T_{sea} = absolute temperature of surface, °R = °F + 460; T_{skya} = average absolute temperature of the sky, °R = °F + 460; and $\overline{T}_{esa}$ = average absolute temperature of the surface of the earth underneath the pipe (see Fig. 8.3), °R = °F + 460.

Eq. 8.6 was derived by assuming that radiant-heat transfer occurs between the surface of the insulation (or outer pipe surface if not insulated) and the sky from the upper half of the surface. Radiating-heat loss also occurs from the lower half of the insulation surface to the Earth. The temperature of the sky, T_{skya}, ranges from 414°R for a clear sky to 515°R when the sky is warm and cloudy (Incropera and DeWitt 1990). Consequently, an average value (460°R) is used in heat-transfer calculations.

This analysis neglects the contact resistance between the insulation and the distribution line. Because the thermal conductivity of aluminum, commonly used to cover the insulation, is high, there is little temperature drop across the aluminum cover. Consequently, T_{is} is equal to the temperature of the outside surface of the aluminum protective covering.

Example 8.1 illustrates the computation of the heat-loss rate from a steam-distribution line.

Example 8.1—Heat Loss From a Steam-Distribution Line. Steam at 600°F is distributed to injection wells through 3-in. line pipe. The lines are bare but will be insulated by adding 1 in. of calcium silicate insulation. Compare the heat-loss rate, in Btu/(hr-ft), for the bare line with that of the insulated line. The calcium silicate insulation will be protected with a thin aluminum covering (ε_{Al} = 0.76). The estimate will be made for average annual conditions. Consider the wind velocity to be 15 miles/hr and the average air temperature to be 70°F. The mean temperature of the earth surface is 55°F. The average sky temperature will be taken as 0°F (460°R).

Solution. Part 1: Heat Loss From Uninsulated Pipe. We begin by computing the heat loss from the bare line pipe. To compute the forced convection heat-transfer coefficient, h_c, from Eqs. 8.4 and 8.5, it is necessary to estimate the density and the viscosity of the air at the average film temperature.

$$\bar{T}_f = \left(T_s + \bar{T}_a\right)/2$$
$$= \left(600 + 70\right)/2$$
$$= 335°F.$$

The following values were obtained from references at 335°F and 14.696 psia. μ_a = 0.0246 cp and k_{ha} = 0.0211 Btu/(hr-ft-°F). The density of air is evaluated at the average ambient temperature of 70°F and 14.696 psia. The value of ρ_a is 0.0750 lbm/ft³.
The Reynolds number is computed from Eq. 8.5.

$$N_{Re} = 4,365 \frac{r_{ec}\rho_a v_w}{\mu_a}$$
$$= \frac{\left(4,365\right)\left(3.5/24 \text{ ft}\right)\left(0.0750 \text{ lbm/ft}^3\right)\left(15 \text{ miles/hr}\right)}{0.246 \text{ cp}}$$
$$= 29,111.$$

Because this is within the range of $1,000 < N_{Re} < 50,000$, the following correlation is valid.

$$h_{fc} = \frac{0.12 k_{ha} N_{Re}^{0.6}}{r_{ec}}$$
$$= \frac{\left(0.12\right)\left(0.021 \text{ Btu/hr-ft-°F}\right)\left(29,111\right)^{0.6}}{\left(3.5/24 \text{ ft}\right)}$$
$$= 8.28 \text{ Btu/}\left(\text{hr-ft}^2\text{-°F}\right).$$

The heat-transfer rate resulting from forced convection is

$$\dot{Q}_{lc} = 2\pi r_{do} h_{fc}\left(T_{se} - T_a\right)$$
$$= \left(2\right)\left(3.14159\right)\left(\frac{3.5}{24} \text{ ft}\right)\left(\frac{8.28 \text{ Btu}}{\text{hr-ft}^2\text{-°F}}\right)\left(600 - 70\right)$$
$$= 4,021 \text{ Btu/}\left(\text{hr-ft}\right).$$

Eq. 8.6 is used to compute the heat-transfer rate caused by radiation from the surface of the pipe. Assume that $T_{isa} = T_{sea}$; therefore,

$$\dot{Q}_r = \pi r_{do} \varepsilon \sigma \left[\left(T_{sea}^4 - T_{skya}^4\right) + \left(T_{sea}^4 - \bar{T}_{esa}^4\right)\right]$$
$$= \pi \left(3.5/24 \text{ ft}\right)\left(0.8\right)\left(1.713 \times 10^{-9} \text{ Btu/hr-ft}^2\text{-°R}^4\right)$$
$$\times \left\{\left[\left(1,060\right)^4 - \left(460\right)^4\right] + \left[\left(1,060\right)^4 - \left(515\right)^4\right]\right\}$$
$$= \left(6.278 \times 10^{-10}\right)\left(2.41 \times 10^{12}\right)$$
$$= 1,513 \text{ Btu/}\left(\text{hr-ft}\right).$$

Therefore, the total heat-loss rate is 4,021 + 1,513 or 5,534 Btu/(hr-ft). Approximately 73% of the heat loss is a result of forced convection.

Part 2: Heat Loss From Insulated Pipe. When 1 in. of calcium silicate insulation is added to the pipe, the heat loss is reduced substantially because of the low thermal conductivity of calcium silicate. The thermal conductivity of calcium silicate insulation varies with temperature, as shown by

$$k_{hins} = 0.0256 + \left(\overline{T}_i - 50\right)\left(3.67 \times 10^{-5}\right) \text{ Btu/(hr-ft)}, \dots\dots\dots\dots\dots\dots\dots\dots\dots\dots \quad (8.7)$$

where $\overline{T}_i$ = average temperature of the insulation, °F.
  Heat loss from the insulated line is given by Eq. 8.1.

$$\dot{Q}_\ell = \frac{2\pi k_{hins}\left(T_s - T_{is}\right)}{\ln\left(r_{is}/r_{do}\right)}.$$

The surface temperature is determined by setting $\dot{Q}_\ell = \dot{Q}_{\ell c} + \dot{Q}_{\ell r}$ and solving iteratively for T_{is}. This procedure is easily performed with a spreadsheet or a small computer program. The basic approach will be illustrated here.
  For purposes of computing k_{hins}, assume that $\overline{T}_i \approx \left(T_s + \overline{T}_a\right)/2 = 335°F$. Then,

$$k_{hins} = 0.0256 + (335 - 50)\left(3.65 \times 10^{-5}\right)$$

$$= 0.0356 \text{ Btu}/(\text{hr-ft-°F}).$$

Also note that r_{do} = 3.5/24 = 0.1458 ft and r_{is} = 4.5/24 = 0.1875 ft; thus,

$$\dot{Q}_\ell = \frac{(2)(3.14159)(0.0356)\left(600 - T_{is}\right)}{\ln\ (0.1875/0.1458)}$$

$$= 0.89\left(600 - T_{is}\right).$$

For this initial calculation, assume that h_c = 8.28 Btu/(hr-ft²-°F) as determined in Part 1. This will be somewhat in error because the viscosity and thermal conductivity are evaluated at $\overline{T}_f = \left(T_{is} + \overline{T}_a\right)/2$, which will decrease because of the reduced heat flow caused by the insulation.
  The heat flow owing to radiation will be reduced substantially because T_{is} will decrease. Applying Eq. 8.6, we obtain

$$\dot{Q}_r = \pi r_{iA\ell}\sigma\left[\left(T_{isa}^4 - T_{skya}^4\right) + \left(T_{isa}^4 - \overline{T}_{esa}^4\right)\right]$$

$$= (3.14159)(0.1875)(0.76)\left(1.713 \times 10^{-9}\right)$$

$$\times\left[2T_{isa}^4 - (460)^4 - (515)^4\right]$$

$$= 7.6687 \times 10^{-10}\left[2T_{isa}^4 - \left(1.1512 \times 10^{11}\right)\right].$$

The value of T_{is} that satisfies the following heat balance at the surface of the insulation is

$$\dot{Q}_\ell = Q_{\ell c} + \dot{Q}_{\ell r}$$

$$0.89\left(600 - T_{is}\right) = \left\{(2\pi)(0.1875)(8.28)\left(T_{is} - 70\right)\right\}$$

$$+ \left\{7.6687 \times 10^{-10}\left[2T_{isa}^4 - \left(1.1512 \times 10^{11}\right)\right]\right\}.$$

The solution to this equation was found by iterative techniques with a spreadsheet to be T_{is} = 107.64°F. This solution can be checked by substitution into Eq. 8.2.

$$\dot{Q}_\ell = 0.89(600 - 107.64)$$

$$= 438.4 \text{ Btu}/(\text{hr-ft}).$$

$$\dot{Q}_{\ell c} = 9.755(107.64 - 70)$$

$$= 367.2 \text{ Btu}/(\text{hr-ft}).$$

$$\dot{Q}_{\ell r} = 7.6687 \times 10^{-10}$$

$$\times\left[2(460 + 107.64)^4 - \left(1.1512 \times 10^{11}\right)\right]$$

$$= 70.95 \text{ Btu}/(\text{hr-ft}).$$

$$\dot{Q}_\ell + \dot{Q}_{\ell r} = 438.15 \text{ Btu}/(\text{hr-ft}),$$

which is close enough.

Therefore, adding 1 in. of calcium silicate insulation will reduce the heat loss from 5,534 to 438.4 Btu/(hr-ft).

It is recognized that the optimum thickness of the insulation is a tradeoff between the cost of heat losses and the capital cost of the insulation. An optimum insulation can be determined if desired.

In extremely cold environments, it is necessary to bury steam-injection lines or to protect insulated lines in utility boxes. Methods are available to estimate heat losses under these situations.

8.2.2 Wellbore Heat-Loss Rates. Fig. 8.4 shows a typical well completion for steam injection into a shallow sand. Tubing is run above the injection interval set on a packer. The annulus is boiled dry during the initial stages of steam injection, leaving it filled with a mixture of air and water vapor. Heat loss to the formation occurs by a series of heat-transfer mechanisms that includes conduction through the tubing, radiation and natural convection across the annulus, and conduction through the casing steel and cement into the formation. **Fig. 8.5** depicts the temperature distribution in this situation.

Heat transfer to the formation is a transient process. Ramey (1962) demonstrated that the heat-transfer rate at the drilled-hole radius could be approximated by

$$\dot{Q}_\ell = \left[2\pi k_{hf}\left(T_h - T_e\right)\right] / \left[\mathrm{f}(t)\right], \dotfill (8.8)$$

where k_{hf} = thermal conductivity of the formation, Btu/(hr-ft²-°F/ft); T_h = temperature at the cement/formation interface, °F; and f(t) = transient time function, dimensionless. Ramey's model assumes that heat flow occurs only in the radial direction.

For long injection times ($t > 1$ week), Ramey found that

$$\mathrm{f}(t) = \ln\left(2\sqrt{\alpha_f t} / r_{hd}\right) - 0.29. \dotfill (8.9)$$

Willhite (1967) integrated Ramey's model for heat loss into a general wellbore-heat-transfer model to predict casing temperatures and wellbore heat losses. In this model, heat transfer in the region between the flowing fluid and the cement/formation interface is assumed to be quasisteady. That is, any temperature change in the flowing fluid is propagated instantaneously across the annulus. Heat transfer in this region is approximated by a series of steady-state mechanisms and is given by

$$\dot{Q}_\ell = 2\pi r_{to} U_{to}\left(T_s - T_h\right), \dotfill (8.10)$$

where U_{to} = overall heat-transfer coefficient between the fluid and the cement/drilled-hole interface based on the outside tubing surface area and the temperature difference $T_s - T_h$, Btu/(hr-ft²-°F), and

$$T_h = \left\{T_s \mathrm{f}(t) + \left\lfloor k_{hf} / (r_{to} U_{to})\right\rfloor T_e\right\} / \left\{\mathrm{f}(t) + \left[k_{hf} / (r_{to} U_{to})\right]\right\}. \dotfill (8.11)$$

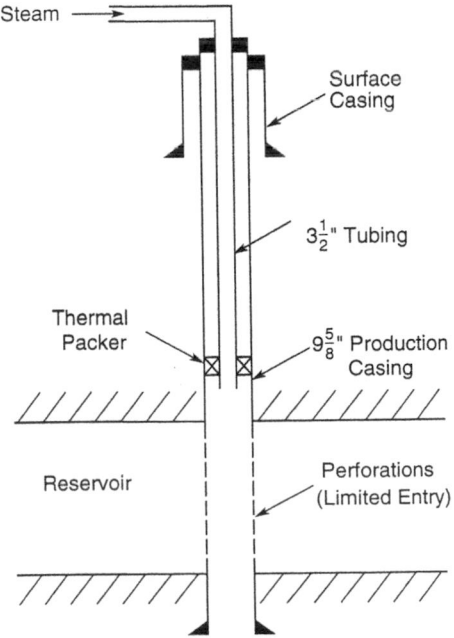

Fig. 8.4—Well completion for steam injection through tubing set on a packer.

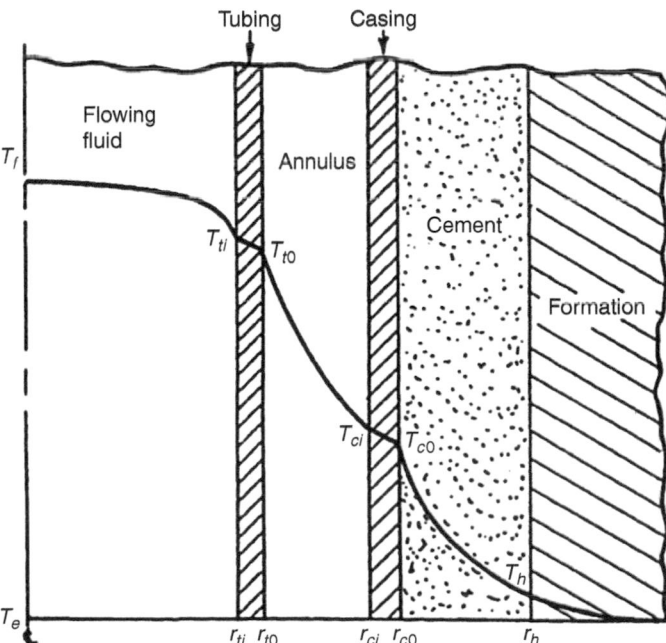

Fig. 8.5—Temperature distribution in an annular completion (Willhite 1967).

$$\frac{r_{to}U_{to}}{k_e} = \alpha t/r_{hd}^2$$

	0.01	0.02	0.05	0.1	0.2	0.5	1.0	2.0	5.0	10	20	50	100	∞
0.1	0.313	0.313	0.314	0.316	0.318	0.323	0.330	0.345	0.373	0.396	0.417	0.433	0.438	0.445
0.2	0.423	0.423	0.424	0.427	0.430	0.439	0.452	0.473	0.511	0.538	0.568	0.572	0.578	0.588
0.5	0.616	0.617	0.619	0.623	0.629	0.644	0.666	0.698	0.745	0.772	0.790	0.802	0.806	0.811
1.0	0.802	0.803	0.806	0.811	0.820	0.842	0.872	0.910	0.958	0.984	1.00	1.01	1.01	1.02
2.0	1.02	1.02	1.03	1.04	1.05	1.08	1.11	1.15	1.20	1.22	1.24	1.24	1.25	1.25
5.0	1.36	1.37	1.37	1.38	1.40	1.44	1.48	1.52	1.56	1.57	1.58	1.59	1.59	1.59
10.0	1.65	1.66	1.66	1.67	1.69	1.73	1.77	1.81	1.84	1.80	1.86	1.87	1.87	1.88
20.0	1.96	1.97	1.97	1.99	2.00	2.05	2.09	2.12	2.15	2.16	2.16	2.17	2.17	2.17
50.0	2.39	2.39	2.40	2.42	2.44	2.48	2.51	2.54	2.56	2.57	2.57	2.57	2.58	2.58
100.0	2.73	2.73	2.74	2.75	2.77	2.81	2.84	2.86	2.88	2.89	2.89	2.89	2.89	2.90

Table 8.2—Time function f(t) for the radiation boundary condition model.

For short times, the value of $f(t)$ is a function of U_{to} and can be obtained from **Table 8.2** (Willhite 1967) when the value of U_{to} is known.

An expression for the overall heat-transfer coefficient for the annulus configuration shown in Fig. 8.5, developed by Willhite (1967), is given by

$$U_{to} = \left[\frac{r_{to}}{r_{ti}h_f} + \frac{r_{to}\ln\frac{r_{to}}{r_{ti}}}{k_{htub}} + \frac{1}{(h_{nc}+h_r)} + \frac{r_{to}\ln\frac{r_{co}}{r_{ci}}}{k_{hcas}} + \frac{r_{to}\ln\frac{r_{hd}}{r_{ro}}}{k_{hcem}} \right]^{-1}, \dots\dots\dots\dots(8.12)$$

where h_f = film heat-transfer coefficient or condensation heat-transfer coefficient between flowing fluid and the inside of the tubing, Btu/(hr-ft²-°F); h_{nc} = heat-transfer coefficient for natural convection in the tubing/casing annulus based on the outside tubing surface and the temperature differences between the outside tubing and inside casing surfaces, Btu/(hr-ft²-°F); h_r = heat-transfer coefficient for radiation based on the outside tubing surface and the temperature difference between the outside tubing and inside casing surfaces, Btu/(hr-ft²-°F); k_{hcas} = thermal conductivity of the casing material at the average casing temperature, Btu/(hr-ft-°F); k_{hcem} = thermal conductivity of the cement at the average temperature and pressure of the cement, Btu/(hr-ft-°F); k_{htub} = thermal conductivity of the tubing, Btu/(hr-ft²-°F/ft); r_{ti} = inside radius of tubing, ft; r_{to} = outside radius of tubing, ft; r_{ci} = inside radius of casing, ft; r_{co} = outside radius of casing, ft; and r_{hd} = radius of drilled hole, ft.

In Prats (1982), Eq. 8.10 is written as

$$\dot{Q}_\ell = (T_s - T_h)/R, \dots\dots\dots\dots\dots\dots\dots\dots\dots (8.13)$$

where $R = 1/2\pi r_{to} U_{to}$ = thermal resistance per unit length of pipe, [Btu/(ft-D-°F)]⁻¹.

In most cases, h_f, k_{htub}, and k_{hcas} are so large that Eq. 8.12 can be approximated by

$$U_{to} = \left[\frac{1}{h_{nc}+h_r} + r_{to}\frac{\ln(r_{hd}/r_{co})}{k_{hcem}} \right]^{-1} \dots\dots\dots\dots\dots\dots (8.14)$$

Evaluation of U_{to} with Eq. 8.14 requires estimation of h_r and h_{nc}. The heat-transfer coefficient for radiation, h_r, is derived in Willhite (1967) and is given by

$$h_r = \sigma F_{tci}(T_{to}^2 + T_{ci}^2)(T_{to} + T_{ci}), \dots\dots\dots\dots\dots\dots (8.15)$$

where

$$\frac{1}{F_{tci}} = \frac{1}{\varepsilon_{to}} + \frac{r_{ti}}{r_{ci}}\left(\frac{1}{\varepsilon_{ci}} - 1\right); \dots\dots\dots\dots\dots\dots (8.16)$$

ε_{to} = emissivity of the external tubing surface, dimensionless; ε_{ci} = emissivity of the internal tubing surface, dimensionless; T = absolute temperature, °R; and σ = Stefan-Boltzman constant [1.713 × 10⁻⁹ Btu/(ft²-hr-°R⁴)].

Emissivities of materials are found in standard heat-transfer texts. The emissivity of mill scale may be taken as 0.9.

The value of h_r can be computed if T_{to} and T_{ci} are known. The casing temperature is given by

$$T_{ci} = T_h + \frac{r_{to}U_{to}}{k_{hcem}}\ln\left(\frac{r_{hd}}{r_{co}}\right)(T_s - T_h). \qquad \text{...} \quad (8.17)$$

Note that T_h and T_{ci} depend on U_{to} and time.

The heat-transfer coefficient for natural convection, h_{nc}, is given by

$$h_{nc} = k_{hc}/\left[r_{to}\ln(r_{ci}/r_{to})\right], \qquad \text{.......................................} \quad (8.18)$$

where $k_{hc}/k_{ha} = 0.049\left(N_{Gr}N_{Pr}\right)^{0.333}N_{Pr}^{0.074}$, $\qquad \text{.................................} \quad (8.19)$

and $N_{Gr} = \dfrac{(r_{ci} - r_{to})^3\,g\rho_{an}^2\,\beta\,(T_{to} - T_{ci})}{\mu_{an}^2}. \quad \text{...} \quad (8.20)$

$$N_{Pr} = C_{an}\mu_{an}/k_{ha}, \qquad \text{.............} \quad (8.21)$$

where N_{Gr} = Grashof number, dimensionless; N_{Pr} = Prandtl number, dimensionless; C_{an} = heat capacity of the fluid in the annulus at the average annulus temperature, Btu/(lbm-°F); k_{ha} = thermal conductivity of the air in the annulus at the average temperature and pressure of the annulus, Btu/(hr-ft-°F); k_{hc} = equivalent thermal conductivity of the annulus fluid with natural convection effects evaluated at the average temperature and pressure of the annulus, Btu/(hr-ft-°F); g = acceleration caused by gravity, 4.17×10^8 ft/hr²; $\overline{T}_{an}$ = average temperature of the fluid in the annulus, $(T_{ci}$ and $T_{to})/2.0$, °F; and β = thermal volumetric expansion coefficient of the fluid in the annulus, $°R^{-1} = \overline{T}_{an}$ for an ideal gas, or generally $= -(1/\rho_{an})(\partial\rho_{an}/\partial T)$, where p is the annulus pressure, and μ_{an} = viscosity of the fluid in the annulus at $\overline{T}_{an}$ and p, lbm/ft-hr.

Eq. 8.19 is valid when the product of the Grashof number and the Prandtl number $(N_{Gr}N_{Pr})$ is between 5×10^4 and 7.2×10^8. Extrapolation to less than 5×10^4 should be performed with caution. The limiting value of $k_{hc}k_{ha}$ is 1.0 at low values of $N_{Gr}N_{Pr}$. At atmospheric pressure (14.7 psia), $N_{Pr} \approx 0.71$ for the range of temperatures encountered in wellbore heat-loss calculations.

Fig. 8.6 plots U_{to} vs. tubing temperature for several modifications of an annulus completion. Steam was injected down 2⅞-in. tubing in 7-in. casing for 14 days. The casing was cemented to the surfaces in a 9⅝-in. hole with a high-temperature-resistance cement. **Table 8.3** summarizes parameters used in these calculations.

The five completions in Fig. 8.6 illustrate heat-transfer mechanisms in the annuli of steam- and hot-water-injection wells. A standard completion is 2⅞-in. tubing (mill scale surface) set on a packer with annulus at 0 psig. Radiation is the dominant heat-transfer mechanism in this case. In principle, radiation can be reduced by reducing the emissivity of the tubing with aluminum paint. The overall heat-transfer coefficient decreases by a factor of two or more in this example. However, in practice, it is difficult to keep the aluminum surface clean (oil free) so this method has limited effectiveness. When it is not possible to isolate the annulus with a packer,

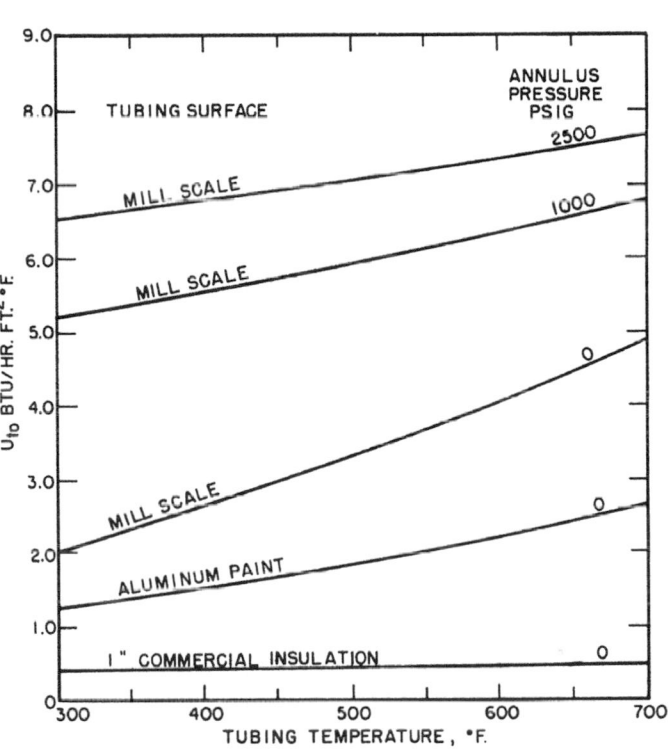

Fig. 8.6—Variation of U_{to} with tubing temperature for parameters of Table 8.3 (Willhite 1967).

Hole size (in.)	9.625
Casing (7 in., 26 lbm, J-55)	
Outer diameter (OD) (in.)	7.000
Inner diameter (ID) (in.)	6.276
Tubing (2⅞ in., 6.4 lbm, J-55)	
OD (in.)	2.875
ID (in.)	2.441
k_h (Btu/hr-ft-°F)	1.4
k_{hcem} (Btu/hr-ft-°F)	0.51
k_{hins} (Btu/hr-ft-°F)	$0.0256 + (T - 50)(3.67 \times 10^{-5})$
α (ft²/hr)	0.04
ε_{to} = (mill scale)	0.9
ε_{to} (aluminum paint)	0.4
T_e (°F)	80

Table 8.3—Parameters for calculation of U_{to} (Fig. 8.6) and casing temperature (Fig. 8.8).

high-pressure gas with a small purge has been used. Increasing the annulus pressure to 1,000 and 2,500 psig by injection of nitrogen causes large increases in the overall heat transfer because natural convection in the tubing/casing/annulus dominates wellbore heat transfer.

Lowest overall heat-transfer coefficients occur when the tubing string is insulated. Eq. 8.22 is the corresponding expression for U_{to} when insulated tubing is used. In Eq. 8.22, the thickness of the insulation is $r_{is} - r_{to}$ and the thermal resistance of the outer cover of the insulated tubing string is neglected. Heat-transfer coefficients h_{nc}' and h'_r are based on the outside radius of the insulation, r_{is}.

$$U_{to} = \left[\frac{r_{to}}{r_{ti}h_f} + \frac{r_{to}\ln\frac{r_{to}}{r_{ti}}}{k_{htub}} + \frac{r_{to}\ln\frac{r_{is}}{r_{to}}}{k_{hins}} + \frac{r_{to}}{r_{is}\left(h_{nc}' + h_r'\right)} + \frac{r_{to}\ln\frac{r_{co}}{r_{ci}}}{k_{hcas}} + \frac{r_{to}\ln\frac{r_{hd}}{r_{ro}}}{k_{hcem}} \right]^{-1} \quad \dots \dots \dots \dots \dots \dots \dots \dots \dots \dots \dots \quad (8.22)$$

In Fig. 8.6, 1 in. of calcium silicate insulation was used to insulate the tubing string. Heat losses in a dry annulus are significantly lower than in the other cases. The low effective thermal conductivity of the insulation dominates heat flow. Insulated injection strings are available with effective thermal conductivities as low as 0.006 to 0.0097 Btu/(hr-ft-°F). **Fig. 8.7** is a schematic of a high-performance insulated tubing string for steam injection (Kawasaki Thermal Systems 1989).

Computation of wellbore heat losses is complicated by the fact that radiation and natural convection coefficients, h_r and h_{nc}, respectively, vary with casing temperature, T_{ci}, which also varies with time and U_{to}. A trial-and-error or iterative solution is required to compute U_{to}, T_{ci}, and T_h, for a particular value of time.

The following procedure is suggested.

1. Select a value of U_{to} corresponding to the temperature of the fluid or tubing from Fig. 8.6 for the corresponding well completion, or estimate a value of U_{to} for the completion.
2. Determine f(t) from Eq. 8.9 or Table 8.2.
3. Calculate T_h using Eq. 8.11.
4. Calculate T_{ci} from Eq. 8.17.
5. Estimate h_r (Eq. 8.15) and h_{nc} (Eqs. 8.19 through 8.21).
6. Determine a new value of U_{to} from Eq. 8.12 or 8.22.
7. Compare the calculated value of U_{to} with the value used in Steps 2 through 5 and repeat Steps 2 through 6 until satisfactory agreement is obtained between successive trials. Convergence is rapid, and three iterations are usually sufficient. This is well-suited for computer or spreadsheet computation.
8. Compute heat loss.

Example 8.2 illustrates the computation of the overall heat-transfer coefficient and the heat loss for steam injection into a well.

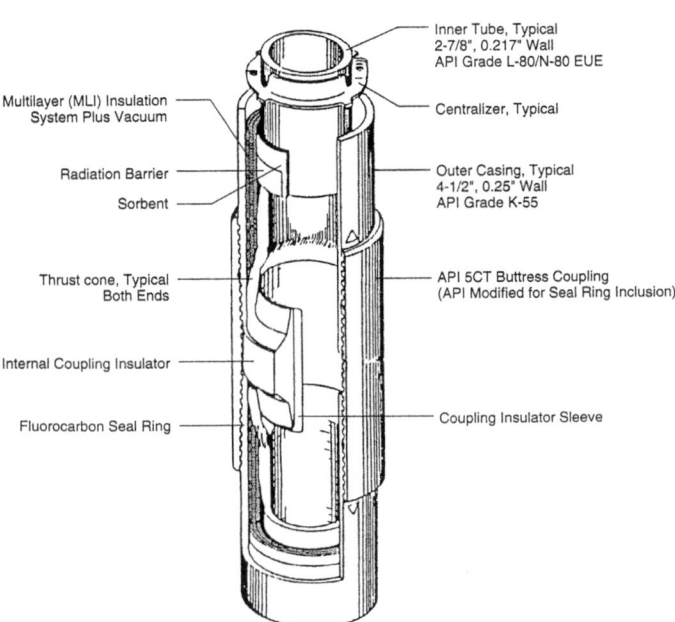

Fig. 8.7—Schematic of high-performance insulated tubing (Kawasaki Thermal Systems 1989).

Example 8.2—Wellbore Heat Loss During Steam Injection. Steam at 600°F is injected down 3½-in. tubing set on a packer in 9⅝-in., 53.5-lbm/ft, N-80 casing. The annulus contains a stagnant gas at 14.7 psia, and the casing is cemented to surface in a 12-in. hole. A temperature survey in the well indicates a mean subsurface temperature of 100°F. The reservoir is at 3,000 ft. Estimate the overall heat-transfer coefficient, average casing temperature, and wellbore heat loss after 21 days of continuous injection.

Data. $r_{to} = 0.146$ ft, $r_{ci} = 0.355$ ft, $r_{co} = 0.400$ ft, $r_h = 0.500$ ft, $a_f = 0.0286$ ft²/hr, $k_{hf} = 1.0$ Btu/(hr-ft²-°F/ft), $\varepsilon_{to} = \varepsilon_{ci} = 0.9$, and $k_{hcem} = 0.2$ Btu/(hr-ft²-°F/ft).

Solution.

1. Estimate U_{to} from Fig. 8.6 for an injection temperature of 600°F and the low-pressure annulus: $U_{to} = 4.05$ Btu/(hr-ft²-°F).

2. Calculate f(t). Because $t = 21$ days, Eq. 8.9 can be used.

$$f(t) = \left[\ln\left(\sqrt{\alpha_f t}/r_{hd} \right) \right] - 0.29$$

$$= \left\{ \ln\left[2\sqrt{(0.0286)(504)}/0.5 \right] \right\} - 0.29 = 2.43.$$

3. Calculate T_h (Eq. 8.11).

$$T_h = \left\{ T_s f(t) + \left[k_{hf}/(r_{to}U_{to}) \right] T_e \right\} / \left\{ f(t) + \left[k_{hf}/(r_{to}U_{to}) \right] \right\}$$

$$= \frac{(600)(2.43) + \left\{ 1.0/[(0.146)(4.05)] \right\}(100)}{2.43 + \left\{ 1.0/[(0.146)(4.05)] \right\}} = 395°F.$$

4. Calculate T_{ci}, neglecting casing and surface resistances (Eq. 8.17).

$$T_{ci} = T_h + \left(r_{to}U_{to}/k_{hcem} \right) \ln\left(r_{hd}/r_{co} \right)(T_s - T_h)$$

$$= 395 + \frac{(0.146)(4.05)}{0.2} \ln\frac{0.5}{0.4}(600 - 395) = 530°F.$$

5. Estimate h_r (Eqs. 8.15 and 8.16).

$$h_r = \sigma F_{tci}\left(T_{toa}^2 + T_{cia}^2 \right)\left(T_{toa} + T_{cia} \right),$$

$$\frac{1}{F_{tci}} = \frac{1}{\varepsilon_{to}} + \frac{r_{to}}{r_{ci}}\left(\frac{1}{\varepsilon_{ci}} - 1 \right),$$

and $F_{tci} = \dfrac{1}{\dfrac{1}{0.9} + \dfrac{0.146}{0.355}\left(\dfrac{1}{0.9} - 1.0 \right)} = 0.865.$

$$h_r = (0.865)(1.713 \times 10^{-9})\left[(600 + 460)^2 + (530 + 460)^2 \right]$$

$$\times \left[(600 + 460) + (530 + 460) \right] = 6.39 \text{ Btu}/\left(\text{hr-ft}^2\text{-°F} \right).$$

Estimate h_{nc} (Eqs. 8.18 through 8.21) using the following parameters: $\overline{T}_{an} = 565$°F, $\rho_{an} = 0.0388$ lbm/ft³, $\mu_{an} = 0.069$ lbm/ft-hr, $C_{an} = 0245$ Btu/lbm-°F, $k_{ha} = 0.0255$ Btu/hr-ft²-°F/ft, $\beta = 1/\overline{T}_{an}$, and $\beta = 9.75 \times 10^{+4}$ °R⁻¹.
Calculate N_{Pr} with Eq. 8.21.

$$N_{Pr} = C_{an}\mu_{an}/k_{ha}$$

$$= \left[(0.245)(0.069) \right]/0.0255 = 0.66.$$

$r_{ci} - r_{to} = 0.209$ ft.

Calculate N_{Gr} with Eq. 8.20.

$$N_{Gr} = \left[(r_{ci} - r_{to})^3 g\rho_{an}^2 \beta(T_{to} - T_{ci}) \right]/\mu_{an}^2$$

$$= \left[(0.209)^3 (4.17 \times 10^8)(0.0388)^2 (9.75 \times 10^{-4}) \times (600 - 530) \right]/(0.069)^2$$

$$= 8.26 \times 10^4.$$

And calculate k_{hc} from Eq. 8.19.

$$k_{hc}/k_{ha} = 0.049 \left(N_{Gr} N_{Pr} \right)^{0.333} N_{Pr}^{0.074}$$
$$= (0.049) \left[\left(8.26 \times 10^4 \right) (0.66) \right]^{0.333} (0.66)^{0.074}$$
$$= 1.81.$$

$$k_{hc} = 0.046 \text{ Btu/} \left(\text{hr-ft-°F} \right).$$

Then, $h_{nc} = k_{hc} r_{to} \ln \left(r_{ci}/r_{to} \right)$

$$= 0.046 / \left[0.146 \ln \left(0.355/0.146 \right) \right]$$
$$= 0.36 \text{ Btu/} \left(\text{hr-ft}^2 \text{-°F} \right).$$

6. Calculate U_{to} (Eq. 8.14).

$$U_{to} = \left[\frac{1}{h_{nc} + h_r} + r_{to} \frac{\ln \left(r_h/r_{co} \right)}{k_{cem}} \right]^{-1}$$
$$= \left[\frac{1}{6.39 + 0.36} + \frac{0.146 \ln \left(0.5/0.4 \right)}{0.2} \right]^{-1}$$
$$= 3.22 \text{ Btu/} \left(\text{hr-ft}^2 \text{-°F} \right).$$

7. Because the assumed and calculated values of U_{to} do not agree, repeat Steps 2 through 6 until agreement is obtained between two successive trials. **Table 8.4** presents results of successive iterations.

The wellbore heat loss can be calculated from Eq. 8.10.

$$\dot{Q}_{\ell wb} = (2\pi)(0.146)(3.14)(600 - 364)(3,000)$$
$$= 2,040,000 \text{ Btu/hr.}$$

If the steam quality at the wellhead, f_{sh}, is 0.75, the steam quality entering the formation, f_{sd}, can be estimated provided that the steam temperature does not vary with depth and the steam-injection rate is specified. Suppose that the steam-injection rate is 1,000 B/D cold-water equivalent (CWE). Then, the wellbore heat loss is equivalent to

$$\frac{2,040,000 \text{ Btu/hr}}{(1,000 \text{ B/D})(350 \text{ lbm bbl})(\text{D/24/hr})} = 139.9 \text{ Btu/lbm.}$$

The latent heat of vaporization at 600°F is approximately 548.4 Btu/lbm. Therefore, the change in quality would be

$$f_{sh} - f_{sd} = 139.9/548.4$$
$$= 0.26$$

and $f_{sd} = 0.49$.

Insulated injection tubing set on a packer is used in deep steam wells and in wells where injection temperatures are high (> 550°F). When the tubing/casing annulus is dry, heat losses can be reduced significantly over those with uninsulated tubing. However, if the annulus cannot be boiled dry, refluxing may occur as a result of boiling in the vicinity of the packer and condensation on the casing walls (Willhite and Griston 1987) and can reduce the effectiveness of the insulation.

Casing Temperatures. Casing temperatures increase during steam injection and are estimated as part of the wellbore-heat-loss calculations. Predicted casing temperatures for the five well completions shown in Fig. 8.6, are shown in **Fig. 8.8**

	Assumed			Calculated		
	U_{to}	T_h	T_{ci}	h_r	h_{nc}	U_{to}
Trial	[Btu/(hr-ft²-°F)]	(°F)	(°F)	[Btu/(hr-ft²-°F)]	[Btu/(hr-ft²-°F)]	(Btu/(hr-ft²-°F))
1	4.05	395	530	6.39	0.36	3.22
2	3.22	367	487	6.00	0.42	3.15
3	3.15	364	485	5.97	0.42	3.14

Table 8.4—Summary of calculations to find U_{to}.

for injection times ranging from 0 to 14 days. When casing temperatures become high, thermal stresses may cause casing failure (Willhite and Dietrich 1967). Methods of completing wells to prevent casing failure are discussed in Prats (1982), Boberg (1988), and Willhite and Dietrich (1967).

8.2.3 Pressure Drop in Two-Phase Flow in Steam Injection.
The heat-loss relationships presented in Sections 8.2.1 and 8.2.2 were developed by assuming that the temperature of the flowing steam was constant. This is a good first approximation for many steam-injection projects. However, friction losses occur when steam flows through a line or injection tubing, leading to changes in pressure with length or depth. The weight of the fluid must be included when estimating pressure in injection wells. Because the temperature of saturated steam is a direct function of the pressure, heat losses may change with the length of the line or injection tubing. The quality of the flowing steam may also change.

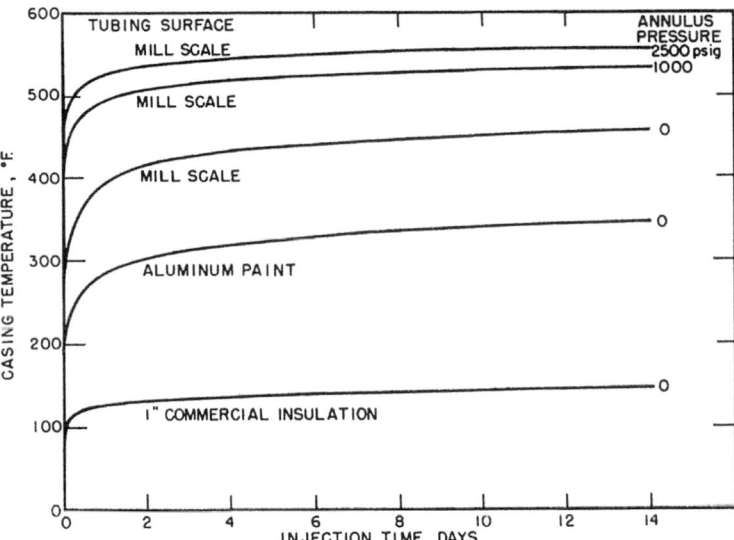

Fig. 8.8—Calculated casing temperatures when 650°F steam is injected down tubing (Willhite 1967).

Changes in pressure and quality of steam flowing in a line can be estimated by solving the conservation equations for thermal and mechanical energy in the flowing fluid over the length of flow (Satter 1965; Earlougher 1969; Pacheco and Farouq Ali 1972; Herrera et al. 1978; Farouq Ali 1981; Griston and Willhite 1987). Eqs. 8.23 and 8.24 describe the conservation of thermal energy and mechanical energy for the flowing fluid. In Eqs. 8.23 and 8.24, it is assumed that the coordinate z is in the direction of flow. When z is horizontal, the component of the gravitational acceleration in the z direction (horizontal), g_z, is zero.

$$\frac{dH}{dz} + \frac{G_m^2 v_f}{g_c J}\frac{dv_f}{dz} - \frac{g_z}{g_c J} + \frac{1}{\dot{m}_s}\dot{Q}_\ell = 0 \quad \dots \quad (8.23)$$

$$\text{and } \frac{G_m^2 v_f}{g_c}\frac{dv_f}{dz} - \frac{g_z}{g_c} + v_f\frac{dp}{dz} + \frac{f_{tp}G^2 vf^2}{2g_c D} = 0, \quad \dots \quad (8.24)$$

where H = enthalpy of flowing steam, Btu/lbm; $\dot{Q}_\ell$ = rate of heat loss from the pipe or tubing to the surroundings, Btu/(hr-ft); v_f = specific volume of flowing fluid, ft³/lbm; and f_{tp} = two-phase friction factor, dimensionless.

In Eq. 8.23, the first term represents the change in enthalpy of the flowing fluid with length, the second term represents the change of kinetic energy of the flowing fluid with length, the third term represents the change of potential energy of the flowing fluid with length, and the fourth term is the rate of heat loss per unit length per unit mass of flowing fluid.

In Eq. 8.24, the mechanical energy equation, the first term represents the change in kinetic energy with length; the second term represents the change in potential energy with length; the third term is the change in the pressure volume work with length; and the fourth term represents the loss in mechanical energy by the irreversible conversion of thermal energy to mechanical energy, commonly called friction loss. Specific terms in Eqs. 8.23 and 8.24 are defined in the Nomenclature.

In steam lines where there is no change in cross section, the change in kinetic energy with length is small and the term $\left(G_m^2 v_f/g_c\right)/\left(dv_f/dz\right)$ is usually omitted when Eqs. 8.23 and 8.24 are solved.

To solve Eqs. 8.23 and 8.24, it is necessary to replace the derivative of the enthalpy with respect to distance with its equivalent expression in terms of steam quality, f_s, and pressure, p (Griston and Willhite 1987). Because

$$H = H\left(f_s, p\right), \quad \dots \quad (8.25)$$

$$\frac{dH}{dz} = \left(\frac{\partial H}{\partial f_s}\right)_p\frac{df_s}{dz} + \left(\frac{\partial H}{\partial p}\right)_{fs}\frac{dp}{dz}, \quad \dots \quad (8.26)$$

where $H = \left(1 - f_s\right)H_{wT} + f_s H_v. \quad \dots \quad (8.27)$

Eqs. 8.28 and 8.29 are obtained when dH/dz is eliminated from Eqs. 8.23 and 8.24.

$$\frac{dp}{dz} = \frac{g_z}{v_f g_c} - \frac{f_{tp} G_m^2 v_f^2}{2 g_c D}. \dots\dots\dots\dots\dots\dots\dots\dots\dots\dots\dots\dots\dots\dots (8.28)$$

$$\frac{df_s}{dz} = \frac{\dfrac{g_z}{J g_c} - \dfrac{\dot{Q}_{twb}}{\dot{m}_s} - \left(\dfrac{\partial H}{\partial p}\right)_{fs}\left(\dfrac{dp}{dz}\right)}{\left(\partial H / \partial f_s\right)_p}. \dots\dots\dots\dots\dots\dots\dots\dots\dots (8.29)$$

The derivatives $(\partial H/\partial p)_{fs}$ and $(\partial H/\partial f_s)$ are obtained from thermodynamic tables or correlations. For example, Farouq Ali (1970) developed correlations for H_{wT} and H_v as a function of pressure. These correlations are given in Eqs. 8.30 and 8.31. When these relationships are used, the derivatives of enthalpy with respect to pressure and quality are given by Eqs. 8.32 and 8.33.

$$H_{wT} = 91 p^{0.2574}, \dots\dots\dots\dots\dots\dots\dots\dots\dots\dots\dots\dots\dots\dots\dots\dots\dots (8.30)$$

with an error of 0.3% for 15 to 1,000 psia, and

$$H_{vT} = 1,119 p^{0.01267}, \dots\dots\dots\dots\dots\dots\dots\dots\dots\dots\dots\dots\dots\dots\dots\dots (8.31)$$

with an error of 0.3% for 15 to 1,000 psia.

$$\left(\partial H / \partial f_s\right)_p = H_{vT} - H_{wT} \dots\dots\dots\dots\dots\dots\dots\dots\dots\dots\dots\dots\dots\dots\dots (8.32)$$

$$\left(\frac{\partial H}{\partial p}\right)_{f_s} = \frac{23.423(1 - f_s)}{p^{0.7426}} + \frac{14.178 f_s}{p^{0.98733}}. \dots\dots\dots\dots\dots\dots\dots\dots\dots (8.33)$$

Tortike and Farouq Ali (1989) present a comprehensive set of polynomial expressions for steam properties as a function of saturation temperature and pressure.

The last factor to be discussed is the nature of two-phase flow in pipes. When gas and liquid flow simultaneously in a pipe, the gas flows faster than the liquid phase in most cases. This phenomenon is called slip or liquid holdup. The effect of slip or liquid holdup is compensated for in two-phase-flow calculations by developing correlations for the specific volume of the flowing fluid and the two phase friction factor as a function of process variables. The specific volume of the flowing steam is given by

$$v_f = \left\{ \left(y_\ell / v_{sw}\right) + \left[(1 - y_\ell)/v_{sv}\right] \right\}^{-1}, \dots\dots\dots\dots\dots\dots\dots\dots\dots\dots\dots\dots (8.34)$$

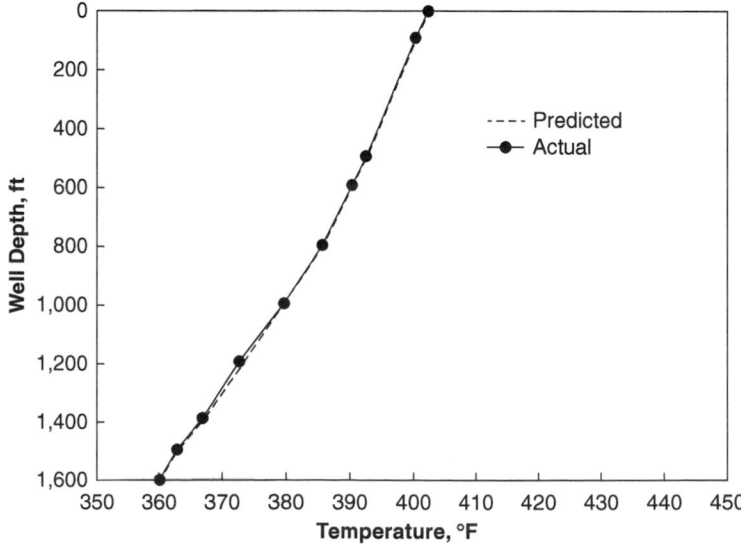

Fig. 8.9—Predicted vs. actual steam temperature, Home Stake data (Griston and Willhite 1987).

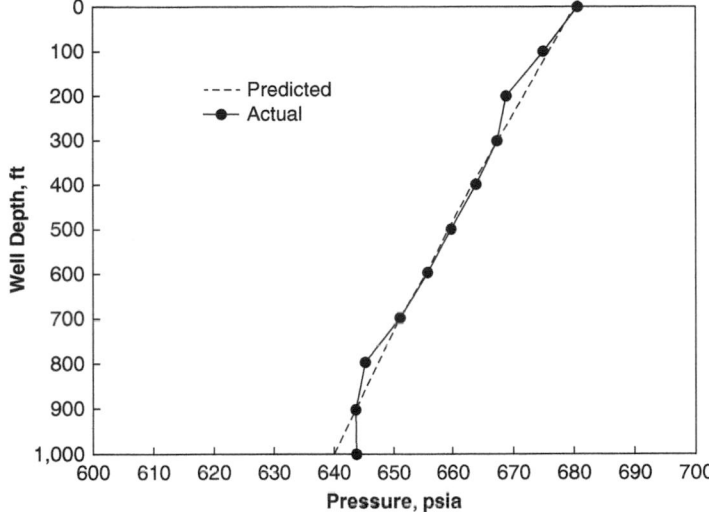

Fig. 8.10—Predicted vs. actual steam pressure, Chevron data (Griston and Willhite 1987).

where v_{sw} = specific volume of the saturated water, ft³/lbm; v_{sv} = specific volume of the saturated vapor, ft³/lbm; y_l = liquid holdup, fraction; and v_f = specific volume of the flowing fluid, ft³/lbm. When slip is neglected, the flowing stream is treated as if it were homogeneous and the holdup, y_l, is replaced by the steam quality, f_s.

There are many two-phase-flow correlations available in the literature that permit computation of the specific volume and the two-phase friction factor from the process variables (e.g., Orkiszewski 1967; Gould et al. 1974; Beggs and Brill 1973; Spedding and Chen 1984; Mukherjee and Brill 1985). Correlations were developed primarily for horizontal flow and vertical upward flow. However, correlations for horizontal flow and vertical upward flow are often used for vertical downward flow because there are limited data for downward flow. The Mukherjee and Brill (1985) correlation was developed from a database including downward flow data and is preferable for two-phase flow in injection wells.

The pressure and quality at any distance, z, is obtained by solving Eqs. 8.28 and 8.29 simultaneously by use of a method such as the fourth-order Runge-Kutta method (Carnahan et al. 1969). The temperature at z is obtained from thermodynamic tables at pressure p or may be calculated from the correlation developed by Farouq Ali:

$$T_s = 115.1 p^{0.225}, \dots\dots\dots\dots\dots\dots\dots\dots\dots\dots\dots\dots (8.35)$$

with an error of 1% for 15 to 3,000 psia.

In solving Eqs. 8.28 and 8.29, the distance is divided into equal increments of distance, Δz. Computations are made for a particular value of time. Solution of the set of equations begins at $z = 0$ and proceeds in increments of Δz until the specified length (or depth of the well) is reached. If all steam is condensed, the computation ends at the depth where the quality becomes zero. The remainder of the injection is treated as hot-water injection. Computation procedures are outlined by Pacheco and Farouq Ali (1972) and Griston and Willhite (1987).

Figs. 8.9 and 8.10 show typical results. In Fig. 8.9, predicted and measured temperatures are compared after 117 hours of injection of steam in a 1,600-ft well completed with 4-in. casing and 2⅜-in. tubing set on a packer (Bleakley 1964). Steam was injected at an average rate of 320 B/D and an injection pressure of 260 psia. Steam quality was assumed to be 70% at the wellhead. The data in Fig. 8.10 (Kamilos 1985) were obtained from a steam-injection well in the Kern River field, Kern County, California. Steam was injected into a 1,000-ft well completed with 5½-in. casing and 2⅜-in. tubing set on a packer. Average wellhead conditions were 500 B/D, 680 psia, and 75% steam. Deviation of the computed results from the measured results below 900 ft is a result of the presence of a downhole choke.

The interactions of two-phase flow, pressure drop, and heat losses are particularly important in concentric steam-injection wells where simultaneous injection of steam into two zones is desired (Fig. 8.11). Griston and Willhite (1987) present an extensive analysis of the concentric injection string in Fig. 8.11, including limitations derived from simulation of fluid flow and heat transfer in this completion.

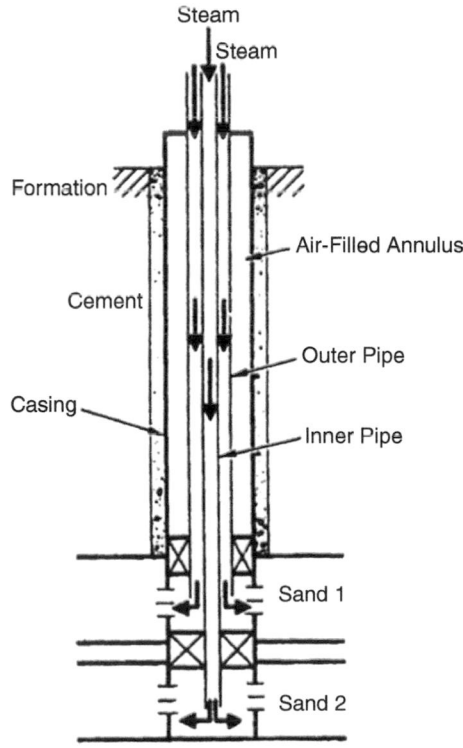

Fig. 8.11—Concentric steam-injection well (Griston and Willhite 1987).

8.3 Cyclic Steam Stimulation

Historically, primary production from heavy-oil reservoirs before the development of thermal recovery techniques was 5% of OOIP or less. Production rates were low, declining with time as the reservoir energy was depleted. In some wells, production declines were known to be caused by deposition of solids, paraffins, or asphalts in the region around the wellbore. Hot-oil squeezes or swabbing the well with a diluent, such as kerosene, would increase the production rate, sometimes by a factor of two or more, and the well would resume its decline. Attempts to use downhole heaters to increase oil production have been attempted and largely discarded.

CSS was discovered by accident in the Mene Grande field in Venezuela in 1959 when steam broke out behind casing in a steam-injection well (de Haan and van Lookeren 1969). This well, which had produced no oil previously, flowed oil at rates of 100 to 200 B/D when the well was blown down. **Fig. 8.12** illustrates the response of a well in the Midway Sunset field to cyclic steam injection Burns (1969). The discovery that steam injection into a heavy-oil reservoir could increase production rates by factors of 5 to 10 was a historic point in the development of thermal recovery techniques. CSS spread rapidly to California, and, by 1965, projects were under way in most major heavy-oil reservoirs in California.

In CSS, steam is injected into a production well for a period of 2 to 4 weeks. The well is shut in and allowed to "soak" before returning to production. The initial oil rate is high because of the reduced oil viscosities at the increased reservoir temperatures. There is also some acceleration from increased reservoir pressure near the wellbore. With time, the heated-zone temperature declines as a result of heat removed with the produced fluids and conduction losses to over- and underlying formations. Oil rates decline as the heated-zone temperature decreases and oil viscosity increases. When the production declines to a predetermined level, another cycle of steam injection is initiated. In some reservoirs, up to 20 cycles have been carried out.

CSS is a precursor to steamdrive in most reservoirs. When the pattern spacing is small, interference between wells may be noticed during steam injection. This is an indication that the heated regions from adjacent wells may overlap or that the steam-slug volume is too large. If interference occurs after several cycles have been conducted, consideration of conversion to steamdrive may be appropriate. CSS is preferred when the natural reservoir energy has not been depleted. Steamdrive is used when the reservoir energy is depleted.

This section introduces fundamental concepts of cyclic steam injection. Simple models will be used to illustrate the process. We recognize that numerical models that simulate CSS are widely used to match field results and to estimate results when cycles are used, but use of such simulators is beyond the scope of this text.

8.3.1 Production Mechanisms—CSS. Heavy-oil reservoirs are characterized by oil viscosities at reservoir temperatures on the order of 100 to 10,000 cp. When a reservoir has a source of natural energy to displace oil from the reservoir to the production wells, the oil rate is controlled primarily by the flow resistance in the immediate vicinity of the wellbore. Cyclic steam

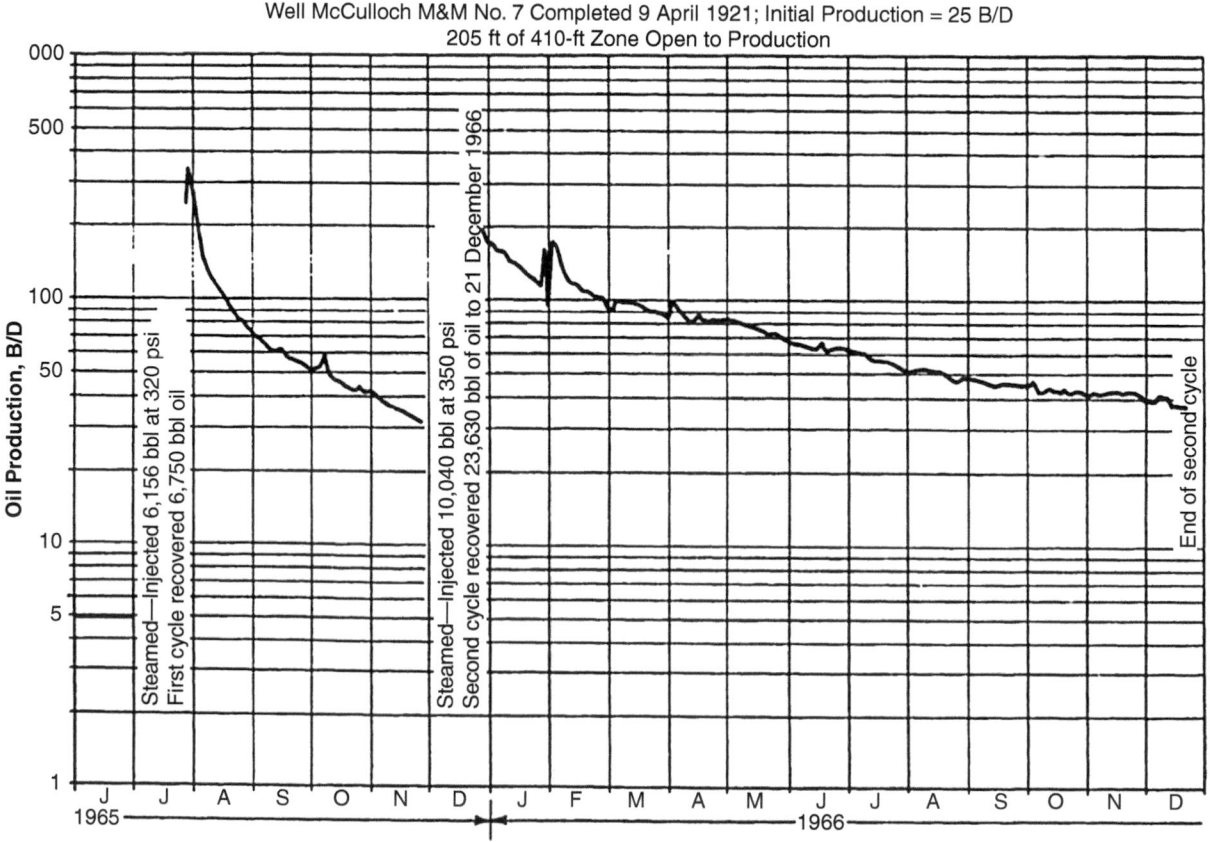

Fig. 8.12—Response of production well to CSS (Burns 1969).

injection heats the reservoir rock around the wellbore and permits this region to remain at an elevated temperature for long periods of time.

The essential requirement for successful CSS is a source of natural reservoir energy. Reservoir energy may be available in the form of (1) fluid expansion by solution-gas drive or reduction in reservoir pressure, (2) natural waterdrive, (3) gravity drainage, or (4) compaction. Examples of each of these sources of reservoir energy are found in the literature describing successful cyclic steam-injection projects (de Haan and van Lookeren 1969; Burns 1969; Boberg and Lantz 1966; Jones 1977).

CSS is examined by first considering a depletion-type reservoir. Consider a heavy-oil primary production rate from a reservoir. The reservoir is developed on 10-acre spacing. Permeability is 3 darcies, and thickness is 100 ft. **Table 8.5** summarizes reservoir rock and fluid properties.

The oil rate at reservoir temperature can be estimated provided that there is sufficient reservoir energy to maintain the reservoir pressure at 500 psi at the effective drainage radius of the well, defined by

$$r_e = \sqrt{(A/\pi)\left(43,560\,\text{ft}^2/\text{acre}\right)}, \quad \dots\dots\dots\dots\dots\dots\dots\dots\dots\dots\dots\dots\dots\dots\dots (8.36)$$

where A = well spacing (10 acres) and r_e = effective drainage radius (372 ft). The oil production rate when the wellbore pressure is maintained at p_w is given by Eq. 8.37, where oilfield units (stock-tank barrels, days, darcies, feet, centipoise, and psi) are used.

$$q_{oc} = \frac{7.082\,k_{oc}h\left(p_e - p_w\right)}{B_{oc}\mu_{oc}\ln\left(r_e/r_w\right)}. \quad \dots\dots\dots\dots\dots\dots\dots\dots\dots\dots\dots\dots\dots\dots\dots\dots (8.37)$$

If the reservoir pressure was 500 psi at discovery and the viscosity at reservoir temperature is 1,000 cp, the initial oil rate is computed as

$$q_{oc} = \frac{(7.082)(3.0\,\text{darcies})(100\,\text{ft})(500-100)\,\text{psi}}{(1.1\,\text{bbl/STB})(1,000\,\text{cp})\ln(372/0.5)}$$

$$= 116.6\ \text{STB/D}.$$

As this reservoir is depleted, the reservoir pressure falls and the production rate declines. For example, when the effective reservoir pressure at the boundary is 50 psi and the bottomhole pressure (BHP) in the production well is maintained at 10 psi, the production rate will be

$$q_{oh} = 116.6\left[(50-10)/(500-100)\right]$$

$$= 11.7\,\text{STB/D}.$$

Next, consider the effects of heating a small region around the wellbore by injecting steam. We assume that the steam zone moves radially from the injection well and is located at r_h, as depicted in **Fig. 8.13.** The steam zone is at T_s, and the region beyond the steam zone is at T_r. If fluid flow is radial, incompressible, and at steady state, the pressure drop between r_e and r_w is given by Eq. 8.38. When the volumetric flow rate at reservoir conditions is equal in heated and cold regions, $q_{oh}B_{oh} = q_{oc}B_{oc}$.

$$p_e - p_w = \frac{q_{oh}B_{oh}\mu_{oh}}{7.082k_{oh}h}\ln\left(\frac{r_h}{r_w}\right) + \frac{q_{oc}B_{oc}\mu_{oc}}{7.082k_{oc}h}\ln\left(\frac{r_e}{r_h}\right). \quad \dots\dots\dots\dots\dots\dots\dots\dots\dots (8.38)$$

Thickness (ft)	100
Permeability (md)	3
Well spacing (acres)	10
Oil viscosity at T_r (cp)	1,000
Reservoir pressure (psi)	500
Initial oil FVF (bbl/STB)	1.1
Wellbore radius (ft)	0.5
Backpressure in pressure well (psi)	100

FVF = formation volume factor.

Table 8.5—Reservoir rock and fluid properties for estimating the effect of CSS on production data.

Thus, $q_{oh} = \dfrac{7.082\,h\left(p_e - p_w\right)}{B_{oh}\left[\dfrac{\mu_{oh}}{k_{oh}}\ln\left(\dfrac{r_h}{r_w}\right) + \dfrac{\mu_{oc}}{k_o}\ln\left(\dfrac{r_e}{r_h}\right)\right]}.$ . . . .(8.39)

The impact of increasing the temperature to T_s for the radius r_h can be determined if the steam zone is assumed to be 50 ft in radius and the oil viscosity is 2 cp at the steam temperature. Then, when $B_{oh} = 1.1$ bbl/STB,

$$q_o = \left[7.082(100\,\text{ft})(50 - 10)\,\text{psi}\right]$$
$$\div \left\{ \begin{array}{l} \left[(1.1\,\text{bbl/STB})(2\,\text{cp})/3\,\text{darcies}\right]\left[\ln(50/0.5)\right] \\ +\left[(1.1\,\text{bbl/STB})(1{,}000\,\text{cp})/3\,\text{darcies}\right]\left[\ln(372/50)\right] \end{array} \right\}$$
$$= 28{,}328/(3.377 + 735.85)$$
$$= 38.3\,\text{STB/D}.$$

In this illustration, heating the region around the wellbore for a distance of 50 ft led to an increase in flow rate from 11.7 to 38.3 STB/D, a factor of approximately three.

If steam stimulation were applied before the reservoir energy was depleted, the initial production rate after the soak period would be

$$q_o = 38.3\left[(500 - 100)/(50 - 10)\right]$$
$$= 383.3\,\text{STB/D}.$$

This calculation illustrates that the response to CSS is determined in large measure by the natural reservoir energy.

This model does not account for the decline in the temperature of the heated zone with time, which occurs because of heat losses and removal of heat with the produced fluids. In Section 8.3.3, the Boberg and Lantz (1966) model is introduced to account for heat losses.

The second important factor in the response of wells stimulated with steam is the removal of wellbore damage. Consider the radial-flow model shown in **Fig. 8.14,** which contains a damaged zone of radius r_d (Boberg 1988) and permeability k_d with $k_d < k$. If we assume steady radial flow in each segment, the pressure drop across the region between r_w and r_e is given by

$$p_e - p_w = \frac{q_{oh}B_{oh}\mu_{oh}}{7.082\,k_d h}\ln\left(\frac{r_d}{r_w}\right) + \frac{q_{oh}B_{oh}\mu_{oh}}{7.082\,k_{oh}h}\ln\left(\frac{r_h}{r_d}\right) + \frac{q_{oc}B_{oc}\mu_{oc}}{7.082\,k_{oc}h}\left[\ln\left(\frac{r_e}{r_h}\right)\right]. \qquad (8.40)$$

The volumetric rate at reservoir conditions is the same in all regions, so that $q_{oh}B_{oh} = q_{oc}B_{oc}$. Terms involving the wellbore damage region can be also rewritten as

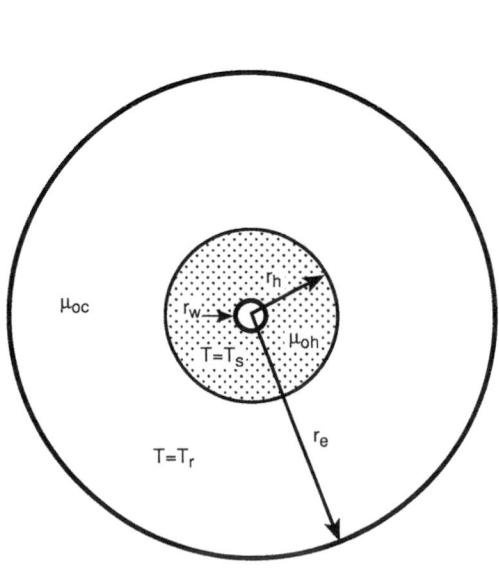

Fig. 8.13—Hot and cold regions surrounding a production well after steam injection to radius r_h.

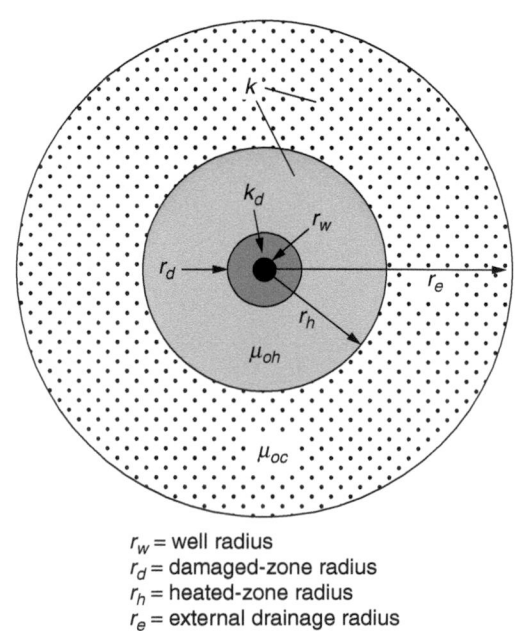

r_w = well radius
r_d = damaged-zone radius
r_h = heated-zone radius
r_e = external drainage radius

Fig. 8.14—Radial flow model. [*Thermal Methods of Oil Recovery*, T. C. Boberg, 1988. Reprinted by permission of John Wiley & Sons Inc.]

$$\frac{q_o B_{oh}}{7.082 \mu_{oh} h} \left[\frac{\ln(r_d/r_w)}{k_d} + \frac{\ln(r_h/r_d)}{k_{oh}} \right] = \frac{q_{oh} B_{oh}}{7.082 \mu_{oh} h} \left[\frac{\ln(r_d/r_w)}{k_d} + \frac{\ln r_w - \ln r_w + \ln(r_h/r_d)}{k_{oh}} \right]$$

$$= \frac{q_{oh} B_{oh} \mu_{oh}}{7.082 h} \left[\frac{\ln(r_d/r_w)}{k_d} - \frac{\ln(r_d/r_w)}{k_{oh}} + \frac{\ln(r_h/r_w)}{k_{oh}} \right]. \quad \dots\dots\dots\dots (8.41)$$

Defining the skin factor for the heated reservoir as S_h,

$$s_h = \left[(k_{oh}/k_d) - 1 \right] \ln(r_d/r_w), \quad \dots\dots\dots\dots\dots\dots\dots\dots\dots\dots (8.42)$$

and rearranging give

$$p_e - p_w = \frac{q_{oh} B_{oh} \mu_{oh}}{7.082 h} \left(\frac{s_h}{k_{oh}} + \frac{\ln r_h/r_w}{k_{oh}} \right) + \frac{q_{oc} B_{oc} \mu_{oc}}{7.082 k_{oc} h} \ln\left(\frac{r_e}{r_h} \right); \quad \dots\dots\dots\dots (8.43)$$

thus, $q_o = \dfrac{7.082 h (p_e - p_w)}{B_{oh} \left[\dfrac{\mu_{oh}}{k_{oh}} \left(s_h + \ln \dfrac{r_h}{r_w} \right) + \dfrac{\mu_{oc}}{k_{oc}} \ln \dfrac{r_e}{r_h} \right]}. \quad \dots\dots\dots\dots\dots\dots (8.44)$

If the effect of temperature on k_o is neglected ($k_{oh} = k_{oc}$), then

$$q_{oh} = \frac{7.082 k h (p_e - p_w)}{B_{oh} \left\{ \mu_{oh} \left[s_h + \ln(r_h/r_w) \right] + \mu_{oc} \ln(r_e/r_h) \right\}}. \quad \dots\dots\dots\dots (8.45)$$

Eq. 8.46 is the corresponding expression for the cold oil rate when there is wellbore damage.

$$q_{oc} = \frac{7.082 k h (p_e - p_w)}{B_{oc} \mu_{oc} \left[s_c + \ln(r_e/r_w) \right]}. \quad \dots\dots\dots\dots\dots\dots (8.46)$$

It is convenient to introduce the productivity index (PI) to investigate the effects of CSS on wellbore damage. The PI is defined by

$$J = q_o / \Delta p. \quad \dots\dots\dots\dots\dots\dots\dots\dots\dots\dots\dots\dots\dots\dots\dots (8.47)$$

The ratio of the PIs for a stimulated well compared with the same well unstimulated is obtained by dividing Eq. 8.45 by Eq. 8.46 and dividing by μ_{oc}.

$$\frac{J_h}{J_c} = \frac{(B_{oc}/B_{oh}) \left[s_c + \ln(r_e/r_w) \right]}{(\mu_{oh}/\mu_{oc}) \left[s_h + \ln(r_h/r_w) \right] + \ln(r_e/r_w)}. \quad \dots\dots\dots\dots\dots\dots (8.48)$$

Table 8.6 (Boberg 1988) presents a range of PIs in which the viscosity ratio in the heated region decreases by a factor of 100. PIs are presented for three wellbore damage scenarios. These results show that damaged wells will respond more favorably to reservoir heating than undamaged wells, even if the wellbore damage is not removed. In cases where wellbore damage is removed ($s_h = 0$), there will be an increase in production rate above q_{oc} that persists well after the formation has cooled. Such an example is shown in **Fig. 8.15** for an average well stimulated with steam in the Duri field in Indonesia (Atmosudiro 1977). Increased production rate was observed in this well long after temperature effects on the crude oil viscosity should have dissipated.

8.3.2 Estimating the Radius of the Heated Zone—The Marx and Langenheim Model.
The radius of the heated region can be estimated using a model developed by Marx and Langenheim (1959). Other reservoir heating models are presented in Section 8.4. We begin by determining the energy requirements for heating porous rock.

Energy Requirements for Heating Porous Rock. The amount of energy required to increase the temperature of a porous rock is easily calculated from thermodynamic tables and from heat capacity data at constant pressure. Eq. 8.49 gives the energy (in Btu) required to increase the temperature of 1 ft³ of reservoir rock from an initial value, T_r, to a higher value, T (in °F).

$$Q = M(T - T_r). \quad \dots\dots\dots\dots\dots\dots\dots\dots\dots\dots\dots\dots\dots (8.49)$$

Amount of Damage			Stimulation Ratio, Heated $(\mu_{oh} = 0.01\mu_{oc})$			Stimulation Ratio, Unheated		
	Permeability Damage Ratio, k_d/k	Skin Factor, S_c	Damage Not Removed	50% Removal*	100% Removal**	Damage Not Removed	50% Removal**	100% Removal**
Undamaged	1	0	2.94	2.94	2.94	1	1	1
Moderately damaged	0.1	21	10.8	11.3	11.8	1	1.6	4.0
Extremely damaged	0.01	228	50.3	67.4	100.0	1	1.94	34.0

*Well radius r_w = 0.3 ft, damaged radius r_d = 3.0 ft, heated radius r_h = 30 ft, drainage radius r_c = 300 ft, s_c = skin before heating, s_h = skin after heating.

$$J_h/J_c = \left[s_c + \ln\left(r_c/r_w\right)\right]\big/\left\{\left(\mu_{oh}/\mu_{oc}\right)\left[s_h + \ln\left(r_h/r_w\right)\right] + \ln\left(r_c/r_h\right)\right\}.$$

**For 50% removal, $s_h = 0.5s_c$; for 100% removal, $s_h = 0$.

$$s_c = \left[\left(k/k_d\right) - 1\right]\ln\left(r_d/r_w\right).$$

Table 8.6—Calculated stimulation ratios for flow model shown in Fig. 8.31 of Boberg (1988) [*Thermal Methods of Oil Recovery*, T. C. Boberg, 1988. Reprinted by permission of John Wiley & Sons Inc.].

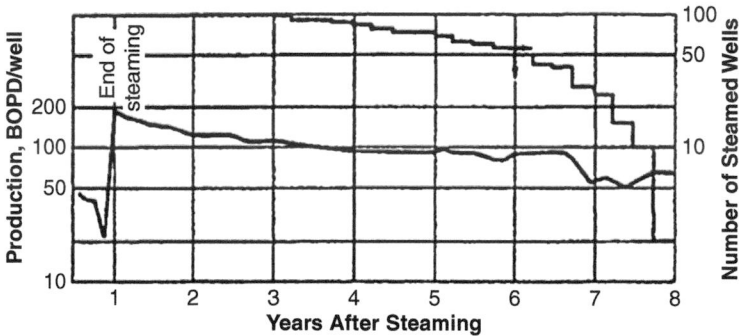

Fig. 8.15—Production response from a well in the Duri field, Indonesia, showing wellbore damage removal (Atmosudiro 1977).

In Eq. 8.49, M is the average volumetric heat capacity of the fluid-saturated rock over the temperature range T_r to T. If T is the temperature of injected steam, T_s, then,

$$M = \left(1 - \phi\right)\rho_r C_r + \phi S_o \rho_o C_o + \phi S_w \rho_w C_w + \phi S_g \rho_s C_s, \quad\quad\quad\quad\quad\quad\quad (8.50)$$

where C_r, C_o, C_w, and C_s are mean heat capacities at constant pressure (Btu/lbm-°F) for the rock, oil, water and steam if present. The mean heat capacities of each component are based on the temperature difference, $T_s - T_r$. Other terms in Eq. 8.50 are defined in the Nomenclature.

Appendix E contains heat capacity data and correlations for rock and oil. Included in Appendix E are thermodynamic tables for steam. Example 8.3 illustrates the calculation of M for a representative reservoir rock when fluid saturations are known.

Example 8.3—Calculation of Volumetric Heat Capacity. A sandstone with a porosity of 25% contains an oil saturation of 0.2 and a water saturation of 0.8. Determine the energy that must be added to the rock to increase its temperature from 80 to 470.9°F (boiling point of saturated steam at 500 psig). The rock is confined, and no vapor phase forms within the pore space as a result of reservoir heating.

Solution. The mean heat capacities must be determined for each fluid and the porous rock for the temperature interval from 80 to 470.9°F. For this example, properties of the rock and oil are C_r = 0.21 Btu/lbm-°F, C_o = 0.50 Btu/lbm-°F, ρ_r = 167.0 lbm/ft³, and ρ_o = 50.0 lbm/ft³. The mean heat capacity for saturated water is defined by

$$C_w = \frac{H_{wT} - H_{wr}}{T_s - T_r}, \quad\quad\quad\quad\quad\quad\quad\quad\quad\quad\quad\quad\quad\quad\quad\quad\quad\quad\quad\quad\quad (8.51)$$

where H_{wT} = enthalpy of saturated water at T_s, Btu/lbm, and H_{wr} = enthalpy of water at T_r, Btu/lbm. Values of the enthalpy can be interpolated from the steam tables in Appendix B. At 80°F, H_{wr} = 48 Btu/lbm. At 470.9°F, H_{wT} = 452.9 Btu/lbm. Thus,

$$C_w = (452.9 - 48)/(470.9 - 80)$$
$$= 1.036 \, \text{Btu/lbm-}°\text{F}.$$

From Table E-2, $\rho_w = 50.6$ lbm/ft³ at 470.9°F. The value of M is found from Eq. 8.50.

$$M = (0.75)(167 \, \text{lbm/ft}^3)(0.21 \, \text{Btu/lbm-}°\text{F})$$
$$+ (0.25)(0.2)(50 \, \text{lbm/ft}^3)(0.5 \, \text{Btu/lbm-}°\text{F})$$
$$+ (0.25)(0.8)(50.6 \, \text{lbm/ft}^3)(1.036 \, \text{Btu/lbm-}°\text{F})$$
$$= 26.3 + 1.25 + 10.48$$
$$= 38.03 \, \text{Btu/ft}^3\text{-}°\text{F}.$$

Approximately 70% of the energy is used to heat the rock matrix.

If the rock contained 40% water saturation, 40% saturated water vapor, and an oil saturation of 20% when heated to 470.9°F, the following changes would be made. From the steam tables at 500 psig, $H_v = 1204.3$ Btu/lbm. Thus, $L_v = 751.4$ Btu/lbm and

$$C_s = C_w + L_v/(T_s - T_r)$$
$$= 1.036 + 1.922$$
$$= 2.96 \, \text{Btu/lbm-}°\text{F}.$$

The density of saturated vapor at 470.9°F is 1.11 lbm/ft³, and

$$M = (0.75)(167 \, \text{lbm/ft}^3)(0.21 \, \text{Btu/lbm-}°\text{F})$$
$$+ (0.25)(0.2)(50 \, \text{lbm/ft}^3)(0.5 \, \text{Btu/lbm-}°\text{F})$$
$$+ (0.25)(0.4)(50.6 \, \text{lbm/ft}^3)(1.036 \, \text{Btu/lbm-}°\text{F})$$
$$+ (0.25)(0.4)(1.11 \, \text{lbm/ft}^3)(2.96 \, \text{Btu/lbm-}°\text{F})$$
$$= 26.3 + 1.25 + 5.25 + 0.33$$
$$= 33.13 \, \text{Btu/ft}^3\text{-}°\text{F}.$$

In this case, approximately 80% of the energy is stored in the rock matrix.

Marx and Langenheim Model for Reservoir Heating. When reservoirs are heated by hot-fluid injection, a significant fraction of the injected energy is lost to the surrounding formations. The reservoir-heating model developed by Marx and Langenheim (1959) retains many of the essential mechanisms of the reservoir-heating process. The reservoir is considered to have uniform thickness and fluid and rock properties. Steam is uniformly distributed in the vertical cross section throughout the heated region so that the temperature is uniform through the vertical cross section. Steam and condensate do not segregate under the influence of gravity. The heated zone advances into the reservoir as a step-function change in temperature, as depicted in **Fig. 8.16.**

The steam zone forms as soon as sufficient steam is injected to increase the reservoir temperature from T_r to T_s by condensation of the steam. Heat is lost to the over- and underlying formations by conduction. The model assumes that heat flow in the over- and underburden occurs only in the direction perpendicular to the direction of fluid flow. This displacement model is piston-like. At the displacement front, the oil saturation is reduced from S_o to S_{ors} and the interstitial water/gas saturations are displaced by condensate and steam. Thus, the heated zone consists of the rock matrix filled with oil (S_{ors}), water (S_w), and saturated water vapor (S_g). The model assumes that there is no heat loss to the cold rock ahead of the moving steam front by conduction or convection.

The Marx and Langenheim model is derived by making an energy balance over the heated region depicted in Fig. 8.16. The model assumes that there is sufficient pressure drop and mobility of the reservoir fluids to allow steam injection at a rate of m_s, lbm/hr. Eq. 8.52 describes the differential energy balance for expansion of the steam zone by an incremental volume element ΔV in a time increment Δt.

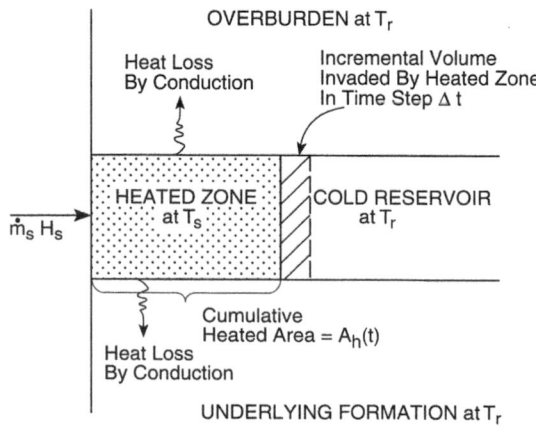

Fig. 8.16—Marx and Langenheim reservoir heating model.

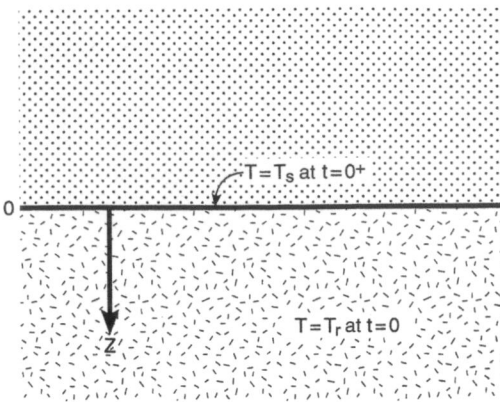

Fig. 8.17—Heat conduction model into a semi-infinite medium.

$$\dot{m}_s H_s = M_R\left(T_s - T_r\right)\frac{\Delta V}{\Delta t} + 2\int_o^{A_h} q(t-\lambda)\,dA_h, \dots \dots (8.52)$$

where H_s = energy content of the injected steam relative to the reservoir temperature, Btu/lbm; $\dot{m}_s$ = steam-injection rate, lbm/hr; M_R = volumetric heat capacity of heated region, Btu/(ft^3-°F); A_h = heated area, ft^2; h = thickness of heated zone, ft; t = total time since start of injection, hours; and λ = time of arrival of heated zone at a specific location (i.e., out to a specific area), hours.

In Eq. 8.52, the first term on the right side represents the heat required to raise the temperature of the reservoir rock from T_r to T_s when the steam zone expands by an increment of bulk volume ΔV. The second term represents the heat-loss rate from the heated zone to the over- and underlying formations. In this second term, t is total time since start of injection and λ is the time of arrival of the heated zone at a specific area from the injection point. Thus, $(t - \lambda)$ is the time that an increment of over- and underburden area, dA, has been in contact with the heated zone. The term $q(t - \lambda)dA$ [meaning q as a function of $(t - \lambda)$] is the instantaneous rate of conduction to overburden or underburden. When $q(t - \lambda)dA$ is integrated over the total heated area, A, and multiplied by two to account for both overburden and underburden, the integral is the instantaneous total rate of heat conduction away from the heated zone. As noted earlier, the model assumes that there is no heat loss from the steam zone in the direction of flow.

To obtain an expression for the conduction term, the rate of heat loss from the steam zone is approximated by assuming that heat flows into a semi-infinite medium by conduction, as depicted in **Fig. 8.17.** When the temperature is changed from T_r to T_s at time = 0 at the boundary of a semi-infinite medium, the temperature distribution is given by Eq. 8.53 (Carslaw and Jaeger 1959) and the heat flux at the heated-zone boundary is given by Eqs. 8.54 and 8.55.

$$T = T_r + \left(T_s - T_r\right)\mathrm{erfc}\left(z/2\sqrt{\alpha t}\right), \dots \dots (8.53)$$

$$q\big|_{z=0} = -k_h\left(\partial T/\partial z\right)\big|_{z=0}, \dots \dots (8.54)$$

$$\text{and } q\big|_{z=0} = \left[k_h\left(T_s - T_r\right)\right]/\sqrt{\pi\alpha t}, \dots \dots (8.55)$$

where k_h = thermal conductivity of the overburden, Btu/hr-ft-°F; α = thermal diffusivity of the overburden = $K_h/(\rho_{ob}C_{ob})$ ft^2/hr; and erfc(x) is the complementary error function defined by

$$\mathrm{erfc}\left(x\right) = \frac{2}{\sqrt{\pi}}\int_x^\infty e^{-t^2}\,dt$$

$$= 1 - \frac{2}{\sqrt{\pi}}\int_o^x e^{-t^2}\,dt. \dots \dots (8.56)$$

$$= 1 - \mathrm{erf}\left(x\right). \dots \dots (8.57)$$

Values of erfc(x) are presented in **Table 8.7** for selected values of x. Eq. 8.58 is a series approximation of erf(x) given by Abramowitz and Stegan (1964):

$$\mathrm{erf}\left(x\right) = 1 - \left(b_1 u + b_2 u^2 + b_3 u^3 + b_4 u^4 + b_5 u^5\right)e^{-x^2} + \varepsilon\left(x\right), \dots \dots (8.58)$$

where

$$\left|\varepsilon\left(x\right)\right| \le 1.5\times10^{-7},$$
$$u = 1/\left(1 + b_6 x\right),$$
$$b_1 = 0.254829592,$$
$$b_2 = -0.284496736,$$
$$b_3 = 1.421413741,$$
$$b_4 = -1.453152027,$$
$$b_5 = 1.061405429, \text{ and}$$
$$b_6 = 0.3274811.$$

Substituting Eq. 8.55 into Eq. 8.52 yields

$$\dot{m}_s H_s = M_R h\left(T_s - T_r\right)\frac{dA_h}{dt} + 2\int_o^t \frac{k_h\left(T_s - T_r\right)}{\sqrt{\pi\alpha\left(t - \lambda\right)}}\left(\frac{dA_h}{d\lambda}\right)d\lambda. \dots\dots\dots\dots\dots\dots\dots\dots\text{(8.59)}$$

It is convenient to express the solution to Eq. 8.57 in dimensionless variables. Following the practice of Prats (1982), the dimensionless time for heat transfer is defined by

$$t_D = 4\left(M_s/M_R\right)^2\left(\alpha/h^2\right)t, \dots\dots\dots\dots\dots\dots\dots\dots\dots\dots\dots\dots\text{(8.60)}$$

where consistent units are used and where M_s = volumetric heat capacity of the over- and underburden.

t_D	erfc(t_D)	t_D	erfc(t_D)
0.00	1.00000	1.05	0.13756
0.02	0.97744	1.1	0.11979
0.04	0.95489	1.15	0.10388
0.06	0.93238	1.2	0.08969
0.08	0.90992	1.25	0.0771
0.1	0.88754	1.3	0.06599
0.12	0.86524	1.35	0.05624
0.14	0.84305	1.4	0.04771
0.16	0.82099	1.45	0.0403
0.18	0.79906	1.5	0.03389
0.2	0.7773	1.55	0.02838
0.22	0.7557	1.6	0.02365
0.24	0.7343	1.65	0.01962
0.26	0.7131	1.7	0.01621
0.28	0.69212	1.75	0.01333
0.3	0.67137	1.8	0.01091
0.32	0.65087	1.85	0.00889
0.34	0.63064	1.9	0.00721
0.36	0.61067	1.95	0.00582
0.38	0.59099	2	0.00468
0.4	0.57161	2.05	0.00374
0.42	0.55253	2.1	0.00298
0.44	0.53377	2.15	0.00236
0.46	0.51534	2.2	0.00186
0.48	0.49725	2.25	0.00146
0.5	0.4795	2.3	0.00114
0.52	0.4621	2.35	0.00089
0.54	0.44506	2.4	0.00069
0.56	0.42838	2.45	0.00053
0.58	0.41208	2.5	0.00041
0.6	0.39614	2.55	0.00031
0.62	0.38059	2.6	0.00024
0.64	0.36541	2.65	0.00018
0.66	0.35062	2.7	0.00013
0.68	0.33622	2.75	0.0001
0.7	0.3222	2.8	0.00008
0.72	0.30857	2.85	0.00006
0.74	0.29532	2.9	0.00004
0.76	0.28246	2.95	0.00003
0.78	0.26999	3	0.00002
0.8	0.2579	3.1	0.00001
0.82	0.24619	3.2	0.00001

Table 8.7—Values of erfc(x) for selected values of x.

t_D	$\mathrm{erfc}(t_D)$	t_D	$\mathrm{erfc}(t_D)$
0.84	0.23486	–	–
0.86	0.2239	–	–
0.88	0.21331	–	–
0.9	0.20309	–	–
0.92	0.19323	–	–
0.94	0.18373	–	–
0.96	0.17458	–	–
0.98	0.16577	–	–
1.00	0.1573	–	–

Table 8.7—Values of erfc(x) for selected values of x (continued).

The solution of Eq. 8.59 gives an expression for the heated area as a function of dimensionless time as

$$A_h = \frac{\dot{m}_s H_s M_R h}{4\left(T_s - T_r\right)\alpha M_s^2}\left[e^{t_D}\mathrm{erfc}\left(\sqrt{t_D}\right)+2\sqrt{\frac{t_D}{\pi}}-1\right]. \quad \text{(8.61)}$$

The term $\left[e^{t_D}\mathrm{erfc}\left(\sqrt{t_D}\right)+2\sqrt{t_D/\pi}-1\right]$ is tabulated as a function of the $\sqrt{t_D}$ in the original paper by Marx and Langenheim (1959). Prats introduced the notation

$$G\left(t_D\right)= e^{t_D}\mathrm{erfc}\sqrt{t_D}+2\sqrt{t_D/\pi}-1, \quad \text{(8.62)}$$

so that

$$A_h = \left\{\dot{m}_s H_s M_R h/\left[4\left(T_s - T_r\right)\alpha M_s^2\right]\right\}G\left(t_D\right). \quad \text{(8.63)}$$

This notation is convenient for further derivations and computations.

The rate of growth of the heated zone is given by

$$dA_h/dt = \left\{\dot{m}_s H_s/M_R\left(T_s - T_r\right)h\right\}e^{t_D}\mathrm{erfc}\sqrt{t_D}. \quad \text{(8.64)}$$

$$= \left\{\dot{m}_s H_s/\left[M_R\left(T_s - T_r\right)h\right]\right\}G_1\left(t_D\right), \quad \text{(8.65)}$$

where $G_1\left(t_D\right)= e^{t_D}\mathrm{erfc}\left(\sqrt{t_D}\right)$. \quad \text{(8.66)}

Table 8.8 contains values of $G(t_D)$ and $G_1(t_D)$ at selected values of t_D. This function is easily calculated for other values of t_D with the series approximation for erf(x) given in Eq. 8.58. Appendix F contains programs to compute erfc(x), $G(t_D)$, and $G_1(t_D)$.

Example 8.4 illustrates use of the Marx and Langenheim model to predict the radius of the steam zone with time when steam is injected at a constant injection rate and the steam zone is assumed to grow radially, as shown in **Fig. 8.18.**

t_D	$G(t_D)$	$E_h(t_D)$	$G_1(t_D)$	t_D	$G(t_D)$	$E_h(t_D)$	$G_1(t_D)$	t_D	$G(t_D)$
0.0001	0.0001	0.99145	0.98882	0.0051	0.00484	0.94869	0.92426	0.02	0.01806
0.0002	0.0002	0.98905	0.98424	0.0052	0.00493	0.94821	0.92356	0.04	0.0347
0.0003	0.0003	0.98605	0.98075	0.0053	0.00502	0.94773	0.92288	0.06	0.05051
0.0004	0.00039	0.98463	0.97783	0.0054	0.00512	0.94727	0.9222	0.08	0.06571
0.0005	0.00049	0.98314	0.97526	0.0055	0.00521	0.94682	0.92152	0.1	0.0804
0.0006	0.00059	0.98161	0.97295	0.0056	0.0053	0.94634	0.92086	0.12	0.09467
0.0007	0.00069	0.98016	0.97083	0.0057	0.00539	0.94591	0.9202	0.14	0.10857
0.0008	0.00078	0.97898	0.96887	0.0058	0.00548	0.94547	0.91955	0.16	0.12214
0.0009	0.00088	0.97789	0.96703	0.0059	0.00558	0.94502	0.9189	0.18	0.13541
0.001	0.00098	0.97646	0.96529	0.006	0.00567	0.94457	0.91826	0.2	0.14841
0.0011	0.00107	0.97536	0.96365	0.0061	0.00576	0.94412	0.91763	0.22	0.16117

Table 8.8—$G(t_D)$, $E_h(t_D)$, and $G_1(t_D)$ for selected values of t_D.

t_D	$G(t_D)$	$E_h(t_D)$	$G_1(t_D)$	t_D	$G(t_D)$	$E_h(t_D)$	$G_1(t_D)$	t_D	$G(t_D)$
0.0012	0.00117	0.97446	0.96208	0.0062	0.00585	0.94369	0.917	0.24	0.1737
0.0013	0.00127	0.97347	0.96058	0.0063	0.00594	0.94329	0.91638	0.26	0.18601
0.0014	0.00136	0.97235	0.95914	0.0064	0.00603	0.94284	0.91576	0.28	0.19813
0.0015	0.00146	0.97148	0.95776	0.0065	0.00613	0.94244	0.91515	0.3	0.21006
0.0016	0.00155	0.97063	0.95642	0.0066	0.00622	0.942	0.91455	0.32	0.22181
0.0017	0.00165	0.96971	0.95512	0.0067	0.00631	0.94162	0.91395	0.34	0.2334
0.0018	0.00174	0.96889	0.95387	0.0068	0.0064	0.94118	0.91335	0.36	0.24483
0.0019	0.00184	0.96804	0.95265	0.0069	0.00649	0.94079	0.91276	0.38	0.25611
0.002	0.00193	0.96719	0.95147	0.007	0.00658	0.94037	0.91218	0.4	0.26726
0.0021	0.00203	0.96649	0.95032	0.0071	0.00667	0.93997	0.91159	0.42	0.27826
0.0022	0.00212	0.96577	0.9492	0.0072	0.00677	0.93959	0.91102	0.44	0.28914
0.0023	0.00222	0.96498	0.9481	0.0073	0.00686	0.93917	0.91045	0.46	0.29989
0.0024	0.00231	0.96428	0.94704	0.0074	0.00695	0.9388	0.90988	0.48	0.31052
0.0025	0.00241	0.96349	0.94599	0.0075	0.00704	0.93841	0.90932	0.5	0.32104
0.0026	0.0025	0.96288	0.94497	0.0076	0.00713	0.93802	0.90876	0.52	0.33145
0.0027	0.0026	0.96217	0.94397	0.0077	0.00722	0.93762	0.9082	0.54	0.34175
0.0028	0.00269	0.9615	0.94298	0.0078	0.00731	0.93723	0.90765	0.56	0.35195
0.0029	0.00279	0.96089	0.94202	0.0079	0.0074	0.93687	0.90711	0.58	0.36206
0.003	0.00288	0.96016	0.94108	0.008	0.00749	0.9365	0.90657	0.6	0.37206
0.0031	0.00297	0.95952	0.94015	0.0081	0.00758	0.93611	0.90603	0.62	0.38198
0.0032	0.00307	0.95897	0.93924	0.0082	0.00767	0.93574	0.90549	0.64	0.3918
0.0033	0.00316	0.95833	0.93834	0.0083	0.00776	0.93539	0.90496	0.66	0.40154
0.0034	0.00326	0.95774	0.93746	0.0084	0.00785	0.93502	0.90444	0.68	0.4112
0.0035	0.00335	0.95711	0.93659	0.0085	0.00794	0.93466	0.90391	0.7	0.42077
0.0036	0.00344	0.95658	0.93574	0.0086	0.00803	0.93429	0.90339	0.72	0.43027
0.0037	0.00354	0.95597	0.9349	0.0087	0.00813	0.93394	0.90288	0.74	0.43969
0.0038	0.00363	0.95543	0.93407	0.0088	0.00822	0.93359	0.90236	0.76	0.44903
0.0039	0.00372	0.95485	0.93326	0.0089	0.00831	0.93323	0.90185	0.78	0.4583
0.004	0.00382	0.95429	0.93245	0.009	0.0084	0.93288	0.90135	0.8	0.4675
0.0041	0.00391	0.95375	0.93166	0.0091	0.00849	0.93252	0.90085	0.82	0.47663
0.0042	0.004	0.95324	0.93088	0.0092	0.00858	0.93217	0.90035	0.84	0.48569
0.0043	0.0041	0.9527	0.9301	0.0093	0.00867	0.93183	0.89985	0.86	0.49469
0.0044	0.00419	0.9522	0.92934	0.0094	0.00876	0.93149	0.89936	0.88	0.50362
0.0045	0.00428	0.95168	0.92859	0.0095	0.00885	0.93116	0.89887	0.9	0.5125
0.0046	0.00438	0.95118	0.92785	0.0096	0.00894	0.93082	0.89838	0.92	0.52131
0.0047	0.00447	0.95066	0.92711	0.0097	0.00903	0.93047	0.89789	0.94	0.53006
0.0048	0.00456	0.95016	0.92638	0.0098	0.00912	0.93014	0.89741	0.96	0.53875
0.0049	0.00465	0.94967	0.92567	0.0099	0.00921	0.9298	0.89693	0.98	0.54738
0.005	0.00475	0.9492	0.92496	0.01	0.00929	0.92949	0.89646	1	0.55596

Table 8.8—$G(t_D)$, $E_h(t_D)$, and $G_1(t_D)$ for selected values of t_D (continued).

$E_h(t_D)$	$G_1(t_D)$	t_D	$G(t_D)$	$E_h(t_D)$	$G_1(t_D)$	t_D	$G(t_D)$	$E_h(t_D)$	$G_1(t_D)$
0.90283	0.85848	1.05	0.57717	0.54969	0.42093	4.1	1.53757	0.37502	0.25278
0.86738	0.80902	1.1	0.59806	0.54369	0.41461	4.2	1.56272	0.37208	0.25023
0.84184	0.77412	1.15	0.61864	0.53795	0.40859	4.3	1.58762	0.36921	0.24776
0.82135	0.74655	1.2	0.63892	0.53244	0.40285	4.4	1.61227	0.36643	0.24537
0.80403	0.72358	1.25	0.65893	0.52714	0.39736	4.5	1.63669	0.36371	0.24304

Table 8.8—$G(t_D)$, $E_h(t_D)$, and $G_1(t_D)$ for selected values of t_D (continued).

$E_h(t_D)$	$G_1(t_D)$	t_D	$G(t_D)$	$E_h(t_D)$	$G_1(t_D)$	t_D	$G(t_D)$	$E_h(t_D)$	$G_1(t_D)$
0.78894	0.70379	1.3	0.67866	0.52205	0.39211	4.6	1.66088	0.36106	0.24078
0.7755	0.68637	1.35	0.69814	0.51714	0.38709	4.7	1.68485	0.35848	0.23858
0.76337	0.67079	1.4	0.71738	0.51241	0.38226	4.8	1.7086	0.35596	0.23645
0.75229	0.65668	1.45	0.73637	0.50784	0.37762	4.9	1.73214	0.3535	0.23437
0.74207	0.64379	1.5	0.75514	0.50343	0.37317	5	1.75548	0.3511	0.23235
0.73259	0.63191	1.55	0.77369	0.49916	0.36887	5.5	1.86925	0.33986	0.22297
0.72374	0.62091	1.6	0.79203	0.49502	0.36473	6	1.97862	0.32977	0.21466
0.71543	0.61065	1.65	0.81017	0.49101	0.36074	6.5	2.08405	0.32062	0.20723
0.7076	0.60105	1.7	0.82811	0.48712	0.35688	7	2.18595	0.31228	0.20054
0.70019	0.59202	1.75	0.84586	0.48335	0.35315	7.5	2.28465	0.30462	0.19446
0.69316	0.5835	1.8	0.86342	0.47968	0.34955	8	2.38045	0.29756	0.18891
0.68647	0.57545	1.85	0.88081	0.47612	0.34605	8.5	2.47358	0.29101	0.18382
0.68009	0.5678	1.9	0.89803	0.47265	0.34267	9	2.56425	0.28492	0.17912
0.67399	0.56054	1.95	0.91508	0.46927	0.33939	9.5	2.65267	0.27923	0.17477
0.66814	0.55361	2	0.93197	0.46599	0.3362	10	2.73898	0.2739	0.17073
0.66253	0.54699	2.05	0.9487	0.46278	0.33311	11	2.90584	0.26417	0.16343
0.65713	0.54066	2.1	0.96528	0.45966	0.33011	12	3.06583	0.25549	0.15702
0.65193	0.53459	2.15	0.98172	0.45661	0.32719	13	3.21974	0.24767	0.15131
0.64692	0.52876	2.2	0.998	0.45364	0.32435	14	3.3682	0.24059	0.1462
0.64208	0.52316	2.25	1.01415	0.45073	0.32158	15	3.51177	0.23412	0.14158
0.6374	0.51776	2.3	1.03016	0.4479	0.31889	16	3.6509	0.22818	0.13738
0.63288	0.51257	2.35	1.04604	0.44512	0.31627	17	3.78597	0.2227	0.13355
0.62849	0.50755	2.4	1.06179	0.44241	0.31372	18	3.91733	0.21763	0.13002
0.62423	0.50271	2.45	1.07741	0.43976	0.31122	19	4.04526	0.21291	0.12677
0.62011	0.49802	2.5	1.09292	0.43717	0.30879	20	4.17002	0.2085	0.12376
0.61609	0.49349	2.55	1.1083	0.43463	0.30642	21	4.29184	0.20437	0.12096
0.61219	0.4891	2.6	1.12356	0.43214	0.3041	22	4.41091	0.2005	0.11835
0.6084	0.48484	2.65	1.13871	0.4297	0.30184	23	4.52741	0.19684	0.1159
0.6047	0.48071	2.7	1.15374	0.42731	0.29963	24	4.64151	0.1934	0.11361
0.6011	0.4767	2.75	1.16867	0.42497	0.29747	25	4.75334	0.19013	0.11145
0.59759	0.47281	2.8	1.18349	0.42268	0.29535	26	4.86304	0.18704	0.10942
0.59417	0.46902	2.85	1.19821	0.42042	0.29329	27	4.97072	0.1841	0.10749
0.59083	0.46533	2.9	1.21282	0.41821	0.29126	28	5.07649	0.1813	0.10567
0.58756	0.46174	2.95	1.22733	0.41605	0.28928	29	5.18045	0.17864	0.10395
0.58437	0.45825	3	1.24175	0.41392	0.28734	30	5.28269	0.17609	0.1023
0.58126	0.45484	3.1	1.2703	0.40977	0.28358	35	5.77075	0.16488	0.09517
0.57821	0.45152	3.2	1.29847	0.40577	0.27997	40	6.2259	0.15565	0.0894
0.57522	0.44828	3.3	1.32629	0.40191	0.27649	45	6.65402	0.14787	0.08462
0.5723	0.44511	3.4	1.35377	0.39817	0.27315	50	7.05941	0.14119	0.08057
0.56944	0.44202	3.5	1.38093	0.39455	0.26993	55	7.44537	0.13537	0.07709
0.56664	0.439	3.6	1.40776	0.39105	0.26682	60	7.81443	0.13024	0.07405
0.56389	0.43605	3.7	1.4343	0.38765	0.26382	65	8.16864	0.12567	0.07136
0.5612	0.43317	3.8	1.46053	0.38435	0.26092	70	8.50966	0.12157	0.06897
0.55855	0.43035	3.9	1.48648	0.38115	0.25812	75	8.83887	0.11785	0.06682
0.55596	0.42758	4	1.51216	0.37804	0.2554	80	9.1574	0.11447	0.06488

Table 8.8—$G(t_D)$, $E_h(t_D)$, and $G_1(t_D)$ for selected values of t_D (continued).

Example 8.4—Radius of Steam Zone at Constant Injection Rate. Steam is to be injected into a reservoir at a rate of 500 BWPD CWE. The steam has 80% quality, f_{sd}, at a pressure of 500 psig at the sandface. Properties of the reservoir rock and fluids are identical to those used in Example 8.3, assuming that 40% of the pore volume (PV) in the heated region is steam. Reservoir thickness is 20 ft. The thermal conductivity of the overburden, k_h, is taken to be 1.5 Btu/hr-ft-°F, and the thermal diffusivity of the overburden, α, is 0.0482 ft²/hr. Find the radius of the heated area after 14 days of continuous injection, assuming the area is cylindrical in shape.

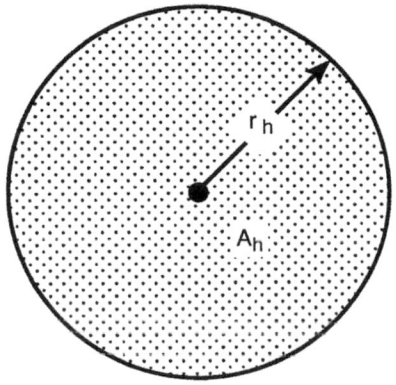

Fig. 8.18—Radial growth of heated area predicted by Marx and Langenheim model.

Solution. The heated area is found from Eq. 8.63.

$$A_h = \left\{ \dot{m}_s H_s M_R h / \left[4\left(T_s - T_r\right)\alpha M_s^2 \right] \right\} G\left(t_D\right).$$

The energy content of the injected steam is determined from the steam tables in Appendix E. In Eq. 8.63, H_s is given by

$$H_s = H_{wT} + f_{sd}L_{vdh} - H_{wr}. \quad\dots\dots\dots\dots\dots\dots\dots\dots\dots\dots\dots\dots\dots(8.67)$$

The saturation temperature of steam at 500 psig (514.7 psia) is 470.9°F. In Example 8.3, the enthalpies of the saturated liquid and vapor were determined to be as follows: $H_{wr} = 48$ Btu/lbm at 80°F, $H_{wT} = 452.9$ Btu/lbm at 470.9°F, $H_s = 1{,}204.3$ Btu/lbm at 470.9°F, $L_{vdh} = 751.4$ Btu/lbm, and $H_s = H_{wT} + f_{sd} L_{vdh} - H_{wr} = 452.9 + 0.8\,(751.4) - 48 = 1{,}006$ Btu/lbm.

The cold-water mass rate is computed assuming 350 lbm/bbl water.

$$\dot{m}_s = \left(500\,\text{B/D}\right)\left(350\,\text{lbm/bbl}\right)/\left(24\,\text{hr/D}\right)$$

$$= 7{,}292\,\text{lbm/hr}.$$

From Example 8.3,

$$M_R = 33.13\,\text{Btu/ft}^3\text{-°F},$$

$$\alpha = k_h / M_s,$$

and $M_s = \left(1.5\,\text{Btu/ft-°F-hr}\right)/\left(0.0482\,\text{ft}^2/\text{hr}\right)$

$$= 31.12\,\text{Btu/ft}^3\text{-°F}.$$

The dimensionless time is computed from Eq. 8.60:

$$t_D = 4\left(M_s/M_R\right)^2 \left(\alpha/h^2\right)t$$

$$= 4\left(31.12/33.13\right)^2 \left[\left(0.0482\,\text{ft}^2/\text{hr}\right)/\left(20\,\text{ft}^2\right)\right]$$

$$\times \left(14\,\text{days}\right)\left(24\,\text{hr/D}\right)$$

$$= 0.143.$$

Interpolating from Table 8.8,

$$G = 0.10857 + \left(\frac{0.143 - 0.14}{0.16 - 0.14}\right)\left(0.12214 - 0.10857\right)$$

$$= 0.111.$$

The heated area can be computed from Eq. 8.63:

$$A_h = \left\{ \left[\left(7.292\,\text{lbm/hr}\right)\left(1006\,\text{Btu/lbm}\right)\left(33.13\,\text{Btu/ft}^3\text{-°F}\right)\right.\right.$$

$$\times\left(20\,\text{ft}\right)\right]/\left[4\left(470.9 - 80\right)\left(0.0482\,\text{ft}^2/\text{hr}\right)\right.$$

$$\left.\left.\times\left(31.12\,\text{Btu/ft}^3\text{°F}\right)^2\right]\right\}\left(0.111\right)$$

$$= \left(66{,}545\right)\left(0.111\right)$$

$$= 7{,}392\,\text{ft}^2.$$

and $r_h \approx \sqrt{A_h/\pi}$

$$= \sqrt{7{,}392\,\text{ft}^2/\pi}$$

$$= 48.5\,\text{ft}.$$

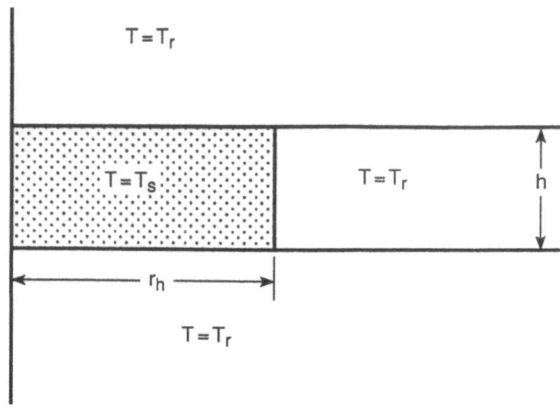

Fig. 8.19—Initial reservoir temperature distribution for Boberg and Lantz model.

8.3.3 Estimating Production Response From Cyclic Steam Injection—Boberg and Lantz Model. Production response to cyclic steam injection can be estimated by use of the CSS model developed by Boberg and Lantz (1966). The model is conceptually simple, retains most of the important features of the process, and will be used to illustrate basic concepts.

The Boberg and Lantz model begins by using the Marx and Langenheim model to compute the radius of the heated zone as shown in Example 8.4. The model assumes that the reservoir is heated "instantaneously" to a temperature of T_s. As a result, the initial temperature distribution around the well is depicted in **Fig. 8.19.** Although heat losses are considered in computing the heated-zone radius, the initial temperature distribution for the model assumes that the region within a disk $0 < r < r_h$ and $0 \leq z \leq h$ is at temperature T_s, while the remainder of the overburden and reservoir is at T_r.

Then, a mathematical model was derived that gives the average temperature of the heated region with time as heat flows from the heated disk into the over- and underburden by conduction. Although colder fluids enter the heated region, the average-temperature model does not explicitly account for this effect. Removal of energy with the produced fluids is accounted for in an approximate manner that will be described later.

The partial-differential equation that describes the energy balance when conduction is the only heat-transfer mechanism is given by

$$\frac{k_h}{r}\frac{\partial}{\partial r}\left(r\frac{\partial T}{\partial r}\right) + k_h\frac{\partial^2 T}{\partial z^2} = \rho C_p\frac{\partial T}{\partial t}. \dots \dots (8.68)$$

Eq. 8.68 is solved for the single-zone case with the following boundary and initial conditions.
Initial Conditions.

At $t = t_i$, $r_w \leq r \leq r_s$, $0 \leq z \leq h$, and $T = T_S$

and $t = t_i$, $0 < r$, $z < 0$, $z > h$, $T = T_r$, $r_s < r$, and $-\infty \leq z \leq \infty$.

Boundary Conditions.
At $r = 0$, $\partial T/\partial r = 0$ (no flow of heat by conduction)

$r \rightarrow \infty \; \partial T/\partial r = 0$,

and at $z = h/2$, $\partial T/\partial z = 0$ (symmetry across centerline of formation)

$z \rightarrow \infty \; (\partial T/\partial z) = 0$ (no flow of heat by conduction).

Solution of Eq. 8.68 is possible by analytical techniques by using the principle of superposition (Carslaw and Jaeger 1959). When the heat-flow problem satisfies certain constraints, the temperature distribution, $T(r,z)$ is a product of solutions derived from the solutions of equations considering heat flow in the r- and z-direction. That is,

$$T(r,z) = T(r)T(z). \dots \dots (8.69)$$

Boberg and Lantz (1966) used this solution technique to develop an analytical solution for the average temperature $\overline{T}(r)$ for $0 \leq r \leq r_h$, and $\overline{T}(z)$ for $0 \leq z \leq h$. **Fig. 8.20** plots these solutions against dimensionless time. Dimensionless average temperature and times are defined by

$$\overline{T}_{Dr} = \frac{\overline{T} - T_r}{\overline{T}_s - T_r} \dots \dots (8.70)$$

and $t_{Dr} = \left[\alpha(t - t_i)\right]/r_h^2$, $\dots \dots (8.71)$

where t_{Dr} = dimensionless time for the radial average heat flow model. Thus, $\overline{T}_{Dr} = 0$ when $\overline{T} = T_r$ and $\overline{T}_{Dr} = 1.0$ when $\overline{T} = T_S$, where $\overline{T}_{Dr}$ = dimensionless temperature averaged over the radius $0 < r < r_h$, $\overline{T}_{Dz}$ = dimensionless average temperature in the region $0 \leq z \leq h$, and t_{Dz} = dimensionless time for the average temperature solution in the z-direction.

$$t_{Dz} = \left[\alpha(t - t_i)\right]/(h/2)^2. \dots \dots (8.72)$$

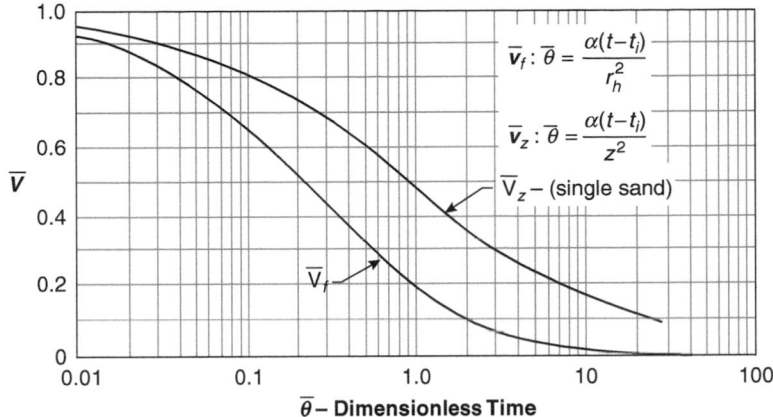

Fig. 8.20—Average dimensionless temperatures vs. dimensionless time for a single zone of radius r_h and thickness h losing heat by conduction (Boberg and Lantz 1966).

Expressions for $\overline{T}_{Dr}$ and $\overline{T}_{Dz}$ are given in Boberg and Lantz (1966) and Bentsen and Donohue (1969). For $1/t_{Dr} \leq 1.0$ (Bentsen and Donohue 1969), Eq. 8.73 may be used to estimate T_{Dr}.

$$\overline{T}_{Dr} = 1 - \sqrt{\frac{t_{Dr}}{\pi}}\left(2 - \frac{t_{Dr}}{2} - \frac{3}{16}t_{Dr}^2 - \frac{15}{16}t_{Dr}^3 - \frac{525}{1,024}t_{Dr}^4 - \ldots\right) \quad \ldots \ldots (8.73)$$

Eq. 8.74 (Boberg and Lantz 1966) is an expansion of the general series, which should be used for $1/t_{Dr} > 0.1$ (Bentsen and Donohue 1969):

$$\overline{T}_{Dr} = \frac{1}{4t_{Dr}}\left(1 - \frac{1}{4t_{Dr}} + \sum_{k=2}^{\infty} S_k\right)$$

$$S_k = S_{k-1}\frac{(k+0.5)\left(-\dfrac{1}{t_{Dr}}\right)}{(k+1)(k+2)}$$

$$s_1 = 1/4,$$

$$s_{k+1} = \left\{-(1/t_{Dr})[(0.5+k)/(1+k)(2+k)]\right\}s_k,$$

and $$\overline{T}_{Dr} = \left(\frac{1}{4t_{Dr}} - \frac{1}{16t_{Dr}^2} + \frac{5}{384t_{Dr}^3} - \frac{1}{439t_{Dr}^4} + \frac{7}{20,480t_{Dr}^5}\right) \quad \ldots \ldots (8.74)$$

$\overline{T}_{Dz}$ is calculated from Eq. 8.75 (Bentsen and Donohue 1969):

$$\overline{T}_{Dr} = \mathrm{erf}\left(1/\sqrt{t_{Dz}}\right) - \left(\sqrt{\frac{t_{Dz}}{\pi}}\right)\left(1 - e^{-1/t_{Dz}}\right). \quad \ldots \ldots (8.75)$$

$$\overline{T}_D = \overline{T}_{Dr}\,\overline{T}_{Dz}. \quad \ldots \ldots 8.76)$$

The Boberg and Lantz model was developed for multiple zones of sand separated by shale. Example 8.5 illustrates the computation of the average temperature of the heated region without consideration of heat removal with produced fluids.

Example 8.5—Estimation of the Average Temperature of a Zone Shut In After Steam Stimulation. A reservoir is heated by steam injection to give a heated radius of 30 ft at a steam temperature of 400°F. The reservoir is 40 ft thick and has an initial temperature of 120°F. Thermal conductivity of the reservoir, k_h, is 1.4 Btu/hr-ft²-°F/ ft, and the average heat capacity of the formation and overburden is 35 Btu/ft³-°F. Using the Boberg and Lantz model, determine the average temperature of the heated zone 100, 200, and 300 days after the reservoir temperature was elevated to 400°F. No fluids are produced from the reservoir during this time.

Solution. The thermal diffusivity, α, must be computed to find $\overline{T}_{Dr}$ and $\overline{T}_{Dz}$ from Eq. 8.77:

$$
\begin{aligned}
\alpha &= k_h/M \\
&= \left(1.4\,\text{Btu/hr-ft°F}\right)\big/\left(35\,\text{Btu/ft}^3\text{-°F}\right) \\
&= 0.04\,\text{ft}^2/\text{hr} \\
&= 0.96\,\text{ft}^2/\text{D}. \qquad\qquad\qquad\qquad\qquad\qquad\qquad\qquad\qquad\qquad (8.77)
\end{aligned}
$$

For the radial dimensionless temperature component, $\overline{T}_{Dr} = \overline{T}_{Dr}\left(t_{Dr}\right)$, where

$$
\begin{aligned}
t_{Dr} &= \left[\alpha\left(t - t_i\right)\right]\big/r_h^2 \\
&= \left[\left(0.96\,\text{ft}^2/\text{D}\right)\left(t - t_i\right)\right]\big/\left(30\,\text{ft}\right)^2 \\
&= 0.001067\left(t - t_i\right).
\end{aligned}
$$

At $t - t_i = 100$ days, $t_{Dr} = 0.1067$ and $\overline{T}_{Dr} = 0.63$; at $t - t_i = 200$ days, $t_{Dr} = 0.2133$ and $\overline{T}_{Dr} = 0.50$; and at $t - t_i, = 300$ days, $t_{Dr} = 0.320$ and $\overline{T}_{Dr} = 0.42$. For the thickness-averaged temperature, $\overline{T}_{Dz}$,

$$
\begin{aligned}
t_{Dz} &= \left[\alpha\left(t - t_i\right)\right]\big/\left(h/2\right)^2 \\
&= \left[\left(0.96\,\text{ft}^2/\text{D}\right)\left(t - t_i\right)\right]\big/\left(20\,\text{ft}\right)^2 \\
&= 0.0024\left(t - t_i\right).
\end{aligned}
$$

At $t - t_i = 100$ days, $t_{Dz} = 0.24$ and $\overline{T}_{Dz} = 0.74$; at $t - t_i = 200$ days, $t_{Dz} = 0.48$ and $\overline{T}_{Dz} = 0.61$; and at $t - t_i = 300$ days, $t_{Dz} = 0.72$ and $\overline{T}_{Dz} = 0.54$. **Table 8.9** summarizes the results on the basis of $\overline{T}_D = \overline{T}_{Dr}\overline{T}_{Dz}$ and $\overline{T} = T_r + \left(T_s - T_r\right)\overline{T}_D$.

The analytical solution to determine the average temperature of the heated zone has two weaknesses that were accounted for by making adjustments in the calculation. The first adjustment is to compensate for the assumption that there is no initial temperature distribution in the shale above or below the heated region. This is achieved by defining a hypothetical thickness, z, which is added to the individual sand thickness, h, to account for all energy lost to over- and underlying formations. The value of z is determined from an overall energy balance

$$
m_s H_s = \pi r_h^2 M\left(T_s - T_r\right)\left(z + h\right), \qquad\qquad\qquad\qquad\qquad\qquad (8.78)
$$

where the term on the left is the total amount of energy injected into the reservoir and the term on the right assumes that all the energy is present in a disk of radius r_h and thickness $z + h$, which is at temperature T_s. Solving for z gives Eq. 8.79. In Eq. 8.78, $m_s = m_s t$ and is the total pounds of water injected as steam.

$$
z = \left\{m_s H_s\big/\left[\pi r_h^2 M\left(T_s - T_r\right)\right]\right\} - h. \qquad\qquad\qquad\qquad\qquad (8.79)
$$

The dimensionless time, t_{Dz}, is modified as

$$
t_{Dz} = \left[\alpha\left(t - t_i\right)\right]\big/\left[\left(z + h\right)/2\right]^2. \qquad\qquad\qquad\qquad\qquad\qquad (8.80)
$$

The second modification accounts for the fact that the average temperature model does not consider removal of heat with the produced fluids. At any time the rate that heat is removed by produced fluids is given by

$$
\dot{Q}_p = 5.615 q_{oh}\left[\rho_o C_o + F_{wo}\left(\rho_w C_w\right)\right]\left(\overline{T}_p - T_r\right), \qquad\qquad\qquad\qquad (8.81)
$$

$t - t_i$ (days)	$\overline{T}_{Dr}$	$\overline{T}_{Dz}$	$\overline{T}_D$	$\overline{T}$ (°F)
100	0.63	0.74	0.466	250.5
200	0.50	0.61	0.305	205.4
300	0.42	0.54	0.227	183.5

Table 8.9—Average heated-zone temperaure after shut-in, initial temperature of heated zone 400°F, initial reservoir temperature 120°F.

where $\dot{Q}_p$ = energy removal rate, Btu/D; $\overline{T}_p$ = average temperature of the heated zone adjusted for heat removal with produced fluids, °F; and F_{wo} = water/oil ratio (WOR). A dimensionless parameter, δ, is defined by Boberg and Lantz to account for energy removed by produced fluids.

$$2\delta = \int_{t_i}^{t} \dot{Q}_p \, d\lambda \Big/ m_s H_s . \quad\dotfill (8.82)$$

The numerator is the cumulative amount of energy removed by the produced fluids in the time interval $t - t_i$ of Eq. 8.82 and the denominator is the amount of energy injected into the reservoir. The factor two is empirical so that

$$\delta = \frac{1}{2} \int_{t_i}^{t} \dot{Q}_p \, d\lambda \Big/ m_s H_s . \quad\dotfill (8.83)$$

With the correction, the average temperature is defined by

$$\overline{T}_{Dp} = \left(\overline{T}_p - T_r \right) \Big/ \left(T_s - T_r \right) = \overline{T}_{Dr} \overline{T}_{Dz} \left(1 - \delta\right) - \delta ; \quad\dotfill (8.84)$$

thus, $\overline{T}_p = T_r + \left(T_s - T_r \right)\left[\overline{T}_{Dr} \overline{T}_{Dz} \left(1 - \delta\right) - \delta \right].$ $\quad\dotfill (8.85)$

When significant gas and water vapor are produced,

$$\dot{Q}_p = q_{oh} \left(H_{ogv} + H_{wrv} \right), \quad\dotfill (8.86)$$

where $H_{ogv} = \left[5.615 \left(\rho_o C_o \right) + F_{go} C_g \right]\left(\overline{T}_p - T_r \right);$ $\quad\dotfill (8.87)$

$$H_{wrv} = 5.615 \left[F_{wo} \left(H_{wT} - H_{wr} \right) + F_{wv} L_v \right]; \quad\dotfill (8.88)$$

H_{wT} = enthalpy of saturated water at $\overline{T}$, Btu/lbm; H_{ogv} = enthalpy of produced oil and gas relative to T_r, Btu/STB; H_{wrv} = enthalpy of produced water, Btu/bbl; F_{go} = gas/oil ratio (GOR), scf/STB; F_{wo} = WOR in produced liquid, bbl/STB; and F_{wv} = WOR for water condensed from the produced gas, bbl/STB.

The WOR for water condensed from the produced gas is given by

$$F_{wv} = \left(0.0001356 \right)\left(\frac{p_{wv}}{p_w - p_{wr}} \right) F_{go} \left(\frac{\text{bbl water at } 60\,°F}{\text{STB}} \right), \quad\dotfill (8.89)$$

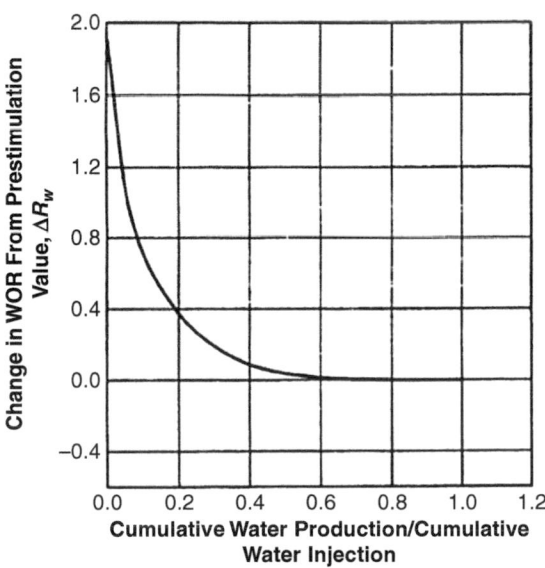

Fig. 8.21—Increase in WOR after steam injection, average for Quiriquire Wells Q-594 and Q-599 (Boberg and Lantz 1966).

where p_{wv} = vapor pressure of water at $\overline{T}$ and P_w = bottomhole pressure (BHP) in the production well. There will always be water vapor produced if there is produced gas. Water vapor will also be produced if $p_{wv} > pw$ because hot water will be flashed to steam in the production wellbore. In this case, Eq. 8.89 is not valid and F_{wv} cannot be estimated without additional calculations. However, F_{wv} cannot exceed F_{wo}.

Water production immediately following steam stimulation is primarily condensate. Water production will decline with time as the condensate is displaced by formation fluids. The Boberg and Lantz model accounts for the production of condensate by use of an empirical correlation of the increase in the WOR of the produced water to the water injected as steam. The correlation, developed for the Quiriquire wells in Venezuela, is presented in **Fig. 8.21.** Because the correlation is empirical, it would be expected to vary from reservoir to reservoir and possibly from well to well. This means that the Boberg and Lantz model must be calibrated against field data before it can be used for prediction.

Estimation of the average fluid temperature is complicated by the fact that Eqs. 8.81 and 8.85 involve, and, in principle, must be solved in, a stepwise manner. Example 8.6 illustrates the estimation of steam stimulation response for a single-zone reservoir with the Boberg and Lantz method.

Thickness (ft)	41
Reservoir temperature (°F)	93
Oil viscosity at reservoir temperature (cp)	2,026
Oil gravity (°API)	13.5
Porosity	0.345
Overburden heat capacity (Btu/ft³-°F)	38.5
Reservoir heat capacity (Btu/ft³-°F)	41
Wellbore radius (ft)	0.292
Effective drainage radius (ft)	233.5
BHP (psi)	25
Formation pressure (psi)	125
Producing GOR	0
Oil heat capacity (Btu/lbm-°F)	0.5
Prestimulation oil rate (B/D)	5.79
WOR	0.25

Table 8.10—Reservoir and fluid properties, Example 8.6.

Example 8.6—Estimation of Steam Stimulation Response With Boberg and Lantz Model. A well in a reservoir with the properties summarized in **Table 8.10** is to be stimulated with steam. The production rate before steam injection is 5.8 STB/D, and the primary production mechanism is depletion drive. There is no evidence of wellbore damage. Estimate the production response of this well to the injection of steam with the Boberg and Lantz method if the temperature/viscosity relationship for the crude oil is given by Eq. 8.90. Steam will be injected for 2 weeks at a rate of 1,000 B/D and 200 psia and then shut in for 4 days before placing the well on production. The viscosity of the crude oil is represented by the following American Society of Testing and Materials viscosity correlation.

$$\log\left\{\log\left[\left(\mu_o/\rho_o\right)+0.6\right]\right\}=a-b\log\left(T+460\right), \quad \ldots \text{(8.90)}$$

where $\rho_o = \rho_{oi}e^{-\beta(T-60)}, \ldots \ldots \ldots \ldots \ldots \ldots \ldots \ldots \ldots$ (8.91)

where $\rho_{oi} = 141.5/(131.5+°API), \ldots \ldots \ldots \ldots \ldots$ (8.92)

$$\beta = 3.5 \times 10^{-4} \text{ °F}^{-1},$$

$$a = 16.1368,$$

$$\text{and } b = 5.6934.$$

Solution. The heated radius owing to steam injection is computed from the Marx and Lagenheim model.

$$A_h = \left\{\dot{m}_s H_s h/\left[4(T_s - T_r)M\right]\right\}G(t_D).$$

From Eq. 8.60,

$$t_D = \left(4k_h M_s/M_R^2 h^2\right)t.$$

In the Boberg and Lantz model, M, the average of the reservoir and shale properties, replaces M_s and Mr; thus,

$$M = \left(M_s + M_R\right)/2$$

$$= (41+38.5)/2$$

$$= 39.75 \text{ Btu/ft}^3\text{-°F},$$

$$k_h = 1.21 \text{ Btu/hr-ft-°F},$$

$$\alpha = 0.0305 \text{ ft}^2/\text{hr}$$

$$= 0.731 \text{ ft}^2/\text{D},$$

and $t_D = 4k_h t/Mh^2$

$$= 4\alpha t/h^2$$

$$= \left[(4)(0.731 \text{ ft}^2/\text{D})(14 \text{ days})\right]/(41.0 \text{ ft})^2$$

$$= 0.0244.$$

From Table 8.8, the value of $G(t_D) \approx 0.0217$. At 200 psia (381.82°F),

$$H_{wT} = 355.4 \text{ Btu/lbm and } L_{vdh} = 843.3 \text{ Btu/lbm}.$$

At $T_r = 93°F$ and $H_{wr} = 61 \text{ Btu/lbm}$,

$$H_s = H_{wT} + f_{sd}L_{vdh} - H_{wr}$$

$$= 355.4 + 0.7(843.3) - 61$$

$$= 884.7 \, \text{Btu/lbm}.$$

$$A_h = \{[(1,000 \, \text{B/D})(5.615 \, \text{ft}^3 \, / \, \text{bbl})(62.4 \, \text{lbm/ft}^3)$$

$$\times (884.7 \, \text{Btu/lbm})(41 \, \text{ft})] \, / \, [4(381.82 - 93°\text{F})$$

$$\times (0.731 \, \text{ft}^2 \, / \, \text{D})(39.75 \, \text{Btu/ft}^3 \text{-}°\text{F})]\}G(t_D)$$

$$= (378,593)(0.0217) \, \text{ft}^2$$

$$= 8,229 \, \text{ft}^2$$

and $r_h = 51.2 \, \text{ft}$.

Next, compute the incremental thickness with Eq. 8.79 to compensate for heat losses to the shale during steam injection.

$$z = \{[(1,000 \, \text{B/D})(5.615 \, \text{ft}^3 \, /\text{bbl})(62.4 \, \text{lbm/ft}^3)$$

$$\times (884.7 \, \text{Btu/lbm})(14 \, \text{days})] \, / \, [\pi (51.2 \, \text{ft})^2$$

$$\times (39.75 \, \text{Btu/ft}^3 \text{-}°\text{F})(381.82 \text{-} 93°\text{F})]\} - 41$$

$$= 45.94 - 41$$

$$= 4.94 \, \text{ft}.$$

Dimensionless times can be computed for radial and vertical dimensionless temperatures.

$$t_{Dr} = \left[\alpha(t - t_i) \right] / r_h^2$$

$$= \left[(0.731 \, \text{ft}^2 \, / \, \text{D})(t - t_r) \right] / (51.2 \, \text{ft}^2)$$

$$= 2.789 \times 10^{-4} (t - t_r).$$

$$t_{Dz} = \left[\alpha(t - t_i) \right] / \left[(z + h)/2 \right]^2$$

$$= \left[(0.731 \, \text{ft}^2 \, / \, \text{D})(t - t_i) \right] / (45.94/2)^2$$

$$= 1.386 \times 10^{-3} (t - t_i).$$

The stimulation performance will be computed in increments of 30 days following shut-in. Thus, at the beginning of production

$$t - t_i = 4 \, \text{days}, \, \delta = 0,$$

$$t_{Dr} = 2.789 \times 10^{-4} (4)$$

$$= 0.0011,$$

and $t_{Dz} = 1.386 \times 10^{-3} (4)$

$$= 0.0055.$$

Values of dimensionless time are not within the range of Fig. 8.20, so the initial temperature cannot be calculated. At $t - t_i = 60$ days, $t_{Dr} = 0.017$, $\overline{T}_{Dr} = 0.88$, $t_{Dz} = 0.084$, and $\overline{T}_{Dz} = 0.83$. Thus,

$$\overline{T}_D = (0.88)(0.83)$$

$$= 0.73$$

and $\overline{T}_D = 93 + (288.82)(0.73)$

$$= 303.84°\text{F}.$$

Values of T uncorrected for energy removed with produced fluids are presented in **Table 8.11.** The procedure used to obtain a value of $\overline{T}_p$ is illustrated below. The production interval is subdivided into time intervals, as done to compute $\overline{T}_D$ in Table 8.9. It is necessary to calculate δ for each timestep. For short timesteps, the integral in Eq. 8.83 is approximated by the summation

$$\int_{t_i}^{t} \dot{Q}_P \, dt = \sum_{i=1}^{n} \dot{Q}_{Pi} \Delta t_{pi}, \quad \dotfill \quad (8.93)$$

$t - t_i$ (days)	t_{Dz}	$\overline{T}_{Dz}$	t_{Dr}	$\overline{T}_{Dr}$	$\overline{T}_D$
4	–	–	–	1.00	–
60	0.083	0.83	0.017	0.88	0.73
90	0.125	0.80	0.025	0.85	0.68
120	0.166	0.75	0.034	0.82	0.64
150	0.208	0.74	0.042	0.78	0.58
180	0.249	0.72	0.050	0.76	0.55
210	0.291	0.69	0.059	0.73	0.50
240	0.333	0.67	0.067	0.72	0.48
260	0.360	0.66	0.073	0.70	0.46
290	0.402	0.64	0.081	0.66	0.44
320	0.443	0.63	0.089	0.66	0.42
350	0.485	0.610	0.098	0.64	0.39
380	0.527	0.59	0.106	0.63	0.370
410	0.568	0.58	0.114	0.61	0.354
440	0.610	0.57	0.123	0.60	0.342
470	0.651	0.55	0.131	0.59	0.325
500	0.693	0.54	0.139	0.58	0.313
530	0.735	0.53	0.148	0.57	0.302
560	0.776	0.525	0.156	0.55	0.289

Table 8.11—Computation of $\overline{T}_{Dr}$, Example 8.6.

where $\dot{Q}_{pi}$ = the energy removal rate at the beginning of Timestep i, Btu/D; Δt_{pi} = number days in Timestep i; and n = number of timesteps. The energy removal rate is determined by assuming that the temperature of the produced fluids is constant during Timestep i and is equal to the average temperature determined from the previous timestep, $\overline{T}_{p,i-1}$.

Computation of $\dot{Q}_{pi}$ requires a correlation between ΔF_{wo} and (W_p/W_{is}), which must be determined from matching the production data. For this example, the following relationship was assumed:

$$\Delta F_{wo} = 1.83\,e^{-7.38\left(w_p/w_{is}\right)}. \dots\dots\dots\dots\dots\dots\dots\dots\dots\dots\dots\dots\dots\dots\dots\dots\dots\dots\dots\dots\dots\dots\dots\dots\dots\dots (8.94)$$

Thus, the WOR following the stimulation is

$$F_{wo} = F_{woc} + \Delta F_{wo}, \dots\dots\dots\dots\dots\dots\dots\dots\dots\dots\dots\dots\dots\dots\dots\dots\dots\dots\dots\dots\dots\dots\dots\dots\dots\dots\dots\dots\dots (8.95)$$

where F_{woc} = WOR before stimulation. Recall that W_{is} = 14,000 bbl water.

For the first timestep, the average temperature is assumed to be the injection temperature, 381.82°F, because the dimensionless times were too low to determine $\overline{T}_{Dr}$ and $\overline{T}_{Dz}$ from Fig. 8.20. From Eqs. 8.90 through 8.92, the oil density at 381.8°F is 0.862 g/cm³ (53.8 lbm/ft³), so that B_{oh} = 1.13 bbl/bbl and μ_{oh} = 1.22 cp. The value of B_{oc} = 1.01 bbl/bbl. Substituting for μ_{oc}, μ_{oh}, r_e, r_w, r_h, B_{oh}, and B_{oc} in Eq. 8.48 yields

$$\overline{J} = \frac{J_h}{J_c} = \frac{(1.01/1.13)\ln(233.5/0.292)}{(1.22/2{,}026)\ln(51.2/0.292) + \ln(233.5/51.2)}$$

$$= 5.974 / (0.00311 + 1.5174)$$

$$= 3.93.$$

The value of J_c is determined from the unstimulated production rate.

$$J_c = q_{oc}/\Delta p$$

$$= (5.79\,\text{STB/D})/(100\,\text{psi})$$

$$= 0.0579\,\text{STB/D-psi}. \dots\dots\dots\dots\dots\dots\dots\dots\dots\dots\dots\dots\dots\dots\dots\dots\dots\dots\dots\dots\dots\dots (8.96)$$

Thus,

$$q_o = \overline{JJ}_c \, \Delta p$$
$$= (3.93)(0.0579)(100)$$
$$= 22.75 \, \text{STB/D}$$

and $q_{oh} = q_o \, B_{oh}$

$$= (22.75 \, \text{STB/D})(1.13 \, \text{bbl/STB})$$
$$= 25.71 \, \text{B/D}.$$

This is not a spectacular production rate but is a four-fold increase in rate over primary production. The rate is low because there is limited reservoir energy.

The average temperature of the produced fluids can now be computed for the second timestep. In this example, the densities and heat capacities were constant. From Eqs. 8.86 through 8.88,

$$\dot{Q}_{pi} = (5.615) q_{oh} \left(\rho_c C_o + F_{wo} \rho_w C_w \right) \left(\overline{T}_p - T_r \right). \quad\quad\quad\quad\quad\quad\quad\quad\quad (8.97)$$

Because $W_p = 0$, $F_{wo} = 2.08$, at 381.8°F, $C_w = 1.016$ Btu/lbm and $p_w = 54.4$ lbm/ft³. For $n = 1$,

$$\dot{Q}_{pi} = (5.615 \, \text{ft}^3/\text{bbl})(25.7 \, \text{B/D}) \{ [(53.8 \, \text{lbm/ft}^3)$$

$$\times (0.5 \, \text{Btu/lbm-°F})] + [(2.08 \, \text{bbl/bbl})(54.4 \, \text{lbm/ft}^3)$$

$$\times (1.016 \, \text{Btu/lbm-°F})] \} (381.8 - 93°\text{F})$$

$$= 5.91 \times 10^6 \, \text{Btu/D}$$

and $\sum_{i=1}^{n} \dot{Q}_{pi} \, \Delta t_{pi} = (5.91 \times 10^6)(56 \, \text{days})$

$$= 3.31 \times 10^8 \, \text{Btu}.$$

Then, δ is found from Eq. 8.83.

$$\delta = \frac{1}{2} \sum_{i=1}^{n} \dot{Q}_{pi} \Delta t_{pi} \, / \, m_s H_s;$$

but $m_s H_s = (1,000 \, \text{B/D})(14 \, \text{days})(5.615 \, \text{ft}^3/\text{bbl})$

$$\times (62.4 \, \text{lbm/ft}^3)(884.7 \, \text{Btu/lbm})$$

$$= 43.5 \times 10^8 \, \text{Btu}.$$

Then, $\delta = \frac{1}{2} (3.31 \times 10^8 \, \text{Btu}) / (43.5 \times 10^8 \, \text{Btu})$

$$= 0.038.$$

The value of $\overline{T}_{Dp}$ must be found where

$$\overline{T}_{Dp} = \overline{T}_D (1 - \delta) - \delta$$
$$= 1.0(1.0 - 0.038) - 0.038$$
$$= 0.924.$$

And the average produced-fluids temperature at the beginning of the second timestep is

$$\overline{T}_p = 93 + (0.924)(381.8 - 93)$$
$$= 359.9°\text{F}.$$

Oil produced during the first timestep ($i = 1$) is

$$\Delta N_{p1} = q_o \Delta t_{p1}$$
$$= (22.75\,\text{B/D})(56\,\text{days})$$
$$= 1{,}274\,\text{STB}.$$

Water produced during the first timestep is

$$W_p = q_o F_{wo} \Delta t_{p1}$$
$$= (22.75\,\text{STD/D})(2.08)(56\,\text{days})$$
$$= 2{,}550\,\text{STB}.$$

Thus, the WOR for the second timestep is obtained from Eq. 8.95.

$$F_{wo} = 0.25 + 1.83 e^{-7.38(2,650/14,000)}$$
$$= 0.703.$$

The effect of temperature on F_{wo} was neglected. The calculations outlined in this section are not difficult, but are time-consuming. However, they are readily adapted to the computer through a FORTRAN program or spreadsheet. The results summarized in **Table 8.12** were obtained in this manner. Small differences between the hand calculations and those in Table 8.12 result from rounding. In computer programs, it is possible to make the timestep increment as small as necessary to make the assumption of constant rates, WORs, and properties valid over the timestep calculated. In Table 8.12, the densities and heat capacities of oil and water were held constant to simplify the computations. These assumptions have little effect on the final results.

Results in Table 8.12 show that the effects of steam stimulation last for more than 1.5 years. There is little decline in oil production rate with time because the viscosity of the oil in the heated region is 66.6 cp 530 days after the well is shut in. This crude oil undergoes a change in viscosity from 2,026 cp at 93°F to 66.6 cp at 152.5°F, the average temperature of the heated region at 530 days. Incremental oil produced 530 days after the stimulation is $12{,}401 - 5.79(530) = 9{,}332$ bbl and the oil/steam ratio (OSR) is approximately 0.67.

The temperature of the heated zone will decrease at a faster rate if the initial WOR is higher. Oil rates following stimulation are shown in **Fig. 8.22** for $F_{woc} = 0.25$, 3.0, and 5.0. High initial WORs have a significant effect on the production responses.

$t - t_i$ (days)	$\overline{T}_p$	μ_{oh} (cp)	q_{oh} (B/D)	F_{wo} (Btu/D)	$\dot{Q}_{pi}$ (Btu)	$\sum_{i=1}^{n} \dot{Q}_{pi}\Delta t_{pi}$	δ	$\overline{T}_{Dp}$	ΔN_{pi} (bbl)	ΔW_{pi} (bbl)	W_i (bbl)	N_p (bbl)
4	381.8	1.22	22.98	2.08	5.29×10^6	2.96×10^8	0.035	0.93	1,287.15	2,677.28	0	0
60	361.67	1.41	23.14	0.7	2.28×10^6	3.65×10^8	0.043	0.656	694.25	483.33	2,677	1,287
90	282.38	3.19	23.73	0.6	1.51×10^6	4.10×10^8	0.048	0.599	711.89	424.17	3,161	1,981
120	265.98	4.01	23.83	0.53	1.30×10^6	4.49×10^8	0.053	0.553	715.05	376.51	3,585	2,693
150	252.81	4.93	23.91	0.48	1.14×10^6	4.83×10^8	0.057	0.49	717.3	341.98	3,961	3,408
180	234.56	6.79	23.99	0.44	9.76×10^5	5.12×10^8	0.06	0.457	719.72	316.21	4,303	4,126
210	224.85	8.21	24.02	0.41	8.81×10^5	5.39×10^8	0.063	0.405	720.49	295.6	4,619	4,845
240	209.93	11.36	24.02	0.39	7.62×10^5	5.54×10^8	0.065	0.383	480.36	185.97	4,915	5,566
260	203.75	13.17	24	0.37	7.10×10^5	5.75×10^8	0.068	0.361	719.97	269.52	5,101	6,046
290	197.29	15.5	23.96	0.36	6.55×10^5	5.95×10^8	0.07	0.339	718.86	257.26	5,371	6,766
320	190.95	18.37	23.9	0.34	6.04×10^5	6.13×10^8	0.072	0.318	717.1	246.82	5,628	7,485
350	184.7	21.94	23.82	0.33	5.55×10^5	6.30×10^8	0.074	0.287	714.53	237.73	5,875	8,202
380	175.88	28.73	23.63	0.32	4.92×10^5	6.44×10^8	0.076	0.266	708.96	228.97	6,112	8,917
410	169.84	35.03	23.45	0.31	4.48×10^5	6.58×10^8	0.077	0.249	703.37	221.33	6,341	9,626
440	164.95	41.5	23.25	0.31	4.12×10^5	6.70×10^8	0.079	0.236	697.42	214.49	6,563	10,329
470	161.19	47.57	23.06	0.3	3.85×10^5	6.82×10^8	0.08	0.219	691.76	208.49	6,777	11,026
500	156.15	57.58	22.75	0.3	3.49×10^5	6.92×10^8	0.081	0.206	682.36	202.01	6,986	11,718
530	152.49	66.57	22.47	0.29	3.23×10^5	7.02×10^8	0.083	0.194	673.96	196.39	7,188	12,401

*For F_{woc} = 0.25.

Table 8.12—Summary of calculations for production rate of the stimulated well, Example 8.6*.

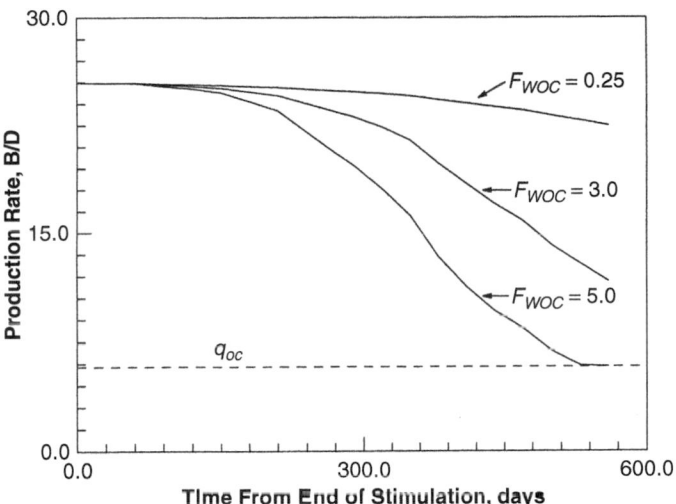

Fig. 8.22—Response of production well to CSS, Example 8.6.

8.3.4 CSS in Gravity-Drainage Reservoirs.

The first description of cyclic steam performance under gravity-dominated producing mechanisms is attributed to Doscher (1966). **Fig. 8.23** depicts a model of the process. Steam is injected into the productive interval, creating a heated zone of radius r_s that extends over the entire oil column. The initial oil content in the heated region is not affected by steam injection. The region beyond the heated region remains at reservoir temperature, and a negligible amount of oil flows from the cold region into the heated region when the well is placed on production.

Oil is produced by draining the oil column in the heated region under the influence of gravity. Because there is no oil influx from the cold reservoir, the heated region acts like a bounded reservoir. Appendix G describes the gravity-drainage process for this case, including the derivation of the fluid-flow equations. The instantaneous production rate at any time is given by (after Eq. G-6)

$$q_o = \frac{\pi \rho_o k_o g \left(h_{os}^2 - h_w^2 \right)}{\mu_o \left[\ln \left(r_s / r_w \right) - \frac{1}{2} \right]}. \quad \ldots \ldots \ldots \ldots \ldots \ldots \ldots (8.98)$$

Oil production declines with time as the height of the oil column at r_s, h_{os}, and the temperature of the heated region decline with time. The well cannot be restimulated until resaturation of the previously heated region occurs.

A second model of steam stimulation, shown in **Fig. 8.24** (Doscher 1966), illustrates the effects of heat transfer by conduction on the temperature and fluid flow in the cold region originally outside of the heated region. In Fig. 8.24b, oil in the heated region is produced by gravity drainage with limited oil influx from the colder part of the reservoir. As time progresses, conduction in the radial direction increases the temperature in the original cold region and oil flows into the heated region under gravity drainage. Figs. 8.24c through 8.24e depict successive stages in this model, which was developed by Seba and Perry (1969).

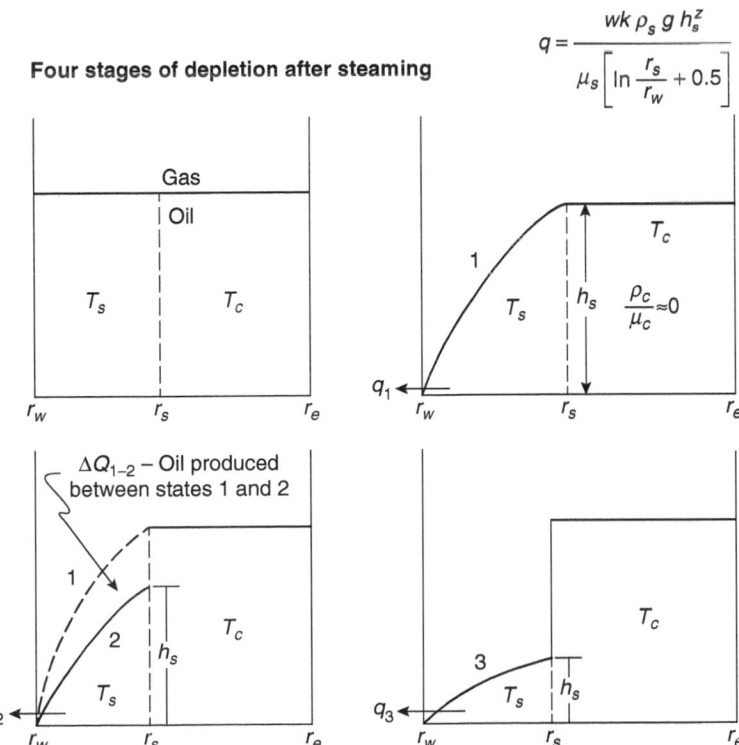

Fig. 8.23—Depletion of heavy-oil reservoir by gravity drainage in heated zone (Doscher 1966).

Outer segment temperature rises with successive cycles

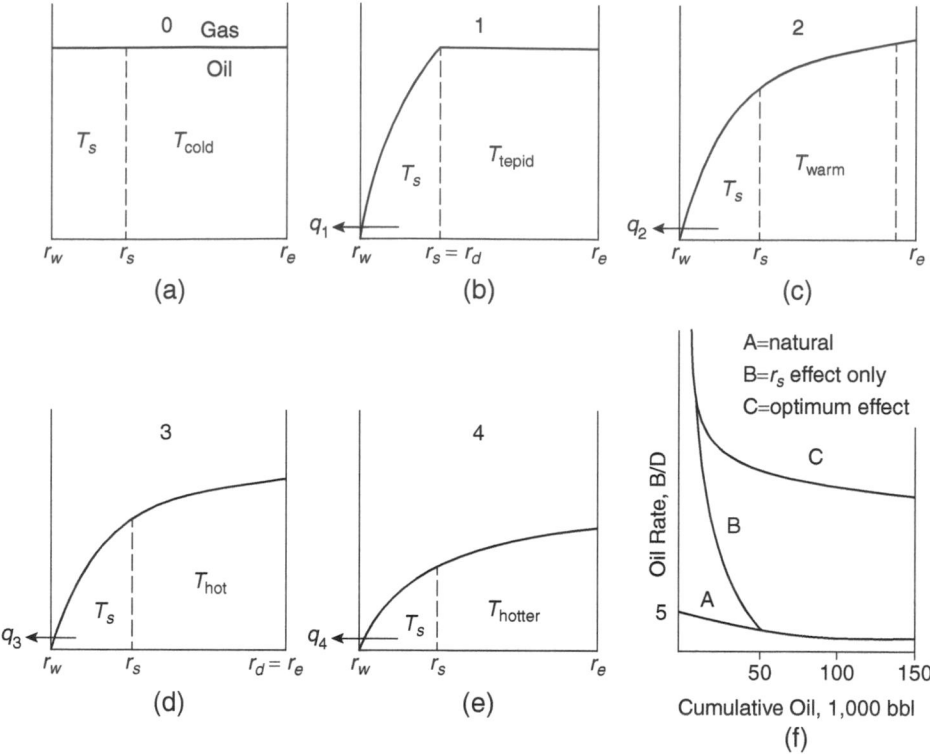

Fig. 8.24—Effect of heat transfer to cold region on the depletion of a heavy-oil reservoir by gravity drainage (Doscher 1966).

In practice, the behavior of CSS in reservoirs where the oil is mobile at reservoir temperature is somewhere between these two examples. Several models have been developed to describe cyclic steam response when gravity drainage is the source of reservoir energy (Jones 1977; Seba and Perry 1969; Towson and Boberg 1967; Gontijo and Aziz 1984; Sylvester and Chen 1988; Gozde et al. 1989; Jones 1992). All the models have at least one adjustable parameter that must be determined from history matching field performance and thus cannot be considered predictive. Some models have proved to be quite useful in studying the effects of operation changes after the parameters have been determined by history matching.

An approximate model, adapted from the model developed by Jones (1977), is introduced in this section. This model retains most of the important features of the process in gravity-drainage reservoirs. In this model, a well is stimulated with steam so that a heated zone of radius r_h is created that extends throughout the entire reservoir thickness, as depicted in **Fig. 8.25.** Oil and interstitial water are displaced from the heated region into the colder portions of the reservoir. The radius of the heated region is estimated with the Marx and Langenheim model described earlier. During the soak period, condensation of steam occurs and the pore space occupied by steam is resaturated with water. Thus, the heated region contains the oil saturation at S_{ors} and water saturation at $S_w = 1 - S_{ors}$.

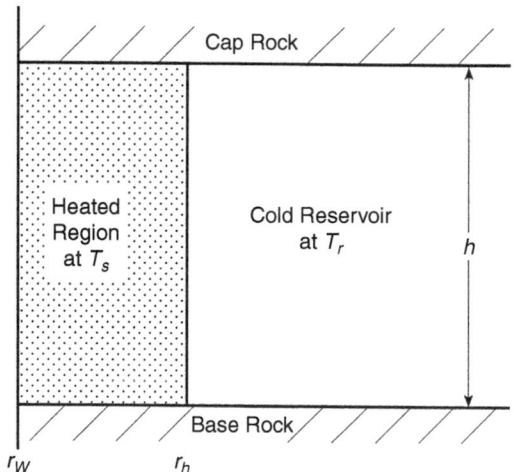

Fig. 8.25—Heated and cold reservoir regions after uniform steam injection into a heavy-oil reservoir.

When the well is placed on production, water will be produced initially at a high rate. Desaturation of the heated region occurs from the top down as water is withdrawn from the well, leaving two distinct regions, as discussed in Appendix G and shown in **Fig. 8.26.** In the unsaturated region, a vapor phase (gas or air) exists, with oil at residual saturation to steam and water at interstitial water saturation. Pressure in the gas phase is controlled by the fluid level that is maintained above the pump and the annulus pressure. Water and vapor phases are in capillary equilibrium in this classic drainage mechanism. The transition between unsaturated and saturated regions is assumed to be sharp with no capillary transition zone. The oil saturation in the saturated region increases with time as oil from the cold region of the reservoir flows into the heated region by gravity drainage.

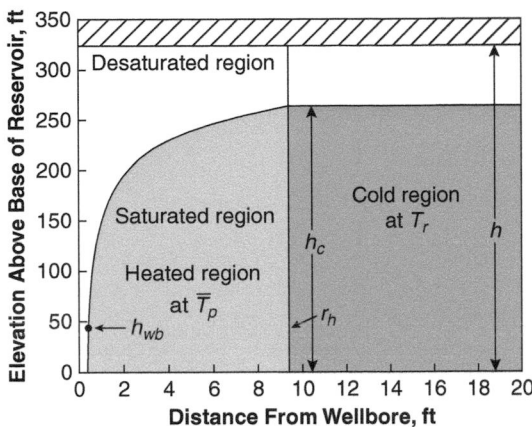

Fig. 8.26—Profile of desaturated and saturated regions in a reservoir producing under gravity drainage when the fluid head is constant at r_h.

Both oil and water flow from the heated zone to the producing wellbore because of the difference in fluid head between the radius of the heated zone and the fluid head maintained in the producing well. The oil content in the saturated region must be re-established by flow from the colder regions of the reservoir. In this model, a constant head is assumed to exist at the heated-zone boundary, which is equivalent to assuming that the reservoir has unlimited capacity to provide cold oil (and water if water is mobile) to the heated region. This is an assumption that might be valid for a single well in an infinite reservoir but cannot be valid for long periods of time for a fully developed reservoir where each well has its prescribed drainage radius. Flow of water in the heated region is restricted as a result of hysteresis in the relative permeability curves (Dietrich 1981), and less water is produced than injected as steam (CWE) during the early cycles.

During the production cycle, the average temperature of the heated zone decreases as heated fluids are produced and heat is lost by conduction to under- and overlying formations as well as by conduction to the reservoir just beyond the region initially heated by steam. This process is identical to that discussed in the previous section for production following CSS in a depleted reservoir. The cycle ends when the oil production rate declines to a specified level.

The calculation procedure is subdivided into two phases. In Phase 1, the desaturated region is created with a corresponding saturated region by removal of hot water from the heated region. In Phase 2, the heated-zone temperature and oil and water flow rates are calculated in a stepwise manner at sequential lengths of time.

At the end of Phase 1, when the desaturated region has formed, fluid flow is quasisteady and the shape of the saturated region is given by (see Appendix G for the derivation)

$$h = \sqrt{\frac{\left(h_h^2 - h_w^2\right)\ln\left(r/r_w\right)}{\ln\left(r_h/r_w\right)} + h_w^2}. \qquad (8.99)$$

The shape of this region is independent of time. Fig. 8.26 shows the saturation profile for this fluid-flow model. Although the shape of the saturated region is fixed, both average temperature and fluid saturations change with time owing to heat losses and production of oil and water.

The PV of the heated region is given by

$$V_{ph} = \left[\pi\left(r_h^2 - r_w^2\right)h\phi\right]/5.615. \qquad (8.100)$$

The PV of the saturated portion of the heated region is given by

$$V_{ps} = \left[\pi\left(r_h^2 - r_w^2\right)\bar{h}\phi\right]/5.615, \qquad (8.101)$$

where $\displaystyle \bar{h} = \int_{r_w}^{r_h} 2\pi r h \, dr \Big/\left[\pi\left(r_h^2 - r_w^2\right)\right]. \qquad (8.102)$

In Eq. 8.102, $\bar{h}$ = average thickness of the oil column in the saturated region and is found by integration of Eq. 8.102 with Eq. 8.99 substituted for h. Integration is performed easily through numerical methods and is described in Appendix G. The PV of the desaturated region is $V_{ph} - V_{ps}$ when the saturation profile shown in Fig. 8.26 is attained.

The volume of water in the heated region of the reservoir at the end of the soak period is $V_{ph}(1 - S_{ors})$, where S_{ors} is the residual oil saturation following steam injection. As the well is pumped off, the water is removed from the upper part of the heated region by gravity drainage and the desaturated region forms. This region contains oil as steamflood residual, S_{ors}, water at interstitial saturation, S_{iw}, and a gas saturation equal to $1 - S_{ors} - S_{iw}$. Thus, the volume of water removed from the heated region during the desaturation process, $W_{ph}^{(1)}$, is given by

$$W_{ph}^{(1)} = \left(V_{ph} - V_{ps}\right)\left(1 - S_{ors} - S_{iw}\right), \qquad (8.103)$$

where the superscript 1 refers to the desaturation period.

Production is essentially 100% water from the heated region during desaturation, and the production rate is often limited by the capacity of the downhole pump. When the maximum pump rate is q_{max}, the time required to form the steady-state shape of the saturated region depicted in Fig. 8.26 is given by

$$t^{(1)} - t_{soak} = W_{ph}^{(1)}/q_{max}, \dotfill (8.104)$$

where t_{soak} = length of soak period, days, and $t^{(1)}$ = time when the desaturated region attains the shape shown in Fig. 8.26, days. Little oil invades the heated region during the desaturation period.

When a stable saturated region is formed, as shown in Fig. 8.26, oil and water production rates are calculated from quasisteady-fluid-flow equations. The calculation procedure involves evaluation of the average water saturation and the average temperature of the saturated portion of the heated region as a function of time. This is done by calculating production rates and produced-fluid temperatures at discrete increments of time, similar to the procedure described earlier for CSS from a depleted reservoir. In the calculations, water saturation and temperature for each time increment are the values known from the end of the previous timestep.

The computations are performed at assumed values of time t^n. The average temperature in the heated region at time t^n is determined from the Boberg and Lantz model with Eqs. 8.84 and 8.85:

$$\bar{T}_{Dp}^{(n)} = \bar{T}_{Dr}^n \bar{T}_{Dz}^n \left[1 - \delta^{(n)}\right] - \delta^{(n)}$$

and $\bar{T}_p^{(n)} = T_r + \bar{T}_{Dp}^{(n)}\left(T_s - T_r\right).$

Oil and water production rates are given by Eqs 8.105 and 8.106. In the heated region ($r_w < r < r_h$), the flow rate of oil is given by Eq. 8.106, where the viscosity and density are evaluated at the average temperature in the heated region. The permeability to oil is evaluated at the average water saturation in the heated region at the previous timestep, S_w^{n-1}.

$$q_{oh} = \left[0.0246\rho_{oh}k_{oh}\left(h_h^2 - h_w^2\right)\right]\Big/\left[a_1\mu_{oh}\ln\left(r_h/r_w\right)\right]. \dotfill (8.105)$$

In Eq. 8.105, a_1 is an empirical parameter used to adjust the production rate to the observed history. By analogy, the production rate for water is given by

$$q_{wh} = \left[0.0246\rho_{wh}k_{wh}\left(h_h^2 - h_w^2\right)\right]\Big/\left[a_2\mu_{wh}\ln\left(r_h/r_w\right)\right], \dotfill (8.106)$$

where a_2 is an empirical parameter used to adjust the water production rate to the observed history.

In the first timestep following desaturation, the water saturation in the saturated region is still at $1 - S_{ors}$ and the oil permeability is small. Water is produced primarily during this timestep, and the production rate is computed by evaluating the permeability to water at $S_w = 1 - S_{ors}$ and the water viscosity, μ_w, at the average temperature of the produced fluids. Cumulative water production during a given timestep is computed with Eq. 8.107:

$$W_{ph}^n - W_{ph}^{n-1} = \bar{q}_{wh}^{-n}\left(t^n - t^{n-1}\right),$$

where $\bar{q}_{wh}^n = q_{wh}^n$ for $n = 1, 2$

and $\bar{q}_{wh}^{-n} = \left(q_{wh}^n + q_{wh}^{n-1}\right)\Big/2$ for $n > 2$. \dotfill (8.107)

Cumulative oil production (at the heated-zone temperature) is given by

$$N_{ph}^n - N_{ph}^{n-1} = \bar{q}_{oh}^n\left(t^n - t^{n-1}\right),$$

where $\bar{q}_{oh}^{(1)} = 0,$

$$\bar{q}_{oh}^{(2)} = q_{oh}^{(2)}\Big/2,$$

and $\bar{q}_{oh}^{-n} = \left(q_{oh}^n + q_{oh}^{n-1}\right)\Big/2$ for $n > 2$. \dotfill (8.108)

The average water saturation in the saturated portion of the heated region at the end of a timestep is found by material balance and is given by Eq. 8.109, where $W_{ph}^{(1)}$ is the volume of water produced during the time interval when the desaturated region is formed.

$$S_w^n = \left\{ \left(1 - s_{ors} \right) V_{ps} - \left[W_{ph}^n - W_{ph}^{(1)} \right] \right\} / V_{ps}. \dotfill (8.109)$$

The value of S_w^n approaches S_{iw} as production continues. Water saturations are often higher than S_{iw} during the production cycle owing to hysteresis of the relative permeability curve for water on the drainage path. If water was not mobile at S_{iw}, water production during the production cycle is limited to the total volume of water injected as steam. Water displaced into the cold region of the reservoir is not accounted for unless some returns to the heated region during the soak period when resaturation occurs.

In this model, cold oil continues to flow into the saturated portion of the heated region at radius r_h. The oil influx rate at r_h must be larger than the oil production rate to enable the oil saturation to increase in the heated region during the production cycle. Oil that flows into the saturated region from the cold region of the reservoir is assumed to be distributed uniformly throughout this region, increasing the average oil saturation and decreasing the water saturation. Oil influx during a particular timestep can be computed as the sum of the oil produced and the net change in oil volume in the heated region with $S_0^n = 1 - S_w^n$ by use of the water saturation computed with Eq. 8.109. The volume of oil entering from the cold region of the reservoir during a timestep is given by

$$V_{oc}^n = \left[\left(N_{ph}^n \quad N_{ph}^{n-1} \right) + \left(S_w^n - S_w^{n-1} \right) V_{ps} \right] \left(B_{oh} / B_{oc} \right). \dotfill (8.110)$$

The last step in the calculation procedure is to determine the value of δ, which will be used to account for the energy removed with the produced fluids in the Boberg and Lantz model. For application to cyclic steam injection in gravity-drainage reservoirs, the expression developed by Boberg and Lantz in Eq. 8.83 was altered to account for the energy that is lost during the soak period. Other adjustments are made to account for energy remaining in the reservoir at the end of a cycle. Eq. 8.111 defines the change in δ under these assumptions.

$$\delta^{n+1} - \delta^{(n)} = \Delta Q^n / \left[2 \left(n_s \right) \left(Q_{\max} \right) \right], \dotfill (8.111)$$

where n_s = number of zones; $Q_{\max}$ = amount of injected energy remaining in the reservoir at the end of the soak period, Btu; and $\Delta Q^n = \dot{Q}_{pn}(t^n - t^{n-1})$, the energy removed with the produced fluids during the current timestep. ΔQ^n can be evaluated with

$$\Delta Q^{(n)} = 5.615 \left[W_{ph}^{(n)} - W_{ph}^{(n-1)} \right] \rho_{wh} \left(H_w^n - H_{wr} \right) + \left[N_{ph}^{(n)} - N_{ph}^{(n-1)} \right] \rho_{oh} \left(H_o^n - H_{or} \right). \dotfill (8.112)$$

$Q_{\max}$ is defined by

$$Q_{\max} = \dot{m}_s H_s t_i - Q_{\text{lost}} \dotfill (8.113)$$

The energy lost during the soak period, Q_{lost}, is computed by use of a model developed by Vogel (1984), which assumes that heat is lost to the formation above and below the heated region by linear flow into a semi-infinite medium. According to this model, the maximum heat loss by conduction during the soak period, Q_{lost}, is given by

$$Q_{\text{lost}} = 4 \pi r_h^2 k \left(T_s - T_r \right) \sqrt{t_{\text{soak}} / \pi a}. \dotfill (8.114)$$

The overall procedure for estimating response to CSS in a gravity-drainage reservoir with a constant head at the heated boundary r_h is summarized in the following.

Initial Calculations—Desaturation of Heated Region.
1. Estimate the radius of the heated region created by steam injection using the Marx and Langenheim equation.
2. Compute the PV of the heated region and the saturated region with Eqs. 8.100 through 8.102.
3. Compute the volume of hot water removed during the formation of the desaturated region with Eq. 8.103.
4. Compute the time required to form the desaturated region at the maximum pump rate with Eq. 8.104.
5. Specify the soak time and estimate the average temperature of the heated region with the Boberg and Lantz method at the soak time given by Eqs. 8.84 and 8.85. This is the average temperature of the produced fluids during the initial desaturation period.
6. Compute the value of δ corresponding to the end of the desaturation period from Eq. 8.111.

Iterative Calculations for All Timesteps After the Desaturation Period.
7. Calculate the average temperature of the heated region with the Boberg and Lantz method with the new time, t^n, and the value of δ^i from the previous timestep. Note that the time t^n corresponds to the time at the beginning of the timestep. The temperature is assumed to be constant through the timestep. The calculated average temperature can be less than the reservoir temperature because of assumptions made in the model. If this occurs, the average temperature is limited to the reservoir temperature.

Well spacing (ft)	160	Overburden thermal diffusivity (ft²/hr)	0.02
Depth to top of zone (ft)	620	Oil density (lbm/ft³)	61.8
Gross section of thickness (ft)	427	Oil specific heat (Btu/lbm-°F)	0.44
Number of zones	4	Oil viscosity (cp)	
Average zone thickness (ft)	80	At 100°F	5,000
Rock porosity	0.32	At 300°F	13
Permeability (darcies)	1.5	Casing inside radius, r_{ci} (ft)	0.31
Initial oil saturation	0.75	Tubing inside radius, r_{ti} (ft)	0.104
Initial water saturation	0.25	Pump shoe depth, h_p (ft)	48
Oil saturation in steam zone	0.20	Steaming method	Open tubing
Geothermal gradient (°F/ft)	0.09	Surface steam quality	0.75
FVF (bbl/bbl)	1.0	Reservoir temperature (°F)	110
Formation thermal conductivity (Btu/ft-hr-°F)	1.0		

Table 8.13—Input data in matching performance of CSS, Example 8.7 (Towson and Boberg 1967; Sylvester and Chen 1988).

8. Calculate the water and oil production rates (other than the first timestep) with Eqs. 8.105 and 8.106 by evaluating the viscosities at the average temperature of the heated region adjusted for production of heated fluids and the permeabilities at the average water saturation at the end of the last timestep.
9. Choose a time increment ($t^{n+1} - t^n$) for calculating oil and water production during the timestep.
10. Calculate the average water saturation in the saturated portion of the heated region by material balance with Eq. 8.109.
11. Calculate the value of $\delta^{(n+1)}$ corresponding to this timestep.
12. Calculate cumulative oil and water production during the time increment, including adjustment for temperature of the produced fluids.
13. Repeat Steps 7 through 12 until the oil production rate reaches the economic limit and another cycle is to be carried out.

A_h [ft² (acres)]	297.7 (0.00683)
r_h (ft)	9.73
V_h (bbl)	16,966
V_{ph} (bbl)	5,429
V_{ps} (bbl)	4,069

Table 8.14—Computed results for heated region, Example 8.7.

The parameters a_1 and a_2 are determined by matching production history to the simulated results, usually a trial-and-error procedure.

Example 8.7 illustrates the use of this model to calculate oil production for the first cycle in a gravity-drainage reservoir. Performance of subsequent cycles can be predicted by adjusting energy- and material-balance calculations for the injected water that was not produced and the energy that remained in the reservoir from previous steam stimulation cycles.

Example 8.7—Estimation of Oil Production After Steam Stimulation of a Reservoir Producing by Gravity Drainage. Jones (1977) presented field data from seven steam stimulation cycles of a test well in the Midway Sunset field. The data have been used by several authors to illustrate steam stimulation models. In this example, we will estimate the CSS behavior using the procedures described in this section for the first cycle.

Table 8.13 gives reservoir fluid and rock data for this well. Gross sand thickness is 427 ft, and there are four zones with an average zone thickness of 80 ft; thus, the net sand thickness is 320 ft. The static fluid level in the well is approximately 260 ft (relative to sea level) and remained near that level throughout the seven cycles reported. The well was pumped off with the pump shoe located at an elevation of 48 ft (relative to sea level). No information was provided on how the nonproductive intervals were distributed within the gross sand. Thus, fluid-flow calculations are performed by treating the net sand as a single zone that is hydraulically connected.

During the first cycle, steam at 485°F was injected at an injection rate of 647 B/D for 6 days followed by a soak period of 5 days. For this example, we will assume that the maximum pump rate is 800 B/D. Steam quality at the wellhead was 0.75. The downhole quality at the sandface is estimated to be 0.71. Estimate the oil and water production rates following the soak period and compare oil production rates with the observed production data.

Solution. The computations required to solve this problem are lengthy because fluid properties are temperature-dependent. Appendix H illustrates calculations for the first two timesteps. The remainder of the solution was achieved with a computer program. A summary of the computed results is presented in this section.

Table 8.14 summarizes the calculations for the heated volume obtained with the Marx and Langenheim model and the PV of the saturated region when the fluid heads stabilize. **Table 8.15** summarizes computed results for key variables. Cumulative oil production following steam stimulation was reported to be 3,800 bbl after 55 days on production. The calculated cumulative oil production in Table 8.15 is 3,783 bbl. The oil production rate is compared with the observed production from the first cycle in **Fig. 8.27.** The parameters a_1 and a_2 that gave the best match with the observed data were 4.0 and 2.0, respectively. These were determined by trial and error. **Figs. 8.28 and 8.29** illustrate the variation of the average temperature and water saturation

Timestep	$t - t_i$ (days)	$\overline{T}_p$ (°F)	S_w	q_o (STB/D)	q_w (STB/D)	q_t (STB/D)	N_p (STB)	W_p (STB)
1	5.00	411.50	0.750	0.00	677.60	677.60	0	571
2	5.84	387.69	0.689	5.38	429.96	435.34	3	786
3	6.34	378.34	0.639	34.57	270.99	305.56	13	962
4	6.84	370.37	0.607	58.54	178.00	236.54	36	1,075
5	7.34	364.24	0.585	76.07	133.72	209.78	70	1,153
6	7.84	359.09	0.568	88.72	108.84	197.56	111	1,213
7	8.34	354.45	0.554	98.31	92.35	190.66	158	1,264
8	8.84	350.14	0.542	105.74	80.37	186.12	209	1,307
9	9.34	346.08	0.531	111.50	71.19	182.69	263	1,345
10	9.84	342.22	0.522	115.90	63.90	179.80	320	1,379
11	10.34	338.54	0.513	119.19	57.96	177.14	379	1,409
12	10.84	335.00	0.483	121.54	53.02	174.56	620	1,520
13	12.84	321.74	0.457	127.38	38.05	165.42	869	1,612
14	14.84	309.96	0.439	127.45	28.69	156.14	1,124	1,679
15	16.04	299.53	0.425	121.76	23.11	144.88	1,374	1,731
16	18.84	290.16	0.414	113.73	19.41	133.14	1,610	1,773
17	20.84	281.70	0.404	105.18	16.73	121.91	1,829	1,810
18	22.84	273.99	0.396	96.83	14.69	111.52	2,031	1,841
19	24.84	266.91	0.388	89.00	13.08	102.08	2,217	1,869
20	26.84	260.40	0.382	81.80	11.77	93.57	2,388	1,894
21	28.84	254.36	0.376	75.25	10.69	85.94	2,545	1,916
22	30.84	248.74	0.370	69.29	9.77	79.07	2,690	1,937
23	32.84	243.49	0.365	63.91	8.99	72.91	2,823	1,956
24	34.84	238.55	0.361	59.03	8.32	67.35	2,946	1,973
25	36.84	233.92	0.357	54.62	7.74	62.35	3,060	1,989
26	38.84	229.54	0.353	50.61	7.22	57.83	3,165	2,004
27	40.84	225.40	0.349	46.98	6.76	53.74	3,263	2,018
28	42.84	221.47	0.346	43.68	6.35	50.03	3,354	2,031
29	44.84	217.74	0.342	40.68	5.99	46.67	3,438	2,043
30	46.84	214.19	0.339	37.94	5.66	43.60	3,517	2,055
31	48.84	210.79	0.336	35.44	5.36	40.79	3,590	2,066
32	50.84	207.56	0.334	33.15	5.08	38.24	3,659	2,077
33	52.84	204.46	0.331	31.06	4.84	35.89	3,723	2,087
34	54.84	201.50	0.329	29.13	4.61	33.74	3,783	2,096
35	56.84	198.65	0.326	27.37	4.40	31.76	3,840	2,105

Table 8.15—Summary of CSS calculations Example 8.7 for a 6-day steam-injection time.

in the heated region with time following steam stimulation. Appendix H contains supporting tables for values of other variables determined in the calculation.

Extension to Multiple Cycles. The CSS model can be modified to account for multiple cycles (Jones 1977; Gontijo and Aziz 1984; Sylvester and Chen 1988). There are three major modifications to the calculation procedure described in this section.

First, the amount of energy injected into each zone in calculating the heated radius with the Marx and Langenheim model is modified to account for the energy in the heated zone at the end of the previous stimulation cycle. The amount of energy remaining in the heated zone is approximated by

$$Q_{\text{last}} = V_{ph}^{(n-1)} M \left[\overline{T}_p^{(n-1)} - T_r \right].$$

Then, $H_t = H_{wT} + f_{sd}L_{vdh} - H_{wr} + \dfrac{Q_{\text{last}}}{n_s \dot{m}_s t_s}.$ . . . . (8.115)

The initial water saturation of the region to be heated is the value calculated at the end of the last cycle. Finally, Eq. 8.113 for the calculation of $\hat{o}$ is altered by adding Q_{last} to Q_{max}.

Figs. 8.30 through 8.32 are representative comparisons of calculated and actual oil production from the first, third, and seventh steam stimulation cycles in the Midway Sunset field developed by Jones (1977). Production data were matched with computed data to find the parameters a_1 and a_2. **Fig. 8.33** shows cumulative production. The match between computed and observed data was quite adequate. Similar matches between observed and computed production for the same set of data were reported by Gontijo and Aziz (1984) and Sylvester and Chen (1988) using different fluid-flow models.

The development of gravity-drainage models continues to evolve as additional production data become available that permit examination of critical assumptions used in earlier models. Jones (1992) observed that steam does not enter into a reservoir uniformly even when the entire interval is open, as assumed in the models discussed earlier in this section. A new model describing the reservoir after steam injection was developed and is shown in **Fig. 8.34.** Steam injection is limited to an interval with zone thickness estimated by the method of van Lookeren (1983, Eq. 38). In the new model, the reservoir is separated into two regions, a hot zone and a cold zone. There is no crossflow between zones. Production and fluid saturations are computed separately for each zone. History matches were made between calculated and observed data.

Fig. 8.35 shows the history match for total oil production from a typical Midway Sunset well. Fig. 8.35 is based on data from 1,500 wells over a period of 20 years. Jones (1992) also presents examples showing how the model is used to examine the effects of operating parameters, such as cycle life, steam-injection rate, steam volume, and well spacing.

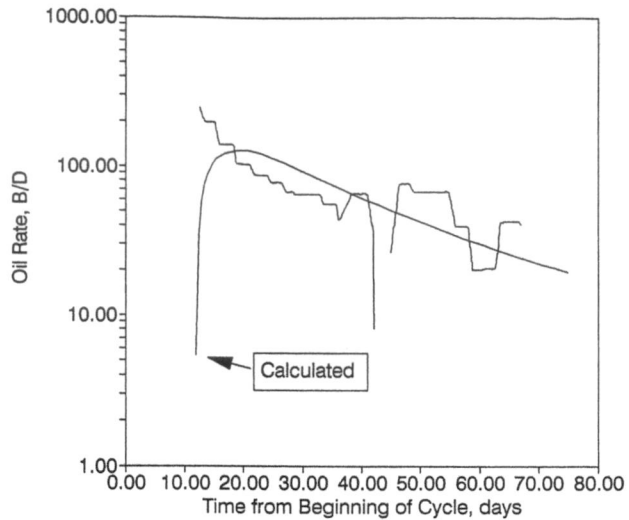

Fig. 8.27—Comparison of calculated and observed oil production rate from first cycle of Midway Sunset test well, Example 8.7.

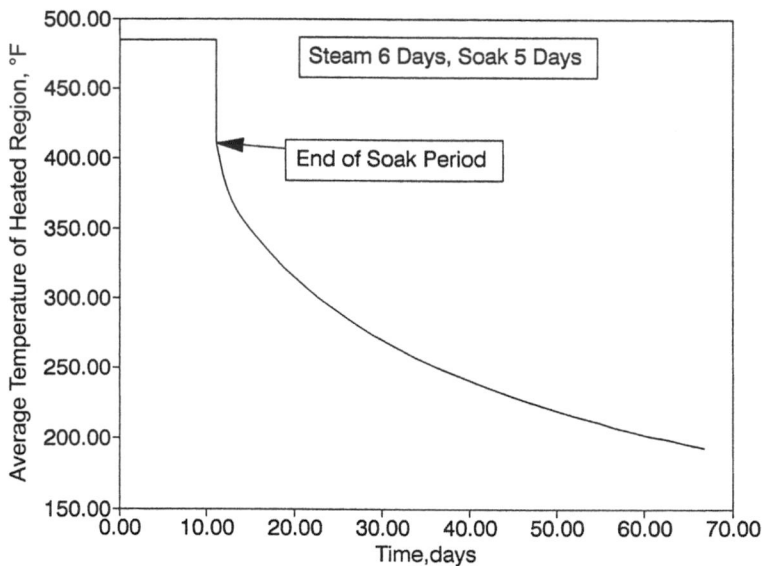

Fig. 8.28—Decline of the average temperature of the heated region with time in Example 8.7.

8.3.5 Field Case Histories. Cyclic steam injection is a proven thermal recovery process. Thousands of wells have been stimulated since the discovery of the process in 1959. As discussed earlier, CSS is usually carried out as a precursor to steamdrive and, in some cases, is necessary before steamdrive can begin.

Cyclic steam injection has enjoyed immense popularity because it is relatively easy to carry out, inexpensive, and the economic payout begins immediately if the well responds to stimulation. In new reservoirs, the process is piloted with minimal design calculations. Although thermal reservoir simulators can be used to predict the response of a well to CSS, it is necessary to match the first cycle results by adjusting parameters in the simulator. That is, CSS cannot be predicted a priori even with the best reservoir simulators. Once a match is obtained, the response to subsequent cycles can be estimated.

Prats (1982) and Boberg (1988) assembled an extensive collection of field results from CSS projects. Their collections are summarized in this section. CSS has been used extensively in heavy-oil fields in California (Burns 1969; Rivero and Heintz 1975; Adams and Khan 1969; Stokes and Doscher 1974), the Bolivar coast of Venezuela (de Haan and van Lookeren 1969; Borregales 1976), Alberta (Buckles 1979; Mainland and Lo 1984), and in Indonesia (Atmosudiro 1977). **Table 8.16** presents the first published results of CSS for first-cycle treatments in 10 California reservoirs. Oil recovered from the stimulation ranged from 2,660 bbl at Poso Creek to 50,000 bbl at San Ardo. Incremental oil recovered varied from 0.21 to 5.0 bbl/bbl of steam. The change in oil production rate following steam injection ranged between 3 and 47 B/D, with an average of 12.8 B/D. Most wells show some evidence of removal of wellbore damage by the steam stimulation.

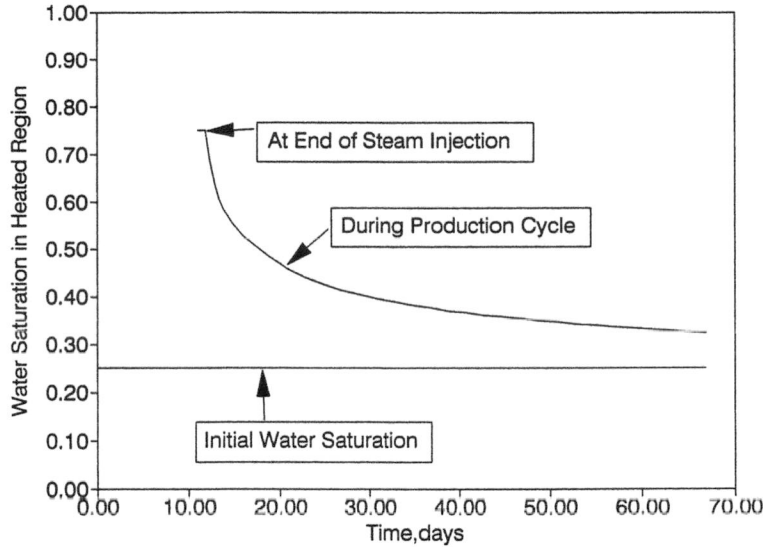

Fig. 8.29—Average water saturation of the heated region during production after steam stimulation, Example 8.7.

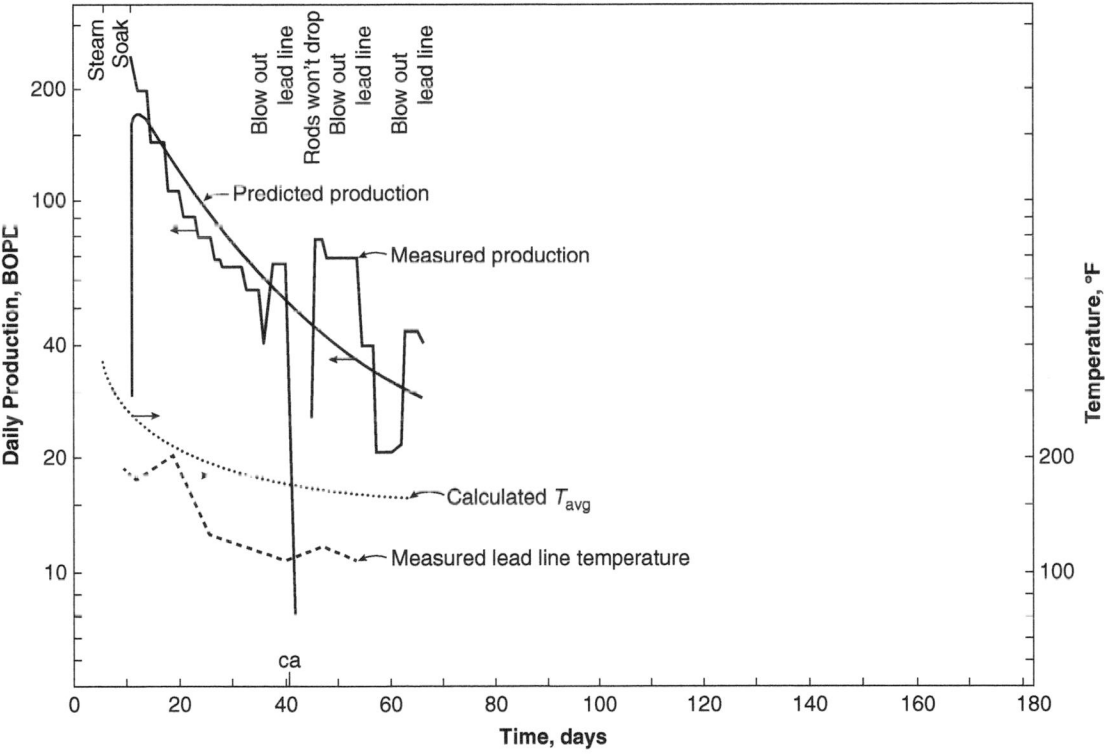

Fig. 8.30—Comparison of calculated and observed data from first cycle of Midway Sunset test well (Jones 1977).

Table 8.17 (Prats 1982) summarizes CSS experience from the Bolivar coast of Venezuela (Borregales 1976). In this area, injection per cycle was approximately 50,000 bbl compared with an average of 8,480 bbl in the California fields presented in Table 8.16. Incremental OSRs are well in excess of 1.0, reflecting the larger volumes of steam injection and the compaction-drive mechanism caused by subsidence. Compaction of the reservoir sands maintains a high level of reservoir energy. **Fig. 8.36** depicts the effects of reservoir drive mechanisms on oil recovery from the Tia Juana field.

Table 8.18 (Boberg 1988) presents performance from the M-6 Project in the Tia Juana field in Venezuela. This project covers 1,831 acres with 145 wells on 757.9-ft spacing. Average net oil sand is 125 ft and the initial oil in place (OIP) was 575 million bbl. Primary production was 67.8 million bbl, or 11.7% OOIP, when cyclic steam injection began in 1969. By December

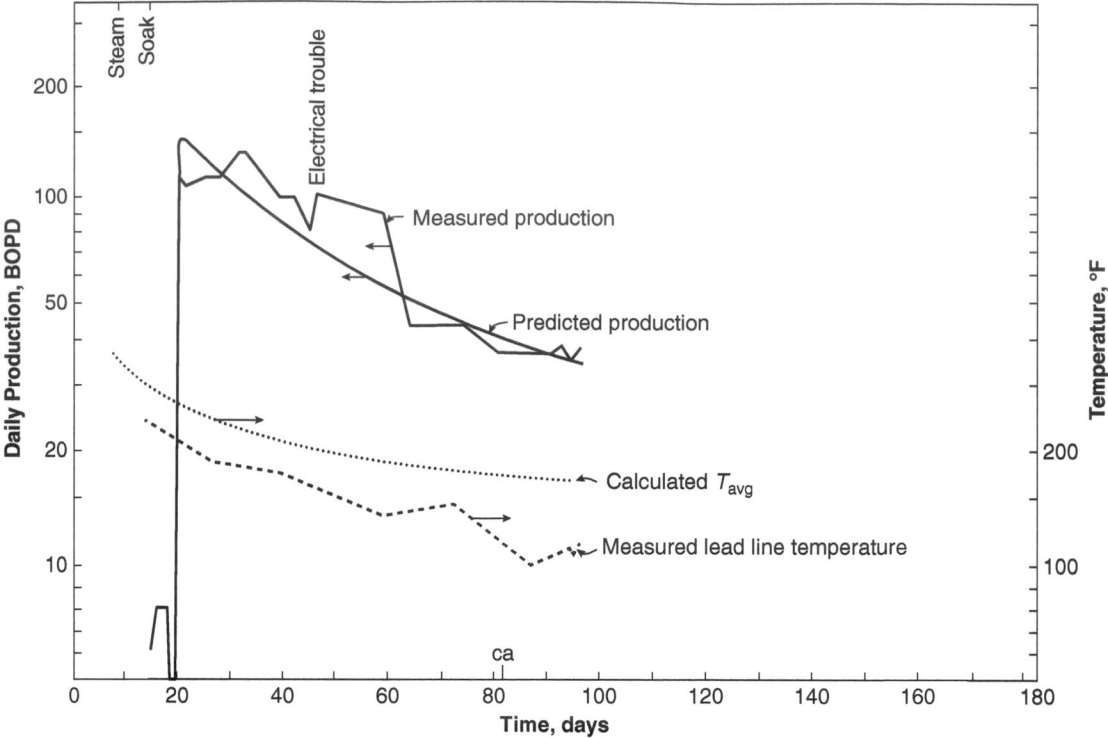

Fig. 8.31—Comparison of calculated and observed data from third cycle of Midway Sunset test well (Jones 1977).

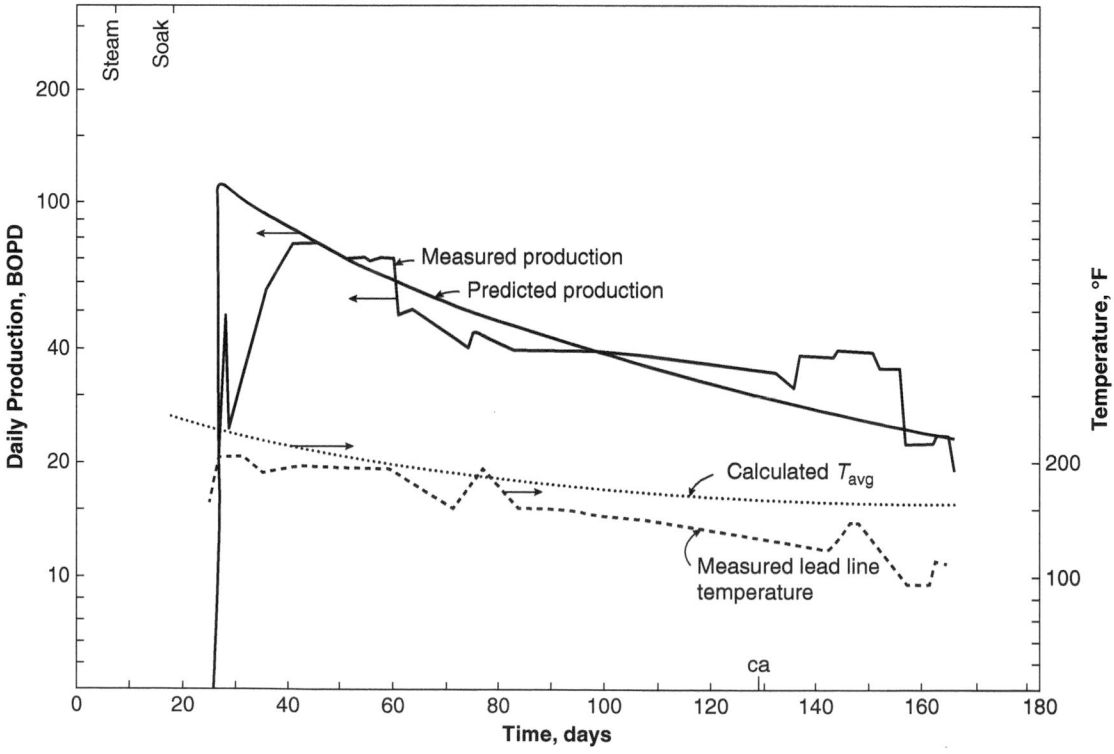

Fig. 8.32—Comparison of calculated and observed data from seventh cycle of Midway Sunset test well (Jones 1977).

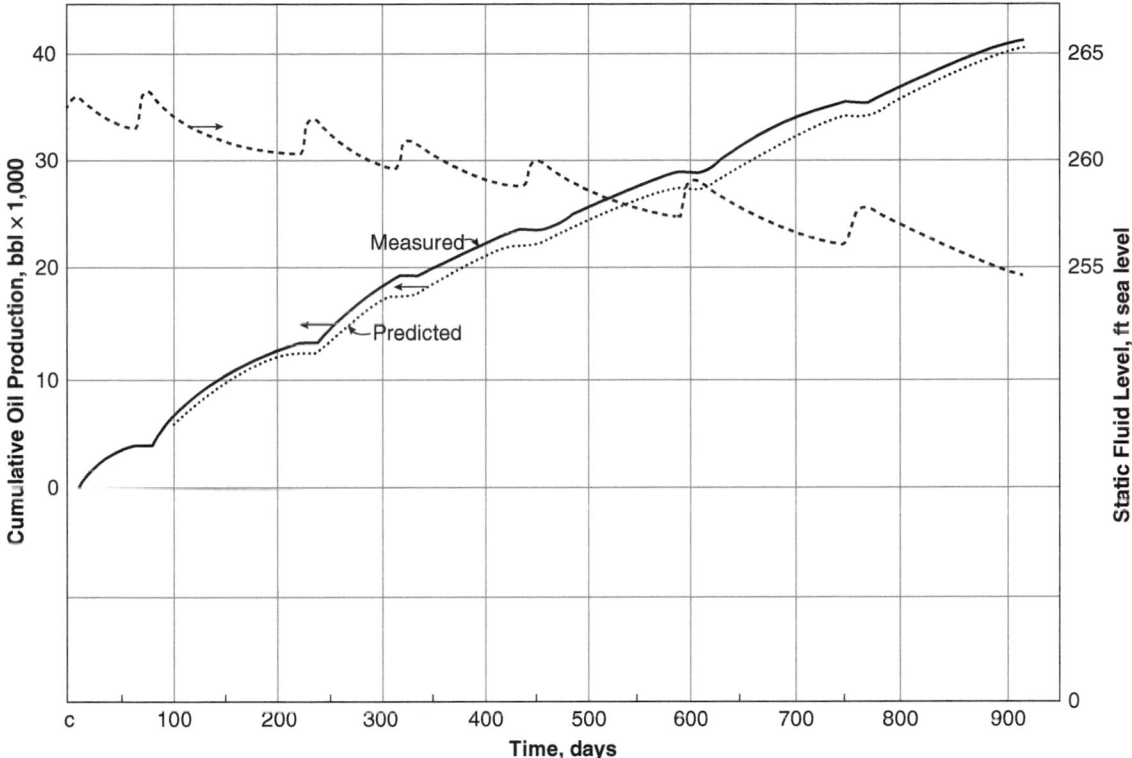

Fig. 8.33—Measured and predicted cumulative oil production vs. time for the first seven cycles of Midway Sunset test well (Jones 1977).

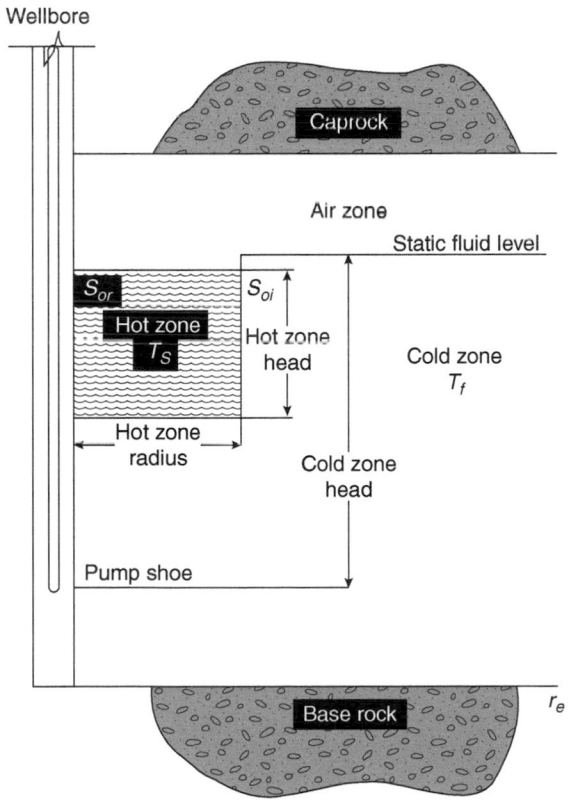

Fig. 8.34—Proposed model of heated zone after steam injection (Jones 1992).

1976, 11.9 million bbl of steam had been injected in 143 wells. Of these, 7 were first cycle, 82 were second cycle, 53 were third cycle, and 1 was fourth cycle. Incremental oil attributed to CSS was 30.7 million bbl for a cumulative OSR of 2.58 bbl oil/bbl steam. A steamdrive was initiated later in this field and is described in Section 8.5.8.

CSS in the Cold Lake oils sands in Alberta (Buckles 1979; Mainland and Lo 1984) has been piloted extensively by Esso Resources Canada Ltd. **Table 8.19** (Boberg 1988) presents characteristics of the Clearwater formation. The oil in this reservoir is properly characterized as a bitumen or oil sand because at reservoir temperature (55°F) the 10.2°API oil has a viscosity of 100,000 cp. Thus, the bitumen is immobile at reservoir temperature even though the reservoir pressure is 450 psig. Porosity (0.35) and oil saturation (0.70) are high, making this reservoir a major hydrocarbon resource.

Reservoir heating is required for any oil movement in the oil sands. Because of the high oil viscosity, essentially all oil production comes from the heated area. The first projects used hydraulic fracturing to create injectivity into the oil sands. Fractures were initiated by injecting steam at pressures greater than 1,300 psi (Boberg 1988). Horizontal or vertical fractures were initiated, depending on the in-situ stress. A southwest/northeast trend in fractures was observed when vertical fracturing occurred.

Three major pilot projects (Ethel, May, and Leming) were operated between 1964 and 1984. The Leming pilot, shown in **Fig. 8.37**, is the most intensive, consisting of 56 wells in eight clusters drilled on 7.2-acre well spacing. Wells were directionally drilled. A major expansion occurred in 1980 with the addition of seven new pads containing 112 wells. The well

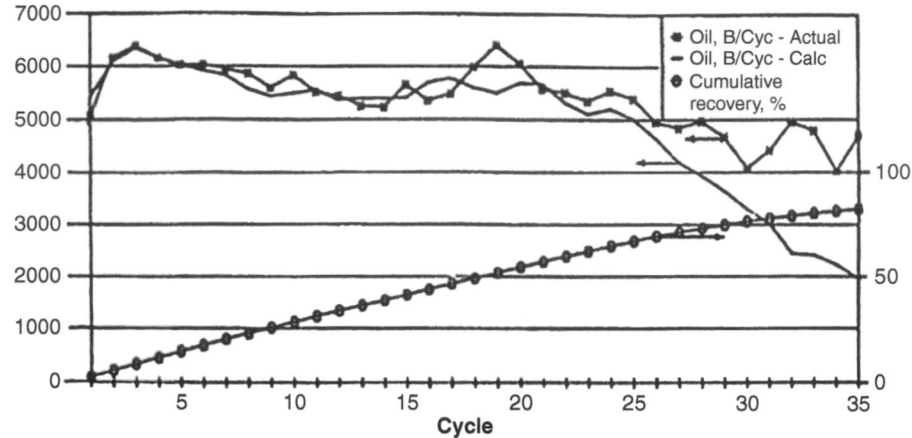

Fig. 8.35—History match of typical Midway Sunset well recovery (Jones 1992).

Field	Zone	Production Rate (B/D)				Net Sand Open (ft)	Steam Injected		Cycle Period (months)	Oil Recovered (bbl)		Additional Oil Recovered/ Barrel Steam (bbl)
		Before Steam, q	After Steam,* q_h	q/q_h	End		Total (bbl)	Per Foot of Sand (bbl)		Total	Per Barrel of Steam	
Huntington Beach	TM	15	160	11	25	40	4,500	112	15	29,000	6.5	5.0
San Ardo	Lombardy	25	360	14	35	220	14,000	64	18	50,000	3.6	2.8
Kern River	China	3	140	47	15	22	4,400	200	6	11,600	2.62	2.5
Midway-Sunset	Potter (A)	10	110	11	25	250	6,000	24	5	9,240	1.54	1.29
Kern River	Kern River	14	65	4.6	20	220	6,500	30	5	4,730	0.73	0.43
Coalinga	Temblor	3	52	17.3	15	107	9,000	84	5	4,300	0.48	0.40
Midway-Sunset	Tulare	5	56	11	10	240	12,000	50	6	4,640	0.38	0.31
Midway-Sunset	Potter (B)	5	35	7	10	250	7,700	31	4	3,000	0.39	0.29
White Wolf	Reef Ridge	30	85	2.4	30	75	14,000	187	4	6,750	0.48	0.23
Poso Creek	Etchegoin	7	20	3	10	80	6,700	84	6	2,660	0.40	0.21

*Average first 30 days.

Table 8.16—Summary of typical well performances: first-cycle steam-soak operations in various fields of California (Prats 1982).

locations were based on fracture orientation to prevent communication between wells in the off-fracture trend. Steam injection was conducted by rows. Oil production from the Leming pilot was 16,350 STB/D in 1984 when a commercial project was developed. Oil recovery in pilot floods exceeded 20% OOIP. **Table 8.20** (Boberg 1988) gives design parameters for the commercial project. In 1990, the Cold Lake Project produced more than 88,060 STB/D (Moritis 1990).

The response of the Cold Lake oil sands to CSS is quite different from that of heavy-oil reservoirs. Cumulative OSRs are on the order of 0.2 to 0.45 bbl oil/bbl steam. First-cycle response is characteristically low, with improvement in subsequent cycles. **Fig. 8.38** (Boberg 1988) compares Cold Lake OSRs for the Leming and May pilots. The increased response of the Leming pilot is attributed to location in a higher-quality area of the reservoir and process design improvements.

The lower OSR for the Cold Lake projects raises an interesting economic question. What is the incremental OSR at the break-even point for the cost of energy? This can be determined easily for an assumed energy content of steam. The following analysis neglects the cost of the water used to make steam. To determine the break-even point, the reference conditions for steam will be taken to be 1,000 Btu/lbm, which is approximately the energy content of 80% quality steam at 800 psi. The energy content per barrel of steam injected is

$$\left(5.615\,\text{ft}^3/\text{bbl}\right)\left(62.4\,\text{lbm/ft}^3\right)\left(1,000\,\text{Btu/lbm}\right) = 350,376\,\text{Btu}, \dots\dots\dots\dots\dots\dots\dots\dots\dots\dots\dots\dots\dots\dots (8.116)$$

Project	Cumulative Steam Injection (10^6 ton)	Number of Wells	Cumulative Oil Productivity (10^6 bbl)	Cumulative Incremental Oil Production (10^6 bbl)	Incremental OSR (bbl/bbl)	Production Resulting From Compaction (% STOOIP*)
Tia Juana						
B/C-3	0.576	63	25.3	18.6	5.14⎫ 3.04⎭	19.4
C-2/3/4	0.268	13	7.02	5.12		–
D-2/E-2	1.96	69	49.8	35.5	2.88	18.9
D/E-3	0.687	69	28.2	16.7	3.86	19.3
G-2/3	0.887	123	50.7	37.7	6.77	22.2
H-6	0.422	29	16.5	12.5	4.71	29.6
H-7	0.227	13	9.23	7.13	5.01	16.1
J-6	0.075	9	4.87	4.33	9.16⎫ 4.85⎭	–
J-7	1.21	125	44.6	36.8		14.6
F-7	1.28	135	46.0	36.2	4.52	10.7
M-6	0.923	95	27.1	20.4	3.51	18.0
D-6	0.982	103	46.7	44.3	7.19	14.1
Lagunillas						
T-6	0.615	31	28.4	18.8	4.86	23.4
V-7	0.685	116	24.4	17.0	3.96	20.4
W-6	1.01	37	22.1	11.2	1.76⎫ 2.51⎭	–
W-6E	1.08	47	24.3	17.1		21.2

*STOOIP = stock-tank OOIP.

Table 8.17—CSS experience in Bolivar Coast of Venezuela (Prats 1982; Borregales 1976).

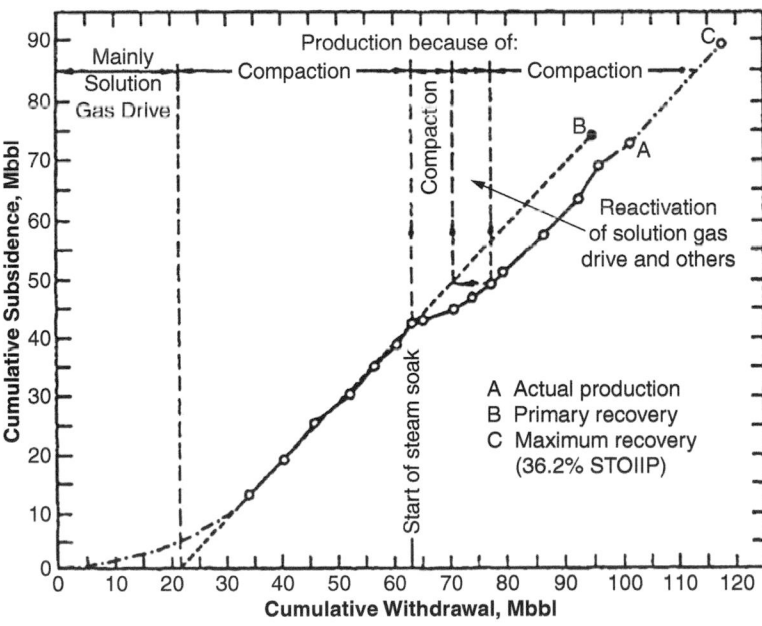

Fig. 8.36—Influence of different reservoir drive mechanisms, D2/E2 project, Tia Juana field (Boberg 1988).

Reservoir Data

Project area (acres)	1,831
Average depth of top sand (ft subsea)	1,600
Average net oil sand (ft)	125
Porosity (%)	38.1
Oil saturation	0.85
Water saturation	0.15
Oil FVF (RB/STB)	1.04
Temperature (°F)	113
Oil gravity (°API)	12
Dead-oil viscosity at 100°F (cp)	1,000 to 9,000
Original reservoir pressure at 1,600 ft (psig)	780
Original STOIP (bbl)	575×10^6

Project Performance, 1 January 1977

Oil-production rate (B/D)	8,714
Cumulative oil production (bbl)	115.7×10^6
Cumulative oil (% STOOIP)	20.1
GOR (instantaneous) (ft³/bbl)	422
Cumulative gas production (ft³)	44.9×10^9
Producing water cut (%)	30.1
Cumulative water production (bbl)	9.9×10^6
Cumulative steam injected (bbl)	11.9×10^6

Project Reserves Estimates

Primary recovery [bbl (% STOOIP)]	105.2×10^6 (18.3)
Steam-soak recovery [bbl (% STOOIP)]	31.7×10^6 (5.5)
Steamdrive recovery [bbl (% STOOIP)]	120.1×10^6 (20.9)
Total recovery [bbl (% STOOIP)]	257.0×10^4 (44.7)

Table 8.18—Reservoir and project performance data—M-6 project, Tia Juana Field (Boberg 1988) [*Thermal Methods of Oil Recovery*, T. C. Boberg, 1988. Reprinted by permission of John Wiley & Sons Inc.].

Reservoir depth (ft)	1,509
Reservoir thickness (ft)	164
Initial reservoir pressure (psi)	450
Initial reservoir temperature (°F)	55
Average reservoir permeability (darcies)	1.5
Oil gravity (°API)	10.2
Viscosity at reservoir temperature (cp)	100,000
Solution GOR (scf/STB)	55
Average oil saturation (%)	70

Table 8.19—Cold Lake Reservoir and fluid properties (Boberg 1988) [*Thermal Methods of Oil Recovery*, T. C. Boberg, 1988. Reprinted by permission of John Wiley & Sons Inc.].

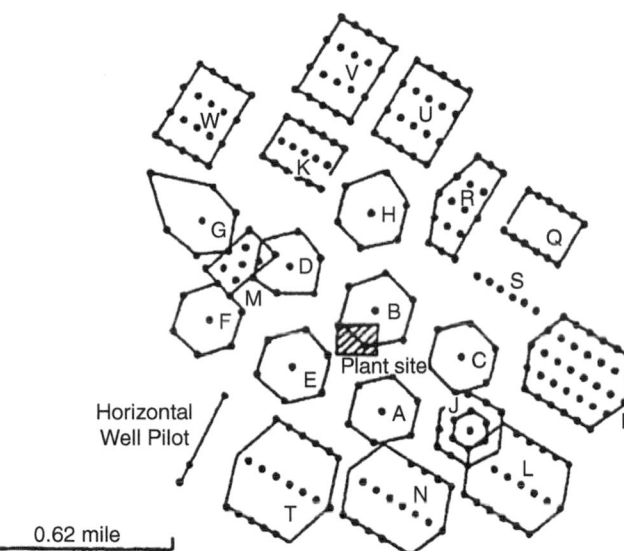

Fig. 8.37—Leming Pilot—Cold Lake thermal recovery project. [*Thermal Methods of Oil Recovery*, T. C. Boberg, 1988. Reprinted by permission of John Wiley & Sons Inc.]

Recovery method	CSS
Number of cycles	8
Well life (years)	8
Well completion	Bottom of good sand
Well spacing (acres)	4
Aspect ratio	1.7
Steam volume* (bbl)	69,200 first cycle (44,000, 50,300, 53,100, 56,600, etc.)**
Average steam rate (B/D)	1,447
Average CDOR (B/D)	55
Steady-state OSR	0.35
Steady-state WOR	2.6
Recovery (%)	20
Produced water recycle (%)	100

CDOR = calendar day oil rate.

*1 m³ = 6.29 bbl.

**For second and subsequent cycles.

Table 8.20—Commercial design parameters for recovery of Cold Lake crude (Boberg 1988) [*Thermal Methods of Oil Recovery*, T. C. Boberg, 1988. Reprinted by permission of John Wiley & Sons Inc.].

which is typically rounded to 350,000 Btu.

Lease crude is used for steam-generator fuel in many projects. Heat release from the combustion of lease crude is approximately 18,000 Btu/lbm. The steam-generator efficiency is 80%. If 10.2 °API oil is burned, the density is 141.5/(131.5 + 10.2), or 0.9986 g/cm³. If F_{OSE} represents the break-even OSR when lease crude is used for steam generation, then

$$F_{OSE} = 350,000 \, \text{Btu/bbl steam}$$

$$\div \left(0.8 \, \text{Btu produced/Btu consumed}\right)$$

$$\times \left(5.615 \, \text{ft}^3/\text{bbl oil}\right)\left[\left(62.4\right)\left(0.9986\right)\text{lbm/ft}^3 \, \text{oil}\right]$$

$$\times \left(18,000 \, \text{Btu/lbm oil}\right)$$

$$= 0.0695 \, \text{bbl oil/bbl stream} \ldots \ldots \ldots (8.117)$$

This ratio will change depending on the gravity of the lease crude oil. Because steam costs are a major expense for CSS projects, this ratio gives some insight into the economics of steam stimulation.

In reservoirs with adequate reservoir energy, several stimulation cycles have been conducted. **Table 8.21** (Rivero and Heintz 1975) summarizes typical response from seven cycles of CSS in the Midway Sunset field. Oil produced per cycle averages 6,531 bbl, with no apparent decline. Production response from this reservoir is sustained by gravity drainage promoted by oil sands that are more than 200 ft thick.

Jones and Cawthon (1990) summarize CSS experience in the Midway Sunset field. By 1987, Santa Fe Energy Co. had performed more than 19,000 steam cycles in 1,500 wells in the field. More than 75 wells had been stimulated 30 or more times, while 350 wells had 20 or more cycles. CSS was found to be superior to conventional steamflooding.

Fig. 8.35 shows cumulative recovery from a Midway Sunset well after 35 cycles. Recovery from this well was more than 80%. **Fig. 8.39** shows the average production rate per well from a lease with approximately 70 wells that was cyclically steamed over a period of 15 years (Jones and Cawthon 1990). Although the average production rate declined with time, the steam/oil ratio (SOR) averaged less than unity for the entire period.

A number of operating practices evolved that improved performance of CSS in the Midway Sunset field. These included infill drilling to ⅝-acre spacing and sequential steaming of production wells. In sequential steaming, wells were stimulated beginning with alternate wells on downdip rows and moving updip, as depicted in **Fig. 8.40**

(Jones and Cawthon 1990b). **Fig. 8.41** shows representative performance from a combination of infill drilling and sequential steaming on Section 27 USL.

Table 8.22 (Yoelin 1971) summarizes production response from four cycles of steam stimulation in the Huntington Beach offshore field. The Huntington Beach OSR is unusually high when compared with most CSS projects in the U.S. **Table 8.23** (Stokes and Doscher 1974) summarizes results from the Yorba Linda field. A slight decline in OSR occurred with successive cycles. Downstructure wells responded more favorably than upstructure wells, probably because of gravity-drainage effects.

8.4 Reservoir Heating by Steam Injection

In CSS and steamdrive, reservoir heating is accomplished by the injection of steam. The Marx and Langenheim model to estimate the radial extent of the heated zone following a steam stimulation treatment, introduced in Section 8.3, was the first of several models. Other models have been developed that allow estimation of the areal and vertical extent of the steam zone from process parameters and fluid and rock properties. Models range from those that can be easily evaluated by hand calculator

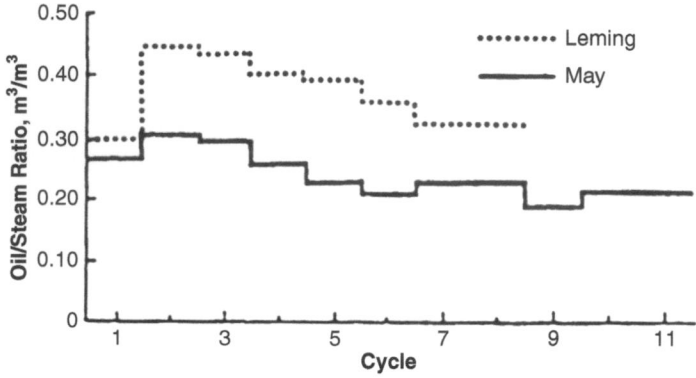

Fig. 8.38—Cold Lake Pilot OSRs (Boberg 1988).

to complex reservoir simulators that run on super-computers. All models assume some distribution of rock and fluid properties in the reservoir. The simple models assume uniform properties, while complex simulators require point-by-point assignment of properties that often are not known and can only be estimated. Simple models assume uniform fluid patterns. In contrast, reservoir simulators incorporate multiphase fluid-flow calculations in the simulation.

This text is aimed at providing an introduction to the fundamentals of EOR processes. Consequently, the focus is on mechanisms of oil recovery and models that represent principal displacement processes. Simple models are used to illustrate these principles. In some cases, these models provide excellent screening tools for evaluation of reservoirs. In the models discussed, the focus is on the amount of energy injected and the fraction of that energy that remains within the reservoir to increase the temperature of reservoir rock and fluids to some desired level at which the displacement process takes place.

8.4.1 Generalization of the Marx and Langenheim Model. The Marx and Langenheim model, given by Eq. 8.63, estimates the heated area as a function of time when the injection rate is specified.

$$A_h = \left[\dot{m}_s H_s M_R h / \left(4 \left(T_s - T_r \right) a M_s^2 \right) \right] G \left(t_D \right).$$

No assumptions are made concerning the shape of the area, but the heated zone has constant thickness. If the reservoir is reasonably homogeneous, then the heated area would be cylindrical in shape, as shown in Fig. 8.18. However, directional

	Cycle						
	First	Second	Third	Fourth	Fifth	Sixth	Seventh
Steam injection (bbl)	6,332	6,179	5,170	4,592	5,618	5,420	5,664
Injection time (days)	11	15	12	8	9	12	7
Injection rate (B/D)	453 to 666	118 to 536	275 to 497	514 to 600	522 to 729	200 to 663	709 to 791
Injection temperature (°F)	350	298 to 334	307 to 338	298 to 327	307 to 324	298 to 307	378 to 415
Injection pressure (psig)	115 to 125	50 to 95	60 to100	50 to 90	60 to 80	50 to 60	175 to 275
Soak time (days)	8	3	3	3	6	3	2
Oil production during cycle (bbl)	5,585	8,784	5,702	5,898	7,218	5,137	7,393
Water production during cycle (bbl)	1,854	3,074	2,800	3,185	4,576	4,382	3,940
Total time (days)	120	216	162	192	298	262	340
Cumulative oil produced (STB)	5,585	14,369	20,071	25,969	33,187	38,324	45,717
Cumulative water injection (bbl)	6,332	12,511	17,681	22,273	27,891	33,311	38,975
Cumulative water production (bbl)	1,854	4,928	7,728	10,913	15,489	19,871	23,811
Water production/water injection	0.293	0.394	0.437	0.490	0.555	0.597	0.611
Cumulative time (days)	120	336	498	694	992	1,254	1,594
Average rate per cycle (B/D)	55.3	44.3	38.8	31.9	25.5	20.8	22.3
OSR (vol/vol)	0.88	1.42	1.10	1.28	1.28	0.95	1.31
SOR (vol/vol)	1.13	0.70	0.91	0.78	0.78	1.06	0.77

Table 8.21—Steam-soak injection/production data—typical well, Buena Fee lease, Midway-Sunset field (Prats 1982; Rivero and Heintz 1975).

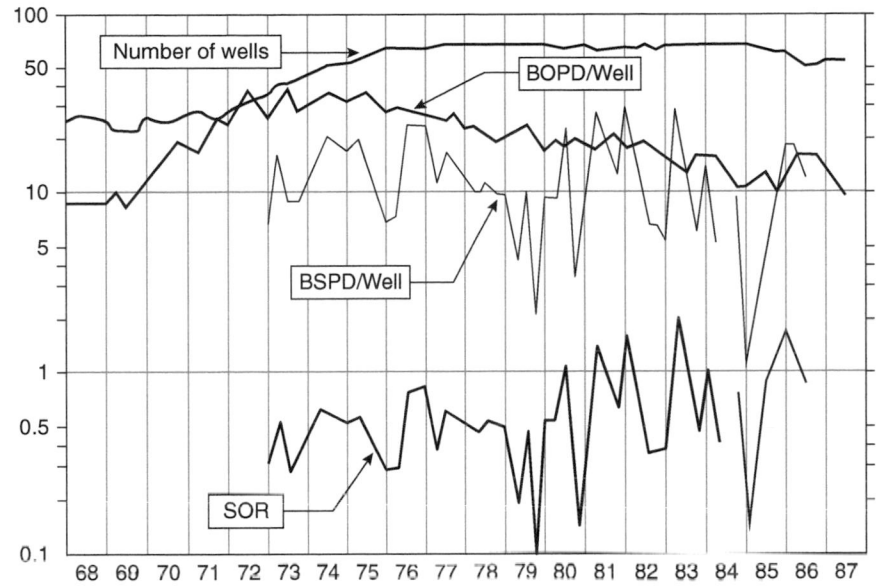

Fig. 8.39—Section 36 performance, conventional cylic steam (Jones and Cawthon 1990).

Fig. 8.40—Schematic of sequential well steaming plan (Jones and Cawthon 1990).

permeability or preferential flow caused by formation dip may lead to a heated region similar to that presented in **Fig. 8.42.** The shape of the heated area is determined by fluid-flow patterns in the reservoir.

Thermal Efficiency of Reservoir Heating Process. The expression for the heated area as a function of time given in Eq. 8.63 can be used to derive relationships for the rate of heat loss and the thermal efficiency of the reservoir heating process.

At any time, the energy stored in the heated region is given by

$$Q = M_R h A_h \left(T_s - T_r \right). \quad\dotfill (8.118)$$

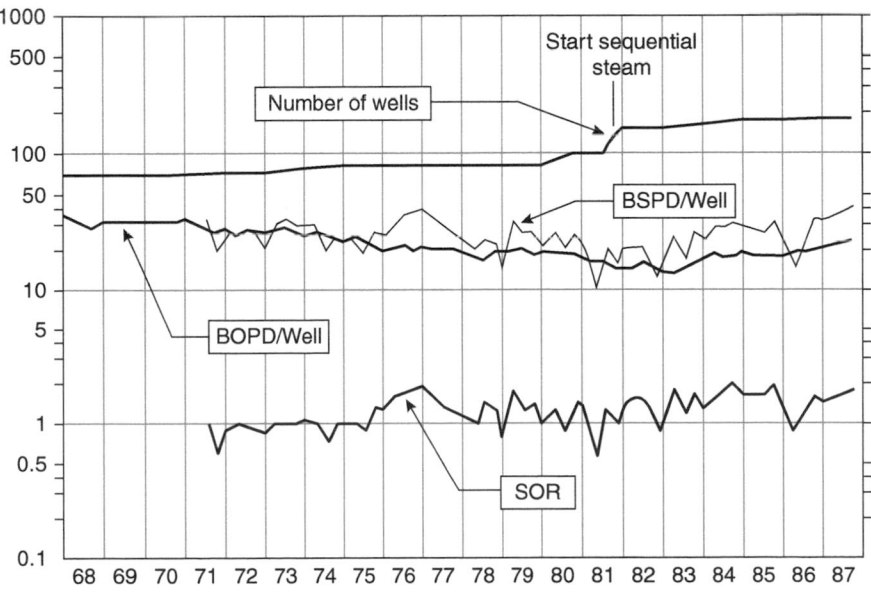

Fig. 8.41—Section USL total lease performance with the impact of infill drilling and sequential steam stimulation (Jones and Cawthon 1990).

	Cycle 1	Cycle 2	Cycle 3	Cycle 4
Number of wells	24	18	11	4
Total oil recovery (STB)	694,000	556,250	271,150	116,900
Average cycle length (months)	14	18	15.3	14.5
Average oil recovery per well (STB)	28,900	30,900	24,650	29,225
Average quality of steam injected (%)	71.4	69.3	75.1	78.5
Average volume of steam injected (bbl)	9,590	8,130	10,190	11,760
Steam injected (bbl/ft)	213.9	191.3	232.1	267.3
Heat input (10^6 Btu/ft)	64	63.8	89.4	98.4
Ratio of oil recovered to steam injected (STB/bbl)	3	3.8	2.4	2.5
Oil recovery (STB/ft)	645	737	560	665
Oil recovery (STB/10^6 Btu)	10.1	11.2	6.3	6.8

Table 8.22—Summary of performance through four huff 'n' puff cycles as of 1 October 1970, TM sand, Huntington Beach offshore field (Prats 1982; Yoelin 1971).

Type of Well and Completion	Cycle	Steam Injected (bbl)	Oil Produced (bbl)	Extrapolated Oil Produced (bbl)	Extrapolated Oil/ Steam Ratio
Upstructure (full)	1	24,500	13,800	19,400	0.79
	2	27,300	10,700	14,100	0.52
	3	19,200	9,100	11,200	0.58
	4	17,200	10,100	12.900	0.75
	5	15,800	1,900	*	—
Upstructure (short)	1	16,500	20,100	21,400	1.30
	2	15,700	10,000	20,600	1.31
	3	15,900	11,800	15,200	0.96
Upstructure (scab)	1	—	—	—	—
	2	—	—	—	—
	3	15,100	11,300	14,200	0.94
	4	14,700	9,900	13,200	0.90
	5	16,400	6,800	9,300	0.57
Downstructure (full)	1	22,300	23,200	13,800	1.42
	2	20,400	18,900	23,300	1.14
	3	15,000	17,300	20,700	1.38
	4	13,500	13,800	18,100	1.34
	5	16,400	17,800	17,100	1.09
Downstructure (full)	1	13,900	18,900	22,200	1.60
	2	15,300	15,100	25,600	1.67
	3	15,100	11,100	23,800	1.48

*Insufficient data to extrapolate the curve.

Table 8.23—Effect of completion type and structure position on well performance, Yorba Linda field (Prats 1982; Stokes and Doscher 1974).

Substituting Eq. 8.63 for A, Eq. 8.118 becomes

$$Q = \left(\dot{m}_s H_s M_R^2 h^2 / 4\alpha M_s^2 \right) G(t_D). \dots\dots\dots\dots\dots\dots\dots\dots\dots\dots\dots\dots\dots\dots\dots\dots \quad (8.119)$$

The thermal efficiency of reservoir heating, E_h, is the fraction of injected energy that remains in the reservoir. The total energy injected is $\dot{m}_s H_t t$. Thus, the thermal efficiency is the ratio of Q to $\dot{m}_s H_t t$ and is given by

$$E_h = G(t_D)/t_D, \dots\dots\dots\dots\dots\dots\dots\dots\dots\dots\dots\dots\dots\dots\dots\dots\dots\dots\dots\dots\dots \quad (8.120)$$

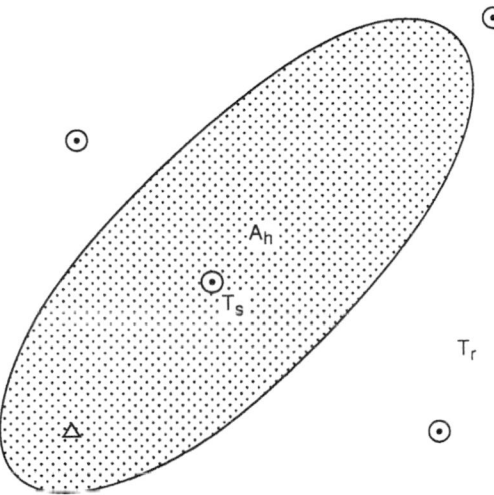

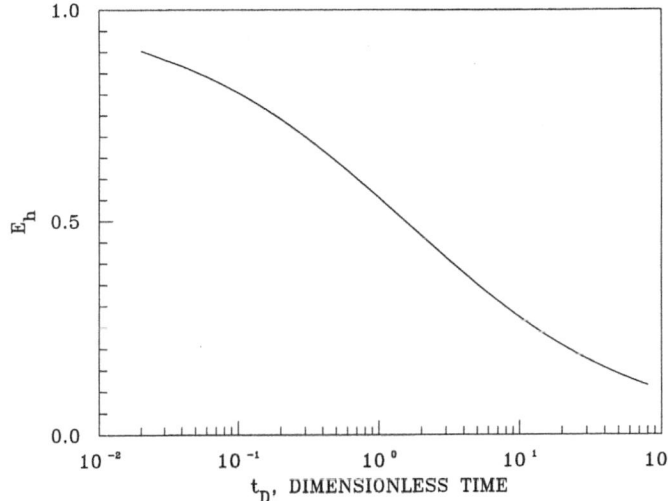

Fig. 8.42—Areal growth predicted by Marx and Langenheim model.

Fig. 8.43—Thermal efficiency for reservoir heating as a function of t_D.

where t_D is given by Eq. 8.60:

$$t_D = 4\left(M_s/M_R\right)^2\left(\alpha/h^2\right)t.$$

Values of E_h, are also included in Table 8.6. **Fig. 8.43** is a graph of E_h vs. t_D.

The energy lost to the overburden and underlying formation at any time is

$$Q_\ell = \dot{m}_s H_s t\left(1 - E_h\right)$$

$$= \dot{m}_s H_s t\left[1 - \left(G/t_D\right)\right]. \quad\ldots\ldots\ldots\ldots\ldots\ldots\ldots\ldots\ldots\ldots\ldots\ldots(8.121)$$

Development of Hot-Water Region. A steam zone can be maintained throughout the heated region predicted by the Marx and Langenheim model as long as there is sufficient latent heat to keep the steam zone at a constant temperature. As the heated zone expands, a time is reached when the heated region is separated into a steam zone and a hot-water zone. The critical time when formation of a distinct hot-water zone occurs was determined by Mandl and Volek (1969) from an energy balance at the condensation front. A derivation is presented in this section that leads to the same expression for the critical time by a less complex derivation.

The rate of heat loss to the over- and underburden is given by rearranging Eq. 8.59 to obtain

$$2\int_o^t \frac{k_h\left(T_s - T_r\right)}{\sqrt{\pi\alpha\left(t-\lambda\right)}}\left(\frac{dA_h}{d\lambda}\right)d\lambda = \dot{m}_s H_s - M_R h\left(T_s - T_r\right)\frac{dA_h}{dt} = \dot{Q}_\ell, \quad\ldots\ldots\ldots\ldots\ldots(8.122)$$

where $\dot{Q}_\ell$ is the rate of heat loss to the overburden and underlying formation. From Eq. 8.65,

$$dA_h/dt = \left[\dot{m}_s H_s / M_R\left(T_s - T_r\right)h\right]e^{t_D}\,\mathrm{erfc}\left(\sqrt{t_D}\right).$$

Substituting into Eq. 8.122 yields

$$\dot{Q}_\ell = \dot{m}_s H_s\left(1 - e^{t_D}\,\mathrm{erfc}\sqrt{t_D}\right). \quad\ldots\ldots\ldots\ldots\ldots\ldots\ldots\ldots\ldots\ldots\ldots(8.123)$$

The entire heated region is filled with steam as long as

$$\dot{m}_s f_{sd} L_{vdh} \geq \dot{m}_s H_s\left(1 - e^{t_D}\,\mathrm{erfc}\sqrt{t_D}\right). \quad\ldots\ldots\ldots\ldots\ldots\ldots\ldots\ldots\ldots\ldots(8.124)$$

The condensation of all steam is just able to satisfy heat losses when $\dot{Q}_\ell = m_{sfsd} L_{vdh}$. The critical dimensionless time, t_{cD}, when this happens is defined by

$$e^{t_{cD}}\,\mathrm{erfc}\left(\sqrt{t_{cD}}\right) = 1 - \left(f_{sd} L_{vdh}/H_s\right)$$

$$= 1 - f_{h,v}. \quad\ldots\ldots\ldots\ldots\ldots\ldots\ldots\ldots\ldots\ldots\ldots\ldots\ldots(8.125)$$

Recall from Eq. 8.66 that

$$G_1(t_D) = e^{t_D} \operatorname{erfc}\left(\sqrt{t_D}\right).$$

Thus, $G_1(t_{cD}) = 1 - f_{h,v}$. The term $f_{h,v}$ is the fraction of the injected energy that is latent heat following the notation introduced by Prats (1982). For $t_D > t_{cD}$, a hot-water region precedes the steam zone. Computation of t_{cD} is illustrated in Example 8.8. **Table 8.24** contains values of t_{cD} for $f_{h,v}$ between 0.05 and 0.90 obtained by solving Eq. 8.98 for t_{cD}.

Example 8.8—Determination of Critical Time for Formation of Hot-Water Zone. Determine the critical time for formation of the hot-water zone for the reservoir heating problem in Example 8.4.

Solution. In Example 8.4,

$$L_{vdh} = 751.4\,\text{Btu/lbm},$$

$$H_s = 1006\,\text{Btu/lbm},$$

$$f_{sd} = 0.8,$$

and $f_{h,v} = \dfrac{(0.8)(751.4)}{1006} = 0.598.$

From Table 8.24, $t_{cD} \approx 1.226$. In Example 8.4, $t_D = 0.0102t$, where t is in days, so that

$t = 1.226/0.0102$
$\quad = 120$ days.

$f_{h,v}$	t_{cD}	$f_{h,v}$	t_{cD}	$f_{h,v}$	t_{cD}
0.05	0.00213	0.51	0.63586	0.77	5.1193
0.1	0.00927	0.52	0.68352	0.78	5.67237
0.15	0.02282	0.53	0.73481	0.79	6.30765
0.2	0.04465	0.54	0.78991	0.8	7.04249
0.25	0.07728	0.55	0.84933	0.81	7.89875
0.3	0.12415	0.56	0.91339	0.82	8.90339
0.31	0.13564	0.57	0.98253	0.83	10.09489
0.32	0.14793	0.58	1.05721	0.84	11.5198
0.33	0.16107	0.59	1.13805	0.85	13.24672
0.34	0.17513	0.6	1.22572	0.86	15.36573
0.35	0.19015	0.61	1.32079	0.87	18.00898
0.36	0.20622	0.62	1.42418	0.88	21.35923
0.37	0.22337	0.63	1.53673	0.89	25.70713
0.38	0.24173	0.64	1.6595	0.9	31.49442
0.39	0.26133	0.65	1.79372	–	–
0.4	0.28227	0.66	1.94055	–	–
0.41	0.30466	0.67	2.10186	–	–
0.42	0.32858	0.68	2.27933	–	–
0.43	0.35416	0.69	2.4751	–	–
0.44	0.38152	0.7	2.69158	–	–
0.45	0.4108	0.71	2.9318	–	–
0.46	0.44212	0.72	3.19908	–	–
0.47	0.47569	0.73	3.49772	–	–
0.48	0.51162	0.74	3.83239	–	–
0.49	0.55017	0.75	4.20937	–	–
0.5	0.05915	0.76	4.6354	–	–

Table 8.24—Critical dimensionless time for formation of hot-water region, Marx and Langenheim model.

Thus, for this set of parameters, the hot-water zone forms a short time after injection starts.

Growth of Steam Zone After the Critical Time. When the injection time exceeds the critical time, a hot-water zone develops that also moves as a zone where there is a step increase from T_r to T_s. It is possible to determine the area exposed to steam by noting that the rate of heat loss can be expressed as the sum of the heat loss rate from the steam zone, $\dot{Q}_{\ell s}$, and the hot-water zone, $\dot{Q}_{\ell w}$.

$$\dot{Q}_\ell = \dot{Q}_{\ell s} + \dot{Q}_{\ell w} \quad\dotfill\quad (8.126)$$

$$\text{and } \dot{Q}_\ell = \dot{m}_s H_s \int_o^{t_D} \frac{e^{\lambda_D}\,\text{erfc}\sqrt{\lambda_D}}{\sqrt{\pi(t_D - \lambda_D)}}\,d\lambda_D, \quad\dotfill\quad (8.127)$$

where λ_D is the dimensionless time when the heated region arrived at a specific location corresponding to a particular area A. The integral in Eq. 8.127 can be written as the sum of two integral terms,

$$\dot{Q}_\ell = \dot{m}_s H_s \int_o^{t_{Ds}} \frac{e^{\lambda_D}\,\text{erfc}\left(\sqrt{\lambda_D}\right)d\lambda_D}{\sqrt{\pi(t_D - \lambda_D)}} + \dot{m}_s H_s \int_{t_{Ds}}^{t_D} \frac{e^{\lambda_D}\,\text{erfc}\left(\sqrt{\lambda_D}\right)d\lambda_D}{\sqrt{\pi(t_D - \lambda_D)}}, \quad\dotfill\quad (8.128)$$

where the first integral is the rate of heat loss from the steam zone to the over- and underburden and the second integral is the rate of heat loss from the hot-water zone to the over- and underburden. Dimensionless time t_{Ds} is the time that the heated front arrived at the position contacted by condensed steam at time t_D. Because the heat loss from the steam zone is supplied totally by condensation of steam,

$$\dot{Q}_{\ell s} = \dot{m}_s f_{sd} L_{vdh} = \dot{m}_s H_s \int_o^{t_{Ds}} \frac{e^{\lambda_D}\,\text{erfc}\left(\sqrt{\lambda_D}\right)}{\sqrt{\pi(t_D - \lambda_D)}}\,d\lambda_D. \quad\dotfill\quad (8.129)$$

It is consistent to designate the integral in Eq. 8.129 as $G_s(t_D, t_{Ds})$. Then,

$$G_s\left(t_D, t_{Ds}\right) = \int_o^{t_{Ds}} \frac{e^{\lambda_D}\,\text{erfc}\left(\sqrt{\lambda_D}\right)}{\sqrt{\pi(t_D - \lambda_D)}}\,d\lambda_D$$

$$\text{and } G_s\left(t_D, t_{Ds}\right) = f_{sd} L_{vdh} / H_s = f_{h,v}, \quad\dotfill\quad (8.130)$$

where $t_{Ds} < t_D$.

Table 8.25 presents values of $G_s\,(t_D, t_{Ds})$ for selected values of $f_{h,v}$ and t_D. A program is also available in Appendix F that calculates t_{Ds} by an iterative procedure when G_s and t_D are specified. When t_D and $f_{h,v}$ are specified, Eq. 8.130 is solved to find t_{Ds}. Then, the steam-zone area and heated volume are determined with

$$A_s = \left[\dot{m}_s H_s M_R h / 4\left(T_s - T_r\right)a_s M_s^2\right]G(t_{Ds}) \quad\dotfill\quad (8.131)$$

$$\text{and } V_s = A_s h \quad\dotfill\quad (8.132)$$

$$= \left[\dot{m}_s H_s M_R h_t^2 / 4\left(T_s - T_r\right)a_s M_s^2\right]G(t_{Ds}). \quad\dotfill\quad (8.133)$$

Example 8.9—Estimation of Steam-Zone Growth When a Hot-Water Region Develops. Steam (200 psig) is injected at a rate of 850 B/D into a reservoir that is 32 ft thick. Steam temperature is 387.9°F at 215 psia, and the formation temperature is 110°F. Estimate the steam-zone area after 4.5 years of injection assuming continuous steam injection and no withdrawal of heated fluids at the production wells.

| $f_{h,v}$ = 0.05000 | | | $f_{h,v}$ = 0.1 | | | $f_{h,v}$ = 0.2 | | |
| t_{cD} = 0.00213 | | | t_{cD} = 0.00927 | | | t_{cD} = 0.04465 | | |
t_D	t_{Ds}	$E_{h,s}$	t_D	t_{Ds}	$E_{h,s}$	t_D	t_{Ds}	$E_{h,s}$
0.01	0.007227	0.678933	0.01	0.009965	0.92634s	–	–	–
0.02	0.011278	0.521821	0.02	0.01817	0.824028	–	–	–
0.04	0.017154	0.389993	0.04	0.030234	0.667313	–	–	–
0.06	0.021762	0.326067	0.06	0.039814	0.575743	0.06	0.059097	0.830185
0.08	0.025707	0.286345	0.08	0.048084	0.51465	0.08	0.075992	0.783826
0.1	0.029225	0.258538	0.1	0.055504	0.470145	0.1	0.091308	0.740731
0.2	0.04338	0.187092	0.2	0.085733	0.349839	0.2	0.15513	0.594317
0.4	0.064244	0.134456	0.4	0.131171	0.256194	0.4	0.254266	0.456253
0.6	0.080854	0.110575	0.6	0.16795	0.212414	0.6	0.336662	0.385794
0.8	0.095225	0.096169	0.8	0.200151	0.185639	0.8	0.410097	0.341035
1	0.108157	0.086271	1	0.229385	0.167075	1	0.47769	0.3093
2	0.161127	0.061447	2	0.351453	0.119982	2	0.767781	0.226322
4	0.241631	0.043677	4	0.542703	0.085784	4	1.242098	0.163946
6	0.307573	0.035755	6	0.7033	0.070391	6	1.65474	0.135313
8	0.366014	0.03103	8	0.847703	0.061146	8	2.034566	0.117944
10	0.419034	0.027773	10	0.981423	0.0548	10	2.392998	0.105959
20	0.644652	0.019704	20	1.564569	0.038953	20	4.012178	0.075763
40	1.009376	0.013999	40	2.543519	0.027658	40	6.872243	0.054006
60	1.317919	0.011428	60	3.413544	0.022625	60	9.514153	0.044252
80	1.592982	0.009868	80	4.225643	0.019614	80	12.04015	0.038402

| $f_{h,v}$ = 0.3 | | | $f_{h,v}$ = 0.4 | | | $f_{h,v}$ = 0.5 | | |
| t_{cD} = 0.12415 | | | t_{cD} = 0.28227 | | | t_{cD} = 0.5915 | | |
t_D	t_{Ds}	$E_{h,s}$	t_D	t_{Ds}	$E_{h,s}$	t_D	t_{Ds}	$E_{h,s}$
0.2	0.194035	0.722814	–	–	–	–	–	–
0.4	0.346292	0.592532	0.4	0.394329	0.660276	–	–	–
0.6	0.475933	0.513949	0.6	0.567338	0.592786	0.6	0.599728	0.619879
0.8	0.593252	0.460871	0.8	0.726219	0.541505	0.8	0.792783	0.580234
1	0.702486	0.421957	1	0.875697	0.501708	1	0.977384	0.546257
2	1.181824	0.315792	2	1.544949	0.385914	2	1.817809	0.434819
4	1.991592	0.232286	4	2.707405	0.28899	4	3.314966	0.332607
6	2.714137	0.192996	6	3.766127	0.241946	6	4.702823	0.280921
8	3.390263	0.168889	8	4.769492	0.212672	8	6.032212	0.24819
10	4.036238	0.15214	10	5.736897	0.192159	10	7.323497	0.225015
20	7.021204	0.10951	20	10.28109	0.139331	20	13.46738	0.164488
40	12.45512	0.078415	40	18.72186	0.10025	40	25.05137	0.118976
60	17.57658	0.064369	60	26.77401	0.082443	60	36.19491	0.098038
80	22.52909	0.055911	80	34.60566	0.071669	80	47.076	0.085311

Table 8.25—Values of $G_s(t_D, t_{Ds})$ for $f_{h,v}$, Marx and Langenheim model.

| $f_{h,v}$ = 0.6 | | | $f_{h,v}$ = 0.7 | | | $f_{h,v}$ = 0.8 | | |
| t_{cD} = 0.122572 | | | t_{cD} = 2.69158 | | | t_{cD} = 7.0425 | | |
t_D	t_{Ds}	$E_{h,s}$	t_D	t_{Ds}	$E_{h,s}$	t_D	t_{Ds}	$E_{h,s}$
2	1.970966	0.461092						
4	3.752328	0.362016	4	3.969072	0.376062			
6	5.430101	0.308938	6	5.875948	0.325311			
8	7.052593	0.274559	8	7.72864	0.2911	8	7.985243	0.297207
10	8.638368	0.249891	10	9.551483	0.266165	10	9.951659	0.273072
20	16.26721	0.184369	20	18.40917	0.198504	20	19.67827	0.206511
40	30.8303	0.134158	40	35.50842	0.14546	40	38.64596	0.152638
60	44.93692	0.110813	60	52.1875	0.120508	60	57.28124	0.126929
80	58.75788	0.096553	80	68.58767	0.105182	80	75.69219	0.111045

$f_{h,v}$ = 0.9

t_{cD} = 31.49442

t_D	t_{Ds}	$E_{h,s}$
40	39.92532	0.155483
60	59.70173	0.129881
80	79.40966	0.114004

Table 8.25—Values of $G_s(t_D,t_{Ds})$ for $f_{h,v}$, Marx and Langenheim model (continued).

Solution. The following values are used in the example.

$L_{vdh} = 837.4\,\text{Btu/lbm}$,

$M_R = 35\,\text{Btu/ft}^3\text{-}°\text{F}$,

$H_s = 870.15\,\text{Btu/lbm}$,

$M_s = 42\,\text{Btu/ft}^3\text{-}°\text{F}$,

$f_{sd} = 0.7$,

$t_D = \dfrac{(35,040)(1.2)(42)}{(32)^2(35)^2}\,t = 1.408t$ (where t is in years)

$\quad = 6.335,$

and $f_{h,v} = \left[(0.7)(837.4)\right]/870.15 = 0.674$.

The critical time t_{cD} corresponding to $f_{h,v}$ = 0.674 is 2.167. Thus, a hot-water region precedes the steam zone. Applying Eq. 8.130, $G_s(t_D,t_{Ds})$ = 0.674. The value of t_{Ds} = 6.095 when t_D = 6.335. The steam-zone area is computed from Eq. 8.131 after evaluating $G(t_{Ds})$. From Table 8.8, the value of $G(t_{Ds})$ is 1.996.

$$A_s = \left\{\dot{m}_s H_s M_R h_t \big/ \left[4(T_s - T_r)a_s M_s^2\right]\right\} G(t_{Ds})$$

$$= \left\{\left[(12,396\,\text{lbm/hr})(870.15\,\text{Btu/lbm})\right.\right.$$

$$\times(35\,\text{Btu/ft}^3\text{-}°\text{F})(32\,\text{ft})]/\left[4(387.9°\text{F} - 110°\text{F})\right.$$

$$\left.\left.\times(0.0286\,\text{ft}^2/\text{hr})(42\,\text{Btu/ft}^3)^2\right]\right\} G(t_{Ds})$$

$$= 215,414(1.996)$$

$$= 429,966\,\text{ft}^2$$

$$= 9.87\,\text{acres}.$$

At $t_D = 6.33$, $G(t_D) = 2.049$ and the total heated area is

$$A = (215,414)(2.049)$$
$$= 441,297 \, \text{ft}^2$$
$$= 10.13 \text{ acres.}$$

In this example, the hot-water region covers an area of $(10.13 - 9.87)$ or 0.26 acres after 4.5 years of injection.

The volume of the steam zone, V_s, is computed directly from Eq. 8.132 as A_{sh}. At $t_D = 6.33$, $V_s = 315.8$ acre-ft or 2,450,287 bbl. The hot-water region occupies a reservoir volume of 8.3 acre-ft. The development of these two regions with time is shown in **Fig. 8.44**, where the heated reservoir volume is plotted as a function of t_D for the parameters of this example.

It should be remembered that the Marx and Langenheim model assumes that there is no heat loss by conduction at the leading edge of the heated region. Thus, the entire region remains at the injection temperature of the steam. However, the hot-water region cannot be maintained at a constant temperature because heat losses must be made up from heat capacity of the flowing fluids rather than latent heat in the steam zone, where condensation can maintain a constant temperature. Because of this, the steam volume computed from the method described in this section is considered a lower bound. Yortos and Gavalas (1981a, b) have studied upper bounds for the growth of the steam zone.

Conduction Heating by Steam Injection.
There are applications of the Marx and Langenheim model to reservoir geometries other than those used in the development of the

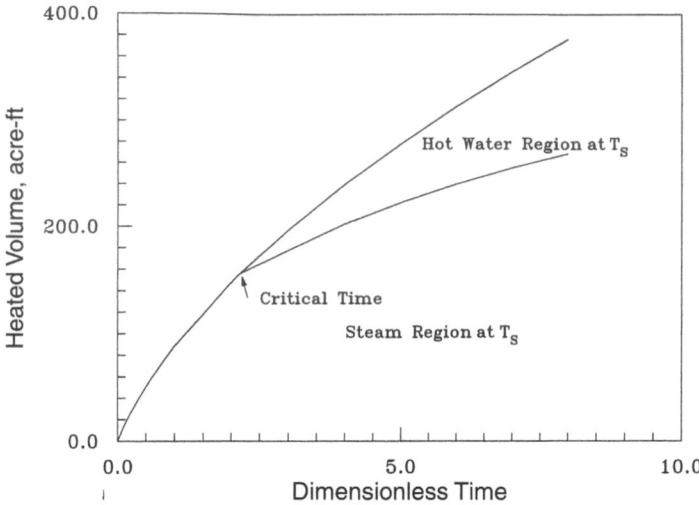

Fig. 8.44—Growth of steam and hot-water regions with dimensionless time, Example 8.9.

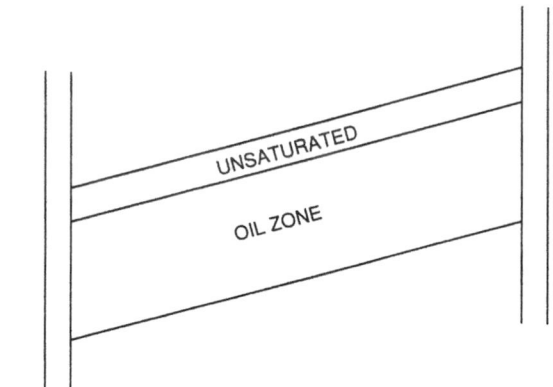

Fig. 8.45—Heavy-oil reservoir with unsaturated zone at top of reservoir.

model. In **Fig. 8.45**, an unsaturated zone of thickness h_{us} is present at the top of a reservoir. This zone is characterized by an initial gas saturation, S_g, and a low oil saturation. When steam is injected into this reservoir, steam will flow preferentially through the unsaturated zone because of the high mobility of the gas phase and the heated zone will develop along the top of the reservoir. Heat loss to the underlying reservoir enhances reservoir heating. The location of the steam front can be computed as a function of time by use of the unsaturated zone thickness, h_{us}, for h in Eqs. 8.60 and 8.61.

It is also possible to compute the temperature profiles at any point in the underlying reservoir formation to assess what fraction of the heat flow from the steam zone is heating the reservoir. To estimate temperature profiles, it is necessary to assume the value of the steam-zone thickness. Example 8.10 illustrates computation of the temperature profile for this case.

Example 8.10—Estimation of Temperature Profile in the Heated Zone and Above and Below the Heated Zone. Steam at 470.9°F (500 psig) is injected into a reservoir that is 50 ft thick at a rate of 500 BWPD. The upper 5 ft of this reservoir has a high gas saturation and, consequently, all the injected steam is assumed to flow through this zone. Determine the temperature profile in the reservoir at a radius of 100 ft from the injection well after 180 days of injection. Thermal properties of the reservoir are identical to those used in Example 8.3. Initial reservoir temperature is 80°F.

Solution. In Example 8.3, the volumetric heat capacity of the reservoir rock in the heated region was determined to be 33.13 Btu/ft³-°F, while the volumetric heat capacity of the over- and underlying formations was 31.12 Btu/ft³-°F. These are not large differences. In this example, the thermal properties of the reservoir zone beneath the steam zone will be assumed to be the same as those of the overburden. This assumption does not introduce a significant error in most situations.

The solution begins by calculating the value of t_D at $t = 180$ days. From Eq. 8.60,

$$t_D = 4 \left(M_s / M_R \right)^2 \left(\alpha / h_s^2 \right) t$$
$$= 4 \left[(31.12 \text{ Btu/ft}^3\text{-}°F) / (33.13 \text{ Btu/ft}^3\text{-}°F) \right]^2$$
$$\times \left[(0.0482 \text{ ft}^3\text{/hr}) / (5 \text{ ft})^2 \right] (180 \text{ days})(24 \text{ hr/D})$$
$$= 29.41.$$

From Table 8.8,

$$G = 5.1905 + \left(\frac{29.41 - 29}{30 - 29} \right)(5.2827 - 5.1805)$$

$$= 5.222.$$

The steam injection rate is 500 BWPD, which is equivalent to 7,292 lbm/hr.

From Eq. 8.63,

$$A_h = \left\{ \dot{m}_s H_s M_R h / \left[4(T_s - T_r) a M_s^2 \right] \right\} G(t_D)$$

$$= \left\{ (7,296 \, \text{lbm/hr})(1006 \, \text{Btu/lbm})(33.13 \, \text{Btu/ft}^3 \text{-}°\text{F})(5 \, \text{ft}) \right.$$

$$\left. \div \left[4(470.9°\text{F} - 80°\text{F})(0.0482 \, \text{ft}^2 / \text{hr})(31.12 \, \text{Btu/ft}^3 - °\text{F})^2 \right] \right\}$$

$$\times G(t_{Ds})$$

$$= (16,648)(5.222)$$

$$= 86,922 \, \text{ft}^2$$

$$= 1.996 \, \text{acres.}$$

If flow is radial,

$$\pi r_h^2 \approx 86,922$$

$$r_h \approx \sqrt{86,922 \, \text{ft}^2 / \pi}$$

$$= 166.3 \, \text{ft.}$$

To determine the temperature profile in the underlying formation at a radial distance of 100 ft from the injection well, it is necessary to determine the arrival time of the heated front at that location.

$$A_h(t_s) = \pi(100)^2$$

$$= 31,416 \, \text{ft}^2.$$

Solve for G using Eq. 8.63. In Eq. 8.63,

$$A = 16,644 \, G,$$

$$= 1.888,$$

and $t_D \approx 5.578$.

For this example, $t_D = 0.163 \, t_s$, where t_s is in days.

$$t_s = 34.1 \, \text{days.}$$

Thus, the zone underlying the reservoir at a radius of 100 ft from the injection well has been exposed to condensing steam at 470.9°F for a total of 180 – 34.1, or 144.4, days. Steam movement in the heated zone is so rapid that the effect of the arrival time on the temperature profile is important, primarily near the steam front.

The temperature profile in the underlying formation is computed assuming that the steam zone is not displaced vertically (i.e., steam does not invade the underlying formation). Applying Eq. 8.53, adjusted for arrival time, A, of the steam zone, the temperature distribution is given by

$$T = T_r + (T_s - T_r) \text{erfc} \left\{ z / \left[2\sqrt{\alpha(t - t_s)} \right] \right\}. \quad \dots \dots \dots \dots \dots \dots \dots \dots \dots (8.134)$$

The boundary between the steam zone and the underlying formation is $z = 0$. At $t = 180$ days, the temperature distribution is computed as illustrated next. Because $a = 0.0482 \, \text{ft}^2/\text{hr}$,

Distance Below Steam Zone (ft)	Temperature (°F)
0	470.90
5	386.60
10	308.35
15	240.92
20	186.99
25	146.94
30	119.33
35	101.66
40	91.17
45	85.39
50	82.43

*Computed temperature distribution under steam zone t = 180 days at r = 100 ft.

Table 8.26—Temperatures* in underlying reservoir owing to conduction heating.

$$T(z) = 80 + (470.9 - 80)$$
$$\times \operatorname{erfc}\left\{z\Big/\left[2\sqrt{(0.0482\ \text{ft}^2/\text{hr})(180 - 34.1\ \text{days})(24\ \text{hr/D})}\right]\right\}$$
$$= 80 + 390.9\ \operatorname{erfc}(0.0385z).$$

At z = 15,

$$T(z) = 80 + 390.9\ \operatorname{erfc}[0.0385(15)]$$
$$= 80 + 390.9\ \operatorname{erfc}(0.577).$$

From Table 8.7, erfc (0.577) = 0.415, so $T(z)$ = 242°F.

Table 8.26 presents temperatures computed at intervals of 5 ft. **Fig. 8.46** shows the temperature profile. At the boundary between the underburden and the reservoir (z = 45 ft), the temperature is 90°F, so this part of the reservoir has not increased much in temperature.

Heat transfer by conduction from the reservoir zone invaded by steam can be an effective means of heating the reservoir. To heat a reservoir in this manner, it is necessary to be able to propagate the steam zone through a portion of the reservoir, as assumed in Example 8.4.

Reservoir heating by injection in a limited undersaturated region at the top of a reservoir is analogous to the process that occurs when a water zone of thickness h_w underlies the reservoir. In this case, the injected steam would flow through the water zone and the reservoir would be heated from the bottom. Nasr and Pierce (1993) describe a study of steamflooding Cold Lake oil reservoir through a bottomwater zone. Example 8.10 can also be used to describe reservoir heating by injection into a bottomwater zone. For example, consider the reservoir in **Fig. 8.47**. As in Example 8.10, the reservoir is 45 ft thick and is underlain by a water zone that is 5 ft thick and has high permeability. All other properties are identical to those in Example 8.10.

The areal advance of the steam zone is identical to that in the undersaturated zone. In practice, there would be some difference in thermal properties of the water zone compared with the unsaturated zone. These effects would be small. The actual temperature profile at the distance of 100 ft from the injection well is just inverted from the profile determined in Example 8.10. **Fig. 8.48** shows this profile for reservoir heating through the bottomwater zone. If the bottom-water zone has sufficient permeability and thickness, a reservoir can be effectively heated from the bottom. Gravity segregation will allow the steam to rise into the reservoir in a relatively short distance if vertical permeability is high.

Variable Injection Rates and Removal of Heat Through Production Wells. In a homogeneous reservoir, the heated region breaks through into the production wells as depicted in **Fig. 8.49**. Energy is removed from the reservoir as hot oil and water. If the drawdown is large in the production wells, water will partially flash to steam. The area contacted by steam at breakthrough is controlled by reservoir fluid-flow paths. Directional permeability in the southwest/northeast directions, for example, will usually lead to low areal sweep at breakthrough, as depicted in Fig. 8.42. Expansion of the heated area and, consequently, the steam-zone area requires continuous injection of steam. However, because heat is removed with produced fluids at the production wells, the rate of areal expansion of the heated region will be reduced.

The amount of energy removed from the reservoir can be estimated if the produced-fluid temperature and oil cut are known. Produced-fluid temperatures cannot be predicted from simple oil-balance models. Eq. 8.135 defines the rate at which energy is removed by produced fluids when oil and water are removed as liquids at temperature T:

$$\dot{Q}_p = q_o\rho_o C_o(T_p - T_r) + q_w\rho_w C_w(T_p - T_r). \quad \ldots (8.135)$$

Additional terms must be added to Eq. 8.135 when the produced fluids contain gas or steam. Section 8.3 gives appropriate expressions for the energy content of the

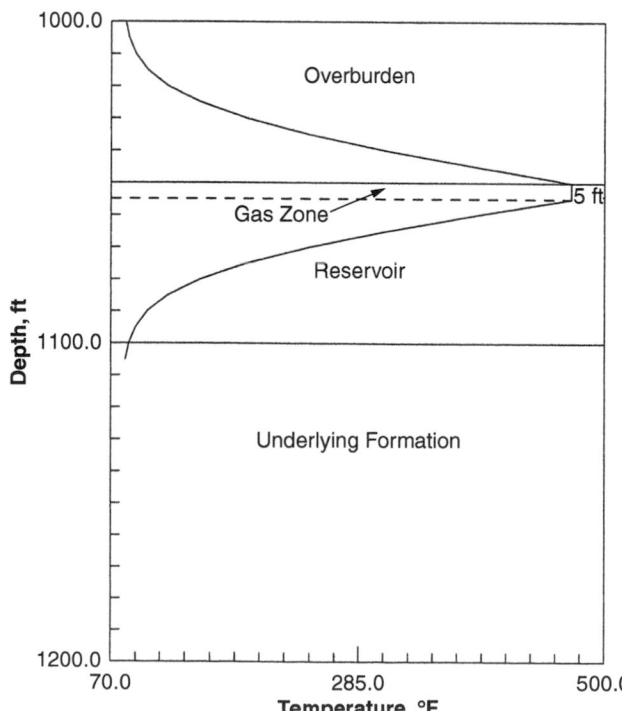

Fig. 8.46—Estimated vertical temperature profile at a radius of 100 ft in Example 8.10 after 180 days of injection.

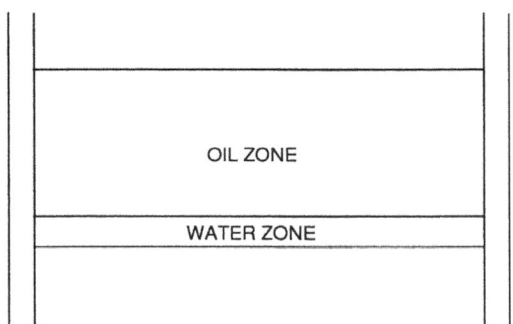

Fig. 8.47—Heavy-oil reservoir with bottomwater zone.

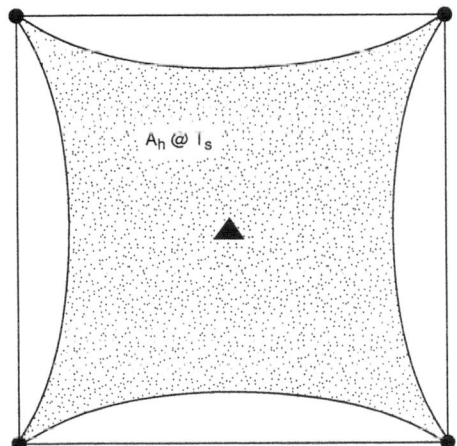

Fig. 8.49—Breakthrough of heated area into a production well, homogeneous reservoir.

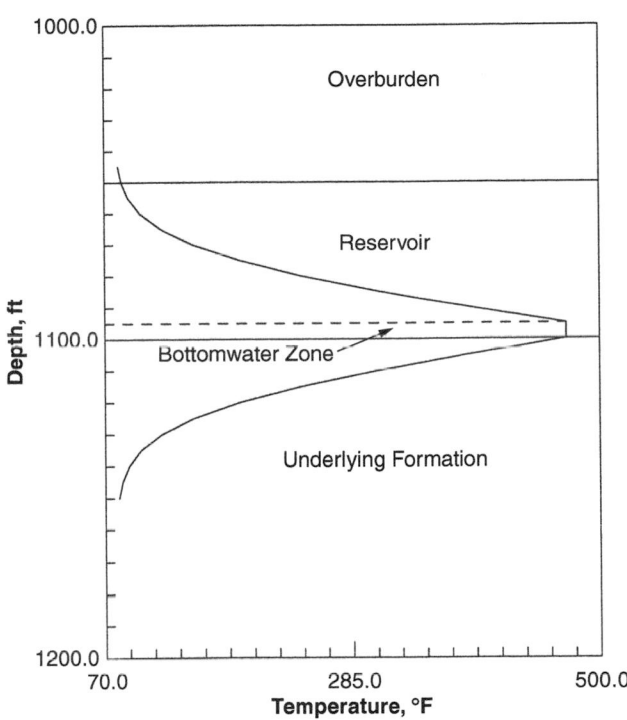

Fig. 8.48—Estimated vertical temperature profile at a radius of 100 ft after 180 days of injection into the bottomwater zone.

produced fluids. Models based on heat injection, such as that of Marx and Langenheim, assume all injected energy is distributed between heat losses and that required for expansion of the heated zones. Removal of energy at production wells is not accounted for explicitly.

The effect of energy removal at production wells on the growth of the heated area can be estimated by use of an approach proposed by Ramey (1959) to account for reservoir heating under variable heat-injection rate. The Marx and Langenheim model assumes that the heat-injection rate, $\dot{m}_s H_s$, is constant. Ramey (1959) extended this model to variable heat-injection rates. If the heat-injection schedule can be represented by a series of step changes,

$$\dot{m}_s H_s = \dot{m}_{s1} H_{s1} \text{ for } 0 \leq t < t_1$$
$$= \dot{m}_{s2} H_{s2} \text{ for } t_1 \leq t < t_2$$
$$= \dot{m}_{sn} H_{sn} \text{ for } t_n \leq t_{n-1} < t.$$

then, $A_h(t) = B_n G(t_D) + \sum_{m-1}^{n-1} \left(B_m - B_{m+1} \right) G\left(t_{D_m} \right),$ . . . .(8.136)

where $B_n = \dot{m}_{s_n} H_{s_n} M_R h / 4 \left(T_s - T_R \right) \alpha M_s^2,$ . . . .(8.137)

where $\dot{m}_{s_n}$ = steam injection rate during time interval n, lbm/D, and H_{s_n} = energy content at the injected steam relative to T_r, Btu/lbm. Example 8.11 illustrates the application of the Ramey extension to estimate the effect of fluid withdrawal at the production well on the heated area.

The variable-rate model can be used to account for heat removal at the production wells. To do this, it is necessary to assume values for the temperature history, fluid production rate, and fluid distribution at the production wells, as illustrated in Example 8.11.

Example 8.11—Removal of Energy Through Production Wells. Consider the reservoir heating problem posed in Example 8.9, where steam was injected into the upper 5 ft of a 50-ft oil-saturated zone at an injection rate of 500 BWPD. Fluid and rock properties are given in Example 8.10. Suppose that hot fluids broke into the production well when the heated area reached

84,559 ft² (1.94 acres). For the purpose of this example, assume the produced fluid is 100% water at T_s. Other temperatures and produced-fluid conditions could be assumed. The fluid withdrawal rate is 200 BWPD (CWE).

From Example 8.4, H_w = 452.9 Btu/lbm at T_s (470.9°F) and H_w = 48 Btu/lbm at 80°F. Thus, $C_w(T - T_r) = H_{w_T} - H_{wr}$ and

$$\dot{Q}_p = q_w \rho_w \left(H_{w_T} - H_{w_{Tr}} \right)$$

$$= 28.34 \times 10^6 \text{ Btu/D or } 1,181,012 \text{ Btu/hr.}$$

Recall from Example 8.10 that $\dot{m}_s H_s$ = 7.125 × 10⁶ Btu/hr and A = 86,922 ft² (2.0 acres) at t = 180 days. The heat-injection schedule corresponding to removal of 548,188 Btu/hr from the production wells beginning at t = 180 days is as follows:

$$\dot{m}_s H_s = 7.336 \times 10^6 \text{ Btu/hr for } 0 \leq t < 180$$

$$= 6.155 \times 10^6 \text{ Btu/hr for } t \geq 180.$$

From Eq. 8.137,

$$B_n = m_{s_n} H_{s_n} M_R h / 4 (T_s - T_R) \alpha M_s^2.$$

From Example 8.10, B_1 = 16,648 ft² and B_2 = 16,648(6.155/7.336) = 13,968 ft². At t_1 = 180 days, t_{D1} = 30.1, and $G(t_{D1})$ = 5.222. The expression for the growth of the heated area becomes

$$A_h(t) = B_2 G(t_D) + (B_1 - B_2) G(t_{D1})$$

$$= 13,968 G(t_D) + (16,648 - 13,968)(5.222)$$

$$= 13,968 G(t_D) + 13,996.$$

Thus, at t_D = 70.0 (t = 70.0/0.163 = 429.4 days), $G(t_D)$ = 8.51 and

$$A_h(t) = 13,968(8.51) + 13,996$$

$$= 132,863 \text{ ft}^2$$

$$= 3.05 \text{ acres.}$$

The effect of hot fluid removal can be examined for the same case by using the above analysis. If no heated fluids are produced,

$$A_h(t) = 16,648 G(t_D)$$

$$= 16,648(8.51)$$

$$= 141,674 \text{ ft}^2$$

$$= 3.05 \text{ acres.}$$

Removal of heat through production wells reduces the growth of the heated region or increases the time required to heat the same area. In this example, the effect was small because produced fluids were removed as saturated liquid at T_s. Larger effects would occur if produced fluids contained steam.

Maintenance of a Heated Region of Constant Areal Extent. Heat losses to over- and underlying formations at a particular location decrease with time, as is evident from Eq. 8.55. Thus, one would expect that the heat-injection rate would decrease with time if the heated area ceases to expand. This situation might occur if the steam zone spread initially over the upper part of a reservoir as described in Example 8.10, where reservoir heating was achieved through the upper 5 ft of a reservoir that was 50 ft thick.

Eq. 8.127 gives the heat-loss rate by conduction to under- and overlying formations. When the heated region is to be limited to that existing at an arbitrary time, t_{D1}, the heat-loss rate is given by

$$\dot{Q}_\ell = \dot{m}_s H_s \int_o^{t_{D1}} \frac{e^{\lambda_D} \text{ erfc} \sqrt{\lambda_D}}{\sqrt{\pi (t_D - \lambda_D)}} d\lambda_D. \quad \dots\dots\dots\dots\dots\dots\dots\dots\dots\dots\dots\dots\dots \quad (8.138)$$

This heat-loss rate can be supplied by reducing the injection rate for $t_D > t_{D1}$. Let $\dot{m}_1(t_D)$ represent the steam injection rate for $t_D > t_{D1}$. Then,

$$\dot{m}_1 H_s = \dot{Q}_\ell \quad \dotfill \quad (8.139)$$

$$\text{or } \dot{m}_1(t_D)/m_s = G_s(t_D, t_{D1}). \quad \dotfill \quad (8.140)$$

Eq. 8.140 can be adjusted easily for removal of heat by produced fluids when the rate of heat removal is known. **Table 8.27** contains selected values of $G_s(t_D, t_{D1})$. A Fortran program, GS(TD,TD1) is included in Appendix F and may be used to compute $G_s(t_D, t_{D1})$.

Example 8.12 Heat Injection Rate To Maintain Constant Heated Area. Consider the reservoir in Example 8.10, where steam was injected at a rate into a reservoir that was 50 ft thick. For Example 8.12, steam will be injected at an initial rate of 1,000 BWPD. Assume that it is desired to limit the heated area to 5.0 acres by reducing the heat-injection rate after the heated region covers 5.0 acres. Properties of the fluids and rocks were given in Example 8.4.

Solution. From Example 8.10, $t_D = 0.163t$, where t is given in days.

$$A_h = \left[(14,583 \text{lbm/hr})(1006 \text{Btu/lbm})(33.12 \text{Btu/ft}^3\text{-}°\text{F})(5\text{ft}) \right.$$
$$\left. \div 4(470.9°\text{F} - 80°\text{F})(0.0482 \text{ft}^2/\text{hr})(31.12 \text{Btu/ft}^3 - °\text{F})^2 \right]$$
$$\times G(t_D)$$
$$= 33,285 G(t_D) \text{ft}^2.$$

When $A_h = 5$ acres,

$$G(t_D) = \left[(5\text{ acres})(43,560 \text{ ft}^2/\text{acres}) \right] / 33,285 \text{ ft}^2$$
$$= 6.543.$$

From Table 8.7, $G(t_D) = 6.226$ at $t_D = 40$ and $G(t_D) = 6.657$ at $t_D = 45$. From linear interpolation,

$$t_D \approx 40 + \left[(6.543 - 6.626)/(6.69 - 6.226) \right](5)$$
$$= 43.71.$$

Thus, the heated zone covers 5 acres after $t = 43.7/0.163$ or 268 days of injection.

It is desired to keep the heated area constant at 5 acres. If all of the injected heat is used to supply heat losses, and none is lost by removal of produced fluids, then Eq. 8.140 gives the fraction of the initial steam injection rate as a function of t_D and t_{D1}.

In this example, $t_{D1} = 43.71$. Values of $G_s(t_D, t_{D1})$ are obtained from Table 8.27 or computed with the FORTRAN program GS(TD,TD1) in Appendix F. **Table 8.28** gives estimated injection rates. The steam injection rates in the fourth column decrease with time because the heat losses decrease with time.

This analysis can be adjusted to take into account withdrawal of heated fluids from the hot region. Let f_{cp} = fraction of the injected energy that is removed with the produced fluids. Then,

$$\dot{m}_1 H_s (1 - f_{cp}) = \dot{Q}_\ell$$
$$= \dot{m}_1 H_s G_s(t_D, t_{D1}); \quad \dotfill \quad (8.141)$$

$$\text{then } \dot{m}_1/m_s = G_s(t_D, t_{D_1})/(1 - f_{cp}). \quad \dotfill \quad (8.142)$$

This example could also be worked in terms of the steam zone. Suppose that it is desired to reduce the injection rate after the steam zone covers 5 acres. The solution is developed by first finding the critical time for the beginning of the hot-water zone, as illustrated in Example 8.8. Then it is necessary to determine how long to inject at the initial injection rate to extend the steam zone to 5 acres.

From Examples 8.4 and 8.8, $L_v = 751.4$ Btu/lbm, $H_s = 1006$ Btu/lbm, $f_{sd} = 0.8$, $t_{cD} \approx 1.396$, and

$$f_{h,v} = 0.8(751.4)/1006$$
$$= 0.598.$$

To follow the development of the steam zone, Eq. 8.130 is solved to find t_{Ds} for each t_D so that $G_s(t_D, t_{Ds}) = f_{h,v}$. Table 8.25 presents values of t_D, t_{Ds} determined by running program TDS from Appendix F.

The steam zone covers 5 acres when $t_{Ds} = 43.71$. Thus, t_D must be approximately 58.9 before the injection rate is reduced if it is desired to maintain the heated region at T_s by condensing steam. To compute the injection rate for $t_D > 58.9$, recall that the heat loss from the region contacted by steam at T_s is given by

$$\dot{Q}_\ell = \dot{m}_s H_s G_s\left(t_D, t_{D1}\right). \dots\dots\dots\dots\dots\dots\dots\dots\dots\dots\dots\dots\dots\dots\dots\dots\dots \text{(8.143)}$$

The rate at which energy is supplied by condensing steam is given by

$$\dot{Q}_{\text{cond}} = \dot{m}_1 f_{sd} L_{vdh}. \dots\dots\dots\dots\dots\dots\dots\dots\dots\dots\dots\dots\dots\dots\dots\dots\dots\dots\dots \text{(8.144)}$$

t_D	t_{D1}									
	0.001	0.002	0.003	0.004	0.005	0.006	0.007	0.008	0.009	0.010
0.001	0.0347									
0.002	0.0144	0.0485								
0.003	0.0111	0.0252	0.0589							
0.004	0.0093	0.0202	0.0342	0.0675						
0.005	0.0082	0.0174	0.0281	0.0420	0.0750					
0.006	0.0074	0.0155	0.0245	0.0352	0.0489	0.0817				
0.007	0.0068	0.0141	0.0221	0.0311	0.0416	0.0552	0.0878			
0.008	0.0064	0.0131	0.0203	0.0282	0.0370	0.0475	0.0610	0.0934		
0.009	0.0060	0.0122	0.0188	0.0260	0.0338	0.0426	0.0530	0.0664	0.0987	
0.010	0.0057	0.0115	0.0177	0.0242	0.0313	0.0391	0.0478	0.0581	0.0715	0.1035
0.020	0.0039	0.0079	0.0120	0.0161	0.0203	0.0246	0.0290	0.0336	0.0384	0.0433
0.030	0.0032	0.0064	0.0096	0.0129	0.0162	0.0195	0.0228	0.0263	0.0297	0.0333
0.040	0.0028	0.0055	0.0083	0.0110	0.0138	0.0166	0.0195	0.0223	0.0252	0.0281
0.050	0.0025	0.0049	0.0074	0.0098	0.0123	0.0148	0.0172	0.0197	0.0222	0.0247
0.060	0.0023	0.0045	0.0067	0.0089	0.0112	0.0134	0.0156	0.0179	0.0201	0.0224
0.070	0.0021	0.0042	0.0062	0.0083	0.0103	0.0124	0.0144	0.0165	0.0185	0.0206
0.080	0.0020	0.0039	0.0058	0.0077	0.0096	0.0115	0.0134	0.0153	0.0172	0.0192
0.090	0.0018	0.0037	0.0055	0.0073	0.0091	0.0108	0.0126	0.0144	0.0162	0.0180
0.100	0.0017	0.0035	0.0052	0.0069	0.0086	0.0103	0.0120	0.0136	0.0153	0.0170
0.200	0.0012	0.0024	0.0036	0.0048	0.0060	0.0072	0.0084	0.0095	0.0107	0.0119
0.300	0.0010	0.0020	0.0030	0.0039	0.0049	0.0059	0.0068	0.0078	0.0087	0.0097
0.400	0.0009	0.0017	0.0026	0.0034	0.0042	0.0051	0.0059	0.0067	0.0075	0.0083
0.500	0.0008	0.0015	0.0023	0.0031	0.0038	0.0045	0.0053	0.0060	0.0067	0.0075
0.600	0.0007	0.0014	0.0021	0.0028	0.0035	0.0041	0.0048	0.0055	0.0061	0.0068
0.700	0.0007	0.0013	0.0019	0.0026	0.0032	0.0038	0.0045	0.0051	0.0057	0.0063
0.800	0.0006	0.0012	0.0018	0.0024	0.0030	0.0036	0.0042	0.0047	0.0053	0.0059
0.900	0.0006	0.0012	0.0017	0.0023	0.0028	0.0034	0.0039	0.0045	0.0050	0.0055

Table 8.27—$G_s(t_D, t_{D1})$.

					t_{D1}					
	0.001	0.002	0.003	0.004	0.005	0.006	0.007	0.008	0.009	0.010
t_D										
1.000	0.0006	0.0011	0.0016	0.0022	0.0027	0.0032	0.0037	0.0042	0.0047	0.0053
2.000	0.0004	0.0008	0.0011	0.0015	0.0019	0.0023	0.0026	0.0030	0.0034	0.0037
3.000	0.0003	0.0006	0.0009	0.0012	0.0015	0.0018	0.0021	0.0024	0.0027	0.0030
4.000	0.0003	0.0005	0.0008	0.0011	0.0013	0.0016	0.0019	0.0021	0.0024	0.0026
5.000	0.0002	0.0005	0.0007	0.0010	0.0012	0.0014	0.0017	0.0019	0.0021	0.0023
6.000	0.0002	0.0004	0.0007	0.0009	0.0011	0.0013	0.0015	0.0017	0.0019	0.0021
7.000	0.0002	0.0004	0.0006	0.0008	0.0010	0.0012	0.0014	0.0016	0.0018	0.0020
8.000	0.0002	0.0004	0.0006	0.0008	0.0009	0.0011	0.0013	0.0015	0.0017	0.0019
9.000	0.0002	0.0004	0.0005	0.0007	0.0009	0.0011	0.0012	0.0014	0.0016	0.0017
10.000	0.0002	0.0003	0.0005	0.0007	0.0008	0.0010	0.0012	0.0013	0.0015	0.0017
20.000	0.0001	0.0002	0.0004	0.0005	0.0006	0.0007	0.0008	0.0009	0.0011	0.0012
30.000	0.0001	0.0002	0.0003	0.0004	0.0005	0.0006	0.0007	0.0008	0.0009	0.0010
40.000	0.0001	0.0002	0.0003	0.0003	0.0004	0.0005	0.0006	0.0007	0.0007	0.0008
50.000	0.0001	0.0002	0.0002	0.0003	0.0004	0.0005	0.0005	0.0006	0.0007	0.0007
60.000	0.0001	0.0001	0.0002	0.0003	0.0003	0.0004	0.0005	0.0005	0.0006	0.0007
70.000	0.0001	0.0001	0.0002	0.0003	0.0003	0.0004	0.0004	0.0005	0.0006	0.0006
80.000	0.0001	0.0001	0.0002	0.0002	0.0003	0.0004	0.0004	0.0005	0.0005	0.0006

Table 8.27—$G_s(t_D, t_{D1})$ (continued).

					t_{D1}				
	0.020	0.030	0.040	0.050	0.060	0.070	0.080	0.090	0.100
t_D									
0.020	0.1415								
0.030	0.0742	0.1689							
0.040	0.0595	0.0989	0.1910						
0.050	0.0512	0.0815	0.1198	0.2096					
0.060	0.0457	0.0712	0.1006	0.1379	0.2259				
0.070	0.0417	0.0641	0.0889	0.1176	0.1541	0.2404			
0.080	0.0385	0.0589	0.0807	0.1049	0.1329	0.1687	0.2534		
0.090	0.0360	0.0547	0.0745	0.0958	0.1194	0.1469	0.1821	0.2654	
0.100	0.0340	0.0514	0.0695	0.0888	0.1096	0.1328	0.1598	0.1944	0.2764
0.200	0.0234	0.0348	0.0461	0.0576	0.0692	0.0810	0.0930	0.1053	0.1180
0.300	0.0189	0.0280	0.0370	0.0459	0.0548	0.0638	0.0727	0.0817	0.0908
0.400	0.0163	0.0241	0.0317	0.0393	0.0468	0.0543	0.0618	0.0692	0.0766
0.500	0.0146	0.0215	0.0282	0.0349	0.0415	0.0481	0.0546	0.0611	0.0676
0.600	0.0133	0.0195	0.0257	0.0317	0.0377	0.0436	0.0495	0.0553	0.0611
0.700	0.0123	0.0181	0.0237	0.0293	0.0348	0.0402	0.0456	0.0509	0.0562
0.800	0.0115	0.0169	0.0222	0.0273	0.0325	0.0375	0.0425	0.0474	0.0523
0.900	0.0108	0.0159	0.0209	0.0257	0.0305	0.0353	0.0400	0.0446	0.0492

Table 8.27—$G_s(t_D, t_{D1})$ (continued).

					t_{D1}				
	0.020	0.030	0.040	0.050	0.060	0.070	0.080	0.090	0.100
t_D									
1.000	0.0102	0.0151	0.0198	0.0244	0.0289	0.0334	0.0378	0.0422	0.0465
2.000	0.0072	0.0106	0.0139	0.0171	0.0203	0.0234	0.0265	0.0295	0.0325
3.000	0.0059	0.0087	0.0113	0.0140	0.0165	0.0191	0.0215	0.0240	0.0264
4.000	0.0051	0.0075	0.0098	0.0121	0.0143	0.0165	0.0186	0.0207	0.0228
5.000	0.0046	0.0067	0.0088	0.0108	0.0128	0.0147	0.0166	0.0185	0.0204
6.000	0.0042	0.0061	0.0080	0.0099	0.0117	0.0134	0.0152	0.0169	0.0186
7.000	0.0039	0.0057	0.0074	0.0091	0.0108	0.0124	0.0141	0.0156	0.0172
8.000	0.0036	0.0053	0.0069	0.0085	0.0101	0.0116	0.0131	0.0146	0.0161
9.000	0.0034	0.0050	0.0065	0.0080	0.0095	0.0110	0.0124	0.0138	0.0152
10.000	0.0032	0.0047	0.0062	0.0076	0.0090	0.0104	0.0117	0.0131	0.0144
20.000	0.0023	0.0033	0.0044	0.0054	0.0064	0.0073	0.0083	0.0092	0.0102
30.000	0.0019	0.0027	0.0036	0.0044	0.0052	0.0060	0.0068	0.0075	0.0083
40.000	0.0016	0.0024	0.0031	0.0038	0.0045	0.0052	0.0059	0.0065	0.0072
50.000	0.0014	0.0021	0.0028	0.0034	0.0040	0.0046	0.0052	0.0058	0.0064
60.000	0.0013	0.0019	0.0025	0.0031	0.0037	0.0042	0.0048	0.0053	0.0059
70.000	0.0012	0.0018	0.0023	0.0029	0.0034	0.0039	0.0044	0.0049	0.0054
80.000	0.0011	0.0017	0.0022	0.0027	0.0032	0.0037	0.0041	0.0046	0.0051

Table 8.27—$G_s(t_D, t_{D1})$ (continued).

					t_{D1}					
	0.200	0.300	0.400	0.500	0.600	0.700	0.800	0.900	1.000	2.000
t_D										
0.200	0.3562									
0.300	0.1910	0.4080								
0.400	0.1536	0.2445	0.4464							
0.500	0.1325	0.2023	0.2867	0.4768						
0.600	0.1183	0.1772	0.2420	0.3214	0.5020					
0.700	0.1080	0.1598	0.2145	0.2754	0.3507	0.5233				
0.800	0.0999	0.1468	0.1950	0.2464	0.3042	0.3761	0.5418			
0.900	0.0935	0.1366	0.1801	0.2254	0.2742	0.3294	0.3984	0.5580		
1.000	0.0881	0.1283	0.1683	0.2093	0.2522	0.2988	0.3517	0.4182	0.5724	
2.000	0.0607	0.0870	0.1121	0.1365	0.1604	0.1840	0.2076	0.2313	0.2552	0.6638
3.000	0.0491	0.0701	0.0899	0.1089	0.1273	0.1452	0.1628	0.1801	0.1973	0.3718
4.000	0.0424	0.0603	0.0772	0.0933	0.1088	0.1238	0.1385	0.1528	0.1668	0.3011
5.000	0.0378	0.0538	0.0687	0.0830	0.0966	0.1098	0.1226	0.1350	0.1472	0.2606
6.000	0.0345	0.0490	0.0625	0.0754	0.0877	0.0996	0.1111	0.1223	0.1332	0.2332
7.000	0.0319	0.0453	0.0578	0.0696	0.0810	0.0919	0.1024	0.1126	0.1226	0.2130
8.000	0.0298	0.0423	0.0539	0.0650	0.0755	0.0857	0.0955	0.1050	0.1142	0.1973
9.000	0.0281	0.0398	0.0508	0.0612	0.0711	0.0806	0.0898	0.0986	0.1073	0.1847

Table 8.27—$G_s(t_D, t_{D1})$ (continued).

					t_{D1}					
	0.200	0.300	0.400	0.500	0.600	0.700	0.800	0.900	1.000	2.000
t_D										
10.000	0.0266	0.0377	0.0481	0.0580	0.0673	0.0763	0.0850	0.0934	0.1015	0.1742
20.000	0.0188	0.0266	0.0339	0.0407	0.0473	0.0535	0.0595	0.0653	0.0709	0.1202
30.000	0.0153	0.0217	0.0276	0.0332	0.0385	0.0436	0.0484	0.0532	0.0577	0.0974
40.000	0.0133	0.0188	0.0239	0.0287	0.0333	0.0377	0.0419	0.0460	0.0499	0.0841
50.000	0.0119	0.0168	0.0214	0.0257	0.0298	0.0337	0.0374	0.0411	0.0446	0.0750
60.000	0.0108	0.0153	0.0195	0.0234	0.0272	0.0307	0.0342	0.0375	0.0406	0.0684
70.000	0.0100	0.0142	0.0180	0.0217	0.0251	0.0284	0.0316	0.0347	0.0376	0.0632
80.000	0.0094	0.0133	0.0169	0.0203	0.0235	0.0266	0.0296	0.0324	0.0352	0.0591

Table 8.27—$G_s(t_D,t_{D1})$ (continued).

					t_{D1}					
	3.000	4.000	5.000	6.000	7.000	8.000	9.000	10.000	20.000	30.000
t_D										
3.000	0.7127									
4.000	0.4453	0.7446								
5.000	0.3714	0.4974	0.7677							
6.000	0.3267	0.4235	0.5369	0.7853						
7.000	0.2954	0.3771	0.4642	0.5682	0.7995					
8.000	0.2718	0.3438	0.4173	0.4972	0.5940	0.8111				
9.000	0.2532	0.3183	0.3831	0.4505	0.5248	0.6155	0.8209			
10.000	0.2380	0.2979	0.3564	0.4158	0.4785	0.5482	0.6340	0.8293		
20.000	0.1620	0.1996	0.2344	0.2675	0.2993	0.3304	0.3610	0.3914	0.8762	
30.000	0.1308	0.1604	0.1876	0.2130	0.2371	0.2603	0.2827	0.3044	0.5145	0.8977
40.000	0.1126	0.1378	0.1609	0.1823	0.2025	0.2218	0.2403	0.2581	0.4200	0.5846
50.000	0.1004	0.1227	0.1431	0.1620	0.1797	0.1965	0.2127	0.2282	0.3648	0.4914
60.000	0.0914	0.1117	0.1302	0.1472	0.1632	0.1784	0.1928	0.2067	0.3272	0.4340
70.000	0.0845	0.1032	0.1202	0.1359	0.1505	0.1644	0.1777	0.1903	0.2993	0.3935
80.000	0.0790	0.0964	0.1122	0.1268	0.1404	0.1533	0.1656	0.1774	0.2776	0.3628

			t_{D1}		
	40.000	50.000	60.000	70.000	80.000
t_D					
40.000	0.9106				
50.000	0.6319	0.9194			
60.000	0.5420	0.6669	0.9260		
70.000	0.4846	0.5806	0.6944	0.9310	
80.000	0.4431	0.5240	0.6115	0.7169	0.9351

Table 8.27—$G_s(t_D,t_{D1})$ (continued).

Because $\dot{Q}_\ell = \dot{Q}_{cond}$ to maintain a stable heated volume at T_s,

$$\dot{m}_1/\dot{m}_s = G_s\left(t_D,t_{D1}\right)\big/f_{h,v}. \dots\dots\dots\dots\dots\dots\dots\dots\dots\dots\dots\dots\dots\dots (8.145)$$

At $t_D = 80$, $G_s(t_D,t_{D1}) = 0.473$. Thus,

$$\dot{m}_1/\dot{m}_s = 0.499/0.598$$
$$= 0.791.$$

t_D	t (days)	$G_s(t_D, t_{D1})$	Steam Injection Rate (B/D)
43.71	268.1	1.000	1000
50	306.7	0.697	697
58.92	361.5	0.598	598
60	368.1	0.585	585
70	429.4	0.519	519
80	490.8	0.473	473
90	552.1	0.437	437
100	613.5	0.409	409

Table 8.28—Injection rate to maintain heated area at 5 acres in Example 8.12.

The injection rate can be reduced to (1,000 B/D)(0.791), or 791 B/D, and still provide sufficient latent heat for heat losses to over- and underlying formations. **Table 8.29** summarizes results of these computations. If maintenance of an expanding steam zone is desired, reducing the injection rate as much as indicated in Table 8.29 will not be possible because all injected energy was used to supply heat losses.

When the steam zone is maintained at T_s by condensation, sensible heat is removed from the heated volume at a rate of $m_1(H_t - f_{sd} L_v)$.

8.4.2 Reservoir Heating by Gravity Override. The reservoir heating models described earlier in this chapter assume that the injected fluid is uniformly distributed through any vertical cross section where fluids flow. This is a convenient assumption for the development of simple heating models but is not the case in practice. The densities of steam and water at the same temperature differ by factors of approximately 10 to 100 in the range of temperature and pressure used in most steamdrive projects. When there are no impermeable barriers to flow that extend between wells, the injected steam will segregate by gravity, with saturated vapor rising to the top of the formation and water flowing in the bottom of the heated region. Reservoir heating occurs as the steam zone spreads horizontally across the top of the reservoir and expands vertically. This process is called gravity override, gravity segregation, or gravity overlay.

Fig. 8.50 compares a gravity-override (or bypass) model for reservoir heating with the frontal-advance model represented by the Marx and Langenheim model discussed earlier in this section (Prats 1982). The bypass, or gravity-override, model is characterized by relatively rapid segregation of the steam in the top of the reservoir, irrespective of where the steam is injected in the formation. Thus, when steam is injected into a viscous-oil reservoir, the vapor phase will migrate to the top of the reservoir, driven by gravity forces that promote the separation of vapor and liquid phases. The rate of migration is controlled by the viscosity of the crude oil flowing countercurrent to the rising steam and the vertical permeability in the strata (Doscher and Ghassemi 1983). Siltstone and shale intervals that are continuous over large distances can be effective barriers to fluid flow and may restrict vertical migration and thus the rate that steam rises within the reservoir.

Ample evidence of gravity override exists from laboratory studies of scaled model experiments (Nasr and Pierce 1993; Baker 1973; van Lookeren 1983), observation-well data (Blevins et al. 1969; Blevins and Billingsley 1975), and numerical simulation of steam injection with multidimensional reservoir simulators (Shutler 1970; Coats 1978). Blevins et al. (1969) first documented gravity override in the Inglewood field. **Fig. 8.51** shows the estimated distribution of steam and hot-water regions in a west/east cross section through the injection well. Steam was injected into the formation through an injection interval that was approximately 10 ft below the bottom of the sand. Steam rose rapidly after leaving the injection interval and reached the top of the reservoir within a distance of approximately 40 ft from the injection well. Other evidence of gravity segregation was reported in the Tia Juana Steamdrive Project (de Haan and Schenk 1969), the Ten Pattern Test in the Kern River field (Blevins and Billingsley 1975), and in the Schoonebeek field (van Lookeren 1983).

Fig. 8.52 shows the temperature profile in Well T6-3 in Chevron's Ten-Pattern steam displacement pilot (Blevins and Billingsley 1975; Neuman 1985) in the Kern River field, California, USA. This observation well is located a distance of

t_D	t_{Ds}	$G_s(t_D, t_{D1})$	$\dot{m}_1/\dot{m}_s$	q (B/D)
2.0	1.968			1,000
4.0	3.744			1,000
6.0	5.414			1,000
8.0	7.028			1,000
10.0	8.605			1,000
20.0	16.176			1,000
40.0	38.572			1,000
58.92	43.71	0.598	1.0	1,000
60.0	43.71	0.585	0.978	978
70.0	43.71	0.519	0.868	868
80.0	43.71	0.473	0.791	791
90.0	43.71	0.437	0.731	731
100.0	43.71	0.409	0.684	684

*$f_{h,v}$ = 0.598, t_{D1} = 0.598, t_{cD} = 1.207.

Table 8.29—Supporting calculations* to maintain steam zone at 5 acres, Example 8.12.

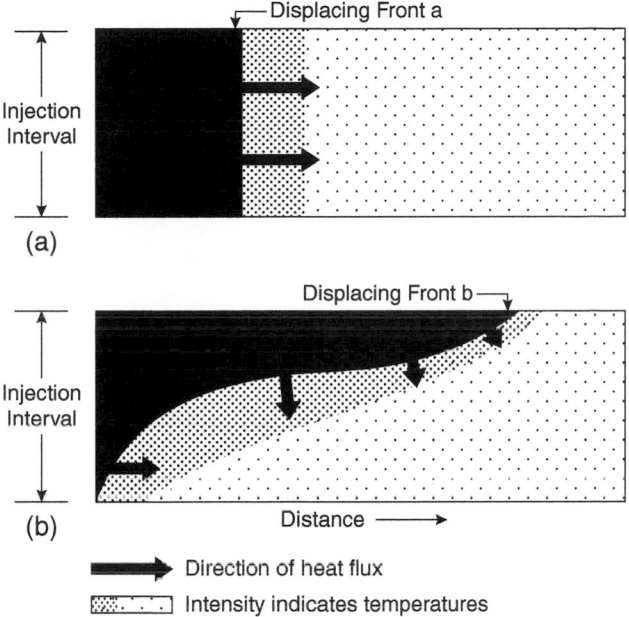

Fig. 8.50—Idealized temperature distribution in (a) frontal-advance model and (b) bypass or override model (Prats 1982).

150 ft from the injection wells. Steam was injected into a 30-ft interval at the bottom of the oil sand in the injection well. After 718 days of continuous injection, a steam zone approximately 30 ft thick was present at the top of the formation in most wells, as depicted in Fig. 8.52. After the steam zone reaches the top of the formation, the reservoir heating process is dominated by conduction to the cooler portions of the reservoir and vertical expansion of the steam zone as heated oil is displaced from the reservoir. **Fig. 8.53** shows calculated and observed steam-zone thicknesses (Neuman 1975) for the 10-pattern steamflood deduced from temperature profiles for several wells in this project. The steam zone is seen to expand vertically with time.

These data show that fluid-flow patterns dominated by gravity segregation affect the shape and growth of the steam zone in many reservoirs. In some reservoirs, flow barriers may exist even though they are not detectable from well logs. Although gravity segregation is evident from the temperature profile in **Fig. 8.54** (Blevins and Billingsley 1975), the initial arrival of the steam zone does not show gravity override, even though this well was 150 ft from the injection well. The delay in onset of gravity segregation is attributed to siltstone lenses that are intermittent between the injection and observation wells. Thus, gravity segregation will occur in steam displacement processes if there is sufficient vertical permeability within the reservoir and injection rates are relatively low.

Gravity override is well-documented in physical scaled models of steam displacement (Doscher and Ghassemi 1983; Baker 1973; Stegemeier e al. 1980). **Fig. 8.55** (Baker 1973) shows vertical temperature distributions in a radial laboratory model during steam injection at two injection rates. The scaling factor for injection rates is 1 lbm/min = 16.4 bbl/(D-ft). In the absence of gravity segregation, the temperature distribution should be constant at T_s across the injection interval. In Fig. 8.55, steam

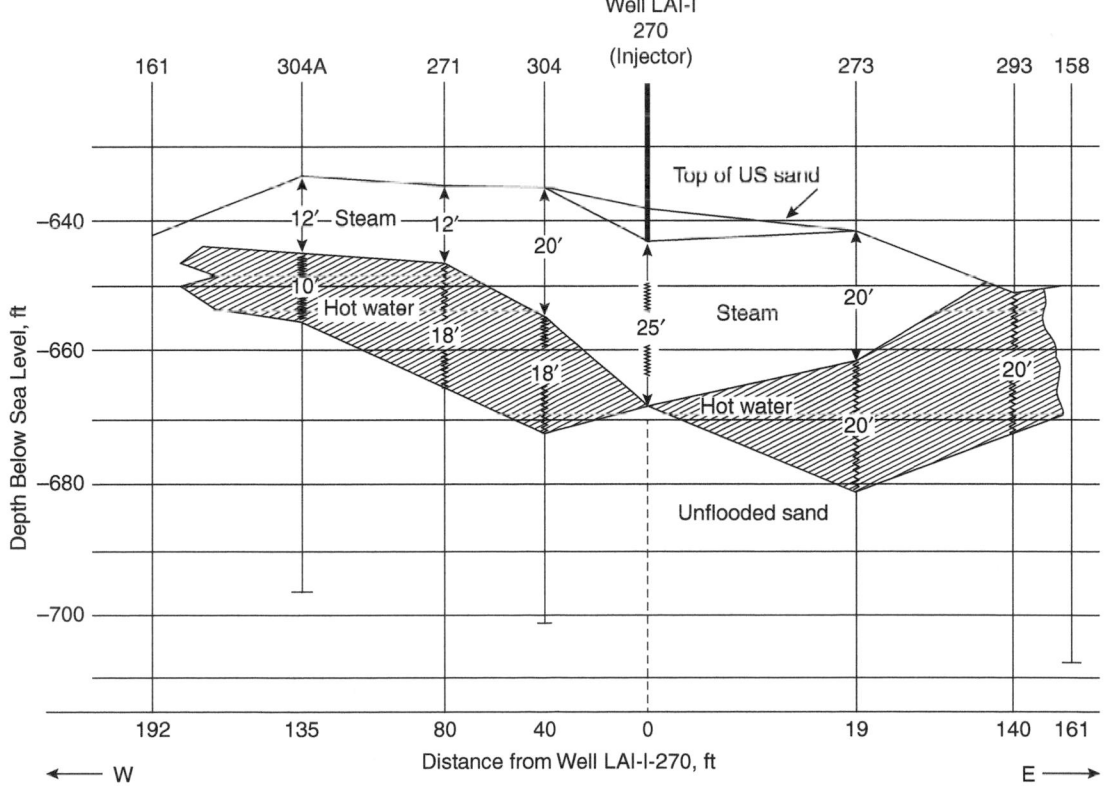

Fig. 8.51—Distribution of steam and hot-water regions in Inglewood steamdrive (Blevins et al. 1969).

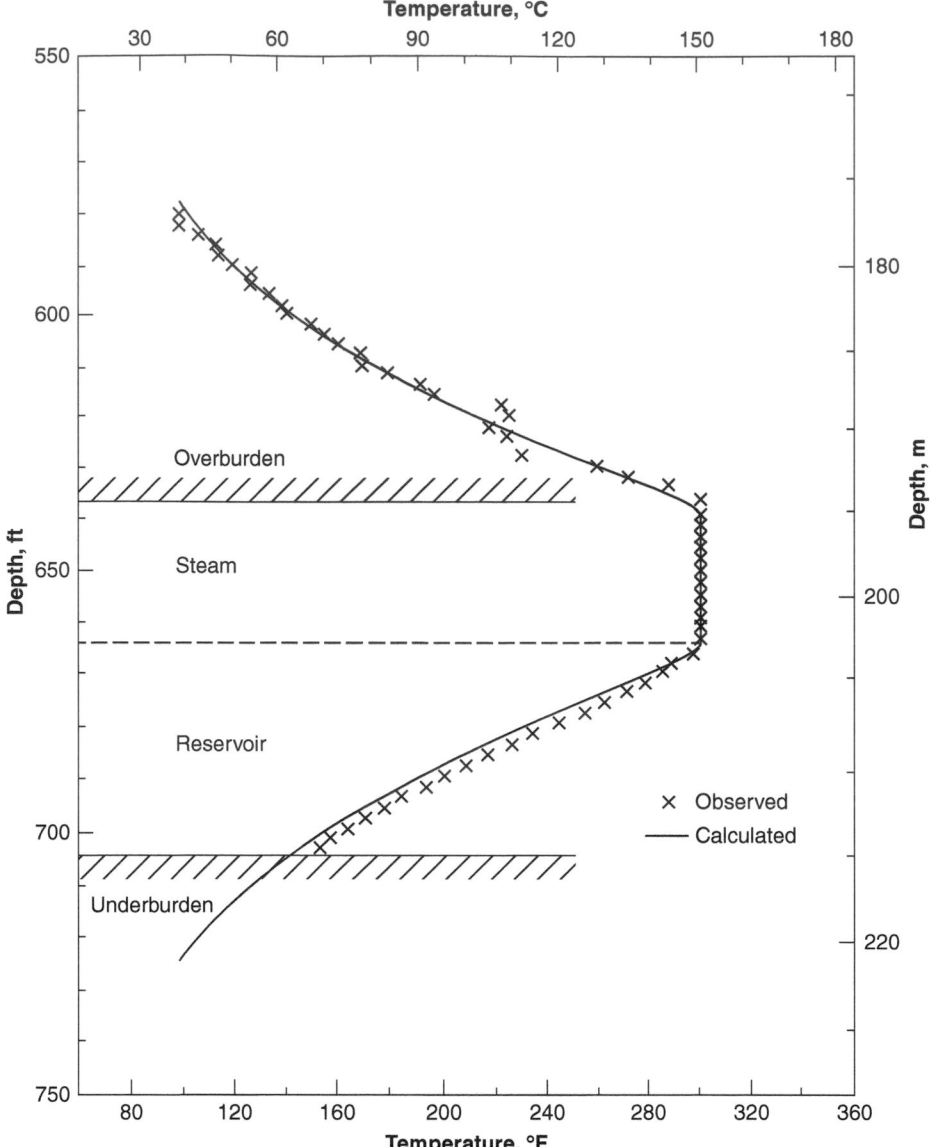

Fig. 8.52—Observed and calculated temperature profiles above and below steam zone, Well T6-3 on September 1971 (*f* = 718 days) (Neuman 1985).

flows at the top of the sand, with hot water segregating at the bottom. There is more segregation at lower injection rates. This model was sufficiently thin that the steam zone was present throughout the vertical cross section except near the heated front where the condensation zone was tilted.

Gravity segregation is a complex process. When gravity segregation is a critical factor, numerical simulation is necessary to predict reservoir heating and oil-displacement performance. Numerical simulators are covered in Coats (1978) and are beyond the scope of this text.

Heat-balance models are still useful for screening purposes as well as for examining heating mechanisms and will be discussed in this section. Some simplifying assumptions must be made to develop reservoir heating models with gravity segregation that are easy to use. Two such reservoir heating models have been developed (Doscher and Ghassemi 1983; Neuman 1975, 1985). Both models assume that gravity segregation occurs rapidly so that reservoir heating occurs primarily when the steam zone expands vertically. Both models assume a sharp interface exists between the heated zone at steam temperature and the unswept reservoir, as depicted in **Fig. 8.56.** The interface moves at a velocity, v, and a constant temperature T_s.

An energy balance is used to develop relationships between the heated area, heated volume, and the heat-injection rate. The two models differ in that the Neuman model was developed from an expression for a variable-steam-interface model. The Doscher-Ghassemi model assumes that the steam interface expands vertically at a constant average velocity.

Neuman Model. The Neuman model (1975, 1985) was developed from an energy and mass balance over a thin slice of reservoir in the vertical cross section shown in **Fig. 8.57.** In the steam zone, condensation of steam supplies the heat loss to the

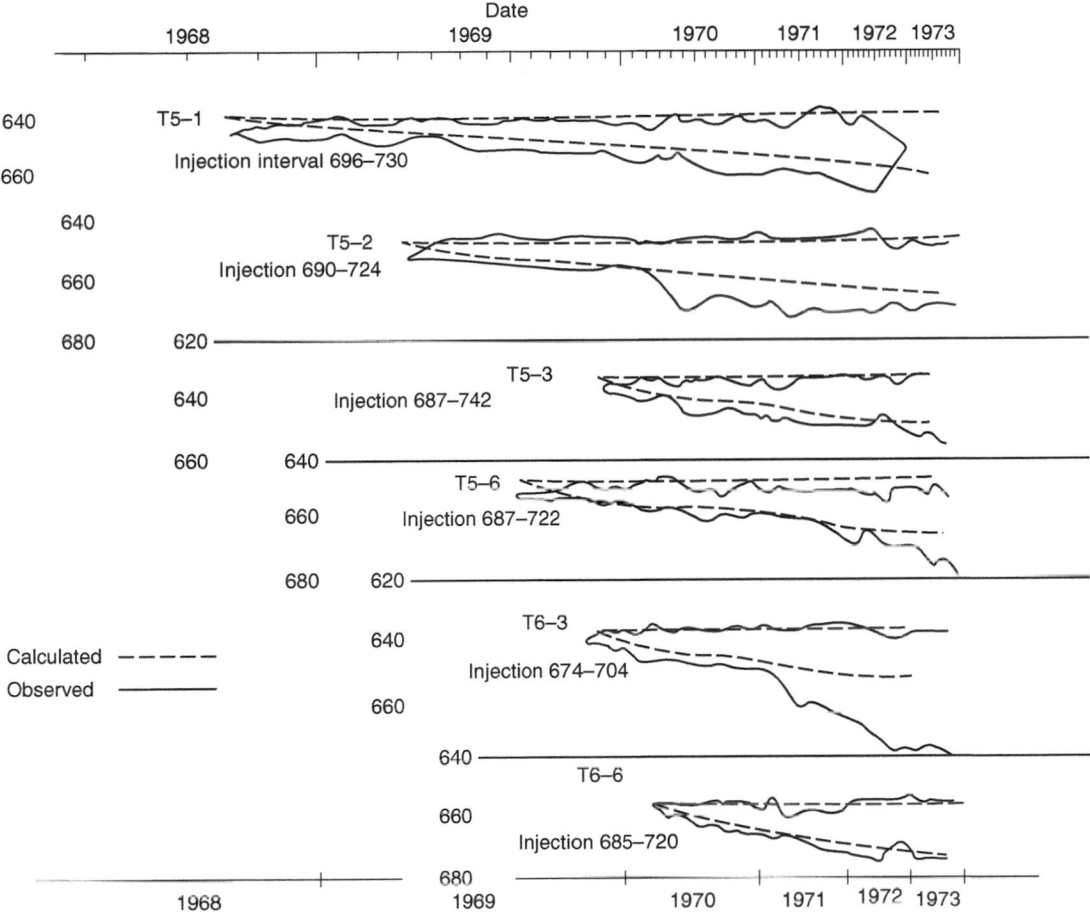

Fig. 8.53—Calculated and observed steam-zone thickness decline from temperature profiles, 10-pattern steamflood, Kern River field, California (Neuman 1975).

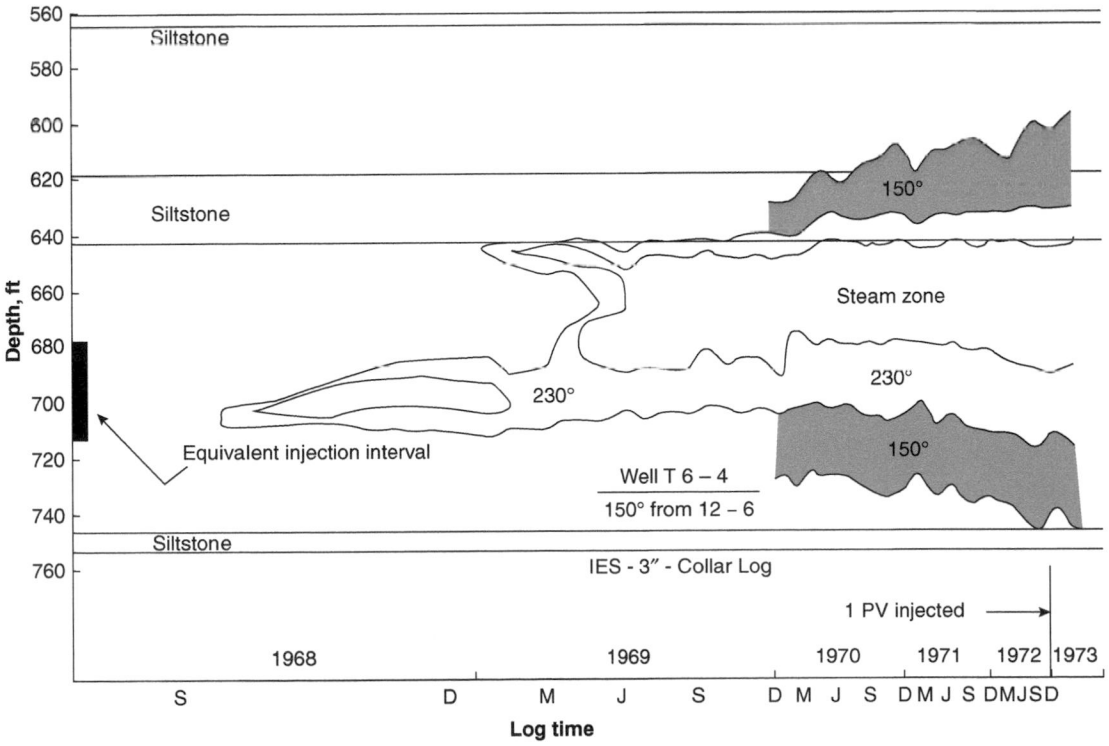

Fig. 8.54—Vertical heat-zone growth or log time, Well T6-4, 10-pattern steamflood, Kern River field, California (Neuman 1975).

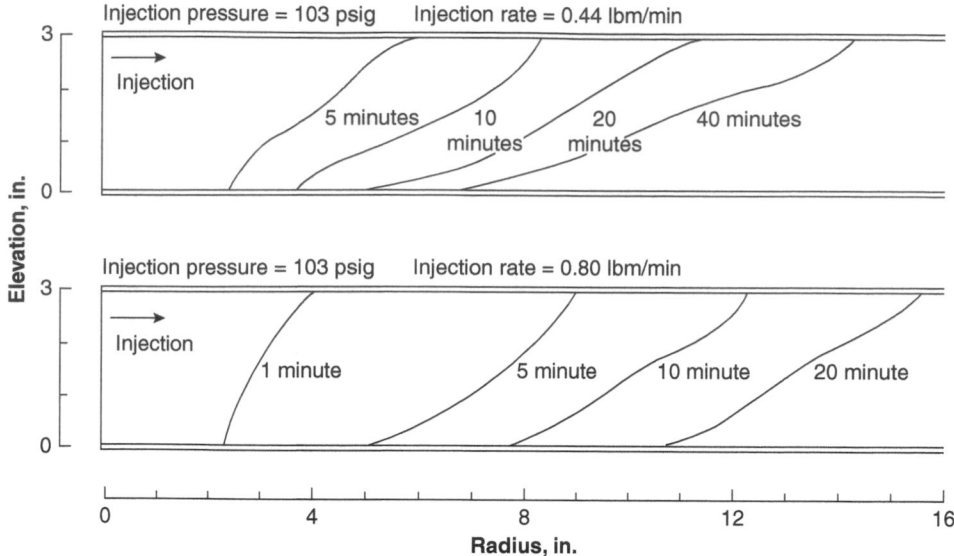

Fig. 8.55—Steam front at different times in two model runs at high injection pressure and different injection rates (Baker 1973).

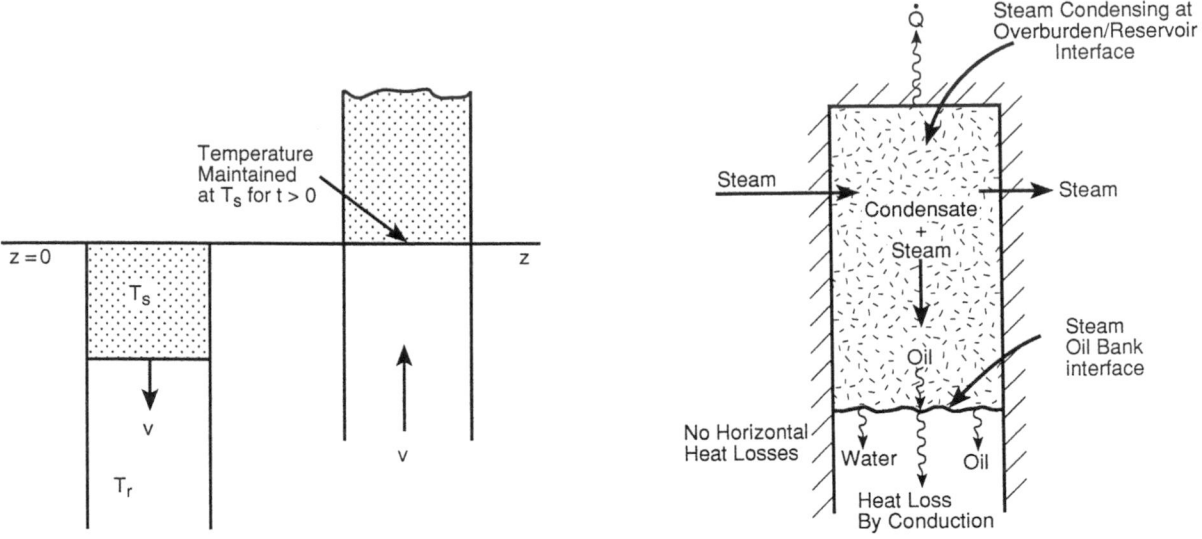

Fig. 8.56—Moving steam-zone interface for uniform-vertical-displacement model of Doscher and Ghassemi.

Fig. 8.57—Schematic of moving steam-zone interface, Neuman model.

overburden. It is assumed that the water produced by condensation of steam flows vertically under the influence of gravity to the steam-zone interface. The rate of heat loss by conduction to the overburden is given by

$$\dot{Q}_{ob} = k_h \left(T_s - T_r \right) \Big/ \sqrt{\pi \alpha \left(t - \lambda \right)}, \quad\dotfill\quad (8.146)$$

where $t = \lambda$ is the time that the steam zone initially arrived at a particular areal position. The rate of condensation, and thus the rate of condensate flow to the steam-zone interface, is given by

$$\dot{m}_{sob} = k_h \Delta T_s \Big/ \left[L_{vdh} \sqrt{\pi \alpha \left(t - \lambda \right)} \right]. \quad\dotfill\quad (8.147)$$

Fig. 8.57 shows the flow of fluids in the same vertical slice in an incremental thickness, Δz, on either side of the expanding steam interface. Temperatures and fluid saturations are assumed to be constant in the steam zone. Oil displacement at the steam displacement front is assumed to be piston-like. Thus, the oil saturation decreases from S_{ow} in the water zone to S_{ors} in the steam zone.

Steam-Interface Velocity. The Neuman model assumes that the temperatures, saturations, and oil and water velocities in the hot-water zone vary with distance below the steam-displacement front. Eq. 8.148 was developed from a mass and energy balance to relate the velocity of the steam zone to the elapsed time $(t - \lambda)$ when the steam zone arrived at that areal location at $t = \lambda$.

$$v(t - \lambda) = 2k_h C_w \Delta T_s \Big/ \left[L_{vdh} M_s \sqrt{\pi \alpha (t - \lambda)} \right]. \dots \dots \dots \dots \dots \dots \dots \dots \dots \dots \dots \dots \quad (8.148)$$

Eq. 8.148 is a phenomenological expression because the derivations in Neuman (1975, 1985) contain inconsistencies. The model is included here because it has been used successfully to correlate steam-zone thickness in several observation wells in the Kern River field.

Steam-Zone Thickness. The thickness of the steam zone at any position is found by integrating the velocity with respect to time.

$$h(t - \lambda) = \int_0^t v(t - \lambda) \, dt \dots \dots \dots \dots \dots \dots \dots \dots \dots \dots \dots \dots \dots \dots \dots \quad (8.149)$$

$$= \left(4k_h C_w \Delta T_s / L_{vdh} M_s \right) \sqrt{(t - \lambda) / \pi \alpha}. \dots \dots \dots \dots \dots \dots \dots \dots \dots \dots \quad (8.150)$$

Steam-Zone Area. The incremental rate that energy, $d\dot{Q}$, is supplied by steam condensation to maintain a vertically expanding steam zone over an incremental area, dA, is given by

$$d\dot{Q} = \left[\frac{4k_h \Delta T_s}{\sqrt{\pi \alpha (t - \lambda)}} + \frac{k_h \Delta T_s}{\sqrt{\pi \alpha (t - \lambda)}} + v M_s \Delta T_s \right] dA_s. \dots \dots \dots \dots \dots \dots \dots \quad (8.151)$$

In Eq. 8.151, the first term on the right side is the rate of heat loss by conduction to the overburden. The second term is the rate of heat loss to the reservoir under the steam interface by conduction, assuming that this region has been exposed to steam temperature for $t - \lambda$. The third term is the rate that energy is required to raise the temperature of the moving interface from T_r to T_s. In the model, the thermal properties of the overburden and the reservoir are assumed to be the same. Thus, $k_h = k_{hu}$, $a_r = a$, and heat losses by conduction to overburden and underlying reservoir are assumed to be equal.

The total rate by which energy is supplied by steam condensation is defined by

$$\dot{Q}_s = \int_0^t \left(\frac{d\dot{Q}_s}{dA_s} \right) \left(\frac{dA_s}{d\lambda} \right) d\lambda. \dots \dots \dots \dots \dots \dots \dots \dots \dots \dots \dots \dots \dots \dots \quad (8.152)$$

The solution of Eq. 8.152 is an expression for the steam area as a function of time.

$$A_s = \frac{\dot{Q}_s}{k_h \Delta T_s} \left(\frac{L_{vdh}}{L_{vdh} + C_w \Delta T_s} \right) \sqrt{\alpha \frac{t}{\pi}}. \dots \dots \dots \dots \dots \dots \dots \dots \dots \dots \dots \quad (8.153)$$

In Eq. 8.153, $\dot{Q}_s$ = rate that energy is delivered to the steam zone by condensation (Neuman 1975, 1985). Before breakthrough of heated fluids into the production wells, $\dot{Q}_s$ is defined by

$$\dot{Q}_s = \dot{m}_s \left(L_{vdh} + C_w \Delta T_s \right) f_{sd}, \dots \dots \dots \dots \dots \dots \dots \dots \dots \dots \dots \dots \dots \dots \quad (8.154)$$

where f_{sd} = quality of the steam at bottomhole conditions. Thus,

$$A_s = \left(\dot{m}_s f_{sd} L_{vdh} / k_h \Delta T_s \right) \sqrt{\alpha t / \pi}. \dots \dots \dots \dots \dots \dots \dots \dots \dots \dots \dots \dots \dots \quad (8.155)$$

Steam-Zone Volume. An expression for steam-zone volume is obtained by integrating the steam-zone thickness at any time over the heated area.

$$V_s = \int_0^{A_s} h(t - \lambda) \, dA_s. \dots \dots \dots \dots \dots \dots \dots \dots \dots \dots \dots \dots \dots \dots \dots \quad (8.156)$$

Eq. 8.156 can be rewritten as

$$V_s = \int_0^t \left(\frac{dA_s}{d\lambda} \right) h(t - \lambda) \, d\lambda. \dots \dots \dots \dots \dots \dots \dots \dots \dots \dots \dots \dots \dots \quad (8.157)$$

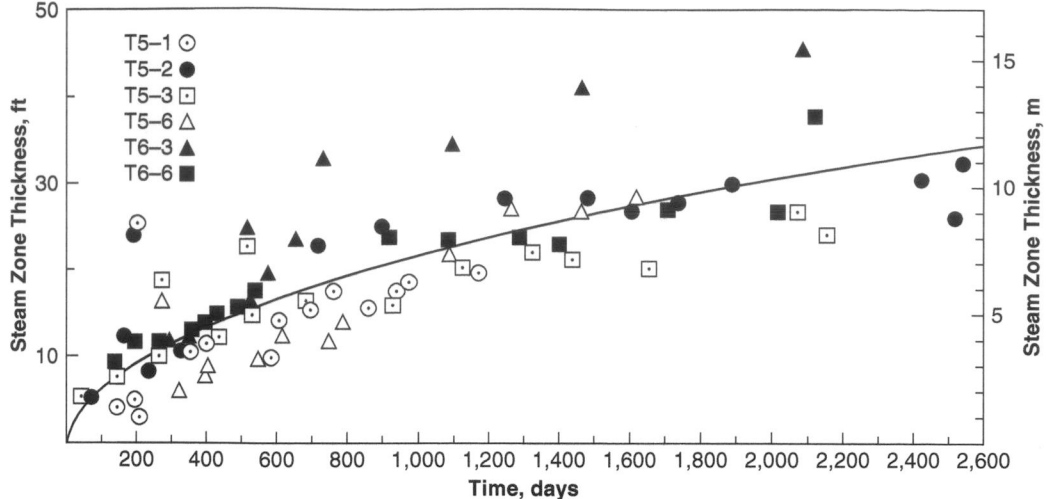

Fig. 8.58—Predicted steam-zone thickness compared with thicknesses inferred from temperature observation wells (Neuman 1985).

Steam injection rate (B/D CWE)	300
L_{vdh} (Btu/lbm)	910
C_w (Btu/lbm-°F)	1
ΔT (°F)	215
k_h (Btu/ft-D-°F)	35.7
M_s (Btu/ft³-°F)	35.2
a (ft²/D)	0.87
f_{sd}	0.8
ρ_w (lbm/bbl)	350
Injection time (days)	500

Table 8.30—Example 8.13, Neuman gravity-override model.

The solution of Eq. 8.157 is

$$V_s = \dot{Q}_s C_w t / \left[M_s \left(L_{vdh} + C_w \Delta T_s \right) \right]. \qquad (8.158)$$

Eq. 8.158 can be simplified by substituting Eq. 8.154 for $\dot{Q}_s$ to obtain

$$V_s = \dot{m}_s C_w f_{sd} t / M_s. \qquad (8.159)$$

Thus, the Neuman model predicts that the steam-zone volume is a linear function of the time before the time that the steam zone breaks into a production well.

The Neuman model can be used to predict the thickness of the steam zone at any particular distance from the injection well. To do this, it is necessary to know when the steam zone arrived at the location. This is the value of λ in Eq. 8.150. **Fig. 8.58** shows a comparison of steam-zone thickness at observation wells in the Kern River field as a function of total time (Neuman 1985). The application of the Neuman model is illustrated in Example 8.13.

Example 8.13—Estimation of Steam-Zone Growth Using the Neuman Model.
Steam is injected into a reservoir at a rate of 300 B/D at a temperature of 300°F. The reservoir is located at a depth of 1,000 ft. Steam quality at the sandface is 0.8. The reservoir is 50 ft thick and has a reservoir temperature of 85°F. **Table 8.30** presents properties representative of the reservoir. Estimate (1) the area heated by steam as a function of time to 2,000 days of continuous injection and (2) the thickness of the steam zone after 500 days of injection.

Solution. The first step in the solution is to find the heated area as a function of time. From Eq. 8.155,

$$A_s = \left(\dot{m}_s f_{sd} L_{vdh} / k_h \Delta T_s \right) \sqrt{\alpha t / \pi}.$$

and $\dot{m}_s = (300 \ \text{B/D})(350 \ \text{lbm/bbl})$

$$= 105,000 \ \text{lbm/D}.$$

From Table 8.27, we obtain values of f_{sd}, L_v, k_h, ΔT_s, and a. Substituting into Eq. 8.155, we obtain the areal extent of the steam zone.

$$A_s = \frac{(105,000 \ \text{lbm/D})(0.8)(910 \ \text{Btu/lbm})}{(35.7 \ \text{Btu/ft-D-°F})(215°F)}$$

$$\times \sqrt{(0.87 \ \text{ft}^2/\text{D})(t/3.14159)}$$

$$= 5,241\sqrt{t}.$$

When $t = 500$ days,

$$A_s = 5,241\sqrt{500}$$

$$= 117,200 \ \text{ft}^2$$

$$= 2.70 \ \text{acres}.$$

Time (days)	Heated Area (ft²)	Heated Area (acres)	Heated Volume (ft³)	Average Steam-Zone Thickness (ft)	Elapsed Time at t = 500 Days (days)	Steam-Zone Thickness at t = 500 Days (ft)
5	11,719	0.27	11,932	1.02	495	12.90
10	16,573	0.38	23,864	1.44	490	12.83
50	37,058	0.85	119,318	3.22	450	12.30
100	52,408	1.20	238,636	4.55	400	11.60
150	64,187	1.47	357,955	5.58	350	10.85
200	74,116	1.70	477,273	6.44	300	10.04
250	82,865	1.90	596,591	7.20	250	9.17
300	90,774	2.08	715,909	7.89	200	8.20
350	98,047	2.25	835,227	8.52	150	7.10
400	104,816	2.41	954,545	9.11	100	5.80
450	111,174	2.55	1,073,864	9.66	50	4.10
500	117,188	2.69	1,193,182	10.18	0	0

Table 8.31—Example 8.13, Neuman gravity-override model.

Thus, the steam zone just covers a heated area of 2.70 acres after 500 days of injection. **Table 8.31** gives locations of the steam zone in increments of 50 days of cumulative injection.

The steam-zone thickness is computed with Eq. 8.150:

$$h(t-\lambda) = \left(4k_h C_w \Delta T_s / L_{vdh} M_s\right)\sqrt{(t-\lambda)/\pi\alpha}.$$

In Eq. 8.150, λ is the time that the steam zone arrived at a particular area, A_s is as noted in Table 8.31, and $t-\lambda$ is the elapsed time that the steam zone has been at a given location A_s.

Compute the steam-zone thickness after 500 days of injection at $A_s = 1$ acre. From the solution above,

$$A_s = 5{,}241\sqrt{t},$$

$$\sqrt{\lambda} - \left[(1 \text{ acre})(43{,}560 \text{ ft}^2/\text{acre})\right]/5{,}241,$$

and $\lambda = 69.1$ days.

Thus, when the total injection time is 500 days, the steam zone at $A_s = 1$ acre has been at steam temperature $500 - 69.1 = 431$ days. Eq. 8.150 can now be used to calculate the thickness of the steam zone:

$$h = \frac{4(35.7\,\text{Btu/ft-°F})(1.0\,\text{Btu/lbm-°F})(215°F)}{(910\,\text{Btu/lbm})(35.2\,\text{Btu/ft}^3\text{-°F})}$$

$$\times \sqrt{431\,\text{days}/\pi(0.87\,\text{ft}^2/\text{D})}$$

$$= (0.9585)(12.56)$$

$$= 12.04\,\text{ft}.$$

Fig. 8.59 shows the distribution of steam-zone thickness with area after 500 days of continuous injection. Table 8.31 presents the average steam-zone thicknesses at other injection times.

The volume heated by the steam zone when $t = 500$ days is computed from Eq. 8.159:

$$V_s = \frac{\dot{m}_s C_w f_{sd} t}{M_s}$$

$$= (105{,}000\,\text{lbm/D})(1.0\,\text{Btu/lbm-°F})$$

$$\times (500\,\text{days})(0.8)$$

$$= 1{,}193{,}000\,\text{ft}^3$$

$$= 27.4\,\text{acre-ft}.$$

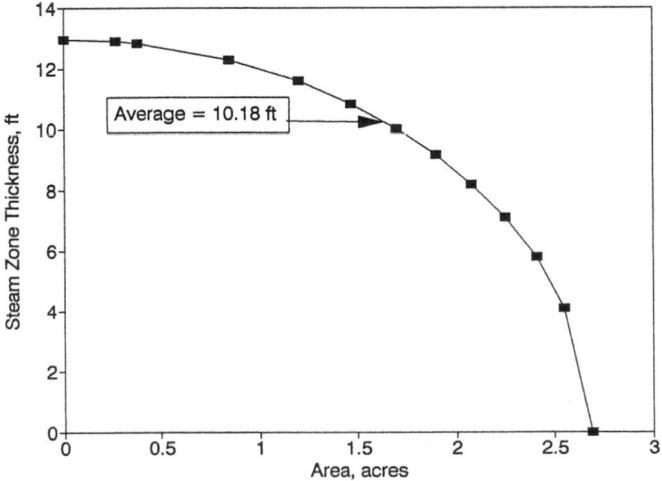

Fig. 8.59—Steam-zone thickness in Example 8.13 after 500 days of injection computed with Neuman model.

The average thickness of the steam zone is

$$\bar{h} = V_s/A_s$$
$$= 27.4 \text{ acre-ft}/2.7 \text{ acres}$$
$$= 10.15 \text{ ft.}$$

The average velocity of the expanding steam zone is 10.15 ft/(500 days) = 0.0203 ft/D.

Constant Vertical Heated-Zone Expansion (Doscher and Ghassemi 1983). Fig. 8.56 shows an interface that moves at a velocity v in the positive z-direction. The region behind (and including) the interface is maintained at constant temperature, T_s, which is analogous to the vertical movement of the heated zone described previously. If the origin is located at the interface, the region of steam temperature is maintained at T_s for $z \leq 0$ and the region $z > 0$ moves toward the origin at a velocity of $-v$. These two situations are equivalent mathematical descriptions.

The mathematical model that describes the flow of heat when the interface is maintained at T_s as in Fig. 8.56 is given by Eq. 8.160 and associated initial and boundary conditions:

$$\left(\partial^2 T/\partial z^2\right) + \left(v/\alpha\right)\left(\partial T/\partial z\right) = \left(1/\alpha\right)\left(\partial T/\partial t\right), \dots\dots\dots\dots\dots\dots\dots\dots\dots\dots\dots\dots(8.160)$$

where at $t = 0$, $z > 0$ and $T = T_r$, and for $t > 0$, $z = 0$, $T = T_s$, $z \geq \infty$, and $T = T_r$.

The temperature distribution corresponding to the solution of Eq. 8.160 with the boundary and initial conditions stated is found in Carslaw and Jaeger (1959) and is expressed here as

$$T = \frac{1}{2}(T_s - T_r)\left[\text{erfc}\left(\frac{z+vt}{2\sqrt{at}}\right) + e^{-\frac{vz}{a}} \text{erfc}\left(\frac{z-vt}{2\sqrt{at}}\right)\right]. \dots\dots\dots\dots\dots\dots\dots\dots\dots(8.161)$$

A reservoir heating model based on uniform vertical expansion of the heated zone is developed by use of the same approach as in the Marx and Langenheim model. Consider the heated region shown in Fig. 8.56, where heat is injected at a constant rate. It is assumed that the steam enters at the top of the formation. Heat is lost by conduction to the overburden at a rate $\dot{Q}_1$ as described by Eq. 8.55:

$$\dot{Q}_1 = k(T_s - T_r)/\sqrt{\pi at}.$$

At the moving interface ($z = 0$), heat flow by conduction through the interface is given by

$$\dot{Q}_2 = -k(\partial T/\partial z)\big|_{z=0}, \dots\dots\dots\dots\dots\dots\dots\dots\dots\dots\dots\dots\dots\dots\dots(8.162)$$

$$= k(T_s - T_r)\left[\left(e^{-a_1^2 t}/\sqrt{\pi at}\right) + \frac{v}{a} + \frac{v}{2a}\text{erfc}\left(a_1\sqrt{t}\right)\right], \dots\dots\dots\dots\dots\dots\dots(8.163)$$

where $a_1 = v/2\sqrt{\alpha}$.

The overall energy balance on the region contacted by the heated fluid is given by

$$\dot{m}_s H_s = \int_0^A \dot{Q}_1(t - \lambda)dA + \int_0^A \dot{Q}_2(t - \lambda)dA. \dots\dots\dots\dots\dots\dots\dots\dots\dots\dots\dots(8.164)$$

In Eq. 8.164, the first integral represents the rate that heat is lost to the overburden by conduction through the heated area A at time t. The second integral is the rate by which heat is lost from the heated area A at time t to the underlying reservoir by conduction through the interface, which is moving at a constant velocity v. The parameter λ is the time that the heated region arrived at a particular area A. In this model, it is necessary to assume that the thermal properties of the reservoir and the overburden are identical.

Each integral in Eq. 8.164 can be expressed in terms of time by recalling that

$$\int_0^A Q_1(t - \lambda)dA = \int_0^t Q_1(t - \lambda)\left(\frac{dA}{d\lambda}\right)d\lambda. \dots\dots\dots\dots\dots\dots\dots\dots\dots\dots\dots(8.165)$$

Thus, the energy balance at any time before breakthrough of heat into a production well is given by

$$\dot{m}_s H_s = k(T_s - T_r)$$

$$\times \left[\int_o^t \frac{1}{\sqrt{\pi a (t - \lambda)}} \left(\frac{dA}{d\lambda} \right) d\lambda + \int_o^t \frac{e^{-a_1^2(t-\lambda)}}{\sqrt{\pi a (t - \lambda)}} \left(\frac{dA}{d\lambda} \right) d\lambda - \frac{v}{2a} \int_o^t \mathrm{erfc} \left(a_1 \sqrt{t - \lambda} \right) \left(\frac{dA}{d\lambda} \right) d\lambda \right]$$

$$+ M v A \left(T_s - T_r \right), \dots \dots \dots \dots \dots \dots \dots \dots \dots (8.166)$$

where $M = \rho C_p$.

Eq. 8.166 was solved[11] for the heated area as a function of time. The solution, developed in Appendix I, is

$$A_h \left(t_{Dv} \right) = \left[m_s H_s / M v \left(T_s - T_r \right) \right] G_3 \left(t_{Dv} \right), \dots \dots \dots \dots \dots \dots \dots \dots (8.167)$$

where $t_{Dv} = v^2 t / 4a$. $\dots \dots \dots \dots \dots \dots \dots \dots \dots (8.168)$

Table 8.32 contains selected values of $G_3(t_{Dv})$. $G_3(t_{Dv})$ is plotted against t_{Dv} in **Fig. 8.60**. Because the velocities of the heated interface are on the order of 0.01 ft/D, dimensionless times of practical interest are in the range of 0.001 to 0.1. Fig. 8.60 indicates that the heated area changes rapidly with t_{Dv} in that time interval. Doscher and Ghassemi (1983) developed an equivalent correlation using numerical integration.

The thickness of the heated region, h, is found by recalling that the distance traveled by the moving interface is $(t - \lambda)v$, where λ is the time when the heated zone arrived at a particular value of the heated area. Therefore, at each areal position, the heated-zone thickness is given by

$$h = (t - \lambda)v. \dots \dots \dots \dots \dots \dots \dots \dots \dots (8.169)$$

At a specified time t, it is possible to compute the thickness of the heated zone. **Fig. 8.61** shows the heated-zone thickness as a function of heated area at a total injection time of 1 year when the heated interface advanced at a rate of 0.01 ft/D. Example 8.14 discusses parameters used to construct this graph.

The heated volume is the integral of the heated thickness over the heated area as defined by

$$V_h = \int_o^{A_h} h \, dA_h. \dots \dots \dots \dots \dots \dots \dots \dots \dots (8.170)$$

This is simply the area between $h = 0$ and the (h-vs.-A) graph in Fig. 8.61. The integral can be evaluated numerically at specific values of time to determine the heated volume, V_h, at those times.

It is possible to determine the heated volume in a more general way. The integral in Eq. 8.170 can also be expressed as

$$V_h = \int_o^t h \left(\frac{dA_h}{d\lambda} \right) d\lambda \dots \dots \dots \dots \dots \dots \dots \dots \dots (8.171)$$

$$= v \int_o^t (t - \lambda) \left(\frac{dA_h}{d\lambda} \right) d\lambda \dots \dots \dots \dots \dots \dots \dots \dots (8.172)$$

The solution to Eq. 8.172 (Appendix I) is

$$V_h = \left[\dot{m}_s H_s / M \left(T_s - T_r \right) a_1^2 \right] G_v \left(t_{Dv} \right), \dots \dots \dots \dots \dots \dots \dots \dots (8.173)$$

where the function $G_v(t_{Dv})$ is defined in Appendix I, including a computer program for evaluation of the function. For many calculations, it is convenient to express $G_v(t_{Dv})$ in terms of the thermal efficiency of the displacement.

The thermal efficiency of the heat injection process is defined as the fraction of the injected energy that remains within the steam zone. For this process,

$$E_h = \left[V_h M \left(T_s - T_r \right) \right] / \dot{m}_s H_s t. \dots \dots \dots \dots \dots \dots \dots \dots (8.174)$$

Substitution of Eq. 8.173 into Eq. 8.174 gives the thermal efficiency for the actual expansion process:

$$E_h = G_v \left(t_{Dv} \right) / t_{Dv}. \dots \dots \dots \dots \dots \dots \dots \dots \dots (8.175)$$

The thermal efficiency is only a function of the dimensionless time t_{Dv}. Values of E_h are given in Table 8.32 at selected values of t_{Dv}. A Fortran program used to determine E_h is included in Appendix I. The heated volume can be expressed in terms of the thermal efficiency by substituting Eq. 8.175 into Eq. 8.174 to obtain

[11]Unpublished notes, G. P. Willhite, University of Kansas, Lawrence (1969).

t_{Dv}	$G_3(t_{Dv})$	$E_h(t_{Dv})$
0.001	0.035182	0.0235
0.002	0.049463	0.0331
0.003	0.060304	0.0405
0.004	0.069366	0.0466
0.005	0.07729	0.0519
0.006	0.084406	0.0568
0.007	0.09091	0.0612
0.008	0.096929	0.0653
0.009	0.102553	0.0691
0.01	0.107844	0.0727
0.02	0.149602	0.1014
0.03	0.180497	0.1228
0.04	0.205775	0.1405
0.05	0.227468	0.1558
0.06	0.246618	0.1693
0.07	0.263843	0.1816
0.08	0.279548	0.1929
0.09	0.294012	0.2033
0.10	0.307439	0.2131
0.15	0.363391	0.2543
0.20	0.407045	0.2872
0.25	0.442939	0.3149
0.30	0.473395	0.3388
0.35	0.499791	0.3600
0.40	0.523022	0.3789
0.45	0.543709	0.3961
0.50	0.562302	0.4118
0.55	0.579139	0.4263
0.60	0.594483	0.4397
0.65	0.60854	0.4521
0.70	0.621477	0.4638
0.75	0.633432	0.4747
0.80	0.644518	0.4850
0.85	0.654831	0.4946
0.90	0.664452	0.5038
0.95	0.673451	0.5125
1.00	0.681887	0.5208

Table 8.32—Selected values of $G_3(t_{Dv})$, $E_h(t_{Dv})$.

$$V_h = \frac{\dot{m}_s H_s t}{M(T_s - T_r)} E_h(t_{Dv}). \dotfill (8.176)$$

Example 8.14 illustrates the prediction of reservoir heating from steam injection when gravity override occurs and the heated region expands at a constant velocity.

Example 8.14—Reservoir Heating by Vertical Expansion of the Heated Region at Constant Velocity. Steam at 470.9°F (500 psig) is injected into a reservoir that is 50 ft thick at a rate of 500 BWPD. Initial temperature of the reservoir is 80°F. Gravity override is expected to occur because there are no vertical barriers to fluid flow within the cross section. Estimate the following when the heated zone expands vertically at a constant velocity of 0.01 ft/D after 1 year of continuous injection: (1) heated area (in acres), (2) thickness of the heated zone as a function of areal position, and (3) volume of the heated region.

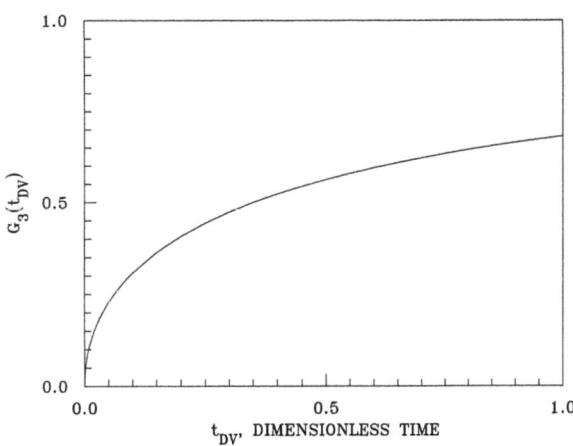

Fig. 8.60—$G_3(t_{Dv})$ vs. t_D for uniform-vertical-displacement model.

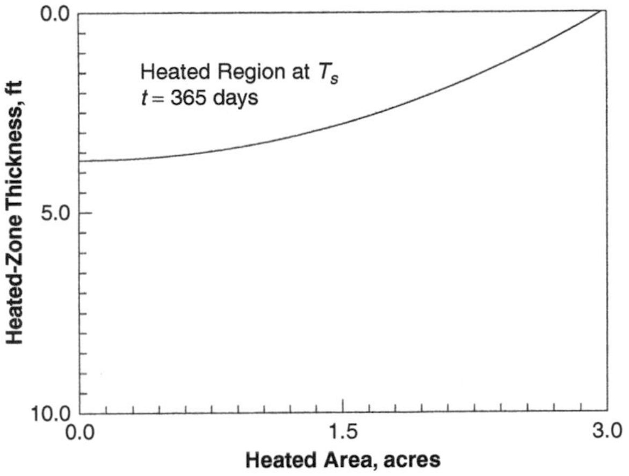

Fig. 8.61—Thickness of heated zone after 1 year of continuous injection, uniform-vertical-displacement model, Example 8.15.

For this example, the volumetric heat capacity of the overburden and reservoir are equal at 32.74 Btu/ft³-°F. Thermal diffusivity of the overburden and reservoir is 0.0482 ft²/hr. The quality of the steam at the sandface is 0.8.

Solution.

1. The heated area for uniform vertical expansion of the heated zone at constant temperature T_s is given as a function of dimensionless time by Eq. 8.167:

$$A_h(t_{Dv}) = \left\{ \dot{m}_s H_s t \big/ \left[Mv(T_s - T_r) \right] \right\} G_3(t_{Dv}),$$

where $t_{Dv} = v^2 t/4\alpha$ is given in Eq. 8.168. At $t = 1$ year,

$$t_{Dv} = \left[(0.01\,\text{ft/D})^2 (365\,\text{days}) \right] \big/ \left[(4)(0.0482\,\text{ft}^2/\text{hr})(24\,\text{hr/D}) \right]$$

$$\approx 0.00789.$$

From Table 8.32, $G_3(0.00789) \approx 0.0952$. In Eq. 8.167,

$$\dot{m}_s H_s \big/ \left[Mv(T_s - T_r) \right] = \left[(500\,\text{B/D})(350\,\text{lbm/bbl})(977\,\text{Btu/lbm}) \right]$$

$$\div \left[(32.74\,\text{Btu.ft}^3\text{-}°\text{F})(0.01\,\text{ft/D}) \right.$$

$$\times (470.9°\text{F} - 80°\text{F})(43,560\,\text{ft}^2/\text{acre}) \Big]$$

$$= 31.6\,\text{acres}.$$

$$A_h = (31.6)(0.0952)$$

$$= 3.01\,\text{acres}.$$

2. The thickness of the heated region, h, is found by recalling from Eq. 8.169 that the distance traveled by the moving interface is $(t - \lambda)v$, where λ is the time the heated zone arrived at each heated area. Therefore, at each areal position, as given by Eq. 8.169,

$$h = (t - \lambda)v.$$

To find the thickness as a function of areal position at $t = 1$ year, it is necessary to find the arrival time for the heated region at area values less than 2.94 acres. From Part 1,

$$A(t_{Dv}) = 30.865 G_3(t_{Dv}),$$

where $t_{Dv} = v^2 t / 4\alpha$

$$= 2.161 \times 10^{-5} t.$$

In this example, values of t_{Dv} between 0 and 0.00789 are chosen and values of t and A are computed for each t_{Dv}. **Table 8.33** summarizes the results.

λ_{Dv}	λ (days)	$t-\lambda$ (days)	h (ft)	A (acres)
0	0	365	3.65	0
0.001	46.3	318.7	3.19	1.11
0.002	92.6	272.4	2.73	1.56
0.003	138.8	226.2	2.26	1.90
0.004	185.1	179.9	1.8	2.19
0.005	231.4	133.6	1.34	2.44
0.006	277.6	87.4	0.87	2.67
0.007	323.9	41.1	0.41	2.87
0.00789	365.0	0	0	3.01

For $v = 0.01$ ft/D.

Table 8.33—Determination of the heated-region thickness after 1 year of continuous injection, Example 8.14.

The heated zone is shown as a function of areal position in Fig. 8.61. If the velocity of the heated zone is constant, it will take 5,000 days before the entire thickness at the injection well sandface is heated to T_s.

3. The heated volume is related to the thermal efficiency through Eq. 8.176. From Table 8.32, at $t_{Dv} = 0.00789$,

$$E_h = 0.061 + \left(\frac{0.00789 - 0.007}{0.008 - 0.007} \right)(0.065 - 0.061)$$

$$= 0.0646$$

and

$$V_h = \frac{(500\,\text{B/D})(350\,\text{lbm/bbl})(1006\,\text{Btu/lbm})(365\,\text{days})(0.0646)}{(32.74\,\text{Btu.ft-°F})(390.7°\text{F})(43,560\,\text{ft/acre})}$$

$$= 7.44\,\text{acre-ft}.$$

The average thickness of the heated region, $\overline{h}$, is

$$\overline{h} = V_h / A_h$$

$$= 7.44 \text{ acre-ft}/3.01 \text{ acres}$$

$$= 2.47\,\text{ft}.$$

Computed results for Example 8.14 are presented in **Table 8.34** for $0.001 \leq t_{Dv} \leq 0.100$, corresponding to a total injection time of 4,626 days (12.8 years). If injection is continuous, the steam zone will be 46.3 ft thick at the injection well after 4,626 days of injection and the heated region ($T = T_s$) will occupy 311.17 acre-ft.

Steam-Zone Volume When Areal Extent Is Limited by Heat Losses. The heated region can be maintained at steam temperature only in the region where heat losses and heat rate needed to maintain the steam-zone expansion at a constant velocity are supplied by condensing steam. An analogous situation was presented in Section 8.4.1 in connection with the Marx and Langenheim model. When the injection rate is constant, the rate that latent heat is injected is $\dot{m}_s f_{sd} L_{vdh}$. If it is assumed that condensing steam is present in the entire vertical cross section behind the temperature front, the amount of heat supplied by condensation is given by

$$\dot{m}_s f_{sd} L_{vdh} = \int_o^{t_1} \dot{Q}_1 (t - \lambda)\left(\frac{dA}{d\lambda}\right)d\lambda + \int_o^{t_1} \dot{Q}_2 (t - \lambda)\left(\frac{dA}{d\lambda}\right)d\lambda. \quad \ldots \ldots \ldots \ldots (8.177)$$

In Eq. 8.177, t_1 is the time that the heated zone ($T = T_s$) arrived at a particular areal location A_1 and t is the time that the steam zone arrives at the same location. The first integral is the heat transferred by condensation to supply heat losses to the overburden. The second integral is the heat transferred by condensation to maintain a constant steam front velocity, v, as the steam zone expands vertically within the area A_1.

The rate at which heat must be supplied within area A_1 to replace heat losses and maintain a constant steam-zone expansion velocity is given by

$$\dot{Q}_{\ell_1} = \dot{m}_s H_s G_e \left(t_{Dv}, t_{Dv1}\right), \quad \ldots \ldots \ldots \ldots \ldots \ldots \ldots \ldots \ldots \ldots \ldots (8.178)$$

	t_{Dv}	t (days)	A_h (acres)	V_h (acre-ft)	$\bar{h}$ (ft)	E_h
1	0.0050	231.3	2.44	3.80	1.56	0.052
2	0.0100	462.6	3.41	10.66	3.13	0.073
3	0.0150	693.9	4.06	19.28	4.74	0.088
4	0.0200	925.2	4.72	29.51	6.25	0.101
5	0.0250	1,156.5	5.21	41.27	7.92	0.113
6	0.0300	1,387.8	5.70	53.91	9.46	0.123
7	0.0350	1,619.1	6.10	67.49	11.07	0.132
8	0.0400	1,850.4	6.50	81.81	12.59	0.140
9	0.0450	2,081.7	6.84	97.30	14.22	0.148
10	0.0500	2,313.1	7.18	113.95	15.86	0.156
11	0.0550	2,544.4	7.49	130.97	17.50	0.163
12	0.0600	2,775.7	7.79	148.13	19.02	0.169
13	0.0650	3,007.0	8.06	167.13	20.74	0.176
14	0.0700	3,238.3	8.33	186.12	22.34	0.182
15	0.0750	3,469.6	8.58	204.89	23.88	0.187
16	0.0800	3,700.9	8.83	225.56	25.55	0.193
17	0.0850	3,932.2	9.06	245.87	27.15	0.198
18	0.0900	4,163.5	9.28	266.91	28.75	0.203
19	0.0950	4,394.8	9.50	288.67	30.40	0.208
20	0.1000	4,626.1	9.71	311.17	32.05	0.213

Table 8.34—Estimated heated volumes for gravity-override model—uniform vertical displacement.

t_{Dv} \ t_{Dv1}	0.0010	0.0020	0.0030	0.0040	0.0050	0.0060	0.0070	0.0080	0.0090	0.0100
0.001	1.0000									
0.002	0.5106	1.0000								
0.003	0.4043	0.6205	1.0000							
0.004	0.3467	0.5151	0.6799	1.0000						
0.005	0.3092	0.4526	0.5807	0.7187	1.0000					
0.006	0.2823	0.4026	0.5186	0.6257	0.747	1.0000				
0.007	0.2617	0.3775	0.4743	0.5655	0.659	0.768	1.0000			
0.008	0.2453	0.3525	0.4406	0.5215	0.601	0.686	0.785	1.0000		
0.009	0.2318	0.3322	0.4137	0.4873	0.558	0.63	0.707	0.799	1.0000	
0.010	0.2205	0.3153	0.3916	0.4597	0.5241	0.5877	0.6532	0.7245	0.8103	1.0000
0.020	0.1605	0.2274	0.2794	0.3241	0.3642	0.4014	0.4364	0.4700	0.5026	0.5345
0.030	0.1345	0.1901	0.2329	0.2692	0.3015	0.3310	0.3585	0.3844	0.4091	0.4328
0.040	0.1192	0.1682	0.2058	0.2376	0.2657	0.2912	0.3148	0.3370	0.3579	0.3779
0.050	0.1089	0.1534	0.1876	0.2164	0.2418	0.2648	0.2860	0.3058	0.3245	0.3423
0.060	0.1013	0.1427	0.1743	0.2010	0.2244	0.2456	0.2652	0.2834	0.3005	0.3168
0.070	0.0954	0.1343	0.1641	0.1891	0.2111	0.2310	0.2492	0.2662	0.2822	0.2974
0.080	0.0907	0.1277	0.1559	0.1796	0.2005	0.2193	0.2366	0.2526	0.2677	0.2820
0.090	0.0868	0.1222	0.1492	0.1718	0.1917	0.2097	0.2262	0.2415	0.2558	0.2694
0.100	0.0835	0.1176	0.1435	0.1653	0.1844	0.2016	0.2175	0.2321	0.2459	0.2589
0.120	0.0783	0.1102	0.1345	0.1549	0.1728	0.1888	0.2036	0.2173	0.2301	0.2422

Table 8.35—$G_e(t_{Dv}, t_{Dv1})$—Gravity-override model.

t_{Dv}	t_{Dv1}									
	0.0010	0.0020	0.0030	0.0040	0.0050	0.0060	0.0070	0.0080	0.0090	0.0100
0.140	0.0743	0.1046	0.1276	0.1469	0.1638	0.1790	0.1930	0.2059	0.2181	0.2295
0.160	0.0711	0.1001	0.1221	0.1405	0.1567	0.1712	0.1845	0.1969	0.2085	0.2194
0.180	0.0685	0.0964	0.1176	0.1353	0.1508	0.1648	0.1776	0.1895	0.2006	0.2111
0.200	0.0663	0.0933	0.1138	0.1309	0.1460	0.1595	0.1719	0.1833	0.1941	0.2042

t_{Dv}	t_{Dv1}								
	0.0200	0.0300	0.0400	0.0500	0.0600	0.0700	0.0800	0.0900	0.1000
0.001									
0.002									
0.003									
0.004									
0.005									
0.006									
0.007									
0.008									
0.009									
0.010									
0.020	1.0000								
0.030	0.6482	1.0000							
0.040	0.5495	0.7095	1.0000						
0.050	0.4911	0.6183	0.7494	1.0000					
0.060	0.4510	0.5612	0.6654	0.7780	1.0000				
0.070	0.4213	0.5207	0.6109	0.7004	0.7998	1.0000			
0.080	0.3981	0.4899	0.5712	0.6487	0.7277	0.8172	1.0000		
0.090	0.3794	0.4654	0.5405	0.6103	0.6787	0.7498	0.8314	1.0000	
0.100	0.3639	0.4453	0.5157	0.5801	0.6418	0.7033	0.7681	0.8433	1.0000
0.120	0.3395	0.4141	0.4777	0.5349	0.5882	0.6394	0.6900	0.7417	0.7971
0.140	0.3210	0.3907	0.4496	0.5020	0.5502	0.5956	0.6394	0.6824	0.7256
0.160	0.3064	0.3724	0.4279	0.4768	0.5215	0.5631	0.6026	0.6407	0.6781
0.180	0.2946	0.3576	0.4104	0.4567	0.4987	0.5376	0.5742	0.6091	0.6429
0.200	0.2847	0.3453	0.3959	0.4402	0.4801	0.5169	0.5514	0.5840	0.6153

Table 8.35—$G_e(t_{Dv}, t_{Dv1})$—Gravity-override model (continued).

where t_{Dv} = dimensionless time from the beginning of injection and t_{Dv1} = dimensionless time when the steam zone reached area A_1. It is understood that $t_{Dv} > t_{D1}$. **Table 8.35** gives values of $G_e(t_{Dv}, t_{Dv1})$ for selected values of t_{Dv} and t_{Dv1}. Appendix I discusses the evaluation of $G_e(t_{Dv}, t_{Dv1})$.

The value of G_e when the heated zone just arrives at t_{Dv1} must be 1.0 to be consistent with the vertical expansion model. As discussed earlier in this section, steam can contact only part of the heated region because a fraction of the injected energy is latent heat.

Arrival time of the steam zone at a given areal location A_1 is determined by Eqs. 8.178 and 8.179. To determine the arrival time of the steam zone, we require that

$$\dot{m}_s f_{sd} L_{vdh} = \dot{Q}_\ell (A_1)$$

$$= \dot{m}_s H_s G_e (t_{Dv}, t_{Dv1}). \dots\dots\dots\dots\dots\dots\dots\dots\dots\dots\dots\dots\dots (8.179)$$

Solving for $G_e(t_{Dv}, t_{Dv1})$ results in

$$G_e\left(t_{Dv}, t_{Dv1}\right) = f_{sd} L_{vdh} / \mathrm{H}_s = f_{h,v}. \dots\dots\dots\dots\dots\dots\dots\dots\dots\dots\dots\dots\dots\dots\dots\dots\dots\dots\dots\dots\dots (8.180)$$

The right side of Eq. 8.180 is the fraction of the injected energy that is latent heat and is designated as $f_{h,v}$. Thus, for a given sandface steam quality, f_{sd}, the steam zone will arrive at A_1 when the value of $G_e(t_{Dv}, t_{Dv1})$ is equal to $f_{sd}(L_{vdh}/H_s)$. Finding t_{Dv} is a root finding process for a fixed value of t_{Dv1}. This process is simplified by constructing a table containing $G_e(t_{Dv}, t_{Dv1})$ at assumed values of t_{Dv} and t_{Dv1}, as in Table 8.35. Then t_{Dv} is found from Table 8.35 for specified values of t_{Dv1} and $G_e(t_{Dv}, t_{Dv1})$ by interpolation, or t_{Dv1} may be found for specified values of t_{Dv} and $G_e(t_{Dv}, t_{Dv1})$.

t_{Dv1}	$f_{h,v} = 0.2$		$f_{h,v} = 0.3$		$f_{h,v} = 0.4$	
	t_{Dv1}	$E_{h,s}$	t_{Dv1}	$E_{h,s}$	t_{Dv1}	$E_{h,s}$
0.00400	–	–	–	–	0.00132	0.03590
0.00500	–	–	–	–	0.00163	0.03995
0.00600	–	–	0.00114	0.03516	0.00192	0.04340
0.00700	–	–	0.00133	0.03795	0.00223	0.04671
0.00800	–	–	0.00151	0.04041	0.00254	0.04978
0.00900	–	–	0.00168	0.04260	0.00283	0.05256
0.01000	–	–	0.00184	0.04458	0.00312	0.05518
0.02000	0.00159	0.04304	0.00346	0.06097	0.00596	0.07594
0.03000	0.00223	0.05092	0.00495	0.07277	0.00863	0.09109
0.04000	0.00285	0.05740	0.00637	0.08236	0.01129	0.10376
0.05000	0.00343	0.06293	0.00771	0.09041	0.01388	0.11466
0.06000	0.00396	0.06757	0.00897	0.09739	0.01620	0.12370
0.07000	0.00450	0.07186	0.01021	0.10375	0.01828	0.13136
0.08000	0.00498	0.07555	0.01155	0.11008	0.02021	0.13809
0.09000	0.00546	0.07905	0.01278	0.11560	0.02240	0.14508
0.10000	0.00590	0.08213	0.01391	0.12046	0.02444	0.15133

Table 8.36—$E_{h,s}$, t_{Dv} at selected values of $f_{h,v}$, gravity-override model.

t_{Dv}	$f_{h,v} = 0.5$		$f_{h,v} = 0.6$		$f_{h,v} = 0.7$		$f_{h,v} = 0.8$	
	t_{Dv1}	$E_{h,s}$	t_{Dv1}	$E_{h,s}$	t_{Dv1}	$E_{h,s}$	t_{Dv1}	$E_{h,s}$
0.00200	–	–	0.00118	0.03074	0.00139	0.03186	0.00159	0.03260
0.00300	0.00144	0.03544	0.00191	0.03818	0.00221	0.03932	0.00247	0.03997
0.00400	0.00191	0.04074	0.00252	0.04387	0.00306	0.04556	0.00338	0.04614
0.00500	0.00237	0.04535	0.00314	0.04892	0.00386	0.05089	0.00429	0.05154
0.00600	0.00283	0.04951	0.00376	0.05345	0.00462	0.05558	0.00521	0.05640
0.00700	0.00328	0.05328	0.00437	0.05756	0.00538	0.05990	0.00614	0.06084
0.00800	0.00373	0.05679	0.00498	0.06140	0.00615	0.06392	0.00707	0.06496
0.00900	0.00418	0.06005	0.00558	0.06496	0.00691	0.06767	0.00801	0.06881
0.01000	0.00463	0.06313	0.00619	0.06832	0.00766	0.07118	0.00888	0.07239
0.02000	0.00892	0.08727	0.01141	0.09353	0.01356	0.09722	0.01570	0.09963
0.03000	0.01312	0.10529	0.01776	0.11447	0.02147	0.11902	0.02432	0.12120
0.04000	0.01711	0.11985	0.02316	0.13037	0.02941	0.13681	0.03312	0.13900
0.05000	0.02070	0.13176	0.02856	0.14422	0.03624	0.15140	0.04202	0.15437
0.06000	0.02445	0.14276	0.03372	0.15631	0.04307	0.16440	0.05099	0.16803
0.07000	0.02792	0.15234	0.03879	0.16722	0.04996	0.17623	0.06001	0.18038

Table 8.36—$E_{h,s}$, t_{Dv} at selected values of $f_{h,v}$, gravity-override model (continued).

t_{Dv}	$f_{h,v}$ = 0.5		$f_{h,v}$ = 0.6		$f_{h,v}$ = 0.7		$f_{h,v}$ = 0.8	
	t_{Dv1}	$E_{h,s}$	t_{Dv1}	$E_{h,s}$	t_{Dv1}	$E_{h,s}$	t_{Dv1}	$E_{h,s}$
0.08000	0.03125	0.16097	0.04371	0.17714	0.05650	0.18690	0.06808	0.19147
0.09000	0.03461	0.16909	0.04853	0.18626	0.06300	0.19679	0.07616	0.20175
0.10000	0.03778	0.17644	0.05323	0.19472	0.06946	0.20602	0.08424	0.21135

Table 8.36—$E_{h,s}$, t_{Dv} at selected values of $f_{h,v}$, gravity-override model (continued).

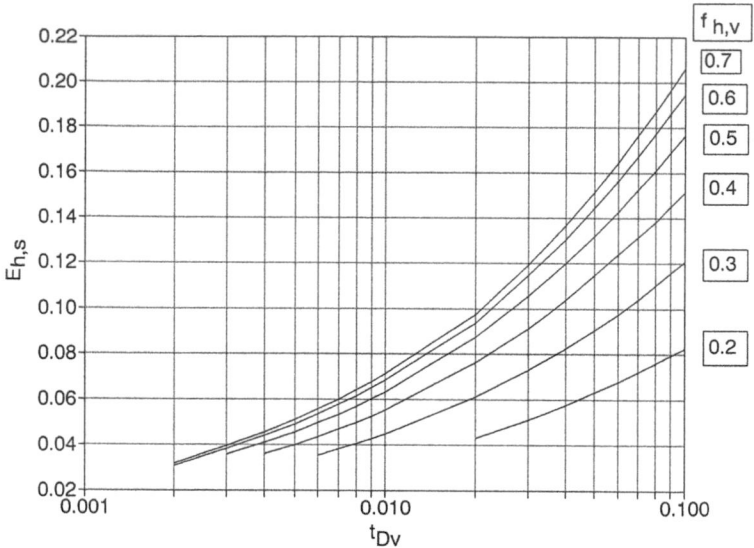

Fig. 8.62—$E_{h,s}$ vs. t_{Dv} for uniform-vertical-displacement model.

The thermal efficiency based on the region contacted by steam is the fraction of the injected energy that is in the heated volume contacted by steam and is defined by

$$E_{h,s} = V_s M \left(T_s - T_r \right) / \left(m_s H_s t \right). \dotfill (8.181)$$

Table 8.36 gives values of $E_{h,s}$, t_{Dv}, and t_{Dv1}. **Fig. 8.62** is a graph of $E_{h,s}$ vs. t_{Dv} for values of $f_{sd}L_{vdh}/H_s$ = 0.2, 0.3, 0.4, 0.5, 0.6, and 0.7. Example 8.15 illustrates calculation of the steam-zone volume after 1 year of continuous steam injection using the parameters from Example 8.14.

Example 8.15—Estimation of Steam-Zone Volume When Steam Zone Expands Vertically at a Constant Velocity. The region contacted by steam is less than the heated volume determined in Part 3 of Example 8.14 because the steam temperature, T_s, can be maintained only by condensation. Estimate the area contacted by steam and the steam-zone volume after 1 year of continuous steam injection at 500 B/D and 500 psig into a 50-ft-thick reservoir. Initial reservoir temperature is 80°F. As in Example 8.14, the steam zone expands vertically at a constant velocity of 0.01 ft/D. The volumetric heat capacities of the over- and underburden are equal at 32.79 Btu/ft³-°F. Thermal diffusivity of the overburden and the reservoir is 0.0482 ft²/hr. The quality of the steam at the sandface is 0.8.

Solution. First, it is necessary to find $f_{h,v}$, the fraction of the injected energy that is condensable. Next, we find the location of the steam zone from Eq. 8.180.

$$G_e \left(t_{Dv}, t_{Dv1} \right) = f_{sd} L_{vdh} / H_s$$
$$= f_{h,v}.$$

$$f_{h,v} = (0.8)(601.1 \, \text{Btu/lbm}) / (1006 \, \text{Btu/lbm})$$
$$= 0.478.$$

The value of t_{Dv} at t = 1 year was determined to be 0.00789 in Example 8.14. Thus, to find out where the steam zone is at t_{Dv} = 0.00789, it is necessary to solve $G_e(0.00789, t_{Dv1})$ = 0.478 for t_{Dv1}. Table 8.35 contains values of G_e as a function of t_{Dv}

and t_{Dv1} at even increments of t_{Dv} and t_{Dv1}. To find t_{Dv1}, it is necessary to interpolate Table 8.35 to find the value of t_{Dv1} at $G_e = 0.478$ when $t_{Dv} = 0.00789$. Referring to Table 8.35, values of G_e that bracket 0.478 are found at $t_{Dv} = 0.00789$. At $t_{Dv} = 0.00789$ and $t_{Dv1} = 0.003$, $G_e = 0.4443$. At $t_{Dv} = 0.00789$ and $t_{Dv1} = 0.004$, $G_e = 0.5255$.

The correct value of t_{Dv1} is found by linear interpolation.

$$t_{Dv1} = 0.003 + \left(\frac{0.478 - 0.4443}{0.5255 - 0.4406} \right)(0.004 - 0.003)$$

$$= 0.0034.$$

This process could be easily performed with a short computer program or on a spreadsheet. Thus, the steam zone is located at the area corresponding to $t_{Dv1} = 0.00356$ when $t = 1$ year.

The area heated by steam is obtained from Eq. 8.167, where $t_{Dv} = t_{Dv1}$.

$$A_s = \left\{ \dot{m}_s H_s / \left[Mv(T_s - T_r) \right] \right\} G_3(t_{Dv1}).$$

From Table 8.32,

$$G_3(t_{Dv1}) = 0.0603 + \left(\frac{0.0034 - 0.003}{0.004 - 0.003} \right)(0.06937 - 0.0603)$$

$$= 0.0639$$

and $A_s = (30.685)(0.0639)$

$$= 1.96 \, \text{acres}.$$

The shape of the steam zone at this time is superimposed on the heated region in **Fig. 8.63.**

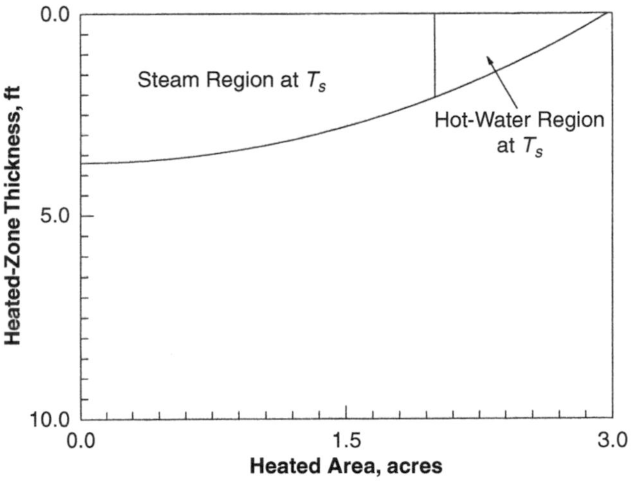

The volume of the steam zone is determined from Eq. 8.181 and Table 8.36. From Eq. 8.181,

$$E_{h,s} = V_s M(T_s - T_r)/\dot{m}_s H_s t.$$

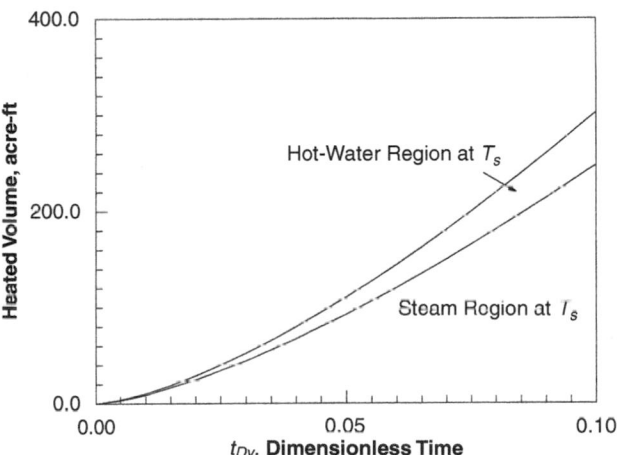

Fig. 8.63—Steam and hot-water zone distribution for uniform-vertical-displacement model.

Fig. 8.64—Heated volumes for containing hot water and steam during steamdrive under uniform-vertical-expansion model, Example 8.15.

Table 8.36 includes values of $E_{h,s}$ as a function of t_{Dv} and $f_{h,v}$. Thus, we want to find $E_{h,s}$ for the following set of parameters: $t_{Dv} = 0.00789$ and $f_{h,v} = 0.0478$. To save double interpolation, assume $t_{Dv} = 0.00789 \approx 0.008$. In Table 8.36, at $t_{Dv} = 0.008$, $f_{h,v} = 0.4$ and 0.5. $E_{h,s} = 0.05679$ at $t_{Dv} = 0.008$, $f_{h,v} = 0.5$, $E_{h,s} = 0.04978$ at $t_{Dv} = 0.008$, and $f_{h,v} = 0.4$.

For $f_{h,v} = 0.478$ and $t_{Dv} = 0.008$,

$$E_{h,s} = 0.04978 + (0.478 - 0.4/0.1)(0.05679 - 0.04978)$$

$$= 0.0552.$$

$$V_s = \dot{m}_s H_s t E_{h,s} / M(T_s - T_r)$$

$$= \frac{(500 \, \text{B/D})(350 \, \text{lbm/bbl})(1006 \, \text{Btu/lbm})(365 \, \text{days})(0.0552)}{(33.13 \, \text{Btu/ft}^3\text{-}°\text{F})(390.7°\text{F})(43,560 \, \text{ft}^2/\text{acre})}$$

$$= 6.29 \, \text{acre-ft}.$$

Recall from Example 8.14 that $V_h = 7.44$ acre-ft at the same time. Therefore, the hot-water zone preceding the steam zone occupies a volume of 1.15 acre-ft after 1 year of injection. **Fig. 8.64** shows the development of the hot-water and steam regions with time for Example 8.15.

Heat Injection Requirement When Heated Area Remains Constant. At some point, heated fluids will break into the production wells. It may be necessary to limit the heat injection rate to the amount needed to maintain vertical expansion of the steam zone. The heat injection schedule can be estimated directly from relationships developed in the previous section. The heat requirement necessary to hold the heated area constant at some value A_1 is given by the right side of Eq. 8.177, which is written in terms of $\dot{Q}_\ell$ as

$$\dot{Q}_\ell = \int_o^{t_1} \dot{Q}_1(t-\lambda)\left(\frac{dA_h}{d\lambda}\right)d\lambda + \int_o^{t_1} \dot{Q}_2(t-\lambda)\left(\frac{dA_h}{d\lambda}\right)d\lambda, \dots\dots\dots\dots\dots\dots\dots\dots\dots\dots \text{(8.182)}$$

where, from Eq. 8.179,

$$\dot{Q}_\ell = \dot{m}_s H_s G_e\left(t_{Dv}, t_{Dv1}\right).$$

Inspection of Eqs. 8.179 and 8.180 reveals that $G_e(t_{Dv}, t_{Dv1})$ is the fraction of the initial heat injection rate that is necessary to maintain vertical expansion of the heated region at a constant velocity v within the area $A(t_{Dv1})$. Thus, to estimate the heat injection schedule once a specified area is heated, it is only necessary to find $G_e(t_{Dv}, t_{Dv1})$. Example 8.16 illustrates the estimation of a heat injection schedule.

Note that the steam zone will not arrive at the heated area until t_{Dv}, where $G_e(t_{Dv}, t_{Dv1}) = f_{h,v}$. Thus, if the area A_1 is to be heated by constant vertical expansion of the steam zone, then heat injection cannot be reduced until $t_{Dv} > t_{Dv1}$, where $G_e(t_{Dv}, t_{Dv1}) = f_{h,v}$.

Example 8.16—Estimation of Heat Injection Schedule for a Uniform Vertical Expansion of the Heated Region in a Specified Area.
This example is a continuation of Example 8.14 and will use all properties and parameters of that example. Consider continuous injection at 500 B/D until the heated area is approximately 5.5 acres. Referring to Table 8.34, A_h = 5.54 acres at t_{Dv} = 0.03 (t = 1,388.4 days). Determine the heat injection schedule for t > 1,388.4 days.

Solution. The heated region ($T = T_s$) arrived at A_h = 5.54 acres after 1,388.4 days of continuous injection. The value of t_{Dv1} = 0.03. Referring to Table 8.35, find values of $G_e(t_{Dv}, t_{Dv1})$ corresponding to $t_{Dv} > t_{Dv1}$ and then compute the injection rates. **Table 8.37** summarizes the results. This solution does not consider energy removed with the produced fluids. This energy would be added to the amount included in Table 8.37.

Estimating Average Steam-Zone Thickness. In heat-balance models of reservoir heating when gravity override is important, two approaches are available. In the first approach, the average vertical thickness of the steam zone is estimated and reservoir heating is computed with the Marx and Langenheim model, as illustrated in Example 8.3. The average steam-zone thickness may be estimated from field data. In **Fig. 8.65** (Blevins and Billingsley 1975), the steam zone is approximately one-half of the total thickness in an observation well located midway between the injection well and producer in a 6-acre pattern. In the Tia Juana steamdrive, de Haan and Schenk (1969) estimated the average steam-zone thickness to be 25% of the average gross interval open, as depicted in **Fig. 8.66.**

Estimation of the steam-zone thickness when gravity segregation occurs is difficult. van Lookeren (1983) developed expressions for the average steam-zone thickness of linear and radial steamdrives using segregated flow theory. The van Lookeren models consider the vertical cross section to consist of a steam zone that is separated by a steam/oil interface from the oil zone, as depicted in **Fig. 8.67.** Both steam and oil zones are assumed to be isothermal at the steam temperature. Two models representing upper and lower bounds on the average steam-zone thickness were developed by assuming relationships between steam flows in the segregated region. For radial flow, the average steam thicknesses are given by Eqs. 8.183 through 8.188.

Case 1.

$$\bar{h}_{st}/h = f\left(A_R\right). \dots\dots\dots\dots\dots\dots\dots\dots\dots\dots\dots\dots\dots\dots\dots \text{(8.183)}$$

Case 2.

$$\bar{h}_{st}/h = A_R\sqrt{\pi/8}\ \text{erf}\left(\sqrt{2}/A_R\right), \dots\dots\dots\dots\dots\dots\dots\dots\dots\dots\dots\dots \text{(8.184)}$$

t_{Dv}	$G_e(tD_V, 0.03)$	Heat Injection Rate [(Btu/D) × 10^6]
0 ≤ 0.03	1.0000	170.98
0.04	0.7095	121.31
0.05	0.6183	105.71
0.06	0.5612	95.95
0.07	0.5207	89.03
0.08	0.4899	83.76
0.09	0.4654	79.57
0.10	0.4453	76.14

Table 8.37—Heat injection schedule to maintain heated area at 5.54 acres with uniform vertical expansion of heated region at 0.01 ft/D, Example 8.16.

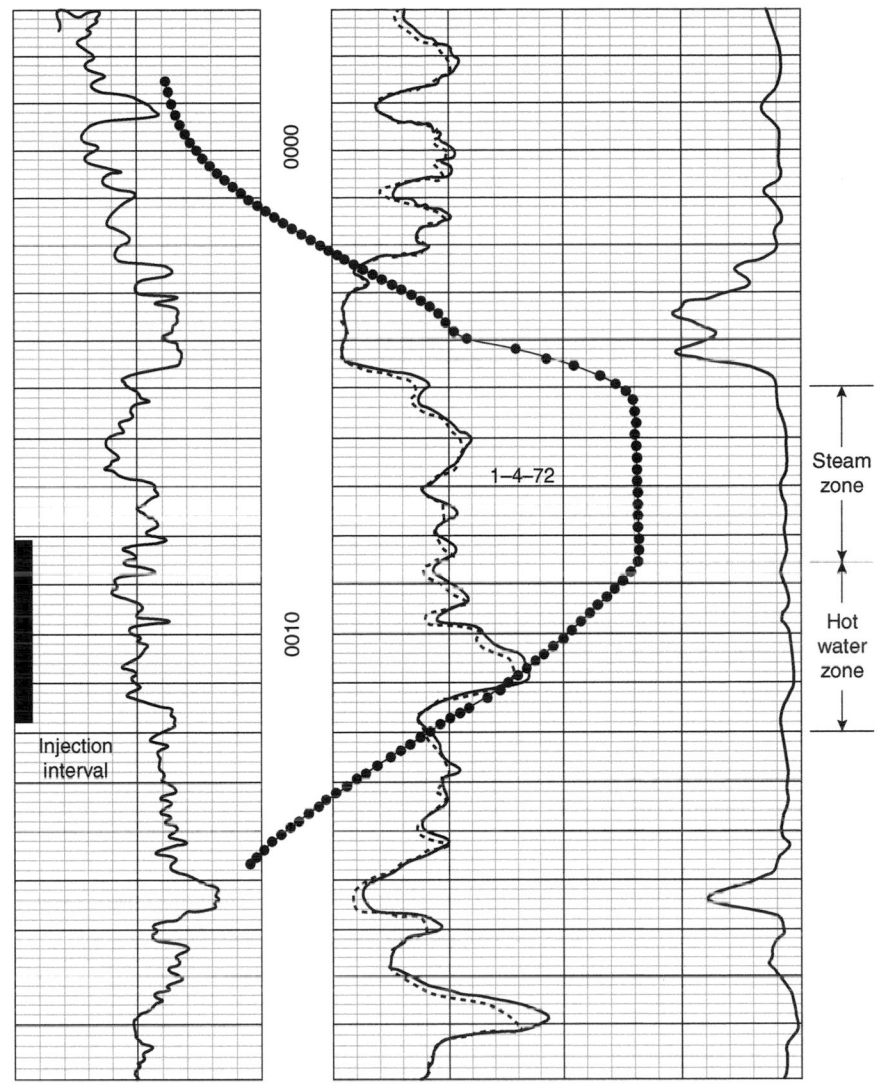

Fig. 8.65—Typical temperature profile showing gravity segregation, Well T6-4 (Blevins and Billingsley 1975).

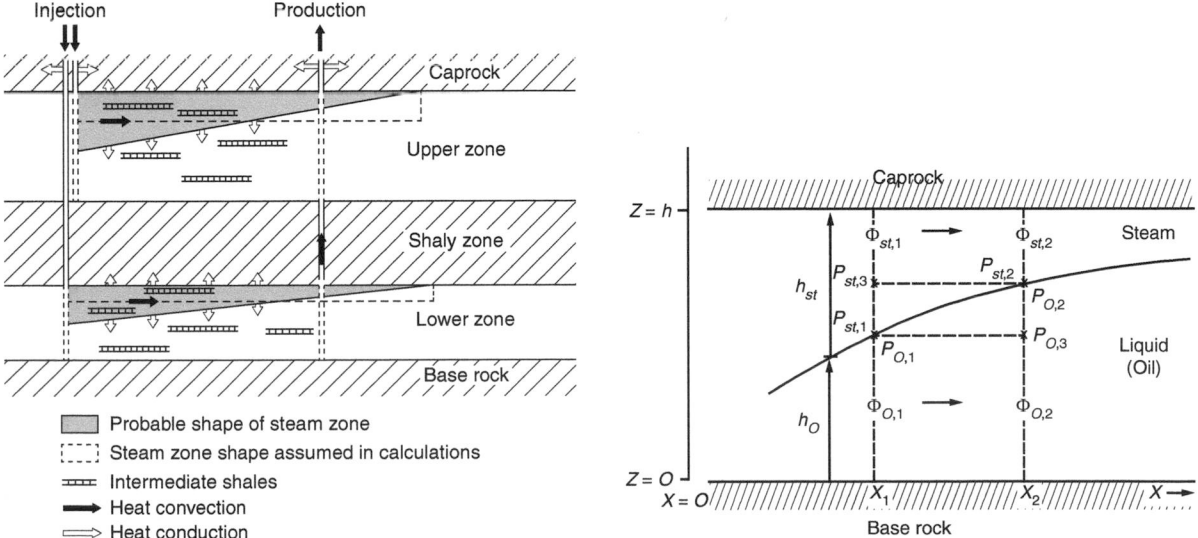

Fig. 8.66—Estimation of steam zone shape, Tia Juana steamdrive (de Haan and Schenk 1969).

Fig. 8.67—Pressures and potentials near steam/liquid interface (van Lookeren 1983).

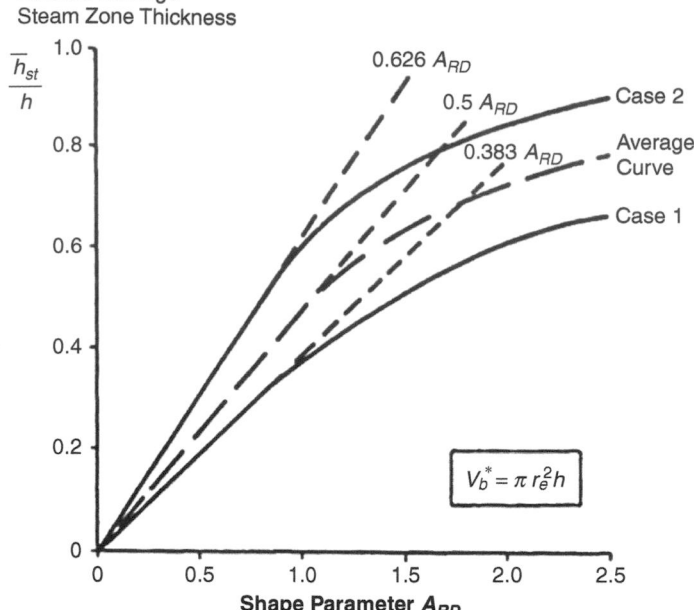

Fig. 8.68—Relative average thickness vs. parameter A_R (van Lookeren 1983).

where $A_R = 7.255 \mu_s \dot{m}_s / (\rho_o - \rho_s) h^2 k_s \rho_s,$ . . . .(8.185)

when oilfield units are used (cp, lbm/ft³, ft, darcies, and lbm/D). **Fig. 8.68** shows correlations representing Cases 1 and 2. For values of $A_R < 1$, the following relationships can be used to compute the average steam-zone thickness.

Case 1.

$\overline{h}_{st}/h = 0.383 A_R.$ . . . . (8.186)

Case 2.

$\overline{h}_{st}/h = 0.626 A_R.$ . . . . (8.187)

Average steam-zone thickness can be determined by averaging Eqs. 8.186 and 8.187 to obtain

$\overline{h}_{st}/h = 0.5 A_R.$ . . . . (8.188)

These correlations may be used to estimate the average steam-zone thickness for subsequent calculations by using the Marx and Langenheim model and to determine the effects of process variables on steam-zone thickness. van Lookeren (1983) reports that steam-zone thicknesses predicted from these models agreed with field data. It should be noted that the assumption of constant temperature used in this model would be valid primarily in the vicinity of the injection well.

Both the Neuman model and the uniform-vertical-expansion model may be used to estimate steam-zone thickness when gravity override occurs. The average steam-zone thickness is defined by

$\overline{h}_s = A_s / A_s.$ . . . . (8.189)

In the Neuman model,

$\overline{h}_s = (C_w \Delta T_s k_h / M_s L_v) \sqrt{\pi t / \alpha}.$ . . . . (8.190)

For the uniform-vertical-expansion model,

$\overline{h}_s = vt \left[E_{h,s}(t_{Dv}, t_{Dv1}) / G_s(t_{Dv1}) \right].$ . . . .(8.191)

8.4.3 Reservoir Heating by Steam Injection Into a Fracture. Some reservoirs have limited or no transmissibility at reservoir temperature as a result of high oil saturations and oil viscosity. Steam can be injected into these reservoirs if a horizontal fracture is created between injection and production wells. Britton et al. (1983) describe a steam displacement process in which a horizontal fracture was created between an injection well and producing wells. Steam was injected at a high rate to heat the San Miguel tar-sand reservoir and subsequently to displace the tar. The San Miguel sand contained –2 °API tar. Although permeabilities of the clean sand were on the order of 250 to 1,000 md, the oil was immobile at reservoir conditions. Powers et al. (1985) used horizontal fractures in a steam displacement project in the Loco field, Stephens County, Oklahoma, USA.

Reservoir heating during the injection of steam into a fracture can be estimated with the models developed in this chapter. The injection rate is assumed for the heated-zone thickness, and then the heated area is computed as a function of time. Steam-zone growth in a fractured reservoir is controlled by the ratio of injection rate to thickness. Example 8.17 illustrates application of the Marx and Langenheim model to predict steam-zone growth in a fracture.

Example 8.17—Estimation of Heated Area in a Horizontal Fracture. A reservoir developed on 10-acre five-spot patterns is to be heated by injecting steam into a horizontal fracture. Reservoir thickness is 50 ft, and the thermal properties of the reservoir and overburden are identical to those used in Example 8.4. Steam of 80% quality is injected into the fracture at a rate of 1,000 BWPD at 500 psig. Assume that the fracture remains open throughout the reservoir heating process.

Although the pattern area is 10 acres, after 9 months (270 days) of continuous injection, hot fluid breaks into the production wells. Determine the heated area when steam reaches the production well.

Solution. The fracture is represented by a thin region of high porosity (40%) and high permeability. For the purposes of this example, the effective fracture width is taken as ¼ in. (0.021 ft). The effective width would increase with time because oil would be displaced from the reservoir through the fracture. Oil displacement will be considered in later sections. Thermal properties of the reservoir and overburden will be assumed to be identical.

For the fracture with 40% porosity filled with water,

$$M_R = (0.6)\left(167\,\text{lbm/ft}^3\right)\left(0.21\,\text{Btu/lbm} - °\text{F}\right)$$
$$+ (0.4)\left(50.6\,\text{lbm/ft}^3\right)\left(1.036\,\text{Btu/lbm-°F}\right)$$
$$= 21.04 + 20.97$$
$$= 42.01\,\text{Btu/ft}^3\text{-°F}.$$

From Example 8.4, $M_s = 31.12$ Btu/ft³-°F.

$$t_D = 4\left(M_s / M_R\right)^2 \left(\alpha / h^2\right) t$$
$$= 4\left(\frac{31.12\,\text{Btu/ft}^3\text{-°F}}{42.01\,\text{Btu/ft}^3\text{-°F}}\right)^2 \frac{\left(0.0482\,\text{ft}^2\,/\text{hr}\right)}{\left(0.021\,\text{ft}\right)^2}$$
$$\times \left(t\,\text{days}\right)\left(24\,\text{hr/D}\right)$$
$$= 5,758t$$
$$= (5,758)(270\ \text{days})$$
$$= 1.555 \times 10^6.$$

$$\dot{m}_s = (1,000\ \text{B/D})(350\ \text{lbm/bbl})(24\ \text{hr/D})$$
$$= 14,583\ \text{lbm/hr}.$$

$$A_h = \left\{\left[\left(14,583\ \text{lbm/hr}\right)\left(1006\ \text{Btu/lbm}\right)\left(42.01\ \text{Btu/ft}^3\text{-°F}\right)\right.\right.$$
$$\times \left(0.021\ \text{ft}\right)\right]/\left[4\left(470.9°\text{F-}80°\text{F}\right)\left(0.0482\,\text{ft}^2\,/\text{hr}\right)\right.$$
$$\left.\left.\times \left(31.12\ \text{Btu/ft}^3\text{-°F}\right)^2\right]\right\}G\left(t_D\right)$$
$$= 177G\left(t_D\right).$$

In principle, it should be possible to determine G for each value of t_D. However, serious computation problems arise immediately. The value of t_D is large for any reasonable value of t, so that evaluation of $e^{t_D}\text{erfc}\sqrt{t_D}$ involves multiplication of a large number by a small number. There is a series approximation to erfc(x) for large values of x that resolves this problem. From Carslaw and Jaeger (1959), for large x,

$$\text{erfc}\left(x\right) \approx \frac{e^{-x^2}}{\sqrt{\pi}}\left[\frac{1}{x} - \frac{1}{2x^3} + \frac{1.3}{2^2 x^5} - \frac{(1.3)(5)}{2^3 x^7} + \cdots\right]. \quad\quad\quad\quad\quad\quad\quad (8.192)$$

Thus, for large x,

$$e^{x^2}\operatorname{erfc}(x) = \frac{1}{\sqrt{\pi}}\left[\frac{1}{x} - \frac{1}{2x^3} + \frac{1.3}{2^2 x^5} - \frac{(1.3)(5)}{2^3 x^7} + \cdots\right]. \quad\quad\quad (8.193)$$

Inspection of Eq. 8.193 shows that $e^{x^2}\operatorname{erfc}(x)$ goes to zero for large x. Thus, for values of t_D typical of steam injection in a fracture,

$$G(t_D) = 2\sqrt{t_D/\pi} - 1$$

$$= 2\sqrt{(1.555\times10^6)/\pi} - 1$$

$$= 14.06. \quad\quad\quad\quad\quad\quad\quad\quad\quad\quad\quad\quad (8.194)$$

Solving for A_h gives

$$A_h = 177G(t_D)$$

$$= (177)(1,406)$$

$$= 248,845 \text{ ft}^2$$

$$= 5.71 \text{ acres.}$$

This example shows that the arrival of the heat front at a production well can be used to estimate the heated area. Heated-area-vs.-time data for other areas are summarized in **Table 8.38.**

Heat losses from the steam zone can be computed with the heat-loss relationships presented earlier. However, if the fracture extends through the middle of the reservoir, most of the heat loss from fluids flowing in the fracture will contribute to reservoir heating. Temperature distributions above and below the fracture are computed as illustrated in Example 8.3.

Example 8.17 shows that it is possible to heat a reservoir through a fracture, with modest steam injection rates. However, long injection times are required as the heated area increases. Remember that there was no consideration in Example 8.17 of oil displacement through the fracture as a result of reservoir heating or other displacement processes. The reservoir heating calculation assumes that the effective fracture width is constant throughout heating and contains water vapor, water, and rock.

8.5 Estimation of Oil Recovery From Steamdrive

Methods of predicting the volume of a reservoir contacted by steam and hot water that were not coupled with displacement calculations were developed in Section 8.2. This can be done because steam is an efficient displacing fluid for many crude oils. Three factors promote high displacement efficiency of a steamdrive: (1) large reduction in oil viscosity with temperature in heavy crude oils, (2) steam distillation of some crude oils, and (3) the stability of the steam zone. These effects combine to give steam recovery efficiencies that are greater than hot waterfloods. Gravity drainage also contributes to the reduction of oil saturation in many reservoirs (Vogel 1984, 1992). This section covers displacement mechanisms and the prediction of steamdrive performance with approximate models.

8.5.1 Displacement Mechanisms. A steamdrive is a complex process. In the simplest representation presented in Section 8.4, the steam zone was assumed to occupy the entire heated region until the critical time was reached. Then, a hot-water zone formed that preceded the steam zone. In these models, oil is assumed to be displaced to residual values in each zone.

Heated Area (acres)	Heated Area (ft²)	Time for Heated-Zone Arrival, t_D	Time for Heated-Zone Arrival (days)
0.2	8,712	2,330	0.38
0.5	21,780	13,926	2.25
1.0	43,560	54,870	8.86
2.0	87,120	217,821	35.2
5.0	217,800	1,355,186	218.9
5.56*	242,112	1,672,000	270.0
10.0	435,600	5,412,495	874.1

1 acre = 43,560 ft².

Table 8.38—Heated-area/time estimates for steam injection in a fracture, Example 8.17.

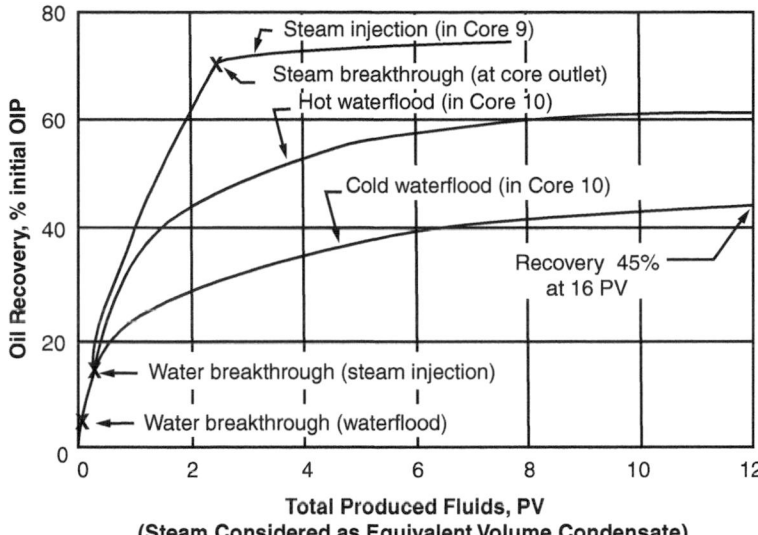

Fig. 8.69—Oil recovery by steam injection, hot waterflood, and cold waterflood in cores initially containing 12.2 °API Ba-chaqueu crude and connate water (Willman et al. 1961).

Willman et al. (1961) identified the principal displacement mechanisms for a steamdrive from the results of a set of laboratory experiments on short and long cores. Crude oil viscosity at ambient temperature varied from 5.4 to 6,500 cp. Waterfloods were conducted at ambient temperature (80°F) and 330°F. Steamfloods were conducted at 330 and 520°F (800 psig). Typical results are shown in **Fig. 8.69,** where oil recoveries by steam injection and hot and cold waterfloods are compared for a heavy crude oil. Recovery from steam injection was significantly higher than that for either hot or cold waterflood. Furthermore, fewer PVs of injected fluid were required for steam injection compared with hot or cold water injection.

In analyzing the results of these tests, Willman et al. (1961) found that viscosity reduction and thermal expansion accounted for the increased recovery in hot waterfloods. The residual oil saturation to waterflooding is not a strong function of temperature in some reservoirs. Consequently, hot waterflood performance can be predicted by applying fractional-flow concepts to conventional waterflooding displacement calculations (Prats 1982; Willman et al. 1961).

In all cases where the oil contained volatile components, recovery from steamdrive was larger than from the corresponding hot waterflood at the same temperature. The additional recovery from steamdrive is primarily the result of steam distillation of oil remaining behind the hot waterflood. Steam distillation occurs because when a vapor phase (steam) is in the presence of two immiscible liquids (water and oil), each liquid phase exerts its own vapor pressure at the temperature of the system. Distillation begins when the sum of the vapor pressures equals the total pressure on the system. As a result, oil will begin distilling at a temperature much lower than the normal boiling point of its constituents. In principle, 100% recovery of a volatile oil is possible by steam distillation if the oil is 100% steam distillable at the temperature of the steamdrive.

An example of the importance of steam distillation is given in **Table 8.39,** where the results are presented of three tests involving steamdrive of a white oil containing 0 and 50% distillable content. Presence of distillable components increased the recovery from 54.8 to 76% when 25% of the oil was steam distillable at injection conditions. Recovery increased from 58% (following the hot waterflood) to 83.9% (from the steamdrive) for an oil that was 50% steam distillable at injection conditions. However recoveries in the steam runs were larger than expected from steam distillation effects. In other experiments, Willman et al. (1961) found that an additional 3% of the OOIP could be attributed to simultaneous gas-/waterdrive in the region no longer containing distillable oil.

Test	Core	Oil	Viscosity at 80°F (cp)	Oil Recovery (%OOIP)		
				Waterflood at 80°F	Hot Waterflood (330°F)	Steamdrive
13	9	Nondistillable white oil	138	49.5	54.8	59.0
15	9	25% distillable white oil	22.5	—	54.8	76.0
16	9	50% distillable white oil	5.4	56.0	58.0	83.9
18	9	Crude B	6,500	—	—	73.8
19	10	Crude B	6,500	45.0	59.8	—
20	9	Crude C	8.2	60.0	67.0	91.8

Table 8.39—Oil recovery by steam injection into torpedo cores containing oil and connate water (Willman et al. 1961).

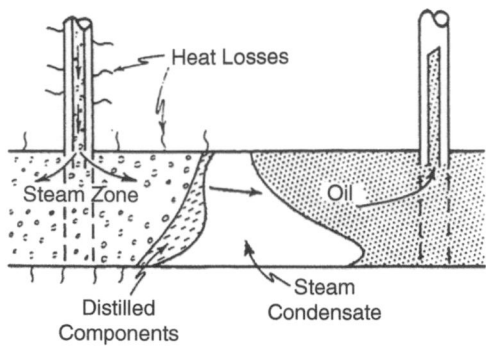

Fig. 8.70—Schematic of the steam distillation process (Volek and Pryor 1972).

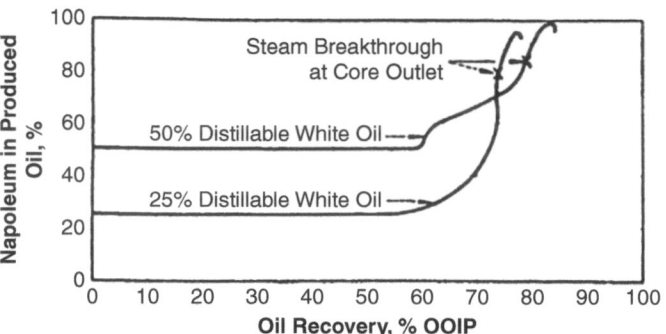

Fig. 8.71—Composition of the oil produced during steam injection into Core 9 initially containing the two steam distillation white oils and connate water (Willman et al. 1961).

	Recovery (% OIP)		
	Nondistillable	23% Distillable	50% Distillable
1. Hot waterflood recovery (includes viscosity reduction and swelling)	54.8	54.8	58.0
2. Extra recovery caused by gasdrive effects	3.0	3.0	3.0
3. Extra recovery owing to distillation*	—	10.5	19.5
4. Predicted recovery based on above mechanisms	57.8	68.3	80.5
5. Actual recovery by steam	59.0	76.0	83.9
6. Unexplained recovery (Line 5 minus Line 4)	1.2	7.7	3.4

*(Percent of oil that is steam-distillable) × [100 − (Line 1 + Line 2)].

Table 8.40—Steamflood recoveries compared with predicted recoveries without solvent extractions (Willman et al. 1961).

The additional recovery in Table 8.39 for distillable oils is thought to be caused by the solvent extraction effects on the residual oil ahead of the steam displacement front. Steam distillation strips the more volatile components from the residual oil. Steam, enriched by hydrocarbons, flows through the steam zone to the condensation front, where both steam and hydrocarbons condense, forming a multiphase region, as depicted in **Fig. 8.70** (Volek and Pryor 1972). The condensed hydrocarbon contains lighter constituents than the oil resident in that region and can displace some of that oil miscibly. Evidence of solvent extraction is shown in **Fig. 8.71** (Willman et al. 1961), where the percentage of the distillable component (napoleum) in the oil is plotted against oil recovery for the two tests presented in **Tables 8.39 and 8.40.** A miscible zone of high distillable component content is clearly present.

Further support for the solvent extraction mechanism was developed by Volek and Pryor (1972). They conducted two displacement experiments on sandpacks saturated with 24 °API oil that had been waterflooded to residual oil saturation. One experiment

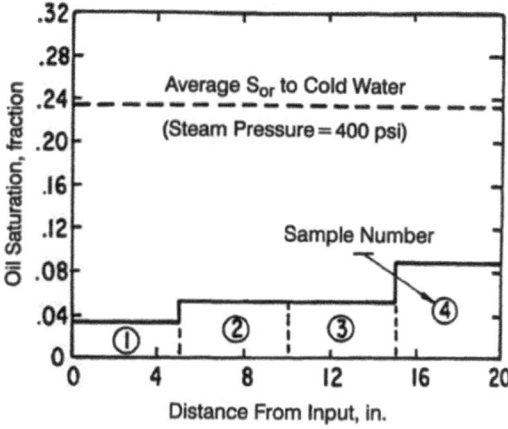

Fig. 8.72—Residual oil saturations after steam breakthrough (Volek and Pryor 1972).

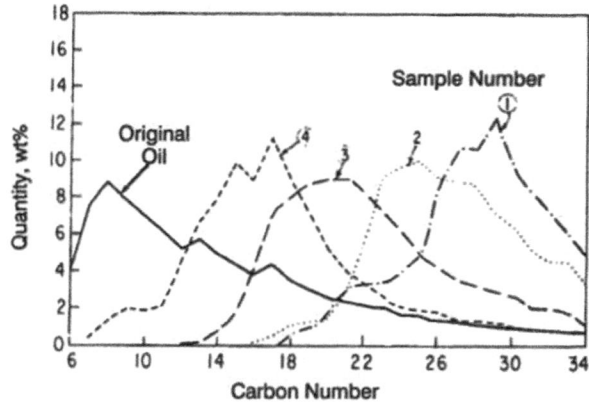

Fig. 8.73—Hydrocarbon component distribution in residual oil after steamdrive (Volek and Pryor 1972). See Fig. 4 of Volek and Pryor (1972) for description of sample numbers.

was terminated at steam breakthrough, and the residual oil saturation was determined at samples taken from the intervals shown in **Fig. 8.72.** Compositions of the oil taken from each interval are shown in **Fig. 8.73.** Fig. 8.73 shows that the residual oil becomes progressively depleted of light constituents (low carbon number) with distance from the inlet. **Fig. 8.74** shows residual oil saturations from the third experiment, which was terminated in the middle of the run. Residual oil saturations averaging less than 2% were found in the steam zone.

Residual oil saturations from cores recovered in steamdrives are often on the order of 0 to 10%. **Table 8.41** presents representative data from several steamdrives. Prats (1982) points out that low residual oil saturations in steamdrives of heavy oils require the formation of an adequate solvent bank. Thus, he concludes that wells near the injection well may not have low oil saturations. Prats also expresses concern about residual oil saturations determined in short laboratory tests.

Gravity override occurs in most steamfloods, with low-density steam rising to the top of the formations. Oil displacement during a steamflood also occurs by gravity drainage. This mechanism may contribute to the low residual oil saturations observed in field cores (Vogel 1992).

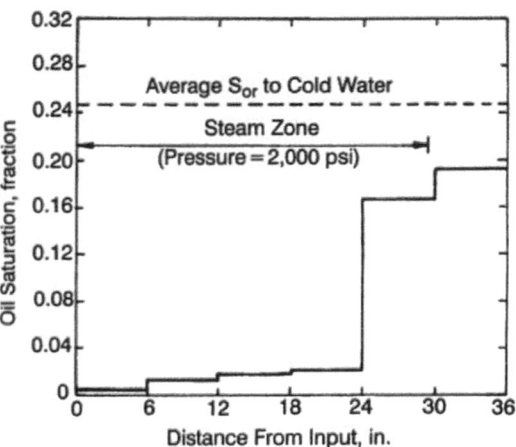

Fig. 8.74—Residual oil saturation after steamdrive (Experiment 3) (Volek and Pryor 1972).

8.5.2 Stability of Steamdrives. The steamdrive process is remarkably stable (Prats 1982; Baker 1973; Harmsen 1971). If the stability of a steamdrive is analyzed with the conventional mobility ratio approach described in Chapter 4, large mobility ratios are obtained because the viscosity of steam is small. However, steamdrives behave as if any viscous fingers that form are limited in extent by condensation and thus are self-healing. When gravity override occurs, the steam front is stabilized by gravity forces.

Prats (1982) includes an excellent discussion of steamdrive stability. An equivalent mobility ratio for steamdrives was defined to explain the stability of a steamdrive. The equivalent mobility ratio is defined in terms of upstream and downstream pressure gradients. His analysis shows that steamdrives are inherently more stable than hot waterfloods. Favorable equivalent mobility ratios have been obtained for steamdrives with temperatures up to 370°F. The stability of steamdrives is not well-understood and undoubtedly will be the subject of further research.

8.5.3 Displacement Performance. Oil displacement from a steamdrive is predicted by use of simple models, such as the Marx and Langenheim or gravity-override models. The region swept by steam is assumed to be at residual oil saturation to steam as if the displacement process is piston-like, as depicted in **Figs. 8.75 and 8.76.** This is justified because of the favorable equivalent mobility ratio for the steamdrive process. The steamdrive residual oil saturation is determined from laboratory displacement tests or field experiments.

At any point in the steamdrive, the steam volume can be estimated with models presented in Sections 8.4.1 and 8.4.2. Then, the amount of oil displaced from the steam zone is given by

$$N_{ps} = \left(\frac{7{,}758 \text{ bbl}}{\text{acre-ft}} \right) \phi \frac{h_n}{h_t} \left(\frac{S_{oi}}{B_{o1}} - \frac{S_{ors}}{B_{os}} \right) V_s, \quad\dotsfill (8.195)$$

where N_{ps} = volume of oil displaced by steam, STB; h_t = reservoir thickness, ft; h_n = net reservoir thickness, ft; S_{oi} = initial oil saturation before steamdrive; S_{ors} = residual oil saturation to steamdrive; V_s = volume of steam zone, acre-ft; B_{o1} = FVF at start of steamdrive, bbl/STB; and B_{os} = FVF in zone swept by steam, bbl/STB.

The steam-swept volume, V_s, is obtained from Eq. 8.61 for the frontal-advance model. When the gravity-override model is used, Eq. 8.159 or 8.174 is used to find V_s. Remember that the dimensionless times are defined differently for each model. From Eq. 8.174,

$$E_{h,s} = M_R V_s (T_s - T_r) / \dot{m}_s H_s t. \quad\dotsfill (8.196)$$

Test	Laboratory	Field Core
Kern	10.9	8.7
San Joaquin	12.4	4.4
Kern A	12.8	5.7
Inglewood	20.0	22.0
Tia Juana	15.0	–
Schoonebeek	–	8.0

Table 8.41—Residual oil saturations for steamdrive from laboratory and field tests.

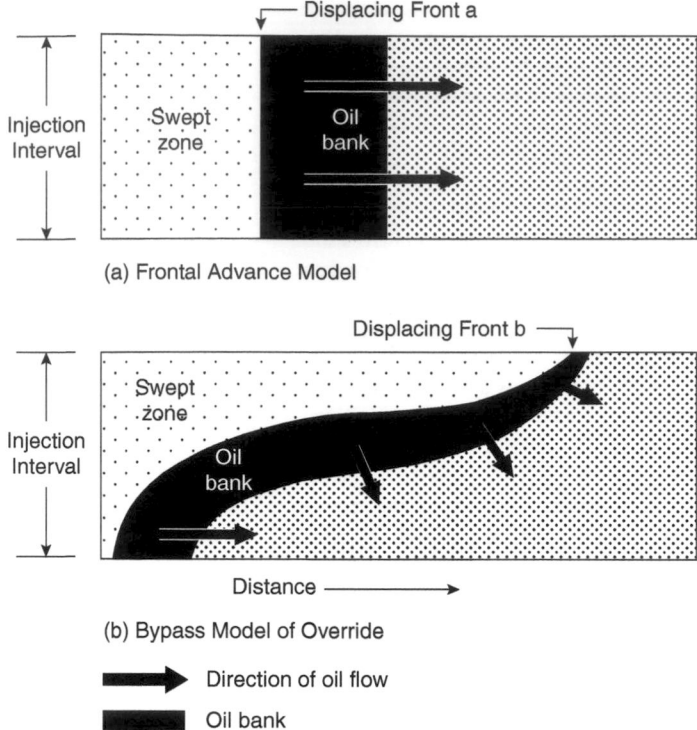

Fig. 8.75—Idealization of oil-bank distribution in frontal-advance and gravity-override models (Prats 1982).

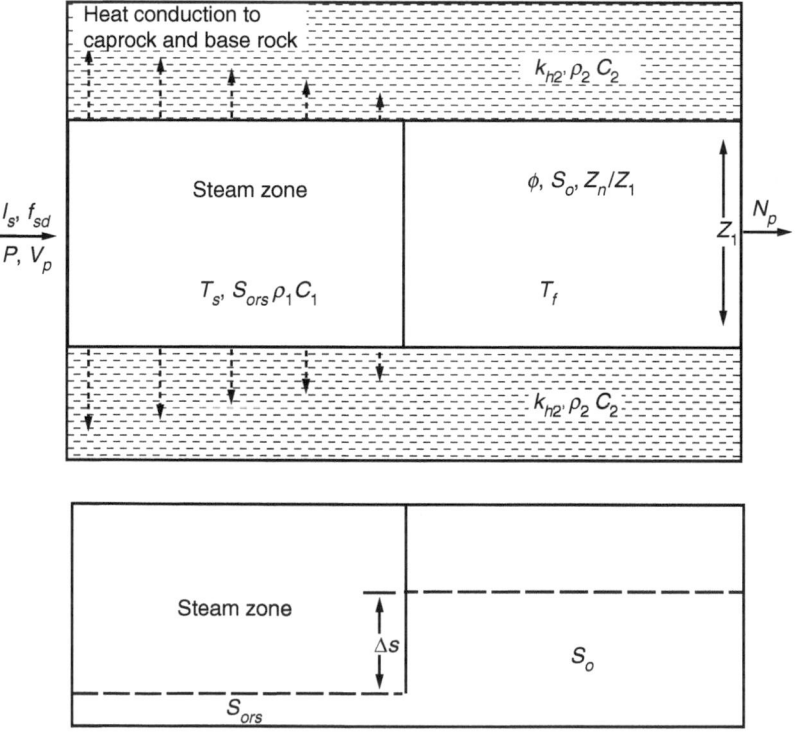

Fig. 8.76—Saturation and temperature distribution for frontal-advance model (Myhill and Stegemeier 1978).

Thus, $V_s = [\dot{m}_s H_s t / M_R (T_s - T_r)] E_{h,s}(t_D).$ .... (8.197)

When the critical time for maintenance of a steam zone is exceeded, part of the heated region is displaced by hot water. Oil displaced from this region may be estimated by assuming an average residual saturation for the hot-water-swept region. Oil displaced from the hot-water-swept region is given by

$$N_{pw} = \left(\frac{7{,}758 \text{ bbl}}{\text{acre-ft}}\right)\phi \frac{h_n}{h_t}\left(\frac{S_{oi}}{B_{o1}} - \frac{S_{ors}}{B_{os}}\right)(V_h - V_s).$$ .... (8.198)

Oil displaced by injection of steam would be the sum of the oil displaced by steam and hot-water regions.

Myhill and Stegemeier Modification to Frontal-Advance Model. Myhill and Stegemeier (1978) account for the combined effects of the steam and hot-water regions in a uniform-displacement model by defining an average steam zone displacement efficiency according to

$$\bar{E}_{h,s} = E_h - (1 - f_{h,v})(E_h - E_{lb})$$ .... (8.199)

and $E_{lb} = \dfrac{1}{t_D \sqrt{\pi}}\left[2\sqrt{t_D} - 2\left(1 - f_{h,v}\right)\sqrt{t_D - t_{cD}}\right.$

$$\left. - \int_0^{t_{cD}} \frac{e^u \text{erfc}\sqrt{u}}{\sqrt{t_D - u}} \, du, \right.$$ .... (8.200)

where E_{lb} = lower bound of steamdrive efficiency based on formation of a hot-water zone when the critical time, t_{cD}, is exceeded during the steamdrive.

Substituting Eqs. 8.120 and 8.200 into Eq. 8.199 gives

$$\bar{E}_{h,s} = \frac{1}{t_D}\left\{G(t_D) + \frac{\left(1 - f_{h,v}\right)U\left(t_D - t_{cD}\right)}{\sqrt{\pi}}\left[2\sqrt{t_D} - 2\left(1 - f_{h,v}\right)\right.\right.$$

$$\left.\left. \times \sqrt{t_D - t_{cD}} - \int_0^{t_{cD}} \frac{e^u \text{erfc}\sqrt{u}}{\sqrt{t_D - u}} \, du - \sqrt{\pi}G(t_D)\right]\right\}.$$ .... (8.201)

The term $U(t_D - t_{cD}) = 1$ for $t_D > t_{cD}$ and 0 for $t_D \le t_{cD}$.

Fig. 8.77 shows the correlation of the average thermal efficiency of the steam zone with dimensionless time (Prats 1982). The weighting factor $(1 - f_{h,v})$ is empirical (Myhill and Stegemeier 1978). The correlation should not be used for low-quality steamdrives ($f_{sd} < 0.2$) because it does not adequately account for oil displacement from the hot-water region (Myhill and Stegemeier 1978).

Eqs. 8.195 and 8.198 give estimates of oil displaced by the steam and hot-water regions. In most cases, more oil is displaced than produced. This may be owing to resaturation of gas-depleted regions or displacement outside of the region being stimulated. As a result, adjustments are made in the displacement models to calibrate field results to predicted performance.

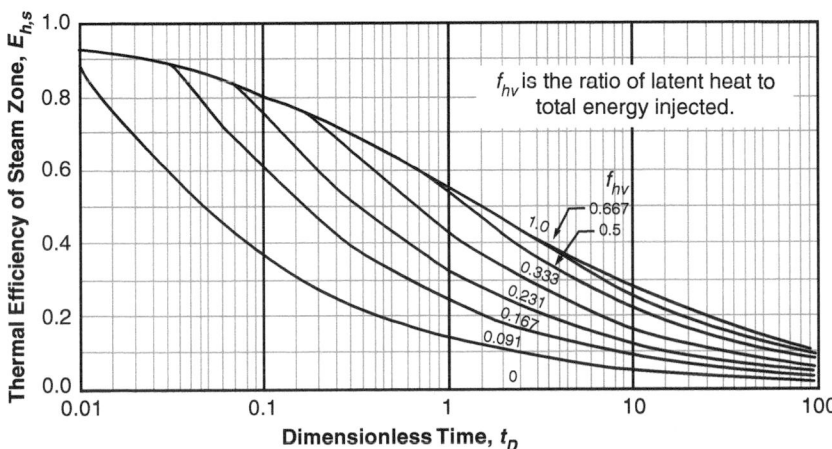

Fig. 8.77—Fraction of heat injected into steamflood remaining in the steam zone (Prats 1982).

One approach is to introduce a capture factor, E_c, into Eqs. 8.195 and 8.198. Capture factors range from 70 to 100%, depending on the field.

Myhill and Stegemeier (1978) used the modified frontal-advance model to predict the ultimate SOR for a steamdrive. In developing this model, they observed that mature steamdrives converge to an OSR determined by reservoir and steam properties and time. The OSR is defined by

$$F_{os} = N_p / W_{s,eq}, \dots\dots\dots\dots\dots\dots\dots\dots\dots\dots\dots\dots\dots\dots\dots\dots \quad (8.202)$$

where $W_{s,eq}$ = volume of steam injected as equivalent barrels of water.

The term $W_{s,eq}$ is defined relative to the average temperature of the boiler feed water, T_A, and the properties of the steam at the boiler outlet by Eq. 8.203. Note that $W_{s,eq}$ assumes the steam leaving the boiler is 1,000 Btu/lbm.

$$(W_{s,eq})(62.4 \text{ lbm/ft}^3)(350 \text{ lbm/bbl})(1,000 \text{ Btu/lbm})$$

$$= m_s \left(H_{wT} - H_{wA} + f_{sb}L_v \right), \dots\dots\dots\dots\dots\dots\dots\dots\dots\dots\dots\dots\dots\dots \quad (8.203)$$

where the subscript b refers to the outlet of the boiler. Thus,

$$W_{s,eq} = 2.854 \times 10^{-6} m_s \left(H_{sbA} \right), \dots\dots\dots\dots\dots\dots\dots\dots\dots\dots\dots\dots\dots \quad (8.204)$$

where H_{sbA}, the enthalpy of the steam at the boiler outlet relative to the feedwater temperature, is equal to $(H_{ws} + f_{sb}L_v - H_{wA})$. Eq. 8.204 is a precise definition of the equivalent volume and permits comparison with other projects on the same basis. The SOR is just the reciprocal of Eq. 8.204:

$$F_{so} = W_{s,eq} / N_p. \dots\dots\dots\dots\dots\dots\dots\dots\dots\dots\dots\dots\dots\dots\dots\dots\dots\dots \quad (8.205)$$

Myhill and Stegemeier computed OSRs for 11 steamdrives with their method. **Table 8.42** presents reservoir, rock, and steamdrive properties, and **Table 8.43** presents the calculated results, which are plotted in **Fig. 8.78**. In general, the predicted SOR is larger than the field equivalent SOR. Application of a capture factor of 71% would bring seven projects into agreement with field equivalent SORs. Whether the capture factor reflects the inability to capture all displaced oil or simply accounts for inadequacies in the calculation technique has not been resolved. However, these comparisons show that it is possible to estimate SORs for mature steamdrives with an uncertainty of approximately 30%. This is adequate for screening studies for many reservoirs.

Example 8.18—Calculation of Ultimate SOR With the Myhill and Stegemeier Model. Data for the Yorba Linda F steamdrive are provided by Myhill and Stegemeier (1978). Steam is injected into a reservoir at a rate of 850 B/D and a pressure of 200 psig. The feedwater temperature entering the boiler is 70°F, and the steam quality leaving the boiler is 0.8. Heat losses in flowlines and the wellbore will reduce the steam quality to 0.7 at the injection sandface. Reservoir temperature is 110°F. Other properties are summarized in the following.

Example 8.18 illustrates the calculation of the ultimate SOR for the Yorba Linda F steamdrive in Table 8.42.

$$
\begin{aligned}
p_s &\approx 215 \text{ psia,} \\
T_s &= 387.9°\text{F,} \\
L_{vdh} &= 837.4 \text{ Btu/lbm,} \\
f_{sd} &= 0.7, \\
H_{wr} &= 77.94 \text{ Btu/lbm at } 110°\text{F,} \\
h &= 32 \text{ ft,} \\
H_{wT} &= 361.91 \text{ Btu/lbm at } 387.9°\text{F,} \\
\phi &= 0.30, \\
H_{wA} &= 38 \text{ Btu/lbm at } 70°\text{F,} \\
\Delta S_o &= 0.31, \\
M_R &= 35 \text{ Btu/ft}^3\text{-}°\text{F,} \\
M_s &= 42 \text{ Btu/ft}^3\text{-}°\text{F,} \\
\alpha &= (1.2 \text{ Btu/hr-ft-}°\text{F})/(42 \text{ Btu/ft}^3\text{-}°\text{F}) \\
&= 0.0286 \text{ ft}^2/\text{hr} \\
&= 0.6857 \text{ ft}^2/\text{D,} \\
k_h &= 1.2 \text{ Btu/hr-ft-}°\text{F.}
\end{aligned}
$$

Solution. From Example 8.6, when t is in days,

Field	Number of Injectors	Thermal Properties* T_r (°F)	Petrophysical Properties					Steam Parameters			A (acres/well)	t (years)
			h_t (ft)	h_n/h_t	ϕ	ΔS**	f_{sb}	f_{sdh}	p_s (psig)	i_s (B/D)		
Brea ("B" sand)[a]	4	175	300	0.63	0.22	0.40	0.75	0.54	2,000	500	10	8
Coalinga (Section 27, Zone 1)[b]	40	96	35	1.0	0.31	0.37	0.7	0.55	400	500	9.2	4
El Dorado (northwest pattern)[c]	4	70	20	0.85	0.26	0.20	0.75	0.45	500	200	1.6	1
Inglewoodd	1	100	43	1.0	0.37	0.40	0.75	0.7	400	1,100	2.6	1
Kern Rivere	85	90	55	1.0	0.32	0.40	0.7	0.5	100	360	2.5	5
Schoonebeek[f]	4	100	83	1.0	0.30	0.70	0.85	0.7	600	1,250	15	6
Slocum (Phase 1)[g]	7	75	40	1.0	0.37	0.34	0.8	0.7	200	1,000	5.65	2.5
Smackover[h]	1	110	50	0.5	0.36	0.55	1.0	0.8	390	2,500	10	1
Tatums (Hefner steamdrive)[i]	4	70	66	0.56	0.28	0.55	0.7	0.6	1,300	685	10	5
Tia Juana[j]	7	113	200	1.0	0.33	0.50	1.0	0.8	300	1,400	12	5.3
Yorba Linda ("F" sand)[k]	2	110	32	1.0	0.30	0.31	0.8	0.7	200	850	35	4.5

*M_R = 35 Btu/ft³-°F, M_S = 42 Btu/ft³-°F, and k_{hs} = 1.2 Btu/ft-hr°F.

**ΔS = Oil saturation at start of steaming minus the change in oil saturation from estimated primary recovery during steamdrive period minus 0.15. The term 0.15 represents an average value of the residual oil saturation after steamdrive.

[a]Volek and Pryor (1972)

[b]Choquette et al. (1993)

[c]Pebdani et al. (1988)

[d]Blevins et al. (1969)

[e]Bursell (1970)

[f]Afoeju (1974)

[g]Hearn (1972)

[h]van Dijk (1978)

[i]Hall and Bowman (1973)

[j]de Haan and Schenk (1969)

[k]Myhill and Stegemeier (1978)

Table 8.42—Steamdrive field projects—summary of conditions (Myhill and Stegemeier 1970).

Field	Quantity of Steam Injected,* W_{sD}	Steam-Zone Size,** V_{sD}	Calculated Additional Equivalent OSR (vol/vol)	Field Additional Equivalent OSR (vol/vol)
Brea	0.5	0.15	0.13	0.14
Coalinga	0.94	0.45	0.16	0.18
El Dorado	1.6	0.315	0.05	0.02
Inglewood	1.26	1.256	0.41	0.28
Kern River	1.92	1.139	0.32	0.26
Schoonebeek	0.95	0.617	0.43	0.35
Slocum	1.41	1.202	0.29	0.18
Smackover	1.23	0.756	0.27	0.21
Tatums	1.54	0.397	0.13	0.10
Tia Juana	0.47	0.551	0.59	0.37
Yorba Linda "F"	0.54	0.280	0.16	0.17

*$W_{sD} = W_{s,eq}/(7758Ah_t)$.

**$V_{sD} = V_s/(7758\,Ah_t)$.

Table 8.43—Comparison of field results.

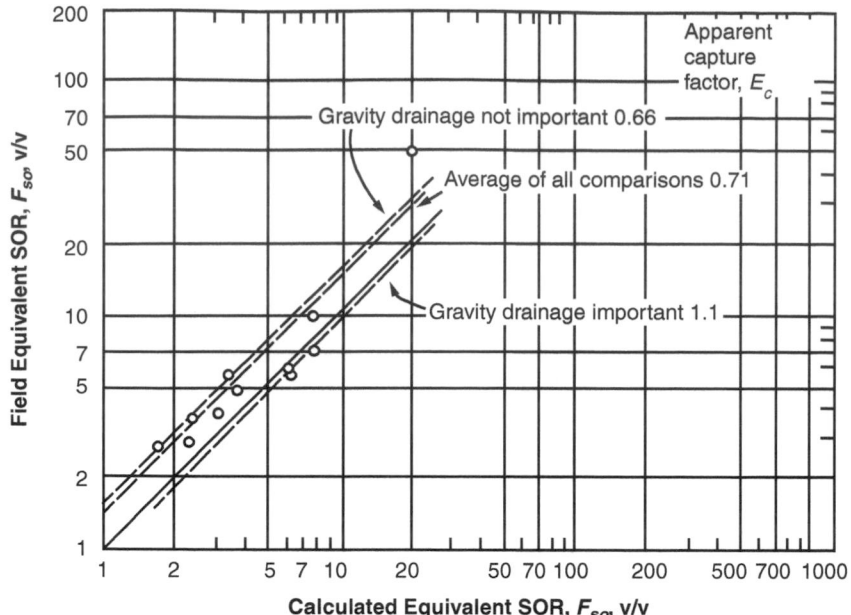

Fig. 8.78—Comparison of calculated and observed SORs for field projects (Prats 1982).

$$t_D = 4(M_s / M_R)^2 (\alpha / h^2)t$$
$$= 4(42 / 35)^2 \left[0.6857 / (32 \text{ ft})^2 \right] t$$
$$= 3.857 \times 10^{-3} t.$$

When t is in years,

$$t_D = 1.408t,$$

so that at $t = 4.5$ years,

$$t_D = 6.335,$$
$$H_s = 361.91 + (0.7)(837.4) - 77.94$$
$$= 870.15 \text{ Btu/lbm,}$$

and $f_{h,v} = (0.7)(837.4) / 870.15$

$$= 0.674.$$

Solution. From Fig. 8.77, $\overline{E}_{h,s} = 0.33$.
Compute the volume of oil displaced from the steam zone from Eq. 8.197.

$$V_s = \left[\dot{m}_s H_s t / M_R (T_s - T)_r \right] \overline{E}_{h,s}$$
$$= \left\{ \left[(850 \text{ B/D})(350.4 \text{ lbm/bbl})(870.15 \text{ Btu/lbm}) \right. \right.$$
$$\times (4.5 \text{ years})(365 \text{ D/yr}) \right] / \left[(35 \text{ Btu/ft}^3\text{-}^\circ\text{F}) \right.$$
$$\times (387.9^\circ\text{F} - 110^\circ\text{F})(43,560 \text{ ft}^2/\text{acre-ft}) \right] \right\} (0.33)$$
$$= 331.5 \text{ acre-ft.}$$

$$N_{ps} = (7,758 \text{ bbl/acre-ft}) \phi \frac{h_n}{h_t} \left(\frac{S_{oi}}{B_{ol}} - \frac{S_{ors}}{B_{os}} \right) V_s$$
$$= (7,758 \text{ bbl/acre-ft})(0.30)(1.0)(0.31)(331.5 \text{ acre-ft})$$
$$= 239,197 \text{ STB.}$$

Finally, the equivalent volume of water injected is determined.

The energy content of the steam relative to the feedwater temperature and the steam leaving the boiler is computed below.

$$H_s = H_{wT} - H_{wA} + f_{sb}L_{vs}$$
$$= 361.91 - 38 + 0.8(837.4)$$
$$= 993.83 \text{ Btu/lbm.}$$

$$W_i = (850 \text{ B/D})(5.615 \text{ ft}^3/\text{bbl})(62.4 \text{ lbm/ft}^3)$$
$$\times (4.5 \text{ years})(365 \text{ D/yr})$$
$$= 489.17 \times 10^6 \text{ lbm.}$$

$$W_{s,eq} = (2.854 \times 10^{-6})(489.17 \times 10^6)(993.83)$$
$$= 1.388 \times 10^6 \text{ bbl.}$$

$$F_{os} = 239{,}198 / 1{,}387{,}500$$
$$= 0.172 \text{ bbl oil/bbl steam,}$$

and $F_{so} = 5.81$ bbl steam/bbl oil.

Oil-Displacement Rates. The oil-displacement rate can be estimated with steamdrive predictive models. In the region swept by the steam zone, piston-like displacement is assumed so that the oil saturation decreases from S_{oi} to S_{ors}, the residual oil saturation to steamdrive. Thus, the oil-displacement rate by the moving steam zone is given by

$$q_o = \left(\frac{7{,}758 \text{ bbl}}{\text{acre-ft}}\right)\phi\left(\frac{S_{oi}}{B_{o1}} - \frac{S_{ors}}{B_{os}}\right)\frac{dV_s}{dt}, \quad\ldots\ldots\ldots\ldots\ldots\ldots\ldots (8.206)$$

where dV_s/dt = rate that the steam-zone volume increases with time and V_s = steam-zone volume in acre-feet. Oil may also be displaced by the hot-water zone that precedes the steam zone.

Frontal-Advance Model. The oil-displacement rate from the steam zone calculated from the frontal-advance model is given by

$$q_o = \left(\frac{7{,}758 \text{ bbl}}{\text{acre-ft}}\right)h\phi\left(\frac{S_{oi}}{B_{o1}} - \frac{S_{ors}}{B_{os}}\right)\frac{dA_s}{dt} \quad\ldots\ldots\ldots\ldots\ldots\ldots\ldots (8.207)$$

when A_s is in acres and the steam-zone volume at any time is $A_s\phi h$ acre-ft.

For times less than the critical time, t_{cD}, Eq. 8.208 gives

$$\frac{dA_s}{dt} = \frac{\dot{m}_s H_s}{M_R(T_s - T_r)h}G_1(t_D). \quad\ldots\ldots\ldots\ldots\ldots\ldots\ldots (8.208)$$

Thus, for $t \leq t_{cD}$, the oil rate can be predicted with

$$q_o = \left(\frac{7{,}758 \text{ bbl}}{\text{acre-ft}}\right)\frac{\dot{m}_s H_s \phi}{M_R(T_s - T)_r}\left(\frac{S_{oi}}{B_{o1}} - \frac{S_{ois}}{B_{os}}\right)G_1(t_D). \quad\ldots\ldots\ldots\ldots\ldots\ldots\ldots (8.209)$$

When $t > t_{cD}$, oil is displaced from both steam and hot-water regions. The oil-displacement rate from the hot-water region is given by Eq. 8.210 with S_{ors} replaced by S_{orh}, the average oil saturation in the hot-water region.

$$q_{oh} = \left(\frac{7{,}758 \text{ bbl}}{\text{acre-ft}}\right)\frac{\dot{m}_s H_s \phi}{M_R(T_s - T)_r}\left(\frac{S_{oi}}{B_{o1}} - \frac{S_{orh}}{B_{oh}}\right)G_1(t_D). \quad\ldots\ldots\ldots\ldots\ldots\ldots\ldots (8.210)$$

Oil-displacement rate from the steam zone is computed with Eq. 8.206, where it is necessary to find the derivative of $G_1(t_{Ds})$ with respect to time to evaluate dA_s/dt, as shown in Eq. 8.208.

$$q_{os} = \left(\frac{7{,}758 \text{ bbl}}{\text{acre-ft}}\right)\frac{\dot{m}_s H_s \phi}{M_R(T_s - T_r)}\left(\frac{S_{orh}}{B_{os}} - \frac{S_{ors}}{B_{os}}\right)\frac{dG_1(t_D)}{dt}. \quad\ldots\ldots\ldots\ldots\ldots\ldots\ldots (8.211)$$

Because dA_s/dt is largest at the beginning of a steamdrive, the highest oil rates are predicted to occur at the instant injection begins. This is contrary to field observations, where the oil production rate increases gradually with time, as noted in **Fig. 8.79** for the Kern River A Project (Jones 1981). Fig. 8.79 also shows the oil production rate predicted by Myhill and Stegemeier

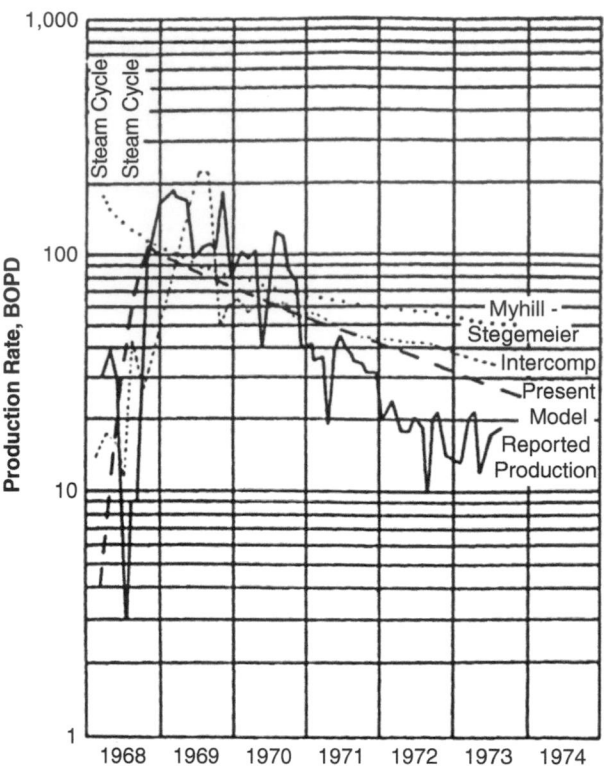

Fig. 8.79—Comparison of oil rate predicted by frontal-advance models with field rate, Kern River A (Jones 1981).

with the frontal-advance model. In general, agreement is poor between oil rates predicted with the frontal-advance models and those observed in the field, particularly in the early periods of displacement. The predicted ultimate OSR is in reasonable agreement with field results because integration of the oil rate/time prediction compensates for over- and underprediction of rates.

Gravity-Override Models. In this subsection, we illustrate concepts using the gravity-override model. Because of gravity segregation, steam rises rapidly to the top of the formation. The heated area expands rapidly, propagating heat to the production wells much faster than frontal-advance models. This model is consistent with the observations of Myhill and Stegemeier, who note that steamdrives do not respond well until the oil bank and heat arrive at the production wells.

Simulation of steamdrive performance with gravity-override models requires a different approach than in frontal-advance models. In a gravity-override model, the steam zone expands both areally and vertically. The process is aided by the heat lost in front of the vertically expanding steam zone, improving the thermal efficiency of the process. The steam zone can expand vertically and horizontally by controlling the injection rate. At long times, most of the expansion is in the vertical direction.

Oil-displacement rate is easily predicted from the gravity-override model when the volume displaced by the steam zone can be estimated. For the Neuman model, the volume of the steam zone is given by Eq. 8.159,

$$V_s = m_s C_w f_{sd} t / M_s,$$

so that $dV_s / dt = \dot{m}_s C_w f_{sd} / M_s$ .... (8.212)

and $q_o = \left(\dfrac{7{,}758 \text{ bbl}}{\text{acre-ft}} \right) \dfrac{\phi \dot{m}_s C_w f_{sd}}{M_s} \left(\dfrac{S_{oi}}{B_{ol}} - \dfrac{S_{ors}}{B_{os}} \right)$ .... (8.213)

After steam breakthrough, Eq. 8.213 is multiplied by the factor $1 - f_p$, where f_p is the fraction of the injected energy that is produced. Note that the Neuman model predicts that the oil-displacement rate from the expansion of the steam zone is constant as long as the steam-injection rate is constant.

The uniform-vertical-displacement model of steam displacement requires an independent estimate of the average vertical velocity of the steam zone to predict oil recovery from a steamdrive. Thus, this model is more useful for correlating performance and understanding displacement mechanisms than for prediction. In this model, the steam zone advances vertically at a constant average velocity throughout the heated zone. Thus, at any time, the volume of the steam zone is expanding at the rate given by Eq. 8.214, and the rate of oil displacement is given by Eq. 8.215.

$$dV_s / dt = A_s v. \qquad (8.214)$$

$$q_o = (7{,}758) A_s v \phi \left[\left(S_{oi} / B_{ol} \right) - \left(S_{ors} / B_{os} \right) \right]. \qquad (8.215)$$

The oil rate increases as the heated volume expands vertically and areally. The oil rate will increase with time until the steam zone reaches the bottom of the productive interval or ceases to grow vertically because of fluid-flow constraints and because areal growth of the steam zone ceases as a result of excessive production of steam and hot fluids. Fig. 8.75 depicts oil displacement during gravity override.

Gravity-override models require that limits be set on the displacement calculations to compute oil displacement and displacement rates. The areal sweep when steam breaks through into the production well is not known but will be assumed to be 80%. This is equivalent to assuming that the steamdrive has a mobility ratio less than 1.0. If fluid withdrawals are limited, the steam zone will continue to expand areally after breakthrough. The heated area can be estimated if the fluid withdrawals are specified by deducting the rate that heat is removed with the produced fluids from the total rate that heat is injected. Otherwise, it is necessary to assume that the heated area remains unchanged after some time. Thereafter, the steam zone expands vertically at a constant velocity.

If the areal sweep is assumed to be 80% at steam breakthrough and does not increase with time, the peak oil-displacement rate occurs when the heated zone arrives at the production well. The oil rate remains constant until the steam zone reaches the bottom of the production interval at the injection well and propagates horizontally throughout the heated area. In practice, fluid-flow considerations may limit the vertical movement of the steam zone to some fraction of the total thickness and the steam zone may expand areally. In this case, the estimated maximum steam-zone thickness is used to determine the time when vertical movement of the steam zone ceases.

Example 8.19 illustrates estimation of steam displacement by gravity override with the same data as used in Example 8.18.

Example 8.19—Calculation of Steamdrive Performance Assuming Uniform Vertical Expansion of the Steam Zone. The Kern River A steamdrive was carried out by injecting steam at 100 psig (115 psia) at a rate of 360 B/D. The drive was conducted on 2.5-acre patterns. Steam was injected for 5 years. Downhole quality was 0.5. The following are the other properties.

$$T_r = 90°F,$$
$$\alpha = 0.0312 \text{ ft}^2/\text{hr},$$
$$\phi = 0.32,$$
$$M_R = 38.5 \text{ Btu/ft}^3\text{-}°F,$$
$$\Delta S_o = 0.40,$$
$$T_s = 336.8°F,$$
$$h = 55 \text{ ft},$$
$$f_{sb} = 0.7,$$
$$f'_{sd} = 0.5.$$

Estimate the oil recovery and oil rate as a function of time for the 5-year period the steamdrive was active.

Solution. The actual velocity of the steam zone is not known but is estimated to be 0.05 ft/D, which is consistent with field experience.

From steam tables,

$$H_{wT} = 307.86 \text{ Btu/lbm at 115 psia},$$
$$H_{wr} = 58 \text{ Btu/lbm at 90°F},$$
$$L_{vdh} = 881.55 \text{ Btu/lbm at 115 psia},$$

and $H_s = 307.86 + 0.5(881.55) - 58$
$$= 690.3 \text{ Btu/lbm}.$$

From Eq. 8.141,

$$t_{Dv} = v^2 t / 4\alpha$$
$$= \left[(0.05 \text{ ft/D})^2 t\right] / \left[4(0.0312 \text{ ft}^2/\text{hr})(24 \text{ hr/D})\right]$$
$$= 8.347 \times 10^{-4} t,$$

where t is in days.

$$f_{h,v} = \left[(0.5)(881.55)\right] / 690.3$$
$$= 0.639.$$

The heated area is found as a function of time by use of Eq. 8.167:

$$A_h(t_{Dv}) = \left[\dot{m}_s H_s / M_v (T_s - T_r)\right] G_3(t_{Dv})$$
$$= \left\{\left[(360 \text{ lbm/D})[350.3 \text{ lbm/bbl}](690.3 \text{ Btu/lbm})\right]\right.$$
$$+ \left[(38.5 \text{ Btu/ft}^3\text{-}°F)(0.05 \text{ ft/D})(336.3 - 90°F)\right.$$
$$\left. \times (43,560 \text{ ft}^2/\text{acre-ft})\right]\right\} G_3(t_D)$$
$$= 4.215 G_3(t_D) \text{ acres}.$$

The heated volume is computed with Eq. 8.176:

$$V_h = \left[\dot{m}_s H_s t / M (T_s - T_r)\right] E_h(t_{Dv})$$
$$= 0.2107 t E_h(t_{Dv}),$$

where t is in days. It is convenient to choose even values of t_D to avoid interpolation. The maximum value of t_D (t = 5 years) is $(8.347 \times 10^{-4})(5)(365) = 1.523$. The initial calculations are made to determine the time for steam breakthrough into the production well.

Assuming $E_A = 0.80$ at steam breakthrough,

$$G_3\left(t_{Dv}\right) = \left[(0.8)(2.5)\right] / 4.215$$
$$= 0.475.$$

From Table 8.32, $t_{Dv} \approx 0.3$. Thus, $t_{BT} = 359$ days. At steam breakthrough, $E_h(t_{Dv}) \approx 0.339$ and

$$V_h = (0.2107)(359)(0.339)$$
$$= 25.67 \text{ acre-ft.}$$

Therefore, the average steam zone thickness is

$$\overline{h} = 25.67 \text{ acre-ft/2 acres}$$
$$= 12.83 \text{ ft.}$$

To compute the oil-displacement rate, recall that

$$q_o = (7{,}758 \text{ bbl/acre-ft}) vA\left[\left(S_{oi} / B_{o1}\right) - \left(S_{ors} / B_{os}\right)\right].$$

At $t_{Dv} = 0.3$,

$$q_o = (7{,}758 \text{ bbl/acre-ft})(0.32)(0.40)(0.05 \text{ ft/D})(2.0 \text{ acres})$$
$$= 99.3 \text{ STB/D.}$$
$$N_{ps} = (7{,}758 \text{ bbl/acre-ft})(0.32)(0.40)(25.7 \text{ acre-ft})$$
$$= 25{,}521 \text{ STB.}$$

Table 8.44 summarizes displacement performance for 0 to 359 days. After steam breakthrough, the oil production rate is constant in this example until the net thickness to be displaced is reached at the injection well. This will occur at a time of approximately 1,100 days (3 years) after injection starts, as indicated next.

$$t_B = (55 \text{ ft})/(0.05 \text{ ft/D})$$
$$= 1{,}100 \text{ days,}$$

where t_B is the time when the steam-zone thickness reaches a maximum value (assumed 55 ft) at the injection well. In this example, the areal sweep efficiency is assumed to remain constant at 80% ($A_s = 2.0$ acres).

The remainder of the displacement calculation is based on excluding area from steam displacement when the steam-zone thickness reaches the maximum value. This will occur for each area in Table 8.44 at a time equal to t + 1,100 days. For example, when the steam zone reaches its maximum thickness at $A = 1.04$ acres, the time when this will occur is (71.9 + 1,100) days, or 1,171.9 days.

Let A_d = area displaced to the maximum steam-zone thickness at time t. Then at time t = 1,172 days,

$$q_o = (7{,}758 \text{ bbl/acre-ft}) v(2.0 - A_d)\phi\left[\left(S_{oi} / B_{o1}\right) - \left(S_{ors} / B_{os}\right)\right]$$
$$= \left(\frac{7{,}758 \text{ bbl}}{\text{acre-ft}}\right)(0.05 \text{ ft})(2.0 - 1.04 \text{ acres})(0.40)$$
$$= 47.67 \text{ STB/D.}$$

The steam volume is given by

$$V_s = V_d + v\left(A_s - A_d\right)\left(t - t_{BT}\right), \quad \dotfill \quad (8.216)$$

where t_{BT} = time for steam breakthrough into the production well.

Average Vertical Steamflood Velocity = 0.05 ft/D

	t_{Dv}	t_{Ds} (days)	A_h (acres)	V_s (acre-ft)	$\bar{h}_s$ (ft)	E_h	q_o (B/D)	N_{ps} (bbl)
	0.0	0.0	0.00	0.00	0.00	0.0000	0.00	0
1	0.0100	12.0	0.45	0.18	0.40	0.0727	22.50	182
2	0.0200	24.0	0.63	0.51	0.81	0.1014	31.22	507
3	0.0300	35.9	0.76	0.93	1.22	0.1228	37.66	921
4	0.0400	47.9	0.86	1.41	1.64	0.1405	42.94	1,405
5	0.0500	59.9	0.96	1.96	2.05	0.1558	47.46	1,947
6	0.0600	71.9	1.04	2.56	2.47	0.1693	51.46	2,540
7	0.0700	83.9	1.11	3.20	2.89	0.1816	55.05	3,178
8	0.0800	95.8	1.17	3.88	3.31	0.1929	58.33	3,858
9	0.0900	107.8	1.24	4.61	3.73	0.2033	61.35	4,575
10	0.1000	119.8	1.29	5.36	4.15	0.2131	64.15	5,327
11	0.1100	131.8	1.34	6.15	4.58	0.2222	66.77	6,111
12	0.1200	143.8	1.39	6.97	5.00	0.2309	69.23	6,926
13	0.1300	155.8	1.44	7.82	5.43	0.2391	71.54	7,770
14	0.1400	167.7	1.49	8.70	5.86	0.2469	73.74	8,640
15	0.1500	179.7	1.53	9.60	6.29	0.2543	75.82	9,536
16	0.1600	191.7	1.57	10.53	6.72	0.2614	77.81	10,456
17	0.1700	203.7	1.61	11.48	7.15	0.2683	79.71	11,400
18	0.1800	215.7	1.64	12.45	7.58	0.2748	81.52	12,366
19	0.1900	227.6	1.68	13.45	8.02	0.2811	83.26	13,353
20	0.2000	239.6	1.71	14.46	8.45	0.2872	84.93	14,361
21	0.2100	251.6	1.74	15.50	8.89	0.2931	86.54	15,388
22	0.2200	263.6	1.77	16.55	9.33	0.2988	88.09	16,434
23	0.2300	275.6	1.80	17.62	9.77	0.3043	89.58	17,499
24	0.2400	287.5	1.83	18.71	10.21	0.3097	91.03	18,581
25	0.2500	299.5	1.86	19.82	10.65	0.3149	92.42	19,680
26	0.2600	311.5	1.89	20.94	11.09	0.3199	93.77	20,795
27	0.2700	323.5	1.91	22.08	11.53	0.3249	95.08	21,926
28	0.2800	335.5	1.94	23.24	11.97	0.3296	96.35	23,073
29	0.2900	347.4	1.97	24.41	12.42	0.3343	97.58	24,235
30	0.3000	359.4	1.99	25.59	12.86	0.3388	98.78	25,411
31	0.3100	371.4	2.01	26.79	13.31	0.3433	99.94	26,602

Table 8.44—Summary of computations, gravity-override, Example 8.19, Kern River A to steam breakthrough.

$$V_d = (1.04 \text{ acre})(55 \text{ ft})$$
$$= 5.72 \text{ acre-ft.}$$
$$V_s = 57.2 + (2.0 - 1.04)(0.05)(1,172 - 359)$$
$$= 96.22 \text{ acre-ft.}$$
$$N_{ps} = (7,758 \text{ bbl/acre-ft})\phi(h_n / h_t)\left[(S_{oi} / B_{o1}) - (S_{ors} / B_{os})\right]v_s$$
$$= (7,758 \text{ bbl/acre-ft})(96.2 \text{ acre-ft})(0.32)(0.40)$$
$$= 95,553 \text{ STB.}$$

Table 8.45 summarizes the remaining displacement calculations.

The prediction of oil production is plotted against time in **Fig. 8.80.** Also shown in Fig. 8.80 are the field data from Fig. 8.79. The equivalent OSR at $t = 1,411.6$ days is computed with Eq. 8.202. Assuming a boiler feedwater temperature of 70°F, H_{wA} = 38 Btu/lbm.

t_{Ds}	t (days)	A_{ds} (acres)	Incremental Oil Rate (B/D)	V_d (acre-ft)	V_s (acre-ft)	N_p (bbl)
47.9	1,147.9	0.86	56.6	47.3	92.2	91,557
71.9	1,171.9	1.04	47.7	57.2	96.2	95,553
95.8	1,195.8	1.17	41.2	64.3	99.1	98,386
131.8	1,231.8	1.34	32.8	73.7	102.5	101,787
167.7	1,267.7	1.49	25.3	81.95	105.1	104,388
203.7	1,303.7	1.61	19.4	88.55	106.97	106,225
239.6	1,339.6	1.71	14.4	94.05	108.3	107,513
275.6	1,375.6	1.80	9.93	99.0	109.2	108,404
311.6	1,411.6	1.89	5.46	103.95	109.7	108,974

Table 8.45—Summary of displacement computations—Example 8.19 after steam break-through, no further expansion of heated area.

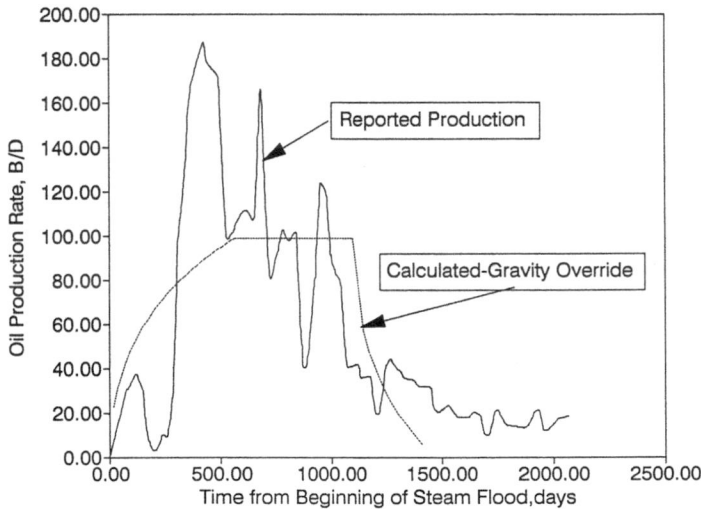

Fig. 8.80—Comparison of oil rate predicted from gravity-override model at $v = 0.05$ ft/D with field data, Kern River A.

$$H_{sbA} = H_{ws} - H_{wA} + f_{sb}L_{vs}$$
$$= 307.86 - 38 + (0.7)(881.55)$$
$$= 886.9 \text{ Btu/lbm.}$$

$$m_s = (360 \text{ B/D})\left(\frac{5.615 \text{ ft}^3}{\text{bbl}}\right)\left(\frac{62.4 \text{ lbm}}{\text{ft}^3}\right)(1,412 \text{ days})$$
$$= 178.10 \times 10^6 \text{ lbm.}$$

$$W_{s,eq} = (2.854 \times 10^{-6})(178.10 \times 10^6 \text{ lbm})(886.9 \text{ Btu/lbm})$$
$$= 450.817 \text{ bbl.}$$

$$F_{os} = 108,974 / 450,817$$
$$= 0.242 \text{ bbl oil/bbl steam.}$$

This SOR is predicted in 3.87 years and is in reasonable agreement with the 0.26 SOR reported from field data at 5 years. This example shows that the gravity-override model correlates steam-drive performance when appropriate parameters are selected. The predicted oil ratio and OSR are consistent with field response.

Estimating Oil Production and Steam-Injection Rates. In the design of steamdrives, it is necessary to estimate the steam-injection rate. Field experience, either from developed drives or from pilot testing is often available to guide the design. Estimation of steam-injection rates in new reservoirs is difficult because simple calculation methods cannot account for complex two- and three-phase flow in a condensing environment. Even numerical models of steamdrive must be adjusted to match injection rates with presumed parameters. This section shows how simple models can be applied to understand the nature of steamdrive projects and, in particular, the importance of reservoir heating on oil-displacement and steam-injection rates.

Frontal-Advance Models. Consider estimating the rate that oil can be displaced from a heavy-oil reservoir at reservoir temperature. For purposes of this discussion, assume that the pattern geometry is a five-spot. When flow is caused by the applied pressure drop (in contrast to gravity drainage), the initial oil rate in a pattern drive is given in oilfield units (barrels per day, darcies, feet, centipoise, and psi) by

$$q_o = \frac{3.54 kh\left(p_i - p_p\right)}{\mu_o\left[\ln\left(d / r_w\right) - 0.619\right]} \dots\dots\dots\dots\dots\dots\dots\dots\dots\dots\dots\dots\dots\dots\dots\dots\dots\dots\dots (8.217)$$

As the steam zone develops, extending to radius r_s, the oil-displacement rate is approximated by (Willhite 1986)

$$q_o = \frac{7.082 h\left(p_i - p_p\right)}{\dfrac{\ln\left(r_s / r_w\right)}{\lambda_s} + \dfrac{\ln\left(d^2 / r_s r_w\right) - 2\ln\sqrt{\pi}}{\lambda_o}}, \dots\dots\dots\dots\dots\dots\dots\dots\dots\dots\dots\dots\dots\dots (8.218)$$

where

d = distance between injection and production wells, ft,
  = $147.6\sqrt{A_p}$;
A_p = pattern area, acres;
P_i = pressure in the injection well, psi;
P_p = pressure in the production well, psi;
h = thickness of the displaced zone, ft;
k = permeability of the formation to oil at initial water saturation, darcies;
μ_o = viscosity of oil at the average reservoir temperature, T_r, °F;
λ_s = mobility of steam in the steam zone, darcies/cp;
λ_o = mobility of oil in the oil zone at reservoir temperature, darcies/cp;
r_s = radius of the steam zone, ft; and
r_w = radius of the wellbore, ft.

In Eq. 8.218, the production well is at reservoir temperature. Eq. 8.218 is easily modified to include a heated region around the production well that might be present as a result of CSS. Eq. 8.219 describes the oil-displacement rate under these conditions:

$$q_o = \frac{7.082 h\left(p_i - p_p\right)}{\dfrac{\ln\dfrac{r_s}{r_w}}{\lambda_s} + \dfrac{\ln\dfrac{d^2}{r_s r_h} - 2\ln\sqrt{\pi}}{\lambda_o} + \dfrac{\ln\dfrac{r_h}{r_w}}{\lambda_{oh}}}, \dots\dots\dots\dots\dots\dots\dots\dots\dots\dots\dots (8.219)$$

where r_h = radius of heated zone around the production well; μ_{oh} = viscosity of oil in the heated zone around the production well, cp; and λ_{oh} = mobility of oil in the heated zone around the production well, darcies/cp.

Exploratory calculations to determine the oil-displacement rate assuming piston-like displacement of oil by steam as in frontal-advance models are instructive. For these calculations, consider steam injection into a five-spot pattern that is on 2.5-acre spacing. Example 8.20 summarizes the calculations.

Example 8.20—Calculation of Oil-Displacement Rates During Steam-Injection, Frontal-Advance Model, Piston-Like Displacement. Estimate the oil-displacement rate in a 2.5-acre, five-spot pattern when steam is injected at 100 psig (115 psia) into a reservoir that is 20 ft thick. Effective permeability of the reservoir to oil is 3 darcies. **Table 8.46** summarizes other properties. **Table 8.47** presents oil viscosities and densities estimated from the American Society of Testing and Materials (ASTM) viscosity correlation (Miller and Leung 1985). Fitting constants for the ASTM correlation were developed from the viscosities at 92 and 300°F in Table 8.46 and correspond to the reported data for the Kern River A crude oil. The oil viscosities decline rapidly with temperature. Also included in Table 8.47 are viscosities of water ("Steamflood Predictive Model," US DOE 1986).

Solution. The initial oil rate is computed from Eq. 8.217:

$$q_o = \left[3.541 kh\left(p_i - p_p\right)\right] / \left\{\mu_o\left[\ln\left(d / r_w\right) - 0.619\right]\right\}.$$

Permeability (darcies)	3
T_r (°F)	92
Oil gravity (°API)	13.5
μ_o at 92°F (cp)	2,200
μ_o at 300°F (cp)	2.6
r_w (ft)	1.0

Table 8.46—Properties of reservoir rock and fluids—Example 8.20, Kern River A.

T (°F)	ρ_o (g/cm³)	μ_o (cp)	μ_w (cp)
92	0.9650	2,200*	–
100	0.9623	1,214	0.681
150	0.9456	75.5	0.435
200	0.9292	14.7	0.305
250	0.9131	5.25	0.235
300*	0.8973	2.6*	0.187
336.8	0.8859	1.8	–

*Data for development of ASTM viscosity correlation (1986) (Miller and Leung 1985).

Table 8.47—Viscosities and densities estimated from ASTM correlation, Example 8.20.

With $A_p = 2.5$ acres,

$$d = 147.6\sqrt{2.5}$$
$$= 233.3 \text{ ft,}$$

and $q_o = \dfrac{(3.541)(3 \text{ darcies})(20 \text{ ft})(100 \text{ psi})}{\mu_o\left[\ln(233.3/1.0) - 0.619\right]}$

$$= \left(4,396 / \mu_o\right) \text{ B/D.}$$

At reservoir temperature of 92°F, $\mu_o = 2,200$ cp and $q_o = 2.0$ B/D.

Suppose the steam injection has extended the radius of the steam zone to a distance of 50 ft from the wellbore. Eq. 8.218 gives the injection rate for piston-like displacement. For the purpose of this example, assume that the steam mobility is large enough that the flow resistance in the steam zone is negligible. Then,

$$q_o = \dfrac{7.082kh\left(p_i - p_p\right)}{\mu_o\left[\ln\left(d^2 / r_s r_w\right) - 2\ln\sqrt{\pi}\right]}$$

$$= \dfrac{(7.082)(3 \text{ darcies})(20 \text{ ft})(100 \text{ psi})}{\mu_o\left\{\ln\left[(233.5)^2 / (50)(1.0)\right] - 1.145\right\}}$$

$$= 7,266 / \mu_o. \quad\ldots\ldots\ldots\ldots\ldots\ldots\ldots\ldots\ldots\ldots\ldots\ldots\ldots\ldots\ldots\ldots\ldots\ldots\ldots (8.220)$$

When the region being displaced by the steam zone is 92°F, $q_o = 3.3$ B/D. Thus, simply injecting steam to reduce the flow resistance in the region around the injection wellbore does not lead to significant increase in oil-displacement rates. Similar results are obtained if a heated region is placed around the production well as a result of cyclic steam injection. However, if the reservoir is heated before displacement by steam, the oil rates shown in **Table 8.48** would be observed if the displacement was represented by the frontal-advance model.

Table 8.48 reveals a fundamental concept in oil displacement by steamdrive in viscous-oil reservoirs. If the displacement is piston-like, as envisioned in frontal-advance models, significant oil-displacement rates in many heavy-oil reservoirs cannot be attained until there is a substantial increase in average reservoir temperature in the flow region between the injection well and production wells. The oil viscosity in the bulk of the pattern controls fluid-flow behavior and displacement rates.

Average Temperature of Oil Zone (°F)	Oil-Displacement Rate, $\Delta p = 100$ psi (B/D)
92	3.3
150	5.99
200	96.2
250	501.1
300	1,384
336.8	4,037

Table 8.48—Estimated oil-displacement rates when steam zone is at 50 ft, Example 8.20.

This observation leads to a conclusion related to steam-injection rates. If displacement were piston-like, injecting steam at the rates observed in field projects would not be possible. The observed steam-injection rates must result from preferential flow in depleted intervals of the reservoir, viscous fingering through a small segment of the reservoir, or displacement of a mobile water saturation. If the oil viscosity were large enough, injecting steam would be impossible because no oil could be displaced. Techniques such as fracturing (Britton et al. 1983; Powers et al. 1985), injection into an underlying water zone, or injection through horizontal wellbores are required to introduce heat into a reservoir that has no permeability at normal reservoir temperature. Clearly, heating the reservoir is necessary before flow resistance can be reduced enough to permit substantial oil-displacement rates by steam injection.

Gravity-Override/Drainage Models. When gravity override occurs, a steam zone overlays the interval between the injection well and the production well, as shown in **Fig. 8.81** (Vogel 1984). A hot oil layer forms in front of the expanding steam zone as a result of heat losses by conduction. Oil and water in the heated region flow to the production well under the influence of gravity because of the difference in fluid head and the pressure drawdown in the production well, as illustrated in **Fig. 8.82**. Gravity drainage is enhanced when the formation is thick and/or steeply dipping (Vogel 1984, 1992). Furthermore, Vogel also points out that the pressure drop between the injection well and the production well in the overlying steam zone is relatively small, probably on the order of a few psi. Thus, increasing injection rates above levels required to maintain the vertical expansion of the steam zone results in excessive steam production and little additional oil recovery. These views lead to substantial differences in operating and design concepts for steamfloods. **Table 8.49** summarizes contrasts between concepts proposed for steamdrive and gravity drainage (Vogel 1992).

Estimating production rates when the steamflood is assumed to override the formation is possible. A significant temperature distribution is present under the moving steam front. The oil in this region becomes mobile and can flow to a production well under the influence of both pressure gradients and differences in fluid head. The temperature distribution in this region is given by Eq. 8.161. With this temperature distribution, it is possible to estimate oil flows in this region. Let $T(z,t)$ represent the temperature distribution in the region underneath the moving steam front, where $z = 0$ at the steam front and z is the distance below the steam front. Because the oil viscosity decreases rapidly with temperature, it is convenient to define a thickness-averaged oil viscosity for this region. The thickness-averaged oil viscosity is given by

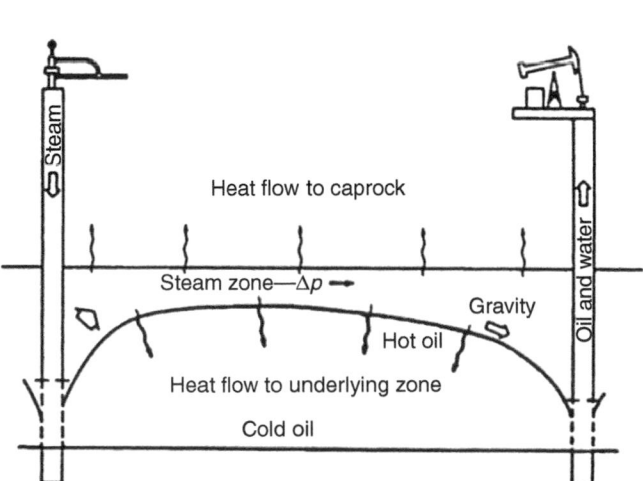

Fig. 8.81—Schematic cross section in continuous steam-injection process where gravity drainage of hot oil and steam drag are important displacement mechanisms (Vogel 1984).

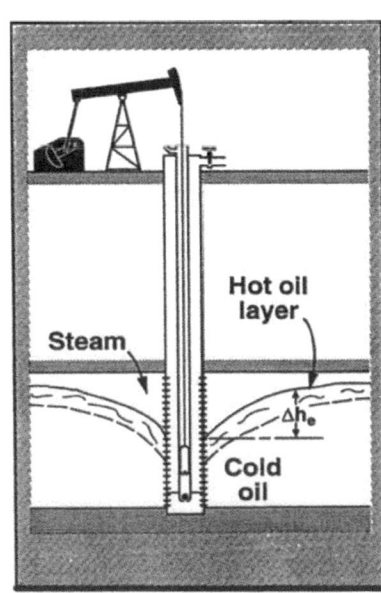

Fig. 8.82—Flow of oil under the influence of a partially effective gravity head (Vogel 1992).

Drive Concepts	Drainage Concepts
1. Steam drives oil to producers.	1. Steam does not drive oil, oil flows mostly by gravity, with some assistance by steam drag.
2. Steam rates determine oil rates.	2. Oil rates largely unaffected by steam rates above minimum required for heating.
3. Steam breakthrough is discouraging.	3. Steam breakthrough is welcomed.
4. Minimize backpressure on producers by freely flowing steam from casing.	4. Do not produce much steam, conserve the heat.
5. Steam soaks are not important.	5. Steam soaks are very important.
6. Most oil production occurs before breakthrough.	6. Most oil production occurs after breakthrough.
7. Constant-rate injection is okay.	7. Diminishing-rate injection is best.
8. Late-life water injection can maintain drive pressure and salvage heat.	8. Late-life water injection won't maintain drive pressures, may chill reservoir, and may interfere with gravity drainage.
9. High recovery efficiency is caused by gasdrive by steam and miscible drive by distilled light ends.	9. High recovery efficiency is a result of gravity drainage.
10. Project producing rates depend on steam rates and little on the number of producers.	10. Project producing rates are directly proportional to number of producers.
11. Injection rates should be proportional to reservoir volumes.	11. Injection rates should be proportional to project areas.
12. Pattern configuration is important.	12. Pattern configurations do not matter much.

Table 8.49—Contrasts between steamdrive and gravity drainage (Vogel 1992).

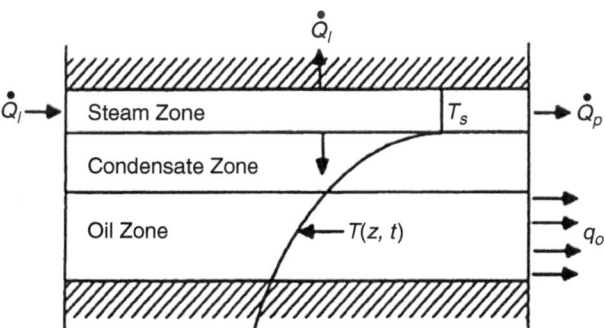

Fig. 8.83—Conceptual schematic of steam-override predictive model (Miller and Leung 1985).

$$\mu_{oz} = \left[\int\limits_{o}^{h-\bar{h}_s} \mu_o \, dz \right] \Big/ \left(h - \bar{h}_s \right), \dots\dots\dots\dots\dots\dots\dots\dots\dots\dots\dots\dots\dots\dots\dots (8.221)$$

where $\mu_o = \mu_o(T)(z,t)$ and $T(z,t)$ is obtained from Eq. 8.161. The term h_s is the average thickness of the steam zone so that $h–h_s$ is the thickness of the oil column.

If flow is primarily driven by pressure gradients between the injection well and the production well, Eq. 8.221 can be used to estimate the oil flow rate in the region below the steam zone.

$$q_{oh} = \frac{3.541 k_o \left(h - \bar{h}_s \right)\left(p_i - p_p \right)}{\mu_{oz} \left[\ln\left(d / r_w \right) - 0.619 \right]}. \dots\dots\dots\dots\dots\dots\dots\dots\dots\dots\dots\dots\dots\dots (8.222)$$

Miller and Leung (1985) developed an approximate method to estimate oil production rates when gravity drainage is the dominant mechanism. In their model, the steam zone is assumed to spread instantaneously across the top of the reservoir as proposed by Vogel (1984). **Fig. 8.83** shows temperature and fluid saturation distributions. The temperature profile is computed from Eq. 8.134. All oil production is assumed to result from conductive heating of the oil zone. The resulting expression for the oil rate is given by

$$\frac{q_{oh}}{q_{oc}} = \left(\frac{\mu_{oc}}{h} \right) \left(\int\limits_{h_c}^{h_c+h_o} \right) \left(\frac{d_z}{\mu_o} \right), \dots\dots\dots\dots\dots\dots\dots\dots\dots\dots\dots\dots\dots\dots\dots (8.223)$$

where q_{oc} = production rate before steamdrive under gravity drainage, BOPD; q_{oh} = stimulated oil production rate, BOPD; h_c = gross condensate zone thickness, ft; and h_o = gross oil zone thickness, ft, and is equal to $h - h_s - h_c$. The size of the steam condensate zone is given by

$$h_c = \frac{\left(1 - f_{cp}\right) 2k_h \left(T_s - T_r\right) \sqrt{t - \tau}}{\rho_{ws} L_v \left(h_n / h\right) \phi \left(S_{oi} - S_{oc}\right) \sqrt{\pi \alpha}}, \dotfill (8.224)$$

where f_{cp} = the fraction of the condensed steam at the interface that is produced from the reservoir.

8.5.4 Advanced Computer Models. A number of computer models are available in the literature for simulation of steamflooding. The US DOE (1986) developed a series of models for screening projects. The US DOE models are in the public domain and are available for use on personal computers and work stations. Emanuel (1993) compared results from the stream-tube model developed by Aydelotte and Pope (1983) with results of several field tests. This model is one of the models in the US DOE package. Gomaa (1980) developed a correlation for prediction of oil recovery from steamflooding from the results of a numerical simulation that used the parameters of the Kern River field. This correlation is also in the US DOE steamflooding programs.

There are many other examples of simulation studies in the literature. Recent studies include the use of numerical simulation to optimize steamflood performance in a California reservoir (Kumar and Ziegler 1993) and the simulation of the Duri steamflood (Pearce and Megginson 1991).

8.5.5 Field Results—Steamdrive. Steamdrive is the principal EOR process used in the world, with production in excess of 700,000 B/D and growing. CSS is an integral part of a steamdrive, and, in some cases, it is difficult to allocate incremental production between the two processes. Growth in steamdrive technology is remarkable in that it occurred over a period of approximately 30 years, with major developments during 1960–70. There is extensive literature on steamdrive projects, which is summarized in several references (Prats 1982; Boberg 1988; Farouq Ali and Meldau 1979; Farouq Ali 1974). This section presents an overview of field results followed by detailed description of selected projects.

Table 8.50 (Farouq Ali and Meldau 1979) gives formation characteristics of 12 major steamdrive projects, and **Table 8.51** presents steamdrive test results. Oil recoveries range from 20% OOIP at Winkleman Dome to 73% OOIP at Charco Redondo. Average recoveries often exceed 50% OOIP. Cumulative OSRs are on the order of 0.2, which is equivalent to an SOR of 5.0. Many of these projects have been under steamdrive for nearly 20 years, and results reflect performance of a mature technology. Steamdrive is applied in a variety of displacement patterns. Most reservoirs had low pressures at the beginning of the steamdrives, with primary recovery averaging less than 15%.

Boberg (1988) identified several characteristics of reservoirs that have had successful steamdrive projects: high oil content, shallow depth, thick sands with low shale content, moderate oil viscosities under cold conditions (so that preheating is not necessary to obtain oil mobility), and high sand permeabilities.

8.5.6 Kern River Field. The Kern River field is a large heavy-oil reservoir 5 miles northeast of Bakersfield, California, USA. OOIP is estimated to exceed 4 billion bbl. The reservoir is shallow and consists of an alternating sequence of unconsolidated sands with interbedded silts and clays (Farouq Ali 1974, Greaser and Shore 1980; Villanueva and Pittman 1970; Bursell 1970). The sands have a high permeability of 1 to 5 darcies and porosities of 28 to 35%. Oil viscosities are on the order of 4,000 cp at reservoir temperature. Reservoir pressure is low at 100 psig.

The Kern River field reservoir consists of four main oil-sand intervals shown in a representative cross section in **Fig. 8.84** (Bursell and Pittman 1975). The main intervals (C, G, K, and R) correlate across the entire field. Main intervals contain subunits with thicknesses between 25 and 125 ft. The reservoir has a slight dip of 3 to 6° to the southwest.

Steamdrive Tests and Expansion. Getty/Texaco Steamdrives. Getty Oil Co. (now Texaco) initiated a pilot steamflood in 1964 in the Kern property shown in **Fig. 8.85**. Additional 9 to 12 pattern pilots were installed in 1968 and 1969, followed by major expansions in 1970 (189 patterns) and 1971 (325 patterns) on 2.5-acre spacing. The Getty property is outlined in Fig. 8.85. Fieldwide expansion has occurred since.

Greaser and Shore (1980) describe results of the 1970 Canfield, 1970 San Joaquin, and 1971 Green and Whittier expansions. Reports 76-34-2/76 and 76-34-2/81 (1976, 1982) describe the Green and Whittier 1971 expansion. **Table 8.52** summarizes the reservoir properties for these three properties, and **Table 8.53** presents estimated oil recoveries. Incremental oil recovery from the steamdrive ranged from 57 to 78% OOIP.

In the Green and Whittier 1971 expansion, the 60-ft-thick K-1 sand was flooded. The 1971 expansion, shown in **Fig. 8.86** (Report 76-34-2/76 1976), consists of 114 patterns on 307 acres. Patterns are inverted five-spots with somewhat irregular shapes. **Fig. 8.87** shows average pattern steamdrive performance. **Fig. 8.88** summarizes production and injection performance from 1972 to 1981. Oil production peaked at approximately 8,000 B/D in June 1973, declining to approximately 4,000 B/D by the end of 1981. Cumulative steamflood recovery through 1981 was approximately 15 million bbl. Cumulative SOR was 5.9.

Kern River sands are separated by silt and clay interbeds (Restine 1983). Individual sands were believed to be isolated, so the development plan was based on starting with the deepest sand and moving uphole to the next zone after the current zone is depleted. This is done by selectively perforating the injection well in the desired zone. Production wells are generally completed with all zones open.

Test	Field, Location (Operator)	Year Started	Formation	Net/Gross Pay (ft)	Depth to Top (ft)	Dip (degrees)	Porosity (%)	Permeability (md)
1	Mount Poso, California (Shell)	1970	Upper Vedder	60/75	1,800	6	33	15,000
2	Midway-Sunset 26C, California (Chevron)	1975	Monarch sand	260/350	1,300	10	27	520
3	Cat Canyon, California (Getty)	1977	S1-B sand	80/80	2,500	10	31	5,000
4	Charco Redondo, Texas (Texaco)	–	–	10/10	200	–	35	2,500
5	Yorba Linda, California (Shell)	1971	Upper Conglomerate	325/–	650	12	30	600
6	Duri Field, Sumatra (Caltex)	1967	Duri sandstone	100/–	525	–	37	–
7	Tia Juana, M6, Venezuela (Maraven)	1975	Unconsolidated sand	125/250	1,624	4	38	2,800
8	Kern River Sec. 3, California (Chevron)	1968	Kern River sand	70/–	705	3	35	7,600
9	South Belridge, California (Mobil)	1969	Tulare D and E	91/210	1,100	7	35	3,000
10	Kern River, California (Getty)	1964	K1,R, R1	60/–	900	4	35	3,000
11	Winkelman Dome, Wyoming (Amoco)	1964	Nugget sand	73/–	1,220	–	25	638
12	Peace River, Alberta (Shell/AOSTRA)	1979	Upper Bullhead	90/40	1,800	0.2	28	1,050
13	Cold Lake, Lemming, Alberta (Esso)	1975	Clearwater	150/155	1,500	–	35	1,500

Table 8.50—Steam-injection test formation characteristics (Farouq Ali and Meldau 1979).

Test	Field, Location (Operator)	Oil Saturation (% PV)	Oil Gravity (°API)	Oil Viscosity at Reservoir Temperature (cp)	Reservoir Temperature (°F)	Reservoir Pressure (psi)	Primary Production (% OOIP)
1	Mount Poso, California (Shell)	58	16	280	110	100	35
2	Midway-Sunset 26C, California (Chevron)	48	14	1,500	105	75	–
3	Cat Canyon, California (Getty)	65	9	25,000	110	–	12
4	Charco Redondo, Texas (Texaco)	34	18	95	72	10	–
5	Yorba Linda, California (Shell)	–	14	6,400	85	–	5
6	Duri Field, Sumatra (Caltex)	62	22	160	98	180	10
7	Tia Juana, M6, Venezuela (Maraven)	85	12	5,000	113	350	11
8	Kern River Sec. 3, California (Chevron)	52	14	2,710	80	140	13
9	S. Belridge, California (Mobil)	76	13	1,600	95	180	9
10	Kern River, California (Getty)	50	14	4,000	95	50	10
11	Winkelman Dome, Wyoming (Amoco)	75	14	900	81	210	–
12	Peace River, Alberta (Shell/AOSTRA)	77	9	200,000	62	530	0
13	Cold Lake, Lemming, Alberta (Esso)	60	10	100,000	55	450	0

Table 8.50—Steam-injection test formation characteristics (continued).

Because steam is injected into the drive zone for several years, some of the heat loss to the overburden will preheat other reservoir zones if they are close enough. Methods discussed in Section 8.4 can be used to estimate the temperature of adjacent zones. This was recognized in the Kern River field, and Restine (1983), who investigated the effects of steamdrive in the R1 zone on a subsequent steamdrive in the R zone, carried out an analysis of preheating effects. **Fig. 8.89** is a cross-sectional view of the 1970 Canfield steamflood with the R1 interval identified. **Fig. 8.90** is a map of the 1970 Canfield steamflood. In 1978, 55 patterns were recompleted in the R zone. **Fig. 8.91** shows temperature profiles in Canfield TW No. 2 before steam injection into the R zone and at selected times during steam injection into the R zone. The effect of conduction heating of the R zone during the previous steamdrive in the R1 zone is apparent; the temperature of the base and the top of the R sand increased from 85 to 165 and 92°F, respectively. An extensive computer study revealed that preheating would not increase ultimate oil recovery in the R zone. However, project life was estimated to decrease 1 to 2 years, combined with a substantial increase in OSR.

Another benefit of preheating was observed in the Kern A-RCA project. In this project, the R sand was thin (22 ft), and economics of steamdrive were considered marginal. **Fig. 8.92** indicates substantial preheating. Preheating this sand improved

Test	Location	Pattern Type	Average Pattern Size/Total Area (acres)	Number of Producer/Injector Wells	Average Injection Pressure (psig)	Average Injection Rate (BWPD/well)
1	Mount Poso	Line	–/2,100	49/10	350	2,000
2	Midway-Sunset	Irregular five-spot	4/23	15/6	400	3,000
3	Cat Canyon	Inverted five-spot	5/20	9/4	1,375	500
4	Charco Redondo	Inverted five-spot	2.5/2.5	4/1	215	580
5	Yorba Linda	Inverted nine-spot	–	74/16	200	850
6	Duri Field	Cyclic	–	239/–	450	1,000
7	Tia Juana M6	Inverted seven-spot	17.1/1,831	131/19	600	3,150
8	Kern River (Chevron)	Inverted seven-spot	6.1/61	32/10	400	600
9	South Belridge	Line	5/204	40/10	—	600
10	Kern River (Getty)	Five-spot	2.5/2,750	2.751/818	200	400
11	Winkleman Dome	Five-spot	2.5/5.3	21/12	1,150	241
12	Peace River	Inverted seven-spot	7/49	24/7	2,500	1,500
13	Cold Lake	Cyclic	6.1/61	70/–	1,300	1,500

Table 8.51— Steam-injection test results (Farouq Ali and Meldau 1979).

Test	Location	Total Oil Rate (BOPD)	Recovery (% OIP)	Cumulative OSR (bbl/bbl)	Cumulative SOR (bbl/bbl)	Source Reference	Notes
1	Mount Poso	13,000	60[a]	0.18	5.6	Bentsen and Donohue (1969); Doscher (1966)	b
2	Midway-Sunset	720	65	0.16	0.25	Kawasaki Thermal Systems (1989)	c
3	Cat Canyon	250	43[a]	0.25[a]	4.[a]	Seba and Perry (1969)	d
4	Charco Redondo	60	73	0.05	20.	Towson and Boberg (1967)	e
5	Yorba Linda	7,000	50	–	–	Willhite and Dietrich (1967)	f
6	Duri Field	34,000	8	0.16	6.10	Gontijo and Aziz (1984)	g
7	Tia Juana M6	–	45	0.34	2.94	Sulvester and Chen (1988)	h
8	Kern River (Chevron)	1,490	63	0.17	5.88	Gozde et al. (1989)	i
9	South Belridge	3,200	26	0.28	3.57	Ramey (1962)	j
10	Kern River (Getty)	65,000	73	0.21	4.76	Jones (1992)	–
11	Winkleman Dome	850	50	0.2	5	Dietrich (1981)	–
12	Peace River	3,500[a]	50[a]	0.25[a]	4.[a]	Boberg (1988)	k
13	Cold Lake	5,000	20	0.4	2.5	Burger et al. (1985)	g, l

Notes

(a) Predicted; (b) strong waterdrive, Phases 1 to 3 only; (c) instantaneous OSR is 0.19 with only 0.105 PV steam injected to date; (d) cyclic stimulation OSR is 1.0; (e) total injection time is 222 days; (f) thick, discontinuous silt barriers in the sand; (g) CSS project; (h) only 3 seven-spots were flooded in 1977; recovery is net owing to steam; (i) response to steamflooding 4 months after start; (j) 1.600 BOPD is peak CSS rate before steamflood; (k) project figures are based on scaled model experiments, steam injection recently started; (l) wells hydraulically fractured during steam injection.

Table 8.51—Steam-injection test results (continued).

the steam displacement response and appears to have been a major factor in being able to steamdrive a thin sand successfully if all the oil came from the R sand.

Kern River steamdrives run by Texaco were developed primarily on 2.5-acre spacing. An infill well program involving 574 wells was developed in previously steamflooded zones, reducing well spacing to 0.625 acres/well (Restine et al. 1987). Subsequently, the wells were recompleted into upper active steamfloods, converting 2½-acre five-spot patterns into 2½-acre nine-spot patterns. The infill program accelerated oil recovery from previously steamflooded zones and apparently increased

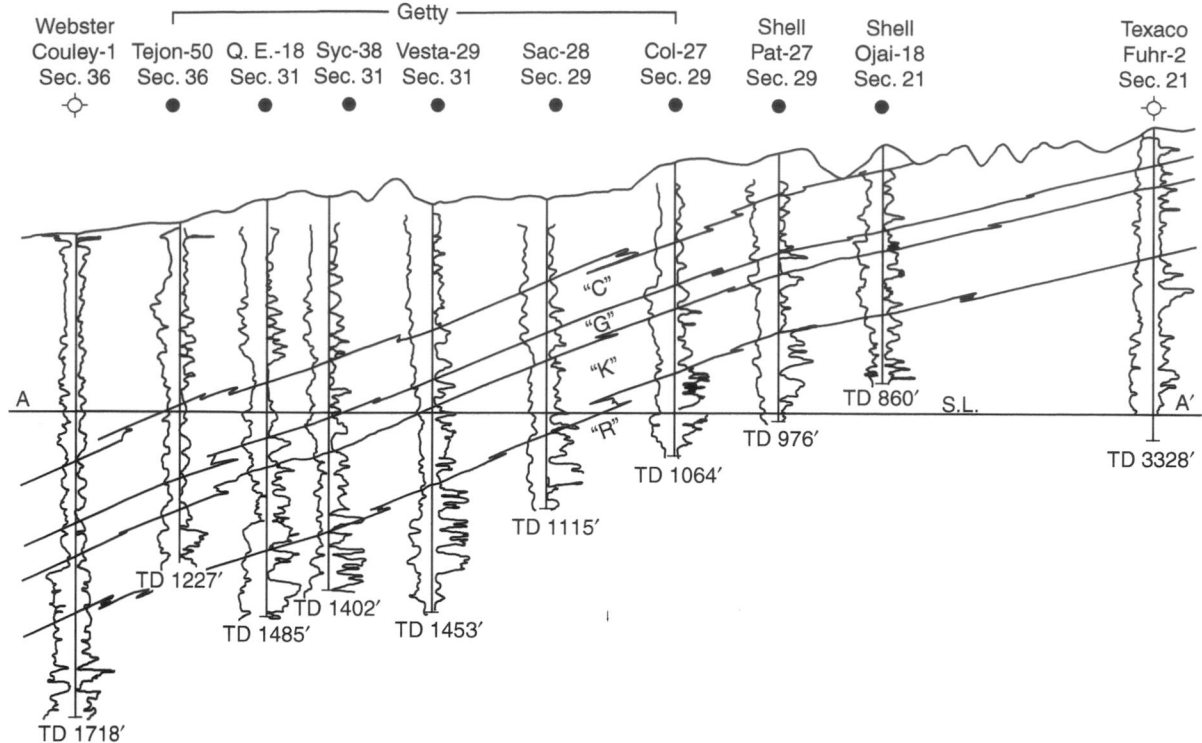

Fig. 8.84—Fieldwide stratigraphic cross section (Bursell and Pittman 1975).

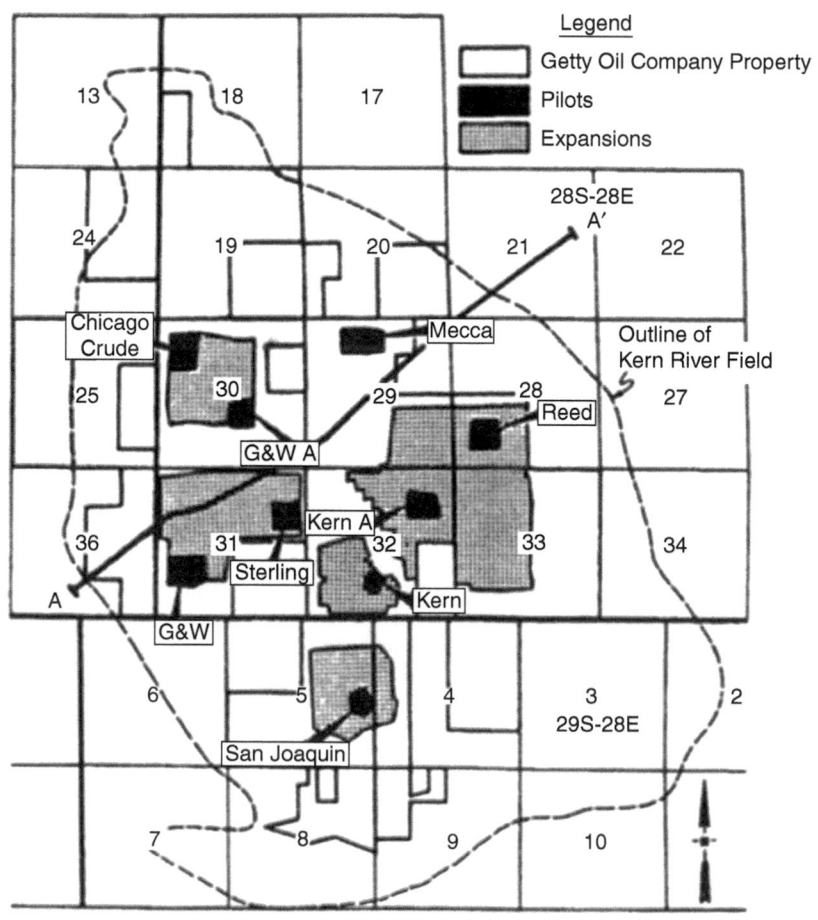

Fig. 8.85—Map of Kern River field (Bursell and Pittman 1975).

Project	Reservoir Properties					
	Injection Zone	Zone Thickness (ft)	Porosity (%)	Initial Oil Saturation (%)	Oil Gravity (°API)	Oil Viscosity at 100°F (cp)
1970 Canfield expansion	R-1	80	31	51	13.5	1,700
1970 San Joaquin expansion	K-1	29	29	52	14.5	1,000
1971 G&W expansion	K-1	60	31	47	13.5	1,780

Table 8.52—Reservoir properties—Kern River steamfloods.

Project	Number of Patterns Installed	Total Acreage (acres)	Project Life (years)	Estimated OIP (10^6 STB)	Cumulative Project Recovery (10^6 STB)	Estimated* Steamflood Interval Recovery (10^6 STB)
1970 Canfield expansion	80	210	6.75	20.61	12.44	11.76
1970 San Joaquin expansion	69	186	5.75	6.15	9.06	4.8
1971 G&W expansion	114	307	8.17+	21.27	16.12	≥ 12.57

*Cumulative production figures exclude estimated nonsteamflood oil production.

Table 8.53—Estimated steamflood recovery—Kern River steamflood.

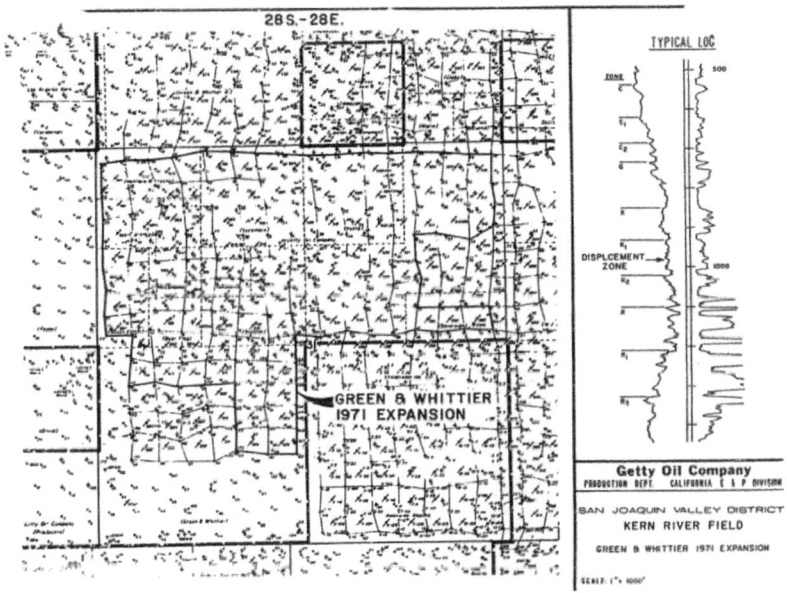

Fig. 8.86—Green and Whittier 1971 expansion (SPE Report 76-34-2/76 1976).

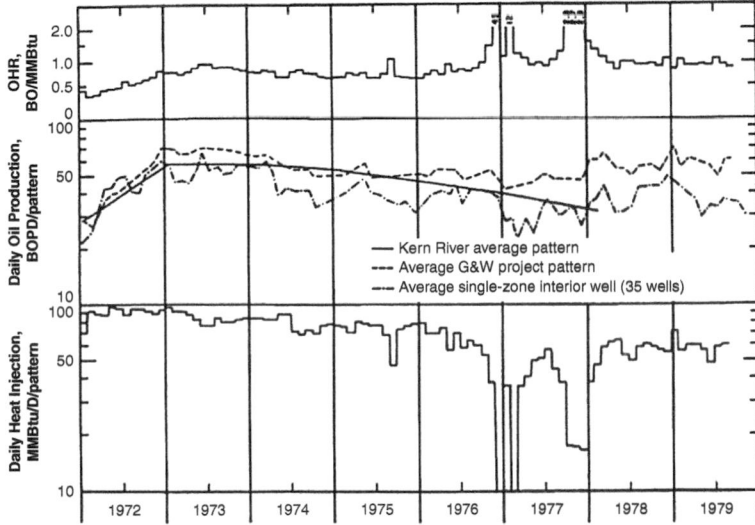

Fig. 8.87—Average pattern performance for the 1971 Green and Whittier steamflood project (Greaser and Shore 1980).

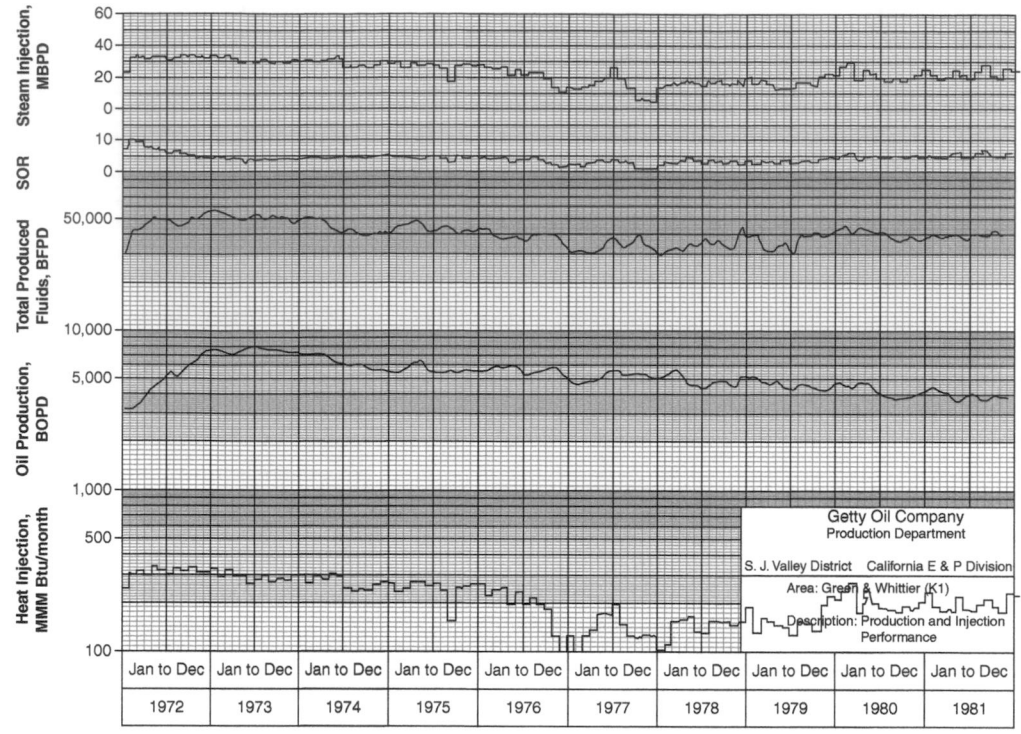

Fig. 8.88—Production and injection performance, Green and Whittier 1971 expansion.

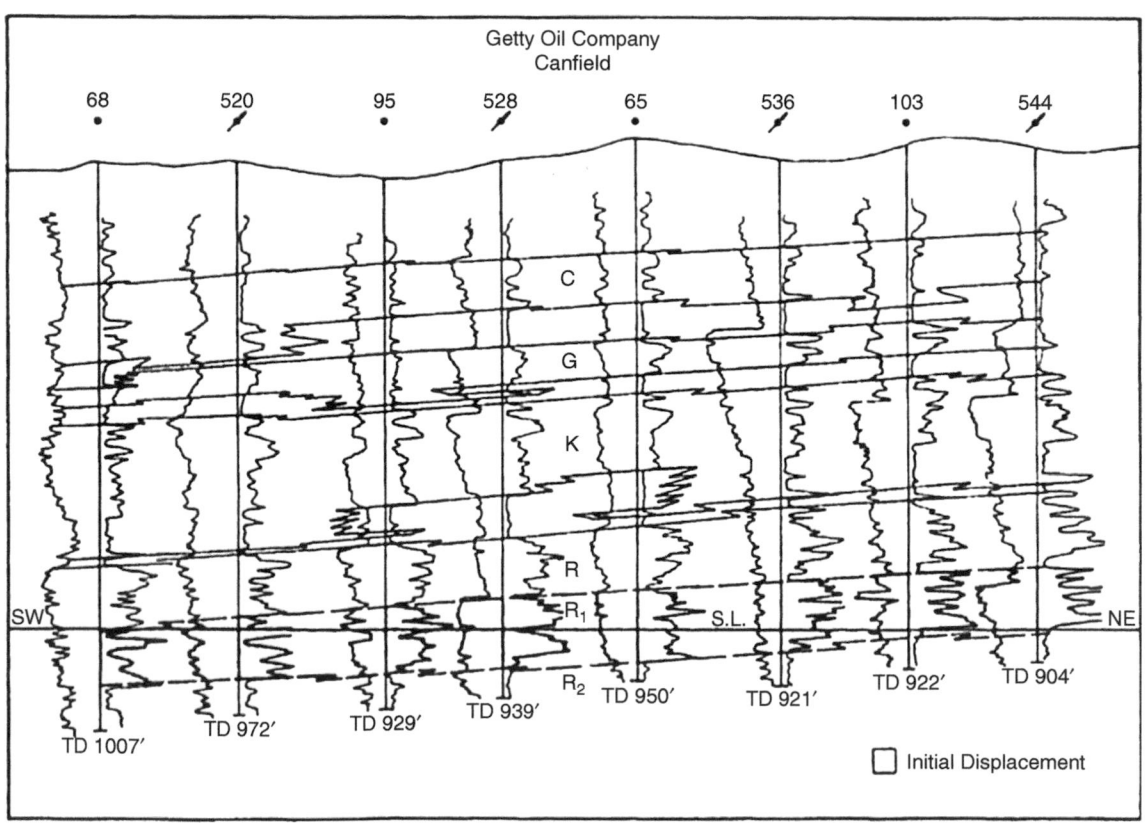

Fig. 8.89—Cross-sectional view through the 1970 Canfield steamflood project (Greaser and Shore 1980).

oil recovery from 50 to 58%, indicating that regions existed that were not effectively drained by five-spot patterns. **Fig. 8.93** presents production response for the 1970 Canfield R1 and R infill projects. **Table 8.54** summarizes steamdrive recoveries for the 1970 Canfield R1 and R and the Sanguinetti R1 steamflood.

The Sanguinetti R infill program occurred during the active drive period in the R zone. Steam injection began in the R zone in June 1984. The infill program began on the same date and was completed by January 1985. **Fig. 8.94** gives average

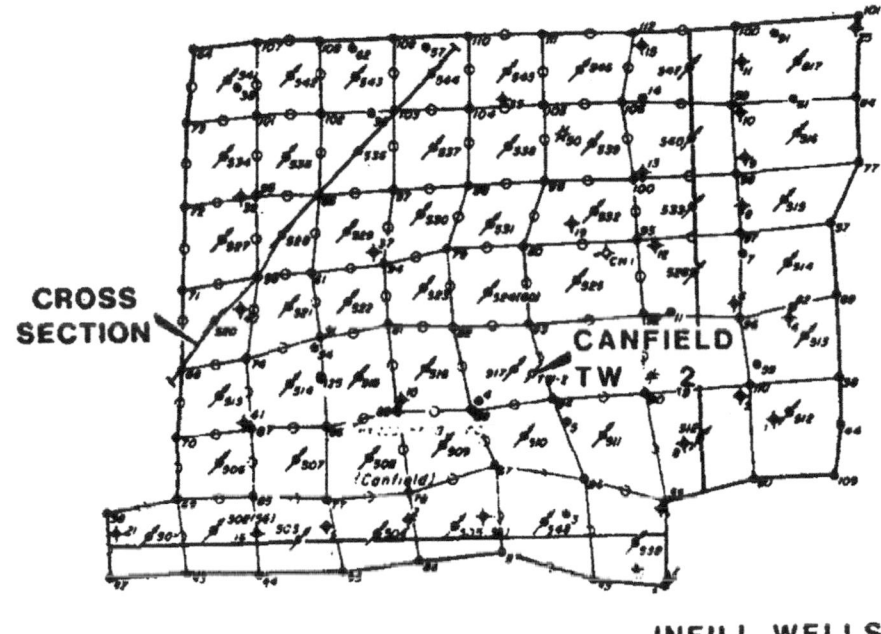

Fig. 8.90—1970 Canfield well location map (Restine et al. 1987).

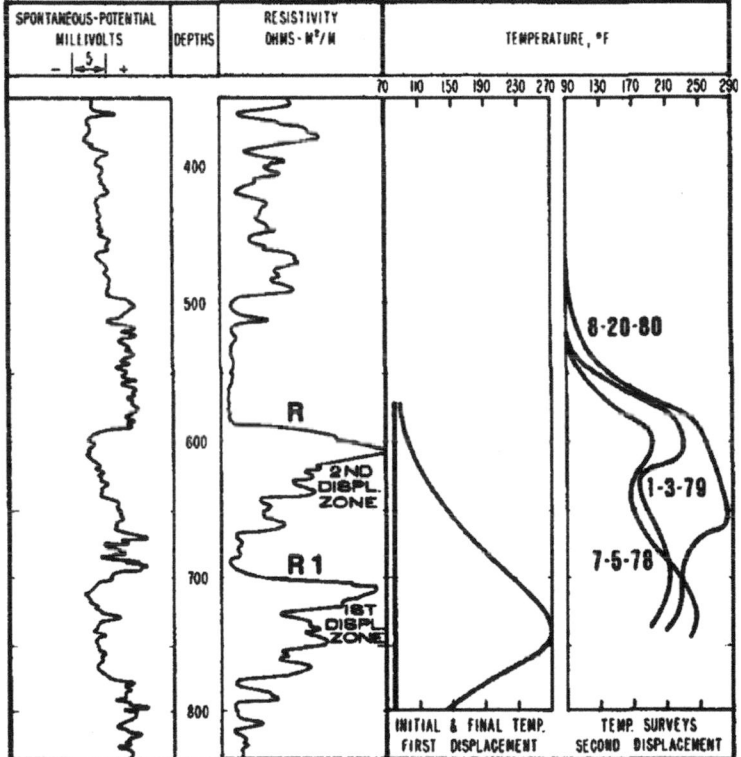

Fig. 8.91—Type log and temperature surveys, Canfield TW No. 2 (Restine et al. 1987).

production response for the Sanguinetti R project pattern. Wells recompleted into zones under active steamdrive showed acceleration of reserves and increased OSRs.

Converting 2½-acre Kern River five-spots to nine-spots by infill drilling accelerates, and may increase, oil recovery at any point in the steamdrive of the five-spot pattern. Infill wells compensate for increased reservoir heterogeneities, such as clay/silt interbeds, that reduce potential recovery by interfering with gravity drainage as well as "dead spots" that received no heat. Infill wells are more effective during the active drive period than when completed in previously steamed zones. These results support the development of steamdrives on more intensive well spacing than the conventional five-spot pattern.

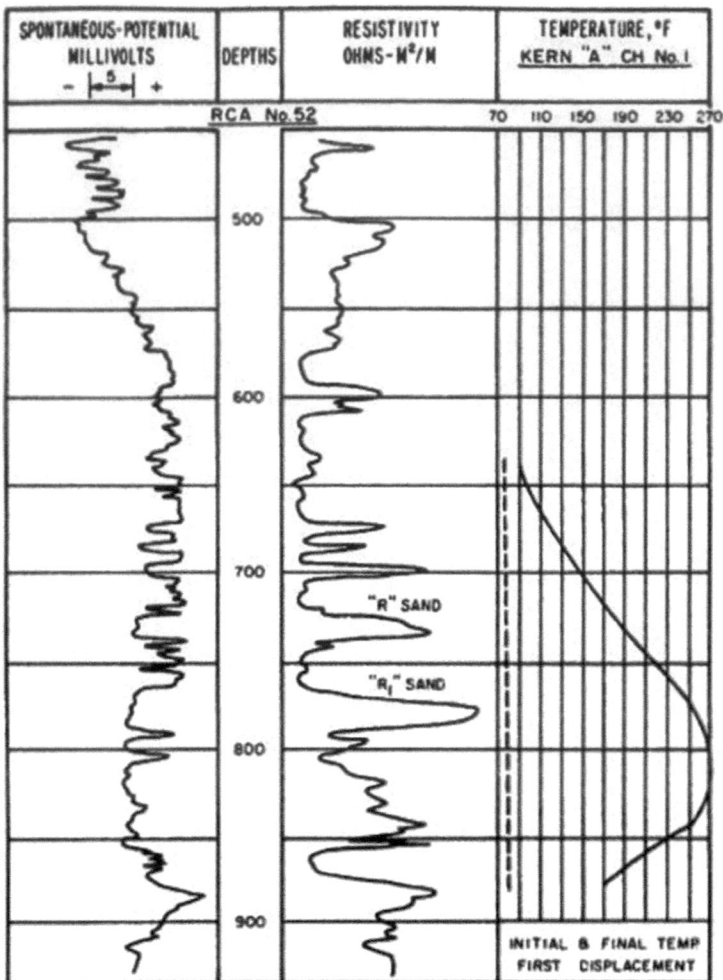

Fig. 8.92—Kern CH No. 1 temperature profile (Restine 1983).

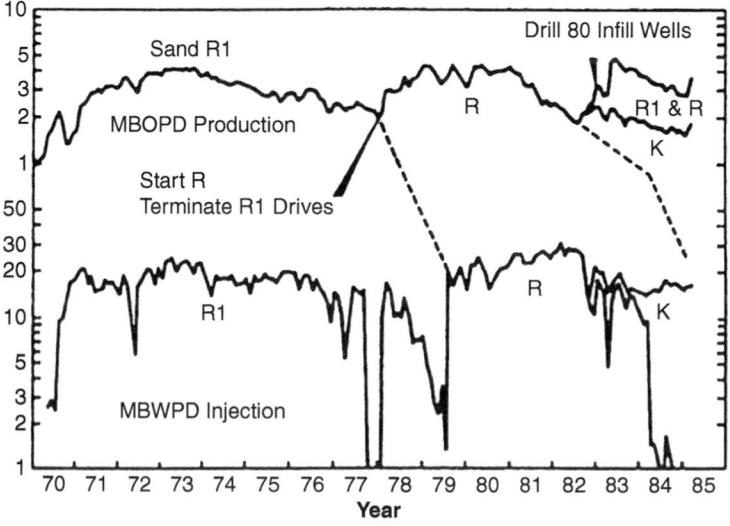

Fig. 8.93—1970 Canfield oil production and steam-injection history (Restine et al. 1987).

Chevron Section 3-10 Pattern Steamflood. Chevron USA is the second major operator in the Kern River field. A 10-pattern steamflood was initiated in September 1968 in Section 3 of the Kern River field (Blevins and Billingsley 1975). The reservoir is 300 to 500 ft thick, with the shallowest reservoir sand located 200 to 300 ft below the surface. The 10-pattern steamflood, shown in **Fig. 8.95,** consists of 6-acre, seven-spot patterns. Steam was injected into the second and third China Grade sands

Project	Patterns Installed	Total Area (acres)	Project Lift (years)	Presteamflood Estimated OIP (10^6 STB)	Cumulative Project Recovery (10^6 STB)	Estimated Steamflood Interval Recovery (10^6 STB)	Original OIP (%)	Presteamflood OIP (%)	Cumulative Fuel/Oil Production Ratio
1970 Canfield									
R1 five-spot	55	148	6.75	14.53	8.77	8.29	48.5	57.0	0.30
Infill					<u>1.22</u>	<u>1.22</u>	<u>7.1</u>	<u>8.4</u>	
Total					9.99	9.51	55.6	65.4	0.26
R five-spot	55	148	4.5	10.89	5.80	5.34	36.8	49.0	0.38
Infill					<u>0.54</u>	<u>0.54</u>	<u>3.7</u>	<u>5.0</u>	
Total					6.34	5.88	40.5	54.0	0.35
Sanguinetti									
R1 five-spot	1	2.7	8	0.274	0.109	0.109	34.5	39.8	0.34
Infill					<u>0.015</u>	<u>0.015</u>	<u>4.7</u>	<u>5.5</u>	
Total					0.124	0.124	39.2	45.3	0.30

Table 8.54—Steamflood recovery following infill drilling.

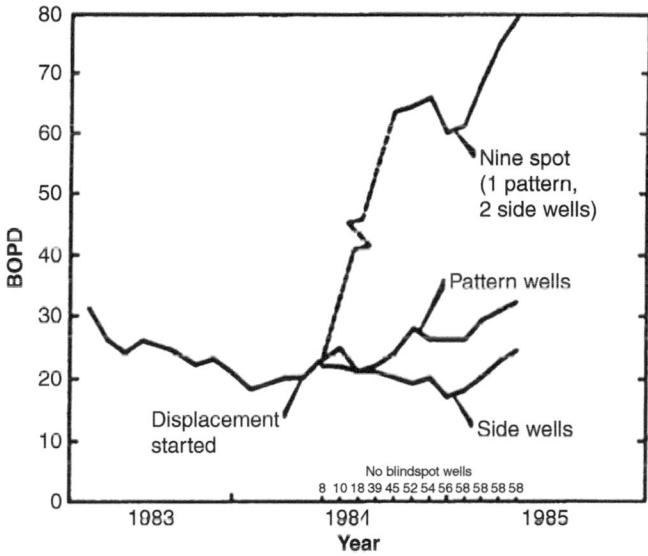

Fig. 8.94—Sanguinetti R project average pattern production (Restine et al. 1987).

at depths of 705 to 765 ft (**Fig. 8.96**). The test was designed with several temperature observation wells and post-steam core wells. Some of these temperature and core data were discussed earlier in this chapter. Blevins and Billingsley (1975) and Oglesby et al. (1982) describe the results of this steamdrive in detail.

Continuous steam injection was preceded by CSS of production wells. Steam was injected at a rate of 6,000 B/D. At the end of the steamdrive, the project was converted to cold-water injection combined with CSS and was waterflooded to the economic limit. **Fig. 8.97** shows response of the pattern to steam injection. An expansion was made in 1971 by adding fifteen 2½-acre patterns. **Table 8.55** summarizes project results. Primary recovery was approximately 11% OOIP. Steamdrive recovery was 36% OOIP, and an additional 15% OOIP had been recovered from the waterflood after the steamdrive was completed, making a total recovery (primary, steam, and post-steam) of 62% OOIP. Cumulative SOR was approximately 6.0.

One finding from the 10-pattern test was verification that gravity override was a dominant factor in the displacement process. At 1.16 PV of steam injected, 73.3% of the thickness was affected by heat. Core analyses from two wells, summarized in **Table 8.56,** show that the steam zone was located at the upper part of the formation, while the bottom 27 to 44% of the net thickness was unaffected by heat injection, depending on when cores were taken and their location.

8.5.7 Mount Poso Field. The Mount Poso field is located approximately 14 miles north of Bakersfield, California, USA, in the San Joaquin Valley. The productive area of the field is 2,100 acres, with Shell Oil Co. owning approximately 97% of the total acreage (Stokes et al. 1978). Production is from four separate horizons in the Vedder sands (**Fig. 8.98**). The Upper Vedder is the largest contributor to production. Structure in the main area is an easterly rising homocline with a dip of approximately 6°.

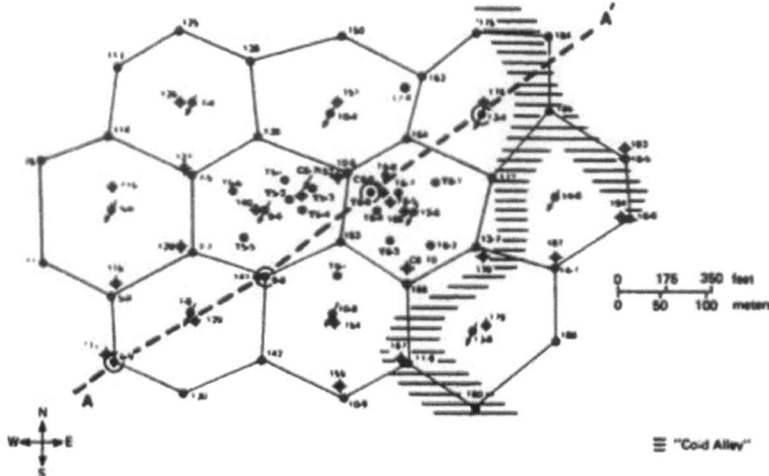

Fig. 8.95—Ten-pattern steamflood, Kern River field, Section 3 (Oglesby et al. 1982).

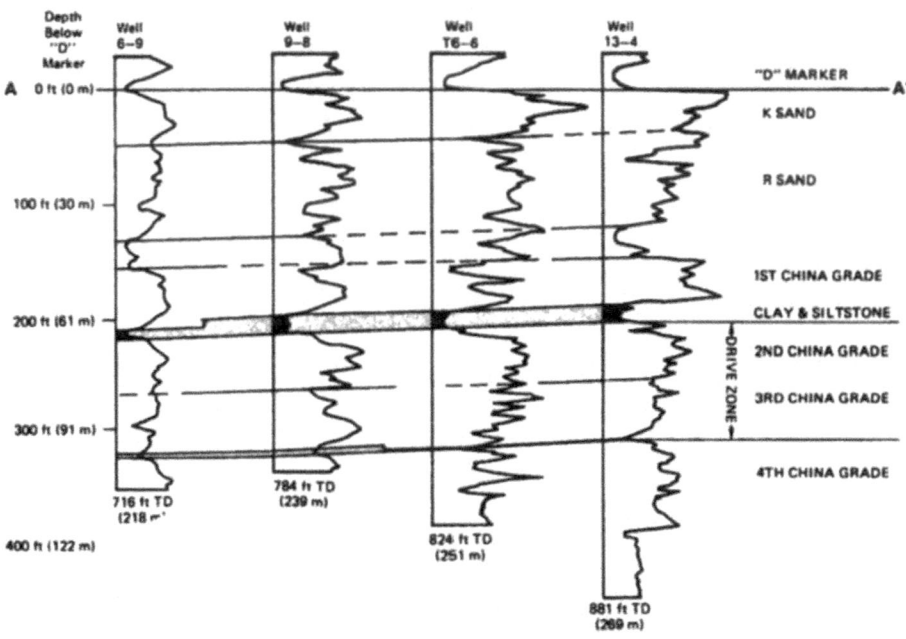

Fig. 8.96—Ten-pattern stratigraphic cross section (Oglesby et al. 1982).

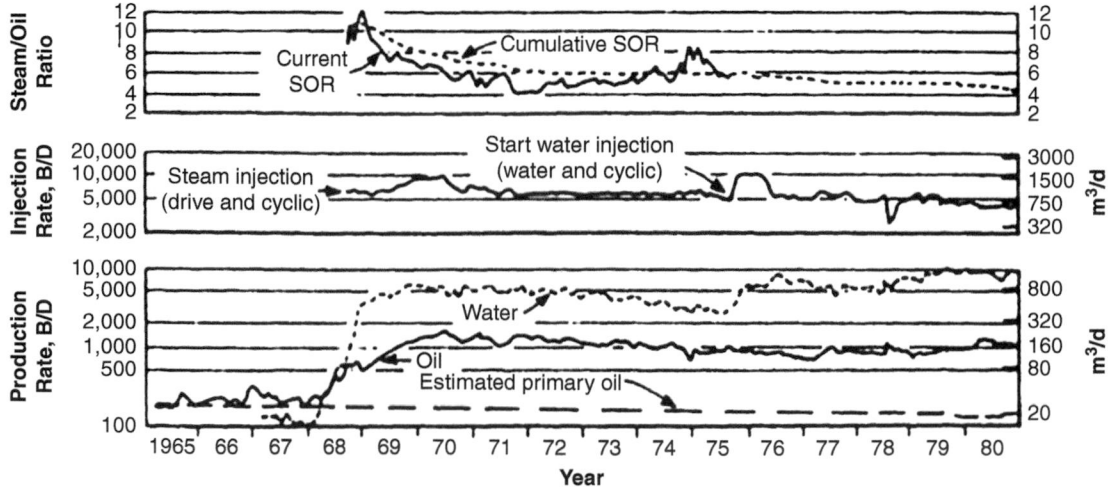

Fig. 8.97—Ten-pattern steamflood performance (Oglesby et al. 1982).

	Primary	Steamflood	Waterflood	Total
Injection				
Total volume (bbl)	0	18,581,829	8,872,858	27,454,687
PV (%)	0	116	55	171
Production				
Total oil volume (STB)	910,000	3,023,241	1,257,444	5,190,685
Total water volume (bbl)	595,000	12,053,576	10,650,971	23,299,547
Total volume (bbl)	1,505,000	15,076,817	11,908,415	28,490,232
PV (%)	9	94	74	177
% OOIP	11	36	15	62
Balance				
Net voidage* (bbl)	1,505,000	(3,505,012)	3,035,557	1,035,545
Pore volumes (%)	9	(22)	19	6

*Produced minus injected volumes.

Parentheses indicate negative value.

Table 8.55—Summary of production data, ten-pattern steamflood, Kern River Field, Section 3.

	Well C6-9 (0.40 PV)	Well C6-10 (1.16 PV)
Steam zone		
Thickness (ft)	18	50
Fraction of total thickness (%)	17.8	47.6
Oil saturation (%)	9.2	14.3
Hot-water zone		
Thickness (ft)	39	27
Fraction of total thickness (%)	38.6	25.7
Oil saturation (%)	22.0	21.5
Unaffected zone		
Thickness (ft)	44.0*	28
Fraction of total thickness (%)	43.6	26.7
Oil saturation (%)	40.0*	30.0
Totals		
Thickness (ft)	101.0*	105
Average oil saturation (%)	27.6	20.3
Displacement efficiency (%)**	43.8	58.7

*Estimated from Well 12-6.

**Based on an average initial oil saturation of 49.1% in Well 12-6.

Table 8.56—Steamflood oil displacement, ten-pattern steamflood, Kern River Field, Section 3.

Fig. 8.99 shows a structure map drawn on top of the Upper Vedder Zone. The Mount Poso fault provides closure to the east and north. Top of the structure is at a depth of 1,600 ft and an oil/water contact (OWC) is at 2,000 ft.

The Upper Vedder reservoir is a clean, unconsolidated sand with an absolute permeability of 20 darcies and a porosity of 33%. Gross thickness is 75 ft, and the average net pay is 55 ft. The sand body has good lateral continuity and contains interbedded, thin, fine-grained silt stringers that are continuous over relatively small areas. Oil gravity is 15 to 16 °API, and the viscosity at the reservoir temperature of 110°F is approximately 280 cp. Reservoir energy is provided by a strong waterdrive with an estimated water influx of 100,000 to 120,000 BWPD. Approximately 90 million bbl of oil was produced on primary production, approximately 35% of the OOIP. **Table 8.57** summarizes reservoir data from the Mount Poso field.

In the early 1960s, water cuts were approaching 99% and the field was nearing the economic limit. A commercial-scale steamdrive project was initiated in 1971 after two steam-injection pilots and physical model experiments (Stegemeier et al. 1980) were used to explore operating parameters and design the commercial-scale operation. This operation was divided into four phases to make optimal use of steam-injection equipment. Phase I involved 300 acres and is shown in Fig. 8.99. **Fig. 8.100** shows the flooding plan for Phase I, and **Fig. 8.101** shows the injection pattern. Steam was injected into five updip wells and

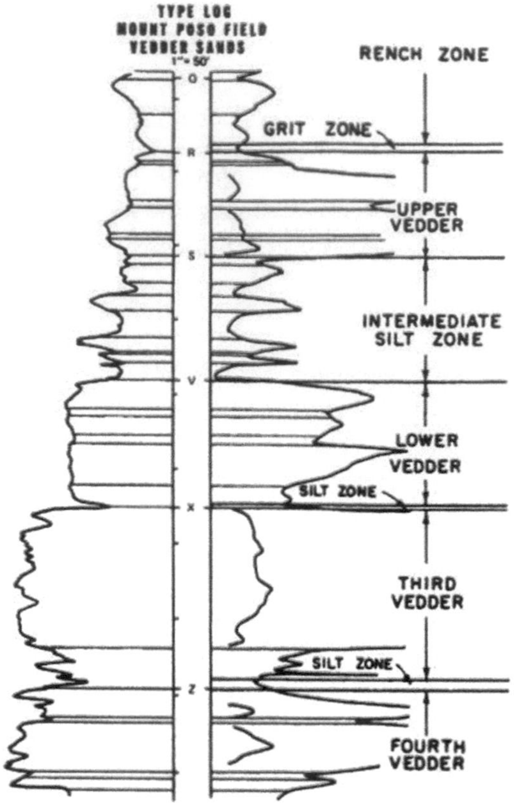

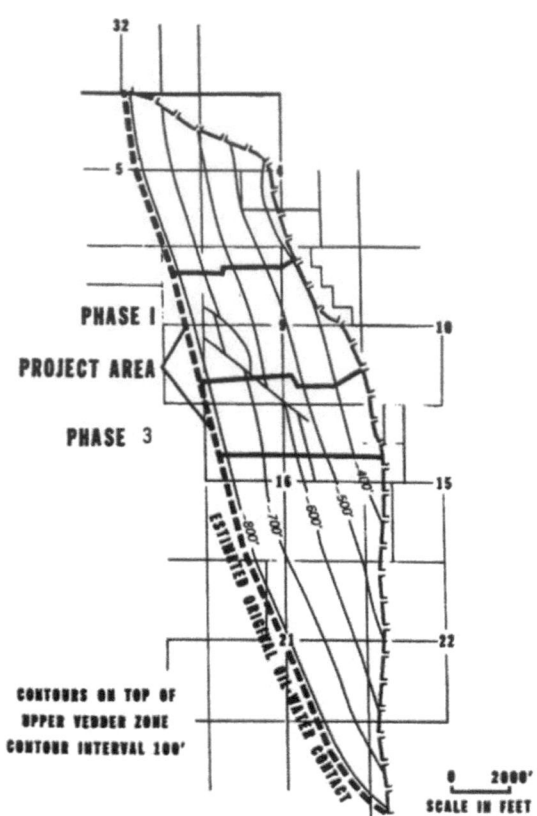

Fig. 8.98—Typical electric log of the Upper Vedder sands, Mount Poso field (Stokes et al. 1978).

Fig. 8.99—Structure map of main area of Mount Poso field (Stokes et al. 1978).

Productive area (acres)	2,100
Average depth (ft)	1,800
Average thickness (ft)	75
Average net thickness (ft)	55
Net/gross ratio	0.73
Dip (degrees)	6
Porosity (%)	33
Permeability (darcies)	20
Oil saturation (%)	
Original	90
Current	58
Residual to steam	15
Original STOIP (million bbl)	260
Estimated cumulative production (1 January 1971) (million bbl)	90
Primary recovery (percent original STOIP)	35
Reservoir temperature (°F)	110
Reservoir pressure (psi) (600-ft datum)	
Original	450
Current	100
Oil gravity (°API)	15 to 16
Viscosity at reservoir temperature (cp)	280

Table 8.57—Mount poso field, Upper Vedder reservoir data.

five downdip wells. The overall plan was to inject steam in the downdip row of wells for a period of 2 to 3 years (0.2 PV) to heat that area of the reservoir and to prevent formation of a cold-oil bank near the OWC. Continuous injection in the updip wells (0.7 PV) would permit efficient contact of the upper and middle parts of the reservoir. Phase 3 injection started in June 1994, following termination of downdip injection in Phase 1 in January 1974. **Table 8.58** summarizes the chronology for development of the steamdrive.

Fig. 8.102 shows project performance for Phases 1 through 4.[12] Overall recovery efficiency is anticipated to be approximately 60% OOIP. Cumulative production in the last published report (Mount Poso Field, Report 75-19-1/86 1986) was 567 STB/acre-ft, or approximately 20.6 million STB, from the 660 acres under steamdrive.

8.5.8 M-6 Steamdrive Project, Tia Juana Field. The M-6 steamdrive project is located in the Tia Juana field, Bolivar Coast, Venezuela. The Tia Juana field is one of several heavy-oil fields on the eastern coast of Lake Maracaibo, which collectively contain OOIP of approximately 33.5 billion bbl (Herrera 1977, 1978). The Tia Juana field has an OOIP of approximately 10.7 billion bbl. The structure is a homocline with a dip of approximately 4°. Four productive intervals are identified on the type log in **Fig. 8.103,** Sands C, Dl, D2, and D3. Sands C and D3 are separated from Sands Dl and D2 by continuous shale units. While Fig. 8.103 indicates that Sands Dl and D2 are separated by shale, they are in direct contact in some parts of the project area as a result of erosion of the intervening shale during deposition. The D sands are difficult to correlate between wells in the M-6 project unit. Some feeling of the reservoir heterogeneity is obtained from the map of the D sands in **Fig. 8.104,** which was prepared by lumping all D units together.

Table 8.59 summarizes reservoir and fluid properties. The sands are unconsolidated and have 38% porosity and permeabilities between 0.5 and 1.0 darcies. Reservoir temperature is 113°F, and the dead-oil viscosity varies with depth between

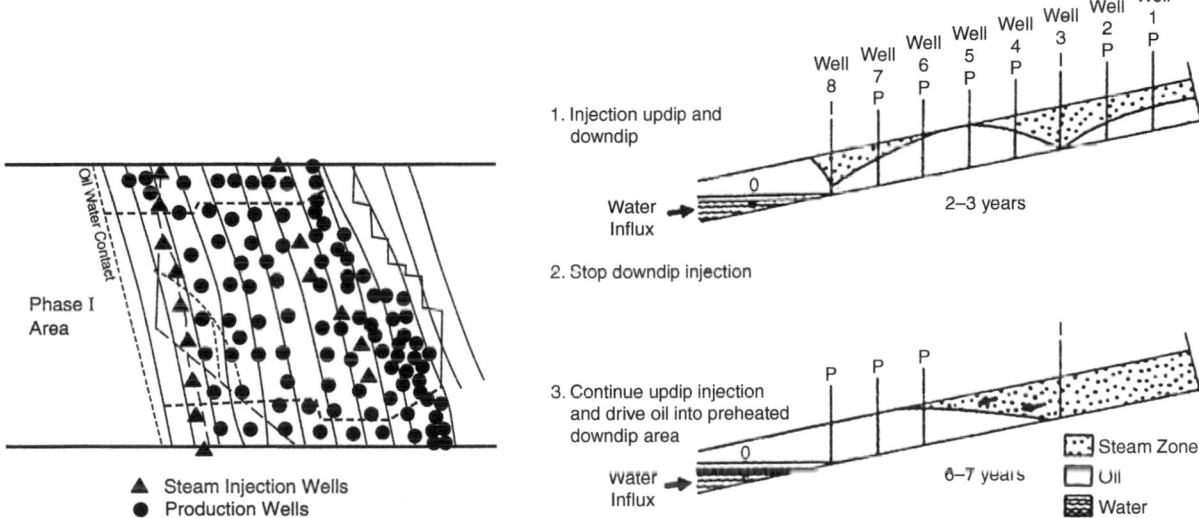

Fig. 8.100—Phase I Steamflood area (SPE Report 75-19-1/86 1986).

Fig. 8.101—Schematic of steamdrive process, Mount Poso field (Stokes et al. 1978).

May 1970—Steam injection in downdip pilot began.
September 1971—Steam injection in Phase 1 began.
January 1974—Downdip steam injection terminated in Phase 1.
June 1974—Steam injection in Phase 3 began.
November 1976—Steam injection in Phase 2 began.
December 1977—Steam injection in Phase 4 began.
June 1978—Downdip steam injection terminated in Phase 3.
July 1979—Downdip steam injection terminated in Phase 2.
November 1979—Began downdip steam injection in Phase 4.
July 1982—Steam injection in Phase 1 reduced by 50%.
March 1983—Downdip steam injection terminated in Phase 4.
November 1983—Steam injection in Phase 2 downdip extension began.
February 1985—Steam injection in Phase 2 downdip extension terminated.

Table 8.58—Chronology of field development, Mount Poso steamflood.

[12]Personal communication with L. Z. Cook, Shell Oil Co., Houston (1992).

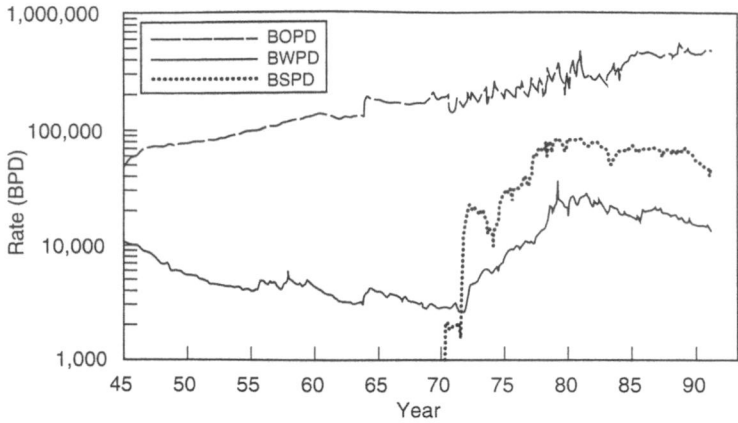

Fig. 8.102—Project performance, Mount Poso field, main area, Upper Vedder zone, Phases 1 through 4.

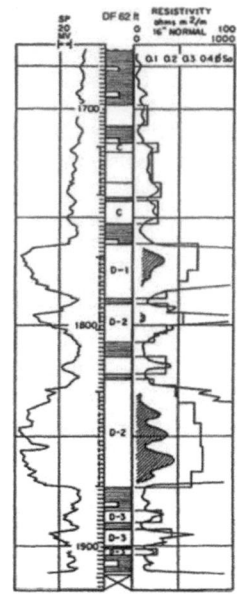

Fig. 8.103—Type log showing sand development, M-6 steamdrive project, Well LSE-1470, Tia Juana field (Schenk 1982).

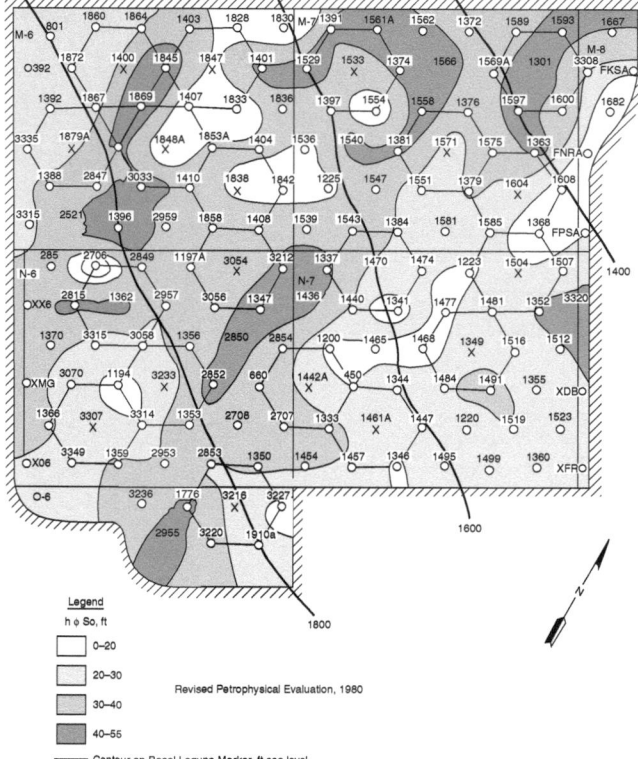

Fig. 8.104—Oil column and structure map of D sands, M-6 steamdrive project (Schenk 1982).

1,000 and 9,000 cp at 100°F and 10 to 20 cp at 350°F. Primary production is by solution-gas drive and compaction resulting from subsidence. Original reservoir pressure was 780 psig at the top sand. At the beginning of the steamdrive, the reservoir pressure had declined to 100 psig at the top sand. Primary recovery, aided by CSS, is approximately 23% stock-tank oil in place (STOIP).

Fig. 8.105 shows the M-6 Project area. The project area had 130 producers and 21 injection wells completed on a triangular grid with a well spacing of 758 ft. The area was chosen for steamdrive because the estimated remaining primary from compaction drive was only 4% STOIP, the area was large enough to conduct a major drive project while avoiding interference from surrounding operations, sands were well-developed in the area and believed to be homogeneous, and few additional wells were required to promote uniform recovery.

A steamdrive began in January 1978 in semiopen seven-spot well patterns and continued until February 1987, when the project was shut in because of low oil prices (Report 78-46-2/88 1988). de Haan and Schenk (1969), Herrera (1977), Kruit et al. (1979), Munoz (1981), Puig and Schenk (1984), and Schenk (1982) describe the steamdrive. Steam-injection rate was approximately 44,000 B/D for most of the project. Average injection rate per well was approximately 2,520 B/D at surface pressures of 400 to 600 psig (Schenk 1982). **Fig. 8.106** shows the response of the project. By the end of 1983, approximately 107 million bbl (17 million tons) of steam had been injected in the steamdrive. **Table 8.60** summarizes project performance. Cumulative oil production during the steamdrive period was more than 41 million bbl. Cumulative SOR is on the order of 2.5, which is unusually low compared with steamdrive projects in California heavy-oil reservoirs. Compaction resulting from subsidence is a dominant source of reservoir energy in the field. **Table 8.61** is a volumetric analysis of performance data (Puig and Schenk 1984). This analysis shows that as much as 15.3 million bbl of the 41.4 million bbl of oil produced during the steamdrive period could be attributed to compaction of the reservoir. The SOR is approximately 4.0 when compaction is taken into consideration.

Analysis of project performance provides basic information about steamdrives. Flowmeter surveys in the injection wells indicated that only the top 45% of the D sand interval was taking steam by 1981. At that time, only 20 of 130 production

Original Conditions

Project area (acres)	1,831
Average depth top sand (ft subsea)	1,800
Formation dip (degress)	3 to 5
Original reservoir pressure at 1,800 ft subsea (psig)	850
Original reservoir temperature at 1,800 ft subsea (°F)	113
Dead-oil viscosity at 100°F (cp)	1,000 to 10,000
In-situ oil viscosity (cp)	2,000
Permeability (darcies)	0.5 to 1.0
Net oil sand (ft)	120
Porosity	0.38
Oil saturation	0.85
Oil FVF (RB/STB)	1.04
STOOIP (10^6 bbl)	525

Conditions at Start of Steamdrive

Primary oil production (depletion enhanced by steam soak), 106 bbl (% STOOIP) 115 (22)	
Reservoir pressure at 1,800 ft subsea (psig)	150
Net oil sand (corrected for 5% sand compaction)	0.35
Oil saturation	0.74
Oil FVF (RB/STB)	1.02
OIP (10^6 bbl)	410

Formation: post-Eocene, D sands only.

Table 8.59—Reservoir data, M-6 steamdrive project area.

wells had shown a strong reaction to the injection of heat similar to that depicted in **Fig. 8.107.** These wells experienced a pronounced increase in oil production, followed shortly by a larger increase in water production and then an increase in wellhead temperature. It was thought that steam was beginning to break through in these wells. Temperature profiles in these wells indicated that heat was confined to an interval of 20 to 30 ft at the top of the formation. Production policy was to shut off this interval and recomplete the lower D sands.

Interpretation of project performance indicated that areal and vertical steam distribution were uneven because of reservoir heterogeneities. The reservoir is so heterogeneous areally and vertically that it was not possible to construct a reservoir model to assist in the development of a reservoir management policy.

8.5.9 Duri Field. The Duri steamflood in east central Sumatra was the world's largest steamflood. In 1992, injection was approximately 700,000 B/D CWE, and production was approximately 200,000 BOPD (Report 91-83-2/92 1992). Field production was expected to peak at more than 300,000 BOPD in the mid-1990s. Production was 190,000 BOPD in 2014 (Koottungal 2014).

Fig. 8.108 (Pearce and Megginson 1991) shows the Duri field, which is approximately 5 miles wide and 11 miles long, covering a surface area of 26,000 acres (Report 91-83-2/91 1991). The Pertama and Kedua sandstones are the principal producing intervals in the Bekasap formation. A type log is shown in **Fig. 8.109.** The intervals are layered, and representative properties of zones within these intervals are shown in **Fig. 8.110.** These sands are separated by a correlatable shale that extends over most of the field but share a common OWC.

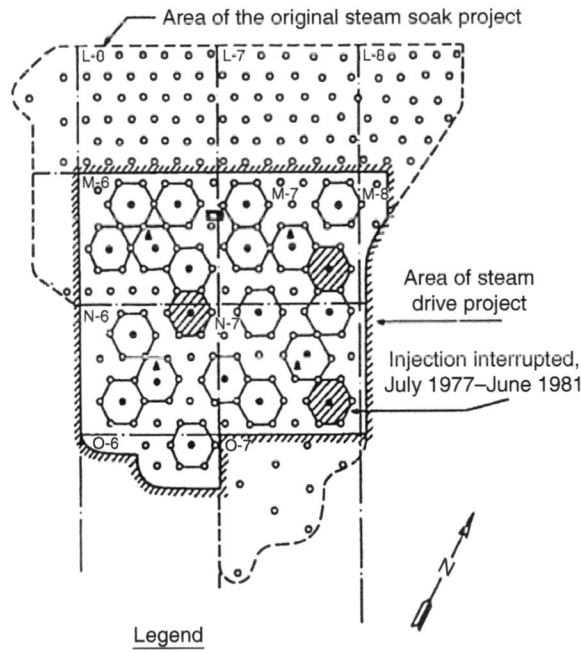

Legend

- • Injection well
- ○ Production well
- ▬ Steam plant
- ▲ Production station
- ⬡ Pilot hexagon

Fig. 8.105—Project layout, M-6 steamdrive project (Schenk 1982).

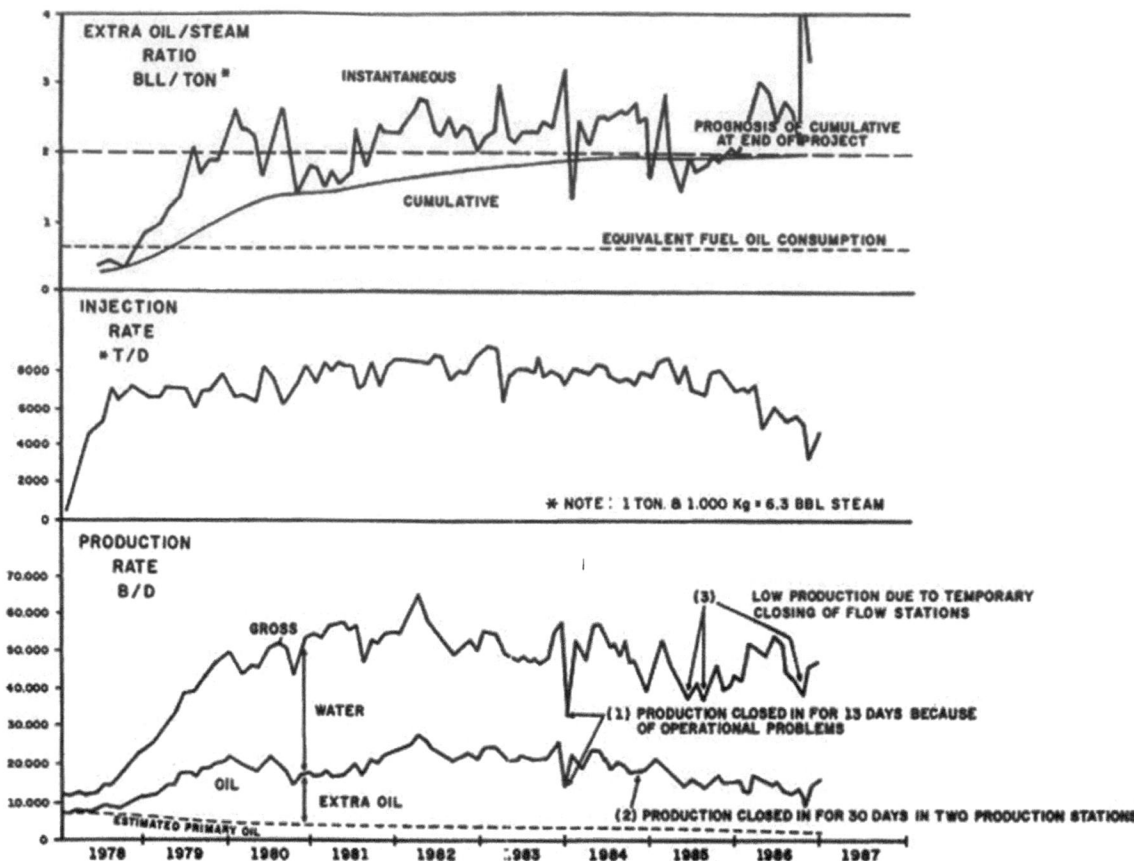

Fig. 8.106—Project performance, M-6 steamdrive (SPE Report 78-46-2/88 1988).

| | | | Oil Production | | | Steam Injection | | |
| | | | | Cumulative | | | Cumulative | |
Exploitation Mode	Period	Number of Years	Average Rate (B/D)	10⁶ bbl	% STOIIP*	Average Rate (ton/D)	10⁶ton	OSR (bbl/ton)
1. Primary depletion	1945–68	24	7,700	67.1	12.8	–	–	–
2. Steam soak	1969–77	9	15,600	51.2	9.8	520	1.7	30
3. Steamdrive**	1978–83	6	18,800	41.1	7.8	7,800	17.0	2.4
TOTAL	–	–	–	159.4	30.4	–	–	–

*The area has a STOIIP of 525×10⁶ bbl.
**Last quarter of 1983 estimated.

Table 8.60—Project performance of M-6 steamdrive to 1983.

The Pertama zone extends over approximately 15,000 acres of the field, while the Kedua is considered floodable for approximately 3,700 acres (Pearce and Megginson 1991). **Table 8.62** summarizes representative properties of the Pertama/Kedua interval.

The field was discovered in 1941 and put on production in 1958. Primary recovery was estimated to be 7.5% OOIP owing to the high oil viscosity (157 cp), small amount of solution gas, and limited natural reservoir energy. CSS of production wells began in 1967 to increase production rates. Two EOR pilots were conducted in 1975 to determine the most efficient way to improve recovery from the field. One test was a caustic waterflood in a 250-acre pilot test area. The caustic waterflood was terminated in 1979 because it failed to produce a significant increase in oil recovery.

The second test was a steamflood pilot consisting of 16 inverted five-spot patterns, approximately 15½ acres in size. There were 27 producers and several observation wells. High-quality steam (65%) was injected into the lower Kedua. Approximately 30% of the OIP was recovered by the end of the test period.

Project expansion began in 1985 in Area 1, shown in **Fig. 8.111,** and has continued following the plan indicated in the figure (Pearce and Megginson 1991). Area 1 was developed with 95 inverted seven-spot patterns (11⅝ acres) with twin wells at each producing location. One of the twin wells produces from the Pertama and the other from the Kedua. Steam is injected into each zone with twin injection wells or through a single string with a downhole choke. New areas have used the twin injection wells.

Primary depletion period	
N_{p1}, cumulative oil production (10^6 bbl)	67.1
W_{p1}, cumulative water production (10^6 bbl)	2.8
ΔV_{s1}, cumulative subsidence volume (10^6 bbl)	48.3
B_{o1}, oil FVF at end of period (RB/STB)	1.63
Steam-soak period	
N_{p2}, cumulative oil production (10^6 bbl)	51.1
W_{p2}, cumulative water production (10^6 bbl)	7.7
ΔV_{s2}, cumulative subsidence volume (10^6 bbl)	31.3
B_{o2}, oil FVF at end of period (RB/STB)	1.02
M_s, cumulative steam injection (10^6 tons)	2.4
[pilot steamdrive hexagons (10^6 tons)]	0.7
Steamdrive period*	
N_{p3}, cumulative oil production (10^6 bbl)	40.6
W_{p3}, cumulative water production (10^6 bbl)	54.7
ΔV_{s3}, cumulative subsidence volume (10^6 bbl)	15.3
W_{p3}, cumulative steam injection (10^6 tons)	16.3

*Includes estimates for August through December 1983.

Table 8.61—M-6 project area, performance data used in volumetric balance analysis.

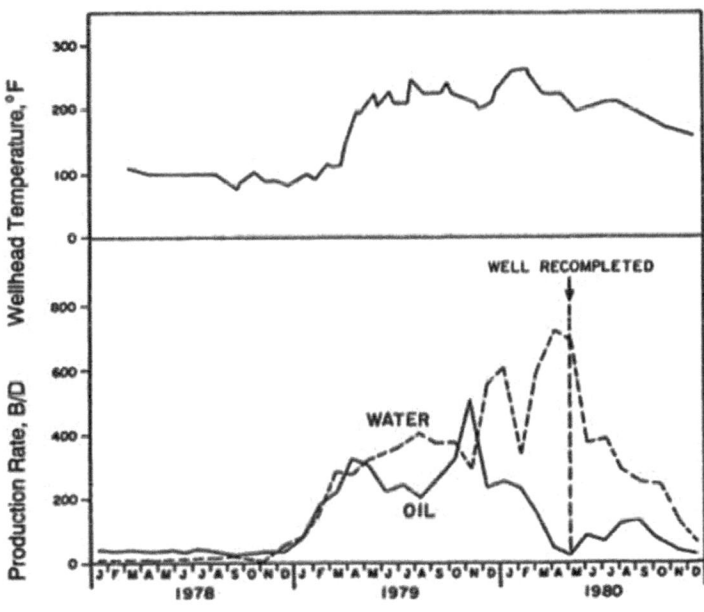

Fig. 8.107—Performance of hexagon with strong response to steamdrive, M-6 project (Schenk 1982).

Fig. 8.108—Duri field steamflood project location, Sumatra (Pearce and Megginson 1991).

Table 8.63 summarizes performance of Area 1. By the beginning of 1991, 244,000 B/D CWE was being injected as steam and the oil production rate was 66,000 B/D. The SOR was in the vicinity of 3.6, compared with the design value of 3.8. Infill wells were drilled at some locations to improve sweep and to increase oil recovery in patterns with poor performance. The infill program resulted in improvements in both injection-to-production and SORs.

Fig. 8.112 summarizes the performance of the Duri steamflood. Expansion of the project is continuing. **Table 8.64** compares the Duri steamflood when fully developed with the Kern River steamflood. Pearce and Megginson (1991) describe the expansion program and the reservoir management program in additional detail. One result of the economic analysis of the simulated expansion program was the discovery that a "hybrid" development scheme with a combination of 15½-acre five- and nine-spot patterns maximized recovery of reserves and the return on capital investment. Changing to the nine-spot geometry increased the producer-to-injector ratio, which permitted more flexibility in balancing steam-injection rates. This pattern was to be used in the expansion into Area 6.

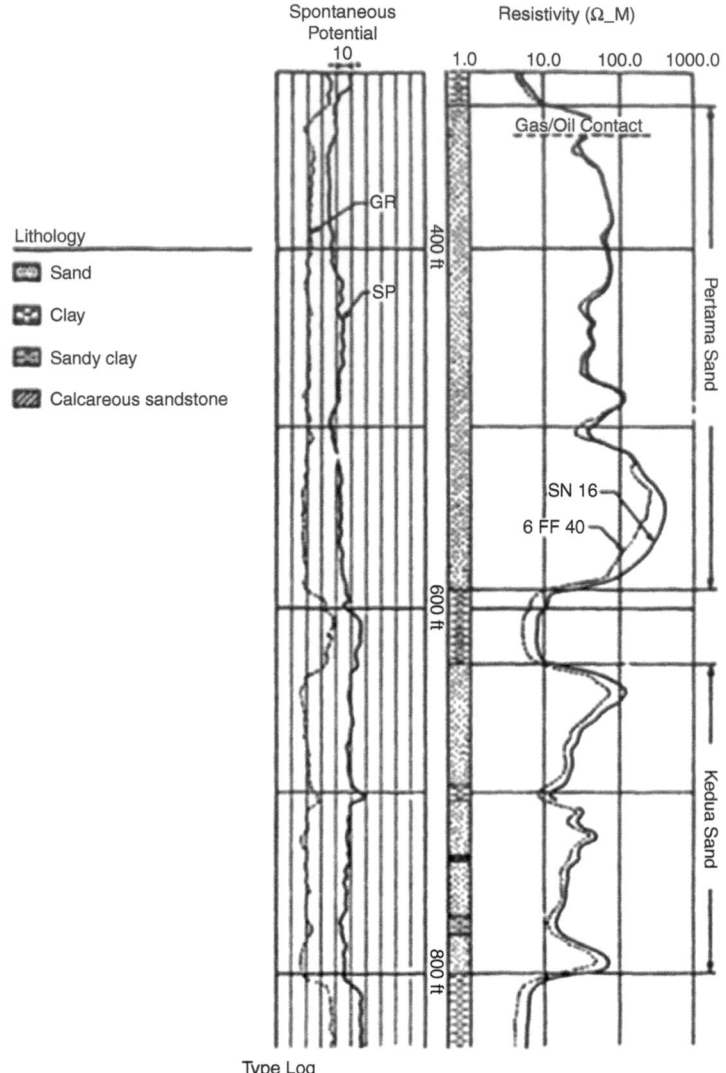

Fig. 8.109—Type log showing Pertama and Kedua sands, Well 522, Duri field; RTE = 94 ft (SPE Report 91-83-2/91 1991).

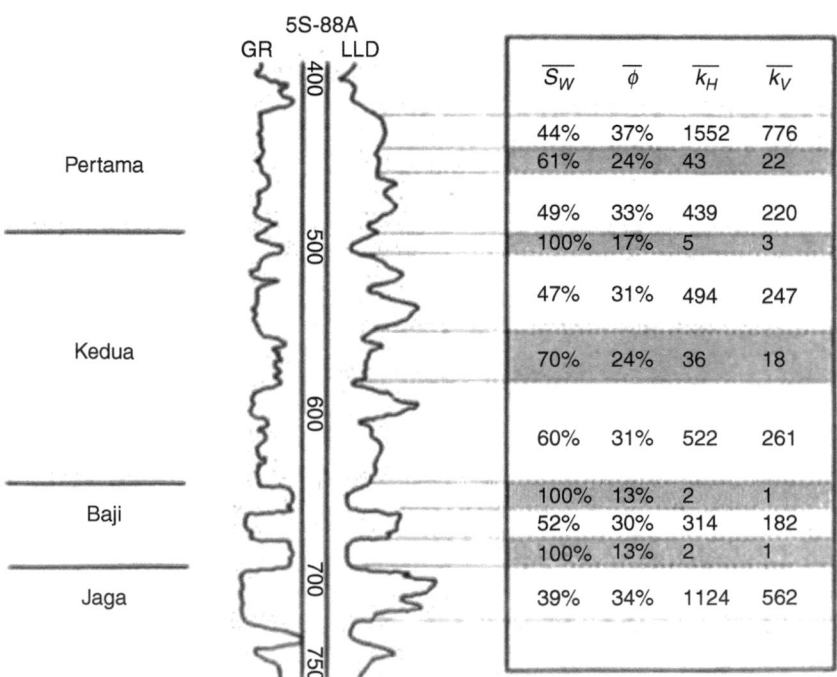

Fig. 8.110—Properties of representative layers in Area 1 of Duri steamflood project from history match (Pearce and Megginson 1991).

Reservoir	Lithology
Pertama/Keduan	Sandstone
Depth (ft)	500 to 800

Reservoir Fluid Parameters	
OOIP (10^6 STB)	4.15
Average porosity (wt%)	36
Horizontal permeability (md)	440 to 1,550
FVF (RB/STB)	1.022
Reservoir temperature (°F)	100
Reservoir pressure (psi)	110
Bubblepoint pressure (psi)	267
Net thickness (ft)	142
Dip (degrees)	0 to 5
Crude type	Paraffin
Stock-tank oil gravity (°API)	22.7
Reservoir oil viscosity (cp)	157
Produced water salinity, NaCl equivalent (ppm)	900 to 4,500
Critical water saturation (%)	28.5
Residual water saturation (%)	19.8

Table 8.62—Summary of initial report data.

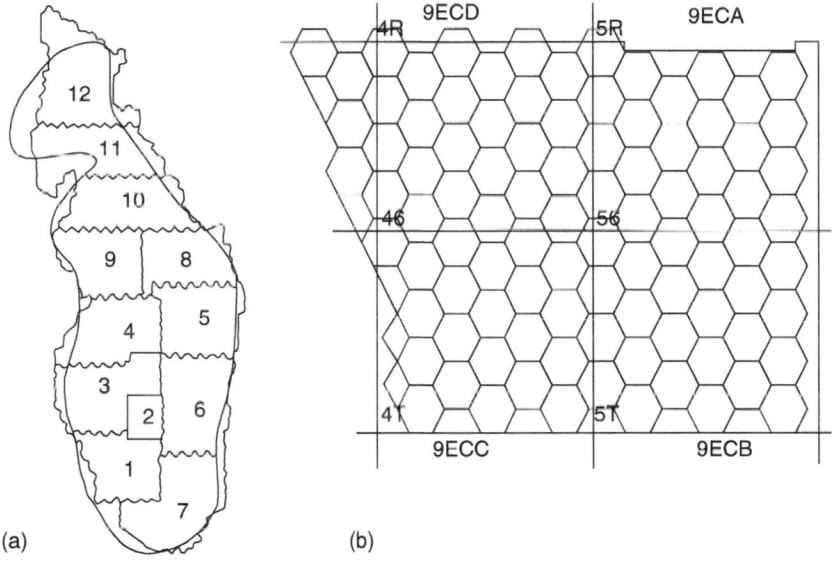

(a)   (b)

Fig. 8.111—Duri steamflood project development; (a) current 12-area development plan and (b) Area 1 (Pearce and Megginson 1991).

Area (acres)	1,137
Patterns	95
Pattern type	Seven-spot
Pattern area (acres)	11⅝
Injectors	106
Pattern producers	397
Extra producers	17
Infill producers	46
Total producers	460
Current production (10^3 BOPD)	66
Current injection (10^3 B/D CWE)	244

Table 8.63—Area 1 performance statistics.

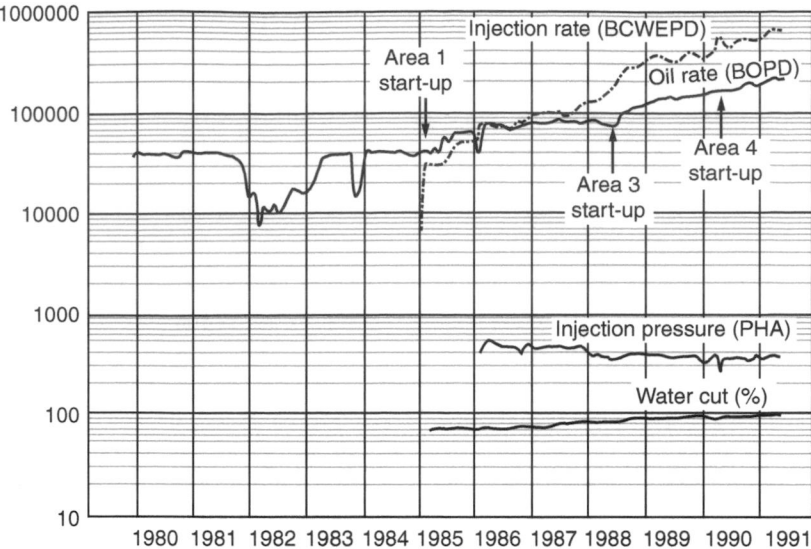

Fig. 8.112—Duri steamflood performance (SPE Report 91-83-2/91 1991).

	Duri	Kern River
Acres to be developed	15,100	9,660
Depth (ft)	600	900
Net pay (ft)	109	60
Pattern configuration	Five-, seven-, and nine-spot	Five- and nine-spot
Well spacing (acres)	3⅝	2½
Permeability (md)	1,550	2,000
Porosity (%)	36	31
Reservoir temperature (°F)	100	90
Oil viscosity at BHT (cp)	157	4,500
Gravity (°API)	23	13
OOIP (10^9 bbl)	4.4	4.0
Ultimate reserves (10^9 bbl)	2.7	2.0
Number of producing wells	3,810*	7,490 total 6,300 active
Number of steam injectors	1,370*	2,640
Gross oil production (10^3 B/D)	266*	125
Net oil production (10^3 B/D)	204*	113
Water production (10^3 B/D)	1,504*	1,200
Steam injection (10^3 B/D)	730*	830
Number of 50×10^6 Btu generators	230*	390

*Ultimate peak forcecast.

BHT = bottomhole temperature.

Table 8.64—Steamflood field comparison.

8.5.10 Reservoir Management of Steamfloods. Steamflooding is a mature EOR method and is practiced in many reservoirs. As the technology matured, strategies for the management of steamfloods have been hypothesized and tested in field tests and expansion of commercial projects. Some of the strategies worked and others did not. A few of the management schemes were discussed in the preceding section. Reservoir management strategies continue to be developed in the Duri steamflood (Gael et al. 1994) as the steamflood is expanded. Similar developments are in progress in other fields.

The management of steamfloods on a field scale is beyond the scope of this text. However, Hong (1994) provides a comprehensive overview of steamflood management and covers many topics of interest to the practicing engineer that could not be included in this text.

8.6 Production of Bitumen by Steam Injection

The bitumen deposits (termed oil sands) in the Canadian province of Alberta are estimated to contain more than 1.845 billion bbl of hydrocarbon (Alberta Energy Regulator 2014). The hydrocarbon resource is found in unconsolidated sands and is immobile at reservoir temperature. Approximately 20% of the bitumen resource is located in shallow formations that are being strip mined. The bitumen is separated from the sand by a hot-water process and is then refined to produce synthetic crude oil, which is transported by pipeline to refineries in Canada and the US.

The remaining 80% of the bitumen resource is deeper than 75 m and remains the target of extensive research and development activities to develop in-situ processes that have the potential to recover this resource economically.

The first challenge to producing bitumen in situ is illustrated by the correlation of bitumen viscosity with temperature shown in **Fig. 8.113.** At reservoir temperatures of approximately 11°C, the viscosity of the bitumen is approximately 1 million cp and is immobile.

Fig. 8.113 indicates that the reservoir temperature must be increased significantly before the viscosity of the bitumen decreases to values that will enable flow through the porous sands to the producing wells. For example, increase in the average reservoir temperature from 11 to 80°C would decrease the viscosity to 800 cp, which is comparable to some heavy-oil reservoirs that produce by fluid expansion and rock compression.

Many processes that are based on reservoir heating were investigated in both laboratory and field tests by various companies. Of the many processes that were investigated, the research and field applications of Butler and colleagues at Imperial/Exxon in 1978 (Butler

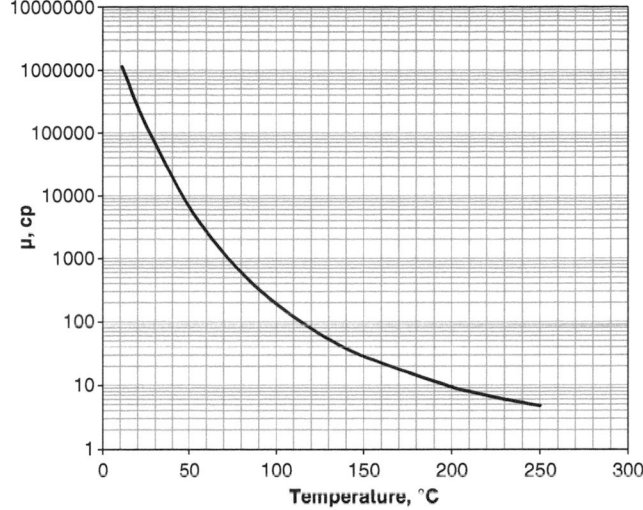

Fig. 8.113—Correlation of bitumen viscosity with temperature [Gates and Chakrabarty (2006) after Mehrotra and Svrcek (1986)].

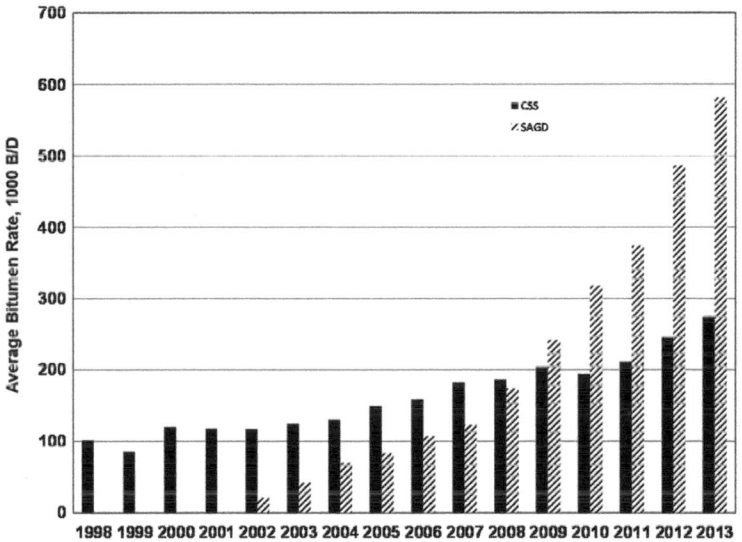

Fig. 8.114—Bitumen production from SAGD and CSS (Alberta Energy Regulator 2014).

et al. 1981; Butler and Stephens 1981; Butler 1985) led to the development of the thermal recovery process known as SAGD (Butler 1991, 1994a, b, c, d, 1998, 2001). A second major process under fieldwide application is CSS. Initial pilot testing of CSS was presented in Section 8.3, and a number of papers describe the early development of CSS (Buckles 1979; Vittoratos et al. 1990; Beattie et al. 1991; Denbina et al. 1991). Full-scale field results are presented in Section 8.6.2, and statistical reports for many SAGD and CSS projects are available through the Alberta Energy Regulator website (http://www.aer. ca/data-and-publications/statistical-reports/st98). The impact of SAGD and CSS on the annual production of bitumen in Alberta is shown in **Fig. 8.114** (Alberta Energy Regulator 2014).

8.6.1 SAGD. SAGD is based on reservoir heating through a set of parallel horizontal wells, as shown in **Fig. 8.115.** In this process, well pairs are completed so that they are approximately 5 m apart in the vertical plane. The upper well is used for steam injection and the lower well is the production well.

In the SAGD process, steam is injected into the injection well and condenses, heating the reservoir. Gravity segregation of steam and condensate occur because of the differences in densities, and the steam eventually rises to the top of the steam chamber, as shown in **Fig. 8.116** (Butler 1994c). As the reservoir heats, bitumen is mobilized and flows by gravity to the production well. Field development consists of horizontal wells spaced approximately 100 m apart, as shown in **Fig. 8.117.**

Wells may be 500 to 1000 m in length, with small variations in distance between wells, and offset from the vertical plane, as shown in **Fig. 8.118.** The horizontal segments are completed with slotted liners.

Horizontal wells are completed with injection, production, and measurement strings that run the entire length of the horizontal section. **Figs. 8.119 and 8.120** show typical completions for an injection well and a production well, respectively.

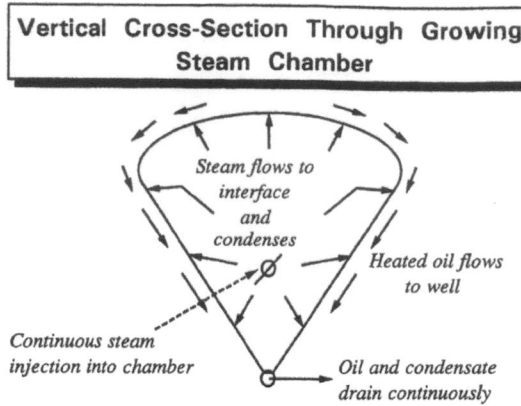

Fig. 8.115—Schematic of the SAGD process (Butler 1994b). *Printed in* Horizontal Wells for the Recovery of Oil, Gas and Bitumen, *Petroleum Society Monograph No. 2, CIM, Calgary. Reproduced with permission from the Canadian Institute of Mining, Metallurgy and Petroleum.*

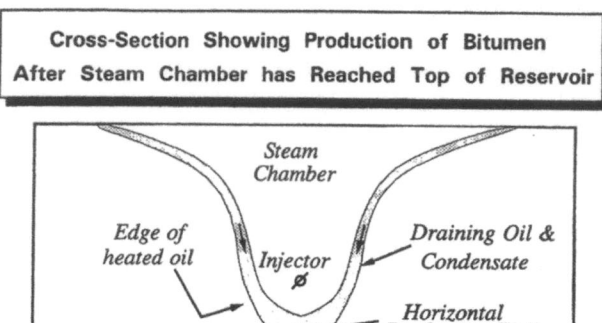

Fig. 8.116—Cross section showing production of bitumen after the steam chamber has reached the top of the reservoir (Butler 1994c). *Printed in* Horizontal Wells for the Recovery of Oil, Gas and Bitumen, *Petroleum Society Monograph No. 2, CIM, Calgary. Reproduced with permission from the Canadian Institute of Mining, Metallurgy and Petroleum.*

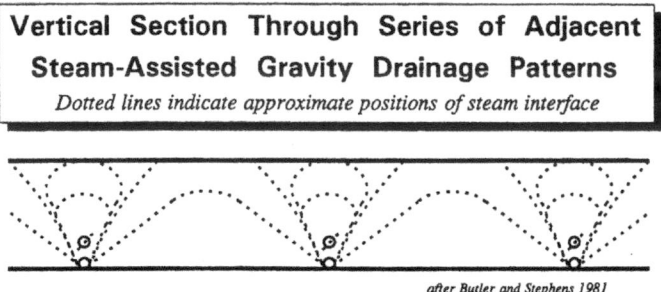

Fig. 8.117—Well placement for fieldwide application of SAGD (Butler 1994c). *Printed in* Horizontal Wells for the Recovery of Oil, Gas and Bitumen, *Petroleum Society Monograph No. 2, CIM, Calgary. Reproduced with permission from the Canadian Institute of Mining, Metallurgy and Petroleum.*

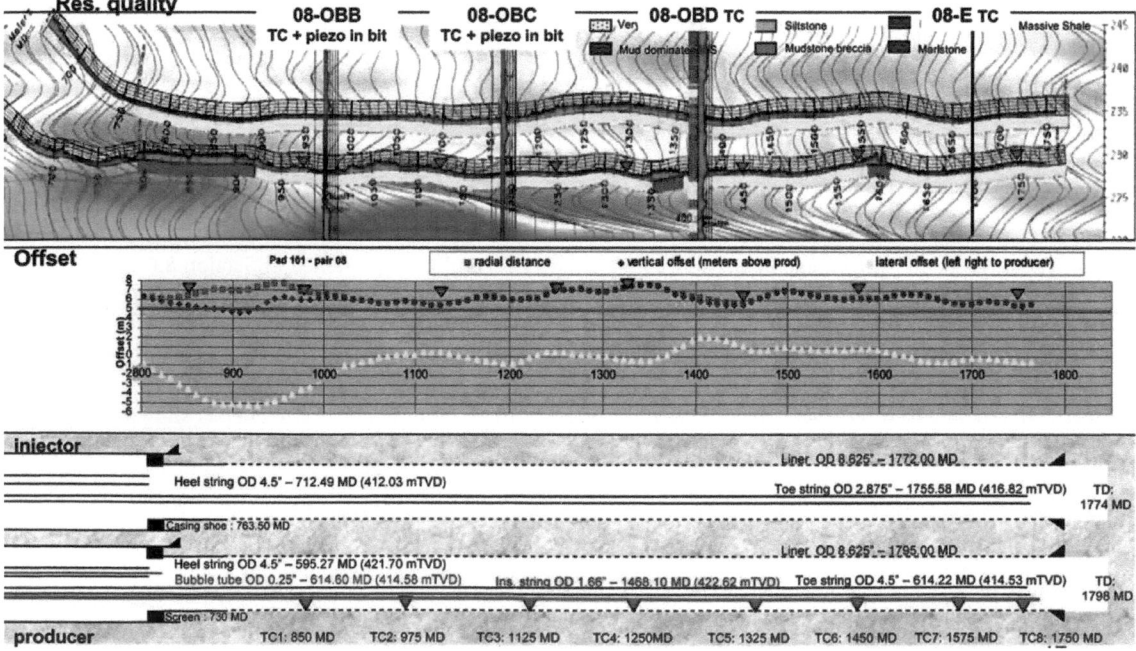

Fig. 8.118—Cross section through Well Pair 101-08(101-17) showing variation of the distance between production and injection wells as the displacement from vertical (ConocoPhillips 2009a).

Reservoir heating begins by injection of steam (220–250°C) in both injection and production wells through the toe string. Because the bitumen is immobile at this point, circulation is necessary in both injection and production wells through the toe string to heat the reservoir. Reservoir heating occurs primarily by conduction from both wells until the bitumen begins to flow. This process takes approximately 90 days (JACOS 2010). During reservoir heating, the pressure in the horizontal wells is maintained below the fracturing pressure of the formation. Fracturing pressure is approximately 6300 kPa (JACOS 2010), which is approximately 21 kPa/m, but must be determined by mini-fracture tests; therefore, steam pressures are limited to approximately 5000 kPa during the initial heating period. **Fig. 8.121** shows a series of vertical temperature profiles in an observation well (ConocoPhillips 2009d).

In Fig. 8.121, reservoir heating continued through May 2008, with conversion to steam injection in the injection well and fluid production from the production well. Temperature

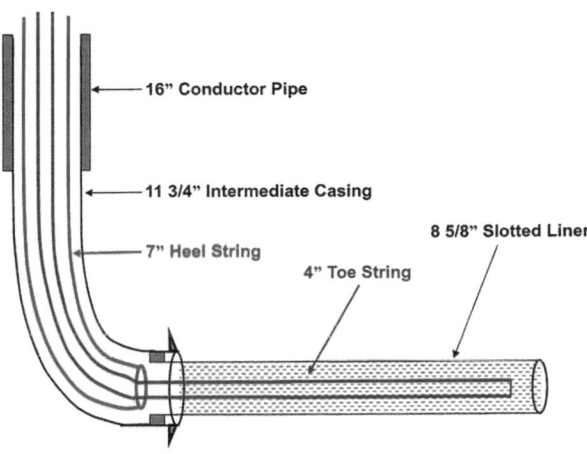

Fig. 8.119—Well schematic of a horizontal injection well showing heel and toe strings (ConocoPhillips 2009b).

profiles show the evolution of the steam chamber, which appears fully developed by August 2008.

After the steam chamber is established, injection pressure at the wellhead in production wells is adjusted to maintain pressures between 1000 and 1500 kPa. Bottomhole temperature in production wells is maintained below the steam temperature to avoid production of steam. Bitumen production increases with time, reaching a plateau, as shown in **Fig. 8.122** for wells from several SAGD projects.

Estimating Performance of the SAGD Process. The SAGD process is conceptually simple to describe but difficult to simulate without simulation packages using commercial software. The SAGD process is difficult to model with simple analytical models because the model must represent the initial formation of the steam chamber, which develops vertically and is eventually limited by the overburden and expansion horizontally as the oil is heated and flows by gravity to the production well. The geometry of the model is further complicated by the heated oil/steam interface, which is constantly expanding horizontally, increasing the area of the interface as the top of the steam chamber expands horizontally. At long times, adjacent steam chambers merge at the top of the steam chamber at a distance that, in theory, is one-half of the distance between adjacent horizontal wells

Butler et al. (1981) proposed a model of the gravity-drainage process to relate the expansion of the steam/oil interface by combining an expression for gravity drainage with a model for heat loss when the interface expanded at a constant rate. This model provided qualitative insight into the expansion of the steam/oil interface and was modified in subsequent papers (Butler and Stephens 1981; Butler 1985, 1994a) to provide a method for estimating oil-drainage rates that has been used widely

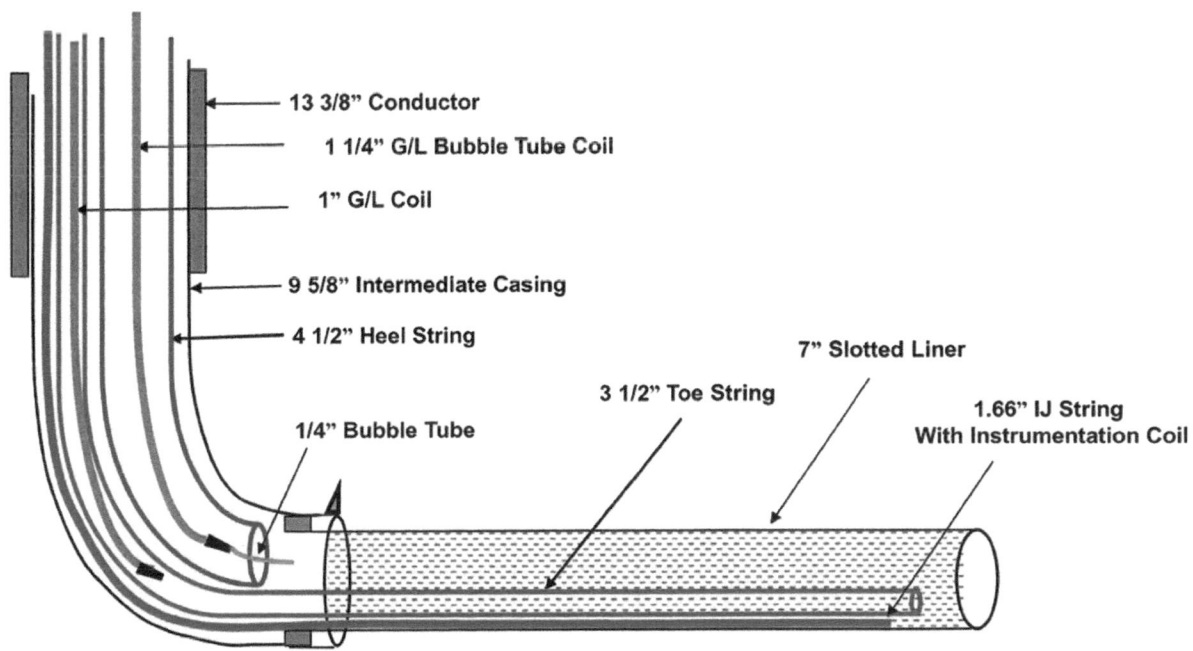

Fig. 8.120—Well schematic for completion of horizontal production well (ConocoPhillips 2009c).

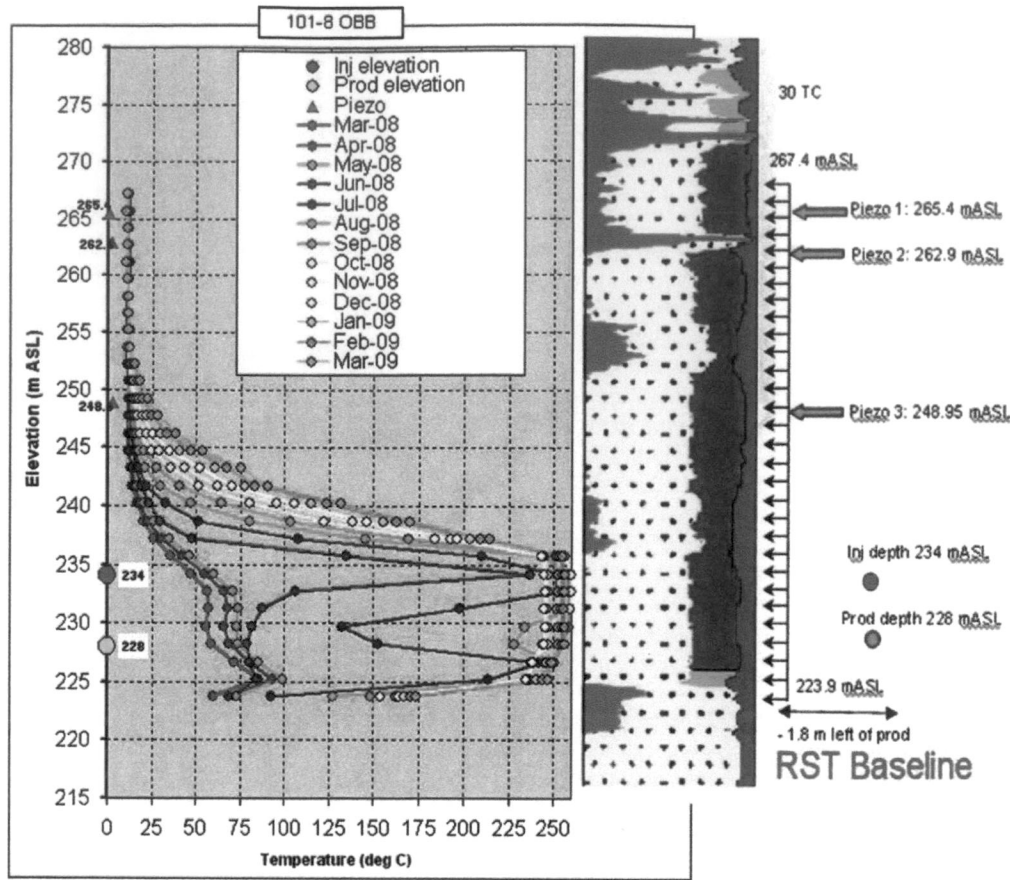

Fig. 8.121—Temperature profiles in Observation Well 101-08 OBC located 1.8 m left of the production well (ConocoPhillips 2009d).

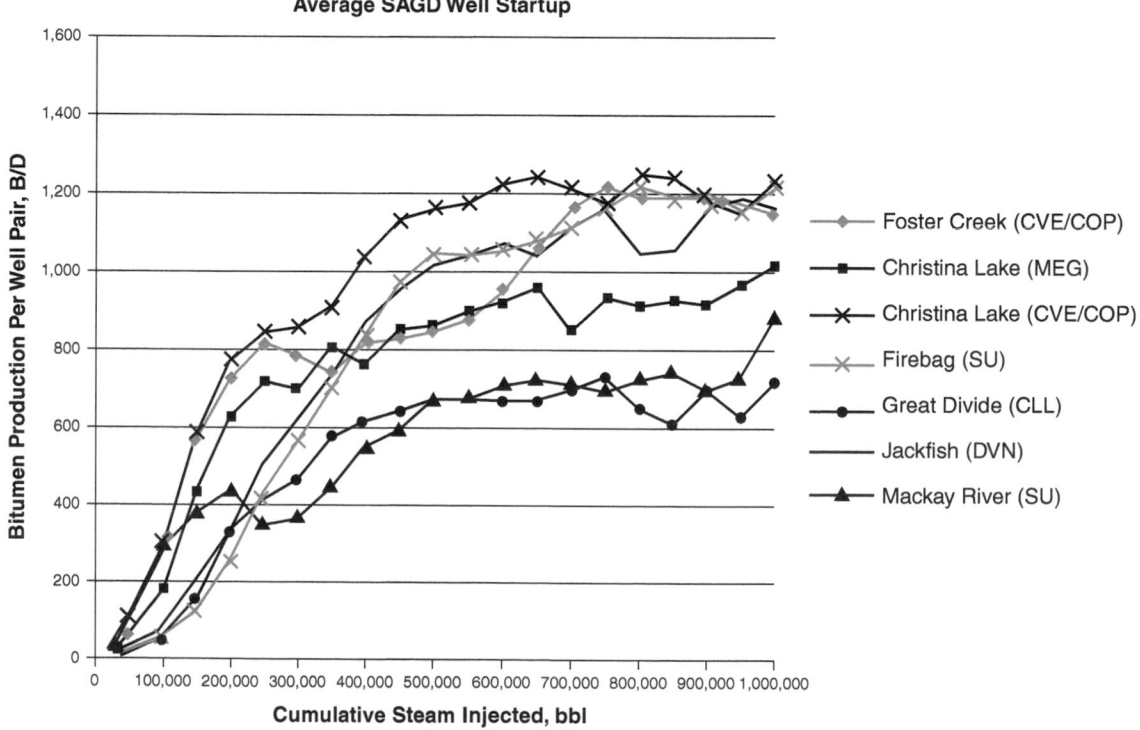

Fig. 8.122—Bitumen production rates during average SAGD well startup (Edmunds 2013).

in Canadian companies. A description of Butler's model, augmented by the impact of no-flow boundaries and the rising steam chamber, is presented in this section and is based on material presented by Butler (1991).

Fig. 8.123 (Butler 1994d) describes a model that is based on the expansion of the steam-chamber interface at a constant velocity U. The steam chamber is at constant temperature and pressure. Heat is transported by conduction through the interface and by convection as the interface advances at a constant velocity U. In this model, oil flows by gravity drainage parallel to the steam/oil interface. Because there is a temperature gradient in the direction ξ, the viscosity of the oil decreases with distance from the interface and is so large that there is no oil flow at some distance δ from the interface.

In Butler's model, Eq. 8.225 is assumed to describe the drainage of oil in the small element $d\xi$ shown in Fig. 8.123 per unit length of the steam chamber at the base of the horizontal well. The direction η is parallel to the interface, which is assumed to be straight.

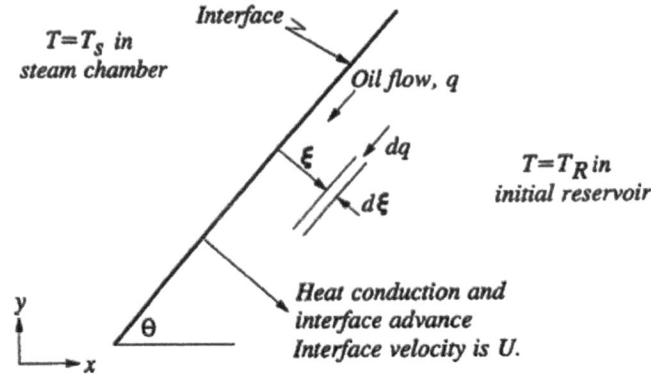

Flows of Oil and Heat in Reservoir Element Close to the Steam Interface

Fig. 8.123—Incremental element of reservoir illustrating gravity drainage (Butler 1994d). *Printed in* Horizontal Wells for the Recovery of Oil, Gas and Bitumen, *Petroleum Society Monograph No. 2, CIM, Calgary. Reproduced with permission from the Canadian Institute of Mining, Metallurgy and Petroleum.*

$$dq_o = \frac{k\rho g \sin\theta}{\mu}\,d\xi \quad\ldots\ldots\ldots\ldots\ldots\ldots\ldots\ldots\ldots\ldots\ldots\ldots\ldots\ldots\ldots\ldots\ldots\ldots \text{(8.225)}$$

$$\text{and } dq = \frac{kg \sin\theta}{\nu}\,d\xi, \quad\ldots\ldots\ldots\ldots\ldots\ldots\ldots\ldots\ldots\ldots\ldots\ldots\ldots\ldots\ldots\ldots\ldots \text{(8.226)}$$

where $\nu = \dfrac{\mu}{\rho} = $ kinematic viscosity.

The oil rate can be obtained by integrating Eq. 8.226, assuming the permeability is constant:

$$q = kg \sin\theta \int_0^{\infty} \frac{d\xi}{\nu}. \quad\ldots\ldots\ldots\ldots\ldots\ldots\ldots\ldots\ldots\ldots\ldots\ldots\ldots\ldots\ldots\ldots\ldots \text{(8.227)}$$

In Eq. 8.227, the kinematic viscosity varies with ξ because ν is a strong function of temperature. Integration of Eq. 8.227 requires a relationship between the temperature and ξ.

In developing this model, flow of heat at the steam interface is assumed to be quasisteady, leading to Eq. 8.228, in which the heat loss at the steam interface is equal to the convective energy transport of the interface moving at a constant velocity U.

$$-K\frac{dT}{d\xi} = U\rho C (T - T_R). \quad\ldots\ldots\ldots\ldots\ldots\ldots\ldots\ldots\ldots\ldots\ldots\ldots\ldots\ldots\ldots\ldots \text{(8.228)}$$

The solution to this differential equation, assuming that U is constant, is

$$\frac{T - T_R}{T_s - T_R} = e^{-\frac{U\xi}{\alpha}}, \quad\ldots\ldots\ldots\ldots\ldots\ldots\ldots\ldots\ldots\ldots\ldots\ldots\ldots\ldots\ldots\ldots\ldots \text{(8.229)}$$

where it is assumed that $T = T_s$ at $\xi = 0$ and $T = T_R$ at $\xi = \infty$.

Eq. 8.229 gives the location ξ of every temperature between T_s and T_R. When the relationship between ν and T is known, the value of ν is known at every value of ξ. The integral in Eq. 8.227 can be evaluated numerically for any relationship between ν and temperature for a constant value of U. Eq. 8.230 is an example of the *ASTM D341* (2015) correlation between the kinematic viscosity and temperature for crude oils and bitumen.

$$\log_{10}\left[\log_{10}(\nu + 0.7)\right] = A\log_{10}(T + 273) + B, \quad\ldots\ldots\ldots\ldots\ldots\ldots\ldots\ldots\ldots\ldots \text{(8.230)}$$

where A and B are determined from fitting experimental data at two different temperatures.

Butler (1985) used an analytical approach to integrating Eq. 8.227. Eq. 8.228 can be expressed in terms of $d\xi$:

$$d\xi = -\frac{K}{U\rho C}\frac{dT}{(T - T_R)}, \qquad \dots \dots \dots (8.231)$$

so that

$$dq = \frac{kg\sin\theta}{v}d\xi = \frac{kg\sin\theta}{v}\left(-\frac{\alpha}{U}\frac{dT}{T - T_R}\right). \qquad \dots \dots \dots (8.232)$$

The relationship proposed by Butler is given by Eq. 8.233:

$$\frac{v_s}{v} = \left(\frac{T - T_R}{T_s - T_R}\right)^m. \qquad \dots \dots \dots (8.233)$$

Note that in Eq. 8.233, $v \rightarrow \infty$ at $T = T_R$, and Eq. 8.234 is the result of the integration:

$$\int_0^\infty \left(\frac{1}{v}\right)d\xi = \left(\frac{\alpha}{Umv_s}\right). \qquad \dots \dots \dots (8.234)$$

$$q = \frac{kg\alpha\sin\theta}{Umv_s}. \qquad \dots \dots \dots (8.235)$$

At this point, U is considered to be a variable, which is not consistent with the assumptions used to derive the temperature profile and causes inconsistencies in the remaining part of the model.

$$U = \frac{d\xi}{dt} = \sin\theta\left(\frac{dx}{dt}\right)_y. \qquad \dots \dots \dots (8.236)$$

Eq. 8.237 is the location of the steam/bitumen interface as a function of y and t on the basis of this assumption, while Eq. 8.238 gives the velocity of the steam/bitumen interface at a constant value of y.

$$x = t\sqrt{\frac{kg\alpha}{2\phi\Delta S_o mv_s(h - y)}}. \qquad \dots \dots \dots (8.237)$$

The velocity of this point (x, y) at a constant value of y is given by Eq. 8.238:

$$\left(\frac{dx}{dt}\right)_y = \sqrt{\frac{kg\alpha}{2\phi\Delta S_o mv_s(h - y)}}. \qquad \dots \dots \dots (8.238)$$

Depletion by Gravity Drainage: Steam Interface Begins as a Vertical Plane. Eq. 8.238 was solved numerically to determine the location of point x as a function of t when y was constant. Details of the numerical solution are presented by Butler and Stephens (1981). In this model, the steam-zone interface began as a vertical plane through the injection well and expanded horizontally. A correlation of the location of the steam-zone interface as a function of dimensionless time and dimensionless distances was developed, as shown in **Fig. 8.124,** in which w is the distance between adjacent well pairs. The vertical line at $x/w = 1.0$ represents the no-flow boundary between adjacent well pairs.

In Fig. 8.124,

$$t^* = \frac{t}{w}\sqrt{\frac{kg\alpha}{\phi\Delta S_o mv_s h}}, \qquad \dots \dots \dots (8.239)$$

$$X = x/w, \qquad \dots \dots \dots (8.240)$$

$$Y = y/H, \qquad \dots \dots \dots (8.241)$$

and $h = H - y_p, \qquad \dots \dots \dots (8.242)$

where w = distance from the production well to the no-flow boundary of the adjacent well pair, y_p = elevation of the production well above the base of the formation, and H = total thickness of the formation.

A dimensionless gravity-drainage-rate correlation was developed from the steam-zone interface correlations for the case of a no-flow boundary at a dimensionless distance of $X = 1$. Eq. 8.243 is the correlation of dimensionless gravity-drainage rate q^* as a function of dimensionless time t^*:

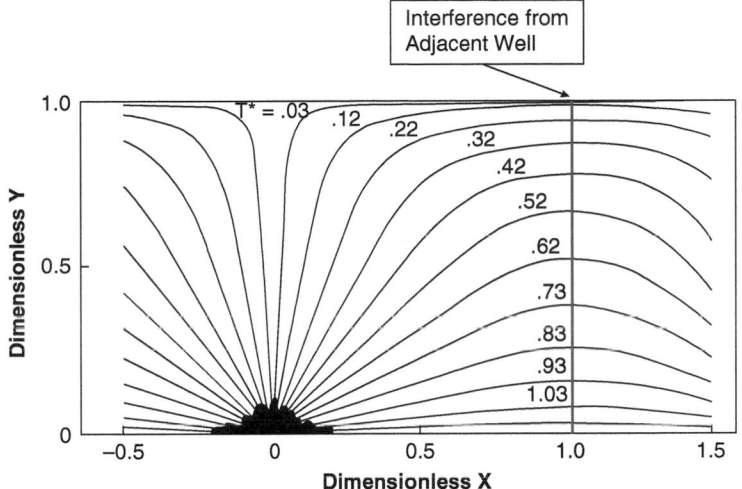

Fig. 8.124—Correlation of steam-zone interface as a function of dimensionless time and dimensionless distances (Butler et al. 1981).

$$q^* = \sqrt{\frac{3}{2} - t^{*2}\sqrt{\frac{2}{3}}} \quad \dotfill \quad (8.243)$$

for one-half of the drainage region, and

$$q^* = \frac{q}{2\sqrt{\dfrac{mv_s}{kg\alpha h\phi \Delta S_o}}}, \quad \dotfill \quad (8.244)$$

where q = drainage rate for the entire region.

The fractional recovery (FR) is obtained by integrating Eq. 8.243:

$$FR = \int_0^{t^*} q^* \, dt^*$$

$$= \sqrt{\frac{3}{2}}t^* - \frac{t^{*3}}{3}\sqrt{\frac{2}{3}}. \quad \dotfill \quad (8.245)$$

Fig. 8.125 illustrates the variation of q^* and FR with dimensionless time for the complete drainage of a steam-contacted region. The dimensionless production rate is maximum at $t = 0$ because the steam interface is initially vertical. As the steam chamber expands with time, the rate decreases slowly until the interface reaches the no-flow boundary of an adjacent well pair and the head for gravity drainage decreases with time. The entire drainage volume is swept by steam to residual oil saturation when $t^* \approx 1.22$ with $FR \approx 1.00$.

The application of the Butler model to estimate gravity-drainage rates requires correlations for m as a function of the steam temperature and properties of the oil. In Appendix 5 of Butler (1991), the following correlation between the kinematic viscosity and temperature for a wide variety of Californian and Canadian oils is presented. The correlation is an adjusted version of the widely used *ASTM D341* correlation.

$$\log_{10}\left[\log_{10}\left(v + 0.7\right)\right] = m_1 \log_{10}\left(T + 273\right) + b, \quad \dotfill \quad (8.246)$$

where b = constant for any oil and $m_1 = 0.3249 - 0.4106b$.

The constant b can be found when one value of the kinematic viscosity is known at one temperature and v can be estimated at other temperatures by use of Eq. 8.246.

The other crude-oil parameter that must be estimated is m. **Fig. 8.126** is a correlation of m with steam temperature and the ratio of $v_{100°C}/v_{200°C}$, which can be used to estimate m. Fig. 8.126 was developed for a reservoir temperature of 13°C. The upper curve gives estimates of m for Athabasca-type bitumen, the middle curve for Cold Lake crude, and the lower curve for Lloydminster oils.

Rising-Steam-Chamber Correlation. It is known that the development of the steam chamber involves heating of both production and injection wells followed by gradual growth of the steam chamber, as depicted in **Fig. 8.127.**

Gravity drainage during this phase of the SAGD process is difficult to model mathematically. Butler developed a correlation of cumulative production with process variables during the rising steam-chamber phase on the basis of experimental data

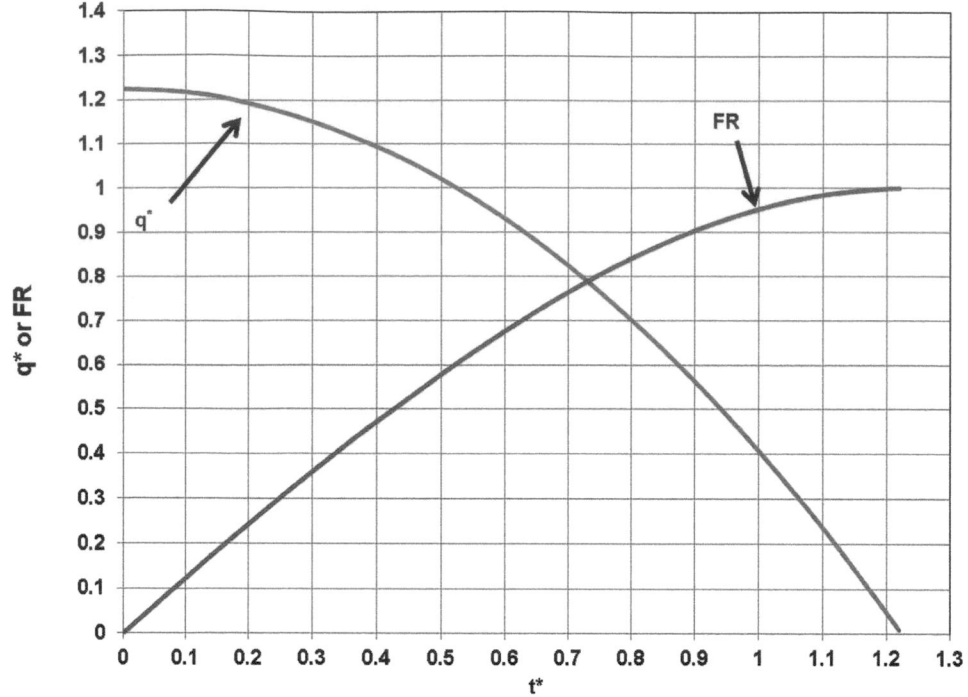

Fig. 8.125—Dimensionless gravity-drainage rate (one side) and FR for depletion model of Butler.

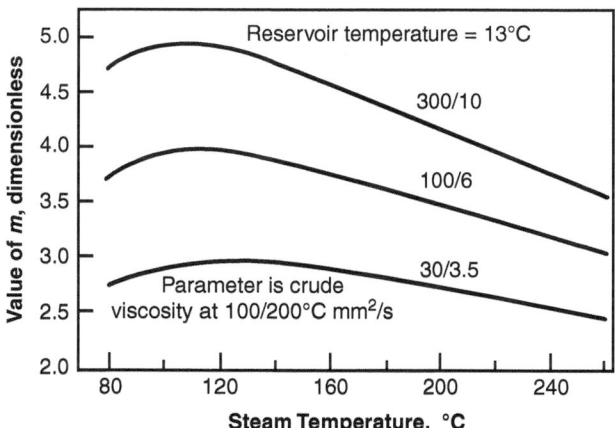

Fig. 8.126—Correlation of m with steam temperature and oil viscosity ratios (Butler 1991).

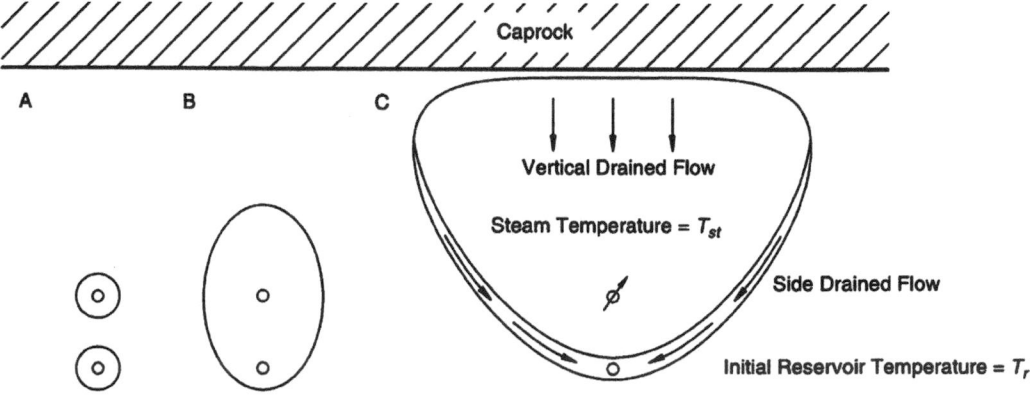

Fig. 8.127—Development of rising steam chamber during initial phase of SAGD (Irani and Ghannadi 2013).

obtained from scaled physical models. Cumulative production during the rising- steam-chamber period was estimated by use of Eq. 8.247:

$$q_{cum} = 2.25\left(\frac{kg\alpha}{mv_s}\right)^{2/3} \left(\phi\Delta S_o\right)^{1/3} t^{4/3} \dots\dots\dots\dots\dots\dots\dots\dots\dots\dots\dots\dots\dots\dots\dots (8.247)$$

$$q = \frac{dq_{cum}}{dt} = 3\left(\frac{kg\alpha}{mv_s}\right)^{2/3} \left(\phi\Delta S_o\right)^{1/3} t^{1/3}. \dots\dots\dots\dots\dots\dots\dots\dots\dots\dots\dots\dots\dots\dots\dots (8.248)$$

The FR is based on the volume of the region drained above the production well. If the production well is located at an elevation y_p above the base of the formation, the volume drained by the SAGD process is given by Eq. 8.249:

$$\text{Mobile oil} = 2wh\phi\left(S_{oi} - S_{or}\right). \dots\dots\dots\dots\dots\dots\dots\dots\dots\dots\dots\dots\dots\dots\dots\dots\dots (8.249)$$

FR = fractional recovery above the production well:

$$FR = \frac{q_{cum}}{2wh\phi\left(S_{oi} - S_{or}\right)}. \dots\dots\dots\dots\dots\dots\dots\dots\dots\dots\dots\dots\dots\dots\dots\dots\dots\dots (8.250)$$

Integration of Gravity-Drainage Depletion Model With Rising-Steam-Chamber Model. When steam injection starts in a well pair, the displacement performance is estimated by use of the rising-steam-chamber model. In time, the steam will reach the top of the steam chamber and the drainage of the steam chamber will be controlled by the gravity-drainage model, referred to as the depletion model.

The transition between the two models was defined by Butler as the point at which the gravity-drainage rates from both models were equal at the same FR. This point will occur at different cumulative injection times in each model. Butler's approach was to adjust the time in the depletion model by the difference between the time determined in the rising-steam-chamber model and the time determined from the depletion model at the point where the intersection of the two models occurred. The remaining gravity drainage was governed by the depletion model, with time adjusted by the difference between the times at the intersection of the two models described in the preceding. Application of this model is illustrated in Example 8.21.

Example 8.21—Application of the Butler SAGD Model. The application of Butler's SAGD model is illustrated in this example by use of the data presented by Butler (1991) in **Table 8.65.**

Estimate the cumulative oil production and oil production rate as a function of time for complete drainage of the volume above the location of the production well. The solution will be developed on the basis of 1 m of horizontal well length.

Solution. Because the production well is located at $y_p = 2.5$ m, the mobile oil volume in this pattern, with a total width of 75 m, is calculated by use of Eq. 8.249.

Reservoir temperature (°C)	15
Oil kinematic viscosity at T_r (cs)	100,000
Oil density (g/cm³)	1.00
Oil kinematic viscosity at 100°C (cs)	80
Reservoir thickness (m)	20
Elevation of the production well above the base of the formation (m)	2.5
Distance between adjacent well pairs (m)	75
Distance from the center of the production well to the no-flow boundary (m)	37.5
Porosity	0.33
Initial oil saturation, S_{oi}	0.75
Residual oil saturation to steam, S_{or}	0.13
Permeability to oil flow (darcies)	0.4
Steam temperature at steam pressure = 1.2 MPa (°C)	188
Thermal diffusivity of the formation (m²/s)	8.1×10^{-7}
Acceleration of gravity (m/s²)	9.81

Table 8.65—Properties of the tar sand reservoir (Butler 1991, page 316).

Mobile Oil (Gravity-Drainage Volume)

$$= \left(h - y_p\right) 2w\phi \left(S_{oi} - S_{or}\right)$$
$$= (20 - 2.5)(75)(0.33)(0.75 - 0.13)$$
$$= 268.5 \ m^3/m.$$

It is necessary to estimate the values of the kinematic viscosity (v_s) of the oil at steam temperature and the value of m. Applying Eq. 8.246,

$$\log_{10}\left[\log_{10}\left(v + 0.7\right)\right] = m_1 \log_{10}\left(T + 273\right) + b.$$
$$m_1 = 0.3249 - 0.4106b.$$

Because $v = 80$ cs at 100°C, m_1 and b can be found and v_s calculated:

$$\log_{10}\left[\log_{10}\left(80 + 0.7\right)\right] = m_1 \log_{10}\left(100 + 273\right) + b.$$
$$0.28032 = 2.5171m_1 + b$$
$$= 2.5171(0.3249 - 0.4106b) + b.$$
$$b = \frac{(0.28032 - 0.83555)}{-0.05594} = 9.9248.$$
$$m_1 = -3.7502.$$

At $T_s = 188$°C,

$$\log_{10}\left[\log_{10}\left(v_s + 0.7\right)\right] = -3.7502 \times \log_{10}\left(188 + 273\right) + 9.9248$$
$$= -0.06468.$$

$$v_s + 0.7 = 7.27.$$
$$v_s = 6.57 \ cs.$$

The value of m is obtained from Fig. 8.126. To use Fig. 8.126, the ratio of kinematic viscosities at 100/200°C is required. By use of the previous correlation, $v = 5.36$ cs at 200°C, so the ratio of $v_{100°C}/v_{200°C} = 14.92$. From Fig. 8.126 at $T_s = 188$°C, $m \approx 3.37$, rounded to 3.4.

Drainage rates estimated from the depletion model are calculated by use of Eqs. 8.243 and 8.244:

$$q^* = \sqrt{\frac{3}{2}} - t^{*2}\sqrt{\frac{2}{3}}$$

for one-half of the drainage region, and

$$q^* = \frac{q}{2\sqrt{\dfrac{mv_s}{kg\alpha h\phi\Delta S_o}}}.$$

$$\sqrt{\frac{mv_s}{kg\alpha h\phi\Delta S_o}} = \sqrt{\frac{(3.4)\left(6.57 \times 10^{-6} \ \dfrac{m^2}{s}\right)}{\left(0.4 \times 10^{-12} \ m^2\right)\left(9.81 \ \dfrac{m}{s^2}\right)\left(8.1 \times 10^{-7} \ \dfrac{m^2}{s}\right)(17.5 \ m)(0.33)(0.75 - 0.13)}}$$

$$= \sqrt{0.1962847 \times 10^{13}\left(\frac{s^2}{m^4}\right)}$$

$$= 1401016 \ \frac{s}{m^2}\left(\frac{d}{86,400s}\right) = 16.215 \ \frac{d}{m^2}.$$

$$q = 2\left(\frac{q^*}{16.215}\right)\frac{m^2}{d}.$$

$$q = 0.12334 q^* \frac{m^2}{d}.$$

The corresponding time is estimated from the dimensionless time by use of Eq. 8.239:

$$t^* = \frac{t}{w}\sqrt{\frac{kg\alpha}{\phi \Delta S_o m v_s h}}$$

$$\frac{1}{w}\sqrt{\frac{kg\alpha}{\phi \Delta S_o m v_s h}} = \frac{1}{37.5\,m}\sqrt{\frac{\left(0.4\times10^{-12}\,m^2\right)\left(9.81\dfrac{m}{s^2}\right)\left(8.1\times10^{-7}\dfrac{m^2}{s}\right)}{(0.33)(0.75-0.13)(3.4)\left(6.57\times10^{-6}\dfrac{m^2}{s}\right)(17.5\,m)}}.$$

$$t^* = \frac{2.004\times10^{-7}\left(\dfrac{m}{s}\right)\left(\dfrac{86,400s}{d}\right)}{37.5\,m} = \frac{0.0004593}{d}t(d).$$

$$q^* = \sqrt{\frac{3}{2}} - t^{*2}\sqrt{\frac{2}{3}}$$

$$= 1.2247 - 0.8165 t^{*2}.$$

$$FR = \sqrt{\frac{3}{2}}t^* - \frac{t^{*3}}{3}\sqrt{\frac{2}{3}}.$$

$$FR = 1.2247 t^* - 0.27217 t^{*3}.$$

Estimated rates and FRs are presented in **Table 8.66.**

Estimation of the Effect of a Rising Steam Chamber on Gravity-Drainage Rates. Cumulative displacement is estimated by use of Eq. 8.247:

t (years)	t (days)	t*	q*	FR	q (m³/d/m)
0	0	0	1.225	0.000	0.1511
0.5	182.5	0.084	1.219	0.103	0.1504
1	365	0.168	1.202	0.204	0.1482
1.054	384.69	0.177	1.199	0.215	0.1479
1.096	400	0.184	1.197	0.223	0.1477
1.233	450	0.207	1.190	0.251	0.1468
1.332	486.17	0.223	1.184	0.270	0.1460
1.368	499.45	0.229	1.182	0.278	0.1458
1.425	520	0.239	1.178	0.289	0.1453
1.5	547.5	0.251	1.173	0.304	0.1447
2	730	0.335	1.133	0.400	0.1397
3	1,095	0.503	1.018	0.581	0.1256
4	1,460	0.671	0.858	0.739	0.1058
5	1,825	0.838	0.651	0.866	0.0803
6	2,190	1.006	0.399	0.955	0.0492
7	2,555	1.174	0.100	0.997	0.0124

Table 8.66—Estimated gravity-drainage rates and FR-depletion model.

$$q_{cum} = 2.25 \left(\frac{kg\alpha}{mv_s} \right)^{2/3} (\phi \Delta S_o)^{1/3} t^{4/3}$$

$$= 2.25 \left[\frac{\left(0.4 \times 10^{-12}\, m\right)\left(9.81 \frac{m}{s^2}\right)\left(8.1 \times 10^{-7} \frac{m^2}{s}\right)\left(\frac{86{,}400 s}{d}\right)^2}{3.4 \left(6.57 \times 10^{-6} \frac{m^2}{s}\right)} \right]^{2/3} \left[(0.33)(0.75 - 0.13)\right]^{1/3} t^{4/3}$$

$$= (2.25)\left(0.01041 \frac{m^3}{d^{4/3} m}\right)(0.589250) t^{4/3}$$

$$= 0.01380 t^{4/3} \left(\frac{m^3}{dm} \right) \text{ when } t \text{ is in days.}$$

$$FR = \frac{0.01380 t^{4/3}\left(\frac{m^3}{dm} \right)}{268.5\ m^3 / m}.$$

$$FR = 5.1397 \times 10^{-5}\, t^{4/3} \text{ when } t \text{ is in days.}$$

$$q = \frac{dq_{cum}}{dt} = 3\left(\frac{kg\alpha}{mv_s} \right)^{2/3} (\phi \Delta S_o)^{1/3} t^{1/3}$$

$$= 3\left(0.010441 \frac{m^2}{d^{4/3}}\right)(0.589250) t^{1/3}$$

$$= 0.018403 t^{1/3} \frac{m^2}{d} \text{ when } t = \text{days.}$$

Table 8.67 presents the estimates of gravity-drainage rate and FR for the rising-steam-chamber model.

Inspection of Tables 8.66 and 8.67 indicates that the displacement rate for the rising-steam-chamber model increases with time, while the gravity-drainage rate for the depletion model decreases with time. These curves will intersect at a transition point where the displacement process transitions from the rising-steam-chamber model to the depletion model. Butler proposed setting the transition point where the drainage rates and FR for both models were equal.

Fig. 8.128 is a graph of the gravity-drainage rate for both models as a function of FR. At the transition point, $FR_{RSC} = FR_{Dep}$ = 0.215 and $q_{RSC} = q_{Dep} = 0.148$ m³/d/m. However, the model times at the transition point are not equal and an adjustment of time must be made in the depletion model after the transition point.

For the rising-steam-chamber model,

$$q = 0.148 \text{ m}^3/d/m$$

$$t^{1/3} = \left(\frac{0.148}{0.018403} \right) = 8.042$$

$$t_{RSC} = 520.12 \text{ days.}$$

t (years)	t (days)	q_{cum} (m³/m)	Fraction Mobile Oil	q (m³/m/d)
0	0	0	0	0
0.5	182.5	14.29	0.053	0.104
1	365	36.00	0.134	0.132
1.426	520.5	57.78	0.215	0.1480
1.5	562.5	64.08	0.239	0.152
2	730	90.71	0.338	0.166
3	1,095	155.75	0.580	0.190
3.458	1,262	188.23	0.701	0.199

Table 8.67—Estimated cumulative displacement and gravity-drainage rate by use of the rising-steam-chamber model.

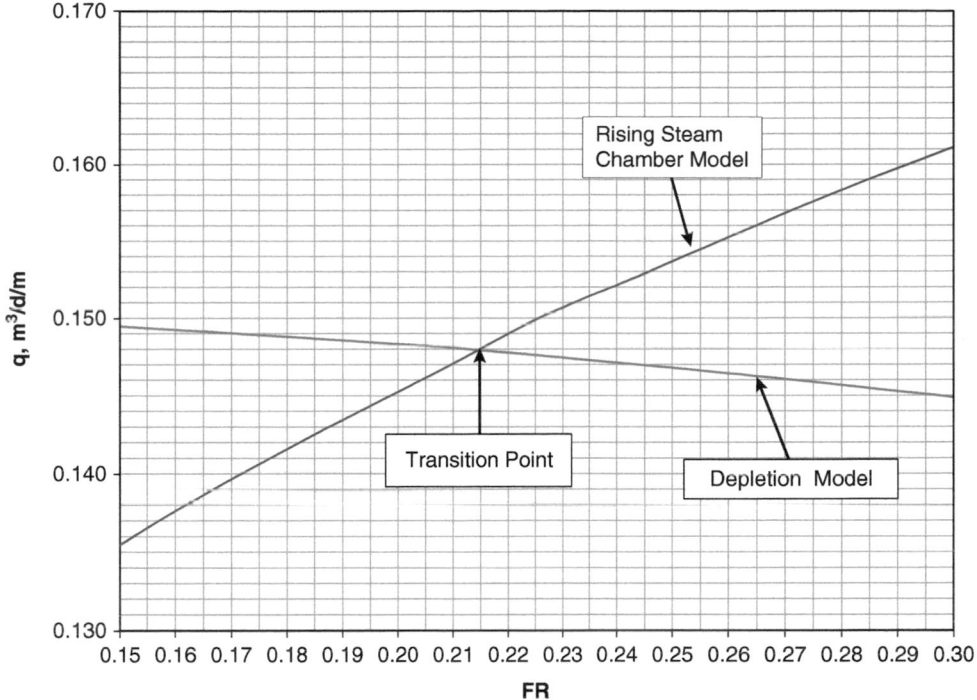

Fig. 8.128—Estimation of transition point from rising-steam-chamber model to depletion model.

For the depletion model,

$$q^* = \frac{(16.215)(0.148)}{2} =$$
$$q^* = 1.1999.$$

From Eq. 8.243,

$$1.1999 = 1.2247 - 0.8165t^{*2}.$$

$$t^* = 0.1771.$$

$$t = \frac{0.1771}{0.0004593} = 385.5 \text{ days.}$$

The adjustment time for the depletion model after the transition point is $520.12 - 385.5 = 134.6$ days. The rising-steam-chamber model is used to estimate the drainage rate and FR from $t = 0$ to $t = 520.12$ days where the transition occurs. The adjusted time is determined by subtracting 134.6 days from the total time. Thereafter, **Table 8.68** shows the adjustment calculations for the depletion model after the transition time.

Table 8.69 summarizes the estimated gravity-drainage rates and FR of the mobile oil above the production well as a function of time. **Fig. 8.129** is a plot of the oil production rate as a function of time.

Butler continued to develop the gravity-drainage model to remove some of the limitations in the first models. Butler (1985) introduced a semianalytical model that was based on the concept of heat penetration ahead of the steam/bitumen interface. This model has been refined by several investigators, including Heidari et al. (2009), who applied the boundary integral method to solve the heat-balance equations at the steam/bitumen interface in a study of the effect of drainage height and permeability on SAGD performance. In subsequent work, Irani et al. (2015) and Irani and Cokar (2014) investigated the impact of a temperature-dependent thermal conductivity on the prediction of the temperature front. The effect of temperature-dependent thermal conductivity and heat capacity on the temperature profile ahead of the steam/bitumen interface and on the movement of this interface was investigated by Heidari et al. (2015).

A common feature of these models is that heat loss from the steam zone to the overburden is not considered. Consequently, while the models provide estimations of the steam/bitumen interface and oil-drainage rate because of gravity drainage, they do not provide estimations of the steam requirement to obtain the predicted drainage rates and oil recovery.

Estimating Performance of the SAGD Process From Heat-Balance Models. The SAGD model presented in this section is based on an approximate energy balance around the steam chamber as the steam chamber develops, but the oil-displacement rate is based on the overall energy balance.

t (years)	t_adj (days)	t (years adjusted)	t*	q*	FR	q (m³/d/m)
0	0					
0.5		Rising-Steam-Chamber Model				
1						
1.426	520	1.426	NA	NA	0.215	0.1479
2	594	1.628	0.250	1.174	0.302	0.145
3	959	2.628	0.404	1.091	0.477	0.135
4	1,324	3.628	0.558	0.970	0.636	0.120
5	1,689	4.628	0.712	0.811	0.774	0.100
6	2,054	5.628	0.866	0.613	0.884	0.076
7	2,419	6.628	1.020	0.376	0.960	0.046
8	2,784	7.628	1.174	0.100	0.997	0.012

Table 8.68—Summary of calculations for the depletion model after the transition time.

	t (years)	t (days)	q (m³/m/d)	Fraction Mobile Oil
	0	0	0	0
Rising steam chamber	0.5	182.5	0.104	0.076
	1	365	0.132	0.191
	1.4260	520	0.1480	0.280
Depletion model	1.628	594	0.1447	0.302
	2.628	959	0.1346	0.477
	3.628	1,324	0.1197	0.636
	4.628	1,689	0.1000	0.774
	5.628	2,054	0.0756	0.884
	6.628	2,419	0.0463	0.960
	7.628	2,784	0.0124	0.997

Table 8.69—Estimated gravity-drainage rates and FR as a function of time.

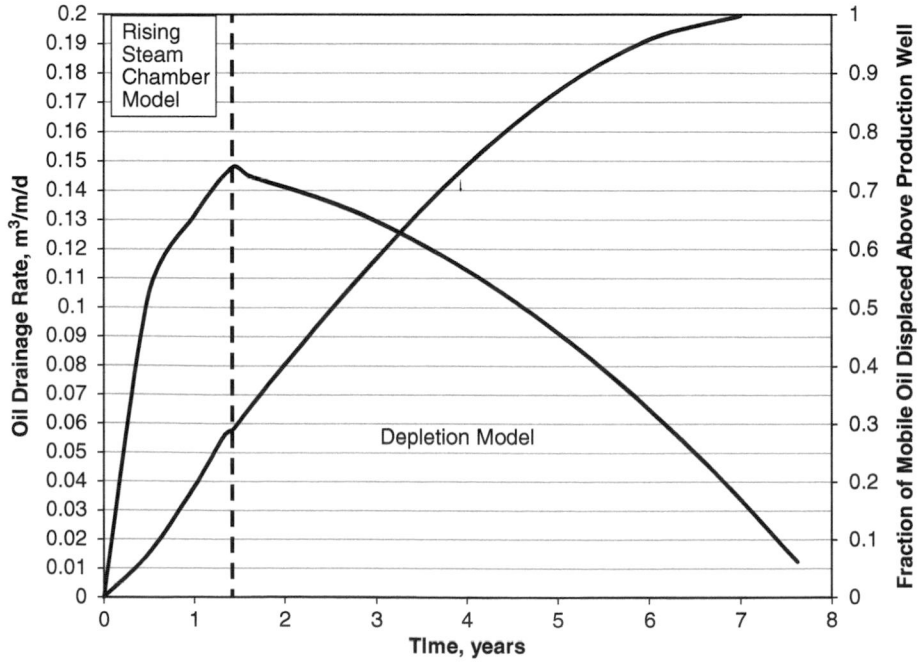

Fig. 8.129—Estimated gravity-drainage rate and FR for SAGD process by use of the Butler models.

Reis (1992, 1993) introduced the concept of approximating the steam chamber by a triangular shape with its apex located at the production well near the bottom of the formation. This model was adapted by Edmunds and Peterson (2007) to develop a heat-balance model to predict the cumulative SOR (CSOR) for an SAGD process. The Edmunds and Peterson model was extended by Miura and Wang (2012).

A conceptual model of SAGD is shown in **Fig. 8.130.** In this model, the steam zone rises instantaneously to the top of the chamber and spreads horizontally along the interface between the reservoir and the overburden. Horizontal expansion of the steam chamber is limited in pattern developments by coalescence with adjacent SAGD well pairs. When this happens, no-flow boundaries exist on the sides of the heated region shown by **Fig. 8.131,** and further expansion of the steam chamber occurs vertically along the oil/water interface. The heated area above the steam chamber is constant, and, for the triangular shape assumed for the steam chamber in this model, coalescence occurs when the steam-chamber volume is 50% of the pattern volume.

Edmunds and Peterson (2007) approximated the expansion of the heated area by use of an empirical expression on the basis of heat loss from the top, sides, and bottom of the steam chamber. **Fig. 8.132** shows the diagram of the Edmunds and Peterson model with a heated area of A in a steam chamber that has thickness h. They developed an approximate relationship for the cumulative SOR (m^3_{CWE}/m^3) given by Eq. 8.251.

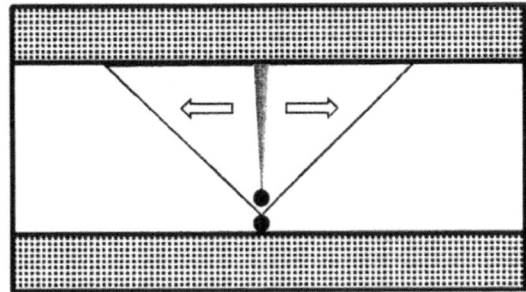

Fig. 8.130—Expansion of steam chamber horizontally after steam reaches the top of the steam chamber (Miura and Wang 2012).

$$CSOR\left(\frac{m^3_{CWE}}{m^3}\right) = \frac{(T_s - T_r)}{L^*_{dv}\phi(S_{oi} - S_{or})}\left(M_R + \frac{\sqrt{k_t M_s t}}{h\eta_s}\right), \quad \ldots (8.251)$$

where

$$\eta_s = \frac{V_{sz}}{Ah} \quad \ldots\ldots\ldots\ldots\ldots\ldots\ldots\ldots\ldots (8.252)$$

and V_{sz} = volume of the steam zone, m^3/m; A = distance between adjacent well pairs, m^2/m; and η_s = effective sweep efficiency, which is constant at 50% = 0.5. With this assumption, Eq. 8.251 becomes

$$CSOR = \frac{(T_s - T_r)}{L^*_{dv}\phi(S_{oi} - S_{or})}\left(M_R + \frac{2\sqrt{k_t M_s t}}{h}\right). \quad \ldots\ldots\ldots (8.253)$$

The Edmunds and Peterson model can be developed without making the assumption that the steam-chamber sweep efficiency is 50% by adapting the Marx and Langenheim model, which was developed for horizontal steamfloods in Section 8.3.

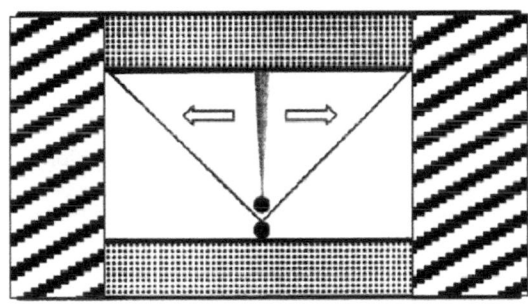

Fig. 8.131—Horizontal expansion of the steam chamber limited by no-flow boundaries for both heat and fluid flow from adjacent SAGD well pairs, assuming balanced injection and withdrawal.

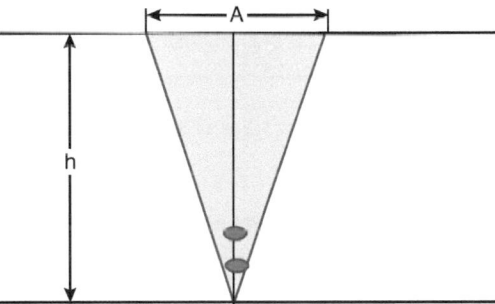

Fig. 8.132—SAGD model of Edmunds and Peterson.

The Marx and Langenheim model for steam injection is a piston-like model that accounts for heat loss to the overburden and underburden in the linear/radial forms of the model. The oil saturation in the region contacted by steam is reduced from S_{oi} to S_{or}. The displaced oil flows ahead of the steam interface at reservoir temperature. The model does not account for heat loss by conduction or convection ahead of the steam interface.

The SAGD adaptation of the Marx and Langenheim model retains two of the model characteristics:

1. Heat loss to the overburden is modeled correctly.
2. There is no heat loss by conduction ahead of the steam interface.

The SAGD adaption differs from the Marx and Langenheim model in that the displaced oil and condensate are removed at steam temperature by gravity drainage to the lower well of the two well systems, as shown in **Fig. 8.133,** so there is no displacement of oil ahead of the steam interface. This is accounted for in the overall energy and material balances.

The second difference is indicated by the expanding steam interface depicted by the red region in Fig. 8.133. Every incremental expansion of the steam zone (assumed to be triangular in shape—this is the assumption of Edmunds and Peterson, which is also used by Miura and Wang) heats the incremental volume from reservoir temperature to steam temperature.

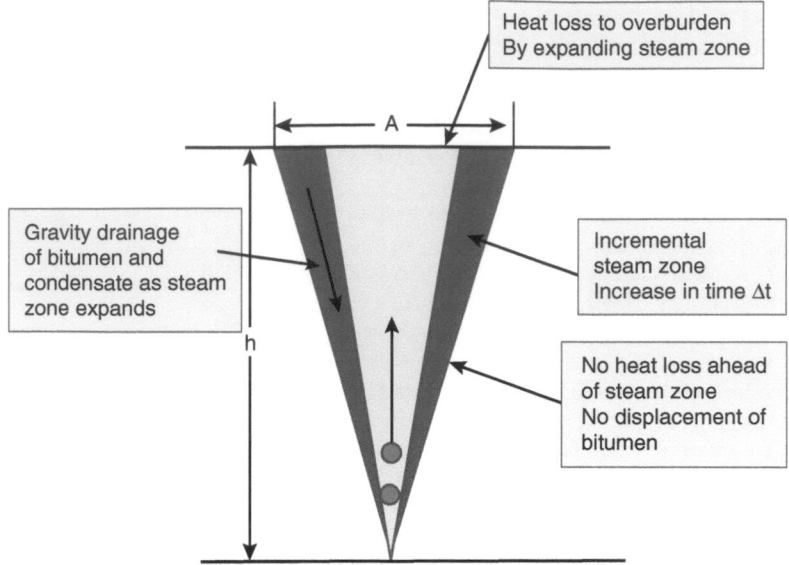

Fig. 8.133—SAGD adaptation of Marx and Langenheim model for steam injection.

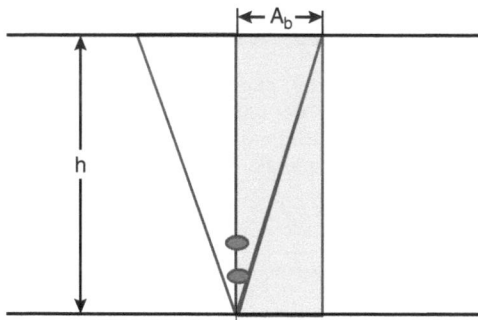

Fig. 8.134—Equivalent linear steam model for Fig. 8.132.

In actuality, there would be heat loss by conduction ahead of the steam zone, which means that the amount of steam required to heat the incremental increase in the steam zone is overestimated. This cannot be avoided in a simple model, so the OSR is underestimated by the model.

Gravity drainage is assumed to occur at a rate such that all oil mobilized by the expanding steam zone and all condensate produced by condensation of steam are removed at the same rate as calculated from the energy and material balances. The SAGD adaptation satisfies both material and energy balances, but assumes that the heat-injection rate controls the process.

In the Marx and Langenheim approximation of SAGD, heat is lost only from the top of the steam chamber. Heat losses ahead of the moving sides and the bottom of the steam chamber are neglected. Some of the heat lost from the sides is actually recovered by the expanding steam zone, but is not included in the model.

When these assumptions are made, heat loss from the heated zone in Fig. 8.133, which is triangular in shape until coalescence, can be approximated by the linear steam zone in **Fig. 8.134,** with heat loss from both the top and the bottom of the rectangular region, but has the same pore volume as in Fig. 8.132. In Fig. 8.132, $V_p = \frac{1}{2}Ah$. In Fig. 8.134, $V_p = A_bh$. Because $A_b = A/2$, the steam chamber volumes are identical.

If steam was injected into the injection well at a constant rate and no fluids were removed from the production well, the growth of the steam zone would be described by Eq. 8.61 as presented in Section 8.3.

In SAGD, after injectivity into the formation is established, steam is injected into the upper well at the same time as condensate and displaced oil are removed from the production well. Gravity segregation causes the steam to rise to the top of the steam chamber, with displaced oil and condensate flowing to the bottom of the interval by gravity. When this happens, the energy balance for the expanding steam chamber is given by Eq. 8.254 when steam is injected at a constant rate and temperature. In Eq. 8.254, the latent heat of the injected steam supplies the expansion of the steam chamber and heat loss from the top and bottom of the steam chamber in Fig. 8.134. This heat balance is the same balance that would be written for the model in Fig. 8.132, when heat loss occurs only from the top of the steam chamber.

$$\dot{m}_s f_{dv} L_{vd} = M_R \left(T_s - T_r \right) h \frac{dA_b}{dt} + 2 \int_0^t \frac{k_h \left(T_s - T_r \right)}{\sqrt{\pi \alpha \left(t - \lambda \right)}} \frac{dA_b}{d\lambda} \, d\lambda. \quad \dots \dots \dots \dots \dots \dots \dots \dots \dots \dots \dots (8.254)$$

In Eq. 8.254,

$$M_R = \left[\rho_r C_r \left(1 - \phi \right) + \rho_w C_w S_{iw} \phi + \rho_o C_o S_{oi} \phi + \rho_s S_s \phi \frac{\left(H_{ss} - H_{ws} \right)}{\left(T_s - T_r \right)} \right]. \quad \dots \dots \dots \dots \dots \dots \dots (8.255)$$

The solution to Eq. 8.254 is Eq. 8.256:

$$A_b = \frac{\dot{m}_s f_{sd} L_{vd} M_R h}{4(T_s - T_r)\alpha_o M_S^2} \left[e^{t_D} \operatorname{erfc}\left(\sqrt{t_D}\right) + 2\sqrt{\frac{t_D}{\pi}} - 1 \right], \dotfill (8.256)$$

$$A_b = \frac{\dot{m}_s f_{sd} L_{vd} M_R h}{4(T_s - T_r)\alpha_o M_S^2} G(t_D), \dotfill (8.257)$$

$$t_D = 4\left(\frac{M_S}{M_R}\right)^2 \frac{\alpha_o}{h^2} t. \dotfill (8.258)$$

It follows that the volume of the steam zone increases with time, as given by Eq. 8.259, as does the sweep efficiency:

$$V_{sz} = \frac{1}{2}(2A_b h)$$

$$V_{sz} = \frac{\dot{m}_s f_{sd} L_{vd} M_R h^2}{4(T_s - T_r)\alpha_o M_S^2} G(t_D), \dotfill (8.259)$$

and that the volume of oil displaced is given by Eq. 8.260,

$$O = V_{sz}\phi(S_{oi} - S_{or}) \dotfill (8.260)$$

$$= \frac{\dot{m}_s f_{sd} L_{vd} M_R h^2 \phi(S_{oi} - S_{or})}{4(T_s - T_r)\alpha_o M_S^2} G(t_D). \dotfill (8.261)$$

The oil-displacement rate is estimated from the Marx and Langenheim model by determining the rate at which the heated volume expands with time. Recall that oil is displaced immediately from the expanding steam zone, with the oil saturation decreasing form S_{oi} to S_{or}.

The rate at which the heated area expands with time at a constant injection rate is given by Eq. 8.262.

$$\frac{dA_b}{dt} = \left[\frac{\dot{m}_s f_{sd} L_{vd}}{M(T_s - T_r)h} \right] G_1(t_D). \dotfill (8.262)$$

Assuming a horizontal thickness of 1 m,

$$q_o\left[\left(\frac{m^3}{m}\right)\frac{1}{y}\right] = h(m)(S_{oi} - S_{or})\frac{dA_b}{dt}\left(\frac{m^2}{m}\right)\frac{1}{y}$$

$$q_o = h(S_{oi} - S_{or})\left[\frac{\dot{m}_s f_{sd} L_{vd}}{M(T_s - T_r)h} \right] G_1(t_D)$$

$$= \frac{m^3}{my}. \dotfill (8.263)$$

$$G_1(t_D) = e^{t_D}\operatorname{erfc}\left(\sqrt{t_D}\right). \dotfill (8.264)$$

The CSOR is used as a measure of efficiency for the SAGD process. Steam consumption S is defined by Eq. 8.265 as the total amount of energy injected divided by the latent heat of vaporization at T_s.

$$S(m^3_{CWE}) = \frac{\dot{m}_s t f_{sd} L_{vd}\left(\dfrac{kJ}{kg}\right)}{L_{vd}^*\left(\dfrac{kJ}{m^3}\right)_{CWE}}; \dotfill (8.265)$$

O = volume of oil displaced, m³;

$$O = A_b h \phi (S_{oi} - S_{or}); \quad \dots\dots\dots\dots\dots\dots\dots\dots\dots\dots\dots\dots\dots\dots\dots\dots\dots\dots (8.266)$$

and $CSOR \left(\dfrac{m^3_{CWE}}{m^3} \right) = \dfrac{S}{O}$ $\dots\dots\dots\dots\dots\dots\dots\dots\dots\dots\dots\dots\dots\dots\dots\dots\dots\dots\dots (8.267)$

From the overall energy balance,

$$CSOR = \frac{M_R (T_s - T_r)}{L^*_{vd} \phi (S_{oi} - S_{or})} \left[\frac{t_D}{G(t_D)} \right]. \quad \dots\dots\dots\dots\dots\dots\dots\dots\dots\dots\dots\dots\dots (8.268)$$

Example 8.22 illustrates the application of this model to estimate the minimum CSOR for JACOS Well Pair O (Miura and Wang 2012) in the JACOS Hangingstone Project. In this project, the distance between well pairs is approximately 100 m, with lengths ranging from 500 to 750 m. Initial OIP, assuming well-pair lengths of 750 m³, is 421 800 m³.

Example 8.22—Application of the Heat-Balance Model To Estimate CSOR for the JACOS Well Pair O. Properties for Well Pair O are presented in **Table 8.70.** Steam was injected into this well pair at pressures ranging from 5000 kPa to approximately 3500 kPa. The production well is assumed to be located 2.5 m above the base of the reservoir.

From Equation 8.251,

$$CSOR = \frac{\left(2350 \dfrac{kJ}{m^3 \, ^\circ C} \right) (263.94 - 11) \, ^\circ C}{\left(1640000 \dfrac{kJ}{m^3_{CWE}} \right) (0.38)(0.8 - 0.15)} \left[\frac{t_D}{G(t_D)} \right].$$

$$CSOR = \frac{594,409}{405,080} \left[\frac{t_D}{G(t_D)} \right].$$

$$CSOR = 1.467 \left[\frac{t_D}{G(t_D)} \right].$$

From Eq. 8.258,

$$t_D = 4 \left(\frac{2380 \dfrac{kJ}{m^3 \, ^\circ C}}{2350 \dfrac{kJ}{m^3 \, ^\circ C}} \right)^2 \frac{35.71 \dfrac{m^2}{y}}{(18.5 - 2.5)^2 \, m^2} t(y).$$

$$t_D = 0.5723t$$

S_{oi}	0.8	0.8
Porosity	0.38	0.38
h (m)	18.5	18.5
M_R [kJ/(m³°C)]	2350	2350
T (°C)	11	11
ρ_w (kg/m³$_{CWE}$)	999.56	999.56
M_S [kJ/(m³°C)]	2380	2380
k (MJ/m·C·a)	85	85
α_o (m²/y)	35.71	35.71
p_s (kPa)	5000	3500
T_s (°C)	264	243
L_{vd}^* (MJ/m³$_{CWE}$)	1640	1753
S_{or}	0.15	0.15
A_{max} (m²/m)	100	100

Table 8.70—Representative properties for JACOS Well Pair O (Miura and Wang 2012).

or $t = 1.747t_D$.

When $t_D = 0.5$, $G(t_D) = 0.32104$, $CSOR = 2.29$, and $t = 0.874$ years.

Table 8.71 presents estimates of CSOR vs. time for steam pressures of 5000 and 3500 kPa. Estimated CSORs from Table 8.71 are presented in **Fig. 8.135** with field data. The estimated CSOR is consistent with field data from Well Pair O approximately 1.5 years after steam injection begins.

This comparison shows that simple analytical models that are based on constant steam-injection rate do not represent the initial phases of an SAGD project. In practice, steam is circulated in injection and production wells for periods of 3 to 6 months before substantial injectivity is established into the formation. Even this may require use of solvent preflush or mini-fractures to create injectivity. In the startup phase, steam injection is performed at a specific pressure. Net steam injection into the formation and oil production are low and CSOR values are large, as indicated in Fig. 8.135.

Eq. 8.258 indicates that the minimum CSOR at any time during the SAGD process is dependent on the steam temperature and the change in oil saturation as the steam chamber expands. CSOR does not depend on the injection rate. **Fig. 8.136** shows the effect of injection pressure on the CSOR for Well Pair O.

Effect of Initial Preheating To Establish Injectivity. The Marx and Langenheim SAGD model can be used to predict cumulative oil displaced and production rates that are consistent with the model assumptions. Because the model assumes that steam

t_D	$G(t_D)$	t (years)	CSOR (m^3_{cwe}/ m^3 Oil), 5000 kPa	CSOR (m^3_{cwe}/m^3 Oil) 3500 kPa
0.001	0.00098	0.002	1.50	1.28
0.002	0.00193	0.003	1.52	1.30
0.003	0.00288	0.005	1.53	1.31
0.004	0.00382	0.007	1.54	1.32
0.005	0.00475	0.009	1.54	1.32
0.006	0.00567	0.010	1.55	1.33
0.007	0.00658	0.012	1.56	1.34
0.008	0.00749	0.014	1.57	1.34
0.009	0.0084	0.016	1.57	1.35
0.01	0.00929	0.017	1.58	1.35
0.02	0.01806	0.035	1.62	1.39
0.04	0.0347	0.070	1.69	1.45
0.06	0.05051	0.105	1.74	1.49
0.08	0.06571	0.140	1.79	1.53
0.1	0.0804	0.175	1.82	1.56
0.2	0.14841	0.349	1.98	1.69
0.3	0.21006	0.524	2.10	1.80
0.4	0.26726	0.699	2.20	1.88
0.5	0.32104	0.874	2.29	1.96
0.6	0.37206	1.048	2.37	2.03
0.7	0.42077	1.223	2.44	2.09
0.8	0.4675	1.398	2.51	2.15
0.9	0.5125	1.572	2.58	2.21
1	0.55596	1.747	2.64	2.26
1.1	0.59806	1.922	2.70	2.31
1.2	0.63892	2.097	2.76	2.36
1.3	0.67866	2.271	2.81	2.41
1.4	0.71738	2.446	2.86	2.45
1.5	0.75514	2.621	2.91	2.50
1.6	0.79203	2.795	2.96	2.54
1.7	0.82811	2.970	3.01	2.58
1.8	0.86342	3.145	3.06	2.62

Table 8.71—Estimates of CSOR for Well Pair O at 3500 and 5000 kPa.

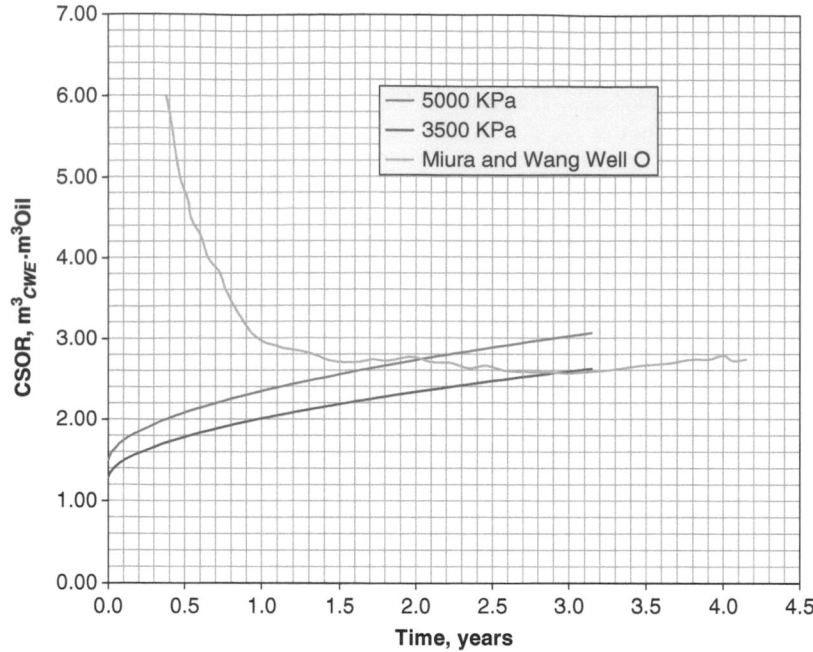

Fig. 8.135—Comparison with field data and estimated CSOR at steam-injection pressures of 3500 and 5000 kPa with time (field data courtesy of Miura and Wang 2012).

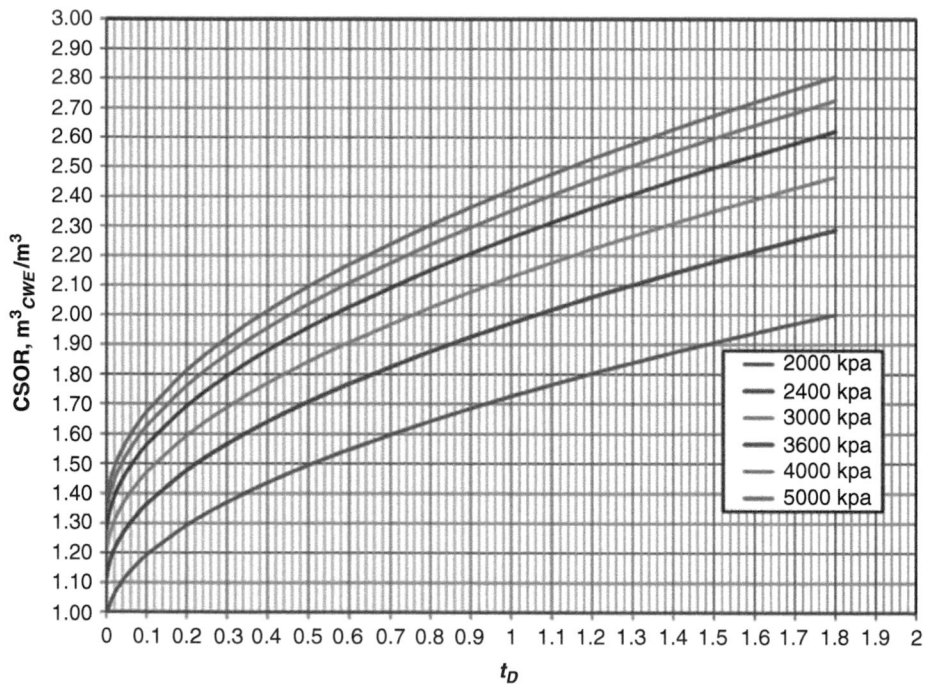

Fig. 8.136—Variation of CSOR with steam-chamber pressure over a range of dimensionless times from 0 to 1.8 for the parameters of Well Pair O.

injection is established immediately, it does not represent the initial period of 3 to 6 months when steam is circulated in both injection and production wells to establish injectivity and production.

Net steam-injection rates begin at low values (≈ 0 in most cases) and are ramped up as injectivity is established. Although the Marx and Langenheim model was based on constant injection rate, as discussed in Section 8.4, Ramey (1959) extended this model to include variable heat-injection rates. Example 8.11 illustrates the application of this model for steam injection in a vertical well. As noted in Section 8.3, if the injection-rate schedule is approximated by the following series of step changes with steam pressure fixed, then

$$\dot{m}_s f_s dL_{vd} = \dot{m}_{s1} f_s dL_{vd}.....0 \leq t \leq t_1$$
$$= \dot{m}_{s2} f_s dL_{vd}.......t_1 \geq t \leq t_2$$
$$= \dot{m}_{sn} f_s dL_{vd}......t_{n-1} \leq t \leq t.$$

From Eq. 8.136,

$$A_h(t) = B_n G(t_D) + \sum_{m=1}^{n-1}(B_m - B_{m+1})G(t_{Dm}),$$

where $B_n = \dfrac{\dot{m}_{s_n} f_{sd} L_{vd} M_R h}{4(T_s - T_r)\alpha_o M_s^2}.$

Eq. 8.136 can be simplified by expanding the summation. In general,

$$A(t_n) = A(t_{n-1}) + B_n\left[G(t_{D_n}) - G(t_{D_n-1})\right]. \dots\dots\dots\dots\dots\dots\dots\dots\dots (8.269)$$

Example 8.23 illustrates the application of the Ramey extension to SAGD, where injection rates are ramped up following a specified schedule.

Example 8.23—Application of the Ramey Extension to SAGD. This example uses the same reservoir, oil, and steam properties as in Example 8.22. For this example, the production well is located 2.5 m above the base of the formation, resulting in a net reservoir thickness of 16 m subject to SAGD. Because the injection rates were not provided in the references on the JACOS Hangingstone Project, oil rates will be estimated as a function of time for a range of injection rates that are in the range of those used in SAGD projects. In SAGD projects operated by JACOS, steam quality at the wellhead averages between 97 and 100% (JACOS). This example is based on injection of steam at 5000 kPa, having a steam quality of 95% entering the formation. In applying the model, it is assumed that sufficient preheating of injection and production wells occurred to establish injectivity, and the injection rate can be maintained constant throughout the flood.

Injection rates on the order of 0.2 to 0.6 $(m^3/d/m)_{CWE}$ appear to be common in ongoing SAGD projects. This is equivalent to 150 to 450 m^3/d in wells that are 750 to 800 m in length. In this example, a 3-month period of steam circulation is assumed to establish injectivity. No net steam injection occurs during this period of time, and heat losses are not accounted for in the Marx and Langenheim model. Steam injection is assumed to begin after 3 months and is ramped up from 0 to 0.6 $m^3/d/m$ over a period of 12 months. **Fig. 8.137** shows the assumed injection-rate profile. The injection-rate schedule is presented in **Table 8.72**, along with computed results.

Calculations for the first two timesteps are illustrated in this section.

Solution. From Eq. 8.258,

$$t_D = 4\left(\frac{M_S}{M_R}\right)^2 \frac{\alpha_o}{(h - z_p)^2} t.$$

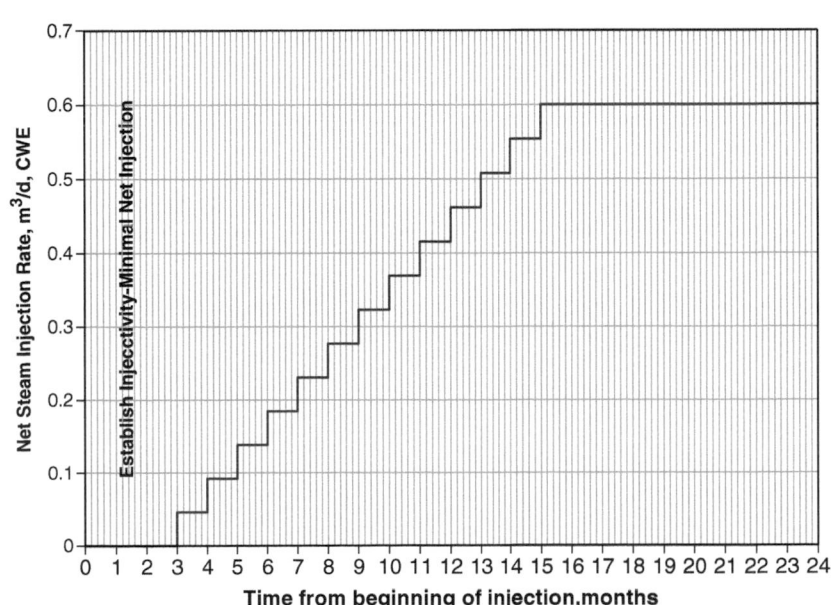

Fig. 8.137—Steam-injection profile for development of steam injectivity over a 1-year period in equal steps/month.

Total Time (months)	t_n (months injected)	t_n (years)	t_D	$G(t_D)$	q_s (m³/m/d)	q_s (m³/m/y)	B_n	$A_b(t_n)$	V_{sz} (m³)	η_s	Cumulative Oil (m³)	Oil Rate (m³/y/m)	Oil Rate (m³/d)
0	0	0	–	–	0	0	–	0	0	0	0	0	0
1	0	0	–	–	0	0	–	0	0	0	0	0	0
2	0	0	–	–	0	0	–	0	0	0	0	0	0
3	0	0	–	–	0	0	–	0	0	0	0	0	0
4	1	0.083	0.048	0.041	0.046	16.85	4.822	0.197	3.15	0.002	0.78	9.35	19.20
5	2	0.167	0.095	0.077	0.092	33.69	9.645	0.546	8.74	0.005	2.16	16.55	34.02
6	3	0.250	0.143	0.111	0.138	50.54	14.467	1.03	16.52	0.010	4.08	23.07	47.41
7	4	0.333	0.191	0.142	0.185	67.38	19.289	1.65	26.33	0.016	6.50	29.06	59.72
8	5	0.417	0.238	0.173	0.231	84.23	24.111	2.38	38.02	0.024	9.39	34.65	71.19
9	6	0.500	0.286	0.202	0.277	101.08	28.934	3.22	51.48	0.032	12.72	39.90	81.99
10	7	0.583	0.334	0.230	0.323	117.92	33.756	4.16	66.62	0.042	16.46	44.88	92.22
11	8	0.667	0.382	0.257	0.369	134.77	38.578	5.21	83.36	0.052	20.59	49.63	101.98
12	9	0.750	0.429	0.283	0.415	151.62	43.400	6.35	101.64	0.064	25.11	54.18	111.32
13	10	0.833	0.477	0.309	0.462	168.46	48.223	7.59	121.39	0.076	29.98	58.54	120.30
14	11	0.917	0.525	0.334	0.508	185.31	53.045	8.91	142.57	0.089	35.21	62.75	128.94
15	12	1.000	0.572	0.358	0.554	202.15	57.867	10.32	165.11	0.103	40.78	66.82	137.29
16	13	1.083	0.620	0.382	0.600	219.00	62.689	11.81	188.98	0.118	46.68	70.75	145.38
17	14	1.167	0.668	0.405	0.600	219.00	62.689	13.27	212.34	0.133	52.45	69.24	142.28
18	15	1.250	0.715	0.428	0.600	219.00	62.689	14.70	235.23	0.147	58.10	67.84	139.40
27	24	2.000	1.145	0616	0.600	219.00	62.689	26.51	424.17	0.265	104.77	62.22	127.86
39	36	3.000	1.717	0.834	0.600	219.00	62.689	40.16	642.54	0.402	158.71	53.94	110.83
48.68	45.68	3.807	2.179	0.991	0.600	219.00	62.689	50.00	800.00	0.500	197.60	48.19	99.03

Table 8.72—Estimated oil displacement from Well Pair O at 5000 kPa until coalescence point with adjacent well pair. Results are based on 1 m of steam-chamber length. Time is based on the beginning of net steam injection.

$$t_D = 4\left(\frac{2380}{2350}\right)^2 \frac{35.71}{\left(18.5 - 2.5\right)^2} t$$

$$= 0.5723t.$$

For Timestep 1, $t_1 = 1$ month $= 0.083$ year.

$$t_{D1} = (0.5723)(0.083)$$

$$= 0.0476.$$

$$G(t_{D1}) = 0.0407.$$

$$B_1 = \frac{q_{s1} L_{vd}^* f_{sd} M_R h}{4\left(T_s - T_r\right)\alpha_o M_s^2} = \frac{q_{s1}\left(1640000\right)\left(0.95\right)\left(2350\right)\left(18.5 - 2.5\right)}{4\left(263.94 - 11\right)\left(35.71\right)\left(2380\right)^2}.$$

$$B_1 = 0.2862 q_{s1}.$$

$$q_{s1} = 16.85 \frac{m^3}{m \cdot a}.$$

$$B_1 = (0.2862)(16.85) = 4.823.$$

$$A(t_1) = B_1 G(t_{D1}).$$

$$A(t_1) = (4.823)(0.0407).$$

$$A(t_1) = 0.1963 \frac{m^2}{m}.$$

For $t_2 = 2$ months $= 0. \ 0.167$ years, $t_{D2} = 0.0956$, $G(t_{D2}) = 0.0772$, $q_{s2} = 33.69$ m³/my, $B_2 = 9.642$,

$$A(t_2) = A(t_1) + B_2\left[G(t_{D2}) - G(t_{D1})\right].$$

$$A(t_2) = 0.1963 + 9.642(0.0772 - 0.0476).$$

$$A(t_2) = 0.5491 \frac{m^2}{m}.$$

The remaining parameters can be calculated after the area is determined.

$$V_{sz} = A(t)\left(h - z_p\right).$$

$$V_o = A\left(h - z_p\right)\phi\left(S_{oi} - S_{or}\right).$$

At $t_1 = 1$ month,

$$V_{sz} = (0.1963)(18.5 - 2.5)$$

$$= 3.1408 \ m^3 /m.$$

$$V_o = (3.1408)(0.38)(0.8 - 0.15)$$

$$= 0.776 \ m^3 /m.$$

$$q_o = \frac{0.776 \ m^3 /m}{0.083 \ a}$$

$$= 9.347 \frac{m^3}{m \cdot a} = 0.0256 \frac{m^3}{m \cdot d}.$$

Results for Example 8.23 are presented in Table 8.72. On the basis of the injection schedule for this example, the coalescence point ($A_b = 50$ m) is reached after 48.7 months of injection (45.7 months net injection after preheating). The coalescence time is the limit of the application of the Marx and Langenheim model for SAGD.

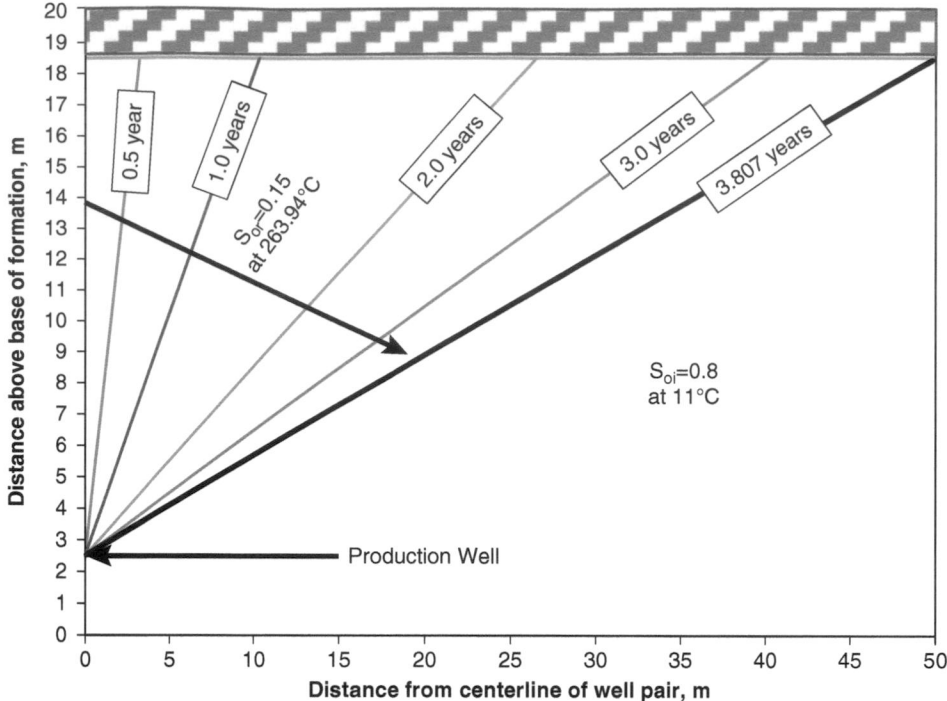

Fig. 8.138—Position of steam/oil interface as a function of net injection time.

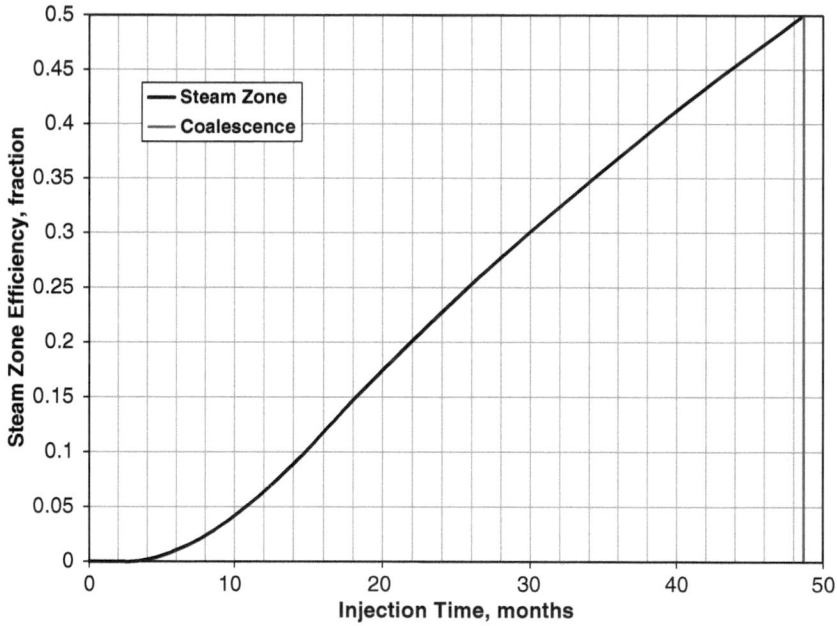

Fig. 8.139—Development of steam-zone efficiency with time for the injection-rate schedule in Table 8.72, Example 8.23.

Fig. 8.138 shows the position of the steam chamber/oil interface at several injections times. **Fig. 8.139** shows the development of the steam-zone efficiency with time for Example 8.23. Steam-zone efficiency reaches 50% when A_b reaches the coalescence point of 50 m on the basis of the spacing between adjacent well pairs of 100 m. The calculations in Table 8.67 were developed on the basis of a steam-chamber length of 1 m. Estimated performance of the well pair was obtained by multiplying values of Table 8.72 by 750, the assumed length of the well pair. The oil displacement rate can be estimated using the Marx and Langenheim model, but it is simpler when the steam rate varies to take the derivative of the cumulative oil displaced with respect to time.

Fig. 8.140 is the estimated oil-displacement rate for Example 8.23 plotted against time from the beginning of the project. Results from Table 8.72 are displaced by 3 months to account for the three periods assumed necessary to establish net injection. Cumulative oil displacement to the coalescence point is 197.6 m³/m or 148 200 m³. This is equivalent to 40.6% of the

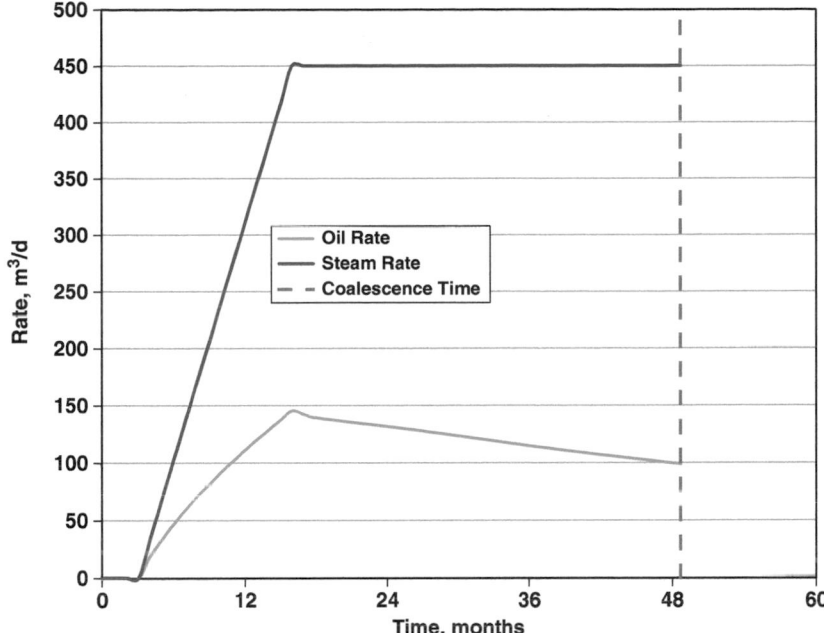

Fig. 8.140—Estimated oil-displacement rate from SAGD using Marx and Langenheim model with a steam-injection ramp from 0 to 450 m³/d for the first 12 months of injection. Well pair length is 750 m.

OOIP in the 16-m thickness above the production well. It is possible to estimate the rate of heat loss when the coalescence point is reached from the overall energy balance.

$$\text{Cumulative energy injected by steam} = S\left(\text{m}^3_{CWE}\right) f_{sd} L^*_{vd}\left(\frac{\text{kJ}}{\text{m}^3_{CWE}}\right). \quad\quad\quad\quad (8.270)$$

$$\text{Cumulative energy stored in the steam chamber} = V_{sz}\left(\text{m}^3\right) M_R\left(\frac{\text{kJ}}{\text{m}^3\,^\circ\text{C}}\right)\left(T_s - T_r\right). \quad\quad (8.271)$$

$$\text{Cumulative energy lost to the overburden} = S f_{sd} L^*_{vd} - V_{sz} M_R\left(T_s - T_r\right). \quad\quad\quad (8.272)$$

Fig. 8.141 shows the cumulative energy lost to the overburden to the coalescence point for the parameters of Example 8.23.

The rate of heat loss to overlying formations can be estimated from the calculations for Example 8.23 by differentiating the cumulative heat-loss data. **Fig. 8.142** shows the heat-loss rate as a function of time. The heat rate approaches a constant rate as the coalescence point is reached.

The heat-loss rate at the coalescence point may be used to estimate the minimum steam-injection rate needed to maintain the steam chamber at the steam-zone temperature. For the conditions of Example 8.23,

$$i_{min}\left(\frac{\text{m}^3_{CWE}}{\text{a}}\right) = \frac{\dot{H}_l\left(\dfrac{\text{kJ}}{\text{a}}\right)}{\left(L^*_{vd}\dfrac{\text{kJ}}{\text{m}^3_{CWE}}\right)f_{sd}}.$$

At $t = 4.07$ years, $\dot{H}_l = 2.252 \times 10^8 \dfrac{\text{kJ}}{\text{a}\cdot\text{m}}$.

$$i_{min}\left(\frac{\text{m}^3_{CWE}}{\text{a}\cdot\text{m}}\right) = \frac{2.252 \times 10^8}{\left(1.64 \times 10^6\right)\left(0.95\right)}.$$

$$i_{min}\left(\frac{\text{m}^3_{CWE}}{\text{a}\cdot\text{m}}\right) = 144.54.$$

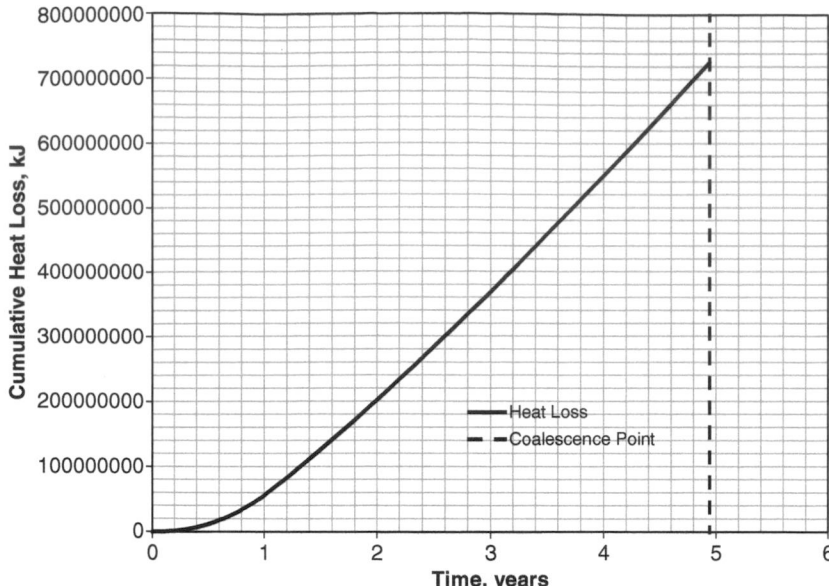

Fig. 8.141—Cumulative heat loss to the overburden for the well pair in Example 8.23.

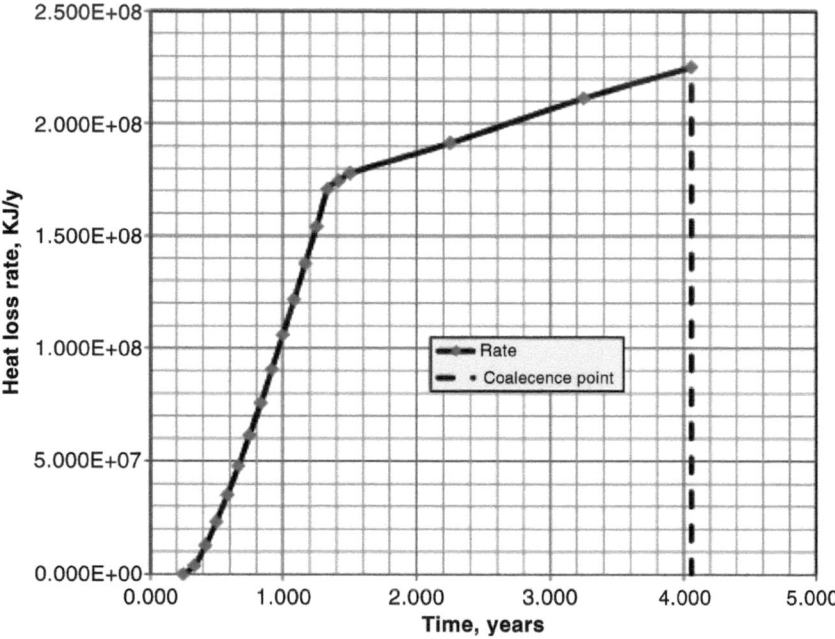

Fig. 8.142—Heat-loss rate during expansion of steam chamber to the coalescence point for Example 8.23.

$$i_{min} = 0.396\frac{m^3_{CWE}}{d \cdot m}$$

$$= 297\frac{m^3_{CWE}}{d}$$

to meet heat losses to the overburden after coalescence. Thus, the steam-injection rate must be 0.396 m^3_{CWE}/d·m to maintain the steam chamber at T_s = 263.94°C. Reduction of the injection pressure would allow maintaining the steam chamber at a lower injection rate.

Approximating Gravity Drainage During the Post-Coalescence Period by Use of Heat-Balance Modeling. At the coalescence point, the lateral expansion of the steam chamber stops at the top of the formation because a no-flow boundary will form with the adjacent steam chamber in a fully developed SAGD process. Volumetric sweep is 50% of the drainage area, corresponding to 50% of the mobile oil. Heat-loss rate to the overburden is a maximum at the coalescence point. Because heat loss by conduction oil is inversely proportional to the square root of time, the heat-loss rate to the overburden decreases with time at every location. The reduction in heat-loss rate can be estimated by use of the methods presented in Section 8.3.

When the steam-injection rate is specified and steam temperature remains unchanged, it is possible to estimate the SAGD performance after the coalescence point. For this model, the heat-loss rate is assumed to be constant at the value determined at the coalescent point Q_{cp}. Continued steam injection will cause the steam zone to expand vertically, as shown in **Fig. 8.143.** The energy balance for the vertically expanding steam zone is given by Eq. 8.273:

$$\dot{m}_s f_{dv} L_{vd} = M_R \left(T_s - T_r \right) \frac{dV_{sz}}{dt} + 2Q_c \dots\dots\dots\dots\dots\dots\dots\dots\dots\dots\dots\dots\dots\dots\dots\dots \text{(8.273)}$$

As the steam zone expands, injected steam may short circuit into the production well, depending on the difference in BHP in the injection well and the drawdown pressure in the production well located a short distance below the injection well. In this model, gravity drainage is assumed to control the oil production rate. The post-coalescence oil-displacement rate can be estimated by assuming that gravity drainage controls the oil production rate by use of the simple model depicted in **Fig. 8.144.** At the coalescence time,

$$q_o = \frac{k_o \rho_o \delta g \left(z_j - z_p \right)}{\mu_o l_{\cdot j}}, \dots\dots\dots\dots\dots\dots\dots\dots\dots\dots\dots\dots\dots\dots\dots\dots\dots\dots\dots\dots\dots\dots\dots\dots \text{(8.274)}$$

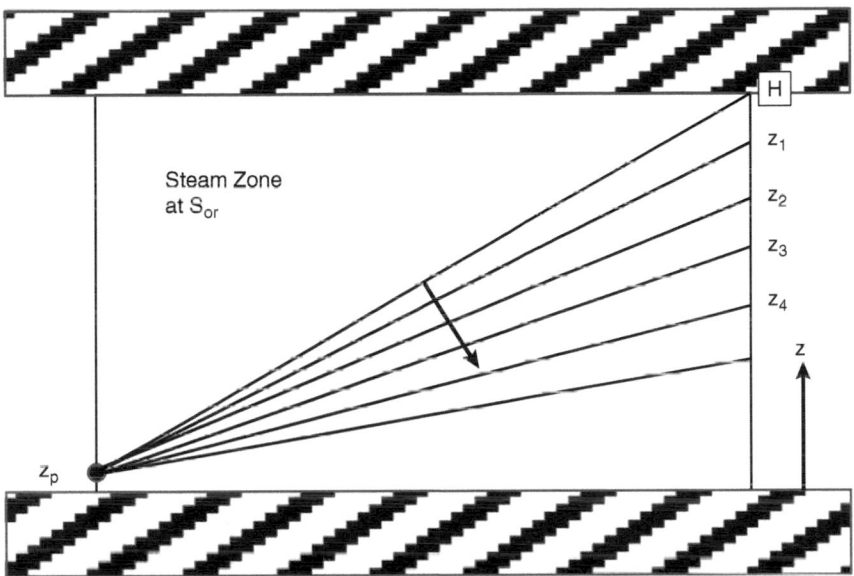

Fig. 8.143—Expansion of the steam zone after the coalescence point is reached.

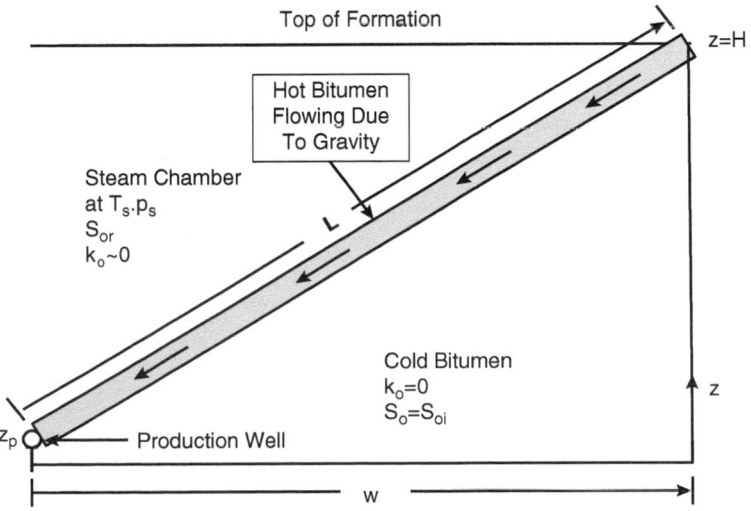

Fig. 8.144—Gravity-drainage model approximating bitumen flow after coalescence of steam chambers.

where z_j = elevation of the steam zone at w at time t_j, m; z_p = elevation of production well above the base of the reservoir, m; L_j = length of the steam interface at time t_j, m; and δ = average thickness of flowing oil stream, m. The term $(k_o \rho_o \delta g / \mu_o)$ is assumed to be constant and determined from the oil rate at the coalescence point.

At the coalescence point,

$$L_{cp} = \sqrt{w^2 + \left(h - z_p\right)^2} \quad \dots\dots\dots\dots\dots\dots\dots\dots\dots\dots \quad (8.275)$$

and $$\frac{k_o \delta g}{\upsilon_o} = \frac{q_{oc} L_{cp}}{\left(h - z_p\right)}. \quad \dots\dots\dots\dots\dots\dots\dots\dots\dots\dots \quad (8.276)$$

In the next timestep, the end of the steam-zone interface is moved from $z = h$ to z_i:

$$L_j = \sqrt{w^2 + \left(z_j - z_p\right)^2} \quad \dots\dots\dots\dots\dots\dots\dots\dots\dots\dots \quad (8.277)$$

and $$q_{oj} = \left(\frac{k_o \delta g}{\upsilon_o}\right) \frac{\left(z_j - z_p\right)}{L_j}, \quad \dots\dots\dots\dots\dots\dots\dots\dots\dots\dots \quad (8.278)$$

when the steam zone is advanced from endpoint elevation z_j to z_{j+1}.

The change in steam-zone volume is determined by a volume balance.

$$\Delta V_{sz_j} = \frac{1}{2} w \left(z_{j-1} - z_j\right), \quad \dots\dots\dots\dots\dots\dots\dots\dots\dots\dots \quad (8.279)$$

where $z_{j-1} > z_j$, as noted in Fig. 8.143 and $z_o = h$.

Incremental time is found by calculating ΔV_{sz} from the volume balance.

$$\Delta V_{oj} = \Delta V_{szj} \phi \left(S_{oi} - S_{or}\right) \quad \dots\dots\dots\dots\dots\dots\dots\dots\dots\dots \quad (8.280)$$

and $$\Delta t_j = \frac{\Delta V_{oj}}{q_{oj}}. \quad \dots\dots\dots\dots\dots\dots\dots\dots\dots\dots \quad (8.281)$$

The steam injection rate is found by applying Eq. 8.273:

$$\dot{m}_{sj} f_{dv} L_{vd} = M_R \left(T_s - T_r\right) \frac{q_{oj}}{\phi \left(S_{oi} - S_{or}\right)} + 2Q_c \quad \dots\dots\dots\dots\dots\dots \quad (8.282)$$

and $$\dot{m}_{sj} = \frac{M_R \left(T_s - T_r\right) \dfrac{q_{oj}}{\phi \left(S_{oi} - S_{or}\right)} + 2Q_c}{f_{dv} L_{vd}}. \quad \dots\dots\dots\dots\dots\dots \quad (8.283)$$

Example 8.24—Estimation of Oil Displacement After Coalescence. This example is a continuation of Example 8.23, in which the steam zone reached the coalescence distance of 50 m after 3.807 years of steam injection. At the coalescence point, the oil production rate (both sides) was 0.1320 m³/m/d, which is equivalent to 0.0660 m³/m/d for one side of the model. The heat-loss rate to the overburden was held constant at 2.252×10^8 kJ/a·m at the coalescence point, as discussed in Example 8.23. The gravity-drainage-rate constant is found by use of Eq. 8.276.

Solution.

$$\frac{k_o \delta g}{\upsilon_o} = \frac{q_{oc} L_{cp}}{\left(h - z_p\right)}.$$

At $z = h = 18.5$ m.

$$L_{cp} = \sqrt{(50)^2 + (18.5 - 2.5)^2}$$
$$= 52.50 \text{ m}.$$

$$\frac{k_o \delta g}{v_o} = \frac{\left(0.0660 \frac{m^3}{d \cdot m}\right)(52.50 \text{ m})}{(18.5 - 2.5)m}$$

$$= 0.217 \left(\frac{m^3}{d \cdot m}\right).$$

The calculation proceeds by selecting values of z_j, beginning from the top of the reservoir in increments of 1 m after the first step of $\Delta z = 0.5$ m.

For $z_1 = 18.0$ m,

$$L_1 = \sqrt{(50)^2 + (18 - 2.5)^2}$$

$$= 52.35 \text{ m}.$$

$$q_{o1} = 0.217 \frac{(18 - 0)}{53.1}$$

$$= 0.128 \frac{m^3}{d \cdot m} \text{(both sides)}$$

$$= 0.064 \frac{m^3}{d \cdot m} \text{(one side)}.$$

$$\Delta V_{sz1} = \frac{1}{2}(50)(18.5 - 18.0)$$

$$= 25 \text{ m}^3/\text{m (two sides)}$$

$$= 12.5 \text{ m}^3/\text{m (one side)}.$$

$$\Delta V_{o1} = 25 \frac{m^3}{m}(0.38)(0.8 - 0.15)$$

$$= 6.175 \frac{m^3}{m}.$$

$$\Delta t_1 = \frac{6.175 \frac{m^3}{m}}{0.128 \frac{m^3}{d \cdot m}} = 48.24 \text{ days}$$

$$= 0.132 \text{ years}.$$

$$\frac{\Delta V_{sz1}}{\Delta t} = \frac{25}{48.24} \frac{m^3}{d \cdot m}$$

$$= 0.518 \frac{m^3}{d \cdot m}.$$

The steam rate to displace the oil is found from

$$\dot{m}_s f_{dv} L_{vd} = M_R (T_s - T_r) \frac{dV_{sz}}{dt}. \dots\dots\dots\dots\dots\dots\dots\dots\dots\dots\dots\dots\dots\dots\dots \text{(8.284)}$$

$$\dot{m}_s = \frac{2350 \frac{kJ}{m^3 \, °C}(263.9 - 11)\left(0.518 \frac{m^3}{m \cdot d}\right)}{(0.95)(1640000) \frac{kJ}{m^3_{CWE}}}$$

$$= 0.198 \frac{m^3_{CWE}}{d \cdot m}$$

$$= 148.2 \frac{m^3_{CWE}}{d}.$$

The final step of the calculations is to estimate the cumulative steam injected and CSOR. For the first timestep, incremental cumulative steam injected to heat the steam zone

$$= \left(148.2 \frac{m^3_{CWE}}{d}\right) \frac{(48.24 \text{ days})}{(750 \text{ m})}$$

$$= 9.532 \frac{m^3_{CWE}}{m}.$$

The incremental steam injected to supply heat losses to the overburden during this time step is

$$= 297.06 \frac{m^3_{CWE}}{d} \frac{(48.27 \text{ days})}{(750 \text{ m})}$$

$$= 19.11 \frac{m^3_{CWE}}{m}.$$

At $t = 3.807$ years of injection, cumulative steam injection

$$= 724.23 + 9.53 + 19.11$$

$$= 752.87 \frac{m^3_{CWE}}{m}.$$

Cumulative oil displaced = 203.78 m³.

$$CSOR = \frac{752.9 \frac{m^3_{CWE}}{m}}{203.78 \frac{m^3}{m}} = 3.695.$$

The instantaneous SOR (ISOR) for the first time step is

$$ISOR = \frac{q_s}{q_o}$$

$$= \frac{445.7 \frac{m^3_{CWE}}{d}}{96.24 \frac{m^3}{d}}$$

$$= 4.63 \frac{m^3_{CWE}}{m^3}.$$

Table 8.73 summarizes the calculations of oil rates for the expanding steam chamber following the triangular shape in Fig. 8.143. **Table 8.74** presents calculated steam-injection rates and heat-loss rates for the post-coalescence period, and **Table 8.75** presents heat-balance and CSOR ratios. **Fig. 8.145** shows the location of the steam/oil interface during post-coalescence displacement by use of this model. **Fig. 8.146** shows the oil production history for this example, including the post-coalescence period.

Other Studies of SAGD. There are a large number of studies of the SAGD process that encompass a wide range of topics including convective heat transfer in the steam chamber (Irani and Ghannadi 2013; Irani and Gates 2013), steam-chamber development (Wei et al. 2014), interpretation of temperature data (Birrell 2001), numerical simulation (Sasaki et al. 2001), and solvent-assisted SAGD (Dickson et al. 2013).

8.6.2 CSS. The CSS process is a major contributor to in-situ bitumen production in the Province of Alberta, Canada. Average bitumen production increased from approximately 100,000 B/D in 1998 to more than 200,000 B/D in 2010, as shown in Fig. 8.114. Approved projected expansions are anticipated to increase production continually over the next several years.

The Cold Lake area shown in **Fig. 8.147** is the location of the world's largest thermal recovery process, operated by Imperial Oil. Cyclic steam injection has been developed as a large-scale commercial project.

The bitumen deposit (Denbina et al. 1991) at Cold Lake is found in the Clearwater formation at an average depth of 1,450 ft (442 m), which contains a clean, well-sorted, unconsolidated sand. Gross pay ranges from 160 ft (49 m) with net pay up to 150

Post-Coalescence Displacement Step	z (m)	L (m)	Scaled Oil Rate, Two Sides (m³/d/m)	Scaled Rate (m³/a/m)	V (m³)	ΔV_{sz} (m³)	V_{sz} (m³)	ΔVo (m³)	Cumulative Oil (m³)	Δt (years)	t (years)	Scaled Rate (m³/d)
Coalescence	18.5	53.3	0.1320	48.19	800	0	800	197.60	197.60	0	4.057	99.03
1	18	53.1	0.1283	46.84	775	25	825	6.175	203.78	0.132	4.189	96.24
2	17	52.8	0.1208	44.09	725	50	875	12.35	216.13	0.280	4.469	90.60
3	16	52.5	0.1131	41.30	675	50	925	12.35	228.48	0.299	4.768	84.85
4	15	52.2	0.1054	38.45	625	50	975	12.35	240.83	0.321	5.089	79.01
5	14	51.9	0.0974	35.57	575	50	1025	12.35	253.18	0.347	5.436	73.08
6	13	51.7	0.0894	32.64	525	50	1075	12.35	265.53	0.378	5.815	67.06
7	12	51.4	0.0813	29.67	475	50	1125	12.35	277.88	0.416	6.231	60.96
8	11	51.2	0.0730	26.66	425	50	1175	12.35	290.23	0.463	6.694	54.78
9	10	51.0	0.0647	23.62	375	50	1225	12.35	302.58	0.523	7.217	48.53
10	9	50.8	0.0563	20.55	325	50	1275	12.35	314.93	0.601	7.818	42.22
11	8	50.6	0.0478	17.44	275	50	1325	12.35	327.28	0.708	8.526	35.84
12	7	50.5	0.0392	14.31	225	50	1375	12.35	339.63	0.863	9.389	29.41
13	6	50.4	0.0306	11.16	175	50	1425	12.35	351.98	1.107	10.496	22.93
14	5	50.2	0.0219	7.99	125	50	1475	12.35	364.33	1.546	12.041	16.42
15	4	50.2	0.0132	4.80	75	50	1525	12.35	376.68	2.572	14.613	9.87
16	3	50.1	0.0044	1.60	25	50	1575	12.35	389.03	7.704	22.318	3.29

Table 8.73—Estimates of oil rate after coalescence with adjacent steam chamber.

t (months)	Oil Recovery, Fraction Mobile Oil	$\Delta V_{sz}/\Delta t$ (m³/d/m)	Steam Rate To Displace Oil (m³$_{CWE}$/d)	Steam Rate To Supply Heat Loss (m³$_{CWE}$/d)	Total Steam Rate (m³$_{CWE}$/d)	ISOR
48.7	0.50	0.535	152.94	297.06	450.0	4.54
50.3	0.52	0.5195	148.64	297.06	445.7	4.63
53.6	0.55	0.489	139.92	297.06	437.0	4.82
57.2	0.58	0.458	131.04	297.06	428.1	5.05
61.1	0.61	0.427	122.03	297.06	419.1	5.30
65.2	0.64	0.395	112.87	297.06	409.9	5.61
69.8	0.67	0.362	103.57	297.06	400.6	5.97
74.8	0.70	0.329	94.15	297.06	391.2	6.42
80.3	0.73	0.296	84.61	297.06	381.7	6.97
86.6	0.77	0.262	74.95	297.06	372.0	7.67
93.8	0.80	0.228	65.20	297.06	362.3	8.58
102.3	0.83	0.193	55.35	297.06	352.4	9.83
112.7	0.86	0.159	45.42	297.06	342.5	11.64
125.9	0.89	0.124	35.42	297.06	332.5	14.50
144.5	0.92	0.089	25.35	297.06	322.4	19.64
175.4	0.95	0.053	15.24	297.06	312.3	31.65
267.8	0.98	0.018	5.09	297.06	302.1	91.73

Table 8.74—Steam-injection and heat-loss rates.

t (months)	Steam To Heat Steam Zone and Displace Oil (m³$_{CWE}$/m)	Heat-Loss Rate (m³$_{CWE}$/a·m)	Heat Loss (m³$_{CWE}$/m)	Cumulative Steam (m³$_{CWE}$/m)	CSOR
48.7	–	–	–	724	3.67
50.3	9.5	0.0	0.00	734	3.60
53.6	19.1	0.0	0.00	753	3.48
57.2	19.1	0.0	0.00	772	3.38
61.1	19.1	0.0	0.00	791	3.28
65.2	19.1	0.0	0.00	810	3.20
69.8	19.1	0.0	0.00	829	3.12
74.8	19.1	0.0	0.00	848	3.05
80.3	19.1	0.0	0.00	867	2.99
86.6	19.1	0.0	0.00	886	2.93
93.8	19.1	0.0	0.00	905	2.88
102.3	19.1	0.0	0.00	924	2.82
112.7	19.1	0.0	0.00	944	2.78
125.9	19.1	0.0	0.00	963	2.73
144.5	19.1	0.0	0.00	982	2.69
175.4	19.1	0.0	0.00	1001	2.66
267.8	19.1	0.0	0.00	1020	2.62

Table 8.75—Heat loss and CSOR for post-coalescence model.

ft (46 m). Porosity is between 30 and 35%, and oil saturation is approximately 70%. The total amount of bitumen in the Cold Lake oil-sands deposit is approximately 150 billion bbl (25×10^9 m³). The deposit covers an area of approximately 2 million acres (800 000 hectares) (Buckles 1979).

　　The bitumen has an °API value of 10.6 and a viscosity of approximately 100,000 cp at the reservoir temperature of 13°C. Reservoir and fluid properties are summarized in **Tables 8.76 and 8.77.** Formation pressure is approximately 450 psi (Buckles 1979), and the amount of solution gas is approximately 50 scf/bbl (8 std. m³/m³) (Stark 2011). Although the absolute in-situ permeability of the sand ranges from 0.5 to 2 darcies, the bitumen is essentially immobile at reservoir temperature, with cold production rates on the order of 1 m³/d or less.

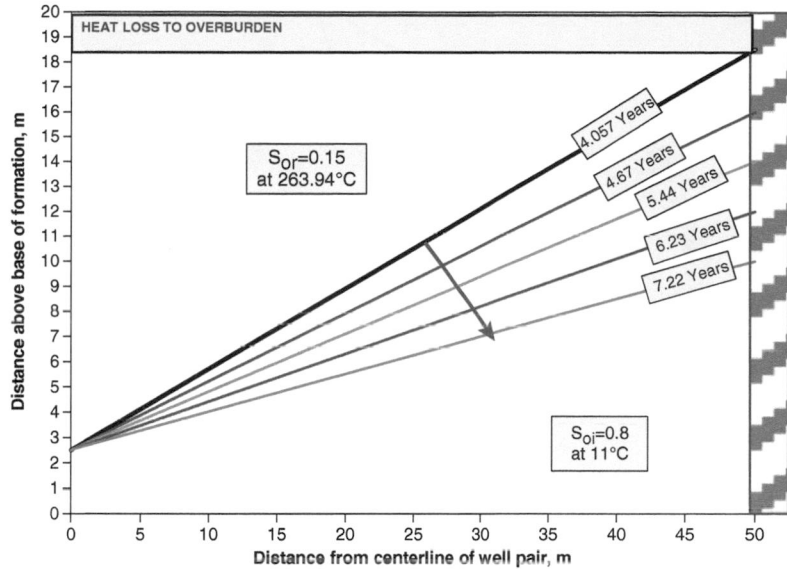

Fig. 8.145—Location of steam/oil interface during post-coalescence displacement.

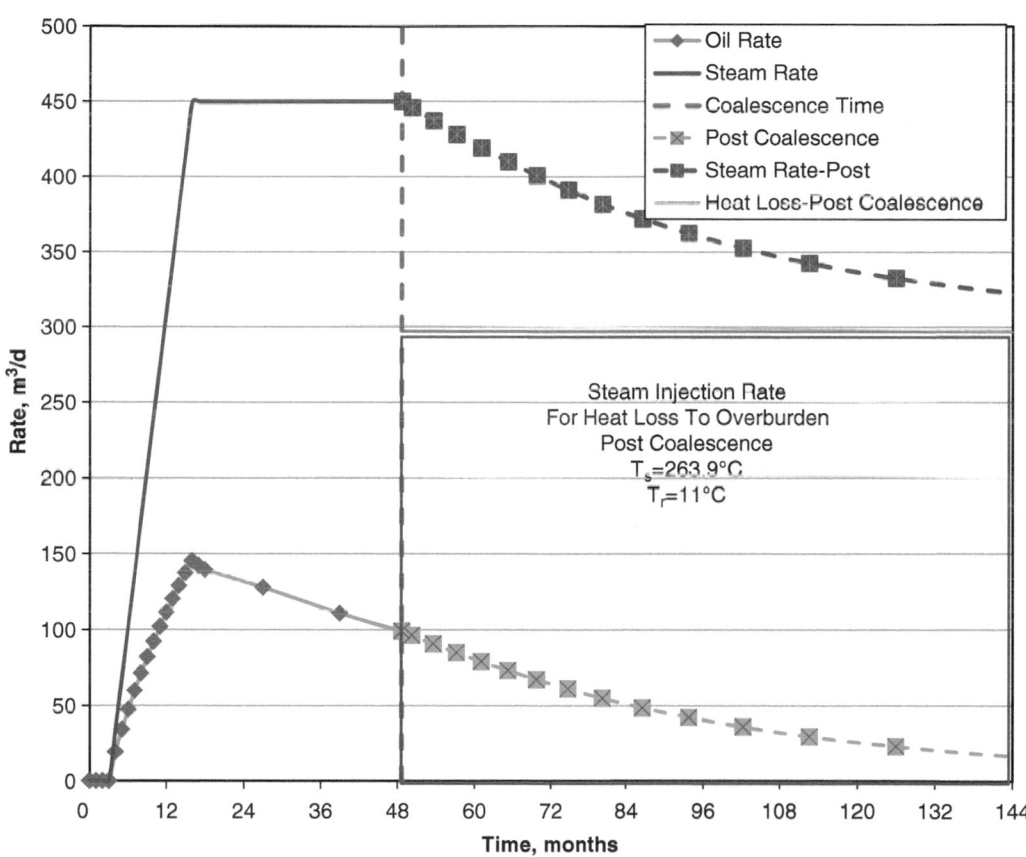

Fig. 8.146—Estimated oil production rates for SAGD using Marx and Langenheim and post-coalescence models.

The viscosity of the Cold Lake bitumen decreases rapidly with temperature, as shown in **Fig. 8.148,** reaching a value of approximately 2 cp at 500°F (260°C).

In the initial development of the process, wells were drilled directionally from a pad with a spacing of approximately 4 acres, as shown in **Fig. 8.149.** Each pad was composed of 20 wells that were steamed and produced as a unit. **Fig. 8.150** shows the design of a typical production well.

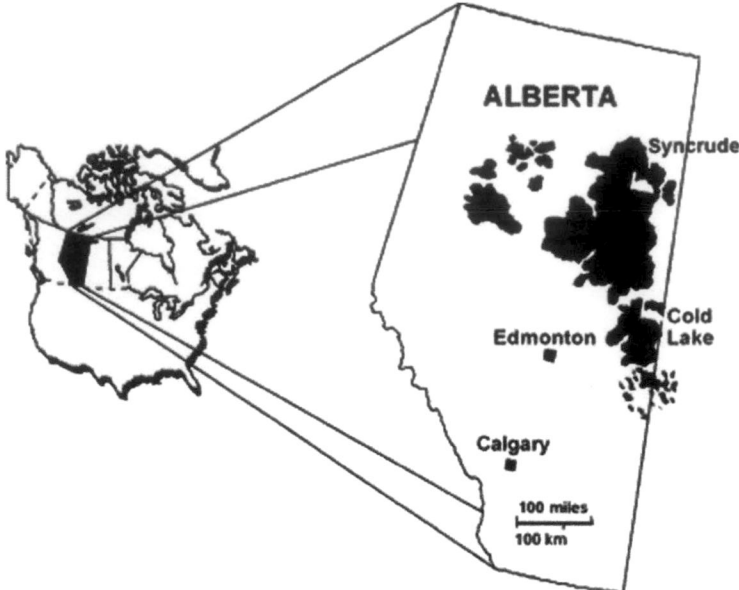

Fig. 8.147—Location of bitumen deposits in Alberta (Williams 2000).

Reservoir heating is required to mobilize the bitumen. At Cold Lake, it is necessary to fracture the reservoir sands to inject steam at rates required for development of a commercial project. Fracture pressures range from 1,050 psi (Denbina et al. 1991] to approximately 9000 kPa (1,300 psi) (Buckles 1979). Initial breakdown pressure may be as high as 14 MPa (2,031 psia) (Buckles 1979). In Cold Lake projects, steam is injected at pressures from 1,450 psi (10 MPa) to as high as 2,031 psi (14 MPa). According to Denbina et al. 1991, most fractures are horizontal in the steam-injection area, but some vertical fractures have been created.

In a typical CSS, steam is injected into each well at a rate of 1,320 B/D (210 m³/d) CWE for periods of 30 to 50 days. The extent of the heated region can be estimated by use of the model developed by Boberg and Lantz (1966), as presented in Section 8.3.3. During steam injection, surface heave (elevation of the ground level) occurs above the region heated by steam. Surface heave on the order of 45 cm (Beattie et al. 1991) has been observed. There is some subsidence after injection ceases, but permanent elevations of 15 cm have been reported many years after the completion of stimulation (Beattie et al. 1991).

Each well is placed on production after a soak period until the production declines to an economic limit determined by the SOR or the OSR. As the reservoir is heated by application of successive stimulation cycles, the length of the production period increases. The length of the production period ranges from approximately 120 days in the first cycle to approximately 400 days in the eighth cycle. Average production rates range from 155 B/D (25 m³/d) for the first cycle to 40 B/D (6 m³/d) for the eighth cycle (Denbina et al. 1991). **Fig. 8.151** shows typical production responses from the application of 10 steam stimulation cycles for a 20-well pad drilled directionally, as shown in Fig. 8.149.

Temperature (°F)	Oil FVF (RB/STB)	Oil Viscosity (cp)
50	0.996	80,700
150	1.033	329
250	1.072	26
450	1.160	2.6
650	1.264	0.8

Table 8.76—Properties of bitumen at Cold Lake (with gas in solution at 450 psia; gas in solution at original reservoir temperature of 55°F is 55 scf/STB). (Denbina et al. 1991)

Reservoir	
Rock volumetric heat capacity [Btu/(ft³-°F)]	35
Thermal conductivity [Btu/(ft°F-D)]	33.6
Normal rock compressibility (psi⁻¹)	7×10^{-6}
Fracture pressure (psi)	1,000
Dilation compressibility* (psi⁻¹)	7×10^{-4}
Recompaction onset pressure* (psi)	600
Residual dilation fraction* (dimensionless)	0.35
Irreducible water saturation (dimensionless)	0.26
Over/Underburden	
Volumetric heat capacity [Btu/(ft³-°F)]	40.0
Thermal conductivity [Btu/(ft-°F-D)]	33.6

*Beattie et al. (1991)

Table 8.77—Formation properties (Denbina et al. 1991).

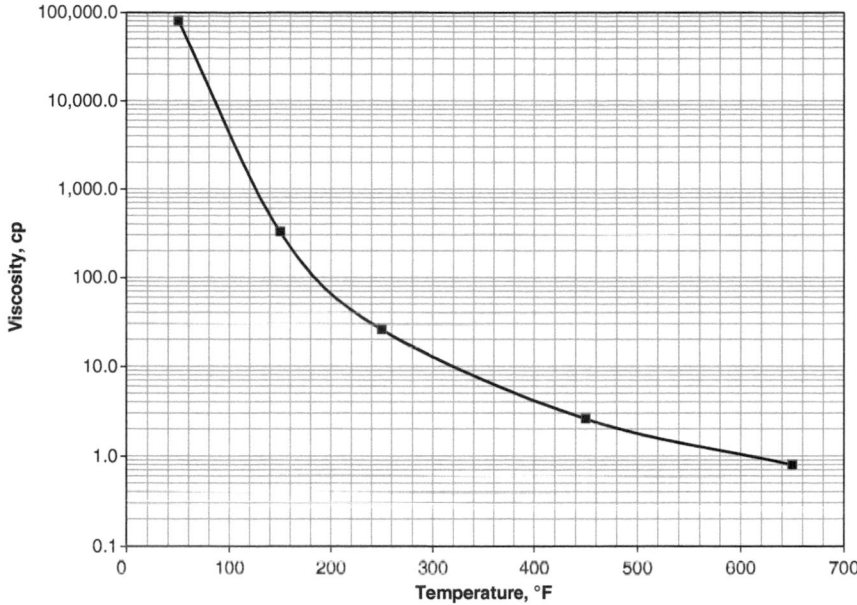

Fig. 8.148—Viscosity of Cold Lake bitumen as a function of temperature (Denbina et al. 1991, Table 1).

Original Pad Design

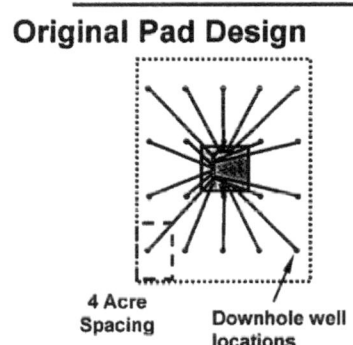

4 Acre Spacing   Downhole well locations

Fig. 8.149—Pad design for CSS at Cold Lake (Fair et al. 2008).

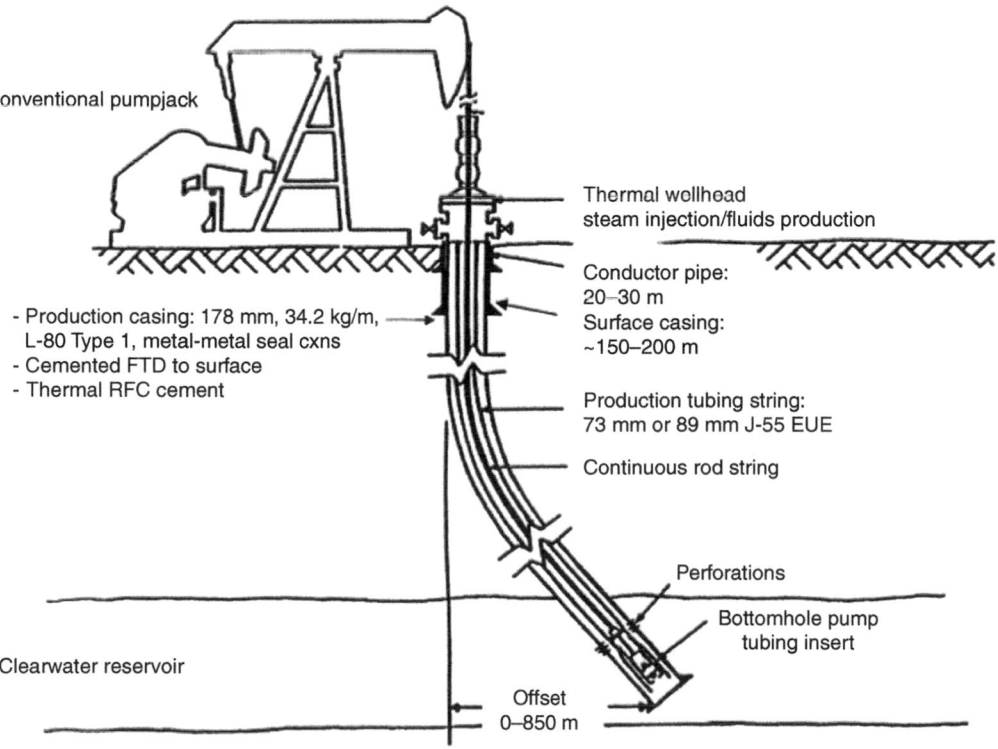

Fig. 8.150—Design of production well for CSS at Cold Lake (Imperial Oil 2012).

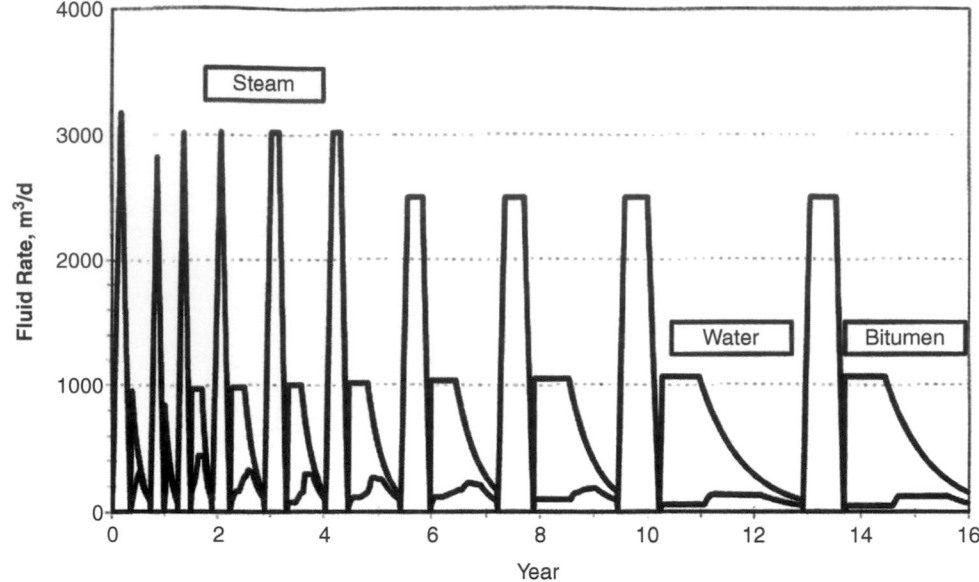

Fig. 8.151—Summary of fluid injection and production history for 10 CSS cycles applied in a 20-well pad (Imperial Oil 2012).

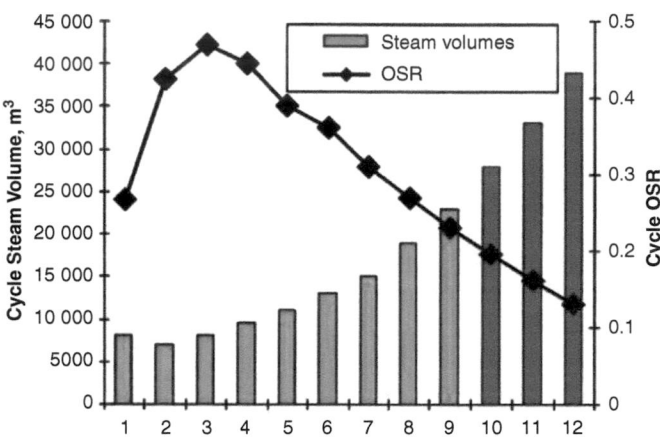

Fig. 8.152—Cyclic steam volume and OSR progression (Stark 2011).

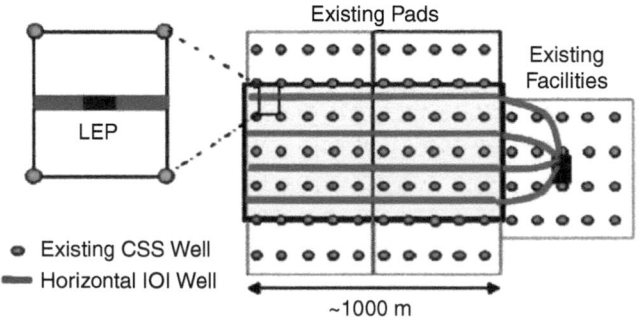

Fig. 8.153—Cold Lake horizontal infill well configuration (Stark 2011).

Initially, there is little water produced in the first 2 to 3 cycles. Water production increases during subsequent cycles with production of water/oil emulsions. Reservoir drive mechanisms change during the life of the CSS project. During the early cycles, Beattie et al. (1991) demonstrated that dilation of the formation during steam injection facilitates the recompaction of formation as pressure decreases after steam injection. Recompaction is the dominant reservoir drive mechanism in early cycles, followed by solution gas drive (Denbina et al. 1991). Fluid expansion and gravity drainage were not found to be important contributors during these periods.

As reservoir heating progresses during successive stimulation cycles, fluid mobility develops between adjacent wells, which causes production performance to decline. Overlapping of heated regions prevents maintenance of high injection pressures, declining reservoir temperature, and reservoir drive because of recompaction. The reservoir drive in the expanding steam chamber around each production well becomes dominated by gravity drainage, leading to decline in CSS response. The key measure of CSS effectiveness is the OSR. **Fig. 8.152** shows the progression of the cyclic steam volume and the OSR (Stark 2011) for nine CSS cycles.

The decrease in OSR as a result of the overlapping of heated volumes leads to the drilling of horizontal infill wells, as shown in **Fig. 8.153** (Stark 2011), to provide reservoir heating to the bypassed regions between production wells. The horizontal infill wells are completed with limited-entry perforations to control steam distribution along the well and are used only for steam injection. Steam is injected at lower pressures and continues to expand reservoir heating.

In 2011, expected recovery of bitumen by use of CSS and steam injection through horizontal infill wells was on the order of 40 to 50% of the effective bitumen in place (Stark 2011). Expansion of the CSS operation to a steamflood is depicted by the schematic of an infill steamflood shown in **Fig. 8.154** (Imperial Oil 2012). Expected injection pressures were 0.5 to 1.5 MPa, with production dominated by gravity drainage.

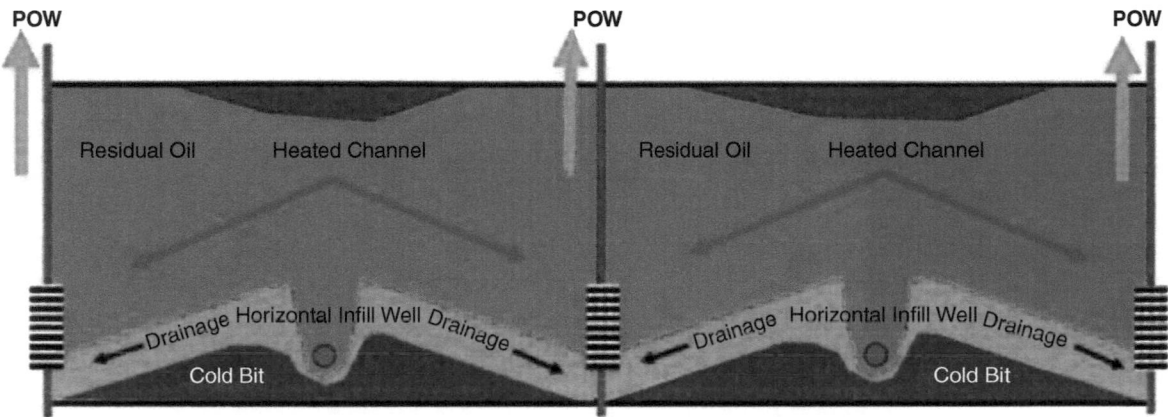

Fig. 8.154—Cross section through the region between production wells for a steamflood conducted from horizontal infill injection wells (Imperial Oil 2012).

Fig. 8.155 shows the layout of a pilot flood (Fig. 8.153) consisting of two pads (H01 and H02) that contained 50 wells (45 producers, five abandoned), and four horizontal infill injection wells. Each infill well had a 1000-m-long horizontal section. Steam injection began in March 2008 and continued for 3 years. Injection pressures of 1 MPa were observed.

Fig. 8.156 shows the performance of Well H01-14, which was representative of the pilot response. The well responded to steam injection with average water rates of 40 m^3/d and stable bitumen rates of 10 m^3/d. Incremental OSR was 0.14, which was more than the economic limit. Incremental recovery from the steamflood was 8%, increasing the total recovery of bitumen from the two pads to 66% of the effective bitumen in place.

The steam chamber expands on subsequent cycles, and it is difficult to determine the shape and areal extent of the heated region without additional information. Kry (1989) recognized that seismic properties of the steam and liquid regions might be identified by use of seismic surveys. **Fig. 8.157** (Fair et al. 2008) shows a series of time-lapsed 3D images of the vapor region (light gray) during a series of three CSS cycles in five wells. The seismic images show the growth of the steam zone from individual wells in Cycle 1, coalesc-

Fig. 8.155—Horizonal steamflood pilot with continuous injection of steam in horizontal infill wells (Stark 2011).

ing with chambers developed by adjacent wells. By the third cycle, the coalesced steam chambers covered a sizeable area of the pad with continued expansion. The authors attributed the cold region at the end of the northern well to a sand bridge that blocked the flow of steam to the toe (Smith and Perepelecta 2002).

Field Results. A review of thermal recovery applications in the Cold Lake region was presented by Jiang et al. (2010). Field data developed by Sandhar (2011) that summarize CSS field production for 2010 are presented in **Table 8.78.**

8.7 In-Situ Combustion

In-situ combustion is a displacement process in which an oxygen-containing gas is injected into a reservoir where it reacts with the crude oil to create a high-temperature combustion front that is propagated through the reservoir. In most cases, the injected gas is air, although the use of 100% oxygen has been reported. The fuel consumed by the combustion front is a residuum produced by a complex process of cracking, coking, and steam distillation that occurs ahead of the combustion front. In-situ combustion is possible if the crude-oil/rock combination produces enough fuel to sustain the combustion front. In-situ combustion field tests have been carried out in reservoirs containing gravities from 9 to 40 °API.

Forward and reverse combustion have been studied in the laboratory and in field tests (Martin et al. 1958; Alexander et al. 1962; Showalter 1963; Penberthy and Ramey 1966; Reed et al. 1960; Craig and Parrish 1974; Gates et al. 1978; Gates and Ramey 1958). Reverse combustion has not been developed commercially and will not be discussed further. Forward

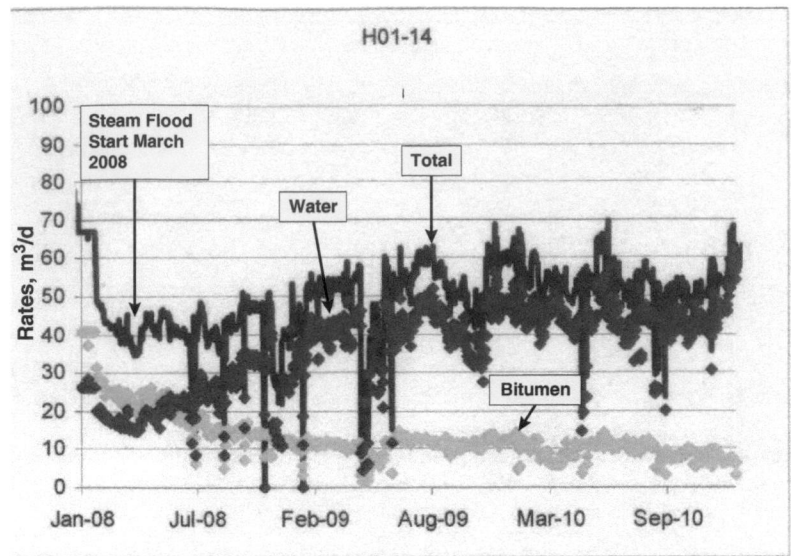

Fig. 8.156—Production rates from a Well H01-14 during steamflood pilot that began in March 2008 and ended in February 2011 (Stark 2011).

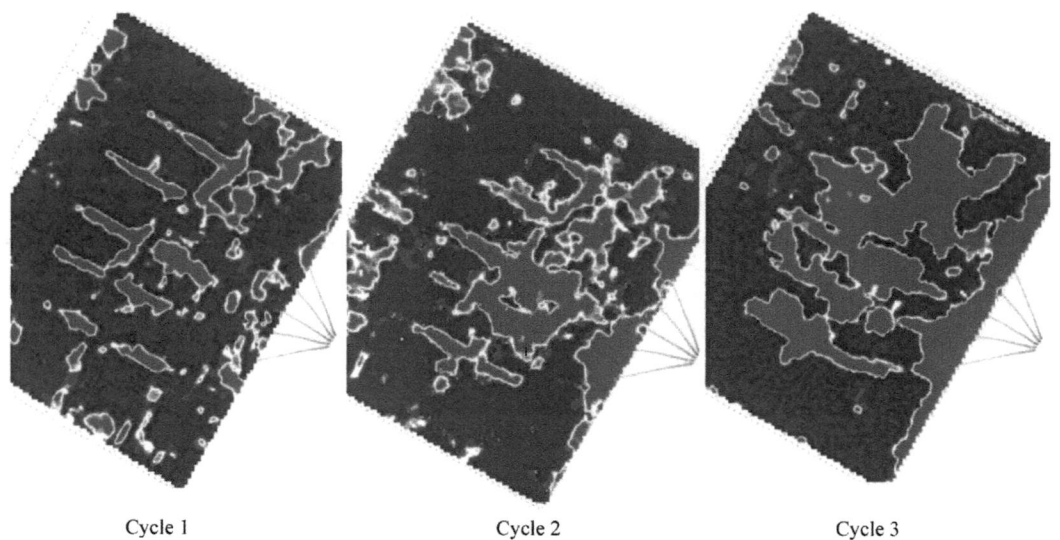

Cycle 1   Cycle 2   Cycle 3

Fig. 8.157—Development of steam conformance during application of three cycles in five wells (Fair et al. 2008).

	Production (B/D)	Usage	ISOR	CSOR
Shell – Peace River	14,800	58%	2.2	3.0
Imperial – Cold Lake	133,200	95%	4.0	3.0
Canadian Natural – Primrose	87,700	69%	5.0	4.9
Median	–	60%	4.0	3.4
Weighted average	–	83%	3.5	3.9
Total	235,700	–	–	–

Table 8.78—Estimated CSS field production for 2010 (Sandhar 2011).

combustion will be considered in this text. There are two types of forward combustion: dry combustion and wet combustion. In wet combustion, water is injected simultaneously with the injected gas or alternated in slugs. Significant improvements in process performance have been obtained in laboratory experiments and field projects with the wet-combustion process.

This section presents fundamental concepts of in-situ combustion beginning with fuel availability and air requirement and ending with a brief description of major field projects for dry-combustion projects. As mentioned in Chapter 1, in-situ combustion is not a widely used recovery process because steamdrive is usually preferred when both processes are technically feasible. However, steamdrive is limited to reservoirs in which pressures are in the range of 2,500 psi or less and at depths on the order of 3,000 ft because of wellbore heat losses. Consequently, in-situ combustion is the only thermal recovery process that is potentially useful for deep, high-pressure reservoirs.

Although there have been many field tests of in-situ combustion, the number of commercial scale projects of in-situ combustion was reported (Turta et al. 2007) to be four, with estimated production of 17,700 B/D. This estimate does not include in-situ combustion projects in the dolomite reservoirs in North Dakota and Montana of approximately 20.590 B/D (Koottungal 2014).

8.7.1 Process Mechanisms—Dry Combustion. In-situ combustion occurs when oxygen reacts with the coke contained within the pore space of a porous rock to create a self-sustaining combustion front. Ignition may be induced through electrical or gas igniters or may be spontaneous (Tadema and Weijdema 1970) if the crude oil has sufficient reactivity. When the reservoir is relatively thin, the displacement process behaves like a frontal-advance process, with the temperature and saturation distribution depicted in **Fig. 8.158** (White and Moss 1965). A narrow combustion zone forms where temperatures may be very high. The injected air is preheated to combustion temperature (650 to 1,200°F) as it flows through the rock behind the combustion zone. Combustion products, primarily water (as water vapor), CO_2, and CO, flow ahead of the slowly moving (0.125 to 1.0 ft/D) front. Oxygen not consumed by the combustion front and nitrogen (if air is injected) also flow with the combustion gases.

In a fully developed front, hot combustion gases strip light ends from the crude oil flowing ahead of the front (because of the high mobility of gas compared with that of liquids). Hydrocarbons stripped by the hot combustion gases and water vapor condense to form a small steam plateau of hot water and light hydrocarbon banks, as depicted in Fig. 8.158. The oil saturation that remains after steam stripping is subjected to thermal cracking as the combustion front approaches, leaving a residual deposit on the sand grains that is rich in carbon. This residuum becomes the fuel for the process. In general, no more than 5 to 6% of the oil is consumed. Hydrocarbon products and other compounds released by the cracking process (SO_2, CO, CH_4, and H_2) join the combustion gases and are either absorbed by crude oil ahead of the front or are produced in the effluent. In dry forward combustion, the fuel must be consumed for the combustion front to advance. Thus, the rate of frontal advance is controlled by fuel availability and the rate that oxygen is delivered to the burning front.

When the reservoir is thick and has good vertical permeability, gravity segregation or override occurs, as discussed with steamdrive. The combustion front migrates to the top of the reservoir where it stretches out across the reservoir, as shown in **Fig. 8.159** (Prats et al. 1968). The combustion front expands horizontally and vertically in a complicated manner. The same process mechanisms are present when gravity segregation occurs, but front movement is influenced by the complex multiphase flow pattern inherent in a combustion front that expands both vertically and horizontally. The reservoir under the combustion front is heated by conduction, which is particularly important in heavy-oil reservoirs where large changes in oil viscosity occur with relatively small changes in temperature, as illustrated in Fig. 8.1.

The impact of gravity drainage on the recovery of oil during in-situ combustion was first observed in the South Belridge thermal recovery experiment (Gates et al. 1978; Gates and Ramey 1958). The project began as an inverted five-spot pattern (2.75 acres) with one air-injection well and four production wells **(Fig. 8.160)**. **Fig. 8.161** shows cross sections through the oil sand in the test pattern. Air injection began on 1 March 1956, and spontaneous ignition occurred in June 1956. The project

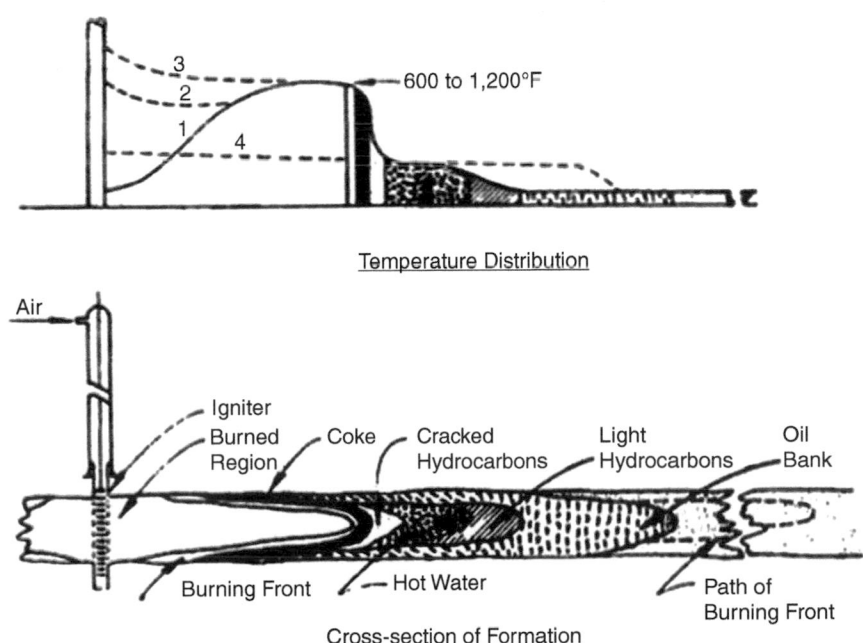

Temperature Distribution

Cross-section of Formation

Fig. 8.158—Temperature distribution and displacement zones in a dry-combustion process (White and Moss 1965).

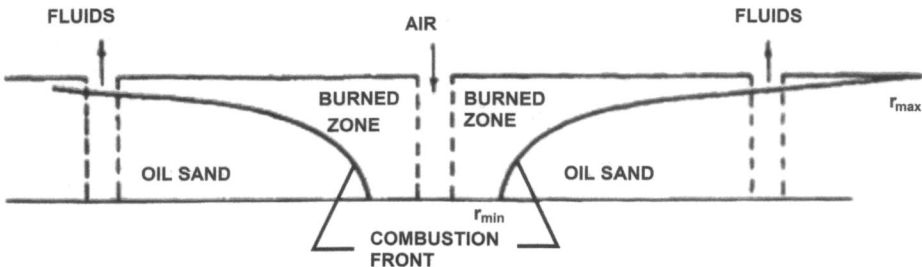

Fig. 8.159—Gravity override of combustion front (Prats et al. 1968).

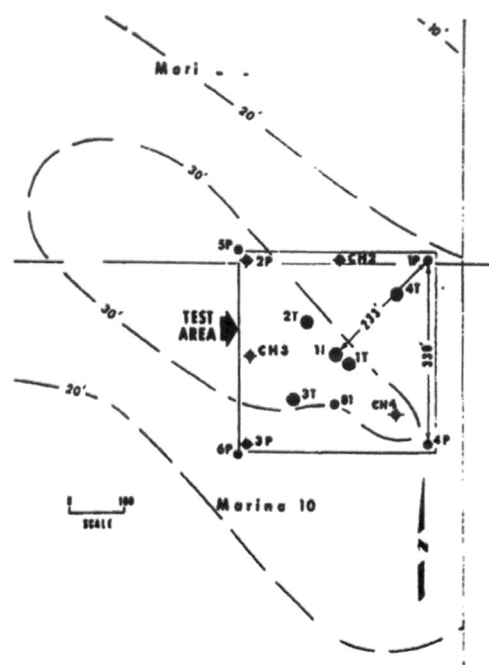

Fig. 8.160—Isopachous map of 700-ft Tulare sand, South Belridge field (Gates and Ramey 1958).

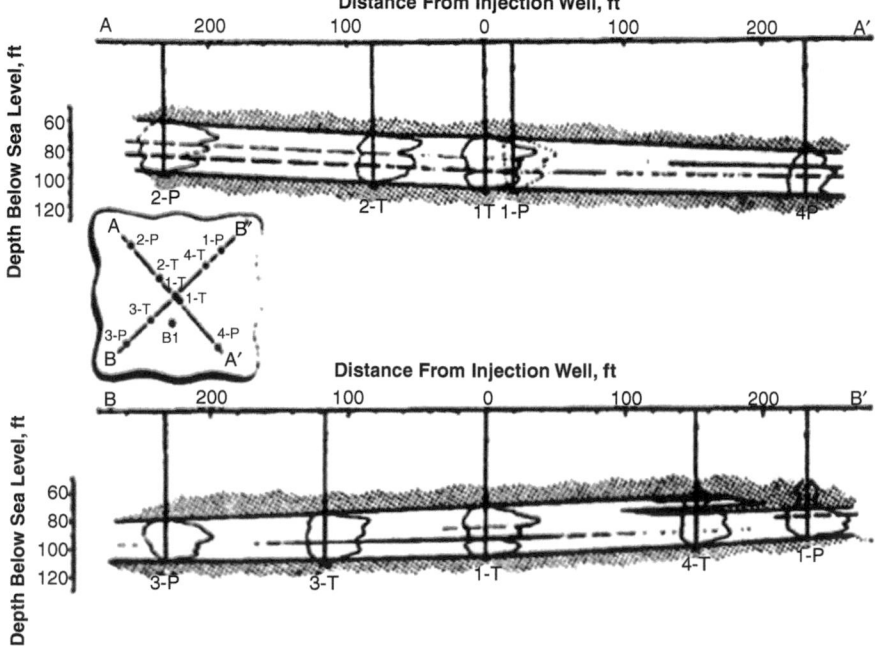

Fig. 8.161—Cross sections of oil sand through test pattern (Gates and Ramey 1958).

was operated through 1959. Several core holes were drilled at various stages of the project to determine the areal and vertical extent of the burned zone.

Fig. 8.162 shows the estimated thickness of the burned zone as of November 1959. Approximately 7.90 acres were burned, an increase of 1.57 acres from November 1957. The burned zone developed along the top of the sand under the influence of gravity segregation. **Fig. 8.163** shows cross sections of the burned-zone thickness between Core Holes 7 and 10 in November 1957 and November 1959. There was significant vertical movement of the combustion front during this period. Although the formation dip was 3°, the burned zone moved preferentially updip, coinciding with the movement of air updip under the influence of gravity. Gravity segregation dominated the combustion process, and the combustion front moved vertically under the influence of gravity segregation throughout the burned area. In the burned zone, oil flowed countercurrent to the air under the influence of gravity, and significant volumes were produced in wells downstructure from the location of the combustion front.

Breakthrough of the combustion front into a production well occurs in most projects. In thin reservoirs, most of the oil displaced by the combustion front will be produced before breakthrough. In reservoirs where the combustion front does not burn the entire vertical cross section, large amounts of oil have been produced from "hot" wells by cooling the production well

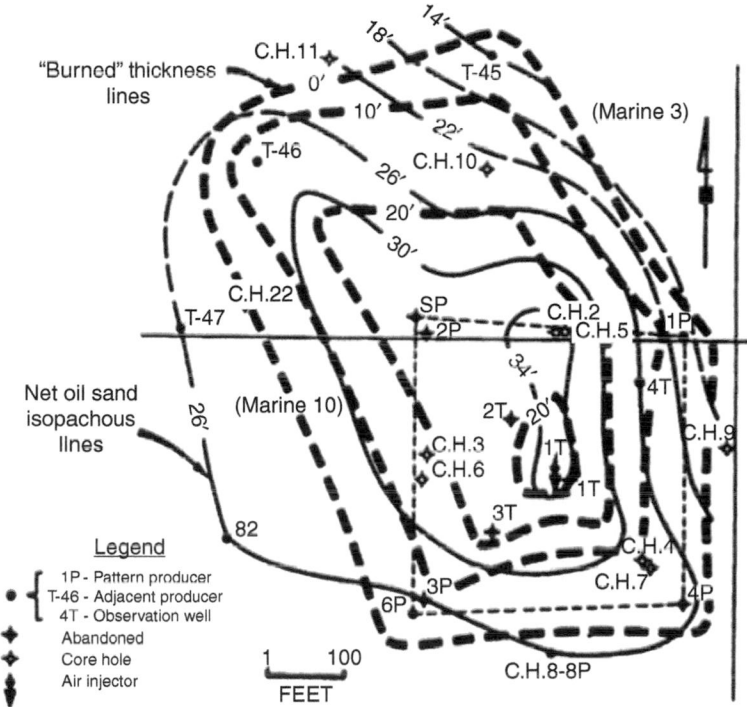

Fig. 8.162—Burned thickness (November 1959), South Belridge thermal recovery experiment (Gates et al. 1978).

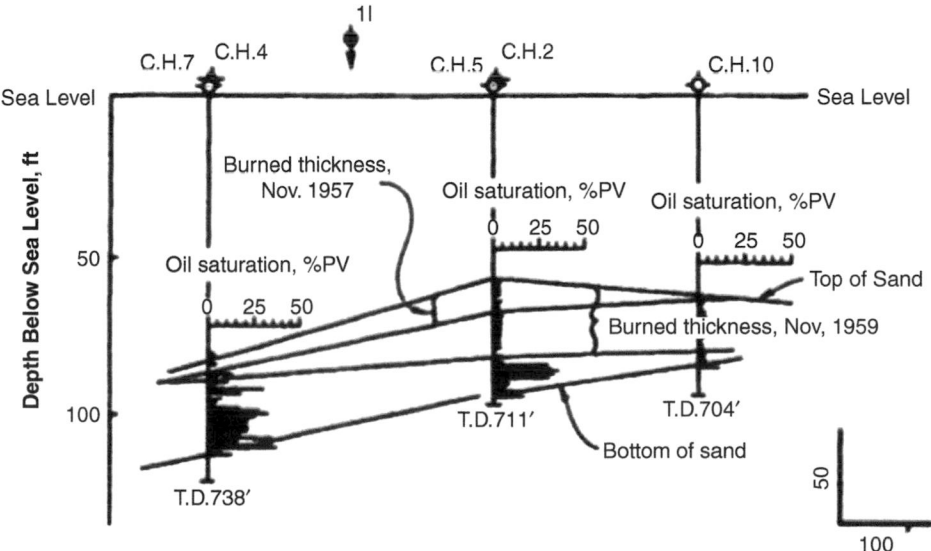

Fig. 8.163—Cross sections showing burned thickness, South Belridge thermal recovery experiment (Gates et al. 1978).

by injecting water down the casing annulus or by shutting the well in for a period of time after the arrival of the combustion front. The latter practice allows heating of the reservoir by conduction from the combustion zone. Production wells may be lost when the combustion front breaks through if not managed properly. Three factors control the in-situ combustion process: fuel availability, air requirement, and air flux.

Fuel Availability. The term "fuel availability" refers to the amount of fuel laid down by the advancing combustion front. Fuel availability is expressed in several different units. Common units found in the literature are lbm fuel (hydrocarbon)/100 lbm rock, lbm fuel (hydrocarbon)/ft^3 rock, and lbm carbon/ft^3 rock. The following definition of fuel availability will be used in this text.

$$m_R = \text{lbm fuel/ft}^3 \text{ reservoir volume burned.} \dots\dots\dots\dots\dots\dots\dots\dots\dots\dots\dots\dots\dots\dots\dots (8.285)$$

Two distinct combustion regimes are known to exist. When combustion temperatures exceed approximately 650°F, the combustion reaction is called high-temperature oxidation (HTO) and the combustion products are CO_2, CO, and water. At temperatures less than 650°F, water and oxygenated hydrocarbons are formed (Burger and Sahuquet 1972). This combustion regime is known as low-temperature oxidation (LTO) and is discussed later in this section.

HTO. Fuel availability under the conditions of HTO is usually determined from the analysis of laboratory combustion experiments with the reservoir rock and crude oil. Several types of experiments have been used (Martin et al. 1958; Alexander et al. 1962; Showalter 1963). The most common experiment is to carry out a material balance on the effluent gases collected during a combustion-tube experiment. **Fig. 8.164** shows a typical combustion tube. It consists of a thin-walled inner tube 2 to 4 in. in diameter that contains the porous material. This tube is mounted inside the larger tube, which also serves as a pressure vessel. The inner tube is instrumented with thermocouples to determine the temperature distribution as the combustion front advances through the tube. Some tubes use a series of guard heaters mounted on the outside of the inner tube to maintain adiabatic conditions. In small-diameter tubes, heat loss is minimized by insulating the tube and operating the combustion front at high velocities. Combustion-tube experiments are run vertically from the top down to minimize gravity effects.

In-situ combustion is initiated by use of an electric heater to heat the air and inlet end of the porous rock to a high temperature. The heater is turned off once a sustained combustion front develops. **Fig. 8.165** illustrates the progression of the combustion front through a sandpack at different times. The air-injection rate is held constant at the inlet of the tube. Inlet and outlet gas rates and compositions are measured as a function of time. A steady rate of combustion is established relatively soon after ignition.

Combustion gases contain N_2, O_2, CO_2, CO, and H_2O. Concentration of O_2 is usually small. To determine the fuel availability, it is necessary to make some assumptions about the combustion reactions. In HTO, it is assumed that all hydrogen in the fuel reacts with oxygen to produce water. All carbon in the fuel is assumed to be converted to CO or CO_2, which are recovered in the produced gases. Carbon dioxide can dissolve in crude oil with substantial reduction in oil viscosity. However, in laboratory combustion-tube tests, the entire porous rock is usually swept by the combustion front so that dissolved CO_2 is

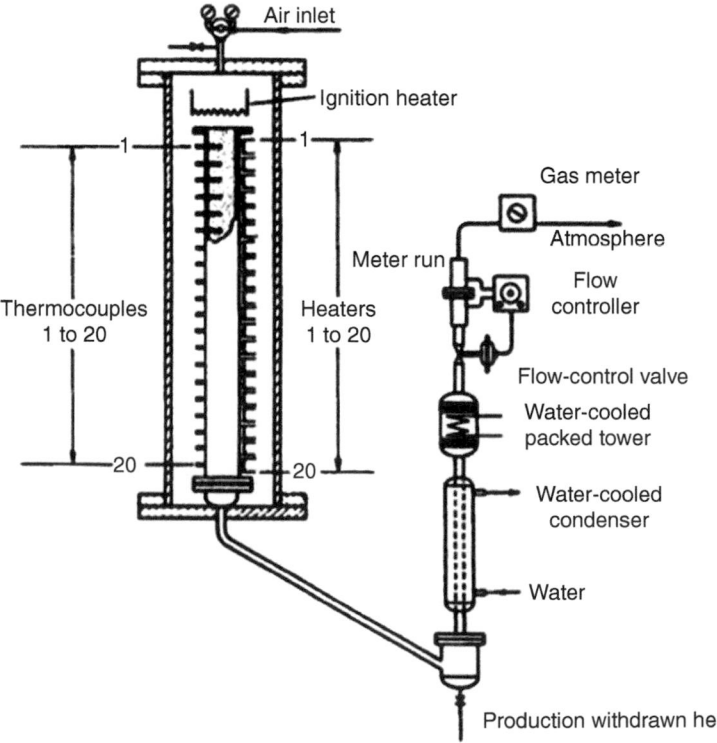

Fig. 8.164—Schematic of combustion tube (Showalter 1963).

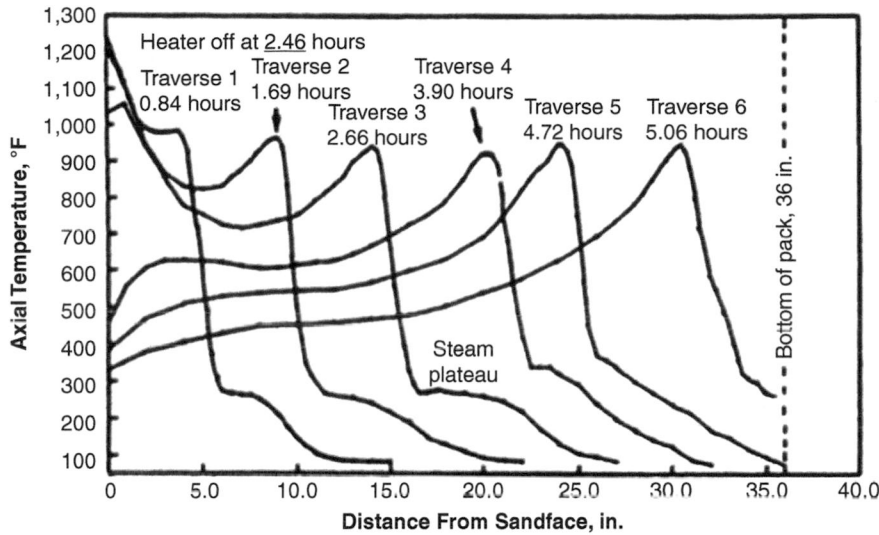

Fig. 8.165—Axial temperatures as functions of distance from sandface and time (Penberthy and Ramey 1966).

effectively displaced by the combustion front. Methane is observed occasionally in the produced gases and can be accounted for in material-balance calculations. In a reservoir environment, hydrocarbon gases would most likely be dissolved in the crude oil ahead of the front and would not be detected.

The determination of fuel availability from the analysis of combustion-tube data begins with an analysis of the stoichiometry of the combustion reaction. The composition of the fuel is not known but is assumed to be represented as a hypothetical hydrocarbon with the chemical formula $CH_{F_{HC}}$, where F_{HC} is the atomic ratio of hydrogen to carbon in the fuel. The high-temperature combustion of $CH_{F_{HC}}$ when the combustion products are water, CO_2, and CO is represented by

$$CH_{F_{HC}} + n_{O_2}O_2 \rightarrow n_{CO_2}CO_2 + n_{CO}CO + (F_{HC}/2)H_2O, \quad \dots\dots\dots\dots\dots\dots\dots\dots\dots\dots\dots\dots\dots \text{(8.286)}$$

where n_{O_2} = moles of oxygen reacting, n_{CO_2} = moles of CO_2 in the combustion gases, n_{CO} = moles of CO in the combustion gases, and F_{HC} = atomic ratio of hydrogen to carbon in the fuel.

Eq. 8.286 can be simplified by defining the parameter m as in Eq. 8.287.

$$m = \frac{\text{moles } CO_2 \text{ produced}}{\text{moles CO produced}} = \frac{n_{CO_2}}{n_{CO}}. \quad \dots\dots\dots\dots\dots\dots\dots\dots\dots\dots\dots\dots\dots\dots\dots \text{(8.287)}$$

This definition is useful because the composition of the produced gas is determined as mole fraction or mole percent. The value of m is computed directly from the produced-gas analysis:

$$m = V_{CO_2}/V_{CO}. \quad \dots\dots\dots\dots\dots\dots\dots\dots\dots\dots\dots\dots\dots\dots\dots\dots\dots\dots\dots\dots\dots\dots \text{(8.288)}$$

A mole balance on carbon shows that

$$n_{CO} + n_{CO_2} = 1$$

and $n_{CO}(m+1) = 1,$

so that $n_{CO_2} = m/(m+1)$ $\dots\dots\dots\dots\dots\dots\dots\dots\dots\dots\dots\dots\dots\dots\dots\dots\dots\dots$ (8.289)

and $n_{CO} = 1/(m+1).$ $\dots\dots\dots\dots\dots\dots\dots\dots\dots\dots\dots\dots\dots\dots\dots\dots\dots\dots\dots$ (8.290)

A mole balance on oxygen yields the following relationships:

$$n_{O_2} = n_{CO_2} + \left(n_{CO_2}/2\right) + \left(F_{HC}/4\right)$$

$$= \frac{m}{m+1} + \frac{1}{2(m+1)} + \frac{F_{HC}}{4}. \quad \dots\dots\dots\dots\dots\dots\dots\dots\dots\dots\dots\dots\dots\dots \text{(8.291)}$$

$$n_{O_2} = \left\{ \left[(2m+1)/(2m+2) \right] + \left(F_{HC}/4 \right) \right\}. \quad \dots \dots \dots \dots \dots \dots \dots \dots \dots \dots \dots \quad (8.292)$$

Substituting Eqs. 8.287, 8.289, 8.290 into Eq. 8.286 gives (Alexander et al. 1962; Poettmann 1964)

$$CH_{F_{HC}} + \left(\frac{2m+1}{2m+2} \frac{F_{HC}}{4} \right) O_2 \rightarrow \left(\frac{m}{m+1} \right) CO_2 + \left[1/(m+1) \right] CO + \left(F_{HC}/2 \right) H_2O. \quad \dots \dots \dots \dots \quad (8.293)$$

When the oxygen is not totally consumed, an oxygen usage efficiency can be introduced to measure the extent of combustion. Let

$$E_{O_2} = \frac{\text{moles of oxygen consumed}}{\text{moles of oxygen injected}} \quad \dots \dots \dots \dots \dots \dots \dots \dots \dots \dots \dots \dots \dots \dots \dots \quad (8.294)$$

The injected gas is usually air but may be enriched with oxygen in some cases. The oxygen usage efficiency is determined from the analysis of produced gases. From Eq. 8.294,

$$E_{O_2} = \left(y_{iO_2} n_i - y_{pO_2} n_p \right) / y_{iO_2} n_i, \quad \dots \dots \dots \dots \dots \dots \dots \dots \dots \dots \dots \dots \dots \quad (8.295)$$

where y_{iO_2} = average mole fraction of oxygen in the injected gas, y_{pO_2} = average mole fraction of oxygen in the produced gas, n_i = moles of gas injected during a specified time interval, and n_p = moles of produced gas during a specified time interval. When the injected gas contains nitrogen, the nitrogen flows through the combustion front as an inert gas and is completely recovered in the produced gas. Consequently, a mole balance on nitrogen produces the relationship between n_i and n_p given by

$$y_{iN_2} n_i = y_{pN_2} n_p, \quad \dots \dots \dots \dots \dots \dots \dots \dots \dots \dots \dots \dots \dots \dots \dots \dots \dots \dots \quad (8.296)$$

where y_{iN_2} = mole fraction of nitrogen in the injected gas and y_{pN_2} = mole fraction of nitrogen in the produced gas. Substituting Eq. 8.296 into Eq. 8.295 gives a general equation for the oxygen usage efficiency:

$$E_{O_2} = 1 - \left(y_{iN_2} / y_{iO_2} \right) \left(y_{pO_2} / y_{pN_2} \right). \quad \dots \dots \dots \dots \dots \dots \dots \dots \dots \quad (8.297)$$

If air is the injected gas,

$$E_{O_2} = 1 - (0.79/0.21) \left(y_{pO_2} / y_{pN_2} \right). \quad \dots \dots \dots \dots \dots \dots \dots \dots \dots \dots \dots \dots \quad (8.298)$$

Naji and Poettmann (1991) show that the oxygen use is given by Eq. 8.299 when the injected gas is 100% oxygen.

$$E_{O_2} = 1 - y_{pO_2}. \quad \dots \dots \dots \dots \dots \dots \dots \dots \dots \dots \dots \dots \dots \dots \dots \dots \dots \quad (8.299)$$

The hydrogen/carbon (H/C) ratio is determined from a material balance on the hydrogen and the carbon oxides, which are produced by HTO. All hydrogen in the fuel reacts with oxygen to produce water, which is not completely recovered or measured. Thus, hydrogen consumption is determined by the material balance on oxygen. The oxygen that is consumed appears as CO_2, CO, and water. Hydrogen consumption is overestimated by this balance because some of the oxygen is consumed in low-temperature reactions where oxygen combines with hydrocarbon to produce oxygenated compounds. For this reason, hydrogen is referred to as apparent hydrogen consumption. Writing a balance that is based on the stoichiometry of the compounds, the H/C ratio is given by (Naji and Poettmann 1991)

$$F_{HC} = \left\{ 4 \left(y_{iO_2} / y_{iN_2} \right) - \left[\left(y_{iO_2} / y_{iN_2} \right) + 1 \right] \left(1 - y_{pN_2} \right) + 2 y_{pCO} \right\} \div y_{pCO_2} + y_{pCO}. \quad \dots \dots \dots \dots \quad (8.300)$$

For the case in which the injected gas is pure oxygen, Eq. 8.300 becomes (Naji and Poettmann 1991)

$$F_{HC} = \frac{4 \left(1 - y_{pO_2} - 0.5 y_{pCO} - y_{pCO_2} \right)}{y_{pCO_2} + y_{pCO}}. \quad \dots \dots \dots \dots \dots \dots \dots \dots \dots \quad (8.301)$$

When air is injected (y_{iO_2} = 0.21 and y_{iN_2} = 0.79), the apparent H/C ratio can be computed directly from combustion-gas analysis with (Dew and Martin 1965)

$$F_{HC} = \frac{106.3 + 2CO - 5.06 \left(CO_2 + CO + O_2 \right)}{CO_2 + CO}, \quad \dots \dots \dots \dots \dots \dots \dots \dots \dots \quad (8.302)$$

where CO_2 = mol% of CO_2 in produced gas, CO = mol% of CO in produced gas, and O_2 = mol% of O_2 in produced gas.

Fuel availability can be computed from the analysis of combustion-tube data over a specified period of time. The fuel availability from the combustion-tube experiment is defined by

$$m_E = w_f / V_b, \quad \dots \dots \dots \dots \dots \dots \dots \dots \dots \dots \dots \dots \dots \dots \dots \dots \dots \dots \dots \dots \dots \dots \dots \dots \dots \dots \dots \dots (8.303)$$

where w_f = mass of carbon and hydrogen consumed when V_b (in cubic feet) of reservoir was y_{pCO} burned. In a combustion-tube run, n_p moles of gas with an average composition of y_{pCO_2}, y_{pCO}, and y_{pO_2} are produced during a time interval Δt when the combustion front is propagated through V_b. It is assumed that the combustion tube is operating under steady-state conditions so that the velocity of the combustion front is constant. Under these conditions,

$$V_b = (x_2 - x_1) A \quad \dots \dots \dots \dots \dots \dots \dots \dots \dots \dots \dots \dots \dots \dots \dots \dots \dots \dots \dots \dots \dots (8.304)$$

$$\text{and } V_b = v_f (t_2 - t_1) A, \quad \dots \dots \dots \dots \dots \dots \dots \dots \dots \dots \dots \dots \dots \dots \dots \dots \dots \dots \dots (8.305)$$

where x_1 and x_2 = locations of the combustion front at times t_1 and t_2, respectively; A = cross-sectional area of the tube; and v_f = average velocity of the combustion front. One may determine v_f by plotting the location of the peak temperature in the combustion tube vs. time. The slope of this graph is the velocity of the combustion front.

The fuel consumed is the sum of the mass of carbon and hydrogen consumed. Applying a material balance to the produced gases, the mass of carbon and hydrogen consumed is given by

$$w_f = (12 + F_{HC})(y_{pCO_2} + y_{pCO}) n_p \quad \dots \dots \dots \dots \dots \dots \dots \dots \dots \dots \dots \dots \dots \dots \dots \dots (8.306)$$

$$\text{and } w_f = (12 + F_{HC})(m+1) y_{pCO} n_p. \quad \dots \dots \dots \dots \dots \dots \dots \dots \dots \dots \dots \dots \dots \dots \dots (8.307)$$

The determination of fuel availability from the analysis of combustion-tube data is illustrated in Example 8.25.

Example 8.25—Computation of Fuel Availability From Results of an In-Situ Combustion Run. A combustion-tube run was carried out with reservoir rock and fluid to determine the fuel availability for an in-situ combustion project (Nelson and McNeil 1961). The combustion tube was 4 in. in diameter and 6 ft long and was packed with formation sand. Porosity of the packed bed was 35%. During the combustion run, the volume of the produced gas was 190 scf (at 60°F and 14.7 psia) measured on a dry basis. **Table 8.79** gives compositions of injected and produced gases. It is assumed that the combustion front propagated through the sandpack at a constant velocity. Determine the fuel availability from the data as lbm fuel/ft³ rock, lbm fuel/100 lbm rock, and lbm fuel/acre-ft of reservoir burned. For this example, the grain density of the sand is 2.65 g/cm³.

Solution. The H/C ratio is computed directly from the analysis of the produced gas. From Eq. 8.302,

$$F_{HC} = \left[106.3 + 2CO - 5.06(CO_2 + CO + O_2) \right] / (CO_2 + CO).$$

Substitution of gas compositions (in mole percent) into Eq. 8.302 gives

$$F_{HC} = \left[106.3 + 2(3.0) - 5.06(11.7 + 3.0 + 1.1) \right] / (11.7 + 3.0)$$
$$- 2.20.$$

The value of m is computed from Eq. 8.288:

Injected air (dry basis) (vol%)	
N_2	79.0
O_2	21.0
Produced gas (dry basis) (vol%)	
N_2	84.2
O_2	1.1
CO_2	11.7
CO	3.0

Table 8.79—Analysis of injected and produced gas, combustion-tube run, Example 8.25.

$$m = V_{CO_2} / V_{CO}$$
$$= 11.7 / 3.0$$
$$= 3.90.$$

The weight of fuel consumed is computed from Eq. 8.307:

$$w_f = (12 + F_{HC})(m+1) y_{pCO} n_p.$$

To use Eq. 8.307, it is necessary to compute the value of n_p from the produced gas data.

The total volume of produced gas (dry basis) is 190 scf. Because volumes are given at standard conditions, it will be necessary to convert volume to moles. The molar volume at 60°F and 14.7 psia is 379 ft³/lbm mol. Thus, during the time interval represented by the data

$$n_p = 190 \, \text{scf} / (379 \, \text{scf/lbm mol})$$
$$= 0.5013 \, \text{lbm mol}.$$

The weight of fuel consumed can be computed from Eq. 8.307, as illustrated in the following.

$$w_f = (12 + 2.20)(3.90 + 1)(0.03)(0.5013)$$
$$= 1.046 \, \text{lbm fuel}.$$

To express fuel consumption in terms of the volume of reservoir rock burned, it is necessary to compute fuel availability in terms of burned volume. For this computation, assume that the entire sandpack was burned during the time the 190 scf of produced gas was measured. Thus,

$$v_b = \pi d_c^2 L / 4$$
$$= \pi (4 / 12 \, \text{ft})^2 (6 \, \text{ft})$$
$$= 0.524 \, \text{ft}^3.$$

In practice, it is preferred to determine the actual burned volume from temperature profiles in the combustion tube during the time the combustion front is moving at a steady velocity. This would give a lower volume than obtained in this example. Fuel availability is computed from Eq. 8.303:

$$m_E = w_f / V_b$$
$$= 1.046 \, \text{lbm} / 0.524 \, \text{ft}^3$$
$$= 2.00 \, \text{lbm/ft}^3.$$

Fuel availability may also be expressed in other units, as discussed earlier. For example, the fuel availability as lbm fuel/100 lbm rock can be computed from the data. First, it is necessary to determine the mass of the reservoir rock in the burned volume.

$$m_R = \rho_r (1 - \phi) V_B$$
$$= (2.65 \, \text{g/cm}^3) \left(\frac{62.4 \, \text{lbm/ft}^3}{\text{g/cm}^3} \right) (1 - 0.35)(0.524 \, \text{ft}^3)$$
$$= 56.28 \, \text{lbm}.$$
$$m_E = 1.046 \, \text{lbm fuel} / 56.28 \, \text{lbm rock}$$
$$= 1.86 \, \text{lbm fuel} / 100 \, \text{lbm rock}.$$

Fuel availability may also be expressed in terms of acre-feet of burned reservoir volume, as indicated in the following.

$$m_E = (2.00 \, \text{lbm fuel})(43,560 \, \text{ft}^3) / (\text{ft}^3 \, \text{rock})(\text{acre-ft})$$
$$= 86,871 \, \text{lbm fuel/acre-ft reservoir burned}.$$

The oxygen usage efficiency for this combustion-tube run may be calculated with Eq. 8.299. From Table 8.79, $y_{pO_2} = 0.011$ and $y_{pN_2} = 0.842$.

$$E_{O_2} = 1 - (0.79/0.21)\left(y_{pO_2}/y_{pN_2}\right)$$
$$= 1 - (0.79/0.21)(0.011/0.842)$$
$$= 0.955.$$

The fuel availability determined from combustion-tube runs, m_E, must be adjusted to reservoir conditions when the porosity of the porous material in the combustion tube is not equal to the porosity of the reservoir rock. Nelson and McNeil (1961) introduced the correction factor given by Eq. 8.308 to account for differences between the reservoir porosity, ϕ_R, and that of the combustion-tube experiment, ϕ_E.

$$m_R = \left[(1-\phi_R)/(1-\phi_E)\right]m_E. \dotfill (8.308)$$

Rearranging Eq. 8.308 gives

$$m_R/(1-\phi_R) = m_E/(1-\phi_E)$$
$$= \left(\text{lbm fuel/ft}^3 \text{ sandpack}\right)\left(\text{ft}^3 \text{ sandpack/ft}^3 \text{ sand grain}\right)$$
$$= \text{lbm fuel/ft}^3 \text{ sand grain.}$$

Thus, this correction factor assumes that the fuel availability per unit volume of sand grain in the matrix is constant.

Several techniques have been developed to determine fuel availability and in-situ combustion reactions that are less complicated than running a combustion tube. A flood-pot technique (Alexander et al. 1962) was developed that uses small samples of reservoir rock. This technique exposes the reservoir rock and crude oil to the same temperature history they would see as the combustion front traveled through a combustion tube. Effects of fluid flow, including solvent extraction, vaporization, and condensation, cannot be simulated in small-sample techniques.

Laboratory data indicate that the amount of fuel formed depends on the type of crude oil, oil gravity, oil saturation, and mineralogy of the rock. Because combustion-tube experiments are tedious, time-consuming, and require special equipment, empirical correlations have been developed between fuel availability and process variables. **Fig. 8.166** (Alexander et al. 1962) shows the variation of fuel availability with oil gravity for the same porous rock. Fuel availability is expressed in lbm carbon/100 lbm rock because it did not appear to correlate with porosity for the same type of reservoir rock. There is considerable scatter in Fig. 8.166, but the trend is correct: fuel availability decreases with increasing oil gravity. Fuel availability for a 10 °API oil is three times the fuel availability for a 30 °API oil. This fact is important to the economics of dry forward combustion because fuel must be consumed for the combustion front to move forward. A similar trend, with much less scatter, was found by Showalter (1963), who conducted a series of combustion-tube runs using the same sand composition (3.6 wt% kaolinite clay). **Fig. 8.167** shows the Showalter correlation; he was unable to attain sustained combustion for the 40 °API oil. Correlations like those in Figs. 8.166 and 8.167 suggest that the data could be extended to other reservoirs and crude oils, thus eliminating the requirement for combustion-tube runs to determine fuel availability. However, experience has shown that this cannot be done.

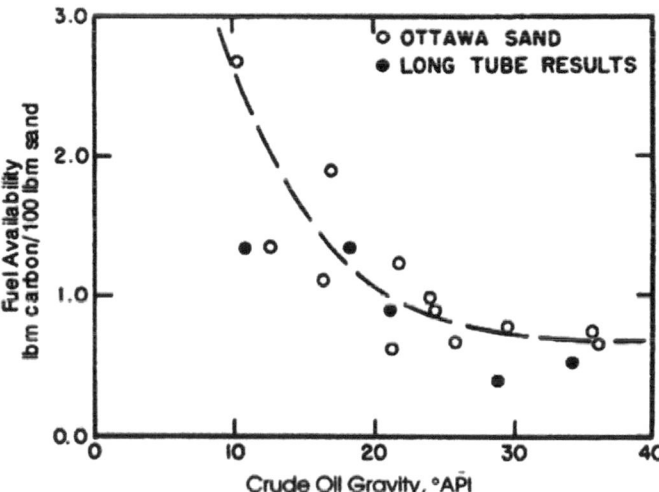

Fig. 8.166—Correlation of fuel burned with crude oil gravity (Alexander et al. 1962).

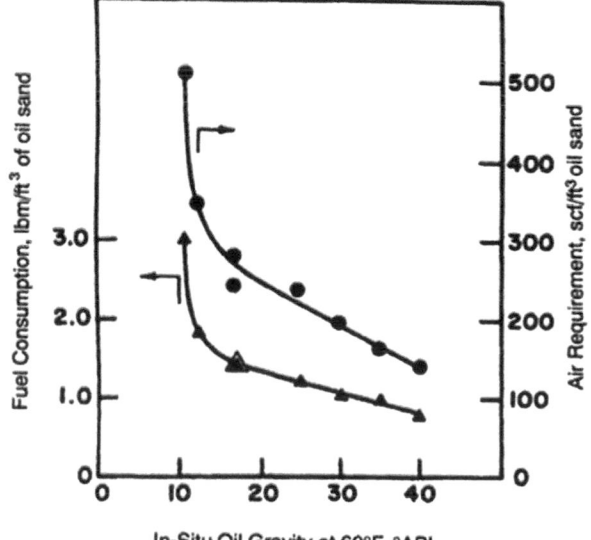

Fig. 8.167—Combustion drive fuel burned and air requirements vs. oil gravity (Showalter 1963).

Fuel availability depends on the porous matrix and the amounts of clay and metallic minerals present in the porous matrix in particular. This was recognized early in in-situ combustion research by Dew and Martin (1965), who pointed out that combustion data must be obtained from reservoir rocks and crude oil. This observation has been verified by several investigations. Vossoughi et al. (1982) demonstrated that small amounts of clay (5 to 15 wt%) were required to obtain a self-sustaining combustion front in sandpacks packed with 35-mesh silica sand. The mineralogy of the May-Libby reservoir rock was considered to be a dominant factor in a successful in-situ combustion project in a reservoir containing 40 °API crude oil (Hardy et al. 1972). Fuel availability may also be determined from the analysis of field data. The procedure is basically the same as that described in Example 8.25 with the exception that it is necessary to estimate the burned volume to compute fuel consumed per unit of rock or burned reservoir volume.

LTO. Combustion-front temperatures are on the order of 650 to 1,000°F when a combustion front is fully developed. When heat losses are large or air fluxes are low, combustion temperatures drop considerably and fuel availability increases. This region is called LTO in contrast to normal combustion (HTO). The importance of oxidation temperature on fuel availability is shown in **Fig. 8.168** (Alexander et al. 1962). Prolonged exposure to combustion temperatures on the order of 300 to 400°F doubled the fuel availability. Abu-Khamsin et al. (1985) discuss the mechanism affecting fuel formation during in-situ combustion.

The fuel formed under conditions of LTO is also quite different in composition from that formed in HTO. **Fig. 8.169** shows the apparent H/C ratio for fuel formed as a function of oxidation temperature. Burger and Sahuquet (1973) provide additional insight into the complex reaction mechanisms involving coke formation. At low temperatures, oxidation of crude oil produces carboxylic acids, aldehydes, and alcohols, which incorporate oxygen into the reaction product instead of CO_2 or CO.

It is possible to assess the extent of LTO in field projects with a method developed by Ramey et al. (1992) if the correct H/C ratio is known for HTO. To use this method, the H/C ratio from a combustion-tube run on native core material is assumed to be the true H/C ratio. The fraction of injected oxygen being consumed in the LTO process when air is injected is given by

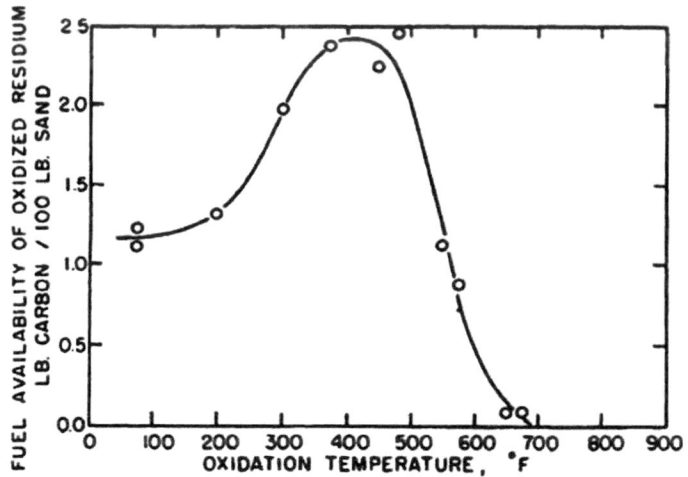

Fig. 8.168—Effect of LTO on fuel burned at 800°F (Alexander et al. 1962).

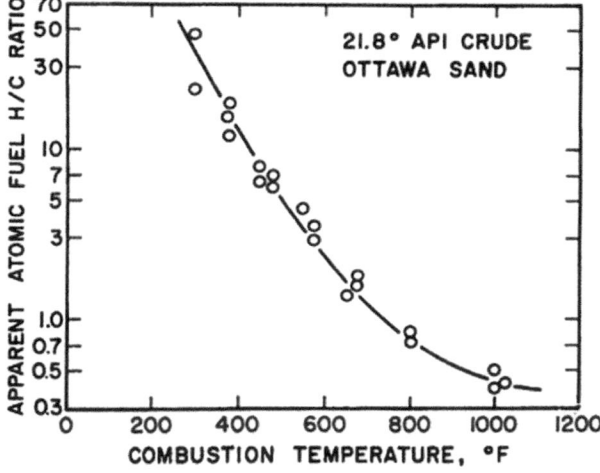

Fig. 8.169—Effect of temperature on apparent H/C ratio of fuel (Alexander et al. 1962).

$$\% O_{2_{LTO}} = \frac{100\left(F_{HC} - F_{HC_{TRUE}}\right)\left(\% CO_2 + \% CO\right)}{4\left(0.266\% N_2 - \% O_2\right)}. \dots\dots\dots\dots\dots\dots\dots\dots\dots\dots\dots\dots\dots\dots (8.309)$$

Ramey et al. (1992) give an example for the South Belridge in-situ combustion project. LTO is a contributor to the extinction of a combustion front, as will be discussed later.

Air Requirement. The combustion front can advance only by consuming fuel. Thus, the air required will be directly proportional to the fuel availability. Oxygen (pure or in a mixture of air) is required to burn fuel. Because air is the common source of oxygen, subsequent discussions and examples will refer to air as the oxygen source. The air requirement for dry forward combustion is defined as the standard volume of air required to burn a unit volume of reservoir. In common oilfield units, the air requirement is defined as

$$a_R = \text{scf/ft}^3 \text{ reservoir volume burned.} \dots\dots\dots\dots\dots\dots\dots\dots\dots\dots\dots\dots\dots\dots\dots (8.310)$$

Combustion Stoichiometry. The air requirement is calculated from combustion stoichiometry from the apparent H/C ratio of the fuel. From Eq. 8.285,

$$m_R = \text{lbm fuel/ft}^3 \text{ reservoir burned.}$$

The fuel has an apparent molecular formula $CH_{F_{HC}}$ and the apparent molecular weight of the fuel is $12 + F_{HC}$. Thus, the moles of fuel burned when 1 ft³ of reservoir rock is burned is given by

$$\text{moles fuel/ft}^3 \text{ reservoir} = m_R / \left(12 + F_{HC}\right). \dots\dots\dots\dots\dots\dots\dots\dots\dots\dots\dots\dots\dots\dots (8.311)$$

According to the combustion stoichiometry in Eq. 8.294, n_{O_2} moles of oxygen are consumed for every mole of fuel consumed. The oxygen requirement is obtained by multiplying Eq. 8.311 by Eq. 8.292. Oxygen is normally supplied by injecting gas that contains y_{iO_2} mole fraction of oxygen. At standard conditions (60°F and 14.7 psia), the molar volume is 379 scf/lbm mol. Thus, the air requirement as scf/ft³ of reservoir volume burned is given by

$$a_R = \frac{379}{y_{iO_2} E_{O_2}} \left(\frac{2m+1}{2m+2} + \frac{F_{HC}}{4}\right)\left(\frac{m_R}{12 + F_{HC}}\right), \dots\dots\dots\dots\dots\dots\dots\dots\dots\dots\dots (8.312)$$

where E_{O_2} = combustion efficiency of oxygen, fraction, and y_{iO_2} is the mole fraction oxygen in the injected gas. For air, y_{iO_2} = 0.21. Fig. 8.122 (Showalter 1963) gives typical values of the air requirement calculated from combustion gas analyses.

Some of the injected air is stored in the burned volume and does not contribute to the combustion efficiency (Prats 1982). This is not a factor when the pressure is low. However, at high pressures, the stored air should be considered. If the burned volume is assumed to be at the injection pressure p_i, then the air requirement at p is approximated by (Naji and Poettmann 1991)

$$a_R^* = a_R + \left(\phi_R / B_{gi}\right), \dots\dots\dots\dots\dots\dots\dots\dots\dots\dots\dots\dots\dots\dots\dots\dots\dots\dots\dots (8.313)$$

where a_R^* = air requirement at p_i, scf/ft³ rock; B_{gi} = FVF for air at injection pressure and the average temperature of the region behind the combustion front, ft³/scf; and p_i = BHP of the air in the injection well, psi. At high pressures, the air requirement is increased considerably by storage in the burned zone.

Material Balance. The air requirement can be determined experimentally during steady-state combustion by measuring the injected air rate and the combustion front velocity. The air requirement is given by

$$a_R^* = u_a / v_f, \dots\dots\dots\dots\dots\dots\dots\dots\dots\dots\dots\dots\dots\dots\dots\dots\dots\dots\dots\dots\dots\dots\dots (8.314)$$

where u_a = injected air flux, scf/ft²-hr = G_i/A.

Note that the air requirement determined from the measured gas-injection rate includes storage in the burned zone. The air requirement computed from stoichiometry should be in reasonable agreement with that determined from combustion stoichiometry when storage in the burned zone is accounted for.

Example 8.26 illustrates computation of the air requirement from combustion-tube data.

Example 8.26—Determination of Air Requirement From a Combustion-Tube Run. Table 8.80 summarizes combustion-tube data for Oil A and Sand S2 (Burger and Sahuquet 1973). Compute the air requirement, in scf/ft³ of reservoir burned, from these data.

Material Balance. The reported air flux density in Table 8.80 during the steady-state part of this combustion run, u_{ai}, is 29.52 scf/ ft²-hr measured at the inlet, and the combustion front velocity was 0.1286 ft/hr. Substituting these values into Eq. 8.314 yields

Run number	A5
Sand	S2
Absolute pressure (bar)	11
Porosity (%)	37.0
Oil saturation (%)	49.0
Oil content (lbm/ft³)	10.1
Air flux density (ft³/ft³-hr)	29.5
Velocity, v_b, of the combustion front (in./hr)	1.54
Velocity, v_c, of the condensation front (in./hr)	1.61
v_c/v_b	1.04
Air requirement, a (scf/ft³)	230
Peak temperature, T_b (°F)	752 to 788
Oil recovery (wt%)	82.5
Average composition of the exhaust gases (vol%)	
O_2	3.0
CO_2	10.0
CO	4.0
Apparent H/C ratio of the fuel	1.96
Fuel burned (lbm/ft³)	1.1
Fuel (wt% of initial oil)	10.7
Oxygen utilization (%)	86

Table 8.80—Summary of dry-combustion test conditions and results (Oil A).

$$a_R^* = u_{ai} / v_f$$

$$= \left(29.52 \text{ scf/ft}^2\text{-hr}\right)\left(\text{hr}/0.1286\text{-ft}\right)$$

$$= 229.6 \text{ scf/ft}^3.$$

Combustion Stoichiometry. Fuel availability is given in SI units, which must be converted to customary oilfield units before the computation is made:

$$m_R = \left(18 \text{ kg/m}^3\right)\left(\text{m}^3/35.315 \text{ ft}^3\right)\left(\text{lbm}/0.454 \text{ kg}\right)$$

$$= 1.123 \text{ lbm fuel/ft}^3 \text{ reservoir burned.}$$

$$m = n_{CO_2}/n_{CO}$$

$$= 10/4$$

$$= 2.5.$$

$$E_{O_2} = 0.86.$$

$$F_{HC} = 1.96.$$

The true air requirement must include the effect of gas storage. In this run, the injection pressure was 159.5 psi and the estimated temperature of the combustion zone was 751.4°F. Because air is essentially ideal at these conditions, Eq. 8.315 can be used to estimate the contribution of gas storage:

$$B_{gi} = \left[\left(T_b + 460\right)/520\right]\left(14.696 / p\right)$$

$$= \left(1,211.4 / 520\right)\left(14.696 / 159.5\right)$$

$$= 0.2146 \text{ ft}^3/\text{scf}. \dotfill (8.315)$$

p (bar)	p (psia)	Average Burned-Zone Temperature					
		170°F			260°F		
		B_{gi} (ft³/scf)	ϕ_R/B_{gi} (scf/ft³)	a_R^* (scf/ft³)	B_{gi} (ft³/scf)	ϕ_R/B_{gi} (scf/ft³)	a_R^* (scf/ft³)
1	14.50	1.219	0.304	227.7	1.393	0.27	227.7
10	145	0.123	3.01	230.4	0.1395	2.65	230.1
20	290	0.0610	6.07	233.5	0.0700	5.39	232.7
40	580	0.0306	12.09	239.5	0.0352	10.51	237.9
60	870	0.0205	18.0	245.4	0.0235	15.7	243.1
80	1160	0.0154	24.0	251.4	0.0178	20.8	248.2
100	1450	0.0124	29.8	257.2	0.0143	25.9	253.3
150	2175	0.0084	44.0	271.4	0.0098	37.8	265.2
200	2900	0.0065	56.9	284.3	0.0075	49.3	276.7
300	4350	0.0041	90.2	317.6	0.0053	69.8	297.2
400	5000	0.0037	100.0	327.4	0.0042	88.1	315.5

$\phi_R = 0.37a$ and $a_R = 227.4$ scf/ft³

Table 8.81—Air requirement at elevated pressures, Example 8.26.

$$a_R^* = a_R + \left(\phi_R / B_{gi}\right)$$
$$= 227.4 + \left(0.36/0.2146\right)$$
$$= 229.1 \text{ scf/ft}^3.$$

From Burger and Sahuquet (1973), the reported air requirement is 230 m³/m³, which is in good agreement.

In Example 8.26, the air requirement owing to gas storage is 1.7 scf/ft³ or 0.7% of the air requirement. **Table 8.81** illustrates the effect of injection pressure on the air requirement determined in Example 8.26 when the average temperatures in the region behind the combustion front are assumed to be 170 and 260°F.

Minimum Air Flux—Extinction of the Combustion Front. A combustion front can be propagated as long as heat is generated at a sufficient rate to compensate for heat losses to over- and underlying formations, to elevate the burned volume from ambient temperature to combustion temperature, and to compensate for heat losses by conduction and convection ahead of the front. Below this burning-front rate, the combustion front will be extinguished.

In laboratory combustion-tube experiments, heat loss to the surroundings is minimized so that the minimum frontal-advance rate required to sustain a combustion front in these experiments approximates a thick burning zone where heat losses above and below the zone are negligible. Combustion zones in reservoirs have finite thickness, and heat losses to the surroundings must be considered. Field experience shows that there is a minimum burning-front advance rate to sustain combustion (Nelson and McNeil 1961). This rate is on the order of 0.125 to 0.5 ft/D for formations 20 to 30 ft thick.

The minimum frontal-advance rate to sustain combustion is normally expressed in terms of a minimum air flux. The air flux, u_{am}, is the volumetric rate of air required per unit cross-sectional area of burning front and is defined by

$$v_{fm} = u_{am}/a_R, \dotfill (8.316)$$

where v_{fm} = minimum burning front velocity, ft/D, and u_{am} = minimum air flux at the combustion front, scf/D-ft². In Example 8.25, the air requirement was 386.5 scf/ft³ of burned reservoir.

If the minimum burning-front velocity is 0.125 ft/D, the minimum air flux can be obtained by rearranging Eq. 8.316:

$$u_{am} = v_{fm}a_R$$
$$= \left(0.125 \text{ ft/D}\right)\left(386.5 \text{ scf/ft}^3 \text{ burned volume}\right)$$
$$= 48.3 \text{ scf/D-ft}^2.$$

The injected air flux would be increased by the amount needed for gas storage in the burned region.

In principle, the minimum frontal-advance rate can be determined with numerical simulators. Chu (1963) studied the propagation of a combustion front in a radial system. The combustion front was assumed to be vertical, and gravity segregation was not considered. A combustion front was initiated at $t = 0$ at the wellbore radius r_b, shown in **Fig. 8.170,** and temperature profiles were simulated as a function of time and radial and vertical position. The combustion front was assumed to extend through the

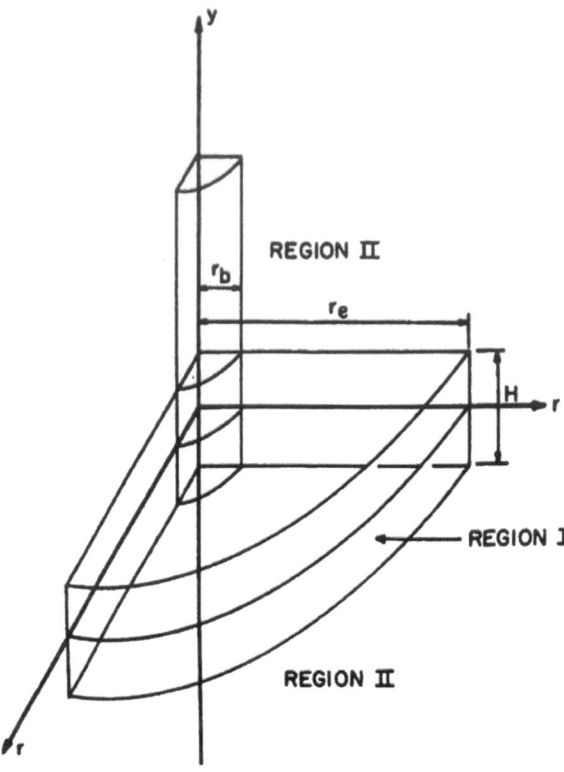

Fig. 8.170—Geometrical model for the reservoir and its bounding formations (Chu 1963).

entire reservoir thickness. In-situ combustion was represented in the simulator by use of a heat-source model, so the kinetics of the combustion reaction were not properly simulated.

Fig. 8.171 shows typical isotherms and corresponding radial and vertical temperature profiles at one instant of time. Reservoir thickness was 36 ft, and the air flux at the wellbore was 1,000 scf/ft sand. A steep temperature profile is present just ahead of the combustion front. Isotherms and temperature profiles show the distribution of the energy generated by in-situ combustion. As the radial combustion front expands, the air flux at the combustion front decreases because the air flux is inversely proportional to the radius of the combustion front. At the same time, heat losses to the over- and underlying formations increase because the combustion zone extends over a larger area. If the injected air flux remains constant, the temperature of the combustion zone will decrease with distance from the injection well. At some radial position, the flux will not be high enough to sustain HTO and the front will begin to extinguish. Because Chu's model did not have reaction kinetics, Chu assumed that HTO could not be maintained if the center-plane peak temperature dropped below 650°F + T_r. This temperature was called the decay temperature, and the radial position where the temperature first decreased below the decay temperature was called the decay point. Although this concept is arbitrary, it serves to illustrate fundamental concepts governing the extinction of the combustion front.

Fig. 8.172 shows the effect of injected air flux on the location of the decay point. Increasing the injected air flux extends the location of the decay point, as expected. At fixed flux,

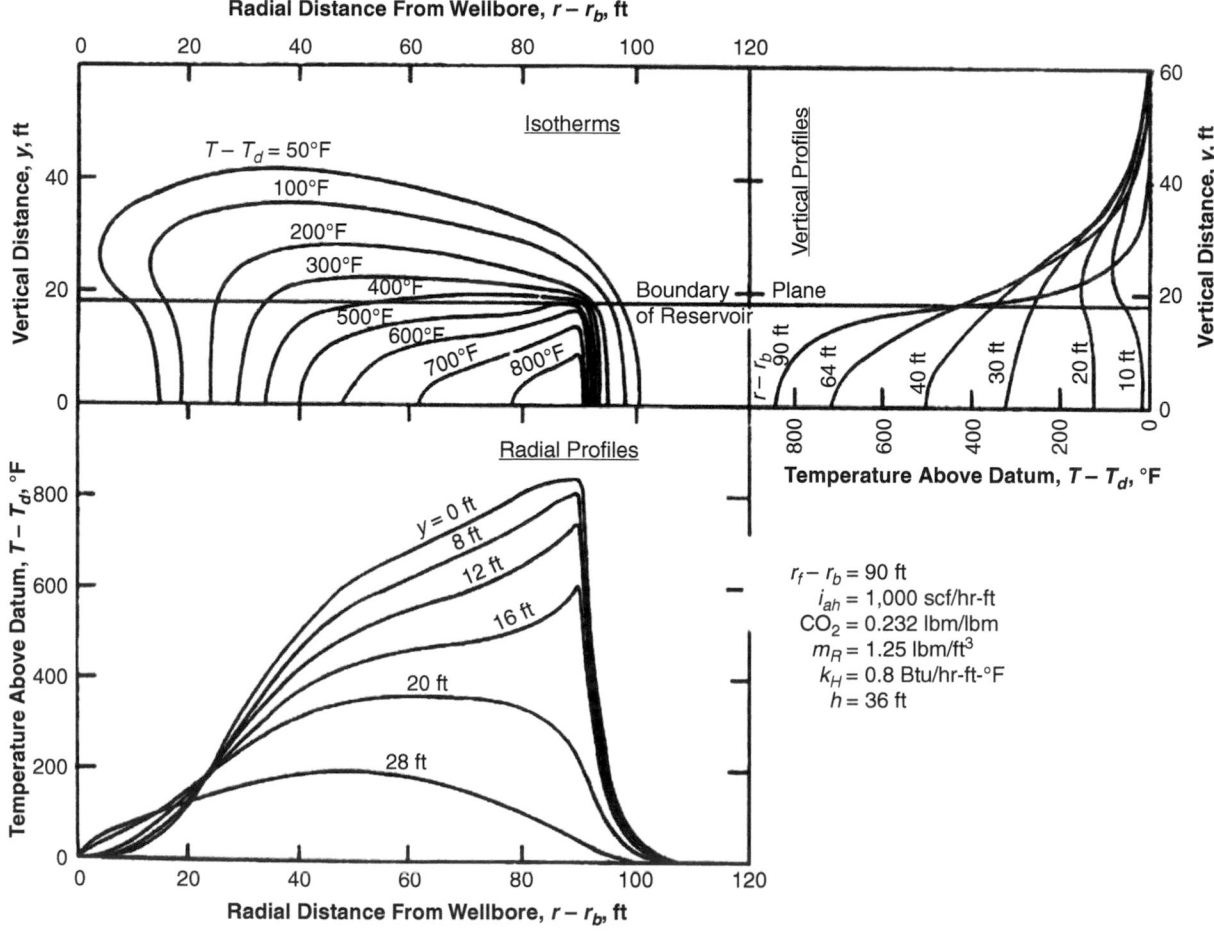

Fig. 8.171—Typical isotherms and corresponding radial and vertical temperature profiles (Chu 1963).

two variables control the location of the decay point: fuel content and thickness of the formation. The location of the decay point is shown as a function of reservoir thickness in **Fig. 8.173** for fuel availabilities of 1.25 and 2.50 lbm/ft³. For the conditions simulated, propagation of a combustion front in a thin reservoir is difficult unless the fuel availability is significantly larger than 1.25 lbm/ft³.

Chu's analysis is not based on kinetics of the combustion process. Consequently, the effects of LTO on the propagation of the combustion front can only be approximated by the onset of the decay point. As noted in the section on LTO, when combustion temperatures decrease to less than 650°F, the nature of the combustion reaction changes significantly. Fuel availability increases markedly, and a fuel that has a higher H/C ratio is deposited. Less heat is released in this process, which contributes to reduction of the temperature and formation of large regions of coke. If the air flux is not adequate to burn this coke in the HTO regime, the combustion front will be extinguished.

The extinguishing of a combustion front was observed in early field tests of in-situ combustion. **Fig. 8.174** (Moss et al. 1959) shows the layout of a five-spot test pattern for an in-situ combustion pilot in southern Oklahoma, USA (White and Moss 1965). **Fig. 8.175** (Moss et al. 1959) shows the estimated distribution of burned and coked sands from that test. Apparently, the air flux in the direction of Well PW-2 was not sufficient to maintain HTO. As **Fig. 8.176** (Moss et al. 1959) shows, an extensive region of coked sand was mapped on the basis of cores taken after the pilot test.

Estimating when extinction will occur is difficult when gravity segregation dominates the displacement process, as observed in thick California reservoirs. When gravity override occurs, the combustion front expands areally and vertically. **Fig. 8.177** shows that fluid flow is complex. In this case, the combustion front extends over a large area and is moving primarily vertically. Because of the large areal extent, lateral heat losses are negligible. The minimum air flux at the combustion front may be estimated by determining the minimum burning rate in an adiabatic combustion tube where heat losses to the surroundings are negligible.

8.7.2 Displacement From Burned Zone. Combustion stoichiometry for HTO may also be used to estimate the volume of oil displaced by the moving combustion front, the volume of water displaced and produced by the combustion process, the air/oil ratio for the burned zone, and the volume of combustion gases produced. However, the rates of fluid production cannot be predicted from stoichiometry.

A combustion front displaces all oil and water that is not consumed by the combustion process. Thus, oil displacement from the burned region can be calculated by a material balance. The material balance requires that the oil displaced equal the oil initially present minus oil burned.

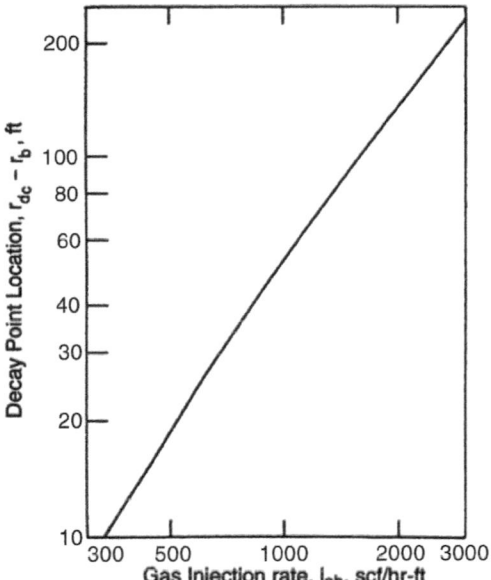

Fig. 8.172—Effect of gas injection rate on decay-point location (Chu 1963).

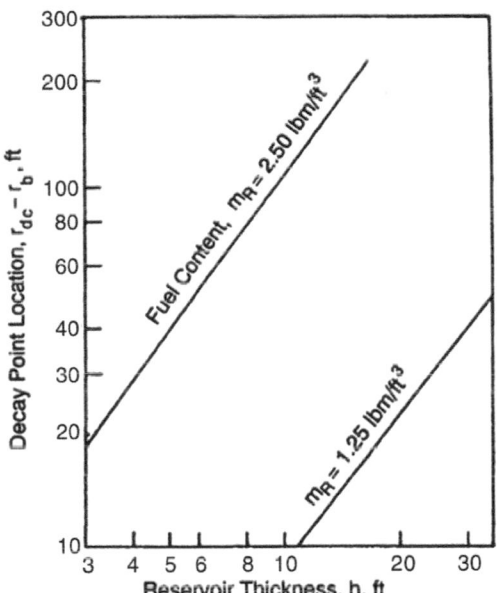

Fig. 8.173—Effect of reservoir thickness on decay-point location (Chu 1963).

$$V_{ob} = \phi V_{Rb} \left(S_{oi} - S_{oF} \right), \quad\quad\quad\quad\quad\quad (8.317)$$

where V_{ob} = oil displaced from the burned volume, ft³; V_{Rb} = bulk volume burned, ft³; S_{oF} = oil saturation equivalent to fuel consumed; and S_{oi} = initial oil saturation. The equivalent oil saturation is given by

$$S_{oF} = m_R / \phi \rho_F, \quad\quad\quad\quad\quad\quad\quad\quad\quad\quad\quad\quad\quad\quad\quad\quad\quad (8.318)$$

where ρ_F = density of fuel, lbm/ft³.

The equivalent oil saturation requires an estimate of the density of the fuel. As discussed earlier, the fuel has a composition that is quite different from that of the original crude oil. Consequently, the density of the fuel is not known accurately. Nelson and McNeil (1961) suggest using a specific gravity of 1.0 (ρ_F = 62.4 lbm/ft³) for the fuel to recognize the change in the crude oil as a result of coking and cracking that occurs during the fuel deposition process. Prats (1982) assumes that the density of the fuel is equal to that of the oil.

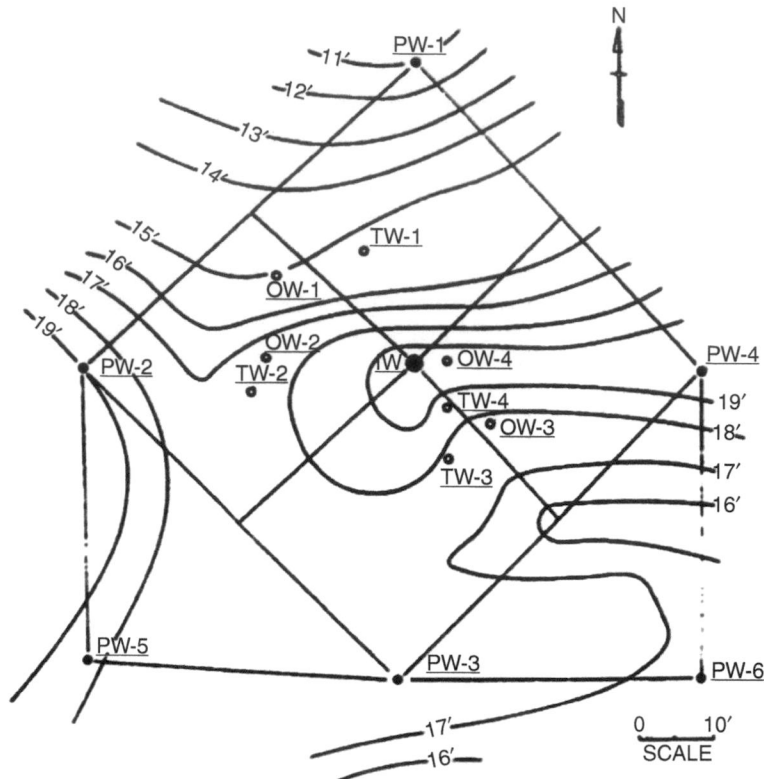

Fig. 8.174—Isopachous map of net oil sand in the five-spot area (Moss et al. 1959).

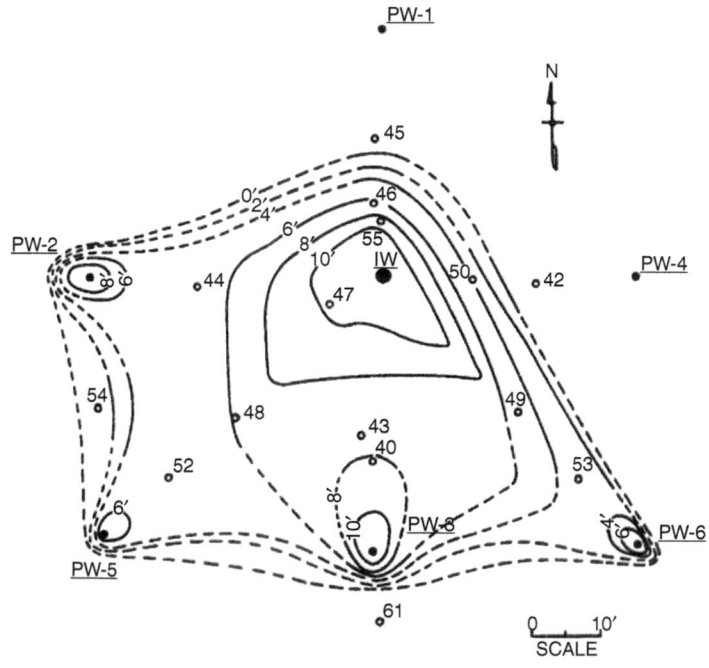

Fig. 8.175—Isopach of burned and coked sand at end of combustion test (Moss et al. 1959).

Water displaced by the in-situ combustion process comes from the initial water saturation and the water produced by the combustion reaction. The volume of water displaced from the burned volume is given by

$$V_{wb} = \phi V_{Rb} \left(S_{iw} - S_{wF} \right), \quad \ldots\ldots\ldots\ldots\ldots\ldots\ldots\ldots\ldots\ldots\ldots\ldots\ldots\ldots\ldots \quad (8.319)$$

where S_{wF} = water saturation equivalent to water produced by the combustion reaction and V_{wb} = water (liquid equivalent) displaced from the burned volume plus water produced in the reaction, ft³.

The water produced by the combustion reaction is derived from the stoichiometry. There is $F_{HC}/2$ mol of water produced per mole of fuel ($CH_{F_{HC}}$) burned. The volume of water produced per cubic foot of reservoir volume burned is given by

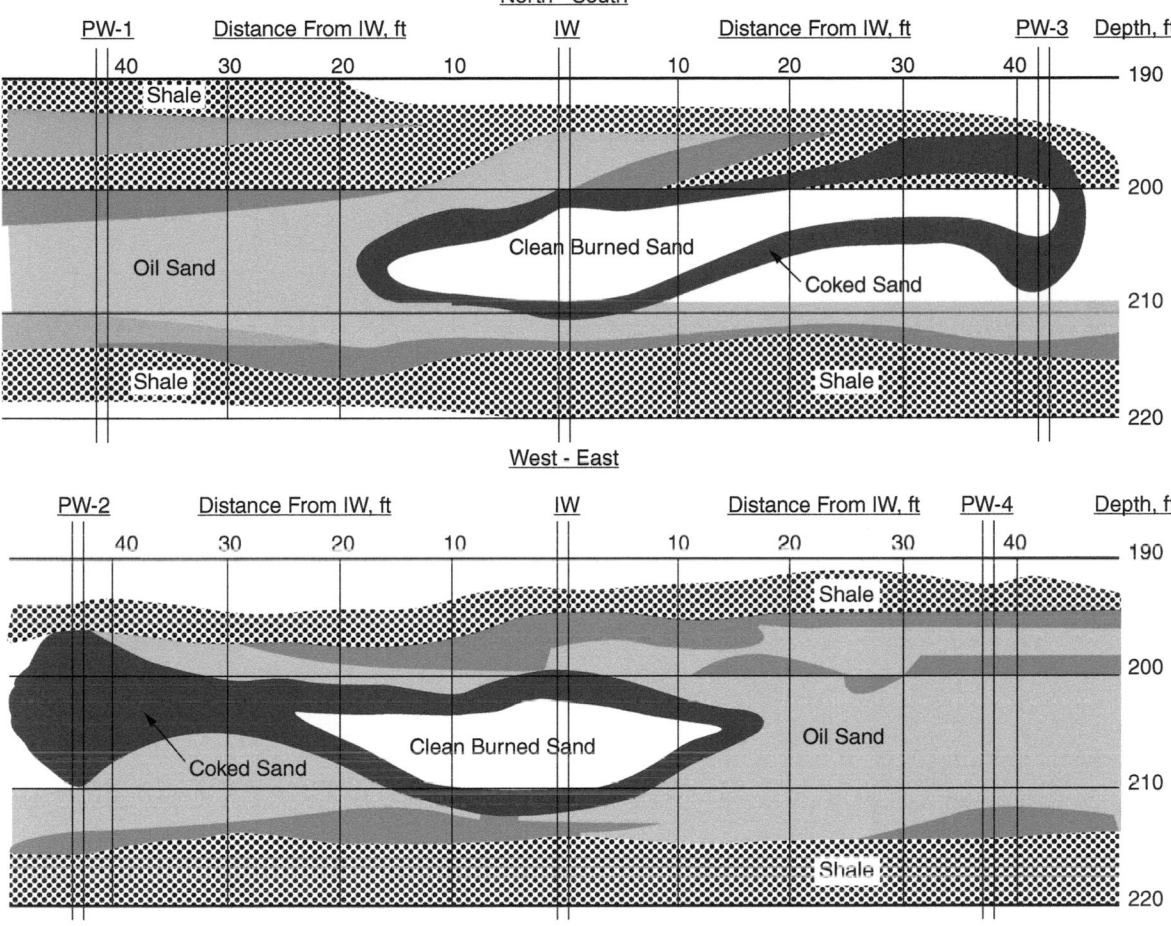

Fig. 8.176—Reservoir cross section after combustion operations (Moss et al. 1959).

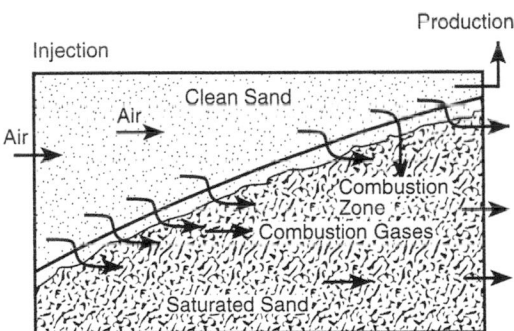

Fig. 8.177—Flow of air and combustion gases, gravity segregation model.

$$V_w = m_R \left(\frac{\text{lbm fuel}}{\text{ft}^3 \ \text{burned}} \right) \left(\frac{1}{12 + F_{HC}} \right) \left(\frac{\text{moles fuel}}{\text{lbm fuel}} \right) \left(\frac{F_{HC}}{2} \right)$$

$$\times \left(\frac{\text{moles H}_2\text{O}}{\text{mole fuel}} \right) \left(\frac{18 \ \text{lbm H}_2\text{O}}{\text{mole water}} \right) \left(\frac{1\,\text{ft}^3 \ \text{H}_2\text{O}}{\rho_w \ \text{lbm}} \right)$$

$$= \frac{9.0 \, m_R F_{HC}}{\rho_w \left(12 + F_{HC} \right)} \frac{\text{ft}^3 \ \text{H}_2\text{O}}{\text{ft}^3 \ \text{reservoir burned}}. \dotfill (8.320)$$

Thus, $S_{wF} = 9.0 m_R F_{HC} / \rho_w \phi \left(12 + F_{HC} \right).$ \dotfill (8.321)

The economics of in-situ combustion is controlled by the cost of air compression. The air/oil ratio is a measure of the effectiveness of the combustion process and can be computed from combustion stoichiometry. Considering only the burned zone, the ratio of the air injected to oil displaced is given by

$$F_{AO_b} = \frac{a_R^*\left(\text{scf/ft}^3 \text{ reservoir burned}\right)\left(5.615\,\text{ft}^3/\text{bbl}\right)}{\phi_R\left(\text{ft}^3\,\text{PV}/\text{ft}^3\text{ reservoir volume}\right)\left(S_{oi}-S_{oF}\right)\left(\text{ft}^3\,\text{oil}/\text{ft}^3\,\text{PV}\right)}, \dotfill (8.322)$$

which simplifies to

$$F_{AO_b} = 5.615\left[a_R^*/\phi_R\left(S_{oi}-S_{oF}\right)\right]\left(\text{scf/bbl}\right). \dotfill (8.323)$$

It is interesting to estimate F_{AOb} for a field project where the parameters in Eq. 8.323 are known. Gates and Ramey (1958, 1980) present data for the South Belridge thermal recovery project. The following values are representative of that project.

$$a_R^* = 385 \ \text{scf/ft}^3,$$
$$m_R = 2.20 \ \text{lbm/ft}^3,$$
$$\phi_R = 0.36,$$
$$\rho_F = 343 \ \text{lbm/bbl},$$
$$S_{oi} = 0.60.$$

From Eq. 8.318, $S_{oF} = 0.10$. Substituting these values into Eq. 8.324 yields

$$F_{AO_b} = 5.615\left[385/0.36\left(0.60-0.10\right)\right]\left(\text{scf/bbl}\right)$$
$$= 12{,}010 \ \text{scf/bbl}.$$

Stoichiometry reveals the volume of oil that is displaced from the burned volume. There is no information on how much of the displaced oil is produced. The actual air/oil ratio will increase if resaturation occurs in the reservoir. Furthermore, as will be discussed in Section 8.7.5, combustion stoichiometry does not account for oil that is displaced from adjacent regions that are either heated by the combustion front or affected by the combustion gases. At South Belridge, the air/oil ratio for the entire project was 3,600 scf/bbl (Ramey et al. 1992). Thus, other mechanisms, such as gravity drainage and gas drive, made a significant contribution to the air/oil ratio observed in the South Belridge project.

The volume of combustion gases is a useful parameter in the evaluation of the operation of a combustion project. Expressions for estimating the volume of combustion gases are derived from the combustion stoichiometry, assuming that the project is operating in the HTO regime. One relationship for computing the volume of combustion gases, expressed as scf/ft^3 burned volume, is developed in this section. The basis for development is Eq. 8.293, which relates the moles of combustion products per mole of fuel consumed. From Eqs. 8.289 through 8.291,

$$\text{CH}_{F_{HC}} + \left(\frac{2m+1}{2m+2}+\frac{F_{HC}}{4}\right)\text{O}_2 \rightarrow \left(\frac{m}{m+1}\right)\text{CO}_2 + \left[1/(m+1)\right]\text{CO} + \left(F_{HC}/2\right)\text{H}_2\text{O}$$

$$n_{pCO_2} + n_{pCO} + n_{pH_2O} = \left(1+\frac{F_{HC}}{2}\right)\left(\frac{\text{moles}}{\text{mole fuel consumed}}\right), \dotfill (8.324)$$

$$n_{pO_2} = \left(1-E_{O_2}\right)\left(\frac{n_{O_2}}{E_{O_2}}\right)\left(\frac{\text{moles}}{\text{mole fuel consumed}}\right), \dotfill (8.325)$$

and $n_{pN_2} = \left(\frac{1-y_{iO_2}}{y_{iO_2}}\right)\left(\frac{n_{O_2}}{E_{O_2}}\right)\left(\frac{\text{moles N}_2}{\text{mole fuel consumed}}\right). \dotfill (8.326)$

Thus, the total amount of combustion gases per mole of fuel consumed is given by

$$n_{pf} = \left[1+\frac{F_{HC}}{2}+\left(\frac{1-y_{iO_2}E_{O_2}}{y_{iO_2}E_{O_2}}\right)n_{O_2}\right]\left(\frac{\text{moles}}{\text{mole fuel consumed}}\right). \dotfill (8.327)$$

Eq. 8.311 gives the fuel consumed per cubic foot of reservoir burned. Multiplying Eq. 8.327 by Eq. 8.311 and converting from moles of gas to standard cubic feet gives the combustion gas volume in scf/ft³ reservoir burned:

$$G_{pf} = \left(\frac{379m_R}{12+F_{HC}}\right)\left[1+\frac{F_{HC}}{2}+\left(\frac{1-y_{iO_2}E_{O_2}}{y_{iO_2}E_{O_2}}\right)n_{O_2}\right] \times \left(\text{scf/ft}^3 \text{ reservoir burned}\right). \quad \dots \dots \dots (8.328)$$

Recall from Eq. 8.292 that

$$n_{O_2} = \left\{[(2m+1)/(2m+2)]+(F_{HC}/4)\right\}.$$

Thus,

$$G_{pf} = \left(\frac{379m_R}{12+F_{HC}}\right)\left[1+\frac{F_{HC}}{2}+\left(\frac{1-y_{iO_2}E_{O_2}}{y_{iO_2}E_{O_2}}\right)\left(\frac{2m+1}{2m+2}+\frac{F_{HC}}{4}\right)\right]. \quad \dots \dots \dots (8.329)$$

Eq. 8.329 may be used to estimate the burned-zone volume from the produced-gas analysis, assuming that all combustion gases are produced. Eq. 8.329 must be modified to account for an increase in gas saturation in the unburned region of the reservoir. Water produced by the combustion reaction condenses and increases the water saturation in the reservoir, as shown in Eq. 8.321. This water is displaced by the combustion front, and most of the water will be produced with the oil bank created by the combustion front. Because the water produced by combustion remains primarily in the liquid phase, the term $F_{HC}/2$ in Eq. 8.329 is deleted in most applications.

Heat Release by In-Situ Combustion. In-situ combustion releases considerable quantities of heat to the reservoir and surrounding formations. The amount of energy released can be estimated from heat of combustion data (Dew and Martin 1965; Burger and Sahuquet 1973). Eq. 8.330 gives the heat of reaction (Prats 1982; Burger and Sahuquet 1972), assuming water produced by the combustion reaction condenses:

$$\Delta h_a = \frac{94.0-67.9m'+31.2F_{HC}}{1-8.5m'+0.25F_{HC}} \text{ Btu/scf air}, \quad \dots \dots \dots (8.330)$$

where $m' = CO(CO + CO_2)$ in the effluent gas.

Eq. 8.330 does not account for the oxygen usage efficiency because $E_{O_2} = 1.0$. **Figs. 8.178 and 8.179** present heat-release data. The parameter $\beta = 1/m$ or $m'/(1-m')$ defines a range of possible operational conditions from no CO ($\beta = 0$) to 100% CO ($\beta = \infty$) in the effluent gases. Over the range of values expected in laboratory and field operations ($0 < \beta < 1$ and $1 < F_{HC} < 2$), a heat release of 100 Btu/scf is a reasonable average value. This is equivalent to an average heat of combustion of 18,000 Btu/lbm fuel burned or 180 scf/lbm of fuel burned (Dew and Martin 1964, 1965).

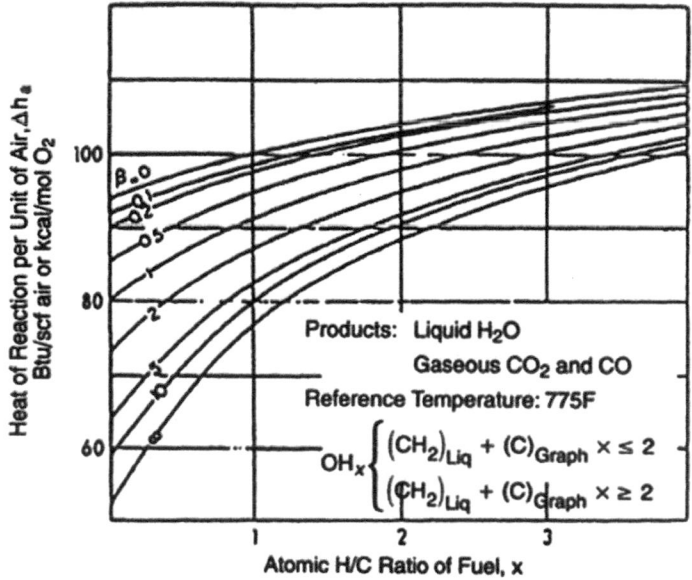

Fig. 8.178—Heat released (kcal/mol O_2 or Btu/scf air) as a function of the H/C ratio, F_{HC}, of the fuel and the CO/CO$_2$ ratio in the produced gases (Burger and Sahuquet 1972).

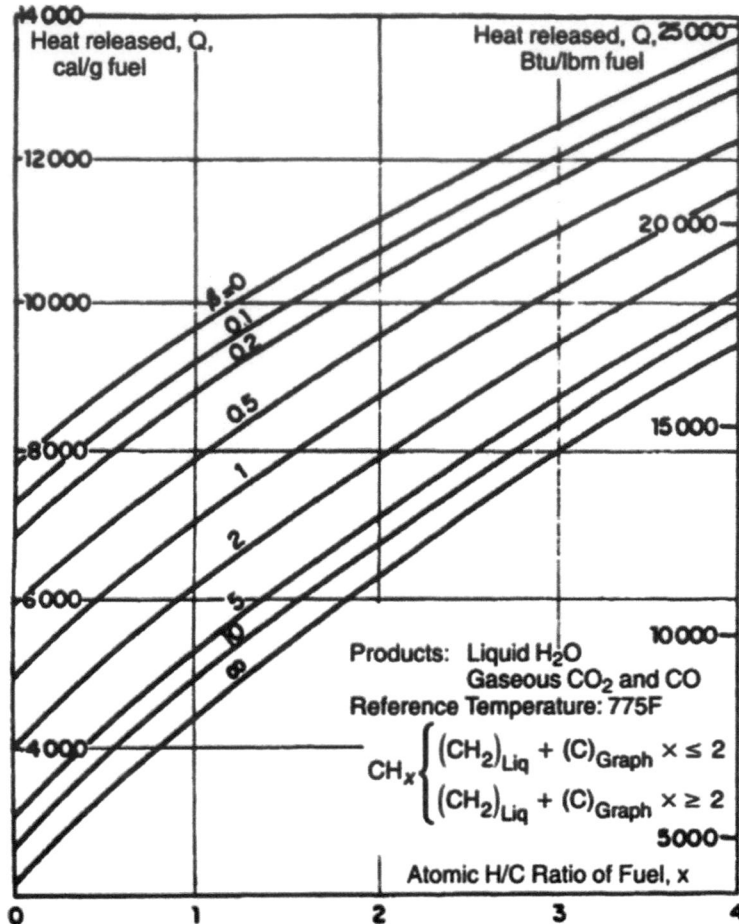

Fig. 8.179—Heat of combustion (cal/g and Btu/lbm CH$_x$) as a function of the H/C ratio, F_{HC}, of the fuel and the CO/CO$_2$ ratio in the produced gases (Burger and Sahuquet 1972).

The heat released by burning fuel is given by (Prats 1982; Burger and Sahuquet 1973)

$$\Delta h_f = \frac{1,800}{12 + F_{HC}}\left(94.0 - 67.9m' + 31.2F_{HC}\right) \text{Btu/lbm fuel.} \quad \dots\dots\dots\dots\dots\dots\dots\dots\dots\dots\dots\dots \quad (8.331)$$

Fig. 8.179 is a graph of Δh_f vs. m' and atomic H/C ratio.

Interpretation of Field Results With Gas Analysis, Injection, and Production Data. The concepts of fuel availability, air requirement, and heat of reaction can be applied to analyze the performance of an in-situ combustion project. In an in-situ combustion field test, the data typically available include (1) gas injection and production rate and (2) gas analyses (CO$_2$, O$_2$, and CO). Evaluating the performance of the combustion project in terms of the following parameters is desired.

1. Average gas analysis.
2. Apparent H/C ratio.
3. Oxygen usage efficiency.
4. Total combustion rate, Mscf/D.
5. Total fuel burned, lbm/D.
6. Heat generation rate, Btu/D.

Application of material- and energy-balance concepts to an in-situ combustion project is illustrated in Example 8.27.

Example 8.27—Interpretation of Field Data, In-Situ Combustion Project. A combustion project is underway in the five-spot pattern depicted in **Fig. 8.180**. The project has been on stream for some time and injection/production rates, as well as gas analyses, are relatively stable. We assume that HTO occurs in this project and that all injected gas not stored in the burned zone passes through the combustion

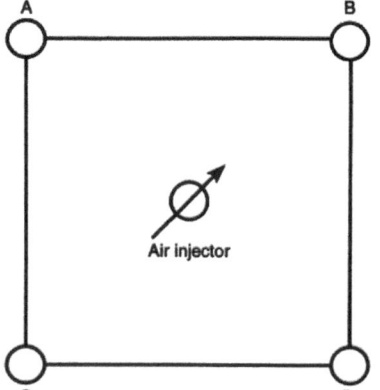

Fig. 8.180—Five-spot pattern for analysis of in-situ combustion performance, Example 8.27.

Well	Rate (Mscf/D)	Gas Analysis (Dry Basis) (%)		
		CO_2	O_2	CO
A	300	12.0	3.0	0.5
B	200	10.0	5.0	0.3
C	100	6.0	10.0	0.1
D	200	8.0	7.0	0.2
E	1,000	0.0	21.0	0.0

Table 8.82—Flow rates and gas analyses, in-situ combustion pilot, Example 8.27.

front. **Table 8.82** gives average data for each well. Find the average gas analysis, apparent H/C ratio, oxygen usage efficiency, combustion rate, total fuel burned, and heat generation rate.

Solution. It is clear from the produced gas rates that the combustion front is not moving at the same rate in all directions and is not radial in shape. This is not unusual because reservoir heterogeneities and dip often determine the shape of the burning zone. All rates are at standard conditions (60°F and 14.7 psia), so a material balance can be performed on each component.

Material Balances.
CO_2 production rate

$$= (0.12)(300) + (0.10)(200) + (0.06)(100) + (0.08)(200)$$
$$= 78\,\text{Mscf/D}.$$

O_2 production rate

$$= (0.03)(300) + (0.05)(200) + (0.10)(100) + (0.07)(200)$$
$$= 43\,\text{Mscf/D}.$$

CO production rate

$$= (0.005)(300) + (0.003)(200) + (0.001)(100) + (0.002)(200)$$
$$= 2.4\,\text{Mscf/D}.$$

Total combustion rate = 800 Mscf/D.

Pattern Average Gas Analysis.

$$CO_2 = (78/800)(100) = 9.750\%$$
$$O_2 = (43/800)(100) = 5.375\%$$
$$CO = (2.4/800)(100) = 0.300\%$$
$$\overline{15.430\%}.$$

Total Combustion Gases.
  $N_2 = 84.57\%.$

Average Apparent H/C Ratio. From Eq. 8.159,

$$F_{HC} = \left[106 + 2CO - 5.06\left(CO_2 + CO + O_2\right)\right] / \left(CO_2 + CO\right)$$
$$= 100 + \left[2(0.3) - 5.06(15.43)\right] / (9.75 + 0.3)$$
$$= 2.84.$$

Recall that F_{HC} is based on the assumption that all oxygen not converted to CO and CO_2 reacts with hydrogen to form water. When HTO is present, the value of F_{HC} is likely to be on the order of 1.5 to 2. The value of 2.84 suggests that some LTO is going on in this pattern.

There is another problem in the analysis of this pattern. If the combustion operation is near steady state, the rate that N_2 is injected must equal the rate that N_2 is produced less the N_2 required for storage as the burned region grows in volume. The effect of gas storage can be estimated if the fuel availability and average injection pressure are known. For this problem, the volume of injected air stored is not considered. In this example, the volume of N_2 injected = (0.79)(100) = 790 scf/D, while the volume of N_2 produced is (0.8457)(800) scf, or 676.6 scf/D. There is a loss of 113.4 scf/D N_2 from the pattern as well as other combustion gases. It is not possible to say much about these gases except to assume that they have the same composition as the produced gases.

Oxygen Consumption Efficiency.

$$E_{O_2} = 1 - (0.79/0.21)(0.0538/0.8457)$$
$$= 76.1\%.$$

Total Combustion Gas Rate. As noted earlier, not all combustion gas is captured. Assume that all injected air passes through the combustion front. A material balance on nitrogen yields

$$N_2 \text{ in injected air} = N_2 \text{ in combustion gas}$$
$$0.79(1{,}000 \text{ Mscf/D}) = 0.8457 \dot{G}_p$$
$$\dot{G}_p = 933 \text{ Mscf/D}.$$

Gases not captured by production wells = 133 Mscf/D. These presumably leave the pattern and migrate to other portions of the reservoir.

Total Fuel Burned. In pounds per day (based on 933 Mscf/D combustion gases),

$$\text{lbm fuel/D} = (\text{lbm carbon/D}) + (\text{lbm H}_2 / D)$$
$$= (\text{lbm carbon/D})[1 + (F_{HC}/12)].$$

$$\frac{\text{lbm carbon}}{D} = \left[\frac{(78 \text{ Mscf/D}) + (2.4 \text{ Mscf/D})}{379 \text{ ft}^3/\text{lbm mol}}\right]$$
$$\times \left(\frac{12 \text{ lbm carbon}}{\text{lbm mol}}\right)\left(\frac{933 \text{ Mscf/D}}{800 \text{ Mscf/D}}\right)$$
$$= 2{,}969 \text{ lbm/D}.$$

$$\text{lbm fuel/D} = 2{,}969[1 + (2.84/12)]$$
$$= 3{,}672 \text{ lbm/D}.$$

Heat Generated. From Eq. 8.330,

$$\Delta h_a = \frac{94.0 - 67.9m' + 31.2F_{HC}}{1 - 0.5m' + 0.25F_{HC}} \text{ Btu/scf air},$$
$$m' = CO/(CO + CO_2)$$
$$= 0.3/(9.75 + 0.3)$$
$$= 0.0299,$$

and
$$\Delta h_a = \frac{94.0 - 67.9(0.0299) + 31.2(2.84)}{1 - 0.5(0.0299) + 0.25(2.84)}$$
$$= 180.58/1.695$$
$$= 106.5 \text{ Btu/scf}.$$

Thus, the heat generated is

$$\dot{Q}_b = (1{,}000 \text{ Mscf/D})(E_{O_2})\Delta h_a$$
$$= (1{,}000 \text{ Mscf/D})(0.744)(106.5 \text{ Btu/scf})$$
$$= 79.3 \times 10^6 \text{ Btu/D}.$$

It is not possible to determine the fuel availability from the analysis of combustion gases in field projects unless additional information on the volume of the burned volume is known. In many in-situ combustion projects, the fuel availability would be known from a combustion-tube run. When these data are available, it is possible to estimate the burned volume and the amount of injected air stored in the burned volume.

8.7.3 Design of Dry Forward In-Situ Combustion Projects. Dry forward in-situ combustion projects can be designed by three methods: the frontal-advance method developed by Nelson and McNeil (1961), the burned-volume method developed by Gates and Ramey (1980), and empirical correlations based on ultimate oil recovery developed by Brigham et al. (1980). There are some common elements in these methods, which are presented before each method is discussed. Methods to estimate compression requirements for air injection are available in other references and are not included in this text.

First, we assume that fuel availability and the air requirement have been determined from a combustion-tube run or other correlation. From combustion stoichiometry, it is possible to estimate several parameters related to the burned volume. The volume of oil displaced by the burning front is given by

$$N_{pb} = \left(7{,}758\,\frac{\text{bbl}}{\text{acre-ft}}\right)\frac{\phi_R V_{Rb}\left(S_o - S_{oF}\right)}{B_o}, \quad \dots\dots\dots\dots\dots\dots\dots\dots\dots\dots\dots\dots\dots \text{(8.332)}$$

where N_{pb} = oil displaced from the burned volume, STB, and V_{rb} = bulk volume of reservoir burned, acre-ft.

The volume of water displaced from the burned volume is given by

$$W_{pb} = \left(7{,}758\ \text{bbl/acre-ft}\right)\phi_R V_{Rb}\left(S_{iw} + S_{wF}\right), \quad \dots\dots\dots\dots\dots\dots\dots\dots\dots\dots\dots\dots \text{(8.333)}$$

where S_{wF} is given by Eq. 8.321 and W_{pb} = water (liquid equivalent) displaced from the burned volume plus water produced in the reaction, bbl.

At this point in design calculations, the models are identical. Major departures come when the effects of the burning zone on oil recovery from the reservoir above, below, and ahead of the combustion front are estimated. Laboratory combustion-tube runs indicate that more oil is displaced than is calculated directly from the burned volume except at 100% burn. This is because of the combined effects of the combustion front, gas drive, steam distillation, and waterdrives. Core analyses from field tests show that substantial oil is displaced (although not always captured) from unburned sections of the reservoir.

Fig. 8.181 shows the relationship between total recovery and volume burned from the Moco-T in-situ combustion project described in Section 8.7.4 (Curtis 1989). A straight line through the origin with a slope of $\phi(S_{oi} - S_{oF})$ would represent oil displacement from the burned zone. Production response from the initiation of in-situ combustion occurs soon after the initiation of combustion and exceeds the amount of oil displaced by the burning front. Thus, Eq. 8.332 underestimates the amount of oil displaced by an in-situ combustion project. The methods of Nelson and McNeil, Gates and Ramey, and Brigham et al. represent empirical approaches to estimate this additional oil displacement.

Nelson and McNeil Model. The Nelson and McNeil model is based on interpretation of field data available at the time the model was developed (circa 1960). On the basis of these data, they assumed that the average recovery efficiency for the unburned portion of the reservoir was 40%. Thus, the total oil and water displaced by the combustion front are given by

$$N_p = \left(7{,}758\ \text{bbl/acre-ft}\right)\phi\left[V_{Rb}\left(S_{oi} - S_{oF}\right) + 0.4\left(V_T - V_{Rb}\right)S_{oi}\right] \dots\dots\dots\dots\dots\dots\dots\dots \text{(8.334)}$$

and $W_p = \left(7{,}758\ \text{bbl/acre-ft}\right)\phi V_{Rb}\left(S_{wi} + S_{wF}\right),$ $\dots\dots\dots\dots\dots\dots\dots\dots\dots\dots\dots\dots\dots \text{(8.335)}$

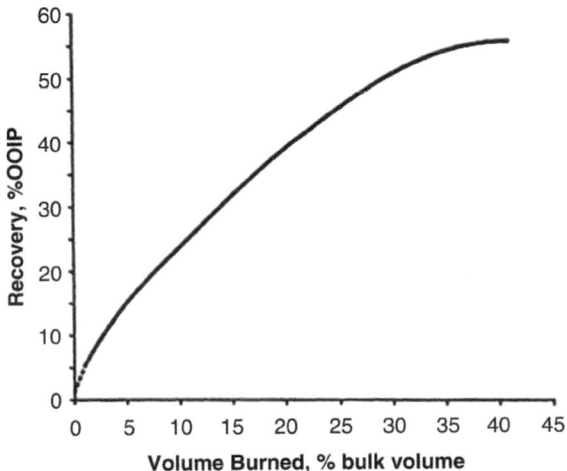

Fig. 8.181—Recovery vs. volume burned from Moco-T in-situ combustion project (Curtis 1989).

where V_T = volume associated with the well pattern, acre-ft. The total amount of air required is determined directly from the burned volume:

$$G_T = a_R^* \left(\frac{\text{scf}}{\text{ft}^3 \, \text{burned reservoir}} \right) \left(\frac{43,560 \, \text{ft}^3}{\text{acre-ft}} \right) V_{Rb} \, (\text{acre-ft})$$

$$= (43,560) a_R^* V_{Rb}. \quad \dots \dots \dots \dots \dots \dots \dots \dots \dots \dots \dots \dots (8.336)$$

Note that, although oil is displaced from the unburned regions of the pattern by combustion gases, there is no water displacement. This implies that the initial water saturation was immobile at reservoir conditions.

The remainder of the Nelson and McNeil method involves estimating the burned volume, the air-injection rate, oil- and water-displacement rates, injection pressure, and compressor requirements. The maximum burned volume is estimated from the volumetric sweep efficiency of the combustion front when the combustion front breaks into a production well. Oil production is assumed to cease at breakthrough. The volumetric sweep efficiency of the combustion front is defined by

$$E_{vb} = E_{Ab} E_{hb}, \quad \dots \dots \dots \dots \dots \dots \dots \dots \dots \dots \dots \dots \dots \dots \dots \dots \dots \dots (8.337)$$

where E_{Ab} = the fraction of the pattern area (A_b/A_p) swept by the combustion front; A_b = maximum area burned within the pattern, acres; A_p = pattern area, acres; and E_{hb} = the vertical sweep efficiency, which is the volumetric sweep efficiency within the burned area A_b. If the combustion zone has an average thickness of $\overline{h}_b$ within the burned area, then

$$E_{hb} = \overline{h}_b / h. \quad \dots \dots \dots \dots \dots \dots \dots \dots \dots \dots \dots \dots \dots \dots \dots \dots \dots \dots \dots (8.338)$$

Thus, $V_{Rb} = E_{vb} V_p. \quad \dots \dots \dots \dots \dots \dots \dots \dots \dots \dots \dots \dots \dots \dots \dots \dots \dots (8.339)$

The Nelson and McNeil method was developed for a five-spot pattern. The areal sweep efficiency of a combustion front in a five-spot pattern is correlated with the constant dimensionless injection rate, as shown in **Table 8.83.** This correlation was developed using potentiometric modeling. The dimensionless injection rate is defined by

$$i_D = (i_a)_{\text{max}} / u_{am} d \overline{h}_b, \quad \dots \dots \dots \dots \dots \dots \dots \dots \dots \dots \dots \dots \dots (8.340)$$

where $(i_a)_{\text{max}}$ = maximum air-injection rate, scf/D; u_{am} = minimum air flux required to sustain the combustion front, scf/(D-ft^2) burning front area; d = distance between the injection and production well in a five-spot pattern; and $\overline{h}_b$ = average burned zone thickness, ft. The average burned thickness must be estimated and cannot be predicted in advance. Table 8.83 shows that it is possible, within limits, to increase the areal sweep efficiency at breakthrough by increasing the injection rate.

For a given areal sweep efficiency, the maximum air rate is obtained by rearranging Eq. 8.340 and solving for $(i_a)_{\text{max}}$:

$$(i_a)_{\text{max}} = i_D u_{am} d \overline{h}_b. \quad \dots \dots \dots \dots \dots \dots \dots \dots \dots \dots \dots \dots \dots (8.341)$$

Oil and water production rates from the expanding burned volume are direct functions of the rate that burned volume changes with time. It is possible to estimate oil and water displacement rates. Eq. 8.342 defines the rate that the burned volume increases, which is directly proportional to the injection rate:

$$\frac{dV_{Rb}}{dt} = \frac{i_a \, (\text{scf/D})}{a_R^* \left(\text{scf/ft}^3 \, \text{reservoir burned} \right) \left(43,560 \, \text{ft}^3 / \text{acre-ft} \right)}$$

$$= (2.296 \times 10^{-5}) (i_a / a_R^*). \quad \dots \dots \dots \dots \dots \dots \dots \dots \dots \dots \dots (8.342)$$

i_D	Areal Sweep Efficiency at Breakthrough (%)
3.39	50.0
4.77	55.0
6.06	57.5
∞	62.6

Table 8.83—Correlation of dimensionless maximum injection rate with areal sweep efficiency at breakthrough.

The oil displacement rate from the burned zone is obtained by differentiating Eq. 8.332 with respect to time and substituting Eq. 8.342 for dV_{Rb}/dt to obtain

$$q_{ob} = (0.1781)\left(i_a\phi/a_R^*\right)\left(S_{oi} - S_{oF}\right). \dotfill (8.343)$$

The oil displacement rate from the unburned region cannot be computed directly from the burned volume rate. Only total oil displaced from the unburned zone at the end of the project and the total air required are determined from the design calculations. Displacement from the unburned region is estimated empirically. Nelson and McNeil assume that oil is displaced from the unburned region as given by

$$F_{oAu} = \left(7{,}758 \text{ bbl/acre-ft}\right)\frac{0.4\phi S_{oi}\left(V_T - V_{Rb}\right)}{G_T}\frac{\text{acre-ft}}{\text{scf}}, \dotfill (8.344)$$

where F_{oAu} = barrels of oil displaced from unburned region per unit volume of air injected, bbl/scf. The rate that oil is displaced from the unburned region is assumed to be given by

$$q_{ou} = F_{oAu}\left(\text{bbl/scf}\right)i_a\left(\text{scf/D}\right)$$

$$= 7{,}758\left[0.4\phi i_a S_{oi}\left(V_T - V_{Rb}\right)/G_T\right]; \dotfill (8.345)$$

thus, $q_o = 0.1781\left(i_a/\phi a_R^*\right)\left(S_{oi} - S_{oF}\right)$

$$+ 7{,}758\left[0.4\phi i_a S_{oi}\left(V_T - V_{Rb}\right)/G_T\right]. \dotfill (8.346)$$

The producing air/oil ratio is defined as the ratio of the air injected to the oil displaced, and is given by

$$F_{AO} = i_a / q_o. \dotfill (8.347)$$

The water displacement rate is obtained by differentiating Eq. 8.335 ($q_w = dW_t/dt$) and substituting Eq. 8.342 for dV_{Rb}/dt and is given by

$$q_w = (0.1781)\left(i_a\phi/a_R^*\right)\left(S_{wi} + S_{wF}\right). \dotfill (8.348)$$

If the air-injection rate is maintained constant at $(i_a)_{max}$, both oil and water displacement rates will be constant. Oil and water rates would not be constant if there was an initial gas saturation because it would be necessary to resaturate some fraction of the reservoir volume initially occupied by gas before oil could be displaced from the region affected by the combustion front.

It is necessary to relate air, oil, and water rates to time for economic evaluation. If the injection rate is maintained at $(i_a)_{max}$, the total time would be computed from the total volume of air required. Because

$$G_T = (i_a)_{max}\left(\text{scf/D}\right)t_{min}\left(\text{days}\right), \dotfill (8.349)$$

$$t_{min} = 43{,}560 a_R^* V_{Rb} / (i_a)_{max}. \dotfill (8.350)$$

Usually, it is not possible to start at the maximum injection rate. Furthermore, Nelson and McNeil point out that scheduling of air compressors for multipattern development precludes running every pattern at the maximum rate during the burn period. They suggest an air-injection schedule based on increasing the rate of injection linearly with time until u_{min} is reached. In this approach, the velocity of the burning front leaving the injection well is v_1; where $v_1 > v_b$, the minimum frontal-advance rate necessary to sustain combustion v_1 is taken to be 0.5 ft/D. Then, the time required for the increasing rate period is given by

$$t_1 = (i_a)_{max}/2\pi\bar{h}_b a_R^* v_1^2. \dotfill (8.351)$$

At the end of the burn, the injection rate is assumed to decline linearly with time, as depicted in **Fig. 8.182.** The injection rate schedule is as follows:

$$0 \leq t_1 \, i_a = (i_a)_{max}\left(t/t_1\right), \dotfill (8.352)$$

$$t_1 \leq t \leq t_2 \, i_a = (i_a)_{max}, \dotfill (8.353)$$

$$\text{and } t_2 < t \leq t_3 \, i_a = (i_a)_{max}\left[\left(t_3 - t\right)/\left(t_3 - t_2\right)\right], \dotfill (8.354)$$

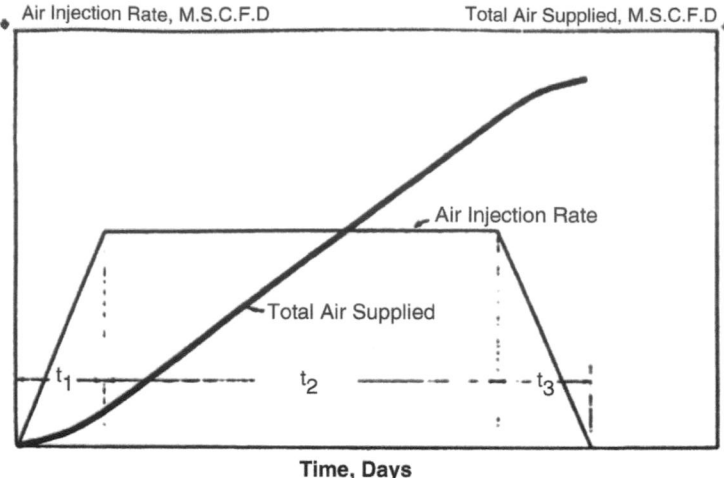

Fig. 8.182—Air requirements for inverted five-spot well pattern (Nelson and McNeil 1961).

where $t_3 = t_2 + t_1$.

The remaining parameter to estimate in the design of an in-situ combustion project is the compressor requirements. It is not possible to fix both the maximum injection rate and the maximum BHP. However, the injection pressure can be estimated for a five-spot pattern if certain assumptions are made. In the Nelson and McNeil method, the maximum injection rate will occur when the combustion front has been propagated to a radial distance of r_1 at an average frontal advance rate of v_1. Eq. 8.355 gives the injection rate in a five-spot pattern as a function of injection pressure; assuming all air passes through the burned zone of average thickness $\overline{h}_b$, flow is radial in the burning zone, flow resistance is negligible in the burned zone, and there are no large changes in the volume of gases as a result of in-situ combustion.

$$\left(i_a\right)_{max} = 0.703 \frac{k_g \overline{h}_b \left(p_i^2 - p_p^2\right)}{\mu_a T_r \left[\ln\left(d^2 / r_w r_f\right) - 1.238\right]}, \dots\dots\dots\dots\dots\dots\dots\dots\dots\dots\dots\dots\dots (8.355)$$

where r_f = radius of combustion front, ft = $v_1 t_1$ when the injection rate is maximum; k_g = effective permeability to air, md; μ_a = viscosity of air at T_r, cp; i_a = air-injection rate, scf/D; r_w = radius of the production well, ft; p_i = BHP of the injection well, psia; p_p = BHP of the production well, psia; and T_r = initial reservoir temperature, °R.

When the injection rate is specified, the pressure is found by solving Eq. 8.355 for p_i:

$$p_i^2 = p_p^2 + \frac{\left(i_a\right)_{max} \mu_a T_r}{0.703 k_g \overline{h}_b}\left(\ln\frac{d^2}{r_w r_f} - 1.238\right)\dots\dots\dots\dots\dots\dots\dots\dots\dots\dots\dots\dots (8.356)$$

Compressor requirements can be determined from standard compressor horsepower charts when the injection rate and desired injection pressure are known.

The Nelson and McNeil correlation is empirical and applies to the type of reservoirs used to develop the correlation (i.e., reservoirs where there is no initial gas saturation that might be resaturated by oil displaced by the combustion front). Prats (1982) has an extensive discussion of the uncertainties associated with this model. Example 8.28 illustrates the use of the Nelson and McNeil method to design an in-situ combustion project.

Example 8.28—Application of Nelson and McNeil Model. An in-situ combustion project is to be evaluated for the field reservoir described in **Table 8.84.** A 30% volumetric sweep of the burned zone when air injection is terminated is desired. Use combustion-rate data for the reservoir rock and oil from Example 8.18.

Solution. The fuel availability and air requirement determined in Example 8.26 were m_R = 1.994 lbm fuel/ft³ reservoir burned, a_R = 386.5 scf/ft³ reservoir burned, and F_{HC} = 2.20.

The pattern volume is 150 acre-ft. Because the burned volume is assumed to be 30% of the pattern volume at completion of the project,

$$V_{Rb} = E_v V_p$$

$$= (0.30)(0.50)$$

$$= 45 \text{ acre-ft.}$$

The equivalent oil saturation consumed as fuel is determined from Eq. 8.318:

Pattern area (acres)	5
Distance between injection and production wells (ft)	330
Formation thickness (ft)	30
Formation temperature (°F)	85
Production well BHP (psia)	14.7
Porosity	0.35
Permeability (md)	500
S_{oi}	0.55
S_{wi}	0.40
Radius of production well (ft)	0.276

Table 8.84—Field data for Example 8.28.

$$S_{oF} = m_R/\phi\rho_F$$
$$= \left(1.994\,\text{lbm/ft}^3\right)\Big/\left[(0.35)\left(62.4\,\text{lbm/ft}^3\right)\right]$$
$$= 0.091,$$

which is approximately 16.5% of the OOIP.

The equivalent water saturation resulting from the combustion process is given by Eq. 8.321:

$$S_{wF} = 9.0 m_R F_{HC}/\rho_w\phi\left(12 + F_{HC}\right)$$
$$= \left[(9)(1.994)(2.20)\right]\Big/\left[(62.4)(0.35)(12 + 2.220)\right]$$
$$= 0.127.$$

Eq. 8.334 gives the total oil displaced by the combustion process:

$$N_p = (7{,}758\text{ bbl/acre ft})\phi\left[V_{Rb}\left(S_{oi} - S_{oF}\right) + 0.4\left(V_I - V_{Rb}\right)S_{oi}\right]$$
$$= (7{,}758\text{ bbl/acre-ft})(0.35)\left[(45\text{ acre-ft})(0.55 - 0.091)\right.$$
$$\left. + 0.4(150 - 45)(0.55)\text{ acre-ft}\right]$$
$$= (7{,}758\text{ bbl/acre-ft})(0.35)(20.66 + 23.1)$$
$$= 118{,}822\text{ bbl.}$$

Total water displaced by the combustion process is obtained from Eq. 8.335:

$$W_p = (7{,}758\text{ bbl/acre-ft})\phi V_{Rb}\left(S_{wi} + S_{wF}\right)$$
$$= (7{,}758\text{ bbl/acre-ft})(0.35)(45\text{ acre-ft})(0.40 + 0.13)$$
$$= 64{,}760\text{ bbl.}$$

The maximum amount of air injected is computed from Eq. 8.336:

$$G_r = 43{,}560 a_R^* V_{Rb}$$
$$= (43{,}560)\left(386.5\text{ scf/ft}^3\text{ burned reservoir}\right)(45\text{ acre-ft})$$
$$= 757.6 \times 10^6\text{ scf.}$$

Referring to Table 8.79, there are several values of E_{Ab} that could be chosen. In practice, each case could be analyzed for economic evaluation to determine which case gave the best economics. For this example, E_{Ab} is assumed to be 0.55 and the value of $i_D = 4.77$. Because $E_{vb} = 0.30$, the vertical sweep efficiency is computed from Eq. 8.337:

$$E_{hb} = E_{vb}/E_{Ab}$$
$$= 0.30 / 0.55$$
$$= 0.545.$$

Referring to Eq. 8.338, this means that, on the average, slightly more than one-half of the thickness will be burned within the area covered by the combustion front. Thus,

$$\bar{h}_b = 0.545h$$
$$= 16.4 \, \text{ft}.$$

There is no way to estimate how the thickness varies within the burned volume. It is likely that the burned thickness approaches h near the injection well when sand thickness is on the order of 20 to 30 ft (Nelson and McNeil 1961).

When $E_{hb} < 1.0$, it is necessary to consider what the effective thickness should be in computing the maximum injection rate. Nelson and McNeil used the total thickness of the formation. This can be valid only when $E_{hb} = 1.0$. Thus, for design purposes, h_b is chosen here as the effective thickness in computing $(i_a)_{max}$ and p_i.

The maximum injection rate is given by Eq. 8.341:

$$\left(i_a\right)_{max} = i_D u_a d\bar{h}_b.$$

From Eq. 8.316, assuming that $v_b = 0.125$ ft/D,

$$u_a = v_f a_R^*$$
$$= (0.125 \, \text{ft/D})(386.5 \, \text{scf/ft}^3 \, \text{reservoir burned})$$
$$= 48.3 \, \text{scf/ft}^2\text{-D}.$$

Then, $\left(i_a\right)_{max} = (4.77)(48.3 \, \text{scf/ft}^2\text{-D})(330 \, \text{ft})(16.4 \, \text{ft})$

$$= 1.25 \times 10^6 \, \text{scf/D}.$$

For a constant injection rate, the total time required to burn the pattern would be

$$t_{min} = G_T / \left(i_a\right)_{max}$$
$$= 757.6 \times 10^6 \, \text{scf}/1.25 \times 10^6 \, \text{scf/D}$$
$$= 606 \, \text{days}.$$

For a multipattern burn, the rate schedule shown in Fig. 8.181 will be used assuming $v_1 = 0.5$ ft/D. From Eq. 8.350,

$$t_1 = \left(i_a\right)_{max} / 2\pi \bar{h}_b a_R^* v_1^2$$

$$= \frac{1.25 \times 10^6 \, \text{scf/D}}{2\pi (16.4 \, \text{ft})(386.5 \, \text{scf/ft}^3 \, \text{reservoir burned})(0.5 \, \text{ft/D})^2}$$

$$= 125.5 \, \text{days}.$$

Then, $G_T = \left(i_a\right)_{max}\left(t_1 / 2\right) + \left(t_2 - t_1\right)\left(i_a\right)_{max} + \left(i_a\right)_{max}\left(t_1 / 2\right)$

$$= \left(i_a\right)_{max} t_2,$$

$$t_2 = 757.6 \times 10^6 / 1.25 \times 10^6$$
$$= 606 \, \text{days},$$

and $t_3 = 731.5$ days. Thus, the injection rate schedule is as follows. For $0 < t \le 125.5$ days,

$$i_a = 1.25 \times 10^6 \left(t / 125.5\right) \text{scf/D};$$

for $125.5 \le t \le 606$ days,

$i_a = 1.25 \times 10^6 \, \text{scf/D};$

and for $606 \le t \le 731.5$ days,

$$i_a = 1.25 \times 10^6 \, \text{scf/D} \left(\frac{t - 606}{731.5 - 606} \right).$$

During the period of constant air-injection rate, the oil displacement rate is obtained from Eq. 8.346:

$$
\begin{aligned}
q_o &= 0.1781 \frac{i_a \phi}{a_R} \left(S_{oi} - S_{oF} \right) + 7{,}758 \frac{0.4 \phi i_a S_{oi} \left(V_p - V_{Rb} \right)}{G_T} \\
&= (0.1781)\left(1.25 \times 10^6\right)(0.35)(0.55 - 0.091) / 386.5 \\
&\quad + \frac{(7{,}758)(0.4)(0.35)\left(1.25 \times 10^6\right)(0.55)(150 - 15)}{757 \times 10^6 \, \text{scf}} \\
&= 92.54 + 103.49 \\
&= 196 \, \text{B/D}.
\end{aligned}
$$

The water displacement rate is obtained from Eq. 8.348:

$$
\begin{aligned}
q_w &= 0.178 \left(i_a \phi / a_R \right) \left(S_{wi} + S_{wF} \right) \\
&= (0.178) \left[\left(1.25 \times 10^6 \right)(0.35) / 386.5 \right] (0.40 + 0.13) \\
&= 106.85 \, \text{B/D}.
\end{aligned}
$$

The cumulative air/oil ratio is

$$
\begin{aligned}
F_{AO} &= 757.6 \times 10^6 \, \text{scf} / 118{,}822 \, \text{bbl} \\
&= 6{,}376 \, \text{scf/bbl}.
\end{aligned}
$$

Finally, the BHP in the injection well must be computed to estimate compressor requirements. From Eq. 8.356,

$$p_i^2 = p_p^2 + \left[\left(i_a \right)_{\max} \mu_a T_r / 0.703 k_g \overline{h_b} \right] \left[\ln \left(d^2 / r_w r_f \right) - 1.238 \right].$$

The air permeability is assumed to be 5% of the absolute permeability, and the air viscosity is 0.0186 cp at 85°E. Because $t_1 = 125.5$ days, $r_f = (0.5 \, \text{ft/D})(125.5 \, \text{days}) = 62.75$ days. The producing well is assumed to be pumped off so that $p_p = 14.7$ psia. Substituting parameters into Eq. 8.356 yields

$$
\begin{aligned}
p_i^2 &= (14.7)^2 + \frac{\left(1.25 \times 10^6\right)(0.0186)(460 + 85)}{(0.703)(0.05)(500)(16.4)} \\
&\quad \times \left\{ \ln \left[(330)^2 / (0.276)(62.75) \right] - 1.238 \right\} \\
&= 216 + 330{,}087 \\
&= 330{,}302
\end{aligned}
$$

and $p_i = 575$ psia.

This example maintains consistency of material balances throughout and assumes that all the air passes through the combustion zone, where combustion efficiency is identical to that determined in the combustion-tube runs.

In designing in-situ combustion projects, it is often necessary to consider the possibility of air bypassing the combustion front. Nelson and McNeil incorporated this factor into their analysis by assuming that the volumetric sweep efficiency at breakthrough of the combustion front was 62.6% with 100% vertical sweep efficiency. This assumption is inconsistent with Table 8.83. The effect of this assumption is to increase the volume of gas required by a factor of 0.626/0.30 = 2.087 in this example. Injection rates are computed from the correlation in Table 8.83, assuming that the total thickness is burned. Thus, $(i_a)_{\max} = 1.25 \times 10^6 (30/16.4) = 2.29 \times 10^6$ scf/D and 45% of the injected air is not burned in the combustion zone.

Curtis (1989) illustrates how the Nelson and McNeil method was adjusted to obtain a history match of the oil production from burned and unburned regions in the Moco-T in-situ combustion project, which is discussed in Section 8.7.4.

Gates and Ramey Method. The Gates and Ramey method is based on interpretation of laboratory combustion-rate data and field results from the South Belridge thermal recovery experiment (Gates and Ramey 1980). The method uses a correlation

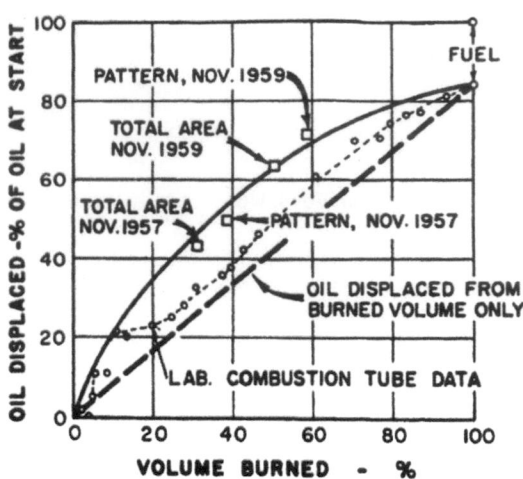

Fig. 8.183—Correlation of oil displaced with volume burned (Gates and Ramey 1980).

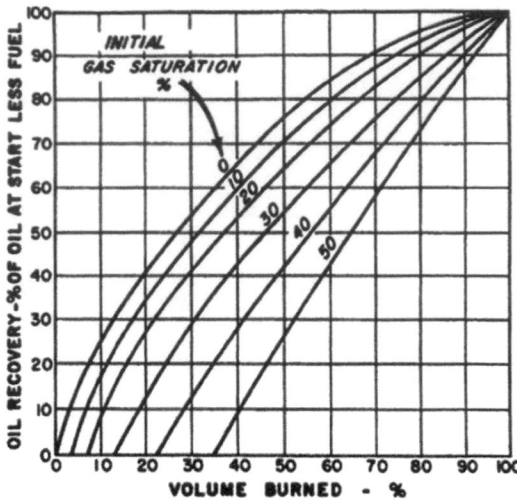

Fig. 8.184—Estimated oil recovery vs. volume burned (Gates and Ramey 1980).

of oil recovery and volume of reservoir burned as the design basis. **Fig. 8.183** illustrates such a correlation from the South Belridge thermal recovery experiment. The dashed line in Fig. 8.183 represents the oil displaced from the burned volume by the combustion front and is a linear function of burned volume. The solid curved line is a correlation of field data, which illustrates that much more oil was displaced than computed from the burned volume. This is analogous to the assumption of Nelson and McNeil that at the end of an in-situ combustion project, 40% of the oil was displaced from unburned portions of the pattern. Also shown on Fig. 8.183 is a dashed curve representing the oil displaced as a function of burned volume from the combustion-tube run. The rapid rise in initial oil displacement and the endpoint oil recovery are believed to be representative of field performance.

Combustion-tube runs carried out over a range of initial gas saturations demonstrated that oil recovery can be correlated with volume burned by use of the empirical relationship shown in **Fig. 8.184.** The effect of an initial gas saturation is to delay oil recovery because of resaturation of the gas region. Furthermore, less oil is recovered with increasing initial gas saturation because the initial oil saturation is lower. From Fig. 8.184, oil recovery at any initial gas saturation can be estimated as a function of V_{Rb}.

Fassihi et al. (1981) developed a set of equations that represents the oil-recovery/burned-volume relationships in Fig. 8.184. These equations are

$$N_{p1} = 100x_R + y_R \left(26.8229 - 0.4678 S_{gi} \right), \dotfill (8.357)$$

where N_{p1} = oil recovered, initial oil less fuel, %;

$$x_R = \left(V_{Rb}^* - V_{Rbo}^* \right) / \left(100 - V_{Rbo}^* \right), \dotfill (8.358)$$

where $V_{Rb}^* = 0.14714 S_{gi} + 0.01071 S_{qo}^2 \dotfill (8.359)$

and S_{gi} = initial gas saturation; and

$$y_R = 6.77526x_R - 15.947794x_R^2 + 16.187187x_R^3 - 7.014659x_R^4. \dotfill (8.360)$$

Thus,

$$N_p = \left(N_{p1} / 100 \right) \underbrace{V_T \left[7,758\phi \left(S_{oi} - S_{oF} \right) \right]}_{\text{ultimate oil recovery, bbl.acre-ft}}. \dotfill (8.361)$$

In principle, the rate of oil displacement can be obtained from Fig. 8.184 or the corresponding correlations when the air-injection rate is known.

$$q_o = \left(\frac{dN_{p1}}{dV_{Rb}^*} \right) \left(\frac{dV_{Rb}^*}{dt} \right) \frac{V_T}{100} \left[7,758\phi \left(S_{oi} - S_{oF} \right) \right]. \dotfill (8.362)$$

The derivative dN_{p1}/dV^*_{Rb} is the slope of the oil-displacement/burned-volume curve in Fig. 8.184 at constant S_{gi} and a particular value of V^*_{Rb}. Note that

$$dN_{p1}/dV^*_{Rb} = \left(dN_{p1}/dx_R\right)\left(dx_R/dV^*_{Rb}\right) \dotfill (8.363)$$

and $dN_{p1}/dx_R = 100 + \left(26.8229 - 0.4678 S_{gi}\right)\left(dy_R/dx_R\right).$ $\dotfill (8.364)$

From Eq. 8.358,

$$dx_R/dV^*_{Rb} = 1/\left(100 - V^*_{Rbo}\right). \dotfill (8.365)$$

Eq. 8.360 yields

$$dy_R/dx_R = 6.775267 - 31.8965 x_R + 48.562 x_R^2 - 28.059 x_R^3. \dotfill (8.366)$$

Recall from Eq. 8.342 that

$$dV_{Rb} / dt = \left(2.296 \times 10^{-5}\right)\left(i_a/a^*_R\right)\left(\text{acre-ft burned/D}\right),$$

so that

$$dV^*_{Rb} / dt = \left(2.296 \times 10^{-5}\right)\left[\frac{i_a(100)}{a^*_R V_T}\right]\frac{\text{acre-ft burned/D}}{\text{acre-ft pattern volume}}.$$

Therefore,

$$q_o = 2.296 \times 10^{-5}\left(dN_{p1}/dV^*_{Rb}\right)\left(i_a/a_R\right)\left[7,758\phi\left(S_{oi} - S_{oF}\right)\right]. \dotfill (8.367)$$

Eq. 8.367 overestimates the oil rate when dN_{p1}/dV^*_{Rb} is evaluated from Eqs. 8.364 through 8.366, possibly because the empirical curve-fitting functions represent the values of the parameters reasonably well at each point but not the derivatives. Therefore, an alternative method is preferred to compute oil rates. The displacement calculation is subdivided into increments of V_{Rb}, and the cumulative oil production is computed at each value of V^*_{Rb}. Total injection time is computed for each V^*_{Rb}. Then, the average oil production rate is computed for the time interval by dividing the incremental oil production by the time interval:

$$\bar{q}_o^{n+1} = \left(N_p^{n+1} - N_p^n\right)/\left(t^{n+1} - t^n\right), \dotfill (8.368)$$

where it is understood that $\bar{q}_o^{n+1}$ = average oil displacement rate for the time interval $(t_n + 1 - t_n)$. The water rate is computed from Eq. 8.348:

$$q_w = 0.1781\left(i_a\phi/a^*_R\right)\left(S_{wi} + S_{wF}\right).$$

The Gates and Ramey model accounts for oil displacement after the arrival of the combustion front at a production well. This is one major difference from the Nelson and McNeil model, where oil displacement is completed at combustion-front breakthrough. Air that bypasses the combustion front is called excess air. Excess air is determined from the analysis of combustion gases as outlined next.

At low pressures, the volume fraction of O_2 and N_2 are equal to their mole fractions. Let y_{pO_2} = volume fraction of O_2 in the combustion gases produced from the pattern; y_{pN_2} = volume fraction of N_2 in the combustion gases from the pattern; and $\dot{G}_p$ = rate of gas production, scf/D.

$$\begin{aligned}\text{percent excess air} &= \frac{\text{produced oxygen}}{\text{oxygen consumed}}(100)\\[6pt] &= \frac{\text{unused air}}{\text{air used to burn fuel}}\\[6pt] &= \frac{y_{pO_2}\dot{G}_p}{0.21 i_a - y_{pO_2}\dot{G}_p}(100). \dotfill (8.369)\end{aligned}$$

When all the injected nitrogen is produced, $y_{pN_2}\dot{G}_p = 0.79 i_a.$

$$\text{percent excess air} = \frac{0.79\left(y_{pO_2} / y_{pN_2}\right)(100)}{0.21 - 0.79\left(y_{pO_2} / y_{pN_2}\right)} \quad\dots\dots\dots\dots\dots\dots\dots\dots\dots\dots\dots\dots\dots\dots\dots (8.370)$$

When injection pressures are high, it is necessary to account for the volume of air that is stored in the burned volume.

Excess air affects the air requirement as well as the air/oil ratio. Gates and Ramey correlated excess air with the percentage of the oil recovered from the pattern in **Fig. 8.185** on the basis of data from the South Belridge thermal recovery experiment. Eq. 8.371 is the corresponding correlation developed by Fassihi et al. (1981)

$$\text{Excess air} = \left[0.9\left(N_{p1}\right) - 15.85\right]/100, \quad\dots\dots\dots\dots\dots\dots\dots\dots\dots\dots\dots\dots\dots\dots\dots (8.371)$$

where $N_{p1} > 17.61$.

Example 8.29 illustrates the application of the Gates and Ramey (1958) method to estimate the performance of the South Belridge thermal recovery project.

Example 8.29—Application of the Gates and Ramey Method. The South Belridge thermal recovery project was a cooperative field test of in-situ combustion that began in 1956. **Table 8.85** presents reservoir properties at the start of the test. The test pattern was a 2.5-acre inverted five-spot. Distance from the injection well to each of the four production wells was 233 ft. The fuel availability, based primarily on interpolation of field data, is 280 bbl/acre-ft, while the combustion-tube data indicated a fuel availability of 262 bbl/acre-ft. A value of 280 bbl/acre-ft is used in this example. The apparent atomic H/C ratio in the fuel, F_{HC}, is estimated to be 1.6 (Alexander et al. 1962), and the density of the fuel is taken as 343 lbm/bbl. The initial injection rate was 1×10^6 scf/D, increasing linearly to 3.5×10^6 scf/D over a period of 7 months and held constant thereafter. Estimate the oil production as a function of time using the Gates and Ramey method.

Solution. The volume of the pattern can be determined from the data in Table 8.85. Note that the initial gas saturation is 3%.

$$V_T = (141,000 + 87,000) / (0.97)(0.36)$$
$$= 652,921 \text{ bbl}$$
$$= 84.2 \text{ acre-ft.}$$

For this example, the design calculation will be done on the basis of oil content per acre-ft. Thus, the initial oil content is

$$(7,758 \text{ bbl acre-ft})\phi S_{oi} = (7,758)(0.36)(0.60)$$
$$= 1,676 \text{ bbl/acre-ft.}$$

The oil burned as fuel is given as 280 bbl/acre-ft, so that the oil less fuel is

$$\frac{\text{Initial oil} - \text{fuel}}{\text{acre-ft}} = \frac{1,626 \text{ bbl}}{\text{acre-ft}} - \frac{280 \text{ bbl}}{\text{acre-ft}}$$
$$= 1,396 \text{ bbl/acre-ft.}$$

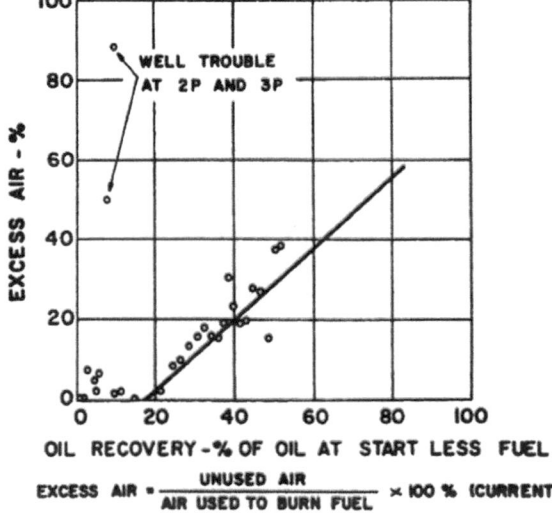

Fig. 8.185—Excess air vs. oil recovery, South Belridge thermal recovery experiment (Gates and Ramey 1980).

Table 8.85—Reservoir properties at start of South Belridge thermal recovery project.	
Porosity (%)	36
Oil saturation (%)	60
Water saturation (%)	37
Gas saturation (%)	3
Total oil in test pattern* (bbl)	141,000
Total water in test pattern* (bbl)	87,000
Reservoir temperature (°F)	87
Oil gravity (°API)	12.9
Oil viscosity at 87°F (cp)	2,700
Air permeability (darcies)	8
*Test Patterns 1P, 4P, 6P, and 5P.	

The oil burned is equivalent to $S_{oF} = 0.10$ and $m_R = 2.20$ lbm/ft^3 of reservoir burned.

S_{wF} is computed from Eq. 8.321:

$$S_{wF} = 9.0 m_R F_{HC} / \left[\rho_w \phi \left(12 + F_{HC} \right) \right]$$

$$= 0.104.$$

The air requirement was estimated to be 175 scf/lbm fuel (Showalter 1963), which is equivalent to 385 scf/ft^3 of reservoir burned. Because the injection pressure was 330 psi (Gates and Ramey 1958), the correction for gas storage is small and we assume that $a_R = a_R^*$. The air required per acre-foot of volume burned (in Mcf/acre-ft) is

$$F_a = a_R^* \left(\text{scf/ft}^3 \right) \left(43,560 \text{ft}^3 / \text{acre-ft} \right) \left(10^{-3} \text{Mcf/scf} \right)$$

$$= (385)(43,560)\left(10^{-3} \right)$$

$$= 16,770 \text{ Mcf/acre-ft.}$$

The following calculations are for the first step of 5% volume burned. From Eq. 8.359,

$$V_{Rbo}^* = 0.14714 S_{gi} + 0.01071 S_{gi}^2$$

$$= (0.1471)(3) + (0.01071)(3)^2$$

$$= 0.538\%$$

and $v_{Rb}^* = 5\%$.

$$x_R = \left(V_{Rb}^* - V_{Rbo}^* \right) / \left(100 - V_{Rbo}^* \right)$$

$$= (5 - 0.54)/(100 - 0.54)$$

$$= 0.045.$$

The value of y_R is calculated from Eq. 8.360.

$$y_R = 6.775 x_R - 15.948 x_R^2 + 16.187 x_R^3 - 7.015 x_R^4$$

$$= 6.775(0.045) - 15.948(0.045)^2 + 16.187(0.045)^3$$

$$- 7.015(0.045)^4$$

$$= 0.273.$$

Oil recovery as a percentage of the initial oil less fuel is computed with Eq. 8.357.

$$N_{p1} = 100 x_R + y_R \left(26.8229 - 0.4678 S_{gi} \right)$$

$$= 100(0.045) + 0.273 \left[26.823 - 0.4678(3) \right]$$

$$= 11.44\% \text{ of (initial oil} - \text{fuel).}$$

From Eq. 8.361,

$$N_p / V_T = \left(N_{p1} / 100 \right) \left[7,758 \phi \left(S_{oi} - S_{oF} \right) \right]$$

$$= 0.1144 \text{ (initial oil} - \text{fuel)}$$

$$= 0.114 \left(1,396 \text{ bbl/acre-ft} \right)$$

$$= 159.7 \text{ bbl/acre-ft.}$$

$$N_p = (159.7 \text{ bbl/acre-ft})(84.2 \text{ acre-ft})$$

$$= 13,447 \text{ bbl,}$$

and $F_a = \left(V_{Rb}^* / 100 \right) (16,770 \text{ Mcf/acre-ft})$

$$= (0.05)(16,770)$$
$$= 839 \text{ Mcf/acre-ft}$$

The cumulative volume of air injected is

$$(839 \text{ Mcf/acre-ft})(84.2 \text{ acre-ft}) = 70.64 \times 10^6 \text{ scf}.$$

From the problem statement, the initial injection rate was 1×10^6 scf/D with a linear increase in rate of 11.9 Mscf/D for 7 months (210 days). The time required to inject 70.64×10^6 scf is computed as follows:

$$70.64 \times 10^6 = \left[\left(1 \times 10^6 + 11.9 \times 10^3 t \right) / 2 \right] t.$$

This is a quadratic equation, which can be solved to yield $t = 74.8$ days and $i_a = 1.89 \times 10^6$ scf/D. The average oil displacement rate for the first 74.8 days of air injection is computed from Eq. 8.368:

$$\bar{q}_o^{n+1} = \left(N_p^{n+1} - N_p^n \right) / \left(t^{n+1} - t^n \right)$$
$$= 13,447 \text{ B} / 74.8 \text{ D}$$
$$= 179.8 \text{ B/D}$$
$$q_w = 0.1781 \left\{ \left[\left(1.89 \times 10^6 \right)(0.36) \right] / 385 \right\} (0.30 + 0.10)$$
$$= 127.1 \text{ B/D}.$$

The instantaneous injected-air/produced-oil ratio is 10.512 Mscf/bbl.

The remainder of the displacement performance was computed with a spreadsheet. **Table 8.86** summarizes the results. The air/oil ratio (AOR) is adjusted for excess air with Eq. 8.368 when $N_{p1} > 17.61$.

In-situ combustion continued in the South Belridge field with the initiation of the Section 12 in-situ combustion project in 1962 (Ramey et al. 1992; Gates et al. 1978). **Fig. 8.186** compares oil recovery vs. volume burned from this project. The dashed curve in Fig. 8.186 is the design curve based on Fig. 8.184 for an initial gas saturation of 7%. Section 12 performance was much better than predicted from the design curve. This is attributed to the presence of a large gravity-drainage component in the 30-ft-thick sand, which had only a few degrees of dip (Ramey et al. 1992).

V_{Rb} (%)	x	y	N_{p1} (%)	N_p (bbl/ acre-ft)	N_p (bbl)	Excess Air	Air Required (Mcf/acre-ft)	Cumulative AOR (Mcf/bbl)	Cumulative Air (Mcf)
1	0.004648	0.031148	1.26	17.54	1,477	0.00	168	9.57	14,129
5	0.044864	0.273289	11.43	159.61	13,439	0.00	839	5.26	70,644
7.69	0.07191	0.410553	17.63	246.07	20,719	0.00	1,291	5.24	108,666
10	0.095135	0.513561	22.57	315.05	26,527	0.04	1,753	5.56	147,591
15	0.145405	0.694563	32.20	449.46	37,844	0.13	2,847	6.34	239,750
20	0.195675	0.826061	40.57	566.30	47,682	0.21	4,049	7.15	340,953
21	0.205729	0.847204	42.11	587.83	49,496	0.22	4,301	7.32	362,120
21.26	0.208343	0.852443	42.50	593.34	49,959	0.22	4,367	7.36	367,670
25	0.245946	0.916746	47.90	668.65	56,301	0.27	5,338	7.98	449,499
30	0.296216	0.974234	54.39	759.23	63,927	0.33	6,700	8.82	564,151
40	0.396756	1.0147	65.47	913.94	76,954	0.43	9,603	10.51	808,572
50	0.497297	0.986882	74.82	1,044.43	87,941	0.51	12,710	12.17	1,070,142
60	0.597838	0.912995	82.99	1,158.56	97,551	0.59	15,992	13.80	1,346,548
70	0.698378	0.798054	90.12	1,258.13	105,935	0.65	19,412	15.43	1,634,459
80	0.798919	0.62987	95.90	1,338.80	112,727	0.70	22,883	17.09	1,926,740
90	0.899459	0.37905	99.58	1,390.15	117,051	0.74	26,243	18.88	2,209,678
100	1	0	100.00	1,396.00	117,543	0.74	29,222	20.93	2,460,524

Design of in-situ combustion project: Gates and Ramey method; initial fuel = 1,396 bbl/acre-ft; air requirement = 16,780 Mcf/acre-ft burned or 385 scf/ft^3 reservoir burned; S_{gi} = 3%; V_{rbo} = 0.53769; i_a = 3.5 MMscf/D; and pattern volume = 84.2 acre-ft.

Table 8.86—Summary of design calculations, Example 8.29.

V_{Rb} (%)	Cumulative Air (Mcf)	Time (days)	Injection Rate (Mcf/D)	Oil Displacement During Time Treatment (bbl)	Average Oil Rate (B/D)	Instantaneous AOR (Mcf/bbl)	Water Rate (B/D)
1	14,129	22.33	1,266	1,477	66.16	19.13	84.31
5	70,644	74.77	1,890	11,962	228.11	8.28	125.88
7.69	108,666	99.51	2,184	7,280	294.28	7.42	145.49
10	147,591	120.99	2,440	5,808	270.35	9.02	162.52
15	239,750	163.07	2,941	11,317	268.95	10.93	195.88
20	340,953	201.02	3,392	9,833	259.20	13.09	225.97
21	362,120	208.23	3,478	1,814	251.46	13.83	231.68
21.26	367,670	210.09	3,500	464	249.77	14.01	233.16
25	449,499	233.47	3,500	6,341	271.23	12.90	233.15
30	564,151	266.23	3,500	7,627	232.82	15.03	233.15
40	808,572	336.06	3,500	13,027	186.54	18.76	233.15
50	1,070,142	410.80	3,500	10,987	147.01	23.81	233.15
60	1,346,548	489.77	3,500	9,610	121.69	28.76	233.15
70	1,634,459	572.03	3,500	8,384	101.92	34.34	233.15
80	1,926,740	655.54	3,500	8,793	81.34	43.03	233.15
90	2,209,678	736.38	3,500	4,324	53.48	65.44	233.15
100	2,460,524	808.05	3,500	492	6.87	509.57	233.15

Table 8.86—Summary of design calculations, Example 8.29 (continued).

Correlations of Brigham et al. In the late 1970s, Brigham et al. (1980) reviewed data from 40 field tests to develop a correlation between oil recovery from dry in-situ combustion and the volume of air injected. Data from 12 field tests were used in a multiple regression program to correlate oil recovery as a function of oil saturation S_o, thickness h, and oil viscosity μ_o. The correlating groups were developed to represent groups that had physical meaning. Two correlating equations were developed that depended on the viscosity of the oil at reservoir temperature. When the oil viscosity is 10 cp or smaller, Eq. 8.372 should be used.

$$y_B / 36.53 = (2.00 S_o - 0.001h - 0.00082 \mu_o) x_B, \qquad (8.372)$$

where $y_B = \left[\left(\Delta N_p + \Delta N_b \right) / N \right] 100 \quad \dots \dots \dots \dots \dots \dots (8.373)$

and $x_B = G_i E_{O_2} / \left[\left(N_c / \phi S_o \right) (1 - \phi) \right]. \quad \dots \dots \dots \dots \dots \dots (8.374)$

In this correlation, N_c = OIP at the start of combustion, STB; N = OOIP, STB; ΔN_p = cumulative oil produced, STB; ΔN_b = cumulative oil burned as fuel, STB; G_i = cumulative air injected, Mscf; E_{O_2} = oxygen utilization efficiency, fraction; and S_o = oil saturation at the start of combustion, fraction.

Fig. 8.187 shows the correlation of data from 12 in-situ combustion field projects with the correlation functions given in Eq. 8.372. The correlation represented by Eq. 8.372 is a smooth curve drawn through the data in Fig. 8.187. The correlation is limited to the following ranges of oil saturation, reservoir thicknesses, and oil viscosity, respectively: S_o =

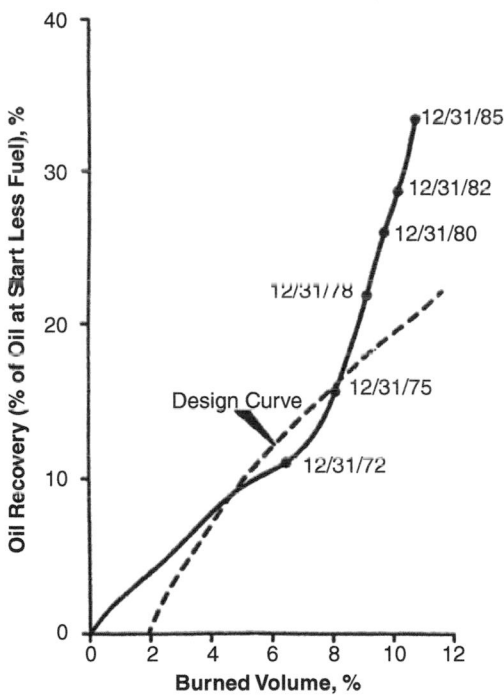

Fig. 8.186—Oil recovery vs. volume burned for Section 12 in-situ combustion project, South Belridge field (Ramey et al. 1992).

0.36 to 0.79, h = 4.4 to 150 ft, and μ_o = 10 to 700 cp. Use of the correlation outside these ranges may give unreasonable results.

A second correlating function was developed to improve the influence of oil viscosity on the recovery predictions. The functional relationship that gave the best fit to the field data is represented by Eqs. 8.375 and 8.376. **Fig. 8.188** shows the correlation of field data with the second correlation. This correlation should be used when oil viscosity is 700 cp or more and when there has been considerable recovery before the onset of combustion (Brigham et al.1980). Use of the second correlation is restricted to parameters in the range of those used to define the correlation.

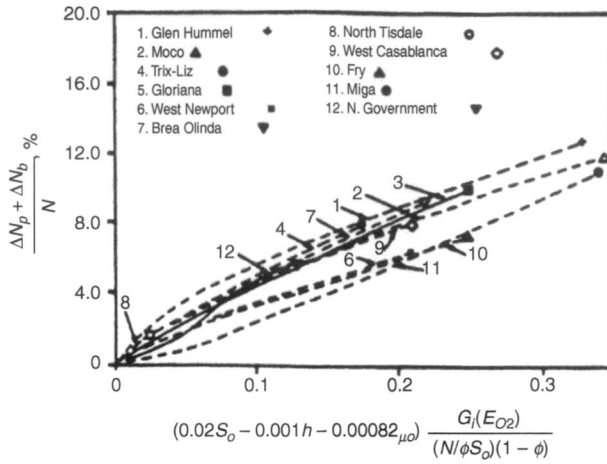

Fig. 8.187—Effect of multiple linear regression analysis on dry in-situ combustion data, low-viscosity correlation (Fassihi et al. 1981).

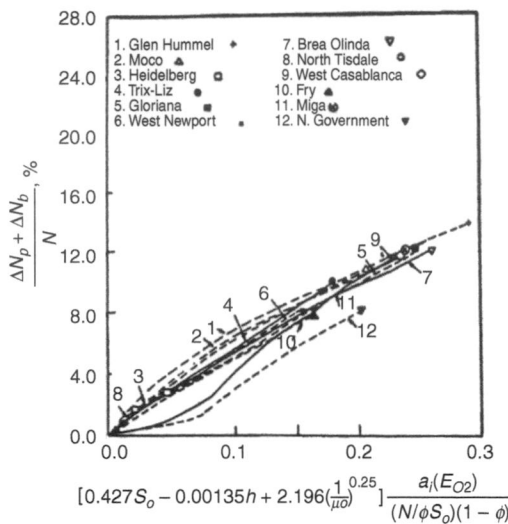

Fig. 8.188—Data for development of high-viscosity correlation (Fassihi et al. 1981).

$$y_B/47.00 = \left[0.427 S_o - 0.00135 h + 2.196 \left(1/\mu_o \right)^{0.25} \right] x_B. \quad \dots \dots \dots \dots \dots \dots \dots \dots (8.375)$$

In the second correlation, y_B is given by Eq. 8.376.

$$y_B = \left[\left(\Delta N_p + \Delta N_b \right) / N_c \right] 100. \quad \dots \dots \dots \dots \dots \dots \dots \dots \dots \dots \dots \dots \dots \dots (8.376)$$

The parameter x_B is obtained from Eq. 8.374.

The Brigham et al. correlation was found to correlate dry in-situ combustion field data with a standard deviation of 9% when the value of the abscissa (right side of Eq. 8.375) was 0.20, and a maximum error of ±14% when compared to the field data. This correlation can be used to make rapid estimates of ultimate recovery by in-situ combustion vs. air injected for dry in-situ combustion projects. Parameters in the correlation that depend on air requirement and fuel availability can be estimated by use of the concepts introduced earlier in this section.

The Brigham et al. correlation was used by the National Petroleum Council in the 1982 study of EOR, and a Fortran program of the in-situ combustion model developed by the US DOE (1986) is in the public domain. Example 8.30 illustrates the application of the Brigham et al. correlation to the Fry in-situ combustion project.

Example 8.30—Estimation of Oil Recovery by In-Situ Combustion—Fry Combustion Project. The Fry in-situ combustion project in the Robinson field, Illinois, (Howell and Peterson 1979) is described in Section 8.7.4. An in-situ combustion project was initiated in 1965, and by 1 January 1978, approximately 18,295,612 Mscf of air had been injected into three injection wells creating three active burn zones. **Table 8.87** summarizes other characteristics of this project. Using the Brigham et al. method, estimate the oil recovery by in-situ combustion as of 1 January 1978.

S_o	0.522
ϕ	0.197
h (ft)	22.25
μ_o (cp)	35
N (STB)	13,888,464
N_c (STB)	9,085,426
ΔN_b (STB)	243,341
G_i (Mcf)	18,295,612
E_{o_2}	0.90

Table 8.87—Reservoir and fluid properties, Fry in-situ combustion project, Robinson Field, Illinois, 1 January 1978 (Howell and Peterson 1979).

Solution. The oil viscosity at reservoir temperature is 35 cp. According to Brigham et al., the correlation given by Eq. 8.375 should be used when the oil viscosity is greater than 10 cp.

$$y_B/47.00 = \left[0.427 S_o - 0.00135h + 2.196\left(1/\mu_o\right)^{0.25}\right]x_B.$$

Substituting appropriate values into Eq. 8.375 gives

$$\frac{y_B}{47.00} = \left[0.427(0.522) - (0.00135)(22.25) + 2.196\left(\frac{1}{35}\right)^{0.25}\right]x_B$$

$$= (0.2229 - 0.03 + 0.9028)x_B$$

$$= 1.0957 x_B.$$

The value of x_B is calculated from Eq. 8.374; recalling that G_i in Eq. 8.374 is in Mscf,

$$x_B = G_i E_{O_2}/\left[\left(N_c/\phi S_o\right)(1-\phi)\right]$$

$$= \frac{\left(18.295 \times 10^6\right)(0.9)}{\left[9,085,426/(0.197)(0.522)\right](1-0.197)}$$

$$= 0.232.$$

The value of y_B can now be calculated.

$$y_B = (1.0957)(0.232)(47)$$

$$= 11.95.$$

From Eq. 8.376, the parameter y is the percentage of the OIP at the start of combustion recovered as a result of in-situ combustion and burned as fuel.

$$y_B = \left[\left(\Delta N_p + \Delta N_b\right)/N_c\right]100,$$

$$\left(\Delta N_p + \Delta N_b\right) = (0.1195)N_c$$

$$= (0.1195)(9,085,426)$$

$$= 1,085,708 \text{ STB},$$

$$\Delta N_b = 243,341 \text{ STB},$$

and $\Delta N_p = 842,367$ STB.

Actual production attributed to in-situ combustion was 1,056,167 STB, so the second Brigham et al. correlation underestimates recovery by 20%.

Closer agreement was reported by Howell and Peterson (1979), who used the correlation given by Eq. 8.372 to estimate oil recovery from in-situ combustion.

$$\frac{y_B}{36.53} = \left(2.00 S_o - 0.001h - 0.00082\mu_o\right)x_B$$

$$= \left[(2.00)(0.522) - (0.001)(25.25) - (0.00082)(35)\right](0.232)$$

$$= 0.23.$$

Solving for y_B and ΔN_p yields $y_B = 8.39$ and $\Delta N_p = 926,670$ STB.

The estimated oil production resulting from in-situ combustion is within 12.7% of the actual value obtained with the first correlation (Eq. 8.372). This example illustrates that there are uncertainties in applying the Brigham et al. models to actual field data even though the data from the Fry project was used to develop the correlations. In this case, the estimate was conservative, which would support good projects and eliminate marginal prospects.

Selection of Pattern—In-Situ Combustion Projects. The first in-situ combustion projects were designed on a five-spot pattern. Results of many field tests, including several discussed in the next section, show that few projects can be considered pattern floods after combustion is initiated. The directions that a combustion front will move and, thus, the shape of the front are determined by the distributions of permeability and fluid saturations in a reservoir coupled with gravity segregation. There has been little success in trying to control the presumed orientation of the flood pattern by controlling rates of surrounding producing wells.

The in-situ combustion front goes where the air flows. Formation dip, aided by gravity segregation, has a powerful influence on the development and stability of a combustion front. A combustion front will move updip if there are no flow barriers, enhanced by gravity effects. Thus, where possible, the selection of a structurally high well for injection allows gravity to help in stabilizing the combustion front as well as in driving the oil downdip into the production wells. Once established, a combustion front will move vertically, stabilized by gravity segregation, if there is sufficient vertical permeability.

8.7.4 Field Cases—Dry In-Situ Combustion. In-situ combustion has been extensively field tested in a variety of reservoirs. Chu (1982a, b) provides an extensive survey of in-situ combustion projects. As noted earlier in this chapter, steamdrive has been preferred to in-situ combustion in most cases where the two processes are applicable. In 1989, there were 10 active in-situ combustion projects operated on a fieldwide scale in the US and Canada producing approximately 11,805 B/D. Another large ongoing in-situ combustion project was in Romania, with an estimated production rate of 10,400 B/D, giving a total in-situ combustion production rate of approximately 22,200 B/D. **Table 8.88** gives fields and operators. Although there have been many field tests of in-situ combustion, the number of commercial scale projects of in-situ combustion was reported (Turta et al. 2007) to be four with estimated production of 17,700 B/D. This estimate does not include in-situ combustion projects in the dolomite reservoirs in North Dakota and Montana of approximately 20,590 B/D (Koottungal 2014).

Some existing in-situ combustion field projects have been in progress since the early 1960s or 1970s. There have been numerous pilot tests. The only new commercial projects initiated in heavy-oil reservoirs were the projects in the Balol/Santhal fields in India (Ursenbach et al. 2010), which began in the 1990s and became commercial in 2000. Field-scale projects have been applied to a wide range of reservoirs with oil gravities from 10 to 40 °API. These include reservoirs that are too thin and/or too deep to use the steamdrive process. Many of the in-situ combustion projects were applied to reservoirs that were not ideal candidates. Consequently, there have been many technical and economic failures. There are several field projects that have been exceptionally successful. Four in-situ combustion projects are reviewed in this section to demonstrate the technology. In addition, the South Belridge Section 12 thermal recovery project (Ramey et al. 1992; Gates et al. 1978) is adequately described in the literature. Other projects in Table 8.88 and not described in this section, such as the Battram projects in Canada, have not been presented in the technical literature.

Moco-T Midway-Sunset Field. The Moco zone reservoir in the Midway-Sunset field, Kern County, California, is a small reservoir (150 acres) with six major sands separated by interbedded shales (Curtis 1989; Fassihi et al. 1981; Howell and Peterson 1979; Chu 1982a, b; Gates and Sklar 1971). The reservoir is located on a small anticline. **Fig. 8.189** shows sand distribution and structure. Cross Sections X-X' and E-E' in **Figs. 8.190 and 8.191** show that Sands M_1 through M_6 are correlated across the field, separated by shale zones. Dip is up to 45° to the north and 20° to the south. Some of the sands appear to merge on the west side of the reservoir. **Table 8.89** gives reservoir characteristics. The reservoir originally contained 38 million bbl of 14.5 °API oil.

In-situ combustion was initiated in January 1960 by injecting air at a rate of 2,000 Mscf/D into two wells (Wells 504 and 592) at the top of the structure. All zones were open. Spontaneous ignition occurred approximately 18 days after the start of air injection. Production response was immediate following air injection, with production rates increasing to 2,850 BOPD in

Field	Operator	Province	Start Date	Area (acres)	Thermal Production (B/D)
US					
Bellevue	Bayou State	LA	1970	200	420
Bellevue	Texaco	LA	1963	385	850
Forest Hills	Greenwich Oil	TX	1976	1,900	1,050
Lost Hills	Mobil	CA	1961	164	520
Midway-Sunset	Mobil	CA	1960	150	1,100
West Newport	Mobil	CA	1958	300	980
West Heidelberg	Chevron	MS	1971	362	470
Canada					
Battram	Mobil	Sask.	1966	4,920	3,300
Battram	Mobil	Sask.	1967	2,400	1,635
Battram	Mobil	Sask.	1965	680	1,400
Fosterton, NW	Mobil	Sask.	1960	200	50
TOTAL					11,805
Suplacu de Barcau		Romania	1964		10,400* 22,205

– – Data not available.

*1987

Table 8.88—Summary of active in-situ combustion projects.

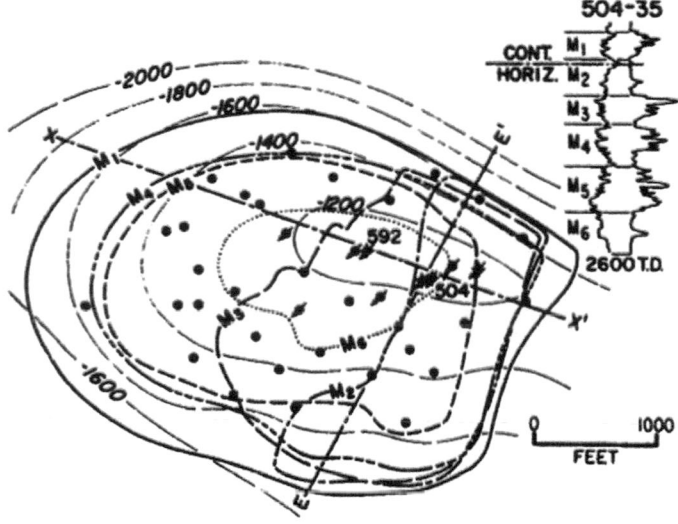

Fig. 8.189—Moco zone sand distribution and structure (Gates and Sklar 1971).

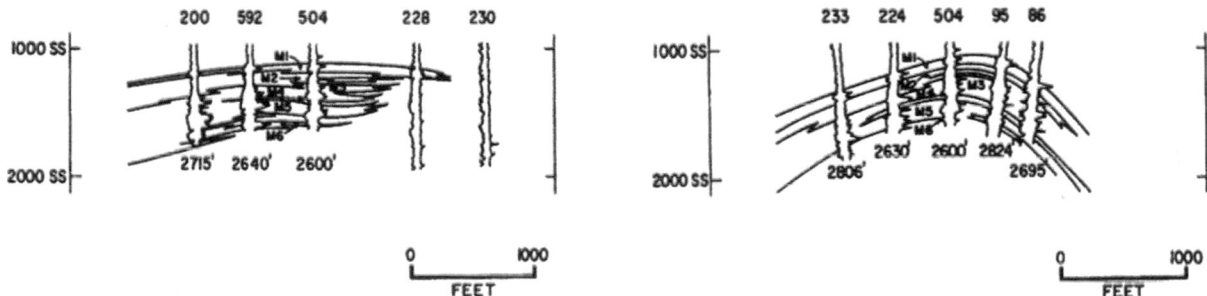

Fig. 8.190—Cross Section X-X', Moco zone (Gates and Sklar 1971).

Fig. 8.191—Cross Section E-E', Moco zone (Gates and Sklar 1971).

Productive area (acres)	150
Average depth (ft)	2,100 to 2,700
Gross formation thickness (ft)	500
Average net sand thickness (ft)	129
Porosity (%)	75
Oil saturation (%)	75
Water saturation (%)	25
FVF (RB/STB)	1.06
Initial OIP (bbl/acre-ft)	1,980
Initial OIP [total bbl (10^6)]	38
Specific permeability (md)	1,575
Initial formation pressure (psi)	1,000
Reservoir temperature (°F)	125
Oil gravity (°API)	14.5
Oil viscosity at formation temperature (cp)	110
Sand character	Unconsolidated

Table 8.89—Reservoir and fluid properties, Moco Zone.

February 1960 in response to the high volume of air injection. Peak oil rate was reached in 1964 at 4,200 BOPD. Air injection was increased to an average rate of 8,000 Mscf/D in early 1964. In 1976, the air-injection rate was reduced to 5,500 Mscf/D and remained at this rate thereafter. Declines in oil production rates correlate with declines in air injection in 1963, 1967, and 1973 (Curtis 1989). Four additional injection wells were added to improve distribution of the air between zones.

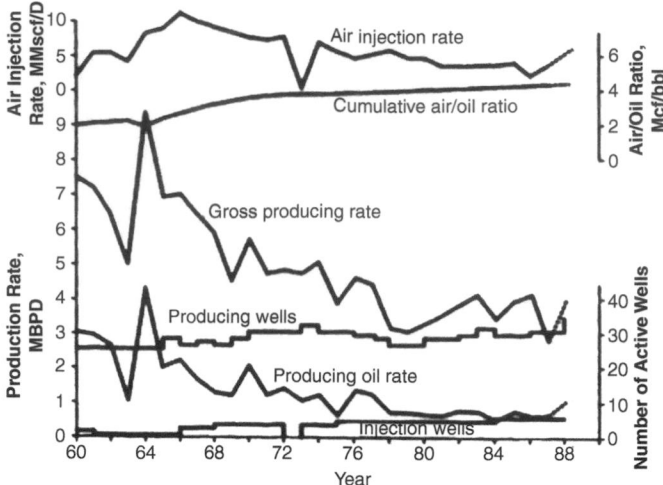

Fig. 8.192—Moco zone performance (Curtis 1989).

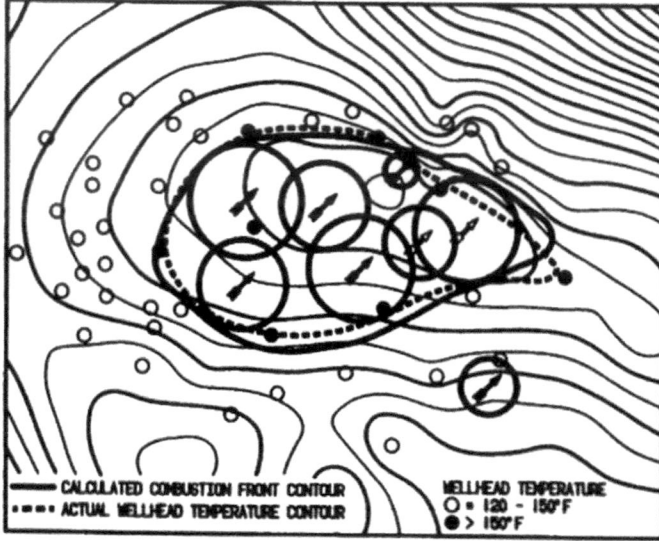

Fig. 8.193—Individual injection well burned radii, Moco T in-situ combustion project (Curtis 1989).

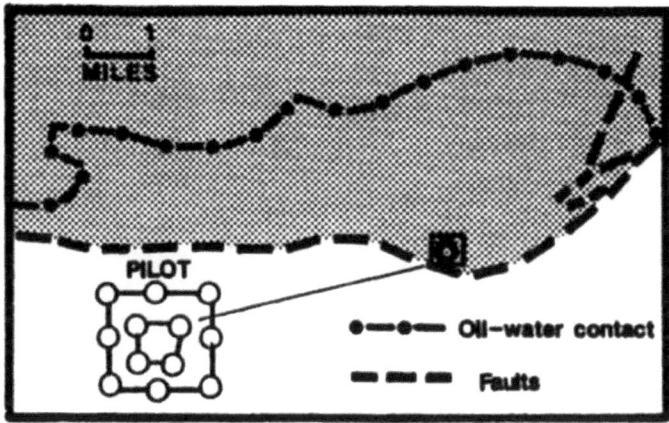

Fig. 8.194—Suplacu de Barcau field (Carcoana 1990).

Fig. 8.192 shows the Moco zone performance and air-injection rates over the life of the project. Also shown as a dashed line in Fig. 8.192 is the extrapolated primary decline rate resulting from air injection if in-situ combustion had not occurred. As of 1989 (Curtis 1989), the cumulative oil produced was 14 million bbl and the cumulative air injected was 59.5 Bscf. Oil production rate as of 1989 was 1,040 BOPD from 34 active wells. The project's AOR has remained at 4 to 4.3 Mscf/bbl, with a current producer to injector ratio of 6:1. This project is the longest sustained dry-combustion project known to exist and the most successful.

The combustion fronts tend to move along the top of each sand, strongly supported by gravity segregation and the steep dip of the reservoir. Consequently, the combustion process resembles a gravity-stabilized frontal-advance displacement (Curtis 1989). Location of the burning front was estimated by several methods, including estimating the burned radius around each injection well with material balances. **Fig. 8.193** shows the composite burning-front contour estimated as of spring 1987. The dashed contour in Fig. 8.193 is an estimate of burning-front location based on producing wellhead temperatures in excess of 150°F. The two estimates are in reasonable agreement. Curtis (1989) reported that eight new wells drilled just outside the estimated burning front were among the highest producers, with rates greater than 60 BOPD. Oxygen usage was 100% with negligible oxygen produced in the combustion gases.

The ultimate recovery is estimated to reach 50 to 57% of OOIP at a cumulative air-injection volume of 85 to 90 Bscf (Curtis 1989). The remaining producing life was estimated by Curtis to be 12 to 15 years at an average air-injection rate of 5,000 Mscf/D. This is an exceptional result for a thermal recovery project.

Suplacu de Barcau. The Suplacu de Barcau field, located in northwestern Romania, is the site of the world's largest in-situ combustion project (Cadelle et al. 1981; Carcoana 1990). The Suplacu de Barcau oil reservoir, shown in **Fig. 8.194,** is a monocline with an average dip of 5° north. It contains unconsolidated, slightly shaly sands and is located at a depth of 165 ft at the south end of the monocline and 656 ft at the north end. **Table 8.90** presents reservoir and fluid properties. Net-pay thickness is 33 ft, with an average porosity of 32%. Permeability of the sand ranges between 1,700 and 2,000 md. The reservoir oil has a viscosity of 2,000 cp at the reservoir temperature of 64.4°F. OOIP was estimated to be 295 million bbl. Primary recovery was estimated to be 9% OOIP.

In-situ combustion was selected as a recovery process for this field because steam generators and well equipment were not available. A dry-combustion pilot was initiated in the area at the top of the structure, as indicated in Fig. 8.194. Eight 9.9-acre nine-spot patterns were pilot tested between 1967 and 1971. These were located directly east and south of the initial pilot pattern. Expansion continued in the upper part of the reservoir in the east and west directions until the patterns merged to form the linedrive shown in **Fig. 8.195. Fig. 8.196** shows

Average depth (ft)	400
Average net pay (ft)	33
Porosity (%)	32
Average permeability (md)	1,850
Interstitial water (%)	15
Oil gravity (°API)	16
Oil viscosity (cp)	2,000
OOIP (10^6 STB)	295

Table 8.90—Reservoir and fluid properties, Suplacu de Barcau reservoir.

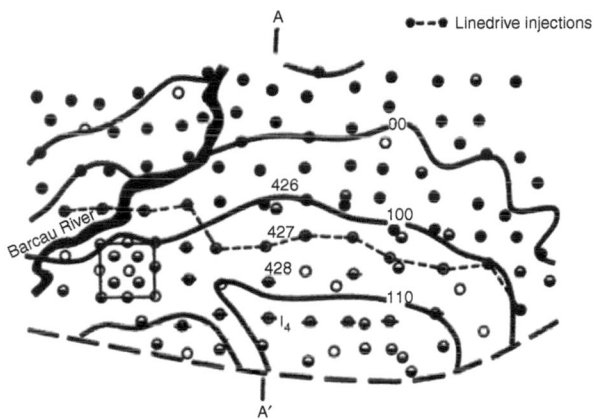

Fig. 8.195—Linedrive development of combustion front (Carcoana 1990).

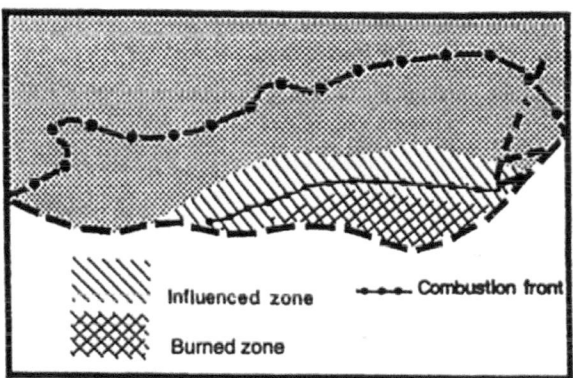

Fig. 8.196—Estimated location of combustion front and the area affected by combustion (Carcoana 1990).

the combustion front and the burned region. The combustion front was estimated to be 5 miles in length by 1990. At that time, 120 MMscf/D of air was being injected with 100 injection wells.

The exploitation plan is to displace the combustion front downdip. The linedrive is maintained by converting production wells to injectors when the combustion front breaks into the production well. The conversion process is depicted in **Fig. 8.197,** which is a cross section between Wells 426 and 428 showing conversion of Well 428 into an air injection well after the combustion front arrives. Observation wells drilled behind the combustion front revealed that the top half of the pay zone was burned. This region was separated from the lower unburned zone by a coke layer. Gravity segregation is prominent in this field, as indicated by core analyses and the depiction of the combustion process in Fig. 8.197. Infill drilling and CSS ahead of the combustion front are used to maintain the air pressure below the fracturing pressure of the formation and to promote a uniform advance of the combustion front downstructure. A pilot test of wet combustion was conducted and expanded to 20 wells during 1976–79. However, the project remains primarily a dry-combustion project.

Fig. 8.198 (Aldea and Turta 1988) shows performance of the in-situ combustion project. Oil production in 1987 was 10,400 B/D with approximately 600 wells affected by in-situ combustion. The average AOR ranged between 9.5 and 11.3 Mscf/bbl during 1973–79 and increased to 14.2 Mscf/bbl after 1985. Ultimate oil recovery is expected to be 52% of OOIP. In 2010, the production from Suplacu de Barcau was reported to be 8,000 to 10,000 B/D (Ursenbach et al. 2010).

Fry Project, Robinson Field. The Fry project is an example of an

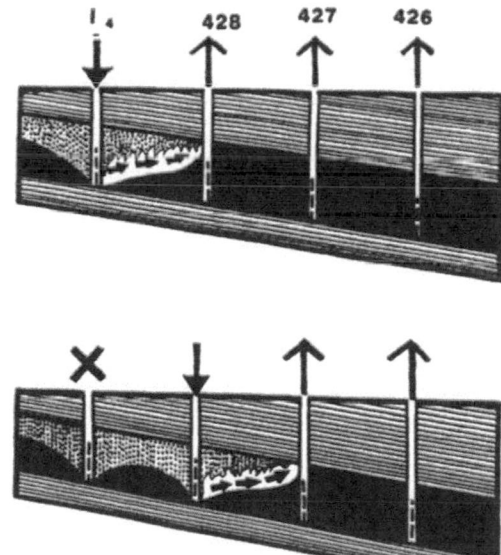

Fig. 8.197—Maintenance of linedrive combustion front by conversion of a production well to air injection after breakthrough of the combustion front (Carcoana 1990).

in-situ combustion project in a light-oil (26.5 to 30.3 °API) reservoir. Production in the Fry reservoir was from the Robinson sandstone, which is believed to have been formed by stream deposition along a series of migrating point bars (Howell and

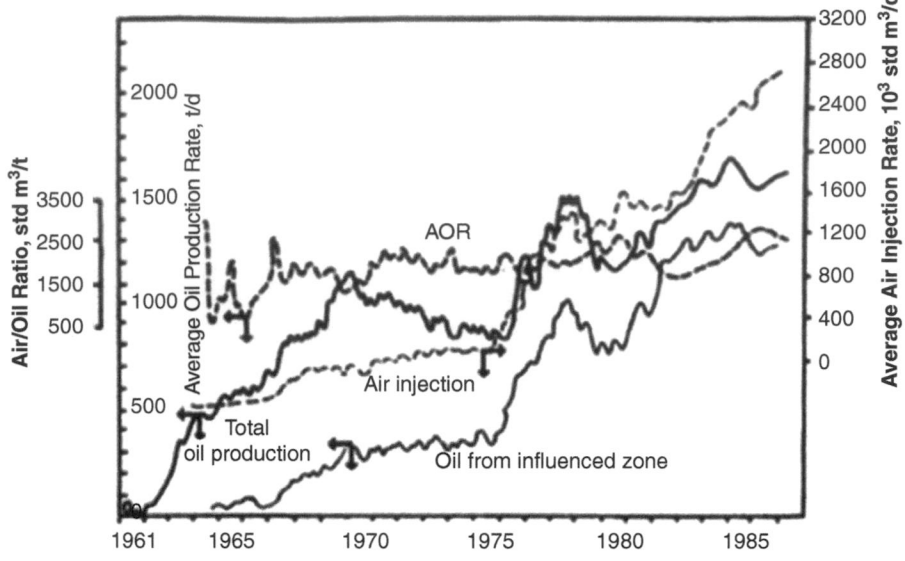

Fig. 8.198—Field performance, Suplacu de Barcau in-situ combustion project (Carcoana 1990).

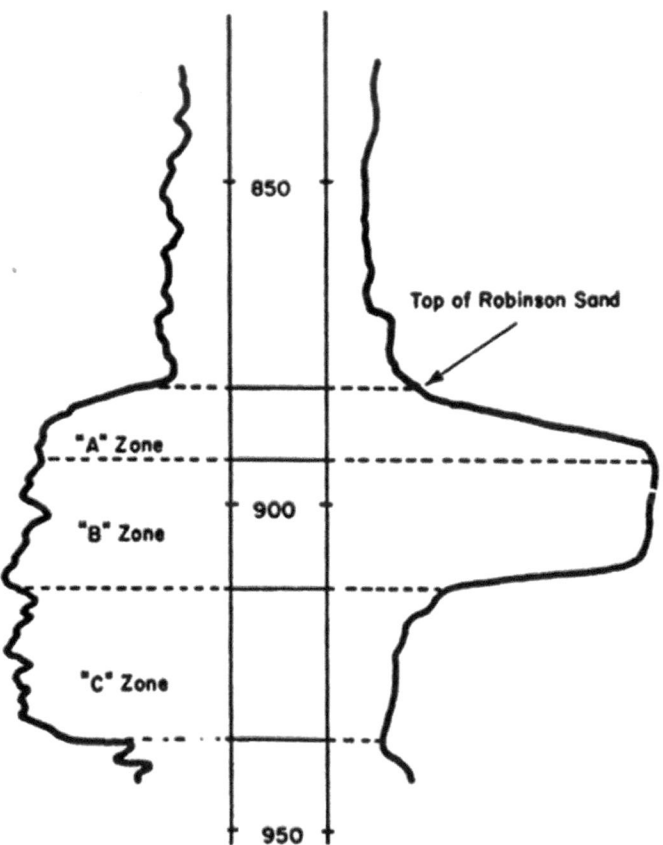

Fig. 8.199—Type log, Robinson sand, L.B. Wampler AI-1 (Howell and Peterson 1979).

Peterson 1979). The productive interval contains four distinct zones, Zones A, B, C, and M. Zones A, B, and C correlate across the field. **Fig. 8.199** shows a type log. Zone M is present only around Well 1-2. **Table 8.91** summarizes properties of the three main zones. The reservoir trends from southwest to northeast, with net thickness varying from 0 to 55 ft. **Fig. 8.200** is an isopach map showing the net thickness of the Robinson sand. Average porosity was 19.7%, and the average permeability was 320 md in the project area.

The Fry reservoir contains oil with viscosities ranging from 10 to 80 cp. Primary production was by solution-gas drive. Following depletion of the reservoir energy in the early 1900s, the field was placed under vacuum in 1917 to increase production with little success. Repressuring with air was attempted during 1945–51. Prewaterflood recovery was estimated to be 287 STB/acre-ft.

Waterflooding was initiated between 1948 and 1952, continuing in nonproject areas until 1967. Waterflooding produced mixed response in this reservoir, with best response in the west portion of the unit (Hughes Unit) where oil viscosities were on the order of 10 to 20 cp. Poorest responses were observed in the east (Weirich Heirs Community), where the oil viscosity was as high as 80 cp. **Table 8.92** summarizes the waterflood response.

The Fry in-situ combustion project was initiated in 1961 on the Fry Unit, where waterflood response had been poor. The initial project was a 3.3-acre five-spot pilot test to evaluate the technical feasibility of in-situ combustion. The reservoir description, geology, and operations of the project are summarized in Chu (1982a), Hewitt and Morgan (1965), Clark et al. (1965), Poettmann et al. (1967), and Earlougher et al. (1970). An electric ignitor was used to initiate combustion. Air injection in Well AI-1 averaged 1.52 MMscf/D at an average pressure of 310 to 315 psia. The pilot project continued until 1 January 1964, when the project was evaluated and determined to be a technical success.

The project was expanded to 517 productive acres (797 surface acres) in 1965 by drilling air-injection wells AI-2, AI-3, and AI-4, as shown in Fig. 8.200, and named Fry Project 144-R. Well AI-2 was ignited in March 1965 and Well AI-4 in October

Zone	Lithology	Average Thickness (ft)	Average Porosity (%)	Average Permeability	Permeability Variation
Upper (A)	Ripple laminated sandstone	7	18	50	0.79
Middle (B)	Cross stratified sandstone	24	19	300	0.60
Lower (C)	Laminated sandstone (85%) and shale (15%)	19	20	425	0.57
Total Reservoir		50	19.7	320	0.68

Table 8.91—Reservoir properties, Fry project, Robinson field.

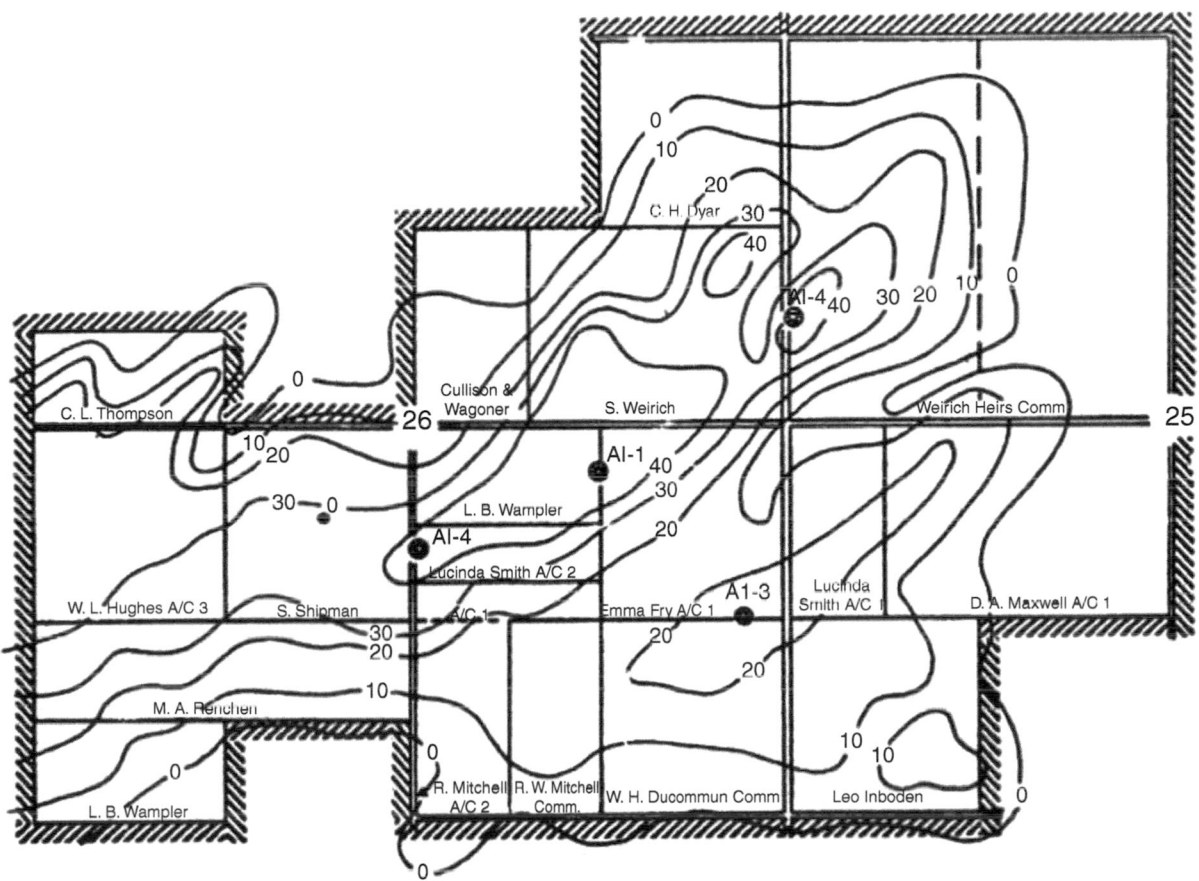

Fig. 8.200—Isopach of net sand (feet), Robinson sand (Howell and Peterson 1979).

Unit	Waterflood Recovery (bbl/acre-ft)	Waterflood Pattern Used	Approximate Crude Viscosity Range* (cp)
Fry unit	73	Peripheral	20 to 40
Hughes unit	233	Five-spot	10 to 20
Weirich Heirs	67	Peripheral	40 to 80
*At 70°F.			

Table 8.92—Waterflood data, Fry in-situ combustion project.

1966. Well AI-3 was not ignited because of insufficient net sand. The air-injection capacity by the end of 1965 was approximately 7 MMscf/D at 600 psi. Oxygen usage averaged 90% from the beginning of the project to 1 January 1978. **Fig. 8.201** summarizes oil production and air-injection data for the project. Cumulative oil production was 1,056,167 STB, and the cumulative air injected was 18.295 MMscf through 1977. This corresponds to a cumulative AOR of 17.32 Mscf/bbl. Cumulative oil production was a linear function of cumulative air injection during most of the project, as shown in **Fig. 8.202.** Water was injected simultaneously with air in Wells AI-2 and AI-4 from late 1970 until late 1972 in an attempt to initiate wet combustion, as discussed in Section 8.7.3. Injection into the burned zone continued through 1978 by use of burned-out wells. No notable changes were observed as a result of water injection (Howell and Peterson 1979).

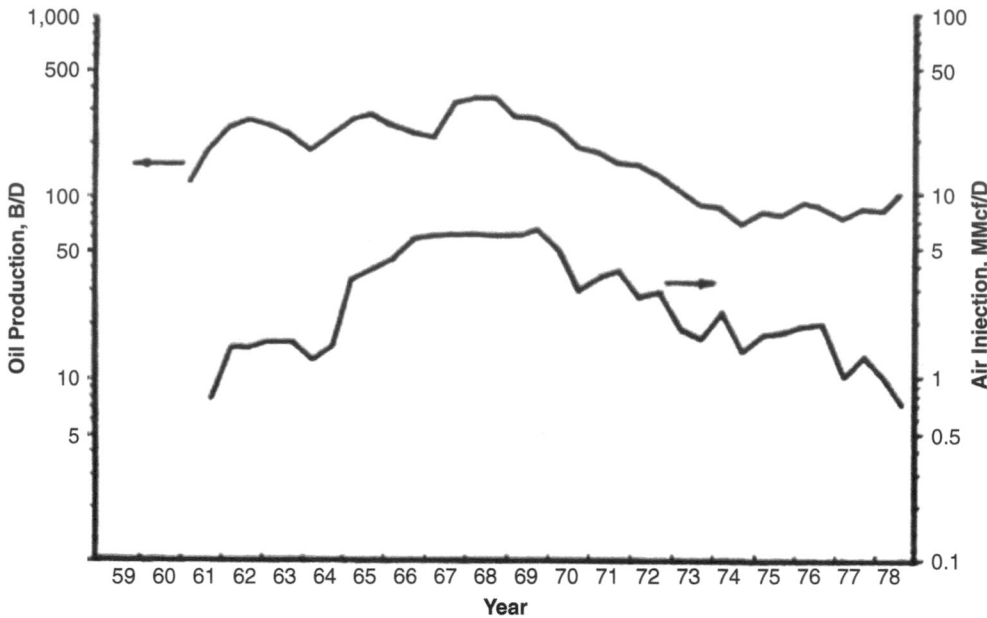

Fig. 8.201—Oil production and air injection vs. time, Fry in-situ combustion project (Howell and Peterson 1979).

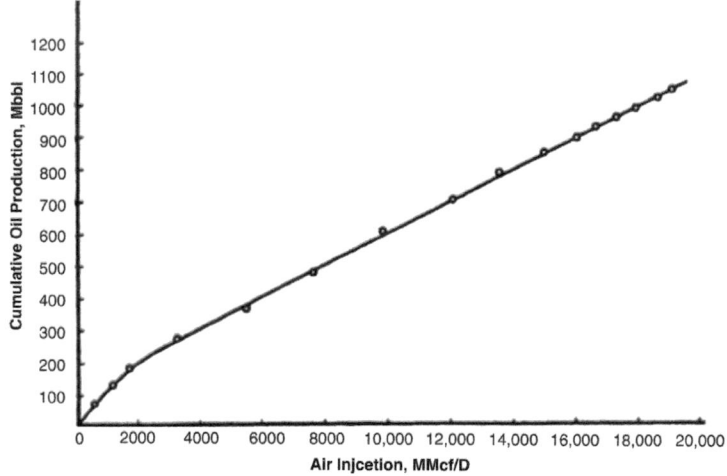

Fig. 8.202—Cumulative oil production vs. cumulative air injection, Fry in-situ combustion project (Howell and Peterson 1979).

Fig. 8.203 shows estimates of burned-zone locations. These locations were estimated from volumetric calculations on the basis of material balances from produced-gas analyses and permeability distributions. Pressure-transient analyses (van Poolen 1965) were also used to confirm the locations of burning fronts. Arrival of the combustion front at a production well usually meant that the well was nearing the end of its productive life.

Recovery for the project was estimated to be 555 bbl/acre-ft through 1977 on the basis of the volume burned. However, four cores from the burned-out region of the pilot area were interpreted to estimate an overall volumetric sweep efficiency of 59.7%. This corresponds to a swept volume that is 1.675 times the burned volume in each of the burned areas. These results are consistent with the discussions in Section 8.7.2, which indicated that more oil is produced from an in-situ combustion process than is displaced by the burning front.

May-Libby Reservoir. The May-Libby reservoir, shown in **Fig. 8.204,** is located in Richland Parish, Louisiana, USA. The reservoir is a sand lens that appears to be isolated by shale (Hardy et al. 1972). The reservoir is approximately 2.5 miles long and 1 mile wide and has an average thickness of 4.4 ft. Reservoir dip is approximately 250 ft/mile in the direction perpendicular to the major axis of the reservoir. **Fig. 8.205** is an isopach map of the reservoir. **Table 8.93** summarizes properties of the reservoir. The May-Libby reservoir is unique in that it is thin and contains a light oil (40 °API). At one time, it was thought that reservoirs containing light oil could not be candidates for in-situ combustion because not enough fuel could be laid down during the combustion process.

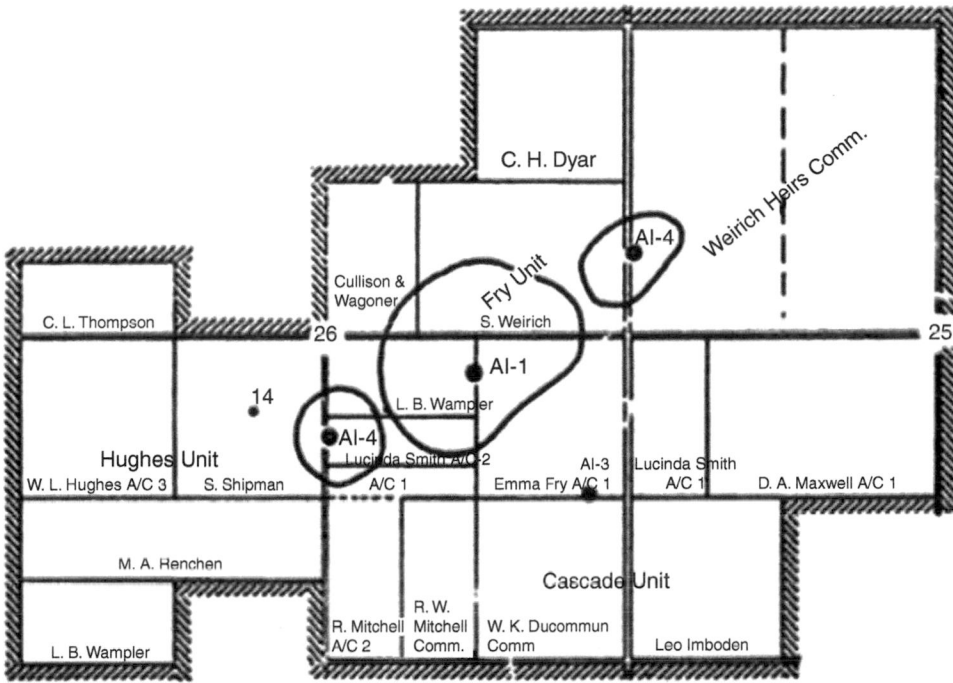

Fig. 8.203—Burned zone location, Fry in-situ combustion project (Howell and Peterson 1979).

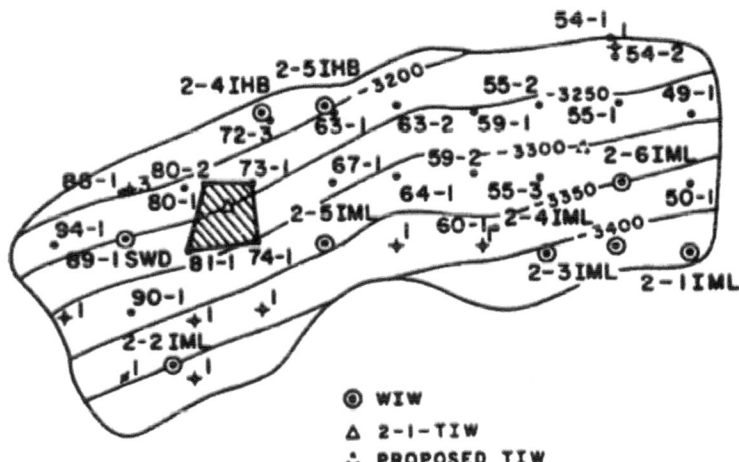

Fig. 8.204—Structure map, May-Libby reservoir, Delhi field, Louisiana, USA (Hardy et al. 1972).

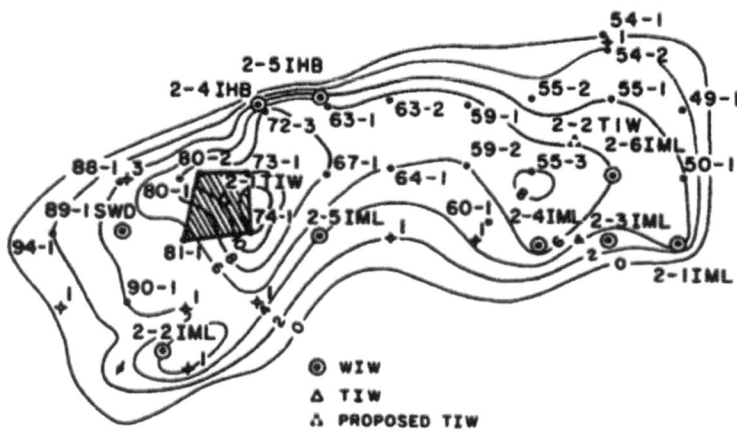

Fig. 8.205—Isopachous map of net-pay thickness, May-Libby oil sand, Delhi field, Louisiana, USA (Hardy et al. 1972).

Reservoir	
Date of discovery	12 January 1945
Type of formation	Tuscaloosa sandstone
Number of producing wells	32
Depth of pay (ft)	3,400
Size of reservoir (miles)	1 × 2½
Area of reservoir (acres)	1,362
Volume of reservoir (acre-ft)	5,986
Well spacing (acres)	40
Net-pay thickness (ft)	2 to 12
Average net pay (ft)	4.4
Permeability range (md)	20 to 2,900
Average permeability (md)	1,069
Oil gravity (°API)	40
FVF at 1,556 psig (RB/STB)	1.216
Original BHP (psig)	1,556
Original bottomhole temperature (°F)	135
Solution GOR (scf/STB)	446
Connate water saturation (%)	30
Oil content (STB/acre-ft)	1,395
OOIP (STB)	8,350,000
Primary Production	
Oil (STB)	2,676,411
Water (bbl)	177,519
Gas (Mcf)	2,364,674
Waterflood Recovery	
Oil (STB)	1,245,800
Water (bbl)	2,273,538
Gas (Mcf)	714,060

Table 8.93—Reservoir and fluid properties, in-situ combustion pilot, May-Libby reservoir.

This field had excellent primary recovery (2.7 million bbl or 32% OOIP) and was waterflooded. Oil viscosity at reservoir temperature (135°F) was 3 cp, so the mobility ratio was probably favorable for the waterflood, as indicated when relative permeability effects are considered. Waterflood recovery was 1,245 million bbl, or 14.9% of OOIP, for a total recovery (primary and secondary) of 47% of OOIP. Thus, 53% of the OOIP remained in the reservoir. This oil was the target for the in-situ combustion pilot conducted in this reservoir.

Combustion-tube runs with crushed, natural core material were made, and no difficulty was encountered in establishing a vigorous combustion front. **Table 8.94** summarizes combustion-tube data. Fuel availability was 0.94 lbm/ft^3 rock, and the air requirement was 239 scf/ft^3 rock. It was established that the mineral composition of the porous matrix played a dominant factor in the fuel availability of the 40 °API oil.

In-situ combustion was initiated in August 1966 and maintained as a dry forward-combustion project until October 1967 when slugs of water were alternated with air to move some of the heat behind the combustion front to regions ahead of the front. This procedure is discussed in Section 8.7.5. Injection of water was performed by alternating injection of 1,000 bbl of water with 8 to 10 MMscf of air. **Tables 8.95 and 8.96** summarize performance of the in-situ combustion pilot through January 1970. Oil recovery was 705 bbl/acre-ft, which was approximately 68% of the OIP at the beginning of the pilot test. Because the reservoir was thin (average thickness was 8.3 ft in the pilot area), gravity override was not a factor and the vertical sweep efficiency was 100%. Estimated areal sweep efficiency is shown in **Fig. 8.206,** where the combustion front was estimated by a computer model. The combustion front arrived nearly simultaneously at Wells DU 80-2 and DU 73-1, but had not reached Wells DU 81-1 and DU 74-1 when this analysis was completed.

West Heidelberg Unit. The West Heidelberg Cotton Valley Sands Unit is located in Jasper County, Mississippi, USA. (Huffman et al. 1983). The Cotton Valley sands range in depth from 10,900 to 11,800 ft. Two sands, Sands 4 and 5 shown in the type log in **Fig. 8.207,** are the major oil reservoirs in this unit. As **Fig. 8.208** shows, the Cotton Valley sands lie on the flank of

Combination Tube Test (CTT) Data	
Air requirement (scf/ft^3 of rock)	239
Fuel requirement (lbm/ft^3 of rock)	0.94
Oxygen utilization efficiency (%)	94.5
Actual H/C atomic ratio	0.9
Apparent H/C atomic ratio (gas analysis)	3.1
Heating value of fuel (Btu/100 lbm rock)	21,000
Pressure of CTT (psi)	500
Air flux CTT (scf/ft^3-hr)	25.3
Combustion-zone advance (ft/D)	2.5
Oil Properties	
Combustion index (°F-cp)	411
Viscosity (cp)	
At 80°F	6.2
At 135°F	3.0
At 350°F	0.79

Distillation	
Amount Distilled (%)	Temperature (°F)
Initial	150
10	255
20	325
30	412
40	503
50	570
60	612
70	635
80	650
90	658

Table 8.94—Combustion tube test data and oil properties, May-Libby reservoir.

Pilot acreage	40
Pilot volume (acre-ft)	333
Average thickness (ft)	8.3
Average water saturation (%)	30.0
Porosity (%)	31.24
Cumulated air injected (1 January 1970) (Mcf)	2,327,060
Air requirement (scf/ft^3)	240
Volume burned (acre-ft)	205
Fuel consumed (DU 73-1 and DU 80-2 envelopes) (bbl oil/acre-ft)	120
Average oxygen utilization (%)	92.1
Vertical sweep efficiency (%)	100

Table 8.95—Summary of in-situ combustion performance, May-Libby reservoir.

a piercement salt dome, with a dip of approximately 8° from east to west. The updip reservoir limit is the salt intrusion, and the downdip reservoir limit is an asphalt deposit believed to be immobile.

Table 8.97 summarizes reservoir and fluid properties for Sands 4 and 5, which have similar properties. Average net-pay thickness is 62 ft, porosity is approximately 14%, and permeability averages 85 md. The OOIP was estimated to be 10 million

Well	Drainage Volume (acre-ft)	Gas Saturation (% PV)	OOIP (bbl)	Oil Recovery (bbl)
DU 73-1	65.78	41.37	45,651	37,871
DU 74-1	56.71	0.00	96,214	88,205
DU 81-1	63.99	16.91	82,339	27,235
DU 80-2	44.61	55.61	15,553	9,773
Total (1 January 1970)				163,084
Unit Oil Recovery (bbl oil/acre-ft)				
Burned zone				795
Drainage volume				705
DU 73-1 drainage volume				574
DU 80-2 drainage volume				219

Table 8.96—Summary of displacement performance by drainage area.

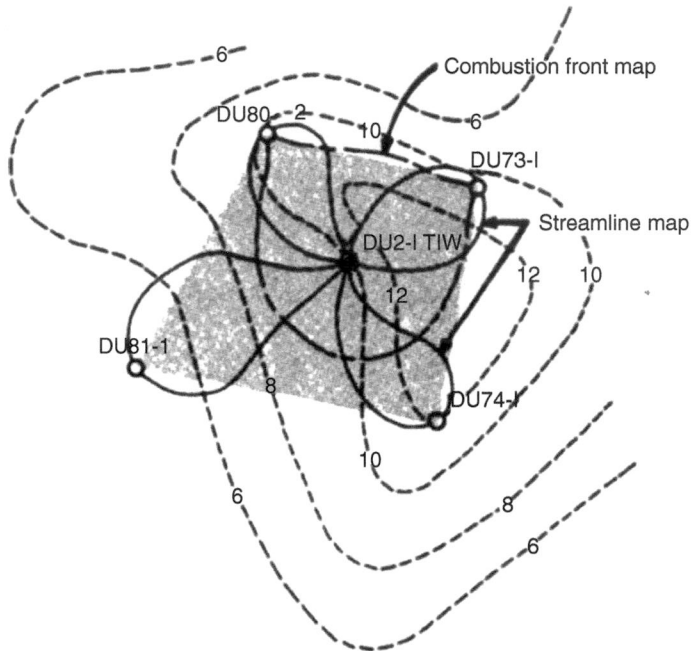

Fig. 8.206—Estimated location of combustion front on overlay of isopach and streamline maps on May-Libby in-situ combustion pilot project (Hardy et al. 1972).

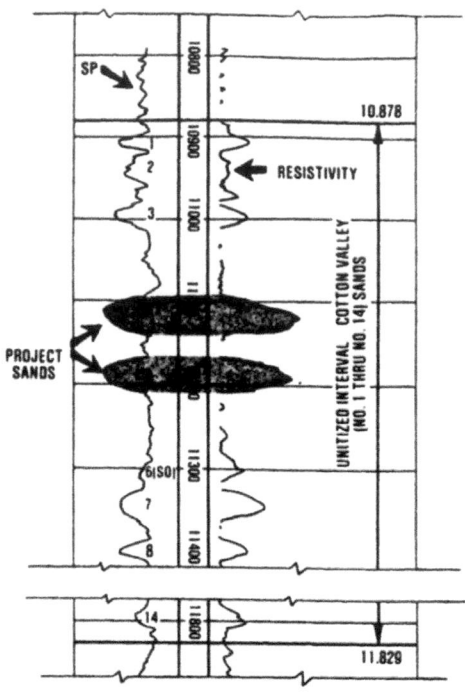

Fig. 8.207—Type log (Well 5-11-1) West Heidelberg Unit (Huffman et al. 1983).

bbl for Sand 4 and 8 million bbl for Sand 5. The reservoir had an initial reservoir pressure of approximately 5,100 psi and a bubblepoint pressure of 930 psi, indicating a highly undersaturated reservoir. A solution GOR of 108 scf/STB was reported between 5,100 psi and the bubblepoint pressure of 930 psi, possibly because of the wide range of oil gravities in the reservoir. Oil gravity varied from 15 °API near the asphalt region to 27 °API at the top of the structure. Oil viscosity is 6 cp at the reservoir temperature of 221°F.

Primary production was typical of an undersaturated reservoir and was characterized by rapid decline of both reservoir pressure and production rates. Primary recovery was estimated as 3% of OOIP for Sand 4 and 10% of OOIP for Sand 5, for an average of approximately 6% of OOIP for the combined sands. Air injection was selected for pressure maintenance because the mobility ratio was unfavorable for a waterflood and lifting costs were projected to be high. Preparations were made for pressure maintenance in Sand 5 by squeeze cementing all zones except Sand 5. Subsequently, it was determined that, although Sands 4 and 5 are separated by impermeable shale, there was communication behind pipe in several wells; therefore, the two zones are now treated as a single unit. Well 5-6-2 was converted to injection, and air injection began in December 1971. Ignition was spontaneous, and flue-gas production began in early 1972, rising rapidly, as **Fig. 8.209** shows. Injection began in 1977 into Sand 5 at Well 5-11-1. **Table 8.98** contains a typical flue-gas analysis. In addition to CO_2, which indicates in-situ combustion, substantial amounts of C_1 through C_{7+} were in the flue gas, possibly stripped from the reservoir oil by the gas.

The project was expanded between 1978 and 1982 by adding one injector and eight producers. As **Fig. 8.210** shows, air/flue-gas injection rates increased significantly (SPE Report 75-25-2/90 1990). Oil production rates from 1970 to mid-1990 are also shown in Fig. 8.210. Incremental oil was being produced at a rate of approximately 550 STB/D in mid-1990. The incremental oil recovery resulting from pressure maintenance was more than 4 million STB. Anticipated ultimate recovery is 30% of OOIP (SPE Report 75-25-2/90 1990). This project illustrates how in-situ combustion can be used to provide reservoir energy (pressure maintenance) as well as to displace oil.

8.7.5 Wet Combustion—Process Fundamentals. Dry in-situ combustion generates a large amount of heat that is either stored in the clean porous rock behind the combustion front or lost to the surroundings. A small amount of the energy in the hot rock behind the front is transported to the combustion front by the injected air as it is preheated from injection temperature to near combustion temperature. About one-fifth of the heat is picked up by the incoming air (SPE Report 75-25-2/75 1975). A large quantity of energy remains in this region because the heat capacity of air is relatively small (0.2 Btu/lbm-°F).

Practitioners of in-situ combustion recognized that much of the energy stored in the hot rock would eventually be lost to the over- and underlying formations. Several approaches were developed to use this energy to displace oil. In one method, air injection was stopped at some point in the combustion project and then converted to water injection to pick up some of the heat behind the combustion front.

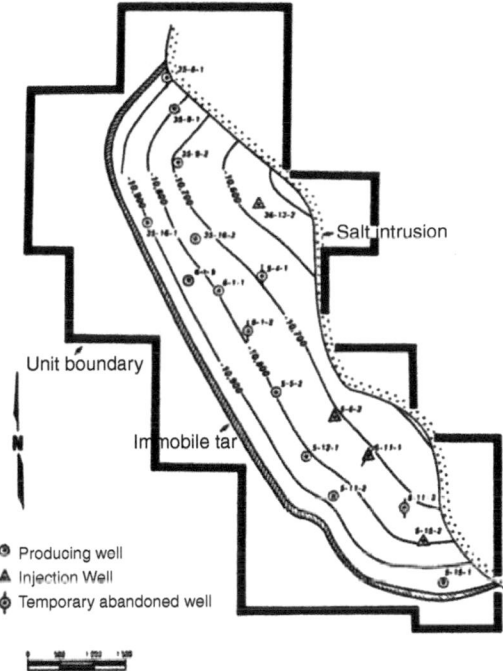

Fig. 8.208—Top of Sand 4, West Heidelberg Unit (SPE Report 75-25-2/90 1990).

This created an in-situ hot waterflood. In other projects, water was injected into wells behind the combustion front in an attempt to control air flow and to salvage some of the energy behind the front.

A process known as wet combustion was developed to improve the efficiency of forward combustion by simultaneous (or alternate) injection of air and water during the combustion process. This process is also referred to as combination of forward combustion and water (COFCAW). When water and air are injected simultaneously, the water initially fills part of the region

Formation	Sandstone
Depth (ft)	11,300
Productive acres	352
Average net-pay thickness (ft)	62
Average porosity (%)	14
Average air permeability (md)	85
Initial water saturation (% PV)	15
Initial oil saturation (% PV)	85
Original reservoir pressure (psi)	5,100
Current reservoir pressure (psi)	2,500
Bubblepoint pressure (psi)	930
Solution GOR above 930 psi (scf/STB)	108
FVF (RB/STB)	1,1397
Reservoir temperature (°F)	221
Oil viscosity at original reservoir conditions (cp)	6.0
Oil gravity (°API)	18 to 27
OOIP (10^6 STB)	18
Primary recovery (10^6 STB)	1.1
Primary recovery factor (% OOIP)	6.1
Recovery as of 1 August 1982 (10^6 STB)	4.0
Recovery factor as of 1 August 1982 (% OOIP)	22.2

Table 8.97—Reservoir characteristics of Cotton Valley Sands 4 and 5, West Heidelberg unit.

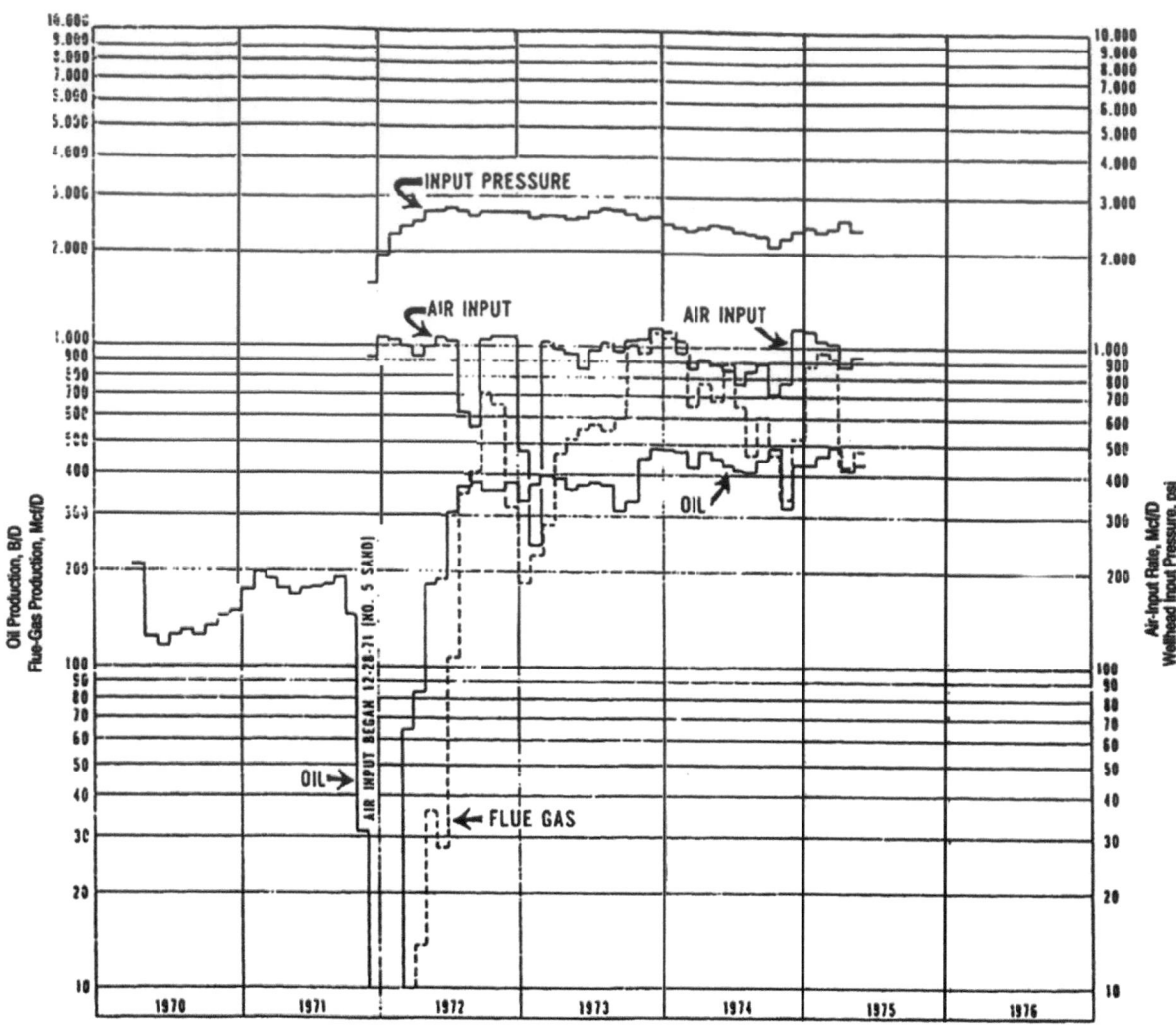

Fig. 8.209—Performance curves, West Heidelberg Unit, 1970–75 (SPE Report 75-25-2/90 1990).

Methane (vol%)	3.29
Ethane (vol%)	0.65
Propane (vol%)	0.89
i-butane (vol%)	0.15
n-butane (vol%)	0.51
i-pentane (vol%)	0.05
n-pentane (vol%)	0.22
Hexanes plus (vol%)	0.14
Carbon dioxide (vol%)	14.61
Carbon monoxide (vol%)	0.07
Oxygen/argon (vol%)	1.02
Nitrogen (vol%)	78.40
Gross heating value (Btu/scf)	100.00

Table 8.98—Typical flue-gas analysis, West Heidelberg unit.

behind the combustion front. As water saturations increase, the water is displaced into the heated region where it is converted to superheated steam. It then flows through the combustion front. The additional steam created by water injection mixes with the combustion gases and volatile hydrocarbons downstream of the combustion zone, where the steam forms an enlarged condensation zone. Under ideal conditions, the enlarged condensation zone travels up to three times faster than the combustion

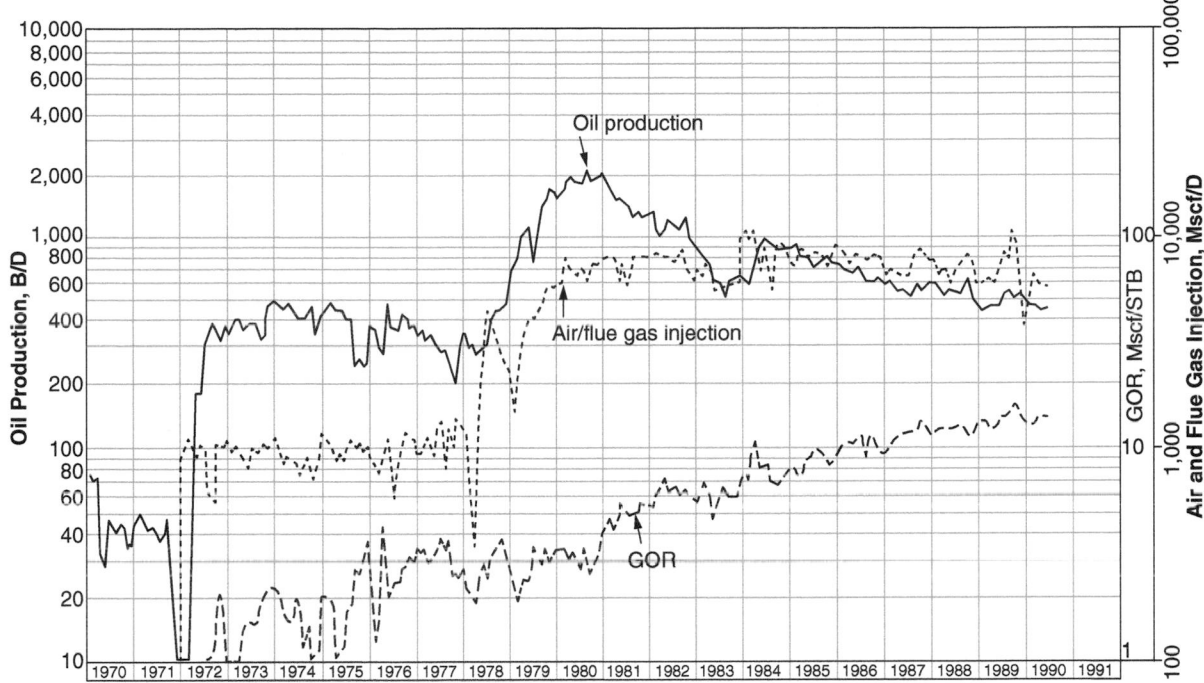

Fig. 8.210—Performance of Composite Sands 4 and 5, West Heidelberg Unit (SPE Report 75-25-2/90 1990).

zone (Parrish and Craig 1969), thereby creating an extended region of steam distillation ahead of the combustion front. The length of the steam zone is determined by the amount of heat recovered from the region behind the combustion front.

The concept of wet combustion is illustrated in a series of temperature profiles (**Figs. 8.211 through 8.213**) measured along the length of a combustion tube in which wet combustion was initiated. The combustion tube was equipped with thermocouples at the center and edge of the porous medium and in the annulus between the core holder and the external pressure vessel. Fig. 8.211 shows the typical temperature profile just after stable dry in-situ combustion was initiated in a long combustion tube. A small steam plateau is emerging in the vicinity of 400°F.

Fig. 8.212 is the temperature profile at steam breakthrough after water was injected alternately with air for the remainder of the displacement. Injected air/water ratios were on the order of 1,800 to 2,000 scf/bbl. Under the conditions of this run, the combustion front is still burning at the same temperature as under dry combustion. A steep temperature gradient has formed behind the combustion front owing to vaporization of water flowing into this region. Vaporization occurs over a range of temperatures because the partial pressure of water changes

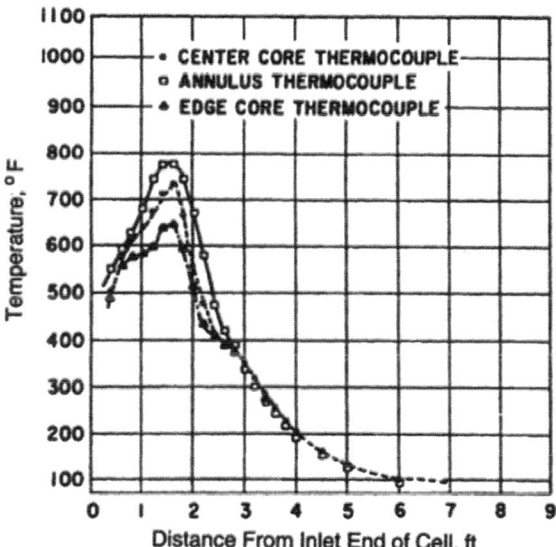

Fig. 8.211—Temperature profile, start of wet combustion (Parrish and Craig 1969).

continually owing to the temperature gradient. A long steam plateau with a constant temperature that corresponds to the pressure in that region forms ahead of the front. Fig. 8.213 shows the temperature profiles along the combustion tube as the combustion zone breaks out of the tube. The back of the combustion zone resembles a heat wave that is moving through the porous rock.

Table 8.99 summarizes the wet-combustion test conditions of Parrish and Craig (1969). A wide range of crude oil properties were evaluated, including reservoirs that had been waterflooded before initiation of in-situ combustion. **Table 8.100** summarizes displacement results. Oil recovery was between 80 and 94% of OOIP in all runs except Run 1. However, the air requirement was reduced significantly because less oil was burned as fuel. Dietz and Weijdema (1968) show that, under the conditions of optimal wet combustion depicted in **Fig. 8.214,** an in-situ steamdrive is created and the air requirement is reduced to approximately one-third of that required for dry combustion. A small amount of unburned coke was found in the region behind the front. Thus, the combustion front was able to advance without consuming all the coke.

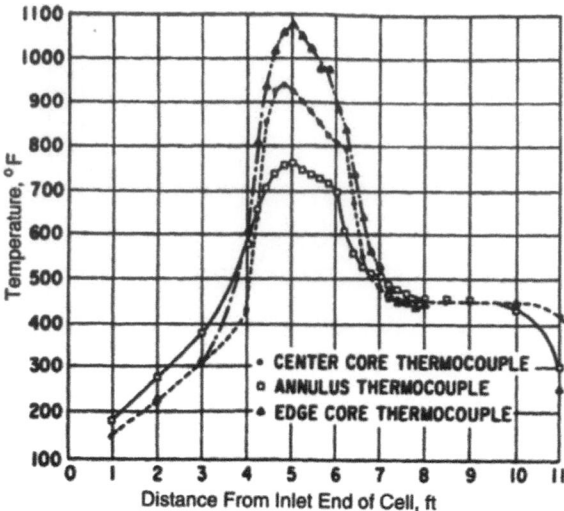

Fig. 8.212—Temperature profile, steam breakthrough in wet-combustion test (Parrish and Craig 1969).

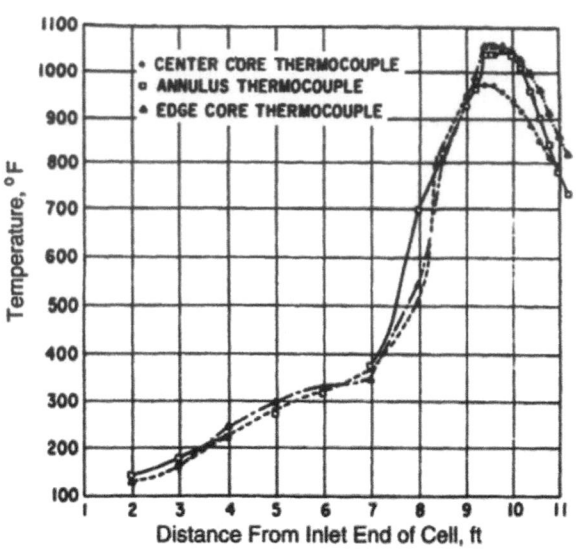

Fig. 8.213—Temperature profile, combustion-zone breakthrough in wet-combustion test (Parrish and Craig 1969).

Run	Crude Source	Oil Gravity (°API)	Oil Viscosity at 100°F (cp)	Initial Cell Temperature (°F)	Stage of Depletion	Fluid Saturation at Start of Test (% PV)			Injected Air/ Water Ratio (scf/bbl)
						Oil	Water	Gas	
1	Rocky Mountains	38.9	3.6	200	Waterflooded	24.0	42.8	33.2	2,522
2	Rocky Mountains	13.5	2,900.0	130	Nonwaterflooded	71.2	0.0	22.8	2,435
3	Rocky Mountains	13.5	2,900.0	130	Nonwaterflooded	68.0	0.0	32.0	1,505
4	West Texas	30.5	5.7	100	Nonwaterflooded	61.2	22.6	16.2	1,680
5	Rocky Mountains	25.2	28.8	100	Nonwaterflooded	79.0	0.0	21.0	1,636
6	West Texas	19.9	60.0	100	Waterflooded	75.3	8.2	16.5	1,077
7	Rocky Mountains	18.4	67.2	120	Waterflooded	41.7	29.8	28.7	1,485
8	US gulf coast	29.2	3.8	200	Waterflooded	29.0	28.6	42.4	1,660
9	US gulf coast	35.2	2.0	177	Waterflooded	30.0	23.1	46.9	2,750
10	Rocky Mountains	19.2	244.0	100	Nonwaterflooded	54.8	23.6	21.6	1,385
11	West Texas	40.9	3.5	100	Nonwaterflooded	43.5	31.8	24.7	2,430

Table 8.99—Summary of COFCAW test conditions.

If water is injected at a high rate, the evaporation zone at the trailing edge of the combustion zone overruns the combustion front (Dietz and Weijdema 1968) and the temperature of the combustion zone declines. This variation of wet combustion is called partially quenched combustion or superwet combustion. Dietz and Weijdema found that it was not necessary to maintain the combustion temperature at 752°F to propagate a combustion zone through a porous rock. Under conditions of partially quenched combustion, the combustion-zone temperature declines to approximately 392°F, the same temperature as the steam plateau. Combustion is maintained because oxygen will react completely with hot oil at 392°F. Heat supplied by the combustion of the fuel maintains the combustion front. **Fig. 8.215** shows experimental temperature profiles in partially quenched combustion obtained in a well-insulated combustion tube. Other experiments demonstrated that the velocities of the combined combustion/evaporation front were determined by the specific water throughput (correlated in **Fig. 8.216).** In all

Run	Oil Gravity (°API)	Air/Water Ratio (scf/bbl) Injected	Air/Water Ratio (scf/bbl) At Combustion Zone	Oil Burned (% PV)	Non-condensable Light Ends in Produced Gas (% PV)	Oil Equivalent to Carbon Residue (% PV)	Total Unrecovered Oil (% PV)	Oil Recovery (% OIP)	Injected Air to Produced Oil (scf/bbl)	Maximum Produced Oil Gravity (°API)
1	38.9	2,522	6,011	4.39	1.92	2.18	8.49	64.6	11,730	40.8
2	13.5	2,435	4,875	5.61	1.41	1.55	8.57	87.9	6,020	25.1
3	13.5	1,505	2,360	4.54	0.62	0.97	5.95	84.4	5,040	24.8
4	30.5	1,680	2,997	4.49	0.60	0.24	5.33	91.4	3,380	44.0
5	25.2	1,636	2,885	4.32	0.44	0.33	5.09	93.8	4,100	31.3
6	19.9	1,077	1,499	3.60	3.04	1.34	6.97	90.9	3,360	27.2
7	18.4	1,485	2,650	6.00	1.34	0.58	7.92	81.0	10,900	31.9
8	29.2	1,660	2,670	4.01	0.24	0.13	4.38	84.9	8,240	36.6
9	35.2	2,750	5,820	1.59	0.75	0.04	1.71	94.3	3,500	43.1
10	19.2	1,385	1,710	3.84	0.33	0.06	4.23	92.3	1,870	36.8
11	40.9	2,430	4,430	3.37	0.57	0.01	3.95	90.9	4,340	64.8

Table 8.100—Summary of COFCAW test results.

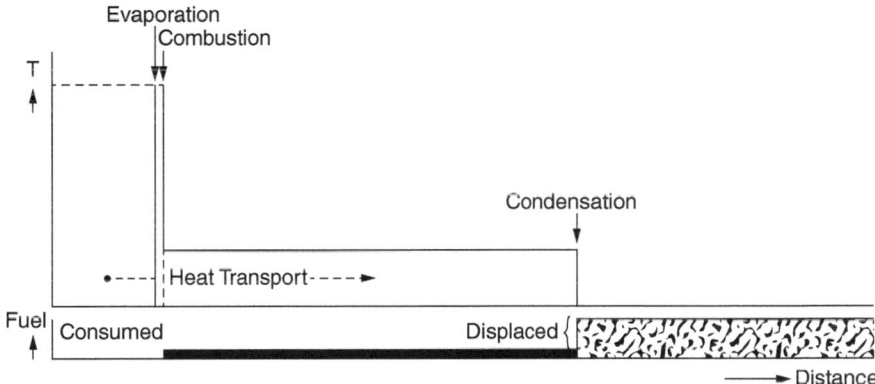

Fig. 8.214—Optimal normal wet combustion (Dietz and Weijdema 1968).

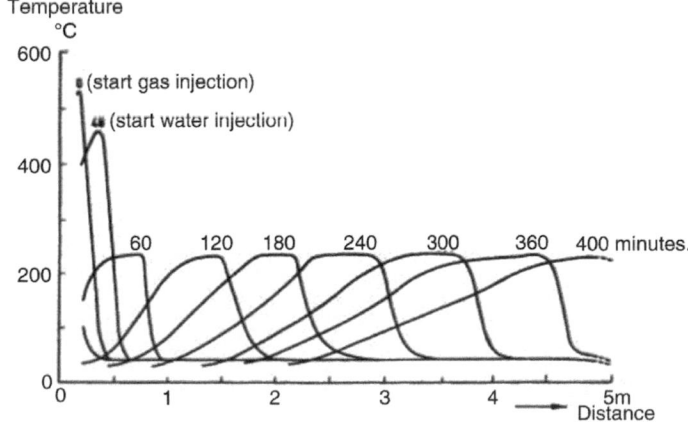

Fig. 8.215—Experimental temperature profiles in partially quenched combustion (Dietz and Weijdema 1968).

cases, unburned fuel remained on the sand as a result of partially quenched combustion. The unburned fuel consisted primarily of carbon, indicating preferential oxidation of hydrogen. Dietz and Weijdema also observed less production of acids, which are emulsion-forming constituents during partially quenched combustion.

Burger and Sahuquet (1973) carried out a comprehensive set of wet-combustion experiments that confirmed much of the work of Parrish and Craig. In an extensive set of experiments, they observed that no coke remained in the burned zone as long

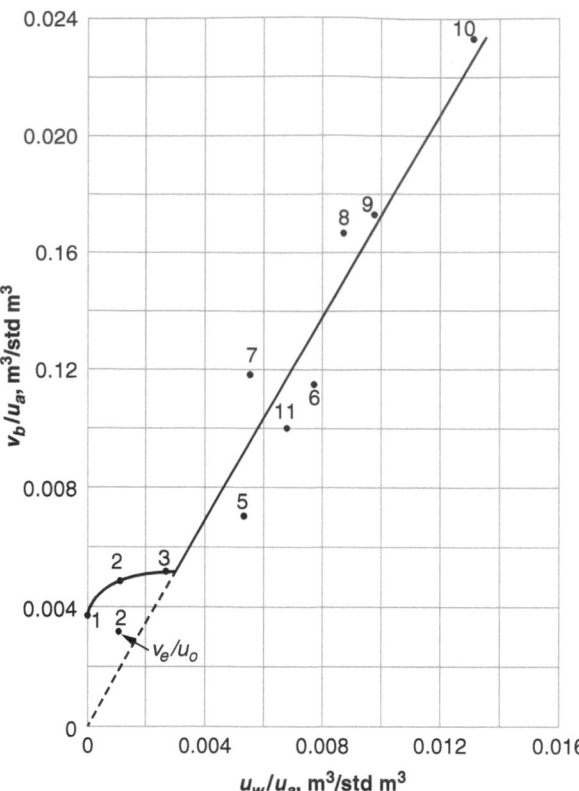

Fig. 8.216—Combustion front velocity as a function of specific water throughput (Dietz and Weijdema 1968).

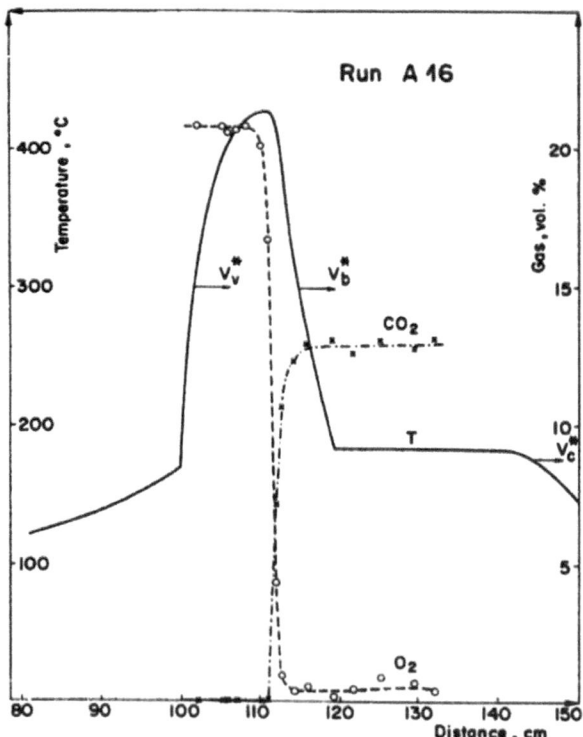

Fig. 8.217—Typical temperature and concentration profiles for Run A16, wet-combustion phase at t = 45 hours (Burger and Sahuquet 1973).

as a high-temperature zone existed. Less fuel was consumed under wet combustion, which they attributed to the effects of the prolonged steam plateau on the oil saturation remaining to be used as fuel.

Burger and Sahuquet introduced a systematic method to characterize the wet-combustion process. The wet-combustion process was described in terms of three characteristic velocities (shown on **Fig. 8.217**): v_v^*, the velocity of the vaporization front; v_b^*, the velocity of the combustion front; and v_c^*, the velocity of the condensation front. Velocities are determined from the interpretation of temperature profiles (as in **Fig. 8.218**). Such an interpretation for Run B2 is presented in **Fig. 8.219,** where the locations of the vaporization front, combustion front, and condensation front are plotted as functions of time.

The slope of each line is the velocity of that zone at a specific time. Inspection of Fig. 8.219 shows that the velocities of the combustion and condensation fronts were constant during the dry-combustion period ($t \leq 22.5$ hours) with v_c^* nearly the same as v_b^*. The velocity of the condensation front increases shortly after the initiation of wet combustion, and a vaporization front is formed that also moves at a constant velocity. The combustion-front velocity increases at a slower rate, eventually reaching what appears to be a steady-state value.

As **Fig. 8.220** shows, the vaporization front overruns the combustion front when the water injection rate is large, and high-temperature combustion is extinguished. This occurs because $v_v^* > v_b^*$. Extinction was observed at an air/water ratio as low as approximately 1,400 scf/bbl water. Water injection was effective at air/water ratios of approximately 5,600 scf/bbl.

8.7.6 Design of Wet-Combustion Projects. The design of wet-combustion projects follows the same approaches used for dry combustion except that the air requirement is reduced and AORs decrease. The critical variable is air/water ratio. Burger and Sahuquet (1973) proposed methods of estimating the minimum and maximum air/water ratio for wet combustion from a steady-state analysis of saturation banks and an energy balance. However, in most cases, the air/water ratio is determined from combustion-tube data.

When combustion-tube data are unavailable, estimates may be made with data from correlations. Garon and Wygal (1974) conducted a series of 131 combustion-tube tests covering a wide range of process variables. The air requirements for wet combustion were always less than for dry in-situ combustion. Prats (1982) correlated the data of Garon and Wygal, Dietz and Weijdema, Parrish and Craig, and Burger and Sahuquet, as shown in **Fig. 8.221.** For most of the data, there is a strong relationship between a_R^* and F_{wa}. The data of Garon and Wygal were correlated in US DOE (1986) to estimate a_R^* with

$$a_R^* = 73 \log(\mu_o) + 200 - 100 F_{wa}, \ldots\ldots\ldots (8.377)$$

where μ_o = dead oil viscosity at 60°F, 14.7 psia, cp, and F_{wa} = water/air ratio, bbl/Mscf.

Oil recovery under conditions of wet combustion may be estimated with the air requirement determined for the conditions of wet combustion in the correlations discussed in Section 8.7.4 for dry combustion. This approach was selected for the National Petroleum Council Study of EOR in 1982. The FORTRAN program simulating in-situ combustion incorporates wet combustion by altering the air requirement (US DOE 1986).

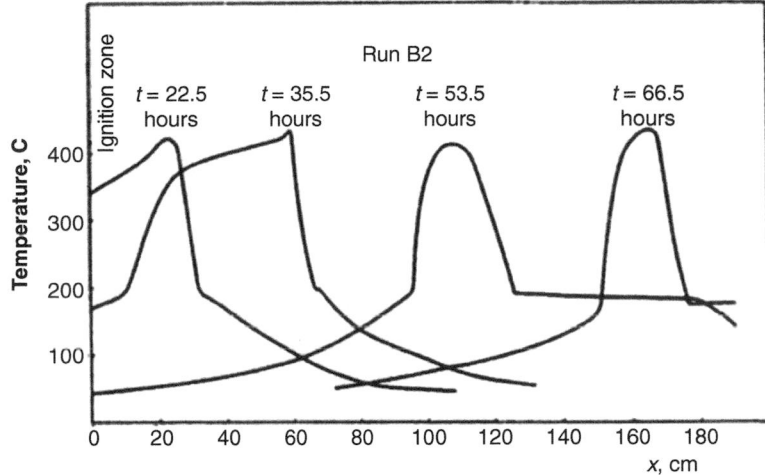

Fig. 8.218—Temperature profiles vs. distance for Run B2, wet combustion (Burger and Sahuquet 1973).

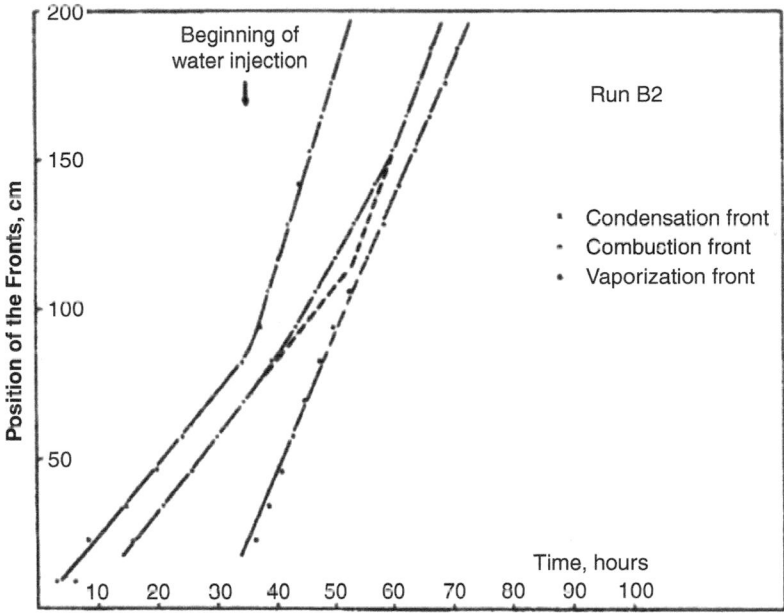

Fig. 8.219—Positions of the fronts along Run B2 (Burger and Sahuquet 1973).

The wet-combustion process depends on the flow of water into the heated rock behind the combustion front, where the water vaporizes and flows through the combustion front. Laboratory combustion-tube tests are usually run under conditions where gravity segregation is negligible. Gravity segregation of air and water may cause field results to differ from those observed in laboratory combustion-tube experiments. Ideal wet-combustion conditions observed in laboratory combustion tubes are more likely in thin reservoirs ($h \leq 30$ ft), where gravity segregation may not be a significant factor.

8.7.7 Field Cases—Wet Combustion. The discovery that the air requirement for forward combustion could be reduced by simultaneous or alternate injection of water, thereby reducing the air compression costs, stimulated many pilot tests of wet combustion and conversion of some dry-combustion projects to wet combustion. Craig and Parrish (1974) describe an extensive series of tests in which wet combustion was pilot tested in seven different fields to define the potential of the process. These tests provided experience about operating problems in the field and identified types of reservoirs where application of the process was unlikely. Dietz (1970) also describes operating problems in wet-combustion processes.

Evaluation of wet combustion in the field was difficult because operators sometimes used water injection to control air channeling. In a few cases, water was injected into wells behind the combustion front, possibly to fill the void space or to control the movement of the combustion front. As a result, few well-documented field-scale wet-combustion projects are in the published literature and much speculation exists as to how much wet combustion will improve the performance of in-situ combustion processes. This section reviews three field tests.

Bellevue Field. The Bellevue field is located in Bossier Parish, Louisiana, USA. Oil is produced from the Nacatoch sands at depths ranging from 300 to 420 ft. The field is on a dome that has closure on top of the Nacatoch sands. The Nacatoch is

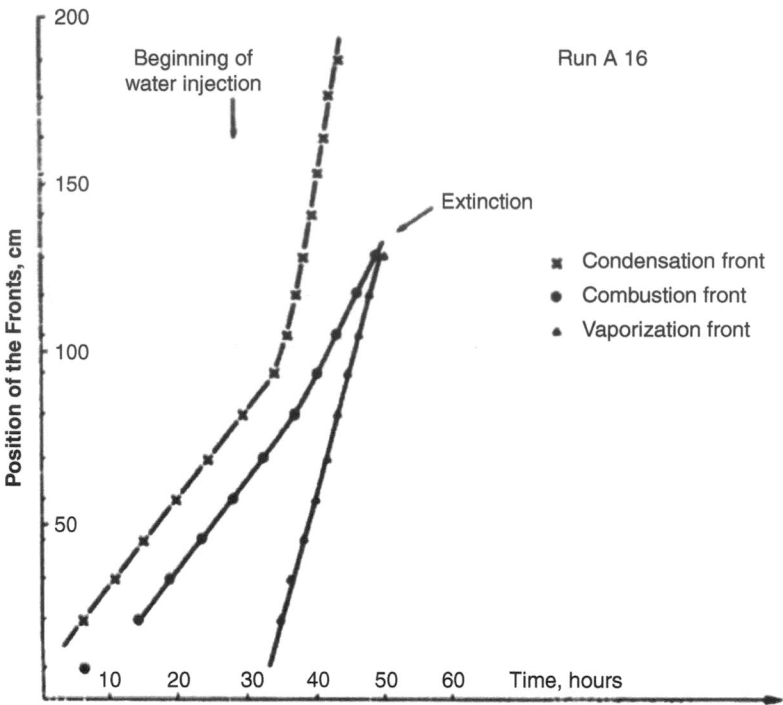

Fig. 8.220—Positions of the fronts along Run A16, showing extinction of the combustion front (Burger and Sahuquet 1973).

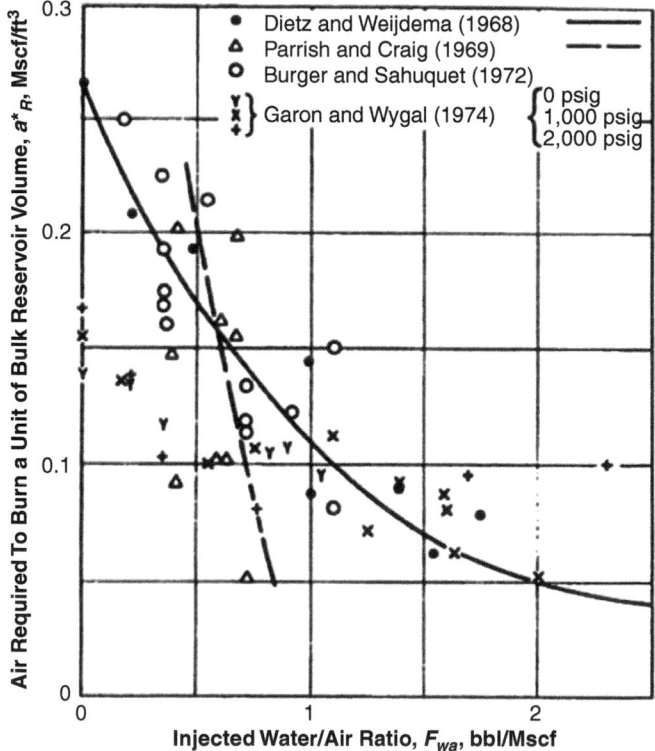

Fig. 8.221—Effect of injected water/air ratio on air required to burn unit reservoir volume (Prats 1982).

broken by numerous faults. A lime caprock overlays the upper sands, and thin lime sections separate zones of the sand. The upper and lower Nacatoch are separated by a lime zone approximately 4 ft thick that appears to be continuous over much of the field. There is vertical communication between the two zones.

Thickness of the sands varies from 20 to 90 ft. Porosity varies from 34 to 39%, and permeabilities range from 540 to 1,100 md. The reservoir oil gravity is 19.5 °API, and viscosity is 675 cp at the reservoir temperature of 75°F. Average oil saturation is 52%. The OOIP in the region covered by this test was approximately 35 million bbl.

Texaco (formerly Getty Oil Co.) initiated a pilot in-situ combustion project in 1963 in a 1.25-acre inverted five-spot. This pilot was expanded and converted to an inverted nine-spot in 1966. Air injection began at 1 MMscf/D, increasing to 2 MMscf/D after 30 months of operation. Air injection was stopped in 1966, and water injection was used to scavenge the heat remaining in the combustion zone. Water injection continued until WORs became excessive. By the end of 1966, the pilot was producing 145 B/D, compared with 6 B/D before the initiation of in-situ combustion. Cumulative oil production from the pilot as of July 1969 was 184,000 bbl, and at the end of 1981, approximately 372,000 bbl had been produced from the pilot area. The project was expanded several times; **Fig. 8.222** shows the pattern development plan (SPE Report 76-33-2/90 1990), and **Fig. 8.223** shows rates of air injection, water injection, and oil production for the period from 1968 to 1986. By 1989, production was 850 B/D from 200 production wells and 30 injection wells (Moritis 1990).

The Bellevue project began as a dry in-situ combustion project with an operating plan to burn approximately 50% of the pattern volume before converting to water injection. Wet combustion was introduced in 1969, when four patterns were converted from dry to wet combustion. Subsequently, the development plan included a sequence of dry combustion, wet combustion, and water injection. The decision to include a period of wet combustion in the development plan was based on comparison of recoveries from two groups of patterns having comparable areas and thicknesses (Long and Nuar 1982). **Fig. 8.224** compares annual oil production as percent of OIP as a function of years after ignition for the wet-combustion group (dry combustion, wet combustion, and water injection) and the nonwet-combustion group (dry combustion and water injection). The comparison

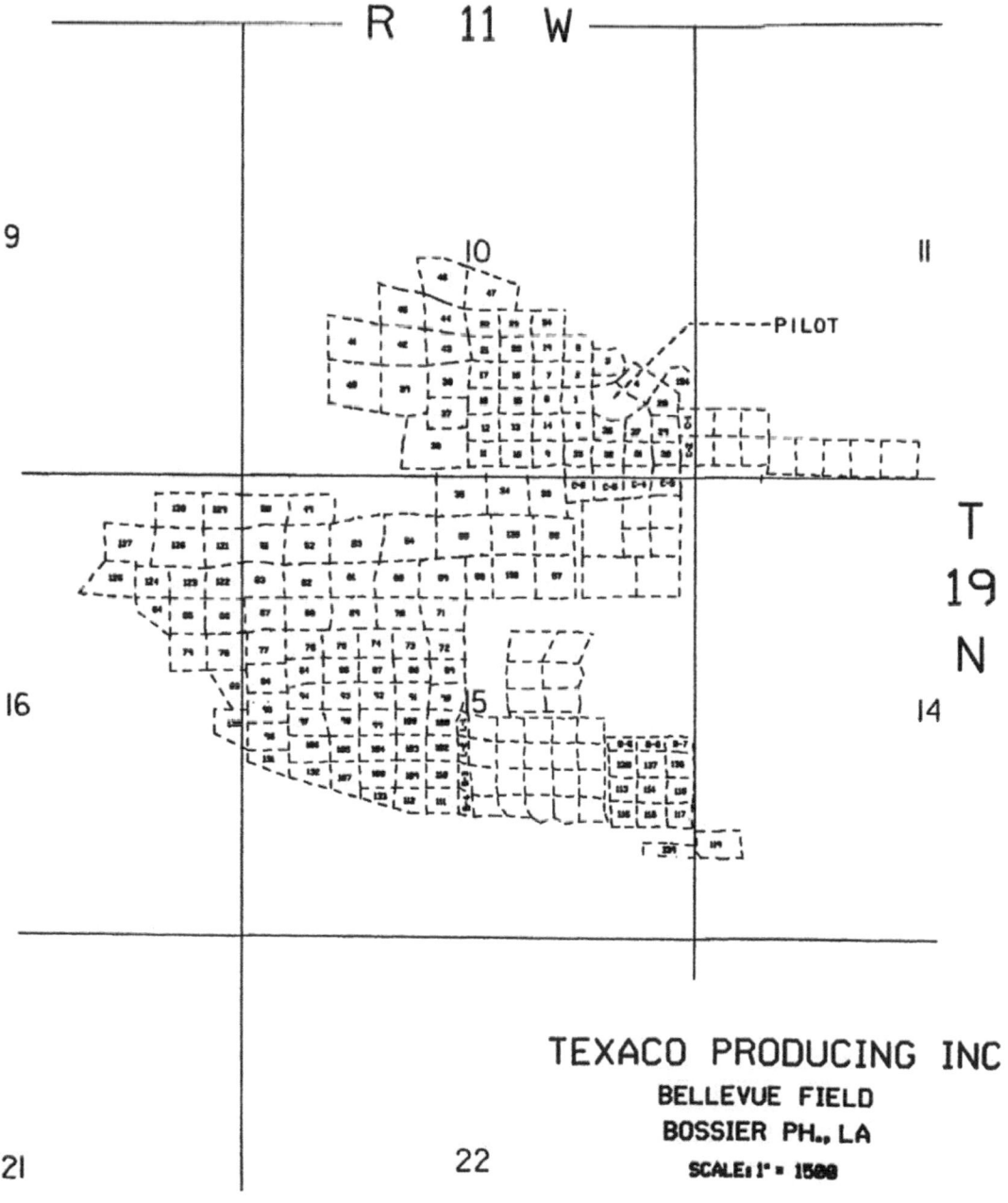

Fig. 8.222—Layout of patterns in development of in-situ combustion project, Bellevue field (SPE Report 76-33-2/90 1990).

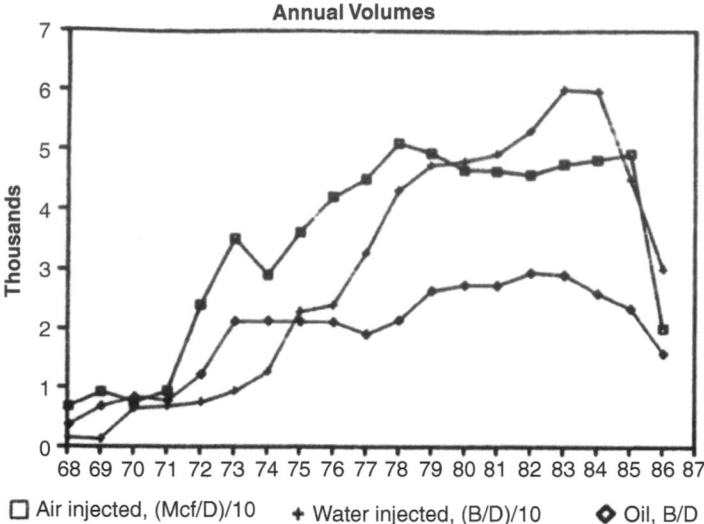

Fig. 8.223—Summary of air injection, water injection, and oil production rates from 1968 to 1986, Bellevue in-situ combustion project (SPE Report 76-33-2/90 1990).

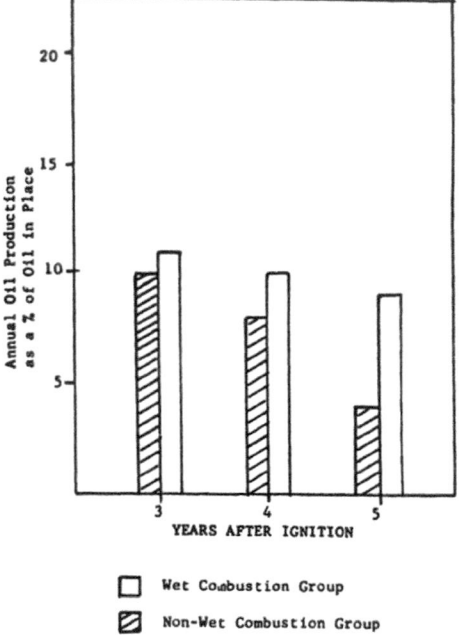

Fig. 8.224—Comparison of recovery vs. time for dry-combustion and wet-combustion development sequences (Long and Nuar 1982).

shows that patterns with a wet-combustion period declined at a slower rate than those that were directly converted to water injection.

Patterns in the Bellevue field are developed as inverted nine-spots with pattern volumes of approximately 185 acre-ft. Thus, patterns with thicknesses greater than 50 ft were drilled with areas as low as 2.2 acres, while sands with thicknesses of 20 ft had areas as large as 8.5 acres. Patterns with large areas were slow to respond to the combustion drive. An analysis of production was made for three groups of six patterns that had been ignited for at least 5 years to determine the effect of pattern area on displacement performance in areas of thin sand. Patterns were grouped into three categories: Group A (area < 3 acres), Group B (3.1 to 3.8 acres), and Group C (4.2 to 5.8 acres). Performance of these groups (**Fig. 8.225**) shows that recovery was significantly better for the smaller patterns.

The Bellevue field is not a continuous-wet-combustion project; therefore, it is not possible to compare dry and wet combustion directly. The combination of dry combustion, wet combustion, and water injection developed for this field has led to a successful project in which more than 9 million bbl of oil had been produced by 1982. A wet-combustion field test was carried out by another operator in another section of the Bellevue field. Additional information can be found in the literature (Joseph and Pusch 1980). Production from the Bellevue field was reported to be 300 BOPD in 2007 (Turta et al. 2007).

East Tia Juana Field. The East Tia Juana field is in Venezuela (Dietz 1970). The reservoir consists of two zones of unconsolidated sand separated by shale. Net sand thickness is 39.4 ft in the upper sand and 88.6 ft in the lower sand. A major steamdrive project was conducted in another part of the field, and de Haan and Schenk (1969) and de Haan and van Lookeren (1969) give extensive reservoir description data on the field. **Table 8.101** gives reservoir data for the East Tia Juana pilot test. The pilot area is a 11.4-acre inverted seven-spot (**Fig. 8.226**).

The wet-combustion design was based on vertical combustion and displacement fronts assuming an areal sweep efficiency of 66.7%. Water/air ratio was 0.007 ft³/scf (0.00136 bbl/scf). Wet combustion was initiated spontaneously in two injection wells after each well had been treated with sufficient steam to create a high-temperature region around the well. Production wells were also given steam stimulation treatments. Simultaneous air and water injection began in 1966 and was terminated in March 1968. **Fig. 8.227** summarizes injection and production data. By March 1968, 1.5 million bbl of oil had been produced. However, there was some uncertainty about how to allocate oil to the wet-combustion process because of the effects of steam soaks on the six pattern producing wells. A conservative estimate of the oil attributed to wet combustion would be 220,000 bbl. The AOR was 938 scf/bbl, which is significantly lower than reported from dry-combustion tests and probably has a high degree of uncertainty owing to the uncertainty in the oil attributed to wet combustion.

Sloss Wet-Combustion Project. The Sloss field is in Kimball County, Nebraska, USA, and produces light (38.8 °API) oil from a stratigraphic trap in the Muddy "J" sandstone. Two reservoirs are in the field, the J1, at a depth of 6,200 ft, and the J2, which is slightly deeper. These reservoirs are separated by shale. The wet-combustion project was conducted in the J1 reservoir shown in **Fig. 8.228.** The Sloss field was produced by solution-gas drive until 1958, when a waterflood was initiated. Most of the reservoir had been waterflooded when the wet-combustion project was initiated.

A full-scale wet-combustion project was initiated in Sand J1 after results from a pilot test were encouraging (Parrish et al. 1974). The test started with the six 80-acre five-spots (**Fig. 8.229**) and was later expanded to include 960 acres. **Table 8.102** summarizes reservoir data for the area affected by the wet-combustion process. The viscosity of the Sloss J1 reservoir oil was 0.8 cp at reservoir temperature; therefore, the wet-combustion project relied primarily on displacement by the combustion front rather than viscosity reduction to increase oil recovery. Furthermore, the project was initiated in a waterflood area where the average residual oil saturation after waterflood was 0.30 ± 0.10.

Air injection into the six Phase I wells began in 1967. Ignition was spontaneous, assisted by preinjection of 150 bbl of reactive oil into each injection well in the Phase I area. Air was injected with water from 1967 to 1971 at a ratio of 1,271 scf/bbl. Water injection continued for 2.5 years after termination of air injection. **Figs. 8.230 through 8.232** show fluid injection, production, air/water ratio, and AOR. **Table 8.103** summarizes project performance. Approximately 646,776 bbl of oil was produced during the wet-combustion phase, and another 189,000 bbl was produced to 1 January 1974. The cumulative oil recovered to 1

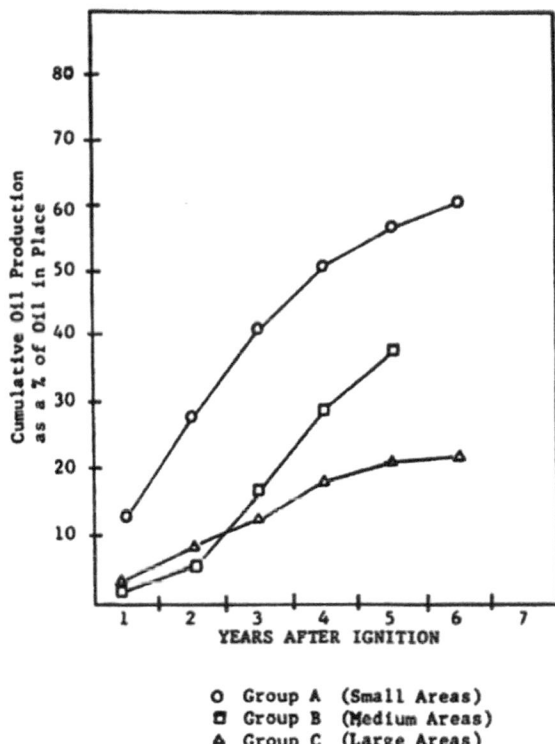

Fig. 8.225—Effect of pattern area on rate of oil recovery, Bellevue in-situ combustion project (Long and Nuar 1982).

January 1974, was 916,000 bbl (including 80,000 bbl from the pilot area), or 796,000 bbl incremental over waterflooding. The produced gas contained recoverable hydrocarbons, but was vented. Of the 340,000 bbl estimated to be vented, approximately 210,000 bbl were considered to be recoverable. Adding the potentially recoverable hydrocarbons would increase the recoverable oil to 1,126,000 bbl. AORs were 21,226 scf/bbl to 1 July 1971, considerably greater than expected from laboratory and pilot test results.

Parrish et al. (1974) present a detailed analysis of the operating problems associated with the Sloss wet-combustion project and a well-documented appraisal of the results. The project was judged to be economically unattractive for three reasons: low initial oil content, poor volumetric sweep, and high air/water ratio resulting from injectivity problems. Evidence of poor volumetric sweep during the project is shown in **Fig. 8.233,** where approximate shapes of the combustion fronts are shown assuming 100% vertical sweep efficiency. Because subsequent core holes showed less than 100% vertical sweep efficiency, the areal sweep is larger than indicated in Fig. 8.233. However, Fig. 8.233 shows the dominance of reservoir heterogeneity in determining the direction that the combustion front moved. Parrish et al. (1974) reported the results of extensive coring in one five-spot pattern. Their data revealed that combustion took place in the more-permeable top part of the pay, the average vertical sweep of the heated zone (T > 350°F) was approximately 28%, and the areal sweep of the heated zone was approximately 50%. Thus, the volumetric sweep efficiency of the wet-combustion process was approximately 14%.

Area (acres)	11.4
Net thickness, upper sand complex (ft)	39.36
Net thickness, lower sand complex (ft)	88.56
Dip (degrees)	4
OIP at start of project (bbl)	3,244,873
Permeability (darcies)	5
Oil viscosity (cp)	6,000
Well spacing (ft)	436.24

Table 8.101—Reservoir data, East Tia Juana wet-combustion pilot test.

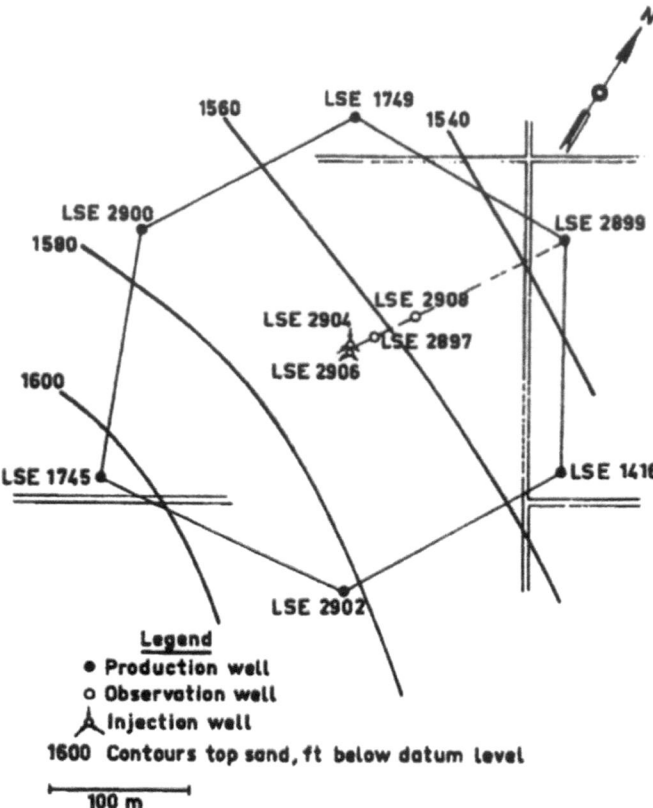

Fig. 8.226—East Tia Juana wet underground combustion test site (Dietz 1970).

No obvious relationship exists between the estimated shape of the front and controllable operating parameters. In the Sloss project, production wells were essentially ineffective after heat breakthrough. It was thought that a well pattern other than a five-spot might have been more effective in increasing the volumetric sweep efficiency of the process. Although the Sloss project was well designed and executed, the wet-combustion process was not applicable to the Sloss field.

8.7.8 Enriched Oxygen Injection. In-situ combustion was developed with air used as the oxygen-containing gas. Because air contains 79% nitrogen, which acts as an inert component in the process, it has long been recognized that the minimum air flux could be reduced by a factor of 0.79/0.21 if pure oxygen were injected. Other benefits of oxygen enrichment include reduced compression costs per unit of injected oxygen (Buxton and Pollock 1974) and gases with increased CO_2 content, which can dissolve in the reservoir oil and reduce the viscosity. There were also concerns about the hazards of use of pure oxygen in laboratory and field experiments.

Combustion-tube experiments (Buxton and Pollock 1974; Hansel et al. 1984) with a light crude oil (31 °API) verified the benefits of enriched oxygen injection. Hansel et al. (1984) report that most combustion characteristics (fuel availability, H/C ratio, peak temperatures, CO_2/CO ratio, and oxygen usage) between 40 and 95% oxygen were similar to those

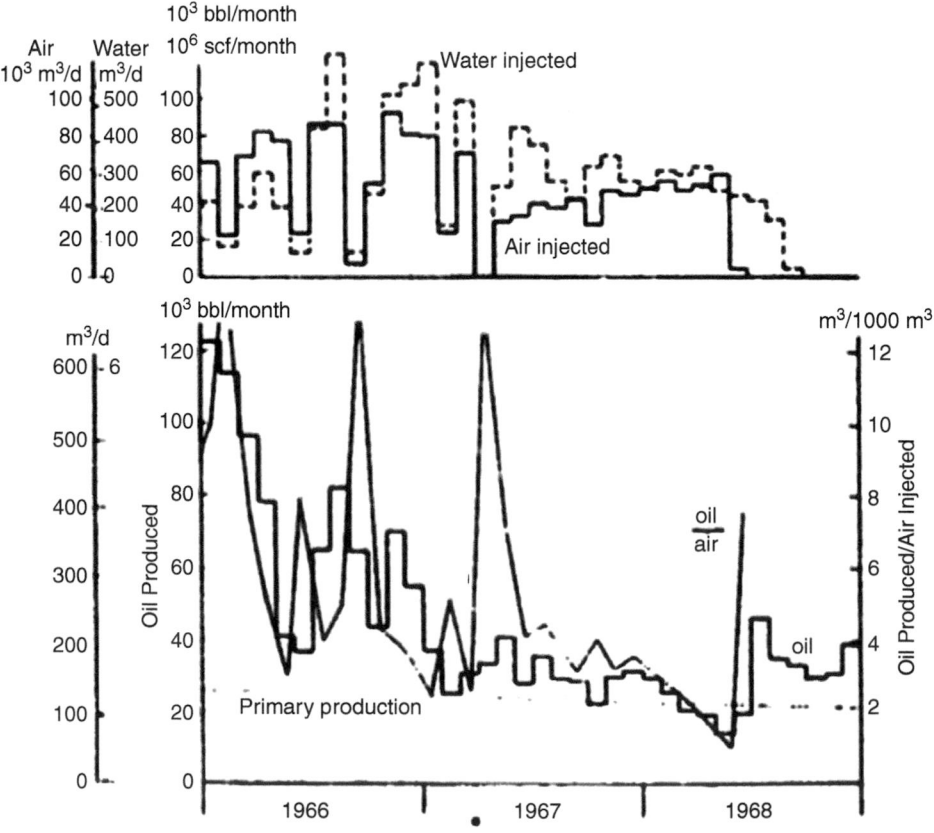

Fig. 8.227—East Tia Juana wet underground combustion reservoir performance (Dietz 1970).

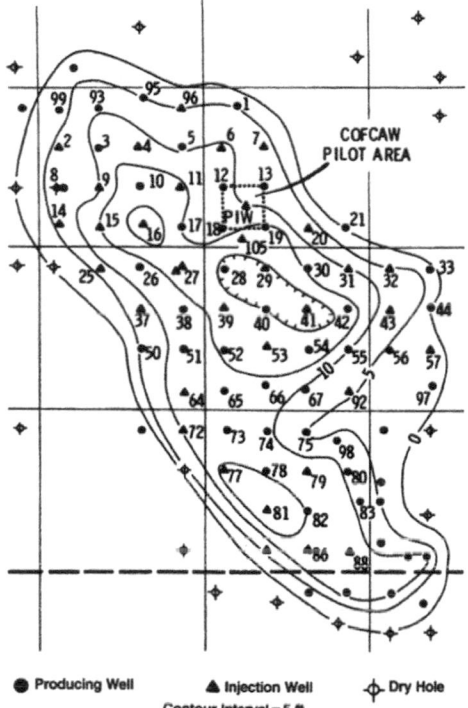

Fig. 8.228—Isopach map of the Muddy J1 reservoir, Sloss field, Nebraska (Parrish et al. 1974).

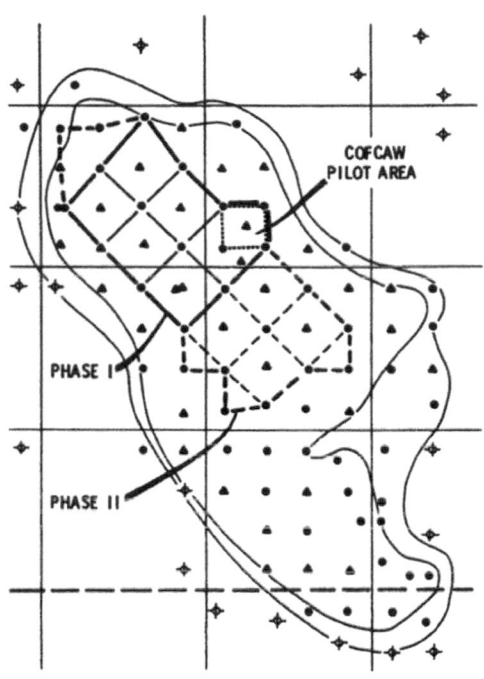

Fig. 8.229—Wet-combustion project area (Parrish et al. 1974).

Formation	Muddy J1 sandstone
Structure	Stratigraphic
Producing mechanism	
Primary	Solution-gas drive
Secondary	Waterflood
Average depth (ft)	9,200
Average porosity (%)	19.3
Average air permeability (md)	191
Oil saturation after waterflood (% PV)	30 ± 10
BHP (psig)	
1968 survey	2,274
1970 survey	3,158
Bottomhole temperature (°F)	200
Crude gravity before COFCAW (°API)	38.8
Oil viscosity before COFCAW (cp)	0.8
Reservoir volume factor (vol/vol)	1.05
Full-scale project	
Area (acres)	960
Average net pay (ft)	14.3
OIP*	
bbl/acre-ft	427
Million bbl	5.9

*Oil saturation = 30%.

Table 8.102—Pertinent reservoir data, Sloss field COFCAW area.

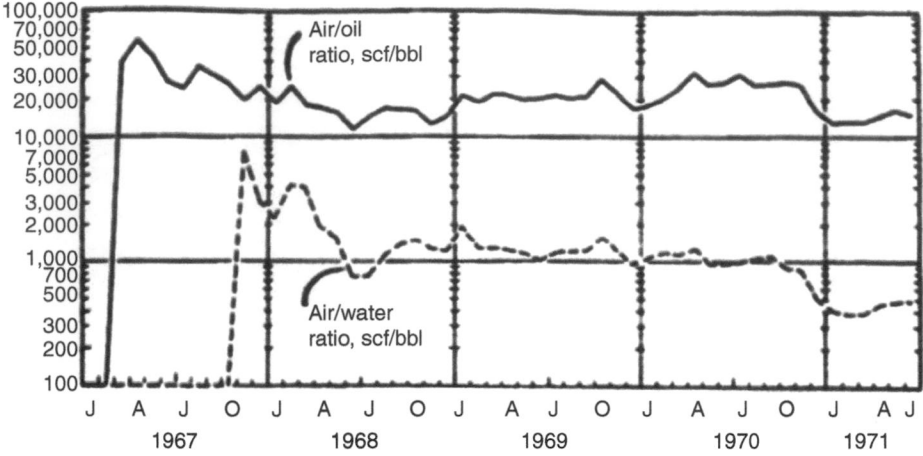

Fig. 8.230—Air/water and air/oil ratios, Sloss wet-combustion project (Parrish et al. 1974).

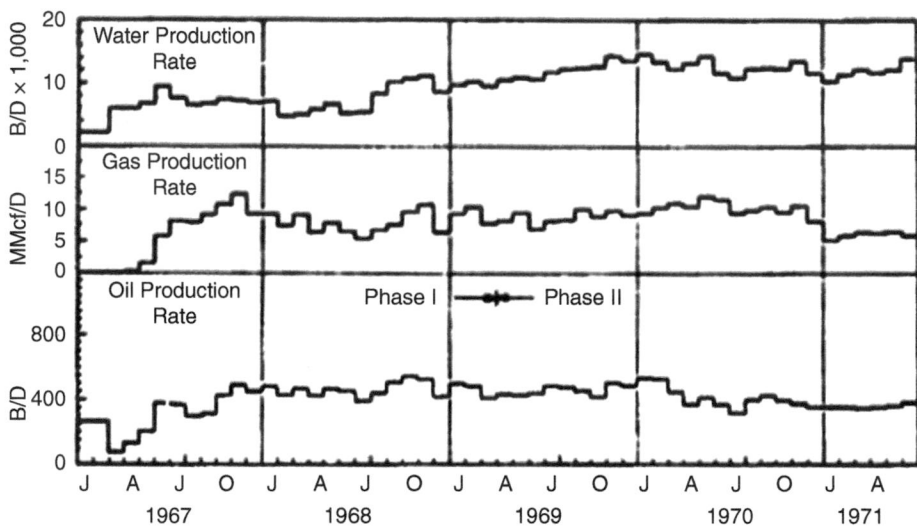

Fig. 8.231—Fluid production during Sloss wet-combustion project (Parrish et al. 1974).

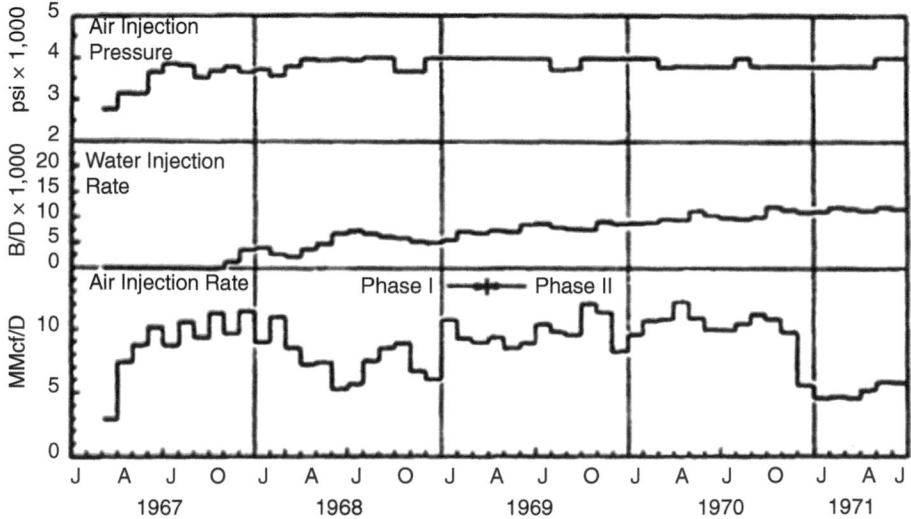

Fig. 8.232—Fluid injection during the Sloss wet-combustion project (Parrish et al. 1974).

Fluid	Injection 27 February 1967 to 1 July 1971	Production 27 February 1967 to 1 July 1971
Air (gas) (Mcf)	13,754,077	(15,604,035)
Water (bbl)	10,818,307	15,919,374
Oil (bbl)	—	646,776
Incremental oil captured (bbl)	646,776 − 120,000 = 527,000	
Hydrocarbons vented in produced gas (equivalent bbl)		340,000
Recoverable hydrocarbons vented (equivalent bbl)		210,000
Additional oil captured to 1 January 1974 (bbl)		189,000
Overall air/water ratio (scf/bbl)		1,271
Overall air/oil ratio to 1 July 1971 (scf air/bbl)		21,266
Over-all produced WOR		24.6

Table 8.103—Fluid injection and production data, Sloss wet-combustion project.

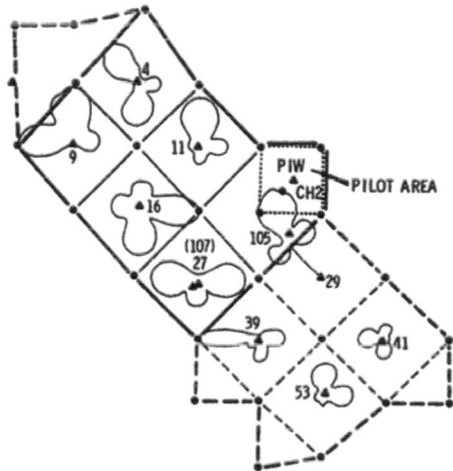

Fig. 8.233—Estimated burned area, Sloss wet-combustion project (Parrish et al. 1974).

obtained with air injection. The oxygen requirement was 463 scf/ft^3 reservoir volume. No evidence of partial oxidation of the displaced oil was observed.

Two pilot tests with oxygen enrichment have been completed (Petit et al. 1992; Hvizdos et al. 1983). Equipment and procedures for handling oxygen-enriched gas and pure oxygen injection have been developed. These tests show that field application is possible if the process is economically feasible. Petit et al. (1992) presented a comparison of the economics of oxygen and air fireflooding.

Esperson Dome Pilot Test. A pilot test with oxygen was conducted in the Esperson Dome field, Liberty County, Texas, USA (Choquette et al. 1993; Pebdani et al. 1988). The test was conducted in a watered-out, medium-gravity (21 °API) oil reservoir. A total of 200 MMscf of O_2 and 55 MMscf N_2 was injected between April 1984 and August 1987 at O_2 concentrations up to 100% (Choquette et al. 1993).

Fig. 8.234 shows the pilot area and well loca-

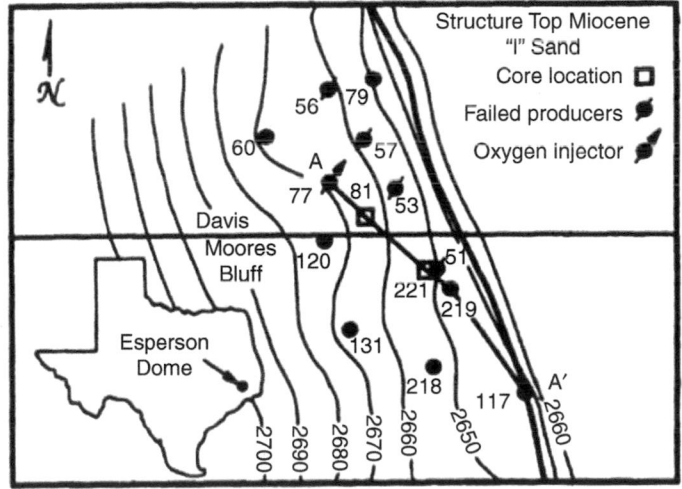

Fig. 8.234—Esperson Dome oxygen pilot test, project wells and corehole location (Choquette et al. 1993).

tions. The pilot area is part of a Miocene sand body bounded on the east by a fault. **Table 8.104** summarizes properties of the reservoir and fluid. The sand ranges from 50 to 80 ft thick with a porosity of approximately 31% and permeabilities varying from 200 to > 6,000 md. Reservoir pressure is 1,190 psi and is maintained by a strong aquifer. Water cut was 95% at the beginning of the project.

Depth (ft)	2,670
Porosity (%)	31
Permeability (md)	
Reservoir	200 to 6,300
Upper zone	200 to 2,500
Sand thickness (ft)	
Reservoir	50 to 80
Upper zone	15 to 25
Pressure (psi)	1,190
Temperature (°F)	125
Oil gravity (°API)	21
Oil viscosity at reservoir temperature (cp)	90
Estimated oil saturation at start (%)	30
Atomic H/C ratio of oil	1.78

Table 8.104—Esperson Dome field O_2 pilot reservoir and fluid properties of Miocene sand.

Oxygen was injected into Well D-77 through perforations in the bottom 10 ft of a 36-ft oil-sand interval. Interpretation of production and post-combustion cores indicated that the injected oxygen rose to the top of the reservoir near the injection well and apparently continued to rise to the top of the structure, accumulating updip near the eastern fault. Combustion occurred only in a portion of the upper zone. The burned zone was clean. Oil saturations in adjacent zones ranged from 11 to 34%. **Fig. 8.235** shows the estimated burned-zone isochore.

Approximately 96,000 bbl oil was recovered from the pilot area, which is approximately 49,000 bbl above the preproject baseline. Most of the oil production was from wells south of the injection well. The apparent injected-oxygen/produced-oil ratio was 4.1 Mscf/STB. Approximately 36,000 STB was displaced from the burned zone, with an equivalent fuel consumption of 9 saturation percent.

8.8 Comparison of Steam and In-Situ Combustion

An operator with a reservoir that is amenable to thermal recovery techniques is faced with a dilemma as to which process to pursue if both steam and in-situ combustion are technically feasible. This dilemma occurs because most commercial thermal recovery projects are either steamfloods, CSS, or a combination of these processes. However, when the thermal efficiency of each process is analyzed, in-situ combustion shows a clear advantage over steamflooding. **Fig. 8.236** shows this. The figure plots the ratio of energy in the produced oil to the energy consumed to produce the oil (coefficient of performance) vs. pressure for AORs and OSRs that are in the range of those observed for successful projects. Two factors contribute to the results. One is that the fuel consumed by the moving combustion front is about the same as the residual oil saturation left by the steam front. The second factor is that the fuel required to run the air compressor is less than the energy needed to generate steam on the surface.

Ramey et al. (1992) compared the thermal efficiency of steam and in-situ combustion in an analysis of the energy requirements for thermal oil recovery in the South Belridge field. **Table 8.105** (Ramey et al. 1992) summarizes the analysis. On the basis of observed field performance, in-situ combustion required approximately 4 Mscf/bbl oil, which corresponds to a fuel requirement of approximately 288,000 Btu fuel/bbl oil. At an SOR of 0.35 for South Belridge, the energy requirement is 1,400,000 Btu fuel/bbl oil. Thus, approximately 4.9 times as much surface energy is required for steamflooding as

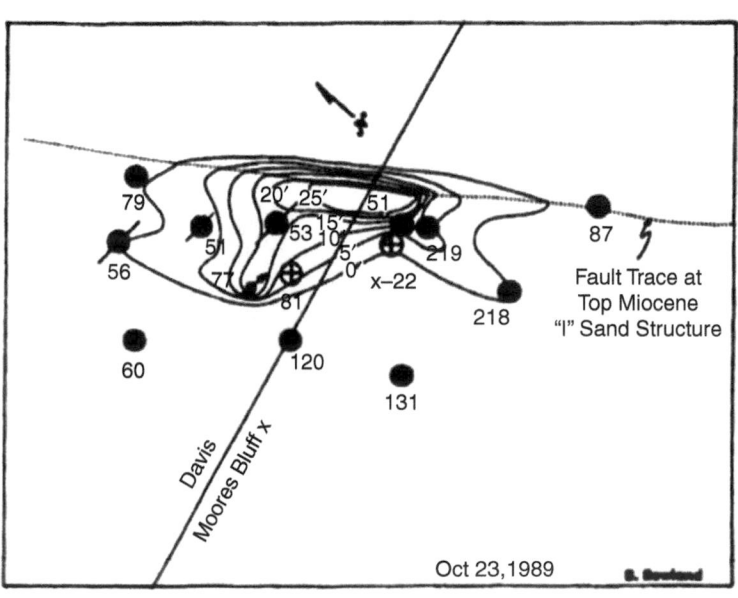

Fig. 8.235—Estimated burned-zone isochore, Esperson Dome project (Choquette et al. 1993).

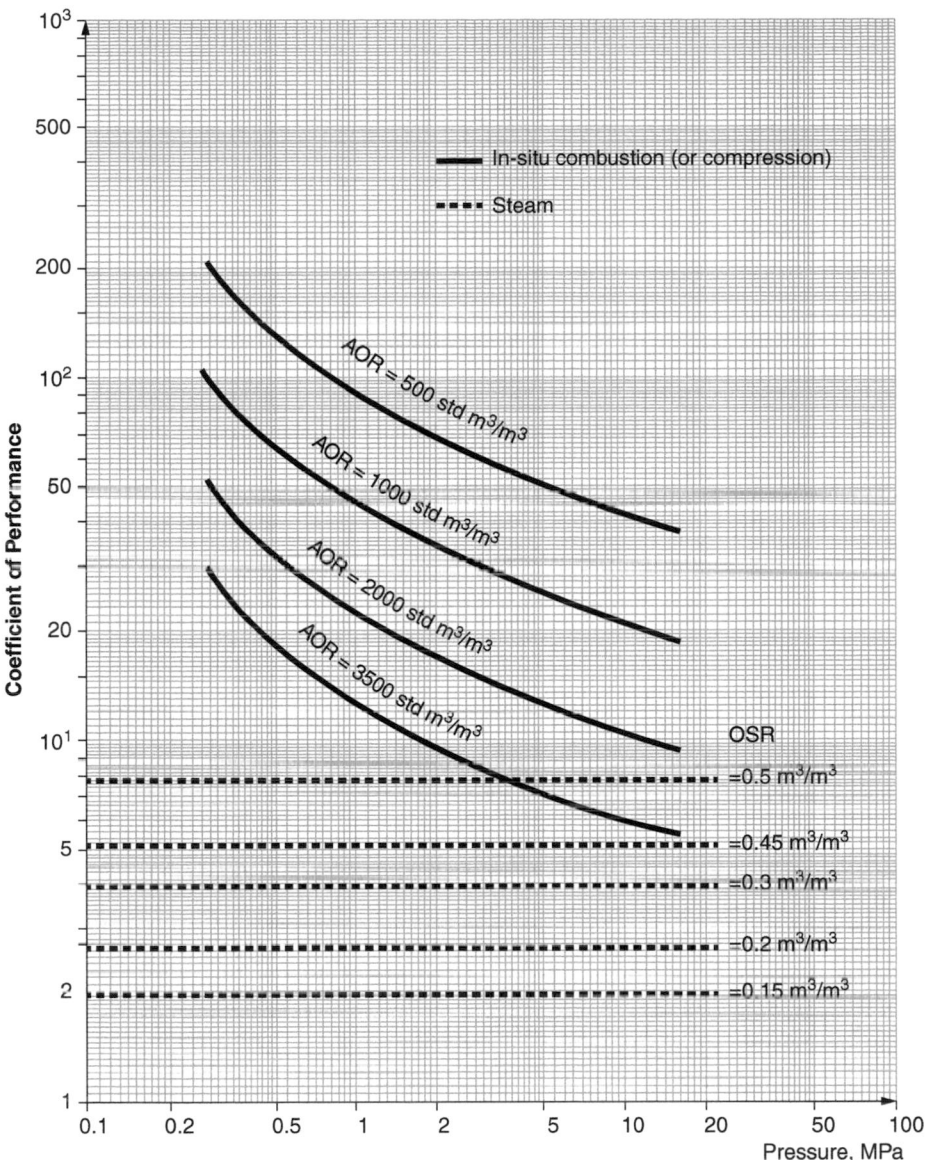

Fig. 8.236—Comparison of the coefficients of performance (ratio of energy in produced oil to energy consumed to produce oil) for air fireflooding and steamdrive (Burger et al. 1985).

1. In-Situ Combustion

Fuel: 216 Mcf/D fuel gas will provide 3,000 Mcf/D air at 500 psi from a 900-hp compressor. Btu fuel/scf air = 216/3 =72

Injected air/bbl oil (Mcf/bbl)	5	10	15
Btu fuel/bbl oil	360,000	720,000	1,080,000

2. Steam Injection

Fuel: 0.08 bbl oil will generate 1 bbl steam (feedwater), and the heat of combustion of oil is 6,000,000 Btu/bbl. Btu fuel/bbl steam = 6,000,000 (0.08) = 480,000

OSR	0.1	0.2	0.3	0.4
Btu/bbl oil	4,800,000	2,400,000	1,600,000	1,200,000

3. At South Belridge, in-situ combustion requires approximately 4 Mcf air/bbl oil, or 288,000 Btu/bbl oil. Steam injection has resulted in an OSR of 0.35, or an energy requirement of 1,400,000 Btu fuel/bbl oil. Thus, approximately 4.9 times as much energy per barrel oil is required for steam injection as for in-situ combustion.

4. At South Belridge, the total combustion products exhausted may be computed by adding the fuel burned in in-situ combustion to the surface energy requirement. There is 1 Mcf oxygen in 5 Mcf air, and approximately 500 Btu is generated per 1 scf oxygen. Thus, 500,000 Btu/bbl oil should be added to 288,000 Btu/bbl oil, for a total of 788,000 Btu/bbl oil. Thus, the ratio of combustion products from steam injection and in-situ combustion is 1,400/788, or 1.8. About twice as much total fuel per barrel oil is required for steam injection at South Belridge as was required for in-situ combustion.

Table 8.105—Energy requirements for thermal oil recovery (Ramey et al. 1992).

for in-situ combustion in the South Belridge field. Table 8.105 also indicates that steam injection requires about twice the total fuel compared with in-situ combustion.

Although thermal efficiency favors in-situ combustion, field-scale in-situ combustion projects represent approximately 3% of the oil produced by thermal recovery processes. Steamflooding and cyclic steam injection are widely used and are usually the processes of choice in reservoirs where both processes are technically feasible. Thus, there must be reasons other than thermal efficiency for the selection process that has developed.

Many early in-situ combustion projects were carried out in five-spot patterns in thin reservoirs as pilot tests. Most of these projects were economic failures because of high AORs. Ramey et al. (1992) claim that many of these projects actually demonstrated LTO instead of the HTO observed in successful California in-situ combustion tests and in combustion-tube runs used to design field tests. Location and type of air-injection pattern were not considered to be important variables in early field tests. However, field experience shows that in-situ combustion projects appear to be aided by a significant component of gravity drainage. Finally, most successful in-situ combustion projects have been developed by starting the combustion at the top of the structure and expanding the project as a linedrive rather than as a multipattern flood.

Operating experience with proven equipment is often a deciding factor in selecting the thermal recovery process. Equipment for generating steam is readily available in a wide range of sizes. Personnel can be easily trained to operate the equipment, and there is extensive operating experience in the industry. In contrast, there appears to be less confidence in selecting and operating air compressors for in-situ combustion processes. Thus, practical considerations appear to have favored the development of steamflooding as a preferred displacement process.

Finally, as noted earlier in this chapter, in-situ combustion is feasible in reservoirs too deep for steam injection. In some cases, in-situ combustion is being used for pressure maintenance by creating a combustion front at the top of the formation, where gravity acts to sustain a combustion front that displaces oil effectively in the vertical direction over large areas.

Problems

8.1   The quality of steam coming from a generator can be determined by condensing the wet steam into a measured quantity of water at given temperature and pressure. The increase in temperature and the weight of water resulting from the condensation of steam are then used to calculate the enthalpy of the wet steam. If the enthalpy of the wet steam was determined to be 1,013.6 Btu/lbm, calculate the quality of the wet steam for a generator pressure of 450 psia.

8.2   Injection of 80% quality steam at 400°F at the wellhead is desired in a steam-displacement project. The saturated steam passes through 200 ft of 3-in.-diameter pipe insulated with 1.5-in. magnesia insulation from the generator to the wellhead. The air temperature is 80°F. Under these conditions, the heat-loss rate is 0.552 million Btu/D. The injection rate is 200 B/D CWE.

    1.   What is the quality of the steam generated?
    2.   If bare metal pipe were used, the heat-loss rate would be 3.732 million Btu/D. What would the quality of the steam at the generator need to be? Compare with Part 1.

8.3   **Fig. 8.237** depicts a steam-distribution system. The distribution line is 3-in. line pipe (3.5-in. OD) and is insulated with 2 in. of magnesia silicate. The insulation is protected with a thin cover of aluminum ($E_{A1} = 0.76$). Injection rate is to be maintained at 300 B/D, injection pressure is 1,800 psia, and injection temperature is 621.03°F. Average ambient temperature is 100°F, average wind velocity is 10 miles/hr, and subsurface temperature is 65°F. Assume that the steam temperature is constant in the distribution system; this neglects pressure drop caused by friction/two-phase flow.

    1.   What is the rate of heat loss from the distribution line?
    2.   What is the steam quality at the wellhead if the quality at the steam generator is 0.8?

8.4   What would the steam quality be at the wellhead in Problem 8.3 if the distribution line was not insulated?

8.5   The design of a steam-distribution line involves determination of the amount of insulation needed. This requires comparing the cost of insulation with the savings in energy cost over the life of the project. You are to determine the heat loss from a 100-ft-long steam-distribution line. Steam at 500°F is to be injected at a rate of 1,000 B/D through 2.5-in (nominal pipe size) Schedule 80 line pipe. The line will be insulated with calcium silicate insulation covered with an aluminum cover, as described in Example 8.1. The average ambient temperature is 65°F, and the average wind velocity is 15 miles/hr in the direction normal to the injection line. Determine the heat loss from this line (in Btu/D) as a function of insulation thickness. Use insulation thickness from 0.5 to 3 in. in 0.5-in. increments. Plot heat-loss rate (in Btu/D) vs. insulation thickness. Example 8.1 gives the thermal properties of the insulation. Note that the OD of the 2.5-in. pipe is 2.875 in. Other properties needed to solve this problem can be found in the Appendices or in heat-transfer texts. How would you select the proper insulation thickness using your heat-loss estimates?

8.6   The pressure loss resulting from friction for the flow of wet steam in a distribution line may be estimated with the following empirical equation (Hong 1994).

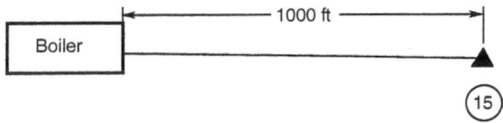

Fig. 8.237—Steam-distribution system.

Mean subsurface temperature (°F)	70
Geothermal gradient (°F/ft)	0.02
Overall heat-transfer coefficient, U_{ti} (Btu/D-ft-°F)	30
Tubing ID (in.)	2
Drillhole diameter (in.)	7
Thermal conductivity of Earth (Btu/D-ft-°F)	36
Thermal diffusivity of Earth (ft²/D)	0.96

Table 8.106—Steam-injection conditions.

$$\Delta p_f = \left(3.628 \times 10^{-8}\right)\left[1+\left(3.6/D_{pi}\right)\right] \times \left[f_s v_{sv} + \left(1-f_s\right)v_{sl}\right]\left(Lm_s^2/D_{pi}^5\right), \quad\quad\quad\quad (8.378)$$

where Δp_f = pressure drop caused by friction in psi and D_{pi} = ID of pipe in inches. Estimate the pressure and steam quality at the wellhead of Well 15 in Fig. 8.237 if frictional pressure drop is included. Assume that the distribution line is horizontal and that kinetic energy effects are negligible. The distribution line is 3-in. Schedule 160 line pipe (wall thickness = 0.438 in.). The distribution-line ID is 2.124 in.

8.7   Estimate the wellbore-heat-loss rate for steam injection through 2-in. tubing at 100 days injection time for the conditions listed in **Table 8.106.** Saturated steam of 80% quality is being injected at the wellhead at 400°F at an equivalent feedwater rate of 1,000 B/D. The formation is 1,000 ft deep.

1. What is the heat-loss rate from the tubing in million Btu/D?
2. What is the heat-injection rate at the surface in million Btu/D?
3. What is the percentage of wellbore heat loss?
4. What is the quality of the saturated steam being injected into the formation?

8.8   Steam at 300 psia (417.33°F) is injected into a reservoir at a depth of 1,000 ft. Injection is down casing at a rate of 300 B/D CWE. The well is completed with 7-in. casing (J-55, 26 lbm/ft) cemented to surface in a 10.75-in. drilled hole. Mean subsurface temperature is 100°F, and the geothermal gradient is 0.015°F/ft. Thermal conductivity of the Earth is assumed to be 1.0 Btu/(hr-ft-°F), and the mean heat capacity is 35 Btu/ft-°F. Assume that the thermal conductivity of the cement is 0.6 Btu/(hr-ft²-°F/ft). Quality of the steam at the wellhead is 0.8. Estimate the wellbore-heat-loss rate and the downhole steam quality after 6 months of injection. Neglect pressure changes in the casing.

8.9   In Problem 8.8, consider the alternative of injecting down 2.875-in. tubing (6.4 lbm/ft) set on a packer. The tubing/casing annulus will be full of water when the packer is set, but the annulus is vented to the atmosphere to permit the water to be boiled out. Assume that the tubing and casing surfaces are mill scale.

1. Estimate the wellbore-heat-loss rate after 6 months of injection.
2. Estimate the casing temperature after 6 months of injection. Neglect the effect of boiloff of the annulus on the casing temperature. Why can you consider the annulus to be dry?
3. Estimate the downhole steam quality after 6 months of injection assuming that the annulus boils dry.

8.10 Suppose the injection tubing in Problem 8.8 is prestressed insulated tubing. Prestressed insulated tubing is constructed from 2.875-in. tubing prestressed in a 4.5-in. casing string (K-55, 11.6 lbm/ft) to protect the insulation. The ID of the protective casing is 4.00 in. The insulation has an effective thermal conductivity of 0.002 Btu/(hr-ft²-°F/ft). The insulated tubing is run on a thermal packer, and the annulus is filled with water when the packer is set. Estimate the wellbore-heat-loss rate, the downhole steam quality, and the casing temperature after 6 months of injection if the casing/tubing annulus is boiled dry by a heat transfer from a bare expansion joint just above the thermal packer. In solving this problem, note that Eq. 8.22 is modified to obtain Eq. 8.379 to account for the finite thickness of the outer casing of the insulated tubing string. Assume that the collars have the same effective thermal conductivity as the body of the insulated tubing string.

$$U_{to} = \left[\frac{r_{to}}{r_t h_f} + \frac{r_{to}\ln\left(r_{to}/r_{ti}\right)}{k_{htub}} + \frac{r_{to}\ln\left(r_{ins}/r_{to}\right)}{k_{hins}} + \frac{r_{to}\ln\left(r_{inso}/r_{ins}\right)}{k_{hcas}}\right.$$

$$\left.+ \frac{r_{to}}{r_{inso}\left(h_c' + h_r'\right)}, \frac{r_{to}\ln\left(r_{co}/r_{ci}\right)}{k_{hcas}} + \frac{r_{to}\ln\left(r_{hd}/r_{co}\right)}{k_{hcem}}\right]^{-1}, \quad\quad\quad\quad (8.379)$$

where r_{inso} = radius of outer casing in feet.

8.11 **Table 8.107** presents properties of a well that is to be used for steam injection. Injection rate is 1,000 B/D CWE. The table shows thermal properties of the formation and cement, the surface temperature, and the geothermal gradient.

Well completion data	
Injection tubing, ID (in.)	2.99
Injection tubing, OD (in.)	3.50
Casing OD (in.)	7.0
Casing ID (in.)	6.276
Hole (in.)	10.75
Depth (ft)	3500
Thermal properties	
Injection tubing (Btu/hr-ft-°F)	24.84
Casing (Btu/hr-ft-°F)	24.84
Cement (Btu/hr-ft-°F)	0.3
Formation (Btu/hr-ft-°F)	1.4
Thermal diffusivity of formation (0.04 ft²/hr)	0.04
Mean subsurface temperature (°F)	77
Geothermal gradient (°F/ft)	0.017

Table 8.107—Properties of steam-injection well.

Evaluate the heat-loss rate and casing temperature when this well is completed with tubing set on a packer. The casing annulus is assumed to boil dry during steam injection.

1. Estimate heat-loss rate and casing temperature after 30 days of continuous steam injection at a pressure of 2,500 psi. Steam quality at the wellhead is 89.7%.
2. Find the steam quality at the sandface if the pressure change in the tubing is neglected.

8.12 An oil well in a reservoir containing viscous oil (μ_{oc} = 1,500 cp) has been stimulated by steam injection at 400°F. Estimated heated-zone radius is 30 ft. The heated interval is 100 ft thick. Pressure at the drainage boundary at the beginning of the production period is 100 psig, and the well is to be pumped so that the BHP is 20 psig. Estimate the initial oil production rate following stimulation when the properties listed in **Table 8.108** are believed to be representative of the reservoir.

8.13 A petroleum reservoir at a depth of 1,500 ft contains an oil with a viscosity of 500 cp at reservoir temperature (120°F). The reservoir is 50 ft thick and produces by solution-gas drive. **Table 8.109** gives properties of the reservoir, rock, and fluid. The initial solution gas was ≈ 200 scf/STB, and the initial reservoir pressure was 1,500 psia. Reservoir pressure declined to 1,000 psia during primary production. Wells are on 5-acre spacing, and a fluid head equivalent to 50 psia is maintained in the wellbore. Average production rate is 166 B/D. There is negligible water production. Consider the possibility of stimulating wells in this field by CSS. The oil viscosity decreases to 14 cp at 300°F.

1. Estimate the production rate from this reservoir under the current conditions using the data from Table 8.109. Is the average production rate consistent with your estimates? Why or why not?
2. Estimate the maximum oil production rate following CSS if wells are stimulated by injecting steam at 1,500 psia long enough to create a heated zone with a 25-ft radius. Assume that the entire productive interval takes steam at the same rate so that the heated zone has a uniform radius. To solve this problem, it is necessary to develop relationships for B_o and μ_o as functions of temperature. Use Eq. E-20 (Appendix E) to find B_o at different temperatures and Eq. E-23 to estimate the oil viscosity as a function of temperature. Remember that, in Eq. E-23, μ_o is in centipoise and p_o is in grams per cubic centimeter.

k_{oc} (md)	3,000
k_{oh} (md)	2,000
μ_{oh} (cp)	10
B_{oh} (bbl/STB)	1.3
B_{oc} (bbl/STB)	1.1
r_w (ft)	0.5
r_e (ft)	372

Table 8.108—Representative reservoir properties.

Porosity (fraction)	0.35
Permeability (md)	1,500
Initial oil saturation (fraction)	0.65
Initial water saturation (fraction)	0.35
Thickness (ft)	50
Reservoir temperature (°F)	120
Drill-hole diameter (in.)	10¾
Gravel-pack diameter (in.)	16
Well spacing (acres)	5
Initial reservoir pressure (psig)	1,500
Current reservoir pressure (psig)	1,000
Oil gravity (°API)	12.7

Table 8.109—Reservoir, rock, and oil properties.

8.14 Steam is to be injected at a rate of 500 B/D CWE into a well for CSS. Steam quality at the sandface is 0.75. Neglect pressure loss down the injection tubing. Estimate the length of the steam-injection period (in days) required to heat the reservoir described in Problem 8.13 uniformly to a radius of 25 ft. **Table 8.110** gives thermal properties of the rock matrix and the overburden/underlying formation; other properties are given in Table 8.109 or are to be estimated from correlations in Appendix E. Use the Marx and Langenheim (1959) model to estimate the injection time. Estimate the volume of oil (in reservoir barrels) displaced from this zone during steam injection. The residual oil saturation (ROS) to steam is 0.15.

8.15 Steam is to be injected into the reservoir described in Problems 8.13 and 8.14 for 21 days. Estimate the radius of the heated zone and the volume of oil displaced during steam injection. Is any water displaced during steam injection? How would you estimate the volume of water that was displaced, if any?

8.16 A 90-ft-thick zone in a heavy-oil reservoir is to be stimulated by steam injection. The zone produces under primary depletion. The current production rate is 5 B/D with no water production. The stimulation design includes injection of 70% quality steam (at the sandface) at 500°F into the zone for 14 days at a rate of 500 B/D (CWE at 70°F) followed by a 14-day shut-in period to allow the injected steam to condense. **Table 8.111** gives other properties of the reservoir, reservoir rock, and fluids. Reservoir temperature is 80°F.

1. Estimate the radius of the steam zone (in feet) at the end of the steam-injection period using the Marx and Langenheim (1959) model.
2. Estimate the volume of oil (in reservoir barrels) displaced from the heated region during steam injection.
3. Estimate the average temperature of the heated region at the end of the soak period.
4. Estimate the volume of condensate and interstitial water (at 80°F) displaced from the hot region into the cold region of the reservoir at the end of the shut-in period. For the purposes of this calculation, you may assume that displaced interstitial water and condensate resaturates the heated region during the shut-in period. If you were unable to calculate the average temperature at the end of the shut-in period in Part 3, use 300°F.

8.17 Use the Boberg and Lantz (1966) model to estimate the oil-production history following CSS of a well in the field described in Problem 8.13. Use the WOR/cumulative-production correlation in Example 8.6 for this problem.

8.18 A thick reservoir with 2.5-acre well spacing produces oil by gravity drainage. **Table 8.112** summarizes the properties of the reservoir and the oil. The oil viscosity is 6,400 cp at reservoir temperature and decreases to 11 cp at 300°F. Wells have been pumped off, with 50 ft of head (above the bottom of the reservoir) maintained at the pump shoe. An estimate is desired of the effect of steam stimulation on the production in this reservoir assuming that the fluid head (195 ft) remains constant at the heated radius. Estimate the oil and water production rates as functions of time following a steam-stimulation treatment where steam is injected into the well for 30 days at a rate of 750 B/D CWE. Injection pressure is 200 psig, and steam quality downhole is 0.70. For this problem, use Eqs. H-4 and H-5 (Appendix H) to calculate the relative permeability of the oil and water following CSS. Assume that the values of a_1 and a_2 in Eqs. 8.105 and 8.106 are 1.0. The pumping equipment is capable of maintaining 50 ft of fluid at the pump shoe following stimulation.

Mean heat capacity of rock (Btu/lbm-°F)	0.255
Density of rock (lbm/ft³)	165.4
Mean heat capacity of rock matrix (Btu/ft³-°F)	42.18
Thermal diffusivity of overburden (ft²/hr)	0.0482
Thermal conductivity of overburden (Btu/hr-ft-°F)	1.5

Table 8.110—Properties of reservoir rock and surrounding formations.

8.19   In reservoirs that produce by gravity drainage for some period of time, a zone develops at the top of the reservoir with a substantial gas saturation. Consider steam stimulation of a well in the reservoir shown in **Fig. 8.238.** During injection, most of the steam will enter the unsaturated zone at the top of the reservoir as depicted, leaving substantial portions of the reservoir in the lower interval uncontacted except in the immediate vicinity of the wellbore. The lower interval is heated by conduction from the heated region. In this problem, steam (at 200 psig) is injected into the well at a rate of 1,000 B/D for 14 days. Steam quality at the sandface is 0.7. Reservoir temperature is 75°F.

1. Estimate the radius of the heated zone if steam enters the top 10 ft of the formation.
2. Determine the temperature distribution (as a function of vertical distance from the heated zone) in the underlying reservoir at a distance of one-half of the heated radius. Calculate the temperature at increments of 1 ft from the boundary of the upper and lower zones until the temperature is within 0.05°F of the initial reservoir temperature. Plot the temperature as a function of distance from the boundary between the two zones. Thermal properties are M_R = 35 Btu/ ft³-°F, M_s = 35 Btu/ft³-°F, k_{hs} = 1.2 Btu/(hr-ft-°F). For this problem, assume that the thermal properties of the lower zone are the same as those of the overburden.

8.20   The well in **Fig. 8.239** penetrates a reservoir that has two noncommunicating zones separated by a thin permeability barrier. Zone 1 has a high water saturation, while Zone 2 has such a high oil saturation and viscosity that the mobility of the oil is limited at reservoir temperature (80°F). Stimulation of the well by steam injection is proposed. Because the Zone 2 oil saturation is high, it is believed that most of the steam will enter Zone 1. Zone 2 would be heated by conduction from heat losses in Zone 1. Steam injection into Zone 1 will continue until the average temperature of Zone 2 within the heated areas covered by the steam front in Zone 1 increases to 280°F. Develop a model of this process [based on the Marx and Langenheim (1959) model] to estimate the injection time required to increase the temperature in the heated area of Zone 2 by 200°F. For the purposes of this problem, assume that one-half of the heat lost from the Zone 1 steam zone flows into Zone 2 and that there is no heat loss from Zone 2. Thus, the amount of energy stored in Zone 2 is $M_2 h_2 A_h (\overline{T}_2 - T_r)$. **Table 8.113** gives the reservoir and rock properties. Assume that the overburden has the same thermal properties as Zone 2.

8.21   Consider steam injection into a thick reservoir (h = 200 ft) like that shown in **Fig. 8.240.** An unsaturated interval (50 ft) with a high gas saturation occurs at the top of the reservoir. Steam enters this interval and displaces oil to production wells but does not expand vertically. The underlying formation is heated by conduction. The steam-injection rate is 15 × 10⁶ Btu/hr at 500 psig. **Table 8.114** gives the reservoir properties.

Porosity	0.25
Initial water saturation	0.20
Initial oil saturation	0.80
Initial formation temperature (°F)	80
ROS to steam	0.10
Average heat capacity of overburden (Btu/ft³-°F)	31.13
Average heat capacity of the reservoir rock (Btu/ft³-°F)	34.43
Overburden thermal conductivity (Btu/ft-hr-°F)	1.5
Overburden thermal diffusivity (ft²/hr)	0.0482

Useful constants: 1 acre = 43,560 ft², 1 bbl = 5.615 ft³, and 1 bbl of water at 70°F = 350 lbm.

Table 8.111—Reservoir, reservoir rock, and fluid properties.

Reservoir properties	
Gross thickness (ft)	335
Net thickness (ft)	195
Porosity	0.33
Permeability (md)	1750
Well spacing (acres)	2.5
Wellbore radius (ft)	0.5
Initial oil saturation	0.83
ROS to steam	0.20
Reservoir temperature (°F)	85
Fluid properties	
Oil gravity (°API)	14
Oil viscosity at 85°F (cp)	6400
Oil viscosity at 300°F (cp)	11
Thermal Properties	
Thermal diffusivity (ft²/D)	1.00

Table 8.112—Reservoir rock and fluid properties.

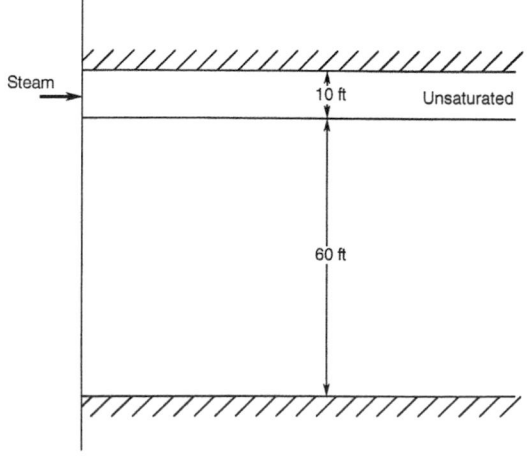

Fig. 8.238—Steam injection into an unsaturated zone during CSS.

1. Calculate the rate of heat conduction (in Btu/hr) into the oil sand below the steam zone after 1 year of continuous injection.

2. Consider an overall energy balance over the first year of injection. Calculate (a) total amount of energy injected (in Btu), (b) energy retained in the steam zone (in Btu), (c) total energy conducted to the overburden (in Btu), and (d) total energy conducted into the oil sand below the steam-out zone (in Btu).

3. Calculate the bulk volume of the steam zone after 1 year of continuous injection.

8.22 Compare the fraction of heat lost by vertical conduction from a 100-ft-thick formation with that from a 10-ft-thick formation for steam injection over a period of 1,000 days. All other conditions are the same for both thicknesses. Thermal diffusivity of the overburden is 0.96 ft²/D.

8.23 A cold-waterflood, hot-waterflood, and steam displacement are run on linear Dexter sandstone cores

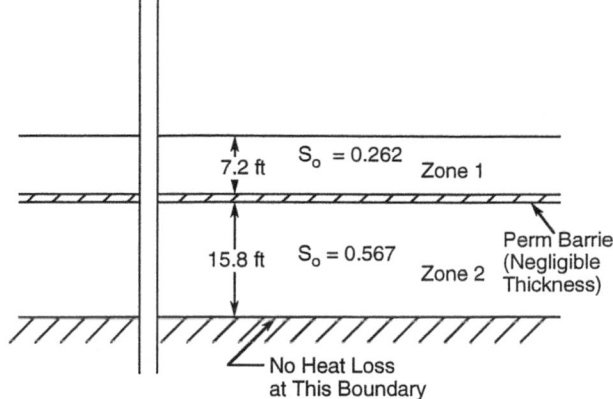

Fig 8.239—Reservoir cross section.

initially saturated with water and Crude Oil A. **Fig. 8.241** shows the displacement curves. All runs are conducted with the system at the same initial oil saturation. Separate distillation tests conducted on the crude oil, at the same pressure and temperature as used for the displacement runs, indicate that 30% of the oil is distillable. Analyze the results to determine the percent recovery (as percent OOIP) that can be attributed to each recovery mechanism that Willman et al. (1961) specify for steam displacement (see Section 8.5.1). Be as specific as possible, and indicate the basis for your answer.

8.24 Two separate steam-displacement experiments are to be conducted in the laboratory with Dexter sandstone linear cores and Crude Oil B. Each experiment is to be conducted with the same initial oil saturation in the core (S_{oi} = 70%). In Experiment 1, the steam displacement is conducted at 250°F (29.8 psia); in Experiment 2, the steam displacement is conducted at 400°F (247.3 psia). Both experiments are stopped after 0.75 PV of steam (as CWE) has been injected. Which experiment would be expected to give the greater oil recovery? State the reasons for your answer.

8.25 A steam displacement is to be conducted in a reservoir with the properties given in **Table 8.115.** The steam-injection rate will be 3,000 lbm/hr, and steam temperature will be 500°F. The steam quality as it enters the reservoir is estimated to be 85%. Assume that the steam will invade the entire thickness of the oil sand [as assumed in the Marx and Langenheim (1959) model].

1. Calculate the point in time (i.e., the number of hours after start of steam injection) when the rate of heat loss to the overburden and the underlying formation will be exactly equal to one-half the rate of energy injection.

2. Determine the economic limit (time when the value of displaced oil equals the net revenue) for steam injection if the steam cost is USD 2.00/million Btu and the net oil price to the operator is USD 12.00/bbl. Assume a capture efficiency factor of 80% (0.8 bbl produced/bbl displaced).

Mean heat capacity of rock (Btu/lbm-°F)	0.21
Mean heat capacity of oil (Btu/lbm-°F)	0.5
Mean heat capacity of water (Btu/lbm-°F)	1.0
Density of rock (lbm/ft³)	167
Density of water (lbm/ft³)	62.4
Density of oil (lbm/ft³)	50
Thermal conductivity of overburden (Btu/hr-ft-°F)	1.5

Table 8.113—Reservoir and rock properties.

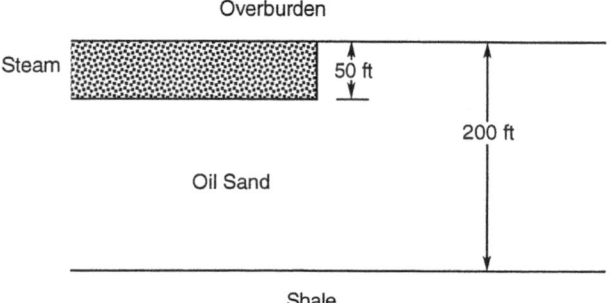

Fig 8.240—Conduction heating by steam injection into an unsaturated zone.

Mean heat capacity of reservoir rock (Btu/ft³-°F)	33.2
Thermal conductivity of overburden and underlying formations (Btu/ft-°F)	1.5
Thermal diffusivity of overburden and underlying formations (ft²/hr)	0.0482
Porosity	0.26
Initial oil saturation	0.60
Initial water saturation	0.20
ROS after steam injection	0.10
Reservoir temperature (°F)	80

Table 8.114—Properties of rock and formations.

3. Calculate the oil displacement rate and the cumulative oil production as a function of time from the beginning of steam injection to the economic limit.

8.26 Use the Neuman (1975, 1985) model to estimate the following (using the data from Problem 8.25) when steam is injected at 3,000 lbm/hr for 3 years (1,095 days).

1. Area of steam zone in acre-ft.
2. Average thickness of steam zone in feet.
3. Steam-zone thickness as a function of heated area at t = 3 years.
4. Volume of oil displaced by steam zone.
5. Oil displacement rate: Compare the oil displacement rate with the estimated oil displacement rate obtained in Problem 8.25.
6. Rate at which the steam zone is expanding vertically at the wellbore and at a radius midway between the wellbore and the front of the steam zone.

8.27 Assume that gravity override dominates the displacement process described in Problem 8.25 and that the average vertical rate of steam-zone advance is 0.02 ft/D. Use the uniform-vertical-expansion model to estimate the following after 3 years (1,095 days) of steam injection.

1. Heated area in acre-ft.
2. Volume of reservoir contacted by steam and hot water in acre-ft.
3. Thickness of the heated zone as a function of heated area.
4. Average thickness of heated zone in feet.
5. Average thickness of steam zone in feet.
6. Volume of oil displaced by steam and hot water. Assume that the ROS for the hot-water zone is 0.35.
7. Estimated oil displacement rate after 3 years of steam injection.
8. Estimated fraction of injected energy remaining in the reservoir.

8.28 A reservoir with the properties listed in **Table 8.116** is to be steamflooded. The heat-injection rate is 15 million Btu/hr with a steam temperature of 500°F. The initial reservoir temperature is 100°F. Use the Marx and Langenheim (1959) model to answer the following questions.

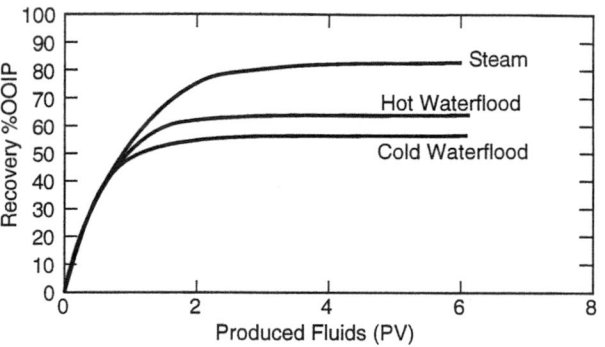

Fig 8.241—Recovery from cold waterflood, hot waterflood, and steamflood in linear Dexter sandstone cores.

Thickness (ft)	40
Reservoir temperature (°F)	120
Initial oil saturation	0.65
ROS to steamflood	0.15
Porosity	0.30
Thermal conductivity of overburden/underlying formation (Btu/hr-ft-°F)	1.2
Thermal diffusivity of overburden/underlying formation (ft²/hr)	0.04
Average heat capacity of oil sand [Btu/(ft³-°F)]	35

Table 8.115—Reservoir and rock properties.

Thickness (ft)	40
Reservoir temperature (°F)	120
Initial oil saturation	0.65
ROS to steamflood	0.10
Porosity	0.22
Thermal conductivity of overburden/underlying formation (Btu/hr-ft-°F)	1.5
Thermal diffusivity of overburden/underlying formation (ft²/hr)	0.05
Average heat capacity of oil sand [Btu/(ft³-°F)]	32

Table 8.116—Reservoir rock and fluid properties.

1. Assume that the net oil value to the company is USD 10.00/bbl. For a single injection well, calculate the daily oil-production income (in USD/D) 1 year after the start in injection. Assume that the capture efficiency is 100% (i.e., oil produced equals oil displaced).
2. Calculate the rate of heat loss (Btu/hr) to the overburden and the underlying formation after 1 year of injection.

8.29 In **Fig. 8.242**, Bursell and Pittman (1975) present a correlation of the oil recovery (percent PV) as a function of viscosity for the Kern River field. Assume a pilot flood is to be conducted in an area of the field where the crude oil viscosity is 1,000 cp and the initial oil saturation is 65%. Use their correlation to estimate the oil recovery in this new area as %OOIP. Assume a sweep efficiency of 1.0 (100%) for the calculation.

8.30 **Fig. 8.243** is a correlation developed by Bursell and Pittman (1975) to optimize the injection rate for steamflooding the Kern River field on 2.5-acre spacing (i.e., the area of each pattern is 2.5 acres). It is desired to use this correlation in an adjacent region where the average thickness of the reservoir is 33 ft. In this region, the current steam-injection rate is 206 B/D of steam (BSPD) for each pattern. The oil production rate is approximately 61 BOPD for each pattern.

1. Are the present operating conditions in agreement with the correlation? Plot the present operating conditions on the figure.
2. Assume that you want to operate at the optimum steam-injection rate. What is your recommended new steam-injection rate in BSPD?
3. What would the oil recovery rate be at your recommended steam-injection rate? Give your answer in BOPD per pattern.
4. Tables 8.42 and 8.43 give the properties of the Kern River field. Estimate the average rate of steam-front movement in the vertical direction using the uniform-vertical-displacement model if the areal sweep efficiency is 100% and gravity override dominates the displacement process.

8.31 Estimate the ultimate SOR for a steamdrive in the Kern River reservoir using Myhill and Stegemeier's method. Tables 8.42 and 8.43 give the properties for the Kern River reservoir. Compare your estimate with the value in Table 8.43.

8.32 The hydrocarbon in the San Miguel-4 sand in Maverick County, Texas, is properly identified as a tar. It is immobile at the reservoir temperature of 95°F and has a pour point of 180°F. The tar viscosity decreases to approximately 870 cp at 300°F, suggesting that production could occur if the reservoir were heated. Although the permeability of the San Miguel-4 sand ranges from 250 to 1,000 md, the mobility of the tar is so low that steam injection is not possible without fracturing the formation.

The reservoir is 52 ft thick and is at a depth of 1,500 ft. **Fig. 8.244** shows the reservoir properties for a pilot test of steam injection into a fracture to heat the reservoir before steam displacement (Britton et al. 1983). The pattern is an inverted 5-acre five-spot with injection into the center well (Well JSR-5). Fig. 8.245 shows the well locations for the test region. A 4-ft-thick impermeable limestone streak is in the middle of the formation and separates the reservoir into two zones of equal thickness (26 ft). In the test-pattern area, sand development in the lower zone was absent; therefore, steam injection was targeted at the upper zone. Fig. 8.245 shows net sand after deducting thickness of dense streaks in the net sand.

Production wells (Wells JSR-1 through JSR-4) were fractured hydraulically, creating horizontal fractures. Injection Well JSR-5 was also fractured hydraulically, allowing connection to each production well through the network of horizontal fractures. For the purposes of this problem, we assume that the fracture is in the middle of the upper zone of the reservoir. This is a simplification because, in practice, it is usually not possible to confine a fracture to a particular horizontal plane, even if the fracture is horizontal. As in Example 8.17, assume that the horizontal fracture has a width of 0.25 in. and a porosity of 0.40. Other parameters are $M_R = 40.51$ Btu/ft³-°F, $M_s = 31.12$ Btu/ft³-°F, and $a = 0.0482$ ft²/hr. Steam was injected into the injection well at 1,850 psig at a rate of 3,000 B/D for 146 days to preheat the upper zone before steam displacement. Injection pressure was slightly greater than the fracture pressure during the preheating phase. Although tar was produced during the preheating phase, assume that this had little effect on the fracture width.

1. What is the area (in acres) heated by steam injection during the preheating phase?
2. Estimate the temperature distribution in the upper zone above the fracture for the path between Wells JSR-5 and JSR-4 at a distance midway between the two wells. Plot temperature vs. distance from the fracture in 1-ft increments to the top of the upper reservoir.
3. Estimate the average temperature of the upper reservoir zone at the end of the preheating period at the location in Part 2.

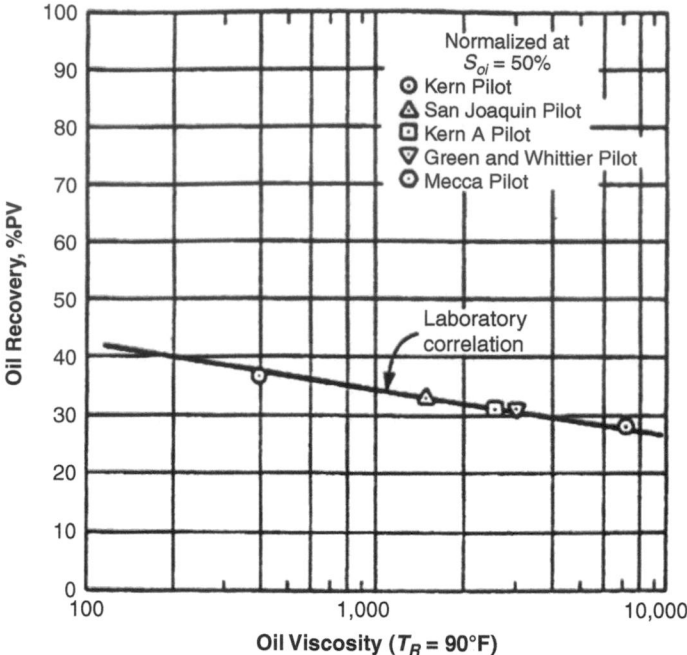

Fig 8.242—Comparison of field steam-displacement performance and laboratory correlation (Bursell and Pittman 1975).

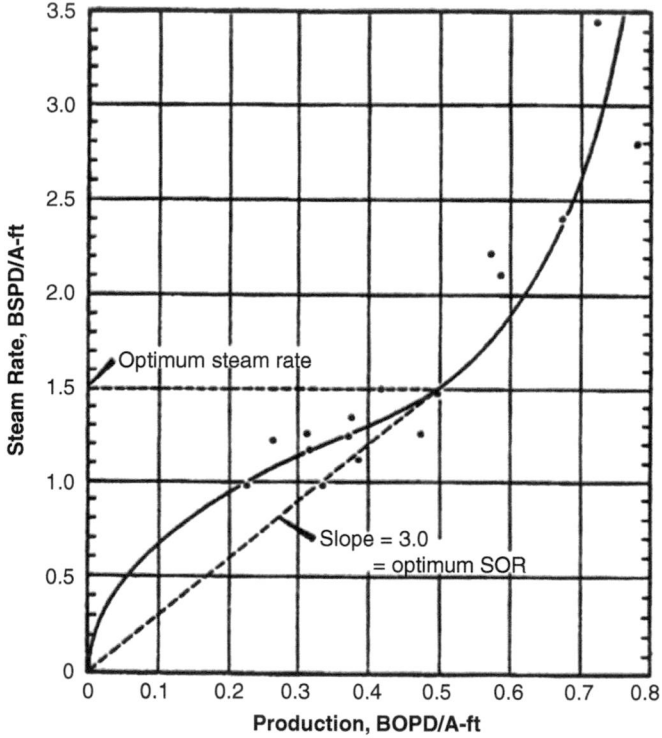

Fig 8.243—Optimum volumetric injection rate (Bursell and Pittman 1975).

8.33 The Kern River field contains a reservoir with up to seven sands that are separated by silt and clay interbeds (Restine 1983) and are considered to be impermeable. The development plan was to inject steam into the lowest layer and move up hole to the next zone after the lower zone was steamed out. A single zone is perforated in the injection wells; all zones are open in the production wells. In one project, steam was injected into the R-1 zone (as shown in **Fig. 8.246**) at a rate of 330 B/D. The injection pattern was 2.5-acre five-spot spacing, injection pressure was 100 psig, and steam quality at the sandface was estimated to be 0.5. Steam was injected for approximately 6.75 years into the R-1

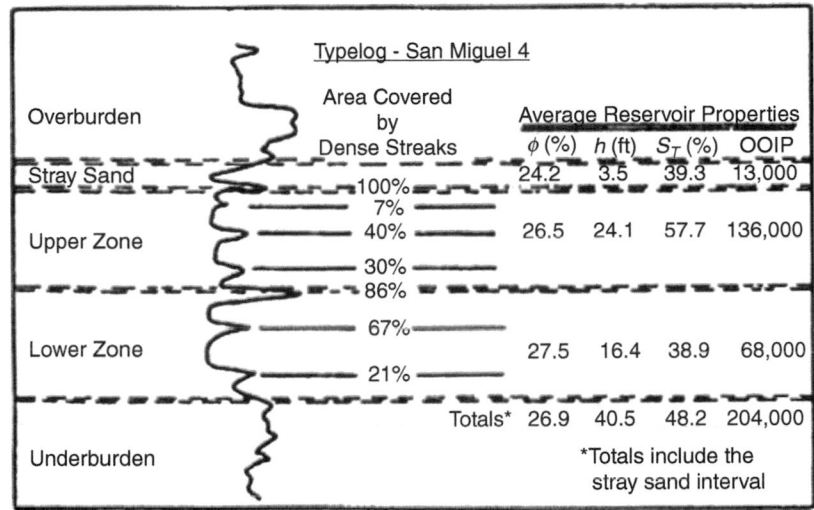

Fig 8.244—Reservoir properties in the test pattern, Street Ranch Pilot (Britton et al. 1983).

zone. During this time, heat was lost by conduction to the upper and lower formations. A beneficial effect of heat loss to the upper formation is the warming of the R zone.

Estimate the temperature distribution in the R zone in Fig. 8.246 when steam injection was discontinued in the R-1 zone ($r = 6.75$ years). The R zone is 68 ft thick and is separated from the lower R-1 zone by a 30-ft-thick silt/clay layer. The temperature distribution in the R zone is desired for two locations. At each location, plot the temperature distribution as a function of distance from the midpoint of the R-1 zone. Use the following parameters in your calculations: $M_R = 35$ Btu/ft³-°F, $M_S = 42$ Btu/ft³-°F, and $k_{hs} = 1.2$ Btu/ (hr-ft-°F). Neglect the difference in thermal properties between the shale and R zone for these calculations.

1. Estimate the temperature distribution 50 ft from the injection wellbore. Neglect the effect of conduction from the wellbore (wellbore heat loss) on the temperature distribution.
2. Estimate the temperature distribution at a distance halfway between the injection and production wells. Assume that the steam zone grows radially around the injection well until the steam zone reaches this midway point.
3. What would be the effect of "preheating" the upper layer on the steam displacement of the R zone? Compare your conclusions with the results of Restine (1983).

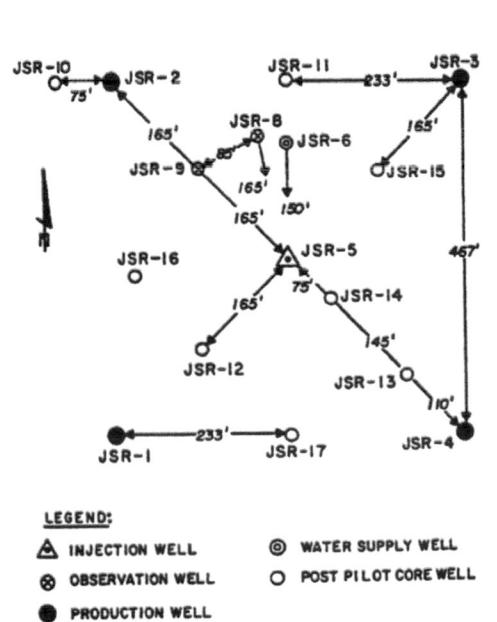

Fig. 8.245—Well locations in Street Ranch Pilot (Britton et al. 1983).

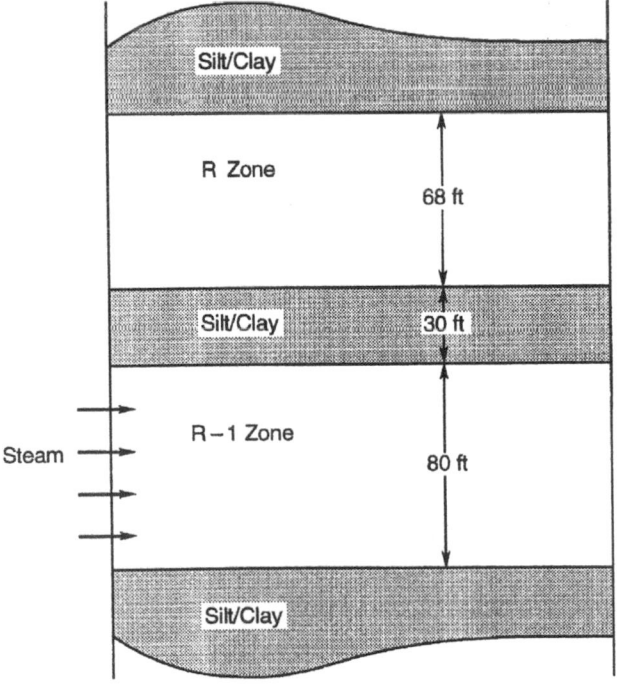

Fig. 8.246—Conduction heating of R zone during steam injection into R-1 zone, Kern River field.

8.34 Steam is being injected at an equivalent feedwater rate of 1,000 B/D. The quality of the steam entering the reservoir is 0.73 at a temperature of 565°F. **Table 8.117** gives pertinent reservoir and fluid properties. Estimate the recovery history in terms of cumulative oil production and SOR at 100, 250 and 500 days of injection. Use a capture factor of $E_c = 0.7$.

8.35 What is the composition of the effluent gas on a water-free basis in an in-situ-combustion operation if the CO_2/CO mole ratio is 4 and the atomic H/C ratio is 1.6? Air is used as the combustion gas, and the oxygen-usage efficiency is 94%.

8.36 1. What would be the composition of the effluent gas on a water-free basis in an in-situ-combustion operation if all the oxygen is consumed and the H/C atomic ratio is 1.6. Air was injected into the well.

2. What would be the composition of the effluent gas on a water free basis if an enriched oxygen stream consisting of 40% O and 60% nitrogen is injected instead of air?

8.37 Start with the general stoichiometric equation for burning one unit of coke, which is

$$CH_{F_{HC}} = \left(\frac{2m+1}{2m+2} + \frac{F_{HC}}{4}\right)O_2 \rightarrow \left(\frac{m}{m+1}\right)CO_2$$

$$+ \left[1/(m+1)\right]CO + \left(F_{HC}/2\right)H_2O. \quad\dots\dots\dots\dots\dots\dots\dots\dots\dots\dots\dots\dots (8.294)$$

Use Eq. 8.294 for the following questions.

1. Derive the general relationship for the ratio of moles of gas produced (or cubic feet) per mole (or cubic feet) of oxygen-enriched gas passing through the combustion front. Designate the concentration (mole fraction) of the oxygen in the injected gas by y_{iO_2}. Check your relationship by substituting $y_{iO_2} = 0.21$ for the case when air is injected.

2. What would be the ratio for pure-oxygen injection?

3. What would be the ratio for pure-oxygen injection with 100% oxygen usage?

8.38 The data in **Table 8.118** were obtained from a combustion-tube-test run. Determine the following from analysis of the data using stoichiometric equations.

1. Burning-front velocity.
2. Air/fuel ratio.
3. Air necessary to burn out 1 acre-ft of reservoir.
4. Air necessary to displace 1 bbl of oil.

The combustion-tube results from either direct measurement of material balance were burning-front velocity of 5.1 in./hr, air/fuel ratio of 162.5 scf/lbm, and air/sand ratio of 12.8 MMscf/acre-ft. Compare your calculated results with the tube results.

8.39 **Table 8.119** summarizes results from a combustion-tube run with a 13.4 °API California heavy oil and sands from the same reservoir used by Shu and Lu (1984). The oil has a viscosity of 6,600 cp at 75°F and an atomic H/C ratio of 1.52. The combustion-tube run was conducted by preflooding with N_2 and then injecting gas (21% O_2, 79% N_2). Table 8.119 summarizes the properties of the sandpack, operating conditions, and combustion characteristics. Evaluate the data and determine the following.

1. Average fuel H/C ratio.
2. Oxygen/fuel ratio in scf/lbm.
3. Oxygen-usage efficiency.
4. Oil saturation equivalent to the fuel consumed.
5. Water saturation equivalent to the fuel consumed.
6. Volume of water displaced/burned volume (ft³/ft³).
7. Ratio of air injected to oil displaced, scf/bbl.

8.40 **Table 8.120** presents properties of three reservoirs that are candidates for in-situ combustion. The table includes estimates of coke lay-down based on properties of the oil and rocks. Reservoir A contains a heavy oil with a viscosity of 100,000 cp at reservoir temperature. Owing to the low oil mobility, the reservoir was stimulated with six cycles of steam stimulation. Approximately 5.1% of the OOIP was recovered in 4 years. CSS consisted of injection of steam for 15 to 30 days followed by 7 days of soaking time; 9,200 tons of steam was injected in each well, and the cumulative OSR was 0.774 bbl/ton. Reservoir B is a deep reservoir (4,800 ft) containing a moderately viscous oil. Reservoir

Reservoir properties	
Porosity	0.18
Net thickness (ft)	34.0
Gross thickness (ft)	41.0
S_{oi}	0.66
S_{or} (steamflood residual)	0.17
Reservoir temperature (°F)	93
M_R (Btu/ft³-°F)	38.1
a (ft²/D)	0.70
Fluid properties	
Injection temperature (°F)	565
L_{dhv} (Btu/lbm)	615.4
f_{sd}	0.73
C_w (Btu/lbm-°F)	1.08

Table 8.117—Reservoir and fluid properties.

Injected air (L/hr)	167
Tube diameter (in.)	2.968
Fuel concentration from tube results (lbm/ft³)	1.87
Oil gravity (°API)	11.2
Porosity (fraction)	0.47
S_o	0.29
S_w	0
S_g	0.71
Combustion gas analysis	
CO_2	14.5
CO	5.2
O_2	1.4
N_2	78.9

Table 8.118—Combustion-tube data.

Sandpack properties	
Porosity (%)	40.8
Air permeability (darcies)	5.7
Oil saturation (%)	58.0
Water saturation (%)	16.2
Gas saturation (%)	25.8
Operating conditions	
Initial pack temperature (°F)	150
Injection pressure (psig)	150
Injection-gas flux (scf/hr-ft²⁾	90
Stabilized burning characteristics	
Maximum combustion temperature (°F)	900
Combustion-front velocity (ft/hr)	0.325
Fuel availability (lbm/ft³⁾	1.5
Average product gas (vol%)	
CO_2	13.37
CO	3.33
O_2	1.30
N_2	82.0

Table 8.119—Combustion-tube results (from Run 2, Farouq Ali 1981).

thickness is approximately 15 ft. Primary production was 5.4% OOIP. Reservoir C is located at a depth of 11,400 ft, which is clearly too deep for steam injection. The reservoir oil has a viscosity of 6 cp at a reservoir temperature of 221°F. Using the Brigham et al. (1980) correlation, estimate the ultimate recovery of oil (%OOIP) by in-situ combustion in each reservoir assuming a 90% oxygen-usage efficiency for each.

8.41 A combustion pilot was run on an inverted 1.5-acre five-spot with air as the injected gas (Naji and Poettmann 1991). Gas production was 350 MMscf. Table 8.121 lists the compositions obtained from analysis of the produced gases after steady-state operations were achieved. The fuel content was 1.7 lbm/ft³ of bulk volume. Table 8.122 gives properties of the reservoir and rock.

1. Estimate the volumetric sweep efficiency of the combustion front assuming that all the combustion gas was produced.
2. Calculate the total air necessary to displace the oil from the pilot pattern assuming a 70% volumetric-displacement efficiency. The reservoir oil is a dead oil with an oil gravity of 25 °API. Neglect the air stored in the burned volume.
3. In Part 2, the average reservoir pressure behind the combustion front was 500 psia and the average temperature was 150°F. Estimate the volume of air stored in the burned volume and the air requirement. Assume a Z factor of 0.96 at these conditions. Compare the air requirement computed in Parts 2 and 3.

8.42 A reservoir that is a candidate for in-situ combustion has the properties given in Table 8.123. Using the Brigham et al. (1980) correlation, predict the oil recovery vs. cumulative air injected. Base your calculations on 1 acre-ft of reservoir.

8.43 In his Table 5, Showalter (1963) presents production well data on the South Belridge thermal recovery experiment. Analyze these data. For South Belridge, fuel content is 1.9 lbm/ft³, porosity is 0.30, oil gravity is 17 °API, and oil saturation is 0.60.

1. Calculate the well-effluent-gas composition for Months A and F, including the oxygen-usage efficiency.
2. With the gas composition calculated in Part 1 and the stoichiometric equations, calculate the air required to burn 1 lbm of fuel for Months A and F. Calculate H/C atomic ratio from the gas analysis.
3. How does the H/C atomic ratio calculated from your gas analysis compare with Showalter's ratio? (Note: Showalter measured the C/H atomic weight ratio; you will need to convert.) Compare for Month A and Month F.
4. Compare your values calculated in Part 2 with the experimental values in Showalter's Table 5. (Note: His values were measured at 32°F and 1 atm; correct them to your standard values.)
5. Calculate the standard cubic feet of air necessary to displace 1 bbl of oil for Months A and F.

8.44 Table 8.124 presents the composition of the produced gas in the example problem and auxiliary data of Nelson and McNeil (1961). Use stoichiometric equations to calculate the following.

1. Air injected per cubic foot of reservoir burned.
2. Air flux for a burning-front-advance rate of 0.125 ft/D.
3. Oil displaced per acre-foot of reservoir burned.

	Reservoir		
	A	B	C
Depth (ft)	1,476	4,800	11,400
Thickness (ft)	75	15	62
Pattern size (five-spot) (acres)	2.5	15.4	15.4
Oil gravity (°API)	11	10	24
Viscosity at T_r (cp)	100,000	1,000	6
Porosity (%)	30	27.7	14
Permeability (md)	1,500	626	85
Initial water saturation (%)	36	36	15
Reservoir temperature (°F)	60	185	221
Fuel consumed (lbm/ft³)	2.62	2.99	1.37

Table 8.120—Reservoir and oil characteristics (from Hansel et al. 1984).

CO_2	11.95
CO	1.45
O_2	3.20
N_2	83.40

Table 8.121—Produced-gas analysis (vol%).

Porosity (%)	30
Thickness (ft)	25.83
Initial oil saturation	0.6
Interstitial water saturation	0.2

Table 8.122—Reservoir and rock properties.

4. Oil displaced per million cubic feet of air injected. This value should approach the oil produced per million cubic feet of air injected at the end of the project.
5. Barrels of water produced as a result of combustion per acre-foot of reservoir burned.

Compare your results from Parts 1 through 3 with the results calculated by Nelson and McNeil with their tank mass-balance method. Compare your results from Part 4 with their calculated results.

8.45 During the life of an in-situ-combustion project, the composition of the produced gases were monitored constantly. **Table 8.125** gives the gas composition on a water-free basis. A carbon balance on the produced gases indicated that 4,716,000 lbm of carbon was consumed as fuel. If it can be assumed that the oil produced in the tank battery is equal to the oil moved by the combustion zone, then the produced oil plus that consumed as fuel must equal the OOIP in the reservoir that was burned out. Oil produced from the combustion project was 92,030 STB. The density of the oil is 55.2 lbm/ft³ (28.6 °API). The original oil saturation was 0.68, and the porosity was 0.21. Assume that B_o = 1.0. Evaluate this project by carrying out the following calculations.

1. Calculate the total pounds of fuel consumed.
2. Calculate the volume (in cubic feet) of reservoir that was burned out.
3. On the basis of Parts 1 and 2, what was the fuel content?
4. On the basis of your answer to Part 3 and the stoichiometric equation, what is the air required to burn out 1 ft³ of reservoir?
5. Estimate the fuel content from the stoichiometrically derived equation relating cubic feet of air necessary to displace 1 bbl of oil. The total air injected was corrected for the air stored in the burned-out portion of the reservoir. The corrected value was 1.027 Bscf of air. How does this value of fuel content calculated from the stoichiometric equation compare with the fuel content estimated in Part 3?

Thickness (ft)	30
Porosity	0.36
Oil viscosity at T_r (cp)	2,700
Oil gravity (°API)	12.9
S_o	0.60
S_w	0.37
N (bbl/acre-ft)	1,645
m_R (lb/ft³)	2.02
e_o (%)	100

Table 8.123—Reservoir and fluid properties.

Produced-gas composition (vol%)	
N_2	84.2
O_2	1.1
CO_2	11.7
CO	3.0
Total	100.0
Auxiliary data	
Fuel (lbm/ft³)	2.0
Porosity	0.35
S_o	0.55
Oil gravity (°API)	10

Table 8.124—Produced-gas analysis and auxilliary data.

O$_2$	2.94
CO$_2$	13.80
CO	0.41
N$_2$	82.85
Total	100.00

Table 8.125—Produced-gas composition (vol%).

N$_2$	83.40
O$_2$	3.20
CO$_2$	11.95
CO	1.45
Total	100.00

Table 8.126—Combustion-gas analysis (mol%).

Well A	81.551
Well B	43.095
Well C	23.555
Well D	20.641
Total	168.842

Table 8.127—Combustion-gas volumes (MMcf).

8.46 An in-situ-combustion project was conducted on the 0.592-acre inverted five-spot shown in **Fig. 8.247. Table 8.126** gives the produced-gas composition averaged over a period of time of steady operation. **Table 8.127** gives gas production from the four producing wells from ignition to shutdown. The reservoir oil is a dead oil with an oil gravity of 25.8 °API (or a specific gravity of 0.8996). The fuel content is 1.69 lbm/ft^3, and the average porosity is 0.25. **Table 8.128** gives the post-combustion reservoir model derived from core holes. Total air injected was 179.751 MMscf. Actual oil recovery was 2,692 bbl. No air traversed the upper unburned zone. Because of the large vertical permeability, any air entering this zone at the wellbore was diverted into the lower zone.

1. What is the amount of air passing through the combustion front (a) for each quadrant and (b) for the total pattern?
2. How much air is stored behind the combustion front?
3. How much oil is *displaced* in each zone in combined Quadrants A, B, and D and in each zone in Quadrant C? How much total oil is displaced?
4. Calculate the OIP for combined Quadrants A, B, and D and for Quadrant C for the cleanly burned zone and for the partially burned zone. For calculation purposes assume that all quadrants are of equal area. What is the total OIP in the cleanly burned and partially burned zones?
5. Calculate the *displaceable* OIP for combined Quadrants A, B, and D and for quadrant C for the cleanly burned zone and for the partially burned zone. For calculation purposes, assume that all quadrants are of equal area. What is the total displaceable oil? Estimate the volumetric- or oil-displacement efficiency for the cleanly burned zone, for the partially burned zone, and for the total pilot.
6. Calculate the oil-recovery efficiency on the basis of total OOIP in the cleanly burned zone and in the partially burned zone assuming that all the produced oil came from these zones. Assuming that all the oil in the cleanly burned and partially burned zones was completely swept by the combustion front, what is the maximum potential percent oil recovery based on the OOIP in these zones? What is the actual percent oil recovery based on the maximum possible oil that could be recovered from the cleanly burned and partially burned zones?
7. Actual production from the pilot was 2,692 bbl. Compare this with the displaced oil, and offer possible explanations for the difference.

8.47 A reservoir is considered to be a candidate for thermal recovery. A pilot test of in-situ combustion is to be conducted in a 2.5-acre pattern. **Table 8.129** summarizes the properties of the reservoir, and **Table 8.130** presents the combustion-tube data. The anticipated air-injection rate is 3.5 MMscf/D. Estimate the oil recovery for this pilot test as a function

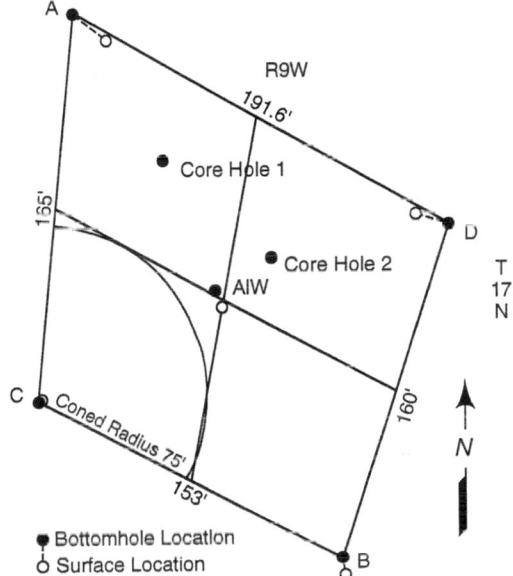

Fig. 8.247—Pilot pattern well location (after Struna and Poettmann 1988).

Zone	h (ft)	Initial S_o Quadrants A, B, and D	Quadrant C	kh (md-ft)
Upper unburned zone	8	0.474	0.313	4,768
Cleanly burned zone	18	0.286	0.183	26,478
Lower partially burned zone	14	0.157	0.157	1,484

Table 8.128—Post-combustion analysis.

of time using the Gates and Ramey method (Gates and Ramey 1980). Assume that the correlations developed from the South Belridge tests can be used for this calculation and that the entire pattern volume can be burned. Estimate the oil recovery if the volume burned is 30% of the pattern volume. Assume that the recovery efficiency from the unburned volume is 20%.

8.48 The Moco T in-situ-combustion project has been in operation since 1962 (Curtis 1989). In 1987, cumulative oil produced from the project was 14 million bbl and cumulative water produced was 35.44 million bbl. All the oil production is attributed to in-situ combustion, while part of the water production is the result of water influx. Cumulative air injection was 59 Bcf. Combustion-tube data are not available, but the fuel requirement is estimated to be 2.5 lbm/ft³. The air requirement can be estimated with the Chu (1982a) correlation:

$$a_R^* = 0.963\left(4.72 + 0.03656h + 9.9965S_{oi} + 0.0006\right), \quad \dots\dots\dots\dots\dots\dots\dots\dots\dots\dots\dots\dots (8.380)$$

where a_R^* is in MMscf/acre-ft and k is in millidarcies. **Table 8.131** gives rock and fluid properties.

1. Estimate the burned volume at the end of 1987. Assume that all the air is consumed in the burned volume.
2. Estimate the volume of oil displaced from the burned volume and the volume of oil displaced from other regions assuming a capture efficiency of 100%.
3. Estimate the volume of produced water that is attributed to water influx if F_{HC} is assumed to be 2.80.

Thickness (ft)	30
Porosity (fraction)	0.37
Original oil saturation (fraction)	0.60
bbl/acre-ft	1,720
Water saturation (fraction)	0.37
bbl/acre-ft	1,060
Gas saturation	0.03
Total OOIP (bbl)	145,000
Total water (bbl)	90,000
OIP at start of pilot test (bbl/acre-ft)	1,545
Oil gravity (°API)	12.9
Oil viscosity (cp at 87°F)	2,700
Pattern area (acres)	2.5

Table 8.129—Formation properties (from Williams et al. 1980).

Porosity	0.35
Radius (in.)	1.25
Air flux (scf/hr)	4.0
Burning-front velocity (in./hr)	35.
CO (wt%)	1.1
CO₂ (wt%)	15.2
O₂ (wt%)	0.2
N₂ (wt%)	83.2
Fuel consumed (bbl/acre-ft)	287.5
Air/fuel ratio (scf/bbl)	184.1
Air required (MMscf/acre-ft)	18.171

Table 8.130—Combustion-tube data.

Productive area (acres)	150
Gross formation thickness (ft)	500
Net formation thickness (ft)	129
Porosity (fraction)	0.36
Oil saturation (fraction)	0.71 to 0.75
Water saturation (fraction)	0.25
Gas saturation (fraction)	0.04
FVF	1.06
Initial OIP (bbl/acre-ft)	1,980
OOIP (million bbl)	38
Specific permeability (md)	1,575
Initial formation pressure (psi)	1,000
Initial formation temperature (°F)	125
Oil gravity (°API)	14.5
Oil viscosity at formation temperature (cp)	110

Table 8.131—Rock and fluid properties, Moco T in-situ combustion project (from Curtis 1989).

Nomenclature

a_R = air consumed in in-situ combustion, L^3/m, scf/ft^3 burned volume

a_R^* = air requirement at pressure p_{inj}, L^3/L^3, scf/ft^3

A = area, L^2, ft^2; distance between adjacent well pairs, m^2/m

A_b = maximum area burned in in-situ combustion process, L^2, acres

A_d = area displaced to the maximum steam-zone thickness at time t in uniform vertical displacement model, L^2, acres

A_h = heated area at temperature T_s, L^2, ft^2 or acres

A_p = area of well pattern, L^2, acres

A_r = dimensionless parameter defined by Eq. 8.185

A_s = area of reservoir zone invaded by steam, L^2, ft^2

b = constant for any oil

$b_1 \ldots b_6$ = parameters in infinite-series approximation for the error function of x

B_{gi} = FVF for air at injection pressure and the average temperature of the region behind the combustion front, L^3/L^3, ft^3/scf

B_n = dimensionless parameter defined by Eq. 8.137 for Timestep n

B_o = FVF, L^3/L^3, RB/STB

B_{oc} = FVF in cold region of reservoir, L^3/L^3, RB/STB

B_{oh} = FVF in the heated region, L^3/L^3, RB/STB

B_{oi} = FVF at start of steamdrive, L^3/L^3, RB/STB

B_{os} = FVF in zone swept by steam, L^3/L^3, RB/STB

B_{ow} = FVF for oil in the hot-water-swept region in a steamflood, L^3/L^3, bbl/STB

c_{O_2} = oxygen concentration in injected air, m/m, lbm O_2/lbm air

C_{an} = heat capacity of fluid in annulus at average annulus temperature, L^2/t^2T, Btu/lbm-°F

C_g = mean heat capacity of gas at constant pressure, L^2/t^2T, Btu/lbm-°F

C_i = mean heat capacity at constant pressure, L^2/t^2T, Btu/lbm-°F

C_o = mean heat capacity of oil at constant pressure, L^2/t^2T, Btu/lbm-°F

C_{ob} = mean heat capacity of overburden at T_s, L^2/t^2T, Btu/lbm-°F

C_p = heat capacity of reservoir rock, saturated with fluids, ahead of the steam zone when the steam zone expands at a constant vertical velocity, L^2/t^2T, Btu/lbm-°F

C_r = mean heat capacity of the matrix material at constant pressure, L^2/t^2T, Btu/lbm-°F

C_s = mean heat capacity of saturated water vapor at T_s, L^2/t^2T, Btu/lbm-°F

C_w = mean heat capacity of water at constant pressure, L^2/t^2T, Btu/lbm-°F

d = distance between injection and production well, L, ft

d_c = diameter of laboratory core, L, ft

D = diameter of tubing, L, ft

D_{pi} = ID of pipe, L, ft

E_A = areal sweep efficiency at steam breakthrough, fraction

E_{Ab} = fraction of a pattern area swept by combustion front, L^3/L^3, volume fraction

F_c = capture efficiency, fraction of the oil displaced by steam and hot-water regions that is actually produced, fraction

E_h = thermal efficiency of reservoir heating, fraction of injected energy remaining in heated reservoir region

E_{hb} = vertical sweep efficiency within the area invaded by combustion front, L^3/L^3, volume fraction

$E_{h,s}$ = thermal efficiency, fraction of injected energy in steam zone

$\bar{E}_{h,s}$ = average steam-zone displacement efficiency defined by Myhill and Stegemeier (1978), dimensionless

E_{lb} = lower bound of steamdrive efficiency based on formation of a hot-water zone when the critical time, t_{cD}, is exceeded during the steamdrive

E_l = energy lost to over- and underburden, mL^2/t^2, Btu

E_{O_2} = oxygen-usage efficiency, fraction

E_{vb} = volumetric sweep efficiency of the combustion front

E_V = volumetric sweep efficiency, volume fraction of volume burned by combustion front, L^3/L^3

f = fraction of gas saturation that is noncondensible, L^3/L^3, volume fraction

f_{cp} = fraction of injected energy that is removed with produced fluids

$f_{h,v}$ = fraction of injected energy that is latent heat

f_p = fraction of the injected energy that is produced

f_s = steam quality, m/m, mass fraction

f_{sb} = steam quality at outlet of boiler, mass fraction

f_{sd} = steam quality at sandface, m/m, mass fraction

f_{sh} = steam quality at wellhead, m/m, mass fraction

$f(t)$ = transient time function, dimensionless

f_{tp} = two-phase friction factor, dimensionless

F_a = air required for in-situ combustion, L^3/L^2-L, Mcf/acre-ft

F_{AO} = air/oil ratio based on the burned zone, L^3/L^3, scf/bbl

F_{go} = GOR in produced fluid, L^3/L^3, scf/STB

F_{HC} = atomic H/C ratio in a fuel for in-situ combustion

F_{oAu} = oil displaced from unburned region per volume of air injected, L^3/L^3, scf/bbl

F_{os} = OSR, L^3/L^3, STB/bbl

F_{OSE} = economic breakeven OSR when lease crude is used for steam generation, Eq. 8.117, L^3/L^3, bbl oil/bbl steam

F_{so} = SOR, barrels of steam injected (CWE volume) as steam/barrels of oil produced, L^3/L^3, bbl/STB

F_{tci} = view factor based on outside tubing and inside casing surfaces, dimensionless

F_{wa} = produced water/air ratio, L^3/L^3, bbl/scf

F_{wo} = producing WOR, L^3/L^3, bbl/STB

F_{woc} = producing WOR before steam stimulation, L^3/L^3, bbl/STB

F_{wv} = WOR for water condensed from produced gas, L^3/L^3, bbl/STB

ΔF_{wo} = change in producing WOR after CSS from the prestimulation WOR, L^3/L^3, bbl/STB

g = acceleration owing to gravity, L/t^2, 4.17 × 10^8 ft/hr²

g_c = gravitational constant, 32.2 ft lbm/sec²-lbf

g_z = acceleration of gravity in z-direction ($g_z = 0$ when z is horizontal), 32.2 ft/sec²

G = dimensionless compilation of terms, Eq. 8.62

G_e = dimensionless compilation of terms, Eq. 8.178

G_i = cumulative air injected, L^3, scf

G_m = mass flux of steam, m/L^2t, lbm/ft²-hr

$\dot{G}_p$ = combustion gas rate, L^3/t, scf/D

G_{pf} = volume of combustion gases per volume of reservoir burned, L^3/L^3, scf/ft³

G_s = integral expression defined by Eq. 8.129

G_T = amount of air required in in-situ combustion process, L^3, scf

G_v = integral expression defined by Eq. 8.173

G_1 = dimensionless compilation of terms, Eq. 8.66

G_3 = dimensionless compilation of terms, Eq. 8.167

h = thickness of formation, L, ft

$\overline{h}$ = volumetric average thickness to the oil column in the saturated portion of the region, L, ft

$\overline{h}_b$ = average burned-zone thickness, L, ft

h_c = heat-transfer coefficient for convection based on the outer surface area of the insulation ($2\pi r_{is}L$) and the temperature difference between surface of insulation and ambient air, m/t^3T, Btu/hr-ft²-°F

h'_c = heat-transfer coefficient for natural convection based on the outside radius of the insulated tubing, r_{ins}, m/t^3T, Btu/hr-ft²-°F

h_{cz} = gross thickness of condensate zone, L, ft

h_f = heat-transfer coefficient between flowing fluid and inside surface of tubing, m/t^3T, Btu/hr-ft²-°F

h_{fc} = heat-transfer coefficient for forced convection, m/t^3T, Btu/hr-ft²-°F

h_h = thickness of saturated region in a reservoir being produced by gravity drainage at the radius r_h, L, ft

h_n = net reservoir thickness, L, ft

h_{nc} = heat-transfer coefficient for natural convection in the tubing/casing annulus based on the outside tubing surface and the temperature differences between the outside tubing and the inside casing surfaces, m/t^3T, Btu/hr-ft²-°F

h'_{nc} = heat-transfer coefficient for natural convection in the insulation/casing annulus based on the outside insulation surface and the temperature differences between the outside insulation and the inside casing surfaces, m/t^3T, Btu/hr-ft²-°F

h_o = gross thickness of oil zone, L, ft

h_{os} = thickness of oil column at radius r_s, L, ft

h_r = heat-transfer coefficient for radiation based on the outside tubing surface and the temperature difference between the outside tubing and the inside casing surfaces, m/t^3T, Btu/hr-ft²-°F

h'_r = heat-transfer coefficient for radiation based on the outside insulation surface and the temperature difference between the outside insulation and the inside casing surfaces, m/t^3T, Btu/hr-ft²-°F

h_s = thickness of reservoir zone that contains initial gas saturation, L, ft

$\overline{h}_s$ = average steam-zone thickness, L, ft

$\overline{h}_{st}$ = average steam-zone thickness, van Lookeren model, L, ft

h_t = reservoir thickness, L, ft

h_{us} = thickness of unsaturated zone at the top of a reservoir, L, ft

h_w = thickness of oil-saturated zone at r_w, L, ft

h_{wp} = fluid level in production well, L, ft

Δh_a = heat released by combustion reaction assuming that water produced by combustion condenses and $E_{O_2}w = 1.0$, m/t²L, Btu/scf air

Δh_f = heat released by burning fuel during in-situ combustion, L²/t², Btu/lbm

H = specific enthalpy, L²/t², Btu/lbm

H_o^n = enthalpy of oil at time t^n, L²/t², Btu/lbm

H_{ogv} = enthalpy of produced oil and gas relative to T_r, L²/t², Btu/STB

H_{or} = enthalpy of oil at reservoir temperature, L²/t², Btu/lbm

H_{oT} = enthalpy of oil at temperature T, L²/t², Btu/lbm

H_s = energy content of injected steam relative to initial reservoir temperature, Eq. 8.4, L²/t², Btu/lbm

H_{sbA} = enthalpy of steam at boiler outlet relative to the feed water temperature, TA, L²/t², Btu/lbm

H_{Sn} = energy content of injected steam relative to the initial reservoir temperature (Eq. 8.67) at time steam n, L²/t², Btu/lbm

H_t = energy content of injected steam relative to reservoir temperature adjusted for the energy remaining in the reservoir at the end of the previous cyclic steam-injection period, L²/t², Btu/lbm

H_v = enthalpy of saturated vapor, L²/t², Btu/lbm

H_{vT} = enthalpy of saturated water vapor at temperature T, L²/t², Btu/lbm

H_w = enthalpy of produced water and water vapor, mL²/t², Btu

H_w^n = enthalpy of water at time t^n, mL²/t², Btu/lbm

H_{wA} = enthalpy of water at feed water temperature, T_A, L²/t², Btu/lbm

H_{wr} = enthalpy of water at reservoir temperature, L²/t², Btu/lbm

H_{wrv} = enthalpy of produced water relative to T_r, L²/L³, Btu/bbl

H_{ws} = enthalpy of water at steam temperature, L²/t², Btu/lbm

H_{wT} = enthalpy of water at temperature T_s, L²/t², Btu/lbm

i_a = air-injection rate, L³/t, scf/D

i_{ah} = air injection per foot of thickness, L²/t, scf/hr-ft

$(i_a)_{max}$ = maximum air-injection rate, L³/t, scf/D

i_D = dimensionless injection rate, Eq. 8.182

I_0 = modified Bessel function of zero order

I_1 = modified Bessel function of first order

J = ratio of PI of stimulated well to the PI of the well before stimulation

$\bar{J}$ = ratio of PI after CSS to PI before CSS

J_c = PI, flow rate per unit of pressure drawdown before reservoir has been heated by cyclic steam injection, L⁴ t/m, STB/(D-psi)

J_h = PI, flow rate per unit of pressure drawdown after reservoir has been heated by cyclic steam injection, L⁴t/m, STB/(D-psi)

J_Q = mechanical equivalent of heat, 778 ft-lbf/Btu

K = permeability of porous rock, L², md

k_d = permeability in damaged zone of reservoir, L², md

k_g = effective permeability to gas, L², md

k_h = thermal conductivity, mL/t³ T, Btu/(hr-ft-°F)

k_{ha} = thermal conductivity of air in annulus at average temperature and pressure of annulus or at the film temperature, m/Lt³T, Btu/hr-ft-°F

k_{hc} = equivalent thermal conductivity of annulus fluid with natural convection effects evaluated at average temperature and pressure of annulus, m/Lt³T, Btu/hr-ft-°F

k_{hcas} = thermal conductivity of casing material at average casing temperature, m/Lt³T, Btu/hr-ft-°F

k_{hcem} = thermal conductivity of cement at average temperature and pressure of cement, m/Lt³T, Btu/hr-ft-°F

k_{hf} = thermal conductivity of formation, m/Lt³T, Btu/hr-ft²-°F/ft

k_{hins} = thermal conductivity of insulation, m/Lt³T, Btu/hr-ft-°F

k_{hr} = thermal conductivity of the reservoir rock at initial saturation, m/Lt³T, Btu/hr-ft-°F

k_{htub} = thermal conductivity of tubing, m/Lt³T, Btu/hr-ft-°F

k_o = effective permeability to oil, L², md

k_{oc} = effective oil permeability in cold region of reservoir, L², md

k_{oh} = effective permeability to oil in heated region, L², md

k_s = effective permeability of steam, L², md

k_{wh} = effective permeability to water in the heated region following CSS, L², md

L = length of laboratory core, L, ft

L_j = length of steam interface at time t_j, m

L_v = latent heat of vaporization, L²/t², Btu/lbm

L_{vdh} = latent heat of vaporization at sandface, L^2/t^2, Btu/lbm

m = moles of CO_2 produced per mole of CO produced during in-situ combustion

m' = ratio of $CO/(CO + CO_2)$ in effluent gas

m_E = fuel consumption in combustion-tube run, m/L^3, lbm/ft^3

m'_{ob} = rate of steam condensation in Neuman model, m/t, lbm/hr

m_R = fuel consumed per area of reservoir burned for in-situ combustion, m/L^3, lbm/ft^3

m_s = mass of water injected as steam, m, lbm

$\dot{m}_s$ = steam-injection rate, m/t, lbm/hr, lbm/D

$\dot{m}_{s_n}$ = steam-injection rate during Timestep n, m/t, lbm/hr, lbm/D

$\dot{m}_{sob}$ = rate of condensate flow to the steam-zone interface in Neuman model, m/t, lbm/hr, lbm/D

M = average volumetric heat capacity of rock, Eq. 8.50, m/Lt^2T, Btu/(ft^3-°F)

M_R = average volumetric heat capacity of heated region, m/Lt^2T, Btu/(ft^3-°F)

M_s = average volumetric heat capacity of overburden rock, m/Lt^2T, Btu/(ft^3-°F)

n = number of moles

n_i = moles of gas injected during a specified time interval, m/M_w, moles

n_p = moles of produced gas, moles

n_{pf} = amount of combustion gases produced per fuel consumed, mole/mole

n_s = number of zones in the gross interval that receive steam

N = OOIP, L^3, STB

N_c = OIP at start of combustion, L^3, STB

N_{Gr} = Grashof number, dimensionless

N_p = volume of oil produced by steamdrive, L^3, STB

N_{pb} = oil displaced from burned volume of reservoir, L^3, bbl

$N_{ph}^{(n)}$ = cumulative oil produced from heated zone at the end of Timestep n, L^3, bbl

N_{Pr} = Prandtl number, dimensionless

N_{ps} = volume of oil displaced by steam, L^3, bbl

N_{pw} = volume of oil displaced from region swept by hot water, L^3, STB

N_{p1} = percent oil recovered by in-situ combustion, initial oil less oil consumed as fuel, Eq. 8.357

N_{Re} = modified Reynolds number for air flow normal to the pipe

ΔN_b = cumulative volume of oil burned as fuel from the start of combustion, L^3, STB

ΔN_p = cumulative volume of oil produced from the start of combustion, L^3, STB

ΔN_{pi} = volume of oil produced during Timestep i following CSS, L^3, STB

ΔN_{p1} = volume of oil produced during the first timestep following CSS, L^3, STB

O = volume of oil displaced, m^3

p = pressure, m/Lt^2, psia

p_e = pressure at effective radius of drainage of well, m/Lt^2, psi

P_i = BHP in injection well, m/Lt^2, psia

P_{inj} = injection pressure, m/Lt^2, psi

P_p = BHP in a production well, m/Lt^2, psi

p_s = steam-injection pressure, m/Lt^2, psia

p_{sc} = pressure at standard conditions, m/Lt^2, psi

P_w = BHP in the production well, m/Lt^2, psi

P_{wr} = vapor pressure of water at reservoir temperature T_r, m/Lt^2, psi

P_{wv} = vapor pressure of water, m/Lt^2, psi

Δp = pressure drop between the production well and the effective drainage radius of the well, $P_w - P_e$, m/Lt^2, psia

Δp_f = pressure drop caused by friction, m/Lt^2, psia

q = fluid production rate of single fluid, L^3/t, B/D; or drainage rate for the entire region

$q*$ = gravity-drainage rate, dimensionless

q_g = rate of gas production, L^3/t, scf/D

q_{max} = maximum pumping rate after soak period, L^3/t, B/D

q_o = oil production rate, L^3/t, B/D

$\bar{q}_o^{n+1}$ = average oil displacement rate for the time interval $t^{n+1} - t^n$, L^3/t, STB/D

q_{ob} = oil production rate produced by combustion front, L^3/t, B/D

q_{oc} = oil production rate when well is cold, L^3/t, B/D

q_{oh} = oil flow rate in heated zone following CSS, L^3/t, B/D

q_{oh}^n = oil production rate in heated zone at time t^n following CSS, L^3/t, B/D

$\bar{q}_{oh}^n$ = average oil production rate following CSS at Timestep n, L^3/t, B/D

q_{oh}^{n-1} = oil production rate in heated zone at time t^{n-1} following CSS, L^3/t, B/D

q_{os} = oil displacement rate in the steam zone when $t > t_{cD}$, L^3/t, STB/D

q_{ou} = oil production rate from unburned region in in-situ combustion process, L^3/t, B/D

q_t = total fluid production rate, L^3/t, B/D

q_w	=	water production rate as CWE, or rate that water is displaced by combustion front, L^3/t, B/D
q_w^n	=	water production rate during Timestep n, L^3/t, B/D
q_{wh}	=	water production rate following CSS, L^3/t, B/D
$\overline{q}_{wh}^n$	=	water production rate during Timestep n, L^3/t, B/D
Q	=	energy stored in heated region, mL^2/t^2, Btu
$\dot{Q}$	=	rate that energy is supplied by steam condensation to maintain a vertically expanding steam zone, m/t^3L, Btu/hr or Btu/D
$\dot{Q}_b$	=	rate of heat generation during in-situ combustion, m/t^3L, Btu/D
$\dot{Q}_{cond}$	=	rate that energy is supplied to the steam zone by condensation, m/t^3L, Btu/hr or Btu/D
Q_f	=	heat released by burning fuel, L^2/t^2, Btu/lbm
Q_l	=	energy lost to overburden and underlying formations during steam injection, mL^2/t^2, Btu
$\dot{Q}_\ell$	=	rate of energy loss to over- and underburden rock, mL^2/t^3, Btu/hr
$\dot{Q}_{\ell c}$	=	heat loss rate per unit length of pipe owing to convection, mL/t^3, Btu/hr-ft
$\dot{Q}_{\ell p}$	=	heat loss rate per unit length of pipe, mL/t^3, Btu/hr-ft
$\dot{Q}_{\ell r}$	=	heat loss rate per unit length of pipe owing to radiation, mL/t^3, Btu/hr-ft
$\dot{Q}_{\ell s}$	=	rate of heat loss from steam zone, mL^2/t^3, Btu/hr
$\dot{Q}_{\ell w}$	=	rate of heat loss from hot-water zone, mL^2/t^3, Btu/hr
$\dot{Q}_{\ell wb}$	=	wellbore heat-loss rate, mL^2/t^3, Btu/hr
$\dot{Q}_{\ell_1}$	=	rate by which heat must be supplied in Area A_1 to supply heat losses and maintain a constant steam-zone expansion velocity, mL/t^3, Btu/hr-ft
Q_{last}	=	energy remaining in heated region, mL^2/t^2, Btu
Q_{lost}	=	energy lost during soak period, mL^2/t^2, Btu
Q_{max}	=	amount of injected energy remaining in the reservoir at the end of the soak period, mL^2/t^2, Btu
$\dot{Q}_{ob}$	=	rate of heat loss by conduction to overburden in Neuman model, mL/t^3, Btu/hr
$\dot{Q}_p$	=	rate of energy removed from reservoir by produced fluids, Eq. 8.81, mL^2/t^2, Btu/D
$\dot{Q}_{pf}$	=	rate that energy is removed with produced fluids during Timestep n, mL^2/t^3, Btu/D
$\dot{Q}_{pi}$	=	energy removal rate at beginning of Timestep i, mL^2/t^3, Btu/D

$\dot{Q}_{pn}$	=	rate that energy is removed from reservoir by produced fluids during Timestep n following CSS, mL^2/t^3, Btu/D
$\dot{Q}_r$	=	heat-transfer rate by conduction, $m\,L^2/t^3$, Btu/hr
$\dot{Q}_r(t-\lambda)$	=	heat-loss rate from heated zone, Eq. 8.52, mL^2/t^3, Btu/hr
$\dot{Q}_s$	=	rate that energy is delivered to the steam zone by condensation, mL^2/t^3, Btu/hr or Btu/D
$\dot{Q}_1$	=	heat-loss rate by conduction to the overburden from a condensing steam zone, mL^2/t^3, Btu/hr or Btu/D
$\dot{Q}_2$	=	rate that heat is lost by conduction through the interface of a steam zone expanding at a velocity of v, mL^2/t^3, Btu/D
ΔQ^n	=	energy removed from the reservoir with the produced fluids in the time period n, $t^n - t^{n-1}$
r	=	radius, L, ft
r_b	=	wellbore radius in Chu (1963) combustion model, L, ft
r_c	=	radius of cold region in Table 8.6, same as r_e, L, ft
r_{ci}	=	inside radius of casing, L, ft
r_{co}	=	outside radius of casing, L, ft
r_d	=	radius of damaged zone, L, ft
r_{dc}	=	radius where combustion front begins to decay, L, ft
r_{do}	=	radius of outside surface of pipe, L, ft
r_e	=	effective drainage radius of well, L, ft
r_{ev}	=	external radius of conduit exposed to air, L, ft
r_f	=	radius of location of combustion front, L, ft
r_h	=	radius of heated zone, L, ft
r_{hd}	=	radius of drilled hole, L, ft
r_{is}	=	radius of outside surface of insulation, L, ft
r_r	=	radius of surface that is radiating, L, ft
r_s	=	radius of steam zone around injection well, L, ft
r_{ti}	=	inside radius of tubing, L, ft
r_{to}	=	outside radius of tubing, L, ft
r_w	=	wellbore radius, L, ft
R	=	thermal resistance defined by Eq. 8.13, Tt^3/m, °F-hr-ft²/Btu
R_g	=	solution-gas/water ratio, L^3/L^3, bbl/STB
s_c	=	skin factor for cold reservoir before cyclic steam injection
S	=	steam consumption, $m^3{}_{CWE}$
S_h	=	skin factor in heated reservoir
$s_1...s_k$	=	parameters in the infinite-series approximation in Eq. 8.73

S_g	=	gas saturation, volume fraction
S_{gi}	=	initial gas saturation, L^3/L^3, volume fraction
S_{iw}	=	interstitial water saturation, fraction
S_o	=	oil saturation, L^3/L^3, volume fraction
S_o^n	=	oil saturation at time t^n, L^3/L^3, volume fraction
S_{oc}	=	oil saturation in the condensate zone, Miller and Leung (1985) model, fraction
S_{oF}	=	oil saturation equivalent to fuel consumed, L^3/L^3, volume fraction
S_{oi}	=	initial oil saturation, L^3/L^3, volume fraction
S_{orh}	=	oil saturation in the hot-water-swept region of a steamflood, L^3/L^3, volume fraction
S_{ors}	=	ROS after displacement by steam, L^3/L^3, volume fraction
S_{ow}	=	oil saturation in Neuman model in the water zone, fraction
S_w	=	water saturation, fraction
S_w^n	=	water saturation at time t^n, fraction
S_w^{n-1}	=	water saturation at time t^{n-1}, fraction
S_{wF}	=	water saturation equivalent to water, (liquid equivalent) produced by combustion reaction, L^3/L^3, volume fraction
S_{wi}	=	initial water saturation, volume fraction
ΔS_o	=	change in oil saturation in steam zone, L^3/L^3, volume fraction
t	=	time, t, hours
t^*	=	dimensionless time defined by Eq. 8.239 for gravity-drainage model
t_B	=	time that net thickness is steamed out at the injection well, t, days
t_{BT}	=	time of steam breakthrough into a production well, t, days
t_{cD}	=	critical dimensionless time, Eq. 8.125
t_D	=	dimensionless time, Eq. 8.60
t_{Dm}	=	dimensionless time in Eq. 8.136 corresponding to time t_m
t_{Dr}	=	modified dimensionless time, Eq. 8.71
t_{Ds}	=	dimensionless time at which heated front arrives at position contacted by condensed steam at time t_D
t_{Dv}	=	dimensionless time for constant vertical velocity model defined by Eq. 8.168
t_{Dv1}	=	dimensionless time when the steam zone reaches area A_1 when the steam zone is expanding at a constant velocity v
t_{Dz}	=	modified dimensionless time, Eq. 8.72
t_i	=	initial time of process, t, hours
t_{min}	=	minimum time of injection of air corresponding to maximum air-injection rate, t, days

t_{soak}	=	length of soak period (shut in) following steam injection into a production well, t, days
$t^{(n)}$	=	time following CSS at Timestep n, t, days
$t^{(1)}$	=	time when the desaturated region attains the shape shown in Fig. 8.26, t, days
Δt	=	incremental time, t, hours
Δt_{pi}	=	number of days in Timestep i following CSS
Δt_{p1}	=	number of days in Timestep 1 following CSS
T	=	temperature, T, °F
$\bar{T}$	=	average temperature of heated region, T, °F
$\bar{T}_a$	=	average ambient air temperature, T, °F
$\bar{T}_{an}$	=	average temperature of the fluid in the annulus, $(T_{ci} + T_{to})/2.0$, T, °F or °R
T_b	=	temperature of burned zone behind the combustion front, T, °F
T_{ci}	=	inside surface temperature of casing, T, °F
T_{cia}	=	absolute inside surface temperature of casing, T, °R
$\bar{T}_D$	=	dimensionless average temperature of the heated zone, $0 \le r \le r_h$ and $0 \le z \le h$, T, °F
$\bar{T}_{Dp}$	=	dimensionless average temperature of the produced fluid following CSS, T, °F
$\bar{T}_{Dp}^n$	=	average dimensionless temperature of the produced fluid following CSS at time t^n
T_{Dr}	=	dimensionless temperature as function of radial position
$\bar{T}_{Dr}$	=	dimensionless average temperature in the radial direction, $0 \le r \le r_h$
$\bar{T}_{Dr}^n$	=	dimensionless average temperature in the radial direction, $0 \le r \le r_h$, at time t^n
T_{Dz}	=	dimensionless temperature as function of vertical position
$\bar{T}_{Dz}$	=	dimensionless average temperature in the z-direction, $0 \le z \le h$
$\bar{T}_{Dz}^n$	=	dimensionless average temperature in the z-direction, $0 \le z \le h$ at time t^n
T_e	=	temperature of Earth at depth $z = az + b$, where b = mean subsurface temperature and a = geothermal gradient, T, °F
T_{ea}	=	absolute temperature of Earth at depth $z = az + b$, T, °R
T_{es}	=	temperature of surface of Earth under distribution line, T, °F
$\bar{T}_{esa}$	=	average absolute temperature of surface of Earth underneath pipe (Fig. 8.3), T, °R = °F + 460
$\bar{T}_f$	=	average film temperature of air flowing around circular conduit = $(T_{se} + T_a)/2$, T, °F

T_h = temperature at cement/formation interface, T, °F

$\overline{T}_i$ = average temperature of insulation, T, °F

T_{is} = temperature of the surface of the insulation, °R = °F + 460, T, °F

T_{isa} = absolute temperature of surface of insulation, °R = °F + 460, T, °F

$T_o(t)$ = surface temperature at time t, T, °F

T_p = temperature of produced fluids, T, °F

$\overline{T}_p^{(n-1)}$ = average temperature of heated zone adjusted for removal of energy by produced fluids at Timestep n_1, T, °F

T_r = reservoir temperature, T, °F

T_s = temperature of steam, T, °F

T_{se} = temperature of surface losing heat by convection, T, °F

T_{gea} = absolute temperature of surface, T, °R = °F + 460

T_{sky} = temperature of sky, T, °F

T_{skya} = average absolute temperature of sky, T, °R = °F + 460

T_{to} = outside surface temperature of tubing, T, °F

T_{toa} = absolute outside surface temperature of tubing, T, °R

T_z = temperature as function of vertical position

ΔT_s = $T_s - T_r$, T, °F

u = dummy variable in Eq. 8.58 for infinite-series approximation of error function

u_a = minimum air flux to sustain in-situ combustion, L^3/t-L^2, scf/D-ft^2

u_{ai} = air flux density measured at the inlet of a combustion tube, L^3/L^2t, scf/ft^2-hr

u_{am} = minimum air flux required to sustain combustion front, L/t, scf/(D-ft^2)

U_{ti} = overall heat-transfer coefficient between fluid and cement/drilled hole interface based on inside tubing surface area and temperature difference $T_s - T_h$, mL/t^{3}T, Btu/hr-ft^2-°F

U_{to} = overall heat-transfer coefficient between fluid and cement/drilled hole interface based on outside tubing surface area and temperature difference $T_s - T_h$, mL/t^{3}T, Btu/hr-ft^2-°F

v = velocity of steam-zone interface, L/t, ft/D

v_b = minimum front velocity of burning zone during in-situ combustion, L, ft/D

v_b^* = velocity of combustion front, L/t, ft/D

v_c^* = velocity of condensation front, L/t, ft/D

v_f = specific volume of flowing fluid, L^3/m, ft/lbm

v_{fm} = minimum burning front velocity to sustain a combustion front, L/t, ft/D

v_{sv} = specific volume of saturated vapor, L^3/m, ft/lbm

v_{sw} = specific volume of saturated water, L^3/m, ft^3/lbm

v_v^* = velocity of vaporization front, L/t, ft/D

v_w = wind velocity normal to the pipe, L/t, mile/hr

v_1 = velocity of burning front, L/t, ft/D

V = volume, L^3, ft^3

V_b = volume of reservoir burned, L^3, ft^3

V_d = volume of steam zone displaced to maximum thickness at time t in the gravity override model for steam displacement, L^3, ft^3

V_h = volume of heated region, L^3, ft^3

V_i = volume of injected gas, L^3, scf

V_{ob} = oil displaced from burned volume, L^3, ft^3

V_{oc}^n = volume of oil entering the heated region from the cold region of the reservoir during Timestep n, L^3, bbl

V_p = volume of produced gas, L^3, scf

V_{ph} = pore volume of heated region following steam injection, L^3, ft^3

V_{ps} = pore volume of saturated portion of heated region following cyclic steam injection, L^3, ft^3

V_{Rb} = bulk volume of reservoir burned, L^3, acre-ft

V_{Rb}^* = volume of reservoir burned, % pattern volume

V_{Rbo}^* = volume of reservoir burned before oil recovery begins, %

V_s = volume of reservoir zone invaded by steam, L^3, ft^3

V_{sz} = volume of the steam zone, m^3/m

V_T = volume associated with well pattern, L^3, acre-ft

V_w = volume of water (liquid equivalent) produced by combustion reaction, L^3, bbl

V_{wb} = water displaced from burned volume, L^3, ft^3

ΔV = incremental volume element heated by the expansion of the steam zone in the time increment Δt, L^3, ft^3

w = distance from production well to no-flow boundary of the adjacent well pair

w_f = carbon and hydrogen burned when V_b (ft^3) of reservoir is burned, m, lbm

W_i = cumulative volume of cold water injected as steam, L^3, bbl

W_{is} = volume of water injected as steam during CSS CWE, L^3, bbl

W_p	=	volume of water produced following CSS, L^3, STB
W_{pb}	=	water (liquid equivalent) displaced from burned volume plus water (liquid equivalent) produced in reaction, L^3/L^3, bbl
$W_{ph}^{(n)}$	=	volume of water removed from the heated region following CSS during Timestep n, L^3, bbl
$W_{s,eq}$	=	volume of steam injected as equivalent barrels of water at the average temperature of the boiler feed water, L^3, bbl
x	=	argument of error function, Eq. 8.56, or distance along flow path or in direction of heat transfer, L, ft
x_B	=	correlating parameter for Brigham correlation of in-situ combustion performance
x_r	=	correlating parameter for Gates and Ramey (1980) model developed by Fassihi et al. (1981)
X	=	steam quality, weight fraction of vapor phase
y	=	mole fraction of component in gas
y_B	=	correlating parameter for Brigham correlation of in-situ combustion performance
y_{iN_2}	=	mole fraction of N_2 in the injected gas
y_{iO_2}	=	mole fraction of O_2 in the injected gas
y_l	=	liquid holdup, fraction
y_p	=	elevation of production well above the base of the formation
y_{pCO}	=	mole fraction of CO in the combustion gases produced from pattern, volume fraction
y_{pCO_2}	=	mole fraction of CO_2 in the combustion gases produced from pattern, volume fraction
y_{pN_2}	=	mole fraction of N_2 in combustion gases produced from pattern, volume fraction
y_{pO_2}	=	mole fraction of O_2 in combustion gases produced from pattern, L^3/L^3, volume fraction
y_R	=	correlating parameter for Gates and Ramey (1980) model developed by Fassihi et al. (1981)
z	=	distance in vertical direction, L, ft
z_j	=	elevation of the steam zone at w at time t_j, m
z_p	=	elevation of the production well above the base of the reservoir, m
Δz	=	increment of distance in z-direction, L, ft
α	=	thermal diffusivity, L^2/t, ft²/hr
α_f	=	thermal diffusivity of formation, L^2/t, ft²/hr
α_r	=	thermal diffusivity of reservoir rock saturated with fluids, L^2/t, ft²/hr or ft²/D

α_s	=	thermal diffusivity of shale, L^2/t, ft²/hr or ft²/D
β	=	thermal volumetric expansion coefficient of fluid in annulus, $°R^{-1} = 1/\bar{T}_{an}$ for an ideal gas, or generally $= -(1/p_{an})(\partial p_{an}/\partial T)_p$, where p_{an} is the annulus pressure, 1/T
δ	=	dimensionless parameter defined in Eq. 8.82 or average thickness of flowing oil stream, m
$\delta^{(n)}$	=	dimensionless parameter defined in Eq. 8.82 at Timestep n
$\Delta\delta^t$	=	change in dimensionless parameter defined in Eq. 8.82 during Timestep n
ε	=	emissivity of surface, dimensionless
ε_{al}	=	emissivity of aluminum surface, dimensionless
ε_{ci}	=	emissivity of internal tubing surface, dimensionless
ε_{to}	=	emissivity of external tubing surface, dimensionless
η_s	=	effective sweep efficiency
λ	=	time of arrival of heated region at specific location, Eqs. 8.52 and 8.59, t, hours
λ	=	mobility, k/μ, L^3t/m, md/cp
λ_D	=	dimensionless time of arrival of heated region at specified location, defined in a manner analogous to that for t_D
λ_{Dv}	=	dimensionless time of arrival of heated region at a specific location defined in a manner analogous to that for t_{Dv} for the constant vertical velocity steam-displacement model
λ_o	=	mobility of oil in oil zone at reservoir temperature, L^3t/m, md/cp
λ_{oh}	=	mobility of oil in heated zone, k_{oh}/μ_{oh}, L^3t/m, md/cp
λ_s	=	mobility of steam in steam zone, k_s/μ_s, L^3t/m, md/cp
μ	=	viscosity, m/Lt, cp
μ_a	=	viscosity of air, m/Lt, cp
μ_{an}	=	viscosity of fluid in annulus at T_{an} and p, m/Lt, lbm/ft-hr
μ_o	=	viscosity of oil, m/Lt, cp
μ_{oc}	=	viscosity of oil in cold region of reservoir, m/Lt, cp
μ_{oh}	=	viscosity of oil in heated region of reservoir, m/Lt, cp
μ_{oz}	=	thickness averaged viscosity of oil defined by Eq. 8.221, m/Lt, cp
μ_s	=	viscosity of steam, m/Lt, cp
μ_w	=	viscosity of water, m/Lt, cp

μ_{wh}	=	viscosity of water in the heated region following CSS, m/Lt, cp		ρ_{ob}	=	density of overburden, m/L^3, lbm/ft^3
v	=	kinematic viscosity		ρ_{oh}	=	density of oil produced from heated region following CSS, m/L^3, lbm/ft^3
v_s	=	kinematic viscosity at steam temperature		ρ_{oi}	=	density of oil at 60°F, m/L^3, lbm/ft^3
ϕ	=	porosity, L^3/L^3, volume fraction		ρ_r	=	density of rock (matrix), m/L^3, lbm/ft^3
ϕ_E	=	porosity of rock in the combustion-tube test		ρ_s	=	density of steam, m/L^3, lbm/ft^3
ϕ_R	=	porosity of reservoir rock		ρ_w	=	density of water, m/L^3, lbm/ft^3
ρ	=	density, m/L^3, lbm/ft^3		ρ_{wh}	=	density of water in the heated region following CSS, m/L^3, lbm/ft^3
ρ_a	=	density of air evaluated at ambient air temperature, m/L^3, lbm/ft^3		ρ_{ws}	=	density of water at T_s, m/L^3, lbm/ft^3
ρ_{an}	=	density of fluid in annulus at average temperature and pressure of annulus, m/L^3, lbm/ft^3		σ	=	Stefan-Boltzman constant, 1.713×10^{-9}, m/t^3T^4, Btu/hr-ft^2-°R^4
ρ_F	=	density of fuel, m/L^3, lbm/ft^3		τ	=	time when areal sweep efficiency of steamdrive is 100% in Miller and Leung (1985) model, t, days
ρ_o	=	density of oil, m/L^3, lbm/ft^3				

Subscripts

a	=	air phase property		o	=	oil property
c	=	property at reservoir temperature		ob	=	overburden rock property
CO	=	property of carbon monoxide		O_2	=	property of oxygen
CO_2	=	property of carbon dioxide		r	=	rock property
g	=	gas property		T_r	=	property at temperature T_r
h	=	property in the heated zone		T_s	=	property at steam temperature T_s
H_2O	=	property of water		w	=	water property
i	=	physical state		1, 2...	=	particular time level, position, or location

References

Abramowitz, M. and Stegan, L.A. 1964. *Handbook of Mathematical Functions*, Natl. Bureau of Standards, Appl. Math Series, 55, 104.

Abu-Khamsin, S., Ramey, H. J. Jr., Pettit, P. et al. 1985. The Reaction Kinetics of Fuel Formation for In-Situ Combustion. SUPRI-46 prepared by the Stanford U. Petroleum Research Inst., Report DOE/SF/11564-12, US DOE (July 1985).

Adams, R. H. and Khan, A. M. 1969. Cyclic Steam Injection Project Performance Analysis and Some Results of a Continuous Steam Displacement Pilot. *J Pet Technol* **21** (1): 95–100. SPE-1916-PA. https://doi.org/10.2118/1916-PA.

Afoeju, B. I. 1974. Conversion of Steam Injection to Waterflood, East Coalinga Field. *J Pet Technol* **26** (11): 1227–1232. SPE-4502-PA. https://doi.org/10.2118/4502-PA.

Alberta Energy Regulator. 2014. ST98-2014: Alberta's Energy Reserves 2013 and Supply/Demand Outlook 2014–2023. Report, Alberta Energy Regulator, Calgary (May 2014).

Aldea, G. and Turta, A. 1988. The In-Situ Combustion Industrial Exploitation of Suplacu de Barcau Panonian Field, Romania. Paper presented at the UNITAR/UNDP International Conference on Heavy Crude and Tar Sands, Edmonton, 7–8 August.

Alexander, J. D., Martin, W. L., and Dew, J. N. 1962. Factors Affecting Fuel Availability and Composition During In-Situ Combustion. *J Pet Technol* **14** (10): 1154–1164. SPE-296-PA. https://doi.org/10.2118/296-PA.

ASTM D341-09(2015), Standard Practice for Viscosity-Temperature Charts for Liquid Petroleum Products 2015. West Conshohocken, Pennsylvania: ASTM International. https://doi.org/10.1520/D0341-09R15.

Atmosudiro, H. W. 1977. Steam Soak Increases Recovery in Indonesia. *Oil & Gas J.* (1 August 1977): 104–108.

Aydelotte, S. R. and Pope, G. A. 1983. A Simplified Predictive Model for Steamdrive Performance. *J Pet Technol* **35** (5): 991–1002. SPE-10748-PA. https://doi.org/10.2118/10748-PA.

Baker, P. E. 1973. Effect of Pressure and Rate of Steam Zone Development in Steamflooding. *SPE J.* **13** (5): 274–284. SPE-4141-PA. https://doi.org/10.2118/4141-PA.

Beattie, C. I., Boberg, T. C., and McNab, G. S. 1991. Reservoir Simulation of Cyclic Steam Stimulation in the Cold Lake Oil Sands. *SPE Res Eng* **6** (2): 200–206. SPE-18752-PA. https://doi.org/10.2118/18752-PA.

Beggs, H. D. and Brill, J. P. 1973. A Study of Two-Phase Flow in Inclined Pipes. *J Pet Technol* **25** (5): 607–617. SPE-4007-PA. https://doi.org/10.2118/4007-PA.

Bellevue Field. 1990. Report 76-33-2/90, *SPE Enhanced Oil Recovery Field Reports*, Vol. 15, No. 2, 3025.

Bentsen, R. G. and Donohue, D. A. T. 1969. A Dynamic Programming Model of the Cyclic Steam Injection Process. *J Pet Technol* **21** (12): 1582–1596. SPE-2032-PA. https://doi.org/10.2118/2032-PA.

Birrell, G. 2001. Heat Transfer Ahead of a SAGD Steam Chamber: A Study of Thermocouple Data from Phase B of the Underground Test Facility (Dover Project). Presented at the Canadian International Petroleum Conference, Calgary, 12–14 June. PETSOC-2001-088. https://doi.org/10.2118/2001-088.

Bleakley, W. B. 1964. Here Are Case Histories of Two Thermal Properties. *Oil & Gas J.* **62:** 123–30.

Blevins, T. R. and Billingsley, R. H. 1975. The Ten-Pattern Steamflood, Kern River Field, California. *J Pet Technol* **27** (12): 1505–1514. SPE-4756-PA. https://doi.org/10.2118/4756-PA.

Blevins, T. R., Aseltine, R. J., and Kirk, R. S. 1969. Analysis of a Steam Drive Project, Inglewood Field, California. *J Pet Technol* **21** (9): 1141–1150. SPE-2291-PA. https://doi.org/10.2118/2291-PA.

Boberg, T. C. 1988. *Thermal Methods of Oil Recovery*. New York City: John Wiley & Sons.

Boberg, T. C. and Lantz, R. B. 1966. Calculation of the Production Rate of a Thermally Stimulated Well. *J Pet Technol* **18** (12): 1613–1623. SPE-1578-PA. https://doi.org/10.2118/1578-PA.

Borregales, C. 1976. Inyeccion Alternada de Vapor en la Costa Bolivar. Presented at the Simposio de Crudos Extra-Pesados, Petroleos de Venezuela, Maracay, 13–15 October.

Braden, W. B. 1966. A Viscosity-Temperature Correlation at Atmospheric Pressure for Gas-Free Oils. *J Pet Technol* **18** (11): 1487–1490. SPE-1580-PA. https://doi.org/10.2118/1580-PA.

Brigham, W. E., Satman, A., and Soliman, M. Y. 1980. Recovery Correlations for In-Situ Combustion Field Projects and Application to Combustion Pilots. *J Pet Technol* **32** (12): 2132–2138. SPE-7130-PA. https://doi.org/10.2118/7130-PA.

Britton, M. W., Martin, W. L., Leibrecht, R. J. et al. 1983. The Street Ranch Pilot Test of Fracture-Assisted Steamflood Technology. *J Pet Technol* **35** (3): 511–522. SPE-10707-PA. https://doi.org/10.2118/10707-PA.

Buckles, R. S. 1979. Steam Stimulation Heavy Oil Recovery at Cold Lake, Alberta. Presented at the SPE California Regional Meeting, Ventura, California, USA, 18–20 April. SPE-7994-MS. https://doi.org/10.2118/7994-MS.

Burger, J. G. and Sahuquet, B. C. 1972. Chemical Aspects of In-Situ Combustion—Heat of Combustion and Kinetics. *SPE J.* **12** (5): 410–422. SPE-3599-PA. https://doi.org/10.2118/3599-PA.

Burger, J. G. and Sahuquet, B. C. 1973. Laboratory Research on Wet Combustion. *J Pet Technol* **25** (10): 1137–1146. SPE-4144-PA. https://doi.org/10.2118/4144-PA.

Burger, J. G., Sourieau, P., and Combarnous, M. 1985. *Thermal Methods of Oil Recovery*. Houston: Gulf Publishing Co.

Burns, J. 1969. A Review of Steam Soak Operation in California. *J Pet Technol* **21** (1): 25–34. SPE-2117-PA. https://doi.org/10.2118/2117-PA.

Bursell, C. G. 1970. Steam Displacement—Kern River Field. *J Pet Technol* **22** (10): 1225–1231. SPE-2738-PA. https://doi.org/10.2118/2738-PA.

Bursell, C. G. and Pittman, G. M. 1975. Performance of Steam Displacement in the Kern River Field. *J Pet Technol* **27** (8): 997–1004. SPE-5017-PA. https://doi.org/10.2118/5017-PA.

Butler, R. M. 1985. A New Approach to the Modeling of Steam-Assisted Gravity Drainage. *J Can Pet Technol* **24** (3): 42–51. PETSOC-85-03-01. https://doi.org/10.2118/85-03-01.

Butler, R. M. 1991. *Thermal Recovery of Oil and Bitumen*. Englewood Cliffs, New Jersey: Prentice Hall (rep. GravDrain, 1997)

Butler, R. M. 1994a. Steam-Assisted Gravity Drainage: Concept, Development, Performance and Future. *J Can Pet Technol* **33** (2): 44–50. PETSOC-94-02-05. https://doi.org/10.2118/94-02-05.

Butler, R. M. 1994b. *Horizontal Wells for the Recovery of Oil, Gas and Bitumen*, Petroleum Society Monograph No. 2, CIM, Calgary, 171.

Butler, R. M. 1994c. *Horizontal Wells for the Recovery of Oil, Gas and Bitumen*, Petroleum Society Monograph No. 2, CIM, Calgary, 172.

Butler, R. M. 1994d. *Horizontal Wells for the Recovery of Oil, Gas and Bitumen*, Petroleum Society Monograph No. 2, CIM, Calgary, 173.

Butler, R. M. 1998. SAGD Comes of Age! Distinguished Author Series, *J Can Pet Technol* **37** (7): 9–12. PETSOC-98-07-DA. https://doi.org/10.2118/98-07-DA.

Butler, R. M. 2001. Some Recent Developments in SAGD. Distinguished Author Series, *J Can Pet Technol* **40** (1): 18–22. PETSOC-01-01-DAS. https://doi.org/10.2118/01-01-DAS.

Butler, R. M. and Stephens, D. J. 1981. The Gravity Drainage of Steam-Heated Heavy Oil to Parallel Horizontal Wells. *J Can Pet Technol* **20** (2): 90–96. PETSOC-81-02-07. https://doi.org/10.2118/81-02-07.

Butler, R. M., McNab, G. S, and Lo, H. Y. 1981. Theoretical Studies on the Gravity Drainage of Heavy Oil During In-Situ Steam Heating. *Can. J. Chem. Eng.* **59** (4): 455-460. https://doi.org/10.1002/cjce.5450590407.

Buxton, T. S. and Pollock, C. B. 1974. The Sloss COFCAW Project—Further Evaluation of Performance During and After Air Injection. *J Pet Technol* **26** (12): 1439–1448. SPE-4766-PA. https://doi.org/10.2118/4766-PA.

Cadelle, C. P., Burger, J. G., Bardon, C. P. et al. 1981. Heavy-Oil Recovery by In-Situ Combustion—Two Field Cases in Romania. *J Pet Technol* **33** (11): 2057–2066. SPE-8905-PA. https://doi.org/10.2118/8905-PA.

Carcoana, A. 1990. Results and Difficulties of the World's Largest In-Situ Combustion Process: Suplacu de Barcau Field, Romania. Presented at the SPE/DOE Enhanced Oil Recovery Symposium, Tulsa, 22–25 April. SPE-20248-MS. https://doi.org/10.2118/20248-MS.

Carnahan, B., Luther, H. A., and Wilkes, J. O. 1969. *Applied Numerical Methods*, 361–79. New York City: John Wiley & Sons.

Carslaw, H. S. and Jaeger, J. C. 1959. *Conduction of Heat in Solids*, second edition, 388. London: Oxford Press.

Choquette, S. P., Sampath, K., Northrop, P. S. et al. 1993. Esperson Dome Oxygen Combustion Pilot Test: Postburn Coring Results. *SPE Res Eng* **8** (2): 85–93. SPE-21774-PA. https://doi.org/10.2118/21774-PA.

Chu, C. 1963. Two-Dimensional Analysis of a Radial Heat Wave. *J Pet Technol* **15** (10): 1137–1144. SPE-560-PA. https://doi.org/10.2118/560-PA.

Chu, C. 1982a. State-of-the-Art Review of Fireflood Field Projects (includes associated papers 10901 and 10918). *J Pet Technol* **34** (1): 19–36. SPE-9772-PA. https://doi.org/10.2118/9772-PA.

Chu, C. 1982b. Author's Reply to Discussion of State-of-the-Art Review of Fireflood Field Projects. *J Pet Technol* **34** (4): 861–62.

Clark, G. A., Jones, R. G., Kinney, W. L. et al. 1965. The Fry In Situ Combustion Test—Performance. *J Pet Technol* **17** (3): 348–353. SPE-956-PA. https://doi.org/10.2118/956-PA.

Coats, K. H. 1978. A Highly Implicit Steamflood Model. *SPE J.* **18** (5): 369–383. SPE-6105-PA. https://doi.org/10.2118/6105-PA.

ConocoPhillips. 2009a. Surmont Oil Sands Commercial Project Approval 9426B: ERCB Annual Performance Review, Slide 17, Subsection 3.1.1(5) Appendix (29 April 2009) http://www.aer.ca/documents/oilsands/insitu-presentations/2009Athaba scaConocoPhillipsSurmont9483_9460.pdf.

ConocoPhillips. 2009b. Surmont Oil Sands Commercial Project Approval 9426B: ERCB Annual Performance Review, Slide 38, Subsection 3.1.1(3c) (29 April 2009) http://www.aer.ca/documents/oilsands/insitu-presentations/2009AthabascaCono coPhillipsSurmont9483_9460.pdf.

ConocoPhillips. 2009c. Surmont Oil Sands Commercial Project Approval 9426B: ERCB Annual Performance Review, Slide 37 Subsection 3.1.1(3c) (29 April 2009) http://www.aer.ca/documents/oilsands/insitu-presentations/2009AthabascaConoc oPhillipsSurmont9483_9460.pdf.

ConocoPhillips. 2009d. Surmont Oil Sands Commercial Project Approval 9426B: ERCB Annual Performance Review, Slide 25, Subsection 3.1.1(5) Appendix (29 April 2009) http://www.aer.ca/documents/oilsands/insitu-presentations/2009Athaba scaConocoPhillipsSurmont9483_9460.pdf.

Craig, F. F. Jr. and Parrish, D. R. 1974. A Multipilot Evaluation of the COFCAW Process. *J Pet Technol* **26** (6): 659–666. SPE-3778-PA. https://doi.org/10.2118/3778-PA.

Curtis, J. H. 1989. Performance Evaluation of the MOCO T In-Situ Combustion Project, Midway-Sunset Field. Presented at the SPE California Regional Meeting, Bakersfield, California, USA, 5–7 April. SPE-18809-MS. https://doi.org/10.2118/18809-MS.

de Haan, H. J. and Schenk, L. 1969. Performance and Analysis of a Major Steam Drive Project in the Tia Juana Field, Western Venezuela. *J Pet Technol* **21** (1): 111–19. SPE-1915-PA. https://doi.org/10.2118/ 1915-PA.

de Haan, H. J. and van Lookeren, J. 1969. Early Results of the First Large-Scale Steam Soak Project in the Tia Juana Field, Western Venezuela. *J Pet Technol* **21** (1): 101–110. SPE-1913-PA. https://doi.org/10.2118/ 1913-PA.

Denbina, E. S., Boberg, T. C., and Rotter. M. B. 1991. Evaluation of Key Reservoir Drive Mechanisms in the Early Cycles of Steam Stimulation at Cold Lake. *SPE Res Eng* **6** (2): 207–211. SPE-16737-PA. https://doi.org/10.2118/16737-PA.

Dew, J. N. and Martin, W. L. 1964. Air Requirements for In-Situ Combustion, Part 1. *Pet. Eng.* (December): 82–86.

Dew, J. N. and Martin, W. L. 1965. Air Requirements for In-Situ Combustion, Part 2. *Pet. Eng.* (January): 81–85.

Dickson, J. L., Dittaro, L. M., and Boone, T. J. 2013. Integrating the Key Learnings From Laboratory, Simulation, and Field Tests To Assess the Potential for Solvent Assisted -Steam Assisted Gravity Drainage. Presented at the SPE Heavy Oil Conference–Canada, Calgary, 11–13, June. SPE-165485-MS. https://doi.org/10.2118/165485-MS.

Dietrich, J. K. 1981. Relative Permeability During Cyclic Steam Stimulation of Heavy-Oil Reservoirs. *J Pet Technol* **33** (10): 1987–1989. SPE-7968-PA. https://doi.org/10.2118/7968-PA.

Dietz, D. N. 1970. Wet Underground Combustion, State of Art. *J Pet Technol* **22** (5): 605–617. SPE-2518-PA. https://doi.org/10.2118/2518-PA.

Dietz, D. N. and Weijdema, J. 1968. Wet and Partially Quenched Combustion. *J Pet Technol* **20** (4): 411–415. SPE-1899-PA. https://doi.org/10.2118/1899-PA.

Doscher, T. M. 1966. Factors That Spell Success in Steaming Viscous Crudes. *Oil & Gas J.* (11 July 1966): 95–100.

Doscher, T. M. and Ghassemi, F. 1983. The Influence of Oil Viscosity and Thickness on the Steam Drive. *J Pet Technol* **35** (2): 291–298. SPE-9897-PA. https://doi.org/10.2118/9897-PA.

Duri Field. 1991. Report 91-83-2/91, *SPE Enhanced Oil Recovery Field Reports* **16** (2): 3491.

Duri Field. 1992. Report 91-83-2/92, *SPE Enhanced Oil Recovery Field Reports* **17** (2): 3791.

Earlougher, R. C. Jr. 1969. Some Practical Considerations in the Design of Steam Injection Wells. *J Pet Technol* **21** (1): 79–86. SPE-2202-PA. https://doi.org/10.2118/2202-PA.

Earlougher, R. C., Galloway, J. R., and Parsons, R. W. 1970. Performance of the Fry In-Situ Combustion Project. *J Pet Technol* **22** (5): 551–557. SPE-2409-PA. https://doi.org/10.2118/2409-PA.

Edmunds, N. 2013. SAGD Technology: First the Fundamentals, Then the Advancements, Slide 20. Presented at the 2013 APEGA SAGD Forum, Edmonton, Alberta, Canada, 4 September. http://www.laricinaenergy.com/uploads/corporate/APEGA%20SAGD%20Forum_09_04_13.pdf.

Edmunds, N. and Peterson, J. 2007. A Unified Model for Prediction of CSOR in Steam-Based Bitumen Recovery. Presented at the Canadian International Petroleum Conference, Calgary, 12–14 June. PETSOC-2007-027. https://doi.org/10.2118/2007-027.

Emanuel, A. S. 1993. Development of an Analytical Streamtube Model for Estimating Steam-Drive Performance. *SPE Advanced Technology Series* **1** (l): 81–89. SPE-21756-PA. https://doi.org/10.2118/21756-PA.

Fair, D. R., Trudell, C. L., Boone, T. J. et al. 2008. Cold Lake Heavy Oil Development—A Success Story in Technology Application. Presented at the International Petroleum Technology Conference, Kuala Lumpur, Malaysia, 3–5 December. IPTC-12361-MS. https://doi.org/10.2523/IPTC-12361-MS.

Farouq Ali, S. M. 1970. *Oil Recovery by Steam Injection*, 4–6. Bradford, Pennsylvania: Producers Publishing Co. Inc.

Farouq Ali, S. M. 1974. Current Status of Steam Injection as a Heavy Oil Recovery Method. *J Can Pet Technol* **13** (1): 1–15. PETSOC-74-01-06. https://doi.org/10.2118/74-01-06.

Farouq Ali, S. M. 1981. A Comprehensive Wellbore Steam/Water Flow Model for Steam Injection and Geothermal Applications. *SPE J.* **21** (5): 527–534. SPE-7966-PA. https://doi.org/10.2118/7966-PA.

Farouq Ali, S. M. and Meldau, R. F. 1979. Current Steamflood Technology. *J Pet Technol* **31** (10): 1332–1342. SPE-7183-PA. https://doi.org/10.2118/7183-PA.

Fassihi, M. R., Gobran, B. D., and Ramey, H. J. Jr. 1981. Algorithm Calculates Performance of In-Situ Combustion. *Oil & Gas J.* (16 November): 90–98.

Gael, B. T., Putro, E. S., Masykur, A. et al. 1994. Reservoir Management in the Duri Steamflood. Presented at the SPE/DOE Improved Oil Recovery Symposium, Tulsa, 17–20 April. SPE-27764-MS. https://doi.org/10.2118/27764-MS.

Garon, A. M. and Wygal, R. J. Jr. 1974. A Laboratory Investigation of Fire-Water Flooding. *SPE J.* **14** (6): 537–544. SPE-4762-PA. https://doi.org/10.2118/4762-PA.

Gates, C. F. and Ramey, H. J. Jr. 1958. Field Results of South Belridge Thermal Recovery Experiment. In *Petroleum Transactions, AIME*, Vol. 213, 236–244. SPE-1179-G.

Gates, C. F. and Ramey, H. J. Jr. 1980. A Method for Engineering In-Situ Combustion Oil Recovery Projects. *J Pet Technol* **32** (2): 285–294. SPE-7149-PA. https://doi.org/10.2118/7149-PA.

Gates, C. F. and Sklar, I. 1971. Combustion as a Primary Recovery Process—Midway Sunset Field. *J Pet Technol* **23** (8): 981–86. SPE-3054-PA. https://doi.org/10.2118/3054-PA.

Gates, C. F., Jung, K. D., and Surface, R. A. 1978. In-Situ Combustion in the Tulare Formation, South Belridge Field, Kern County, California. *J Pet Technol* **30** (5): 798–806. SPE-6554-PA. https://doi.org/10.2118/ 6554-PA.

Gates, I. D. and Chakrabarty, N. 2006. Design of the Steam and Solvent Injection Strategy in Expanding-Solvent Steam-Assisted Gravity Drainage. Presented at the Canadian International Petroleum Conference, Calgary, 13–15 June. PETSOC-2006-023. https://doi.org/10.2118/2006-023.

Gomaa, E. E. 1980. Correlations for Prediction Oil Recovery by Steamflood. *J Pet Technol* **32** (2): 325–332. SPE-6169-PA. https://doi.org/10.2118/6169-PA.

Gontijo, J. E. and Aziz, K. 1984. A Simple Analytical Model for Simulating Heavy Oil Recovery by Cyclic Steam in Pressure-Depleted Reservoirs. Presented at the SPE Annual Technical Conference and Exhibition, Houston, 16–19 September. SPE-13037-MS. https://doi.org/10.2118/13037-MS.

Gould, T. L., Tek, M. R., and Katz, D. L. 1974. Two-Phase Flow Through Vertical, Inclined, or Curved Pipe. *J Pet Technol* **26** (8): 915–926. SPE-4487-PA. https://doi.org/10.2118/4487-PA.

Gozde, S., Chhina, H. S., and Best, D. A. 1989. An Analytical Cyclic Steam Stimulation Model for Heavy-Oil Reservoirs. Presented at the SPE California Regional Meeting, Bakersfield, California, USA, 5–7 April. SPE-18807-MS. https://doi.org/10.2118/18807-MS.

Greaser, G. R. and Shore, R. A. 1980. Steamflood Performance in the Kern River Field. Presented at the SPE/DOE Enhanced Oil Recovery Symposium, Tulsa, 20–23 April. SPE-8834-MS. https://doi.org/10.2118/8834-MS.

Griston, S. and Willhite, G. P. 1987. Numerical Model for Evaluating Concentric Steam Injection Wells. Presented at the SPE California Regional Meeting, Ventura, California, USA, 8–10 April. SPE-16337-MS. https://doi.org/10.2118/16337-MS.

Hall, A. L. and Bowman, R. W. 1973. Operation and Performance of the Slocum Thermal Recovery Project. *J Pet Technol* **25** (4): 402–408. SPE-2843-PA. https://doi.org/10.2118/2843-PA.

Hansel, J. G., Benning, M. A., and Fernbacher, J. M. 1984. Oxygen In-Situ Combustion for Oil Recovery: Combustion Tube Tests. *J Pet Technol* **36** (7): 1139–1144. SPE-11253-PA. https://doi.org/10.2118/ 11253-PA.

Hardy, W. C., Fletcher, P. B., Shepard, J. C. et al. 1972. In-Situ Combustion in a Thin Reservoir Containing High-Gravity Oil. *J Pet Technol* **24** (2): 199–208. SPE-3053-PA. https://doi.org/10.2118/3053-PA.

Harmsen, G. J. 1971. Oil Recovery by Hot-Water and Steam Injection. *Proc*, Eighth World Petroleum Congress, Applied Science Publishers, London, 3241–51.

Hearn, C. L. 1972. The El Dorado Steam Drive—A Pilot Tertiary Recovery Test. *J Pet Technol* **24** (11): 1377–1384. SPE-3780-PA. https://doi.org/10.2118/3780-PA.

Heidari, M., Hejazi, S. H., and Farouq Ali, S. M. 2015. SAGD Performance With Temperature Dependent Properties—An Analytical Approach. Presented at the SPE Annual Technical Conference and Exhibition, Houston, 28–30 September. SPE-175036-MS. https://doi.org/10.2118/175036-MS.

Heidari, M., Pooladi-Darvish, M., Azaiez, J. et al. 2009. Effect of Drainage Height and Permeability on SAGD Performance. *Journal of Petroleum Science and Engineering* **68** (1–2): 99–106. https://doi.org/10.1016/j.petrol.2009.06.020.

Herrera, A. J. 1977. The M6 Steam Drive Project Design and Implementation. *J Can Pet Technol* **16** (3): 62–71. PETSOC-77-03-04. https://doi.org/10.2118/77-03-04.

Herrera, A. J. 1978. Steam Drive Recovery Being Used on Lake Maracaibo Coastal Field. *Oil & Gas J.* (17 July 1978): 74–80.

Herrera, J. D. Jr., George, W. D., Birdwell, B. F. et al. 1978. Wellbore Heat Losses in Deep Steam Injection Wells, S1-B Zone, Cat Canyon Field. Presented at the SPE California Regional Meeting, San Francisco, 12–14 April. SPE-7117-MS. https://doi.org/10.2118/7117-MS.

Hewitt, C. H. and Morgan, J. T. 1965. The Fry In-Situ Combustion Test-Reservoir Characteristics. *J Pet Technol* **17** (3): 337–342. SPE-954-PA. https://doi.org/10.2118/954-PA.

Hong, K. C. 1994. *Steamflood Reservoir Management: Thermal Oil Recovery*. Tulsa: PennWell Publishing.

Howell, J. C. and Peterson, M. E. 1979. The Fry In-Situ Combustion Project Performance and Economic Status. Presented at the SPE Annual Technical Conference and Exhibition, Las Vegas, 23–26 September. SPE-8381-MS. https://doi.org/10.2118/8381-MS.

Huffman, G. A., Benton, J. P., El-Messidi, A. E. et al. 1983. Pressure Maintenance by In-Situ Combustion, West Heidelberg Unit, Jasper County, Mississippi. *J Pet Technol* **35** (10): 1877–1883. SPE-10247-PA. https://doi.org/10.2118/10247-PA.

Hvizdos, L. J., Howard, J. V., and Roberts, G. W. 1983. Enhanced Oil Recovery Through Oxygen-Enriched In-Situ Combustion: Test Results From the Forest Hill Field in Texas. *J Pet Technol* **35** (6): 1061–1070. SPE-11218-PA. https://doi.org/10.2118/11218-PA.

Imperial Oil. 2012. Cold Lake Approvals 8558 and 4510: 2012 Annual Performance Review. Report, Imperial Oil, Calgary. https://www.aer.ca/documents/oilsands/insitu-presentations/2012ColdLakeImperialColdLakeCSS8558.pdf.

Incropera, E. P. and DeWitt, D. P. 1990. *Introduction to Heat Transfer*, second edition, 707–714. New York City: John Wiley & Sons.

In-Situ Combustion Predictive Model. 1986. Report DOE/BC-86/7/SP (DE 86000284), US DOE, Washington, DC (December 1986).

Irani, M. and Cokar, M. 2014. Understanding the Impact of Temperature-Dependent Thermal Conductivity on the Steam-Assisted Gravity-Drainage (SAGD) Process. Part 1: Temperature Front Prediction. Presented at the SPE Heavy Oil Conference–Canada, Calgary, 10–12 June. SPE-170064-MS. https://doi.org/10.2118/ 170064-MS.

Irani, M. and Gates, I. D. 2013. Understanding the Convection Heat-Transfer Mechanism in the Steam-Assisted-Gravity-Drainage Process. *SPE J.* **18** (6): 1202–1215. SPE-167258-PA. https://doi.org/10.2118/167258-PA.

Irani, M. and Ghannadi. S. 2013. Understanding the Heat-Transfer Mechanism in the Steam-Assisted-Gravity-Drainage (SAGD) Process and Comparing the Conduction and Convection Flux in Bitumen Reservoirs. *SPE J.* **18** (1): 134–145. SPE-163079-PA. https://doi.org/10.2118/163079-PA.

Irani, M., Ghannadi, S., and Shu, H. 2015. Discussion on the Effects of Temperature on Thermal Properties in the Steam-Assisted Gravity-Drainage (SAGD) Process. Part 2: Heat Capacity. Presented at World Heavy Oil Congress, Edmonton, Alberta, Canada, 24–26 March.

Japan Canada Oil Sands Limited (JACOS). 2010. Volume 1 Project Description: JACOS Hangingstone Expansion Project. Report, JACOS, Calgary (April 2010).

Jiang, Q., Thornton, B., Russel-Houston, J. et al. 2010. Review of Thermal Recovery Technologies for the Clearwater and Lower Grand Rapids Formations in the Cold Lake Area in Alberta. *J Can Pet Technol* **49** (9): 57–68. SPE-140118-PA. https://doi.org/10.2118/140118-PA.

Jones, J. 1977. Cyclic Steam Reservoir Model for Viscous Oil, Pressure Depleted, Gravity Drainage Reservoirs. Presented at the SPE California Regional Meeting, Bakersfield, California, USA, 13–15 April. SPE-6544-MS. https://doi.org/10.2118/6544-MS.

Jones, J. 1981. Steam Drive Model for Hand-Held Programmable Calculators. *J Pet Technol* **33** (9): 1583–1558.SPE-8882-PA. https://doi.org/10.2118/8882-PA.

Jones, J. 1992. Why Cyclic Steam Predictive Models Get No Respect. *SPE Res Eng* **7** (1): 67–74. SPE-20022-PA. https://doi.org/10.2118/20022-PA.

Jones, J. and Cawthon, G. J. 1990. Sequential Steam: An Engineered Cyclic Steaming Method. *J Pet Technol* **42** (7): 848–901. SPE-17421-PA. https://doi.org/10.2118/17421-PA.

Joseph, C. and Pusch, W. H. 1980. A Field Comparison of Wet and Dry Combustion. *J Pet Technol* **32** (9): 1523–1528. SPE-7992-PA. https://doi.org/10.2118/7992-PA.

Kamilos, G. N. 1985. Downhole Steam Chokes, An Improved Control. Presented at the Chevron-Western Northern California Div. Quarterly Petroleum Engineering Meeting, Bakersfield, California, USA, 19 March.

Kawasaki Thermal Systems. 1989. Thermocase Insulated Tubulars. Kawasaki Thermal Systems, Tacoma, Washington, USA.

Kern River Field. 1976. Report 76-34-2/76, *SPE Improved Oil Recovery Field Reports* **2** (2): 285 (September 1976).

Kern River Field. 1982. Report 76-34-2/81, *SPE Enhanced Oil Recovery Field Reports* **7** (2): 595 (March 1982).

Koottungal, L. 2014. 2014 Worldwide EOR Survey. *Oil & Gas Journal* (7 April): 79–91.

Kruit, C, Meyer, R. F., and Steele, C. T. 1979. Sedimentary-Geological Study of a Steam Drive Project in Deltaic River Sands. *Proc*, International Conference on the Future of Heavy Crude and Tar Sands, Edmonton, Canada, 321-326. New York City: McGraw-Hill Inc.

Kry, P. R. 1989. Field Observations of Steam Distribution During Injection to the Cold Lake Reservoir. In *Proc.*, Rocks at Great Depth: Rock Mechanics and Rock Physics at Great Depth, ISRM International Symposium, Pau, France, 30 August–2 September, 853–881.

Kumar, M. and Ziegler, V. M. 1993. Injection Schedules and Production Strategies for Optimizing Steamflood Performance. *SPE Res Eng* **8** (2): 101–107. SPE-20763-PA. https://doi.org/10.2118/20763-PA.

Long, R. E. AND Nuar, M. F. 1982. A Study of Getty Oil Co.'s Successful In-Situ Combustion Project in the Bellevue Field. Presented at the SPE Enhanced Oil Recovery Symposium, Tulsa, 4–7 April. SPE-10708-MS. https://doi.org/10.2118/10708-MS.

Mainland, G. G. and Lo, H. Y. 1984. Technological Basis for Commercial In-Situ Recovery of Cold Lake Bitumen. *Proc*, Eleventh World Petroleum Congress, London, July, 235–242.

Mandl, G. and Volek, C. W. 1969. Heat and Mass Transport in Steam-Drive Processes. *SPE J.* **9** (1): 59–79. SPE-2049-PA. https://doi.org/10.2118/2049-PA.

Martin, W. L., Alexander, J. D., and Dew, J. N. 1958. Process Variables of In-Situ Combustion. In *Petroleum Transactions, AIME*, Vol. 213, 28–35. SPE-914-G.

Marx, J. W. and Langenheim, R. H. 1959. Reservoir Heating by Hot Fluid Injection. In *Petroleum Transactions, AIME*, Vol. 216, 312–315. SPE-1266-G.

McAdams, W. H. 1954. *Heat Transmission*, third edition, 260. New York City: McGraw-Hill Book Co. Inc.

Mehrotra, A. K. and Svrcek, W. Y. 1986. Viscosity of Compressed Athabasca Bitumen. *Can. J. Chem. Eng.* **64** (5): 844–847. https://doi.org/10.1002/cjce.5450640520.

Miller, M. A. and Leung, W. K. 1985. A Simple Gravity Drainage Model of Steamdrive. Presented at the SPE Annual Technical Conference and Exhibition, Las Vegas, 22–25 September. SPE-14241-MS. https://doi.org/10.2118/14241-MS.

Miura, K. and Wang, J. 2012. An Analytical Model To Predict Cumulative Steam/Oil Ratio (CSOR) in Thermal-Recovery SAGD Process. *J Can Pet Technol* **51** (4): 268–275. SPE-137604-PA. https://doi.org/10.2118/137604-PA.

Moritis, G. 1990. CO_2 and HC Injection Lead EOR Production Increase. *Oil & Gas J.* (23 April 1990): 49–82.

Moss, J. T, White, P. D., and McNeil, J. S. 1959. In-Situ Combustion Process—Results of a Five-Well Field Experiment in Southern Oklahoma. In *Petroleum Transactions, AIME*, Vol. 216, 55–64. SPE-1102-G.

Mount Poso Field. 1986. Report 75-19-1/86, *SPE Enhanced Oil Recovery Field Reports* **12** (1), 1971 (September 1986).

Mukherjee, H. and Brill, J. P. 1985. Empirical Equations To Predict Flow Patterns in Two-Phase Inclined Flow. *Intl. J. Multiphase Flow* **11** (3): 299–315. https://doi.org/10.1016/0301-9322(85)90060-6.

Munoz, J. D. 1981. Numerical Simulation of Steam Drive in the Tia Juana M-6 Project. *Proc*, 1979 International Conference on the Future of Heavy Crude and Tar Sands, Edmonton, Canada. New York City: McGraw-Hill Inc.

Myhill, N. A. and Stegemeier, G. L. 1978. Steam-Drive Correlation and Prediction. *J Pet Technol* **30** (2): 173–182. SPE-5572-PA. https://doi.org/10.2118/5572-PA.

Naji, H. S. A. and Poettmann, F. H. 1991. Reservoir Engineering Equations for In-Situ Combustion. *In Situ* **15** (2): 175–194.

Nasr, T. N. and Pierce, G. E. 1993. Steamflooding Cold Lake Oil Reservoirs Through a Bottomwater Zone: A Scaled Physical Model Study. *SPE Res Eng* **8** (2): 94–100. SPE-21772-PA. https://doi.org/10.2118/ 21772-PA.

Nelson, T. W. and McNeil, J. S. Jr. 1961. How To Engineer an In-Situ Combustion Project. *Oil & Gas J.* (5 June): 58–65.

Neuman, C. H. 1975. A Mathematical Model of the Steam Drive Process Derivation. Paper SPE-5495-MS available from SPE, Richardson, Texas, USA.

Neuman, C. H. 1985. A Gravity Override Model of Steamdrive. *J Pet Technol* **37** (1): 163–169. SPE-13348-PA. https://doi.org/10.2118/13348-PA.

Oglesby, K. D., Blevins, T. R., Rogers, E. E. et al. 1982. Status of the 10-Pattern Steamflood, Kern River Field, California. *J Pet Technol* **34** (10): 2251–2257. SPE-8833-PA. https://doi.org/10.2118/8833-PA.

Orkiszewski, J. 1967. Predicting Two-Phase Pressure Drops in Vertical Pipe. *J Pet Technol* **19** (6): 829–838. SPE-1546-PA. https://doi.org/10.2118/1546-PA.

Pacheco, E. F. and Farouq Ali, S. M. 1972. Wellbore Heat Losses and Pressure Drop in Steam Injection. *J Pet Technol* **24** (2): 139–144. SPE-3428-PA. https://doi.org/10.2118/3428-PA.

Parrish, D. R. and Craig, F. F. Jr. 1969. Laboratory Study of a Combination of Forward Combustion and Waterflooding—The COFCAW Process. *J Pet Technol* **21** (6): 753–761. SPE-2209-PA. https://doi.org/10.2118/2209-PA.

Parrish, D. R., Pollock, C. B., and Craig, F. F. Jr. 1974. Evaluation of COFCAW as a Tertiary Recovery Method, Sloss Field, Nebraska. 1974. *J Pet Technol* **26** (6): 676–686. SPE-3777-PA. https://doi.org/10.2118/3777-PA.

Pearce, J. C. and Megginson, E. A. 1991. Current Status of the Duri Steamflood Project, Sumatra, Indonesia. Presented at the SPE International Thermal Operations Symposium, Bakersfield, California, USA, 7–8 February. SPE-21527-MS. https:// doi.org/10.2118/21527-MS.

Pebdani, F. N., Longoria, R., Wilkerson, D. N. et al. 1988. Enhanced Oil Recovery by Wet In-Situ Oxygen Combustion: Esperson Dome Field, Liberty County, Texas. Presented at the SPE Annual Technical Conference and Exhibition, Houston, 2–5 October. SPE-18072-MS. https://doi.org/10.2118/18072-MS.

Penberthy, W. L. Jr. and Ramey, H. J. Jr. 1966. Design and Operation of Laboratory Combustion Tubes. *SPE J.* **6** (2): 183–198. SPE-1290-PA. https://doi.org/10.2118/1290-PA.

Petit, H. J-M., Valentin, E. P., and Desmarquest, J-P. 1992. Air/0_2 Fireflood: Comparison of Field Scale Performances and Economics. Presented at the SPE Western Regional Meeting, Bakersfield, California, USA, 30 March–1 April. SPE-24082-MS. https://doi.org/10.2118/24082-MS.

Poettmann, F. H. 1964. In Situ Combustion: A Current Appraisal. *World Oil*, Part 1, 124–128; Part 2, 95–98.

Poettmann, F. H., Schilson, R. E., and Surkalo, H. 1967. Philosophy and Technology of In-Situ Combustion in Light Oil Reservoirs. *Proc*, Seventh World Petroleum Congress, Mexico City, Vol. 3, 487–498.

Powers, M. L., Dodson, C. J., Ghassemi, F. et al. 1985. Commercial Application of Steamflooding in an Oilfield Comprising Multiple Thin Sand Reservoirs. *J Pet Technol* **37** (9): 1707–1715. SPE-13035-PA. https://doi.org/10.2118/13035-PA.

Prats, M. 1982. *Thermal Recovery*, SPE Monograph Series, 7. Richardson, Texas: SPE.

Prats, M., Jones, R. F., and Truitt, N. E. 1968. In Situ Combustion Away From Thin, Horizontal Gas Channels. *J Pet Technol* **8** (1): 18–32. SPE-1898-PA. https://doi.org/10.2118/1898-PA.

Puig, F. and Schenk, L. 1984. Analysis and Performance of the M-6 Area of the Tia Juana Field, Venezuela, Under Primary, Steam-Soak, and Steamdrive Conditions. Presented at the SPE/DOE Enhanced Oil Recovery Symposium, Tulsa, 15–18 April. SPE-12656-MS. https://doi.org/10.2118/12656-MS.

Ramey, H. J. Jr. 1959. Discussion of Reservoir Heating by Hot Fluid Injection. In *Petroleum Transactions, AIME*, Vol. 216, 364–365. SPE-1266-G (discussion follows paper).

Ramey, H. J. Jr. 1962. Wellbore Heat Transmission. *J Pet Technol* **14** (4): 427–435. SPE-96-PA. https://doi.org/10.2118/96-PA.

Ramey, H. J. Jr., Stamp, V. W., Pebdani, F. N. et al. 1992. Case History of South Belridge, California, In-Situ Combustion Oil Recovery. Presented at the SPE/DOE Enhanced Oil Recovery Symposium, Tulsa, 22–24 April. SPE-24200-MS. https:// doi.org/10.2118/24200-MS.

Reed, R. L., Reed, D. W., and Tracht, J. H. 1960. Experimental Aspects of Reverse Combustion in Tar Sands. In *Petroleum Transactions, AIME*, Vol. 219, 99–108. SPE-1313-G1.

Reis, J. C. 1992. A Steam-Assisted Gravity Drainage Model for Tar Sands: Linear Geometry. *J Can Pet Technol* **31** (10): 14–20. PETSOC-92-10-01. https://doi.org/10.2118/92-10-01.

Reis, J. C. 1993. A Steam Assisted Gravity Drainage Model for Tar Sands: Radial Geometry. *J Can Pet Technol* **32** (8): 43–48. PETSOC-93-08-05. https://doi.org/10.2118/93-08-05.

Restine, J. L. 1983. Effect of Preheating on Kern River Field Steam Drive. *J Pet Technol* **35** (3): 523–529. SPE-10750-PA. https://doi.org/10.2118/10750-PA.

Restine, J. L., Graves, W. G., and Elias, R. Jr. 1987. Infill Drilling in a Steamflood Operation: Kern River Field. *SPE Res Eng* **2** (2): 243–248. SPE-14337-PA. https://doi.org/10.2118/14337--PA.

Rivero, R. T. and Heintz, R. C. 1975. Resteaming Time Determination—Case History of a Steam-Soak Well in Midway Sunset. *J Pet Technol* **27** (6): 665–671. SPE-4892-PA. https://doi.org/10.2118/4892-PA.

Sandhar. K. 2011. Opportunities in the Canadian Oil Sands. Technical Report, prepared for SPE by Peters & Company Limited (1 February 2001).

Sasaki, K., Akibayahsi, S., Yazawa, N. et al. 2001. Numerical and Experimental Modelling of the Steam Assisted Gravity Drainage (SAGD) Process. *J Can Pet Technol* **40** (1): 44–50. PETSOC-01-01-04. https://doi.org/10.2118/01-01-04.

Satter, A. 1965. Heat Losses During Flow of Steam Down a Wellbore. *J Pet Technol* **17** (7): 845–851. SPE-1071-PA. https://doi.org/10.2118/1071-PA.

Schenk, L. 1982. Analysis of the Early Performance of the M-6 Steam-Drive Project, Venezuela. Presented at the SPE/DOE Enhanced Oil Recovery Symposium, Tulsa, 4–7 April. SPE-10710-MS. https://doi.org/10.2118/10710-MS.

Seba, R. D. and Perry, G. E. 1969. A Mathematical Model of Repeated Steam Soaks of Thick Gravity Drainage Reservoirs. *J Pet Technol* **21** (1): 87–94. SPE-1894-PA. https://doi.org/10.2118/1894-PA.

Showalter, W. E. 1963. Combustion-Drive Tests. *SPE J.* **3** (1): 53–58. SPE-456-PA. https://doi.org/10.2118/ 456-PA.

Shu, W. R. and Lu, H. S. 1984. Potential Benefit of CO_2 in Oxygen Combustion. *J Pet Technol* **36** (7): 1137–1138. SPE-11687-PA. https://doi.org/10.2118/11687-PA.

Shutler, N. D. 1970. Numerical Three-Phase Model of the Two-Dimensional Steamflood Process. *SPE J.* **10** (4): 405–417. SPE-2798-PA. https://doi.org/10.2118/2798-PA.

Smith, R. J. and Perepelecta, K. R. 2002. Steam Conformance Along Horizontal Wells at Cold Lake. Presented at the SPE/PS-CIM/CHOA International Thermal Operations and Heavy Oil Symposium and International Horizontal Well Technology Conference, Calgary, 4–7 November. SPE-79009-MS. https://doi.org/10.2118/79009-MS.

Spedding, P. L. and Chen, J. J. J. 1984. Holdup in Two-Phase Flow. *Intl. J. Multiphase Flow* **10** (3): 307–339. https://doi.org/10.1016/0301-9322(84)90024-7.

Stark, S. D. 2011. Increasing Cold Lake Recovery by Adapting Steamflood Principles to a Bitumen Reservoir. Presented at the SPE Enhanced Oil Recovery Conference, Kuala Lumpur, Malaysia, 19–21 July. SPE-145052-MS. https://doi.org/10.2118/145052-MS.

Steamflood Predictive Model. 1986. Report DOE/BC-8616/SP (DE87001219), US DOE, Washington, DC (December 1986).

Stegemeier, G. L., Laumbach, D. D., and Volek, C. W. 1980. Representing Steam Processes With Vacuum Models. *SPE J.* **20** (3): 151–174. SPE-6787-PA. https://doi.org/10.2118/6787-PA.

Stokes, D. D. and Doscher, T. M. 1974. Shell Makes a Success of Steam Hood at Yorba Linda. *Oil & Gas J.* (2 September 1974): 71–76.

Stokes, D. D., Brew, J. R., Whitten, D. G. et al. 1978. Steam Drive as a Supplemental Recovery Process in an Intermediate-Viscosity Reservoir, Mount Poso Field, California. *J Pet Technol* **30** (1): 125–131. SPE-6522-PA. https://doi.org/10.2118/6522-PA.

Struna, S. M. and Poettmann, F. H. 1988. In-Situ Combustion in the Lower Hospah Formation, McKinley County, New Mexico. *SPE Res Eng* **3** (2): 440–448. SPE-14917-PA. https://doi.org/10.2118/14917-PA.

Sylvester, N. B. and Chen, H. L. 1988. An Improved Cyclic Steam Stimulation Model for Pressure-Depleted Reservoirs. Presented at the SPE California Regional Meeting, Long Beach, California, USA, 23–25 March. SPE-17420-MS. https://doi.org/10.2118/17420-MS.

Tadema, H. J. and Weijdema, J. 1970. Spontaneous Ignition in Oil Sands. *Oil & Gas J.* (14 December): 77–80.

Tia Juana Este Field. 1988. Report 78-46-2/88, *SPE Enhanced Oil Recovery Field Reports* **13** (2): 2559 (September 1988).

Tortike, W. S. and Farouq Ali, S. M. 1989. Saturated-Steam-Property Functional Correlations for Fully Implicit Thermal Reservoir Simulation. *SPE Res Eng* **4** (4): 471–474. SPE-17094-PA. https://doi.org/10.2118/ 17094-PA.

Towson, D. E. and Boberg, T. C. 1967. Gravity Drainage in Thermally Stimulated Wells. *J Can Pet Technol* **6** (4): 130–135. PETSOC-67-04-06. https://doi.org/10.2118/67-04-06.

Turta, A. T., Chattopadhay, S. K., Bhattacharya, R. N. et al. 2007. Current Status of Commercial In Situ Combustion Projects Worldwide. *J Can Pet Technol* **46** (11): 1–7. PETSOC-07-11-GE. https://doi.org/10.2118/07-11-GE.

Ursenbach, M. G., Moore, R. G., and Mehta, S. A. 2010. Air Injection in Heavy Oil Reservoirs—A Process Whose Time Has Come (Again). *J Can Pet Technol* **49** (1): 48–54. SPE-132487-PA. https://doi.org/10.2118/132487-PA.

van Dijk, C. 1978. Steam-Drive Project in the Schoonebeek Field, The Netherlands. *J Pet Technol* **20** (3): 295–302. SPE-1917-PA. https://doi.org/10.2118/1917-PA.

van Lookeren, J. 1983. Calculation Methods for Linear and Radial Steam Flow in Oil Reservoirs. *SPE J.* **23** (3): 427–439. SPE-6788-PA. https://doi.org/10.2118/6788-PA.

van Poolen, H. K. 1965. Transient Tests Find Fire Front in an In-Situ Combustion Project. *Oil & Gas J.* (1 February): 78–80.

Villanueva, L. F. and Pittman, G. M. 1970. Geology of the Kern River Oil Field with a Case History of Its Development and Rejuvenation by Steam. Presented at the AIME Pacific Southwest Mineral Industry Conference, San Francisco, 27–29 May.

Vittoratos, E., Scott, G. R., and Beattie, C. I. 1990. Cold Lake Cyclic Steam Stimulation: A Multiwell Process. *SPE Res Eng* **5** (1): 19–24. SPE-17422-PA. https://doi.org/10.2118/17422-PA.

Vogel, J. V. 1984. Simplified Heat Calculations for Steamfloods. *J Pet Technol* **36** (7): 1127–1136. SPE-11219-PA. https://doi.org/10.2118/11219-PA.

Vogel, J. V. 1992. Gravity Drainage Vital Factor for Understanding Steam Floods. *Oil & Gas J.* (30 November 1992): 42–47.

Volek, C. W. and Pryor, J. A. 1972. Steam Distillation Drive—Brea Field, California. *J Pet Technol* **24** (8): 899–906. SPE-3441-PA. https://doi.org/10.2118/3441-PA.

Vossoughi, S., Willhite, G. P., Kritikos, W. P. et al. 1982. Automation of In-Situ Combustion Tube and Study of the Effect of Clay on the In-Situ Combustion Process. *SPE J.* **22** (4): 493–502. SPE-10320-PA. https://doi.org/10.2118/10320-PA.

Wei, S., Cheng, L.-S., Huang, S. et al. 2014. Steam Chamber Development and Production Performance Prediction of Steam Assisted Gravity Drainage. Presented at the SPE Heavy Oil Conference–Canada, Calgary, 10–12-June. SPE-170002-MS. https://doi.org/10.2118/170002-MS.

West Heidelberg Field, Cotton Valley Unit. 1975. Report 75-25-2/75, *SPE Improved Oil Recovery Field Reports*, Vol. 1, No. 2, 351.

West Heidelberg Field, Cotton Valley Unit. 1990. Report 75-25-2/90, *SPE Enhanced Oil Recovery Field Reports*, Vol. 15, No. 2, 3019.

White, P. D. and Moss, J. T. 1965. High-Temperature Thermal Techniques for Stimulating Oil Recovery. *J Pet Technol* **17** (9): 1007–1011. SPE-1119-PA. https://doi.org/10.2118/1119-PA.

Willhite, G. P. 1967. Overall Heat Transfer Coefficients in Steam and Hot Water Injection Wells. *J Pet Technol* **19** (5): 607–615. SPE-1449-PA. https://doi.org/10.2118/1449-PA.

Willhite, G. P. 1986. *Waterflooding*, Vol. 3. Richardson, Texas: Textbook Series, SPE.

Willhite, G. P. and Dietrich, W. K. 1967. Design Criteria for Completion of Steam Injection Wells. *J Pet Technol* **19** (1): 15–21. SPE-1560-PA. https://doi.org/10.2118/1560-PA.

Willhite, G. P. and Griston, S. 1987. Wellbore Refluxing in Steam-Injection Wells. *J Pet Technol* **39** (3): 353–362. SPE-15056-PA. https://doi.org/10.2118/15056-PA.

Williams, K. C. 2000. Technology Evolution and Commercial Development at Imperial's Cold Lake Production Project. Presented at the 16th World Petroleum Congress, Calgary, 11–15 June. WPC-30140.

Williams, R. L., Ramey, H. J. Jr., Brown, S. C. et al. 1980. An Engineering Economic Model for Thermal Recovery Methods. Presented at the SPE California Regional Meeting, Los Angeles, 9–11 April. SPE-8906-MS. https://doi.org/10.2118/8906-MS.

Willman, B. T., Valleroy, V. V., Runberg, G. W. et al. 1961. Laboratory Studies of Oil Recovery by Steam Injection. *J Pet Technol* **13** (7): 681–690. SPE-1537-G-PA. https://doi.org/10.2118/1537-G-PA.

Yoelin, S. D. 1971. The TM Sand Steam Stimulation Project. *J Pet Technol* **23** (8): 987–994. SPE-3104-PA. https://doi.org/10.2118/3104-PA.

Yortsos, Y. C. and Gavalas, G. R. 1981a. Analytical Modeling of Oil Recovery by Steam Injection: Part 1—Upper Bounds. *SPE J.* **21** (2): 162–178. SPE-8148-PA. https://doi.org/10.2118/8148-PA.

Yortsos, Y. C. and Gavalas, G. R. 1981b. Analytical Modeling of Oil Recovery by Steam Injection: Part 2—Asymptotic and Approximate Solutions (includes associated papers 10943 and 10951). *SPE J.* **21** (2): 179–190. SPE-8149-PA. https://doi.org/10.2118/8149-PA.

General References

Benham, A. L. and Poettmann, F. H. 1958. The Thermal Recovery Process—An Analysis of Laboratory Combustion Tube Data. *J Pet Technol* **10** (9): 83–85. SPE-1022-G. https://doi.org/10.2118/1022-G.

Buchwald, R. W. Jr., Hardy, W. C., and Neinast, G. S. 1973. Case Histories of Three In-Situ Combustion Projects. *J Pet Technol* **25** (7): 784–792. SPE-3781-PA. https://doi.org/10.2118/3781-PA.

Butler, R. M. 1985. A New Approach to the Modeling of Steam-Assisted Gravity Drainage. *J Can Pet Technol* **24** (3): 42–51. PETSOC-85-03-01. https://doi.org/10.2118/85-03-01.

Butler, R. M. and Stephens, D. J. 1981. The Gravity Drainage of Steam-Heated Heavy Oil to Parallel Horizontal Wells. *J Can Pet Technol* **20** (2): 90–96. PETSOC-81-02-07. https://doi.org/10.2118/81-02-07.

Butler, R. M., McNab, G. S., and Lo, H. Y. 1981. Theoretical Studies on the Gravity Drainage of Heavy Oil During In-Situ Steam Heating. *Canadian Journal of Chemical Engineering* **59** (4): 455–460. https://doi.org/10.1002/cjce.5450590407.

Chu, C. 1985. State-of-the-Art Review of Steamflood Field Projects. *J Pet Technol* **37** (10): 1887–1902. SPE-11733-PA. https://doi.org/10.2118/11733-PA.

Farouq Ali, S. M. 1975. Current Status of In-Situ Recovery From the Tar Sands of Alberta. *J Can Pet Technol* **14** (1): 51–58. PETSOC-75-01-05. https://doi.org/10.2118/75-01-05.

Hong, K. C. 1987. Guidelines for Converting Steamflood to Waterflood. *SPE Res Eng* **2** (1): 67–76. SPE-13605-PA. https://doi.org/10.2118/13605-PA.

Kern River Field, SOC AL. 1975. Report 75-8-1, *SPE Improved Oil Recovery Reports*.

Kuo, C. H., Shain, S. A., and Phocas, D. M. 1970. A Gravity Drainage Model for the Steam-Soak Process. *SPE J.* **10** (2): 119–126. SPE-2329-PA. https://doi.org/10.2118/2329-PA.

Matthews, C. S. and Lefkovits, H. C. 1956. Gravity Drainage Performance of Depletion-Type Reservoirs in the Stripper Stage. In *Petroleum Transactions, AIME*, Vol. 207, 265–274.

Matthews, C. W and Gorrill, R. G. 1985. Shell Plans Peace River Steamflood. *Pet. Eng.* (July): 34–42.

Mikkelsen, P. L., Cook, W. C., and Ostapovich, G. 1988. Fosterton Northwest: An In-Situ Combustion Case History. Presented at the SPE/DOE Enhanced Oil Recovery Symposium, Tulsa, 17–20 April. SPE-17391-MS. https://doi.org/10.2118/17391-MS.

Mukherjee, H. and Brill, J.P. 1981. Design Manual: Mukerjee and Brill Inclined Two-Phase Row Correlations. University of Tulsa Fluid Flow Projects (April 1981).

Ostapovich, E. and Ross, M. F. 1991. Fosterton Northwest: Post In-Situ Combustion Corehole Analysis. Presented at the SPE International Thermal Operations Symposium, Bakersfield, California, USA, 7–8 February.

Petit, H. J. M. 1987. In-Situ Combustion With Oxygen-Enriched Air. Presented at the SPE Annual Technical Conference and Exhibition, Dallas, Texas, USA, 27–30 September. SPE-16741-MS. https://doi.org/10.2118/16741-MS.

Thermal Recovery Processes. 1985. Reprint Series, 7, Richardson, Texas: Society of Petroleum Engineers.

Vittoratos, E., Scott, G. R., and Beattie, C. I. 1990. Cold Lake Cyclic Steam Stimulation: A Multiwell Process. *SPE Res Eng* **5** (1): 19–24. SPE-17422-PA. https://doi.org/10.2118/17422-PA.

SI Metric Conversion Factors

acre	×	4.046 873	E–01	=	ha
acre-ft	×	1.233 489	E–01	=	ha·m
°API		141.5/(131.5 + °API)		=	g/cm^3
bar	×	1.0*	E+05	=	Pa
bbl	×	1.589 873	E–01	=	m^3
Btu	×	1.055 056	E+00	=	kJ
Btu/lbm mol	×	2.236	E+03	=	J/mol
cp	×	1.0*	E–03	=	Pa·s
cSt	×	1.0*	E–06	=	m^2/s
cycle/sec	×	1.0*	E+00	=	Hz
dyne	×	1.0*	E–02	=	mN
dyne/cm	×	1.0*	E+00	=	mN/m
ft	×	3.048*	E–01	=	m
ft^2	×	9.290 304*	E–02	=	m^2
ft^3	×	2.831685	E–02	=	m^3
°F		(°F–32)/1.8		=	°C
°F		(°F+459.67)/1.8		=	K
gal	×	3.785412	E–03	=	m^3
gauss	×	1.0*	E–04	=	T
g mol	×	1.0*	E–03	=	kmol
hp	×	7.46043	E–01	=	kW
in.	×	2.54*	E+00	=	cm
in.	×	2.54*	E+01	=	mm
in.^{2}	×	6.4516*	E+00	=	cm^2
in.^{3}	×	1.638 706	E+01	=	cm^3
kip	×	4.448 222	E+03	=	N
lbf	×	4.448 222	E+00	=	N
lbf-ft	×	1.355 818	E+00	=	N·m
lbm	×	4.535 924	E–01	=	kg
lbm mol	×	4.535 924	E–01	=	kmol
md	×	9.869 233	E–04	=	μm^2
mile	×	1.609 344*	E+00	=	km
psf	×	4.788 026	E–02	=	kPa
psi	×	6.894 757	E+00	=	kPa
psi^{-1}	×	1.450 377	E–01	=	kPa^{-1}
°R		5/9		=	K
sq mile	×	2.589 988	E+00	=	km^2
ton	×	9.071847	E–01	=	Mg
tonne	×	1.0*	E+00	=	Mg

*Conversion factor is exact.

Appendix A

Formation of a Viscous Shock

In Chapter 3, the formation of a viscous shock or a chemical shock was assumed to be instantaneous. Saturation profiles shown in Figs. 3.13 and 3.14 also were assumed to form immediately after the injection of viscous water or water containing a chemical species that might increase the viscosity of the injected fluid or reduce the interfacial tension. Appendix A describes how the viscous shock and the saturation profiles develop.

Formation of the Viscous Shock

A conceptual model of the displacement process is developed by considering the injection of viscous water at some time, t_{Do}, after a waterflood has started. For this development, the waterflood parameters in Example 3.1 and the viscous solution parameters in Example 3.2 are used. When a waterflood has been in operation before the injection of a viscous fluid, a saturation profile exists such as that depicted in Fig. 3.6. As discussed in Section 3.2, the saturation profile contains all saturations $S_{wf} < S_w < 1 - S_{or}$.

When viscous water enters the linear system at $x_D = 0$ and $t_D = t_{Do}$ and dispersion is neglected, a miscible boundary forms between the injected waterflood water and the viscous water. These fluids are called injected water and viscous water in this section. Because the two fluids are miscible, their specific velocities (Eq. 3.39) must be equal. That is, at the miscible boundary between the two fluids,

$$v_{Dv}^* = v_{Dv} = f_w/S_w = f_w^*/S_w^*. \dots\dots\dots\dots\dots\dots\dots\dots\dots\dots\dots\dots\dots\dots\dots\dots\dots\dots(A\text{-}1)$$

As this boundary between the viscous water and injected water propagates into the linear system, the saturations that satisfy Eq. A-1 are found by drawing a straight line from the origin of the fractional-flow curve through S_w^* to intersect the $f_w - S_w$ curve at (f_w, S_w).

Fig. A-1 is an enlargement of the upper portion of Fig. 3.12. Lines a through f are drawn from the origin to $f_w^* - S_w^*$, which intersects the $f_w - S_w$ curve at f_w, S_w. **Table A-1** contains pairs of saturations that satisfy Eq. A-1 for the parameters of Example 3.2. Note that $S_w > S_w^*$ for every pair of points except at $1 - S_{or}$. Table A-1 also includes the specific velocity of each saturation (i.e., $f_w'^*$ and f_w') from the respective fractional-flow curves.

The boundary between the viscous fluid and the injected water is a viscous shock because there is a viscosity discontinuity across this boundary. In Fig. A-1, the viscosity of the viscous water is 4 cp while the viscosity of the injected waterflood water is 1.0 cp. A saturation discontinuity also develops across the viscous shock, as seen in Table A-1. This discontinuity is small at high S_w^*, but becomes noticeable as $S_w^* \to S_{w3}^*$. The viscous shock must satisfy the continuity equation for a saturation discontinuity, as discussed in Section 3.3.1. Thus, the velocity of the viscous shock is given by

$$v_{Dv} = \left(f_w - f_w^*\right)/\left(S_w - S_w^*\right). \dots\dots\dots (A\text{-}2)$$

The development of a viscous shock can be visualized with the assistance of Fig. A-1 and

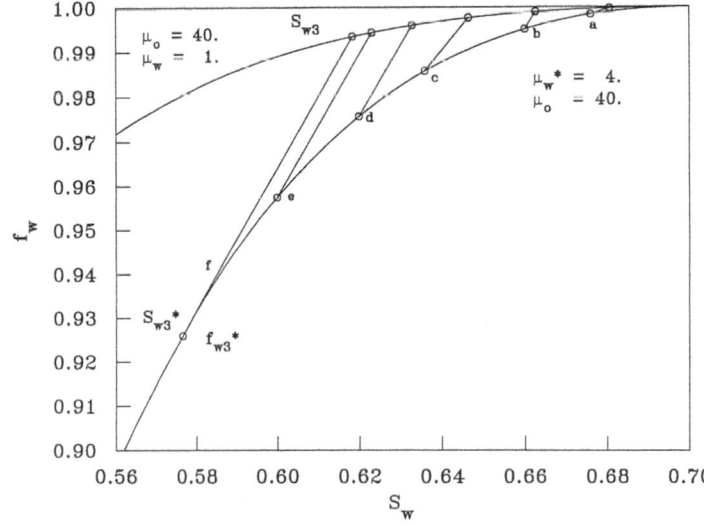

Fig. A-1—Fractional-flow diagram showing saturation of the developing viscous flood front; Example A-1 with parameters from Examples 3.1 and 3.2.

Pair	S_w^*	f_w^*	$f_w''^*$	$v_{Dv} - f_w^*/S_w^*$	S_w	f_w	f_w	$v_{Dv} = f_w/S_w$	$v_{Dv} = (f_w - f_w^*)/(S_w - S_w^*)$
—	0.70	1.0000	0.00000	1.4286	0.700	1.0000	0.0000	1.4286	1.4286
a	0.68	0.9989	0.11640	1.4690	0.6806	0.9997	0.0282	1.4690	1.4690
b	0.66	0.9951	0.2717	1.5077	0.6625	0.9989	0.0627	1.5077	1.5077
c	0.64	0.9877	0.4766	1.5433	0.6464	0.9976	0.1026	1.5433	1.5433
d	0.62	0.9756	0.7436	1.5736	0.6329	0.9960	0.1442	1.5736	1.5736
e	0.60	0.9575	1.0865	1.5958	0.6231	0.9944	0.1802	1.5958	1.5958
f	0.5764	0.9259	1.6064	1.6064	0.6184	0.9934	0.1995	1.6064	1.6064

Table A-1—Development of viscous flood front, Example A-1, Computed values corresponding to selected values of S_w^*.

Table A-1. When viscous water is injected into the linear system at $x_D = 0$ and $t_D = t_{Do}$, the initial specific velocity of the viscous shock fluid is

$$v_{Dv} = 1/(1 - S_{or})$$
$$= 1/(1 - 0.30)$$
$$= 1.429. \dots\dots\dots\dots\dots\dots\dots\dots\dots\dots\dots\dots\dots\dots\dots\dots\dots\dots\dots\dots\dots\dots\dots\dots\dots\dots(A-3)$$

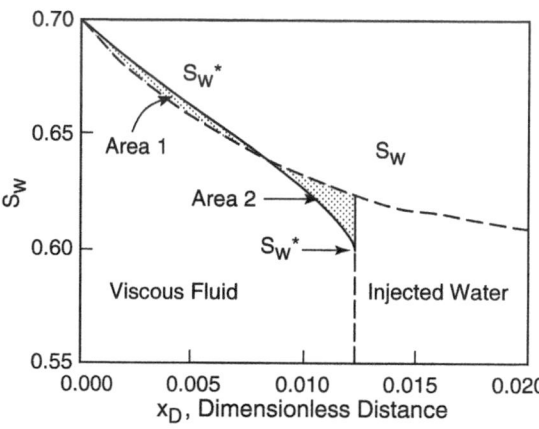

Fig. A-2—Development of viscous saturation discontinuity at $t_D = 0.06805$ when viscous waterflood begins at $t_{Do} = 0.06$; Example A-1 with parameters from Examples 3.1 and 3.2.

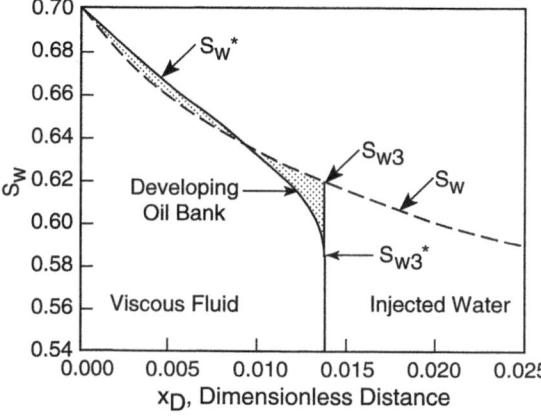

Fig. A-3—Saturation profile when the velocity of the viscous saturation discontinuity reaches v_{D3}^*; Example A-1 with parameters from Examples 3.1 and 3.2.

Inspection of Table A-1 or Fig. A-1 reveals that $v_{Dv} > f_w'$ and $v_{Dv} > f_w''^*$ until $S_w^* = S_{w3}^*$, where $v_{Dv} = f_{w3}''^*$. Thus, the velocity of the viscous shock increases until it is equal to f_{Sw3}' at S_{w3}^*, where it becomes stabilized at its maximum value. Smaller S_w values are encountered as the viscous fluid penetrates the rear of the original waterflood saturation profile. S_w^* must decrease from $1 - S_{or}$ along the path from Line a to Line f to S_w^*, while S_w must decrease from $1 - S_{or}$ to S_{w3} along the same path.

Figs. A-2 through A-4 illustrate the development of the saturation profile. These figures depict the saturation profiles at three dimensionless times when a viscous waterflood is initiated for the parameters of Examples 3.1 and 3.2 at $t_{Do} = 0.06$. The saturation profile in Fig. A-2 corresponds to $t_D = 0.06805$ or 0.00805 pore volumes (PV) of viscous water injection. The dashed saturation profile is the saturation profile of the waterflood if no viscous water is injected. The solid curve is the saturation profile caused by the injection of viscous water. A viscous shock is present at $x_D = 0.0123$, where saturations S_w^* and S_w satisfy Eqs. A-1 and A-2. The oil bank beginning to develop represents oil that was displaced by the viscous water from the region $x_D < 0.0123$. Shaded Area 1 between the solid and dashed curves in Fig. A-2 is the oil displaced by the viscous saturation profile. This oil has accumulated in an oil bank shown as Shaded Area 2. Note that the region where $x_D = 0.0123$ is a normal waterflood and that this region is not affected by the fact that viscous water was injected at $t_D = 0.06$. Also note that all saturations in the viscous water greater than S_w^* evolved from the saturation profile as the viscous fluid overtook locations previously contacted by injected water. Similarly, all water saturations greater than S_w^* have been removed from the profile by the advancing viscous water.

Fig. A-3 shows the development of the viscous waterflood at $t_D = 0.06896$ or 0.00896 PV of viscous water injected. As in Fig. A-2, the oil in the oil bank was displaced by the viscous fluid from the shaded area between the viscous saturation profile and the dashed line representing the profile if the original waterflood was in place and viscous fluid had not been injected. This value of t_D corresponds to the dimensionless time when the viscous shock reaches its maximum velocity. The values of the saturations at this point are S_{w3}^* and S_{w3}. Thereafter, the viscous shock moves through the linear system at a constant velocity given by Eq. 3.42.

Fig. A-4 shows the saturation profile at $t_D = 0.10035$. An oil bank of constant saturation, S_{w1}, has formed. S_{w1} evolves from the viscous shock at $t_D = 0.06896$ because the saturation discontinuity from S_{w3}^* to S_{w1} also has the same velocity as the viscous shock $S_{w3}^* - S_{w3}$, which corresponds to Fluid Pair f in Table A-1. As noted in Figs. A-2 and A-3, the oil in this bank was displaced from the shaded area between the two saturation profiles. Another saturation discontinuity is created when the oil bank is formed. This saturation discontinuity is at the front of the oil bank. At this discontinuity, the water saturation increases from S_{w1} to S_{wr}. The velocity of this saturation discontinuity is

$$v_{Dr} = \left(f_{wr} - f_{w1} \right) / \left(S_{wr} - S_{w1} \right), \quad \dots\dots\dots\dots\dots(A\text{-}4)$$

where v_{Dr} = velocity of the oil-bank shock at the front of the oil bank, f_{wr} = fractional flow of water at the rear of the water bank, and S_{wr} = water saturation at the rear of the water bank. Because $v_{Dr} > f_{wr}'$, water saturations (in the saturation profile of the previous waterflood) are overtaken by the advancing oil bank. The front of the oil bank is also the rear of the water bank. S_{wr} eventually becomes equal to S_{w1}, and the discontinuity at the rear of the water bank disappears.

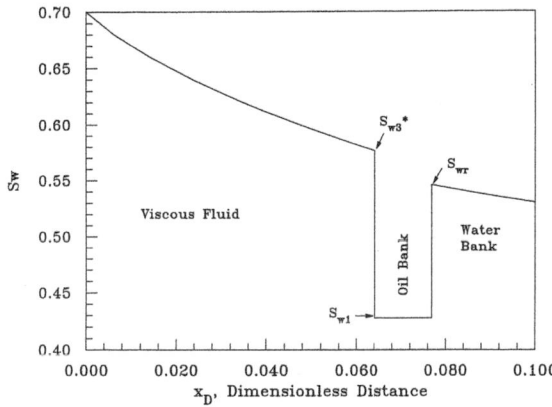

Fig. A-4—Saturation profile illustrating the formation of an oil bank between the viscous shock and the flood front; Example A-1 with parameters from Examples 3.1 and 3.2.

The location of the discontinuity between the viscous water and the interstitial water can be determined at any t_D by integrating Eq. A-2 to obtain

$$x_{Dv} = \int_0^{t_D} v_{Dv}\, dt_D$$

$$= \int_0^{t_D} \left(\frac{f_w - f_w^*}{S_w - S_w^*} \right) dt_D. \quad \dots\dots\dots\dots\dots\dots\dots\dots\dots\dots\dots\dots\dots\dots\dots\dots(A\text{-}5)$$

Because v_{Dv} varies with saturation, it is possible to integrate Eq. A-5 by averaging the velocity over small time increments. Thus,

$$x_{Dv}^{n+1} = x_{Dv}^n + \overline{v}_{Dv}^{n+1} \left(t_D^{n+1} - t_D^n \right), \quad \dots\dots\dots\dots\dots\dots\dots\dots\dots\dots\dots\dots\dots\dots\dots\dots\dots\dots(A\text{-}6)$$

$$\text{where } \overline{v}_{Dv}^{n+1} = \tfrac{1}{2} \left[\left(\frac{f_w - f_w^*}{S_w - S_w^*} \right)^{n+1} + \left(\frac{f_w - f_w^*}{S_w - S_w^*} \right)^n \right]. \quad \dots\dots\dots\dots\dots\dots\dots\dots\dots\dots\dots(A\text{-}7)$$

Because the shock intersects the path traveled by saturation S_w^{n+1}, the frontal advance equation gives

$$x_{Dv}^{n+1} = t_D^{n+1} f_w'^{n+1}. \quad \dots\dots\dots\dots\dots\dots\dots\dots\dots\dots\dots\dots\dots\dots\dots\dots\dots\dots\dots\dots\dots\dots\dots\dots\dots\dots\dots\dots(A\text{-}8)$$

Substituting into Eq. A-6 and solving for the time when the shock intersects the path of S_w^{n+1} yield

$$t_n^{n+1} = \left(x_{Dv}^n - \overline{v}_{Dv}^{n+1} t_D^n \right) / \left(f_w'^{n+1} - \overline{v}_{Dv}^{n+1} \right). \quad \dots\dots\dots\dots(A\text{-}9)$$

The evolution of the viscous shock is depicted in the distance/time diagram in **Fig. A-5.** The solid line shows the trajectory of the viscous discontinuity. The dashed line is the trajectory of the viscous shock when it intersects the next line of constant saturation S_w. Note that the specific velocity of the viscous shock is larger than the specific velocity of S_w or S_w^* and is increasing. However, the specific velocity of the viscous shock becomes constant when S_{w3}^* evolves from the saturation profile. As the saturation discontinuity advances from the origin, saturations $S_{wa}^*, S_{wb}^* \to S_{w3}^*$ evolve from the moving shock, while saturations $S_{wa}, S_{wb} \to S_{w3}$ disappear.

Saturation profiles can be constructed once the trajectory of the viscous shock has been determined; t_{Do}^* is introduced to indicate the time when S_w^* evolves from the moving saturation

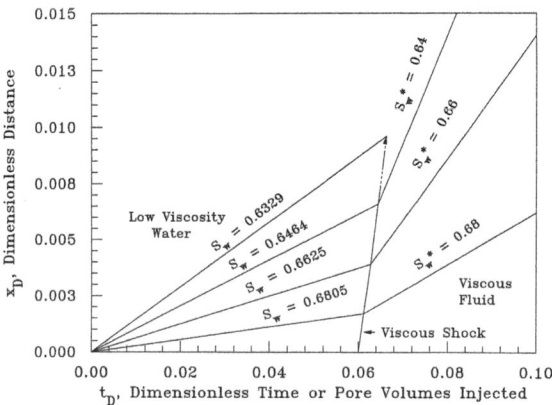

Fig. A-5—Evolution of the viscous shock when a viscous waterflood begins at $t_{Do} = 0.06$ after a conventional waterflood began; Example A-1 with parameters from Examples 3.1 and 3.2.

discontinuity. The location of S_w^* at subsequent t_D is given by the frontal-advance solution in Eq. A-10, with appropriate adjustments for the origin of the saturation.

$$x_D^* = \left(t_D - t_{Do}^*\right)f_w^{'*} + x_{Do}^*, \dots\dots\dots\dots\dots\dots\dots\dots\dots\dots\dots\dots\dots\dots\dots\dots\dots\dots\text{(A-10)}$$

where x_{Do}^* = location where S_w^* evolves from the saturation discontinuity.

Fig. A-2 shows the saturation profile at t_D = 0.06805 when the viscous waterflood in Example 3.2 begins at t_{Do} = 0.06. Fig. A-3 shows the saturation profile when t_D = 0.06896 or after 0.00896 PV of viscous water has been injected. This saturation profile is unstable because the specific velocity of S_{w3}^* is greater than that for S_{w3}. Recall from Eq. 3.44 that

$$v_{D3}^* = \frac{f_{w3} - f_{w3}^*}{S_{w3} - S_{w3}^*} = \frac{f_{w3}^* - f_{w1}}{S_{w3}^* - S_{w1}}. \dots\dots\dots\dots\dots\dots\dots\dots\dots\dots\dots\dots\dots\dots\dots\text{(A-11)}$$

The water saturation S_{w1} evolves from this saturation profile because $v_{D3}^* = v_{D1}$ from Eq. 3.44 and all saturations $S_{w3} > S_w > S_{w1}$ travel at slower velocities than S_{w1} in the water bank ahead of the viscous shock. The small region of increased oil saturation behind the discontinuity expands into the oil bank with constant water saturation S_{w1}, as shown in Fig. A-4. S_{w1} at the base of the viscous shock is displaced through the system at a constant specific velocity, v_{D1}, given by Eq. 3.44.

The small oil bank that forms between the viscous shock and the water bank, depicted in Fig. A-4, creates a third discontinuity at the boundary between the front of the oil bank and the rear of the water bank. Eq. A-4 gives the specific velocity of this shock.

Following the Oil-Bank Shock by Front Tracking

The front of the oil bank, called the oil shock, can be followed in the same manner used to track the evolution of the viscous shock. As the oil shock moves, it intersects the paths of saturations in the water bank after the flood front forms (i.e, $S_{w3} > S_w > S_{wf}$). A schematic of this path is shown on the x_D/t_D diagram in **Fig. A-6.** The oil bank is formed at Point A when S_{w3}^* evolves from the viscous shock. The dashed line in Fig. A-6 shows the trajectory of the oil bank and the intersection of this shock with the next path of constant S_w. The intersection of the oil shock with a path of constant saturation S_w is a point where the respective x_D and t_D for each path coincide. This point can be computed from the frontal-advance equation and the position of the oil shock.

Consider the movement of the oil-bank shock from the S_{wr}^n saturation line to the S_{wr}^{n+1} saturation line. The distance traversed by the oil-bank shock is given by

$$x_r^{n+1} = x_r^n + \int_{t_n}^{t_{n+1}} v_r \, dt$$

$$= x_r^n + \overline{v}_r^{n+1}\left(t^{n+1} - t^n\right), \dots\dots\dots\dots\dots\dots\dots\dots\dots\dots\dots\dots\dots\dots\dots\dots\dots\dots\text{(A-12)}$$

where v_r = velocity of oil-bank shock, x_r^n = location of the shock at t^n, x_r^{n+1} = location of the shock at t^{n+1}, and $\overline{v}_r^{n+1}$ = average velocity of the shock between t^n and t^{n+1}. The average velocity, $\overline{v}_r^{n+1}$, may be approximated for small timesteps by

$$\overline{v}_r^{n+1} = \frac{q_t}{2A\phi}\left[\left(\frac{f_{wr}^{n+1} - f_{w1}}{S_{wr}^{n+1} - S_{w1}}\right) + \left(\frac{f_{wr}^n - f_{w1}}{S_{wr}^n - S_{w1}}\right)\right]. \dots\dots\dots\dots\dots\dots\dots\dots\dots\dots\text{(A-13)}$$

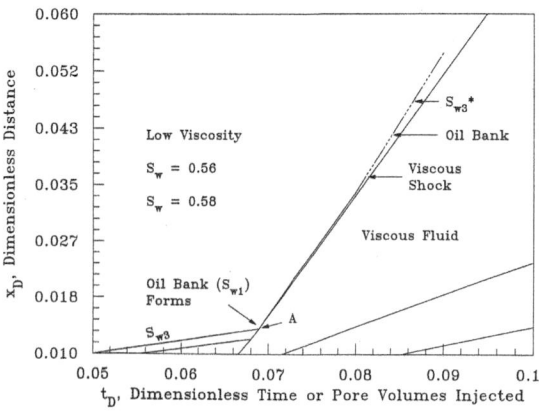

Fig. A-6—Path followed by oil-bank shock when viscous waterflood begins at t_{Do} = 0.06 after a conventional waterflood began. Stabilized oil-bank saturation, S_{w1}, forms at Point A. Example A-1 with parameters from Examples 3.1 and 3.2.

Eq. A-13 can be converted into dimensionless form as follows. Let

$$\overline{v}_{Dr}^{n+1} = \tfrac{1}{2}\left[\left(\frac{f_{wr}^{n+1} - f_{w1}}{S_{wr}^{n+1} - S_{w1}}\right) + \left(\frac{f_{wr}^n - f_{w1}}{S_{wr}^n - S_{w1}}\right)\right]. \dots\dots\dots\text{(A-14)}$$

Then, $\dfrac{x_r^{n+1}}{L} = \dfrac{x_r^n}{L} + \dfrac{q_t \overline{v}_{Dr}^{n+1}}{A\phi L}\left(t^{n+1} - t^n\right). \dots\dots\dots\dots\text{(A-15)}$

or $x_{Dr}^{n+1} = x_{Dr}^n + \overline{v}_{Dr}^{n+1}\left(t_D^{n+1} - t_D^n\right), \dots\dots\dots\dots\dots\text{(A-16)}$

where x_{Dr} = location of the rear of the water bank and $\overline{v}_{Dr}$ = average specific velocity of the saturation discontinuity at the rear of the water bank. Because x_{Dr}^n and x_{Dr}^{n+1} can also be expressed in terms of the frontal-advance equation (i.e., $x_{Dr} = t_D f_{wr}'$),

$$t_{Dr}^{n+1} f_{wr}^{'n+1} = t_D^n f_{wr}^{'n} + \overline{v}_{Dr}^{n+1}\left(t_D^{n+1} - t_D^n\right) \dots\dots\dots\dots\text{(A-17)}$$

and $t_D^{n+1} = t_D^n \dfrac{\left(f_{wr}'^n - \overline{v}_{Dr}^{n+1}\right)}{\left(f_{wr}'^{n+1} - \overline{v}_{Dr}^{n+1}\right)}$ ....(A-18)

The path followed by the front of the oil bank is computed by determining the value of v_{Dr} in Eq. A-4 as S_{wr} decreases. This path may be visualized by constructing the series of lines from S_{w1} to the f_w/S_w curve. The slope of each line is the specific velocity of the oil bank shock at S_{wr}.

Fig. A-7 is the distance/time diagram for a viscous waterflood $\left(\mu_w^* = 4 \text{ cp}\right)$ that began after injecting 0.06 PV of water into a linear system. Supporting calculations for this figure are included in Example A-1. Formation of the viscous shock and the oil bank, depicted in Figs. A-2 through A-4, can be visualized in Fig. A-7. The path of the viscous shock appears to be linear because the maximum specific velocity of the shock, v_{D3}, is reached at $t_{Do}^* = 0.06896$. However, the specific velocity of the front of the oil bank continuously increases with t_D. In this example, the oil bank does not overtake the flood front, S_{wf}, before it leaves the

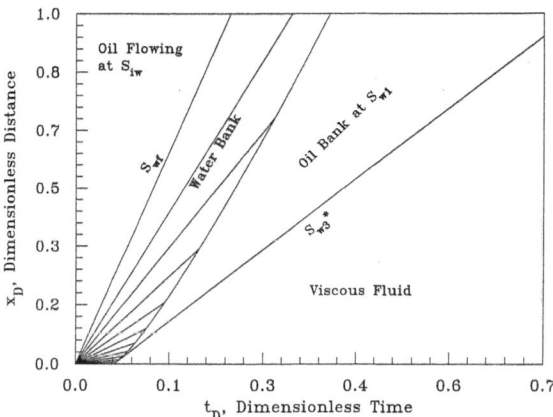

Fig. A-7—Time/distance diagram for a viscous waterflood which began at t_{Do} = 0.06; Example A-1 with parameters from Examples 3.1 and 3.2.

system at $x_d = 1.0$. Consequently, the viscous waterflood would perform like the regular waterflood until breakthrough of the oil bank at $t_D = 0.33$ in Fig. A-7. It is clear from this example that a viscous fluid injected late in the life of a linear waterflood in a homogeneous reservoir would have little effect on the oil recovery. Example A-1 illustrates the computation of the oil-bank advance by the front-tracking method.

Example A.1—Application of Front-Tracking Method. The waterflood described in Example 3.1 has been in operation for $t_D = 0.06$ or 0.06 PV of water injected. At this time, a viscous waterflood is initiated by injecting a nonadsorbing aqueous solution with a viscosity of 4 cp.

1. Determine the trajectory followed by the viscous shock until the viscous shock arrives at the end of the system.
2. Determine the path followed by the leading edge of the oil bank.
3. Prepare an x_D/t_D diagram from the beginning of the waterflood until the viscous shock arrives at the end of the system.

Solution.

1. *Trajectory of the Viscous Shock.* The path of the viscous shock is found in two parts. The maximum velocity of the viscous shock is determined from Eq. 3.44. From Example 3.2, $f_{w3}^* = 0.9289$ and $S_{w3}^* = 0.5764$. Thus,

$$v_{D3}^* = f_{w3}^*/S_{w3}^*$$
$$= 0.9259/0.5764$$
$$= 1.606.$$

The location of this shock is given by the relationships $x_{D3} = \left(t_D - t_{Do}^*\right) + x_{Do}^*$ and $x_{Dv} = t_D f_w'$. The values of t_{Do}^* and x_{Do}^* are found by use of Eq. A-9 to find t_D^{n+1} as S_w^* decreases from $S_w^* = 1 - S_{or}$ to S_{w3}^*. The saturation interval from S_{w3}^* to $S_w^* = 1 - S_{or}$ in Fig. A.1 was split into six increments by choosing values of $S_w^* w$. For each value of S_w^*, a corresponding point on the f_w, S_w curve is sought to satisfy

$$f_w^*/S_w^* = f_w/S_w.$$

The corresponding values of f_w and S_w are determined from the intersection of a straight line from the origin through the Point f_w^*, S_w^*, or by an equivalent root-finding scheme. Table A-1 presents pairs of $\left(f_w^*, S_w^*\right) - \left(f_w, S_w\right)$ determined for this example. Table A-1 also includes values of v_{Dv} and f_w'.

In the first timestep, the viscous shock advances from $S_w^* = 0.7$ to $S_w^* = 0.68$. The average specific velocity for this timestep is computed with Eq. A-14. For the first timestep, $n = 0$ and $t_D(0) = t_{Do} = 0.06$,

$$\overline{v}_D^{(1)} = \tfrac{1}{2}\left[v_{Dv}^{(0)} + v_{Dv}^{(1)}\right]$$
$$= \tfrac{1}{2}(1.4286 + 1.4690)$$
$$= 1.4488.$$

Substituting into Eq. A-10 gives

$$t_D^{(1)} = \left[x_{Dv}^{(0)} - v_{Dv}^{(1)} t_{Do} \right] / \left[f_w^{\prime(1)} - \overline{v}_{Dv}^{(1)} \right]$$
$$= \left[0 - (1.4488)(0.06) \right] / (0.0282 - 1.4488)$$
$$= 0.0612.$$

The location of the shock at this saturation, $x_{Dv}^{(1)}$, is obtained from Eq. A-8:

$$x_{Do}^{(1)} = t_D^{(1)} f_w^{\prime(1)}$$
$$= (0.0612)(0.0282)$$
$$= 0.0017.$$

Table A-2 summarizes values of t_{Do}^* and x_{Do}^* from the computations.

The viscous shock forms rapidly and is fully developed within 0.009 PV $\left(t_{Do}^* = 0.069, x_{Do}^* = 0.0138 \right)$ after the beginning of viscous fluid injection. For most cases, little error would be caused by assuming that the shock started at $x_D = 0$. For $t_D > t_{Do}^*$, the location of the viscous shock is

$$x_{D3}^* = \left(t_D - t_{Do}^* \right) f_{w3}^{\prime *} + x_{Do}^*$$
$$= \left(t_D - 0.069 \right)(1.606) + 0.0138.$$

When $x_{D3}^* = 1.0$, $t_D = 0.683$.

All saturations S_w^* travel at a constant velocity after they evolve from the viscous shock.
The paths for saturations $S_w^* > S_{w3}$ can be found from Eq. A-10:

$$x_D^* = \left(t_D - t_{Do}^* \right) f_w^{\prime *} + x_{Do}^*.$$

For example, the location of $S_w^* = 0.6$ at $t_D = 1.0$ is

$$x_D^* = (1.0 - 0.068)(1.0865) + 0.0123$$
$$= 1.0249.$$

Table A-3 presents locations of other values of S_w^* with parameters used to compute x_D^*. When $t_D = 1.0$, the last two saturations in Table A-3 have left the system.

2. *Leading Edge of Oil Bank.* The path followed by the leading edge of the oil bank is computed with Eq. A-18. Recall that a saturation discontinuity develops as soon as the injection of viscous fluid begins and a region of increased oil saturation is shown ahead of the viscous shock in Figs. A-2 and A-3. The region of constant water saturation, S_{w1}, begins at $t_{Do}^* = 0.069$ and $x_{Dr}^* = 0.0138$. This is called the leading edge of the oil bank. The computation proceeds by selecting saturations S_{wr} for $S_{w1} < S_{wr} < S_{w3}$ and computing the time when the front of the oil bank moves from S_{wr}^n to the intersection with the path of constant saturation at S_{wr}^{n+1}. Values of (f_{w3}, S_{w3}) determined from the extension of the line from (0, 0) through (f_{w1}, S_{w1}) and (f_{w3}^*, S_{w3}^*) are $f_{w3} = 0.9935$ and $S_{w3} = 0.6185$. Recall that $f_{w1} = 0.6869$ and $S_{w1} = 0.4276$.

From Eq. A-18,

$$t_D^{n+1} = t_D^n \frac{\left(f_{wr}^{\prime n} - \overline{v}_{Dr}^{n+1} \right)}{\left(f_{wr}^{\prime n+1} - \overline{v}_{Dr}^{n+1} \right)}.$$

	S_w^*	t_{Do}^*	x_{Do}^*
(Fig. A-1)	0.70	0.06	0
A	0.68	0.0612	0.0017
B	0.66	0.0627	0.0039
C	0.64	0.0644	0.0066
D	0.62	0.0663	0.0096
E	0.60	0.0680	0.0123
F	0.5764	0.0690	0.0138

Table A-2—Location of viscous flood front when viscous water injection begins at $t_{Do} = 0.06$, Example A-1.

S_w^*	t_{Do}^*	$t_D - t_{Do}^*$	$f_w'^*$	$\dfrac{\left(t_D - t_{Do}^*\right)}{f_w'^*}$	x_D^*
0.70	0.06	0.94	0	0	0
0.68	0.0612	0.9388	0.1164	0.1093	0.1110
0.66	0.0627	0.9373	0.2717	0.2547	0.2586
0.64	0.0644	0.9356	0.4766	0.4459	0.4525
0.62	0.0663	0.9337	0.7436	0.6943	0.7039
0.60	0.0680	0.9320	1.0865	1.0126	1.0249
0.5764	0.0690	0.9310	1.6064	1.4956	1.5094

Table A-3—Summary of supporting calculations, Example A-1; location of viscous fluid saturations at $t_D = 1.0$.

For the first timestep, S_{wr} moves from $S_{w3} = 0.6184$ to $S_{wr} = 0.60$, where $f_{wr} = 0.9890$. Applying Eq. A-14 gives

$$v_{Dr}^{(2)} = \tfrac{1}{2}\left(\frac{0.9935 - 0.6869}{0.6185 - 0.4276} + \frac{0.9890 - 0.6869}{0.60 - 0.4276}\right)$$
$$= \tfrac{1}{2}(1.6069 + 1.7523)$$
$$= 1.6792.$$

Then, $t_D^{(2)} = (0.069)\left(\dfrac{0.1995 - 1.6792}{0.2898 - 1.6792}\right)$

$$= 0.0735$$

and $x_D^{(2)} = (0.0735)(0.2898)$

$$= 0.0213.$$

Table A-4 summarizes results of the remaining computations from S_{w3} to S_{w1}.

As discussed earlier, the x_D values in Table A-4 represent locations where S_{wr} is overtaken by the front of the oil bank and disappears from the saturation profile. The path followed by each S_{wr} is linear from the origin to (x_D, t_D). In this example, the oil bank arrives at the end of the system ($x_D = 1.0$) when $S_{wr} = 0.47$.

3. *Preparation of x_D/t_D Diagram.* The x_D/t_D graph in Fig. A-7 was obtained by plotting results presented in Tables A-2 and A-4. The saturation profile shown in **Fig. A-8** is at $t_D = 0.10035$. As t_{Do} becomes smaller, the size of the oil bank decreases. If $S_{w1} > S_{wf}$, the saturation at the front of the oil bank approaches S_{w1} asymptotically as $S_{wr} \rightarrow S_{w1}$,

$$v_{Dr} = \left(\partial f_w / \partial S_w\right)_{S_{w1}} \dots\dots\dots\dots\dots\dots\dots\dots\dots\dots\dots\dots\dots\dots\dots\dots\dots\dots\text{(A-19)}$$

S_{wr}	f_{wr}	f_{wr}'	t_D	x_{Dr}
0.6184	0.9935	0.1995	0.069	0.0138
0.6000	0.9890	0.2898	0.073	0.0213
0.5800	0.9820	0.4217	0.080	0.0339
0.5600	0.9718	0.6018	0.090	0.0544
0.5400	0.9575	0.8488	0.106	0.0899
0.5200	0.9373	1.1878	0.132	0.1569
0.5000	0.9091	1.6529	0.182	0.3005
0.4800	0.8700	2.2844	0.297	0.6783
0.4600	0.8163	3.1237	0.699	2.1822
0.4400	0.7599	4.1908	1.693	7.0952

Table A-4—Supporting calculations, Example A-1; determining of the oil-bank path.

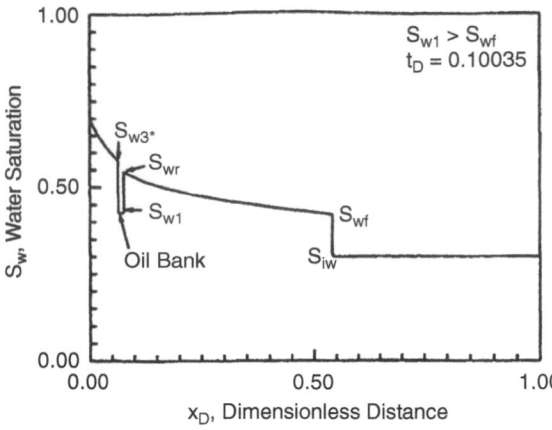

Fig. A-8—Saturation profile at $t_D = 0.10035$ for a viscous waterflood which began at $t_{Do} = 0.06$; Example A-1 with parameters from Examples 3.1 and 3.2.

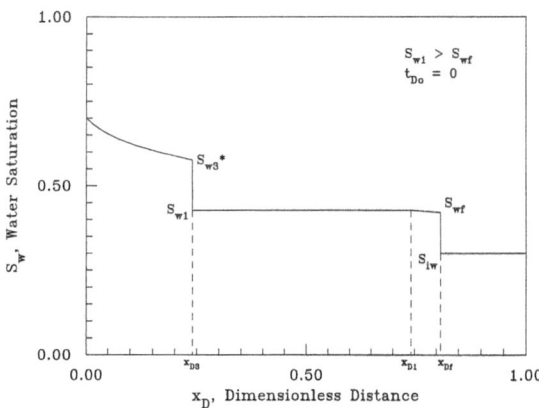

Fig. A-9—Saturation profile during a viscous waterflood at interstitial water saturation ($t_{Do} = 0$) when $S_{w1} > S_{wf}$.

The remainder of the saturation profile (i.e., $S_{wf} < S_w < S_{w1}$) is unaffected by the viscous flood, and the saturation profile is shown in **Fig. A-9**. Note in Example 3.2 that $S_{w1} = 0.428$ and $S_{wf} = 0.4206$ (i.e., the two values are approximately equal).

In many cases, $S_{w1} < S_{wf}$. If t_{Do} is small, S_{w1} can overtake S_{wf} leading to the saturation profile in **Fig. A-10.** The saturation profiles in Figs. A-9 and A-10 form instantaneously when $t_{Do} = 0$. Computation of the displacement performance under this assumption is discussed in Section 3.4.

Following the Oil-Bank Shock by Material Balance
The oil-bank shock may be located by a material balance on the interstitial water and the water injected during the waterflood. Because the viscous solution displaces the resident water miscibly, the only resident water ahead of the viscous boundary is injected and interstitial water, as shown in **Fig. A-11.** In this case, $S_{w1} > S_{wf}$, as in Example A-1.

A volume balance on the injected and resident water at time t when the original waterflood ran to t_o before starting the viscous water injection is given by

$$q_t t_0 + x_f A\phi S_{iw} = (x_r - x_3) A\phi S_{w1} + (x_f - x_r) A\phi \overline{S}_{wrf}, \quad ..(A-20)$$

where x_f = location of the flood-front saturation, S_{wf}; x_r = location of the front at the oil bank; x_3 = location of the viscous shock; and $\overline{S}_{wrf}$ = average water saturation in the region between x_r and x_f.

Dividing by $A\phi L$ and converting to dimensionless variables give

$$t_{Do} + x_{Df} S_{iw} = (x_{Dr} - x_{Dv}) S_{w1} + (x_{Df} - x_{Dr}) \overline{S}_{wrf}. \quad ....(A-21)$$

Recall that

$$x_{Df} = t_D f'_{wf}, \quad ....(A-22)$$

$$x_{Dr} = t_D f'_{wr}, \quad ....(A-23)$$

and $x_{D3} = (t_D - t^*_{Do})(f_{w1}/S_{w1}) + x^*_{D3o}, \quad ....(A-24)$

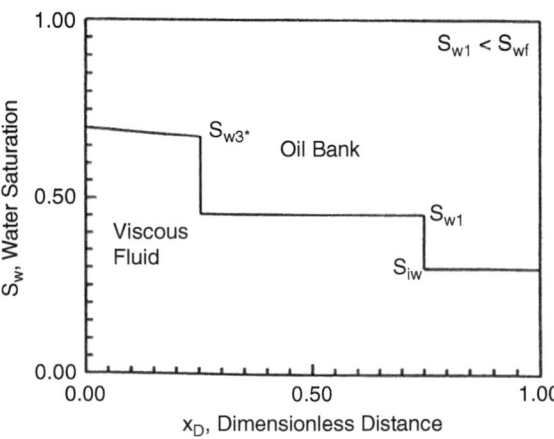

Fig. A-10—Saturation profile during a viscous waterflood at interstitial water saturation $(t_{Do} = 0)$ when $S_{w1} < S_{wf}$.

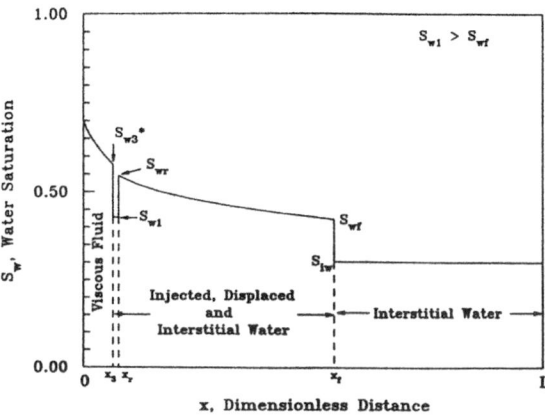

Fig. A-11—Saturation profile used to determine the location of the oil-bank shock by material balance when $S_{w1} > S_{wf}$.

where t^*_{Do} = time when the oil bank was initiated and x^*_{D3o} = distance from the origin where oil bank started.

In addition, from the Welge equation,

$$\overline{S}_{wrf} = \frac{x_{Df}S_{wf} - x_{Dr}S_{wr}}{x_{Df} - x_{Dr}} - t_D\frac{\left(f_{wf} - f_{wr}\right)}{\left(x_{Df} - x_{Dr}\right)}. \dots\dots\dots\dots\dots\dots\dots\dots\dots\dots\dots\dots\dots\text{(A-25)}$$

Substituting Eqs. A-22 through A-25 into Eq. A-21 and solving for t_D give

$$t_D = \frac{t_{Do} - t^*_{Do}S_{w1} - x^*_{D3o}S_{w1}}{f_{wr} - f_{w1} - f_{wf} + f_{wf}\left(S_{wf} - S_{iw}\right) - f_{wr}\left(S_{wr} - S_{w1}\right)}, \dots\dots\dots\dots\dots\dots\dots\text{(A-26)}$$

where t_D = dimensionless time when the fractional flow of water at the rear of the water bank is f_{wr}.

Under some conditions, the flood front may be overtaken by the oil bank before the flood front reaches the end of the system. The time required for this to occur can be obtained by allowing $f_{wr} = f_{wf}$ and $S_{wr} = S_{wf}$ in

$$t_{D1} = \frac{t_{Do} - t^*_{Do}S_{w1} - x^*_{D3o}S_{w1}}{f'_{w1}\left(S_{w1} - S_{iw}\right) - f_{w1}}. \dots\dots\dots\dots\dots\dots\dots\dots\dots\dots\dots\dots\dots\text{(A-27)}$$

The location where $x_{Dr} = x_{Df}$ is computed by substituting t_{D1} into the frontal-advance equation. Thus,

$$x_{Df} = x_{Dr} = t_{D1}f'_{wf}. \dots\dots\dots\dots\dots\dots\dots\dots\dots\dots\dots\dots\dots\dots\dots\dots\dots\text{(A-28)}$$

For this to occur within the linear system,

$$t_{D1} < 1/f'_{wf}. \dots\dots\dots\dots\dots\dots\dots\dots\dots\dots\dots\dots\dots\dots\dots\dots\dots\dots\dots\text{(A-29)}$$

This method is equivalent to the incremental approach developed earlier when the saturation increments are small.

Nomenclature

A = cross-sectional area available for flow, L^2, ft^2

f'_{w3} = derivative of f^*_w with respect to S^*_w at S_{w3}

f_w = fractional flow of water

f'_w = derivative of water fractional-flow curve with respect to S_w

f^*_w = fractional flow of viscous or chemical solution having a different fractional-flow curve from that of water

f'^*_w = derivative of f^*_w with respect to S^*_w

f'_{wf} = derivative of f_w with respect to S_w at S_{wf}

f_{wr} = fractional flow of water in oil-bank shock

f'_{wr} = derivative of water fractional-flow curve at S_{wr}

f_{w1} = fractional flow of water in oil bank preceding viscous or chemical flood saturation discontinuity

f_{w3} = fractional flow of water before stable oil bank forms

f^*_{w3} = fractional flow of viscous water at S^*_{w3}

L = length, L, ft

q_t = total injection rate, L^3/t, B/D

S_{iw} = interstitial water saturation, L^3/L^3, volume fraction

S_{or} = residual oil saturation, L^3/L^3, volume fraction

S_w = water saturation, L^3/L^3, volume fraction

S^*_w = water saturation in region where flowing fluid system is viscous fluid or chemical solution, L^3/L^3, volume fraction

S_{wa} = water saturation on fractional flow curve for resident water corresponding to Pair a

S^*_{wa} = water saturation on viscous fractional-flow curve corresponding to Pair a

S_{wb} = water saturation on fractional-flow curve for resident water corresponding to Pair b

S^*_{wb} = water saturation on viscous fractional-flow curve corresponding to Pair b

S_{wf} = flood-front saturation, L^3/L^3, volume fraction

S_{wr} = water saturation in the water bank at the oil-bank shock, L^3/L^3, volume fraction

$\overline{S}_{wrf}$ = average water saturation between x_r and x_f, fraction

S_{w1} = water saturation in oil bank, L^3/L^3, volume fraction

S_{w3} = water saturation on μ_w fractional-flow curve in developing viscous or chemical shock when chemical flood begins after waterflood was under way, L^3/L^3, volume fraction

S_{w3}^* = water saturation of shock front associated with f_w^*, L^3/L^3, volume fraction

t = time, t, days

t_D = dimensionless time, PVs injected

t_{Do} = dimensionless time when a viscous water-flood or chemical injection starts

t_{Do}^* = dimensionless time when S_w^* evolves from saturation profile

t_{D1} = dimensionless time when S_{w1} evolves from saturation profile

t_o = time when viscous water injection begins

v_{Dr} = specific velocity of front of oil-bank shock

$\bar{v}_{Dr}$ = average specific velocity of oil-bank shock

v_{Dv} = specific velocity of shock for injected water

v_{Dv}^* = specific velocity of shock for viscous water

v_{D1} = specific velocity of water saturation S_{w1}

v_{D3}^* = specific velocity of S_{w3}^*

v_r = velocity of oil-bank shock

$\bar{v}_r$ = average velocity of oil-bank shock

x_D = dimensionless distance in x-direction

x_D^* = location of saturation S_w^* in x-direction, dimensionless

x_{Df} = location of flood front saturation, S_{wf}, dimensionless

x_{Do}^* = location where S_{w3} evolved from saturation profile, dimensionless

x_{Dr} = location of front of oil bank, dimensionless

x_{Dv} = location of saturation discontinuity between viscous water and injected water, dimensionless

x_{D3} = location of viscous-flood-front saturation, S_{w3}^*, dimensionless

x_{D3o}^* = location where S_{w3}^* evolved from saturation profile, dimensionless

x_f = location of flood-front saturation in x-direction, L, ft

x_r = location of front of oil-bank shock (rear of water bank), L, ft

x_3 = location of S_{w3}^*, L, ft

μ_w = viscosity of water, m/Lt, cp

μ_w^* = viscosity of viscous water, m/Lt, cp

ϕ = porosity, fraction

Superscripts

$n,(0),(1),(2)$ = timestep indices

$^-$ = average

SI Metric Conversion Factor

cp $\times 1.0^*$ E–03 = Pa·s

*Conversion factor is exact.

Appendix B

Error-Function Tabulation

Appendix B comprises error-function tabulation in the form of **Table B-1.**

x	erf(x)	x	erf(x)	x	erf(x)	x	erf(x)
0.00	0.00000	0.50	0.52049	1.00	0.84270	1.50	0.96610
0.01	0.01128	0.51	0.52924	1.01	0.84681	1.51	0.96727
0.02	0.02256	0.52	0.53789	1.02	0.85083	1.52	0.96841
0.03	0.03384	0.53	0.54646	1.03	0.85478	1.53	0.96951
0.04	0.04511	0.54	0.55493	1.04	0.85864	1.54	0.97058
0.05	0.05637	0.55	0.56332	1.05	0.86243	1.55	0.97162
0.06	0.06762	0.56	0.57161	1.06	0.86614	1.56	0.97262
0.07	0.07885	0.57	0.57981	1.07	0.86977	1.57	0.97360
0.08	0.09007	0.58	0.58792	1.08	0.87332	1.58	0.97454
0.09	0.10128	0.59	0.59593	1.09	0.87680	1.59	0.97546
0.10	0.11246	0.60	0.60385	1.10	0.88020	1.60	0.97634
0.11	0.12362	0.61	0.61168	1.11	0.88353	1.61	0.97720
0.12	0.13475	0.62	0.61941	1.12	0.88678	1.62	0.97803
0.13	0.14586	0.63	0.62704	1.13	0.88997	1.63	0.97884
0.14	0.15694	0.64	0.63458	1.14	0.89308	1.64	0.97962
0.15	0.16799	0.65	0.64202	1.15	0.89612	1.65	0.98037
0.16	0.17901	0.66	0.64937	1.16	0.89909	1.66	0.98110
0.17	0.18999	0.67	0.65662	1.17	0.90200	1.67	0.98181
0.18	0.20093	0.68	0.66378	1.18	0.90483	1.68	0.98249
0.19	0.21183	0.69	0.67084	1.19	0.90760	1.69	0.98315
0.20	0.22270	0.70	0.67780	1.20	0.91031	1.70	0.98379
0.21	0.23352	0.71	0.68466	1.21	0.91295	1.71	0.98440
0.22	0.24429	0.72	0.69143	1.22	0.91553	1.72	0.98500
0.23	0.25502	0.73	0.69810	1.23	0.91805	1.73	0.98557
0.24	0.26570	0.74	0.70467	1.24	0.92050	1.74	0.98613
0.25	0.27632	0.75	0.71115	1.25	0.92290	1.75	0.98667
0.26	0.28689	0.76	0.71753	1.26	0.92523	1.76	0.98719
0.27	0.29741	0.77	0.72382	1.27	0.92751	1.77	0.98769
0.28	0.30788	0.78	0.73001	1.28	0.92973	1.78	0.98817
0.29	0.31828	0.79	0.73610	1.29	0.93189	1.79	0.98864
0.30	0.32862	0.80	0.74210	1.30	0.93400	1.80	0.98909
0.31	0.33890	0.81	0.74800	1.31	0.93606	1.81	0.98952
0.32	0.34912	0.82	0.75381	1.32	0.93806	1.82	0.98994
0.33	0.35927	0.83	0.75952	1.33	0.94001	1.83	0.99034
0.34	0.36936	0.84	0.76514	1.34	0.94191	1.84	0.99073
0.35	0.37938	0.85	0.77066	1.35	0.94376	1.85	0.99111
0.36	0.38932	0.86	0.77610	1.36	0.94556	1.86	0.99147
0.37	0.39920	0.87	0.78143	1.37	0.94731	1.87	0.99182
0.38	0.40900	0.88	0.78668	1.38	0.94901	1.88	0.99215

Table B-1—Error-function tabulation.

x	erf(x)	x	erf(x)	x	erf(x)	x	erf(x)
0.39	0.41873	0.89	0.79184	1.39	0.95067	1.89	0.99247
0.40	0.42839	0.90	0.79690	1.40	0.95228	1.90	0.99279
0.41	0.43796	0.91	0.80188	1.41	0.95385	1.91	0.99308
0.42	0.44746	0.92	0.80676	1.42	0.95537	1.92	0.99337
0.43	0.45683	0.93	0.81156	1.43	0.95685	1.93	0.99365
0.44	0.46622	0.94	0.81627	1.44	0.95829	1.94	0.99392
0.45	0.47543	0.95	0.82089	1.45	0.95969	1.95	0.99417
0.46	0.48465	0.96	0.82542	1.46	0.96105	1.96	0.99442
0.47	0.49374	0.97	0.82987	1.47	0.96237	1.97	0.99466
0.48	0.50274	0.98	0.83425	1.48	0.96365	1.98	0.99489
0.49	0.51166	0.99	0.83850	1.49	0.96489	1.99	0.99511
–	–	–	–	–	–	2.00	0.99532

Table B-1—Error-function tabulation (continued).

Auxiliary Programs To Compute Location of Polymer Bank When Polymer Slug Is Displaced by Drive Water

Appendix C contains a program named PREAR that computes the location of the rear of the polymer bank when a polymer slug of size D_p is displaced through a linear system by drive water. Included is the situation when the slug is not large enough to satisfy retention on the rock and the polymer slug is overtaken by the drive water. In this case, the program computes the location of the rear of the oil bank after the polymer flood dissipates and the flood becomes a normal waterflood. The method of calculation is described in Chapter 5 and illustrated in Examples 5.8 through 5.10.

At the beginning of water injection, the polymer bank contains all water saturations from $1 - S_{or}$ to S_{w3}^*. As drive water is injected, the water cuts into the rear of the polymer bank, generating two regions. The first region contains drive water with saturations between $1 - S_{or}$ and S_{wr}. The second region contains polymer saturations ranging from S_{w3}^* at the rear of the polymer bank to S_{w3}^*. If the volume of the polymer slug is not sufficient to satisfy retention, the polymer region will be overtaken by the drive water and the oil bank formed by the previous polymer flood contains saturations S_{wr} to S_{w1} as it is displaced through the linear system. The calculation procedure involves a main program, PREAR, and two subprograms, SWR and PROP. Listings of these programs are included in this appendix.

To calculate the movement of the various regions created during a polymer flood, it is necessary to estimate flood-front saturation for a waterflood, S_{wf}, the flood-front saturation for the polymer flood, S_{w3}^*, and the water saturation, S_{w1}, in the oil bank that forms between the flood-front saturation and the polymer flood front. These saturations are used in PREAR to determine the movement of the rear of the polymer bank and the oil bank. Programs are included in this appendix to compute these saturations.

Program SW3S determines the flood-front saturation, S_{w3}^*, for a polymer flood when the effective inaccessible PV, ϕ_e, and the polymer retention factor, D_p, are known. Program SW1FW1 computes the values of S_{w1} and f_{w1} for the oil bank. Program SWF computes the flood-front saturation, S_{wf}, for the waterflood that precedes the polymer flood.

All programs are based on the general form of the relative permeability relationships given by Eqs. C-1 through C-7.

$$k_{ro} = \alpha_1 \left(1 - S_{wD}\right)^m, \dots\dots\dots\dots\dots\dots\dots\dots\dots\dots\dots\dots\dots\dots\dots\dots \text{(C-1)}$$

$$\text{and } k_{rw} = \alpha_2 S_{wD}^n, \dots\dots\dots\dots\dots\dots\dots\dots\dots\dots\dots\dots\dots\dots\dots\dots\dots \text{(C-2)}$$

$$\text{where } S_{wD} = \left(S_w - S_{iw}\right)/\left(1 - S_{or} - S_{iw}\right), \dots\dots\dots\dots\dots\dots\dots\dots\dots\dots\dots \text{(C-3)}$$

$$f_w = S_{wD}^n \Big/ \left[S_{wD}^n + A\left(1 - S_{wD}\right)^m\right], \dots\dots\dots\dots\dots\dots\dots\dots\dots\dots\dots\dots \text{(C-4)}$$

$$\frac{\partial f_w}{\partial S_w} = \frac{AB\left[nS_{wD}^{n-1}\left(1 - S_{wD}\right)^m + mS_{wD}^n\left(1 - S_{wD}\right)^{m-1}\right]}{\left[S_{wD}^n + A\left(1 - S_{wD}\right)^m\right]^2}, \dots\dots\dots\dots\dots\dots \text{(C-5)}$$

$$A = \alpha_1 \mu_w / \alpha_2 \mu_o, \dots\dots\dots\dots\dots\dots\dots\dots\dots\dots\dots\dots\dots\dots\dots\dots\dots\dots \text{(C-6)}$$

$$\text{and } B = 1/\left(1 - S_{or} - S_{iw}\right). \dots\dots\dots\dots\dots\dots\dots\dots\dots\dots\dots\dots\dots\dots\dots \text{(C-7)}$$

Program Descriptions

SW3S. Program SW3S calculates the flood-front saturation, S^*_{w3}, for a polymer flood by determining the intersection of a tangent drawn from the location $(0, -D_s)$ to the fractional flow curve $f^*_w - S^*_w$ based on the apparent viscosity of the polymer solution and the viscosity of the oil at reservoir conditions. The parameter $D_s = D_p - \phi_e$. Data entry is interactive, and **Table C-1** lists the input data for Example 5.8, with the interactive input values in italics.

The input data and computed values of S^*_{w3}, f^*_{w3}, and f'^*_{w3} are saved in a file named POLS.DAT. SW3S also subdivides the saturation interval from $1 - S_{or}$ to S^*_{w3} into 100 increments, computes the fractional flow data corresponding to the 101 saturations in the polymer region, and stores the computed results in a data file GSW2P.DAT in the following order:

Line	S^*_w	f^*_w	f'^*_{wr}
1	S^*_{w3}	f^*_{w3}	f'^*_{w3}
.			
.			
.			
101	$1 - S_{or}$	1.00	0.

File GSW2P.DAT is required as an input file by PREAR. **Program C-1** is a listing of SW3S.FOR.

SW1FW1. Program SW1FW1 calculates the intersection of a tangent drawn from the location $(0, -D_s)$ to the fractional flow curve, f^*_w/S^*_w, with the fractional curve for the waterflood, f_w/S_w. This intersection is f_{w1}, S_{w1}. **Program C-2** is a listing of SW1FW1. The program runs interactively in a similar manner to SW3S. The values of S^*_{w3} and f^*_{w3} determined from SW3S are also read in interactively. Computed values of f_{w1} and S_{w1} are saved in file SWT.OUT for further reference. **Table C-2** contains the data in SWT.OUT for Examples 5.8 through 5.10.

SWF (Willhite 1986). Program SWF calculates the flood-front saturation for a waterflood by finding the tangent drawn from $0, S_{iw}$ to the fractional curve for the waterflood, f_w/S_w. The program runs interactively. Values of S_{wf}, f_{wf}, and f'_{wf} are saved in file SWF.OUT. **Program C-3** is a listing of SWF, and **Table C-3** gives SWF.OUT for Examples 5.8 through 5.10.

PREAR. Program PREAR calculates the location of the rear of the polymer bank, x_{Dr}, by subdividing the saturation interval from $1 - S_{or}$ to S^*_{w3} into 100 increments and computes the location of each saturation S^*_{wr} using Eq. 5.112. The program assumes that saturations and fractional-flow data corresponding to the 101 saturations in the polymer region have been computed and are stored in a data file GSW2P.DAT in the following order:

Initial water saturation = *0.30*
Interstitial water saturation = *0.30*
Retention Factor-DS (fraction) = *0.174*
Residual oil saturation = *0.30*
KRO = ALPHA1*(1 .-SWD)**M
M = *2.00*
ALPHA1 = *0.80*
KRW = ALPHA2*(SWD)**N
N = *2.00*
ALPHA2 = *0.20*
Oil viscosity (cp) = *40.0*
Polymer or water viscosity (cp) = *4.00*
FW3* = 0.948714
SW3* = 0.592478
FPRIME from computed tangent = 1.23776
FPRIME from equation at SW3S = 1.23776
FP3WW2 from Eq. G3P at SW3S = 1.23776
Note: Interactive input values are in italics.

Table C-1—Sample interactive input and output for SW3S.FOR.

Line	S_w^*	f_w^*	$f_{wr}^{*\prime}$
1	$1 - S_{or}$	1.00	0.
.			
.			
.			
101	S_{w3}^*	f_{w3}^*	$f_{w3}^\prime$

This file is prepared by running SW3S before PREAR is executed.

When PREAR is executed, data files PREARI.DAT and GSW2P.DAT are read. The file PREARI.DAT contains the variable ALPHA1,ALPHA2,M,N,SWC,SOR,DS,SWI,VISO,VISW,SWI, FW1,FPSW1,SWF,FWF,FPSWF.

Initial water saturation = *0.30*

Interstitial water saturation = *0.30*

Retention Factor-DS (fraction) = *0.174*

Residual oil saturation = *0.30*

Fractional flow from polymer tangent, FW3* – *0.948714*

Flood front saturation from polymer tangent, SW3* = *0.592478*

KRO = ALPHA1*(1.-SWD)**M

M = *2.00*

ALPHA1= *0.80*

KRW = ALPHA2*(SWD)**N

N = *2.00*

ALPHA2 = 0.20

Mobility ratio based on endpoints of relative permeability curves = 10.0000

FW1 = 0.767291

SW1 = 0.445905

FINT from polymer tangent = 0.767292

FPRIME at SW1 = 3.85298

Note: Interactive input values are in italics.

Table C-2—Sample interactive input and output from SW1FW1.FOR.

Initial water saturation = *0.30*

Connate (interstitial) water saturation = *0.30*

Residual oil saturation = *0.30*

KRO = ALPHA1*(1.-SWD)**M

Exponent (M) of oil relative permeability relationship = *2.00*

ALPHA1 = *0.80*

KRW = ALPHA2*(SWD)**N

Exponent (N) of water relative permeability relationship = *2.00*

ALPHA2 = *0.20*

Oil viscosity (cp) = *40.0*

Water viscosity (cp) =*1.00*

Mobility ratio based on endpoints of relative permeability curves = 10.0000

FWSF = 0.650756

SWF = 0.420605

FPRIME from FW curve = 5.39578

FPSWW2 from analytical expression = 5.39578

Mobility ratio based on average flood-front saturation = 2.14670

Interactive input values are in italics.

Table C-3—Sample interactive input and output from SWF.FOR.*

0.8	0.2	2.0	2.0	0.3	0.3	−0.25	0.3	40.0	1.0
0.4459049		0.7672911		3.852982					
0.4206		0.6508		5.3958					

Table C-4—PREARI.DAT.

Table C-4 contains the data in PREARI.DAT for Example 5.10. Three variables are read in interactively. These are ϕ_e, D_p, and t_{Dp}. **Program C-4** is a listing of PREAR.

In PREAR, the values of x_{Dr} are computed for each saturation in the polymer bank ($1 - S_{or}$ to S^*_{w3}) starting with $1 - S_{or}$, which is always located at $x_{Dr} = 0$ for the relative permeability relationships used in this model. PREAR calls subroutine function program SWR to compute the values of S_{wr} corresponding to S^*_{w3} at the interface between the drive water and the rear of the polymer bank. Subroutine subprogram prop calculates the values of f_w, f'_w, k_{ro}, and k_{rw} when the water saturation, S_w, is specified.

When the value of x_{Dr} corresponding to S^*_{w3} is < 1.0, the polymer bank is overtaken by the drive water and the remainder of the flood is a waterflood. Recall from Chapter 5 that the water saturation in the oil bank at this time is S_{w1} and that there is still a saturation discontinuity at the interface between the drive water and the oil bank.

The second part of PREAR computes the location of the rear of the oil bank as the drive water cuts into the oil bank. The method is based on a material balance that is given in Eq. 5.111. In the program, the saturation interval between the value of S_{wr} when the polymer bank was overtaken and S_{w1} is subdivided into 50 increments. The location of the rear of the oil bank and the time corresponding to this location are computed from material balance, as described in Example 5.10. PREAR uses the same material-balance calculation to check the location of the rear of the polymer slug as it is displaced by the drive water. The locations of the rear of the polymer bank and the rear of the oil bank are written to file PSLUG.OUT. **Table C-5** contains selected results from the PREAR for Example 5.10.

SWR. SWR is a function subprogram that computes the value of S_{wr} and f_{wr} when the values of S^*_{wr} and f^*_{wr} are specified. Saturations in the drive-water region evolve as the drive water cuts into the rear of the polymer bank. That is, as values of S^*_{wr} disappear at the rear of the polymer bank, a corresponding saturation emerges from the drive-water bank denoted as S_{wr}. The water saturation at the interface between the drive-water and polymer regions is determined from the intersection of a line from $(0, -D_s)$ drawn through S^*_{wr}, f^*_{wr} to the fractional-flow curve for the waterflood at f_{wr}, S_{wr}, as discussed in Chapter 5.

SWR is accessed through the FORTRAN statement

SWRS = SWR(FWS,SWS,FWR).

Program C-5 is a listing of SWR.

PROP (Willhite 1986). PROP is a subroutine subprogram that computes k_{ro}, k_{rw}, f_w, and f'_w from generalized relative permeability relations given as Eqs. C-1 through C-7 when S_w and water and oil viscosities are given.

PROP is accessed through the FORTRAN call statement

CALL PROP(VISO,VISW,SW1,KRO1,KRW1,FW1,FPSW1).

t_D	t_{Dm}	x_{Dr}	S^*_{wr}
0.21200	0.21199	0.00000	0.69998
0.21703	0.21701	0.01130	0.69040
0.22372	0.22371	0.02668	0.67993
0.23187	0.23185	0.04585	0.66973
0.24185	0.24184	0.06992	0.65979
0.25423	0.25421	0.10045	0.65016
0.26976	0.26974	0.13965	0.64087
0.28958	0.28956	0.19079	0.63199
0.31542	0.31539	0.25888	0.62358
0.35003	0.35000	0.35197	0.61571
0.39812	0.39809	0.48385	0.60846
0.40381**	0.40381	0.50000	0.59248

IPV = 0.250000, DP = 0.424000, and PVs polymer injected = 0.212000.
*Headings added after run.
**Polymer bank is overtaken by drive water.

Table C-5—Selected results for Example 5.10 from PREAR (file PSLUG.OUT).*

A call to this subroutine gives values for only the value of S_{w1} specified in the calling sequence. Thus, PROP must be called for every saturation where values of k_{ro}, k_{rw}, f_w, and f'_w are desired. The values of α_1, α_2, m, n, S_{wc}, S_{or}, D_s, S_{wi}, μ_o, and μ_w are made available to PROP through the common label PDATA. These variables are read into PREAR. **Program C-6** is a listing of PROP.

Nomenclature

A	=	parameter defined by Eq. C-6
B	=	parameter defined by Eq. C-7
D_p	=	retention factor, dimensionless
D_s	=	$D_p - \phi_e$, dimensionless
f_w	=	fractional flow of water, L^3/L^3, volume fraction
f_w^*	=	fractional flow of water in polymer bank, L^3/L^3, volume fraction
f'_w	=	derivative of f_w with respect to S_w, L^3/L^3, volume fraction
f_{wf}	=	fractional flow of water at S_{wf}, L^3/L^3, volume fraction
f'_{wf}	=	derivative of f_{wf} with respect to S_w at S_{wf}
f_{wr}	=	fractional flow of drive water at rear of polymer bank, L^3/L^3, volume fraction
f_{wr}^*	=	fractional flow of water at rear of polymer bank, L^3/L^3, volume fraction
$f_{wr}^{*'}$	=	derivative of f_{wr}^* with respect to S_w^* at rear of polymer bank
f_{w1}	=	fractional flow of water in at S_{w1}, L^3/L^3, volume fraction
f_{w3}^*	=	fractional flow of water at the polymer-flood front, L^3/L^3, volume fraction
$f_{w3}^{'*}$	=	derivative of fractional flow of water with respect to S_w^* at the polymer-flood front
k_{ro}	=	relative permeability to oil, dimensionless
k_{rw}	=	relative permeability to water, dimensionless
m	=	exponent in relative permeability correlation for oil, dimensionless
n	=	exponent in relative permeability correlation for water, dimensionless

S_{iw}, S_{wc}	=	interstitial water saturation, L^3/L^3, volume fraction
S_{or}	=	residual oil saturation, L^3/L^3, volume fraction
S_w	=	water saturation, L^3/L^3, volume fraction
S_w^*	=	water saturation in polymer region, L^3/L^3, volume fraction
S_{wD}	=	dimensionless water saturation defined by Eq. C-3
S_{wf}	=	flood-front saturation, L^3/L^3, volume fraction
S_{wi}	=	initial water saturation
S_{wr}	=	drive-water saturation at rear of polymer bank, L^3/L^3, volume fraction
S_{wr}^*	=	water saturation at rear of polymer bank, L^3/L^3, volume fraction
S_{w1}	=	water saturation in oil bank, L^3/L^3, volume fraction
S_{w3}^*	=	water saturation at polymer front, L^3/L^3, volume fraction
t_D	=	dimensionless time, PVs injected
t_{Dm}	=	dimensionless time estimated from material balance on total volume of water injected
t_{Dp}	=	PVs of polymer injected
x_{Dr}	=	dimensionless location of S_{wr}
α_1	=	coefficient in Eq. C-1 for oil relative permeability correlation, relative permeability to oil at S_{wi}
α_2	=	coefficient in Eq. C-2 for water relative permeability correlation, relative permeability to water at S_{or}
μ_o	=	viscosity of oil, M/Lt, cp
μ_w	=	viscosity of water, M/Lt, cp
ϕ_e	=	inaccessible fraction of porosity

Reference

Willhite, G. P. 1986. *Waterflooding,* Vol. 3. Richardson, Texas: Textbook Series, SPE.

PROGRAM C-1—SW3S.FOR

```
********************************************************************************************************
*       SW3S.FOR
*       PROGRAM TO COMPUTE TANGENT TO FRACTIONAL FLOW CURVE
*       FOR POLYMER/VISCOUS WATER DISPLACEMENT MODEL
*
*       KRO = ALPHA1*(1.-SWD)**M
*
*       KRW = ALPHA2*SWD**N
*
*       CALCULATES THE FLOOD FRONT SATURATION, SW3S,
*       FRACTIONAL FLOW FW3S,
*       AND SLOPE OF THE FRACTIONAL FLOW CURVE, FPW3S
*       FOR A POLYMER FLOOD WHEN THE RETENTION FACTOR IS DS
*       AND VISP IS THE APPARENT VISCOSITY OF THE POLYMER SOLUTION
*
*       SUBDIVIDES SATURATION INTERVAL BETWEEN 1-SOR AND SWFS
*       INTO 101 INCREMENTS AND CALCULATES FWS AND FPS FOR
*       EACH SATURATION-SWS.
*
*       OUTPUT OF SWS,FWS,FPS SAVED ON FILE GWS2PI.DAT
*       STORED IN ORDER FROM SWFS TO 1-SOR
*       SOR IS ASSUMED TO BE THE SAME FOR POLYMER FLOOD
*       AS FOR WATERFLOOD
*
*       WHEN DS = 0, THE FLOOD FRONT SATURATION FOR A VISCOUS
*       WATERFLOOD IS DETERMINED.. I.E. SWF, FWF.FPWF WITH
*       NO INACCESSIBLE PORE VOLUME AND NO RETENTION


********************************************************************************************************


        REAL KO,KW,M,N,MOBEND
        DIMENSION SW(101),FW(101),FP(101)
        COMMON/PDATA/ALPHA1,ALPHA2,M,N,SIW1,SOR1


********************************************************************************************************


*       FUNCTIONS USED IN CALCULATIONS
*
        Y(SW1) = (SW1-SIW)*B
        KW(Y) = ALPHA2*Y**N
        KO(Y) = ALPHA1*(1.-Y)**M
        FWF(SWD1) = SWD1**N/(SWD1**N +A*(1.-SWD1 )**M)

        B1(SWD1) = N*SWD1**(N-1 )*(1.-SWD1 )**M
        B2(SWD1) = M*SWD1**N*(1.-SWD1 )**(M-1)
        B3(SWD1) = SWD1**N+A*(1.-SWD1 )**M

        C1(SWD1) = N*SWD1**(N-1 ) +A*M*(1.-SWD1)**(M-1)
        C2(SWD1) = ((N-1)/SWD1 +M/(1.-SWD1 ))*B1 (SWD1)
        C3(SWD1) = (M/(1.-SWD1) +M*(M-1)*SWD1/(N*(1.-SWD1)**2))*B1(SWD1)

        G1(SWD1) = (FWF(SWD1)-FWF(SWDI))/(SWD1-SWDI)
        G2(SWD1) = A*(B1(SWD1) +B2(SWD1))/B3(SWD1)**2
        G1S(SWD1) = FWF(SWD1)/(SWD1 +SWO)
        G3(SWD1) = G1S(SWD1)-G2(SWD1)

        G1SP(SWD1) = (G2(SWD1)-G1S(SWD1))/(SWD1 +SWO)
        G1P(SWD1) = (G2(SWD1)-G1(SWD1))/(SWD1-SWDI)
        G2P(SWD1) = -2.*G2(SWD1)*C1(SWD1)/B3(SWD1)
      : + A*B1(SWD1)*(C2(SWD1)+C3(SWD1))/B3(SWD1)**2
        G3P(SWD1) = G1SP(SWD1)-G2(SWD1)
        SWF(SWD1) = SWD1/B + SIW
```

```
***********************************************************************************************************
*          OPEN FILES AND READ INPUT DATA INTERACTIVELY
*
           OPEN(8,FILE = 'POLS. DAT')
           OPEN(9,FILE = 'GSW2P.DAT')
*
           PRINT*,' INITIAL WATER SATURATION'
           READ*, SWI
           WRITE(8,*)'INITIAL WATER SATURATION = ',SWI
           PRINT*,'INTERSTITIAL WATER SATURATION'
           READ*,SIW
           WRITE(8,*)'INTERSTITIAL WATER SATURATION = ',SIW
*          SIW < = SWI
C
           PRINT*,'RETENTION FACTOR–DS, FRACTION'
           READ*, DS
           WRITE(8,*) 'RETENTION FACTOR–DS, FRACTION = ',DS
           PRINT*,'RESIDUAL OIL SATURATION'
           READ*, SOR
           WRITE(8,*)'RESIDUAL OIL SATURATION =', SOR


***********************************************************************************************************

C2345678901234567890123456789012345678901234567890123456789012345678901234567890123456789012
           PRINT*,'KRO = ALPHA1*(1.–SWD)**M'
           PRINT*,'M = '
           READ*,M
           WRITE(8,*)'KRO = ALPHA1*(1.–SWD)**M'
           WRITE(8,*)'M = ',M
           PRINT*,'SPECIFY ALPHA1'
           READ*,ALPHA1
           WRITE(8,*)'ALPHA1 = ',ALPHA1
C
           PRINT*,'KRW=ALPHA2*(SWD)**N'
           PRINT*,'N = '
           WRITE(8,*)'KRW = ALPHA2*(SWD)**N'
           READ*, N
           WRITE(8,*)'N =',N

           PRINT*,'ALPHA2 ='
           READ',ALPHA2
           WRITE(8,*)'ALPHA2 =',ALPHA2
C2345678901234567890123456789012345678901234567890123456789012345678901234567890123456789012
           PRINT*,'OIL VISCOSITY(CP)'
           READ*,VISO
           WRITE(8,*)'OIL VISCOSITY(CP) = ',VISO
C
           PRINT*,'POLYMER OR WATER VISCOSITY(CP) = '
           READ*,VISP
           WRITE(8,*)'POLYMER OR WATER VISCOSITY(CP) =',VISP
C

***********************************************************************************************************
*          INITIALIZE CONSTANTS
*
           A = ALPHA1*VISP/(ALPHA2*VISO)
           B = 1./(1.–SOR–SIW)
           SWO = B*(DS+SIW)
           SWDI = Y(SWI)
*          SWDF = SQRT(A/(1.+A))

***********************************************************************************************************

*          COMPUTE ENDPOINT MOBILITY RATIO
*
           MOBEND = 1./A
           PRINT*,'MOBILITY RATIO BASED ON ENDPOINTS OF RELATIVE',
```

```
                :' PERMEABILITY CURVES = ',MOBEND
*
*
*               COMPUTE BREAKTHROUGH SATURATION FOR GENERAL
*               FORM OF RELATIVE PERMEABILITY CURVES BY NEWTONIAN ITERATION
*               AND INTERVAL HALVING
*
                X1=SWDI
                SDMAX = Y(1.-SOR)
                DSD = (SDMAX-SWDI)/49
                PRINT*,'SDMAX = ',SDMAX,' DELSD = ',DSD
                SWX = X1+DSD
                G3X = G3(SWX)
                DO 10 I = 1,100
                SWX = SWX+DSD
                G3XP1 = G3(SWX)
C               PRINT*,SWX,G3X,G3XP1
                IF((G3X.LT.0.) .AND.G3XP1.GT.0.) GO TO 90
                G3X = G3XP1
                GO TO 10
        90      SWX = SWX-DSD
                DSD = DSD/2.
                IF(ABS(G3X) .LE.1.E-06) GO TO 20
        10      CONTINUE
                PRINT*,'N0 CONVERGENCE IN 50 ITERATIONS'
                STOP
        20      SWDFS = SWX
*
                PRINT*,'SWDFS = ', SWDFS,' FOR MOBILITY RATIO = ',MOBEND
                FW3S = FWF(SWDFS)
C
                PRINT*,'FW3* =',FW3S
                WRITE(8,*)'FW3* = ', FW3S
                SW3S = SWF(SWDFS)

                PRINT*,'SW3* = ', SW3S
                WRITE(8,*)'SW3* = ', SW3S
*
*               CHECK SLOPE OF TANGENT TO FRACTIONAL FLOW CURVE
*               WITH EXPRESSIONS BASED ON FRACTIONAL FLOW
*
                FPRIME = FW3S/(SW3S+DS)
                PRINT*,'FPRIME FROM COMPUTED TANGENT = ',FPRIME
                WRITE(8,*)'FPRIME FROM COMPUTED TANGENT = ',FPRIME
                FP3S = B*G2(SWDFS)
                PRINT*,'FPRIME FROM EQUATION AT SW3S = ',FP3S
                WRITE(8,*)'FPRIME FROM EQUATION AT SW3S = ',FP3S
                FP3WW2 = B*G1S(SWDFS)
                PRINT*,'FP3WW2 FROM EQUATION G3P AT SW3S = ',FP3WW2
                WRITE(8,*)'FP3WW2 FROM EQUATION G3P AT SW3S = ',FP3WW2
*               END OF FLOODFRONT CALCULATIONS

***************************************************************************************************************************
C               GENERATE FRACTIONAL FLOW DATA FOR BREAKTHROUGH SATURATION
C               THROUGH 1-SOR
C
                DELS = (1-SOR-SW3S)/100.
                SW(1) =SW3S
                FP(1) = FPRIME
                FW(1) = FW3S
                SWB = SW3S
*
                DO 30 I = 2,101
                SWB = SWB+DELS
                SWD = Y(SWB)
                SW(I) = SWF(SWD)
                FW(I) = FWF(SWD)
                FP(I) = B*G2(SWD)
        30      CONTINUE
*
*               OUTPUT ON FILE8 FOR SUBSEQUENT CALCULATIONS
*
```

```
        WRITE(9,43)(SW(I),FW(I),FP(I),I = 1,101)
43      FORMAT(",F8.6,2X,F8.6,2X,F10.7)
        STOP
        END
```

PROGRAM C-2—SW1FW1.FOR

```
*********************************************************************************************
*       SW1FW1.FOR
*
*       PROGRAM TO COMPUTE INTERSECTION OF TANGENT TO
*       VISCOUS/POLYMER FRACTIONAL FLOW CURVE WITH THE
*       FRACTIONAL FLOW CURVE FOR A NORMAL WATERFLOOD
*       THE INTERSECTION IS(FW1,SW1) ON THE FRACTIONAL
*       FLOW CURVE FOR WATER
*
*********************************************************************************************
*       RELATIVE PERMEABILITY DATA ARE CORRELATED IN THE
*       FOLLOWING FORM FOR BOTH WATERFLOOD AND POLYMER FLOOD
*       KRO = ALPHA1*(1.–SWD)**M
*
*       KRW = ALPHA2*SWD**N
*
        REAL KO,KW,M,N,MOBEND
*234567
        COMMON/PDATA/ALPHA1,ALPHA2, M, N, SIW1, SOR1
        DATA IOUT1/18/

*********************************************************************************************
*       DEFINE STATEMENT FUNCTIONS
*
        Y(SW1) = (SW1–SIW)*B
        KW(Y) = ALPHA2*Y**N
        KO(Y) = ALPHA1*(1.–Y)**M
        FWF(SWD1) = SWD1**N/(SWD1**N +A*(1.–SWD1 )**M)
        FWS(SWD1) = A1*(DS+SIW+SWD1/B)
        B1(SWD1) = N*SWD1**(N–1)*(1.–SWD1)**M
        B2(SWD1) = M*SWD1**N*(1.–SWD1)**(M–1)
        B3(SWD1) = SWD1**N+A*(1.–SWD1)**M

        C1(SWD1) = N*SWD1**(N–1) +A*M*(1.–SWD1 )**(M–1)
        C2(SWD1) = ((N–1)/SWD1 +M/(1.–SWD1))*B1(SWD1)
        C3(SWD1) = (M/(1.–SWD1 ) +M*(M–1)*SWD1/(N*(1 –SWD1)**2))*B1(SWD1)

        H3(SWD1) = FWS(SWD1)–FWF(SWD1)
        G1(SWD1) = (FWF(SWD1)–FWF(SWDI))/(SWD1–SWDI)
        G2(SWD1) = A*(B1(SWD1)+B2(SWD1))/B3(SWD1)**2

        SWF(SWD1) = SWD1/B+SIW

*********************************************************************************************
*       READ INPUT DATA INTERACTIVELY AND SAVE ON FILE SWT.OUT
*
        OPEN(IOUT1, FILE = 'SWT.OUT')
        PRINT*,'INITIAL WATER SATURATION'
        READ*, SWI
        WRITE(IOUT1 ,*)'INITIAL WATER SATURATION = ',SWI
        PRINT*,'INTERSTITIAL WATER SATURATION'
        READ*, SIW
        WRITE(IOUT1 ,*)'INTERSTITIAL WATER SATURATION = ',SIW
*2345678790
        PRINT*,'RETENTION FACTOR–DS, FRACTION'
        READ*, DS
        WRITE(IOUT1,*)'RETENTION FACTOR–DS, FRACTION = ',DS
```

```
          PRINT*,'RESIDUAL OIL SATURATION'
          READ*,SOR
          WRITE(IOUT1 ,*)'RESIDUAL OIL SATURATION = ',SOR
          B = 1./(1.-SOR-SIW)
          SWO = B*(DS+SIW)
          PRINT*,'FRACTIONAL FLOW FROM POLYMER TANGENT-FW3* = '
          READ*, FW3S
          WRITE(IOUT1,*)'FRACTIONAL FLOW FROM POLYMER TANGENT-FW3*',
         :' = ',FW3S
          PRINT*,'FLOOD FRONT SATURATION FROM POLYMER TANGENT-SW3*'
          READ*, SW3S
          WRITE(IOUT1,*)'FLOOD FRONT SATURATION FROM POLYMER',
         : ' TANGENT-SW3* = ',SW3S
          A1 = FW3S/(DS+SW3S)
*
          SWDI = Y(SWI)
*
          PRINT*,'KRO = ALPHA1*(1.-SWD)**M'
          PRINT*,'SPECIFY M'
          READ*, M
          PRINT*,'SPECIFY ALPHA1'
          READ*,ALPHA1
*234567
          PRINT*,'KRW = ALPHA2*(SWD)**N'
          PRINT*,'SPECIFY N'
          READ*, N
          PRINT*,'SPECIFY ALPHA2'
          READ*,ALPHA2
*
          PRINT*,'OIL VISCOSITY IN CP'
          READ*,VISO
*
          PRINT*,'WATER VISCOSITY IN CP'
          READ*,VISW
*
          WRITE(IOUT1 ,*)'KRO = ALPHA1*(1.-SWD)**M'
          WRITE(IOUT1,*)'M = ', M
          WRITE(IOUT1,*)'ALPHA1 =',ALPHA1
*234567
          WRITE(IOUT1 ,*)'KRW = ALPHA2*(SWD)**N'
          WRITE(IOUT1,*)'N = ',N
          WRITE(IOUT1 ,*)'ALPHA2 = ',ALPHA2
*
          PRINT*,'OIL VISCOSITY IN CP =,'VISO
          PRINT*,'WATER VISCOSITY IN CP = ,'VISW
*
*         A = ALPHA1*VISW/(ALPHA2*VISO)
          SWDF = SQRT(A/(1+A))
          MOBEND = 1./A
          PRINT*,'MOBILITY RATIO BASED ON ENDPOINTS OF',
         :' RELATIVE PERMEABILITY CURVES = ', MOBEND

****************************************************************************************************************
*         COMPUTE INTERSECTION OF TANGENT FROM(0,-DS) TO FW*
*         WITH FRACTIONAL FLOW CURVE FOR WATER AT(FW1,SW1)
*         BY FINDING THE INTERSECTION OF A LINE
*         FROM(0,-DS) TO(FW3S,SW3S) WITH FW(SW) FOR THE WATERFLOOD
*
*         INTERSECTION IS FOUND BY INTERVAL HALVING
*234567
          X1 = SWDI
          SDMAX = Y(1.-SOR)
          DSD = (SDMAX-SWDI)/49
          SWX = X1+DSD
          H3X = H3(SWX)
          DO 10 1 = 1,100
          SWX = SWX+DSD
          H3XP1 = H3(SWX)
*         PRINT*, SWX, H3X, H3XP1
          IF((H3X.GT.0.).AND.H3XP1.LT.0.) GO TO 90
```

```
         H3X = H3XP1
         GO TO 10
90       SWX = SWX–DSD
         DSD = DSD/2.
         IF(ABS(H3X).LE.1.E–06) GO TO 20
10       CONTINUE
         PRINT*,'NO CONVERGENCE IN 50 ITERATIONS'
         STOP
20       SWDF = SWX
         PRINT*,'MOBILITY RATIO BASED ON ENDPOINTS OF RELATIVE',
         :' PERMEABILITY CURVES = ',MOBEND
         FW1 = FWF(SWDF)
*
         PRINT*,'FW1 = ', FW1SW1 = SWF(SWDF)
         PRINT*,'SW1 = ',SW1
*****************************************************************************************************
*        CHECK FOR CONSISTENCY IN VALUES OF FW1 ,SW1
         FWINT = FWS(SWDF)
         PRINT*,'FWINT FROM POLYMER TANGENT, FWINT
         FP1 = B*G2(SWDF)
         PRINT*,'FP1 =', FP1
         WRITE(IOUT1,*)'MOBILITY RATIO BASED ON ENDPOINTS OF RELATIVE',
         :' PERMEABILITY CURVES =',MOBEND
         WRITE(IOUT1,*)'FW1 = ', FW1
         WRITE(IOUT1,*)'SW1 = ', SW1
         WRITE(IOUT1,*)'FINT FROM POLYMER TANGENT, FWINT
         WRITE(IOUT1,*)'FPRIME @ SW1 = ', FP1
         STOP
         END
```

PROGRAM C-3—SWF.FOR
****

```
C        SWF.FOR
C        PROGRAM TO COMPUTE FRACTIONAL FLOW INFORMATION FOR
C        RELATIVE PERMEABILITY CURVES INCLUDING MOBILE CONNATE WATER
C
         KRO = ALPHA*(1-SWD)**M
C
         KRW = ALPHA2*SWD**N
C
         COMMON/PDATA/ALPHA1,ALPHA2,M, N, SWC1, SOR1
         REAL KO,KW,MOCGM,MOBEND,M,N

*****************************************************************************************************
*        DEFINE FUNCTIONS
*
         Y(SW1) = (SW1–SWC)*BKW(Y) = ALPHA2*Y**N
         KO(Y) = ALPHA1*(1.–Y)**M
         FWF(SWD1) = SWD1**N/(SWD1**N+A*(1.–SWD1)**M)
         SWF1(SWD1) = SWD1/B+SWC

         B1(SWD1) = N*SWD1**(N–1)*(1.–SWD1)**M
         B2(SWD1) = M*SWD1**N*(1.–SWD1)**(M–1)
         B3(SWD1) = SWD1**N+A*(1.–SWD1)**M

         C1(SWD1) = N*SWD1**(N–1)+A*M*(1.–SWD1)**(M–1)
         C2(SWD1) = ((N–1)/SWD1 +M/(1.–SWD1))*B1(SWD1)
         C3(SWD1) = (M/(1.–SWD1)+M*(M–1)*SWD1/(N*(1.–SWD1)**2))*B1(SWD1)

         G1(SWD1) = (FWF(SWD1)–FWF(SWDI))/(SWD1–SWDI)
         G2(SWD1) = A*(B1(SWD1)+B2(SWD1))/B3(SWD1)**2
         G3(SWD1) = G1(SWD1)–G2(SWD1)

         G1 P(SWD1) = (G2(SWD1)–G1(SWD1))/(SWD1–SWDI)
```

```
          G2P1(SWD1) = -2.*G2(SWD1)*C1(SWD1)/B3(SWD1)
          G2P(SWD1) = G2P1(SWD1) +A*B1(SWD1)*(C2(SWD1)+C3(SWD1))/B3(SWD1)**2
          G3P(SWD1) = G1P(SWD1)-G2P(SWD1)
***************************************************************************************************************
*         DATA INPUT IS INTERACTIVE
*         INPUT DATA AND COMPUTED RESULTS WRITTEN TO FILE SWF OUT
*
          OPEN(9,FILE = 'SWF.OUT)
          WRITE(*,*)'INITIAL WATER SATURATION = '
          READ(*,*) SWI
          WRITE(9,*)'INITIAL WATER SATURATION = ',SWI
*
          WRITE(*,*)'CONNATE(INTERSTITIAL) WATER SATURATION = '
          READ(*,*) SWC
          WRITE(9,*)'CONNATE(INTERSTITIAL) WATER SATURATION = ', SWC
*
          WRITE(*,*)'RESIDUAL OIL SATURATION = '
          READ(*,*) SOR
          WRITE(9,*)'RESIDUAL OIL SATURATION = ',SOR

          B = 1./(1.-SOR-SWC)
          SWDI = Y(SWI)
C
C
C
          WRITE(*,*)'KRO = ALPHA1*(1.-SWD)**M'
          WRITE(9,*)'KRO = ALPHA1*(1.-SWD)**M'
C
          WRITE(*,*)'M = '
          READ(*,*) M
          WRITE(9,*)'M = '
          WRITE(*,*)'ALPHA1 = '
          READ(*,*) ALPHA1
          WRITE(9,*)'ALPHA1 = ', ALPHA1
C
C
          WRITE(*,*)'KRW = ALPHA2*(SWD)**N'
          WRITE(9,*)'KRW = ALPHA2*(SWD)**N'

          WRITE(*,*)'N = '
          READ(*,*) N
          WRITE(9,*)'N = ',N
C
          WRITE(*,*)'ALPHA2 = '
          READ(*,*) ALPHA2
          WRITE(9,*)'ALPHA2 = ,'ALPHA2
C
C
C
          WRITE(*,*)'OIL VISCOSITY(CP) = '
          READ(*,*) VISO
          WRITE(9,*)'OIL VISCOSITY(CP) = ,'VISO
C
          WRITE(*,*)'WATER VISCOSITY(CP) ='
          READ(*,*) VISW
          WRITE(9,*)'WATER VISCOSITY(CP) =,'VISW
*
*         END OF DATA INPUT

***************************************************************************************************************
          A = ALPHA1*VISW/(ALPHA2*VISO)
          SWDF = SQRT(A/(1+A))
          MOBEND = 1./A
          WRITE(*,*)'MOBILITY RATIO BASED ON ENDPOINTS OF RELATIVE',
```

```
     &' PERMEABILITY CURVES = ', MOBEND
     WRITE(9,*)'MOBILITY RATIO BASED ON ENDPOINTS OF RELATIVE',
     &' PERMEABILITY CURVES = ', MOBEND
C
C        COMPUTE BREAKTHROUGH SATURATION FOR GENERAL
C        FORM OF RELATIVE PERMEABILITY CURVES BY NEWTONIAN ITERATION
C        USING INTERVAL HALVING
C
         X1 = SWDI
         SDMAX = Y(1.-SOR)
         DSD = (SDMAX-SWDI)/49
C        WRITE(*,*)'SDMAX = '.SDMAX.'DELSD = ',DSD
         SWX = X1+DSD
         G3X = G3(SWX)
         DO 10 I = 1,100
         SWX = SWX+DSD
         G3XP1 = G3(SWX)
C        WRITE(*,*) SWX.G3X.G3XP1
         IF((G3X.LT.0.).AND.G3XP1.GT.0.) GO TO 90
         G3X = G3XP1
         GO TO 10
90       SWX = SWX-DSD
         DSD = DSD/2.
         IF(ABS(G3X).LE.1.e-06) GO TO 20
10       CONTINUE
         WRITE(*,*)'NO CONVERGENCE IN 100 ITERATIONS'
         STOP
20       SWDF = SWX

*************************************************************************************************************************

*        OUTPUT OF COMPUTED RESULTS
*
         FWSF = FWF(SWDF)
         WRITE(*,*)'FWSF = ,' FWSF
         WRITE(9,*)'FWSF = ', FWSF
         SWF = SWF1(SWDF)
         WRITE(*,*)'SWF = ', SWF
         WRITE(9,*)'SWF = ',SWF

         FSWF = FWF(SWDF)

         FPSWF = B*G2(SWDF)
         WRITE(*,*)'FPRIME FROM FW CURVE = ',FPSWF
         WRITE(9,*)'FPRIME FROM FW CURVE = ',FPSWF

         FPSWW2 = B*G1(SWDF)
         WRITE(*,*)'FPSWW2 FROM ANALYTICAL EXPRESSION = ',FPSWW2
         WRITE(9,*)'FPSWW2 FROM ANALYTICAL EXPRESSION = ',FPSWW2
C
         SOR1 = SOR
         SWC1 = SWC
C
C        CALCULATE MOBILITY RATIO BASED ON SWF
C
         SBWF = SWF+(1.-FSWF)/FPSWF
         MOCGM = KW(Y(SBWF))*VISO/(KO(Y(SWC))*VISW)
         WRITE(*,*)'MOBILITY RATIO BASED ON',
         &'AVERAGE FLOOD FRONT SATURATION =',MOCGM
         WRITE(9,*)'MOBILITY RATIO BASED ON',
         &'AVERAGE FLOOD FRONT SATURATION = ',MOCGM
         STOP
         END
```

PROGRAM C-4—PREAR.FOR

```
**********************************************************************************************
*       PREAR. FOR
*       PROGRAM TO COMPUTE LOCATION OF REAR OF THE POLYMER
*       BANK BY MATERIAL BALANCE WHEN A POLYMER SLUG EQUAL
*       TO TDP PORE VOLUMES IS FOLLOWED BY DRIVE WATER
*       TDP IS LESS THAN DP
*
        REAL IPV,M,N,KR01,KRW1
        DATA INFILE, IDFILE, IPLOT/8, 9,10/
        DIMENSION SW(151),SW3AVG(151),TD(151),XDPR(151),FP(151)
      : ,FW(151),SWRS(151),FWR(151),FPR(151),FPWR(151 ),SBAR1 (151)
      :,VDR(151),VDRBAR(151)
      : ,SBARP(151),SBARW1 (151),SBAR1 F(151),SBARIW(151),TDM(151)
        COMMON/PDATA/ALPHA1,ALPHA2,M,N, SWC, SOR, DS, SWI,VISO,VISW
*
        XDM(J1,IT) = XDPR(J1)+(TD(IT)−TD(J1))*FPWR(J1)
*
        OPEN(INFILE, FILE = 'GSW2PI.DAT', STATUS = 'OLD')
        OPEN(IDFILE, FILE = 'PREARI.DAT, STATUS = 'OLD')
        OPEN(IPLOT, FILE = ' PSLUG.DAT')
*
**********************************************************************************************
*       READ PARAMETERS FOR RELATIVE PERMEABILITY RELATIONSHIPS,      *
*       FLUID PROPERTIES AND OTHER PARAMETERS FOR POLYMER FLOOD        *
**********************************************************************************************
*
        READ(IDFILE,*) ALPHA1 ,ALPHA2,M,N,SWC,SOR,DS,SWI,VISO,VISW
      :,SW1, FW1, FPSW1, SWF, FWF, FPSWF
*
*       KRO = ALPHA1*(1.−SWD)**M
*
*       KRW = ALPHA2*SWD**N
*
**********************************************************************************************
*       ALPHA1 = COEFFICIENT OF OIL RELATIVE PERMEABILITY             *
*       CORRELATION                                                   *
*       ALPHA2 = COEFFICIENT OF WATER RELATIVE PERMEABILITY           *
*       CORRELATION                                                   *
*       M = EXPONENT IN OIL RELATIVE PERMEABILITY                     *
*       CORRELATION                                                   *
*       N = EXPONENT IN WATER RELATIVE PERMEABILITY                   *
*       CORRELATION                                                   *
*       SWD = (SW−SWC)/(1−SOR−SWC)                                    *
*       SWC = INTERSTITAL WATER SATURATION                           *
*       SOR = RESIDUAL OIL SATURATION                                *
*       −DS = COORDINATE ON X AXIS(FW = 0)FOR CONSTRUCTION           *
*       OF A LINE FROM(−DS,0) THROUGH FWS.SWS THAT                    *
*       INTERSECTS THE FRACTIONAL FLOW CURVE FOR A                    *
*       NORMAL WATERFLOOD AT SWR.FWR                                  *
*       VISO = OIL VISCOSITY, CP                                      *
*       VISW = WATER VISCOSITY, CP                                    *
*       SW1 = WATER SATURATION IN OIL BANK                            *
*       FW1 = FRACTIONAL FLOW OF WATER IN OIL BANK                    *
*       FPSW1 = DERIVATIVE OF THE FRACTIONAL FLOW CURVE FOR           *
*       THE NORMAL WATERFLOOD AT SW1                                  *
*       SWF = FLOOD FRONT SATURATION                                  *
*       FWF = FRACTIONAL FLOW OF WATER AT FLOOD FRONT                 *
*       SATURATION                                                    *
*       FPSWF = DERIVATIVE OF THE FRACTIONAL FLOW CURVE FOR           *
*       THE NORMAL WATERFLOOD AT SWF                                  *
*
**********************************************************************************************
```

```
*
        SIW = SWC
        SBARWF1 = (FPSWF*SWF–FPSW1*SW1–FWF+FW1)/(FPSWF–FPSW1)
*
C2345678901234567890123456789012345678901234567890123456789012345
*
*
**************************************************************************************************
*       READ FRACTIONAL FLOW DATA FOR THE POLYMER BANK              *
*       FROM BREAKTHROUGH SATURATION(SW3*) TO 1–SOR                 *
*       PREVIOUSLY COMPUTED AND STORED IN FILE GSW2PI.DAT           *
*       STORE IN REVERSE ORDER OF INPUT DATA FILE-I.E.              *
*       1–SOR TO SW3*                                               *
*
**************************************************************************************************
*
        READ(INFILE,43)(SW(102–I),FW(102–I),FP(102–I),I = 1,101)
    43  FORMAT(F8.6,2X,F8.6,2X,F10.7)
*
*       INPUT DATA SPECIFIC TO THE POLYMER SLUG
*       WHEN IT WAS INJECTED AND IPV
*
*       IPV = INACCESSIBLE PORE VOLUME
*       DP = POLYMER RETENTION FACTOR
*       TDP = PORE VOLUMES OF POLYMER INJECTED IN A SLUG

        PRINT*,'SPECIFY VALUE OF IPV
        READ*,IPV
        WRITE(IPLOT,*)IPV = ',IPV
        PRINT*,'SPECIFY VALUE OF DP'
        READ*,DP
        WRITE(IPLOT,*)'DP = ',DP
        PRINT*,'SPECIFY PORE VOLUMES OF POLYMER INJECTED, TDP'
        READ*,TDP
        WRITE(IPLOT,*)'PORE VOLUMES OF POLYMER INJ = ',TDP
*
C2345678901234567890123456789012345678901234567890123456789012345
*
        ICOUNT = 0

**************************************************************************************************
*       POLYMER FLOOD SECTION
**************************************************************************************************
        DO 10 I = 1,100
        SW3AVG(I) = (FP(101)*SW(101)—FP(I)*SW(I)–FW(101)+FW(I)) /
       :(FP(101)–FP(I))
        TD(I) = TDP/(FP(101)*(DP+SW3AVG(I)–IPV)–FP(I)
       :*(SW3AVG(I)–IPV))
        XDPR(I) = FP(I)*TD(I)
        SWRS(I) = SWR(FW(I),SW(I),FWR(I))
        FPR(I) = FW(I)/(SW(I)–IPV)
    10  CONTINUE
*
*
        XDPR(101) = TDP/DP
        TD(101) = XDPR(101)/FP(101)
        SW3AVG(101) = SW(101)
        SWRS(101) = SW(101)
        CALL PROP(VISO,VISW,SWRS(101),KRO1,KRW1,FWR(101),FPWR(101))
        FPR(101) = FWR(101)/SWRS(101)
        VDR(101) = (FWR(101)–FW1 )/(SWRS(101)–SW1)

**************************************************************************************************
*       CHECK ON MATERIAL BALANCE OF WATER BEFORE POLYMER FRONT
*       IS OVERTAKEN
*       TDM(IT) = DIMENSIONLESS TIME COMPUTED FROM MATERIAL
*       BALANCE–POLYMER FRONT IS OVERTAKEN IN THE LINEAR
*       SYSTEM FOR TDP μ DP
```

```
********************************************************************************
*
      DO 20 1 = 1,100
  20  CALL PROP(VISO,VISW,SWRS(I),KRO1,KRW1,FWR(I),FPWR(I))
      DO 39 IT = 1,101
      SBAR1(IT) = 0.
      DO 38 J = 1,IT-1
      SBAR = 0.5*(SWRS(J) +SWRS(J+1))
      DELXDM = XDM(J+1, IT)-XDM(J, IT)
  38  SBAR1(IT) = SBAR1(IT)+SBAR*(XDM(J+1,IT)-XDM(J,IT))
      SBARP(IT) = SW3AVG(IT)*TD(IT)*(FP(101)-FP(IT))
      SBARW1(IT) = SW1*TD(IT)*(FPSW1-FP(101))
      SBAR1F(IT) = SBARWF1*TD(IT)*(FPSWF-FPSW1)
      SBARIW(IT) = SIW*TD(IT)*FPSWF
      TDM(IT) = SBAR1(IT)+SBARP(IT)+SBARW1(IT)+SBAR1F(IT)
      :-SBARIW(IT)

  39  CONTINUE
*


********************************************************************************

*     END MATERIAL BALANCE SECTION BEFORE POLYMER FRONT IS            *
*     OVERTAKEN                                                        *

********************************************************************************

*
*     POLYMER FLOOD FRONT IS OVERTAKEN BY DRIVE WATER
*     IF XDPR(101)< 1.0
*

********************************************************************************

      IF(XDPR(101).GE. 1)THEN
      PRINT*,'PR0GRAM TERMINATED AFTER POLYMER SECTION'
      PRINT*,'AFTER STATEMENT 39 BECAUSE TDP > = DP'
      STOP
      ELSE

********************************************************************************

*     POLYMER BANK IS OVERTAKEN BY DRIVE WATER                        *
*     CALCULATE PROPERTIES AT END OF POLYMER REGION                   *
*     THESE ARE STARTING VALUES FOR BEGINNING OF OIL BANK             *
*     CALCULATIONS                                                    *

********************************************************************************

      CALL PROP(VISO,VISW,SWRS(101),KRO1,KRW1,FWR(101),FPWR(101))
      FPR(101) = FWR(101)/SWRS(101)
      VDR(101) = (FWR(101)-FW1)/(SWRS(101)-SW1)
      ENDIF

********************************************************************************

*     COMPUTE SPECIFIC VELOCITY OF THE REAR OF THE OIL BANK           *
*     FOR EACH TIME INCREMENT                                         *

********************************************************************************

*
*
*     COMPUTE TIME AND POSITION OF REAR OF OIL BANK AFTER            *
*     DRIVE WATER OVERTAKES THE POLYMER FRONT                         *
*     SMALL MATERIAL BALANCE ERRORS DUE TO TRUNCATION                 *
*     SET TD(101) = TDM(101)                                          *

********************************************************************************

*

********************************************************************************

*     DETERMINE SATURATIONS AND FRACTIONAL PROPERTIES AFTER           *
*     POLYMER BANK IS OVERTAKEN BY DRIVE WATER                        *
*     SATURATIONS SWRS(101)—> TO SW1 EVOLVE AFTER THE POLYMER         *
```

```
*          BANK IS OVERTAKEN—>SATURATION INTERVAL BETWEEN SWRS(101)          *
*          AND SW1 IS SUBDIVIDED INTO 50 INCREMENTS                          *

*****************************************************************************************************

           DELS = (SWRS(101)–SW1)/50.
*
           DO 30 1 = 102,151
           SWRS(I) = SWRS(I–1)–DELS
           CALL PROP(VISO,VISW,SWRS(I),KRO1,KRW1,FWR(I),FPWR(I))
     30    VDR(I) = (FWR(I)–FW1)/(SWRS(I)–SW1)
           TD(101) = TDM(101)
           PRINT*,XDPR(101)
           DELTD = TDM(101)–TDM(100)
           DO 50 IT = 102,151
           TDM(IT) = 0.
           VDRBAR(IT) = 0.5*(VDR(IT)+VDR(IT–1))
           IF(IT.GT.102) DELTD = TD(IT–1)–TD(IT–2)
           TD(IT)=TD(IT–1)+DELTD
           ICOUNT = 0.
     46    IF(TDM(IT).EQ.O.) GO TO 47
           TD(IT) = TD(IT)+0.5*(TD(IT)–TDM(IT))
     47    CONTINUE
           XDPR(IT) = XDPR(IT–1)+VDRBAR(IT)*(TD(IT)–TD(IT–1))
           SBAR1(IT) = 0.
           DO 49 J = 1,IT–1
           SBAR = 0.5*(SWRS(J) +SWRS(J+1))
     49    SBAR1(IT) = SBAR1(IT)+SBAR*(XDM(J+1,IT)–XDM(J,IT))
           SBARW1(IT) = SW1*(TD(IT)*FPSW1–XDPR(IT))
           SBAR1F(IT) = SBARWF1*TD(IT)*(FPSWF–FPSW1)
           SBARIW(IT) = SIW*TD(IT)*FPSWF
           TDM(IT) = SBAR1(IT)+SBARW1(IT)+SBAR1F(IT)–SBARIW(IT)
*
*          CONVERGENCE CHECK ON MATERIAL BALANCE
*
           ICOUNT = ICOUNT+1
           PRINT*,ICOUNT,' = ICOUNT'
           IF(ICOUNT .GE. 100) GO TO 50
           IF(ABS(TD(IT)–TDM(IT)).LE.1.5E–04) GO TO 60
           GO TO 46
     60    PRINT*,'IT = ',IT,' ICOUNT AT CONVERGENCE = ',ICOUNT
     50    CONTINUE
*

*****************************************************************************************************

*          COMPLETE CALCULATIONS TO LOCATE REAR OF OIL BANK                  *
*          OIL BANK LEAVES THE SYSTEM WHEN XDPR(IT) = 1.0 AND IT             *
*          IS GREATER THAN 101 WHEN TDP < DP                                 *
*                                                                           *
*          LINEAR INTERPOLATION IS USED TO FIND THE VALUES OF               *
*          PARAMETERS IF THE OIL BANK LEAVES BEFORE IT = 151                *

*****************************************************************************************************

*
    105    DO 110 IT = 101,150
           IF(XDPR(IT).LT1.0 .AND. XDPR(IT+1).GE. 1.0) GOTO 115
    110    CONTINUE
           PRINT*,'LARGEST VALUE OF XDPR IS LESS THAN 1.0'
           PRINT*,'XDPR(151) = ',XDPR(151)
           PRINT*,'PROGRAM TERMINATED AFTER STATEMENT NO. 110'
           STOP
    115    NTR1 = IT+1
           TDSWR = TD(IT)+(TD(IT+1)–TD(IT))*(1.0–XDPR(IT))/
          :(XDPR(IT+1)–XDPR(IT))
           TDMSWR = TDM(IT)+(TDM(IT+1)–TDM(IT))*(1.0–XDPR(IT))/
          :(XDPR(IT+1)–XDPR(IT))
           SWROB = SWRS(IT)+(SWRS(IT+1)–SWRS(IT))*(1.0–XDPR(IT))/
```

```
      :(XDPR(IT+1)-XDPR(IT))
      SBAROB = SBAR1(IT)+(SBAR1(IT+1)-SBAR1(IT))*(1.0-XDPR(IT))/
      :(XDPR(IT+1)-XDPR(IT))
      FWROB = FWR(IT) +(FWR(IT+1)-FWR(IT))*(1.0-XDPR(IT))/
      (XDPR(IT+1)-XDPR(IT))
      TD(NTR1) = TDSWR
      TDM(NTR1) = TDMSWR
      SWRS(NTR1) = SWROB
      SBAR1(NTR1) = SBAROB
      FWR(NTR1) = FWROB
      XDPR(NTR1) = 1.0
*


******************************************************************************************************

*         OUTPUT OF COMPUTED RESULTS

******************************************************************************************************

*
      WRITE(IPLOT,65)(TD(IT),TDM(IT),XDPR(IT),SWRS(IT),IT = 1,NTR1)
  65  FORMAT(' ',F7.5,5X,7.5,5X,F7.5,5X,F7.5)
      STOP
      END
```

PROGRAM C-5—SWR.FOR

****

```
*         SWR. FOR
          REAL FUNCTION SWR(FWFS, SWFS, FWR)
*
*         PROGRAM TO COMPUTE INTERSECTION OF A LINE TO
*         VISCOUS/POLYMER FRACTIONAL FLOW CURVE WITH THE
*         FRACTIONAL FLOW CURVE FOR A NORMAL WATERFLOOD
*
*         KRO = ALPHA1*(1.-SWD)**M
*
*         KRW = ALPHA2*SWD**N

          REAL M,N
*234567
          COMMON/PDATA/ALPHA1,ALPHA2,M,N, SWC, SOR,DS, SWI,VISO,VISW
          Y(SW1) = (SW1-SWC)*B
          FWF(SWD1) = SWD1**N/(SWD1**N+A*(1.-SWD1)**M)
          FWS(SWD1) = A1*(DS+SWC+SWD1/B)

          H3(SWD1) = FWS(SWD1)-FWF(SWD1)

          SWF(SWD1) = SWD1/B+SWC
*
          B = 1./(1.-SOR-SWC)
          A1 = FWFS/(DS+SWFS)
*
*234567
          A = ALPHA1*VISW/(ALPHA2*VISO)
          SWDF = SQRT(A/(1.+A))

*         COMPUTE BREAKTHROUGH SATURATION FOR GENERAL
*         FORM OF RELATIVE PERMEABILITY CURVES BY NEWTONIAN ITERATION
*         AND INTERVAL HALVING
          X1 = SWFS
          SDMAX = Y(1.-SOR)
          DSD = (SDMAX-SWFS)/49
*         PRINT*,'SDMAX = ',SDMAX,' DELSD = ', DSD
          SWX = X1+DSD
          H3X = H3(SWX)
          DO 10 1 = 1,200
          SWX = SWX+DSD
          H3XP1 = H3(SWX)
          IF((H3X.LT.0.).AND.H3XP1.GT.0.) GO TO 90
          H3X = H3XP1
```

```
            GO TO 10
     90     SWX = SWX–DSD
            DSD = DSD/2.
            IF(ABS(H3X).LE..5E–04) GO TO 20
     10     CONTINUE
            PRINT*,'NO CONVERGENCE IN 200 ITERATIONS'
            RETURN
     20     SWDF = SWX
*
            FWR = FWF(SWDF)
*
            SWR = SWF(SWDF)
            RETURN
            END
```

PROGRAM C-6—PROP.FOR

****

```
C          PROP.FOR
C          SUBPROGRAM TO COMPUTE KRO,KRW,FW,FPSW FOR PREAR.FOR
C
*234567890
            SUBROUTINE PROP(VISO,VISW, SW1, KRO1, KRW1, FW1, FPSW1)
            REAL KRO1,KRW1,KO,KW,M,N
            COMMON/PDATA/ALPHA1,ALPHA2,M,N,SWC1,SOR1,DS,SWI,VISO1,VISW1
            Y(SW1) = (SW1–SWC1)*B
            KW(Y) = ALPHA2*Y**N
            KO(Y) = ALPHA1*(1.–Y)**M
            B1(SWD1) = N*SWD1**(N–1)*(1.–SWD1)**M
            B2(SWD1) = M*SWD1**N*(1.–SWD1)**(M–1)
            B3(SWD1) = SWD1**N+A*(1.–SWD1)**M
            G2(SWD1) = A*(B1(SWD1)+B2(SWD1))/B3(SWD1)**2
            FWF(SWD1) = SWD1**N/(SWD1**N+A*(1.–SWD1)**M)
            B = 1./(1.–SOR1–SWC1)
            A = ALPHA1*VISW/(ALPHA2*VISO)
            SWD = Y(SW1)
            IF(ABS(SWD–1.0).LE.9.E–07) GO TO 10
            KRO1 = KO(SWD)
            KRW1 = KW(SWD)
            FW1 = FWF(SWD)
            FPSW1 = B*G2(SWD)
            RETURN
     10     KRO1 = 0.
            KRW1 = ALPHA2
            FW1 = 1.0
            FPSW1 = 0.
            RETURN
            END
```

Computer Programs—Stalkup Displacement Models

```
        PROGRAM TERTIARY
        REAL M,MKO,K,PHI,MAX,N,NP,KRW,KRS,INCREMENT
        OPEN(UNIT = 8, FILE = TERT.DAT', STATUS = 'NEW')
C
C   THIS PROGRAM USES THE STALKUP MODEL TO CALCULATE OIL RECOVERY
C   FROM A MISCIBLE FLOOD INITIATED AFTER A WATERFLOOD WHEN THE
C   OIL SATURATION IS AT ITS RESIDUAL VALUE TO WATER.
C
C   A CORRELATION FROM CLARIDGE IS USED TO CALCULATE THE AREAL SWEEP
C   EFFICIENCY OF THE RESPECTIVE FLOOD FRONTS. CONTINUOUS SOLVENT INJECTION IS
    ASSUMED.
C
C   GET VARIABLES FROM USER
C
10      PRINT*,'INPUT THE FOLLOWING VARIABLES'
        PRINT*,'WELL SPACING (ACRES)'
        READ*,SPC
        PRINT*,'FORMATION THICKNESS (FT)'
        READ*,H
        PRINT*,'FORMATION POROSITY (decimal) AND PERMEABILITY (MD)'
        READ*,PHI,K
        PRINT*,'OIL VISCOSITY(CP), WATER VISCOSITY(CP), AND'
        PRINT*,'SOLVENT VISCOSITY (CP)'
        READ*,UO,UW,US
        PRINT*,'OIL FORMATION VOLUME FACTOR (RES BBL/STB)'
        READ*,BO
        PRINT*,'ENTER THE FOLLOWING SATURATIONS'
        PRINT*,'SWC, SORM, SORW, SOB, SWT'
        READ*,SWC,SORM,SORW,SOB,SWT
        PRINT*,'ENTER RELATIVE PERMEABILITY TO WATER AT SORW'
        READ*,KRW
        PRINT*,'ENTER RELATIVE PERMEABILITY OF SOLVENT AT SORM'
        READ*,KRS
        PRINT*,'ENTER MAXIMUM DISPLACEABLE PORE VOLUMES INJECTED'
        READ*,MAX
        PRINT*,'ENTER SOLVENT INJECTION INCREMENT (PORE VOLUMES)'
        READ*,INCREMENT
C
C   PRINT OUT VARIABLES FOR USER CHECK
C
        PRINT*,'SPACING = ',SPC,' ACRES'
        PRINT*,'FORMATION THICKNESS = ',H,' FT'
        PRINT*,'POROSITY = ',PHI
        PRINT*,'PERMEABILITY = ',K,' MD'
        PRINT*,'OIL VISCOSITY = ',UO,' CP'
        PRINT*,'SOLVENT VISCOSITY = ',US,' CP'
        PRINT*,'WATER VISCOSITY = ',UW,' CP'
        PRINT*,'FORMATION VOLUME FACTOR = ',BO
        PRINT*,'SWC = ',SWC
        PRINT*,'SORM = ',SORM
        PRINT*,'SORW = ',SORW
        PRINT*,'SOB = ',SOB
        PRINT*,'SWT = ',SWT
        PRINT*,'REL PERM. OF WATER @ SORW = ',KRW
        PRINT*,'REL. PERM. OF SOLVENT @ SORM = ',KRS
        PRINT*,'MAX DISPLACEABLE VOLUMES OF SOLVENT = ',MAX
```

```
              PRINT*,'PV INJECTION INCREMENT = '.INCREMENT
              PRINT*,ENTER 1 IF DATA IS INCORRECT.'
              PRINT*,ENTER 0 IF DATA IS CORRECT.'
              READ*,ICHECK
              IF(ICHECK .EQ. 1) GOTO 10
C
C  ASSUME THE MOBILITY RATIO OF THE PROCESS IS GOVERNED BY THE
C  SOLVENT FRONT AND THE OIL BANK TO CALCULATE DISPLACEABLE PORE
C  VOLUMES OF SOLVENT INJECTED AT BREAKTHROUGH (FIBT)
C
              M = (KRS/US)/(KRW/UW)
              VP = SPC*H*PHI*4.356E4/(5.615)
              FIBT = (0.9/(M + 1.1))**0.5
              FIOBT = FIBT/(1. +(SORW-SORM)/(SOB-SORW))
C  HYDROCARBONS IN PLACE AT START OF FLOOD
              N = VP*SORW/BO
              WRITE(8,*)'MOBILITY RATIO = ',M
              WRITE(8,*)'SOLVENT BREAKTHROUGH = ',FIBT,' DIS. P.V.'
              WRITE(8,*)'OIL BANK BREAKTHROUGH = ',FIOBT,' DIS P.V.'
              WRITE(8,*)
              MKO = (0.78 + 0.22*M**0.25)**4
              R1 = (1.6/MKO**0.61)
              FI = 0.0
15            CONTINUE
C
C  START SOLVENT INJECTION BY THE SPECIFIED INCREMENT
C
              FI = FI +INCREMENT
              HCPVI = FI*(1.-SORM-SWT)/(1-SWC)
C
C  CHECK TO SEE IF ENOUGH SOLVENT HAS BEEN INJECTED TO GET TO BREAKTHROUGH
C  IF NOT SET THE SOLVENT SWEEPOUT EQUAL TO PORE VOLUMES OF SOLVENT INJECTED.
C  IF SO, USE CLARIDGE CORRELATION TO CALCULATE ES AND THEN CALCULATE THE
C  PSEUDO DISPLACEABLE PORE VOLUMES OF SOLVENT INJECTED (DVOB). DVOB
C  ACCOUNTS FOR THE VOLUME OF OIL MOBILIZED BY THE SOLVENT.
C
              IF(FI .GE. FIBT) GOTO 20
              ES = FI
              GOTO 30
20            R2 = ((FI–FIBT)/(1–FIBT))**(1.28*MKO**(–0.26))
              ES = (R1*R2 + FIBT)/(1+R1*R2)
30            DVOB = FI + (ES*(SORW–SORM)/(SOB–SORW))
              IF(DVOB .GE. FIBT) GOTO 40
              EOB = DVOB
              GOTO 50
40            R3 = ((DVOB – FIBT)/(1 – FIBT))* *(1.28*MKO* *(–0.26))
              EOB = (R1*R3 + FIBT)/(1 + R1*R3)
C
C  NOW CALCULATE OIL RECOVERY USING STALKUP'S TECHNIQUE
C
50            NP = (VP*ES*(SORW – SORM) – VP*(EOB – ES)*(SOB – SORW))/BO
              RECFAC = NP/N
C
C  PRINT OUT RESULTS
              IF(FI .EQ. INCREMENT) THEN
              WRITE(8,62)
62            FORMAT(' ',2X'DVS',7X,'HCPV',3X,'SOLVENT',3X,'OIL BANK',5X,
     &        'OIL',3X,'RECOVERY')
              WRITE(8,63)
              FORMAT(' ','INJECTED',2X,'INJECTED',2X,'SWEEPOUT',2X,'SWEEPOUT'
     &        ,2X,'RECOVERY',2X,'FACTOR')
              END IF
              WRITE(8,66)FI,HCPVI,ES,EOB,NP,RECFAC
66            FORMAT(5X,4(F6.4,2X),F10.1,2X,F6.4)
              IF (FI .GT. MAX) STOP
              GOTO 15
              END
              PROGRAM SECONDARY
```

```
C
C   THIS PROGRAM ESTIMATES OIL RECOVERY FROM A MISCIBLE FLOOD INITIATED
C   AS A SECONDARY RECOVERY PROCEDURE. THE STALKUP MODEL IS USED AND
C   SWEEPOUT EFFICIENCY IS CALCULATED FROM THE CLARIDGE CORRELATION.
C   THIS MODEL ASSUMES NO FREE GAS SATURATION AND IMMOBILE WATER AT THE
C   START OF THE FLOOD. AN INFINITE SOLVENT SLUG IS INJECTED.
C
          REAL M,SPC,H,K,PHI,SOI,SWC,SORM,UO,US,BO,MAX,N,NP,INCREMENT
10        PRINT*,'INPUT THE FOLLOWING VARIABLES'
C
C   HAVE USER INPUT VARIABLES C
          PRINT*,'PATTERN SPACING (ACRES)'
          READ*,SPC
          PRINT*,'FORMATION THICKNESS (FT)'
          READ*,H
          PRINT*,'FORMATION POROSITY (DEC) AND PERMEABILITY (MD)'
          READ*,PHI,K
          PRINT*,'OIL AND SOLVENT VISCOSITY (CP)'
          READ*,UO,US
          PRINT*,'OIL FORMATION VOLUME FACTOR'
          READ*,BO
          PRINT*,'ENTER THE FOLLOWING SATURATIONS'
          PRINT*,'SOI,SWC,SORM'
          READ*,SOI,SWC,SORM
          PRINT*,'ENTER THE MAXIMUM DISPLACEABLE VOLUME OF SOLVENT'
          PRINT*,'TO BE INJECTED'
          READ*,MAX
          PRINT*,'ENTER THE SOLVENT INJECTION INCREMENT (DIS. PORE VOL.)'
          READ*, INCREMENT
          PRINT*,' '
          PRINT*,'SPACING = ', SPC,'ACRES'
          PRINT*,'FORMATION THICKNESS = ',H,' FT'
          PRINT*,'POROSITY = ',PHI
          PRINT*,'PERMEABILITY = ',K,' MD'
          PRINT*,'SOI = ',SOI
          PRINT*, 'SWC = ',SWC
          PRINT*, 'SORM = ',SORM
          PRINT*,'OIL VISCOSITY = ',UO,' CP'
          PRINT*,'SOLVENT VISCOSITY =',US,' CP'
          PRINT*,'BO = ',BO
          PRINT*,'MAX DISPL. PV OF SOLVENT TO INJECT = ',MAX
          PRINT*,'INJECTION INCREMENT = ',INCREMENT
          PRINT*,' '
          PRINT*,'ENTER 1 IF DATA IS INCORRECT'
          PRINT*,'ENTER 0 TO CONTINUE'
          READ*,'CHECK
          IF(ICHECK .EQ. 1) GOTO 10
          M = UO/US
C   CALCULATE AMOUNT OF SOLVENT INJECTED AT BREAKTHROUGH
          FIBT = (0.9/(M + 1.1))**0.5
C   CALCULATE OOIP
          N = SPC*H*PHI*SOI*4.356E4/BO/5.615
          PRINT*,' MOBILITY RATIO = ',M
          MKO = (0.78 + 0.22*M**0.25)**4
          R1=(1.6/MKO**0.61)
C
C   UNTIL SOLVENT BREAKTHROUGH OIL PRODUCTION IS EQUAL TO SOLVENT INJECTION
C   ASSUMING IMMOBILE WATER SATURATION. START STALKUPS MODEL AT SOLVENT
C   BREAKTHROUGH.
C
          FI = FIBT + 0.0001
          WRITE(*,65)
65        FORMAT(' ',6X,'DIS. PV',5X,'HCPV',7X,'SWEEP',7X,'NP',10X,'RECOVERY')
          WRITE(*,66)
66        FORMAT(' ',5X,'INJECTED',3X,'INJECTED',3X,'EFFICIENCY',17X,'FACTOR')
60        CONTINUE
          FI = FI + INCREMENT
          HCPV = FI*(1.-SORM-SWC)/(1.-SORM)
```

```fortran
C
C  USE CLARIDGE CORRELATION TO CALCULATE SWEEP EFFICIENCY
C
          R2 = ((FI – FIBT)/(1 – FIBT))* *(1.28*MKO* *(– 0.26))
          E = (R1*R2 + FIBT)/(1 + R1*R2)
C
C  CALCULATE AMOUNT OF OIL PRODUCED
C
          NP = SPC*H*PHI*(SOI–SORM)*E*4.356E4/(5.615*BO)
          RECFAC = NP/N
C
C  PRINT OUT RESULTS
C
          WRITE(*,70)FI,HCPV,E,NP,RECFAC
70        FORMAT(' ',5(F12.4))
          IF(FI .GE. MAX)STOP
          GOTO 60
          END
          PROGRAM SLUG2
C
C  THIS PROGRAM CALCULATES THE TERTIARY RECOVERY OF OIL BY A SOLVENT
C  SLUG DRIVEN BY WATER
C
          REAL MS,MO,K,PHI,MAX,N,NP,KRW,KRS,INCREMENT,KRO
          REAL MKO,HCPV
          OPEN(UNIT = 8.FILE = 'OUT.DAT',STATUS = 'NEW')
C
C
C
C
C  GET INPUT DATA FROM USER
C
          PRINT*,'INPUT THE FOLLOWING VARIABLES'
10        PRINT*,'PATTERN SPACING (ACRES)'
          READ* SPC
          PRINT*,'FORMATION THICKNESS (FT)'
          READ*,H
          PRINT*,'FORMATION POROSITY (FRAC) AND PERMEABILITY (MD)'
          READ*,PHI,PERM
          PRINT*,'OIL, WATER AND SOLVENT VISCOSITY (CP)'
          READ*,UO,UW,US
          PRINT*,'OIL FORMATION VOLUME FACTOR'
          READ*, BO
          PRINT*,'INPUT THE FOLLOWING SATURATIONS'
          PRINT*,'SORM,SORW,SOB,SWC,SWT,SSRW'
          READ*,SORM,SORW,SOB,SWC,SWT,SSRW
          PRINT*,'INPUT THE FOLLOWING RELATIVE PERMEABILITIES'
          PRINT*,'REL. PERM. OF OIL AT SWC
          READ*,KRO
          PRINT*,'REL PERM. OF SOLVENT AT SWC
          READ*,KRS
          PRINT*,'REL. PERM. OF WATER AT SOR'
          READ*,KRW
          PRINT*,'INPUT SLUG SIZE (HCPV)'
          READ*, MAX
          PRINT*,'INPUT INJECTION INCREMENT (DIS. PV INJ)'
          READ*,INCREMENT
C
C  PRINT DATA OUT FOR CHECK BY USER
C
          PRINT*,'SPACING = ',SPC,' ACRES'
          PRINT*,'FORMATION THICKNESS = ',H,' FT'
          PRINT*,'POROSITY = ',PHI
          PRINT*,'PERMEABILITY = ',K,' MD'
          PRINT*,'OIL VISCOSITY = ',UO,' CP'
          PRINT*,'WATER VISCOSITY = ',UW,' CP'
          PRINT*,'SOLVENT VISCOSITY = ',US,' CP'
          PRINT*,'BO = ',BO
          PRINT*,'SWC = ',SWC
```

```
          PRINT*,'SORM = ',SORM
          PRINT*,'SORW = ',SORW
          PRINT*,'SOB = ',SOB
          PRINT*,'SWT = ',SWT
          PRINT*,'SSRW = ',SSRW
          PRINT*,'KRW = ',KRW
          PRINT*,'KRO = ',KRO
          PRINT*,'KRW = ',KRS
          PRINT*,'SLUG SIZE = ',MAX,' HCPV
          PRINT*,'INCREMENT = ',INCREMENT
          PRINT*,'ENTER 1 TO CHANGE DATA'
          PRINT*,'ENTER 0 TO CONTINUE'
          READ*,CHECK
          IF(ICHECK .EQ. 1) GOTO 10
C
C    CALCULATE MOBILITY RATIO AS THE RATIO OF THE MOBILITY BETWEEN
C    THE SOLVENT FRONT AND THE WATERFLOOD WATER
C
          MS = (KRS/US)/(KRW/UW)
C    CALCULATE AMOUNT OF SOLVENT INJECTED TO GET SOLVENT BREAKTHROUGH
C    AND THE "PSEUDO" DISPLACEABLE VOLUMES TO GET OIL BREAKTHROUGH
C    BY CALCULATING THE AMOUNT OF OIL DISPLACED BY THE SOLVENT AND
C    ADJUSTING FIBT
          FIBT = (0.9/(MS + 1.1))**0.5
          FIOBT = FIBT/(1. + (SORW–SORM)/(SOB–SORW))
C    CALCULATE PORE VOLUME AND OOIP
          VP = SPC*H*PHI*4.356E4/(5.615)
          N = SPC*H*PHI*SORW*4.356E4/(BO*5.615)
          MKO = (0.78 + 0.22*MS**0.25)**4
          R1=(1.6/MKO**0.61)
          FI = 0.0
          SLUGINDVS = MAX*(1–SWC)/(1–SWT–SORM)
          SLUGMAX = SLUGINDVS
          WRITE(8,*)'SLUG SIZE = ',SLUGMAX,' DIS. P.V'
          WRITE(8,*)'SOLVENT/WATER MOBILITY RATIO = ',MS
          WRITE(8,*)'FIBT = ',FIBT,' DIS. P.V.'
          WRITE(8,*)'OIL BANK B.T. = ',FIOBT,' DIS. P.V.'
          WRITE(8,*)
          WRITE(8,*)'  DIS. PV  HCPV  ES  EOB  NP  NP/N'
          WRITE(8,*)'  INJ  INJ'
17        CONTINUE
C
C            ––––––CONTINUOUS SOLVENT INJECTION––––––
C
C    DURING SLUG INJECTION, TREAT THE PROCESS AS A CONTINUOUS
C    SOLVENT INJECTION, FI EQUALS DVS IN STALKUP.
C
          FI = FI + INCREMENT
          HCPV = F1*(1.–SWT–SORM)/(1.–SWC)
C
C    IF AMOUNT INJECTED IS LESS THAN AMOUNT TO BREAKTHROUGH, SET THE
C    SWEEP EFFICIENCY EQUAL TO THE DISPLACEABLE PORE VOLUMES INJECTED.
C
          IF (FI .GE. FIBT) GOTO 20
          ES = FI
          GOTO 30
C
C    USE CLARIDGE CORRELATION TO CALCULATE SWEEP EFFICIENCY OF SOLVENT (ES)
C    AND IF APPLICABLE OF THE OIL BANK USING THE PSEUDO DISPLACEABLE VOLUME
C    INJECTED (DVOB). DVOB ACCOUNTS FOR THE VOLUME THE MOBILIZED OIL OCCUPIES.
20        R2 = ((FI–FIBT)/(1–FIBT))* *(1.28*MKO**(–0.26))
          ES = (R1 *R2 + FIBT)/(1 + R1 *R2)
30        DVOB = FI + (ES*(SORW – SORM)/(SOB – SORW))
          IF(DVOB .GE. FIBT) GOTO 40
          EOB = DVOB
          GOTO 50
40        R3 = ((DVOB – FIBT)/(1 – FIBT))* *(1.28*MKO* *(– 0.26))
          EOB = (R1 *R3 + FIBT)/(1 + R1 *R3)
```

```
50        NP = (VP*ES*(SORW – SORM) – VP*(EOB – ES)*(SOB – SORW))/BO
          RECFAC = NP/N
          WRITE(8,66)FI,HCPV,ES,EOB,NP,RECFAC
66        FORMAT(5X,4(F6.4,2X),F7.0,2X,F6.4)
          IF(FI + INCREMENT .GT. SLUGMAX .AND. L .EQ. 0)THEN
          FI = SLUGMAX – INCREMENT
          L = 1
          GOTO 17
          END IF
          IF(FI .LT. SLUGMAX)GOTO 17
C
C    ––––––END SOLVENT INJECTION––––––
C
          FI = SLUGMAX
C
C    FIT = TOTAL SLUG SIZE IN DISPLACEABLE PORE VOLUMES
C
          FIT = SLUGMAX
          WRITE(8,*)'END OF SLUG INJECTION'
C
C    ––––––START DRIVE WATER INJECTION––––––
C
          DVW = 0.0
60        CONTINUE
          DVW = DVW + INCREMENT
C
C    FI IS INCREASED BY ADDING WATER INJECTION AND AMOUNT OF
C    SOLVENT DISPLACED BY WATER
C.
          FI = FIT + DVW*((1 – SORM – SWT – SSRW)/(1 – SORM – SWT))
          HCPV = FI*(1.–SWT–SORM)/(1.–SWC)
          EW = DVW
C
C    ONCE AGAIN, IF THE FRONT IS NOT AT BREAKTHROUGH SET THE SOLVENT
C    SWEEP EFFICIENCY EQUAL TO THE DISPLACEABLE PORE VOLUMES INJECTED
C    AND USE THE CLARIDGE CORRELATION TO CALCULATE THE SWEEP EFFICIENCY
C    OF THE SOLVENT DRIVEN BY WATER AND OF THE OIL BANK
C
          IF(FI .GE. FIBT) GOTO 70
          ES = FI
          GOTO 80
70        R2 = ((FI–FIBT)/(1–FIBT))**(1.28*MKO**(–0.26))
          ES = (R1*R2 + FIBT)/(1 + R1*R2)
80        DVOB = FI + (ES*(SORW – SORM)/(SOB – SORW))
          IF(DVOB .GE. FIBT)GOTO 100
          EOB = DVOB
          GOTO 110
100       R3 = ((DVOB - FIBT)/(1 - FIBT))* *(1.28*MKO* *(- 0.26))
          EOB = (R1 *R3 + FIBT)/(1 + R1 *R3)
C
C    CALCULATE AMOUNT OF OIL PRODUCED DURING PERIOD WHEN
C    SLUG IS DRIVEN BY WATER
C
110       NP = (VP*ES*(SORW – SORM) – VP*(EOB – ES)*(SOB – SORW))/BO
          RECFAC = NP/N
          WRITE(8,66)FI,HCPV,ES,EOB,NP,RECFAC
C
C    CONTINUE THIS LOOP UNTIL THE SWEEP EFFICIENCY OF THE WATER
C    IS GREATER THAN THE SWEEP EFFICIENCY OF THE SOLVENT BANK
C    AT WHICH ALL OF THE SOLVENT IS ASSUMED LEFT AS A RESIDUAL SATURATION
C
          IF(EW .LT. ES) GOTO 60
C
C         ––––––END OF SOLVENT DRIVEN BY WATER––––––
C
          WRITE(8,*)'ALL SOLVENT LEFT AS RESIDUAL'
C
```

```
C   NOW THE REMAINDER OF THE FLOOD IS BETWEEN THE DRIVE WATER BANK
C   AND THE OIL BANK. THE RESIDUAL OIL SATURATION REVERTS BACK TO
C   SORW WHILE IT IS PUSHED WITH WATER.
C
C
C   CALCULATE THE NEW MOBILITY RATIO AND THE INJECTION UNTIL
C   WATER BREAKTHROUGH
C
        M = (KRW/UW)/(KRO/UO)
        FIBT = (0.9/(M + 1.1))**0.5
        MKO = (0.78 + 0.22*M**0.25)**4
        R1=(1.6/MKO**0.61)
C
C   DEPENDING ON THE SOLVENT SLUG SIZE, SET ARBITRARY VALUES OF
C   ADDED DISPLACEABLE VOLUMES OF WATER (F) UNTIL THE PROCESS IS
C   OIL DRIVEN BY WATER.
C
        IF(MAX .LE. .10)THEN
          F = 1
        ELSE IF(MAX .LE. .25)THEN
          F = 1.5
        ELSE IF(MAX .LE. .50) THEN
          F = 2
        ELSE
          F = 2.5
        END IF
C
C   CALCULATE THE END POINTS WHERE THE PROCESS IS OIL DRIVEN BY WATER
C
        RVOB = DVOB + F
        RVW = DVW + F
C   USE CLARIDGE CORRELATION TO CALCULATE THE ENDPOINT SWEEP EFFICIENCIES
C   OF THE OIL AND WATER BANK (ROB) AND (RW)
        R2 = ((RVW - FIBT)/(1 - FIBT))* *(1.28*MKO* *(- 0.26))
        RW = (R1*R2 + FIBT)/(1 + R1*R2)
        R3 = ((RVOB-FIBT)/(1-FIBT))* *(1.28*MKO**(- 0.26))
        ROB = (R1 * R3 + FIBT)/(1 + R1 * R3)
C   CALCULATE THE END POINT OIL PRODUCED
        RP = (VP*ES*(SORW-SORM)-VP*(ROB-RW)*(SOB-SORW))/BO
        RECFAC = RP/N
C   SET THE OTHER POINTS FOR LINEAR INTERPOLATION AT THE EFFICIENCIES
C   WITH THE LAST SOLVENT PROPERTIES
        EOBO = EOB
        EWO = EW
        DVWO = DVW
        DVOBO = DVOB
C
C   BEGIN STRAIGHT LINE APPROXIMATION OF MOBILITY RATIO GOING FROM
C   SOLVENT/OIL TO DRIVE WATER/OIL.
C
150     CONTINUE
        FI = FI + INCREMENT
        HCPV = FI*(1.-SWT-SORM)/(1.-SWC)
        DVOB = DVOB + INCREMENT
        DVW = DVW + INCREMENT
        EW = (RW-EWO)*((DVW-DVWO)/(RVW-DVWO)) + EWO
        EOB = (ROB-EOBO)* ((DVOB-DVOBO)/(RVOB-DVOBO)) + EOBO
        NP = (VP*ES*(SORW-SORM) - VP*(EOB - EW)*(SOB - SORW))/BO
        RECFAC = NP/N
        WRITE(8,66)FI,HCPV,ES,EOB,NP,RECFAC
        IF(DVW .LT. RVW) GOTO 150
        WRITE(8,*)'END STRAIGHT LINE APPROXIMATION'
C
C   NOW THE PROCESS IS STRICTLY AN OIL BANK BEING DISPLACED BY WATER
200     CONTINUE
        FI = FI + INCREMENT
        HCPV = FI*(1.-SWT-SORM)/(1.-SWC)
```

```
            DVOB = DVOB + INCREMENT
            DVW = DVW + INCREMENT
C
C   CALCULATE SWEEP EFFICIENCIES OF THE WATER FRONT AND THE OIL BANK
C   FROM THE CLARIDGE CORRELATION
C
            R3 = ((DVOB-FIBT)/(1–FIBT))* *(1.28*MKO**(–0.26))
            EOB = (R1*R3 + FIBT)/(1+R1*R3)
            R2 = ((DVW – FIBT)/(1 – FIBT))* *(1.28*MKO* *(– 0.26))
            EW = (R1*R2 + FIBT)/(1 + R1*R2)
            NP = (VP*ES*(SORW – SORM) – VP*(EOB – EW)*(SOB – SORW))/BO
            RECFAC = NP/N
            WRITE(8,66)FI,HCPV,EW,EOB,NP,RECFAC
            IF(FI .LT. 3.) GOTO 200
            STOP
            END
```

Fluid and Rock Property Data and Supporting Data for Thermal Recovery Calculations

It is always desirable to have actual data for engineering calculations. However, in most cases, values of key parameters are not available and must be estimated. Appendix E contains methods for estimating fluid and rock properties needed to calculate the performance of thermal recovery processes. We also include casing and tubing specifications of materials commonly used in the completion of steam-injection wells. The data and methods presented here are based on extensive collections in the literature, including several sections of Appendix B from the monograph *Thermal Oil Recovery* by Prats (1982) and several tables from Boberg (1988). The appendix is organized by fluid type, with rock properties combined under one heading.

Water

The thermodynamic properties of water are available in Keenan and Keyes (1936). **Table E-1** contains values of the enthalpy and specific volume of saturated liquid and vapor at selected temperatures. **Table E-2** contains the same data at selected pressures.

Tabulated data are not convenient to use in computer models. Consequently, empirical correlations of the data have been developed. Farouq Ali (1970) presents several useful formulas to calculate steam properties for the pressure range of 15 to $\approx$ 1,000 psia. Eqs. E-1 through E-5 are the correlations, with average absolute error indicated in parenthesis.

$$T_s = 115.1 p^{0.225} \left(\text{error } 1\%, 15 \text{ to } 3{,}000 \text{ psia} \right), \dotfill \text{(E-1)}$$

$$L_v = 1{,}318 p^{-0.08774} \left(\text{error } 1.9\% \right), \dotfill \text{(E-2)}$$

$$H_w = 91 p^{0.2574} \left(\text{error } 0.3\% \right), \dotfill \text{(E-3)}$$

$$H_v = 1{,}119 p^{0.01267} \left(\text{error } 0.3\% \right), \dotfill \text{(E-4)}$$

$$\text{and } v_{sv} = 363.9 p^{-0.9588} \left(\text{error } 1.2\% \right). \dotfill \text{(E-5)}$$

The density of water can be estimated with the Gros (1984) correlation.

$$\rho_w = \left[0.01602 + 0.000023 \left(-6.6 + 0.0325T + 0.000657T^2 \right) \right]^{-1}. \dotfill \text{(E-6)}$$

Sylvester and Chen (1988) report that Eq. E-6 has a difference of less than 0.5% with steam tables for $T < 400°F$. They report that Eq. E-7 is accurate to within 1% for $r \geq 400°F$ (Burger et al. 1985).

$$\rho_w = \frac{9.97 - 0.046 \left(\dfrac{T-32}{1.8} \right) - 0.306 \times 10^{-2} \left(\dfrac{T-32}{1.8} \right)^2}{16}. \dotfill \text{(E-7)}$$

The viscosity of water is estimated from Eq. E-8 (Yao 1985).

$$\mu_w = 159.5T^{-1.182} \dotfill \text{(E-8)}$$

in units of °F, psia, Btu/lbm, and cp. Dissolved solids affect the viscosity of water. **Fig. E-1** is a correlation of water viscosity with temperature, indicating the corrections that should be applied to account for dissolved NaCl in the water (Matthews and Russell 1967).

Tortike and Farouq Ali (1989) present an extensive set of correlations of saturated-steam properties as functions of simple polynomials. Correlations were developed for both SI and customary units. Correlations for customary units are included in this appendix. All temperatures are given in °R (°F + 459.6). Eqs. E-9 through E-19 are the polynomial correlations.

Temperature T (°F)	Absolute Pressure p (psia)	Specific Volume			Enthalpy		
		Saturated Liquid v_{sw}	Evaporate v_{swv}	Saturated Vapor v_{sv}	Saturated Liquid H_w	Evaporate L_v	Saturated Vapor H_v
70	0.3631	0.01606	867.8	867.9	38.04	1,054.3	1,092.3
80	0.5069	0.01608	633.1	633.1	48.02	1,048.6	1,096.6
90	0.6982	0.01610	468.0	468.0	57.99	1,042.9	1,100.9
95	0.8153	0.01612	404.3	404.3	62.98	1,040.1	1,103.1
100	0.9492	0.01613	350.3	350.4	67.97	1,037.2	1,105.2
105	1.1016	0.01615	304.5	304.5	72.95	1,034.3	1,107.3
110	1.2748	0.01617	265.3	265.4	77.94	1,031.6	1,109.5
115	1.4709	0.01618	231.9	231.9	82.93	1,028.7	1,111.6
120	1.6924	0.01620	203.25	203.27	87.92	1,025.8	1,113.7
125	1.9420	0.01622	178.59	178.61	92.91	1,022.9	1,115.8
130	2.2225	0.01625	157.32	157.34	97.90	1,020.2	1,117.9
135	2.5370	0.01627	138.93	138.95	102.90	1,017.0	1,119.9
140	2.8886	0.01629	122.99	123.01	107.89	1,014.1	1,122.0
145	3.281	0.01632	109.13	109.15	112.89	1,011.2	1,124.1
150	3.718	0.01634	97.06	97.07	117.89	1,008.2	1,126.1
155	4.203	0.01637	86.51	86.52	122.89	1,005.2	1,128.1
160	4.741	0.01639	77.27	77.29	127.89	1,002.3	1,130.2
165	5.335	0.01642	69.17	69.19	132.89	999.3	1,132.2
170	5.992	0.01645	62.04	62.06	137.90	996.3	1,134.2
175	6.715	0.01648	55.76	55.78	142.91	993.3	1,136.2
180	7.510	0.01651	50.21	50.23	147.92	990.2	1,138.1
185	8.383	0.01654	45.29	45.31	152.93	987.2	1,140.1
190	9.339	0.01657	40.94	40.96	157.95	984.1	1,142.0
195	10.385	0.01660	37.07	37.09	162.97	981.0	1,144.0
200	11.526	0.01633	33.62	33.64	167.99	977.9	1,145.9
210	14.123	0.01670	27.80	27.82	178.05	971.6	1,149.7
220	17.186	0.01677	23.13	23.15	188.13	965.2	1,153.4
230	20.780	0.01684	19.365	19.382	198.23	958.8	1,157.0
240	24.969	0.01692	16.306	16.323	208.34	952.2	1,160.5
250	29.825	0.01700	13.804	13.821	218.48	945.5	1,164.0
260	35.429	0.01709	11.746	11.763	228.64	938.7	1,167.3
270	41.858	0.01717	10.044	10.061	238.84	931.8	1,170.6
280	49.203	0.01726	8.628	8.645	249.06	924.7	1,173.8
290	57.556	0.01735	7.444	7.461	259.31	917.5	1,176.8
300	67.013	0.01745	6.449	6.466	269.59	910.1	1,179.7
310	77.68	0.01755	5.609	5.626	279.92	902.6	1,182.5
312	79.96	0.01757	5.457	5.474	281.99	901.0	1,183.1
314	82.30	0.01759	5.310	5.327	284.06	899.5	1,183.6
316	84.70	0.01761	5.167	5.185	286.13	898.0	1,184.1
318	87.15	0.01763	5.030	5.047	288.20	896.5	1,184.7
320	89.66	0.01765	4.896	4.914	290.28	894.9	1,185.2
322	92.22	0.01768	4.767	4.785	292.36	893.3	1,185.7
324	94.84	0.01770	4.642	4.660	294.43	891.8	1,186.2
326	97.52	0.01772	4.521	4.538	296.52	890.2	1,186.7
328	100.26	0.01774	4.403	4.421	298.60	888.6	1,187.2
330	103.06	0.01776	4.289	4.307	300.68	887.0	1,187.7

Table E-1—Thermodynamic data for water at selected saturation temperatures (Keenan and Keyes 1936).

Temperature T (°F)	Absolute Pressure p (psia)	Specific Volume			Enthalpy		
		Saturated Liquid v_{sw}	Evaporate v_{swv}	Saturated Vapor v_{sv}	Saturated Liquid H_w	Evaporate L_v	Saturated Vapor H_v
332	105.92	0.01778	4.179	4.197	302.77	885.4	1,188.2
334	108.85	0.01781	4.072	4.090	304.86	883.8	1,188.7
336	111.84	0.01783	3.968	3.986	306.95	882.2	1,189.2
338	114.89	0.01785	3.868	3.886	309.04	880.6	1,189.6
340	118.01	0.01787	3.770	3.788	311.13	879.0	1,190.1
342	121.20	0.01790	3.675	3.693	313.23	877.4	1,190.6
344	124.45	0.01792	3.584	3.602	315.33	875.7	1,191.0
346	127.77	0.01794	3.495	3.513	317.43	874.1	1,191.5
348	131.17	0.01797	3.408	3.426	319.53	872.4	1,191.9
350	134.63	0.01799	3.324	3.342	321.63	870.7	1,192.3
352	138.16	0.01801	3.243	3.261	323.74	869.1	1,192.8
354	141.77	0.01804	3.164	3.182	325.85	867.3	1,193.2
356	145.45	0.01806	3.087	3.105	327.96	865.6	1,193.6
358	149.21	0.01808	3.012	3.030	330.07	863.9	1,194.0
360	153.04	0.01811	2.939	2.957	332.18	862.2	1,194.4
362	156.95	0.01813	2.869	2.887	334.30	860.5	1,194.8
364	160.93	0.01816	2.801	2.819	336.42	858.8	1,195.2
366	165.00	0.01818	2.734	2.752	338.54	857.1	1,195.6
368	169.15	0.01821	2.669	2.687	340.66	855.3	1,196.0
370	173.37	0.01823	2.606	2.625	342.79	853.5	1,196.3
372	177.68	0.01826	2.545	2.564	344.91	851.8	1,196.7
374	182.07	0.01829	2.486	2.504	347.04	850.0	1,197.0
376	186.55	0.01831	2.428	2.446	349.18	848.2	1,197.4
378	191.12	0.01834	2.372	2.390	351.31	846.4	1,197.7
380	195.77	0.01836	2.317	2.335	353.45	844.6	1,198.1
382	200.50	0.01839	2.264	2.282	355.59	842.8	1,198.4
384	205.33	0.01842	2.212	2.231	357.73	841.0	1,198.7
386	210.25	0.01844	2.162	2.180	359.88	839.1	1,199.0
388	215.26	0.01847	2.113	2.131	362.02	837.3	1,199.3
390	220.37	0.01850	2.0651	2.0836	364.17	835.4	1,199.6
392	225.56	0.01853	2.0187	2.0372	366.33	833.6	1,199.9
394	230.85	0.01855	1.9734	1.9920	368.48	831.7	1,200.2
396	236.24	0.01858	1.9293	1.9479	370.64	829.9	1,200.5
398	241.73	0.01861	1.8864	1.9050	372.80	827.9	1,200.7
400	247.31	0.01864	1.8447	1.8633	374.97	826.0	1,201.0
405	261.71	0.01871	1.7448	1.7635	380.39	821.2	1,201.6
410	276.75	0.01878	1.6512	1.6700	385.83	816.3	1,202.1
415	292.45	0.01886	1.5635	1.5823	391.29	811.3	1,202.6
420	308.83	0.01894	1.4811	1.5000	396.77	806.3	1,203.1
425	325.92	0.01902	1.4036	1.4226	402.27	801.2	1,203.5
430	343.72	0.01910	1.3308	1.3499	407.79	796.0	1,203.8
435	362.27	0.01918	1.2623	1.2815	413.34	790.8	1,204.1
440	381.59	0.01926	1.1979	1.2171	418.90	785.4	1,204.3

Keenan, J. H. and Keyes, F. G. 1936. *Thermodynamic Properties of Steam*, first edition, 28–33. New York City: John Wiley & Sons Inc. Reprinted by permission of John Wiley & Sons Inc.

Table E-1—Thermodynamic data for water at selected saturation temperatures (Keenan and Keyes 1936) (continued).

Temperature T (°F)	Absolute Pressure p (psia)	Specific Volume			Enthalpy		
		Saturated Liquid v_{sw}	Evaporate v_{swv}	Saturated Vapor v_{sv}	Saturated Liquid H_w	Evaporate L_v	Saturated Vapor H_v
445	401.68	0.01935	1.1371	1.1565	424.49	780.0	1,204.5
450	422.6	0.0194	1.0799	1.0993	430.1	774.5	1,204.6
455	444.3	0.0195	1.0258	1.0453	435.7	768.9	1,204.6
460	466.9	0.0196	0.9748	0.9944	441.4	763.2	1,204.6
465	490.3	0.0197	0.9266	0.9463	447.1	757.4	1,204.5
470	514.7	0.0198	0.8811	0.9009	452.8	751.7	1,204.3
475	539.9	0.0199	0.8380	0.8579	458.6	745.4	1,204.0
480	566.1	0.0200	0.7972	0.8172	464.4	739.4	1,203.7
485	593.3	0.0201	0.7586	0.7787	470.2	733.1	1,203.3
490	621.4	0.0202	0.7221	0.7423	476.0	726.8	1,202.8
495	650.6	0.0203	0.6874	0.7077	481.9	720.4	1,202.3
500	680.8	0.0204	0.6545	0.6749	487.8	713.9	1,201.7
505	712.0	0.0205	0.6233	0.6438	493.8	707.1	1,200.9
510	744.3	0.0207	0.5935	0.6142	499.8	700.3	1,200.1
515	777.8	0.0208	0.5653	0.5861	505.8	693.4	1,199.2
520	812.4	0.0209	0.5385	0.5594	511.9	686.4	1,198.2
525	848.1	0.0210	0.5130	0.5340	518.0	679.1	1,197.1
530	885.0	0.0212	0.4886	0.5098	524.1	671.8	1,195.9
535	923.2	0.0213	0.4655	0.4868	530.3	664.3	1,194.6
540	962.5	0.0215	0.4434	0.4649	536.6	656.6	1,193.2
545	1,003.2	0.0216	0.4224	0.4440	542.9	648.8	1,191.7
550	1,045.2	0.0218	0.4022	0.4240	549.3	640.8	1,190.0
555	1,088.5	0.0219	0.3831	0.4050	555.7	632.6	1,188.3
560	1,133.1	0.0221	0.3647	0.3868	562.2	624.2	1,186.4
565	1,179.1	0.0222	0.3472	0.3694	568.8	615.5	1,184.3
570	1,226.5	0.0224	0.3304	0.3528	575.4	606.7	1,182.1
575	1,275.4	0.0226	0.3143	0.3369	582.1	597.7	1,179.8
580	1,325.8	0.0228	0.2989	0.3217	588.9	588.4	1,177.3
585	1,377.7	0.0230	0.2841	0.3071	595.8	578.8	1,174.6
590	1,431.2	0.0232	0.2700	0.2931	602.8	569.0	1,171.8
595	1,486.2	0.0234	0.2563	0.2797	609.8	558.9	1,168.7
600	1,542.9	0.0236	0.2432	0.2668	617.0	548.5	1,165.5
605	1,601.2	0.0239	0.2306	0.2545	624.3	537.7	1,162.0
610	1,661.2	0.0241	0.2185	0.2426	631.6	526.7	1,158.4
615	1,723.0	0.0244	0.2068	0.2312	639.1	515.3	1,154.4
620	1,786.6	0.0247	0.1955	0.2201	646.7	503.6	1,150.3
625	1,852.0	0.0250	0.1895	0.2095	654.4	491.4	1,145.8
630	1,919.3	0.0253	0.1740	0.1992	662.3	478.8	1,141.1
635	1,988.5	0.0256	0.1637	0.1893	670.4	465.6	1,136.0
640	2,059.7	0.0260	0.1538	0.1798	678.6	452.0	1,130.5
645	2,132.9	0.0264	0.1441	0.1705	687.0	437.7	1,124.7
650	2,208.2	0.0268	0.1348	0.1616	695.7	422.8	1,118.5
655	2,285.7	0.0273	0.1256	0.1528	704.8	406.9	1,111.7
660	2,365.4	0.0278	0.1165	0.1442	714.2	390.2	1,104.4
665	2,447.4	0.0283	0.1076	0.1359	724.1	372.4	1,096.4
670	2,531.8	0.0290	0.0987	0.1277	734.4	353.2	1,087.7

Table E-1—Thermodynamic data for water at selected saturation temperatures (Keenan and Keyes 1936) (continued).

Temperature T (°F)	Absolute Pressure p (psia)	Specific Volume			Enthalpy		
		Saturated Liquid v_{sw}	Evaporate v_{swv}	Saturated Vapor v_{sv}	Saturated Liquid H_w	Evaporate L_v	Saturated Vapor H_v
675	2,618.7	0.0297	0.0899	0.1196	754.4	332.6	1,078.0
680	2,708.1	0.0305	0.0810	0.1115	757.3	309.9	1,067.2
685	2,800.2	0.0315	0.0719	0.1034	770.1	284.7	1,054.8
690	2,895.1	0.0328	0.0625	0.0953	784.4	256.0	1,040.4
695	2,992.9	0.0344	0.0520	0.0864	801.2	220.7	1,021.9
700	3,093.7	0.0369	0.0392	0.0761	823.3	172.1	995.4
702	3,134.9	0.0385	0.0325	0.0710	823.3	172.1	995.4
704	3,176.7	0.0410	0.0234	0.0645	852.7	106.0	958.7
705	3,197.7	0.0438	0.0152	0.0589	869.2	69.1	938.4
705.4	3,602.2	0.0503	0	0.0503	902.7	0	902.7

Keenan, J. H. and Keyes, F. G. 1936. *Thermodynamic Properties of Steam,* first edition, 28–33. New York City: John Wiley & Sons Inc. Reprinted by permission of John Wiley & Sons Inc.

Table E-1—Thermodynamic data for water at selected saturation temperatures (Keenan and Keyes 1936) (continued).

Absolute Pressure p (psia)	Temperature T (°F)	Specific Volume		Enthalpy		
		Saturated Liquid v_{sw}	Saturated Vapor v_{sv}	Saturated Liquid H_w	Evaporate L_v	Saturated Vapor H_v
14.696	212.00	0.01672	26.80	180.07	970.3	1,150.4
20	227.96	0.01683	20.089	196.16	960.1	1,156.3
30	250.33	0.01701	13.746	218.82	945.3	1,164.1
40	267.25	0.01715	10.498	236.03	933.7	1,169.7
50	281.01	0.01727	8.515	250.09	924.0	1,174.1
60	292.71	0.01738	7.175	262.09	915.5	1,177.6
70	302.92	0.01748	6.206	272.61	907.9	1,180.6
80	312.03	0.01757	5.472	282.02	901.1	1,183.1
90	320.27	0.01766	4.896	290.56	894.7	1,185.3
100	327.81	0.01774	4.432	298.40	888.8	1,187.2
110	334.77	0.01782	4.049	305.66	883.2	1,188.9
120	341.25	0.01789	3.728	312.44	877.9	1,190.4
130	347.32	0.01796	3.455	318.81	872.9	1,191.7
140	353.02	0.01802	3.220	324.82	868.2	1,193.0
150	358.42	0.01809	3.015	330.51	863.6	1,194.1
160	363.53	0.01815	2.834	335.93	859.2	1,195.1
170	368.41	0.01822	2.675	341.09	854.9	1,196.0
180	373.06	0.01827	2.532	346.03	850.8	1,196.9
190	377.51	0.01833	2.404	350.79	846.8	1,197.6
200	381.79	0.01839	2.288	355.36	843.0	1,198.4
205	383.86	0.01842	2.234	357.58	841.1	1,198.7
210	385.90	0.01844	2.183	359.77	839.2	1,199.0
215	387.89	0.01847	2.134	361.91	837.4	1,199.3
220	389.86	0.01850	2.087	364.02	835.6	1,199.6
225	391.79	0.01852	2.0422	366.09	833.8	1,199.9

Keenan, J. H. and Keyes, F. G. 1936. *Thermodynamic Properties of Steam,* first edition, 28–33. New York City: John Wiley & Sons Inc. Reprinted by permission of John Wiley & Sons Inc.

Table E-2—Thermodynamic data for water at selected saturation pressures (Keenan and Keyes 1936).

Absolute Pressure p (psia)	Temperature T (°F)	Specific Volume		Enthalpy		
		Saturated Liquid v_{sw}	Saturated Vapor v_{sv}	Saturated Liquid H_w	Evaporate L_v	Saturated Vapor H_v
230	393.68	0.01854	1.9992	368.13	832.0	1,200.1
235	395.54	0.01857	1.9579	370.14	830.3	1,200.4
240	397.37	0.01860	1.9183	372.12	828.5	1,200.6
245	399.18	0.01863	1.8803	374.08	826.8	1,200.9
250	400.95	0.01865	1.8438	376.00	825.1	1,201.1
255	402.70	0.01868	1.8086	377.89	823.4	1,201.3
260	404.42	0.01870	1.7748	379.76	821.8	1,201.5
265	406.11	0.01873	1.7422	381.60	820.1	1,201.7
270	407.78	0.01875	1.7107	383.42	818.5	1,201.9
275	409.43	0.01878	1.6804	385.21	816.9	1,202.1
280	411.05	0.01880	1.6511	386.98	815.3	1,202.3
285	412.65	0.01883	1.6228	388.73	813.7	1,202.4
290	414.23	0.01885	1.5954	390.46	812.1	1,202.6
295	415.79	0.01887	1.5689	392.16	810.5	1,202.7
300	417.33	0.01890	1.5433	393.84	809.0	1,202.8
310	420.35	0.01894	1.4944	397.15	806.0	1,203.1
320	423.29	0.01899	1.4485	400.39	803.0	1,203.4
330	426.16	0.01904	1.4053	403.56	800.0	1,203.6
340	428.97	0.01908	1.3645	406.66	797.1	1,203.7
350	431.72	0.01913	1.3260	409.69	794.2	1,203.9
360	434.40	0.01917	1.2895	412.67	791.4	1,204.1
370	437.03	0.01921	1.2550	415.59	788.6	1,204.2
380	439.60	0.01925	1.2222	418.45	785.8	1,204.3
390	442.12	0.01930	1.1910	421.27	783.1	1,204.4
400	444.59	0.0193	1.1613	424.0	780.5	1,204.5
410	447.01	0.0194	1.1330	426.8	777.7	1,204.5
420	449.39	0.0194	1.1061	429.4	775.2	1,204.6
430	451.73	0.0194	1.0803	432.1	772.5	1,204.6
440	454.02	0.0195	1.0556	434.6	770.0	1,204.6
450	456.28	0.0195	1.0320	437.2	767.4	1,204.6
460	458.50	0.0196	1.0094	439.7	764.9	1,204.6
470	460.68	0.0196	0.9878	442.2	762.4	1,204.6
480	462.82	0.0197	0.9670	444.6	759.9	1,204.5
490	464.93	0.0197	0.9470	447.0	757.5	1,204.5
500	467.01	0.0197	0.9278	449.4	755.0	1,204.4
520	471.07	0.0198	0.8915	454.1	750.1	1,204.2
540	475.01	0.0199	0.8578	458.6	745.4	1,204.0
560	478.85	0.0200	0.8265	463.0	740.8	1,203.8
580	482.58	0.0201	0.7973	467.4	736.1	1,203.5
600	486.21	0.0201	0.7698	471.6	731.6	1,203.2
620	489.75	0.0202	0.7440	475.7	727.2	1,202.9
640	493.21	0.0203	0.7198	479.8	722.7	1,202.5
660	496.58	0.0204	0.6971	483.8	718.3	1,202.1
680	499.88	0.0204	0.6757	487.7	714.0	1,201.7
700	503.10	0.0205	0.6554	491.5	709.7	1,201.2
720	506.25	0.0206	0.6362	495.3	705.4	1,200.7

Table E-2—Thermodynamic data for water at selected saturation pressures (Keenan and Keyes 1936) (continued).

Absolute Pressure p (psia)	Temperature T (°F)	Specific Volume		Enthalpy		
		Saturated Liquid v_{sw}	Saturated Vapor v_{sv}	Saturated Liquid H_w	Evaporate L_v	Saturated Vapor H_v
740	509.34	0.0207	0.6180	499.0	701.2	1,200.2
760	512.36	0.0207	0.6007	502.6	697.1	1,199.7
780	515.33	0.0208	0.5843	506.2	692.9	1,199.1
800	518.23	0.0209	0.5687	509.7	688.9	1,198.6
820	521.08	0.0209	0.5538	513.2	684.8	1,198.0
840	532.88	0.0210	0.5396	516.6	680.8	1,197.4
860	526.63	0.0211	0.5260	520.0	676.8	1,196.8
880	529.33	0.0212	0.5130	523.3	672.8	1,196.1
900	531.98	0.0212	0.5006	526.6	668.8	1,195.4
920	534.59	0.0213	0.4886	529.8	664.9	1,194.7
940	537.16	0.0214	0.4772	533.0	661.0	1,194.0
960	539.68	0.0214	0.4663	536.2	657.1	1,193.3
980	542.17	0.0215	0.1557	539.3	653.3	1,192.6
1,000	544.61	0.0216	0.4456	542.4	649.4	1,191.8
1,050	550.57	0.0218	0.4218	550.0	639.9	1,189.9
1,100	556.31	0.0220	0.4001	557.4	630.4	1,187.8
1,150	561.86	0.0221	0.3802	564.6	621.0	1,185.6
1,200	567.22	0.0223	0.3619	571.7	611.7	1,183.4
1,250	572.42	0.0225	0.3450	578.6	602.4	1,181.0
1,300	577.46	0.0227	0.3293	585.4	593.2	1,178.6
1,350	582.35	0.0229	0.3148	592.1	584.0	1,176.1
1,400	587.10	0.0231	0.3012	598.7	574.7	1,173.4
1,450	591.73	0.0233	0.2884	605.2	565.5	1,170.7
1,500	596.23	0.0235	0.2765	611.6	556.3	1,167.9
1,600	604.90	0.0239	0.2548	624.1	538.0	1,162.1
1,700	613.15	0.0243	0.2354	636.3	519.6	1,155.9
1,800	621.03	0.0247	0.2179	648.3	501.1	1,149.4
1,900	628.58	0.0252	0.2021	660.1	482.4	1,142.4
2,000	635.82	0.0257	0.1878	671.7	463.4	1,135.1
2,100	642.77	0.0262	0.1746	683.3	444.1	1,127.4
2,200	649.46	0.0268	0.1625	694.8	424.4	1,119.2
2,300	655.91	0.0274	0.1513	706.5	403.9	1,110.4
2,400	662.12	0.0280	0.1407	718.4	328.7	1,101.1
2,500	668.13	0.0287	0.1307	730.6	360.5	1,091.1
2,600	673.94	0.0295	0.1213	743.0	337.2	1,080.2
2,700	679.55	0.0305	0.1123	756.2	312.1	1,068.3
2,800	684.99	0.0315	0.1035	770.1	284.7	1,054.8
2,900	690.26	0.0329	0.0947	785.4	253.6	1,039.0
3,000	695.36	0.0346	0.0858	802.5	217.8	1,020.3
3.100	700.31	0.0371	0.0753	825.0	168.1	993.1
3,200	705.11	0.0444	0.0580	872.4	62.0	934.4
3,206.2	705.40	0.0503	0.0503	902.7	0	902.7

Keenan, J. H. and Keyes, F. G. 1936. *Thermodynamic Properties of Steam*, first edition, 28–33. New York City: John Wiley & Sons Inc. Reprinted by permission of John Wiley & Sons Inc.

Table E-2—Thermodynamic data for water at selected saturation pressures (Keenan and Keyes 1936) (continued).

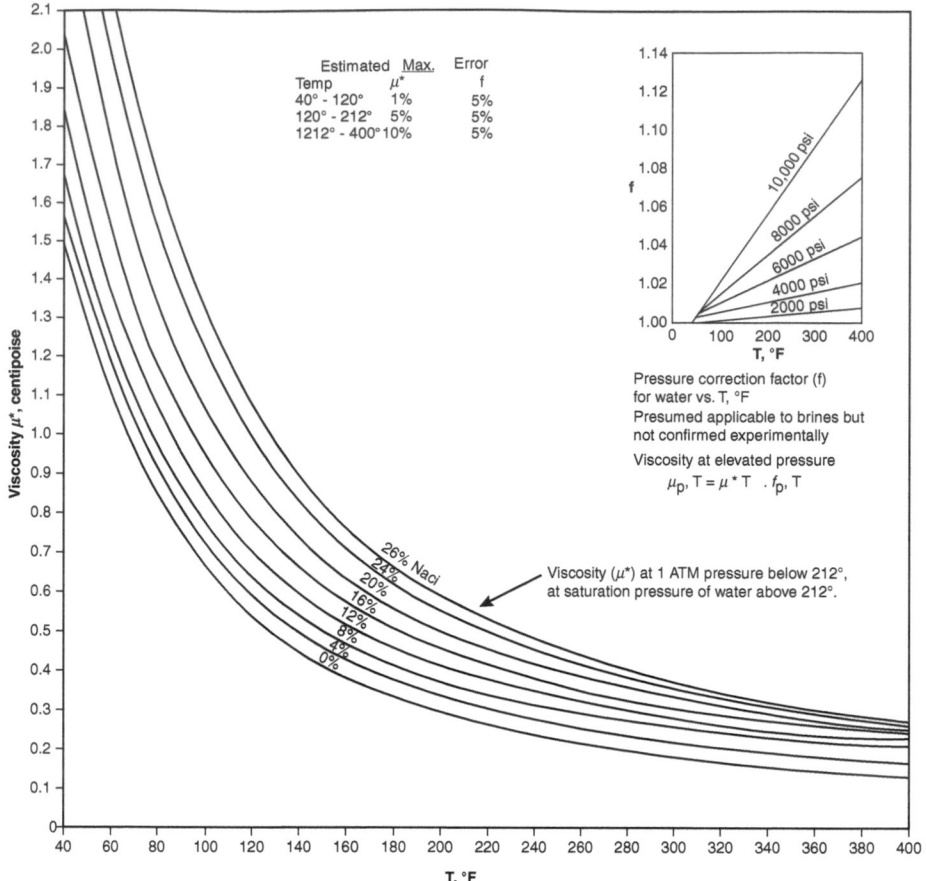

Fig. E-1—Water viscosities for various salinities and temperatures, from Chesnut, unpublished, Shell Development Company (Yao 1985).

Saturation Temperature ($0.089 < p < 3{,}208$ psia).

$$T_s = 561.435 + 33.8866 \ln p + 2.18893(\ln p)^2$$

$$+ 0.0808998(\ln p)^3 + 0.0342030(\ln p)^4 \dots \dots \dots \dots \dots \dots \dots \dots \dots \dots \dots \text{(E-9)}$$

Saturation Pressure.

$$p_s = \big(-66.9421 + 0.485086T - 1.33944 \times 10^{-3} T^2$$
$$+ 1.71599 \times 10^{-6} T^3 - 9.93039 \times 10^{-10} T^4$$
$$+ 2.29394 \times 10^{-13} T^5 \big)^2 \dots \dots \dots \dots \dots \dots \dots \dots \dots \dots \dots \dots \dots \dots \text{(E-10)}$$

Enthalpy of Saturated Water ($492 < T < 1{,}161°$R).

$$H_w = 10{,}174.2 - 87.4729T + 0.301147T^2 - 5.38409 \times 10 - 4T^3$$
$$+ 5.33392 \times 10^{-7} T^4 - 2.77814 \times 10^{-10} T^5$$
$$+ 5.95201 \times 10^{-14} T^6 \dots \dots \dots \dots \dots \dots \dots \dots \dots \dots \dots \dots \dots \dots \text{(E-11)}$$

Density of Saturated Water ($492 < T < 1{,}152°$R).

$$\rho_w = 236.372 - 1.19187T + 0.00378125T^2$$
$$- 5.4258 \times 10^{-6} T^3 + 3.74277 \times 10^{-9} T^4$$
$$- 1.01916 \times 10^{-12} T^5 \dots \dots \dots \dots \dots \dots \dots \dots \dots \dots \dots \dots \dots \dots \text{(E-12)}$$

Viscosity of Saturated Water (492 <T< 1,152°R).

$$\mu_w = -12.3274 + \frac{48,786.8}{T} - \frac{7.62292 \times 10^7}{T^2} + \frac{5.91509 \times 10^{10}}{T^3}$$

$$-\frac{2.28157 \times 10^{13}}{T^4} + \frac{3.53226 \times 10^{15}}{T^5}. \dots\dots\dots\dots\dots\dots\text{(E-13)}$$

Thermal Conductivity of Saturated Water (492 < T < 1,161°R).

$$k_{hw} = 2.02892 - 0.0142394T + 4.30191 \times 10^{-5}T^2$$

$$-5.99485 \times 10^{-8}T^3 + 3.97811 \times 10 - 11T^4$$

$$-1.02089 \times 10^{-14}T^5. \dots\dots\dots\dots\dots\dots\dots\text{(E-14)}$$

Latent Heat of Vaporization (492 <T< 1,161°R).

$$L_v = \left(1,327,940 + 1,134.53T - 5.04327T^2\right.$$

$$\left. +5.15204 \times 10^{-3}T^3 - 2.13711 \times 10^{-6}T^4\right)^{0.5}. \dots\dots\dots\text{(E-15)}$$

Enthalpy of Saturated Vapor (492 <T< 1,161°R).

$$H_v = -9,469.85 + 87.2545T - 0.299668T^2$$

$$+5.43610 \times 10^{-4}T^3 - 5.46484 \times 10^{-7}T^4$$

$$+2.88759 \times 10^{-10}T^5 - 6.28068 \times 10^{-14}T^6. \dots\dots\dots\text{(E-16)}$$

Density of Saturated Vapor (492 < T < 1,161°R).

$$\ln \rho_s = -96.4809 + 0.463301T - 9.90153 \times 10^{-4}T^2$$

$$+1.12766 \times 10^{-6}T^3 - 6.60862 \times 10^{-10}T^4$$

$$+1.57286 \times 10^{-13}T^5. \dots\dots\dots\dots\dots\dots\text{(E-17)}$$

Viscosity of Saturated Vapor (492 < T < 1,161°R).

$$\mu_s = -0.546807 + 3.83050 \times 10^{-3}T - .04938 \times 10^{-5}T^2$$

$$+1.42291 \times 10^{-8}T^3 - 9.49798 \times 10^{-12}T^4$$

$$+2.49747 \times 10^{-15}T^5. \dots\dots\dots\dots\dots\dots\text{(E-18)}$$

Thermal Conductivity of Saturated Vapor (492 < T < 1,161°R).

$$k_{hs} = -1.36235 + 9.54729 \times 10^{-3}T - 2.61945 \times 10^{-5}T^2$$

$$+3.54448 \times 10^{-8}T^3 - 2.36542 \times 10^{-11}T^4$$

$$+6.25351 \times 10^{-15}T^5. \dots\dots\dots\dots\dots\dots\text{(E-19)}$$

Oil

Phase-behavior data are correlated in a number of references. We begin with correlations of density, specific heat, and viscosity with temperatures that are useful in thermal-recovery applications. These correlations require the oil gravity and the temperature and measured oil viscosities at two temperatures. The effects of pressure are ignored.

The density of oil can be estimated with the Gros (1984) correlation, which is given by

$$\rho_o = \rho_{oR} - C_1(T - 60) + C_2(T - 60)^2, \dots\dots\dots\dots\dots\text{(E-20)}$$

where

$$\rho_{oR} = 62.4278\left[141.5/(131.5 + °API)\right],$$

$$C_1 = 0.0133 + 152.4\rho_{oR}^{-2.45},$$

and $C_2\ 0.0000081 - 0.0622 \times 10^{-(0.0764\rho_{oR})}$.

The specific heat of oil is estimated with the Gambill (1957) correlation, which is given by

$$C_{po} = (0.3881 + 0.00045T)/\sqrt{\gamma_o}, \dotfill \text{(E-21)}$$

where γ_o = specific gravity of the oil at 60°F and is given by

$$\gamma_o = 141.5/(131.5 + °API). \dotfill \text{(E-22)}$$

The oil viscosity is correlated with the two-point equation developed by Wright (1974) (Eq. E-23). Density and viscosity data at two temperatures are necessary to determine constants a and b in Eq. E-23.

$$\log\left\{\log\left[(\mu_o/\rho_o) + 0.6\right]\right\} = a - b\left[\log(T + 460)\right]. \dotfill \text{(E-23)}$$

Units in Eqs. E-20 through E-23 are cp and g/cm³.

Thermal conductivities of hydrocarbon liquids decrease with temperature. For petroleum fractions and hydrocarbon mixtures, Cragoe (1929) proposed Eq. E-24 to estimate the thermal conductivity.

$$k_{ho} = \left\{1.62\left[1 - 3(T - 32)\times 10^{-4}\right]\right\}/\gamma_o \dotfill \text{(E-24)}$$

Physical-Property Correlations for Crude Oil

Experimental phase-behavior data frequently are unavailable. However, empirical correlations of physical properties, such as oil-formation volume factor (FVF), solution-gas/oil ratio (GOR), bubblepoint pressure, viscosity of gas-free oil, viscosity of gas-saturated oil, viscosity of undersaturated oil, isothermal compressibility of saturated oil, and gas-gravity correction factor, and conversion factors for changing flash oil FVFs to differential oil FVFs have been developed from experimental data. There are many correlations available in the petroleum engineering literature.

Kartoatmodjo and Schmidt (1994) present the most recent correlations. These correlations give physical properties as a function of parameters that are typically measured in the field, such as temperature, pressure, separator-gas gravity, and stock-tank-oil gravity. The separator-gas gravity must be converted to a pressure of 100 psi if measured at a different pressure. These correlations are presented as Eqs. E-25 through E-41.

Bubblepoint Pressure (When R_{sf} Is Known). For °API < 30,

$$p_b = \left\{\frac{R_{sf}}{0.05958\left[\gamma_{g100}^{0.7972} \times 10^{13.1405\ API/(T+460)}\right]}\right\}^{0.9986} \dotfill \text{(E-25)}$$

For °API > 30,

$$p_b = \left\{\frac{R_{sf}}{0.03150\left[\gamma_{g100}^{0.7587} \times 10^{13.2895\ API/(T+460)}\right]}\right\}^{0.9143} \dotfill \text{(E-26)}$$

Gas Specific Gravity.

$$\gamma_{g100} = \gamma_{gsep}\left[1 + 0.1596\ °API^{0.4076}T_{sep}^{-0.2466}\log\left(p_{sep}/114.7\right)\right]. \dotfill \text{(E-27)}$$

Oil FVF. *Below the Bubblepoint.*

$$B_{of} = 0.98496 + 0.0001F_B^{1.50}, \dotfill \text{(E-28)}$$

where $F_B = R_{sf}^{0.755}\gamma_{g100}^{0.25}\gamma_o^{-1.50} + 0.45T \dotfill \text{(E-29)}$

Above the Bubblepoint.

$$B_{of} = B_{ofb}e^{c_o(p_b-p)}, \dots\dots\dots\dots\dots\dots\dots\dots\dots\dots\dots\dots\dots\dots\dots\dots\dots\dots\dots \text{(E-30)}$$

where c_o = isothermal oil compressibility and is given by

$$c_o = 6.8257 \times 10^{-6}\frac{R_{sf}^{0.5002}\,{}^\circ\text{API}^{0.3613}T^{0.76606}}{p\gamma_{g100}^{0.35505}} \dots\dots\dots\dots\dots\dots\dots\dots\dots\dots\dots\dots \text{(E-31)}$$

Solution GOR (Solved from Eqs. E-25 and E-26).

$$R_{sf} = 0.5958\gamma_{g100}^{0.7972}\left[p^{1.0014}\,10^{13.1405\,{}^\circ\text{API}/(T+460)}\right]. \dots\dots\dots\dots\dots\dots\dots\dots\dots\dots \text{(E-32)}$$

For $^\circ$API > 30,

$$R_{sf} = 0.03150\gamma_{g100}^{0.7587}\left[p^{1.0937}\,10^{11.2895\,{}^\circ\text{API}/(T+460)}\right]. \dots\dots\dots\dots\dots\dots\dots\dots\dots \text{(E-33)}$$

Oil Viscosity. *Gas Free.*

$$\mu_{od} = \left(16 \times 10^8\right)T^{-2.8177}\left(\log{}^\circ\text{API}\right)^{5.7526\log(T)-26.9718} \dots\dots\dots\dots\dots\dots\dots\dots \text{(E-34)}$$

Gas Saturated (p ≤ p_b).

$$\mu_o = -\,0.06821 + 0.9824f + 0.0004034f_{ov}^{\;2}, \dots\dots\dots\dots\dots\dots\dots\dots\dots\dots\dots\dots\dots \text{(E-35)}$$

$$f_{ov} = \left[0.2001 + \left(0.8428 \times 10^{-0.000845\,R_{sf}}\right)\right]\mu_{od}^{(0.43+0.5165y_{ov})}, \dots\dots\dots\dots\dots\dots\dots\dots \text{(E-36)}$$

and $y_{ov} = 10^{-0.00081R_{sf}}$. $\dots\dots\dots\dots\dots\dots\dots\dots\dots\dots\dots\dots\dots\dots\dots\dots\dots\dots\dots$ (E-37)

Gas Undersaturated (p < pb).

$$\mu_o = 1.00081\mu_{ob} + 0.001127\left(p-p_b\right)\left(-0.006517\mu_{ob}^{1.18148}\right.$$
$$\left. +10.038\mu_{ob}^{1.590}\right), \dots\dots\dots\dots\dots\dots\dots\dots\dots\dots\dots\dots\dots\dots\dots\dots\dots\dots\dots\dots\dots\dots\dots\dots \text{(E-38)}$$

Conversion From Flash to Differential Data. *Oil FVF.*

$$B_{od} = B_{of}\left(B_{odb}/B_{ofb}\right). \dots\dots\dots\dots\dots\dots\dots\dots\dots\dots\dots\dots\dots\dots\dots\dots\dots\dots\dots \text{(E-39)}$$

Solution GOR.

$$R_{sd} = R_{sdb} + \left(R_{sf} - R_{sfb}\right)\left(B_{odb}/B_{ofb}\right), \dots\dots\dots\dots\dots\dots\dots\dots\dots\dots\dots\dots\dots \text{(E-40)}$$

where

$$B_{ofb}/B_{odb} = 0.7264\gamma_o^{0.3202} - 0.3126\gamma_{gsep}^{-0.02087} + 0.6459$$
$$\times\left[\left(T_{\text{sep}} + 460\right)/T + 460\right]^{0.5596} \dots\dots\dots\dots\dots\dots\dots\dots\dots\dots\dots\dots\dots \text{(E-41)}$$

These correlations were developed from a large data bank that includes oils from throughout the world. **Table E-3** shows the range of variables in the data bank.

Reservoir Rocks

Thermal properties of reservoir rocks have been studied extensively, and data have been correlated empirically that allow estimation of thermal properties, such as heat capacity and effective thermal conductivity.

Heat Capacity. Fig. E-2 shows the heat capacity of dry reservoir rock, C_m, for several reservoir rocks as a function of temperature (Somerton 1958). The heat capacity for a porous rock saturated with a fluid can be calculated from the properties of the solid matrix and the porosity with

Parameter	Units	Minimum	Mean	Maximum
°API	–	14.4	37.2	58.9
B_{od}	bbl/bbl residual oil	1.007	1.307	2.747
B_{of}	bbl/STB	1.007	1.237	2.144
B_{odb}	bbl/STB	1.027	1.347	2.485
B_{ofb}	bbl/STB	1.004	1.303	2.410
p_{sep}	psia	14.7	116.4	514.7
p_b	psia	14.7	1,317.1	6,054.7
R_{sf}	scf/STB	0.0	426.2	2,890.0
R_{sfb}	scf/STB	1.4	484.4	2,473.0
T	°F	75.0	186.7	320.0
T_{sep}	°F	65.0	107.4	186.0
γ_{gsep}	–	0.379	0.841	1.709
μ_o	cp	0.2	32.2	517.0
μ_{ob}	cp	0.2	23.6	184.8
μ_{od}	cp	0.5	6.3	682.0
μ_{ol}	cp	0.1	3.7	586.0

Table E-3—Range of variables for oil property correlations (Kartoatmodjo and Schmidt 1994).

$$C_m = \left\{\left[(1-\phi)\rho_r C_r\right]/\rho_m\right\} + \left(\phi\rho_f C_f / \rho_m\right), \dots\dots\dots\dots\dots\dots\dots\dots\dots\dots\dots\text{(E-42)}$$

where ρ_r = density of the solid and C_r = heat capacity of the solid.

When the rock is filled with air at atmospheric pressure, the density is small and the contribution by the fluid is negligible. The heat capacity of the porous rock is given by

$$C_m = \left[(1-\phi)\rho_r C_r\right]/\rho_m \dots\dots\dots\dots\dots\dots\dots\dots\dots\dots\dots\dots\dots\dots\dots\dots\dots\text{(E-43)}$$

Consequently, the variation of C_m with temperature is caused by the variation of the heat capacities of the mineral constituents of the solid matrix with temperature. This observation is supported by a comparison of calculated heat capacities for a limestone ($\phi = 0.38$) and a sandstone ($\phi = 0.196$) based on the properties of pure quartz and pure calcite. **Fig. E-3** shows the results of this comparison. The data in Fig. E-2 suggest that the heat capacity increases approximately 10%/100°F.

Effective Thermal Conductivity of Porous Rocks. Somerton and Boozer (1960), Somerton et al. (1974), and Messmer (1985) report extensive data for the effective thermal conductivity of reservoir rock. The term "effective thermal conductivity" is used to describe the thermal conductivity of rocks because the thermal properties are determined by the mineral constituents, the porosity, and the fluids saturating the pore space.

Unconsolidated Oil Sands

Fig. E-4 illustrates the effect of porosity and the type of fluid saturating the pore space for unconsolidated oil sands. In Fig. E-4, the fluid (air, Stoddard solvent/oil, and water) thermal conductivities vary from 0.0165 (air) to 0.375 Btu/ft-hr-°F (water) and are the primary factors controlling the magnitude of the effective thermal conductivity. The effect of porosity is smaller.

Fig. E-5 shows the variation of the effective thermal conductivity of Kern River oil sands with brine saturation over a porosity range of 0.28 to 0.37. These sands contained various combinations of brine, oil, and air saturations at an average temperature of 125°F. The experimental data were fitted with

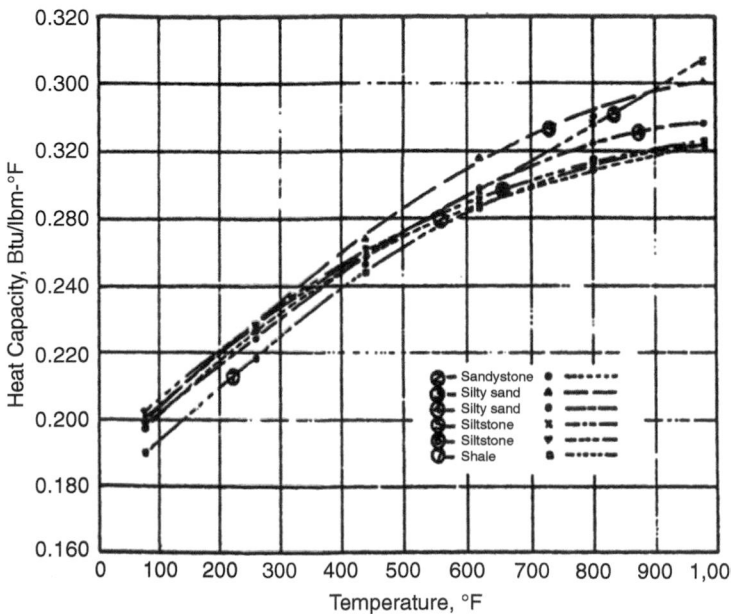

Fig. E-2—Heat capacities of some reservoir rocks (Somerton 1958).

$$k_{hR} = 0.735 - 1.30\phi + \sqrt{S_w}. \dots\dots\text{(E-44)}$$

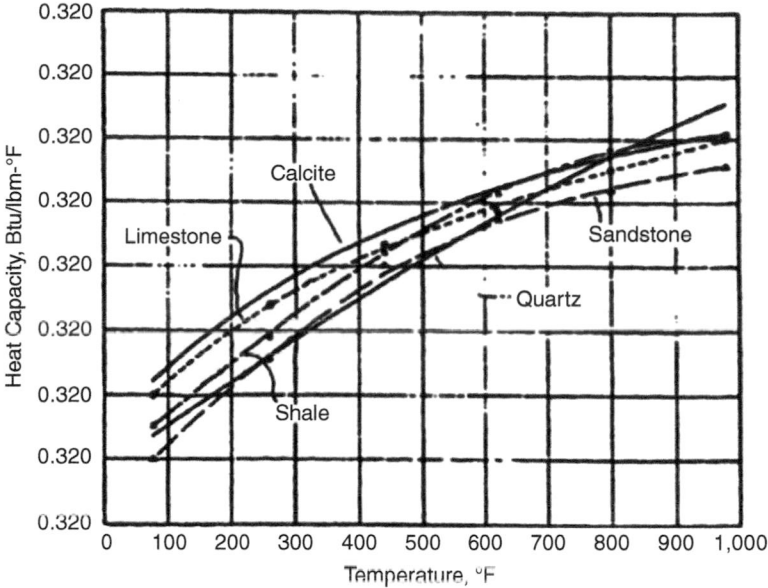

Fig. E-3—Comparison of calculated heat capacities of reservoir rocks and principal constituents (Somerton 1958).

Eq. E-44 was modified to correlate data for quartzitic sands by introducing the thermal conductivity of the solid (quartz thermal conductivity = 4.45 Btu/ft-hr°F) into the relationship to account for conduction in the solid phase. Eq. E-45 represents the correlation for quartzitic oil sands.

$$k_{hR} = 0.735 - 1.30\phi + 0.39k_{hs}\sqrt{S_w}, \dots\dots\dots\dots\dots\dots\dots\dots\dots\dots\dots\dots\dots\dots\dots\dots\dots\dots\dots \text{(E-45)}$$

with the thermal conductivity of the solid phase, k_{hs}, determined from

$$k_{hs} = 4.45f_q + 1.65\left(1 - f_q\right), \dots\dots\dots\dots\dots\dots\dots\dots\dots\dots\dots\dots\dots\dots\dots\dots\dots \text{(E-46)}$$

where f_q = volume fraction of quartz in the solid phase.

The effective thermal conductivity of liquid-saturated unconsolidated quartzitic oil sands decreases as temperature increases because the thermal conductivity of the quartz decreases with temperature. Although the thermal conductivity of brine increases with temperature to 260°F, the increase is not sufficient to offset the decrease in the thermal conductivity of the mineral constituents. Changes in unconsolidated oil sands are small and may be estimated with

$$k_h = k_{hR} - 0.28 \times 10^{-3}(T - 125)\left(k_{hR} - 0.82\right), \dots \text{(E-47)}$$

where k_{hR} is the effective thermal conductivity at 125°F determined from Eq. E-45.

The effect of pressure on effective thermal conductivities of liquid-saturated sands is small (Anand et al. 1973). For highly compressible formations, Somerton et al. (1974) suggest the following expression.

$$\Delta k_{hR} = \Delta p \times 10^{-5}$$

$$\left(0.5\rho_b\phi + 5.75\phi - 0.37k^{0.10} + 0.12\right), \dots \text{(E-48)}$$

where ρ_b is bulk density, g/cm³.

As noted earlier, fluid saturations and the type of fluid determine the magnitude of the effective thermal conductivity of unconsolidated oil sands. Although there are a number of correlations that may be used to estimate the effective thermal conductivity of unconsolidated sand saturated with a single liquid, there is no general correlation that applies when there are two or more phases present.

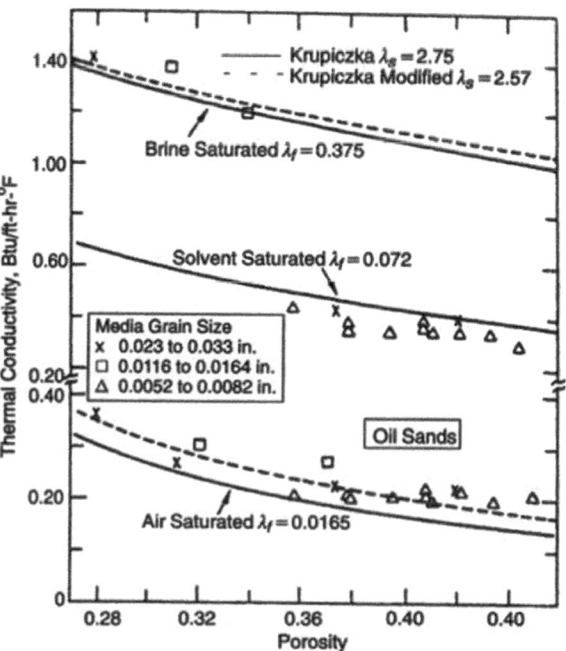

Fig. E-4—Thermal conductivity of unconsolidated oil sands (Somerton et al. 1974).

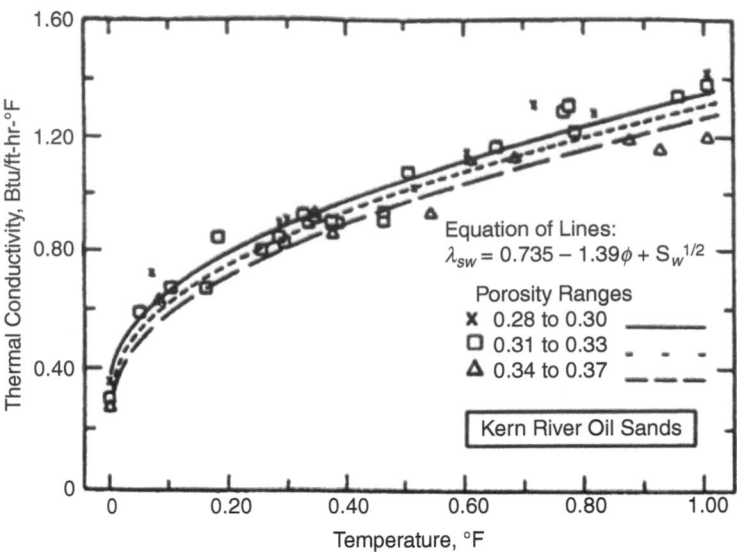

Fig. E-5—Thermal conductivity of Kern River oil sands (Somerton et al. 1974).

The correlation given by Eq. E-45 was developed for Kern River oil sands. Other oil sands may have different relationships between the effective thermal conductivity and brine saturation.

Consolidated Rocks

The thermal properties of consolidated rocks differ from those of unconsolidated oil sands primarily because the solid-phase contribution to the effective thermal conductivity is considerably larger because of cementation. This is illustrated in **Fig. E-6,** which represents data for the effects of fluid type and temperature on the effective thermal conductivity of Berea sandstone (Anand et al. 1973).

For consolidated sandstones, the following correlation was developed by Anand et al. (1973) to estimate the effective thermal conductivity of a liquid-saturated sandstone when the effective thermal conductivity of the dry (saturated with air) sandstone and the thermal conductivity of the liquid are known at 68°F.

$$k_{hR,l}/k_{hR,d} = 1.00 + 0.3\left[\left(k_{hl}/k_{ha}\right) - 1.0\right]^{1/3}$$
$$+ 4.57\left(\frac{\phi}{1-\phi}\frac{k_{hl}}{k_{hR,d}}\right)^{0.48m}\left(\frac{\rho_{R,l}}{\rho_{R,d}}\right)^{-4.30} \dots \dots \dots (E-49)$$

In Eq. E-49, the subscript R,l refers to the rock saturated with liquid, R,d refers to the dry rock saturated with air, l is liquid, and a is air. The parameter m is the Archie cementation factor, which is nominally taken as 2.15 for consolidated sandstones. The thermal conductivity of the dry sandstone at 68°F is estimated from Eq. E-50.

$$k_{hR,d} = 0.34\rho_{R,d} - 3.20\phi + 0.53k^{0.10} + 0.0310F - 0.031. \dots \dots \dots (E-50)$$

Prats (1982) reports that the agreement between values of effective thermal conductivity reported in the literature and those computed with Eq. E-50 is within 15% for 85% of the values when the ratio of $k_{hR,l}/k_{hR,d}$ is between the values of 1.20 and 2.30.

At low moderate temperatures (< 250°F), the effect of temperature on the effective thermal conductivity of liquid-saturated sandstones is given by Eq. E-51.

$$k_{hR,l}(T) = k_{hR,l} - 0.71\times10^{-3}(T-68)\left(k_{hR,l} - 0.80\right)$$
$$\times\left[k_{hR,l}(T+459.6)\times10^{-3}\right]^{-0.55k_{hR,l}} + 0.74. \dots \dots \dots (E-51)$$

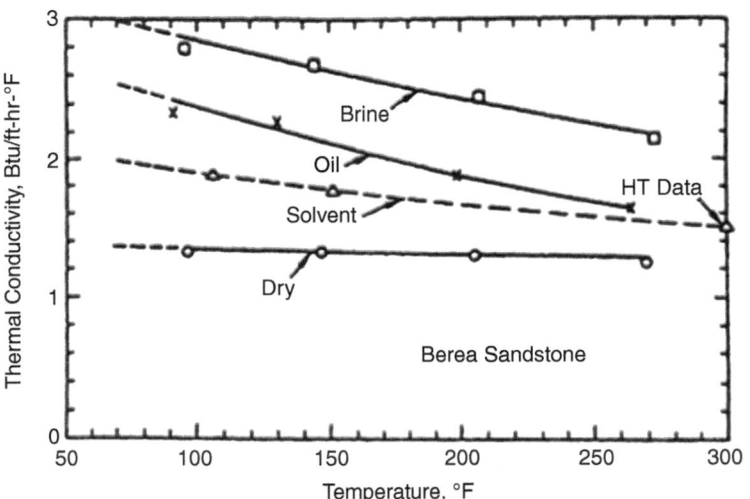

Fig. E-6—Thermal conductivity of Berea sandstone: effects of temperature and fluid saturant (Anand et al. 1973).

At elevated temperatures, the effective thermal conductivity of consolidated rocks decreases markedly with temperature, as shown in **Fig. E-7.** Thermal alteration of clays and other mineral constituents may occur at elevated temperatures with the result that samples exposed to elevated temperatures often exhibit hysteresis in the effective thermal conductivity. Effects of hysteresis are shown in **Table E-4,** which presents the data used to prepare Fig. E-7 relative to measurements at 200°F. **Table E-5** is a description of the samples used in Table E-4 (Somerton and Boozer 1960). Messmer (1985) reported similar data (shown in **Fig. E-8**) for a Teapot sandstone. Less sensitivity to temperature is indicated after exposure to elevated temperatures, such as might be encountered when a combustion front passes through the rock.

Messmer (1985) determined the effect of temperature on the effective thermal conductivity of air-saturated consolidated sandstones over the temperature range of 50 to 1,000°F and was able to correlate the data with Eq. E-52. The effect of pressure on the effective thermal conductivity of liquid-saturated porous rocks is negligible.

$$k_h(T) = k_{hf}^{\phi} k_{hs}^{(1-\phi)} \dots \dots \dots \dots \dots \dots (E-52)$$

As noted earlier, the type of fluid saturating a porous rock has a significant effect on the effective thermal conductivity of the rock. This is illustrated in **Fig. E-9** (Messmer 1985), where effective thermal conductivity data are presented for three systems—air/water, air/Soltrol, and water (S_w = 19.2%)/Soltrol/air—for the same sandstone. Significant differences in the effective thermal conductivity are indicated. In Fig. E-9, the effective thermal conductivity is a linear function of liquid saturation for liquid saturations > 20%. The variation with saturation is less than 20% over the range of liquid saturations from 50 to 100%. Thus, Messmer concludes that effective thermal conductivities for water/oil systems can be approximated from water/gas and oil/gas effective thermal conductivities.

Tikhomirov (1968) developed the empirical relationship given by Eq. E-53 from a statistical analysis of 274 data points for sandstones saturated with water and gas.

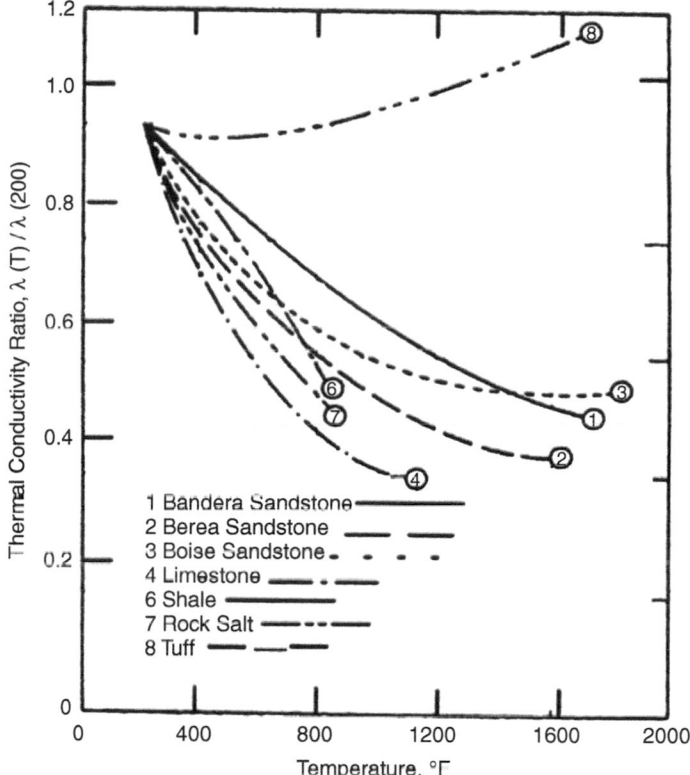

Fig. E-7—Thermal conductivity relative to that at 200°F—samples subjected to increasing temperatures (Somerton and Boozer 1960).

$$k_h(T) = 8.787 \left[e^{0.6(\rho_b + S_w)} / (T + 459.6)^{0.55} \right]. \dots \dots \dots \dots \dots \dots \dots \dots \dots \dots \dots \dots \dots \dots \dots (E-53)$$

Few data are available on the effective thermal conductivities of consolidated porous rocks containing more than a single phase, such as oil/water, oil/gas, or water/gas. One might expect that wettability would influence the effective thermal conductivity of rocks containing multiple phases. Further research is needed to determine these effects. In the absence of experimental data, Prats (1982) suggests the following averaging procedure for estimating the effective thermal conductivity of consolidated porous rocks containing oil, water, and gas saturations.

$$k_{hR} = k_{hR,w} \left(k_{ho} / k_{hR,w} \right)^{\phi S_o} \left(k_{hg} / k_{hR,w} \right)^{\phi S_g}. \dots \dots \dots \dots \dots \dots \dots \dots \dots \dots \dots \dots \dots \dots \dots \dots (E-54)$$

Thermal Diffusivity. Transient heat transfer is governed by the thermal diffusivity of the porous rocks defined by Eq. E-55.

$$\alpha = k_h / M. \dots \dots \dots \dots \dots \dots \dots \dots \dots \dots \dots \dots \dots \dots \dots \dots \dots \dots \dots \dots \dots \dots \dots \dots \dots \dots \dots \dots \dots \dots \dots \dots \dots (E-55)$$

Sample	Bulk Density (lbm/ft³)		Thermal Diffusivity Unsteady State (ft²/D at 200°F)		Thermal Diffusivity Steady State (ft²/D)		Thermal Conductivity Unsteady State (Btu/D-ft-°F at 200°F)	
	Initial	Repeat	Initial	Repeat	At 90°F	At 275°F	Initial	Repeat
Bandera sandstone	134.2	131.8 1	0.816	0.660	0.895	0.660	21.6	16.6
Berea sandstone	134.8	26.0	0.821	0.581	0.948	0.665	21.8	15.7
Boise sandstone	118.9	116.0	0.833	0.492	0.694	0.624	19.5	11.8
Limestone	140.2	78.5*	0.780	0.497	0.780	0.643	21.7	13.9
CaO*	78.5	–	0.924	–	–	–	14.6	–
Shale	137.1	128.2	0.936	0.516	0.950	0.698	26.2	13.9
Rock salt	135.0	128.0	2.95	–	1.7	1.5	8.30	8.30
Tuffaceous sandstone	115.3	170.2	0.444	–	0.504	–	9.55	9.55

*After reaction

Table E-4—Thermal characteristics of test samples (Somerton and Boozer 1960).

Sample	Description	Principal Minerals Quartz (%)	Feldspar (%)	Other	Porosity
Bandera sandstone	Well-consolidated, very fine grained	35	25	Calcite, clay	0.200
Berea sandstone	Well-consolidated, fine grained	65	10	Calcite, sericite, clay	0.205
Boise sandstone	Well-consolidated, medium grained	40	35	Clay, sericite	0.265
Limestone	Small vugs, medium-course grained	–	–	CaCO₃	0.186
Shale	Hard, laminated, very fine grained	50	–	Clay, iron oxides, biotite	0.170
Rock salt	Crystalline	–	–	Halite	0.010
Tuffaceous sandstone	Well-consolidated, very fine grained	10	60	Clay, pumice lapilli, calcite	0.280

Table E-5—Description of samples used in Table E-4 (Somerton and Boozer 1960).

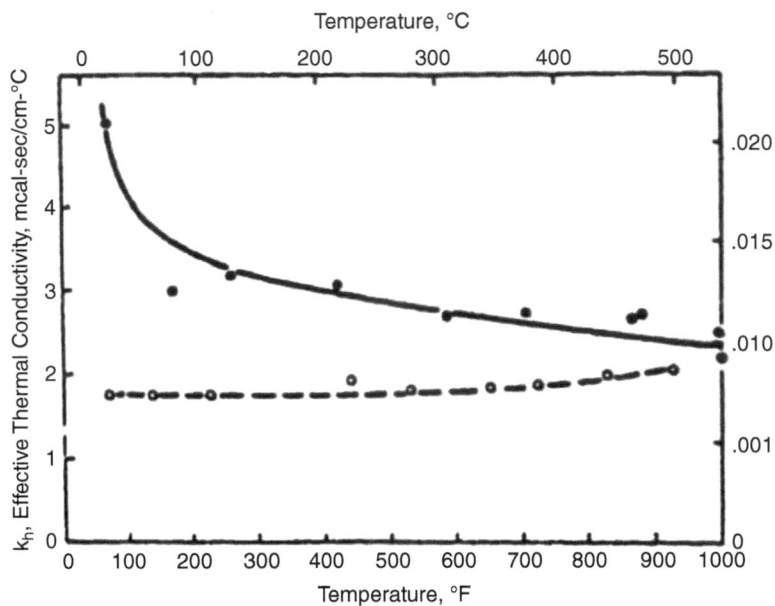

Fig. E-8—The effect of temperature on the thermal conductivity of a Teapot sandstone (Messmer 1985).

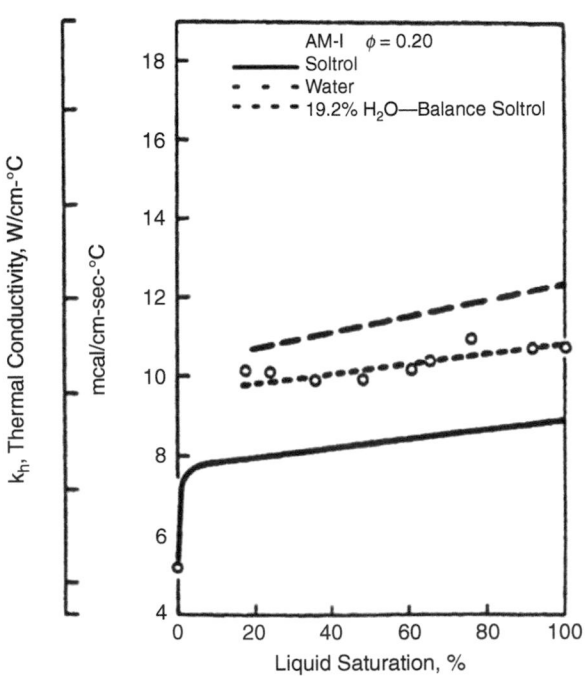

Fig. E-9—The effective thermal conductivity of sandstone (Messmer 1985).

Table E-6 summarizes representative properties of porous rocks (Boberg 1988; Somerton 1958), and **Table E-7** gives a description of the porous rocks (Somerton 1958).

Air

Properties of air are used in wellbore heat-transfer calculations when the casing annulus is dry. The following correlations represent the density, heat capacity, thermal conductivity, and viscosity of air over the range of temperatures expected in wellbores.

Density. (Determined assuming an ideal gas.)

$$\rho_a = 2.71(p+14.7)/(T+459.7). \dots\dots\dots\dots\dots(E\text{-}56)$$

Heat Capacity (Himmelblau 1974).

$$C_a = a + bT + cT^2 + dT^3,$$

$$a = 0.2391,$$

$$b = 9.99 \times 10^{-6},$$

$$c = 8.42 \times 10^{-9},$$

$$\text{and } d = 2.79 \times 10^{-12}. \dots\dots\dots\dots\dots\dots\dots(E\text{-}57)$$

Sample	Bulk Density (lbm/ft³)		Heat Capacity (Btu/lbm-°F)		Thermal Conductivity (Btu/hr-ft-°F)		Thermal Diffusivity (ft²/hr)	
	Air	Water	Air	Water	Air	Water	Air	Water
1. Sandstone	130	142	0.183	0.252	0.507	1.592	0.0213	0.0445
2. Sandstone	90	115	0.200	0.374	0.285	1.050	0.0158	0.0244
(original)	109	126	0.200	0.308	(0.34)	(1.32)	(0.0156)	(0.0340)
3. Silty sand	90	115	0.202	0.376	0.285	(1.05)	0.0157	(0.0242)
(original)	119	132	0.202	0.288	(0.40)	(1.50)	(0.0167)	(0.0394)
4. Silty sand	85	112	0.200	0.393	0.261	1.110	0.0153	0.0249
(original)	116	130	0.200	0.286	(0.47)	(1.51)	(0.0202)	(0.0406)
5. Siltstone	96	118	0.202	0.351	0.338	1.036	0.0174	0.0250
(original)	105	123	0.202	0.318	(0.44)	(1.29)	(0.0207)	(0.0329)
6. Siltstone	120	132	0.204	0.276	0.396	(1.51)	0.0162	(0.0414)
7. Shale	145	149	0.192	0.213	0.603	0.975	0.0216	0.0307
8. Limestone	137	149	0.202	0.266	0.983	2.050	0.0355	0.0517
9. Sand (fine)	102	126	0.183	0.339	0.362	1.590	0.0194	0.0372
10. Sand (coarse)	109	130	0.183	0.315	0.322	1.775	0.0161	0.0433

Values in parentheses estimated.

Table E-6—Calculated thermal diffusivities (Somerton 1958).

Sample	Description	Porosity	Principal Minerals		
			Quartz (%)	Clay Mineral	Other
1. Sandstone	Well consolidated, medium-coarse grain	0.196	80	Tr. kaolinite	Tr. Pyrite
2. Sandstone	Poorly consolidated, medium-fine grain	0.273	40	Illite (?)	Feldspars
3. Silty sand	Poorly consolidated, poorly sorted	0.207	20	Kaolinite type	Feldspars
4. Silty sand	Medium hard, poorly sorted	0.225	20	Kaolinite type	Feldspars
5. Siltstone	Medium hard, broken	0.296	20	Kaolinite type	Feldspars
6. Siltstone	Hard	0.199	25	Illite	Feldspars
7. Shale	Hard, laminated	0.071	40	Illite-Kaolinite	–
8. Limestone	Granular, uniform texture	0.186	–	–	Calcium carbonate
9. Sand	Unconsolidated, fine-grained	0.38	100	–	–
10. Sand	Unconsolidated, coarse-grained	0.34	100	–	–

Table E-7—Description of test samples used in Table E-6 (Somerton 1958).

Thermal Conductivity.

$$k_{ha} = 0.0135 + 2.38 \times 10^{-5}T \dots\dots\dots\dots\dots\dots\dots\dots\dots\dots\dots\dots\dots\dots\dots\dots\dots\dots\dots\text{(E-58)}$$

Viscosity.

$$\mu_a = 0.01709\left[(T + 459.7)/273.1\right]^{0.768} \dots\dots\dots\dots\dots\dots\dots\dots\dots\dots\dots\dots\dots\dots\dots\text{(E-59)}$$

Thermal Properties of Other Materials (Prats 1982; Boberg 1988). Numerous situations are encountered in designing and evaluating thermal recovery processes in which heat-transfer calculations are made involving metals and insulating materials. **Table E-8** includes thermophysical properties of selected metals and alloys (Prats 1982; Rosenow and Choi 1961). **Table E-9** contains thermophysical properties of selected nonmetallic materials (Prats 1982; Rosenow and Choi 1961).

    Table E-10 contains thermal conductivities of insulating materials at high temperatures (Boberg 1988; Marks 1978). These data are useful in evaluating the feasibility of reducing heat losses from surface equipment and lines. **Table E-11** presents thermal conductivities of various materials at different temperatures (McAdams 1954). Thermal radiation is a significant heat-transfer mechanism in dry and in many steam-injection wells. **Table E-12** gives emissivities of various surfaces (Boberg 1988; McAdams 1954).

Metals and Alloys	Density ρ at 68°F (lbm/ft³)	Isobaric Heat Capacity C at 68°F (Btu/lbm-°F)	Thermal Conductivity k_h (Btu/D-ft-°F)			Thermal Diffusivity α at 68°F (ft²/D)
			At 68°F	At 212°F	At 1,112°F	
Aluminum, pure	169	0.214	2,830	2,860	–	87.96
Brass (70% Cu, 30% Zn)	532	0.092	1,500	1,800	–	31.73
Constantin (60% Cu, 40 Ni)	557	0.098	310	310	–	5.69
Copper, pure	559	0.0915	5,350	5,260	4,900	104.5
Iron						
Pure	493	0.108	1,000	940	550	18.8
Cast (C ≈ 4%)	454	0.10	720	–	–	16.0
Wrought (C < 0.5%)	490	0.11	820	790	500	15.2
Lead, pure	710	0.031	480	463	–	22.2
Magnesium, pure	109	0.242	2,400	2,300	–	90.29
Molybdenum	638	0.060	1,700	1,600	1,500	49.78
Nickel						
Pure (99.9%)	556	0.1065	1,200	1,200	–	21.1
Impure (99.2%)	556	0.106	960	890	770	16.2
Silver, pure	657	0.056	5,800	5,760	–	158.4
Steel, mild, 1% C	487	0.113	600	600	460	10.8
Stainless steel (18 Cr, 8 Ni)	488	0.11	230	240	310	4.13
Tin, pure	456	0.054	890	820	–	36.12
Tungsten	1,208	0.032	2,300	2,100	1,600	58.32
Zinc, pure	446	0.092	1,560	1,500	–	38.18
		516				

Reprinted by permission of Prentice-Hall Inc.

Table E-8—Thermophysical properties of selected metals and alloys (Rosenow and Choi 1961).

Material	Temperature T (°F)	Density ρ (lbm/ft³)	Isobaric Heat Capacity C (Btu/lbm-°F)	Thermal Conductivity k_h (Btu/P-ft-°F)	Thermal Diffusivity α (ft²/D)
Aerogel, silica	100	5.3	0.205	0.31	0.29
Asbestos	32	36	0.25	2.1	0.24
	800	36	–	3.1	–
Brick					
Common	68	100	0.20	2.4 to 4.8	0.2 to 0.5
Fire Clay	1,472	145	0.23	19	0.58
Bakelite	68	79.5	0.38	3.2	0.11
Concrete	68	119 to 144	0.21	11 to 19	0.45 to 0.65
Corkboard	100	10	0.4	0.60	0.1
Diatomaceous earth, powdered	100	14	0.21	0.72	0.2
Fiber insulating board	100	14.3	–	0.58	–
Glass, window	68	162	0.16	12	0.48
Glass wool					
Fine	100	1.5	–	0.74	–
Packed	100	6.0	–	0.52	–
Ice	32	57	0.46	30.7	1.2
Magnesia, 85%	100	17	–	0.94	–

Table E-9—Thermophysical properties of selected nonmetallic materials (Rosenow and Choi 1961).

Material	Temperature T (°F)	Density ρ (lbm/ft³)	Isobaric Heat Capacity C (Btu/lbm-°F)	Thermal Conductivity k_h (Btu/P-ft-°F)	Thermal Diffusivity α (ft²/D)
Marble	68	156-169	0.193	38	1.3
Paper	–	–	–	1.8	–
Rock wool	100	12	–	0.55	–
Rubber, hard	32	74.8	0.48	2.1	0.058
Wood, oak, $\perp$ to grain	70	51	0.57	2.9	0.1
Wood, oak, $\parallel$ to grain	70	51	0.57	5.5	0.17

Reprinted by permission of Prentice-Hall Inc.

Table E-9—Thermophysical properties of selected nonmetallic materials (Rosenow and Choi 1961) (continued).

Material*	Bulk Density (lbm/ft³)	Maximum Temperature (°F)	Thermal Conductivities at Various Temperatures					
			100°F	300°F	500°F	1,000°F	1,500°F	2,000°F
Asbestos paper, laminated	22	400	0.038	0.042	–	–	–	–
Asbestos paper, corrugated	16	300	0.031	0.042	–	–	–	–
Diatomaceous earth, silica, powder	18.7	1,500	0.037	0.045	0.053	0.074	–	–
Diatomaceous earth, asbestos and bonding material	18	1,600	0.045	0.049	0.053	0.065	–	–
Fiberglas block, PF612	2.5	500	0.023	0.039	–	–	–	–
Fiberglas block, PF614	4.25	500	0.021	0.033	–	–	–	–
Fiberglas block, PF617	9	500	0.020	0.033	–	–	–	–
Fiberglas, metal mesh blanket, #900	–	1,000	0.020	0.030	0.040	–	–	–
Cellular glass blocks, average value	8.5	900	0.033	0.045	0.062	–	–	–
Hydrous calcium silicate, "Kaylo"	11	1,200	0.032	0.038	0.045	–	–	–
85% magnesia	12	600	0.029	0.035	–	–	–	–
Micro-quartz fiber, blanket	3	3,000	0.021	0.028	0.042	0.075	0.108	0.142
Potassium titanate, fibers	71.5	–	–	0.022	0.024	0.030	–	–
Rock wool, loose	8 to 12	–	0.027	0.038	0.049	0.078	–	–
Zirconia grain	113	3,000	–	–	0.108	0.129	0.163	0.217

*[k = Btu/hr-ft-°F]. Marks, L. S. 1978. *Standard Handbook for Mechanical Engineers*. New York City: McGraw-Hill. Reproduced with permission of McGraw-Hill Co.

Table E-10—Thermal conductivities of insulating materials at high temperatures (Marks 1978).

	At 32°F	At 212°F	At 392°F		At 572°F	At 752°F
Aluminum*	117	119	124	–	133	144
Stainless steel**	–	9.4	10.2	–	10.9	11.7
Mild steel*	–	26	26	–	25	23
Cement, dry†	–	0.2	–	0.4	–	–
Cement, wet†	–	0.5	–	0.6	–	–

*McAdams, W. H. 1954. *Heat Transmission*. New York City: McGraw-Hill. Reproduced with permission of McGraw-Hill Co.
**304, 316 Stainless Steel.
†From Boberg, ©1988 John Wiley Publishing.

Table E-11—Thermal conductivities of various materials (McAdams 1954).

Iron	
Rusted	0.69
Dark grey	0.31
Steel	
Oxidized	0.74
Rough oxide layer	0.80
Rough plate	0.94 to 0.97
Stainless steel	
316	0.56
304	0.36 to 0.44
Aluminum, rough polish	0.18

McAdams, W.H. 1954. *Heat Transmission.* New York City: McGraw-Hill. Reproduced with permission of McGraw-Hill Co.

Table E-12—Emissivities of various surfaces (McAdams 1954).

Casing and Tubing Properties (Prats 1982; Allen and Roberts 1978). Tables E-13 and E-14 contain standard data on casing and tubing commonly used for well completions in thermal-recovery projects.

Nomenclature

a,b = constants in Eq. E-23

a,b,c,d = constants in Eq. E-56

B_{od} = volume of oil at some reservoir pressure other than bubblepoint pressure required to yield 1 bbl of residual oil at 60°F when differentially liberated from pressure p to atmospheric pressure, L^3/L^3, bbl/bbl residual oil

B_{odb} = FVF for oil at the bubblepoint pressure, p_b, by differential liberation from the bubblepoint pressure to atmospheric pressure, L^3/L^3, bbl/bbl residual oil

B_{of} = FVF for oil at pressure, p, when flashed through separator to stock-tank conditions, L^3/L^3, bbl/bbl

B_{ofb} = FVF for oil at bubblepoint pressure when flashed through separator to stock-tank conditions, L^3/L^3, bbl/bbl

c_o = isothermal compressibility of oil, m/Lt^2, 1/psi

C_a = mean heat capacity of air at constant pressure, L^2/t^2-T, Btu/lbm-°F

C_f = mean heat capacity of fluid at constant pressure, L^2/t^2-T, Btu/lbm-°F

C_m = mean heat capacity of a saturated porous rock, L^2/t^2-T, Btu/lbm-°F

C_{po} = heat capacity of oil at constant pressure, L^2/t^2-T, Btu/lbm-°F

C_r = mean heat capacity of matrix material at constant pressure, L^2/t^2-T, Btu/lbm-°F

C_s = mean heat capacity of steam or of saturated water vapor at T_s, L^2/t^2-T, Btu/lbm-°F

C_1 = term in oil density correlation in Eq. E-20

C_2 = term in oil density correlation in Eq. E-20

f = fraction of gas saturation that is noncondensible, L^3/L^3, volume fraction

f_{ov} = correlating factor for calculating viscosity of gas-saturated oil in Eq. E-36

f_q = volume fraction of quartz in solid matrix, L^3/L^3, ft^3/ft^3

F = formation factor

F_B = correlating factor for calculating oil FVF

H_v = enthalpy of saturated vapor, L^2/t^2, Btu/lbm

H_w = enthalpy of saturated water or enthalpy of produced water and water vapor, L^2/t^2, Btu/lbm

k = permeability of porous rock, L^2, md

k_h = thermal conductivity, mL/t^3T, Btu/(hr-ft-°F)

k_{ha} = thermal conductivity of air in annulus at average temperature and pressure of annulus or at the film temperature, mL/t^3T, Btu/(hr-ft-°F)

k_{hf} = thermal conductivity of fluid or of formation, mL/t^3T, Btu/(hr-ft-°F)

k_{hg} = thermal conductivity of gas, mL/t^3T, Btu/(hr-ft-°F)

| Tubing Size | | Nominal Weight | | | | | | | Threaded and Coupled | | | | Integral Joint | |
Nom. (in.)	OD (in.)	Nonupset (lbm/ft)	T&C Upset (lbm/ft)	Internal Joint (lbm/ft)	Grade	Wall Thickness (in.)	ID (in.)	Drift Diameter (in.)	Coupled OD (in.) Nonupset	Upset Regular	Upset Special	Drift Diameter (in.)	Box OD (in.)
¾	1.050	1.14	1.20	–	H-40	0.113	0.824	0.730	1.313	1.660	–	–	–
	1.050	1.14	1.20	–	J-55	0.113	0.824	0.730	1.313	1.660	–	–	–
	1.050	1.14	1.20	–	C-75	0.113	0.824	0.730	1.313	1.660	–	–	–
	1.050	1.14	1.20	–	N-80	0.113	0.824	0.730	1.313	1.660	–	–	–
1	1.315	1.70	1.80	1.72	H-40	0.133	1.049	0.955	1.660	1.900	–	0.955	1.550
	1.315	1.70	1.80	1.72	J-55	0.133	1.049	0.955	1.660	1.900	–	0.955	1.550
	1.315	1.70	1.80	1.72	C-75	0.133	1.049	0.955	1.660	1.900	–	0.955	1.550
	1.315	1.70	1.80	1.72	N-80	0.133	1.049	0.955	1.660	1.900	–	0.955	1.550
1¼	1.660	–	–	2.10	H-40	0.125	1.410	–	–	–	–	1.286	1.880
	1.660	2.30	2.40	2.33	H-40	0.140	1.380	1.286	2.054	2.200	–	1.286	1.880
	1.660	–	–	2.10	J-55	0.125	1.410	–	–	–	–	1.286	1.880
	1.660	2.30	2.40	2.33	J-55	0.140	1.380	1.286	2.054	2.200	–	1.286	1.880
	1.660	2.30	2.40	2.33	C-75	0.140	1.380	1.286	2.054	2.200	–	1.286	1.880
	1.660	2.30	2.40	2.33	N-80	0.140	1.380	1.286	2.054	2.200	–	1.286	1.880
1½	1.900	–	–	2.40	I-40	0.125	1.650	–	–	–	–	1.516	2.110
	1.900	2.75	2.90	2.76	H-40	0.145	1.610	1.516	2.200	2.500	–	1.516	2.110
	1.900	–	–	2.40	J-55	0.125	1.650	–	–	–	–	1.516	2.110
	1.900	2.75	2.90	2.76	J-55	0.145	1.610	1.516	2.200	2.500	–	1.516	2.110
	1.900	2.75	2.90	2.76	C-75	0.145	1.610	1.516	2.200	2.500	–	1.516	2.110
	1.900	2.75	2.90	2.76	N-80	0.145	1.610	1.516	2.200	2.500	–	1.516	2.110
2¹/₁₆	2.063	–	–	3.25	H-40	0.156	1.751	–	–	–	–	1.657	2.325
	2.063	–	–	3.25	J-55	0.156	1.751	–	–	–	–	1.657	2.325
	2.063	–	–	3.25	C-75	0.156	1.751	–	–	–	–	1.657	2.325
	2.063	–	–	3.25	N-80	0.156	1.751	–	–	–	–	1.657	2.325
2⅜	2.375	4.00	–	–	H-40	0.167	2.041	1.947	2.875	–	–	–	–
	2.375	4.60	4.70	–	H-40	0.190	1.995	1.901	2.875	3.063	2.910	–	–
	2.375	4.00	–	–	J-55	0.167	2.041	1.947	2.873	–	–	–	–

Table E-13—Tubing minimum performance properties (Allen and Roberts 1978).

| Tubing Size | | Nominal Weight | | | Grade | Wall Thickness (in.) | ID (in.) | Drift Diameter (in.) | Threaded and Coupled — Coupled OD (in.) | | | Integral Joint | |
Nom. (in.)	OD (in.)	Nonupset (lbm/ft)	T&C Upset (lbm/ft)	Internal Joint (lbm/ft)					Nonupset	Upset Regular	Upset Special	Drift Diameter (in.)	Box OD (in.)
	2.375	4.60	4.70	–	J-55	0.190	1.995	1.901	2.875	3.063	2.910	–	–
	2.375	4.00	–	–	C-75	0.167	2.041	1.947	2.875	–	–	–	–
	2.375	4.60	4.70	–	C-75	0.190	1.995	1.901	2.875	3.063	2.910	–	–
	2.375	5.80	5.95	–	C-75	0.254	1.867	1.773	2.875	3.063	2.910	–	–
	2.375	4.00	–	–	N-80	0.167	2.041	1.947	2.875	–	–	–	–
	2.375	4.60	4.70	–	N-80	0.190	1.995	1.901	2.875	3.063	2.910	–	–
	2.375	5.80	5.95	–	N-80	0.254	1.867	1.773	2.875	3.063	2.910	–	–
	2.375	4.60	4.70	–	P-105	0.190	1.995	1.901	2.875	3.063	2.910	–	–
	2.375	5.80	5.95	–	P-105	0.254	1.867	1.773	2.875	3.063	2.910	–	–
2⅞	2.875	6.40	6.50	–	H-40	0.217	2.441	2.347	3.500	3.668	3.460	–	–
	2.875	6.40	6.50	–	J-55	0.217	2.441	2.347	3.500	3.668	3.460	–	–
	2.875	6.40	6.50	–	C-75	0.217	2.441	2.347	3.500	3.668	3.460	–	–
	2.875	8.60	8.70	–	C-75	0.308	2.259	2.165	3.500	3.668	3.460	–	–
	2.875	6.40	6.50	–	N-80	0.217	2.441	2.347	3.500	3.668	3.460	–	–
	2.875	8.60	8.70	–	N-80	0.308	2.259	2.165	3.500	3.668	3.460	–	–
	2.875	6.40	6.50	–	P-105	0.217	2.441	2.347	3.500	3.668	3.460	–	–
	2.875	8.60	8.70	–	P-105	0.308	2.259	2.165	3.500	3.668	3.460	–	–
3½	3.500	7.70	–	–	H-40	0.216	3.068	2.943	4.250	–	–	–	–
	3.500	9.20	9.30	–	H-40	0.254	2.992	2.867	4.250	4.500	4.180	–	–
	3.500	10.20	–	–	H-40	0.289	2.922	2.797	4.250	–	–	–	–
	3.500	7.70	–	–	J-55	0.216	3.068	2.943	4.250	–	–	–	–
	3.500	9.20	9.30	–	J-55	0.254	2.992	2.867	4.250	4.500	4.180	–	–
	3.500	10.20	–	–	J-55	0.289	2.922	2.797	4.250	–	–	–	–
	3.500	7.70	–	–	C-75	0.216	3.068	2.943	4.250	–	–	–	–
	3.500	9.20	9.30	–	C-75	0.254	2.992	2.867	4.250	4.500	4.180	–	–
	3.500	10.20	–	–	C-75	0.289	2.922	2.797	4.250	–	–	–	–
	3.500	12.70	12.95	–	C-75	0.375	2.750	2.625	4.250	4.500	4.180	–	–
	3.500	7.70	–	–	N-80	0.216	3.068	2.943	4.250	–	–	–	–

Table E-13—Tubing minimum performance properties (Allen and Roberts 1978) (continued).

| Tubing Size | | Nominal Weight | | | | | | | Threaded and Coupled | | | | Integral Joint | |
Nom. (in.)	OD (in.)	Nonupset (lbm/ft)	T&C Upset (lbm/ft)	Internal Joint (lbm/ft)	Grade	Wall Thickness (in.)	ID (in.)	Drift Diameter (in.)	Coupled OD (in.) Nonupset	Coupled OD (in.) Upset Regular	Coupled OD (in.) Upset Special	Drift Diameter (in.)	Box OD (in.)
	3.500	9.20	9.30	–	N-80	0.254	2.992	2.867	4.250	4.500	4.180	–	–
	3.500	10.20	–	–	N-80	0.289	2.922	2.797	4.250	–	–	–	–
	3.500	12.70	12.95	–	N-80	0.375	2.750	2.625	4.250	4.500	4.180	–	–
	3.500	9.20	9.30	–	P-105	0.254	2.992	2.867	4.250	4.500	4.180	–	–
	3.500	12.70	12.95	–	P-105	0.375	2.750	2.625	4.250	4.500	4.180	–	–
4	4.000	9.50	–	–	I-40	0.225	3.548	3.423	4.750	–	–	–	–
	4.000	–	11.00	–	I-40	0.262	3.476	3.351	–	5.000	–	–	–
	4.000	9.50	–	–	J-55	0.226	3.548	3.423	4.750	–	–	–	–
	4.000	–	11.00	–	J-55	0.262	3.476	3.351	–	5.000	–	–	–
	4.000	9.50	–	–	C-75	0.226	3.548	3.423	4.750	–	–	–	–
	4.000	–	11.00	–	C-75	0.262	3.476	3.351	–	5.000	–	–	–
	4.000	9.50	–	–	N-80	0.226	3.548	3.423	4.750	–	–	–	–
	4.000	–	11.00	–	N-80	0.262	3.376	3.351	–	5.000	–	–	–
4½	4.500	12.60	12.75	–	H-40	0.271	3.958	3.833	5.200	5.563	–	–	–
	4.500	12.60	12.75	–	J-55	0.271	3.958	3.833	5.200	5.563	–	–	–
	4.500	12.60	12.75	–	C-75	0.271	3.958	3.833	5.200	5.563	–	–	–
	4.500	12.60	12.75	–	N-80	0.271	3.958	3.833	5.200	5.563	–	–	–

Table E-13—Tubing minimum performance properties (Allen and Roberts 1978) (continued).

Tubing Size		Nominal Weight								Joint Yield Strength		
Nom. (in.)	OD (in.)	Nonupset (lbm/ft)	T&C Upset (lbm/ft)	Internal Joint (lbm/ft)	Grade	Well Thickness (in.)	ID (in.)	Collapse Resistance (psi)	Internal Yield Pressure (psi)	T&C Nonupset (lbm)	T&C Upset (lbm)	Internal Joint (lbm)
¾	1.050	1.14	1.20	—	H-40	0.113	0.824	7,680	7,530	6,360	13,300	—
	1.050	1.14	1.20	—	J-55	0.113	0.824	10,560	10,360	8,740	18,290	—
	1.050	1.14	1.20	—	C-75	0.113	0.824	14,410	14,120	11,920	24,940	—
	1.050	1.14	1.20	—	N-80	0.113	0.824	15,370	15,070	12,710	26,610	—
1	1.315	1.70	1.80	1.72	H-40	0.133	1.049	7,270	7,080	10,960	19,760	15,970
	1.315	1.70	1.80	1.72	J-55	0.133	1.049	10,000	9,730	15,060	27,160	21,960
	1.315	1.70	1.80	1.72	C-75	0.133	1.049	13,640	13,270	20,540	37,040	29,940
	1.315	1.70	1.80	1.72	N-80	0.133	1.049	14,550	14,160	21,910	39,510	31,940
1¼	1.660	—	—	2.10	H-40	0.125	1.410	5,570	5,270	—	—	22,180
	1.660	2.30	2.40	2.33	H-40	0.140	1.380	6,180	5,900	15,530	26,740	22,180
	1.660	—	—	2.10	J-55	0.125	1.410	7,660	7,250	—	—	30,500
	1.660	2.30	2.40	2.33	J-55	0.140	1.380	8,500	8,120	21,360	36,770	30,500
	1.660	2.30	2.40	2.33	C-75	0.140	1.380	11,580	11,070	29,120	50,140	41,600
	1.660	2.30	2.40	2.33	N-80	0.140	1.380	12,360	11,810	31,060	53,480	44,370
1½	1.900	—	—	2.40	H-40	0.125	1.650	4,920	4,610	—	—	26,890
	1.900	2.75	2.90	2.76	H-40	0.145	1.610	5,640	5,340	19,090	31,980	26,890
	1.900	—	—	2.40	J-55	0.125	1.650	6,640	6,330	—	—	36,970
	1.900	2.75	2.90	2.76	J-55	0.145	1.610	7,750	7,350	26,250	43,970	36,970
	1.900	2.75	2.90	2.76	C-75	0.145	1.610	10,570	10,020	35,800	59,960	50,420
	1.900	2.75	2.90	2.76	N-80	0.145	1.610	11,280	10,680	38,180	63,960	53,780
2¹/₁₆	2.063	—	—	3.25	H-40	0.156	1.751	5,590	5,290	—	—	35,690
	2.063	—	—	3.25	J-55	0.156	1.751	7,690	7,280	—	—	49,070
	2.063	—	—	3.25	C-75	0.156	1.751	10,480	9,920	—	—	66,910
	2.063	—	—	3.25	N-80	0.156	1.751	11,180	10,590	—	—	71,370
2⅜	2.375	4.00	—	—	H-40	0.167	2.041	5,230	4,920	30,130	—	—
	2.375	4.60	4.70	—	H-40	0.190	1.995	5,890	5,600	35,960	52,170	—
	2.375	4.00	—	—	J-55	0.167	2.041	7,190	6,770	41,430	—	—
	2.375	4.60	4.70	—	J-55	0.190	1.995	8,100	7,700	49,450	71,730	—
	2.375	4.00	—	—	C-75	0.167	2.041	9,520	9,230	56,500	—	—
	2.375	4.60	4.70	—	C-75	0.190	1.995	11,040	10,500	67,430	97,820	—
	2.375	5.80	5.95	—	C-75	0.254	1.867	14,330	14,040	96,560	126,940	—
	2.375	4.00	—	—	N-80	0.167	2.041	9,980	9,840	60,260	—	—
	2.375	4.60	4.70	—	N-80	0.190	1.995	11,780	11,200	71,930	104,340	—
	2.375	5.80	5.95	—	N-80	0.254	1.867	15,280	14,970	102,990	135,400	—
	2.375	4.60	4.70	—	P-105	0.190	1.995	15,460	14,700	94,410	136,940	—
	2.375	5.80	5.95	—	P-105	0.254	1.867	20,060	19,650	135,180	177,710	—

Table E-13—Tubing minimum performance properties (Allen and Roberts 1978) (continued).

| Tubing Size | | Nominal Weight | | | | Well | | | Collapse | Internal Yield | Joint Yield Strength | | |
Nom. (in.)	OD (in.)	Nonupset (lbm/ft)	T&C Upset (lbm/ft)	Internal Joint (lbm/ft)	Grade	Thickness (in.)	ID (in.)	Resistance (psi)	Pressure (psi)	T&C Nonupset (lbm)	T&C Upset (lbm)	Internal Joint (lbm)
2⅞	2.875	6.40	6.50	—	H-40	0.217	2.441	5,580	5,280	52,780	72,480	—
	2.875	6.40	6.50	—	J-55	0.217	2.441	7,680	7,260	72,580	99,660	—
	2.875	6.40	6.50	—	C-75	0.217	2.441	10,470	9,910	98,970	135,900	—
	2.875	8.60	8.70	—	C-75	0.308	2.259	14,350	14,060	149,360	186,290	—
	2.875	6.40	6.50	—	N-80	0.217	2.441	11,160	10,570	105,570	144,960	—
	2.875	8.60	8.70	—	N-80	0.308	2.259	15,300	15,000	159,310	198,710	—
	2.875	6.40	6.50	—	P-105	0.217	2.441	14,010	13,870	138,560	190,260	—
	2.875	8.60	8.70	—	P-105	0.308	2.259	20,090	19,690	209,100	260,810	—
3½	3.500	7.70	—	—	H-40	0.216	3.068	4,630	4,320	65,070	—	—
	3.500	9.20	9.30	—	H-40	0.254	2.992	5,380	5,080	79,540	103,610	—
	3.500	10.20	—	—	H-40	0.289	2.922	6,060	5,780	92,550	—	—
	3.500	7.70	—	—	J-55	0.216	3.068	5,970	5,940	89,470	—	—
	3.500	9.20	9.30	—	J-55	0.254	2.992	7,400	6,980	109,370	142,460	—
	3.500	10.20	—	—	J-55	0.289	2.922	8,330	7,950	127,250	—	—
	3.500	7.70	—	—	C-75	0.216	3.068	7,540	8,100	122,010	—	—
	3.500	9.20	9.30	—	C-75	0.254	2.992	10,040	9,520	149,140	194,260	—
	3.500	10.20	—	—	C-75	0.289	2.922	11,360	10,840	173,530	—	—
	3.500	12.70	12.95	—	C-75	0.375	2.750	14,350	14,060	230,990	276,120	—
	3.500	7.70	—	—	N-80	0.216	3.068	7,870	8,640	130,140	—	—
	3.500	9.20	9.30	—	N-80	0.254	2.992	10,530	10,160	159,090	207,220	—
	3.500	10.20	—	—	N-80	0.289	2.922	12,120	11,560	185,100	—	—
	3.500	12.70	12.95	—	N-80	0.375	2.750	15,310	15,000	246,390	294,530	—
	3.500	9.20	9.30	—	P-105	0.254	2.992	13,050	13,330	208,800	271,970	—
	3.500	12.70	12.95	—	P-105	0.375	2.750	20,090	19,690	323,390	386,570	—
4	4.000	9.50	—	—	H-40	0.226	3.548	4,060	3,950	72,000	—	—
	4.000	—	11.00	—	H-40	0.262	3.476	4,900	4,580	—	123,070	—
	4.000	9.50	—	—	J-55	0.226	3.548	5,110	5,440	99,010	—	—
	4.000	—	11.00	—	J-55	0.262	3.476	6,590	6,300	—	169,220	—
	4.000	9.50	—	—	C-75	0.226	3.548	6,350	7,420	135,010	—	—
	4.000	—	11.00	—	C-75	0.262	3.476	8,410	8,630	—	230,750	—
	4.000	9.50	—	—	N-80	0.226	3.548	6,590	7,910	144,010	—	—
	4.000	—	11.00	—	N-80	0.262	3.376	8,800	9,170	—	246,140	—
4½	4.500	12.60	12.75	—	H-40	0.271	3.958	4,500	4,220	104,360	144,020	—
	4.500	12.60	12.75	—	J-55	0.271	3.958	5,720	5,830	143,500	198,030	—
	4.500	12.60	12.75	—	C-75	0.271	3.958	7,200	7,900	195,680	270,040	—
	4.500	12.60	12.75	—	N-80	0.271	3.958	78,500	8,430	208,730	288,040	—

Table E-13—Tubing minimum performance properties (Allen and Roberts 1978) (continued).

OD Size (in.)	Weight/ Ft Nom. (lbm)	Wall (in.)	ID (in.)	Drift (in.)	Coupling OD (in.)	Minimum Collapse Pressure (psi)				Minimum Internal Yield Pressure (psi)			
						H40	J55	N80	P110	H40	J55	N80	P110
	45.50	0.400	9.950	9.794	11.750	–	2,090	–	–	–	3,580	–	–
	51.00	0.450	9.850	9.694	11.750	–	2,700	3,220	3,670	–	4,030	5,860	8,060
	55.50	0.495	9.760	9.604	11.750	–	–	4,020	4,630	–	–	6,450	8,860
	60.70	0.545	9.660	9.504	11.750	–	–	–	5,860	–	–	–	9,760
	65.70	0.595	9.560	9.404	11.750	–	–	–	7,490	–	–	–	10,650
11¾	42.00	0.333	11.084	10.928	12.750	1,070	–	–	–	1,980	–	–	–
	47.00	0.375	11.000	10.844	12.750	–	1,510	–	–	–	3,070	–	–
	54.00	0.435	10.880	10.724	12.750	–	2,070	–	–	–	3,560	–	–
	60.00	0.489	10.772	10.616	12.750	–	2,660	3,180	–	–	4,010	5,830	–
13⅜	48.00	0.330	12.715	12.559	14.375	740	–	–	–	1,730	–	–	–
	54.50	0.380	12.615	12.459	14.375	–	1,130	–	–	–	2,730	–	–
	61.00	0.430	12.515	12.359	14.375	–	1,540	–	–	–	3,090	–	–
	68.00	0.480	12.415	12.259	14.375	–	1,950	–	–	–	3,450	–	–
	72.00	0.514	12.347	12.191	14.375	–	–	2,670	–	–	–	5,380	–
16	65.00	0.375	15.250	15.062	17.000	630	–	–	–	1,640	–	–	–
	75.00	0.438	15.124	14.936	17.000	–	1,020	–	–	–	2,630	–	–
	84.00	0.495	15.010	14.822	17.000	–	1,410	–	–	–	2,980	–	–
18⅝	87.50	0.435	17.755	17.567	19.625	630	630	–	–	1,630	2,250	–	–
20	94.00	0.438	19.124	18.936	21.000	520	520	–	–	1,530	2,110	–	–
	106.50	0.500	19.000	18.812	21.000	–	770	–	–	–	2,410	–	–
	133.00	0.635	18.730	18.542	21.000	–	1,500	–	–	–	3,060	–	–

Table E-14—Casing minimum performance properties (Allen and Roberts 1978).

OD (in.)	Weight/ Ft Nom. (lbm)	Wall (in.)	ID (in.)	Drift (in.)	Coupling OD (in.)	Joint Yield Strength (1,000 lbm)						
						Short Thread				Long Thread		
						H40	J55	N80	P110	J55	N80	P110
4½	9.50	0.205	4.090	3.965	5.000	77	101	–	–	–	–	–
	11.60	0.250	4.000	3.875	5.000	–	154	–	–	162	223	279
	13.50	0.290	3.920	3.795	5.000	–	–	–	–	–	270	338
	15.10	0.337	3.826	3.701	5.000	–	–	–	–	–	–	406
5	11.50	0.220	4.560	4.435	5.563	–	133	–	–	–	–	–
	13.00	0.253	4.494	4.369	5.563	–	169	–	–	182	–	–
	15.00	0.296	4.408	4.283	5.563	–	207	–	–	223	311	388
	18.00	0.362	4.276	4.151	5.563	–	–	–	–	–	396	495
5½	14.00	0.244	5.012	4.887	6.050	130	172	–	–	–	–	–
	15.50	0.275	4.950	4.825	6.050	–	202	–	–	217	–	–
	17.00	0.304	4.892	4.767	6.050	–	229	–	–	247	348	445
	20.00	0.361	4.778	4.653	6.050	–	–	–	–	–	428	548
	23.00	0.415	4.670	4.545	6.050	–	–	–	–	–	502	643
6⅝	20.00	0.288	6.049	5.924	7.390	184	245	–	–	266	–	–
	24.00	0.352	5.921	5.796	7.390	–	314	–	–	340	481	641
	28.00	0.417	5.791	5.666	7.390	–	–	–	–	–	586	781
	32.00	0.475	5.675	5.550	7.390	–	–	–	–	–	677	904

Table E-14—Casing minimum performance properties (Allen and Roberts 1978) (continued).

| | Dimensions | | | | | Joint Yield Strength (1,000 lbm) | | | | | | |
| | | | | | | Short Thread | | | | Long Thread | | |
OD (in.)	Weight/ Ft Nom. (lbm)	Wall (in.)	ID (in.)	Drift (in.)	Coupling OD (in.)	H40	J55	N80	P110	J55	N80	P110
7	17.00	0.231	6.538	6.413	7.656	122	–	–	–	–	–	–
	20.00	0.272	6.456	6.331	7.656	176	234	–	–	–	–	–
	23.00	0.317	6.366	6.241	7.656	–	284	–	–	313	442	–
	26.00	0.362	6.276	6.151	7.656	–	334	–	–	367	519	693
	29.00	0.408	6.184	6.059	7.656	–	–	–	–	–	597	797
	32.00	0.453	6.094	5.969	7.656	–	–	–	–	–	672	897
	35.00	0.498	6.004	5.879	7.656	–	–	–	–	–	746	996
	38.00	0.540	5.920	5.795	7.656	–	–	–	–	–	814	1,087
7⅝	24.00	0.300	7.025	6.900	8.500	212	–	–	–	–	–	–
	26.40	0.328	6.969	6.844	8.500	–	315	–	–	346	490	–
	29.70	0.375	6.875	6.750	8.500	–	–	–	–	–	575	769
	33.70	0.430	6.765	6.640	8.500	–	–	–	–	–	674	901
	39.00	0.500	6.625	6.500	8.500	–	–	–	–	–	798	1,066
8⅝	24.00	0.264	8.097	7.972	9.625	–	244	–	–	–	–	–
	28.00	0.304	8.017	7.892	9.625	233		–	–	–	–	–
	32.00	0.352	7.921	7.796	9.625	279	372	–	–	417	–	–
	36.00	0.400	7.825	7.700	9.625	–	434	–	–	486	688	–
	40.00	0.450	7.725	7.600	9.625	–	–	–	–	–	788	1,055
	44.00	0.500	7.625	7.500	9.625	–	–	–	–	–	887	1,186
	49.00	0.557	7.511	7.386	9.625	–	–	–	–	–	997	1,335
9⅝	32.30	0.312	9.001	8.845	10.625	254	–	–	–	–		
	36.00	0.352	8.921	8.765	10.625	294	394	–	–	453	–	–
	40.00	0.395	8.835	8.679	10.625	–	452	–	–	520	737	–
	43.50	0.435	8.755	8.599	10.625	–	–	–	–	–	825	1,106
	47.00	0.472	8.681	8.525	10.625	–	–	–	–	–	905	1,213
	53.50	0.545	8.535	8.379	10.625	–	–	–	–	–	1,062	1,422
10¾	32.75	0.279	10.192	10.036	11.750	205	–	–	–	–	–	–
	40.50	0.350	10.050	9.894	11.750	314	420	–	–	–	–	–
	45.50	0.400	9.950	9.794	11.750	–	493	–	–	–	–	–
	51.00	0.450	9.850	9.694	11.750	–	565	804	1,080	–	–	–
	55.50	0.495	9.760	9.604	11.750	–	–	895	1,203	–	–	–
	60.70	0.545	9.660	9.504	11.750	–	–	–	1,338	–	–	–
	65.70	0.595	9.560	9.404	11.750	–	–	–	1,472	–	–	–
11¾	42.00	0.333	11.084	10.928	12.750	307	–	–	–	–	–	–
	47.00	0.375	11.000	10.844	12.750	–	477	–	–	–	–	–
	54.00	0.435	10.880	10.724	12.750	–	568	–	–	–	–	–

Table E-14—Casing minimum performance properties (Allen and Roberts 1978) (continued).

	Dimensions					Joint Yield Strength (1,000 lbm)						
						Short Thread				Long Thread		
	60.00	0.489	10.772	10.616	12.750	–	649	924	–	–	–	–
13⅜	48.00	0.330	12.715	12.559	14.375	322	–	–	–	–	–	–
	54.50	0.380	12.615	12.459	14.375	–	514	–	–	–	–	–
	61.00	0.430	12.515	12.359	14.375	–	595	–	–	–	–	–
	68.00	0.480	12.415	12.259	14.375	–	675	–	–	–	–	–
	72.00	0.514	12.347	12.191	14.375	–	–	1040	–	–	–	–
16	65.00	0.375	15.250	15.062	17.000	439	–	–	–	–	–	–
	75.00	0.438	15.124	14.936	17.000	–	710	–	–	–	–	–
	84.00	0.495	15.010	14.822	17.000	–	817	–	–	–	–	–
18⅝	87.50	0.435	17.755	17.567	19.625	559	754	–	–	–	–	–
20	94.00	0.438	19.124	18.936	21.000	581	784	–	–	907	–	–
	106.50	0.500	19.000	18.812	21.000	–	913	–	–	1,057	–	–
	133.00	0.635	18.730	18.542	21.000	–	1192	–	–	1,380	–	–

Table E-14—Casing minimum performance properties (Allen and Roberts 1978) (continued).

k_{hl}	=	thermal conductivity of liquid, mL/ t^3T, Btu/(hr-ft-°F)
k_{ho}	=	thermal conductivity of oil, mL/t^3T, Btu/(hr-ft-°F)
k_{hR}	=	effective thermal conductivity of oil sands at 125°F, mL/t^3T, Btu/(hr-ft-°F)
$k_{hR,d}$	=	effective thermal conductivity of dry sandstone at 68°F, mL/t^3T, Btu/(hr-ft-°F)
$k_{hR,l}$	=	effective thermal conductivity of liquid-saturated sandstone at 68°F, mL/ t^3T, Btu/(hr-ft-°F)
$k_{hR,w}$	=	effective thermal conductivity of water-saturated sandstone at 68°F, mL/t^3T, Btu/(hr-ft-°F)
k_{hs}	=	thermal conductivity of steam, mL/ t^3T, Btu/(hr-ft-°F)
k_{ht}	=	effective thermal conductivity of porous rock at temperature T, mL/t^3T, Btu/(hr-ft-°F)
k_{hw}	=	thermal conductivity of saturated water, mL/t^3T, Btu/(hr-ft-°F)
k_{hws}	=	thermal conductivity of saturated water vapor, mL/t^3T, Btu/(hr-ft-°F)
L_v	=	latent heat of vaporization for water, L^2/t^2, Btu/lbm
m	=	Archie cementation factor taken as 2.15 for sandstones, or moles of CO_2 produced per mole of CO produced during in-situ combustion
M	=	average volumetric heat capacity of porous rock, Eq. 8.50, m/Lt^2T, Btu/(ft³-°F)
P	=	reservoir pressure, m/Lt^2, psia
p_b	=	bubblepoint pressure, m/Lt^2, psia
p_s	=	saturation pressure of water or steam injection pressure, m/Lt^2, psia
p_{sep}	=	separator pressure, m/Lt^2, psia
R_{sd}	=	gas in solution at any pressure less than bubblepoint pressure in 1 bbl of residual oil when measured by differential liberation, scf/bbl of residual oil
R_{sdb}	=	gas in solution at the bubblepoint pressure in 1 bbl of residual oil when measured by differential liberation, scf/bbl of residual oil
R_{sf}	=	separator and stock-tank gas at any pressure less than the bubblepoint pressure, scf/bbl
R_{sfb}	=	separator and stock-tank gas at the bubblepoint pressure, L^3/L^3, scf/bbl
S_g	=	gas saturation, L^3/L^3, volume fraction
S_o	=	oil saturation, L^3/L^3, volume fraction
S_w	=	water saturation, volume fraction
T	=	reservoir temperature, T, °F
T_s	=	temperature of steam, T, °F
T_{sep}	=	separator temperature, T,°F
y_{ov}	=	correlating parameter for calculating live-oil viscosity in Eq. E-37
α	=	thermal diffusivity of porous rock, L^2/t, ft²/hr
γ_{gsep}	=	gas specific gravity at T_{sep} and p_{sep} (air = 1)
γ_{g100}	=	gas specific gravity at T_{sep} and separation pressure of 100 psig (air = 1)
γ_o	=	stock-tank oil specific gravity at 60°F (water = 1)
μ_a	=	viscosity of air, m/Lt, cp
μ_o	=	viscosity of oil, m/Lt, cp
μ_{ob}	=	viscosity of oil at bubblepoint pressure, m/Lt, cp
μ_{od}	=	viscosity of oil that has no dissolved gas (dead oil), m/Lt, cp
μ_{of}	=	viscosity of gas-saturated oil when $p > p_b$, m/Lt, cp
μ_s	=	viscosity of steam, m/Lt, cp
μ_w	=	viscosity of water, m/Lt, cp
v_{sf}	=	change in specific volume of saturated water upon vaporization, L^3/m, ft³/lbm
v_{sv}	=	specific volume of saturated water vapor, L^3/m, ft³/lbm
v_{swv}	=	change in specific volume of water when vaporized, L^3/m, ft³/lbm
v_{sw}	=	specific volume of saturated water, L^3/m, ft³/lbm
ρ_a	=	density of air, m/L^3, lbm/ft³
ρ_b	=	density of dry porous rock, m/L^3, lbm/ft³
ρ_f	=	density of fluid in pore space, m/L^3, lbm/ft³
ρ_m	=	density of saturated porous rock, m/L^3, lbm/ft³
ρ_o	=	density of oil, m/L^3, lbm/ft³
ρ_{oR}	=	density of oil at 60°F, m/L^3, lbm/ft³
ρ_r	=	density of porous rock (matrix), m/L^3, lbm/ft³
$\rho_{R,d}$	=	density of dry (air-saturated) sandstone at 68°F, m/L^3, lbm/ft³
$\rho_{R,l}$	=	density of liquid-saturated sandstone at 68°F, m/L^3, lbm/ft³
ρ_s	=	density of steam, m/L^3, lbm/ft³
ρ_w	=	density of water, m/L^3, lbm/ft³
ϕ	=	porosity, L^3/L^3, fraction

References

Allen, T. O. and Roberts, A. P. 1978. *Production Operations,* Vol. 1. Tulsa: Oil & Gas Consultants Intl. Inc.

Anand, J., Somerton, W. H., and Gomaa, E. E. 1973. Predicting Thermal Conductivities of Formations From Other Known Properties. *SPE J.* **13** (5): 267–273. SPE-4171-PA. https://doi.org/10.2118/4171-PA.

Boberg, T. C. 1988. *Thermal Methods of Oil Recovery*, 355–391. New York City: John Wiley & Sons.

Burger, J. G., Sourieau, P., and Combarnous, M. 1985. *Thermal Methods of Oil Recovery,* 47. Houston: Gulf Publishing Co.

Cragoe, C. S. 1929. *Thermal Properties of Petroleum Products,* Vol. NBS Miscellaneous Publication No. 97. Washington, DC: US Department of Commerce, Bureau of Standards.

Farouq Ali, S. M. 1970. *Oil Recovery by Steam Injection,* 4–6. Bradford, Pennsylvania: Producers Publishing Co. Inc.

Gambill, W. R. 1957. You Can Predict Heat Capacities. *Chemical Engineering* (June): 88–91.

Gros, R. P. 1984. *Steam Soak Predictive Model.* MS thesis, University of of Texas at Austin, Austin, Texas.

Himmelblau, D. M. 1974. *Basic Principles and Calculations in Chemical Engineering,* third edition, 494. Englewood Cliffs, New Jersey: Prentice-Hall Inc..

Kartoatmodjo, T. and Schmidt, Z. 1994. Large Data Bank Improves Crude Physical Property Correlations. *Oil & Gas J.* **92** (27): 51–55.

Keenan, J. H. and Keyes, E. G. 1936. *Thermodynamic Properties of Steam,* first edition, 28–33, 35–39. New York City: John Wiley and Sons Inc.

Marks, L. S. 1978. *Standard Handbook for Mechanical Engineers,* eds. E. A. Avallone and T. Baumeister III, ninth edition, 4–86. New York City: McGraw-Hill.

Matthews, C. S. and Russell, D. G. 1967. *Pressure Buildup and Flow Tests in Wells*, Vol. 1. Richardson, Texas: Monograph Series, Society of Petroleum Engineers.

McAdams, W. H. 1954. *Heat Transmission,* third edition. New York City: McGraw-Hill.

Messmer, J. H. 1985. The Effective Thermal Conductivity of Quartz Sands and Sandstones. In *Thermal Recovery,* Vol. 7, 30–39. Richardson, Texas: Reprint Series, Society of Petroleum Engineers.

Prats, M. 1982. *Thermal Recovery,* SPE Monograph Series, 7. Richardson, Texas: SPE.

Rosenow, W. M. and Choi, H. Y. 1961. *Heat, Mass and Momentum Transfer*, 516–517. Englewood Cliffs, New Jersey: Prentice-Hall Inc.

Somerton, W. H. 1958. Some Thermal Characteristics of Porous Rocks. In *Petroleum Tranactions,* AIME, Vol. 213, 375–378, SPE-965-G. Richardson, Texas: SPE.

Somerton, W. H. and Boozer, G. D. 1960. Thermal Characteristics of Porous Rocks at Elevated Temperatures. *J Pet Technol* **12** (6): 77–81. SPE-1372-G. https://doi.org/10.2118/1372-G.

Somerton, W. H., Keese, J. A., and Chu, S. L. 1974. Thermal Behavior of Unconsolidated Oil Sands. *SPE J.* **14** (5): 513–521. SPE-4506-PA. https://doi.org/10.2118/4506-PA.

Sylvester, N. B. and Chen, H. L. 1988. An Improved Cyclic Steam Stimulation Model for Pressure-Depleted Reservoirs. Presented at the SPE California Regional Meeting, Long Beach, California, 23–25 March. SPE-17420-MS. https://doi.org/10.2118/17420-MS.

Tiknomirov, V. M. 1968. Thermal Conductivity of Rock Samples and Its Relation to Liquid Saturation, Density, and Temperature. *Neftyanoe Khozaistov* **46** (4): 36.

Tortike, W. S. and Farouq Ali, S. M. 1989. Saturated-Steam-Property Functional Correlations for Fully Implicit Thermal Reservoir Simulation. *SPE Res Eng* **4** (4): 471–474. SPE-17094-PA. https://doi.org/10.2118/17094-PA.

Wright, W. A. 1974. An Improved Viscosity-Temperature Chart for Hydrocarbons. *J. Materials* **4** (1): 19–25.

Yao, S. C. 1985. *Fluid Mechanics and Heat Transfer in Steam Injection Wells.* MS thesis, University of Tulsa, Tulsa, Oklahoma.

SI Metric Conversion Factors

°API		141.5/(131.5 + °API)		=	g/cm^3
bbl	×	1.589 873	E–01	=	m^3
Btu	×	1.055 056	E+00	=	kJ
cp	×	1.0*	E–03	=	Pa.s
ft	×	3.048*	E–01	=	m
ft^2	×	9.290 304*	E–02	=	m^2
ft^3	×	2.831 685	E–02	=	m^3
°F		(°F–32)/1.8		=	°C
in.	×	2.54*	E+00	=	cm
in.^{2}	×	6.4516*	E+00	=	cm^2
lbm	×	4.535 924	E–01	=	kg
psi	×	6.894 757	E+00	=	kPa

*Conversion factor is exact.

Programs To Evaluate Functions Derived From Marx and Langenheim Model

G(t_D)

$$G\left(t_D\right) = e^{t_D}\,\mathrm{erfc}\sqrt{t_D} + 2\sqrt{t_D/\pi} - 1. \quad\dots\dots\dots\dots\dots\dots\dots\dots\dots\dots\dots (8.62)$$

The function subprogram $G(t_D)$ is listed next. Included in the listing is a function subprogram erfc(x), as described in Section 8.3.

```
C       G.FOR
C       FUNCTION SUBPROGRAM TO COMPUTE G(TD) FOR MARX
C       LANGENHEIM MODEL GPW 12-11-1990
C
        REAL FUNCTION G(TD)
        REALTD.PI
        DATA PI/3.141596/
        G=EXP(TD)*ERFC(SQRT(TD))+2*SQRT(TD/PI) – 1.
        RETURN
        END
C
C       PROGRAM TO COMPUTE ERFC(X)
C       BASED ON SERIES REPRESENTATION—ABRAMOWITZ AND STEGUN
C       HANDBOOK OF MATHEMATICAL FUNCTIONS
C       NATIONAL BUREAU OF STANDARDS
C       APPLIED MATH SERIES 55, P 299(1964)
C       EQUATION 7.1.26 FOR ERF
C       ERFC = 1 – ERF
C
        REAL FUNCTION ERFC(X)
        REAL A1,A2,A3,A4,A5,P,X,U
        DATA A1,A2,A3,A4,A5,P/0.254829592, – 0.284496736,1.421413741,
        : –1.453152027,1.061405429,0.3275911/
        U=1/(1+P*X)
        ERFC=EXP(– X*X)*(A1*U+A2*U*U+A3*U**3+A4*U**4+A5*U**5)
        RETURN
        END
```

G₁(t_D)

$$G_1\left(t_D\right) = e^{t_D}\,\mathrm{erfc}\left(\sqrt{t_D}\right). \quad\dots\dots\dots\dots\dots\dots\dots\dots\dots\dots\dots\dots\dots\dots\dots\dots\dots (8.66)$$

The function program $G_1(t_D)$ is

```
C       G1.FOR
C       FUNCTION SUBPROGRAM TO COMPUTE G1 (TD) FOR MARX
C       LANGENHEIM MODEL GPW 12-11 -1990
C
        REAL FUNCTION G 1(TD)
        REALTD.PI
        DATA PI/3.141596/
        G1 =EXP(TD)*ERFC(SQRT(TD))
        RETURN
        END
```

t_{cD} ($f_{h,v}$). A hot-water zone forms for $t_D > t_{cD}$. The value of t_{cD} is found as the root of Eq. 8.125:

$$e^{t_{cD}} \mathrm{erfc}\left(\sqrt{t_{cD}}\right) = 1 - \left(f_{sd} L_{vdh} / H_s\right) \dots\dots\dots\dots\dots\dots\dots\dots\dots\dots\dots\dots\dots\dots\dots (8.125)$$

and $G_1\left(t_{cD}\right) = 1 - f_{h,v}$.

The function subprogram t_{cD} that finds the root t_{cD} by interval halving is listed next. The program finds roots for $f_{h,v} > 0.26$.

```
C     TCD.FOR
C
C     FUNCTION SUBPROGRAM TO COMPUTE THE CRITICAL TIME
C     FOR THE ONSET OF THE HOT WATER REGION
C     MARX AND LANGENHEIM MODEL
C
C     GPW 12-12-1990
C
      REAL FUNCTION TCD(FHV)
      REAL FHV,X,XD1
      G(X) = 1. – FHV–EXP(X)*ERFC(SQRT(X))
C
C     FIND ROOT BY INTERVAL HALVING
C     G(X) IS MONOTONICALLY INCREASING
C     FROM X = 0 TO X = TDC
C
      XD1 = 0.
      DELX = FHV/10.
      ICOUNT = 0
10    XD1A=XD1+DELX
      ICOUNT = ICOUNT+1
      IF(ICOUNT.GT.10000) STOP
      G1 = G(XD1)
      G1A = G(XD1A)
      IF(G1.LT.0..AND.G1A.LT.0.) GO TO 20
C
C     ROOT IS BRACKETED-BEGIN INTERVAL HALVING
C
      DIFF = G1A – G1
      IF(ABS(DIFF).LT.1.E–05) GO TO 30
      DELX = DELX/2.
      GO TO 10
20    XD1 = XD1A
      GO TO 10
30    TCD=XD1A
      RETURN
      END
```

t_{Ds}. t_{Ds} is the time when the heated zone arrives at the area contacted by condensing steam at t_D. The value of t_{Ds} at t_D satisfies the relationship

$$G_s\left(t_D, t_{D_s}\right) = f_{sd} L_{vdh} / H_s = f_{h,v}$$

$$= \int_o^{t_{Ds}} \frac{e^{\lambda_D} \mathrm{erfc}\left(\sqrt{\lambda_D}\right)}{\sqrt{\pi\left(t_D - \lambda_D\right)}} \, d\lambda_D, \dots\dots\dots\dots\dots\dots\dots\dots\dots\dots\dots\dots\dots\dots (8.130)$$

where $t_D > t_{Ds}$.

The function subprograms $G_s(t_D, t_{Ds})$ and t_{Ds} are listed in the following section.

G_s(t_D, t_{Ds}). The function $G_s(t_D, t_{Ds})$ is defined by Eq. F-l where $t_D > t_{Ds}$:

$$G_s\left(t_D, t_{Ds}\right) = \int_o^{t_{Ds}} \frac{e^\lambda D \, \mathrm{erfc}\left(\sqrt{\lambda_D}\right)}{\sqrt{\pi\left(t_D - \lambda_D\right)}} \, d\lambda_D. \dots\dots\dots\dots\dots\dots\dots\dots\dots\dots\dots\dots (F\text{-}1)$$

$G_s(t_D, t_{D1})$ is given by Eq. F-l, with t_{DI} replacing t_{Ds}. The function subprogram $G_s(t_D, t_{Ds})$ is

```
C     GS.FOR
C     GPW12-7-90
C
C     FUNCTION SUBPROGRAM TO COMPUTE HEAT LOSS INTEGRAL
C     TRAPEZODIAL RULE INTEGRATION
C
      REAL FUNCTION GS (TD,TDS)
      REAL TD,TDS,LD1,LDB,LDE,INTEG
      INTEG(LD1 )=EXP(LD1)*ERFC(SQRT(LD1))/SQRT(PI*(TD – LD1))
      DATA PI/3.141596/
      IF (TD.GT.TDS) GO TO 5
      WRITE(*,1)TD,TDS
    1 FORMAT('PROGRAM STOPPED—VALUE OF TD LE TDS–TD=
     :',F10.7,;TDS = ',F10.7)
      STOP
    5 DELTD=TDS/1000.
      SUM=0.
      DO 10 1=1,1000
      LDB=(I–1)*DELTD
      LDE=I*DELTD
   10 SUM=SUM+DELTD/2.*(INTEG(LDB)+INTEG(LDE))
      GS=SUM
      RETURN
      END
```

$t_{Ds}(t_{Ds}, f_{h,v})$. t_{Ds} is the time when the heated zone $(T = T_S)$ first arrived at the area contacted by condensing steam at t_D. The initial heated zone will be hot water if $t_D > t_{cD}$.

As discussed in Chapter 8, the value of t_{Ds} at t_D satisfies the relationship

$$G_s\left(t_D, t_{Ds}\right) = f_{sd} L_{vdh} / H_t = f_{h,v} \dots\dots\dots\dots\dots\dots\dots\dots\dots\dots\dots\dots\dots (8.130)$$

t_{Ds} is computed by an iterative procedure by finding the value of t_{Ds} that satisfies Eq. 8.130. A function subprogram $t_{Ds}(t_D, f_{h,v})$ for computing t_{Ds} is

```
C     TDS.FOR
C
C     FUNCTION SUBPROGRAM TO COMPUTE THE ARRIVAL TIME
C     OF THE STEAM ZONE IN THE MARX AND LANGENHEIM MODEL
C     FOR TD>TDS
C
C     GPW 12-11-1990
C
      REAL FUNCTION TDS (TD.FHV)
      REALTD,FVALUE(100),TTDS(100)
      DELTDS=TD/100.
      TTDS(100)=TD
      FVALUE(100)=1.-EXP(TD)*ERFC(SQRT(TD))
      DO 10 I=1,99
      TTDS(I)=I*DELTDS
   10 FVALUE(I)=HLF(TD,TTDS(I))
C
C     TABLE LOOKUP TO FIND GS CORRESPONDING TO VALUE OF
      FHV
C     FVALUE IS MONOTONIC INCREASING FUNCTION OF TTDS
C
      DO 20 1=1,100
      IF (FHV.LE.FVALUE(I)) GO TO 30
   20 CONTINUE
      GO TO 40
   30 TDS1=TTDS(I –1)+(FHV – FVALUE(I – 1))/(FVALUE(I) – FVALUE
     : (I – 1))*:(TTDS(I) – TTDS(I – 1))
      TDS=TDS1
      RETURN
   40 PRINT*,' FHV IS ',FHV,' FVALUE(100) = ', FVALUE(100)
```

```
            PRINT*,' PROGRAM HALTED IN TDS'
               STOP
               END

C      HLF.FOR
C      GPW 12-7/90
C
C      FUNCTION SUBPROGRAM TO COMPUTE HEAT LOSS INTE-
C      GRAL
C      TRAPEZODIAL RULE INTEGRATION
C
       REAL FUNCTION HLF (TD,LD)
       REAL TD,LD,LD1,LDB,LDE,INTEG
       INTEG(LD1)=EXP(LD1)*ERFC(SQRT(LD1))/SQRT(PI*(TD – LD1))
       DATA PI/3.141596/
       IF(TD.GT.LD) GO TO 5
       WRITE(*,1)TD,LD
1      FORMAT(' PROGRAM STOPPED, VALUE OF TD .LE. LD, TD= ',F10.7,
       :'LD= ',F10.7)
       STOP
5      DELTD=LD/1000.
       SUM=0.
       DO 10 I=1,1000
       LDB=(I–1)*DELTD
       LDE=I*DELTD
10     SUM=SUM+DELTD/2.*(INTEG(LDB)+INTEG(LDE))
       HLF=SUM
       RETURN
       END
```

Nomenclature

$f_{h,v}$ = fraction of injected energy that is latent heat

f_{sd} = steam quality at sandface, m/m, mass fraction

G = dimensionless compilation of terms, Eq. 8.62

G_s = integral expression defined by Eq. 8.130

G_1 = dimensionless compilation of terms, Eq. 8.66

H_s = energy content of injected steam relative to initial reservoir temperature, Eq. 8.4, L^2/t^2, Btu/lbm

L_{vdh} = latent heat of vaporization at sandface, L^2/t^2, Btu/lbm

t_{cD} = critical dimensionless time, Eq. 8.125

t_D = dimensionless time, Eq. 8.60

t_{Ds} = dimensionless time at which heated front arrives at position contacted by condensed steam at time t_D

λ_D = dimensionless time of arrival of heated region at specified location, defined in an analogous manner to t_D

Appendix G

An Introduction to Gravity Drainage

When gravity forces dominate during production from a reservoir, a gas/oil interface develops at the top of the interval, as depicted in **Fig. G-1** (Matthews and Lefkovits 1956). Pressure in the gas zone is uniform (unless there is appreciable production of gas) and is often low. Oil enters the wellbore from the saturated interval at the bottom of the formation and flows to the bottom of the well along the wellbore. If there is sufficient solution gas, the void space created by oil draining toward the bottom of the zone is filled with gas coming out of solution and migrating to the top of the zone as the average pressure in the gas zone decreases. When the casing annulus is open to the atmosphere, air enters the formation through the production well to form a gas/air zone.

Fluid flow during gravity drainage can be described by approximate models. Consider gravity drainage in a reservoir that is horizontal with no dip. A gas zone forms as the liquid saturation is reduced by gravity drainage. This desaturated zone contains oil at residual saturation, S_{or}; water at interstitial saturation, S_{iw}; and a gas saturation. Although a capillary transition zone would exist between the desaturated and saturated zones, the thickness of this zone is neglected in this analysis; therefore, the interface is depicted as a sharp saturation discontinuity that becomes a free surface separating gas and liquid regions. The oil saturation is assumed to be at initial saturation, S_{oi}, in the oil zone.

If the slope of the surface representing the gas/oil interface is small, fluid flow is essentially horizontal (Dupuit 1863). This assumption is good except in the immediate vicinity of the wellbore. Vertical fluid flow is treated as if all changes in the vertical direction occur instantaneously with no driving force. Under these assumptions, the flow rate in the radial (horizontal) direction is given by Eq. G-1. Note that in Eq. G-1, the driving force for fluid flow is the potential difference.

$$q_r = -\left[2\pi k\rho r h(r)/\mu\right](\partial\Phi/\partial r), \dots\dots\dots\dots\dots\dots\dots\dots\dots\dots\dots(G-1)$$

where $h(r)$ = vertical distance from the base of the reservoir to the free surface (the height of the fluid column) and Φ = fluid potential at radius r as defined by

$$\Phi = \int_{p_d}^{p} \frac{dp}{\rho} + gZ. \dots\dots\dots\dots\dots\dots\dots\dots\dots\dots\dots\dots\dots\dots\dots\dots\dots(G-2)$$

In Eq. G-2, p_d = pressure at an arbitrary datum and Z = vertical elevation above the datum, which is usually chosen as the base of the reservoir.

The assumption of horizontal flow means that isopotentials are vertical and pressure is in hydrostatic equilibrium. When the pressure in the free-gas space, p, is uniform at a particular time and does not vary with radius r, $\Phi(r) = (p - p_d)p + h(r)$. Thus, Eq. G-1 can be written as

$$q_r = -\left[2\pi k\rho r h(r)/\mu\right](\partial h/\partial r). \dots\dots\dots\dots\dots\dots\dots\dots\dots\dots\dots(G-3)$$

Eq. G-3 can be solved analytically for two cases of interest in oil production. In the first case, the well is producing from a region that has a no-flow boundary at radius r_e. This case is analogous to producing from a developed field with each well producing from its own drainage region. This is called a bounded-flow region. In the second case, the fluid head is known at some radius r_e and remains constant throughout the production period.

Bounded Region. When a well has a finite drainage volume, the fluid head at the exterior boundary, h_e, must decline as fluid is produced because there will be no fluid flow across the exterior boundary. Matthews and Lefkovits (1956) developed one solution to the problem of gravity drainage from a bounded region. They visualized the bounded region as a cylinder with radius r_e and assumed that, within the drainage volume, the flow rate varied with radius as given by

$$q_r = q\left[1 - \left(r^2/r_e^2\right)\right], \dots\dots\dots\dots\dots\dots\dots(G-4)$$

where $q = q_r$ at $r = r_w$. At the exterior boundary ($r = r_e$), the flow rate is zero. Substitution of Eq. G-4 into Eq. G-3 and integrating gives Eq. G-5 for the oil production rate corresponding to this assumption when h_e is known at r_e.

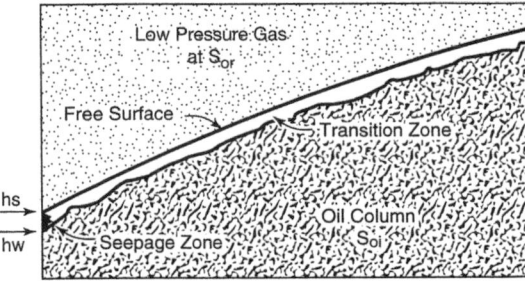

Fig. G-1—Gas-/oil-free surface in gravity-dominated reservoirs.

$$q_o = \frac{0.0246\rho_o k_o \left(h_e^2 - h_w^2\right)}{\mu_o \left\{\ln\left(r_e/r_w\right) - \left[\left(r_e^2 - r_w^2\right)/2r_e^2\right]\right\}} \dotfill \text{(G-5)}$$

When $r_e^2 \gg r_w^2$, Eq. G-5 can be approximated by

$$q_o = \frac{0.0246\rho_o k_o \left(h_e^2 - h_w^2\right)}{\mu_o \left[\ln\left(r_e/r_w\right) - (1/2)\right]} \dotfill \text{(G-6)}$$

The head at the exterior boundary, h_e, decreases with time as fluid is removed by gravity drainage. Eq. G-5 or Eq. G-6 can also be used to find the location of the free surface (gas/liquid interface), $h(r)$, at all radii, $r_w < r < r_e$. The expression for $h(r)$ when $r_e^2 \gg r_w^2$ is given by

$$h = \sqrt{h_w^2 + \left(h_e^2 - h_w^2\right) \frac{\left[\ln\left(r/r_w\right) - \left(r^2 - r_w^2/2r_e^2\right)\right]}{\ln\left(r_e/r_w\right) - (1/2)}} \dotfill \text{(G-7)}$$

Eq. G-7 defines the saturation profile for specific values of h_e and h_w. Recall that the gas/liquid interface separates the desaturated region where the saturations are S_{or}, S_{iw}, and S_g from the liquid region where the saturations are S_o and S_w. The saturation profile is generated by plotting h vs. r, as illustrated in **Fig. G-2**, which shows the height of the oil column as a function of radius when $h_w = 48$ ft, $r_w = 0.31$ ft, and $h_e = 260$ ft at $r_e = 90$ ft.

The pore volume (PV) of the saturated region in the reservoir is given by

$$V_{ps} = \pi\left(r_e^2 - r_w^2\right)\bar{h}\phi/5.615, \dotfill \text{(8.103)}$$

where $\bar{h}$, the average thickness of the oil column, is defined by

$$\bar{h} = \left(\int_{r_w}^{r_e} 2\pi r h \, dr\right)\Bigg/\left[\pi\left(r_e^2 - r_w^2\right)\right]. \dotfill \text{(8.104)}$$

Integration of Eq. 8.104 can be performed by substituting Eq. G-7 for $h(r)$. Matthews and Leftkovits (1956) give an approximate expression for the average thickness of the oil column derived from analytical solutions. The examples presented in this section were obtained by integrating Eq. 8.104 to obtain $\bar{h}$ by use of Simpson's rule and a short computer program.

As the reservoir is produced, the oil column drops, as shown by the second line in Fig. G-2, the saturation profile when the head at r_e has dropped to 250 ft. A third oil/gas surface is shown in Fig. G-2 when the head at r_e has declined to 160 ft. Oil production by gravity drainage can be estimated by assuming successively smaller values of h_e and computing the average thickness of the oil column for each value of h_e. Different values of h_e correspond to different times during the production process. Oil displaced by gravity drainage is given by

$$N_p^{n+1} - N_p^n = \left(V_{ps}^{n+1} - V_{ps}^n\right)\left(S_{oi} - S_{or}\right). \dotfill \text{(G-8)}$$

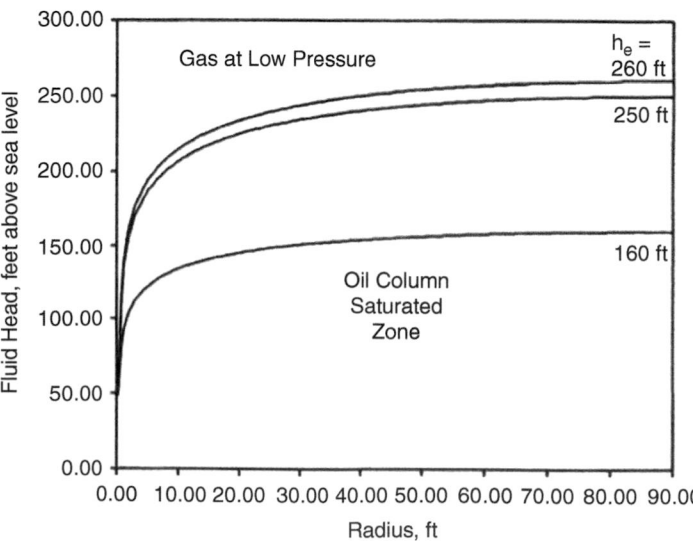

Fig. G-2—Saturation profiles in a bounded reservoir producing by gravity drainage.

The time required for the reservoir to drain by gravity from h_e^n to h_e^{n+1} is given by

$$t^{n+1} - t^n = N_p^{n+1} - N_p^n \Big/ \left[\left(q_o^n + q_o^{n+1}\right)/2\right], \quad \text{(G-9)}$$

where the superscripts n and $n+1$ indicate evaluation of terms at values of h_e^n and h_e^{n+1} respectively. Example G-1 illustrates the prediction of production rates in a bounded viscous oil reservoir producing under gravity drainage.

Example G-1—Estimation of Production Rate for a Bounded Heavy Oil Reservoir Producing Under Gravity Drainage. Reservoir data from the Midway-Sunset field (Jones 1977) presented in Example 8.7 are used to estimate primary production under gravity drainage in this example. The reservoir has four zones with 320 ft of net sand, and wells are drilled so that the effective drainage radius is assumed to be 90 ft (well spacing is 160 ft).

Static head when the wells are shut in is 260 ft above the pump, and the head is maintained at 48 ft above the pump when on production. Estimate the production history from the time that the exterior boundary is reached to a total time of 10 years. Reservoir and rock properties are given in **Table G-1.**

Table G-2 summarizes the production history from the time the head is 260 ft at r_e = 90 ft. Sample calculations for the first timestep are outlined in the following. The first time increment begins with the assumption that h_e = 260 ft at r_e = 90 ft. Under these conditions, the oil production rate is given by Eq. G-6.

$$q_o = \left[0.0246\rho_o k_o \left(h_e^2 - h_w^2\right)\right]/\left\{\mu_o\left[\ln\left(r_e / r_w\right) - (1/2)\right]\right\}$$

$$= \frac{(0.0246)(61.8)(1.5)\left[(260)^2 - (48)^2\right]}{(5,000)\left[\ln\left(90/0.31\right) - (1/2)\right]}$$

$$= 5.76 \text{ B/D}.$$

The saturation profile for h_e = 260 ft was computed with Eq. G-7 and is shown in Fig. G-1. Average thickness determined from Eq. 8.104 was found to be 253.65 ft. The PV of the oil-saturated region ($S_o = S_{oi}$) is computed from Eq. 8.103:

$$V_{ps}^{(1)} = \left\{(3.1416)\left[(90)^2 - (0.31)^2\right](0.32)(253.65)\right\}/5.615$$

$$= 367,845 \text{ bbl}.$$

Porosity	0.32
Oil viscosity at reservoir temperature (cp)	5,000
Initial oil saturation	0.75
r_w (ft)	0.31
Residual oil saturation	0.25
Oil permeability at S_{oi} (darcies)	1.5

Table G-1—Fluid and rock properties–Midway-Sunset Field (Jones 1977).

Timestep	h_e (ft)	$\bar{h}$ (ft)	V_{ps} (bbl)	ΔN_p (bbl)	N_p (bbl)	q_o (B/D)	Time (days)	Time (years)
1	260	253.65	367,850	0	0	5.76	0	0
2	250	243.92	353,729	7,061	7,061	5.31	1,276	3.50
3	240	234.18	339,609	7,060	14,121	4.88	2,662	7.29
4	230	224.44	325,490	7,059	21,180	4.46	4,174	11.43
5	220	214.71	311,373	7,059	28,239	4.07	5,829	15.97
6	210	204.98	297,257	7,058	35,297	3.69	7,650	20.96
7	200	195.24	283,143	7,057	42,354	3.32	9,663	26.47
8	190	185.51	269,031	7,056	49,409	2.98	11,901	32.61
9	180	175.78	254,922	7,055	56,464	2.65	14,405	39.46
10	170	166.06	240,816	7,053	63,517	2.35	17,226	47.19
11	160	156.33	226,714	7,051	70,568	2.05	20,430	55.97
12	150	146.61	212,615	7,049	77,618	1.78	24,106	66.04
13	140	136.89	198,522	7,047	84,664	1.53	28,368	77.72
14	130	127.18	184,435	7,044	91,708	1.29	33,376	91.44
15	120	117.47	170,356	7,040	98,747	1.07	39,356	107.82
16	110	107.77	156,286	7,035	105,782	0.86	46,643	127.79
17	100	98.08	142,229	7,028	112,810	0.68	55,754	152.75
18	90	88.39	128,189	7,020	119,831	0.51	67,552	185.07
19	80	78.73	114,171	7,009	126,839	0.36	83,619	229.09
20	70	69.08	100,185	6,993	133,833	0.23	107,315	294.01

Table G-2—Summary of gravity-drainage calculations for a bounded reservoir, Example G-1; after saturated region formed as depicted in Fig. G-2.

When h_e = 250 ft, the corresponding values are $\bar{h}$ = 243.92 ft, $V_{ps}^{(2)}$ = 353,734 bbl, and q_o = 5.31 B/D. The volume of oil produced when h_e declines from 260 to 250 ft is given by Eq. G-8.

$$Np^{n+1} - N_p^n = \left(V_{ps}^{n+1} - V_{ps}^n\right)\left(S_{oi} - S_{or}\right)$$
$$= (367,845 - 353,734)(0.75 - 0.25)$$
$$= 7,056 \text{ bbl.}$$

The time required to produce this volume of oil is given by Eq. G-9:

$$t^{n+1} - t^n = \left[N_p^{(2)} - N_p^1\right]\big/\left[\left(q_o^n + q_o^{n+1}\right)\big/2\right]$$
$$= 7,056\big/\left[(5.75 + 5.31)\big/2\right]$$
$$= 1,276 \text{ days,}$$

where q_o^n is the oil rate when $h_e = h_e^n$ and q_o^{n+1} is the oil rate when $h_e = h_e^{n+1}$. The remainder of the gravity-drainage calculations are summarized in Table G-2.

Gravity drainage is a slow process, and this reservoir would produce at low rates for a long period of time unless stimulated by cyclic steam injection.

Constant Flow Rate. When fluid flow is quasisteady and the flow rate, q_r, is constant, Eq. G-3 can be integrated to obtain (Muskat 1949)

$$q = \left(0.0246\rho k/\mu\right)\left[\left(h^2 - h_w^2\right)\big/\ln\left(r/r_w\right)\right], \dots\dots\dots\dots\dots\dots\dots\dots\dots\dots\dots\dots\dots\dots\text{(G-10)}$$

where the fluid head is maintained constant with values of h_w at the production well with radius r_w. If the fluid head is held constant at h_e at radius r_e, then the flow rate is constant at every value of r and can be estimated from Eq. G-10. For oil and water production, Eq. G-10 can be written as Eqs. G-11 and G-12 in this case. Oilfield units (darcies, STB/D, feet, centipoise, and lbm/ft³) are used in these equations.

$$q_o = \left(0.0246\rho_o k_o/\mu_o B_o\right)\left[\left(h_e^2 - h_w^2\right)\big/\ln\left(r_e/r_w\right)\right] \dots\dots\dots\dots\dots\dots\dots\dots\dots\dots\dots\text{(G-11)}$$

and $$q_w = \left(0.0246\rho_w k_w/\mu_w B_w\right)\left[\left(h_e^2 - h_w^2\right)\big/\ln\left(r_e/r_w\right)\right]. \dots\dots\dots\dots\dots\dots\dots\dots\dots\dots\text{(G-12)}$$

The fluid head can be estimated as a function of radius r with Eq. G-11 or Eq. G-12. The graph of fluid head vs. radius is the saturation profile in the region affected by gravity drainage. Thus, for $z < h$, the oil saturation is S_{oi}, and for $z > h$, the oil saturation is S_{or}. The shape of the saturated region is defined by

$$h = \sqrt{\left[\left(h_e^2 - h_w^2\right)\ln\left(r/r_w\right)\big/\ln\left(r_e/r_w\right)\right] + h_w^2}. \dots\dots\dots\dots\dots\dots\dots\dots\dots\dots\dots\text{(G-13)}$$

The average thickness of the desaturated region can be calculated by substituting Eq. G-13 for h in Eq. 8.104.

When the head is constant at radius r_e, an adjoining region must exist that has a large source of reservoir energy to maintain the constant head. In this model, oil flows continuously from this region into the saturated region.

Nomenclature

B_o	=	oil formation volume factor (FVF), L³/L³, bbl/STB
B_w	=	water FVF, L³/L³, bbl/STB
g	=	acceleration of gravity, L/t, ft/sec²
$\bar{h}$	=	volumetric average thickness of oil column, L, ft
h_e	=	fluid head at exterior boundary, L, ft
h_w	=	fluid head at wellbore, L, ft
k	=	permeability of porous rock, L², md
k_o	=	effective permeability to oil, L², md
k_w	=	effective permeability to water, L², md
N_p	=	cumulative oil production, L³, STB
p	=	pressure in the free gas space, m/Lt², psi
p_d	=	pressure at arbitrary datum, m/Lt², psi
q	=	fluid production rate of single fluid, L³/t, B/D
q_o	=	oil production rate, L³/t, B/D
q_r	=	flow rate at radius r, L³/t, B/D
q_w	=	water production rate, L³/t, STB/D
r	=	radius, L, ft

r_e	=	effective drainage radius, L, ft
r_w	=	wellbore radius, L, ft
S_g	=	gas saturation, volume fraction
S_{iw}	=	interstitial water saturation, volume fraction
S_o	=	oil saturation, volume fraction
S_{oi}	=	initial oil saturation, volume fraction
S_{or}	=	residual oil saturation, volume fraction
S_w	=	water saturation, volume fraction
t	=	time, t, days
V_{ps}	=	PV of saturated portion of heated region following cyclic steam injection, L^3, ft^3

z	=	vertical elevation above base of reservoir, L, ft
Z	=	vertical elevation above datum, L, ft
ρ	=	density, m/L^3, lbm/ft^3
ρ_o	=	density of oil, m/L^3, lbm/ft^3
ρ_w	=	density of water, m/L^3, lbm/ft^3
μ	=	viscosity, m/Lt, cp
μ_o	=	viscosity of oil, m/Lt, cp
μ_w	=	viscosity of water, m/Lt, cp
ϕ	=	porosity, fraction
Φ	=	fluid potential at radius r defined in Eq. G-2

Superscripts

n = timestep number

References

Dupuit, J. 1863. *Études Théoriques et Pratiques sur le Mouvement des Eaux,* second edition. Paris: Dalmont.

Jones, J. 1977. Cyclic Steam Reservoir Model for Viscous Oil, Pressure Depleted, Gravity Drainage Reservoirs. Presented at the SPE California Regional Meeting, Bakersfield, California, 13–15 April. SPE-6544-MS. https://doi.org/10.2118/6544-MS.

Matthews, C. S. and Lefkovits, H. C. 1956. Gravity Drainage Performance of Depletion-Type Reservoirs in the Stripper Stage. In *Petroleum Transactions,* AIME, Vol. 207, 265–74. SPE-665-G-P. Richardson, Texas: SPE.

Muskat, M. 1949. *Physical Principles of Oil Production.* New York City: McGraw-Hill Book Co. Inc.

Solution to Example 8.7—Cyclic Steam Stimulation of a Gravity-Drainage Reservoir

The solution of Example 8.7 involves estimating the radius of the heated region created by steam injection. Then, the temperature of the heated region and the water and oil production rates are calculated at discrete time increments, as Section 8.3.4 describes. When the time interval is subdivided into timesteps, the temperature of the heated region is assumed to be constant at the value estimated for the beginning of each timestep. Water saturation for a timestep is also taken as the value from the previous timestep.

The calculation procedure is not difficult, but it is lengthy because the temperature changes with time and temperature-dependent properties must be evaluated at each timestep. The solution presented in this appendix includes detailed calculations for the first two timesteps. The desaturated region is assumed to form in the first timestep. In the second timestep and all subsequent timesteps, water and oil production are estimated from the saturated region. Table 8.12 presents data for these calculations.

Estimation of Heated Radius

The heated radius is estimated by use of the Marx and Langenheim model, assuming that the injected steam is equally distributed between the four zones. From Eq. 8.63,

$$A_h = \left\{ \dot{m}_s H_s M_R h / \left[4(T_s - T_r) \alpha M_s^2 \right] \right\} G(t_D),$$

where $\dot{m}_s$ = injection rate per zone and h = thickness of each zone.

$$\dot{m}_s = (646 \text{ B/D})(350 \text{ lbm/bbl})\left[1/(4 \text{ zones}) \right]$$
$$= 56{,}525 \text{ lbm/D-zone.}$$

The net energy content of the injected steam is found by evaluating H_s with Eq. 8.67:

$$H_s = H_w + f_{sd} L_{vdh} - H_{wr}.$$

Enthalpies of water are obtained from Table E-1. At 485°F, H_w = 470.2 Btu/lbm and L_{vdh} = 733.1 Btu/lbm. At 110°F, H_{wr} = 77.94 Btu/lbm and

$$H_s = 470.2 + (0.71)(733.1) - 77.94$$
$$= 912.8 \text{ Btu/lbm}$$

and $\dot{m}_s H_s = 51.6 \times 10^6$ Btu/D-zone.

In this example, the values of M_r and M_s are assumed to be equal. The value of M is calculated from Eq. 8.50:

$$M = (1 - \phi)\rho_r C_r + \phi S_o \rho_o C_o + \phi S_w \rho_w C_w.$$

To evaluate M, the heat capacities and densities of water are required.

Correlations for fluid properties of oil and water are discussed in Appendix E. Correlations used in this example are summarized below. Units are lbm/ft^3 and Btu/lbm-°F with temperatures in °F.

Heat Capacity of Water.

$$C_w = (H_w - H_{wr})/(T_s - T_r)$$
$$= (470.2 - 77.94)/(485 - 110)$$
$$= 1.046 \text{ Btu/lbm-°F.}$$

Heat Capacity of Oil. The heat capacity of oil between T_s and T_r is found with

$$C_o = (H_o - H_{or})/(T_s - T_r), \quad \dots\dots\dots\dots\dots\dots\dots\dots\dots\dots\dots\dots\dots\dots\dots\dots\dots\dots\dots\dots\dots\text{(H-1)}$$

where H_o = enthalpy of oil at T_s, Btu/lbm-°F, and H_{or} = enthalpy of oil at T_r, Btu/lbm-°F.

The enthalpy of oil relative to an arbitrary base temperature, T_b, is found by integrating Eq. E-21 from T_b, to temperature T as in Eq. H-2 to obtain Eq. H-3.

$$H_o = \int_{T_b}^{T} C_{po}\, dT. \quad \dots\dots\dots\dots\dots\dots\dots\dots\dots\dots\dots\dots\dots\dots\dots\dots\dots\dots\dots\dots\dots\dots\dots\text{(H-2)}$$

$$= \int_{T_b}^{T} \left[(0.3881 + 0.00045\, T)/\sqrt{\gamma_o} \right] dT$$

$$= \frac{0.3881(T - T_b) + 0.000225(T^2 - T_b^2)}{\sqrt{\gamma_o}}. \quad \dots\dots\dots\dots\dots\dots\dots\dots\dots\dots\dots\dots\dots\dots\text{(H-3)}$$

In Example 8.7, °API = 11.43 and $\gamma_o = 0.99$. At $T_s = 485°F$ and $T_b = 60°F$,

$$H_o = \frac{0.3881(485 - 60) + 0.000225\left[(485)^2 - (60)^2 \right]}{\sqrt{0.99}}$$

$$= 165.77 + 52.38$$

$$= 218.15 \text{ Btu/lbm.}$$

At $T_r = 110°F$, $H_{or} = 17.05$ Btu/lbm

and $C_o = (218.15 - 17.05)/(485 - 110)$

$$= 0.54 \text{ Btu/lbm-°F.}$$

Calculation of M.

$$M = (1 - 0.32)(165 \text{ lbm/ft}^3)(0.20 \text{ Btu/lbm-°F})$$

$$+ (0.32)(0.75)(54.75 \text{ lbm-ft}^3)$$

$$+ (0.32)(0.25)(50.54 \text{ lbm-ft}^3)(1.046 \text{ Btu/lbm-°F})$$

$$= 22.40 + 7.10 + 4.23$$

$$= 33.73 \text{ Btu/ft}^3\text{-°F.}$$

All values needed to compute A_h are known except $G(t_D)$. The injection time is 6 days. Thus,

$$t_D = 4\alpha t/h^2$$

$$= \left[(4)(0.02 \text{ ft}^2/\text{hr})(24 \text{ hr/D})(6 \text{ days}) \right]/(80 \text{ ft})^2$$

$$= 0.0018.$$

From Table 8.8, the value of $G(t_D) = 0.00175$ at $t_D = 0.0018$. Substituting into Eq. 8.63,

$$A_h = \left\{ \left[\dot{m}_s H_s M h 4 (T_s - T_r) \right]/\alpha M^2 \right\} G(t_D)$$

$$= \left\{ \left[(51.6 \times 10^6 \text{ Btu/D-zone})(33.7 \text{ Btu/ft}^3\text{-°F})(80 \text{ ft}) \right] \right.$$

$$\div \left[(4)(485 - 110)(°F)(0.02 \text{ ft}^2/\text{hr})(24 \text{ hr/D}) \right.$$

$$\left. \times \left(33.7 \text{ Btu/ft}^3\text{-°F} \right)^2 \right] \right\}(0.00175)$$

$$= 297.7 \text{ ft}^2$$

and $r_h = 9.73$ ft.

Thus, if steam enters the entire net thickness uniformly, the heated zone will extend to a distance of 9.73 ft from the injection wellbore, a relatively small fraction of the distance between wells.

The volume of the heated zone is found by material balance:

$$V_h = (4 \text{ zones})\left[A_h \left(\text{ft}^2 \right) \right] (h \text{ ft/zone})$$

$$= (4)(297.7)(80)$$

$$= 95,264 \text{ ft}^3$$

$$= 16,966 \text{ bbl.}$$

The heated pore volume (PV), V_{ph}, is $V_p \phi = (0.32)(16,966) = 5,429$ bbl. In comparison, the volume of water injected as steam was 3,882 bbl cold water equivalent (CWE) at ambient temperature.

Boberg and Lantz Method To Compute the Heated Region Average Temperature at the End of Soak Period. First, compute the hypothetical thickness z in Eq. 8.79 to account for the heat loss by conduction during steam injection.

$$z = \left\{ (\dot{m}_s H_s t_i) / \left[\pi r_h^2 M (T_s - T_r) \right] \right\} - h$$

$$= \left\{ \left[(51.6 \times 10^6 \text{ Btu/D-zone})(6 \text{ days}) \right] / \left[\pi (9.73 \text{ ft})^2 \right. \right.$$

$$\times \left. \left. (33.7 \text{ Btu/ft}^3 \text{-}°F)(485 - 110)(°F) \right] \right\} - 80$$

$$= 82.3 - 80$$

$$= 2.3 \text{ ft.}$$

The soak time for the first cycle is 5 days. The dimensionless average temperature at the end of the soak period is given by Eq. 8.76:

$$\bar{T}_D = \bar{T}_{Dr} \bar{T}_{Dz}$$

$$\bar{T}_{Dr} = \bar{T}_{Dr} (t_{Dr})$$

$$\bar{T}_{Dz} = \bar{T}_{Dz} (t_{Dz})$$

$$t_{Dr} = \left[\alpha (t - t_i) \right] / r_h^2$$

$$= \left[(0.02 \text{ ft}^2 \text{/hr})(24 \text{ hr/D})(5 \text{ days}) \right] / (9.73 \text{ ft})^2$$

$$= 0.0253.$$

From Eq. 8.73, $\bar{T}_{Dr} = 0.824$.

Recall that t_{Dz} is evaluated at $(z + h)/2$ in Eq. 8.72. Thus,

$$t_{Dz} = \left[\alpha (t - t_i) \right] / \left[(z + h)/2 \right]^2$$

$$= \left[(0.02 \text{ ft}^2 \text{/hr})(24 \text{ hr/D})(5 \text{ days}) \right] / (82.3/2 \text{ ft})^2$$

$$= 0.00142.$$

From Eq. 8.75, $\bar{T}_{Dz} = 0.979$.

The average temperature of the produced fluid can be computed from Eq. 8.84, noting that there is no fluid production during the soak period and, consequently, $\delta^{(1)} = 0$.

$$\bar{T}_{Dp}^{(n)} = \bar{T}_{Dr} \bar{T}_{Dz} \left[1 - \delta^{(n)} \right] - \delta^{(n)}.$$

But $\delta^{(1)} = 0$, giving

$$\bar{T}_{Dp}^{(1)} = (0.824)(0.979)$$

$$= 0.807$$

and $\bar{T}_p^{(1)} = T_r + \bar{T}_D \left(T_s - T_r \right)$

$$= 110 + (0.807)(485 - 110)$$

$$= 412.6°F.$$

$\bar{T}_p^{(1)} = 412.6°F$ is the average temperature of the heated region at the end of the soak period and the temperature used to compute the heat removed during the first timestep when the desaturated region is formed.

Fluid properties required in subsequent calculations were evaluated with correlations presented in Appendix E. The following summarizes the values at 412.6°F.

ρ_w	=	53.31bm/ft³,
ρ_o	=	55.77 lbm/ft³,
μ_w	=	0.129 cp,
μ_o	=	3.10 cp,
B_w	=	1.17 bbl/bbl,
B_o	=	1.11 bbl/bbl,
H_{wT}	=	388.7 Btu/lbm, and
H_{oT}	=	175.18 Btu/lbm.

First Timestep: Formation of the Desaturated Region. *Compute the Volume of Hot Water Produced During the Desaturation Period.* The net thickness of the heated region is 320 ft. At the end of the soak period, all steam has condensed and this region is assumed to be saturated with hot water and oil at residual oil saturation (ROS) to steam, S_{ors}. The hydrostatic head at the boundary of the heated region is assumed to be constant at 260 ft, and the well will be placed on pump with a fluid level maintained at 48 ft.

A significant volume of water will be removed from the heated region when the well is placed on pump. As described earlier, the first step involves desaturation of the heated region so that the shape of the saturated region conforms to that illustrated in Fig. 8.26. In this example, the average thickness of the saturated region was found to be 240.3 ft from Eq. 8.102. The volume of the saturated zone after the initial desaturation is found from Eq. 8.101:

$$V_{ps} = \left[\phi \bar{h} \pi \left(r_h^2 - r_w^2 \right) \right] / 5.615$$

$$= \left\{ (0.32)(240.3 \text{ ft})(\pi) \left[(9.73 \text{ ft})^2 - (0.31 \text{ ft})^2 \right] \right\} / 5.615$$

$$= 4,069 \text{ bbl.}$$

During the initial desaturation, hot water is removed from the top of the heated zone and replaced by water vapor, gas, or air. The volume of water removed is found by a volume balance assuming that the saturations of the desaturated region are S_{ors}, S_{iw}, and S_g, where $S_g = 1 - S_{ors} - S_{iw}$. Thus, the volume of water removed during the initial desaturation period is

$$W_{ph}^{(1)} = \left(V_{ph} - V_{ps} \right) \left(1 - S_{ors} - S_{iw} \right)$$

$$= (5,429 - 4,069)(1 - 0.25 - 0.25)$$

$$= 680 \text{ bbl.}$$

The temperature of the produced water is 412.6°F, and the volume of water produced at stock-tank conditions is

$$W_p^{(1)} = W_{ph}^{(1)} / B_w$$

$$= 680 \text{ bbl} / (1.17 \text{ bbl/bbl})$$

$$= 582.8 \text{ bbl.}$$

In this example, the pump rate is limited to 800 B/D. Therefore, the time required for the initial desaturation is estimated from Eq. 8.106:

$$t^{(1)} - t_{\text{soak}} = W_{ph}^{(1)} / q_{\max}$$

$$= 680 \text{ bbl}/(800 \text{ B/D})$$

$$= 0.85 \text{ days,}$$

giving $t^{(1)} = 5.85$ days.

Because only water is removed from the heated region, the average water saturation in the saturated region is $S_w^{(1)} = 1 - S_{ors} = 0.75$.

Compute the Value of $\delta^{(2)}$ at the End of the Timestep. Next, it is necessary to compute the value of δ from the current timestep. This value of δ will be used to compute the average temperature of the heated zone at the beginning of the next timestep. The average temperature of the heated zone during this timestep is 412.6°F. Because only water is produced during the desaturation period, Eq. 8.113 gives the energy removed from the reservoir with the water produced during the current timestep:

$$\Delta Q^{(1)} = 5.615 W_{ph}^{(1)} \rho_{wh} \left(H_{wT} - H_{wr} \right).$$

Substituting appropriate values into Eq. 8.113 yields the amount of energy removed during the initial timestep when the desaturated region forms:

$$\Delta Q^{(1)} = 5.615 (680 \text{ bbl}) \left(53.3 \text{ lbm/ft}^3 \right)$$

$$\times \left[(388.7 \text{ Btu/lbm}) - (77.9 \text{ Btu/lbm}) \right]$$

$$= 63.2 \times 10^6 \text{ Btu.}$$

The last step in the computation is to determine the value of $\delta^{(2)}$. Eq. 8.111 defines the change in δ under the assumption

$$\delta^{n+1} - \delta^n = \Delta Q^n / \left[2(n_s)(Q_{\max}) \right],$$

where $Q_{\max}$ is defined by Eq. 8.114 as

$$Q_{\max} = \dot{m}_s H_s t_i - Q_{\text{lost}}$$

and Q_{lost} is given by Eq. 8.115 as

$$Q_{\text{lost}} = 4 \pi r_h^2 k_h \left(T_s - T_r \right) \sqrt{t_{\text{soak}} / \pi \alpha}.$$

For the current timestep,

$$Q_{\text{lost}} = (4\pi)(9.73 \text{ ft})^2 (1.0 \text{ Btu/ft-}°\text{F-hr})(485 - 110)(°\text{F})$$

$$\times \sqrt{(5 \text{ days})(24 \text{ hr/D}) / \pi \left(0.02 \text{ ft}^2 / \text{hr} \right)}$$

$$= 19.5 \times 10^6 \text{ Btu}$$

and $Q_{\max} = \left(51.6 \times 10^6 \text{ Btu/D-zone} \right)(6 \text{ days})$

$$-19.5 \times 10^6 \text{ Btu}$$

$$= 290.1 \times 10^6 \text{ Btu/zone.}$$

Substituting into Eq. 8.111,

$$\delta^{(2)} - \delta^{(1)} = \Delta Q / \left[2(n_s)(Q_{\max}) \right]$$

$$= \frac{63.2 \times 10^6 \text{ Btu}}{(2)(4 \text{ zone}) \left(290.1 \times 10^6 \text{ Btu/zone} \right)}$$

$$= 0.0273,$$

giving $\delta^{(2)} = \delta^{(1)} + 0.0273$, $\delta^{(1)} = 0$, and $\delta^{(2)} = 0.0273$.

This value of δ is used to compute the average temperature of the heated zone for the next timestep. Calculations for the first timestep have been completed.

Calculations After the Desaturation Period—Second and Subsequent Timesteps. Production performance after the desaturation period is computed by following the procedure outlined earlier. A timestep increment is selected for the calculations. The temperature of the heated region is assumed to be constant throughout this timestep and is estimated with the method of Boberg and Lantz, taking into account the energy removed with the produced fluids.

Second Timestep. The time increment for the desaturation period was 0.85 days, so the elapsed time at the beginning of the second timestep is 5.85 days. The values of t_{Dr}, t_{Dz}, $\overline{T}_{Dr}$, and $\overline{T}_{Dz}$, summarized next, were calculated by the same procedure used in the first timestep.

$$t_{Dr} = 0.0296,$$

$$\overline{T}_{Dr} = 0.807,$$

$$t_{Dz} = 0.00166, \text{ and}$$

$$\overline{T}_{Dr} = 0.977.$$

The average temperature of the heated region is computed as

$$\overline{T}_{Dp}^{(2)} = \overline{T}_{Dr}\overline{T}_{Dz}\left[1 - \delta^{(2)}\right] - \delta^{(2)}$$

$$= (0.807)(0.977)(1 - 0.0273) - 0.0273$$

$$= 0.74$$

and $\overline{T}_p^{(2)} = T_r + \overline{T}_{Dp}^{(2)}(T_s - T_r)$

$$= 110 + (0.740)(485 - 110)$$

$$= 387.4°F.$$

A number of fluid properties are evaluated at the average temperature of the heated region. The following properties were calculated with the correlations illustrated in the first timestep. At 387.4°F,

μ_w	=	0.139 cp,
μ_o	=	2.97 cp,
ρ_w	=	54.3 lbm/ft³,
ρ_o	=	56.3 lbm/ft³,
H_w	=	361.5 Btu/lbm,
H_o	=	160.8 Btu/lbm,
B_w	=	1.15 bbl/bbl, and
B_o	=	1.10 bbl/bbl.

The water saturation in the heated region is $1 - S_{ors} = 0.75$.

Calculation of Water and Oil Production Rates. The water production rate (at 387.4°F) is given by Eq. 8.106, where fluid properties are evaluated at the average temperature of the heated region:

$$q_{wh} = \left[0.0246\rho_{wh}k_{wh}\left(h_h^2 - h_w^2\right)\right]/\left[a_2\mu_{wh}\ln\left(r_h/r_w\right)\right].$$

In this model, water and oil production flow through the saturated region shown on Fig. 8.26. Water and oil relative permeabilities are evaluated at the water saturation at the end of the previous timestep. Thus, $k_{rw}^{(n)} = k_{rw}\left[S_w^{(n-1)}\right]$ and $k_{ro}^{(n)} = k_{ro}\left[S_w^{(n-1)}\right]$. In the second timestep, $S_w^{(1)} = 1 - S_{ors}$, or 0.75 in this region. Permeability of water is estimated by use of the Dietrich (1981) correlation of relative permeability given by*

$$k_{rw} = A + \left(B/\left\{1 + e^{-\left[(10S_w - C)/D\right]}\right\}\right), \dotfill \text{(H-4)}$$

---

*Personal communication with M. J. Michnick. 1994. University of Kansas: Tertiary Oil Recovery Project.

where $A = -2.6856 \times 10^{-5}$, $B = 0.012026113$, $C = 8.353369$, $D = 1.045583$, and k_{rw} is set equal to 0 at $S_w = 0.25$. At $S_w = 0.75$, $k_{rw} = 0.00366$. The water rate at reservoir temperature is calculated with a value of $a_2 - 2.0$.

$$q_{wh} = \{(0.0246)(54.3 \text{ lbm/ft}^3)(0.00366)(1.5 \text{ darcies})$$
$$\times \left[(260 \text{ ft})^2 - (48 \text{ ft})^2 \right]\}$$
$$\div \left[(2.0)(0.139 \text{ cp}) \ln (9.73/0.31) \right]$$
$$= 499.8 \text{ B/D}.$$

The oil production rate (at 387.4°F) is estimated with Eq. 8.105:

$$q_{oh} = \left[0.0246 \rho_{oh} k_{oh} \left(h_h^2 - h_w^2 \right) \right] / \left[a_1 \mu_{oh} \ln (r_h / r_w) \right].$$

The oil relative permeability curve presented by Jones (1977) for Midway Sunset oil was digitized and correlated with Eq. H-5 for $S_w > 0.1175$.

$$k_{ro} = A_1 + B_1 X_s + C_1 X_s^2 + D_1 X_s^3 + E_1 X_s^4 + F_1 X_s^5$$
$$+ G_1 X_s^6 + H_1 X_s^7 + I_1 X_s^8 + J_1 X_s^9 + K_1 X_s^{10}, \quad \dots\dots\dots\dots\dots\dots\dots\dots\dots\dots\text{(H-5)}$$

where $X_s = \ln(S_w)$ and $A_1 \dots K_1$ are correlating constants. **Table H-1** gives values of $A_1 \dots K_1$.

The correlating equation is complex but is easily evaluated with a short computer program or spreadsheet. At $S_w^{(1)} = 0.75$, $k_{ro} = 0.00182$. The oil production rate is calculated for this example with a value of $a_1 = 4.0$ for the empirical adjustment factor. For the second timestep,

$$q_{oh}^{(2)} = \{(0.0246)(56.1 \text{ lbm/ft}^3)(0.00182)(1.5 \text{ darcies})$$
$$\times \left[(260 \text{ ft})^2 - (48 \text{ ft})^2 \right]\} / \left[(4.0)(2.97 \text{ cp}) \ln (9.73/0.31) \right]$$
$$= 6.01 \text{ B/D}.$$

Calculate Average Water Saturation at the End of Timestep. Water production during the current timestep is found with Eq. 8.107:

$$W_{ph}^{(2)} - W_{ph}^{(1)} = q_{wh}^{(2)} \left[t^{(2)} - t^{(1)} \right]$$
$$= q_{wh}^{(2)} \left[t^{(2)} - t^{(1)} \right]$$
$$= (499.8 \text{ B/D})(0.5 \text{ days})$$
$$= 249.9 \text{ bbl}.$$

A_1	−0.1243664
B_1	−0.98316363
C_1	−2.8201304
D_1	−3.2248968
E_1	0.80827171
F_1	2.9771635
G_1	−0.17679942
H_1	−3.0213165
I_1	−2.330329
J_1	−0.73862929
K_1	−0.088007532

Table H-1—Constants for relative permeability correlation in Eq. H-5.

$W_{ph}^{(2)} - W_{ph}^{(1)}$ is the cumulative volume of water produced from the saturated region because water produced during the first timestep was assumed to result from desaturation of the heated region, as discussed earlier. Note: Average oil and water rates are used for Timesteps 3,....

The water saturation in the saturated region at the end of the second timestep can be calculated by substituting the value of $W_{ph}^{(2)} - W_{ph}^{(1)}$ into Eq. 8.109 to obtain

$$S_w^{(2)} = \left\{ (1 - S_{ors}) V_{ps} - \left[W_{ph}^{(2)} - W_{ph}^{(1)} \right] \right\} / V_{ps}$$
$$= \left[(1 - 0.25)(4,069) - 249.9 \right] / 4,069$$
$$= 0.689.$$

The volume of oil produced during the second timestep can be calculated from Eq. 8.108:

$$N_{ph}^{(2)} - N_{ph}^{(1)} = q_{oh}^{(2)} \left[t^{(2)} - t^{(1)} \right]$$
$$= q_{oh}^{(2)} \left[t^{(2)} - t^{(1)} \right]$$
$$= (6.01 \text{ B/D})(0.5 \text{ days})$$
$$= 3.00 \text{ bbl.}$$

Because $N_{ph}^{(1)} = 0$, $N_{ph}^{(2)} = 3.00$ bbl.

Calculate the Value of $\delta^{(3)}$ Corresponding to This Timestep. The amount of energy removed with the produced fluids is computed from Eq. 8.113:

$$\Delta Q^{(2)} = 5.615 \left[W_{ph}^{(2)} - W_{ph}^{(1)} \right] \rho_{wh} \left(H_{wT} - H_{wr} \right)$$
$$+ \left[N_{ph}^{(2)} - N_{ph}^{(1)} \right] \rho_{oh} \left(H_{oT} - H_{or} \right)$$
$$= 5.615 \text{ ft}^3/\text{bbl} \left[(249.5 \text{ bbl})(54.3 \text{ lbm/ft}^3) \right.$$
$$\times (361.5 \text{ Btu/lbm} - 78 \text{ Btu/lbm}) + (3.0 \text{ bbl})$$
$$\left. \times (56.1 \text{ lbm/ft}^3)(160.8 \text{ Btu/lbm} - 21.4 \text{ Btu/lbm}) \right.$$
$$= 21.73 \times 10^6 \text{ Btu.}$$

The value of $\delta^{3)}$ can be calculated by

$$\delta^{(3)} - \delta^{(2)} = \Delta Q / \left[2(n_s)(Q_{\max}) \right]$$
$$= \frac{21.73 \times 10^6 \text{ Btu}}{(2)(4 \text{ zones})(290.1 \times 10^6 \text{ Btu/zone})}$$
$$= 0.00935$$

and $\delta^{(3)} = 0.0273 + 0.00935$

$$= 0.0366.$$

Convert Production to Stock-Tank Quantities. All computed volumes and rates are determined at the average temperature of the heated region for the prevailing timestep. To convert these quantities to stock-tank volumes, the following conversion is made.

$$W_p^{(2)} - W_p^{(1)} = \left[W_{ph}^{(2)} - W_{ph}^{(1)} \right] / B_{wh}$$
$$= (249.9 \text{ bbl}) / (1.147 \text{ bbl/bbl})$$
$$= 218 \text{ bbl.}$$

By analogy,

$$N_p^{(2)} - N_p^{(1)} = \left[N_{ph}^{(2)} - N_{ph}^{(1)} \right] B_{oh}$$
$$= (3.00 \text{ bbl}) / (1.10 \text{ bbl/bbl})$$
$$= 2.73 \text{ bbl.}$$

Because $N_{ph}^{(1)} = 0$, $N_p^{(2)} = 2.73$ bbl. This completes the calculations for the second timestep.

This example illustrates that estimation of performance of a cyclic steam stimulation treatment with a model analogous to those developed by Jones (1977), Gontijo and Aziz (1984), and Sylvester and Chen (1988) is not difficult. Because many calculations involve temperature-dependent properties, this analysis has been performed with a computer program. **Tables H-2 and H-3** contain selected output from the computer program used to complete this example. Slight differences between computer output and results included in this appendix are the result of roundoff and truncation. Table 8.15 presents the computed results. Fig. 8.27 shows the oil and water rates calculated as functions of time when the well was produced for 55 days after stimulation.

Timestep	T_p (°F)	S_w	k_{ro}	ρ_o (lbm/ft³)	k_{rw}	ρ_w (lbm/ft³)	μ_o (cp)	μ_w (cp)	δ
1	411.5	0.75	0.0018	55.79	0.0037	53.38	2.134	0.13	0
2	387.69	0.689	0.0133	56.14	0.0023	54.26	2.959	0.139	0.0269
3	378.34	0.639	0.0252	56.28	0.0016	54.6	3.383	0.143	0.0363
4	370.37	0.607	0.0358	56.4	0.0012	54.89	3.802	0.147	0.0438
5	364.24	0.585	0.045	56.5	0.001	55.11	4.166	0.15	0.0488
6	359.09	0.568	0.0534	56.57	0.0008	55.29	4.505	0.152	0.0525
7	354.45	0.554	0.0613	56.65	0.0007	55.46	4.838	0.155	0.0557
8	350.14	0.542	0.0687	56.71	0.0007	55.61	5.173	0.157	0.0584
9	346.08	0.531	0.0758	56.78	0.0006	55.75	5.515	0.159	0.0609
10	342.22	0.522	0.0826	56.84	0.0005	55.89	5.865	0.161	0.0633
11	338.54	0.513	0.089	56.89	0.0005	56.01	6.224	0.163	0.0655
12	335	0.483	0.1156	56.95	0.0004	56.13	6.593	0.165	0.0676
13	321.74	0.457	0.141	57.16	0.0003	56.58	8.228	0.173	0.0755
14	309.96	0.439	0.1616	57.35	0.0002	56.97	10.096	0.181	0.0824
15	299.53	0.425	0.1787	57.52	0.0002	57.31	12.182	0.189	0.0882
16	290.16	0.414	0.1934	57.67	0.0002	57.61	14.5	0.196	0.0932
17	281.7	0.404	0.2065	57.81	0.0002	57.88	17.055	0.203	0.0975
18	273.99	0.396	0.2182	57.94	0.0001	58.11	19.861	0.21	0.1012
19	266.91	0.388	0.2288	58.05	0.0001	58.33	22.926	0.216	0.1044
20	260.4	0.382	0.2385	58.16	0.0001	58.52	26.255	0.223	0.1072
21	254.36	0.376	0.2475	58.26	0.0001	58.7	29.857	0.229	0.1096
22	248.74	0.37	0.2558	58.36	0.0001	58.86	33.754	0.235	0.1117
23	243.49	0.365	0.2636	58.45	0.0001	59.01	37.944	0.241	0.1136
24	238.55	0.361	0.2709	58.53	0.0001	59.15	42.453	0.247	0.1152
25	233.92	0.357	0.2777	58.61	0.0001	59.28	47.273	0.253	0.1167
26	229.54	0.353	0.2842	58.69	0.0001	59.4	52.437	0.258	0.118
27	225.4	0.349	0.2903	58.76	0.0001	59.51	57.95	0.264	0.1192
28	221.47	0.346	0.2961	58.83	0.0001	59.61	63.815	0.27	0.1202
29	217.74	0.342	0.3016	58.89	0.0001	59.71	70.046	0.275	0.1212
30	214.19	0.339	0.3069	58.95	0.0001	59.8	76.664	0.28	0.122
31	210.79	0.336	0.3119	59.01	0.0001	59.89	83.68	0.286	0.1228
32	207.56	0.334	0.3167	59.07	0.0001	59.97	91.091	0.291	0.1235
33	204.46	0.331	0.3213	59.12	0.0001	60.05	98.917	0.296	0.1241
34	201.5	0.329	0.3258	59.17	0.0001	60.12	107.163	0.301	0.1247
35	198.65	0.326	0.33	59.22	0.0001	60.19	115.854	0.306	0.1252

Table H-2—Summary of calculations for the heated region, Example 8.7.

Timestep	$T - t_{io}$ (days)	T_p (°F)	S_w	q_{wh} (B/D)	q_{oh} (B/D)	q_t (B/D)	W_{ph} (bbl)	N_{ph} (bbl)
1	5.00	411.5	0.750	800	0	0	674	0
2	5.84	387.69	0.689	499.36	6.03	505.39	924	3
3	6.34	378.34	0.639	312.76	38.69	351.45	1,127	14
4	6.84	370.37	0.607	204.36	65.39	269.75	1,256	40
5	7.34	364.24	0.585	152.91	84.82	237.73	1,345	78
6	7.84	359.09	0.568	124.05	98.79	222.83	1,415	124
7	8.34	354.45	0.554	104.95	109.33	214.27	1,472	176
8	8.84	350.14	0.542	91.08	117.46	208.54	1,521	232
9	9.34	346.08	0.531	80.47	123.72	204.19	1,564	293
10	9.84	342.22	0.522	72.06	128.47	200.52	1,602	356
11	10.34	338.54	0.513	65.21	131.97	197.18	1,636	421
12	10.84	335.00	0.483	59.52	134.45	193.97	1,761	687
13	12.84	321.74	0.457	42.37	140.39	182.76	1,863	962
14	14.84	309.96	0.439	31.74	140	171.74	1,937	1,242
15	16.84	299.53	0.425	25.41	133.37	158.78	1,994	1,516
16	18.84	290.16	0.414	21.23	124.24	145.46	2,041	1,773
17	20.84	281.7	0.404	18.21	114.62	132.83	2,080	2,012
18	22.84	273.99	0.396	15.93	105.29	121.22	2,114	2,232
19	24.84	266.91	0.388	14.13	96.58	110.71	2,144	2,434
20	26.84	260.4	0.382	12.67	88.60	101.27	2,171	2,619
21	28.84	254.36	0.376	11.47	81.37	92.84	2,195	2,789
22	30.84	248.74	0.370	10.46	74.80	85.27	2,217	2,945
23	32.84	243.49	0.365	9.61	68.89	78.49	2,237	3,089
24	34.84	238.55	0.361	8.87	63.53	72.40	2,256	3,222
25	36.84	233.92	0.357	8.22	58.71	66.93	2,273	3,344
26	38.84	229.54	0.353	7.66	54.33	61.99	2,289	3,457
27	40.84	225.4	0.349	7.16	50.37	57.53	2,304	3,562
28	42.84	221.47	0.346	6.72	46.78	53.49	2,318	3,659
29	44.84	217.74	0.342	6.32	43.52	49.84	2,331	3,749
30	46.84	214.19	0.339	5.96	40.54	46.50	2,343	3,833
31	48.84	210.79	0.336	5.64	37.83	43.47	2,354	3,911
32	50.84	207.56	0.334	5.34	35.36	40.70	2,365	3,985
33	52.84	204.46	0.331	5.07	33.09	38.17	2,376	4,053
34	54.84	201.5	0.329	4.83	31.02	35.85	2,386	4,117
35	56.84	198.65	0.326	4.6	29.11	33.71	2,395	4,177

Table H-3—Summary of flow-rate calculations, Example 8.7.

Nomenclature

a_1, a_2 = empirical parameters to adjust oil and water production rates after cyclic steam stimulation

A, B, C, D = constants in Eq. H-4

$A_1 ... K_1$ = correlating constants in Eq. H-5

A_h = heated area at temperature T_s, L², ft²

B_o = FVF, L³/L³, RB/STB

B_{oh} = FVF in the heated region, L³/L³, RB/STB

B_w = FVF for water, L³/L³, RB/STB

B_{wh} = FVF for water in the heated region, L³/L³, RB/STB

C_o = mean heat capacity of oil at constant pressure, L²/t²T, Btu/lbm-°F

C_{po} = specific heat capacity of oil at constant pressure, L²/t²T, Btu/lbm-°F

C_r = mean heat capacity of the matrix material at constant pressure, L²/t²T, Btu/lbm-°F

C_w = mean heat capacity of water at constant pressure, L^2/t^2T, Btu/lbm-°F

f_{sd} = steam quality at sandface, m/m, mass fraction

$G(t_D)$ = dimensionless compilation of terms, Eq. 8.62

h = thickness of formation, L, ft

$\overline{h}$ = volumetric average thickness to the oil column in the saturated portion of the region, L, ft

h_h = thickness of saturated region in a reservoir being produced by gravity drainage at the radius r_h, L, ft

h_w = thickness of oil-saturated zone at r_w, L, ft

H_o = enthalpy of oil at temperature T, L^2/t^2, Btu/lbm

H_{or} = enthalpy of oil at reservoir temperature, L^2/t^2, Btu/lbm

H_{oT} = enthalpy of oil at temperature $\overline{T}_p$, L^2/t^2, Btu/lbm

H_s = energy content of injected steam relative to initial reservoir temperature, Eq. 8.4, L^2/t^2, Btu/lbm

H_w = enthalpy of water at temperature T_s, L^2/t^2, Btu/lbm

H_{wr} = enthalpy of water at reservoir temperature, L^2/t^2, Btu/lbm

H_{wT} = enthalpy of water at temperature T_p, L^2/t^2, Btu/lbm

k = permeability of porous rock, L^2, md

k_{oh} = effective permeability of oil in the heated region, L^2, md

k_r = relative permeability, dimensionless

k_{ro} = relative permeability to oil, dimensionless

k_{rw} = relative permeability to water, dimensionless

k_{wh} = permeability of water in the heated region, L^2, md

L_{vdh} = latent heat of vaporization at sandface, L^2/t^2, Btu/lbm

$\dot{m}_s$ = steam-injection rate, m/t, lbm/D, lbm/hr

M = average volumetric heat capacity of rock, Eq. 8.50, m/Lt^2T, Btu/(ft³-°F)

M_R = average volumetric heat capacity of heated region, Eq. 8.4, m/Lt^2T, Btu/(ft³-°F)

M_s = average volumetric heat capacity of overburden rock, m/Lt^2T, Btu/(ft³-°F)

n_s = number of zones in the gross interval that receive steam

N_p = cumulative oil produced, L^3, bbl

$N_p^{(n)}$ = cumulative oil produced following cyclic steam stimulation at the end of Timestep n, L^3, STB

$N_{ph}^{(n)}$ = cumulative oil produced from heated zone at the end of Timestep n, L^3, STB

q = production rate or fluid production rate of single fluid, L^3/t, B/D

q_{max} = maximum pumping rate after soak period, L^3/t, B/D

q_{oh} = oil flow rate in the heated zone following cyclic steam injection, L^3/t, B/D

q_{wh} = water production rate following cyclic steam stimulation, L^3/t, B/D

Q_{lost} = energy lost during soak period, mL^2/t^2, Btu

Q_{max} = amount of injected energy remaining in the reservoir at the end of the soak period, mL^2/t^2, Btu

$\Delta Q^{(n)}$ = energy removed from the reservoir with the produced fluids during Timestep n, mL^2/t^2, Btu

r = radius, L, ft

r_h = radius of heated region, L, ft

r_w = wellbore radius, L, ft

S = saturation, L^3/L^3, volume fraction

S_g = gas saturation, volume fraction

S_{ors} = ROS after displacement by steam, L^3/L^3, volume fraction

S_w = water saturation, volume fraction

S_{iw} = interstitial water saturation, volume fraction

t = time, t, days

t_D = dimensionless time, Eq. 8.60

t_{Dr} = modified dimensionless time, Eq. 8.71

t_{Dz} = modified dimensionless time, Eq. 8.72

t_i = time that steam was injected during cyclic steam stimulation treatment or initial time of process, t, days

t_{soak} = length of soak period (shut in) following steam injection into a production well, t, days

T = temperature, T, °F

$\overline{T}_D$ = dimensionless average temperature of heated region, $0 \leq r \ll r_h$, and $0 \leq z \leq h$, T, °F

$\overline{T}_{Dp}$ = dimensionless average temperature adjusted for removal of energy by produced fluids

$\overline{T}_{Dp}^{(n)}$ = dimensionless average temperature of the produced fluid following cyclic steam stimulation at Time t^n

$\overline{T}_{Dr}$ = dimensionless average temperature in the radial direction, $0 \leq r \leq r_h$

$\overline{T}_{Dz}$ = dimensionless average temperature in the z-direction, $0 \leq z \leq h$

$\overline{T}_p$ = average heated-zone temperature adjusted for removal of energy by produced fluids, T, °F

$\overline{T}_p^{(n)}$ = average temperature of produced fluid (and heated zone) at the end of Timestep n following cyclic steam stimulation adjusted for removal of energy by produced fluids, T, °F

T_r = reservoir temperature, T, °F

T_s = temperature of steam, T, °F

V = volume, L^3, ft^3

V_h = volume of heated region, L^3, ft^3

V_{ph} = PV of heated region following steam injection, L^3, ft^3

V_{ps} = PV of saturated portion of heated region following cyclic steam injection, L^3, ft^3

$W_p^{(n)}$ = cumulative volume of water produced at Timestep n following cyclic steam stimulation, L^3, STB

$W_{ph}^{(n)}$ = cumulative volume of heated water removed from the heated region following cyclic steam stimulation at Timestep n, L^3, bbl

X_s = term representing ln S_w in Eq. H-5

z = hypothetical thickness in Boberg model to account for heat loss by conduction during steam stimulation, L, ft

α = thermal diffusivity of formation, L^2/t, ft^2/hr

γ_o = specific gravity of oil

δ = dimensionless parameter defined in Eq. 8.82

μ_o = viscosity of oil, m/Lt, cp

μ_{oh} = viscosity of oil in heated region of reservoir, m/Lt, cp

μ_w = viscosity of water, m/Lt, cp

μ_{wh} = viscosity of water in the heated region following cyclic steam stimulation, m/Lt, cp

ρ = density, m/L^3, lbm/ft^3

ρ_o = density of oil, m/L^3, lbm/ft^3

ρ_{oh} = density of oil produced from heated region following cyclic steam stimulation, m/L^3, lbm/ft^3

ρ_r = density of rock (matrix), m/L^3, lbm/ft^3

$\rho_{w,}$ = density of water, m/L^3, lbm/ft^3

ρ_{wh} = density of water in the heated region following cyclic steam stimulation, m/L^3, lbm/ft^3

ϕ = porosity, L^3/L^3, volume fraction

Subscripts

o = oil

w = water

g = gas

D = dimensionless

h = heated zone

Superscripts

n	=	timestep number
1,2,3	=	particular time level, position, location

References

Dietrich, J. K. 1981. Relative Permeability During Cyclic Steam Stimulation of Heavy-Oil Reservoirs. *J Pet Technol* **33** (10): 1987–1989. SPE-7968-PA. https://doi.org/10.2118/7968-PA.

Gontijo, J. E. and Aziz, K. 1984. A Simple Analytical Model for Simulating Heavy Oil Recovery by Cyclic Steam in Pressure-Depleted Reservoirs. Presented at the SPE Annual Technical Conference and Exhibition, Houston, 16–19 September. SPE-13037-MS. https://doi.org/10.2118/13037-MS.

Jones, J. 1977. Cyclic Steam Reservoir Model for Viscous Oil, Pressure Depleted, Gravity Drainage Reservoirs. Presented at the SPE California Regional Meeting, Bakersfield, California, 13–15 April. SPE-6544-MS. https://doi.org/10.2118/6544-MS.

Sylvester, N. B. and Chen, H. L. 1988. An Improved Cyclic Steam Stimulation Model for Pressure-Depleted Reservoirs. Presented at the SPE California Regional Meeting, Long Beach, California, USA, 23–25 March. SPE-17420-MS. https://doi.org/10.2118/17420-MS.

SI Metric Conversion Factors

°API		$141.5/(131.5+°API)$		=	g/cm^3
bbl	×	1.589 873	E–01	=	m^3
Btu	×	1.055 056	E+00	=	kJ
cp	×	1.0*	E–03	=	Pa·s
ft	×	3.048*	E–01	=	m
ft^2	×	9.290 304*	E–02	=	m^2
ft^3	×	2.831685	E–02	=	m^3
°F		(°F–32)/1.8		=	°C
lbm	×	4.535 924	E–01	=	kg

*Conversion factor is exact.

Appendix I

Development of Constant-Vertical-Velocity Model for Gravity Override

Appendix I presents development of the constant-vertical-velocity model and the computer programs to evaluate the functions that describe the solutions developed with this model.

The energy balance at any time before breakthrough of heat into a production well is given by

$$\dot{m}_s H_s = k_h (T_s - T_r)$$
$$\times \left[\int_0^t \frac{1}{\sqrt{\pi a(t-\lambda)}} \left(\frac{dA}{d\lambda} \right) d\lambda + \int_0^t \frac{e^{-u_1^2(t-\lambda)}}{\sqrt{\pi a(t-\lambda)}} \left(\frac{dA}{d\lambda} \right) d\lambda \right.$$
$$\left. - \frac{v}{2a} \int_0^t \mathrm{erfc}\left(a_1 \sqrt{t-\lambda} \right) \left(\frac{dA}{d\lambda} \right) d\lambda \right]$$
$$+ \rho c_p v A (T_s - T_r). \dots\dots\dots\dots\dots\dots\dots\dots \text{(8.166)}$$

Eq. 8.166 was solved* for the heated area as a function of time. The solution is Eq. I-1 where I_0 and I_1 are modified Bessel functions of the zero and the first order. Abramowitz and Stegun (1970) define these functions mathematically, and they can be computed from series expressions given by

$$A(t) = \dot{m}_s H_s / \rho C_p v (T_s - T_r)$$
$$\times \left\{ 2a_1 \sqrt{1/\pi} - e^{-a_1^2} t/2 \left[(1 + a_1^2 t) I_0 \left(a_1^2 t/2 \right) \right. \right.$$
$$\left. \left. + a_1^2 t I_1 \left(a_1^2 t \right) \right] + 1 \right\}. \dots\dots\dots\dots\dots \text{(I-1)}$$

In Eq. I-1,

$$I_0(x) = 1 + A_0 y^2 + B_0 y^4 + C_0 y^6 + D_0 y^8 + E_0 y^{10} + F_0 y^{12}, \dots\dots\dots \text{(I-2)}$$

where $y = x/3.75$, $-3.75 < x < 3.75$, $A_0 = 3.5156229$, $B_0 = 3.0899424$, $C_0 = 1.2067492$, $D_0 = 0.2659732$, $E_0 = 0.0360768$, and $F_0 = 0.0045813$, and

$$I_1(x) = \left(0.5 + A_1 y^2 + B_1 y^4 + C_1 y^6 + D_1 y^8 + E_1 y^{10} + F_1 y^{12} \right), \dots\dots\dots \text{(I-3)}$$

where $A_1 = 0.87890594$, $B_1 = 0.51498869$, $C_1 = 0.15084934$, $D_1 = 0.02658733$, $E_1 = 0.00301532$, and $F_1 = 0.00032411$.

It is convenient for computation to introduce a dimensionless time defined by

$$t_{Dv} = v^2 t / 4a. \dots\dots\dots\dots\dots\dots\dots\dots\dots \text{(I-4)}$$

Introducing t_{Dv} into Eq. I-1 gives

$$A(t_{Dv}) = \dot{m}_s H_s / \rho C_p v (T_s - T_r)$$
$$\times \left\{ 2\sqrt{t_{Dv}/\pi} - e^{-t_{Dv}/2} \left[(1 + t_{Dv}) I_0 (t_{Dv}/2) \right. \right.$$
$$\left. \left. + t_{Dv} I_1 (t_{Dv}/2) \right] + 1 \right\}. \dots\dots\dots\dots\dots \text{(I-5)}$$

---

*G. P. Willhite. 1969. Unpublished notes, University of Kansas, Lawrence, Kansas.

The term within the braces is a function of t_{Dv} and can be evaluated separately. Let

$$G_3\left(t_{Dv}\right)=2\sqrt{t_{Dv}/\pi}-e^{-t_{Dv}/2}\left[\left(1+t_{Dv}\right)I_0\left(t_{Dv}/2\right)\right.$$
$$\left.+t_{Dv}I_1\left(t_{Dv}/2\right)\right]+1.\dots\dots\dots\dots\dots\dots\dots\dots\dots\dots\text{(I-6)}$$

Then,

$$A\left(t_{Dv}\right)=\frac{\dot{m}_sH_s}{\rho C_pv\left(T_s-T_r\right)}G_3\left(t_{Dv}\right).\dots\dots\dots\dots\dots\dots\dots\dots\text{(8.167)}$$

Doscher and Ghassemi (1983) developed an equivalent relationship using numerical integration. The following is the function subprogram $G_3(t_{Dv})$.

$G_3(t_{Dv})$

```
C       G3.FOR
C       PROGRAM TO COMPUTE HEATED AREA WHEN STEAM ZONE
        GROWS
C       VERTICALLY
C       AT A CONSTANT FRONTAL ADVANCE RATE
C       GPW 3-3-1991
C
        REAL FUNCTION G3(TD) DATA PI/3.141596/
        ARGI=TD/2.
        TERM1 = 2.*SQRT(TD/PI)
        TERM2 = (1 + TD)*BES0(ARG1) + TD*BES1(ARG1)
        TERM3 = EXP( – ARG1)*TERM2
        G3(I)=TERM1 – TERM3 + 1
        RETURN
        END
```

The heated volume is expressed as

$$v_h=\int_0^t v\left(t-\lambda\right)\left(dA/d\lambda\right)d\lambda.\dots\dots\dots\dots\dots\dots\dots\dots\dots\text{(8.172)}$$

The derivative of A with respect to t is given by

$$dA/dt=\left[\dot{m}_sH_s\rho/C_pv\left(T_s-T_r\right)\right]\left\{a_1/\sqrt{\pi t}\right.$$
$$\left.-\left(a_1^2/2\right)e^{-a_1^{2t/2}}\left[I_0\left(a_1^{2t}/2\right)+I_1\left(a_1^{2t}/2\right)\right]\right\},\dots\dots\dots\dots\dots\text{(I-7)}$$

where $a_1=v/2\sqrt{a}$.

Substituting Eq. 1-7 into Eq. 8.172 and integrating give

$$V_h=vtA-\left(2vCa_{1/}3\sqrt{\pi}\right)t^{3/2}+\left(vCa_1^2/2\right)$$
$$\times\int_0^t\lambda\left[I_0\left(a_1^2\lambda/2\right)+I_1\left(a_1^2\lambda/2\right)\right]d\lambda,\dots\dots\dots\dots\dots\dots\dots\text{(I-8)}$$

where $C=\dot{m}_sH_s/\rho C_pv\left(T_s-T_r\right).\dots\dots\dots\dots\dots\dots\dots\dots\dots\text{(I-9)}$

The integral in Eq. I-8 can be expressed in dimensionless time and evaluated numerically:

$$\int_0^t\lambda\left[I_0\left(a_1^2/\lambda2\right)+I_1\left(a_1^2/\lambda2\right)\right]d\lambda$$
$$=\left(4/a_1^4\right)\int_0^{t_{Dv}/2}\lambda_D\left[I_0\left(\lambda_D\right)+I_1\left(\lambda_D\right)\right]d\lambda_D\dots\dots\dots\dots\dots\dots\text{(I-10)}$$

and $G_4\left(t_{Dv}/2\right)=\int_0^{t_{Dv}/2}\lambda_D\left[I_0\left(\lambda_D\right)+I_1\left(\lambda_D\right)\right]d\lambda_D.\dots\dots\dots\dots\text{(I-11)}$

Then,

$$V_h=vtA-\frac{2vC}{3\sqrt{\pi}}t^{3/2}+\frac{2vC}{a_1^2}G_4\left(\frac{t_{Dv}}{2}\right).\dots\dots\dots\dots\dots\dots\dots\text{(I-12)}$$

Eq. I-12 can be written in terms of t_{Dv} by substituting Eq. 8.167 for A and Eq. I-4 for t. The resulting expression is

$$V_h = \frac{\dot{m}_s H_s}{\rho C_p \left(T_s - T_r\right)a_1^2}\left[t_{Dv}G_3\left(t_{Dv}\right) - \frac{2}{3}\frac{t_{Dv}^{3/2}}{\sqrt{\pi}} + 2G_4\left(\frac{t_{Dv}}{2}\right)\right]. \quad \dots\dots\dots\dots\dots\dots\dots\dots \text{(8.168)}$$

The term in the brackets is a function only of t_{Dv}. However, it is convenient to define a new function $G_v(t_{Dv})$ by

$$G_v\left(t_{Dv}\right) = t_{Dv}G_3\left(t_{Dv}\right) - \frac{2}{3}\frac{t_{Dv}^{3/2}}{\sqrt{\pi}} + 2G_4\left(\frac{t_{Dv}}{2}\right) \quad \dots\dots\dots\dots\dots\dots\dots\dots \text{(I-13)}$$

Then,

$$V_h = \left[\dot{m}_s H_s / \rho C_p \left(T_s - T_r\right)\right]a_1^2 G_v\left(t_{Dv}\right). \quad \dots\dots\dots\dots\dots\dots\dots\dots\dots \text{(8.173)}$$

The following subprogram is used to calculate $G_v(t_{Dv})$.

$G_v(t_{Dv})$

```
C
C      GV.FOR
C
C      COMPUTES HEATED VOLUME DIMENSIONLESS GROUP IN UNIFORM C
C      VERTICAL EXPANSION MODEL
C      GPW 3-29-1991
C
C2345678
       REAL FUNCTION GV(TD,NINT)
       DATA PI/3.141596/
       GV = TD*G3(TD) − 2./3.*TD**1.5/SQRT(PI) + 2.*G4(TD/2.,NINT)
       RETURN
       END
```

$G_4(t_D/2)$

```
C      G4.FOR
C
C234567
C
C      FUNCTION SUBPROGRAM USED TO EVALUATE INTEGRAL FOR
C      COMPUTATION OF HEATED VOLUME IN VERTICAL EXPANSION
C      MODEL FOR STEAM DISPLACEMENT
C      GPW 3-15-1991
C
       REAL FUNCTION G4(TD02,NINT)
       DIMENSION F(501),TD2(501)
       DTD=TD02/NINT
       DO10I=I,NINT+I
       TD2(I)=(I-1)*DTD
       10
F(I) = TD2(I)*EXP(−TD2(I))*(BES0(TD2(I))+BES1(TD2(I)))
       SUM = 0.
C      INTEGRATION USING TRAPEZOIDAL RULE
C
       DO 20 I=1,NINT
       20
       SUM=SUM+0.5*(F(I)+F(I+1))*DTD
       G4 = SUM
       RETURN
       END
```

$G_e(t_{Dv}, t_{Dv1})$

The term $G_e(t, t_1)$ is the fraction of the initial heat-injection rate necessary to maintain vertical expansion of the heated region at a constant velocity of v within the area $A(t_{Dv1})$ and is defined by

$$G_e\left(t, t_1\right) = \frac{1}{\dot{m}_s H_s}\int_0^{t_1} q_1\left(t - \lambda\right)\left(dA/d\lambda\right)d\lambda$$

$$+ \int_0^{t_1} q_2\left(t - \lambda\right)\left(dA/d\lambda\right)d\lambda, \quad \dots\dots\dots\dots\dots\dots\dots\dots \text{(I-14)}$$

where $q_1(t) = k(T_s - T_r)/\sqrt{\pi a t}$,

$$q_2 = k(T_s - T_r)\left[\left(e^{-a_1^2 t}/\sqrt{\pi a t}\right) + \frac{v}{a} - \frac{v}{2a}\,\mathrm{erfc}\left(a_1\sqrt{t}\right)\right], \quad\ldots\ldots\ldots\ldots\ldots\ldots\ldots\ldots\ldots \text{(8.163)}$$

and $\dfrac{dA}{dt} = \dfrac{\dot{m}_s H_s}{M_R v(T_s - T_r)}\left\{\dfrac{a_1}{\sqrt{\pi t}} - \dfrac{a_1^2}{2}\left[I_0\left(\dfrac{t_D}{2} + \dfrac{I_1}{2}\right)\right]\right\}.$

In Eq. I-14, t_1 is the time that the heated zone arrives at an area A_1 and t is the time that the steam zone arrives at the same location. Thus, $t \geq t_1$.

The integral in Eq. I-14 can be evaluated by introducing the dimensionless time t_{Dv} defined by Eqs. 8.168 and I-4, $t_{Dv} = v^2 t/4a$, and subdividing the integral into four parts as in

$$\begin{aligned}
G_e(t_{Dv}, t_{Dv1}) &= \left(1/2\sqrt{\pi}\right)\left[G_5(t_{Dv}, t_{Dv1}) + G_6(t_{Dv}, t_{Dv1})\right] \\
&\quad + (1/2)\left[G_7(t_{Dv}, t_{Dv1}) + G_8(t_{Dv}, t_{Dv1})\right], \quad\ldots\ldots\ldots\ldots\ldots\ldots \text{(I-15)}
\end{aligned}$$

where $G_5(t_{Dv}, t_{Dv1}) = \displaystyle\int_0^{t_{Dv1}} \frac{d\lambda_D}{\sqrt{t_{Dv} - \lambda_D}\,\sqrt{\pi \lambda_D}}, \quad\ldots\ldots\ldots\ldots\ldots\ldots\ldots\ldots\ldots \text{(I-16)}$

$$G_6(t_{Dv}, t_{Dv1}) = \int_0^{t_{Dv1}} \frac{e^{-\lambda_D/2}\left[I_0(\lambda_D/2) + I_1(\lambda_D/2)\right]}{2\sqrt{t_{Dv} - \lambda_D}}\,d\lambda_D, \quad\ldots\ldots\ldots\ldots\ldots\ldots \text{(I-17)}$$

$$\begin{aligned}
G_7(t_{Dv}, t_{Dv1}) &= \int_0^{t_{Dv1}} \left[\frac{e^{-(t_{Dv} - \lambda_D)}}{\sqrt{\pi(t_{Dv} - \lambda_D)}} + 2 - \mathrm{erfc}\sqrt{t_{Dv} - \lambda_D}\right] \\
&\quad \times \left\{-\left[\left(e^{-\lambda_D}/2\right)/2\right]\right. \\
&\quad \left. \times \left[I_0(\lambda_D/2) + I_1(\lambda_D/2)\right]\right\}d\lambda_D, \quad\ldots\ldots\ldots\ldots\ldots\ldots \text{(I-18)}
\end{aligned}$$

$$\begin{aligned}
G_8(t_{Dv}, t_{Dv1}) &= \int_0^{t_{Dv1}} \left(1/\sqrt{\pi \lambda_D}\right)\left\{\left[e^{-(t_{Dv} - \lambda_D)}/\sqrt{\pi(t_{Dv} - \lambda_D)}\right]\right. \\
&\quad \left. + 2 - \mathrm{erfc}\sqrt{t_{Dv} - \lambda_D}\right\}d\lambda_D. \quad\ldots\ldots\ldots\ldots\ldots\ldots\ldots\ldots \text{(I-19)}
\end{aligned}$$

These integrals were evaluated numerically in the following programs with the trapezoidal rule to obtain values of $G_e(t_{Dv}, t_{Dv1})$ given in Table 8.35 for selected values of t_{Dv} and t_{Dv1}. Remember that $G_e(t_{Dv}, t_{Dv1}) = f_{h,v}$, the fraction of the injected energy that is latent heat, and $t_{Dv1} < t_{Dv}$.

$G_5(t_{Dv}, t_{Dv1})$

```
C     G5.FOR
C234567890
C     PROGRAM TO COMPUTE THE FIRST INTEGRAL FOR
C     DETERMINATION
C     OF THE HEAT INJECTION RATE TO MAINTAIN VERTICAL
C     EXPANSION
C     OF A HEATED ZONE AT CONSTANT RATE
C
C     GPW 3-18-1991
C
      REAL FUNCTION G5(TD,TD1M,NINT)
      DATA PI/3.141596/
      DIMENSION TD1(501),F(501)
      IF(TI.LE.TD1M) GO TO 30
      DTD = SQRT(TD1M)/NINT
      DO 10 I = I,NINT+I
      TD1(I) = (I–1)*DTD
10    F(I) = 1./SQRT(TD–TD1(I)**2)
      SUM = 0.
```

```
C
C     INTEGRATION USING TRAPEZOIDAL RULE
C
      DO 20 I = 1,NINT
20    SUM = SUM+0.5*(F(I)+F(I +1))*DTD
      G5 = 2./SQRT(PI)*SUM
      RETURN
30    PRINT VTI.LE. TD1–PROGRAM STOPPED IN G5.FOR'
      PRINT *,'TD = ',TD,' TD1 = ',TD1M
      STOP
      END
```

$G_6(t_{Dv}, t_{Dv1})$

```
C     G6.FOR
C234567890
C     PROGRAM TO COMPUTE THE FIRST INTEGRAL FOR
C     DETERMINATION
C     OF THE HEAT INJECTION RATE TO MAINTAIN VERTICAL
C     EXPANSION
C     OF A HEATED ZONE AT CONSTANT RATE
C
C     GPW 3-18-1991
C
      REAL FUNCTION G6(TD,TD1M,NINT)
      DATA PI/3.141596/
      DIMENSION TD1(501),F(501)
      IF(TI.LE.TD1M) GO TO 30
      DTD = TD1M/NINT
      DO 10 I = 1,NINT+1
      TD1(I) = (I–1)*DTD
      TERM1 = 1./SQRT(TD–TD1(I))
      TERM2 = –0.5*EXP(–TDI(I)/2.)*(BES0(TDI(I)/2.)+ :BESI(TDI(I)/2.))
10    F(I) = TERM1*TERM2
      SUM = 0.
C
C     INTEGRATION USING TRAPEZOIDAL RULE
C
      DO 20 I = 1,NINT
20    SUM = SUM + 0.5*(F(I) + F(I + 1))*DTD
      G6 = SUM
      RETURN
30    PRINT VTLLE. TD1–PROGRAM STOPPED IN G6.FOR'
      PRINT *,'TD –',TD,' TD1 = ',TD1M
      STOP
      END
```

$G_7(t_{Dv}, t_{Dv1})$

```
C     G7.FOR
C234567890
C     PROGRAM TO COMPUTE THE SECOND INTEGRAL FOR
C     DETERMINATION
C     OF THE HEAT INJECTION RATE TO MAINTAIN VERTICAL
C     EXPANSION
C     OF A HEATED ZONE AT CONSTANT RATE
C
C     GPW 3-18-1991
C
      REAL FUNCTION G7(TD,TD1M,NINT)
      DATA PI/3.141596/
      DIMENSION TD1(501),F(501)
      IF(TI.LE.TD1M) GO TO 30
      DTD = TD1M/NINT
      DO 10 I = 1,NINT+1
      TD1(I) = (I–1)*DTD
```

```
        TD2 = TD–TD1(I)
        TERM1 = EXP(–TD2)/SQRT(PI*TD2)+2 –
        : ERFC(SQRT(TD2))
        TERM2 = –0.5*EXP(–TDI(I)/2.)*(BES0(TDI(I)/2.)+
        : BESI(TDI(I)/2.))
10      F(I) = TERM1*TERM2
        SUM = 0.
C
C       INTEGRATION USING TRAPEZOIDAL RULE
C
        DO 20 I = 1,NINT
20      SUM = SUM + 0.5*(F(I) + F(I + 1))*DTD
        G7 = SUM
        RETURN
30      PRINT VTLLE. TD1–PROGRAM STOPPED IN G7.FOR'
        PRINT *,'TD =',TD,' TD1E',TD1M
        STOP
        END
```

$G_8(t_{Dv}, t_{Dv1})$

```
C       G8.FOR
C234567890
C       PROGRAM TO COMPUTE THE SECOND INTEGRAL FOR
C       DETERMINATION
C       OF THE HEAT INJECTION RATE TO MAINTAIN VERTICAL
C       EXPANSION
C       OF A HEATED ZONE AT CONSTANT RATE
C
C       GPW 3-19-1991
C
        REAL FUNCTION G8(TD,TD1M,NINT)
        DATA PI/3.141596/
        DIMENSION TD1(501),F(501)
        IF(TI.LE.TD1M) GO TO 30
        DTD = SQRT(TD1M)/NINT
        DO 10 I = 1,NINT+1
        TD1(I) = (I–1)*DTD
        TD2 = TD–TD1(I)**2
10      F(I) = EXP(–TD2)/SQRT(PI*TD2)+2–
        : ERFC(SQRT(TD2))
        SUM = 0.
C
C       INTEGRATION USING TRAPEZOIDAL RULE
C
        DO 20 I = 1,NINT
20      SUM = SUM + 0.5*(F(I)+F(I+1))*DTD
        G8 = 2.*SUM/SQRT(PI)
        RETURN
30      PRINT VTI.LE. TD1–PROGRAM STOPPED IN G8.FOR'
        PRINT *,'TD = ',TD,' TD1 = ',TD1M
        STOP
        END
```

Nomenclature

A	=	area, L^2, ft^2	C	=	group of terms defined by Eq. I-9
a_1	=	$v/2\sqrt{a}$ in Eq. 8.172	C_p	=	heat capacity of reservoir rock, saturated with fluids, ahead of the steam zone when the steam zone expands at a constant vertical velocity, L^2/t^2-T, Btu/lbm-°F
$A_0, B_0,$ $C_0, D_0,$ E_0, F_0	=	constants in Eq. I-2			
$A_1, B1,$ $C1, D_1,$ E_1, F_1	=	constants in Eq. 1-3	$f_{h,v}$	=	fraction of injected energy that is latent heat

G_e = dimensionless compilation of terms, Eq. 8.178

G_v = integral expression defined by Eq. 8.173

G_3 = dimensionless compilation of terms, Eq. 8.167

G_4 = dimensionless compilation of terms, Eq. I-11

G_5 = dimensionless compilation of terms, Eq. I-16

G_6 = dimensionless compilation of terms, Eq. I-17

G_7 = dimensionless compilation of terms, Eq. I-18

G_8 = dimensionless compilation of terms, Eq. I-19

H_s = energy content of injected steam relative to initial reservoir temperature, Eq. 8.4, L^2/t^2, Btu/lbm

I_0, I_1 = modified Bessel functions of zero and first order, respectively

k = permeability of porous rock, L^2, md

$\dot{m}_s$ = steam injection rate, m/t, lbm/hr

M_r = average volumetric heat capacity of heated region, m/Lt^2T, Btu/(ft^3-°F)

q_1 = rate of heat loss to overburden by conduction, Btu/hr-ft^2

q_2 = rate of heat loss by conduction at the front of the steam zone that is moving at a velocity at specific heated area, t, hours

t = time, t, hours

t_1 = time when heated zone arrived at a specific heated area, t, hours

t_D = dimensionless time

t_{Dv} = modified dimensionless time, Eq. 8.168

t_{Dv1} = dimensionless time when the steam zone reaches Area A_1 when the steam zone is expanding at a constant velocity v

T_r = reservoir temperature, T, °F

T_s = temperature of steam, T, °F

v = velocity of steam zone interface, L/t, ft/hr

V_h = volume of heated zone, L^3, ft^3

x = argument of Bessel functions I_0, I_1

y = correlating parameter in Eqs. I-2 and I-3 = $x/3.75$

α = thermal diffusivity, L^2/t, ft^2/hr

λ = time of arrival of heated region at specific location, t, hours

λ_D = dimensionless time of arrival of heated region at specific location

ρ = density, m/L^3, lbm/ft^3

References

Abramowitz, M. and Stegun, L. A. 1970. *Handbook of Mathematical Functions,* National Bureau of Standards Applied Mathematics Series 55, ninth printing, Washington, DC, 378.

Doscher, T. M. and Ghassemi, F. 1983. The Influence of Oil Viscosity and Thickness on the Steam Drive. *J Pet Technol* 35 (2): 291–298. SPE-9897-PA. https://doi.org/10.2118/9897-PA.

AUTHOR INDEX

SUBJECT INDEX